# Problem Books in Mathematics

*Series Editor:*
Peter Winkler
Department of Mathematics
Dartmouth College
Hanover, NH 03755
USA

More information about this series at http://www.springer.com/series/714

Vladimir V. Tkachuk

# A $C_p$-Theory Problem Book

## Functional Equivalencies

Springer

Vladimir V. Tkachuk
Departamento de Matematicas
Universidad Autonoma Metropolitana-Iztapal
Mexico, Mexico

ISSN 0941-3502          ISSN 2197-8506   (electronic)
Problem Books in Mathematics
ISBN 978-3-319-24383-2      ISBN 978-3-319-24385-6   (eBook)
DOI 10.1007/978-3-319-24385-6

Library of Congress Control Number: 2015958805

Mathematics Subject Classification (2010): 54C35

Springer Cham Heidelberg New York Dordrecht London

Springer International Publishing AG Switzerland is part of Springer Science+Business Media (www.
springer.com)

# Preface

This is the fourth and the last volume of the series of books of problems in $C_p$-theory entitled *A $C_p$-Theory Problem Book*, i.e., this book is a continuation of the three volumes subtitled *Topological and Function Spaces, Special Features of Function Spaces* and *Compactness in Function Spaces*. The series was conceived as an introduction to $C_p$-theory with the hope that each volume could also be used as a reference guide for specialists.

The first volume provides a self-contained introduction to general topology and $C_p$-theory and contains some highly nontrivial state-of-the-art results. For example, Section 1.4 presents Shapirovsky's theorem on existence of a point-countable $\pi$-base in any compact space of countable tightness, and Section 1.5 brings the reader to the frontier of the modern knowledge about realcompactness in the context of function spaces.

The second volume covers a wide variety of topics in $C_p$-theory and general topology at the professional level bringing the reader to the frontiers of modern research. It presents, among other things, a self-contained introduction to Descriptive Set Theory and Advanced Set Theory providing a basis for working with most popular axioms independent of ZFC.

The third volume introduces the reader to compactness and its generalizations in the context of function spaces. It continues dealing with topology and $C_p$-theory at a professional level. The main objective is to develop from the very beginning the theory of compact spaces most used in functional analysis, i.e., Corson compacta, Eberlein compacta, and Gul'ko compacta.

This volume presents a reasonably complete and up-to-date information on preservation of topological properties by homeomorphisms of function spaces. An exhaustive theory of $t$-equivalent, $u$-equivalent and $l$-equivalent spaces is developed from scratch. Since the policy of the author is to make this book self-contained, the reader will find here an introduction to the theory of uniform spaces, the theory of locally convex spaces as well as to the theory of inverse systems and dimension theory. The above-mentioned policy also made it necessary to include Kolmogorov's solution of Hilbert's Problem 13 since it is needed for the presentation of the theory of $l$-equivalent spaces.

The author's intention was to include in this volume all classical results on functional equivalencies. In particular, we present the famous theorem of Gul'ko and Khmyleva on non-preservation of compactness by $t$-equivalence as well as Okunev's results on $t$-invariance of spread, hereditary density, hereditary Lindelöf number and $\sigma$-compactness. Of course, it was impossible to omit the fundamental result of Gul'ko on preservation of the dimension dim by $u$-equivalence, a deep theorem of Marciszewski which states that $\mathbb{I}$ and $\mathbb{I}^\omega$ are not $t$-equivalent as well as the Bouziad's result on preservation of the Lindelöf number by $l$-equivalence.

We apply here all topological methods developed in the first three volumes, so we refer to their problems and solutions when necessary; the abbreviation for the first volume is *TFS*, and we will use the expressions *SFFS* and *CFS* to refer to the second and third volumes, respectively. For example, TFS-273 refers to Problem 273 of the book TFS. The references to the solutions are not that straightforward: the abbreviation S.115 means "solution of Problem 115 of the book TFS," while T.025 stands for "solution of Problem 025 of the book SFFS." The expression U.249 abbreviates the phrase "solution of Problem 249 of the book CFS," and, finally, V.411 denotes the solution of Problem 411 of this volume. The author did his best to keep *every* solution as independent as possible, so a short argument could be repeated several times in different places.

The author wants to emphasize that if a postgraduate student mastered the material of the first three volumes, it will be more than sufficient to understand every problem and solution of this book. However, for a concrete topic much less might be needed. Finally, the author outlines some points which show the potential usefulness of this work.

- *The only background needed is some knowledge of set theory and real numbers; any reasonable course in calculus covers everything needed to understand this book.*
- *The student can learn all of general topology required without recurring to any textbook or papers; the amount of general topology is strictly minimal and is presented in such a way that the student works with the spaces $C_p(X)$ from the very beginning.*
- *What is said in the previous paragraph is true as well if a mathematician working outside of topology (e.g., in functional analysis) wants to use results or methods of $C_p$-theory; he (or she) will find them easily in a concentrated form or with full proofs if there is such a need.*
- *The material we present here is up to date and brings the reader to the frontier of knowledge in a reasonable number of important areas of $C_p$-theory.*
- *This book seems to be the first self-contained introduction to $C_p$-theory. Although there is an excellent textbook written by Arhangel'skii (1992a), it heavily depends on the reader's good knowledge of general topology.*

Mexico City, Mexico                                                        Vladimir V. Tkachuk

# Contents

# Detailed Summary of Exercises

## 1.1. Equivalencies That Arise From Homeomorphisms of $C_p(X)$.

## 1.2. Uniformities, Dimension, and u-Equivalence.

## 1.3. Linear Topological Spaces and l-Equivalence.

## 1.4. Metrizable Spaces and l-Equivalence.

## 1.5. The Last-Minute Updates. Yet More on l-Equivalence.

# Introduction

The term "$C_p$-theory" was invented to abbreviate the phrase "the theory of function spaces endowed with the topology of pointwise convergence." The credit for the creation of $C_p$-theory must undoubtedly be given to Alexander Vladimirovich Arhangel'skii. The author is proud to say that Arhangel'skii also was the person who taught him general topology and directed his PhD thesis. Arhangel'skii was the first to understand the need to unify and classify a bulk of heterogeneous results from topological algebra, functional analysis and general topology. He was the first to obtain crucial results that made this unification possible. He was also the first to formulate a critical mass of open problems which showed this theory's huge potential for development.

Later, many mathematicians worked hard to give $C_p$-theory the elegance and beauty it boasts nowadays. The author hopes that the work he presents for the reader's judgement will help to attract more people to this area of mathematics.

The main text of this volume consists of 500 statements formulated as problems; it constitutes Chapter 1. These statements provide a gradual development of many popular topics of $C_p$-theory to bring the reader to the frontier of the present-day knowledge. A complete solution is given to every problem of the main text.

The material of Chapter 1 is divided into five sections with 100 problems in each one. The sections start with an introductory part where the definitions and concepts to be used are given. The introductory part of any section *never exceeds two pages and covers everything that was not defined previously*. Whenever possible, we try to save the reader the effort of ploughing through various sections, chapters and volumes, so we give the relevant definitions in the current section not caring much about possible repetitions.

Chapter 1 ends with some bibliographical notes to give the most important references related to its results. The selection of references is made according to the author's preferences and by no means can be considered complete. However, a complete list of contributors to the material of this book can be found in our bibliography of 800 items. It is the author's pleasant duty to acknowledge that he consulted the paper of Arhangel'skii (1998a) to include quite a few of its 375 references in his bibliography.

Sometimes, as we formulate a problem, we use without reference definitions and constructions introduced in other problems. The general rule is to try to find the relevant definition *not more than ten problems before*.

The complete solutions of all problems of Chapter 1 are given in Chapter 2. Chapter 3 begins with a selection of 80 statements which were proved as auxiliary facts in the solutions of the problems of the main text. This material is split into 7 sections to classify the respective results and make them easier to find. Chapter 4 consists of 100 open problems presented in 7 sections with the same idea: to classify this bulk of problems and make the reader's work easier.

Chapter 4 also witnesses an essential difference between the organization of our text and the book by Arhangel'skii and Ponomarev (1974): *we never put unsolved problems in the main text as is done in their book.* All problems formulated in Chapter 1 are given complete solutions in Chapter 2, and the unsolved ones are presented in Chapter 4.

There is little to explain about how to use this book as a reference guide. In this case the methodology is not that important, and the only thing the reader wants is to find the results he (or she) needs as fast as possible. To help with this, the titles of chapters and sections give the first approximation. To better see the material of a chapter, one can consult the second part of the "Contents" section where a detailed summary is given; it is supposed to cover all topics presented in each section. Besides, the index can also be used to find necessary material.

To sum up the main text, the author believes that the coverage of $C_p$-theory will be reasonably complete and many of the topics can be used by postgraduate students who want to specialize in $C_p$-theory. Formally, this book can also be used as an introduction to general topology. However, it would be a somewhat biased introduction, because the emphasis is always given to $C_p$-spaces and the topics are only developed when they have some applications in $C_p$-theory.

To conclude, let the author quote an old saying which states that the best way for one to learn a theorem is to prove it oneself. This text provides a possibility to do this. If the reader's wish is to read the proofs, there they are concentrated immediately after the main text.

# Chapter 1
# Properties Preserved by Homeomorphisms of Function Spaces

The reader who has found his (or her) way through the first fifteen hundred problems of this book is fully prepared to enjoy working professionally in $C_p$-theory. Such a work implies choosing a topic, reading the papers with the most recent progress thereon, and attacking the unsolved problems. Now the first two steps are possible without doing heavy library work, because Chapter 1 provides information on the latest advances in all areas of $C_p$-theory, where functional equivalencies are concerned. Here, many ideas, results, and constructions came from functional analysis and the theory of uniform spaces giving a special flavor to this part of $C_p$-theory, but at the same time making it more difficult to master. I must warn the reader that most topics, outlined in the forthcoming bulk of 500 problems, constitute the material of important research papers—in many cases very difficult ones. The proofs and solutions, given in Chapter 2, are complete, but sometimes they require a very high level of understanding of the matter. The reader should not be discouraged if some proofs seem to be unfathomable. We still introduce new themes in general topology and formulate, after a due preparation, some nontrivial results which might be later used in $C_p$-theory.

This volume presents a very popular line of research in $C_p$-theory. The objective here is to find common features of the spaces $X$ and $Y$ knowing that $C_p(X)$ and $C_p(Y)$ are similar in some way. Theorem of Nagata (Problem TFS-200) gives a complete solution if the rings $C_p(X)$ and $C_p(Y)$ are topologically isomorphic because in this case the spaces $X$ and $Y$ must be homeomorphic. It turns out that the existence of a linear homeomorphism between $C_p(X)$ and $C_p(Y)$ need not imply that the spaces $X$ and $Y$ are homeomorphic. Nevertheless, $X$ and $Y$ have to share quite a few important properties even if there exists a nonlinear homeomorphism between $C_p(X)$ and $C_p(Y)$.

This volume requires a much broader vision of general topology than the previous ones because the general scheme of research here is to take some property (which might come from any area of general topology or even from outside) and check whether it is shared by the spaces $X$ and $Y$ in case $C_p(X)$ is (linearly or

© Springer International Publishing Switzerland 2016
V.V. Tkachuk, *A Cp-Theory Problem Book*, Problem Books in Mathematics,
DOI 10.1007/978-3-319-24385-6_1

uniformly) homeomorphic to $C_p(Y)$. As a consequence, we had to introduce some theories which could be the subject of a separate book. In particular, we had to develop the basic techniques of dealing with uniformities, linear topological spaces, dimension theory, and topological games. Once accomplished, this made it possible to present in a self-contained way almost all important results in the area.

If nothing is said about the separation axioms of a space $X$ then $X$ is assumed to be Tychonoff. Section 1.1 deals with pairs of spaces $X, Y$ for which $C_p(X)$ is homeomorphic to $C_p(Y)$ (such spaces are called $t$-equivalent). One of the outstanding results presented here is Gul'ko and Khmyleva's theorem on non-preservation of compactness by $t$-equivalence (Problem 027). The theorems on invariance of spread, hereditary density, hereditary Lindelöf number, and $\sigma$-compactness (Problems 043, 068–070) constitute a breakthrough due to Okunev. Another gem of this collection is a result of W. Marciszewski which states that $\mathbb{I}$ and $\mathbb{I}^\omega$ are not $t$-equivalent (Problem 099).

Section 1.2 is devoted to the study of pairs of spaces $X, Y$ for which $C_p(X)$ is uniformly homeomorphic to $C_p(Y)$ (such spaces are called $u$-equivalent). We give the reader a glimpse of the theory of uniform spaces. Inverse systems and dimension theory are also developed to some extent to make it possible to present a famous result of S. Gul'ko on invariance of the dimension dim under $u$-equivalence (Problem 180). Another important result is a theorem of W. Marciszewski on preservation of absolute Borel classes by $u$-equivalence (Problems 197–198). We conclude this section with another beautiful result of S. Gul'ko: if $X$ and $Y$ are infinite countable compact spaces then they are $u$-equivalent (Problem 200).

Section 1.3 starts with a short introduction to linear topological spaces. The results that follow are intended to give information on pairs of spaces $X, Y$ for which $C_p(X)$ is linearly homeomorphic to $C_p(Y)$ (such spaces are called $l$-equivalent). We present a general method of Okunev for constructing pairs of $l$-equivalent spaces (Problem 257) and a classification of spaces $l$-equivalent to some standard ones (Problems 293, 295, 297–300).

Section 1.4 presents another portion of deep results on $l$-equivalence. A fact which could not be omitted here is a theorem of Baars, de Groot, and Pelant on preservation of Čech-completeness by $l$-equivalence in metrizable spaces (Problem 366). It is a nontrivial theorem of Dranishnikov that any nonempty open subspace of $\mathbb{R}^n$ is $l$-equivalent to $\mathbb{R}^n$ (Problem 394); this Section concludes with a very difficult example of Marciszewski of an infinite compact space $K$ for which there exists no continuous linear surjection of $C_p(K)$ onto $C_p(K) \times \mathbb{R}$ (Problem 400).

This book was in preparation for almost ten years and six years passed in the process of publishing the first three volumes; so quite a few new results in $C_p$-theory emerged during this period. The author included them in the book where it was possible to avoid violation of the existing classification scheme. However, at the moment of writing Section 1.5 (which was originally planned to cover what was left from the theory of $l$-equivalent spaces) many new fundamental results appeared and they did not fit into any classification at all. That is why Section 1.5 stands completely apart: it contains the most recent results which could not be left out and continues the study of functional equivalences.

One of the main results here is Bouziad's theorem on preservation of the Lindelöf number by $l$-equivalence (Problem 500). Another one is a theorem of Leiderman, Levin, and Pestov stating that $C_p(\mathbb{I})$ can be linearly and continuously mapped onto $C_p(X)$ for any finite-dimensional metrizable compact space $X$ (Problem 499). Problem 494 gives a complete classification (due to Gul'ko and Os'kin) of $l$-equivalent countable compact spaces. A fundamental result of Marciszewski and Pol states that if $X$ is a linearly orderable separable compact space or a separable dyadic compact space, then $C_p(X)$ has a $\sigma$-discrete network (Problems 484 and 485).

## 1.1   Equivalences that arise from homeomorphisms of $C_p(X)$

All spaces are assumed to be Tychonoff. Given a space $X$, the family $\tau(X)$ is its topology, $\tau^*(X) = \tau(X)\backslash\{\emptyset\}$ and $\tau(x, X) = \{U \in \tau(X) : x \in U\}$ for any $x \in X$. The expression $X \hookrightarrow Y$ says that the space $X$ can be embedded in the space $Y$. If we write $X \simeq Y$, this means that $X$ is homeomorphic to $Y$. The spaces $X$ and $Y$ are called *t-equivalent (which is denoted by $X \overset{t}{\sim} Y$), if $C_p(X) \simeq C_p(Y)$.* A topological property $\mathcal{P}$ (a cardinal invariant $\varphi$) is called *t-invariant* if it is preserved by $t$-equivalence, i.e., if $X \vdash \mathcal{P}$ (or $\varphi(X) \leq \kappa$) and $X \overset{t}{\sim} Y$ imply $Y \vdash \mathcal{P}$ (or $\varphi(Y) \leq \kappa$ respectively). A subspace $Y$ of a space $X$ is *t-embedded (l-embedded) in $X$,* if there is a (linear) continuous map $\varphi : C_p(Y) \to C_p(X)$ such that $\varphi(f)|Y = f$ for any $f \in C_p(Y)$. If $X$ is a space, then $\mathbf{0}_X \in C_p(X)$ is the function equal to zero at all points of $X$. The space $\mathbb{I}$ is the set $[-1, 1]$ with the topology inherited from $\mathbb{R}$. A set $F \subset C_p(X, \mathbb{I})$ is called *D-separating,* if $\mathbf{0}_X \in F$ and, given $\varepsilon > 0$, a finite $A \subset X$ and a closed $P \subset X\backslash A$, there is $f \in F$ such that $f(A) \subset (-\varepsilon, \varepsilon)$ and $f(P) \subset [\frac{3}{4}, 1]$. Now, if $F \subset C_p(X, \mathbb{I})$ is any set which contains $\mathbf{0}_X$, let $Z_F(X) = \{\varphi : F \to \mathbb{I} : \varphi(\mathbf{0}_X) = 0$ and $\varphi(V) \subset [-\frac{1}{2}, \frac{1}{2}]$ for some $V \in \tau(\mathbf{0}_X, F)\}$.

Given a cardinal invariant $\varphi$, we define $\varphi^*(X) = \sup\{\varphi(X^n) : n \in \mathbb{N}\}$. Call a class of spaces $\mathcal{C}$ *complete* if $\mathcal{C}$ is invariant under finite products, countable unions, closed subspaces, and continuous images. *Only in this Section, we denote by $\mathcal{K}(X)$* the smallest complete class which contains the space $X$ and all compact spaces.

Any homeomorphism $f : X \to X$ is called *autohomeomorphism.* A trivial example of an autohomeomorphism is the map $\mathrm{id}_X : X \to X$ defined by $\mathrm{id}_X(x) = x$ for any $x \in X$. Given maps $f, g \in C(X, Y)$, we say that $f$ and $g$ are *homotopic* if there exists a continuous map $F : X \times [0, 1] \to Y$ such that $F(x, 0) = f(x)$ and $F(x, 1) = g(x)$ for any $x \in X$. The map $F$ is called *the homotopy which connects the mappings $f$ and $g$.* If $F : X \times [0, 1] \to Y$ is a homotopy, it generates the family $\{F_t : t \in [0, 1]\}$ of functions defined by $F_t(x) = F(x, t)$ for any $x \in X$ and $t \in [0, 1]$. By $\exp(X)$ we denote the family of all subsets of a set $X$. If $\kappa$ is a cardinal which can be finite, we will need the families $[X]^\kappa = \{A \subset X : |A| = \kappa\}$ and $[X]^{\leq \kappa} = \{A \subset X : |A| \leq \kappa\}$ together with $[X]^{<\kappa} = \{A \subset X : |A| < \kappa\}$. In particular, $[X]^{<\omega}$ is the family of all finite subsets of $X$.

If $X$ is a space denote by $\mathcal{F}(X)$ the family of all closed subsets of $X$. For every open subset $U$ of the space $X$ consider the families $P(U) = \{F \in \mathcal{F}(X) : F \subset U\}$ and $Q(U) = \{F \in \mathcal{F}(X) : F \cap U \neq \emptyset\}$; let $\mathcal{S} = \bigcup\{P(U) \cup Q(U) : U \in \tau(X)\}$. The topology $\tau_V(X)$ generated by $\mathcal{S}$ as a subbase is called *the Vietoris topology on the family of all closed subset of $X$.* Let $\mathcal{CL}(X) = (\mathcal{F}(X), \tau_V(X))$; if $\mathcal{G} \subset \mathcal{F}(X)$, by *the Vietoris topology on $\mathcal{G}$* we mean the topology on $\mathcal{G}$ inherited from $\mathcal{CL}(X)$.

Given sets $X$ and $Y$, we say that $p$ is *a multi-valued map from $X$ to $Y$* if $p : X \to \exp(Y)$. The map $p$ is called *finite-valued* if $p(X) \subset [Y]^{<\omega}$ and *single-valued* if $|p(x)| = 1$ for any $x \in X$. Any single-valued map $p : X \to \exp(Y)$ is identified with the map $p' : X \to Y$ such that $p(x) = \{p'(x)\}$ for any $x \in X$. Now, if $X$ and $Y$ are spaces, a map $p : X \to \exp(Y)$ is called *lower semicontinuous* if $p_l^{-1}(U) = \{x \in X : p(x) \cap U \neq \emptyset\}$ is open in $X$ for any open $U \subset Y$. The map $p$ is called *upper semicontinuous* if $p_u^{-1}(U) = \{x \in X : p(x) \subset U\}$ is open in $X$ for any open $U \subset Y$. Given two maps $p, q : X \to \exp(Y)$, we say that $q$ *is lower semicontinuous with respect to* $p$ if, for any $x \in X$, we have $p(x) \subset q(x)$ and $q_l^{-1}(U)$ is a neighborhood of $p_l^{-1}(U)$ for any open $U \subset Y$. A finite-valued map $p : X \to \exp(Y)$ is called *almost lower semicontinuous* if there exists a space $Z \supset Y$ and a finite-valued map $q : X \to \exp(Z)$ such that $q$ is lower semicontinuous with respect to $p$. We say that a single-valued map $p : X \to Y$ *has defect* $\leq n$ if there exists a space $Z \supset Y$ and a map $q : X \to [Z]^{\leq n+1}$ which is lower semicontinuous with respect to $p$. The map $p$ *has finite defect* if it has defect $\leq n$ for some $n \in \omega$.

Given a natural number $n$ and set-valued maps $p_i : X_i \to \exp(Y_i)$ for every $i < n$ let $(\prod_{i<n} p_i)(x_0, \ldots, x_{n-1}) = p_0(x_0) \times \ldots \times p_{n-1}(x_{n-1})$ for any point $(x_0, \ldots, x_{n-1}) \in \prod_{i<n} X_i$. Thus, $\prod_{i<n} p_i : \prod_{i<n} X_i \to \exp(\prod_{i<n} Y_i)$.

Let $\Sigma_*(A) = \{x \in \mathbb{R}^A :$ for any $\varepsilon > 0$, the set $\{a \in A : |x(a)| \geq \varepsilon\}$ is finite$\}$. Given $x \in \Sigma_*(A)$, let $\|x\| = \sup\{|x(a)| : a \in A\}$. The space $\mathbb{D}$ is the two-point set $\{0, 1\}$ endowed with the discrete topology. The symbol $\mathbb{P}$ denotes the irrational numbers with the usual topology. If $X$ is a space and $A, B \subset X$ are disjoint closed subsets of $X$, we say that a (closed) set $P \subset X$ is *a partition between $A$ and $B$* if there are open $U, V \subset X$ such that $A \subset U$, $B \subset V$, $U \cap V = \emptyset$ and $U \cup V = X \setminus P$.

For every $n \in \omega$, the set $S^n = \{x \in \mathbb{R}^{n+1} : (x(0))^2 + \ldots + (x(n))^2 = 1\}$ with the topology induced from $\mathbb{R}^{n+1}$ is called *the $n$-dimensional sphere*. A set $F \subset \mathbb{I}^n$ is *a face of the cube* $\mathbb{I}^n$ if $F = \{x \in \mathbb{I}^n : x(i) = 1\}$ or $F = \{x \in \mathbb{I}^n : x(i) = -1\}$ for some $i < n$. A set $\{a_0, \ldots, a_m\} \subset \mathbb{R}^n$ is called *independent* if, for any $\lambda_0, \ldots, \lambda_m \in \mathbb{R}$ such that $\lambda_0 + \ldots + \lambda_m = 0$ and $\lambda_0 \cdot a_0 + \ldots + \lambda_m \cdot a_m = \mathbf{0}$ we have $\lambda_i = 0$ for all $i \leq m$; here $\mathbf{0}$ is the vector of $\mathbb{R}^n$ whose all coordinates are zeros. If $\{a_0, \ldots, a_m\} \subset \mathbb{R}^n$ is an independent set, let $[a_0, \ldots, a_m] = \{\lambda_0 \cdot a_0 + \ldots + \lambda_m \cdot a_m : \lambda_i \geq 0$ for all $i \leq m$ and $\lambda_0 + \ldots + \lambda_m = 1\}$. The set $[a_0, \ldots, a_m]$ is called *an $m$-dimensional simplex with the vertices $a_0, \ldots, a_m$*.

It is easy to see that if $x \in [a_0, \ldots, a_m]$ then there are unique $\lambda_0(x), \ldots, \lambda_m(x)$ such that $x = \lambda_0(x) \cdot a_0 + \ldots + \lambda_m(x) \cdot a_m$, where $\lambda_i(x) \geq 0$ for all $i \leq m$ and $\lambda_0(x) + \ldots, + \lambda_m(x) = 1$. The numbers $\lambda_0(x), \ldots, \lambda_m(x)$ are called *the barycentric coordinates of the point $x$*. *A face of the simplex* $[a_0, \ldots, a_m]$ is any simplex $[a_{i_0}, \ldots, a_{i_k}]$ for which $a_{i_j} \neq a_{i_l}$ if $j \neq l$. *The barycenter of a simplex* $S = [a_0, \ldots, a_m]$ is the point $b(S) = \frac{1}{m+1} a_0 + \ldots + \frac{1}{m+1} a_m$. *A simplicial subdivision of a simplex $S$* is a finite family $\mathcal{P} = \{S_i : 0 \leq i < k\}$ of simplexes such that $\bigcup \mathcal{P} = S$, for any $i, j < k$, all faces of $S_i$ belong to $\mathcal{P}$ and $S_i \cap S_j$ is either empty or is a common face of the simplexes $S_i$ and $S_j$. *The mesh of a simplicial subdivision $\mathcal{P}$ of a simplex $S$* is the number $\delta(\mathcal{P}) = \max\{\mathrm{diam}(T) : T \in \mathcal{P}\}$.

For any $\xi \in \omega_1$ the class $\mathcal{M}_\xi$ (or $\mathcal{A}_\xi$) of absolute Borel sets of multiplicative (additive) class $\xi$ consists of metrizable spaces $X$ such that for any embedding of $X$ into a metrizable space $Y$ the set $X$ belongs to $\Pi^0_\xi(Y)$ (or to $\Sigma^0_\xi(Y)$ respectively).

**001.** Prove that cardinality, network weight, $i$-weight, as well as density are $t$-invariant.

**002.** Prove that if $C_p(Y) \hookrightarrow C_p(X)$ then $nw(Y) \leq nw(X)$, $d(Y) \leq d(X)$ and $|Y| \leq |X|$. Give an example showing that the inequality $iw(Y) \leq iw(X)$ is not necessarily true.

**003.** Prove that $p(Y) \leq p(X)$ whenever $C_p(Y) \hookrightarrow C_p(X)$. As a consequence, point-finite cellularity is $t$-invariant.

**004.** Suppose that $X$ and $Y$ are $t$-equivalent Baire spaces. Prove that $c(X) = c(Y)$. In particular, the Souslin numbers of $t$-equivalent pseudocompact spaces coincide.

**005.** Let $\kappa$ be a caliber of $X$. Knowing that $C_p(Y)$ embeds in $C_p(X)$, prove that $\kappa$ is a caliber of $Y$. In particular, calibers are $t$-invariant.

**006.** Suppose that $C_p(Y)$ embeds into $C_p(X)$. Prove that $l^*(Y) \leq l^*(X)$. As a consequence, $l^*$ is $t$-invariant.

**007.** Suppose that $C_p(Y)$ embeds into $C_p(X)$. Prove that $\varphi(Y) \leq \varphi(X)$ for any $\varphi \in \{hl^*, hd^*, s^*\}$ and hence $\varphi$ is $t$-invariant.

**008.** Prove that, if $X \overset{t}{\sim} Y$ then $t_m(X) = t_m(Y)$. Give an example of spaces $X$ and $Y$ such that $C_p(Y)$ embeds into $C_p(X)$ and $t_m(Y) > t_m(X)$.

**009.** Prove that $X \overset{t}{\sim} Y$ implies $q(X) = q(Y)$. In particular, realcompactness is $t$-invariant.

**010.** Give an example of spaces $X$ and $Y$ such that $C_p(Y)$ embeds into $C_p(X)$, the space $X$ is realcompact and $Y$ is not realcompact.

**011.** Suppose that $X$ is a $P$-space and $X \overset{t}{\sim} Y$. Prove that $Y$ is also a $P$-space.

**012.** Prove that discreteness is $t$-invariant.

**013.** Suppose that $X$ and $Y$ are compact spaces such that $C_p(Y) \hookrightarrow C_p(X)$. Prove that $Y$ is scattered whenever $X$ is scattered. In particular, if $X \overset{t}{\sim} Y$ then $X$ is scattered if and only if so is $Y$.

**014.** Suppose that $X^n$ is a Hurewicz space for each $n \in \mathbb{N}$ and $X \overset{t}{\sim} Y$. Prove that $Y^n$ is also a Hurewicz space for each $n \in \mathbb{N}$.

**015.** Suppose that $X \overset{t}{\sim} Y$ and $X$ is a $\sigma$-compact space with a countable network. Prove that $Y$ is also $\sigma$-compact. As a consequence, if $X$ is a metrizable compact space and $X \overset{t}{\sim} Y$ then $Y$ is $\sigma$-compact.

**016.** Given an arbitrary number $\varepsilon > 0$ prove that there exists a homeomorphism $u : \mathbb{R} \times \Sigma_*(\omega) \to \Sigma_*(\omega)$ for which we have the inequality $|\, \|u(r, x)\| - \|x\| \,| \leq \varepsilon$ for any $r \in \mathbb{R}$ and $x \in \Sigma_*(\omega)$.

**017.** Prove that $\Sigma_*(\omega)$ is homeomorphic to $\mathbb{R}^\omega \times \Sigma_*(\omega)$.

**018.** Suppose that $X$ is a pseudocompact space. Given any function $f \in C_p(X)$, let $\|f\| = \sup\{|f(x)| : x \in X\}$. Prove that $C_*(X) \simeq C_*(X) \times (C_p(X))^\omega$, where $C_*(X) = \{\varphi \in (C_p(X))^\omega : \|\varphi(n)\| \to 0\}$.

**019.** Let $X$ be a pseudocompact space. As usual, for any $f \in C_p(X)$, we define $\|f\| = \sup\{|f(x)| : x \in X\}$. Supposing that the space $C_p(X)$ is homeomorphic to $C_*(X) = \{\varphi \in (C_p(X))^\omega : \|\varphi(n)\| \to 0\}$, prove that $C_p(X) \simeq (C_p(X))^\omega$.

**020.** Prove that $\Sigma_*(\omega)$ is homeomorphic to $(\Sigma_*(\omega))^\omega$.

**021.** Prove that, for every infinite space $X$, the space $\mathbb{R}^\omega$ embeds into $C_p(X)$ as a closed subspace.

**022.** Prove that a space $X$ is not pseudocompact if and only if $\mathbb{R}^\omega$ embeds in $C_p(X)$ as a linear subspace.

**023.** Prove that, if either $X$ or $C_p(X)$ is Lindelöf, then $\mathbb{R}^{\omega_1}$ does not embed into $C_p(X)$.

**024.** Prove that there exists a space $X$ such that $c(X) = \omega$ and $\mathbb{R}^{\omega_1}$ embeds in $C_p(X)$ as a closed linear subspace.

**025.** Prove that if $\omega + 1 \hookrightarrow X$ then $C_p(X) \simeq C_p(X) \times \mathbb{R}^\omega$. Deduce from this fact that pseudocompactness, countable compactness, and compactness are not $t$-invariant.

**026.** Prove that $C_p(\mathbb{R})^\omega \simeq C_p(\mathbb{R})$ and $C_p(\mathbb{I})^\omega \simeq C_p(\mathbb{I})$.

**027.** Prove that $\mathbb{R}$ is $t$-equivalent to $[0, 1]$.

**028.** Prove that $\upsilon X \overset{t}{\sim} \upsilon Y$ whenever $X \overset{t}{\sim} Y$. Give an example which shows that $X \overset{t}{\sim} Y$ does not necessarily imply $\beta X \overset{t}{\sim} \beta Y$.

**029.** Give an example of spaces $X$ and $Y$ such that $\upsilon X \simeq \upsilon Y$ (and hence $\upsilon X \overset{t}{\sim} \upsilon Y$) while $X$ is not $t$-equivalent to $Y$.

**030.** Prove that $\kappa$-monolithity and $\kappa$-stability are $t$-invariant for any infinite cardinal $\kappa$.

**031.** Given spaces $X$ and $Y$ such that $X \overset{t}{\sim} Y$ prove that $X$ is functionally perfect if and only if so is $Y$.

**032.** Give an example of spaces $X$ and $Y$ such that $X$ is functionally perfect, $C_p(Y)$ embeds into $C_p(X)$ while $Y$ is not functionally perfect.

**033.** Suppose that compact spaces $X$ and $Y$ are $t$-equivalent. Prove that $X$ is Eberlein (Corson or Gul'ko) compact if and only if so is $Y$.

**034.** Suppose that $F \subset C_p(X, \mathbb{I})$ is a $D$-separating set (and hence $\mathbf{0}_X \in F$). For each $x \in X$, let $e_x(f) = f(x)$ for any $f \in F$. Prove that $\tilde{X} = \{e_x : x \in X\}$ is a closed subset of the space $Z_F(X) = \{\varphi : F \to \mathbb{I} : \varphi(\mathbf{0}_X) = 0$ and $\varphi(V) \subset [-\frac{1}{2}, \frac{1}{2}]$ for some $V \in \tau(\mathbf{0}_X, F)\}$, and the map $x \to e_x$ is a homeomorphism between $X$ and $\tilde{X}$. In other words, $X$ is canonically homeomorphic to a closed subset of $Z_F(X)$.

**035.** Knowing that $\mathbf{0}_X \in F \subset C_p(X, \mathbb{I})$ and $\mathbf{0}_Y \in G \subset C_p(Y, \mathbb{I})$, suppose that there is an embedding $i : G \to F$ with $i(\mathbf{0}_Y) = \mathbf{0}_X$. Prove that $Z_F(X)$ maps continuously onto $Z_G(Y)$.

**036.** Given a space $X$ prove that if $\mathbf{0}_X \in F \subset C_p(X, \mathbb{I})$ then $Z_F(X)$ belongs to the class $\mathcal{K}(X)$.

**037.** Let $G$ be a $D$-separating subspace of $C_p(Y)$. Prove that, if $G$ embeds into $C_p(X)$ then $Y \in \mathcal{K}(X)$.

**038.** Given spaces $X, Y$ and a subspace $Z \subset Y$ suppose that the space $C_p(Z|Y) = \{f \in C_p(Z) : f = g|Z$ for some $g \in C_p(Y)\}$ embeds in $C_p(X)$. Prove that $Z \in \mathcal{K}(X)$.

**039.** Let $X$ be a $\sigma$-compact space. Prove that any space $Y \in \mathcal{K}(X)$ is also $\sigma$-compact.

**040.** Let $X$ be a Lindelöf $\Sigma$-space. Prove that any $Y \in \mathcal{K}(X)$ is also a Lindelöf $\Sigma$-space.

**041.** Let $X$ be a $K$-analytic space. Prove that any $Y \in \mathcal{K}(X)$ is also a $K$-analytic space.

**042.** Prove that $ext^*(Y) \leq ext^*(X)$ for any $Y \in \mathcal{K}(X)$.

**043.** Suppose that $C_p(Y)$ embeds into $C_p(X)$. Prove that

(1) if $X$ is $\sigma$-compact then $Y$ is $\sigma$-compact.
(2) if $X$ is Lindelöf $\Sigma$-space then $Y$ is Lindelöf $\Sigma$.
(3) if $X$ is $K$-analytic then $Y$ is also $K$-analytic.
(4) $ext^*(Y) \leq ext^*(X)$.

As a consequence, if $X \overset{t}{\sim} Y$ then $ext^*(X) = ext^*(Y)$ and, for any property $\mathcal{P} \in \{\sigma$-compactness, Lindelöf $\Sigma$-property, $K$-analyticity$\}$, we have $X \vdash \mathcal{P}$ if and only if $Y \vdash \mathcal{P}$.

**044.** Suppose that $X$ is an analytic space and $C_p(Y)$ embeds into $C_p(X)$. Prove that $Y$ is also analytic. In particular, analyticity is $t$-invariant, i.e., if $X \overset{t}{\sim} Y$ then $X$ is analytic if and only if so is $Y$.

**045.** Suppose that $X \overset{t}{\sim} Y$. Prove that $X$ is $\sigma$-bounded if and only if so is $Y$.

**046.** Given a zero-dimensional space $Y$, suppose that $C_p(Y, \mathbb{D})$ embeds in $C_p(X)$. Prove that, for any property $\mathcal{P} \in \{\sigma$-compactness, Lindelöf $\Sigma$-property, analyticity, $K$-analyticity$\}$, we have $Y \vdash \mathcal{P}$ whenever $X \vdash \mathcal{P}$.

**047.** Let $X$ be a zero-dimensional space. Prove that $l^*(X) = t(C_p(X, \mathbb{D}))$.

**048.** Let $X$ and $Y$ be zero-dimensional spaces with $C_p(X, \mathbb{D}) \simeq C_p(Y, \mathbb{D})$. Prove that $X$ is pseudocompact if and only if so is $Y$. Deduce from this fact that $X$ is compact if and only if so is $Y$.

**049.** Prove that there exist zero-dimensional spaces $X$ and $Y$ such that $C_p(X)$ is homeomorphic to $C_p(Y)$ and $C_p(X, \mathbb{D})$ is not homeomorphic to $C_p(Y, \mathbb{D})$.

**050.** Prove that $\chi(C_p(X, \mathbb{D})) = w(C_p(X, \mathbb{D})) = |X|$ for any zero-dimensional space $X$.

**051.** Prove that a zero-dimensional compact space $X$ is scattered if and only if $C_p(X, \mathbb{D})$ is Fréchet–Urysohn.

**052.** Suppose that $X$ is not $\sigma$-compact and $w(X) \leq \kappa$. Prove that there is a subspace $Z \subset C_p(X)$ such that $|Z| \leq \kappa$ and $Z$ is not embeddable into $C_p(Y)$ for any $\sigma$-compact $Y$. In particular, there is a countable subspace of $C_p(\mathbb{P})$ which cannot be embedded into $C_p(Y)$ for any $\sigma$-compact $Y$.

**053.** Prove that, for any $X$, the space $C_p(X)$ embeds in $C_p(\mathbb{P})$ if and only if $C_p(X)$ is homeomorphic to a linear subspace of $C_p(\mathbb{P})$.

**054.** Suppose that a space $X$ is compact and there exists a homeomorphism $\varphi : \mathbb{R}^X \to \mathbb{R}^Y$ such that $\varphi(C_p(X)) \supset C_p(Y)$. Prove that $Y$ is compact.

**055.** Suppose that $p, q : X \to \exp(Y)$ are finite-valued mappings such that $q$ is lower semicontinuous with respect to $p$. Prove that, for any nonempty set $A \subset X$, the map $q|A : A \to \exp(\bigcup q(A))$ is lower semicontinuous with respect to $p|A : A \to \exp(\bigcup q(A))$.

**056.** Suppose that $p : X \to \exp(Y)$ and $q : X \to \exp(Y)$ are finite-valued mappings such that $q$ is lower semicontinuous with respect to $p$. Given an open set $U \subset Y$, let $p_U(x) = p(x) \cap U$ and $q_U(x) = q(x) \cap U$ for every $x \in X$. Prove that the map $q_U : X \to \exp(U)$ is lower semicontinuous with respect to $p_U : X \to \exp(U)$.

**057.** Given a number $n \in \mathbb{N}$ and spaces $X_i, Y_i$, for every $i < n$ suppose that $p_i, q_i : X_i \to \exp(Y_i)$ are set-valued maps such that $q_i$ is lower semicontinuous with respect to $p_i$. Prove that $\prod_{i<n} q_i$ is lower semicontinuous with respect to $\prod_{i<n} p_i$. As a consequence, any finite product of almost lower semicontinuous maps is an almost lower semicontinuous map.

**058.** Suppose that, for finite-valued mappings $p, r, q : X \to \exp(Y)$, we have $p(x) \subset r(x) \subset q(x)$ for any $x \in X$. Prove that, if $q$ is lower semicontinuous with respect to $r$ then $q$ is lower semicontinuous with respect to $p$.

**059.** Suppose that $X$ is a nonempty space and $p : X \to \exp(Y)$ is an almost lower semicontinuous map such that $\bigcup p(X) = Y$. Prove that $Y$ is a countable union of images of subspaces of $X$ under single-valued almost lower semicontinuous maps of finite defect.

**060.** Given nonempty spaces $X, Z$ and $k \in \mathbb{N}$, suppose that we have maps $p : X \to Z$ and $q : X \to \exp(Z)$ such that $p(x) \in q(x)$ and $|q(x)| \leq k$ for each $x \in X$ while $q(x) \cap q(y) = \emptyset$ if $x \neq y$ and $q$ is lower semicontinuous with respect to $p$. Prove that, if $p(X)$ is discrete then $X = \bigcup_{j<k} D_j$ where every $D_j \subset X$ is discrete and hence there exists a discrete subspace $S \subset X$ such that $|S| = |X|$.

**061.** Given nonempty spaces $X, Z$ and $k \in \mathbb{N}$, suppose that we have maps $p : X \to Z$ and $q : X \to \exp(Z)$ such that $p(x) \in q(x)$ and $|q(x)| \leq k$ for each $x \in X$ while $q(x) \cap q(y) = \emptyset$ if $x \neq y$ and $q$ is lower semicontinuous with respect to $p$. Prove that, if $p(X)$ is left-separated then $X = \bigcup_{j<k} D_j$ where every $D_j \subset X$ is left-separated and hence there exists a left-separated subspace $S \subset X$ such that $|S| = |X|$.

**062.** Given nonempty spaces $X, Z$ and $k \in \mathbb{N}$, suppose that we have maps $p : X \to Z$ and $q : X \to \exp(Z)$ such that $p(x) \in q(x)$ and $|q(x)| \leq k$ for each $x \in X$ while $q(x) \cap q(y) = \emptyset$ if $x \neq y$ and $q$ is lower semicontinuous with respect to $p$. Prove that, if $p(X)$ is right-separated then $X = \bigcup_{j<k} D_j$ where every $D_j \subset X$ is right-separated and hence there exists a right-separated subspace $S \subset X$ such that $|S| = |X|$.

**063.** Suppose that an infinite space $Y$ is an image of a space $X$ under a finite-valued almost lower semicontinuous map, i.e., there exists an almost lower semicontinuous map $p : X \to \exp(Y)$ such that $Y = \bigcup p(X)$. Prove that for each $n \in \mathbb{N}$, we have $s(Y^n) \leq s(X^n)$, $hl(Y^n) \leq hl(X^n)$ and $hd(Y^n) \leq hd(X^n)$.

**064.** Let $h : C_p(Y) \to C_p(X)$ be an embedding such that $h(0_Y) = 0_X$. Given a point $x \in X$ and $\varepsilon > 0$, a point $y \in Y$ is called $\varepsilon$-inessential for $x$ if there is $U \in \tau(y, Y)$ such that $|h(g)(x)| \leq \varepsilon$ whenever $g(Y \setminus U) \subset \{0\}$. The point $y$ is $\varepsilon$-essential for $x$ if it is not $\varepsilon$-inessential for $x$. Denote by $\text{supp}_\varepsilon(x)$ the set of all points which are $\varepsilon$-essential for $x$. Prove that $\text{supp}_\varepsilon(x)$ is finite for any $x \in X$ and $\varepsilon > 0$.

**065.** Let $h : C_p(Y) \to C_p(X)$ be an embedding such that $h(\mathbf{0}_Y) = \mathbf{0}_X$. Denote by $\text{supp}_\varepsilon(x)$ the set of all points which are $\varepsilon$-essential for $x$. Prove that $Y = \bigcup\{\text{supp}_{1/n}(x) : x \in X, \ n \in \mathbb{N}\}$.

**066.** Let $h : C_p(Y) \to C_p(X)$ be an embedding such that $h(\mathbf{0}_Y) = \mathbf{0}_X$. Denote by $\text{supp}_\varepsilon(x)$ the set of all points which are $\varepsilon$-essential for $x$. Prove that, for any $\varepsilon > 0$, the finite-valued map $\text{supp}_\varepsilon : X \to \exp(Y)$ is almost lower semicontinuous.

**067.** Suppose that $C_p(Y)$ embeds in $C_p(X)$. Prove that $Y$ is a countable union of images of $X$ under finite-valued almost lower semicontinuous maps.

**068.** Suppose that $C_p(Y)$ embeds in $C_p(X)$. Prove that, for any $n \in \mathbb{N}$, we have $s(Y^n) \leq s(X^n)$. In particular, if $X$ and $Y$ are $t$-equivalent then $s(Y^n) = s(X^n)$ for any $n \in \mathbb{N}$.

**069.** Suppose that $C_p(Y)$ embeds in $C_p(X)$. Prove that, for any $n \in \mathbb{N}$, we have $hd(Y^n) \leq hd(X^n)$. In particular, if the spaces $X$ and $Y$ are $t$-equivalent then $hd(Y^n) = hd(X^n)$ for any $n \in \mathbb{N}$.

**070.** Suppose that $C_p(Y)$ embeds in $C_p(X)$. Prove that, for any $n \in \mathbb{N}$, we have $hl(Y^n) \leq hl(X^n)$. In particular, if the spaces $X$ and $Y$ are $t$-equivalent then $hl(Y^n) = hl(X^n)$ for any $n \in \mathbb{N}$.

**071.** Suppose that $f : X \to Y$ is a closed continuous onto map such that $t(Y) \leq \kappa$ and $t(f^{-1}(y)) \leq \kappa$ for any $y \in Y$. Prove that $t(X) \leq \kappa$. Deduce from this fact that, for every infinite compact space $X$ we have $t(X^n) = t(X)$ for any $n \in \mathbb{N}$.

**072.** Given countably compact sequential spaces $X_1, \ldots, X_m$ prove that the space $X_1 \times \ldots \times X_m$ is countably compact and sequential.

**073.** Assuming that a space $X$ maps continuously onto a compact space $K$ prove that $t(K) \leq l(X) \cdot t(X)$.

**074.** Suppose that $K$ is a compact space and $K = \bigcup_{n \in \omega} K_n$ where every $K_n$ is a sequential closed subspace of $K$. Prove that if either Martin's Axiom or Luzin's Axiom ($2^{\omega_1} > 2^\omega$) holds then the space $K$ is sequential.

**075.** Suppose that we have a space $X$, a compact space $K$ and a compact-valued upper semicontinuous map $p : X \to \exp(K)$ such that $\bigcup p(X) = K$. Knowing that $l(X) \cdot t(X) \leq \kappa$ and $t(p(x)) \leq \kappa$ for any $x \in X$ prove that $t(K) \leq \kappa$.

**076.** Suppose that $K$ is a compact sequential space and $L$ is a compact space for which there exists a finite-valued upper semicontinuous map $p : K \to \exp(L)$ such that $\bigcup p(K) = L$. Prove that $L$ is also sequential.

**077.** Suppose that there exists an open continuous map of a subspace of $C_p(X)$ onto $C_p(Y)$. Prove that, for any $n \in \mathbb{N}$ there is a finite-valued upper semicontinuous map $\varphi_n : X^n \to \exp(Y)$ such that $\bigcup\{\bigcup \varphi_n(X^n) : n \in \mathbb{N}\} = Y$.

**078.** Suppose that there is an open continuous map of a subspace of $C_p(X)$ onto $C_p(Y)$. Prove that $t(K) \leq t^*(X) \cdot l^*(X)$ for any compact $K \subset Y$. Deduce from this fact that if $K$ and $L$ are $t$-equivalent compact spaces then $t(K) = t(L)$.

**079.** Suppose that $X$ is a compact sequential space and $Y$ is a compact space such that there is an open continuous map of some subspace of $C_p(X)$ onto $C_p(Y)$. Prove that $Y$ is a countable union of its compact sequential subspaces. As a consequence, under Martin's Axiom if $K$ and $L$ are compact $t$-equivalent spaces and $K$ is sequential then $L$ is also sequential.

**080.** Let $X$ and $Y$ be metrizable $t$-equivalent spaces. Prove that we have $Y = \bigcup\{Y_n : n \in \omega\}$, where each $Y_n$ is a $G_\delta$-subspace of $Y$, homeomorphic to some $G_\delta$-subspace of $X$.

**081.** Let $X$ and $Y$ be metrizable spaces such that $C_p^*(X) \simeq C_p^*(Y)$. Prove that $Y = \bigcup\{Y_n : n \in \omega\}$, where each $Y_n$ is a $G_\delta$-subspace of $Y$, homeomorphic to some $G_\delta$-subspace of $X$.

**082.** Let $X$ and $Y$ be metrizable $t$-equivalent spaces. Prove that $X$ is a countable union of zero-dimensional subspaces if and only if so is $Y$.

**083.** Let $X$ and $Y$ be metrizable $t$-equivalent spaces. Prove that $X$ is a countable union of its Čech-complete subspaces if and only if so is $Y$.

**084.** Suppose that $\{a_0, \dots, a_m\} \subset \mathbb{R}^n$ is an independent set. Prove that

   (i)   the simplex $S = [a_0, \dots, a_m]$ is a compact subset of $\mathbb{R}^n$;
   (ii)  any two $m$-dimensional simplexes are homeomorphic;
   (iii) the barycentric coordinates are continuous functions from $S$ to $[0, 1]$.

**085.** Let $S$ be a simplex in $\mathbb{R}^n$ and suppose that $S_0 \supset S_1 \supset \dots \supset S_k$ are distinct faces of $S$. Prove that the points $\{b(S_0), \dots, b(S_k)\}$ are independent. Here $b(S_i)$ is the barycenter of the simplex $S_i$ for all $i \leq k$.

**086.** For a simplex $S = [a_0, \dots, a_m]$, consider the family $\mathcal{B}(S)$ of all simplexes of the form $[b(S_0), \dots, b(S_k)]$, where $S_0 \supset S_1 \supset \dots \supset S_k$ are distinct faces of $S$. Prove that $\mathcal{P}$ is a simplicial subdivision of $S$ such that any $(m - 1)$-dimensional simplex $T \in \mathcal{B}(S)$ is a face of one or two $m$-dimensional members of $\mathcal{B}(S)$ depending on whether $T$ is contained in an $(m - 1)$-dimensional face of $S$. The subdivision $\mathcal{B}(S)$ is called *the barycentric subdivision of the simplex $S$*.

**087.** Given a simplex $S$, let $\mathcal{B}_1(S)$ be the barycentric subdivision $\mathcal{B}(S)$ of the simplex $S$. If we have a simplicial subdivision $\mathcal{B}_n(S)$ of the simplex $S$, let $\mathcal{B}_{n+1}(S) = \bigcup\{\mathcal{B}(T) : T \in \mathcal{B}_n(S)\}$. The family $\mathcal{B}_n(S)$ is called *the n-th barycentric subdivision of the simplex $S$*. Prove that, for any simplex $S$ and any $\varepsilon > 0$, there exists a natural number $n$ such that the mesh of the $n$-th barycentric subdivision of the simplex $S$ is less than $\varepsilon$.

**088.** (Sperner's lemma). Given a number $l \in \mathbb{N}$ and an $m$-dimensional simplex $[a_0, \dots, a_m]$ let $V$ be the set of all vertices of simplexes in $\mathcal{B}_l([a_0, \dots, a_m])$. Suppose that, for a function $h : V \to \{0, 1, \dots, m\}$, we have $h(v) \in \{i_0, \dots, i_k\}$ whenever $v \in [a_{i_0}, \dots, a_{i_k}]$. Prove that the family of simplexes in $\mathcal{B}_l([a_0, \dots, a_m])$, on vertices of which $h$ takes all values from 0 to $m$, has an odd cardinality.

**089.** (Brower's fixed-point theorem). Prove that, for any $n \in \mathbb{N}$, if $S$ is an $n$-dimensional simplex and $f : S \to S$ is a continuous function then there exists a point $x \in S$ such that $f(x) = x$.

**090.** Prove that, for any $n \in \mathbb{N}$, there is no retraction of the cube $\mathbb{I}^n$ onto its boundary $\partial \mathbb{I}^n = \{x \in \mathbb{I}^n : |x(i)| = 1 \text{ for some } i < n\}$.

**091.** Given spaces $X$ and $Y$ and functions $f, g \in C(X, Y)$, let $f \sim g$ denote the fact that $f$ and $g$ are homotopic. Prove that $\sim$ is an equivalence relation on $C(X, Y)$.

**092.** Given a space $X$, let $f, g : X \to \partial \mathbb{I}^n$ be continuous maps such that the points $f(x)$ and $g(x)$ belong to the same face of $\mathbb{I}^n$ for any $x \in X$. Prove that $f$ and $g$ are homotopic.

**093.** (Mushroom lemma). Let $X$ be a normal countably paracompact space. Suppose that $F \subset X$ is closed and we have continuous homotopic mappings $f_0, f_1$ of $F$ to the $n$-dimensional sphere $S^n = \{x \in \mathbb{R}^{n+1} : x(0)^2 + \ldots + x(n)^2 = 1\}$. Prove that, if there exists a continuous map $g_0 : X \to S^n$ with $g_0|F = f_0$ then there is a continuous map $g_1 : X \to S^n$ such that $g_1|F = f_1$ and $g_1$ is homotopic to $g_0$.

**094.** For each $i < n$, consider the faces $F_i = \{x \in \mathbb{I}^n : x(i) = -1\}$ and $G_i = \{x \in \mathbb{I}^n : x(i) = 1\}$ of the $n$-dimensional cube $\mathbb{I}^n$. Prove that, if $C_i$ is a partition between $F_i$ and $G_i$ then $\bigcap\{C_i : i < n\} \neq \emptyset$.

**095.** For each $i \in \omega$, consider the subsets $F_i = \{x \in \mathbb{I}^\omega : x(i) = -1\}$ and $G_i = \{x \in \mathbb{I}^\omega : x(i) = 1\}$ of the cube $\mathbb{I}^\omega$. Prove that, if $C_i$ is any partition between $F_i$ and $G_i$ then $\bigcap\{C_i : i \in \omega\} \neq \emptyset$.

**096.** Prove that, for any $n \in \mathbb{N}$, the space $\mathbb{I}^n$ is the finite union of its zero-dimensional subspaces.

**097.** Prove that, for any $n \in \mathbb{N}$, the space $\mathbb{I}^n$ cannot be represented as the union of $\leq n$-many of its zero-dimensional subspaces.

**098.** Prove that the cube $\mathbb{I}^\omega$ cannot be represented as the countable union of its zero-dimensional subspaces. Prove (in ZFC) that there exist zero-dimensional spaces $\{X_\alpha : \alpha < \omega_1\}$ such that $\mathbb{I}^\omega = \bigcup\{X_\alpha : \alpha < \omega_1\}$.

**099.** Prove that, for any $n \in \mathbb{N}$, the spaces $\mathbb{I}^n$ and $\mathbb{I}^\omega$ are not $t$-equivalent.

**100.** Suppose that $X$ is one of the spaces $\omega_1$ or $\omega_1 + 1$. Prove that, for any distinct $m, n \in \mathbb{N}$, the spaces $(C_p(X))^n$ and $(C_p(X))^m$ are not homeomorphic. In particular, $X$ is not $t$-equivalent to $X \oplus X$.

## 1.2 Uniformities, Dimension, and u-Equivalence

Given a set $X$ and $A \subset X^2$, let $A^{-1} = \{(x, y) \in X \times X : (y, x) \in A\}$. The set $A$ is *symmetric* if $A^{-1} = A$. If $A, B \subset X \times X$, then $A \circ B = \{(x, y) \in X^2 :$ there exists a point $z \in X$ such that $(x, z) \in B$ and $(z, y) \in A\}$. Here, as usual, $\Delta = \Delta_X = \{(x, x) : x \in X\}$. For any point $x \in X$ and $U \subset X^2$ let $U(x) = \{y \in X : (x, y) \in U\}$; if $A \subset X$ and $U \subset X^2$, then $U(A) = \bigcup\{U(x) : x \in A\}$. A nonempty family $\mathcal{U} \subset \exp(X \times X)$ is *a uniformity* on a set $X$ if it has the following properties:

(U1) $\bigcap \mathcal{U} = \Delta$, besides, if $U, V \in \mathcal{U}$ then $U^{-1} \in \mathcal{U}$ and $U \cap V \in \mathcal{U}$;

(U2) if $U \in \mathcal{U}$ then $W \in \mathcal{U}$ whenever $U \subset W$; besides, $V \circ V \subset U$ for some $V \in \mathcal{U}$.

The pair $(X, \mathcal{U})$ is called *a uniform space*. A family $\mathcal{B} \subset \mathcal{U}$ is called *a base of the uniformity* $\mathcal{U}$ if, for any $U \in \mathcal{U}$, there is $V \in \mathcal{B}$ such that $V \subset U$. A family $\mathcal{S} \subset \mathcal{U}$ is called *a subbase of the uniformity* $\mathcal{U}$ if the family of all finite intersections of the elements of $\mathcal{S}$ is a base for $\mathcal{U}$. As usual, if $f : X \to Y$ then $f \times f : X \times X \to Y \times Y$ is defined by $(f \times f)(x_1, x_2) = (f(x_1), f(x_2))$.

If $(X, \mathcal{U})$ is a uniform space, let $\tau_\mathcal{U} = \{\emptyset\} \cup \{O \subset X : \text{for any } x \in O \text{ there}$ is $U \in \mathcal{U}$ such that $U(x) \subset O\}$. The family $\tau_\mathcal{U}$ is a topology; it is called *the topology generated by the uniformity* $\mathcal{U}$. If a topological notion (closure, interior, base, continuity, etc.) is mentioned for a uniform space $(X, \mathcal{U})$, the topology in question is always $\tau_\mathcal{U}$. Given uniform spaces $(X, \mathcal{U})$ and $(Y, \mathcal{V})$, a map $f : X \to Y$ is called *uniformly continuous with respect to* $\mathcal{U}$ *and* $\mathcal{V}$ if, for any $V \in \mathcal{V}$, we have $U = (f \times f)^{-1}(V) = \{(x, y) \in X^2 : (f(x), f(y)) \in V\} \in \mathcal{U}$. If the uniformities $\mathcal{U}$ and $\mathcal{V}$ are clear, we will just say that $f$ is uniformly continuous. The mapping $f$ is called *a uniform isomorphism* (or *uniform homeomorphism)* if it is a bijection and both maps $f$ and $f^{-1}$ are uniformly continuous. If there exists a uniform isomorphism $f : X \to Y$, the spaces $(X, \mathcal{U})$ and $(Y, \mathcal{V})$ are called *uniformly isomorphic* (or *uniformly homeomorphic*). It is impossible to distinguish two uniformly isomorphic uniform spaces because their properties, which can be expressed in terms of their uniformities, are identical.

An arbitrary function $d : X \times X \to \mathbb{R}$ is called *a pseudometric on the set* $X$ if $d(x, x) = 0$, $d(x, y) \geq 0$ for any $x, y \in X$ and $d(x, z) \leq d(x, y) + d(y, z)$ for any $x, y, z \in X$. If $(X, \mathcal{U})$ is a uniform space and $Y \subset X$, then the family $\mathcal{U}_Y^X = \{U \cap (Y \times Y) : U \in \mathcal{U}\}$ is a uniformity on $Y$; the pair $(Y, \mathcal{U}_Y^X)$ is called *a uniform subspace of the uniform space* $(X, \mathcal{U})$ and $\mathcal{U}_Y^X$ is *the uniformity induced on* $Y$ *from* $X$. If $X$ is a set, $(Y, \mathcal{V})$ is a uniform space and $f : X \to Y$ then $f^{-1}(\mathcal{V}) = \{(f \times f)^{-1}(V) : V \in \mathcal{V}\}$. If we have a family $\{(X_t, \mathcal{U}_t) : t \in T\}$ of uniform spaces then $\bigcup\{p_t^{-1}(\mathcal{U}_t) : t \in T\}$ is a subbase of a (uniquely defined) uniformity $\mathcal{U}$ on the set $X = \prod\{X_t : t \in T\}$ (check it, please). The uniform space $(X, \mathcal{U})$ is called *the uniform product* of the uniform spaces $\{(X_t, \mathcal{U}_t) : t \in T\}$. Given a uniform space $(X, \mathcal{U})$, a family $\mathcal{F} \subset \exp(X)$ is *Cauchy* if, for any $U \in \mathcal{U}$, we have $F \times F \subset U$ for some $F \in \mathcal{F}$. The uniform space $(X, \mathcal{U})$ is called *complete* if, for any Cauchy filter $\mathcal{F}$ on $X$, we have $\bigcap\{\overline{A} : A \in \mathcal{F}\} \neq \emptyset$. A uniform space $(X, \mathcal{U})$ is called *totally bounded* if, for any $U \in \mathcal{U}$, there is a finite $P \subset X$ such that $U(P) = X$.

Given a linear topological space $L$ let $G(U) = \{(x, y) \in L \times L : x - y \in U\}$ for any $U \in \tau(\mathbf{0}, L)$ where $\mathbf{0}$ is the zero vector of $L$. Then $\{G(U) : U \in \tau(\mathbf{0}, L)\}$ is a base for the *linear uniformity* on the set $L$. From now on, if a uniformity notion is used in a linear topological space $L$ (in particular, when $L = \mathbb{R}$ or $L = C_p(X)$), then the relevant uniformity is assumed to be linear. If $X$ and $Y$ are Tychonoff spaces, they are called *u-equivalent (which is denoted by $X \overset{u}{\sim} Y$),* if $C_p(X)$ is uniformly isomorphic to $C_p(Y)$. A set $E \subset C_p(X)$ is *a QS-algebra* for $X$ if, given $x \in X$ and closed $F \subset X$ with $x \notin F$, there is $f \in E$ with $f(x) = 1$ and $f(F) \subset \{0\}$ and, besides, for any $f, g \in E$ and $q \in \mathbb{Q}$, we have $f + g \in E$, $f \cdot g \in E$ and $q \cdot f \in E$.

A set $P \subset X$ is called *functionally closed (open)* if there exists $f \in C(X)$ and a closed (open) $Q \subset \mathbb{R}$ such that $P = f^{-1}(Q)$. A family $\mathcal{C}$ of sets is called closed (open, functionally open, functionally closed) if all elements of $\mathcal{C}$ are closed (open, functionally open, functionally closed). If $\mathcal{U} \subset \exp(X)$, we say that *the order of $\mathcal{U}$ is $\leq k$ (denoting it by $\mathrm{ord}(\mathcal{U}) \leq k$)*, if every $x \in X$ belongs to at most $k$ elements of $\mathcal{U}$. If $\mathcal{A} = \{A_s : s \in S\} \subset \exp(X)$, a family $\{B_s : s \in S\} \subset \exp(X)$ is called *a swelling* of $\mathcal{A}$ if $A_s \subset B_s$ for each $s \in S$ and $B_{s_0} \cap \ldots B_{s_m} = \emptyset$ if and only if $A_{s_0} \cap \ldots A_{s_m} = \emptyset$ for any $s_0, \ldots s_m \in S$. If $\bigcup \mathcal{A} = X$, a family $\mathcal{B} = \{B_s : s \in S\} \subset \exp(X)$ is called *a shrinking* of $\mathcal{A}$ if $B_s \subset A_s$ for each $s \in S$ and $\bigcup \mathcal{B} = X$. Given $X \in T_{3\frac{1}{2}}$, let $\dim X = -1$ if and only if $X = \emptyset$. If $X \neq \emptyset$, we say that $\dim X \leq n \in \omega$ if any finite functionally open cover $\mathcal{U}$ of the space $X$ has a finite functionally open refinement $\mathcal{V}$ with $\mathrm{ord}(\mathcal{V}) \leq n + 1$. It is said that $\dim X = n$ if $\dim X \leq n$ and $\dim X \leq n - 1$ does not hold. Finally, $\dim X = \infty$ if $\dim X \leq n$ is false for all $n \in \omega$.

A partial order $\leq$ on a set $T$ is called *a direction* on $T$ if, for any $s, t \in T$, there is $w \in T$ such that $s \leq w$ and $t \leq w$. A set with a direction is called *a directed set*. A subset $T'$ of a directed set $T$ is *cofinal* in $T$ if, for any $t \in T$, there is $t' \in T'$ such that $t \leq t'$. Suppose that, for every element $t$ of a directed set $T$, we have a topological space $X_t$ and, for every pair $t, s \in T$ with $s \leq t$, there is a continuous map $\pi_s^t : X_t \to X_s$. If, for all $r, s, t \in T$ with $r \leq s \leq t$, we have $\pi_s^t = id_{X_t}$ and $\pi_r^s \circ \pi_s^t = \pi_r^t$ then the family $\mathcal{S} = \{X_t, \pi_s^t : t, s \in T\}$ is called *an inverse system* and the maps $\pi_s^t$ are called *projections*. An inverse system $\mathcal{S} = \{X_m, \pi_n^m : m, n \in \omega\}$ is called *an inverse sequence*. An element $x \in X = \prod\{X_t : t \in T\}$ is called *a thread* of $\mathcal{S}$ if $\pi_s^t(x(t)) = x(s)$ whenever $t, s \in T$ and $s < t$. The set $\varprojlim \mathcal{S}$ of all threads of $\mathcal{S}$ (with the topology inherited from $X$) is called *the limit of the inverse system $\mathcal{S}$*. The natural projection $p_t : X \to X_t$, restricted to $\varprojlim \mathcal{S}$, is called *the limit projection* and is denoted by $\pi_t$. If we have inverse systems $\mathcal{S} = \{X_t, \pi_s^t : t, s \in T\}$ and $\mathcal{T} = \{Y_t, p_s^t : t, s \in T\}$ with the same directed set $T$, a family $\{f_t : t \in T\}$ is a *topological isomorphism (embedding)* between $\mathcal{S}$ and $\mathcal{T}$ (of $\mathcal{S}$ in $\mathcal{T}$), if $f_t : X_t \to Y_t$ is a homeomorphism (a topological embedding) such that $p_s^t \circ f_t = f_s \circ \pi_s^t$ for any $s, t \in T$ with $s \leq t$.

**101.** Prove that a nonempty family $B \subset \exp(X \times X)$ is a base for some uniformity on $X$ if and only if it has the following properties:

(1) $\bigcap B = \Delta$;
(2) for any $U \in B$, there is $V \in B$ such that $V^{-1} \subset U$;
(3) for any $U \in B$, there is $V \in B$ such that $V \circ V \subset U$;
(4) if $U, V \in B$ then there is $W \in B$ such that $W \subset U \cap V$.

**102.** Suppose that a nonempty family $S \subset \exp(X \times X)$ has the following properties:

(1) $\bigcap S = \Delta$;
(2) for any $U \in S$, there is $V \in S$ such that $V^{-1} \subset U$;
(3) for any $U \in S$, there is $V \in S$ such that $V \circ V \subset U$.

Prove that $S$ is a subbase for some uniformity on $X$. As a consequence, the union of any family of uniformities on $X$ is a subbase of some uniformity on $X$.

**103.** Let $(X, \mathcal{U})$ be a uniform space. Prove that

(1) $\mathrm{Int}(A) = \{x : U(x) \subset A \text{ for some } U \in \mathcal{U}\}$ for any set $A \subset X$; in particular, $x \in \mathrm{Int}(U(x))$ for any $U \in \mathcal{U}$;
(2) if $B$ is a base of the uniformity $\mathcal{U}$ then, for any $x \in X$ and $O \in \tau(x, X)$, there is $B \in B$ such that $B(x) \subset O$. In particular, the family $\{\mathrm{Int}(B(x)) : B \in B\}$ is a local base of the space $X$ at $x$.
(3) if $S$ is a subbase of $\mathcal{U}$ then, for any $x \in X$ and $O \in \tau(x, X)$, there is a finite $S' \subset S$ such that $\bigcap \{S(x) : S \in S'\} \subset O$.
(4) for any $U \in \mathcal{U}$, the interior (in $X \times X$) of the set $U$ also belongs to $\mathcal{U}$. As a consequence, the family of all open symmetric elements of $\mathcal{U}$ is a base of $\mathcal{U}$;
(5) $\overline{A} = \bigcap \{U(A) : U \in \mathcal{U}\}$ for any $A \subset X$;
(6) $\overline{B} = \bigcap \{U \circ B \circ U : U \in \mathcal{U}\}$ for any $B \subset X \times X$;
(7) the family of all closed symmetric elements of $\mathcal{U}$ is a base of $\mathcal{U}$.

**104.** Given uniform spaces $(X, \mathcal{U})$ and $(Y, \mathcal{V})$, prove that every uniformly continuous map $f : X \to Y$ is continuous. In particular, every uniform isomorphism is a homeomorphism.

**105.** Suppose that $(X_t, \mathcal{U}_t)$ is a uniform space for every $t \in T$ and consider the set $X = \prod_{t \in T} X_t$. Let $p_t : X \to X_t$ be the natural projection for every $t \in T$; prove that

(1) the family $S = \bigcup \{p_t^{-1}(\mathcal{U}_t) : t \in T\}$ is a subbase of a unique uniformity on $X$, i.e., the uniform product $(X, \mathcal{U})$ of the spaces $\{(X_t, \mathcal{U}_t) : t \in T\}$ is well defined;
(2) every map $p_t : (X, \mathcal{U}) \to (X_t, \mathcal{U}_t)$ is uniformly continuous;
(3) $\tau_{\mathcal{U}}$ coincides with the topology of the product $\prod \{(X_t, \tau_{\mathcal{U}_t}) : t \in T\}$.

**106.** Let $(X, \mathcal{U})$ be the uniform product of the family $\{(X_t, \mathcal{U}_t) : t \in T\}$. Given a uniform space $(Y, \mathcal{V})$, prove that a map $f : Y \to X$ is uniformly continuous if and only if $p_t \circ f : Y \to X_t$ is uniformly continuous for any $t \in T$. Here, as always, the map $p_t : X \to X_t$ is the natural projection.

**107.** Prove that, for any uniform space $(X, \mathcal{U})$, a pseudometric $d : X \times X \to \mathbb{R}$ is uniformly continuous with respect to the uniform square of $(X, \mathcal{U})$ if and only if the set $O_r = \{(x, y) \in X \times X : d(x, y) < r\}$ belongs to $\mathcal{U}$ for any $r > 0$. Such pseudometrics will be called *uniformly continuous on* $(X, \mathcal{U})$.

**108.** Suppose that a sequence $\{U_n : n \in \omega\}$ of subsets of $X \times X$ has the following properties:

(1) $U_0 = X \times X$ and $\Delta \subset U_n$ for any $n \in \omega$;
(2) the set $U_n$ is symmetric and $U_{n+1} \circ U_{n+1} \circ U_{n+1} \subset U_n$ for any $n \in \omega$.

Prove that there exists a pseudometric $d$ on the set $X$ such that, for any $n \in \mathbb{N}$, we have $U_n \subset \{(x, y) : d(x, y) \le 2^{-n}\} \subset U_{n-1}$.

**109.** Given a uniform space $(X, \mathcal{U})$ and $U \in \mathcal{U}$, prove that there is a uniformly continuous pseudometric $\rho$ on $(X, \mathcal{U})$ such that $\{(x, y) \in X \times X : \rho(x, y) < 1\} \subset U$.

**110.** Prove that a topological space $X$ is Tychonoff if and only if there exists a uniformity $\mathcal{U}$ on the set $X$ such that $\tau(X) = \tau_{\mathcal{U}}$.

**111.** Given a metric $\rho$ on a set $X$ and a number $r > 0$, consider the set $U_r = \{(x, y) \in X \times X : \rho(x, y) < r\}$. Prove that the family $\mathcal{B} = \{U_r : r > 0\}$ forms a base of some uniformity $\mathcal{U}_\rho$ on the set $X$. The uniformity $\mathcal{U}_\rho$ is called *the uniformity induced by the metric* $\rho$. A uniform space $(X, \mathcal{U})$ is called *uniformly metrizable* if $\mathcal{U} = \mathcal{U}_\rho$ for some metric $\rho$ on the set $X$.

**112.** Prove that a uniform space $(X, \mathcal{U})$ is uniformly metrizable if and only if $\mathcal{U}$ has a countable base.

**113.** Prove that every uniform space is uniformly isomorphic to a subspace of a product of uniformly metrizable spaces.

**114.** Given a uniform space $(X, \mathcal{U})$ prove that the following conditions are equivalent:

(i) the space $(X, \mathcal{U})$ is complete;
(ii) if $\mathcal{C}$ is a centered Cauchy family of closed subsets of $X$ then $\bigcap \mathcal{C} \ne \emptyset$;
(iii) if $\mathcal{C}$ is a centered Cauchy family of subsets of $X$ then $\bigcap \{\overline{C} : C \in \mathcal{C}\} \ne \emptyset$;
(iv) if $\mathcal{B}$ is a Cauchy filter base on $X$ then $\bigcap \{\overline{B} : B \in \mathcal{B}\} \ne \emptyset$;
(v) if $\mathcal{B}$ is a Cauchy filter base of closed subsets of $X$ then $\bigcap \mathcal{B} \ne \emptyset$;
(vi) any Cauchy filter $\mathcal{F}$ on $X$ converges to a point $x \in X$, i.e., $\tau(x, X) \subset \mathcal{F}$;
(vii) if $\mathcal{E}$ is a Cauchy ultrafilter on $X$ then $\bigcap \{\overline{E} : E \in \mathcal{E}\} \ne \emptyset$;
(viii) if $\mathcal{E}$ is a Cauchy ultrafilter on $X$ then it converges to a point $x \in X$.

**115.** Prove that any closed uniform subspace of a complete uniform space is complete.

**116.** Let $(X, \mathcal{U})$ be a uniform space such that some $Y \subset X$ is complete considered as a uniform subspace of $(X, \mathcal{U})$. Prove that $Y$ is closed in $X$.

**117.** Prove that, for any family $\mathcal{A} = \{(X_t, \mathcal{U}_t) : t \in T\}$ of complete uniform spaces, the uniform product $(X, \mathcal{U})$ of the family $\mathcal{A}$ is complete.

**118.** Let $(X, \mathcal{U})$ be a uniform space such that the uniformity $\mathcal{U}$ is induced by a metric $\rho$. Prove that $(X, \mathcal{U})$ is complete if and only if the metric space $(X, \rho)$ is complete.

**119.** Let $A$ be a dense subspace of a uniform space $(X, \mathcal{U})$. Suppose that $f : A \to Y$ is a uniformly continuous map of $(A, \mathcal{U}_A^X)$ to a complete uniform space $(Y, \mathcal{V})$. Prove that there is a uniformly continuous map $F : X \to Y$ such that $F|A = f$.

**120.** Let $(X, \mathcal{U})$ and $(Y, \mathcal{V})$ be complete uniform spaces. Suppose that $A$ is a dense subspace of $X$ and $B$ is a dense subspace of $Y$. Prove that every uniform isomorphism between the uniform spaces $(A, \mathcal{U}_A^X)$ and $(B, \mathcal{V}_B^Y)$ is extendable to a uniform isomorphism between $(X, \mathcal{U})$ and $(Y, \mathcal{V})$.

**121.** Prove that every uniform space $(X, \mathcal{U})$ is uniformly isomorphic to a dense subspace of a complete uniform space $(X^*, \mathcal{U}^*)$. The space $(X^*, \mathcal{U}^*)$ is called *the completion* of the space $(X, \mathcal{U})$. Prove that the completion of $(X, \mathcal{U})$ is unique in the sense that, if $(Y, \mathcal{V})$ is a complete uniform space such that $(X, \mathcal{U})$ is a dense uniform subspace of $(Y, \mathcal{V})$ then there is a uniform isomorphism $\Phi : X^* \to Y$ such that $\Phi(x) = x$ for any $x \in X$.

**122.** Let $X$ be a paracompact Tychonoff space. Prove that the family of all neighborhoods of the diagonal of $X$ is a uniformity on $X$ which generates $\tau(X)$.

**123.** Suppose that $X$ is a Tychonoff space such that the family of all neighborhoods of the diagonal of $X$ is a uniformity on $X$ which generates $\tau(X)$. Prove that $X$ is collectionwise normal.

**124.** Give an example of a Tychonoff countably compact non-compact (and hence non-paracompact) space $X$ such that the family of all neighborhoods of the diagonal of $X$ is a uniformity on $X$ which generates $\tau(X)$.

**125.** Prove that, for any compact uniform space $(X, \mathcal{U})$, the uniformity $\mathcal{U}$ coincides with the family of all neighborhoods of the diagonal of $X$.

**126.** Suppose that $(X, \mathcal{U})$ a compact uniform space. Prove that, for any uniform space $(Y, \mathcal{V})$, any continuous map $f : X \to Y$ is uniformly continuous.

**127.** Let $(X, \mathcal{U})$ be a uniform space such that the uniformity $\mathcal{U}$ is induced by a metric $\rho$. Prove that $(X, \mathcal{U})$ is totally bounded if and only if the metric space $(X, \rho)$ is totally bounded.

**128.** Prove that a uniform space $(X, \mathcal{U})$ is totally bounded if and only if every ultrafilter on $X$ is a Cauchy family with respect to $\mathcal{U}$.

**129.** Prove that any uniform product of totally bounded uniform spaces is totally bounded.

**130.** Prove that a uniform space is compact if and only if it is complete and totally bounded. Deduce from this fact that a uniform space is totally bounded if and only if its completion is compact.

**131.** Prove that a Tychonoff space $X$ is pseudocompact if and only if every uniformity $\mathcal{U}$ on the set $X$ with $\tau_{\mathcal{U}} = \tau(X)$ is totally bounded.

**132.** For any Tychonoff space $X$ let $\mathbb{U}_X$ be the family of all uniformities on the set $X$ which generate $\tau(X)$. Note that $\bigcup \mathbb{U}_X$ can be considered a subbase of a uniformity $\mathcal{N}_X$ (called *the universal uniformity*) on the set $X$. Prove that

(i) the topology generated by $\mathcal{N}_X$ coincides with $\tau(X)$ and hence $\mathcal{N}_X \in \mathbb{U}_X$;

(ii) if $Y$ is a Tychonoff space and $f : X \to Y$ is a continuous map then the map $f : (X, \mathcal{N}_X) \to (Y, \mathcal{V})$ is uniformly continuous for any uniformity $\mathcal{V} \in \mathbb{U}_Y$.

**133.** Let $X$ be a Tychonoff space. Prove that the following are equivalent:

(i) there exists a complete uniformity $\mathcal{U}$ on the set $X$ such that $\tau_{\mathcal{U}} = \tau(X)$;

(ii) the universal uniformity on the space $X$ is complete;

(iii) the space $X$ is Dieudonné complete.

**134.** For any linear topological space $L$ denote by $\mathbf{0}_L$ its zero vector and let $G(U) = \{(x, y) \in L \times L : x - y \in U\}$ for any $U \in \tau(\mathbf{0}_L, L)$. Prove that

(i) the family $\mathcal{B}_L = \{G(U) : U \in \tau(\mathbf{0}_L, L)\}$ forms a base for a uniformity $\mathcal{U}_L$ on the set $L$ (called *the linear uniformity* on $L$).

(ii) If $M$ is a linear subspace of $L$ then the linear uniformity $\mathcal{U}_M$ on the set $M$ coincides with the subspace uniformity induced on $M$ from $L$.

(iii) If $L'$ is a linear topological space then a map $f : L \to L'$ is uniformly continuous if and only if, for any $U' \in \tau(\mathbf{0}_{L'}, L')$ there exists $U \in \tau(\mathbf{0}_L, L)$ such that $f(x) - f(y) \in U'$ for any $x, y \in L$ with $x - y \in U$.

(iv) If $L'$ is a linear topological space then any linear continuous map $f : L \to L'$ is uniformly continuous if $L$ and $L'$ are considered with their linear uniformities. In particular, any linear isomorphism between $L$ and $L'$ is a uniform isomorphism.

**135.** Prove that the linear uniformity of $\mathbb{R}^X$ coincides with the uniform product of the respective family of real lines. Deduce from this fact that $\mathbb{R}^X$ is the completion of $C_p(X)$ for any space $X$ so $C_p(X)$ is complete as a uniform space if and only if $X$ is discrete.

**136.** Prove that $C_p(X)$ is $\sigma$-totally bounded as a uniform space if and only if $X$ is pseudocompact. More formally, $X$ is pseudocompact if and only if there exists a family $\{C_n : n \in \omega\} \subset \exp(C_p(X))$ such that $C_p(X) = \bigcup\{C_n : n \in \omega\}$ and each $C_n$ is totally bounded considered as a uniform subspace of $C_p(X)$. In particular, if $C_p(X)$ is uniformly isomorphic to $C_p(Y)$ then the space $X$ is pseudocompact if and only if so is $Y$.

**137.** Observe that $X \overset{u}{\sim} Y$ implies $X \overset{t}{\sim} Y$. Given an example of $t$-equivalent spaces $X$ and $Y$ which are not $u$-equivalent.

**138.** Suppose that $C_p(X)$ is uniformly isomorphic to $C_p(Y)$. Prove that $X$ is compact if and only if so is $Y$.

**139.** Suppose that the spaces $X$ and $Y$ are $u$-equivalent. Prove that there exists a homeomorphism $\varphi : \mathbb{R}^X \to \mathbb{R}^Y$ such that $\varphi(C_p(X)) = C_p(Y)$.

**140.** Let $\mathcal{F} = \{F_1, \ldots, F_k\}$ be a family of functionally closed subsets of a Tychonoff space $X$. Suppose that $\{U_1, \ldots, U_k\}$ is a family of functionally open subsets of $X$ such that $F_i \subset U_i$ for each $i$. Prove that the family $\mathcal{F}$ has a functionally open swelling $\{W_1, \ldots, W_k\}$ such that $F_i \subset W_i \subset \overline{W}_i \subset U_i$ for each $i \leq k$.

**141.** Let $\mathcal{F} = \{F_1, \ldots, F_k\}$ be a family of closed subsets of a normal space $X$. Suppose that $\{U_1, \ldots, U_k\}$ is a family of open subsets of $X$ such that $F_i \subset U_i$ for each $i \leq k$. Prove that the family $\mathcal{F}$ has an open swelling $\{W_1, \ldots, W_k\}$ such that $F_i \subset W_i \subset \overline{W}_i \subset U_i$ for each $i \leq k$.

**142.** Let $\mathcal{U} = \{U_1, \ldots, U_k\}$ be a functionally open cover of a Tychonoff space $X$. Prove that $\mathcal{U}$ has shrinkings $\mathcal{F} = \{F_1, \ldots, F_k\}$ and $\mathcal{W} = \{W_1, \ldots, W_k\}$ such that $\mathcal{F}$ is functionally closed, $\mathcal{W}$ is functionally open and $F_i \subset W_i \subset \overline{W}_i \subset U_i$ for every $i \leq k$.

**143.** Let $\mathcal{U} = \{U_1, \ldots, U_k\}$ be an open cover of a normal space $X$. Prove that $\mathcal{U}$ has shrinkings $\mathcal{F} = \{F_1, \ldots, F_k\}$ and $\mathcal{W} = \{W_1, \ldots, W_k\}$ such that $\mathcal{F}$ is closed, $\mathcal{W}$ is open, and $F_i \subset W_i \subset \overline{W}_i \subset U_i$ for every $i \leq k$.

**144.** Prove that, for any Tychonoff space $X$, the following conditions are equivalent:

   (i) $\dim X \leq n$;

  (ii) every finite functionally open cover of $X$ has a finite functionally closed refinement of order $\leq n + 1$;

 (iii) every finite functionally open cover of $X$ has a functionally closed shrinking of order $\leq n + 1$;

 (iv) every finite functionally open cover of $X$ has a functionally open shrinking of order $\leq n + 1$.

**145.** Prove that, for any normal $X$, the following conditions are equivalent:

   (i) $\dim X \leq n$;

  (ii) every finite open cover of $X$ has a finite open refinement of order $\leq n+1$;

 (iii) every finite open cover of $X$ has a finite closed refinement of order $\leq n + 1$;

 (iv) every finite open cover of $X$ has a closed shrinking of order $\leq n + 1$;

  (v) every finite open cover of $X$ has an open shrinking of order $\leq n + 1$.

**146.** Suppose that $X$ is a Tychonoff space and $Y$ is a $C^*$-embedded subset of $X$. Prove that $\dim Y \leq \dim X$. In particular, if $X$ is normal then $\dim F \leq \dim X$ for any closed $F \subset X$.

**147.** Prove that $\dim X = \dim \beta X$ for any Tychonoff space $X$. Deduce from this fact that $\dim X = \dim Y$ for any $Y$ with $X \subset Y \subset \beta X$.

**148.** Prove that a Tychonoff space $X$ is strongly zero-dimensional if and only if $X$ is normal and $\dim X = 0$. Give an example of a Tychonoff space $X$ such that $\dim X = 0$ while $X$ is not strongly zero-dimensional.

**149.** Prove that $\dim X = 0$ implies that $X$ is zero-dimensional. Give an example of a zero-dimensional space $Y$ such that $\dim Y > 0$.

**150.** (The countable sum theorem for normal spaces). Given $n \in \omega$, suppose that a normal space $X$ has a countable closed cover $\mathcal{F}$ such that $\dim F \leq n$ for every $F \in \mathcal{F}$. Prove that $\dim X \leq n$.

**151.** (General countable sum theorem). Given $n \in \omega$, suppose that we have a closed countable cover $\mathcal{F}$ of a Tychonoff space $X$ such that

(i) every $F \in \mathcal{F}$ is $C^*$-embedded in $X$;

(ii) $\dim F \leq n$ for each $F \in \mathcal{F}$.

Prove that $\dim X \leq n$; give an example of a Tychonoff non-normal space $Y$ such that $\dim Y > 0$ and $Y = \bigcup\{Y_i : i \in \omega\}$, where $Y_i$ is closed in $Y$ and $\dim Y_i = 0$ for every $i \in \omega$.

**152.** Give an example of a compact (and hence normal) space $X$ such that $\dim X = 0$ while $\dim Y > 0$ for some $Y \subset X$.

**153.** Give an example of a Tychonoff space $X$ such that $\dim X = 0$ and $\dim Y > 0$ for some closed $Y \subset X$.

**154.** Let $X$ be a normal space with $\dim X \leq n$. Given a subspace $Y \subset X$, suppose that, for every open $U \supset Y$, there exists an $F_\sigma$-set $P$ such that $Y \subset P \subset U$. Prove that $\dim Y \leq n$.

**155.** Prove that, for any perfectly normal space $X$, we have $\dim Y \leq \dim X$ for any $Y \subset X$. In particular, $\dim Y \leq \dim X$ for any subspace $Y$ of a metrizable space $X$.

**156.** Given $n \in \omega$ and a Tychonoff space $X$, prove that $\dim X \leq n$ if and only if, for any family $\{(A_0, B_0), \ldots, (A_n, B_n)\}$ of $n + 1$ pairs of disjoint functionally closed sets, it is possible to choose, for each $i \leq n$, a functionally closed partition $C_i$ between $A_i$ and $B_i$ in such a way that $L_0 \cap \ldots \cap L_n = \emptyset$.

**157.** Given a natural $n \geq 0$ and a normal space $X$, prove that $\dim X \leq n$ if and only if, for any family $\{(A_0, B_0), \ldots, (A_n, B_n)\}$ of $n+1$ pairs of disjoint closed sets, it is possible to choose, for each $i \leq n$, a partition $C_i$ between $A_i$ and $B_i$ in such a way that $L_0 \cap \ldots \cap L_n = \emptyset$.

**158.** Let $X$ be a normal space. Prove that $\dim X \leq n$ if and only if, for any closed $F \subset X$ and any continuous map $f : F \to S^n$, there exists a continuous map $g : X \to S^n$ such that $g|F = f$. Here $S^n = \{(x_0, \ldots, x_n) \in \mathbb{R}^{n+1} : x_0^2 + \ldots + x_n^2 = 1\}$ is the $n$-dimensional sphere with the topology inherited from $\mathbb{R}^{n+1}$.

**159.** Prove that $\dim(\mathbb{I}^n) = \dim(\mathbb{R}^n) = \dim(S^n) = n$ for any $n \in \mathbb{N}$. Here $S^n = \{(x_0, \ldots, x_n) \in \mathbb{R}^{n+1} : x_0^2 + \ldots + x_n^2 = 1\}$ is the $n$-dimensional sphere with the topology inherited from $\mathbb{R}^{n+1}$.

**160.** Given $n \in \mathbb{N}$ prove that, for any set $X \subset \mathbb{R}^n$, we have $\dim X = n$ if and only if the interior of $X$ in $\mathbb{R}^n$ is nonempty.

**161.** Prove that, for any Tychonoff space $X$ and $n \in \omega$, we have $\dim X \leq n$ if and only if, for any second countable space $Y$ and any continuous $f : X \to Y$, there exists a second countable space $M$ and continuous maps $g : X \to M$, $h : M \to Y$ such that $\dim M \leq n$ and $f = h \circ g$.

**162.** Prove that, for any $n \in \omega$, there exists a compact second countable space $U_n$ such that $\dim U_n \leq n$ and any second countable $X$ with $\dim X \leq n$ can be embedded in $U_n$.

**163.** Suppose that $X$ is a second countable space, $Y \subset X$ and $\dim Y \leq n$. Prove that there exists a $G_\delta$-set $Y'$ of the space $X$ such that $Y \subset Y'$ and $\dim Y' \leq n$.

**164.** Given $n \in \omega$ and a second countable Tychonoff space $X$ prove that $\dim X \leq n$ if and only there exist $X_0, \ldots, X_n \subset X$ such that $X = X_0 \cup \ldots \cup X_n$ and $\dim X_i \leq 0$ for each $i \leq n$.

**165.** Let $\mathcal{S} = \{X_t, \pi_s^t : t, s \in T\}$ be an inverse system of Hausdorff topological spaces. Prove that the set $\varprojlim \mathcal{S}$ is closed in $\prod\{X_t : t \in T\}$. Therefore the limit of an inverse system of Hausdorff compact spaces is a Hausdorff compact space.

**166.** Suppose that $\mathcal{S} = \{X_t, \pi_s^t : t, s \in T\}$ is an inverse system in which $X_t$ is a nonempty compact Hausdorff space for each $t$. Prove that $\varprojlim \mathcal{S} \neq \emptyset$.

**167.** Let $\mathcal{S} = \{X_t, \pi_s^t : t, s \in T\}$ be an inverse system. Suppose that, for a space $Y$, a continuous map $f_t : Y \to X_t$ is given for every $t \in T$ and, besides, $\pi_s^t \circ f_t = f_s$ for any $s, t \in T$ with $s \leq t$. Prove that the diagonal product $f = \Delta_{t \in T} f_t$ maps $Y$ continuously into $\varprojlim \mathcal{S}$.

**168.** Let $X$ be a topological space. Suppose that, for a nonempty directed set $T$, a subspace $X_t \subset X$ is given for each $t \in T$ in such a way that $X_t \subset X_s$ whenever $s \leq t$. Given $s, t \in T$ with $s \leq t$, let $\pi_s^t(x) = x$ for each $x \in X_t$. Prove that the inverse system $\mathcal{S} = \{X_t, \pi_s^t : t, s \in T\}$ is well defined and the limit of $\mathcal{S}$ is homeomorphic to $\bigcap\{X_t : t \in T\}$.

**169.** Give an example of an inverse sequence $\mathcal{S} = \{X_n, \pi_m^n : n, m \in \omega\}$ such that every $X_n$ is a nonempty second countable Tychonoff space while $\varprojlim \mathcal{S} = \emptyset$.

**170.** Given an inverse system $\mathcal{S} = \{X_t, \pi_s^t : t, s \in T\}$ of topological spaces, prove that the family $\mathcal{B} = \{\pi_t^{-1}(U) : t \in T, \ U \in \tau(X_t)\}$ is a base of the space $L = \varprojlim \mathcal{S}$. Here $\pi_t : L \to X_t$ is the limit projection for every $t \in T$.

**171.** Suppose that $\mathcal{S} = \{X_t, \pi_s^t : t, s \in T\}$ is an inverse system of topological spaces. Prove that, for any closed set $F \subset \varprojlim \mathcal{S}$, the subspace $F$ is the limit of the inverse system $\mathcal{S}_F = \{\pi_t(F), \pi_s^t|\pi_t(F) : t, s \in T\}$. Here $\pi_t : \varprojlim \mathcal{S} \to X_t$ is the limit projection for every $t \in T$.

**172.** Given an inverse system $\mathcal{S} = \{X_t, \pi_s^t : t, s \in T\}$ and a cofinal set $T' \subset T$, prove that the limit of the inverse system $\mathcal{S}' = \{X_t, \pi_s^t : t, s \in T'\}$ is homeomorphic to $\varprojlim \mathcal{S}$.

**173.** Suppose that an inverse system $\mathcal{S} = \{X_t, \pi_s^t : t, s \in T\}$ consists of compact Hausdorff spaces $X_t$ and all projections $\pi_s^t$ are onto. Prove that all limit projections $\pi_t$ are also surjective maps.

**174.** Suppose that $n \in \omega$ and $\mathcal{S} = \{X_t, \pi_s^t : t, s \in T\}$ is an inverse system in which all spaces $X_t$ are compact and Hausdorff. Knowing additionally that $\dim X_t \leq n$ for each $t \in T$ prove that $\dim(\varprojlim \mathcal{S}) \leq n$.

**175.** Suppose that $n \in \omega$ and $\mathcal{S} = \{X_l, \pi_m^l : l, m \in \omega\}$ is an inverse sequence in which every $X_l$ is a Lindelöf $\Sigma$-space. Knowing additionally that $\dim X_l \leq n$ for every $l \in \omega$ prove that $\dim(\varprojlim \mathcal{S}) \leq n$.

**176.** Prove that if $X$ is second countable and $A \subset C_p(X)$ is countable then there exists a countable QS-algebra $E \subset C_p(X)$ such that $A \subset E$.

**177.** Let $K = \{x_0, \ldots, x_n\} \subset X$ be a finite subset of a space $X$. Suppose additionally that $U \in \tau(K, X)$ and $q_0, \ldots, q_n \in \mathbb{Q}$. Prove that, for any QS-algebra $E$ on a space $X$, there is $f \in E$ such that $f(X \setminus U) \subset \{0\}$ and $f(x_i) = q_i$ for each $i \leq n$.

**178.** Given second countable Tychonoff spaces $X$ and $Y$, suppose that some QS-algebras $E(X)$ and $E(Y)$ are chosen in $C_p(X)$ and $C_p(Y)$ respectively. Prove that, if $E(X)$ is uniformly homeomorphic to $E(Y)$ then $X$ is representable as a countable union of closed subspaces each one of which embeds in $Y$.

**179.** Given second countable Tychonoff spaces $X$ and $Y$, suppose that some QS-algebras $E(X)$ and $E(Y)$ are chosen in $C_p(X)$ and $C_p(Y)$ respectively. Prove that, if $E(X)$ is uniformly homeomorphic to $E(Y)$ then $\dim X = \dim Y$.

**180.** Suppose that $X$ and $Y$ are Tychonoff spaces such that $C_p(X)$ is uniformly homeomorphic to $C_p(Y)$. Prove that $\dim X = \dim Y$.

**181.** Let $X$ be a zero-dimensional compact space. Prove that $Y$ is also a zero-dimensional compact space whenever $Y \overset{u}{\sim} X$.

**182.** Suppose that $X$ is a zero-dimensional Lindelöf space and $Y \overset{u}{\sim} X$. Prove that $Y$ is also zero-dimensional.

**183.** Given a countable ordinal $\xi \geq 1$, prove that a metrizable space $X$ is an absolute Borel set of multiplicative class $\xi$ (i.e., $X \in \mathcal{M}_\xi$) if and only if there exists a completely metrizable space $Z$ such that $X$ is homeomorphic to some $Y \in \Pi_\xi^0(Z)$.

**184.** Given a countable ordinal $\xi \geq 2$, prove that a metrizable space $X$ is an absolute Borel set of additive class $\xi$ (i.e., $X \in \mathcal{A}_\xi$) if and only if there exists a completely metrizable space $Z$ such that $X$ is homeomorphic to some $Y \in \Sigma_\xi^0(Z)$.

**185.** Suppose that $n \in \mathbb{N}$ and a space $X_i$ is metrizable for every $i \leq n$. Prove that, for any countable ordinal $\xi \geq 2$,

(i) if $X_i \in \mathcal{A}_\xi$ for all $i \leq n$ then $X_1 \times \ldots \times X_n \in \mathcal{A}_\xi$;
(ii) if $X_i \in \mathcal{M}_\xi$ for all $i \leq n$ then $X_1 \times \ldots \times X_n \in \mathcal{M}_\xi$.

**186.** Given ordinals $\alpha, \beta \in \omega_1$ such that $\alpha \geq 2$ and $\beta < \alpha$ suppose that $X$ is a metrizable space and $X = \bigcup \{X_n : n \in \omega\}$ where $X_n \in \Sigma_\beta^0(X) \cap \mathcal{M}_\alpha$ for every $n \in \omega$. Prove that $X \in \mathcal{M}_\alpha$.

**187.** Prove that a metrizable space $X$ is a Borel set of absolute additive class $\xi \geq 2$ (i.e., $X \in \mathcal{A}_\xi$) if and only if there exists a sequence $\{\xi_n : n \in \omega\} \subset \xi$ such that $X = \bigcup \{X_n : n \in \omega\}$ and $X_n \in \mathcal{M}_{\xi_n}$ for every $n \in \omega$.

**188.** Given a countable ordinal $\xi \geq 2$, let $\mathcal{M}_\xi$ be the class of absolute Borel sets of multiplicative class $\xi$. Prove that the following conditions are equivalent for any metrizable $X$:

(i) the space $X$ belongs to $\mathcal{M}_\xi$;
(ii) there is a complete sequence $\{\mathcal{U}_n : n \in \omega\}$ of $\sigma$-discrete covers of $X$ such that, for any $n \in \omega$, there is $\xi_n < \xi$ with $\mathcal{U}_n \subset \Sigma_{\xi_n}^0(X)$;

(iii) there is a complete sequence $\{\mathcal{V}_n : n \in \omega\}$ of $\sigma$-discrete covers of $X$ such that, for any $n \in \omega$, there is $\xi_n < \xi$ with $\mathcal{V}_n \subset \bigcup\{\Pi_\alpha^0(X) : \alpha < \xi_n\}$.

**189.** Given a countable ordinal $\xi \geq 2$ prove that the following conditions are equivalent for any second countable $X$:

(i) the space $X$ belongs to $\mathcal{M}_\xi$;

(ii) there is a complete sequence $\{\mathcal{U}_n : n \in \omega\}$ of countable covers of $X$ such that, for any $n \in \omega$, there is $\xi_n < \xi$ with $\mathcal{U}_n \subset \Sigma_{\xi_n}^0(X)$;

(iii) there is a complete sequence $\{\mathcal{V}_n : n \in \omega\}$ of countable covers of $X$ such that, for any $n \in \omega$, there is $\xi_n < \xi$ with $\mathcal{V}_n \subset \bigcup\{\Pi_\alpha^0(X) : \alpha < \xi_n\}$.

**190.** Prove that any analytic space has a complete sequence of countable covers. Show that in metrizable spaces the converse is also true, i.e., a metrizable space $X$ is analytic if and only if there exists a complete sequence of countable covers of $X$.

**191.** For any metrizable space $X$ and $n \in \mathbb{N}$ define a map $e : X^n \to [X]^{\leq n}$ by $e((x_1, \ldots, x_n)) = \{x_1, \ldots, x_n\}$ for every $(x_1, \ldots, x_n) \in X^n$. Prove that there exists an $F_\sigma$-set $G$ in the space $X^n$ such that $e(G) = [X]^n$ and the map $e|G : G \to [X]^n$ is a bijection.

**192.** Given a metrizable space $X$ and $n \in \mathbb{N}$ consider the set $[X]^n$ together with its Vietoris topology. Prove that there exists a family $\{Y_m : m \in \omega\}$ of closed subsets of $[X]^n$ such that $[X]^n = \bigcup\{Y_m : m \in \omega\}$ and every $Y_m$ is homeomorphic to some closed subspace of $X^n$.

**193.** Suppose that there exists a uniformly continuous surjection of $C_p(X)$ onto $C_p(Y)$. Prove that if $X$ is pseudocompact then $Y$ is also pseudocompact. Deduce from this fact that if $X$ is a metrizable compact space and there exists a uniformly continuous surjection of $C_p(X)$ onto $C_p(Y)$ then $Y$ is also compact. Give an example of a (non-metrizable!) compact space $X$ such that there is a non-compact space $Y$ and a uniformly continuous surjection of $C_p(X)$ onto $C_p(Y)$.

**194.** Assume that $X$ and $Y$ are metrizable spaces and there exists either a uniformly continuous surjection of $C_p(X)$ onto $C_p(Y)$ or a uniformly continuous surjection of $C_p^*(X)$ onto $C_p^*(Y)$. Prove that there exists a family $\{Y_n : n \in \omega\}$ of closed subspaces of $Y$ such that $Y = \bigcup_{n \in \omega} Y_n$ and each $Y_n$ can be perfectly mapped onto a closed subspace of $[X]^{k_n}$ (with the Vietoris topology) for some $k_n \in \mathbb{N}$.

**195.** Let $\mathcal{P}$ be a class of metrizable spaces with the following properties:

(1) $\mathcal{P}$ contains all complete metrizable spaces;

(2) $\mathcal{P}$ is invariant under finite products and closed subspaces;

(3) if $M$ is a metrizable space with $M = \bigcup\{M_n : n \in \omega\}$, where $M_n$ is closed in $M$ and $M_n \in \mathcal{P}$ for each $n \in \omega$, then $M \in \mathcal{P}$.

Suppose that $X \in \mathcal{P}$ and $Y$ is a metrizable space. Prove that, if there exists a uniformly continuous surjection of $C_p(X)$ onto $C_p(Y)$ (or $C_p^*(X)$ onto $C_p^*(Y)$), then $Y \in \mathcal{P}$.

**196.** Let $\mathcal{P}$ be a class of second countable spaces such that

  (1) every compact metrizable space belongs to $\mathcal{P}$;
  (2) $\mathcal{P}$ is invariant under finite products and closed subspaces;
  (3) if $M$ is a second countable space with $M = \bigcup\{M_n : n \in \omega\}$, where $M_n$ is closed in $M$ and $M_n \in \mathcal{P}$ for each $n \in \omega$, then $M \in \mathcal{P}$.

  Suppose that $X \in \mathcal{P}$ and $Y$ is a metrizable space. Prove that, if there exists a uniformly continuous surjection of $C_p(X)$ onto $C_p(Y)$ (or $C_p^*(X)$ onto $C_p^*(Y)$), then $Y \in \mathcal{P}$.

**197.** Given a countable ordinal $\xi$, let $\mathcal{M}_\xi$ be the class of absolute Borel sets of multiplicative class $\xi$. Suppose that $X$ is a metrizable space such that $X \in \mathcal{M}_\alpha$ for some $\alpha \geq 2$. Let $Y$ be a metrizable space such that $C_p(Y)$ (or $C_p^*(Y)$) is a uniformly continuous image of $C_p(X)$ (or $C_p^*(X)$ respectively). Prove that $Y \in \mathcal{M}_\alpha$. In particular, if $X \overset{u}{\sim} Y$ then $X$ belongs to $\mathcal{M}_\alpha$ if and only if so does $Y$.

**198.** Given a countable ordinal $\xi$, let $\mathcal{A}_\xi$ be the class of absolute Borel sets of additive class $\xi$. Suppose that $X$ is a metrizable space such that $X \in \mathcal{A}_\alpha$ for some $\alpha \geq 2$. Let $Y$ be a metrizable space such that $C_p(Y)$ (or $C_p^*(Y)$) is a uniformly continuous image of $C_p(X)$ (or $C_p^*(X)$ respectively). Prove that $Y \in \mathcal{A}_\alpha$. In particular, if $X \overset{u}{\sim} Y$ then $X$ belongs to $\mathcal{A}_\alpha$ if and only if so does $Y$.

**199.** Prove that every nonempty countable compact space $X$ is homeomorphic to the space $\alpha + 1 = \{\beta : \beta \leq \alpha\}$ for some countable ordinal $\alpha$. Here, as usual, the set $\alpha + 1$ is considered with the topology generated by the well ordering on $\alpha + 1$.

**200.** Let $X$ and $Y$ be infinite countable compact spaces. Prove that $X \overset{u}{\sim} Y$, i.e., the spaces $C_p(X)$ and $C_p(Y)$ are uniformly homeomorphic.

## 1.3 Linear Topological Spaces and l-Equivalence

Given a space $X$, consider the family $\mathcal{L}$ of all continuous maps of $X$ into locally convex linear topological spaces of cardinality not exceeding $|X| \cdot 2^\omega$. If $\varphi \in \mathcal{L}$, we denote by $L_\varphi$ the relevant linear topological space. The map $i = \Delta\{\varphi : \varphi \in \mathcal{L}\}$ is an embedding of $X$ into $L = \prod\{L_f : f \in \mathcal{L}\}$. The linear span $L(X)$ of $i(X)$ in $L$ is called *the free locally convex space of the space* $X$. We will identify $X$ and $i(X)$. A space $X$ is said to be *L-equivalent* to a space $Y$ (or $X \overset{L}{\sim} Y$), if $L(X)$ is linearly homeomorphic to $L(Y)$. The spaces $X$ and $Y$ are *l*-equivalent (or $X \overset{l}{\sim} Y$) if $C_p(X)$ is linearly homeomorphic to $C_p(Y)$. A topological property $\mathcal{P}$ (a cardinal invariant $\varphi$) is called *l-invariant* if it is preserved by *l*-equivalence, i.e., if $X \vdash \mathcal{P}$ (or $\varphi(X) \le \kappa$) and $X \overset{l}{\sim} Y$ imply $Y \vdash \mathcal{P}$ (or $\varphi(Y) \le \kappa$ respectively).

The space $L_p(X)$ is the linear span of $X$ in $C_p(C_p(X))$. If $L_t$ is a linear space for all $t \in T$, the set $L = \prod\{L_t : t \in T\}$ carries a natural structure of linear space: let $(x + y)(t) = x(t) + y(t)$ and $(\alpha x)(t) = \alpha x(t)$ for any $t \in T$, $x, y \in L$ and $\alpha \in \mathbb{R}$. The set $L$ with the operations defined above is called *the product of linear spaces* $\{L_t : t \in T\}$. If each $L_t$ is a linear topological space, the linear space $L$ is always considered with the respective product topology.

If $f : X \to Y$ is a map of a space $X$ onto a set $Y$, consider the families $\mu = \{U \subset Y : f^{-1}(U) \in \tau(X)\}$ and $\nu = \bigcup\{\{g^{-1}(U) : U \in \tau(\mathbb{R})\} : g \in \mathbb{R}^Y$ and $g \circ f \in C(X)\}$. The family $\mu$ is called *the quotient topology induced by* $f$. The map $f : X \to (Y, \mu)$ is quotient and the space $(Y, \mu)$ may fail to be a Tychonoff space. The topology $\nu'$ generated by $\nu$ as a subbase is called *the $\mathbb{R}$-quotient topology induced by the map* $f$. The map $f : X \to (Y, \nu')$ is $\mathbb{R}$-quotient and $(Y, \nu')$ is a completely regular (but, maybe, not Tychonoff) space. If $X$ is a space and $F$ is a nonempty closed subset of $X$, let $X_F = (X \backslash F) \cup \{F\}$. For any $y \in X \backslash F$, let $p_F(y) = y$ and $p_F(x) = F$ for any $x \in F$. The set $X_F$ with the $\mathbb{R}$-quotient topology induced by the *contraction map* $p_F : X \to X_F$ is called *the $\mathbb{R}$-quotient space* $X_F$. Given two retractions $r, s : X \to X$ in a space $X$, we say that $r$ is *parallel to* $s$ if $r \circ s = r$ and $s \circ r = s$. Two retracts of a space $X$ are called *parallel* if they are images of $X$ under parallel retractions. If $\tau \subset \exp(X)$ and $Y \subset X$ then $\tau|Y = \{U \cap Y : U \in \tau\}$.

A subset $B$ of a linear space $L$ is *balanced* if $tB = \{tx : x \in B\} \subset B$ for any $t \in \mathbb{I}$. The set $B$ is *absorbing* if, for any $x \in L$, there exists $\delta > 0$ such that $tx \in B$ for all $t \in (-\delta, \delta)$. Recall that $\mathbf{0}_L$ denotes the zero vector of $L$. If $L$ is clear, we write $\mathbf{0}$ instead of $\mathbf{0}_L$. The set $B$ is *linearly bounded or l-bounded* if, for any $U \in \tau(\mathbf{0}, L)$, there exists $s > 0$ such that $B \subset tU$ for all $t \ge s$. A subset $A$ of a space $X$ is called *bounded* if $f(A)$ is bounded in $\mathbb{R}$ for any $f \in C(X)$. If $L$ and $M$ are linear spaces, a map $f : L \to M$ is called *linear* if $f(\alpha x + \beta y) = \alpha f(x) + \beta f(y)$ for all $x, y \in L$ and $\alpha, \beta \in \mathbb{R}$. The linear topological spaces $M$ and $L$ are *linearly homeomorphic* if there exists a linear map $f : L \to M$ which is a homeomorphism. Given a linear space $L$, a function $p : L \to \mathbb{R}$ is *a seminorm* on $L$, if $p(x + y) \le p(x) + p(y)$ and $p(\alpha x) = |\alpha| p(x)$ for all $x, y \in L$ and $\alpha \in \mathbb{R}$. If, additionally, $p(x) \ne 0$ for any

$x \neq \mathbf{0}$, then $p$ is called *a norm*; in this case we write $\|x\|$ instead of $p(x)$. A family $\mathcal{P}$ of seminorms on a linear space $L$ is called *separating* if, for any $x \neq \mathbf{0}$, there is $p \in \mathcal{P}$ such that $p(x) \neq 0$. Given an absorbing set $A$ in a linear space $L$, the *Minkowski functional* $\mu_A$ is defined on $L$ as follows: $\mu_A(x) = \inf\{t > 0 : \frac{x}{t} \in A\}$ for any $x \in L$.

If $L$ is a linear topological space then $H \subset L$ is *a Hamel basis* of $L$ if $H$ is linearly independent (i.e., for any distinct $x_1, \ldots, x_n \in H$ and $\lambda_1, \ldots, \lambda_n \in \mathbb{R}$, the equality $\lambda_1 x_1 + \ldots + \lambda_n x_n = \mathbf{0}$ implies $\lambda_i = 0$ for all $i \leq n$) and the linear span of $H$ is equal to $L$. Now, $L^* = \{f \in C(L) : f$ is a linear functional$\}$ and $L' = \{f \in \mathbb{R}^L : f$ is a linear functional$\}$; the space $L^*$ is called the *dual* of $L$. The topology on $L$ with the subbase $\{\varphi^{-1}(U) : U \in \tau(\mathbb{R}), \varphi \in L^*\}$ is called *the weak topology of L*. The set $L^*$ with the topology inherited from $C_p(L)$ is *the weak dual* of $L$. Given a closed linear subspace $N \subset L$, let $\pi(x) = x + N$ for any $x \in L$. Denote the set $\{\pi(x) : x \in L\}$ by $L/N$. For any $x, y \in L$ let $\pi(x) + \pi(y) = \pi(x + y)$ and $\alpha\pi(x) = \pi(\alpha x)$. Then $\pi(x) = \pi(x')$ and $\pi(y) = \pi(y')$ imply $\pi(x) + \pi(y) = \pi(x') + \pi(y')$ and $\alpha\pi(x) = \alpha\pi(x')$; thus $L/N$, with the zero element $N = \pi(\mathbf{0})$, is a linear space called *the quotient space of L over N*. If $L/N$ is dealt with as a topological space, it is assumed to carry the quotient topology $\mu = \{U \subset L/N : \pi^{-1}(U) \in \tau(L)\}$, induced by $\pi$.

A linear topological space $L$ is *(completely) normable* if there exists a norm $\|\cdot\|$ on $L$ such that the metric $d(x, y) = \|x - y\|$ (is complete and) generates $\tau(L)$. The space $L$ is *barreled* if any convex closed balanced and absorbing subset of $L$ is a neighborhood of $\mathbf{0}$ in $L$. Suppose that $X$ is a topological space and $(Y, \mathcal{U})$ is a uniform space. A family $\mathcal{F} \subset C(X, Y)$ is called *equicontinuous at a point $x \in X$* if, for any $U \in \mathcal{U}$, there exists an open $V \subset X$ such that $x \in V$ and $(f(x), f(y)) \in U$ for any $y \in V$ and $f \in \mathcal{F}$. If the family $\mathcal{F}$ is equicontinuous at every $x \in X$, it is called *equicontinuous*. Note that a set $\mathcal{F} \subset C(X)$ is equicontinuous at $x \in X$ if and only if, for any $\varepsilon > 0$, there is $U \in \tau(x, X)$ such that $|f(y) - f(x)| < \varepsilon$ for any $y \in U$ and $f \in \mathcal{F}$. If $L$ and $M$ are linear topological spaces and $\mathcal{F}$ is a set of linear continuous maps from $L$ to $M$, then $\mathcal{F}$ is equicontinuous if and only if, for any $U \in \tau(\mathbf{0}_M, M)$, there exists $V \in \tau(\mathbf{0}_L, L)$ such that $f(V) \subset U$ for any $f \in \mathcal{F}$.

Suppose that $X$ is a space and $\mathcal{C}$ is a cover of $X$. If $\varepsilon > 0$ and $P_1, \ldots, P_n \in \mathcal{C}$ then $[P_1, \ldots, P_n, \varepsilon] = \{f \in C(X) : f(P_i) \subset (-\varepsilon, \varepsilon)$ for all $i \leq n\}$. The family $\tau_\mathcal{C} = \{\emptyset\} \cup \{U \subset C(X) :$ for any $f \in U$, there are $n \in \mathbb{N}$, $\varepsilon > 0$ and $P_1, \ldots, P_n \in \mathcal{C}$ such that $f + [P_1, \ldots, P_n, \varepsilon] \subset U\}$ is a topology on $C(X)$ and, if every $P \in \mathcal{C}$ is bounded in $X$, then $(C(X), \tau_\mathcal{C})$ is a linear topological space. The topology $\tau_\mathcal{C}$ is called *the topology of uniform convergence on the elements of $\mathcal{C}$*. If $\mathcal{C} = \{K \subset X : K$ is compact$\}$ then $(C(X), \tau_\mathcal{C})$ is denoted by $C_k(X)$ and $\tau_\mathcal{C}$ is called *the compact-open topology*. Denote by $\mathcal{BD}_X$ the family of all bounded subsets of $X$. The set $C(X)$ with the topology of uniform convergence on the elements of $\mathcal{BD}_X$ is denoted by $C_b(X)$. A function $f : X \to \mathbb{R}$ is called *b-continuous* if, for any $B \in \mathcal{BD}_X$, there is $g \in C(X)$ with $g|B = f$. The space $X$ is called *a $b_f$-space* if every $b$-continuous function on $X$ is continuous. A set $P \subset C(X)$ is *pointwise bounded* if the set $\{f(x) : f \in P\}$ is bounded in $\mathbb{R}$ for every $x \in X$.

**201.** Prove that the topology of any linear topological $T_0$-space is Tychonoff.

**202.** Let $L$ be a linear topological Tychonoff space. Prove that, for any local base $\mathcal{B}$ of the space $L$ at $\mathbf{0}$, the following properties hold:

(1) for any $U, V \in \mathcal{B}$, there is $W \in \mathcal{B}$ such that $W \subset U \cap V$;
(2) every $B \in \mathcal{B}$ is an absorbing set and $\bigcap \mathcal{B} = \{\mathbf{0}\}$;
(3) for any $U \in \mathcal{B}$, there exists $V \in \mathcal{B}$ such that $V + V \subset U$;
(4) for any $U \in \mathcal{B}$ and $x \in U$, there exists $V \in \mathcal{B}$ such that $x + V \subset U$;
(5) for any $U \in \mathcal{B}$ and $\varepsilon > 0$ there is $V \in \mathcal{B}$ such that $\lambda \cdot V \subset U$ for any $\lambda \in (-\varepsilon, \varepsilon)$.

Prove that, if $L$ is a linear space without topology and $\mathcal{B}$ is a family of subsets of $L$ which has the properties (1)–(5) then there exists a unique Tychonoff topology $\tau$ on $L$ such that $(L, \tau)$ is a linear topological space and $\mathcal{B}$ is a local base of $\tau$ at $\mathbf{0}$.

**203.** Let $L$ be a linear topological space. Prove that

(1) for any local base $\mathcal{B}$ of $L$ at $\mathbf{0}$ and any $A \subset L$, we have $\overline{A} = \bigcap \{A + V : V \in \mathcal{B}\}$;
(2) for any $A, B \subset L$, we have $\overline{A} + \overline{B} \subset \overline{A + B}$;
(3) if $M$ is a linear subspace of $L$ then $\overline{M}$ is also a linear subspace of $L$;
(4) if $C$ is a convex subset of $L$ then the sets $\overline{C}$ and $\text{Int}(C)$ are also convex;
(5) if $B$ is a balanced subset of $L$ then $\overline{B}$ is also balanced; if, additionally, we have $\mathbf{0} \in \text{Int}(B)$ then $\text{Int}(B)$ is also balanced;
(6) if $E$ is an $l$-bounded subset of $L$ then $\overline{E}$ is also $l$-bounded.

**204.** Let $L$ be a linear topological space. Prove that

(1) every neighborhood of $\mathbf{0}$ contains an open balanced neighborhood of $\mathbf{0}$;
(2) every convex neighborhood of $\mathbf{0}$ contains an open convex balanced neighborhood of $\mathbf{0}$.

Deduce from (2) that any locally convex linear topological space has a local base $\mathcal{B}$ at $\mathbf{0}$ such that each $U \in \mathcal{B}$ is convex and balanced.

**205.** Let $L$ be a linear topological space. Given a nontrivial linear functional $f : L \to \mathbb{R}$, prove that the following properties are equivalent:

(i) $f$ is continuous;
(ii) $f^{-1}(0)$ is closed in $L$;
(iii) $f^{-1}(0)$ is not dense in $L$;
(iv) there exists $U \in \tau(\mathbf{0}, L)$ such that $f(U)$ is a bounded subset of $\mathbb{R}$.

**206.** Suppose that $L$ is a locally convex linear topological space which has a countable local base at $\mathbf{0}$. Prove that there exists a metric $d$ on the set $L$ with the following properties:

(i) $d$ generates the topology of $L$;
(ii) all $d$-open balls are convex and all balls with the center at $\mathbf{0}$ are balanced;
(iii) the metric $d$ is invariant, i.e., $d(x+z, y+z) = d(x, y)$ for all $x, y, z \in L$.

As a consequence, a locally convex space is metrizable if and only if it has countable character.

**207.** Let $p$ be a seminorm on a linear space $L$. Prove that

(1) $p(\mathbf{0}) = 0$ and $p(x) \geq 0$ for any $x \in L$;
(2) $|p(x) - p(y)| \leq p(x - y)$ for any $x, y \in L$;
(3) $\{x \in L : p(x) = 0\}$ is a linear subspace of $L$;
(4) the set $B = \{x : p(x) < 1\}$ is convex, balanced, absorbing, and $p = \mu_B$.

**208.** Let $A$ be a convex absorbing set in a linear space $L$. Prove that

(1) $\mu_A(x + y) \leq \mu_A(x) + \mu_A(y)$ for any $x, y \in L$;
(2) $\mu_A(tx) = t\mu_A(x)$ for any $x \in L$ and $t \geq 0$;
(3) if $A$ is balanced then $\mu_A$ is a seminorm;
(4) if $B = \{x \in L : \mu_A(x) < 1\}$ and $C = \{x \in L : \mu_A(x) \leq 1\}$ then $B \subset A \subset C$ and $\mu_A = \mu_B = \mu_C$.

**209.** Given a locally convex linear topological space $L$, take any local base $\mathcal{B}$ at $\mathbf{0}$ such that all elements of $\mathcal{B}$ are convex and balanced. Prove that $\{\mu_V : V \in \mathcal{B}\}$ is a separating family of continuous seminorms on $L$.

**210.** Let $\mathcal{P}$ be a separating family of seminorms on a linear space $L$. Given $p \in \mathcal{P}$ and $n \in \mathbb{N}$, let $O(p, n) = \{x \in L : p(x) < \frac{1}{n}\}$. Prove that the family $\mathcal{B} = \{O(p_1, n) \cap \ldots \cap O(p_n, n) : n \in \mathbb{N}, p_1, \ldots, p_n \in \mathcal{P}\}$ is a convex balanced local base at $\mathbf{0}$ for some topology $\tau$ on $L$ such that $(L, \tau)$ is a locally convex space in which all elements of $\mathcal{P}$ are continuous and any $E \subset L$ is $l$-bounded if and only if $p(E)$ is bounded for any $p \in \mathcal{P}$.

**211.** Prove that a linear topological space is normable if and only if it has a convex $l$-bounded neighborhood of zero.

**212.** Let $N$ be a closed subspace of a linear topological space $L$. Prove that

(1) the quotient topology of $L/N$ makes $L/N$ a linear topological space;
(2) the quotient map $\pi : L \to L/N$ is linear, open, and continuous;
(3) If $\mathcal{P} \in \{\text{metrizability, local convexity, normability, complete normability}\}$ and $L$ has $\mathcal{P}$ then $L/N$ also has $\mathcal{P}$.

**213.** Prove that any product of locally convex spaces is a locally convex space.

**214.** Suppose that $L$ and $M$ are linear topological spaces and $\Phi$ is an equicontinuous family of linear maps from $L$ to $M$. Prove that, for any $l$-bounded set $A \subset L$ there is an $l$-bounded set $B \subset M$ such that $f(A) \subset B$ for all $f \in \Phi$.

**215.** Suppose that $L$ and $M$ are linear topological spaces and $\Phi$ is a family of linear continuous maps from $L$ to $M$. Let $\Phi(x) = \{f(x) : f \in \Phi\}$ for every $x \in L$ and assume that the set $B = \{x \in L : \Phi(x) \text{ is } l\text{-bounded in } M\}$ is of second category in $L$. Prove that $B = L$ and the family $\Phi$ is equicontinuous.

**216.** (Hahn–Banach theorem) Let $L$ be a linear space (without topology). Suppose that we are given a map $p : L \to \mathbb{R}$ such that $p(x + y) \leq p(x) + p(y)$ and $p(tx) = tp(x)$ for all $x, y \in L$ and $t \geq 0$. Prove that, for any linear subspace $M$ of the linear space $L$ and any linear functional $f : M \to \mathbb{R}$ such that $f(x) \leq p(x)$ for any $x \in M$, there exists a linear functional $F : L \to \mathbb{R}$ such that $F|M = f$ and $-p(-x) \leq F(x) \leq p(x)$ for any $x \in L$.

**217.** Let $L$ be a linear space (without topology). Suppose that we are given a seminorm $p : L \to \mathbb{R}$, a linear subspace $M \subset L$ and a linear functional $f : M \to \mathbb{R}$ such that $|f(x)| \leq p(x)$ for any $x \in M$. Prove that there exists a linear functional $F : L \to \mathbb{R}$ such that $F|M = f$ and $|F(x)| \leq p(x)$ for any $x \in L$.

**218.** Given a linear topological space $L$ prove that any nontrivial continuous linear functional $f : L \to \mathbb{R}$ is an open map.

**219.** Let $L$ be a linear topological space and suppose that $A$ and $B$ are nonempty disjoint convex subsets of $L$ and $A$ is open. Prove that there exists a continuous linear functional $f : L \to \mathbb{R}$ such that, for some $r \in \mathbb{R}$, we have $f(x) < r \leq f(y)$ for any $x \in A$ and $y \in B$.

**220.** Let $L$ be a locally convex linear topological space. Suppose that $A$ and $B$ are disjoint convex subsets of $L$ such that $A$ is compact and $B$ is closed. Prove that there exists a continuous linear functional $f : L \to \mathbb{R}$ such that, for some $r, s \in \mathbb{R}$, we have $f(x) < r < s < f(y)$ for any $x \in A$ and $y \in B$.

**221.** Let $L$ be a locally convex linear topological space. Prove that $L^*$ separates the points of $L$.

**222.** Let $M$ be a linear subspace of a locally convex linear topological space $L$ and $x_0 \notin \overline{M}$. Prove that there exists $f \in L^*$ such that $f(x_0) = 1$ and f(M)={0}.

**223.** Let $B$ be a closed convex balanced subset of a locally convex space $L$. Prove that, for any $x \in L \backslash B$, there exists a continuous linear functional $f : L \to \mathbb{R}$ such that $f(B) \subset [-1, 1]$ and $f(x) > 1$.

**224.** Let $L$ be a locally convex linear topological space. Given a linear subspace $M$ of the linear space $L$ and a continuous linear functional $f : M \to \mathbb{R}$, prove that there exists a functional $g \in L^*$ such that $g|M = f$.

**225.** Given a linear space $L$ (without topology) denote by $L'$ the family of all linear functionals on $L$. Suppose that $M \subset L'$ is a linear subspace of $L'$ (i.e., $\alpha f + \beta g \in M$ whenever $f, g \in M$ and $\alpha, \beta \in \mathbb{R}$) and $M$ separates the points of $L$; let $\mu$ be the topology generated by the set $M$. Then $L_M = (L, \mu)$ is a locally convex space and $(L_M)^* = M$. Deduce from this fact that if $L$ is a locally convex space and $L_w$ is the set $L$ with the weak topology of the space $L$ then $L_w$ is a locally convex space such that $(L_w)^* = L^*$.

**226.** Let $E$ be a convex subset of a locally convex space $L$. Prove that the closure of $E$ in $L$ coincides with the closure of $E$ in the weak topology of $L$.

**227.** Let $V$ be a neighborhood of $\mathbf{0}$ in a locally convex space $L$. Prove that the set $P(V) = \{f \in L^* : f(V) \subset [-1, 1]\}$ is compact if considered with the topology induced from $C_p(L)$.

**228.** Given $n \in \mathbb{N}$ suppose that $L$ is a linear topological space and $M$ is a linear subspace of $L$ of linear dimension $n$. Prove that $M$ is closed in $L$ and every linear isomorphism $\varphi : \mathbb{R}^n \to M$ is a homeomorphism.

**229.** Given a linear topological space $L$ prove that the following conditions are equivalent:

  (i) $L$ has a finite Hamel basis, i.e., the linear dimension of $L$ is finite;
  (ii) $\dim L \leq n$ for some $n \in \mathbb{N}$;
  (iii) $L$ is locally compact.

**230.** Suppose that $L$ is a finite-dimensional linear topological space. Prove that any linear functional $f : L \to \mathbb{R}$ is continuous on $L$, i.e., $L' = L^*$. Give an example of an infinite-dimensional locally convex space $M$ such that $M' = M^*$.

**231.** Let $L$ be a locally convex space. Denote by $L' \subset \mathbb{R}^L$ the set of all (not necessarily continuous) linear functionals on $L$ with the topology induced from $\mathbb{R}^L$. Prove that $L^*$ is dense in $L'$.

**232.** Given a linear space $L$ let $L' \subset \mathbb{R}^L$ be the set of all linear functionals on $L$ with the topology induced from $\mathbb{R}^L$. Prove that $L'$ is linearly homeomorphic to $\mathbb{R}^B$ for some $B$.

**233.** For any linear topological space $L$ denote by $\tau_w(L)$ the weak topology of the space $L$. Prove that

(1) if $\tau_w(L) = \tau(L)$ and $M$ is a linear subspace of $L$ then $\tau_w(M) = \tau(M)$;
(2) for any space $X$, the topology of $C_p(X)$ coincides with its weak topology;
(3) for any space $X$, the topology of $L_p(X)$ coincides with its weak topology.

**234.** Suppose that $L$ is a locally convex space with its weak topology and $X \subset L$ is a Hamel basis in $L$. Prove that the following conditions are equivalent:

(i) there exists a linear homeomorphism $h : L \to L_p(X)$ such that $h(x) = x$ for all $x \in X$;
(ii) for every $f \in C(X)$ there exists a continuous linear functional $\varphi : L \to \mathbb{R}$ such that $\varphi|X = f$;
(iii) for every continuous map $f : X \to M$ from $X$ to a locally convex space $M$ with its weak topology, there exists a continuous linear map $\Phi : L \to M$ such that $\Phi|X = f$.

**235.** Given a space $X$ let $(L_p(X))^* = \{\varphi \in C_p(L_p(X)) : \varphi$ is a linear functional on $L_p(X)\}$ and consider the restriction map $\pi : (L_p(X))^* \to C_p(X)$. Prove that $(L_p(X))^*$ is a closed linear subspace of $C_p(L_p(X))$ and $\pi$ is a linear homeomorphism. Deduce from this fact that the operation of extending continuous real-valued functions on $X$ to continuous linear functionals on $L_p(X)$ is also a linear homeomorphism between $C_p(X)$ and $(L_p(X))^*$.

**236.** Prove that any $L_p(X)$ is homeomorphic to a dense subspace of $\mathbb{R}^A$ for some $A$. Deduce from this fact that every uncountable regular cardinal is a precaliber of $L_p(X)$. In particular, $c(L_p(X)) = \omega$ for any space $X$.

**237.** Given spaces $X$ and $Y$ prove that

(i) there exists a linear continuous map of $C_p(X)$ onto $C_p(Y)$ if and only if $L_p(Y)$ is linearly homeomorphic to a linear subspace of $L_p(X)$;
(ii) there exists a linear continuous open map of $C_p(X)$ onto $C_p(Y)$ if and only if $L_p(Y)$ is linearly homeomorphic to a closed linear subspace of $L_p(X)$;
(iii) the space $C_p(X)$ linearly condenses onto $C_p(Y)$ if and only if $L_p(Y)$ is linearly homeomorphic to a dense linear subspace of $L_p(X)$;
(iv) $C_p(X)$ is linearly homeomorphic to $C_p(Y)$ if and only if $L_p(Y)$ is linearly homeomorphic to $L_p(X)$.

**238.** Given spaces $X$ and $Y$ prove that

    (i) there is a linear continuous map of $C_p(X)$ onto $C_p(Y)$ if and only if $Y$ is homeomorphic to a subspace $Y' \subset L_p(X)$ such that every $f \in C(Y')$ extends to a continuous linear functional on $L_p(X)$;

    (ii) the space $C_p(X)$ linearly condenses onto $C_p(Y)$ if and only if $Y$ is homeomorphic to a subspace $Y' \subset L_p(X)$ such that every $f \in C(Y')$ extends to a uniquely determined continuous linear functional on $L_p(X)$;

    (iii) $C_p(X)$ is linearly homeomorphic to $C_p(Y)$ if and only if $Y$ is homeomorphic to some $Y' \subset L_p(X)$ whose linear hull coincides with $L_p(X)$ and every $f \in C(Y')$ extends to a continuous linear functional on $L_p(X)$.

**239.** Let $\mathcal{P}$ be a class of spaces which have the following properties:

    (1) if $Y \in \mathcal{P}$ and $Z$ is a continuous image of $Y$ then $Z \in \mathcal{P}$;

    (2) if $Y = \bigcup \{Y_i : i \in \omega\}$, $Y_i \subset Y_{i+1}$, $Y_i \in \mathcal{P}$ and $Y_i$ closed in $Y$ for every $i \in \omega$, then $Y \in \mathcal{P}$;

    (3) if $Y \in \mathcal{P}$ and $n \in \mathbb{N}$ then $Y^n \times \mathbb{R}^n \in \mathcal{P}$;

    Prove that if a space $X$ belongs to $\mathcal{P}$ then $L_p(X) \in \mathcal{P}$.

**240.** Prove that $iw(L_p(X)) = \psi(L_p(X)) = \Delta(L_p(X)) = iw(X)$ for any space $X$; show that we also have $nw(X) = nw(L_p(X))$ and $d(X) = d(L_p(X))$.

**241.** Prove that, for any space $X$, we have the following equalities.

    (i) $s^*(X) = s(L_p(X)) = s^*(L_p(X))$;

    (ii) $hl^*(X) = hl(L_p(X)) = hl^*(L_p(X))$;

    (iii) $hd^*(X) = hd(L_p(X)) = hd^*(L_p(X))$.

**242.** Given a space $X$ prove that $l^*(X) = l(L_p(X)) = l^*(L_p(X))$ and $ext^*(X) = ext(L_p(X)) = ext^*(L_p(X))$.

**243.** Prove that an uncountable regular cardinal $\kappa$ is a caliber of $X$ if and only if $\kappa$ is a caliber of $L_p(X)$.

**244.** Denote by $\mathcal{L}$ the following collection of classes of Tychonoff spaces: {analytic spaces, $K$-analytic spaces, $\sigma$-compact spaces, Lindelöf $\Sigma$-spaces, realcompact spaces}. Prove that, for any class $\mathcal{P}$ from the list $\mathcal{L}$, a space $X$ belongs to $\mathcal{P}$ if and only if $L_p(X)$ belongs to $\mathcal{P}$.

**245.** Given $w = \lambda_1 x_1 + \ldots + \lambda_n x_n \in L_p(X)$, where $x_1, \ldots, x_n \in X$ and $\lambda_1, \ldots, \lambda_n \in \mathbb{R} \backslash \{0\}$, let $\mathrm{supp}(w) = \{x_1, \ldots, x_n\}$. If $w = \mathbf{0}$, then $\mathrm{supp}(w) = \emptyset$. Say that a set $B \subset L_p(X)$ is *weakly bounded* if $\xi(B)$ is a bounded subset of $\mathbb{R}$ for any continuous linear functional $\xi : L_p(X) \to \mathbb{R}$. Observe that any bounded subset of $L_p(X)$ is weakly bounded and prove that, for any weakly bounded set $B \subset L_p(X)$, the set $\bigcup \{\mathrm{supp}(w) : w \in B\}$ is bounded in the space $X$.

**246.** Prove that, for any Dieudonné complete space $X$, if $A$ is a bounded subset of $L_p(X)$ then $\overline{A}$ is compact.

**247.** Suppose that a space $X$ has a weaker metrizable topology and $A$ is a bounded subset of $L_p(X)$. Prove that $\overline{A}$ is compact and metrizable.

**248.** Prove that, for any infinite pseudocompact space $X$, there exists an infinite closed discrete set $D$ in the space $L_p(X)$ which is weakly bounded in $L_p(X)$. Therefore, even for a metrizable compact space $X$, the closure of a weakly bounded subset of $L_p(X)$ can fail to be compact.

**249.** Give an example of a space $X$ in which all compact subspaces are metrizable while there are non-metrizable compact subspaces in $L_p(X)$.

**250.** Given spaces $X$ and $Y$ and a continuous map $\varphi : X \to Y$ observe that there exists a unique continuous linear map $u_\varphi : L_p(X) \to L_p(Y)$ such that $u_\varphi|X = \varphi$. Prove that the following conditions are equivalent for any continuous onto map $\varphi : X \to Y$.

  (i)  The map $\varphi$ is $\mathbb{R}$-quotient.
  (ii)  The map $u_\varphi$ is $\mathbb{R}$-quotient.
  (iii).  The map $u_\varphi$ is quotient.
  (iv)  The map $u_\varphi$ is open.

**251.** Let $f : X \to Y$ be an $\mathbb{R}$-quotient map. Prove that, for any open $U \subset Y$, the map $f|(f^{-1}(U)) : f^{-1}(U) \to U$ is also $\mathbb{R}$-quotient.

**252.** Let $X$ be a Tychonoff space. Prove that, for any nonempty closed set $F \subset X$, the $\mathbb{R}$-quotient space $X_F$ is also Tychonoff and if $p_F : X \to X_F$ is the contraction map then $p_F|(X\backslash F) : X\backslash F \to X_F\backslash\{F\}$ is a homeomorphism.

**253.** Suppose that $X$ is a space and $F$ is a nonempty closed subspace of $X$; in the $\mathbb{R}$-quotient space $X_F$ denote by $a_F$ the point represented by the set $F$. Say that $F$ is *deeply inside* a set $U \in \tau(X)$ if there exists a zero-set $G$ in the space $X$ such that $F \subset G \subset U$. For the family $\mathcal{U} = \{U : U$ is a cozero subset of $X$ and $F$ is deeply inside the set $U\}$ prove that $\mathcal{V} = \{\{a_F\} \cup (U\backslash F) : U \in \mathcal{U}\}$ is a local base of the space $X_F$ at the point $a_F$.

**254.** Suppose that $X$ is a normal space and $F$ is a nonempty closed subspace of $X$; in the $\mathbb{R}$-quotient space $X_F$ denote by $a_F$ the point represented by the set $F$. Prove that $U \in \tau(a_F, X_F)$ if and only if $(U\backslash\{a_F\}) \cup F$ is an open neighborhood of $F$ in the space $X$.

**255.** Suppose that $X$ is a space and $K$ is a nonempty compact subspace of $X$; in the $\mathbb{R}$-quotient space $X_K$ denote by $a_K$ the point represented by the set $K$. Prove that $U \in \tau(a_K, X_K)$ if and only if $(U\backslash\{a_K\}) \cup K$ is an open neighborhood of $K$ in the space $X$.

**256.** Given a nonempty space $X$ prove that closed sets $P, Q \subset X$ are parallel retracts of $X$ if and only if there exists a retraction $r : X \to P$ such that $r|Q : Q \to P$ is a homeomorphism.

**257.** (Okunev's method of constructing $l$-equivalent spaces). Suppose that $P$ and $Q$ are parallel retracts of a nonempty space $X$. Prove that the completely regular quotient spaces $X_P$ and $X_Q$ are $l$-equivalent.

**258.** Suppose that $K$ is a nonempty $l$-embedded subspace of a space $X$ and fix a point $a \notin X$. Prove that the spaces $X \oplus \{a\}$ and $X_K \oplus K$ are $l$-equivalent. Deduce from this fact that if $K$ is a retract of the space $X$ then $X \oplus \{a\} \overset{l}{\sim} X_K \oplus K$. Here $X_K$ is the $\mathbb{R}$-quotient space obtained by contracting $K$ to a point.

**259.** Given a space $X_i$ and a point $x_i \in X_i$ for any $i = 1, \ldots, n$ consider the space $X = X_1 \oplus \ldots \oplus X_n$ and the set $F = \{x_1, \ldots, x_n\} \subset X$. The $\mathbb{R}$-quotient space $X_F$ is denoted by $(X_1, x_1) \vee \ldots \vee (X_n, x_n)$ and called *a bunch* of spaces $X_1, \ldots, X_n$ with respect to the points $x_1, \ldots, x_n$. Prove that if we choose any point $y_i \in X_i$ for every $i = 1, \ldots, n$ then the spaces $(X_1, x_1) \vee \ldots \vee (X_n, x_n)$ and $(X_1, y_1) \vee \ldots \vee (X_n, y_n)$ are $l$-equivalent.

**260.** Let $K$ be a retract of a nonempty space $X$ and fix any point $z \in K$. Denote by $a_K$ the point of the space $X_K$ represented by the set $K$. Prove that the space $X$ is $l$-equivalent to the bunch $(X_K, a_K) \vee (K, z)$ of the spaces $X_K$ and $K$ with respect to the points $a_K$ and $z$.

**261.** Assume that $K$ and $L$ are retracts of a nonempty space $X$ and there exists a retraction $r : X \to L$ such that $r(K) = K \cap L = \{a\}$ for some point $a \in L$; let $M = K \cup L$. Prove that the space $X$ is $l$-equivalent to the bunch $(X_M, c_0) \vee (K, c_1) \vee (L, c_2)$ where the points $c_0 \in X_M$, $c_1 \in K$ and $c_2 \in L$ are chosen arbitrarily.

**262.** Given spaces $Y$ and $Z$ consider the space $X = Y \times Z$; choose arbitrary points $y_0 \in Y$, $z_0 \in Z$ and let $M = (Y \times \{z_0\}) \cup (\{y_0\} \times Z)$. Prove that, for any $x_0 \in X_M$, the space $X$ is $l$-equivalent to the bunch $(X_M, x_0) \vee (Y, y_0) \vee (Z, z_0)$.

**263.** Let $a = 0$ and $a_n = \frac{1}{n+1}$ for all $n \in \omega$; then $S = \{a_n : n \in \omega\} \cup \{a\}$ is a faithfully indexed convergent sequence with limit $a$. Given an infinite cardinal $\kappa$ consider the discrete space $D(\kappa)$ of cardinality $\kappa$ and let $E = D(\kappa) \times S$. Observe that $F = D(\kappa) \times \{a\}$ is a retract of $E$; as usual let $E_F$ be the $\mathbb{R}$-quotient space obtained by contracting $F$ to a point. The space $E_F$ will be denoted by $V(\kappa)$; it is often called the *Fréchet–Urysohn $\kappa$-fan*. The space $V(\omega)$ is called *the Fréchet–Urysohn fan*. Prove that $V(\kappa)$ is $l$-equivalent to $D(\kappa) \times S$ for any infinite cardinal $\kappa$. Deduce from this fact that

  (i) there exist $l$-equivalent spaces $X$ and $Y$ with $w(X) \neq w(Y)$ and $\chi(X) \neq \chi(Y)$;

  (ii) metrizability is not preserved by $l$-equivalence;

  (iii) a space $l$-equivalent to a locally compact space need not be Čech-complete.

**264.** Given infinite cardinals $\kappa_1, \ldots, \kappa_n$ prove that $A(\kappa_1) \oplus \ldots \oplus A(\kappa_n)$ is $l$-equivalent to the space $A(\kappa)$ where $\kappa = \max\{\kappa_1, \ldots, \kappa_n\}$.

**265.** Given a family of spaces $\{X_t : t \in T\}$ let $X = \bigoplus_{t \in T} X_t$ and prove that $C_p(X)$ is linearly homeomorphic to $\prod_{t \in T} C_p(X_t)$. Deduce from this fact that if $X_t \overset{l}{\sim} Y_t$ for any $t \in T$ then $X \overset{l}{\sim} Y = \bigoplus_{t \in T} Y_t$.

**266.** Suppose that a space $J_i$ is homeomorphic to $\mathbb{I}$ for any $i = 1, \ldots, n$ and let $J = \bigoplus\{J_i : 1 \leq i \leq n\}$. Prove that the space $J \oplus D$ is $l$-equivalent to $\mathbb{I}$ for any finite space $D$. Deduce from this fact that

  (i) connectedness is not preserved by $l$-equivalence;

  (ii) for any cardinal $\kappa$ there exist $l$-equivalent spaces $X$ and $Y$ such that $X$ has no isolated points and $Y$ has $\kappa$-many isolated points.

**267.** Let $X$ be a compact space with $|X| = \kappa \geq \omega$. Prove that $AD(X)$ is $l$-equivalent to $X \oplus A(\kappa)$. Here $AD(X)$ is the Alexandroff double of the space $X$ and $A(\kappa)$ is the one-point compactification of a discrete space of cardinality $\kappa$.

**268.** Let $X$ be a compact space such that $|X| = \kappa \geq \omega$. Prove that $AD(X)$ is $l$-equivalent to $AD(X) \oplus A(\kappa)$. Here $AD(X)$ is the Alexandroff double of the space $X$.

**269.** Prove that there exist $l$-equivalent compact spaces $X$ and $Y$ such that $\chi(X) \neq \chi(Y)$. As a consequence, pseudocharacter is not $l$-invariant.

**270.** Prove that there exist $l$-equivalent compact spaces $X$ and $Y$ such that $X$ has nontrivial convergent sequences and $Y$ does not have any.

**271.** Prove that for any uncountable regular cardinal $\kappa$ with its usual order topology we have the equivalencies $\kappa \overset{l}{\sim} \kappa \oplus A(\kappa)$ and $(\kappa+1) \overset{l}{\sim} (\kappa+1) \oplus A(\kappa)$. Therefore there exist $l$-equivalent spaces $X$ and $Y$ such that all compact subspaces of $X$ are metrizable while $Y$ has non-metrizable compact subspaces.

**272.** Suppose that compact spaces $X$ and $Y$ are $l$-equivalent. Prove that $X \times Z$ is $l$-equivalent to $Y \times Z$ for any space $Z$.

**273.** Given a family $\{X_1, \ldots, X_n\}$ of compact spaces assume that $X_i \overset{l}{\sim} Y_i$ for all $i \in \{1, \ldots, n\}$. Prove that the spaces $X = X_1 \times \ldots \times X_n$ and $Y = Y_1 \times \ldots \times Y_n$ are $l$-equivalent.

**274.** Give an example of $l$-equivalent spaces $X$ and $Y$ such that $X \times Z$ is not $t$-equivalent to $Y \times Z$ for some space $Z$.

**275.** Give an example of $l$-equivalent spaces $X$ and $Y$ such that $X \times X$ is not $t$-equivalent to $Y \times Y$.

**276.** Given infinite cardinals $\kappa_1, \ldots, \kappa_n$ prove that $A(\kappa_1) \times \ldots \times A(\kappa_n)$ is $l$-equivalent to the space $A(\kappa)$ where $\kappa = \max\{\kappa_1, \ldots, \kappa_n\}$.

**277.** Prove that there exist $l$-equivalent spaces $X$ and $Y$ such that $X$ is hereditarily paracompact while $Y$ is not hereditarily normal.

**278.** Prove that $L_p(D)$ is $l$-equivalent to $L_p(D) \oplus D$ for any infinite discrete space $D$. Deduce from this fact that the Souslin property is not $l$-invariant.

**279.** Given spaces $X$ and $Y$ suppose that $\varphi : C_p(X) \to C_p(Y)$ is a continuous linear surjection. Prove that,

(i) for any point $y \in Y$, there exist uniquely determined $n = n(y) \in \mathbb{N}$, distinct points $x_1(y), \ldots, x_n(y) \in X$ and numbers $\lambda_1(y), \ldots, \lambda_n(y) \in \mathbb{R}\backslash\{0\}$ such that $\varphi(f)(y) = \sum_{i=1}^{n(y)} \lambda_i(y) f(x_i(y))$ for any $f \in C_p(X)$; for further reference denote the set $\{x_1(y), \ldots, x_n(y)\}$ by $\operatorname{supp}(\varphi, y)$.

(ii) if $\varphi^* : C_p(C_p(Y)) \to C_p(C_p(X))$ is the dual map of $\varphi$ then $\varphi^*$ embeds $L_p(Y)$ in $L_p(X)$ and $\varphi^*(y) = \lambda_1(y)x_1(y) + \ldots + \lambda_{n(y)}(y)x_{n(y)}(y)$ for any $y \in Y$.

**280.** Suppose that $\varphi : C_p(X) \to C_p(Y)$ is a continuous linear surjection and let $\xi(y) = \operatorname{supp}(\varphi, y)$ for any $y \in Y$. Prove that the map $\xi : Y \to \exp(X)$ is lower semicontinuous.

**281.** Given a continuous linear surjection $\varphi : C_p(X) \to C_p(Y)$ prove that, for any bounded subset $A$ of the space $Y$, the set $\text{supp}(A) = \bigcup\{\text{supp}(\varphi, y) : y \in A\}$ is bounded in $X$.

**282.** Suppose that $\varphi : C_p(X) \to C_p(Y)$ is a continuous linear surjection. Prove that, for any bounded subset $B \subset X$, the set $C = \{y \in Y : \text{supp}(\varphi, y) \subset B\}$ is bounded in $Y$.

**283.** Say that $X$ is a *$\mu$-space* if $\overline{A}$ is compact for any bounded set $A \subset X$. Prove that $X$ is a $\mu$-space if and only if $L_p(X)$ is a $\mu$-space. As a consequence, $\mu$-property is preserved by $l$-equivalence.

**284.** Given spaces $X$ and $Y$ assume that $X$ is a $\mu$-space and there exists a linear surjection $\varphi : C_p(X) \to C_p(Y)$ which is an $\mathbb{R}$-quotient map. Prove that $Y$ is also a $\mu$-space. Give an example of a compact space $X$ (which is, automatically, a $\mu$-space) such that there exists a continuous linear surjection of $C_p(X)$ onto $C_p(Y)$ for some $Y$ which is not a $\mu$-space.

**285.** Given $\mu$-spaces $X$ and $Y$, let $\varphi : C_p(X) \to C_p(Y)$ be a continuous linear surjection. Prove that, if $X$ is compact then $Y$ is also compact. Observe that the same conclusion about $Y$ may be false if $Y$ is not a $\mu$-space.

**286.** Given $\mu$-spaces $X$ and $Y$, let $\varphi : C_p(X) \to C_p(Y)$ be a continuous linear surjection. Prove that, if $X$ is $\sigma$-compact then $Y$ is also $\sigma$-compact. Observe that the same conclusion about $Y$ may be false if $Y$ is not a $\mu$-space.

**287.** For any space $X$ let $\mathcal{K}(X)$ be the family of all compact subspaces of $X$. Prove that a second countable space $X$ is Čech-complete if and only if there exists a Polish space $M$ and a map $\varphi : \mathcal{K}(M) \to \mathcal{K}(X)$ such that, for any $F, G \in \mathcal{K}(M)$ the inclusion $F \subset G$ implies $\varphi(F) \subset \varphi(G)$ and, for any $P \in \mathcal{K}(X)$, there exists $F \in \mathcal{K}(M)$ such that $\varphi(F) \supset P$.

**288.** Let $X$ and $Y$ be second countable spaces for which there is a continuous linear surjection of $C_p(X)$ onto $C_p(Y)$. Prove that, if $X$ is Čech-complete then $Y$ is also Čech-complete. In particular, if two second countable spaces $X$ and $Y$ are $l$-equivalent then $X$ is Čech-complete if and only if so is $Y$.

**289.** Give an example of second countable $l$-equivalent spaces $X$ and $Y$ such that $X$ is pseudocomplete and $Y$ is not Baire. As a consequence, having a dense Čech-complete subspace is not an $l$-invariant property in the class of second countable spaces.

**290.** Prove that there exist $l$-equivalent $\sigma$-compact second countable spaces $X$ and $Y$ such that $X$ can be condensed onto a compact space and $Y$ doesn't have such a condensation.

**291.** Prove that a countable second countable space is scattered if and only if it is Čech-complete. Deduce from this fact that if $X$ and $Y$ are countable second countable $l$-equivalent spaces then $X$ is scattered if and only if $Y$ is scattered.

**292.** Let $X$ and $Y$ be metrizable spaces such that $C_p(X)$ is linearly homeomorphic to $C_p(X) \times C_p(X)$ and $C_p(Y)$ is linearly homeomorphic to $C_p(Y) \times C_p(Y)$. Prove that if $X$ embeds in $Y$ as a closed subspace and $Y$ embeds in $X$ as a closed subspace then $X$ and $Y$ are $l$-equivalent.

**293.** Let $X$ be a countable second countable space. Prove that the following properties are equivalent:

   (i)  $X$ is $l$-equivalent to $\mathbb{Q}$;

  (ii)  $X$ is not scattered;

 (iii)  $X$ has a subspace homeomorphic to $\mathbb{Q}$;

 (iv)  $X$ has a closed subspace homeomorphic to $\mathbb{Q}$.

**294.** Prove that, for any infinite cardinal $\kappa$ there exist $l$-equivalent spaces $X$ and $Y$ such that $X$ is dense-in-itself and $Y$ has a dense set of $\kappa$-many isolated points.

**295.** Suppose that a second countable space $X$ is Čech-complete and has a closed subspace homeomorphic to $\mathbb{R}^\omega$. Prove that $X$ is $l$-equivalent to $\mathbb{R}^\omega$.

**296.** Let $K$ be an uncountable metrizable compact space. Prove that $K$ is $l$-equivalent to $K \oplus E$ for any metrizable zero-dimensional compact space $E$.

**297.** Prove that a compact space $X$ is $l$-equivalent to the Cantor set $\mathbb{K}$ if and only if $X$ is metrizable, zero-dimensional, and uncountable. As a consequence, any two zero-dimensional metrizable uncountable compact spaces are $l$-equivalent.

**298.** Prove that a second countable space $X$ is $l$-equivalent to space $\mathbb{P}$ of the irrational numbers if and only if $X$ is non-$\sigma$-compact, zero-dimensional, and Čech-complete.

**299.** Given any $n \in \mathbb{N}$ prove that a compact set $K \subset \mathbb{R}^n$ is $l$-equivalent to $\mathbb{I}^n$ if and only if $\mathbb{I}^n$ embeds in $K$. Deduce from this fact that $K \overset{l}{\sim} \mathbb{I}^n$ if and only if $\dim K = n$.

**300.** Prove that a space $X$ is $l$-equivalent to $\mathbb{I}^\omega$ if and only if $X$ is compact, metrizable and $\mathbb{I}^\omega$ embeds in $X$.

## 1.4 Metrizable Spaces and l-Equivalence

All topological spaces are assumed to be Tychonoff. Given a space $X$, let $\tau^*(X) = \tau(X)\backslash\{\emptyset\}$ and $X^+ = X \oplus \{0\}$; by $\mathcal{K}(X)$ we denote the family of all compact subsets of $X$. If $L$ and $M$ are linear topological spaces, the expression $L \approx M$ says that $L$ is linearly homeomorphic to $M$; the space $M$ is *a linear topological factor of* $L$ if there exists a linear topological space $N$ such that $L \approx M \times N$. If, for a compact $X$, we treat $C(X)$ as a Banach space, then the respective norm is defined by $\|f\| = \sup\{f(x) : x \in X\}$ for any $f \in C(X)$. A sequence $\{x_n\} \subset L$ is called *linearly Cauchy* if, for any $U \in \tau(\mathbf{0}, L)$, there is $m \in \omega$ such that $x_n - x_k \in U$ for any $n, k \geq m$. Given spaces $X$ and $Y$, a map $f : X \to Y$ is called *compact-covering* if, for any compact $K \subset Y$, there is a compact $K' \subset X$ such that $f(K') = K$.

A space $X$ is called *hemicompact* if there exists a sequence $\{K_n : n \in \omega\}$ of compact subsets of $X$ such that, for any compact $K \subset X$, we have $K \subset K_n$ for some $n \in \omega$. The space $X$ is *an $\aleph_0$-space* if there exists a countable family $\mathcal{N} \subset \exp(X)$ which is *a network for* $\mathcal{K}(X)$, i.e., for any compact $K \subset X$ and any open $U \supset K$, we have $K \subset P \subset U$ for some $P \in \mathcal{N}$. The space $X$ is *a q-space* if, for any $x \in X$, there is a sequence $\{U_n : n \in \omega\} \subset \tau(x, X)$ such that the sequence $\{x_n : n \in \omega\}$ has an accumulation point whenever $x_n \in U_n$ for each $n \in \omega$. A space $X$ is called *a μ-space* if $\overline{A}$ is compact for any bounded $A \subset X$. A space $X$ is *σ-metacompact* if any open cover of $X$ has a $\sigma$-point-finite open refinement, i.e., a refinement which is a countable union of point-finite families.

A family $\mathcal{K}$ of subsets of a space $X$ *has a discrete open expansion* if there is a discrete family $\{U_K : K \in \mathcal{K}\} \subset \tau(X)$ such that $K \subset U_K$ for any $K \in \mathcal{K}$. A family $\mathcal{K}$ of nonempty compact subsets of $X$ is called *a moving off collection* if, for any compact $L \subset X$, there is $K \in \mathcal{K}$ such that $K \cap L = \emptyset$. A space $X$ has *the moving off property* if every moving off collection contains an infinite subcollection which has a discrete open expansion. For a locally compact non-compact $X$, let $\alpha(X)$ be its one-point compactification. A family $\mathcal{A}$ of subsets of $X$ is $T_1$-*separating* if, for any distinct $x, y \in X$, there are $A, B \in \mathcal{A}$ such that $A \cap \{x, y\} = \{x\}$ and $B \cap \{x, y\} = \{y\}$.

*The Gruenhage–Ma game* is played on a space $X$ by players $I$ and $II$. For every $n \in \mathbb{N}$ the $n$-th move of player $I$ is to choose a set $K_n \in \mathcal{K}(X)$ while the player $II$ responds with a set $L_n \in \mathcal{K}(X\backslash K_n)$. The player $I$ wins if the collection $\{L_n : n \in \mathbb{N}\}$ chosen by $II$ has a discrete open expansion. Let $\mathcal{G} = \bigcup\{(\mathcal{K}(X))^n : n \in \mathbb{N}\}$; *a strategy of player $II$ in Gruenhage–Ma game on a space $X$* is a map $s : \mathcal{G} \to \mathcal{K}(X)$, such that $s(K_1, \ldots, K_n) \in \mathcal{K}(X\backslash K_n)$ for every $(K_1, \ldots, K_n) \in \mathcal{G}$. A play $\{K_i, L_i : i \in \mathbb{N}\}$ of a Gruenhage–Ma game is said to have been *played by $II$ applying a strategy s* if $L_n = s(K_1, \ldots, K_n)$ for all $n \in \mathbb{N}$. A strategy $s$ of player $II$ in Gruenhage–Ma game is called *winning on a space $X$* if $II$ wins in every play on $X$, in which he (or she!) applies the strategy $s$.

A *Banach–Mazur game* on a space $X$ is a two-person game in which players $E$ (for empty) and $NE$ (for nonempty) take turns picking a nonempty open subset of $X$ contained in the opponent's previous move (if any). Thus, if $U_1 \in \tau^*(X)$ is the first move of $E$, player $NE$ has to respond with a set $V_1 \in \tau^*(U_1)$. The second move of $E$ has to be some $U_2 \in \tau^*(V_1)$. Then $NE$ has to choose $V_2 \in \tau^*(U_2)$ and so on. The Banach–Mazur game as described above, where $E$ makes the first move, is called *an E-game*. Now, if $V_1 \in \tau^*(X)$ is the first move of $NE$, player $E$ has to respond with a set $U_1 \in \tau^*(V_1)$. The second move of $NE$ has to be some $V_2 \in \tau^*(U_1)$. Then $E$ has to choose $U_2 \in \tau^*(V_2)$ and so on. The Banach–Mazur game, where player $NE$ makes the first move, is called *an NE-game*. In both games player $E$ wins after $\omega$ moves if the intersection of the moves is empty. Otherwise the winner is $NE$. The sequence of moves is called *a play* of the relevant game.

A *strategy of player E in E-game on a space X* is a map $s$ defined inductively as follows. First we have to choose a set $U_1 = s(\emptyset) \in \tau^*(X)$. If the strategy $s$ is defined for first $n$ moves then an $n$-tuple $(V_1, \ldots, V_n) \in (\tau^*(X))^n$ is called *admissible* if $V_1 \subset U_1$ and $V_i \subset U_i = s(V_1, \ldots, V_{i-1})$ for any $i \in \{2, \ldots, n\}$. For any admissible $n$-tuple $(V_1, \ldots, V_n)$ we have to choose a set $U_{n+1} = s(V_1, \ldots, V_n) \in \tau^*(V_n)$. We say that $E$ *applies the strategy s* in a play $\{U_i, V_i : i \in \mathbb{N}\}$ of an $E$-game if $U_1 = s(\emptyset)$ and $U_{k+1} = s(V_1, \ldots, V_k)$ for all $k \in \mathbb{N}$. A strategy $s$ of player $E$ in $E$-game is *winning on a space X* if $E$ wins in every play on $X$, in which he (or she!) applies $s$.

To define inductively *a strategy s of player E in NE-game on a space X* we have to choose a set $s(V_1) \in \tau^*(V_1)$ for any $V_1 \in \tau^*(X)$. If the strategy $s$ is defined for the first $n$ moves then say that an $(n + 1)$-tuple $(V_1, \ldots, V_{n+1}) \in (\tau^*(X))^{n+1}$ is *admissible* if $V_{i+1} \subset s(V_1, \ldots, V_i)$ for all $i = 1, \ldots, n$. If $(V_1, \ldots, V_{n+1}) \in (\tau^*(X))^{n+1}$ is admissible then we have to choose $s(V_1, \ldots, V_{n+1}) \in \tau^*(V_{n+1})$. We say that $E$ *applies the strategy s* in a play $\{V_i, U_i : i \in \mathbb{N}\}$ of an $NE$-game if $U_n = s(V_1, \ldots, V_n)$ for all $n \in \mathbb{N}$. A strategy $s$ of player $E$ in $NE$-game is *winning on X* if $E$ wins in every play on $X$, in which he (or she!) applies $s$.

Given a linear space $L$ and a norm $|| \cdot ||$ on $L$ let $d(x, y) = ||x - y||$ for any $x, y \in L$. It is easy to see that $d$ is a metric on $L$; if $\tau$ is the topology generated by the metric $d$ then $\tau$ is called *the topology generated by the norm* $|| \cdot ||$. A linear topological space is *normable* if its topology is generated by a norm.

**301.** Prove that there exist $l$-equivalent spaces $X$ and $Y$ such that $X$ is hereditarily paracompact while $Y$ is not collectionwise normal.

**302.** Prove that there exist $l$-equivalent spaces $X$ and $Y$ such that $X$ is collection-wise normal while $Y$ is not normal.

**303.** Prove that there exist $l$-equivalent spaces $X$ and $Y$ such that $X$ is hereditarily normal while $Y$ is not normal.

**304.** Prove that $\pi$-weight is not preserved by $l$-equivalence neither in the class of compact spaces nor in the class of countable spaces.

**305.** Give an example of $l$-equivalent spaces $X$ and $Y$ with $ext(X) \neq ext(Y)$.

**306.** Prove that there exist $l$-equivalent spaces $X$ and $Y$ such that $X$ is Fréchet–Urysohn while $t(Y) > \omega$ and there is a non-closed set $A \subset Y$ such that $B \cap A$ is finite whenever $B$ is a bounded subset of $Y$. As a consequence, Fréchet–Urysohn property, $k$-property, sequentiality, and countable tightness are not $l$-invariant.

**307.** Show that Fréchet–Urysohn property is not preserved by $l$-equivalence in the class of compact spaces.

**308.** Let $Y$ be a space in which every closed subspace has the Baire property. Suppose that $Y$ is $l$-equivalent to a space $X$ and a nonempty set $Z \subset X$ also has the Baire property. Prove that there is a nonempty $W \subset Z$ which is open in $Z$ and homeomorphic to a subspace of $Y$.

**309.** Let $X$ and $Y$ be Čech-complete $l$-equivalent spaces. Prove that every nonempty subspace of $X$ has a $\pi$-base whose elements are embeddable in $Y$. Deduce from this fact that if $X$ and $Y$ are nonempty Čech-complete $l$-equivalent spaces and $X$ is scattered then $Y$ is also scattered.

**310.** Let $X$ and $Y$ be Čech-complete $l$-equivalent spaces such that $\chi(X) \leq \kappa$. Prove that $Y$ has a dense open subspace $D$ such that $\chi(x, Y) \leq \kappa$ for each $x \in D$. In particular, $\chi(D) \leq \kappa$.

**311.** Prove that $X$ is a closed Hamel basis of $L(X)$ for any space $X$.

**312.** Prove that, for any space $X$ and any continuous map $f : X \to L$ of $X$ to a locally convex space $L$, there exists a unique continuous linear map $\xi_f : L(X) \to L$ such that $\xi_f | X = f$. Observe that this makes it possible to consider that $L(X)$, as a linear space, coincides with $L_p(X) \subset C(C_p(X))$ while the topology of $L(X)$ is stronger than $\tau(L_p(X))$. In all problems that follow we use this observation identifying the underlying set of $L(X)$ with $L_p(X)$.

**313.** Suppose that $L$ is a locally convex space such that $X$ is embedded as a Hamel basis in $L$. Prove that the following conditions are equivalent:

(i) there exists a linear homeomorphism $i : L \to L(X)$ such that $i(x) = x$ for all $x \in X$;

(ii) every continuous function $f : X \to M$ of the space $X$ to a locally convex space $M$ can be extended to a continuous linear functional $\xi_f : L \to M$.

**314.** Prove that, for any space $X$, the set $(L(X))^*$ coincides with the set $(L_p(X))^*$. Deduce from this fact that the weak topology of the space $L(X)$ coincides with the topology of $L_p(X)$.

**315.** Given a space $X$ let $E$ be the weak dual of $L(X)$, i.e., $E = (L(X))^*$ and the topology of $E$ is induced from $C_p(L(X))$. For every $f \in E$ let $\pi(f) = f|X$, i.e., $\pi : E \to C_p(X)$ is a restriction map. Prove that $\pi$ is a linear homeomorphism and hence $E$ is linearly homeomorphic to $C_p(X)$.

**316.** Observe that $l$-equivalence implies $u$-equivalence, i.e., for any spaces $X$ and $Y$, if $X \overset{l}{\sim} Y$ then $X \overset{u}{\sim} Y$. Prove that $L$-equivalence implies $l$-equivalence.

**317.** Prove that $C_b(X)$ is complete (as a uniform space with its linear uniformity) if and only if $X$ is a $b_f$-space.

**318.** Given a space $X$ call a set $P \subset C(X)$ *equicontinuous at a point* $x \in X$ if, for any $\varepsilon > 0$ there exists $U \in \tau(x, X)$ such that $f(U) \subset (f(x)-\varepsilon, f(x)+\varepsilon)$ for any $f \in P$. The family $P$ is called *equicontinuous* if it is equicontinuous at every point $x \in X$. Say that $P$ is *pointwise bounded* if the set $\{f(x) : f \in P\}$ is bounded in $\mathbb{R}$ for any $x \in X$. Prove that, for any equicontinuous pointwise bounded set $P \subset C(X)$, the closure of $P$ in the space $C_b(X)$ is compact. In particular, if $X$ is pseudocompact and $P \subset C(X)$ is equicontinuous and pointwise bounded then the closure of $P$ in $C_u(X)$ is compact.

**319.** Prove that, for any $b_f$-space $X$, a set $P \subset C(X)$ is equicontinuous and pointwise bounded if and only if the closure of $P$ in the space $C_b(X)$ is compact.

**320.** Given a space $X$, let $[P, \varepsilon] = \{\varphi \in L(X) : \varphi(P) \subset (-\varepsilon, +\varepsilon)\}$ for every $P \subset C(X)$ and $\varepsilon > 0$. Prove that a set $U \subset L(X)$ is open in $L(X)$ if and only if, for any $\eta \in U$ there exists an equicontinuous pointwise bounded set $P \subset C(X)$ and $\varepsilon > 0$ such that $\eta+[P, \varepsilon] \subset U$. In other words, the topology of $L(X)$ coincides with the topology of uniform convergence on equicontinuous pointwise bounded subsets of $C(X)$.

**321.** Given a space $X$ and $P \subset X$ let $I_P = \{f \in C(X) : f(P) \subset \{0\}\}$. Prove that for any linear continuous functional $\varphi : C_k(X) \to \mathbb{R}$ which is not identically zero on $C_k(X)$, there exists a compact subspace $K \subset X$ (called the support of $\varphi$ and denoted by supp$(\varphi)$) such that $\varphi(I_K) = \{0\}$ and $\varphi(I_{K'}) \neq \{0\}$ whenever $K'$ is a proper compact subset of $K$.

**322.** Recall that a set $B$ is *a barrel* in a locally convex space $L$ if $B$ is closed, convex, balanced, and absorbing in $L$. The space $L$ is *barreled* if any barrel in $L$ is a neighborhood of $\mathbf{0}$. Prove that a locally convex space $L$ is barreled whenever it has the Baire property.

**323.** Prove that $C_k(X)$ is a barreled space if and only if $X$ is a $\mu$-space, i.e., the closure of any bounded subspace of $X$ is compact. Deduce from this fact that $C_k(X)$ is barreled for any realcompact space $X$.

**324.** Prove that $C_p(X)$ is barreled if and only if all bounded subspaces of $X$ are finite.

**325.** Give an example of a space $X$ such that $C_p(X)$ barreled but does have the Baire property.

**326.** Given a point $z$ in a space $Z$, say that a family $\mathcal{B}$ of subsets of $Z$ is *a local base of neighborhoods of $Z$ at the point $z$*, if $z \in \text{Int}(U)$ for any $U \in \mathcal{B}$, and $z \in V \in \tau(Z)$ implies $U \subset V$ for some $U \in \mathcal{B}$. Prove that the family of all barrels in $C_k(X)$ constitutes a local base of neighborhoods of $\mathbf{0}$ in $C_b(X)$. Deduce from this fact that the family of all barrels in $C_p(X)$ is also a local base of neighborhoods of $\mathbf{0}$ in $C_b(X)$.

**327.** Let $\varphi : C_k(X) \to C_p(Y)$ be a linear continuous map. Prove that $\varphi$ is continuous considered as a map from $C_b(X)$ to $C_b(Y)$.

**328.** Assuming that a space $Y$ is $l$-equivalent to a $b_f$-space $X$ prove that $Y$ is also a $b_f$-space. In other words, $b_f$-property is $l$-invariant.

**329.** Let $\varphi : L_p(X) \to L_p(Y)$ be a linear homeomorphism. Prove that, if $X$ is a $b_f$-space then $\varphi$ is a linear homeomorphism of $L(X)$ onto $L(Y)$.

**330.** Let $X$ and $Y$ be spaces one of which is a $b_f$-space. Prove that $X$ is $L$-equivalent to $Y$ if and only if $X$ and $Y$ are $l$-equivalent.

**331.** Prove that there exist $l$-equivalent spaces which are not $L$-equivalent.

**332.** Prove that a space $X$ has a weaker metrizable topology if and only if $L(X)$ has a weaker metrizable topology. In particular, if $X$ and $Y$ are $L$-equivalent and $X$ can be condensed onto a metrizable space then $Y$ can also be condensed onto a metrizable space.

**333.** Suppose that $X$ and $Y$ are $l$-equivalent spaces. Prove that, if $X$ is metrizable, then $Y$ can be condensed onto a metrizable space.

**334.** Say that a space $X$ is *$\sigma$-metrizable* if $X$ is the countable union of its closed metrizable subspaces. Prove that a space $X$ is $\sigma$-metrizable and paracompact if and only if $L(X)$ is $\sigma$-metrizable and paracompact.

**335.** Suppose that a space $X$ is $l$-equivalent to a metrizable space. Prove that $X$ is $\sigma$-metrizable and paracompact.

**336.** For an arbitrary space $X$, prove that $X$ is hemicompact if and only if $C_k(X)$ is first countable.

**337.** Prove that hemicompactness is preserved by $l$-equivalence.

**338.** Given a space $X$ prove that

    (i) if $X$ is a $k$-space then a sequence $\{f_n : n \in \omega\} \subset C_k(X)$ is convergent whenever it is linearly Cauchy;

    (ii) for a hemicompact space $X$ the converse is true, i.e., if any linearly Cauchy sequence in $C_k(X)$ is convergent then $X$ is a $k$-space.

**339.** Let $X$ be an arbitrary space. Prove that $X$ is a hemicompact space with $k$-property if and only if $C_k(X)$ is metrizable by a complete metric.

**340.** Let $X$ and $Y$ be $l$-equivalent spaces. Prove that $X$ is a hemicompact $k$-space if and only if $Y$ is a hemicompact $k$-space.

**341.** Prove that any subspace of an $\aleph_0$-space is an $\aleph_0$-space and any countable product of $\aleph_0$-spaces is an $\aleph_0$-space.

**342.** Observe that a compact-covering continuous image of an $\aleph_0$-space is an $\aleph_0$-space. Prove that a space $X$ is an $\aleph_0$-space if and only if $X$ is a compact-covering continuous image of a second countable space.

**343.** Prove that a space $X$ is an $\aleph_0$-space with the $k$-property if and only if it is a quotient image of a second countable space.

**344.** Prove that any $\aleph_0$-space of countable character is second countable.

**345.** Let $\varphi : X \to Y$ be a continuous map. Recall that the dual map $\varphi^* : C_k(Y) \to C_k(X)$ is defined by the formula $\varphi^*(f) = f \circ \varphi$ for every $f \in C_k(Y)$. Assuming that $\varphi$ is compact-covering, prove that $\varphi^*$ is an embedding.

**346.** Given a compact subspace $K$ of a space $X$ let $v(f, x) = f(x)$ for every $f \in C_k(X)$ and $x \in K$. Prove that the map $v : C_k(X) \times K \to \mathbb{R}$ is continuous.

**347.** Prove that the following properties are equivalent for any space $X$:

  (i) $X$ is an $\aleph_0$-space;
  (ii) $C_k(X)$ is an $\aleph_0$-space.
  (iii) $C_k(X)$ has a countable network.

**348.** Let $X$ and $Y$ be $l$-equivalent spaces. Prove that $X$ is an $\aleph_0$-space if and only if so is $Y$. In particular, if some space $Z$ is $l$-equivalent to a second countable space then $Z$ is an $\aleph_0$-space. Deduce from this fact that any first countable space $l$-equivalent to a second countable space must be second countable.

**349.** Suppose that a space $X$ has a countable network and $Y$ is an $\aleph_0$-space. Prove that $C_p(X, Y)$ has a countable network.

**350.** Given spaces $X, Y$ and a function $u : X \times Y \to \mathbb{R}$ let $u_x(y) = u(x, y)$ for all $y \in Y$; then $u_x : Y \to \mathbb{R}$ for every $x \in X$. Analogously, let $u^y(x) = u(x, y)$ for all $x \in X$; then $u^y : X \to \mathbb{R}$ for every $y \in Y$. Say that the function $u$ is *separately continuous* if the functions $u_x$ and $u^y$ are continuous (on $Y$ and $X$ respectively) for all $x \in X$ and $y \in Y$. Let $C_p^s(X \times Y)$ be the set of all separately continuous functions on $X \times Y$ with the topology induced from $\mathbb{R}^{X \times Y}$. Observe that $C_p^s(X \times Y)$ is a locally convex space and let $\xi(\varphi)(x, y) = \varphi(x)(y)$ for any continuous function $\varphi : X \to C_p(Y)$. Prove that $\xi(\varphi) \in C_p^s(X \times Y)$ for every $\varphi \in C_p(X, C_p(Y))$ and $\xi : C_p(X, C_p(Y)) \to C_p^s(X \times Y)$ is a linear homeomorphism.

**351.** Prove that the space $C_p(X, C_p(X))$ has a countable network if and only if $X$ is countable. Deduce from this fact that $C_p(X)$ is an $\aleph_0$-space if and only if $X$ is countable.

**352.** Prove that a space $X$ is of second category in itself if and only if the player $E$ has no winning strategy in the Banach–Mazur $NE$-game on $X$.

**353.** Prove that a space $X$ has the Baire property if and only if the player $E$ has no winning strategy in the Banach–Mazur $E$-game on $X$.

**354.** Prove that

  (i) any pseudocompact space with the moving off property is compact.
  (ii) any paracompact locally compact space has the moving off property.

**355.** Let $X$ be a $q$-space. Prove that, if $X$ has the moving off property then it is locally compact.

**356.** Prove that the following conditions are equivalent for any space $X \neq \emptyset$:

    (i) $X$ has the moving off property;

    (ii) given a sequence $\{\mathcal{K}_i : i \in \omega\}$ of moving off collections in $X$, we can choose $K_i \in \mathcal{K}_i$ for each $i \in \omega$, such that the family $\{K_i : i \in \omega\}$ has a discrete open expansion;

    (iii) the player $II$ has no winning strategy in the Gruenhage–Ma game on $X$.

**357.** Prove that, if $C_k(X)$ has the Baire property then the space $X$ has the moving off property.

**358.** Prove that, for any $q$-space $X$, the following conditions are equivalent:

    (i) $C_k(X)$ has the Baire property;

    (ii) $X$ has the moving off property;

    (iii) the player $II$ has no winning strategy in Gruenhage–Ma game on the space $X$.

**359.** Let $X$ be a paracompact $q$-space. Prove that $C_k(X)$ has the Baire property if and only if $X$ is locally compact.

**360.** Let $X$ be a paracompact $q$-space. Prove that, if $X$ is $l$-equivalent to a locally compact paracompact space then $X$ is also locally compact. In particular, any first countable paracompact space $l$-equivalent to a locally compact paracompact space is locally compact. Deduce from this fact that

    (i) if $X$ and $Y$ are metrizable $l$-equivalent spaces then $X$ is locally compact if and only if $Y$ is locally compact;

    (ii) if a first countable space $X$ is $l$-equivalent to a second countable locally compact space then $X$ is also locally compact and second countable.

**361.** Suppose that a $q$-space $X$ is $l$-equivalent to a locally compact metrizable space. Prove that $X$ is metrizable and locally compact.

**362.** Suppose that $X$ is $l$-equivalent to a metrizable space, $Y \subset X$ and $Y$ is Čech-complete. Prove that $Y$ is metrizable. In particular, if a Čech-complete space $X$ is $l$-equivalent to a metrizable space then $X$ is metrizable.

**363.** Suppose that $X$ is $l$-equivalent to a metrizable space. Prove that $\overline{A}$ is an $\aleph_0$-space for any countable set $A \subset X$.

**364.** For any space $X$ let $dc(X) = \sup\{|\mathcal{U}| : \mathcal{U} \subset \tau^*(X)$ and $\mathcal{U}$ is a discrete family$\}$. Prove that if $X$ is $l$-equivalent to a metrizable space then $nw(X) = dc(X)$. In particular, if $X$ is $l$-equivalent to a metrizable space then both the Souslin property of $X$ and $ext(X) \leq \omega$ imply that $X$ has a countable network.

**365.** Given a space $X$ and a first countable space $Y$ assume that there exists a continuous linear surjection $\varphi : C_p(X) \to C_p(Y)$. For any $y \in Y$ there exist $x_1, \ldots, x_n \in X$ and $\lambda_1, \ldots, \lambda_n \in \mathbb{R}\backslash\{0\}$ such that $\varphi(f)(y) = \sum_{i=1}^{n} \lambda_i f(x_i)$ for any $f \in C_p(X)$; denote the set $\{x_1, \ldots, x_n\}$ by $\text{supp}(y)$. Suppose that $\mathcal{U}$ is a locally finite open cover of $X$ and let $T(U) = \{y \in Y : \text{supp}(y) \cap U \neq \emptyset\}$ for every $U \in \mathcal{U}$. Prove that the family $\{T(U) : U \in \mathcal{U}\}$ is a locally finite open cover of $Y$.

**366.** Let $X$ and $Y$ be metrizable spaces for which there exists a continuous linear surjection of $C_p(X)$ onto $C_p(Y)$. Prove that, if $X$ is Čech-complete then $Y$ is also Čech-complete.

**367.** Prove that if a metrizable space $X$ is $l$-equivalent to a Čech-complete space then $X$ is also Čech-complete. As a consequence, if $X$ and $Y$ are $l$-equivalent metrizable spaces then $X$ is metrizable by a complete metric if and only if so is $Y$.

**368.** Show that there exist first countable $l$-equivalent spaces $X$ and $Y$ such that $X$ is locally compact and $Y$ is not locally compact.

**369.** Given a space $Z$ let $Z'$ be the set of non-isolated points of $Z$. Suppose that $X$ and $Y$ are normal first countable $l$-equivalent spaces. Prove that if $X'$ is countably compact then $Y'$ is also countably compact. Show that this statement can be false if we omit first countability of $X$ and $Y$.

**370.** Given a nonempty closed subspace $F$ of a normal space $X$ suppose that $F$ is a retract of some neighborhood of $F$. Prove that $F$ is $l$-embedded in $X$ and hence $C_p(X) \approx C_p(F) \times I$, where $I = \{f \in C_p(X) : f(F) = \{0\}\}$.

**371.** Given a space $X$ assume that $X_0$ and $X_1$ are closed subspaces of $X$ such that $X = X_0 \cup X_1$ and the set $F = X_0 \cap X_1$ is $l$-embedded in $X$; suppose additionally that $C_p(F) \approx C_p(F) \times C_p(F)$. Prove that $C_p(X) \approx C_p(X_0) \times C_p(X_1)$.

**372.** Suppose that a space $X$ has a nontrivial convergent sequence. Prove that $X$ is $l$-equivalent to $X \oplus (\omega + 1)$ and $X^+ \overset{l}{\sim} X$. Deduce from this fact that $X^+$ is $l$-equivalent to $X$ for every infinite metrizable space $X$.

**373.** Suppose that a space $X$ has a nontrivial convergent sequence and $Y$ is $l$-embedded in $X$. Prove that $X$ is $l$-equivalent to $X_Y \oplus Y$. Consequently, for any infinite metrizable space $X$, if $Y$ is closed in $X$ then $X \overset{l}{\sim} X_Y \oplus Y$. Here $X_Y$ is the $\mathbb{R}$-quotient space obtained from $X$ by contracting $Y$ to a point.

**374.** Suppose that $X$ is a compact space and $F$ is $l$-embedded in $X$. Prove that $X^+$ is $l$-equivalent to $F \oplus \alpha(X \setminus F)$. In particular, if $X$ is an infinite metrizable compact space then $X \overset{l}{\sim} F \oplus \alpha(X \setminus F)$ for any closed $F \subset X$.

**375.** Let $X$ and $Y$ be metrizable compact spaces. Suppose that $F$ and $G$ are closed subspaces of $X$ and $Y$ respectively such that $F \overset{l}{\sim} G$ and $X \setminus F$ is homeomorphic to $Y \setminus G$. Prove that $X$ is $l$-equivalent to $Y$.

**376.** Let $X$ and $Y$ be nonempty compact metrizable spaces such that either $Y \times (\omega + 1)$ or $\alpha(Y \times \omega)$ embeds in $X$. Prove that $C_p(X) \approx C_p(X) \times (C_p(Y))^n$ for every $n \in \mathbb{N}$.

**377.** Say that a metrizable compact space $K$ is *universal in the dimension* $n \in \omega$ if $\dim K = n$ and any metrizable compact space of dimension at most $n$ embeds in $K$. Prove that if $X$ and $Y$ are metrizable compact spaces universal in the dimension $n$, then $X \overset{l}{\sim} Y$.

**378.** Suppose that a space $X$ is $l$-equivalent to $Y$ and $Y$ is a metrizable compact space universal in the dimension $n$. Prove that $X$ is also a metrizable compact space universal in the dimension $n$.

**379.** Prove that, for any $n_1, \ldots, n_k \in \mathbb{N}$, the space $C_p(\mathbb{I}^{n_1}) \times \ldots \times C_p(\mathbb{I}^{n_k})$ is linearly homeomorphic to $C_p(\mathbb{I}^n)$ where $n = \max\{n_1, \ldots, n_k\}$.

**380.** Given $n \in \mathbb{N}$ and a closed subset $F$ of the space $\mathbb{I}^n$ such that $\emptyset \neq F \neq \mathbb{I}^n$ prove that $(\mathbb{I}^n)_F$ is $l$-equivalent to $\mathbb{I}^n$. Here $(\mathbb{I}^n)_F$ is the $\mathbb{R}$-quotient image of $\mathbb{I}^n$ obtained by contracting $F$ to a point.

**381.** Prove that, for any $n \in \mathbb{N}$, if $U$ is a nonempty open subset of the space $\mathbb{R}^n$, then the space $\alpha(U)$ is $l$-equivalent to $\mathbb{I}^n$.

**382.** Given a space $X$ with $\dim X = n \in \mathbb{N}$ assume that $X$ is homeomorphic to a finite union of Euclidean cubes, i.e., there is a finite family $\mathcal{F}$ of subsets of $X$ such that $\bigcup \mathcal{F} = X$ and every $F \in \mathcal{F}$ is homeomorphic to $\mathbb{I}^k$ for some $k \in \mathbb{N}$. Prove that $X$ is $l$-equivalent to $\mathbb{I}^n$.

**383.** For any $n \in \mathbb{N}$ prove that both spaces $\mathbb{I}^n \times (\omega + 1)$ and $\alpha(\mathbb{I}^n \times \omega)$ are $l$-equivalent to $\mathbb{I}^n$.

**384.** Prove that $\mathbb{I}^n \times \mathbb{D}^\omega$ is not $l$-equivalent to $\mathbb{I}^n$ for any $n \in \mathbb{N}$.

**385.** Suppose that $K$ is a compact space and there exists a continuous bijective map of $[0, +\infty)$ onto $K$. Prove that $K$ is $l$-equivalent to $\mathbb{I}$. Deduce from this fact that if there is a continuous bijection of $\mathbb{R}$ onto a compact space $L$ then $L$ is also $l$-equivalent to $\mathbb{I}$.

**386.** Assume that $X$ is a second countable space and $\dim X = n \in \omega$. Let $\mathcal{O} = \{U \in \tau(X) : \dim U < n\}$ and $O = \bigcup \mathcal{O}$. The set $K(X) = X \backslash O$ is called *the dimensional kernel of $X$*. Prove that $\dim O < n$ and $\dim W = n$ for any nonempty open subset $W$ of the space $K(X)$.

**387.** Prove that if $n \in \mathbb{N}$ and a space $X$ is $l$-equivalent to $\mathbb{I}^n$ then the dimensional kernel $K(X)$ of the space $X$ is also $l$-equivalent to $\mathbb{I}^n$.

**388.** Call a second countable space $Y$ *weakly $n$-Euclidean* if $\dim Y = n$ and every $n$-dimensional subspace of $Y$ has nonempty interior and contains a homeomorphic copy of $\mathbb{I}^n$. Prove that a compact space $X$ is $l$-equivalent to $\mathbb{I}^n$ if and only if its dimensional kernel $K(X)$ has a nonempty open weakly $n$-Euclidean subspace and every $U \in \tau^*(K(X))$ contains a subset which is $l$-equivalent to $X$.

**389.** Given (linear) topological spaces $X$ and $Y$ the expression $X \sim Y$ says that they are (linearly) homeomorphic. Suppose that $X^\omega \sim X$ and there exist (linear) topological spaces $E$ and $F$ such that $X \sim Y \times F$ and $Y \sim X \times E$. Prove that $X \sim Y$.

**390.** Suppose that $L$ is a linear topological space, $M$ is a linear subspace of $L$ and there exists a linear retraction $r : L \to M$. Prove that $L \approx M \times r^{-1}(\mathbf{0})$. Deduce from this fact that for any linear topological spaces $L$ and $E$ there exists a linear topological space $N$ such that $L \approx E \times N$ (i.e., $E$ is a linear topological factor of $L$) if and only if there exists a linear retract $E'$ of the space $L$ such that $E' \approx E$.

**391.** Given linear topological spaces $L, M$, and $N$ prove that $L \approx M \times N$ if and only if there exist linear subspaces $M'$ and $N'$ of the space $L$ for which $M' \approx M$, $N' \approx N$ and there exist linear retractions $r : L \to M'$ and $s : L \to N'$ such that $r(x) + s(x) = x$ for any $x \in L$.

**392.** For any linear topological space $L$ denote by $L^*$ the set of all continuous linear functionals on $L$ with the topology inherited from $C_p(L)$. Prove that, for any locally convex spaces $M$ and $N$, we have $(M \times N)^* \approx M^* \times N^*$. In particular, $C_p(Y)$ is a linear topological factor of $C_p(X)$ if and only if $L_p(Y)$ is a linear topological factor of $L_p(X)$.

**393.** Suppose that $X$ is a second countable non-compact space, $n \in \mathbb{N}$ and there exists a locally finite cover $\mathcal{I}$ of the space $X$ such that $\bigcup \{\text{Int}(I) : I \in \mathcal{I}\} = X$ and every $I \in \mathcal{I}$ is homeomorphic to $\mathbb{I}^n$. Prove that $X$ is $l$-equivalent to $\mathbb{I}^n \times \omega$.

**394.** Prove that, for any $n \in \mathbb{N}$, every nonempty open subspace of $\mathbb{R}^n$ is $l$-equivalent to $\mathbb{I}^n \times \omega$. In particular, if $U \in \tau^*(\mathbb{R}^n)$ then $U \overset{l}{\sim} \mathbb{R}^n$.

**395.** Given spaces $X$ and $Y$ assume that $nw(X) = \omega$ and $C_p(Y)$ is a linear topological factor of $C_p(X)$. Prove that $\dim Y \leq \dim X$.

**396.** Assuming that $X$ is a compact space and $C_p(Y)$ is a linear topological factor of $C_p(X)$ prove that $\dim Y \leq \dim X$.

**397.** (Open Mapping Theorem) Suppose that $L$ is a Banach space, $M$ is a linear topological space which is of second category in itself, and $f : L \to M$ is a surjective continuous linear map. Prove that $f$ is open. In particular, any continuous linear onto map between two Banach spaces is open.

**398.** (Closed Graph Theorem) Suppose that $L$ and $M$ are Banach spaces and $f : L \to M$ is a linear map such that its graph $G = \{(x, f(x)) : x \in L\}$ is closed in $L \times M$. Prove that the map $f$ is continuous.

**399.** Prove that, for any nonempty space $X$, there exists a continuous surjection of the space $C_p(X)$ onto $C_p(X) \times \mathbb{R}$.

**400.** Prove that there exists an infinite compact space $K$ for which there is no linear continuous surjection of $C_p(K)$ onto $C_p(K) \times \mathbb{R}$. In particular, $K^+$ is not $l$-equivalent to the space $K$.

## 1.5 The Last-Minute Updates. Yet More on l-Equivalence

All spaces are assumed to be Tychonoff; the space $\mathbb{D}$ is the doubleton $\{0, 1\}$ with the discrete topology. Given an infinite cardinal $\kappa$, say that a set $P \subset X$ is *an $F_\kappa$-subset of $X$* if $P$ is a union of $\leq \kappa$ closed subsets of $X$; if $Q$ is the intersection of at most $\kappa$-many open subsets of $X$ then $Q$ is called *a $G_\kappa$-subset of $X$*. For any space $X$ let $\rho(X) = \min\{\kappa : \text{there is no continuous surjective map of } X \text{ onto } \mathbb{I}^{\kappa^+}\}$. The cardinal $\rho(X)$ is called *the dyadicity index* of the space $X$. Let $\Psi(X) = \min\{\kappa \geq \omega : \text{every closed subset of } X \text{ is a } G_\kappa\text{-set in } X\}$. If $X$ is a space then $x \in X$ is called *a P-point in $X$* if $x \in \text{Int}(F)$ for any $G_\delta$-subset $F$ of the space $X$ such that $x \in F$.

Given a point $\xi \in \beta\omega \setminus \omega$ let $\mathcal{U}_\xi = \{U \cap \omega : U \in \tau(\xi, \beta\omega)\}$; it is an easy exercise to see that $\mathcal{U}_\xi$ is an ultrafilter on $\omega$. We will follow the usual practice to identify $\xi$ with $\mathcal{U}_\xi$ for any $\xi \in \beta\omega \setminus \omega$. In particular, the elements of $\beta\omega \setminus \omega$ will be called ultrafilters on $\omega$. Observe that, for each ultrafilter $\xi \in \beta\omega \setminus \omega$ and any set $A \subset \omega$ we have $\xi \in \overline{A}$ if and only if $A \in \xi$; let $\omega_\xi$ be the set $\omega \cup \{\xi\}$ endowed with the topology inherited from $\beta\omega$. Recall that $\sigma(\mathbb{R}^A) = \{x \in \mathbb{R}^A : |\{a \in A : x(a) \neq 0\}| < \omega\}$ for any set $A$.

Given an infinite cardinal $\kappa$, a space $X$ is called *$\kappa$-cosmic* if $nw(X) \leq \kappa$; the space $X$ is *strongly $\kappa$-cosmic* if there exists a family $\{X_\alpha : \alpha < \kappa\}$ of subspaces of $X$ such that $X = \bigcup\{X_\alpha : \alpha < \kappa\}$ and $nw(X_\alpha) < \kappa$ for any $\alpha < \kappa$. If $P \subset X$ then a set $A \subset X$ is said to be *concentrated around $P$* if $A \setminus U$ is countable for any $U \in \tau(P, X)$. If $X$ is a space and $n > 1$ is a natural number then the set $\Delta_n(X) = \{x \in X^n : x(i) = x(j) \text{ for some distinct } i, j < n\}$ is called the *$n$-diagonal of $X$*. For technical reasons it is convenient to define the 1-diagonal $\Delta_1(X)$ to be the empty set for any space $X$. If $\mathcal{A} = \{A_n : n \in \omega\}$ is a sequence of sets then $\underline{\lim}\,\mathcal{A} = \bigcup_{n \in \omega} \bigcap_{i \geq n} A_i$.

A game of two players $E$ and $NE$ on a nonempty space $X$ is called *a Choquet game* if at the $n$-th move $E$ chooses an open set $U_n$ and a point $x_n \in U_n$. The player $NE$ responds by choosing an open set $V_n \ni x_n$ with $V_n \subset U_n$. At the move $(n + 1)$ the player $E$ takes a point $x_{n+1} \in V_n$ and an open set $U_{n+1}$ such that $x_{n+1} \in U_{n+1} \subset V_n$. The player NE has to respond with a set $V_{n+1} \in \tau(X)$ such that $x_{n+1} \in V_{n+1} \subset U_{n+1}$. The game ends after the moves $(x_n, U_n), V_n$ are made for all $n \in \omega$ and the player $E$ wins if $\bigcap\{U_n : n \in \omega\} = \emptyset$; otherwise, player $NE$ is the winner. In the *point-open game* $\mathcal{PO}$ on a nonempty space $X$ at the $n$-th move the player $P$ chooses a point $x_n$ and the player $O$ takes an open set $U_n$ which contains $x_n$. The game ends after the moves $x_n, U_n$ are made for all $n \in \omega$ and the player $P$ wins if $\bigcup_{n \in \omega} U_n = X$; otherwise $O$ is the winner.

A *neighborhood assignment* on a space $X$ is any function $N : X \to \tau(X)$ such that $x \in N(x)$ for every $x \in X$; if $A \subset X$ then $N(A) = \bigcup\{N(x) : x \in A\}$. Say that $X$ is a *D-space* if for any neighborhood assignment $N$ on $X$ there exists a closed discrete set $D \subset X$ such that $N(D) = X$.

If $\beta$ is an ordinal then a family $\{A_\alpha : \alpha < \beta\}$ of subsets of a space $X$ is called *increasing* if $\alpha < \gamma < \beta$ implies $A_\alpha \subset A_\gamma$. If $X$ is a space and $P \subset X$ say that a family $\mathcal{N} \subset \exp(X)$ is *an external network (base)* of $P$ in $X$ if ($\mathcal{N} \subset \tau(X)$ and) for any $x \in P$ and $U \in \tau(x, X)$ there exists $N \in \mathcal{N}$ such that $x \in N \subset U$.

Given sets $X$ and $Y$ and an infinite cardinal $\kappa$ suppose that we have families $\mathcal{A} \subset \exp(X), \mathcal{B} \subset \exp(Y)$ and a map $\varphi : \mathcal{A} \to \mathcal{B}$. Say that $\varphi$ is $\kappa$-*monotone* if

(1) $|\varphi(A)| \leq \max\{|A|, \omega\}$ whenever $A \in \mathcal{A}$ and $|A| \leq \kappa$;
(2) if $A \subset B$ and $A, B \in \mathcal{A}$, then $\varphi(A) \subset \varphi(B)$;
(3) if $\lambda \leq \kappa$ is an infinite cardinal, $\{A_\alpha : \alpha < \lambda\} \subset \mathcal{A}$ and $\alpha < \alpha'$ implies $A_\alpha \subset A_{\alpha'}$
   then we have the equality $\varphi(\bigcup_{\alpha < \lambda} A_\alpha) = \bigcup_{\alpha < \lambda} \varphi(A_\alpha)$.

Given an infinite cardinal $\kappa$, say that a space $X$ is *(strongly) monotonically* $\kappa$-*monolithic* if, for any $A \subset X$ with $|A| \leq \kappa$, there exists an external network (base) $\mathcal{O}(A)$ of the set $\overline{A}$ in $X$ such that the assignment $A \to \mathcal{O}(A)$ is $\kappa$-monotone. If $X$ is a space and $f : X \to Y$ is a continuous map, then a family $\mathcal{N} \subset \exp(X)$ is *a network for* $f$ if for any $x \in X$ and $U \in \tau(f(x), Y)$ there exists $N \in \mathcal{N}$ such that $x \in N$ and $f(N) \subset U$. Say that a space $X$ is *monotonically* $\kappa$-*stable* if to any set $B \subset C_p(X)$ with $|B| \leq \kappa$ we can assign a network $\mathcal{N}(B)$ for the map $e_{\overline{B}} : X \to C_p(\overline{B})$ in such a way that the correspondence $B \to \mathcal{N}(B)$ is $\kappa$-monotone. Recall that $e_{\overline{B}}(x)(f) = f(x)$ for any $f \in \overline{B}$.

A space $X$ is *(strongly) monotonically monolithic* if, for any $A \subset X$, there exists an external network (base) $\mathcal{O}(A)$ of the set $\overline{A}$ in $X$ such that $|\mathcal{O}(A)| \leq \max\{|A|, \omega\}$ while $A \subset B$ implies $\mathcal{O}(A) \subset \mathcal{O}(B)$ and $\mathcal{O}(\bigcup\{A_\alpha : \alpha < \beta\}) = \bigcup\{\mathcal{O}(A_\alpha) : \alpha < \beta\}$ whenever $\beta$ is an ordinal and $\{A_\alpha : \alpha < \beta\}$ is an increasing family of subsets of $X$.

Say that a space $X$ is *monotonically retractable* if we can assign to any countable $A \subset X$ a retraction $r_A : X \to X$ and a countable network $\mathcal{N}(A)$ for the map $r_A$ such that $A \subset r_A(X)$ and the assignment $\mathcal{N}$ is $\omega$-monotone. Say that a space $X$ is *monotonically Sokolov* if we can assign to any countable family $\mathcal{F}$ of closed subsets of $X$ a continuous retraction $r_\mathcal{F} : X \to X$ and a countable external network $\mathcal{N}(\mathcal{F})$ for $r_\mathcal{F}(X)$ in $X$ such that $r_\mathcal{F}(F) \subset F$ for each $F \in \mathcal{F}$ and the assignment $\mathcal{N}$ is $\omega$-monotone.

Say that $X$ is a $\sigma$-*space* if the space $X$ has a $\sigma$-discrete network. A space $X$ is *a strong $\Sigma$-space* if there exists a family $\mathcal{C}$ of compact subsets of $X$ such that $\bigcup \mathcal{C} = X$ and there exists a $\sigma$-discrete family $\mathcal{N} \subset \exp(X)$ which is a network modulo $\mathcal{C}$, i.e., for any $C \in \mathcal{C}$ and $U \in \tau(C, X)$ there exists $N \in \mathcal{N}$ with $C \subset N \subset U$. Call a space $Z$ *hereditarily Baire* if every closed subspace of $Z$ has the Baire property. A space $X$ is called $d$-*separable* if it has a dense $\sigma$-discrete subspace.

Given any ordinal $\alpha$ let $\alpha + 0 = \alpha$; if $\beta$ is an ordinal and we defined $\alpha + \beta$ then $\alpha + (\beta + 1) = (\alpha + \beta) + 1$. If $\beta$ is a limit ordinal and we defined $\alpha + \gamma$ for any $\gamma < \beta$ then $\alpha + \beta = \sup\{\alpha + \gamma : \gamma < \beta\}$. This defines an ordinal $\alpha + \beta$ for any ordinals $\alpha$ and $\beta$. If $\alpha$ is an ordinal then let $\alpha \cdot 0 = 0$. If $\beta$ is an ordinal and we have defined $\alpha \cdot \beta$ then $\alpha \cdot (\beta + 1) = \alpha \cdot \beta + \alpha$. If $\beta$ is a limit ordinal and we defined $\alpha \cdot \gamma$ for every $\gamma < \beta$ then $\alpha \cdot \beta = \sup\{\alpha \cdot \gamma : \gamma < \beta\}$. This defines an ordinal $\alpha \cdot \beta$ for any ordinals $\alpha$ and $\beta$. For each ordinal $\alpha$ let $\alpha^0 = 1$. If $\beta$ is an ordinal and we have defined $\alpha^\beta$ then $\alpha^{\beta+1} = (\alpha^\beta) \cdot \alpha$. If $\beta$ is a limit ordinal and we defined $\alpha^\gamma$ for every $\gamma < \beta$ then $\alpha^\beta = \sup\{\alpha^\gamma : \gamma < \beta\}$. This defines an ordinal $\alpha^\beta$ for any ordinals $\alpha$ and $\beta$.

Say that a space $X$ is *finite-dimensional* if $\dim X \leq n$ for some $n \in \omega$. The space $X$ is *$\sigma$-zero-dimensional* if $X = \bigcup_{i \in \omega} X_i$ and every $X_i$ is zero-dimensional.

**401.** Suppose that $\kappa$ is an infinite cardinal and $X$ is a Lindelöf $\Sigma$-space such that $t(X) \leq \kappa$ and $\pi\chi(X) \leq \kappa$. Prove that $X$ has a $\pi$-base of order $\leq \kappa$. In particular, if $X$ is a Lindelöf $\Sigma$-space with $t(X) = \pi\chi(X) = \omega$ then $X$ has a point-countable $\pi$-base. Deduce from this fact that any Lindelöf $\Sigma$-space $X$ with $\chi(X) \leq \kappa$ has a $\pi$-base of order $\leq \kappa$. In particular, any first countable Lindelöf $\Sigma$-space has a point-countable $\pi$-base.

**402.** Given a space $X$ and an infinite cardinal $\kappa$ suppose that $\pi\chi(X) \leq \kappa$ and $d(X) \leq \kappa^+$. Prove that $X$ has a $\pi$-base of order $\leq \kappa$. In particular, if $\pi\chi(X) \leq \omega$ and $d(X) \leq \omega_1$ then $X$ has a point-countable $\pi$-base.

**403.** Assuming CH prove that

    (i) any Lindelöf first countable space has a point-countable $\pi$-base;
    (ii) any space $X$ with $\chi(X) = c(X) = \omega$ has a point-countable $\pi$-base;
    (iii) if $\omega_1$ is a caliber of $X$ and $\chi(X) \leq \omega$ then $X$ is separable.

**404.** Give an example of a first countable space which has no point-countable $\pi$-base.

**405.** Let $X$ be a space for which we can find a family of sets $\{A_m : m \in \omega\}$ and a sequence $\{k_m : m \in \omega\} \subset \mathbb{N}\backslash\{1\}$ such that $\sup\{|A_m| : m \in \omega\} = |X|$ while $A_m \subset X^{k_m}\backslash\Delta_{k_m}(X)$ and $A_m$ is concentrated around $\Delta_{k_m}(X)$ for every $m \in \omega$. Prove that the space $C_p(X)$ has a point-countable $\pi$-base. In particular, the space $C_p(X)$ has a point-countable $\pi$-base if there is a set $A$ with $|A| = |X|$ such that either $A \subset X^n\backslash\Delta_n(X)$ and $A$ is concentrated around $\Delta_n(X)$ or $A \subset X$ and $A$ is concentrated around some point of $X$.

**406.** Prove that

    (a) $C_p(\alpha)$ has a point-countable $\pi$-base whenever $\alpha$ is an ordinal with its order topology;
    (b) $C_p(AD(X))$ has a point-countable $\pi$-base for any countably compact space $X$. Here $AD(X)$ is the Alexandroff double of $X$.

**407.** Prove that the following conditions are equivalent for any infinite space $X$ with $l^*(X) \leq \omega$:

    (i) $C_p(X)$ has a point-countable $\pi$-base;
    (ii) there is a family of sets $\{A_m : m \in \omega\}$ and a sequence $\{k_m : m \in \omega\} \subset \mathbb{N}\backslash\{1\}$ such that $\sup\{|A_m| : m \in \omega\} = |X|$ while $A_m \subset X^{k_m}\backslash\Delta_{k_m}(X)$ and $A_m$ is concentrated around $\Delta_{k_m}(X)$ for every $m \in \omega$.

**408.** Given a space $X$ suppose that the cardinality of $X$ is regular and uncountable while $l^*(X) = \omega$, i.e., all finite powers of $X$ are Lindelöf. Prove that the space $C_p(X)$ has a point-countable $\pi$-base if and only if there exists a natural number $n > 1$ such that some set $A \subset X^n\backslash\Delta_n(X)$ is concentrated around $\Delta_n(X)$ and $|A| = |X|$.

**409.** Given a metrizable space $X$ prove that $C_p(X)$ has a point-countable $\pi$-base if and only if $X$ is countable.

**410.** Suppose that $X$ is an infinite space with $l^*(X) = \omega$. Prove that if $C_p(X)$ has a point-countable $\pi$-base then $|X| = \Delta(X)$. Here $\Delta(X) = \min\{\kappa : \Delta_2(X)$ is a $G_\kappa$-subset of $X \times X\}$ is the diagonal number of the space $X$. Deduce from this fact that if $X$ is compact and $C_p(X)$ has a point-countable $\pi$-base then $w(X) = |X|$.

**411.** Prove that if $K$ is a scattered Corson compact space then $C_p(K)$ has a point-countable $\pi$-base.

**412.** Prove that if $C_p(X)$ embeds in a $\Sigma$-product of first countable spaces then it has a point-countable $\pi$-base.

**413.** Say that a space $X$ is $P$-favorable (for the point-open game) if the player $P$ has a winning strategy in the point-open game on $X$. Prove that

   (i) any countable space is $P$-favorable;

   (ii) a continuous image of a $P$-favorable space is $P$-favorable;

  (iii) the countable union of $P$-favorable space is $P$-favorable;

  (iv) any nonempty closed subspace of a $P$-favorable space is $P$-favorable;

   (v) any nonempty Lindelöf scattered space is $P$-favorable.

**414.** Prove that

  (a) if $X$ is $P$-favorable for the point-open game and $\psi(X) \leq \omega$ then $X$ is countable;

  (b) a compact space $X$ is $P$-favorable for the point-open game if and only if $X$ is scattered.

**415.** Given a nonempty space $X$ define a game $\mathcal{PO}'$ on the space $X$ as follows: at the $n$-th move the player $P$ chooses a finite set $F_n \subset X$ and the player $O$ takes a set $U_n \in \tau(X)$ such that $F_n \subset U_n$. The game ends after the moves $F_n, U_n$ are made for all $n \in \omega$. The player $P$ wins if $\bigcup_{n \in \omega} U_n = X$; otherwise the victory is assigned to $O$. Prove that the player $P$ has a winning strategy in the game $\mathcal{PO}'$ on a space $X$ if and only if $X$ is $P$-favorable for the point-open game.

**416.** Given a nonempty space $X$ define a game $\mathcal{PO}''$ on the space $X$ as follows: at the $n$-th move the player $P$ chooses a finite set $F_n \subset X$ and the player $O$ takes a set $U_n \in \tau(X)$ such that $F_n \subset U_n$. After all moves $\{F_n, U_n : n \in \omega\}$ are made let $\mathcal{U} = \{U_n : n \in \omega\}$. The player $P$ wins if $\varprojlim \mathcal{U} = X$; otherwise the victory is assigned to $O$. Prove that the player $P$ has a winning strategy in the game $\mathcal{PO}''$ on a space $X$ if and only if $X$ is $P$-favorable for the point-open game.

**417.** Prove that a space $X$ is $P$-favorable for the point-open game if and only if $C_p(X)$ is a $W$-space.

**418.** Assuming that $2^\omega = \omega_1$ and $2^{\omega_1} = \omega_2$ prove that there exists a scattered Lindelöf $P$-space $X$ such that $hd^*(X) = \omega_1$ and $|X| = \omega_2$. Show that $C_p(X)$ is a $W$-space with no point-countable $\pi$-base. In particular, $C_p(X)$ is a $W$-space which cannot be embedded into a $\Sigma$-product of first countable spaces.

**419.** Prove that if $X_t$ is a $d$-separable space for each $t \in T$ then the product space $\prod_{t \in T} X_t$ is $d$-separable. In other words, any product of $d$-separable spaces must be $d$-separable.

**420.** Given an infinite cardinal $\kappa$ and a space $X$ prove that $X^\kappa$ is $d$-separable if and only if there exists a family $\mathcal{D} = \{D_n : n \in \omega\}$ of discrete subspaces of $X^\kappa$ such that $\sup\{|D_n| : n \in \omega\} \geq d(X)$. In particular, if $X^\kappa$ has a discrete subspace of cardinality $d(X)$ then $X^\kappa$ is $d$-separable. Deduce from this fact that $X^{d(X)}$ is $d$-separable for any space $X$.

**421.** Prove that

(a) if $K$ is a compact space then $K^\omega$ is $d$-separable;
(b) there exists a compact space $K$ such that $K^n$ is not $d$-separable for any $n \in \mathbb{N}$. Thus $K^\omega$ is $d$-separable but no finite power of $K$ is $d$-separable.

**422.** Prove that if $C_p(X)$ is $d$-separable then there is a discrete subspace $D \subset C_p(X)$ such that $|D| = d(C_p(X))$.

**423.** Given a space $X$ and $n \in \mathbb{N}$ say that a discrete subspace $D \subset X^n$ is *essential* if $\overline{D} \cap \Delta_n(X) = \emptyset$ and $|D| = iw(X)$. Prove that if, for some $n \in \mathbb{N}$, there exists an essential discrete set $D \subset X^n$ then $C_p(X)$ is $d$-separable. In particular, if there exists a discrete subspace of $X$ of cardinality $iw(X)$ then $C_p(X)$ is $d$-separable.

**424.** Assume that $X$ is a space for which there exists a discrete subspace $D \subset X \times X$ such that $|D| = iw(X)$. Prove that $C_p(X)$ is $d$-separable.

**425.** Let $X$ be a space such that the cardinal $\kappa = iw(X)$ has uncountable cofinality. Prove that the following conditions are equivalent:

(i) $(C_p(X))^n$ is $d$-separable for all $n \geq 2$;
(ii) $(C_p(X))^n$ is $d$-separable for some $n \geq 2$;
(iii) $C_p(X) \times C_p(X)$ is $d$-separable;
(iv) for some $m \in \mathbb{N}$ there is a discrete set $D \subset X^m$ with $|D| = \kappa$.

**426.** Prove that

(a) if $\sup\{s(X^n) : n \in \mathbb{N}\} > iw(X)$ then $C_p(X)$ is $d$-separable;
(b) if $K$ is a Corson compact space then $C_p(K)$ is $d$-separable;
(c) if $X$ is a metrizable space then $C_p(X)$ is $d$-separable.

**427.** Prove that if $C_p(X)$ is a Lindelöf $\Sigma$-space then it is $d$-separable.

**428.** Prove that if $C_p(X)$ is a Lindelöf $\Sigma$-space then the space $X$ must be hereditarily $d$-separable.

**429.** Prove that under CH,

(a) there exists a compact space $K$ such that $C_p(K)$ is not $d$-separable;
(b) there exists a space $X$ such that $X \times X$ is $d$-separable while $X$ is not $d$-separable.

**430.** Suppose that $\kappa$ is an infinite cardinal, $T \neq \emptyset$ is a set, and $N_t$ is a space such that $nw(N_t) \leq \kappa$ for all $t \in T$. Assume that $D$ is a dense subspace of $N = \prod_{t \in T} N_t$ and $f : D \to K$ is a continuous map of $D$ onto a compact space $K$.

Prove that if $\rho(K) \leq \kappa$ then $w(K) \leq \kappa$. Deduce from this fact that if a compact space $K$ is a continuous image of a dense subspace of a product of cosmic spaces then $w(K) = t(K) = \rho(K)$.

**431.** Suppose that $\kappa$ is a cardinal of uncountable cofinality, $T \neq \emptyset$ is a set, and $N_t$ is a space such that $nw(N_t) \leq \omega$ for all $t \in T$. Assume that $D$ is a dense subspace of $N = \prod_{t \in T} N_t$ and $f : D \to K$ is a continuous map of $D$ onto a compact space $K$ with $w(K) = \kappa$. Prove that $K$ maps continuously onto $\mathbb{I}^{\kappa}$.

**432.** Given an infinite cardinal $\kappa$ suppose that $nw(N_t) \leq \kappa$ for any $t \in T$ and $C \subset N = \prod_{t \in T} N_t$ is a dense subspace of $N$. Assume additionally that we have a continuous (not necessarily surjective) map $\varphi : C \to L$ of $C$ into a compact space $L$. Prove that if $y \in C' = \varphi(C)$ and $h\pi\chi(y, L) \leq \kappa$ then $\psi(y, C') \leq \kappa$. Here the cardinal $h\pi\chi(y, L)$ is the hereditary $\pi$-character of the space $L$ at the point $y$, i.e., $h\pi\chi(y, L) = \sup\{\pi\chi(y, Z) : y \in Z \subset L\}$.

**433.** Suppose that $C$ is a dense subspace of a product $N = \prod_{t \in T} N_t$ such that $nw(N_t) \leq \kappa$ for each $t \in T$. Assume that $K$ is a compact space with $t(K) \leq \kappa$ and $\varphi : C \to K$ is a continuous (not necessarily surjective) map; let $C' = \varphi(C)$. Prove that every closed subspace of $C'$ is a $G_{\kappa}$-set; in particular, $\psi(C') \leq \kappa$.

**434.** Suppose that $C$ is a dense subspace of a product $N = \prod_{t \in T} N_t$ such that $nw(N_t) \leq \kappa$ for each $t \in T$. Assume additionally that $l(C) \leq \kappa$ and $K$ is a compact space with $t(K) \leq \kappa$ such that there exists a continuous (not necessarily surjective) map $\varphi : C \to K$. Prove that if $C' = \varphi(C)$ then $hl(C') \leq \kappa$.

**435.** Prove that if $C$ is a dense subspace of a product of cosmic spaces and $K$ is a compact space then, for any continuous map $\varphi : C \to K$, we have $\Psi(\varphi(C)) \leq t(K)$.

**436.** Suppose that $C$ is a dense subspace of a product of cosmic spaces and $\varphi : C \to K$ is a continuous (not necessarily surjective) map into a compact space $K$ of countable tightness; let $Y = \varphi(C)$. Prove that $Y$ is a perfect space and $\pi w(Y) \leq \omega$. In particular, if $\varphi : C_p(X) \to K$ is a continuous map of $C_p(X)$ into a compact space $K$ with $t(K) \leq \omega$ then $\varphi(C_p(X))$ is a perfect space of countable $\pi$-weight.

**437.** Suppose that $C$ is a dense Lindelöf $\Sigma$-subspace of a product of cosmic spaces and $\varphi : C \to K$ is a continuous (not necessarily surjective) map into a compact space $K$ of countable tightness. Prove that $nw(\varphi(C)) \leq \omega$. In particular, if $C_p(X)$ is a Lindelöf $\Sigma$-space and $\varphi : C_p(X) \to K$ is a continuous map of $C_p(X)$ into a compact space $K$ of countable tightness then $\varphi(C_p(X))$ is cosmic.

**438.** Given an infinite cardinal $\kappa$ observe that the property $\mathcal{P}_{\kappa}$ of being strongly $\kappa$-cosmic is stronger than being $\kappa$-cosmic. Besides, $\mathcal{P}_{\kappa}$ is preserved by subspaces and continuous images. Prove that if $\mu < \mathrm{cf}(\kappa)$ and $X_{\alpha}$ is strongly $\kappa$-cosmic for any $\alpha < \mu$ then $X = \prod_{\alpha < \mu} X_{\alpha}$ is also strongly $\kappa$-cosmic.

**439.** Given an infinite cardinal $\kappa$ prove that $\mathbb{D}^{\kappa}$ is not strongly $\kappa$-cosmic and hence no strongly $\kappa$-cosmic space can be continuously mapped onto $\mathbb{I}^{\kappa}$.

**440.** Assume that $\kappa$ is an infinite cardinal and $K$ is a strongly $\kappa$-cosmic compact space. Prove that there exists $x \in K$ such that $\chi(x, K) < \kappa$.

**441.** Suppose that $\kappa$ is an uncountable cardinal and $X$ is a space such that $w(X) \leq \kappa$ and $l(X) < \mathrm{cf}(\kappa)$. Prove that $C_p(X)$ is strongly $\kappa$-cosmic. In particular, if $\mathrm{cf}(\kappa) > \omega$ and $X$ is a Lindelöf space with $w(X) = \kappa$ then $C_p(X)$ is strongly $\kappa$-cosmic.

**442.** Let $\kappa$ be a cardinal of uncountable cofinality. Prove that if $X$ is a Lindelöf $\Sigma$-space with $nw(X) \leq \kappa$ then $C_p(X)$ is strongly $\kappa$-cosmic.

**443.** Given an uncountable cardinal $\kappa$ suppose that a space $X$ is strongly $\kappa$-monolithic, i.e., $w(\overline{A}) \leq \kappa$ for every $A \subset X$ with $|A| \leq \kappa$ and, additionally, $l(X) < \mathrm{cf}(\kappa)$. Prove that $w(K) < \kappa$ for any compact continuous image $K$ of the space $C_p(X)$. Deduce from this fact that if $X$ is a Lindelöf strongly $\omega_1$-monolithic space (in particular, if $l(X) = \omega$ and $w(X) \leq \omega_1$) then every compact continuous image of $C_p(X)$ is metrizable.

**444.** Given a cardinal $\kappa$ with $\mathrm{cf}(\kappa) > \omega$ assume that $X$ is a $\kappa$-monolithic Lindelöf $\Sigma$-space. Prove that if $K$ is a compact continuous image of $C_p(X)$ then $w(K) < \kappa$. In particular, if $X$ is an $\omega_1$-monolithic Lindelöf $\Sigma$-space then every compact continuous image of $C_p(X)$ is metrizable.

**445.** Let $\kappa$ be a cardinal and denote by $D_\kappa$ a discrete space of cardinality $\kappa$. For the compact space $K = \beta D_\kappa$ prove that $C_p(K)$ maps continuously onto $\mathbb{I}^\kappa$.

**446.** Prove that every one of the following statements is equivalent to Luzin's axiom $(2^{\omega_1} > \mathfrak{c})$:

    (i)   for every separable compact space $K$ any compact continuous image of $C_p(K)$ is metrizable;

    (ii)   any compact continuous image of $C_p(\beta\omega)$ is metrizable;

    (iii)   any compact continuous image of $C_p(\mathbb{I}^\mathfrak{c})$ is metrizable;

    (iv)   for every compact space $K$ with $w(K) \leq \mathfrak{c}$, any compact continuous image of $C_p(K)$ is metrizable.

**447.** Under Luzin's Axiom prove that

    (a)   if $X$ is Lindelöf and $w(X) \leq \mathfrak{c}$ then every compact continuous image of $C_p(X)$ is metrizable.

    (b)   if $X$ is Lindelöf and first countable then every compact continuous image of $C_p(X)$ is metrizable.

**448.** Prove that, for every hereditarily Lindelöf first countable space $X$, any compact continuous image of $C_p(X)$ is metrizable. In particular, for each perfectly normal compact space $X$, any compact continuous image of $C_p(X)$ is metrizable.

**449.** Suppose that $X$ is a space such that $l(C_p(X)) = t(C_p(X)) = \omega$. Prove that every compact continuous image of $C_p(X)$ is metrizable.

**450.** Prove that if $X$ is a space such that $C_p(X)$ is a Lindelöf $\Sigma$-space then every compact continuous image of $C_p(X)$ is metrizable.

**451.** For the double arrow space $K$ prove that the space $C_p(K)$ can be condensed onto $\mathbb{I}^\omega$.

**452.** Prove that $C_p(\mathbb{D}^{\mathfrak{c}})$ condenses onto $\mathbb{I}^{\omega}$.

**453.** Suppose that $X$ is a nonempty $\sigma$-compact second countable space. Prove that $C_p(X)$ condenses onto a compact space.

**454.** Prove that there exists a set $X \subset \mathbb{R}$ such that $C_p(X)$ does not condense onto an analytic space.

**455.** Given a nonempty space $X$ such that $|C_p(X)| < 2^{d(X)}$ prove that $C_p(X)$ cannot be condensed onto a compact space. Deduce from this fact that neither one of the spaces $C_p(\beta\omega\backslash\omega)$ and $C_p(\Sigma(\mathbb{D}^{\mathfrak{c}}))$ condenses onto a compact space. Here $\Sigma(\mathbb{D}^{\mathfrak{c}}) = \{x \in \mathbb{D}^{\mathfrak{c}} : |x^{-1}(1)| \leq \omega\}$ is the $\Sigma$-product of $\mathbb{D}^{\mathfrak{c}}$.

**456.** Prove that if $2^{\omega} = \omega_1$ and $2^{\omega_1} > \omega_2$ then $C_p(\mathbb{D}^{\omega_2})$ does not condense onto a compact space.

**457.** Suppose that $2^{\omega} < 2^{\omega_1}$ and $C_p(X)$ condenses onto a compact space. Prove that $X$ is separable if and only if $|C_p(X)| \leq \mathfrak{c}$.

**458.** Suppose that $l(X^n) = \omega$ for all $n \in \mathbb{N}$ and $C_p(X)$ is Lindelöf. Observe that if $C_p(X)$ condenses onto a compact space $K$ then $K$ is metrizable and $X$ is separable. Prove that if $C_p(X)$ condenses onto a $\sigma$-compact space $Y$ then $X$ is separable and $\psi(Y) = \omega$. Deduce from this fact that if $X$ is a non-metrizable Corson compact space then $C_p(X)$ does not condense onto a $\sigma$-compact space.

**459.** Given a space $X$ prove that the space $C_p(X)$ condenses onto a space embeddable in a compact space of countable tightness if and only if $C_p(X)$ condenses onto a second countable space.

**460.** Assuming MA+¬CH prove that if $K$ is a compact space such that $C_p(K)$ is Lindelöf and condensable onto a $\sigma$-compact space then $K$ is metrizable.

**461.** Assume that $C_p(X)$ is a Lindelöf $\Sigma$-space and there exists a condensation of $C_p(X)$ onto a $\sigma$-compact space $Y$. Prove that $nw(X) \leq \omega$ and $nw(Y) \leq \omega$.

**462.** Prove that, for any compact space $X$ such that $w(X) \leq \omega_1$, the space $C_p(X)$ is hereditarily metalindelöf.

**463.** Given an uncountable discrete space $X$ prove that the space $C_p(\beta X, \mathbb{D})$ is not metalindelöf; therefore $C_p(X)$ is not metalindelöf either.

**464.** Prove that

(i) every strong $\Sigma$-space is a $D$-space. In particular, any Lindelöf $\Sigma$-space is a $D$-space;

(ii) every space with a point-countable base is a $D$-space.

**465.** Prove that,

(i) for any infinite cardinal $\kappa$, if a space is monotonically $\kappa$-monolithic then it is $\kappa$-monolithic;

(ii) for any infinite cardinal $\kappa$, if $X$ is a monotonically $\kappa$-monolithic space and $Y \subset X$ then $Y$ is also monotonically $\kappa$-monolithic;

(iii) if $\kappa$ is an infinite cardinal, then any countable product of monotonically $\kappa$-monolithic spaces is monotonically $\kappa$-monolithic;

(iv) for any infinite cardinal $\kappa$, every closed continuous image of a monotonically $\kappa$-monolithic space is monotonically $\kappa$-monolithic;

(v) a space $X$ is monotonically monolithic if and only if $X$ is monotonically $\kappa$-monolithic for any infinite cardinal $\kappa$;

(vi) for any infinite cardinal $\kappa$, a space $X$ is monotonically $\kappa$-monolithic if and only if to any finite set $F \subset X$ we can assign a countable family $\mathcal{O}(F) \subset \exp(X)$ in such a way that for any set $A \subset X$ with $|A| \leq \kappa$, the family $\bigcup\{\mathcal{O}(F) : F \in [A]^{<\omega}\}$ is an external network at all points of $\overline{A}$;

(vii) a space $X$ is monotonically monolithic if and only if to any finite set $F \subset X$ we can assign a countable family $\mathcal{O}(F) \subset \exp(X)$ in such a way that for any $A \subset X$, the family $\bigcup\{\mathcal{O}(F) : F \in [A]^{<\omega}\}$ is an external network at all points of $\overline{A}$;

(viii) if $X$ is a space and the set $X'$ of non-isolated points of $X$ has a countable network then $X$ is monotonically monolithic. In particular, if a space $X$ is cosmic or has a unique non-isolated point then $X$ is monotonically monolithic;

(ix) given an infinite cardinal $\kappa$, if a space $X$ is monotonically $\kappa$-monolithic and $t(X) \leq \kappa$, then $X$ is monotonically monolithic;

(x) for any infinite cardinal $\kappa$, every $\sigma$-product of monotonically $\kappa$-monolithic spaces is monotonically $\kappa$-monolithic. In particular, any $\sigma$-product of monotonically monolithic spaces is monotonically monolithic.

**466.** Prove that,

(i) a space $X$ is strongly monotonically monolithic if and only if $X$ is strongly monotonically $\omega$-monolithic;

(ii) every space with a point-countable base must be strongly monotonically monolithic;

(iii) if a space $X$ is strongly monotonically monolithic, then it must be strongly monolithic and monotonically monolithic;

(iv) every subspace of a strongly monotonically monolithic space is strongly monotonically monolithic;

(v) any countable product of strongly monotonically monolithic spaces must be strongly monotonically monolithic;

(vi) if $X$ is strongly monotonically monolithic and $f : X \to Y$ is an open map such that $d(f^{-1}(y)) \leq \omega$ for any $y \in Y$ then $Y$ is strongly monotonically monolithic;

(vii) every countably compact strongly monotonically monolithic space is compact and metrizable.

**467.** For any infinite cardinal $\kappa$ prove that

(a) a space $X$ is monotonically $\kappa$-stable if and only if $C_p(X)$ is monotonically $\kappa$-monolithic.

(b) a space $X$ is monotonically $\kappa$-monolithic if and only if $C_p(X)$ is monotonically $\kappa$-stable.

**468.** Prove that if $X$ is a Lindelöf $\Sigma$-space then $C_p(X)$ is monotonically mono-lithic. In particular, if $C_p(Y)$ is a Lindelöf $\Sigma$-space then $Y$ is monotonically monolithic.

**469.** Prove that if $\upsilon X$ is a Lindelöf $\Sigma$-space then $C_p(X)$ is monotonically $\omega$-monolithic. In particular, if $\upsilon(C_p(X))$ is a Lindelöf $\Sigma$-space then $C_p(X)$ is monotonically $\omega$-monolithic. Deduce from these facts that if $X$ is pseudo-compact then $C_p(X)$ is monotonically $\omega$-monolithic.

**470.** Let $X$ be a monotonically $\kappa$-monolithic space with $ext(X) = \lambda \leq \kappa$. Prove that

(i) if $\lambda < \kappa$ then $l(X) \leq \lambda$;
(ii) if $\lambda = \kappa$ and $t(X) \leq \kappa$ then $l(X) \leq \kappa$.

**471.** Prove that every subspace of a monotonically monolithic space is a $D$-space. As a consequence,

(i) if $X$ is a Lindelöf $\Sigma$-space then every subspace of $C_p(X)$ is a $D$-space.
(ii) Observe that $ext(Y) = l(Y)$ for any $D$-space $Y$. Therefore (i) general-izes Baturov's theorem (SFFS-269).

**472.** Given an infinite cardinal $\kappa$, suppose that a space $X$ is monotonically $\kappa$-monolithic and $nw(X) \leq \kappa^+$. Prove that every subspace of $X$ is a $D$-space. Deduce from this fact that for any pseudocompact space $Z$ with $nw(Z) \leq \omega_1$, every subspace of the space $C_p(Z)$ is a $D$-space.

**473.** Prove that any countably compact monotonically $\omega$-monolithic space is Corson compact.

**474.** Give an example of a Corson compact space that is not monotonically $\omega$-monolithic.

**475.** Observe that if $X$ is a monotonically $\omega$-monolithic compact space and $\omega_1$ is a caliber of $X$, then the space $X$ is metrizable. Prove that there exists a monotonically $\omega_1$-monolithic (and hence monotonically $\omega$-monolithic) non-compact pseudocompact space $X$ such that $\omega_1$ is a caliber of $X$.

**476.** Suppose that $X$ is a monotonically Sokolov space. Prove that

(a) $X$ is monotonically $\omega$-monolithic;
(b) any $F_\sigma$-subset of $X$ is monotonically Sokolov.

**477.** Suppose that a space $X$ is monotonically retractable and fix, for any countable set $A \subset X$, a retraction $r_A : X \to X$ and a network $\mathcal{N}(A)$ for the map $r_A$ that witness monotone retractability of $X$. Prove that, for any countable family $\mathcal{G}$ of closed subsets of $X$ we can find a countable set $P(\mathcal{G}) \subset X$ such that $r_{P(\mathcal{G})}(G) \subset G$ for any $G \in \mathcal{G}$ and the assignment $\mathcal{G} \to P(\mathcal{G})$ is $\omega$-monotone.

**478.** Given a space $Y$ denote by $\mathcal{CL}(Y)$ the family of all closed subsets of $Y$ and let $\mathcal{CL}^*(Y) = \bigcup\{\mathcal{CL}(Y^n) : n \in \mathbb{N}\}$. Prove that, for any space $X$, the following conditions are equivalent:

(a) $X$ is monotonically retractable;
(b) for any countable family $\mathcal{G}$ of closed subsets of $X$, there exists a retraction $s_\mathcal{G} : X \to X$ and a countable network $\mathcal{O}(\mathcal{G})$ for the map $s_\mathcal{G}$ such that $s(G) \subset G$ for any $G \in \mathcal{G}$ and the assignment $\mathcal{O}$ is $\omega$-monotone;

(c) for any countable family $\mathcal{G} \subset \mathcal{CL}^*(X)$ there exists a retraction $t_\mathcal{G} : X \to X$ and a countable network $\mathcal{P}(\mathcal{G})$ for the map $t_\mathcal{G}$ such that, for any $n \in \mathbb{N}$, we have the inclusion $t_\mathcal{G}^n(G) \subset G$ for any $G \in \mathcal{G} \cap \mathcal{CL}(X^n)$ and the assignment $\mathcal{P}$ is $\omega$-monotone.

**479.** Given a monotonically retractable space $X$, prove that

(a) every $F_\sigma$-subset of $X$ is monotonically retractable;
(b) the space $X$ is Sokolov; in particular, $X$ is $\omega$-stable, collectionwise normal, and $ext(X) \leq \omega$;
(c) $C_p(X)$ is a Lindelöf $D$-space.

**480.** Prove that

(a) any $\sigma$-product of monotonically retractable spaces is monotonically retractable;
(b) any $\Sigma$-product of monotonically retractable spaces is monotonically retractable and hence every countable product of monotonically retractable spaces is monotonically retractable;
(c) any closed subspace of a $\Sigma$-product of cosmic spaces is monotonically retractable.

**481.** For any space $X$ prove that

(a) $X$ is monotonically retractable if and only if $C_p(X)$ is monotonically Sokolov.
(b) $X$ is monotonically Sokolov if and only if $C_p(X)$ is monotonically retractable.

**482.** Prove that

(a) any monotonically Sokolov space is Lindelöf and Sokolov;
(b) any countable product of monotonically Sokolov spaces is monotonically Sokolov;
(c) every $\mathbb{R}$-quotient image of a monotonically Sokolov space is monotonically Sokolov;
(d) every $\mathbb{R}$-quotient image of a monotonically retractable space $X$ must be monotonically retractable;
(e) a monotonically retractable space must be $\omega$-monolithic but it can fail to be monotonically $\omega$-monolithic.

**483.** Prove that a compact space $X$ is monotonically retractable if and only if $X$ is Corson compact.

**484.** Prove that a monotonically retractable space $X$ is monotonically Sokolov if and only if $X$ is monotonically $\omega$-monolithic. In particular, a compact space is monotonically Sokolov if and only if it is monotonically $\omega$-monolithic.

**485.** Suppose that $X$ is a first countable countably compact subspace of an ordinal with its order topology. Prove that $X$ is monotonically retractable and hence $C_p(X)$ is a Lindelöf $D$-space.

**486.** Prove that

(a) the set $E$ of all ordinals $\alpha < \omega_2$ of uncountable cofinality is monotonically $\omega$-monolithic but is neither a $D$-space nor monotonically $\omega_1$-monolithic;

(b) for set $Y$ of all ordinals $\alpha < \omega_2$ of countable cofinality, the space $X = Y \cup \{\omega_2\}$ is countably compact and $ext(C_p(X)) = \omega$ while $l(C_p(X)) = \omega_2$. In particular, $X$ is a countably compact space such that $C_p(X)$ is not a $D$-space.

**487.** Suppose that $X$ and $C_p(X)$ are Lindelöf $\Sigma$-spaces. Prove that both $X$ and $C_p(X)$ must be monotonically retractable and monotonically Sokolov. In particular, $X$ and $C_p(X)$ have the Sokolov property.

**488.** Prove that

(a) any Eberlein compact space $X$ has a $\sigma$-closure-preserving local base at every point, i.e., for any $x \in X$ there exists a local base $\mathcal{B}_x$ at the point $x$ such that $\mathcal{B}_x = \bigcup_{n \in \omega} \mathcal{B}_x^n$ and every $\mathcal{B}_x^n$ is closure-preserving;

(b) there exists an Eberlein compact space $K$ that does not have a closure-preserving local base at some point;

(c) there exists a non-metrizable compact space $Y$ such that $C_p(Y)$ has a closure-preserving local base at every point.

**489.** Given a compact space $K$ let $\|f\| = \sup\{|f(x)| : x \in K\}$ for any $f \in C(K)$; if $f, g \in C(K)$ then $\rho_K(f, g) = \|f - g\|$. As usual, $C_u(K)$ is the set $C(K)$ with the topology generated by the metric $\rho_K$. If $A \subset C_p(K)$ and $A \neq \emptyset$ then $\operatorname{diam}(A) = \sup\{\|f - g\| : f, g \in A\}$. For each $\varepsilon > 0$ say that a family $\mathcal{A}$ of subsets of $C(K)$ is $\varepsilon$-small if $\operatorname{diam}(A) < \varepsilon$ for any $A \in \mathcal{A}$. The space $C_p(K)$ is said to have the property JNR if for every $\varepsilon > 0$ we can find a family $\{M_n : n \in \omega\}$ of closed subsets of $C_p(K)$ such that $C_p(K) = \bigcup_{n \in \omega} M_n$ and, for every $n \in \omega$ there exists an $\varepsilon$-small family $\mathcal{U}_n \subset \tau(M_n)$ such that $\bigcup \mathcal{U}_n = M_n$. Prove that the following conditions are equivalent:

(i) the space $C_p(K)$ has the property JNR;

(ii) there exists a $\sigma$-discrete family $\mathcal{N}$ in the space $C_p(K)$ such that $\mathcal{N}$ is a network in both spaces $C_p(K)$ and $C_u(K)$.

In particular, if $C_p(K)$ has the property JNR then it has a $\sigma$-discrete network.

**490.** Prove that

(a) if $L$ is a linearly ordered separable compact space then $C_p(L)$ is a $\sigma$-space, i.e., it has a $\sigma$-discrete network;

(b) for any separable dyadic compact space $K$, the space $C_p(K)$ has a $\sigma$-discrete network.

**491.** Prove that there exists a scattered compact space $K$ with a countable dense set of isolated points such that $C_p(K, \mathbb{D})$ is not perfect. Deduce from this fact that $C_p(\beta\omega, \mathbb{D})$ is not perfect and, in particular, the space $C_p(\beta\omega)$ does not have a $\sigma$-discrete network.

**492.** Prove that, for any ultrafilter $\xi \in \beta\omega\backslash\omega$, the following conditions are equivalent:

(i) $\xi$ is not a $P$-point in $\beta\omega\backslash\omega$;
(ii) $\sigma(\mathbb{R}^\omega)$ is homeomorphic to a closed subspace of $C_p(\omega_\xi)$;
(iii) $C_p(\omega_\xi)$ is not hereditarily Baire.

**493.** Let $\xi, \eta \in \beta\omega\backslash\omega$. Prove that the spaces $\omega_\xi$ and $\omega_\eta$ are $l$-equivalent if and only if there exists a bijection $b : \omega \to \omega$ such that $b(\xi) = \{b(U) : U \in \xi\} = \eta$. In particular, $\omega_\xi$ and $\omega_\eta$ are $l$-equivalent if and only if they are homeomorphic.

**494.** Suppose that $\alpha$ and $\beta$ are ordinals such that $\omega \le \alpha \le \beta < \omega_1$. Prove that $(\alpha + 1)$ is $l$-equivalent to $(\beta + 1)$ if and only if $\beta < \alpha^\omega$. Deduce from this fact that there exist countable compact $u$-equivalent spaces which are not $l$-equivalent.

**495.** Suppose that $X$ is a metrizable zero-dimensional compact space and there exists a continuous linear surjection of $C_p(X)$ onto $C_p(Y)$. Prove that $Y$ is a metrizable compact zero-dimensional space. In particular, there exists no continuous linear surjection of $C_p(\mathbb{K})$ onto $C_p(\mathbb{I})$. Here, as usual, $\mathbb{K}$ is the Cantor set and $\mathbb{I} = [-1, 1] \subset \mathbb{R}$.

**496.** Suppose that a space $X$ is metrizable, compact, and $\sigma$-zero-dimensional. Prove that if there exists a continuous linear surjection of $C_p(X)$ onto $C_p(Y)$ then $Y$ is also a compact metrizable $\sigma$-zero-dimensional space. In particular, if $X$ is compact, metrizable, and finite-dimensional then there is no continuous linear surjection of $C_p(X)$ onto $C_p(\mathbb{I}^\omega)$.

**497.** Prove that a space $K$ is $l$-equivalent to the Cantor set $\mathbb{K}$ if and only if there exists a continuous linear surjection of $C_p(K)$ onto $C_p(\mathbb{K})$ as well as a continuous linear surjection of $C_p(\mathbb{K})$ onto $C_p(K)$.

**498.** Let $I$ be the closed interval $[0, 1] \subset \mathbb{R}$. Given any $n \in \omega$ suppose that $K$ is a metrizable compact space and dim $K \le n$. Prove that there exists a compact subspace $K' \subset I^{2n+1}$ with the following properties:

(i) $K'$ is homeomorphic to $K$;
(ii) for any function $\varphi \in C(K')$, we can choose $\varphi^1, \ldots, \varphi^{2n+1} \in C(I)$ such that $\varphi(y) = \varphi^1(y_1) + \ldots + \varphi^{2n+1}(y_{2n+1})$ for any $y = (y_1 \ldots, y_{2n+1}) \in K'$.

Deduce from this fact that if $X$ is a second countable space and dim $X \le n$ then $X$ embeds in $\mathbb{I}^{2n+1}$.

**499.** Prove that, for any finite-dimensional metrizable compact space $K$, there exists a surjective continuous linear mapping of $C_p(\mathbb{I})$ onto $C_p(K)$.

**500.** Prove that $l(X) = l(Y)$ for any $l$-equivalent spaces $X$ and $Y$. In particular, if $X \overset{l}{\sim} Y$ then $X$ is Lindelöf if and only if so is $Y$.

## 1.6   Bibliographic notes to Chapter 1

The material of Chapter 1 consists of problems of the following types:

(1)  textbook statements which give a gradual development of some topic;
(2)  folkloric statements that might not be published but are known by specialists;
(3)  famous theorems cited in textbooks and well-known surveys;
(4)  comparatively recent results which have practically no presence in textbooks.

*We will almost never cite the original papers for statements of the first three types.* We are going to cite them for a very small sample of results of the fourth type. The selection of theorems to cite is made according to the preferences of the author and *does not mean that all statements of the fourth type are mentioned.* I bring my apologies to readers who might think that I did not cite something more important than what is cited. The point is that a selection like that has to be biased because it is impossible to mention all contributors. As a consequence, *there are quite a few statements of the main text, published as results in papers, which are never mentioned in our bibliographic notes.* A number of problems of the main text cite published or unpublished results of the author. However, those are treated exactly like the results of others: some are mentioned and some aren't. On the other hand, the Bibliography contains (to the best knowledge of the author) the papers and books of *all contributors to the material of this book.*

Chapter 1 seems to be the first complete and systematic presentation of the material on functional equivalences. Section 1.1 contains many results on $t$-invariance which are simple consequences of previously proved results on duality. Gul'ko and Khmyleva were the first ones to prove in (1986) that compactness is not $t$-invariant (Problems 016–027). Okunev proved in (1990) that $\sigma$-compactness and similar properties are preserved by $t$-equivalence (Problems 034–045). The $t$-invariance of spread, hereditary density, and hereditary Lindelöf number (Problems 055–070) was proved in Okunev's paper (1997a). Dobrowolski et al. established in (1990) that all metrizable countable non-discrete spaces are $t$-equivalent. Since they applied very deep results and methods of infinite-dimensional topology, it was impossible to present their theorem in this book. Another fundamental result of Section 1.1 is Marciszewski's theorem (2000) which states that $\mathbb{I}$ is not $t$-equivalent to $\mathbb{I}^\omega$ (Problem 099).

Section 1.2 gives a brief introduction to the theory of uniform spaces. This material is best covered in Engelking's book (1977). Some knowledge of uniformities is necessary to tackle Gul'ko's theorem (1993) on $u$-invariance of the dimension dim (Problems 176–180). There are also some deep results on the behavior of the dimension dim, which are used in Gul'ko's proof. This made it necessary to develop some methods used in dimension theory and the theory of inverse systems. The reader can find an exhaustive information on the mentioned topics in Engelking's books (1977) and (1978). The $u$-invariance of absolute Borel sets (Problems 197 and 198) was proved by Marciszewski and Pelant in (1997). The last group of problems of Section 1.2 presents another theorem of Gul'ko (1988) which states that all infinite countable compact spaces are $u$-equivalent (Problem 200).

Section 1.3 develops the most important tools to deal with linear topological spaces. A good reference is W. Rudin's book (1973). Okunev's method of constructing pairs of $l$-equivalent spaces (Problems 257–263) was published in (1986) in a stronger form.

Section 1.4 presents quite a few nontrivial results on $l$-equivalence. Uspenskij's theorem on spaces which are $l$-equivalent to metrizable ones (Problems 329–335) was proved in (1983a). The theorem of Baars, de Groot, and Pelant on $l$-invariance of Čech-completeness in metrizable spaces was established in (1993) (Problems 365–367). A very difficult example of an infinite compact space which is not $l$-equivalent to itself with an added isolated point was constructed by Marciszewski in (1997a) (Problem 400).

Section 1.5 contains some more material on $l$-equivalence and the latest results in $C_p$-theory which did not appear in previous parts of the book. Marciszewski and Pol constructed in (2009) examples of non-cosmic spaces $X$ such that the space $C_p(X)$ has a $\sigma$-discrete network (Problems 489–490). Gruenhage proved in (2012) that there exists a Corson compact space which is not monotonically $\omega$-monolithic (Problem 474). The exhaustive classification of countable compact $l$-equivalent spaces (Problem 494) was obtained in a paper of Gul'ko and Os'kin (1975). The first readable text of the solution of Hilbert's Problem 13 (Problem 498) was published in Kolmogorov (1957). Leiderman et al. (1997) used this result to prove that $C_p(\mathbb{I})$ can be linearly and continuously mapped onto $C_p(X)$ for any finite-dimensional metrizable compact $X$ (Problem 499). Velichko proved $l$-invariance of the Lindelöf property in (1998a). Later Bouziad established in (2001) that Lindelöf number is $l$-invariant (Problem 500).

# Chapter 2
# Solutions of problems 001–500

It took the author six long years to come to finally start writing this last portion of his work. In this period he had a sabbatical stay, celebrated the arrival of the new century and the new millennium, changed his citizenship, became a grandfather, and published about thirty papers. Also, he understood much better the material of this book.

Since it could take just as much (or even more) time to read this book and/or to solve its problems, it is now almost impossible for the author to write the standard phrase: "the reader who mastered the previous material is now able to ...." How many readers will repeat this accomplishment? Well, at the present moment there is at least one (who, evidently, coincides with the author), so, formally, the previous three volumes show that it is possible to solve the problems of this book at least, when one already knows the solutions.

So, left behind are 1500 solutions and about 700 statements proved as auxiliary facts; some of these facts are quite famous and highly nontrivial theorems. As in the previous volume, the treatment of topology and $C_p$-theory is professional. When you read a solution of a problem of the main text, it has more or less the same level of exposition as a published paper on a similar topic.

The author hopes, however, that reading our solutions is more helpful than ploughing through the proofs in published papers; the reason is that we are not so constrained by the amount of the available space as a journal contributor, so we take much more care about all details of the proof. It is also easier to work with the references in our solutions than with those in research papers because in a paper the author does not need to bother about whether the reference is accessible for the reader whereas we only refer to what we have proved in this book apart from some very simple facts of calculus and set theory.

This volume has the same policy about the references as the third one; we use the textbook facts from general topology without giving a reference to them. This book is self-contained, so all necessary results are proved in the previous volumes but the references to standard things have to stop sometime. This makes it difficult

© Springer International Publishing Switzerland 2016
V.V. Tkachuk, *A Cp-Theory Problem Book*, Problem Books in Mathematics,
DOI 10.1007/978-3-319-24385-6_2

for a beginner to read this volume's results without some knowledge of the previous material. However, a reader who mastered the first four chapters of R. Engelking's book (1977) will have no problem with this.

We also omit references to some standard facts of $C_p$-theory. The reader can easily find the respective proofs using the index. Our reference omission rule can be expressed as follows: we omit references to textbook results from topology and $C_p$-theory *proved in the previous volumes*. There are quite a few phrases like "it is easy to see" or "it is an easy exercise"; the reader should trust the author's word and experience that the statements like that are *really* easy to prove as soon as one has the necessary background. On the other hand, the highest percentage of errors comes exactly from omissions of all kinds, so my recommendation is that, even though you should trust the author's claim that the statement is easy to prove or disprove, you shouldn't take just his word for the truthfulness of *any* statement. Verify it yourself and if you find any errors communicate them to me to correct the respective parts.

**V.001.** *Prove that cardinality, network weight, i-weight, as well as density are t-invariant.*

**Solution.** Suppose that $X$ is a space and $C_p(X)$ is homeomorphic to $C_p(Y)$. Then $|X| = w(C_p(X)) = w(C_p(Y)) = |Y|$ (see TFS-169); this proves that cardinality is $t$-invariant. Furthermore, $nw(X) = nw(C_p(X)) = nw(C_p(Y)) = nw(Y)$ (see TFS-172) and hence network weight is also $t$-invariant. Applying TFS-174 we conclude that the equalities $iw(X) = d(C_p(X)) = d(C_p(Y)) = iw(Y)$ show that $i$-weight is $t$-invariant as well. Finally observe that $d(X) = iw(C_p(X)) = iw(C_p(Y)) = d(Y)$ (see TFS-173), so density is $t$-invariant too.

**V.002.** *Prove that if $C_p(Y) \hookrightarrow C_p(X)$ then $nw(Y) \leq nw(X)$, $d(Y) \leq d(X)$ and $|Y| \leq |X|$. Give an example showing that the inequality $iw(Y) \leq iw(X)$ is not necessarily true.*

**Solution.** We have $nw(Y) = nw(C_p(Y)) \leq nw(C_p(X)) = nw(X)$ (see TFS-159 and TFS-172) which settles the first inequality. It follows from TFS-159 and TFS-173 that we have $d(Y) = iw(C_p(Y)) \leq iw(C_p(X)) = d(X)$. Now, TFS-159 together with TFS-169 imply that $|Y| = w(C_p(Y)) \leq w(C_p(X)) = |X|$.

Finally, consider the spaces $X = D(\mathfrak{c})$ and $Y = A(\mathfrak{c})$; the space $X$ is discrete and $|X| = |Y|$, so $X$ maps continuously onto $Y$ and hence $C_p(Y) \hookrightarrow C_p(X)$ by TFS-163. Since $|\mathbb{I}| = \mathfrak{c}$, any bijection between $X$ and $\mathbb{I}$ is a condensation of $X$ onto $\mathbb{I}$, so $iw(X) = \omega$. The space $Y$ being compact we have $iw(Y) = w(Y) = \mathfrak{c} > \omega = iw(X)$, so $X$ and $Y$ is a pair of spaces such that $C_p(Y) \hookrightarrow C_p(X)$ while $iw(Y) > iw(X)$.

**V.003.** *Prove that $p(Y) \leq p(X)$ whenever $C_p(Y) \hookrightarrow C_p(X)$. As a consequence, point-finite cellularity is t-invariant.*

**Solution.** We have $p(Y) = a(C_p(Y)) \leq a(C_p(X)) = p(X)$ by Problem TFS-178; as a consequence, $p(Y) \leq p(X)$. If $X$ is $t$-equivalent to $Y$ then $C_p(X) \hookrightarrow C_p(Y)$ and $C_p(Y) \hookrightarrow C_p(X)$ so $p(X) = p(Y)$ and hence point-finite cellularity is $t$-invariant.

**V.004.** *Suppose that $X$ and $Y$ are $t$-equivalent Baire spaces. Prove that $c(X) = c(Y)$. In particular, the Souslin numbers of $t$-equivalent pseudocompact spaces coincide.*

**Solution.** The spaces $X$ and $Y$ being Baire we can apply TFS-282 to see that we have $p(X) = c(X)$ and $p(Y) = c(Y)$. It follows from TFS-178 and our observation that $c(X) = p(X) = a(C_p(X)) = a(C_p(Y)) = p(Y) = a(Y)$ and hence $c(X) = c(Y)$. Any pseudocompact space is Baire (see TFS-274), so the Souslin numbers of $t$-equivalent pseudocompact spaces coincide.

**V.005.** *Let $\kappa$ be a caliber of $X$. Knowing that $C_p(Y)$ embeds in $C_p(X)$, prove that $\kappa$ is a caliber of $Y$. In particular, calibers are $t$-invariant.*

**Solution.** Apply SFFS-290 to see that the diagonal of $C_p(X)$ is $\kappa$-small; it follows from $C_p(Y) \hookrightarrow C_p(X)$ that the diagonal of $C_p(Y)$ is $\kappa$-small as well (it is an easy exercise to prove that having a $\kappa$-small diagonal is a hereditary property). Applying SFFS-290 again we conclude that $\kappa$ is a caliber of $Y$. If $C_p(X) \simeq C_p(Y)$ then $C_p(X) \hookrightarrow C_p(Y)$ and $C_p(Y) \hookrightarrow C_p(X)$ which shows that $\kappa$ is a caliber of $X$ if and only if it is a caliber of $Y$; therefore calibers are $t$-invariant.

**V.006.** *Suppose that $C_p(Y)$ embeds into $C_p(X)$. Prove that $l^*(Y) \leq l^*(X)$. As a consequence, $l^*$ is $t$-invariant.*

**Solution.** We have $l^*(Y) = t(C_p(Y)) \leq t(C_p(X)) = l^*(X)$ (see TFS-149 and TFS-159). If $X \overset{t}{\sim} Y$ then $C_p(Y) \hookrightarrow C_p(X)$ and $C_p(X) \hookrightarrow C_p(Y)$ so $l^*(X) = l^*(Y)$ and hence $l^*$ is $t$-invariant.

**V.007.** *Suppose that $C_p(Y)$ embeds into $C_p(X)$. Prove that $\varphi(Y) \leq \varphi(X)$ for any $\varphi \in \{hl^*, hd^*, s^*\}$ and hence $\varphi$ is $t$-invariant.*

**Solution.** We have $hl^*(Y) = hd^*(C_p(Y)) \leq hd^*(C_p(X)) = hl^*(X)$ (see SFFS-026), so the promised inequality is true for $\varphi = hl^*$. Apply SFFS-027 to convince ourselves that $hd^*(Y) = hl^*(C_p(Y)) \leq hl^*(C_p(X)) = hd^*(X)$ which shows that $\varphi(Y) \leq \varphi(X)$ also in case when $\varphi = hd^*$. Besides, $s^*(Y) = s^*(C_p(Y)) \leq s^*(C_p(X)) = s^*(X)$ by SFFS-025 and hence the inequality $\varphi(Y) \leq \varphi(X)$ is true for $\varphi = s^*$ as well. An immediate consequence is that any $\varphi \in \{hd^*, hl^*, s^*\}$ is $t$-invariant.

**V.008.** *Prove that, if $X \overset{t}{\sim} Y$ then $t_m(X) = t_m(Y)$. Give an example of spaces $X$ and $Y$ such that $C_p(Y)$ embeds into $C_p(X)$ and $t_m(Y) > t_m(X)$.*

**Solution.** The spaces $C_p(X)$ and $C_p(Y)$ being homeomorphic we can apply TFS-429 to conclude that $t_m(X) = q(C_p(X)) = q(C_p(Y)) = t_m(Y)$. Now if $X$ is a discrete space of cardinality $\omega_1$ and $Y = L(\omega_1)$ is the one-point lindelöfication of $X$ then any bijection between $X$ and $Y$ maps $X$ continuously onto $Y$, so $C_p(Y)$ embeds in $C_p(X)$ by TFS-163. Observe that $t_m(X) \leq t(X) = \omega$ (see TFS-419) and let $a \in Y$ be the unique non-isolated point of $Y$. It is easy to see that $a \in \overline{Y \setminus \{a\}}$ and $a \notin \overline{B}$ for any countable $B \subset Y \setminus \{a\}$. This, together with Fact 1 of S.419 shows that $t_m(Y) > \omega$. Therefore $C_p(Y) \hookrightarrow C_p(X)$ and $t_m(Y) > t_m(X)$.

**V.009.** *Prove that* $X \overset{t}{\sim} Y$ *implies* $q(X) = q(Y)$. *In particular, realcompactness is* *t-invariant.*

**Solution.** The spaces $C_p(X)$ and $C_p(Y)$ being homeomorphic we can apply TFS-434 to conclude that $q(X) = t_m(C_p(X)) = t_m(C_p(Y)) = q(Y)$. Since a space $X$ is realcompact if and only if $q(X) = \omega$ (see TFS-401), realcompactness is a $t$-invariant property.

**V.010.** *Give an example of spaces* $X$ *and* $Y$ *such that* $C_p(Y)$ *embeds into* $C_p(X)$, *the space* $X$ *is realcompact and* $Y$ *is not realcompact.*

**Solution.** Let $X$ be a discrete space of cardinality $\mathfrak{c}$; if $Y$ is a Mrowka space (see TFS-142) then $Y$ is pseudocompact and non-compact which implies that $Y$ is not realcompact (see TFS-407). Any bijection between $X$ and $\mathbb{I}$ is a condensation of $X$ onto a second countable space, so $iw(X) \leq \mathfrak{c}$ and hence $X$ is realcompact by TFS-446. Furthermore, it follows from $|Y| \leq \mathfrak{c}$ that $X$ maps continuously onto $Y$ and hence $C_p(Y)$ embeds in $C_p(X)$ by TFS-163.

**V.011.** *Suppose that* $X$ *is a P-space and* $X \overset{t}{\sim} Y$. *Prove that* $Y$ *is also a P-space.*

**Solution.** Every countable subset of $X$ is closed and $C$-embedded in $X$ by Fact 1 of S.479, so $C_p(X)$ is pseudocomplete by TFS-485. Since $C_p(Y) \simeq C_p(X)$, the space $C_p(Y)$ is also pseudocomplete and hence $C_p(Y, \mathbb{I})$ is pseudocompact by TFS-476.

*Fact 1.* If $Z$ is a $P$-space then $\overline{A} \subset C_p(Z)$ for any countable $A \subset C_p(Z)$ (the bar denotes the closure in $\mathbb{R}^Z$).

*Proof.* Take a countable $A \subset C_p(Z)$ and $f \in \overline{A}$. Given any $z \in Z$, we will prove that $f$ is continuous at the point $z$. Note that the set $\{h(z)\}$ is a $G_\delta$-set in the space $\mathbb{I}$ for any $h \in A$, so $W = \bigcap\{h^{-1}(h(z)) : h \in A\}$ is a $G_\delta$-set in $Z$. Since $Z$ is a $P$-space, the set $W$ is an open neighborhood of $z$; we claim that $f(W) = \{f(z)\}$. To see this, suppose that $w \in W$ and $|f(w) - f(z)| > \varepsilon$ for some $\varepsilon > 0$. Since $f \in \overline{A}$, there is $h \in A$ such that $|h(z) - f(z)| < \frac{\varepsilon}{2}$ and $|h(w) - f(w)| < \frac{\varepsilon}{2}$. However, $h(w) = h(z)$ and hence

$$|f(w) - f(z)| \leq |f(w) - h(w)| + |h(z) - f(z)| < \frac{\varepsilon}{2} + \frac{\varepsilon}{2} = \varepsilon$$

which is a contradiction. Therefore, $f(W) = \{f(z)\} \subset (f(z) - \varepsilon, f(z) + \varepsilon)$, for any $\varepsilon > 0$, i.e., $f$ is continuous at $z$. Therefore $f \in C_p(Z)$ and Fact 1 is proved.

Returning to our solution fix a homeomorphism $\varphi : C_p(Y) \to C_p(X)$ and assume that $K = C_p(Y, \mathbb{I})$ is not countably compact. Then there is a countably infinite closed discrete set $B \subset K$; it is evident that $B$ is also closed in $C_p(Y)$, so $A = \varphi(B)$ is closed and discrete in $C_p(X)$. The set $F = \varphi(K)$ is pseudocompact, so $L = \overline{F}$ is also pseudocompact by Fact 18 of S.351 (the bar denotes the closure in $\mathbb{R}^X$). However, every closed pseudocompact subspace of $\mathbb{R}^X$ is compact (see TFS-401 and TFS-415), so $L$ is compact. Since $A \subset L$, the set $\overline{A}$ has to be compact;

however, $\overline{A} \subset C_p(X)$ by Fact 1, so $A$ is an infinite closed discrete subspace of a compact space $\overline{A}$. This contradiction shows that $C_p(Y, \mathbb{I})$ is countably compact and hence $Y$ is a $P$-space by TFS-397.

**V.012.** *Prove that discreteness is $t$-invariant.*

**Solution.** Suppose that $X$ is a discrete space and $Y \overset{t}{\sim} X$. Then $C_p(X) = \mathbb{R}^X$ is pseudocomplete (see TFS-470 and TFS-467), so $C_p(Y)$ is pseudocomplete as well. Besides, $C_p(X)$ is realcompact by TFS-401, so the space $C_p(Y)$ is also realcompact and hence $t_m(Y) = q(C_p(Y)) = \omega$ (see TFS-429). If some point $y \in Y$ is not isolated in $Y$ then we can apply Fact 1 of S.419 to see that there is a countable set $A \subset Y \backslash \{y\}$ such that $y \in \overline{A}$ and hence $y \in \overline{A} \backslash A$. However, every countable subset of $Y$ is closed in $Y$ by TFS-485; this contradiction shows that all points of $Y$ are isolated, i.e., $Y$ is discrete.

**V.013.** *Suppose that $X$ and $Y$ are compact spaces such that $C_p(Y) \overset{}{\hookrightarrow} C_p(X)$. Prove that $Y$ is scattered whenever $X$ is scattered. In particular, if $X \overset{t}{\sim} Y$ then $X$ is scattered if and only if so is $Y$.*

**Solution.** If $X$ is scattered then $C_p(X)$ is a Fréchet–Urysohn space by SFFS-134. Therefore the space $C_p(Y) \overset{}{\hookrightarrow} C_p(X)$ also has the Fréchet–Urysohn property; applying SFFS-134 once more we conclude that $Y$ is scattered. If $X \overset{t}{\sim} Y$ then $C_p(X) \overset{}{\hookrightarrow} C_p(Y)$ and $C_p(Y) \overset{}{\hookrightarrow} C_p(X)$ so $X$ is scattered if and only if so is $Y$.

**V.014.** *Suppose that $X^n$ is a Hurewicz space for each $n \in \mathbb{N}$ and $X \overset{t}{\sim} Y$. Prove that $Y^n$ is also a Hurewicz space for each $n \in \mathbb{N}$.*

**Solution.** Apply CFS-057 to see that $vet(C_p(X)) = \omega$; since $C_p(Y) \simeq C_p(X)$, we also have $vet(C_p(Y)) = \omega$, so we can apply CFS-057 once more to conclude that $Y^n$ is a Hurewicz space for any $n \in \mathbb{N}$.

**V.015.** *Suppose that $X \overset{t}{\sim} Y$ and $X$ is a $\sigma$-compact space with a countable network. Prove that $Y$ is also $\sigma$-compact. As a consequence, if $X$ is a metrizable compact space and $X \overset{t}{\sim} Y$ then $Y$ is $\sigma$-compact.*

**Solution.** We have $X = \bigcup_{n \in \omega} K_n$ where every set $K_n$ is compact; since also $nw(K_n) = \omega$, the space $K_n$ is metrizable (see Fact 4 of S.307) and hence analytic for any $n \in \omega$. It follows from SFFS-337 that $X$ is analytic; since $C_p(Y) \simeq C_p(X)$, we can apply Fact 12 of T.250 to see that the space $Y$ is $K$-analytic. Furthermore, $nw(Y) = nw(C_p(Y)) = nw(C_p(X)) = nw(X) \leq \omega$, so the space $Y$ is analytic by SFFS-346.

It is an easy exercise that every $\sigma$-compact space is Hurewicz; since every finite power of $X$ is $\sigma$-compact, the space $X^n$ is Hurewicz for any $n \in \mathbb{N}$ and therefore $vet(C_p(X)) = \omega$ (see CFS-057). As a consequence, $vet(C_p(Y)) = \omega$ and hence $Y$ is Hurewicz as well. Thus $Y$ is analytic and Hurewicz, so we can apply CFS-053 to conclude that $Y$ is $\sigma$-compact.

**V.016.** *Given an arbitrary number $\varepsilon > 0$ prove that there exists a homeomorphism* $u : \mathbb{R} \times \Sigma_*(\omega) \to \Sigma_*(\omega)$ *such that* $|\, ||u(r, x)|| - ||x|| \,| \leq \varepsilon$ *for any* $r \in \mathbb{R}$ *and* $x \in \Sigma_*(\omega)$.

**Solution.** Denote by $I$ the subspace $[0, 1]$ of the real line $\mathbb{R}$. Given a space $X$, the function $\mathrm{id}_X : X \to X$ is the identity on $X$, i.e., $\mathrm{id}_X(x) = x$ for any $x \in X$. A family $\{f_t : t \in I\}$ is called *an autoisotopy* of $X$ if, for any $t \in I$, the map $f_t : X \to X$ is a homeomorphism and both maps $f_+, f_- : I \times X \to X$ defined by $f_+(t, x) = f_t(x)$ and $f_-(t, x) = f_t^{-1}(x)$ respectively for any $t \in I$ and $x \in X$ are continuous.

*Fact 1.* Fix a number $a > 0$ and let $a_n = (1 - 2^{-n})a$ for any $n \in \mathbb{N}$. Given a point $z = (z_0, z_1) \in \mathbb{R}^2$ let $||z|| = \max\{|z_0|, |z_1|\}$. Then, for any $n \in \mathbb{N}$, there exists an autoisotopy $\{f_t^n : t \in I\}$ of the plane $\mathbb{R}^2$ with the following properties:

(1) $f_0^n(z) = z$ for any $z \in \mathbb{R}^2$;
(2) $f_t^n(z) = z$ for any $z \in [-a_n, a_n] \times \mathbb{R}$ and $t \in I$;
(3) $f_1^n((-a_m, a_m) \times \mathbb{R}) = \mathbb{R} \times (-\infty, \frac{1}{2}a_m)$ for any $m > n$;
(4) $||f_t^n(z)|| = ||z||$ for any $z \in \mathbb{R}^2$ and $t \in I$.

*Proof.* To obtain the property (1) we have to define $f_0^n$ to be the identity on $\mathbb{R}^2$, so let $f_0^n(z) = z$ for any $z \in \mathbb{R}^2$. To define the homeomorphism $f_1^n : \mathbb{R}^2 \to \mathbb{R}^2$ we will need the set $A_m = [0, a_m] \times \mathbb{R}$ for all $m \in \mathbb{N}$. To satisfy (2) let $f_1^n(z) = z$ for any $z \in A_n$; we will next define $f_1^n$ on the set $[a_n, +\infty) \times \mathbb{R}$ meaning to extend it symmetrically over the whole plane. First, let $f_1^n(z) = z$ for every $z = (z_0, z_1) \in [a_n, +\infty) \times \mathbb{R}$ such that $z_0 < z_1$. Thus we defined $f_1^n$ on the set $B = A_n \cup \{(z_0, z_1) \in \mathbb{R}^2 : a_n \leq z_0 < z_1\}$. To complete our construction of $f_1^n$ on the half-plane $P_0 = [0, +\infty) \times \mathbb{R}$, we must define it on the set $Q = \{(z_0, z_1) : a_n \leq z_0 \text{ and } z_1 \leq z_0\}$.

Observe first that $Q = \bigcup\{Q_b : b \in (-\infty, -a_n]\}$ where this union is disjoint and $Q_b = ([a_n, -b] \times \{b\}) \cup (\{-b\} \times [b, -b])$ for any $b \leq -a_n$. In other words, every set $Q_b$ is the union of two line segments: the horizontal segment $[a_n, -b] \times \{b\}$ and the vertical one $\{-b\} \times [b, -b]$. Observe that $||z|| = |b|$ for any $z \in Q_b$, so, to satisfy the condition (4), it suffices to assure that $f_1^n(Q_b) \subset Q_b$ for every $b \leq -a_n$. Another requirement we must meet is not to move the "endpoints" of the set $Q_b$, i.e., $f_1^n((a_n, b)) = (a_n, b)$ and $f_1^n((-b, -b)) = (-b, -b)$.

Fix any number $b \leq -a_n$; to define the function $f_1^n | Q_b$ we will consider an auxiliary interval $J_b = [a_n, -3b] \subset \mathbb{R}$ which, intuitively, is obtained by "straightening up" the vertical part of $Q_b$. We will construct a piecewise linear increasing function $h_b : J_b \to J_b$ which gives $f_1^n | Q_b$ if we identify $Q_b$ and $J_b$. If we construct the function $f_1^n$ according to the above plan then, to meet the condition (3), it suffices to guarantee that $f_1^n((a_m, b)) = (-b, \frac{1}{2}a_m)$ whenever $a_n < a_m \leq -b$. If $-b$ is sufficiently large then $a \in (a_n, -b]$; in this case we will also require that $f_1^n((a, b)) = (-b, \frac{1}{2}a)$. In the terms of the function $h_b$ this is equivalent to saying that $h_b(a_n) = a_n$, $h_b(-3b) = -3b$ while $h_b(a_m) = -2b + \frac{1}{2}a_m$ for any $m$ with $a_n < a_m \leq -b$. Besides, $h_b(a) = -2b + \frac{1}{2}a$ whenever $a \leq -b$.

It is easy to see that these conditions uniquely determine every function $h_b$ for all $b \leq -a_{n+1}$. To visualize it better, take in consideration the key points in the square

$J_b \times J_b$ which belong to its graph $\Gamma_b$. It is obligatory for the points $C_n = (a_n, a_n)$ and $D_b = (-3b, -3b)$ to belong to $\Gamma_b$. If $C_m^b = (a_m, -2b + \frac{1}{2}a_m)$ for every $m > n$ then $C_m^b \in \Gamma_b$ whenever $a_m \leq -b$. Finally, the point $C^b = (a, -2b + \frac{1}{2}a)$ also belongs to $\Gamma_b$ if $a \leq -b$.

If $-a_{n+2} < b \leq -a_{n+1}$ then the set $\Gamma_b$ consists of two line segments: the one which connects the points $C_n$ and $C_{n+1}^b$ and the second one which connects the points $C_{n+1}^b$ and $D_b$. In the case when $b \leq -a_{n+2}$, the set $\Gamma_b$ is the union of three line segments because the set $\{C^b\} \cup \{C_m^b : m > n\}$ is contained in the straight line determined by the points $C_{n+1}^b$ and $C_{n+2}^b$.

To construct $h_b$ for $-a_{n+1} < b \leq -a_n$ observe that the function $h_{-a_n}$ has to be an identity on the interval $[a_n, 3a_n]$ and hence $h_{-a_{n+1}}$ has to be continuously transformed into identity as $b$ runs from $-a_{n+1}$ to $-a_n$. To do this consider the function $\delta(x) = h_{-a_{n+1}}(x) - x$ on the set $[a_n, 3a_{n+1}]$. Then $h_{-a_{n+1}}(x) = x + \delta(x)$ and $\delta(a_n) = \delta(3a_{n+1}) = 0$.

Given $b \in [-a_{n+1}, -a_n]$ let $\lambda(b, x)$ be the unique linear function such that $\lambda(b, a_n) = a_n$ and $\lambda(b, -3b) = 3a_{n+1}$. If $\mu(b, x) = x + \frac{b + a_n}{a_n - a_{n+1}} \delta(\lambda(b, x))$ then $\mu(-a_{n+1}, x) = x + \delta(x) = h_{-a_{n+1}}(x)$ for any $x \in [a_n, 3a_{n+1}]$. It is also easy to see that $\mu(-a_n, x) = x$ for any $x \in [a_n, 3a_n]$. The only problem with the function $\mu(b, x)$ could be its range, so let $h_b(x) = \backslash n\{\max\{\mu(b, x), a_n\}, -3b\}$. Then the function $h_b$ maps the interval $[a_n, -3b]$ onto itself while $h_b(a_n) = a_n$ and $h_b(-3b) = -3b$.

Consider the set $E = \{(b, x) : b \leq -a_n \text{ and } a_n \leq x \leq -3b\} \subset \mathbb{R}^2$; it follows from our construction of the family $\{h_b : b \leq -a_n\}$ that the function $H : E \to E$ defined by $H((b, x)) = (b, h_b(x))$ is continuous. Given $b \leq -a_n$ define $f_1^n | Q_b$ by applying $h_b$ on $\{b\} \times J_b$ and "bending back" the interval $\{b\} \times J_b$. More formally, for any point $z = (b, x) \in E$ let $\varphi(z) = z$ if $x \leq -b$ and $\varphi(z) = (-b, x + 2b)$ otherwise. Then $\varphi : E \to Q$ is a homeomorphism, so the map $\varphi \circ H \circ \varphi^{-1} : Q \to Q$ is continuous; let $f_1^n | Q = \varphi \circ H \circ \varphi^{-1}$. This consistently defines the function $f_1^n | P_0$. The sets $F = ([0, a_n] \times \mathbb{R}) \cup \{(x_0, x_1) : a_n \leq x_0 \leq x_1\}$ and $Q$ are closed in $P_0$ and $P_0 = F \cup Q$. The function $f_1^n | F$ is continuous being an identity on $F$; we already saw that $f_1^n | Q$ is also continuous, so $f_1^n | P_0$ is also continuous by Fact 2 of T.354.

Since $f_1^n | P_0$ is an identity on the $y$-axis, its symmetric extension over the whole plane gives us a continuous map $f_1^n : \mathbb{R}^2 \to \mathbb{R}^2$ (to see this apply again Fact 2 of T.354) for which the conditions (1)–(4) are satisfied.

Let $K_n = \{x \in \mathbb{R}^2 : \|x\| \leq n\}$ for any $n \in \mathbb{N}$. It is evident that every $K_n$ is compact and $\bigcup \{\text{Int}(K_n) : n \in \mathbb{N}\} = \mathbb{R}^2$. Since $f_1^n(K_n) = K_n$ and $f_1^n$ is a bijection, the map $(f_1^n)^{-1} | K_n : K_n \to K_n$ is continuous. Next apply Fact 1 of S.472 to see that $(f_1^n)^{-1}$ is continuous and hence the map $f_1^n$ is a homeomorphism.

Fix $t \in I$; as before, we will first construct the map $f_t^n$ on the half-plane $P_0$. First of all declare $f_t^n$ to be an identity on the set $B$. To define $f_t^n | Q$ we will again guarantee that $f_t^n(Q_b) = Q_b$ for any $b \leq -a_n$.

To this end consider, for any $b \leq -a_n$, the function $h_b^t(x) = x + t(h_b(x) - x)$ which maps the interval $J_b$ into itself. Since $h_b$ is an increasing function, the function $h_b^t$ is increasing as well.

It easily follows from our definition of the family $\{h_b^t : b \le -a_n, \; t \in I\}$ that the function $G : I \times E \to E$ defined by $G(t, (b, x)) = (b, h_b^t(x))$ is continuous. Define a map $G_t : E \to E$ by $G_t((b, x)) = G(t, (b, x)) = (b, h_b^t(x))$ for any $t \in I$ and $(b, x) \in E$. Then the mapping $f_t^n | Q = \varphi \circ G_t \circ \varphi^{-1} : Q \to Q$ is continuous. This gives us a function $f_t^n | P_0$. As before, $P_0 = F \cup Q$ and the function $f_t^n | F$ is continuous being an identity; the function $f_t^n | Q$ is continuous by our construction, so $f_t^n | P_0$ is continuous.

Since every $f_t^n | (\{0\} \times \mathbb{R})$ is an identity, we can symmetrically extend it over the whole plane obtaining a continuous map $f_t^n : \mathbb{R}^2 \to \mathbb{R}^2$ for which the condition (4) is satisfied.

To show that every $f_t^n$ is a homeomorphism replace $f_1^n$ by $f_t^n$ in the proof for $t = 1$. Now that we have the family $\{f_t^n : t \in I\}$ of autohomeomorphisms of $\mathbb{R}^2$ let $f_+^n(t, x) = f_t^n(x)$ and $f_-^n(t, x) = (f_t^n)^{-1}(x)$ for any $t \in I$ and $x \in \mathbb{R}^2$. To see that the map $f_+^n : I \times \mathbb{R}^2 \to \mathbb{R}^2$ is continuous observe that the symmetry of our situation and Fact 2 of T.354 show that it suffices to prove that $f_+^n | (I \times P_0) \to P_0$ is continuous. The fact that $f_+^n | (I \times F)$ is an identity which does not depend on the first coordinate shows that it suffices to prove that $f_+^n | (I \times Q) : I \times Q \to Q$ is continuous. To do this note that $f_+^n(t, z) = f_t^n(z) = \varphi(G(t, \varphi^{-1}(z)))$ for any $t \in I$ and $z \in Q$ which, together with continuity of $G$ and $\varphi$, implies that $f_+^n | (I \times Q)$ is continuous and hence the function $f_+^n$ is, indeed, continuous.

The proof that the function $f_-^n$ is continuous is analogous if we consider the map $G_- : I \times E \to E$ defined by $G_-(t, (b, x)) = (b, (h_b^t)^{-1}(x))$ for any $t \in I$ and $(b, x) \in E$. Once we observe that $G_-$ is continuous and $f_-^n(t, x) = \varphi(G_-(t, \varphi^{-1}(z)))$ for any $t \in I$ and $z \in Q$, we can make the necessary changes in the proof of continuity of $f_+^n$ to convince ourselves that $f_-^n$ is continuous and hence the family $\{f_t^n : t \in I\}$ is an autoisotopy on $\mathbb{R}^2$ with the properties (1)–(4). Fact 1 is proved.

*Fact 2.* Given spaces $X$ and $Y$ suppose that $\{\varphi_t : t \in I\}$ is an autoisotopy on $X$ and $s : Y \to I$ is a continuous function. If $u(x, y) = (\varphi_{s(y)}(x), y)$ for any $(x, y) \in X \times Y$ then the map $u : X \times Y \to X \times Y$ is an autohomeomorphism of the space $X \times Y$ and $u^{-1}(x, y) = (\varphi_{s(y)}^{-1}(x), y)$ for all $x \in X$ and $y \in Y$.

*Proof.* It is straightforward from the definition that both maps $u \circ u^{-1}$ and $u^{-1} \circ u$ are identities on $X \times Y$, so $u$ is a bijection and $u^{-1}$ is its inverse. Let $\Phi_+ : I \times X \to X$ and $\Phi_- : I \times X \to X$ be defined by $\Phi_+(t, x) = \varphi_t(x)$ and $\Phi_-(t, x) = (\varphi_t)^{-1}(x)$ respectively for any $t \in I$ and $x \in X$. Both maps $\Phi_+$ and $\Phi_-$ are continuous because $\{\varphi_t : t \in I\}$ is an isotopy on $X$. The map $\mu = s \times \mathrm{id}_X : Y \times X \to I \times X$ is continuous and hence so is the map $\Phi_+ \circ \mu$. Since $u$ is the diagonal product of $\Phi_+ \circ \mu$ and $\mathrm{id}_Y$, it has to be continuous. Analogously, the map $u^{-1}$ is continuous being the diagonal product of $\Phi_- \circ \mu$ and $\mathrm{id}_Y$. This shows that $u$ is a homeomorphism, so Fact 2 is proved.

*Fact 3.* Given $\varepsilon > 0$ suppose that $\{G_n : n \in \omega\}$ is an increasing sequence of open subsets of $\Sigma_*(\omega)$ and $\{v_n : n \in \omega\}$ is a set of autohomeomorphisms of $\Sigma_*(\omega)$ with the following properties:

(1)  $v_0|G_0 = \mathrm{id}_{G_0}$ and $v_{n+1}|G_{n+1} = v_n|G_{n+1}$ for any $n \in \omega$;
(2)  $v_n(G_{n+1}) \supset \bigcup_{1 \leq k \leq n+1} \{x \in \Sigma_*(\omega) : |x(k)| < \varepsilon\}$ for all $n \in \omega$;
(3)  $||v_n(x)|| = ||x||$ for all $x \in \Sigma_*(\omega)$ and $n \in \omega$.

Then, for every $x \in G = \bigcup_{n \in \omega} G_n$ the sequence $\{v_n(x) : n \in \omega\}$ converges to a point $v(x) \in \Sigma_*(\omega)$ and the respective map $v : G \to \Sigma_*(\omega)$ is a homeomorphism between $G$ and $\Sigma_*(\omega)$ such that $||v(x)|| = ||x||$ for any $x \in G$.

*Proof.* Using the property (1) it is easy to prove by induction that

(∗)  for any $m \in \omega$, if $n \geq m$ then $v_n|G_m = v_m|G_m$.

If $x \in G$ then there is $m \in \omega$ such that $x \in G_m$. It follows from (∗) that $v_n(x) = v_m(x)$ for any $n \geq m$ and hence the sequence $\{v_n(x) : n \in \omega\}$ is convergent because it is eventually constant. Thus the element $v(x) = \lim_{n \to \infty} v_n(x)$ is well defined for any $x \in G$ and

(∗∗)  $v|G_m = v_m|G_m$ for any $m \in \omega$.

Given distinct points $x, y \in G$ there is $m \in \omega$ with $x, y \in G_m$ which shows that $v(x) = v_m(x)$ and $v(y) = v_m(y)$. The map $v_m$ being a homeomorphism, we have $v(x) = v_m(x) \neq v_m(y) = v(y)$, i.e., $v$ is injective. If $y \in \Sigma_*(\omega)$ then there is $k \in \mathbb{N}$ such that $|y(k)| < \varepsilon$; the property (2) shows that $y \in v_k(G_{k+1})$. Take $x \in G_{k+1}$ such that $v_k(x) = y$; by (1) and (∗∗), we have $v(x) = v_{k+1}(x) = v_k(x) = y$ and hence the map $v$ is a bijection between $G$ and $\Sigma_*(\omega)$.

Fact 1 of S.472 and the property (∗∗) easily imply that $v$ is a continuous map. To prove that $v^{-1}$ is also continuous fix a point $y \in \Sigma_*(\omega)$ and $x \in G$ with $v(x) = y$. There is $m \in \omega$ such that $x \in G_m$ and therefore $v(x) = v_m(x)$. The map $v_m$ being a homeomorphism, the set $v_m(G_m)$ is open in $\Sigma_*(\omega)$ and it follows from (∗∗) that $v^{-1}|v_m(G_m) = v_m^{-1}|v_m(G_m)$. Thus $\{v_n(G_n) : n \in \omega\}$ is an open cover of the space $\Sigma_*(\omega)$ such that $v^{-1}|v_n(G_n)$ is continuous for any $n \in \omega$; this makes it possible to apply Fact 1 of S.472 again to conclude that $v^{-1}$ is continuous and hence $v : G \to \Sigma_*(\omega)$ is a homeomorphism.

Finally, take any $x \in G$; there is $m \in \omega$ with $x \in G_m$, so we can apply (∗∗) once more to convince ourselves that $v(x) = v_m(x)$ and hence $||v(x)|| = ||v_m(x)|| = ||x||$ by (3). This proves that $||v(x)|| = ||x||$ for any $x \in G$, so Fact 3 is proved.

*Fact 4.* Suppose that $a > 0$ and let $G = \{x \in \Sigma_*(\omega) : |x(0)| < a\}$. Then there exists a homeomorphism $\varphi : G \to \Sigma_*(\omega)$ such that $||\varphi(x)|| = ||x||$ for any $x \in G$.

*Proof.* Let $a_n = (1 - 2^{-n-1})a$ and $G_n = \{x \in \Sigma_*(\omega) : |x(0)| < a_n\}$ for every $n \in \omega$; then $\{G_n : n \in \omega\}$ is an increasing sequence of open subsets of $\Sigma_*(\omega)$ and $\bigcup_{n \in \omega} G_n = G$. We are going to construct a sequence $\{v_n : n \in \omega\}$ of autohomeomorphisms of $\Sigma_*(\omega)$ which satisfy the premises of Fact 3.

To that end we will produce a sequence $\{u_n : n \in \omega\}$ of autohomeomorphisms of $\Sigma_*(\omega)$ such that $v_n = u_n \circ \ldots \circ u_0$ for any $n \in \omega$. Besides, every map $u_n$ will be two-dimensional, i.e., it will change at most two coordinates of any point of $\Sigma_*(\omega)$.

Let $s_0(x) = 1$ for all $x \in \Sigma_*(\omega)$; to construct the functions $s_n$ for $n > 0$ we will need the function $z_n$ defined by $z_n(x) = \setminus n\{x(1), \ldots, x(n)\}$ for any $n \in \mathbb{N}$ and

$x \in \Sigma_*(\omega)$. It is clear that $z_n : \Sigma_*(\omega) \to \mathbb{R}$ is a continuous function, so the sets $Z_n^0 = \{x \in \Sigma_*(\omega) : z_n(x) \leq \frac{1}{2}a_n\}$ and $Z_n^1 = \{x \in \Sigma_*(\omega) : z_n(x) \geq \frac{1}{2}a_{n+1}\}$ are closed and disjoint in $\Sigma_*(\omega)$. Take a continuous function $s_n' : \mathbb{R} \to I$ such that $s_n'((-\infty, \frac{1}{2}a_n]) = \{0\}$ and $s_n'([\frac{1}{2}a_{n+1}, +\infty)) = \{1\}$ and let $s_n = s_n' \circ z_n$. Then $s_n(Z_n^0) = \{0\}$ and $s_n(Z_n^1) = \{1\}$.

For any $n \in \omega$ and $x \in \Sigma_*(\omega)$ let $u_n(x)(i) = x(i)$ for all $i \in \omega \backslash \{0, n+1\}$; this guarantees that the map $u_n$ can only change the coordinates $0$ and $n+1$. Fact 1 provides an autoisotopy $\{f_t^{n+1} : t \in I\}$ of the plane $\mathbb{R}^2$; observe that every map $f_t^{n+1} : \mathbb{R}^2 \to \mathbb{R}^2$ has its two components, i.e., $f_t^{n+1}((a,b)) = (f_{t,0}^{n+1}(a,b), f_{t,1}^{n+1}(a,b))$ for any $(a,b) \in \mathbb{R}^2$. To finish our definition of the mapping $u_n$ let

$$u_n(x)(0) = f_{s_n(x),0}^{n+1}(x(0), x(n+1)) \text{ and } u_n(x)(n+1) = f_{s_n(x),1}^{n+1}(x(0), x(n+1)).$$

To see that every $u_n$ is an autohomeomorphism of $\Sigma_*(\omega)$ observe that we can consider that $\Sigma_*(\omega) = X \times Y$ where $X$ is the plane determined by the coordinates $0$ and $n+1$ and $Y = \Sigma_*(\omega \backslash \{0, n+1\})$. The function $s_n(x)$ is formally defined on the whole space $\Sigma_*(\omega)$ but it does not depend on the coordinates $0$ and $n+1$, so we can consider that it is defined on $Y$. Thus we can apply Fact 2 to conclude that $u_n$ is an autohomeomorphism for any $n \in \omega$.

An immediate consequence of Fact 1 is that $\|u_n(x)\| = \|x\|$ for any $x \in \Sigma_*(\omega)$ and $n \in \omega$. Let $v_n = u_n \circ \ldots \circ u_0$ for any $n \in \omega$. We will prove by induction that

(i) $v_n(G_m) = G_m \cup (\bigcup_{k=1}^{n+1} \{x \in \Sigma_*(\omega) : {}'x(k) < \frac{1}{2}a_m\})$ for any $n \in \omega$ and $m > n$.

If $n = 0$ then $v_0 = u_0$ and the property (i) is an immediate consequence of Fact 1. Assume that (i) is proved for $n = l$ and consider the set $v_{l+1}(G_m)$ for some $m > l+1$. Observe that $v_{l+1}(G_m) = u_{l+1}(v_l(G_m))$ and let $L = \bigcup_{k=1}^{n+1} \{x \in \Sigma_*(\omega) : x(k) < \frac{1}{2}a_m\}$. The map $u_{l+1}$ only changes coordinates $0$ and $l+2$ while $L$ is determined by the coordinates from $1$ to $l+1$; this implies $u_{l+1}(L) = L$. Now, if $x \in G_m \backslash L$ then $z_{l+1}(x) \geq \frac{1}{2}a_m \geq \frac{1}{2}a_{l+2}$ and therefore $s_{l+1}(x) = 1$. Thus $s_{l+1}(x) = 1$ for any point $x \in G_m \backslash L$, so $u_{l+1}(x)(n+1) = f_{1,1}^{l+2}(x(0), x(l+2))$. Recalling the properties of the function $f_1^{l+2}$ (see Fact 1), it is easy to see that $u_{l+1}(G_m \backslash L) = (G_m \cup \{x \in \Sigma_*(\omega) : x_{l+2} < \frac{1}{2}a_m\}) \backslash L$. This shows that the formula (i) is true if we substitute $l$ by $l+1$, so (i) is proved.

An immediate consequence of (i) is that our autohomeomorphisms $v_n$ have the property (2) of Fact 3 for $\varepsilon = \frac{1}{4}a$ because $\frac{1}{2}a_m \geq \frac{1}{2}a_0 = \frac{1}{4}a$ for all $m \in \omega$. It is evident that we also have the property (3) of Fact 3, so the last thing we must check is the property (1). The property (2) of Fact 1 shows that we have the equalities $v_0|G_0 = u_0|G_0 = \text{id}_{G_0}$. Now, take $n \in \omega$; we must show that $v_{n+1}|G_{n+1} = v_n|G_{n+1}$, so fix any point $x \in G_{n+1}$. If $y = v_n(x) \in G_{n+1}$ then $|x(0)| < a_{n+1}$, so $u_{n+1}(y) = y$, i.e., $v_{n+1}(x) = u_{n+1}(y) = y = v_n(x)$ by the property (2) of Fact 1. If $y \notin G_{n+1}$ then we can apply (i) for $m = n+1$ to conclude that $y(k) < \frac{1}{2}a_m$ for some number $k \in \{1, \ldots, n+1\}$ and hence $z_{n+1}(y) \leq \frac{1}{2}a_{n+1}$ which implies $s_{n+1}(y) = 0$. Recalling the definition of $u_{n+1}$ and the condition (1) of Fact 1, we conclude that again $u_{n+1}(y) = y$ and hence $v_{n+1}(x) = v_n(x)$.

Thus all premises of Fact 3 are satisfied for our sequences $\{G_n : n \in \omega\}$ and $\{v_n : n \in \omega\}$, so Fact 3 is applicable to see that there is a homeomorphism $\varphi : G \to \Sigma_*(\omega)$ such that $||\varphi(x)|| = ||x||$ for any $x \in G$. Fact 4 is proved.

Returning to our solution, fix a homeomorphism $h : \mathbb{R} \to (-\varepsilon, \varepsilon)$. Given a point $(t, x) \in \mathbb{R} \times \Sigma_*(\omega)$ let $\mu(t, x)(0) = h(t)$ and $\mu(t, x)(n) = x(n-1)$ for any $n \in \mathbb{N}$. It is straightforward that, for the set $G = \{x \in \Sigma_*(\omega) : |x(0)| < \varepsilon\}$, the map $\mu : \mathbb{R} \times \Sigma_*(\omega) \to G$ is a homeomorphism such that $| \,||\mu(t, x)|| - ||x|| \,| \leq \varepsilon$ for any $t \in \mathbb{R}$ and $x \in \Sigma_*(\omega)$. By Fact 4 there is a homeomorphism $\varphi : G \to \Sigma_*(\omega)$ such that $||\varphi(x)|| = ||x||$ for any $x \in G$. Therefore the map $u = \varphi \circ \mu$ is a homeomorphism between $\mathbb{R} \times \Sigma_*(\omega)$ and $\Sigma_*(\omega)$ such that $| \,||u(t, x)|| - ||x|| \,| \leq \varepsilon$, so our solution is complete.

**V.017.** *Prove that $\Sigma_*(\omega)$ is homeomorphic to $\mathbb{R}^\omega \times \Sigma_*(\omega)$.*

**Solution.** Fix a disjoint family $\mathcal{A} = \{A_n : n \in \omega\}$ of infinite subsets of $\omega$ such that $\omega = \bigcup \mathcal{A}$ and let $\pi_n : \mathbb{R}^\omega \to \mathbb{R}^{A_n}$ be the natural projection of $\mathbb{R}^\omega$ onto is face $\mathbb{R}^{A_n}$ for any $n \in \omega$. We will need the space $E_n = \Sigma_*(A_n)$ for any $n \in \omega$; let $||x||_n = \sup\{|x(k)| : k \in A_n\}$ for each $x \in E_n$. It is easy to see that there is a homeomorphism $h_n : E_n \to \Sigma_*(\omega)$ such that $||h(x)|| = ||x||_n$ for any $x \in E_n$.

Consider the set $E = \{x \in \mathbb{R}^\omega : \pi_n(x) \in E_n$ for any $n \in \omega\}$. It is straightforward that $E \simeq \prod_{n \in \omega} E_n$ and $\Sigma_*(\omega) = \{x \in E :$ the sequence $\{||\pi_n(x)||_n : n \in \omega\}$ converges to zero$\}$. Apply Problem 016 to find a homeomorphism $\varphi_n : \mathbb{R} \times E_n \to E_n$ such that

(1)  $| \,||\varphi_n(s, y)||_n - ||y||_n \,| \leq 2^{-n}$ for any $s \in \mathbb{R}$, $y \in E_n$ and $n \in \omega$.

Given $t \in \mathbb{R}^\omega$ and $x \in E$ let $\varphi(t, x)$ be the unique point of $E$ such that $\pi_n(\varphi(t, x)) = \varphi_n(t(n), \pi_n(x))$ for every $n \in \omega$. It follows from Fact 1 of S.271 that the product map $\varphi' = \prod_{n \in \omega} \varphi_n : \prod_{n \in \omega} (\mathbb{R} \times E_n) \to \prod_{n \in \omega} E_n$ is a homeomorphism. It is easy to see that there exist homeomorphisms $\delta : \mathbb{R}^\omega \times E \to \prod_{n \in \omega} (\mathbb{R} \times E_n)$ and $\mu : \prod_{n \in \omega} E_n \to E$ such that $\varphi = \mu \circ \varphi' \circ \delta$. Therefore the map $\varphi : \mathbb{R}^\omega \times E \to E$ is a homeomorphism.

If $t \in \mathbb{R}^\omega$ and $x \in E$ then $\pi_n(x) \in E_n$ for any $n \in \omega$; it follows from (1) that the sequence $\{||\pi_n(x)||_n : n \in \omega\}$ converges to zero if and only if the sequence $\{||\varphi_n(t(n), \pi_n(x))||_n : n \in \omega\}$ converges to zero. An immediate consequence is that $\varphi(\mathbb{R}^\omega \times \Sigma_*(\omega)) = \Sigma_*(\omega)$ and therefore $\varphi|(\mathbb{R}^\omega \times \Sigma_*(\omega))$ is a homeomorphism between $\mathbb{R}^\omega \times \Sigma_*(\omega)$ and $\Sigma_*(\omega)$.

**V.018.** *Suppose that $X$ is a pseudocompact space. Given any function $f \in C_p(X)$, let $||f|| = \sup\{|f(x)| : x \in X\}$. Prove that $C_*(X) \simeq C_*(X) \times (C_p(X))^\omega$, where $C_*(X) = \{\varphi \in (C_p(X))^\omega : ||\varphi(n)|| \to 0\}$.*

**Solution.** Fix a disjoint family $\mathcal{A} = \{A_n : n \in \omega\}$ of infinite subsets of $\omega$ such that $\omega = \bigcup \mathcal{A}$ and let $\pi_n : \mathbb{R}^\omega \to \mathbb{R}^{A_n}$ be the natural projection of $\mathbb{R}^\omega$ onto is face $\mathbb{R}^{A_n}$ for any $n \in \omega$. We will need the space $E_n = \Sigma_*(A_n)$ for any $n \in \omega$; let $||x||_n = \sup\{|x(k)| : k \in A_n\}$ for each $x \in E_n$. It is easy to see that there is a homeomorphism $h_n : E_n \to \Sigma_*(\omega)$ such that $||h(x)|| = ||x||_n$ for any $x \in E_n$.

From now on the norm symbol $||\cdot||$ is applied only to the functions on $X$ with the meaning defined in the formulation of this problem.

Consider the set $E = \{x \in \mathbb{R}^\omega : \pi_n(x) \in E_n \text{ for any } n \in \omega\}$. It is straightforward that $E \simeq \prod_{n\in\omega} E_n$ and $\Sigma_*(\omega) = \{x \in E : \text{the sequence } \{||\pi_n(x)||_n : n \in \omega\}$ converges to zero$\}$. Apply Problem 016 to find a homeomorphism $u_n : \mathbb{R} \times E_n \to E_n$ such that

(1)  $|\,||u_n(s, y)||_n - ||y||_n\,| \leq 2^{-n}$ for any $s \in \mathbb{R}$, $y \in E_n$ and $n \in \omega$.

Given $t \in \mathbb{R}^\omega$ and $x \in E$ let $u(t, x)$ be the unique point of $E$ such that $\pi_n(u(t, x)) = u_n(t(n), \pi_n(x))$ for every $n \in \omega$. It follows from Fact 1 of S.271 that the product map $u' = \prod_{n\in\omega} u_n : \prod_{n\in\omega}(\mathbb{R} \times E_n) \to \prod_{n\in\omega} E_n$ is a homeomorphism. It is easy to see that there exist homeomorphisms $\delta : \mathbb{R}^\omega \times E \to \prod_{n\in\omega}(\mathbb{R} \times E_n)$ and $\mu : \prod_{n\in\omega} E_n \to E$ such that $u = \mu \circ u' \circ \delta$. Therefore the map $u : \mathbb{R}^\omega \times E \to E$ is a homeomorphism.

For any $n \in \omega$ let $p_n : \mathbb{R}^\omega \to \mathbb{R}$ be the natural projection of $\mathbb{R}^\omega$ onto the $n$-th factor of $\mathbb{R}^\omega$; recall that $p_n(x) = x(n)$ for any $x \in \mathbb{R}^\omega$ and $n \in \omega$. Let $q_0 : \mathbb{R}^\omega \times E \to \mathbb{R}^\omega$ and $q_1 : \mathbb{R}^\omega \times E \to E$ be the natural projections. Consider the sets $H = \{f \in C_p(X, \mathbb{R}^\omega \times E) : ||p_n \circ (q_1 \circ f)|| \to 0\}$ and $G = \{f \in C_p(X, E) : ||p_n \circ f|| \to 0\}$ and define a map $\varphi : C_p(X, \mathbb{R}^\omega \times E) \to C_p(X, E)$ by $\varphi(f) = u \circ f$ for any $f \in C_p(X, \mathbb{R}^\omega \times E)$. Since $u$ is a homeomorphism, it follows easily from TFS-091 that the map $\varphi$ is also a homeomorphism. Our next step is to establish that

(2)  $\varphi(H) = G$ and hence the spaces $H$ and $G$ are homeomorphic.

To prove (2) take any $f \in H$; since the sequence $\{||p_k \circ (q_1 \circ f)|| : k \in \omega\}$ converges to zero, there exists $a \in \Sigma_*(\omega)$ such that $|q_1(f(x))(k)| \leq |a(k)|$ for any $x \in X$ and $k \in \omega$. If $r_n = \sup\{|a(k)| : k \in A_n\}$ then the sequence $\{r_n : n \in \omega\}$ converges to zero and $||\pi_n(q_1(f(x)))||_n \leq r_n$ for every $n \in \omega$.

The property (1) and the equality $\pi_n(u(f(x))) = u_n(q_0(f(x))(n), \pi_n(q_1(f(x))))$ imply that $|\,||\pi_n(q_1(f(x)))||_n - ||\pi_n(u(f(x)))||_n\,| \leq 2^{-n}$ and hence we have the inequality $||\pi_n(u(f(x)))||_n \leq r_n + 2^{-n}$ for any point $x \in X$ and $n \in \omega$. Given $k \in \omega$ let $s_k = r_n + 2^{-n}$ if $n \in \omega$ is the unique number with $k \in A_n$. It is straightforward that the sequence $\{s_k : k \in \omega\}$ converges to zero and $|u(f(x))(k)| \leq s_k$ for any $x \in X$ and $k \in \omega$. Therefore $||p_k(\varphi(f))|| \to 0$ which shows that $\varphi(f) \in G$. This proves that $\varphi(H) \subset G$.

Now assume that $g = \varphi(f) = u \circ f \in G$; by the definition of $G$ we can choose an element $a \in \Sigma_*(\omega)$ such that $|u(f(x))(k)| \leq |a(k)|$ for any $x \in X$ and $k \in \omega$. If $s_n = \sup\{|a(k)| : k \in A_n\}$ then the sequence $\{s_n : n \in \omega\}$ converges to zero and $||\pi_n(u(f(x)))||_n \leq s_n$ for every $n \in \omega$.

The property (1) and the equality $\pi_n(u(f(x))) = u_n(q_0(f(x))(n), \pi_n(q_1(f(x))))$ imply $|\,||\pi_n(q_1(f(x)))||_n - ||\pi_n(u(f(x)))||_n\,| \leq 2^{-n}$, so $||\pi_n(q_1(f(x)))||_n \leq s_n + 2^{-n}$ for any point $x \in X$ and $n \in \omega$. Given $k \in \omega$ let $r_k = s_n + 2^{-n}$ if $n \in \omega$ is the unique number with $k \in A_n$. It is straightforward that the sequence $\{r_k : k \in \omega\}$ converges to zero and $|q_1((f(x))(k)| \leq r_k$ for any $x \in X$ and $k \in \omega$. Therefore $||p_k \circ q_1 \circ f|| \to 0$ which shows that $f \in H$. This proves that $\varphi(H) \supset G$ and hence $\varphi(H) = G$, i.e., (2) is proved.

Consider the map $\mu_0 : C_p(X, \mathbb{R}^\omega \times E) \to C_p(X, \mathbb{R}^\omega)$ defined by the formula $\mu_0(f) = q_0 \circ f$ for any $f \in C_p(X, \mathbb{R}^\omega \times E)$. If we let $\mu_1(f) = q_1 \circ f$ for each $f \in C_p(X, \mathbb{R}^\omega \times E)$ then we obtain a map $\mu_1 : C_p(X, \mathbb{R}^\omega \times E) \to C_p(X, E)$. It follows from TFS-091 that $\mu_0$ and $\mu_1$ are continuous and it is left to the reader as an exercise (consisting in the extraction of the relevant part of the proof of TFS-112) that the map $\mu = \mu_0 \Delta \mu_1 : C_p(X, \mathbb{R}^\omega \times E) \to C_p(X, \mathbb{R}^\omega) \times C_p(X, E)$ is a homeomorphism.

It is straightforward that $\mu(H) = C_p(X, \mathbb{R}^\omega) \times G$, so we can apply (2) to see that $C_p(X, \mathbb{R}^\omega) \times G \simeq G$. Observe that $E \subset \mathbb{R}^\omega$, so $G \subset C_p(X, \mathbb{R}^\omega)$. For any $n \in \omega$ let $\delta_n(f) = p_n \circ f$ for any $f \in C_p(X, \mathbb{R}^\omega)$; then $\delta_n : C_p(X, \mathbb{R}^\omega) \to C_p(X)$ is a continuous map. It is an easy exercise (being again an extraction the relevant part of the proof of TFS-112) to show that the map $\delta = \Delta_{n \in \omega} \delta_n : C_p(X, \mathbb{R}^\omega) \to (C_p(X))^\omega$ is a homeomorphism. We also omit a simple checking that $\delta(G) = C_*(X)$ and hence $G \simeq C_*(X)$. We already saw that $C_p(X, \mathbb{R}^\omega) \simeq (C_p(X))^\omega$ (another way to see it is to apply TFS-112 directly), so $(C_p(X))^\omega \times G \simeq (C_p(X))^\omega \times C_*(X) \simeq C_*(X)$ which shows that our solution is complete.

**V.019.** *Let $X$ be a pseudocompact space. As usual, for any $f \in C_p(X)$, we define $\|f\| = \sup\{|f(x)| : x \in X\}$. Supposing that the space $C_p(X)$ is homeomorphic to $C_*(X) = \{\varphi \in (C_p(X))^\omega : \|\varphi(n)\| \to 0\}$, prove that $C_p(X) \simeq (C_p(X))^\omega$.*

**Solution.** It follows from Problem 018 that $C_*(X) \simeq C_*(X) \times (C_p(X))^\omega$. Since we also have $C_p(X) \simeq C_*(X)$, we conclude that $C_p(X) \simeq C_p(X) \times (C_p(X))^\omega \simeq (C_p(X))^\omega$.

**V.020.** *Prove that $\Sigma_*(\omega)$ is homeomorphic to $(\Sigma_*(\omega))^\omega$.*

**Solution.** Given any function $f \in C_p(\omega + 1)$ let $\|f\| = \sup\{|f(x)| : x \in (\omega + 1)\}$. Take a disjoint family $\mathcal{A} = \{A_n : n \in \omega\}$ of countably infinite subsets of $\omega$ such that $\bigcup \mathcal{A} = \omega$ and fix a bijection $\mu_n : (\omega + 1) \to A_n$ for every $n \in \omega$. For any $x \in \Sigma_*(A_n)$ let $\|x\|_n = \sup\{|x(k)| : k \in A_n\}$.

For each $f \in C_p(\omega + 1)$ let $\varphi_n(f)(\mu_n(m)) = f(m) - f(\omega)$ for any $m \in \omega$ and $\varphi_n(f)(\mu_n(\omega)) = f(\omega)$. It is easy to see that $\varphi_n(f) \in \Sigma_*(A_n)$ for all $f \in C_p(\omega+1)$ and $\varphi_n : C_p(\omega + 1) \to \Sigma_*(A_n)$ is a homeomorphism. Besides,

(1)  $\frac{1}{2}\|f\| \leq \|\varphi_n(f)\|_n \leq 2\|f\|$ for any $f \in C_p(\omega + 1)$ and $n \in \omega$.

The map $\varphi = \prod_{n \in \omega} \varphi_n : (C_p(\omega + 1))^\omega \to \prod_{n \in \omega} \Sigma_*(A_n)$ is a homeomorphism. If $f \in (C_p(\omega + 1))^\omega$ then it follows from (1) that the sequence $\{\|f(n)\| : n \in \omega\}$ converges to zero if and only if the sequence $\{\|\varphi_n(f(n))\|_n : n \in \omega\}$ converges to zero.

For any $x \in P = \prod_{n \in \omega} \Sigma_*(A_n)$ let $\mu(x)(k) = x(n)(k)$ where $n \in \omega$ is the unique number for which $k \in A_n$. Again, it is straightforward that $\mu : P \to \mu(P) \subset \mathbb{R}^\omega$ is a homeomorphism and $\mu(x) \in \Sigma_*(\omega)$ if and only if $\|x(n)\|_n \to 0$. This shows that, for any $f \in (C_p(\omega + 1))^\omega$, the sequence $\{\|f(n)\| : n \in \omega\}$ converges to zero if and only if the sequence $\{\mu(\varphi(f))(n) : n \in \omega\}$ converges to zero. In other words, $\mu(\varphi(C_*(\omega + 1))) = \Sigma_*(\omega)$ and therefore $C_*(\omega + 1) \simeq \Sigma_*(\omega)$. Since $(\omega + 1) \simeq A(\omega)$, we can apply CFS-105 to convince ourselves that

$C_*(\omega+1) \simeq \Sigma_*(\omega) \simeq C_p(\omega+1)$. This makes it possible to apply Problem 019 to conclude that $C_p(\omega+1) \simeq (C_p(\omega+1))^\omega$ and hence $(\Sigma_*(\omega))^\omega \simeq (C_p(\omega+1))^\omega \simeq C_p(\omega+1) \simeq \Sigma_*(\omega)$.

**V.021.** *Prove that, for every infinite space $X$, the space $\mathbb{R}^\omega$ embeds into $C_p(X)$ as a closed subspace.*

**Solution.** If $X$ is not pseudocompact then $\mathbb{R}^\omega$ embeds in $C_p(X)$ as a closed subspace by Fact 6 of T.132. If $X$ is pseudocompact apply Fact 7 of T.132 to find a function $\varphi \in C(X)$ such that $\varphi(X) \subset \mathbb{R}$ is infinite; let $K = \varphi(X)$. The function $\varphi : X \to K$ is $\mathbb{R}$-quotient (see Fact 3 of 6.154), so the space $C_p(K)$ embeds in $C_p(X)$ as a closed subspace.

The space $K$ is infinite, compact, and metrizable, so there exists a subspace $S \subset K$ with $S \simeq (\omega+1)$. Now apply Fact 2 of U.216 to see that $C_p(\omega+1)$ embeds in $C_p(K)$ as a closed subspace. Furthermore, $C_p(\omega+1) \simeq \Sigma_*(\omega) \simeq \Sigma_*(\omega) \times \mathbb{R}^\omega$ (see CFS-105 and Problem 017) which shows that $\mathbb{R}^\omega$ embeds in $C_p(\omega+1)$ as a closed subspace. Therefore $\mathbb{R}^\omega$ embeds in $C_p(K)$ as a closed subspace; an immediate consequence is that $\mathbb{R}^\omega$ also embeds in $C_p(X)$ as a closed subspace.

**V.022.** *Prove that a space $X$ is not pseudocompact if and only if $\mathbb{R}^\omega$ embeds in $C_p(X)$ as a linear subspace.*

**Solution.** We will need some notions of the theory of linear topological spaces. If $L$ is a linear space and $A, B \subset L$ then $A + B = \{x + y : x \in A, \ y \in B\}$; if $A = \{x\}$ then we write $x + B$ instead of $\{x\} + B$. If $L$ is a linear topological space and $0 \in L$ is its zero vector, a set $K \subset L$ is called *totally bounded* if for any $U \in \tau(0, L)$ there is a finite $A \subset L$ such that $K \subset A + U$; a set $P \subset L$ is $\sigma$-*totally bounded* if $P = \bigcup_{n \in \omega} K_n$ where $K_n$ is totally bounded for any $n \in \omega$. A set $P \subset L$ is called *symmetric* if $P = -P = \{-x : x \in P\}$.

*Fact 1.* Suppose that $L$ is a linear topological space and $M \subset L$ is a linear subspace of $L$. If a set $K \subset L$ is totally bounded in $L$ then $K' = K \cap M$ is totally bounded in $M$.

*Proof.* Let $U' \in \tau(0, M)$ and take a set $U \in \tau(L)$ such that $U \cap M = U'$. Using continuity of the addition operation in the space $L$ it is easy to find a symmetric set $V \in \tau(0, L)$ such that $V + V \subset U$; let $V' = V \cap M$. The set $K$ being totally bounded in $L$ there is a finite set $A \subset L$ such that $K \subset A + V$. For any $a \in A$ choose a point $v(a) \in (a + V) \cap M$ if $(a + V) \cap M \neq \emptyset$. If $(a + V) \cap M = \emptyset$ then let $v(a) = 0$. It is evident that $A' = \{v(a) : a \in A\}$ is a finite subset of $M$.

Fix a point $x \in K'$; there is $a \in A$ such that $x \in (a + V)$ and, in particular, $(a + V) \cap M \neq \emptyset$. Therefore $b = v(a) \in A'$, so there is $v_0 \in V$ with $b = a + v_0$. Besides, there is $v_1 \in V$ such that $x = a + v_1$; an immediate consequence is that $x = b - v_0 + v_1$. Observe that $w = -v_0 + v_1 \in V + V \subset U$ and, since also $w = x - b$, the point $w$ belongs to $M$ because $M$ is a linear subspace of $L$. Thus $w \in U \cap M = U'$ and we proved that $x = b + w \in b + U' \subset A' + U'$. The point $x \in K'$ was chosen arbitrarily, so we established that $K' \subset A' + U'$, i.e., $K'$ is totally bounded in $M$. Fact 1 is proved.

*Fact 2.* Suppose that $L$ and $L'$ are linear topological spaces and $f : L \to L'$ is a continuous linear map. If a set $K \subset L$ is totally bounded in $L$ then $K' = f(K)$ is totally bounded in $L'$.

*Proof.* If $U'$ is an open neighborhood of zero in $L'$ then $U = f^{-1}(U')$ is an open neighborhood of zero in $L$, so there exists a finite $A \subset L$ with $K \subset A + U$. It is straightforward that $A' = f(A) \subset L'$ is a finite set such that $K' \subset A' + U'$, so $K'$ is totally bounded in $L'$ and hence Fact 2 is proved.

*Fact 3.* Suppose that $L$ is a linear topological space and $K \subset L$ is totally bounded in $L$. Then

  (i)   any $K' \subset K$ is totally bounded;
 (ii)   the set $\overline{K}$ is totally bounded;
(iii)   the set $z + K$ is totally bounded for any $z \in L$.

*Proof.* Since (i) is evident, let us prove (ii). Fix a set $U \in \tau(\mathbf{0}, L)$; using continuity of the addition operation in $L$ it is easy to find a symmetric set $V \in \tau(\mathbf{0}, L)$ such that $V + V \subset U$. Since $K$ is totally bounded, we can find a finite set $A \subset L$ with $K \subset A + V$. If $x \in \overline{K}$ then $W = x + V \in \tau(x, L)$, so there is $y \in W \cap K$; this means that there is $v_0 \in V$ such that $y = x + v_0$. Since $y \in A + V$, there are $a \in A$ and $v_1 \in V$ for which $y = a + v_1$.

Thus $x = a + v_1 - v_0 \in (a + V + V) \subset a + U \subset A + U$. Since the point $x \in \overline{K}$ was chosen arbitrarily, we proved that $\overline{K} \subset A + U$ and hence $\overline{K}$ is totally bounded. This settles (ii).

To prove (iii) take a set $U \in \tau(\mathbf{0}, L)$; since $K$ is totally bounded, there is a finite set $A' \subset L$ such that $K \subset A' + U$. The set $A = z + A'$ is finite and it is easy to see that $(z + K) \subset A + U$, so $z + K$ is totally bounded; this shows that (iii) is also true, so Fact 3 is proved.

*Fact 4.* The space $\mathbb{R}^{\omega}$ is not $\sigma$-totally bounded (as a linear topological space).

*Proof.* Suppose that $\mathbb{R}^{\omega} = \bigcup_{n \in \omega} K_n$ and every $K_n$ is totally bounded. Fact 3 shows that every $F_n = \overline{K}_n$ is totally bounded as well; the space $\mathbb{R}^{\omega}$ has the Baire property, so there is $n \in \omega$ for which the interior $V$ of the set $\overline{K}_n$ is nonempty. Apply again Fact 3 to see that $V$ is totally bounded.

Any nonempty open subset of $\mathbb{R}^{\omega}$ contains a standard open subset of $\mathbb{R}^{\omega}$, so we can fix $n \in \omega$ and $W_i \in \tau^*(\mathbb{R})$ for any $i < n$ such that $W = (\prod_{i < n} W_i) \times \mathbb{R}^{\omega \setminus n} \subset V$. Let $\pi : \mathbb{R}^{\omega} \to \mathbb{R}$ be the projection of $\mathbb{R}^{\omega}$ onto its $n$-th factor. Then $\pi$ is a continuous linear map; since $W$ is totally bounded (here we used Fact 3 once more), we can apply Fact 2 to see that the set $\pi(W) = \mathbb{R}$ is totally bounded in $\mathbb{R}$.

However, $\mathbb{R}$ is not totally bounded in itself because, for the open neighborhood $H = (-1, 1)$ of zero in $\mathbb{R}$ there is no finite $A \subset \mathbb{R}$ with $A + H = \mathbb{R}$ (this is an easy exercise which can be left to the reader). This contradiction shows that $\mathbb{R}^{\omega}$ is not $\sigma$-totally bounded and hence Fact 4 is proved.

*Fact 5.* A space $Z$ is pseudocompact if and only if the space $C_p(Z)$ (considered as a linear topological space) is $\sigma$-totally bounded.

*Proof.* If $Z$ is not pseudocompact then there exists a countably infinite discrete family of nonempty open subsets of $Z$; an easy consequence of Fact 1 of T.217 is that $\mathbb{R}^\omega$ is linearly homeomorphic to a closed linear subspace of $C_p(Z)$. If $C_p(Z)$ is $\sigma$-totally bounded then it follows from Fact 1 that $\mathbb{R}^\omega$ is also $\sigma$-totally bounded which, together with Fact 4, gives a contradiction. This shows that if $C_p(Z)$ is $\sigma$-totally bounded then the space $Z$ is pseudocompact, i.e., we proved sufficiency.

Now, if $Z$ is pseudocompact then $C_p(Z) = \bigcup_{n\in\omega} C_p(Z,[-n,n])$. For any $n \in \omega$ the set $[-n,n]^Z$ is compact and hence totally bounded in $\mathbb{R}^Z$. The space $C_p(Z)$ is a linear subspace of $\mathbb{R}^Z$, so we can apply Fact 2 to convince ourselves that $C_p(Z,[-n,n]) = [-n,n]^Z \cap C_p(Z)$ is totally bounded in $C_p(Z)$ for any $n \in \omega$. Thus $C_p(Z)$ is $\sigma$-totally bounded and hence we established necessity. Fact 5 is proved.

Returning to our solution observe that if the space $X$ is not pseudocompact then there exists a countably infinite discrete family of nonempty open subsets of $X$; an easy consequence of Fact 1 of T.217 is that $\mathbb{R}^\omega$ is linearly homeomorphic to a closed linear subspace of $C_p(X)$. This proves necessity.

Now, assume that $X$ is pseudocompact and hence $C_p(X)$ is $\sigma$-totally bounded by Fact 5. If $\mathbb{R}^\omega$ embeds in $C_p(X)$ as a linear subspace then it follows from Fact 1 that $\mathbb{R}^\omega$ is also $\sigma$-totally bounded which contradicts Fact 4. Thus $\mathbb{R}^\omega$ does not embed in $C_p(X)$ as a linear subspace; this proves sufficiency and makes our solution complete.

**V.023.** *Prove that, if either $X$ or $C_p(X)$ is Lindelöf, then $\mathbb{R}^{\omega_1}$ does not embed into $C_p(X)$.*

**Solution.** If $Z$ is a space and $Y \subset Z$ then $e(Y,Z)$ (called *the relative extent of $Y$ in $Z$*) is the supremum of cardinalities of discrete subspaces of $Y$ which are closed in $Z$ (and hence in $Y$). In the same spirit, $l(Y,Z)$ (called *the relative Lindelöf number of $Y$ in $Z$*) is the minimal cardinal number $\kappa$ such that every open cover of $Z$ contains a subfamily of cardinality at most $\kappa$ which covers $Y$.

*Fact 1.* Given an infinite cardinal $\kappa$, if $hd(Y) \le \kappa$ and $w(Z) \le \kappa$ then $hd(Y \times Z) \le \kappa$. In particular, the product of a second countable space and a hereditarily separable space is hereditarily separable.

*Proof.* Fix a base $\mathcal{B}$ in the space $Z$ such that $|\mathcal{B}| \le \kappa$ and assume, toward a contradiction, that some $P = \{p_\alpha : \alpha < \kappa^+\} \subset Y \times Z$ is left-separated, i.e., every set $P_\alpha = \{p_\beta : \beta < \alpha\}$ is closed in $P$. For any $\alpha < \kappa^+$ there are $q_\alpha \in Y$ and $r_\alpha \in Z$ such that $p_\alpha = (q_\alpha, r_\alpha)$; choose $B_\alpha \in \mathcal{B}$ and $U_\alpha \in \tau(Z)$ such that $q_\alpha \in U_\alpha$, $r_\alpha \in B_\alpha$ and $(U_\alpha \times B_\alpha) \cap P_\alpha = \emptyset$. There exist $L \subset \kappa^+$ and $B \in \mathcal{B}$ such that $|L| = \kappa^+$ and $B_\alpha = B$ for all $\alpha \in L$.

Suppose that $\alpha, \beta \in L$ and $\beta < \alpha$. If $q_\beta \in U_\alpha$ then the point $p_\beta = (q_\beta, r_\beta)$ belongs to the set $U_\alpha \times B_\beta = U_\alpha \times B = U_\alpha \times B_\alpha$ which is a contradiction with the choice of the sets $B_\alpha$ and $U_\alpha$. Consequently, $q_\beta \notin U_\alpha$ whenever $\beta < \alpha$ and therefore $\{q_\alpha : \alpha \in L\}$ is a left-separated subspace of $Y$ of cardinality $\kappa^+$; this contradiction with $hd(Y) \leq \kappa$ (see SFFS-004) shows that there are no left-separated subspaces of $Y \times Z$ of cardinality $\kappa^+$, so $hd(Y \times Z) \leq \kappa$, i.e., Fact 1 is proved.

*Fact 2.* Suppose that $X$ is a space with $l^*(X) \leq \omega$. If $X$ can be perfectly mapped onto a space $M$ with $hd^*(M) \leq \omega$ then $l(Y, Z) = ext(Y, Z)$ for any $Y \subset Z \subset C_p(X)$.

*Proof.* Suppose that we have spaces $Z, T$ and a map $u : Z \to T$. Given any $n \in \mathbb{N}$ let $u^n(z_1, \ldots, z_n) = (u(z_1), \ldots, u(z_n))$ for any $z = (z_1, \ldots, z_n) \in Z^n$. Thus $u^n : Z^n \to T^n$. If $P$ is a set then $\mathrm{Fin}(P)$ is the family of all nonempty finite subsets of $P$. For the family $\mathcal{O} = \{(a, b) : a < b, a, b \in \mathbb{Q}\}$ and every $n \in \mathbb{N}$ let $\mathcal{O}^n = \{O_1 \times \ldots \times O_n : O_i \in \mathcal{O}$ for any $i = 1, \ldots, n\}$. Choose some enumeration $\{O_k : k \in \omega\}$ of the family $\bigcup\{\mathcal{O}^n : n \in \mathbb{N}\}$. Thus, for each $k \in \omega$ there is $m_k \in \mathbb{N}$ and $O_1^k, \ldots, O_{m_k}^k \in \mathcal{O}$ such that $O_k = O_1^k \times \ldots \times O_{m_k}^k$. For any $x = (x_1, \ldots, x_{m_k}) \in X^{m_k}$ let $[x, O_k] = \{f \in C_p(X) : f^{m_k}(x) \in O^k\}$. If $\mathcal{B}_k = \{[x, O_k] : x \in X^{m_k}\}$ then $\mathcal{B} = \bigcup\{\mathcal{B}_k : k \in \omega\}$ is a base of the space $C_p(X)$.

Fix a perfect map $p : X \to M$; then $p_k = p^{m_k}$ maps $X^{m_k}$ perfectly onto the space $M^{m_k}$ for every $k \in \omega$ (see Fact 4 of S.271).

To prove that $ext(Y, Z) = l(Y, Z)$ is suffices to show that $l(Y, Z) \leq ext(Y, Z)$, so assume the contrary; then there is an infinite cardinal $\kappa \geq ext(Y, Z)$ and a family $\mathcal{U} \subset \tau(C_p(X))$ such that $Z \subset \bigcup \mathcal{U}$ and no subfamily of $\mathcal{U}$ of cardinality $\leq \kappa$ covers $Y$. It is easy to see that, without loss of generality, we can assume that $\mathcal{U} \subset \mathcal{B}$. We will need the set $A_k = \{x \in X^{m_k} : [x, O_k] \in \mathcal{U}\}$ for any $k \in \omega$; observe that the fact that $\mathcal{U}$ covers $Z$ implies that

(1) for any $f \in Z$ there is $k \in \omega$ and $x \in A_k$ such that $f^{m_k}(x) \in O_k$.
On the other hand, no subfamily of $\mathcal{U}$ of cardinality $\leq \kappa$ covers $Y$, so we have

(2) if $B_k \subset A_k$ and $|B_k| \leq \kappa$ for any $k \in \omega$ then there is $f \in Y$ such that $f^{m_k}(B_k) \cap O_k = \emptyset$ for every $k \in \omega$.
Choose a function $f_0 \in Y$ arbitrarily and let $B(k, 0) = \emptyset$ for all $k \in \omega$. Assume that $0 < \alpha < \kappa^+$ and we have chosen a set $\{f_\beta : \beta < \alpha\} \subset Y$ and a family $\{B(k, \beta) : \beta < \alpha, k \in \omega\}$ with the following properties:

(3) $B(k, \beta) \subset A_k$ and $|B(k, \beta)| \leq \kappa$ for all $k \in \omega$ and $\beta < \alpha$;
(4) if $\gamma < \beta < \alpha$ then $B(k, \gamma) \subset B(k, \beta)$ for every $k \in \omega$;
(5) for any $\beta < \alpha$, $k \in \omega$ and any $H \in \mathrm{Fin}(\{f_\gamma : \gamma < \beta\})$ the set $u_H(B(k, \beta))$ is dense in $u_H(A_k)$ where $u_H = p_k \Delta(\Delta\{f^{m_k} : f \in H\}) : X \to M^{m_k} \times \mathbb{R}^{m_k \cdot |H|}$;
(6) $f_\beta^{m_k}(B(k, \beta)) \cap O_k = \emptyset$ for all $\beta < \alpha$ and $k \in \omega$.

To get $f_\alpha$, let $F_\alpha = \{f_\beta : \beta < \alpha\}$ and fix any $k \in \omega$; for every $H \in \mathrm{Fin}(F_\alpha)$ let $u_H = p_k \Delta(\Delta\{f^{m_k} : f \in H\}) : X \to M^{m_k} \times \mathbb{R}^{m_k \cdot |H|}$. The space $u_H(A_k)$ being hereditarily separable by Fact 1, there is a countable $B(H, k) \subset A_k$ such that $u_H(B(H, k))$ is dense in $u_H(A_k)$. The set

$$B(k, \alpha) = (\bigcup\{B(k, \beta) : \beta < \alpha\}) \cup (\bigcup\{B(H, k) : H \in \mathrm{Fin}(F_\alpha)\})$$

has cardinality $\leq \kappa$. Once we have a set $B(k, \alpha)$ for every $k \in \omega$ apply (2) to find a function $f_\alpha \in Y$ such that $f_\alpha^{m_k}(B(k, \alpha)) \cap O_k = \emptyset$ for all $k \in \omega$. It is immediate that the properties (3)–(6) still hold for the set $\{f_\beta : \beta \leq \alpha\}$ and the family $\{B(k, \beta) : \beta \leq \alpha, k \in \omega\}$ and therefore our inductive construction can be continued to give us a set $D = \{f_\alpha : \alpha < \kappa^+\}$ and a family $\{B(k, \beta) : \beta < \kappa^+, k \in \omega\}$ such that the properties (3)–(6) hold for all $\alpha < \kappa^+$.

Assume that $\beta < \alpha < \kappa^+$; it follows from (5) that $(p_k \Delta f_\beta^{m_k})(B(k, \alpha))$ is dense in $(p_k \Delta f_\beta^{m_k})(A_k)$ and hence $f_\beta^{m_k}(B(k, \alpha))$ is dense in $f_\beta^{m_k}(A_k)$ for all $k \in \omega$. The property (1) shows that $f_\beta^{m_k}(A_k) \cap O_k \neq \emptyset$ and therefore $f_\beta^{m_k}(B(k, \alpha)) \cap O_k \neq \emptyset$ for some $k \in \omega$. On the other hand, $f_\alpha^{m_k}(B(k, \alpha)) \cap O_k = \emptyset$ for all $k \in \omega$ by the property (6). Consequently, $f_\alpha \neq f_\beta$ and therefore $|D| = \kappa^+$.

Our purpose is to prove that $D$ is closed and discrete in $Z$, so assume, toward a contradiction, that $g$ is an accumulation point in $Z$ for the set $D$. Recall that $t(C_p(X)) = \omega$ by TFS-149; so $g$ is also an accumulation point for some countable subset of $D$ and hence the ordinal $\alpha = \setminus n\{\beta < \kappa^+ : g$ is an accumulation point for $F_\beta\}$ is well defined. It is evident that $\alpha$ is a limit ordinal. There is $k \in \omega$ and $y \in A_k$ such that $g \in [y, O_k]$; it is evident that $g$ is also an accumulation point for the set $G = F_\alpha \cap [y, O_k]$. The set $K = \bigcap\{(f^{m_k})^{-1}(f^{m_k}(y)) : f \in G\}$ is nonempty because $y \in K$.

Let $W = (g^{m_k})^{-1}(O_k)$ and assume that $K \setminus W \neq \emptyset$. Take any $x \in K \setminus W$ and observe that $g^{m_k}(x) \notin O_k$ while $g^{m_k}(y) \in O_k$ and therefore $g^{m_k}(x) \neq g^{m_k}(y)$. On the other hand, $f^{m_k}(x) = f^{m_k}(y)$ for all $f \in G$ which contradicts $g \in \overline{G}$. We proved that the case $K \setminus W \neq \emptyset$ is impossible, i.e., $K \subset W$.

Let $K_f = (f^{m_k})^{-1}(f^{m_k}(y))$ for all $f \in G$; the set $N = p_k^{-1}(p_k(y))$ is compact because the map $p_k$ is perfect. Therefore $N \cap K_f$ is a nonempty compact set for all $f \in G$ and $y \in \bigcap\{N \cap K_f : f \in G\} \subset K \subset W$. Now we can apply Fact 1 of S.326 to conclude that there is a finite $H \subset G$ such that $Q = \bigcap\{N \cap K_f : f \in H\} \subset W$. Observe that, for the map $u_H = p_k \Delta(\Delta\{f^{m_k} : f \in H\})$ we have $Q = u_H^{-1}(u_H(y))$. Now, if $Y = u_H(X)$ then the map $u_H : X \to Y$ is perfect because $p_k$ is perfect (see Fact 1 of T.266).

Therefore Fact 1 of S.226 is applicable to conclude that there is $U \in \tau(Y)$ such that $u_H(y) \in U$ and $u_H^{-1}(U) \subset W$. Let $\gamma = \max\{\beta : f_\beta \in H\}$. Then $\gamma < \mu = \gamma + 1 < \alpha$ because $\alpha$ is a limit ordinal. We have $H \in \mathrm{Fin}(F_\mu)$ and therefore $u_H(B(k, \mu))$ is dense $u_H(A_k)$ by (5). Furthermore, $u_H(y) \in U \cap u_H(A_k)$ which shows that $U \cap u_H(A_k)$ is a nonempty open subset of $u_H(A_k)$. The set $u_H(B(k, \mu))$ being dense in $u_H(A_k)$ by (5), we have $u_H(B(k, \mu)) \cap U \neq \emptyset$ and therefore there is $z \in B(k, \mu)$ for which $u_H^{-1}(u_H(z)) \subset W$ and, in particular, $z \in W$. This implies that $g^{m_k}(z) \in O_k$.

On the other hand, the condition (6) implies that $f_\mu^{m_k}(B(k, \mu)) \cap O_k = \emptyset$; the conditions (4) and (6) show that, for any ordinal $\beta$ with $\mu < \beta < \alpha$ we have $f_\beta^{m_k}(B(k, \mu)) \cap O_k \subset f_\mu^{m_k}(B(k, \beta)) \cap O_k = \emptyset$. Consequently, $f_\beta^{m_k}(z) \notin O_k$ whenever $\mu \leq \beta < \alpha$ which shows that $g \notin \overline{G \setminus F_\mu}$ and hence $g$ is an accumulation point for the set $F_\mu$ which is a contradiction with $\mu < \alpha$ and the choice of $\alpha$.

This contradiction proves that $D$ is a closed discrete subspace of $Z$. We already saw that $|D| = \kappa^+ > ext(Y, Z)$; this final contradiction shows that $l(Y, Z) \leq ext(Y, Z)$ and hence Fact 2 is proved.

*Fact 3.* Suppose that $X$ is a space such that $C_p(X)$ is Lindelöf. Given a set $A \subset X$ assume that every countable subset of $A$ is $C$-embedded in $X$. Then $A$ is $C$-embedded in $X$.

*Proof.* Fix a continuous function $\varphi : A \to \mathbb{R}$; for any countable $B \subset A$ the set $F_B = \{f \in C_p(X) : f|B = \varphi\}$ is nonempty and closed in $C_p(X)$. It is easy to see that the family $\mathcal{F} = \{F_B : B$ is a countable subset of $A\}$ is countably centered, so the Lindelöf property of $C_p(X)$ implies that $\bigcap \mathcal{F} \neq \emptyset$. Any element of $\bigcap \mathcal{F}$ is a continuous extension of $\varphi$ over the whole space $X$, so $A$ is $C$-embedded in $X$ and hence Fact 3 is proved.

*Fact 4.* Given spaces $X, Y$ and a map $\varphi : X \to Y$ suppose that $D \subset X$ is discrete, $\varphi|D$ is injective and $D' = \varphi(D)$ is a discrete subspace of $Y$. If $D'$ is $C$-embedded (or $C^*$-embedded) in $Y$ then the set $D$ is $C$-embedded (or $C^*$-embedded respectively) in the space $X$.

*Proof.* Suppose that $f : D \to \mathbb{R}$ is a (bounded) function (observe that $f$ is automatically continuous because $D$ is discrete). Then $g = f \circ (\varphi|D)^{-1} : D' \to \mathbb{R}$ is a continuous (bounded) function because $D'$ is also discrete. The set $D'$ being $C$-embedded ($C^*$-embedded) in $Y$, there is a continuous (bounded) function $h : Y \to \mathbb{R}$ such that $h|D' = g$. It is straightforward that $f_1 = h \circ \varphi : X \to \mathbb{R}$ is a continuous (bounded) function such that $f_1|D = f$, so Fact 4 is proved.

*Fact 5.* If $D \subset C_p(\omega_1 + 1)$ is closed, discrete, and $|D| \leq \omega$ then $D$ is $C$-embedded in $C_p(\omega_1 + 1)$.

*Proof.* Let $\pi : C_p(\omega_1 + 1) \to C_p(\omega_1)$ be the restriction map. Since $\omega_1$ is countably compact and $\omega_1 + 1 = \beta\omega_1$ (see TFS-314), the space $\omega_1 + 1$ is the Hewitt extension of $\omega_1$ (see TFS-417 and Fact 3 of S.309). Therefore $\pi|A : A \to \pi(A)$ is a homeomorphism for any countable $A \subset C_p(\omega_1 + 1)$ (see TFS-437). As a consequence, the set $D' = \pi(D)$ is closed and discrete in $C_p(\omega_1)$. The space $C_p(\omega_1)$ being Lindelöf (see TFS-316), the set $D'$ is $C$-embedded in $C_p(\omega_1)$, so we can apply Fact 4 to see that $D$ is also $C$-embedded in $C_p(\omega_1 + 1)$. Fact 5 is proved.

*Fact 6.* Given an infinite cardinal $\kappa$ if a set $Z \subset C_p(\kappa^+ + 1)$ separates the points of $\kappa^+ + 1$ then there is a discrete subspace $D \subset Z$ which is closed in $C_p(\kappa^+ + 1)$ and $|D| = \kappa^+$.

*Proof.* For each $\alpha < \kappa^+$ fix rational numbers $s_\alpha, t_\alpha$ and a function $f_\alpha \in Z$ such that $f_\alpha(\alpha) < s_\alpha < t_\alpha < f_\alpha(\kappa^+)$ or $f_\alpha(\alpha) > s_\alpha > t_\alpha > f_\alpha(\kappa^+)$. Since each $f_\alpha$ is continuous, there exists $\beta_\alpha < \alpha$ such that $f_\alpha(\gamma) < s_\alpha$ or $f_\alpha(\gamma) > s_\alpha$ for every $\gamma \in (\beta_\alpha, \alpha]$.

The map $r : \kappa^+ \to \kappa^+$ defined by $r(\alpha) = \beta_\alpha$ for all $\alpha < \kappa^+$ satisfies the hypothesis of Fact 3 of U.074, so there is $\beta < \kappa^+$ and a stationary set $R \subset \kappa^+$ such that $\beta_\alpha = \beta$ for all $\alpha \in R$. There is a set $R' \subset R$ with $|R'| = \kappa^+$ for which there are $s, t \in \mathbb{Q}$ such that $s_\alpha = s$ and $t_\alpha = t$ for all $\alpha \in R'$; let $E = \{f_\alpha : \alpha \in R'\}$. Passing, if necessary, to a subset of $E$ of cardinality $\kappa^+$, we can assume that either $f_\alpha(\alpha) < s < t < f_\alpha(\kappa^+)$ or $f_\alpha(\alpha) > s > t > f_\alpha(\kappa^+)$ for all $\alpha \in R'$. Since these two cases are analogous, we will only consider the first one.

For every function $f \in C_p(\kappa^+ + 1)$ let $O_f = C_p(\kappa^+ + 1) \backslash \overline{E}$ if $f \notin \overline{E}$. Then $O_f$ is an open neighborhood of $f$ in $C_p(\kappa^+ + 1)$ such that $O_f \cap E = \emptyset$ and hence $B_f = \{\alpha \in R' : f_\alpha \in O_f\} = \emptyset$. If $f \in \overline{E}$ then $f(\kappa^+) \geq t$ because $g(\kappa^+) > t$ for all $g \in E$. Choose any $s' \in (s, t)$ and observe that, by continuity of $f$, there is $\gamma > \beta$ such that $f(\gamma) > s' > s$. The set $O_f = \{g \in C_p(\kappa^+ + 1) : g(\gamma) > s'\}$ is an open neighborhood of $f$ in $C_p(\kappa^+ + 1)$. If $\alpha > \gamma$ and $\alpha \in R'$ then $\gamma \in (\beta, \alpha] = (\beta_\alpha, \alpha]$ which implies, by the choice of $\beta_\alpha$, that $f_\alpha(\gamma) < s < s'$ whence $f_\alpha \notin O_f$. Thus $O_f \cap E \subset \{f_\alpha : \alpha \leq \gamma\}$ and therefore, for the set $B_f = \{\alpha \in R' : f_\alpha \in O_f\}$, we have $|B_f| \leq |\gamma| \leq \kappa$.

The family $\mathcal{U} = \{O_f : f \in Z\}$ is an open cover of the space $C_p(\kappa^+ + 1)$ such that $|U \cap E| \leq \kappa$ for any $U \in \mathcal{U}$. If $\mathcal{U}' \subset \mathcal{U}$ and $|\mathcal{U}'| \leq \kappa$ then $|(\bigcup \mathcal{U}') \cap E| \leq \kappa$, so the set $\bigcup \mathcal{U}'$ does not cover the set $E \subset Z$ and hence $Z$ is not contained in $\bigcup \mathcal{U}'$. This shows that $l(Z, C_p(\kappa^+ + 1)) \geq \kappa^+$, so we have $e(Z, C_p(\kappa^+ + 1)) \geq \kappa^+$ (see Fact 2) and hence there exists a discrete $D \subset Z$ which is closed in $C_p(\kappa^+ + 1)$ and $|D| = \kappa^+$. Fact 6 is proved.

*Fact 7.* If $X$ is a space and $\omega_1 + 1$ embeds in $C_p(X)$ then there is a closed discrete uncountable $D \subset X$ such that every countable $A \subset D$ is $C$-embedded in $X$.

*Proof.* Let $\Omega \subset C_p(X)$ be a subspace homeomorphic to $\omega_1 + 1$. For any $x \in X$ and $f \in \Omega$ let $\varphi(x)(f) = f(x)$. Then $\varphi : X \to C_p(\Omega)$ is a continuous map such that $Y = \varphi(X)$ separates the points of $\Omega$. By Fact 6, there is an uncountable discrete $D' \subset Y$ which is closed in $C_p(\Omega)$. For any $y \in D'$ choose a point $a(y) \in \varphi^{-1}(y)$; then the set $D = \{a(y) : y \in D'\} \subset X$ is uncountable, discrete, and closed in $X$. Since the map $\varphi|D$ is an injection, we can apply Fact 4 and Fact 5 to conclude that every countable $A \subset D$ is $C$-embedded in $X$, so Fact 7 is proved.

*Fact 8.* If $C_p(X)$ is Lindelöf then $\omega_1$ does not embed in $C_p(X)$.

*Proof.* Assume that there is a subspace $Z \subset C_p(X)$ which is homeomorphic to $\omega_1$. Since $\overline{Z}$ is pseudocompact and Lindelöf, it has to be compact, so $K = \overline{Z}$ is a compact extension of $Z$. Observe that $Z$ is not closed in $C_p(X)$ because $\omega_1$ is not Lindelöf. Since $\omega_1 + 1$ is canonically homeomorphic to $\beta\omega_1$ (see TFS-314), there is a continuous onto map $\varphi : (\omega_1 + 1) \to \overline{Z}$ such that $\varphi|\omega_1 : \omega_1 \to Z$ is a homeomorphism.

Apply Fact 3 of S.261 to see that $\varphi((\omega_1 + 1) \backslash \omega_1) = \overline{Z} \backslash Z$ and hence $\overline{Z} \backslash Z$ is a singleton. An immediate consequence is that $\varphi$ is a bijection and hence homeomorphism. Thus $K$ is a subspace of $C_p(X)$ homeomorphic to $\omega_1 + 1$. By Fact 7, there is a closed discrete uncountable $D \subset X$ such that every countable $A \subset D$ is $C$-embedded in $X$. Apply Fact 3 to conclude that $D$ is $C$-embedded in $X$. If $\pi : C_p(X) \to C_p(D)$ is the restriction map then $\pi(C_p(X)) = C_p(D) = \mathbb{R}^D$.

This, together with the Lindelöf property of $C_p(X)$, implies that $\mathbb{R}^D$ is Lindelöf which is a contradiction (see e.g., Fact 2 of S.215). Thus $\omega_1$ cannot be embedded in $C_p(X)$, so Fact 8 is proved.

Returning to our solution observe that if $X$ is Lindelöf then $\omega_1 + 1$ does not embed in $C_p(X)$ by Fact 1 of U.089. Since $\omega_1 + 1$ embeds in $\mathbb{R}^{\omega_1}$, the space $\mathbb{R}^{\omega_1}$ cannot be embedded in $C_p(X)$ either. Now, if $C_p(X)$ is Lindelöf then even the space $\omega_1$ is not embeddable in $C_p(X)$ by Fact 8; as before, this implies that $\mathbb{R}^{\omega_1}$ is not embeddable in $C_p(X)$ as well.

**V.024.** *Prove that there exists a space $X$ such that $c(X) = \omega$ and $\mathbb{R}^{\omega_1}$ embeds in $C_p(X)$ as a closed linear subspace.*

**Solution.** All spaces in this solution are considered to be nonempty. We will often use without explicit reference the fact that any linear space has a Hamel basis (see Fact 1 of S.489).

*Fact 1.* Suppose that $Z$ is a space and $H$ is a Hamel basis in $C_p(Z)$. If $A \subset H$ is a finite nonempty set then, for any $u \in \mathbb{R}^A$, there is a continuous linear functional $\varphi : C_p(Z) \to \mathbb{R}$ such that $\varphi | A = u$.

*Proof.* For any $x \in Z$ let $e_x(f) = f(x)$ for any $f \in C_p(Z)$; then $e_x$ is a continuous linear functional on $C_p(Z)$ (see TFS-196).

We will show first by induction on $n \in \mathbb{N}$ that

(1) for any $f_1, \ldots, f_n \in H$ there is a set $\{x_1, \ldots, x_n\} \subset Z$ such that the family $\{(f_i(x_1), \ldots, f_i(x_n)) : i = 1, \ldots, n\}$ of vectors of $\mathbb{R}^n$ is linearly independent.

If $n = 1$ then it follows from $f_1 \in H$ that $f_1$ is not identically zero, so there is $x_1 \in Z$ such that $f_1(x_1) \neq 0$. It is clear that the vector $(f_1(x_1))$ forms an independent family in $\mathbb{R}$, so (1) is proved for $n = 1$.

Assume that the property (1) is proved for all $n \leq m$ and fix any functions $f_1, \ldots, f_{m+1} \in H$. By the induction hypothesis there is a set $\{x_1, \ldots, x_m\} \subset Z$ such that the family $\{(f_i(x_1), \ldots, f_i(x_m)) : i = 1, \ldots, m\}$ of vectors of $\mathbb{R}^m$ is linearly independent; let $a_i = (f_i(x_1), \ldots, f_i(x_m))$ for all $i \leq m + 1$. It is clear that the family $\{a_i : i \leq m + 1\}$ cannot be linearly independent in $\mathbb{R}^m$, so we can find $\mu_1, \ldots, \mu_m \in \mathbb{R}$ such that $a_{m+1} = \mu_1 a_1 + \ldots + \mu_m a_m$. However, the functions $f_1, \ldots, f_{m+1}$ are linearly independent, so $f_{m+1} \neq \mu_1 f_1 + \ldots + \mu_m f_m$; therefore there is a point $x_{m+1} \in Z$ such that $f_{m+1}(x_{m+1}) \neq \mu_1 f_1(x_{m+1}) + \ldots + \mu_m f_m(x_{m+1})$. We leave it to the reader to verify that the family $\{(f_i(x_1), \ldots, f_i(x_{m+1})) : i \leq m + 1\}$ of vectors of $\mathbb{R}^{m+1}$ is linearly independent, so (1) is proved.

Now choose a faithful enumeration $\{f_1, \ldots, f_n\}$ of the set $A$ and let $b_i = u(f_i)$ for any $i = 1, \ldots, n$. Apply the property (1) to find a set $\{x_1, \ldots, x_n\} \subset Z$ such that the family $\{(f_i(x_1), \ldots, f_i(x_n)) : i \leq n\}$ of vectors of the space $\mathbb{R}^n$ is linearly independent. A well-known theorem of algebra shows that there exist $\lambda_1, \ldots, \lambda_n \in \mathbb{R}$ such that $\lambda_1 f_i(x_1) + \ldots + \lambda_n f_i(x_n) = b_i$ for each $i = 1, \ldots, n$. It is evident that $\varphi = \lambda_1 e_{x_1} + \ldots + \lambda_n e_{x_n}$ is a continuous linear functional on $C_p(Z)$ and $\varphi(f_i) = b_i$ for every $i \leq n$, i.e., $\varphi | A = u$, so Fact 1 is proved.

*Fact 2.* Suppose that $Z$ is a space and $H \subset C_p(Z)$ is a Hamel basis in $C_p(Z)$. Let $\mathcal{J} = \{(a,b) : a,b \in \mathbb{Q} \text{ and } a < b\}$. Given functions $f_1, \ldots, f_n \in C_p(Z)$ and $O_1, \ldots, O_n \in \mathcal{J}$ let $[f_1, \ldots, f_n; O_1, \ldots, O_n] = \{\varphi \in L_p(Z) : \varphi(f_i) \in O_i \text{ for all } i \le n\}$. Then the family $\mathcal{O} = \{[f_1, \ldots, f_n; O_1, \ldots, O_n] : n \in \mathbb{N}, \ f_1, \ldots, f_n \in H$ and $O_1, \ldots, O_n \in \mathcal{J}\}$ is a base in the space $L_p(Z)$.

*Proof.* It is clear that all elements of $\mathcal{O}$ are open in $L_p(Z)$. To prove that $\mathcal{O}$ is a base in $L_p(Z)$ take a point $\varphi \in L_p(Z)$ and a set $W \in \tau(\varphi, L_p(Z))$. There are $g_1, \ldots, g_k \in C_p(Z)$ and $\varepsilon > 0$ such that $\varphi \in V = \{u \in L_p(Z) : |u(g_i) - \varphi(g_i)| < \varepsilon$ for each $i \le k\} \subset W$.

The set $H$ being a Hamel basis in the space $C_p(Z)$ there are $f_1, \ldots, f_n \in H$ such that every function $g_i$ is a linear combination of $f_1, \ldots, f_n$. Thus for any $i \in \{1, \ldots, k\}$ there are $\lambda_1^i, \ldots, \lambda_n^i \in \mathbb{R}$ such that $g_i = \lambda_1^i f_1 + \ldots + \lambda_n^i f_n$. For the number $\lambda = \max\{|\lambda_j^i| : i \le k, \ j \le n\} + 1$ choose $\delta > 0$ such that $\lambda n \delta < \varepsilon$ and take a set $O_i = (a,b) \in \mathcal{J}$ such that $\varphi(f_i) \in O_i \subset (\varphi(f_i) - \delta, \varphi(f_i) + \delta)$ for every $i \le n$. The set $U = [f_1, \ldots, f_n; O_1, \ldots, O_n]$ belongs to $\mathcal{O}$ and $\varphi \in U$. Given any $u \in U$ observe that $|u(g_i) - \varphi(g_i)| = |\sum_{j=1}^n \lambda_j^i (u(f_j) - \varphi(f_j))| \le \lambda n \delta < \varepsilon$ for any $i \le k$ and therefore $u \in V$. This proves that $U \subset V$ and hence $\varphi \in U \subset V \subset W$. Thus $\mathcal{O}$ is a base in $L_p(Z)$ and Fact 2 is proved.

*Fact 3.* The space $L_p(Z)$ has the Souslin property for any space $Z$.

*Proof.* If $c(L_p(Z)) > \omega$ then there exists a disjoint family $\mathcal{U} \subset \tau^*(L_p(Z))$ with $|\mathcal{U}| = \omega_1$. Since the family $\mathcal{O}$ from Fact 2 is a base in $L_p(Z)$, we can assume, without loss of generality, that $\mathcal{U} \subset \mathcal{O}$. For any $U = [f_1, \ldots, f_n; O_1, \ldots, O_n] \in \mathcal{U}$ let $k_U = n$, $\text{supp}(U) = \{f_1, \ldots, f_n\}$ and $O(U) = (O_1, \ldots, O_n) \in \mathcal{J}^n$. The family $\mathcal{J}$ being countable we can consider, without loss of generality (passing to an appropriate uncountable family of $\mathcal{U}$ if necessary), that there are $n \in \mathbb{N}$ and $O_1, \ldots, O_n \in \mathcal{J}$ such that, for any $U \in \mathcal{U}$, we have $k_U = n$ and $U = [f_1^U, \ldots, f_n^U; O_1, \ldots, O_n]$ for some $f_1^U, \ldots, f_n^U \in H$.

Apply the Delta-lemma (SFFS-038) to see that there exists an uncountable $\mathcal{U}' \subset \mathcal{U}$ and a finite set $D \subset H$ such that $\text{supp}(U) \cap \text{supp}(V) = D$ for any distinct $U, V \in \mathcal{U}'$. If $\{f_1, \ldots, f_m\}$ is a faithful enumeration of $D$ then changing the respective order in the sets $\{f_1^U, \ldots, f_n^U\}$ and $\{O_1, \ldots, O_n\}$ if necessary, we can assume, without loss of generality, that for any $U \in \mathcal{U}'$ we have $U = [f_1, \ldots, f_m, f_{m+1}^U, \ldots, f_n^U; O_1, \ldots, O_n]$ while the sets $\{f_{m+1}^V, \ldots, f_n^V\}$ and $\{f_{m+1}^U, \ldots, f_n^U\}$ are disjoint whenever $U, V \in \mathcal{U}'$ and $U \ne V$.

Take distinct $U, V \in \mathcal{U}'$ and apply Fact 1 to find a linear continuous functional $\varphi : C_p(Z) \to \mathbb{R}$ such that $\varphi(f_i) \in O_i$ for all $i \le m$ (observe that the set $D$ can be empty; in this case $m = 0$) while $\varphi(f_i^U) \in O_i$ and $\varphi(f_i^V) \in O_i$ for all $i \in \{m+1, \ldots, n\}$. It is immediate that $\varphi \in U \cap V$; this contradiction shows that $c(L_p(Z)) = \omega$ and hence Fact 3 is proved.

Returning to our solution let $D$ be a discrete space with $|D| = \omega_1$ and let $X = L_p(D)$. There exists an $l$-embedding of $D$ in $X$ by CFS-466, so $C_p(D) = \mathbb{R}^D \simeq \mathbb{R}^{\omega_1}$ embeds in $C_p(X)$ as a closed linear subspace by CFS-448. Finally, apply Fact 3 to see that $c(X) = \omega$ and hence our solution is complete.

**V.025.** *Prove that if $\omega + 1 \hookrightarrow X$ then $C_p(X) \simeq C_p(X) \times \mathbb{R}^\omega$. Deduce from this fact that pseudocompactness, countable compactness, and compactness are not t-invariant.*

**Solution.** Let $K \subset X$ be a subspace homeomorphic to $\omega + 1$ and consider the space $I = \{f \in C_p(X) : f(K) = \{0\}\}$. The set $K$ is $l$-embedded in $X$ (see CFS-482) and hence $C_p(X) \simeq I \times C_p(K)$ by CFS-448. Furthermore, $C_p(K) \simeq \Sigma_*(\omega)$ (see CFS-105), so we can apply Problem 017 to conclude that $C_p(X) \simeq I \times \Sigma_*(\omega) \simeq I \times \Sigma_*(\omega) \times \mathbb{R}^\omega \simeq C_p(X) \times \mathbb{R}^\omega$.

We proved, in particular, that $C_p(\omega + 1) \simeq C_p(\omega + 1) \times \mathbb{R}^\omega \simeq C_p((\omega + 1) \oplus \omega)$ and therefore $(\omega + 1) \overset{t}{\sim} (\omega + 1) \oplus \omega$. The space $\omega + 1$ is compact while $(\omega + 1) \oplus \omega$ is not even pseudocompact. This shows that pseudocompactness, countable compactness, and compactness are not $t$-invariant.

**V.026.** *Prove that $C_p(\mathbb{R})^\omega \simeq C_p(\mathbb{R})$ and $C_p(\mathbb{I})^\omega \simeq C_p(\mathbb{I})$.*

**Solution.** It follows from TFS-177 that $C_p(\mathbb{R}) \simeq (C_p(\mathbb{R}))^\omega$. To prove that the space $(C_p(\mathbb{I}))^\omega$ is homeomorphic to $C_p(\mathbb{I})$ it suffices to establish the same for the space $I = [0, 1]$ because $I \simeq \mathbb{I}$.

The space $K = \{0\} \cup \{2^{-n} : n \in \omega\} \subset I$ is homeomorphic to $\omega + 1$; since it is $l$-embedded in $I$ (see CFS-482), for the set $M = \{f \in C_p(I) : f(K) = \{0\}\}$, we have $C_p(I) \simeq M \times C_p(K)$. Recalling that $C_p(K) \simeq \Sigma_*(\omega)$ (see CFS-105) we conclude that $C_p(I) \simeq M \times \Sigma_*(\omega)$. Let $a_n = 2^{-n-1}$, $b_n = 2^{-n}$ and $I_n = [a_n, b_n]$; given a function $f \in C_p(I_n)$ let $\|f\|_n = \{\sup |f(x)| : x \in I_n\}$ be its usual norm in the space $C_p(I_n)$ for any $n \in \omega$.

We will also need the space $M_n = \{f \in C_p(I_n) : f(a_n) = f(b_n) = 0\}$; let $\pi_n : C_p(I) \to C_p(I_n)$ be the restriction map and observe that $\pi_n(M) = M_n$ for each $n \in \omega$. We claim that

(1) the map $\pi = \Delta_{n\in\omega} \pi_n : M \to M' = \pi(M) \subset \prod_{n\in\omega} M_n$ is a homeomorphism.

Let $p_n : \prod_{n\in\omega} M_n \to M_n$ be the projection for every $n \in \omega$. It is an easy exercise that $\pi$ is a continuous bijection; to see that the map $\pi^{-1} : M' \to M$ is continuous, let $q_x(f) = f(x)$ for any $x \in I$ and $f \in M$. We also need an analogous map in every space $C_p(I_n)$: let $q_x^n(f) = f(x)$ for any $x \in I_n$ and $f \in C_p(I_n)$. If $x \in K$ then $(q_x \circ \pi^{-1})(g) = 0$ for any $g \in M'$, so the map $q_x \circ \pi^{-1}$ is continuous. If $x \in I \setminus K$ then fix the unique $n \in \omega$ with $x \in I_n$. Given a function $g \in M'$ observe that $q_x(\pi^{-1}(g)) = g(n)(x)$; this shows that $q_x \circ \pi^{-1} = q_x^n \circ p_n$ is continuous being the composition of two continuous maps. Therefore we can apply TFS-102 to see that $\pi^{-1}$ is continuous and hence $\pi$ is, indeed, a homeomorphism.

Next observe that $M' = \{f \in \prod_{n\in\omega} M_n : \|f(n)\|_n \to 0\}$. If $\varphi_n : I_n \to I$ is a homeomorphism then the dual map $\varphi_n^* : C_p(I) \to C_p(I_n)$ is a homeomorphism which preserves the norm, i.e., $\|\varphi_n^*(f)\|_n = \|f\|$ for any $f \in C_p(I)$. As a consequence, the map $\varphi = \prod_{n\in\omega} \varphi_n^* : (C_p(I))^\omega \to \prod_{n\in\omega} C_p(I_n)$ is a homeomorphism such that $E = \varphi(C_*(I)) = \{f \in \prod_{n\in\omega} C_p(I_n) : \|f(n)\|_n \to 0\}$ (see Problem 018 for the definition of $C_*(X)$ for any space $X$).

For any $n \in \omega$ we will need an extender $e_n$ from the doubleton $S_n = \{a_n, b_n\}$ to $I_n$. Define $e_n$ by letting $e_n(f)(t) = \frac{f(b_n) - f(a_n)}{b_n - a_n} \cdot (t - a_n) + f(a_n)$ for any $f \in \mathbb{R}^{S_n}$ and $t \in I_n$. In other words, $e(f)$ is the linear function whose graph is obtained connecting the points $(a_n, f(a_n))$ and $(b_n, f(b_n))$ by a line segment. It is easy to check that the map $e : \mathbb{R}^{S_n} \to C_p(I_n)$ is a linear continuous extender such that $\|e(f)\| = \max\{|f(a_n)|, |f(b_n)|\}$ for any $f \in \mathbb{R}^{S_n}$.

Let $r_n^0 : \mathbb{R}^{S_n} \times M_n \to \mathbb{R}^{S_n}$ and $r_n^1 : \mathbb{R}^{S_n} \times M_n \to M_n$ be the natural projections for each $n \in \omega$. It follows from CFS-448 that every map $u_n : \mathbb{R}^{S_n} \times M_n \to C_p(I_n)$ defined by $u_n(f, g) = e(f) + g$ for any $(f, g) \in \mathbb{R}^{S_n} \times M_n$, is a homeomorphism; denote by $v_n$ its inverse. By the choice of our extender $e$ we can see that

(2)  $\|u_n(f, g)\|_n \leq \max\{|f(a_n)|, |f(b_n)|\} + \|g\|_n$ for any $(f, g) \in \mathbb{R}^{S_n} \times M_n$.

Besides, for any $h \in C_p(I_n)$ we have

(3)  $\|r_n^1(v_n(h))\|_n \leq 2\|h\|_n$ and $\max\{|r_n^0(v_n(h))(a_n)|, |r_n^0(v_n(h))(b_n)|\} \leq \|h\|_n$.

The product map $v = \prod_{n \in \omega} v_n : \prod_{n \in \omega} C_p(I_n) \to P = \prod_{n \in \omega}(\mathbb{R}^{S_n} \times M_n)$ is, evidently, a homeomorphism. Let $w : P \to R = (\prod_{n \in \omega} \mathbb{R}^{S_n}) \times (\prod_{n \in \omega} M_n)$ be the homeomorphism obtained by an evident coordinate permutation. It follows from the properties (2) and (3) that $\mu = w \circ v : \prod_{n \in \omega} C_p(I_n) \to R$ is a homeomorphism such that, for any $f \in \prod_{n \in \omega} C_p(I_n)$, if $\mu(f) = (g, h) \in R$ then the sequence $\{\|f(n)\|_n : n \in \omega\}$ converges to zero if and only if both sequences $S[g] = \{\max\{|g(n)(a_n)|, |g(n)(b_n)|\} : n \in \omega\}$ and $\{\|h(n)\|_n : n \in \omega\}$ converge to zero.

In other words, $\mu(E) = A \times M'$ where $A = \{g \in \prod_{n \in \omega} \mathbb{R}^{S_n} : S[g] \to 0\}$ and hence $E \simeq A \times M'$. It is evident that $A \simeq \Sigma_*(\omega)$, so $E \simeq M' \times \Sigma_*(\omega)$. We already saw that $E \simeq C_*(I)$ and $M \simeq M'$, so $C_*(I) \simeq M' \times \Sigma_*(\omega) \simeq C_p(I)$. This makes it possible to apply Problem 019 to conclude that $C_p(I) \simeq (C_p(I))^\omega$ and hence our solution is complete.

**V.027.** *Prove that $\mathbb{R}$ is $t$-equivalent to $[0, 1]$.*

**Solution.** For any $n \in \mathbb{Z}$ let $I_n = [n, n + 1] \subset \mathbb{R}$ and $S_n = \{n, n + 1\}$; we will also need the set $M = \{f \in C_p(\mathbb{R}) : f(\mathbb{Z}) = \{0\}\}$. It is easy to see that the family $\{(n - \frac{1}{3}, n + \frac{1}{3}) : n \in \mathbb{Z}\} \subset \tau^*(\mathbb{R})$ is discrete so $C_p(\mathbb{R})$ is homeomorphic to the space $\mathbb{R}^{\mathbb{Z}} \times M \simeq \mathbb{R}^\omega \times M$ (see Fact 1 of T.217). Every restriction map $\pi_n : C_p(\mathbb{R}) \to C_p(I_n)$ is continuous; let $M_n = \{f \in C_p(I_n) : f(S_n) = \{0\}\}$ for each $n \in \mathbb{Z}$.

The diagonal product map $\pi = \Delta\{\pi_n : n \in \mathbb{Z}\} : C_p(\mathbb{R}) \to \prod\{C_p(I_n) : n \in \mathbb{Z}\}$ is continuous; it is straightforward to check that $\pi(M) = \prod\{M_n : n \in \mathbb{Z}\}$ and the map $\pi|M : M \to \prod\{M_n : n \in \mathbb{Z}\}$ is a homeomorphism. As a consequence, the space $C_p(\mathbb{R})$ is homeomorphic to the space $H = \mathbb{R}^\omega \times \prod\{M_n : n \in \mathbb{Z}\}$; an evident permutation of coordinates shows that $H \simeq \prod\{\mathbb{R}^2 \times M_n : n \in \mathbb{Z}\}$. Furthermore, $\mathbb{R}^2 \times M_n \simeq C_p(I_n)$ (see Fact 1 of S.409) for any $n \in \mathbb{Z}$, so the space $C_p(\mathbb{R})$ is homeomorphic to $\prod\{C_p(I_n) : n \in \mathbb{Z}\}$; recalling that every $C_p(I_n)$ is homeomorphic to $C_p(\mathbb{I})$ and $\mathbb{Z}$ is countable, we can see that $C_p(\mathbb{R}) \simeq (C_p(\mathbb{I}))^\omega$. Finally, observe that $(C_p(\mathbb{I}))^\omega \simeq C_p(\mathbb{I})$ by Problem 026 and $\mathbb{I} \simeq [0, 1]$, so $C_p(\mathbb{R}) \simeq C_p(\mathbb{I}) \simeq C_p([0, 1])$ which shows that $\mathbb{R}$ is $t$-equivalent to $[0, 1]$.

**V.028.** *Prove that* $\upsilon X \overset{t}{\sim} \upsilon Y$ *whenever* $X \overset{t}{\sim} Y$. *Give an example which shows that* $X \overset{t}{\sim} Y$ *does not necessarily imply* $\beta X \overset{t}{\sim} \beta Y$.

**Solution.** If $Z$ is a space and $\mu \subset \exp(Z)$ then $\mu | A = \{M \cap A : M \in \mu\}$ for any $A \subset Z$; let $\mathcal{M} = \{\tau : \tau \text{ is a Tychonoff topology on } Z \text{ and } \tau | A = \tau(Z) | A \text{ for any countable } A \subset Z\}$. The family $\mathcal{M}$ is nonempty because $\tau(Z) \in \mathcal{M}$. Therefore the family $\bigcup \mathcal{M}$ can be considered a subbase for a topology $\mu_Z$ on the set $Z$. It is easy to see that $\mu_Z \in \mathcal{M}$ and, if $Z \simeq Y$ then the spaces $(Z, \mu_Z)$ and $(Y, \mu_Y)$ are homeomorphic.

Suppose that $X \overset{t}{\sim} Y$ and hence $C_p(X) \simeq C_p(Y)$; since the restriction map $\pi : C_p(\upsilon X) \to C_p(X)$ is a condensation, we can identify the sets $C(X)$ and $C(\upsilon X)$ and consider that the topology of $C_p(\upsilon X)$ is given on the set $C(X)$. Reformulating TFS-437 we can see that the topology of $C_p(\upsilon X)$ on $C(X)$ coincides with $\mu_{C_p(X)}$. Analogously, the topology of $C_p(\upsilon Y)$ on $C(Y)$ coincides with $\mu_{C_p(Y)}$. As a consequence, $C_p(\upsilon X) \simeq (C(X), \mu_{C_p(X)}) \simeq (C(Y), \mu_{C_p(Y)}) \simeq C_p(\upsilon Y)$ which shows that the spaces $\upsilon X$ and $\upsilon Y$ are $t$-equivalent.

Next observe that the spaces $X = A(\omega)$ and $Y = A(\omega) \oplus \omega$ are $t$-equivalent by Problem 017 and CFS-105. We claim that the spaces $\beta X = X$ and $\beta Y$ are not $t$-equivalent. Indeed, it easily follows from Fact 2 of S.451 that the space $\beta Y$ contains the space $\beta \omega$ and therefore $nw(\beta Y) \geq nw(\beta \omega) = w(\beta \omega) > \omega$ (see TFS-368). However, $nw(\beta X) = nw(A(\omega)) = \omega$, so $nw(\beta X) \neq nw(\beta Y)$ and hence the spaces $\beta X$ and $\beta Y$ cannot be $t$-equivalent by Problem 001.

**V.029.** *Give an example of spaces $X$ and $Y$ such that $\upsilon X \simeq \upsilon Y$ (and hence $\upsilon X \overset{t}{\sim} \upsilon Y$) while $X$ is not $t$-equivalent to $Y$.*

**Solution.** Let $X = \omega_1$ and $Y = \omega_1 + 1$; then $Y = \upsilon Y \simeq \upsilon X$ (see TFS-314, Fact 3 of S.309 and TFS-417). However, $X$ is not $t$-equivalent to $Y$ because $C_p(X)$ is Lindelöf and $C_p(Y)$ is not (see TFS-316 and TFS-320).

**V.030.** *Prove that $\kappa$-monolithity and $\kappa$-stability are $t$-invariant for any infinite cardinal $\kappa$.*

**Solution.** If $X \overset{t}{\sim} Y$ and $X$ is $\kappa$-monolithic then $C_p(X)$ is $\kappa$-stable (see SFFS-152) and hence so is $C_p(Y)$. Applying SFFS-152 again we conclude that $Y$ is $\kappa$-monolithic. This proves that $\kappa$-monolithity is $t$-invariant.

Now, if $X \overset{t}{\sim} Y$ and $X$ is $\kappa$-stable then $C_p(X)$ is $\kappa$-monolithic (see SFFS-154) and hence $C_p(Y)$ is $\kappa$-monolithic. Applying SFFS-154 again we conclude that $Y$ is $\kappa$-stable. This proves that $\kappa$-stability is also $t$-invariant.

**V.031.** *Suppose that $X \overset{t}{\sim} Y$. Prove that $X$ is functionally perfect if and only if so is $Y$.*

**Solution.** If $X$ is functionally perfect, then $C_p(X)$ has a dense $\sigma$-compact subspace (see CFS-301); therefore $C_p(Y)$ also has a dense $\sigma$-compact subspace, so we can apply CFS-301 once more to conclude that $Y$ is functionally perfect. Analogously, if $Y$ is functionally perfect then so is $X$ and hence the space $X$ is functionally perfect if and only if $Y$ is functionally perfect.

**V.032.** *Give an example of spaces $X$ and $Y$ such that $X$ is functionally perfect, $C_p(Y)$ embeds into $C_p(X)$ while $Y$ is not functionally perfect.*

**Solution.** Let $X$ be a discrete space of cardinality $\omega_1$ and $Y = \omega_1 + 1$. Then $Y$ is not an Eberlein compact because $t(Y) > \omega$. Therefore $Y$ is not functionally perfect. However, $X$ is functionally perfect (see CFS-316) and $X$ maps continuously onto $Y$; this implies that $C_p(Y)$ embeds in $C_p(X)$ (see TFS-163).

**V.033.** *Suppose that compact spaces $X$ and $Y$ are $t$-equivalent. Prove that $X$ is Eberlein (Corson or Gul'ko) compact if and only if so is $Y$.*

**Solution.** If $X$ is Eberlein compact then $X$ is functionally perfect, so $Y$ is also functionally perfect by Problem 031 and hence $Y$ is Eberlein compact. Analogously, if $Y$ is Eberlein compact then so is $X$.

Now, if $X$ is Corson compact then $C_p(X)$ is primarily Lindelöf (see CFS-150), so $C_p(Y)$ is also primarily Lindelöf and hence $Y$ is Corson compact. Analogously, if $Y$ is Corson compact then so is $X$.

Finally, if $X$ is Gul'ko compact then $C_p(X)$ is a Lindelöf $\Sigma$-space so $C_p(Y)$ is also a Lindelöf $\Sigma$-space and hence $Y$ is Gul'ko compact. Analogously, if $Y$ is Gul'ko compact then so is $X$.

**V.034.** *Suppose that $F \subset C_p(X, \mathbb{I})$ is a D-separating set (and hence $\mathbf{0}_X \in F$). For each $x \in X$, let $e_x(f) = f(x)$ for any $f \in F$. Prove that $\widetilde{X} = \{e_x : x \in X\}$ is a closed subset of the space $Z_F(X) = \{\varphi : F \to \mathbb{I} : \varphi(\mathbf{0}_X) = 0 \text{ and } \varphi(V) \subset [-\frac{1}{2}, \frac{1}{2}]$ for some $V \in \tau(\mathbf{0}_X, F)\}$, and the map $x \to e_x$ is a homeomorphism between $X$ and $\widetilde{X}$. In other words, $X$ is canonically homeomorphic to a closed subset of $Z_F(X)$.*

**Solution.** For an arbitrary space $Z$ and $p \in Z$ let $D(Z, p) = \{f \in \mathbb{I}^Z : f(p) = 0$ and there is $U \in \tau(p, Z)$ such that $f(U) \subset [-\frac{1}{2}, \frac{1}{2}]\}$. We consider that $D(Z, p)$ is a space with the topology induced from $\mathbb{I}^Z$.

Given any point $x \in X$ let $e(x) = e_x$; then $e(x) \in C_p(F, \mathbb{I})$ and the map $e : X \to C_p(F, \mathbb{I})$ is continuous (see TFS-166). Since $e(x)(\mathbf{0}_X) = \mathbf{0}_X(x) = 0$ for any $x \in X$ and $e(x)$ is continuous at $\mathbf{0}_X$, we have $e(x) \in D(F, \mathbf{0}_X)$ for any $x \in X$, i.e., $e(X) \subset D(F, \mathbf{0}_X)$. It is an evident consequence of the fact that $F$ is $D$-separating that $F$ separates the points and closed sets, i.e., for any $x \in X$ and any closed $P \subset X$ with $x \notin P$, there is $f \in F$ such that $f(x) \notin \overline{f(P)}$. Therefore $e$ is an embedding by TFS-166 and we only must prove that $\widetilde{X} = e(X)$ is closed in $Z_F(X) = D(F, \mathbf{0}_X)$.

Take any element $\varphi \in D(F, \mathbf{0}_X) \backslash e(X)$. There exists $O \in \tau(\mathbf{0}_X, F)$ such that $\varphi(O) \subset [-\frac{1}{2}, \frac{1}{2}]$. By definition of the pointwise convergence topology there is a finite $K \subset X$ and $\varepsilon > 0$ such that $\mathbf{0}_X \in W = \{f \in F : f(K) \subset (-\varepsilon, \varepsilon)\} \subset O$ and hence $\varphi(W) \subset [-\frac{1}{2}, \frac{1}{2}]$. Since $\varphi \notin e(K)$, there is $U \in \tau(K, X)$ such that $\varphi \notin \overline{e(U)}$. The family $F$ being $D$-separating, there is $g \in F$ such that $g(K) \subset (-\varepsilon, \varepsilon)$ and $g(X \backslash U) \subset [\frac{3}{4}, 1]$ and, in particular, $g \in W$. This implies $e(x)(g) = g(x) \in [\frac{3}{4}, 1]$ for every point $x \in X \backslash U$ while we have $\varphi(g) \in \varphi(W) \subset [-\frac{1}{2}, \frac{1}{2}]$. Consequently, the set $G = \{\delta \in D(F, \mathbf{0}_X) : \delta(g) < \frac{3}{4}\}$ is an open neighborhood of $\varphi$ in $D(F, \mathbf{0}_X)$

such that $G \cap e(X \setminus U) = \emptyset$ and therefore $\varphi \notin \overline{e(X \setminus U)}$. It is easy to see that this implies $\varphi \notin \overline{e(U)} \cup \overline{e(X \setminus U)} = \overline{e(X)}$. The function $\varphi \in D(F, \mathbf{0}_X) \setminus e(X)$ was chosen arbitrarily, so $\widetilde{X} = e(X)$ is closed in the space $D(F, \mathbf{0}_X) = Z_F(X)$.

**V.035.** *Knowing that* $\mathbf{0}_X \in F \subset C_p(X, \mathbb{I})$ *and* $\mathbf{0}_Y \in G \subset C_p(Y, \mathbb{I})$, *suppose that there is an embedding* $i : G \to F$ *with* $i(\mathbf{0}_Y) = \mathbf{0}_X$. *Prove that* $Z_F(X)$ *maps continuously onto* $Z_G(Y)$.

**Solution.** For an arbitrary space $Z$ and $p \in Z$ let $D(Z, p) = \{f \in \mathbb{I}^Z : f(p) = 0$ and there is $U \in \tau(p, Z)$ such that $f(U) \subset [-\frac{1}{2}, \frac{1}{2}]\}$. We consider that $D(Z, p)$ is a space with the topology induced from $\mathbb{I}^Z$.

Let $H = i(G)$; for any $f \in \mathbb{I}^H$ the function $i^*(f) = f \circ i$ belongs to $\mathbb{I}^G$ and it is easy to see, using TFS-163, that the map $i^* : \mathbb{I}^H \to \mathbb{I}^G$ is a homeomorphism. Besides, it follows from $i(\mathbf{0}_Y) = \mathbf{0}_X$ that $i^*(D(H, \mathbf{0}_X)) = D(G, \mathbf{0}_Y)$. Since $\mathbf{0}_X \in H \subset F$, the restriction map $\pi : \mathbb{I}^F \to \mathbb{I}^H$ maps $D(F, \mathbf{0}_X)$ onto $D(H, \mathbf{0}_X)$ by Fact 8 of T.250. Therefore $i^* \circ \pi$ maps $Z_F(X) = D(F, \mathbf{0}_X)$ continuously onto $Z_G(Y) = D(G, \mathbf{0}_Y)$.

**V.036.** *Given a space* $X$ *prove that if* $\mathbf{0}_X \in F \subset C_p(X, \mathbb{I})$ *then* $Z_F(X)$ *belongs to the class* $\mathcal{K}(X)$.

**Solution.** This was proved in Fact 10 of T.250.

**V.037.** *Let* $G$ *be a* $D$-separating subspace of $C_p(Y)$. *Prove that, if* $G$ *embeds into* $C_p(X)$ *then* $Y \in \mathcal{K}(X)$.

**Solution.** It is easy to see that there exists an embedding $e : C_p(X) \to C_p(X, \mathbb{I})$ such that $e(\mathbf{0}_X) = \mathbf{0}_X$. The space $C_p(X)$ being homogeneous (i.e., for any functions $f, g \in C_p(X)$ there is a homeomorphism $\varphi : C_p(X) \to C_p(X)$ such that $\varphi(f) = g$ (see TFS-079)), there is an embedding $w : G \to C_p(X)$ such that $w(\mathbf{0}_Y) = \mathbf{0}_X$. Therefore $i = e \circ w$ embeds $G$ in $C_p(X, \mathbb{I})$ in such a way that $i(\mathbf{0}_Y) = \mathbf{0}_X$. This shows that, for the set $F = C_p(X, \mathbb{I})$, the space $Z_G(Y)$ is a continuous image of $Z_F(X)$ by Problem 035. The space $Z_F(X)$ belongs to the class $\mathcal{K}(X)$ by Problem 036 and hence $Z_G(Y)$ also belongs to $\mathcal{K}(X)$. The space $Y$ embeds in $Z_G(Y)$ as a closed subspace (see Problem 034), so $Y \in \mathcal{K}(X)$.

**V.038.** *Given arbitrary spaces* $X, Y$ *and a subspace* $Z \subset Y$ *suppose that the space* $C_p(Z|Y) = \{f \in C_p(Z) : f = g|Z$ *for some* $g \in C_p(Y)\}$ *embeds in* $C_p(X)$. *Prove that* $Z \in \mathcal{K}(X)$.

**Solution.** Suppose that $A \subset Z$ is finite, $F \subset Z$ is closed in $Z$ and $A \cap F = \emptyset$. Then $G = \mathrm{cl}_Y(F)$ is closed in $Y$ and $A \cap G = \emptyset$. By the Tychonoff property of $Y$, for any $a \in A$, there is a function $f_a \in C(Y, [0, 1])$ such that $f_a(a) = 1$ and $f_a(G) \subset \{0\}$. The function $g = \prod_{a \in A}(1 - f_a) \in C(Y, [0, 1])$ is equal to zero on $A$ and $g(G) \subset \{1\}$. Consequently, $h = g|Z \in C_p(Z|Y)$ while $h(F) \subset \{1\}$ and $h(A) \subset \{0\}$. This proves that the set $E = \{f|Z : f \in C_p(Y, \mathbb{I})\}$ is $D$-separating. Since $C_p(Z|Y)$ embeds in $C_p(X)$, the space $E \subset C_p(Z|Y)$ also embeds in $C_p(X)$, so we can apply Problem 037 to conclude that $Z$ belongs to the class $\mathcal{K}(X)$.

**V.039.** *Let* $X$ *be a* $\sigma$-*compact space. Prove that any space* $Y \in \mathcal{K}(X)$ *is also* $\sigma$-*compact.*

**Solution.** It is easy to see that the class $\mathcal{SK}$ of $\sigma$-compact spaces is complete and contains all compact spaces. Therefore $X \in \mathcal{SK}$ implies $\mathcal{K}(X) \subset \mathcal{SK}$ because $\mathcal{K}(X)$ is the minimal complete class which contains $X$ and all compact spaces. Thus every $Y \in \mathcal{K}(X)$ belongs to $\mathcal{SK}$, i.e., $Y$ is $\sigma$-compact.

**V.040.** *Let* $X$ *be a Lindelöf* $\Sigma$-*space. Prove that any* $Y \in \mathcal{K}(X)$ *is also a Lindelöf* $\Sigma$-*space.*

**Solution.** It is easy to see that the class $\mathcal{LS}$ of Lindelöf $\Sigma$-spaces is complete and contains all compact spaces. Therefore $X \in \mathcal{LS}$ implies $\mathcal{K}(X) \subset \mathcal{LS}$ because $\mathcal{K}(X)$ is the minimal complete class which contains $X$ and all compact spaces. Thus every $Y \in \mathcal{K}(X)$ belongs to $\mathcal{LS}$, i.e., $Y$ is a Lindelöf $\Sigma$-space.

**V.041.** *Let* $X$ *be a* $K$-*analytic space. Prove that any space* $Y \in \mathcal{K}(X)$ *is also* $K$-*analytic.*

**Solution.** It is easy to see that the class $\mathcal{KA}$ of $K$-analytic spaces is complete and contains all compact spaces. Therefore $X \in \mathcal{KA}$ implies $\mathcal{K}(X) \subset \mathcal{KA}$ because $\mathcal{K}(X)$ is the minimal complete class which contains $X$ and all compact spaces. Thus every $Y \in \mathcal{K}(X)$ belongs to $\mathcal{KA}$, i.e., $Y$ is a $K$-analytic space.

**V.042.** *Prove that* $ext^*(Y) \leq ext^*(X)$ *for any* $Y \in \mathcal{K}(X)$.

**Solution.** Assume that $ext^*(X) = \kappa$ and consider the class $\mathcal{E}$ of spaces $Z$ such that $ext^*(Z) \leq \kappa$. We leave to the reader a simple verification of the fact that $\mathcal{E}$ is a complete class and all compact spaces are in $\mathcal{E}$. Therefore $\mathcal{K}(X) \subset \mathcal{E}$ by minimality of the class $\mathcal{K}(X)$, so any $Y \in \mathcal{K}(X)$ belongs to $\mathcal{E}$, i.e., $ext^*(Y) \leq \kappa = ext^*(X)$.

**V.043.** *Suppose that* $C_p(Y)$ *embeds into* $C_p(X)$. *Prove that*

*(1) if* $X$ *is* $\sigma$-*compact then* $Y$ *is* $\sigma$-*compact;*
*(2) if* $X$ *is Lindelöf* $\Sigma$-*space then* $Y$ *is Lindelöf* $\Sigma$.
*(3) if* $X$ *is* $K$-*analytic then* $Y$ *is also* $K$-*analytic.*
*(4)* $ext^*(Y) \leq ext^*(X)$.

*As a consequence, if* $X \overset{t}{\sim} Y$ *then* $ext^*(X) = ext^*(Y)$ *and, for any property* $\mathcal{P} \in \{\sigma$-*compactness, Lindelöf* $\Sigma$-*property, K-analyticity*$\}$, *we have* $X \vdash \mathcal{P}$ *if and only if* $Y \vdash \mathcal{P}$.

**Solution.** It follows from $C_p(Y) \hookrightarrow C_p(X)$ that $C_p(Y, \mathbb{I}) \hookrightarrow C_p(X)$; since $C_p(Y, \mathbb{I})$ is a $D$-separating subset of $C_p(Y)$, we can apply Problem 037 to see that $Y \in \mathcal{K}(X)$.

Now, the statement (1) follows from Problem 039, the property (2) is an immediate consequence of Problem 040, the property (3) is implied by Problem 041 and (4) can be deduced from Problem 042.

**V.044.** *Suppose that $X$ is an analytic space and $C_p(Y)$ embeds into $C_p(X)$. Prove that $Y$ is also analytic. In particular, analyticity is $t$-invariant, i.e., if $X \overset{t}{\sim} Y$ then $X$ is analytic if and only if so is $Y$.*

**Solution.** The space $X$ being $K$-analytic (see SFFS-346), it follows from Problem 043 that $Y$ is also $K$-analytic. Besides, $nw(X) \leq \omega$ because $X$ is analytic and network weight is not increased by continuous images. Therefore

$$nw(Y) = nw(C_p(Y)) \leq nw(C_p(X)) = nw(X) \leq \omega$$

and hence we can apply SFFS-346 again to conclude that $Y$ is also analytic.

**V.045.** *Suppose that $X \overset{t}{\sim} Y$. Prove that $X$ is $\sigma$-bounded if and only if so is $Y$.*

**Solution.** The space $X$ is $\sigma$-bounded if and only if $\upsilon X$ is $\sigma$-compact (see TFS-416); since also $\upsilon X \overset{t}{\sim} \upsilon Y$ (see Problem 028), the space $\upsilon Y$ is $\sigma$-compact if and only if the space $\upsilon X$ is $\sigma$-compact (see Problem 043). Finally, $\upsilon Y$ is $\sigma$-compact if and only if $Y$ is $\sigma$-bounded (here we used TFS-416 again). This shows that $X$ is $\sigma$-bounded if and only if so is $Y$.

**V.046.** *Given a zero-dimensional space $Y$, suppose that $C_p(Y, \mathbb{D})$ embeds in $C_p(X)$. Prove that, for any property $\mathcal{P} \in \{\sigma\text{-compactness, Lindelöf } \Sigma\text{-property,}$ analyticity, $K$-analyticity\}, we have $Y \vdash \mathcal{P}$ whenever $X \vdash \mathcal{P}$.*

**Solution.** We leave it to the reader to verify that the set $C_p(Z, \mathbb{D})$ is $D$-separating in $C_p(Z)$ for any zero-dimensional space $Z$. Thus it follows from $C_p(Y, \mathbb{D}) \hookrightarrow C_p(X)$ that $Y \in \mathcal{K}(X)$ (see Problem 037). Now, if $X$ is $\sigma$-compact then so is $Y$ by Problem 039. If $X$ is Lindelöf $\Sigma$ then $Y$ is Lindelöf $\Sigma$ by Problem 040. If $X$ is $K$-analytic then $Y$ is also $K$-analytic by Problem 041.

If the space $X$ is analytic then $Y$ is $K$-analytic by our above observations and $nw(C_p(Y, \mathbb{D})) \leq nw(C_p(X)) = \omega$. Since $C_p(Y, \mathbb{D})$ is $D$-separating, it separates the points from closed sets in $Y$, so $Y$ embeds in $C_p(C_p(Y, \mathbb{D}))$ by TFS-166. Therefore $nw(Y) \leq nw(C_p(C_p(Y, \mathbb{D}))) = nw(C_p(Y, \mathbb{D})) \leq \omega$ which shows that we can apply SFFS-346 to conclude that $Y$ is analytic.

**V.047.** *Let $X$ be a zero-dimensional space. Prove that $l^*(X) = t(C_p(X, \mathbb{D}))$.*

**Solution.** Denote by Fin($X$) the family of all finite subsets of the space $X$ and observe that $t(C_p(X, \mathbb{D})) \leq t(C_p(X)) = l^*(X)$ (see TFS-149), so we only have to prove that $l^*(X) \leq \kappa = t(C_p(X, \mathbb{D}))$. To do so fix an open $\omega$-cover $\mathcal{U}$ of the space $X$. For any finite $F \subset X$ choose a set $U_F \in \mathcal{U}$ with $F \subset U_F$; since $X$ is zero-dimensional, we can find a clopen set $O_F$ such that $F \subset O_F \subset U_F$. Let $h_F$ be the characteristic function of the set $O_F$, i.e., $h_F(x) = 1$ for all $x \in O_F$ and $h_F(x) = 0$ if $x \in X \setminus O_F$.

The function $u \in C_p(X, \mathbb{D})$ with $u(x) = 1$ for all $x \in X$ belongs to the closure of the set $\{h_F : F \in \text{Fin}(X)\} \subset C_p(X, \mathbb{D})$, so there is a family $\mathcal{F} \subset \text{Fin}(X)$ such that $|\mathcal{F}| \leq \kappa$ and $u \in \overline{\{h_F : f \in \mathcal{F}\}}$; let $\mathcal{U}' = \{U_F : F \in \mathcal{F}\}$. Given a finite $K \subset X$ the set $G = \{f \in C_p(X, \mathbb{D}) : f(K) \subset \{1\}\}$ is an open neighborhood of $u$ in $C_p(X, \mathbb{D})$, so there is $F \in \mathcal{F}$ with $h_F \in G$; this, evidently, implies $K \subset O_F \subset U_F \in \mathcal{U}'$, so we proved that $\mathcal{U}'$ is an $\omega$-cover of $X$ of cardinality $\leq \kappa$. Thus every $\omega$-cover of $X$ has an $\omega$-subcover of cardinality at most $\kappa$, so we can apply TFS-148 to conclude that $l^*(X) \leq \kappa$ and hence $l^*(X) = t(C_p(X, \mathbb{D}))$.

**V.048.** *Let $X$ and $Y$ be zero-dimensional spaces with $C_p(X, \mathbb{D}) \simeq C_p(Y, \mathbb{D})$. Prove that $X$ is pseudocompact if and only if so is $Y$. Deduce from this fact that $X$ is compact if and only if so is $Y$.*

**Solution.** As usual, given a space $Z$ and $A \subset Z$, the characteristic function $\chi_A$ of the set $A$ is defined by $\chi_A(x) = 1$ for all $x \in A$ and $\chi_A(x) = 0$ for all $x \in Z \backslash A$.

Suppose that $X$ is pseudocompact and take a countable set $A \subset C_p(X, \mathbb{D})$. Since the space $C_p(X)$ is $\omega$-monolithic (see Fact 9 of S.351), we have $nw(\overline{A}) \leq \omega$. For any $x \in X$ let $\varphi(x)(f) = f(x)$ for any $f \in B = \overline{A}$. Then $\varphi : X \to C_p(B, \mathbb{D})$ is a continuous map (see TFS-166); let $X' = \varphi(X)$. We have

$$nw(X') \leq nw(C_p(B, \mathbb{D})) \leq nw(C_p(B)) = nw(B) = \omega;$$

this, together with pseudocompactness of $X'$, shows that the space $X'$ is compact and metrizable.

For any $u \in B$ let $\mu(u)(f) = f(u)$ for any $f \in X'$; since $X'$ generates the topology of $B$ (see TFS-166), the map $\mu$ embeds $B$ in $C_p(X', \mathbb{D})$. Apply Fact 1 of U.077 to conclude that $|B| \leq |C_p(X', \mathbb{D})| = \omega$ and therefore the closure of every countable subset of $C_p(X, \mathbb{D})$ is countable. Since $C_p(Y, \mathbb{D}) \simeq C_p(X, \mathbb{D})$, we conclude that

(1) the closure of every countable subset of $C_p(Y, \mathbb{D})$ is countable.

Assume toward a contradiction that $Y$ is not pseudocompact and fix a discrete family $\{U_n : n \in \omega\} \subset \tau^*(Y)$; pick a point $x_n \in U_n$, a clopen set $O_n$ such that $x_n \in O_n \subset U_n$ and let $f_n$ be the characteristic function of $O_n$ for any $n \in \omega$. Let $D = \{x_n : n \in \omega\}$; if $g \in \mathbb{D}^D$ then the function $e(g) = \sum \{f_n : n \in g^{-1}(1)\}$ belongs to $C_p(Y, \mathbb{D})$. Apply Fact 5 of T.132 to see that the map $e : \mathbb{D}^D \to C_p(Y, \mathbb{D})$ is continuous; it is straightforward that $e$ is injective, so $K = e(\mathbb{D}^D)$ is homeomorphic to $\mathbb{D}^D \simeq \mathbb{D}^\omega$.

The space $K$ being second countable, we can choose a countable dense $A \subset K$. Since the set $K = \overline{A}$ is uncountable, we obtain a contradiction with (1) which shows that $Y$ has to be pseudocompact. This shows that pseudocompactness of $X$ implies pseudocompactness of $Y$; changing the roles of $X$ and $Y$ in the above proof we can derive pseudocompactness of $X$ from pseudocompactness of $Y$. Thus $X$ is pseudocompact if and only if so is $Y$.

Finally, assume that the space $X$ is compact; then $Y$ is pseudocompact by the above result. Furthermore, $t(C_p(X, \mathbb{D})) = l^*(X) = \omega$ (see Problem 047) and hence $l^*(Y) = t(C_p(Y, \mathbb{D})) = \omega$. This shows that $Y$ is compact being pseudocompact and Lindelöf. Therefore compactness of $X$ implies compactness of $Y$. Analogously, compactness of $Y$ implies compactness of $X$, so $X$ is compact if and only if so is $Y$.

**V.049.** *Prove that there exist zero-dimensional spaces $X$ and $Y$ such that $C_p(X)$ is homeomorphic to $C_p(Y)$ and $C_p(X, \mathbb{D})$ is not homeomorphic to $C_p(Y, \mathbb{D})$.*

**Solution.** Let $X = \omega + 1$ and $Y = (\omega + 1) \oplus \omega$. It is trivial that $X$ and $Y$ are zero-dimensional; it follows from CFS-105 and Problem 017 that $C_p(X)$ is homeomorphic to $C_p(Y)$. However, $C_p(X, \mathbb{D})$ is not homeomorphic to $C_p(Y, \mathbb{D})$ because $X$ is compact and $Y$ is not even pseudocompact (see Problem 048).

**V.050.** *Prove that $\chi(C_p(X, \mathbb{D})) = w(C_p(X, \mathbb{D})) = |X|$ for any zero-dimensional space $X$.*

**Solution.** We have $\chi(C_p(X, \mathbb{D})) \leq w(C_p(X, \mathbb{D})) \leq w(C_p(X)) = |X|$ (see TFS-169), so it suffices to show that $|X| \leq \kappa = \chi(C_p(X, \mathbb{D}))$. To do this let $u \in C_p(X, \mathbb{D})$ be the function which is identically zero on $X$ and fix a local base $\mathcal{B}$ at the point $u$ in the space $C_p(X, \mathbb{D})$ such that $|\mathcal{B}| \leq \kappa$. Taking smaller neighborhoods of $u$ if necessary we can assume that all elements of $\mathcal{B}$ belong to the standard base of the space $C_p(X, \mathbb{D})$, i.e., for any $B \in \mathcal{B}$ there is a finite set $K_B \subset X$ such that $B = \{f \in C_p(X, \mathbb{D}) : f(K_B) \subset \{0\}\}$.

The set $Y = \bigcup\{K_B : B \in \mathcal{B}\}$ has cardinality at most $\kappa$; if $Y \neq X$ then pick a point $x \in X \setminus Y$. The set $U = \{f \in C_p(X, \mathbb{D}) : f(x) = 0\}$ is an open neighborhood of $u$ in $C_p(X, \mathbb{D})$. Therefore there is $B \in \mathcal{B}$ with $B \subset U$; however, $x \notin K_B$, so zero-dimensionality of $X$ shows that there is a clopen set $G \subset X$ such that $x \in G \subset X \setminus K_B$. Let $f(y) = 1$ for all $y \in G$ and $f(y) = 0$ whenever $y \in X \setminus G$. It is straightforward that $f \in B \setminus U$; this contradiction shows that $Y = X$ and hence $|X| \leq \kappa$. We already saw that this implies $\chi(C_p(X, \mathbb{D})) = w(C_p(X, \mathbb{D})) = |X|$.

**V.051.** *Prove that a zero-dimensional compact space $X$ is scattered if and only if $C_p(X, \mathbb{D})$ is Fréchet–Urysohn.*

**Solution.** Recall that an $\omega$-cover of a space $Z$ is a family $\mathcal{U} \subset \exp(Z)$ such that, for any finite $A \subset Z$, there is $U \in \mathcal{U}$ with $A \subset U$. If $B_n \in \exp(Z)$ for each $n \in \omega$ then $B_n \to Z$ says that, for any $z \in Z$, there is $m \in \omega$ such that $z \in B_n$ for all $n \geq m$.

If the space $X$ is scattered then $C_p(X)$ is Fréchet–Urysohn (see SFFS-134) and hence $C_p(X, \mathbb{D}) \subset C_p(X)$ is also Fréchet–Urysohn; this proves necessity.

To prove sufficiency denote by $\mathrm{Fin}(X)$ the family of all finite subsets of $X$ and assume that $C_p(X, \mathbb{D})$ is a Fréchet–Urysohn space; take an open $\omega$-cover $\mathcal{U}$ of the space $X$. For any finite $K \subset X$ fix a set $U_K \in \mathcal{U}$ with $K \subset U_K$ and choose, using zero-dimensionality of $X$, a clopen set $O_K$ such that $K \subset O_K \subset U_K$; let $h_K(x) = 1$ for all $x \in O_K$ and $h_K(x) = 0$ if $x \in X \setminus O_K$. We will also need the function $u : X \to \mathbb{R}$ such that $u(x) = 1$ for all $x \in X$; it is straightforward that $u \in C_p(X, \mathbb{D})$ belongs to the closure of the set $P = \{h_K : K \in \mathrm{Fin}(X)\} \subset C_p(X, \mathbb{D})$. The space $C_p(X, \mathbb{D})$ being Fréchet–Urysohn, we can choose a sequence $\{K_n : n \in \omega\}$ such that the sequence $\{h_{K_n} : n \in \omega\}$ converges to $u$; let $U_n = U_{K_n}$ for each $n \in \omega$.

Given a point $x \in X$ the set $G = \{f \in C_p(X, \mathbb{D}) : f(x) = 1\}$ is an open neighborhood of $u$ in $C_p(X, \mathbb{D})$, so there is $m \in \omega$ such that $h_{K_n} \in G$ for all $n \geq m$. This implies $x \in O_{K_n} \subset U_n$ for all $n \geq m$ and hence $U_n \to X$. It turns out that any open $\omega$-cover $\mathcal{U}$ of $X$ has a subfamily $\{U_n : n \in \omega\} \subset \mathcal{U}$ with $U_n \to X$. This implies that $C_p(X)$ is a Fréchet–Urysohn space (see TFS-144), so we can apply SFFS-134 once more to conclude that $X$ is scattered and complete our solution.

**V.052.** *Suppose that $X$ is not $\sigma$-compact and $w(X) \le \kappa$. Prove that there is a subspace $Z \subset C_p(X)$ such that $|Z| \le \kappa$ and $Z$ is not embeddable into $C_p(Y)$ for any $\sigma$-compact $Y$. In particular, there is a countable subspace of $C_p(\mathbb{P})$ which cannot be embedded into $C_p(Y)$ for any $\sigma$-compact $Y$.*

**Solution.** Choose a base $\mathcal{B}$ of the space $X$ such that $|\mathcal{B}| \le \kappa$ and call a pair $p = (U, V) \in \mathcal{B} \times \mathcal{B}$ *adequate* if $U \subset V$ and there is a function $f_p \in C(X, [0, 1])$ for which $f_p(U) \subset \{0\}$ and $f_p(X \backslash V) \subset \{1\}$; let $W_p = V$ and denote by $P$ the set of all adequate pairs of elements of $\mathcal{B}$.

It is evident that the family $\mathcal{P} = \{Q : Q \subset P$ is finite and the collection $\{W_q : q \in Q\}$ is disjoint$\}$ has cardinality at most $\kappa$; let $h_Q = \prod\{f_q : q \in Q\}$ for any $Q \in \mathcal{P}$. The set $Z = \{h_Q : Q \in \mathcal{P}\} \cup \{\mathbf{0}_X\}$ also has cardinality $\le \kappa$; we claim that $Z$ is $D$-separating in $C_p(X)$.

To prove this take a finite $K \subset X$ and a closed set $F \subset X \backslash K$. It is easy to find a finite disjoint family $\mathcal{B}' = \{B_x : x \in K\} \subset \mathcal{B}$ such that $x \in B_x \subset X \backslash F$ for every $x \in K$. Using the Tychonoff property of $X$ choose, for any $x \in K$, a function $g_x \in C(X, [0, 1])$ for which $g_x(x) = 0$ and $g_x(X \backslash B_x) \subset \{1\}$; pick a set $C_x \in \mathcal{B}$ such that $x \in C_x \subset g_x^{-1}([0, \frac{1}{2}))$. There exists a continuous function $\varphi : [0, 1] \to [0, 1]$ such that $\varphi([0, \frac{1}{2}]) = \{0\}$ and $\varphi(1) = 1$. Consequently, $(\varphi \circ g_x)(C_x) \subset \{0\}$ and $(\varphi \circ g_x)(X \backslash B_x) \subset \{1\}$ which shows that every $p_x = (C_x, B_x)$ is an adequate pair and hence the set $Q = \{p_x : x \in K\}$ belongs to $\mathcal{P}$.

Since $f_{p_x}(X \backslash B_x) \subset \{1\}$ and $F \subset X \backslash B_x$, we have $f_{p_x}(F) \subset \{1\}$ for any $x \in K$; as a consequence, $h_Q(F) \subset \{1\}$. Furthermore, $f_{p_x}(x) = 0$ for any $x \in K$, so $h_Q(K) \subset \{0\}$; recalling that $h_Q \in Z$ we convince ourselves that $Z$ is indeed, a $D$-separating family. If $Z$ embeds in the space $C_p(Y)$ for some $\sigma$-compact $Y$ then $X \in \mathcal{K}(Y)$ (see Problem 037) and hence $X$ is $\sigma$-compact by Problem 039 which is a contradiction. Thus $Z \subset C_p(X)$ is a set of cardinality at most $\kappa$ which cannot be embedded in $C_p(Y)$ for any $\sigma$-compact space $Y$.

**V.053.** *Prove that, for any $X$, the space $C_p(X)$ embeds in $C_p(\mathbb{P})$ if and only if $C_p(X)$ is homeomorphic to a linear subspace of $C_p(\mathbb{P})$.*

**Solution.** Since sufficiency is evident, assume that $C_p(X) \hookrightarrow C_p(\mathbb{P})$. Then $X$ is analytic by Problem 044, so let $\varphi : \mathbb{P} \to X$ be a continuous onto map. The dual map $\varphi^* : C_p(X) \to C_p(\mathbb{P})$ is an embedding (see TFS-163) and it is straightforward that $\varphi^*(C_p(X))$ is a linear subspace of $C_p(\mathbb{P})$.

**V.054.** *Suppose that a space $X$ is compact and there exists a homeomorphism $\varphi : \mathbb{R}^X \to \mathbb{R}^Y$ such that $\varphi(C_p(X)) \supset C_p(Y)$. Prove that $Y$ is compact.*

**Solution.** It follows from $\varphi(C_p(X)) \supset C_p(Y)$ that $C_p(Y)$ is embeddable in the space $C_p(X)$, so $Y$ is $\sigma$-compact by Problem 043. The space $X$ being compact, there exists a $\sigma$-compact $P \subset \mathbb{R}^X$ such that $C_p(X) \subset P$ (see CFS-204); then $P' = \varphi(P) \subset \mathbb{R}^Y$ is a $\sigma$-compact space such that $C_p(Y) \subset P'$. Apply CFS-204 again to conclude that $Y$ is pseudocompact and hence compact.

**V.055.** *Suppose that* $p, q : X \to \exp(Y)$ *are finite-valued mappings such that* $q$ *is lower semicontinuous with respect to* $p$. *Prove that, for any nonempty set* $A \subset X$, *the map* $q|A : A \to \exp(\bigcup q(A))$ *is lower semicontinuous with respect to* $p|A :$ $A \to \exp(\bigcup q(A))$.

**Solution.** Let $Z = \bigcup q(A)$ and denote the mappings $p|A : A \to \exp(Z)$ and $q|A : A \to \exp(Z)$ by $p_A$ and $q_A$ respectively. Given a set $U \in \tau(Z)$ there is $U' \in \tau(Y)$ such that $U = U' \cap Z$; observe that, for any $a \in A$, we have $q(a) \subset Z$, so $q(a) \cap U' \neq \emptyset$ implies $q(a) \cap U \neq \emptyset$. The set $q_l^{-1}(U')$ being a neighborhood of the set $p_l^{-1}(U')$ in $X$, for any $a \in A$ with $p(a) \cap U' \neq \emptyset$, there is $U_a \in \tau(a, X)$ such that $q(x) \cap U' \neq \emptyset$ for any $x \in U_a$.

Now, if we are given a point $a \in (p_A)_l^{-1}(U)$ then $p_A(a) \cap U = p(a) \cap U \neq \emptyset$ and hence $p(a) \cap U' \neq \emptyset$. The set $V_a = U_a \cap A$ is an open neighborhood of $a$ in $A$; for any point $x \in V_a$ we have $q(x) \cap U' \neq \emptyset$ and hence $q_A(x) \cap U = q(x) \cap U \neq \emptyset$ by the above observation. This shows that $V_a \subset (q_A)_l^{-1}(U)$ for any $a \in (p_A)_l^{-1}(U)$ and hence the set $\bigcup \{V_a : a \in (p_A)_l^{-1}(U)\}$ is an open neighborhood (in $A$) of $(p_A)_l^{-1}(U)$ contained in $(q_A)_l^{-1}(U)$, i.e., $q_A$ is lower semicontinuous with respect to $p_A$.

**V.056.** *Suppose that* $p : X \to \exp(Y)$ *and* $q : X \to \exp(Y)$ *are finite-valued mappings such that* $q$ *is lower semicontinuous with respect to* $p$. *Given an open set* $U \subset Y$, *let* $p_U(x) = p(x) \cap U$ *and* $q_U(x) = q(x) \cap U$ *for every* $x \in X$. *Prove that the map* $q_U : X \to \exp(U)$ *is lower semicontinuous with respect to* $p_U : X \to \exp(U)$.

**Solution.** Take a set $V \in \tau(U)$; then $V$ is open in $Y$, so the set $q_l^{-1}(V)$ is a neighborhood of the set $p_l^{-1}(V)$. This shows that, for any $x \in X$ such that $p(x) \cap V \neq \emptyset$, there is $O_x \in \tau(x, X)$ such that $q(y) \cap V \neq \emptyset$ for each $y \in O_x$.

The set $O = \bigcup \{O_x : x \in (p_U)_l^{-1}(V)\}$ is an open neighborhood of $(p_U)_l^{-1}(V)$. If $y \in O$ then $y \in O_x$ for some $x \in (p_U)_l^{-1}(V)$; we have $p(y) \cap V = p_U(y) \cap V \neq \emptyset$, so $q(y) \cap V \neq \emptyset$ and hence $q_U(y) \cap V \neq \emptyset$ because $V \subset U$. Thus $O \subset (q_U)_l^{-1}(V)$, i.e., $(q_U)_l^{-1}(V)$ is a neighborhood of $(p_U)_l^{-1}(V)$ which shows that $q_U$ is lower semicontinuous with respect to $p_U$.

**V.057.** *Given a number* $n \in \mathbb{N}$ *and spaces* $X_i, Y_i$, *for every* $i < n$ *suppose that* $p_i, q_i : X_i \to \exp(Y_i)$ *are set-valued maps such that* $q_i$ *is lower semicontinuous with respect to* $p_i$. *Prove that* $\prod_{i<n} q_i$ *is lower semicontinuous with respect to* $\prod_{i<n} p_i$. *As a consequence, any finite product of almost lower semicontinuous maps is an almost lower semicontinuous map.*

**Solution.** Let $p = \prod_{i<n} p_i$ and $q = \prod_{i<n} q_i$; fix an open set $U \subset \prod_{i<n} Y_i$ and a point $a = (a_0, \ldots, a_{n-1}) \in p_l^{-1}(U)$. Since $p(a) \cap U \neq \emptyset$, we can find $U_i \in \tau(Y_i)$ for any $i < n$ such that $U' = U_0 \times \ldots \times U_{n-1} \subset U$ and $p(a) \cap U' \neq \emptyset$. This implies that, for every $i < n$, we have $p(a_i) \cap U_i \neq \emptyset$ and hence there exists $W_i \in \tau(a_i, X_i)$ such that $q_i(y) \cap U_i \neq \emptyset$ for any $y \in W_i$.

Now, if $W(a) = W_0 \times \ldots \times W_{n-1}$ then, for any $b = (b_0, \ldots, b_{n-1}) \in W(a)$, the set $q(b)$ meets $U' \subset U$ and hence $q(b) \cap U \neq \emptyset$ which shows that $W(a) \subset$

$q_l^{-1}(U)$. Thus the set $W = \bigcup\{W(a) : a \in p_l^{-1}(U)\}$ is an open neighborhood of $p_l^{-1}(U)$; since also $W \subset q_l^{-1}(U)$, we proved that the map $q = \prod_{i<n} q_i$ is lower semicontinuous with respect to $p = \prod_{i<n} p_i$.

**V.058.** *Suppose that, for finite-valued mappings $p, r, q : X \to \exp(Y)$, we have $p(x) \subset r(x) \subset q(x)$ for any $x \in X$. Prove that, if $q$ is lower semicontinuous with respect to $r$ then $q$ is lower semicontinuous with respect to $p$.*

**Solution.** If $U \in \tau(Y)$ then $q_l^{-1}(U)$ is a neighborhood of the set $r_l^{-1}(U)$; since also $p_l^{-1}(U) \subset r_l^{-1}(U)$, the set $q_l^{-1}(U)$ is a neighborhood of $p_l^{-1}(U)$, so $q$ is lower semicontinuous with respect to $p$.

**V.059.** *Suppose that $X$ is a nonempty space and $p : X \to \exp(Y)$ is an almost lower semicontinuous map such that $\bigcup p(X) = Y$. Prove that $Y$ is a countable union of images of subspaces of $X$ under single-valued almost lower semicontinuous maps of finite defect.*

**Solution.** Fix a space $Z \supset Y$ and a finite-valued map $q : X \to \exp(Z)$ such that $q$ is lower semicontinuous with respect to $p$. For any $m, n \in \mathbb{N}$ consider the set $X_{mn} = \{x \in X : |p(x)| = m \text{ and } |q(x)| = n\}$; if $Y_{mn} = \bigcup p(X_{mn})$ and $Z_{mn} = \bigcup q(X_{mn})$ then $Y = \bigcup\{Y_{mn} : m, n \in \mathbb{N}\}$. Apply Problem 055 to see that, for any $m, n \in \mathbb{N}$, the map $q|X_{mn} : X_{mn} \to \exp(Z_{mn})$ is lower semicontinuous with respect to $p|X_{mn} : X_{mn} \to \exp(Y_{mn})$, so it suffices to prove our statement for every $p|X_{mn}$. Thus we can assume, without loss of generality, that there exist $m, n \in \mathbb{N}$ such that $|p(x)| = m$ and $|q(x)| = n$ for each $x \in X$.

Fix a faithful enumeration $\{p^1(x), \ldots, p^m(x)\}$ of every set $p(x)$; this gives a single-valued map $p^i : X \to Y$; let $Y_i = p^i(X)$ for any $i \in \{1, \ldots, m\}$. By Problem 058, the map $q$ is lower semicontinuous with respect to every $p^i : X \to Y_i$; since $|q(x)| = n$ for any point $x \in X$, the defect of the map $p^i$ does not exceed $n - 1$. Finally, it follows from $Y = \bigcup\{Y_i : i \in \{1, \ldots, m\}\}$ that we obtained the promised representation of the space $Y$.

**V.060.** *Given nonempty spaces $X, Z$ and $k \in \mathbb{N}$, suppose that we have maps $p : X \to Z$ and $q : X \to \exp(Z)$ such that $p(x) \in q(x)$ and $|q(x)| \leq k$ for each $x \in X$ while $q(x) \cap q(y) = \emptyset$ if $x \neq y$ and $q$ is lower semicontinuous with respect to $p$. Prove that, if $p(X)$ is discrete then $X = \bigcup_{j<k} D_j$ where every $D_j \subset X$ is discrete and hence there exists a discrete subspace $S \subset X$ such that $|S| = |X|$.*

**Solution.** If $Z$ is a space then let $Z^{(0)} = Z$ and $I_0(Z) = \emptyset$. If $n \in \omega$ and we have a set $Z^{(n)} \subset Z$, let $I_{n+1}(Z)$ be the set of all isolated points of the space $Z^{(n)}$ and $Z^{(n+1)} = Z^{(n)} \backslash I_{n+1}(Z)$. If $Z^{(n)} = \emptyset$ for some $n \in \omega$ then we say that *dispersion index of $Z$* does not exceed $n$. If the dispersion index of the space $Z$ does not exceed $n$ then, evidently, $Z$ can be represented as the union of at most $n$ discrete subspaces.

If $X$ is finite then it is discrete and hence the sets $D_0 = \ldots = D_{k-1} = S = X$ do the job; this shows that we can assume, without loss of generality, that $X$ is infinite. Given an open set $U \subset X$, a point $x \in U$ is isolated in $U$ if and only it is isolated in $X$. Using this fact it is easy to prove by induction that $U^{(n)} = X^{(n)} \cap U$ for any $n \in \omega$. This implies that

(1) if $U \cap X^{(n+1)} \neq \emptyset$ for some $n \in \omega$ then the set $U \cap X^{(n)}$ is infinite.

For any $x \in X$ fix a set $O_x \in \tau(p(x), Z)$ such that $\overline{O}_x \cap p(X) = \{p(x)\}$. It suffices to show that $X^{(k)} = \emptyset$, so assume toward a contradiction, that there is a point $x_0 \in X^{(k)}$. Let $V_0 = O_{x_0}$; since $q$ is lower semicontinuous with respect to $p$, there is a set $U_0 \in \tau(x_0, X)$ such that $q(y) \cap V_0 \neq \emptyset$ for any $y \in U_0$. Proceeding by induction suppose that $m < k$ and we have points $x_0, \dots, x_m \in X$ and sets $U_0, V_0, \dots, U_m, V_m$ with the following properties:

(2) $x_i \in X^{(k-i)} \setminus \{x_0, \dots, x_{i-1}\}$ for any $i \leq m$;
(3) $V_i \in \tau(p(x_i), Z)$ and $\overline{V}_i \cap p(X) = \{p(x_i)\}$ for any $i \leq m$;
(4) the family $\{\overline{V}_0, \dots, \overline{V}_m\}$ is disjoint;
(5) if $i \leq m$ then $U_i \in \tau(x_i, X)$ and $q(y) \cap V_i \neq \emptyset$ for any $y \in U_i$;
(6) $U_0 \supset \dots \supset U_m$.

It follows from the properties (1) and (2) that the set $U_m \cap X^{(k-m-1)}$ is infinite, so we can choose a point $x_{m+1} \in X^{(k-m-1)} \setminus \{x_0, \dots, x_m\}$. The map $p$ is easily seen to be injective, so it follows from (3) that $p(x_{m+1}) \notin G = \overline{V}_0 \cup \dots \cup \overline{V}_m$. Choose a set $V_{m+1} \in \tau(p(x_{m+1}), Z)$ such that $V_{m+1} \subset O_{x_{m+1}}$ and $\overline{V}_{m+1} \cap G = \emptyset$. The map $q$ being lower semicontinuous with respect to $p$, we can choose a set $U_{m+1} \in \tau(x_{m+1}, X)$ such that $U_{m+1} \subset U_m$ and $q(y) \cap V_{m+1} \neq \emptyset$ for any $y \in U_{m+1}$. It is straightforward that the conditions (2)–(6) are still satisfied if we substitute $m$ by $m+1$, so our inductive procedure can be continued to construct points $x_0, \dots, x_k$ and sets $U_0, V_0, \dots, U_k, V_k$ for which the properties (2)–(6) hold for any $i \leq k$.

It follows from (5) and (6) that $q(x_k) \cap V_i \neq \emptyset$ for any $i = 0, \dots, k$; an immediate consequence of (4) is that $|q(x_k)| \geq k+1$; this contradiction shows that $X^{(k)} = \emptyset$, so we can let $D_j = I_j(X)$ for any $j < k$. Since $X$ is infinite, some $D_j$ has the same cardinality as $X$; letting $S = D_j$ we obtain a discrete subspace $S \subset X$ such that $|S| = |X|$.

**V.061.** *Given nonempty spaces $X, Z$ and $k \in \mathbb{N}$, suppose that we have maps $p : X \to Z$ and $q : X \to \exp(Z)$ such that $p(x) \in q(x)$ and $|q(x)| \leq k$ for each $x \in X$ while $q(x) \cap q(y) = \emptyset$ if $x \neq y$ and $q$ is lower semicontinuous with respect to $p$. Prove that, if $p(X)$ is left-separated then $X = \bigcup_{j<k} D_j$ where every $D_j \subset X$ is left-separated and hence there exists a left-separated subspace $S \subset X$ such that $|S| = |X|$.*

**Solution.** If $Z$ is a space and $<$ is a well-order on $Z$ say that a point $z \in Z$ is *left-separated in* $Z$ if there exists $U \in \tau(z, Z)$ such that $z$ is the minimal point of $U$, i.e., $x \leq y$ for any $y \in U$. Let $Z^{(0)} = Z$ and $L_0(Z) = \emptyset$. If $n \in \omega$ and we have a set $Z^{(n)} \subset Z$, let $L_{n+1}(Z) = \{z \in Z^{(n)} : z$ is left-separated in $Z^{(n)}\}$ and $Z^{(n+1)} = Z^{(n)} \setminus L_{n+1}(Z)$. If $Z^{(n)} = \emptyset$ for some $n \in \omega$ then we say that *left-separation index of* $Z$ *does not exceed* $n$. It is clear that every $L_i(Z)$ is a left-separated space with respect to the order inherited from $Z$; this shows that if the left-separation index of the space $Z$ does not exceed $n$ then $Z$ can be represented as the union of at most $n$ left-separated subspaces.

If the space $X$ is finite then it is discrete and hence left-separated, so the sets $D_0 = \ldots = D_{k-1} = S = X$ do the job; this shows that we can assume, without loss of generality, that $X$ is infinite. Let $<$ be a well-order on $p(X)$ which witnesses that $p(X)$ is left-separated. It is straightforward that $p : X \to p(X)$ is a bijection, so we obtain a well-order $\prec$ on $X$ if we declare that $x \prec y$ if and only if $p(x) < p(y)$.

Given an open set $U \subset X$, a point $x \in U$ is left-separated in $U$ if and only it is left-separated in $X$. Using this fact it is easy to prove by induction that $U^{(n)} = X^{(n)} \cap U$ for any $n \in \omega$. This implies that

(1) if $n \in \omega$ and $x \in U \cap X^{(n+1)}$ then the set $\{y \in U \cap X^{(n)} : y \prec x\}$ is infinite.

For any $x \in X$ fix a set $O_x \in \tau(p(x), Z)$ such that $\overline{O}_x \cap p(X)$ does not meet the set $\{y \in p(X) : y < p(x)\}$ and hence $p(x)$ is the minimal element of $\overline{O}_x \cap p(X)$. It suffices to show that $X^{(k)} = \emptyset$, so assume toward a contradiction, that there is a point $x_0 \in X^{(k)}$. Let $V_0 = O_{x_0}$; since $q$ is lower semicontinuous with respect to $p$, there is a set $U_0 \in \tau(x_0, X)$ such that $q(y) \cap V_0 \neq \emptyset$ for any $y \in U_0$. Proceeding by induction suppose that $m < k$ and we have points $x_0, \ldots, x_m \in X$ and sets $U_0, V_0, \ldots, U_m, V_m$ with the following properties:

(2) $x_0 \succ \ldots \succ x_m$ and $x_i \in X^{(k-i)}$ for any $i \leq m$;
(3) $V_i \in \tau(p(x_i), Z)$ and $p(x_i)$ is the minimal element of $\overline{V}_i \cap p(X)$ for any $i \leq m$;
(4) the family $\{\overline{V}_0, \ldots, \overline{V}_m\}$ is disjoint;
(5) if $i \leq m$ then $U_i \in \tau(x_i, X)$ and $q(y) \cap V_i \neq \emptyset$ for any $y \in U_i$;
(6) $U_0 \supset \ldots \supset U_m$.

It follows from (1) and (2) that the set $\{y \in U_m \cap X^{(k-m-1)} : y \prec x_m\}$ is infinite, so we can choose a point $x_{m+1} \in X^{(k-m-1)} \cap U_m$ such that $x_{m+1} \prec x_m$. Since $p$ is injective, it follows from the property (3) that $p(x_{m+1}) \notin G = \overline{V}_0 \cup \ldots \cup \overline{V}_m$. Choose a set $V_{m+1} \in \tau(p(x_{m+1}), Z)$ such that $V_{m+1} \subset O_{x_{m+1}}$ and $\overline{V}_{m+1} \cap G = \emptyset$. Since $q$ is lower semicontinuous with respect to $p$, we can find a set $U_{m+1} \in \tau(x_{m+1}, X)$ such that $U_{m+1} \subset U_m$ and $q(y) \cap V_{m+1} \neq \emptyset$ for any $y \in U_{m+1}$. It is straightforward that the conditions (2)–(6) are still satisfied if we substitute $m$ by $m + 1$, so our inductive procedure can be continued to construct points $x_0, \ldots, x_k$ and sets $U_0, V_0, \ldots, U_k, V_k$ for which the properties (2)–(6) hold for any $i \leq k$.

It follows from (5) and (6) that $q(x_k) \cap V_i \neq \emptyset$ for any $i = 0, \ldots, k$; an immediate consequence of (4) is that $|q(x_k)| \geq k + 1$; this contradiction shows that $X^{(k)} = \emptyset$, so we can let $D_j = L_j(X)$ for any $j < k$. Since $X$ is infinite, some $D_j$ has the same cardinality as $X$; letting $S = D_j$ we obtain a left-separated subspace $S \subset X$ such that $|S| = |X|$.

**V.062.** *Given nonempty spaces $X, Z$ and $k \in \mathbb{N}$, suppose that we have maps $p : X \to Z$ and $q : X \to \exp(Z)$ such that $p(x) \in q(x)$ and $|q(x)| \leq k$ for each $x \in X$ while $q(x) \cap q(y) = \emptyset$ if $x \neq y$ and $q$ is lower semicontinuous with respect to $p$. Prove that, if $p(X)$ is right-separated then $X = \bigcup_{j<k} D_j$ where every $D_j \subset X$ is right-separated and hence there exists a right-separated subspace $S \subset X$ such that $|S| = |X|$.*

**Solution.** If $Z$ is a space and $<$ is a well-order on $Z$ say that a point $z \in Z$ is *right-separated in $Z$* if there exists $U \in \tau(z, Z)$ such that $z$ is the maximal point of $U$, i.e., $y \leq x$ for any $y \in U$. Let $Z^{(0)} = Z$ and $M_0(Z) = \emptyset$. If $n \in \omega$ and we have a set $Z^{(n)} \subset Z$, let $M_{n+1}(Z) = \{z \in Z^{(n)} : z$ is right-separated in $Z^{(n)}\}$ and $Z^{(n+1)} = Z^{(n)} \backslash M_{n+1}(Z)$. If $Z^{(n)} = \emptyset$ for some $n \in \omega$ then we say that *right-separation index of $Z$ does not exceed $n$*. It is clear that every $M_i(Z)$ is a right-separated space with respect to the order inherited from $Z$; this shows that if the right-separation index of the space $Z$ does not exceed $n$ then $Z$ can be represented as the union of at most $n$ right-separated subspaces.

If the space $X$ is finite then it is discrete and hence right-separated, so the sets $D_0 = \ldots = D_{k-1} = S = X$ do the job; this shows that we can assume, without loss of generality, that $X$ is infinite. Let $<$ be a well-order on $p(X)$ which witnesses that $p(X)$ is right-separated. It is straightforward that $p : X \to p(X)$ is a bijection, so we obtain a well-order $\prec$ on $X$ if we declare that $x \prec y$ if and only if $p(x) < p(y)$.

Given an open set $U \subset X$, a point $x \in U$ is right-separated in $U$ if and only it is right-separated in $X$. Using this fact it is easy to prove by induction that $U^{(n)} = X^{(n)} \cap U$ for any $n \in \omega$. This implies that

(1) if $n \in \omega$ and $x \in U \cap X^{(n+1)}$ then the set $\{y \in U \cap X^{(n)} : x \prec y\}$ is infinite.

For any $x \in X$ fix a set $O_x \in \tau(p(x), Z)$ such that $\overline{O}_x \cap p(X)$ does not meet the set $\{y \in p(X) : p(x) < y\}$ and hence $p(x)$ is the maximal element of $\overline{O}_x \cap p(X)$. It suffices to show that $X^{(k)} = \emptyset$, so assume toward a contradiction that there is a point $x_0 \in X^{(k)}$. Let $V_0 = O_{x_0}$; since $q$ is lower semicontinuous with respect to $p$, there is a set $U_0 \in \tau(x_0, X)$ such that $q(y) \cap V_0 \neq \emptyset$ for any $y \in U_0$. Proceeding by induction suppose that $m < k$ and we have points $x_0, \ldots, x_m \in X$ and sets $U_0, V_0, \ldots, U_m, V_m$ with the following properties:

(2) $x_0 \prec \ldots \prec x_m$ and $x_i \in X^{(k-i)}$ for any $i \leq m$;
(3) $V_i \in \tau(p(x_i), Z)$ and $p(x_i)$ is the maximal element of $\overline{V}_i \cap p(X)$ for any $i \leq m$;
(4) the family $\{\overline{V}_0, \ldots, \overline{V}_m\}$ is disjoint;
(5) if $i \leq m$ then $U_i \in \tau(x_i, X)$ and $q(y) \cap V_i \neq \emptyset$ for any $y \in U_i$;
(6) $U_0 \supset \ldots \supset U_m$.

It follows from (1) and (2) that the set $\{y \in U_m \cap X^{(k-m-1)} : x_m \prec y\}$ is infinite, so we can choose a point $x_{m+1} \in X^{(k-m-1)} \cap U_m$ such that $x_m \prec x_{m+1}$. Since $p$ is injective, it follows from the property (3) that $p(x_{m+1}) \notin G = \overline{V}_0 \cup \ldots \cup \overline{V}_m$. Choose a set $V_{m+1} \in \tau(p(x_{m+1}), Z)$ such that $V_{m+1} \subset O_{x_{m+1}}$ and $\overline{V}_{m+1} \cap G = \emptyset$. Since $q$ is lower semicontinuous with respect to $p$, we can find a set $U_{m+1} \in \tau(x_{m+1}, X)$ such that $U_{m+1} \subset U_m$ and $q(y) \cap V_{m+1} \neq \emptyset$ for any $y \in U_{m+1}$. It is straightforward that the conditions (2)–(6) are still satisfied if we substitute $m$ by $m + 1$, so our inductive procedure can be continued to construct points $x_0, \ldots, x_k$ and sets $U_0, V_0, \ldots, U_k, V_k$ for which the properties (2)–(6) hold for any $i \leq k$.

It follows from (5) and (6) that $q(x_k) \cap V_i \neq \emptyset$ for any $i = 0, \ldots, k$; an immediate consequence of (4) is that $|q(x_k)| \geq k + 1$; this contradiction shows that $X^{(k)} = \emptyset$, so we can let $D_j = M_j(X)$ for any $j < k$. Since $X$ is infinite, some $D_j$ has the same cardinality as $X$; letting $S = D_j$ we obtain a right-separated subspace $S \subset X$ such that $|S| = |X|$.

**V.063.** *Suppose that an infinite space $Y$ is an image of a space $X$ under a finite-valued almost lower semicontinuous map, i.e., there exists an almost lower semicontinuous map $p : X \to \exp(Y)$ such that $Y = \bigcup p(X)$. Prove that for each $n \in \mathbb{N}$, we have $s(Y^n) \leq s(X^n)$, $hl(Y^n) \leq hl(X^n)$ and $hd(Y^n) \leq hd(X^n)$.*

**Solution.** Assume for a moment that

$(*)$  for any infinite spaces $Z$ and $T$, if there exists an almost lower semicontinuous map $\varphi : Z \to \exp(T)$ such that $\bigcup \varphi(Z) = T$ then $s(T) \leq s(Z)$, $hd(T) \leq hd(Z)$ and $hl(T) \leq hl(Z)$.

For any $n \in \mathbb{N}$ and $x = (x_0, \ldots, x_{n-1}) \in X^n$ let $p^n(x) = \prod_{i<n} p(x_i)$. Then $p^n : X^n \to \exp(Y^n)$ is an almost lower semicontinuous map (see Problem 057) such that $\bigcup p^n(X^n) = Y^n$. This shows that, letting $Z = X^n$, $T = Y^n$, $\varphi = p^n$ and applying $(*)$, we obtain the promised inequalities for every $n \in \mathbb{N}$. Therefore it suffices to establish $(*)$.

Since spread, hereditary density, and hereditary Lindelöf number are hereditary and countably additive, we can apply Problem 059 to see that we can assume, without loss of generality, that $\varphi$ is a single-valued almost lower semicontinuous map of finite defect. Thus there exists a space $H \supset T$, a number $n \in \mathbb{N}$ and a finite-valued map $q : Z \to \exp(H)$ such that $|q(z)| \leq n$ for any $z \in Z$ and $q$ is lower semicontinuous with respect to $\varphi$.

To deal with spread assume that $s(Z) = \kappa$ and $s(T) > \kappa$; therefore we can find a discrete subspace $D \subset T$ such that $|D| = \kappa^+$. It is easy to find a set $E \subset Z$ such that $\varphi|E : E \to D$ is a bijection. Apply the Delta-lemma (SFFS-038) to see that there exists a set $E' \subset E$ and a set $K \subset Z$ such that $|E'| = \kappa^+$ and $q(x) \cap q(y) = K$ for any distinct $x, y \in E'$.

Throwing away a finite subset of $E'$ if necessary, we can assume, without loss of generality, that $\varphi(E') \cap K = \emptyset$. Let $r(x) = q(x) \backslash K$ for any $x \in E'$; it follows from Problems 055 and 056 that $r : E' \to Z$ is still lower semicontinuous with respect to $\varphi|E'$. Since we also have $r(x) \cap r(y) = \emptyset$ for distinct $x, y \in E'$, we can apply Problem 060 to conclude that there is a discrete subspace $A \subset E'$ such that $|A| = |E'| = \kappa^+$. This contradiction with $s(Z) \leq \kappa$ shows that $s(T) \leq \kappa = s(Z)$.

To prove our inequality for hereditary density, assume that $hd(Z) = \kappa$ and $hd(T) > \kappa$; then there exists a left-separated subspace $D \subset T$ such that $|D| = \kappa^+$ (see SFFS-004). It is easy to find a set $E \subset Z$ such that $\varphi|E : E \to D$ is a bijection. Apply the Delta-lemma to see that there exists a set $E' \subset E$ and a set $K \subset Z$ such that $|E'| = \kappa^+$ and $q(x) \cap q(y) = K$ for any distinct $x, y \in E'$.

As in the previous case, we can assume that $\varphi(E') \cap K = \emptyset$. Let $r(x) = q(x) \backslash K$ for any $x \in E'$; it follows from Problems 055 and 056 that $r : E' \to Z$ is still lower semicontinuous with respect to $\varphi|E'$. Since we also have $r(x) \cap r(y) = \emptyset$

for distinct $x, y \in E'$, we can apply Problem 061 to conclude that there is a left-separated subspace $A \subset E'$ such that $|A| = |E'| = \kappa^+$. The obtained contradiction with $hd(Z) \leq \kappa$ shows that $hd(T) \leq \kappa = hd(Z)$.

To prove our inequality for hereditary Lindelöf number assume that $hl(Z) = \kappa$ and $hl(T) > \kappa$; therefore we can find a right-separated subspace $D \subset T$ such that $|D| = \kappa^+$ (see SFFS-005). It is easy to find a set $E \subset Z$ such that $\varphi|E : E \to D$ is a bijection. Apply the Delta-lemma to see that there exists a set $E' \subset E$ and a set $K \subset Z$ such that $|E'| = \kappa^+$ and $q(x) \cap q(y) = K$ for any distinct $x, y \in E'$.

As before, we can assume, without loss of generality, that $\varphi(E') \cap K = \emptyset$. Let $r(x) = q(x) \backslash K$ for any $x \in E'$; then $r : E' \to Z$ is still lower semicontinuous with respect to $\varphi|E'$. Since we also have $r(x) \cap r(y) = \emptyset$ for distinct $x, y \in E'$, we can apply Problem 062 to conclude that there is a right-separated subspace $A \subset E'$ such that $|A| = |E'| = \kappa^+$ which is a contradiction with $hl(Z) \leq \kappa$. This shows that $hl(T) \leq \kappa = hl(Z)$ finishing the proof of $(*)$ and completing our solution.

**V.064.** *Let $h : C_p(Y) \to C_p(X)$ be an embedding such that $h(0_Y) = 0_X$. Given $x \in X$ and $\varepsilon > 0$, a point $y \in Y$ is called $\varepsilon$-inessential for $x$ if there is $U \in \tau(y, Y)$ such that $|h(g)(x)| \leq \varepsilon$ whenever $g(Y \backslash U) \subset \{0\}$. The point $y$ is $\varepsilon$-essential for $x$ if it is not $\varepsilon$-inessential for $x$. Denote by $\mathrm{supp}_\varepsilon(x)$ the set of all points which are $\varepsilon$-essential for $x$. Prove that $\mathrm{supp}_\varepsilon(x)$ is finite for any $x \in X$ and $\varepsilon > 0$.*

**Solution.** Given a set $T$ let $\mathrm{Fin}(T)$ be the family of all finite subsets of $T$. If $Z$ is a space and $A \subset Z$ then $O_Z(A, \varepsilon) = \{f \in C_p(Z) : |f(z)| < \varepsilon$ for any $z \in A\}$. It is evident that family $\{O_Z(A, \varepsilon) : A \in \mathrm{Fin}(Z), \varepsilon > 0\}$ is a local base of $C_p(Z)$ at $0_Z$. If $A = \{x_1, \ldots, x_n\}$ then we write $O_Z(x_1, \ldots, x_n, \varepsilon)$ instead of $O_Z(\{x_1, \ldots, x_n\}, \varepsilon)$.

*Fact 1.* Given spaces $Z$ and $T$ suppose that $\varphi : C_p(T) \to C_p(Z)$ is an embedding such that $\varphi(0_T) = 0_Z$. Assume also that $z \in Z$, $\varepsilon > 0$, $K \in \mathrm{Fin}(T)$ and $\varphi(O_T(K, \delta)) \subset O_Z(z, \varepsilon)$ for some $\delta > 0$. Then $\mathrm{supp}_\varepsilon(z) \subset K$ and hence the set $\mathrm{supp}_\varepsilon(z)$ is finite.

*Proof.* If $t \in T \backslash K$ then $U = T \backslash K$ is an open neighborhood of the point $z$; assume that $f(T \backslash U) = f(K) \subset \{0\}$ for some $f \in C_p(T)$. Then $f \in O_T(K, \delta)$ and hence $\varphi(f) \in O_Z(z, \varepsilon)$, i.e., $|\varphi(f)(z)| < \varepsilon$. Thus the set $U$ witnesses that $t$ is $\varepsilon$-inessential for $z$. Since $t \in T \backslash K$ was chosen arbitrarily, we established that no point of $T \backslash K$ can be $\varepsilon$-essential for $z$, so $\mathrm{supp}_\varepsilon(z) \subset T$ and Fact 1 is proved.

Returning to our solution, fix a point $x \in X$ and $\varepsilon > 0$. Since $O_X(x, \varepsilon)$ is an open neighborhood of $0_X$ in $C_p(X)$, by continuity of $h$, there is a finite $K \subset Y$ and $\delta > 0$ such that $h(O_Y(K, \delta)) \subset O_X(z, \varepsilon)$. Now apply Fact 1 to see that $\mathrm{supp}_\varepsilon(x) \subset K$ and hence $\mathrm{supp}_\varepsilon(x)$ is finite.

**V.065.** *Let $h : C_p(Y) \to C_p(X)$ be an embedding such that $h(0_Y) = 0_X$. Denote by $\mathrm{supp}_\varepsilon(x)$ the set of all points which are $\varepsilon$-essential for $x$. Prove that $Y = \bigcup \{\mathrm{supp}_{1/n}(x) : x \in X, n \in \mathbb{N}\}$.*

**Solution.** Given a set $T$ let $\mathrm{Fin}(T)$ be the family of all finite subsets of $T$. If $Z$ is a space and $A \subset Z$ then $O_Z(A, \varepsilon) = \{f \in C_p(Z) : |f(z)| < \varepsilon$ for any $z \in A\}$. It is evident that family $\{O_Z(A, \varepsilon) : A \in \mathrm{Fin}(Z), \varepsilon > 0\}$ is a local base of $C_p(Z)$ at $\mathbf{0}_Z$. If $A = \{x_1, \ldots, x_n\}$ then we write $O_Z(x_1, \ldots, x_n, \varepsilon)$ instead of $O_Z(\{x_1, \ldots, x_n\}, \varepsilon)$.

Fix $y \in Y$; since $h(O_Y(y, 1))$ is an open neighborhood of $\mathbf{0}_X$ in $h(C_p(Y))$, there is $K \in \mathrm{Fin}(X)$ and $\varepsilon > 0$ such that $O_X(K, \varepsilon) \cap h(C_p(Y)) \subset h(O_Y(y, 1))$. Choose $n \in \mathbb{N}$ such that $\frac{1}{n} < \varepsilon$; we claim that $y \in \mathrm{supp}_{1/n}(x)$ for some $x \in K$.

Indeed, if this is not so then, for any $x \in K$, there is $U_x \in \tau(y, Y)$ which witnesses that $y$ is $\frac{1}{n}$-inessential for $x$; let $U = \bigcap\{U_x : x \in K\}$. Take a function $f \in C_p(Y)$ such that $f(y) = 1$ and $f(Y \setminus U) \subset \{0\}$. Then $f(Y \setminus U_x) \subset \{0\}$ and hence $|h(f)(x)| \le \frac{1}{n} < \varepsilon$ for any $x \in K$. This shows that $h(f) \in O_X(K, \varepsilon) \cap h(C_p(Y))$ and hence $h(f) \in h(O_Y(y, 1))$. The map $h$ being injective we have $f \in O_Y(y, 1)$, i.e., $|f(y)| < 1$ which is a contradiction. Thus $y$ is $\frac{1}{n}$-essential for some $x \in K$; since the point $y \in Y$ was chosen arbitrarily, we proved that $\bigcup\{\mathrm{supp}_{1/n}(x) : x \in X\} = Y$.

**V.066.** *Let $h : C_p(Y) \to C_p(X)$ be an embedding such that $h(\mathbf{0}_Y) = \mathbf{0}_X$. Denote by $\mathrm{supp}_\varepsilon(x)$ the set of all points which are $\varepsilon$-essential for $x$. Prove that, for any $\varepsilon > 0$, the finite-valued map $\mathrm{supp}_\varepsilon : X \to \exp(Y)$ is almost lower semicontinuous.*

**Solution.** Given a set $T$ let $\mathrm{Fin}(T)$ be the family of all finite subsets of $T$. If $Z$ is a space and $A \subset Z$ then $O_Z(A, \varepsilon) = \{f \in C_p(Z) : |f(z)| < \varepsilon$ for any $z \in A\}$. It is evident that family $\{O_Z(A, \varepsilon) : A \in \mathrm{Fin}(Z), \varepsilon > 0\}$ is a local base of $C_p(Z)$ at $\mathbf{0}_Z$. If $A = \{x_1, \ldots, x_n\}$ then we write $O_Z(x_1, \ldots, x_n, \varepsilon)$ instead of $O_Z(\{x_1, \ldots, x_n\}, \varepsilon)$.

Fix $\varepsilon > 0$; for any $x \in X$ there is a finite set $K_x \subset Y$ and $\delta_x > 0$ such that $h(O_Y(K_x, \delta_x)) \subset O_X(x, \varepsilon)$. Let $q(x) = K_x$ for any $x \in X$; then $q : X \to \exp(Y)$ is a finite-valued map. By Fact 1 of V.064, we have $\mathrm{supp}_\varepsilon(x) \subset K_x = q(x)$ for any $x \in X$, so we only need to prove that $q$ is lower semicontinuous with respect to $\mathrm{supp}_\varepsilon$.

To do so take an arbitrary set $U \in \tau(Y)$ and consider a point $x \in X$ such that $\mathrm{supp}_\varepsilon(x) \cap U \ne \emptyset$. Pick a point $y \in \mathrm{supp}_\varepsilon(x) \cap U$; since $y$ is $\varepsilon$-essential for $x$, there is a function $f \in C_p(Y)$ such that $f(Y \setminus U) \subset \{0\}$ and $|h(f)(x)| > \varepsilon$. The set $W_x = \{z \in X : |h(f)(z)| > \varepsilon\}$ is an open neighborhood of $x$ in $X$. If $z \in W_x$ and $q(z) \cap U = \emptyset$ then $f(q(z)) \subset \{0\}$; recalling that $q(z) = K_z$ we conclude that $f \in O_Y(K_z, \delta_z)$ and therefore $h(f) \in O_X(z, \varepsilon)$, i.e., $|h(f)(z)| < \varepsilon$ which contradicts $z \in W_x$. Thus $q(z) \cap U \ne \emptyset$ and hence $z \in q_l^{-1}(U)$ for any $z \in W_x$ which shows that $W_x \subset q_l^{-1}(U)$. An immediate consequence is that

$$(\mathrm{supp}_\varepsilon)_l^{-1}(U) \subset \bigcup\{W_x : x \in (\mathrm{supp}_\varepsilon)_l^{-1}(U)\} \subset q_l^{-1}(U);$$

this proves that $q_l^{-1}(U)$ is a neighborhood of $(\mathrm{supp}_\varepsilon)_l^{-1}(U)$, i.e., $q$ is lower semicontinuous with respect to $\mathrm{supp}_\varepsilon$ and therefore $\mathrm{supp}_\varepsilon$ is almost lower semicontinuous.

**V.067.** *Suppose that $C_p(Y)$ embeds in $C_p(X)$. Prove that $Y$ is a countable union of images of $X$ under finite-valued almost lower semicontinuous maps.*

**Solution.** Using homogeneity of the space $C_p(X)$ it is easy to find an embedding $h : C_p(Y) \to C_p(X)$ such that $h(\mathbf{0}_Y) = \mathbf{0}_X$. For any $\varepsilon > 0$ and $x \in X$ let $\operatorname{supp}_\varepsilon(x)$ be the set of points of the space $Y$ which are $\varepsilon$-essential for $x$. Given $n \in \mathbb{N}$ let $p_n(x) = \operatorname{supp}_{1/n}(x)$ for any $x \in X$. Then every $p_n : X \to \exp(Y)$ is an almost lower semicontinuous finite-valued map (see Problems 064 and 066) and the space $Y_n = \bigcup p_n(X) \subset Y$ is the image of $X$ under $p_n$. Since $Y = \bigcup\{Y_n : n \in \mathbb{N}\}$ by Problem 065, the space $Y$ is a countable union of images of $X$ under finite-valued almost lower semicontinuous maps.

**V.068.** *Suppose that $C_p(Y)$ embeds in $C_p(X)$. Prove that, for any $n \in \mathbb{N}$, we have $s(Y^n) \le s(X^n)$. In particular, if $X$ and $Y$ are $t$-equivalent then $s(Y^n) = s(X^n)$ for any $n \in \mathbb{N}$.*

**Solution.** Fix $n \in \mathbb{N}$ and let $s(X^n) = \kappa$; there is a sequence $\{p_m : m \in \omega\}$ such that every $p_m : X \to \exp(Y)$ is a finite-valued almost lower semicontinuous map and $\bigcup\{\bigcup p_m(X) : m \in \omega\} = Y$ (see Problem 067). Let $Y_m = \bigcup p_m(X)$ for each $m \in \omega$; for any $n$-tuple $\xi = (m_0, \ldots, m_{n-1}) \in \omega^n$ consider the space $Y_\xi = Y_{m_0} \times \ldots \times Y_{m_{n-1}}$ and a finite-valued map $q_\xi : X^n \to \exp(Y_\xi)$ defined by $q_\xi(x) = \prod_{i<n} p_{m_i}(x_i)$ for any $x = (x_0, \ldots, x_{n-1}) \in X^n$. Every $q_\xi$ is almost lower semicontinuous by Problem 057; since also $Y_\xi = \bigcup q_\xi(X^n)$, we can apply Problem 063 to see that $s(Y_\xi) \le \kappa$ for any $\xi \in \omega^n$. Finally observe that $Y^n = \bigcup\{Y_\xi : \xi \in \omega^n\}$; the spread being countably additive, we conclude that $s(Y^n) \le \kappa = s(X^n)$.

**V.069.** *Suppose that $C_p(Y)$ embeds in $C_p(X)$. Prove that, for any $n \in \mathbb{N}$, we have $hd(Y^n) \le hd(X^n)$. In particular, if the spaces $X$ and $Y$ are $t$-equivalent then $hd(Y^n) = hd(X^n)$ for any $n \in \mathbb{N}$.*

**Solution.** Fix $n \in \mathbb{N}$ and let $hd(X^n) = \kappa$; there is a sequence $\{p_m : m \in \omega\}$ such that every $p_m : X \to \exp(Y)$ is a finite-valued almost lower semicontinuous map and $\bigcup\{\bigcup p_m(X) : m \in \omega\} = Y$ (see Problem 067). Let $Y_m = \bigcup p_m(X)$ for each $m \in \omega$; for any $n$-tuple $\xi = (m_0, \ldots, m_{n-1}) \in \omega^n$ consider the space $Y_\xi = Y_{m_0} \times \ldots \times Y_{m_{n-1}}$ and a finite-valued map $q_\xi : X^n \to \exp(Y_\xi)$ defined by $q_\xi(x) = \prod_{i<n} p_{m_i}(x_i)$ for any $x = (x_0, \ldots, x_{n-1}) \in X^n$. Every $q_\xi$ is almost lower semicontinuous by Problem 057; since also $Y_\xi = \bigcup q_\xi(X^n)$, we can apply Problem 063 to see that $hd(Y_\xi) \le \kappa$ for any $\xi \in \omega^n$. Finally observe that $Y^n = \bigcup\{Y_\xi : \xi \in \omega^n\}$; the hereditary density being countably additive, we conclude that $hd(Y^n) \le \kappa = hd(X^n)$.

**V.070.** *Suppose that $C_p(Y)$ embeds in $C_p(X)$. Prove that, for any $n \in \mathbb{N}$, we have $hl(Y^n) \le hl(X^n)$. In particular, if the spaces $X$ and $Y$ are $t$-equivalent then $hl(Y^n) = hl(X^n)$ for any $n \in \mathbb{N}$.*

**Solution.** Fix $n \in \mathbb{N}$ and let $hl(X^n) = \kappa$; there is a sequence $\{p_m : m \in \omega\}$ such that every $p_m : X \to \exp(Y)$ is a finite-valued almost lower semicontinuous map and $\bigcup\{\bigcup p_m(X) : m \in \omega\} = Y$ (see Problem 067). Let $Y_m = \bigcup p_m(X)$ for each $m \in \omega$; for any $n$-tuple $\xi = (m_0, \ldots, m_{n-1}) \in \omega^n$ consider the space $Y_\xi = Y_{m_0} \times \ldots \times Y_{m_{n-1}}$ and a finite-valued map $q_\xi : X^n \to \exp(Y_\xi)$ defined by $q_\xi(x) = \prod_{i<n} p_{m_i}(x_i)$ for any $x = (x_0, \ldots, x_{n-1}) \in X^n$. Every $q_\xi$ is almost lower semicontinuous by Problem 057; since also $Y_\xi = \bigcup q_\xi(X^n)$, we can apply Problem 063 to see that $hl(Y_\xi) \leq \kappa$ for any $\xi \in \omega^n$. Finally observe that $Y^n = \bigcup\{Y_\xi : \xi \in \omega^n\}$; the hereditary Lindelöf number being countably additive, we conclude that $hl(Y^n) \leq \kappa = hl(X^n)$.

**V.071.** *Suppose that $f : X \to Y$ is a closed continuous onto map such that $t(Y) \leq \kappa$ and $t(f^{-1}(y)) \leq \kappa$ for any $y \in Y$. Prove that $t(X) \leq \kappa$. Deduce from this fact that, for every infinite compact space $X$, we have $t(X^n) = t(X)$ for any $n \in \mathbb{N}$.*

**Solution.** Given a space $Z$ say that a set $A \subset Z$ is $\kappa$-*closed* in $Z$ if $\overline{B} \subset A$ for any $B \subset A$ with $|B| \leq \kappa$. It is straightforward that

(1) if $A$ is $\kappa$-closed in $Z$ then $A \cap F$ is $\kappa$-closed both in $Z$ and in $F$ for any closed $F \subset Z$.

Now assume that $A$ is a $\kappa$-closed subset of $X$. Let $P = f(A)$ and take any $Q \subset P$ such that $|Q| \leq \kappa$. It is easy to find a set $B \subset A$ for which $|B| \leq \kappa$ and $f(B) = Q$. The map $f$ being closed, we have $\overline{Q} = f(\overline{B}) \subset f(A) = P$ and hence $P$ is $\kappa$-closed in $Y$; since $t(Y) \leq \kappa$, the set $P$ is closed in $Y$, so we proved that

(2) if $A$ is $\kappa$-closed in $X$ then $f(A)$ is closed in $Y$.

Suppose that a set $A \subset X$ is $\kappa$-closed but not closed in $X$; fix a point $x \in \overline{A} \backslash A$ and let $F = f^{-1}(f(x))$. The set $E = F \cap A$ is $\kappa$-closed in $F$ by (1); recalling that $t(F) \leq \kappa$, we conclude that $E$ is closed in $F$ and hence in $X$. Choose a set $U \in \tau(x, X)$ such that $\overline{U} \cap E = \emptyset$; then $A' = A \cap \overline{U}$ is $\kappa$-closed in $X$ while $A' \cap F = \emptyset$ and $x \in \overline{A'}$. Finally observe that the set $P = f(A')$ is closed in $Y$ by (2); since also $x \in \overline{A'}$, we have $f(x) \in \overline{P} = P$ which is a contradiction with $A' \cap f^{-1}(f(x)) = \emptyset$. This contradiction shows that every $\kappa$-closed subset of $X$ is closed in $X$, so $t(X) \leq \kappa$ by Lemma from S.162.

Finally, assume that $X$ is an infinite compact space and $t(X) \leq \lambda$. To prove by induction that $t(X^n) \leq \lambda$ for any $n \in \mathbb{N}$ suppose that $k \in \mathbb{N}$ and we proved that $t(X^k) \leq \lambda$. Let $\pi : X^{k+1} \to X$ be the projection onto the first factor of $X^{k+1}$. For any $x \in X$, the space $\pi^{-1}(x)$ is homeomorphic to $X^k$, so $t(\pi^{-1}(x)) \leq \lambda$ by the induction hypothesis; since the map $\pi$ is closed, we can apply our above result to see that $t(X^{k+1}) \leq \lambda$. Therefore $t(X^n) \leq t(X)$ for any $n \in \mathbb{N}$; since $X$ embeds in every $X^n$, we also have the opposite inequality and hence $t(X^n) = t(X)$ for any $n \in \mathbb{N}$.

**V.072.** *Given countably compact sequential spaces $X_1, \ldots, X_m$ prove that the space $X_1 \times \ldots \times X_m$ is countably compact and sequential.*

**Solution.** A space $Z$ is called *sequentially compact* if every sequence in $Z$ has a convergent subsequence. Say that a set $A \subset Z$ is *sequentially closed* in the space $Z$ if $\overline{S} \subset A$ for any convergent sequence $S \subset A$. It is straightforward that

(1)  if $A$ is sequentially closed in $Z$ then $A \cap F$ is sequentially closed both in $Z$ and in $F$ for any closed $F \subset Z$.

*Fact 1.*  Any sequentially compact space is countably compact; if $Z$ is countably compact and sequential then $Z$ is sequentially compact.

*Proof.* If a space $Y$ is not countably compact then there is a countably infinite closed discrete $D \subset Y$; taking a faithful enumeration $\{d_n : n \in \omega\}$ of the set $D$ we obtain a sequence in $Y$ which has no convergent subsequence. This proves the first part of our Fact.

To prove the second statement, suppose that $Z$ is countably compact and sequential and take a sequence $S \subset Z$; if the set $S$ is finite then it is easy to extract a constant (and hence convergent) subsequence from $S$. If $S$ is infinite then, applying Fact 4 of S.382 and passing, if necessary, to a subsequence of $S$ we can assume, without loss of generality, that $S$ is an infinite discrete subspace of $Z$. Since $Z$ is countably compact, the set $S$ cannot be closed; the space $Z$ being sequential, there is a sequence $S' \subset S$ which converges to a point $x \in Z \backslash S$. It is clear that $S'$ is the promised convergent subsequence of $S$, so Fact 1 is proved.

*Fact 2.*  Any finite product of sequentially compact spaces is sequentially compact and hence countably compact.

*Proof.* Suppose that the spaces $Z_1, \ldots, Z_n$ are sequentially compact and take a sequence $S \subset Z = Z_1 \times \ldots \times Z_n$. Denote by $\pi_i : Z \to Z_i$ the projection of $Z$ onto its $i$-th factor for every $i \leq n$. Use sequential compactness of the factors to pass $n$ times to a subsequence of $S$ to finally obtain a sequence $S' \subset S$ such that $\pi_i(S')$ converges to a point $z_i \in Z_i$ for any $i \in \{1, \ldots, n\}$. It is easy to check that the sequence $S'$ converges to the point $z = (z_1, \ldots, z_n)$; this proves that $Z$ is sequentially compact and hence countably compact by Fact 1, so Fact 2 is proved.

Returning to our solution observe that an easy induction argument shows that there is no loss of generality to assume that $m = 2$; thus it suffices to prove that $X = X_1 \times X_2$ is countably compact and sequential. Let $\pi_i : X \to X_i$ be the natural projection of $X$ onto $X_i$ for every $i \in \{1, 2\}$. Fact 1 and Fact 2 imply that the space $X$ is sequentially compact and hence countably compact.

Take a sequentially closed set $A \subset X_1 \times X_2$; if $B = \pi_1(A)$ is not sequentially closed in $X_1$ then fix a point $t \in X_1 \backslash B$ and a sequence $T = \{t_n : n \in \omega\} \subset B$ which converges to $t$. Choose a point $s_n \in \pi_1^{-1}(t_n) \cap A$ for each $n \in \omega$; by sequential compactness of $X$ some subsequence $S'$ of the sequence $S = \{s_n : n \in \omega\}$ converges to a point $x \in X$. Then $T' = \pi_1(S')$ has to converge to $\pi_1(x)$; since $T'$ is a subsequence of $T$, it converges to $t$ and therefore $\pi_1(x) = t$. Consequently, $x \notin A$ and hence $S' \subset A$ is a sequence which converges to a point outside of $A$. This contradiction with the set $A$ being sequentially closed proves that

(2) the set $\pi_1(A)$ is sequentially closed and hence closed in $X_1$ for any sequentially closed $A \subset X$.

Now assume that $A \subset X$ is a sequentially closed non-closed set and fix a point $x \in \overline{A} \backslash A$. The subspace $F = \pi_1^{-1}(\pi_1(x))$ is closed in $X$ and sequential being homeomorphic to $X_2$, so the set $E = A \cap F$ is sequentially closed in $F$ by (1); an immediate consequence is that $E$ is closed in $F$ and hence in $X$.

Therefore we can take $U \in \tau(x, X)$ such that $\overline{U} \cap E = \emptyset$. Apply (1) again to see that the set $A' = A \cap \overline{U}$ is sequentially closed in $X$; it is evident that $A' \cap F = \emptyset$. Furthermore, we have $x \in \overline{A'}$ and hence $t = \pi_1(x) \in \pi_1(A')$. However, $\pi_1(A')$ is closed in $X_1$ by (2) which shows that $\pi_1(x) \in \pi_1(A')$ and hence $\pi_1^{-1}(\pi_1(x)) \cap A' \neq \emptyset$. This contradiction shows that every sequentially closed subset of $X$ is closed and hence $X$ is sequential; we already saw that $X$ is countably compact, so our solution is complete.

**V.073.** *Assuming that a space $X$ maps continuously onto a compact space $K$ prove that $t(K) \leq l(X) \cdot t(X)$.*

**Solution.** Given a space $Z$ and a cardinal $\kappa$ say that a set $A \subset Z$ is $\kappa$-*closed* in $Z$ if $\overline{B} \subset A$ for any $B \subset A$ with $|B| \leq \kappa$. For the cardinal $\kappa = t(X) \cdot l(X)$ we must prove that $t(K) \leq \kappa$; fix a continuous onto map $f : X \to K$.

*Fact 1.* Suppose that $Z$ is a space and $\lambda$ is a cardinal such that $l(Z) \leq \lambda$ and $\overline{A}$ is compact for any $A \subset Z$ with $|A| \leq \lambda$. Then $Z$ is compact.

*Proof.* To obtain a contradiction assume that $\mathcal{U}$ is an open cover of $Z$ which has no finite subcover. It follows from $l(Z) \leq \lambda$ that we can assume, without loss of generality, that $|\mathcal{U}| \leq \lambda$. For any finite $\mathcal{V} \subset \mathcal{U}$ pick a point $z(\mathcal{V}) \in Z \backslash (\bigcup \mathcal{V})$; then the set $A = \{z(\mathcal{V}) : \mathcal{V} \text{ is a finite subfamily of } \mathcal{U}\}$ has cardinality at most $\lambda$ and hence $F = \overline{A}$ is compact. The family $\mathcal{U}$ being an open cover of $F$, there exists a finite $\mathcal{V} \subset \mathcal{U}$ such that $F \subset \bigcup \mathcal{V}$. However, $z(\mathcal{V}) \in F \backslash (\bigcup \mathcal{V})$; this contradiction shows that $Z$ is compact so Fact 1 is proved.

Returning to our solution, assume toward a contradiction that $t(K) > \kappa$; then there is a non-closed set $A \subset K$ which is $\kappa$-closed in $K$ (see Lemma from S.162) and hence $\overline{P} = \text{cl}_A(P)$ is compact for any $P \subset A$ with $|P| \leq \kappa$. It is straightforward that $B = f^{-1}(A)$ is $\kappa$-closed in $X$; since $t(X) \leq \kappa$, the set $B$ is closed in $X$. An immediate consequence is that $l(B) \leq l(X) \leq \kappa$; the set $A$ being a continuous image of $B$, we have $l(A) \leq \kappa$. This makes it possible to apply Fact 1 to conclude that $A$ is compact and hence closed in $K$; this contradiction shows that $t(K) \leq \kappa = l(X) \cdot t(X)$.

**V.074.** *Suppose that $K$ is a compact space and $K = \bigcup_{n \in \omega} K_n$ where every $K_n$ is a sequential closed subspace of $K$. Prove that if either Martin's Axiom or Luzin's Axiom ($2^{\omega_1} > 2^\omega$) holds then the space $K$ is sequential.*

**Solution.** Given a space $Z$ and $A \subset Z$ the set $A$ is *sequentially closed in $Z$* if $\overline{S} \subset A$ for any convergent sequence $S \subset A$. It is easy to see that $Z$ is sequential if and only if every sequentially closed subset of $Z$ is closed in $Z$.

Assume toward a contradiction that there exists a sequentially closed non-closed set $A \subset K$; observe that every set $A_n = A \cap K_n$ is sequentially closed and hence closed in $K_n$. Therefore the set $A = \bigcup_{n \in \omega} A_n$ is $\sigma$-compact. We assumed that $A$ is not compact, so it cannot be countably compact; fix a countably infinite set $D \subset A$ which is closed and discrete in $A$. The set $F = \overline{D}$ is compact and $G = F \setminus D \subset K \setminus A$.

Since $F$ is separable, we can apply Fact 2 of S.368 to see that $w(F) \leq \mathfrak{c}$. If $F_n = F \cap K_n$ then $w(F_n) \leq w(F) \leq \mathfrak{c}$, so we can take a set $H_n \subset F_n$ such that $|H_n| \leq \mathfrak{c}$ and $F_n = \overline{H}_n$ for any $n \in \omega$. The space $K_n \supset F_n$ being sequential, we have $|F_n| \leq \mathfrak{c}$ (see Fact 2 of T.041) for each $n \in \omega$, so $|F| \leq \omega \cdot \mathfrak{c} = \mathfrak{c}$.

Now assume that Luzin's Axiom holds. If $\chi(x, G) \geq \omega_1$ for any point $x \in G$ then we can apply TFS-330 to see that $|G| \geq 2^{\omega_1} > \mathfrak{c}$ which is contradiction. Therefore, there is a point $x \in G$ with $\chi(x, G) \leq \omega$. Since $G$ is a $G_\delta$-subset of $F$, the point $x$ is also a $G_\delta$-subset of $F$ (see Fact 2 of S.358), so $\chi(x, F) \leq \omega$ by TFS-327. An immediate consequence is that there is a sequence $S \subset D$ which converges to $x$; since $x \notin A$, we obtained a contradiction which proves that $K$ is sequential.

Finally assume that Martin's Axiom takes place; since Luzin's Axiom is already taken care of and CH implies Luzin's Axiom, we can assume that MA$+\neg$CH holds. If $\chi(x, G) \geq \mathfrak{c}$ for any point $x \in G$ then we can apply TFS-330 to see that $|G| \geq 2^{\mathfrak{c}} > \mathfrak{c}$ which is contradiction. Therefore, there is a point $x \in G$ with $\chi(x, G) < \mathfrak{c}$. Since $\chi(G, F) = \psi(G, F) \leq \omega$, we can apply Fact 2 of S.358 again to conclude that $\chi(x, F) = \psi(x, F) < \mathfrak{c}$. Consequently, $\chi(x, \{x\} \cup D) < \mathfrak{c}$ and hence there is a sequence $S \subset D$ which converges to $x$ (see SFFS-054); since $x \notin A$, we again obtained a contradiction which shows that $K$ is sequential.

**V.075.** *Suppose that we have a space $X$, a compact space $K$, and a compact-valued upper semicontinuous map $p : X \to \exp(K)$ such that $\bigcup p(X) = K$. Knowing that $l(X) \cdot t(X) \leq \kappa$ and $t(p(x)) \leq \kappa$ for any $x \in X$ prove that $t(K) \leq \kappa$.*

**Solution.** For the space $Y = X \times K$ let $\pi_X : Y \to X$ and $\pi_K : Y \to K$ be the natural projections. We will next prove that the set $Z = \bigcup\{\{x\} \times p(x) : x \in X\}$ is closed in $Y$. To do so take an arbitrary point $u = (a, b) \notin Z$; then $b \notin p(a)$ and hence there exist disjoint sets $U \in \tau(p(a), K)$ and $V \in \tau(b, K)$. By upper semicontinuity of $p$, the set $W = \{x \in X : p(x) \subset U\}$ is an open neighborhood of $a$ in $X$ and hence $O = W \times V \in \tau(u, Y)$. If $(x, y) \in O$ then $p(x) \subset U$ and $y \in V$ which shows that $y \notin p(x)$ and hence $(x, y) \notin Z$. Thus $O \cap Z = \emptyset$ and we proved that every $u \in Y \setminus Z$ has an open neighborhood $O$ in $Y$ with $O \cap Z = \emptyset$. Therefore $Z$ is closed in $Y$.

Let $H = \{x \in X : p(x) = \emptyset\}$; by upper semicontinuity of $p$ the set $H$ is open in $X$, so $X_0 = X \setminus H$ is closed in $X$ and hence $l(X_0) \leq l(X) \leq \kappa$. The projection $\pi_X$ is perfect by Fact 3 of S.288, so the map $\varphi = \pi_X | Z : Z \to X_0$ is perfect as well. We have $\varphi^{-1}(x) = \{x\} \times p(x) \simeq p(x)$, so $t(\varphi^{-1}(x)) \leq \kappa$ for any $x \in X_0$. Besides, $t(X_0) \leq t(X) \leq \kappa$, so we can apply Problem 071 to see that $t(Z) \leq \kappa$. We also have $l(Z) \leq \kappa$ by Fact 5 of S.271; since the map $\pi_K | Z : Z \to K$ is continuous and onto, we can apply Problem 073 to conclude that $t(K) \leq \kappa$.

**V.076.** *Suppose that $K$ is a compact sequential space and $L$ is a compact space for which there exists a finite-valued upper semicontinuous map $p : K \to \exp(L)$ such that $\bigcup p(K) = L$. Prove that $L$ is also sequential.*

**Solution.** Given a space $Z$ say that $A \subset Z$ is *sequentially closed* if $\overline{S} \subset A$ for any convergent sequence $S \subset A$.

It turns out that the set $F = \{(a, b) \in K \times L : b \in p(a)\} \subset K \times L$ is closed in $K \times L$ and hence compact. To see this take any point $z = (a, b) \in (K \times L) \backslash F$; then $b \notin p(a)$ and hence there exist disjoint sets $U \in \tau(b, L)$ and $V \in \tau(p(a), L)$. By upper semicontinuity of $p$, the set $W = \{x \in K : p(x) \subset V\}$ is an open neighborhood of $a$ in $K$, so $z \in W \times U \in \tau(z, K \times L)$. If $t = (x, y) \in W \times U$ then $y \in U$ and $p(x) \subset V$ which shows that $y \notin p(x)$; this proves that $(W \times V) \cap F = \emptyset$ and hence every point $z \in (K \times L) \backslash F$ has an open neighborhood disjoint from $F$. This implies that the set $(K \times L) \backslash F$ is open in $K \times L$, so $F$ is, indeed, closed in $K \times L$. Let $\pi_K : K \times L \to K$ and $\pi_L : K \times L \to L$ be the natural projections.

Our next step is to prove that

(1) for any sequentially closed set $A \subset F$ the set $\pi_K(A)$ is sequentially closed and hence closed in $K$.

Indeed, if the set $B = \pi_K(A)$ is not closed in $K$ then there is a faithfully indexed sequence $S = \{s_n : n \in \omega\} \subset B$ which converges to a point $x \in K \backslash B$; let $S' = S \cup \{x\}$ and choose a set $O_n \in \tau(s_n, K)$ such that $O_n \cap S = \{s_n\}$ for every $n \in \omega$. Take a point $t_n \in \pi_K^{-1}(s_n) \cap A$ for every $n \in \omega$. The sequence $T = \{t_n : n \in \omega\}$ is an infinite discrete subspace of $F$, so it cannot be closed in $F$. Observe that $Z = \pi_K^{-1}(S') \cap F$ is a closed subset of $F$; since $\pi_K^{-1}(s) \cap F$ is finite for any $s \in S'$, the set $Z$ is countable. Being compact and countable, $Z$ is metrizable so there is a sequence $T' \subset T$ which converges to a point $y \in Z \backslash T$. Since $\pi_K^{-1}(O_n) \cap T = \{t_n\}$, the sequence $T'$ cannot converge to a point of $\pi_K^{-1}(s_n)$ for any $n \in \omega$. Thus $\pi_K(y) \in \overline{S} \backslash S = \{x\}$, i.e., $\pi_K(y) = x$; as a consequence, $\pi_K(y) \notin \pi_K(A)$ and therefore $y \in F \backslash A$. This gives a contradiction with the set $A$ being sequentially closed, so (1) is proved.

Now assume that $A \subset F$ is sequentially closed and not closed in $F$; fix a point $z \in \overline{A} \backslash A$. The set $B = \pi_K(A)$ is closed in $K$ by (1). We have $\pi_K(z) \in \overline{B} = B$, so $z \in \pi_K^{-1}(B)$. Since $z$ cannot be in the closure of the set $E = \pi_K^{-1}(\pi_K(z)) \cap A$ we can find $W \in \tau(z, K \times L)$ such that $\overline{W} \cap E = \emptyset$. It is straightforward that the set $A' = A \cap \overline{W}$ is also sequentially closed; besides, $z \in \overline{A'}$ and hence $\pi_K(z) \in \overline{\pi_K(A')}$. However, $\pi_K(A')$ is closed in $K$ by (1); since $\pi_K(z) \notin \pi_K(A')$, it is impossible that $\pi_K(z) \in \overline{\pi_K(A')}$. This contradiction shows that every sequentially closed subset of $F$ is closed in $F$, i.e., $F$ is a sequential space.

It follows from $\bigcup p(K) = L$ that $\pi_L(F) = L$, so $L$ is a continuous (and hence closed) image of the compact space $K$. Applying Fact 1 of S.224 we conclude that $L$ is also sequential, so our solution is complete.

**V.077.** *Suppose that there exists an open continuous map of a subspace of $C_p(X)$ onto $C_p(Y)$. Prove that, for any $n \in \mathbb{N}$ there is a finite-valued upper semicontinuous map $\varphi_n : X^n \to \exp(Y)$ such that $\bigcup \{\bigcup \varphi_n(X^n) : n \in \mathbb{N}\} = Y$.*

**Solution.** The following easy fact will be useful for this solution and future references.

*Fact 1.* Given spaces $Z, T$ and an open continuous onto map $f : Z \to T$, the map $f_A = f|f^{-1}(A) : f^{-1}(A) \to A$ is open for any $A \subset T$.

*Proof.* It is clear that the map $f_A$ is continuous and onto. If $U$ is an open subset of $f^{-1}(A)$ then take $U' \in \tau(Z)$ such that $U' \cap f^{-1}(A) = U$; it is straightforward that $f_A(U) = f(U) = f(U') \cap A$, so $f_A(U)$ is open in $A$. Fact 1 is proved.

Returning to our solution observe that, by homogeneity of $C_p(X)$ and $C_p(Y)$, there exists a subspace $C_0 \subset C_p(X)$ for which there is a continuous open onto map $\xi_0 : C_0 \to C_p(Y)$ such that $\mathbf{0}_X \in C_0$ and $\xi_0(\mathbf{0}_X) = \mathbf{0}_Y$. Let $C = \xi_0^{-1}(C_p(Y, \mathbb{I}))$; then the map $\xi = \xi_0|C : C \to C_p(Y, \mathbb{I})$ is open by Fact 1 and $\xi(\mathbf{0}_X) = \mathbf{0}_Y$. For any function $f \in C_p(Y, \mathbb{I})$ there is $\tilde{f} \in C_p(\beta Y, \mathbb{I})$ such that $\tilde{f}|Y = f$. Given points $x_1, \ldots, x_k \in X$ and $\varepsilon > 0$ let $O_X(x_1, \ldots, x_k, \varepsilon) = \{f \in C : |f(x_i)| < \varepsilon$ for each $i \leq k\}$. It is clear that the family $\{O_X(x_1, \ldots, x_k, \frac{1}{n}) : k, n \in \mathbb{N}, \ x_1, \ldots, x_k \in X\}$ is a local base at $\mathbf{0}_X$ in the space $C$.

Analogously, let $O_Y(y_1, \ldots, y_k, \varepsilon) = \{f \in C_p(Y, \mathbb{I}) : |\tilde{f}(y_i)| < \varepsilon$ for all $i \leq k\}$ for any $y_1, \ldots, y_k \in \beta Y$ and $\varepsilon > 0$. A set $O_Y(y_1, \ldots, y_k, \varepsilon)$ is not necessarily open in $C_p(Y, \mathbb{I})$; however, the family $\{O_Y(y_1, \ldots, y_k, \frac{1}{n}) : k, n \in \mathbb{N}$ and $y_1, \ldots, y_k \in Y\}$ is a local base at $\mathbf{0}_Y$ in $C_p(Y, \mathbb{I})$. For any points $y_1, \ldots, y_k \in \beta Y$ and $\varepsilon > 0$ let $H_Y(y_1, \ldots, y_k, \varepsilon) = \{f \in C_p(Y, \mathbb{I}) : |\tilde{f}(y_i)| \leq \varepsilon$ for all $i \leq k\}$.

Consider, for every point $x = (x_1, \ldots, x_n) \in X^n$, the set

$$\varphi_n(x) = \{y \in \beta Y : \xi(O_X(x_1, \ldots, x_n, \frac{1}{n})) \subset H_Y(y, \frac{1}{2})\}.$$

Our first observation is that

(1) the set $\varphi_n(x)$ is finite and contained in $Y$ for any $x \in X^n$.

To prove (1) recall that the map $\xi$ is open, so there are $y_1, \ldots, y_m \in Y$ and $\varepsilon > 0$ such that $\xi(O_X(x_1, \ldots, x_n, \frac{1}{n})) \supset O_Y(y_1, \ldots, y_m, \varepsilon)$. If $y \in \beta Y \backslash \{y_1, \ldots, y_m\}$ then there exists a function $g \in C_p(Y, \mathbb{I})$ such that $\tilde{g}(y) = 1$ and $g(y_i) = 0$ for each $i \leq m$. Pick a function $f \in O_X(x_1, \ldots, x_n, \frac{1}{n})$ with $\xi(f) = g$. It follows from $g \notin H_Y(y, \frac{1}{2})$ that $\xi(O_X(x_1, \ldots, x_n, \frac{1}{n}))$ is not contained in $H_Y(y, \frac{1}{2})$ and hence $y \notin \varphi_n(x)$. Thus $\varphi_n(x) \subset \{y_1, \ldots, y_m\}$ is a finite subset of $Y$, so (1) is proved.

To see that the map $\varphi_n : X^n \to \exp(Y)$ is upper semicontinuous fix a set $U \in \tau(Y)$ and $x = (x_1, \ldots, x_n) \in X^n$ with $\varphi_n(x) \subset U$. Choose $U' \in \tau(\beta Y)$ such that $U' \cap Y = U$. For any $y \in \beta Y \backslash U'$ we have $y \notin \varphi_n(x)$, so there is a function $f_y \in O_X(x_1, \ldots, x_n, \frac{1}{n})$ such that $|\xi(f_y)(y)| > \frac{1}{2}$; let $g_y = \xi(f_y)$ and $F_y = (\tilde{g}_y)^{-1}([-\frac{1}{2}, \frac{1}{2}])$. Since $y \notin F_y$ for any $y \in \beta Y \backslash U'$, the set $\bigcap \{F_y : y \in \beta Y \backslash U'\}$ is contained in $U'$. By Fact 1 of S.326, there are $y_1, \ldots, y_k \in \beta U \backslash U'$ such that $F_{y_1} \cap \ldots \cap F_{y_k} \subset U'$.

The set $W_x = \{z = (z_1, \ldots, z_n) \in X^n : |f_{y_i}(z_i)| < \frac{1}{n}$ for any $i \leq k\}$ is an open neighborhood of $x$ in $X^n$; we claim that $\bigcup \{\varphi_n(z) : z \in W_x\} \subset U$. Indeed, take any $z = (z_1, \ldots, z_n) \in W_x$; if $y \in \beta Y \setminus U'$ then there is $i \in \{1, \ldots, k\}$ such that $y \notin F_{y_i}$ and hence $|\tilde{g}_{y_i}(y)| > \frac{1}{2}$. An immediate consequence is that the function $g_{y_i} = \xi(f_{y_i}) \in \xi(O_X(z_1, \ldots, z_n, \frac{1}{n})) \setminus H_Y(y, \frac{1}{2})$ witnesses that $y \notin \varphi_n(z)$ and hence $\varphi_n(z) \subset U'$; the property (1) shows that $\varphi_n(z) \subset Y$, so $\varphi_n(z) \subset Y \cap U' = U$. Thus $(\varphi_n)_u^{-1}(U) = \{x \in X^n : \varphi_n(x) \subset U\} = \bigcup \{W_x : x \in U\}$ is an open set and hence the map $\varphi_n$ is upper semicontinuous.

Finally, for any $y \in Y$ the set $O_Y(y, \frac{1}{2})$ is an open neighborhood of $\mathbf{0}_Y$ in $C_p(Y, \mathbb{I})$; the map $\xi$ is continuous and $\xi(\mathbf{0}_X) = \mathbf{0}_Y$, so there are $z_1, \ldots, z_m \in X$ and $\varepsilon > 0$ such that $\xi(O_X(z_1, \ldots, z_m, \varepsilon)) \subset O_Y(y, \frac{1}{2})$. Take $n \in \mathbb{N}$ such that $m \leq n$ and $\frac{1}{n} < \varepsilon$. If $x = (x_1, \ldots, x_n) \in X^n$ and $\{x_1, \ldots, x_n\} \supset \{z_1, \ldots, z_m\}$ then $O_X(x_1, \ldots, x_n, \frac{1}{n}) \subset O_X(z_1, \ldots, z_m, \varepsilon)$ and hence $y \in \varphi_n(x)$. This proves that $\bigcup \{\bigcup \varphi_n(X^n) : n \in \mathbb{N}\} = Y$ and shows that our solution is complete.

**V.078.** *Suppose that there is an open continuous map of a subspace of $C_p(X)$ onto $C_p(Y)$. Prove that $t(K) \leq t^*(X) \cdot l^*(X)$ for any compact $K \subset Y$. Deduce from this fact that if $K$ and $L$ are $t$-equivalent compact spaces then $t(K) = t(L)$.*

**Solution.** Let $\kappa = t^*(X) \cdot l^*(X)$; there exists a subspace $C \subset C_p(X)$ and an open continuous onto map $\varphi : C \to C_p(Y)$. Given a compact subspace $K \subset Y$ let $\pi : C_p(Y) \to C_p(K)$ be the restriction map; since $\pi$ is open continuous and surjective, the map $\pi \circ \varphi : C_p(X) \to C_p(K)$ is also open continuous and onto. Apply Problem 077 to see that we can choose for any $n \in \mathbb{N}$, an upper semicontinuous finite-valued map $p_n : X^n \to \exp(K)$ such that $K = \bigcup \{\bigcup p_n(X) : n \in \mathbb{N}\}$.

Identify every $X^n$ with the respective clopen subspace of $\widetilde{X} = \bigoplus \{X^n : n \in \mathbb{N}\}$. If $x \in \widetilde{X}$ then there is a unique $n \in \mathbb{N}$ such that $x \in X^n$; let $p(x) = p_n(x)$. Then $p : \widetilde{X} \to \exp(K)$ is a finite-valued map and it follows from the choice of the sequence $\{p_n : n \in \mathbb{N}\}$ that $\bigcup p(\widetilde{X}) = K$.

Fix a point $x \in X^n$ and a set $U \in \tau(p(x), K)$; by upper semicontinuity of $p_n$ there is a set $V \in \tau(x, X^n)$ such that $\bigcup p_n(V) \subset U$. The set $V$ is also open in $\widetilde{X}$ and $\bigcup p(V) = \bigcup p_n(V) \subset U$. This, together with Fact 1 of T.346, shows that the map $p$ is upper semicontinuous. Since $t(\widetilde{X}) \cdot l(\widetilde{X}) = t^*(X) \cdot l^*(X) = \kappa$, we can apply Problem 075 to conclude that $t(K) \leq \kappa$.

Finally, if $K$ and $L$ are compact $t$-equivalent spaces then $C_p(L)$ is homeomorphic to $C_p(K)$ and hence it is an open continuous image of the space $C_p(K)$, so $t(L) \leq t^*(K) \cdot l^*(K) = t^*(K) = t(K)$ (see Problem 071). Analogously, $t(K) \leq t(L)$, so $t(K) = t(L)$.

**V.079.** *Suppose that $X$ is a compact sequential space and $Y$ is a compact space such that there is an open continuous map of some subspace of $C_p(X)$ onto $C_p(Y)$. Prove that $Y$ is a countable union of its compact sequential subspaces. As a consequence, under Martin's Axiom if $K$ and $L$ are compact $t$-equivalent spaces and $K$ is sequential then $L$ is also sequential.*

**Solution.** Apply Problem 077 to fix, for any $n \in \mathbb{N}$, an upper semicontinuous finite-valued map $p_n : X^n \to \exp(Y)$ such that $Y = \bigcup \{\bigcup p_n(X) : n \in \mathbb{N}\}$. Consider the space $Y_n = \bigcup p_n(X)$; it is easy to see that $p_n : X^n \to \exp(Y_n)$ is also an upper semicontinuous map, so we can apply SFFS-241 to convince ourselves that $Y_n$ is compact for any $n \in \mathbb{N}$. Apply Problem 076 to see that every $Y_n$ is sequential, so $Y = \bigcup \{Y_n : n \in \mathbb{N}\}$ is a countable union of compact sequential subspaces.

If $K$ and $L$ are compact $t$-equivalent spaces then $C_p(L)$ is homeomorphic to $C_p(K)$ and hence it is an open continuous image of the space $C_p(K)$. The space $K$ being sequential, $L$ is a countable union of compact sequential subspaces which, together with Martin's Axiom and Problem 074, implies that $L$ is sequential.

**V.080.** *Suppose that $X$ and $Y$ are metrizable $t$-equivalent spaces. Prove that we have $Y = \bigcup \{Y_n : n \in \omega\}$, where each $Y_n$ is a $G_\delta$-subspace of $Y$, homeomorphic to some $G_\delta$-subspace of $X$.*

**Solution.** For any set $P$ denote by $\mathrm{Fin}(P)$ the family of all finite subsets of $P$; if $n \in \mathbb{N}$ then $[P]^n = \{B \subset P : |B| = n\}$. Given a space $Z$ the map $\mathrm{id}_Z : Z \to Z$ is the identity, i.e., $\mathrm{id}_Z(z) = z$ for any $z \in Z$. If $A$ is a finite subset of $Z$ and $n \in \mathbb{N}$ then $O_Z(A, n) = \{f \in C_p(Z) : |f(x)| < \frac{1}{n}$ for any $x \in A\}$. For all $z \in Z$ and $n \in \mathbb{N}$ let $G_Z(z, n) = \{f \in C_p(Z) : |f(z)| \leq \frac{1}{n}\}$.

Choose metrics $\rho_X$ and $\rho_Y$ which generate the topologies of the spaces $X$ and $Y$ respectively; the spaces $C_p(X)$ and $C_p(Y)$ being homogeneous, it is easy to convince ourselves that there exists a homeomorphism $\Phi : C_p(X) \to C_p(Y)$ such that $\Phi(\mathbf{0}_X) = \mathbf{0}_Y$. Fix $m, n \in \mathbb{N}$ and consider, for every point $y \in Y$, the family $\mathcal{F}_Y(y, m, n) = \{A \in \mathrm{Fin}(X) : \Phi(O_X(A, m)) \subset G_Y(y, n)\}$; if $\mathcal{F}_Y(y, m, n) \neq \emptyset$ then we can define $a_{m,n}(y) = \setminus n\{|A| : A \in \mathcal{F}_Y(y, m, n)\}$. Likewise, for all $x \in X$ let $\mathcal{F}_X(x, m, n) = \{B \in \mathrm{Fin}(Y) : \Phi^{-1}(O_Y(B, m)) \subset G_X(x, n)\}$; if $\mathcal{F}_X(x, m, n) \neq \emptyset$ then let $b_{m,n}(x) = \setminus n\{|B| : B \in \mathcal{F}_X(x, m, n)\}$. In what follows, when we deal with $a_{m,n}(y)$ and/or $b_{m,n}(x)$ we assume, without mentioning it explicitly, that the respective family $\mathcal{F}_Y(y, m, n)$ and/or $\mathcal{F}_X(x, m, n)$ is nonempty.

Observe that $O_X(\emptyset, m) = C_p(X)$ and hence $\Phi(O_X(\emptyset, m)) = C_p(Y) \not\subset G(y, n)$ for any $y \in Y$; this shows that if $\mathcal{F}_Y(y, m, n) \neq \emptyset$ then all elements of the family $\mathcal{F}_Y(y, m, n)$ are nonempty and hence $a_{m,n}(y) \in \mathbb{N}$ for all $m, n \in \mathbb{N}$; let

$$\mathcal{E}_Y(y, m, n) = \{A \in \mathcal{F}_Y(y, m, n) : |A| = a_{m,n}(y)\}.$$

The same reasoning demonstrates that if $\mathcal{F}_X(x, m, n) \neq \emptyset$ then all elements of $\mathcal{F}_X(x, m, n)$ are nonempty; let $\mathcal{E}_X(x, m, n) = \{B \in \mathcal{F}_X(x, m, n) : |B| = b_{m,n}(x)\}$.

*Fact 1.* Suppose that $M$ and $L$ are metrizable spaces and we have $A \subset M$ and $B \subset L$ which are $G_\delta$-sets in $M$ and $L$ respectively. Assume additionally that there are continuous maps $f : A \to L$ and $g : B \to A$ such that $f \circ g = \mathrm{id}_B$. Then the map $g : B \to g(B)$ is a homeomorphism and $g(B)$ is a $G_\delta$-subset of $M$.

*Proof.* Since $f|g(B)$ is a continuous inverse of $g$, the map $g : B \to g(B)$ is a homeomorphism. To see that $g(B)$ is a $G_\delta$-subset of $X$ take complete metric spaces $\widetilde{M}$ and $\widetilde{L}$ such that $M \subset \widetilde{M}$ and $L \subset \widetilde{L}$. Apply Fact 2 of T.333 to see that there are $G_\delta$-sets $A'$ and $B'$ in $\widetilde{M}$ and $\widetilde{L}$ respectively such that $g(B) \subset A'$, $B \subset B'$ and there is a homeomorphism $h : B' \to A'$ with $h|B = g$; note that $A \cap A'$ is a $G_\delta$-subset of $M$. Since $h^{-1}|g(B) = f|g(B)$, the set $E = \{x \in A \cap A' : h^{-1}(x) = f(x)\}$ contains $g(B)$; it is straightforward that $E$ is a $G_\delta$-subset of $A \cap A'$, so it is also a $G_\delta$-subset of $M$. If $E' = f(E)$ then $f_0 = f|E : E \to E'$ is a homeomorphism; since $E' \subset L$, the subspace $B \subset E'$ is a $G_\delta$-subset of $E'$, so $g(B) = f_0^{-1}(B)$ is a $G_\delta$-subset in $E$ and hence in $M$. Fact 1 is proved.

*Fact 2.* Suppose that $M$ and $L$ are metrizable spaces, $n \in \mathbb{N}$ and $p : M \to [L]^n$ is a lower semicontinuous map. Then there exists a family $\mathcal{U} = \{U_i : i \in \omega\} \subset \tau(M)$ such that $\bigcup \mathcal{U} = M$ and, for any $i \in \omega$, there are continuous maps $f_1^i, \ldots, f_n^i$ such that $f_j^i : U_i \to L$ for every $j \le n$ and $p(x) = \{f_1^i(x), \ldots, f_n^i(x)\}$ for any $x \in U_i$.

*Proof.* Fix a point $x \in M$ and let $\{y_1, \ldots, y_n\}$ be an enumeration of $p(x)$. Choose disjoint sets $W_1, \ldots, W_n \in \tau(L)$ such that $y_i \in W_i$ for all $i \le n$. By lower semicontinuity of $p$, the set $O_x = \{z \in M : p(z) \cap W_i \ne \emptyset$ for every $i \le n\}$ is an open neighborhood of $x$ in $M$ and $p(z) \cap W_i$ is a singleton for any $z \in O_x$ and $i \le n$. For any $z \in O_x$ let $h_i^x(z)$ be the unique point of the set $p(z) \cap W_i$; it is an easy exercise that the map $h_i^x : O_x \to W_i$ is continuous for any $i \le n$. Besides, $p(z) = \{h_1^x(z), \ldots, h_n^x(z)\}$ for any $z \in O_x$.

The space $M$ being metrizable, we can find a $\sigma$-discrete refinement $\mathcal{U}$ of the cover $\{O_x : x \in M\}$. Pick a sequence $\{\mathcal{U}_i : i \in \omega\}$ of discrete subfamilies of $\mathcal{U}$ such that $\mathcal{U} = \bigcup_{n \in \omega} \mathcal{U}_i$. Fix $i \in \omega$; for every $U \in \mathcal{U}_i$ there is $x(U) \in M$ such that $U \subset O_{x(U)}$. Let $f_j^i(z) = h_j^{x(U)}(z)$ for any $z \in U$ and $j \le n$. This defines a continuous map $f_j^i$ on the set $U_i = \bigcup \mathcal{U}_i$ for any $j \le n$. It is evident that $M = \bigcup_{n \in \omega} U_i$ and the maps $f_1^i, \ldots, f_n^i$ are as promised for any $i \in \omega$, so Fact 2 is proved.

*Fact 3.* Suppose that $m, n, k \in \mathbb{N}$ and a sequence $\{y_i : i \in \omega\} \subset Y$ converges to a point $y \in Y$. If $A_i \in \mathcal{F}_Y(y_i, m, n)$ and $|A_i| = k$ for any $i \in \omega$ then there is a subsequence $\{A_{i_j} : j \in \omega\}$ of the sequence $\mathcal{A} = \{A_i : i \in \omega\}$ and a number $q \le k$ for which we can choose a faithful enumeration $\{a_1^j, \ldots, a_k^j\}$ of every set $A_{i_j}$ in such a way that the sequence $\{a_p^j : j \in \omega\}$ converges to a point $x_p \in X$ for all $p \le q$ and $A = \{x_1, \ldots, x_q\} \in \mathcal{F}_Y(y, m, n)$.

*Proof.* We will pass several times to a subsequence of the sequence $\mathcal{A} = \{A_i : i \in \omega\}$; since our aim is to find a certain subsequence of $\mathcal{A}$, at each step we will identify the obtained subsequence with $\mathcal{A}$ considering that all elements of $\mathcal{A}$ have the property we have found in a subsequence.

The first step is to use Fact 2 of U.337 to choose a subsequence $\mathcal{A}' \subset \mathcal{A}$ for which there is a set $D = \{d_1, \ldots, d_r\} \subset X$ such that $A \cap A' = D$ for distinct $A, A' \in \mathcal{A}'$ (observe that it is possible that $r = 0$ in which case $D = \emptyset$). According to the above-mentioned politics we can consider that, for any $i \in \omega$, we have $A_i = \{d_1, \ldots, d_r, a_1^i, \ldots, a_{k-r}^i\}$ and the family $\{A_i \setminus D : i \in \omega\}$ is disjoint.

An evident property of metric spaces is that any sequence contains either a convergent subsequence or an infinite closed discrete subspace. This makes it possible to pass to a subsequence of $\mathcal{A}$ once more to guarantee that, for any

$j \in \{1,\ldots,k-r\}$, the sequence $S_j = \{a_j^i : i \in \omega\}$ is either convergent or constitutes a closed discrete subspace of $X$. If $S_j$ is convergent then denote by $x_j$ its limit. Renumbering every $A_i$ if necessary we can assume that $A_i = \{d_1,\ldots,d_r,a_1^i,\ldots,a_l^i,a_{l+1}^i,\ldots,a_{k-r}^i\}$ while the set $Q = \{a_{l+j}^i : i \in \omega, 1 \leq j \leq k-r-l\}$ is closed and discrete in $X$ and the sequence $S_j$ converges to $x_j$ for any $j \in \{1,\ldots,l\}$.

It turns out that the set $B = \{d_1,\ldots,d_r,x_1,\ldots,x_l\}$ belongs to $\mathcal{F}_Y(y,m,n)$. To prove this assume toward a contradiction that there is a function $f \in C_p(X)$ such that $f(B) \subset (-\frac{1}{m},\frac{1}{m})$ and $|\Phi(f)(y)| > \frac{1}{n}$. By continuity of $\Phi$ there is a finite set $E \supset B$ and $\delta > 0$ such that,

(1) for any $g \in C_p(X)$, if $|g(x) - f(x)| < \delta$ for all $x \in E$ then $|\Phi(g)(y)| > \frac{1}{n}$.

It is easy to find $W \in \tau(B,X)$ such that $\overline{W} \cap Q$ is finite and $f(W) \subset (-\frac{1}{m},\frac{1}{m})$. Every sequence $S_j$ is eventually in $W$, so there is $p \in \mathbb{N}$ such that $a_j^i \in W$ for all $j \leq l$ and $i \geq p$. Therefore we can pass to a subsequence of $\mathcal{A}$ once more to assume, without loss of generality, that $S_j \subset W$ for all $j \leq l$ and $(\overline{W} \cup E) \cap Q = \emptyset$. Now we can find a function $g \in C_p(X)$ such that $g|(\overline{W} \cup E) = f|(\overline{W} \cup E)$ and $g(Q) \subset \{0\}$. Then $g(A_i) \subset (-\frac{1}{m},\frac{1}{m})$ and hence $|\Phi(g)(y_i)| \leq \frac{1}{n}$ for every $i \in \omega$. The sequence $\{y_i : i \in \omega\}$ converges to $y$, so $|\Phi(g)(y)| \leq \frac{1}{n}$ by continuity of $\Phi(g)$. However $g|E = f|E$, so (1) implies that $|\Phi(g)(y)| > \frac{1}{n}$; this contradiction shows that $B \in \mathcal{F}_Y(y,m,n)$. Letting $q = l + r$ and $x_{l+i} = d_i$ for all $i \in \{1,\ldots,r\}$ we obtain the promised set $A = \{x_1,\ldots,x_q\} \in \mathcal{F}_Y(y,m,n)$; now it is evident how to enumerate the sets $A_i$ to obtain the convergence we need, so Fact 3 is proved.

*Fact 4.* For any $k,m,n \in \mathbb{N}$, the set $C_Y(k,m,n) = \{y \in Y : a_{m,n}(y) \leq k\}$ is closed in $Y$. Analogously, every set $C_X(k,m,n) = \{x \in X : b_{m,n}(x) \leq k\}$ is closed in $X$.

*Proof.* We have a symmetric situation with $C_X(k,m,n)$ and $C_Y(k,m,n)$, so it suffices to show that $C_Y(k,m,n)$ is closed in $Y$. To do so suppose that $y_i \in C_Y(k,m,n)$ for every $i \in \omega$ and the sequence $S = \{y_i : i \in \omega\}$ converges to a point $y \in Y$. Fix a set $A_i \in \mathcal{F}_Y(y_i,m,n)$ such that $|A_i| \leq k$ for each $i \in \omega$. Passing to a subsequence of $S$ if necessary, we can assume that there is $k_0 \leq k$ such that $|A_i| = k_0$ for all $i \in \omega$. It follows from Fact 3 that there is $k_1 \leq k_0$ and a set $A \in \mathcal{F}_Y(y,m,n)$ with $|A| \leq k_1$. Consequently, $a_{m,n}(y) \leq k_1 \leq k$ and hence $y \in C_Y(k,m,n)$. This shows that $C_Y(k,m,n)$ is, indeed, closed in $Y$, so Fact 4 is proved.

*Fact 5.* We have $Y = \bigcup_{k,m \in \mathbb{N}} C_Y(k,m,n)$ and $X = \bigcup_{k,m \in \mathbb{N}} C_X(k,m,n)$ for any $n \in \mathbb{N}$.

*Proof.* Fix any $n \in \mathbb{N}$; again, by the symmetry of the situation, it suffices to prove the first equality. To do so, take any point $y \in Y$. Since the set $O = \{f \in C_p(Y) : |f(y)| < \frac{1}{n}\} \subset G(y,n)$ is an open neighborhood of $\mathbf{0}_Y$, there is a set $A \in \mathrm{Fin}(X)$ and $m \in \mathbb{N}$ such that $\Phi(O_X(A,m)) \subset O$. We have $A \in \mathcal{F}_Y(y,m,n)$; if $|A| = k$ then $a_{m,n}(y) \leq k$ which shows that $y \in C_Y(k,m,n)$. Fact 5 is proved.

*Fact 6.* For any $m, n \in \mathbb{N}$ and $y \in Y$ the family $\mathcal{E}_Y(y, m, n)$ is finite. Analogously, every family $\mathcal{E}_X(x, m, n)$ is finite.

*Proof.* Once more we have a symmetric situation, so it suffices to show that the family $\mathcal{E}_Y(y, m, n)$ is finite. Assume the contrary, let $k = a_{m,n}(y)$ and fix an infinite family $\mathcal{A} \subset \mathcal{E}_Y(y, m, n)$. Apply Fact 2 of U.337 to find a set $D \subset X$ and an infinite subfamily $\mathcal{A}' \subset \mathcal{A}$ such that $A \cap A' = D$ for any distinct $A, A' \in \mathcal{A}'$; since also $|A| = |A'| = k$ for any $A, A' \in \mathcal{A}'$, we have $A \backslash D \neq \emptyset$ for all $A \in \mathcal{A}'$ and hence $|D| < k$.

We claim that $D \in \mathcal{F}_Y(y, m, n)$; to prove it suppose that there is a function $f \in O_X(D, m)$ such that $|\Phi(f)(y)| > \frac{1}{n}$. By continuity of $\Phi$, there is a finite set $E \supset D$ and $\delta > 0$ such that

(2)  if $|g(x) - f(x)| < \delta$ for all $x \in E$ then $|\Phi(g)(y)| > \frac{1}{n}$.

The family $\{A \backslash D : A \in \mathcal{A}'\}$ being disjoint and infinite, there is $A \in \mathcal{A}'$ such that $(A \backslash D) \cap E = \emptyset$. Pick $h \in C_p(X)$ such that $h(A \backslash D) = \{0\}$ and $h|E = f|E$. Then $h \in O_X(A, m)$; this, together with $A \in \mathcal{F}_Y(y, m, n)$ implies that $|\Phi(h)(y)| \leq \frac{1}{n}$. On the other hand, it follows from $f|E = h|E$ and (2) that $|\Phi(h)(y)| > \frac{1}{n}$ which is a contradiction. Therefore the family $\mathcal{A}$ is finite and hence Fact 6 is proved.

*Fact 7.* Fix $m, n \in \mathbb{N}$; we will need the set $D_Y(1, m, n) = C_Y(1, m, n)$; if $k > 1$ then let $D_Y(k, m, n) = C_Y(k, m, n) \backslash C_Y(k - 1, m, n)$. Now fix $k \in \mathbb{N}$ and define a map $\varphi_{m,n} : D_Y(k, m, n) \rightarrow \exp(X)$ by the equality $\varphi_{m,n}(y) = \bigcup \mathcal{E}_Y(y, m, n)$ for all $y \in D_Y(k, m, n)$. Next, for any $p, q \in \mathbb{N}$ such that $p \geq k$ consider the set $E_Y(k, m, n, p, q) = \{y \in D_Y(k, m, n) : |\varphi_{m,n}(y)| = p \text{ and } \rho_X(x, x') \geq \frac{1}{q} \text{ for distinct } x, x' \in \varphi_{m,n}(y)\}$. Then every $E_Y(k, m, n, p, q)$ is a $G_\delta$-subset of $Y$ and the map $\varphi_{m,n}|E_Y(k, m, n, p, q) : E_Y(k, m, n, p, q) \rightarrow [X]^p$ is lower semicontinuous.

*Proof.* The set $D_Y(k, m, n)$ is open in the closed set $C_Y(k, m, n)$; an immediate consequence is that $D_Y(k, m, n)$ is a $G_\delta$-subset of $Y$. For fixed $p \geq k$ and $q \in \mathbb{N}$ consider the set $K(p, q) = \{y \in D_Y(k, m, n) : \text{there exists a set } A \subset \varphi_{m,n}(y) \text{ such that } |A| = p \text{ and } \rho_X(x, x') \geq \frac{1}{q} \text{ for any distinct } x, x' \in A\}$.

To see that the set $K(p, q)$ is closed in $D_Y(k, m, n)$ suppose that a sequence $\{y_i : i \in \omega\} \subset K(p, q)$ converges to a point $y \in D_Y(k, m, n)$ and fix, for any $i \in \omega$, a set $A_i \subset \varphi_{m,n}(y_i)$ such that $|A_i| = p$ and $\rho_X(x, x') \geq \frac{1}{q}$ for any distinct $x, x' \in A_i$.

Take an arbitrary point $z_i \in A_i$; since $A_i \subset \bigcup \mathcal{E}_Y(y_i, m, n)$, we can pick a set $B_i \in \mathcal{E}_Y(y_i, m, n)$ with $z_i \in B_i$ for every $i \in \omega$. Apply Fact 3 to find a subsequence $\{B_{i_j} : j \in \omega\}$ of the sequence $\{B_i : i \in \omega\}$ for which we can choose an enumeration $\{b_1^j, \ldots, b_k^j\}$ of every $B_{i_j}$ in such a way that there is $k_0 \leq k$ for which every sequence $\{b_l^j : j \in \omega\}$ converges to a point $b_l$ and the set $\{b_1, \ldots, b_{k_0}\}$ belongs to $\mathcal{F}_Y(y, m, n)$. However, $a_{m,n}(y) = k$, so $k_0 = k$ and hence $z_{i_j} \in \{b_1^j, \ldots, b_{k_0}^j\}$ and therefore the sequence $\{z_{i_j} : j \in \omega\}$ converges to a point of $\varphi_{m,n}(y)$.

This shows that, passing several times to a subsequence of $\{A_i : i \in \omega\}$ if necessary, we can assume that there is an enumeration $\{a_1^i, \ldots, a_p^i\}$ of every $A_i$

such that the sequence $\{a_l^i : i \in \omega\}$ converges to a point $a_l \in \varphi_{m,n}(y)$ for any $l \leq p$. Given distinct numbers $l, l' \in \{1, \ldots, p\}$ we have $\rho_X(a_l^i, a_{l'}^i) \geq \frac{1}{q}$ for any $i \in \omega$; this, evidently, implies that $\rho_X(a_l, a_{l'}) \geq \frac{1}{q}$ and hence the set $\{a_1, \ldots, a_p\} \subset \varphi_{m,n}(y)$ witnesses that the point $y$ belongs to $K(p, q)$. This proves that every set $K(p, q)$ is closed in a $G_\delta$-set $D_Y(k, m, n)$, so it is a $G_\delta$-set in $Y$ as well. Now, it follows from the equality $E_Y(k, m, n, p, q) = K(p, q)\backslash(\bigcup\{K(p+1, r) : r \in \mathbb{N}\})$ that $E_Y(k, m, n, p, q)$ is a $G_\delta$-subset of $Y$.

To finally see that the map $\varphi = \varphi_{m,n}|E_Y(k, m, n, p, q)$ is lower semicontinuous let $E = E_Y(k, m, n, p, q)$ and fix a set $U \in \tau(X)$ such that $\varphi(y) \cap U \neq \emptyset$ for some $y \in E$. It suffices to show that there is a set $V \in \tau(y, E)$ such that $\varphi(z) \cap U \neq \emptyset$ for any $z \in V$. If such a set $V$ does not exist then it is easy to find a sequence $S = \{y_i : i \in \omega\} \subset E$ such that $S \to y$ and $\varphi(y_i) \cap U = \emptyset$ for any $i \in \omega$. Using the observation in the third paragraph of this proof we can pass to a subsequence of $S$ if necessary to be able to assume, without loss of generality, that $\varphi(y_i) = \{z_1^i, \ldots, z_p^i\}$ for every $i \in \omega$ and the sequence $\{z_l^i : i \in \omega\}$ is convergent to a point $z_l \in \varphi(y)\backslash U$ for any $l \leq p$. Given distinct numbers $l, l' \in \{1, \ldots, p\}$ we have $\rho_X(z_l^i, z_{l'}^i) \geq \frac{1}{q}$ for any $i \in \omega$; this, evidently, implies that $\rho_X(z_l, z_{l'}) \geq \frac{1}{q}$ and hence $z_l \neq z_{l'}$. Thus the set $\{z_1, \ldots, z_p\} \subset \varphi(y)\backslash U$ has cardinality $p$; since also $\varphi(y) \cap U \neq \emptyset$, we conclude that $|\varphi(y)| \geq p + 1$ which is a contradiction. Therefore the map $\varphi$ is lower semicontinuous and Fact 7 is proved.

*Fact 8.* Fix $m, n \in \mathbb{N}$; we will need the set $D_X(1, m, n) = C_X(1, m, n)$; if $k > 1$ then let $D_X(k, m, n) = C_X(k, m, n)\backslash C_X(k-1, m, n)$. Now fix $k \in \mathbb{N}$ and define a map $\xi_{m,n} : D_X(k, m, n) \to \exp(Y)$ by the equality $\xi_{m,n}(x) = \bigcup \mathcal{E}_X(x, m, n)$ for all $x \in D_X(k, m, n)$. Next, for any $p, q \in \mathbb{N}$ such that $p \geq k$ consider the set $E_X(k, m, n, p, q) = \{y \in D_X(k, m, n) : |\xi_{m,n}(x)| = p$ and $\rho_Y(y, y') \geq \frac{1}{q}$ for distinct $y, y' \in \xi_{m,n}(x)\}$. Then every $E_X(k, m, n, p, q)$ is a $G_\delta$-subset of $X$ and the map $\xi_{m,n}|E_X(k, m, n, p, q) : E_X(k, m, n, p, q) \to [Y]^p$ is lower semicontinuous.

*Proof.* Left as an exercise to the reader; one only needs to substitute the notions we used for $Y$ by their analogues for $X$ and vice versa.

*Fact 9.* If $k, m \in \mathbb{N}$ and $y \in D_Y(k, m, 1)$ then there are numbers $l, s \in \mathbb{N}$ and a point $x \in \varphi_{m,1}(y) \cap D_X(s, l, m+1)$ such that $y \in \xi_{l,m+1}(x)$.

*Proof.* Assume the contrary; by definition of $\varphi_{m,1}$ we can find a set $A \subset \varphi_{m,1}(y)$ such that $\Phi(O_X(A, m)) \subset G_Y(y, 1)$. Since $\bigcup\{C_X(p, q, m+1) : p, q \in \mathbb{N}\} = X$ by Fact 5, for each $x \in A$ there are $p, q_x \in \mathbb{N}$ with $x \in C_X(p, q_x, m+1)$. If $p_x = \backslash n\{p \in \mathbb{N} : x \in C_X(p, q_x, m+1)\}$ then $x \in D_X(p_x, q_x, m+1)$.

By our assumption, $y \notin P = \bigcup\{\xi_{q_x, m+1}(x) : x \in A\}$, so we can take $f \in C_p(Y)$ such that $f(y) = 2$ and $f|P \equiv 0$. Every $\xi_{q_x, m+1}(x)$ contains a set $B_x$ such that $\Phi^{-1}(O_Y(B_x, q_x)) \subset G_X(x, m+1)$. It is evident that $f \in O_Y(B_x, q_x)$ for any $x \in A$, so $|\Phi^{-1}(f)(x)| \leq \frac{1}{m+1} < \frac{1}{m}$ which shows that $g = \Phi^{-1}(f) \in O_X(A, m)$ and therefore $f = \Phi(g) \in G_Y(y, 1)$, i.e., $|f(y)| \leq 1$ which is a contradiction. As a consequence, $y \in \xi_{q_x, m+1}(x)$ for some $x \in \varphi_{m,1}(y) \cap D_X(p_x, q_x, m+1)$, so the numbers $s = p_x$ and $l = q_x$ are as promised. Fact 9 is proved.

Returning to our solution observe that $Y = \bigcup_{k,m \in \mathbb{N}} C_Y(k,m,1)$ by Fact 5; it follows from the definition of the sets $D_Y(k,m,1)$ that $Y = \bigcup_{k,m \in \mathbb{N}} D_Y(k,m,1)$. Another easy observation is that $D_Y(k,m,1) = \bigcup\{E_Y(k,m,1,p,q) : p \geq k, q \in \mathbb{N}\}$ which shows that $Y$ is a countable union of its $G_\delta$-subspaces $E_Y(k,m,1,p,q)$, so it suffices to show that every set $E_Y(k,m,1,p,q)$ is a countable union of subspaces homeomorphic to their respective $G_\delta$-subspaces of $X$. To that end fix $k, m, p, q \in \mathbb{N}$ with $k \leq p$.

Apply Fact 7 and Fact 2 to see that $E_Y(k,m,1,p,q) = \bigcup_{r \in \mathbb{N}} G_r$ where every $G_r$ is open in $E_Y(k,m,1,p,q)$ (and hence is a $G_\delta$-subset of $Y$) and there are continuous functions $f_1, \ldots, f_p : G_r \to X$ such that $\varphi_{m,1}(y) = \{f_1(y), \ldots, f_p(y)\}$ for any point $y \in G_r$. It suffices to show that every $G_r$ is a countable union of its $G_\delta$-subsets which are homeomorphic to their respective $G_\delta$-subspaces of $X$, so we fix $r \in \mathbb{N}$ and $f_1, \ldots, f_p$ as above.

Analogously, $X = \bigcup\{E_X(k',m',m+1,p',q') : k',m',p',q' \in \mathbb{N}, \ p' \geq k'\}$, so we can apply Fact 2 and Fact 7 again to convince ourselves that $X = \bigcup\{H_s : s \in \mathbb{N}\}$ where every $H_s$ is a $G_\delta$-subset of $X$, contained in some $E_X(k',m',m+1,p',q')$, for which there exist continuous maps $g_1^s, \ldots, g_{l_s}^s : H_s \to Y$ such that $l_s = p'$ and $\xi_{m',m+1}(x) = \{g_1^s(x), \ldots, g_{l_s}^s(x)\}$ for any $x \in H_s$.

Applying Fact 9 we conclude that

(3) for any $y \in G_r$ with $f_i(y) \in H_{s_i}$ for all $i \leq p$ there exist $i \leq p$ and $j \leq l_{s_i}$ such that $g_j^{s_i}(f_i(y)) = y$.

Given any $\theta = (s_1, \ldots, s_p) \in \mathbb{N}^p$ the set $G_r[\theta] = \{y \in G_r : f_i(y) \in H_{s_i}$ for each $i \leq p\}$ is a $G_\delta$-subspace of $Y$; it is evident that $G_r = \bigcup\{G_r[\theta] : \theta \in \mathbb{N}^p\}$ and hence again it suffices to show that every set $G_r[\theta]$ is the countable union of its $G_\delta$-subsets which are homeomorphic to their respective $G_\delta$-subspaces of $X$, so we fix an arbitrary $\theta = (s_1, \ldots, s_p) \in \mathbb{N}^p$.

The property (3) implies that $G_r[\theta] = \bigcup_{i \leq p} \bigcup_{j \leq l_{s_i}} \{y \in G_r[\theta] : g_j^{s_i}(f_i(y)) = y\}$; every set $Q(i,j) = \{y \in G_r[\theta] : g_j^{s_i}(f_i(y)) = y\}$ is a $G_\delta$-subset of $Y$ being closed in $G_r[\theta]$; therefore $Q(i,j)$ is homeomorphic to a $G_\delta$-subspace of $X$ by Fact 1. This completes the desired representation of $Y$ as the countable union of its $G_\delta$-subspaces each one of which is homeomorphic to a $G_\delta$-subspace of $X$.

**V.081.** *Let $X$ and $Y$ be metrizable spaces such that $C_p^*(X) \simeq C_p^*(Y)$. Prove that $Y = \bigcup\{Y_n : n \in \omega\}$, where each $Y_n$ is a $G_\delta$-subspace of $Y$, homeomorphic to some $G_\delta$-subspace of $X$.*

**Solution.** For any set $P$ denote by $\mathrm{Fin}(P)$ the family of all finite subsets of $P$; if $n \in \mathbb{N}$ then $[P]^n = \{B \subset P : |B| = n\}$. If $A$ is a finite subset of $Z$ and $n \in \mathbb{N}$ then $O_Z(A,n) = \{f \in C_p^*(Z) : |f(x)| < \frac{1}{n}$ for any $x \in A\}$. For all $z \in Z$ and $n \in \mathbb{N}$ let $G_Z(z;n) = \{f \in C_p^*(Z) : |f(z)| \leq \frac{1}{n}\}$.

Choose metrics $\rho_X$ and $\rho_Y$ which generate the topologies of the spaces $X$ and $Y$ respectively; the spaces $C_p^*(X)$ and $C_p^*(Y)$ being homogeneous, it is easy to convince ourselves that there exists a homeomorphism $\Phi : C_p^*(X) \to C_p^*(Y)$ such that $\Phi(0_X) = 0_Y$. Fix $m, n \in \mathbb{N}$ and consider, for every point $y \in Y$, the family $\mathcal{F}_Y(y,m,n) = \{A \in \mathrm{Fin}(X) : \Phi(O_X(A,m)) \subset G_Y(y,n)\}$; if $\mathcal{F}_Y(y,m,n) \neq \emptyset$

then we can define $a_{m,n}(y) = \setminus n\{|A| : A \in \mathcal{F}_Y(y,m,n)\}$. Likewise, for all $x \in X$ let $\mathcal{F}_X(x,m,n) = \{B \in \text{Fin}(Y) : \Phi^{-1}(O_Y(B,m)) \subset G_X(x,n)\}$; if $\mathcal{F}_X(x,m,n) \neq \emptyset$ then let $b_{m,n}(x) = \setminus n\{|B| : B \in \mathcal{F}_X(x,m,n)\}$. In what follows, when we deal with $a_{m,n}(y)$ and/or $b_{m,n}(x)$ we assume, without mentioning it explicitly, that the respective family $\mathcal{F}_Y(y,m,n)$ and/or $\mathcal{F}_X(x,m,n)$ is nonempty.

Observe that $O_X(\emptyset, m) = C_p^*(X)$ and hence $\Phi(O_X(\emptyset,m)) = C_p^*(Y) \not\subset G(y,n)$ for any $y \in Y$; this shows that if $\mathcal{F}_Y(y,m,n) \neq \emptyset$ then all elements of the family $\mathcal{F}_Y(y,m,n)$ are nonempty and hence $a_{m,n}(y) \in \mathbb{N}$ for all $m,n \in \mathbb{N}$; let

$$\mathcal{E}_Y(y,m,n) = \{A \in \mathcal{F}_Y(y,m,n) : |A| = a_{m,n}(y)\}.$$

The same reasoning demonstrates that if $\mathcal{F}_X(x,m,n) \neq \emptyset$ then all elements of $\mathcal{F}_X(x,m,n)$ are nonempty; let $\mathcal{E}_X(x,m,n) = \{B \in \mathcal{F}_X(x,m,n) : |B| = b_{m,n}(x)\}$.

*Fact 1.* Suppose that $m,n,k \in \mathbb{N}$ and a sequence $\{y_i : i \in \omega\} \subset Y$ converges to a point $y \in Y$. If $A_i \in \mathcal{F}_Y(y_i,m,n)$ and $|A_i| = k$ for any $i \in \omega$ then there is a subsequence $\{A_{i_j} : j \in \omega\}$ of the sequence $\mathcal{A} = \{A_i : i \in \omega\}$ and a number $q \leq k$ for which we can choose a faithful enumeration $\{a_1^j, \ldots, a_k^j\}$ of every set $A_{i_j}$ in such a way that the sequence $\{a_p^j : j \in \omega\}$ converges to a point $x_p \in X$ for all $p \leq q$ and $A = \{x_1, \ldots, x_q\} \in \mathcal{F}_Y(y,m,n)$.

*Proof.* We will pass several times to a subsequence of the sequence $\mathcal{A} = \{A_i : i \in \omega\}$; since our aim is to find a certain subsequence of $\mathcal{A}$, at each step we will identify the obtained subsequence with $\mathcal{A}$ considering that all elements of $\mathcal{A}$ have the property we have found in a subsequence.

The first step is to use Fact 2 of U.337 to choose a subsequence $\mathcal{A}' \subset \mathcal{A}$ for which there is a set $D = \{d_1, \ldots, d_r\} \subset X$ such that $A \cap A' = D$ for distinct $A, A' \in \mathcal{A}'$ (observe that it is possible that $r = 0$ in which case $D = \emptyset$). According to the above-mentioned politics we can consider that, for any $i \in \omega$, we have $A_i = \{d_1, \ldots, d_r, a_1^i, \ldots, a_{k-r}^i\}$ and the family $\{A_i \setminus D : i \in \omega\}$ is disjoint.

An evident property of metric spaces is that any sequence contains either a convergent subsequence or a closed discrete subspace. This makes it possible to pass to a subsequence of $\mathcal{A}$ once more to guarantee that, for any $j \in \{1, \ldots, k-r\}$, the sequence $S_j = \{a_j^i : i \in \omega\}$ is either convergent or constitutes a closed discrete subspace of $X$. If $S_j$ is convergent then denote by $x_j$ its limit. Renumbering every $A_i$ if necessary we can assume that $A_i = \{d_1, \ldots, d_r, a_1^i, \ldots, a_l^i, a_{l+1}^i, \ldots, a_{k-r}^i\}$ while the set $Q = \{a_{l+j}^i : i \in \omega, 1 \leq j \leq k-r-l\}$ is closed and discrete in $X$ and the sequence $S_j$ converges to $x_j$ for any $j \in \{1, \ldots, l\}$.

It turns out that the set $B = \{d_1, \ldots, d_r, x_1, \ldots, x_l\}$ belongs to $\mathcal{F}_Y(y,m,n)$. To prove this assume toward a contradiction that there is a function $f \in C_p^*(X)$ such that $f(B) \subset (-\frac{1}{m}, \frac{1}{m})$ and $|\Phi(f)(y)| > \frac{1}{n}$. By continuity of $\Phi$ there is a finite set $E \supset B$ and $\delta > 0$ such that,

(1) for any $g \in C_p^*(X)$, if $|g(x) - f(x)| < \delta$ for all $x \in E$ then $|\Phi(g)(y)| > \frac{1}{n}$.

It is easy to find $W \in \tau(B, X)$ such that $\overline{W} \cap Q$ is finite and $f(W) \subset (-\frac{1}{m}, \frac{1}{m})$. Every sequence $S_j$ is eventually in $W$, so there is $p \in \mathbb{N}$ such that $a_j^i \in W$ for all $j \leq l$ and $i \geq p$. Therefore we can pass to a subsequence of $\mathcal{A}$ once more to assume, without loss of generality, that $S_j \subset W$ for all $j \leq l$ and $(\overline{W} \cup E) \cap Q = \emptyset$. Now we can find a function $g \in C_p^*(X)$ such that $g|(\overline{W} \cup E) = f|(\overline{W} \cup E)$ and $g(Q) \subset \{0\}$. Then $g(A_i) \subset (-\frac{1}{m}, \frac{1}{m})$ and hence $|\Phi(g)(y_i)| \leq \frac{1}{n}$ for every $i \in \omega$. The sequence $\{y_i : i \in \omega\}$ converges to $y$, so $|\Phi(g)(y)| \leq \frac{1}{n}$ by continuity of $\Phi(g)$. However $g|E = f|E$, so (1) implies that $|\Phi(g)(y)| > \frac{1}{n}$; this contradiction shows that $B \in \mathcal{F}_Y(y, m, n)$. Letting $q = l + r$ and $x_{l+i} = d_i$ for all $i \in \{1, \dots, r\}$ we obtain the promised set $A = \{x_1, \dots, x_q\} \in \mathcal{F}_Y(y, m, n)$; now it is evident how to enumerate the sets $A_i$ to obtain the convergence we need, so Fact 1 is proved.

*Fact 2.* For any $k, m, n \in \mathbb{N}$, the set $C_Y(k, m, n) = \{y \in Y : a_{m,n}(y) \leq k\}$ is closed in $Y$. Analogously, every set $C_X(k, m, n) = \{x \in X : b_{m,n}(x) \leq k\}$ is closed in $X$.

*Proof.* We have a symmetric situation with $C_X(k, m, n)$ and $C_Y(k, m, n)$, so it suffices to show that $C_Y(k, m, n)$ is closed in $Y$. To do so suppose that $y_i \in C_Y(k, m, n)$ for every $i \in \omega$ and the sequence $S = \{y_i : i \in \omega\}$ converges to a point $y \in Y$. Fix a set $A_i \in \mathcal{F}_Y(y_i, m, n)$ such that $|A_i| \leq k$ for each $i \in \omega$. Passing to a subsequence of $S$ if necessary, we can assume that there is $k_0 \leq k$ such that $|A_i| = k_0$ for all $i \in \omega$. It follows from Fact 1 that there is $k_1 \leq k_0$ and a set $A \in \mathcal{F}_Y(y, m, n)$ with $|A| \leq k_1$. Consequently, $a_{m,n}(y) \leq k_1 \leq k$ and hence $y \in C_Y(k, m, n)$. This shows that $C_Y(k, m, n)$ is, indeed, closed in $Y$, so Fact 2 is proved.

*Fact 3.* We have $Y = \bigcup_{k,m \in \mathbb{N}} C_Y(k, m, n)$ and $X = \bigcup_{k,m \in \mathbb{N}} C_X(k, m, n)$ for any $n \in \mathbb{N}$.

*Proof.* Fix any number $n \in \mathbb{N}$; again, by the symmetry of the situation, it suffices to prove only the first equality. To do so, take any point $y \in Y$. Since the set $O = \{f \in C_p^*(Y) : |f(y)| < \frac{1}{n}\} \subset G(y, n)$ is an open neighborhood of $0_Y$, there is a set $A \in \text{Fin}(X)$ and $m \in \mathbb{N}$ such that $\Phi(O_X(A, m)) \subset O$. We have $A \in \mathcal{F}_Y(y, m, n)$; if $|A| = k$ then $a_{m,n}(y) \leq k$ which shows that $y \in C_Y(k, m, n)$. Fact 3 is proved.

*Fact 4.* For any $m, n \in \mathbb{N}$ and $y \in Y$ the family $\mathcal{E}_Y(y, m, n)$ is finite. Analogously, every family $\mathcal{E}_X(x, m, n)$ is finite.

*Proof.* Once more we have a symmetric situation, so it suffices to show that the family $\mathcal{E}_Y(y, m, n)$ is finite. Assume the contrary, let $k = a_{m,n}(y)$ and fix an infinite family $\mathcal{A} \subset \mathcal{E}_Y(y, m, n)$. Apply Fact 2 of U.337 to find a set $D \subset X$ and an infinite subfamily $\mathcal{A}' \subset \mathcal{A}$ such that $A \cap A' = D$ for any distinct $A, A' \in \mathcal{A}'$; since also $|A| = |A'| = k$ for any $A, A' \in \mathcal{A}'$, we have $A \backslash D \neq \emptyset$ for all $A \in \mathcal{A}'$ and hence $|D| < k$.

We claim that $D \in \mathcal{F}_Y(y, m, n)$; to prove it suppose that there is a function $f \in O_X(D, m)$ such that $|\Phi(f)(y)| > \frac{1}{n}$. By continuity of $\Phi$, there is a finite set $E \supset D$ and $\delta > 0$ such that

(2) if $|g(x) - f(x)| < \delta$ for all $x \in E$ then $|\Phi(g)(y)| > \frac{1}{n}$.

The family $\{A\backslash D : A \in \mathcal{A}'\}$ being disjoint and infinite, there is $A \in \mathcal{A}'$ such that $(A\backslash D) \cap E = \emptyset$. Pick $h \in C_p^*(X)$ such that $h(A\backslash D) = \{0\}$ and $h|E = f|E$. Then $h \in O_X(A, m)$; this, together with $A \in \mathcal{F}_Y(y, m, n)$ implies $|\Phi(h)(y)| \leq \frac{1}{n}$. On the other hand, it follows from $f|E = h|E$ and (2) that $|\Phi(h)(y)| > \frac{1}{n}$ which is a contradiction. Therefore the family $\mathcal{A}$ is finite and hence Fact 4 is proved.

*Fact 5.* Fix $m, n \in \mathbb{N}$; we will need the set $D_Y(1, m, n) = C_Y(1, m, n)$; if $k > 1$ then let $D_Y(k, m, n) = C_Y(k, m, n)\backslash C_Y(k - 1, m, n)$. Now fix $k \in \mathbb{N}$ and define a map $\varphi_{m,n} : D_Y(k, m, n) \to \exp(X)$ by the equality $\varphi_{m,n}(y) = \bigcup \mathcal{E}_Y(y, m, n)$ for all $y \in D_Y(k, m, n)$. Next, for any $p, q \in \mathbb{N}$ such that $p \geq k$ consider the set $E_Y(k, m, n, p, q) = \{y \in D_Y(k, m, n) : |\varphi_{m,n}(y)| = p$ and $\rho_X(x, x') \geq \frac{1}{q}$ for distinct $x, x' \in \varphi_{m,n}(y)\}$. Then every $E_Y(k, m, n, p, q)$ is a $G_\delta$-subset of $Y$ and the map $\varphi_{m,n}|E_Y(k, m, n, p, q) : E_Y(k, m, n, p, q) \to [X]^p$ is lower semicontinuous.

*Proof.* The set $D_Y(k, m, n)$ is open in the closed set $C_Y(k, m, n)$; an immediate consequence is that $D_Y(k, m, n)$ is a $G_\delta$-subset of $Y$. For fixed $p \geq k$ and $q \in \mathbb{N}$ consider the set $K(p, q) = \{y \in D_Y(k, m, n) :$ there exists a set $A \subset \varphi_{m,n}(y)$ such that $|A| = p$ and $\rho_X(x, x') \geq \frac{1}{q}$ for any distinct $x, x' \in A\}$.

To see that the set $K(p, q)$ is closed in $D_Y(k, m, n)$ suppose that a sequence $\{y_i : i \in \omega\} \subset K(p, q)$ converges to a point $y \in D_Y(k, m, n)$ and fix, for any $i \in \omega$, a set $A_i \subset \varphi_{m,n}(y_i)$ such that $|A_i| = p$ and $\rho_X(x, x') \geq \frac{1}{q}$ for any distinct $x, x' \in A_i$.

Take an arbitrary point $z_i \in A_i$; since $A_i \subset \bigcup \mathcal{E}_Y(y_i, m, n)$, we can pick a set $B_i \in \mathcal{E}_Y(y_i, m, n)$ with $z_i \in B_i$ for every $i \in \omega$. Apply Fact 1 to find a subsequence $\{B_{i_j} : j \in \omega\}$ of the sequence $\{B_i : i \in \omega\}$ for which we can choose an enumeration $\{b_1^j, \ldots, b_k^j\}$ of every $B_{i_j}$ in such a way that there is $k_0 \leq k$ for which every sequence $\{b_l^j : j \in \omega\}$ converges to a point $b_l$ and the set $\{b_1, \ldots, b_{k_0}\}$ belongs to $\mathcal{F}_Y(y, m, n)$. However, $a_{m,n}(y) = k$, so $k_0 = k$ and hence $z_{i_j} \in \{b_1^j, \ldots, b_{k_0}^j\}$ and therefore the sequence $\{z_{i_j} : j \in \omega\}$ converges to a point of $\varphi_{m,n}(y)$.

This shows that, passing several times to a subsequence of $\{A_i : i \in \omega\}$ if necessary, we can assume that there is an enumeration $\{a_1^i, \ldots, a_p^i\}$ of every $A_i$ such that the sequence $\{a_l^i : i \in \omega\}$ converges to a point $a_l \in \varphi_{m,n}(y)$ for any $l \leq p$. Given distinct numbers $l, l' \in \{1, \ldots, p\}$ we have $\rho_X(a_l^i, a_{l'}^i) \geq \frac{1}{q}$ for any $i \in \omega$; this, evidently, implies that $\rho_X(a_l, a_{l'}) \geq \frac{1}{q}$ and hence the set $\{a_1, \ldots, a_p\} \subset \varphi_{m,n}(y)$ witnesses that the point $y$ belongs to $K(p, q)$. This proves that every set $K(p, q)$ is closed in a $G_\delta$-set $D_Y(k, m, n)$, so it is a $G_\delta$-set in $Y$ as well. Now, it follows from the equality $E_Y(k, m, n, p, q) = K(p, q)\backslash(\bigcup\{K(p + 1, r) : r \in \mathbb{N}\})$ that $E_Y(k, m, n, p, q)$ is a $G_\delta$-subset of $Y$.

To finally see that the map $\varphi = \varphi_{m,n}|E_Y(k, m, n, p, q)$ is lower semicontinuous let $E = E_Y(k, m, n, p, q)$ and fix a set $U \in \tau(X)$ such that $\varphi(y) \cap U \neq \emptyset$ for some $y \in E$. It suffices to show that there is a set $V \in \tau(y, E)$ such that $\varphi(z) \cap U \neq \emptyset$ for any $z \in V$. If such a set $V$ does not exist then it is easy to find a sequence $S = \{y_i : i \in \omega\} \subset E$ such that $S \to y$ and $\varphi(y_i) \cap U = \emptyset$ for any $i \in \omega$. Using the observation in the third paragraph of this proof we can pass to a subsequence of $S$ if necessary to be able to assume, without loss of generality, that $\varphi(y_i) = \{z_1^i, \ldots, z_p^i\}$ for every $i \in \omega$ and the sequence $\{z_l^i : i \in \omega\}$ is convergent to a point $z_l \in \varphi(y) \backslash U$ for any $l \leq p$. Given distinct numbers $l, l' \in \{1, \ldots, p\}$ we have $\rho_X(z_l^i, z_{l'}^i) \geq \frac{1}{q}$ for any $i \in \omega$; this, evidently, implies that $\rho_X(z_l, z_{l'}) \geq \frac{1}{q}$ and hence $z_l \neq z_{l'}$. Thus the set $\{z_1, \ldots, z_p\} \subset \varphi(y) \backslash U$ has cardinality $p$; since also $\varphi(y) \cap U \neq \emptyset$, we conclude that $|\varphi(y)| \geq p+1$ which is a contradiction. Therefore the map $\varphi$ is lower semicontinuous and Fact 5 is proved.

*Fact 6.* Fix $m, n \in \mathbb{N}$; we will need the set $D_X(1, m, n) = C_X(1, m, n)$; if $k > 1$ then let $D_X(k, m, n) = C_X(k, m, n) \backslash C_X(k-1, m, n)$. Now fix $k \in \mathbb{N}$ and define a map $\xi_{m,n} : D_X(k, m, n) \to \exp(Y)$ by the equality $\xi_{m,n}(x) = \bigcup \mathcal{E}_X(x, m, n)$ for all $x \in D_X(k, m, n)$. Next, for any $p, q \in \mathbb{N}$ such that $p \geq k$ consider the set $E_X(k, m, n, p, q) = \{y \in D_X(k, m, n) : |\xi_{m,n}(x)| = p$ and $\rho_Y(y, y') \geq \frac{1}{q}$ for distinct $y, y' \in \xi_{m,n}(x)\}$. Then every $E_X(k, m, n, p, q)$ is a $G_\delta$-subset of $X$ and the map $\xi_{m,n}|E_X(k, m, n, p, q) : E_X(k, m, n, p, q) \to [Y]^p$ is lower semicontinuous.

*Proof.* Left as an exercise to the reader; one only needs to substitute the notions we used for $Y$ by their analogues for $X$ and vice versa.

*Fact 7.* If $k, m \in \mathbb{N}$ and $y \in D_Y(k, m, 1)$ then there are numbers $l, s \in \mathbb{N}$ and a point $x \in \varphi_{m,1}(y) \cap D_X(s, l, m+1)$ such that $y \in \xi_{l, m+1}(x)$.

*Proof.* Assume the contrary; by definition of $\varphi_{m,1}$ we can find a set $A \subset \varphi_{m,1}(y)$ such that $\Phi(O_X(A, m)) \subset G_Y(y, 1)$. Since $\bigcup \{C_X(p, q, m+1) : p, q \in \mathbb{N}\} = X$ by Fact 3, for each $x \in A$ there are $p, q_x \in \mathbb{N}$ with $x \in C_X(p, q_x, m+1)$. If $p_x = \backslash n\{p \in \mathbb{N} : x \in C_X(p, q_x, m+1)\}$ then $x \in D_X(p_x, q_x, m+1)$.

By our assumption, $y \notin P = \bigcup \{\xi_{q_x, m+1}(x) : x \in A\}$, so we can take $f \in C_p^*(Y)$ such that $f(y) = 2$ and $f|P \equiv 0$. Every $\xi_{q_x, m+1}(x)$ contains a set $B_x$ such that $\Phi^{-1}(O_Y(B_x, q_x)) \subset G_X(x, m+1)$. It is evident that $f \in O_Y(B_x, q_x)$ for any $x \in A$, so $|\Phi^{-1}(f)(x)| \leq \frac{1}{m+1} < \frac{1}{m}$ which shows that $g = \Phi^{-1}(f) \in O_X(A, m)$ and therefore $f = \Phi(g) \in G_Y(y, 1)$, i.e., $|f(y)| \leq 1$ which is a contradiction. As a consequence, $y \in \xi_{q_x, m+1}(x)$ for some $x \in \varphi_{m,1}(y) \cap D_X(p_x, q_x, m+1)$, so the numbers $s = p_x$ and $l = q_x$ are as promised. Fact 7 is proved.

Returning to our solution observe that $Y = \bigcup_{k, m \in \mathbb{N}} C_Y(k, m, 1)$ by Fact 3; it follows from the definition of the sets $D_Y(k, m, 1)$ that $Y = \bigcup_{k, m \in \mathbb{N}} D_Y(k, m, 1)$. Another easy observation is that $D_Y(k, m, 1) = \bigcup \{E_Y(k, m, 1, p, q) : p \geq k, q \in \mathbb{N}\}$ which shows that $Y$ is a countable union of its $G_\delta$-subspaces $E_Y(k, m, 1, p, q)$, so it suffices to show that every set $E_Y(k, m, 1, p, q)$ is a countable union of subspaces homeomorphic to their respective $G_\delta$-subspaces of $X$. To that end fix $k, m, p, q \in \mathbb{N}$ with $k \leq p$.

Apply Fact 5 and Fact 2 of V.080 to see that $E_Y(k, m, 1, p, q) = \bigcup_{r \in \mathbb{N}} G_r$ where every $G_r$ is open in $E_Y(k, m, 1, p, q)$ (and hence is a $G_\delta$-subset of $Y$) and there are continuous functions $f_1, \ldots, f_p : G_r \to X$ such that $\varphi_{m,1}(y) = \{f_1(y), \ldots, f_p(y)\}$ for any point $y \in G_r$. It suffices to show that every $G_r$ is a countable union of its $G_\delta$-subsets which are homeomorphic to their respective $G_\delta$-subspaces of $X$, so we fix $r \in \mathbb{N}$ and $f_1, \ldots, f_p$ as above.

Analogously, $X = \bigcup \{E_X(k', m', m + 1, p', q') : k', m', p', q' \in \mathbb{N}, \ p' \geq k'\}$, so we can apply Fact 5 and Fact 2 of V.080 again to convince ourselves that $X = \bigcup \{H_s : s \in \mathbb{N}\}$ where every $H_s$ is a $G_\delta$-subset of $X$, contained in some $E_X(k', m', m+1, p', q')$, for which there exist continuous maps $g_1^s, \ldots, g_{l_s}^s : H_s \to Y$ such that $l_s = p'$ and $\xi_{m', m+1}(x) = \{g_1^s(x), \ldots, g_{l_s}^s(x)\}$ for any $x \in H_s$.

Applying Fact 7 we conclude that

(3) for any $y \in G_r$ with $f_i(y) \in H_{s_i}$ for all $i \leq p$ there exist $i \leq p$ and $j \leq l_{s_i}$ such that $g_j^{s_i}(f_i(y)) = y$.

Given any $\theta = (s_1, \ldots, s_p) \in \mathbb{N}^p$ the set $G_r[\theta] = \{y \in G_r : f_i(y) \in H_{s_i}$ for each $i \leq p\}$ is a $G_\delta$-subspace of $Y$; it is evident that $G_r = \bigcup \{G_r[\theta] : \theta \in \mathbb{N}^p\}$ and hence again it suffices to show that every set $G_r[\theta]$ is the countable union of its $G_\delta$-subsets which are homeomorphic to their respective $G_\delta$-subspaces of $X$, so we fix an arbitrary $\theta = (s_1, \ldots, s_p) \in \mathbb{N}^p$.

The property (3) implies that $G_r[\theta] = \bigcup_{i \leq p} \bigcup_{j \leq l_{s_i}} \{y \in G_r[\theta] : g_j^{s_i}(f_i(y)) = y\}$; every set $Q(i, j) = \{y \in G_r[\theta] : g_j^{s_i}(f_i(y)) = y\}$ is a $G_\delta$-subset of $Y$ being closed in $G_r[\theta]$; therefore $Q(i, j)$ is homeomorphic to a $G_\delta$-subspace of $X$ by Fact 1 of V.080. This completes the desired representation of $Y$ as the countable union of its $G_\delta$-subspaces each one of which is homeomorphic to a $G_\delta$-subspace of $X$.

**V.082.** *Let $X$ and $Y$ be metrizable $t$-equivalent spaces. Prove that $X$ is a countable union of zero-dimensional subspaces if and only if so is $Y$.*

**Solution.** Say that a space $Z$ belongs to the class $\mathcal{P}$ if $Z$ is the countable union of its zero-dimensional subspaces. Since every subspace of a zero-dimensional space is zero-dimensional, we conclude that

(1) if $Z \in \mathcal{P}$ then each $Y \subset Z$ also belongs to $\mathcal{P}$.

Suppose that $X$ is the countable union of its zero-dimensional subspaces, i.e., $X \in \mathcal{P}$. By Problem 080, there exists a family $\mathcal{F} = \{Y_n : n \in \omega\}$ of subspaces of $Y$ such that $Y = \bigcup \mathcal{F}$ and every $Y_n$ is homeomorphic to a subspace of $X$. Apply (1) to see that $Y_n \in \mathcal{P}$ for any $n \in \omega$; the class $\mathcal{P}$ is, evidently, $\sigma$-additive, so $Y \in \mathcal{P}$. Analogously, if $Y \in \mathcal{P}$ then $X \in \mathcal{P}$, so $X$ is the countable union of its zero-dimensional subspaces if and only if so is $Y$.

**V.083.** *Let $X$ and $Y$ be metrizable $t$-equivalent spaces. Prove that $X$ is a countable union of its Čech-complete subspaces if and only if so is $Y$.*

**Solution.** Say that a space $Z$ belongs to the class $\mathcal{P}$ if $Z$ is the countable union of its Čech-complete subspaces. Since every $G_\delta$-subspace of a Čech-complete space is Čech-complete (see TFS-260), we conclude that

(1) if $Z \in \mathcal{P}$ then each $G_\delta$-subspace $Y \subset Z$ also belongs to $\mathcal{P}$.

Suppose that $X$ is the countable union of its Čech-complete subspaces, i.e., $X \in \mathcal{P}$. By Problem 080, there exists a family $\mathcal{F} = \{Y_n : n \in \omega\}$ of subspaces of $Y$ such that $Y = \bigcup \mathcal{F}$ and every $Y_n$ is homeomorphic to a $G_\delta$-subspace of $X$.

Apply (1) to see that $Y_n \in \mathcal{P}$ for any $n \in \omega$; the class $\mathcal{P}$ is, evidently, $\sigma$-additive, so $Y \in \mathcal{P}$. Analogously, if $Y \in \mathcal{P}$ then $X \in \mathcal{P}$, so $X$ is the countable union of its Čech-complete subspaces if and only if so is $Y$.

**V.084.** *Suppose that $\{a_0, \ldots, a_m\} \subset \mathbb{R}^n$ is an independent set. Prove that*

*(i)   the simplex $S = [a_0, \ldots, a_m]$ is a compact subset of $\mathbb{R}^n$;*
*(ii)  any two $m$-dimensional simplexes are homeomorphic;*
*(iii) the barycentric coordinates are continuous functions from $S$ to $[0, 1]$.*

**Solution.** Consider the set $\Delta_m = \{\lambda = (\lambda_0, \ldots, \lambda_m) \in \mathbb{R}^{m+1} : \lambda_i \geq 0$ for all $i \leq m$ and $\lambda_0 + \ldots + \lambda_m = 1\}$. The $i$-th projection $p_i : \mathbb{R}^{m+1} \to \mathbb{R}$ defined by $p_i(\lambda) = \lambda_i$ for any $\lambda = (\lambda_0, \ldots, \lambda_m) \in \mathbb{R}^{m+1}$ is continuous for any $i \leq m$; an easy consequence is that the set $\Delta_m$ is closed in $\mathbb{R}^{m+1}$. Since also $0 \leq \lambda_i \leq 1$ for any $\lambda = (\lambda_0, \ldots, \lambda_m) \in \Delta_m$ and $i \leq m$, we have $\Delta_m \subset \mathbb{I}^{m+1}$, so $\Delta_m$ is compact being closed in the compact space $\mathbb{I}^{m+1}$.

For any $\lambda = (\lambda_0, \ldots, \lambda_m) \in \Delta_m$ let $\varphi(\lambda) = \lambda_0 \cdot a_0 + \ldots + \lambda_m \cdot a_m$; since $\mathbb{R}^n$ is a linear topological space, the map $\varphi : \Delta_m \to \mathbb{R}^n$ is continuous. Besides, $S = \varphi(\Delta_m)$, so $S$ is compact and hence (i) is proved.

Suppose that $\lambda = (\lambda_0, \ldots, \lambda_m)$, $\mu = (\mu_0, \ldots, \mu_m)$, $\lambda, \mu \in \Delta_m$ and $\varphi(\lambda) = \varphi(\mu)$. If $\mathbf{0} \in \mathbb{R}^n$ is the zero vector then $(\lambda_0 - \mu_0) \cdot a_0 + \ldots + (\lambda_m - \mu_m) \cdot a_m = \mathbf{0}$ and $(\lambda_0 - \mu_0) + \ldots + (\lambda_m - \mu_m) = 0$ which implies, together with independency of $\{a_0, \ldots, a_m\}$, that $\lambda_i = \mu_i$ for every $i \leq m$, i.e., $\lambda = \mu$; this shows that the map $\varphi$ is injective. Since $\varphi(\Delta_m) = S$, the map $\varphi$ is a homeomorphism, i.e., we proved that any $m$-dimensional simplex is homeomorphic to $\Delta_m$; it follows that any two $m$-dimensional simplexes are homeomorphic, so we verified (ii). Finally observe that if $b_i : S \to \mathbb{R}$ is the $i$-th barycentric coordinate function then $b_i = p_i \circ \varphi^{-1}$, so every $b_i$ is continuous; this settles (iii) and completes our solution.

**V.085.** *Let $S$ be a simplex in $\mathbb{R}^n$ and suppose that $S_0 \supset S_1 \supset \ldots \supset S_k$ are distinct faces of $S$. Prove that the points $\{b(S_0), \ldots, b(S_k)\}$ are independent. Here $b(S_i)$ is the barycenter of the simplex $S_i$ for all $i \leq k$.*

**Solution.** Choose an enumeration $\{a_0, \ldots, a_m\}$ of the vertices of $S$ in such a way that if $T_0 = [a_0, \ldots, a_m], \ldots, T_i = [a_i, \ldots, a_m], \ldots, T_m = [a_m]$ then every simplex $S_i$ is listed in the sequence $\{T_0, \ldots, T_m\}$. Since any subset of an independent set is independent, it is sufficient to prove that the set $\{b(T_0), \ldots, b(T_m)\}$ is independent.

Let $\mathbf{0} \in \mathbb{R}^n$ be the zero vector and suppose that we are given $\lambda_0, \ldots, \lambda_m \in \mathbb{R}$ such that $\sum_{i=0}^m \lambda_i = 0$ and $\lambda_0 b(T_0) + \ldots + \lambda_m b(T_m) = \mathbf{0}$. Recalling that, for any $i \leq m$, we have $b(T_i) = \frac{1}{m-i+1} \sum_{j=i}^m a_j$, we conclude that $\sum_{i=0}^m \frac{\lambda_i}{m-i+1} \sum_{j=i}^m a_j = \mathbf{0}$ which shows that $\mu_0 a_0 + \ldots + \mu_m a_m = \mathbf{0}$ where $\mu_i = \frac{1}{m+1} \lambda_0 + \ldots + \frac{1}{m+1-i} \lambda_i$ for every $i \leq m$.

Note that $\sum_{i=0}^m \mu_i = (m+1) \cdot (\frac{1}{m+1} \lambda_0) + \ldots + (m+1-i) \cdot (\frac{1}{m+1-i} \lambda_i) + \ldots + \lambda_m$ and hence $\sum_{i=0}^m \mu_i = \sum_{i=0}^m \lambda_i = 0$ which, together with independency of the set $\{a_0, \ldots, a_m\}$, implies that $\mu_i = 0$ for each $i \leq m$. An immediate consequence is that $\lambda_i = 0$ for all $i \leq m$ and hence the set $H = \{b(T_0), \ldots, b(T_m)\}$ is independent. We already saw that this implies that the set $\{b(S_0), \ldots, b(S_k)\} \subset H$ is also independent.

**V.086.** *For a simplex* $S = [a_0, , \ldots, a_m]$, *consider the family* $\mathcal{B}(S)$ *of all simplexes of the form* $[b(S_0), \ldots, b(S_k)]$, *where* $S_0 \supset S_1 \supset \ldots \supset S_k$ *are distinct faces of* $S$. *Prove that* $\mathcal{P}$ *is a simplicial subdivision of* $S$ *such that any* $(m-1)$-*dimensional simplex* $T \in \mathcal{B}(S)$ *is a face of one or two m-dimensional members of* $\mathcal{B}(S)$ *depending on whether* $T$ *is contained in an* $(m-1)$-*dimensional face of* $S$. *The subdivision* $\mathcal{B}(S)$ *is called the barycentric subdivision of the simplex* $S$.

**Solution.** For any $x \in S$ we denote by $\lambda_i(x)$ its $i$-th barycentric coordinate in $S$ for all $i = 0, \ldots, m$. Take a permutation $\mathcal{P} = \{a_{i_0}, \ldots, a_{i_m}\}$ of the set $\{a_0, \ldots, a_m\}$ and let $T_0 = [a_{i_0}, a_{i_1}, \ldots, a_{i_m}], \ldots, T_l = [a_{i_l}, a_{i_{l+1}}, \ldots, a_{i_m}], \ldots, T_m = [a_{i_m}]$. Say that the sequence $\mathcal{T} = \{T_0, \ldots, T_m\}$ and the simplex $C = [b(T_0), \ldots, b(T_m)]$ are generated by the permutation $\mathcal{P}$.

Given a point $x \in C$ there are nonnegative $\alpha_0, \ldots, \alpha_m$ such that $\sum_{i=0}^m \alpha_i = 1$ and $\alpha_0 b(T_0) + \ldots + \alpha_m b(T_m) = x$. Recalling that $b(T_l) = \frac{1}{m-l+1} \sum_{j=l}^m a_{i_j}$ for any $l \leq m$, we conclude that $\sum_{l=0}^m \frac{\alpha_l}{m-l+1} \sum_{j=l}^m a_{i_j} = x$; an immediate consequence is that $x = \beta_0 a_{i_0} + \ldots + \beta_m a_{i_m}$ where $\beta_l = \frac{1}{m+1}\alpha_0 + \ldots + \frac{1}{m+1-l}\alpha_l$ for every $l \leq m$. Note that $\sum_{l=0}^m \beta_l = (m+1) \cdot (\frac{1}{m+1}\alpha_0) + \ldots + (m+1-l) \cdot (\frac{1}{m+1-l}\alpha_l) + \ldots + \alpha_m$ and hence $\sum_{l=0}^m \beta_l = \sum_{l=0}^m \alpha_l = 1$ which shows that every $\beta_l$ is the $i_l$-th barycentric coordinate of $x$ in $S$ and, in particular, $x \in S$. Thus

(1) for any permutation $\mathcal{P} = \{a_{i_0}, \ldots, a_{i_m}\}$ of the set $\{a_0, \ldots, a_m\}$ if the sequence $\{T_0, \ldots, T_m\}$ and the simplex $C = [b(T_0), \ldots, b(T_m)]$ are generated by $\mathcal{P}$ then $C \subset S$ and, for any $x \in C$, if $\alpha_0, \ldots, \alpha_m$ are the barycentric coordinates of the point $x$ in $C$ then $\lambda_{i_l}(x) = \frac{1}{m+1}\alpha_0 + \ldots + \frac{1}{m+1-l}\alpha_l$ for every $l \leq m$.

Now take any $B \in \mathcal{B}(S)$; there are distinct faces $S_0 \supset S_1 \supset \ldots \supset S_k$ of the simplex $S$ such that $B = [b(S_0), \ldots, b(S_k)]$. It is easy to find a permutation $\mathcal{P} = \{a_{i_0}, \ldots, a_{i_m}\}$ of the set $\{a_0, \ldots, a_m\}$ such that every $S_i$ is listed in the sequence $\{T_0, \ldots, T_m\}$ generated by $\mathcal{P}$ and hence $B \subset C = [b(T_0), \ldots, b(T_m)]$. Applying (1) we conclude that $B \subset C \subset S$. Since the simplex $B \in \mathcal{B}(S)$ was chosen arbitrarily, we proved that $\bigcup \mathcal{B}(S) \subset S$. To verify the rest of the properties of $\mathcal{B}(S)$ let us show that

(2) if $\mathcal{P} = \{a_{i_0}, \ldots, a_{i_m}\}$ is a permutation of the set $\{a_0, \ldots, a_m\}$ and $C$ is the element of $\mathcal{B}(S)$ generated by $\mathcal{P}$ then, a point $x \in S$ belongs to $C$ if and only if $\lambda_{i_0}(x) \leq \lambda_{i_1}(x) \leq \ldots \leq \lambda_{i_m}(x)$.

If $x \in C$ then it is easy to see that (1) implies the inequalities in (2). If, on the other hand, we have a point $x \in S$ such that $\lambda_{i_0}(x) \leq \lambda_{i_1}(x) \leq \ldots \leq \lambda_{i_m}(x)$ then let $\alpha_0 = (m+1)\lambda_{i_0}(x)$ and $\alpha_l = (m+1-l)(\lambda_{i_l}(x) - \lambda_{i_{l-1}}(x))$ for any $l \in \{1, \ldots, m\}$. It is easy to check that $\alpha_l \geq 0$ for all $l \leq m$ and $\sum_{l=0}^m \alpha_l = 1$ while $x = \sum_{l=0}^m \alpha_l b(T_l)$ which shows that $x \in C$ and $\alpha_l$ is the $l$-th barycentric coordinate of the point $x$ for every $l \leq m$, so (2) is proved.

Now, if $x \in S$ then choose a permutation $\mathcal{P} = \{a_{i_0}, \ldots, a_{i_m}\}$ of the set $\{a_0, \ldots, a_m\}$ such that $\lambda_{i_0}(x) \leq \lambda_{i_1}(x) \leq \ldots \leq \lambda_{i_m}(x)$ and apply (2) to see that $x$ belongs to the simplex $C$ generated by $\mathcal{P}$. Since $C \in \mathcal{B}(S)$, we established that $S \subset \bigcup \mathcal{B}(S)$ and hence $S = \bigcup \mathcal{B}(S)$. It is immediate from the definition of $\mathcal{B}(S)$ that if $B \in \mathcal{B}(S)$ then any face of $B$ also belongs to $\mathcal{B}(S)$.

Take a permutation $\mathcal{P} = \{a_{i_0}, \ldots, a_{i_m}\}$ of the set $\{a_0, \ldots, a_m\}$ and let $C$ be the simplex generated by $\mathcal{P}$. We have $C = \{x \in S : \lambda_{i_0}(x) \leq \lambda_{i_1}(x) \leq \ldots \leq \lambda_{i_m}(x)\}$ by the property (2). For any $x \in C$ let $\alpha_0(x), \ldots, \alpha_m(x)$ be the barycentric coordinates of $x$ in $C$. If $B$ is a face of $C$ then there is a set $E \subset \{0, \ldots, m\}$ such that $B = \{x \in C : \alpha_l(x) = 0 \text{ for all } l \in E\}$. The property (1) shows that $\alpha_0(x) = 0$ is equivalent to $\lambda_{i_0}(x) = 0$; if $l > 0$ then $\alpha_l(x) = 0$ is equivalent to $\lambda_{i_l}(x) = \lambda_{i_{l-1}}(x)$. Letting $\lambda_{i_{-1}}(x) = 0$ for any $x \in S$ we conclude that $B = \{x \in C_1 : \lambda_{i_l}(x) = \lambda_{i_{l-1}}(x) \text{ for all } l \in E\}$. Therefore

(3) a simplex $B$ is a face of $C$ if and only if there is a set $E \subset \{0, \ldots, m\}$ such that $B = \{x \in C : \lambda_{i_l}(x) = \lambda_{i_{l-1}}(x) \text{ for any } l \in E\}$.

Now assume that $\mathcal{P}_1 = \{a_{i_0}, \ldots, a_{i_m}\}$ and $\mathcal{P}_2 = \{a_{j_0}, \ldots, a_{j_m}\}$ are permutations of the set $\{a_0, \ldots, a_m\}$. If $C_i$ is the simplex generated by $\mathcal{P}_i$ for every $i \in \{1, 2\}$ then $C_1 = \{x \in S : \lambda_{i_{l-1}}(x) \leq \lambda_{i_l}(x) \text{ for each } l \leq m\}$ and $C_2 = \{x \in S : \lambda_{j_{l-1}}(x) \leq \lambda_{j_l}(x) \text{ for each } l \leq m\}$.

Given a face $B_2$ of the simplex $C_2$, an immediate consequence of (3) is that there exists a set $E' \subset \{0, \ldots m\}$ for which $B_2 = \{x \in C_2 : \lambda_{j_{l-1}}(x) = \lambda_{j_l}(x) \text{ for all } l \in E'\}$.

Fix a point $x \in C_1$ and $l \leq m$ and take $u, v \leq m$ with $j_{l-1} = i_u$ and $j_l = i_v$; it is clear that $u \neq v$. If $u < v$ then $\lambda_{j_{l-1}}(x) \leq \lambda_{j_l}(x)$ is fulfilled automatically; the property $\lambda_{j_{l-1}}(x) = \lambda_{j_l}(x)$ holds if and only if $\lambda_{i_{l-1}} = \lambda_{i_l}$ for all $l \in \{u + 1, \ldots, v\}$.

If $v < u$ then both conditions $\lambda_{j_{l-1}}(x) \leq \lambda_{j_l}(x)$ and $\lambda_{j_{l-1}}(x) = \lambda_{j_l}(x)$ are equivalent to the equalities $\lambda_{i_{l-1}} = \lambda_{i_l}$ for all $l \in \{v + 1, \ldots, u\}$. Carrying out this procedure for every $l \leq m$ we conclude that

(4) there is a set $E \subset \{0, \ldots, m\}$ such that $C_1 \cap B_2 = \{x \in C_1 : \lambda_{i_l}(x) = \lambda_{i_{l-1}}(x)$ for all $l \in E\}$ and, in particular, $C_1 \cap B_2$ is a face of $C_1$.

For the general case suppose that $B_1, B_2 \in \mathcal{B}(S)$ have nonempty intersection and choose permutations $\mathcal{P}_1 = \{a_{i_0}, \ldots, a_{i_m}\}$ and $\mathcal{P}_2 = \{a_{j_0}, \ldots, a_{j_m}\}$ of the set $\{a_0, \ldots, a_m\}$ such that $B_i$ is a face of the simplex $C_i$ generated by $\mathcal{P}_i$ for each $i \in \{1, 2\}$. The condition (2) implies that $C_1 = \{x \in S : \lambda_{i_0}(x) \leq \ldots \leq \lambda_{i_m}(x)\}$. Apply the property (4) to see that we can find a set $E_2 \subset \{0, \ldots, m\}$ such that $C_1 \cap B_2 = \{x \in C_1 : \lambda_{i_l}(x) = \lambda_{i_{l-1}}(x) \text{ for all } l \in E_2\}$.

The property (3) demonstrates that there exists a set $E_1 \subset \{0, \ldots, m\}$ for which $B_1 = \{x \in C_1 : \lambda_{i_l}(x) = \lambda_{i_{l-1}}(x) \text{ for all } l \in E_1\}$. It is straightforward to check that $B_1 \cap B_2 = \{x \in C_1 : \lambda_{i_l}(x) = \lambda_{i_{l-1}}(x) \text{ for all } l \in E_1 \cup E_2\}$, so we can apply (3) to conclude that $B_1 \cap B_2$ is a face of $C_1$. Since $C_1$ and $C_2$ are in a symmetric situation, the set $B_1 \cap B_2$ is also a face of $C_2$. It is evident that any face of $C_i$ contained in $B_i$ is a face of $B_i$ for every $i \in \{1, 2\}$, so $B_1 \cap B_2$ is a common face of $B_1 \cap B_2$. This proves that $\mathcal{B}(S)$ is a simplicial subdivision of $S$.

To check the last promise for the family $\mathcal{B}(S)$ take an $(m - 1)$-dimensional $T \in \mathcal{B}(S)$ such that $T = [b(S_0), \ldots, b(S_{m-1})]$ for some faces $S_0 \supset \ldots \supset S_{m-1}$ of the simplex $S$; fix a permutation $\{a_{i_0}, \ldots, a_{i_m}\}$ of the set $\{a_0, \ldots, a_m\}$ such that every $S_l$ is listed in the sequence $\{[a_{i_l}, \ldots, a_{i_m}] : l \in \{0, \ldots, m\}\}$

If $S_0 \neq S$ then $S_l = [a_{i_{l+1}}, \ldots, a_{i_m}]$ for every $l \in \{0, \ldots, m-1\}$; in particular, $T$ is contained in $S_0$ which is an $(m-1)$-dimensional face of $S$. Since no simplex other than $S$ can have $S_0$ as its proper face, $T$ is a face of the unique $m$-dimensional simplex $B = [b(S), b(S_0), \ldots, b(S_{m-1})]$ of the family $\mathcal{B}(S)$.

Now, if $S_0 = S$ then $b(S) = b(S_0) \in T$; it is easy to see that $b(S)$ does not belong to any $(m-1)$-dimensional face of $S$, so the simplex $T$ is not contained in any $(m-1)$-dimensional face of $S$. We have two cases to consider.

1) $S_{m-1} = [a_{m-1}, a_m]$ and hence $S_l = [a_{i_l}, \ldots, a_{i_m}]$ for any $l \leq m-1$. It is clear that the simplexes $[b(S_0), \ldots, b(S_{m-1}), b([a_{m-1}])]$ and $[b(S_0), \ldots, b(S_{m-1}), b([a_m])]$ are two distinct $m$-dimensional elements of $\mathcal{B}(S)$ such that $T$ is their common face. It is straightforward that there are no other possibilities to add to the sequence $\mathcal{T} = \{S_0, \ldots, S_{m-1}\}$ a face $F$ of $S$ in such a way that the sequence $\mathcal{T} \cup F$ be decreasing.

2) There exists a number $l \in \{0, \ldots, m-2\}$ such that $S_l = [a_{i_l}, \ldots, a_{i_m}]$ and $S_{l+1} = [a_{i_{l+2}}, \ldots, a_{i_m}]$. If $S' = [a_{i_l}, a_{i_{l+2}}, \ldots, a_{i_m}]$ and $S'' = [a_{i_{l+1}}, a_{i_{l+2}}, \ldots, a_{i_m}]$ then it is evident that the simplexes $[b(S_0), \ldots, b(S_l), b(S'), b(S_{l+1}), \ldots, b(S_{m-1})]$ and $[b(S_0), \ldots, b(S_l), b(S''), b(S_{l+1}), \ldots, b(S_{m-1})]$ are the $m$-dimensional elements of $\mathcal{B}(S)$ whose common face is $T$. It is easy to see that there are no other possibilities to add to the sequence $\mathcal{T} = \{S_0, \ldots, S_{m-1}\}$ a face $F$ of $S$ in such a way that the sequence $\mathcal{T} \cup F$ be decreasing.

This proves that an $(m-1)$-dimensional simplex $T \in \mathcal{B}(S)$ is contained in an $(m-1)$-dimensional face of $S$ (we saw that this happens if and only if the first element of the sequence that determines $T$ is distinct from $S$) if and only if $T$ is a face of exactly one $m$-dimensional element of $\mathcal{B}(S)$. If $S$ is not contained in an $(m-1)$-dimensional face of $S$ (we saw that this happens if and only if the first element of the sequence that determines $T$ coincides with $S$) then $T$ is the common face of exactly two distinct elements of $\mathcal{B}(S)$.

**V.087.** *Given a simplex $S$, let $\mathcal{B}_1(S)$ be the barycentric subdivision of $S$. If $\mathcal{B}_n(S)$ is a simplicial subdivision of $S$, let $\mathcal{B}_{n+1}(S) = \bigcup\{\mathcal{B}_1(T) : T \in \mathcal{B}_n(S)\}$. The family $\mathcal{B}_n(S)$ is called the n-th barycentric subdivision of the simplex $S$. Prove that, for any simplex $S$ and any $\varepsilon > 0$, there exists a natural number $n$ such that the mesh of the n-th barycentric subdivision of the simplex $S$ is less than $\varepsilon$.*

**Solution.** For any $n \in \mathbb{N}$ and $x = (x_1, \ldots, x_n) \in \mathbb{R}^n$ let $||x||_n = \sqrt{x_1^2 + \ldots + x_n^2}$. If $n$ is clear we will write $||x||$ instead of $||x||_n$. Given $x, y \in \mathbb{R}^n$, it is evident that $||x - y||_n$ is the distance between the points $x$ and $y$. For any set $A \subset \mathbb{R}^n$ let $\text{diam}(A) = \sup\{||x - y||_n : x, y \in A\}$. If $T$ is a simplex then $b(T)$ is its barycenter.

*Fact 1.* Let $T = [a_0, \ldots, a_m] \subset \mathbb{R}^n$ be an $m$-dimensional simplex. Then for any points $x, y$ such that $x \in \mathbb{R}^n$ and $y \in T$ there is $k \leq m$ for which $||x - y|| \leq ||x - a_k||$. In particular, $\text{diam}(T) = \max\{||a_i - a_j|| : i, j \in \{0, \ldots, m\}\}$.

*Proof.* Take $k \in \{0, \ldots, m\}$ such that $||x - a_i|| \leq ||x - a_k||$ for any $i \leq m$ and choose numbers $\lambda_0, \ldots, \lambda_m \in [0, 1]$ such that $\sum_{i=0}^{m} \lambda_i = 1$ and $y = \sum_{i=0}^{m} \lambda_i a_i$. Then $||x - y|| = ||x - \sum_{i=0}^{m} \lambda_i a_i|| = ||\sum_{i=0}^{m} (\lambda_i x - \lambda_i a_i)|| \leq \sum_{i=0}^{m} \lambda_i \cdot ||x - a_i||$; therefore $||x - y|| \leq \sum_{i=0}^{m} \lambda_i ||x - a_i|| \leq \sum_{i=0}^{m} \lambda_i ||x - a_k|| \leq ||x - a_k||$, so $k$ is as promised. Applying twice this part of our Fact we conclude that, for any $x, y \in T$, there are $k, l \in \{0, \ldots, m\}$ such that $||x - y|| \leq ||a_l - a_k||$; as a consequence, $\text{diam}(T) \leq d = \max\{||a_i - a_j|| : i, j \in \{0, \ldots, m\}\}$. The opposite inequality is evident, so $\text{diam}(T) = d$ and hence Fact 1 is proved.

*Fact 2.* Given an $m$-dimensional simplex $T = [a_0, \ldots, a_m] \subset \mathbb{R}^n$, the mesh of its barycentric subdivision $\mathcal{B}(T)$ does not exceed $\frac{m}{m+1} \cdot \text{diam}(T)$.

*Proof.* Take any simplex $B \in \mathcal{B}(T)$; there are faces $T_0 \supset \ldots \supset T_k$ of the simplex $T$ such that $B = [b(T_0), \ldots, b(T_k)]$. Fix $j, l \in \{0, \ldots, k\}$ with $j < l$. Since $T_l \subset T_j$, there is a permutation $\{a_{i_0}, \ldots, a_{i_m}\}$ of the set $\{a_0, \ldots, a_m\}$ and $p, q \in \{0, \ldots, m\}$ such that $p < q$ while $T_l = [a_{i_0}, \ldots, a_{i_p}]$ and $T_j = [a_{i_0}, \ldots, a_{i_q}]$.

For any $r \leq p$ the equality $||a_{i_r} - b(T_j)|| = ||a_{i_r} - \frac{1}{q+1}(a_{i_0} + \ldots + a_{i_q})||$ shows that $||a_{i_r} - b(T_j)|| = \frac{1}{q+1} ||\sum_{h=0}^{q} (a_{i_r} - a_{i_h})|| \leq \frac{1}{q+1} \cdot q \cdot \text{diam}(T)$ (the last inequality holds because the summand for $h = r$ is equal to zero). This shows that, for any $r \leq p$, we have $||a_{i_r} - b(T_j)|| \leq \frac{m}{m+1} \cdot \text{diam}(T)$. Now apply Fact 1 to see that there is $r \leq q$ such that $||b(T_j) - b(T_l)|| \leq ||a_{i_r} - b(T_l)|| \leq \frac{m}{m+1} \cdot \text{diam}(T)$. Since $j, l \in \{0, \ldots, k\}$ were chosen arbitrarily, we can apply Fact 1 once more to conclude that $\text{diam}(B) \leq \frac{m}{m+1} \cdot \text{diam}(T)$. Thus every element of $\mathcal{B}(T)$ has diameter not exceeding $\frac{m}{m+1} \cdot \text{diam}(T)$ and hence the mesh of $\mathcal{B}(T)$ does not exceed $\frac{m}{m+1} \cdot \text{diam}(T)$, i.e., Fact 2 is proved.

Returning to our solution suppose that $S = [a_0, \ldots, a_m]$ and $\varepsilon > 0$. It is a consequence of Fact 2 that the mesh of $\mathcal{B}_n(S)$ does not exceed $(\frac{m}{m+1})^n \cdot \text{diam}(T)$ for any $n \in \mathbb{N}$. Since $(\frac{m}{m+1})^n \cdot \text{diam}(T) \to 0$ as $n \to \infty$, there exists $n \in \mathbb{N}$ such that $(\frac{m}{m+1})^n \cdot \text{diam}(T) < \varepsilon$ and hence the mesh of $\mathcal{B}_n(S)$ is less than $\varepsilon$.

**V.088 (Sperner's lemma).** *Given a number $l \in \mathbb{N}$ and an $m$-dimensional simplex $[a_0, \ldots, a_m]$ let $V$ be the set of all vertices of simplexes in $\mathcal{B}_l([a_0, \ldots, a_m])$. Suppose that, for a function $h : V \to \{0, 1, \ldots, m\}$, we have $h(v) \in \{i_0, \ldots, i_k\}$ whenever $v \in [a_{i_0}, \ldots, a_{i_k}]$. Prove that the family of simplexes in $\mathcal{B}_l([a_0, \ldots, a_m])$, on vertices of which $h$ takes all values from 0 to $m$, has an odd cardinality.*

**Solution.** All simplexes are subspaces of some $\mathbb{R}^p$; in most cases, the power $p$ is not mentioned since we make no use of it. The symbol $\mathbf{0}$ is used to denote the respective zero vector of $\mathbb{R}^p$. Given a simplex $S$, the point $b(S)$ is its barycenter and $\mathcal{B}(S)$ is the barycentric subdivision of $S$. We also let $\mathcal{B}_1(S) = \mathcal{B}(S)$ and, if $\mathcal{B}_k(S)$ is defined, then $\mathcal{B}_{k+1}(S) = \bigcup \{\mathcal{B}(T) : T \in \mathcal{B}_k(S)\}$.

*Fact 1.* Assume that $S = [d_0, \ldots, d_n]$ is an $n$-dimensional simplex. If a simplex $T = [c_0, \ldots, c_k]$ is contained in $S$ then $k \leq n$.

*Proof.* Suppose that $k > n$ and let $\lambda_0^i, \ldots, \lambda_n^i$ be the barycentric coordinates of $c_i$ in $S$ for any $i \leq k$. Considering the $(n + 1)$-tuple $v_i = (\lambda_0^i, \ldots, \lambda_n^i)$ to be a vector of the space $\mathbb{R}^{n+1}$ for every $i \le k$, observe that the family $M = \{v_i : i \le k\}$ has $k + 1$ vectors; since $k + 1 > n + 1$, the family $M$ cannot be linearly independent, so we can choose $\beta_0, \ldots, \beta_k \in \mathbb{R}$ such that $\sum_{i=0}^{k} \beta_i^2 \ne 0$ and $\sum_{i=0}^{k} \beta_i v_i = \mathbf{0}$.

Thus $\sum_{i=0}^{k} \beta_i \lambda_j^i = 0$ for every $j \le n$ and hence $A = \sum_{j=0}^{n} \sum_{i=0}^{k} \beta_i \lambda_j^i = 0$. It is immediate that $A = \sum_{i=0}^{k} \beta_i (\sum_{j=0}^{n} \lambda_j^i) = \sum_{i=0}^{k} \beta_i = 0$ (we used the fact that $\sum_{j=0}^{n} \lambda_j^i = 1$ for every $i \le k$), so the set $\{c_0, \ldots, c_k\}$ is not independent; since only independent sets can span a simplex, we obtained a contradiction which shows that Fact 1 is proved.

*Fact 2.* Given an arbitrary simplex $S = [d_0, \ldots, d_n]$ and a point $x \in S$, denote by $\lambda_i(x)$ the $i$-th barycentric coordinate of $x$ in $S$ for all $i = 0, \ldots, n$. Fix a permutation $\mathcal{P} = \{d_{i_0}, \ldots, d_{i_n}\}$ of the set $\{d_0, \ldots, d_n\}$ and consider the simplexes $T_0 = [d_{i_0}, d_{i_1}, \ldots, d_{i_n}], \ldots, T_l = [d_{i_l}, d_{i_{l+1}}, \ldots, d_{i_n}], \ldots, T_n = [d_{i_n}]$. Then the simplex $C = [b(T_0), \ldots, b(T_n)]$ is contained in $S$ and, for any $x \in C$, if $\alpha_0, \ldots, \alpha_n$ are the barycentric coordinates of the point $x$ in $C$ then $\lambda_{i_l}(x) = \frac{1}{n+1}\alpha_0 + \ldots + \frac{1}{n+1-l}\alpha_l$ for every $l \le n$. We will say that the sequence $\{T_0, \ldots, T_n\}$ and the simplex $C$ are generated by the permutation $\mathcal{P}$.

*Proof.* Given a point $x \in C$ there are nonnegative $\alpha_0, \ldots, \alpha_n$ such that $\sum_{i=0}^{n} \alpha_i = 1$ and $\alpha_0 b(T_0) + \ldots + \alpha_n b(T_n) = x$. Recalling that $b(T_l) = \frac{1}{n-l+1} \sum_{j=l}^{n} d_{i_j}$ for any $l \le n$, we conclude that $\sum_{l=0}^{n} \frac{\alpha_l}{n-l+1} \sum_{j=l}^{n} d_{i_j} = x$; an immediate consequence is that $x = \beta_0 d_{i_0} + \ldots + \beta_n d_{i_n}$ where $\beta_l = \frac{1}{n+1}\alpha_0 + \ldots + \frac{1}{n+1-l}\alpha_l$ for every $l \le n$.

Note that $\sum_{l=0}^{n} \beta_l = (n+1) \cdot (\frac{1}{n+1}\alpha_0) + \ldots + (n+1-l) \cdot (\frac{1}{n+1-l}\alpha_l) + \ldots + \alpha_n$ and hence $\sum_{l=0}^{n} \beta_l = \sum_{l=0}^{n} \alpha_l = 1$ which shows that $\beta_l = \lambda_{i_l}(x)$ for every $l \le n$, so $x \in S$ and hence Fact 2 is proved.

*Fact 3.* Given a simplex $S = [d_0, \ldots, d_n]$ and a point $x \in S$, denote by $\lambda_i(x)$ the $i$-th barycentric coordinate of $x$ in $S$ for all $i = 0, \ldots, n$. If $\mathcal{P} = \{d_{i_0}, \ldots, d_{i_n}\}$ is a permutation of the set $\{d_0, \ldots, d_n\}$ and $C$ is the element of $\mathcal{B}(S)$ generated by $\mathcal{P}$ then a point $x \in S$ belongs to $C$ if and only if $\lambda_{i_0}(x) \le \lambda_{i_1}(x) \le \ldots \le \lambda_{i_n}(x)$.

*Proof.* If $x \in C$ then it is easy to see that Fact 1 implies our inequalities. If, on the other hand, we have a point $x \in S$ such that $\lambda_{i_0}(x) \le \lambda_{i_1}(x) \le \ldots \le \lambda_{i_n}(x)$ then let $\alpha_0 = (n+1)\lambda_{i_0}(x)$ and $\alpha_l = (n+1-l)(\lambda_{i_l}(x) - \lambda_{i_{l-1}}(x))$ for any $l \in \{1, \ldots, n\}$. It is easy to check that $\alpha_l \ge 0$ for all $l \le n$ and $\sum_{l=0}^{n} \alpha_l = 1$ while $x = \sum_{l=0}^{n} \alpha_l b(T_l)$ which shows that $x \in C$ and $\alpha_l$ is the $l$-th barycentric coordinate of the point $x$ in $C$ for every $l \le n$, so Fact 3 is proved.

*Fact 4.* Given a simplex $S = [d_0, \ldots, d_n]$ and a point $x \in S$, denote by $\lambda_i(x)$ the $i$-th barycentric coordinate of $x$ in $S$ for all $i = 0, \ldots, n$. Take a permutation $\mathcal{P} = \{d_{i_0}, \ldots, d_{i_n}\}$ of the set $\{d_0, \ldots, d_n\}$ and let $C$ be the simplex generated by $\mathcal{P}$. Then a simplex $B$ is a face of $C$ if and only if there is a set $E \subset \{0, \ldots, n\}$ such that $B = \{x \in C : \lambda_{i_l}(x) = \lambda_{i_{l-1}}(x)$ for any $l \in E\}$; here $\lambda_{i_{-1}}(x) = 0$ for any $x \in S$.

*Proof.* Fact 3 shows that $C = \{x \in S : \lambda_{i_0}(x) \le \lambda_{i_1}(x) \le \ldots \le \lambda_{i_n}(x)\}$. For any point $x \in C$ let $\alpha_0(x), \ldots, \alpha_n(x)$ be the barycentric coordinates of $x$ in $C$.

A set $B$ is a face of $C$ if and only if there exists a set $E \subset \{0, \ldots, n\}$ such that $B = \{x \in C : \alpha_l(x) = 0 \text{ for all } l \in E\}$. It follows from Fact 2 that $\alpha_0(x) = 0$ is equivalent to $\lambda_{i_0}(x) = 0$; if $l > 0$ then $\alpha_l(x) = 0$ is equivalent to $\lambda_{i_l}(x) = \lambda_{i_{l-1}}(x)$. Therefore $B = \{x \in C : \lambda_{i_l}(x) = \lambda_{i_{l-1}}(x) \text{ for all } l \in E\}$ and Fact 4 is proved.

*Fact 5.* Given a simplex $S = [d_0, \ldots, d_n]$ and $B \in \mathcal{B}(S)$ there exists a permutation $\mathcal{P} = \{d_{i_0}, \ldots, d_{i_n}\}$ of the set $\{d_0, \ldots, d_n\}$ such that $B$ is a face of the simplex generated by $\mathcal{P}$. In particular, every element of $\mathcal{B}(S)$ is a face of an $n$-dimensional element of $\mathcal{B}(S)$.

*Proof.* By definition of the family $\mathcal{B}(S)$ there exist faces $S_0 \supset \ldots \supset S_k$ of the simplex $S$ such that $B = [b(S_0), \ldots, b(S_k)]$. Take a permutation $\mathcal{P} = \{d_{i_0}, \ldots, d_{i_n}\}$ of the set $\{d_0, \ldots, d_n\}$ such that every simplex $S_i$ is listed in the sequence

$$T_0 = [d_{i_0}, d_{i_1}, \ldots, d_{i_n}], \ldots, T_l = [d_{i_l}, d_{i_{l+1}}, \ldots, d_{i_n}], \ldots, T_n = [d_{i_n}].$$

It is clear that $B$ is a face of the $n$-dimensional simplex $C$ generated by the permutation $\mathcal{P}$, so Fact 5 is proved.

*Fact 6.* For any simplex $S$, suppose that $B \in \mathcal{B}(S)$ and $S'$ is a face of $S$ such that $B \cap S' \neq \emptyset$; then $B \cap S'$ is a face of $B$ and hence $B \cap S' \in \mathcal{B}(S)$.

*Proof.* By Fact 5, there is a permutation $\mathcal{P} = \{d_{i_0}, \ldots, d_{i_n}\}$ of the set $\{d_0, \ldots, d_n\}$ such that $B$ is a face of the simplex $C$ generated by $\mathcal{P}$. By Fact 4 we can find a set $E \subset \{0, \ldots n\}$ for which $B = \{x \in C : \lambda_{i_{l-1}}(x) = \lambda_{i_l}(x) \text{ for all } l \in E\}$.

There is a set $E_1 \subset \{0, \ldots, n\}$ such that $S' = \{x \in S : \lambda_{i_l}(x) = 0 \text{ for all } l \in E_1\}$. If $p$ is the maximal element of $E_1$ then $B \cap S' = \{x \in C : \lambda_{i_l}(x) = \lambda_{i_{l-1}}(x)$ for all $l \in \{0, \ldots, p\} \cup E\}$. Applying Fact 4 again we conclude that $B \cap S'$ is a face of $C$; any face of $C$ contained in $B$ is, evidently, a face of $B$, so Fact 6 is proved.

*Fact 7.* For any simplex $S$ and $k \in \mathbb{N}$, the family $\mathcal{B}_k(S)$ is a simplicial subdivision of $S$.

*Proof.* For $k = 1$ this was proved in Problem 086. Proceeding inductively, assume that $\mathcal{B}_k(S)$ of $S$ is simplicial. It is immediate from the definition that $\bigcup \mathcal{B}_{k+1}(S) = S$ and, for any $B \in \mathcal{B}_{k+1}(S)$, all faces of $B$ also belong to $\mathcal{B}_{k+1}(S)$.

Now, if $B_1, B_2 \in \mathcal{B}_{k+1}(S)$ and $B_1 \cap B_2 \neq \emptyset$ then there are $P_1, P_2 \in \mathcal{B}_k(S)$ such that $B_i \in \mathcal{B}(P_i)$ for each $i \in \{1, 2\}$. If $P_1 = P_2$ then $B_1 \cap B_2$ is a common face of $B_1$ and $B_2$ by Problem 086. If $P_1 \neq P_2$ then $P = P_1 \cap P_2$ is a common face of $P_1$ and $P_2$ by the induction hypothesis. It follows from Fact 6 that $B_i \cap P \in \mathcal{B}(P)$ and $B_i \cap P$ is a face of $B_i$ for every $i = 1, 2$. Therefore we can apply Problem 086 once more to see that $(B_1 \cap P) \cap (B_2 \cap P) = B_1 \cap B_2$ is a common face of $B_1 \cap P$ and $B_2 \cap P$. It is clear that the common face of $B_1 \cap P$ and $B_2 \cap P$ is also a common face of $B_1 \cap B_2$, so we proved that $\mathcal{B}_{k+1}(S)$ is also a simplicial subdivision of $S$. Therefore $\mathcal{B}_k(S)$ is a simplicial subdivision of $S$ for any $k \in \mathbb{N}$, i.e., Fact 7 is proved.

*Fact 8.* If $S$ is a simplex and $k \in \mathbb{N}$ then for any face $S'$ of the simplex $S$ and any $P \in \mathcal{B}_k(S)$, the set $P \cap S'$ is a face of $P$ and hence $P \cap S' \in \mathcal{B}_k(S')$.

*Proof.* Fact 6 says that our statement is true for $k = 1$. Proceeding by induction assume that we proved (6) for any $k \leq l$ and take a simplex $P \in \mathcal{B}_{l+1}(S)$ and a face $S'$ of the simplex $S$. There is $B \in \mathcal{B}_l(S)$ such that $P \in \mathcal{B}(B)$. By the induction hypothesis, the set $B' = B \cap S'$ is a face of $B$, so we can apply Fact 6 again to see that $P \cap S' = P \cap B'$ is a face of $P$. An immediate consequence of Fact 7 is that $P \cap S' \in \mathcal{B}_{l+1}(S)$, so our induction procedure shows that our statement is true for any $k \in \mathbb{N}$, i.e., Fact 8 is proved.

*Fact 9.* Given $k \in \mathbb{N}$ and a simplex $S$ if a simplex $B \in \mathcal{B}_k(S)$ is contained in an $(n-1)$-dimensional face of $S$ then there is a unique $n$-dimensional simplex $C \in \mathcal{B}_k(S)$ such that $B$ is a face of $C$.

*Proof.* For $k = 1$ this statement was proved in Problem 086. Proceeding inductively assume that we proved our Fact for all $k \leq l$ and take an $(n-1)$-dimensional simplex $B \in \mathcal{B}_{l+1}(S)$ which is contained in an $(n-1)$-dimensional face $S'$ of the simplex $S$. By Fact 5, there is an $n$-dimensional simplex $E \in \mathcal{B}_l(S)$ such that $B \in \mathcal{B}(E)$. By Fact 8, the set $E' = E \cap S'$ is a face of $E$; since $B \subset E'$, it follows from Fact 1 that the dimension of $E'$ is at least $n - 1$. Apply Fact 1 again to see that the simplex $E'$ cannot be $n$-dimensional because $S'$ is $(n-1)$-dimensional and $E' \subset S'$.

Therefore $E'$ is an $(n-1)$-dimensional face of $E$; by the induction hypothesis it is not a face of any other $n$-dimensional element of $\mathcal{B}_l(S)$. Since $B \subset E'$, we can apply Problem 086 to convince ourselves that there is a unique $n$-dimensional simplex $B' \in \mathcal{B}(E)$ such that $B$ is a face of $B'$. We must show that $B$ cannot be a face of any other $n$-dimensional element of $\mathcal{B}_{l+1}(S)$.

Assume, toward a contradiction, that there exists an $n$-dimensional simplex $C \in \mathcal{B}_{l+1}(S)\backslash\{B'\}$ such that $B$ is a face of $C$. We saw already that $C \notin \mathcal{B}(E)$, so there is an $n$-dimensional simplex $D \in \mathcal{B}_l(S)\backslash\{E\}$ such that $C \in \mathcal{B}(D)$. As a consequence, $\emptyset \neq B \subset D \cap E$ and hence the simplex $D' = D \cap E$ is the common $(n-1)$-dimensional face of $D$ and $E$ (see Fact 7). Since $E'$ is not a face of $D$, the simplexes $D'$ and $E'$ are distinct. We have $\emptyset \neq B \subset E' \cap D'$, so $E' \cap D'$ is the common face of $E'$ and $D'$ by Fact 7. Apply Fact 1 once more to see that $E' \cap D'$ is an $(n-1)$-dimensional face of both $E'$ and $D'$. Since the unique $p$-dimensional face of a $p$-dimensional simplex is the simplex itself, we have $D' = E'$ which is a contradiction. Thus $B$ is a face of a unique $n$-dimensional element of $\mathcal{B}_{l+1}(S)$, so the induction step is accomplished and hence Fact 9 is proved.

*Fact 10.* Given $k \in \mathbb{N}$ and a simplex $S$ if a simplex $B \in \mathcal{B}_k(S)$ is not contained in any $(n-1)$-dimensional face of $S$ then there are exactly two $n$-dimensional simplexes $C_1, C_2 \in \mathcal{B}_k(S)$ such that $B$ is a common face of $C_1$ and $C_2$.

*Proof.* For $k = 1$ this statement was proved in Problem 086. Proceeding inductively assume that we proved our Fact for all $k \leq l$ and take an $(n-1)$-dimensional simplex $B \in \mathcal{B}_{l+1}(S)$ which is not contained in any $(n-1)$-dimensional face $S'$ of the simplex $S$. By Fact 5, there is an $n$-dimensional simplex $E \in \mathcal{B}_l(S)$ such that $B \in \mathcal{B}(E)$. We have two possible cases.

*Case 1.* The simplex $B$ is contained in an $(n-1)$-dimensional face $E'$ of the simplex $E$. By Problem 086, there is a unique $n$-dimensional simplex $C_1 \in \mathcal{B}(E)$ such that $B$ is a face of $C_1$. The simplex $E' \supset B$ is not contained in any $(n-1)$-dimensional face of $S$, so we can apply the induction hypothesis to see that it is a common face of exactly two $n$-dimensional elements of $\mathcal{B}_l(S)$. Fix an $n$-dimensional simplex $D \in \mathcal{B}_l(S) \setminus \{E\}$ such that $E'$ is also a face of $D$.

Since our simplex $B$ is also contained in the $(n-1)$-dimensional face $E'$ of the simplex $D$, there is a unique $n$-dimensional simplex $C_2 \in \mathcal{B}(D)$ such that $B$ is a face of $C_2$. Observe that $C_1 \neq C_2$ because otherwise $C_1 = C_1 \cap C_2 \subset E'$ which contradicts Fact 1. Thus $C_1$ and $C_2$ are distinct $n$-dimensional elements of $\mathcal{B}_{l+1}(S)$ such that $B$ is the common face of $C_1$ and $C_2$.

Assume, toward a contradiction, that there exists an $n$-dimensional simplex $C_3 \in \mathcal{B}_{l+1}(S) \setminus \{C_1, C_2\}$ such that $B$ is also a face of $C_3$. We saw already that $C \notin \mathcal{B}(E) \cup \mathcal{B}(D)$, so there is an $n$-dimensional simplex $F \in \mathcal{B}_l(S) \setminus \{E \cup D\}$ such that $C_3 \in \mathcal{B}(F)$. As a consequence, $\emptyset \neq B \subset F \cap E$ and hence $F' = F \cap E$ is the common $(n-1)$-dimensional face of $F$ and $E$ (see Fact 7 and Fact 1). Analogously, $F'' = F \cap D$ is the common $(n-1)$-dimensional face of $F$ and $D$. It follows from $B \subset F' \cap F''$ and Fact 1 that $F' \cap F''$ is a common $(n-1)$-dimensional face of $F'$ and $F''$. Therefore $F' = F' \cap F'' = F''$ which shows that $F'$ is the common $(n-1)$-dimensional face of the three distinct simplexes $E, D, F \in \mathcal{B}_l(S)$ which is a contradiction with the induction hypothesis. Thus our induction step is accomplished if Case 1 takes place.

*Case 2.* The simplex $B$ is not contained in any $(n-1)$-dimensional face of the simplex $E$. By Problem 086, there are exactly two $n$-dimensional simplexes $C_1 \in \mathcal{B}(E)$ such that $B$ is the common face of $C_1$ and $C_2$.

Assume, toward a contradiction, that there exists an $n$-dimensional simplex $C_3 \in \mathcal{B}_{l+1}(S) \setminus \{C_1, C_2\}$ such that $B$ is also a face of $C_3$. We saw already that $C_3 \notin \mathcal{B}(E)$, so there exists an $n$-dimensional simplex $F \in \mathcal{B}_l(S) \setminus \{E\}$ such that $C_3 \in \mathcal{B}(F)$. As a consequence, $\emptyset \neq B \subset F \cap E$ and hence $F' = F \cap E$ is the common $(n-1)$-dimensional face of $F$ and $E$ (see Fact 7 and Fact 1). It turns out that the simplex $B \subset F'$ is contained in an $(n-1)$-dimensional face $F'$ of the simplex $E$; this contradiction shows that our induction step is accomplished in Case 2 as well, so Fact 10 is proved.

Returning to our solution let $S = [a_0, \ldots, a_m]$; assume first that $m = 1$ and hence $S = [a_0] = \{a_0\}$. Then $\mathcal{B}_l(S) = \{S\}$; the simplex $S$ has the unique vertex $a_0$ and $h(a_0) = a_0$ by our assumption about the function $h$. Therefore the relevant number of simplexes is 1, i.e., our solution is carried out for $m = 1$.

Proceeding by induction assume that we proved our statement for all $m < n$ and take an $n$-dimensional simplex $S = [a_0, \ldots, a_n]$ and a function $h$ with values in $\{0, \ldots, n\}$ defined on the set $V$ of all vertices of $\mathcal{B}_l(S)$ such that $h(v) \in \{i_0, \ldots, i_k\}$ whenever $v \in [a_{i_0}, \ldots, a_{i_k}]$.

Let $S = \{S_1, \ldots, S_p\}$ be a faithfully enumerated family of all $n$-dimensional simplexes from $\mathcal{B}_l(S)$. For any number $i \in \mathbb{N}$ let $M_i = \{1, \ldots, i\}$; for any simplex $T \in \mathcal{B}_l(S)$ let $V(T)$ be the set of its vertices. Call a simplex $T \in S$ *marked* if $h(V(T)) = \{0, \ldots, n\}$. We must prove that the cardinality $q$ of the family of marked simplexes from $S$ is an odd number.

We will say that an $(n-1)$-dimensional simplex $T \in \mathcal{B}_l(S)$ is *adequate* if $h(V(T)) = \{0, \ldots, n-1\}$. Let $\mathcal{T}$ be the family of all $(n-1)$-dimensional simplexes from $\mathcal{B}_l(S)$. If we consider the face $S' = [a_0, \ldots, a_{n-1}]$ of the simplex $S$ then it takes a trivial induction to prove that $\mathcal{B}_i(S') \subset \mathcal{B}_i(S)$ for any $i \in \mathbb{N}$, so $\mathcal{B}_l(S') \subset \mathcal{B}_l(S)$. It follows from Fact 8 that if $T \in \mathcal{B}_l(S)$ and $T \subset S'$ then $T \in \mathcal{B}_l(S')$. Thus the family $\mathcal{T}' = \{T \in \mathcal{T} : T \subset S'\}$ consists of all $(n-1)$-dimensional elements of $\mathcal{B}_l(S')$.

Let $g$ be the restriction of our function $h$ to the set $W$ of all vertices of the simplexes from $\mathcal{B}_l(S')$. Then $g : W \to \{0, \ldots, n-1\}$ and $g(v) \in \{i_0, \ldots, i_k\}$ whenever $v \in [a_{i_0}, \ldots, a_{i_k}]$. Therefore we can apply the induction hypothesis to see that the cardinality $s$ of the family $\mathcal{E} = \{T \in \mathcal{T}' : g(V(T)) = \{0, \ldots, n-1\}\}$ is an odd number.

Given $i \in M_p$ let $w_i$ be the number of adequate faces of $S_i$. If $S_i$ is marked then it has only one adequate face, i.e., $w_i = 1$. If a non-marked simplex $S_i$ has an adequate face then we can consider that $S_i = [d_0, \ldots, d_n]$ and the enumeration $\{d_0, \ldots, d_n\}$ of the vertices of $S_i$ is chosen in such a way that $h(d_i) = i$ for any $i \leq n-1$. The simplex $S_i$ being non-marked, we have $h(d_n) = j \in \{0, \ldots, n-1\}$ and hence the face $[d_0, \ldots, d_{j-1}, d_{j+1}, \ldots, d_n]$ of the simplex $S_i$ is also adequate. It is easy to see that $S_i$ has no other adequate faces, so $w_i = 2$. This proves that

(1) if a simplex $S_i$ is marked then $w_i = 1$; if $S_i$ is not marked then either $w_i = 0$
    or $w_i = 2$.

An immediate consequence of (1) is that the parity of the number $w = \sum_{i=1}^p w_i$ coincides with the parity of $q$, i.e., $w - q = 2t$ for some $t \in \mathbb{N}$.

Next observe that it follows from our assumption about the function $h$ that the unique $(n-1)$-dimensional face of $S$ that can contain an adequate simplex is $S'$. Therefore an adequate simplex $T$ is either contained in $S'$ or not contained in any $(n-1)$-dimensional face of $S$. In the first case $T$ is a face of a unique element of $S$ by Fact 9; in the second case $T$ is the common face of exactly two simplexes from $S$ (see Fact 10).

Let $u$ be the number of adequate simplexes $T \in \mathcal{T}$ which are not contained in $S'$. For any $i \in M_p$ let $w_i'$ be the number of the faces of $S_i$ which belong to $\mathcal{E}$. It is clear that either $w_i' = 1$ or $w_i' = 0$ and hence $s = \sum_{i=1}^p w_i'$. Since every adequate simplex $T \in \mathcal{T} \setminus \mathcal{E}$ belongs to exactly two elements of $S$, in the sum $\sum_{i=1}^p (w_i - w_i') = w - s$ every adequate $T \in \mathcal{T} \setminus \mathcal{E}$ is counted twice, so $w - s = 2u$. Therefore $q = w - 2t = s + 2(u - t)$ is an odd number. This accomplishes the induction step of our proof and completes our solution.

**V.089 (Brower's fixed-point theorem).** *Prove that, for any* $n \in \mathbb{N}$, *if* $S$ *is an* $n$-*dimensional simplex and* $f : S \to S$ *is a continuous function then there exists a point* $x \in S$ *such that* $f(x) = x$.

**Solution.** All simplexes are subspaces of some $\mathbb{R}^p$ with the respective metric; the power $p$ is not mentioned since we make no use of it. If $T$ is a simplex then $V(T)$ is the set of vertices of $T$ and $\mathcal{B}(T)$ is the barycentric subdivision of $T$. We also let $\mathcal{B}_1(T) = \mathcal{B}(T)$ and, if $\mathcal{B}_k(T)$ is defined, then $\mathcal{B}_{k+1}(T) = \bigcup \{\mathcal{B}(T) : T \in \mathcal{B}_k(T)\}$.

*Fact 1.* Given a simplex $T = [a_0, \ldots, a_m]$ suppose that $F_i$ is a closed subset of $T$ for any $i = 0, \ldots, m$. If, additionally, $[a_{i_0}, \ldots, a_{i_k}] \subset F_{i_0} \cup \ldots \cup F_{i_k}$ for any $i_0, \ldots, i_k \in \{0, \ldots, m\}$ then $F_0 \cap \ldots \cap F_m \neq \emptyset$.

*Proof.* Consider the set $U_i = T \backslash F_i$ for any $i \leq m$; if $F_0 \cap \ldots \cap F_n = \emptyset$ then the family $\mathcal{U} = \{U_i : i \leq m\}$ is an open cover of the compact space $T$ (see Problem 084), so we can apply TFS-244 to find a number $\delta > 0$ such that any set of diameter not exceeding $\delta$ is contained in one of the elements of $\mathcal{U}$. Apply Problem 087 to find $k \in \mathbb{N}$ such that every element of the $k$-th barycentric subdivision $\mathcal{B}_k(T)$ of the simplex $T$ has diameter less than $\delta$. Let $V$ be the set of vertices of the elements of $\mathcal{B}_k(T)$.

For each $v \in V$ let $T_v$ be the intersection of all faces of $T$ which contain $v$. It is evident that $T_v$ is still a face of $T$; therefore there are $a_{i_0}, \ldots, a_{i_l} \in \{a_0, \ldots, a_m\}$ such that $T_v = [a_{i_0}, \ldots, a_{i_l}]$. Since $T_v \subset F_{i_0} \cup \ldots \cup F_{i_l}$, there is $j \leq l$ such that $v \in F_{i_j}$; let $h(v) = i_j$.

This gives us a map $h : V \to \{0, \ldots, m\}$ such that $v \in F_{h(v)}$ and $a_{h(v)}$ is one of the vertices of $T_v$ for any $v \in V$. Consequently, if $T'$ is a face of $T$ and $v \in T'$ then $T_v \subset T'$ and hence $a_{h(v)} \in V(T_v)$ is also a vertex of $T'$. This shows that we can apply Problem 088 to find a simplex $B \in \mathcal{B}_k(T)$ such that $h(V(B)) = \{0, \ldots, m\}$. For any $i \leq m$ there is $v \in V(B)$ such that $h(v) = i$ and hence $v \in F_i$; therefore $B \cap F_i \neq \emptyset$ and hence $B$ is not contained in $U_i$ for each $i \leq m$. This contradiction with the choice of $\delta$ shows that $F_0 \cap \ldots \cap F_m \neq \emptyset$, i.e., Fact 1 is proved.

Returning to our solution let $a_0, \ldots, a_n$ be the vertices of the simplex $S$; then $S = [a_0, \ldots, a_n]$. Denote by $\lambda_i(x)$ the $i$-th barycentric coordinate of $x$ in $S$ for any $i \leq n$. The set $F_i = \{x \in S : \lambda_i(f(x)) \leq \lambda_i(x)\}$ is closed in $S$ for any $i \leq n$; let us check that the family $\{F_0, \ldots, F_n\}$ satisfies the premises of Fact 1.

Indeed, if $x \in [a_{i_0}, \ldots, a_{i_m}]$ then $\lambda_k(x) = 0$ for all $k \notin \{i_0, \ldots, i_m\}$ and hence $\sum_{j=0}^{m} \lambda_{i_j}(x) = 1$ which shows that $\sum_{j=0}^{m} \lambda_{i_j}(f(x)) \leq \sum_{j=0}^{m} \lambda_{i_j}(x)$ and hence there exists $j \leq m$ such that $\lambda_{i_j}(f(x)) \leq \lambda_{i_j}(x)$, i.e., $x \in F_{i_j}$. This proves that $[a_{i_0}, \ldots, a_{i_m}] \subset F_{i_0} \cup \ldots \cup F_{i_m}$ and therefore we can apply Fact 1 to conclude that $F = F_0 \cap \ldots \cap F_n \neq \emptyset$. If $x \in F$ then $\lambda_i(f(x)) \leq \lambda_i(x)$ for any $i \leq n$; this, together with the equalities $\sum_{i=1}^{n} \lambda_i(x) = \sum_{i=1}^{n} \lambda_i(f(x)) = 1$ shows that $\lambda_i(x) = \lambda_i(f(x))$ for every $i \leq n$, so $f(x) = x$ and hence the point $x \in S$ is as promised.

**V.090.** *Prove that, for any* $n \in \mathbb{N}$, *there is no retraction of the cube* $\mathbb{I}^n$ *onto its boundary* $\partial \mathbb{I}^n = \{x \in \mathbb{I}^n : |x(i)| = 1 \text{ for some } i < n\}$.

**Solution.** For any $m \in \mathbb{N}$ denote by $\mathbf{0}_m$ the zero point of $\mathbb{R}^m$ and let $e_i^m \in \mathbb{R}^m$ be the point defined by $e_i^m(i) = 1$ and $e_i^m(j) = 0$ for all $j < m$ with $j \neq i$. We consider the space $\mathbb{R}^m$ with the usual metric $d_m$ defined by $d_m(x, y) = \sqrt{\sum_{i=0}^{m-1}(x(i) - y(i))^2}$ for any points $x, y \in \mathbb{R}^m$. Besides, $B_m(x, s) = \{y \in \mathbb{R}^m : d_m(x, y) < s\}$ is the ball of radius $s$ centered at $x$.

The set $E_m = \{x \in \mathbb{R}^m : x(i) \geq 0$ for all $i < m$ and $\sum_{i=0}^{m-1} x(i) \leq 1\}$ is easily seen to be the simplex spanned by the points $\mathbf{0}_m, e_0^m, \ldots, e_{m-1}^m$. For any $m$-dimensional simplex $S$ we denote by $\partial S$ the union of all $(m-1)$-dimensional faces of $S$. Then $\partial E_m = \{x \in E_m : x(0) \cdot \ldots \cdot x(m-1) = 0$ or $\sum_{i=0}^{m-1} x(i) = 1\}$.

*Fact 1.* For any $m \in \mathbb{N}$ there exists a homeomorphism $h : E_m \to \mathbb{I}^m$ such that $h(\partial E_m) = \partial \mathbb{I}^m$.

*Proof.* Define a point $v_0 \in \mathbb{R}^m$ by requiring that $v_0(i) = -1$ for all $i < m$; the map $f : \mathbb{R}^m \to \mathbb{R}^m$, defined by $f(x) = 2x + v_0$ for any $x \in \mathbb{R}^m$, is a homeomorphism. Let $I = \{x \in \mathbb{R}^m : 0 \leq x(i) \leq 1$ for all $i < m\}$. Then $f(I) = \mathbb{I}^m$ and, for the set $\partial I = \{x \in I : \prod_{i=0}^{m-1} x(i)(x(i) - 1) = 0\}$, we have $f(\partial I) = \partial \mathbb{I}^m$. Thus it suffices to construct a homeomorphism $g : E^m \to I$ such that $g(\partial E_m) = \partial I$.

The geometric idea of construction of $g$ is very simple: for any point $u \in \mathbb{R}^m$ with $u(i) \geq 0$ for all $i < m$ we define $g$ on the ray $R_u = \{tu : t \in \mathbb{R}, t \geq 0\}$ as the unique linear map for which $g(A_u) = B_u$ where $A_u$ and $B_u$ are the intersection points of $R_u$ with $\partial E_m$ and $\partial I$ respectively.

To do it formally, let $s(x) = \sum_{i<m} x(i)$ and $m(x) = \max_{i<m} x(i)$ for any $x \in \mathbb{R}^m$. Now, define $g$ by requiring that $g(\mathbf{0}_m) = \mathbf{0}_m$ and, if $x \in E_m \backslash \{\mathbf{0}_m\}$ then $g(x) = \frac{s(x)}{m(x)} \cdot x$. It is trivial that $g : E_m \to I$ is a bijection which is continuous at every point $x \in E_m \backslash \{\mathbf{0}_m\}$ and $g(\partial E_m) = \partial I$. However, the map $g$ is also continuous at $\mathbf{0}_m$ because $g(B_m(\mathbf{0}_m, \frac{s}{m})) \subset B_m(\mathbf{0}_m, s)$ for any $s > 0$. Thus $g$ is a homeomorphism, so $h = f \circ g : E_m \to \mathbb{I}^m$ is the promised homeomorphism. Fact 1 is proved.

*Fact 2.* Given a simplex $S = [a_0, \ldots, a_m]$ denote by $S_i$ the $(m-1)$-dimensional face of $S$ with $a_i \notin S_i$ and let $F_i$ be a closed subspace of $S$ such that $F_i \cap S_i = \emptyset$ for any $i \leq m$. If, additionally, $S = \bigcup \{F_i : i \in \{0, \ldots, m\}\}$ then $F_0 \cap \ldots \cap F_m \neq \emptyset$.

*Proof.* Take a face $B = [a_{i_0}, \ldots, a_{i_k}]$ of the simplex $S$. If $F_i \cap B \neq \emptyset$ then we must have $i \in \{i_0, \ldots, i_k\}$ because otherwise $B \subset S_i$ and hence $B \cap F_i = \emptyset$. As a consequence, $B \subset F_{i_0} \cup \ldots \cup F_{i_k}$, so we can apply Fact 1 of V.089 to conclude that $F_0 \cap \ldots \cap F_m \neq \emptyset$ and hence Fact 2 is proved.

*Fact 3.* Given a simplex $S = [a_0, \ldots, a_m]$ there is no retraction $r : S \to \partial S$.

*Proof.* Suppose that $r : S \to \partial S$ is a retraction, i.e., $r$ is continuous and $r(x) = x$ for any $x \in \partial S$. For any $i \in \{0, \ldots, m\}$ denote by $S_i$ the $(m-1)$-dimensional face of $S$ for which $a_i \notin S_i$. It is easy to see that $S_0 \cap \ldots \cap S_m = \emptyset$, so the family $\{\partial S \backslash S_i : i \leq m\}$ is an open cover of $\partial S$; since also $S_0 \cup \ldots \cup S_m = \partial S$, we have $\bigcap_{i \leq m}(\partial S \backslash S_i) = \emptyset$. Let $U_i = r^{-1}(\partial S \backslash S_i)$ for every $i \leq m$; then $\{U_0, \ldots, U_m\}$ is an open cover of $S$ such that $U_0 \cap \ldots \cap U_m = \emptyset$. The map $r$ being a retraction, the equality $U_i \cap S_i = \emptyset$ holds for any $i \leq m$.

Apply Fact 2 of S.226 to choose a family $F_0, \ldots, F_m$ of closed subsets of $S$ such that $S = \bigcup \{F_i : i \le m\}$ and $F_i \subset U_i$ for every $i \le m$. Thus $F_i \cap S_i \subset U_i \cap S_i = \emptyset$ for every $i \le m$, so we can apply Fact 2 to convince ourselves that $\bigcap_{i \le m} F_i \ne \emptyset$ and hence $\bigcap_{i \le m} U_i \ne \emptyset$; this contradiction shows that Fact 3 is proved.

Returning to our solution suppose that there is a retraction $r : \mathbb{I}^n \to \partial \mathbb{I}^n$. Apply Fact 1 to find a simplex $S$ and a homeomorphism $h : S \to \mathbb{I}^n$ such that $h(\partial S) = \partial \mathbb{I}^n$. It is straightforward that $r_0 = h^{-1} \circ r \circ h : S \to \partial S$ is a retraction; this contradiction with Fact 3 shows that there exists no retraction of $\mathbb{I}^n$ onto $\partial \mathbb{I}^n$.

**V.091.** *Given spaces $X$ and $Y$ and functions $f, g \in C(X, Y)$, let $f \sim g$ denote the fact that $f$ and $g$ are homotopic. Prove that $\sim$ is an equivalence relation on $C(X, Y)$.*

**Solution.** For any function $f \in C(X, Y)$ let $F(x, t) = f(x)$ for any $t \in [0, 1]$. It is clear that $F : X \times [0, 1] \to Y$ is continuous; since $F(x, 0) = F(x, 1) = f(x)$ for any $x \in X$, we proved that $f \sim f$.

Now assume that $f, g \in C(X, Y)$ and $f \sim g$; there exists a continuous map $F : X \times [0, 1] \to Y$ such that $F(x, 0) = f(x)$ and $F(x, 1) = g(x)$ for any $x \in X$. Define a function $G : X \times [0, 1] \to Y$ by the equality $G(x, t) = F(x, 1 - t)$ for any $x \in X$ and $t \in [0, 1]$. It is immediate that $G$ is continuous while $G(x, 0) = g(x)$ and $G(x, 1) = f(x)$ for any $x \in X$. Therefore $g \sim f$.

Finally assume that $f, g, h \in C(X, Y)$ and $f \sim g \sim h$. There exists continuous functions $F, G : X \times [0, 1] \to Y$ such that $F(x, 0) = f(x)$, $F(x, 1) = G(x, 0) = g(x)$ and $G(x, 1) = h(x)$ for any $x \in X$. Define a map $H_1 : X \times [0, \frac{1}{2}] \to Y$ by the formula $H_1(x, t) = F(x, 2t)$ for any $x \in X$ and $t \in [0, \frac{1}{2}]$. We will also need the map $H_2 : X \times [\frac{1}{2}, 1] \to Y$ defined by $H_2(x, t) = G(x, 2t - 1)$ for any $x \in X$ and $t \in [\frac{1}{2}, 1]$. It is clear that both maps $H_1$ and $H_2$ are continuous.

Observe that $H_1(x, \frac{1}{2}) = H_2(x, \frac{1}{2}) = g(x)$ for any $x \in X$; this, together with Fact 2 of T.354 shows that the map $H = H_1 \cup H_2$ is continuous. Recall that the map $H : X \times [0, 1] \to Y$ is defined by $H(x, t) = H_1(x, t)$ if $x \in X$, $t \le \frac{1}{2}$ and $H(x, t) = H_2(x, t)$ whenever $x \in X$ and $t \ge \frac{1}{2}$. Since $H(x, 0) = f(x)$ and $H(x, 1) = G(x, 1) = h(x)$ for any $x \in X$, we proved that $f \sim h$, so being homotopic is an equivalence relation.

**V.092.** *Given a space $X$, let $f, g : X \to \partial \mathbb{I}^n$ be continuous maps such that the points $f(x)$ and $g(x)$ belong to the same face of $\mathbb{I}^n$ for any $x \in X$. Prove that $f$ and $g$ are homotopic.*

**Solution.** Let $F(x, t) = (1 - t)f(x) + tg(x)$ for any $x \in X$ and $t \in [0, 1]$. It is clear that the map $F : X \times [0, 1] \to \mathbb{R}^n$ is continuous while $F(x, 0) = f(x)$ and $F(x, 1) = g(x)$ for any $x \in X$. Now, if $x \in X$ then there is a face $B$ of $\mathbb{I}^n$ such that $\{f(x), g(x)\} \subset B$. Therefore there is $i < n$ such that either $f(x)(i) = g(x)(i) = 1$ or $f(x)(i) = g(x)(i) = -1$. In both cases

$$F(x, t)(i) = (1 - t)f(x)(i) + tg(x)(i) = f(x)(i)(1 - t + t) = f(x)(i)$$

which shows that $F(x,t) \in B$ for any $t \in [0,1]$. Therefore $F : (X \times [0,1]) \subset \partial \mathbb{I}^n$, so the map $F : X \times [0,1] \to \partial \mathbb{I}^n$ witnesses the fact that the functions $f$ and $g$ are homotopic.

**V.093 (Mushroom lemma).** *Let $X$ be a normal countably paracompact space. Suppose that $F \subset X$ is closed and we have continuous homotopic mappings $f_0, f_1$ of $F$ to the n-dimensional sphere $S^n = \{x \in \mathbb{R}^{n+1} : x(0)^2 + \ldots + x(n)^2 = 1\}$. Prove that, if there exists a continuous map $g_0 : X \to S^n$ with $g_0|F = f_0$ then there is a continuous map $g_1 : X \to S^n$ such that $g_1|F = f_1$ and $g_1$ is homotopic to $g_0$.*

**Solution.** If $m \in \mathbb{N}$ then $|x|_m = \sqrt{(x(0))^2 + \ldots + (x(m-1))^2}$ for any $x \in \mathbb{R}^m$.

*Fact 1.* Given a number $m \in \mathbb{N}$, if $Y$ is a normal space, $G$ is closed in $Y$ and we have a continuous map $f : G \to S^m$ then there exists a set $U \in \tau(G, Y)$ and a continuous map $g : U \to S^m$ such that $g|G = f$.

*Proof.* We can consider that $f$ maps $G$ into $\mathbb{R}^{m+1}$; let $\pi_i(x) = x(i)$ for any $x \in \mathbb{R}^{m+1}$ and $i \in \{0, \ldots, m\}$. The map $\pi_i \circ f : Y \to \mathbb{R}$ is continuous, so there exists a continuous function $p_i : Y \to \mathbb{R}$ such that $p_i|G = \pi_i \circ f$ for any $i \leq m$.

Letting $p(y)(i) = p_i(y)$ for any index $i \leq m$ we obtain a continuous map $p : Y \to \mathbb{R}^{m+1}$ such that $p|G = f$. The set $O = \{x \in \mathbb{R}^{m+1} : |x|_{m+1} \neq 0\}$ is open in $\mathbb{R}^{m+1}$; since $S^m \subset O$, the set $U = p^{-1}(O)$ is open in $Y$ and $G \subset U$.

For any point $y \in O$ let $r(y) = \frac{y}{|y|_{m+1}}$; it is evident that the map $r : O \to S^m$ is continuous and $r(y) = y$ whenever $y \in S^m$. Thus the map $g = r \circ (p|U) : U \to S^m$ is continuous as well. Given an arbitrary point $y \in G$ we have $p(y) = f(y) \in S^m$ which shows that $g(y) = r(p(y)) = r(f(y)) = f(y)$ and hence $g$ is the desired continuous extension of $f$. Fact 1 is proved.

Returning to our solution fix a continuous map $f : F \times [0,1] \to S^n$ such that $f(x,0) = f_0(x)$ and $f(x,1) = f_1(x)$; let $h(x,0) = g_0(x)$ for any $x \in X$. Since $g_0 : X \to S^n$ is a continuous map, the map $h : X \times \{0\} \to S^n$ is also continuous. Furthermore, $h|(F \times \{0\}) = f|(F \times \{0\})$, so the map $q = f \cup h$ is continuous by Fact 2 of T.354.

The space $X \times [0,1]$ is normal (see TFS-288) and the set $P = (X \times \{0\}) \cup (F \times [0,1])$ is closed in $X \times [0,1]$, so we can apply Fact 1 to find a set $U \in \tau(P, X \times [0,1])$ and a continuous map $g : U \to S^n$ such that $g|P = q$. Since $[0,1]$ is compact, the projection $\pi : X \times [0,1] \to X$ is a perfect map (see Fact 3 of S.288), so $H = \pi((X \times [0,1])\backslash U)$ is a closed subset of $X$ disjoint from $F$. Therefore the set $V = X\backslash H$ is an open neighborhood of $F$ in $X$ and $\pi^{-1}(V) = V \times [0,1] \subset U$.

By normality of $X$ there is a continuous function $\varphi : X \to [0,1]$ such that $\varphi(x) = 1$ for any $x \in F$ and $\varphi(x) = 0$ whenever $x \in X\backslash V$. Let $\Phi(x,t) = g(x, t\varphi(x))$ for any $(x,t) \in X \times [0,1]$. It is easy to see that $\Phi$ is well defined on $X \times [0,1]$ and $\Phi : X \times [0,1] \to S^n$ is a continuous map. Besides, $x \in F$ implies $t\varphi(x) = t$ and hence $\Phi(x,t) = g(x,t) = f(x,t)$ for any $t \in [0,1]$. This shows that $\Phi|(F \times [0,1]) = f$; observe that also $\Phi(x,0) = g(x,0) = g_0(x)$ for every $x \in X$.

Finally, let $g_1(x) = \Phi(x, 1)$ for any $x \in X$; then $g_1 : X \to S^n$ is a continuous function and $\Phi$ witnesses that $g_0 \sim g_1$. Since also $g_1(x) = \Phi(x, 1) = f(x, 1) = f_1(x)$ for any $x \in F$, the function $g_1$ is a continuous extension of $f_1$, so our solution is complete.

**V.094.** *For each $i < n$, consider the faces $F_i = \{x \in \mathbb{I}^n : x(i) = -1\}$ and $G_i = \{x \in \mathbb{I}^n : x(i) = 1\}$ of the $n$-dimensional cube $\mathbb{I}^n$. Prove that, if $C_i$ is a partition between $F_i$ and $G_i$ then $\bigcap \{C_i : i < n\} \neq \emptyset$.*

**Solution.** If $m \in \mathbb{N}$ then $u_m \in \mathbb{R}^m$ is the zero point of $\mathbb{R}^m$, i.e., $u_m(i) = 0$ for every $i < m$. Furthermore, $|x|_m = \sqrt{(x(0))^2 + \ldots + (x(m-1))^2}$ for any $x \in \mathbb{R}^m$; the set $B_m = \{x \in \mathbb{I}^{m+1} : |x(i)| = 1$ for some $i \leq m\}$ is the boundary of the $(m+1)$-dimensional cube $\mathbb{I}^{m+1}$.

*Fact 1.* Given $m \in \mathbb{N}$ let $\varphi(x) = \frac{x}{|x|_{m+1}}$ for any point $x \in B_m$. Then the map $\varphi : B_m \to S^m$ is a homeomorphism.

*Proof.* It is clear that the map $\varphi$ is continuous and $\varphi(B_m) \subset S^m$. If $y \in S^m$ then $t = \max\{|y(i)| : i \leq m\} > 0$ and the point $x = \frac{y}{t}$ belongs to $B_m$. It is straightforward that $\varphi(x) = y$, so $\varphi$ is an onto map.

If $x_0, x_1 \in B_m$ and $\varphi(x_0) = \varphi(x_1)$ then $x_1 = tx_0$ where $t = \frac{|x_1|_{m+1}}{|x_0|_{m+1}}$. Assume first that $t < 1$; since $|x_0(i)| \leq 1$, we have $|tx_0(i)| < 1$ for all $i \leq m$. Therefore $tx_0 \notin B_m$ and hence $tx_0 \neq x_1$. If $t > 1$ then take a coordinate $i \leq m$ such that $|x_0(i)| = 1$; then $|tx_0(i)| > 1$ and hence $tx_0 \notin B_m$ which shows that $x_1 \neq tx_0$. This proves that $t = 1$ and therefore $x_0 = x_1$, i.e., $\varphi$ is a bijection; the space $B_m$ being compact, $\varphi$ is a homeomorphism, so Fact 1 is proved.

Returning to our solution assume that $\bigcap \{C_i : i < n\} = \emptyset$ and fix any $i < n$. Take disjoint sets $U_i, V_i \in \tau(\mathbb{I}^n)$ such that $F_i \subset U_i$, $G_i \subset V_i$ and $\mathbb{I}^n \setminus (U_i \cup V_i) = C_i$. The space $\mathbb{I}^n$ being metrizable we can choose continuous functions $a_i$ and $b_i$ from $\mathbb{I}^n$ to $I = [0, 1]$ such that $F_i = a_i^{-1}(0)$ and $\mathbb{I}^n \setminus U_i = b_i^{-1}(0)$. Analogously, there exist continuous functions $c_i, d_i : \mathbb{I}^n \to I$ such that $G_i = c_i^{-1}(0)$ and $\mathbb{I}^n \setminus V_i = d_i^{-1}(0)$.

It is easy to check that the function $g_i = \frac{d_i}{c_i + d_i} - \frac{b_i}{a_i + b_i}$ is continuous, $g_i : \mathbb{I}^n \to \mathbb{I}$ while $g_i(F_i) = \{-1\}$ and $g_i(G_i) = \{1\}$; besides, $g_i^{-1}(0) = C_i$. Let $g(x)(i) = g_i(x)$ for any $i < n$ and $x \in \mathbb{I}^n$; then the function $g : \mathbb{I}^n \to \mathbb{I}^n$ is continuous being the diagonal product of continuous functions $g_0, \ldots, g_{n-1}$. It is straightforward that

(1) $g(F_i) \subset F_i$ and $g(G_i) \subset G_i$ for any $i < n$.

For any $x \in \mathbb{I}^n \setminus \{u_n\}$ let $m(x) = \max\{|x(0)|, \ldots, |x(n-1)|\}$; then the point $p(x) = \frac{x}{m(x)}$ is well defined and belongs to $B_{n-1}$. It is evident that the map $p : \mathbb{I}^n \setminus \{u_n\} \to B_{n-1}$ is continuous. If $g(x) = u_n$ for some $x \in \mathbb{I}^n$ then $g_i(x) = 0$ and hence $x \in C_i$ for any $i < n$, i.e., $x \in \bigcap_{i < n} C_i$ which is a contradiction. Therefore $g(\mathbb{I}^n) \subset \mathbb{I}^n \setminus \{u_n\}$ and hence $\Phi = p \circ g : \mathbb{I}^n \to B_{n-1}$ is a continuous map. The function $p$ being an identity on $B_{n-1}$, it follows from (1) that

(2) $\Phi(F_i) \subset F_i$ and $\Phi(G_i) \subset G_i$ for any $i < n$.

Let id : $B_{n-1} \to B_{n-1}$ be the identity map, i.e., $\mathrm{id}(x) = x$ for any point $x \in B_{n-1}$. It follows from (2) that $h = \Phi|B_{n-1} : B_{n-1} \to B_{n-1}$; apply Problem 092 to see that the maps id and $h$ are homotopic. For any $m \in \mathbb{N}$ the spaces $S^m$ and $B_m$ are homeomorphic by Fact 1, so Mushroom lemma (Problem 093) is applicable for the respective maps into $B_m$. Since $\Phi : \mathbb{I}^n \to B_{n-1}$ is a continuous extension of $h$, we can apply Problem 093 to the maps $h$ and id to conclude that there exists a map $r : \mathbb{I}^n \to B_{n-1}$ such that $r|B_{n-1} = \mathrm{id}$, i.e., $r$ is a retraction of $\mathbb{I}^n$ onto $B_{n-1}$; this contradiction with Problem 090 shows that our solution is complete.

**V.095.** *For each $i \in \omega$, consider the subsets $F_i = \{x \in \mathbb{I}^\omega : x(i) = -1\}$ and $G_i = \{x \in \mathbb{I}^\omega : x(i) = 1\}$ of the cube $\mathbb{I}^\omega$. Prove that, if $C_i$ is any partition between $F_i$ and $G_i$ then $\bigcap\{C_i : i \in \omega\} \neq \emptyset$.*

**Solution.** For any $m \in \omega$ let $\pi_m : \mathbb{I}^\omega \to \mathbb{I}^m$ be the projection of $\mathbb{I}^\omega$ onto its face $\mathbb{I}^m$. Say that a set $U \subset \mathbb{I}^\omega$ is *standard* if there is $m \in \omega$ and a set $V \in \tau(\mathbb{I}^m)$ such that $\pi_m^{-1}(V) = U$; in this case let $j(U) = m$.

Assume that $\bigcap_{i<\omega} C_i = \emptyset$; it follows from compactness of $\mathbb{I}^\omega$ that there is $m \in \omega$ such that $\bigcap\{C_i : i < m\} = \emptyset$. Thus $\{\mathbb{I}^\omega \setminus C_i : i < m\}$ is an open cover of the space $\mathbb{I}^\omega$. The standard sets form a base in $\mathbb{I}^\omega$, so we can choose, for any $x \in \mathbb{I}^\omega$ a standard set $O_x \ni x$ such that $O_x \subset \mathbb{I}^\omega \setminus C_i$ for some $i < m$. Fix a finite set $A \subset \mathbb{I}^\omega$ for which $\bigcup\{O_x : x \in A\} = \mathbb{I}^\omega$.

It is easy to see that, for $n = \sum\{j(O_x) : x \in A\} + m$, we have

(1) $\bigcap\{C_i : i < n\} = \emptyset$ and $\pi_n^{-1}(\pi_n(O_x)) = O_x$ for any $x \in A$.

Consider the sets $F_i' = \{x \in \mathbb{I}^n : x(i) = -1\}$ and $G_i' = \{x \in \mathbb{I}^n : x(i) = 1\}$ for every $i < n$. It turns out that

(2) the set $D_i = \pi_n(C_i)$ is a partition between $F_i'$ and $G_i'$ for any $i < n$.

To prove this, fix a number $i < n$ and choose disjoint sets $U_i, V_i \in \tau(\mathbb{I}^\omega)$ such that $F_i \subset U_i$, $G_i \subset V_i$ and $\mathbb{I}^\omega \setminus (U_i \cup V_i) = C_i$. The map $\pi_n$ being closed the sets $U_i' = \mathbb{I}^n \setminus \pi_n(\mathbb{I}^\omega \setminus U_i)$ and $V_i' = \mathbb{I}^n \setminus \pi_n(\mathbb{I}^\omega \setminus V_i)$ are open in $\mathbb{I}^n$; it is straightforward that $U_i' \cap V_i' = \emptyset$ while $F_i' \subset U_i'$ and $G_i' \subset V_i'$. Given a point $x \in \mathbb{I}^n \setminus D_i$, the space $\pi_n^{-1}(x) \subset U_i \cup V_i$ is connected because $\pi_n^{-1}(x) \simeq \mathbb{I}^{\omega \setminus n}$ (see Fact 1 of U.493), so $\pi_n^{-1}(x)$ cannot meet both sets $U_i$ and $V_i$; thus either $\pi_n^{-1}(x) \subset U_i$ or $\pi_n^{-1}(x) \subset V_i$. This shows that $x \in U_i' \cup V_i'$, i.e., $\mathbb{I}^n \setminus (U_i' \cup V_i') = D_i$, so (2) is proved.

Given a point $y \in \mathbb{I}^n$ take any $z \in \mathbb{I}^\omega$ such that $\pi_n(z) = y$. There is $x \in A$ such that $z \in O_x$; pick $i < n$ for which $O_x \subset \mathbb{I}^\omega \setminus C_i$. It follows from (1) that $\pi_n^{-1}(y) = \pi_n^{-1}(\pi_n(z)) \subset O_x \subset \mathbb{I}^\omega \setminus C_i$ and hence $y \notin D_i$. Therefore $y \notin \bigcap_{i<n} D_i$; the point $y \in \mathbb{I}^n$ was chosen arbitrarily, so we proved that $\bigcap_{i<n} D_i = \emptyset$. Since $D_i$ is a partition in the cube $\mathbb{I}^n$ between its faces $F_i'$ and $G_i'$ for any $i < n$, we obtained a contradiction with Problem 094.

**V.096.** *Prove that, for any $n \in \mathbb{N}$, the space $\mathbb{I}^n$ is the finite union of its zero-dimensional subspaces.*

**Solution.** Let $Z_0 = \mathbb{Q} \cap \mathbb{I}$ and $Z_1 = \mathbb{I} \setminus Z_0$. It is clear that $Z_0$ and $Z_1$ are zero-dimensional subspaces of $\mathbb{I}$ such that $\mathbb{I} = Z_0 \cup Z_1$. Now if $n > 1$ then consider the set $P_\sigma = \prod\{Z_{\sigma(i)} : i < n\}$ for any $\sigma \in \mathbb{D}^n$. Every space $P_\sigma$ is zero-dimensional by

SFFS-302; it is straightforward that $\mathbb{I}^n = \bigcup\{P_\sigma : \sigma \in \mathbb{D}^n\}$, so $\{P_\sigma : \sigma \in \mathbb{D}^n\}$ is a finite family of zero-dimensional subspaces of $\mathbb{I}^n$ whose union is $\mathbb{I}^n$.

**V.097.** *Prove that, for any $n \in \mathbb{N}$, the space $\mathbb{I}^n$ cannot be represented as the union of $\leq n$-many of its zero-dimensional subspaces.*

**Solution.** A space is *cosmic* if has a countable network. Given a space $X$ we say that sets $A, B \subset X$ are *separated* in $X$ if $\overline{A} \cap B = A \cap \overline{B} = \emptyset$.

*Fact 1.* Suppose that $X$ is a cosmic space such that $X \setminus \{a\}$ is zero-dimensional for some point $a \in X$. Then $X$ is zero-dimensional. In other words, adding a point to a cosmic zero-dimensional space gives a zero-dimensional space.

*Proof.* Since $\psi(a, X) \leq nw(X) = \omega$, there exists a family $\{U_n : n \in \omega\} \subset \tau(a, X)$ such that $\{a\} = \bigcap\{U_n : n \in \omega\}$. It follows from $hl(X \setminus \{a\}) = \omega$ that any subspace of $X \setminus \{a\}$ is strongly zero-dimensional (see SFFS-301 and SFFS-306); this implies that the set $F_n = X \setminus U_n \subset X \setminus \{a\}$ is closed in $X$ and strongly zero-dimensional for any $n \in \omega$. The family $\mathcal{F} = \{\{a\}\} \cup \{F_n : n \in \omega\}$ is countable and consists of closed strongly zero-dimensional subsets of $X$; the space $X$ is normal and $X = \bigcup \mathcal{F}$, so we can apply SFFS-311 to conclude that $X$ is strongly zero-dimensional and hence zero-dimensional. Fact 1 is proved.

*Fact 2.* Suppose that $X$ is a second countable space and $F, G \subset X$ are disjoint closed subsets of $X$. Then, for any zero-dimensional $Z \subset X$, there is a partition $C$ between the sets $F$ and $G$ such that $C \cap Z = \emptyset$.

*Proof.* Observe first that

(1) there exists a second countable compact space $K$ and an embedding $e : X \to K$ for which $\overline{e(F)} \cap \overline{e(G)} = \emptyset$.
Indeed, the space $X$ is normal, so there is a continuous function $f : X \to \mathbb{I}$ such that $f(F) \subset \{0\}$ and $f(G) \subset \{1\}$. If $\varphi : X \to \mathbb{I}^\omega$ is an arbitrary embedding then the map $e = \varphi \Delta f : X \to K = \mathbb{I}^\omega \times \mathbb{I}$ is still an embedding such that $\overline{g(F)} \cap \overline{g(G)} = \emptyset$.
To simplify the notation we will identify $X$ and $e(X)$; thus $X \subset K$ and the sets $F' = \overline{F}$ and $G' = \overline{G}$ are disjoint (the bar denotes the closure in $K$). Call a set $U \in \tau(K)$ *adequate* if $U \cap Z = \overline{U} \cap Z$ and, in particular, the set $U \cap Z$ is clopen in $Z$. We claim that

(2) adequate sets form a base in $K$.

To prove it fix a point $x \in K$ and $U \in \tau(x, K)$. Since $\{x\} \cup Z$ is zero-dimensional by Fact 1, there is a clopen subset $A$ of the space $\{x\} \cup Z$ such that $x \in A \subset U \cap (\{x\} \cup Z)$. The sets $A$ and $B = Z \setminus A$ are disjoint and clopen in $\{x\} \cup Z$; an easy consequence is that they are separated in $K$.

Apply Fact 1 of S.291 to find disjoint sets $V_A, V_B \in \tau(K)$ such that $A \subset V_A$ and $B \subset V_B$. We leave to the reader a simple verification of the fact that the set $V = V_A \cap U$ is adequate; since also $x \in V \subset U$, the property (2) is proved.

Now apply (2) to find, for any point $x \in F'$, an adequate set $O_x \in \tau(x, K)$ such that $\overline{O}_x \cap G' = \emptyset$. The set $F'$ being compact, there is a finite $P \subset F'$ for which $F' \subset O = \bigcup\{O_x : x \in P\}$. It is evident that $O$ is adequate so $C' = \overline{O}\backslash O$ is a closed set such that $C' \cap Z = \emptyset$. The sets $O \supset F'$ and $W = K\backslash \overline{O}$ witness that $C'$ is a partition between $F'$ and $G'$; an immediate consequence is that $C = C' \cap X$ is the promised partition in $X$ between the sets $F$ and $G$. Fact 2 is proved.

Returning to our solution assume that $\mathbb{I}^n = \bigcup\{Z_i : i < n\}$ and every $Z_i$ is zero-dimensional. The faces $F_i = \{x \in \mathbb{I}^n : x(i) = -1\}$ and $G_i = \{x \in \mathbb{I}^n : x(i) = 1\}$ are disjoint and closed in $\mathbb{I}^n$; by Fact 2 there is a partition $C_i$ between the sets $F_i$ and $G_i$ such that $C_i \cap Z_i = \emptyset$ for any $i < n$. The set $C = \bigcap_{i<n} C_i$ is nonempty by Problem 094 and $C \cap Z_i = \emptyset$ for any $i < n$; therefore no point of $C$ belongs to $\bigcup\{Z_i : i < n\} = \mathbb{I}^n$. This contradiction shows that the above representation of $\mathbb{I}^n$ is impossible and hence our solution is complete.

**V.098.** *Prove that the cube $\mathbb{I}^\omega$ cannot be represented as the countable union of its zero-dimensional subspaces. Prove (in ZFC) that there exist zero-dimensional spaces $\{X_\alpha : \alpha < \omega_1\}$ such that $\mathbb{I}^\omega = \bigcup\{X_\alpha : \alpha < \omega_1\}$.*

**Solution.** Assume that $\mathbb{I}^\omega = \bigcup\{Z_i : i < \omega\}$ and every $Z_i$ is zero-dimensional. The faces $F_i = \{x \in \mathbb{I}^\omega : x(i) = -1\}$ and $G_i = \{x \in \mathbb{I}^\omega : x(i) = 1\}$ are disjoint and closed in $\mathbb{I}^\omega$; by Fact 2 of V.097 there is a partition $C_i$ between the sets $F_i$ and $G_i$ such that $C_i \cap Z_i = \emptyset$ for any $i < \omega$. The set $C = \bigcap_{i<\omega} C_i$ is nonempty by Problem 095 and $C \cap Z_i = \emptyset$ for any $i < \omega$; therefore no point of $C$ belongs to $\bigcup\{Z_i : i < \omega\} = \mathbb{I}^\omega$. This contradiction shows that the above representation of $\mathbb{I}^\omega$ is impossible.

Next observe that $|\mathfrak{c} \times \omega_1| = \mathfrak{c}$, so there is a bijection $\varphi : \mathfrak{c} \to \mathfrak{c} \times \omega_1$. Let $A_\alpha = \varphi^{-1}(\mathfrak{c} \times \{\alpha\})$ for any $\alpha < \omega_1$. Then

(1) the family $\mathcal{A} = \{A_\alpha : \alpha < \omega_1\}$ is disjoint, consists of nonempty subsets of $\mathfrak{c}$ and $\bigcup \mathcal{A} = \mathfrak{c}$.

It is easy to see that

(2) if $Y \subset \mathbb{I}$ and $\mathbb{I}\backslash Y$ is dense in $\mathbb{I}$ then $Y$ is zero-dimensional.

Apply Fact 1 of S.480 to find a disjoint family $\mathcal{E} = \{E_\alpha : \alpha < \mathfrak{c}\}$ of subsets of $\mathbb{R}$ such that every $E_\alpha$ is dense in $\mathbb{R}$. Let $H_0 = (E_0 \cap \mathbb{I}) \cup (\mathbb{I}\backslash \bigcup \mathcal{E})$; if $0 < \alpha < \mathfrak{c}$ then let $H_\alpha = E_\alpha \cap \mathbb{I}$. Then $\mathcal{H} = \{H_\alpha : \alpha < \mathfrak{c}\}$ is a disjoint family of dense subsets of $\mathbb{I}$ such that $\bigcup \mathcal{H} = \mathbb{I}$.

Now, if $G_\alpha = \bigcup\{H_\beta : \beta \in A_\alpha\}$ for any $\alpha < \omega_1$ then it follows from (1) that $\mathcal{G} = \{G_\alpha : \alpha < \omega_1\}$ is still a family of disjoint dense subsets of $\mathbb{I}$ such that $\bigcup \mathcal{G} = \mathbb{I}$. The property (2) shows that the space $P_\alpha = \bigcup\{G_\beta : \beta < \alpha\}$ is zero-dimensional because $\mathbb{I}\backslash P_\alpha$ contains a dense set $G_\alpha$ for any $\alpha < \omega_1$. Therefore

(3) the family $\mathcal{P} = \{P_\alpha : \alpha < \omega_1\}$ is increasing, consists of zero-dimensional subsets of $\mathbb{I}$ and $\bigcup \mathcal{P} = \mathbb{I}$.

Finally let $X_\alpha = (P_\alpha)^\omega$ for every $\alpha < \omega_1$. Then $\{X_\alpha : \alpha < \omega_1\}$ is a family of zero-dimensional subsets of $\mathbb{I}^\omega$ (see SFFS-302). Given a point $x \in \mathbb{I}^\omega$, for any

$i \in \omega$, there is $\alpha_i < \omega_1$ with $x(i) \in P_{\alpha_i}$. If $\alpha > \sup\{\alpha_i : i < \omega\}$ then $x(i) \in P_\alpha$ for any $i < \omega$ and hence $x \in (P_\alpha)^\omega = X_\alpha$. This proves that $\mathbb{I}^\omega = \bigcup\{X_\alpha : \alpha < \omega_1\}$, so our solution is complete.

**V.099.** *Prove that, for any $n \in \mathbb{N}$, the spaces $\mathbb{I}^n$ and $\mathbb{I}^\omega$ are not $t$-equivalent.*

**Solution.** Take any $n \in \mathbb{N}$ and assume that $\mathbb{I}^n$ is $t$-equivalent to $\mathbb{I}^\omega$. Since the space $\mathbb{I}^n$ is representable as the finite union of its zero-dimensional subspaces (see Problem 096), we can apply Problem 082 to conclude that $\mathbb{I}^\omega$ is also the countable union of its zero-dimensional subspaces. This, however, gives a contradiction with Problem 098 and shows that the spaces $\mathbb{I}^n$ and $\mathbb{I}^\omega$ are not $t$-equivalent.

**V.100.** *Suppose that $X$ is one of the spaces $\omega_1$ or $\omega_1 + 1$. Prove that, for any distinct $m, n \in \mathbb{N}$, the spaces $(C_p(X))^n$ and $(C_p(X))^m$ are not homeomorphic. In particular, $X$ is not $t$-equivalent to $X \oplus X$.*

**Solution.** Let $I = [0, 1] \subset \mathbb{R}$ and $\mathbb{P} = \mathbb{R}\backslash\mathbb{Q}$; for any set $X$ the map $\mathrm{id}_X : X \to X$ is the identity on $X$, i.e., $\mathrm{id}_X(x) = x$ for any $x \in X$. Given spaces $X$ and $Y$, a map $h : X \to Y$ is called *constant* if there is a point $b \in Y$ such that $f(x) = b$ for any $x \in X$; if we have maps $f, g \in C(X, Y)$ then the expression $f \sim g$ says that $f$ and $g$ are homotopic. Say that spaces $X$ and $Y$ are *homotopically equivalent* if there exist maps $f : X \to Y$ and $g : Y \to X$ such that $f \circ g \sim \mathrm{id}_Y$ and $g \circ f \sim \mathrm{id}_X$. If $n \in \mathbb{N}$ say that a space $X$ has the *$n$-partition property* if for any family $\{(F_0, G_0), \ldots, (F_{n-1}, G_{n-1})\}$ of pairs of disjoint closed subsets of $X$, for every $i < n$, there exists a partition $C_i$ between the sets $F_i$ and $G_i$ such that $\bigcap\{C_i : i < n\} = \emptyset$.

If $L$ is a linear topological space then a family $\{\varphi_1, \ldots, \varphi_n\}$ of continuous linear functionals on $L$ is called *independent* if, for any $a_1, \ldots, a_n \in \mathbb{R}$ with $\sum_{i=1}^n a_i^2 \neq 0$ the functional $a_1\varphi_1 + \ldots + a_n\varphi_n$ is not identically zero. A linear subspace $L' \subset L$ is said to have *codimension* $n \in \mathbb{N}$ in $L$ if there is an independent family $\{\varphi_1, \ldots, \varphi_n\}$ of continuous linear functionals on $L$ such that $L' = \varphi_1^{-1}(0) \cap \ldots \cap \varphi_n^{-1}(0)$. Given ordinals $\alpha$ and $\beta$ the interval $[\alpha, \beta]$ consists of the ordinals $\gamma$ such that $\alpha \leq \gamma \leq \beta$; analogously, $(\alpha, \beta) = \{\gamma : \alpha < \beta < \gamma\}$.

*Fact 1.* Suppose that we have spaces $X, Y, Z$ and maps $f, g, h$ with $f, g \in C(X, Y)$ and $h \in C(Y, Z)$. If, additionally, $f \sim g$ then $h \circ f \sim h \circ g$.

*Proof.* Let $F : X \times I \to Y$ be a continuous map such that $F(x, 0) = f(x)$ and $F(x, 1) = g(x)$ for any $x \in X$. Then $G = h \circ F : X \times I \to Z$ is a continuous map such that $G(x, 0) = h(F(x, 0)) = (h \circ f)(x)$ and $G(x, 1) = h(F(x, 1)) = (h \circ g)(x)$ for any $x \in X$, i.e., $G$ witnesses that $h \circ f$ and $h \circ g$ are homotopic. Fact 1 is proved.

*Fact 2.* Suppose that we have spaces $X, Y, Z$ and maps $f, g, h$ with $f, g \in C(Y, Z)$ and $h \in C(X, Y)$. If, additionally, $f \sim g$ then $f \circ h \sim g \circ h$.

*Proof.* Since $f$ is homotopic to $g$, we can find a continuous map $F : Y \times I \to Z$ such that $F(y, 0) = f(y)$ and $F(y, 1) = g(y)$ for any $y \in Y$. Let $H(x, t) = (h(x), t)$ for any $x \in X$ and $t \in I$; then $H : X \times I \to Y \times I$ is a continuous map.

It is evident that the map $G = F \circ H : X \times I \to Z$ is also continuous and $G(x, 0) = F(h(x), 0) = (f \circ h)(x)$ and $G(x, 1) = F(h(x), 1) = (g \circ h)(x)$ for any $x \in X$, i.e., $G$ witnesses that $f \circ h$ and $g \circ h$ are homotopic. Fact 2 is proved.

*Fact 3.* Homotopical equivalence is an equivalence relation on the class of topological spaces.

*Proof.* It is clear from the definition that any space is homotopically equivalent to itself and $X$ is homotopically equivalent to $Y$ if and only if $Y$ is homotopically equivalent to $X$. Now assume that $X$ is homotopically equivalent to $Y$ and $Y$ is homotopically equivalent to $Z$. To witness homotopical equivalence of $X$ and $Y$ choose continuous maps $f : X \to Y$ and $g : Y \to X$ such that $f \circ g \sim \mathrm{id}_Y$ and $g \circ f \sim \mathrm{id}_X$. Analogously, there exist continuous maps $u : Y \to Z$ and $v : Z \to Y$ such that $u \circ v \sim \mathrm{id}_Z$ and $v \circ u \sim \mathrm{id}_Y$.

Then $h = u \circ f : X \to Z$ and $w = g \circ v : Z \to X$ are continuous maps; it follows from Fact 1 and Fact 2 that $h \circ w = u \circ f \circ g \circ v \sim u \circ \mathrm{id}_Y \circ v = u \circ v \sim \mathrm{id}_Z$. Analogously, $w \circ h = g \circ v \circ u \circ f \sim g \circ \mathrm{id}_Y \circ f = g \circ f = \mathrm{id}_X$ and hence the maps $h$ and $w$ witness homotopical equivalence of $X$ and $Z$. Fact 3 is proved.

*Fact 4.* If a second countable space is representable as the union of at most $n$-many zero-dimensional subspaces then it has the $n$-partition property.

*Proof.* Take a second countable space $X$ such that $X = \bigcup\{Z_i : i < n\}$ and every $Z_i$ is zero-dimensional. Assume that $F_i$ and $G_i$ are disjoint closed subsets of $X$ for any $i < n$. By Fact 2 of V.097 there is a partition $C_i$ between the sets $F_i$ and $G_i$ such that $C_i \cap Z_i = \emptyset$ for any $i < n$. Then $C = \bigcap_{i<n} C_i$ does not meet the set $\bigcup_{i<n} Z_i = X$ which shows that $C = \emptyset$ and hence Fact 4 is proved.

*Fact 5.* Suppose that $X$ is a normal space. If $F \subset X$ is closed and $U \in \tau(F, X)$ then there exists a cozero set $V$ such that $F \subset V \subset \overline{V} \subset U$.

*Proof.* By normality of $X$ there is a continuous function $f : X \to I$ such that $f(F) \subset \{0\}$ and $f(X \backslash U) \subset \{1\}$. Then $V = f^{-1}([0, \frac{1}{2}))$ is a cozero set by Fact 1 of T.252. We have $F \subset V \subset \overline{V} \subset f^{-1}([0, \frac{1}{2}]) \subset U$, so Fact 5 is proved.

*Fact 6.* Suppose that $X$ is a normal space and $\{U_1, \ldots, U_n\}$ is an open cover of $X$ for some $n \in \mathbb{N}$. If $F_i$ is a closed subset of $X$ with $F_i \subset U_i$ for any $i \leq n$ then there exists a cozero set $V_i$ such that $F_i \subset V_i \subset \overline{V}_i \subset U_i$ for each $i \leq n$ and $\{V_1, \ldots, V_n\}$ is a cover of $X$.

*Proof.* It is evident that the set $F = F_1 \cup (X \backslash (U_2 \cup \ldots \cup U_n))$ is closed in $Z$ and $F \subset U_1$; besides $F \cup U_2 \cup \ldots \cup U_n = X$. Since $X$ is normal, we can apply Fact 5 to find a cozero set $V_1$ such that $F \subset V_1 \subset \overline{V}_1 \subset U_1$; it is clear that $V_1 \cup U_2 \cup \ldots \cup U_n = X$.

Assume that $1 \leq k < n$ and we have constructed cozero sets $V_1, \ldots, V_k$ such that $F_i \subset V_i \subset \overline{V}_i \subset U_i$ and $V_1 \cup \ldots \cup V_i \cup U_{i+1} \cup \ldots \cup U_n = X$ for any $i \leq k$. The set $F = F_{k+1} \cup (X \backslash (V_1 \cup \ldots \cup V_k \cup U_{k+2} \cup \ldots \cup U_n))$ is closed in $X$ and $F \subset U_{k+1}$. Since $X$ is normal, we can apply Fact 5 to find a cozero set $V_{k+1}$ such that $F \subset V_{k+1} \subset \overline{V}_{k+1} \subset U_{k+1}$; it is clear that $F_{k+1} \subset V_{k+1} \subset \overline{V}_{k+1} \subset U_{k+1}$ and $V_1 \cup \ldots \cup V_{k+1} \cup U_{k+2} \cup \ldots \cup U_n = X$, so our inductive construction can be continued to obtain a family $\{V_1, \ldots, V_n\}$ of cozero subsets of $X$ such that $F_i \subset V_i \subset \overline{V}_i \subset U_i$ for every $i \leq n$ and $V_1 \cup \ldots \cup V_n = X$. Fact 6 is proved.

*Fact 7.* Suppose that, for some $n \in \omega$, a normal countably paracompact space $X$ has the $(n+1)$-partition property. Then for any closed set $F \subset X$ and any function $f \in C(F, S^n)$ there is a continuous map $h : X \to S^n$ such that $h|F = f$.

*Proof.* Denote by $u$ the zero point of $\mathbb{R}^{n+1}$, i.e., $u(i) = 0$ for every $i \leq n$. As usual, the set $B_n = \{x \in \mathbb{I}^{n+1} : |x(i)| = 1$ for some $i \leq n\}$ is the boundary of the $(n+1)$-dimensional cube $\mathbb{I}^{n+1}$. Since $B_n \simeq S^n$ by Fact 1 of V.094, it suffices to prove our Fact replacing $S^n$ with $B_n$, so we consider from now on that $f : F \to B_n$. For any $i \leq n$, let $F_i' = \{x \in B_n : x(i) = -1\}$ and $G_i' = \{x \in B_n : x(i) = 1\}$; the sets $F_i = f^{-1}(F_i')$ and $G_i = f^{-1}(G_i')$ are disjoint and closed in $F$ and hence in $X$, so we can find a partition $C_i'$ in $X$ between the sets $F_i$ and $G_i$ such that $\bigcap_{i \leq n} C_i' = \emptyset$. For every partition $C_i'$ take disjoint sets $U_i', V_i' \in \tau(X)$ such that $F_i \subset U_i'$, $G_i \subset V_i'$ and $X \backslash (U_i' \cup V_i') = C_i'$.

The family $\{U_0', V_0', \ldots, U_n', V_n'\}$ is an open cover of the space $X$. By Fact 6 there are cozero sets $U_0, V_0, \ldots, U_n, V_n$ such that $F_i \subset U_i \subset U_i'$, $G_i \subset V_i \subset V_i'$ for every $i \leq n$ and $\bigcup \{U_i \cup V_i : i \leq n\} = X$. Therefore $C_i = X \backslash (U_i \cup V_i)$ is a zero-set which is a partition between $F_i$ and $G_i$ for every $i \leq n$. The family $\{U_0, V_0, \ldots, U_n, V_n\}$ being a cover of $X$, we have $\bigcap_{i \leq n} C_i = \emptyset$.

Fix $i \leq n$ and apply Fact 1 of S.499 and Fact 1 of T.252 to find continuous functions $a_i$ and $b_i$ from $X$ to $I$ such that $a_i(F_i) \subset \{0\}$, $a_i(X \backslash U_i) \subset \{1\}$ and $X \backslash U_i = b_i^{-1}(0)$. Analogously, there exist functions $c_i, d_i \in C_p(X, I)$ such that $c_i(G_i) \subset \{0\}$, $c_i(X \backslash V_i) \subset \{1\}$ and $X \backslash V_i = d_i^{-1}(0)$. It is easy to check that the function $g_i = \frac{d_i}{c_i + d_i} - \frac{b_i}{a_i + b_i}$ is continuous, $g_i : X \to \mathbb{I}$ while $g_i(F_i) \subset \{-1\}$ and $g_i(G_i) \subset \{1\}$; besides, $g_i^{-1}(0) = C_i$.

Let $g(x)(i) = g_i(x)$ for any $i \leq n$ and $x \in X$; then the function $g : X \to \mathbb{I}^{n+1}$ is continuous being the diagonal product of continuous functions $g_0, \ldots, g_n$. It is straightforward that

(∗)  $g(F_i) \subset F_i'$ and $g(G_i) \subset G_i'$ for any $i \leq n$.

For any $x \in \mathbb{I}^{n+1} \backslash \{u\}$, let $m(x) = \max\{|x(0)|, \ldots, |x(n)|\}$; then the point $p(x) = \frac{x}{m(x)}$ is well defined and belongs to $B_n$. It is evident that the mapping $p : \mathbb{I}^{n+1} \backslash \{u\} \to B_n$ is continuous. If $g(x) = u$ for some $x \in X$ then $g_i(x) = 0$ and hence $x \in C_i$ for any $i \leq n$, i.e., $x \in \bigcap_{i \leq n} C_i$ which is a contradiction. Therefore $g(X) \subset \mathbb{I}^{n+1} \backslash \{u\}$ and hence $\Phi = p \circ g : X \to B_n$ is a continuous map. The function $p$ being an identity on $B_n$, it follows from (∗) that

(∗∗)  $\Phi(F_i) \subset F_i'$ and $\Phi(G_i) \subset G_i'$ for any $i \leq n$.

Given $x \in F$, it follows from $\bigcup_{i \leq n} (F_i' \cup G_i') = B_n$ that $f(x) \in F_i' \cup G_i'$ for some $i \leq n$. If $f(x) \in F_i'$ then $x \in F_i$ and hence $\Phi(x) \in F_i'$ by (∗∗); analogously, if $f(x) \in G_i'$ then $\Phi(x) \in G_i'$ which shows that $\Phi(x)$ and $f(x)$ belong to the same face of $B_n$ for any $x \in F$. Apply Problem 092 to see that the maps $f$ and $\Phi|F$ are homotopic. For any $m \in \mathbb{N}$ the spaces $S^m$ and $B_m$ are homeomorphic by Fact 1 of V.094, so Mushroom lemma (Problem 093) is applicable for the respective maps into $B_m$. Since $\Phi : X \to B_n$ is a continuous extension of $\Phi|F$, we can apply Problem 093 to the maps $\Phi|F$ and $f$ to conclude that there is a continuous map $h : X \to S^n$ with $h|F = f$, so Fact 7 is proved.

*Fact 8.* For any $n \in \mathbb{N}$ both spaces $\mathbb{R}^n$ and $S^n$ are representable as the union of at most $(n+1)$-many of their zero-dimensional subspaces.

*Proof.* Consider the set $Z(k,m) = \{x \in \mathbb{R}^m : |\{i < m : x(i) \in \mathbb{P}\}| = k\}$ for every $m \in \mathbb{N}$ and $k \leq m$. We will show that every $Z(k,m)$ is zero-dimensional. To do this, fix $k \leq m$ and let $H(r_1,\ldots,r_{m-k},i_1,\ldots,i_{m-k}) = \{x \in Z(k,m) : x(i_j) = r_j$ for all $j \leq m-k\}$ for any $r_1,\ldots,r_{m-k} \in \mathbb{Q}$ and distinct $i_1,\ldots,i_{m-k} \in \{0,\ldots,m-1\}$.

Every set $H(r_1,\ldots,r_{m-k},i_1,\ldots,i_{m-k})$ is zero-dimensional being homeomorphic to $\mathbb{P}^k$. The family $\mathcal{H} = \{H(r_1,\ldots,r_{m-k},i_1,\ldots,i_{m-k}) : r_i \in \mathbb{Q}$ for each $i \leq m-k$ and $i_1,\ldots,i_{m-k}$ are distinct elements of $\{0,\ldots,m-1\}\}$ is countable and $\bigcup \mathcal{H} = Z(k,m)$. It turns out that every set $H = H(r_1,\ldots,r_{m-k},i_1,\ldots,i_{m-k})$ is closed in $Z(k,m)$.

Indeed, if $x \in Z(k,m)$ is an accumulation point of $H$ then $x(i_j) = r_j$ for any $j \leq m-k$. Therefore $x(i) \in \mathbb{P}$ for any $i \notin \{i_1,\ldots,i_{m-k}\}$ which shows that $x \in H$, i.e., $H$ is closed in $Z(k,m)$. Finally, apply SFFS-311 and SFFS-306 to conclude that $Z(k,m)$ is zero-dimensional for every $k \leq m$. We have $\mathbb{R}^n = Z(0,n) \cup \ldots \cup Z(n,n)$, so $\mathbb{R}^n$ as the union of $(n+1)$-many zero-dimensional subspaces.

Recall that the boundary $B_n = \{x \in \mathbb{I}^{n+1} : |x(i)| = 1$ for some $i \leq n\}$ of the $(n+1)$-dimensional cube $\mathbb{I}^{n+1}$ is homeomorphic to $S^n$ by Fact 1 of V.094, so it suffices to show that $B_n$ is representable as the union of at most $(n+1)$-many of its zero-dimensional subspaces. Now, $\mathbb{R}^{n+1} = Z(0,n+1) \cup \ldots \cup Z(n+1,n+1)$; since every point of $B_n$ has a rational coordinate (either 1 or $-1$), the set $B_n$ does not meet $Z(n+1,n+1)$ and therefore $B_n = (Z(0,n+1) \cap B_n) \cup \ldots \cup (Z(n,n+1) \cap B_n)$ is the desired representation of $B_n$ as the union of at most $(n+1)$-many zero-dimensional spaces. Fact 8 is proved.

*Fact 9.* If $n < m$ then any continuous function $f : S^n \to S^m$ is homotopic to a constant map.

*Proof.* Fix a map $f \in C(S^n, S^m)$ and let $W = \{x \in \mathbb{R}^{n+1} : \sum_{i=0}^{n}(x(i))^2 \leq 1\}$ be the $(n+1)$-dimensional ball in $\mathbb{R}^{n+1}$ whose boundary is $S^n$. There are zero-dimensional spaces $Z_0,\ldots,Z_{n+2}$ such that $\mathbb{R}^{n+1} = Z_0 \cup \ldots \cup Z_{n+2}$ (see Fact 8); every $Z_i' = Z_i \cap W$ is zero-dimensional and $W = Z_0' \cup \ldots \cup Z_{n+2}'$ which shows that the metrizable space $W$ has the $(n+2)$-partition property by Fact 4. This, together with $n+2 \leq m+1$ implies that $W$ also has $(m+1)$-partition property, so we can apply Fact 7 to the space $W$, its closed subspace $S^n$ and the map $f$ to obtain a continuous map $g : W \to S^m$ such that $g|S^n = f$.

Now let $F(x,t) = g(tx)$ for any $x \in S^n$ and $t \in I$. The map $F : S^n \times I \rightarrow S^m$ is continuous, $F(x,1) = g(x) = f(x)$ and $F(x,0) = g(0)$ for any $x \in S^n$ which shows that the map $f$ is homotopic to a constant map $h : S^n \rightarrow \{g(0)\}$ and hence Fact 9 is proved.

*Fact 10.* For any $n \in \omega$ and $p \in \mathbb{R}^{n+1}$ the spaces $\mathbb{R}^{n+1}\backslash\{p\}$ and $S^n$ are homotopically equivalent.

*Proof.* For any $x \in \mathbb{R}^{n+1}$ let $|x| = \sqrt{(x(0))^2 + \ldots + (x(n))^2}$ and denote by $u$ the zero point of $\mathbb{R}^{n+1}$. The space $\mathbb{R}^{n+1}\backslash\{u\}$ and $\mathbb{R}^{n+1}\backslash\{p\}$ are easily seen to be homeomorphic, so it suffices to show that $\mathbb{R}^{n+1}\backslash\{u\}$ is homotopically equivalent to $S^n$ (see Fact 3). Define a map $g : S^n \rightarrow \mathbb{R}^{n+1}\backslash\{u\}$ by $g(x) = x$ for any $x \in S^n$; it is evident that $g$ is continuous.

Letting $f(x) = \frac{x}{|x|}$ for any $x \in \mathbb{R}^{n+1}\backslash\{u\}$ we also obtain a continuous map $f : \mathbb{R}^{n+1}\backslash\{u\} \rightarrow S^n$. To see that the maps $f$ and $g$ witness homotopical equivalence of $\mathbb{R}^{n+1}\backslash\{u\}$ and $S^n$ observe first that $f \circ g = \mathrm{id}_{S^n}$ because $f$ is a retraction.

For any $x \in \mathbb{R}^{n+1}\backslash\{u\}$ and $t \in I$ let $F(x,t) = (1-t)x + tf(x)$; it is clear that $F(x,0) = x$ and $F(x,1) = f(x)$ for any $x \in \mathbb{R}^{n+1}\backslash\{u\}$. If $F(x,t) = u$ then $x(1 - t + \frac{t}{|x|}) = u$ which implies that $1 - t + \frac{t}{|x|} = 0$ and hence $|x|(1-t) + t = 0$. Since the last equality cannot hold for $t \in I$ and $x \neq u$, we have obtained a contradiction; therefore $F : (\mathbb{R}^{n+1}\backslash\{u\}) \times I \rightarrow \mathbb{R}^{n+1}\backslash\{u\}$ is a continuous map which witnesses that $g \circ f = f$ is homotopic to $\mathrm{id}_{\mathbb{R}^{n+1}\backslash\{u\}}$. Fact 10 is proved.

*Fact 11.* For any distinct $n, m \in \mathbb{N}$ the spaces $S^n$ and $S^m$ are not homotopically equivalent.

*Proof.* There is no loss of generality to assume that $n < m$; suppose that the spheres $S^n$ and $S^m$ are homotopically equivalent. There exist continuous maps $f : S^n \rightarrow S^m$ and $g : S^m \rightarrow S^n$ such that $f \circ g$ is homotopic to $\mathrm{id}_{S^m}$. There is a point $a \in S^m$ such that the map $f$ is homotopic to the constant map $d : S^n \rightarrow \{a\}$ (see Fact 9). The map $d \circ g$ is also constant and $d \circ g \sim f \circ g$ by Fact 2. Being homotopic is a transitive relation (see Problem 091), so $\mathrm{id}_{S^m}$ is homotopic to a constant map. It is easy to see that this implies that $S^m$ is homotopically equivalent to a one-point space.

Let $B_m = \{x \in \mathbb{I}^{m+1} : |x(i)| = 1$ for some $i \leq m\}$ be the boundary of the $(m + 1)$-dimensional cube $\mathbb{I}^{m+1}$. It was proved in Fact 1 of V.094 that $B_m$ is homeomorphic to $S^m$, so $B_m$ is also homotopically equivalent to a one-point space (see Fact 3). Let $E$ be a one-point space for which there are (automatically continuous) maps $r : B_m \rightarrow E$ and $s : E \rightarrow B_m$ such that $s \circ r \sim \mathrm{id}_{B_m}$.

The map $w = s \circ r : B_m \rightarrow B_m$ is constant, so let $b \in B_m$ be the point such that $w(B_m) = \{b\}$. Letting $q(x) = b$ for any $x \in \mathbb{I}^{m+1}$ we obtain a continuous extension $q : \mathbb{I}^{m+1} \rightarrow B_m$ of the map $w$. The space $B_m$ being homeomorphic to $S^m$, we can apply Mushroom lemma (Problem 093) to the space $\mathbb{I}^{m+1}$, its closed subset $B_m$ and the pair of homotopic maps $w, \mathrm{id}_{B_m} : B_m \rightarrow B_m$. Since $q$ is a continuous extension of the map $w$, there exists a continuous map $\rho : \mathbb{I}^{m+1} \rightarrow B_m$ such that $\rho|B_m = \mathrm{id}_{B_m}$, i.e., $\rho$ is a retraction of $\mathbb{I}^{m+1}$ onto $B_m$. This, however, contradicts Problem 090 and shows that Fact 11 is proved.

*Fact 12.* Assume that $P$ and $Q$ are spaces with the following properties:

(i) $P = \bigcup\{P_\alpha : \alpha < \omega_1\}$ and $Q = \bigcup\{Q_\alpha : \alpha < \omega_1\}$;

(ii) the sets $P_\alpha$ and $Q_\alpha$ are countable for any $\alpha < \omega_1$;

(iii) $P_\alpha \subset P_\beta$ and $Q_\alpha \subset Q_\beta$ whenever $\alpha < \beta < \omega_1$;

(iv) $P_\alpha = \bigcup\{P_\beta : \beta < \alpha\}$ and $Q_\alpha = \bigcup\{Q_\beta : \beta < \alpha\}$ if $\alpha < \omega_1$ is a limit ordinal.

Suppose additionally that sets $R \subset C_p(P)$ and $S \subset C_p(Q)$ are dense in $C_p(P)$ and $C_p(Q)$ respectively and there exists a homeomorphism $\varphi : R \to S$. Then the set $A = \{\alpha < \omega_1 :$ for any $f, g \in R$ we have $f|P_\alpha = g|P_\alpha$ if and only if $\varphi(f)|Q_\alpha = \varphi(g)|Q_\alpha\}$ is closed and unbounded in $\omega_1$.

*Proof.* Let $p_\alpha : R \to C_p(P_\alpha)$ and $q_\alpha : S \to C_p(Q_\alpha)$ be the restriction maps for any $\alpha < \omega_1$. Suppose that $\{\alpha_n : n \in \omega\} \subset A$ and $\alpha_n < \alpha_{n+1}$ for any $n \in \omega$. Let $\alpha = \sup_{n \in \omega} \alpha_n$ and take any $f, g \in R$.

Observe that the properties (iii) and (iv) imply that $f|P_\alpha = g|P_\alpha$ if and only if $f|P_{\alpha_n} = g|P_{\alpha_n}$ for any $n \in \omega$. By our choice of $A$ this happens if and only if $\varphi(f)|Q_{\alpha_n} = \varphi(g)|Q_{\alpha_n}$ for any $n \in \omega$ and the last statement is equivalent (due to (iii) and (iv) again) to $\varphi(f)|Q_\alpha = \varphi(g)|Q_\alpha$. Therefore $\alpha \in A$, so we proved that the set $A$ is closed in $\omega_1$.

To see that the set $A$ is unbounded take an arbitrary ordinal $\alpha \in \omega_1$ and let $\alpha_0 = \alpha$. Then $q_{\alpha_0} \circ \varphi : R \to C_p(Q_{\alpha_0})$ is a continuous map of $R$ to a second countable space, so we can apply TFS-299 together with (i) and (iii) to find an ordinal $\beta_0 \in (\alpha_0, \omega_1)$ such that $p_{\beta_0}(f) = p_{\beta_0}(g)$ implies $q_{\alpha_0}(\varphi(f)) = q_{\alpha_0}(\varphi(g))$ for any $f, g \in R$. Analogously, there is an ordinal $\alpha_1 \in (\beta_0, \omega_1)$ such that the equality $q_{\alpha_1}(f) = q_{\alpha_1}(g)$ implies $p_{\beta_0}(\varphi^{-1}(f)) = p_{\beta_0}(\varphi^{-1}(g))$ for any $f, g \in S$.

Continuing this construction inductively, we obtain sequences $\{\alpha_n : n \in \omega\}$ and $\{\beta_n : n \in \omega\}$ of countable ordinals with the following properties:

(1) $\alpha = \alpha_0$ and $\alpha_n < \beta_n < \alpha_{n+1}$ for any $n \in \omega$;

(2) for any $n \in \omega$ and $f, g \in R$, if $p_{\beta_n}(f) = p_{\beta_n}(g)$ then $q_{\alpha_n}(\varphi(f)) = q_{\alpha_n}(\varphi(g))$;

(3) for any $n \in \omega$ and $f, g \in S$, if $q_{\alpha_{n+1}}(f) = q_{\alpha_{n+1}}(g)$ then $p_{\beta_n}(\varphi^{-1}(f)) = p_{\beta_n}(\varphi^{-1}(g))$;

It is evident that $\beta = \sup_{n \in \omega} \beta_n = \sup_{n \in \omega} \alpha_n > \alpha$; to prove that $\beta \in A$ take any $f, g \in R$. If $p_\beta(f) = p_\beta(g)$ then $p_{\beta_n}(f) = p_{\beta_n}(g)$ and hence we have the equality $q_{\alpha_n}(\varphi(f)) = q_{\alpha_n}(\varphi(g))$ (see (2)) for any $n \in \omega$. It follows from the properties (iii) and (iv) that $q_\beta(\varphi(f)) = q_\beta(\varphi(g))$.

Now let $f_0 = \varphi(f)$, $g_0 = \varphi(g)$ and assume that $q_\beta(f_0) = q_\beta(g_0)$. Then we have $q_{\alpha_{n+1}}(f_0) = q_{\alpha_{n+1}}(g_0)$ and hence $p_{\beta_n}(\varphi^{-1}(f_0)) = p_{\beta_n}(\varphi^{-1}(g_0))$ for any $n \in \omega$ by (3). Apply (iv) to conclude that $p_\beta(f) = p_\beta(\varphi^{-1}(f_0)) = p_\beta(\varphi^{-1}(g_0)) = p_\beta(g)$, so $p_\beta(f) = p_\beta(g)$ if and only if $q_\beta(\varphi(f)) = q_\beta(\varphi(g))$ and therefore $\beta \in A$. This shows that $A$ is unbounded and finishes the proof of Fact 12.

*Fact 13.* Assume that $R$ and $S$ are spaces with the following properties:

(i) $R = \bigcup\{R_\alpha : \alpha < \omega_1\}$ and $S = \bigcup\{S_\alpha : \alpha < \omega_1\}$;

(ii) $R_\alpha$ is a separable closed subspace of $R$ and $S_\alpha$ is a separable closed subspace of $S$ for any $\alpha < \omega_1$;

(iii) $R_\alpha \subset R_\beta$ and $S_\alpha \subset S_\beta$ whenever $\alpha < \beta < \omega_1$;

(iv) $R_\alpha = \bigcup\{R_\beta : \beta < \alpha\}$ and $S_\alpha = \bigcup\{S_\beta : \beta < \alpha\}$ if $\alpha < \omega_1$ is a limit ordinal.

Suppose additionally that $\varphi : R \to S$ is a homeomorphism. Then $B = \{\alpha < \omega_1 : \varphi(R_\alpha) = S_\alpha\}$ is closed and unbounded in $\omega_1$.

*Proof.* Take a sequence $\{\alpha_n : n \in \omega\} \subset B$ such that $\alpha_n < \alpha_{n+1}$ for any $n \in \omega$ and let $\alpha = \sup_{n \in \omega} \alpha_n$. The map $\varphi$ being a homeomorphism, it is closed, so we have

$$\varphi(R_\alpha) = \varphi\left(\overline{\bigcup\{R_{\alpha_n} : n \in \omega\}}\right) = \overline{\bigcup\{\varphi(R_{\alpha_n}) : n \in \omega\}} = \overline{\bigcup\{S_{\alpha_n} : n \in \omega\}} = S_\alpha$$

and hence $\alpha \in B$, i.e., the set $B$ is closed in $\omega_1$.

To see that $B$ is unbounded take any $\alpha < \omega_1$ and let $\alpha_0 = \alpha$. The space $S_{\alpha_0}$ being separable, it follows from (i) and (iii) that there is an ordinal $\beta_0 > \alpha_0$ such that $\varphi(R_{\beta_0}) \cap S_{\alpha_0}$ is dense in $S_{\alpha_0}$; since $\varphi(R_{\beta_0})$ is closed in $S$, we have $S_{\alpha_0} \subset \varphi(R_{\beta_0})$.

Analogously, there is an ordinal $\alpha_1 > \beta_0$ such that $R_{\beta_0} \subset \varphi^{-1}(S_{\alpha_1})$. Continuing this construction inductively we obtain sequences $\{\alpha_n : n \in \omega\}$ and $\{\beta_n : n \in \omega\}$ with the following properties:

(4) $\alpha = \alpha_0$ and $\alpha_n < \beta_n < \alpha_{n+1}$ for any $n \in \omega$;

(5) $S_{\alpha_n} \subset \varphi(R_{\beta_n})$ for any $n \in \omega$;

(6) $R_{\beta_n} \subset \varphi^{-1}(S_{\alpha_{n+1}})$ and hence $\varphi(R_{\beta_n}) \subset S_{\alpha_{n+1}}$ for any $n \in \omega$.

If $\beta = \sup_{n \in \omega} \alpha_n = \sup_{n \in \omega} \beta_n$ then it follows from the properties (4) and (5) that

$$\varphi(R_\beta) = \varphi\left(\overline{\bigcup\{R_{\beta_n} : n \in \omega\}}\right) = \overline{\bigcup\{\varphi(R_{\beta_n}) : n \in \omega\}} \supset \overline{\bigcup\{S_{\alpha_n} : n \in \omega\}} = S_\beta,$$

so $\varphi(R_\beta) \supset S_\beta$. Analogously, it follows from the properties (4) and (6) that

$$\varphi(R_\beta) = \varphi\left(\overline{\bigcup\{R_{\beta_n} : n \in \omega\}}\right) = \overline{\bigcup\{\varphi(R_{\beta_n}) : n \in \omega\}} \subset \overline{\bigcup\{S_{\alpha_{n+1}} : n \in \omega\}} = S_\beta,$$

so $\varphi(R_\beta) = S_\beta$ and hence $\beta \in B$. Thus $B$ is also unbounded, i.e., Fact 13 is proved.

*Fact 14.* Suppose that $L$ is a linear space and $f_0, \ldots, f_{n-1}$ are linear functionals on $L$. Then a linear functional $f : L \to \mathbb{R}$ is a linear combination of $f_0, \ldots, f_{n-1}$ if and only if $f_0^{-1}(0) \cap \ldots \cap f_{n-1}^{-1}(0) \subset f^{-1}(0)$.

*Proof.* If there are $\lambda_0, \ldots, \lambda_{n-1} \in \mathbb{R}$ such that $f = \lambda_0 f_0 + \ldots + \lambda_{n-1} f_{n-1}$ then for any point $x \in K = f_0^{-1}(0) \cap \ldots \cap f_{n-1}^{-1}(0)$ we have $f_i(x) = 0$ for all $i < n$, so $f(x) = 0$, i.e., $x \in f^{-1}(0)$ and therefore $K \subset f^{-1}(0)$. This proves necessity.

Now assume that $K \subset f^{-1}(0)$ and consider the map $\Phi : L \to \mathbb{R}^n$ defined by $\Phi(x) = (f_0(x), \ldots, f_{n-1}(x))$ for any $x \in L$. It is straightforward that $\Phi$ is a linear map; let $e_i(j) = 0$ for any distinct $i, j < n$ and $e_i(i) = 1$ whenever $i < n$. Then $e_0, \ldots, e_{n-1}$ is a basis in $\mathbb{R}^n$; we claim that

(7) there exists a linear functional $h : \mathbb{R}^n \to \mathbb{R}$ such that $h \circ \Phi = f$.

To prove this observe first that if $u$ is the zero vector of $\mathbb{R}^n$ then $\Phi^{-1}(u) = K$; given any $y \in \Phi(L)$ take any $x \in \Phi^{-1}(y)$ and let $h(y) = f(x)$. This definition must be consistent, of course, so take any $x_0, x_1 \in \Phi^{-1}(y)$. Then $\Phi(x_0) = \Phi(x_1)$ which shows that $\Phi(x_0 - x_1) = u$ and hence $x_0 - x_1 \in K$; an immediate consequence is that $x_0 - x_1 \in f^{-1}(0)$ and therefore $f(x_0) = f(x_1)$, i.e., this definition is, indeed, consistent. So far, the function $h$ is defined on $\Phi(L)$ which is a linear subspace of $\mathbb{R}^n$; besides, $h$ is a linear functional on $\Phi(L)$. Using Fact 1 of S.489 and Fact 3 of S.489 it is easy to see that $h$ can be extended linearly over the whole $\mathbb{R}^n$, so (7) is proved.

Finally let $\lambda_i = h(e_i)$ for each $i < n$ and take any point $x \in L$. We have $\Phi(x) = f_0(x)e_0 + \ldots + f_{n-1}(x)e_{n-1}$, so we can apply the property (7) to conclude that

$$f(x) = h(\Phi(x)) = f_0(x)h(e_0) + \ldots + f_{n-1}(x)h(e_{n-1}) = \lambda_0 f_0(x) + \ldots + \lambda_{n-1} f_{n-1}(x)$$

which shows that $f = \lambda_0 f_0 + \ldots + \lambda_{n-1} f_{n-1}$, so Fact 14 is proved.

*Fact 15.* Suppose that $L$ is a linear topological space and $K \subset L$ is a linear subspace of $L$ of codimension $n$. Then $L \setminus K$ is homotopically equivalent to $S^{n-1}$.

*Proof.* Denote by $u$ the zero vector of the space $L$ and fix independent continuous linear functionals $f_1, \ldots, f_n$ on the space $L$ such that $K = \bigcap_{i=1}^{n} f_i^{-1}(0)$. Then, for any $i \leq n$ there is a vector $x_i \in \bigcap \{f_j^{-1}(0) : j \in \{1, \ldots, n\} \setminus \{i\}\}$ such that $f_i(x_i) \neq 0$ (see Fact 14); now, if $\xi_i = \frac{1}{f(x_i)} x_i$ then $f_i(\xi_j) = 0$ if $i \neq j$ while $f_i(\xi_i) = 1$ for any $i \leq n$.

The vectors $\xi_1, \ldots, \xi_n$ are easily seen to be linearly independent, so their linear hull $M$ is an $n$-dimensional linear subspace of $L$. Let $r(x) = f_1(x)\xi_1 + \ldots + f_n(x)\xi_n$ for any $x \in L$. Then $r : L \to M$ is a continuous linear map (this is an easy exercise left to the reader) such that $r(L \setminus K) = M \setminus \{u\}$ and $r(x) = x$ for any $x \in M$. Let $r_0 = r|(L \setminus K) : L \setminus K \to M \setminus \{u\}$.

Define a map $e : M \setminus \{u\} \to L$ to be the identity, i.e., $e(x) = x$ for any point $x \in M \setminus \{u\}$. It is clear that $r_0 \circ e = \mathrm{id}_{M \setminus \{u\}}$. Let $F(x, t) = (1 - t)x + t r_0(x)$ for any $x \in L \setminus K$ and $t \in I$. Then $F : (L \setminus K) \times I \to L \setminus K$ is a continuous map such that $F(x, 0) = x$ and $F(x, 1) = r_0(x)$ for any $x \in L \setminus K$; as a consequence, the map $e \circ r_0 = r_0$ is homotopic to $\mathrm{id}_{L \setminus K}$, i.e., the maps $e$ and $r_0$ witness homotopic equivalence of $L \setminus K$ and $M \setminus \{u\}$.

For any $i \leq n$ and $a \in \mathbb{R}$ let $\mu_i(a) = a\xi_i$; it follows from the axioms of linear topological space that the map $\mu_i : \mathbb{R} \to L$ is continuous. Consequently, the map $\mu : \mathbb{R}^n \to L$ defined by $\mu(a_1, \ldots, a_n) = a_1\xi_1 + \ldots + a_n\xi_n$ for any $(a_1, \ldots, a_n) \in \mathbb{R}^n$ is continuous as well. It is clear that, actually, $\mu : \mathbb{R}^n \to M$; now if we define a map $\Phi : M \to L$ by $\Phi(x) = (f_1(x), \ldots, f_n(x))$ then $\Phi$ is continuous because so is every $f_i$. Since $\Phi$ is the inverse of $\mu$, the spaces $M$ and $\mathbb{R}^n$ are homeomorphic and hence $M \setminus \{u\} \simeq \mathbb{R}^n \setminus \{p\}$ where $p = \Phi(u)$ is the zero point of $\mathbb{R}^n$. Thus $L \setminus K$ is homotopically equivalent to $\mathbb{R}^n \setminus \{p\}$ by Fact 3; since $\mathbb{R}^n \setminus \{p\}$ is homotopically equivalent to $S^{n-1}$ by Fact 10, we can apply Fact 3 once more to conclude that $L \setminus K$ is homotopically equivalent to $S^{n-1}$, so Fact 15 is proved.

*Fact 16.* Let $L$ and $L'$ be linear topological spaces and suppose that $K \subset L$ and $K' \subset L'$ are linear subspaces of $L$ and $L'$ respectively. Assume also that $K$ is of codimension $m$ in $L$ while $K'$ is of codimension $n$ in $L'$. If there exists a homeomorphism $h : L \to L'$ such that $h(K) = K'$ then $m = n$.

*Proof.* Since $h(L \setminus K) = L' \setminus K'$, the spaces $L \setminus K$ and $L' \setminus K'$ are homeomorphic and hence homotopically equivalent. By Fact 15, the space $L \setminus K$ is homotopically equivalent to $S^{m-1}$ while $L' \setminus K'$ is homotopically equivalent to $S^{n-1}$. By Fact 3, the spaces $S^{m-1}$ and $S^{n-1}$ are homotopically equivalent so $m - 1 = n - 1$ by Fact 11. This, of course, implies $m = n$, so Fact 16 is proved.

Returning to our solution let $X = \omega_1$ and fix distinct $m, n \in \mathbb{N}$. Since $(C_p(X))^k \simeq C_p(X \times k)$ for any $k \in \mathbb{N}$, it suffices to show that the spaces $C_p(X \times n)$ and $C_p(X \times m)$ are not homeomorphic. Let $u$ and $v$ be the zero functions on $X \times n$ and $X \times m$ respectively. Assume toward a contradiction that there is a homeomorphism $\varphi : C_p(X \times n) \to C_p(X \times m)$. It follows from homogeneity of the spaces $C_p(X \times n)$ and $C_p(X \times m)$ that we can assume that $\varphi(u) = v$.

Let $P_\alpha = \alpha \times n$ and $Q_\alpha = \alpha \times m$ for any $\alpha < \omega_1$. The families $\{P_\alpha : \alpha < \omega_1\}$ and $\{Q_\alpha : \alpha < \omega_1\}$ satisfy the premises of Fact 12 if $P = X \times n$ and $Q = X \times m$, so there is a closed unbounded $A \subset \omega_1$ such that,

(8) for any $\alpha \in A$ and $f, g \in C_p(X \times n)$ we have $f | P_\alpha = g | P_\alpha$ if and only if $\varphi(f) | Q_\alpha = \varphi(g) | Q_\alpha$.

For any $\alpha \in \omega_1$ denote by $E_\alpha$ the set of functions from $C_p(X \times n)$ which are constant "starting from $\alpha$," i.e., $E_\alpha = \{f \in C_p(X \times n) : f((\beta, i)) = f((\alpha, i))$ for any $\beta \geq \alpha$ and $i < n\}$. Analogously, $F_\alpha = \{f \in C_p(X \times m) : f((\beta, i)) = f((\alpha, i))$ for any $\beta \geq \alpha$ and $i < m\}$.

It is straightforward to check that every set $E_\alpha$ is closed in $C_p(X \times n)$ and $F_\alpha$ is closed in $C_p(X \times m)$; besides, $E_\alpha \subset E_\beta$ and $F_\alpha \subset F_\beta$ whenever $\alpha < \beta < \omega_1$. It follows from TFS-314 that $\bigcup \{E_\alpha : \alpha < \omega_1\} = C_p(X \times n)$ and we have the equality $\bigcup \{F_\alpha : \alpha < \omega_1\} = C_p(X \times m)$.

Fix $\alpha < \omega_1$ and let $\pi_\alpha : C_p(X \times n) \to C_p(P_{\alpha+1})$ be the restriction map. Then $p_\alpha = \pi_\alpha | E_\alpha : E_\alpha \to C_p(P_{\alpha+1})$ is a condensation. For any $O_1, \ldots, O_k \in \tau(\mathbb{R})$ and $x_1, \ldots, x_k \in X \times n$ let $[x_1, \ldots, x_k; O_1, \ldots, O_k] = \{f \in C_p(X \times n) : f(x_i) \in O_i$ for any $i \leq k\}$. The family $\mathcal{B} = \{[x_1, \ldots, x_k; O_1, \ldots, O_k] : k \in \mathbb{N}, x_1, \ldots, x_k \in X \times n$ and $O_1, \ldots, O_k \in \tau(\mathbb{R})\}$ is a base in the space $C_p(X \times n)$. It is evident that the family $\mathcal{B}' = \{[x_1, \ldots, x_k; O_1, \ldots, O_k] \in \mathcal{B} : (\alpha, i) \in \{x_1, \ldots, x_k\}$ for any $i < n\}$ is also a base in $C_p(X \times n)$.

Observe that, for any $U = [x_1, \ldots, x_k; O_1, \ldots, O_k] \in \mathcal{B}'$, the set $p_\alpha(U)$ is open in $C_p(P_{\alpha+1})$ because $p_\alpha(U) = \{f \in C_p(P_{\alpha+1}) : f(x_{i_j}) \in O_{i_j}, j = 1, \ldots, l\}$ where $\{x_{i_1}, \ldots, x_{i_l}\} = \{x_1, \ldots, x_k\} \cap P_{\alpha+1}$. Thus $p_\alpha$ is a homeomorphism being an open condensation (see TFS-155 and Fact 2 of S.491).

This proves that every $E_\alpha$ is second countable and hence separable. Analogously, the space $F_\alpha$ is separable for any $\alpha < \omega_1$. We leave it to the reader to check that $E_\alpha = \bigcup \{E_\beta : \beta < \alpha\}$ and $F_\alpha = \bigcup \{F_\beta : \beta < \alpha\}$ for any limit ordinal $\alpha$. Therefore we can apply Fact 13 to the families $\{E_\alpha : \alpha < \omega_1\}$ and $\{F_\alpha : \alpha < \omega_1\}$ to conclude that there exists a closed unbounded set $A' \subset \omega_1$ such that $\varphi(E_\alpha) = F_\alpha$ for any $\alpha \in A'$. The set $B = A \cap A'$ is also closed and unbounded in $\omega_1$.

Fix an ordinal $\alpha \in B$ and let $\alpha^+ = \setminus n\{\beta \in B : \alpha < \beta\}$; consider the sets $L_\alpha = \{x \in E_{\alpha^+} : x|P_\alpha = u|P_\alpha\}$ and $M_\alpha = \{y \in F_{\alpha^+} : y|Q_\alpha = v|Q_\alpha\}$. An immediate consequence of (8) is that $\varphi(L_\alpha) = M_\alpha$ for any $\alpha \in B$; besides, both $L_\alpha$ and $M_\alpha$ are closed linear subspaces of $C_p(X \times n)$ and $C_p(X \times m)$ respectively. For any $i < n$ define a linear functional $d_i^\alpha : L_\alpha \to \mathbb{R}$ by the equality $d_i^\alpha(f) = f((\alpha^+, i))$ for any $f \in L_\alpha$. Analogously, let $e_i^\alpha(g) = g((\alpha^+, i))$ for any $i < m$ and $f \in M_\alpha$.

Then $\{d_i^\alpha : i < n\}$ and $\{e_i^\alpha : i < m\}$ are independent families of continuous linear functionals on $L_\alpha$ and $M_\alpha$ respectively; let $K_\alpha = \bigcap\{(d_i^\alpha)^{-1}(0) : i < n\}$ and $N_\alpha = \bigcap\{(e_i^\alpha)^{-1}(0) : i < m\}$.

Since $n \neq m$, we can apply Fact 16 to see that $\varphi(K_\alpha) \neq N_\alpha$ and hence either $\varphi(K_\alpha) \not\subset N_\alpha$ or $\varphi^{-1}(N_\alpha) \not\subset K_\alpha$ for any $\alpha \in B$. Thus either $\varphi(K_\alpha) \not\subset N_\alpha$ or $\varphi^{-1}(N_\alpha) \not\subset K_\alpha$ for uncountably many $\alpha$. These two cases being analogous, we can assume, without loss of generality, that there is an uncountable $B' \subset B$ such that $\varphi(K_\alpha) \not\subset N_\alpha$ for all $\alpha \in B'$. Thus there is a function $f_\alpha \in K_\alpha$ such that $|\varphi(f_\alpha)((\alpha^+, j_\alpha))| > 0$ for any $\alpha \in B'$.

There exist: $\varepsilon > 0$, an uncountable set $B'' \subset B'$ and $j < m$ such that $j_\alpha = j$ and $|\varphi(f_\alpha)((\alpha^+, j))| \geq \varepsilon$ for any $\alpha \in B''$. Choose a sequence $\{\alpha_i : i \in \omega\} \subset B''$ such that $\alpha_{i+1} > \alpha_i^+$ for any $i \in \omega$. Since $f_{\alpha_i}^{-1}(\mathbb{R}\setminus\{0\}) \subset [\alpha_i, \alpha_i^+] \times n$ for any $i \in \omega$, the family $\{f_{\alpha_i}^{-1}(\mathbb{R}\setminus\{0\}) : i \in \omega\}$ is disjoint and hence the sequence $\{f_{\alpha_i} : i \in \omega\}$ converges to $u$ in $L_\alpha$. Every $\varphi(f_{\alpha_i})$ is constant on each $\omega_1 \times \{k\}$ starting from $\alpha_i^+$, so if $\beta > \sup\{\alpha_i^+ : i \in \omega\}$ then $|\varphi(f_{\alpha_i})((\beta, j))| \geq \varepsilon$ for every $i \in \omega$. Therefore the sequence $\{\varphi(f_{\alpha_i}) : i \in \omega\}$ does not converge to zero which is a contradiction with continuity of $\varphi$. Thus the spaces $C_p(X \times n)$ and $C_p(X \times m)$ are not homeomorphic.

To settle the case when $X = \omega_1 + 1$ we will first prove that

(9) the space $C_p(X)$ is linearly isomorphic to the subspace $\Omega = \{f \in C_p(\omega_1) :$ there exists $\alpha < \omega_1$ such that $f(\beta) = 0$ for any $\beta \geq \alpha\}$ of the space $C_p(\omega_1)$.

The space $\omega_1 + 1$ being homeomorphic to $[1, \omega_1]$ it suffices to construct a linear homeomorphism between $\Omega$ and $C_p([1, \omega_1])$. To do so take any function $f \in \Omega$ and let $\eta(f)(\alpha) = f(0) + f(\alpha)$ for any $\alpha \in [1, \omega_1]$; then $\eta(f) \in C_p([1, \omega_1])$ and the map $\eta : \Omega \to C_p([1, \omega_1])$ is a linear homeomorphism (this is an easy exercise which is left to the reader), so (9) is proved.

Now assume that there is a homeomorphism $\mu : \Omega^n \to \Omega^m$; by homogeneity of $\Omega^n$ and $\Omega^m$ we can consider, without loss of generality, that $\mu(u) = v$. We consider that $\Omega^n$ and $\Omega^m$ are the respective subsets of $C_p(\omega_1 \times n)$ and $C_p(\omega_1 \times m)$. Since $\Omega$ is dense in $C_p(\omega_1)$, the sets $\Omega^n$ and $\Omega^m$ are dense in $C_p(\omega_1 \times n)$ and $C_p(\omega_1 \times m)$ respectively. We use the same notation as in the first part of our proof and, in particular, $P_\alpha = \alpha \times n$ and $Q_\alpha = \alpha \times m$ for any $\alpha < \omega_1$. Since Fact 12 is also applicable to the homeomorphism $\mu$ and the families $\{P_\alpha : \alpha < \omega_1\}$ and $\{Q_\alpha : \alpha < \omega_1\}$, we conclude that there exists a closed unbounded set $J \subset \omega_1$ such that

(10) for any $\alpha \in J$ and $f, g \in \Omega^n$ the equality $f|P_\alpha = g|P_\alpha$ holds if and only if $\mu(f)|Q_\alpha = \mu(g)|Q_\alpha$.

Recall that $E_\alpha = \{f \in C_p(\omega_1 \times n) : f((\beta, i)) = f((\alpha, i))$ for any $\beta \geq \alpha$ and $i < n\}$ and $F_\alpha = \{f \in C_p(\omega_1 \times m) : f((\beta, i)) = f((\alpha, i))$ for any $\beta \geq \alpha$ and $i < m\}$ for every $\alpha < \omega_1$. We saw that $E_\alpha$ and $F_\alpha$ are closed second countable subsets of $C_p(\omega_1 \times n)$ and $C_p(\omega_1 \times m)$ respectively; therefore $E'_\alpha = E_\alpha \cap \Omega^n$ and $F'_\alpha = F_\alpha \cap \Omega^m$ are second countable closed subsets of $\Omega^n$ and $\Omega^m$ respectively. It is easy to check that, for the families $\{E'_\alpha : \alpha < \omega_1\}$ and $\{F'_\alpha : \alpha < \omega_1\}$, the premises of Fact 13 are satisfied, so there exists a closed unbounded subset $J' \subset \omega_1$ such that $\mu(E'_\alpha) = F'_\alpha$ for all $\alpha \in J'$.

The set $J \cap J'$ being also closed and unbounded in $\omega_1$, we can find a limit ordinal $\alpha \in J \cap J'$. Consider the sets $K = \{f \in \Omega^n : f((\alpha, i)) = 0$ for all $i < n\}$ and $M = \{g \in \Omega^m : g((\alpha, i)) = 0$ for all $i < m\}$. It is easy to see that $K$ is a linear subspace of $\Omega^n$ of codimension $n$ and $M$ is a linear subspace of $\Omega^m$ of codimension $m$. Therefore $\mu(K) \neq M$ by Fact 16.

Now fix any function $f \in K$ and let $q_f((\beta, i)) = f((\beta, i))$ for any $\beta < \alpha$. If $\beta \geq \alpha$ then let $q_f((\beta, i)) = 0$ for all $i < n$. It follows from $q_f \in E'_\alpha$ and $\alpha \in J'$ that $h = \mu(q_f) \in F'_\alpha$ and, in particular, $h((\alpha, i)) = 0$ for any $i < m$. We also have $q_f|\alpha \times n = f|\alpha \times n$, so $\alpha \in J$ implies that $h|\alpha \times m = \mu(f)|\alpha \times m$. The ordinal $\alpha$ is a limit while the functions $h$ and $\mu(f)$ are continuous on $\alpha \times n$; this, evidently, implies that $\mu(f)((\alpha, i)) = h((\alpha, i)) = 0$ for any $i < m$ and therefore $\mu(f) \in M$. An identical proof shows that $\mu^{-1}(g) \in K$ for any $g \in M$, so $\mu(K) = M$; this contradiction proves that the spaces $(C_p(\omega_1 + 1))^n$ and $(C_p(\omega_1 + 1))^m$ are not homeomorphic and hence our solution is complete.

**V.101.** *Prove that a nonempty family $\mathcal{B} \subset \exp(X \times X)$ is a base for some uniformity on $X$ if and only if it has the following properties:*

*(1) $\bigcap \mathcal{B} = \Delta$;*
*(2) for any $U \in \mathcal{B}$, there is $V \in \mathcal{B}$ such that $V^{-1} \subset U$;*
*(3) for any $U \in \mathcal{B}$, there is $V \in \mathcal{B}$ such that $V \circ V \subset U$;*
*(4) if $U, V \in \mathcal{B}$ then there is $W \in \mathcal{B}$ such that $W \subset U \cap V$.*

**Solution.** Suppose that $(X, \mathcal{U})$ is a uniform space and $\mathcal{B} \subset \mathcal{U}$ is a base of $\mathcal{U}$. Then $\Delta \subset \bigcap \mathcal{B}$ by (U1); now if $z \in X^2 \backslash \Delta$ then there is $U \in \mathcal{U}$ such that $z \notin U$. Since $\mathcal{B}$ is a base of $\mathcal{U}$, there is $B \in \mathcal{B}$ with $B \subset U$. Therefore $z \notin B$ and hence no point of $X^2 \backslash \Delta$ belongs to $\bigcap \mathcal{B}$, i.e., $\bigcap \mathcal{B} = \Delta$, so we proved that $\mathcal{B}$ has (1).

Now, if $U \in \mathcal{B}$ then $V' = U^{-1} \in \mathcal{U}$; choose $V \subset V'$ with $V \in \mathcal{B}$. Then $V^{-1} \subset (V')^{-1} = U$, so (2) is proved. Analogously, if $U \in \mathcal{B}$ then there is $V' \in \mathcal{U}$ such that $V' \circ V' \subset U$; choose $V \subset V'$ with $V \in \mathcal{B}$. Then $V \circ V \subset V' \circ V' \subset U$ and hence we also have (3).

To see that (4) holds too, assume that $U, V \in \mathcal{B}$; then $U \cap V \in \mathcal{U}$ by (U1), so there is $W \in \mathcal{B}$ such that $W \subset U \cap V$; this finishes the proof of necessity.

To establish sufficiency, take a nonempty $\mathcal{B} \subset \exp(X \times X)$ with the properties (1)–(4) and consider the family $\mathcal{U} = \{U \subset X \times X :$ there is $B \in \mathcal{B}$ such that

$B \subset U\}$. Given $U \in \mathcal{U}$, we have $U \supset B \supset \Delta$ for some $B \in \mathcal{B}$ and therefore $\Delta \subset \bigcap \mathcal{U}$. It is clear that $\mathcal{B} \subset \mathcal{U}$, so $\bigcap \mathcal{U} \subset \bigcap \mathcal{B} = \Delta$ and hence $\bigcap \mathcal{U} = \Delta$.

Next, take any $U \in \mathcal{U}$ and fix $B \in \mathcal{B}$ with $B \subset U$. The property (2) shows that there is $V \in \mathcal{B}$ such that $V^{-1} \subset B$. This implies that $V = (V^{-1})^{-1} \subset B^{-1} \subset U^{-1}$ and hence $U^{-1} \in \mathcal{U}$. Analogously, if $U, V \in \mathcal{U}$ then there are $B, C \in \mathcal{B}$ with $B \subset U$ and $C \subset V$. It follows from (4) that we can find a set $D \in \mathcal{B}$ such that $D \subset B \cap C \subset U \cap V$; thus $U \cap V \in \mathcal{U}$ and hence the condition (U1) is satisfied for the family $\mathcal{U}$.

Given $U \in \mathcal{U}$ and $V \subset X \times X$ with $V \supset U$ there is $B \in \mathcal{B}$ such that $B \subset U$; then $B \subset V$ and hence $V \in \mathcal{U}$. This proves that the first part of (U2) holds for $\mathcal{U}$. Finally, take any $U \in \mathcal{U}$ and pick $B \in \mathcal{B}$ with $B \subset U$. The property (3) shows that there is a set $V \in \mathcal{B}$ such that $V \circ V \subset B$; then $V \in \mathcal{U}$ and $V \circ V \subset U$, so (U2) is also fulfilled for $\mathcal{U}$ and hence $\mathcal{U}$ is a uniformity on $X$. Our definition of $\mathcal{U}$ implies that $\mathcal{B}$ is a base of $\mathcal{U}$, so we established sufficiency and hence our solution is complete.

**V.102.** *Suppose that a nonempty family $\mathcal{S} \subset \exp(X \times X)$ has the following properties:*

*(1) $\bigcap \mathcal{S} = \Delta$;*
*(2) for any $U \in \mathcal{S}$, there is $V \in \mathcal{S}$ such that $V^{-1} \subset U$;*
*(3) for any $U \in \mathcal{S}$, there is $V \in \mathcal{S}$ such that $V \circ V \subset U$.*

*Prove that $\mathcal{S}$ is a subbase for some uniformity on $X$. As a consequence, the union of any family of uniformities on $X$ is a subbase of some uniformity on $X$.*

**Solution.** Let $\mathcal{B}$ be the family of all finite intersections of the elements of $\mathcal{S}$. Then $\mathcal{B} \neq \emptyset$ and $\mathcal{S} \subset \mathcal{B}$. It is clear that $\Delta \subset B$ for any $B \in \mathcal{B}$ and hence we have $\Delta \subset \bigcap \mathcal{B} \subset \bigcap \mathcal{S} = \Delta$ which shows that $\bigcap \mathcal{B} = \Delta$.

Take any $B \in \mathcal{B}$ and fix $S_1, \ldots, S_n \in \mathcal{S}$ such that $B = S_1 \cap \ldots \cap S_n$. It follows from (2) that there exist $T_1, \ldots, T_n \in \mathcal{S}$ with $T_i^{-1} \subset S_i$ for any $i \leq n$. Then $C = T_1 \cap \ldots \cap T_n \in \mathcal{B}$ and $C^{-1} = T_1^{-1} \cap \ldots \cap T_n^{-1} \subset S_1 \cap \ldots \cap S_n = B$; this proves that the properties (1) and (2) of Problem 101 are fulfilled for $\mathcal{B}$. It is clear that $U \cap V \in \mathcal{B}$ whenever $U, V \in \mathcal{B}$, so the condition (4) of Problem 101 is also satisfied for $\mathcal{B}$.

To prove that the property (3) of Problem 101 holds as well take any $B \in \mathcal{B}$ and pick $S_1, \ldots, S_n \in \mathcal{S}$ for which $B = S_1 \cap \ldots \cap S_n$. Our property (3) guarantees existence of $T_1, \ldots, T_n \in \mathcal{S}$ such that $T_i \circ T_i \subset S_i$ for any $i \leq n$. The set $C = T_1 \cap \ldots \cap T_n$ belongs to $\mathcal{B}$; if $(x, y) \in C \circ C$ then there exists $z \in X$ such that $(x, z) \in C$ and $(z, y) \in C$. As a consequence, $(x, z) \in T_i$ and $(z, y) \in T_i$, i.e., $(x, y) \in T_i \circ T_i \subset S_i$ for each $i \leq n$. This proves that $(x, y) \in \bigcap_{i \leq n} S_i = B$ and hence $C \circ C \subset B$, so the condition (3) of Problem 101 is satisfied.

Thus we can apply Problem 101 to the family $\mathcal{B}$ to conclude that there is a uniformity $\mathcal{U}$ on $X$ such that $\mathcal{B}$ is a base of $\mathcal{U}$; it is evident that $\mathcal{S}$ is a subbase of $\mathcal{U}$. Finally, observe that our properties (1)–(3) are fulfilled for any uniformity and hence for any union of uniformities. Therefore any union of uniformities on $X$ is a base of some uniformity of $X$.

**V.103.** *Let $(X, \mathcal{U})$ be a uniform space. Prove that*

*(1)* $\text{Int}(A) = \{x : U(x) \subset A \text{ for some } U \in \mathcal{U}\}$ *for any set* $A \subset X$*; in particular,* $x \in \text{Int}(U(x))$ *for any* $U \in \mathcal{U}$*;*

*(2) if* $\mathcal{B}$ *is a base of the uniformity* $\mathcal{U}$ *then, for any* $x \in X$ *and* $O \in \tau(x, X)$ *there is* $B \in \mathcal{B}$ *such that,* $B(x) \subset O$*. In particular, the family* $\{\text{Int}(B(x)) : B \in \mathcal{B}\}$ *is a local base of the space* $X$ *at* $x$*.*

*(3) if* $\mathcal{S}$ *is a subbase of* $\mathcal{U}$ *then, for any* $x \in X$ *and* $O \in \tau(x, X)$*, there is a finite* $\mathcal{S}' \subset \mathcal{S}$ *such that* $\bigcap\{S(x) : S \in \mathcal{S}'\} \subset O$*.*

*(4) for any* $U \in \mathcal{U}$*, the interior (in* $X \times X$*) of the set* $U$ *also belongs to* $\mathcal{U}$*. As a consequence, the family of all open symmetric elements of* $\mathcal{U}$ *is a base of* $\mathcal{U}$*;*

*(5)* $\overline{A} = \bigcap\{U(A) : U \in \mathcal{U}\}$ *for any* $A \subset X$*;*

*(6)* $\overline{B} = \bigcap\{U \circ B \circ U : U \in \mathcal{U}\}$ *for any* $B \subset X \times X$*;*

*(7) the family of all closed symmetric elements of* $\mathcal{U}$ *is a base of* $\mathcal{U}$*.*

**Solution.** To prove (1) let $O_A = \{x : U(x) \subset A \text{ for some } U \in \mathcal{U}\}$. If $x \in \text{Int}(A)$ then, by definition of the topology generated by $\mathcal{U}$, there is $U \in \mathcal{U}$ such that $x \in U(x) \subset \text{Int}(A) \subset A$, so $x \in O_A$. This proves that $\text{Int}(A) \subset O_A$.

Now, if $x \in O_A$ then fix $U, V \in \mathcal{U}$ for which $U(x) \subset A$ and $V \circ V \subset U$. Take a point $y \in V(x)$; if $z \in V(y)$ then it follows from $(x, y) \in V$ and $(y, z) \in V$ that $(x, z) \in V \circ V$, so $z \in (V \circ V)(x) \subset U(x) \subset A$. Thus $V(y) \subset A$ for any $y \in V(x)$ which shows that $V(x) \subset O_A$. It turns out that, for any $x \in O_A$, there is $V \in \mathcal{U}$ such that $V(x) \subset O_A$; therefore $O_A \subset A$ is an open set and hence $O_A \subset \text{Int}(A)$. We checked that $O_A = \text{Int}(A)$, so (1) is proved.

To deal with (2) observe that, by definition of the topology generated by $\mathcal{U}$, there is a set $U \in \mathcal{U}$ with $U(x) \subset O$; pick $B \in \mathcal{B}$ such that $B \subset U$. Then $B(x) \subset U(x) \subset O$; since also $x \in \text{Int}(B(x)) \subset B(x) \subset O$ (see (1)), we proved that the family $\{\text{Int}(B(x)) : B \in \mathcal{B}\}$ is a local base at $x$.

As to (3), take a set $U \in \mathcal{U}$ for which $U(x) \subset O$. The family $\mathcal{S}$ being a subbase of $\mathcal{U}$ there are $n \in \omega$ and $S_1, \ldots, S_n \in \mathcal{S}$ such that $S = S_1 \cap \ldots \cap S_n \subset U$. If $y \in S_1(x) \cap \ldots \cap S_n(x)$ then $(x, y) \in S_i$ for any $i \leq n$ and therefore $(x, y) \in S \subset U$ which implies that $y \in U(x) \subset O$. Thus $S_1(x) \cap \ldots \cap S_n(x) \subset O$, so the family $\mathcal{S}' = \{S_1, \ldots, S_n\}$ is as promised.

To settle (4) take any $U \in \mathcal{U}$; there exists a set $V_0 \in \mathcal{U}$ such that $V_0 \circ V_0 \subset U$. Applying the property (U2) again we can find $V_1 \in \mathcal{U}$ with $V_1 \circ V_1 \subset V_0$. Therefore $(V_1 \circ V_1) \circ (V_1 \circ V_1) \subset V_0 \circ V_0 \subset U$. It follows from $\Delta \subset V_1$ that $V_1 \subset V_1 \circ V_1$, so $V_1 \circ (V_1 \circ V_1) \subset (V_1 \circ V_1) \circ (V_1 \circ V_1) \subset U$. It is easy to see that the composition is associative, i.e., $A \circ (B \circ C) = (A \circ B) \circ C$ for any $A, B, C \subset X \times X$; this makes it possible to omit parenthesis in the expressions that involve composition, so we will write $V_1 \circ V_1 \circ V_1$ instead of $V_1 \circ (V_1 \circ V_1)$.

The set $V = V_1 \cap (V_1)^{-1}$ is a symmetric element of $\mathcal{U}$ and it follows from $V \subset V_1$ that $V \circ V \circ V \subset V_1 \circ V_1 \circ V_1 \subset U$. If $(x, y) \in V$ then $x \in G = \text{Int}(V(x))$ and $y \in H = \text{Int}(V(y))$ (see (1)). The set $Q = G \times H$ is open in $X \times X$ and $(x, y) \in Q$. If $(z, t) \in Q$ then $(x, z) \in V$ and $(y, t) \in V$; the set $V$ being symmetric we have $(z, x) \in V$; this, together with $(x, y) \in V$ and $(y, t) \in V$ implies that $(z, t) \in V \circ V \circ V \subset U$.

The point $(z, t) \in Q$ was taken arbitrarily, so $Q \subset U$ and hence the point $(x, y) \in Q \subset U$ belongs to the interior of $U$. Thus $V \subset \text{Int}(U)$ which shows that $\text{Int}(U) \in \mathcal{U}$. This proves that open elements of $\mathcal{U}$ form a base of $\mathcal{U}$. Now, if $U \in \mathcal{U}$ then $V = \text{Int}(U)$ is open in $X \times X$; it is an easy exercise that $V^{-1}$ is open in $X \times X$ as well. Therefore $W = V \cap V^{-1}$ is an open symmetric element of $\mathcal{U}$ with $W \subset U$, so open symmetric elements of $\mathcal{U}$ also constitute a base of $\mathcal{U}$, i.e., (4) is proved.

For (5), fix a set $A \subset X$; given $U \in \mathcal{U}$, it follows from (U2) and (4) that there is a symmetric $V \in \mathcal{U}$ with $V \subset U$. For any $x \in \overline{A}$, the set $V(x)$ is a neighborhood of $x$ by (1), so $V(x) \cap A \neq \emptyset$. Therefore there is $a \in A$ with $(x, a) \in V$ and hence $(a, x) \in V$, i.e., $x \in V(a) \subset U(a)$ which shows that $x \in U(A)$. The point $x \in \overline{A}$ and the set $U \in \mathcal{U}$ were chosen arbitrarily, so $\overline{A} \subset \bigcap\{U(A) : U \in \mathcal{U}\}$.

To establish the opposite inclusion take any point $x \in X \backslash \overline{A}$. There is $U \in \mathcal{U}$ with $U(x) \subset X \backslash \overline{A}$; apply (U2) and (4) to choose a symmetric $V \in \mathcal{U}$ such that $V \subset U$. If $x \in V(a)$ for some $a \in A$ then $(a, x) \in V$ and hence $(x, a) \in V$; this implies $a \in V(x)$ while $V(x) \subset U(x) \subset X \backslash A$. This contradiction shows that $x \notin V(A)$ and therefore $x \notin \bigcap\{U(A) : U \in \mathcal{U}\}$ whence $\overline{A} = \bigcap\{U(A) : U \in \mathcal{U}\}$, i.e., (5) is proved.

Passing to (6), fix a set $B \subset X \times X$ and let $P = \bigcap\{U \circ B \circ U : U \in \mathcal{U}\}$. Given $(x, y) \in \overline{B}$ fix any set $U \in \mathcal{U}$ and apply (4) to choose a symmetric set $V \in \mathcal{U}$ such that $V \subset U$. Then $V(x) \times V(y)$ is a neighborhood of $(x, y)$ (see (1)), so there is $(z, t) \in (V(x) \times V(y)) \cap B$. The set $V$ being symmetric, it follows from $(y, t) \in V$ that $(t, y) \in V$ which implies, together with $(x, z) \in V$ and $(z, t) \in B$, that $(x, y) \in V \circ B \circ V \subset U \circ B \circ U$. This shows that $\overline{B} \subset U \circ B \circ U$ for any $U \in \mathcal{U}$, i.e., $\overline{B} \subset P$.

To obtain the opposite inclusion take any point $(x, y) \in (X \times X) \backslash \overline{B}$. It follows from (2) and (U1) that there is $U \in \mathcal{U}$ such that $(U(x) \times U(y)) \cap \overline{B} = \emptyset$. Take a symmetric $V \in \mathcal{U}$ with $V \subset U$ (see (4)). If $(x, y) \in V \circ B \circ V$ then there are $z, t \in X$ for which $(x, z) \in V$, $(z, t) \in B$ and $(t, y) \in V$. The set $V$ being symmetric we have $t \in V(y)$ and therefore $(z, t) \in (V(x) \times V(y)) \cap B$ which is a contradiction. Thus $P \subset \overline{B}$ and hence $P = \overline{B}$, i.e., (6) is proved.

To prove (7) take any $U \in \mathcal{U}$ and find a set $V_0 \in \mathcal{U}$ with $V_0 \circ V_0 \subset U$. Apply (U2) again to obtain a set $V_1 \in \mathcal{U}$ such that $V_1 \circ V_1 \subset V_0$. It is easy to see that $V_1 \circ V_1 \circ V_1 \subset U$. It follows from (4) that there is a symmetric $V \in \mathcal{U}$ with $V \subset V_1$; consequently, $V \circ V \circ V \subset V_1 \circ V_1 \circ V_1 \subset U$. Now, (6) implies that $\overline{V} \subset V \circ V \circ V \subset U$. To see that $\overline{V}$ is symmetric, it suffices to show that $(\overline{V})^{-1} \subset \overline{V}$, so assume toward a contradiction that $(x, y) \in \overline{V}$ and $(y, x) \notin \overline{V}$. It follows from (6) that there is $W' \in \mathcal{U}$ such that $(y, x) \notin W' \circ V \circ W'$. Choose a symmetric $W \in \mathcal{U}$ such that $W \subset W'$; it is straightforward that $R = W \circ V \circ W$ is a symmetric set for which $(x, y) \in R$ and $(y, x) \notin R$; this contradiction shows that $G = \overline{V}$ is a closed symmetric set contained in $U$. Finally, observe that $G \in \mathcal{U}$ because $V \subset G$ and $V \in \mathcal{U}$. Thus every $U \in \mathcal{U}$ contains a closed symmetric $G \in \mathcal{U}$, so closed symmetric elements of $\mathcal{U}$ constitute a base of $\mathcal{U}$; this proves (7) and makes our solution complete.

**V.104.** *Given uniform spaces $(X, \mathcal{U})$ and $(Y, \mathcal{V})$, prove that every uniformly contin-*
*uous map $f : X \to Y$ is continuous. In particular, every uniform isomorphism is a*
*homeomorphism.*

**Solution.** Fix an arbitrary point $x \in X$ and let $y = f(x)$. To prove that $f$ is
continuous at $x$ take a set $O \in \tau(y, Y)$. There is $V \in \mathcal{V}$ with $V(y) \subset O$;
by uniform continuity of $f$, there exists a set $U \in \mathcal{U}$ such that $(x, y) \in U$
implies $(f(x), f(y)) \in V$. The set $G = \mathrm{Int}(U(x))$ is open in $X$ and $x \in G$
(see Problem 103). If $t \in G$ then $t \in U(x)$ and hence $(x, t) \in U$; consequently,
$(y, f(t)) \in V$ which shows that $f(t) \in V(y) \subset O$. The point $t \in G$ was chosen
arbitrarily, so $f(G) \subset V(x) \subset O$, i.e., the set $G$ witnesses continuity of the map $f$
at the point $x$.

**V.105.** *Suppose that $(X_t, \mathcal{U}_t)$ is a uniform space for every $t \in T$ and consider the*
*set $X = \prod_{t \in T} X_t$. Let $p_t : X \to X_t$ be the natural projection for every $t \in T$;*
*prove that*

*(1)  the family $\mathcal{S} = \bigcup \{ p_t^{-1}(\mathcal{U}_t) : t \in T \}$ is a subbase of a unique uniformity on $X$,*
*i.e., the uniform product $(X, \mathcal{U})$ of the spaces $\{(X_t, \mathcal{U}_t) : t \in T\}$ is well defined;*
*(2)  every map $p_t : (X, \mathcal{U}) \to (X_t, \mathcal{U}_t)$ is uniformly continuous;*
*(3)  $\tau_{\mathcal{U}}$ coincides with the topology of the product $\prod \{(X_t, \tau_{\mathcal{U}_t}) : t \in T\}$.*

**Solution.** Let $\Delta = \Delta(X)$ be the diagonal of $X$. Denote by $\Delta_t$ the diagonal $\Delta(X_t)$
of the space $X_t$ and let $\mathcal{S}_t = p_t^{-1}(\mathcal{U}_t)$; it follows from $\Delta \subset (p_t \times p_t)^{-1}(\Delta_t)$ that
$\Delta \subset \bigcap \mathcal{S}_t$ for every $t \in T$.

Fix $t \in T$, a set $U \in \mathcal{S}_t$ and $V \in \mathcal{U}_t$ such that $U = (p_t \times p_t)^{-1}(V)$. A point
$(x, y)$ from $X \times X$ belongs to the set $U^{-1}$ if and only if $(y, x) \in U$ which happens
if and only if $(p_t(y), p_t(x)) \in V$ which occurs if and only if $(p_t(x), p_t(y)) \in V^{-1}$.
This shows that $U^{-1} = (p_t \times p_t)^{-1}(V^{-1})$ and hence

(i)  $\Delta \subset \bigcap \mathcal{S}_t$ and, for any $U \in \mathcal{S}_t$, the set $U^{-1}$ also belongs to $\mathcal{S}_t$.

Next, choose a set $V' \in \mathcal{U}_t$ such that $V' \circ V' \subset V$. The set $W = (p_t \times p_t)^{-1}(V')$
belongs to $\mathcal{S}_t$. If $(x, z) \in W \circ W$ then there is $y \in X$ such that $(x, y) \in W$ and
$(y, z) \in W$. This implies that $(p_t(x), p_t(y)) \in V'$ and $(p_t(y), p_t(z)) \in V'$ and
therefore $(p_t(x), p_t(z)) \in V' \circ V' \subset V$. Thus $(x, z) \in U$ for any $(x, z) \in W \circ W$,
i.e., $W \circ W \subset U$. This proves that

(ii)  for any $U \in \mathcal{S}_t$ there is $W \in \mathcal{S}_t$ such that $W \circ W \subset U$.

It is straightforward that the family $\mathcal{S} = \bigcup_{t \in T} \mathcal{S}_t$ also has the properties (i) and
(ii). Given a point $(x, y) \in (X \times X) \backslash \Delta$ we have $x \neq y$, so there is $t \in T$ such
that $p_t(x) \neq p_t(y)$ and hence $(p_t(x), p_t(y)) \in (X_t \times X_t) \backslash \Delta_t$. The family $\mathcal{U}_t$ being
a uniformity on $X_t$, there is $V \in \mathcal{U}_t$ such that $(p_t(x), p_t(y)) \notin V$ and therefore
$(x, y) \notin (p_t \times p_t)^{-1}(V) \in \mathcal{S}_t \subset \mathcal{S}$. This implies that

(iii)  $\bigcap \{ S : S \in \mathcal{S} \} = \Delta$,

so the family $\mathcal{S}$ satisfies the premises of Problem 102 and hence there exists a
uniformity for which $\mathcal{S}$ is a subbase. The family $\mathcal{B}$ of all finite intersections of

the elements of $S$ has to be a base of any uniformity $\mathcal{U}$ for which $S$ is a subbase. Besides, $\mathcal{U} = \{E \subset X \times X : \text{there is } B \in \mathcal{B} \text{ with } B \subset E\}$, so $\mathcal{U}$ is uniquely determined by $\mathcal{B}$ and hence by $S$. This settles (1).

Given $V \in \mathcal{U}_t$ the set $(p_t \times p_t)^{-1}(V)$ belongs to $S_t$ and hence to $\mathcal{U}$. This shows that every $p_t$ is uniformly continuous, i.e., we established (2).

Denote by $\mu$ the topology of the product $\prod_{t \in T}(X_t, \tau_{\mathcal{U}_t})$ and let $e : X \to X$ be the identity map. To see that $e : (X, \tau_{\mathcal{U}}) \to (X, \mu)$ is continuous, observe that every map $p_t \circ e = p_t : (X, \tau_{\mathcal{U}}) \to (X_t, \mathcal{U}_t)$ is uniformly continuous being the natural projection of the uniform product $(X, \tau_{\mathcal{U}})$ onto the factor $(X_t, \mathcal{U}_t)$ (see (2)); thus $p_t$ is continuous by Problem 104. Therefore $e$ is continuous by TFS-102; as a consequence, $e^{-1}(V) = V \in \tau_{\mathcal{U}}$ for any $V \in \mu$, i.e., $\mu \subset \tau_{\mathcal{U}}$.

To show that $e : (X, \mu) \to (X, \tau_{\mathcal{U}})$ is also continuous fix a point $x \in X$ and a set $O \in \tau_{\mathcal{U}}$ with $x \in O$. There is $U \in \mathcal{U}$ such that $U(x) \subset O$; the family $S$ being a subbase of $(X, \mathcal{U})$ we can choose a finite $S \subset T$ and a set $U_t \in \mathcal{U}_t$ for any $t \in S$ in such a way that $\bigcap\{(p_t \times p_t)^{-1}(U_t) : t \in S\} \subset U$.

For every $t \in T$ let $S_t = (p_t \times p_t)^{-1}(U_t)$; it follows from Problem 103 that there is $V_t \in \tau_{\mathcal{U}_t}$ such that $p_t(x) \in V_t \subset U_t(p_t(x))$. The set $V = \bigcap\{p_t^{-1}(V_t) : t \in S\}$ belongs to $\mu$ and $x \in V$. If $y \in V$ then $p_t(y) \in V_t$; thus $(p_t(x), p_t(y)) \in U_t$ and hence $(x, y) \in S_t$ for any $t \in S$. This shows that $(x, y) \in \bigcap\{S_t : t \in S\} \subset U$; an immediate consequence is that $y \in U(x) \subset O$. Since the point $y \in V$ was chosen arbitrarily, we proved that $V = e(V) \subset O$ and hence the map $e : (X, \mu) \to (X, \tau_{\mathcal{U}})$ is continuous at every point $x \in X$. Finally, observe that $e^{-1}(V) = V \in \mu$ for any $V \in \tau_{\mathcal{U}}$, so $\tau_{\mathcal{U}} \subset \mu$ and hence $\tau_{\mathcal{U}} = \mu$, i.e., (3) is proved, so our solution is complete.

**V.106.** *Let $(X, \mathcal{U})$ be the uniform product of the family $\{(X_t, \mathcal{U}_t) : t \in T\}$. Given a uniform space $(Y, \mathcal{V})$, prove that a map $f : Y \to X$ is uniformly continuous if and only if $p_t \circ f : Y \to X_t$ is uniformly continuous for any $t \in T$. Here, as always, the map $p_t : X \to X_t$ is the natural projection.*

**Solution.** It is an easy exercise that a composition of two uniformly continuous maps is a uniformly continuous map. Therefore, if $f : Y \to X$ is uniformly continuous then so is every map $p_t \circ f$ (see Problem 105). This proves necessity.

To establish sufficiency assume that the map $p_t \circ f$ is uniformly continuous for any $t \in T$ and fix a set $U \in \mathcal{U}$. Since the family $S = \bigcup\{p_t^{-1}(\mathcal{U}_t) : t \in T\}$ is a subbase of $\mathcal{U}$, there exists a finite $S \subset T$ and a set $U_t \in \mathcal{U}_t$ for every $t \in S$ such that $\bigcap\{(p_t \times p_t)^{-1}(U_t) : t \in S\} \subset U$; let $S_t = (p_t \times p_t)^{-1}(U_t)$ for each $t \in S$. The set $V_t = (f \times f)^{-1}(S_t) = ((p_t \circ f) \times (p_t \circ f))^{-1}(U_t)$ belongs to $\mathcal{V}$ for every $t \in S$ because the map $p_t \circ f$ is uniformly continuous. Consequently, the set $V = \bigcap_{t \in S} V_t$ belongs to $\mathcal{V}$; if $(z, t) \in V$ then $(z, t) \in V_t$ and hence $(f(z), f(t)) \in S_t$ for every $t \in S$.

This implies that $(f(z), f(t)) \in \bigcap_{t \in S} S_t \subset U$ for any $(z, t) \in V$ and therefore $(f \times f)(V) \subset U$, i.e., $V \subset (f \times f)^{-1}(U)$ which shows that $(f \times f)^{-1}(U) \in \mathcal{V}$. The set $U \in \mathcal{U}$ was chosen arbitrarily so the map $f$ is uniformly continuous and hence we completed the proof of sufficiency.

**V.107.** *Prove that, for any uniform space* $(X, \mathcal{U})$, *a pseudometric* $d : X \times X \to \mathbb{R}$ *is uniformly continuous with respect to the uniform square of* $(X, \mathcal{U})$ *if and only if the set* $O_r = \{(x, y) \in X \times X : d(x, y) < r\}$ *belongs to* $\mathcal{U}$ *for any* $r > 0$. *Such pseudometrics will be called uniformly continuous on* $(X, \mathcal{U})$.

**Solution.** Let $(X \times X, \mathcal{V})$ be the uniform square of the space $(X, \mathcal{U})$. For any point $z = (x, y) \in X \times X$ let $p_0(z) = x$ and $p_1(z) = y$. The maps $p_0, p_1 : X \times X \to X$ are the natural projections of $X \times X$ onto its factors so the family $p_0^{-1}(\mathcal{U}) \cup p_1^{-1}(\mathcal{U})$ is a subbase of $(X \times X, \mathcal{V})$.

Assume that the pseudometric $d$ is uniformly continuous and fix $r > 0$. There exists $V \in \mathcal{V}$ such that $|d(z_1) - d(z_2)| < r$ for any point $(z_1, z_2) \in V$. We can find finite families $\mathcal{S}_0 \subset p_0^{-1}(\mathcal{U})$ and $\mathcal{S}_1 \subset p_1^{-1}(\mathcal{U})$ such that $S = (\bigcap \mathcal{S}_0) \cap (\bigcap \mathcal{S}_1) \subset V$. It is easy to see that $S_0 = \bigcap \mathcal{S}_0 \in p_0^{-1}(\mathcal{U})$ and $S_1 = \bigcap \mathcal{S}_1 \in p_1^{-1}(\mathcal{U})$, so fix $W_0, W_1 \in \mathcal{U}$ for which $(p_0 \times p_0)^{-1}(W_0) = S_0$ and $(p_1 \times p_1)^{-1}(W_1) = S_1$.

Since $W = W_0 \cap W_1 \in \mathcal{U}$, the set $G = (p_0 \times p_0)^{-1}(W) \cap (p_1 \times p_1)^{-1}(W)$ belongs to $\mathcal{V}$; besides, $G \subset S_0 \cap S_1 \subset V$ and therefore $|d(z_0) - d(z_1)| < r$ for any $(z_0, z_1) \in G$. Now, if $z = (x, y) \in W$ then let $z_0 = z$ and $z_1 = (y, y)$. It is clear that $(x, y) = (p_0(z_0), p_0(z_1)) \in W$ and $(y, y) = (p_1(z_0), p_1(z_1)) \in W$, so $(z_0, z_1) \in G$ and therefore $|d(x, y) - d(y, y)| = d(x, y) < r$; an immediate consequence is that $G \subset O_r$ and hence $O_r \in \mathcal{U}$. Thus $O_r \in \mathcal{U}$ for any $r > 0$, i.e., we have established necessity.

For sufficiency, assume that $O_r \in \mathcal{U}$ for any $r > 0$ and fix $\varepsilon > 0$. Then $W = O_{\varepsilon/2} \in \mathcal{U}$ and hence $V = (p_0 \times p_0)^{-1}(W) \cap (p_1 \times p_1)^{-1}(W) \in \mathcal{V}$. Take any point $(z_0, z_1) \in V$; then $z_0 = (x_0, y_0)$ and $z_1 = (x_1, y_1)$ where $\{(x_0, x_1), (y_0, y_1)\} \subset W$, so it follows from the triangle inequality that

$$d(x_0, y_0) \leq d(x_0, x_1) + d(x_1, y_1) + d(y_1, y_0) < \frac{\varepsilon}{2} + d(x_1, y_1) + \frac{\varepsilon}{2} = d(x_1, y_1) + \varepsilon,$$

so $d(x_0, y_0) - d(x_1, y_1) < \varepsilon$. Analogously,

$$d(x_1, y_1) \leq d(x_1, x_0) + d(x_0, y_0) + d(y_0, y_1) < \frac{\varepsilon}{2} + d(x_0, y_0) + \frac{\varepsilon}{2} = d(x_0, y_0) + \varepsilon,$$

which shows that $d(x_1, y_1) - d(x_0, y_0) < \varepsilon$ and therefore $|d(x_1, y_1) - d(x_0, y_0)| < \varepsilon$. We proved that $|d(z_0) - d(z_1)| < \varepsilon$ for any point $(z_0, z_1) \in V$, so the function $d : (X \times X, \mathcal{V}) \to \mathbb{R}$ is uniformly continuous, i.e., we settled sufficiency.

**V.108.** *Suppose that a sequence* $\{U_n : n \in \omega\}$ *of subsets of* $X \times X$ *has the following properties:*

*(1)* $U_0 = X \times X$ *and* $\Delta \subset U_n$ *for any* $n \in \omega$;
*(2)* *the set* $U_n$ *is symmetric and* $U_{n+1} \circ U_{n+1} \circ U_{n+1} \subset U_n$ *for any* $n \in \omega$.

*Prove that there exists a pseudometric* $d$ *on the set* $X$ *such that, for any* $n \in \mathbb{N}$, *we have* $U_n \subset \{(x, y) : d(x, y) \leq 2^{-n}\} \subset U_{n-1}$.

**Solution.** Given $x, y \in X$, a set $C = \{x_0, x_1, \ldots, x_n\} \subset X$ is *a chain which connects the points $x$ and $y$* if $x_0 = x$ and $x_n = y$; we will say that the number of links of $C$ is equal to $n$. It is an easy exercise, that for any $A \subset X \times X$ with $\Delta \subset A$,

(∗) a point $(x, y) \in X \times X$ belongs to $A \circ A \circ A$ if and only if there exist $z, t \in X$ such that $\{(x, z), (z, t), (t, y)\} \subset A$.

An immediate consequence of (∗) is that $U_{n+1} \subset U_n$ for each $n \in \omega$. Given any point $(x, y) \in X \times X$ let $\sigma(x, y) = 0$ if $(x, y) \in U = \bigcap\{U_n : n \in \omega\}$. If the point $(x, y)$ belongs to $(X \times X) \setminus U$ then there is a unique $n \in \omega$ such that $(x, y) \in U_n \setminus U_{n+1}$; let $\sigma(x, y) = 2^{-n}$. For each chain $C = \{x_0, \ldots, x_n\}$ the number $l(C) = \sum_{i=0}^{n-1} \sigma(x_i, x_{i+1})$ can be thought of as the length of the chain $C$. If $x, y \in X$ then the formula $d(x, y) = \inf\{l(C) : C$ is a chain that connects $x$ and $y\}$ defines a function $d : X \times X \to [0, +\infty)$; we will show that $d$ is the promised pseudometric.

Since $\{x, y\}$ is a chain which connects $x$ and $y$, we have $d(x, y) \leq \sigma(x, y)$ for any $x, y \in X$. Therefore $d(x, x) \leq \sigma(x, x) = 0$ and hence $d(x, x) = 0$ for any $x \in X$.

To prove that $d$ is symmetric observe that $\sigma(x, y) = \sigma(y, x)$ for any $x, y \in X$ because every $U_n$ is a symmetric set. Now if $x, y \in X$ and $C = \{x_0, x_1, \ldots, x_n\}$ is a chain which connects $x$ and $y$ then the chain $C' = \{x_n, x_{n-1}, \ldots, x_1, x_0\}$ connects $y$ and $x$; it follows from the equalities $\sigma(x_i, x_{i+1}) = \sigma(x_{i+1}, x_i)$, $i = 0, \ldots, n-1$, that $l(C) = l(C')$. Thus, for every chain $C$ connecting $x$ and $y$ there is a chain $C'$ which connects $y$ and $x$ such that $l(C') = l(C)$. An evident consequence is that $d(x, y) = d(y, x)$ for any $x, y \in X$.

To see that the triangle inequality also holds for $d$ take any points $x, y, z \in X$ and $\varepsilon > 0$. There exist chains $C' = \{x_0, \ldots, x_n\}$ and $C'' = \{y_0, \ldots, y_k\}$ such that $l(C') < d(x, y) + \frac{\varepsilon}{2}$ and $l(C'') < d(y, z) + \frac{\varepsilon}{2}$ while $C'$ connects $x$ and $y$ and $C''$ connects $y$ and $z$. The set $C = \{x_0, \ldots, x_n, y_0, \ldots, y_k\}$ is a chain connecting the points $x$ and $z$, so $d(x, z) \leq l(C) = l(C') + l(C'') < d(x, y) + d(y, z) + \varepsilon$. Thus $d(x, z) < d(x, y) + d(y, z) + \varepsilon$ for any $\varepsilon > 0$ and hence $d(x, z) \leq d(x, y) + d(y, z)$, i.e., we indeed have the triangle inequality for $d$, so $d$ is a pseudometric on the set $X$.

Our next step is to establish that

(i) $\frac{1}{2}\sigma(x, y) \leq d(x, y) \leq \sigma(x, y)$ for any $x, y \in X$.

We have already seen that the rightmost inequality in (i) is true, so let us prove that $d(x, y) \geq \frac{1}{2}\sigma(x, y)$ for any $x, y \in X$. It suffices to show that,

(ii) for any chain $C$ which connects the points $x$ and $y$, we have $l(C) \geq \frac{1}{2}\sigma(x, y)$.

We will prove (ii) by induction on the number $m$ of links of $C$. If $m = 1$ then $C = \{x, y\}$, so $l(C) = \sigma(x, y) \geq \frac{1}{2}\sigma(x, y)$. Now assume that $m > 1$ and we have established that $l(C) \geq \frac{1}{2}\sigma(x, y)$ whenever $C$ is a chain with $< m$-many links connecting $x$ and $y$.

To accomplish the inductive step take any chain $C = \{x_0, \ldots, x_m\}$ with $x_0 = x$ and $x_m = y$. Letting $a_0 = 0$ and $a_k = \sigma(x_0, x_1) + \ldots + \sigma(x_{k-1}, x_k)$ for any $k \in \{1, \ldots, m\}$ we obtain a sequence $a_0 \leq \ldots \leq a_m$ in which $a = a_m$ is equal to the length of the chain $C$. It is easy to see that there exists a number $k \in \{0, \ldots, m-1\}$ such that $a_k \leq \frac{a}{2}$ and $a_{k+1} \geq \frac{a}{2}$.

If $(x, y) \in U$ then $\sigma(x, y) = 0$, so $l(C) \geq \frac{1}{2}\sigma(x, y)$. If the point $(x, y)$ does not belong to $U$ then $(x, y) \in U_n \backslash U_{n+1}$ for some $n \in \omega$ and hence $\sigma(x, y) = 2^{-n}$. If $H = \{(x, x_k), (x_k, x_{k+1}), (x_{k+1}, y)\} \subset U_{n+2}$ then $(x, y) \in U_{n+2} \circ U_{n+2} \circ U_{n+2} \subset U_{n+1}$ (see (*)) which is a contradiction. Thus one of the elements of $H$ does not belong to $U_{n+2}$.

If $(x_k, x_{k+1}) \notin U_{n+2}$ then $l(C) \geq \sigma(x_k, x_{k+1}) \geq 2^{-n-1} = \frac{1}{2}\sigma(x, y)$. In the case when $(x, x_k) \notin U_{n+2}$ apply the induction hypothesis to the chain $C' = \{x_0, \ldots, x_k\}$ to see that $a_k = l(C') \geq \frac{1}{2}\sigma(x, x_k) \geq \frac{1}{2} \cdot 2^{-n-1} = 2^{-n-2}$. By our choice of $k$, we have $l(C) = a \geq 2a_k \geq 2^{-n-1} = \frac{1}{2}\sigma(x, y)$. Finally, if $(x_{k+1}, x_m) \notin U_{n+2}$ then we can apply the induction hypothesis to the chain $C'' = \{x_{k+1}, \ldots, x_m\}$. This gives $\frac{1}{2}\sigma(x_{k+1}, x_m) \leq l(C'') = l(C) - a_{k+1} \leq \frac{a}{2}$, so it follows from $\sigma(x_{k+1}, x_m) \geq 2^{-n-1}$ that $\frac{a}{2} \geq 2^{-n-2}$, i.e., $a \geq 2^{-n-1} = \frac{1}{2}\sigma(x, y)$, so we proved (ii). The leftmost inequality of (i) is an immediate consequence of (ii), so (i) is proved as well.

Given an arbitrary $n \in \omega$, we have $\sigma(x, y) \leq 2^{-n}$ for any $(x, y) \in U_n$, so $d(x, y) \leq \sigma(x, y) \leq 2^{-n}$; this shows that $U_n \subset \{(x, y) : d(x, y) \leq 2^{-n}\}$. Now if $d(x, y) \leq 2^{-n}$ then $\sigma(x, y) \leq 2^{-n-1}$ by the property (i), so $(x, y) \in U_{n-1}$; this implies that $\{(x, y) : d(x, y) \leq 2^{-n}\} \subset U_{n-1}$ and hence our solution is complete.

**V.109.** *Given a uniform space $(X, \mathcal{U})$ and $U \in \mathcal{U}$, prove that there is a uniformly continuous pseudometric $\rho$ on $(X, \mathcal{U})$ such that $\{(x, y) \in X \times X : \rho(x, y) < 1\} \subset U$.*

**Solution.** We will first prove the following easy statement.

(i) Suppose that $A, B, C \subset X \times X$ and $\Delta \subset A \cap B \cap C$. If $B \circ B \subset A$ and $C \circ C \subset B$ then $C \circ C \circ C \subset A$.

To show that (i) holds take a point $(x, y) \in C \circ C \circ C$; there exist $z, t \in X$ such that $\{(x, z), (z, t), (t, y)\} \subset C$. As a consequence, $(x, t) \in C \circ C \subset B$; it follows from $\{(t, y), (y, y)\} \subset C$ that $(t, y) \in C \circ C \subset B$, so $\{(x, t), (t, y)\} \subset B$ and therefore $(x, y) \in B \circ B \subset A$, so (i) is proved.

It follows from Problem 103 that there exists a symmetric set $U_1 \in \mathcal{U}$ with $U_1 \subset U$. The property (i) and Problem 103 make it possible to construct a sequence $\{U_n : n \in \mathbb{N}\backslash\{1\}\}$ of symmetric elements of $\mathcal{U}$ such that $U_{n+1} \circ U_{n+1} \circ U_{n+1} \subset U_n$ for any $n \in \mathbb{N}$. Letting $U_0 = X \times X$ we obtain a sequence $\{U_n : n \in \omega\}$ which satisfies the premises of Problem 108. Thus there exists a pseudometric $d : X \times X \to [0, +\infty)$ such that

(ii) $U_n \subset \{(x, y) \in X \times X : d(x, y) \leq 2^{-n}\} \subset U_{n-1}$ for any $n \in \mathbb{N}$.

It is straightforward that $\rho = 4d$ is also a pseudometric on $X$. If $\rho(x, y) < 1$ then $d(x, y) < \frac{1}{4} = 2^{-2}$, so (ii) shows that $(x, y) \in U_1 \subset U$. Therefore we have the inclusion $\{(x, y) \in X \times X : d(x, y) < 1\} \subset U$.

Finally, take any $r > 0$ and fix $n \in \omega$ such that $2^{-n-2} < r$. If $\rho(x, y) \leq 2^{-n}$ then $d(x, y) \leq \frac{1}{4} \cdot 2^{-n} = 2^{-n-2} < r$ which shows, together with (ii), that
$$U_n \subset \{(x, y) \in X \times X : d(x, y) \leq 2^{-n}\} \subset \{(x, y) \in X \times X : d(x, y) < r\}.$$

An immediate consequence is that $\{(x, y) \in X \times X : d(x, y) < r\} \in \mathcal{U}$ for any $r > 0$, so the pseudometric $\rho$ is uniformly continuous on $(X, \mathcal{U})$ by Problem 107.

**V.110.** *Prove that a topological space $X$ is Tychonoff if and only if there exists a uniformity $\mathcal{U}$ on the set $X$ such that $\tau(X) = \tau_{\mathcal{U}}$.*

**Solution.** Suppose first that there exists a uniformity $\mathcal{U}$ on the set $X$ such that $\tau_{\mathcal{U}} = \tau(X)$. Given $x \in X$ pick any $y \in X \backslash \{x\}$; the point $(y, x)$ does not belong to the diagonal $\Delta$ of the space $X$, so there is $U \in \mathcal{U}$ such that $(y, x) \notin U$ and therefore $x \notin U(y)$. It follows from Problem 103 that $V = \mathrm{Int}(U(y)) \in \tau(y, X)$; since $x \notin V$, we proved that every $y \in X \backslash \{x\}$ has a neighborhood $V \subset X \backslash \{x\}$. Thus $X \backslash \{x\}$ is open for every $x \in X$, i.e., $X$ is a $T_1$-space.

Now fix a point $a \in X$ and $O \in \tau(a, X)$. There is a set $U \in \mathcal{U}$ such that $U(a) \subset O$; by Problem 109 there exists a uniformly continuous pseudometric $\rho : X \times X \to \mathbb{R}$ such that $G = \{(x, y) : \rho(x, y) < 1\} \subset U$. Since $\rho$ is continuous on $X \times X$ by Problem 104, the function $g : X \to \mathbb{R}$, defined by $g(x) = \rho(a, x)$ for any $x \in X$, is continuous as well.

It is clear that $g(a) = 0$; given $x \in X \backslash O$, it follows from $U(a) \subset O$ that $(a, x) \notin U$ and hence $g(x) = \rho(a, x) \geq 1$. Thus the function $f : X \to \mathbb{R}$ defined by the formula $f(x) = 1 - \backslash n\{g(x), 1\}$ for any $x \in X$, is continuous on $X$ and $f(a) = 1$ while $f(X \backslash O) \subset \{0\}$. This shows that $X$ is a Tychonoff space, i.e., we proved sufficiency.

To establish necessity, suppose that $X$ is a Tychonoff space and consider the set $O(f, r) = \{(x, y) \in X \times X : |f(x) - f(y)| < r\}$ for each $f \in C_p(X)$. Clearly, every element of the family $\mathcal{S} = \{O(f, r) : f \in C_p(X), r > 0\}$ contains $\Delta$. If $(x, y) \notin \Delta$ then $x \neq y$, so there is $f \in C_p(X)$ for which $f(x) = 0$ and $f(y) = 1$; an immediate consequence is that $(x, y) \notin O(f, 1) \in \mathcal{S}$, so $\bigcap \mathcal{S} = \Delta$.

Suppose that $U = O(f, r) \in \mathcal{S}$; then $V = O(f, \frac{r}{2}) \in \mathcal{S}$. If $\{(x, y), (y, z)\} \subset V$ then $|f(x) - f(z)| \leq |f(x) - f(y)| + |f(y) - f(z)| < \frac{r}{2} + \frac{r}{2} = r$ and hence $(x, z) \in U$. This proves that $V \circ V \subset U$, so the family $\mathcal{S}$ satisfies the conditions (1) and (3) of Problem 102. Since every element of $\mathcal{S}$ is a symmetric subset of $X \times X$, the condition (2) is satisfied as well, so there is a uniformity $\mathcal{U}$ on the set $X$ such that $\mathcal{S}$ is a subbase of $\mathcal{U}$.

Take any $O \in \tau(X)$ and $x \in O$. There is $f \in C_p(X)$ such that $f(x) = 1$ and $f(X \backslash O) \subset \{0\}$. The set $U = O(f, 1)$ belongs to the family $\mathcal{U}$; if $y \in U(x)$ then $|f(y) - f(x)| = |f(y) - 1| < 1$ which shows that $f(y) > 0$ and hence $y \in O$. Therefore $U(x) \subset O$, i.e., we proved that, for every $x \in O$ there is $U \in \mathcal{U}$ for which $U(x) \subset O$. Consequently, $O \in \tau_{\mathcal{U}}$, so we proved that $\tau(X) \subset \tau_{\mathcal{U}}$.

Finally, fix a set $O \in \tau_{\mathcal{U}}$; given a point $x \in O$ there is $U \in \mathcal{U}$ with $U(x) \subset O$. The family $\mathcal{S}$ being a subbase of $\mathcal{U}$ we can find $k \in \mathbb{N}$ and $V_1, \ldots, V_k \in \mathcal{S}$ for which $V = V_1 \cap \ldots \cap V_k \subset U$. Choose $f_i \in C_p(X)$ and $r_i > 0$ such that $V_i = O(f_i, r_i)$ for all $i \leq k$. The function $f_i$ is continuous so the set $V_i(x) = \{y \in X : |f_i(y) - f_i(x)| < r_i\}$ is open in $X$ for every $i \in \{1, \ldots, k\}$; thus $V(x) = V_1(x) \cap \ldots \cap V_k(x) \in \tau(x, X)$ and $V(x) \subset U(x) \subset O$. It turns out that, for every point $x \in O$, there is a set $V(x) \in \tau(x, X)$ with $V(x) \subset O$; this shows that $O \in \tau(X)$ and proves that $\tau_{\mathcal{U}} \subset \tau(X)$ and hence $\tau_{\mathcal{U}} = \tau(X)$, i.e., we settled necessity.

**V.111.** *Given a metric $\rho$ on a set $X$ and a number $r > 0$, consider the set $U_r = \{(x, y) \in X \times X : \rho(x, y) < r\}$. Prove that the family $\mathcal{B} = \{U_r : r > 0\}$ forms a base of some uniformity $\mathcal{U}_\rho$ on the set $X$. The uniformity $\mathcal{U}_\rho$ is called the uniformity induced by the metric $\rho$. A uniform space $(X, \mathcal{U})$ is called uniformly metrizable if $\mathcal{U} = \mathcal{U}_\rho$ for some metric $\rho$ on the set $X$.*

**Solution.** It is evident that the diagonal $\Delta$ of the space $X$ is contained in every element of $\mathcal{B}$. If $(x, y) \in (X \times X) \setminus \Delta$ then $x \neq y$ and hence $r = \rho(x, y) > 0$; it is clear that $(x, y) \notin U_r$, so $\bigcap \mathcal{B} = \Delta$.

An immediate consequence of the definition is that every $U_r$ is a symmetric subset of $X \times X$. Now if $U, V \in \mathcal{B}$ then $U = U_r$ and $V = U_s$ for some positive numbers $r$ and $s$. If $t = \backslash n\{r, s\}$ then $U \cap V = U_t \in \mathcal{B}$, so the properties (1), (2), and (4) of Problem 101 are fulfilled for $\mathcal{B}$.

To see that the property (3) of Problem 101 also holds for $\mathcal{B}$ take any $U \in \mathcal{B}$; then $U = U_r$ for some $r > 0$. The set $V = U_{r/2}$ also belongs to $\mathcal{B}$; if $\{(x, y), (y, z)\} \subset V$ then $\rho(x, z) \leq \rho(x, y) + \rho(y, z) < \frac{r}{2} + \frac{r}{2} = r$, so $(x, z) \in U_r = U$. Thus $V \circ V \subset U$, so all conditions (1)–(4) of Problem 101 are satisfied for $\mathcal{B}$ and hence there exists a uniformity $\mathcal{U}_\rho$ such that $\mathcal{B}$ is a base of $\mathcal{U}_\rho$.

**V.112.** *Prove that a uniform space $(X, \mathcal{U})$ is uniformly metrizable if and only if $\mathcal{U}$ has a countable base.*

**Solution.** Assume that the space $(X, \mathcal{U})$ is uniformly metrizable and hence there exists a metric $\rho$ such that the family $\mathcal{B} = \{U_r : r > 0\}$ is a base for $\mathcal{U}$; here $U_r = \{(x, y) \in X \times X : \rho(x, y) < r\}$ for any $r > 0$. The family $\mathcal{B}_0 = \{U_r : r > 0 \text{ and } r \in \mathbb{Q}\}$ is countable and it is straightforward that $\mathcal{B}$ is a base for $\mathcal{U}$; this proves necessity.

Now assume that there exists a countable base $\mathcal{B}$ for the uniformity $\mathcal{U}$ and let $\{B_n : n \in \mathbb{N}\}$ be an enumeration of $\mathcal{B}$. Using Problems 101 and 103 it is easy to construct by induction a family $\{U_n : n \in \omega\}$ of symmetric elements of $\mathcal{U}$ with the following properties:

(i)  $U_0 = X \times X$ and $U_n \subset B_n$ for each $n \in \mathbb{N}$;
(ii) $U_{n+1} \circ U_{n+1} \circ U_{n+1} \subset U_n$ for any $n \in \omega$.

By Problem 108, there exists a pseudometric $d$ on the set $X$ such that
(iii) $U_n \subset \{(x, y) : d(x, y) \leq 2^{-n}\} \subset U_{n-1}$ for every $n \in \mathbb{N}$.

Given distinct $x, y \in X$ the point $(x, y)$ does not belong to the diagonal $\Delta$ of the space $X$, so there is $U \in \mathcal{U}$ with $(x, y) \notin U$. The family $\mathcal{B}$ being a base of $\mathcal{U}$ we can find $n \in \mathbb{N}$ for which $B_n \subset U$ and therefore $(x, y) \notin B_n$. It follows from (i) that $(x, y) \notin U_n$ and hence $d(x, y) > 2^{-n-1} > 0$ by (iii). Thus $d$ is actually a metric on $X$; let us show that $\mathcal{U}$ is induced by $d$.

Let $O_r = \{(x, y) \in X \times X : d(x, y) < r\}$ for any $r > 0$; we must prove that the family $\mathcal{O} = \{O_r : r > 0\}$ is a base of $\mathcal{U}$. If $r > 0$ then there is $n \in \omega$ with $2^{-n} < r$. The property (iii) implies that $U_n \subset \{(x, y) \in X \times X : d(x, y) \leq 2^{-n}\} \subset O_r$, so $O_r \in \mathcal{U}$ for any $r > 0$, i.e., $\mathcal{O} \subset \mathcal{U}$.

Finally, if $U \in \mathcal{U}$ then pick $n \in \mathbb{N}$ with $B_n \subset U$. Then $U_n \subset U$ by (i); it follows from the property (iii) that, for $r = 2^{-n-1} > 0$ we have the inclusions $O_r \subset \{(x, y) \in X \times X : d(x, y) \leq r\} \subset U_n \subset U$, so $\mathcal{O}$ is, indeed, a base of $\mathcal{U}$, i.e., $\mathcal{U}$ is induced by the metric $d$. This proves sufficiency and completes our solution.

**V.113.** *Prove that every uniform space is uniformly isomorphic to a subspace of a product of uniformly metrizable spaces.*

**Solution.** The following statement will help us to show that, for any uniform space, there is a sufficiently large family of uniformly continuous mappings of this space into uniformly metrizable spaces.

*Fact 1.* If $(X, \mathcal{U})$ is a uniform space and $d$ is a uniformly continuous pseudometric on $(X, \mathcal{U})$ then there exists a metric space $(Y, \rho)$ and a map $\varphi : X \to Y$ such that $d(x, y) = \rho(\varphi(x), \varphi(y))$ for any $x, y \in X$. In particular, the map $\varphi$ is uniformly continuous if we consider $Y$ with the uniformity $\mathcal{V}$ induced by the metric $\rho$.

*Proof.* For any $x, y \in X$, let $x \sim y$ if and only if $d(x, y) = 0$. Then $\sim$ is an equivalence relationship on $X$. Indeed, it follows from $d(x, x) = 0$ that $x \sim x$; now, if $x \sim y$ then $d(y, x) = d(x, y) = 0$, so $y \sim x$. To check transitivity, observe that $x \sim y$ and $y \sim z$ imply that $d(x, y) = d(y, z) = 0$, so $d(x, z) \leq d(x, y) + d(y, z) = 0$ which shows that $x \sim z$. Let $P_x = \{y \in X : y \sim x\}$ for any $x \in X$.

The space $X$ is the disjoint union of the classes of equivalence with respect to $\sim$, i.e., for the family $\{P_x : x \in X\}$ we have either $P_x = P_y$ or $P_x \cap P_y = \emptyset$ for any $x, y \in X$. Choose a set $Y \subset X$ such that $\bigcup \{P_y : y \in Y\} = X$ and $P_y \cap P_z = \emptyset$ for distinct $y, z \in Y$.

Given $y, z \in Y$ let $\rho(y, z) = d(y, z)$; the function $\rho : Y \times Y \to \mathbb{R}$ is a metric on $Y$. To prove it, assume that $\rho(y, z) = 0$ for some $y, z \in Y$. If $y \neq z$ then $P_y \cap P_z = \emptyset$ and hence $z \notin P_y$ which shows that $\rho(y, z) = d(y, z) > 0$. Therefore $\rho(y, z) = 0$ implies $y = z$; the other axioms of metric are fulfilled for $\rho$ because it coincides with the pseudometric $d$ on $Y$. Thus $(Y, \rho)$ is metric space.

For each $x \in X$ there is a unique $y \in Y$ with $x \in P_y$; let $\varphi(x) = y$. To see that the map $\varphi : X \to Y$ has the promised properties take $x, y \in X$. If $a = \varphi(x)$ and $b = \varphi(y)$ then $x \in P_a$ and $y \in P_b$; therefore $d(x, a) = d(y, b) = 0$. Consequently, $\rho(a, b) = d(a, b) \leq d(a, x) + d(x, y) + d(y, b) = d(x, y)$. On the other hand, we have the inequality $d(x, y) \leq d(x, a) + d(a, b) + d(b, y) = d(a, b) = \rho(a, b)$ which implies that $\rho(a, b) = d(x, y)$, i.e., $d(x, y) = \rho(\varphi(x), \varphi(y))$ for any $x, y \in X$.

Let $\mathcal{V}$ be the uniformity induced on $Y$ by the metric $\rho$. To see that the map $\varphi : (X, \mathcal{U}) \to (Y, \mathcal{V})$ is uniformly continuous take any $V \in \mathcal{V}$. There is $r > 0$ such that $W_r = \{(y, z) \in Y \times Y : \rho(y, z) < r\} \subset V$. It is straightforward that $(\varphi \times \varphi)^{-1}(W_r) \supset O_r = \{(x, y) \in X \times X : d(x, y) < r\}$. The pseudometric $d$ being uniformly continuous, we have $O_r \in \mathcal{U}$ (see Problem 107), so it follows from the equality $U = (\varphi \times \varphi)^{-1}(V) \supset O_r$ that $U \in \mathcal{U}$, so the map $\varphi$ is uniformly continuous and Fact 1 is proved.

Returning to our solution fix a uniform space $(X, \mathcal{U})$ and apply Problem 109 to construct, for any $U \in \mathcal{U}$, a uniformly continuous pseudometric $d_U$ on the space $(X, \mathcal{U})$ such that $O_U = \{(x, y) \in X \times X : d_U(x, y) < 1\} \subset U$. Fact 1 guarantees us that there exists a metric space $(Y_U, \rho_U)$ and a map $\varphi_U : X \to Y_U$ such that $d_U(x, y) = \rho_U(\varphi_U(x), \varphi_U(y))$ for any $x, y \in X$ and, in particular, $\varphi_U$ is uniformly continuous if we consider the set $Y$ with the uniformity $\mathcal{V}_U$ induced by $\rho_U$.

Let $Y = \prod\{Y_U : U \in \mathcal{U}\}$ be the uniform product of the family $\{Y_U : U \in \mathcal{U}\}$; since every $Y_U$ is uniformly metrizable, it suffices to construct a uniform isomorphism of $(X, \mathcal{U})$ onto a subspace of $Y$. For every $x \in X$ let $\varphi(x)(U) = \varphi_U(x)$ for each $U \in \mathcal{U}$; then $\varphi : X \to Y$ is the diagonal product of the family $\{\varphi_U : U \in \mathcal{U}\}$.

To show that the map $\varphi : X \to X' = \varphi(X)$ is a uniform isomorphism consider the natural projection $p_U : Y \to Y_U$ for each $U \in \mathcal{U}$. It is straightforward that $p_U \circ \varphi = \varphi_U$, so the map $p_U \circ \varphi$ is uniformly continuous for every $U \in \mathcal{U}$. It follows from Problem 106 that the map $\varphi$ is uniformly continuous.

Given distinct points $x, y \in X$ there is a set $U \in \mathcal{U}$ such that $(x, y) \notin U$ and hence $d_U(x, y) \geq 1$. Then $\rho_U(\varphi_U(x), \varphi_U(y)) \geq 1$ which implies that $\varphi_U(x) \neq \varphi_U(y)$, so $\varphi(x) \neq \varphi(x)$, i.e., the map $\varphi$ is a bijection.

To finally see that $\varphi^{-1} : X' \to X$ is uniformly continuous fix a set $U \in \mathcal{U}$. Then $W = \{(s, t) \in Y_U \times Y_U : \rho_U(s, t) < 1\}$ belongs to the uniformity $\mathcal{V}_U$ and hence the set $V = \{(a, b) \in X' \times X' : (p_U(a), p_U(b)) \in W\}$ is an element of the uniformity $\mathcal{V}'$ induced on $X'$ from $Y$. It is an easy exercise that $(\varphi^{-1} \times \varphi^{-1})(V)$ coincides with the set $O_U \subset U$, so $(\varphi^{-1} \times \varphi^{-1})^{-1}(U) = (\varphi \times \varphi)(U) \supset (\varphi \times \varphi)(O_U) = V$ belongs to $\mathcal{V}'$ and therefore the map $\varphi^{-1}$ is uniformly continuous. This proves that $\varphi : X \to X'$ is a uniform isomorphism and completes our solution.

**V.114.** *Given a uniform space $(X, \mathcal{U})$ prove that the following conditions are equivalent:*

  (i) *the space $(X, \mathcal{U})$ is complete;*
 (ii) *if $\mathcal{C}$ is a centered Cauchy family of closed subsets of $X$ then $\bigcap \mathcal{C} \neq \emptyset$;*
(iii) *if $\mathcal{C}$ is a centered Cauchy family of subsets of $X$ then $\bigcap\{\overline{C} : C \in \mathcal{C}\} \neq \emptyset$;*
 (iv) *if $\mathcal{B}$ is a Cauchy filter base on $X$ then $\bigcap\{\overline{B} : B \in \mathcal{B}\} \neq \emptyset$;*
  (v) *if $\mathcal{B}$ is a Cauchy filter base of closed subsets of $X$ then $\bigcap \mathcal{B} \neq \emptyset$;*
 (vi) *any Cauchy filter $\mathcal{F}$ on $X$ converges to a point $x \in X$, i.e., $\tau(x, X) \subset \mathcal{F}$;*
(vii) *if $\mathcal{E}$ is a Cauchy ultrafilter on $X$ then $\bigcap\{\overline{E} : E \in \mathcal{E}\} \neq \emptyset$;*
(viii) *if $\mathcal{E}$ is a Cauchy ultrafilter on $X$ then it converges to a point $x \in X$.*

**Solution.** Assume that $(X, \mathcal{U})$ is complete and fix a centered Cauchy family $\mathcal{C}$ of closed subsets of $X$. There exists a filter $\mathcal{F} \subset \exp(X)$ such that $\mathcal{C} \subset \mathcal{F}$ (see TFS-117). It is evident that $\mathcal{F}$ is still a Cauchy family, so $P = \bigcap\{\overline{F} : F \in \mathcal{F}\} \neq \emptyset$. Then $P \subset \bigcap\{\overline{C} : C \in \mathcal{C}\} = \bigcap \mathcal{C}$, so $\bigcap \mathcal{C} \neq \emptyset$, i.e., we proved that (i)$\Longrightarrow$(ii).

Now, if the property (ii) holds and $\mathcal{C} \subset \exp(X)$ is a centered Cauchy family then $\mathcal{C}' = \{\overline{C} : C \in \mathcal{C}\}$ is a centered family of closed subsets of $X$; take any $U \in \mathcal{U}$. By Problem 103, there is a closed set $V \subset X \times X$ such that $V \in \mathcal{U}$ and $V \subset U$. The family $\mathcal{C}$ being Cauchy there is $C \in \mathcal{C}$ with $C \times C \subset V$. An immediate consequence is that $\overline{C} \times \overline{C} \subset \overline{V} = V \subset U$; this proves that $C'$ is also a Cauchy family. Applying (ii) we conclude that $\bigcap C' \neq \emptyset$, i.e., (iii) holds and hence we settled (ii)$\Longrightarrow$(iii).

Every Cauchy filter base on $X$ is a centered Cauchy family of subsets of $X$, so (iii)$\Longrightarrow$(iv). If (iv) holds and $\mathcal{B}$ is a Cauchy filter base of closed subsets of $X$ then $\bigcap \mathcal{B} = \bigcap \{\overline{B} : B \in \mathcal{B}\} \neq \emptyset$ by (iv); this shows that (iv)$\Longrightarrow$(v).

Next assume that (v) holds and take a Cauchy filter $\mathcal{F}$ on the set $X$. Given $F, G \in \mathcal{F}$ we have $\overline{F \cap G} \subset \overline{F} \cap \overline{G}$; this implies that the family $\mathcal{F}' = \{\overline{F} : F \in \mathcal{F}\}$ is a filter base of closed subsets of $X$. Take any $U \in \mathcal{U}$; by Problem 103, there is a closed set $V \subset X \times X$ such that $V \in \mathcal{U}$ and $V \subset U$. The family $\mathcal{F}$ being Cauchy there is $F \in \mathcal{F}$ with $F \times F \subset V$. An immediate consequence is that $\overline{F} \times \overline{F} \subset \overline{V} = V \subset U$; this proves that $\mathcal{F}'$ is also a Cauchy family.

Thus we can pick a point $x \in \bigcap \mathcal{F}'$; we claim that the filter $\mathcal{F}$ converges to $x$. To see this take an arbitrary set $O \in \tau(x, X)$; there exists $U \in \mathcal{U}$ with $U(x) \subset O$. Apply Problem 103 to find a closed $V \subset X \times X$ such that $V \subset U$ and $V \in \mathcal{U}$. The family $\mathcal{F}$ being Cauchy, we can find $F \in \mathcal{F}$ with $F \times F \subset V$. Then $\overline{F} \times \overline{F} \subset \overline{V} = V \subset U$; given any $y \in F$ it follows from $\{x, y\} \subset \overline{F}$ that $(x, y) \in \overline{F} \times \overline{F} \subset U$, so $y \in U(x)$. Thus $F \subset U(x) \subset O$ and hence $O \in \mathcal{F}$; this proves that $\tau(x, X) \subset \mathcal{F}$ and settles the implication (v)$\Longrightarrow$(vi).

To establish (vi)$\Longrightarrow$(vii), assume that (vi) is true and take a Cauchy ultrafilter $\mathcal{E} \subset \exp(X)$. Since any ultrafilter is a filter, we can apply (vi) to find a point $x \in X$ such that $\tau(x, X) \subset \mathcal{E}$. If $x \notin \overline{E}$ for some $E \in \mathcal{E}$ then $O = X \backslash \overline{E} \in \tau(x, X)$ and hence $O \in \mathcal{E}$ which, together with $O \cap E = \emptyset$ gives a contradiction. Therefore $x \in \bigcap \{\overline{E} : E \in \mathcal{E}\}$, so the implication (vi)$\Longrightarrow$(vii) is proved.

For (vii)$\Longrightarrow$(viii) suppose that (vii) is fulfilled and take any Cauchy ultrafilter $\mathcal{E} \subset \exp(X)$; there exists a point $x \in \bigcap \{\overline{E} : E \in \mathcal{E}\}$. If $O \notin \mathcal{E}$ for some $O \in \tau(x, X)$ then $E = X \backslash O \in \mathcal{E}$ (see TFS-117), so $x \notin E = \overline{E}$ which is a contradiction. Thus the ultrafilter $\mathcal{E}$ converges to $x$ and (vii)$\Longrightarrow$(viii) is proved.

Finally assume that (viii) is satisfied and take any Cauchy filter $\mathcal{F}$ on the set $X$. There exists an ultrafilter $\mathcal{E} \subset \exp(X)$ with $\mathcal{F} \subset \mathcal{E}$ (see TFS-117); it is straightforward that $\mathcal{E}$ is also Cauchy, so it converges to a point $x \in X$. If there is $F \in \mathcal{F}$ with $x \notin \overline{F}$ then $O = X \backslash \overline{F} \in \tau(x, X)$ which shows that $O \in \mathcal{E}$ which, together with $F \in \mathcal{E}$ and $O \cap F = \emptyset$, gives a contradiction. Therefore $x \in \bigcap \{\overline{F} : F \in \mathcal{F}\}$; this proves that (viii)$\Longrightarrow$(i) and hence our solution is complete.

**V.115.** *Prove that any closed uniform subspace of a complete uniform space is complete.*

**Solution.** Suppose that $(X, \mathcal{U})$ is a complete uniform space and $F$ is a closed subspace of $X$; denote by $\mathcal{V}$ the uniformity induced on $F$ from $(X, \mathcal{U})$. Take any centered family $\mathcal{C}$ of closed subsets of $F$ which is Cauchy in $(F, \mathcal{V})$; all elements of $\mathcal{C}$ are also closed in $X$. If $U \in \mathcal{U}$ then $V = U \cap (F \times F) \in \mathcal{V}$; the family $\mathcal{C}$ being Cauchy in $(F, \mathcal{V})$ there is $C \in \mathcal{C}$ such that $C \times C \subset V \subset U$. An immediate consequence is that $\mathcal{C}$ is also a Cauchy family in $(X, \mathcal{U})$; thus $\bigcap \mathcal{C} \neq \emptyset$ and hence the space $(F, \mathcal{V})$ is complete by Problem 114.

**V.116.** *Let $(X, \mathcal{U})$ be a uniform space such that some $Y \subset X$ is complete considered as a uniform subspace of $(X, \mathcal{U})$. Prove that $Y$ is closed in $X$.*

**Solution.** Let $\mathcal{V}$ be the uniformity induced on $Y$ from $(X, \mathcal{U})$. If $Y$ is not closed in $X$ then fix a point $y \in \overline{Y} \backslash Y$ and let $\mathcal{B} = \{O' \cap Y : O' \in \tau(y, X)\}$. Obviously, $\mathcal{B}$ is a filter base of subsets of $Y$. Given a set $V \in \mathcal{V}$ fix $U \in \mathcal{U}$ such that $U \cap (Y \times Y) = V$ and choose a symmetric $W \in \mathcal{U}$ for which $W \circ W \subset U$.

Apply Problem 103 to find an open neighborhood $O'$ of the point $y$ in $X$ such that $O' \subset W(y)$; then $O = O' \cap Y \in \mathcal{B}$. If $z, t \in O$ then $\{(y, z), (y, t)\} \subset W$; the set $W$ being symmetric, we have $\{(z, y), (y, t)\} \subset W$ and therefore $(z, t) \in W \circ W \subset U$. This shows that $O \times O \subset U \cap (Y \times Y) = V$, so $\mathcal{B}$ is a Cauchy family in $(Y, \mathcal{V})$.

The space $(Y, \mathcal{V})$ is complete, so there exists a point $x \in \bigcap \{cl_Y(O) : O \in \mathcal{B}\}$; the points $x$ and $y$ are distinct, so we can choose disjoint sets $G, H \in \tau(X)$ such that $x \in G$ and $y \in H$. Then $O = G \cap Y \in \mathcal{B}$ and $x \notin cl_Y(O)$; this contradiction shows that $Y$ is closed in $X$.

**V.117.** *Prove that, for any family $\mathcal{A} = \{(X_t, \mathcal{U}_t) : t \in T\}$ of complete uniform spaces, the uniform product $(X, \mathcal{U})$ of the family $\mathcal{A}$ is complete.*

**Solution.** Let $p_t : X \to X_t$ be the natural projection for any $t \in T$. Suppose that $\mathcal{F}$ is a Cauchy filter on the space $(X, \mathcal{U})$. It is easy to verify that the family $\mathcal{F}_t = \{p_t(F) : F \in \mathcal{F}\}$ is a Cauchy filter on $(X_t, \mathcal{U}_t)$; by completeness of $(X_t, \mathcal{U}_t)$ we can apply Problem 114 to fix a point $a_t \in X_t$ with $\tau(a_t, X_t) \subset \mathcal{F}_t$ for every $t \in T$.

Define a point $x \in X$ by letting $x(t) = a_t$ for each $t \in T$ and assume that there exists a set $F \in \mathcal{F}$ with $x \notin \overline{F}$. Then there is a finite $S \subset T$ and a set $O_t \in \tau(a_t, X_t)$ for any $t \in S$ such that the set $O = \bigcap \{p_t^{-1}(O_t) : t \in S\}$ does not meet $F$. However, $O_t \in \mathcal{F}_t$, so there is $W_t \in \mathcal{F}$ such that $p_t(W_t) = O_t$; an immediate consequence is that $W_t \subset p_t^{-1}(O_t)$ and hence $p_t^{-1}(O_t) \in \mathcal{F}$ for any $t \in S$. Therefore $O$ and $F$ are two disjoint elements of $\mathcal{F}$; this contradiction shows that $x \in \bigcap \{\overline{F} : F \in \mathcal{F}\}$, so $\bigcap \{\overline{F} : F \in \mathcal{F}\} \neq \emptyset$ for any Cauchy filter $\mathcal{F}$ on $(X, \mathcal{U})$, i.e., the space $(X, \mathcal{U})$ is complete.

**V.118.** *Let $(X, \mathcal{U})$ be a uniform space such that the uniformity $\mathcal{U}$ is induced by a metric $\rho$. Prove that $(X, \mathcal{U})$ is complete if and only if the metric space $(X, \rho)$ is complete.*

**Solution.** Assume first that the space $(X, \mathcal{U})$ is complete and take a sequence $\mathcal{F} = \{F_n : n \in \omega\}$ of nonempty closed subsets of $X$ such that $F_{n+1} \subset F_n$ for any $n \in \omega$ and $diam(F_n) \to 0$. It is clear that $\mathcal{F}$ is a filter base in $X$. Given $U \in \mathcal{U}$ there exists $r > 0$ for which $O_r = \{(x, y) \in X \times X : \rho(x, y) < r\} \subset U$. Pick $n \in \omega$ such that $d = diam(F_n) < r$; then $\rho(x, y) \leq d < r$ for any $x, y \in F_n$ which implies that $F_n \times F_n \subset O_r \subset U$. Thus $\mathcal{F}$ is a Cauchy family; by completeness of $(X, \mathcal{U})$ we can apply Problem 114 to see that $\bigcap \mathcal{F} \neq \emptyset$. Therefore the metric space $(X, \rho)$ is complete by TFS-236; this proves necessity.

Now assume that the metric space $(X, \rho)$ is complete and take a centered Cauchy family $\mathcal{F}$ of closed subsets of $X$. Given any $r > 0$, the set $O_{r/2}$ belongs to $\mathcal{U}$, so there exists $F \in \mathcal{F}$ such that $F \times F \subset O_{r/2}$; then $d(x, y) < \frac{r}{2}$ for any $x, y \in F$ and hence $\mathrm{diam}(F) \leq \frac{r}{2} < r$. This shows that the family $\mathcal{F}$ has sets of arbitrarily small diameter, so we can apply TFS-236 once more to see that $\bigcap \mathcal{F} \neq \emptyset$. Finally, apply Problem 114 to conclude that $(X, \mathcal{U})$ is complete; this settles sufficiency.

**V.119.** *Let $A$ be a dense subspace of a uniform space $(X, \mathcal{U})$. Suppose that $f : A \to Y$ is a uniformly continuous map of $(A, \mathcal{U}_A^X)$ to a complete uniform space $(Y, \mathcal{V})$. Prove that there is a uniformly continuous map $F : X \to Y$ such that $F|A = f$.*

**Solution.** If $x \in X$ then the set $O \cap A$ is nonempty for any $O \in \tau(x, X)$, so the family $\mathcal{O}_x = \{O \cap A : O \in \tau(x, X)\}$ is a filter base of subsets of $A$. An easy consequence is that the family $\mathcal{U}_x = \{f(P) : P \in \mathcal{O}_x\}$ is a filter base of subsets of $Y$ for every $x \in X$.

Fix a point $x \in X$ and $V \in \mathcal{V}$; by uniform continuity of $f$ there is a set $U' \in \mathcal{U}$ such that $(f(a), f(b)) \in V$ whenever $(a, b) \in U' \cap (A \times A)$. Apply Problem 103 to find a symmetric $U \in \mathcal{U}$ for which $U \circ U \subset U'$. The set $O = \mathrm{Int}(U(x))$ is an open neighborhood of $x$ in $X$ by Problem 103, so $P = O \cap A \in \mathcal{O}_x$ and therefore $Q = f(P) \in \mathcal{U}_x$. Now if $z, t \in Q$ then pick $a, b \in P$ with $f(a) = z$, $f(b) = t$. Since $\{a, b\} \subset U(x)$, we have $\{(x, a), (x, b)\} \subset U$ and hence $(a, x) \in U$ by the symmetry of $U$. Consequently, $(a, b) \in U \circ U \subset U'$ which shows that $(z, t) = (f(a), f(b)) \in V$. The points $z, t \in Q$ were chosen arbitrarily, so $Q \times Q \subset V$ and hence $\mathcal{U}_x$ is a Cauchy filter base in $(Y, \mathcal{V})$. By Problem 114 there is a point $y \in \bigcap\{\overline{Q} : Q \in \mathcal{U}_x\}$; let $F(x) = y$. This defines a map $F : X \to Y$.

Given $a \in A$ observe that $f(a) \in Q$ for any $Q \in \mathcal{U}_a$, so $f(a) \in \bigcap\{\overline{Q} : Q \in \mathcal{U}_a\}$. If $z \in Y \setminus \{f(a)\}$ then there is $V \in \mathcal{V}$ with $(f(a), z) \notin V$; by Problem 103, we can find a closed $W \in \mathcal{V}$ such that $W \subset V$. The family $\mathcal{U}_a$ being Cauchy, there is $Q \in \mathcal{U}_a$ for which $Q \times Q \subset W$. The set $W$ being closed we also have $\overline{Q} \times \overline{Q} \subset W \subset V$ which shows that $(f(a), z) \notin \overline{Q} \times \overline{Q}$, i.e., $z \notin \overline{Q}$. Therefore $z \notin R = \bigcap\{\overline{Q} : Q \in \mathcal{U}_a\}$ and hence $f(a)$ is the unique element of $R$; this proves that $F(a) = f(a)$ for each $a \in A$, i.e., $F|A = f$.

To finally see that the map $F : (X, \mathcal{U}) \to (Y, \mathcal{V})$ is uniformly continuous fix any $V \in \mathcal{V}$ and apply Problem 103 to find a closed set $V_0 \in \mathcal{V}$ with $V_0 \subset V$. The map $f$ being uniformly continuous, we can find $U \in \mathcal{U}$ such that $(f(a), f(b)) \in V_0$ whenever $(a, b) \in U \cap (A \times A)$. Apply Problem 103 to find a symmetric set $W \in \mathcal{U}$ with $W \circ W \circ W \subset U$. Now, if $(x, y) \in W$ then $H = \mathrm{Int}(W(x)) \cap A \in \mathcal{O}_x$ and $G = \mathrm{Int}(W(y)) \cap A \in \mathcal{O}_y$, so $Q_0 = f(H) \in \mathcal{U}_x$ and $Q_1 = f(G) \in \mathcal{U}_y$. If $a \in H$ and $b \in G$ then $\{(a, x), (x, y), (y, b)\} \subset W$ which shows that $(a, b) \in W \circ W \circ W \subset U$ and therefore $(f(a), f(b)) \in V_0$. This proves that $Q_0 \times Q_1 \subset V_0$; the set $V_0$ being closed, we conclude that $\overline{Q}_0 \times \overline{Q}_1 \subset V_0 \subset V$. By definition of $F$, we have $F(x) \in \overline{Q}_0$ and $F(y) \in \overline{Q}_1$. Consequently, $(F(x), F(y)) \in \overline{Q}_0 \times \overline{Q}_1 \subset V$; the point $(x, y) \in W$ was chosen arbitrarily so $(F \times F)(W) \subset V$ and hence $F$ is a uniformly continuous map.

**V.120.** *Let $(X, \mathcal{U})$ and $(Y, \mathcal{V})$ be complete uniform spaces. Suppose that $A$ is a dense subspace of $X$ and $B$ is a dense subspace of $Y$. Prove that every uniform isomorphism between the uniform spaces $(A, \mathcal{U}_A^X)$ and $(B, \mathcal{V}_B^Y)$ is extendable to a uniform isomorphism between $(X, \mathcal{U})$ and $(Y, \mathcal{V})$.*

**Solution.** Let $\mathrm{id}_X : X \to X$ and $\mathrm{id}_Y : Y \to Y$ be the identity maps on the spaces $X$ and $Y$ respectively and assume that a map $\varphi : (A, \mathcal{U}_A^X) \to (B, \mathcal{V}_B^Y)$ is a uniform isomorphism. It is straightforward that the maps $\varphi : (A, \mathcal{U}_A^X) \to (Y, \mathcal{V})$ and $\varphi^{-1} : (B, \mathcal{V}_B^Y) \to (X, \mathcal{U})$ are uniformly continuous, so we can apply Problem 119 to find uniformly continuous maps $\Phi : (X, \mathcal{U}) \to (Y, \mathcal{V})$ and $\Psi : (Y, \mathcal{V}) \to (X, \mathcal{U})$ such that $\Phi|A = \varphi$ and $\Psi|B = \varphi^{-1}$.

Since $(\Phi \circ \Psi)(y) = \varphi(\varphi^{-1}(y)) = y$ for any $y \in B$, the continuous maps $\mathrm{id}_Y$ and $\Phi \circ \Psi$ coincide on a dense set $B$ of the space $Y$. By Fact 0 of S.351, we have $\Phi \circ \Psi = \mathrm{id}_Y$. Analogously, $(\Psi \circ \Phi)(x) = \varphi^{-1}(\varphi(x)) = x$ for any $x \in A$, i.e., the continuous maps $\Psi \circ \Phi$ and $\mathrm{id}_X$ coincide on a dense subset $A$ of the space $X$. Therefore $\Psi \circ \Phi = \mathrm{id}_X$ which shows that $\Psi = \Phi^{-1}$ and hence both maps $\Phi$ and $\Psi$ are uniform isomorphisms.

**V.121.** *Prove that every uniform space $(X, \mathcal{U})$ is uniformly isomorphic to a dense subspace of a complete uniform space $(X^*, \mathcal{U}^*)$. The space $(X^*, \mathcal{U}^*)$ is called the completion of the space $(X, \mathcal{U})$. Prove that the completion of $(X, \mathcal{U})$ is unique in the sense that, if $(Y, \mathcal{V})$ is a complete uniform space such that $(X, \mathcal{U})$ is a dense uniform subspace of $(Y, \mathcal{V})$ then there is a uniform isomorphism $\Phi : X^* \to Y$ such that $\Phi(x) = x$ for any $x \in X$.*

**Solution.** By Problem 113, we can assume that $X$ is a uniform subspace of the uniform product $(M, \mathcal{U}) = \prod_{t \in T}(M_t, \mathcal{U}_t)$ of some family $\{(M_t, \mathcal{U}_t) : t \in T\}$ of uniformly metrizable spaces. Let $\rho_t : M_t \times M_t \to \mathbb{R}$ be a metric which induces the uniformity $\mathcal{U}_t$ for every $t \in T$. It follows from TFS-237 that there exists a complete metric space $(N_t, d_t)$ such that $M_t$ is a dense subspace of $N_t$ and $d_t|(M_t \times M_t) = \rho_t$ for each $t \in T$.

If $\mathcal{W}_t$ is the uniformity on $N_t$ induced by the metric $d_t$ then it is easy to see that the uniformity induced by $\mathcal{W}_t$ on $M_t$ coincides with $\mathcal{U}_t$ for any $t \in T$. Another easy exercise is that if $(N, \mathcal{W}) = \prod_{t \in T}(N_t, \mathcal{W}_t)$ is the uniform product of the family $\{(N_t, \mathcal{W}_t) : t \in T\}$ then the uniformity $\mathcal{W}$ induces the uniformity $\mathcal{U}$ on the set $M$.

Thus $(X, \mathcal{U})$ is a uniform subspace of the space $(N, \mathcal{W})$; apply Problem 117 to see that the space $(N, \mathcal{W})$ is complete. Let $X^*$ be the closure of $X$ in $N$ and denote by $\mathcal{U}^*$ the uniformity induced on $X^*$ from $(N, \mathcal{W})$. Clearly, $(X, \mathcal{U})$ is a dense uniform subspace of the space $(X^*, \mathcal{U}^*)$; it follows from Problem 115 that $(X^*, \mathcal{U}^*)$ is complete.

Finally, assume that $(X, \mathcal{U})$ is a dense uniform subspace of some complete uniform space $(Y, \mathcal{V})$. Letting $\varphi(x) = x$ for any $x \in X$ we obtain a uniform isomorphism of $(X, \mathcal{U})$ onto itself. Considering $\varphi$ to be an isomorphism between the respective dense subspaces of $(X^*, \mathcal{U}^*)$ and $(Y, \mathcal{V})$ (both of which coincide with $X$), we can apply Problem 120 to find a uniform isomorphism $\Phi : (X^*, \mathcal{U}^*) \to (Y, \mathcal{V})$ such that $\Phi|X = \varphi$, i.e., $\Phi(x) = x$ for every $x \in X$.

**V.122.** *Let $X$ be a paracompact Tychonoff space. Prove that the family of all neighborhoods of the diagonal of $X$ is a uniformity on $X$ which generates $\tau(X)$.*

**Solution.** Let $\mathcal{U}$ be the family of all neighborhoods of the diagonal $\Delta$ of the space $X$. Since $X$ is Tychonoff, any point $(x, y)$ is closed in $X \times X$, so the set $(X \times X) \backslash \{(x, y)\}$ belongs to $\mathcal{U}$ for any $(x, y) \in (X \times X) \backslash \Delta$. Thus $\Delta = \bigcap \mathcal{U}$.

The map $\varphi : X \times X \to X \times X$ defined by $\varphi(x, y) = (y, x)$ for any $(x, y) \in X \times X$ is easily seen to be a homeomorphism for which $\varphi(\Delta) = \Delta$. Therefore, for any $U \in \mathcal{U}$, the set $\varphi(U) = U^{-1}$ is still a neighborhood of $\Delta$, so $U^{-1} \in \mathcal{U}$. It is evident that the intersection of two neighborhoods of $\Delta$ is still a neighborhood of $\Delta$, so the axiom (U1) is fulfilled for $\mathcal{U}$.

If $U \in \mathcal{U}$ then $U$ is a neighborhood of $\Delta$, so every $W \supset U$ is also a neighborhood of $\Delta$, i.e., $W \in \mathcal{U}$. To check the second part of (U2) fix a set $U \in \mathcal{U}$; then $U_0 = \text{Int}(U) \supset \Delta$. For each $x \in X$ fix a set $O_x \in \tau(x, X)$ such that $O_x \times O_x \subset U_0$; then $\mathcal{O} = \{O_x : x \in X\}$ is an open cover of the space $X$. By paracompactness of $X$ we can find an open star refinement $\mathcal{H}$ of the cover $\mathcal{O}$ (see TFS-230). The set $V = \bigcup \{H \times H : H \in \mathcal{H}\}$ is an open neighborhood of $\Delta$, so $V \in \mathcal{U}$.

Next, assume that $\{(a, b), (b, c)\} \subset V$ for some $a, b, c \in X$; then there are $H, G \in \mathcal{H}$ with $(a, b) \in H \times H$ and $(b, c) \in G \times G$. This implies $\{a, b\} \subset H$ and $\{b, c\} \subset G$ which shows that $b \in H \cap G$; by our choice of the cover $\mathcal{H}$ there is $x \in X$ for which $G \cup H \subset O_x$. Thus $(a, c) \in O_x \times O_x \subset U_0 \subset U$; this proves that $V \circ V \subset U_0 \subset U$, so the property (U2) also holds for $\mathcal{U}$. Therefore the family $\mathcal{U}$ of all neighborhoods of the diagonal of $X$ is a uniformity on $X$.

To see that $\mathcal{U}$ generates the topology of $X$ take any $U \in \mathcal{U}$ and fix $U_0 \in \tau(X \times X)$ with $\Delta \subset U_0 \subset U$. Given any $x \in X$ the set $O = U_0 \cap (\{x\} \times X)$ is open in $\{x\} \times X$; there is an evident homeomorphism between $\{x\} \times X$ and $X$ which takes $O$ onto $U_0(x)$, so $U_0(x) \subset U(x)$ is open in $X$. Therefore $U(x)$ is a neighborhood of $x$ for any $x \in X$ and $U \in \mathcal{U}$. If $H \in \tau_\mathcal{U}$ then, for every $x \in H$, there is $U \in \mathcal{U}$ with $U(x) \subset H$; we already saw that every $U(x)$ is a neighborhood of $x$, so $H$ is open in $X$ and hence we proved that $\tau_\mathcal{U} \subset \tau(X)$.

Finally, if $O \in \tau(X)$ and $x \in O$ then pick a set $W \in \tau(x, X)$ with $\overline{W} \subset O$ and let $G = X \backslash \overline{W}$; the set $U = (O \times O) \cup (G \times G)$ is an open neighborhood of the diagonal, i.e., $U \in \mathcal{U}$. If $(x, y) \in U$ then $(x, y) \notin G \times G$, so $y \in O$ and hence $U(x) \subset O$. This shows that $O \in \tau_\mathcal{U}$; since we have chosen $O \in \tau(X)$ arbitrarily, we proved that $\tau(X) \subset \tau_\mathcal{U}$ and hence $\tau(X) = \tau_\mathcal{U}$, i.e., the topology generated by $\mathcal{U}$ coincides with $\tau(X)$.

**V.123.** *Suppose that $X$ is a Tychonoff space such that the family of all neighborhoods of the diagonal of $X$ is a uniformity on $X$ which generates $\tau(X)$. Prove that $X$ is collectionwise normal.*

**Solution.** Let $\mathcal{U}$ be the family of all neighborhoods of the diagonal $\Delta$ of the space $X$. Given a discrete family $\mathcal{F}$ of closed subsets of $X$, we can fix a set $O_x \in \tau(x, X)$ which intersects at most one element of $\mathcal{F}$ for every point $x \in X$. It is clear that $U = \bigcup \{O_x \times O_x : x \in X\}$ is an element of $\mathcal{U}$, so we can find a symmetric set $V \in \mathcal{U}$ with $V \circ V \subset U$.

For every $F \in \mathcal{F}$, apply Problem 103 to see that $O_F = \bigcup \{V(x) : x \in F\}$ is a neighborhood of the set $F$. Suppose that $F$ and $G$ are distinct elements of $\mathcal{F}$ and $a \in O_F \cap O_G$. There are $y \in F$ and $z \in G$ such that $a \in V(y) \cap V(z)$; it follows from $\{(y,a),(a,z)\} \subset V$ that $(y,z) \in V \circ V \subset U$. As a consequence, there exists $x \in X$ with $(y,z) \in O_x \times O_x$ and hence $\{y,z\} \subset O_x$, i.e., $O_x$ intersects both $F$ and $G$ which is a contradiction. Thus the family $\{\mathrm{Int}(O_F) : F \in \mathcal{F}\}$ is disjoint, so we can apply Fact 1 of S.302 to conclude that $X$ is collectionwise normal.

**V.124.** *Give an example of a Tychonoff countably compact non-compact (and hence non-paracompact) space $X$ such that the family of all neighborhoods of the diagonal of $X$ is a uniformity on $X$ which generates $\tau(X)$.*

**Solution.** Our space $X$ will be the ordinal $\omega_1$ with its order topology; given ordinals $\alpha, \beta \le \omega_1$ let $(\alpha, \beta] = \{\gamma : \alpha < \gamma \le \beta\}$. We will prove that the family

$$\mathcal{O} = \{O \subset X \times X : O \text{ is a neighborhood of } \Delta_X \text{ in } X \times X\}$$

is a uniformity on $X$ which generates $\tau(X)$. Observe first that the space $X$ is countably compact and non-compact being a proper dense subspace of the space $K = \omega_1 + 1$ (see TFS-314). The space $K$ being compact, it follows from Problem 122 that the family $\mathcal{V}$ of all neighborhoods of the diagonal $\Delta_K$ in the space $K \times K$ is a uniformity on $K$ which generates $\tau(K)$. Therefore $\mathcal{U} = \{V \cap (X \times X) : V \in \mathcal{V}\}$ is a uniformity on $X$ which generates the topology of $X$. It is evident that every $U \in \mathcal{U}$ is a neighborhood of the diagonal $\Delta_X$ in the space $X \times X$, i.e., $\mathcal{U} \subset \mathcal{O}$.

To see that $\mathcal{O} \subset \mathcal{U}$ fix any $O \in \mathcal{O}$. For any limit ordinal $\alpha \in X$ there is $\varphi(\alpha) < \alpha$ such that the set $W_\alpha = (\varphi(\alpha), \alpha] \times (\varphi(\alpha), \alpha]$ is contained in $O$; let $W_\alpha = \{(\alpha, \alpha)\}$ for any successor ordinal $\alpha$. By Pressing-Down Lemma (SFFS-067) there is $\gamma < \omega_1$ and an uncountable $A \subset X$ such that $\varphi(\alpha) = \gamma$ for every $\alpha \in A$.

The set $H = \bigcup\{W_\alpha : \alpha \le \gamma\}$ is open in $K$ while $G = \bigcup\{W_\alpha : \alpha \in A\}$ is easily seen to coincide with $((\gamma, \omega_1] \times (\gamma, \omega_1]) \cap (X \times X)$. The set $V = H \cup ((\gamma, \omega_1] \times (\gamma, \omega_1])$ belongs to the family $\mathcal{V}$ being a neighborhood of the set $\Delta_K$ in the space $K \times K$. Since $V \cap (X \times X) \subset H \cup G \subset O$, we conclude that $O \in \mathcal{U}$; this proves that $\mathcal{O} = \mathcal{U}$ and hence the family $\mathcal{O}$ of all neighborhoods of $\Delta_X$ in $X \times X$ is a uniformity which generates $\tau(X)$.

**V.125.** *Prove that, for any compact uniform space $(X, \mathcal{U})$, the uniformity $\mathcal{U}$ coincides with the family of all neighborhoods of the diagonal of $X$.*

**Solution.** Let $\mathcal{O}$ be the family of all neighborhoods in $X \times X$ of the diagonal $\Delta$ of the space $X$. It follows from Problem 103 that every $U \in \mathcal{U}$ is a neighborhood of $\Delta$, i.e., $\mathcal{U} \subset \mathcal{O}$.

To see that $\mathcal{O} \subset \mathcal{U}$ fix a set $O \in \mathcal{O}$. Apply Problem 103 again to observe that the family $\mathcal{F} = \{U \in \mathcal{U} : U \text{ is closed in } X \times X\}$ is a base of $\mathcal{U}$ and therefore $\bigcap \mathcal{F} = \Delta$. Since all elements of $\mathcal{F}$ are compact, we can apply Fact 1 of S.326 to conclude that there is a finite $\mathcal{F}' \subset \mathcal{F}$ with $U = \bigcap \mathcal{F}' \subset O$. The set $U \subset O$ is an element of $\mathcal{U}$, so $O \in \mathcal{U}$; this proves that $\mathcal{O} \subset \mathcal{U}$ and hence $\mathcal{O} = \mathcal{U}$, i.e., the uniformity $\mathcal{U}$ coincides with the family of all neighborhoods of $\Delta$ in $X \times X$.

**V.126.** *Suppose that $(X, \mathcal{U})$ a compact uniform space. Prove that, for any uniform space $(Y, \mathcal{V})$, any continuous map $f : X \to Y$ is uniformly continuous.*

**Solution.** Denote by $\Delta_X$ and $\Delta_Y$ the diagonals of the spaces $X$ and $Y$ respectively. Fix a set $V \in \mathcal{V}$; by Problem 103 we can find $W \in \tau(\Delta_Y, Y \times Y)$ with $W \subset V$. Since $f$ is continuous, the map $f \times f$ is continuous as well, so the set $U = (f \times f)^{-1}(W)$ is open in $X \times X$ and $\Delta_X \subset U$. The space $X$ being compact, we can apply Problem 125 to see that $U \in \mathcal{U}$; it follows from $U \subset (f \times f)^{-1}(V)$ that $(f \times f)^{-1}(V) \in \mathcal{U}$, so the map $f$ is uniformly continuous.

**V.127.** *Let $(X, \mathcal{U})$ be a uniform space such that the uniformity $\mathcal{U}$ is induced by a metric $\rho$. Prove that $(X, \mathcal{U})$ is totally bounded if and only if the metric space $(X, \rho)$ is totally bounded.*

**Solution.** Assume first that the space $(X, \mathcal{U})$ is totally bounded; given $\varepsilon > 0$, the set $U = \{(x, y) \in X \times X : \rho(x, y) < \varepsilon\}$ belongs to $\mathcal{U}$, so there is a finite set $P \subset X$ such that $U(P) = X$. Thus, for any $x \in X$, there is $p \in P$ with $x \in U(p)$ and hence $d(x, p) < \varepsilon$. Therefore $\bigcup\{B_\rho(p, \varepsilon) : p \in P\} = X$ which shows that $(X, \rho)$ is totally bounded, i.e., we established necessity.

Now, if $(X, \rho)$ is totally bounded, then take any $U \in \mathcal{U}$; there is $\varepsilon > 0$ such that $V = \{(x, y) \in X \times X : \rho(x, y) < \varepsilon\} \subset U$. There exists a finite set $P \subset X$ for which $\bigcup\{B_\rho(p, \varepsilon) : p \in P\} = X$. Given any point $x \in X$ there is $p \in P$ with $\rho(x, p) < \varepsilon$, so $(p, x) \in V \subset U$ and hence $x \in U(p)$. This proves that $X = U(P)$ and hence $(X, \mathcal{U})$ is totally bounded, i.e., we settled sufficiency.

**V.128.** *Prove that a uniform space $(X, \mathcal{U})$ is totally bounded if and only if every ultrafilter on $X$ is a Cauchy family with respect to $\mathcal{U}$.*

**Solution.** Suppose first that $(X, \mathcal{U})$ is totally bounded and take an ultrafilter $\mathcal{F}$ on the set $X$. Fix a set $U \in \mathcal{U}$ and apply Problem 103 to find a symmetric $V \in \mathcal{U}$ such that $V \circ V \subset U$. There exists a finite set $A \subset X$ with $V(A) = X$.

Given a point $x \in A$ with $V(x) \notin \mathcal{F}$ we have $X \backslash V(x) \in \mathcal{F}$, so if $V(x) \notin \mathcal{F}$ for every $x \in A$ then $\emptyset = \bigcap\{X \backslash V(x) : x \in A\} \in \mathcal{F}$ which is a contradiction. Therefore $V(a) \in \mathcal{F}$ for some $a \in A$. The set $V$ being symmetric, for any points $x, y \in V(a)$ we have $\{(x, a), (a, y)\} \subset V$ and hence $(x, y) \in V \circ V \subset U$; this shows that $V(a) \times V(a) \subset U$. Recalling that $V(a) \in \mathcal{F}$ we conclude that $\mathcal{F}$ is a Cauchy ultrafilter on $X$, i.e., we proved necessity.

Now assume that every ultrafilter on $X$ is a Cauchy family with respect to $\mathcal{U}$ and fix a set $U \in \mathcal{U}$. If $U(A) \neq X$ for any finite $A \subset X$ then the family $\mathcal{E} = \{X \backslash U(x) : x \in X\}$ is centered, so there exists an ultrafilter $\mathcal{F}$ on the set $X$ such that $\mathcal{E} \subset \mathcal{F}$ (see TFS-117). By our assumption, $\mathcal{F}$ is a Cauchy family, so there is $F \in \mathcal{F}$ with $F \times F \subset U$.

Fix an arbitrary point $a \in F$ and observe that, for any $x \in F$, it follows from $(a, x) \in F \times F \subset U$ that $x \in U(a)$; this proves that $F \subset U(a)$. An immediate consequence is that $U(a) \in \mathcal{F}$; since also $X \backslash U(a) \in \mathcal{F}$ by our choice of $\mathcal{F}$, we obtained a contradiction which shows that $U(A) = X$ for some finite $A \subset X$ and hence $(X, \mathcal{U})$ is totally bounded. This settles sufficiency.

**V.129.** *Prove that any uniform product of totally bounded uniform spaces is totally bounded.*

**Solution.** Suppose that a uniform space $(X_t, \mathcal{U}_t)$ is totally bounded for any $t \in T$ and let $(X, \mathcal{U})$ be the uniform product of the family $\{(X_t, \mathcal{U}_t) : t \in T\}$. Then $X = \prod_{t \in T} X_t$; let $p_t : X \to X_t$ be the natural projection for any $t \in T$.

Fix an ultrafilter $\mathcal{F}$ on the set $X$ and any $U \in \mathcal{U}$; by definition of the uniform product, there is a finite $S \subset T$ and a family $\{W_t : t \in S\}$ such that $W_t \in \mathcal{U}_t$ for each $t \in S$ and $\bigcap \{(p_t \times p_t)^{-1}(W_t) : t \in S\} \subset U$.

It is straightforward that $\mathcal{F}_t = \{p_t(F) : F \in \mathcal{F}\}$ is an ultrafilter on the set $X_t$; the space $(X_t, \mathcal{U}_t)$ being totally bounded, $\mathcal{F}_t$ is a Cauchy family with respect to $\mathcal{U}_t$ for every $t \in S$. Consequently, there is $F_t \in \mathcal{F}$ such that $p_t(F_t) \times p_t(F_t) \subset W_t$ for each $t \in S$. The set $F = \bigcap_{t \in S} F_t$ belongs to $\mathcal{F}$; take any $x, y \in F$.

We have $p_t(F) \times p_t(F) \subset p_t(F_t) \times p_t(F_t) \subset W_t$, so $(p_t(x), p_t(y)) \in W_t$ for every $t \in S$ and therefore $(x, y) \in \bigcap \{(p_t \times p_t)^{-1}(W_t) : t \in S\} \subset U$. The points $x, y \in F$ were taken arbitrarily, so we proved that $F \times F \subset U$ and hence $\mathcal{F}$ is a Cauchy ultrafilter on $X$. Finally apply Problem 128 to conclude that $(X, \mathcal{U})$ is totally bounded.

**V.130.** *Prove that a uniform space is compact if and only if it is complete and totally bounded. Deduce from this fact that a uniform space is totally bounded if and only if its completion is compact.*

**Solution.** If a uniform space $(X, \mathcal{U})$ is compact then each centered family of closed subsets of $X$ (which does not even need to be a Cauchy family) has a nonempty intersection; this, together with Problem 114 implies that $X$ is complete. Given any $U \in \mathcal{U}$, the family $\{\text{Int}(U(x)) : x \in X\}$ is an open cover of $X$ by Problem 103. Therefore there is a finite $A \subset X$ with $X = \bigcup \{\text{Int}(U(x)) : x \in A\}$; an immediate consequence is that $U(A) = X$ and hence $X$ is totally bounded, i.e., we proved necessity.

Now assume that a uniform space $(X, \mathcal{U})$ is complete and totally bounded. Given an ultrafilter $\mathcal{F}$ on the set $X$ it is a Cauchy family by Problem 128; by completeness of $(X, \mathcal{U})$ we have $\bigcap \{\overline{F} : F \in \mathcal{F}\} \neq \emptyset$ (see Problem 114). Finally, apply TFS-118 to conclude that $X$ is compact; this settles sufficiency and shows that a uniform space is compact if and only if it is complete and totally bounded.

Next suppose that $(X, \mathcal{U})$ is a uniform space such that its completion $(X^*, \mathcal{U}^*)$ is compact. Fix an arbitrary ultrafilter $\mathcal{F}$ on the set $X$ and $U \in \mathcal{U}$. Since $\mathcal{F}$ can be considered a filter base on $X^*$, there exists an ultrafilter $\mathcal{F}^*$ on the set $X^*$ such that $\mathcal{F} \subset \mathcal{F}^*$. Choose a set $V \in \mathcal{U}^*$ with $V \cap (X \times X) = U$. The uniform space $(X^*, \mathcal{U}^*)$ being totally bounded, the ultrafilter $\mathcal{F}^*$ is a Cauchy family by Problem 128, so there is $F^* \in \mathcal{F}^*$ with $F^* \times F^* \subset V$. Observe that $X \in \mathcal{F}^*$, so $F = F^* \cap X \in \mathcal{F}^*$. If $F \notin \mathcal{F}$ then $G = X \setminus F \in \mathcal{F}$ and hence $G \in \mathcal{F}^*$, so $F$ and $G$ are two disjoint elements of $\mathcal{F}^*$. This contradiction shows that $F \in \mathcal{F}$; besides, $F \times F \subset (F^* \times F^*) \cap (X \times X) \subset V \cap (X \times X) = U$. Thus every ultrafilter on $X$ is a Cauchy family and hence $(X, \mathcal{U})$ is totally bounded by Problem 128.

Finally assume that $(X, \mathcal{U})$ is totally bounded and let $(X^*, \mathcal{U}^*)$ be the completion of $(X, \mathcal{U})$. Fix an arbitrary set $V \in \mathcal{U}^*$ and apply Problem 103 to find a closed $W \in \mathcal{U}^*$ with $W \subset V$. The set $G = W \cap (X \times X)$ belongs to $\mathcal{U}$, so there is a finite $A \subset X$ such that $G(A) = X$.

For any $a \in A$ the set $P_a = W \cap (\{a\} \times X^*)$ is closed in $\{a\} \times X^*$; an evident homeomorphism from $\{a\} \times X^*$ onto $X^*$ takes $P_a$ onto $W(a)$, so $W(a)$ is closed in $X^*$ for any $a \in A$. Thus $W(A) = \bigcup \{W(a) : a \in A\}$ is a closed subset of $X^*$ such that $X = G(A) \subset W(A)$. The set $X$ being dense in $X^*$, we have $W(A) = X^*$; this proves that $(X^*, \mathcal{U}^*)$ is totally bounded. Since $(X^*, \mathcal{U}^*)$ is also complete, it has to be compact, so we have finally established that a space $(X, \mathcal{U})$ is totally bounded if and only if its completion $(X^*, \mathcal{U}^*)$ is compact.

**V.131.** *Prove that a Tychonoff space $X$ is pseudocompact if and only if every uniformity $\mathcal{U}$ on the set $X$ with $\tau_{\mathcal{U}} = \tau(X)$ is totally bounded.*

**Solution.** Suppose that $X$ is pseudocompact and some uniformity $\mathcal{U}$ generates the topology of $X$. Given $U \in \mathcal{U}$ suppose that $U(A) \neq X$ for any finite $A \subset X$ and find a symmetric set $V \in \mathcal{U}$ such that $V \circ V \circ V \circ V \subset U$ (see Problem 103). Construct by induction a set $Y = \{x_n : n \in \omega\} \subset X$ such that $x_{n+1} \notin \bigcup \{U(x_i) : i \leq n\}$ for every $n \in \omega$.

To see that the family $\{V(x_i) : i \in \omega\}$ is discrete fix a point $x \in X$ and suppose that there are distinct $i, j \in \omega$ such that $V(x) \cap V(x_i) \neq \emptyset$ and $V(x) \cap V(x_j) \neq \emptyset$; we can assume, without loss of generality, that $i < j$. If $y \in V(x) \cap V(x_i)$ and $z \in V(x) \cap V(x_j)$ then $\{(x_i, y), (y, x), (x, z), (z, x_j)\} \subset V$ (we used the fact that $V$ is symmetric). Thus $(x_i, x_j) \in V \circ V \circ V \circ V \subset U$, i.e., $x_j \in U(x_i)$ which is a contradiction with the choice of the set $Y$. This shows that $\operatorname{Int}(V(x))$ is an open neighborhood of the point $x$ (see Problem 103) which intersects at most one element of the family $\{V(x_i) : i \in \omega\}$. Therefore $\{\operatorname{Int}(V(x_i)) : i \in \omega\}$ is an infinite discrete family of nonempty open subsets of $X$. This contradiction with pseudocompactness of $X$ proves that there is a finite $A \subset X$ with $U(A) = X$, i.e., $(X, \mathcal{U})$ is totally bounded and hence we established necessity.

For sufficiency, assume that every uniformity on the set $X$ is totally bounded whenever it generates the topology of $X$. The space $X$ being Tychonoff, we can fix a uniformity $\mathcal{U}$ on the set $X$ such that $\tau_{\mathcal{U}} = \tau(X)$. If $X$ is not pseudocompact then there exists a continuous unbounded function $f : X \to \mathbb{R}$; consider the set $O_r = \{(x, y) \in X \times X : |f(x) - f(y)| < r\}$ for each $r > 0$. It is clear that $O_r = O_r^{-1}$ and $O_r \circ O_r \subset O_{2r}$ for any $r > 0$. An easy consequence is that the family $\mathcal{U} \cup \{O_r : r > 0\}$ generates a uniformity $\mathcal{V}$ on $X$ as a subbase (see Problem 102); let $\mathcal{O} = \{O_r : r > 0\}$.

Given a set $W \in \tau_{\mathcal{V}}$ fix a point $x \in W$; there is $V \in \mathcal{V}$ with $V(x) \subset W$. There exist finite families $\mathcal{U}' \subset \mathcal{U}$ and $\mathcal{O}' \subset \mathcal{O}$ such that $(\bigcap \mathcal{U}') \cap (\bigcap \mathcal{O}) \subset V$. It is clear that $U = \bigcap \mathcal{U}' \in \mathcal{U}$ and $O = \bigcap \mathcal{O}' \in \mathcal{O}$, so fix $r > 0$ such that $O = O_r$. It follows from continuity of the function $f$ that the set $O_r(x)$ is open in $X$; thus the set $\operatorname{Int}(U(x)) \cap O_r(x)$ is an open neighborhood of $x$ contained in $V(x) \subset W$. The point $x \in W$ was chosen arbitrarily, so we proved that every point of $W$ has a neighborhood contained in $W$; thus $W$ is open in $X$, i.e., $\tau_{\mathcal{V}} \subset \tau(X)$. It follows from $\mathcal{U} \subset \mathcal{V}$ that $\tau(X) = \tau_{\mathcal{U}} \subset \tau_{\mathcal{V}}$, so we have $\tau_{\mathcal{V}} = \tau(X)$ and therefore the space $(X, \mathcal{V})$ is totally bounded.

Since $O_1 \in \mathcal{V}$, there exists a finite set $A \subset X$ such that $O_1(A) = X$; let $q = \max\{|f(x)| : x \in A\} + 2$. Given any $y \in X$ there is $x \in A$ with $y \in O_1(x)$ and hence $|f(x) - f(y)| < 1$; an immediate consequence is that $|f(y)| \le |f(x)| + 1 \le q$. It turns out that $|f(y)| \le q$ for any $y \in X$, i.e., the function $f$ is bounded. This contradiction shows that $X$ is pseudocompact, so we settled sufficiency.

**V.132.** *For any Tychonoff space $X$ let $\mathbb{U}_X$ be the family of all uniformities on the set $X$ which generate $\tau(X)$. Note that $\bigcup \mathbb{U}_X$ can be considered a subbase of a uniformity $\mathcal{N}_X$ (called the universal uniformity) on the set $X$. Prove that*

(i)  *the topology generated by $\mathcal{N}_X$ coincides with $\tau(X)$ and hence $\mathcal{N}_X \in \mathbb{U}_X$;*
(ii)  *if $Y$ is a Tychonoff space and $f : X \to Y$ is a continuous map then the map $f : (X, \mathcal{N}_X) \to (Y, \mathcal{V})$ is uniformly continuous for any uniformity $\mathcal{V} \in \mathbb{U}_Y$.*

**Solution.** It is straightforward that all conditions of Problem 102 are satisfied for the family $\bigcup \mathbb{U}_X$, so it can, indeed, be considered a subbase for a uniformity $\mathcal{N}_X$ on the set $X$. Apply Problem 110 to fix a uniformity $\mathcal{U}$ on the set $X$ such that $\tau_{\mathcal{U}} = \tau(X)$. It follows from $\mathcal{U} \subset \mathcal{N}_X$ that $\tau(X) = \tau_{\mathcal{U}} \subset \tau_{\mathcal{N}_X}$.

To prove the opposite inclusion take any $O \in \tau_{\mathcal{N}_X}$ and a point $x \in O$; there is $N \in \mathcal{N}_X$ with $N(x) \subset O$. The family $\bigcup \mathbb{U}_X$ being a subbase of $\mathcal{N}_X$ we can choose $\mathcal{U}_1, \ldots, \mathcal{U}_n \in \mathbb{U}_X$ and a set $U_i \in \mathcal{U}_i$ for each $i \le n$ in such a way that $U = U_1 \cap \ldots \cap U_n \subset N$. It follows from Problem 103 that $x \in W_i = \mathrm{Int}(U_i(x))$ for every $i \le n$, so $W = W_1 \cap \ldots \cap W_n \in \tau(x, X)$; since also $W \subset N(x) \subset O$, we proved that every point of $O$ has a neighborhood contained in $O$, i.e., $O$ is open in $X$ and hence $\tau_{\mathcal{N}_X} = \tau(X)$ which shows that we have verified (i).

Now fix a space $Y$, a continuous map $f : X \to Y$ and a uniformity $\mathcal{V}$ on $Y$ such that $\tau_{\mathcal{V}} = \tau(Y)$. Consider the family $\mathcal{V}' = \{(f \times f)^{-1}(V) : V \in \mathcal{V}\}$. Given $W = (f \times f)^{-1}(V) \in \mathcal{V}'$ there is a symmetric $H \in \mathcal{V}$ with $H \subset V$. It is straightforward that the set $G = (f \times f)^{-1}(H)$ is symmetric and $G^{-1} = G \subset W$. Furthermore, there is $P \in \mathcal{V}$ such that $P \circ P \subset V$; an immediate consequence is that, for the set $Q = (f \times f)^{-1}(P)$, we have $Q \circ Q \subset W$.

This shows that the family $\mathcal{M} = \mathcal{N}_X \cup \mathcal{V}'$ satisfies the conditions of Problem 102 and hence there is a uniformity $\mathcal{W}$ on the set $X$ for which $\mathcal{M}$ is a subbase. It is clear that $\tau(X) = \tau_{\mathcal{N}_X} \subset \tau_{\mathcal{M}}$. To show that also $\tau_{\mathcal{M}} \subset \tau(X)$ take any $O \in \tau_{\mathcal{M}}$ and fix a point $x \in O$; there is a set $W \in \mathcal{W}$ with $W(x) \subset O$. The family $\mathcal{M}$ being a subbase of $\mathcal{W}$ there are $N_1, \ldots, N_k \in \mathcal{N}_X$ and $H_1, \ldots, H_l \in \mathcal{V}'$ such that, for the sets $N = \bigcap_{i \le k} N_i$ and $H = \bigcap_{i \le l} H_i$, we have $N \cap H \subset W$.

It is evident that $N \in \mathcal{N}_X$ and $H \in \mathcal{V}'$; since $\mathcal{N}_X$ generates the topology of $X$, there is $Q \in \tau(x, X)$ with $Q \subset N(x)$ (see Problem 103). Take a set $G \in \mathcal{V}$ such that $H = (f \times f)^{-1}(G)$; since the set $G(y)$ is a neighborhood of $y = f(x)$ (see Problem 103), there is a set $P \in \tau(x, X)$ for which $f(P) \subset G(y)$; this implies that $P \subset H(x)$ and therefore $Q \cap P$ is an open neighborhood of the point $x$ contained in the set $N(x) \cap H(x) \subset W(x) \subset O$. This proves that $O$ is open in $X$ and hence $\tau_{\mathcal{W}} \subset \tau(X)$, i.e., $\tau_{\mathcal{W}} = \tau(X)$.

As a consequence, $\mathcal{W} \in \mathbb{U}_X$ which shows that $\mathcal{W} \subset \mathcal{N}_X$ and, in particular, $\mathcal{V}' \subset \mathcal{N}_X$. In other words, $(f \times f)^{-1}(V) \in \mathcal{N}_X$ for any $V \in \mathcal{V}$; thus the map $f : (X, \mathcal{N}_X) \rightarrow (Y, \mathcal{V})$ is uniformly continuous and hence we settled (ii).

**V.133.** *Let $X$ be a Tychonoff space. Prove that the following are equivalent:*

(i)  *there exists a complete uniformity $\mathcal{U}$ on the set $X$ such that $\tau_{\mathcal{U}} = \tau(X)$;*
(ii)  *the universal uniformity on the space $X$ is complete;*
(iii)  *the space $X$ is Dieudonné complete.*

**Solution.** It is evident that (ii)$\Longrightarrow$(i). Assume that there exists a complete uniformity $\mathcal{U}$ on the set $X$ such that $\tau_{\mathcal{U}} = \tau(X)$ and apply Problem 113 to find a family $\mathcal{F} = \{(M_t, \mathcal{M}_t) : t \in T\}$ of uniformly metrizable spaces such that $(X, \mathcal{U})$ is a uniform subspace of the uniform product $(M, \mathcal{M})$ of the family $\mathcal{F}$. By, Problem 116, the set $X$ is closed in the product $M = \prod_{t \in T} M_t$, so $X$ is Dieudonné complete and hence we established that (i)$\Longrightarrow$(iii).

*Fact 1.* Suppose that $Z$ is a set and $\mathcal{V}, \mathcal{V}'$ are uniformities on $Z$ such that $\mathcal{V} \subset \mathcal{V}'$ and $\tau_{\mathcal{V}} = \tau_{\mathcal{V}'}$. If $(Z, \mathcal{V})$ is complete then the space $(Z, \mathcal{V}')$ is also complete.

*Proof.* Fix a filter $\mathcal{E}$ on the set $Z$ which is a Cauchy family in $(Z, \mathcal{V}')$. It is clear that $\mathcal{E}$ is also a Cauchy family in $(Z, \mathcal{V})$, so $\bigcap \{\overline{E} : E \in \mathcal{E}\} \neq \emptyset$ (see Problem 114; the closure is taken in the topology $\tau = \tau_{\mathcal{V}} = \tau_{\mathcal{V}'}$). Applying Problem 114 again we conclude that $(X, \mathcal{V}')$ is also complete so Fact 1 is proved.

Returning to our solution assume that there is a complete uniformity $\mathcal{U}$ on the set $X$ for which $\tau_{\mathcal{U}} = \tau(X)$. If $\mathcal{N}$ is the universal uniformity on $X$ then $\tau_{\mathcal{N}} = \tau(X)$ and $\mathcal{U} \subset \mathcal{N}$, so we can apply Fact 1 to see that $(X, \mathcal{N})$ is complete as well. This settles (i)$\Longrightarrow$(ii), so (i) $\Longleftrightarrow$ (ii).

Finally, if $X$ is Dieudonné complete then we can consider that $X$ is a closed subspace of a product $M = \prod_{t \in T} M_t$ of complete metric spaces (see TFS-459). If $\rho_t$ is the respective complete metric on $M_t$ then the uniformity $\mathcal{M}_t$ on the set $M_t$, generated by $\rho_t$, is complete for any $t \in T$ (see Problems 111 and 118). Thus the uniform product $(M, \mathcal{M})$ of the family $\{(M_t, \mathcal{M}_t) : t \in T\}$ is a complete uniform space such that $\tau(M) = \tau_{\mathcal{M}}$ (see Problems 105 and 117). The set $X$ being closed in $M$, the uniformity $\mathcal{V}$ generated on $X$ from $(M, \mathcal{M})$ is complete by Problem 115. Since $X$ is a subspace of $M$, the topology $\tau_{\mathcal{V}}$ coincides with $\tau(X)$. Thus, any Dieudonné complete space $X$ has a complete uniformity $\mathcal{V}$ with $\tau_{\mathcal{V}} = \tau(X)$, i.e., we showed that (iii)$\Longrightarrow$(i); this finishes our solution.

**V.134.** *For any linear topological space $L$ denote by $\mathbf{0}_L$ its zero vector and let $G(U) = \{(x, y) \in L \times L : x - y \in U\}$ for any $U \in \tau(\mathbf{0}_L, L)$. Prove that*

(i)  *the family $\mathcal{B}_L = \{G(U) : U \in \tau(\mathbf{0}_L, L)\}$ forms a base for a uniformity $\mathcal{U}_L$ on the set $L$ (called the linear uniformity on $L$).*
(ii)  *If $M$ is a linear subspace of $L$ then the linear uniformity $\mathcal{U}_M$ on the set $M$ coincides with the subspace uniformity induced on $M$ from $L$.*

(iii) *If $L'$ is a linear topological space then a map $f : L \to L'$ is uniformly continuous if and only if, for any $U' \in \tau(\mathbf{0}_{L'}, L')$ there exists $U \in \tau(\mathbf{0}_L, L)$ such that $f(x) - f(y) \in U'$ for any $x, y \in L$ with $x - y \in U$.*

(iv) *If $L'$ is a linear topological space then any linear continuous map $f : L \to L'$ is uniformly continuous if $L$ and $L'$ are considered with their linear uniformities. In particular, any linear isomorphism between $L$ and $L'$ is a uniform isomorphism.*

**Solution.** (i) Fix a set $U \in \tau(\mathbf{0}_L, L)$; it follows from continuity of arithmetical operations in $L$ that $-U = \{-x : x \in U\}$ is an open neighborhood of $\mathbf{0}_L$. Therefore $V = U \cap (-U) \in \tau(\mathbf{0}_L, L)$ and it is straightforward that $G(V) = (G(V))^{-1} \subset G(U)$.

Apply continuity of operations in $L$ once more to see that there exists a set $W \in \tau(\mathbf{0}_L, L)$ such that $W + W = \{x + y : x, y \in W\} \subset U$. If $\{(x, y), (y, z)\} \subset G(W)$ then $x - y \in W$ and $y - z \in W$, so $x - z \in W + W \subset U$, i.e., $(x, z) \in G(U)$. This shows that $G(W) \circ G(W) \subset G(U)$.

Now, if $x, y \in L$ and $x \neq y$ then $x - y \neq \mathbf{0}_L$, so there is $U \in \tau(\mathbf{0}_L, L)$ such that $x - y \notin U$ and hence $(x, y) \notin G(U)$. This proves that the set $\bigcap \mathcal{B}_L$ coincides with the diagonal of the space $L$. Since also $G(U \cap V) = G(U) \cap G(V)$ for any $U, V \in \tau(\mathbf{0}_L, L)$, the family $\mathcal{B}_L$ satisfies all premises of Problem 101, so it indeed forms a base of a uniformity on $L$, i.e., (i) is proved.

(ii) Let $G_M(U) = \{(x, y) \in M \times M : x - y \in U\}$ for any $U \in \tau(\mathbf{0}_L, M)$. It is immediate that $G(U) \cap (M \times M) = G_M(U \cap M)$ for any $U \in \tau(\mathbf{0}_L, L)$; this shows that $\{W \cap (M \times M) : W \in \mathcal{B}_L\} = \mathcal{B}_M$. Given $Q \in \mathcal{U}_L$ there is $W \in \mathcal{B}_L$ with $W \subset Q$; thus $W' = W \cap (M \times M) \in \mathcal{B}_L$ and hence $W' \subset Q \cap (M \times M)$. This proves that $Q \cap (M \times M)$ belongs to $\mathcal{U}_M$ for any $Q \in \mathcal{U}_L$, i.e., the uniformity induced on $M$ from $L$ is contained in the linear uniformity of $L$.

Next take any $Q \in \mathcal{U}_M$; there is $U \in \tau(\mathbf{0}_L, M)$ such that $G_M(U) \subset Q$. Fix $U' \in \tau(L)$ such that $U' \cap M = U$: then $W = G(U') \cup Q \in \mathcal{U}_L$ and $W \cap (M \times M) = Q$. Thus the subspace uniformity induced on $M$ from $L$ coincides with $\mathcal{U}_M$, i.e., (ii) is proved.

(iii) Suppose that a map $f : L \to L'$ is uniformly continuous and fix a set $U' \in \tau(\mathbf{0}_{L'}, L')$. The set $W' = G'(U') = \{(z, t) \in L' \times L' : z - t \in U'\}$ belongs to $\mathcal{U}_{L'}$, so there is $W \in \mathcal{U}_L$ such that $(f \times f)(W) \subset W'$. The family $\mathcal{B}_L$ being a base of $\mathcal{U}_L$ we can find $U \in \tau(\mathbf{0}_L, L)$ for which $G(U) \subset W$. Now, if $x, y \in L$ and $x - y \in U$ then $(x, y) \in G(U) \subset W$ and therefore $(f(x), f(y)) \in W'$ which shows that $f(x) - f(y) \in U'$ and hence we settled necessity.

For sufficiency assume that the second condition of (iii) is satisfied. Given any $W' \in \mathcal{U}_{L'}$ there is $U' \in \tau(\mathbf{0}_{L'}, L')$ with $G'(U') \subset W'$; fix a set $U \in \tau(\mathbf{0}_L, L)$ such that $f(x) - f(y) \in U'$ whenever $x - y \in U$; it is easy to see that this implies that $(f \times f)(G(U)) \subset G'(U') \subset W'$. Since $W = G(U)$ belongs to $\mathcal{U}_L$, we established that $f$ is uniformly continuous, so (iii) is proved.

(iv) If $f : L \to L'$ is a continuous linear map then fix a set $W \in \mathcal{U}_{L'}$ and any $U' \in \tau(\mathbf{0}_{L'}, L')$ such that $G'(U') \subset W'$. The map $f$ being continuous at $\mathbf{0}_L$ there is $U \in \tau(\mathbf{0}_L, L)$ for which $f(U) \subset U'$; let $W = G(U)$.

If $(x, y) \in W$ then $x - y \in U$ and hence $f(x) - f(y) = f(x - y) \in f(U) \subset U'$ which shows that $(f(x), f(y)) \in W'$. Thus $(f \times f)(W) \subset W'$ and hence the map $f$ is uniformly continuous. This settles (iv) and makes our solution complete.

**V.135.** *Prove that the linear uniformity of $\mathbb{R}^X$ coincides with the uniform product of the respective family of real lines. Deduce from this fact that $\mathbb{R}^X$ is the completion of $C_p(X)$ for any space $X$, so $C_p(X)$ is complete as a uniform space if and only if $X$ is discrete.*

**Solution.** Denote by $\mathcal{L}$ the linear uniformity on the space $\mathbb{R}^X$ and let $\mathcal{U}$ be the uniform product of the $|X|$-many real lines; the symbol $\mathbf{0}$ stands for the zero function on $X$, i.e., $\mathbf{0}$ is the zero vector of $\mathbb{R}^X$. Recall that the natural projection $p_x : \mathbb{R}^X \to \mathbb{R}$ of $\mathbb{R}^X$ onto its factor determined by $x$ is defined by $p_x(f) = f(x)$ for any $f \in \mathbb{R}^X$. Besides, $O_\varepsilon = \{(x, y) \in \mathbb{R}^2 : |x - y| < \varepsilon\}$ for all $\varepsilon > 0$ and $\mathcal{O} = \{O_\varepsilon : \varepsilon > 0\}$ is the standard base of the linear uniformity on $\mathbb{R}$.

Given an arbitrary set $L \in \mathcal{L}$ there exists a set $U \in \tau(\mathbf{0}, \mathbb{R}^X)$ such that

$$G(U) = \{(f, g) \in (C_p(X))^2 : f - g \in U\} \subset L.$$

Making the set $U$ smaller if necessary, we can assume that there are $x_1, \dots, x_n \in X$ and $\varepsilon > 0$ such that $U = [x_1, \dots, x_n, \varepsilon] = \{f \in \mathbb{R}^X : |f(x_i)| < \varepsilon$ for all $i \leq n\}$. Then $G(U) = \bigcap\{(p_{x_i} \times p_{x_i})^{-1}(O_\varepsilon) : i \leq n\}$ is an element of the standard base of the uniform product (see Problem 105), so $G(U) \in \mathcal{U}$ and hence $L \in \mathcal{U}$; this proves that $\mathcal{L} \subset \mathcal{U}$.

To prove the opposite inclusion take an arbitrary set $W \in \mathcal{U}$; by the definition of the uniform product, there exist $x_1, \dots, x_n \in X$ and $\varepsilon_1, \dots, \varepsilon_n > 0$ such that $W' = \bigcap\{(p_{x_i} \times p_{x_i})^{-1}(O_{\varepsilon_i}) : i \leq n\} \subset W$. Now, if $\varepsilon = \setminus n\{\varepsilon_1, \dots, \varepsilon_n\}$ and $U = [x_1, \dots, x_n, \varepsilon]$ then it is straightforward that $G(U) \subset W' \subset W$ and hence $W \in \mathcal{L}$, i.e., we proved that $\mathcal{U} = \mathcal{L}$.

By Problem 134, the space $C_p(X)$ with its linear uniformity is a dense uniform subspace of $\mathbb{R}^X$; the space $\mathbb{R}^X$ being complete by Problem 117, we can apply Problem 121 to conclude that $\mathbb{R}^X$ is canonically isomorphic to the completion of the uniform space $C_p(X)$.

Finally, if $X$ is discrete then $C_p(X) = \mathbb{R}^X$ is complete by Problem 117; if, on the other hand, the uniform space $C_p(X)$ is complete then its closed in $\mathbb{R}^X$ by Problem 116, so $C_p(X) = \mathbb{R}^X$ and hence $X$ is discrete (see Fact 1 of S.265).

**V.136.** *Prove that $C_p(X)$ is $\sigma$-totally bounded as a uniform space if and only if $X$ is pseudocompact. More formally, $X$ is pseudocompact if and only if there exists a family $\{C_n : n \in \omega\} \subset \exp(C_p(X))$ such that $C_p(X) = \bigcup\{C_n : n \in \omega\}$ and each $C_n$ is totally bounded considered as a uniform subspace of $C_p(X)$. In particular, if $C_p(X)$ is uniformly isomorphic to $C_p(Y)$ then the space $X$ is pseudocompact if and only if so is $Y$.*

**Solution.** Suppose first that the space $X$ is pseudocompact and consider the set $C_n = \{f \in C_p(X) : f(x) \in [-n, n]\}$ for any $n \in \omega$. It follows from pseudocompactness of $X$ that $C_p(X) = \bigcup\{C_n : n \in \omega\}$; besides, every $C_n$ is dense in the compact set $K_n = [-n, n]^X$.

Let $\mathcal{U}_n$ (or $\mathcal{V}_n$ respectively) be the subspace uniformity induced on $K_n$ (or $C_n$ respectively) from $\mathbb{R}^X$. Then $(C_n, \mathcal{V}_n)$ is a dense uniform subspace of $(K_n, \mathcal{U}_n)$. The space $(K_n, \mathcal{U}_n)$ is compact and hence complete by Problem 130; apply Problem 121 to see that $(K_n, \mathcal{U}_n)$ is the completion of $(C_n, \mathcal{V}_n)$, so $(C_n, \mathcal{V}_n)$ is totally bounded by Problem 130. It is easy to see that $\mathcal{V}_n$ is also the subspace uniformity induced on $C_n$ from $C_p(X)$ (see Problems 134 and 135). Therefore every $C_n$ is a totally bounded uniform subspace of $C_p(X)$, so $C_p(X)$ is $\sigma$-totally bounded. This proves sufficiency.

*Fact 1.* Given uniform spaces $(Y, \mathcal{V})$ and $(Z, \mathcal{W})$ suppose that $f : Y \to Z$ is a uniformly continuous surjective map and $(Y, \mathcal{V})$ is totally bounded. Then $(Z, \mathcal{W})$ is totally bounded as well.

*Proof.* If $W \in \mathcal{W}$ then there is $V \in \mathcal{V}$ such that $(f \times f)(V) \subset W$. The space $(Y, \mathcal{V})$ being totally bounded there is a finite set $A \subset Y$ for which $V(A) = Y$. The set $B = f(A) \subset Z$ is finite; given any $z \in Z$ fix $y \in Y$ with $f(y) = z$. There exists $a \in A$ such that $y \in V(a)$ and hence $(a, y) \in V$. Let $b = f(a)$; then $b \in B$ and $(f(a), f(y)) = (b, z) \in W$, so $z \in W(b)$. This shows that $W(B) = Z$ and hence $(Z, \mathcal{W})$ is totally bounded, so Fact 1 is proved.

*Fact 2.* The space $\mathbb{R}^\omega$ with its linear uniformity is not $\sigma$-totally bounded.

*Proof.* For every $n \in \omega$ the natural projection $p_n : \mathbb{R}^\omega \to \mathbb{R}$ onto the $n$-th factor is uniformly continuous (see Problems 106 and 135; each factor is also considered with its linear uniformity).

Assume that $\mathbb{R}^\omega = \bigcup_{n \in \omega} C_n$ and every subspace $C_n$ is totally bounded. If $r_n \in \mathbb{R} \backslash p_n(C_n)$ for each $n \in \omega$ then define a point $x \in \mathbb{R}^\omega$ by letting $x(n) = r_n$ for all $n \in \omega$. Then $x \in \mathbb{R}^\omega \backslash (\bigcup_{n \in \omega} C_n)$ which is a contradiction. Thus there exists $m \in \omega$ such that $p_m(C_m) = \mathbb{R}$. The map $p_m|C_m : C_m \to \mathbb{R}$ is also uniformly continuous, so $\mathbb{R}$ is totally bounded by Fact 1. Since $\mathbb{R}$ is also complete, it has to be compact (see Problem 130); this contradiction shows that $\mathbb{R}^\omega$ is not $\sigma$-totally bounded, i.e., Fact 2 is proved.

Returning to our solution assume that the uniform space $C_p(X)$ is $\sigma$-totally bounded. If $X$ is not pseudocompact then there is a closed discrete faithfully indexed set $D = \{d_n : n \in \omega\} \subset X$ which is $C$-embedded in $X$. The restriction map $\pi : C_p(X) \to \mathbb{R}^D$ is continuous, surjective, linear and hence uniformly continuous (see Problem 134). This, together with Fact 1, implies that $\mathbb{R}^D$ is $\sigma$-totally bounded. The space $\mathbb{R}^D$ is linearly (and hence uniformly) isomorphic to $\mathbb{R}^\omega$, so $\mathbb{R}^\omega$ is $\sigma$-totally bounded. This contradiction with Fact 2 shows that $X$ is pseudocompact and hence we proved necessity.

Finally note that if $C_p(X)$ and $C_p(Y)$ are uniformly isomorphic then $C_p(X)$ is $\sigma$-totally bounded if and only if so is $C_p(Y)$; therefore the space $X$ is pseudocompact if and only if so is $Y$.

**V.137.** *Observe that* $X \overset{u}{\sim} Y$ *implies* $X \overset{t}{\sim} Y$. *Given an example of t-equivalent spaces $X$ and $Y$ which are not u-equivalent.*

**Solution.** If $\varphi : C_p(X) \to C_p(Y)$ is a uniform isomorphism then it is also a homeomorphism by Problem 104. Therefore $X \overset{t}{\sim} Y$, i.e., we proved that $u$-equivalence implies $t$-equivalence.

Next observe that $\mathbb{R}$ is $t$-equivalent to $[0, 1]$ by Problem 027; every space $u$-equivalent to the compact (and hence pseudocompact) space $[0, 1]$ has to be pseudocompact by Problem 136, so $X = \mathbb{R}$ and $Y = [0, 1]$ are $t$-equivalent spaces which are not $u$-equivalent.

**V.138.** *Suppose that $C_p(X)$ is uniformly isomorphic to $C_p(Y)$. Prove that $X$ is compact if and only if so is $Y$.*

**Solution.** Assume that $X$ is compact; we have $X \overset{t}{\sim} Y$ by Problem 137, so the space $Y$ is $\sigma$-compact by Problem 043. It follows from Problem 136 that $Y$ is pseudocompact and hence compact. Analogously, compactness of $Y$ also implies compactness of $X$, so $X$ is compact if and only if so is $Y$.

**V.139.** *Suppose that the spaces $X$ and $Y$ are u-equivalent. Prove that there exists a homeomorphism $\varphi : \mathbb{R}^X \to \mathbb{R}^Y$ such that $\varphi(C_p(X)) = C_p(Y)$.*

**Solution.** Let $\mu : C_p(X) \to C_p(Y)$ be a uniform isomorphism. The uniform spaces $\mathbb{R}^X$ and $\mathbb{R}^Y$ being the completions of $C_p(X)$ and $C_p(Y)$ respectively (see Problem 135) we can apply Problem 120 to conclude that there exists a uniform isomorphism $\varphi : \mathbb{R}^X \to \mathbb{R}^Y$ such that $\varphi|C_p(X) = \mu$ and hence $\varphi(C_p(X)) = \mu(C_p(X)) = C_p(Y)$. It follows from Problem 104 that the map $\varphi$ is a homeomorphism.

**V.140.** *Let $\mathcal{F} = \{F_1, \dots, F_k\}$ be a family of functionally closed subsets of a Tychonoff space $X$. Suppose that $\{U_1, \dots, U_k\}$ is a family of functionally open subsets of $X$ such that $F_i \subset U_i$ for each $i$. Prove that the family $\mathcal{F}$ has a functionally open swelling $\{W_1, \dots, W_k\}$ such that $F_i \subset W_i \subset \overline{W}_i \subset U_i$ for each $i \leq k$.*

**Solution.** Observe first that a functionally open set is the same as a cozero set (see Fact 1 of T.252) and the concept of a functionally closed set coincides with the concept of a zero-set by Fact 1 of S.499.

*Fact 1.* Given disjoint functionally closed sets $F$ and $G$ in a space $Z$ there exists a function $f \in C(Z, [0, 1])$ such that $f(F) \subset \{0\}$ and $f(G) \subset \{1\}$. In particular, there exist functionally open sets $O(F)$ and $O(G)$ and disjoint functionally closed sets $P(F)$, $P(G)$ such that $F \subset O(F) \subset P(F)$ and $G \subset O(G) \subset P(G)$.

*Proof.* Take $u_F, u_G \in C(Z, [0, 1])$ such that $F = u_F^{-1}(0)$ and $G = u_G^{-1}(0)$. Then $u_F + u_G$ is strictly positive on $Z$, so the function $f = \frac{u_F}{u_F + u_G}$ is well defined and continuous on $Z$. It is evident that $f(x) = 1$ for any $x \in G$ and $f(x) = 0$ whenever $x \in F$. Therefore the sets $O(F) = f^{-1}([0, \frac{1}{3}))$ and $P(F) = f^{-1}([0, \frac{1}{3}])$ together with the sets $O(G) = f^{-1}((\frac{2}{3}, 1])$ and $P(G) = f^{-1}([\frac{2}{3}, 1])$ constitute the promised functionally open (closed) neighborhoods of the sets $F$ and $G$ respectively, so Fact 1 is proved.

**Fact 2.** Suppose that $F$ is functionally closed in a space $Z$ and $O$ is a functionally open subset of $Z$ for which $F \subset O$. Then there is a functionally open set $G \subset Z$ such that $F \subset G \subset \overline{G} \subset O$.

**Proof.** The sets $F$ and $X \setminus O$ are functionally closed and disjoint; this makes it possible to apply Fact 1 to find disjoint functionally open sets $G$ and $G'$ such that $F \subset G$ and $X \setminus O \subset G'$. It follows from $\overline{G} \cap G' = \emptyset$ that $\overline{G} \subset O$, so $F \subset G \subset \overline{G} \subset O$ and hence Fact 2 is proved.

Returning to our solution consider the set $\Phi_1 = \bigcup \{P$ : there exist distinct numbers $i_1, \ldots, i_m \in \{2, \ldots, k\}$ such that $P = F_{i_1} \cap \ldots \cap F_{i_m}$ and $P \cap F_1 = \emptyset\}$. The set $\Phi_1$ is functionally closed being a finite union of functionally closed sets (see Fact 1 of S.499); since also $\Phi_1 \cap F_1 = \emptyset$, we can find a function $f_1 \in C(X, [0, 1])$ for which $f_1(F_1) \subset \{0\}$ and $f(\Phi_1) \subset \{1\}$ (see Fact 1). The set $K_1 = f_1^{-1}([0, \frac{1}{2}])$ is functionally closed and the family $\mathcal{F}_1 = \{K_1, F_2, \ldots, F_k\}$ is a swelling of $\mathcal{F}$.

Proceeding by induction assume that $1 \leq n < k$ and we have constructed functions $f_1, \ldots, f_n \in C(X, [0, 1])$ with the following properties:

(1)  $f_i(F_i) \subset \{0\}$ for any $i \leq n$;
(2)  if $K_i = f_i^{-1}([0, \frac{1}{2}])$ for each $i \leq n$ then the family $\{K_1, \ldots, K_n, F_{n+1}, \ldots, F_k\}$ is a swelling of $\mathcal{F}$.

For the family $\mathcal{F}_n = \{K_1, \ldots, K_n, F_{n+1}, \ldots, F_k\}$ let $\Phi_{n+1} = \bigcup \{P$ : there exist $P_1, \ldots, P_m \in \mathcal{F}_n \setminus \{F_{n+1}\}$ such that $P = P_1 \cap \ldots \cap P_m$ and $P \cap F_{n+1} = \emptyset\}$. The set $\Phi_{n+1}$ being functionally closed and disjoint from $F_{n+1}$ we can apply Fact 1 again to find $f_{n+1} \in C(X, [0, 1])$ such that $f_{n+1}(F_{n+1}) \subset \{0\}$ and $f_{n+1}(\Phi_{n+1}) \subset \{1\}$.

If $K_{n+1} = f_{n+1}^{-1}([0, \frac{1}{2}])$ then $\mathcal{F}_{n+1} = \{K_1, \ldots, K_n, K_{n+1}, F_{n+2}, \ldots, F_k\}$ is easily seen to be a swelling of $\mathcal{F}_n$; since $\mathcal{F}_n$ is a swelling of $\mathcal{F}$, the family $\mathcal{F}_{n+1}$ is a swelling of $\mathcal{F}$ as well, so our inductive procedure can be continued to construct functions $f_1, \ldots, f_k \in C(X, [0, 1])$ such that the properties (1) and (2) are fulfilled for $n = k$. In particular, $\mathcal{K} = \{K_1, \ldots, K_k\}$ is a swelling of $\mathcal{F}$.

If $H_i = f_i^{-1}([0, \frac{1}{2}))$ then the set $H_i$ is functionally open for every $i \leq k$ and the family $\mathcal{H} = \{H_1, \ldots, H_n\}$ is also a swelling of $\mathcal{F}$. Apply Fact 2 to find a functionally open set $W_i$ for which $F_i \subset W_i \subset \overline{W}_i \subset H_i \cap U_i$ for every $i \leq k$; then $\{W_1, \ldots, W_k\}$ is the promised swelling of $\mathcal{F}$, so our solution is complete.

**V.141.** *Let $\mathcal{F} = \{F_1, \ldots, F_k\}$ be a family of closed subsets of a normal space $X$. Suppose that $\{U_1, \ldots, U_k\}$ is a family of open subsets of $X$ such that $F_i \subset U_i$ for each $i \leq k$. Prove that the family $\mathcal{F}$ has an open swelling $\{W_1, \ldots, W_k\}$ such that $F_i \subset W_i \subset \overline{W}_i \subset U_i$ for each $i \leq k$.*

**Solution.** Consider the set $\Phi_1 = \bigcup\{P :$ there exist $i_1, \ldots, i_m \in \{2, \ldots, k\}$ such that $P = F_{i_1} \cap \ldots \cap F_{i_m}$ and $P \cap F_1 = \emptyset\}$. The set $\Phi_1$ is closed and disjoint from $F_1$; by normality of the space $X$ we can find a function $f_1 \in C(X, [0, 1])$ for which $f_1(F_1) \subset \{0\}$ and $f(\Phi_1) \subset \{1\}$. The set $K_1 = f_1^{-1}([0, \frac{1}{2}])$ is closed and the family $\mathcal{F}_1 = \{K_1, F_2, \ldots, F_k\}$ is a swelling of $\mathcal{F}$.

Proceeding by induction assume that $1 \leq n < k$ and we have constructed functions $f_1, \ldots, f_n \in C(X, [0, 1])$ with the following properties:

(1) $f_i(F_i) \subset \{0\}$ for any $i \leq n$;
(2) if $K_i = f_i^{-1}([0, \frac{1}{2}])$ for each $i \leq n$ then the family $\{K_1, \ldots, K_n, F_{n+1}, \ldots, F_k\}$ is a swelling of $\mathcal{F}$.

For the family $\mathcal{F}_n = \{K_1, \ldots, K_n, F_{n+1}, \ldots, F_k\}$ let $\Phi_{n+1} = \bigcup\{P :$ there exist $P_1, \ldots, P_m \in \mathcal{F}_n \backslash \{F_{n+1}\}$ such that $P = P_1 \cap \ldots \cap P_m$ and $P \cap F_{n+1} = \emptyset\}$. The set $\Phi_{n+1}$ is closed and disjoint from $F_{n+1}$; the space $X$ being normal we can find a function $f_{n+1} \in C(X, [0, 1])$ such that $f_{n+1}(F_{n+1}) \subset \{0\}$ and $f_{n+1}(\Phi_{n+1}) \subset \{1\}$.

If $K_{n+1} = f_{n+1}^{-1}([0, \frac{1}{2}])$ then $\mathcal{F}_{n+1} = \{K_1, \ldots, K_n, K_{n+1}, F_{n+2}, \ldots, F_k\}$ is easily seen to be a swelling of $\mathcal{F}_n$; since $\mathcal{F}_n$ is a swelling of $\mathcal{F}$, the family $\mathcal{F}_{n+1}$ is a swelling of $\mathcal{F}$ as well, so our inductive procedure can be continued to construct functions $f_1, \ldots, f_k \in C(X, [0, 1])$ such that the properties (1) and (2) are fulfilled for $n = k$. In particular, $\mathcal{K} = \{K_1, \ldots, K_k\}$ is a swelling of $\mathcal{F}$.

If $H_i = f_i^{-1}([0, \frac{1}{2}))$ then the set $H_i$ is open for every $i \leq k$ and the family $\mathcal{H} = \{H_1, \ldots, H_n\}$ is also a swelling of $\mathcal{F}$. By normality of $X$ we can find an open set $W_i$ for which $F_i \subset W_i \subset \overline{W}_i \subset H_i \cap U_i$ for all $i \leq k$; then $\{W_1, \ldots, W_k\}$ is the promised swelling of $\mathcal{F}$.

**V.142.** *Let $\mathcal{U} = \{U_1, \ldots, U_k\}$ be a functionally open cover of a Tychonoff space $X$. Prove that $\mathcal{U}$ has shrinkings $\mathcal{F} = \{F_1, \ldots, F_k\}$ and $\mathcal{W} = \{W_1, \ldots, W_k\}$ such that $\mathcal{F}$ is functionally closed, $\mathcal{W}$ is functionally open and $F_i \subset W_i \subset \overline{W}_i \subset U_i$ for every $i \leq k$.*

**Solution.** Consider the set $P_i = X \backslash U_i$ for each $i \leq k$. Then all elements of the family $\mathcal{P} = \{P_1, \ldots, P_k\}$ are functionally closed, so we can apply Problem 140 to find a functionally open swelling $\mathcal{O} = \{O_1, \ldots, O_k\}$ of the family $\mathcal{P}$. If $F_i = X \backslash O_i$ then $F_i \subset U_i$ for all $i \leq k$; it follows from $\bigcap \mathcal{P} = \emptyset$ that $\bigcap \mathcal{O} = \emptyset$ and hence the family $\mathcal{F} = \{F_1, \ldots, F_k\}$ covers $X$, i.e., $\mathcal{F}$ is a functionally closed shrinking of $\mathcal{U}$. By Fact 2 of V.140 there exists a functionally open set $W_i$ such that $F_i \subset W_i \subset \overline{W}_i \subset U_i$ for each $i \leq k$; it is evident that the functionally open family $\mathcal{W} = \{W_1, \ldots, W_k\}$ is also a shrinking of $\mathcal{U}$.

**V.143.** *Let $\mathcal{U} = \{U_1, \ldots, U_k\}$ be an open cover of a normal space $X$. Prove that $\mathcal{U}$ has shrinkings $\mathcal{F} = \{F_1, \ldots, F_k\}$ and $\mathcal{W} = \{W_1, \ldots, W_k\}$ such that $\mathcal{F}$ is closed, $\mathcal{W}$ is open and $F_i \subset W_i \subset \overline{W}_i \subset U_i$ for every $i \leq k$.*

**Solution.** Consider the set $P_i = X \backslash U_i$ for each $i \leq k$. Then all elements of the family $\mathcal{P} = \{P_1, \ldots, P_k\}$ are closed in $X$, so we can apply Problem 141 to find an open swelling $\mathcal{O} = \{O_1, \ldots, O_k\}$ of the family $\mathcal{P}$. If $F_i = X \backslash O_i$ then $F_i \subset U_i$ for all $i \leq k$; it follows from $\bigcap \mathcal{P} = \emptyset$ that $\bigcap \mathcal{O} = \emptyset$ and hence the family

$\mathcal{F} = \{F_1, \ldots, F_k\}$ covers $X$, i.e., $\mathcal{F}$ is a closed shrinking of $\mathcal{U}$. By normality of $X$ there exists an open set $W_i$ such that $F_i \subset W_i \subset \overline{W}_i \subset U_i$ for each $i \leq k$; it is evident that the family $\mathcal{W} = \{W_1, \ldots, W_k\}$ is also a shrinking of $\mathcal{U}$.

**V.144.** *Prove that, for any Tychonoff space $X$, the following conditions are equivalent:*

(i) $\dim X \leq n$;

(ii) *every finite functionally open cover of $X$ has a finite functionally closed refinement of order $\leq n + 1$;*

(iii) *every finite functionally open cover of $X$ has a functionally closed shrinking of order $\leq n + 1$;*

(iv) *every finite functionally open cover of $X$ has a functionally open shrinking of order $\leq n + 1$.*

**Solution.** Suppose that $\dim X \leq n$ and take a finite functionally open cover $\mathcal{U}$ of the space $X$. There exists a finite functionally open refinement $\mathcal{V}$ of the family $\mathcal{U}$ of order at most $n + 1$; take a faithful enumeration $\{V_1, \ldots, V_k\}$ of the family $\mathcal{V}$. It follows from Problem 142, that $\mathcal{V}$ has a functionally closed shrinking $\mathcal{F} = \{F_1, \ldots, F_k\}$. It is straightforward that the order of $\mathcal{F}$ is at most $n + 1$; the family $\mathcal{F}$ being a refinement of $\mathcal{V}$, it is also a refinement of $\mathcal{U}$, so we proved that (i)$\Longrightarrow$(ii).

Now, if (ii) holds then take a functionally open cover $\mathcal{U} = \{U_1, \ldots, U_k\}$ of the space $X$. There exists a functionally closed refinement $\mathcal{F}$ of the family $\mathcal{U}$ of order at most $n + 1$. For any $F \in \mathcal{F}$ fix a number $i(F)$ such that $F \subset U_{i(F)}$ and let $G_j = \bigcup \{F \in \mathcal{F} : i(F) = j\}$ for every $j \leq k$. The family $\mathcal{G} = \{G_1, \ldots, G_k\}$ is a functionally closed shrinking of $\mathcal{U}$. If there are distinct $j_1, \ldots, j_{n+2}$ such that $P = G_{j_1} \cap \ldots \cap G_{j_{n+2}} \neq \emptyset$ then pick a point $x \in P$. There are $F_1, \ldots, F_{n+2} \in \mathcal{F}$ such that $x \in F_1 \cap \ldots \cap F_{n+2}$ and $i(F_m) = j_m$ for every $m \leq n + 2$. Therefore $F_1, \ldots, F_{n+2}$ are distinct elements of $\mathcal{F}$ with nonempty intersection. This contradiction with $\mathrm{ord}(\mathcal{F}) \leq n + 1$ shows that $\mathrm{ord}(\mathcal{G}) \leq n + 1$, so $\mathcal{G}$ is a functionally closed shrinking of $\mathcal{U}$ of order at most $n + 1$, i.e., we established that (ii)$\Longrightarrow$(iii).

Next assume that the condition (iii) is satisfied and take a functionally open cover $\mathcal{U} = \{U_1, \ldots, U_k\}$ of the space $X$. There exists a functionally closed shrinking $\mathcal{F} = \{F_1, \ldots, F_k\}$ of the family $\mathcal{U}$ such that $\mathrm{ord}(\mathcal{F}) \leq n + 1$. By Problem 140 there exists a functionally open swelling $\mathcal{V} = \{V_1, \ldots, V_n\}$ of the family $\mathcal{F}$ such that $V_i \subset \overline{V}_i \subset U_i$ for each $i \leq k$. Thus $\mathcal{V}$ is a functionally open shrinking of $\mathcal{U}$ of order at most $n + 1$, i.e., we settled (iii)$\Longrightarrow$(iv). The implication (iv)$\Longrightarrow$(i) is evident, so our solution is complete.

**V.145.** *Prove that, for any normal $X$, the following conditions are equivalent:*

(i) $\dim X \leq n$;

(ii) *every finite open cover of $X$ has a finite open refinement of order $\leq n + 1$;*

(iii) *every finite open cover of $X$ has a finite closed refinement of order $\leq n + 1$;*

(iv) *every finite open cover of $X$ has a closed shrinking of order $\leq n + 1$;*

(v) *every finite open cover of $X$ has an open shrinking of order $\leq n + 1$.*

**Solution.** Suppose that $\dim X \leq n$ and $\mathcal{U} = \{U_1, \ldots, U_k\}$ is an open cover of $X$. It follows from Problem 143 that $\mathcal{U}$ has a closed shrinking $\mathcal{F} = \{F_1, \ldots, F_k\}$. By Fact 5 of V.100 we can choose a functionally open set $V_i$ such that $F_i \subset V_i \subset U_i$ for every $i \leq k$. It is evident that $\mathcal{V} = \{V_1, \ldots, V_k\}$ is a functionally open refinement of $\mathcal{U}$. Since $\dim X \leq n$, the family $\mathcal{V}$ has a finite functionally open refinement $\mathcal{W}$ with $\mathrm{ord}(\mathcal{W}) \leq n + 1$; it is clear that $\mathcal{W}$ is also an open refinement of $\mathcal{U}$, so (i)$\Longrightarrow$(ii) is proved.

Now assume that (ii) holds and take a finite open cover $\mathcal{U}$ of the space $X$. There exists a finite open refinement $\mathcal{V}$ of the family $\mathcal{U}$ of order at most $n + 1$; take a faithful enumeration $\{V_1, \ldots, V_k\}$ of the family $\mathcal{V}$. It follows from Problem 143, that $\mathcal{V}$ has a closed shrinking $\mathcal{F} = \{F_1, \ldots, F_k\}$. It is straightforward that the order of $\mathcal{F}$ is at most $n + 1$; the family $\mathcal{F}$ being a refinement of $\mathcal{V}$, it is also a refinement of $\mathcal{U}$, so we proved that (ii)$\Longrightarrow$(iii).

Next suppose that (iii) holds and take an open cover $\mathcal{U} = \{U_1, \ldots, U_k\}$ of the space $X$. There exists a closed refinement $\mathcal{F}$ of the family $\mathcal{U}$ of order at most $n + 1$. For any set $F \in \mathcal{F}$ fix a number $i(F)$ such that $F \subset U_{i(F)}$ and consider the set $G_j = \bigcup\{F \in \mathcal{F} : i(F) = j\}$ for every $j \leq k$. The family $\mathcal{G} = \{G_1, \ldots, G_k\}$ is a closed shrinking of $\mathcal{U}$. If there are distinct $j_1, \ldots, j_{n+2}$ such that $P = G_{j_1} \cap \ldots \cap G_{j_{n+2}} \neq \emptyset$ then pick a point $x \in P$. There are $F_1, \ldots, F_{n+2} \in \mathcal{F}$ such that $x \in F_1 \cap \ldots \cap F_{n+2}$ and $i(F_m) = j_m$ for every $m \leq n + 2$. Therefore $F_1, \ldots, F_{n+2}$ are distinct elements of $\mathcal{F}$ with nonempty intersection. This contradiction with $\mathrm{ord}(\mathcal{F}) \leq n + 1$ shows that $\mathrm{ord}(\mathcal{G}) \leq n + 1$, so $\mathcal{G}$ is a closed shrinking of $\mathcal{U}$ of order at most $n + 1$, i.e., we established that (iii)$\Longrightarrow$(iv).

If the property (iv) holds then take an open cover $\mathcal{U} = \{U_1, \ldots, U_k\}$ of the space $X$. There exists a closed shrinking $\mathcal{F} = \{F_1, \ldots, F_k\}$ of the family $\mathcal{U}$ such that $\mathrm{ord}(\mathcal{F}) \leq n + 1$. By Problem 141 there exists an open swelling $\mathcal{V} = \{V_1, \ldots, V_n\}$ of the family $\mathcal{F}$ such that $V_i \subset \overline{V}_i \subset U_i$ for each $i \leq k$. Thus $\mathcal{V}$ is an open shrinking of $\mathcal{U}$ of order at most $n + 1$, i.e., we settled (iv)$\Longrightarrow$(v).

Finally, assume that the condition (v) is satisfied and fix a functionally open cover $\mathcal{U} = \{U_1, \ldots, U_k\}$ of the space $X$. Take an open shrinking $\mathcal{V} = \{V_1, \ldots, V_k\}$ of the family $\mathcal{U}$ such that $\mathrm{ord}(\mathcal{V}) \leq n + 1$; it follows from Problem 143 that we can find a closed shrinking $\mathcal{F} = \{F_1, \ldots, F_k\}$ of the family $\mathcal{V}$. Apply Fact 5 of V.100 again to find a functionally open set $W_i$ such that $F_i \subset W_i \subset V_i$ for all $i \leq k$. It is easy to see that $\mathcal{W} = \{W_1, \ldots, W_k\}$ is a functionally open shrinking of $\mathcal{U}$ with $\mathrm{ord}(\mathcal{W}) \leq n + 1$, so $\dim X \leq n$ by Problem 144; this settles (v)$\Longrightarrow$(i) and hence our solution is complete.

**V.146.** *Suppose that $X$ is a Tychonoff space and $Y$ is a $C^*$-embedded subset of $X$. Prove that $\dim Y \leq \dim X$. In particular, if $X$ is normal then $\dim F \leq \dim X$ for any closed $F \subset X$.*

**Solution.** Our statement is clear if $\dim X = \infty$, so assume $\dim X \leq n \in \omega$. Take a functionally open cover $\mathcal{U} = \{U_1, \ldots, U_m\}$ of the space $Y$; by Problem 142 the family $\mathcal{U}$ has a functionally closed shrinking $\mathcal{F} = \{F_1, \ldots, F_m\}$. Apply Fact 1 of V.140 to find a function $f_i \in C(Y, [0, 1])$ such that $f_i(F_i) \subset \{1\}$ and $f_i(Y \setminus U_i) \subset \{0\}$ for every $i \leq m$. The set $Y$ being $C^*$-embedded in $X$ there exist $g_1, \ldots, g_m \in$

$C(X)$ for which $g_i|Y = f_i$ for every $i \leq m$. Let $u(x) = 1$ for all $x \in X$; if $h_i = \max(u - 1, \backslash n(g_i, u))$ then $h_i \in C(X, [0, 1])$ and $h_i|Y = f_i$ for each $i \leq m$.

Every set $O_i = h_i^{-1}((0, 1])$ is open in $X$; if $O = \bigcap\{h_i^{-1}([0, \frac{1}{2})) : i \leq m\}$ then $\mathcal{O} = \{O, O_1, \ldots, O_m\}$ is a functionally open cover of $X$; observe that $O_i \cap Y \subset U_i$ for all $i \leq m$. By Problem 144 we can find a functionally closed shrinking $\mathcal{P} = \{P, P_1, \ldots, P_m\}$ of the cover $\mathcal{O}$ such that $\operatorname{ord}(\mathcal{P}) \leq n + 1$. However, $O \cap Y = \emptyset$ which shows that $P \cap Y = \emptyset$ and therefore $\{P_1 \cap Y, \ldots, P_m \cap Y\}$ is a functionally closed shrinking of $\mathcal{U}$ of order at most $n + 1$. Applying Problem 144 again we conclude that $\dim(Y) \leq n$, so $\dim Y \leq \dim X$. Finally, if $X$ is normal and $F$ is closed in $X$ then $F$ is even $C$-embedded in $X$, so $\dim F \leq \dim X$.

**V.147.** *Prove that* $\dim X = \dim \beta X$ *for any Tychonoff space* $X$. *Deduce from this fact that* $\dim X = \dim Y$ *for any* $Y$ *with* $X \subset Y \subset \beta X$.

**Solution.** Since $X$ is $C^*$-embedded in $\beta X$, the inequality $\dim X \leq \dim \beta X$ is a consequence of Problem 146.

*Fact 1.* Given a Tychonoff space $Z$, suppose that $F_1, \ldots, F_m$ are functionally closed subsets of $Z$ such that $F_1 \cap \ldots \cap F_m = \emptyset$. Then $\operatorname{cl}_{\beta Z}(F_1) \cap \ldots \cap \operatorname{cl}_{\beta Z}(F_m) = \emptyset$.

*Proof.* Suppose that there exists a point $z \in \operatorname{cl}_{\beta Z}(F_1) \cap \ldots \cap \operatorname{cl}_{\beta Z}(F_m)$ and fix a function $f_i \in C_p(Z, [0, 1])$ such that $F_i = f_i^{-1}(0)$ for every $i \leq m$. It follows from $F_1 \cap \ldots \cap F_m = \emptyset$ that the function $f = f_1 + \ldots + f_m$ is strictly positive at any point of $Z$, so $g_i = \frac{f_i}{f} \in C_p(Z, [0, 1])$ and $g_i(F_i) \subset \{0\}$ for each $i \leq m$. Besides, $\sum_{i=1}^{m} g_i(x) = 1$ for any $x \in Z$.

Let $u(x) = 1$ for any $x \in \beta Z$. There exists a function $h_i \in C_p(\beta Z, [0, 1])$ such that $h_i|Z = g_i$; it follows from $g_i(F_i) \subset \{0\}$ that $h_i(\operatorname{cl}_{\beta Z}(F_i)) \subset \{0\}$ and hence $h_i(z) = 0$ for every $i \leq m$. The function $h = \sum_{i=1}^{m} h_i$ is continuous on $\beta Z$ and $h|Z = \sum_{i=1}^{m} g_i = u|Z$; the set $Z$ being dense in $\beta Z$ we have $h = u$ and, in particular, $h(z) = 1$. This contradiction with $h_i(z) = 0$ for every $i \leq m$ shows that $\operatorname{cl}_{\beta Z}(F_1) \cap \ldots \cap \operatorname{cl}_{\beta Z}(F_m) = \emptyset$ and hence Fact 1 is proved.

To prove that $\dim \beta X \leq \dim X$ suppose that $\dim X \leq n$; if $n = \infty$ then there is nothing to prove. If $n \in \omega$, take an open cover $\mathcal{U} = \{U_1, \ldots, U_m\}$ of the space $\beta X$. By Problem 143, there exists a functionally open shrinking $\mathcal{W} = \{W_1, \ldots, W_m\}$ of the cover $\mathcal{U}$ such that $\overline{W}_i \subset U_i$ for every $i \leq m$ (the bar denotes the closure in $\beta X$). The cover $\{X \cap W_i : i \leq m\}$ of the space $X$ has a functionally open (in $X$) shrinking $\{V_1, \ldots, V_m\}$ of order at most $n + 1$ (see Problem 144).

Consider the set $O_i = \beta X \backslash \overline{X \backslash V_i}$; it is straightforward that $O_i$ is open in $\beta X$ and $O_i \cap X = V_i$ for each $i \leq m$. Therefore $\overline{O}_i = \overline{V}_i \subset \overline{W}_i \subset U_i$ for every $i \leq m$. The family $\mathcal{F} = \{X \backslash V_i : i \leq m\}$ consists of functionally closed subsets of $X$ and $\bigcap \mathcal{F} = \emptyset$. Apply Fact 1 to conclude that $\bigcap\{\overline{X \backslash V_i} : i \leq m\} = \emptyset$ and hence the family $\mathcal{O} = \{O_1, \ldots, O_m\}$ is a cover of $\beta X$.

Next, assume that $i_1, \ldots, i_{n+2}$ are distinct elements of $\{1, \ldots, m\}$ and the set $H = O_{i_1} \cap \ldots \cap O_{i_{n+2}}$ is nonempty. The space $X$ being dense in $\beta X$, we have $H \cap X = V_{i_1} \cap \ldots V_{i_{n+2}} \neq \emptyset$, i.e., the order of the family $\{V_1, \ldots, V_m\}$ is at least

$n + 2$ which is a contradiction. Thus $\mathcal{O}$ is an open shrinking of $\mathcal{U}$ of order $\leq n + 1$, so $\dim \beta X \leq n$ by Problem 145. Therefore $\dim \beta X \leq \dim X$ and hence we proved that $\dim X = \dim \beta X$ for any space $X$.

Finally, observe that if $X \subset Y \subset \beta X$ then $\beta Y = \beta X$ by Fact 1 of S.393, so $\dim Y = \dim \beta Y = \dim \beta X = \dim X$ and hence our solution is complete.

**V.148.** *Prove that a Tychonoff space $X$ is strongly zero-dimensional if and only if $X$ is normal and $\dim X = 0$. Give an example of a Tychonoff space $X$ such that $\dim X = 0$ while $X$ is not strongly zero-dimensional.*

**Solution.** It is evident that, $\dim X = 0$ for any strongly zero-dimensional space $X$; besides, $X$ has to be normal by SFFS-308. This proves necessity. Now if $X$ is normal and $\dim X = 0$ then take any open cover $\mathcal{U} = \{U_1, \ldots, U_m\}$ of the space $X$. By Problem 145, there exists an open shrinking $\mathcal{V}$ of order 1 of the cover $\mathcal{U}$; this, evidently, implies that the family $\mathcal{V}$ is an open disjoint refinement of $\mathcal{U}$, i.e., $X$ is strongly zero-dimensional.

*Fact 1.* Given an uncountable cardinal $\kappa$ let $\Sigma = \{x \in \mathbb{D}^\kappa : |x^{-1}(1)| \leq \omega\}$. Then $P = \mathbb{D}^\kappa \setminus \Sigma$ is a pseudocompact non-countably compact (and hence non-normal) dense subspace of $\mathbb{D}^\kappa$.

*Proof.* We will need the zero element $u \in \mathbb{D}^\kappa$ defined by $u(\alpha) = 0$ for all $\alpha < \kappa$. For any $A \subset \kappa$ let $\chi_A \in \mathbb{D}^\kappa$ be the characteristic function of $A$, i.e., $\chi_A(\alpha) = 1$ for every $\alpha \in A$ and $\chi_A(\alpha) = 0$ whenever $\alpha \notin A$. Let $\Sigma_1 = \{x \in \mathbb{D}^\kappa : |x^{-1}(0)| \leq \omega\}$. It is easy to see that $\Sigma_1$ covers all countable faces of $\mathbb{D}^\kappa$, so it is pseudocompact by Fact 2 of S.433. Furthermore, $\Sigma_1 \subset P$ is a dense subspace of $\mathbb{D}^\kappa$, so $P$ is dense in $\mathbb{D}^\kappa$ and pseudocompact by Fact 18 of S.351.

To see that $P$ is not countably compact choose a disjoint family $\{A_n : n \in \omega\}$ of subsets of $\kappa$ such that $|A_n| = \kappa$ and let $x_n = \chi_{A_n}$ for any $n \in \omega$. The set $\{x_n : n \in \omega\} \subset P$ is closed and discrete in $P$ because the sequence $\{x_n : n \in \omega\}$ converges to $u \notin P$. Therefore $P$ is not countably compact and hence not normal by TFS-137 which shows that Fact 1 is proved.

Returning to our solution apply Fact 1 to take a pseudocompact non-normal dense subspace $X \subset \mathbb{D}^{\omega_1}$. It follows from Fact 2 of S.309 that $\beta X = \mathbb{D}^{\omega_1}$. Apply SFFS-303 and SFFS-306 to see that the space $\mathbb{D}^{\omega_1}$ is strongly zero-dimensional and hence $\dim(\mathbb{D}^{\omega_1}) = 0$. As a consequence, $\dim X = \dim \beta X = \dim(\mathbb{D}^{\omega_1}) = 0$ (see Problem 147); however, $X$ is not strongly zero-dimensional because it is not normal.

**V.149.** *Prove that $\dim X = 0$ implies that $X$ is zero-dimensional. Give an example of a zero-dimensional space $Y$ such that $\dim Y > 0$.*

**Solution.** If $\dim X = 0$ then $\dim \beta X = \dim X = 0$ (see Problem 147); the space $\beta X$ being compact and hence normal, we can apply Problem 148 to see that $\beta X$ is strongly zero-dimensional and hence zero-dimensional by SFFS-309. Therefore $X$ is also zero-dimensional by SFFS-301. To give the required example observe that it follows from SFFS-309 that there exists a normal zero-dimensional space $Y$ which is not strongly zero-dimensional; it follows from Problem 148 that $\dim Y > 0$.

**V.150 (The countable sum theorem for normal spaces).** *Given $n \in \omega$, suppose that a normal space $X$ has a countable closed cover $\mathcal{F}$ such that $\dim F \leq n$ for every $F \in \mathcal{F}$. Prove that $\dim X \leq n$.*

**Solution.** Take an enumeration $\{F_i : i \in \mathbb{N}\}$ of the family $\mathcal{F}$ and fix an open cover $\mathcal{U} = \{U_1, \ldots, U_m\}$ of the space $X$. Let $U_k^0 = U_k$ for every $k \leq m$; then $\mathcal{U}_0 = \{U_1^0, \ldots, U_m^0\}$ is an open cover of $X$. Suppose that $l \in \omega$ and we have constructed open covers $\mathcal{U}_0, \ldots, \mathcal{U}_l$ of the space $X$ with the following properties:

(1) $\mathcal{U}_i = \{U_1^i, \ldots, U_m^i\}$ for every $i \leq l$;
(2) $\overline{U_k^{i+1}} \subset U_k^i$ for any $i < l$ and $k \leq m$;
(3) if $i \in \{1, \ldots, l\}$ then the order of the family $\{U_1^i \cap F_i, \ldots, U_m^i \cap F_i\}$ does not exceed $n + 1$.

Apply Problem 143 to find an open shrinking $\{V_1, \ldots, V_m\}$ of the family $\mathcal{U}_l$ such that $\overline{V}_k \subset U_k^l$ for every $k \leq m$. The family $\mathcal{V} = \{V_1 \cap F_{l+1}, \ldots, V_m \cap F_{l+1}\}$ is an open cover of the space $F_{l+1}$, so there exists an open shrinking $\mathcal{W} = \{W_1, \ldots, W_m\}$ (in the space $F_{l+1}$) of the cover $\mathcal{V}$ such that $\mathrm{ord}(\mathcal{W}) \leq n + 1$ (see Problem 145).

It is easy to check that $U_k^{l+1} = (V_k \backslash F_{l+1}) \cup W_k$ is an open subset of $X$ for each $k \leq m$ and $\mathcal{U}_{l+1} = \{U_1^{l+1}, \ldots, U_m^{l+1}\}$ is a cover of $X$ such that the properties (1)–(3) hold for all $i \leq l + 1$. Therefore our inductive procedure can be continued to construct a sequence $\{\mathcal{U}_i : i \in \omega\}$ of open covers of $X$ such that the conditions (1)–(3) are satisfied for all $i \in \omega$.

It follows from (2) that the set $H_k = \bigcap_{i \in \omega} U_k^i = \bigcap_{i \in \omega} \overline{U_k^i}$ is closed in $X$ and $H_k \subset U_k$ for every $k \leq m$. An easy consequence of (3) is that the order of the family $\mathcal{H} = \{H_1, \ldots, H_m\}$ does not exceed $n + 1$.

Given any point $x \in X$ and $i \in \omega$ there is $k_i \leq m$ such that $x \in U_{k_i}^i$, so we can choose $k \leq m$ and an infinite $A \subset \omega$ such that $k_i = k$ for all $i \in A$. This, together with the property (2), implies that $x \in U_k^i$ for all $i \in \omega$, i.e., $x \in H_k$. Thus $\mathcal{H}$ is a closed shrinking of the cover $\mathcal{U}$ of order at most $n + 1$; applying Problem 145 we conclude that $\dim X \leq n$.

**V.151 (General countable sum theorem).** *Given $n \in \omega$, suppose that we have a closed countable cover $\mathcal{F}$ of a Tychonoff space $X$ such that*

*(i) every $F \in \mathcal{F}$ is $C^*$-embedded in $X$;*
*(ii) $\dim F \leq n$ for each $F \in \mathcal{F}$.*

*Prove that $\dim X \leq n$; give an example of a Tychonoff non-normal space $Y$ such that $\dim Y > 0$ and $Y = \bigcup\{Y_i : i \in \omega\}$, where $Y_i$ is closed in $Y$ and $\dim Y_i = 0$ for every $i \in \omega$.*

**Solution.** For any $F \in \mathcal{F}$ let $H(F) = \mathrm{cl}_{\beta X}(F)$; since $H(F)$ is homeomorphic to $\beta F$ (see Fact 2 of S.451), we have $\dim(H(F)) = \dim F \leq n$ (see Problem 147). The space $Z = \bigcup\{H(F) : F \in \mathcal{F}\}$ is $\sigma$-compact and hence normal while the family $\mathcal{H} = \{H(F) : F \in \mathcal{F}\}$ is a countable closed cover of $Z$ and $\dim H \leq n$ for every $H \in \mathcal{H}$. Therefore we can apply Problem 150 to see that $\dim Z \leq n$; since $X \subset Z \subset \beta X$, we conclude that $\dim X = \dim Z \leq n$ (see Problem 147).

Finally, recall that it was proved in SFFS-312 that there exists a space $Y$ such that $Y = \bigcup\{Y_i : i \in \omega\}$ and every $Y_i$ is closed in $Y$ and strongly zero-dimensional while $Y$ is not zero-dimensional and hence $\dim Y > 0$ (see Problem 149). It follows from Problem 148 that $\dim Y_i = 0$ for each $i \in \omega$, so $Y$ is our promised example.

**V.152.** *Give an example of a compact (and hence normal) space $X$ such that $\dim X = 0$ while $\dim Y > 0$ for some $Y \subset X$.*

**Solution.** There exists a zero-dimensional space $Y$ such that $\dim Y > 0$ (see Problem 149). By SFFS-303, we can assume that $Y \subset \mathbb{D}^\kappa$ for some cardinal $\kappa$. If $X = \mathbb{D}^\kappa$ then $X$ is compact and $\dim X = 0$ (see SFFS-303, Problem 148 and SFFS-306), so $X$ is as promised.

**V.153.** *Give an example of a Tychonoff space $X$ such that $\dim X = 0$ and there exists a closed set $Y \subset X$ with $\dim Y > 0$.*

**Solution.** Fix a zero-dimensional space $Y$ such that $\dim Y > 0$ (see Problem 149). By SFFS-303 and SFFS-306 we can assume that $Y \subset \mathbb{D}^\kappa$ for some uncountable cardinal $\kappa$. Let $\Sigma = \{x \in \mathbb{D}^\kappa : |x^{-1}(1)| \le \omega\}$ and take a set $A \subset \kappa$ such that $|A| = |\kappa \backslash A| = \kappa$. Define a point $u \in \mathbb{D}^{\kappa \backslash A}$ by letting $u(\alpha) = 1$ for every $\alpha \in \kappa \backslash A$. It is straightforward that the set $H = \{u\} \times \mathbb{D}^A$ is homeomorphic to $\mathbb{D}^\kappa$ and $H \subset \mathbb{D}^\kappa \backslash \Sigma$. Therefore we can assume that $Y \subset H$.

The set $\Sigma$ is dense in $\mathbb{D}^\kappa$ and covers all countable faces of $\mathbb{D}^\kappa$, so $\beta \Sigma = \mathbb{D}^\kappa$ (see Fact 2 of S.433 and Fact 2 of S.309). If $X = \Sigma \cup Y$ then $\Sigma \subset X \subset \beta \Sigma$, so $\beta X = \beta \Sigma = \mathbb{D}^\kappa$ by Fact 1 of S.393. Therefore $\dim X = \dim(\mathbb{D}^\kappa) = 0$ (see Problem 147); the set $H$ being closed in $\mathbb{D}^\kappa$, the equality $Y = H \cap X$ shows that $Y$ is closed in $X$. Thus $X$ is a space with $\dim X = 0$ such that $\dim Y > 0$ for some closed subspace $Y \subset X$.

**V.154.** *Let $X$ be a normal space with $\dim X \le n$. Given a subspace $Y \subset X$, suppose that, for every open $U \supset Y$, there exists an $F_\sigma$-set $P$ such that $Y \subset P \subset U$. Prove that $\dim Y \le n$.*

**Solution.** Fix an open cover $\mathcal{U} = \{U_1, \ldots, U_k\}$ of the space $Y$ and choose, for every $i \in \{1, \ldots, k\}$ a set $V_i \in \tau(X)$ such that $V_i \cap Y = U_i$. Then $Y$ is contained in the set $V = V_1 \cup \ldots \cup V_k$, so we can find an $F_\sigma$-set $P$ in the space $X$ such that $Y \subset P \subset V$.

The space $P$ is normal by Fact 1 of S.289. Besides, $P = \bigcup_{i \in \omega} P_i$ where every $P_i$ is closed in $X$ and hence $\dim P_i \le n$ (see Problem 146); therefore $\dim P \le n$ by Problem 150. Apply Problem 143 to find a closed shrinking $\{F_1, \ldots, F_k\}$ of the open cover $\{V_1 \cap P, \ldots, V_k \cap P\}$ of the space $P$. There exist functionally open sets $W_1, \ldots, W_k$ in the space $P$ such that $F_i \subset W_i \subset V_i \cap P$ for every $i \le k$ (see Fact 6 of V.100).

The functionally open cover $\{W_1, \ldots, W_k\}$ of the space $P$ has a functionally open shrinking $\mathcal{O} = \{O_1, \ldots, O_k\}$ such that $\text{ord}(\mathcal{O}) \le n + 1$ (see Problem 144). It is evident that $\{O_1 \cap Y, \ldots, O_k \cap Y\}$ is a functionally open refinement of $\mathcal{U}$ of order at most $n + 1$, so $\dim Y \le n$.

**V.155.** *Prove that, for any perfectly normal space* $X$, *we have* $\dim Y \leq \dim X$ *for any* $Y \subset X$. *In particular,* $\dim Y \leq \dim X$ *for any subspace* $Y$ *of a metrizable space* $X$.

**Solution.** If $\dim X = \infty$ then there is nothing to prove, so assume that $n \in \omega$ and $\dim X = n$. If $Y \subset X$ then any $U \in \tau(Y, X)$ is an $F_\sigma$-subset of $X$ because $X$ is perfect. Therefore Problem 154 is applicable to conclude that $\dim Y \leq n = \dim X$.

**V.156.** *Given* $n \in \omega$ *and a Tychonoff space* $X$, *prove that* $\dim X \leq n$ *if and only if, for any family* $\{(A_0, B_0), \ldots, (A_n, B_n)\}$ *of* $n + 1$ *pairs of disjoint functionally closed sets, it is possible to choose, for each* $i \leq n$, *a functionally closed partition* $C_i$ *between* $A_i$ *and* $B_i$ *in such a way that* $L_0 \cap \ldots \cap L_n = \emptyset$.

**Solution.** Say that a space $Z$ has $(n + 1)$-partition property if, for any family $\{(P_0, Q_0), \ldots, (P_n, Q_n)\}$ of pairs of disjoint functionally closed subsets of $Z$ there exist functionally closed sets $R_0, \ldots, R_n$ such that $\bigcap_{i \leq n} R_i = \emptyset$ while $R_i$ is a partition between $P_i$ and $Q_i$ for every $i \leq n$. We must prove that $\dim X \leq n$ if and only if our space $X$ has $(n + 1)$-partition property.

*Fact 1.* Given a space $Z$ and a natural number $m$ we have $\dim Z \leq m$ if and only if any functionally open cover $\mathcal{U} = \{U_0, \ldots, U_{m+1}\}$ of the space $Z$ has a functionally open shrinking $\mathcal{V} = \{V_0, \ldots, V_{m+1}\}$ such that $\bigcap_{i \leq m+1} V_i = \emptyset$.

*Proof.* If $\dim Z \leq m$ and $\mathcal{U} = \{U_0, \ldots, U_{m+1}\}$ is a functionally open cover of $Z$ then it has a functionally open shrinking $\mathcal{V} = \{V_0, \ldots, V_{m+1}\}$ with $\mathrm{ord}(\mathcal{V}) \leq m + 1$ (see Problem 144). It is evident that $\bigcap_{i \leq m+1} V_i = \emptyset$, so we proved necessity.

To establish sufficiency assume that every functionally open cover of $Z$ of cardinality $m + 2$ has a functionally open shrinking with empty intersection and take an arbitrary functionally open cover $\mathcal{O} = \{O_0, \ldots, O_k\}$ of the space $Z$. Given a functionally open cover $\mathcal{W} = \{W_0, \ldots, W_k\}$ of $Z$ a set $B \subset \{0, \ldots, k\}$ is $\mathcal{W}$-*irregular* if $\bigcap_{i \in B} W_i \neq \emptyset$ but there exists a functionally open shrinking $\mathcal{W}' = \{W_0', \ldots, W_k'\}$ of the family $\mathcal{W}$ such that $\bigcap_{i \in B} W_i' = \emptyset$. Let $r_0$ be the number of $\mathcal{O}$-irregular subsets. Proceeding inductively, assume that $p \geq 0$ and we have a sequence $\mathcal{O}_0, \ldots, \mathcal{O}_p$ of functionally open covers of the space $Z$ with the following properties:

(1) $\mathcal{O}_0 = \mathcal{O}$ and $\mathcal{O}_{i+1}$ is a shrinking of $\mathcal{O}_i$ for all $i < p$;
(2) if $r_i$ is the number of $\mathcal{O}_i$-irregular sets then $r_{i+1} < r_i$ for all $i < p$.

If $r_p > 0$ then take an $\mathcal{O}_p$-irregular set $B$ and a functionally open shrinking $\mathcal{O}_{p+1} = \{O_0', \ldots, O_k'\}$ of the cover $\mathcal{O}_p$ such that $\bigcap_{i \in B} O_i' = \emptyset$. It is evident that $B$ is not $\mathcal{O}_{p+1}$-irregular set and any $\mathcal{O}_{p+1}$-irregular set is also $\mathcal{O}_p$-irregular. This shows that the number $r_{p+1}$ of $\mathcal{O}_{p+1}$-irregular sets is strictly less than $r_p$, so our inductive procedure can be continued as long as $r_p > 0$. Since there are only finitely many $\mathcal{O}_0$-irregular sets, at some step we will obtain a functionally open shrinking $\mathcal{V} = \{V_0, \ldots, V_k\}$ of the family $\mathcal{O}$ for which there are no $\mathcal{V}$-irregular sets. In other words, the family $\mathcal{V}$ is a swelling of any of its functionally open shrinkings.

If $\operatorname{ord}(\mathcal{V}) \leq m+1$ then we already have the needed refinement of the family $\mathcal{O}$. If not, assume without loss of generality that $V_0 \cap \ldots \cap V_{m+1} \neq \emptyset$ and consider the set $G = \bigcup\{V_i : m+1 \leq i \leq k\}$. By our assumption about $Z$ there exists a functionally open shrinking $\mathcal{H} = \{H_0, \ldots, H_{m+1}\}$ of the cover $\{V_0, \ldots, V_m, G\}$ such that $\bigcap \mathcal{H} = \emptyset$. Then the family $\mathcal{H}' = \{H_0, \ldots, H_m, H_{m+1} \cap V_{m+1}, \ldots, H_{m+1} \cap V_k\}$ is a functionally open shrinking of $\mathcal{V}$; since $\mathcal{V}$ is a swelling of $\mathcal{H}'$, we must have $H_0 \cap \ldots \cap H_m \cap (H_{m+1} \cap V_{m+1}) \neq \emptyset$ which contradicts $\bigcap \mathcal{H} = \emptyset$.

Thus any finite functionally open cover of $Z$ has a functionally open shrinking of order at most $m+1$, so $\dim Z \leq m$; this settles sufficiency and shows that Fact 1 is proved.

*Fact 2.* Given a space $Z$ suppose that $C \subset Z$ is a functionally closed set such that $Z \backslash C = U \cup V$ where $U, V \in \tau(Z)$ and $U \cap V = \emptyset$. Then both sets $U$ and $V$ are functionally open.

*Proof.* Fix a function $f \in C(Z)$ such that $C = f^{-1}(0)$ and define a function $g : Z \to \mathbb{R}$ as follows: $g(x) = f(x)$ for any $x \in U$ and $g(x) = 0$ whenever $x \in Z \backslash U$. If $f_0 = g|(U \cup C)$ and $f_1 = G|(V \cup C)$ then $f_0$ and $f_1$ are continuous because $f_0 = f|(U \cup C)$ and $f_1$ is identically zero on $V \cup C$. Since $\operatorname{dom}(f_0) \cap \operatorname{dom}(f_1) = C$ and $f_0|C = f_1|C$, we can apply Fact 2 of T.354 to see that $g$ is continuous. Since $U = Z \backslash g^{-1}(0)$, the set $U$ is functionally open in $Z$. An analogous reasoning shows that the set $V$ is also functionally open, so Fact 2 is proved.

*Fact 3.* For any space $Z$ if $A$ and $B$ are functionally closed disjoint subsets of $Z$ then there exists a functionally closed partition $C$ between the sets $A$ and $B$.

*Proof.* By Fact 1 of V.140 there exist disjoint functionally open sets $U$ and $V$ such that $A \subset U$ and $B \subset V$; it is evident that the set $C = X \backslash (U \cup V)$ is as promised, so Fact 3 is proved.

Returning to our solution assume that $X$ has $(n+1)$-partition property and take a functionally open cover $\mathcal{U} = \{U_0, \ldots, U_{n+1}\}$ of the space $X$. Fix a functionally closed shrinking $\mathcal{F} = \{F_0, \ldots, F_{n+1}\}$ of the cover $\mathcal{U}$ (see Problem 142). There exists a family $\{C_0, \ldots, C_n\}$ of functionally closed sets such that $C_i$ is a partition between the sets $F_i$ and $X \backslash U_i$ for every $i \leq n$ and $\bigcap_{i \leq n} C_i = \emptyset$. For every $i \leq n$ take disjoint sets $V_i \in \tau(F_i, X)$ and $W_i \in \tau(X \backslash U_i, X)$ such that $X \backslash (V_i \cup W_i) = C_i$. The sets $W_i$ and $V_i$ are functionally open for every $i \leq n$ (see Fact 2) and $\bigcup_{i \leq n}(W_i \cup V_i) = X$.

If $W = \bigcup_{i \leq n} W_i$ then $\mathcal{V} = \{V_0, \ldots, V_n, U_{n+1} \cap W\}$ is a functionally open shrinking of $\mathcal{U}$. Indeed, if $x \notin V_0 \cup \ldots \cup V_n$ then $x \notin \bigcup_{i \leq n} F_i$, so $x \in F_{n+1}$ because $\mathcal{F}$ is a cover of $X$; besides, $x \in W$ and hence $x \in W \cap \overline{F}_{n+1} \subset W \cap U_{n+1}$. Now, if $x \in \bigcap \mathcal{V}$ then $x \in W$ and hence $x \in W_i$ for some $i \leq n$; since also $x \in V_i$, we have $V_i \cap W_i \neq \emptyset$ which is a contradiction. Therefore $\bigcap \mathcal{V} = \emptyset$ and hence $\dim X \leq n$ by Fact 1; this settles sufficiency.

Now assume that $\dim X \leq n$ and take a family $\{(A_0, B_0), \ldots, (A_n, B_n)\}$ of pairs of disjoint functionally closed sets. If $B = \bigcup_{i \leq n} B_i$ then $B \cap A_0 \cap \ldots \cap A_n = \emptyset$, so the sets $O_0 = X \backslash A_0, \ldots, O_n = X \backslash A_n, O_{n+1} = X \backslash B$ constitute a functionally open cover of $X$. Choose a functionally open shrinking $\mathcal{O}' = \{O_0', \ldots, O_{n+1}'\}$ of the cover $\mathcal{O} = \{O_0, \ldots, O_{n+1}\}$ with $\bigcap \mathcal{O}' = \emptyset$ (see Fact 1) and take a functionally closed shrinking $\{P_0, \ldots, P_{n+1}\}$ of the family $\mathcal{O}$ (see Problem 142). Letting $U_i = X \backslash P_i$ for all $i \leq n+1$ we obtain a functionally open cover $\mathcal{U} = \{U_0, \ldots, U_{n+1}\}$ of the space $X$ such that $A_i \subset U_i$ for all $i \leq n$ and $\bigcap \mathcal{U} = \emptyset$; let $V_i = U_i \backslash B_i$ for each $i \leq n$.

It is easy to check that $\mathcal{V} = \{V_0, \ldots, V_n, U_{n+1}\}$ is a functionally open cover of $X$; take a functionally closed shrinking $\mathcal{F} = \{F_0, \ldots, F_{n+1}\}$ of the cover $\mathcal{V}$. For every $i \leq n$ consider the functionally closed sets $A_i' = A_i \cup F_i$ and $B_i' = B_i \cup (F_{n+1} \backslash U_i)$; it is easy to check that $A_i \subset A_i'$, $B_i \subset B_i'$ and $A_i' \cap B_i' = \emptyset$. Given a point $x \in X$, if $x \notin \bigcup_{i \leq n} A_i'$ then $x \in F_{n+1}$; it follows from $\bigcap \mathcal{U} = \emptyset$ that $(\bigcap_{i \leq n} U_i) \cap F_{n+1} = \emptyset$, so $x \in F_{n+1} \backslash U_i \subset B_i'$ for some $i \leq n$. This proves that $\bigcup_{i \leq n}(A_i' \cup B_i') = X$. Fix a functionally closed partition $C_i$ between the sets $A_i'$ and $B_i'$ (see Fact 3); then $C_i$ is also a partition between the sets $A_i$ and $B_i$ for every $i \leq n$. Besides, $\bigcap_{i \leq n} C_i \subset X \backslash \bigcup_{i \leq n}(A_i' \cup B_i') = \emptyset$. Therefore $X$ has $(n+1)$-partition property, i.e., we have established necessity and hence our solution is complete.

**V.157.** *Given a natural $n \geq 0$ and a normal space $X$, prove that $\dim X \leq n$ if and only if, for any family $\{(A_0, B_0), \ldots, (A_n, B_n)\}$ of $n+1$ pairs of disjoint closed sets, it is possible to choose, for each $i \leq n$, a partition $C_i$ between $A_i$ and $B_i$ in such a way that $L_0 \cap \ldots \cap L_n = \emptyset$.*

**Solution.** Say that a space $Z$ has $(n+1)$-partition property if, for any family $\{(P_0, Q_0), \ldots, (P_n, Q_n)\}$ of pairs of disjoint closed subsets of $Z$ there exist closed sets $R_0, \ldots, R_n$ such that $\bigcap_{i \leq n} R_i = \emptyset$ while $R_i$ is a partition between $P_i$ and $Q_i$ for every $i \leq n$. We must prove that $\dim X \leq n$ if and only if our space $X$ has $(n+1)$-partition property.

*Fact 1.* Given a normal space $Z$ and a natural number $m$ we have $\dim Z \leq m$ if and only if any open cover $\mathcal{U} = \{U_0, \ldots, U_{m+1}\}$ of the space $Z$ has an open shrinking $\mathcal{V} = \{V_0, \ldots, V_{m+1}\}$ such that $\bigcap_{i \leq m+1} V_i = \emptyset$.

*Proof.* If $\dim Z \leq m$ and $\mathcal{U} = \{U_0, \ldots, U_{m+1}\}$ is an open cover of $Z$ then it has an open shrinking $\mathcal{V} = \{V_0, \ldots, V_{m+1}\}$ with $\text{ord}(\mathcal{V}) \leq m + 1$ (see Problem 145). It is evident that $\bigcap_{i \leq m+1} V_i = \emptyset$, so we proved necessity.

To establish sufficiency assume that every open cover of $Z$ of cardinality $m + 2$ has an open shrinking with empty intersection and take an arbitrary open cover $\mathcal{O} = \{O_0, \ldots, O_k\}$ of the space $Z$. Given an open cover $\mathcal{W} = \{W_0, \ldots, W_k\}$ of $Z$, a set $B \subset \{0, \ldots, k\}$ will be called $\mathcal{W}$-*irregular* if $\bigcap_{i \in B} W_i \neq \emptyset$ but there exists an open shrinking $\mathcal{W}' = \{W_0', \ldots, W_k'\}$ of the family $\mathcal{W}$ such that $\bigcap_{i \in B} W_i' = \emptyset$. Let $r_0$ be the number of $\mathcal{O}$-irregular subsets. Proceeding inductively, assume that $p \geq 0$ and we have a sequence $\mathcal{O}_0, \ldots, \mathcal{O}_p$ of open covers of the space $Z$ with the following properties:

(1) $\mathcal{O}_0 = \mathcal{O}$ and $\mathcal{O}_{i+1}$ is a shrinking of $\mathcal{O}_i$ for all $i < p$;
(2) if $r_i$ is the number of $\mathcal{O}_i$-irregular sets then $r_{i+1} < r_i$ for all $i < p$.

If $r_p > 0$ then we can find an $\mathcal{O}_p$-irregular set $B \subset \{0, \ldots, k\}$ and an open shrinking $\mathcal{O}_{p+1} = \{O'_0, \ldots, O'_k\}$ of the cover $\mathcal{O}_p$ such that $\bigcap_{i \in B} O'_i = \emptyset$. It is clear that $B$ is not $\mathcal{O}_{p+1}$-irregular set and any $\mathcal{O}_{p+1}$-irregular set is also $\mathcal{O}_p$-irregular. This shows that the number $r_{p+1}$ of $\mathcal{O}_{p+1}$-irregular sets is strictly less than $r_p$, so our inductive procedure can be continued as long as $r_p > 0$. Since there are only finitely many $\mathcal{O}_0$-irregular sets, at some step we will obtain an open shrinking $\mathcal{V} = \{V_0, \ldots, V_k\}$ of the family $\mathcal{O}$ for which there are no $\mathcal{V}$-irregular sets. In other words, the family $\mathcal{V}$ is a swelling of any of its open shrinkings.

If $\operatorname{ord}(\mathcal{V}) \leq m + 1$ then we already have the needed refinement of the family $\mathcal{O}$. If not, assume without loss of generality that $V_0 \cap \ldots \cap V_{m+1} \neq \emptyset$ and consider the set $G = \bigcup\{V_i : m + 1 \leq i \leq k\}$. By our assumption about $Z$ there exists an open shrinking $\mathcal{H} = \{H_0, \ldots, H_{m+1}\}$ of the cover $\{V_0, \ldots, V_m, G\}$ such that $\bigcap \mathcal{H} = \emptyset$. The family $\mathcal{H}' = \{H_0, \ldots, H_m, H_{m+1} \cap V_{m+1}, \ldots, H_{m+1} \cap V_k\}$ is an open shrinking of $\mathcal{V}$; since $\mathcal{V}$ is a swelling of $\mathcal{H}'$, we have $H_0 \cap \ldots \cap H_m \cap (H_{m+1} \cap V_{m+1}) \neq \emptyset$ which contradicts $\bigcap \mathcal{H} = \emptyset$.

Thus any finite open cover of $Z$ has an open shrinking of order at most $m + 1$, so $\dim Z \leq m$; this settles sufficiency and shows that Fact 1 is proved.

*Fact 2.* For any normal space $Z$ if $A$ and $B$ are closed disjoint subsets of $Z$ then there exists a closed partition $C$ between the sets $A$ and $B$.

*Proof.* By normality of $Z$ there exist disjoint open sets $U$ and $V$ such that $A \subset U$ and $B \subset V$; it is evident that the set $C = X \setminus (U \cup V)$ is as promised, so Fact 2 is proved.

Returning to our solution assume that $X$ has $(n + 1)$-partition property and take an open cover $\mathcal{U} = \{U_0, \ldots, U_{n+1}\}$ of the space $X$. Fix a closed shrinking $\mathcal{F} = \{F_0, \ldots, F_{n+1}\}$ of the cover $\mathcal{U}$ (see Problem 143). There exists a family $\{C_0, \ldots, C_n\}$ of closed sets such that $C_i$ is a partition between the sets $F_i$ and $X \setminus U_i$ for every $i \leq n$ and $\bigcap_{i \leq n} C_i = \emptyset$. For every $i \leq n$ take disjoint sets $V_i \in \tau(F_i, X)$ and $W_i \in \tau(X \setminus U_i, X)$ such that $X \setminus (V_i \cup W_i) = C_i$; then $\bigcup_{i \leq n} (W_i \cup V_i) = X$.

If $W = \bigcup_{i \leq n} W_i$ then $\mathcal{V} = \{V_0, \ldots, V_n, U_{n+1} \cap W\}$ is an open shrinking of $\mathcal{U}$. Indeed, if $x \notin V_0 \cup \ldots \cup V_n$ then $x \notin \bigcup_{i \leq n} F_i$, so $x \in F_{n+1}$ because $\mathcal{F}$ is a cover of $X$; besides, $x \in W$ and hence $x \in W \cap F_{n+1} \subset W \cap U_{n+1}$. Now, if $x \in \bigcap \mathcal{V}$ then $x \in W$ and hence $x \in W_i$ for some $i \leq n$; since also $x \in V_i$, we have $V_i \cap W_i \neq \emptyset$ which is a contradiction. Therefore $\bigcap \mathcal{V} = \emptyset$ and hence $\dim X \leq n$ by Fact 1; this settles sufficiency.

Now assume that $\dim X \leq n$ and take a family $\{(A_0, B_0), \ldots, (A_n, B_n)\}$ of pairs of disjoint closed sets. If $B = \bigcup_{i \leq n} B_i$ then $B \cap A_0 \cap \ldots \cap A_n = \emptyset$, so the sets $O_0 = X \setminus A_0, \ldots, O_n = X \setminus A_n, O_{n+1} = X \setminus B$ constitute an open cover of $X$. Choose an open shrinking $\mathcal{O}' = \{O'_0, \ldots, O'_{n+1}\}$ of the cover $\mathcal{O} = \{O_0, \ldots, O_{n+1}\}$ with $\bigcap \mathcal{O}' = \emptyset$ (see Fact 1) and take a closed shrinking $\{P_0, \ldots, P_{n+1}\}$ of the family $\mathcal{O}$ (see Problem 143). Letting $U_i = X \setminus P_i$ for all $i \leq n + 1$ we obtain an open cover $\mathcal{U} = \{U_0, \ldots, U_{n+1}\}$ of the space $X$ such that $A_i \subset U_i$ for all $i \leq n$ and $\bigcap \mathcal{U} = \emptyset$; let $V_i = U_i \setminus B_i$ for each $i \leq n$.

It is easy to check that $\mathcal{V} = \{V_0, \ldots, V_n, U_{n+1}\}$ is an open cover of $X$; take a closed shrinking $\mathcal{F} = \{F_0, \ldots, F_{n+1}\}$ of the cover $\mathcal{V}$. For every $i \leq n$ consider the closed sets $A_i' = A_i \cup F_i$ and $B_i' = B_i \cup (F_{n+1} \backslash U_i)$; it is easy to check that $A_i \subset A_i'$, $B_i \subset B_i'$ and $A_i' \cap B_i' = \emptyset$. Given a point $x \in X$, if $x \notin \bigcup_{i \leq n} A_i'$ then $x \in F_{n+1}$; it follows from $\bigcap \mathcal{U} = \emptyset$ that $(\bigcap_{i \leq n} U_i) \cap F_{n+1} = \emptyset$, so $x \in F_{n+1} \backslash U_i \subset B_i'$ for some $i \leq n$. This proves that $\bigcup_{i \leq n} (A_i' \cup B_i') = X$. Fix a closed partition $C_i$ between the sets $A_i'$ and $B_i'$ (see Fact 2); then $C_i$ is also a partition between the sets $A_i$ and $B_i$ for every $i \leq n$. Besides, $\bigcap_{i \leq n} C_i \subset X \backslash \bigcup_{i \leq n} (A_i' \cup B_i') = \emptyset$. Therefore $X$ has $(n + 1)$-partition property, i.e., we have established necessity and hence our solution is complete.

**V.158.** *Let $X$ be a normal space. Prove that* $\dim X \leq n$ *if and only if, for any closed $F \subset X$ and any continuous map $f : F \to S^n$, there exists a continuous map $g : X \to S^n$ such that $g|F = f$. Here $S^n = \{(x_0, \ldots, x_n) \in \mathbb{R}^{n+1} : x_0^2 + \ldots + x_n^2 = 1\}$ is the $n$-dimensional sphere with the topology inherited from $\mathbb{R}^{n+1}$.*

**Solution.** Given $m \in \omega$ say that a space $Z$ has $(m + 1)$-partition property if, for any family $\{(P_0, Q_0), \ldots, (P_m, Q_m)\}$ of pairs of disjoint closed subsets of $Z$ there exist closed sets $R_0, \ldots, R_m$ such that $\bigcap_{i \leq m} R_i = \emptyset$ while $R_i$ is a partition between $P_i$ and $Q_i$ for every $i \leq m$. A space $Z$ will be called an $S^m$-*extensor* if, for every closed $F \subset Z$ and any continuous map $f : F \to S^m$ there exists a continuous map $g : Z \to S^m$ such that $g|F = f$. The set $\mathbb{B}_m = \{x \in \mathbb{I}^{m+1} : |x(i)| = 1$ for some $i \leq m\}$ is the boundary of the $(m + 1)$-dimensional cube $\mathbb{I}^{m+1}$. Let $P_i^m = \{x \in \mathbb{I}^{m+1} : x(i) = -1\}$ and $Q_i^m = \{x \in \mathbb{I}^{m+1} : x(i) = 1\}$ be the respective faces of $\mathbb{I}^{m+1}$ for every $i \leq m$. We must prove that, for a normal space $X$, we have $\dim X \leq n$ if and only if $X$ is an $S^n$-extensor.

*Fact 1.* For any $m \in \omega$, a normal space $Z$ is an $S^m$-extensor if and only if $\beta Z$ is an $S^m$-extensor.

*Proof.* Suppose first that $Z$ is an $S^m$-extensor and fix a closed set $K \subset \beta Z$. If $f : K \to S^m$ is a continuous map then there exists a set $O \in \tau(K, \beta Z)$ and a continuous map $f_0 : O \to S^m$ such that $f_0|K = f$ (see Fact 1 of V.093). The space $\beta Z$ being normal we can find an open subset $W$ of the space $\beta Z$ for which $K \subset W \subset \overline{W} \subset O$ (the bar denotes the closure in $\beta Z$).

The set $F = \overline{W} \cap Z$ is closed in $Z$ and the map $f_1 = f_0|F : F \to S^m$ is continuous; the space $Z$ being an $S^m$-extensor, we can choose $g_0 \in C(Z, S^m)$ such that $g_0|F = f_1$. Since the space $S^m$ is compact, there exists a continuous map $g : \beta Z \to S^m$ for which $g|Z = g_0$.

The functions $g_1 = g|\overline{W}$ and $h = f_0|\overline{W}$ are continuous; besides, the set $F$ is dense in $\overline{W}$ and $g_1|F = g_0|F = f_1 = f_0|F = h|F$. An immediate consequence is that $g_1 = h$; recalling that $K \subset \overline{W}$, we conclude that $g|K = g_1|K = h|K = f$, so $g$ is a continuous extension of $f$ and hence $\beta Z$ is an $S^m$-extensor; this proves necessity.

Now, if $\beta Z$ is an $S^m$-extensor then fix a closed set $F$ in the space $Z$ and a continuous function $f : F \to S^m$. The set $\overline{F}$ is canonically homeomorphic to $\beta F$ (see Fact 2 of S.451), so there exists a continuous map $f_0 : \overline{F} \to S^m$ such that $f_0|F = f$. The space $\beta Z$ being an $S^m$-extensor, there is a continuous map $g_0 : \beta Z \to S^m$ with $g_0|\overline{F} = f_0$. Therefore $g = g_0|Z : Z \to S^m$ is a continuous map such that $g|F = f$; this settles sufficiency and shows that Fact 1 is proved.

*Fact 2.* Given a continuous map $f : Y \to Z$ suppose that $A, B \subset Y$ and a set $C' \subset Z$ is a partition between $f(A)$ and $f(B)$. Then $C = f^{-1}(C')$ is a partition between $A$ and $B$.

*Proof.* There exist disjoint sets $U' \in \tau(f(A), Y)$ and $V' \in \tau(f(B), Y)$ such that $Y \backslash (U' \cup V') = C'$. If $U = f^{-1}(U')$ and $V = f^{-1}(V')$ then the sets $U$ and $V$ are disjoint while $U \in \tau(A, X)$, $V \in \tau(B, X)$ and $C = X \backslash (U \cup V)$, so $C$ is a partition between $A$ and $B$, i.e., Fact 2 is proved.

Returning to our solution suppose that $\dim X \leq n$; then $\dim \beta X \leq n$ by Problem 147, so the space $\beta X$ has the $(n+1)$-partition property (see Problem 157). This makes it possible to apply Fact 7 of V.100 to see that $\beta X$ is an $S^n$-extensor, so $X$ is also an $S^n$-extensor by Fact 1. This proves necessity.

Now assume that $X$ is an $S^n$-extensor and fix a family $\{(A_0, B_0), \ldots, (A_n, B_n)\}$ of pairs of disjoint closed subsets of $X$. Using normality of $X$ it is easy to find a continuous function $\varphi_i : X \to \mathbb{I}$ such that $\varphi_i(A_i) \subset \{-1\}$ and $\varphi_i(B_i) \subset \{1\}$ for all $i \leq n$. The diagonal product $\varphi = \Delta_{i \leq n}\varphi_i : X \to \mathbb{I}^{n+1}$ is continuous while $\varphi(A_i) \subset P_i^n$ and $\varphi(B_i) \subset Q_i^n$ for each $i \leq n$.

Therefore $\bigcup_{i \leq n}(A_i \cup B_i) \subset F = \varphi^{-1}(\mathbb{B}_n)$. The space $\mathbb{B}_n$ being homeomorphic to $S^n$ (see Fact 1 of V.094) there exists a continuous map $h : X \to \mathbb{B}_n$ such that $h|F = \varphi|F$. The set $C_i' = \{x \in \mathbb{I}^{n+1} : x(i) = 0\}$ is, evidently, a partition between the sets $P_i^n$ and $Q_i^n$, so $C_i'$ is also a partition between $\varphi(A_i) = h(A_i)$ and $\varphi(B_i) = h(B_i)$; by Fact 1, the set $C_i = h^{-1}(C_i')$ is a partition between the sets $A_i$ and $B_i$ for all $i \leq n$. If $x \in \bigcap_{i \leq n} C_i$ then $h(x)(i) = 0$ for all $i \leq n$; this contradiction with $h(x) \in \mathbb{B}_n$ shows that $\bigcap_{i \leq n} C_i = \emptyset$ and hence we proved that $X$ has $(n+1)$-partition property. This, together with Problem 157, shows that $\dim X \leq n$, so we established sufficiency and hence our solution is complete.

**V.159.** *Prove that* $\dim(\mathbb{I}^n) = \dim(\mathbb{R}^n) = \dim(S^n) = n$ *for any* $n \in \mathbb{N}$. *Here* $S^n = \{(x_0, \ldots, x_n) \in \mathbb{R}^{n+1} : x_0^2 + \ldots + x_n^2 = 1\}$ *is the n-dimensional sphere with the topology inherited from* $\mathbb{R}^{n+1}$.

**Solution.** Given $m \in \omega$ say that a space $Z$ has $(m+1)$-partition property if, for any family $\{(P_0, Q_0), \ldots, (P_m, Q_m)\}$ of pairs of disjoint closed subsets of $Z$ there exist closed sets $R_0, \ldots, R_m$ such that $\bigcap_{i \leq m} R_i = \emptyset$ while $R_i$ is a partition between $P_i$ and $Q_i$ for every $i \leq m$. The set $\mathbb{B}_m = \{x \in \mathbb{I}^{m+1} : |x(i)| = 1$ for some $i \leq m\}$ is the boundary of the $(m+1)$-dimensional cube $\mathbb{I}^{m+1}$.

Observe first that the set $G = \{x \in \mathbb{I}^{n+1} : x(n) = 1\}$ is a subspace of $\mathbb{B}_n$ homeomorphic to $\mathbb{I}^n$, so $\mathbb{I}^n$ embeds in $\mathbb{B}_n$; the spaces $S^n$ and $\mathbb{B}_n$ are homeomorphic by Fact 1 of V.094, so $\mathbb{I}^n$ embeds in $S^n$ and hence $\dim(\mathbb{I}^n) \leq \dim(S^n)$ (see Problem 155).

The space $(-1, 1)^n \subset \mathbb{I}^n$ is homeomorphic to $\mathbb{R}^n$, so $\mathbb{R}^n$ embeds in $\mathbb{I}^n$ and therefore $\dim(\mathbb{R}^n) \leq \dim(\mathbb{I}^n)$. It follows from $\mathbb{I}^n \subset \mathbb{R}^n$ that $\dim(\mathbb{I}^n) \leq \dim(\mathbb{R}^n)$, so $\dim(\mathbb{I}^n) = \dim(\mathbb{R}^n) \leq \dim(S^n)$.

There exist zero-dimensional spaces $A_0, \ldots, A_n$ such that $S^n = A_0 \cup \ldots \cup A_n$ (see Fact 8 of V.100), so the space $S^n$ has $(n + 1)$-partition property by Fact 4 of V.100. Now apply Problem 157 to see that $\dim(S^n) \leq n$. It was proved in Problem 094 that the space $\mathbb{I}^n$ does not have $n$-partition property; this, together with Problem 157, implies that $\dim(\mathbb{I}^n) > n - 1$. Therefore $n - 1 < \dim(\mathbb{I}^n) = \dim(\mathbb{R}^n) \leq \dim(S^n) \leq n$ which shows that $\dim(\mathbb{I}^n) = \dim(\mathbb{R}^n) = \dim(S^n) = n$.

**V.160.** *Given $n \in \mathbb{N}$ prove that, for any set $X \subset \mathbb{R}^n$, we have $\dim X = n$ if and only if the interior of $X$ in $\mathbb{R}^n$ is nonempty.*

**Solution.** For any number $m \in \mathbb{N}$ and $x, y \in \mathbb{R}^m$ such that $x = (x_1, \ldots, x_m)$ and $y = (y_1, \ldots, y_m)$, let $x \pm y = (x_1 \pm y_1, \ldots, x_m \pm y_m)$ and $(x, y) = x_1 y_1 + \ldots + x_m y_m$; as usual, we let $|x|_m = \sqrt{x_1^2 + \ldots + x_m^2}$ and $tx = (tx_1, \ldots, tx_m)$ for any $t \in \mathbb{R}$. Denote by $\mathbf{0}_m$ the point of $\mathbb{R}^m$ whose all coordinates are equal to zero. We use the symbol $\mathbb{P}$ to denote the set of irrational numbers. If $L$ is a linear space then $\mathbf{0}_L$ is the zero vector of $L$.

*Fact 1.* Suppose that $A, B \subset \mathbb{R}$ are countable dense subsets and $f : A \to B$ is a bijection such that $a < b$ implies $f(a) < f(b)$. Then there exists a unique homeomorphism $h : \mathbb{R} \to \mathbb{R}$ such that $h|A = f$ and $x < y$ implies $h(x) < h(y)$.

*Proof.* If $h_0, h_1 \in C(\mathbb{R})$ and $h_0|A = f = h_1|A$ then $h_0 = h_1$ by Fact 0 of S.351; this proves that the promised homeomorphism is unique (if it exists).

To establish existence, for any $x \in \mathbb{R}$ let $h(x) = \sup\{f(a) : a \in A \text{ and } a < x\}$. Observe first that $h(x)$ is well defined for if $q > x$ and $q \in A$ then $f(q)$ is an upper bound for the set $\{f(a) : a \in A, a < x\}$. It is immediate from the definitions that $h(a) \leq f(a)$ for any $a \in A$. If $h(a) < f(a)$ then, by density of $B = f(A)$ there is $a' \in A$ such that $h(a) < f(a') < f(a)$; then $a' < a$ and therefore $f(a') \leq h(a)$ which is a contradiction. Thus $h(a) = f(a)$ for every $a \in A$. If $x, y \in \mathbb{R}$ and $x < y$ then take $a, b \in A$ such that $x < a < b < y$. Then $h(x) \leq f(a) < f(b) \leq h(y)$, so we proved that

(1) $h|A = f$ and $h(x) < h(y)$ whenever $x < y$ and, in particular, $h$ is injective.

Now, take any point $t \in \mathbb{R}$ and consider the set $P = \{b \in B : b < t\}$. Since $f$ is a bijection, there is $a \in A$ for which $f(a) > t$; then $x < a$ for any $x \in f^{-1}(P)$. Therefore the point $y = \sup(f^{-1}(P))$ is well defined; if $x < y$ and $x \in A$ then there is $z \in f^{-1}(P)$ with $x < z$, so $f(x) < f(z) < t$ which shows that $f(x) < t$ and therefore $h(y) \leq t$. If $h(y) < t$ then there is $c \in A$ such that $h(y) < f(c) < t$. Thus $c \in f^{-1}(P)$ and hence $c \leq \sup(f^{-1}(P)) = y$, so $f(c) \leq h(y)$, a contradiction. This proves that $h(y) = t$ and hence $h : \mathbb{R} \to \mathbb{R}$ is a bijection. It follows from (1) that $h((a, b)) = (h(a), h(b))$ and $h^{-1}((a, b)) = (h^{-1}(a), h^{-1}(b))$ for any $a, b \in \mathbb{R}$ with $a < b$. Therefore $h$ is a homeomorphism, so Fact 1 is proved.

*Fact 2.* If $k \in \mathbb{N}$ then any linear subspace of $\mathbb{R}^k$ is closed in $\mathbb{R}^k$; besides, if $L \subset \mathbb{R}^k$ is a linear subspace and $\mathrm{Int}(L) \neq \emptyset$ then $L = \mathbb{R}^k$.

*Proof.* If $L$ is a linear subspace of $\mathbb{R}^k$ then there are linearly independent vectors $w_1, \ldots, w_l \in \mathbb{R}^k$ such that $L$ is the linear hull of $\{w_1, \ldots, w_l\}$. There exist vectors $\{w_{l+1}, \ldots, w_k\}$ such that $W = \{w_1, \ldots, w_k\}$ is a linear basis in $\mathbb{R}^k$ (see Fact 1 of S.489). For any $i = 1, \ldots, k$, let $e_i = (e_1^i, \ldots, e_k^i) \in \mathbb{R}^k$ be the point for which $e_i^i = 1$ and $e_j^i = 0$ whenever $i \neq j$. Take a set $\{a_j^i : i, j \in \{1, \ldots, k\}\} \subset \mathbb{R}$ such that $e_i = a_1^i w_1 + \ldots + a_k^i w_k$ for every $i \leq k$. It is straightforward that $L = \{x = (x_1, \ldots, x_k) \in \mathbb{R}^k : \sum_{i=1}^{k} x_i a_m^i = 0 \text{ for all } m = l + 1, \ldots, k\}$, so $L$ is closed being a finite intersection of inverse images of zero under linear (and hence continuous) functions on $\mathbb{R}^k$.

Now suppose that $p \in L$ belongs to the interior of $L$ in $\mathbb{R}^k$ and let $\varphi(x) = x - p$ for all $x \in \mathbb{R}^k$. It is clear that $\varphi : \mathbb{R}^k \to \mathbb{R}^k$ is a homeomorphism such that $\varphi(L) \subset L$ and $\varphi(p) = \mathbf{0}_k$. Therefore $\mathbf{0}_k$ also belongs to the interior of $L$ and hence there is $\varepsilon > 0$ such that $B = \{x \in \mathbb{R}^k : |x|_k < \varepsilon\} \subset L$. Thus $\frac{\varepsilon}{2} e_i \in B \subset L$ for every $i \leq k$; the family $\{\frac{\varepsilon}{2} e_i : i \leq k\} \subset L$ being a basis in $\mathbb{R}^k$, we have $L = \mathbb{R}^k$, so Fact 2 is proved.

*Fact 3.* Suppose that $m \geq 2$ is a natural number and $L$ is an $m$-dimensional linear space. If $L_k$ is an $(m-1)$-dimensional linear subspace of $L$ for any $k \in \omega$ then $L \backslash \bigcup_{k \in \omega} L_k \neq \emptyset$.

*Proof.* Fix a linear basis $\{e_1, \ldots, e_m\}$ in the space $L$. If $x = (x_1, \ldots, x_m) \in \mathbb{R}^m$ let $\varphi(x) = x_1 e_1 + \ldots + x_m e_m$; then the map $\varphi : \mathbb{R}^m \to L$ is an isomorphism, so $M_k = \varphi^{-1}(L_k)$ is an $(m-1)$-dimensional linear subspace of $\mathbb{R}^m$ for every $k \in \omega$. Every set $M_k$ is nowhere dense in $\mathbb{R}^m$ by Fact 2; since the space $\mathbb{R}^m$ has the Baire property, we can find a point $y \in \mathbb{R}^k \backslash \bigcup_{i \in \omega} M_i$. Then $x = \varphi(y) \in L \backslash \bigcup_{k \in \omega} L_k$, so Fact 3 is proved.

*Fact 4.* Suppose that $m \geq 2$ is a natural number and $L$ is an $m$-dimensional linear space. If $P \subset L \backslash \{\mathbf{0}_L\}$ is a countable set then there exists an $(m-1)$-dimensional linear subspace $M \subset L$ such that $M \cap P = \emptyset$.

*Proof.* Fix a linear basis $\{e_1, \ldots, e_m\}$ in the space $L$. If $x = (x_1, \ldots, x_m) \in \mathbb{R}^m$ let $\varphi(x) = x_1 e_1 + \ldots + x_m e_m$; then the map $\varphi : \mathbb{R}^m \to L$ is an isomorphism, so $Q = \varphi^{-1}(P)$ is a countable subset of $\mathbb{R}^m$. The set $M[c] = \{x \in \mathbb{R}^m : (x, c) = 0\}$ is a proper linear subspace of $\mathbb{R}^m$, so it is closed and nowhere dense in $\mathbb{R}^m$ for any $c \in Q$ (see Fact 2). By Fact 3 there is a vector $x \in \mathbb{R}^m \backslash \bigcup\{M[c] : c \in Q\}$; then $R = M[x]$ is an $(m-1)$-dimensional linear subspace of $\mathbb{R}^m$ with $R \cap Q = \emptyset$. Therefore $M = \varphi(R)$ is an $(m-1)$-dimensional linear subspace of $L$ such that $M \cap P = \emptyset$, so Fact 4 is proved.

*Fact 5.* Suppose that $m \in \mathbb{N}$ and $L$ is an $m$-dimensional linear space. Then, for any countable $P \subset L \backslash \{\mathbf{0}_L\}$ there exists a linear basis $E = \{e_1, \ldots, e_m\}$ in the space $L$ such that for every $p \in P$ all coordinates of $p$ with respect to $E$ are nonzero, i.e., there are $p_1, \ldots, p_m \in \mathbb{R} \backslash \{0\}$ such that $p = p_1 e_1 + \ldots + p_m e_m$.

*Proof.* If $m = 1$ then any basis will do so assume that $m > 1$. Apply Fact 4 to find an $(m - 1)$-dimensional linear subspace $M \subset L$ such that $M \cap P = \emptyset$; let $E' = \{e_1, \ldots, e_{m-1}\}$ be a linear basis in $M$. Denote by $H_i(p)$ the linear hull of the set $(E' \backslash \{e_i\}) \cup \{p\}$; then $H_i(p)$ is an $(m - 1)$-dimensional linear subspace of $L$ for any $i \le m - 1$ and $p \in P$. Apply Fact 3 to see that there exists a point $e_m \in L \backslash (M \cup \bigcup \{H_i(p) : i \le m - 1, \ p \in P\})$.

It is evident that $E = \{e_1, \ldots, e_m\}$ is a linear basis in $L$. Take any $p \in P$; then $p = p_1 e_1 + \ldots + p_m e_m$ for some $p_1, \ldots, p_m \in \mathbb{R}$. If $p_j = 0$ then $1 \le j \le m - 1$ because $p$ does not belong to $M$. Therefore $e_m = \frac{1}{p_m}(p_1 e_1 + \ldots + p_{m-1} e_{m-1} - p)$ belongs to the linear hull of the set $(E' \backslash \{e_j\}) \cup \{p\}$ which is a contradiction with the choice of $e_m$. Therefore $p_i \ne 0$ for any $i \le m$, i.e., the basis $E$ is as promised, so Fact 5 is proved.

*Fact 6.* Given $m \in \mathbb{N}$ suppose that $L$ is an $m$-dimensional linear space and $P \subset L$ is a countable set. Then there exists a linear basis $E = \{e_1, \ldots, e_m\}$ in the space $L$ such that $P$ is *in general position with respect to* $E$, i.e., for any distinct $p, q \in P$ if $p_1, q_1, \ldots, p_m, q_m \in \mathbb{R}$ and $p = \sum_{i \le m} p_i e_i$, $q = \sum_{i \le m} q_i e_i$ then $p_i \ne q_i$ for all $i \le m$.

*Proof.* The set $A = \{p - q : p, q \in P$ and $p \ne q\} \subset L \backslash \{0_L\}$ is countable, so we can apply Fact 5 to find a linear basis $E = \{e_1, \ldots, e_m\}$ in $L$ such that all coordinates with respect to $E$ of every element of $A$ are distinct from zero. It is clear that $E$ is as promised, so Fact 6 is proved.

*Fact 7.* If $n \in \mathbb{N}$ and $P \subset \mathbb{R}^n$ is a countable set then there is homeomorphism $\varphi : \mathbb{R}^n \to \mathbb{R}^n$ such that $\varphi(P)$ is in *general position*, i.e., for any distinct $p, q \in \varphi(P)$ if $p = (p_1, \ldots, p_n)$ and $q = (q_1, \ldots, q_n)$ then $p_i \ne q_i$ for all $i \le n$.

*Proof.* By Fact 6, there is a linear basis $E = \{w_1, \ldots, w_n\}$ in the space $\mathbb{R}^n$ such that $P$ is in general position with respect to $E$. Let $\{e_1, \ldots, e_n\}$ be the standard linear basis of $\mathbb{R}^n$, i.e., $e_i = (e_1^i, \ldots, e_n^i)$ while $e_i^i = 1$ and $e_j^i = 0$ whenever $i$ and $j$ are distinct elements of $\{1, \ldots, n\}$.

Choose a set $\{a_j^i : i, j \in \{1, \ldots, n\}\} \subset \mathbb{R}$ such that $e_i = a_1^i w_1 + \ldots + a_n^i w_n$ for every $i \le n$. For each point $x = (x_1, \ldots, x_n) \in \mathbb{R}^n$ let $\varphi(x) = (y_1, \ldots, y_n)$ where $y_j = x_1 a_j^1 + \ldots + x_n a_j^n$ for all $j \le n$. The map $\varphi : \mathbb{R}^n \to \mathbb{R}^n$ is a linear homeomorphism and the coordinates of every $\varphi(p)$ coincide with the coordinates of $p$ with respect to $E$. Therefore $\varphi(P)$ is as promised, so Fact 7 is proved.

*Fact 8.* For any $n \in \mathbb{N}$, if $A$ and $B$ are countable dense subspaces of $\mathbb{R}^n$ then there exists a homeomorphism $h : \mathbb{R}^n \to \mathbb{R}^n$ such that $h(A) = B$.

*Proof.* By Fact 7, there is no loss of generality to assume that both $A$ and $B$ are in general position; let $\{a_k : k \in \omega\}$ and $\{b_k : k \in \omega\}$ be some faithful enumerations of $A$ and $B$ respectively. We will choose inductively new faithful enumerations $\{c_k = (c_1^k, \ldots, c_n^k) : k \in \omega\}$ and $\{d_k = (d_1^k, \ldots, d_n^k) : k \in \omega\}$ of the sets $A$ and $B$ respectively in such a way that

(2) $\{a_0, \ldots, a_k\} \subset \{c_0, \ldots, c_{2k}\}$ and $\{b_0, \ldots, b_k\} \subset \{d_0, \ldots, d_{2k}\}$ for any $k \in \omega$;
(3) $(c_i^k - c_i^l)(d_i^k - d_i^l) > 0$ for any distinct $k, l \in \omega$ and $i \leq n$.

To start off, let $c_0 = a_0$ and $d_0 = b_0$ and suppose that we have chosen points $\{c_0, \ldots, c_{2m}\} \subset A$ and $\{d_0, \ldots, d_{2m}\} \subset B$, so that the property (2) is fulfilled for all $k \leq m$ and the property (3) holds for all distinct $k, l \leq 2m$.

Take the minimal $j \in \mathbb{N}$ for which $a_j \notin \{c_0, \ldots, c_{2m}\}$ and let $c_{2m+1} = a_j$. For every $i \leq n$ the set $O_i = \{t \in \mathbb{R} : (c_i^{2m+1} - c_i^k)(t - d_i^k) > 0$ for all $k \leq 2m\}$ is a nonempty open interval of $\mathbb{R}$, so $O = O_1 \times \ldots \times O_n$ is a nonempty open subset of $\mathbb{R}^n$ which shows that we can choose a point $d_{2m+1} \in (B \backslash \{d_0, \ldots, d_{2m}\}) \cap O$.

Take the minimal $l \in \mathbb{N}$ for which $b_l \notin \{d_0, \ldots, d_{2m+1}\}$ and let $d_{2m+2} = b_l$. The set $U_i = \{t \in \mathbb{R} : (t - c_i^k)(d_i^{2m+2} - d_i^k) > 0$ for all $k \leq 2m + 1\}$ is a nonempty open interval of $\mathbb{R}$, so $U = U_1 \times \ldots \times U_n$ is a nonempty open subset of $\mathbb{R}^n$; choose a point $c_{2m+2} \in A \backslash \{a_0, \ldots, a_{2m+1}\}) \cap U$. It is straightforward that the condition (2) is fulfilled for $k \leq m + 1$ and (3) holds for all distinct $k, l \leq 2m + 2$, so our inductive procedure can be continued to construct sequences $\{c_k : k \in \omega\}$ and $\{d_k : k \in \omega\}$ with the properties (2) and (3). It follows from (2) and (3) that these sequences constitute a faithful enumeration of the sets $A$ and $B$ respectively.

Let $C_i = \{c_i^k : k \in \omega\}$ and $D_i = \{d_i^k : k \in \omega\}$; letting $f_i(c_i^k) = d_i^k$ for any $k \in \omega$ we obtain a bijection $f_i : C_i \to D_i$ for every $i \leq n$. An easy consequence of (3) is that $t < s$ implies $f_i(t) < f_i(s)$ for any $i \leq n$ and $t, s \in C_i$. Therefore Fact 1 is applicable to find a homeomorphism $h_i : \mathbb{R} \to \mathbb{R}$ such that $h_i$ is strictly increasing and $h_i | C_i = f_i$ for all $i \leq n$. The diagonal product $h = \Delta_{i \leq n} h_i : \mathbb{R}^n \to \mathbb{R}^n$ is a homeomorphism and it is immediate that $h(A) = B$, so Fact 8 is proved.

*Fact 9.* The set $Z(k, m) = \{x = (x_1, \ldots, x_m) \in \mathbb{R}^m : |\{i \leq m : x_i \in \mathbb{P}\}| = k\}$ is zero-dimensional for all $m \in \mathbb{N}$ and $k \in \{0, \ldots, m\}$.

*Proof.* Let $H(r_1, \ldots, r_{m-k}, i_1, \ldots, i_{m-k}) = \{(x_1, \ldots, x_m) \in Z(k, m) : x_{i_j} = r_j$ for all $j \leq m - k\}$ for any $r_1, \ldots, r_{m-k} \in \mathbb{Q}$ and distinct $i_1, \ldots, i_{m-k} \in \{1, \ldots, m\}$. Every set $H(r_1, \ldots, r_{m-k}, i_1, \ldots, i_{m-k})$ is zero-dimensional being homeomorphic to the space $\mathbb{P}^k$. Besides, the family $\mathcal{H} = \{H(r_1, \ldots, r_{m-k}, i_1, \ldots, i_{m-k}) : r_i \in \mathbb{Q}$ for each $i \leq m-k$ and $i_1, \ldots, i_{m-k}$ are distinct elements of $\{1, \ldots, m\}\}$ is countable and $\bigcup \mathcal{H} = Z(k, m)$.

Now, if $H = H(r_1, \ldots, r_{m-k}, i_1, \ldots, i_{m-k})$ is any element of $\mathcal{H}$ then the set $F = \{x \in \mathbb{R}^m : x_{i_j} = r_j$ for all $j \leq m - k\}$ is closed in $\mathbb{R}^m$; it is easy to see that $F \cap Z(k, m) = H$ and hence every $H \in \mathcal{H}$ is closed in $Z(k, m)$. Finally apply SFFS-311 and SFFS-306 to conclude that $Z(k, m)$ is zero-dimensional. Fact 9 is proved.

Returning to our solution assume that the interior of $X$ is nonempty; then there is a point $a \in X$ and $\varepsilon > 0$ such that the ball $G = \{x \in \mathbb{R}^n : |x - a|_n < \varepsilon\}$ is contained in $X$. The ball $G$ is homeomorphic to the set $G_0 = \{x \in \mathbb{R}^n : |x|_n < \varepsilon\}$. If $\varphi(x) = \frac{\varepsilon}{2n} \cdot x$ for any $x \in \mathbb{R}^n$ then map $\varphi : \mathbb{R}^n \to \mathbb{R}^n$ is a homeomorphism; given any $x \in \mathbb{I}^n$ we have $|x|_n \leq \sqrt{n}$ and hence $|\varphi(x)|_n \leq \frac{\varepsilon}{2n} \cdot \sqrt{n} = \frac{\varepsilon}{2\sqrt{n}} < \varepsilon$ which shows that $\varphi(\mathbb{I}^n) \subset G_0$ and hence $\dim G_0 \geq \dim(\mathbb{I}^n) = n$ (see Problem 155 and Problem 159). Since $G_0 \subset \mathbb{R}^n$, we have $\dim G_0 \leq \dim(\mathbb{R}^n) = n$, so $\dim G_0 =$

$\dim G = n$. It follows from $G \subset X \subset \mathbb{R}^n$ that we can apply Problem 155 again to conclude that $n = \dim G \leq \dim X \leq \dim(\mathbb{R}^n) = n$ and hence $\dim X = n$. This proves sufficiency.

For necessity observe first that $\mathbb{R}^n \backslash \mathbb{Q}^n = Z(1,n) \cup \ldots \cup Z(n,n)$, so Fact 9 together with Fact 4 of V.100 imply that $P = \mathbb{R}^n \backslash \mathbb{Q}^n$ has the $n$-partition property and hence $\dim P \leq n - 1$ by Problem 157. If $X \subset \mathbb{R}^n$ and $\text{Int}(X) = \emptyset$ then $\mathbb{R}^n \backslash X$ is dense in $\mathbb{R}^n$; choose a countable dense set $B \subset \mathbb{R}^n \backslash X$. Then $B$ is dense in $\mathbb{R}^n$ and $X \subset \mathbb{R}^n \backslash B$. Apply Fact 8 to find a homeomorphism $h : \mathbb{R}^n \to \mathbb{R}^n$ such that $h(B) = \mathbb{Q}^n$. Then $h|(\mathbb{R}^n \backslash B)$ is a homeomorphism between $\mathbb{R}^n \backslash B$ and $P$; as a consequence, $\dim(\mathbb{R}^n \backslash B) = \dim P \leq n - 1$. Finally, apply Problem 155 to conclude that $\dim X \leq \dim(\mathbb{R}^n \backslash B) \leq n - 1$; this settles necessity and completes our solution.

**V.161.** *Prove that, for any Tychonoff space $X$ and $n \in \omega$, we have $\dim X \leq n$ if and only if, for any second countable space $Y$ and any continuous $f : X \to Y$, there exists a second countable space $M$ and continuous maps $g : X \to M$, $h : M \to Y$ such that $\dim M \leq n$ and $f = h \circ g$.*

**Solution.** Given a space $Z$, a set $U \subset Z$ and a family $\mathcal{A}$ of subsets of $Z$ let $\text{St}(U, \mathcal{A}) = \bigcup \{A \in \mathcal{A} : A \cap U \neq \emptyset\}$ be the star of the set $U$ with respect to $\mathcal{A}$. As usual, we write $\text{St}(z, \mathcal{A})$ instead of $\text{St}(\{z\}, \mathcal{A})$.

*Fact 1.* Given $m \in \omega$ suppose that $K$ is a compact space with $\dim K \leq m$ and $N$ is a second countable space. Then, for any continuous map $p : K \to N$ there exists a second countable space $L$ such that $\dim L \leq m$ and there are continuous maps $q : K \to L$ and $r : L \to N$ for which $r \circ q = p$.

*Proof.* Fix a countable base $\mathcal{B}$ in the space $N$; let $\mathcal{B}_K = \{p^{-1}(B) : B \in \mathcal{B}\}$ and denote by $\mathbb{V}_0$ the family of all finite covers of $K$ with the elements of $\mathcal{B}_K$. For every finite $\mathcal{S} \subset \mathbb{V}_0$ apply TFS-230 to find a finite open cover $\mathcal{V}$ of the space $K$ such that $\mathcal{V}$ is a star refinement of every $\mathcal{U} \in \mathcal{S}$. Since $\dim K \leq m$, we can find a finite open refinement $\mathcal{V}_\mathcal{S}$ of the cover $\mathcal{V}$ such that $\text{ord}(\mathcal{V}_\mathcal{S}) \leq m + 1$. It is evident that the family $\mathbb{V}_1 = \{\mathcal{V}_\mathcal{S} : \mathcal{S} \text{ is a finite subfamily of } \mathbb{V}_0\}$ is countable.

Proceeding by induction assume that $k \in \mathbb{N}$ and we have constructed a sequence $\mathbb{V}_0, \ldots, \mathbb{V}_k$ of countable families of finite open covers of $K$ such that

(1) if $i \in \{1, \ldots, k\}$ then $\text{ord}(\mathcal{V}) \leq m + 1$ for every $\mathcal{V} \in \mathbb{V}_i$;
(2) for any $j \in \{1, \ldots, k\}$ and any finite family $\mathcal{S} \subset \bigcup_{i=0}^{j-1} \mathbb{V}_i$ there is $\mathcal{V} \in \mathbb{V}_j$ which is a star refinement of every $\mathcal{U} \in \mathcal{S}$.

For every finite $\mathcal{S} \subset \bigcup \{\mathbb{V}_i : i \in \{0, \ldots, k\}\}$ apply TFS-230 to find a finite cover $\mathcal{V}$ of the space $K$ such that $\mathcal{V}$ is a star refinement of every $\mathcal{U} \in \mathcal{S}$. Since $\dim K \leq m$, we can find a finite open refinement $\mathcal{V}_\mathcal{S}$ of the cover $\mathcal{V}$ such that $\text{ord}(\mathcal{V}_\mathcal{S}) \leq m + 1$. It is evident that the family $\mathbb{V}_{k+1} = \{\mathcal{V}_\mathcal{S} : \mathcal{S} \text{ is a finite subfamily of } \bigcup_{i=0}^{k} \mathbb{V}_i\}$ is countable and the conditions (1)–(2) are satisfied for the sequence $\{\mathbb{V}_0, \ldots, \mathbb{V}_{k+1}\}$. Therefore our inductive procedure can be continued to construct a sequence $\{\mathbb{V}_i : i \in \omega\}$ such that (1) and (2) are satisfied for all $k \in \omega$; let $\mathbb{V} = \bigcup \{\mathbb{V}_i : i \in \mathbb{N}\}$.

Given a point $x \in K$ let $[x] = \{y \in K : \text{for every } \mathcal{V} \in \mathbb{V} \text{ there is } V \in \mathcal{V} \text{ such that } \{x, y\} \subset V\}$; we will call the set $[x]$ *the equivalence class of* $x$. It is clear that

(3) for any $x, y \in K$ we have $x \in [x]$ and $x \in [y]$ if and only if $y \in [x]$.

It turns out that we also have transitivity for the equivalence classes, i.e.,

(4) for any $x, y, z \in K$ if $y \in [x]$ and $z \in [y]$ then $z \in [x]$.

Take any family $\mathcal{V} \in \mathbb{V}$; it follows from the property (2) that there exists $\mathcal{W} \in \mathbb{V}$ such that $\mathcal{W}$ is a star refinement of $\mathcal{V}$. There exist $W_0, W_1 \in \mathcal{W}$ such that $\{x, y\} \subset W_0$ and $\{y, z\} \subset W_1$. There is $V \in \mathcal{V}$ such that $\text{St}(W_0, \mathcal{W}) \subset V$; it is clear that $\{x, y, z\} \subset W_0 \cup W_1 \subset \text{St}(W_0, \mathcal{W})$, so $\{x, z\} \subset V$ and hence $z \in [x]$, i.e., (4) is proved.

An immediate consequence of (4) is that

(5) for any $x, y \in K$ if $y \in [x]$ then $[x] = [y]$; if $y \notin [x]$ then $[x] \cap [y] = \emptyset$.

Let $L = \{[x] : x \in K\}$; we will need the set $O(P) = \bigcup P$ for any $P \subset L$. If $q(x) = [x]$ for any $x \in K$ then it follows from the property (5) that the map $q : X \to L$ is well defined. Consider the family $\tau_L = \{U \subset L : O(U) \in \tau(K)\}$; it is evident that $\emptyset \in \tau_L$ and $L \in \tau_L$. If $U, V \in \tau_L$ then it follows from (5) that $O(U \cap V) = O(U) \cap O(V) \in \tau(K)$, so $U \cap V \in \tau_L$. Now, if $\mathcal{U} \subset \tau_L$ then $O(\bigcup \mathcal{U}) = \bigcup \{O(U) : U \in \mathcal{U}\} \in \tau(K)$ which shows that $\bigcup \mathcal{U} \in \tau_L$ and hence $\tau_L$ is a topology on $L$; from now on we will identify $L$ with the space $(L, \tau_L)$.

Observe that $q^{-1}(U) = O(U)$ for any $U \subset L$, so the map $q$ is quotient; since $q(K) = L$, the space $L$ is compact. Our next step is to show that

(6) $[x] = \bigcap \{\text{St}(x, \mathcal{V}) : \mathcal{V} \in \mathbb{V}\} = \bigcap \{\overline{\text{St}(x, \mathcal{W})} : \mathcal{W} \in \mathbb{V}\}$ and, in particular, the set $[x]$ is closed in $K$ for any $x \in K$.

Fix a point $x \in K$ and observe that the equality $[x] = \bigcap \{\text{St}(x, \mathcal{V}) : \mathcal{V} \in \mathbb{V}\}$ is immediate from the definition. Now, if $\mathcal{V} \in \mathbb{V}$ then there is $\mathcal{W} \in \mathbb{V}$ which is a star refinement of $\mathcal{V}$. If $W \in \mathcal{W}$ then, for any $H \in \mathcal{W}$ we have $H \cap W \neq \emptyset$ if and only if $H \cap \overline{W} \neq \emptyset$; this implies that $\text{St}(W, \mathcal{W}) = \text{St}(\overline{W}, \mathcal{W})$. There exists $V_W \in \mathcal{V}$ with $\text{St}(W, \mathcal{W}) \subset V_W$, so $\text{St}(\overline{W}, \mathcal{W}) \subset V_W$. The families $\mathcal{V}$ and $\mathcal{W}$ being finite, we conclude that $\overline{\text{St}(x, \mathcal{W})} \subset \bigcup \{V_W : x \in W \in \mathcal{W}\} \subset \text{St}(x, \mathcal{V})$. The cover $\mathcal{V} \in \mathbb{V}$ was chosen arbitrarily so $[x] \subset \bigcap \{\overline{\text{St}(x, \mathcal{W})} : \mathcal{W} \in \mathbb{V}\} \subset \bigcap \{\text{St}(x, \mathcal{V}) : \mathcal{V} \in \mathbb{V}\} = [x]$ and hence (6) is proved.

We will also need the following property of our covers.

(7) For any $x \in K$ and $U \in \tau([x], K)$ there exists a cover $\mathcal{V} \in \mathbb{V}$ and $V \in \tau(K)$ such that $[x] \subset V \subset \bigcup \{[y] : y \in V\} \subset \text{St}(V, \mathcal{V}) \subset U$.

It follows from (6) that there are $\mathcal{W}_1, \dots, \mathcal{W}_k$ such that $\bigcap_{i \leq k} \overline{\text{St}(x, \mathcal{W}_i)} \subset U$ (see Fact 1 of S.326). Apply (2) to take a cover $\mathcal{W} \in \mathbb{V}$ which is a star refinement of every $\mathcal{W}_i$; there exists $\mathcal{V} \in \mathbb{V}$ such that $\mathcal{V}$ is a star refinement of $\mathcal{W}$. Then $\text{St}(x, \mathcal{W}) \subset U$; let $V = \text{St}(x, \mathcal{V})$ and take any set $H \in \mathcal{V}$ with $x \in H$. There exists $W_H \in \mathcal{W}$ such that $\text{St}(H, \mathcal{V}) \subset W_H$; it follows from $x \in W_H$ that $W_H \subset \text{St}(x, \mathcal{W})$, so $\text{St}(V, \mathcal{V}) \subset \bigcup \{\text{St}(H, \mathcal{V}) : x \in H \in \mathcal{V}\} \subset \bigcup \{W_H : x \in H \in \mathcal{V}\} \subset \text{St}(x, \mathcal{W}) \subset U$.

The inclusion $[x] \subset V = \mathrm{St}(x, V)$ is an immediate consequence of (6); the second inclusion in (7) is evident and the third one follows from the fact that $[y] \subset \mathrm{St}(y, V) \subset \mathrm{St}(V, V)$ for any $y \in V$, so (7) is proved.

For any $U \in \tau(K)$ let $U^* = \{a \in L : a \subset U\}$. Given any $z \in O(U^*)$ there is $x \in K$ such that $z = [x]$. It follows from (7) that there exists a set $V \in \tau(K)$ such that $[x] \subset V$ and $\bigcup\{[t] : t \in V\} \subset U$. Thus $y \in O(U^*)$ for any $y \in V$ which shows that every point of $O(U^*)$ is contained in $O(U^*)$ together with some open neighborhood and hence

(8) the set $q^{-1}(U^*) = O(U^*)$ is open in $K$ and hence $U^*$ is open in $L$ for any $U \in \tau(K)$.

If $a$ and $b$ are distinct points of the space $L$ then there are $x, y \in K$ such that $a = [x]$, $b = [y]$ and $[x] \cap [y] = \emptyset$. By normality of $K$ there exist disjoint sets $U \in \tau([x], K)$ and $V \in \tau([y], K)$. It is clear that $a \in U^*$, $b \in V^*$ and $U^* \cap V^* = \emptyset$. This proves, together with (8) and TFS-124, that

(9) the space $L$ is Hausdorff and hence Tychonoff.

Fix an arbitrary point $x \in X$ and take any $z \neq f(x)$. It is easy to see that the family $\mathcal{B}' = \{B \in \mathcal{B} : |B \cap \{f(x), z\}| \leq 1\}$ is a cover of the space $N$, so there is a finite $\mathcal{C} \subset \mathcal{B}'$ such that $K = \bigcup\{f^{-1}(B) : B \in \mathcal{C}\}$. There exists $V \in \mathbb{V}$ which is a star refinement of the cover $\mathcal{C}' = \{f^{-1}(B) : B \in \mathcal{C}\}$. In particular, $[x] \subset \mathrm{St}(x, V) \subset f^{-1}(B)$ for some $B \in \mathcal{C}$. Then $f(x) \in B$, so $z \notin B$ which, together with $f([x]) \subset B$ shows that $z \notin f([x])$. The point $z \neq f(x)$ was chosen arbitrarily and hence we proved that

(10) $f([x]) = \{f(x)\}$ for any $x \in X$.

Now, given any $a \in L$ take $x \in X$ with $[x] = a$ and let $r(a) = f(x)$. The property (10) shows that the map $r : L \to N$ is well defined and $r \circ q = f$. The map $q$ is quotient and $r \circ q = f$ is continuous, so we can apply Fact 1 of T.268 to see that the map $r$ is also continuous.

Now take a family $V \in \mathbb{V}$ and choose $W \in \mathbb{V}$ such that $W$ is a star refinement of $V$. Given a point $a \in L$ there is $x \in K$ with $a = [x]$; take $W \in W$ with $x \in W$. Since there is $V \in V$ for which $\mathrm{St}(W, W) \subset V$, it follows from (6) that $[x] \subset \mathrm{St}(x, W) \subset \mathrm{St}(W, W) \subset V$ and hence $a = [x] \in V^*$. This proves that

(11) the family $V^* = \{V^* : V \in V\}$ is an open cover of $L$ for any $V \in \mathbb{V}$.

Now take $a \in L$ and $H \in \tau(a, L)$. The set $U = q^{-1}(H)$ is open in $K$; pick a point $x \in X$ with $a = [x]$. Then $[x] \subset U$, so we can apply (6) to find $V \in \mathbb{V}$ with $\mathrm{St}(x, V) \subset U$. It follows from (11) that there is $V \in V$ such that $a \in V^*$ and hence $[x] \subset V$; since $x \in V$, the inclusion $V \subset \mathrm{St}(x, V) \subset U$ shows that $V^* \subset U^* = H$ and therefore $a \in V^* \subset H$. As a consequence,

(12) the family $\{V^* : V \in V \in \mathbb{V}\}$ is a base of $L$ and hence $L$ is second countable.

To finally prove that $\dim L \leq n$ fix a finite open cover $\mathcal{H}$ of the space $L$. Then $\mathcal{G} = \{q^{-1}(H) : H \in \mathcal{H}\}$ is an open cover of $K$, so we can apply (7) to find, for every $x \in K$, a set $V_x \in \tau(x, K)$ such that $\mathrm{St}(V_x, V_x) \subset q^{-1}(H_x)$ for some $H_x \in \mathcal{H}$ and $V_x \in \mathbb{V}$.

There is a finite set $A \subset K$ such that $\bigcup\{V_x : x \in A\} = K$; apply (2) to find a cover $V \in \mathbb{V}$ which is a star refinement of $V_x$ for every $x \in A$. To show that $\mathcal{V}$ refines $\mathcal{G}$ take any $V \in \mathcal{V}$. If $V = \emptyset$ then there is nothing to prove; otherwise, there is $x \in A$ such that $V_x \cap V \neq \emptyset$. Choose $W \in \mathcal{V}_x$ such that $\mathrm{St}(V, \mathcal{V}) \subset W$. Then $W \cap V_x \neq \emptyset$ and therefore $W \subset \mathrm{St}(V_x, \mathcal{V}_x) \subset q^{-1}(H_x)$. As a consequence, $V \subset \mathrm{St}(V, \mathcal{V}) \subset W \subset q^{-1}(H_x)$, so $\mathcal{V}$ refines $\mathcal{G}$. Thus $\mathcal{V}^*$ is a refinement of $\mathcal{H}$; it easily follows from $\mathrm{ord}(\mathcal{V}) \leq n+1$ that $\mathrm{ord}(\mathcal{V}^*) \leq n+1$ as well so any finite open cover of $L$ has an open refinement of order at most $n+1$. Applying Problem 145 we conclude that $\dim L \leq n$, so Fact 1 is proved.

Returning to our solution assume that $\dim X \leq n$ and take a continuous map $f : X \to Y$ for some second countable space $Y$. We can consider that $Y \subset K$ for some metrizable compact space $K$, so there is a continuous map $f_0 : \beta X \to K$ such that $f_0|X = f$. Since $\dim(\beta X) = \dim X \leq n$ (see Problem 147), we can apply Fact 1 to find a second countable space $Z$ and continuous maps $g_0 : \beta X \to Z$ and $h_0 : Z \to K$ such that $\dim Z \leq n$ and $h_0 \circ g_0 = f_0$. Let $M = g_0(X)$, $h = h_0|M$ and $g = g_0|X$. The space $M$ is second countable and $\dim M \leq \dim Z \leq n$ (see Problem 155); besides, $g : X \to M$, $h : M \to Y$ and $h \circ g = f$, so we proved necessity.

Finally, assume that, for any second countable space $Y$, if $f : X \to Y$ is a continuous map then there exists a second countable space $M$ and continuous maps $g : X \to M$, $h : M \to Y$ such that $\dim M \leq n$ and $h \circ g = f$. Given a functionally open cover $\mathcal{U} = \{U_0, \ldots, U_k\}$ of the space $X$ take a continuous function $f_i : X \to \mathbb{R}$ such that $X \backslash U_i = f_i^{-1}(0)$ for each $i \leq k$. The diagonal product $f = \Delta_{i \leq k} f_i : X \to \mathbb{R}^{k+1}$ is continuous; let $Y = f(X)$ and consider the natural projection $\pi_i : \mathbb{R}^{k+1} \to \mathbb{R}$ of the space $\mathbb{R}^{k+1}$ onto its $i$-th factor for all $i \leq k$. If $V_i = \pi_i^{-1}(\mathbb{R} \backslash \{0\}) \cap Y$ then $V_i$ is an open subset of $Y$ and $f^{-1}(V_i) = U_i$ for every $i \leq k$.

By our assumption there is a second countable space $M$ and continuous maps $g : X \to M$, $h : M \to Y$ such that $\dim M \leq n$ and $h \circ g = f$. The family $\mathcal{V} = \{h^{-1}(V_i) : i \leq k\}$ is a functionally open cover of $M$, so it has a functionally open refinement $\mathcal{W}$ with $\mathrm{ord}(\mathcal{W}) \leq n+1$. It is straightforward that the family $\mathcal{U}' = \{g^{-1}(W) : W \in \mathcal{W}\}$ is a functionally open refinement of $\mathcal{U}$ and $\mathrm{ord}(\mathcal{U}') \leq n+1$. Thus $\dim X \leq n$, i.e., we proved sufficiency and hence our solution is complete.

**V.162.** *Prove that, for any $n \in \omega$, there exists a compact second countable space $U_n$ such that $\dim U_n \leq n$ and any second countable $X$ with $\dim X \leq n$ can be embedded in $U_n$.*

**Solution.** The space $S^n = \{x = (x_0, \ldots, x_n) \in \mathbb{R}^{n+1} : \sum_{i=0}^{n} x_i^2 = 1\}$ is the usual $n$-dimensional sphere with the topology inherited from $\mathbb{R}^{n+1}$. Let $\mathcal{A} = \{Y : Y \subset \mathbb{I}^\omega$ and $\dim Y \leq n\}$; choose a family $\{Y_t : t \in T\}$ of spaces with the following properties:

(1) $w(Y_t) \leq \omega$ and $\dim Y_t \leq n$ for every $t \in T$;
(2) a homeomorphism $e_t : Y_t \to A_t \in \mathcal{A}$ is fixed for every $t \in T$;
(3) for any $A \in \mathcal{A}$ there is $t \in T$ such that $A = A_t$.

In the space $G = \bigoplus \{Y_t : t \in T\}$ we identify every $Y_t$ with the respective clopen subset of $G$. It is an easy exercise to see that $G$ is a normal space. If $F \subset G$ is closed and $f : F \to S^n$ is a continuous map then $F_t = F \cap Y_t$ is closed in $Y_t$, so there exists a continuous map $g_t : Y_t \to S^n$ such that $g_t | F_t = f | F_t$ for every $t \in T$ (see Problem 158). Given any $x \in G$ there is a unique $t \in T$ with $x \in Y_t$; let $g(x) = g_t(x)$. It is straightforward that $g : G \to S^n$ is continuous and $g | F = f$, so $\dim G \leq n$ by Problem 158.

For every $x \in G$ there is a unique $t \in T$ with $x \in Y_t$; let $e(x) = e_t(x)$. It is easy to check that the map $e : G \to \mathbb{I}^\omega$ is continuous and $e | Y_t = e_t$ for each $t \in T$. Take a continuous map $p : \beta G \to \mathbb{I}^\omega$ such that $p | G = e$; since $\dim(\beta G) \leq n$ (see Problem 147), we can apply Problem 161 to find a second countable space $M$ such that $\dim M \leq n$ and there exist continuous maps $q : \beta G \to M$ and $r : M \to \mathbb{I}^\omega$ with $r \circ q = p$.

The space $U_n = q(\beta G)$ is compact; it follows from $U_n \subset M$ that $U_n$ is second countable and $\dim U_n \leq \dim M \leq n$ (see Problem 155). If $X$ is a second countable compact space with $\dim X \leq n$ then we can consider that $X \subset \mathbb{I}^\omega$ and hence $X \in \mathcal{A}$; consequently, there is $t \in T$ such that $X = A_t$. The map $p | Y_t = e | Y_t = e_t$ is a homeomorphism of $Y_t$ onto $A_t = X$. If $q_t = q | Y_t$ and $r_t = r | q(Y_t)$ then $e_t = r_t \circ q_t$. The map $e_t$ being a homeomorphism, it is an easy exercise that both maps $r_t$ and $q_t$ are condensations so they are homeomorphisms by Fact 2 of S.337. Thus the space $X = e_t(Y_t) = r_t(q_t(Y_t))$ is homeomorphic to the subspace $q_t(Y_t)$ of the space $U_n$, so every second countable space $X$ with $\dim X \leq n$ embeds in $U_n$.

**V.163.** *Suppose that $X$ is a second countable space, $Y \subset X$ and $\dim Y \leq n$. Prove that there exists a $G_\delta$-set $Y'$ of the space $X$ such that $Y \subset Y'$ and $\dim Y' \leq n$.*

**Solution.** If $n = \infty$ then we can take $Y' = X$, so assume that $n \in \omega$. We can consider that $X \subset K$ for some metrizable compact space $K$. It follows from Problem 162 that there exists a compact metrizable space $M$ such that $\dim M \leq n$ and $Y$ embeds in $M$, so fix a subspace $Z \subset M$ and a homeomorphism $h : Z \to Y$.

Apply Fact 2 of T.333 to find a $G_\delta$-subset $Z' \subset M$ and a $G_\delta$-subset $H \subset K$ such that $Z \subset Z'$, $Y \subset H$ and there exists a homeomorphism $g : Z' \to H$ with $g | Z = h$. We have $\dim Z' \leq \dim M \leq n$ (see Problem 155), so $\dim H \leq n$. It is clear that $Y' = H \cap X$ is a $G_\delta$-subset of $X$ such that $Y \subset Y'$ and $\dim Y' \leq \dim H \leq n$.

**V.164.** *Given $n \in \omega$ and a second countable Tychonoff space $X$ prove that $\dim X \leq n$ if and only there exist $X_0, \ldots, X_n \subset X$ such that $X = X_0 \cup \ldots \cup X_n$ and $\dim X_i \leq 0$ for each $i \leq n$.*

**Solution.** Given spaces $Y$ and $Z$ we say that a map $f : Y \to Z$ is an *embedding* if $f : Y \to f(Y)$ is a homeomorphism. If $(Z, d)$ is a metric space and $\mathcal{A} \subset \exp(Z)$ then $\mathrm{mesh}(\mathcal{A}) = \sup\{\mathrm{diam}_d(A) : A \in \mathcal{A}\}$ and $\mathrm{ord}(x, \mathcal{A}) = |\{A \in \mathcal{A} : x \in A\}|$ for any $x \in Z$; if $T \subset Z$ then $\mathcal{A} | T = \{A \cap T : A \in \mathcal{A}\}$. Given sets $A, B \subset Z$ let $d(A, B) = \inf\{d(a, b) : a \in A,\ b \in B\}$.

*Fact 1.* Suppose that $M$ is a second countable space and $F, G \subset M$ are closed and disjoint. Then

(i)  there exists a metrizable compact space $K$ and an embedding $e : M \to K$ such that $\mathrm{cl}_K(e(F)) \cap \mathrm{cl}_K(e(G)) = \emptyset$;

(ii) if, additionally, $\dim M \leq m$ for some $m \in \omega$ then there exists a compact metrizable space $K$ and an embedding $e : M \to K$ such that $\dim K \leq m$ and $\operatorname{cl}_K(e(F)) \cap \operatorname{cl}_K(e(G)) = \emptyset$.

*Proof.* (i) By normality of $M$ we can find a continuous function $f : M \to [0, 1]$ such that $f(F) \subset \{0\}$ and $f(G) \subset \{1\}$. There exists a metrizable compact space $K'$ such that $M$ embeds in $K'$; fix an embedding $h : M \to K'$. It is easy to check that $e = h \Delta f : M \to K' \times [0, 1]$ is still an embedding; let $M' = e(M)$ and $K = \overline{M'}$.

If $\pi : K \to [0, 1]$ is the restriction of the natural projection of $K' \times [0, 1]$ onto its factor $[0, 1]$, then $\pi(e(F)) = f(F) \subset \{0\}$ and $\pi(e(G)) = f(G) \subset \{1\}$. Therefore $\operatorname{cl}_K(e(F)) \cap \operatorname{cl}_K(e(G)) \subset \operatorname{cl}_K(\pi^{-1}(0)) \cap \operatorname{cl}_K(\pi^{-1}(1)) = \emptyset$, so $e$ is the promised embedding of $M$.

(ii) If, additionally, we have $\dim M \leq m$ then apply (i) to find a metrizable compact space $K'$ for which there exists an embedding $h : M \to K'$ such that $\operatorname{cl}_{K'}(h(F)) \cap \operatorname{cl}_{K'}(h(G)) = \emptyset$. Take a continuous function $g : \beta M \to K'$ such that $g|M = h$; since $\dim(\beta M) = \dim M \leq m$ (see Problem 147), we can apply Problem 161 to find a metrizable compact space $K$ for which $\dim K \leq m$ and there exist continuous maps $p : \beta M \to K$ and $q : K \to K'$ such that $g = q \circ p$.

If $e = p|M$ and $r = q|p(M)$ then $e : M \to K$ and $r : p(M) \to K'$ are continuous maps such that $r \circ e = h$. It is an easy exercise to prove, using Fact 2 of S.337, that both maps $e$ and $r$ are embeddings. It follows from $q(e(F)) = h(F)$ and $q(e(G)) = h(G)$ that $\operatorname{cl}_K(e(F)) \cap \operatorname{cl}_K(e(G)) \subset q^{-1}(\operatorname{cl}_{K'}(h(F)) \cap q^{-1}(\operatorname{cl}_{K'}(h(G))) = \emptyset$. This, together with $\dim K \leq m$, shows that we settled (ii) and hence Fact 1 is proved.

*Fact 2.* If $m \in \omega$ and $(K, d)$ is a metric compact space then $\dim K \leq m$ if and only if there is a sequence $\{\mathcal{U}_k : k \in \omega\}$ of finite covers of $K$ such that $\operatorname{ord}(\mathcal{U}_k) \leq m + 1$ for any $k \in \omega$ and the sequence $\{\operatorname{mesh}(\mathcal{U}_k) : k \in \omega\}$ converges to zero.

*Proof.* Suppose first that $\dim K \leq m$ and fix an arbitrary $k \in \omega$. Since the family $\mathcal{B} = \{B_d(x, 2^{-k-1}) : x \in K\}$ is an open cover of the space $K$, we can find a finite subcover $\mathcal{V} = \{V_1, \ldots, V_l\}$ of the cover $\mathcal{B}$; then $\operatorname{mesh}(\mathcal{V}) \leq 2^{-k}$. Since $\dim K \leq m$, there exists a refinement $\mathcal{U}_k$ of the cover $\mathcal{V}$ such that $\operatorname{ord}(\mathcal{U}_k) \leq m + 1$. After constructing $\mathcal{U}_k$ for every $k \in \omega$ we obtain the promised sequence $\{\mathcal{U}_k : k \in \omega\}$.

Now assume that we have a sequence $\{\mathcal{U}_k : k \in \omega\}$ of finite open covers of $K$ such that $\operatorname{mesh}(\mathcal{U}_k) \to 0$ and $\operatorname{ord}(\mathcal{U}_k) \leq m + 1$ for every $k \in \omega$. If $\mathcal{V}$ is an open finite cover of $K$ then we can apply TFS-244 to find a number $\delta > 0$ such that every subset of $K$ of diameter less than $\delta$ is contained in an element of $\mathcal{V}$. There exists $k \in \omega$ with $\operatorname{mesh}(\mathcal{U}_k) < \delta$ and hence the diameter of every $U \in \mathcal{U}_k$ is less than $\delta$; this implies that every element of $\mathcal{U}_k$ is contained in an element of $\mathcal{V}$, i.e., $\mathcal{U}_k$ is a finite open refinement of $\mathcal{V}$ of order $\leq m + 1$. Therefore $\dim K \leq m$ (see Problem 145), i.e., we settled sufficiency and hence Fact 2 is proved.

*Fact 3.* Given $m \in \omega$ and a second countable space $Z$ with $\dim Z \leq m$, for any closed disjoint sets $A, B \subset Z$ there exists a partition $C$ between the sets $A$ and $B$ such that $\dim C \leq m - 1$.

*Proof.* Apply Fact 2 to find a compact space $K$ such that $\dim K \leq m$ and there is an embedding $e : Z \to K$ with $\mathrm{cl}_K(e(A)) \cap \mathrm{cl}_K(e(B)) = \emptyset$. To simplify the notation we identify $Z$ and $e(Z)$; then $Z \subset K$ and $\overline{A} \cap \overline{B} = \emptyset$ (the bar denotes the closure in $K$). Fix a metric $d$ which generates the topology of $K$.

Proceeding inductively, let $F_0 = \overline{A}$, $G_0 = \overline{B}$ and assume that $k \in \omega$ and we have a family $\{F_0, G_0, \ldots, F_k, G_k\}$ of closed subsets of $K$ with the following properties:

(1)  $F_i \subset \mathrm{Int}(F_{i+1})$ and $G_i \subset \mathrm{Int}(G_{i+1})$ for all $i < k$;
(2)  $F_i \cap G_i = \emptyset$ for every $i \leq k$;
(3)  if $C_i = K \backslash (F_i \cup G_i)$ then there is a finite $\mathcal{U}_i \subset \tau(K)$ such that $C_i \subset \bigcup \mathcal{U}_i$ while $\mathrm{ord}(x, \mathcal{U}_i) \leq m$ for any $x \in C_i$ and $\mathrm{mesh}(\mathcal{U}_i) \leq 2^{-i}$ for every $i \in \{1, \ldots, k\}$.

It follows from compactness of $K$ that $\delta = d(F_k, G_k) > 0$; it is easy to construct a finite open cover $\mathcal{V}$ of the space $K$ such that $\mathrm{mesh}(\mathcal{V}) \leq \backslash n\{\frac{\delta}{2}, 2^{-k-1}\}$. By $\dim K \leq m$ there is a finite refinement $\mathcal{H}$ of the cover $\mathcal{V}$ such that $\mathrm{ord}(\mathcal{H}) \leq m+1$; the union of the families $\mathcal{W}_0 = \{U \in \mathcal{H} : \overline{U} \cap F_k = \emptyset\}$ and $\mathcal{W}_1 = \{U \in \mathcal{H} : \overline{U} \cap F \neq \emptyset\}$ is equal to $\mathcal{H}$, so the open sets $W_0 = \bigcup \mathcal{W}_0$ and $W_1 = \bigcup \mathcal{W}_1$ cover the space $K$.

Since the closure of any element of $\mathcal{H}$ cannot intersect both $F_k$ and $G_k$, the set $G_{k+1} = K \backslash W_1$ is closed and $G_k \subset G_{k+1}$. Besides, $G_k \subset K \backslash \overline{W_1} \subset G_{k+1}$, so $G_k \subset \mathrm{Int}(G_{k+1})$. The set $F_{k+1} = K \backslash W_0$ is also closed and $F_k \subset K \backslash \overline{W_0} \subset F_{k+1}$, so $F_k \subset \mathrm{Int}(F_{k+1})$. It follows from $W_0 \cup W_1 = K$ that the sets $F_{k+1}$ and $G_{k+1}$ are disjoint, so the properties (1) and (2) hold if $k$ is replaced with $k + 1$.

Now, if $C_{k+1} = K \backslash (F_{k+1} \cup G_{k+1}) = W_0 \cap W_1$, so both families $\mathcal{W}_0$ and $\mathcal{W}_1$ cover $C_{k+1}$. As a consequence, the family $\mathcal{U}_{k+1} = \mathcal{W}_0$ covers the set $C_{k+1}$ and it follows from $\mathcal{W}_0 \cap \mathcal{W}_1 = \emptyset$ that $\mathrm{ord}(x, \mathcal{U}_{k+1}) < m+1$ for every $x \in C_{k+1}$ because, from at most $(m + 1)$-many elements of $\mathcal{V}$ that contain $x$, at least one element of $\mathcal{W}_1$ was removed. Thus the condition (3) is also satisfied for all $i \in \{1, \ldots, k + 1\}$ and hence our inductive procedure can be continued to construct families $\{F_i : i \in \omega\}$ and $\{G_i : i \in \omega\}$ such that the conditions (1)–(3) are satisfied for every $k \in \omega$.

It follows from (1) and (2) that $O_F = \bigcup_{i \in \omega} F_i$ and $O_G = \bigcup_{i < \omega} G_i$ are disjoint open sets such that $A \subset O_F$ and $B \subset O_G$. If $C' = K \backslash (O_F \cup O_G)$ then $C'$ is a partition between the sets $A$ and $B$ in $K$. The family $\mathcal{V}_k = \mathcal{U}_k | C'$ is a finite open cover of $C' = \bigcap \{C_i : i \in \mathbb{N}\}$ and the property (3) implies that $\mathrm{ord}(\mathcal{V}_k) \leq m$ for each $k \in \mathbb{N}$; it is evident that the sequence $\{\mathrm{mesh}(\mathcal{V}_k) : k \in \mathbb{N}\}$ converges to zero, so we can apply Fact 2 to conclude that $\dim(C') \leq m - 1$. The set $C = C' \cap Z$ is a partition between $A$ and $B$ in $Z$ and $\dim C \leq \dim(C') \leq m - 1$ (see Problem 155), so Fact 3 is proved.

Returning to our solution assume that $X = X_0 \cup \ldots \cup X_n$ and $\dim X_i = 0$ for all $i \leq n$. Then every $X_i$ is zero-dimensional by Problem 149, so we can apply Fact 4 of V.400 to see that $X$ has $(n + 1)$-partition property. This, together with Problem 157 implies that $\dim X \leq n$, so we proved sufficiency.

We will prove necessity by induction on $n$. If $n = 0$ then $X$ is zero-dimensional by Problem 149, so $X_0 = X$ gives the desired decomposition of $X$ into the union of zero-dimensional subspaces. Now assume that $n \in \omega$, $\dim X \leq n$ and we have proved that every second countable space of dimension $m < n$ is the union of at most $(m + 1)$-many zero-dimensional subspaces.

Let $\mathrm{Bd}(A) = \overline{A} \backslash A$ for any $A \subset X$ and call a set $U \in \tau(X)$ *adequate* if we have the inequality $\dim(\mathrm{Bd}(U)) \leq n - 1$. It turns out that

(4)  the adequate sets form a base in $X$.

Indeed, fix a point $x \in X$ and $W \in \tau(x, X)$. By Fact 3 there exist disjoint open sets $U$ and $V$ such that $x \in U$, $X \backslash W \subset V$ and $\dim(X \backslash (U \cup V)) \leq n - 1$. Observe that $\overline{U} \cap V = \emptyset$ and hence $\mathrm{Bd}(U) \subset C = X \backslash (U \cup V)$; an immediate consequence is that $\dim(\mathrm{Bd}(U)) \leq \dim C \leq n - 1$ (see Problem 155), i.e., the set $U$ is adequate. Since also $x \in U \subset W$, the property (4) is proved.

Next apply Claim of S.088 to conclude that there exists a countable base $\mathcal{B}$ in the space $X$ such that every set $B \in \mathcal{B}$ is adequate. The family $\mathcal{C} = \{\mathrm{Bd}(B) : B \in \mathcal{B}\}$ is countable while every $C \in \mathcal{C}$ is closed and $\dim C \leq n - 1$. Let $F = \bigcup \mathcal{C}$ and apply Problem 150 to convince ourselves that $\dim F \leq n - 1$.

Let $X_0 = X \backslash F$; the family $\mathcal{B}_0 = \mathcal{B}|X_0$ is a base in $X_0$. It is easy to see that $B \cap X_0 = \overline{B} \cap X_0$ for every $B \in \mathcal{B}$, so all elements of $\mathcal{B}_0$ are clopen in $X_0$ and hence $X_0$ is zero-dimensional. By the induction hypothesis, there are zero-dimensional spaces $X_1, \ldots, X_n$ such that $F = X_1 \cup \ldots \cup X_n$. Thus $X = X_0 \cup \ldots \cup X_n$ is the promised decomposition of $X$, so we accomplished our inductive step proving necessity for all $n \in \omega$ and hence our solution is complete.

**V.165.** *Let $\mathcal{S} = \{X_t, \pi_s^t : t, s \in T\}$ be an inverse system of Hausdorff topological spaces. Prove that the set $\varprojlim \mathcal{S}$ is closed in $\prod\{X_t : t \in T\}$. Therefore the limit of an inverse system of Hausdorff compact spaces is a Hausdorff compact space.*

**Solution.** Let $X = \prod_{t \in T} X_t$; observe first that the Hausdorff property is preserved by products and subspaces, so the space $Y = \varprojlim \mathcal{S}$ is Hausdorff. To prove that $Y$ is closed in $X$ take any point $x \in X \backslash Y$. Since $x$ is not a thread, there are $s, t \in T$ such that $s < t$ and $\pi_s^t(x(t)) \neq x(s)$. Take disjoint sets $U, V \in \tau(X_s)$ such that $\pi_s^t(x(t)) \in U$ and $x(s) \in V$. By continuity of $\pi_s^t$ there is a set $W \in \tau(x(t), X_t)$ for which $\pi_s^t(W) \subset U$. The set $G = \{y \in X : y(t) \in W$ and $y(s) \in V\}$ is open in $X$. If $y \in G$ then $y(t) \in W$ and hence $\pi_s^t(y(t)) \in U$, which, together with $y(s) \in V$ shows that $\pi_s^t(y(t)) \neq y(s)$. Thus every point $x \in X \backslash Y$ has an open neighborhood $G \subset X \backslash Y$, so $X \backslash Y$ is open in $X$ and hence $Y$ is a closed subset of $X$. If every $X_t$ is compact then $X$ is also compact, so $Y$ has to be compact as well.

**V.166.** *Suppose that $\mathcal{S} = \{X_t, \pi_s^t : t, s \in T\}$ is an inverse system in which $X_t$ is a nonempty compact Hausdorff space for each $t$. Prove that $\varprojlim \mathcal{S} \neq \emptyset$.*

**Solution.** The space $X = \prod_{t \in T} X_t$ is compact. For any $t \in T$ consider the set $A_t = \{s \in T : s \leq t\}$ and say that a point $x \in X$ is a *thread* on $A_t$ if $\pi_v^u(x(u)) = x(v)$ for any $u, v \in A_t$ with $v \leq u$. Denote by $F_t$ the set of all threads on $A_t$ for each $t \in T$.

Fix $t \in T$ and take any point $x \in X \backslash F_t$. Since $x$ is not a thread on $A_t$, there are $u, v \in A_t$ such that $v < u$ and $\pi_v^u(x(u)) \neq x(v)$. Take disjoint sets $U, V \in \tau(X_v)$ such that $\pi_v^u(x(u)) \in U$ and $x(v) \in V$. By continuity of $\pi_v^u$ there is a set $W \in \tau(x(u), X_u)$ for which $\pi_v^u(W) \subset U$. The set $G = \{y \in X : y(u) \in W$ and $y(v) \in V\}$ is open in $X$. If $y \in G$ then $y(u) \in W$ and hence $\pi_v^u(y(u)) \in U$, which, together with $y(v) \in V$ shows that $\pi_v^u(y(u)) \neq y(v)$. Thus every point $x \in X \backslash F_t$ has an open neighborhood $G \subset X \backslash F_t$, so $F_t$ is a closed subset of $X$.

Given $t \in T$ fix a point $x_s \in X_s$ for any $s \notin A_t$; take a point $y \in X_t$ and let $x_s = \pi_s^t(y)$ for any $s \in A_t$. Letting $x(s) = x_s$ for any $s \in T$ we obtain a point $x \in X$ and it follows from the definition of projections of an inverse system that $\pi_v^u(x(u)) = \pi_v^u(\pi_u^t(y)) = \pi_v^t(y) = x(v)$ for any $u, v \in A_t$ with $v \leq u$, so $x \in F_t$. This proves that

(1) the set $F_t$ is nonempty and closed in $X$ for any $t \in T$.

If $t_1, \ldots, t_n \in T$ then there exists $s \in T$ such that $t_i \leq s$ for each $i \leq n$. It is straightforward that $F_s \subset F_{t_i}$ for every $i \leq n$, so $F_s \subset \bigcap_{i \leq n} F_{t_i}$ and therefore

(2) $\bigcap \mathcal{F}' \neq \emptyset$ for any finite family $\mathcal{F}' \subset \mathcal{F} = \{F_t : t \in T\}$.

The properties (1) and (2) show that $\mathcal{F}$ is a centered family of closed subsets of a compact space $X$. Therefore $\bigcap \mathcal{F} \neq \emptyset$; it is an easy exercise that $\bigcap \mathcal{F} = \varprojlim S$, so $\varprojlim S \neq \emptyset$.

**V.167.** *Let $S = \{X_t, \pi_s^t : t, s \in T\}$ be an inverse system. Suppose that, for a space $Y$, a continuous map $f_t : Y \to X_t$ is given for every $t \in T$ and, besides, $\pi_s^t \circ f_t = f_s$ for any $s, t \in T$ with $s \leq t$. Prove that the diagonal product $f = \Delta_{t \in T} f_t$ maps $Y$ continuously into $\varprojlim S$.*

**Solution.** Let $X = \prod_{t \in T} X_t$; recall that the diagonal product $f : Y \to X$ is defined by letting $f(y)(t) = f_t(y)$ for any $y \in Y$ and $t \in T$. If $p_t : X \to X_t$ is the natural projection then $p_t \circ f = f_t$ for any $t \in T$, so $f$ is continuous by TFS-102.

To see that $f(Y) \subset \varprojlim S$ fix a point $y \in Y$. For any $s, t \in T$ with $s \leq t$ we have $\pi_s^t(f(y)(t)) = \pi_s^t(f_t(y)) = f_s(y) = f(y)(s)$, so the point $f(y)$ is a thread and hence $f(y) \in \varprojlim S$. Thus $f$ is a continuous map from $Y$ to $\varprojlim S$.

**V.168.** *Let $X$ be a topological space. Suppose that, for a nonempty directed set $T$, a subspace $X_t \subset X$ is given for each $t \in T$ in such a way that $X_t \subset X_s$ whenever $s \leq t$. Given $s, t \in T$ with $s \leq t$, let $\pi_s^t(x) = x$ for each $x \in X_t$. Prove that the inverse system $S = \{X_t, \pi_s^t : t, s \in T\}$ is well defined and the limit of $S$ is homeomorphic to $\bigcap \{X_t : t \in T\}$.*

**Solution.** It is evident that every map $\pi_s^t$ is continuous and $\pi_t^t = \mathrm{id}_{X_t}$. Given $s \leq t \leq u$ we have $\pi_s^t(\pi_t^u(x)) = \pi_s^t(x) = x = \pi_s^u(x)$ for any $x \in X_u$, so the inverse system $S$ is well defined.

Let $P = \prod_{t \in T} X_t$ and denote the limit of $S$ by $L$; let $h_t(x) = x$ for any $x \in Y = \bigcap \{X_t : t \in T\}$ and $t \in T$. Then $h_t : Y \to X_t$ is a continuous map and it is straightforward that $\pi_s^t \circ h_t = h_s$ for any $s, t \in T$ with $s \leq t$. Consequently, the diagonal product $h = \Delta_{t \in T} h_t$ maps $Y$ continuously into $L$ (see Problem 167).

If $x, y$ are distinct points of $Y$ then $h_t(x) = x \neq y = h_t(y)$ for any $t \in T$, so $h(x) \neq h(y)$ and hence $h$ is an injection. Given $a \in L$ take any $s, t \in T$ and choose $u \in T$ such that $s \leq u$ and $t \leq u$. Then $a(u) = \pi_t^u(a(u)) = a(t)$ and $a(u) = \pi_s^u(a(u)) = a(s)$ which shows that $a(s) = a(t)$, i.e., there is $x \in X$ such that $a(t) = x$ for all $t \in T$. Since $a(t) \in X_t$, we must have $x \in X_t$ for all $t \in T$ and hence $x \in Y$. It is immediate that $h(x) = a$, so $h : Y \to L$ is an onto map whence $h$ is a condensation.

Finally, fix an index $s \in T$ and let $p_s : P \to X_s$ be the natural projection. The map $p = p_s|L : L \to X_s$ is continuous. We already saw that, for every $a \in L$, we have $a(s) = a(t) \in X_t$ for any $t \in T$, so $p(a) = a(s) \in Y$. Therefore $p$ maps continuously the space $L$ in $Y$; it is easy to see that $p$ is the inverse of $h$, so $h$ is a homeomorphism between the spaces $L = \lim_{\leftarrow} S$ and $Y = \bigcap_{t \in T} X_t$.

**V.169.** *Give an example of an inverse sequence $S = \{X_n, \pi_m^n : n, m \in \omega\}$ such that every $X_n$ is a nonempty second countable Tychonoff space while $\lim_{\leftarrow} S = \emptyset$.*

**Solution.** The space $X_n = (n, +\infty) \subset \mathbb{R}$ is second countable, Tychonoff and nonempty for any $n \in \omega$. Given $m, n \in \omega$ with $m \leq n$ let $\pi_m^n(x) = x$ for every $x \in X_n$; this defines a continuous map $\pi_m^n : X_n \to X_m$. Since $X_n \subset X_m$ whenever $m \leq n$, we can apply Problem 168 to see that the inverse sequence $S = \{X_n, \pi_m^n : n, m \in \omega\}$ is well defined and $\lim_{\leftarrow} S$ is homeomorphic to $\bigcap_{n \in \omega} X_n = \emptyset$, so the limit of $S$ is empty.

**V.170.** *Given an inverse system $S = \{X_t, \pi_s^t : t, s \in T\}$ of topological spaces, prove that the family $\mathcal{B} = \{\pi_t^{-1}(U) : t \in T, U \in \tau(X_t)\}$ is a base of the space $L = \lim_{\leftarrow} S$. Here $\pi_t : L \to X_t$ is the limit projection for every $t \in T$.*

**Solution.** It is evident that all elements of the family $\mathcal{B}$ are open in $L$. For the space $X = \prod_{t \in T} X_t$ let $p_t : X \to X_t$ be the natural projection for every $t \in T$. If $x \in L$ and $G \in \tau(x, L)$ then there is $U \in \tau(X)$ such that $U \cap L = G$. By definition of the product topology there exist $t_1, \ldots, t_n \in T$ and $U_i \in \tau(X_{t_i})$ for all $i \leq n$ such that $x \in \bigcap_{i \leq n} p_{t_i}^{-1}(U_i) \subset U$.

The set $T$ being directed we can find an index $t \in T$ such that $t_i \leq t$ for all $i \leq n$. It follows from $x \in L$ that $\pi_{t_i}^t(x(t)) = x(t_i) \in U_i$, so we can find a set $W_i \in \tau(x(t), X_t)$ such that $\pi_{t_i}^t(W_i) \subset U_i$ for every $i \leq n$. The set $W = \bigcap_{i \leq n} W_i$ is an open neighborhood of $x(t)$ in $X_t$ and $\pi_{t_i}^t(W) \subset U_i$ for each $i \leq n$.

The set $O = \pi_t^{-1}(W) = p_t^{-1}(W) \cap L$ is an element of $\mathcal{B}$ and $x \in O$. If $y \in O$ then $y(t) \in W$ and $y \in L$, so $y(t_i) = \pi_{t_i}^t(y(t)) \in U_i$ for all $i \leq n$. As a consequence, $y \in \bigcap_{i \leq n} p_{t_i}^{-1}(U_i) \cap L \subset U \cap L = G$ and therefore $O \subset G$. We proved that, for any $x \in L$ and $G \in \tau(x, L)$ there is $O \in \mathcal{B}$ such that $x \in O \subset G$, so $\mathcal{B}$ is a base in $L$.

**V.171.** *Suppose that $S = \{X_t, \pi_s^t : t, s \in T\}$ is an inverse system of topological spaces. Prove that, for any closed set $F \subset \lim_{\leftarrow} S$, the subspace $F$ is the limit of the inverse system $S_F = \{\pi_t(F), \pi_s^t|\pi_t(F) : t, s \in T\}$. Here $\pi_t : \lim_{\leftarrow} S \to X_t$ is the limit projection for every $t \in T$.*

**Solution.** Let $L = \varprojlim S$; if $t \in T$ and $x \in \pi_t(F)$ then there is $y \in F$ with $\pi_t(y) = y(t) = x$. For any $s \leq t$ we have $\pi_s^t(x) = \pi_s^t(y(t)) = y(s) = \pi_s(y) \in \pi_s(F)$, so $\pi_s^t(\pi_t(F)) \subset \pi_s(F)$ and hence the inverse system $S_F$ is well defined.

If $L_F = \varprojlim S_F$ then every element of $L_F$ has to be a thread of $S$, so $L_F \subset L$. Given a point $x \in F$ we have $\pi_t(x) \in \pi_t(F)$ for any $t \in T$, so $x$ is also a thread of the inverse system $S_F$ and therefore $x \in L_F$; this proves that $F \subset L_F$.

Finally, take any point $x \in L \backslash F$; the set $F$ being closed in $L$ we can apply Problem 170 to find $t \in T$ and $U \in \tau(X_t)$ such that $x \in \pi_t^{-1}(U) \subset L \backslash F$. This implies that $U \cap \pi_t(F) = \emptyset$ and hence $\pi_t(x) \notin \pi_t(F)$ which shows that $x \notin L_F$; therefore $F = L_F = \varprojlim S_F$.

**V.172.** *Given an inverse system $S = \{X_t, \pi_s^t : t, s \in T\}$ and a cofinal set $T' \subset T$, prove that the limit of the inverse system $S' = \{X_t, \pi_s^t : t, s \in T'\}$ is homeomorphic to $\varprojlim S$.*

**Solution.** Let $X = \prod_{t \in T} X_t$ and $X' = \prod_{t \in T'} X_t$; denote the limit of $S$ by $L$ and the limit of $S'$ by $L'$. We will also need the natural projections $p_t : X \to X_t$ and $q_s : X' \to X_s$ for any $t \in T$ and $s \in T'$.

For any $t \in T'$ let $w(t) = t$; if $t \in T \backslash T'$ then fix an index $w(t) \in T'$ such that $t \leq w(t)$. Given a point $x \in X$ let $\pi(x) = x|T'$; the resulting map $\pi : X \to X'$ is continuous. It is evident that $\pi(x)$ is a thread of $S'$ if $x$ is a thread of $S$, so $\pi(L) \subset L'$. If $x$ and $y$ are distinct points of $L$ then there is $t \in T$ such that $x(t) \neq y(t)$. If $x(w(t)) = y(w(t))$ then $x(t) = \pi_t^{w(t)}(x(w(t))) = \pi_t^{w(t)}(y(w(t))) = y(t)$; this contradiction shows that $x(w(t)) \neq y(w(t))$ and hence $\pi(x) \neq \pi(y)$. Thus the map $h = \pi|L : L \to L'$ is injective.

Take any point $y \in L'$ and let $\varphi(y)(t) = \pi_t^{w(t)}(y(w(t)))$ for any $t \in T$. This gives us a map $\varphi : L' \to X$ such that $\pi(\varphi(y)) = y$ for any $y \in L'$. It turns out that $\varphi(L') \subset L$; to see it take any $y \in L'$, let $x = \varphi(y)$ and fix $s, t \in T$ with $s \leq t$. There exists $v \in T'$ such that $w(s) \leq v$ and $w(t) \leq v$. Since $y$ is a thread of $S'$, we have $y(w(t)) = \pi_{w(t)}^v(y(v))$ and $y(w(s)) = \pi_{w(s)}^v(y(v))$ which shows that $\pi_s^t(x(t)) = \pi_s^t(\pi_t^{w(t)}(\pi_{w(t)}^v(y(v)))) = \pi_s^v(y(v))$, which together with the equalities

$$x(s) = \pi_s^{w(s)}(y(w(s))) = \pi_s^{w(s)}(\pi_{w(s)}^v(y(v))) = \pi_s^v(y(v))$$

shows that $\pi_s^t(x(t)) = x(s)$ and hence $x \in L$.

Thus we can consider that $\varphi : L' \to L$; since $\varphi(h(x)) = x$ for any $x \in L$ and $h(\varphi(y)) = y$ for every $y \in L'$, the maps $h$ and $\varphi$ are mutually inverse bijections. We already saw that $h$ is continuous; fix $t \in T$ and take any $y \in L'$. The equality $\varphi(y)(t) = \pi_t^{w(t)}(y(w(t)))$ implies that $(p_t \circ \varphi)(y) = (\pi_t^{w(t)} \circ q_{w(t)})(y)$ for any $y \in L'$, i.e., the map $p_t \circ \varphi = \pi_t^{w(t)} \circ q_{w(t)}$ is continuous for any $t \in T$, so $\varphi$ is continuous by TFS-102. Therefore $h : L \to L'$ is a homeomorphism.

**V.173.** *Suppose that an inverse system* $\mathcal{S} = \{X_t, \pi_s^t : t, s \in T\}$ *consists of compact Hausdorff spaces* $X_t$ *and all projections* $\pi_s^t$ *are onto. Prove that all limit projections* $\pi_t$ *are also surjective maps.*

**Solution.** The space $X = \prod_{t \in T} X_t$ is compact; fix $t \in T$ and a point $z \in X_t$. For any $s \in T$ consider the set $A_s = \{u \in T : u \leq s\}$ and say that a point $x \in X$ is a *z-thread* on $A_s$ if $x(t) = z$ and $\pi_v^u(x(u)) = x(v)$ for any $u, v \in A_s$ with $v \leq u$. Denote by $F_s$ the set of all $z$-threads on $A_s$ for each $s \in T$.

Fix an index $s \in T$ and take any point $x \in X \backslash F_s$. If $x(t) \neq z$ then the set $G = \{y \in X : y(t) \neq z\}$ is open in $X$ and $x \in G \subset X \backslash F_s$. If $x(t) = z$ then there are $u, v \in A_s$ such that $v < u$ and $\pi_v^u(x(u)) \neq x(v)$. Take disjoint sets $U, V \in \tau(X_v)$ such that $\pi_v^u(x(u)) \in U$ and $x(v) \in V$. By continuity of $\pi_v^u$ there is a set $W \in \tau(x(u), X_u)$ for which $\pi_v^u(W) \subset U$. The set $G = \{y \in X : y(u) \in W$ and $y(v) \in V\}$ is open in $X$. If $y \in G$ then $y(u) \in W$ and hence $\pi_v^u(y(u)) \in U$, which, together with $y(v) \in V$ shows that $\pi_v^u(y(u)) \neq y(v)$. Thus every point $x \in X \backslash F_s$ has an open neighborhood $G \subset X \backslash F_s$, so $F_s$ is a closed subset of $X$.

Given $s \in T$ choose $s' \in T$ such that $t \leq s'$ and $s \leq s'$. Fix a point $x_u \in X_u$ for any $u \notin A_{s'}$; since the map $\pi_t^{s'}$ is surjective, there exists $y \in X_{s'}$ such that $\pi_t^{s'}(y) = z$. Let $x_u = \pi_u^{s'}(y)$ for any $u \in A_{s'}$. Letting $x(u) = x_u$ for any $u \in T$ we obtain a point $x \in X$ such that $x(t) = \pi_t^{s'}(y) = z$; it follows from the definition of projections of an inverse system that $\pi_v^u(x(u)) = \pi_v^u(\pi_u^{s'}(y)) = \pi_v^{s'}(y) = x(v)$ for any $u, v \in A_{s'}$ with $v \leq u$, so $x \in F_{s'}$. Since also $F_{s'} \subset F_s$, we proved that $x \in F_s$ and therefore

(1) the set $F_s$ is nonempty and closed in $X$ for any $s \in T$.

  If $s_1, \ldots, s_n \in T$ then there exists $s \in T$ such that $s_i \leq s$ for each $i \leq n$. It is straightforward that $F_s \subset F_{s_i}$ for every $i \leq n$, so $F_s \subset \bigcap_{i \leq n} F_{s_i}$ and hence

(2) $\bigcap \mathcal{F}' \neq \emptyset$ for any finite family $\mathcal{F}' \subset \mathcal{F} = \{F_s : s \in T\}$.

The properties (1) and (2) show that $\mathcal{F}$ is a centered family of closed subsets of a compact space $X$. Thus $\bigcap \mathcal{F} \neq \emptyset$; take a point $x \in \bigcap \mathcal{F}$. It is an easy exercise that $x \in L = \varprojlim \mathcal{S}$ and $\pi_t(x) = x(t) = z$; the point $z \in X_t$ was chosen arbitrarily, so $\pi_t(L) = X_t$, i.e., the map $\pi_t$ is surjective for any $t \in T$.

**V.174.** *Suppose that* $n \in \omega$ *and* $\mathcal{S} = \{X_t, \pi_s^t : t, s \in T\}$ *is an inverse system in which all spaces* $X_t$ *are compact and Hausdorff. Knowing also that* $\dim X_t \leq n$ *for each* $t \in T$ *prove that* $\dim(\varprojlim \mathcal{S}) \leq n$.

**Solution.** The space $L = \varprojlim \mathcal{S}$ is compact (see Problem 165) and hence normal; let $\pi_t : L \to X_t$ be the limit projection for any $t \in T$. Take a finite open cover $\mathcal{U}$ of the space $L$; it follows from Problem 170 that, for every $x \in L$ there exists $t(x) \in T$ and $U_x \in \tau(X_{t(x)})$ such that $x \in V_x = \pi_{t(x)}^{-1}(U_x)$ and the set $V_x$ is contained in some element of $\mathcal{U}$.

Take a finite set $A \subset L$ for which $L = \bigcup \{V_x : x \in A\}$; the set $T$ being directed there exists $t \in T$ such that $t(x) \leq t$ for any $x \in A$. If $x \in A$ then the set $O_x = (\pi_{t(x)}^t)^{-1}(U_x)$ is open in $X_t$. It follows from the definition of the limit of an inverse system that $\pi^t_{t(x)} \circ \pi_t = \pi_{t(x)}$ and hence $\pi^{-1}_t(O_x) = V_x$ for all $x \in A$. Therefore $L = \bigcup\{\pi^{-1}_t(O_x) : x \in A\}$ and hence $\pi_t(L) \subset \bigcup\{O_x : x \in A\}$.

The set $\pi_t(L)$ being closed in $X_t$, we have $\dim(\pi_t(L)) \leq \dim X_t \leq n$ (see Problem 146), so there is a finite closed refinement $\mathcal{F}$ of the cover $\{O_x \cap \pi_t(L) : x \in A\}$ of the space $\pi_t(L)$ such that $\mathrm{ord}(\mathcal{F}) \leq n + 1$. It is straightforward that $\{\pi^{-1}_t(F) : F \in \mathcal{F}\}$ is a finite closed refinement of $\mathcal{U}$ of order $\leq n + 1$, so we can apply Problem 145 to conclude that $\dim L \leq n$.

**V.175.** *Suppose that $n \in \omega$ and $S = \{X_l, \pi^l_m : l, m \in \omega\}$ is an inverse sequence in which every $X_l$ is a Lindelöf $\Sigma$-space. Knowing additionally that $\dim X_l \leq n$ for every $l \in \omega$ prove that $\dim(\varprojlim S) \leq n$.*

**Solution.** The space $L = \varprojlim S$ is Lindelöf being closed in a Lindelöf space $\prod_{m \in \omega} X_m$ (see Problem 165 and SFFS-256); let $K_l = \beta X_l$ for every $l \in \omega$. For any $m, l \in \omega$ with $m \leq l$ there exists a continuous map $q^l_m : K_l \to K_m$ such that $q^l_m | X_l = \pi^l_m$. If $k \leq m \leq l$ then $(q^m_k \circ q^l_m)| X_l = \pi^m_k \circ \pi^l_m = \pi^l_k = q^l_k | X_l$; since $X_l$ is dense in $K_l$, we conclude that $q^m_k \circ q^l_m = q^l_k$ by Fact 0 of S.351. Thus $\mathcal{K} = \{K_l, q^l_m, m, l \in \omega\}$ is an inverse system of compact spaces with $\dim K_l = \dim X_l \leq n$ for any $l \in \omega$ (see Problem 147). If $K = \varprojlim \mathcal{K}$ then $\dim K \leq n$ by Problem 174.

Every thread of $S$ is also a thread of $\mathcal{K}$, so $L \subset K$. Take any $U \in \tau(L, K)$ and fix a set $V_x \in \tau(x, K)$ such that $\overline{V}_x \subset U$ for every $x \in L$. The space $L$ being Lindelöf there is a countable $A \subset L$ for which $L \subset \bigcup\{V_x : x \in A\}$. As a consequence, $L \subset H = \bigcup\{\overline{V}_x : x \in A\} \subset U$ which shows that $H$ is an $F_\sigma$-subset of $K$ such that $L \subset H \subset U$. This makes it possible to apply Problem 154 to convince ourselves that $\dim L \leq n$. ·

**V.176.** *Prove that if $X$ is second countable and $A \subset C_p(X)$ is countable then there exists a countable QS-algebra $E \subset C_p(X)$ such that $A \subset E$.*

**Solution.** Fix a countable base $\mathcal{B}$ in the space $X$ and consider the family

$$\mathcal{P} = \{(U, V) : U, V \in \mathcal{B} \text{ and } \overline{U} \subset V\};$$

for every $P = (U, V) \in \mathcal{P}$ choose a function $f_P \in C_p(X)$ such that $f_P(\overline{U}) \subset \{1\}$ and $f_P(X \backslash V) \subset \{0\}$. It is evident that the set $B = A \cup \{f_P : P \in \mathcal{P}\}$ is countable, so $C = \{f_1 \cdot \ldots \cdot f_n : n \in \mathbb{N}, f_i \in B \text{ for all } i \leq n\}$ is also countable and hence the set $E = \{p_1 g_1 + \ldots + p_m g_m : m \in \mathbb{N}, p_i \in \mathbb{Q} \text{ and } g_i \in C \text{ for all } i \leq m\}$ is countable as well. It is straightforward that $f, g \in E$ and $q \in \mathbb{Q}$ imply that $f + g \in E$, $f \cdot g \in E$ and $qf \in E$. Given a point $x \in X$ and a closed set $F \subset X$ with $x \notin F$ there exists $V \in \mathcal{B}$ such that $x \in V \subset X \backslash F$. Pick $U \in \mathcal{B}$ for which $x \in U \subset \overline{U} \subset V$; then $P = (U, V) \in \mathcal{P}$ and hence $f = f_P \in E$. Then $f(x) = 1$ and $f(F) \subset \{0\}$, so $E$ is a countable QS-algebra in $C_p(X)$ such that $A \subset E$.

**V.177.** *Let* $K = \{x_0, \ldots, x_n\} \subset X$ *be a finite subset of a space* $X$. *Suppose that* $U \in \tau(K, X)$ *and* $q_0, \ldots, q_n \in \mathbb{Q}$. *Prove that, for any QS-algebra* $E$ *on a space* $X$, *there is* $f \in E$ *such that* $f(X \backslash U) \subset \{0\}$ *and* $f(x_i) = q_i$ *for each* $i \leq n$.

**Solution.** The set $F_i = (X \backslash U) \cup (K \backslash \{x_i\})$ is closed in $X$ and $x_i \notin F_i$, so there exists $f_i \in E$ such that $f_i(x_i) = 1$ and $f_i(F_i) \subset \{0\}$ for each $i \leq n$. It is immediate that the function $f = q_0 f_0 + \ldots + q_n f_n \in E$ is as promised.

**V.178.** *Given second countable Tychonoff spaces* $X$ *and* $Y$, *suppose that some QS-algebras* $E(X)$ *and* $E(Y)$ *are chosen in* $C_p(X)$ *and* $C_p(Y)$ *respectively. Prove that, if* $E(X)$ *is uniformly homeomorphic to* $E(Y)$ *then* $X$ *is representable as a countable union of closed subspaces each one of which embeds in* $Y$.

**Solution.** Given a set $A$ the family $\mathrm{Fin}(A)$ consists of all finite subsets of $A$; for each $n \in \omega$ let $[A]^n = \{B \in \mathrm{Fin}(A) : |B| = n\}$. If $f$ is a function then $\mathrm{dom}(f)$ is its domain; for any space $Z$ the function $\mathbf{0}_Z \in C_p(Z)$ is defined by $\mathbf{0}_Z(z) = 0$ for all $z \in Z$.

Choose a uniformly continuous isomorphism $T : E(X) \to E(Y)$; it is evident that $\mathbf{0}_X \in E(X)$ and $\mathbf{0}_Y \in E(Y)$, so there is no loss of generality to assume that $T(\mathbf{0}_X) = \mathbf{0}_Y$. For every $x \in X$ let $a_x(f) = f(x)$ for all $f \in E(X)$; if $y \in Y$ then $b_y(f) = f(y)$ for all $f \in E(Y)$. For any $K \in \mathrm{Fin}(X)$ and $\varepsilon > 0$ the set $O_X(K, \varepsilon) = \{f \in E(X) : f(K) \subset (-\varepsilon, \varepsilon)\}$ is an open neighborhood of $\mathbf{0}_X$ and the family $\{O_X(K, \varepsilon) : K \in \mathrm{Fin}(X) \text{ and } \varepsilon > 0\}$ is a local base of the space $E(X)$ at the point $\mathbf{0}_X$.

*Fact 1.* If $x \in X$ then $a_x : E(X) \to \mathbb{R}$ is a uniformly continuous unbounded map; therefore the map $\alpha_x = a_x \circ T^{-1} : E(Y) \to \mathbb{R}$ is uniformly continuous and unbounded. Analogously, $b_y : E(Y) \to \mathbb{R}$ is a uniformly continuous unbounded map and hence the map $\beta_y = b_y \circ T : E(X) \to \mathbb{R}$ is also uniformly continuous and unbounded for every $y \in Y$.

*Proof.* If we let $\varphi(x)(f) = f(x)$ for any $f \in C_p(X)$ then $\varphi : C_p(X) \to \mathbb{R}$ is linear and continuous; an evident consequence is that $\varphi$ is uniformly continuous. Therefore $a_x = \varphi | E(X)$ is uniformly continuous. Analogously, the map $b_y : E(Y) \to \mathbb{R}$ is uniformly continuous. It follows from Problem 177 that, for any $n \in \omega$ there is a function $f \in E(X)$ such that $f(x) = n$. This implies $a_x(f) = n$, so $a_x$ is unbounded and hence $\alpha_x$ is also uniformly continuous and unbounded. The analogous statement being also true for $\beta_y$, Fact 1 is proved.

*Fact 2.* Given a point $y \in Y$ and a finite subset $K$ of the space $X$ let

$$u(y, K) = \sup\{|\beta_y(f) - \beta_y(g)| : f, g \in E(X) \text{ and } |f(x) - g(x)|$$

$$< 1 \text{ for every } x \in K\}.$$

Then the family $\mathcal{E}(y) = \{K \in \mathrm{Fin}(X) : u(y, K) < \infty\}$ is nonempty. Analogously, let $v(x, L) = \sup\{|\alpha_x(f) - \alpha_x(g)| : f, g \in E(Y) \text{ and } |f(y) - g(y)| < 1 \text{ for all } y \in L\}$ for any $x \in X$ and finite $L \subset Y$. Then the family $\mathcal{D}(x) = \{L \in \mathrm{Fin}(Y) : v(x, L) < \infty\}$ is nonempty.

*Proof.* It follows from uniform continuity of $\beta_y$ that there is a finite set $K \subset X$ and $\varepsilon > 0$ such that $f - g \in O_X(K, \varepsilon)$ implies $|\beta_y(f) - \beta_y(g)| < 1$. Fix $n \in \mathbb{N}$ with $n\varepsilon > 1$ and take any $f, g \in E(X)$ such that $|f(x) - g(x)| < 1$ for all $x \in K$. The function $h_i = f + \frac{i}{n}(g - f)$ belongs to $E(X)$ for every $i \leq n$ and it is straightforward that $|h_{i+1}(x) - h_i(x)| < \frac{1}{n} < \varepsilon$ for all $x \in K$ and hence $|\beta_y(h_{i+1}) - \beta_y(h_i)| < 1$ for all $i < n$. Now, $h_0 = f$ and $h_n = g$, so $|\beta_y(g) - \beta_y(f)| \leq \sum_{i<n} |\beta_y(h_{i+1}) - \beta_y(h_i)| < n$ which shows that $u(y, K) \leq n$ and hence $K \in \mathcal{E}(y)$. An analogous reasoning shows that $\mathcal{D}(x) \neq \emptyset$, so Fact 2 is proved.

*Fact 3.* For any point $y \in Y$ the empty set does not belong to $\mathcal{E}(y)$; besides, if $K_0, K_1 \in \mathcal{E}(y)$ then $u(y, K_0 \cap K_1) \leq u(y, K_0) + u(y, K_1)$ and hence $K_0 \cap K_1 \in \mathcal{E}(y)$. Analogously, if $x \in X$ then all elements of $\mathcal{D}(x)$ are nonempty and $L_0 \cap L_1 \in \mathcal{D}(x)$ whenever $L_0, L_1 \in \mathcal{D}(x)$; besides, $v(x, L_0 \cap L_1) \leq v(x, L_0) + v(x, L_1)$.

*Proof.* Observe that $O_X(\emptyset, 1) = E(X)$; since $\beta_y$ is unbounded on $E(X)$ by Fact 1, the set $\{|\beta_y(f)| = |\beta_y(f) - \beta_y(0_X)| : f \in E(X)\}$ is unbounded so $u(y, \emptyset) = \infty$ and hence $\emptyset \notin \mathcal{E}(y)$. Now, if $K_0, K_1 \in \mathcal{E}(y)$ then let $K = K_0 \cap K_1$ and consider any $f, g \in E(X)$ such that $|f(x) - g(x)| < 1$ for all $x \in K$.

The set $I_x = (f(x) - 1, f(x) + 1) \cap (g(x) - 1, g(x) + 1)$ is open and nonempty, so we can choose a rational number $q_x \in I_x$ for any $x \in K$. For every $x \in K_0 \backslash K$ choose a rational number $q_x \in (f(x) - 1, f(x) + 1)$; if $x \in K_1 \backslash K$ then we can pick $q_x \in (g(x) - 1, g(x) + 1) \cap \mathbb{Q}$. Apply Problem 177 to construct a function $h \in E(X)$ for which $h(x) = q_x$ for every $x \in K_0 \cup K_1$; it is immediate that $f - h \in O_X(K_0, 1)$ and $h - g \in O_X(K_1, 1)$. Therefore $|\beta_y(f) - \beta_y(h)| \leq u(y, K_0)$ and $|\beta_y(h) - \beta_y(g)| \leq u(y, K_1)$ whence $|\beta_y(f) - \beta_y(g)| \leq u(y, K_0) + u(y, K_1)$. This shows that $u(y, K) \leq u(y, K_0) + u(y, K_1)$, so we proved all statements formulated for $\mathcal{E}(y)$. An analogous reasoning settles the case of $\mathcal{D}(x)$, so Fact 3 is proved.

*Fact 4.* For any $y \in Y$ there exists a unique minimal element $K(y)$ in the family $\mathcal{E}(y)$ with respect to inclusion; let $u(y) = u(y, K(y))$. Analogously, there is a unique minimal element $L(x)$ in the family $\mathcal{D}(x)$; let $v(x) = v(x, L(x))$ for each $x \in X$.

*Proof.* The elements of $\mathcal{E}(y)$ are finite, so there exists a minimal element $K \in \mathcal{E}(y)$; if $K_0$ is another minimal element of $\mathcal{E}(y)$ then $K \cap K_0$ is strictly smaller than $K$; since $K \cap K_0 \in \mathcal{E}(y)$ by Fact 3, we have a contradiction with minimality of $K$. Therefore there is a unique minimal element of $\mathcal{E}(y)$; the same argument proves the relevant fact for $\mathcal{D}(x)$, so Fact 4 is proved.

*Fact 5.* For any $p, q \in \mathbb{N}$ let $Y(p, q) = \{y \in Y :$ there exists $K \in \mathcal{E}(y)$ such that $|K| \leq q$ and $u(y, K) \leq p\}$; we will also need the set

$$X(p, q) = \{x \in X : \text{ there is } L \in \mathcal{D}(x) \text{ such that } |L| \leq q \text{ and } v(x, L) \leq p\}.$$

Given $p \in \mathbb{N}$ consider the sets $M(p, 1) = Y(p, 1)$ and $N(p, 1) = X(p, 1)$; for every natural $q > 1$ let $M(p, q) = Y(p, q) \backslash Y(2p, q - 1)$ and $N(p, q) = X(p, q) \backslash X(2p, q - 1)$. If $M(p) = \bigcup\{M(p, q) : q \in \mathbb{N}\}$ and $N(p) = \bigcup\{N(p, q) : q \in \mathbb{N}\}$ then

(i) for any $y \in Y$ and $p \in \mathbb{N}$ such that $u(y) \leq p$ we have $y \in M(p)$; in particular, $Y = \bigcup \{M(p) : p \in \mathbb{N}\}$.

(ii) For any $x \in X$ and $p \in \mathbb{N}$ such that $v(x) \leq p$ we have $y \in N(p)$; in particular, $X = \bigcup \{N(p) : p \in \mathbb{N}\}$.

(iii) If we have $p \in \mathbb{N}$ and distinct $q_0, q_1 \in \mathbb{N}$ then $M(p, q_0) \cap M(p, q_1) = \emptyset$ and $N(p, q_0) \cap N(p, q_1) = \emptyset$.

*Proof.* If $q = |K(y)|$ then $u(y) = u(y, K(y)) \leq p$, so $y \in Y(p, q)$. The set $K(y)$ being minimal in $\mathcal{E}(y)$, the point $y$ does not belong to $Y(2p, q - 1)$, so $y \in M(p, q) \subset M(p)$; this proves (i). An identical argument shows that (ii) is also true.

Now assume that $p \in \mathbb{N}$ and take distinct $q_0, q_1 \in \mathbb{N}$; there is no loss of generality to assume that $q_0 < q_1$ and hence $q_0 \leq q_1 - 1$. It follows from the relevant definitions that $M(p, q_0) \subset Y(p, q_0) \subset Y(p, q_1 - 1) \subset Y(2p, q_1 - 1) \subset Y \setminus M(p, q_1)$ and therefore $M(p, q_0) \cap M(p, q_1) = \emptyset$. Analogously, $N(p, q_0) \cap N(p, q_1) = \emptyset$, so we settled (iii) and hence Fact 5 is proved.

*Fact 6.* Suppose that $p \in \mathbb{N}$ and $y \in M(p)$; denote by $q(p)$ the unique natural number such that $y \in M(p, q(p))$. Then there exists a unique set $K_p(y) \subset X$ such that $|K_p(y)| = q(p)$ and $u(y, K_p(y)) \leq p$. Analogously, for any $x \in L(p)$ there is a unique set $L_p(x) \subset Y$ for which $|L_p(x)| = q$ and $v(x, L_p(x)) \leq p$ (here $q \in \mathbb{N}$ is the unique number for which $x \in L(p, q)$).

*Proof.* Observe that $q(p)$ is unique because the family $\{M(p, q) : q \in \mathbb{N}\}$ is disjoint by Fact 5. Let $q = q(p)$; by the definition of $Y(p, q)$ there exists a set $K \subset X$ such that $|K| \leq q$ and $a(y, K) \leq p$. It follows from $y \notin Y(2p, q - 1) \supset Y(p, q - 1)$ that $|K| = q$. Now, if $K' \neq K$, $a(y, K') \leq p$ and $|K'| = q$ then $K'' = K \cap K'$ has at most $(q-1)$-many elements and $a(y, K'') \leq 2p$ (see Fact 3) which is a contradiction with $y \notin Y(2p, q - 1)$. This proves uniqueness of the set $K_p(y) = K$. The proof of existence and uniqueness of the set $L_p(x)$ is analogous, so Fact 6 is proved.

*Fact 7.* Given $p \in \mathbb{N}$ assume that $\{y_n : n \in \omega\} \subset Y$ is a sequence which converges to a point $y \in M(p)$; if $Q_n \subset X$ is a finite set such that $u(y_n, Q_n) \leq p$ for every $n \in \omega$ then for every $U \in \tau(X)$ with $U \cap K_p(y) \neq \emptyset$ there exists $m \in \omega$ such that $U \cap Q_n \neq \emptyset$ for all $n \geq m$. Analogously, if $\{x_n : n \in \omega\} \subset Y$ is a sequence which converges to a point $x \in N(p)$ and $R_n \subset X$ is a finite set such that $u(y_n, R_n) \leq p$ for all $n \in \omega$ then for every $V \in \tau(Y)$ with $V \cap L_p(x) \neq \emptyset$ there exists $m \in \omega$ such that $V \cap R_n \neq \emptyset$ for all $n \geq m$.

*Proof.* If the set $A = \{n \in \omega : Q_n \cap U = \emptyset\}$ is infinite then we can pass to the subsequence $\{y_n : n \in A\}$ to see that we can assume, without loss of generality, that $Q_n \cap U = \emptyset$ for all $n \in \omega$. Fix $q \in \mathbb{N}$ such that $y \in M(p, q)$; then $|K_p(y)| = q$. If $K = K_p(y) \setminus U$ then $|K| < q$, so there exist functions $f, g \in E(X)$ such that $|f(x) - g(x)| < 1$ for all $x \in K$ while $|\beta_y(f) - \beta_y(g)| > 2p$. Using Problem 177 it is easy to find a function $h \in E(X)$ such that $h|(X \setminus U) = f|(X \setminus U)$ and $|h(x) - g(x)| < 1$ for all $x \in U \cap K_p(y)$. Then $|h(x) - g(x)| < 1$ for all $x \in K_p(y)$, so $|\beta_y(h) - \beta_y(g)| \leq p$. An immediate consequence is that $|\beta_y(h) - \beta_y(f)| > p$.

On the other hand, it follows from $h|Q_n = f|Q_n$ that

$(*)$  $|\beta_{y_n}(h) - \beta_{y_n}(f)| \leq p$ for all $n \in \omega$.

The function $T(h)$ is continuous on $Y$, so the sequence $\{T(h)(y_n) : n \in \omega\}$ converges to $T(h)(y) = \beta_y(h)$. Since $\beta_{y_n}(h) = T(h)(y_n)$ for every $n \in \omega$, the sequence $\{\beta_{y_n}(h) : n \in \omega\}$ converges to $\beta_y(h)$. Analogously, $\beta_{y_n}(f) \to \beta_y(f)$, so we can apply $(*)$ to conclude that $|\beta_y(h) - \beta_y(f)| \leq p$; this contradiction demonstrates that only finitely many elements of the sequence $\{Q_n : n \in \omega\}$ miss the set $U$. An analogous reasoning shows that only finitely many elements of the sequence $\{R_n : n \in \omega\}$ miss the set $V$, so Fact 7 is proved.

*Fact 8.* For any $p, q \in \mathbb{N}$ the map $K_p : M(p, q) \to [X]^q$ is lower semicontinuous. Analogously, the map $L_p : N(p, q) \to [Y]^q$ is lower semicontinuous.

*Proof.* Fix any $U \in \tau(X)$; if the set $V = \{y \in M(p, q) : K_p(y) \cap U \neq \emptyset\}$ is not open in $M(p, q)$ then there is a sequence $\{y_n : n \in \omega\} \subset M(p, q) \backslash V$ which converges to some $y \in V$. Since $K_p(y_n) \cap U = \emptyset$ and $u(y_n, K_p(y_n)) \leq p$ for every $n \in \omega$, we obtained a contradiction with Fact 7. Therefore the map $K_p$ is lower semicontinuous. An analogous reasoning shows that $L_p$ is lower semicontinuous as well, so Fact 8 is proved.

*Fact 9.* The set $Y(p, q)$ is closed in $Y$ and the set $X(p, q)$ is closed in $X$ for any $p, q \in \mathbb{N}$.

*Proof.* Suppose that $y_n \in Y(p, q)$ for all $n \in \omega$ and $y_n \to y$. We will pass several times to a subsequence of the sequence $S = \{y_n : n \in \omega\}$; since our aim is to prove that $y \in Y(p, q)$, at each step we will identify the obtained subsequence with $S$ considering that all elements of $S$ have the property we have found in a subsequence. Fix a set $Q_n \subset X$ such that $|Q_n| \leq q$ and $u(y_n, Q_n) \leq p$ for every $n \in \omega$. Passing to a subsequence of $S$ if necessary, we can assume that $|Q_n| = k \leq q$ for all $n \in \omega$.

Next, use Fact 2 of U.337 to choose an infinite $A \subset \omega$ for which there is a set $D = \{d_1, \ldots, d_r\} \subset X$ such that $Q_n \cap Q_m = D$ for distinct $n, m \in A$ (observe that it is possible that $r = 0$ in which case $D = \emptyset$). According to the above mentioned politics we can consider that, for any $i \in \omega$, we have $Q_i = \{d_1, \ldots, d_r, a_1^i, \ldots, a_{k-r}^i\}$ and the family $\{Q_i \backslash D : i \in \omega\}$ is disjoint.

An evident property of metric spaces is that any sequence contains either a convergent subsequence or an infinite closed discrete subspace. This makes it possible to pass to a subsequence of $S$ once more to guarantee that, for any $j \in \{1, \ldots, k - r\}$, the sequence $S_j = \{a_j^i : i \in \omega\}$ is either convergent or constitutes a closed discrete subspace of $X$. If $S_j$ is convergent then denote by $x_j$ its limit. Renumbering every $Q_i$ if necessary we can assume that $Q_i = \{d_1, \ldots, d_r, a_1^i, \ldots, a_l^i, a_{l+1}^i, \ldots, a_{k-r}^i\}$ while the set $A = \{a_{l+j}^i : i \in \omega, 1 \leq j \leq k - r - l\}$ is closed and discrete in $X$ and the sequence $S_j$ converges to $x_j$ for any $j \in \{1, \ldots, l\}$.

Consider the set $Q = \{d_1, \ldots, d_r, x_1, \ldots, x_l\}$; since $|Q| \leq q$, it suffices to show that $u(y, Q) \leq p$. To do this, fix $f_0, g_0 \in E(X)$ such that $|f_0(x) - g_0(x)| < 1$ for every $x \in Q$. Given an arbitrary $\varepsilon > 0$ there exists a finite set $E \supset Q$ and $\delta \in (0, 1)$ such that $f - g \in O_X(E, \delta)$ implies $|\beta_y(f) - \beta_y(g)| < \varepsilon$. The set $A$ being closed and discrete in $X$ we can find $U \in \tau(E, X)$ such that $U \cap A$ is a finite set.

Apply Problem 177 to find $h \in E(X)$ such that $h(E) \subset \{1\}$ and $h(X \setminus U) \subset \{0\}$; consider the functions $f_1 = h f_0$ and $g_1 = h g_0$. It follows from $f_0 | E = f_1 | E$ and $g_0 | E = g_1 | E$ that

$(**)\quad |\beta_y(f_0) - \beta_y(f_1)| < \varepsilon$ and $|\beta_y(g_0) - \beta_y(g_1)| < \varepsilon$.

We also have $f_1 | Q = f_0 | Q$ and $g_1 | Q = g_0 | Q$, so $|f_1(x) - g_1(x)| < 1$ for every $x \in Q$. Therefore $W = \{x \in X : |f_1(x) - g_1(x)| < 1\}$ is an open neighborhood of the set $Q$ and so is the set $W' = W \cap U$. There is $m \in \omega$ such that, for all $i \geq m$, we have $a_j^i \in W'$ for all $j \in \{1, \ldots, l\}$ and $a_j^i \notin U$ for all $j \in \{l + 1, \ldots, k - r\}$. As a consequence, for each $i \geq m$ we have $|f_1(x) - g_1(x)| < 1$ for every point $x \in D \cup \{a_1^i, \ldots, a_l^i\}$; besides, $f_1(x) = g_1(x) = 0$ whenever $x \in \{a_{l+1}^i, \ldots, a_{k-r}^i\}$, so $|f_1(x) - g_1(x)| < 1$ for all $x \in Q_i$.

By the choice of the sets $Q_i$, we have $|\beta_{y_i}(f_1) - \beta_{y_i}(g_1)| \leq p$ for all $i \geq m$. We already saw that $\beta_{y_i}(f_1) \to \beta_y(f_1)$ and $\beta_{y_i}(g_1) \to \beta_y(g_1)$ as $i \to \infty$. Passing to the limit in the last inequality, we conclude that $|\beta_y(f_1) - \beta_y(g_1)| \leq p$. This, together with $(**)$, implies that $|\beta_y(f_0) - \beta_y(g_0)| \leq p + 2\varepsilon$. The number $\varepsilon > 0$ was taken arbitrarily so $|\beta_y(f_0) - \beta_y(g_0)| \leq p$ and hence $y \in Y(p, q)$, i.e., we established that the set $Y(p, q)$ is closed in $X$. An analogous reasoning shows that $X(p, q)$ is closed in $Y$, so Fact 9 is proved.

Returning to our solution call a set $A \subset Y$ *adequate* if it is closed in $Y$ and embeds in $X$; the set $A$ is called $\sigma$-adequate if it is representable as the countable union of adequate sets. Given any $p, q \in \mathbb{N}$, say that a set $A \subset Y$ is $(p, q)$-*small* if $A \subset M(p, q)$ and there exist continuous maps $f_1, \ldots, f_q : A \to X$ such that $K_p(y) = \{f_1(y), \ldots, f_q(y)\}$ for all $y \in A$. Analogously, a set $B \subset X$ will be called $(p, q)$-*small* if $B \subset N(p, q)$ and there are continuous maps $g_1, \ldots, g_q : B \to Y$ such that $L_p(x) = \{g_1(x), \ldots, g_q(x)\}$ for all $x \in B$.

Given $p, q \in \mathbb{N}$, we can apply Fact 2 of V.080 together with Fact 8 to see that there exists a family $\{U_n : n \in \omega\}$ of open subsets of $M(p, q)$ such that $M(p, q) = \bigcup_{n \in \omega} U_n$ and every $U_n$ is $(p, q)$-small. By Fact 9 each $M(p, q)$ is an $F_\sigma$-set in $Y$, so there exists a family $\{F_n(p, q) : n \in \omega\}$ of closed subsets of $Y$ such that $M(p, q) = \bigcup_{n \in \omega} F_n(p, q)$ and every $F_n(p, q)$ is $(p, q)$-small. The family $\mathcal{F} = \{F_n(p, q) : n \in \omega, \ p, q \in \mathbb{N}\}$ is a countable closed cover of $Y$ such that every $F \in \mathcal{F}$ is $(p, q)$-small for some $p, q \in \mathbb{N}$; let $S_Y(F) = \{f_1, \ldots, f_q\}$ be the respective set of continuous functions from $F$ to $X$. Analogously, there exists a countable closed cover $\mathcal{G}$ of the space $X$ such that every $G \in \mathcal{G}$ is $(p, q)$-small for some $p, q \in \mathbb{N}$; denote by $S_X(G)$ the set $\{g_1, \ldots, g_q\}$ which witnesses that $G$ is $(p, q)$-small. Next, observe that

(1)  $y \in \bigcup \{L(x) : x \in K(y)\}$ for any $y \in Y$.

Let $A = \bigcup \{L(x) : x \in K(y)\}$ and assume that $y \notin A$. Choose a number $n \in \mathbb{N}$ such that $v(x) < n$ for all $x \in K(y)$ and apply Problem 177 to choose a function

$f \in E(Y)$ such that $f(y) > nu(y)$ and $f(A) \subset \{0\}$. It follows from $f|L(x) = 0_Y|L(x)$ that $|\alpha_x(f) - \alpha_x(0_Y)| = |\alpha_x(f)| \le v(x) < n$ for any $x \in K(y)$. If $g = T^{-1}(f)$ then $|g(x)| = |\alpha_x(f)| < n$ for every $x \in K(y)$.

Consider the function $h_i = \frac{i}{n}g$ for all $i = 0, \ldots, n$. Given any natural number $i < n$ we have $|h_{i+1}(x) - h_i(x)| = \frac{1}{n}|g(x)| < 1$ for all points $x \in K(y)$; therefore $|\beta_y(h_{i+1}) - \beta_y(h_i)| \le u(y)$. Now, it follows from the equalities $h_0 = 0_X$ and $h_n = g$ that $|\beta_y(g)| = |\beta_y(g) - \beta_y(0_X)| \le \sum_{i=0}^{n-1} |\beta_y(h_{i+1}) - \beta_y(h_i)| \le nu(y)$. Recalling that $\beta_y(g) = T(g)(y) = f(y)$ we conclude that $|f(y)| \le nu(y)$; this contradiction with our choice of $f$ shows that (1) is proved.

The sets $S_X = \bigcup\{S_X(B) : B \in \mathcal{G}\}$ and $S_Y = \bigcup\{S_Y(A) : A \in \mathcal{F}\}$ are countable; our next step is to prove that

(2) the set $Q(f, g) = \{y \in \text{dom}(f) : f(y) \in \text{dom}(g) \text{ and } g(f(y)) = y\}$ is adequate for any $f \in S_Y$ and $g \in S_X$.

Suppose that a sequence $\{y_n : n \in \omega\} \subset Q(f, g)$ converges to a point $y \in Y$. The set $\text{dom}(f)$ being closed in $Y$, we have $y \in \text{dom}(f)$; by continuity of $f$, the sequence $\{f(y_n) : n \in \omega\} \subset \text{dom}(g)$ converges to the point $f(y)$. The set $\text{dom}(g)$ is closed in $X$, so $f(y) \in \text{dom}(g)$; the map $g$ being continuous, the sequence $\{g(f(y_n)) : n \in \omega\} = \{y_n : n \in \omega\}$ converges to $g(f(y))$ whence $y = g(f(y))$ which proves that the set $Q = Q(f, g)$ is closed in $Y$. Note that $g$ is the inverse of $f$ on the set $f(Q)$, so $f|Q : Q \to f(Q)$ is a homeomorphism and hence $Q$ is, indeed, an adequate subset of $Y$, i.e., (2) is proved.

The family $\mathcal{F}' = \{F \cap Q(f, g) : F \in \mathcal{F}, f \in S_Y \text{ and } g \in S_X\}$ consists of countably many adequate sets (see (2)). Fix a point $y \in Y$ and take $F \in \mathcal{F}$ such that $y \in F$; the set $F$ is $(p, q)$-small for some $p, q \in \mathbb{N}$, so $S_Y(F) = \{f_1, \ldots, f_q\}$ for some continuous functions $f_1, \ldots, f_q$ from $F$ to $X$.

The set $K(y)$ being the minimal element of the family $\mathcal{E}(y)$, we have the inclusion $K(y) \subset K_p(y) = \{f_1(y), \ldots, f_q(y)\}$, so it follows from (1) that there is $i \le q$ for which $y \in L(f_i(y))$. The family $\mathcal{G}$ being a cover of $X$ we can pick $G \in \mathcal{G}$ with $x = f_i(y) \in G$. By our choice of $\mathcal{G}$, there are $r, s \in \mathbb{N}$ such that $G \subset N(r, s)$ and $S_X(G) = \{g_1, \ldots, g_s\}$. The set $L(x)$ being the minimal element of $\mathcal{D}(x)$, we have $L(x) \subset L_r(x) = \{g_1(x), \ldots, g_s(x)\}$. Thus $y = g_j(x) = g_j(f_i(y))$ for some $j \le s$ and therefore $y \in F \cap Q(f_i, g_j) \in \mathcal{F}'$. The point $y \in Y$ was taken arbitrarily, so we proved that $Y = \bigcup \mathcal{F}'$ is $\sigma$-adequate; analogously, $X$ is $\sigma$-adequate, so our solution is complete.

**V.179.** *Given second countable Tychonoff spaces $X$ and $Y$, suppose that some QS-algebras $E(X)$ and $E(Y)$ are chosen in $C_p(X)$ and $C_p(Y)$ respectively. Prove that, if $E(X)$ is uniformly homeomorphic to $E(Y)$ then $\dim X = \dim Y$.*

**Solution.** Let $m = \dim Y$ and apply Problem 178 to find a family $\mathcal{F} = \{F_n : n \in \omega\}$ of closed subsets of $X$ such that $\bigcup \mathcal{F} = X$ and every $F_n$ embeds in $Y$. If $m = \infty$ then $\dim X \le m$, so assume that $m \in \omega$. It follows from Problem 155 that $\dim F_n \le m$ for all $n \in \omega$. The space $X$ being normal we can apply Problem 150 to see that $\dim X \le m$ which proves that $\dim X \le \dim Y$. Since the spaces $X$ and $Y$ are in a symmetric situation, we also have $\dim Y \le \dim X$ and hence $\dim X = \dim Y$.

**V.180.** *Suppose that $X$ and $Y$ are Tychonoff spaces such that $C_p(X)$ is uniformly homeomorphic to $C_p(Y)$. Prove that* dim $X =$ dim $Y$.

**Solution.** Given a space $Z$ and $P \subset C_p(Z)$ let $e_P^Z(z)(f) = f(z)$ for any $z \in Z$ and $f \in P$. Then $Z[P] = \{e_P^Z(z) : z \in Z\} \subset C_p(P)$ and $e_P^Z : Z \to Z[P]$ is a continuous map by TFS-166. If $P \subset Q \subset C_p(Z)$ then $\pi_{Q,P}^Z : Z[Q] \to Z[P]$ is the restriction map; it is evident that $e_P^Z = \pi_{Q,P}^Z \circ e_Q^Z$. Recall that if $\varphi : Z \to T$ is a continuous onto map then the dual map $\varphi^* : C_p(T) \to C_p(Z)$ is defined by $\varphi^*(f) = f \circ \varphi$ for any $f \in C_p(T)$.

If $Z$ is a space and $P \subset C_p(Z)$ then $(e_P^Z)^* : C_p(Z[P]) \to C_p(Z)$ is an embedding (see TFS-163) such that $P \subset C[Z, P] = (e_P^Z)^*(C_p(Z[P]))$ (see Fact 5 of U.086). Let $d_P^Z : C[Z, P] \to C_p(Z[P])$ be the inverse of $(e_P^Z)^*$, i.e., $d_P^Z(f) = [(e_P^Z)^*]^{-1}(f)$ for every $f \in C[Z, P]$.

*Fact 1.* Given spaces $Z$ and $T$ suppose that $\varphi : Z \to T$ is a continuous onto map and $A \subset C_p(T)$. If and $A' = \varphi^*(A)$ and $\xi = \varphi^*|A : A \to A'$ then $e_A^T \circ \varphi = \xi^* \circ e_{A'}^Z$.

*Proof.* If $z \in Z$ then let $y = \varphi(z)$ and observe that $e_A^T(y)$ is an element of $C_p(A)$ such that $e_A^T(y)(f) = f(y)$ for any $f \in C_p(T)$. Now, it follows from the equalities $\xi^*(e_{A'}^Z(z))(f) = (e_{A'}^Z(z) \circ \varphi^*)(f) = (f \circ \varphi)(z) = f(y) = e_A^T(y)(f)$ that the functions $e_A^T(\varphi(z))$ and $\xi^*(e_{A'}^Z(z))$ coincide for every $z \in Z$, i.e., $\xi^* \circ e_{A'}^Z = e_A^T \circ \varphi$, so Fact 1 is proved.

*Fact 2.* Suppose that $Z$ is a space, $M$ is a second countable space and we have a continuous onto map $\varphi : Z \to M$. Then there exists a countable $P \subset C_p(Z)$ and a homeomorphism $u : M' = e_P^Z(Z) \to M$ such that $u \circ e_P^Z = \varphi$.

*Proof.* Let $A \subset C_p(M)$ be a countable $QS$-algebra (which exists by Problem 176). Then $A$ separates the points from the closed sets of the space $M$, so the reflection map $e_A^M : M \to M_0 = e_A^M(M)$ is a homeomorphism by TFS-166.

The set $P = \varphi^*(A) \subset C_p(Z)$ is countable; let $\xi = \varphi^*|A : A \to P$. The dual map $\xi^* : C_p(P) \to C_p(A)$ is a homeomorphism and $\xi^*(e_P^Z(z)) = e_A^M(\varphi(z))$ for any $z \in Z$ by Fact 1. Therefore $((e_A^M)^{-1} \circ (\xi^*|M')) \circ e_P^Z = \varphi$ and hence $u = (e_A^M)^{-1} \circ (\xi^*|M')$ is the promised homeomorphism, i.e., Fact 2 is proved.

*Fact 3.* Given a space $Z$ and a countable $P \subset C_p(Z)$ let $M_0 = e_P^Z(Z)$. Suppose that $M$ is a second countable space and we have continuous onto maps $\varphi : Z \to M$ and $r : M \to M_0$ such that $r \circ \varphi = e_P^Z$. Then there is a countable set $Q \subset C_p(Z)$ such that $P \subset Q$ and $M_1 = e_Q^Z(Z)$ is homeomorphic to $M$.

*Proof.* Let $\eta : C_p(M_0) \to C_p(Z)$ be the dual map of $e_P^Z$, i.e., $\eta(f) = f \circ e_P^Z$ for any $f \in C_p(M_0)$. It was proved in Fact 5 of U.086 that $P \subset \eta(C_p(M_0))$; the set $P' = \{\eta^{-1}(g) : g \in P\}$ is countable and hence $P'' = r^*(P') \subset C_p(M)$ is countable as well, so there is a countable $QS$-algebra $A \subset C_p(M)$ such that $P'' \subset A$ (see Problem 176). The set $Q = \varphi^*(A) \subset C_p(Z)$ is countable; besides, $e_A^M : M \to e_A^M(M) \subset C_p(A)$ is a homeomorphism and it follows from Fact 1 that $e_A^M \circ \varphi = \xi^* \circ e_Q^Z$ where $\xi = \varphi^*|A$.

Thus, for the space $M_1 = e_Q^Z(Z)$ the map $u = (e_A^M)^{-1} \circ \xi^* : M_1 \to M$ is a homeomorphism. To see that $P \subset Q$ fix any $f \in P$ and let $g = \eta^{-1}(f)$, $h = r^*(g)$. Then $v = \varphi^*(h) \in Q$; to see that $f = v$ take any point $z \in Z$. Then $f(z) = g(e_P^Z(z))$. On the other hand, it follows from $h = g \circ r$ and $v = h \circ \varphi$ that $v = g \circ (r \circ \varphi) = g \circ e_P^Z$, so $v(z) = g(e_P^Z(z)) = f(z)$. Thus $f = v \in Q$, so $Q \subset P$ and hence Fact 3 is proved.

*Fact 4.* Suppose that $Z$ is a space and we have a sequence $\{A_n : n \in \omega\}$ of subsets of $C_p(Z)$ such that, for every $n \in \omega$ there is a $QS$-algebra $Q_n$ in the space $C_p(Z[A_n])$ such that $A_n \subset Q_n' = (e_{A_n}^Z)^*(Q_n) \subset A_{n+1}$. Then, for $A = \bigcup_{n \in \omega} A_n$, the set $d_A^Z(A)$ is a $QS$-algebra in $C_p(Z[A])$.

*Proof.* The maps $(e_A^Z)^*$ and $d_A^Z$ are easily seen to preserve products and linear combinations, i.e., $d_A^Z(u \cdot v) = d_A^Z(u) \cdot d_A^Z(v)$ and $d_A^Z(\alpha u + \beta v) = \alpha d_A^Z(u) + \beta d_A^Z(v)$ for any $u, v \in A$ and $\alpha, \beta \in \mathbb{R}$. Of course, analogous equalities hold for $(e_A^Z)^*$ which easily implies that $f \cdot g \in Q_n'$ and $pf + qg \in Q_n'$ for any $f, g \in Q_n'$ and $p, q \in \mathbb{Q}$.

If $f, g \in d_A^Z(A)$ then $f = d_A^Z(f_0)$ and $g = d_A^Z(g_0)$ for some $f_0, g_0 \in A$. There is $n \in \omega$ for which $f_0, g_0 \in A_n$ and hence $h = f_0 \cdot g_0 \in Q_n' \subset A_{n+1} \subset A$; analogously, $h_0 = f_0 + g_0 \in A$. Consequently, $d_A^Z(h) = d_A^Z(f_0) \cdot d_A^Z(g_0) = f \cdot g \in d_A^Z(A)$ and $d_A^Z(h_0) = d_A^Z(f_0) + d_A^Z(g_0) = f + g \in d_A^Z(A)$. Furthermore, for any $q \in \mathbb{Q}$ the function $qf_0$ also belongs to $Q_n' \subset A$, so $d_A^Z(qf_0) = qd_A^Z(f_0) = qf \in d_A^Z(A)$.

Finally take a closed subset $F \subset Z[A]$ and a point $y_0 \in Z[A] \setminus F$. There exist $f_1, \ldots, f_k \in A$ and $\varepsilon > 0$ such that the set $O = \{y \in Z[A] : |y(f_i) - y_0(f_i)| < \varepsilon$ for all $i \leq k\}$ is disjoint from $F$. Choose $n \in \omega$ with $\{f_1, \ldots, f_k\} \subset A_n$. If $z_0 = y_0 | A_n$ then $O' = \{z \in Z[A_n] : |z(f_i) - z_0(f_i)| < \varepsilon$ for all $i \leq k\}$ is an open subset of $Z[A_n]$ such that $\pi_{A,A_n}^Z(O) \subset O'$ and $z_0 \in O'$. It is easy to see that the set $F' = \pi_{A,A_n}^Z(F)$ is disjoint from $O'$, so there is a function $g \in Q_n$ such that $g(z_0) = 1$ and $g(F') \subset \{0\}$. If $f = g \circ \pi_{A,A_n}^Z$ then $f(y_0) = 1$ and $f(F) \subset \{0\}$.

Now the function $h = f \circ e_A^Z = g \circ (\pi_{A,A_n}^Z \circ e_A^Z) = g \circ e_{A_n}^Z$ belongs to $Q_n'$ and hence $h \in A$. Therefore $f = d_A^Z(h) \in d_A^Z(A)$, so $d_A^Z(A)$ is a $QS$-algebra and hence Fact 4 is proved.

*Fact 5.* Given a space $Z$ suppose that a set $A_i \subset C_p(Z)$ is countable, $A_i \subset A_{i+1}$ and $\dim(Z[A_i]) \leq n$ for all $i \in \omega$. Then, for the set $A = \bigcup_{i \in \omega} A_i$, we have $\dim(Z[A]) \leq n$.

*Proof.* The family $\{Z[A_i], \pi_{A_j,A_i}^Z : i < j < \omega\}$ is an inverse system; denote by $L$ its limit. If $r_i = \pi_{A,A_i}^Z$ then $\pi_{A_j,A_i}^Z \circ r_j = r_i$ for any $i, j \in \omega$ with $i < j$, so the diagonal product $r$ of the family $\{r_i : i \in \omega\}$ maps $Z[A]$ continuously into $L$ by Problem 167. If $y_0, y_1 \in Z[A]$ and $y_0 \neq y_1$ then there is $f \in A$ such that $y_0(f) \neq y_1(f)$; since $f \in A_i$ for some $i \in \omega$, we have $r_i(y_0) \neq r_i(y_1)$ and hence $r(y_0) \neq r(y_1)$, i.e., $r : Z[A] \to L$ is an injection.

Let $T = r(Z[A])$ and fix a point $t_0 \in T$; there is a unique point $y_0 \in Z[A]$ such that $r(y_0) = t_0$. To see that the map $r^{-1}$ is continuous at the point $t_0$ take any set $U \in \tau(y_0, Z[A])$. There are $f_1, \ldots, f_k \in A$ and $\varepsilon > 0$ such that $O = \{y \in Z[A] : |y(f_i) - y_0(f_i)| < \varepsilon$ for all $i \leq k\} \subset U$. There exists $m \in \omega$ for which $\{f_1, \ldots, f_k\} \subset A_m$. The set $W = \{t \in T : |t(m)(f_i) - t_0(m)(f_i)| < \varepsilon$ for all $i \leq k\}$ is an open neighborhood of $t_0$ in $T$. If $t \in W$ and $y = r^{-1}(t)$ then $y|A_m = t(m)$ and $y_0|A_m = t_0(m)$ which shows that $|y(f_i) - y_0(f_i)| = |t(m)(f_i) - t_0(m)(f_i)| < \varepsilon$ for every $i \leq k$. As a consequence, $y = r^{-1}(t) \in O$; the point $y \in W$ was chosen arbitrarily, so we proved that $r^{-1}(W) \subset O \subset U$.

Thus the map $r^{-1}$ is continuous at every point of $T$, so $r : Z[A] \to T$ is a homeomorphism, i.e., $Z[A]$ embeds in $L$. Finally apply Problem 175 to see that $\dim L \leq n$ and hence $\dim(Z[A]) = \dim T \leq \dim L \leq n$ by Problem 155, so Fact 5 is proved.

Returning to our solution let $u : C_p(X) \to C_p(Y)$ be a uniform homeomorphism. The equality $\dim X = \dim Y$ will be established if we prove that $\dim X \leq \dim Y$ and $\dim Y \leq \dim X$. The spaces $X$ and $Y$ are in a symmetric situation, so it suffices to show that $\dim Y \leq \dim X$. If $\dim X = \infty$ then there is nothing to prove, so assume that $\dim X = n \in \omega$.

Suppose that $M$ is a second countable space and $r : Y \to M$ is a continuous map; if $M' = r(M)$ then $r : Y \to M'$ is surjective. Apply Fact 2 to find a countable $A_0 \subset C_p(Y)$ and a homeomorphism $h : Y[A_0] \to M'$ such that $h \circ e_{A_0}^Y = r$. The space $Y[A_0]$ being second countable, there is a countable $QS$-algebra $Q_0 \subset C_p(Y[A_0])$ with $d_{A_0}^Y(A_0) \subset Q_0$; then the set $Q_0' = (e_{A_0}^Y)^*(Q_0)$ is countable and contains $A_0$.

The set $P = u^{-1}(Q_0') \subset C_p(X)$ is countable and the map $e_P^X : X \to X[P]$ is continuous, so there exists a second countable space $Z$ such that $\dim Z \leq n$ and there are maps $\xi : X \to Z$ and $\eta : Z \to X[P]$ such that $\eta \circ \xi = e_P^X$ (see Problem 161). Passing, if necessary, to $Z' = \xi(Z)$ and applying Problem 155 we can assume, without loss of generality, that $\xi(X) = Z$. By Fact 3 there exists a countable set $B_0 \supset P$ such that $X[B_0]$ is homeomorphic to $Z$ and hence $\dim X[B_0] \leq n$.

Proceeding by induction assume that $m \in \omega$ and we have constructed countable sets $A_0, \ldots, A_m$ and $B_0, \ldots, B_m$ with the following properties:

(1) $A_i \subset A_{i+1}$ and $B_i \subset B_{i+1}$ for any $i < m$;
(2) $u^{-1}(A_i) \subset B_i$ for each $i \leq m$ and $u(B_i) \subset A_{i+1}$ for every $i < m$;
(3) $\dim(X[B_i]) \leq n$ for any $i \leq m$;
(4) there is a $QS$-algebra $Q_i \subset C_p(Y[A_i])$ such that $A_i \subset Q_i' = (e_{A_i}^Y)^*(Q_i)$ and $u^{-1}(Q_i') \subset B_i$ for all $i \leq m$;
(5) there is a $QS$-algebra $R_i \subset C_p(Y[B_i])$ such that $B_i \subset R_i' = (e_{B_i}^X)^*(R_i)$ and $u(R_i') \subset A_{i+1}$ for all $i < m$.

Apply Problem 176 to find a countable $QS$-algebra $R_i \subset C_p(X[B_i])$ with $d_{B_i}^X(B_i) \subset R_i$. Then $B_i \subset R_i' = (e_{B_i}^X)^*(R_i)$ and the set $A_{i+1} = u(R_i') \supset A_i$ is countable, so there exists a countable $QS$-algebra $Q_{i+1} \subset C_p(Y[A_{i+1}])$ such that $d_{A_{i+1}}^Y(A_{i+1}) \subset Q_{i+1}$. Then $A_{i+1} \subset Q_{i+1}' = (e_{A_{i+1}}^Y)^*(Q_{i+1})$ and the set $P = u^{-1}(Q_{i+1}') \supset R_i'$ is countable, so there exists a second countable space $Z$

such that dim $Z \leq n$ and there are maps $\xi : X \to Z$ and $\eta : Z \to X[P]$ such that $\eta \circ \xi = e_P^X$ (see Problem 161). Passing, if necessary, to $Z' = \xi(Z)$ and applying Problem 155 we can assume, without loss of generality, that $\xi(X) = Z$. By Fact 3 there exists a countable set $B_{m+1} \supset P$ such that $X[B_{m+1}]$ is homeomorphic to $Z$ and hence dim $X[B_{m+1}] \leq n$.

Now that we have the sets $A_{m+1}$ and $B_{m+1}$ is straightforward that the properties (1)–(5) are fulfilled if we substitute $m$ by $m + 1$, so our inductive procedure can be continued to construct sequences $\{A_i : i \in \omega\}$ and $\{B_i : i \in \omega\}$ for which the conditions (1)–(5) are satisfied for all $m \in \omega$. If $A = \bigcup_{i \in \omega} A_i$ and $B = \bigcup_{i \in \omega} B_i$ then it follows from (1) and (2) that $u(B) = A$, so $A$ and $B$ are uniformly homeomorphic. The properties (4) and (5), together with Fact 4 imply that the set $A' = d_A^Y(A)$ is a $QS$-algebra in $C_p(Y[A])$ and $B' = d_B^X(B)$ is a $QS$-algebra in $C_p(X[B])$. The maps $(e_B^X)^*$ and $d_A^Y$ are linear, so they are uniform homeomorphisms; an immediate consequence is that $v = d_A^Y \circ u \circ (e_B^X)^* : B' \to A'$ is a uniform homeomorphism. We have dim$(X[B]) \leq n$ by the property (3) and Fact 5. Now apply Problem 179 to conclude that dim$(Y[A]) = \dim(X[B]) \leq n$. Finally, for the map $g = h \circ \pi_{A,A_0}^Y$, we have $g \circ e_A^Y = r$, so we can apply Problem 161 again to see that dim $Y \leq n = \dim X$. Since $X$ and $Y$ are in a symmetric situation, we also have dim $X \leq \dim Y$, so dim $X = \dim Y$ and hence our solution is complete.

**V.181.** *Let $X$ be a zero-dimensional compact space. Prove that $Y$ is also a zero-dimensional compact space whenever $Y \overset{u}{\sim} X$.*

**Solution.** It follows from Problem 138 that $Y$ is compact. Apply SFFS-306 to see that the space $X$ is strongly zero-dimensional and hence dim $X = 0$ by Problem 148. Therefore dim $Y = \dim X = 0$ by Problem 180, so we can apply Problem 149 to conclude that the space $Y$ is zero-dimensional.

**V.182.** *Suppose that $X$ is a zero-dimensional Lindelöf space and $Y \overset{u}{\sim} X$. Prove that $Y$ is also zero-dimensional.*

**Solution.** Apply SFFS-306 to see that the space $X$ is strongly zero-dimensional and hence dim $X = 0$ by Problem 148. Therefore dim $Y = \dim X = 0$ by Problem 180, so we can apply Problem 149 to conclude that the space $Y$ is also zero-dimensional.

**V.183.** *Given a countable ordinal $\xi \geq 1$, prove that a metrizable space $X$ is an absolute Borel set of multiplicative class $\xi$ (i.e., $X \in \mathcal{M}_\xi$) if and only if there exists a completely metrizable space $Z$ such that $X$ is homeomorphic to some $Y \in \Pi_\xi^0(Z)$.*

**Solution.** Apply TFS-237 to find a complete metric space $Z$ such that $X$ is homeomorphic to a subspace $Y \subset Z$. If $X \in \mathcal{M}_\xi$ then $Y \in \Pi_\xi^0(Z)$; this proves necessity.

Assume that $X$ is homeomorphic to a subspace $Y$ of a complete metric space $Z$ such that $Y \in \Pi_\xi^0(Z)$ and take an arbitrary metrizable space $T$ which contains a homeomorphic copy $X'$ of the space $X$. It follows from TFS-237 that we can assume, without loss of generality, that $T \subset M$ for some complete metric space $M$.

Fix a homeomorphism $h : Y \to X'$ and apply Fact 2 of T.333 to find $G_\delta$-sets $P$ and $Q$ in the spaces $Z$ and $M$ respectively such that $Y \subset P$, $X' \subset Q$ and there exists a homeomorphism $f : P \to Q$ with $f|Y = h$. It follows from $Y \in \Pi_\xi^0(Z)$ that $Y = Y \cap P \in \Pi_\xi^0(P)$ and hence $X' \in \Pi_\xi^0(Q)$. By Fact 1 of T.319 there exists a set $R \in \Pi_\xi^0(M)$ such that $R \cap Q = X'$; it follows from $\xi \geq 1$ and $Q \in \Pi_1^0(M)$ that $Q \in \Pi_\xi^0(M)$ (see Fact 1 of T.331), so $X' \in \Pi_\xi^0(M)$. Finally apply Fact 1 of T.319 to conclude that $X' = X' \cap T \in \Pi_\xi^0(T)$; this settles sufficiency.

**V.184.** *Given a countable ordinal $\xi \geq 2$, prove that a metrizable space $X$ is an absolute Borel set of additive class $\xi$ (i.e., $X \in \mathcal{A}_\xi$) if and only if there exists a completely metrizable space $Z$ such that $X$ is homeomorphic to some $Y \in \Sigma_\xi^0(Z)$.*

**Solution.** Apply TFS-237 to find a complete metric space $Z$ such that $X$ is homeomorphic to a subspace $Y \subset Z$. If $X \in \mathcal{A}_\xi$ then $Y \in \Sigma_\xi^0(Z)$; this proves necessity.

Now assume that the space $X$ is homeomorphic to a subspace $Y$ of a complete metric space $Z$ such that $Y \in \Sigma_\xi^0(Z)$ and take an arbitrary metrizable space $T$ which contains a homeomorphic copy $X'$ of the space $X$. It follows from TFS-237 that we can assume, without loss of generality, that $T \subset M$ for some complete metric space $M$.

Fix a homeomorphism $h : Y \to X'$ and apply Fact 2 of T.333 to find $G_\delta$-sets $P$ and $Q$ in the spaces $Z$ and $M$ respectively such that $Y \subset P$, $X' \subset Q$ and there exists a homeomorphism $f : P \to Q$ with $f|Y = h$. It follows from $Y \in \Sigma_\xi^0(Z)$ that $Y = Y \cap P \in \Sigma_\xi^0(P)$ and hence $X' \in \Sigma_\xi^0(Q)$. By Fact 1 of T.319 there exists a set $R \in \Sigma_\xi^0(M)$ such that $R \cap Q = X'$; it follows from $\xi \geq 2$ and $Q \in \Pi_1^0(M)$ that $Q \in \Sigma_\xi^0(M)$ (see Fact 1 of T.331), so $X' \in \Sigma_\xi^0(M)$ by Fact 1 of T.341. Finally apply Fact 1 of T.319 to conclude that $X' = X' \cap T \in \Sigma_\xi^0(T)$; this settles sufficiency.

**V.185.** *Suppose that $n \in \mathbb{N}$ and a space $X_i$ is metrizable for every $i \leq n$. Prove that, for any countable ordinal $\xi \geq 2$,*

(i) *if $X_i \in \mathcal{A}_\xi$ for all $i \leq n$ then $X_1 \times \ldots \times X_n \in \mathcal{A}_\xi$;*
(ii) *if $X_i \in \mathcal{M}_\xi$ for all $i \leq n$ then $X_1 \times \ldots \times X_n \in \mathcal{M}_\xi$.*

**Solution.** The following fact is crucial for this solution.

*Fact 1.* Given $\alpha \in \omega_1$ and $m \in \mathbb{N}$ suppose that we have spaces $Y_1, \ldots, Y_m$ and sets $P_i \in \Sigma_\alpha^0(Y_i)$, $Q_i \in \Pi_\alpha^0(Y_i)$ for every $i \leq m$. If $Y = \prod_{i \leq n} Y_i$, $P = \prod_{i \leq n} P_i$ and $Q = \prod_{i \leq n} Q_i$ then $P \in \Sigma_\alpha^0(Y)$ and $Q \in \Pi_\alpha^0(Y)$.

*Proof.* If $\alpha = 0$ then $P_i$ is open in $Y_i$ and $Q_i$ is closed in $Y_i$ for each $i \leq m$, so $P$ is open and $Q$ is closed in $Y$. Proceeding by induction assume that $\beta \geq 1$ is a countable ordinal and we proved our Fact for all $\alpha < \beta$.

Let $\pi_i : Y \to Y_i$ be the natural projection for each $i \leq m$. If $P_i \in \Sigma_\beta^0(Y_i)$ then there exist families $\{F_k^i : k \in \omega\}$ and $\{\alpha_k^i : k \in \omega\} \subset \beta$ such that $F_k^i \in \Pi_{\alpha_k^i}^0(Y_i)$

for every $k \in \omega$ and $P_i = \bigcup_{k \in \omega} F_k^i$ for all $i \leq m$. For any $k_1, \ldots, k_m \in \omega$ if $\alpha(k_1, \ldots, k_m) = \max\{\alpha_{k_i}^i : i \leq m\}$ then the set $F(k_1, \ldots, k_m) = F_{k_1}^1 \times \ldots \times F_{k_m}^m$ belongs to $\Pi_{\alpha(k_1, \ldots, k_m)}^0(Y)$ by the induction hypothesis and Fact 1 of T.331. It is straightforward that $P = \bigcup\{F(k_1, \ldots, k_m) : k_i \in \omega$ for all $i \leq m\}$; since also $\alpha(k_1, \ldots, k_m) < \beta$ for any $k_1, \ldots, k_m \in \omega$, we conclude that $P \in \Sigma_\beta^0(Y)$.

Now, if $Q_i \in \Pi_\beta^0(Y_i)$ then $Y_i \setminus Q_i \in \Sigma_\beta^0(Y_i)$ and hence $\pi_i^{-1}(Y_i \setminus Q_i)$ belongs to $\Sigma_\beta^0(Y)$ for all $i \leq m$ (see Fact 1 of T.318). Since $Y \setminus Q = \bigcup\{\pi_i^{-1}(Y_i \setminus Q_i) : i \leq m\}$, we can apply Fact 1 of T.341 to see that $Y \setminus Q \in \Sigma_\beta^0(Y)$ and hence $Q \in \Pi_\beta^0(Y)$. Thus our statement is verified for $\alpha = \beta$, so our inductive procedure can be continued to guarantee that it holds for all $\alpha < \omega_1$, i.e., Fact 1 is proved.

Returning to our solution take a completely metrizable space $M_i$ such that $X_i$ is homeomorphic to some $Y_i \subset M_i$ for all $i \leq n$ (see TFS-237); let $Y = \prod_{i \leq n} Y_i$ and $M = \prod_{i \leq n} M_i$. If $X_i \in \mathcal{A}_\xi$ then $Y_i \in \Sigma_\xi^0(M_i)$ for every $i \leq n$, so $Y \in \Sigma_\xi^0(M)$ by Fact 1. The space $X = X_1 \times \ldots \times X_n$ is homeomorphic to $Y$, so we can apply Problem 184 to see that $X \in \mathcal{A}_\xi$. Finally, if $X_i \in \mathcal{M}_\xi$ then $Y_i \in \Pi_\xi^0(M_i)$ for every $i \leq n$, so $Y \in \Pi_\xi^0(M)$ by Fact 1. The space $X$ being homeomorphic to $Y$ we can apply Problem 183 to see that $X \in \mathcal{M}_\xi$.

**V.186.** *Given ordinals $\alpha, \beta \in \omega_1$ such that $\alpha \geq 2$ and $\beta < \alpha$ suppose that $X$ is a metrizable space and $X = \bigcup\{X_n : n \in \omega\}$ where $X_n \in \Sigma_\beta^0(X) \cap \mathcal{M}_\alpha$ for every $n \in \omega$. Prove that $X \in \mathcal{M}_\alpha$.*

**Solution.** Take a complete metric space $M$ such that $X \subset M$ (see TFS-237). It follows from Fact 1 of T.319 that there exists $Y_n \in \Sigma_\beta^0(M)$ such that $Y_n \cap X = X_n$ for any $n \in \omega$; thus the set $Y = \bigcup_{n \in \omega} Y_n \in \Sigma_\beta^0(M) \subset \Pi_\alpha^0(M)$ (see Fact 1 of T.331) belongs to the class $\Pi_\alpha^0(M)$. Now, $X_n \in \mathcal{M}_\alpha$ implies that $X_n \in \Pi_\alpha^0(M)$ and therefore $Y_n \setminus X_n \in \Sigma_\alpha^0(M)$ for every $n \in \omega$. The class $\Sigma_\alpha^0(M)$ being $\sigma$-additive (see Fact 1 of T.341), we convince ourselves that $Z = \bigcup_{n \in \omega}(Y_n \setminus X_n)$ also belongs to the class $\Sigma_\alpha^0(M)$. Consequently, $X = Y \setminus Z \in \Pi_\alpha^0(M)$, so we can apply Problem 183 to conclude that $X \in \mathcal{M}_\alpha$.

**V.187.** *Prove that a metrizable space $X$ is a Borel set of absolute additive class $\xi \geq 2$ (i.e., $X \in \mathcal{A}_\xi$) if and only if there exists a sequence $\{\xi_n : n \in \omega\} \subset \xi$ such that $X = \bigcup\{X_n : n \in \omega\}$ and $X_n \in \mathcal{M}_{\xi_n}$ for every $n \in \omega$.*

**Solution.** Suppose that $X \in \mathcal{A}_\xi$ and fix a complete metric space $M$ such that $X \subset M$ (see TFS-237). We have $X \in \Sigma_\xi^0(M)$, so there exist sequences $\{\beta_n : n \in \omega\} \subset \xi$ and $\{X_n : n \in \omega\}$ such that $X = \bigcup_{n \in \omega} X_n$ and $X_n \in \Pi_{\beta_n}^0(M)$ for all $n \in \omega$. Let $\xi_n = \beta_n$ if $\beta_n \geq 1$ and $\xi_n = 1$ if $\beta_n = 0$; it is clear that $\{\xi_n : n \in \omega\} \subset \xi$ and, besides, $\beta_n \leq \xi_n$, so $X_n \in \Pi_{\xi_n}^0(M)$ for every $n \in \omega$. Apply Problem 183 to see that $X_n \in \mathcal{M}_{\xi_n}$ for each $n \in \omega$; this settles necessity.

Now assume that there exist sequences $\{\xi_n : n \in \omega\} \subset \xi$ and $\{X_n : n \in \omega\}$ such that $X_n \in \mathcal{M}_{\xi_n}$ for every $n \in \omega$ and $X = \bigcup_{n \in \omega} X_n$. If $X$ is a subspace of a metric space $M$ then $X_n \subset M$ and hence $X_n \in \Pi^0_{\xi_n}(M)$ for all $n \in \omega$. Therefore $X \in \Sigma^0_\xi(M)$ and hence $X$ is a Borel set of absolute additive class $\xi$, i.e., we proved sufficiency.

**V.188.** *Given a countable ordinal $\xi \geq 2$, let $\mathcal{M}_\xi$ be the class of absolute Borel sets of multiplicative class $\xi$. Prove that the following conditions are equivalent for any metrizable $X$:*

(i) *the space $X$ belongs to $\mathcal{M}_\xi$;*
(ii) *there is a complete sequence $\{\mathcal{U}_n : n \in \omega\}$ of $\sigma$-discrete covers of $X$ such that, for any $n \in \omega$, there is $\xi_n < \xi$ with $\mathcal{U}_n \subset \Sigma^0_{\xi_n}(X)$;*
(iii) *there is a complete sequence $\{\mathcal{V}_n : n \in \omega\}$ of $\sigma$-discrete covers of $X$ such that, for any $n \in \omega$, there is $\xi_n < \xi$ with $\mathcal{V}_n \subset \bigcup\{\Pi^0_\alpha(X) : \alpha < \xi_n\}$.*

**Solution.** Recall that a sequence $\{\mathcal{W}_n : n \in \omega\}$ of covers of a space $Z$ is called *complete* if, for any filter $\mathcal{F}$ on the set $Z$ such that $\mathcal{F} \cap \mathcal{U}_n \neq \emptyset$ for all $n \in \omega$ we have $\bigcap\{\overline{F} : F \in \mathcal{F}\} \neq \emptyset$; the points of the set $\bigcap\{\overline{F} : F \in \mathcal{F}\}$ are called *the cluster points* of $\mathcal{F}$. Given a metric space $(M, d)$, a family $\mathcal{U}$ of subsets of $M$ is called *uniformly discrete* if there exists $\varepsilon > 0$ such that the ball $B_d(x, \varepsilon) = \{y \in M : d(x, y) < \varepsilon\}$ meets at most one element of $\mathcal{U}$ for every $x \in X$; the family $\mathcal{U}$ is *$\sigma$-uniformly discrete* if it is a countable union of uniformly discrete families. For any nonempty set $A \subset M$ let $\mathrm{diam}_d(A) = \sup\{d(x, y) : x, y \in A\}$.

Given a space $Z$ and families $\mathcal{A}_1, \ldots, \mathcal{A}_n$ of subsets of $Z$ we will need the family $\bigwedge^n_{i=1} \mathcal{A}_i = \mathcal{A}_1 \wedge \ldots \wedge \mathcal{A}_n = \{A_1 \cap \ldots \cap A_n : A_i \in \mathcal{A}_i$ for all $i \leq n\}$. For any $\mathcal{A} \subset \exp(Z)$ let $\mathcal{A}|Y = \{A \cap Y : A \in \mathcal{A}\}$.

*Fact 1.* Given a collectionwise normal space $Z$ and $\alpha \in \omega_1$ suppose that $\{P_t : t \in T\}$ is a discrete family of subsets of $Z$. If $P_t \in \Sigma^0_\alpha(Z)$ ($P_t \in \Pi^0_\alpha(Z)$) for each $t \in T$ then $P = \bigcup_{t \in T} P_t \in \Sigma^0_\alpha(Z)$ (or $P \in \Pi^0_\alpha(Z)$ respectively).

*Proof.* If $\alpha = 0$ then every $P_t$ is open (closed) in $Z$, so $P$ is open in $Z$ (or closed in $Z$ respectively, because the union of a discrete family of closed sets is closed); this proves our statement for $\alpha = 0$.

Proceeding inductively, assume that $\beta > 0$ and our statement is proved for all $\alpha < \beta$. If $\{P_t : t \in T\} \subset \Sigma^0_\beta(Z)$ then, for each $t \in T$, there are sequences $\{\beta_n(t) : n \in \omega\} \subset \beta$ and $\{P^n_t : n \in \omega\}$ such that $P_t = \bigcup_{n \in \omega} P^n_t$ and $P^n_t \in \Pi^0_{\beta_n(t)}(Z)$ for every $n \in \omega$. For every $\lambda < \beta$ and $n \in \omega$ let $T^n_\lambda = \{t \in T : \beta_n(t) \leq \lambda\}$.

Take any $n \in \omega$; the family $\{P^n_t : t \in T^n_\lambda\} \subset \Pi^0_\lambda(Z)$ is discrete, so we can apply the induction hypothesis to conclude that the set $Q^n_\lambda = \bigcup_{t \in T^n_\lambda} P^n_t$ belongs to $\Pi^0_\lambda(Z)$ for each ordinal $\lambda < \beta$. The family $\mathcal{Q} = \{Q^n_\lambda : n \in \omega, \lambda < \beta\}$ is countable, every element of $\mathcal{Q}$ belongs to a multiplicative class $\lambda < \beta$, so $\bigcup \mathcal{Q} \in \Sigma^0_\beta(Z)$. It is easy to see that $P = \bigcup \mathcal{Q}$, so $P \in \Sigma^0_\beta(Z)$, i.e., we completed the induction step for the additive class $\beta$.

Now, if $\{P_t : t \in T\} \subset \Pi^0_\beta(Z)$ then choose a discrete family $\{O_t : t \in T\}$ such that $P_t \subset O_t$ for every $t \in T$. Each set $O_t \backslash P_t$ belongs to the family $\Sigma^0_\beta(Z)$, so we can use what we proved for the additive class $\beta$ to see that $W = \bigcup\{O_t \backslash P_t : t \in T\}$

is an element of $\Sigma_\beta^0(Z)$. The set $O = \bigcup_{t \in T} O_t$ being open, the closed set $F = Z \setminus O$ belongs to $\Sigma_\beta^0(Z)$ (recall that $\beta > 0$); as a consequence, $Z \setminus P = W \cup F \in \Sigma_\beta^0(Z)$ which shows that $P \in \Pi_\beta^0(Z)$ and completes our induction step. Thus our statement is true for every $\alpha < \omega_1$, i.e., Fact 1 is proved.

*Fact 2.* Assume that $\alpha \geq 2$ is a countable ordinal,$\mathcal{U} \subset \Sigma_\alpha^0(M)$ is a $\sigma$-discrete cover of a metric space $(M, d)$ and $r > 0$. Then there is a $\sigma$-uniformly discrete refinement $\mathcal{V} \subset \Sigma_\alpha^0(M)$ of the family $\mathcal{U}$ such that $\operatorname{diam}_d(V) < r$ for any $V \in \mathcal{V}$.

*Proof.* By our assumption, $\mathcal{U} = \bigcup_{n \in \omega} \mathcal{U}_n$ and every $\mathcal{U}_n$ is discrete, so we can choose an open cover $\mathcal{G}_n$ of the space $M$ such that every set $G \in \mathcal{G}_n$ meets at most one element of $\mathcal{U}_n$ and $\operatorname{diam}_d(G) < r$. Fix $n \in \omega$ and apply Fact 1 of T.373 to find an open refinement $\mathcal{W}_n$ of the cover $\mathcal{G}_n$ such that $\mathcal{W}_n = \bigcup_{i \in \omega} W_i^n$ and every $W_i^n$ is uniformly discrete. Let $\mathcal{V}(n, i) = W_i^n \wedge \mathcal{U}_n$ for all $n, i \in \omega$.

Fix any $n, i \in \omega$ and take $\varepsilon > 0$ such that $B_d(x, \varepsilon)$ meets at most one element of $W_i^n$ for any $x \in X$. Suppose that we have distinct $V_0, V_1 \in \mathcal{V}(n, i)$ such that $B_d(x, \varepsilon) \cap V_j \neq \emptyset$ for $j = 0, 1$. Take $W_0, W_1 \in W_i^n$ and $U_0, U_1 \in \mathcal{U}_n$ such that $V_j = W_j \cap U_j$ and hence $B_d(x, \varepsilon) \cap (W_j \cap U_j) \neq \emptyset$ if $j \in \{0, 1\}$. It follows from $B_d(x, \varepsilon) \cap W_0 \neq \emptyset \neq B_d(x, \varepsilon) \cap W_1$ that $W_0 = W_1$ and hence $U_0 \neq U_1$; thus the set $W_0 \in W_i^n$ meets both sets $U_0$ and $U_1$. There is $G \in \mathcal{G}_n$ with $W_0 \subset G$; an immediate consequence is that $G$ also meets both sets $U_0$ and $U_1$ which contradicts the choice of the family $\mathcal{G}_n$. This proves that the family $\mathcal{V}(n, i)$ is uniformly discrete for any $n, i \in \omega$. Thus the family $\mathcal{V} = \bigcup\{\mathcal{V}(n, i) : n, i \in \omega\}$ is a $\sigma$-uniformly discrete refinement of $\mathcal{U}$ such that $\operatorname{diam}_d(V) < r$ for every $V \in \mathcal{V}$, so Fact 2 is proved.

Returning to our solution observe that

(1) if $\{\mathcal{U}_n : n \in \omega\}$ is a complete sequence of covers of a space $Z$ and $\mathcal{V}_n$ is a refinement of $\mathcal{U}_n$ for every $n \in \omega$ then the sequence $\{\mathcal{V}_n : n \in \omega\}$ is also complete.

Apply TFS-237 to find a complete metric space $(M, d)$ such that $X \subset M$; we will prove by transfinite induction that (i)$\Longrightarrow$(ii) for every $\xi \geq 1$. If $\xi = 1$ then $X \in \Pi_1^0(M)$, i.e., $X$ is a $G_\delta$-subset of $M$, so $X$ is Čech-complete (see TFS-260 and TFS-269) and hence $X$ has a complete sequence $\{\mathcal{U}_n' : n \in \omega\}$ of open covers by TFS-268. Since $X$ is metrizable, we can choose a $\sigma$-discrete open refinement $\mathcal{U}_n$ of the cover $\mathcal{U}_n'$ for every $n \in \omega$. Then $\mathcal{U}_n \subset \Sigma_0^0(X)$ for every $n \in \omega$ and the sequence $\{\mathcal{U}_n : n \in \omega\}$ is complete by (1), so we proved (i)$\Longrightarrow$(ii) for $\xi = 1$.

Now assume that $\alpha > 1$ is a countable ordinal and we proved (i)$\Longrightarrow$(ii) for all $\xi < \alpha$. If $X \in \mathcal{M}_\alpha$ ; then $X \in \Pi_\alpha^0(M)$, so $M \setminus X \in \Sigma_\alpha^0(M)$ and hence there are sequences $\{\xi_n : n \in \omega\} \subset \alpha$ and $\{H_n : n \in \omega\}$ such that $1 \leq \xi_n \leq \xi_{n+1}$ and $H_n \in \Pi_{\xi_n}^0(M)$ for every $n \in \omega$ while $M \setminus X = \bigcup_{n \in \omega} H_n$. The set $D_n = M \setminus H_n$ belongs to the class $\Sigma_{\xi_n}^0(M)$ for each $n \in \omega$ and $X = \bigcap_{n \in \omega} D_n$.

For each $n \in \omega$ fix sequences $\{\beta_i^n : i \in \omega\} \subset \xi_n$ and $\{D_i^n : i \in \omega\}$ such that $D_i^n \in \Pi_{\beta_i^n}^0(M)$ for every $i \in \omega$ and $D_n = \bigcup_{i \in \omega} D_i^n$; if $E_i^n = D_i^n \setminus \bigcup_{j < i} D_j^n$ for all $i \in \omega$ then the family $\{E_i^n : i \in \omega\}$ is disjoint, $D_n = \bigcup_{i \in \omega} E_i^n$ and it follows from Fact 1 of T.341 and Fact 1 of T.331 that every set $E_i^n$ belongs to the family $\Sigma^0_{\xi_n}(M) \cap \Pi^0_{\xi_n}(M)$.

Apply Problem 183 to see that $E_i^n \in \mathcal{M}_{\xi_n}$; by the induction hypothesis, for every set $E_i^n$ we can find a complete sequence $\{\mathcal{C}(n,i,k) : k \in \omega, k \geq n\}$ of $\sigma$-discrete covers of the space $E_i^n$ such that $\mathcal{C}(n,i,k) \subset \Sigma^0_{\xi_n}(E_i^n)$ for every $k \geq n$ (observe that the induction hypothesis could be used to guarantee that every family $\mathcal{C}(n,i,k)$ is in the class $\Sigma^0_{\beta_k}(E_i^n)$ for some $\beta_k < \xi_n$ but we won't need that). Applying Fact 1 of T.341 and Fact 1 of T.319 we conclude that $\mathcal{C}(n,i,k) \subset \Sigma^0_{\xi_n}(M)$ for all $n,i,k \in \omega$, $k \geq n$. It follows from (1) and Fact 2 that we can additionally assume, without loss of generality, that every $\mathcal{C}(n,i,k)$ is $\sigma$-uniformly discrete in $E_i^n$ and $\mathrm{diam}_d(U) \leq 2^{-k}$ for every $U \in \mathcal{C}(n,i,k)$; then every $\mathcal{C}(n,k,i)$ is also $\sigma$-uniformly discrete in $M$ by Fact 2 of T.373.

For every $p \in \omega$ consider the family $\mathcal{U}_p = \{F : F \neq \emptyset$ while there exist sets $U_0, \ldots, U_p$ and $i_0, \ldots, i_p \in \omega$ such that $F = (\bigcap_{n=0}^{p} U_n) \cap X$ and $U_n \in \mathcal{C}(n,i_n,p)$ for all $n \leq p\}$. It follows from $\xi_0 \leq \ldots \leq \xi_p$ that $\mathcal{U}_p \subset \Sigma^0_{\xi_p}(X)$ for every $p \in \omega$ (see Fact 1 of T.331 and Fact 1 of T.341). Fact 3 of T.373 implies that the family $\mathcal{E}(i_0, \ldots, i_p) = \mathcal{C}(0,i_0,p) \wedge \ldots \wedge \mathcal{C}(p,i_p,p)$ is $\sigma$-uniformly discrete for any $i_0, \ldots, i_p \in \omega$. Therefore the family $\mathcal{E} = \bigcup\{\mathcal{E}(i_0, \ldots, i_p) : i_0, \ldots, i_p \in \omega\}$ is also $\sigma$-uniformly discrete. As a consequence, $\mathcal{U}_p \subset \mathcal{E}|X$ is $\sigma$-uniformly discrete for every $p \in \omega$, so all is left is to prove that the sequence $\{\mathcal{U}_p : p \in \omega\}$ is complete in $X$.

Assume that $\mathcal{F}$ is a filter in $X$ such that $\mathcal{F} \cap \mathcal{U}_p \neq \emptyset$ and hence we can pick a set $U_p \in \mathcal{U}_p \cap \mathcal{F}$ for each $p \in \omega$. The sequence $\{\mathrm{diam}_d(U_p) : p \in \omega\}$ converges to zero, so we have the following property:

(2) the filter $\mathcal{F}$ can have at most one cluster point in $M$.

By the choice of $\mathcal{U}_p$, we have $U_p = (\bigcap_{n=0}^{p} U_p^n) \cap X$ where $U_p^n \in \mathcal{C}(n,i(n,p),p)$ for every $p \in \omega$. Given any $n \in \omega$ and $p, q \geq n$ it follows from $U_p \cap U_q \neq \emptyset$ that $U_p^n \cap U_q^n \neq \emptyset$; the family $\{E_i^n : i \in \omega\}$ being disjoint we must have $i(n,p) = i(n,q)$; thus the number $i_n = i(n,p)$ does not depend on $p$.

Let $\mathcal{F}_n = \mathcal{F}|E_{i_n}^n$ for every $n \in \omega$; it follows from $U_p \subset U_p^n$ that $\mathcal{F}_n$ is a filter on $E_{i_n}^n$ and $U_p^n \in \mathcal{F}_n \cap \mathcal{C}(n,i_n,p)$ for each $p \geq n$. The sequence $\{\mathcal{C}(n,i_n,p) : n \leq p < \omega\}$ being complete in $E_{i_n}^n$, the filter $\mathcal{F}_n$ has a cluster point $a_n \in E_{i_n}^n$. It is clear that every $a_n$ is also a cluster point of $\mathcal{F}$, so we can apply the property (2) to convince ourselves that there is $a \in M$ such that $a_n = a$ for all $n \in \omega$. Now, $a \in \bigcap\{E_{i_n}^n : n \in \omega\} \subset \bigcap_{n \in \omega} D_n = X$, so $a$ is a cluster point of $\mathcal{F}$ in $X$. This shows that the sequence $\{\mathcal{U}_p : p \in \omega\}$ is complete and hence we proved that (i)$\Longrightarrow$(ii).

Assume, toward proving (ii)$\Longrightarrow$(i), that $\mathcal{S} = \{\mathcal{U}_n : n \in \omega\}$ is a complete sequence given in (ii); we are still considering that $X$ is a subspace of a complete metric space $(M,d)$. Applying Fact 2 we can assume, without loss of generality, that $\mathrm{diam}_d(U) \leq 2^{-n}$ for all $n \in \omega$ and $U \in \mathcal{U}_n$ while $\mathcal{U}_n = \bigcup_{i \in \omega} \mathcal{U}_i^n$ where every $\mathcal{U}_i^n$ is uniformly discrete. For each $n \in \omega$ and $U \in \mathcal{U}_n$ we can apply Fact 1 of T.319 to find a set $E(U) \subset \mathrm{cl}_M(U)$ such that $E(U) \cap X = U$ and $E(U) \in \Sigma^0_{\xi_n}(M)$.

Every family $\mathcal{U}_i^n$ is uniformly discrete in $M$ (see Fact 2 of T.373); it is easy to see that the collection $\{\mathrm{cl}_M(U) : U \in \mathcal{U}_i^n\}$ is also uniformly discrete in $M$, so the family $\mathcal{V}_i^n = \{E(U) : U \in \mathcal{U}_i^n\} \subset \Sigma_{\xi_n}^0(M) \subset \Sigma_\xi^0(M)$ is uniformly discrete in $M$ as well. If $m \in \omega$ and we are given arbitrary $k_0 \dots, k_m \in \omega$ let

$$Q(k_0, \dots, k_m) = \{Q \in \mathcal{V}_{k_0}^0 \wedge \dots \wedge \mathcal{V}_{k_m}^m : Q \cap X = \emptyset\}.$$

It follows from Fact 3 of T.373 that $\mathcal{Q}(k_0, \dots, k_m)$ is uniformly discrete and hence the set $Q(k_0, \dots, k_m) = \bigcup \mathcal{Q}(k_0, \dots, k_m)$ belongs to $\Sigma_\xi^0(M)$ (see Fact 1) for any $k_0 \dots, k_m \in \omega$. Consequently, the set $Q = \bigcup\{Q(k_0, \dots, k_m) : k_0, \dots, k_m \in \omega\}$ also belongs to $\Sigma_\xi^0(M)$ whence $R = M \setminus Q$ belongs to $\Pi_\xi^0(M)$ and $X \subset R$. The set $R$ has the following important property:

(3)  for any number $n \in \omega$ if $U_i \in \mathcal{U}_i$ for all $i \leq n$ then $\bigcap_{i \leq n} U_i \neq \emptyset$ if and only if $(\bigcap_{i \leq n} E(U_i)) \cap R \neq \emptyset$.

Apply Fact 1 again to see that every set $V_i^n = \bigcup \mathcal{V}_i^n$ belongs to $\Sigma_{\xi_n}^0(M)$; therefore the set $V_n = \bigcup_{i \in \omega} V_i^n$ belongs to $\Sigma_{\xi_n}^0(M)$ for each $n \in \omega$. This implies that the set $V = \bigcap_{n \in \omega} V_n \supset X$ belongs to $\Pi_\xi^0(M)$. Therefore the set $X' = V \cap R$ belongs to $\Pi_\xi^0(M)$ and $X \subset X'$.

Fix any $x \in X'$; there is a sequence $\{U_n : n \in \omega\}$ such that $U_n \in \mathcal{U}_n$ and $x \in E(U_n)$ for every $n \in \omega$. Since $x \in (\bigcap_{i \leq n} E(U_i)) \cap R$, we can apply (3) to convince ourselves that $\bigcap_{i \leq n} U_i \neq \emptyset$ for every $n \in \omega$, so the family $\mathcal{U} = \{U_n : n \in \omega\}$ is centered; let $\mathcal{F}$ be a filter which contains $\mathcal{U}$. The sequence $S$ being complete, the filter $\mathcal{F}$ has a cluster point $a \in X$. Since the sequence $\{\mathrm{diam}_d(\mathrm{cl}_M(U_i)) : i \in \omega\}$ converges to zero, it follows from $\{x, a\} \subset \bigcap\{\mathrm{cl}_M(U_n) : n \in \omega\}$ that $x = a$ and hence $x \in X$. Therefore $X' \subset X$, i.e., $X = X'$ is an element of $\Pi_\xi^0(M)$, so we can apply Problem 183 to conclude that $X \in \mathcal{M}_\xi$; this proves that (ii)$\Longrightarrow$(i).

Now, if (ii) holds and $\{\mathcal{U}_n : n \in \omega\}$ is the respective complete sequence of covers of $X$ then, for every $n \in \omega$ and $U \in \mathcal{U}_n$ pick a sequence $\mathcal{T}(U) = \{O_n(U) : n \in \omega\}$ such that $U = \bigcup_{n \in \omega} O_n(U)$ and $O_n(U) \in \bigcup\{\Pi_\alpha^0(X) : \alpha < \xi_n\}$ for each $n \in \omega$. If $\mathcal{V}_n = \bigcup\{\mathcal{T}(U) : U \in \mathcal{U}_n\}$ then $\mathcal{V}_n$ is a $\sigma$-discrete refinement of $\mathcal{U}_n$ for every $n \in \omega$, so the sequence $\{\mathcal{V}_n : n \in \omega\}$ is complete by (1). Since also $\mathcal{V}_n \subset \bigcup\{\Pi_\alpha^0(X) : \alpha < \xi_n\}$ for all $n \in \omega$, we settled (ii)$\Longrightarrow$(iii).

Finally, if (iii) holds then take the respective complete sequence $\{\mathcal{V}_n : n \in \omega\}$ of $\sigma$-discrete covers of $X$ and let $\mathcal{U}_n = \mathcal{V}_n$ for every $n \in \omega$. By Fact 1 of T.331 we have $\mathcal{U}_n \subset \Sigma_{\xi_n}^0(X)$ for each $n \in \omega$; this shows that (iii)$\Longrightarrow$(ii) and makes our solution complete.

**V.189.** *Given a countable ordinal $\xi \geq 2$ prove that the following conditions are equivalent for any second countable $X$:*

(i)  *the space $X$ belongs to $\mathcal{M}_\xi$;*
(ii)  *there is a complete sequence $\{\mathcal{U}_n : n \in \omega\}$ of countable covers of $X$ such that, for any $n \in \omega$, there is $\xi_n < \xi$ with $\mathcal{U}_n \subset \Sigma_{\xi_n}^0(X)$;*

*(iii)  there is a complete sequence $\{\mathcal{V}_n : n \in \omega\}$ of countable covers of X such that,
    for any $n \in \omega$, there is $\xi_n < \xi$ with $\mathcal{V}_n \subset \bigcup\{\Pi_\alpha^0(X) : \alpha < \xi_n\}$.*

**Solution.** If $X \in \mathcal{M}_\xi$ then it follows from Problem 188 that $X$ has a complete
sequence $\{\mathcal{U}_n : n \in \omega\}$ of $\sigma$-discrete covers such that, for every $n \in \omega$, we have
$\mathcal{U}_n \subset \Sigma_{\xi_n}^0(X)$ for some $\xi_n < \xi$. The space $X$ being second countable, every $\mathcal{U}_n$ is
countable; this proves that (i)$\Longrightarrow$(ii).

Now, if $X$ has a complete sequence $\{\mathcal{U}_n : n \in \omega\}$ of countable covers as in
(ii) then every $\mathcal{U}_n$ is $\sigma$-discrete, so we can apply Problem 188 to see that there is
a complete sequence $\{\mathcal{V}_n : n \in \omega\}$ of $\sigma$-discrete covers of $X$ such that, for every
$n \in \omega$ there exists $\xi_n < \xi$ for which $\mathcal{V}_n \subset \bigcup\{\Pi_\alpha^0(X) : \alpha < \xi_n\}$. Since $X$ is second
countable, every $\mathcal{V}_n$ is countable, so we settled (ii)$\Longrightarrow$(iii).

Finally, if $\{\mathcal{V}_n : n \in \omega\}$ is a complete sequence of countable covers as in (iii)
then each $\mathcal{V}_n$ is $\sigma$-discrete, so we can apply Problem 188 once more to conclude that
$X \in \mathcal{M}_\xi$ and hence (iii)$\Longrightarrow$(i).

**V.190.** *Prove that any analytic space has a complete sequence of countable covers.
Show that in metrizable spaces the converse is also true, i.e., a metrizable space X
is analytic if and only if there exists a complete sequence of countable covers of X.*

**Solution.** Suppose that a space $X$ is analytic and fix a continuous onto map $\varphi$ :
$\omega^\omega \to X$. For any $x \in X$ choose a point $y_x \in \varphi^{-1}(x)$ and let $Y = \{y_x : x \in X\}$.
Then $\varphi|Y : Y \to X$ is a bijection. Given $n \in \omega$ and $k_0, \ldots, k_n \in \omega$ the set
$O(k_0, \ldots, k_n) = \{f \in \omega^\omega : f(i) = k_i$ for all $i \leq n\}$ is open in $\omega^\omega$, so the family
$\mathcal{U}_n = \{O(k_0, \ldots, k_n) : k_i \in \omega$ for all $i \leq n\}$ is an open disjoint cover of $\omega^\omega$.

The family $\mathcal{V}_n = \{U \cap Y : U \in \mathcal{U}_n\}$ is a cover of the space $Y$, so the family
$\mathcal{W}_n = \{\varphi(V) : V \in \mathcal{V}_n\}$ is a cover of $X$ for every $n \in \omega$. To show that the
sequence $\mathcal{S} = \{\mathcal{W}_n : n \in \omega\}$ is complete in $X$ take a filter $\mathcal{F}$ on the set $X$ such that
$\mathcal{F} \cap \mathcal{W}_n \neq \emptyset$ and hence there is $V_n \in \mathcal{V}_n$ such that $\varphi(V_n) \in \mathcal{F}$; pick a set $U_n \in \mathcal{U}_n$
with $U_n \cap Y = V_n$ for each $n \in \omega$.

Since $\varphi|Y : Y \to X$ is a bijection and $\{\varphi(V_n) : n \in \omega\} \subset \mathcal{F}$ is centered, the
family $\{V_n : n \in \omega\}$ is centered too; fix a point $f_n \in V_0 \cap \ldots \cap V_n \subset U_0 \cap \ldots \cap U_n$
for each $n \in \omega$. By the definition of the family $\mathcal{U}_n$ there are $k_0^n, \ldots, k_n^n \in \omega$ such
that $U_n = O(k_0^n, \ldots, k_n^n)$ for every $n \in \omega$. Take any $m, n, i \in \omega$ with $i \leq m \leq n$.
It follows from $\{f_m, f_n\} \subset U_m$ and $f_n \in U_n$ that $k_i^m = f_m(i) = f_n(i) = k_i^n$ and
hence $k_i^n = k_i^m$, i.e., the number $k_i^n = k_i$ does not depend on $n$.

Let $f(i) = k_i$ for every $i \in \omega$; then $f \in \omega^\omega$. Fix any element $F \in \mathcal{F}$; if
$x = \varphi(f) \notin \overline{F}$ then use continuity of $\varphi$ to find a set $W \in \tau(f, \omega^\omega)$ such that
$\varphi(W) \cap F = \emptyset$. The family $\{U_n : n \in \omega\}$ is easily seen to be a local base at $f$,
so there is $n \in \omega$ with $U_n \subset W$. Then $\varphi(V_n) \subset \varphi(U_n) \subset \varphi(W)$ and therefore
$\varphi(V_n) \cap F = \emptyset$. This contradiction with $\{\varphi(V_n), F\} \subset \mathcal{F}$ shows that $x \in \overline{F}$; the
set $F \in \mathcal{F}$ was chosen arbitrarily, so $x \in \bigcap\{\overline{F} : F \in \mathcal{F}\}$. Consequently, $\mathcal{S}$ is a
complete sequence in $X$.

Now assume that $X$ is metrizable and $\mathcal{S} = \{\mathcal{U}_n : n \in \omega\}$ is a complete sequence
of countable covers of $X$. If $w(X) > \omega$ then there exists a closed discrete $D \subset X$
with $|D| = \omega_1$. Since $\mathcal{U}_0$ is countable, we can choose $U_0 \in \mathcal{U}_0$ such that $D_0 =$

$U_0 \cap D$ is uncountable. Suppose that $n \in \omega$ and we have chosen $U_i \in \mathcal{U}_i$ for all $i \leq n$ in such a way that the set $D_i = D \cap (\bigcap_{j \leq i} U_j)$ is uncountable for any $i \leq n$. The set $D_n$ is uncountable while the family $\mathcal{U}_{n+1}$ is countable, so there is $U_{n+1} \in \mathcal{U}_{n+1}$ such that set $D_{n+1} = D_n \cap U_{n+1}$ is uncountable.

Thus, our inductive procedure shows that we can choose a sequence $\{U_i : i \in \omega\}$ such that $U_i \in \mathcal{U}_i$ and the set $D_i = D \cap (\bigcap_{j \leq i} U_j)$ is uncountable for every $i \in \omega$. This makes it possible to pick a point $d_i \in D_i$ for each $i \in \omega$, so that $d_i \neq d_j$ whenever $i \neq j$. The family $\mathcal{D} = \{\{d_i : i \geq n\} : n \in \omega\}$ is a filterbase in $X$; let $\mathcal{F}$ be any filter containing $\mathcal{D}$. By our choice of the set $\{d_i : i \in \omega\}$ we have $\{d_i : i \geq n\} \subset U_n$ and hence $U_n \in \mathcal{F}$ for every $n \in \omega$. The sequence $\mathcal{S}$ being complete there is a point $x \in \bigcap\{\overline{F} : F \in \mathcal{F}\}$, so $x \in \bigcap \mathcal{D}$ because all elements of $\mathcal{D} \subset \mathcal{F}$ are closed in $X$. This contradiction with $\bigcap \mathcal{D} = \emptyset$ shows that $X$ is second countable. The following property is evident.

(1) if $\mathcal{U}'_n$ is a refinement of $\mathcal{U}_n$ for every $n \in \omega$ then the sequence $\mathcal{S}' = \{\mathcal{U}'_n : n \in \omega\}$ is also complete.

We will pass several times from the sequence $\mathcal{S}$ to a sequence $\mathcal{S}'$ as in (1); to simplify the notation, we will then consider that $\mathcal{S}' = \mathcal{S}$ and hence the sequence $\mathcal{S}$ has the properties we found in $\mathcal{S}'$.

Fix a metric $\rho$ on the set $X$ which generates $\tau(X)$ and let $\mathcal{V}_n$ be a countable subcover of the cover $\{B_\rho(x, 2^{-n-1}) : x \in X\}$ for any $n \in \omega$. Given any $n \in \omega$ the cover $\mathcal{U}'_n = \{U \cap V : U \in \mathcal{U}_n$ and $V \in \mathcal{V}_n\}$ is a countable refinement of $\mathcal{U}_n$ and $\text{diam}_\rho(U) \leq 2^{-n}$ for any $U \in \mathcal{U}'_n$. According to the policy described above, we consider that $\text{diam}_\rho(U) \leq 2^{-n}$ for every $U \in \mathcal{U}_n$ and $n \in \omega$. Since there is nothing to prove in the case of an empty $X$ (and neither is it interesting to decide whether the empty space is analytic or not), we consider that $X \neq \emptyset$, so we can throw away all empty elements of every $\mathcal{U}_n$ obtaining its refinement with only nonempty elements. Therefore we consider that $U \neq \emptyset$ for any $U \in \mathcal{U}_n$.

Take an enumeration $\{O(i) : i \in \omega\}$ of the family $\mathcal{U}_0$ (with possible repetitions) and let $\mathcal{O}_0 = \{O(i) : i \in \omega\}$. Proceeding inductively, assume that $n \in \omega$ and we have families $\mathcal{O}_0, \ldots, \mathcal{O}_n$ with the following properties:

(2) $\mathcal{O}_i$ is a refinement of $\mathcal{U}_i$ and no element of $\mathcal{O}_i$ is empty for all $i \leq n$;
(3) $\mathcal{O}_i = \{O(k_0, \ldots, k_i) : k_j \in \omega$ for all $j \leq i\}$ for each $i \leq n$;
(4) $O(k_0, \ldots, k_i) = \bigcup\{\mathcal{O}(k_0, \ldots, k_i, k_{i+1}) : k_{i+1} \in \omega\}$ for all $k_0, \ldots, k_i \in \omega$ and $i < n$;

Given any $U = O(k_0, \ldots, k_n) \in \mathcal{O}_n$ the family $\mathcal{W} = \{U \cap V : V \in \mathcal{U}_{n+1}\}$ covers $U$ which is nonempty so we can assume that every $W \in \mathcal{W}$ is nonempty; take an enumeration $\{W_i : i \in \omega\}$ of the family $\mathcal{W}$ (with possible repetitions). Let $O(k_0, \ldots, k_n, k_{n+1}) = W_{k_{n+1}}$ for every $k_{n+1} \in \omega$.

This gives us a family $\mathcal{O}_{n+1} = \{O(k_0, \ldots, k_{n+1}) : k_i \in \omega$ for all $i \leq n + 1\}$ and it is straightforward that the properties (2)–(4) still hold if we substitute $n$ by $n + 1$. Thus our inductive procedure provides a sequence $\mathcal{S}_0 = \{\mathcal{O}_n : n \in \omega\}$ of covers of $X$ for which the conditions (2)–(4) are satisfied for all $n \in \omega$; applying

(1) once more we conclude that $S_0$ is a complete sequence. It follows from (2) and our choice of the sequence $S$ that

(5) if $n \in \omega$ then $\mathrm{diam}_\rho(O) \leq 2^{-n}$ for any $O \in \mathcal{O}_n$.

Given any element $f \in \omega^\omega$ let $U_n = O(f(0), \ldots, f(n))$; it follows from the property (4) that $U_{n+1} \subset U_n$ for every $n \in \omega$. This gives us a decreasing sequence $\mathcal{T} = \{U_n : n \in \omega\}$ of nonempty sets such that $U_n \in \mathcal{O}_n$ for all $n \in \omega$. Since $\mathcal{T}$ is a filterbase, we can find a filter $\mathcal{F} \supset \mathcal{T}$. Then $U_n \in \mathcal{F} \cap \mathcal{O}_n$ for every $n \in \omega$; the sequence $S_0$ being complete, we have $\bigcap\{\overline{U}_n : n \in \omega\} \supset \bigcap\{\overline{F} : F \in \mathcal{F}\} \neq \emptyset$. It follows from (5) that the sequence $\{\mathrm{diam}_\rho(U_n) : n \in \omega\}$ converges to zero, so there is $x \in X$ such that $\bigcap\{\overline{U}_n : n \in \omega\} = \{x\}$; let $\varphi(f) = x$. This gives us a map $\varphi : \omega^\omega \to X$.

Fix a point $f \in \omega^\omega$ and an open neighborhood $U$ of the point $x = \varphi(f)$ in $X$; by our choice of $\varphi$, we have $\{x\} = \bigcap\{\overline{U}_n : n \in \omega\}$ where $U_n = O(f(0), \ldots, f(n))$ for all $n \in \omega$. It follows from the property (5) that $\mathrm{diam}_\rho(U_n) \to 0$, so there is $n \in \omega$ such that $\overline{U}_n \subset U$. If $g \in V = \{h \in \omega^\omega : h(i) = f(i) \text{ for all } i \leq n\}$ then $\varphi(g) \in \overline{O(g(0), \ldots, g(n))} = \overline{U}_n \subset U$. This proves that $\varphi(V) \subset U$ and hence $V \in \tau(f, \omega^\omega)$ witnesses continuity of $\varphi$ at the point $f$; thus the map $\varphi$ is continuous.

Finally, take any $x \in X$; the property (4) shows that there exists a sequence $\{k_i : i \in \omega\} \subset \omega$ such that $x \in U_n = O(k_0, \ldots, k_n)$ for any $n \in \omega$. If $f(i) = k_i$ for all $i \in \omega$ then $f \in \omega^\omega$ and $x \in \bigcap_{n \in \omega} U_n \subset \bigcap_{n \in \omega} \overline{U}_n$. The set $\bigcup_{n \in \omega} \overline{U}_n$ being a singleton by (5) we conclude that $\bigcap_{n \in \omega} \overline{U}_n = \{x\}$ and hence $\varphi(f) = x$; this shows that $X$ is analytic being a continuous image of $\omega^\omega$. Thus we have proved the converse for the class of metric spaces, i.e., our solution is complete.

**V.191.** *For any metrizable space $X$ and $n \in \mathbb{N}$ define a map $e : X^n \to [X]^{\leq n}$ by $e((x_1, \ldots, x_n)) = \{x_1, \ldots, x_n\}$ for every $(x_1, \ldots, x_n) \in X^n$. Prove that there exists an $F_\sigma$-set $G$ in the space $X^n$ such that $e(G) = [X]^n$ and the map $e|G : G \to [X]^n$ is a bijection.*

**Solution.** Let $T = \{(x_1, \ldots, x_n) \in X^n : x_i \neq x_j \text{ whenever } i \neq j\}$. Given any bijection $\sigma : \{1, \ldots, n\} \to \{1, \ldots, n\}$ let $h_\sigma((x_1, \ldots, x_n)) = (x_{\sigma(1)}, \ldots, x_{\sigma(n)})$ for any $(x_1, \ldots, x_n) \in T$. It is evident that $h_\sigma : T \to T$ is a homeomorphism. Denote by $S_n$ the set of all bijections $\sigma : \{1, \ldots, n\} \to \{1, \ldots, n\}$. Call a set $Y \subset T$ *adequate* if $e|Y$ is injective.

For any $a = (a_1, \ldots, a_n) \in T$ choose disjoint sets $O_1, \ldots, O_n \in \tau(X)$ such that $a_i \in O_i$ for every $i \leq n$. It is easy to see that the set $O = O_1 \times \ldots \times O_n \subset T$ is adequate, open and $a \in O$; this shows that $T$ has a cover which consists of open adequate sets. The space $T$ is metrizable and hence paracompact so we can choose an open locally finite cover $\mathcal{U} = \{U_\alpha : \alpha < \kappa\}$ of the space $T$ such that $U_\alpha$ is adequate for all $\alpha < \kappa$. Let $V_\alpha = \bigcup\{h_\sigma(V) : \sigma \in S_n\}$ and $G_\alpha = U_\alpha \backslash (\bigcup\{V_\beta : \beta < \alpha\})$ for every $\alpha < \kappa$. The family $\{G_\alpha : \alpha < \kappa\}$ is disjoint and locally finite. Since in the space $T$ every open set is $F_\sigma$, every $G_\alpha$ is an $F_\sigma$-subset of $T$; let $G = \bigcup\{G_\alpha : \alpha < \kappa\}$. It is an easy exercise that the union of a locally finite family of $F_\sigma$-sets is an $F_\sigma$-set, so $G$ is an $F_\sigma$-subset of $T$; the set $T$ being open in $X^n$, we conclude that $G$ is also an $F_\sigma$-set in $X^n$.

Take any set $y = \{x_1, \ldots, x_n\} \in [X]^n$; the point $x = (x_1, \ldots, x_n)$ is an element of $T$, so $e^{-1}(y) = \{h_\sigma(x) : \sigma \in S_n\}$. Take the minimal ordinal $\alpha < \kappa$ such that $U_\alpha \cap e^{-1}(y) \neq \emptyset$ and fix $\sigma \in S_n$ with $h_\sigma(x) \in U_\alpha$. If $h_\sigma(x) \in V_\beta$ for some $\beta < \alpha$ then there is $\mu \in S_n$ and $z \in U_\beta$ for which $h_\mu(z) = h_\sigma(x)$. It is straightforward that $z = h_\mu^{-1}(h_\sigma(x)) = h_{\sigma \circ \mu^{-1}}(x) \in e^{-1}(y) \cap U_\beta$ which is a contradiction. Therefore $a = h_\sigma(x) \in G_\alpha$ and hence $e(a) = y$. The point $y \in [X]^n$ was chosen arbitrarily so we proved that $e(G) = [X]^n$.

Finally take distinct points $a$ and $b$ in the set $G$ and fix $\alpha, \beta \in \kappa$ with $a \in G_\alpha$ and $b \in G_\beta$; we can assume, without loss of generality, that $\alpha \leq \beta$. If $\alpha < \beta$ then $b \notin V_\alpha$ and hence $b \notin e^{-1}(a)$ which shows that $e(a) \neq e(b)$. If $\alpha = \beta$ then $e(a) \neq e(b)$ because $G_\alpha$ is an adequate set. Thus $G$ is an $F_\sigma$-subset of $X^n$ such that $e(G) = [X]^n$ and $e|G : G \to [X]^n$ is a bijection.

**V.192.** *Given a metrizable space $X$ and $n \in \mathbb{N}$ consider the set $[X]^n$ together with its Vietoris topology. Prove that there exists a family $\{Y_m : m \in \omega\}$ of closed subsets of $[X]^n$ such that $[X]^n = \bigcup\{Y_m : m \in \omega\}$ and every $Y_m$ is homeomorphic to some closed subspace of $X^n$.*

**Solution.** For any space $Z$ denote by $\mathcal{K}(Z)$ the family of all compact subsets of $Z$ and let $\mathcal{V}_Z$ be the Vietoris topology on $\mathcal{K}(Z)$. For each $U \in \tau(Z)$ consider the families $O_Z[U] = \{K \in \mathcal{K}(Z) : K \cap U \neq \emptyset\}$ and $O_Z\langle U \rangle = \{K \in \mathcal{K}(Z) : K \subset U\}$. If $\mathcal{S}_0(Z) = \{O_Z[U] : U \in \tau(Z)\}$ and $\mathcal{S}_1(Z) = \{O_Z\langle U \rangle : U \in \tau(Z)\}$ then the family $\mathcal{S}(Z) = \mathcal{S}_0(Z) \cup \mathcal{S}_1(Z)$ is a subbase of the space $(\mathcal{K}(Z), \mathcal{V}_Z)$.

*Fact 1.* For any space $Z$ if $T \subset Z$ then $\mathcal{V}_T = \{V \cap \mathcal{K}(T) : V \in \mathcal{V}_Z\}$ and hence $(\mathcal{K}(T), \mathcal{V}_T)$ is a subspace of $(\mathcal{K}(Z), \mathcal{V}_Z)$.

*Proof.* Since $\mathcal{K}(T) \subset \mathcal{K}(Z)$, we must only check that $\mathcal{V}_T$ is induced by $\mathcal{V}_Z$ on $\mathcal{K}(T)$. It is easy to verify that

(1) $O_Z[U] \cap \mathcal{K}(T) = O_T[U']$ and $O_Z\langle U \rangle \cap \mathcal{K}(T) = O_T\langle U' \rangle$ for any $U \in \tau(Z)$ and $U' \in \tau(T)$ such that $U' = U \cap T$.

An immediate consequence of (1) is that $G \cap \mathcal{K}(T) \in \mathcal{S}(T)$ for any $G \in \mathcal{S}(Z)$. If $G' \in \mathcal{S}_0(T)$ then take $U' \in \tau(T)$ such that $G' = O_T[U']$ and choose $U \in \tau(Z)$ with $U \cap T = U'$. It follows from (1) that $G' = O_Z[U] \cap \mathcal{K}(T)$. If $G' \in \mathcal{S}_1(T)$ then there exists $U' \in \tau(T)$ such that $G' = O_T\langle U' \rangle$; pick $U \in \tau(Z)$ with $U \cap T = U'$ and apply (1) to convince ourselves that $O_Z\langle U \rangle \cap \mathcal{K}(Z) = O_T\langle U' \rangle$. This proves that $\mathcal{S}(T) = \{S \cap \mathcal{K}(T) : S \in \mathcal{S}(Z)\}$. Recalling that $\mathcal{S}(Z)$ is a subbase of $(\mathcal{K}(Z), \mathcal{V}_Z)$ and $\mathcal{S}(T)$ is a subbase of $(\mathcal{K}(T), \mathcal{V}_T)$ and representing the respective open sets as unions of finite intersections of the elements of their corresponding subbases, we conclude that $\mathcal{V}_T = \{V \cap \mathcal{K}(T) : V \in \mathcal{V}_Z\}$, i.e., Fact 1 is proved.

*Fact 2.* For any space $Z$ the space $(\mathcal{K}(Z), \mathcal{V}_Z)$ is Tychonoff.

*Proof.* The family $\mathcal{F}$ of all closed subsets of $\beta Z$ coincides with $\mathcal{K}(\beta Z)$, so we can apply Fact 1 of T.372 to conclude that $(\mathcal{K}(\beta Z), \mathcal{V}_{\beta Z})$ is a compact Hausdorff (and hence Tychonoff) space. It follows from Fact 1 that $(\mathcal{K}(Z), \mathcal{V}_Z)$ is a subspace of $(\mathcal{K}(\beta Z), \mathcal{V}_{\beta Z})$, so $(\mathcal{K}(Z), \mathcal{V}_Z)$ is a Tychonoff space, i.e., Fact 2 is proved.

*Fact 3.* Suppose that $n \in \mathbb{N}$ and $f : Z \to T$ is an open map such that $|f^{-1}(t)| = n$ for any $t \in T$. Then

(a)  $f$ is perfect;
(b)  $f$ is a local homeomorphism, i.e., for any $z \in Z$ there is $W \in \tau(z, Z)$ such that $f|W : W \to f(W)$ is a homeomorphism.

*Proof.* Recall that we consider that all open maps are surjective. The inverse images of all points of $T$ are finite and hence compact so, to prove (a), it suffices to show that $f$ is a closed map. Given $t \in T$ and $U \in \tau(f^{-1}(t), Z)$, we have $f^{-1}(t) = \{z_1, \ldots, z_n\}$, so we can choose disjoint open sets $U_1, \ldots, U_n$ such that $U' = U_1 \cup \ldots \cup U_n \subset U$ and $z_i \in U_i$ for all $i \leq n$. The set $V = f(U_1) \cap \ldots \cap f(U_n)$ is an open neighborhood of the point $t$.

For any $a \in V$ and $i \leq n$ there is $b_i \in U_i$ with $f(b_i) = a$; an immediate consequence is that $B = \{b_1, \ldots, b_n\} \subset f^{-1}(a)$. Since the set $f^{-1}(a)$ has exactly $n$ elements, we conclude that $f^{-1}(a) = B \subset U' \subset U$. The point $a \in V$ was chosen arbitrarily, so we proved that $f^{-1}(V) \subset U' \subset U$ and hence the map $f$ is closed by Fact 2 of S.271; this settles (a).

Fix any point $z \in Z$ and let $t = f(z)$; then $f^{-1}(t) = \{z, z_1, \ldots, z_{n-1}\}$, so we can choose disjoint sets $U, U_1, \ldots, U_{n-1} \in \tau(Z)$ such that $z \in U$ and $z_i \in U_i$ for all $i \leq n - 1$. Then $V = f(U) \cap f(U_1) \cap \ldots \cap f(U_{n-1}) \in \tau(t, T)$ and hence $W = U \cap f^{-1}(V)$ is an open neighborhood of the point $z$.

If $a \in W$ then, for any $i \leq n - 1$ there is $a_i \in U_i$ such that $f(a_i) = f(a)$, so $A = \{a, a_1, \ldots, a_{n-1}\} \subset f^{-1}(f(a))$. Recalling that $|f^{-1}(f(a))| = n$ we conclude that $A = f^{-1}(f(a))$ and hence $f^{-1}(f(a)) \cap W = \{a\}$ for any $a \in W$. This shows that $f|W : W \to f(W)$ is an injection; since $f|W$ is an open map, we conclude that $f|W$ is a homeomorphism so we established (b), i.e., Fact 3 is proved.

*Fact 4.* Given a space $Z$ and $n \in \mathbb{N}$ let $Z[n] = \{(z_1, \ldots, z_n) \in Z^n : z_i \neq z_j$ whenever $i \neq j\}$. For any $z = (z_1, \ldots, z_n) \in Z[n]$ let $\varphi(z) = \{z_1, \ldots, z_n\}$. Then the map $\varphi : Z[n] \to [Z]^n$ is open if $[Z]^n$ is considered with it Vietoris topology.

*Proof.* It is evident that $\varphi$ is surjective. If $y = \{z_1, \ldots, z_n\} \in [Z]^n$ and $U_i \in \tau(z_i, Z)$ for all $i \leq n$ then consider the set $[U_1, \ldots, U_n] = \{a \in [Z]^n : a \cap U_i \neq \emptyset$ for all $i \leq n$ and $a \subset U_1 \cup \ldots \cup U_n\}$. It is straightforward that the family

$$\mathcal{C}_y = \{[U_1, \ldots, U_n] : U_i \in \tau(z_i, Z) \text{ for all } i \leq n \text{ and } U_i \cap U_j = \emptyset \text{ whenever } i \neq j\}$$

is a local base of the space $[Z]^n$ at the point $y$.

For any point $z = (z_1, \ldots, z_n) \in Z[n]$ the family $\mathcal{B}_z = \{U_1 \times \ldots \times U_n : z_i \in U_i$ for all $i \leq n$ and $U_i \cap U_j = \emptyset$ whenever $i \neq j\}$ is easily seen to be a local base of $Z[n]$ at the point $z$. It is an easy exercise that $\varphi(U_1 \times \ldots \times U_n) = [U_1, \ldots, U_n]$ if the family $\{U_1, \ldots, U_n\}$ is disjoint. An immediate consequence is that, for the point $y = \varphi(z)$ we have $\mathcal{C}_y = \{\varphi(G) : G \in \mathcal{B}_z\}$, so we can apply Fact 2 of S.491 to conclude that $\varphi$ is an open continuous map. Fact 4 is proved.

Returning to our solution consider the map $\varphi : X[n] \to [X]^n$ defined by the formula $\varphi(x) = \{x_1, \ldots, x_n\}$ for any $x = (x_1, \ldots, x_n) \in X[n]$. Apply Fact 4 to see that $\varphi$ is open; since also $|\varphi^{-1}(a)| = n!$ for any $a \in [X]^n$, we can apply Fact 3 to convince ourselves that $\varphi$ is also a closed map.

Apply Problem 191 to find a family $\{F_m : m \in \omega\}$ of closed subsets of the space $X[n]$ such that the restriction of the map $\varphi$ to the set $F = \bigcup_{m \in \omega} F_n$ is injective and $\varphi(F) = [X]^n$. For every $m \in \omega$ the set $Y_m = \varphi(F_m)$ is closed in $[X]^n$; the map $\varphi | F_m : F_m \to Y_m$ being closed and injective, it has to be a homeomorphism, so $[X]^n = \bigcup_{m \in \omega} Y_m$ is the promised representation of $[X]^n$ and hence our solution is complete.

**V.193.** *Suppose that there exists a uniformly continuous surjection of $C_p(X)$ onto $C_p(Y)$. Prove that if $X$ is pseudocompact then $Y$ is also pseudocompact. Deduce from this fact that if $X$ is a metrizable compact space and there exists a uniformly continuous surjection of $C_p(X)$ onto $C_p(Y)$ then $Y$ is also compact. Give an example of a (non-metrizable!) compact space $X$ such that there is a non-compact space $Y$ and a uniformly continuous surjection of $C_p(X)$ onto $C_p(Y)$.*

**Solution.** Recall that a uniform space $(Z, \mathcal{U})$ is called $\sigma$-*totally bounded* if there exists a family $\{Z_n : n \in \omega\}$ of totally bounded uniform subspaces of $(Z, \mathcal{U})$ such that $Z = \bigcup_{n \in \omega} Z_n$.

Suppose that $\varphi : C_p(X) \to C_p(Y)$ is a uniformly continuous surjection. If $X$ is pseudocompact then $C_p(X)$ is $\sigma$-totally bounded by Problem 136; fix a family $\{C_n : n \in \omega\}$ of totally bounded uniform subspaces of $C_p(X)$ such that $C_p(X) = \bigcup_{n \in \omega} C_n$. By Fact 1 of V.136, the uniform space $D_n = \varphi(C_n)$ is totally bounded for every $n \in \omega$. The map $\varphi$ being surjective, we have $C_p(Y) = \bigcup_{n \in \omega} D_n$, i.e., the uniform space $C_p(Y)$ is $\sigma$-totally bounded, so $Y$ is pseudocompact by Problem 136.

If $X$ is compact and metrizable then $Y$ is pseudocompact by what we proved before. Besides, $nw(C_p(X)) = nw(X) = w(X) \le \omega$; since network weight is not raised even by continuous maps, we have $nw(C_p(Y)) = \omega$ and hence $nw(Y) = \omega$. Therefore $Y$ is compact (and metrizable).

Finally consider the compact space $X = \omega_1 + 1$ and let $Y = \omega_1 \subset X$. Then the restriction map $\pi : C_p(X) \to C_p(Y)$ is continuous and linear, so $\pi$ is uniformly continuous. If follows from TFS-314 that every bounded continuous function on $Y$ can be continuously extended to $X$; since the space $Y$ is countably compact, this implies $\pi(C_p(X)) = C_p(Y)$. Therefore $Y$ is a non-compact space such that $C_p(Y)$ is a linear (and hence uniform) continuous image of $C_p(X)$ for a compact space $X$.

**V.194.** *Assume that $X$ and $Y$ are metrizable spaces and there exists either a uniformly continuous surjection of $C_p(X)$ onto $C_p(Y)$ or a uniformly continuous surjection of $C_p^*(X)$ onto $C_p^*(Y)$. Prove that there exists a family $\{Y_n : n \in \omega\}$ of closed subspaces of $Y$ such that $Y = \bigcup_{n \in \omega} Y_n$ and each $Y_n$ can be perfectly mapped onto a closed subspace of $[X]^{k_n}$ (with the Vietoris topology) for some $k_n \in \mathbb{N}$.*

**Solution.** Given a set $A$ the family $\mathrm{Fin}(A)$ consists of all finite subsets of $A$; for each $n \in \omega$ let $[A]^n = \{B \in \mathrm{Fin}(A) : |B| = n\}$. For any space $Z$ the function $\mathbf{0}_Z \in C_p(Z)$ is defined by $\mathbf{0}_Z(z) = 0$ for all $z \in Z$. For each $n \in \mathbb{N}$ the space $[Z]^n$ is assumed to carry the Vietoris topology.

Suppose that either $T : C_p(X) \to C_p(Y)$ or $T : C_p^*(X) \to C_p^*(Y)$ and $T$ is a uniformly continuous surjection. Our proof will be valid for both cases, so let us denote by $E(X)$ and $E(Y)$ the respective domain and range of $T$. Since $\mathbf{0}_X \in E(X)$ and $\mathbf{0}_Y \in E(Y)$, there is no loss of generality to assume that $T(\mathbf{0}_X) = \mathbf{0}_Y$. For every $y \in Y$ let $b_y(f) = f(y)$ for all $f \in E(Y)$. For any $K \in \mathrm{Fin}(X)$ and $\varepsilon > 0$ the set $O_X(K, \varepsilon) = \{f \in E(X) : f(K) \subset (-\varepsilon, \varepsilon)\}$ is an open neighborhood of $\mathbf{0}_X$ and the family $\{O_X(K, \varepsilon) : K \in \mathrm{Fin}(X) \text{ and } \varepsilon > 0\}$ is a local base of the space $E(X)$ at the point $\mathbf{0}_X$.

It follows from Fact 1 of V.178 that

(1) $b_y : E(Y) \to \mathbb{R}$ is a uniformly continuous unbounded map and hence the map $\beta_y = b_y \circ T$ is also uniformly continuous and unbounded for every $y \in Y$.

*Fact 1.* Given a point $y \in Y$ and a finite subset $K$ of the space $X$ let

$$u(y, K) = \sup\{|\beta_y(f) - \beta_y(g)| : f, g \in E(X) \text{ and } |f(x) - g(x)| < 1 \text{ for every } x \in K\}.$$

Then the family $\mathcal{E}(y) = \{K \in \mathrm{Fin}(X) : u(y, K) < \infty\}$ is nonempty.

*Proof.* It follows from uniform continuity of $\beta_y$ that there is a finite set $K \subset X$ and $\varepsilon > 0$ such that $f - g \in O_X(K, \varepsilon)$ implies $|\beta_y(f) - \beta_y(g)| < 1$. Fix $n \in \mathbb{N}$ with $n\varepsilon > 1$ and take any $f, g \in E(X)$ such that $|f(x) - g(x)| < 1$ for all $x \in K$. The function $h_i = f + \frac{i}{n}(g - f)$ belongs to $E(X)$ for every $i \leq n$ and it is straightforward that $|h_{i+1}(x) - h_i(x)| < \frac{1}{n} < \varepsilon$ for all $x \in K$ and hence $|\beta_y(h_{i+1}) - \beta_y(h_i)| < 1$ for all $i < n$. Now, $h_0 = f$ and $h_n = g$, so $|\beta_y(g) - \beta_y(f)| \leq \sum_{i<n} |\beta_y(h_{i+1}) - \beta_y(h_i)| < n$ which shows that $u(y, K) \leq n$ and hence $K \in \mathcal{E}(y)$, i.e., Fact 1 is proved.

*Fact 2.* For any point $y \in Y$ the empty set does not belong to $\mathcal{E}(y)$; besides, if $K_0, K_1 \in \mathcal{E}(y)$ then $u(y, K_0 \cap K_1) \leq u(y, K_0) + u(y, K_1)$ and hence $K_0 \cap K_1 \in \mathcal{E}(y)$.

*Proof.* Observe that $O_X(\emptyset, 1) = E(X)$; since $\beta_y$ is unbounded on $E(X)$ by (1), the set $\{|\beta_y(f)| = |\beta_y(f) - \beta_y(\mathbf{0}_X)| : f \in E(X)\}$ is unbounded so $u(y, \emptyset) = \infty$ and hence $\emptyset \notin \mathcal{E}(y)$. Now, if $K_0, K_1 \in \mathcal{E}(y)$ then let $K = K_0 \cap K_1$ and consider any $f, g \in E(X)$ such that $|f(x) - g(x)| < 1$ for all $x \in K$.

The set $I_x = (f(x)-1, f(x)+1) \cap (g(x)-1, g(x)+1)$ is open and nonempty, so choose $q_x \in I_x$ for any $x \in K$. For every $x \in K_0 \setminus K$ take $q_x \in (f(x)-1, f(x)+1)$; if $x \in K_1 \setminus K$ then pick $q_x \in (g(x) - 1, g(x) + 1)$. It is easy to find a function $h \in E(X)$ for which $h(x) = q_x$ for every $x \in K_0 \cup K_1$; it is immediate that $f - h \in O_X(K_0, 1)$ and $h - g \in O_X(K_1, 1)$. Therefore $|\beta_y(f) - \beta_y(h)| \leq u(y, K_0)$ and $|\beta_y(h) - \beta_y(g)| \leq u(y, K_1)$ whence $|\beta_y(f) - \beta_y(g)| \leq u(y, K_0) + u(y, K_1)$. This shows that $u(y, K) \leq u(y, K_0) + u(y, K_1)$, so we proved all statements formulated for $\mathcal{E}(y)$ and hence Fact 2 is proved.

*Fact 3.* For any $y \in Y$ there exists a unique minimal element $K(y)$ in the family $\mathcal{E}(y)$ with respect to inclusion; let $u(y) = u(y, K(y))$.

*Proof.* The elements of $\mathcal{E}(y)$ are finite, so there exists a minimal element $K \in \mathcal{E}(y)$; if $K_0$ is another minimal element of $\mathcal{E}(y)$ then $K \cap K_0$ is strictly smaller than $K$; since $K \cap K_0 \in \mathcal{E}(y)$ by Fact 2, we have a contradiction with minimality of $K$. Therefore there is a unique minimal element of $\mathcal{E}(y)$, so Fact 3 is proved.

*Fact 4.* For any $p, q \in \mathbb{N}$ let $Y(p, q) = \{y \in Y :$ there exists $K \in \mathcal{E}(y)$ such that $|K| \leq q$ and $u(y, K) \leq p\}$. Given $p \in \mathbb{N}$ consider the set $M(p, 1) = Y(p, 1)$; for every natural $q > 1$ let $M(p, q) = Y(p, q) \backslash Y(2p, q - 1)$. If $M(p) = \bigcup\{M(p, q) : q \in \mathbb{N}\}$ then

(i) for any $y \in Y$ and $p \in \mathbb{N}$ such that $u(y) \leq p$ we have $y \in M(p)$; in particular, $Y = \bigcup\{M(p) : p \in \mathbb{N}\}$.
(ii) If we have $p \in \mathbb{N}$ and distinct $q_0, q_1 \in \mathbb{N}$ then $M(p, q_0) \cap M(p, q_1) = \emptyset$.

*Proof.* If $q = |K(y)|$ then $u(y) = u(y, K(y)) \leq p$, so $y \in Y(p, q)$. The set $K(y)$ being minimal in $\mathcal{E}(y)$, the point $y$ does not belong to $Y(2p, q - 1)$, so $y \in M(p, q) \subset M(p)$; this proves (i).

Now assume that $p \in \mathbb{N}$ and take distinct $q_0, q_1 \in \mathbb{N}$; there is no loss of generality to assume that $q_0 < q_1$ and hence $q_0 \leq q_1 - 1$. It follows from the relevant definitions that $M(p, q_0) \subset Y(p, q_0) \subset Y(p, q_1 - 1) \subset Y(2p, q_1 - 1) \subset Y \backslash M(p, q_1)$ and therefore $M(p, q_0) \cap M(p, q_1) = \emptyset$, so we settled (ii) and hence Fact 4 is proved.

*Fact 5.* Suppose that $p \in \mathbb{N}$ and $y \in M(p)$; denote by $q(p)$ the unique natural number such that $y \in M(p, q(p))$. Then there exists a unique set $K_p(y) \subset X$ such that $|K_p(y)| = q(p)$ and $u(y, K_p(y)) \leq p$.

*Proof.* Observe that $q(p)$ is unique because the family $\{M(p, q) : q \in \mathbb{N}\}$ is disjoint by Fact 4. Let $q = q(p)$; by the definition of $Y(p, q)$ there exists a set $K \subset X$ such that $|K| \leq q$ and $a(y, K) \leq p$. It follows from $y \notin Y(2p, q-1) \supset Y(p, q-1)$ that $|K| = q$. Now, if $K' \neq K$, $a(y, K') \leq p$ and $|K'| = q$ then $K'' = K \cap K'$ has at most $(q-1)$-many elements and $a(y, K'') \leq 2p$ (see Fact 2) which is a contradiction with $y \notin Y(2p, q-1)$. This establishes uniqueness of the set $K_p(y) = K$, so Fact 5 is proved.

*Fact 6.* Given $p \in \mathbb{N}$ assume that $\{y_n : n \in \omega\} \subset Y$ is a sequence which converges to a point $y \in M(p)$; if $Q_n \subset X$ is a finite set such that $u(y_n, Q_n) \leq p$ for every $n \in \omega$ then for every $U \in \tau(X)$ with $U \cap K_p(y) \neq \emptyset$ there exists $m \in \omega$ such that $U \cap Q_n \neq \emptyset$ for all $n \geq m$.

*Proof.* If the set $A = \{n \in \omega : Q_n \cap U = \emptyset\}$ is infinite then we can pass to the subsequence $\{y_n : n \in A\}$ to see that we can assume, without loss of generality, that $Q_n \cap U = \emptyset$ for all $n \in \omega$. Fix $q \in \mathbb{N}$ such that $y \in M(p, q)$; then $|K_p(y)| = q$. If $K = K_p(y) \backslash U$ then $|K| < q$, so there exist functions $f, g \in E(X)$ such that $|f(x) - g(x)| < 1$ for all $x \in K$ while $|\beta_y(f) - \beta_y(g)| > 2p$. Take a function

$h \in E(X)$ such that $h|(X \setminus U) = f|(X \setminus U)$ and $|h(x) - g(x)| < 1$ for all $x \in U \cap K_p(y)$. Then $|h(x) - g(x)| < 1$ for all $x \in K_p(y)$, so $|\beta_y(h) - \beta_y(g)| \le p$. An immediate consequence is that $|\beta_y(h) - \beta_y(f)| > p$.

On the other hand, it follows from $h|Q_n = f|Q_n$ that

$(*)$   $|\beta_{y_n}(h) - \beta_{y_n}(f)| \le p$ for all $n \in \omega$.

The function $T(h)$ is continuous on $Y$, so the sequence $\{T(h)(y_n) : n \in \omega\}$ converges to $T(h)(y) = \beta_y(h)$. Since $\beta_{y_n}(h) = T(h)(y_n)$ for every $n \in \omega$, the sequence $\{\beta_{y_n}(h) : n \in \omega\}$ converges to $\beta_y(h)$. Analogously, $\beta_{y_n}(f) \to \beta_y(f)$, so we can apply $(*)$ to conclude that $|\beta_y(h) - \beta_y(f)| \le p$; this contradiction demonstrates that only finitely many elements of the sequence $\{Q_n : n \in \omega\}$ miss the set $U$, so Fact 6 is proved.

*Fact 7.* The set $Y(p, q)$ is closed in $Y$ for any $p, q \in \mathbb{N}$.

*Proof.* Suppose that $y_n \in Y(p, q)$ for all $n \in \omega$ and $y_n \to y$. We will pass several times to a subsequence of the sequence $S = \{y_n : n \in \omega\}$; since our aim is to prove that $y \in Y(p, q)$, at each step we will identify the obtained subsequence with $S$ considering that all elements of $S$ have the property we have found in a subsequence. Fix a set $Q_n \subset X$ such that $|Q_n| \le q$ and $u(y_n, Q_n) \le p$ for every $n \in \omega$. Passing to a subsequence of $S$ if necessary, we can assume that $|Q_n| = k \le q$ for all $n \in \omega$.

Next, use Fact 2 of U.337 to choose an infinite $A \subset \omega$ for which there is a set $D = \{d_1, \ldots, d_r\} \subset X$ such that $Q_n \cap Q_m = D$ for distinct $n, m \in A$ (observe that it is possible that $r = 0$ in which case $D = \emptyset$). According to the above mentioned politics we can consider that, for any $i \in \omega$, we have $Q_i = \{d_1, \ldots, d_r, a_1^i, \ldots, a_{k-r}^i\}$ and the family $\{Q_i \setminus D : i \in \omega\}$ is disjoint.

An evident property of metric spaces is that any sequence contains either a convergent subsequence or an infinite closed discrete subspace. This makes it possible to pass to a subsequence of $S$ once more to guarantee that, for any $j \in \{1, \ldots, k - r\}$, the sequence $S_j = \{a_j^i : i \in \omega\}$ is either convergent or constitutes a closed discrete subspace of $X$. If $S_j$ is convergent then denote by $x_j$ its limit. Renumbering every $Q_i$ if necessary we can assume that $Q_i = \{d_1, \ldots, d_r, a_1^i, \ldots, a_l^i, a_{l+1}^i, \ldots, a_{k-r}^i\}$ while the set $A = \{a_{l+j}^i : i \in \omega, 1 \le j \le k - r - l\}$ is closed and discrete in $X$ and the sequence $S_j$ converges to $x_j$ for any $j \in \{1, \ldots, l\}$.

Consider the set $Q = \{d_1, \ldots, d_r, x_1, \ldots, x_l\}$; since $|Q| \le q$, it suffices to show that $u(y, Q) \le p$. To do this, fix $f_0, g_0 \in E(X)$ such that $|f_0(x) - g_0(x)| < 1$ for every $x \in Q$. Given an arbitrary $\varepsilon > 0$ there exists a finite set $E \supset Q$ and $\delta \in (0, 1)$ such that $f - g \in O_X(E, \delta)$ implies $|\beta_y(f) - \beta_y(g)| < \varepsilon$. The set $A$ being closed and discrete in $X$ we can find $U \in \tau(E, X)$ such that $U \cap A$ is a finite set.

Choose $h \in E(X)$ such that $h(E) \subset \{1\}$ and $h(X \setminus U) \subset \{0\}$; consider the functions $f_1 = hf_0$ and $g_1 = hg_0$. It follows from $f_0|E = f_1|E$ and $g_0|E = g_1|E$ that

$(**)$   $|\beta_y(f_0) - \beta_y(f_1)| < \varepsilon$ and $|\beta_y(g_0) - \beta_y(g_1)| < \varepsilon$.

We also have $f_1|Q = f_0|Q$ and $g_1|Q = g_0|Q$, so $|f_1(x) - g_1(x)| < 1$ for every $x \in Q$. Therefore $W = \{x \in X : |f_1(x) - g_1(x)| < 1\}$ is an open neighborhood of the set $Q$ and so is the set $W' = W \cap U$. There is $m \in \omega$ such that, for all $i \geq m$, we have $a_j^i \in W'$ for all $j \in \{1, \ldots, l\}$ and $a_j^i \notin U$ for all $j \in \{l+1, \ldots, k-r\}$. As a consequence, for each $i \geq m$ we have $|f_1(x) - g_1(x)| < 1$ for every point $x \in D \cup \{a_1^i, \ldots, a_l^i\}$; besides, $f_1(x) = g_1(x) = 0$ whenever $x \in \{a_{l+1}^i, \ldots, a_{k-r}^i\}$, so $|f_1(x) - g_1(x)| < 1$ for all $x \in Q_i$.

By the choice of the sets $Q_i$, we have $|\beta_{y_i}(f_1) - \beta_{y_i}(g_1)| \leq p$ for all $i \geq m$. We already saw that $\beta_{y_i}(f_1) \to \beta_y(f_1)$ and $\beta_{y_i}(g_1) \to \beta_y(g_1)$ as $i \to \infty$. Passing to the limit in the last inequality, we conclude that $|\beta_y(f_1) - \beta_y(g_1)| \leq p$. This, together with (**), implies that $|\beta_y(f_0) - \beta_y(g_0)| \leq p + 2\varepsilon$. The number $\varepsilon > 0$ was taken arbitrarily so $|\beta_y(f_0) - \beta_y(g_0)| \leq p$ and hence $y \in Y(p,q)$, i.e., we established that the set $Y(p,q)$ is closed in $X$, so Fact 7 is proved.

*Fact 8.* The map $K_p : M(p,q) \to [X]^q$ is continuous for any $p, q \in \mathbb{N}$.

*Proof.* Given a set $U \in \tau(X)$ consider the families $I(U) = \{K \in [X]^q : K \subset U\}$ and $J(U) = \{K \in [X]^q : K \cap U \neq \emptyset\}$. Then $\mathcal{S} = \bigcup\{I(U) \cup J(U) : U \in \tau(X)\}$ is a subbase of $[X]^q$, so it suffices to show that $(K_p)^{-1}(W)$ is open in $M(p,q)$ for any $W \in \mathcal{S}$.

Fix and open set $U \subset X$, let $W = I(U)$ and assume that $G = (K_p)^{-1}(W)$ is not open in $M(p,q)$. Then there is a sequence $\{y_n : n \in \omega\} \subset M(p,q)\backslash G$ which converges to a point $y \in G$. Let $K_p(y) = \{x_1, \ldots, x_q\}$ and choose disjoint sets $U_1, \ldots, U_q \in \tau(X)$ such that $x_i \in U_i$ for all $i \leq q$ and $U' = U_1 \cup \ldots \cup U_q \subset U$.

We have $|K_p(y_n)| = q$ and $u(y_n, K_p(y_n)) \leq p$ for each $n \in \omega$ (see Fact 5). For every $i \leq q$ it follows from $K_p(y) \cap U_i \neq \emptyset$ that we can apply Fact 6 to find $m_i \in \omega$ such that $U_i \cap K_p(y_n) \neq \emptyset$ for all $n \geq m_i$. If $m = m_1 + \ldots + m_q$ then, for any $n \geq m$ the set $K_p(y_n)$ meets $U_i$ for all $i \leq q$. This, together with $|K_p(y_n)| = q$ implies that $K_p(y_n) \subset U' \subset U$ for all $n \geq m$, i.e., $y_n \in G$ which is a contradiction. Thus $(K_p)^{-1}(I(U))$ is open in $M(p,q)$ for every $U \in \tau(X)$.

Now assume that $U \in \tau(X)$ and $W = J(U)$ while $G = (K_p)^{-1}(W)$ is not open in $M(p,q)$. Then there is a sequence $\{y_n : n \in \omega\} \subset M(p,q)\backslash G$ which converges to a point $y \in G$; let $K_p(y) = \{x_1, \ldots, x_q\}$. We have $|K_p(y_n)| = q$ and $u(y_n, K_p(y_n)) \leq p$ for each $n \in \omega$ (see Fact 5); it follows from $y \in G$ that $K_p(y) \cap U \neq \emptyset$, so Fact 6 can be applied to see that there is $m \in \omega$ such that $K_p(y_n) \cap U \neq \emptyset$ and hence $y_n \in G$ for all $n \geq m$; this contradiction shows that $(K_p)^{-1}(J(U))$ is open in $M(p,q)$ for all $U \in \tau(X)$, i.e., the map $K_p : M(p,q) \to [X]^q$ is continuous and hence Fact 8 is proved.

*Fact 9.* Given metrizable spaces $Z$ and $T$ and a continuous map $f : Z \to T$ the following conditions are equivalent:

(a) $f$ is an almost perfect map, i.e., $f(F)$ is closed in $T$ for any closed $F \subset Z$ and $f^{-1}(t)$ is compact for any $t \in T$ (recall that an almost perfect continuous map is perfect if and only if it is surjective);

(b) a sequence $S = \{z_n : n \in \omega\} \subset Z$ has a convergent subsequence if and only if the sequence $S' = \{f(z_n) : n \in \omega\} \subset T$ has a convergent subsequence.

*Proof.* If $\{z_{n_k} : k \in \omega\}$ is a convergent subsequence of $S$ then $\{f(z_{n_k}) : k \in \omega\}$ is a convergent subsequence of $S'$ by continuity of $f$, so (a)$\Longrightarrow$(b) (observe that we don't need the map $f$ to be almost perfect to obtain this implication).

Now assume that (b) holds and fix $t \in T$. If $f^{-1}(t)$ is not compact then there is an infinite set $D \subset f^{-1}(t)$ which is closed and discrete in $Z$; let $\{d_n : n \in \omega\}$ be a faithful enumeration of $D$. Since $f(d_n) = t$ for every $n \in \omega$, the sequence $\{f(d_n) : n \in \omega\}$ is convergent while the sequence $\{d_n : n \in \omega\}$ has no convergent subsequences. The obtained contradiction shows that $f^{-1}(t)$ is compact for any $t \in T$.

To see that $f$ is closed take a closed subset $F \subset Z$ and assume that $G = f(F)$ is not closed in $T$. Then there is a sequence $\{t_n : n \in \omega\} \subset G$ which converges to a point $t \in T \backslash G$. Pick $z_n \in F$ with $f(z_n) = t_n$ for each $n \in \omega$. Since the property (b) holds, there exists a subsequence $\{z_{n_k} : k \in \omega\}$ of the sequence $\{z_n : n \in \omega\}$ which converges to a point $z$. The set $F$ being closed in $Z$ we have $z \in F$. By continuity of $f$ the sequence $\{f(z_{n_k}) : k \in \omega\} = \{t_{n_k} : k \in \omega\}$ converges to $f(z) \in G$. However, $\{t_{n_k} : k \in \omega\}$ has to converge to $t \neq f(z)$ being a subsequence of the sequence $\{t_n : n \in \omega\}$. This contradiction demonstrates that $f(F)$ is closed in $T$ for any closed $F \subset Z$, so Fact 9 is proved.

*Fact 10.* Suppose that $F$ is a closed subset of $Y$ such that $F \subset M(p,q)$ for some $p, q \in \mathbb{N}$. Then the map $K_p : F \to [X]^q$ is almost perfect.

*Proof.* Fix a sequence $S = \{y_n : n \in \omega\} \subset F$; if $S$ has a convergent subsequence, then the sequence $S' = \{K_p(y_n) : n \in \omega\}$ also has a convergent subsequence by continuity of $K_p$ (see Fact 8).

Now assume that $S'$ has a convergent subsequence and $S$ does not. Passing to the relevant subsequence of $S$ we can assume, without loss of generality, that $S'$ converges to a point $K = \{x_1, \dots, x_q\} \in [X]^q$ while $S$ has no convergent subsequences and hence the set $S$ is closed and discrete in $F$; therefore $S$ is also closed and discrete in $Y$. There is no loss of generality to assume that $y_n \neq y_m$ if $n \neq m$.

Let $\mathcal{A}$ be an uncountable almost disjoint family of infinite subsets of $\omega$ (see TFS-141). The space $Y$ being metrizable, the set $S$ is $C$-embedded in $Y$, so we can choose, for any $A \in \mathcal{A}$, a function $g_A \in E(Y)$ such that $g_A(y_n) = p + 1$ for all $n \in A$ and $g_A(y_n) = 0$ whenever $n \in \omega \backslash A$. Fix a function $f_A \in E(X)$ such that $T(f_A) = g_A$ for every $A \in \mathcal{A}$.

Let $r_A = (f_A(x_1), \dots, f_A(x_q)) \in \mathbb{R}^q$ for every $A \in \mathcal{A}$. The family $\mathcal{A}$ being uncountable the set $\{r_A : A \in \mathcal{A}\}$ cannot be discrete of cardinality $|\mathcal{A}|$, so there are distinct $A, B \in \mathcal{A}$ such that the distance in $\mathbb{R}^q$ between the points $r_A$ and $r_B$ is strictly less than 1 and hence $|f_A(x_i) - f_B(x_i)| < 1$ for all $i \leq q$.

By continuity of $f_A - f_B$, the set $W = \{x \in X : |f_A(x) - f_B(x)| < 1\}$ is an open neighborhood of $x_i$ for every $i \leq q$, so we can choose a disjoint family $\{O_1, \dots, O_q\} \subset \tau(X)$ such that $x_i \in O_i \subset W$ for all $i \leq q$. The sequence $S'$ being convergent to $K$ there is $m \in \omega$ such that $K_p(y_n) \cap O_i \neq \emptyset$ for all $i \leq q$ and

$n \geq m$; recalling that $|K_p(y_n)| = q$, we conclude that $K_p(y_n) \subset \bigcup_{i \leq q} O_i \subset W$ whence $|f_A(x) - f_B(x)| < 1$ for all $x \in K_p(y_n)$ and $n \geq m$.

As a consequence, $|g_A(y_n) - g_B(y_n)| = |\beta_{y_n}(f_A) - \beta_{y_n}(f_B)| \leq p$ which shows that $\{y_n : n \geq m\} \subset A \cap B$ because $|g_A(y_n) - g_B(y_n)| = p + 1$ for all numbers $n \in (A \backslash B) \cup (B \backslash A)$. This contradiction with $|A \cap B| < \omega$ shows that $S$ have a convergent subsequence and hence we can apply Fact 9 to conclude that $K_p$ is almost perfect map, i.e., Fact 10 is proved.

Returning to our solution observe that it follows from Fact 7 that $M(p,q)$ is an $F_\sigma$-subset of $Y$ for all $p,q \in \mathbb{N}$. Apply Fact 4 to see that $Y = \bigcup\{M(p,q) : p,q \in \mathbb{N}\}$, so we can find a family $\{Y_n : n \in \omega\}$ of closed subsets of $Y$ such that $Y = \bigcup_{n \in \omega} Y_n$ and every $Y_n$ is contained in some $M(p,q)$.

Fix $n \in \omega$ and $p,q \in \mathbb{N}$ such that $Y_n \subset M(p,q)$; the map $K_p : Y_n \to [X]^q$ is almost perfect by Fact 10 and, in particular, the set $F_n = K_p(Y_n)$ is closed in $[X]^q$. If $k_n = q$ then $K_p : Y_n \to F_n$ is a perfect map of $Y_n$ onto a closed subspace $F_n$ of the space $[X]^{k_n}$, so our solution is complete.

**V.195.** *Let $\mathcal{P}$ be a class of metrizable spaces with the following properties:*

*(1)  $\mathcal{P}$ contains all complete metrizable spaces;*
*(2)  $\mathcal{P}$ is invariant under finite products and closed subspaces;*
*(3)  if $M$ is a metrizable space with $M = \bigcup\{M_n : n \in \omega\}$, where $M_n$ is closed in $M$ and $M_n \in \mathcal{P}$ for each $n \in \omega$, then $M \in \mathcal{P}$.*

*Suppose that $X \in \mathcal{P}$ and $Y$ is a metrizable space. Prove that, if there exists a uniformly continuous surjection of $C_p(X)$ onto $C_p(Y)$ (or $C_p^*(X)$ onto $C_p^*(Y)$), then $Y \in \mathcal{P}$.*

**Solution.** Fix a number $n \in \mathbb{N}$ and consider the space $[X]^n$ to carry the Vietoris topology. There exists a family $\{F_m : m \in \omega\}$ of closed subsets of $[X]^n$ such that $[X]^n = \bigcup_{m \in \omega} F_m$ and every $F_m$ is homeomorphic to a closed subspace of $X^n$ (see Problem 192).

It follows from $X \in \mathcal{P}$ that every closed subspace of $X^n$ belongs to $\mathcal{P}$, so each $F_m$ belongs to $\mathcal{P}$ and hence $[X]^n \in \mathcal{P}$ for every $n \in \mathbb{N}$.

Apply Problem 194 to find a family $\{Y_n : n \in \omega\}$ of closed subspaces of $Y$ such that $Y = \bigcup_{n \in \omega} Y_n$ and, for every $n \in \omega$, there exist $k_n \in \mathbb{N}$ and a perfect map $f_n : Y_n \to F_n$ for some closed $F_n \subset [X]^{k_n}$; we can consider that $Y_n \subset M_n$ for some complete metric space $M_n$ (see TFS-237).

Given any $n \in \omega$ the graph $G = \{(y, f_n(y)) : y \in Y_n\}$ is closed in $Y_n \times F_n$ and homeomorphic to $Y_n$ (see Fact 4 of S.390). It turns out that $G$ is closed in $M_n \times F_n$; to prove this assume toward a contradiction that a point $(x, y) \in (M_n \times F_n) \backslash G$ belongs to the closure of $G$. Since $G$ is closed in $Y_n \times F_n$, the point $(x, y)$ cannot belong to $Y_n \times F_n$, so $x \notin Y_n$. There exists a set $\{x_m : m \in \omega\} \subset Y_n$ such that the sequence $\{(x_m, f_n(x_m)) : m \in \omega\}$ converges to $(x, y)$. Thus $x_m \to x \notin Y_n$, so the set $D = \{x_m : m \in \omega\}$ is closed and discrete in $Y_n$. Since the sequence $D$

converges to a point outside of $D$, the set $D$ cannot be finite, so we can pass to a subsequence to see that we can assume, without loss of generality, that $x_m \neq x_k$ whenever $m \neq k$.

Given any $z \in F_n$ the set $f_n^{-1}(z)$ is compact, so $|D \cap f_n^{-1}(z)| < \omega$; an immediate consequence is that the set $f(D)$ is infinite. By continuity of the projection onto $F_n$, the sequence $S = \{f_n(x_m) : m \in \omega\}$ must converge to $y$. The set $S$ being infinite, it is a nontrivial convergent sequence. However, $S = f_n(D)$ must be closed and discrete in $F_n$ because the map $f_n$ is closed; this contradiction shows that $G$ is closed in $M_n \times F_n$. Observe that $M_n \in \mathcal{P}$ and $F_n \in \mathcal{P}$, so $G \in \mathcal{P}$ and hence $Y_n \in \mathcal{P}$ for any $n \in \omega$. Recalling that countable unions of closed subsets preserve the property $\mathcal{P}$ we conclude that $Y \in \mathcal{P}$.

**V.196.** *Let $\mathcal{P}$ be a class of second countable spaces such that*

*(1) every compact metrizable space belongs to $\mathcal{P}$;*
*(2) $\mathcal{P}$ is invariant under finite products and closed subspaces;*
*(3) if $M$ is a second countable space with $M = \bigcup\{M_n : n \in \omega\}$, where $M_n$ is closed in $M$ and $M_n \in \mathcal{P}$ for each $n \in \omega$, then $M \in \mathcal{P}$.*

*Suppose that $X \in \mathcal{P}$ and $Y$ is a metrizable space. Prove that, if there exists a uniformly continuous surjection of $C_p(X)$ onto $C_p(Y)$ (or $C_p^*(X)$ onto $C_p^*(Y)$), then $Y \in \mathcal{P}$.*

**Solution.** Fix a number $n \in \mathbb{N}$ and consider the space $[X]^n$ to carry the Vietoris topology. There exists a family $\{F_m : m \in \omega\}$ of closed subsets of $[X]^n$ such that $[X]^n = \bigcup_{m \in \omega} F_m$ and every $F_m$ is homeomorphic to a closed subspace of $X^n$ (see Problem 192). It follows from $X \in \mathcal{P}$ that every closed subspace of $X^n$ belongs to $\mathcal{P}$, so each $F_m$ belongs to $\mathcal{P}$ and hence $[X]^n \in \mathcal{P}$ for every $n \in \mathbb{N}$.

If there exists a (uniformly) continuous surjection $T : C_p(X) \to C_p(Y)$ then $nw(C_p(Y)) \leq nw(C_p(X)) = nw(X) = \omega$, so $nw(Y) = nw(C_p(Y)) = \omega$ and hence $Y$ is second countable. If there is a continuous surjection $T : C_p^*(X) \to C_p^*(Y)$ then $nw(C_p^*(Y)) \leq nw(C_p^*(X)) \leq nw(X) = \omega$. Since $C_p(Y, (0, 1)) \subset C_p^*(Y)$ is homeomorphic to $C_p(Y)$ we conclude that $nw(C_p(Y)) \leq nw(C_p^*(Y)) = \omega$, so $nw(Y) = nw(C_p(Y)) = \omega$ and hence $Y$ is second countable. Thus $Y$ is second countable in all possible cases.

Apply Problem 194 to find a family $\{Y_n : n \in \omega\}$ of closed subspaces of $Y$ such that $Y = \bigcup_{n \in \omega} Y_n$ and, for every $n \in \omega$, there exist $k_n \in \mathbb{N}$ and a perfect map $f_n : Y_n \to F_n$ for some closed $F_n \subset [X]^{k_n}$; we can consider that $Y_n \subset M_n$ for some compact metric space $M_n$ (see TFS-209).

Given any $n \in \omega$ the graph $G = \{(y, f_n(y)) : y \in Y_n\}$ is closed in $Y_n \times F_n$ and homeomorphic to $Y_n$ (see Fact 4 of S.390). It turns out that $G$ is closed in $M_n \times F_n$; to prove this assume toward a contradiction that a point $(x, y) \in (M_n \times F_n) \setminus G$ belongs to the closure of $G$. Since $G$ is closed in $Y_n \times F_n$, the point $(x, y)$ cannot belong to $Y_n \times F_n$, so $x \notin Y_n$. There exists a set $\{x_m : m \in \omega\} \subset Y_n$ such that the sequence $\{(x_m, f_n(x_m)) : m \in \omega\}$ converges to $(x, y)$. Thus $x_m \to x \notin Y_n$, so the set $D = \{x_m : m \in \omega\}$ is closed and discrete in $Y_n$. Since the sequence $D$

converges to a point outside of $D$, the set $D$ cannot be finite, so we can pass to a subsequence to see that we can assume, without loss of generality, that $x_m \neq x_k$ whenever $m \neq k$.

Given any $z \in F_n$ the set $f_n^{-1}(z)$ is compact, so $|D \cap f_n^{-1}(z)| < \omega$; an immediate consequence is that the set $f(D)$ is infinite. By continuity of the projection onto $F_n$, the sequence $S = \{f_n(x_m) : m \in \omega\}$ must converge to $y$. The set $S$ being infinite, it is a nontrivial convergent sequence. However, $S = f_n(D)$ must be closed and discrete in $F_n$ because the map $f_n$ is closed; this contradiction shows that $G$ is closed in $M_n \times F_n$. Observe that $M_n \in \mathcal{P}$ and $F_n \in \mathcal{P}$, so $G \in \mathcal{P}$ and hence $Y_n \in \mathcal{P}$ for any $n \in \omega$. Recalling that countable unions (of closed subsets) preserve the property $\mathcal{P}$ we conclude that $Y \in \mathcal{P}$.

**V.197.** *Given a countable ordinal $\xi$, let $\mathcal{M}_\xi$ be the class of absolute Borel sets of multiplicative class $\xi$. Suppose that $X$ is a metrizable space such that $X \in \mathcal{M}_\alpha$ for some $\alpha \geq 2$. Let $Y$ be a metrizable space such that $C_p(Y)$ (or $C_p^*(Y)$) is a uniformly continuous image of $C_p(X)$ (or $C_p^*(X)$ respectively). Prove that $Y \in \mathcal{M}_\alpha$. In particular, if $X \overset{u}{\sim} Y$ then $X$ belongs to $\mathcal{M}_\alpha$ if and only if so does $Y$.*

**Solution.** If $M$ is completely metrizable then $M$ is closed in $M$, so $M \in \Pi_\alpha^0(M)$ which shows that we can apply Problem 183 to conclude that $M \in \mathcal{M}_\alpha$; this proves that the class $\mathcal{M}_\alpha$ contains all completely metrizable spaces.

Suppose that $Z \in \mathcal{M}_\alpha$ and $F$ is a closed subset of $Z$. There exists a complete metric space $M$ such that $Z$ is homeomorphic to some $Z' \in \Pi_\alpha^0(M)$ (see Problem 183). Then $F$ is homeomorphic to some closed subspace $F'$ of the space $Z'$ and therefore we can find a closed subset $P$ of the space $M$ such that $P \cap Z' = F'$. Since $P \in \Pi_\alpha^0(M)$, we can apply Fact 1 of T.341 to see that $F' \in \Pi_\alpha^0(M)$ and hence $F \in \mathcal{M}_\alpha$ by Problem 183. Thus the class $\mathcal{M}_\alpha$ is invariant under closed subspaces; by Problem 185, it is also invariant under finite products.

Now suppose that $M$ is a metrizable space, $M = \bigcup_{n \in \omega} M_n$ while every $M_n$ is closed in $M$ and belongs to $\mathcal{M}_\alpha$. If $\beta = 1 < \alpha$ then $M_n \in \Sigma_\beta^0(M)$ for every $n \in \omega$, so we can apply Problem 186 to see that $M \in \mathcal{M}_\alpha$. Therefore the class $\mathcal{M}_\alpha$ satisfies all conditions we need to apply Problem 195 and conclude that if $X$ belongs to $\mathcal{M}_\alpha$ and there exists a uniformly continuous surjection $\varphi : C_p(X) \to C_p(Y)$ or a uniformly continuous surjection $\xi : C_p^*(X) \to C_p^*(Y)$ then $Y \in \mathcal{M}_\alpha$.

**V.198.** *Given a countable ordinal $\xi$, let $\mathcal{A}_\xi$ be the class of absolute Borel sets of additive class $\xi$. Suppose that $X$ is a metrizable space such that $X \in \mathcal{A}_\alpha$ for some $\alpha \geq 2$. Let $Y$ be a metrizable space such that $C_p(Y)$ (or $C_p^*(Y)$) is a uniformly continuous image of $C_p(X)$ (or $C_p^*(X)$ respectively). Prove that $Y \in \mathcal{A}_\alpha$. In particular, if $X \overset{u}{\sim} Y$ then $X$ belongs to $\mathcal{A}_\alpha$ if and only if so does $Y$.*

**Solution.** If $M$ is completely metrizable then $M$ is closed in $M$, so $M \in \Sigma_\alpha^0(M)$ which shows that we can apply Problem 184 to conclude that $M \in \mathcal{A}_\alpha$; this proves that the class $\mathcal{A}_\alpha$ contains all completely metrizable spaces.

Suppose that $Z \in \mathcal{A}_\alpha$ and $F$ is a closed subset of $Z$. There exists a complete metric space $M$ such that $Z$ is homeomorphic to some $Z' \in \Sigma_\alpha^0(M)$ (see Problem 184). Then $F$ is homeomorphic to some closed subspace $F'$ of the space $Z'$

and therefore we can find a closed subset $P$ of the space $M$ such that $P \cap Z' = F'$. Since $P \in \Sigma_\alpha^0(M)$, we can apply Fact 1 of T.341 to see that $F' \in \Sigma_\alpha^0(M)$ and hence $F \in \mathcal{A}_\alpha$ by Problem 184. Thus the class $\mathcal{A}_\alpha$ is invariant under closed subspaces; by Problem 185, it is also invariant under finite products.

Now suppose that $M$ is a metrizable space and $M = \bigcup_{n\in\omega} M_n$ while every $M_n$ is closed in $M$ and belongs to the class $\mathcal{A}_\alpha$. If $e : M \to N$ is an arbitrary embedding of $M$ in a metrizable space $N$ then $e(M_n) \in \Sigma_\alpha^0(N)$ for every $n \in \omega$, so it follows from Fact 1 of T.341 that $e(M) = \bigcup_{n\in\omega} e(M_n) \in \Sigma_\alpha^0(N)$. This implies that $M \in \mathcal{A}_\alpha$ and therefore the class $\mathcal{A}_\alpha$ satisfies all conditions we need to apply Problem 195 and conclude that if $X$ belongs to $\mathcal{A}_\alpha$ and there exists a uniformly continuous surjection $\varphi : C_p(X) \to C_p(Y)$ or a uniformly continuous surjection $\xi : C_p^*(X) \to C_p^*(Y)$ then $Y \in \mathcal{A}_\alpha$.

**V.199.** *Prove that every nonempty countable compact space $X$ is homeomorphic to the space $\alpha + 1 = \{\beta : \beta \le \alpha\}$ for some countable ordinal $\alpha$. Here, as usual, the set $\alpha + 1$ is considered with the topology generated by the well-ordering on $\alpha + 1$.*

**Solution.** The expression $Y \simeq Z$ says that the spaces $Y$ and $Z$ are homeomorphic. In every discrete union $Z = \bigoplus\{Z_t : t \in T\}$ we identify every set $Z_t$ with the respective clopen subspace of $Z$. If $\alpha$ and $\beta$ are ordinals such that $\alpha < \beta$ then $(\alpha, \beta) = \{\gamma : \alpha < \gamma < \beta\}$ is the usual interval of ordinals. Given two linearly ordered spaces $L$ and $M$ we say that they are *canonically homeomorphic* if there exists a homeomorphism $f : L \to M$ which is also an order isomorphism between $L$ and $M$.

If $K \ne \emptyset$ is a countable compact space then $K$ is scattered (see SFFS-133); therefore the set $I(K) = \{x \in K : x$ is an isolated point of $K\}$ is nonempty. Let $K[0] = K$; proceeding inductively assume that $\alpha < \omega_1$ and we have a family $\{K[\beta] : \beta < \alpha\}$ of closed subspaces of $K$ with the following properties:

(1) $K[\beta + 1] = K[\beta] \backslash I(K[\beta])$ whenever $0 \le \beta < \beta + 1 < \alpha$;
(2) if $\beta < \alpha$ is a limit ordinal then $K[\beta] = \bigcap\{K[\gamma] : \gamma < \beta\}$.

If $\alpha$ is a limit ordinal, let $K[\alpha] = \bigcap\{K[\beta] : \beta < \alpha\}$. If $\alpha = \beta + 1$ then let $K[\alpha] = K[\beta] \backslash I(K[\beta])$. It is clear that the properties (1)–(2) still hold for the family $\{K[\beta] : \beta \le \alpha\}$, so our inductive procedure can be carried out for all $\alpha < \omega_1$. Observe that if $K[\alpha] \ne \emptyset$ then $I(K[\alpha]) \ne \emptyset$ and hence the set $K[\alpha + 1]$ is strictly smaller than $K[\alpha]$. Since $K$ is countable, there is $\alpha < \omega_1$ for which $K[\alpha] = \emptyset$; let $\beta$ be the minimal such $\alpha$. Observe that every $K[\alpha]$ is compact, so it follows from (2) that $\beta$ is always a non-limit ordinal; let $\gamma$ be the predecessor of $\beta$. The ordinal $\gamma$ is called *the dispersion index* of $K$ and we will denote it by $di(K)$; it is evident that $K[\gamma]$ is a nonempty finite set.

*Fact 1.* For any ordinals $\alpha$ and $\beta$ there exists an ordinal number $\gamma > \alpha$ such that $(\gamma + 1) \backslash (\alpha + 1)$ is canonically homeomorphic to $\beta + 1$ and hence $\gamma + 1 \simeq (\alpha + 1) \oplus (\beta + 1)$.

*Proof.* Let $\gamma' = (\alpha + 1) + (\beta + 1)$; it follows from the definition of the sum of ordinals that the set $B = \gamma' \backslash (\alpha + 1)$ is canonically homeomorphic to $\beta + 1$.

Furthermore, $\alpha + 1 = \{\mu \in \gamma' : \mu < \alpha + 1\}$ and $B = \{\mu \in \gamma' : \mu > \alpha\}$, so $\alpha + 1$ and $B$ are clopen disjoint subspaces of the ordinal $\gamma'$ which shows that $\gamma'$ is homeomorphic to $(\alpha + 1) \oplus B \simeq (\alpha + 1) \oplus (\beta + 1)$. The set $B$ being isomorphic to $\beta + 1$ it has a maximal element $\gamma$; it is evident that $\gamma$ is also the maximal element of $\gamma'$, so $\gamma' = \gamma + 1$ and hence Fact 1 is proved.

*Fact 2.* If a compact space $X_i$ is homeomorphic to an ordinal $\alpha_i + 1 < \omega_1$ for all $i = 0, \ldots, n$ then there exists an ordinal $\alpha < \omega_1$ such that $X = X_0 \oplus \ldots \oplus X_n$ is homeomorphic to $\alpha + 1$.

*Proof.* Proceeding by induction observe that our statement is trivially true for $n = 0$. Assume that it is true for $n \leq k \in \omega$ and consider countable compact spaces $X_0, \ldots, X_k, X_{k+1}$ such that $X_i$ is homeomorphic to some ordinal $\alpha_i + 1$ for every $i \leq k + 1$. By the induction hypothesis the space $Y = X_0 \oplus \ldots \oplus X_k$ is homeomorphic to an ordinal $\beta + 1 < \omega_1$. Apply Fact 1 to find an ordinal $\alpha$ (which is, evidently, countable) such that $\alpha + 1$ is homeomorphic to $(\beta + 1) \oplus (\alpha_{k+1} + 1) \simeq Y \oplus X_{k+1} \simeq X$. Therefore the space $X$ is homeomorphic to $\alpha + 1$ which shows that we completed the inductive step and hence our statement is true for all $n \in \omega$, i.e., Fact 2 is proved.

*Fact 3.* If a nonempty compact space $X_i$ is homeomorphic to a countable ordinal $\alpha_i + 1$ for all $i \in \omega$ then there exists an ordinal $\alpha < \omega_1$ such that the one-point compactification $X$ of the space $\bigoplus_{i \in \omega} X_i$ is homeomorphic to $\alpha + 1$.

*Proof.* Using Fact 1 it easy to construct by induction a sequence $\{\beta_i : i \in \omega\}$ of countable ordinals with the following properties:

(3) $\beta_i < \beta_{i+1}$ for every $i \in \omega$;
(4) $\beta_0 = \alpha_0$ and $B_{i+1} = (\beta_{i+1} + 1) \backslash (\beta_i + 1)$ is homeomorphic to $\alpha_{i+1} + 1$ for all $i \in \omega$.

Let $\alpha = \sup\{\beta_i : i \in \omega\}$; it follows from $(\beta_i, \beta_{i+1} + 1) \cap \alpha = B_i$ that $B_i$ is an open subset of $\alpha$ for all $i \in \mathbb{N}$. If $B_0 = \alpha_0 + 1$ then the family $\{B_i : i \in \omega\}$ is a disjoint open cover of $\alpha$, so every $B_i$ is a clopen subset of $\alpha$. Recalling that $B_i \simeq X_i$ for each $i \in \omega$ we convince ourselves that $\alpha$ is homeomorphic to $\bigoplus_{i \in \omega} X_i$. The compact space $\alpha + 1$ is obtained from $\alpha$ by adding one point; therefore $\alpha + 1$ is the one-point compactification of $\alpha \simeq \bigoplus_{i \in \omega} X_i$. Thus $\alpha + 1$ is homeomorphic to $X$ and hence Fact 3 is proved.

Returning to our solution call a space $X$ *adequate* if it is homeomorphic to $\alpha + 1$ for some countable ordinal $\alpha$. We will prove that every countable compact space $X$ is adequate by induction on the dispersion index of $X$. If $di(X) = 0$ then $X$ is finite, so $X$ is homeomorphic to the ordinal $\alpha + 1$ for $\alpha = |X| - 1$. Now assume that $\mu = di(X) > 0$ and we proved that every nonempty compact countable space $Y$ is adequate whenever $di(Y) < \mu$.

Let $\{a_1, \ldots, a_k\}$ be a faithful enumeration of the set $X[\mu]$; the space $X$ being zero-dimensional, it is easy to find disjoint clopen subsets $O_1, \ldots, O_k$ of the space $X$ such that $a_i \in O_i$ for all $i \leq k$ and $O_1 \cup \ldots \cup O_k = X$. It is immediate that $X$

is homeomorphic to $O_1 \oplus \ldots \oplus O_k$, so it suffices to prove that every $O_i$ is adequate (see Fact 2). It is straightforward that $O_i[\mu] = X[\mu] \cap O_i$, so $O_i[\mu] = \{a_i\}$ for every $i \leq k$. Therefore we can assume, without loss of generality, that $X = O_1$ and hence $X[\mu] = \{a\}$ for some $a \in X$.

Observe that the point $a$ is not isolated in $X$ because $\mu > 0$, so we can construct a family $\{U_n : n \in \omega\}$ of clopen neighborhoods of the point $a$ such that $U_0 = X$, $a \in U_{n+1} \subset U_n$ and $K_n = U_n \setminus U_{n+1} \neq \emptyset$ for all $n \in \omega$. It follows from the inclusion $K_n[\mu] \subset X[\mu] \cap K_n = \emptyset$ that $di(K_n) < \mu$ and hence we can apply the induction hypothesis to see that $K_n$ is adequate for every $n \in \omega$. Every $K_n$ is clopen in $X$ and hence in $X \setminus \{a\}$; since $\{K_n : n \in \omega\}$ is a disjoint cover of $X \setminus \{a\}$, we conclude that $X \setminus \{a\}$ is homeomorphic to $\bigoplus_{n \in \omega} K_n$. Therefore $X$ is homeomorphic to the one-point compactification of the space $\bigoplus_{n \in \omega} K_n$, so we can apply Fact 3 to convince ourselves that $X$ is adequate. This completes our inductive proof and shows that every countable compact space is homeomorphic to $\alpha + 1$ for a countable ordinal $\alpha$.

**V.200.** *Let $X$ and $Y$ be infinite countable compact spaces. Prove that $X \overset{u}{\sim} Y$, i.e., the spaces $C_p(X)$ and $C_p(Y)$ are uniformly homeomorphic.*

**Solution.** The expression $X \simeq Y$ says that the spaces $X$ and $Y$ are homeomorphic; the function $\mathbf{0}_X$ is defined on $X$ by letting $\mathbf{0}_X(x) = 0$ for any $x \in X$. If $x, y \in \mathbb{R}^2$ then $[x, y] = \{(1 - t)x + ty : t \in [0, 1]\}$ is the segment in $\mathbb{R}^2$ that connects the points $x$ and $y$ whereas for any $\delta > 0$ we let $B_\delta = \{(x_0, x_1) \in \mathbb{R}^2 : x_0^2 + x_1^2 < \delta^2\}$. Given ordinals $\alpha, \beta$ with $\alpha \leq \beta$ we will need the intervals $[\alpha, \beta] = \{\gamma : \alpha \leq \gamma \leq \beta\}$ and $[\alpha, \beta) = \{\gamma : \alpha \leq \gamma < \beta\}$. Recall that if we are given a nonempty set $A$ then $\Sigma_*(A) = \{x \in \mathbb{R}^A : |\{a \in A : |x(a)| \geq \varepsilon\}| < \omega$ for any $\varepsilon > 0\}$.

We will often use norms on different spaces using the same symbol $\| \cdot \|$. This won't lead to a confusion because we never use distinct norms on the same space. If $K$ is a compact space then $\|f\| = \max\{|f(x)| : x \in K\}$ for any $f \in C_p(K)$. Given a nonempty set $A$ let $\|x\| = \max\{|x(a)| : a \in A\}$ for any $x \in \Sigma_*(A)$. For any $n \in \mathbb{N}$ and $x = (x_0, \ldots, x_{n-1}) \in \mathbb{R}^n$ let $\|x\| = \max\{|x_i| : i < n\}$. Now, if we have norms on spaces $M$ and $L$ then $\|(p, q)\| = \max\{\|p\|, \|q\|\}$ for any $(p, q) \in M \times L$.

*Fact 1.* If $K$ is a countably infinite compact space then for any point $a \notin K$ the space $K \oplus \{a\}$ is homeomorphic to $K$.

*Proof.* By Problem 199 there exists a countable ordinal $\alpha$ such that $K \simeq [0, \alpha]$, so there is no loss of generality to assume that $K = [0, \alpha]$; the space $K$ being infinite we have $\alpha \geq \omega$ and hence $[0, \omega]$ is a clopen subspace of $K$. Thus $K \simeq [0, \omega] \oplus (K \setminus [0, \omega])$. The space $[0, \omega]$ is a convergent sequence with its limit, so $[0, \omega] \oplus \{a\}$ is homeomorphic to $[0, \omega]$. Thus $K \oplus \{a\} \simeq ([0, \omega] \oplus \{a\}) \oplus (K \setminus [0, \omega]) \simeq [0, \omega] \oplus (K \setminus [0, \omega]) \simeq K$, so Fact 1 is proved.

*Fact 2.* For any $\varepsilon > 0$ there exist functions $u_\varepsilon : \mathbb{R}^2 \to \mathbb{R}$ and $v_\varepsilon : \mathbb{R}^2 \to \mathbb{R}$ with the following properties:

(a) there exists a constant $C(\varepsilon) > 0$ such that $|u_\varepsilon(x) - u_\varepsilon(y)| \leq C(\varepsilon)||x - y||$ and
$|v_\varepsilon(x) - v_\varepsilon(y)| \leq C(\varepsilon)||x - y||$ for any $x, y \in \mathbb{R}^2$;

(b) if $h(x_0, x_1) = (x_0, u_\varepsilon(x_0, x_1))$ for any $(x_0, x_1) \in \mathbb{R}^2$ then $h : \mathbb{R}^2 \to \mathbb{R}^2$ is a
uniform homeomorphism;

(c) if $g(x_0, x_1) = (x_0, v_\varepsilon(x_0, x_1))$ for any $(x_0, x_1) \in \mathbb{R}^2$ then $g : \mathbb{R}^2 \to \mathbb{R}^2$ is a
uniform homeomorphism which is the inverse of $h$;

(d) $u_\varepsilon(t, t) = 0$ for any $t \in \mathbb{R}$;

(e) $(1 + \varepsilon)^{-1}||(x_0, x_1)|| \leq ||h(x_0, x_1)|| \leq ||(x_0, x_1)||$ for any $(x_0, x_1) \in \mathbb{R}^2$.

*Proof.* For any $(x_0, x_1) \in \mathbb{R}^2$ let

$$u_\varepsilon(x_0, x_1) = \begin{cases} x_1, & \text{if } (x_1 + x_0)(x_1 - (1 + \varepsilon)x_0) \geq 0, \\ \frac{1}{2}(x_1 - x_0), & \text{if } |x_1| \leq |x_0|, \\ (1 + \frac{1}{\varepsilon})(x_1 - x_0), & \text{if } (x_1 - x_0)(x_1 - (1 + \varepsilon)x_0) \leq 0. \end{cases}$$

Observe that $u_\varepsilon$ is a union of three continuous functions defined on closed
subspaces of $\mathbb{R}^2$. It is easy to check that any pair of these functions coincide on
the intersections of their domains, so $u_\varepsilon$ is continuous (see Fact 2 of T.354).

Now, if

$$v_\varepsilon(x_0, x_1) = \begin{cases} x_1, & \text{if } (x_1 + x_0)(x_1 - (1 + \varepsilon)x_0) \geq 0, \\ 2x_1 + x_0, & \text{if } (x_1 + x_0)x_1 \leq 0, \\ \frac{\varepsilon}{1+\varepsilon}x_1 + x_0, & \text{if } x_1(x_1 - (1 + \varepsilon)x_0) \leq 0 \end{cases}$$

for any $(x_0, x_1) \in \mathbb{R}^2$ then the same reasoning as in the case of $u_\varepsilon$ shows that the
function $v_\varepsilon : \mathbb{R}^2 \to \mathbb{R}$ is continuous. It is also straightforward that the functions $h$
and $g$ defined in (b) and (c) are mutually inverse, so $h$ is a homeomorphism. The
property (d) is evident from the definition of $u_\varepsilon$.

If $u_\varepsilon(x_0, x_1) = x_1$ or $u_\varepsilon(x_0, x_1) = \frac{1}{2}(x_1 - x_0)$ then it is evident that we have the
inequality $|u_\varepsilon(x_0, x_1)| \leq ||(x_0, x_1)||$.

To prove the same when $u_\varepsilon(x_0, x_1) = (1 + \frac{1}{\varepsilon})(x_1 - x_0)$ assume that $x_1 \geq x_0$;
then $0 \leq x_0 \leq x_1 \leq (1 + \varepsilon)x_0$ and therefore $(1 + \varepsilon)(x_1 - x_0) \leq \varepsilon x_1$ which implies
that $|u_\varepsilon(x_0, x_1)| = \frac{1+\varepsilon}{\varepsilon}(x_1 - x_0) \leq x_1 = ||(x_0, x_1)||$. Now if $x_1 \leq x_0$ is the case,
then $(1 + \varepsilon)x_0 \leq x_1 \leq x_0 \leq 0$ and hence $(1 + \varepsilon)(x_1 - x_0) \geq \varepsilon x_1$; an immediate
consequence is that $|u_\varepsilon(x_0, x_1)| = \frac{1+\varepsilon}{\varepsilon}(x_0 - x_1) \leq -x_1 = ||(x_0, x_1)||$. This proves
that

(1) $|u_\varepsilon(x_0, x_1)| \leq ||(x_0, x_1)||$ and hence $||h(x_0, x_1)|| \leq ||(x_0, x_1)||$ for any point
$(x_0, x_1) \in \mathbb{R}^2$.

To prove the second inequality in (e) we will establish that

(2) $|v_\varepsilon(x_0, x_1)| \leq (1 + \varepsilon)||(x_0, x_1)||$ for any $(x_0, x_1) \in \mathbb{R}^2$.

If $(x_0 + x_1)(x_1 - (1 + \varepsilon)x_0) \geq 0$ then $|v_\varepsilon(x_0, x_1)| = |x_1| = ||(x_0, x_1)||$, so we
have $|v_\varepsilon(x_0, x_1)| \leq (1 + \varepsilon)||(x_0, x_1)||$.

Next assume that $x_1 + x_0 \geq 0$ and $x_1 \leq 0$. Then $||(x_0, x_1)|| = x_0$ and $v_\varepsilon(x_0, x_1) = 2x_1 + x_0 \leq x_0 \leq (1 + \varepsilon)x_0$. Besides, $2x_1 + x_0 \geq -x_0 \geq -x_0(1 + \varepsilon)$ which shows that $-(1 + \varepsilon)x_0 \leq v_\varepsilon(x_0, x_1) \leq (1 + \varepsilon)x_0$ and hence (2) holds in this case.

If $x_1 + x_0 \leq 0$ and $x_1 \geq 0$ then $||(x_0, x_1)|| = |x_0| = -x_0$, so we obtain the inequality $2x_1 + x_0 \leq -x_0 \leq (1 + \varepsilon)(-x_0)$. On the other hand,

$$2x_1 + x_0 \geq x_0 \geq (1 + \varepsilon)x_0 = -(1 + \varepsilon)|x_0|;$$

this demonstrates that $-(1 + \varepsilon)|x_0| \leq v_\varepsilon(x_0, x_1) \leq (1 + \varepsilon)|x_0|$, so we proved (2) for the case when $v_\varepsilon(x_0, x_1) = 2x_1 + x_0$.

Next assume that $0 \leq x_1 \leq (1 + \varepsilon)x_0$; in this case $\frac{\varepsilon}{1+\varepsilon}x_1 \leq \varepsilon x_0$ and therefore $|v_\varepsilon(x_0, x_1)| = \frac{\varepsilon}{1+\varepsilon}x_1 + x_0 \leq (1 + \varepsilon)x_0 \leq (1 + \varepsilon)||(x_0, x_1)||$. Now, if $(1 + \varepsilon)x_0 \leq x_1 \leq 0$ then $\frac{\varepsilon}{1+\varepsilon}x_1 \geq \varepsilon x_0$ and hence $\frac{\varepsilon}{1+\varepsilon}x_1 + x_0 \geq (1 + \varepsilon)x_0$, so we convince ourselves that $|v_\varepsilon(x_0, x_1)| = -(\frac{\varepsilon}{1+\varepsilon}x_1 + x_0) \leq -(1 + \varepsilon)x_0 \leq (1 + \varepsilon)||(x_0, x_1)||$, i.e., (2) is proved.

It follows from (2) that $||g(y)|| \leq (1 + \varepsilon)||(y)||$ for any $y \in \mathbb{R}^2$. In particular, $||h(g(x_0, x_1))|| \leq (1 + \varepsilon)||h(x_0, x_1)||$; recalling that $h$ is the inverse of $g$ we conclude that $||(x_0, x_1)|| \leq (1+\varepsilon)||h(x_0, x_1)||$ for any $(x_0, x_1) \in \mathbb{R}^2$. This, together with (1) shows that (e) is proved.

Given a set $P \subset \mathbb{R}^2$ say that a function $\varphi : P \to \mathbb{R}$ is *Lipschitz* with constant $C > 0$ if we have $|\varphi(x) - \varphi(y)| \leq C||x - y||$ for any $x, y \in P$.

Let $A_0 = \{(x_0, x_1) \in \mathbb{R}^2 : (x_1 + x_0)(x_1 - (1 + \varepsilon)x_0) \geq 0\}$. Given two points $x, y \in A_0$, $x = (x_0, x_1)$, $y = (y_0, y_1)$ we have $|u_\varepsilon(x) - u_\varepsilon(y)| = |x_1 - y_1| \leq ||x - y||$, so $u_\varepsilon$ is Lipschitz with constant $C_0 = 1$ on $A_0$. Let $B_0 = \{(x_0, x_1) \in \mathbb{R}^2 : |x_1| \leq |x_0|\}$; if $x, y \in B_0$ then $|u_\varepsilon(x) - u_\varepsilon(y)| = \frac{1}{2}|(x_1 - y_1) - (x_0 - y_0)| \leq ||x - y||$, so the function $u_\varepsilon$ is also Lipschitz on $B_0$ with constant $C_1 = 1$. If

$$x, y \in D_0 = \{(x_0, x_1) \in \mathbb{R}^2 : (x_1 - x_0)(x_1 - (1 + \varepsilon)x_0) \leq 0\}$$

then

$$|u_\varepsilon(x) - u_\varepsilon(y)| = (1 + \frac{1}{\varepsilon})|(x_1 - y_1) - (x_0 - y_0)| \leq 2(1 + \frac{1}{\varepsilon})||x - y||$$

so the function $u_\varepsilon$ is Lipschitz on $D_0$ with constant $C_2 = 2 + \frac{2}{\varepsilon}$. Since $C_0 = C_1 \leq C_2$, the function $u_\varepsilon$ is Lipschitz with constant $C_2$ on each one of the sets $A_0$, $B_0$ and $D_0$.

Now take two arbitrary points $x, y \in \mathbb{R}^2$, $x = (x_0, x_1)$, $y = (y_0, y_1)$. We have $A_0 \cup B_0 \cup D_0 = \mathbb{R}^2$ and the boundaries of the sets $A_0$, $B_0$ and $D_0$ are contained in the straight lines $l_1, l_2, l_3$ given respectively by the equations $x_1 = (1 + \varepsilon)x_0$, $x_1 = x_0$ and $x_1 = -x_0$. The intersection of the segment $[x, y]$ with the lines $l_1, l_2$ and $l_3$ gives at most three points on $[x, y]$, so we can choose $t_1, t_2, t_3 \in [0, 1]$ such that $t_1 < t_2 < t_3$ and if $a_i = (1 - t_i)x + t_i y$ for all $i \in \{1, 2, 3\}$ then every interval $[x, a_1]$, $[a_1, a_2]$, $[a_2, a_3]$ and $[a_3, y]$ is contained in one of the sets $A_0, B_0, D_0$.

Letting $a_0 = x$ and $a_4 = y$ we obtain the inequality

$$|u_\varepsilon(x) - u_\varepsilon(y)| \le \sum_{i=0}^{3} |u_\varepsilon(a_{i+1}) - u_\varepsilon(a_i)| \le C_2 \sum_{i=0}^{3} ||a_{i+1} - a_i|| \le 4C_2 ||x - y||,$$

which proves that $u_\varepsilon$ is Lipschitz with constant $C = 4C_2$ on $\mathbb{R}^2$. An analogous reasoning shows that $v_\varepsilon$ is also Lipschitz with some constant $C'$ on $\mathbb{R}^2$. Letting $C(\varepsilon) = \max\{C, C'\}$ we conclude that both functions $u_\varepsilon$ and $v_\varepsilon$ are Lipschitz on $\mathbb{R}^2$ with constant $C(\varepsilon)$, so (a) is proved.

To finally establish (b) and (c) it is sufficient to prove that both functions $g$ and $h$ are uniformly continuous. We will do this simultaneously; let $\varphi$ be one of the functions $u_\varepsilon, v_\varepsilon$, so it suffices to show that the function defined by the formula $f(x_0, x_1) = (x_0, \varphi(x_0, x_1))$ for any $(x_0, x_1) \in \mathbb{R}^2$, is uniformly continuous.

For any $x = (x_0, x_1) \in \mathbb{R}^2$ let $|x|_2 = \sqrt{x_0^2 + x_1^2}$ and fix a set $O \in \tau((0,0), \mathbb{R}^2)$; there exists $\delta > 0$ such that $B_\delta \subset O$. If $\mu = \frac{\delta}{\sqrt{1+(C(\varepsilon))^2}}$ then $B_\mu$ is also an open neighborhood of the point $(0, 0)$.

Given any points $x, y \in \mathbb{R}^2$, $x = (x_0, x_1)$, $y = (y_0, y_1)$ suppose that $x - y \in B_\mu$. Then $||x - y|| < \mu$ and hence $|\varphi(x) - \varphi(y)| \le C(\varepsilon)||x - y||$ which implies the inequality $|f(x) - f(y)|_2 \le \sqrt{(x_0 - y_0)^2 + (C(\varepsilon))^2 ||x - y||^2}$. Observe that $(x_0 - y_0)^2 \le ||x - y||^2$, so $|f(x) - f(y)|_2 \le \sqrt{1 + (C(\varepsilon))^2} ||x - y|| < \delta$ which shows that $f(x) - f(y) \in B_\delta \subset O$ and hence the function $f$ is uniformly continuous. Thus $g$ and $h$ are both uniform homeomorphisms, i.e., we verified (b) and (c), so Fact 2 is proved.

*Fact 3.* Given a compact space $K$ and $a \in K$ let $I(a) = \{f \in C_p(K) : f(a) = 0\}$. For any $\varepsilon > 0$ define a map $\varphi_\varepsilon : C_p(K) \to I(a)$ by the formula $\varphi_\varepsilon(f)(x) = u_\varepsilon(f(a), f(x))$ for all $x \in K$ (see Fact 2). If $h_\varepsilon(f) = (f(a), \varphi_\varepsilon(f))$ for every function $f \in C_p(K)$ then the map $h_\varepsilon : C_p(K) \to \mathbb{R} \times I(a)$ is a uniform homeomorphism such that $(1 + \varepsilon)^{-1} ||f|| \le ||h_\varepsilon(f)|| \le ||f||$ for all $f \in C_p(K)$.

*Proof.* The function $\varphi_\varepsilon(f)$ is continuous being the composition of continuous functions; observe also that $\varphi_\varepsilon(f)(a) = u_\varepsilon(f(a), f(a)) = 0$ for any $f \in C_p(K)$, so $\varphi_\varepsilon$ indeed, maps $C_p(K)$ in $I(a)$. For any $x \in K$ the map $f \to \varphi_\varepsilon(f)(x) = u_\varepsilon(f(a), f(x))$ is continuous being the composition of continuous maps, so $\varphi_\varepsilon$ is continuous by TFS-102; therefore $h_\varepsilon$ is also a continuous map.

For any point $(t, f) \in \mathbb{R} \times I(a)$ let $\mu_\varepsilon(t, f)(x) = v_\varepsilon(t, f(x))$ for any $x \in K$. Then the map $\mu_\varepsilon : \mathbb{R} \times I(a) \to C_p(K)$ is continuous: this is proved in the same way as for $h_\varepsilon$. It follows from (b) and (c) of Fact 2 that

(3)  $u_\varepsilon(x_0, v_\varepsilon(x_0, x_1)) = x_1$ and $v_\varepsilon(x_0, u_\varepsilon(x_0, x_1)) = x_1$ for any $(x_0, x_1) \in \mathbb{R}^2$.

Fix $(t, g) \in \mathbb{R} \times I(a)$ and let $f = \mu_\varepsilon(t, g)$. Observe first that

$$f(a) = v_\varepsilon(t, g(a)) = v_\varepsilon(t, 0) = v_\varepsilon(t, u_\varepsilon(t, t)) = t$$

(we used (3) and (d) of Fact 2). For every $x \in K$ we have $\varphi_\varepsilon(f)(x) = u_\varepsilon(t, f(x))$. Recalling that $f(x) = v_\varepsilon(t, g(x))$ and applying (3) again we convince ourselves that $\varphi_\varepsilon(f)(x) = g(x)$ and hence $\varphi_\varepsilon(f) = g$. Thus $h_\varepsilon(f) = (t, g)$ which shows that the map $h_\varepsilon$ is surjective and $h_\varepsilon \circ \mu_\varepsilon$ is the identity on $\mathbb{R} \times I(a)$.

Now if $f \in C_p(K)$ and $g = \varphi_\varepsilon(f)$ then $\mu_\varepsilon(f(a), g)(x) = v_\varepsilon(f(a), g(x))$; recalling that $g(x) = u_\varepsilon(f(a), f(x))$ we can apply (3) once more to see that we have the equality $\mu_\varepsilon(f(a), g)(x) = f(x)$ for every $x \in K$, i.e., $\mu_\varepsilon(f(a), g) = f$. This proves that $h_\varepsilon$ and $\mu_\varepsilon$ are mutually inverse homeomorphisms.

Once more fix any function $f \in C_p(K)$; then $h_\varepsilon(f) = (f(a), \varphi_\varepsilon(f))$ and, evidently, $|f(a)| \leq \|f\|$. Besides, it follows from (e) of Fact 2 that we have the inequalities

(4) $(1 + \varepsilon)^{-1}\|(f(a), f(x))\| \leq \|(f(a), u_\varepsilon(f(a), f(x)))\| \leq \|(f(a), f(x))\|$ for any point $x \in K$.

An immediate consequence of (4) is that $(1 + \varepsilon)^{-1}\|f\| \leq \|h_\varepsilon(f)\| \leq \|f\|$ for any $f \in C_p(K)$.

Given a finite set $P \subset K$ and $\delta > 0$ let $[P, \delta] = \{f \in C_p(K) : |f(x)| < \delta$ for any $x \in P\}$ and $\langle P, \delta \rangle = [P, \delta] \cap I(a)$. All possible sets $[P, \delta]$ (or $\langle P, \delta \rangle$ respectively) constitute a local base of $C_p(K)$ (of $I(a)$ respectively) at the point $\mathbf{0}_K$. To show that the map $h_\varepsilon$ is uniformly continuous take any open neighborhood $W$ of the point $(0, \mathbf{0}_K)$ in the space $\mathbb{R} \times I(a)$. Choose $\delta > 0$ and a finite set $P \subset K$ such that $(-\delta, \delta) \times \langle P, \delta \rangle \subset W$.

If $\delta' = \backslash n\{\delta, \frac{\delta}{C(\varepsilon)}\}$ and $P' = P \cup \{a\}$ then $U = [P', \delta']$ is an open neighborhood of the point $\mathbf{0}_K$ in $C_p(K)$; take any functions $f, g \in C_p(K)$ such that $f - g \in U$ and let $h_\varepsilon(f) = (t, f_0)$, $h_\varepsilon(g) = (s, g_0)$. Then $t = f(a)$ and $s = g(a)$, so $|t - s| = |f(a) - g(a)| < \delta$. Given any $x \in P$ we have $f_0(x) = u_\varepsilon(t, f(x))$ and $g_0(x) = u_\varepsilon(s, g(x))$; as a consequence,

$$|f_0(x) - g_0(x)| \leq C(\varepsilon)\|(t - s, f(x) - g(x))\| < C(\varepsilon)\delta' = \delta$$

(here we applied the property (a) of Fact 2), so $f_0 - g_0 \in \langle P, \delta \rangle$ and therefore $h_\varepsilon(f) - h_\varepsilon(g) \in (-\delta, \delta) \times \langle P, \delta \rangle \subset W$. This proves that the map $h_\varepsilon$ is uniformly continuous.

Finally, take any open neighborhood $W$ of the point $\mathbf{0}_K$ in the space $C_p(K)$; choose $\delta > 0$ and a finite set $P \subset K$ such that $[P, \delta] \subset W$. If $\delta' = \backslash n\{\delta, \frac{\delta}{C(\varepsilon)}\}$ then $U = \langle P, \delta' \rangle$ is an open neighborhood of the point $\mathbf{0}_K$ in $I(a)$, so $V = (-\delta', \delta') \times U$ is an open neighborhood of $(0, \mathbf{0}_K)$ is $\mathbb{R} \times I(a)$; take any points $(t, f), (s, g) \in \mathbb{R} \times I(a)$ such that $(t, f) - (s, g) \in V$ and let $\mu_\varepsilon(t, f) = f_0$, $\mu_\varepsilon(s, g) = g_0$. Given any $x \in P$ we have $f_0(x) = v_\varepsilon(t, f(x))$ and $g_0(x) = v_\varepsilon(s, g(x))$; as a consequence, $|f_0(x) - g_0(x)| \leq C(\varepsilon)\|(t - s, f(x) - g(x))\| < C(\varepsilon)\delta' = \delta$ (here we applied the property (a) of Fact 2), so $f_0 - g_0 \in [P, \delta]$ and therefore $\mu_\varepsilon(f) - \mu_\varepsilon(g) \in [P, \delta] \subset W$. This demonstrates that the map $\mu_\varepsilon$ is also uniformly continuous and shows that Fact 3 is proved.

Returning to our solution observe that any countably infinite compact space is homeomorphic to a space $[0, \alpha]$ for some countable infinite ordinal $\alpha$ (see Problem 199). Therefore it suffices to prove that $C_p([0, \alpha])$ is uniformly homeomorphic to $\Sigma_*(\omega)$ for any countable infinite ordinal $\alpha$. To avoid confusion, from now on any ordinal $\alpha$ *is considered to be a point only*; if we want to see it as a set then we write $[0, \alpha)$. Evidently, $[0, \alpha]$ is the set given by $\alpha + 1$. Applying transfinite induction we will prove that, for any countable ordinal $\alpha \geq \omega$ and any $\varepsilon > 0$,

(5) there exists is a uniform homeomorphism $\varphi : C_p([0, \alpha]) \to \Sigma_*([0, \omega))$ such that $(1 + \varepsilon)^{-1}||f|| \leq ||\varphi(f)|| \leq ||f||$ for every $f \in C_p([0, \alpha])$.

To establish (5) we will often use the following assertion; its proof is straightforward and can be left to the reader.

(6) for any $(t, x) \in \mathbb{R} \times \Sigma_*([0, \omega))$ let $e(t, x)(0) = t$ and $e(t, x)(n + 1) = x(n)$ for every $n < \omega$. Then $e : \mathbb{R} \times \Sigma_*([0, \omega)) \to \Sigma_*([0, \omega))$ is a uniform homeomorphism such that $||e(t, x)|| = ||(t, x)||$ for any $(t, x) \in \mathbb{R} \times \Sigma_*([0, \omega))$.

Our first step is to prove (5) for $\alpha = \omega$; let $I = \{f \in C_p([0, \omega]) : f(\omega) = 0\}$. The restriction map $\pi : I \to C_p([0, \omega))$ is continuous and $\pi(I) = \Sigma_*([0, \omega))$, so we can and will consider that $\pi : I \to \Sigma_*([0, \omega))$. Then $\pi$ is a uniform homeomorphism such that $||\pi(f)|| = ||f||$ for any $f \in I$. Apply Fact 3 to see that there exists a uniform homeomorphism $\varphi_0 : C_p([0, \omega]) \to \mathbb{R} \times I$ such that $(1 + \varepsilon)^{-1}||f|| \leq ||\varphi_0(f)|| \leq ||f||$ for any $f \in C_p([0, \omega])$.

Let $\pi_0(t, f) = (t, \pi(f))$ for every $(t, f) \in \mathbb{R} \times I$. Then $\pi_0 : \mathbb{R} \times I \to \mathbb{R} \times \Sigma_*([0, \omega))$ is a uniform homeomorphism such that $||\pi_0(z)|| = ||z||$ for any $z \in \mathbb{R} \times I$. It follows from (6) that $\varphi = e \circ \pi_0 \circ \varphi_0 : C_p([0, \omega]) \to \Sigma_*([0, \omega))$ is a uniform homeomorphism. Given $f \in C_p([0, \omega])$ the equality $||e(\pi_0(\varphi_0(f)))|| = ||\pi_0(\varphi_0(f))|| = ||\varphi_0(f)||$ shows that $(1 + \varepsilon)^{-1}||f|| \leq ||\varphi(f)|| \leq ||f||$ for any $f \in C_p([0, \omega])$, i.e., (5) is proved for $\alpha = \omega$.

Now assume that $\beta < \omega_1$ and (5) is proved for all $\alpha < \beta$. If $\beta = \beta_0 + 1$ then the space $[0, \beta] = [0, \beta_0] \cup \{\beta\}$ is obtained from $[0, \beta_0]$ by adding an isolated point, so $[0, \beta] \simeq [0, \beta_0]$ by Fact 1; the property (5) being true for $[0, \beta_0]$ by the induction hypothesis, it is also true for $[0, \beta]$.

If $\beta$ is a limit ordinal then choose a sequence $\{\alpha_i : i < \omega\} \subset [0, \beta)$ such that $\alpha_i < \alpha_{i+1}$ for each $i < \omega$ and $\sup\{\alpha_i : i < \omega\} = \beta$; let $H_0 = [0, \alpha_0]$ and $H_{i+1} = [\alpha_i + 1, \alpha_{i+1}]$ for all $i < \omega$. It is easy to find a disjoint family $\{S_i : i < \omega\}$ of subsets of $[0, \omega)$ such that $\bigcup_{i<\omega} S_i = [0, \omega)$ and there exists a bijection between $H_i$ and $S_i$ (we won't need this bijection but we must assure that $H_i$ and $S_i$ are always of the same cardinality) for every $i < \omega$.

For any $i < \omega$ and $f \in \Sigma_*(S_i)$ let $||f||_i = \max\{f(n) : n \in S_i\}$. Observe that $\Sigma_*(S_i) = \mathbb{R}^{S_i}$ if $S_i$ is finite. Every space $H_i$ is compact and countable so there is $\gamma_i < \beta$ such that $H_i \simeq [0, \gamma_i]$ (see Problem 199). Choose a positive number $\delta$ such that $(1 + \delta)^{-2} > (1 + \varepsilon)^{-1}$; by the induction hypothesis, if $H_i$ is infinite, then there exists a uniform homeomorphism $\varphi_i : C_p(H_i) \to \Sigma_*(S_i)$ such that

(7) $(1 + \delta)^{-1}||f|| \leq ||\varphi_i(f)||_i \leq ||f||$ for every $f \in C_p(H_i)$.

If $H_i$ is finite then the uniform homeomorphism $\varphi_i$ can even be chosen in such a way that $||\varphi_i(f)||_i = ||f||$ for every $f \in C_p(H_i) = \mathbb{R}^{H_i}$, so (7) still holds for $\varphi_i$.

If we have a function $f_i \in \mathbb{R}^{S_i}$ for every $i < \omega$ then there is a unique $f \in \mathbb{R}^{[0,\omega)}$ such that $f|S_i = f_i$ for all $i < \omega$; this function $f$ will be referred to as $\bigcup_{i<\omega} f_i$.

Let $I = \{f \in C_p([0,\beta]) : f(\beta) = 0\}$; then $\pi_i : I \to C_p(H_i)$ is the restriction map for every $i < \omega$. For every $f \in I$ let $\mu(f) = \bigcup_{i<\omega} \varphi_i(\pi_i(f))$; then $\mu : I \to \mathbb{R}^{[0,\omega)}$ is a well-defined map. The sets $H_i$ are eventually contained in any given neighborhood of $\beta$, so it follows from $f(\beta) = 0$ that $||\pi_i(f)|| \to 0$ for every $f \in I$; it follows from (7) that $||\varphi_i(\pi_i(f))||_i \to 0$, so $\mu$ actually maps $I$ in $\Sigma_*([0,\omega))$. Let us show that $\mu : I \to \Sigma_*([0,\omega))$ is a uniform homeomorphism.

If $f, g \in I$ and $f \neq g$ then it follows from $\bigcup_{i<\omega} H_i = [0,\beta)$ that $f|H_i \neq g|H_i$ for some $i < \omega$. Then $\varphi_i(\pi_i(f)) \neq \varphi_i(\pi_i(g))$ and hence $\mu(f) \neq \mu(g)$, i.e., $\mu$ is injective. Given any $x \in \Sigma_*([0,\omega))$ let $h_i = x|S_i$; then $h_i \in \Sigma_*(S_i)$, so we can find $f_i \in C_p(H_i)$ such that $\varphi_i(f_i) = h_i$ for each $i < \omega$. It follows from (7) that $||f_i|| \leq (1+\delta)||h_i||_i$ for all $i < \omega$, so it follows from $||h_i||_i \to 0$ that $||f_i|| \to 0$. If we let $f(\beta) = 0$ and $f|H_i = f_i$ for all $i < \omega$ then it is straightforward that $f \in I$ and $\mu(f) = x$; this proves that $\mu$ is a bijection.

To see that $\mu$ is uniformly continuous let $\mathbf{0}$ be the zero point of $\Sigma_*([0,\omega))$ and fix a set $O \in \tau(\mathbf{0}, \Sigma_*([0,\omega)))$. There are $r > 0$ and $m < \omega$ such the set $O' = \{x \in \Sigma_*([0,\omega)) : |x(i)| < r$ for all $i \leq m\}$ is contained in $O$. Take $k < \omega$ for which $A = \{0, \ldots, m\} \subset S_0 \cup \ldots \cup S_k$. For any $i \leq k$, by uniform continuity of $\varphi_i$ there is a finite set $B_i \subset H_i$ and $s_i > 0$ such that if $f, g \in C_p(H_i)$ and $|f(\alpha) - g(\alpha)| < s_i$ for all $\alpha \in B_i$ then $|\varphi_i(f)(l) - \varphi_i(g)(l)| < r$ for all $l \in A \cap S_i$. Let $s = \backslash n\{s_i : i \leq k\}$ and $B = \bigcup\{B_i : i \leq k\}$. It is immediate that if $f, g \in I$ and $|f(\alpha) - g(\alpha)| < s$ for all $\alpha \in B$ then $|\mu(f)(i) - \mu(g)(i)| < r$ for all $i \leq m$ and hence $\mu(f) - \mu(g) \in O' \subset O$. In other words, the open set $U = \{f \in I : |f(\alpha)| < s$ for all $\alpha \in B\}$ witnesses that $\mu$ is uniformly continuous.

To establish that $\mu^{-1}$ is uniformly continuous take an open neighborhood $O$ of the zero function of $I$. Pick $s > 0$ and a finite set $B \subset [0,\beta)$ such that the set $O' = \{f \in I : |f(\alpha)| < s$ for every $\alpha \in B\}$ is contained in $O$. There is $m < \omega$ such that $B \subset H_0 \cup \ldots \cup H_m$. For any $i \leq k$, by uniform continuity of $\varphi_i^{-1}$ there exists a finite set $A_i \subset S_i$ and $r_i > 0$ such that if $f, g \in \Sigma_*(S_i)$ and $|f(l) - g(l)| < r_i$ for all $l \in A_i$ then $|\varphi_i^{-1}(f)(\alpha) - \varphi_i^{-1}(g)(\alpha)| < r$ for all $\alpha \in B \cap H_i$. Let $r = \backslash n\{r_i : i \leq m\}$ and $A = \bigcup\{A_i : i \leq m\}$. It is immediate that if $f, g \in \Sigma_*([0,\omega))$ and $|f(l) - g(l)| < s$ for all $l \in A$ then $|\mu^{-1}(f)(\alpha) - \mu^{-1}(g)(\alpha)| < s$ for all $\alpha \in B$ and hence $\mu^{-1}(f) - \mu^{-1}(g) \in O' \subset O$. In other words, the open set $U = \{x \in \Sigma_*([0,\omega)) : |x(l)| < r$ for all $l \in A\}$ witnesses that $\mu^{-1}$ is uniformly continuous.

Given $f \in I$ and $i < \omega$ we have $||\pi_i(f)|| \leq ||f||$ and hence $||\varphi_i(\pi_i(f))||_i \leq ||f||$ by (7). An immediate consequence is that $||\mu(f)|| \leq ||f||$. The family $\{H_i : i \in \omega\}$ covers $[0,\beta)$, so there exists $m < \omega$ such that $||f|| = ||\pi_m(f)||$. Apply (7) once more to see that $||\varphi_m(\pi_m(f))||_m \geq (1+\delta)^{-1}||\pi_m(f)|| = (1+\delta)^{-1}||f||$. Thus $||\mu(f)|| \geq ||\varphi_m(\pi_m(f))||_m \geq (1+\delta)^{-1}||f||$, so we proved that

(8) $(1+\delta)^{-1}||f|| \leq ||\mu(f)|| \leq ||f||$ for every $f \in I$.

Apply Fact 3 to find a uniform homeomorphism $\nu : C_p([0,\beta]) \to \mathbb{R} \times I$ such that $(1+\delta)^{-1}||f|| \leq ||\nu(f)|| \leq ||f||$ for every $f \in C_p([0,\beta])$. For every $(t, f) \in \mathbb{R} \times I$

let $\xi(t, f) = (t, \mu(f))$. The reader who understood this proof up to the present point, will have no difficulty to check, using (8), that $\xi : \mathbb{R} \times I \rightarrow \mathbb{R} \times \Sigma_*([0, \omega))$ is a uniform homeomorphism such that $(1 + \delta)^{-1}||z|| \le ||\xi(z)|| \le ||z||$ for every $z \in \mathbb{R} \times I$. It follows from (6) that the map $\varphi = e \circ \xi \circ \upsilon : C_p([0, \beta]) \rightarrow \Sigma_*([0, \omega))$ is a uniform homeomorphism such that $(1 + \delta)^{-2}||f|| \le ||\varphi(f)|| \le ||f||$. Recalling that $(1 + \varepsilon)^{-1} < (1 + \delta)^{-2}$ we conclude that $(1 + \varepsilon)^{-1}||f|| \le ||\varphi(f)|| \le ||f||$ for any $f \in C_p([0, \beta])$. Thus we checked (5) for all $\alpha \le \beta$, so our inductive procedure guarantees that (5) holds for all infinite ordinals $\alpha < \omega_1$.

This implies that, for any infinite ordinals $\alpha, \beta < \omega_1$ the spaces $C_p([0, \alpha])$ and $C_p([0, \beta])$ are uniformly homeomorphic; we already saw that this is the same as saying that $C_p(X)$ and $C_p(Y)$ are uniformly homeomorphic for any countably infinite compact spaces $X$ and $Y$, so our solution is complete.

**V.201.** *Prove that the topology of any linear topological $T_0$-space is Tychonoff.*

**Solution.** Fix a linear topological $T_0$-space $L$. Given any $a \in L$ define a map $\varphi_a : L \rightarrow L$ by the formula $\varphi_a(x) = x + a$ for any $x \in L$. The operations in $L$ being continuous, the map $\varphi_a$ is continuous for any $a \in L$. Since $\varphi_{-a}$ is the inverse of $\varphi_a$, every $\varphi_a$ is a homeomorphism. If $i(x) = -x$ for every $x \in L$ then the map $i : L \rightarrow L$ is also a homeomorphism inverse to itself.

Now take any point $a \in L\backslash\{0\}$; there exists an open set $U \subset L$ such that $P = U \cap \{a, 0\}$ is a singleton. If $P = \{a\}$ then $U$ is a neighborhood of $a$ missing $0$. If $P = \{0\}$ then let $V = U \cap i(U)$; since $i(0) = 0$, the set $V$ is an open neighborhood of $0$. Therefore $W = \varphi_a(V)$ is an open neighborhood of $a$. If $0 \in W$ then there is $x \in V$ with $x + a = 0$, i.e., $x = -a \in V$. However, $i(V) = V$, so $a = i(-a) \in V \subset U$ which is a contradiction. Thus $W$ is an open neighborhood of $a$ which does not contain $0$.

We proved that every $a \in L\backslash\{0\}$ has a neighborhood which misses $0$, so the set $\{0\}$ is closed in $L$. For any $x \in L$ the homeomorphism $\varphi_x$ maps $\{0\}$ onto $\{x\}$, so $\{x\}$ is closed in $L$ and hence $L$ is a $T_1$-space.

For any $U \in \tau(0, L)$ let $[U] = \{(x, y) \in L \times L : x - y \in U\}$ and consider the family $\mathcal{B} = \{[U] : U \in \tau(0, L)\}$. It follows from $0 + 0 = 0$ and continuity of addition in $L$ that

(1) for any $U \in \tau(0, L)$ there exists $V \in \tau(0, L)$ such that $V + V \subset U$.

Call a set $U \in \tau(0, L)$ *symmetric* if $-U = i(U) = U$. It is easy to check that the set $U \cap (-U)$ is symmetric for any $U \in \tau(0, L)$, so we have

(2) for any $U \in \tau(0, L)$ there exists a symmetric set $V \in \tau(0, L)$ with $V \subset U$.

For any $A, B \subset L \times L$ we will need the sets $A^{-1} = \{(x, y) : (y, x) \in A\}$ and $A \circ B = \{(x, y) : \text{there is } z \in L \text{ such that } (x, z) \in B \text{ and } (z, y) \in A\}$; besides, we let $A(x) = \{y \in L : (x, y) \in A\}$ for any $x \in L$.

If $x$ and $y$ are distinct points of $L$ then $z = x - y \ne 0$, so $U = L\backslash\{z\}$ is an open neighborhood of $0$ such that $(x, y) \notin [U]$ and hence $(x, y) \notin \bigcap \mathcal{B}$. It is evident that $\Delta = \{(x, x) : x \in L\} \subset \bigcap \mathcal{B}$, so we proved that $\bigcap \mathcal{B} = \Delta$. Given any $U \in \tau(0, L)$ use (2) to find a symmetric $V \in \tau(0, L)$ with $V \subset U$. Then

$[V]^{-1} = [-V] = [V] \subset [U]$, so we proved that for any $B \in \mathcal{B}$ there is $C \in \mathcal{B}$ such that $C^{-1} \subset B$. Given $U_0, U_1 \in \tau(\mathbf{0}, L)$ if $U = U_0 \cap U_1$ then $[U] \subset [U_0] \cap [U_1]$ which shows that, for any $A, B \in \mathcal{B}$ there exists $C \in \mathcal{B}$ such that $C \subset A \cap B$.

Now, if $U \in \tau(\mathbf{0}, L)$ then apply (1) and (2) to find a set $V \in \tau(\mathbf{0}, L)$ such that $V + V \subset U$. If $(x, y) \in [V] \circ [V]$ then there is $z \in L$ such that $x - z \in V$ and $z - y \in V$ which shows that $x - y \in V + V \subset U$, so $[V] \circ [V] \subset [U]$. An immediate consequence is that, for any $B \in \mathcal{B}$ there exists $C \in \mathcal{B}$ such that $C \circ C \subset B$. We finally checked that all conditions of Problem 101 are satisfied for the family $\mathcal{B}$, so there exists a unique uniformity $\mathcal{U}$ on the set $L$ such that $\mathcal{B}$ is a base of $\mathcal{U}$; let $\tau$ be the topology generated by the uniformity $\mathcal{U}$.

Fix a set $U \in \tau(L)$ and a point $x \in U$. The set $U' = -x + U = \varphi_{-x}(U)$ is an open neighborhood of $\mathbf{0}$, so we can apply (2) to find a symmetric $V \in \tau(\mathbf{0}, L)$ such that $V \subset U'$. It is clear that $x \in x + V \subset U$; if $y \in [V](x)$ then $x - y \in V$, so $y - x \in V$ because the set $V$ is symmetric. This shows that $y \in x + V \subset U$, so we proved that $[V](x) \subset U$, i.e., for any $x \in U$ there exists $B \in \mathcal{B}$ (and hence $B \in \mathcal{U}$) with $B(x) \subset U$. Therefore the set $U$ belongs to $\tau$, i.e., $\tau(L) \subset \tau$.

Now take any set $U \in \tau$ and $x \in U$. There exists $A \in \mathcal{U}$ such that $A(x) \subset U$; the family $\mathcal{B}$ being a base of $\mathcal{U}$ we can find $V \in \tau(\mathbf{0}, L)$ such that $[V] \subset A$. Apply (2) to find a symmetric $W \in \tau(\mathbf{0}, L)$ for which $W \subset V$. If $y \in x + W$ then $y = x' + w$ for some $w \in W$. The set $W$ being symmetric we have $-w \in W$ and hence $x - y = -w \in W$, i.e., $y \in [W](x) \subset V(x) \subset U$. Thus $\varphi_x(W) = x + W \subset [W](x) \subset U$ which shows that, for every point $x \in U$, there is $G \in \tau(x, L)$ with $G \subset U$. Therefore $U \in \tau(L)$; this proves that $\tau \subset \tau(L)$ and hence $\tau(L) = \tau$, i.e., the topology of $L$ is generated by the uniformity $\mathcal{U}$, so $\tau(L)$ is Tychonoff by Problem 110.

**V.202.** *Let $L$ be a linear topological Tychonoff space. Prove that, for any local base $\mathcal{B}$ of the space $L$ at $\mathbf{0}$, the following properties hold:*

*(1) for any $U, V \in \mathcal{B}$, there is $W \in \mathcal{B}$ such that $W \subset U \cap V$;*
*(2) every $B \in \mathcal{B}$ is an absorbing set and $\bigcap \mathcal{B} = \{\mathbf{0}\}$;*
*(3) for any $U \in \mathcal{B}$, there exists $V \in \mathcal{B}$ such that $V + V \subset U$;*
*(4) for any $U \in \mathcal{B}$ and $x \in U$, there exists $V \in \mathcal{B}$ such that $x + V \subset U$;*
*(5) for any $U \in \mathcal{B}$ and $\varepsilon > 0$ there is $V \in \mathcal{B}$ such that $\lambda \cdot V \subset U$ for any $\lambda \in (-\varepsilon, \varepsilon)$.*

*Prove that, if $L$ is a linear space without topology and $\mathcal{B}$ is a family of subsets of $L$ which has the properties (1)–(5) then there exists a unique Tychonoff topology $\tau$ on $L$ such that $(L, \tau)$ is a linear topological space and $\mathcal{B}$ is a local base of $\tau$ at $\mathbf{0}$.*

**Solution.** Suppose that $L$ is a linear topological space and $\tau(L)$ is Tychonoff. The property (1) holds because, for any $U, V \in \mathcal{B}$ the set $U \cap V$ is a neighborhood of $\mathbf{0}$, so there is $W \in \mathcal{B}$ with $W \subset U \cap V$. It is clear that $\mathbf{0} \in \bigcap \mathcal{B}$; if $x \in L \backslash \{\mathbf{0}\}$ then $L \backslash \{x\}$ is an open neighborhood of $\mathbf{0}$, so there is $U \in \mathcal{B}$ such that $U \subset L \backslash \{x\}$ and hence $x \notin \bigcap \mathcal{B}$. This proves that $\bigcap \mathcal{B} = \{\mathbf{0}\}$. Consider the map $m : \mathbb{R} \times L \to L$ defined by $m(t, x) = tx$ for any $t \in \mathbb{R}$ and $x \in L$. If $B \in \mathcal{B}$ and $x \in L$ then we have $m(0, x) = \mathbf{0} \in B$, so, by continuity of $m$, there exist $\delta > 0$ and $W \in \tau(x, L)$ such that $m((-\delta, \delta), W) \subset B$. In particular, $m((-\delta, \delta), \{x\}) \subset B$ which says precisely that $tx \in B$ for any $t \in (-\delta, \delta)$. Therefore the set $B$ is absorbing and hence we proved the property (2).

Now fix any $U \in \mathcal{B}$; since $\mathbf{0} + \mathbf{0} = \mathbf{0}$, it follows from continuity of the addition in $L$ that there exists $W \in \tau(\mathbf{0}, L)$ with $W + W \subset U$. Take $V \in \mathcal{B}$ with $V \subset W$; then $V + V \subset U$, so (3) is proved.

Given $U \in \mathcal{B}$ and $x \in U$ apply again continuity of addition together with the equality $x + \mathbf{0} = x$ to find $G \in \tau(x, L)$ and $W \in \tau(\mathbf{0}, L)$ such that $G + W \subset U$. Then $x + W \subset G + W \subset U$; if $V \in \mathcal{B}$ and $V \subset W$ then $x + V \subset U$, so we settled (4).

Next take any $U \in \mathcal{B}$ and $\varepsilon > 0$. The map $m$ is continuous so it follows from $m(0, \mathbf{0}) = \mathbf{0} \in U$ that there exist $W \in \tau(\mathbf{0}, L)$ and $G \in \tau(0, \mathbb{R})$ for which $m(G, W) \subset U$. Choose $\delta > 0$ with $(-\delta, \delta) \subset G$ and let $V = \frac{\delta}{\varepsilon}W$. If $x \in V$ and $|\lambda| < \varepsilon$ then there is $w \in W$ such that $x = \frac{\delta}{\varepsilon}w$; let $\lambda' = \lambda \cdot \frac{\delta}{\varepsilon}$. Then $\lambda' \in (-\delta, \delta)$ and $\lambda x = \lambda' w = m(\lambda', w) \in m(G, W) \subset U$. This shows that $\lambda x \in U$ for any $x \in V$ and $\lambda \in (-\varepsilon, \varepsilon)$, so (5) holds as well and hence we checked (1)–(5) for any linear topological Tychonoff space.

To prove the second part of our statement assume that $L$ is a linear space and we have a family $\mathcal{B}$ of subsets of $L$ with the properties (1)–(5). Consider the family $\tau = \{\emptyset\} \cup \{U : U \subset L$ and for any $x \in U$ there is $V \in \mathcal{B}$ such that $x + V \subset U\}$. It is immediate that $\emptyset \in \tau$ and $L \in \tau$. Given $U_0, U_1 \in \tau$ suppose that $U = U_0 \cap U_1 \neq \emptyset$ and take any $x \in U$. There exist $V_0, V_1 \in \mathcal{B}$ such that $x + V_0 \subset U_0$ and $x + V_1 \subset U_1$. The property (1) enables us to find $W \in \mathcal{B}$ with $W \subset V_0 \cap V_1$. Then $x + W \subset U_0 \cap U_1 = U$, so $U \in \tau$; this shows that the intersection of any two elements of $\tau$ belongs to $\tau$.

Given any $\mathcal{U} \subset \tau$ if $G = \bigcup \mathcal{U} = \emptyset$ then $G \in \tau$. If $G \neq \emptyset$ then fix a point $x \in G$; there is $U \in \mathcal{U}$ with $x \in U$, so we can find $V \in \mathcal{B}$ such that $x + V \subset U$. It follows from $U \subset G$ that $x + V \subset G$, so $G \in \tau$ and hence $\tau$ is a topology on $L$.

Fix any point $x \in L$; if $B \in \mathcal{B}$ and $y \in x + B$ then $y - x \in B$, so we can apply (4) to find a set $C \in \mathcal{B}$ such that $y - x + C \subset B$ and hence $y + C \subset x + B$, i.e., $x + B \in \tau$. Now, if $x \in U \in \tau$ then, by the choice of $\tau$ there is $W \in \mathcal{B}$ with $x + W \subset U$. This proves that

(a) for any $x \in L$ the family $\{x + B : B \in \mathcal{B}\}$ is a local base at the point $x$ in the space $(L, \tau)$.

As before, we let $m(t, x) = tx$ for any $t \in \mathbb{R}$ and $x \in L$. To prove that the operation $m : \mathbb{R} \times L \to L$ is continuous fix a point $z = (t_0, x_0) \in \mathbb{R} \times L$ and an open neighborhood $G$ of $t_0 x_0$ in the space $(L, \tau)$. The property (a) shows that there is $U \in \mathcal{B}$ such that $t_0 x_0 + U \subset G$. Using the property (3) of the family $\mathcal{B}$ it is easy to find a set $W \in \mathcal{B}$ such that $W + W + W \subset U$. The set $W$ being absorbing there is $\delta > 0$ such that $M_0 = m((-\delta, \delta), \{x_0\}) \subset W$. Applying the property (5) for $\varepsilon = \delta + |t_0| + 1$ we can find a set $V \in \mathcal{B}$ such that $M_1 = m((-\varepsilon, \varepsilon), V) \subset W$. The set $H = (t_0 - \delta, t_0 + \delta) \times (x_0 + V)$ is an open neighborhood of $(t_0, x_0)$ in $\mathbb{R} \times L$. If $(t, x) \in H$ then we can pick $v \in V$ and $s \in (-\delta, \delta)$ such that $t = t_0 + s$ and $x = x_0 + v$.

Observe that $tx = t_0x_0 + t_0v + sx_0 + sv$ while $t_0v \in M_1$, $sv \in M_1$ and $sx_0 \in M_0 \subset W$. As a consequence, $tx \in t_0x_0 + W + W + W \subset t_0x_0 + U \subset G$. This shows that $m(H) \subset G$, so the set $H$ witnesses continuity of the scalar multiplication $m$ at the point $(t_0, x_0)$.

Let $a(x, y) = x + y$ for any $x, y \in L$. To see that the addition map $a : L \times L \to L$ is continuous fix a point $(x_0, y_0) \in L \times L$ and $G \in \tau(x_0 + y_0, L)$. The property (a) shows that there exists $U \in \mathcal{B}$ such that $x_0 + y_0 + U \subset G$. Apply (3) to find a set $V \in \mathcal{B}$ such that $V + V \subset U$. The set $O = (x_0 + V) \times (y_0 + V)$ is an open neighborhood of $(x_0, y_0)$ in $L \times L$ and

$$a(O) = (x_0 + V) + (y_0 + V) = x_0 + y_0 + V + V \subset x_0 + y_0 + U \subset G$$

so the set $O$ witnesses continuity of the addition $a$ at the point $(x_0, y_0)$.

Thus $(L, \tau)$ is a linear topological space and it follows from (a) that $\mathcal{B}$ is a local base of $(L, \tau)$ at $\mathbf{0}$. If $x$ and $y$ are distinct points of $L$ then $z = x - y \neq \mathbf{0}$, so we can apply (2) to see that there exists $U \in \mathcal{B}$ with $z \notin U$. Consequently, $x \notin y + U$, i.e., every $y \in L \backslash \{x\}$ has a neighborhood $y + U$ contained in $L \backslash \{x\}$. Therefore $L \backslash \{x\}$ is open in $(L, \tau)$, so $\{x\}$ is closed in $(L, \tau)$ which proves that $(L, \tau)$ is a $T_1$-space. Applying Problem 201 we conclude that $(L, \tau)$ is a Tychonoff space.

Finally, assume that $\mu$ is a topology on $L$ such that $(L, \mu)$ is a linear topological space for which $\mathcal{B}$ is a local base at $\mathbf{0}$. Given $x \in L$ let $\varphi_x(y) = y + x$ for any $y \in L$. Then the map $\varphi_x : L \to L$ is continuous both for $(L, \tau)$ and $(L, \mu)$; since $\varphi_{-x}$ is the inverse of $\varphi_x$, every $\varphi_x$ is a homeomorphism with respect to both topologies. An immediate consequence is that

(b) the family $\{x + B : B \in \mathcal{B}\}$ is a local base at $x$ in $(L, \mu)$ for every $x \in L$.

Now, if $U \in \tau$ and $x \in U$ then, by (a), there is $B \in \mathcal{B}$ with $U_x = x + B \subset U$; every set $U_x$ belongs to $\mu$ by (b), so $\bigcup \{U_x : x \in U\} = U$ also belongs to $\mu$. This proves that $\tau \subset \mu$. If $U \in \mu$ and $x \in U$ then apply (b) to find $B \in \mathcal{B}$ such that $U_x = x + B \subset U$. The property (a) shows that $U_x \in \tau$ for every $x \in U$, so $\bigcup \{U_x : x \in U\} = U$ also belongs to $\tau$. Therefore $\mu \subset \tau$ and hence $\mu = \tau$. This proves uniqueness of the topology $\tau$ and makes our solution complete.

**V.203.** *Let $L$ be a linear topological space. Prove that*

*(1) for any local base $\mathcal{B}$ of $L$ at $\mathbf{0}$ and any $A \subset L$, we have $\overline{A} = \bigcap \{A + V : V \in \mathcal{B}\}$;*

*(2) for any $A, B \subset L$, we have $\overline{A} + \overline{B} \subset \overline{A + B}$;*

*(3) if $M$ is a linear subspace of $L$ then $\overline{M}$ is also a linear subspace of $L$;*

*(4) if $C$ is a convex subset of $L$ then the sets $\overline{C}$ and $\mathrm{Int}(C)$ are also convex;*

*(5) if $B$ is a balanced subset of $L$ then $\overline{B}$ is also balanced; if, additionally, we have $\mathbf{0} \in \mathrm{Int}(B)$ then $\mathrm{Int}(B)$ is also balanced;*

*(6) if $E$ is an $l$-bounded subset of $L$ then $\overline{E}$ is also $l$-bounded.*

**Solution.** (1) Take any point $x \in \overline{A}$; if $V \in \mathcal{B}$ then the set $W = V \cap (-V)$ is easily seen to be symmetric, i.e., $W = -W$. Since $x + W$ is an open neighborhood of $x$, we can choose a point $a \in (x + W) \cap A$. There is $w \in W$ such that $a = x + w$ and hence $x = a - w \in a + (-W) = a + W \subset a + V \subset A + V$. This gives us the inclusion $\overline{A} \subset P = \bigcap \{A + V : V \in \mathcal{B}\}$.

Now, if $x \in P$ then take any $G \in \tau(x, L)$. It is an easy exercise that there exists $U \in \mathcal{B}$ such that $x + U \subset G$. The set $W = U \cap (-U)$ is a symmetric neighborhood of $\mathbf{0}$, so we can find $V \in \mathcal{B}$ with $V \subset W$. It follows from $x \in P$ that $x \in A + V$, so there are $a \in A$ and $v \in V$ such that $x = a + v$. Then $a = x - v \in x + (-V) \subset x + (-W) = x + W \subset x + U \subset G$, i.e., $a \in G \cap A$ and hence $G \cap A \neq \emptyset$ for any $G \in \tau(x, L)$ which shows that $x \in \overline{A}$. As a consequence, $P = \overline{A}$, i.e., (1) is proved.

(2) Let $a(x, y) = x + y$ for any $x, y \in L$. The map $a : L \times L \to L$ is continuous, so $a(\overline{P}) \subset \overline{a(P)}$ for any $P \subset L \times L$. In particular, if $A, B \subset L$ and $P = A \times B$ then $\overline{P} = \overline{A} \times \overline{B}$, so $a(\overline{A}, \overline{B}) = \overline{A} + \overline{B} \subset \overline{a(A \times B)} = \overline{A + B}$; this settles (2).

(3) Fix any $\alpha, \beta \in \mathbb{R}$ and let $\varphi(x, y) = \alpha x + \beta y$ for any $x, y \in L$. The map $\varphi : L \times L \to L$ is continuous and $\varphi(M \times M) \subset M$ because $M$ is a linear subspace of $L$. It follows from continuity of $\varphi$ that $\varphi(\overline{P}) \subset \overline{\varphi(P)}$ for any $P \subset L \times L$. In particular, for the set $P = M \times M$ we have $\overline{P} = \overline{M} \times \overline{M}$, so $\varphi(\overline{M} \times \overline{M}) \subset \overline{\varphi(M \times M)} \subset \overline{M}$. In other words, $\alpha x + \beta y \in \overline{M}$ for any $x, y \in \overline{M}$ and $\alpha, \beta \in \mathbb{R}$ which shows that $\overline{M}$ is a linear subspace of $L$ and hence we are done with (3).

(4) Observe first that, for any $A \subset L$ if a set $U$ is open in the space $L$ then $A + U = \bigcup\{a + U : a \in A\}$ is an open subset of $L$. Fix any $t \in [0, 1]$ and let $\xi_t(x, y) = tx + (1 - t)y$ for any $x, y \in L$. The map $\xi_t : L \times L \to L$ is continuous and $\xi_t(C \times C) \subset C$ because $C$ is convex. Given an open $U \subset L$ at least one of the sets $tU$ or $(1-t)U$ is open in $L$ because $t$ and $1-t$ cannot be both equal to zero. Thus we can apply our above observation to see that $\xi_t(U \times U) = tU + (1-t)U$ is an open subset of $L$ for any $U \in \tau(L)$. In particular, $\xi_t(\mathrm{Int}(C) \times \mathrm{Int}(C))$ is an open set contained in $C$, so $\xi_t(\mathrm{Int}(C) \times \mathrm{Int}(C)) \subset \mathrm{Int}(C)$ for any $t \in [0, 1]$ which is a reformulation of the fact that $\mathrm{Int}(C)$ is a convex set.

It follows from continuity of $\xi_t$ that $\xi_t(\overline{P}) \subset \overline{\xi_t(P)}$ for any $P \subset L \times L$. In particular, for the set $P = C \times C$ we have $\overline{P} = \overline{C} \times \overline{C}$, so $\xi_t(\overline{C} \times \overline{C}) \subset \overline{\xi_t(C \times C)} \subset \overline{C}$. In other words, $tx + (1 - t)y \in \overline{C}$ for any $x, y \in \overline{C}$ and $t \in [0, 1]$ which shows that $\overline{C}$ is a convex subset of $L$, i.e., we finished the proof of (4).

(5) Fix any $t \in \mathbb{I}$ and let $\mu_t(x) = tx$ for any $x \in L$; by continuity of the operations in $L$, the map $\mu_t : L \to L$ is continuous. The set $B$ being balanced, we have the inclusion $\mu_t(B) \subset B$. Furthermore, $\mu_t(\overline{B}) \subset \overline{\mu_t(B)} \subset \overline{B}$ which shows that $\mu_t(\overline{B}) \subset \overline{B}$ for any $t \in \mathbb{I}$ and hence $\overline{B}$ is a balanced set.

Next observe that $\mu_t$ is a homeomorphism if $t \neq 0$; therefore $\mu_t(\mathrm{Int}(B))$ is an open subset of $B$; this, evidently, implies that $\mu_t(\mathrm{Int}(B)) \subset \mathrm{Int}(B)$ for any $t \in \mathbb{I}\backslash\{0\}$. Since $\mathbf{0} \in \mathrm{Int}(B)$, we also have the inclusion $\mu_0(\mathrm{Int}(B)) = \{\mathbf{0}\} \subset \mathrm{Int}(B)$ which shows that $\mu_t(\mathrm{Int}(B)) \subset \mathrm{Int}(B)$ for any $t \in \mathbb{I}$, i.e., $\mathrm{Int}(B)$ is balanced and hence we showed that (5) is true.

(6) Take any $U \in \tau(\mathbf{0}, L)$; the space $L$ being Tychonoff, we can find $W \in \tau(\mathbf{0}, L)$ such that $\overline{W} \subset U$. The set $E$ is $l$-bounded, so there exists $r > 0$ such that $E \subset tW$ for any $t \geq r$. The map $\mu_t : L \to L$ is a homeomorphism, so $\overline{E} \subset \overline{tW} = t\overline{W} \subset tU$ for any $t \geq r$ and hence $\overline{E}$ is $l$-bounded; this settles (6) and completes our solution.

**V.204.** *Let L be a linear topological space. Prove that*

*(1) every neighborhood of* **0** *contains an open balanced neighborhood of* **0***;*
*(2) every convex neighborhood of* **0** *contains an open convex balanced neighborhood of* **0***.*

Deduce from (2) that any locally convex linear topological space has a local base $\mathcal{B}$ at **0** such that each $U \in \mathcal{B}$ is convex and balanced.

**Solution.** (1) If $G$ is a neighborhood of **0** then there is $U \in \tau(\mathbf{0}, L)$ with $U \subset G$. The family $\mathcal{B} = \tau(\mathbf{0}, L)$ is a local base of $L$ at **0**, so we can apply Problem 202 to find a set $V \in \tau(\mathbf{0}, L)$ such that $tV \subset U$ for any $t \in (-1, 1)$. The set $tV$ is open in $L$ and $\mathbf{0} \in tV$ for every $t \in (-1, 1)\backslash\{0\}$. Therefore the set $W = \bigcup\{tV : t \in (-1, 1)\backslash\{0\}\}$ is open in $L$. Observe also that, for $t = 0$, the set $tV = \{\mathbf{0}\}$ is contained in $W$, so we have the equality $W = \bigcup\{tV : t \in (-1, 1)\}$. For any $s \in \mathbb{I}$ and $x \in W$ there are $t \in (-1, 1)$ and $v \in V$ such that $x = tv$. Then $sx = stv \in stV \subset W$ because $st \in (-1, 1)$. As a consequence, $sW \subset W$ for any $s \in \mathbb{I}$, i.e., $W \subset U \subset G$ is an open balanced neighborhood of **0**, so we proved (1).

(2) Suppose that $G$ is a convex neighborhood of **0**. Then $U = \text{Int}(G)$ is an open convex neighborhood of **0** by Problem 203. It is an easy exercise to see that the set $-U$ is also open and convex, so $V = U \cap (-U)$ is an open symmetric convex neighborhood of **0** with $V \subset G$ (it is again an easy exercise that any intersection of convex subsets of $L$ is convex). It turns out that the set $V$ is also balanced; to see that, take any $t \in \mathbb{I}$ and $x \in V$. If $t \geq 0$ then $tx = tx + (1-t)\mathbf{0} \in V$ because $V$ is convex and $\{\mathbf{0}, x\} \subset V$. If $t < 0$ then $tx = |t|(-x) + (1 - |t|)\mathbf{0} \in V$; we used again convexity of $V$ and the fact that $\{-x, \mathbf{0}\} \subset V$. Thus $tx \in V$ for any $t \in \mathbb{I}$ and $x \in V$, so $V$ is the promised open balanced and convex neighborhood of **0** contained in $G$; this proves (2).

Finally, if $L$ is locally convex then it follows from (2) that the family $\mathcal{B}$ of all open convex balanced neighborhoods of **0** is a local base at **0**, so our solution is complete.

**V.205.** *Let L be a linear topological space. Given a nontrivial linear functional $f : L \to \mathbb{R}$, prove that the following properties are equivalent:*

*(i) f is continuous;*
*(ii) $f^{-1}(0)$ is closed in L;*
*(iii) $f^{-1}(0)$ is not dense in L;*
*(iv) there exists $U \in \tau(\mathbf{0}, L)$ such that $f(U)$ is a bounded subset of $\mathbb{R}$.*

**Solution.** The implication (i)$\Longrightarrow$(ii) holds because $f^{-1}(0)$ is the inverse image of a closed set $\{0\}$. The functional $f$ being nontrivial the set $f^{-1}(0)$ does not coincide with $L$, so if $f^{-1}(0)$ is closed then it is not dense in $L$. This proves (ii)$\Longrightarrow$(iii).

Now suppose that $f^{-1}(0)$ is not dense in $L$ and choose a nonempty open set $G \subset L$ such that $G \cap f^{-1}(0) = \emptyset$. Pick a point $z \in G$; then $K = |f(z)| > 0$ and $V = -z + G$ is an open neighborhood of **0**. By Problem 204 there exists an open balanced set $U \in \tau(\mathbf{0}, L)$ such that $U \subset V$. If $|f(x)| \geq K$ for some

$x \in U$ then $r = \frac{K}{f(x)} \in \mathbb{I}$, so $y = rx \in U$ because the set $U$ is balanced. Then $f(y) = rf(x) = K = f(z)$ and hence $f(-y + z) = 0$. However, $-y \in U \subset V$, so $t = -y + z \in z + V = G$, so $t \in f^{-1}(0) \cap G$, a contradiction. Therefore $|f(x)| < K$ for any $x \in U$ which shows that the set $f(U) \subset (-K, K)$ is bounded in $\mathbb{R}$, i.e., we established (iii)$\Longrightarrow$(iv).

Finally, assume that there exists $U \in \tau(\mathbf{0}, L)$ such that $f(U)$ is bounded in $\mathbb{R}$, i.e., we can choose a number $K > 0$ such that $f(U) \subset (-K, K)$ and hence $|f(y)| < K$ for any $y \in U$. Given any open neighborhood $G$ of $f(\mathbf{0}) = 0$ in $\mathbb{R}$ there exists $\varepsilon > 0$ with $(-\varepsilon, \varepsilon) \subset G$. The set $V = \frac{\varepsilon}{K}U$ is an open neighborhood of $\mathbf{0}$; if $x \in V$ then there is $y \in U$ such that $x = \frac{\varepsilon}{K}y$. An immediate consequence is that $|f(x)| = |f(\frac{\varepsilon}{K}y)| = \frac{\varepsilon}{K}|f(y)| < \varepsilon$. Thus $f(V) \subset (-\varepsilon, \varepsilon) \subset G$ and hence the set $V$ witnesses continuity of $f$ at the point $\mathbf{0}$; applying Fact 2 of S.496 we conclude that $f$ is continuous. This settles (iv)$\Longrightarrow$(i) and shows that our solution is complete.

**V.206.** *Suppose that $L$ is a locally convex linear topological space which has a countable local base at $\mathbf{0}$. Prove that there exists a metric $d$ on the set $L$ with the following properties:*

  *(i)  $d$ generates the topology of $L$;*
  *(ii)  all $d$-open balls are convex and all balls with the center at $\mathbf{0}$ are balanced;*
  *(iii)  the metric $d$ is invariant, i.e., $d(x + z, y + z) = d(x, y)$ for all $x, y, z \in L$.*

  *As a consequence, a locally convex space is metrizable if and only if it has countable character.*

**Solution.** Using Problems 202 and 204 we can find a local base $\{B_n : n \in \mathbb{N}\}$ of the space $L$ at $\mathbf{0}$ such that $B_n$ is convex, balanced and $B_{n+1} + B_{n+1} \subset B_n$ for every $n \in \mathbb{N}$; let $B_0 = L$.

Consider the set $D_n = \{\frac{k}{2^n} : k = 0, \ldots, 2^n\}$; it is clear that $D_n \subset D_{n+1} \subset [0, 1]$ for every $n \in \omega$; let $D = \bigcup_{n \in \omega} D_n$. It is an easy exercise that

(1)  the set $D$ is dense in $[0, 1]$ and, for any $n \in \omega$, if $r + s \in D_n$ and $r + s \leq 1$ then $r + s \in D_n$ and $|r - s| \in D_n$.

  We will define inductively a set $A(r) \subset L$ for all $r \in D$. To start off, let $A(0) = \{\mathbf{0}\}$ and $A(1) = L$; this gives us a set $A(r)$ for every $r \in D_0$. Suppose that $n \in \mathbb{N}$ and we have the set $A(r)$ for every $r \in D_{n-1}$. If $r \in D_n \setminus D_{n-1}$ then $r = \frac{2k+1}{2^n}$ for some $k \in \{0, \ldots, 2^{n-1} - 1\}$ and hence $r = r' + \frac{1}{2^n}$ where $r' = \frac{k}{2^{n-1}} \in D_{n-1}$; let $A(r) = A(r') + B_n$. This inductive procedure gives us a set $A(r)$ for each $r \in D$; if $r > 0$ then $A(r)$ is an open subset of $L$.

  Let us prove first that

(2)  if $r, s \in D$ and $r + s \leq 1$ then $A(r) + A(s) \subset A(r + s)$.

  If $r, s \in D_0$ and $rs = 0$ then $A(r) + A(s) = A(r + s)$; if $r = s = \frac{1}{2}$ then $A(r) + A(s) \subset L = A(r + s)$, so (2) is true. Proceeding inductively, assume that $n \in \mathbb{N}$ and (2) is proved for any $r, s \in D_{n-1}$. Given any $s, r \in D_n$ with $s + r \leq 1$ we have several cases to consider.

(a) If $r, s \in D_{n-1}$ then (2) holds by the induction hypothesis.

(b) If $r \in D_{n-1}$ and $s \in D_n \backslash D_{n-1}$ then there is $s' \in D_{n-1}$ with $s = s' + 2^{-n}$, so $A(r) + A(s) = A(r) + A(s') + B_n \subset A(r + s') + B_n = A(r + s' + 2^{-n}) = A(r + s)$. The inclusion in the above formula is true by the induction hypothesis and the equalities follow from (1) and the definition of the family $\{A(t) : t \in D\}$.

(c) The case when $r \in D_n$ and $s \in D_{n-1}$ is considered as in (b).

(d) If $r, s \in D_n \backslash D_{n-1}$ then there are $r', s' \in D_{n-1}$ such that $r = r' + 2^{-n}$ and $s = s' + 2^{-n}$. Then $A(r) = A(r') + B_n$ and $A(s) = A(s') + B_n$, so

$$A(r) + A(s) = A(r') + A(s') + B_n + B_n \subset A(r' + s') + B_{n-1} = A(r' + s') + A(2^{-n+1}).$$

Since $r' + s' \in D_{n-1}$, $2^{-n+1} \in D_{n-1}$ and $r' + s' + 2^{-n+1} = r + s \le 1$, we can apply the induction hypothesis again to see that $A(r' + s') + A(2^{-n+1}) \subset A(r + s)$, so $A(r) + A(s) \subset A(r + s)$ for any $r, s \in D_n$. Therefore our inductive procedure can be continued to establish that, for any $n \in \omega$, the condition (2) is satisfied for all $r, s \in D_n$. Since the family $\{D_n : n \in \omega\}$ is increasing, the property (2) is proved. An immediate consequence is that

(3) given $t, s \in D$, if $t \le s$ then $A(t) \subset A(t) + A(s - t) \subset A(s)$.

Let $\xi(x) = \inf\{r \in [0, 1] : x \in A(r)\}$ for every $x \in L$. The function $\xi : L \to \mathbb{R}$ is well defined and $\xi(L) \subset [0, 1]$. If we let $d(x, y) = \xi(x - y)$ then it is straightforward that we have (iii) for the function $d$; besides, $d(x, x) = 0$ and $d(x, y) \ge 0$ for any $x, y \in L$. If $z \in L \backslash \{0\}$ then $z \notin B_n = A(2^{-n})$ for some $n \in \omega$, so it follows from (3) that $\xi(z) \ge 2^{-n} > 0$. Therefore $d(x, y) > 0$ whenever $x \ne y$.

If $P, Q$ are balanced (convex) subsets of $L$ then $P + Q$ is also balanced (or convex respectively). Using this fact it takes an easy induction to prove that $A(r)$ is balanced and convex for every $r \in D$. Thus, $x \in A(r)$ if and only if $-x \in A(r)$ and hence $\xi(x) = \xi(-x)$ for any $x \in L$. This implies that $d(x, y) = d(y, x)$ for all $x, y \in L$.

To prove the triangle inequality for $d$ we will show first that

(4) $\xi(x + y) \le \xi(x) + \xi(y)$ for any $x, y \in L$.

Since (4) is trivially true if $\xi(x) + \xi(y) \ge 1$, we can assume that $\xi(x) + \xi(y) < 1$. Fix an arbitrary number $\varepsilon > 0$; there exist $r, s \in D$ such that $\xi(x) < r < \xi(x) + \frac{\varepsilon}{2}$ and $\xi(y) < s < \xi(y) + \frac{\varepsilon}{2}$ while $r + s < 1$. It follows from (3) and the definition of the function $\xi$ that $x \in A(r)$ and $y \in A(s)$. Therefore $x + y \in A(r + s)$ by the property (2); as a consequence, $\xi(x + y) \le r + s < \xi(x) + \xi(y) + \varepsilon$. The inequality $\xi(x + y) < \xi(x) + \xi(y) + \varepsilon$ being true for any $\varepsilon > 0$, the property (4) is proved.

Thus $d(x, z) = \xi(x - z) = \xi(x - y + y - z) \le d(x, y) + d(y, z)$ for any $x, y, z \in L$, so $d$ is a metric on the set $L$; let $B(x, \delta) = \{y \in L : d(x, y) < \delta\}$ for all $x \in L$ and $\delta > 0$. It is an easy exercise to see that

(5) $B(0, \delta) = \bigcup\{A(r) : r \in D, \ r < \delta\}$ for any $\delta > 0$;

this implies that every $B(0, \delta)$ is balanced. Given $x, y \in B(0, \delta)$ and $t \in [0, 1]$ it follows from (3) that there exists $r \in D$, $r < \delta$ such that $x, y \in A(r)$; since $A(r)$ is convex, the point $tx + (1-t)y$ belongs to $A(r)$ and hence to $B(0, \delta)$. This proves that $B(0, \delta)$ is a convex balanced set for any $\delta > 0$.

It is a consequence of (iii) that $B(x, \delta) = x + B(0, \delta)$ for all $x \in L$ and $\delta > 0$; this shows that all $d$-open balls are convex, i.e., (ii) is proved.

The property (5) shows that every $B(0, \delta)$ is open in $L$; since every $d$-open ball is a shift of some $B(0, \delta)$, all $d$-open balls are open in $L$. If $\tau$ is the topology generated by $d$ then the family $\mathcal{B} = \{B(x, \delta) : x \in L, \ \delta > 0\}$ is a base of $(L, \tau)$, so it follows from $\mathcal{B} \subset \tau(L)$ that $\tau \subset \tau(L)$.

Since $B_n = A(2^{-n})$ it follows from (3) that $A(r) \subset B_n$ for every $r < 2^{-n}$, so $B(0, 2^{-n}) \subset B_n$ for each $n \in \mathbb{N}$. Therefore the family $\{B(0, \delta) : \delta > 0\}$ is a local base at $0$, so the family $\{B(x, \delta) : \delta > 0\}$ is a local base at $x$ for every $x \in L$. This shows that the family $\{B(x, \delta) : x \in L, \ \delta > 0\} \subset \tau$ is a base of $\tau(L)$, so $\tau(L) \subset \tau$ and hence $\tau(L) = \tau$. This settles (i) and demonstrates that our solution is complete.

**V.207.** *Let $p$ be a seminorm on a linear space $L$. Prove that*

*(1)  $p(0) = 0$ and $p(x) \geq 0$ for any $x \in L$;*
*(2)  $|p(x) - p(y)| \leq p(x - y)$ for any $x, y \in L$;*
*(3)  $\{x \in L : p(x) = 0\}$ is a linear subspace of $L$;*
*(4)  the set $B = \{x : p(x) < 1\}$ is convex, balanced, absorbing and $p = \mu_B$.*

**Solution.** It follows from $p(0) = p(0 \cdot 0) = 0 \cdot p(0) = 0$ that $p(0) = 0$. Given any $x \in L$ observe that $0 = p(x - x) \leq p(x) + p(-x) = 2p(x)$ and hence $p(x) \geq 0$; this proves (1).

Given any points $x, y \in L$ we have the inequalities $p(x) \leq p(y) + p(x - y)$ and $p(y) \leq p(x) + p(y - x)$ which, together with the fact that $p(y - x) = p(x - y)$ show that $-p(x - y) \leq p(x) - p(y) \leq p(x - y)$, i.e., $|p(x) - p(y)| \leq p(x - y)$, so (2) is proved.

To prove that $M = \{x \in L : p(x) = 0\}$ is a linear subspace of $L$, take any $x, y \in M$ and $\alpha, \beta \in \mathbb{R}$. Since $0 \leq p(\alpha x + \beta y) \leq |\alpha|p(x) + |\beta|p(y) = 0$, we conclude that $\alpha x + \beta y \in M$; this settles (3).

To prove (4) take any $x, y \in B$ and $t \in [0, 1]$. Then

$$p(tx + (1-t)y) \leq tp(x) + (1-t)p(y) < t + (1-t) = 1$$

so $tx + (1-t)y \in B$ for any $x, y \in B$ and $t \in [0, 1]$, i.e., $B$ is convex. If $\alpha \in \mathbb{I}$ and $x \in B$ then $p(\alpha x) = |\alpha|p(x) \leq p(x) < 1$, so $\alpha x \in B$ and hence $B$ is balanced.

Now fix any $x \in L$ and let $\delta = \frac{1}{p(x)+1}$. If $|t| < \delta$ then $p(tx) = |t|p(x) < 1$, so $tx \in B$ for any $t \in (-\delta, \delta)$ which shows that $B$ is an absorbing set.

To finally prove that $p$ coincides with the Minkowski functional of the set $B$ take any $x \in L$. If $p(x) = 0$ then $p(\frac{x}{t}) = 0$ and hence $\frac{x}{t} \in B$ for any $t > 0$; therefore $\mu_B(x) = 0$, so we have the equality $p(x) = \mu_B(x)$ in this case.

If $p(x) > 0$ and $0 < t < p(x)$ then $p(\frac{x}{t}) > 1$, so $\frac{x}{t} \notin B$ and hence $\mu_B(x) \geq p(x)$. If $t > p(x)$ then $p(\frac{x}{t}) < 1$ which implies that $\frac{x}{t} \in B$, so $\mu_B(x) \leq t$; the last inequality being true for all $t > p(x)$ we conclude that $\mu_B(x) \leq p(x)$ and hence $\mu_B(x) = p(x)$ for every $x \in L$, i.e., we completed the proof of (4).

**V.208.** *Let $A$ be a convex absorbing set in a linear space $L$. Prove that*

*(1)* $\mu_A(x + y) \leq \mu_A(x) + \mu_A(y)$ *for any* $x, y \in L$;
*(2)* $\mu_A(tx) = t\mu_A(x)$ *for any* $x \in L$ *and* $t \geq 0$;
*(3)* *if $A$ is balanced then $\mu_A$ is a seminorm;*
*(4)* *if* $B = \{x \in L : \mu_A(x) < 1\}$ *and* $C = \{x \in L : \mu_A(x) \leq 1\}$ *then* $B \subset A \subset C$
*and* $\mu_A = \mu_B = \mu_C$.

**Solution.** (1) Fix any $x, y \in L$; given $t > \mu_A(x) + \mu_A(y)$ it is possible to choose $t_0 > \mu_A(x)$ and $t_1 > \mu_A(y)$ such that $t_0 + t_1 = t$. There exist $s_0, s_1$ such that $\mu_A(x) \leq s_0 < t_0$ and $\mu_A(y) \leq s_1 < t_1$ while $\frac{x}{s_0} \in A$ and $\frac{y}{s_1} \in A$. The set $A$ is absorbing, so $\mathbf{0} \in A$. It follows from convexity of $A$ that

$(*)$   $z \in A$ implies $sz + (1 - s)\mathbf{0} = sz \in A$ for any $s \in [0, 1]$.

Therefore $\frac{x}{t_0} \in A$ and $\frac{y}{t_1} \in A$. If $s = \frac{t_0}{t}$ then $1 - s = \frac{t_1}{t}$, so we can use convexity of $A$ again to see that $\frac{x}{t_0} \cdot \frac{t_0}{t} + \frac{y}{t_1} \cdot \frac{t_1}{t} = \frac{x+y}{t} \in A$. This shows that $\mu_A(x + y) \leq t$ for any $t > \mu_A(x) + \mu_A(y)$, so $\mu_A(x + y) \leq \mu_A(x) + \mu_A(y)$, i.e., (1) is proved.

(2) Since $\frac{0}{s} = \mathbf{0} \in A$ for any $s > 0$, we have $\mu_A(\mathbf{0}) = 0$ which implies that $\mu_A(0 \cdot x) = 0 \cdot \mu_A(x)$ for any $x \in L$. Next, fix $x \in L$ and $t > 0$; if $s > t\mu_A(x)$ then $\mu_A(x) < \frac{s}{t}$. This, together with $(*)$, shows that $\frac{tx}{s} \in A$ and therefore $\mu_A(tx) \leq s$. The number $s > t\mu_A(x)$ was chosen arbitrarily, so $\mu_A(tx) \leq t\mu_A(x)$.

Now, if $s > \mu_A(tx)$ then apply $(*)$ once more to see that $\frac{tx}{s} = \frac{x}{s/t} \in A$ and hence $\mu_A(x) \leq \frac{s}{t}$, i.e., $t\mu_A(x) \leq s$. As before, this implies $t\mu_A(x) \leq \mu_A(tx)$, so $\mu_A(tx) = t\mu_A(x)$ and hence we proved (2).

(3) Suppose, additionally, that $A$ is balanced. Then $\frac{x}{t} \in A$ if and only if $\frac{-x}{t} \in A$ for any $x \in L$ and $t > 0$; consequently, $\mu_A(x) = \mu_A(-x)$ for any $x \in L$. If $\alpha \geq 0$ then we can apply (2) to conclude that $\mu_A(\alpha x) = \alpha\mu_A(x) = |\alpha|\mu_A(x)$. If $\alpha < 0$ then $\mu_A(\alpha x) = \mu_A(|\alpha|(-x)) = |\alpha|\mu_A(-x) = |\alpha|\mu_A(x)$ for any $x \in L$; this, together with what was proved in (1), completes the proof of (3).

(4) If $x \in B$ then $\mu_A(x) < 1$, so $(*)$ implies that $x = \frac{x}{1} \in A$. If $x = \frac{x}{1} \in A$ then $\mu_A(x) \leq 1$, i.e., $x \in C$. This proves that $B \subset A \subset C$; an immediate consequence of the definition of the Minkowski functional is that $\mu_C(x) \leq \mu_A(x) \leq \mu_B(x)$ for any $x \in L$.

Now, assume that $x \in L$ and $\mu_C(x) < t$; there is $s \in [\mu_C(x), t)$ such that $\frac{x}{s} \in C$ and hence $1 \geq \mu_A(\frac{x}{s}) = \frac{1}{s}\mu_A(x)$, so $\mu_A(x) \leq s < t$. The number $t > \mu_C(x)$ was chosen arbitrarily, so we proved that $\mu_A(x) \leq \mu_C(x)$ for any $x \in L$, i.e., $\mu_A = \mu_C$.

If $x \in L$ and $\mu_A(x) < t$ then $\mu_A(\frac{x}{t}) < 1$ by (2); thus $\frac{x}{t} \in B$, so $\mu_B(x) \leq t$ for any $t > \mu_A(x)$. Therefore $\mu_B(x) \leq \mu_A(x)$ for any $x \in L$ and hence $\mu_A = \mu_B$; this settles (4) and finishes our solution.

**V.209.** *Given a locally convex linear topological space L, take any local base $\mathcal{B}$ at* **0** *such that all elements of $\mathcal{B}$ are convex and balanced. Prove that $\{\mu_V : V \in \mathcal{B}\}$ is a separating family of continuous seminorms on L.*

**Solution.** Observe first a local base $\mathcal{B}$ at the point **0** such that all elements of $\mathcal{B}$ are convex and balanced exists in any locally convex space by Problem 204. All elements of $\mathcal{B}$ are also absorbing by Problem 202, so we can apply Problem 208 to see that $\mu_V$ is a seminorm on $L$ for any $V \in \mathcal{B}$.

Fix $V \in \mathcal{B}$ and take any $x \in L$, $\varepsilon > 0$. The set $W = x + \frac{\varepsilon}{2}V$ is an open neighborhood of $x$; if $y \in W$ then $z = y - x \in \frac{\varepsilon}{2}V$ and hence $\frac{z}{\varepsilon/2} \in V$, so $\mu_V(z) \leq \frac{\varepsilon}{2} < \varepsilon$. Applying Problem 207 we conclude that $|\mu_V(y) - \mu_V(x)| \leq \mu_V(z) < \varepsilon$ for any $y \in W$ and therefore the set $W$ witnesses that $\mu_V$ is continuous at the point $x$. Thus every $\mu_V$ is continuous on $L$.

Finally, take any $x \in L\backslash\{0\}$; the family $\mathcal{B}$ being a local base at **0**, there exists $V \in \mathcal{B}$ with $x = \frac{x}{1} \notin V$. If $t < 1$ and $\frac{x}{t} \in V$ then $x = t \cdot \frac{x}{t} \in V$ because the set $V$ is balanced. This contradiction shows that $\frac{x}{t} \notin V$ for any $t < 1$ and hence $\mu_V(x) \geq 1 > 0$. Therefore $\{\mu_V : V \in \mathcal{B}\}$ is a separating family of continuous seminorms on $L$.

**V.210.** *Let $\mathcal{P}$ be a separating family of seminorms on a linear space L. Given $p \in \mathcal{P}$ and $n \in \mathbb{N}$, let $O(p,n) = \{x \in L : p(x) < \frac{1}{n}\}$. Prove that the family $\mathcal{B} = \{O(p_1,n) \cap \ldots \cap O(p_n,n) : n \in \mathbb{N}, \ p_1, \ldots, p_n \in \mathcal{P}\}$ is a convex balanced local base at **0** for some topology $\tau$ on L such that $(L, \tau)$ is a locally convex space in which all elements of $\mathcal{P}$ are continuous and any $E \subset L$ is $l$-bounded if and only if $p(E)$ is bounded for any $p \in \mathcal{P}$.*

**Solution.** Observe first that $O(p,n) = \frac{1}{n}O(p,1)$, so $O(p,n)$ is convex, balanced and absorbing for any $p \in \mathcal{P}$ and $n \in \mathbb{N}$ (see Problem 207). We leave it to the reader to verify that any finite intersection of convex, balanced and absorbing sets is a convex, balanced and absorbing set; therefore every element of $\mathcal{B}$ is convex, balanced and absorbing.

Given $U, V \in \mathcal{B}$ there are $k, n \in \mathbb{N}$ and $p_1, \ldots, p_n, q_1, \ldots, q_k \in \mathcal{P}$ such that $U = O(p_1,n) \cap \ldots \cap O(p_n,n)$ and $V = O(q_1,k) \cap \ldots \cap O(q_k,k)$. Let $r_i = p_i$ for all $i \leq n$ and $r_i = q_{i-n}$ for all $i = n+1, \ldots, n+k$; then, for $m = n + k$, the set $W = O(r_1,m) \cap \ldots \cap O(r_m,m)$ belongs to $\mathcal{B}$ and $W \subset U \cap V$.

If $x \in L\backslash\{0\}$ then there is $p \in \mathcal{P}$ such that $p(x) \neq 0$ and hence there is $n \in \mathbb{N}$ for which $p(x) \geq \frac{1}{n}$. This shows that $x \notin O(p,n) \in \mathcal{B}$, so $x \notin \bigcap \mathcal{B}$ and therefore $\bigcap \mathcal{B} = \{0\}$.

Given $U = O(p_1,n) \cap \ldots \cap O(p_n,n) \in \mathcal{B}$ let $r_i = p_i$ for all $i \leq n$ and $r_i = p_1$ for all $i = n+1, \ldots, 2n$. Then $\{r_1, \ldots, r_{2n}\} \subset \mathcal{P}$ and $V = O(r_1, 2n) \cap \ldots \cap O(r_{2n}, 2n)$ belongs to $\mathcal{B}$. If $x, y \in V$ and $i \leq n$ then $r_i(x) = p_i(x) < \frac{1}{2n}$ and $r_i(y) = p_i(y) < \frac{1}{2n}$; an immediate consequence is that $p_i(x + y) \leq p_i(x) + p_i(y) < \frac{1}{n}$ for all $i \leq n$ and therefore $x + y \in U$. This proves that $V + V \subset U$.

Now, if $U = O(p_1,n) \cap \ldots \cap O(p_n,n) \in \mathcal{B}$ and $x \in U$ then there exists $m \in \mathbb{N}$ such that $p_i(x) < \frac{1}{n} - \frac{1}{m}$ for all $i \leq n$. For $k = m + n$ let $q_i = p_i$ for all $i \leq n$ and $q_i = p_1$ for all $i = n+1, \ldots, k$. Then $\{q_1, \ldots, q_k\} \subset \mathcal{P}$, so

$V = O(q_1, k) \cap \ldots \cap O(q_k, k)$ is an element of $\mathcal{B}$. If $y \in V$ then $p_i(x + y) \le p_i(x) + p_i(y) = p_i(x) + q_i(y) < \frac{1}{n} - \frac{1}{m} + \frac{1}{k} \le \frac{1}{n}$ for every $i \le n$, so $x + y \in U$ for any $y \in V$. This proves that, for any $x \in U$ there is $V \in \mathcal{B}$ such that $x + V \subset U$.

Once more, if $U = O(p_1, n) \cap \ldots \cap O(p_n, n) \in \mathcal{B}$ and $\varepsilon > 0$ then choose $m \in \mathbb{N}$ such that $m \ge n$ and $\frac{1}{m} < \frac{1}{\varepsilon n}$. Let $q_i = p_i$ for all $i \le n$ and $q_i = p_1$ for all $i = n + 1, \ldots, m$. Then $\{q_1, \ldots, q_m\} \subset \mathcal{P}$, so $V = O(q_1, m) \cap \ldots \cap O(q_m, m)$ is an element of $\mathcal{B}$. For any $\lambda \in (-\varepsilon, \varepsilon)$ and $x \in V$ we have $p_i(\lambda x) = |\lambda| p_i(x) < \varepsilon \cdot \frac{1}{m} < \frac{1}{n}$, so $\lambda x \in U$.

Thus we checked that all conditions of Problem 202 for $\mathcal{B}$ are satisfied, so there exists a unique linear space topology $\tau$ on $L$ such that $\mathcal{B}$ is a local base at $\mathbf{0}$ for the space $(L, \tau)$. For every $x \in L$ the family $\mathcal{B}_x = \{x + B : B \in \mathcal{B}\}$ is a local base at $x$ in $(L, \tau)$. Since all elements of $\mathcal{B}_x$ are convex for each $x \in L$, the family $\bigcup\{\mathcal{B}_x : x \in L\}$ is a base of $(L, \tau)$ which consists of convex sets. Therefore $(L, \tau)$ is a locally convex space.

Take any $p \in \mathcal{P}$ and fix $x \in L$, $\varepsilon > 0$. Choose $n \in \mathbb{N}$ with $\frac{1}{n} < \varepsilon$; the set $U = O(p, n)$ belongs to $\mathcal{B}$, so $V = x + U$ is an open neighborhood of $x$. For any $y \in V$ we have $p(y - x) < \frac{1}{n}$, so $|p(y) - p(x)| \le p(y - x) < \frac{1}{n}$ (see Problem 207); therefore the set $V$ witnesses continuity of $p$ at the point $x$. This proves that every $p \in \mathcal{P}$ is continuous on $(L, \tau)$.

Assume that $E \subset L$ is $l$-bounded in $(L, \tau)$. Given $p \in \mathcal{P}$ the set $U = O(p, 1)$ is an open neighborhood of $\mathbf{0}$, so there is $r > 0$ such that $E \subset rU$. Therefore, for any $x \in E$, there is $y \in U$ with $x = ry$ and hence $p(x) = rp(y) < r$; this shows that $p(E) \subset [0, r)$ is a bounded subset of $\mathbb{R}$.

Finally suppose that $E \subset L$ and the set $p(E)$ is bounded in $\mathbb{R}$ for any $p \in \mathcal{P}$. Given any $U \in \tau$ with $\mathbf{0} \in U$ there exist $n \in \mathbb{N}$ and seminorms $p_1, \ldots, p_n \in \mathcal{P}$ such that $V = O(p_1, n) \cap \ldots \cap O(p_n, n) \subset U$. Take $r > 0$ such that $p_i(E) \subset [0, r)$ for every $i \le n$ and let $s = nr$. If $t \ge s$ and $x \in E$ then $p_i(\frac{x}{t}) = \frac{1}{t} \cdot p_i(x) < \frac{r}{t} \le \frac{r}{s} = \frac{1}{n}$ for all $i \le n$, so $\frac{x}{t} \in V$ and therefore $x \in tV \subset tU$. We proved that $E \subset tU$ for any $t \ge s$, so the set $E$ is $l$-bounded in $(L, \tau)$. Thus a set $E \subset L$ is $l$-bounded in $(L, \tau)$ if and only if $p(E)$ is bounded in $\mathbb{R}$ for all $p \in \mathcal{P}$. We have finally proved all that was promised for $(L, \tau)$, so our solution is complete.

**V.211.** *Prove that a linear topological space is normable if and only if it has a convex $l$-bounded neighborhood of zero.*

**Solution.** Suppose that $L$ is a normable linear topological space and fix a norm $||\cdot||$ such that the metric $d : L \times L \to \mathbb{R}$ defined by $d(x, y) = ||x - y||$ for all $x, y \in L$, generates the topology of $L$. Since every norm is also a seminorm, we can apply Problem 207 to see that the set $B = \{x \in L : ||x|| < 1\}$ is convex.

We also have the equality $B = \{x \in L : d(x, \mathbf{0}) < 1\}$, so the set $B$ is open in $L$ being a ball in the metric space $(L, d)$. To see that $B$ is $l$-bounded, take any $U \in \tau(\mathbf{0}, L)$; there exists $\varepsilon > 0$ such that $C = \{x \in L : ||x|| < \varepsilon\} \subset U$. Let $s = \frac{1}{\varepsilon}$ and take any $t \ge s$. If $x \in B$ then $||x|| < 1$, so $||\frac{x}{t}|| = \frac{1}{t} ||x|| < \frac{1}{t} \le \frac{1}{s} = \varepsilon$. Therefore $y = \frac{x}{t} \in C$, so $x = ty \in tC \subset tU$ and hence $B \subset tU$ for any $t \ge s$. This proves that $B$ is an $l$-bounded convex neighborhood of $\mathbf{0}$ and hence we have established necessity.

To verify sufficiency, assume that $L$ is a linear topological space and there exists an $l$-bounded convex neighborhood $B$ of the point $\mathbf{0}$. Apply Problem 204 to find an open convex balanced neighborhood $V$ of the point $\mathbf{0}$ such that $V \subset B$; it follows from $V \subset B$ that $V$ is also $l$-bounded. The set $V$ is absorbing by Problem 202 (applied to $\mathcal{B} = \tau(\mathbf{0}, L)$), so its Minkowski functional $\mu_V$ is a seminorm on $L$ by Problem 208.

If $x \neq \mathbf{0}$ then $W = L\backslash\{x\}$ is an open neighborhood of $\mathbf{0}$, so there is $t > 0$ such that $V \subset tW$ and hence $\frac{1}{t}V \subset W$ which shows that $x \notin \frac{1}{t}V$, i.e., $tx = \frac{x}{1/t} \notin V$. Suppose that $0 < s < \frac{1}{t}$ and $\frac{x}{s} \in V$; since $0 < st < 1$ and $x \in sV$, we have $tx \in tsV \subset V$ the last inclusion being true because the set $V$ is balanced. This contradiction with $tx \notin V$ demonstrates that $\frac{x}{s} \notin V$ for any $s \in (0, \frac{1}{t})$ and therefore $\mu_V(x) \geq \frac{1}{t} > 0$.

We showed that $\mu_V(x) > 0$ for any $x \neq \mathbf{0}$, so $\mu_V$ is a norm on $L$; let $\|x\| = \mu_V(x)$ for every $x \in L$. We must prove that the metric $d : L \times L \to \mathbb{R}$ defined by $d(x, y) = \|x - y\|$ for any $x, y \in L$, generates the topology of $L$; denote temporarily by $\tau$ the topology generated by the metric $d$. For any $r > 0$ and $x \in L$ the set $B(x, r) = \{y \in L : d(x, y) < r\}$ is the ball of radius $r$ centered at $x$.

Take any point $x = 1 \cdot x \in V$; the multiplication by scalars is continuous in $L$, so we can find $\varepsilon > 0$ and $W \in \tau(x, L)$ such that $tw \in V$ for any $t \in (1 - \varepsilon, 1 + \varepsilon)$ and $w \in W$. In particular, $(1 + \frac{\varepsilon}{2})x = \frac{x}{(1+\varepsilon/2)^{-1}} \in V$, so $\mu_V(x) \leq (1+\varepsilon/2)^{-1} < 1$. This shows that $V \subset B(\mathbf{0}, 1)$; applying Problem 208 we conclude that $V = B(\mathbf{0}, 1)$ and therefore $B(\mathbf{0}, r) = rB(\mathbf{0}, 1) = rV$ is an open subset of $L$ for any $r > 0$. It follows from the equality $B(x, r) = x + B(\mathbf{0}, r)$ that every ball $B(x, r)$ is open in $L$. The family $\mathcal{B} = \{B(x, r) : x \in L, \ r > 0\} \subset \tau(L)$ is a base of $\tau$; an easy consequence is that $\tau \subset \tau(L)$.

Take any $U \in \tau(\mathbf{0}, L)$; the set $V$ being $l$-bounded, there is $t > 0$ such that $V \subset tU$ and hence $\frac{1}{t}V \subset U$. Therefore $B(\mathbf{0}, \frac{1}{t}) = \frac{1}{t}B(\mathbf{0}, 1) = \frac{1}{t}V \subset U$ which shows that the family $\{B(\mathbf{0}, r) : r > 0\}$ is a local base of $L$ at $\mathbf{0}$. Consequently, $\{B(x, r) : r > 0\}$ is a local base of $L$ at $x$ for any $x \in L$. This implies that the family $\mathcal{B} \subset \tau$ is a base of $L$, so $\tau(L) \subset \tau$ and hence $\tau = \tau(L)$. We proved that $\mu_V$ is a norm on $L$ which generates the topology of $L$; this settles sufficiency and makes our solution complete.

**V.212.** *Let $N$ be a closed subspace of a linear topological space $L$. Prove that*

*(1)  the quotient topology of $L/N$ makes $L/N$ a linear topological space;*
*(2)  the quotient map $\pi : L \to L/N$ is linear, open and continuous;*
*(3)  If $\mathcal{P} \in \{metrizability,\ local\ convexity,\ normability,\ complete\ normability\}$ and $L$ has $\mathcal{P}$ then $L/N$ also has $\mathcal{P}$.*

**Solution.** Observe that $\pi(x) = \pi(y)$ if and only if $x - y \in N$. If $\pi(x) = \pi(x')$ and $\pi(y) = \pi(y')$ then $x - x' \in N$ and $y - y' \in N$; since $N$ is a subspace of $L$, the point $x - x' + y - y' = (x + y) - (x' + y')$ belongs to $N$ and hence $\pi(x+y) = \pi(x'+y')$. Therefore the addition is well defined in $L/N$. If $\alpha \in \mathbb{R}$ and $\pi(x) = \pi(x')$ then $x - x' \in N$ implies that $\alpha(x - x') \in N$ and hence $\alpha x - \alpha x' \in N$, i.e., $\pi(\alpha x) = \pi(\alpha x')$, so the definition of the scalar multiplication is also consistent.

It is straightforward that $a + b = b + a$ and $(a + b) + c = a + (b + c)$ for any $a, b, c \in L/N$. For the set $e = N = \mathbf{0} + N \in L/N$ we have $a + e = a$ and $a + (-1)a = e$ for any $a \in L/N$, so $e$ is the zero element of $L/N$. We leave it to the reader to verify the scalar multiplication properties for $L/N$, i.e., that the equalities $\alpha(\beta a) = (\alpha\beta)a$, $(\alpha + \beta)a = \alpha a + \beta a$, $\alpha(a + b) = \alpha a + \alpha b$ and $1a = a$ are fulfilled for all $a, b \in L/N$ and $\alpha, \beta \in \mathbb{R}$. Thus $L/N$ is a linear space.

It follows from the definition that the map $\pi$ is linear and quotient, so $\pi$ is continuous. If $U$ is open in $L$ then $\pi^{-1}\pi(U) = U + N$ is open in $L$, so, by definition of quotient topology, the set $\pi(U)$ is open in $L/N$. This proves that the map $\pi$ is open.

To see that the topology of $L/N$ is compatible with its algebraic structure fix a point $a_0 \in L/N$ and $t_0 \in \mathbb{R}$; there is $x_0 \in L$ such that $a_0 = \pi(x_0)$. If $U \in \tau(t_0 a_0, L/N)$ then the set $V = \pi^{-1}(U)$ is open in $L$ and $t_0 x_0 \in V$. By continuity of the scalar multiplication in $L$ we can find $\varepsilon > 0$ and $W \in \tau(x_0, L)$ such that $tw \in V$ whenever $|t - t_0| < \varepsilon$ and $w \in W$.

The set $W' = \pi(W)$ is open in $L/N$ and contains $a_0$; take any $t \in (t_0 - \varepsilon, t_0 + \varepsilon)$ and $a \in W'$. There is $x \in W$ such that $a = \pi(x)$. Then $tx \in V$, so $ta = \pi(tx) \in U$. Thus $ta \in U$ for any $t \in (t_0 - \varepsilon, t_0 + \varepsilon)$ and $a \in W'$; this proves that the scalar multiplication is continuous in $L/N$.

Next, fix $a_0, b_0 \in L/N$ and $U \in \tau(a_0 + b_0, L/N)$; there are $x_0, y_0 \in L$ such that $\pi(x_0) = a_0$ and $\pi(y_0) = b_0$. By linearity of $\pi$ we have $\pi(x_0 + y_0) = a_0 + b_0$, so the set $V = \pi^{-1}(U)$ is an open neighborhood of $x_0 + y_0$ in $L$. By continuity of the addition in $L$ we can find $G \in \tau(x_0, L)$ and $H \in \tau(y_0, L)$ such that $G + H \subset V$. Recalling again that the map $\pi$ is open we conclude that $H' = \pi(H) \in \tau(a_0, L/N)$ and $G' = \pi(G) \in \tau(b_0, L/N)$. Given any $a \in H'$ and $b \in G'$ pick $x \in G$ and $y \in H$ with $\pi(x) = a$ and $\pi(y) = b$. Then $a + b \in V$, so $x + y = \pi(a + b) \in U$; this proves that $G' + H' \subset U$ and hence the addition is continuous in $L/N$, so $L/N$ is a linear topological space, i.e., we settled (1) and (2).

*Fact 1.* Suppose that $M$ is a normed linear space with a norm $|| \cdot ||$. Given a closed linear subspace $F \subset M$ let $\mu(a) = \inf\{||x|| : x \in a\}$ for any $a \in M/F$. Then $\mu$ is a norm on the space $M/F$ which generates the quotient topology on $M/F$ and, if $|| \cdot ||$ is complete, then $\mu$ is a complete norm on $M/F$.

*Proof.* Let $q : M \to M/F$ be the quotient map and denote by $z_f$ the zero element of $M/F$ (which coincides with the set $F$); the symbol $\mathbf{0}$ stands for the zero element of $M$. For any $x \in M$ and $r > 0$ let $B(x, r) = \{y \in M : ||y - x|| < r\}$. Since $\mathbf{0} \in F = z_f$ and $||\mathbf{0}|| = 0$, we conclude that $\mu(z_f) = 0$. It is evident that $\mu(a) \geq 0$ for any $a \in M/F$. Now, if $a \neq z_f$ then $M \backslash a$ is an open neighborhood of $\mathbf{0}$, so there is $\varepsilon > 0$ such that $B(\mathbf{0}, \varepsilon) \cap a = \emptyset$; an immediate consequence is that $||x|| \geq \varepsilon$ for any $x \in a$ and hence $\mu(a) \geq \varepsilon > 0$. Thus $\mu(a) > 0$ for any $a \in M/F \backslash \{z_f\}$.

Next fix $a \in M/F$ and $t \in \mathbb{R}$; then $a = x_0 + F$ for some $x_0 \in a$ and $ta = tx_0 + F$. If $t = 0$ then $ta = z_f$, so $\mu(ta) = 0 = |t| \cdot \mu(a)$, so the second axiom of norm is fulfilled in this case. If $t \neq 0$ and $\varepsilon > 0$ then choose a point $x \in ta$ such that $||x|| < \mu(ta) + \varepsilon$. Since $ta = tx_0 + F = t(x_0 + F)$, we can pick $f \in F$

with $t(x_0 + f) = x$. It follows from $x_0 + f \in a$ that $||x_0 + f|| \geq \mu(a)$, so $||x|| = |t|||x_0 + f|| \geq |t|\mu(a)$ which shows that $|t|\mu(a) \leq \mu(ta) + \varepsilon$ for any $\varepsilon > 0$. Therefore $|t|\mu(a) \leq \mu(ta)$.

Now, let us choose a point $y \in a$ such that $||y|| < \mu(a) + \frac{\varepsilon}{|t|}$. Then we have $|t|||y|| = ||ty|| < |t|\mu(a) + \varepsilon$ and $ty \in ta$, so $\mu(ta) \leq ||ty|| < |t|\mu(a) + \varepsilon$. The inequality $\mu(ta) < |t|\mu(a) + \varepsilon$ being true for any $\varepsilon > 0$ we conclude that $\mu(ta) \leq |t|\mu(a)$ and hence $\mu(ta) = |t|\mu(a)$ for any $a \in M/F$ and $t \in \mathbb{R}$.

To check the triangle inequality for $\mu$ fix any $a, b \in M/F$ and $\varepsilon > 0$. There are $x, y \in M$ such that $q(x) = a$, $q(y) = b$ while $||x|| \leq \mu(a) + \frac{\varepsilon}{2}$ and $||y|| \leq \mu(b) + \frac{\varepsilon}{2}$. Then $||x + y|| \leq ||x|| + ||y|| \leq \mu(a) + \mu(b) + \varepsilon$; it follows from $x + y \in a + b$ that $\mu(a + b) \leq \mu(a) + \mu(b) + \varepsilon$. The number $\varepsilon > 0$ was taken arbitrarily, so $\mu(a + b) \leq \mu(a) + \mu(b)$ for any $a, b \in M/F$ and hence $\mu$ is a norm on $M/F$; denote by $\tau$ the topology generated by $\mu$ on the set $M/F$.

For any $a \in M/F$ and $\varepsilon > 0$ the set $C(a, \varepsilon) = \{b \in M/F : \mu(b - a) < \varepsilon\}$ is the $\varepsilon$-ball with respect to $\mu$ centered at $a$. It turns out that

$(*)$  $q(B(x, \varepsilon)) = C(q(x), \varepsilon)$ for any $x \in M$ and $\varepsilon > 0$.

Given any $y \in B(x, \varepsilon)$ we have $||y - x|| < \varepsilon$; it follows from $y - x \in q(y - x)$ that $\mu(q(y - x)) = \mu(q(y) - q(x)) \leq ||y - x|| < \varepsilon$. Therefore $q(y) \in C(q(x), \varepsilon)$, so we proved that $q(B(x, \varepsilon)) \subset C(q(x), \varepsilon)$.

Now, if $a \in C(q(x), \varepsilon)$ then $\mu(a - q(x)) < \varepsilon$, so there exists $y_0 \in a - q(x)$ with $||y_0|| < \varepsilon$. If $y = y_0 + x$ then $y \in B(x, \varepsilon)$ and $q(y) = q(y_0) + q(x) = a - q(x) + q(x) = a$. The point $a \in C(q(x), \varepsilon)$ was chosen arbitrarily, so $C(q(x), \varepsilon) \subset q(B(x, \varepsilon))$ and therefore $C(q(x), \varepsilon) = q(B(x, \varepsilon))$, i.e., $(*)$ is proved.

Since the open balls form a base in any metric space, we conclude that the map $q : M \to (M/F, \tau)$ is open and continuous (see Fact 2 of S.491). Any open map being quotient, the topology $\tau$ on $M/F$ is the quotient one with respect to $q$, so $\tau$ coincides with the quotient space topology on $M/F$.

Finally, assume that $|| \cdot ||$ is a complete norm on $M$ and take a Cauchy sequence $\{a_n : n \in \omega\} \subset M/F$ with respect to $\mu$. It is easy to construct by induction a sequence $\{n_i : i \in \omega\} \subset \omega$ such that $n_i < n_{i+1}$ and $\mu(a_{n_i} - a_{n_{i+1}}) < 2^{-i}$ for each $i \in \omega$. Pick $x_0 \in q^{-1}(a_{n_0})$ arbitrarily; proceeding by induction assume that $j \in \omega$ and we have chosen $x_0, \ldots, x_j \in M$ in such a way that $q(x_i) = a_{n_i}$ and $||x_i - x_{i+1}|| < 2^{-i}$ for every $i < j$. It follows from $\mu(a_{n_j} - a_{n_{j+1}}) < 2^{-j}$ that there exists $y \in a_{n_{j+1}} - a_{n_j}$ with $||y|| < 2^{-j}$; let $x_{j+1} = x_j + y$. Then $||x_{j+1} - x_j|| = ||y|| < 2^{-j}$ and $q(x_{j+1}) = q(x_j) + q(y) = a_{n_j} + (a_{n_{j+1}} - a_{n_j}) = a_{n_{j+1}}$, so we accomplished our inductive procedure and hence we can construct a sequence $\{x_i : i \in \omega\}$ such that $q(x_i) = a_{n_i}$ and $||x_i - x_{i+1}|| < 2^{-i}$ for every $i \in \omega$.

It is easy to see that $\{x_i : i \in \omega\}$ is a Cauchy sequence so there is $x \in M$ with $x_i \to x$. By continuity of $q$ the sequence $\{q(x_i) : i \in \omega\} = \{a_{n_i} : i \in \omega\}$ converges to $q(x)$, i.e., $\{a_{n_i} : i \in \omega\}$ is a convergent subsequence of our Cauchy sequence $S = \{a_n : n \in \omega\}$. Therefore $S$ is convergent and hence we established that $\mu$ is a complete norm on $M/F$, so Fact 1 is proved.

Returning to our solution observe that if $L$ is (completely) normable then $L/N$ is also (completely) normable by Fact 1. If $L$ is metrizable then $L/N$ is first countable being an open continuous image of $L$ by (2). Therefore we can apply Problem 206 to conclude that $L/N$ is also metrizable.

If $L$ is locally convex then take any $a \in L/N$ and $U \in \tau(a, L/N)$; fix a point $x \in \pi^{-1}(a)$. The set $U' = \pi^{-1}(U)$ is an open neighborhood of $x$, so there exists a convex $V \in \tau(x, L)$ with $V \subset U'$ The set $W = \pi(V)$ is an open neighborhood of $a$ in $L/N$ and $W \subset U$. Take any $b, c \in W$, $t \in [0, 1]$ and pick $y, z \in V$ such that $\pi(y) = b$ and $\pi(z) = c$. The set $V$ being convex, the point $v = ty + (1 - t)z$ belongs to $V$ and therefore $tb + (1 - t)c = \pi(v) \in W$. This shows that $W$ is a convex open subset of $L/N$ such that $a \in W \subset U$, so convex open subsets of $L/N$ form a base in $L/N$, i.e., the space $L/N$ is locally convex. Thus we have proved all statements in (3), so our solution is complete.

**V.213.** *Prove that any product of locally convex spaces is a locally convex space.*

**Solution.** We will need the following general statement.

*Fact 1.* Suppose that $L_t$ is a linear space for any $t \in T$ and $L = \prod_{t \in T} L_t$. If $C_t \subset L_t$ is a convex set for every $t \in T$ then $C = \prod_{t \in T} C_t$ is a convex subset of $L$.

*Proof.* Take any $x, y \in C$, $\alpha \in [0, 1]$ and let $z = \alpha x + (1 - \alpha)y$; for any $t \in T$, it follows from $x(t), y(t) \in C_t$ that $z(t) = \alpha x(t) + (1 - \alpha)y(t) \in C_t$, so $z \in \prod_{t \in T} C_t = C$. Therefore $C$ is convex and Fact 1 is proved.

Now assume that $L_t$ is a locally convex space for all $t \in T$ and fix, for any $t \in T$ and $x \in L_t$, a local base $\mathcal{B}_x^t$ of $L_t$ at $x$ such that all elements of $\mathcal{B}_x^t$ are convex. Given any point $a \in L = \prod_{t \in T} L_t$ the family $\mathcal{U}_a = \{\prod_{t \in T} W_t : W_t = L_t$ for all but finitely many $t$ and if $W_t \neq L_t$ then $W_t \in \mathcal{B}_{a(t)}^t\}$ is a local base at $a$ in the space $L$. All elements of $\mathcal{U}_a$ are convex being the product of convex sets (see Fact 1). Therefore open convex subsets of $L$ constitute a base of $L$, so $L$ is a locally convex space.

**V.214.** *Suppose that $L$ and $M$ are linear topological spaces and $\Phi$ is an equicontinuous family of linear maps from $L$ to $M$. Prove that, for any $l$-bounded set $A \subset L$ there is an $l$-bounded set $B \subset M$ such that $f(A) \subset B$ for all $f \in \Phi$.*

**Solution.** Let $B = \bigcup\{f(A) : f \in \Phi\}$; then $f(A) \subset B$ for any $f \in \Phi$. Take any open set $W$ in the space $M$ such that $0_M \in W$. The family $\Phi$ being equicontinuous there is $V \in \tau(0_L, L)$ such that $f(V) \subset W$ for any $f \in \Phi$. Since $A$ is $l$-bounded, there exists $s > 0$ such that $A \subset tV$ for any $t \geq s$. If $y \in B$ and $t \geq s$ pick $f \in \Phi$ and $x \in A$ with $f(x) = y$; it follows from $x \in tV$ that $y = f(x) \in tf(V) \subset tW$. Thus $B \subset tW$ for all $t \geq s$, so the set $B$ is $l$-bounded in $M$.

**V.215.** *Suppose that $L$ and $M$ are linear topological spaces and $\Phi$ is a family of linear continuous maps from $L$ to $M$. Let $\Phi(x) = \{f(x) : f \in \Phi\}$ for every $x \in L$ and assume that the set $B = \{x \in L : \Phi(x)$ is $l$-bounded in $M\}$ is of second category in $L$. Prove that $B = L$ and the family $\Phi$ is equicontinuous.*

**Solution.** Take any $W \in \tau(\mathbf{0}_M, M)$ and use Problem 202 to find a set $W' \in \tau(\mathbf{0}_M, M)$ with $W' + W' \subset W$. By regularity of $M$ and Problem 204 there exists a balanced set $V \in \tau(\mathbf{0}_M, M)$ such that $\overline{V} \subset W'$. Observe that the set $\overline{V}$ is also balanced by Problem 203 and, in particular, $-\overline{V} \subset \overline{V}$.

All elements of $\Phi$ are continuous, so $E = \bigcap \{f^{-1}(\overline{V}) : f \in \Phi\}$ is a closed subset of $L$. If $x \in B$ then there is $n \in \mathbb{N}$ such that $\Phi(x) \subset nV$ because $\Phi(x)$ is $l$-bounded in $M$. Therefore $f(x) \in nV$ which shows that $f(\frac{x}{n}) \subset V$ and hence $\frac{x}{n} \in f^{-1}(V)$ for any $f \in \Phi$; consequently, $\frac{x}{n} \in \bigcap \{f^{-1}(V) : f \in \Phi\} \subset E$. This proves that, for any $x \in B$ there exists $n \in \mathbb{N}$ such that $\frac{x}{n} \in E$, i.e., $x \in nE$.

We proved the inclusion $B \subset \bigcup \{nE : n \in \mathbb{N}\}$, so $nE$ is of second category in $L$ for some $n \in \mathbb{N}$. The multiplication by $n$ being a homeomorphism of $L$ onto $L$, we conclude that $E$ is of second category in $L$. Recalling that $E$ is closed we convince ourselves that the interior of $E$ is nonempty, so we can find a set $U \in \tau(\mathbf{0}_L, L)$ and $a \in E$ such that $a + U \subset E$.

Given any $u \in U$ and $f \in \Phi$ it follows from $f(u) = f(a + u) - f(a)$ that $f(u) \in f(E) + f(-E) \subset \overline{V} + (-\overline{V}) \subset \overline{V} + \overline{V} \subset W' + W' \subset W$. This shows that $f(U) \subset W$; the set $W \in \tau(\mathbf{0}_M, M)$ was chosen arbitrarily, so we proved that $\Phi$ is an equicontinuous family. Given any $x \in L$ the set $\{x\}$ is $l$-bounded in $L$ (this is an easy consequence of the fact that every $G \in \tau(\mathbf{0}_L, L)$ is absorbing), so the set $\Phi(x)$ is $l$-bounded in $M$ by Problem 214. Thus every $x \in L$ belongs to $B$, i.e., $B = L$ as promised.

**V.216 (Hahn–Banach theorem).** *Let $L$ be a linear space (without topology). Suppose that we are given a map $p : L \to \mathbb{R}$ such that $p(x+y) \leq p(x) + p(y)$ and $p(tx) = tp(x)$ for all $x, y \in L$ and $t \geq 0$. Prove that, for any linear subspace $M$ of the linear space $L$ and any linear functional $f : M \to \mathbb{R}$ such that $f(x) \leq p(x)$ for any $x \in M$, there exists a linear functional $F : L \to \mathbb{R}$ such that $F|M = f$ and $-p(-x) \leq F(x) \leq p(x)$ for any $x \in L$.*

**Solution.** Consider the family $\mathcal{L} = \{(N, \varphi) : N$ is a linear subspace of $L$ with $M \subset N$ and $\varphi : N \to \mathbb{R}$ is a linear functional such that $\varphi|M = f$ and $\varphi(x) \leq p(x)$ for any $x \in N\}$. Given $(N, \varphi) \in \mathcal{L}$ and $(N', \varphi') \in \mathcal{L}$ say that $(N, \varphi) \prec (N', \varphi')$ if $N \subset N'$ and $\varphi \subset \varphi'$., i.e., $\varphi'|N = \varphi$. It is easy to see that $(\mathcal{L}, \prec)$ is a partially ordered set; for any $\alpha = (N, \varphi) \in \mathcal{L}$ let $S(\alpha) = N$ and $\Phi(\alpha) = \varphi$.

Suppose that $\mathcal{C}$ is a chain in $(\mathcal{L}, \prec)$ and let $N = \bigcup \{S(\alpha) : \alpha \in \mathcal{C}\}$. The family $\{S(\alpha) : \alpha \in \mathcal{C}\}$ is a chain of linear subspaces of $L$, so $N$ is a linear subspace of $L$ as well. We leave to the reader the checking that $\varphi = \bigcup \{\Phi(\alpha) : \alpha \in \mathcal{C}\}$ is a well-defined linear functional on $N$. Since $\Phi(\alpha)|M = f$ for any $\alpha \in \mathcal{C}$, we have $\varphi|M = f$. Given $x \in N$ there is $\alpha \in \mathcal{C}$ such that $|\varphi(x)| = |\Phi(\alpha)(x)| \leq p(x)$, so $\varphi(x) \leq p(x)$ for any $x \in N$. This proves that $(N, \varphi) \in \mathcal{L}$; it is evident that $\alpha \prec (N, \varphi)$ for any $\alpha \in \mathcal{C}$, so $(N, \varphi)$ is an upper bound for the chain $\mathcal{C}$.

We proved that any chain of $(\mathcal{L}, \prec)$ has an upper bound which belongs to $\mathcal{L}$, so Zorn's lemma is applicable to conclude that $(\mathcal{L}, \prec)$ has a maximal element $(N, \varphi)$. If $N = L$ then letting $F = \varphi$ we obtain a linear functional $F : L \to \mathbb{R}$ such that $F|M = f$ and $F(x) \leq p(x)$ for any $x \in L$.

Assume, toward a contradiction, that $N \neq L$ and fix a vector $v \in L \backslash N$. It is evident that $N' = \{x + tv : x \in N, \ t \in \mathbb{R}\}$ is a linear subspace of $L$ such that $N \subset N'$ and $N \neq N'$. Observe also that, for any $w \in N'$ there are unique $x \in N$ and $t \in \mathbb{R}$ such that $w = x + tv$.

The formula $\varphi(x) + \varphi(y) = \varphi(x + y) \leq p(x + y) \leq p(x - v) + p(v + y)$ implies that

(1) $\varphi(x) - p(x - v) \leq -\varphi(y) + p(v + y)$ for any $x, y \in N$.

It follows from (1) that, for $a = \sup\{\varphi(x) - p(x - v) : x \in N\}$, we have

(2) $\varphi(x) - p(x - v) \leq a \leq -\varphi(y) + p(v + y)$ for all $x, y \in N$.

For any $x \in N$ and $t \in \mathbb{R}$ let $\xi(x + tv) = \varphi(x) + ta$; it is straightforward that $\xi : N' \to \mathbb{R}$ is a linear functional such that $\xi|N = \varphi$. Fix any $w \in N'$; there are unique $x \in N$ and $t \in \mathbb{R}$ such that $w = x + tv$. If $t = 0$ then $w \in N$, so $\xi(w) = \varphi(w) \leq p(w)$. Now, assume that $t > 0$ and let $y = \frac{x}{t}$ in the second inequality of (2); then $a + \varphi(\frac{x}{t}) \leq p(v + \frac{x}{t})$ and hence $\varphi(x) + ta \leq p(x + tv)$, i.e., $\xi(w) \leq p(w)$.

If $t < 0$ then replace $x$ with $\frac{x}{-t}$ in the first inequality of (2). This gives us the formula $\varphi(\frac{x}{-t}) - a \leq p(\frac{x}{-t} - v)$ which can be multiplied by $-t$ to conclude that $\varphi(x) + ta \leq p(x + tv)$, i.e., $\xi(w) \leq p(w)$ in this case as well.

We have proved that $\xi(w) \leq p(w)$ for any $w \in N'$ and hence the pair $(N', \xi)$ belongs to $\mathcal{L}$. We also have $(N, \varphi) \prec (N', \xi)$ and $(N, \varphi) \neq (N', \xi)$; this contradiction with maximality of $(N, \varphi)$ shows that the case of $N \neq L$ is impossible. Thus $N = L$ and the linear functional $F = \varphi$ is an extension of $f$ such that $F(x) \leq p(x)$ for any $x \in L$. Multiplying the last inequality by $(-1)$ we convince ourselves that $-F(x) \geq -p(x)$; replacing $x$ with $-x$ we conclude that $F(x) = -F(-x) \geq -p(-x)$, i.e., $-p(-x) \leq F(x) \leq p(x)$ for all $x \in L$, so our solution is complete.

**V.217.** *Let $L$ be a linear space (without topology). Suppose that we are given a seminorm $p : L \to \mathbb{R}$, a linear subspace $M \subset L$ and a linear functional $f : M \to \mathbb{R}$ such that $|f(x)| \leq p(x)$ for any $x \in M$. Prove that there exists a linear functional $F : L \to \mathbb{R}$ such that $F|M = f$ and $|F(x)| \leq p(x)$ for any $x \in L$.*

**Solution.** We can apply Problem 216 to find a linear functional $F : L \to \mathbb{R}$ such that $F|M = f$ and $-p(-x) \leq F(x) \leq p(x)$ for every $x \in L$. Since $p$ is a seminorm, we have $p(-x) = p(x)$ and hence $-p(x) \leq F(x) \leq p(x)$, i.e., $|F(x)| \leq p(x)$ for every $x \in L$.

**V.218.** *Given a linear topological space $L$ prove that any nontrivial continuous linear functional $f : L \to \mathbb{R}$ is an open map.*

**Solution.** Take any $U \in \tau(\mathbf{0}, L)$ and apply Problem 204 to find an open balanced set $V$ such that $\mathbf{0} \in V \subset U$. Suppose for a moment that $f(V) = \{0\}$ and take any $x \in L$; the set $V$ is absorbing by Problem 202, so there is $t > 0$ such that $x \in tV$, i.e., $\frac{x}{t} \in V$ and therefore $f(\frac{x}{t}) = 0$ which implies that $f(x) = tf(\frac{x}{t}) = t \cdot 0 = 0$. This shows that $f(x) = 0$ for any $x \in L$, i.e., $f$ is a trivial functional which is a contradiction. Therefore we can find a point $x_0 \in V$ with $f(x_0) \neq 0$. The interval

$G = (-|f(x_0)|, |f(x_0)|)$ is an open neighborhood of 0; since $V$ is balanced, the set $H = \{tx_0 : t \in (-1, 1)\}$ is contained in $V$ and hence $G = f(H) \subset f(V) \subset f(U)$. This shows that for any open neighborhood $U$ of the point $\mathbf{0}$ the set $f(U)$ contains an open neighborhood of 0, so the map $f$ is open by Fact 3 of S.496.

**V.219.** *Let $L$ be a linear topological space and suppose that $A$ and $B$ are nonempty disjoint convex subsets of $L$ and $A$ is open. Prove that there exists a continuous linear functional $f : L \to \mathbb{R}$ such that, for some $r \in \mathbb{R}$, we have $f(x) < r \le f(y)$ for any $x \in A$ and $y \in B$.*

**Solution.** Pick any $a \in A$, $b \in B$ and consider the set $C = A - B + (b - a)$. We leave it to the reader to verify that $\mathbf{0} \in C$ and $C$ is a convex open (and hence absorbing) subset of $L$; let $p : L \to \mathbb{R}$ be the Minkowski functional of $C$. Observe that $x_0 = b - a$ does not belong to $C$ for otherwise there exist $x \in A$, $y \in B$ such that $x - y + b - a = b - a$ which implies $x = y$ and hence $x \in A \cap B = \emptyset$ which is a contradiction. It follows from $x_0 \notin C$ and Problem 208 that $p(x_0) \ge 1$.

The set $M = \{tx_0 : t \in \mathbb{R}\}$ is a linear subspace of $L$; let $\varphi(tx_0) = t$ for any $t \in \mathbb{R}$. It is straightforward that $\varphi : M \to \mathbb{R}$ is a linear functional. If $t \ge 0$ then $\varphi(tx_0) = t \le tp(x_0) = p(tx_0)$; if $t < 0$ then $\varphi(tx_0) = t < 0 \le p(tx_0)$, so $\varphi(y) \le p(y)$ for any $y \in M$. By Problem 216 there is a linear functional $f : L \to \mathbb{R}$ such that $f|M = \varphi$ and $f(x) \le p(x)$ for any $x \in L$. In particular, $f(x) \le p(x) \le 1$ for any $x \in C$ (see Problem 208) and hence $f(x) \ge -1$ for any $x \in -C$; thus $W = C \cap (-C)$ is a neighborhood of $\mathbf{0}$ such that $f(W) \subset [-1, 1]$ is a bounded subset of $\mathbb{R}$. Therefore the functional $f$ is continuous by Problem 205.

Given $x \in A$ and $y \in B$ the point $z = x - y + x_0$ belongs to $C$. By continuity of scalar multiplication in $L$ there exists $t > 1$ such that $tz = \frac{z}{1/t} \in C$, so $p(z) \le \frac{1}{t} < 1$. Therefore $f(z) = f(x) - f(y) + 1 \le p(z) < 1$ which shows that $f(x) < f(y)$ for any $x \in A$ and $y \in B$. As a consequence, for the number $r = \sup\{f(x) : x \in A\}$, we have $f(x) \le r \le f(y)$ whenever $x \in A$ and $y \in B$. Observe that $f(x_0) = \varphi(x_0) = 1$, so the functional $f$ is nontrivial and hence open by Problem 218. This implies that $f(A)$ is an open subset of $\mathbb{R}$, so $r = \sup f(A)$ cannot belong to $f(A)$. Consequently, $f(x) < r \le f(y)$ for any $x \in A$ and $y \in B$.

**V.220.** *Let $L$ be a locally convex linear topological space. Suppose that $A$ and $B$ are disjoint convex subsets of $L$ such that $A$ is compact and $B$ is closed. Prove that there exists a continuous linear functional $f : L \to \mathbb{R}$ such that, for some $r, s \in \mathbb{R}$, we have $f(x) < r < s < f(y)$ for any $x \in A$ and $y \in B$.*

**Solution.** For any $a \in A$ there exists an open neighborhood $U_a$ of the point $\mathbf{0}$ such that $(a + U_a) \cap B = \emptyset$; choose a convex open neighborhood $V_a$ of $\mathbf{0}$ with $V_a + V_a \subset U_a$. The set $A$ being compact we can find a finite set $K \subset A$ such that $A \subset \bigcup\{a + V_a : a \in K\}$. The set $V = \bigcap\{V_a : a \in K\}$ is an open convex neighborhood of $\mathbf{0}$. If $x \in (A + V) \cap B$ then $x = y + v$ for some $y \in A$ and $v \in V$; pick a point $a \in K$ such that $y \in a + V_a$. Then $x \in y + V \subset y + V_a \subset a + V_a + V_a \subset a + U_a$, so $x \in (a + U_a) \cap B$ which is a contradiction. Therefore $W = A + V$ is a convex open set (it is an easy exercise that the sum of two convex sets is a convex set) disjoint from $B$.

Applying Problem 219 we conclude that there exists a continuous linear functional $f : L \to \mathbb{R}$ such that, for some $t \in \mathbb{R}$, we have $f(x) < t \leq f(y)$ for all $x \in W$ and $y \in B$. The set $f(A) \subset f(W) \subset (-\infty, t)$ being compact, there exists $r < t$ for which $f(A) \subset (-\infty, r)$; if $s = \frac{r+t}{2}$ then $f(A) \subset (-\infty, r)$ and $f(B) \subset (s, +\infty)$ which is equivalent to saying that $f(x) < r < s < f(y)$ for any $x \in A$ and $y \in B$.

**V.221.** *Let L be a locally convex linear topological space. Prove that $L^*$ separates the points of L.*

**Solution.** If $a$ and $b$ are distinct points of $L$ then $A = \{a\}$ and $B = \{b\}$ are disjoint convex compact subsets of $L$, so we can apply Problem 220 to see that there is a continuous linear functional $f : L \to \mathbb{R}$ (i.e., $f \in L^*$) such that, for some $r, s \in \mathbb{R}$, we have $f(x) < r < s < f(y)$ whenever $x \in A$ and $y \in B$. This is the same as saying that $f(a) < r < s < f(b)$, so $f(a) \neq f(b)$ and hence $L^*$ separates the points of $L$.

**V.222.** *Let M be a linear subspace of a locally convex linear topological space L and $x_0 \notin \overline{M}$. Prove that there exists $f \in L^*$ such that $f(x_0) = 1$ and $f(M) = \{0\}$.*

**Solution.** The set $N = \overline{M}$ is also a linear subspace of $L$ by Problem 203. Every linear subspace of $L$ is clearly convex so $A = \{x_0\}$ and $N$ are closed disjoint convex subsets of $L$ and $A$ is compact. This makes it possible to apply Problem 220 to find a functional $g \in L^*$ such that, for some $r, s \in \mathbb{R}$, we have

(1) $g(x_0) < r < s < g(y)$ and, in particular, $g(y) \neq g(x_0)$ for any $y \in N$.

If $g(x) \neq 0$ for some $x \in N$ then $t = \frac{g(x_0)}{g(x)}$ is well defined and $y = tx \in N$, so $g(y) = tg(x) = g(x_0)$ which is a contradiction with (1). Therefore $g(x) = 0$ for any $x \in N$; letting $f = \frac{1}{g(x_0)} \cdot g$ we obtain a functional $f \in L^*$ such that $f(N) = \{0\}$ and $f(x_0) = 1$ as promised.

**V.223.** *Let B be a closed convex balanced subset of a locally convex space L. Prove that, for any $x \in L \setminus B$, there exists a continuous linear functional $f : L \to \mathbb{R}$ such that $f(B) \subset [-1, 1]$ and $f(x) > 1$.*

**Solution.** If $B = \emptyset$ then we can apply Problem 221 to find a functional $g \in L^*$ with $g(x) \neq 0$. It is evident that, for the functional $f = \frac{2}{g(x)} \cdot g \in L^*$, we have $f(x) = 2 > 1$ and $f(B) = \emptyset \subset [-1, 1]$.

Now assume that $B \neq \emptyset$ and fix a point $b \in B$; the sets $A = \{x\}$ and $B$ are closed, convex and disjoint; besides, $A$ is compact. Therefore Problem 220 is applicable to convince ourselves that there exists a functional $g \in L^*$ with the following property:

(1) there is $r \in \mathbb{R}$ such that $g(x) < r < g(y)$ for any $y \in B$.

Since $B$ is balanced, the point $0 = 0 \cdot b$ belongs to $B$ and therefore $r < g(0) = 0$. Take any $z \in B$ and suppose that $g(z) > -r = |r|$; the point $y = -z$ also belongs to $B$ and hence $g(y) = -g(z) < r$ which is a contraction with (1). This shows that $g(B) \subset [-|r|, |r|]$. The functional $f = \frac{1}{r} \cdot g$ is continuous on $L$ and we have $|f(y)| = \frac{1}{|r|}|g(y)| \leq 1$ for any $y \in B$, i.e., $f(B) \subset [-1, 1]$. Finally, observe that $g(x) < r < 0$ implies that $f(x) = \frac{1}{r} \cdot g(x) > 1$.

**V.224.** *Let $L$ be a locally convex linear topological space. Given a linear subspace $M$ of the linear space $L$ and a continuous linear functional $f : M \to \mathbb{R}$, prove that there exists a functional $g \in L^*$ such that $g|M = f$.*

**Solution.** If $f$ is trivial then the trivial functional on $L$ is the desired extension of $f$. Now assume that $f$ is nontrivial and fix a point $x_0 \in M$ such that $f(x_0) \neq 0$. If $N = f^{-1}(0)$ then $N \subset M$ is a linear subspace of $L$; it follows from continuity of $f$ that $x_0 \notin \overline{N}$, so we can apply Problem 222 to find a functional $h \in L^*$ such that $h(x_0) = 1$ and $h(N) = \{0\}$. Then $g = f(x_0) \cdot h$ is a continuous linear functional on $L$ such that $g(N) = \{0\}$ and $g(x_0) = f(x_0)$; take any point $x \in M$. If $y = x - \frac{f(x)}{f(x_0)} \cdot x_0$ then it is immediate that $f(y) = 0$ and hence $y \in N$. Therefore $g(y) = g(x) - \frac{f(x)}{f(x_0)} g(x_0) = 0$; recalling that $g(x_0) = f(x_0)$ we conclude that $g(x) - f(x) = 0$, i.e., $g(x) = f(x)$. We proved that $g(x) = f(x)$ for any $x \in M$, so $g|M = f$ as required.

**V.225.** *Given a linear space $L$ (without topology) denote by $L'$ the family of all linear functionals on $L$. Suppose that $M \subset L'$ is a linear subspace of $L'$ (i.e., $\alpha f + \beta g \in M$ whenever $f, g \in M$ and $\alpha, \beta \in \mathbb{R}$) and $M$ separates the points of $L$; let $\mu$ be the topology generated by the set $M$. Then $L_M = (L, \mu)$ is a locally convex space and $(L_M)^* = M$. Deduce from this fact that if $L$ is a locally convex space and $L_w$ is the set $L$ with the weak topology of the space $L$ then $L_w$ is a locally convex space such that $(L_w)^* = L^*$.*

**Solution.** For any $f \in M$ let $O(f, n) = \{x \in L : |f(x)| < \frac{1}{n}\}$ for each $n \in \mathbb{N}$; observe also that the function $p_f = |f|$ is a seminorm on $L$ and $\{p_f : f \in M\}$ is a separating family of seminorms on $L$. Therefore we can apply Problem 210 to see that there exists a unique locally convex linear space topology $\tau$ on the set $L$ such that $\mathcal{B} = \{O(f_1, n) \cap \ldots \cap O(f_n, n) : n \in \mathbb{N}, \ f_1, \ldots, f_n \in M\}$ is a local base at $\mathbf{0}$; it is evident that $\tau \subset \mu$. If $f \in M$ and $\varepsilon > 0$ then pick $n \in \mathbb{N}$ with $\frac{1}{n} < \varepsilon$. The set $O(f, n) \ni \mathbf{0}$ belongs to $\tau$ and $f(O(f, n)) \subset (-\varepsilon, \varepsilon)$; this proves that every $f \in M$ is continuous at $\mathbf{0}$ and hence continuous on $(L, \tau)$ (see Fact 2 of S.496). An immediate consequence is that $\mu \subset \tau$ and hence $\mu = \tau$ is a locally convex linear space topology on $L$.

It is straightforward that $M \subset (L_M)^*$; to prove the opposite inclusion take any functional $f \in (L_M)^*$. By continuity of $f$ at $\mathbf{0}$ there exist $n \in \mathbb{N}$ and $f_1, \ldots, f_n \in M$ such that $f(O(f_1, n) \cap \ldots \cap O(f_n, n)) \subset (-1, 1)$. Let $N = f_1^{-1}(0) \cap \ldots \cap f_n^{-1}(0)$; we have $f_i^{-1}(0) \subset O(f_i, n)$ for every $i \leq n$, so $N \subset \bigcap_{i=1}^{n} O(f_i, n)$ and therefore $f(N) \subset (-1, 1)$. If $x \in N$ and $r = f(x) \neq 0$ then $y = \frac{1}{r} \cdot x \in N$ (because $N$ is a linear subspace of $L$), so $f(y) = \frac{1}{r} \cdot f(x) = 1 \notin (-1, 1)$ which is a contradiction. Thus $f(N) = \{0\}$, so we can apply Fact 14 of V.100 to find $t_1, \ldots, t_n \in \mathbb{R}$ such that $f = t_1 f_1 + \ldots + t_n f_n$; since $M$ is a linear subspace of $L'$, we proved that $f \in M$ and hence $(L_M)^* = M$.

Finally observe that if $L$ is a locally convex space then $L^*$ separates the points of $L$ by Problem 221; since $\tau(L_w)$ is generated by $L^*$, the space $L_w$ is locally convex and $(L_w)^* = L^*$.

**V.226.** *Let $E$ be a convex subset of a locally convex space $L$. Prove that the closure of $E$ in $L$ coincides with the closure of $E$ in the weak topology of $L$.*

**Solution.** Denote by $\mu$ the weak topology of $L$; given $A \subset L$ the set $\overline{A}$ is the closure of $A$ in $L$ and $\mathrm{cl}_\mu(A)$ is the closure of $A$ in $(L, \mu)$. It follows from $\mu \subset \tau(L)$ that $\overline{A} \subset \mathrm{cl}_\mu(A)$ for any $A \subset L$, so it suffices to show that $\mathrm{cl}_\mu(E) \subset \overline{E}$.

Fix a point $a \notin \overline{E}$; the sets $A = \{a\}$ and $\overline{E}$ are convex (see Problem 203), closed, disjoint and $A$ is compact so we can apply Problem 220 to find a functional $f \in L^*$ and $r \in \mathbb{R}$ such that $f(a) < r < f(y)$ for any $y \in \overline{E}$. The set $U = f^{-1}((-\infty, r))$ belongs to $\mu$ while $a \in U$ and $U \cap E = \emptyset$; an immediate consequence is that $a \notin \mathrm{cl}_\mu(E)$ and hence $\mathrm{cl}_\mu(E) = \overline{E}$.

**V.227.** *Let $V$ be a neighborhood of $\mathbf{0}$ in a locally convex space $L$. Prove that the set $P(V) = \{f \in L^* : f(V) \subset [-1, 1]\}$ is compact if considered with the topology induced from $C_p(L)$.*

**Solution.** The topology induced on $P(V)$ from $C_p(L) \subset \mathbb{R}^L$ is the same as the topology induced from $\mathbb{R}^L$, so it suffices to show that $P(V)$ is closed in $\mathbb{R}^L$ and there is a compact $K \subset \mathbb{R}^L$ such that $P(V) \subset K$.

Denote by $L'$ the set of all (not necessarily continuous) linear functionals on $L$ and take any point $f \in \overline{P(V)}$ (the bar denotes the closure in $\mathbb{R}^L$). The set $L' \subset \mathbb{R}^L$ is closed in $\mathbb{R}^L$ (see CFS-393) and $P(V) \subset L'$, so $f \in L'$, i.e., $f$ is a linear functional on $L$. If there is $x \in L$ such that $f(x) \notin \mathbb{I}$ then $G = \{g \in \mathbb{R}^L : g(x) \notin \mathbb{I}\}$ is an open neighborhood of $f$ in $\mathbb{R}^L$ such that $G \cap P(V) = \emptyset$; this contradiction shows that $f(x) \in \mathbb{I}$ for any $x \in V$, i.e., $f(V) \subset \mathbb{I}$ and hence $f(V)$ is a bounded subset of $\mathbb{R}$. Applying Problem 205 we conclude that $f$ is continuous and hence $f \in P(V)$. Thus $P(V)$ is closed in $\mathbb{R}^L$.

Now fix a point $x \in L$; the set $V$ is absorbing by Problem 202, so there is $r(x) \in \mathbb{R}$ such that $r(x) > 0$ and $x \in r(x)V$, i.e., there exists $v \in V$ with $x = r(x) \cdot v$. If $f \in P(V)$ then $|f(x)| = r(x) \cdot |f(v)| \leq r(x)$; therefore $f(x) \in [-r(x), r(x)]$ for any $f \in P(V)$. The point $x \in L$ was chosen arbitrarily, so we have proved the inclusion $P(V) \subset K = \prod\{[-r(x), r(x)] : x \in L\}$. Consequently, $P(V)$ is compact being a closed subset of a compact set $K$.

**V.228.** *Given $n \in \mathbb{N}$ suppose that $L$ is a linear topological space and $M$ is a linear subspace of $L$ of linear dimension $n$. Prove that $M$ is closed in $L$ and every linear isomorphism $\varphi : \mathbb{R}^n \to M$ is a homeomorphism.*

**Solution.** Our first step is to prove that locally compact linear subspaces are always closed in a linear topological space.

*Fact 1.* Suppose that $N$ is a linear topological space and $G \subset N$ is a locally compact linear subspace of $N$. Then $G$ is closed in $N$.

*Proof.* Fix a compact set $K \subset G$ which contains a neighborhood of $\mathbf{0}$ in the space $G$ and pick a set $O \in \tau(\mathbf{0}, N)$ such that $O \cap G \subset K$. Use Problems 202 and 204 to find a symmetric set $V \in \tau(\mathbf{0}, N)$ with $V + V \subset O$ and take any point $x \in \overline{G}$.

The set $U = x + V$ is an open neighborhood of the point $x$ in $N$; an easy consequence is that $x \in \overline{U \cap G}$. Fix a point $x_0 \in P = U \cap G$ and take any $y \in P$. There exist points $v_0, v_1 \in V$ such that $x_0 = x + v_0$ and $y = x + v_1$. Then $y - x_0 = v_1 - v_0 \in V + V \subset O$; since $G$ is a linear subspace of $N$, the point $y - x_0$ belongs to $G$. Therefore $y - x_0 \in O \cap G \subset K$ which shows that $y \in x_0 + K$ for any $y \in P$ and hence $P \subset K' = x_0 + K \subset G$. The space $K'$ being compact we conclude that $\overline{P} \subset K'$, so $x \in \overline{P} \subset G$ and hence $x \in G$. Thus $\overline{G} \subset G$, i.e., $G$ is closed in $N$, so Fact 1 is proved.

Returning to our solution consider, for any $n \in \mathbb{N}$, the following statement:

($S_n$) if $L$ is a linear topological space and $M$ is a linear subspace of $L$ of linear dimension $n$ then every isomorphism $\varphi : \mathbb{R}^n \to M$ is a homeomorphism.

To prove inductively that ($S_n$) is true for each $n \in \mathbb{N}$ assume first that $n = 1$ and $\varphi : \mathbb{R} \to M$ is an isomorphism. If $x_0 = \varphi(1)$ then $\varphi(t) = t \cdot x_0$ for any $t \in \mathbb{R}$. It follows from continuity of multiplication by scalars in $L$ that the map $\varphi$ is continuous. The map $\xi = \varphi^{-1} : M \to \mathbb{R}$ is linear, i.e., $\xi$ is a nontrivial linear functional on $M$. Besides, $\xi^{-1}(0) = \{\mathbf{0}\}$ is a closed subspace of $M$, so $\xi$ is also continuous by Problem 205. Thus $\varphi$ is a homeomorphism, so ($S_1$) is proved.

Suppose that $n \in \mathbb{N}$, $n > 1$, the statement ($S_k$) is proved for any $k < n$ and take any linear isomorphism $\varphi : \mathbb{R}^n \to M$. For each $i = 1, \ldots, n$ let $e_i = (e_1^i, \ldots, e_n^i)$ where $e_i^i = 1$ and $e_j^i = 0$ whenever $i \neq j$. Then $\{e_1, \ldots, e_n\}$ is a linear basis in $\mathbb{R}^n$, so $G = \{\varphi(e_1), \ldots \varphi(e_n)\}$ is a linear basis in $M$. If $g_i = \varphi(e_i)$ for all $i \leq n$ then $\varphi(t_1, \ldots, t_n) = t_1 g_1 + \ldots + t_n g_n$ for any point $(t_1, \ldots, t_n) \in \mathbb{R}^n$. An easy consequence of continuity of the operations in $L$ is that the map $\varphi$ is continuous.

For any point $x \in M$ there are uniquely determined $t_1, \ldots, t_n \in \mathbb{R}$ such that $x = t_1 g_1 + \ldots + t_n g_n$; let $p_i(x) = t_i$ for each $i \leq n$. It is easy to see that every $p_i : M \to \mathbb{R}$ is a linear functional and $\varphi^{-1} = p_1 \Delta \ldots \Delta p_n$, so, to prove continuity of $\varphi^{-1}$, it suffices to show that every $p_i$ is continuous. Observe that $p_i^{-1}(0)$ coincides with the linear hull of the set $G \backslash \{g_i\}$, so $F = p_i^{-1}(0)$ is an $(n - 1)$-dimensional linear subspace of $M$. By the induction hypothesis, the space $F$ is homeomorphic to $\mathbb{R}^{n-1}$ and hence locally compact. Therefore $F$ is closed in $M$ by Fact 1. This makes it possible to apply Problem 205 again to conclude that every $p_i$ is continuous and hence $\varphi^{-1}$ is also continuous, i.e., $\varphi$ is a homeomorphism.

Our inductive step having been accomplished, we have proved that ($S_n$) holds for every $n \in \mathbb{N}$. It is an easy exercise that if a linear space $M$ has linear dimension $n$ then there exists an isomorphism between $M$ and $\mathbb{R}^n$; consequently, the subspace $M$ in ($S_n$) is homeomorphic to $\mathbb{R}^n$ which, in turn, shows that $M$ is locally compact and hence closed in $L$ by Fact 1. Therefore, in the statement ($S_n$), the set $M$ is automatically closed in $L$, so our solution is complete.

**V.229.** *Given a linear topological space $L$ prove that the following conditions are equivalent:*

*(i)  L has a finite Hamel basis, i.e., the linear dimension of L is finite;*
*(ii)  dim $L \leq n$ for some $n \in \mathbb{N}$;*
*(iii)  L is locally compact.*

**Solution.** If $L$ has linear dimension $n$ for some $n \in \mathbb{N}$ then $L$ is locally compact being homeomorphic to $\mathbb{R}^n$ (see Problem 228); this settles (i)$\Longrightarrow$(iii). Besides, it follows from Problem 159 that dim $L = n$, so we also have the implication (i)$\Longrightarrow$(ii).

Next, assume that $n \in \mathbb{N}$ and dim $L \leq n$. If there are some linearly independent vectors $e_1, \ldots, e_{n+1} \in L$ then the linear hull $M$ of the set $\{e_1, \ldots, e_{n+1}\}$ is a linear subspace of $L$ of linear dimension $n + 1$, so it follows from Problem 228 that $M$ is homeomorphic to $\mathbb{R}^{n+1}$. As a consequence, $L$ has a subspace $K$ homeomorphic to $\mathbb{I}^{n+1}$. A compact subspace is $C$-embedded in any space (see Fact 1 of T.218), so dim $K \leq$ dim $L \leq n$ by Problem 146. However, dim $K = \dim(\mathbb{I}^{n+1}) = n + 1$ by Problem 159; this contradiction shows that the linear dimension of $L$ does not exceed $n$ and hence (ii)$\Longrightarrow$(i) is proved.

To establish that (iii)$\Longrightarrow$(i) assume that $L$ is locally compact and hence we can find a balanced set $O \in \tau(\mathbf{0}, L)$ such that $K = \overline{O}$ is compact; let $W_n = 2^{-n} \cdot O$ for any $n \in \omega$. If $U$ is an open neighborhood of $\mathbf{0}$ then pick an open balanced neighborhood $V$ of $\mathbf{0}$ such that $V \subset U$. Since $V$ is an absorbing subset of $L$ by Problem 202, we can find, for every $x \in K$ a number $t_x > 0$ such that $x \in t_x \cdot V$. The set $K$ being compact there is a finite $A \subset K$ for which $K \subset \bigcup\{t_x \cdot V : x \in A\}$; let $t = \sum\{t_x : x \in A\}$.

Since the set $V$ is balanced, we have $t_x \cdot V \subset t \cdot V$ for any $x \in A$ and hence $K \subset t \cdot V$. This implies $O \subset t \cdot V$ and therefore $\frac{1}{t} \cdot O \subset V$; pick $n \in \omega$ with $2^{-n} < \frac{1}{t}$. Then $W_n = 2^{-n} \cdot O \subset \frac{1}{t} \cdot O \subset V \subset U$. We proved that, for any $U \in \tau(\mathbf{0}, L)$ there is $n \in \omega$ with $W_n \subset U$ and hence the family $\mathcal{W} = \{W_n : n \in \omega\}$ is a local base of $L$ at $\mathbf{0}$.

Apply once more compactness of $K$ to find a finite set $B \subset K$ such that $K \subset \bigcup\{x + \frac{1}{2} \cdot O : x \in B\}$. If $M$ is the linear hull of $B$ then $M$ has finite linear dimension and $O \subset K \subset M + \frac{1}{2} \cdot O$. Proceeding inductively assume that $n \in \mathbb{N}$ and $O \subset M + 2^{-n} \cdot O$. The set $M$ being a linear subspace of $L$ we have $2^{-n} \cdot M = M$, so $2^{-n} \cdot O \subset 2^{-n} \cdot (M + \frac{1}{2} \cdot O) = M + 2^{-n-1} \cdot O$. An immediate consequence is that $O \subset M + M + 2^{-n-1} \cdot O = M + 2^{-n-1} \cdot O$. Thus our inductive procedure shows that $O \subset M + W_n$ for any $n \in \omega$. Since $\mathcal{W}$ is a local base of $L$ at $\mathbf{0}$, we can apply Problem 203 to see that $O \subset \bigcap\{M + W_n : n \in \omega\} = \overline{M}$. By Problem 228, the set $M$ is closed in $L$, so $\overline{M} = M$ and hence $O \subset M$. The set $O$ is absorbing, so, for any $x \in L$ there is $t \in \mathbb{R}$ with $x \in t \cdot O \subset t \cdot M = M$; therefore $L = M$ and hence linear dimension of $L$ is finite; this settles (iii)$\Longrightarrow$(i) and completes our solution.

**V.230.** *Suppose that $L$ is a finite-dimensional linear topological space. Prove that any linear functional $f : L \to \mathbb{R}$ is continuous on $L$, i.e., $L' = L^*$. Give an example of an infinite-dimensional locally convex space $M$ such that $M' = M^*$.*

**Solution.** The set $N = f^{-1}(0) \subset L$ is a finite-dimensional linear subspace of $L$, so $N$ is closed in $L$ by Problem 228. Therefore $f$ is continuous by Problem 205 and hence $L' = L^*$.

Let $D$ be an infinite discrete space. The space $M = L_p(D)$ is locally convex being a linear subspace of a locally convex space $C_p(C_p(D))$ (see Fact 1 of T.131). We can consider that $D \subset L_p(D)$ and $D$ is a Hamel basis of $L_p(D)$ (see Fact 5 of S.489); therefore the space $M$ is infinite-dimensional. If $f : M \to \mathbb{R}$ is a linear functional then $f$ is a linear extension of the function $g = f|D$; the space $D$ being discrete, the function $g$ is continuous, so there exists a continuous linear functional $h : M \to \mathbb{R}$ such that $h|D = g$. The set $D$ is a Hamel basis of $M$, so there is only one linear extension of $g$ over $M$ by Fact 3 of S.489; therefore $f = h$ is a continuous linear functional on $M$. Thus $M$ is an infinite-dimensional locally convex space such that $M' = M^*$.

**V.231.** *Let $L$ be a locally convex space. Denote by $L' \subset \mathbb{R}^L$ the set of all (not necessarily continuous) linear functionals on $L$ with the topology induced from $\mathbb{R}^L$. Prove that $L^*$ is dense in $L'$.*

**Solution.** Fix a linear functional $f : L \to \mathbb{R}$ and let $A$ be a finite subset of $L$. The linear span $M$ of the set $A$ is a finite-dimensional linear subspace of $L$, so the functional $f_0 = f|M$ is continuous by Problem 230. Apply Problem 224 to find a continuous linear functional $g : L \to \mathbb{R}$ such that $g|M = f_0$; since $A \subset M$, we also have $g|A = f_0|A = f|A$. This shows that, for any finite $A \subset L$ there is $g \in L^*$ with $g|A = f|A$. An immediate consequence is that $f$ belongs to the closure of the set $L^*$. The point $f \in L'$ was chosen arbitrarily, so $L^*$ is dense in $L'$.

**V.232.** *Given a linear space $L$ let $L' \subset \mathbb{R}^L$ be the set of all linear functionals on $L$ with the topology induced from $\mathbb{R}^L$. Prove that $L'$ is linearly homeomorphic to $\mathbb{R}^B$ for some $B$.*

**Solution.** Apply Fact 1 of S.489 to fix a Hamel basis $B \subset L$ and consider the restriction map $\pi : L' \to \mathbb{R}^B$. It is clear that $\pi$ is linear and continuous; Fact 3 of S.489 implies that $\pi(L') = \mathbb{R}^B$ and $\pi$ is an injection, i.e., $\pi$ is a condensation. We will denote by $\mathbf{0}_L$ the zero function on $L$ (which is also the zero element of $L'$); let $\mathbf{0}_B = \pi(0)$ be the zero function on $B$.

Given a finite set $A \subset L$ and $\varepsilon > 0$ let $O(A, \varepsilon) = \{f \in L' : f(A) \subset (-\varepsilon, \varepsilon)\}$. It is clear that the family $\mathcal{O} = \{O(A, \varepsilon) : A$ is a finite subset of $L$ and $\varepsilon > 0\}$ is a local base of $L'$ at $\mathbf{0}_L$. Analogously, if $C \subset B$ is a finite set and $\varepsilon > 0$ then $W(C, \varepsilon) = \{f \in \mathbb{R}^B : f(C) \subset (-\varepsilon, \varepsilon)\}$; the family $\mathcal{W} = \{W(C, \varepsilon) : C$ is a finite subset of $B$ and $\varepsilon > 0\}$ is a local base of $\mathbb{R}^B$ at $\mathbf{0}_B$.

Take an arbitrary set $U \in \tau(\mathbf{0}_L, L')$; we can fix $A \subset L$ and $\varepsilon > 0$ such that $O(A, \varepsilon) \subset U$. The set $B$ being a Hamel basis of $L$ there exist $b_0, \dots, b_n \in B$ such that $A$ is contained in the linear span of the set $C = \{b_0, \dots, b_n\}$. Thus, for every point $x \in A$ there exist $t_0^x, \dots, t_n^x \in \mathbb{R}$ such that $x = t_0^x b_0 + \dots + t_n^x b_n$. We will need the number $\delta = \frac{\varepsilon}{K} > 0$ where $K = \sum\{|t_i^x| : x \in A, \ i \le n\} + 1$.

Take any $f \in W(C, \delta)$ and let $\varphi = \pi^{-1}(f)$. Since the functional $\varphi$ is a linear extension of $f$, for any point $x \in A$ we have the equality $\varphi(x) = \sum_{i=0}^{n} t_i^x f(b_i)$, so $|\varphi(x)| \leq |t_0^x||f(b_0)| + \ldots + |t_n^x||f(b_n)| \leq \delta \cdot \sum_{i=0}^{n} |t_i^x| < K\delta = \varepsilon$ and therefore $\varphi \in O(A, \varepsilon)$. An immediate consequence is that $\pi(U) \supset \pi(O(A, \varepsilon)) \supset W(C, \delta)$, so we proved that the image under $\pi$ of every $U \in \tau(\mathbf{0}_L, L')$ contains an open neighborhood of $\mathbf{0}_B$ in $\mathbb{R}^B$. Applying Fact 3 of S.496 we convince ourselves that the map $\pi$ is open, i.e., $\pi$ is the desired linear homeomorphism between the spaces $L'$ and $\mathbb{R}^B$.

**V.233.** *For any linear topological space $L$ denote by $\tau_w(L)$ the weak topology of the space $L$. Prove that*

*(1) if $\tau_w(L) = \tau(L)$ and $M$ is a linear subspace of $L$ then $\tau_w(M) = \tau(M)$;*
*(2) for any space $X$, the topology of $C_p(X)$ coincides with its weak topology;*
*(3) for any space $X$, the topology of $L_p(X)$ coincides with its weak topology.*

**Solution.** The following fact is useful for establishing coincidence of topologies on linear spaces.

*Fact 1.* Given a linear space $L$ suppose that $\tau$ and $\mu$ are linear space topologies on $L$ such that, for any $U \in \tau$ with $\mathbf{0}_L \in U$ there exists $V \in \mu$ such that $\mathbf{0} \in V \subset U$. Then $\tau \subset \mu$.

*Proof.* Let $i : (L, \mu) \to (L, \tau)$ be the identity map, i.e., $i(x) = x$ for any $x \in L$. Our hypothesis is a reformulation of the fact that the map $i$ is continuous at $\mathbf{0}_L$. Since $i$ is a linear map, we can apply Fact 2 of S.496 to conclude that $i$ is continuous. Therefore $U \in \tau$ implies that $U = i^{-1}(U) \in \mu$, so $\tau \subset \mu$ and hence Fact 1 is proved.

Returning to our solution fix a linear topological space $L$ with $\tau(L) = \tau_w(L)$ and take any linear subspace $M \subset L$. It is evident that $\tau_w(M) \subset \tau(M)$, so take any $U \in \tau(\mathbf{0}_L, M)$; there exists $V \in \tau(\mathbf{0}_L, L)$ such that $V \cap M = U$. It follows from the equality $\tau(L) = \tau_w(L)$ that there exist $f_0, \ldots, f_n \in L^*$ and $\varepsilon > 0$ such that the set $G = \bigcap\{f_i^{-1}((-\varepsilon, \varepsilon)) : i \leq n\}$ is contained in $V$. The functional $g_i = f_i|M$ is continuous on $M$, so the set $W_i = g_i^{-1}((-\varepsilon, \varepsilon))$ belongs to $\tau_w(M)$ for each $i \leq n$. Thus the set $H = \bigcap\{W_i : i \leq n\}$ also belongs to $\tau_w(M)$; it is straightforward that $\mathbf{0}_L \in H \subset G \cap M \subset V \cap M = U$. This shows that, for every $U \in \tau(\mathbf{0}_L, M)$, there exists $H \in \tau_w(M)$ such that $\mathbf{0}_L \in H \subset U$ and hence $\tau(M) \subset \tau_w(M)$ by Fact 1. Therefore $\tau(M) = \tau_w(M)$; this settles (1).

(2) Let $\tau$ be the topology of $C_p(X)$ and denote by $\tau_w$ its weak topology; it is evident that $\tau_w \subset \tau$. Fix a set $U \in \tau$ with $\mathbf{0}_X \in U$. There exists a finite set $A \subset X$ such that $U_0 = \{f \in C_p(X) : f(A) \subset (-\varepsilon, \varepsilon)\} \subset U$.

For any $x \in A$ let $e_x(f) = f(x)$ for any $f \in C_p(X)$; then $e_x : C_p(X) \to \mathbb{R}$ is a continuous linear functional on $C_p(X)$ and hence the set $V_x = e_x^{-1}((-\varepsilon, \varepsilon))$ belongs to $\tau_w$. It is immediate that $U_0 = \bigcap\{V_x : x \in A\}$, so $U_0 \in \tau_w$. This shows that, for every $U \in \tau$ with $\mathbf{0}_X \in U$ there is $U_0 \in \tau_w$ such that $\mathbf{0} \in U_0 \subset U$. Therefore $\tau \subset \tau_w$ (see Fact 1) and hence $\tau = \tau_w$.

(3) Recall that $L_p(X)$ is a linear subspace of $C_p(C_p(X))$; since the weak topology of $C_p(C_p(X))$ coincides with its topology by (2), we can apply (1) to conclude that the weak topology of $L_p(X)$ coincides with the topology of $L_p(X)$.

**V.234.** *Suppose that $L$ is a locally convex space with its weak topology and $X \subset L$ is a Hamel basis in $L$. Prove that the following conditions are equivalent:*

(i) *there exists a linear homeomorphism $h : L \to L_p(X)$ such that $h(x) = x$ for all $x \in X$;*

(ii) *for every $f \in C(X)$ there exists a continuous linear functional $\varphi : L \to \mathbb{R}$ such that $\varphi|X = f$;*

(iii) *for every continuous map $f : X \to M$ from $X$ to a locally convex space $M$ with its weak topology, there exists a continuous linear map $\Phi : L \to M$ such that $\Phi|X = f$.*

**Solution.** Assume that (i) holds and take a continuous function $f : X \to \mathbb{R}$. Apply Fact 6 of S.489 to find a continuous linear functional $\varphi_0 : L_p(X) \to \mathbb{R}$ such that $\varphi_0|X = f$. Then $\varphi = \varphi_0 \circ h$ is a continuous linear functional on $L$. If $x \in X$ then $\varphi(x) = \varphi_0(h(x)) = \varphi_0(x) = f(x)$, so $\varphi|X = f$ and hence we proved that (i)$\Longrightarrow$(ii).

Next, assume (ii) and suppose that we have a continuous map $f : X \to M$ for some locally convex space $M$ with its weak topology. It follows from Problem 225 that the set $M^* \subset C_p(M)$ generates the topology of $M$; let $e(a)(\xi) = \xi(a)$ for any $a \in M$ and $\xi \in M^*$. If $M_0 = \{e(a) : a \in M\} \subset C_p(M^*)$ then $M_0$ is a linear subspace of $C_p(M^*)$ and $e : M \to M_0$ is a linear homeomorphism (see TFS-166).

For every $\xi \in M^*$, there exists a continuous linear functional $r_\xi : L \to \mathbb{R}$ such that $r_\xi|X = \xi \circ f$. For any point $y \in L$ let $r(y)(\xi) = r_\xi(y)$ for any $\xi \in M^*$; this gives us a continuous linear map $r : L \to \mathbb{R}^{M^*}$. Since $X$ is a Hamel basis of $L$, the set $r(L)$ is contained in a linear hull $N$ of the set $r(X)$. For any $x \in X$ and $\xi \in M^*$ we have the equalities $r(x)(\xi) = r_\xi(x) = \xi(f(x)) = e(f(x))(\xi)$ which show that $r(x) = e(f(x))$, i.e., $r = e \circ f$. Therefore $r(X) = e(f(X))$ is contained in $M_0$ and hence $N \subset M_0$. Thus we can consider that $r : L \to M_0$ and therefore the map $\Phi = e^{-1} \circ r : L \to M$ is linear and continuous. Besides, $\Phi(x) = e^{-1}(r(x)) = f(x)$ for any $x \in X$; this shows that we obtained a continuous linear map $\Phi : L \to M$ such that $\Phi|X = f$, i.e., (ii)$\Longrightarrow$(iii) is proved.

Finally assume that (iii) is fulfilled for $L$ and denote by $\mathbf{0}_L$ and $\mathbf{0}$ the zeros in $L$ and $L_p(X)$ respectively. Observe that the property (ii) holds for $L_p(X)$ by Fact 6 of S.489, so $L_p(X)$ satisfies (iii) as well. Let id $: X \to X$ be the identity map, i.e., $\text{id}(x) = x$ for any $x \in X$; we can consider that id $: X \to L_p(X)$. In the space $L_p(X)$ the original topology coincides with its weak topology by Problem 233, so there exists a continuous linear map $h : L \to L_p(X)$ such that $h|X = \text{id}$. The set $h(L)$ is a linear subspace of $L_p(X)$ with $X \subset h(L)$; since $X$ is a Hamel basis in $L_p(X)$ (see Fact 5 of S.489), we conclude that $h(L) = L_p(X)$.

To simplify the notation we denote in the same way the operations in $L$ and $L_p(X)$. Since $X \subset L$ and $X \subset L_p(X)$, for any $x_1, \ldots, x_n \in X$ and $t_1, \ldots, t_n \in \mathbb{R}$, the linear combination $z = t_1x_1 + \ldots + t_nx_n$ can be viewed either as an element of $L$ or as an element of $L_p(X)$; the space to which the point $z$ belongs will be always clear from the context.

Given distinct $x_1, \ldots, x_n \in X$ and $t_1, \ldots, t_n \in \mathbb{R} \backslash \{0\}$ suppose that a point $z = t_1x_1 + \ldots + t_nx_n$ is an element of the space $L$ and $h(z) = \mathbf{0}$; by linearity of $h$, we have $h(z) = t_1h(x_1) + \ldots + t_nh(x_n)$. Since $h(x_i) = \mathrm{id}(x_i) = x_i$ for any $i \leq n$, it turns out that $h(z) = t_1x_1 + \ldots + t_nx_n = \mathbf{0}$ (the operations in the last equality are carried out in $L_p(X)$), which is a contradiction with the fact that $X$ is linearly independent in $L_p(X)$. Therefore $h$ is a continuous linear isomorphism.

If we consider that id : $X \to L$ then apply the property (iii) for $L_p(X)$ to find a continuous linear map $\mu : L_p(X) \to L$ such that $\mu|X = \mathrm{id}$. As before, it is easy to check that, for any $z = t_1x_1 + \ldots + t_nx_n \in L_p(X)$ the point $\mu(z)$ coincides with $w = t_1x_1 + \ldots + t_nx_n$ considered to be an element of $L$. Therefore the map $h^{-1}$ coincides with a continuous map $\mu$ and hence $h : L \to L_p(X)$ is the required linear homeomorphism, i.e., settled (iii)$\Longrightarrow$(i) thus completing our solution.

**V.235.** *Given a space $X$ let $(L_p(X))^* = \{\varphi \in C_p(L_p(X)) : \varphi$ is a linear functional on $L_p(X)\}$ and consider the restriction map $\pi : (L_p(X))^* \to C_p(X)$. Prove that $(L_p(X))^*$ is a closed linear subspace of $C_p(L_p(X))$ and $\pi$ is a linear homeomorphism. Deduce from this fact that the operation of extending continuous real-valued functions on $X$ to continuous linear functionals on $L_p(X)$ is also a linear homeomorphism between $C_p(X)$ and $(L_p(X))^*$.*

**Solution.** Let $(L_p(X))'$ be the set of all (not necessarily continuous) linear functionals on $L_p(X)$. Then $(L_p(X))^* = (L_p(X))' \cap C_p(L_p(X))$; since $(L_p(X))'$ is closed in $\mathbb{R}^{L_p(X)}$ by CFS-393, the set $(L_p(X))^*$ is closed in $C_p(L_p(X))$. It is evident that $(L_p(X))^*$ is a linear subspace of $C_p(L_p(X))$.

To see that the restriction map $\pi : (L_p(X))^* \to C_p(X)$ is well defined recall that TFS-167 and Fact 5 of S.489 make it possible to consider that $X \subset L_p(X)$ identifying every $x \in X$ with the linear functional $e_x : C_p(X) \to \mathbb{R}$ defined by $e_x(f) = f(x)$ for any $f \in C_p(X)$. It follows from Fact 6 of S.489 that the map $\pi$ is a continuous linear bijection.

For any $f \in C_p(X)$ and $\xi \in L_p(X)$ let $u(f)(\xi) = \xi(f)$; it follows from TFS-166 that the map $u : C_p(X) \to C_p(L_p(X))$ is continuous. Fix a function $f \in C_p(X)$; given any $\xi_1, \xi_2 \in L_p(X)$ and $\alpha, \beta \in \mathbb{R}$ it follows from the equalities

$$u(f)(\alpha\xi_1 + \beta\xi_2) = (\alpha\xi_1 + \beta\xi_2)(f) = \alpha\xi_1(f) + \beta\xi_2(f) = \alpha u(f)(\xi_1) + \beta u(f)(\xi_2)$$

that $u(f)$ is a linear functional on the space $L_p(X)$ for any $f \in C_p(X)$ and hence we can consider that $u : C_p(X) \to (L_p(X))^*$. Now if $f \in C_p(X)$ and $x \in X$ then $u(f)(x) = x(f) = f(x)$ which shows that $\pi(u(f)) = f$, so the continuous map $u$ coincides with $\pi^{-1}$. Therefore both $\pi$ and $u$ are linear isomorphisms between the spaces $C_p(X)$ and $(L_p(X))^*$. Since $f \to u(f)$ is precisely the operation of extending a function $f \in C(X)$ to a continuous linear functional $u(f)$ on the space $L_p(X)$, this operation is also a linear homeomorphism between $C_p(X)$ and $(L_p(X))^*$.

**V.236.** *Prove that any $L_p(X)$ is homeomorphic to a dense subspace of $\mathbb{R}^A$ for some A. Deduce from this fact that every uncountable regular cardinal is a precaliber of $L_p(X)$. In particular, $c(L_p(X)) = \omega$ for any space X.*

**Solution.** Let $L = (C_p(X))' \subset \mathbb{R}^{C_p(X)}$ be the set of all linear functionals on $C_p(X)$; apply Problem 232 to see that $L$ is homeomorphic to $\mathbb{R}^A$ for some set $A$. The space $C_p(X)$ is locally convex (see Problem 225 and Problem 233), so the set $(C_p(X))^*$ is dense in $L$ by Problem 231. It follows from TFS-197 that $(C_p(X))^* = L_p(X)$, so $L_p(X)$ is dense in $L$. If $\kappa$ is an uncountable regular cardinal then it is a caliber of $\mathbb{R}^A$ (and hence of $L$) by SFFS-282; therefore $\kappa$ is a precaliber of $L_p(X)$ by SFFS-278.

**V.237.** *Given spaces X and Y prove that*

 (i)   *there exists a linear continuous map of $C_p(X)$ onto $C_p(Y)$ if and only if $L_p(Y)$ is linearly homeomorphic to a linear subspace of $L_p(X)$;*
 (ii)  *there exists a linear continuous open map of $C_p(X)$ onto $C_p(Y)$ if and only if $L_p(Y)$ is linearly homeomorphic to a closed linear subspace of $L_p(X)$;*
 (iii) *the space $C_p(X)$ linearly condenses onto $C_p(Y)$ if and only if $L_p(Y)$ is linearly homeomorphic to a dense linear subspace of $L_p(X)$;*
 (iv)  *$C_p(X)$ is linearly homeomorphic to $C_p(Y)$ if and only if $L_p(Y)$ is linearly homeomorphic to $L_p(X)$.*

**Solution.** Given any set $P \subset L_p(X)$ we denote by $\langle P \rangle$ the linear hull of $P$ in $L_p(X)$. If $i \in \omega$ and we consider a set $A = \{a_1, \ldots, a_i\} \subset L_p(X)$ then $i = 0$ says that $A = \emptyset$. By $\mathbf{0}_X$ we denote the element of $C_p(X)$ which is identically zero on $X$; the zero vector of $L_p(X)$ is denoted by $\mathbf{0}$.

We will simultaneously prove sufficiency in (i)–(iv). Assume that we have a linear homeomorphism $\varphi : L_p(Y) \to L$ for some $L \subset L_p(X)$ (which is closed in $L_p(X)$ or dense in $L_p(X)$ or coincides with $L_p(X)$ respectively) and let $Y' = \varphi(Y)$. For every function $f \in C_p(X)$ there exists a unique continuous linear functional $u(f)$ on $L_p(X)$ such that $u(f)|X = f$ (see Problem 234); it follows from Problem 235 that the map $u : C_p(X) \to (L_p(X))^*$ is a linear homeomorphism. If $\pi : (L_p(X))^* \to C_p(Y')$ is the restriction map then it is continuous and linear; besides, $\pi$ is a homeomorphism (see Problem 235) if we are proving sufficiency in (iv). In the case when $L$ is dense in $L_p(X)$ observe that $\pi = \pi_1 \circ \pi_0$ where $\pi_0 : (L_p(X))^* \to L^*$ and $\pi_1 : L^* \to C_p(Y')$ are the restriction maps. The map $\pi_1$ is injective by Problem 235 and $\pi_0$ is injective because $L$ is dense in $L_p(X)$ (see TFS-152). Therefore the map $\pi$ is injective if we are dealing with the property (iii).

As a consequence, the map $\mu = \pi \circ u : C_p(X) \to C_p(Y')$ is continuous and linear (and homeomorphic respectively), so the proof of sufficiency in (iv) ends here. In the remaining three cases apply Problem 234 to see that, for every $g \in C_p(Y')$ there is a continuous linear functional $e : L \to \mathbb{R}$ such that $e|Y' = g$. By Problem 224 there exists $w \in (L_p(X))^*$ such that $w|L = e$. Pick $f \in C_p(X)$ such that $w = u(f)$. Then $\mu(f) = (\pi \circ u)(f) = \pi(w) = e|Y' = g$ and hence $\mu$ maps $C_p(X)$ (injectively) onto $C_p(Y')$; the space $Y$ being homeomorphic to $Y'$, there exists a continuous linear (injective or homeomorphic respectively) map of $C_p(X)$ onto $C_p(Y)$, so we settled sufficiency in (i), (iii), and (iv).

Let us show that the map $\mu$ is open if the set $L$ is closed in $L_p(X)$. Fix any open set $U \subset C_p(X)$ such that $\mathbf{0}_X \in U$. There is a finite set $A \subset X$ such that $W = \{f \in C_p(X) : f(A) \subset (-\varepsilon, \varepsilon)\} \subset U$ for some $\varepsilon > 0$. It is easy to construct inductively a set $A' = \{a_1, \ldots, a_k\} \subset A$ such that $A \subset \langle L \cup A' \rangle$ and $a_{i+1} \notin \langle L \cup \{a_1, \ldots, a_i\} \rangle$ whenever $0 \le i < k$.

Suppose that $Y' \cup A'$ is not independent; since both sets $Y'$ and $A'$ are independent, the number $k$ has to be positive and there exists $n \in \mathbb{N}$ such that $\{y_1, \ldots, y_n, a_1, \ldots, a_k\}$ is a dependent set for some distinct points $y_1, \ldots, y_n \in Y'$. If $\lambda_1 y_1 + \ldots \lambda_n y_n + t_1 a_1 + \ldots + t_k a_k = \mathbf{0}$ then not all $t_i$'s are equal to zero, so let $j = \max\{i : t_i \neq 0\}$. It follows that $a_j \in \langle Y' \cup \{a_1, \ldots, a_{j-1}\} \rangle$; this contradiction with our choice of $A'$ shows that $Y' \cup A'$ is a linearly independent set. Choose a finite set $B \subset Y'$ such that $A \subset \langle B \cup A' \rangle$. Observe that

(1)  there exists $\delta \in (0, \varepsilon)$ such that, for any $\varphi \in (L_p(X))^*$, if $\varphi(B \cup A') \subset (-\delta, \delta)$ then $\varphi(A) \subset (-\varepsilon, \varepsilon)$.

Indeed, for each $x \in A \setminus A'$ we can choose a number $\lambda_y \in \mathbb{R}$ for every $y \in B \cup A'$ such that $x = \sum\{\lambda_y \cdot y : y \in B \cup A'\}$; let $K_x = \sum\{|\lambda_y| : y \in B \cup A'\}$. If $K = \sum\{K_x : x \in A \setminus A'\} + 1$ then $\delta = \frac{\varepsilon}{K}$ is easily seen to be as promised.

. The set $V = \{f \in C_p(Y') : f(B) \subset (-\delta, \delta)\}$ is open in $C_p(Y')$; fix any $f \in V$. If $f(Y) = \{0\}$ then $\mu(\mathbf{0}_X) = f$, so we can assume that $r = f(y) \neq 0$ for some $y \in Y'$.

There exists a continuous linear functional $\nu : L \to \mathbb{R}$ such that $\nu|Y' = f$. The set $Q = \nu^{-1}(0)$ is closed in $L$ and hence in $L_p(X)$. Since $Y \cup A'$ is independent, the set $G = \langle Q \cup A' \rangle$ does not contain the point $y$. Besides, $G = Q + \langle A' \rangle$, so $G$ is closed in $L_p(X)$ by Fact 2 of V.250. Apply Problem 222 to see that there exists a continuous linear functional $\varphi : L_p(X) \to \mathbb{R}$ such that $\varphi(y) = r$ and $\varphi(G) = \{0\}$. It follows from Fact 14 of V.100 that the functional $\varphi|L$ is proportional to $\nu$, so $\varphi(y) = \nu(y)$ shows that $\varphi|L = \nu$ and hence $\varphi|Y = f$. Recalling that $\varphi(A') \subset \{0\}$ and $\varphi(B) = f(B) \subset (-\delta, \delta)$ we convince ourselves that $\varphi(B \cup A') \subset (-\delta, \delta)$ and hence (1) implies that $\varphi(A) \subset (-\varepsilon, \varepsilon)$. If $g = \varphi|X$ then $f = \mu(g)$ and $g \in W \subset U$; since $f \in V$ was chosen arbitrarily, this proves that $\mu(U) \supset V$, i.e., the image under $\mu$ of every neighborhood of $\mathbf{0}_X$ contains a neighborhood of the zero function of $C_p(Y')$. By Fact 3 of S.496, the map $\mu$ is an open surjection of $C_p(X)$ onto $C_p(Y')$. Therefore there exists an open continuous linear surjection of $C_p(X)$ onto $C_p(Y)$ and hence we settled sufficiency in (ii).

Suppose that $\xi : C_p(X) \to C_p(Y)$ is a linear surjective map and let $\xi^*(f) = f \circ \xi$ for any $f \in (C_p(Y))^* = L_p(Y)$ (see TFS-197). Then $\xi^* : L_p(Y) \to (C_p(X))^* = L_p(X)$; if $L = \xi^*(L_p(Y))$ then $L$ is a linear subspace of $L_p(X)$ and $\xi^* : L_p(Y) \to L$ is a linear homeomorphism (see Fact 4 of S.489). This proves necessity in (i). If $\xi$ is an open map then $\xi^*(C_p(C_p(Y)))$ is closed in $C_p(C_p(X))$ (see TFS-163); since $L_p(Y)$ is closed in $C_p(C_p(Y))$, the set $L$ is closed in $C_p(C_p(X))$ and hence in $L_p(X)$. This settles necessity in (ii).

Now, if $\xi$ is a linear condensation (homeomorphism) then $L$ is dense in $L_p(X)$ (or coincides with $L_p(X)$ respectively). To see this take any functional $w \in L_p(X)$ and a finite set $A \subset C_p(X)$; denote by $M$ the linear hull of $A$. It is evident that

$w' = w \circ \xi^{-1}$ is a (continuous) linear functional on $C_p(Y)$ (and $\xi^*(w') = w$); in the case when $\xi$ is a linear homeomorphism we conclude that $\xi^*(L_p(Y)) = L_p(X)$, so $\xi^*$ is a linear homeomorphism between $L_p(Y)$ and $L_p(X)$, i.e., the proof of necessity in (iv) ends here.

Since $M' = \xi(M)$ is a finite-dimensional linear subspace of $C_p(Y)$, the functional $w'|M'$ is continuous on $M'$ (see Problem 230). Apply Problem 224 to find a continuous linear functional $v' : C_p(Y) \to \mathbb{R}$ such that $v'|M' = w'$. Then $v = v' \circ \xi \in L$; given any $x \in M$ we have $v(x) = v'(\xi(x))$. It follows from $\xi(x) \in M'$ that $v(x) = v'(\xi(x)) = w'(\xi(x)) = (w \circ \xi^{-1})(\xi(x)) = w(x)$; an immediate consequence is that $v|M = w|M$ and hence $v|A = w|A$. Thus, for any $w \in L_p(X)$ and any finite $A \subset C_p(X)$ there exists $v \in L$ with $v|A = w|A$; this implies that $L$ is dense in $L_p(X)$, so we established necessity in (iii) and completed our solution.

**V.238.** *Given spaces $X$ and $Y$ prove that*

(i) *there is a linear continuous map of $C_p(X)$ onto $C_p(Y)$ if and only if $Y$ is homeomorphic to a subspace $Y' \subset L_p(X)$ such that every $f \in C(Y')$ extends to a continuous linear functional on $L_p(X)$;*

(ii) *the space $C_p(X)$ linearly condenses onto $C_p(Y)$ if and only if $Y$ is homeomorphic to a subspace $Y' \subset L_p(X)$ such that every $f \in C(Y')$ extends to a uniquely determined continuous linear functional on $L_p(X)$;*

(iii) *$C_p(X)$ is linearly homeomorphic to $C_p(Y)$ if and only if $Y$ is homeomorphic to some $Y' \subset L_p(X)$ whose linear hull coincides with $L_p(X)$ and every $f \in C(Y')$ extends to a continuous linear functional on $L_p(X)$.*

**Solution.** We will simultaneously prove sufficiency in (i)–(iii). Assume that we have a homeomorphism $\varphi : Y \to Y' \subset L_p(X)$ such that every $f \in C(Y')$ extends to a continuous linear functional $w(f)$ on the space $L_p(X)$ where the functional $w(f)$ is unique if we are dealing with (ii) and the linear hull of $Y'$ is $L_p(X)$ if we are proving sufficiency in (iii). For every $f \in C_p(X)$ there exists a unique continuous linear functional $u(f)$ on $L_p(X)$ such that $u(f)|X = f$ (see Problem 234); besides, the map $u : C_p(X) \to (L_p(X))^*$ is a linear homeomorphism (see Problem 235). If $\pi : (L_p(X))^* \to C_p(Y')$ is the restriction map then it is continuous, linear and surjective; besides, $\pi$ is injective in case when we deal with (ii). Therefore $\mu = \pi \circ u$ is a continuous linear map of $C_p(X)$ onto $C_p(Y')$ which is injective if we are dealing with (ii); the space $Y$ being homeomorphic to $Y'$ there exists a linear continuous (injective) map of $C_p(X)$ onto $C_p(Y)$, so the proof of sufficiency in (i) and (ii) ends here.

For the remaining part of the proof of sufficiency we are considering that the linear hull of $Y'$ coincides with $L_p(X)$. Our next step is to show that

(1) the set $Y'$ is linearly independent in $L_p(X)$.

Assume that $a_1, \ldots, a_n \in \mathbb{R}$ and $y_1, \ldots, y_n$ are distinct points of $Y'$ such that $\sum_{i=1}^{n} a_i y_i = \mathbf{0}$. For every $i \leq n$ there exists a continuous function $f_i : Y' \to \mathbb{R}$ such that $f_i(y_i) = 1$ and $f_i(y_j) = 0$ for all $j \neq i$. If $\mu_i : L_p(X) \to \mathbb{R}$ is the continuous linear functional which extends $f_i$ then

$$0 = \mu(\mathbf{0}) = \mu_i\left(\sum_{k=1}^{n} a_k y_k\right) = \sum_{k=1}^{n} a_k \mu_i(y_k) = a_i.$$

Thus $a_i = 0$ for each $i \leq n$ and hence we proved (1).

The property (1) shows that the set $Y'$ is a Hamel basis of $L_p(X)$, so we can apply Problem 234 to see that $L_p(Y')$ is linearly homeomorphic to $L_p(X)$ and hence $L_p(Y)$ is linearly homeomorphic to $L_p(X)$ as well. By Problem 237, the space $C_p(X)$ is linearly homeomorphic to $C_p(Y)$ and hence we completed the proof of sufficiency in (i)–(iii).

Suppose that $\xi : C_p(X) \to C_p(Y)$ is a linear surjective map and let $\xi^*(f) = f \circ \xi$ for any $f \in (C_p(Y))^* = L_p(Y)$ (see TFS-197). Then $\xi^* : L_p(Y) \to (C_p(X))^* = L_p(X)$; if $L = \xi^*(L_p(Y))$ then $L$ is a linear subspace of $L_p(X)$ and $\xi^* : L_p(Y) \to L$ is a linear homeomorphism (see Fact 4 of S.489). If $Y' = \xi^*(Y)$ then, for every $f \in C(Y')$ there is a continuous linear functional $u : L \to \mathbb{R}$ such that $u|Y' = f$ (see Problem 234). Apply Problem 224 to find a continuous linear functional $w : L_p(X) \to \mathbb{R}$ with $w|L = u$. Then $w|Y' = u|Y' = f$, so we proved necessity in (i).

Now, if $\xi$ is a linear condensation (or linear homeomorphism respectively) then there exists a linear homeomorphism $\varphi : L_p(Y) \to L \subset L_p(X)$ such that $L$ is dense in $L_p(X)$ (or $L = L_p(X)$ respectively) by Problem 237. If $Y' = \varphi(Y)$ then, for every $f \in C(Y')$ there is a continuous linear functional $u : L \to \mathbb{R}$ such that $u|Y' = f$. Apply Problem 224 to find a continuous linear functional $w : L_p(X) \to \mathbb{R}$ with $w|L = u$. Then $w|Y' = u|Y' = f$; if $w' \in (L_p(X))^*$ and $w'|Y' = f$ then $w'|L = u$ because the linear extension of $f$ from $Y'$ to $L$ is unique (see Problem 235). The set $L$ being dense in $L_p(X)$ it follows from $w'|L = u = w|L$ that $w = w'$ and hence the linear extension of $f$ over $L_p(X)$ is unique, i.e., we established necessity in (ii). If we deal with (iii) then $Y'$ is a Hamel basis in $L_p(X)$ because $Y$ is a Hamel basis in $L_p(Y)$ and Hamel bases are preserved by linear homeomorphisms. This proves necessity in (iii) and shows that our solution is complete.

**V.239.** *Let $\mathcal{P}$ be a class of spaces which have the following properties:*

*(1) if $Y \in \mathcal{P}$ and $Z$ is a continuous image of $Y$ then $Z \in \mathcal{P}$;*
*(2) if $Y = \bigcup\{Y_i : i \in \omega\}$, $Y_i \subset Y_{i+1}$, $Y_i \in \mathcal{P}$ and $Y_i$ closed in $Y$ for every $i \in \omega$, then $Y \in \mathcal{P}$;*
*(3) if $Y \in \mathcal{P}$ and $n \in \mathbb{N}$ then $Y^n \times \mathbb{R}^n \in \mathcal{P}$;*

*Prove that if a space $X$ belongs to $\mathcal{P}$ then $L_p(X) \in \mathcal{P}$.*

**Solution.** For any $n \in \mathbb{N}$ consider the map $\varphi_n : X^n \times \mathbb{R}^n \to L_p(X)$ defined by $\varphi_n(x, r) = \sum_{i=1}^{n} r_i x_i$ for any $x = (x_1, \ldots, x_n) \in X^n$ and $r = (r_1, \ldots, r_n) \in \mathbb{R}^n$. It follows from continuity of operations in $L_p(X)$ that every $\varphi_n$ is continuous; if $Y_n = \varphi(X^n \times \mathbb{R}^n)$ then $Y_n \subset Y_{n+1}$ for every $n \in \mathbb{N}$. Since $X$ is a Hamel basis of $L_p(X)$, we have the equality

$(*)$  $L_p(X) = \bigcup\{Y_n : n \in \mathbb{N}\}$.

It follows from Fact 1 of U.485 that every $Y_n$ is closed in $L_p(X)$. If $X \in \mathcal{P}$ then $X^n \times \mathbb{R}^n \in \mathcal{P}$ by (3); applying (1) we convince ourselves that $Y_n \in \mathcal{P}$ for any $n \in \mathbb{N}$. Finally, apply (2) and (*) to conclude that $L_p(X) \in \mathcal{P}$.

**V.240.** *Prove that* $iw(L_p(X)) = \psi(L_p(X)) = \Delta(L_p(X)) = iw(X)$ *for any space* $X$; *show that also* $nw(X) = nw(L_p(X))$ *and* $d(X) = d(L_p(X))$.

**Solution.** It follows from $X \subset L_p(X)$ that $nw(X) \leq nw(L_p(X))$. To prove the opposite inequality observe that $L_p(X) \subset C_p(C_p(X))$, so we can apply TFS-172 to conclude that $nw(L_p(X)) \leq nw(C_p(C_p(X))) = nw(C_p(X)) = nw(X)$ and hence $nw(X) = nw(L_p(X))$.

Given an infinite cardinal $\kappa$ observe that the class $\mathcal{P}$ of spaces of density $\leq \kappa$ satisfies the conditions (1)–(3) of Problem 239, so $d(X) \leq \kappa$ implies $d(L_p(X)) \leq \kappa$. This proves that $d(L_p(X)) \leq d(X)$.

Now assume that $d(L_p(X)) \leq \kappa$. Then $iw(C_p(L_p(X)) \leq \kappa$ (see TFS-173); the space $C_p(X)$ embeds in $C_p(L_p(X))$ (see Problem 235), so $iw(C_p(X)) \leq \kappa$ and hence we can apply TFS-173 again to convince ourselves that $d(X) = iw(C_p(X)) \leq \kappa$. This shows that $d(X) \leq d(L_p(X))$ and therefore $d(X) = d(L_p(X))$.

To check the rest of our equalities suppose that $\psi(L_p(X)) \leq \kappa$. For any finite $A \subset C_p(X)$ and $\varepsilon > 0$ let $[A, \varepsilon] = \{\xi \in L_p(X) : |\xi(f)| < \varepsilon$ for any $f \in A\}$. The family $\mathcal{B} = \{[A, \varepsilon] : A$ is a finite subset of $C_p(X)$ and $\varepsilon > 0\}$ is a local base of the space $L_p(X)$ at $\mathbf{0}$, so we can find a family $\mathcal{A} = \{[A_\alpha, \varepsilon_\alpha] : \alpha < \kappa\} \subset \mathcal{B}$ such that $\bigcap \mathcal{A} = \{\mathbf{0}\}$.

The set $A = \bigcup\{A_\alpha : \alpha < \kappa\}$ has cardinality at most $\kappa$, so its linear hull $H$ has density at most $\kappa$. Assume for a moment that $H$ is not dense in $C_p(X)$ and fix a function $f \in C_p(X) \setminus \overline{H}$. By Problem 222 and TFS-197, there exists $\xi \in L_p(X)$ such that $\xi(f) = 1$ and $\xi(H) = \{0\}$. Then $\xi \neq \mathbf{0}$ and $\xi \in \bigcap \mathcal{A}$ which is a contradiction. Therefore $H$ is dense in $C_p(X)$ and hence $iw(X) = d(C_p(X)) \leq d(H) \leq \kappa$ which shows that $iw(X) \leq \psi(L_p(X))$.

It follows from $L_p(X) \subset C_p(C_p(X))$ that $iw(L_p(X)) \leq iw(C_p(C_p(X)))$. Furthermore, $iw(C_p(C_p(X))) = d(C_p(X)) = iw(X)$, so $iw(L_p(X)) \leq iw(X)$. The inequalities $iw(X) \leq \psi(L_p(X)) \leq \Delta(L_p(X)) \leq iw(L_p(X)) \leq iw(X)$ imply that $\psi(L_p(X)) = \Delta(L_p(X)) = iw(L_p(X)) = iw(X)$, i.e., our solution is complete.

**V.241.** *Prove that, for any space* $X$, *we have the following equalities.*

(i) $s^*(X) = s(L_p(X)) = s^*(L_p(X))$;
(ii) $hl^*(X) = hl(L_p(X)) = hl^*(L_p(X))$;
(iii) $hd^*(X) = hd(L_p(X)) = hd^*(L_p(X))$.

**Solution.** For any $n \in \mathbb{N}$ the space $X^n$ embeds in $L_p(X)$ by TFS-337; this shows that $s^*(X) \leq s(L_p(X))$, $hl^*(X) \leq hl(L_p(X))$ and $hd^*(X) \leq hd(L_p(X))$.

It follows from SFFS-025 that $s^*(X) = s^*(C_p(X)) = s^*(C_p(C_p(X)))$; since $L_p(X)$ is a subspace of $C_p(C_p(X))$, we have $s^*(L_p(X)) \leq s^*(C_p(C_p(X))) = s^*(X)$. This gives us inequalities $s^*(X) \leq s(L_p(X)) \leq s^*(L_p(X)) \leq s^*(X)$ which show that (i) is proved.

Apply SFFS-026 and SFFS-027 to see that $hl^*(X) = hd^*(C_p(X)) = hl^*(C_p(C_p(X)))$; since $L_p(X) \subset C_p(C_p(X))$, we have $hl^*(L_p(X)) \leq hl^*(C_p(C_p(X))) = hl^*(X)$. Therefore $hl^*(X) \leq hl(L_p(X)) \leq hl^*(L_p(X)) \leq hl^*(X)$, so we settled (ii).

Since SFFS-026 and SFFS-027 also imply $hd^*(X) = hl^*(C_p(X)) = hd^*(C_p(C_p(X)))$, we can recall once more that $L_p(X)$ is a subspace of $C_p(C_p(X))$ to conclude that $hd^*(L_p(X)) \leq hd^*(C_p(C_p(X))) = hd^*(X)$. An immediate consequence is that $hd^*(X) \leq hd(L_p(X)) \leq hd^*(L_p(X)) \leq hd^*(X)$, i.e., (iii) is proved.

**V.242.** *Given a space X prove that we have* $l^*(X) = l(L_p(X)) = l^*(L_p(X))$ *and* $ext^*(X) = ext(L_p(X)) = ext^*(L_p(X))$.

**Solution.** For any $n \in \mathbb{N}$ the space $X^n$ embeds in $L_p(X)$ as a closed subspace (see TFS-337); this shows that $l^*(X) \leq l(L_p(X))$ and $ext^*(X) \leq ext(L_p(X))$.

*Fact 1.* If $Z$ and $T$ are spaces and there exists a perfect map $f : Z \to T$ then

(1) $ext(Z) \leq ext(T)$ and $ext^*(Z) \leq ext^*(T)$;
(2) $ext(Z \times S) \leq ext(Z)$ and $ext^*(Z \times S) \leq ext^*(Z)$ for any $\sigma$-compact space $S$.
(3) $l(Z) \leq l(T)$ and $l^*(Z) \leq l^*(T)$;
(4) $l(Z \times S) \leq l(Z)$ and $l^*(Z \times S) \leq l^*(Z)$ for any $\sigma$-compact space $S$.

*Proof.* Fix an infinite cardinal $\mu$ and assume that $ext(T) \leq \mu$. If there exists a closed discrete subspace $D \subset Z$ with $|D| = \mu^+$ then let $E = f(D)$. The map $f$ is closed, so $E$ is a closed subspace of $T$. By the same reason $f(G)$ is closed in $T$ for every $G \subset D$ and hence $E$ is a discrete subspace of $T$. For any $y \in E$ the set $f^{-1}(y)$ is compact, so the set $F_y = \{x \in D : f(x) = y\}$ is finite for any $y \in E$. If $|E| \leq \mu$ then $D = \bigcup\{F_y : y \in E\}$ is the union of at most $\mu$-many finite sets, i.e., $|D| \leq \mu$ which is a contradiction. Therefore $|E| = \mu^+$; this contradiction with $ext(T) \leq \mu$ shows that $ext(Z) \leq \mu$ and hence $ext(Z) \leq ext(T)$. The inequality $l(Z) \leq l(T)$ was proved in Fact 2 of T.490. It is easy to see that, for each $n \in \mathbb{N}$, there exists a perfect map of $Z^n$ onto $T^n$; therefore $ext(Z^n) \leq ext(T^n) \leq ext^*(T)$ and $l(Z^n) \leq l(T^n) \leq l^*(T)$ for every $n \in \mathbb{N}$ which shows that $ext^*(Z) \leq ext^*(T)$ and $l^*(Z) \leq l^*(T)$, so we settled (1) and (3).

If $K$ is compact then the projection map $Z \times K \to Z$ is perfect by Fact 3 of S.288, so $ext(Z \times K) \leq ext(Z)$ and $l(Z \times K) \leq l(Z)$ by (1) and (3) respectively. Now, if $S = \bigcup_{n \in \omega} K_n$ where $K_n$ is compact and $K_n \subset K_{n+1}$ for every $n \in \omega$ then it follows from the equality $Z \times S = \bigcup_{n \in \omega}(Z \times K_n)$ that $ext(Z \times S) \leq ext(Z)$ and $l(Z \times S) \leq l(Z)$, i.e., we established the first parts of (2) and (4). Apply these inequalities to see that $ext((Z \times S)^n) = ext(Z^n \times S^n) \leq ext(Z^n) \leq ext^*(Z)$ and analogously, $l((Z \times S)^n) = l(Z^n \times S^n) \leq l(Z^n) \leq l^*(Z)$ for all $n \in \mathbb{N}$; this shows that $ext^*(Z \times S) \leq ext^*(Z)$ and $l^*(Z \times S) \leq l^*(Z)$, so Fact 1 is proved.

Returning to our solution fix an infinite cardinal $\kappa$ and consider the class $\mathcal{P}$ (the class $\mathcal{Q}$) of spaces $Y$ such that $l^*(Y) \leq \kappa$ (or $ext^*(Y) \leq \kappa$ respectively). It is an easy exercise that any continuous image of a space from $\mathcal{P}$ (or from $\mathcal{Q}$) belongs to $\mathcal{P}$ (or to $\mathcal{Q}$ respectively).

Next, assume that $Y = \bigcup_{i \in \omega} Y_i$ where every $Y_i$ is closed in $Y$, belongs to $\mathcal{P}$ (or to $\mathcal{Q}$) and the sequence $\{Y_i : i \in \omega\}$ is increasing, i.e., $Y_i \subset Y_{i+1}$ for all $i \in \omega$. Given $n \in \mathbb{N}$ the space $Y^n$ can be represented as the countable union of its closed subspaces $F(i_1, \dots, i_n) = Y_{i_1} \times \dots \times Y_{i_n}$ where $i_1, \dots, i_n \in \omega$. If $j = \max\{i_1, \dots, i_n\}$ then the set $F = F(i_1, \dots, i_n)$ is a closed subspace of the space $Y_j^n$, so it follows from $l(Y_j^n) \le \kappa$ (or $ext(Y_j^n) \le \kappa$ respectively) that $l(F) \le \kappa$ (or $ext(F) \le \kappa$). Therefore $l(Y^n) \le \kappa$ (or $ext(Y^n) \le \kappa$) for every $n \in \mathbb{N}$, i.e., $Y \in \mathcal{P}$ (or $Y \in \mathcal{Q}$ respectively).

It follows from Fact 1 that we have the equalities $ext^*(X^n \times \mathbb{R}^n) = ext^*(X)$ and $l^*(X^n \times \mathbb{R}^n) = l^*(X)$ which shows that $X^n \times \mathbb{R}^n \in \mathcal{P}$ if $X \in \mathcal{P}$ and $X^n \times \mathbb{R}^n \in \mathcal{Q}$ if $X \in \mathcal{Q}$. Therefore the classes $\mathcal{P}$ and $\mathcal{Q}$ satisfy the conditions (1)–(3) of Problem 239 and hence $X \in \mathcal{P}$ implies $L_p(X) \in \mathcal{P}$ while $X \in \mathcal{Q}$ implies $L_p(X) \in \mathcal{Q}$. This gives us the inequalities $l^*(X) \le l(L_p(X)) \le l^*(L_p(X)) \le l^*(X)$; analogously, we conclude that $ext^*(X) \le ext(L_p(X)) \le ext^*(L_p(X)) \le ext^*(X)$. This implies the equalities $l^*(X) = l(L_p(X)) = l^*(L_p(X))$ and $ext^*(X) = ext(L_p(X)) = ext^*(L_p(X))$, i.e., our solution is complete.

**V.243.** *Prove that an uncountable regular cardinal $\kappa$ is a caliber of $X$ if and only if $\kappa$ is a caliber of $L_p(X)$.*

**Solution.** Consider the class $\mathcal{P}$ of spaces $Y$ such that $\kappa$ is a caliber of $Y$. By SFFS-276, SFFS-277 and SFFS-281 the class $\mathcal{P}$ satisfies the conditions (1)–(3) of Problem 239, so it follows from $X \in \mathcal{P}$ that $L_p(X) \in \mathcal{P}$, i.e., we established necessity.

Now, if $\kappa$ is a caliber of $L_p(X)$ then the diagonal of $C_p(L_p(X))$ is $\kappa$-small (see SFFS-290). The space $C_p(X)$ embeds in $C_p(L_p(X))$ by Problem 235, so the diagonal of $C_p(X)$ is also $\kappa$-small. Applying SFFS-290 again we conclude that $\kappa$ is a caliber of $X$; this proves sufficiency.

**V.244.** *Let $\mathcal{L}$ be the following collection of classes of Tychonoff spaces: {analytic spaces, $K$-analytic spaces, $\sigma$-compact spaces, realcompact spaces, Lindelöf $\Sigma$-spaces}. Prove that, for any class $\mathcal{P}$ from the list $\mathcal{L}$, a space $X$ belongs to $\mathcal{P}$ if and only if $L_p(X)$ belongs to $\mathcal{P}$.*

**Solution.** It is an easy exercise to verify that if $\mathcal{P}$ is one of the first four classes then $\mathcal{P}$ satisfies the conditions (1)–(3) of Problem 239, so if $X \in \mathcal{P}$ then $L_p(X) \in \mathcal{P}$. Besides, if $\mathcal{P}$ is the class of realcompact spaces then $X \in \mathcal{P}$ implies $C_p(C_p(X)) \in \mathcal{P}$ by TFS-435; the set $L_p(X)$ being closed in $C_p(C_p(X))$ (see TFS-078) we also have $L_p(X) \in \mathcal{P}$.

Now, if $L_p(X) \in \mathcal{P}$ (where $\mathcal{P}$ is any class from the list $\mathcal{L}$) then $X \in \mathcal{P}$ because $X$ is a closed subspace of $L_p(X)$. Thus $X$ belongs to $\mathcal{P}$ if and only if so does $L_p(X)$.

**V.245.** *Given $w = \lambda_1 x_1 + \dots + \lambda_n x_n \in L_p(X)$, where $x_1, \dots, x_n \in X$ and $\lambda_1, \dots, \lambda_n \in \mathbb{R} \setminus \{0\}$, let $supp(w) = \{x_1, \dots, x_n\}$. If $w = 0$, then $supp(w) = \emptyset$. Say that a set $B \subset L_p(X)$ is weakly bounded if $\xi(B)$ is a bounded subset of $\mathbb{R}$ for any continuous linear functional $\xi : L_p(X) \to \mathbb{R}$. Observe that any bounded subset of $L_p(X)$ is weakly bounded and prove that, for any weakly bounded set $B \subset L_p(X)$, the set $\bigcup\{supp(w) : w \in B\}$ is bounded in the space $X$.*

**Solution.** It is immediate from the definitions that any bounded subset of $L_p(X)$ is weakly bounded in $L_p(X)$, so assume that $B \subset L_p(X)$ is weakly bounded while $S(B) = \bigcup\{\operatorname{supp}(w) : w \in B\}$ is not bounded in $X$.

*Fact 1.* Given a space $Z$, a set $Y \subset Z$ is not bounded in $Z$ if and only if there exists a discrete family $\{U_n : n \in \omega\} \subset \tau(Z)$ such that $U_n \cap Y \neq \emptyset$ for any $n \in \omega$.

*Proof.* If a family $\{U_n : n \in \omega\} \subset \tau(Z)$ is discrete and every $U_n$ meets $Y$, pick a point $y_n \in U_n \cap Y$ for each $n \in \omega$. Apply Fact 1 of T.217 to find a continuous function $f : Z \to \mathbb{R}$ such that $f(y_n) = n$ for every $n \in \omega$. Then $f(Y)$ is not a bounded subset of $\mathbb{R}$, so $Y$ is not bounded in $Z$ and hence we proved sufficiency.

Now assume that $Y$ is not bounded in $Z$ and hence there exists a continuous function $f : Z \to \mathbb{R}$ such that $f(Y)$ is not bounded in $\mathbb{R}$. Let $g(x) = |f(x)|$ for any $x \in Z$; the function $g : Z \to \mathbb{R}$ is continuous and it is clear that $g(Y)$ is not bounded in $\mathbb{R}$. It is easy to choose by induction on $n \in \omega$ a set $D = \{y_n : n \in \omega\} \subset Y$ such that $g(y_n) > n + \sum_{i=0}^{n-1} g(y_i)$ for every $n \in \omega$. The sequence $\{g(y_n) : n \in \omega\} \subset \mathbb{R}$ is strictly increasing and $g(y_n) > n$ for each $n \in \omega$. Consequently, the set $D$ is closed and discrete in $\mathbb{R}$, so we can find a discrete family $\{O_n : n \in \omega\} \subset \tau(\mathbb{R})$ such that $g(y_n) \in O_n$ for every $n \in \omega$. Letting $U_n = g^{-1}(O_n)$ for all $n$ we obtain a discrete family $\{U_n : n \in \omega\} \subset \tau(Z)$ with $y_n \in U_n \cap Y$ for every $n \in \omega$; this settles necessity and shows that Fact 1 is proved.

Returning to our solution apply Fact 1 to find a countably infinite discrete family $\mathcal{U} \subset \tau(X)$ such that $U \cap S(B) \neq \emptyset$ for every $U \in \mathcal{U}$. Proceeding by an evident induction we can choose a family $\{U'_n : n \in \omega\} \subset \mathcal{U}$ and a sequence $\{w_n : n \in \omega\} \subset B$ such that $U'_n \cap \operatorname{supp}(w_n) \neq \emptyset$ and $U'_n \cap \operatorname{supp}(w_k) = \emptyset$ for any $k < n$. For every $n \in \omega$ take a point $x_n \in U'_n \cap \operatorname{supp}(w_n)$ and an open set $U_n \subset U'_n$ such that $U_n \cap \operatorname{supp}(w_n) = \{x_n\}$.

We have $w_n = \lambda_n x_n + \sum_{i=1}^{k_n} \mu_i^n y_i^n$ where $\{\lambda_n, \mu_1^n, \ldots, \mu_{k_n}^n\} \subset \mathbb{R}$, $\lambda_n \neq 0$ and $x_n \notin P_n = \{y_1^n, \ldots, y_{k_n}^n\} \subset X$; let $v_n = \max\{|\mu_i^n| : i \leq k_n\}$ for each $n \in \omega$. Start off with $r_0 = |\lambda_0|^{-1}$ and construct inductively a sequence $R = \{r_n : n \in \omega\}$ of positive numbers such that $r_n > r_{n-1} + \frac{1}{|\lambda_n|}(n + v_n k_n r_{n-1})$ for every $n \in \mathbb{N}$; it is evident that $R$ is strictly increasing. Take a continuous function $f_n : X \to [0, 1]$ such that $f_n(x_n) = 1$ and $f_n(X \setminus U_n) = \{0\}$ for each $n \in \omega$. By Fact 1 of T.217 the function $f = \sum\{r_n f_n : n \in \omega\}$ is continuous, so we can apply Problem 234 to find a continuous linear functional $\xi : L_p(X) \to \mathbb{R}$ such that $\xi | X = f$.

Fix any $n \in \omega$; since the set $P_n$ does not meet $H_n = \bigcup\{U_i : i \geq n\}$, we conclude that $|f(y_i^n)| = f(y_i^n) \leq r_{n-1}$ for every $i \leq k_n$. An immediate consequence is that $\sum_{i=1}^{k_n} |\mu_i^n||f(y_i^n)| \leq v_n k_n r_{n-1}$ which implies, together with the inequality $|\lambda_n f(x_n)| > n + v_n k_n r_{n-1}$, that $|\xi(w_n)| \geq |\lambda_n f(x_n)| - |\sum_{i=1}^{k_n} \mu_i^n f(y_i^n)| \geq n$. Therefore the continuous linear functional $\xi : L_p(X) \to \mathbb{R}$ is not bounded on the set $B$; this contradiction shows that $S(B)$ is bounded in $X$ and makes our solution complete.

**V.246.** *Prove that, for any Dieudonné complete space $X$, if $A$ is a bounded subset of $L_p(X)$ then $\overline{A}$ is compact.*

**Solution.** Our first step is to prove that closures of bounded sets are compact in Dieudonné complete spaces.

*Fact 1.* If $Y$ is a Dieudonné complete space and $B$ is a bounded subset of $Y$ then $\overline{B}$ is compact.

*Proof.* Observe first that $F = \overline{B}$ is also bounded in $Y$ by Fact 2 of S.398. We can consider that $Y$ is a closed subspace of a product $M = \prod\{M_t : t \in T\}$ in which every $M_t$ is a metrizable space; let $\pi_t : M \to M_t$ be the natural projection for every $t \in T$. The set $F_t = \pi_t(F)$ is bounded in $M_t$ (see Fact 1 of S.399); since $P_t = \overline{F}_t$ is also bounded and $C$-embedded in $M_t$, the set $P_t$ is pseudocompact and hence compact (here we used again Fact 2 of S.398). The set $F$ is closed in $Y$ and hence in $M$; since $F \subset P = \prod\{P_t : t \in T\}$, the set $F$ has to be closed in the compact space $P$. This shows that $F$ is compact, so Fact 1 is proved.

*Fact 2.* For any space $Y$, if $Z$ is a closed subspace of $Y$ then the linear hull $H$ of the set $Z$ in $L_p(Y)$ is closed in $L_p(Y)$.

*Proof.* Suppose that a point $w = \lambda_1 y_1 + \ldots + \lambda_n y_n$ belongs to $L_p(Y) \backslash H$ and $y_i \neq y_j$ whenever $i \neq j$. Then $\lambda_k \neq 0$ and $y_k \notin Z$ for some $k \leq n$. By the Tychonoff property of $Y$ we can find a continuous function $f : Y \to [0, 1]$ such that $f(y_k) = 1$, $f(y_i) = 0$ for all $i \neq k$ and $f(Z) \subset \{0\}$. The set $U = \{v \in L_p(Y) : v(f) \neq 0\}$ is an open subset of $L_p(Y)$ and $w \in U$. It is immediate that $u(f) = 0$ for any $u \in H$, so $U \cap H = \emptyset$. Thus every point $w \in L_p(Y) \backslash H$ has an open neighborhood which does not meet $H$. This shows that $H$ is closed in $L_p(X)$, so Fact 2 is proved.

*Fact 3.* Given a space $Y$, if $Z \subset Y$ is $C$-embedded in $Y$ then, for the linear hull $H$ of the set $Z$ in $L_p(Y)$, there exists a linear homeomorphism $\varphi : L_p(Z) \to H$ such that $\varphi(z) = z$ for any $z \in Z$.

*Proof.* Given any continuous function $f : Z \to \mathbb{R}$ there exists $g \in C(Y)$ such that $g|Z = f$. By Problem 234, there is a continuous linear functional $\xi : L_p(Y) \to \mathbb{R}$ for which $\xi|Y = g$. Then $\xi_0 = \xi|H : H \to \mathbb{R}$ is a continuous linear functional such that $\xi_0|Z = \xi|Z = g|Z = f$, so we can apply Problem 234 again to conclude that there exists a linear homeomorphism $\varphi : L_p(Z) \to H$ such that $\varphi(z) = z$ for any $z \in Z$, i.e., Fact 3 is proved.

Returning to our solution recall that $X$ is $C$-embedded in the Hewitt realcompactification $\upsilon X$ of the space $X$, so we can apply Fact 3 to identify $L_p(X)$ with the linear hull of the set $X$ in $L_p(\upsilon X)$. Assume that $A$ is a bounded subset of $L_p(X)$. For any $w = \lambda_1 x_1 + \ldots + \lambda_n x_n \in L_p(\upsilon X)$ where $\lambda_i \in \mathbb{R} \backslash \{0\}$ and $x_i \in \upsilon X$ for all $i \leq n$, let supp$(w) = \{x_1, \ldots, x_n\}$; if $w = \mathbf{0}$ then supp$(w) = \emptyset$. The set $\overline{A}$ is still bounded in $L_p(X)$ by Fact 2 of S.398 (the bar denotes the closure in $L_p(X)$), so we can apply Problem 245 to see that $B = \bigcup\{\text{supp}(w) : w \in \overline{A}\}$ is a bounded subset of $X$, so $K = \text{cl}_X(B)$ is compact by Fact 1.

Let $G$ be the linear hull of the set $K$ in the space $L_p(\upsilon X)$; it is evident that $\overline{A} \subset G \subset L_p(X)$. By Fact 2, the set $G$ is closed in $L_p(\upsilon X)$; besides, $\overline{A}$ is closed in $L_p(X)$ and hence in $G$, so $\overline{A}$ is closed in $L_p(\upsilon X)$. The set $\overline{A}$ is also bounded in $L_p(\upsilon X)$ because $L_p(X) \subset L_p(\upsilon X)$; since $L_p(\upsilon X)$ is realcompact by Problem 244, the set $\overline{A}$ is compact by Fact 1, so our solution is complete.

**V.247.** *Suppose that a space $X$ has a weaker metrizable topology and $A$ is a bounded subset of $L_p(X)$. Prove that $\overline{A}$ is compact and metrizable.*

**Solution.** The space $X$ is Dieudonné complete by TFS-461, so we can apply Problem 246 to see that $K = \overline{A}$ is compact. Fix a condensation $f : X \to M$ of $X$ onto a metrizable space $M$. Given a point $w = \lambda_1 x_1 + \ldots + \lambda_n x_n \in L_p(X)$ with $\lambda_i \in \mathbb{R}\backslash\{0\}$ for all $i \leq n$ let $\mathrm{supp}(w) = \{x_1, \ldots, x_n\}$. The set $B = \bigcup\{\mathrm{supp}(w) : w \in K\}$ is bounded in $X$ by Problem 245, so $P = \overline{B}$ is compact by Fact 1 of T.246. The map $f|P$ condenses $P$ onto a metrizable space $f(P)$, so $P$ is metrizable and second countable.

The set $P$ is $C$-embedded in $X$ (see Fact 1 of T.218), so we can apply Fact 3 of V.246 to convince ourselves that the linear hull $H$ of the set $P$ in $L_p(X)$ is linearly homeomorphic to $L_p(P)$ and hence $nw(H) = nw(P) \leq \omega$ (see Problem 240). It follows from $K \subset H$ that $w(K) = nw(K) \leq \omega$ (see Fact 4 of S.307), so $K$ is metrizable.

**V.248.** *Prove that, for any infinite pseudocompact space $X$, there exists an infinite closed discrete set $D$ in the space $L_p(X)$ which is weakly bounded in $L_p(X)$. Therefore, even for a metrizable compact space $X$, the closure of a weakly bounded subset of $L_p(X)$ can fail to be compact.*

**Solution.** The space $X$ being infinite we can apply Fact 4 of S.382 to find a discrete subspace $A = \{d_n : n \in \omega\} \subset X$ such that $d_n \neq d_m$ for distinct $m, n \in \omega$; it is easy to construct a disjoint family $\{U_n : n \in \omega\}$ of open subsets of $X$ such that $U_n \cap A = \{d_n\}$ for each $n \in \omega$. Let $w_n = \sum_{i=0}^{n} 2^{-i} d_i$ for every $n \in \omega$; we claim that the set $D = \{w_n : n \in \omega\} \subset L_p(X)$ is as promised.

Take an arbitrary point $u = \lambda_1 x_1 + \ldots + \lambda_n x_n \in L_p(X)$ and choose $k \in \omega$ such that $U_k \cap \{x_1, \ldots, x_n\} = \emptyset$. There exists a continuous function $f : X \to \mathbb{R}$ with $f(d_k) = 2^k$ and $f(X\backslash U_k) \subset \{0\}$. The set $W = \{w \in L_p(X) : w(f) < 1\}$ is open in $L_p(X)$ and $u \in W$ because $u(f) = 0$. Furthermore, $w_n(f) = 1$ for any $n \geq k$, so $W \cap D \subset \{d_0, \ldots, d_{k-1}\}$. We proved that every point of $L_p(X)$ has a neighborhood whose intersection with $D$ is finite, so $D$ is a closed discrete subspace of $L_p(X)$. The set $X$ being linearly independent, we have $w_n \neq w_m$ for any distinct $m, n \in \omega$, so $D$ is an infinite subset of $L_p(X)$.

Finally, take a continuous linear functional $\xi : L_p(X) \to \mathbb{R}$. The space $X$ is pseudocompact, so the function $h = \xi|X$ is bounded on $X$; fix a number $K > 0$ such that $|h(x)| \leq K$ for all $x \in X$. Then $|\xi(w_n)| \leq \sum_{i=0}^{n} 2^{-i}|h(d_i)| \leq K \sum_{i=0}^{n} 2^{-i} \leq 2K$ for every $n \in \omega$; this shows that $\xi(D) \subset [-2K, 2K]$ is a bounded subset of $\mathbb{R}$ and hence $D$ is weakly bounded.

**V.249.** *Give an example of a space $X$ in which all compact subspaces are metrizable while there are non-metrizable compact subspaces in $L_p(X)$.*

**Solution.** Denote by $X$ the space $\omega_1$ with its order topology and consider the set $O_\alpha = \{\beta \in X : \beta < \alpha\}$ for every $\alpha < \omega_1$. If $K$ is a compact subspace of $X$ then $\{O_\alpha : \alpha < \omega_1\}$ is an open cover of $K$, so there is a finite set $A \subset \omega_1$ such that $K \subset O = \bigcup\{O_\alpha : \alpha \in A\}$. Every set $O_\alpha$ is countable, so $O$ is countable as well; thus the set $K$ is also countable and hence metrizable.

To avoid confusion with the addition of ordinals, we denote the addition in $L_p(X)$ by the symbol $\oplus$; let $w_\alpha = \alpha \oplus (-1)(\alpha + 1)$ for every $\alpha < \omega_1$ and consider the set $F = \{\mathbf{0}\} \cup \{w_\alpha : \alpha < \omega_1\} \subset L_p(X)$. Take any open set $U$ in the space $L_p(X)$ with $\mathbf{0} \in U$. We can find $\varepsilon > 0$ and a finite set $B \subset C_p(X)$ such that the set $V = \{u \in L_p(X) : |u(f)| < \varepsilon$ for every $f \in B\}$ is contained in $U$. Given any $f \in B$ we have $w_\alpha(f) = f(\alpha) - f(\alpha + 1)$ for each $\alpha < \omega_1$, so there is a finite set $P_f \subset \omega_1$ such that $|w_\alpha(f)| < \varepsilon$ for any $\alpha \notin P_f$ (see Fact 1 of S.334). The set $P = \bigcup\{P_f : f \in B\}$ is finite and $|w_\alpha(f)| < \varepsilon$ for any $f \in B$ and $\alpha \notin P$. Therefore $w_\alpha \in V \subset U$ for any $\alpha \notin P$ and hence any neighborhood of $\mathbf{0}$ in $F$ contains all but finitely many points of $F$. An easy consequence is that $F$ is homeomorphic to $A(\omega_1)$, i.e., $F$ is a non-metrizable compact subspace of $L_p(X)$.

**V.250.** *Given spaces $X$ and $Y$ and a continuous map $\varphi : X \to Y$ observe that there is a unique continuous linear map $u_\varphi : L_p(X) \to L_p(Y)$ such that $u_\varphi | X = \varphi$. Prove that the following conditions are equivalent for any continuous onto map $\varphi : X \to Y$.*

*(i)  The map $\varphi$ is $\mathbb{R}$-quotient.*
*(ii)  The map $u_\varphi$ is $\mathbb{R}$-quotient.*
*(iii)  The map $u_\varphi$ quotient.*
*(iv)  The map $u_\varphi$ is open.*

**Solution.** By Problem 233 the topology of $L_p(Y)$ coincides with its weak topology, so we can apply Problem 234 to see that there is a continuous linear map $u_\varphi : L_p(X) \to L_p(Y)$ such that $u_\varphi | X = \varphi$. Suppose that $u : L_p(X) \to L_p(Y)$ is a linear map with $u | X = \varphi$. Given a point $z \in L_p(X)$ we can find $n \in \mathbb{N}$, $\lambda_1, \ldots, \lambda_n \in \mathbb{R}$ and $x_1, \ldots, x_n \in X$ such that $z = \sum_{i=1}^n \lambda_i x_i$. By linearity of the map $u_\varphi$ we have $u_\varphi(z) = \sum_{i=1}^n \lambda_i \varphi(x_i)$; analogously, $u(z) = \sum_{i=1}^n \lambda_i \varphi(x_i) = u_\varphi(z)$, so $u = u_\varphi$, i.e., our linear extension of the map $\varphi$ is unique.

It is evident that (iv)$\Longrightarrow$(iii)$\Longrightarrow$(ii). Assume that $u_\varphi : L_p(X) \to L_p(Y)$ is an $\mathbb{R}$-quotient map for some continuous onto map $\varphi : X \to Y$ and take a function $f : Y \to \mathbb{R}$ such that $f \circ \varphi$ is continuous. By Problem 234 there exists a continuous linear functional $\xi : L_p(X) \to \mathbb{R}$ such that $\xi | X = f \circ \varphi$. There exists a linear functional $\mu : L_p(Y) \to \mathbb{R}$ such that $\mu | Y = f$. It is evident that $\xi' = \mu \circ u_\varphi$ is a linear functional on $L_p(X)$. If $x \in X$ then $\xi'(x) = \mu(u_\varphi(x)) = \mu(\varphi(x))$; since $\varphi(x) \in Y$, we have $\xi'(x) = \mu(\varphi(x)) = f(\varphi(x)) = \xi(x)$.

We proved that $\xi | X = \xi' | X$, so $\xi = \xi'$ by Fact 3 of S.489. Therefore the functional $\xi' = \mu \circ u_\varphi$ is continuous so $\mu$ is continuous because $u_\varphi$ is $\mathbb{R}$-quotient. Consequently, the map $f = \mu | Y$ is continuous and hence $\varphi$ is $\mathbb{R}$-quotient, i.e., we checked that (ii)$\Longrightarrow$(i).

*Fact 1.* Suppose that $L$ is a linear topological space and $K$ is a compact subset of $L$. Then the set $K + A$ is closed in $L$ for any closed $A \subset L$.

*Proof.* To prove that the set $F = K + A$ is closed take any point $x \in L \backslash F$. For any $y \in K$ the set $y + A$ is closed in $L$ and does not contain $x$, so we can find a balanced open neighborhood $U_y$ of $\mathbf{0}$ such that $(x + U_y) \cap (y + A) = \emptyset$. Fix a balanced open neighborhood $V_y$ of the point $\mathbf{0}$ such that $V_y + V_y \subset U_y$. The family $\{y + V_y : y \in K\}$ is an open cover of $K$, so there is a finite set $P \subset K$ such that $K \subset \bigcup \{y + V_y : y \in P\}$. Then $V = \bigcap \{V_y : y \in P\}$ is an open neighborhood of $\mathbf{0}$, so $O = x + V \in \tau(x, L)$.

If $O \cap F \neq \emptyset$ then we can pick $z \in K$ and $a \in A$ with $z + a \in x + V$ and hence $v = z + a - x \in V$. There is $y \in P$ for which $z \in y + V_y$ and therefore $v_0 = z - y \in V_y$. It follows from $z = v_0 + y = v + x - a$ that $a + y = x + v - v_0 \in x + V + V_y$; since $V \subset V_y$, we conclude that $a + y \in x + V_y + V_y \subset x + U_y$, i.e., $(x + U_y) \cap (y + A) \neq \emptyset$ which is a contradiction. Therefore every $x \in L \backslash F$ has a neighborhood which does not meet $F$, so $L \backslash F$ is open and hence $F$ is closed in $L$, i.e., Fact 1 is proved.

*Fact 2.* Given a locally convex space $L$ suppose that $M$ is a closed linear subspace of $L$ and $G$ is a finite-dimensional linear subspace of $L$. Then $M + G$ is a closed subspace of $L$.

*Proof.* Fix a Hamel basis $H$ in the subspace $M$ and apply Fact 1 of S.489 to find vectors $\{e_1, \ldots, e_n\} \subset M + G$ such that $H \cup \{e_1, \ldots, e_n\}$ is a Hamel basis in $M + G$. Letting $M_0 = M$ it takes an easy induction to construct linear subspaces $M_0 \subset M_1 \subset \ldots \subset M_n$ of the space $L$ in such a way that $M_{i+1}$ is a linear hull of $M_i \cup e_{i+1}$ for all $i \in \{0, \ldots, n-1\}$. It is clear that $M_n = M + G$, so it suffices to show inductively that every $M_i$ is closed in $L$. By our assumption the set $M_0$ is closed in $L$, so assume that $0 \leq i < n$ and we have proved that $M_i$ is closed in $L$.

Fix a point $x \notin M_{i+1}$; we claim that there exists a continuous linear functional $f : L \to \mathbb{R}$ such that $f(x) = 1$ and $f(M_i \cup \{e_{i+1}\}) = \{0\}$. To prove this, apply Problem 222 to find a continuous linear functional $g : L \to \mathbb{R}$ for which $g(x) = 1$ and $g(M_i) = \{0\}$. If $r = g(e_{i+1}) = 0$ then let $f = g$. If $r \neq 0$ then $z = rx - e_{i+1} \notin M_i$ for otherwise $x = \frac{1}{r}(z + e_{i+1}) \in M_{i+1}$ which is a contradiction. Thus we can apply Problem 222 once more to find a continuous linear functional $h : L \to \mathbb{R}$ such that $h(z) = 1$ and $h(M_i) = \{0\}$.

Since the determinant $\begin{vmatrix} g(x) & h(x) \\ g(e_{i+1}) & h(e_{i+1}) \end{vmatrix} = \begin{vmatrix} 1 & h(x) \\ r & h(e_{i+1}) \end{vmatrix} = h(e_{i+1} - rx) = -1$ is not equal to zero, any system of linear equations with this determinant has a solution, so we can find numbers $\alpha, \beta \in \mathbb{R}$ such that $\alpha g(x) + \beta h(x) = 1$ and $\alpha g(e_{i+1}) + \beta h(e_{i+1}) = 0$. The linear functional $f = \alpha g + \beta h$ is continuous on $L$ while $f(x) = 1$ and $f(M_i \cup \{e_{i+1}\}) = 0$. As a consequence, $f(M_{i+1}) = \{0\}$, so $x$ cannot belong to the closure of $M_{i+1}$. Since $x \notin M_{i+1}$ was chosen arbitrarily, we proved that $M_{i+1}$ is closed in $L$, so our inductive procedure shows that $M_n = G + M$ is closed in $L$, i.e., Fact 2 is proved.

Returning to our solution assume that $\varphi : X \to Y$ is an $\mathbb{R}$-quotient map and consider the dual map $\varphi^* : C_p(Y) \to C_p(X)$ defined by $\varphi^*(f) = f \circ \varphi$ for any $f \in C_p(Y)$. It is straightforward that $L = \varphi^*(C_p(Y))$ is a linear subspace of $C_p(X)$ and $\varphi^* : C_p(Y) \to L$ is a linear homeomorphism; by TFS-163, the set $L$ is closed in $C_p(X)$. Apply Fact 1 of S.489 to choose a Hamel basis $H$ in $C_p(X)$ in such a way that $H' = H \cap L$ is a Hamel basis in $L$.

We will first show that

(1) for any finite set $A \subset H \backslash H'$ and $f \in A$ there exists a continuous linear functional $\mu : C_p(X) \to \mathbb{R}$ such that $\mu(f) = 1$ and $\mu(L \cup (A \backslash \{f\})) = \{0\}$.

Let $G$ be the linear hull of the set $A \backslash \{f\}$; since $G$ is a finite-dimensional linear subspace of $C_p(X)$, we can apply Fact 2 to see that $L' = G + L$ is a closed linear subspace of $C_p(X)$. Since $f \notin L'$, we can apply Problem 222 to find a continuous linear functional $\mu : C_p(X) \to \mathbb{R}$ such that $\mu(f) = 1$ and $\mu(L') = \{0\}$; then $\mu(L \cup (A \backslash \{f\})) \subset \mu(L') = \{0\}$, so (1) is proved. Our next step is to establish that

(2) given a finite set $A \subset H \backslash H'$, for any $u \in \mathbb{R}^A$ there exists a continuous linear functional $\mu : C_p(X) \to \mathbb{R}$ such that $\mu(L) = \{0\}$ and $\mu|A = u$.

Apply (1) to fix, for any $f \in A$ a continuous linear functional $\mu_f : C_p(X) \to \mathbb{R}$ such that $\mu_f(L \cup (A \backslash \{f\})) = \{0\}$ and $\mu_f(f) = 1$. It is straightforward that the linear functional $\mu = \sum_{f \in A} u(f) \cdot \mu_f$ is continuous on $C_p(X)$ while $\mu(L) = \{0\}$ and $\mu|A = u$, so (2) is proved.

To finally prove that $u_\varphi$ is an open map take an open neighborhood $U$ of $\mathbf{0}$ in the space $L_p(X)$. Apply Fact 2 of V.024 to find a finite set $A \subset H$ and $\varepsilon > 0$ such that $V = \{z \in L_p(X) : |z(f)| < \varepsilon$ for all $f \in A\} \subset U$. Let $A' = A \cap L$ and choose a finite set $B \subset C_p(Y)$ with $\varphi^*(B) = A'$. The set $W = \{w \in L_p(Y) : |w(g)| < \varepsilon$ for any $g \in B\}$ is an open neighborhood of $\mathbf{0}$ in $L_p(Y)$. We claim that $u_\varphi(V) \supset W$; to prove this, fix an arbitrary element $w \in W$.

The map $(\varphi^*)^{-1} : L \to C_p(Y)$ being linear and continuous, $w' = w \circ (\varphi^*)^{-1}$ is a continuous linear functional on $L$. By Problem 224 there exists a continuous linear functional $w'' : C_p(X) \to \mathbb{R}$ such that $w''|L = w'$. Apply Fact 2 to find a continuous linear functional $\mu : C_p(X) \to \mathbb{R}$ such that $\mu|(A \backslash A') = w''|(A \backslash A')$ and $\mu(L) = \{0\}$. Then $v = w'' - \mu$ is a continuous linear functional on the space $C_p(X)$ such that $v(A \backslash A') \subset \{0\}$ and $v|L = w'$. Since $L_p(X)$ coincides with $(C_p(X))^*$ (see TFS-197), the functional $v$ belongs to $L_p(X)$.

If $f \in A'$ then $f = g \circ \varphi$ for some $g \in B$; therefore $g = (\varphi^*)^{-1}(f)$ and hence $|v(f)| = |w'(f)| = |w(g)| < \varepsilon$ by definition of the set $W$. If $f \in A \backslash A'$ then $v(f) = 0$, so $|v(f)| < \varepsilon$ for all $f \in A$ which shows that $v \in V$.

Given any $g \in C_p(Y)$ the function $f = g \circ \varphi$ belongs to $L$, so $v(f) = w'(f)$; it follows from $g = (\varphi^*)^{-1}(f)$ that $v(f) = w(g)$. Choose $n \in \mathbb{N}$, $\lambda_1, \ldots, \lambda_n \in \mathbb{R}$ and $x_1, \ldots, x_n \in X$ such that $v = \sum_{i=1}^n \lambda_i x_i$; let $y_i = \varphi(x_i)$ for every $i \leq n$. Then $u_\varphi(v) = \sum_{i=1}^n \lambda_i y_i$ and hence

$$u_\varphi(v)(g) = \sum_{i=1}^n \lambda_i g(y_i) = \sum_{i=1}^n \lambda_i f(x_i) = v(f) = w(g),$$

which shows that $u_\varphi(v)(g) = w(g)$ for every $g \in C_p(Y)$, i.e., $u_\varphi(v) = w$. We proved that, for each $w \in W$ there exists $v \in V$ such that $u_\varphi(v) = w$; as a consequence, $u_\varphi(U) \supset u_\varphi(V) \supset W$. Thus, for any $U \in \tau(\mathbf{0}, L_p(X))$ there exists $W \in \tau(\mathbf{0}, L_p(Y))$ such that $W \subset u_\varphi(U)$. Applying Fact 3 of S.496 we conclude that the map $u_\varphi$ is open; this proves that (i)$\Longrightarrow$(iv) and makes our solution complete.

**V.251.** Let $f : X \to Y$ be an $\mathbb{R}$-quotient map. Prove that, for any open $U \subset Y$, the map $f|(f^{-1}(U)) : f^{-1}(U) \to U$ is also $\mathbb{R}$-quotient.

**Solution.** Let $U_0 = f^{-1}(U)$ and denote by $f_0$ the map $f$ restricted to $U_0$.

*Fact 1.* Given a space $Z$ and a set $W \in \tau(Z)$ suppose that $h : W \to \mathbb{R}$ is a bounded continuous function and $g : Z \to \mathbb{R}$ is a continuous function such that $g(Z \backslash W) \subset \{0\}$. Consider a function $w = g * h : Z \to \mathbb{R}$ defined as follows: $w(z) = g(z)h(z)$ for all $z \in W$ and $w(z) = 0$ whenever $z \in Z \backslash W$. Then the function $w$ is continuous.

*Proof.* By our hypothesis, there exists $K > 0$ such that $|h(z)| \leq K$ for all $z \in W$. The function $w|W = h \cdot (g|W)$ being continuous on $W$, it is an easy exercise that $w$ is continuous at every $z \in W$. If $z \in Z \backslash W$ and $\varepsilon > 0$ then we can find a set $V \in \tau(z, Z)$ for which $g(V) \subset (-\frac{\varepsilon}{K}, \frac{\varepsilon}{K})$. If $x \in V \backslash W$ then $w(x) = 0 \in (-\varepsilon, \varepsilon)$; if $x \in W \cap V$ then $|w(x)| = |h(x)||g(x)| < K \cdot \frac{\varepsilon}{K} = \varepsilon$. Therefore $w(V) \subset (-\varepsilon, \varepsilon)$, i.e., $V$ witnesses continuity of $w$ at the point $z$. Thus $w$ is continuous at all points of $Z$, so Fact 1 is proved.

*Fact 2.* A map $g : Z \to T$ is $\mathbb{R}$-quotient if and only if, for any function $h : T \to \mathbb{I}$, it follows from continuity of $h \circ g$ that $h$ is continuous.

*Proof.* If $g$ is $\mathbb{R}$-quotient and $h : T \to \mathbb{I}$ is a function such that $h \circ g$ is continuous then $h$ is continuous by Fact 1 of T.268, i.e., we established necessity. To prove sufficiency assume that our hypothesis about the functions from $T$ to $\mathbb{I}$ is true and take an arbitrary function $u : T \to \mathbb{R}$ such that $u \circ g$ is continuous. Fix a homeomorphism $\xi : \mathbb{R} \to (-1, 1)$ and consider the function $h = \xi \circ u$. It is evident that $h : T \to \mathbb{I}$ and the function $h \circ g = \xi \circ (u \circ g)$ is continuous. Therefore $h$ is continuous and hence $u = \xi^{-1} \circ h$ is continuous as well. This shows that the map $g$ is $\mathbb{R}$-quotient and settles sufficiency so Fact 2 is proved.

Returning to our solution take any function $h : U \to \mathbb{I}$ such that $u = h \circ f_0$ is continuous. To see that $h$ is continuous fix any point $y_0 \in U$. Using the Tychonoff property of the space $Y$ and Fact 1 of S.499 it is easy to find a continuous function $v : Y \to [0, 1]$ such that $v(Y \backslash U) \subset \{0\}$ and $v(W) = \{1\}$ for some open $W \subset U$ with $y_0 \in W$. Then $w = v \circ f$ is a continuous function on $X$ such that $w(X \backslash U_0) \subset \{0\}$. By Fact 1, the function $q = w * (h \circ f_0) : X \to \mathbb{I}$ is continuous. If $p = v * h$ then $p \circ f = q$, so the function $p$ is continuous on $Y$ by Fact 1 of T.268. Since $p|W = h|W$, we proved that every point $y_0 \in U$ has a neighborhood $W$ such that $h|W$ is continuous. Applying Fact 1 of S.472 we conclude that $h$ is continuous. Therefore $f_0$ is $\mathbb{R}$-quotient by Fact 2, so our solution is complete.

**V.252.** *Let $X$ be a Tychonoff space. Prove that, for any nonempty closed set $F \subset X$, the $\mathbb{R}$-quotient space $X_F$ is also Tychonoff and if $p_F : X \to X_F$ is the contraction map then $p_F|(X\setminus F) : X\setminus F \to X_F\setminus\{F\}$ is a homeomorphism.*

**Solution.** If $Y$ is a space and $G \neq \emptyset$ is a closed subset of $Y$ then a function $f \in C(Y)$ is called $G$-constant if there is $c \in \mathbb{R}$ such that $f(G) = \{c\}$. Let $p_G(x) = x$ for any $x \in Y\setminus G$ and $p_G(x) = G$ for each $x \in G$; this gives the map $p_G : Y \to Y_G$. If $f \in C(Y)$ is a $G$-constant function with $f(G) = \{c\}$ then let $u_f(x) = f(x)$ for any $x \in Y_G\setminus\{G\}$ and $u_f(G) = c$; therefore we constructed a function $u_f : Y_G \to \mathbb{R}$ for any $G$-constant $f \in C(Y)$.

*Fact 1.* Given a space $Y$ and a closed nonempty set $G \subset Y$, for any $g \in C_p(Y_G)$ let $p_G^*(g) = g \circ p_G$; then $p_G^* : C_p(Y_G) \to C_p(Y)$ is a linear embedding such that $p_G^*(C_p(Y_G))$ coincides with the set $L_G$ of all $G$-constant functions. Furthermore, a function $g : Y_G \to \mathbb{R}$ is continuous if and only if there is a $G$-constant function $f \in C(Y)$ such that $g = u_f$.

*Proof.* It follows from TFS-163 that $p_G^*$ is an embedding; it is straightforward that it is a linear map. If $g \in C(Y_G)$ then $f = g \circ p_G$ is a continuous $G$-constant function on $X$ and it is immediate that $g = u_f$; this proves that $p_G^*(C_p(Y_G)) \subset L_G$ and, for any $g \in C(Y_G)$ there is a $G$-constant function with $g = u_f$.

If $f \in C(Y)$ is a $G$-constant function then, for the function $g = u_f$ we have $f = g \circ p_G$, so $\{g^{-1}(O) : O \in \tau(\mathbb{R})\} \subset \tau(Y_G)$ by definition of the topology of $Y_G$. Therefore $g$ is continuous on $Y_G$, i.e., $u_f$ is continuous on $Y_G$ for any $G$-constant $f \in C(Y)$ and it follows from $f = u_f \circ p_G = p_G^*(u_f)$ that $L_G \subset p_G^*(C_p(Y_G))$, so Fact 1 is proved.

Returning to our solution recall first that $X_F$ is completely regular by Fact 1 of T.139, so it suffices to show that it is Hausdorff. To distinguish the set $F$ from the point $F \in X_F$ we will denote the point $F$ in the space $X_F$ by $a_F$.

Given distinct points $a, b \in X_F$ we can assume that $a \neq a_F$ and hence $a \notin F$. Consider first the case when $b \neq a_F$ and choose $f \in C(X)$ such that $f(a) = 1$ and $f(\{b\} \cup F) = \{0\}$. It is clear that $f$ is an $F$-constant function so $g = u_f$ is continuous on $X_F$ by Fact 1. We have $g(a) = 1$ and $g(b) = 0$, so $U = g^{-1}((-\infty, 1/2))$ and $V = g^{-1}((1/2, +\infty))$ are disjoint open neighborhoods of $b$ and $a$ respectively.

Now, if $b = a_F$ then there exists $f \in C(X)$ such that $f(a) = 1$ and $f(F) = \{0\}$. It is clear that $f$ is $F$-constant so $g = u_f$ is continuous on $X_F$. We have again $g(a) = 1$ and $g(b) = 0$, so $U = g^{-1}((-\infty, 1/2))$ and $V = g^{-1}((1/2, +\infty))$ are disjoint open neighborhoods of $b$ and $a$ respectively. This shows that $X_F$ is Tychonoff being a completely regular Hausdorff space.

Finally observe that the map $p_F$ is $\mathbb{R}$-quotient and $p_F^{-1}(X_F\setminus\{a_F\}) = X\setminus F$, so we can apply Problem 251 to convince ourselves that $p_F|(X\setminus F)$ is also $\mathbb{R}$-quotient. Since $p_F|(X\setminus F)$ is a continuous bijection between $X\setminus F$ and $X_F\setminus\{a_F\}$, we can apply TFS-155 to see that $p_F|(X\setminus F)$ is a homeomorphism.

**V.253.** *Suppose that $X$ is a space and $F$ is a nonempty closed subspace of $X$; in the $\mathbb{R}$-quotient space $X_F$ denote by $a_F$ the point represented by the set $F$. Say that $F$ is deeply inside a set $U \in \tau(X)$ if there exists a zero-set $G$ in the space $X$ such that $F \subset G \subset U$. For the family $\mathcal{U} = \{U : U$ is a cozero subset of $X$ and $F$ is deeply inside the set $U\}$ prove that $\mathcal{V} = \{\{a_F\} \cup (U \backslash F) : U \in \mathcal{U}\}$ is a local base of the space $X_F$ at the point $a_F$.*

**Solution.** Let $p_F : X \to X_F$ be the contraction map and fix an arbitrary set $V \in \mathcal{V}$; by definition of $\mathcal{V}$ we can find a cozero set $U$ and a zero-set $G$ in the space $X$ such that $F \subset G \subset U$ and $V = (U \backslash F) \cup \{a_F\}$. There exists a continuous function $\varphi : X \to [0,1]$ such that $\varphi(G) \subset \{1\}$ and $\varphi(X \backslash U) \subset \{0\}$ (see Fact 1 of V.140). Since $\varphi$ is constant on $F$, we can find a function $u : X_F \to [0,1]$ such that $u \circ p_F = \varphi$. The map $p_F$ being $\mathbb{R}$-quotient, the function $u$ is continuous; it is straightforward that $u(a_F) = 1$ and $u(X_F \backslash V) \subset \{0\}$. It follows from $a_F \in u^{-1}((0,1]) \subset V$ that $a_F$ belongs to the interior of $V$.

Now, if $x \in V \backslash \{a_F\}$ then $x \in U \backslash F$ and hence $U \backslash F = p_F^{-1}(U \backslash F)$ is an open neighborhood of $x$ in $X_F$ with $U \backslash F \subset V$. This shows that each point of $V$ belongs to the interior of $V$ in $X_F$, i.e., every $V \in \mathcal{V}$ is open in $X_F$ and therefore $\mathcal{V} \subset \tau(X_F)$.

Take an open set $O$ in the space $X_F$ with $a_F \in O$. The space $X_F$ being Tychonoff (see Problem 252) there exists a continuous function $f : X_F \to [0,1]$ such that $f(a_F) = 1$ and $f(X_F \backslash O) \subset \{0\}$. Evidently, $g = f \circ p_F$ is a continuous function on $X$ such that $F \subset G = g^{-1}(1)$; it is clear that $G$ is a zero-set in $X$. Furthermore, $U = g^{-1}((0,1])$ is a cozero subset of $X$ and it follows from $G \subset U$ that $F$ is deeply inside the set $U$, i.e., $U \in \mathcal{U}$. Besides, $V = (U \backslash F) \cup \{a_F\}$ coincides with the set $f^{-1}((0,1])$, so $a_F \in V \subset O$. Since also $V \in \mathcal{V}$, we have verified that $\mathcal{V}$ is a local base of the space $X_F$ at the point $a_F$.

**V.254.** *Suppose that $X$ is a normal space and $F$ is a nonempty closed subspace of $X$; in the $\mathbb{R}$-quotient space $X_F$ denote by $a_F$ the point represented by the set $F$. Prove that $U \in \tau(a_F, X_F)$ if and only if $(U \backslash \{a_F\}) \cup F$ is an open neighborhood of $F$ in the space $X$.*

**Solution.** Let $p_F : X \to X_F$ be the contraction map; apply Problem 252 to see that the mapping $p_F|(X \backslash F) : X \backslash F \to X_F \backslash \{a_F\}$ is a homeomorphism and therefore

(∗)  the set $p_F(O)$ is open in $X_F$ for any $O \in \tau(X)$ such that $O \subset X \backslash F$.

If $U \in \tau(a_F, X_F)$ then $(U \backslash \{a_F\}) \cup F = p_F^{-1}(U)$ is an open subset of $X$ containing $F$; this proves necessity.

Now, if $W = (U \backslash \{a_F\}) \cup F \in \tau(F, X)$ then use normality of $X$ to find a continuous function $f : X \to [0,1]$ such that $f(F) = \{1\}$ and $f(X \backslash W) \subset \{0\}$. Then $V = f^{-1}((0,1])$ is a cozero set such that $F \subset f^{-1}(1) \subset V$, so $H = (V \backslash F) \cup \{a_F\}$ is open in the space $X_F$ by Problem 253. Observe that $V \backslash F \subset W \backslash F = U \backslash \{a_F\}$ and hence $a_F \in H \subset U$; this shows that the point $a_F$ belongs to the interior of the set $U$. If $x \in U \backslash \{a_F\} = W \backslash F = p_F(W \backslash F)$ then apply (∗) to convince ourselves that $W \backslash F \subset U$ is an open neighborhood of $x$ in the space $X_F$. Thus every point of $U$ belongs to the interior of $U$ in $X_F$, so $U$ is open in $X_F$ and hence we settled sufficiency.

**V.255.** *Suppose that $X$ is a space and $K$ is a nonempty compact subspace of $X$; in the $\mathbb{R}$-quotient space $X_K$ denote by $a_K$ the point represented by the set $K$. Prove that $U \in \tau(a_K, X_K)$ if and only if $(U \setminus \{a_K\}) \cup K$ is an open neighborhood of $K$ in the space $X$.*

**Solution.** Let $p_K : X \to X_K$ be the contraction map; apply Problem 252 to see that the mapping $p_K | (X \setminus K) : X \setminus K \to X_K \setminus \{a_K\}$ is a homeomorphism and therefore

$(*)$ the set $p_K(O)$ is open in $X_K$ for any $O \in \tau(X)$ such that $O \subset X \setminus K$.

If $U \in \tau(a_K, X_K)$ then $(U \setminus \{a_K\}) \cup K = p_K^{-1}(U)$ is an open subset of $X$ containing $K$; this proves necessity.

*Fact 1.* Suppose that $Z$ is a space, $F \subset Z$ is closed, $P \subset Z$ is compact and $P \cap F = \emptyset$. Then the sets $F$ and $P$ are functionally separated, i.e., there exists a continuous function $f : X \to [0, 1]$ such that $f(P) \subset \{1\}$ and $f(F) \subset \{0\}$.

*Proof.* The sets $P$ and $G = \text{cl}_{\beta Z}(F)$ are closed in the space $\beta Z$; if $x \in G \cap P$ then $x \in \text{cl}_{\beta Z}(F) \cap Z = \text{cl}_Z(F) = F$ which is a contradiction with $P \cap F = \emptyset$. By normality of $\beta Z$ there exists a continuous function $g : \beta Z \to [0, 1]$ for which $g(P) \subset \{1\}$ and $g(G) \subset \{0\}$. The function $f = g|Z : Z \to [0, 1]$ is continuous on $Z$ while $f(P) \subset \{1\}$ and $f(F) \subset \{0\}$, so Fact 1 is proved.

Returning to our solution assume that $W = (U \setminus \{a_K\}) \cup K \in \tau(K, X)$ and use Fact 1 to find a continuous function $f : X \to [0, 1]$ such that $f(K) = \{1\}$ and $f(X \setminus W) \subset \{0\}$. Then $V = f^{-1}((0, 1])$ is a cozero set such that $K \subset f^{-1}(1) \subset V$, so the set $H = (V \setminus K) \cup \{a_K\}$ is open in the space $X_K$ by Problem 253. Observe that $V \setminus K \subset W \setminus K = U \setminus \{a_K\}$ and hence $a_K \in H \subset U$; this shows that the point $a_K$ belongs to the interior of the set $U$. If $x \in U \setminus \{a_K\} = W \setminus K = p_K(W \setminus K)$ then apply $(*)$ to convince ourselves that $W \setminus K \subset U$ is an open neighborhood of $x$ in the space $X_K$. Thus every point of $U$ belongs to the interior of $U$ in $X_K$, so $U$ is open in $X_K$ and hence we settled sufficiency.

**V.256.** *Given a nonempty space $X$ prove that closed sets $P, Q \subset X$ are parallel retracts of $X$ is and only if there is a retraction $r : X \to P$ such that $r|Q : Q \to P$ is a homeomorphism.*

**Solution.** If $P$ and $Q$ are parallel retracts then there exist parallel retractions $r : X \to P$ and $s : X \to Q$; to see that the map $r_0 = r|Q : Q \to P$ is a homeomorphism let $s_0 = s|P : P \to Q$. By definition of parallel retractions we have the equalities $s_0(r_0(x)) = s(r(x)) = s(x) = x$ for any $x \in Q$; since also $r_0(s_0(y)) = r(s(y)) = r(y) = y$ for any $y \in P$, the maps $r_0$ and $s_0$ are mutually inverse homeomorphisms and hence we proved necessity.

Now assume that there exists a retraction $r : X \to P$ such that the mapping $r_0 = r|Q : Q \to P$ is a homeomorphism; let $s_0 : P \to Q$ be the inverse of $r_0$ and consider the map $s = s_0 \circ r$. It is immediate that $s : X \to Q$ is continuous; given any $x \in Q$ we have $s(x) = s_0(r(x)) = s_0(r_0(x)) = x$, so $s$ is a retraction. Finally observe that, for every $x \in X$, we have $s(r(x)) = s_0(r(r(x))) = s_0(r(x)) = s(x)$

and $r(s(x)) = r(s_0(r(x))) = (r_0 \circ s_0)(r(x)) = r(x)$. Therefore $r$ and $s$ are parallel retractions, i.e., we settled sufficiency.

**V.257.** *(Okunev's method of constructing $l$-equivalent spaces). Suppose that $P$ and $Q$ are parallel retracts of a nonempty space $X$. Prove that the completely regular quotient spaces $X_P$ and $X_Q$ are $l$-equivalent.*

**Solution.** Fix some parallel retractions $r : X \to P$ and $s : X \to Q$; as usual, $\pi_P : C_p(X) \to C_p(P)$ and $\pi_Q : C_p(X) \to C_p(Q)$ are the respective restriction maps. We will also need the $\mathbb{R}$-quotient map $\xi_P : X \to X_P$ defined as follows: $\xi_P(x) = x$ if $x \in X \backslash P$ and $\xi_P(x) = P$ if $x \in P$. Analogously, define a map $\xi_Q : X \to X_Q$ by $\xi_Q(x) = x$ whenever $x \in X \backslash Q$ and $\xi_Q(x) = Q$ for each $x \in Q$. If we have spaces $Z$ and $T$ and a map $\varphi : Z \to T$ then its dual map $\varphi^* : C_p(T) \to C_p(Z)$ is defined by $\varphi^*(f) = f \circ \varphi$ for any $f \in C_p(T)$.

Given a nonempty set $Y \subset X$ say that a function $f \in C(X)$ is $Y$-*constant* if there is $c \in \mathbb{R}$ such that $f(Y) = \{c\}$. The dual maps $\xi_P^* : C_p(X_P) \to C_p(X)$ and $\xi_Q^* : C_p(X_Q) \to C_p(X)$ are linear embeddings (see TFS-163); let $L_P = \xi_P^*(C_p(X_P))$ and $L_Q = \xi_Q^*(C_p(X_Q))$. To prove $l$-equivalence of $X_P$ and $X_Q$ it suffices to show that the spaces $L_P$ and $L_Q$ are linearly homeomorphic. We will actually construct a linear homeomorphism $\mu : C_p(X) \to C_p(X)$ such that $\mu(L_P) = L_Q$.

Given any function $f \in C_p(X)$ let $\mu(f) = \pi_P(f) \circ r + \pi_Q(f) \circ s - f$. It follows from TFS-163 and continuity of operations in $C_p(X)$ that the map $\mu : C_p(X) \to C_p(X)$ is linear and continuous. Fix an arbitrary function $f \in C_p(X)$ and let $g = \mu(f)$. For any point $y \in X$ we have $g(y) = f(r(y)) + f(s(y)) - f(y)$; now if $x \in X$ and $y = r(x)$ then $g(y) = f(r \circ r(x)) + f(s \circ r(x)) - f(r(x))$; recalling that $r(r(x)) = r(x)$ and $s(r(x)) = s(x)$ because the retractions $r$ and $s$ are parallel, we convince ourselves that $g(r(x)) = f(s(x))$. An analogous calculation shows that $g(s(x)) = f(r(x))$. Therefore

$$\mu(g)(x) = g(r(x)) + g(s(x)) - g(x) =$$
$$= f(s(x)) + f(r(x)) - (f(r(x)) + f(s(x)) - f(x)) = f(x).$$

This proves that for any $f \in C_p(X)$ we have the equality $\mu(\mu(f))(x) = f(x)$ for every $x \in X$, i.e., $\mu(\mu(f)) = f$. An immediate consequence is that $\mu : C_p(X) \to C_p(X)$ is a linear homeomorphism which coincides with its inverse.

Apply Fact 1 of V.252 to see that $L_P$ coincides with the set of all $P$-constant functions and $L_Q$ is exactly the set of all $Q$-constant functions on $X$. Take any function $f \in L_P$; there exists $c \in \mathbb{R}$ for which $f(P) = \{c\}$. Now, if $x \in Q$ then $\mu(f)(x) = f(r(x)) + f(s(x)) - f(x)$; since $s(x) = x$ and $r(x) \in P$, we conclude that $f(r(x)) = c$ and hence $\mu(f)(x) = c + f(x) - f(x) = c$. Therefore $\mu(f)(x) = c$ for any $x \in Q$, so $\mu(f)$ is $Q$-constant for any $f \in L_P$; in other words, $\mu(L_P) \subset L_Q$. An analogous proof shows that $\mu(L_Q) \subset L_P$, so $\mu|L_P : L_P \to L_Q$ is a linear homeomorphism between the spaces $L_P$ and $L_Q$. We have already observed that $C_p(X_P)$ and $C_p(X_Q)$ are linearly homeomorphic to $L_P$

and $L_Q$ respectively. Therefore $C_p(X_P)$ is linearly homeomorphic to $C_p(X_Q)$, i.e., the spaces $X_P$ and $X_Q$ are $l$-equivalent.

**V.258.** *Suppose that $K$ is a nonempty $l$-embedded subspace of a space $X$ and fix a point $a \notin X$. Prove that the spaces $X \oplus \{a\}$ and $X_K \oplus K$ are $l$-equivalent. Deduce from this fact that if $K$ is a retract of the space $X$ then $X \oplus \{a\} \overset{l}{\sim} X_K \oplus K$. Here $X_K$ is the $\mathbb{R}$-quotient space obtained by contracting $K$ to a point.*

**Solution.** If $Z$ is a space and $F$ is a nonempty closed subset of $Z$ then $Z_F$ is the $\mathbb{R}$-quotient space obtained by contracting $F$ to a point; say that a function $f : Z \to \mathbb{R}$ is $F$-constant if there exists $c \in \mathbb{R}$ such that $f(F) = \{c\}$. Any spaces $Z$ and $T$ are identified with the respective clopen subspaces of $Z \oplus T$; the expression $Z \simeq T$ says that $Z$ is homeomorphic to $T$. We denote by $L \approx M$ the fact that the linear topological spaces $L$ and $M$ are linearly homeomorphic. It is an easy exercise which we leave to the reader that

(1) $C_p(Y \oplus Z) \approx C_p(Y) \times C_p(Z)$ for any spaces $Y$ and $Z$.

After Section 6.3 was finished, this solution was changed for the sake of optimization. As a result, the following fact is not needed for our solution and is left here only for further references.

*Fact 1.* Given spaces $Z$ and $T$ suppose that $F$ is a nonempty closed subset of $Z$ and let $P = Z \oplus T$. Then the space $P_F$ is homeomorphic to $Z_F \oplus T$.

*Proof.* To avoid confusion denote the point $F$ in both spaces $Z_F$ and $P_F$ by $z_F$. Given an $F$-constant function $f$ on the space $P_F$ there is $c \in \mathbb{R}$ with $f(F) = \{c\}$; let $u_f(x) = f(x)$ if $x \neq z_F$ and $f(z_F) = c$. This gives a function $u_f : P_F \to \mathbb{R}$ for any function $f : P \to \mathbb{R}$. Analogously, if $f : Z \to \mathbb{R}$ is an $F$-constant function then $f(F) = \{c\}$ for some $c \in \mathbb{R}$; let $v_f(x) = f(x)$ if $x \in Z_F \backslash \{z_F\}$ and $v_f(z_F) = c$. Therefore we have constructed a function $v_f : Z_F \to \mathbb{R}$ for any $F$-constant function $f : Z \to \mathbb{R}$.

The spaces $P_F$ and $Z_F \oplus T$ have the same underlying set so it suffices to prove that a function $g : P_F \to \mathbb{R}$ is continuous on $P_F$ if and only if it is continuous on $Z_F \oplus T$. Assume first that a function $g : P_F \to \mathbb{R}$ is continuous. There exists a continuous $F$-constant function $f : P \to \mathbb{R}$ such that $g = u_f$ (see Fact 1 of V.252). The function $f_0 = f|Z$ is also $F$-constant and continuous, so $v_{f_0}$ is continuous on $Z_F$; it is straightforward that $v_{f_0} = g|Z_F$. Besides, $f_1 = f|T = g|T$ is a continuous function on $T$, so $\{Z_F, T\}$ is an open cover of $Z_F \oplus T$ such that the restrictions of $g$ to its elements are continuous. Therefore $g$ is continuous on $Z_F \oplus T$ by Fact 1 of S.472.

Now, if $g : Z_F \oplus T \to \mathbb{R}$ is continuous then $g_0 = g|Z_F$ is also continuous, so there is an $F$-constant function $f_0 \in C(Z)$ such that $g_0 = v_{f_0}$. Let $f(z) = f_0(z)$ for any $z \in Z$ and $f(z) = g(z)$ whenever $z \in T$. It follows from continuity of $f_0$ and $g|T$ that $f : Z \oplus T \to \mathbb{R}$ is continuous. Since $f$ is $F$-constant, the function $u_f : P_F \to \mathbb{R}$ is continuous as well and it is straightforward that $u_f = g$. This shows that $g$ is continuous if considered as a function on $P_F$, so the spaces $P_F$ and $Z_F \oplus T$ are homeomorphic, i.e., Fact 1 is proved.

Let $p_K : X \to X_K$ be the contraction map. Given any function $f \in C_p(X_K)$ let $p_K^*(f) = f \circ p_K$. Then $p_K^* : C_p(X_K) \to C_p(X)$ is a linear embedding and $p_K^*(C_p(X_K))$ coincides with the set $C = \{f \in C_p(X) : f(x) = f(y)$ for any $x, y \in K\}$ of $K$-constant functions on $X$ (see Fact 1 of V.252).

Consider the set $I = \{f \in C_p(X) : f(K) = \{0\}\} \subset C$; for any pair $(f, t) \in I \times \mathbb{R}$ let $\varphi(f, t) = f + t$. Then $\varphi : I \times \mathbb{R} \to C$ is a linear continuous map. Fix a point $z \in K$ and let $\xi(f) = f - f(z)$ for any $f \in C$; it is easy to see that the map $\xi : C \to I$ is linear and continuous. Therefore the map $\mu : C \to I \times \mathbb{R}$ defined by the equality $\mu(f) = (\xi(f), f(z))$ for any $f \in C$, is also linear and continuous.

It is straightforward that the maps $\varphi$ and $\mu$ are mutually inverse, so we proved that the space $C \approx C_p(X_K)$ is linearly homeomorphic to $I \times \mathbb{R}$. Now apply CFS-448 to see that $C_p(X) \approx C_p(K) \times I$ which implies, together with $C_p(X \oplus \{a\}) \approx C_p(X) \times \mathbb{R}$ and the property (1) that

$$C_p(X \oplus \{a\}) \approx C_p(K) \times (I \times \mathbb{R}) \approx C_p(K) \times C_p(X_K) \approx C_p(X_K \oplus K),$$

so the spaces $X \oplus \{a\}$ and $X_K \oplus K$ are $l$-equivalent. If $K$ is a retract of $X$ then it is $l$-embedded in $X$ by CFS-465, so $X \oplus \{a\} \overset{l}{\sim} X_K \oplus K$ is this case as well.

**V.259.** *Given a space $X_i$ and a point $x_i \in X_i$ for any $i = 1, \ldots, n$ consider the space $X = X_1 \oplus \ldots \oplus X_n$ and the set $F = \{x_1, \ldots, x_n\} \subset X$. The $\mathbb{R}$-quotient space $X_F$ is denoted by $(X_1, x_1) \vee \ldots \vee (X_n, x_n)$ and called a bunch of spaces $X_1, \ldots, X_n$ with respect to the points $x_1, \ldots, x_n$. Prove that if we choose any point $y_i \in X_i$ for every $i = 1, \ldots, n$ then the spaces $(X_1, x_1) \vee \ldots \vee (X_n, x_n)$ and $(X_1, y_1) \vee \ldots \vee (X_n, y_n)$ are $l$-equivalent.*

**Solution.** We identify every space $X_i$ with the respective clopen subspace of $X$. Given $x \in X$ there is a unique $i \in \{1, \ldots, n\}$ such that $x \in X_i$; let $r(x) = x_i$. The family $\{X_1, \ldots, X_n\}$ is an open cover of $X$ while $r|X_i$ is constant and hence continuous for every $i \in \{1, \ldots, n\}$. Therefore the map $r : X \to F$ is continuous by Fact 1 of S.472. It follows from $x_i \in X_i$ that $r(x_i) = x_i$ for all $i \leq n$; in other words, $r(x) = x$ for each $x \in F$, so $r$ is a retraction.

If $G = \{y_1, \ldots, y_n\}$ then $r(y_i) = x_i$ for all $i \leq n$, so $r|G : G \to F$ is a homeomorphism which shows that $F$ and $G$ are parallel retracts (see Problem 256). Applying Problem 257 we conclude that the spaces $X_F = (X_1, x_1) \vee \ldots \vee (X_n, x_n)$ and $X_G = (X_1, y_1) \vee \ldots \vee (X_n, y_n)$ are $l$-equivalent.

**V.260.** *Let $K$ be a retract of a nonempty space $X$ and fix any point $z \in K$. Denote by $a_K$ the point of the space $X_K$ represented by the set $K$. Prove that the space $X$ is $l$-equivalent to the bunch $(X_K, a_K) \vee (K, z)$ of the spaces $X_K$ and $K$ with respect to the points $a_K$ and $z$.*

**Solution.** We identify any spaces $Y$ and $Z$ with the respective clopen subspaces of the space $Y \oplus Z$. If $Y$ is a space and $F \neq \emptyset$ is a closed subspace of $Y$ then $Y_F$ is the $\mathbb{R}$-quotient space obtained from $Y$ by contracting $F$ to a point and $p_F : Y \to Y_F$ is the contraction map; to avoid confusion, the point represented by $F$ in the space $Y_F$ is denoted by $a_F$.

*Fact 1.* Suppose that $Y$ is a space and $F$ is a nonempty closed subspace of $Y$. If $r : Y \to Q$ is a retraction such that the set $r(F)$ is a singleton then the map $p = p_F | Q : Q \to p_F(Q)$ is a homeomorphism.

*Proof.* If $F \cap Q = \emptyset$ then apply Problem 252 to see that the map $p_F :$ $Y \backslash F \to Y_F \backslash \{a_F\}$ is a homeomorphism; since $Q \subset Y \backslash F$, the map $p$ is also a homeomorphism.

If $F \cap Q \neq \emptyset$ then there exists a point $a \in Q$ such that $F \cap Q = r(F) = \{a\}$. It follows from continuity of $p_F$ that the map $p$ is continuous. Since $Q \backslash \{a\} \subset Q \backslash F$, the map $p|(Q \backslash \{a\}) : Q \backslash \{a\} \to p_F(Q) \backslash \{a_F\}$ is also a homeomorphism; besides, $p(a) = a_F$, so $p$ is a condensation. Therefore it suffices to show that the map $p^{-1}$ is continuous at the point $a_F$.

Fix a set $U \in \tau(a, Q)$; it is easy to find a cozero set $V$ and a zero-set $H$ in the space $Q$ such that $a \in H \subset V \subset U$. Then $H' = r^{-1}(H)$ is a zero-set and $V' = r^{-1}(V)$ is a cozero set in the space $Y$ (see Fact 1 of S.499); the map $r$ being a retraction, we have $V' \cap Q = V$. It follows from $F \subset H' \subset V'$ that the set $W = (V' \backslash F) \cup \{a_F\}$ is an open neighborhood of $a_F$ in $Y_F$ (see Problem 253). As a consequence, the set $W' = W \cap p_F(Q) = (V \backslash \{a\}) \cup \{a_F\}$ is an open neighborhood of $a_F$ in $p_F(Q)$. If $x \in W' \backslash \{a_F\}$ then $p^{-1}(x) = x \in V$; since $p^{-1}(a_F) = a \in V$, we conclude that $p^{-1}(W') \subset V \subset U$ and hence the map $p^{-1}$ is continuous at the point $a_F$. Therefore $p$ is a homeomorphism, i.e., Fact 1 is proved.

*Fact 2.* Suppose that $Y$ is a space and $Q$ is a retract of $Y$. If $F$ is a nonempty closed subset of $Y$ with $F \subset Q$ and $p_F : Y \to Y_F$ is a contraction map then the topology induced on $Q' = p_F(Q)$ from the space $Y_F$ coincides with the topology of the space $Q_F$.

*Proof.* Since the map $p_F$ is an identity on the set $Y \backslash F$, we will identify the sets $U$, $p_F(U)$ and $p_F^{-1}(U)$ for any $U \subset Y \backslash F$. Note first that the sets $Q'$ and $Q_F$ coincide; let $\tau$ be the topology on $Q'$ induced from $Y_F$ and denote by $\mu$ the topology of $Q_F$. Suppose first that $U \in \mu$ and $a_F \notin U$; then $U \subset Q \backslash F$, so it follows from Problem 252 that $U$ is open in $Q \backslash F$ and hence there is $U' \in \tau(Y)$ such that $U' \cap Q = U$. Apply Problem 252 again to see that $U' \subset Y \backslash F$ is open in $Y_F$ and hence $U = U' \cap Q \in \tau$. This shows that

(1) if $U \in \mu$ and $a_F \notin U$ then $U \in \tau$.

Now, assume that $U \in \tau$ and $a_F \notin U$; there exists an open set $U'$ in the space $Y_F$ such that $U = U' \cap Q'$. Since $U' \subset Y \backslash F$, it follows from Problem 252 that $U'$ is open in $Y \backslash F$ and hence $U$ belongs to the topology of $Q \backslash F$. Applying Problem 252 to the space $Q_F$ we convince ourselves that $U \in \mu$. This proves that

(2) if $a_F \notin U \subset Q'$ then $U \in \mu$ if and only if $U \in \tau$.

Next assume that $a_F \in U \in \mu$; there exist a zero-set $H$ and a cozero set $W$ in the space $Q$ such that $F \subset H \subset W$ and $U' = (W \backslash F) \cup \{a_F\} \subset U$ (see Problem 253). If $H' = r^{-1}(H)$ and $W' = r^{-1}(W)$ then $H'$ is a zero-set and $W'$ is a cozero set in the space $Y$; the map $r$ being a retraction we have $W' \cap Q = W$. Since also $F \subset H' \subset W'$, we can apply Problem 253 again to conclude that the set

$V = (W' \backslash F) \cup \{a_F\}$ is open in $Y_F$, so $U' = V \cap Q'$ belongs to $\tau$. It follows from (2) that $U \backslash \{a_F\} \in \tau$, so $U = U' \cup (U \backslash \{a_F\})$ also belongs to $\tau$. Therefore,

(3) if $a_F \in U \in \mu$ then $U \in \tau$.

Finally suppose that $a_F \in U \in \tau$ and choose an open subset $U'$ of the space $Y_F$ such that $U' \cap Q' = U$. There exist a zero-set $H$ and a cozero set $W$ in the space $Y$ such that $F \subset H \subset W$ and $V = (W \backslash F) \cup \{a_F\} \subset U'$ (see Problem 253). If $H' = H \cap Q$ and $W' = W \cap Q$ then $H'$ is a zero-set and $W'$ is a cozero set in the space $Q$. Since also $F \subset H' \subset W'$, we can apply Problem 253 again to conclude that the set $V' = (W' \backslash F) \cup \{a_F\} \subset U$ is open in $Q_F$. It follows from (2) that $U \backslash \{a_F\} \in \mu$, so $U = V' \cup (U \backslash \{a_F\})$ also belongs to $\mu$. As a consequence,

(4) if $a_F \in U \in \tau$ then $U \in \mu$.

It follows from (1)–(4) that $\tau = \mu$, so Fact 2 is proved.

*Fact 3.* Given a space $Y$ and $a \in Y$ suppose that $Q_1, \ldots, Q_n$ are closed subspaces of $Y$ such that $Y = Q_1 \cup \ldots \cup Q_n$ and $Q_i \cap Q_j = \{a\}$ for any distinct $i, j \in \{1, \ldots, n\}$. Then $Y$ is homeomorphic to the space $(Q_1, a) \vee \ldots \vee (Q_n, a)$.

*Proof.* We will identify every $Q_i$ with the respective clopen subspace of the space $Q = Q_1 \oplus \ldots \oplus Q_n$; the point $a_i \in Q_i$ stands for the copy of the point $a$ in the space $Q_i'$. Let $F = \{a_1, \ldots, a_n\}$; it suffices to show that $Y$ is homeomorphic to the space $Q_F$. Let $h(a) = a_F$; if $y \in Y \backslash \{a\}$ then there is a unique $i \leq n$ with $y \in Q_i$; let $h(y) = y$. This gives us a bijection $h : Y \to Q_F$; an easy consequence of Problem 252 is that $h|(Y \backslash \{a\}) : Y \backslash \{a\} \to Q \backslash F$ is a homeomorphism.

Fix a set $U \in \tau(a_F, Q_F)$; by Problem 253, there is an open set $V$ in the space $Q$ such that $F \subset V$ and $(V \backslash F) \cup \{a_F\} \subset U$. The set $V \backslash F$ is also open in $Y$; furthermore, $G_i = Q_i \backslash V$ is closed in $Q_i$ and hence in $Y$ for each $i \leq n$. Therefore the set $G = G_1 \cup \ldots \cup G_n$ is closed in $Y$, so $W = Y \backslash G$ is an open neighborhood of the point $a$. It is immediate that $h(W) = (V \backslash F) \cup \{a_F\} \subset U$ and hence $h$ is continuous at the point $a$.

Finally take any set $U \in \tau(a, Y)$; it is easy to find a zero-set $H$ and a cozero set $W$ in the space $Y$ such that $a \in H \subset W \subset U$. Then $H_i = H \cap Q_i$ is a zero-set and $W_i = W \cap Q_i$ is a cozero set in $Q_i$ for every $i \leq n$; consider the copy $W_i'$ of $W_i$ and the copy $H_i'$ of $H_i$ in the set $Q_i \subset Q$. It is evident that $H' = H_1' \cup \cdots \cup H_n'$ is a zero-set and $W' = W_1' \cup \cdots \cup W_n'$ is a cozero set in the space $Q$. It follows from $F \subset H' \subset W'$ that $O = (W' \backslash F) \cup \{a_F\}$ is an open neighborhood of $a_F$ in $Q_F$ (see Problem 253). Since $h^{-1}(O) \subset W \subset U$, the set $O$ witnesses continuity of the map $h^{-1}$ at the point $a_F$. Thus $h$ is a homeomorphism between the spaces $Y$ and $Q_F$; the space $Q_F$ is homeomorphic to $(Q_1, a) \vee \ldots \vee (Q_n, a)$, so Fact 3 is proved.

Returning to our solution fix a retraction $r : X \to K$ and a copy $K_0$ of the space $K$ with a homeomorphism $h : K \to K_0$. Let $z_0 = h(z)$ and consider the space $Z = X \oplus K_0$ together with the set $F = \{z, z_0\} \subset Z$; let $p_F : Z \to Z_F$ be the contraction map.

It is trivial that there exist retractions $\mu_0 : Z \to X$ and $\mu_1 : Z \to K_0$ such that $\mu_0(F)$ and $\mu_1(F)$ are singletons, so we can apply Fact 1 to conclude that the maps

$p = p_F|X : X \to p_F(X)$ and $q = p_F|K_0 : K_0 \to p_F(K_0)$ are homeomorphisms; let $X' = p_F(X)$, $K' = p_F(K)$ and $K_1 = p_F(K_0)$. The map $s_1 = p \circ r \circ p^{-1} :$ $X' \to K'$ is easily seen to be a retraction while the map $s_2 = (p|K) \circ h^{-1} \circ q^{-1} :$ $K_1 \to K'$ is a homeomorphism.

The family $\{X', K_1\}$ is a closed cover of $Z_F$ such that $X' \cap K_1 = \{a_F\}$. Since $s_1(a_F) = s_2(a_F) = a_F$, there exists a continuous map $s : Z_F \to K'$ such that $s|X' = s_1$ and $s|K_1 = s_2$ (see Fact 2 of T.354). It is straightforward that $s$ is a retraction such that $s|K_1$ is a homeomorphism, so $K'$ and $K_1$ are parallel retracts of $Z' = Z_F$ by Problem 256.

Letting $\nu(x) = x$ for any $x \in X'$ and $\nu(x) = a_F$ for any $x \in K_1 \setminus \{a_F\}$ we obtain a retraction $\nu : Z' \to X'$ such that $\nu(K_1)$ is a singleton. This makes it possible to apply Fact 1 again to see that the space $Z'_{K_1}$ is homeomorphic to $X'$ which is in turn homeomorphic to $X$.

Observe that $p : X \to X'$ is a homeomorphism such that $p(K) = K'$; an immediate consequence is that $X'_{K'}$ is homeomorphic to $X_K$. Since $X'$ is a retract of the space $Z'$, we can apply Fact 2 to see that $Q = X'_{K'}$ is homeomorphic to the respective subspace of $Z'_{K'}$; let $q_{K'} : Z' \to Z'_{K'}$ be the contraction map. There exists a retraction $u$ of $Z'$ onto $K_1$ such that $u(K')$ is a singleton, so we can apply Fact 1 to convince ourselves that $q_{K'}|K_1$ embeds $K_1$ in the space $Z'_{K'}$; let $P = q_{K'}(K_1)$.

This shows that $Z'_{K'} = P \cup Q$ where $P$ and $Q$ are closed subsets of $Z'_{K'}$ such that $P \cap Q = \{a_{K'}\}$. By Fact 3, the space $Z'_{K'}$ is homeomorphic to $(Q, a_{K'}) \vee (P, a_{K'})$. The spaces $Z'_{K'}$ and $Z'_{K_1}$ are $l$-equivalent by Problem 257; besides, $Z'_{K_1}$ is homeomorphic to $X$ and $Z'_{K'}$ is homeomorphic to $(X_K, b) \vee (K, c)$ for some points $b \in X_K$ and $c \in K$. It follows from Problem 259 that $X$ is $l$-equivalent to $(X_K, a_K) \vee (K, z)$, i.e., our solution is complete.

**V.261.** *Assume that $K$ and $L$ are retracts of a nonempty space $X$ and there exists a retraction $r : X \to L$ such that $r(K) = K \cap L = \{a\}$ for some point $a \in L$; let $M = K \cup L$. Prove that the space $X$ is $l$-equivalent to the bunch $(X_M, c_0) \vee (K, c_1) \vee (L, c_2)$ where the points $c_0 \in X_M$, $c_1 \in K$ and $c_2 \in L$ are chosen arbitrarily.*

**Solution.** We identify any spaces $Y$ and $Z$ with the respective clopen subspaces of the space $Y \oplus Z$. If $Y$ is a space and $F \neq \emptyset$ is a closed subspace of $Y$ then $Y_F$ is the $\mathbb{R}$-quotient space obtained from $Y$ by contracting $F$ to a point.

Denote by $a_K$ the point represented in $X_K$ by the set $K$ and take a homeomorphic copy $K_0$ of the space $K$. Pick a point $z_0 \in K_0$ and let $Q = \{a_K, z_0\}$. Define a map $\beta$ on the space $Y = X_K \oplus K_0$ by letting $\beta(x) = x$ for any $x \in X_K$ and $\beta(x) = a_K$ for every $x \in K_0$. It is evident that $\beta : Y \to X_K$ is a retraction such that $\beta(Q)$ is a singleton. Analogously, we can construct a retraction $\alpha : Y \to K_0$ such that $\alpha(Q)$ is a singleton.

The space $Y_Q = (X_K, a_K) \vee (K_0, z_0)$ is $l$-equivalent to $X$ by Problem 260; denote by $w_Q$ the point of $Y_Q$ represented by the set $Q$. Let $q(x) = x$ for any $x \in X_K \setminus \{a_K\}$ and $q(a_K) = w_Q$; then $q$ is the restriction to $X_K$ of the $\mathbb{R}$-quotient map that contracts $Q$ to a point. By Fact 1 of V.260, the map $q : X_K \to (X \setminus K) \cup \{w_Q\}$ is a homeomorphism so we identify $X_K$ with the subspace $T = (X \setminus K) \cup \{w_Q\}$ of the space $Z = Y_Q$. Analogously, $K' = (K_0 \setminus \{z_0\}) \cup \{w_Q\}$ is a homeomorphic copy of the space $K$.

If $p_K : X \to X_K$ is the contraction map then $p_K|L : L \to L' = p_K(L)$ is a homeomorphism by Fact 1 of V.260. Define a map $r_0 : X_K \to L'$ by letting $r_0(a_K) = a_K$ and $r_0(x) = p_K(r(x))$ for any point $x \in X_K \setminus \{a_K\}$. It follows from $r_0 \circ p_K = p_K \circ r$ that $r_0 \circ p_K$ is continuous; the map $p_K$ being $\mathbb{R}$-quotient, we can apply Fact 1 of T.268 to conclude that $r_0$ is a continuous map.

If $x \in L' \setminus \{a_K\} = L \setminus \{a\}$ then $r(x) = x$, so $r_0(x) = p_K(x) = x$; this shows that the map $r_0 : X_K \to L'$ is a retraction. Recalling that $T$ is a copy of $X_K$, we can consider that $L' \subset T$ is a copy of $L$ and $r_0 : T \to L'$ is a retraction of $T$ onto $L'$. Let $\gamma(t) = t$ for any $t \in T$ and $\gamma(x) = w_Q$ whenever $x \in K' \setminus \{w_Q\}$. It is straightforward that $\gamma : Z \to T$ is a retraction so $r_0 \circ \gamma$ is a retraction of $Z$ onto $L'$.

Denote by $b_{L'}$ the point represented by the set $L'$ in the space $Z_{L'}$ and let $p_{L'} : Z \to Z_{L'}$ be the contraction map. Since $T$ is a retract of $Z$ and $L' \subset T$, it follows from Fact 2 of V.260 that $T' = T_{L'}$ can be identified with the set $p_{L'}(T)$.

Apply Problem 260 again to convince ourselves that the space $Z$ is $l$-equivalent to the bunch $B = (Z_{L'}, b_{L'}) \vee (L', t)$ where the point $t \in L'$ is chosen arbitrarily. Let $\delta(x) = w_Q$ if $x \in T$ and $\delta(x) = x$ whenever $x \in K'$. It is immediate that $\delta : Z \to K'$ is a retraction such that $\delta(L')$ is a singleton, so $p_{L'}|K' : K' \to K_1 = p_{L'}(K')$ is a homeomorphism. It is easy to see that $K_1 \cap T' = \{b_{L'}\}$.

Denote by $u_G$ the point of the space $B$ represented by the set $G = \{b_{L'}, t\}$. It is easy to verify, applying Fact 1 of V.260, that the contraction map $v : Z_{L'} \oplus L' \to B$ is a homeomorphism if restricted to either $Z_{L'}$ or $L'$; let $K_2 = v(K_1)$ and $L_2 = v(L')$. If $T_1 = v(T')$ then the family $\{T_1, K_2, L_2\}$ is a closed cover of the space $B$ and $T_1 \cap K_2 = T_1 \cap L_2 = K_2 \cap L_2 = \{u_G\}$. Therefore we can apply Problem 259 and Fact 3 of V.260 to conclude that $B$ is homeomorphic to the bunch $(T_1, x) \vee (K_2, y) \vee (L_2, z)$ for some $x \in T_1$, $y \in K_2$ and $z \in L_2$.

The spaces $K_2$ and $L_2$ were obtained by applying several homeomorphisms to the spaces $K$ and $L$ respectively. As a consequence, $B$ is homeomorphic to the bunch $(T_1, x) \vee (K, y) \vee (L, z)$ for some $x \in Z_2$, $y \in K$ and $z \in L$.

Consider the space $E = X_K$ and let $h = p_{L'}|E : E \to p_{L'}(E) = E_{L'}$. Our construction shows that the space $T_1$ is homeomorphic to $E_{L'}$. Let $s_M : X \to X_M$ be the contraction map and denote by $z_M$ the point represented in $X_M$ by the set $M = K \cup L$. Let $\xi(b_{L'}) = z_M$; if $x \in E_{L'} \setminus \{b_{L'}\}$ then $x \in X \setminus M$, so we can let $\xi(x) = x$. This gives us a bijection $\xi : E_{L'} \to X_M$ and it is immediate that $\xi \circ (h \circ p_K) = s_M$. The map $s_M$ is continuous and $h \circ p_K$ is $\mathbb{R}$-quotient, so $\xi$ is a continuous map (see Fact 1 of T.268). The map $\xi^{-1} \circ s_M = h \circ p_K$ is also continuous; since $s_M$ is $\mathbb{R}$-quotient, we can apply Fact 1 of T.268 again to see that $\xi^{-1}$ is continuous, i.e., $\xi$ is a homeomorphism. This shows that $T_1$ is homeomorphic to $X_M$ and hence $B$ is homeomorphic to the bunch $(X_M, x) \vee (K, y) \vee (L, z)$ for some $x \in X_M$, $y \in K$ and $z \in L$.

Finally recall that $X$ is $l$-equivalent to $Z$ and $Z$ is $l$-equivalent to $B$, so we can apply Problem 259 to conclude that $X$ is $l$-equivalent to $(X_M, c_0) \vee (K, c_1) \vee (L, c_2)$ for any choice of points $c_0 \in X_M$, $c_1 \in K$ and $c_2 \in L$.

**V.262.** *Given spaces $Y$ and $Z$ consider the space $X = Y \times Z$; choose arbitrary points $y_0 \in Y$, $z_0 \in Z$ and let $M = (Y \times \{z_0\}) \cup (\{y_0\} \times Z)$. Prove that, for any $x_0 \in X_M$, the space $X$ is $l$-equivalent to the bunch $(X_M, x_0) \vee (Y, y_0) \vee (Z, z_0)$.*

**Solution.** If $K = Y \times \{z_0\}$ and $L = \{y_0\} \times Z$ then both $K$ and $L$ are retracts of $X$ and $M = K \cup L$. Letting $r(y, z) = (y_0, z)$ for any $(y, z) \in X$ we obtain a retraction $r : X \to L$ such that $r(K) = \{(y_0, z_0)\} = K \cap L$. This makes it possible to apply Problems 261 and 259 to conclude that $X$ is $l$-equivalent to the bunch $(X_M, x) \vee (K, a) \vee (L, b)$ for any points $x \in X_M$, $a \in K$ and $b \in L$. The space $K$ is homeomorphic to $Y$ and $L$ is homeomorphic to $Z$, so $X$ is also $l$-equivalent to the bunch $(X_M, x_0) \vee (Y, y_0) \vee (Z, z_0)$ where the point $x_0 \in X_M$ can be chosen arbitrarily.

**V.263.** *Let $a = 0$ and $a_n = \frac{1}{n+1}$ for all $n \in \omega$; then $S = \{a_n : n \in \omega\} \cup \{a\}$ is a faithfully indexed convergent sequence with limit $a$. Given an infinite cardinal $\kappa$ consider the discrete space $D(\kappa)$ of cardinality $\kappa$ and let $E = D(\kappa) \times S$. Observe that $F = D(\kappa) \times \{a\}$ is a retract of $E$; as usual let $E_F$ be the $\mathbb{R}$-quotient space obtained by contracting $F$ to a point. The space $E_F$ will be denoted by $V(\kappa)$; it is often called the Fréchet–Urysohn $\kappa$-fan. The space $V(\omega)$ is called the Fréchet–Urysohn fan. Prove that $V(\kappa)$ is $l$-equivalent to $D(\kappa) \times S$ for any infinite cardinal $\kappa$. Deduce from this fact that*

  (i)  *there exist $l$-equivalent spaces $X$ and $Y$ with $w(X) \neq w(Y)$ and $\chi(X) \neq \chi(Y)$;*

 (ii)  *metrizability is not preserved by $l$-equivalence;*

(iii)  *a space $l$-equivalent to a locally compact space, need not be Čech-complete.*

**Solution.** Given spaces $X$ and $Y$ the expression $X \simeq Y$ says that they are homeomorphic. For any point $(x, y) \in E$ let $r(x, y) = (x, a)$; it is evident that the map $r : E \to F$ is a retraction, so $F$ is, indeed, a retract of $E$. If $D$ is a discrete space with $|D| \leq \kappa$ then $D(\kappa) \oplus D$ is also a discrete space of cardinality $\kappa$, so we have the following property:

(1)  $D(\kappa) \oplus D \simeq D(\kappa)$ for any discrete space $D$ with $|D| \leq \kappa$.

Since $a_0$ is an isolated point of $S$, the set $A = D(\kappa) \times \{a_0\}$ is clopen and discrete in $E$. The space $S_0 = S \setminus \{a_0\}$ is homeomorphic to $S$, so $G = D(\kappa) \times S_0$ is a clopen subspace of $E$ homeomorphic to $E$; it follows from $E = G \cup A$ that $E \simeq E \oplus D(\kappa)$. If $D$ is a discrete space of cardinality at most $\kappa$ then $E \oplus D \simeq E \oplus (D(\kappa) \oplus D)$, so we can apply (1) to see that $E \oplus D \simeq E \oplus D(\kappa) \simeq E$. Therefore

(2)  $E \oplus D \simeq E$ for any discrete space $D$ with $|D| \leq \kappa$.

By (2) and Fact 1 of V.258, the space $E_F$ is homeomorphic to $E_F \oplus D(\kappa)$; the set $F$ being a discrete subspace of $E$ of cardinality $\kappa$, we convince ourselves that $E_F \simeq E_F \oplus F$. Fix any point $b \notin E$ and apply Problem 258 to see that

$E_F \oplus F \overset{l}{\sim} E \oplus \{b\}$; it follows from (2) that $E \oplus \{b\} \simeq E$, so $V(\kappa) = E_F$ is $l$-equivalent to $E = D(\kappa) \times S$ as promised.

Therefore the spaces $X = V(\omega)$ and $Y = D(\omega) \times S$ are $l$-equivalent. It is straightforward that $Y$ is a locally compact metrizable space of countable weight. To analyze the space $X$ let $D = \{d_n : n \in \omega\}$ be a faithfully indexed discrete space; we can identify the space $Y$ with $D \times S$. If $F = D \times \{a\}$ then $X$ is homeomorphic to $Y_F$; let $S_n = \{d_n\} \times S$ and $b_n = (d_n, a)$ for each $n \in \omega$.

Denote by $x_F$ the point of $X$ represented by the set $F$ and take an arbitrary family $\mathcal{U} = \{U_n : n \in \omega\}$ of open neighborhoods of $x_F$ in $X$. The space $Y$ being normal, every set $W_n = (U_n \setminus \{x_F\}) \cup F$ is an open neighborhood of $F$ in $Y$ (see Problem 254). In particular, $W_n \in \tau(b_n, Y)$, so $S_n \setminus W_n$ is finite because the sequence $S_n$ converges to $b_n$; pick a point $c_n \in (S_n \setminus \{b_n\}) \cap W_n$ for every $n \in \omega$ and let $C = \{c_n : n \in \omega\}$.

The set $V = Y \setminus C$ is an open neighborhood of $F$ in $Y$, so we can apply Problem 254 once more to see that the set $W = (V \setminus F) \cup \{x_F\}$ is an open neighborhood of $x_F$ in $X$. Since $c_n \in U_n \setminus W$ for every $n \in \omega$, no element of $\mathcal{U}$ is contained in $W$ and hence $\mathcal{U}$ is a not a local base at $x_F$ in the space $X$. Therefore $X$ has no countable local base at $x_F$ and hence $w(X) \geq \chi(X) > \omega = \chi(Y) = w(Y)$, so the property (i) holds for the spaces $X$ and $Y$. The space $X$ cannot be metrizable being a countable space of uncountable weight; this settles (ii). It follows from $nw(X) = \omega < w(X)$ that $X$ cannot be Čech-complete either (see TFS-270); this proves (iii) and completes our solution.

**V.264.** *Given infinite cardinals $\kappa_1, \ldots, \kappa_n$ prove that the space $A(\kappa_1) \oplus \ldots \oplus A(\kappa_n)$ is $l$-equivalent to $A(\kappa)$ where $\kappa = \max\{\kappa_1, \ldots, \kappa_n\}$.*

**Solution.** To model every space $A(\kappa_i)$ fix a number $i \in \{1, \ldots, n\}$ and choose a set $B_i$ of a cardinality $\kappa_i$ together with a point $a_i \notin B_i$. Introduce the topology of $A(\kappa_i)$ on the set $A_i = \{a_i\} \cup B_i$ declaring all points of $B_i$ isolated in $A_i$ and letting $\tau(a_i, A_i) = \{U \subset A_i : a_i \in U$ and $|A_i \setminus U| < \omega\}$.

In the space $A = A_1 \oplus \ldots \oplus A_n$ the set $F = \{a_1, \ldots, a_n\}$ is a retract of $A$. Let $a_F$ be the point represented by $F$ in the $\mathbb{R}$-quotient space $A_F$. It follows from Problem 260 that the space $A$ is $l$-equivalent to the bunch $B = (A_F, a_F) \vee (F, z)$ where the point $z \in F$ is chosen arbitrarily. Recall that $B$ is defined as the $\mathbb{R}$-quotient space of the space $Q = A_F \oplus F$ in which the set $G = \{a_F, z\}$ is contracted to a point $b_G$.

Since $a_F$ is the unique non-isolated point of $A_F$ and $F$ is finite, $b_G$ is the unique non-isolated point of the space $B = Q_G$. The space $B$ is a continuous image of $Q$ while $Q$ is a continuous image of a compact space $A \oplus F$. Therefore $B$ is a compact space of cardinality $\kappa$ with a unique non-isolated point. A trivial consequence is that $B$ is homeomorphic to $A(\kappa)$ and hence $A \overset{l}{\sim} A(\kappa)$. Since $A(\kappa_1) \oplus \ldots \oplus A(\kappa_n)$ is homeomorphic to $A$, we conclude that the space $A(\kappa_1) \oplus \ldots \oplus A(\kappa_n)$ is $l$-equivalent to $A(\kappa)$.

**V.265.**  *Given a family of spaces $\{X_t : t \in T\}$ let $X = \bigoplus_{t \in T} X_t$ and prove that $C_p(X)$ is linearly homeomorphic to $\prod_{t \in T} C_p(X_t)$. Deduce from this fact that if $X_t \overset{l}{\sim} Y_t$ for any $t \in T$ then $X \overset{l}{\sim} Y = \bigoplus_{t \in T} Y_t$.*

**Solution.**  We identify every $X_t$ with the respective clopen subspace of $X$; it is straightforward that the restriction map $\pi_t : C_p(X) \to C_p(X_t)$ is linear and continuous. We omit a trivial proof that any diagonal product of linear maps is a linear map, so $\pi = \Delta\{\pi_t : t \in T\} : C_p(X) \to \prod_{t \in T} C_p(X_t)$ is a linear continuous map; recall that $\pi(f)(t) = \pi_t(f)$ for any $f \in C_p(X)$ and $t \in T$.

If $f$ and $g$ are distinct elements of $C_p(X)$ then $f(x) \neq g(x)$ for some $x \in X$. There is $t \in T$ with $x \in X_t$, so $\pi_t(f) \neq \pi_t(g)$ and hence $\pi(f) \neq \pi(g)$ which shows that $\pi$ is an injective map. Given an arbitrary $g \in \prod_{t \in T} C_p(X_t)$ the function $g(t)$ belongs to $C_p(X_t)$ for each $t \in T$. Since $\{X_t : t \in T\}$ is a clopen disjoint cover of $X$, there is a function $f \in C_p(X)$ such that $\pi_t(f) = g(t)$ for all $t \in T$. It is immediate that $\pi(f) = g$, so we established that $\pi$ is a surjective map.

Recall that $C_p(X) \subset \mathbb{R}^X$ and let $p_x(f) = f(x)$ for all $f \in \mathbb{R}^X$ and $x \in X$. Denote the set $\prod_{t \in T} C_p(X_t)$ by $C$ and consider the projection map $q_t : C \to C_p(X_t)$ for each $t \in T$. Fix any $x \in X$ and the unique $t \in T$ such that $x \in X_t$; if $r_x(f) = f(x)$ for any $f \in C_p(X_t)$ then the map $r_x : C_p(X_t) \to \mathbb{R}$ is continuous. For any $g \in C$ we have $\varphi^{-1}(g)(x) = g(t)(x) = q_t(g)(x) = r_x(q_t(g))$ which shows that $p_x \circ \varphi^{-1} = r_x \circ q_t$ is a continuous map for every $x \in X$. Therefore the map $\varphi^{-1}$ is continuous by TFS-102 and hence $\varphi$ is a linear homeomorphism between the spaces $C_p(X)$ and $\prod_{t \in T} C_p(X_t)$.

Finally, assume that $X_t$ is $l$-equivalent to $Y_t$ and fix a linear homeomorphism $\varphi_t : C_p(X_t) \to C_p(Y_t)$ for each $t \in T$. An easy proof of the fact that the product map $\varphi = \prod_{t \in T} \varphi_t : \prod_{t \in T} C_p(X_t) \to \prod_{t \in T} C_p(Y_t)$ is linear can be left to the reader. It follows from Fact 1 of S.271 and the equality $\varphi^{-1} = \prod_{t \in T} \varphi_t^{-1}$ that $\varphi$ is a linear homeomorphism between the spaces $\prod_{t \in T} C_p(X_t)$ and $\prod_{t \in T} C_p(Y_t)$. Since they are linearly homeomorphic to $C_p(X)$ and $C_p(Y)$ respectively, we conclude that $C_p(X)$ is linearly homeomorphic to $C_p(Y)$, i.e., $X \overset{l}{\sim} Y$.

**V.266.**  *Suppose that a space $J_i$ is homeomorphic to $\mathbb{I}$ for any $i = 1, \ldots, n$ and let $J = \bigoplus\{J_i : 1 \leq i \leq n\}$. Prove that the space $J \oplus D$ is $l$-equivalent to $\mathbb{I}$ for any finite space $D$. Deduce from this fact that*

  *(i)  connectedness is not preserved by $l$-equivalence;*
  *(ii)  for any cardinal $\kappa$ there exist $l$-equivalent spaces $X$ and $Y$ such that $X$ has no isolated points and $Y$ has $\kappa$-many isolated points.*

**Solution.**  In any discrete union $Z = Z_1 \oplus \ldots \oplus Z_n$ we identify every space $Z_i$ with the respective clopen subspace of $Z$. The expression $Z \simeq T$ says that the spaces $Z$ and $T$ are homeomorphic while $M \cong L$ denotes the fact that the linear topological spaces $M$ and $L$ are linearly homeomorphic. We will first prove that

(1)  $\mathbb{I} \oplus D$ is $l$-equivalent to $\mathbb{I}$ for any finite space $D$.

Fix a finite space $D$ and consider the subspace $S = \{0\} \cup \{\frac{1}{n+1} : n \in \omega\}$ of the space $\mathbb{I}$. Since $S$ is a convergent sequence, the space $S \oplus D$ is homeomorphic to $S$ and hence $C_p(S) \times \mathbb{R}^D$ is linearly homeomorphic to $C_p(S)$ (see Problem 265).

Apply Fact 2 of U.216 to see that there is a linear topological subspace $L$ of the space $C_p(\mathbb{I})$ such that $C_p(\mathbb{I})$ is linearly homeomorphic to $L \times C_p(S)$. Therefore $C_p(\mathbb{I} \oplus D) \cong (L \times C_p(S)) \times \mathbb{R}^D \cong L \times (C_p(S) \times \mathbb{R}^D) \cong L \times C_p(S) \cong C_p(\mathbb{I})$, so the property (1) is proved.

*Fact 1.* Given a space $Z$ suppose that $a \in Z$ and $Z_0$, $Z_1$ are closed subsets of $Z$ such that $Z_0 \cup Z_1 = Z$ and $Z_0 \cap Z_1 = \{a\}$. If there exist homeomorphisms $h_0 : \mathbb{I} \to Z_0$ and $h_1 : \mathbb{I} \to Z_1$ such that $h_0(1) = h_1(1) = a$ then $Z$ is homeomorphic to $\mathbb{I}$.

*Proof.* Let $f_0(x) = h_0^{-1}(x)$ for any $x \in Z_0$ and $f_1(x) = 2 - h_1^{-1}(x)$ for each $x \in Z_1$. Then $f_0 : Z_0 \to [-1, 1]$ and $f_1 : Z_1 \to [1, 3]$ are continuous functions such that $f_0(a) = f_1(a)$. By Fact 2 of T.354 there exists a continuous function $f : X \to [-1, 3]$ such that $f|Z_i = f_i$ for every $i \in \{0, 1\}$. It is straightforward that $f$ is a bijection; the space $Z$ is compact, so $f$ is a homeomorphism between $Z$ and $[-1, 3]$. Since $[-1, 3] \simeq \mathbb{I}$, we have $Z \simeq \mathbb{I}$ and hence Fact 1 is proved.

Returning to our solution let us show that

(2)   $\mathbb{I} \times D$ is $l$-equivalent to $\mathbb{I}$ for any nonempty finite space $D$.

Obviously, the property (2) holds when $|D| = 1$. Assume that $D = \{d_0, d_1\}$ and $d_0 \neq d_1$. The set $F = \{(1, d_0), (1, d_1)\}$ is a retract of $Q = \mathbb{I} \times D$; denote by $a_F$ the point of the $\mathbb{R}$-quotient space $Q_F$ represented by $F$. We can apply Problem 260 to see that $Q$ is $l$-equivalent to the bunch $B = (Q_F, a_F) \vee (F, d_0)$. Recall that the space $B$ is obtained as the $\mathbb{R}$-quotient space of $Q \oplus F$ when the set $G = \{a_F, d_0\}$ is contracted to a point; let $b_G$ be the point of $B$ represented by the set $G$. It is easy to see that $d_1$ is an isolated point of $B$. If $Z_i = ([-1, 1) \times \{d_i\}) \cup \{a_F\}$ then Fact 1 of V.260 can be applied to the space $Q_F$ to convince ourselves that there exists a homeomorphism $h_i : \mathbb{I} \to Z_i$ for which $h_i(1) = a_F$ for every $i \in \{0, 1\}$. Since $Z_0 \cup Z_1 = Q_F$ and $Z_0 \cap Z_1 = \{a_F\}$, we can apply Fact 1 to conclude that $Q_F \simeq \mathbb{I}$. Apply Fact 1 of V.260 once more to note that the set $Q' = (Q_F \setminus \{a_F\}) \cup \{b_G\}$ is homeomorphic to $Q_F$, so $Q' \simeq \mathbb{I}$ and therefore $B \simeq \mathbb{I} \oplus \{d_1\}$. The property (1) shows that $B \overset{l}{\sim} \mathbb{I}$, so $Q = \mathbb{I} \times D \overset{l}{\sim} \mathbb{I}$ and hence we established (2) for all two-element sets $D$.

Proceeding by induction assume that $n \geq 3$ and we proved (2) for all sets $D$ with $1 \leq |D| < n$. If $|D| = n$ then $D = D_0 \oplus D_1$ where $|D_0| = 2$ and $|D_1| = n - 2$. It is evident that $\mathbb{I} \times D \simeq (\mathbb{I} \times D_0) \oplus (\mathbb{I} \times D_1)$. Since $\mathbb{I} \times D_0 \overset{l}{\sim} \mathbb{I}$, it follows from Problem 265 that $\mathbb{I} \times D \overset{l}{\sim} \mathbb{I} \oplus (\mathbb{I} \times D_1)$. The space $\mathbb{I} \oplus (\mathbb{I} \times D_1)$ is easily seen to be homeomorphic to $\mathbb{I} \times E$ for some set $E$ with $|E| = |D_1| + 1 = n - 1$, so the induction hypothesis shows that $\mathbb{I} \oplus (\mathbb{I} \times D_1) \overset{l}{\sim} \mathbb{I}$ and hence $\mathbb{I} \times D \overset{l}{\sim} \mathbb{I}$, i.e., (2) is proved.

Next observe that the set $J = \bigoplus\{J_i : 1 \le i \le n\}$ is homeomorphic to $\mathbb{I} \times D_0$ for some finite set $D_0$, so $J \overset{l}{\sim} \mathbb{I}$ by the property (2). Applying (1) we conclude that $J \oplus D$ is $l$-equivalent to $\mathbb{I}$ and hence our first statement is proved. An immediate consequence of (2) is that a disconnected space $\mathbb{I} \oplus \mathbb{I}$ is $l$-equivalent to a connected space $\mathbb{I}$; this settles (i).

Finally, fix a cardinal $\kappa$, a homeomorphic copy $I_\alpha$ of the space $\mathbb{I}$ and a point $b_\alpha \notin I_\alpha$ for every $\alpha < \kappa$. It follows from (1) that $I_\alpha \overset{l}{\sim} E_\alpha = I_\alpha \oplus \{b_\alpha\}$ for all $\alpha < \kappa$, so we can apply Problem 265 to see that the space $X = \bigoplus_{\alpha<\kappa} I_\alpha$ is $l$-equivalent to $Y = \bigoplus_{\alpha<\kappa} E_\alpha$ while $X$ is dense-in-itself and $Y$ has $\kappa$-many isolated points. This proves (ii) and completes our solution.

**V.267.** *Let $X$ be a compact space with $|X| = \kappa \ge \omega$. Prove that the space $AD(X)$ is $l$-equivalent to $X \oplus A(\kappa)$. Here $AD(X)$ is the Alexandroff double of the space $X$ and $A(\kappa)$ is the one-point compactification of a discrete space of cardinality $\kappa$.*

**Solution.** The expression $Y \simeq Z$ says that the spaces $Y$ and $Z$ are homeomorphic. Any spaces $Y$ and $Z$ are identified with the respective clopen subspaces of $Y \oplus Z$. Recall that $\mathbb{D} = \{0, 1\}$ is the standard discrete two-point space and $AD(X) = X \times \mathbb{D}$; all points of $X_1 = X \times \{1\}$ are isolated in $AD(X)$ and a local base at any point $y = (x, 0)$ is given by the family $\{(U \times \mathbb{D})\setminus\{(x, 1)\} : U \in \tau(x, X)\}$. It is evident that the subspace $X_0 = X \times \{0\}$ of the space $AD(X)$ is homeomorphic to $X$. For any $y \in X_0$ let $r(y) = y$; if $y = (x, 1) \in X_1$ then let $r(y) = (x, 0)$. It is straightforward from the definitions that $r : AD(X) \to X_0$ is a retraction. Let $Y$ be the $\mathbb{R}$-quotient space obtained from $AD(X)$ by contracting $X_0$ to a point $a_0$ and fix any point $z_0 \in X_0$.

By Problem 260, the space $AD(X)$ is $l$-equivalent to the bunch $B' = (Y, a_0) \vee (X_0, z_0)$. Apply Problem 259 to see that $B'$ is $l$-equivalent to the bunch $B = (Y, a) \vee (X_0, z_0)$ where $a \in Y\setminus\{a_0\}$. Observe that $Y$ is a compact space of cardinality $\kappa$ in which only the point $a_0$ is not isolated; consequently, $Y$ is homeomorphic to $A(\kappa)$.

Let $G = \{a, z_0\}$ and denote by $a_G$ the point represented by the set $G$ in the space $B$. Recall that $B$ is the $\mathbb{R}$-quotient space of the space $Q = Y \oplus X_0$ when $G$ is contracted to a point. It is routine to verify that we can apply Fact 1 of V.260 to convince ourselves that the contraction map $p_G : Q \to Q_G = B$ is a homeomorphism when restricted to either $Y$ or $X_0$. In particular, the set $X' = (X_0\setminus\{z_0\}) \cup \{a_G\}$ is homeomorphic to $X_0$ and hence to $X$.

By the same reason, $Y_0 = (Y\setminus\{a\}) \cup \{a_G\}$ is homeomorphic to $Y$; since $a$ is an isolated point of $Y$, the point $a_G = p_G(a)$ is isolated in $Y_0$. Therefore $Y_1 = Y_0\setminus\{a_G\}$ is a clopen subset of $Y_0$; a minute's thought shows that $Y_1$ is also clopen in $B$ and hence $B$ is homeomorphic to $Y_1 \oplus X' \simeq Y_1 \oplus X$. Taking away an isolated point from the space $A(\kappa)$ gives a subspace homeomorphic to $A(\kappa)$, so $B \simeq Y_1 \oplus X \simeq X \oplus A(\kappa)$. It follows from $X \overset{l}{\sim} B' \overset{l}{\sim} B \simeq X \oplus A(\kappa)$ that $X \overset{l}{\sim} X \oplus A(\kappa)$, i.e., we proved what was promised.

**V.268.** *Let $X$ be a compact space such that $|X| = \kappa \geq \omega$. Prove that $AD(X)$ is $l$-equivalent to $AD(X) \oplus A(\kappa)$. Here $AD(X)$ is the Alexandroff double of the space $X$.*

**Solution.** Apply Problem 267 to see that $AD(X)$ is $l$-equivalent to $X \oplus A(\kappa)$. Therefore $AD(X) \oplus A(\kappa)$ is $l$-equivalent to $(X \oplus A(\kappa)) \oplus A(\kappa)$ by Problem 265. It follows from Problem 264 that $A(\kappa) \oplus A(\kappa) \overset{l}{\sim} A(\kappa)$, so $AD(X) \oplus A(\kappa) \overset{l}{\sim} X \oplus (A(\kappa) \oplus A(\kappa)) \overset{l}{\sim} X \oplus A(\kappa) \overset{l}{\sim} AD(X)$ which shows that $AD(X) \oplus A(\kappa)$ is $l$-equivalent to $AD(X)$.

**V.269.** *Prove that there exist $l$-equivalent compact spaces $X$ and $Y$ such that $\chi(X) \neq \chi(Y)$. As a consequence, pseudocharacter is not $l$-invariant.*

**Solution.** As usual, $\mathbb{D} = \{0, 1\}$ is the standard two-point discrete space. Recall that the Alexandroff double of $\mathbb{I}$ is the space $AD(\mathbb{I}) = \mathbb{I} \times \mathbb{D}$ in which all points of $I_1 = \mathbb{I} \times \{1\}$ are isolated while the standard local base at any point $y = (t, 0)$ is given by the family $\{(U \times \mathbb{D}) \setminus \{(t, 1)\} : U \in \tau(t, \mathbb{I})\}$. The space $X = AD(\mathbb{I})$ is compact by TFS-364. It clear that $\chi(y, X) \leq \omega$ for any $x \in I_1$; if $y = (t, 0)$ for some $t \in \mathbb{I}$ then the family $\{((t - \frac{1}{n}, t + \frac{1}{n}) \times \mathbb{D}) \setminus \{(t, 1)\} : n \in \mathbb{N}\}$ is easily seen to be a countable local base of $X$ at the point $y$. Therefore $\chi(y, X) \leq \omega$ for any $y \in X$, i.e., $\chi(X) \leq \omega$.

It follows from Problem 267 that $X \overset{l}{\sim} Y = \mathbb{I} \oplus A(\mathfrak{c})$; we identify $A(\mathfrak{c})$ with the respective clopen subspace of $Y$. It is obvious that, for the unique non-isolated point $a$ of the space $A(\mathfrak{c})$, we have $\psi(a, Y) = \mathfrak{c}$. Therefore $X$ and $Y$ are $l$-equivalent compact spaces such that $\psi(X) = \chi(X) = \omega < \mathfrak{c} = \psi(Y)$ which shows that pseudocharacter is not preserved by $l$-equivalence.

**V.270.** *Prove that there exist $l$-equivalent compact spaces $X$ and $Y$ such that $X$ has nontrivial convergent sequences and $Y$ does not have any.*

**Solution.** As usual, $\mathbb{D} = \{0, 1\}$ is the standard two-point discrete space. Recall that the Alexandroff double of $\beta\omega$ is the space $AD(\beta\omega) = \beta\omega \times \mathbb{D}$ in which all points of $B_1 = \beta\omega \times \{1\}$ are isolated while the standard local base at any point $y = (x, 0)$ is given by the family $\{(U \times \mathbb{D}) \setminus \{(x, 1)\} : U \in \tau(x, \beta\omega)\}$. The space $Y = AD(\beta\omega)$ is compact by TFS-364.

Assume that $S = \{s_n : n \in \omega\}$ is a nontrivial convergent sequence in $Y$; we can assume that $S$ is faithfully indexed and its limit $s$ does not belong to $S$. No isolated point can be a limit of a nontrivial convergent sequence, so $s \in B_0 = \beta\omega \times \{0\}$; fix a point $t \in \beta\omega$ such that $s = (t, 0)$. The space $B_0$ is homeomorphic to $\beta\omega$ which does not have nontrivial convergent sequences (see Fact 2 of T.131), so we can assume, without loss of generality, that $S \subset B_1$ and $(t, 1) \notin S$. For every $n \in \omega$ there is $x_n \in \beta\omega$ with $s_n = (x_n, 1)$; let $w_n = (x_n, 0)$. It is a trivial consequence of the definition of the topology of the space $AD(\beta\omega)$ that the nontrivial sequence $\{w_n : n \in \omega\} \subset B_0 \setminus \{s\}$ converges to $s$, i.e., we have found a nontrivial convergent sequence in $B_0 \simeq \beta\omega$ which is a contradiction. Therefore the space $Y$ does not have nontrivial convergent sequences.

It follows from Problem 267 that $Y \overset{l}{\sim} X = \beta\omega \oplus A(2^c)$; we identify $A(2^c)$ with the respective clopen subspace of $X$. It is obvious that, for the unique non-isolated point $a$ of the space $A(2^c)$, if $Q$ is a countably infinite subset of $A(2^c)\backslash\{a\}$ then $Q \cup \{a\}$ is a nontrivial convergent sequence in $X$. As a consequence, $X$ and $Y$ are $l$-equivalent compact spaces such that $X$ has nontrivial convergent sequences and $Y$ does not have any.

**V.271.** *Prove that for any uncountable regular cardinal $\kappa$ with its usual order topology we have the equivalencies $\kappa \overset{l}{\sim} \kappa \oplus A(\kappa)$ and $(\kappa + 1) \overset{l}{\sim} (\kappa + 1) \oplus A(\kappa)$. Therefore there exist $l$-equivalent spaces $X$ and $Y$ such that all compact subspaces of $X$ are metrizable while $Y$ has non-metrizable compact subspaces.*

**Solution.** The expression $X \simeq Y$ says that the spaces $X$ and $Y$ are homeomorphic. For any uncountable regular cardinal $\mu$ let $L(\mu) = \{\alpha < \mu : \alpha$ is a limit ordinal$\}$ and $L^+(\mu) = L(\mu) \cup \{\mu\}$. Suppose that we have a map $f : \mu \to Z$ and an ordinal $\alpha < \mu$. To avoid confusion, we will consider that $f(\alpha)$ is the image of the point $\alpha$ and $f[\alpha] = \{f(\beta) : \beta < \alpha\}$. We will need the interval $(\alpha, \beta] = \{\gamma : \alpha < \gamma \leq \beta\}$ for any ordinals $\alpha$ and $\beta$.

*Fact 1.* The space $L(\mu)$ is homeomorphic to $\mu$ and $L^+(\mu)$ is homeomorphic to $(\mu + 1)$ for any uncountable regular cardinal $\mu$.

*Proof.* Fix an uncountable regular cardinal $\mu$ and let $f(0) = \omega$. If $0 < \alpha \leq \mu$ and we defined an ordinal $f(\beta) \in L(\mu)$ for any $\beta < \alpha$, let

(1) $f(\alpha) = \backslash n\{\gamma \in L(\mu) : f(\beta) < \gamma$ for all $\beta < \alpha\}$.

This inductive procedure defines a point $f(\alpha) \in L(\mu)$ for any $\alpha \in \mu$; letting $f(\mu) = \mu$ we obtain a function $f : (\mu + 1) \to L^+(\mu)$. It is an immediate consequence of (1) that

(2) $\beta < \alpha \leq \mu$ implies $f(\beta) < f(\alpha)$.

If the set $A = \{\alpha \leq \mu : f(\alpha) < \alpha\}$ is nonempty then let $\gamma$ be the minimal element of $A$. The ordinal $\gamma_0 = f(\gamma)$ is strictly less than $\gamma$ and $f(\gamma_0) < f(\gamma) = \gamma_0$, so $\gamma_0 \in A$ which is a contradiction. Therefore the set $A$ is empty, i.e.,

(3) $\alpha \leq f(\alpha)$ for any $\alpha \leq \mu$.

If the set $B = L(\mu)\backslash f[\mu]$ is nonempty then let $\gamma$ be the minimal element of $B$. The set $C = \{\beta < \mu : f(\beta) < \gamma\}$ is nonempty because $0 \in C$. It follows from (3) that $f(\beta) \geq \beta \geq \gamma$ for any $\beta \geq \gamma$, so $C \subset \gamma$ and hence the ordinal $\gamma_0 = \backslash n\{\alpha < \mu : C \subset \alpha\}$ is consistently defined. If $\alpha < \gamma_0$ then $\beta \geq \alpha$ for some $\beta \in C$ and hence $f(\alpha) \leq f(\beta) < \gamma$, i.e., $\alpha \in C$ for any $\alpha < \gamma_0$ which shows that $\gamma_0 = C$.

By minimality of $\gamma$, for any $\beta \in L(\mu) \cap \gamma$ there is $\alpha \in \mu$ such that $f(\alpha) = \beta$ and hence $\alpha \in C$. Therefore $f[\gamma_0] = f(C) \supset L(\mu) \cap \gamma$, so $\gamma$ is the minimal element of

$L(\mu)$ such that $f(\beta) < \gamma$ for any $\beta < \gamma_0$. The definition (1) shows that $f(\gamma_0) = \gamma$; this contradiction proves that

(4) $f[\mu] = L(\mu)$ and hence $f$ is a bijection between $(\mu + 1)$ and $L^+(\mu)$.

The map $f$ is clearly continuous at every $\alpha \in (\mu + 1) \backslash L^+(\mu)$, so assume that $\alpha_0 \leq \mu$ is a limit ordinal and fix an ordinal $\gamma_0 < \beta_0 = f(\alpha_0)$. It follows from (2) that the set $H = \{f(\alpha) : \alpha < \alpha_0\}$ does not have a maximal element. If $H \subset \gamma_0$ then $\nu = \sup(H) \leq \gamma_0 < \beta_0$. Besides, $\nu \in L^+(\mu)$ because $L^+(\mu)$ is closed in $\mu + 1$. This together with (1) shows that $f(\alpha_0) \leq \gamma_0$ which is a contradiction.

As a consequence, there exists an ordinal $\delta < \alpha_0$ such that $f(\delta) \geq \gamma_0$; the set $W = (\delta, \alpha_0]$ is an open neighborhood of $\alpha_0$ and $f(W) \subset (\gamma_0, \beta_0]$, i.e., $W$ witnesses continuity of $f$ at the point $\alpha_0$; this proves that $f$ is continuous at every point of $\mu + 1$, i.e., $f$ is a continuous map. Furthermore, $\mu + 1$ is compact and $f$ is a bijection, so $f$ is a homeomorphism between $\mu + 1$ and $L^+(\mu)$. Since $f[\mu] = L(\mu)$, the spaces $\mu$ and $L(\mu)$ are also homeomorphic and hence Fact 1 is proved.

Returning to our solution denote by $Z$ the space $\kappa$ and let $T = (\kappa + 1)$. The set $F = L(\kappa)$ is a retract of $Z$ and $G = L^+(\kappa)$ is a retract of $T$ (see Fact 5 of U.074). Fix a point $a \notin T$ and observe that $Z \simeq (\omega + 1) \oplus (Z \backslash (\omega + 1))$; since $(\omega + 1) \oplus \{a\} \simeq (\omega + 1)$, we conclude that $Z \oplus \{a\}$ is homeomorphic to $Z$. The spaces $Z \oplus \{a\}$ and $Z_F \oplus F$ are $l$-equivalent by Problem 258, so the space $Z$ is also $l$-equivalent to $Z_F \oplus F \simeq Z_F \oplus \kappa$ (see Fact 1).

The space $\kappa$ is countably compact and hence so is $Z_F$; let $y_F$ be the point of $Z_F$ represented by $F$. Evidently, $y_F$ is the unique non-isolated point of $Z_F$. If $O \in \tau(y_F, Z_F)$ then $Z_F \backslash O$ is finite being discrete and countably compact. An immediate consequence is that the space $Z_F$ is homeomorphic to $A(\kappa)$ which shows that $Z = \kappa \overset{l}{\sim} \kappa \oplus A(\kappa)$.

Analogously, $T \simeq T \oplus \{a\}$; therefore the space $T$ is $l$-equivalent to $T_G \oplus G$. Since $T_G$ is a compact space with a unique non-isolated point, it is homeomorphic to $A(\kappa)$. It follows from Fact 1 that $G \simeq (\kappa + 1)$, so $T = (\kappa + 1) \overset{l}{\sim} (\kappa + 1) \oplus A(\kappa)$.

Finally, observe that, in the space $X = \omega_1$, all compact subspaces are metrizable while $X$ is $l$-equivalent to $Y = X \oplus A(\omega_1)$ in which $A(\omega_1)$ is a non-metrizable compact subspace.

**V.272.** *Suppose that compact spaces $X$ and $Y$ are $l$-equivalent. Prove that $X \times Z$ is $l$-equivalent to $Y \times Z$ for any space $Z$.*

**Solution.** If $K$ is a compact space then $C_u(K)$ is the topology of uniform convergence on $K$; the topology of $C_u(K)$ is generated by the metric $\rho_K$ defined by the formula $\rho_K(f, g) = \sup\{|f(x) - g(x)| : x \in K\}$ for any $f, g \in C(K)$.

*Fact 1.* Given spaces $T$ and $W$ if $\varphi : C_p(T) \to C_p(W)$ is a linear homeomorphism then there exists a linear homeomorphism $\Phi : \mathbb{R}^T \to \mathbb{R}^W$ such that $\Phi|C_p(T) = \varphi$.

*Proof.* For any $w \in W$ and $f \in \mathbb{R}^W$ let $p_w(f) = f(w)$; then $p_w : \mathbb{R}^W \to \mathbb{R}$ is the projection onto the factor determined by $w$. The map $p_w \circ \varphi$ is a continuous linear functional on $C_p(T)$, so we can apply Problem 224 to find a continuous linear functional $\xi_w : \mathbb{R}^T \to \mathbb{R}$ such that $\xi_w | C_p(T) = p_w \circ \varphi$ for each $w \in W$. The diagonal product $\Phi = \Delta\{\xi_w : w \in W\} : \mathbb{R}^T \to \mathbb{R}^W$ is a continuous linear map and it is an easy exercise that $\Phi | C_p(T) = \varphi$. Analogously, there exists a continuous linear map $\Psi : \mathbb{R}^W \to \mathbb{R}^T$ such that $\Psi | C_p(W) = \varphi^{-1}$.

Let $i_T : \mathbb{R}^T \to \mathbb{R}^T$ and $i_W : \mathbb{R}^W \to \mathbb{R}^W$ be the respective identity maps. If $f \in C_p(T)$ then $\varphi(f) \in C_p(W)$, so $\Psi(\Phi(f)) = \Psi(\varphi(f)) = \varphi^{-1}(\varphi(f)) = f$; analogously, $\Phi(\Psi(g)) = \Phi(\varphi^{-1}(g)) = \varphi(\varphi^{-1}(g)) = g$ for any $g \in C_p(W)$. This shows that $(\Psi \circ \Phi) | C_p(T) = i_T$ and $(\Phi \circ \Psi) | C_p(W) = i_W$, so we can apply Fact 0 of S.351 to conclude that $\Phi \circ \Psi = i_W$ and $\Psi \circ \Phi = i_T$. Therefore $\Phi$ and $\Psi$ are mutually inverse linear homeomorphisms and hence Fact 1 is proved.

*Fact 2.* Given spaces $T, U$ and $W$ suppose that $\varphi : \mathbb{R}^T \to \mathbb{R}^U$ is a linear homeomorphism. For any function $f : T \times W \to \mathbb{R}$ let $f_w(t) = f(t, w)$ for any $t \in T$; then $f_w \in \mathbb{R}^T$ for any $w \in W$. Let $\Phi(f)(u, w) = \varphi(f_w)(u)$ for any $f \in \mathbb{R}^{T \times W}$ and $(u, w) \in U \times W$. Then $\Phi : \mathbb{R}^{T \times W} \to \mathbb{R}^{U \times W}$ is a linear homeomorphism.

*Proof.* It is immediate that $\Phi$ is a linear map. For any function $g : U \times W \to \mathbb{R}$ let $g^w(u) = g(u, w)$ for any $(u, w) \in U \times W$; then $g^w \in \mathbb{R}^U$ for any $w \in W$ and we can let $\Psi(g)(t, w) = \varphi^{-1}(g^w)(t)$ for any $(t, w) \in T \times W$. Then $\Psi : \mathbb{R}^{U \times W} \to \mathbb{R}^{T \times W}$ and we omit a straightforward proof that $\Phi$ and $\Psi$ are mutually inverse linear maps.

Fix a point $(u, w) \in U \times W$ and let $p_{(u,w)} : \mathbb{R}^{U \times W} \to \mathbb{R}$ be the projection onto the factor determined by $(u, w)$. If $\xi(f) = f_w$ for any function $f \in \mathbb{R}^{T \times W}$ then $\xi : \mathbb{R}^{T \times W} \to \mathbb{R}^T$ is the composition of the restriction map to $T \times \{w\}$ and an evident homeomorphism which "forgets" $w$; therefore $\xi$ is a continuous map. If $q(f) = f(u)$ for any $f \in \mathbb{R}^U$ then the map $q : \mathbb{R}^U \to \mathbb{R}$ is continuous being a natural projection. Thus $p_{(u,w)} \circ \Phi = q \circ \varphi \circ \xi$ is a continuous map; since the point $(u, w) \in U \times W$ was chosen arbitrarily, we can apply TFS-102 to conclude that $\Phi$ is continuous. An identical proof shows that $\Psi$ is also continuous, so $\Phi$ is, indeed, a linear homeomorphism, i.e., Fact 2 is proved.

Returning to our solution fix a linear homeomorphism $\varphi : C_p(X) \to C_p(Y)$ and a space $Z$; if we consider that $\varphi : C_u(X) \to C_u(Y)$ then $\varphi$ is also a homeomorphism (see Fact 1 of U.481). Apply Fact 1 to find a linear homeomorphism $\mu : \mathbb{R}^X \to \mathbb{R}^Y$ such that $\mu | C_p(X) = \varphi$. Let $f_z(x) = f(x, z)$ for any $f \in \mathbb{R}^{X \times Z}$ and $(x, z) \in X \times Z$. Then $f_z \in \mathbb{R}^X$ whenever $f \in \mathbb{R}^{X \times Z}$ and $z \in Z$. Letting $\Phi(f)(y, z) = \mu(f_z)(y)$ for any function $f \in \mathbb{R}^{X \times Z}$ and $(y, z) \in Y \times Z$ we obtain a linear homeomorphism $\Phi : \mathbb{R}^{X \times Z} \to \mathbb{R}^{Y \times Z}$ (see Fact 2), so it suffices to prove that $\Phi(C(X \times Z)) = C(Y \times Z)$.

Fix a function $p \in C(X \times Z)$; then $p_z \in C(X)$ for any $z \in Z$. To see that $q = \Phi(p)$ is a continuous function on $Y \times Z$ take any point $a_0 = (y_0, z_0) \in Y \times Z$ and $\varepsilon > 0$. The map $\varphi : C_u(X) \to C_u(Y)$ being continuous there exists $\delta > 0$ such that $f \in C(X)$ and $\rho_X(f, p_{z_0}) < \delta$ implies $\rho_Y(\varphi(f), \varphi(p_{z_0})) < \frac{\varepsilon}{2}$. For any $x \in X$ there exist $O_x \in \tau(x, X)$ and $W_x \in \tau(z_0, Z)$ such that $|p(a) - p(x, z_0)| < \frac{\delta}{2}$ for any $a \in O_x \times W_x$. The open cover $\{O_x : x \in X\}$ of the space $X$ has a subcover $\{O_x : x \in F\}$ for some finite set $F \subset X$; let $W = \bigcap\{W_x : x \in F\}$.

Take any points $x \in X$ and $z \in W$; there exists $x' \in F$ such that $x \in O_{x'}$. Then $(x, z) \in O_{x'} \times W \subset O_{x'} \times W_{x'}$ which implies that $|p(x, z) - p(x', z_0)| < \frac{\delta}{2}$ and $|p(x, z_0) - p(x', z_0)| < \frac{\delta}{2}$. As an immediate consequence we obtain the inequality $|p(x, z) - p(x, z_0)| < \delta$; the points $x \in X$ and $z \in W$ were chosen arbitrarily, so we established that $|p_z(x) - p_{z_0}(x)| < \delta$ for any $x \in X$ and $z \in W$. Recalling that $X$ is compact we convince ourselves that

(1) $\rho_X(p_z, p_{z_0}) < \delta$ and hence $\rho_Y(\varphi(p_z), \varphi(p_{z_0})) < \frac{\varepsilon}{2}$ for any $z \in W$.

The function $h_{z_0} = \varphi(p_{z_0})$ being continuous on $Y$ there is $V \in \tau(y_0, Y)$ such that $|h_{z_0}(y) - h_{z_0}(y_0)| < \frac{\varepsilon}{2}$ for any $y \in V$. Take any point $a = (y, z)$ from the set $V \times W$ and let $h_z = \varphi(p_z)$. It follows from (1) that $\rho_Y(h_z, h_{z_0}) < \frac{\varepsilon}{2}$ and therefore $|q(a) - q(a_0)| = |h_z(y) - h_{z_0}(y_0)| \leq |h_z(y) - h_{z_0}(y)| + |h_{z_0}(y) - h_{z_0}(y_0)| < \frac{\varepsilon}{2} + \frac{\varepsilon}{2} = \varepsilon$.

Thus the set $V \times W$ witnesses continuity of $q$ at the point $a_0$. This proves that $\Phi(C(X \times Z)) \subset C(Y \times Z)$. Given any function $g \in \mathbb{R}^{Y \times Z}$ let $g_z(y) = g(y, z)$ for any $y \in Y$; then $g_z \in \mathbb{R}^Y$ for any $z \in Z$. If we let $\Psi(g)(x, z) = \mu^{-1}(g_z)(x)$ for any $(x, z) \in X \times Z$ and $g \in \mathbb{R}^{Y \times Z}$ then $\Psi : \mathbb{R}^{Y \times Z} \to \mathbb{R}^{X \times Z}$ is the inverse map of $\Phi$. Since $\Psi$ is a symmetric version of $\Phi$, the above proof for $\Phi$ can be easily modified to establish that $\Psi(C(Y \times Z)) \subset C(X \times Z)$ and hence $\Phi(C(X \times Z)) = C(Y \times Z)$. This proves that the spaces $C_p(X \times Z)$ and $C_p(Y \times Z)$ are linearly homeomorphic, i.e., the spaces $X \times Z$ and $Y \times Z$ are $l$-equivalent, so our solution is complete.

**V.273.** *Given a family* $\{X_1, \ldots, X_n\}$ *of compact spaces assume that* $X_i \overset{l}{\sim} Y_i$ *for all* $i \in \{1, \ldots, n\}$. *Prove that the spaces* $X = X_1 \times \ldots \times X_n$ *and* $Y = Y_1 \times \ldots \times Y_n$ *are $l$-equivalent.*

**Solution.** Observe first that every $Y_i$ is compact by Problem 134 and 138. For any $i \in \{1, \ldots, n\}$ let $\mathcal{S}_i$ be the statement which asserts that the space $P_i = X_1 \times \ldots X_i$ is $l$-equivalent to $Q_i = Y_1 \times \ldots \times Y_i$. It is clear that $\mathcal{S}_1$ is true; proceeding by induction assume that $1 \leq k < n$ and $\mathcal{S}_k$ holds, i.e., $P_k \overset{l}{\sim} Q_k$. It follows from Problem 272 that $P_k \times X_{k+1}$ is $l$-equivalent to $Q_k \times X_{k+1}$. Since $X_{k+1} \overset{l}{\sim} Y_{k+1}$, we can apply Problem 272 once more to conclude that $Q_k \times X_{k+1} \overset{l}{\sim} Q_k \times Y_{k+1} = Q_{k+1}$. Therefore $P_{k+1} = P_k \times X_{k+1}$ is $l$-equivalent to $Q_{k+1}$, i.e., we established $\mathcal{S}_{k+1}$, so our inductive procedure shows that $\mathcal{S}_i$ is true for all $i \leq n$. In particular $\mathcal{S}_n$ is true, i.e., the spaces $P_n = X_1 \times \ldots \times X_n$ and $Q_n = Y_1 \times \ldots \times Y_n$ are $l$-equivalent as required.

**V.274.** *Give an example of $l$-equivalent spaces* $X$ *and* $Y$ *such that* $X \times Z$ *is not $t$-equivalent to* $Y \times Z$ *for some space* $Z$.

**Solution.** Let $a = 0$ and $a_n = \frac{1}{n+1}$ for any $n \in \omega$; then $S = \{a\} \cup \{a_n : n \in \omega\}$ is a faithfully indexed convergent sequence with its limit $a$. For any infinite cardinal $\kappa$ we denote by $D(\kappa)$ a discrete space of cardinality $\kappa$. In the space $D(\kappa) \times S$ the set $F = D(\kappa) \times \{a\}$ is a retract of $D(\kappa) \times S$; let $V(\kappa)$ be the $\mathbb{R}$-quotient space obtained from $D(\kappa) \times S$ by contracting $F$ to a point. A subset $A$ of a space $T$ will be called $\kappa$-closed in $T$ if $\overline{B} \subset A$ for any $B \subset A$ with $|B| \leq \kappa$. It is immediate that every closed subset of $T$ is $\kappa$-closed and the intersection of every two $\kappa$-closed subsets of $T$ is a $\kappa$-closed subset of $T$. Recall that *weak functional tightness* $t_m(P)$ of a space $P$ is the minimal cardinal $\kappa$ such that every strictly $\kappa$-continuous real-valued function on $P$ is continuous.

*Fact 1.* Given spaces $P, Q$ and an infinite cardinal $\kappa$ suppose that $t(Q) \leq \kappa$ and there exists a continuous closed onto map $f : P \to Q$ such that $t(f^{-1}(q)) \leq \kappa$ for any $q \in Q$. Then $t(P) \leq \kappa$. As a consequence, if $K$ is a compact space with $t(K) \leq \kappa$ then $t(Q \times K) \leq \kappa$.

*Proof.* Suppose that $A \subset P$ is a $\kappa$-closed non-closed subspace of $P$ and fix a point $z \in \overline{A} \backslash A$. The set $F = f^{-1} f(z)$ is closed in $P$, so $G = F \cap A$ is $\kappa$-closed; since $t(F) \leq \kappa$, we can apply Fact 1 of S.328 to conclude that $G$ is closed in $F$ and hence in $P$. Choose an open neighborhood $U$ of the point $z$ in $P$ such that $\overline{U} \cap G = \emptyset$. The set $E = A \cap \overline{U}$ is still $\kappa$-closed while $z \in \overline{E} \backslash E$ and $f^{-1}(f(z)) \cap E = \emptyset$.

If $y = f(z)$ then it follows from continuity of $f$ that $y \in \overline{f(E)}$; by our choice of $E$, the point $y$ does not belong to $f(E)$, so $f(E)$ is not closed in $Q$. It is an easy exercise that the image of a $\kappa$-closed set under a closed map is $\kappa$-closed, so $f(E)$ is a $\kappa$-closed non-closed subset of $Q$. This, together with Fact 1 of S.328, gives a contradiction with $t(Q) \leq \kappa$. Therefore every $\kappa$-closed subset of $P$ is closed, so we can apply Fact 1 of S.328 once more to conclude that $t(P) \leq \kappa$.

Finally, if $K$ is a compact space with $t(K) \leq \kappa$ then the natural projection $\pi : Q \times K \to Q$ is a perfect map (see Fact 3 of S.288) such that $\pi^{-1}(z)$ is homeomorphic to $K$ and hence $t(\pi^{-1}(z)) \leq \kappa$ for every $z \in Q$; this implies that $t(Q \times K) \leq \kappa$ and shows that Fact 1 is proved.

*Fact 2.* Given a space $P$ suppose that $\{U_a : a \in A\}$ is an open cover of $P$ such that $t(U_a) \leq \kappa$ for any $a \in A$. Then $t(P) \leq \kappa$.

*Proof.* Let $i_a : U_a \to P$ be the identity map, i.e., $i_a(x) = x$ for any $x \in U_a$. If we identify every $U_a$ with the respective clopen subspace of $U = \bigoplus\{U_a : a \in A\}$ then there is a continuous map $i : U \to P$ such that $i|U_a = i_a$ for any $a \in A$. It is clear that $i(U) = P$; if $O$ is an open subset of $U$ then $O_a = O \cap U_a$ is an open subset both of $U_a$ and $P$ for each $a \in A$. Therefore $i(O) = \bigcup\{O_a : a \in A\}$ is an open subset of $P$; this shows that the map $i$ is open and hence quotient. It is an easy exercise that $t(U) \leq \kappa$, so we can apply TFS-162 to convince ourselves that $t(P) \leq \kappa$ and hence Fact 2 is proved.

*Fact 3.* Weak functional tightness of the space $P = V(\omega) \times V(\mathfrak{c})$ is uncountable.

*Proof.* To prove that $t_m(P) > \omega$ it suffices to construct a strictly $\omega$-continuous discontinuous function $f : P \to \mathbb{R}$. We will need a more convenient representation of the spaces $V(\omega)$ and $V(\mathfrak{c})$. Denote by $a$ the unique non-isolated point of $V(\omega)$ and let $b$ be the unique non-isolated point of $V(\mathfrak{c})$. Then $V(\omega) \backslash \{a\} = \{a^i_j : i, j \in \omega\}$ where the indexation is faithful and $A_i = \{a^i_j : j \in \omega\}$ is a copy of the subspace $\{d\} \times S$ of $V(\omega)$ for each $i \in \omega$. In this model, a set $U \subset V(\omega)$ is open in $V(\omega)$ if and only if either $U \subset A = V(\omega) \backslash \{a\}$ or $a \in U$ and $A_i \backslash U$ is finite for every $i \in \omega$.

Analogously, $V(\mathfrak{c}) = \{b\} \cup \{b_i^\alpha : i \in \omega,\ \alpha < \mathfrak{c}\}$ where the indexation is faithful and the set $B_\alpha = \{b_i^\alpha : i \in \omega\}$ is a sequence that converges to $b$ for every $\alpha < \mathfrak{c}$. It is easy to see that a set $U \subset V(\mathfrak{c})$ is open in the space $V(\mathfrak{c})$ if and only if either $U \subset B = V(\mathfrak{c})\backslash\{b\}$ or $b \in U$ and $B_\alpha\backslash U$ is finite for each $\alpha < \mathfrak{c}$.

Call a countably infinite set $D \subset A$ *thin* if $|D \cap A_i| \leq 1$ for all $i \in \omega$. Choose an enumeration $\{D_\alpha : \alpha < \mathfrak{c}\}$ of all thin subsets of $A$; let $\xi_\alpha : B_\alpha \to D_\alpha$ be a bijection and consider the set $E_\alpha = \{(\xi_\alpha(y), y) : y \in B_\alpha\} \subset V(\omega) \times V(\mathfrak{c})$ for every $\alpha < \mathfrak{c}$. All points of the set $E = \bigcup\{E_\alpha : \alpha < \mathfrak{c}\}$ are isolated in $P$. If $v = (a, y)$ for some $y \in B$ then there is $\alpha < \mathfrak{c}$ such that $y \in B_\alpha$, so the set $V(\omega) \times \{y\}$ is a clopen neighborhood of $v$ which contains only the point $(\xi_\alpha(y), y)$ of the set $E$. If $v = (x, b)$ for some $x \in A$ then, for any $\alpha < \mathfrak{c}$, it follows from the fact that $\xi_\alpha$ is a bijection that there is a point $y_\alpha \in B_\alpha$ such that $\xi_\alpha(y) \neq x$ for all $y \in B_\alpha\backslash\{y_\alpha\}$. Therefore the set $W' = V(\mathfrak{c})\backslash\{y_\alpha : \alpha < \mathfrak{c}\}$ is an open neighborhood of $b$ in $V(\mathfrak{c})$, so $W = \{x\} \times W'$ is an open neighborhood of $v$ in $P$ such that $W' \cap E = \emptyset$.

This shows that every point $v \in P\backslash(E \cup \{(a, b)\})$ has an open neighborhood which does not meet the set $E' = E \cup \{(a, b)\}$; as a result, $E'$ is closed in $P$. We are going to prove next that the point $u = (a, b)$ belongs to the closure of $E$. Indeed, if $W$ is an open neighborhood of $u$ in $P$ then there exist $G \in \tau(a, V(\omega))$ and $H \in \tau(b, V(\mathfrak{c}))$ such that $G \times H \subset W$. It is immediate that there exists a thin set $D \subset G$; then $D = D_\alpha$ for some $\alpha < \mathfrak{c}$. Since the sequence $B_\alpha$ converges to $b$, we can find $y \in B_\alpha \cap H$; then $(\xi_\alpha(y), y) \in (G \times H) \cap E$, so $E \cap W \neq \emptyset$ for any $W \in \tau(u, P)$ which shows that $u \in \overline{E}$. Let us prove that

(1) $u \notin \overline{C}$ for any countable $C \subset E$.

Fix a countable set $C \subset E$; there exists a sequence $\{\alpha_n : n \in \omega\} \subset \mathfrak{c}$ such that $C \subset L = \bigcup_{n \in \omega} E_{\alpha_n}$, so it suffices to show that $u$ does not belong to the closure of the set $L$. Let $R_i = \xi_{\alpha_i}(B_{\alpha_i})$ for all $i \in \omega$. Since every $R_k$ is a thin set, we can find a number $m_k \in \omega$ such that $\xi_{\alpha_k}(b_n^{\alpha_k}) \notin A_j$ whenever $j \leq k$ and $n \geq m_k$; let $R_i' = R_i\backslash\{\xi_{\alpha_i}(b_j^{\alpha_i}) : j < m_k\}$ for all $i \in \omega$. An immediate consequence is that $A_i \cap R_j' = \emptyset$ whenever $i \leq j$, so there exists a finite set $A_i' \subset A_i$ such that $(A_i\backslash A_i') \cap R_n' = \emptyset$ for all $i, n \in \omega$. The set $O = \{a\} \cup (\bigcup\{A_i\backslash A_i' : i \in \omega\})$ is an open neighborhood of $a$ in $V(\omega)$ while $W = V(\mathfrak{c})\backslash(\bigcup\{b_j^{\alpha_k} : k \in \omega,\ j < m_k\})$ is an open neighborhood of $b$ in $V(\mathfrak{c})$, so the set $G = O \times W$ is an open neighborhood of $u$ in $P$.

Take any point $v \in L$; there are $k, j \in \omega$ such that $v = (\xi_{\alpha_k}(y), y)$ where $y = b_j^{\alpha_k}$. If $j < m_k$ then $y \notin W$, so $v \notin G$. If $j \geq m_k$ then $\xi_{\alpha_k}(y) \in R_k'$ which shows that $\xi_{\alpha_k}(y) \notin O$ and hence $v \notin G$. Thus $v \notin G$ for any $v \in L$ which shows that $G \cap L = \emptyset$ and hence $G \cap C = \emptyset$, so $u \notin \overline{C}$ and hence (1) is proved.

Let $f(v) = 0$ for any $v \in E$ and $f(v) = 1$ whenever $v \in P\backslash E$. The function $f : P \to \mathbb{R}$ is discontinuous because $f(E) = \{0\}$ while $u \in \overline{E}$ and $f(u) = 1$. To see that $f$ is strictly $\omega$-continuous fix any countable set $C \subset P$. If $C' = C \cap E$ then $u \notin \overline{C'}$ by (1), so $C'$ is closed in $E'$ and hence in $P$. Since all points of $C'$ are isolated in $P$, the set $C'$ is clopen in $P$. As a consequence, the function $h$ defined by $h(x) = 0$ for all $x \in C'$ and $h(x) = 1$ whenever $x \in P\backslash C'$ is continuous on $P$. It is immediate that $h|C = f|C$, so $f$ is a strictly $\omega$-continuous discontinuous function; therefore $t_m(P) > \omega$, i.e., Fact 3 is proved.

Returning to our solution let $X = V(\omega)$ and $Y = D(\omega) \times S$; the spaces $X$ and $Y$ are $l$-equivalent by Problem 263. If $Z = V(\mathfrak{c})$ then $t_m(X \times Z) > \omega$ by Fact 3. The space $Y \times Z$ is representable as a countable union of open subspaces homeomorphic to $V(\mathfrak{c}) \times S$. It is an easy exercise that the space $V(\mathfrak{c})$ is Fréchet–Urysohn and hence $t(V(\mathfrak{c})) \leq \omega$. Therefore $t(S \times V(\mathfrak{c})) \leq \omega$ by Fact 1, so we can apply Fact 2 to conclude that $t(Y \times Z) \leq \omega$ and hence $t_m(Y \times Z) \leq \omega$ (see TFS-419). It follows from $t_m(X \times Z) \neq t_m(Y \times Z)$ that the spaces $X \times Z$ and $Y \times Z$ are not $t$-equivalent (see Problem 008), so our solution is complete.

**V.275.** *Give an example of $l$-equivalent spaces $X$ and $Y$ such that $X \times X$ is not $t$-equivalent to $Y \times Y$.*

**Solution.** Let $a = 0$ and $a_n = \frac{1}{n+1}$ for any $n \in \omega$; then $S = \{a\} \cup \{a_n : n \in \omega\}$ is a faithfully indexed convergent sequence with its limit $a$. For any infinite cardinal $\kappa$ we denote by $D(\kappa)$ a discrete space of cardinality $\kappa$. In the space $D(\kappa) \times S$ the set $F = D(\kappa) \times \{a\}$ is a retract of $D(\kappa) \times S$; let $V(\kappa)$ be the $\mathbb{R}$-quotient space obtained from $D(\kappa) \times S$ by contracting $F$ to a point. Recall that *weak functional tightness* $t_m(P)$ of a space $P$ is the minimal cardinal $\kappa$ such that every strictly $\kappa$-continuous real-valued function on $P$ is continuous.

We have $V(\omega) \overset{l}{\sim} D(\omega) \times S$ and $V(\mathfrak{c}) \overset{l}{\sim} D(\mathfrak{c}) \times S$ by Problem 263, so we can apply Problem 265 to see that the spaces $X = V(\omega) \oplus V(\mathfrak{c})$ and $Y = (D(\omega) \times S) \oplus (D(\mathfrak{c}) \times S)$ are $l$-equivalent. The space $Y$ is easily seen to be metrizable, so $Y \times Y$ is also metrizable and hence $t_m(Y \times Y) \leq t(Y \times Y) = \omega$ (see TFS-419).

The space $X \times X$ contains a clopen subspace $G$ homeomorphic to $V(\omega) \times V(\mathfrak{c})$, so $t_m(G) > \omega$ by Fact 3 of V.274. It is an easy exercise that $G$ is a retract of $X \times X$, so fix a retraction $r : X \times X \to G$. The map $r$ is quotient (see Fact 11 of U.074) and hence $\mathbb{R}$-quotient, so we can apply TFS-420 to conclude that $t_m(X \times X) \geq t_m(G) > \omega$. Now, it follows from $t_m(X \times X) \neq t_m(Y \times Y)$ that the spaces $X \times X$ and $Y \times Y$ are not $t$-equivalent (see Problem 008).

**V.276.** *Given infinite cardinals $\kappa_1, \ldots, \kappa_n$ prove that the space $A(\kappa_1) \times \ldots \times A(\kappa_n)$ is $l$-equivalent to $A(\kappa)$ where $\kappa = \max\{\kappa_1, \ldots, \kappa_n\}$.*

**Solution.** Our statement is clearly true if $n = 1$; let us prove it for $n = 2$. Represent $A(\kappa_1)$ as $D \cup \{a\}$ where $a \notin D$ is the unique non-isolated point of $A(\kappa_1)$ and $|D| = \kappa_1$. Analogously, $A(\kappa_2) = \{b\} \cup E$ where $b \notin E$ is the unique non-isolated point of $A(\kappa_2)$ and $|E| = \kappa_2$. In the space $X = A(\kappa_1) \times A(\kappa_2)$ both sets $K = \{a\} \times A(\kappa_2)$ and $L = A(\kappa_1) \times \{b\}$ are retracts. Let $r(x, y) = (a, y)$ for any $(x, y) \in X$; it is straightforward that $r : X \to K$ is a retraction such that $r(L) = L \cap K = \{(a, b)\}$.

Let $M = K \cup L$ and denote by $X_M$ the $\mathbb{R}$-quotient space obtained from $X$ by contracting $M$ to a point $a_M$. It is easy to see that $a_M$ is the unique non-isolated point of the space $X_M$. If $p = (a, b)$ then $p \in K \cap L$ and $p$ is the unique non-isolated point of $K \cup L$. Apply Problem 261 to see that the space $X$ is $l$-equivalent to the bunch $B = (X_M, a_M) \vee (K, p_0) \vee (L, p_1)$ where $p_0$ is the copy of the point $p$ in the space $K$ and $p_1$ is the copy of $p$ in $L$. It is evident that only the point represented by the set $\{a_M, p_0, p_1\}$ is non-isolated in $B$ (recall that $B$ is the $\mathbb{R}$-quotient space of $X_M \oplus K \oplus L$ obtained by contracting the set $\{a_M, p_0, p_1\}$ to a point). Since $B$ is compact and $|B| = \kappa = \max\{\kappa_1, \kappa_2\}$, we conclude that $B$ is homeomorphic to $A(\kappa)$. Therefore $X = A(\kappa_1) \times A(\kappa_2)$ is $l$-equivalent to $A(\kappa)$, so we settled the case when $n = 2$.

Proceeding inductively, assume that $m \geq 2$ and we proved that, for any $i \leq m$ and any infinite cardinals $\lambda_1, \ldots, \lambda_i$ the space $A(\lambda_1) \times \ldots \times A(\lambda_i)$ is $l$-equivalent to $A(\lambda)$ where $\lambda = \max\{\lambda_1, \ldots, \lambda_i\}$. Given infinite cardinals $\kappa_1, \ldots, \kappa_m, \kappa_{m+1}$ consider the cardinals $\kappa' = \max\{\kappa_1, \ldots, \kappa_m\}$ and $\kappa = \max\{\kappa_1, \ldots, \kappa_m, \kappa_{m+1}\}$; by the induction hypothesis the space $P = A(\kappa_1) \times \ldots \times A(\kappa_m)$ is $l$-equivalent to $A(\kappa')$. It follows from Problem 273 that $P \times A(\kappa_{m+1})$ is $l$-equivalent to $A(\kappa') \times A(\kappa_{m+1})$. Again, our induction hypothesis guarantees that $A(\kappa') \times A(\kappa_{m+1}) \overset{l}{\sim} A(\kappa)$, so the space $P \times A(\kappa_{m+1}) = A(\kappa_1) \times \ldots \times A(\kappa_{m+1})$ is $l$-equivalent to $A(\kappa)$ and hence our inductive procedure shows that, for any $n \in \mathbb{N}$ and any infinite cardinals $\kappa_1, \ldots, \kappa_n$, the space $A(\kappa_1) \times \ldots \times A(\kappa_n)$ is $l$-equivalent to $A(\kappa)$ where $\kappa = \max\{\kappa_1, \ldots, \kappa_n\}$.

**V.277.** *Prove that there exist $l$-equivalent spaces $X$ and $Y$ such that $X$ is hereditarily paracompact while $Y$ is not hereditarily normal.*

**Solution.** If $X = A(\omega_1)$ and $Y = A(\omega_1) \times A(\omega_1)$ then $X$ is $l$-equivalent to $Y$ by Problem 276. It is straightforward that any subspace of $X$ is either discrete or compact, so $X$ is hereditarily paracompact. If $a$ is the unique non-isolated point of $A(\omega_1)$ then $\chi(a, A(\omega_1)) = \omega_1$, so $\{a\}$ is not a $G_\delta$-subset of $A(\omega_1)$ (see TFS-327) and hence $A(\omega_1)$ is not perfectly normal.

Assume that $Y$ is hereditarily normal. Since $A(\omega_1) \times A(\omega)$ embeds in $Y$, the space $A(\omega_1) \times A(\omega)$ is also hereditarily normal and hence $A(\omega_1)$ is perfectly normal by Fact 2 of S.292; this contradiction proves that $Y$ is not hereditarily normal.

**V.278.** *Prove that $L_p(D)$ is $l$-equivalent to $L_p(D) \oplus D$ for any infinite discrete space $D$. Deduce from this fact that the Souslin property is not $l$-invariant.*

**Solution.** The expression $M \approx L$ says that the linear topological spaces $M$ and $L$ are linearly homeomorphic. If $I = \{f \in C_p(L_p(D)) : f(D) = \{0\}\}$ then $I$ is a linear subspace of the space $C_p(L_p(D))$; since $D$ is $l$-embedded in $L_p(D)$ by CFS-467, the space $C_p(L_p(D))$ is linearly homeomorphic to $I \times C_p(D) = I \times \mathbb{R}^D$ (see CFS-448). Therefore $C_p(L_p(D)) \times \mathbb{R}^D \approx I \times (\mathbb{R}^D \times \mathbb{R}^D) \approx I \times \mathbb{R}^D \approx C_p(L_p(D))$ which shows that $L_p(D) \overset{l}{\sim} L_p(D) \oplus D$ (see Problem 265).

Finally, let $D$ be an uncountable discrete space and apply Fact 3 of V.024 to see that $c(L_p(D)) = \omega$; however, the space $L_p(D)$ is $l$-equivalent to $L_p(D) \oplus D$ while $c(L_p(D) \oplus D) = c(D) = |D| > \omega$ and hence the Souslin property is not $l$-invariant.

**V.279.** *Given spaces $X$ and $Y$ suppose that $\varphi : C_p(X) \to C_p(Y)$ is a continuous linear surjection. Prove that,*

(i) *for any point $y \in Y$, there exist uniquely determined $n = n(y) \in \mathbb{N}$, distinct points $x_1(y), \ldots, x_n(y) \in X$ and numbers $\lambda_1(y), \ldots, \lambda_n(y) \in \mathbb{R} \setminus \{0\}$ such that $\varphi(f)(y) = \sum_{i=1}^{n(y)} \lambda_i(y) f(x_i(y))$ for any $f \in C_p(X)$; for further reference denote the set $\{x_1(y), \ldots, x_n(y)\}$ by $\operatorname{supp}(\varphi, y)$.*

*(ii) if $\varphi^* : C_p(C_p(Y)) \to C_p(C_p(X))$ is the dual map of $\varphi$ then $\varphi^*$ embeds $L_p(Y)$ in $L_p(X)$ and $\varphi^*(y) = \lambda_1(y)x_1(y) + \ldots + \lambda_{n(y)}(y)x_{n(y)}(y)$ for any $y \in Y$.*

**Solution.** Fix a point $y \in Y$ and recall that we also consider $y$ to be a linear functional on the space $C_p(Y)$. Then $y \circ \varphi$ is a continuous linear functional on $C_p(X)$, i.e., $y \circ \varphi \in L_p(X)$ (see TFS-197). It is easy to see that $y \circ \varphi$ is not identically zero, so we can find $n = n(y) \in \mathbb{N}$ and distinct points $x_1, \ldots, x_n \in X$ such that $y \circ \varphi = \sum_{i=1}^{n} \lambda_i x_i$ for some $\lambda_1, \ldots, \lambda_n \in \mathbb{R}\backslash\{0\}$. Since $X$ is a Hamel basis in $L_p(X)$ (see Fact 5 of S.489), this representation of $y \circ \varphi$ is unique; let $\lambda_i(y) = \lambda_i$ and $x_i(y) = x_i$ for all $i \in \{1, \ldots, n\}$. For any function $f \in C_p(X)$ we have the equalities $\varphi(f)(y) = (y \circ \varphi)(f) = (\sum_{i=1}^{n} \lambda_i x_i)(f) = \sum_{i=1}^{n} \lambda_i f(x_i)$, so (i) is proved. It follows from Fact 4 of S.489 that $\varphi^*$ embeds $L_p(Y)$ in $L_p(X)$. Since we have the equalities $\varphi^*(y) = y \circ \varphi = \sum_{i=1}^{n(y)} \lambda_i(y)x_i(y)$ for any $y \in Y$, we are also done with (ii).

**V.280.** *Suppose that $\varphi : C_p(X) \to C_p(Y)$ is a continuous linear surjection and let $\xi(y) = supp(\varphi, y)$ for any $y \in Y$. Prove that the map $\xi : Y \to exp(X)$ is lower semicontinuous.*

**Solution.** Fix a set $U \in \tau(X)$ and a point $y_0 \in \xi_l^{-1}(U) = \{y \in Y : \xi(y) \cap U \neq \emptyset\}$. Pick a point $x_0 \in \xi(y_0) \cap U$ and take a function $f \in C(X)$ such that $f(x_0) = 1$ and $f(x) = 0$ for any $x \in (X\backslash U) \cup (\xi(y_0)\backslash\{x_0\})$. For any point $x \in \xi(y_0)$ there exists $\lambda(x) \in \mathbb{R}\backslash\{0\}$ such that $\varphi(g)(y_0) = \sum_{x \in \xi(y_0)} \lambda(x)g(x)$ for any $g \in C_p(X)$. In particular, $\varphi(f)(y_0) = \lambda(x_0)f(x_0) = \lambda(x_0) \neq 0$, so the set $V = [\varphi(f)]^{-1}(\mathbb{R}\backslash\{0\})$ is an open neighborhood of the point $y_0$.

If $y \in V$ and $\xi(y) = \{x_1, \ldots, x_n\} \subset X\backslash U$ then there exist $\lambda_1, \ldots, \lambda_n \in \mathbb{R}$ such that $\varphi(f)(y) = \sum_{i=1}^{n} \lambda_i f(x_i)$, so it follows from $f(\xi(y)) = \{0\}$ that $\varphi(f)(y) = 0$ which is a contradiction. Therefore $\xi(y) \cap U \neq \emptyset$ for any $y \in V$, so every point $y_0 \in \xi_l^{-1}(U)$ has an open neighborhood $V$ contained in $\xi_l^{-1}(U)$. This shows that $\xi_l^{-1}(U)$ is open in $Y$ for any $U \in \tau(X)$, i.e., $\xi$ is a lower semicontinuous map.

**V.281.** *Given a continuous linear surjection $\varphi : C_p(X) \to C_p(Y)$ prove that, for any bounded subset $A$ of the space $Y$, the set $supp(A) = \bigcup\{supp(\varphi, y) : y \in A\}$ is bounded in $X$.*

**Solution.** Recall that if $u = \lambda_1 x_1 + \ldots + \lambda_n x_n \in L_p(X)$ for some $x_1, \ldots, x_n \in X$ and $\lambda_1, \ldots, \lambda_n \in \mathbb{R}\backslash\{0\}$ then $supp(u) = \{x_1, \ldots, x_n\}$; if $u = \mathbf{0}$ then $supp(u) = \emptyset$.

Any $v \in L_p(Y)$ is a continuous linear functional on $C_p(Y)$; let $\varphi^*(v) = v \circ \varphi$. Then $\varphi^* : L_p(Y) \to L_p(X)$ is a linear embedding and it follows from Problem 279 that $supp(\varphi^*(y)) = supp(\varphi, y)$ for any $y \in Y$.

If a set $A \subset Y$ is bounded in $Y$ then $A$ is bounded in $L_p(Y)$, so $\varphi^*(A)$ is bounded in $\varphi^*(L_p(Y))$ and hence in $L_p(X)$. Our observation above shows that $supp(A) = \bigcup\{supp(\varphi^*(y)) : y \in A\}$ and hence we can apply Problem 245 to conclude that the set $supp(A)$ is bounded in $X$.

**V.282.** *Suppose that $\varphi : C_p(X) \to C_p(Y)$ is a continuous linear surjection. Prove that, for any bounded subset $B \subset X$, the set $C = \{y \in Y : supp(\varphi, y) \subset B\}$ is bounded in $Y$.*

**Solution.** Assume that the set $C$ is not bounded in $Y$; by Fact 1 of V.245, there exists a discrete family $\{U_n : n \in \omega\}$ of open subsets of $Y$ such that $U_n \cap C \neq \emptyset$ and hence we can fix a point $y_n \in U_n \cap C$ for each $n \in \omega$.

Let $supp(\varphi, y_n) = \{x_1^n, \ldots, x_{k_n}^n\}$ and choose numbers $\lambda_1^n, \ldots, \lambda_{k_n}^n \in \mathbb{R}\backslash\{0\}$ such that $\varphi(f)(y_n) = \sum_{i=1}^{k_n} \lambda_i^n f(x_i^n)$ for any $f \in C_p(X)$ and $n \in \omega$. It follows from Fact 5 of T.132 that we can find a function $g \in C_p(Y)$ such that $g(y_n) = n \cdot (|\lambda_1^n| + \ldots + |\lambda_{k_n}^n|)$ for every $n \in \omega$. Pick $f \in C_p(X)$ with $\varphi(f) = g$; the set $B$ being bounded in $X$ there is $K > 0$ such that $|f(x)| \leq K$ for all $x \in B$; it follows from $supp(\varphi, y_n) \subset B$ that $|f(x_i^n)| \leq K$ for any $n \in \omega$ and $i \leq k_n$. Take any $n \in \omega$ with $K < n$; then

$$|g(y_n)| = |\sum_{i=1}^{k_n} \lambda_i^n f(x_i^n)| \leq K \cdot \sum_{i=1}^{k_n} |\lambda_i^n| < n \cdot \sum_{i=1}^{k_n} |\lambda_i^n| = |g(y_n)|,$$

which is a contradiction. Therefore the set $C$ is bounded in $Y$.

**V.283.** *Say that $X$ is a $\mu$-space if $\overline{A}$ is compact for any bounded set $A \subset X$. Prove that $X$ is a $\mu$-space if and only if $L_p(X)$ is a $\mu$-space. As a consequence, $\mu$-property is preserved by $l$-equivalence.*

**Solution.** Suppose that $L_p(X)$ is a $\mu$-space and take any bounded subset $A$ of the space $X$. It is evident that $A$ is also bounded in $L_p(X)$, so the closure $K$ of the set $A$ in $L_p(X)$ is compact. If $\overline{A}$ is the closure of $A$ in $X$ then $\overline{A}$ is closed in $L_p(X)$ because $X$ is closed in $L_p(X)$. An easy consequence is that the set $\overline{A} = K$ is compact, so $X$ is a $\mu$-space.

To prove necessity assume that $X$ is a $\mu$-space; recall first that $X$ is $C$-embedded in the Hewitt realcompactification $\upsilon X$ of the space $X$, so we can apply Fact 3 of V.246 to identify $L_p(X)$ with the linear hull of the set $X$ in $L_p(\upsilon X)$. Assume that $A$ is a bounded subset of $L_p(X)$. For any $w = \lambda_1 x_1 + \ldots + \lambda_n x_n \in L_p(\upsilon X)$ where $\lambda_i \in \mathbb{R}\backslash\{0\}$ and $x_i \in \upsilon X$ for all $i \leq n$, let $supp(w) = \{x_1, \ldots, x_n\}$; if $w = \mathbf{0}$ then $supp(w) = \emptyset$. The set $\overline{A}$ is still bounded in $L_p(X)$ by Fact 2 of S.398 (the bar denotes the closure in $L_p(X)$), so we can apply Problem 245 to see that $B = \bigcup\{supp(w) : w \in \overline{A}\}$ is a bounded subset of $X$ and hence $K = cl_X(B)$ is compact.

Let $G$ be the linear hull of the set $K$ in the space $L_p(\upsilon X)$; it is evident that $\overline{A} \subset G \subset L_p(X)$. By Fact 2 of V.246, the set $G$ is closed in $L_p(\upsilon X)$; besides, $\overline{A}$ is closed in $L_p(X)$ and hence in $G$, so $\overline{A}$ is closed in $L_p(\upsilon X)$. Besides, $\overline{A}$ is also bounded in $L_p(\upsilon X)$ because $L_p(X) \subset L_p(\upsilon X)$; since $L_p(\upsilon X)$ is realcompact by Problem 244, the set $\overline{A}$ is compact by Fact 1 of V.246, so $L_p(X)$ is a $\mu$-space, i.e., we established necessity and hence our solution is complete.

**V.284.** *Given spaces $X$ and $Y$ assume that $X$ is a $\mu$-space and there exists a linear surjection $\varphi : C_p(X) \to C_p(Y)$ which is an $\mathbb{R}$-quotient map. Prove that $Y$ is also a $\mu$-space. Give an example of a compact space $X$ (which is, automatically, a $\mu$-space) such that there exists a continuous linear surjection of $C_p(X)$ onto $C_p(Y)$ for some $Y$ which is not a $\mu$-space.*

**Solution.** For any $f \in C_p(C_p(Y))$ let $\varphi^*(f) = f \circ \varphi$; this definition gives us the map $\varphi^* : C_p(C_p(Y)) \to C_p(C_p(X))$ which embeds $C_p(C_p(Y))$ in $C_p(C_p(X))$ as a closed subspace (see TFS-163). Besides, $\varphi^*(L_p(Y)) \subset L_p(X)$ by Problem 279; this shows that $\varphi^*(L_p(Y))$ is closed in $L_p(X)$ and hence $\varphi^*(Y)$ is closed in $L_p(X)$. Thus, $Y$ embeds in $L_p(X)$ as a closed subspace. Apply Problem 283 to see that $L_p(X)$ is a $\mu$-space; it is an easy exercise that $\mu$-property is closed-hereditary, so $Y$ is a $\mu$-space as well.

Now if $X = \omega_1 + 1$ and $Y = \omega_1$ then $X$ is compact and $Y$ is a countably compact non-compact space, so $Y$ fails to be a $\mu$-space. We also have $X = \beta Y$ by TFS-314 and hence the restriction map $\pi : C_p(X) \to C_p(Y)$ is a continuous linear surjection.

**V.285.** *Given $\mu$-spaces $X$ and $Y$, let $\varphi : C_p(X) \to C_p(Y)$ be a continuous linear surjection. Prove that, if $X$ is compact then $Y$ is also compact. Observe that the same conclusion about $Y$ may be false if $Y$ is not a $\mu$-space.*

**Solution.** The map $\varphi$ is uniformly continuous by Problem 134, so we can apply Problem 193 to see that the space $Y$ is pseudocompact. Any pseudocompact $\mu$-space is compact, so $Y$ is compact.

Now if $X = \omega_1 + 1$ and $Y = \omega_1$ then $X$ is compact and $Y$ is a countably compact non-compact space, so $Y$ fails to be a $\mu$-space. We also have $X = \beta Y$ by TFS-314 and hence the restriction map $\pi : C_p(X) \to C_p(Y)$ is a continuous linear surjection.

**V.286.** *Given $\mu$-spaces $X$ and $Y$, let $\varphi : C_p(X) \to C_p(Y)$ be a continuous linear surjection. Prove that, if $X$ is $\sigma$-compact then $Y$ is also $\sigma$-compact. Observe that the same conclusion about $Y$ may be false if $Y$ is not a $\mu$-space.*

**Solution.** It follows from $\sigma$-compactness of $X$ that $X = \bigcup_{n\in\omega} X_n$ where $X_n$ is compact and $X_n \subset X_{n+1}$ for any $n \in \omega$. For any point $y \in Y$ there exists a uniquely determined $n \in \mathbb{N}$ together with $x_1, \ldots, x_n \in X$ and $\lambda_1, \ldots, \lambda_n \in \mathbb{R}\backslash\{0\}$ such that $\varphi(f)(y) = \sum_{i=1}^n \lambda_i f(x_i)$ for any $f \in C_p(X)$; let $\operatorname{supp}(\varphi, y) = \{x_1, \ldots, x_n\}$.

The set $Y_n = \{y \in Y : \operatorname{supp}(\varphi, y) \subset X_n\}$ is bounded in $Y$ for any $n \in \omega$ (see Problem 282). If we let $\xi(y) = \operatorname{supp}(\varphi, y)$ for each $y \in Y$ then the map $\xi : Y \to \exp(X)$ is lower semicontinuous by Problem 280. An immediate consequence is that the set $Y_n$ is closed in $Y$ for every $n \in \omega$; since $Y$ is a $\mu$-space, every $Y_n$ is compact. Given any $y \in Y$ it is clear that $\operatorname{supp}(\varphi, y) \subset X_n$ for some $n \in \omega$, so $Y = \bigcup_{n\in\omega} Y_n$ is $\sigma$-compact.

Now if $X = \omega_1 + 1$ and $Y = \omega_1$ then $X$ is compact and $Y$ is a countably compact non-compact space, so $Y$ fails to be $\sigma$-compact. We also have $X = \beta Y$ by TFS-314 and hence the restriction map $\pi : C_p(X) \to C_p(Y)$ is a continuous linear surjection.

**V.287.** *For any space $X$ let $\mathcal{K}(X)$ be the family of all compact subspaces of $X$. Prove that a second countable space $X$ is Čech-complete if and only if there exists a Polish space $M$ and a map $\varphi : \mathcal{K}(M) \to \mathcal{K}(X)$ such that, for any $F, G \in \mathcal{K}(M)$ the inclusion $F \subset G$ implies $\varphi(F) \subset \varphi(G)$ and, for any $P \in \mathcal{K}(X)$, there exists $F \in \mathcal{K}(M)$ such that $\varphi(F) \supset P$.*

**Solution.** Let $\mathbb{P}$ be the space $\omega^{\omega}$ of the irrational numbers; for any $s, t \in \mathbb{P}$ we write $s \leq t$ if $s(i) \leq t(i)$ for all $i \in \omega$. If $K \in \mathcal{K}(\mathbb{P})$ and $i \in \omega$ then the set $\{s(i) : s \in K\}$ is finite being a discrete continuous image of $K$, so the number $s_K(i) = \max\{s(i) : s \in K\}$ is well defined; this gives a point $s_K \in \mathbb{P}$ for every $K \in \mathcal{K}(\mathbb{P})$. It is immediate that $K \subset L$ implies $s_K \leq s_L$ for all $K, L \in \mathcal{K}(\mathbb{P})$. For any $s \in \mathbb{P}$ we will need the set $K_s = \{t \in \mathbb{P} : t \leq s\}$ which is compact being homeomorphic to $\prod_{i \in \omega}\{0, \ldots, s(i)\}$; it is clear that $s_{K_s} = s$ and $s \leq t$ implies $K_s \subset K_t$.

Suppose first that a second countable space $X$ is Čech-complete. By SFFS-365, there exists a map $\xi : \mathbb{P} \to \mathcal{K}(X)$ such that $s \leq t$ implies $\xi(s) \subset \xi(t)$ and $\xi$ "swallows" all compact subsets of $X$ in the sense that, for each $P \in \mathcal{K}(X)$ there exists $s \in \mathbb{P}$ such that $P \subset \xi(s)$. Let $\varphi(K) = \xi(s_K)$ for any $K \in \mathcal{K}(\mathbb{P})$; this gives us a map $\varphi : \mathcal{K}(\mathbb{P}) \to \mathcal{K}(X)$. If $F, G \in \mathcal{K}(\mathbb{P})$ and $F \subset G$ then $s_F \leq s_G$ and hence $\varphi(F) = \xi(s_F) \subset \xi(s_G) = \varphi(G)$. If $P$ is a compact subset of $X$ then there exists $s \in \mathbb{P}$ such that $\xi(s) \supset P$, so $\varphi(K_s) = \xi(s) \supset P$. This shows that for the Čech-complete space $M = \mathbb{P}$ the map $\varphi : \mathcal{K}(M) \to \mathcal{K}(X)$ preserves inclusions and swallows all compact subsets of $X$, i.e., we proved necessity.

Now assume that $M$ is a Čech-complete second countable space and we have a map $\varphi : \mathcal{K}(M) \to \mathcal{K}(X)$ which preserves inclusions and swallows all compact subsets of $X$. Apply SFFS-328 to find an open continuous onto map $h : \mathbb{P} \to M$. For any $s \in \mathbb{P}$ let $\xi(s) = \varphi(h(K_s))$. It is clear that $\xi : \mathbb{P} \to \mathcal{K}(X)$; if $s \leq t$ then $K_s \subset K_t$ and hence $h(K_s) \subset h(K_t)$ which in turn implies that $\varphi(h(K_s)) \subset \varphi(h(K_t))$, i.e., $\xi(s) \subset \xi(t)$.

Now if $P \in \mathcal{K}(X)$ then there exists $F \in \mathcal{K}(M)$ for which $\varphi(F) \supset P$. The map $h_1 = h|h^{-1}(F) : h^{-1}(F) \to F$ is open by Fact 1 of V.077, so there is a closed set $K \subset h^{-1}(F)$ such that $h_1(K) = h(K) = F$ and $h_1|K = h|K : K \to F$ is a perfect map (see SFFS-326). It follows from Fact 2 of S.259 that $K$ is compact; let $s = s_K$. Then $K \subset K_s$, so $\xi(s) \supset \varphi(h(K)) = \varphi(F) \supset P$ and hence we proved that the map $\xi : \mathbb{P} \to \mathcal{K}(X)$ satisfies all premises of SFFS-365; the relevant conclusion of SFFS-365 guarantees that $X$ is Čech-complete. Thus we established sufficiency and completed our solution.

**V.288.** *Let $X$ and $Y$ be second countable spaces for which there is a continuous linear surjection of $C_p(X)$ onto $C_p(Y)$. Prove that, if $X$ is Čech-complete then $Y$ is also Čech-complete. In particular, if two second countable spaces $X$ and $Y$ are $l$-equivalent then $X$ is Čech-complete if and only if so is $Y$.*

**Solution.** Given a space $Z$ let $\mathcal{K}(Z)$ be the family of all compact subsets of $Z$. Fix a continuous linear surjection $\mu : C_p(X) \to C_p(Y)$. For any $y \in Y$ we can find uniquely determined $n \in \mathbb{N}$, distinct points $x_1, \ldots, x_n \in X$ and $\lambda_1, \ldots, \lambda_n \in$

$\mathbb{R}\backslash\{0\}$ such that $\mu(f)(y) = \sum_{i=1}^{n} \lambda_i f(x_i)$ for any $f \in C_p(X)$; let $\text{supp}(\mu, y) = \{x_1, \ldots, x_n\}$. Letting $\xi(y) = \text{supp}(\mu, y)$ for any point $y \in Y$ we obtain a lower semicontinuous map $\xi : Y \to \mathcal{K}(X)$ (see Problem 280). Thus, for any compact subspace $F$ of the space $X$ the set $\varphi(F) = \{y \in Y : \xi(y) \subset F\}$ is closed in $Y$; besides, $\varphi(F)$ is bounded in $Y$ (see Problem 282) and hence compact. It is evident that the map $\varphi : \mathcal{K}(X) \to \mathcal{K}(Y)$ preserves inclusions, i.e., and $F \subset G$ implies $\varphi(F) \subset \varphi(G)$ for any $F, G \in \mathcal{K}(X)$. If $K \in \mathcal{K}(Y)$ then the set $A = \bigcup\{\text{supp}(\mu, y) : y \in K\}$ is bounded in $X$ by Problem 281. Therefore $F = \overline{A}$ is compact and it follows from the definition of $\varphi$ that $\varphi(F) \supset K$.

We checked that all premises of Problem 287 are fulfilled for the map $\varphi$, so it follows from Čech-completeness of $X$ that $Y$ is also Čech-complete.

**V.289.** *Give an example of second countable $l$-equivalent spaces $X$ and $Y$ such that $X$ is pseudocomplete and $Y$ is not Baire. As a consequence, having a dense Čech-complete subspace is not an $l$-invariant property in the class of second countable spaces.*

**Solution.** The expression $P \simeq Q$ says that the spaces $P$ and $Q$ are homeomorphic. Choose a faithful enumeration $\{q_n : n \in \omega\}$ of the set $\mathbb{Q}$ and consider the set $T = (\mathbb{Q} \times \{0\}) \cup \{(q_n, \frac{1}{n+1}) : n \in \omega\} \subset \mathbb{R}^2$ with the topology induced from the plane. Observe that $D = \{(q_n, \frac{1}{n+1}) : n \in \omega\}$ is a discrete subspace of $T$. Indeed, the set $U_n = \mathbb{R} \times (\frac{1}{n+2}, \frac{1}{n})$ is open in $\mathbb{R}^2$ and $U_n \cap D = \{(q_n, \frac{1}{n+1})\}$ for any $n \in \mathbb{N}$; the point $x_0 = (q_0, 1)$ is also isolated in $T$ because the set $U_0 = \mathbb{R} \times (\frac{1}{2}, 2)$ is open in $\mathbb{R}^2$ and $U_0 \cap D = \{x_0\}$.

Take any point $x = (q, 0) \in \mathbb{Q} \times \{0\}$ and $U \in \tau(x, T)$; there exists $\varepsilon > 0$ such that $V \cap T \subset U$ where $V = (q-\varepsilon, q+\varepsilon) \times [0, \varepsilon)$. The set $A = \mathbb{Q} \cap (q-\varepsilon, q+\varepsilon)$ being infinite we can find $n \in \omega$ for which $q_n \in A$ and $\frac{1}{n+1} < \varepsilon$. The point $y = (q_n, \frac{1}{n+1})$ belongs to $D \cap U$; this proves that $D \cap U \neq \emptyset$ for any $U \in \tau(x, T)$ and hence $x \in \overline{D}$. The point $x \in \mathbb{Q} \times \{0\}$ was chosen arbitrarily, so we proved that $D$ is dense in $T$ and, in particular, $T$ has a dense Čech-complete subspace.

For any $z = (x, y) \in T$ let $r(z) = (x, 0)$; we omit a simple proof that $r$ is a retraction of $T$ onto the set $F = \mathbb{Q} \times \{0\}$. Let $T_F$ be the $\mathbb{R}$-quotient space under the map which contracts $F$ to a point. If $a = (\sqrt{2}, \sqrt{2}) \in \mathbb{R}^2$ then the space $T_F \oplus F$ is $l$-equivalent to the space $X = T \oplus \{a\}$ (see Problem 258). It is evident that $w(X) \leq \omega$ and $X$ has a dense Čech-complete subspace, so it is pseudocomplete by TFS-465 and TFS-467.

If $Y = X \oplus \mathbb{Q}$ then $w(Y) \leq \omega$ and we can apply Problem 265 to see that the space $Y$ is $l$-equivalent to $Z = T_F \oplus F \oplus \mathbb{Q}$. The space $F$ being homeomorphic to $\mathbb{Q}$, we have $F \oplus \mathbb{Q} \simeq F$ by SFFS-349, so $Z \simeq T_F \oplus F$ which shows that $Y \overset{l}{\sim} Z \simeq T_F \oplus F \overset{l}{\sim} X$ and hence $Y$ is $l$-equivalent to $X$. The space $Y$ is not Baire because it has a clopen subspace homeomorphic to $\mathbb{Q}$; it follows from TFS-274 and TFS-275 that $Y$ does not have a dense Čech-complete subspace.

**V.290.** *Prove that there exist l-equivalent σ-compact second countable spaces X and Y such that X can be condensed onto a compact space and Y doesn't have such a condensation.*

**Solution.** For any space $Z$ denote by $\mathcal{O}(Z)$ the family of all clopen subsets of $Z$; if $\mathcal{O}(Z) = \{\emptyset, Z\}$ then $Z$ is called *connected*. Call a subset $U$ of the space $Z$ *nontrivial* if $U \neq \emptyset$ and $U \neq Z$. For each point $z \in Z$ let $Q_z = \bigcap \{U : z \in U$ and $U \in \mathcal{O}(Z)\}$; the set $Q_z$ is called *the quasi-component* of the point $z$ in the space $Z$. The set $K_z = \bigcup \{C : z \in C \subset Z$ and $C$ is connected$\}$ is called *the component* of the point $z$ in $Z$.

*Fact 1.* Suppose that $Z$ is a space and, for any points $x, y \in Z$ there exists a connected subspace $C \subset Z$ such that $\{x, y\} \subset C$. Then $Z$ is connected.

*Proof.* Suppose that $U \in \mathcal{O}(Z)$ and $U \neq \emptyset$; fix a point $x \in U$. If $U \neq Z$ then choose a point $y \in Z \backslash U$ and a connected subspace $C \subset Z$ with $\{x, y\} \subset C$. It follows from $x \in U$ and $y \notin U$ that $U' = U \cap C$ is a clopen subspace of $C$ such that $U \neq \emptyset$ and $U \neq C$ which is a contradiction. Therefore $U = Z$ and hence $\mathcal{O}(Z) = \{\emptyset, Z\}$, i.e., $Z$ is connected which shows that Fact 1 is proved.

*Fact 2.* Suppose that $Z$ is a space and $\mathcal{C} = \{C_t : t \in T\}$ is a family of connected subspaces of $Z$ such that $\bigcap \mathcal{C} \neq \emptyset$. Then the set $C = \bigcup \mathcal{C}$ is connected. As a consequence, the component of any point of $Z$ is a closed connected subspace of $Z$.

*Proof.* Fix a point $x \in \bigcap \mathcal{C}$ and suppose that $U$ is a nontrivial clopen subset of $C$. We can assume, without loss of generality, that $x \in U$ for otherwise $C \backslash U$ is a nontrivial clopen subset of $C$ which contains $x$. Take a point $y \in C \backslash U$ and $t \in T$ with $y \in C_t$. Then $x \in U' = C_t \cap U$ and $y \notin U'$, so $U'$ is a nontrivial clopen subspace of a connected space $C_t$. This contradiction shows that $C$ has no nontrivial clopen subsets, i.e., $C$ is connected.

An immediate consequence is that the component $K_z$ of any $z \in Z$ is connected. It follows from Fact 1 of T.312 that $\overline{K_z}$ is also a connected set; since $z \in \overline{K_z}$, we must have $\overline{K_z} \subset K_z$, so $K_z$ is closed in $Z$ and hence Fact 2 is proved.

*Fact 3.* Given any space $Z$ and $z \in Z$ we have $K_z \subset Q_z$, i.e., the component of any point is contained in its quasi-component. If $Z$ is compact then $K_z = Q_z$.

*Proof.* Take any $U \in \mathcal{O}(Z)$ such that $z \in U$. The set $U \cap K_z$ is clopen in $K_z$ and nonempty; since $K_z$ is connected by Fact 2, the set $U \cap K_z$ must coincide with $K_z$ and hence $K_z \subset U$. Therefore $K_z \subset U$ for any $U \in \mathcal{O}(Z)$ such that $z \in U$, i.e., $K_z \subset Q_z$.

Now assume that $Z$ is compact; since $K_z$ is the maximal connected set which contains $z$, to prove that $Q_z \subset K_z$, it suffices to show that $Q_z$ is connected. Assume toward a contradiction that $G$ is a nontrivial clopen subset of $Q_z$; then $H = Q_z \backslash G$ is also a nontrivial clopen subset of $Q_z$, so $G$ and $H$ are nonempty closed disjoint subsets of $Z$. Assume, without loss of generality, that $z \in G$. By normality of the space $Z$ we can find disjoint sets $U, V \in \tau(Z)$ such that $G \subset U$ and $H \subset V$. Since $Q_z = G \cup H \subset U \cup V$, we can apply Fact 1 of S.326 to find a finite family $\mathcal{V}$ of clopen subsets of $Z$ such that $Q_z \subset W = \bigcap \mathcal{V} \subset U \cup V$.

It is evident that $W$ is a clopen subset of $Z$, so $W \cap U$ is open in $Z$. Besides, $U \cap W$ and $V \cap W$ are open complementary subsets of $W$, so they are both closed in $W$ and hence in $Z$. This shows that $W' = W \cap U$ is a clopen subset of $Z$ such that $z \in W'$ and $\emptyset \neq H \subset Q_z \backslash W$ which contradicts the fact that $Q_z$ is contained in *every* clopen set which contains $z$. As a consequence, $Q_z$ is connected, so $Q_z = K_z$, i.e., Fact 3 is proved.

**Fact 4.** Suppose that $Z$ is a connected compact space and $F$ is a nonempty closed subset of $Z$ such that $F \neq Z$. Then, for any $z \in F$, the component $K$ of the point $z$ in the space $F$ intersects the boundary $B = F \backslash \mathrm{Int}(F)$ of the set $F$ in $Z$.

*Proof.* If this is false then $K \subset O = \mathrm{Int}(F)$. By Fact 3, the set $K$ is also the quasi-component of $z$ in $F$, so we can apply Fact 1 of S.326 to find a finite family $\mathcal{V}$ of clopen subsets of $F$ such that $K \subset W = \bigcap \mathcal{V} \subset O$. It is evident that $W$ is a clopen subset of $F$, so $W$ is closed in $Z$. Besides, $W$ is open in $F$ and hence in $O \subset F$. Since $O$ is open in $Z$, the set $W$ is also open in $Z$. It follows from $W \subset F \neq Z$ that $W \neq Z$; we also have $z \in W$, so $W \neq \emptyset$ and hence $W$ is a nontrivial clopen subset of $Z$. The space $Z$ being connected, this is a contradiction which shows that Fact 4 is proved.

**Fact 5.** Suppose that $Z$ is a connected compact space and $\mathcal{A} = \{Z_n : n \in \omega\}$ is a disjoint family of closed subsets of $Z$ such that $Z = \bigcup_{n \in \omega} Z_n$. Then at most one element of $\mathcal{A}$ is nonempty.

*Proof.* Say that a cover $\mathcal{F}$ of a space $E$ is *adequate* if it is disjoint, every $F \in \mathcal{F}$ is closed in $E$ and at least two elements of $\mathcal{F}$ are nonempty. We will first prove that, for any connected compact space $E$, if $\mathcal{F}$ is an adequate cover of $E$ then,

(1) for any $H \in \mathcal{F}$ there exists a connected compact $C \subset E$ such that $C \cap H = \emptyset$
    and $\{F \cap C : F \in \mathcal{F}\}$ is an adequate cover of $C$.

If $H = \emptyset$ then we can take $C = E$, so assume that $H \neq \emptyset$. The cover $\mathcal{F}$ being adequate there is $G \in \mathcal{F}$ such that $G \neq \emptyset$ and $G \cap H = \emptyset$. By normality of $E$ we can find a set $U \in \tau(G, E)$ such that $\overline{U} \cap H = \emptyset$; fix a point $z \in G$. The component $C$ of the point $z$ in the space $\overline{U}$ is connected and compact by Fact 2. The set $C$ meets the set $\overline{U} \backslash U$ (see Fact 4); it follows from $G \subset U$ that $C \backslash G \neq \emptyset$ and hence there is $G' \in \mathcal{F}$ distinct from $G$ such that $G' \cap C \neq \emptyset$, i.e., the family $\{F \cap C : F \in \mathcal{F}\}$ is adequate. Furthermore, $C \subset \overline{U}$ and hence $C \cap H = \emptyset$, so (1) is proved.

Apply (1) to construct by induction a sequence $\{C_i : i \in \omega\}$ of compact connected subsets of $Z$ such that $C_{i+1} \subset C_i$, $C_i \cap Z_i = \emptyset$ while $\{Z_n \cap C_i : n \in \omega\}$ is an adequate family and hence $C_i \neq \emptyset$ for all $i \in \omega$. By compactness of $Z$, the set $C = \bigcap_{i \in \omega} C_i$ is nonempty and no point of $C$ belongs to $\bigcup_{i \in \omega} Z_i$; this contradiction shows that Fact 5 is proved.

Returning to our solution consider the sequence $S = \{\frac{1}{n+1} : n \in \omega\} \subset \mathbb{R}$ which converges to zero and let $u_0 = (0, 0)$ be the zero point of the plane. We will also need the point $u_1 = (0, 1)$ and the subsets $P_0 = \{0\} \times [1, 2)$ and $P_1 = \{1\} \times [1, 2)$ of the plane $\mathbb{R}^2$. Denote the interval $[0, 1] \subset \mathbb{R}$ by $I$ and consider the subspace $Z = \{u_0\} \cup P_0 \cup P_1 \cup (S \times I)$ of the plane. For any point $z = (x, y) \in P_0 \cup P_1$ let $r(z) = (0, y)$; if $w \in Z \setminus (P_0 \cup P_1)$ then $r(w) = u_1$. It is trivial to check that $r : Z \to P_0$ is a retraction such that $r|P_1 : P_1 \to P_0$ is a homeomorphism, so $P_0$ and $P_1$ are parallel retracts of $Z$ by Problem 256.

Denote by $X$ the $\mathbb{R}$-quotient space obtained from $Z$ by contracting the set $P_1$ to a point $p_1$ and let $Y$ be the $\mathbb{R}$-quotient space obtained from $Z$ by contracting $P_0$ to a point $p_0$. The spaces $X$ and $Y$ are $l$-equivalent by Problem 257.

A set $U \subset X$ is an open neighborhood of $p_1$ in $X$ if and only if $(U \setminus \{p_1\}) \cup P_1$ is an open neighborhood of $P_1$ in $Z$ (see Problem 254). It is easy to see that this happens if and only if $(U \setminus \{p_1\}) \cup \{(1, 1)\}$ is an open neighborhood of the point $(1, 1)$ in the space $(Z \setminus P_1) \cup \{(1, 1)\}$. In other words, the space

$$(Z \setminus P_1) \cup \{(1, 1)\} = \{u_0\} \cup P_0 \cup (S \times I)$$

is homeomorphic to $X$ and, in particular, $w(X) \leq \omega$. From now on we will identify $X$ with the subspace $\{u_0\} \cup P_0 \cup (S \times I)$ of the plane $\mathbb{R}^2$.

An analogous reasoning shows that the space $Y$ is homeomorphic to the set $\{u_0\} \cup \{u_1\} \cup P_1 \cup (S \times I)$ with the topology induced from the plane, so we will identify $Y$ with this set from now on. It follows from $Y \subset \mathbb{R}^2$ that $w(Y) \leq \omega$.

The space $K = (S \cup \{0\}) \times I$ is compact and, intuitively, the space $X$ condenses onto $K$ by reflecting the interval $P_0$ symmetrically with respect to the point $u_1$. To define this condensation rigorously let $h(x) = x$ for any $x \in Q = \{u_0\} \cup \{u_1\} \cup (S \times I)$. If $x = (0, t)$ where $t > 1$ then let $h(x) = (0, 2 - t)$. It is evident that $h : X \to K$ is a bijection such that $h|Q$ and $h|P_0$ is continuous; applying Fact 2 of T.354 we conclude that $h$ is continuous and hence $X$ condenses onto the compact space $K$.

Assume toward a contradiction that $Y$ can be condensed onto a compact space $L$. We can consider that $L$ and $Y$ have the same underlying set and the topology of $L$ is contained in $\tau(Y)$. The set $I_n = \{\frac{1}{n}\} \times I$ being compact in $Y$ the topologies induced on $I_n$ from $L$ and $Y$ coincide for all $n \in \mathbb{N}$.

Suppose that $U \in \mathcal{O}(Y)$ and $u_0 \in U$. The set $U$ is also clopen in $Y$, so there is $m \in \omega$ such that $(\frac{1}{n}, 0) \in U$ and hence $U \cap I_n \neq \emptyset$ for all $n \geq m$. The set $I_n$ is connected and $U \cap I_n$ is a nonempty clopen subset of $I_n$, so $I_n \subset U$ for all $n \geq m$. The point $u_1$ belongs to the closure of the set $\bigcup\{I_n : n \geq m\} \subset U$ in $Y$ and hence in $L$. Since $U$ is closed in $L$, we conclude that $u_1 \in U$ which shows that $u_1$ belongs to the quasi-component of the point $u_0$ in the space $L$.

By Fact 2 and Fact 3 the component $C$ of the point $u_0$ in $L$ is a connected compact space with $\{u_0, u_1\} \subset C$. Fix $U_0, U_1 \in \tau(C)$ such that $u_0 \in U_0$, $u_1 \in U_1$ while $\overline{U}_0 \cap \overline{U}_1 = \emptyset$ and $w_0 = (1, 0) \notin \overline{U}_i$ for each $i = 0, 1$. The sets $\overline{U}_0, \overline{U}_1$ being disjoint there exists $j \in \{0, 1\}$ such that $\overline{U}_j$ does not contain the set $\{1\} \times (2 - \varepsilon, 2)$ for any number $\varepsilon \in (0, 1)$. Therefore there exists a strictly increasing sequence $\{t_n : n \in \mathbb{N}\} \subset (1, 2)$ such that $\sup\{t_n : n \in \omega\} = 2$ and $w_n = (1, t_n) \notin \overline{U}_j$ for all $n \in \mathbb{N}$; let $t_0 = 0$. The set $J_n = \{1\} \times [t_n, t_{n+1}]$ is compact both in $Y$ and $L$ for each $n \in \omega$. If $W_n = \{1\} \times (t_n, t_{n+1})$ for all $n \in \omega$ then the family $\{W_n : n \in \omega\}$ is disjoint; it follows from $\{w_i : i \in \omega\} \cap \overline{U}_j = \emptyset$ that $E_n = J_n \cap \overline{U}_j = W_n \cap \overline{U}_j$ is a compact set in $L$ for all $n \in \omega$ and the family $\mathcal{E} = \{E_n : n \in \omega\}$ is disjoint.

Apply Fact 4 to see that the component $D$ of the point $u_j$ in the set $\overline{U}_j$ is a compact set which meets the boundary of $\overline{U}_j$; since the point $u_j$ is in the interior of $\overline{U}_j$, there is a point $z \in D \backslash \{u_j\}$. It is easy to see that the family $\mathcal{H} = \{E_n \cap D : n \in \omega\} \cup \{I_n \cap D : n \in \mathbb{N}\} \cup \{u_j\}$ is a countable disjoint cover of $D$ and all elements of $\mathcal{H}$ are compact. If $H \in \mathcal{H}$ and $z \in H$ then $H$ and $\{u_j\}$ are nonempty distinct elements of $\mathcal{H}$; this contradiction with Fact 5 shows that $Y$ cannot be condensed onto a compact space.

It is straightforward that $Z$ is $\sigma$-compact, so both spaces $X$ and $Y$ are also $\sigma$-compact being continuous images of $Z$. Thus $X$ and $Y$ are $l$-equivalent second countable $\sigma$-compact spaces such that $X$ condenses onto a compact space and $Y$ does not have such a condensation, i.e., our solution is complete.

**V.291.** *Prove that a countable second countable space is scattered if and only if it is Čech-complete. Deduce from this fact that if $X$ and $Y$ are countable second countable $l$-equivalent spaces then $X$ is scattered if and only if $Y$ is scattered.*

**Solution.** Given a space $Z$ denote by $I(Z)$ the set of isolated points of $Z$. Say that a space $Z$ is *locally Čech-complete* if, for any $z \in Z$ there exists a Čech-complete set $V \in \tau(z, Z)$.

*Fact 1.* Any paracompact locally Čech-complete space is Čech-complete.

*Proof.* Suppose that $Z$ is a paracompact locally Čech-complete space and choose a Čech-complete set $O_z \in \tau(z, Z)$ for any $z \in Z$. Fix a homeomorphic copy $W_z$ of every space $O_z$ in such a way that the family $\{W_z : z \in Z\}$ is disjoint; the space $W = \bigoplus \{W_z : z \in Z\}$ is Čech-complete by TFS-262. Fix a homeomorphism $\varphi_z : W_z \to O_z$ for each $z \in Z$; then the union $\varphi : W \to Z$ of the family of maps $\{\varphi_z : z \in Z\}$ is a continuous map by Fact 2 of T.354. Given any $w \in W$ there is $z \in Z$ such that $w \in W_z$; fix a local base $\mathcal{B}$ of the space $W$ at the point $w$ such that $\bigcup \mathcal{B} \subset W_z$. The map $\varphi_z$ being a homeomorphism, the family $\{\varphi_z(B) : B \in \mathcal{B}\}$ is a local base of $Z$ at the point $\varphi(w)$. Since $\varphi(B) = \varphi_z(B)$ for any $B \in \mathcal{B}$, the family $\{\varphi(B) : B \in \mathcal{B}\}$ is also a local base of $Z$ at the point $\varphi(w)$, so we can apply Fact 2 of S.491 to see that $\varphi : W \to Z$ is an open continuous onto map.

Fact 1 of S.491 shows that there exists a closed (and hence Čech-complete) subspace $F \subset W$ such that $\varphi(F) = Z$ and $\varphi_0 = \varphi | F : F \to Z$ is a perfect map. It follows from TFS-261 that $Z$ is Čech-complete, so Fact 1 is proved.

*Fact 2.* Given a paracompact space $Z$ suppose that $D$ is a closed discrete subset of $Z$ and $\chi(d, Z) \le \omega$ for each $d \in D$. If $Z \backslash D$ is Čech-complete then $Z$ is also Čech-complete.

*Proof.* The Čech-complete open set $U = Z \backslash D$ witnesses that $Z$ is locally Čech-complete at any point $z \in U$. Fix an arbitrary point $d \in D$; it is immediate that $U \cup \{d\}$ is an open neighborhood of $d$. If $d$ is not in the closure of $U$ then $U \cup \{d\}$

is Čech-complete being homeomorphic to $U \oplus \{d\}$. Now if $d$ is not an isolated point of $U \cup \{d\}$ then $K = \beta(U \cup \{d\})$ is a compact extension of $U$ and therefore $U$ is a $G_\delta$-subset of $K$. It follows from $\chi(d, \{d\} \cup U) \leq \omega$ that $\chi(d, K) \leq \omega$ (see Fact 1 of S.158); this implies that $\{d\}$ is a $G_\delta$-subset of $K$. It is easy to see that the union of two $G_\delta$-sets is a $G_\delta$-set, so $U \cup \{d\}$ is Čech-complete being a $G_\delta$-subset of $K$. Therefore every point of $Z$ has an open Čech-complete neighborhood, i.e., $Z$ is locally Čech-complete. By Fact 1, the space $Z$ is Čech-complete, so Fact 2 is proved.

Returning to our solution suppose that $X$ is a countable Čech-complete space. If $X$ is not scattered then there exists a subspace $A \subset X$ which has no isolated points. The set $B = \overline{A}$ is also dense-in-itself; besides, $B$ is Čech-complete being closed in $X$. However, the countable family $\mathcal{B} = \{\{b\} : b \in B\}$ consists of nowhere dense subspaces of $B$ and $\bigcup \mathcal{B} = B$, i.e., $B$ does not have the Baire property which is a contradiction with TFS-274. Therefore $X$ is scattered and hence we proved sufficiency.

Now, if $X$ is a scattered countable space then consider the set $F_0 = X \setminus I(X)$; it is clear that $F_0$ is closed in $X$. If $\beta < \omega_1$ and we have a closed subset $F_\beta \subset X$ then let $F_{\beta+1} = F_\beta \setminus I(F_\beta)$. If $\beta < \omega_1$ is a limit ordinal and we have a family $\{F_\alpha : \alpha < \beta\}$ of closed subsets of $X$ then the set $F_\beta = \bigcap \{F_\alpha : \alpha < \beta\}$ is also closed in $X$.

This inductive procedure gives us a decreasing family $\{F_\beta : \beta < \omega_1\}$ of closed subsets of $X$. Observe that every $F_\beta$ is scattered, so $F_\beta \neq \emptyset$ implies that $F_{\beta+1}$ is strictly smaller than $F_\beta$. Since the space $X$ is countable, there exists $\alpha < \omega_1$ such that $F_\alpha = \emptyset$. The set $X_\beta = X \setminus F_\beta$ is open in $X$ for any $\beta \leq \alpha$ and $X_\alpha = X$, so it suffices to prove by induction that $X_\beta$ is Čech-complete for every $\beta \leq \alpha$.

Observe first that $X_0$ is Čech-complete being a discrete space. If $\beta < \alpha$ and we proved that $X_\beta$ is Čech-complete then $X_{\beta+1} = X_\beta \cup E$ where $E$ is the set of isolated points of $F_\beta$. Therefore $E$ is closed and discrete in $X_{\beta+1}$ while $X_{\beta+1} \setminus E = X_\beta$ is Čech-complete. Applying Fact 2 we conclude that $X_{\beta+1}$ is Čech-complete.

If $\beta \leq \alpha$ is a limit ordinal and the space $X_\gamma$ is Čech-complete for all $\gamma < \beta$ then $X_\beta = \bigcup \{X_\gamma : \gamma < \beta\}$ and the family $\{X_\gamma : \gamma < \beta\}$ is an open cover of $X_\beta$ which witnesses that $X_\beta$ is locally Čech-complete. Therefore $X_\beta$ is Čech-complete by Fact 1 and our inductive procedure shows that $X_\beta$ is Čech-complete for every $\beta < \omega_1$. In particular, $X = X_\alpha$ is Čech-complete so we proved necessity.

Finally, assume that $X$ and $Y$ are countable metrizable $l$-equivalent spaces. If $X$ is scattered then it is Čech-complete, so $Y$ is also Čech-complete (see Problem 288) and hence scattered. This proves that being scattered is $l$-invariant in the class of countable metrizable spaces and completes our solution.

**V.292.** *Let $X$ and $Y$ be metrizable spaces such that $C_p(X)$ is linearly homeomorphic to $C_p(X) \times C_p(X)$ and $C_p(Y)$ is linearly homeomorphic to $C_p(Y) \times C_p(Y)$. Prove that if $X$ embeds in $Y$ as a closed subspace and $Y$ embeds in $X$ as a closed subspace then $X$ and $Y$ are $l$-equivalent.*

**Solution.** The expression $L \approx M$ says that the linear topological spaces $L$ and $M$ are linearly homeomorphic. Suppose first that $F$ is a closed subspace of $X$

homeomorphic to $Y$. Since $F$ is $l$-embedded in $X$ (see CFS-469), we can apply CFS-448 to see that $C_p(X) \approx C_p(F) \times I$ where $I = \{f \in C_p(X) : f(F) \subset \{0\}\}$. The set $F$ is a homeomorphic copy of the space $Y$, so we have $C_p(F) \approx C_p(Y)$ and hence $C_p(X) \approx C_p(Y) \times I$; recalling that $C_p(Y)$ is linearly homeomorphic to its square we conclude that $C_p(X) \approx C_p(Y) \times (C_p(Y) \times I) \approx C_p(Y) \times C_p(X)$.

Now assume that $G$ is a closed subspace of $Y$ homeomorphic to $X$. Since the set $G$ is $l$-embedded in $Y$ by CFS-469, we have $C_p(Y) \approx C_p(G) \times J$ (see CFS-448), where $J = \{g \in C_p(Y) : g(G) \subset \{0\}\}$. The set $G$ is a homeomorphic copy of the space $X$, so $C_p(G) \approx C_p(X)$ and hence $C_p(Y) \approx C_p(X) \times J$; recalling that the space $C_p(X)$ is linearly homeomorphic to $C_p(X) \times C_p(X)$ we conclude that $C_p(Y) \approx C_p(X) \times (C_p(X) \times J) \approx C_p(X) \times C_p(Y)$. An immediate consequence is that $C_p(X) \approx C_p(X) \times C_p(Y) \approx C_p(Y)$, so the spaces $X$ and $Y$ are $l$-equivalent.

**V.293.** *Let $X$ be a countable second countable space. Prove that the following properties are equivalent:*

*(i)  $X$ is $l$-equivalent to $\mathbb{Q}$;*
*(ii)  $X$ is not scattered;*
*(iii)  $X$ has a subspace homeomorphic to $\mathbb{Q}$;*
*(iv)  $X$ has a closed subspace homeomorphic to $\mathbb{Q}$.*

**Solution.** The expression $L \approx M$ says that the linear topological spaces $L$ and $M$ are linearly homeomorphic. Given spaces $Y$ and $Z$ say that $Y \simeq Z$ if $Y$ and $Z$ are homeomorphic; we will write $Y \ll Z$ if $Y$ embeds in $Z$ as a closed subspace.

If $X \overset{l}{\sim} \mathbb{Q}$ then $X$ is not scattered because $\mathbb{Q}$ is not scattered (see Problem 291); this proves that (i)$\Longrightarrow$(ii). If $X$ is not scattered then some $Y \subset X$ is dense-in-itself; it follows from SFFS-349 that the space $Y$ is homeomorphic to $\mathbb{Q}$, so (ii)$\Longrightarrow$(iii). Next, assume that some $Y \subset X$ is homeomorphic to $\mathbb{Q}$ and hence $Y$ has no isolated points. Then $Z = \overline{Y}$ has no isolated points either and therefore $Z \subset X$ is a closed homeomorphic copy of $\mathbb{Q}$, so we established that (iii)$\Longrightarrow$(iv).

Now suppose that $\mathbb{Q} \ll X$; apply SFFS-349 again to see that $\mathbb{Q}$ is homeomorphic to $\mathbb{Q} \times \omega$, so we can choose a discrete family $\{F_i : i \in \omega\}$ in the space $X$ such that $F_i$ is closed in $X$ and $F_i \simeq \mathbb{Q}$ for every $i \in \omega$. Apply SFFS-350 to choose a closed set $G_i \subset F_i$ such that $G_i \simeq X$ for any $i \in \omega$.

If $G = \bigcup_{i \in \omega} G_i$ then $G$ is $l$-embedded in $X$ (see CFS-469), so $C_p(X) \approx C_p(G) \times I$ where $I$ is a linear subspace of $C_p(X)$ (see CFS-448). Every $G_i$ is a homeomorphic copy of $X$, so it follows from $G \simeq \bigoplus_{i \in \omega} G_i$ that $C_p(G) \approx (C_p(X))^\omega$. Consequently, $C_p(X) \approx (C_p(X))^\omega \times I \approx (C_p(X))^\omega \times ((C_p(X))^\omega \times I) \approx (C_p(X))^\omega \times C_p(X)$ which shows that $C_p(X) \approx (C_p(X))^\omega \approx (C_p(X))^\omega \times (C_p(X))^\omega \approx (C_p(X))^2$, i.e., the space $C_p(X)$ is linearly homeomorphic to its square.

Apply SFFS-349 once more to see that $\mathbb{Q} \simeq \mathbb{Q} \oplus \mathbb{Q}$, so $C_p(\mathbb{Q}) \approx (C_p(\mathbb{Q}))^2$. We have already observed that $X \ll \mathbb{Q}$ by SFFS-350, so we can apply Problem 292 to conclude that $X \overset{l}{\sim} \mathbb{Q}$. This settles (iv)$\Longrightarrow$(i) and makes our solution complete.

**V.294.** *Prove that, for any infinite cardinal $\kappa$ there exist $l$-equivalent spaces $X$ and $Y$ such that $X$ is dense-in-itself and $Y$ has a dense set of $\kappa$-many isolated points.*

**Solution.** Choose a faithful enumeration $\{q_n : n \in \omega\}$ of the set $\mathbb{Q}$ and consider the subspace $E = (\mathbb{Q} \times \{0\}) \cup \{(q_n, \frac{1}{n}) : n \in \omega\}$ of the plane $\mathbb{R}^2$. It is easy to see that $D = \{(q_n, \frac{1}{n}) : n \in \omega\}$ is a dense set of isolated points of $E$. Since $\mathbb{Q}$ embeds in $E$, we can apply Problem 293 to see that $E \overset{l}{\sim} \mathbb{Q}$. Let $E_\alpha = E$, $D_\alpha = D$ and $Q_\alpha = \mathbb{Q}$ for all $\alpha < \kappa$. If $Y = \bigoplus\{E_\alpha : \alpha < \kappa\}$ and $X = \bigoplus\{Q_\alpha : \alpha < \kappa\}$ then $X \overset{l}{\sim} Y$ by Problem 265 while $X$ is dense-in-itself and $Y$ has a dense set $Q = \bigoplus\{D_\alpha : \alpha < \kappa\}$ of $\kappa$-many isolated points.

**V.295.** *Suppose that a second countable space $X$ is Čech-complete and has a closed subspace homeomorphic to $\mathbb{R}^\omega$. Prove that $X$ is $l$-equivalent to $\mathbb{R}^\omega$.*

**Solution.** The expression $Y \simeq Z$ says that the spaces $Y$ and $Z$ are homeomorphic. We denote by $L \approx M$ the fact that the linear topological spaces $L$ and $M$ are linearly homeomorphic.

The family $\{\mathbb{R}^\omega \times \{n\} : n \in \omega\}$ is discrete in $\mathbb{R}^\omega \times \mathbb{R} \simeq \mathbb{R}^\omega$, so we can find a discrete family $\mathcal{F} = \{F_n : n \in \omega\}$ of closed subsets of $\mathbb{R}^\omega$ such that $F_n \simeq \mathbb{R}^\omega$ for every $n \in \omega$. Since the set $F = \bigcup_{n \in \omega} F_n$ is closed in $\mathbb{R}^\omega$, we can apply CFS-469 and CFS-448 to see that $C_p(\mathbb{R}^\omega) \approx C_p(F) \times I$ where $I = \{f \in C_p(\mathbb{R}^\omega) : f(F) = \{0\}\}$.

Recall that $F \simeq \bigoplus_{n \in \omega} F_n$ where every space $F_n$ is a homeomorphic copy of $\mathbb{R}^\omega$. Therefore

$$C_p(\mathbb{R}^\omega) \approx (C_p(\mathbb{R}^\omega))^\omega \times I \approx (C_p(\mathbb{R}^\omega))^\omega \times (C_p(\mathbb{R}^\omega))^\omega \times I;$$

replacing in the last product the expression $(C_p(\mathbb{R}^\omega))^\omega \times I$ with $C_p(\mathbb{R}^\omega)$ we conclude that $C_p(\mathbb{R}^\omega) \approx (C_p(\mathbb{R}^\omega))^\omega \times C_p(\mathbb{R}^\omega)$, i.e., $C_p(\mathbb{R}^\omega) \approx (C_p(\mathbb{R}^\omega))^\omega$. As an immediate consequence we obtain the formula

$$(C_p(\mathbb{R}^\omega))^2 \approx (C_p(\mathbb{R}^\omega))^\omega \times (C_p(\mathbb{R}^\omega))^\omega \approx (C_p(\mathbb{R}^\omega))^\omega \approx C_p(\mathbb{R}^\omega).$$

Since $\mathbb{R}^\omega$ embeds in $X$ as a closed subspace, taking the respective copies of the elements of the family $\mathcal{F}$ we obtain a discrete family $\{G_n : n \in \omega\}$ of closed subspaces of $X$ such that every $G_n$ is a homeomorphic copy of $\mathbb{R}^\omega$. It follows from TFS-273 that $X$ embeds in $\mathbb{R}^\omega$ as a closed subspace, so we can find a closed set $H_n \subset G_n$ such that $H_n \simeq X$ for each $n \in \omega$. Since the set $H = \bigcup_{n \in \omega} H_n$ is closed in $X$, we can apply CFS-469 and CFS-448 again to see that $C_p(X) \approx C_p(H) \times J$ where $J = \{f \in C_p(X) : f(H) \subset \{0\}\}$.

Furthermore, $H = \bigoplus_{n \in \omega} H_n$, so $C_p(H) \approx \prod_{n \in \omega} C_p(H_n) \approx (C_p(X))^\omega$. It follows from $C_p(X) \approx (C_p(X))^\omega \times J \approx (C_p(X))^\omega \times ((C_p(X))^\omega \times J)$ that the space $C_p(X)$ is linearly homeomorphic to $(C_p(X))^\omega \times C_p(X) \approx (C_p(X))^\omega$. As a consequence, $C_p(X) \times C_p(X) \approx (C_p(X))^\omega \times (C_p(X))^\omega \approx (C_p(X))^\omega \approx C_p(X)$ which shows that the spaces $X$ and $\mathbb{R}^\omega$ satisfy all assumptions of Problem 292 and therefore $X \overset{l}{\sim} \mathbb{R}^\omega$.

**V.296.** *Let $K$ be an uncountable metrizable compact space. Prove that $K$ is $l$-equivalent to $K \oplus E$ for any metrizable zero-dimensional compact space $E$.*

**Solution.** The expression $X \simeq Y$ says that the spaces $X$ and $Y$ are homeomorphic. We denote by $L \approx M$ the fact that the linear topological spaces $L$ and $M$ are linearly homeomorphic.

*Fact 1.* For any infinite compact space $X$ let $\Omega(X)$ be the one-point compactification of the space $X \times \omega$. Then

(1) $w(X) = w(\Omega(X))$; in particular, if $X$ is second countable then $\Omega(X)$ is also second countable;

(2) if $X$ is zero-dimensional then $\Omega(X)$ is also zero-dimensional;

(3) $\Omega(X) \oplus X \simeq \Omega(X)$.

*Proof.* Let $\kappa = nw(X)$; since $X$ embeds in $\Omega(X)$, we have $nw(\Omega(X)) \geq \kappa$. Let $a \in \Omega(X)$ be the point which "compactifies" $X \times \omega$, i.e., $a \notin X \times \omega$. Since network weight is countably additive and $nw(X \times \{n\}) = \kappa$ for each $n \in \omega$, it follows from $\Omega(X) = \bigcup \{X \times \{n\} : n \in \omega\} \cup \{a\}$ that $nw(\Omega(X)) \leq \kappa$ and hence we have the equality $nw(\Omega(X)) = nw(X)$. Applying Fact 4 of S.307 we convince ourselves that $w(\Omega(X)) = nw(\Omega(X)) = nw(X) = w(X)$, so (1) is proved.

Assume that the space $X$ is zero-dimensional and let $F_n = X \times \{n\}$; the set $U_n = \{a\} \cup (\bigcup \{F_i : i \geq n\})$ is an open neighborhood of $a$ for every $n \in \omega$. By the definition of the topology of $\Omega(X)$ at the point $a$, the family $\mathcal{U} = \{U_n : n \in \omega\}$ is a local base at $a$ in the space $\Omega(X)$; it is evident that every element of $\mathcal{U}$ is clopen in $\Omega(X)$. Given any $n \in \omega$ and $x \in F_n$ choose a family $\mathcal{U}_x$ of clopen subsets of $F_n$ which forms a local base at $x$ in the space $F_n$. Since the set $F_n$ is clopen in $\Omega(X)$, the family $\mathcal{U}_x$ is also a local clopen base of $\Omega(X)$ at the point $x$. This shows that $\Omega(X)$ is zero-dimensional, i.e., (2) is proved.

We can consider that $\Omega(X) \oplus X = \Omega(X) \cup G$ where $G$ is a homeomorphic copy of $X$ while the sets $\Omega(X)$ and $G$ are disjoint and clopen in $\Omega(X) \oplus X$. Since every $F_n$ is a homeomorphic copy of $X$, we can fix a homeomorphism $q_n : F_n \to F_{n+1}$ for any $n \in \omega$. Fix a homeomorphism $q : G \to F_0$ and define a map $\varphi : \Omega(X) \oplus X \to \Omega(X)$ by the equalities $\varphi(a) = a$, $\varphi|G = q$ and $\varphi|F_n = q_n$ for all $n \in \omega$. It is clear that $\varphi$ is a bijection which is continuous at all points of $(X \times \omega) \cup G$. Observing that $\varphi(U_n) \subset U_n$ for all $n \in \omega$, we verify continuity of $\varphi$ at the point $a$. This shows that $\varphi$ is a condensation and hence homeomorphism (see TFS-123), so we established (3) and finished the proof of Fact 1.

Returning to our solution observe first that there is a closed $F \subset K$ which is homeomorphic to the Cantor set $\mathbb{K}$ (see SFFS-353). The space $H = \Omega(E)$ is second countable and zero-dimensional by Fact 1, so we can apply Fact 2 of U.003 to find a closed set $D \subset F$ with $D \simeq \Omega(E)$. It follows from CFS-469 and CFS-448 that there is a linear topological space $I$ such that $C_p(K) \approx C_p(D) \times I$. Fact 1 implies that $D \oplus E \simeq D$, so $C_p(D) \approx C_p(D) \times C_p(E)$ and hence the relevant substitution of $C_p(D)$ gives us the formula $C_p(K) \approx C_p(E) \times (C_p(D) \times I) \approx C_p(E) \times C_p(K)$. Since $C_p(K \oplus E) \approx C_p(K) \times C_p(E)$, we conclude that $K \overset{l}{\sim} K \oplus E$.

**V.297.** *Prove that a compact space $X$ is $l$-equivalent to the Cantor set $\mathbb{K}$ if and only if $X$ is metrizable, zero-dimensional and uncountable. As a consequence, any two zero-dimensional metrizable uncountable compact spaces are $l$-equivalent.*

**Solution.** The expression $X \simeq Y$ says that the spaces $X$ and $Y$ are homeomorphic. We denote by $L \approx M$ the fact that the linear topological spaces $L$ and $M$ are linearly homeomorphic.

If $X \overset{l}{\sim} \mathbb{K}$ then $w(X) = nw(X) = nw(\mathbb{K}) = w(\mathbb{K}) = \omega$ (see Problem 001 and Fact 4 of S.307), so $X$ is second countable and hence metrizable. It follows from Problem 180 that $\dim X = \dim \mathbb{K} = 0$, so $X$ is zero-dimensional. Since cardinality is even $t$-invariant, we have $|X| = |\mathbb{K}| > \omega$, i.e., $X$ is uncountable; this proves necessity.

Now assume that $X$ is a metrizable zero-dimensional uncountable compact space. By SFFS-353, the space $\mathbb{K}$ embeds in $X$; it follows from SFFS-303 that $X$ also embeds in $\mathbb{K}$. Next apply Problem 296 to see that $X \oplus X \overset{l}{\sim} X$ and $\mathbb{K} \oplus \mathbb{K} \overset{l}{\sim} \mathbb{K}$ which is the same as saying that $(C_p(X))^2 \approx C_p(X)$ and $(C_p(\mathbb{K}))^2 \approx C_p(\mathbb{K})$. Therefore we can apply Problem 292 to conclude that the spaces $X$ and $\mathbb{K}$ are $l$-equivalent; this proves sufficiency and completes our solution.

**V.298.** *Prove that a second countable space $X$ is $l$-equivalent to space $\mathbb{P}$ of the irrational numbers if and only if $X$ is non-$\sigma$-compact, zero-dimensional and Čech-complete.*

**Solution.** The expression $X \simeq Y$ says that the spaces $X$ and $Y$ are homeomorphic. We denote by $L \approx M$ the fact that the linear topological spaces $L$ and $M$ are linearly homeomorphic.

*Fact 1.* A second countable zero-dimensional space is Čech-complete if and only if it embeds in $\mathbb{P}$ as a closed subspace.

*Proof.* If $F$ is a closed subspace of the space $\mathbb{P}$ then it is Čech-complete because so is $\mathbb{P}$ (see TFS-260); this proves sufficiency. Now, if $F$ is a second countable zero-dimensional space then it embeds in $\mathbb{D}^\omega \subset \omega^\omega = \mathbb{P}$ (see SFFS-303), so we can consider that $F \subset \mathbb{P}$. If $F$ is, additionally, Čech-complete then it is a $G_\delta$-subset of $\mathbb{P}$, so fix a family $\{U_n : n \in \omega\}$ of open subsets of $\mathbb{P}$ such that $F = \bigcap_{n \in \omega} U_n$.

It is clear that every $U_n$ is Polish and zero-dimensional. If $K$ is a compact subset of $U_n$ and the interior of $K$ in $U_n$ is nonempty then the interior of $F$ in $\mathbb{P}$ is nonempty as well which is a contradiction with SFFS-347; therefore we can apply SFFS-347 to the space $U_n$ to convince ourselves that $U_n \simeq \mathbb{P}$ for any $n \in \omega$. By Fact 7 of S.271, the space $F$ embeds in $\prod_{n \in \omega} U_n \simeq \mathbb{P}^\omega$ as a closed subspace. Finally observe that $\mathbb{P}^\omega \simeq (\omega^\omega)^\omega \simeq \omega^{\omega \times \omega} \simeq \omega^\omega \simeq \mathbb{P}$, so $F$ is embeddable in $\mathbb{P}$ as a closed subspace and hence Fact 1 is proved.

Returning to our solution assume that $X$ is a second countable space and $X \overset{l}{\sim} \mathbb{P}$. Then the space $X$ is Čech-complete by Problem 288 and zero-dimensional by Problem 180. Recall that $\sigma$-compactness is even $t$-invariant property (see Problem 043); since $\mathbb{P}$ is not $\sigma$-compact, the space $X$ cannot be $\sigma$-compact either so we proved necessity.

Now assume that $X$ is a second countable zero-dimensional Čech-complete space which is not $\sigma$-compact. Fact 1 shows that $X$ embeds in $\mathbb{P}$ as a closed subspace. It follows from SFFS-352 that $\mathbb{P}$ also embeds in $X$ as a closed subspace. Since $\mathbb{P} \times \omega$ is homeomorphic to $\mathbb{P}$, we can construct a discrete family $\{F_n : n \in \omega\}$ of closed subsets of $X$ such that $F_n \simeq \mathbb{P}$ for any $n \in \omega$. Apply Fact 1 once more to find a closed set $G_n \subset F_n$ such that $G_n \simeq X$ for each $n \in \omega$. Since the set $G = \bigcup_{n\in\omega} G_n$ is closed in $X$, we can apply CFS-469 and CFS-448 to see that $C_p(X) \approx C_p(G) \times I$ where $I = \{f \in C_p(X) : f(G) = \{0\}\}$.

Recall that $G \simeq \bigoplus_{n\in\omega} G_n$ where every space $G_n$ is a homeomorphic copy of $X$; this implies that $C_p(G) \approx \prod_{n\in\omega} C_p(G_n) \approx (C_p(X))^\omega$. Therefore $C_p(X) \approx (C_p(X))^\omega \times I \approx (C_p(X))^\omega \times (C_p(X))^\omega \times I$; replacing in the last product the expression $(C_p(X))^\omega \times I$ with $C_p(X)$ we conclude that $C_p(X) \approx (C_p(X))^\omega \times C_p(X)$, i.e., $C_p(X) \approx (C_p(X))^\omega$. As an immediate consequence we obtain the formula $(C_p(X))^2 \approx (C_p(X))^\omega \times (C_p(X))^\omega \approx (C_p(X))^\omega \approx C_p(X)$. Finally observe that $\mathbb{P} \oplus \mathbb{P} \simeq \mathbb{P}$ and hence $C_p(\mathbb{P}) \approx (C_p(\mathbb{P}))^2$; this makes it possible to apply Problem 292 to the spaces $X$ and $\mathbb{P}$ to conclude that they are $l$-equivalent. This settles sufficiency and makes our solution complete.

**V.299.** *Given any $n \in \mathbb{N}$ prove that a compact set $K \subset \mathbb{R}^n$ is $l$-equivalent to $\mathbb{I}^n$ if and only if $\mathbb{I}^n$ embeds in $K$. Deduce from this fact that $K \overset{l}{\sim} \mathbb{I}^n$ if and only if $\dim K = n$.*

**Solution.** The expression $X \simeq Y$ says that the spaces $X$ and $Y$ are homeomorphic. We denote by $L \approx M$ the fact that the linear topological spaces $L$ and $M$ are linearly homeomorphic. For any infinite compact space $X$ let $\Omega(X)$ be the one-point compactification of the space $X \times \omega$. If we have a point $x = (x_1, \ldots, x_n) \in \mathbb{R}^n$ then $|x|_n = \sqrt{x_1^2 + \ldots + x_n^2}$; we denote by $\mathbf{0}$ the zero point of $\mathbb{R}^n$, i.e., $\mathbf{0} = (u_1, \ldots, u_n)$ where $u_i = 0$ for all $i \in \{1, \ldots, n\}$.

*Fact 1.* For any nonempty open set $U \subset \mathbb{R}^n$ there is an open set $V \subset U$ such that $V \simeq \mathbb{R}^n$.

*Proof.* Fix a point $a \in U$; then, for some $\varepsilon > 0$, the ball $B = \{x \in \mathbb{R}^n : |x - a|_n < \varepsilon\}$ is contained in $U$. Let $\varphi(x) = \frac{1}{\varepsilon} \cdot (x - a)$ for any $x \in \mathbb{R}^n$. It is straightforward that $\varphi : \mathbb{R}^n \to \mathbb{R}^n$ is a homeomorphism such that $\varphi(B) = E = \{x \in \mathbb{R}^n : |x|_n < 1\}$.

If $\delta = \frac{1}{\sqrt{n}}$ then $(0, \delta) \simeq \mathbb{R}$, so the set $G = (0, \delta)^n$ is homeomorphic to $\mathbb{R}^n$. If $x = (x_1, \ldots, x_n) \in G$ then $x_i^2 < \delta^2 = \frac{1}{n}$ for each $i \leq n$, so $|x|_n < \sqrt{n \cdot \frac{1}{n}} = 1$ and hence $x \in E$. This proves that $G \subset E$ and therefore $G$ is an open subset of $E$ homeomorphic to $\mathbb{R}^n$; consequently, $\varphi^{-1}(G) \subset \varphi^{-1}(E) = B \subset U$ is an open subset of $U$ homeomorphic to $\mathbb{R}^n$, so Fact 1 is proved.

*Fact 2.* If a compact space $X$ embeds in $\mathbb{R}^n$ then $\Omega(X)$ also embeds in $\mathbb{R}^n$.

*Proof.* Let $\varepsilon_i = 2^{-i}$ and $B_i = \{x \in \mathbb{R}^n : |x|_n < \varepsilon_i\}$ for all $i \in \omega$. It is evident that $U_i = B_i \setminus \overline{B}_{i+1}$ is a nonempty open subset of $\mathbb{R}^n$, so we can use Fact 1 to find a set $X_i \subset U_i$ with $X_i \simeq X$ for every $i \in \omega$. If $Y = \bigcup_{i \in \omega} X_i$ then $U_i \cap Y = X_i$ for each $i \in \omega$; the family $\{U_i : i \in \omega\}$ being disjoint, the collection $\{X_i : i \in \omega\}$ is also disjoint and consists of clopen subsets of $Y$. Therefore $Y$ is homeomorphic to $\bigoplus_{i \in \omega} X_i \simeq X \times \omega$.

If $G \in \tau(\mathbf{0}, \mathbb{R}^n)$ then there exists $\delta > 0$ such that $W = \{x \in \mathbb{R}^n : |x|_n < \delta\} \subset G$. Choose $m \in \omega$ with $2^{-m} < \delta$; then $U_i \subset B_i \subset W \subset G$ for all $i \geq m$; consequently, $X_i \subset U$ for all $i \geq m$. This shows that any open set $G \ni \mathbf{0}$ contains all but finitely many sets $X_i$; an easy consequence is that $\{\mathbf{0}\} \cup Y$ is a compact space so $\{\mathbf{0}\} \cup Y$ is homeomorphic to $\Omega(X)$ and hence Fact 2 is proved.

Returning to our solution assume that $K \overset{l}{\sim} \mathbb{I}^n$; then $\dim K = \dim(\mathbb{I}^n) = n$ (see Problems 180 and 159). It follows from Problem 160 that the interior $U$ of the set $K$ is nonempty, so $\mathbb{R}^n$ embeds in $U$ (see Fact 1) and hence $\mathbb{I}^n$ embeds in $U \subset K$. This proves necessity.

Now assume that $G \subset K$ is a subset homeomorphic to $\mathbb{I}^n$; since the interior of $\mathbb{I}^n$ in $\mathbb{R}^n$ is nonempty, we can apply Fact 1 again to see that $\mathbb{R}^n$ embeds in $\mathbb{I}^n$ and hence in $G$. This, together with Fact 2, shows that we can choose a set $H \subset G$ with $H \simeq \Omega(K)$.

It follows from CFS-469 and CFS-448 that there is a linear topological space $I$ such that $C_p(K) \approx C_p(H) \times I$. Fact 1 of V.296 implies that $H \oplus K \simeq H$, so $C_p(H)$ is linearly homeomorphic to $C_p(H) \times C_p(K)$ and hence the relevant substitution of $C_p(H)$ gives us the formula $C_p(K) \approx C_p(K) \times (C_p(H) \times I) \approx C_p(K) \times C_p(K)$.

Note that $\mathbb{I} \overset{l}{\sim} \mathbb{I} \oplus \mathbb{I}$ by Problem 266, so if $n = 1$ then $\mathbb{I}^n \overset{l}{\sim} \mathbb{I}^n \oplus \mathbb{I}^n$. If $n > 1$ then applying Problem 272 we can see that $\mathbb{I}^n = \mathbb{I} \times \mathbb{I}^{n-1}$ is $l$-equivalent to $(\mathbb{I} \oplus \mathbb{I}) \times \mathbb{I}^{n-1} \simeq \mathbb{I}^n \oplus \mathbb{I}^n$. As a consequence, $(C_p(\mathbb{I}^n))^2 \approx C_p(\mathbb{I}^n)$, so Problem 292 is applicable to conclude that the spaces $K$ and $\mathbb{I}^n$ are $l$-equivalent; this settles sufficiency.

Finally, if $\mathbb{I}^n$ embeds in $K$ then $n = \dim(\mathbb{I}^n) \leq \dim K$ (see Problems 155 and 159); it follows from $K \subset \mathbb{R}^n$ that $\dim K \leq \dim(\mathbb{R}^n) = n$, so $\dim K = n$. Now, if $\dim K = n$ then the interior $U$ of the set $K$ is nonempty (see Problem 160), so we can apply Fact 1 to convince ourselves that $\mathbb{R}^n$, and hence $\mathbb{I}^n$, embeds in $K$. Therefore the condition $\dim K = n$ is equivalent to $K \overset{l}{\sim} \mathbb{I}^n$, so our solution is complete.

**V.300.** *Prove that a space $X$ is $l$-equivalent to $\mathbb{I}^\omega$ if and only if $X$ is compact, metrizable and $\mathbb{I}^\omega$ embeds in $X$.*

**Solution.** The expression $Y \simeq Z$ says that the spaces $Y$ and $Z$ are homeomorphic. We denote by $L \approx M$ the fact that the linear topological spaces $L$ and $M$ are linearly homeomorphic. For any infinite compact space $K$ let $\Omega(K)$ be the one-point compactification of the space $K \times \omega$.

Suppose that $X$ is a metrizable compact space such that a closed set $G \subset X$ is homeomorphic to $\mathbb{I}^\omega$. The space $\Omega(X)$ being compact and metrizable (see Fact 1 of V.296), there is a set $H \subset G$ which is homeomorphic to $\Omega(X)$.

It follows from CFS-469 and CFS-448 that there is a linear topological space $I$ such that $C_p(X) \approx C_p(H) \times I$. Apply Fact 1 of V.296 again to see that $H \oplus X$ is homeomorphic to $H$, so $C_p(H) \approx C_p(H) \times C_p(X)$ and hence the relevant substitution of $C_p(H)$ gives us the formula $C_p(X) \approx C_p(X) \times (C_p(H) \times I) \approx C_p(X) \times C_p(X)$.

Note that $\mathbb{I} \overset{l}{\sim} \mathbb{I} \oplus \mathbb{I}$ by Problem 266, so it follows from Problem 272 that $\mathbb{I} \times \mathbb{I}^\omega \overset{l}{\sim} (\mathbb{I} \oplus \mathbb{I}) \times \mathbb{I}^\omega$ which, together with $\mathbb{I} \times \mathbb{I}^\omega \simeq \mathbb{I}^\omega$ and $(\mathbb{I} \oplus \mathbb{I}) \times \mathbb{I}^\omega \simeq \mathbb{I}^\omega \oplus \mathbb{I}^\omega$ shows that $\mathbb{I}^\omega \overset{l}{\sim} \mathbb{I}^\omega \oplus \mathbb{I}^\omega$. As a consequence, $(C_p(\mathbb{I}^\omega))^2 \approx C_p(\mathbb{I}^\omega)$, so Problem 292 is applicable to conclude that the spaces $X$ and $\mathbb{I}^\omega$ are $l$-equivalent; this proves sufficiency.

Next assume that $X \overset{l}{\sim} \mathbb{I}^\omega$; then $X$ is compact by Problem 138. Furthermore, we have $nw(X) = nw(\mathbb{I}^\omega) \leq \omega$ (see Problem 001) and hence $w(X) \leq \omega$ by Fact 4 of S.307 so $X$ is metrizable.

Fix a linear homeomorphism $\varphi : C_p(X) \to C_p(\mathbb{I}^\omega)$ and let $\varphi^*(a) = a \circ \varphi$ for any $a \in L_p(\mathbb{I}^\omega)$; it is an easy consequence of Fact 4 of S.489 that $\varphi^* : L_p(\mathbb{I}^\omega) \to L_p(X)$ is a linear isomorphism. The set $H = \varphi^*(\mathbb{I}^\omega)$ is a Hamel basis in the space $L_p(X)$ such that $H \simeq \mathbb{I}^\omega$ and every $f \in C_p(H)$ extends to a continuous linear functional on $L_p(X)$. Given nonempty sets $A_1, \ldots, A_n \subset L_p(X)$ and $O_1, \ldots, O_n \subset \mathbb{R}$ let $O_1 A_1 + \ldots + O_n A_n = \{\lambda_1 x_1 + \ldots + \lambda_n x_n : \lambda_i \in O_i \text{ and } x_i \in A_i \text{ for all } i \leq n\}$.

For any $a \in H$ we can find a uniquely determined $n \in \mathbb{N}$ and distinct points $x_1, \ldots, x_n \in X$ such that $a = \lambda_1 x_1 + \ldots + \lambda_n x_n$ for some $\lambda_1, \ldots, \lambda_n \in \mathbb{R} \setminus \{0\}$; let $\xi_X(a) = \{x_1, \ldots, x_n\}$ and $l_X(a) = n$. Analogously, for any point $x \in X$ it is possible to find a uniquely determined $m \in \mathbb{N}$ and distinct points $a_1, \ldots, a_m \in H$ such that $x = \mu_1 a_1 + \ldots + \mu_m a_m$ for some $\mu_1, \ldots, \mu_m \in \mathbb{R} \setminus \{0\}$; let $l_H(x) = m$ and $\xi_H(x) = \{a_1, \ldots, a_m\}$. It follows from Problem 279 that $\xi_X : H \to \exp(X)$ and $\xi_H : X \to \exp(H)$ are the support maps for $\varphi$ and $\varphi^{-1}$ respectively. Applying Problem 280 we infer that both $\xi_X$ and $\xi_H$ are lower semicontinuous.

The set $X_n = \{x \in X : l_H(x) \leq n\}$ is closed in $X$ for any $n \in \mathbb{N}$ by Fact 1 of U.485; by the same reason every set $H_n = \{a \in H : l_X(a) \leq n\}$ is closed in $H$. Let $Y_1 = X_1$ and $Y_n = X_n \setminus X_{n-1}$ for all $n > 1$; we will also need the sets $G_1 = H_1$ and $G_n = H_n \setminus H_{n-1}$ for all $n \geq 2$.

Since $H = \bigcup \{H_n : n \in \mathbb{N}\}$, an easy consequence of the Baire property of $H$ is that there exists a nonempty open set $U$ in the space $H$ such that $U \subset G_n$ for some $n \in \mathbb{N}$. Pick a point $a \in U$; then $\xi_X(a) = \{x_1, \ldots, x_n\}$ and we can choose a set $W_i \in \tau(x_i, X)$ for every $i \leq n$ such that the family $\{W_1, \ldots, W_n\}$ is disjoint. The set $U' = \{b \in H : \xi_X(b) \cap W_i \neq \emptyset \text{ for all } i \leq n\}$ is open in $H$ and $a \in U'$. Therefore $V = U \cap U'$ is an open neighborhood of $a$ in $H$.

Given any $b \in V$ it follows from $l_X(b) = n$ that for every $i \in \{1, \ldots, n\}$ there is a unique element $y_i \in \xi_X(b) \cap W_i$; let $p_i(b) = y_i$. This gives us a mapping $p_i : V \to W_i$ for each $i \leq n$.

Let $O = \mathbb{R} \setminus \{0\}$; if $W$ is a nonempty open subset of $W_i$ then it is easy to see that $p_i^{-1}(W) = (OW_1 + \ldots + OW_{i-1} + OW + OW_{i+1} + \ldots + OW_n) \cap V$; it follows from Fact 2 of U.485 that $p_i^{-1}(W)$ is an open subset of $V$, so the map $p_i$ is continuous for any $i \leq n$.

Let $V_0 = V$; we will inductively construct nonempty open sets $V_0, \ldots, V_n$ of the space $H$ and $k_1, \ldots, k_n \in \mathbb{N}$ such that

(1) $V_i \subset V_{i-1}$ and $p_i(V_i) \subset Y_{k_i}$ for all $i = 1, \ldots, n$.

Suppose that $j < n$ and we have $V_0, \ldots, V_j \in \tau^*(H)$ and $k_1, \ldots, k_j \in \mathbb{N}$ such that the condition (1) is satisfied for all $i = 1, \ldots, j$. The set $\overline{p_{j+1}(V_j)}$ is compact, so it has the Baire property; this, together with $\overline{p_{j+1}(V_j)} \subset \bigcup\{X_n : n \in \mathbb{N}\}$ implies that there is a nonempty open subset $E$ of the space $\overline{p_{j+1}(V_j)}$ such that $E \subset Y_{k_{j+1}}$ for some $k_{j+1} \in \mathbb{N}$. The set $E' = E \cap p_{j+1}(V_j)$ is nonempty and open in $p_{j+1}(V_j)$, so $V_{j+1} = p_{j+1}^{-1}(E') \cap V_j$ is open in $V_j$ and hence in $H$. It is straightforward that (1) now holds for all $i \leq j + 1$, so our inductive procedure gives us the promised sets $V_0, \ldots, V_n$ and numbers $k_1, \ldots, k_n \in \mathbb{N}$. As a consequence,

(2) $l_H(p_i(b)) = k_i$ for any $b \in V_n$ and $i = 1, \ldots, n$.

Fix a point $b \in V_n$ and $i \in \{1, \ldots, n\}$. Let $y_i = p_i(b)$ and consider the set $\xi_H(y_i) = \{a_1^i, \ldots, a_{k_i}^i\}$; choose a disjoint family $\{U_1^i, \ldots, U_{k_i}^i\} \subset \tau^*(H)$ such that $a_j^i \in U_j^i$ for all $j \leq k_i$. The set $Q_i = \{y \in X : \xi_H(y) \cap U_j^i \neq \emptyset$ for all $j \leq k_i\}$ is open in $X$, so $W_i' = Q_i \cap W_i$ is open in $X$ and nonempty because $y_i \in W_i'$. If $y \in W_i'$ then, for every $j \leq k_i$, there is a unique point $b_j \in U_j^i \cap \xi_H(y)$; let $q_j^i(y) = b_j$. Then $q_j^i : W_i' \to U_j^i$ is a continuous map; this is proved in the same way as we verified continuity of $p_i$.

The set $E = \bigcap\{p_i^{-1}(W_i') : i \leq n\} \cap V_n$ is open in the space $H$ and nonempty due to the fact that $b \in E$. For any $i \in \{1, \ldots, n\}$ and $j \in \{1, \ldots, k_i\}$ the set $S_j^i = \{c \in E : q_j^i(p_i(c)) = c\}$ is closed in $E$. If $c \in E$ and $\xi_X(c) = \{z_1, \ldots, z_n\}$ then there are $\lambda_1, \ldots, \lambda_n \in O$ for which $c = \sum_{i=1}^n \lambda_i z_i$. For every $i \leq n$ let $\xi_H(z_i) = \{d_j^i : j \leq k_i\}$; there are $\mu_1^i, \ldots, \mu_{k_i}^i \in O$ such that $z_i = \sum_{j=1}^{k_i} \mu_j^i d_j^i$. The equality $c = \sum_{i=1}^n \lambda_i (\sum_{j=1}^{k_i} \mu_j^i d_j^i)$ shows that $c = d_j^i$ for some $i \in \{1, \ldots, n\}$ and $j \leq k_i$. By definition of the maps $p_i$ and $q_j^i$ we have $c = q_j^i(p_i(c))$, i.e., $c \in S_j^i$.

Therefore $E \subset \bigcup\{S_j^i : i \leq n, \ j \leq k_i\}$; an immediate consequence is that we can choose $i \leq n$ and $j \leq k_i$ for which there is a set $Q \in \tau^*(H)$ with $Q \subset S_j^i$. Then $q_j^i(p_i(c)) = c$ for every $c \in Q$ which shows that the map $p_i|Q : Q \to p_i(Q)$ is a homeomorphism because $q_j^i|p_i(Q)$ is its continuous inverse.

We proved that a nonempty open set $Q$ of the space $H \simeq \mathbb{I}^\omega$ embeds in $X$. It is an easy exercise that $\mathbb{I}^\omega$ embeds in any nonempty open subspace $B \subset \mathbb{I}^\omega$, so $\mathbb{I}^\omega$ also embeds in $X$. This settles necessity and makes our solution complete.

**V.301.** *Prove that there exist l-equivalent spaces $X$ and $Y$ such that $X$ is hereditarily paracompact while $Y$ is not collectionwise normal.*

**Solution.** If $Z$ is a space and $A \subset Z$ let $\mathcal{U}_A = \tau(Z) \cup \{\{z\} : z \in Z \backslash A\}$ and denote by $\tau_A$ the topology on $Z$ generated by the family $\mathcal{U}_A$ as a subbase; denote the space $(Z, \tau_A)$ by $Z[A]$. The space $Z[A]$ is always Tychonoff and all points of $Z \backslash A$ are isolated in $Z[A]$; besides, $a \in A$ and $a \in U \in \tau_A$ implies that there is $V \in \tau(a, Z)$ with $V \subset U$ (see Fact 1 of S.293).

*Fact 1.* Any space with at most one non-isolated point is hereditarily paracompact.

*Proof.* The property of having at most one non-isolated point is hereditary, so it suffices to show that any space $Z$ with at most one non-isolated point is paracompact. Fix a point $a \in Z$ such that any $x \in Z \setminus \{a\}$ is isolated in $Z$ and take an open cover $\mathcal{U}$ of the space $Z$. If $U \in \mathcal{U}$ and $a \in U$ then $\{U\} \cup \{\{x\} : x \in Z \setminus U\}$ is a disjoint (and hence locally finite) open refinement of $\mathcal{U}$, so $Z$ is paracompact, i.e., Fact 1 is proved.

*Fact 2.* If $Z$ is a space and $A \subset Z$ is $l$-embedded in $Z$ then the set $A$ is also $l$-embedded in $Z[A]$.

*Proof.* It follow from $\tau(Z) \subset \tau(Z[A])$ that $C_p(Z) \subset C_p(Z[A])$ and it is immediate that $C_p(Z)$ is a linear subspace of $C_p(Z[A])$. There exists a continuous linear map $\mu : C_p(A) \to C_p(Z)$ such that $\mu(f)|A = f$ for each $f \in C_p(A)$. Therefore $\mu$ can be considered to be a map from $C_p(A)$ to $C_p(Z[A])$.

Let us consider that $A$ carries the topology induced from $Z$ and denote by $A'$ the set $A$ with the topology inherited from $Z[A]$. By Fact 1 of S.293, these two topologies on $A$ coincide, so we can identify $C_p(A)$ and $C_p(A')$; this implies that $\mu$ is also a continuous linear extender of continuous functions on $A'$ in the space $Z[A]$. Therefore $A$ is $l$-embedded in $Z[A]$, i.e., Fact 2 is proved.

Returning to our solution take a discrete space $D$ of cardinality $\omega_1$; then $D$ is closed and $l$-embedded in the space $Z = L_p(D)$ (see CFS-466). Take a point $a \notin Z$ and let $Y = Z[D] \oplus \{a\}$. The set $D$ is still closed and discrete in $Z[D]$; assume that there is a disjoint family $\{U_d : d \in D\}$ of open subsets of $Z[D]$ such that $d \in U_d$ for every $d \in D$. There exists $V_d \in \tau(d, Z)$ such that $d \in V_d \subset U_d$ for each $d \in D$. Therefore $\{V_d : d \in D\}$ is an uncountable disjoint family of nonempty open subsets of $L_p(D)$ which contradicts Problem 236. Thus the set $D$ witnesses that the space $Z[D]$ is not collectionwise normal, so the space $Y$ is not collectionwise normal either.

Let $P$ be the $\mathbb{R}$-quotient space obtained from $Z[D]$ by contracting the set $D$ to a point $q$. Since $D$ is $l$-embedded in $Z[D]$ by Fact 2, it follows from Problem 258 that $Y$ is $l$-equivalent to the space $X = P \oplus D$. It is straightforward that $q$ is the unique non-isolated point of $X$, so $X$ is hereditarily paracompact by Fact 1.

**V.302.** *Prove that there exist l-equivalent spaces $X$ and $Y$ such that $X$ is collectionwise normal while $Y$ is not normal.*

**Solution.** Given a space $Z$ and $F \subset Z$ we denote by $Z_F$ the $\mathbb{R}$-quotient space obtained by contracting the set $F$ to a point which will be referred to as $a_F$.

*Fact 1.* In the space $N = \omega_1 \times (\omega_1 + 1)$ consider the set $F = \{(\alpha, \omega_1) : \alpha < \omega_1\}$. If $f \in C_p(N)$ and $f(F) = \{0\}$ then there exists an ordinal $\gamma < \omega_1$ such that $\{(\alpha, \beta) : \gamma < \alpha < \omega_1 \text{ and } \gamma < \beta \leq \omega_1\} \subset f^{-1}(0)$.

*Proof.* If such an ordinal does not exist then it is easy to construct an $\omega_1$-sequence $\{q_\alpha = (\eta_\alpha, \xi_\alpha) : \alpha < \omega_1\} \subset \omega_1 \times \omega_1$ such that $f(q_\alpha) \neq 0$ and $\alpha < \ln\{\eta_\alpha, \xi_\alpha\}$ for all $\alpha < \omega_1$ while $\max\{\eta_\alpha, \xi_\alpha\} < \ln\{\eta_\beta, \xi_\beta\}$ whenever $\alpha < \beta < \omega_1$.

There exists $\varepsilon > 0$ such that, for an uncountable $A \subset \omega_1$, we have $|f(q_\alpha)| \geq \varepsilon$ for all $\alpha \in A$. It follows from TFS-314 that, for any $\alpha < \omega_1$ there is $\beta < \omega_1$ such that $f(\alpha, \delta) = 0$ for all $\delta > \beta$. This makes it possible to construct, by induction on $n \in \omega$, sequences $\{\alpha_n : n \in \omega\} \subset A$ and $\{(\mu_n, \nu_n) : n \in \omega\} \subset \omega_1 \times \omega_1$ such that $f(\mu_n, \nu_n) = 0$ and $\max\{\eta_{\alpha_n}, \xi_{\alpha_n}\} < \mu_n < \nu_n < \backslash n\{\eta_{\alpha_{n+1}}, \xi_{\alpha_{n+1}}\}$ for all $n \in \omega$.

Let $\mu = \sup\{\mu_n : n \in \omega\}$ and take any $O \in \tau((\mu, \mu), N)$; pick an ordinal $\nu < \mu$ such that $W = \{(\alpha, \beta) : \nu < \backslash n\{\alpha, \beta\}$ and $\max\{\alpha, \beta\} \leq \mu\} \subset O$. There exists $m \in \omega$ such that $\mu_m > \nu$ and hence $(\mu_i, \nu_i) \in W$ for all $i \geq m$; since also $(\eta_{\alpha_i}, \xi_{\alpha_i}) \in W$ for all $i \geq m + 1$, both sequences $S = \{(\eta_{\alpha_n}, \xi_{\alpha_n}) : n \in \omega\}$ and $T = \{(\mu_n, \nu_n) : n \in \omega\}$ converge to the point $u = (\mu, \mu)$. It follows from $f(T) = \{0\}$ that $f(u) = 0$. However, $f(S) \subset \mathbb{R}\backslash(-\varepsilon, \varepsilon)$, so $|f(u)| \geq \varepsilon$; this contradiction shows that the promised ordinal $\gamma$ exists and hence Fact 1 is proved.

**Fact 2.**  The space $T = ((\omega_1 + 1) \times (\omega_1 + 1))\backslash\{(\omega_1, \omega_1)\}$ is not normal.

*Proof.*  The sets $F = \{(\alpha, \omega_1) : \alpha < \omega_1\}$ and $G = \{(\alpha, \alpha) : \alpha < \omega_1\}$ are disjoint and closed in $T$. If $T$ is normal then there exists a continuous function $f : T \to \mathbb{R}$ such that $f(F) = \{0\}$ and $f(G) = \{1\}$. If $N = \omega_1 \times (\omega_1 + 1)$ then $f|N$ is continuous and identically zero on $F$, so we can apply Fact 1 to find an ordinal $\gamma < \omega_1$ such that $f(\alpha, \beta) = 0$ whenever $\gamma < \backslash n\{\alpha, \beta\}$ and $\max\{\alpha, \beta\} < \omega_1$. In particular, $f(\gamma + 1, \gamma + 1) = 0$ which contradicts $(\gamma + 1, \gamma + 1) \in G \subset f^{-1}(1)$. Therefore the space $T$ cannot be normal, i.e., Fact 2 is proved.

**Fact 3.**  For any ordinal $\xi$ let $f(\alpha, \beta) = \backslash n\{\alpha, \beta\}$ for every $(\alpha, \beta) \in \xi \times \xi$. Then the map $f : \xi \times \xi \to \xi$ is continuous.

*Proof.*  Fix an arbitrary point $(\mu, \nu) \in \xi \times \xi$ and let $\gamma = f(\mu, \nu)$. If $\gamma' < \gamma$ then $\gamma' < \mu$ and $\gamma' < \nu$, so the set $U = \{(\alpha, \beta) : \gamma' < \alpha \leq \mu$ and $\gamma' < \beta \leq \nu\}$ is an open neighborhood of $(\mu, \nu)$ in $\xi \times \xi$. If $(\alpha, \beta) \in U$ then $\gamma' < \backslash n\{\alpha, \beta\} \leq \backslash n\{\mu, \nu\}$, i.e., $\gamma' < f(\alpha, \beta) \leq \gamma$; this shows that $f$ is continuous at the point $(\alpha, \beta)$ for any $(\alpha, \beta) \in \xi \times \xi$, so Fact 3 is proved.

**Fact 4.**  In the space $(\omega_1 + 1) \times (\omega_1 + 1)$ consider the sets $N = \omega_1 \times (\omega_1 + 1)$ and $F = \{(\alpha, \omega_1) : \alpha < \omega_1\}$. Let $G = (\omega_1 + 1) \times \{\omega_1\}$ and fix any set $P \subset \omega_1$. The set $F$ is closed in the space $Z = N \cup (\{\omega_1\} \times P)$ while $G$ is closed in the space $T = Z \cup \{(\omega_1, \omega_1)\}$. For the $\mathbb{R}$-quotient spaces $Z_F$ and $T_G$ define a map $\varphi : Z_F \to T_G$ by letting $\varphi(a_F) = a_G$ and $\varphi(x) = x$ for every $x \in Z\backslash F$. Then $\varphi$ is a homeomorphism; in particular, the spaces $Z_F$ and $T_G$ are homeomorphic.

*Proof.*  It is immediate that $\varphi$ is a bijection. It follows from Problem 252 that the map $\varphi|(Z_F\backslash\{a_F\}) : Z_F\backslash\{a_F\} \to T_G\backslash\{a_G\}$ is a homeomorphism, so it suffices to show $\varphi$ is continuous at the point $a_F$ and $\varphi^{-1}$ is continuous at $a_G$.

Suppose that $W \in \tau(a_G, T_G)$; by Problem 253, there is a cozero set $U$ and a zero-set $E$ in the space $T$ such that $G \subset E \subset U$ and $\{a_G\} \cup (U\backslash G) \subset W$. Then $F \subset E' = E \cap Z \subset U' = U \cap Z$ while $E'$ is a cozero set and $U'$ is a zero-set in $Z$, so we can apply Problem 253 again to see that $W' = \{a_F\} \cup (U'\backslash F)$ is an open neighborhood of the point $a_F$ in the space $Z_F$. It is clear that $\varphi(W') \subset W$, so $\varphi$ is continuous at the point $a_F$.

Now assume that $W' \in \tau(a_F, Z_F)$; by Problem 253, we can find a cozero set $U'$ and a zero-set $E'$ in the space $Z$ such that $F \subset E' \subset U'$ and $\{a_F\} \cup (U' \backslash F) \subset W'$. By Fact 1 of V.140, there is a function $f \in C_p(Z, [0, 1])$ such that $f(E') = \{0\}$ and $f(Z \backslash U') \subset \{1\}$. As a consequence, $f(F) = \{0\}$, so we can apply Fact 1 to find an ordinal $\delta < \omega_1$ such that the set $H = \{(\alpha, \beta) : \delta < \alpha < \omega_1 \text{ and } \delta < \beta < \omega_1\}$ is contained in $f^{-1}(0)$.

Thus $f(\omega_1, \alpha) = 0$ for every ordinal $\alpha > \delta$ with $\alpha \in P$, so the set $f^{-1}(0)$ contains $H' = H \cup \{(\omega_1, \alpha) : \alpha \in P \text{ and } \alpha > \delta\}$. Since $H' \cup \{(\omega_1, \omega_1)\}$ is an open neighborhood of the point $v = (\omega_1, \omega_1)$, letting $g(v) = 0$ and $g|Z = f$ we obtain a function $g : T \to \mathbb{R}$ which is identically zero on a neighborhood of $v$, so $g \in C_p(T)$. Besides, $G \subset E = g^{-1}(0)$ and $E \subset U = g^{-1}([0, 1))$. Since $E$ is a zero-set and $U$ is a cozero set in the space $T$, it follows from Problem 253 that $W = (U \backslash G) \cup \{a_G\}$ is an open neighborhood of $a_G$ in the space $T_G$. It is easy to check that $\varphi^{-1}(W) \subset W'$, so $\varphi^{-1}$ is also continuous and hence Fact 4 is proved.

Returning to our solution consider the subspace $Z = (\omega_1 + 1)^2 \backslash \{(\omega_1, \omega_1)\}$ of the space $(\omega_1 + 1)^2$. The set $F = \{(\alpha, \omega_1) : \alpha < \omega_1\}$ is closed in $Z$; choose a point $a \notin Z$ and let $r(\alpha, \beta) = (\backslash n\{\alpha, \beta\}, \omega_1)$ for every $(\alpha, \beta) \in Z$. It follows from Fact 3 that the map $r : Z \to F$ is a retraction, so we can apply Problem 258 to convince ourselves that $Y = Z \oplus \{a\}$ is $l$-equivalent to the space $X = Z_F \oplus F$. Observe that $Y$ is not normal because $Z$ is a closed subspace of $Y$ and $Z$ not normal by Fact 2.

If $G = \{(\alpha, \omega_1) : \alpha \le \omega_1\}$ then $G$ is a closed subset of the space $T = (\omega_1 + 1)^2$. Applying Fact 4 to the spaces $Z$ and $T$ we conclude that $Z_F$ is homeomorphic to $T_G$; the space $T$ being compact, $T_G$ is also compact which shows that $Z_F$ is compact and hence collectionwise normal. The space $F$ is homeomorphic to the ordinal $\omega_1$, so it is collectionwise normal by Fact 3, Fact 4 and Fact 5 of S.232. Therefore $X$ is also collectionwise normal.

**V.303.** *Prove that there exist l-equivalent spaces $X$ and $Y$ such that $X$ is hereditarily normal while $Y$ is not normal.*

**Solution.** For any space $Z$ and $F \subset Z$ we denote by $Z_F$ the $\mathbb{R}$-quotient space obtained by contracting the set $F$ to a point which will be referred to as $a_F$. Given sets $A, B \subset Z$ say that they are *separated (in $Z$)* if $\overline{A} \cap B = \emptyset = \overline{B} \cap A$. Call the sets $A$ and $B$ *open-separated (in $Z$)* if there exist disjoint sets $U, V \in \tau(Z)$ such that $A \subset U$ and $B \subset V$.

If $Z$ is a space and $A \subset Z$ let $\mathcal{U}_A = \tau(Z) \cup \{\{z\} : z \in Z \backslash A\}$ and consider the topology $\tau_A$ on $Z$ generated by the family $\mathcal{U}_A$ as a subbase; denote the space $(Z, \tau_A)$ by $Z[A]$. The space $Z[A]$ is always Tychonoff and all points of $Z \backslash A$ are isolated in $Z[A]$; besides, $a \in A$ and $a \in U \in \tau_A$ implies that there is $V \in \tau(a, Z)$ with $V \subset U$ (see Fact 1 of S.293).

*Fact 1.* Given a space $Z$ suppose that $K$ is a compact subspace of $Z$ and $Z \backslash K$ is normal. Then $Z$ is also normal.

*Proof.* Assume that $F$ and $G$ are closed disjoint subsets of $Z$. The sets $F_0 = F \cap K$ and $G_0 = G \cap K$ are compact, so they are open-separated in $Z$ by Fact 4 of T.309;

pick disjoint sets $U_0, V_0 \in \tau(Z)$ such that $F_0 \subset U_0$ and $G_0 \subset V_0$. The sets $F_0$ and $G \backslash V_0$ are also open-separated, so we can find disjoint $U_1, W_1 \in \tau(Z)$ for which $F_0 \subset U_1$ and $G \backslash V_0 \subset W_1$. If $U = U_0 \cap U_1$ then $F_0 \subset U$ and $\overline{U} \cap G = \emptyset$.

The set $F_1 = F \backslash U$ is closed in $Z$ and disjoint from the set $G_0$, so we can apply Fact 4 of T.309 again to find disjoint sets $E_0, H_0 \in \tau(Z)$ such that $F_1 \subset E_0$ and $G_0 \subset H_0$. Furthermore, $F_1$ and $G \backslash H_0$ are disjoint closed subsets of $Z \backslash K$, so they are open-separated in $Z \backslash K$ and hence in $Z$. Choose disjoint sets $E_1, H_1 \in \tau(Z)$ for which $F_1 \subset E_1$ and $G \backslash H_0 \subset H_1$. It is immediate that $E = E_0 \cap E_1$ is an open neighborhood of $F_1$ such that $\overline{E} \cap G = \emptyset$. Therefore we found open subsets $U$ and $E$ of the space $Z$ such that $F \subset U \cup E$ and $\overline{U} \cup \overline{E}$ does not meet $G$. An analogous reasoning shows that we can find open sets $H$ and $W$ in the space $Z$ such that $G \subset H \cup W$ and $(\overline{H} \cup \overline{W}) \cap F = \emptyset$. It is straightforward that the sets $O = (U \cup E) \backslash (\overline{H} \cup \overline{W})$ and $Q = (H \cup W) \backslash (\overline{U} \cup \overline{E})$ are disjoint open neighborhoods of $F$ and $G$ respectively, so $Z$ is normal and hence Fact 1 is proved.

*Fact 2.* Suppose that $Z \times A(\kappa)$ is normal for any infinite cardinal $\kappa$. Then $Z$ is collectionwise normal.

*Proof.* The space $Z$ embeds in $Z \times A(\omega)$, so it is normal. Fix a discrete family $\mathcal{F} = \{F_t : t \in T\}$ of closed subsets of $Z$; by normality of the space $Z$ we can assume, without loss of generality, that the set $T$ is infinite. Take a point $a \notin T$ and consider the topology $\tau$ on the set $A = \{a\} \cup T$ such that $\{t\} \in \tau$ for every $t \in T$ and $a \in U \in \tau$ if and only if $a \in U$ and $T \backslash U$ if finite. It is easy to see that $A$ is homeomorphic to $A(\kappa)$ for $\kappa = |T|$. Therefore the space $Z \times A$ is normal.

The set $G = Z \times \{a\}$ is, evidently, closed in $Z \times A$ and disjoint from the set $F = \bigcup \{F_t \times \{t\} : t \in T\}$. If $w = (z, t) \in (Z \times A) \backslash F$ for some $t \in T$ then $(Z \backslash F_t) \times \{t\}$ is an open neighborhood of $w$ that does not meet $F$. If $w = (z, a)$ then there exists $O \in \tau(z, Z)$ such that $O$ meets at most one element of $\mathcal{F}$, i.e., there exists $t \in T$ such that $U \cap F_s = \emptyset$ whenever $s \neq t$. Then $O \times (A \backslash \{t\})$ is an open neighborhood of $w$ with $O \cap F = \emptyset$; this proves that the set $F$ is closed in $Z \times A$.

By normality of $Z \times A$ there exist disjoint open subsets $U$ and $V$ of the space $Z \times A$ such that $F \subset U$ and $G \subset V$. For each $t \in T$ choose a set $V_t \in \tau(F_t, Z)$ such that $V_t \times \{t\} \subset V$. Given a point $z \in Z$ there is a finite $K \subset T$ and $O \in \tau(z, Z)$ such that $O \times (A \backslash K) \subset U$. If the set $O$ meets infinitely many elements of the family $\mathcal{V} = \{V_t : t \in T\}$ then there is $t \in T \backslash K$ with $O \cap V_t \neq \emptyset$ and hence $(O \times (A \backslash K)) \cap (V_t \times \{t\}) \neq \emptyset$; this implies $U \cap V \neq \emptyset$, so we obtained a contradiction.

Therefore $\mathcal{V}$ is locally finite; for every $t \in T$ choose a set $W_t \in \tau(F_t, Z)$ such that $\overline{W}_t \subset V_t$ and $\overline{W}_t \cap F_s = \emptyset$ for all $s \in T \backslash \{t\}$. The family $\{W_t : t \in T\}$ is also locally finite, so the set $H_t = \bigcup \{\overline{W}_s : s \in T \backslash \{t\}\}$ is closed in $Z$ and does not meet $F_t$ for each $t \in T$. It is straightforward to check that the family $\{(Z \backslash H_t) \cap W_t : t \in T\}$ is disjoint and $F_t \subset (Z \backslash H_t) \cap W_t$ for each $t \in T$. It follows from Fact 1 of S.302 that the space $Z$ is collectionwise normal, so Fact 2 is proved.

Returning to our solution take a discrete space $D$ of cardinality $\omega_1$; then $D$ is closed and $l$-embedded in the space $Z = L_p(D)$ (see CFS-466). The set $D$ is still closed and discrete in $Z[D]$; assume that there is a disjoint family $\{U_d : d \in D\}$ of open subsets of $Z[D]$ such that $d \in U_d$ for every $d \in D$. There exists $V_d \in \tau(d, Z)$ such that $d \in V_d \subset U_d$ for each $d \in D$. Therefore $\{V_d : d \in D\}$ is an uncountable disjoint family of nonempty open subsets of $L_p(D)$ which contradicts Problem 236. Thus the set $D$ witnesses that the space $Z[D]$ is not collectionwise normal. By Fact 2 there exists a space $K$ homeomorphic to $A(\kappa)$ for some cardinal $\kappa$ such that $Z[D] \times K$ is not normal. Denote by $a$ the unique non-isolated point of $K$ and choose a point $b \notin T = Z[D] \times K$. Then the space $Y = T \oplus \{b\}$ is not normal.

Assume that $A$ and $B$ are separated subsets of $Z[D]$ and choose a function $f :$ $D \to \mathbb{R}$ such that $f(\overline{A} \cap D) \subset \{0\}$ and $f(D \setminus (\overline{A} \cap D)) \subset \{1\}$. Fix a continuous function $g : L_p(D) \to \mathbb{R}$ with $g|D = f$; then $g$ is also continuous on $Z[D]$, so the sets $U' = g^{-1}((-\frac{1}{3}, \frac{1}{3}))$ and $V' = g^{-1}((\frac{2}{3}, \frac{4}{3}))$ are open in $Z[D]$ and disjoint. Therefore the sets $U = U' \setminus \overline{B}$ and $V = V' \setminus \overline{A}$ are closed and disjoint as well. It is immediate that $A \cup U$ and $B \cup V$ are disjoint open neighborhoods of the sets $A$ and $B$ respectively in the space $Z[D]$. We proved that any two separated subsets of $Z[D]$ are open-separated; thus $Z[D]$ is hereditarily normal by Fact 3 of U.193.

In the space $T$, the set $F = Z \times \{a\}$ is a retract of $T$, so we can apply Problem 258 to see that the spaces $Y$ and $X = T_F \oplus F$ are $l$-equivalent. The space $F$ is hereditarily normal being homeomorphic to $Z[D]$. Take an arbitrary subspace $E$ of the space $T_F$; it follows from Problem 252 that the space $T_F \setminus \{a_F\}$ is homeomorphic to $Z[D] \times (K \setminus \{a\})$, so it is hereditarily normal because $K \setminus \{a\}$ is a discrete space. Therefore the space $E \setminus \{a_F\} \subset T_F \setminus \{a_F\}$ is normal and hence we can apply Fact 1 to conclude that $E$ is normal. This proves that $T_F$ is hereditarily normal, so $X$ is a hereditarily normal space which is $l$-equivalent to a non-normal space $Y$.

**V.304.** *Prove that $\pi$-weight is not preserved by $l$-equivalence neither in the class of compact spaces nor in the class of countable spaces.*

**Solution.** For an arbitrary space $Z$ and a closed set $F \subset Z$ we denote by $Z_F$ the $\mathbb{R}$-quotient space obtained by contracting the set $F$ to a point. For any set $A \subset Z$ let $\mathcal{U}_A = \tau(Z) \cup \{\{z\} : z \in Z \setminus A\}$ and consider the topology $\tau_A$ on $Z$ generated by the family $\mathcal{U}_A$ as a subbase; denote the space $(Z, \tau_A)$ by $Z[A]$. The space $Z[A]$ is always Tychonoff and all points of $Z \setminus A$ are isolated in $Z[A]$; besides, $a \in A$ and $a \in U \in \tau_A$ implies that there is $V \in \tau(a, Z)$ with $V \subset U$ (see Fact 1 of S.293).

*Fact 1.* Suppose that $A$ is a retract of a space $Z$. Then $A$ is also a retract of the space $Z[A]$.

*Proof.* Fix a retraction $r : Z \to A$. The space $Z[A]$ induces the same topology on the set $A$, so the map $r : Z[A] \to A$ is also continuous because the topology of $Z[A]$ is stronger than the topology of $Z$. Thus $r$ is a retraction of $Z[A]$ onto $A$ and hence Fact 1 is proved.

*Fact 2.* If $A$ is a retract of a space $Z$ then $F = \text{cl}_{\beta Z}(A)$ is a retract of $\beta Z$.

*Proof.* Fix a retraction $r : Z \to A$; the space $F$ being compact, there exists a continuous map $s : \beta Z \to F$ (see TFS-258) such that $s|Z = r$. Let $id_F : F \to F$

be the identity map, i.e., $id_F(x) = x$ for all $x \in F$. Observe that $s|A = r|A = id_F|A$; the set $A$ being dense in $F$, we can apply Fact 0 of S.351 to see that $s|F = id_F$. This shows that $s$ is a retraction, so Fact 2 is proved.

Returning to our solution recall that the space $\mathbb{R}^{\omega_1}$ is separable by TFS-108; fix a countable dense subspace $A \subset \mathbb{R}^{\omega_1}$. Since $\mathbb{R}^{\omega_1}$ is homeomorphic to $C_p(D)$ for a discrete space $D$ with $|D| = \omega_1$, we can apply TFS-169 and Fact 1 of T.158 to see that $\pi w(\mathbb{R}^{\omega_1}) > \omega$. Next apply Fact 1 of T.187 to convince ourselves that $\pi w(A) > \omega$.

Take a space $S$ homeomorphic to the convergent sequence $\omega + 1$ and let $a$ be the unique non-isolated point of $S$. In the space $Z = A \times S$, the set $F = A \times \{a\}$ is a retract of $Z$. Therefore the set $F$ is also a retract of the space $T = Z[F]$ by Fact 1; observe that $Z \setminus F$ is a dense set of isolated points of $T$. Choose a point $b \notin T$ and let $X = T \oplus \{b\}$. It is clear that $X$ is a countable space in which the isolated points form a dense set; an immediate consequence is that $\pi w(X) = \omega$.

It follows from Problem 258 that $X$ is $l$-equivalent to the space $Y = T_F \oplus F$. The set $F$ is homeomorphic to $A$, so $A$ embeds in $Y$ as a clopen subspace; as a consequence, $\pi w(Y) \geq \pi w(A) > \omega$. Therefore the spaces $X$ and $Y$ witness that $\pi$-weight is not $l$-invariant in the class of countable spaces.

The space $K = \beta T$ is compact and it follows from Fact 2 that $G = cl_{\beta T}(F)$ is a retract of $K$. The space $K$ still has a dense set of isolated points, so $\pi w(K) \leq \omega$. Take a point $c \notin K$ and let $L = K \oplus \{c\}$. It is clear that $\pi w(L) \leq \omega$. Apply Problem 258 again to see that $L$ is $l$-equivalent to the space $M = K_G \oplus G$. Since $F$ is homeomorphic to a dense subspace of $G$, we have $\pi w(G) = \pi w(F) > \omega$ (see Fact 1 of T.187). Now, $G$ is homeomorphic to a clopen subspace of $M$, so $\pi$-weight of $M$ has to be uncountable. Therefore, $L$ and $M$ constitute an example which shows that $\pi$-weight is not $l$-invariant in the class of compact spaces.

**V.305.** *Give an example of $l$-equivalent spaces $X$ and $Y$ with $ext(X) \neq ext(Y)$.*

**Solution.** For any space $Z$ and $F \subset Z$ we denote by $Z_F$ the $\mathbb{R}$-quotient space obtained by contracting the set $F$ to a point.

Consider the set $P = \{(\omega_1, \alpha + 1) : \alpha < \omega_1\} \subset (\omega_1 + 1) \times (\omega_1 + 1)$. It is evident that $P$ is closed and discrete in the space $N = (\omega_1 \times (\omega_1 + 1)) \cup P$. The set $F = \{(\alpha, \omega_1) : \alpha < \omega_1\}$ is closed in $N$ and if we let $r(\alpha, \beta) = (\backslash n\{\alpha, \beta\}, \omega_1)$ for any $(\alpha, \beta) \in N$ then the map $r : N \to F$ is a retraction (see Fact 3 of V.302). Choose a point $a \notin N$ and let $X = N \oplus \{a\}$. The set $P$ is still closed and discrete in $X$, so $ext(X) \geq |P| > \omega$.

Let $w = (\omega_1, \omega_1)$ and consider the space $M = N \cup \{w\}$; it is clear that the set $G = (\omega_1 + 1) \times \{\omega_1\}$ is compact and hence closed in $M$. If $A \subset \omega_1 \times \omega_1$ is a countable set then there is $\alpha < \omega_1$ such that $A \subset (\alpha + 1) \times (\alpha + 1)$; an immediate consequence is that $\overline{A} \subset (\alpha + 1) \times (\alpha + 1)$ is compact, so $Q = \omega_1 \times \omega_1$ is countably compact by Fact 1 of S.310. If $U \in \tau(w, M)$ then there exists $\beta < \omega_1$ such that $(\omega_1, \alpha) \in U$ for all $\alpha > \beta$. Therefore the set $P \setminus U \subset \{\omega_1\} \times (\beta + 1)$ is countable. Consequently, the space $P' = P \cup \{w\}$ is Lindelöf and hence $ext(P') = \omega$.

Suppose that $D$ is a closed discrete subset of $M$; then the set $D_0 = D \cap Q$ is finite being closed and discrete is a countably compact space $Q$. The set $D_1 = D \cap G$ is finite because $G$ is compact and it follows from $ext(P') = \omega$ that the set $D_2 = D \cap P'$ is countable. This shows that the set $D = D_0 \cup D_1 \cup D_2$ is countable and hence $ext(M) = \omega$.

The $\mathbb{R}$-quotient space $M_G$ also has countable extent being a continuous image of $M$. By Problem 258, the space $X$ is $l$-equivalent to $Y = N_F \oplus F$. The space $F$ is countably compact being homeomorphic to $\omega_1$; therefore $ext(F) = \omega$. It follows from Fact 4 of V.302 that the spaces $N_F$ and $M_F$ are homeomorphic, so $ext(N_F) = ext(M_F) = \omega$. This shows that $ext(Y) = \omega$ and hence we obtained $l$-equivalent spaces $X$ and $Y$ such that $ext(X) \neq ext(Y)$.

**V.306.** *Prove that there exist $l$-equivalent spaces $X$ and $Y$ such that $X$ is Fréchet–Urysohn while $t(Y) > \omega$ and there is a non-closed set $A \subset Y$ such that $B \cap A$ is finite whenever $B$ is a bounded subset of $Y$. As a consequence, Fréchet–Urysohn property, $k$-property, sequentiality and countable tightness are not $l$-invariant.*

**Solution.** Given a space $Z$ and a closed set $F \subset Z$, the symbol $Z_F$ stands for the $\mathbb{R}$-quotient space obtained from $Z$ by contracting the set $F$ to a point which will be denoted by $a_F$.

*Fact 1.* For any space $Z$ its diagonal $\Delta = \{(z, z) : z \in Z\}$ is a retract of $Z \times Z$.

*Proof.* Let $r(x, y) = (x, x)$ for any $(x, y) \in Z \times Z$. It is clear that $r : Z \times Z \to \Delta$ and $r(z) = z$ for any $z \in \Delta$. Letting $r_1(x) = (x, x)$ for every $x \in Z$ we obtain a continuous map $r_1 : Z \to \Delta$ (which is, in fact, a homeomorphism); if $r_0 : Z \times Z \to Z$ is the projection onto the first factor then $r = r_1 \circ r_0$ which shows that the map $r$ is continuous, i.e., $r : Z \times Z \to \Delta$ is a retraction and hence Fact 1 is proved.

Returning to our solution let $c_n = \frac{1}{n+1}$ for all $n \in \omega$. Then the sequence $\{c_n : n \in \omega\}$ converges to the point $c = 0$; let $C = \{c_n : n \in \omega\} \cup \{c\}$. Fix a space $D$ homeomorphic to the one-point compactification $A(\omega_1)$ of a discrete space of cardinality $\omega_1$ and denote by $d$ the unique non-isolated point of $D$. Consider the space $S = (C \times D) \setminus \{(c, d)\}$ and let $X = (S \times S) \oplus \{a\}$ where $a$ is a point which does not belong to the set $T = S \times S$. It is easy to see that $T$ is homeomorphic to a subspace of $D^4$; the latter space being Eberlein compact, it is Fréchet–Urysohn so $T$ is Fréchet–Urysohn and hence $X$ is a Fréchet–Urysohn space as well.

The diagonal $F = \{(s, s) : s \in S\}$ of the space $S$ is a retract of $T$ by Fact 1, so we can apply Problem 258 to see that the space $Y = T_F \oplus F$ is $l$-equivalent to $X$.

The set $E = \{c\} \times (D \setminus \{d\})$ is closed and discrete in $S$, so $A = (E \times E) \setminus F$ is a closed discrete subset of $T$ which does not meet $F$. Suppose that $U$ is an open subspace of $T$ with $F \subset U$; then the point $((c_n, d), (c_n, d))$ belongs to $U$ for each $n \in \omega$. Therefore there is a set $V_n \in \tau(d, D)$ such that $(\{c_n\} \times V_n)^2 \subset U$ for every $n \in \omega$. The set $V = \bigcap_{n \in \omega} V_n$ is uncountable, so we can choose distinct points $d_1, d_2 \in V \setminus \{d\}$. If $x_i = (c, d_i)$ for $i = 1, 2$ then the point $x = (x_1, x_2)$ belongs to $A$.

Given an arbitrary set $W \in \tau(x, T)$ we can find $W_i \in \tau(x_i, S)$ for $i = 1, 2$ such that $W_1 \times W_2 \subset W$. It is easy to see that there is $m \in \omega$ for which $y_i = (c_m, d_i) \in W_i$ for $i = 1, 2$; therefore the point $y = (y_1, y_2)$ belongs to the set $W$. It follows from $y_i \in \{c_m\} \times V \subset \{c_m\} \times V_m$ for every $i \in \{1, 2\}$ that $y \in (\{c_m\} \times V_m)^2 \subset U$; an immediate consequence is that $y \in W \cap U$, so every neighborhood of the point $x$ in $T$ meets the set $U$. In other words, $x \in \overline{U}$; since the set $U \in \tau(F, T)$ was chosen arbitrarily, we proved that

(1) $\overline{U} \cap A \neq \emptyset$ for any set $U \in \tau(F, T)$.

Take an arbitrary neighborhood $W$ of the point $a_F$ in $T_F$. There exist a zero-set $Q$ and a cozero set $U$ in the space $T$ such that $F \subset Q \subset U$ and $U \setminus F \subset W \setminus \{a_F\}$ (see Problem 253). It is possible to find a set $V \in \tau(Q, T)$ such that $\overline{V} \subset U$ (see Fact 2 of V.140), so (1) shows that $U \cap A \neq \emptyset$ and therefore $W \cap A \neq \emptyset$. This proves that $a_F$ belongs to the closure of the set $A$ in the space $T_F$.

It turns out that

(2) for any faithfully indexed set $N = \{x_n : n \in \omega\} \subset E$ there exists a continuous function $f : S \to \mathbb{R}$ such that $f(x_n) = 2^n$ for all $n \in \omega$.

Pick $d_n \in D \setminus \{d\}$ for which $x_n = (c, d_n)$ and let $R_n = \{(c_k, d_n) : k \geq n\} \cup \{x_n\}$ for every $n \in \omega$. For each $n \in \omega$, if $x \in R_n$ then let $f(x) = 2^n$. If $x \notin R = \bigcup_{n \in \omega} R_n$ then let $f(x) = 0$. Observe that the set $R_n$ is an open neighborhood of $x_n$, so $f$ is continuous at every $x_n$. If $x = (c, y) \in E \setminus N$ then $C \times \{y\}$ is an open neighborhood of $x$ on which $f$ is constant, so $f$ is also continuous at $x$. Finally, if $x = (c_k, d)$ for some $k \in \omega$ then $V = \{c_k\} \times (D \setminus \{d_i : i \leq k\})$ is an open neighborhood of the point $x$ on which $f$ is constant and hence $f$ is continuous at $x$. This shows that $f$ is continuous on $S$, so (2) is proved.

All promised properties of $Y$ will be deduced from the following fact.

(3) For any countably infinite subset $N \subset A$ there is a function $\varphi \in C_p(T)$ such that $\varphi(F) = \{0\}$, $\varphi(N) \subset [1, +\infty)$ and $\varphi$ is not bounded on $N$.

Take a faithfully indexed set $M = \{z_n : n \in \omega\} \subset E$ such that $N \subset M \times M$ and apply (2) to construct a function $f \in C_p(S)$ for which $f(z_n) = 2^n$ for every $n \in \omega$. Letting $\varphi(x, y) = |f(x) - f(y)|$ for any $(x, y) \in T$ we obtain a continuous function $\varphi : T \to \mathbb{R}$ and it is evident that $\varphi(F) = \{0\}$. If $u \in N$ then there are distinct $m, n \in \omega$ such that $u = (z_n, z_m)$, so $\varphi(u) = |2^m - 2^n| \geq 1$; this shows that $\varphi(N) \subset [1, +\infty)$.

The set $N$ being infinite there exists a sequence $\{u_k : k \in \omega\} \subset N$ such that $u_k = (z_{n_k}, z_{m_k})$ for all $k \in \omega$ and the sequence $\{l_k = \max\{n_k, m_k\} : k \in \omega\}$ is unbounded. Then $f(u_k) = |2^{n_k} - 2^{m_k}| \geq |2^{l_k} - 2^{l_k - 1}| = 2^{l_k - 1}$, so the sequence $\{f(u_k) : k \in \omega\}$ is unbounded as well and hence $f$ is not bounded on $N$, i.e., (3) is proved.

Let $\pi : T \to T_F$ be the contraction map. We already noted that $a_F$ belongs to the closure of the set $A$ in $T_F$. Suppose that $N \subset A$ is a countably infinite set and take a function $\varphi \in C_p(T)$ as in (3). It follows from Fact 1 of V.252 that there is a function $f : T_F \to \mathbb{R}$ such that $f \circ \pi = \varphi$. In particular, $f(a_F) = 0$

and $f(N) \subset [1, +\infty)$ which shows that $a_F$ cannot belong to the closure of $N$, so $t(T_F) > \omega$ and hence $t(Y) > \omega$. Furthermore, the function $f$ is not bounded on $N$, so $N$ cannot be contained in a bounded subset of $T_F$; the set $T_F$ being clopen in $Y$ no bounded subset of $Y$ can contain $N$ either. Therefore $B \cap A$ is finite for any bounded subset $B$ of the space $Y$.

Consequently, $K \cap A$ is finite and hence closed for any compact $K \subset Y$; since $A$ is not closed in $Y$, the space $Y$ is not a $k$-space and hence it is not sequential. Therefore the spaces $X$ and $Y$ witness the fact that a Fréchet–Urysohn space can be $l$-equivalent to a non-$k$-space of uncountable tightness. This shows that the Fréchet–Urysohn property, $k$-property, sequentiality and countable tightness all fail to be $l$-invariant, i.e., our solution is complete.

**V.307.** *Show that Fréchet–Urysohn property is not preserved by $l$-equivalence in the class of compact spaces.*

**Solution.** Given a space $Z$ and a set $F \subset Z$, the symbol $Z_F$ stands for the $\mathbb{R}$-quotient space obtained from $Z$ by contracting the set $F$ to a point which will be denoted by $a_F$. The expression $X \simeq Y$ says that the spaces $X$ and $Y$ are homeomorphic. If $N$ is a countably infinite set then a family $\mathcal{A}$ of infinite subsets of $N$ is *almost disjoint* if $A \cap A'$ is finite for any distinct $A, A' \in \mathcal{A}$. Given an almost disjoint family $\mathcal{A}$ of subsets of $N$ fix a point $p_A \notin N$ for any $A \in \mathcal{A}$ in such a way that $p_A \neq p_{A'}$ for distinct $A, A' \in \mathcal{A}$. Denote by $M(N, \mathcal{A})$ the set $N \cup \{p_A : A \in \mathcal{A}\}$ with a topology $\tau$ defined by declaring all points of $N$ isolated while $\mathcal{B}_A = \{\{p_A\} \cup (A \setminus K) : K$ is a finite subset of $N\}$ is a local base at every point $p_A$ in the space $M(N, \mathcal{A})$.

It is straightforward that $M(N, \mathcal{A})$ is a first countable locally compact space; denote by $K(N, \mathcal{A})$ its one-point compactification. For any infinite subset $N' \subset N$ let $\mathcal{A}|N' = \{A \cap N' : A \in \mathcal{A}$ and $A \cap N'$ is infinite$\}$. Say that an almost disjoint family $\mathcal{A}$ on the set $N$ is *Fréchet–Urysohn* if there exists no infinite set $N' \subset N$ such that $\mathcal{A}|N'$ is a infinite maximal almost disjoint family on $N'$.

*Fact 1.* For any countably infinite set $N$ and an almost disjoint family $\mathcal{A}$ on $N$, the space $K(N, \mathcal{A})$ fails to be Fréchet–Urysohn if and only if there exists a set $N' \subset N$ such that $\mathcal{A}|N'$ is an infinite maximal almost disjoint family on $N'$. In other words, the space $K(N, \mathcal{A})$ is Fréchet–Urysohn if and only if $\mathcal{A}$ is a Fréchet–Urysohn family on the set $N$.

*Proof.* Let $P = \{p_A : A \in \mathcal{A}\}$ and suppose that $\mathcal{B} = \mathcal{A}|N'$ is an infinite maximal almost disjoint family on $N'$ for some $N' \subset N$. Denote by $a$ the unique point of the set $K(N, \mathcal{A}) \setminus M(N, \mathcal{A})$ and observe that if $A \in \mathcal{A}$ and $A \cap N'$ is infinite then $p_A \in \overline{N'}$. Therefore $C = \overline{N'} \cap P$ is an infinite set; since $P \setminus U$ is finite for any neighborhood $U$ of the point $a$, we have $a \in \overline{C}$ and hence $a \in \overline{N'}$.

If $S \subset N'$ is a sequence which converges to $a$ then $S$ is infinite but $S \cap A$ is finite for any $A \in \mathcal{B}$, so $\mathcal{B}$ is not a maximal almost disjoint family on $N'$. This contradiction shows that no sequence from $N'$ converges to $a$, so the space $K(N, \mathcal{A})$ is not Fréchet–Urysohn and hence we proved sufficiency.

Now assume that the space $K(N, \mathcal{A})$ is not Fréchet–Urysohn. Recall that we have $\chi(x, K(N, \mathcal{A})) \leq \omega$ for every point $x \in K(N, \mathcal{A}) \backslash \{a\}$, so there must exist a set $E \subset K(N, \mathcal{A})$ such that $a \in \overline{E}$ but no sequence from $E$ converges to $a$. If $E \cap P$ is infinite then any countably infinite subset of $E \cap P$ converges to $a$ which is a contradiction, so $E \cap P$ is finite and hence the set $N' = E \backslash P$ is infinite. We still have $a \in \overline{N'}$ while there is no sequence in $N'$ which converges to $a$.

If some infinite set $D \subset N'$ is closed and discrete in $M(N, \mathcal{A})$ then $D$ is a sequence which converges to $a$. Therefore, for any infinite $D \subset N'$ there is $A \in \mathcal{A}$ with $p_A \in \overline{D}$ and hence $D \cap A$ is an infinite set. Assumé for a moment that the family $\mathcal{A}' = \{A \in \mathcal{A} : A \cap N'$ is infinite$\}$ is finite. It is evident that the set $U = K(N, \mathcal{A}) \backslash (\bigcup \{\{p_A\} \cup A : A \in \mathcal{A}'\})$ is an open neighborhood of $a$, so $D = U \cap N'$ is an infinite set which is closed and discrete in $M(N, \mathcal{A})$. This contradiction shows that $\mathcal{A}|N'$ is an infinite maximal almost disjoint family on $N'$, i.e., we established necessity and hence Fact 1 is proved.

*Fact 2.* Suppose that $N$ is a countably infinite set and we are given almost disjoint families $\mathcal{A}$ and $\mathcal{B}$ on the set $N$. If $\mathcal{C} = \mathcal{A} \cup \mathcal{B}$ is an infinite maximal almost disjoint family on $N$ then the space $K(N, \mathcal{A}) \times K(N, \mathcal{B})$ is not Fréchet–Urysohn.

*Proof.* Let $a$ be the unique point of the set $K(N, \mathcal{C}) \backslash M(N, \mathcal{C})$. It is immediate that the space $K_0 = N \cup \{p_A : A \in \mathcal{A}\} \cup \{a\} \subset K(N, \mathcal{C})$ is homeomorphic to $K(N, \mathcal{A})$ and $K_1 = N \cup \{p_B : B \in \mathcal{B}\} \cup \{a\}$ is homeomorphic to $K(N, \mathcal{B})$.

The space $K(N, \mathcal{C})$ is not Fréchet–Urysohn by Fact 1. Since the subspace $M(N, \mathcal{C})$ is first countable, there exists a set $E \subset K(N, \mathcal{C})$ such that $a \in \overline{E}$ but no sequence from $E$ converges to $a$. If $E' = E \cap \{p_C : C \in \mathcal{C}\}$ is infinite then it is easy to see that $E'$ contains a sequence convergent to $a$. Therefore $a \in \overline{E \backslash E'}$; since $E \backslash E' \subset N$, the space $N \cup \{a\}$ is not Fréchet–Urysohn.

Let $\Delta = \{(x, x) : x \in K(N, \mathcal{C})\}$ be the diagonal of the space $K(N, \mathcal{C})$. The set $\Delta' = \Delta \cap (N \cup \{a\})^2$ coincides with the diagonal of $N \cup \{a\}$, so it is not Fréchet–Urysohn being homeomorphic to $N \cup \{a\}$. Now it follows from $\Delta' \subset K_0 \times K_1$ that $K_0 \times K_1$ is not a Fréchet–Urysohn space. Therefore $K(N, \mathcal{A}) \times K(N, \mathcal{B}) \simeq K_0 \times K_1$ is not Fréchet–Urysohn, i.e., Fact 2 is proved.

*Fact 3.* Suppose that $N$ is a countably infinite set and $\mathcal{A}$ is an infinite maximal almost disjoint family on $N$. Call an infinite set $N' \subset N$ *nontrivial* if $\mathcal{A}|N'$ is infinite. Then, for any decreasing family $N_0 \supset N_1 \supset \ldots \supset N_k \supset \ldots$ of nontrivial subsets of $N$ there exists a nontrivial set $H \subset N$ such that $H \backslash N_k$ is finite for every $k \in \omega$.

*Proof.* Say that a set $A \subset N$ is *small* if $A \backslash H_k$ is finite for any $k \in \omega$. Observe that

(1) if $A_k$ is small and $A_k \subset N_k$ for every $k \in \omega$ then $A = \bigcup_{k \in \omega} A_k$ is also small, because $A \backslash N_k = (A_0 \backslash N_k) \cup \ldots \cup (A_{k-1} \backslash N_k)$ is a finite set for every $k \in \omega$.

We are going to construct by induction a family $\{A_k : k \in \omega\}$ of disjoint small sets such that

(2) $A_k \subset N_k$ and there exists $B_k \in \mathcal{A}$ such that $A_k \subset B_k$ for all $k \in \omega$;
(3) the sets $B_k$ and $B_l$ are distinct if $k \neq l$.

Since every $N_i$ is infinite, it is easy to pick a point $x_i \in N_i$ in such a way that the set $A = \{x_i : i \in \omega\}$ is faithfully indexed. The family $\mathcal{A}$ being maximal almost disjoint, there exists $B_0 \in \mathcal{A}$ such that $A_0 = B_0 \cap A$ is infinite. It is clear that $A_0$ is a small set, so (2) and (3) hold for $k = 0$.

Assume that $n \in \omega$ and we have small disjoint sets $A_0, \ldots, A_n$ and sets $B_0, \ldots, B_n \in \mathcal{A}$ such that the properties (2) and (3) are fulfilled for each $k \leq n$. Since $N_k$ is nontrivial, the set $N'_k = N_k \backslash (B_0 \cup \ldots \cup B_n)$ has to be infinite for every $k \in \omega$, so we can choose a point $y_k \in N'_k$ in such a way that the set $A' = \{y_k : k > n\}$ is faithfully indexed. The family $\mathcal{A}$ being maximal almost disjoint, there is $B_{n+1} \in \mathcal{A}$ such that $A_{n+1} = B_{n+1} \cap A'$ is infinite. It follows from $A_{n+1} \cap (B_0 \cup \ldots \cup B_n) = \emptyset$ that the family $\{A_0, \ldots, A_{n+1}\}$ is disjoint and it is evident that $A_{n+1}$ is a small set. Besides, $B_{n+1} \backslash B_i \supset A_{n+1}$ is infinite and hence $B_{n+1} \neq B_i$ for any $i \leq n$. This shows that the sets $A_0, B_0, \ldots, A_{n+1}, B_{n+1}$ satisfy (2) and (3) for each $k \leq n + 1$, so our inductive procedure can be continued to construct the promised sequence $\{A_k : k \in \omega\}$ with the properties (2) and (3).

By (1) the set $H = \bigcup_{k \in \omega} A_k$ is small; the properties (2) and (3) guarantee that $\{B_k \cap H : k \in \omega\}$ is an infinite family contained in $\mathcal{A}|H$, so $H$ is nontrivial and hence Fact 3 is proved.

*Fact 4.* Suppose that $N$ is a countably infinite set and $\mathcal{A}$ is a maximal almost disjoint family on $N$ with $|\mathcal{A}| = \mathfrak{c}$. Then there exists an infinite set $Q \subset N$ such that $\mathcal{A}|Q$ is infinite and $\mathcal{A}|Q = \mathcal{F}_0 \cup \mathcal{F}_1$ for some Fréchet–Urysohn families $\mathcal{F}_0, \mathcal{F}_1$ on the set $Q$. As a consequence, there exists an infinite maximal almost disjoint family on a countably infinite set which is a union of two Fréchet–Urysohn families.

*Proof.* Assume the contrary; then, for any infinite set $Q \subset N$ if $\mathcal{A}|Q$ is infinite and $\mathcal{A}|Q = \mathcal{F}_0 \cup \mathcal{F}_1$ then one of the families $\mathcal{F}_0, \mathcal{F}_1$ is not Fréchet–Urysohn, i.e., there exists an infinite set $A \subset Q$ such that $\mathcal{F}_i|A$ is an infinite maximal almost disjoint family on $A$.

Recall that $\mathbb{D} = \{0, 1\}$ and take a faithful enumeration $\{A_f : f \in \mathbb{D}^\omega\}$ of the family $\mathcal{A}$; let $\mathcal{A}^i_n = \{A_f : f(n) = i\}$ for any $n \in \omega$ and $i \in \mathbb{D}$. Since $\mathcal{A}^0_0 \cup \mathcal{A}^1_0 = \mathcal{A}$, there exists an infinite set $N_0 \subset N$ such that $\mathcal{A}^{i_0}_0 | N_0$ is an infinite maximal almost disjoint family on $N_0$ for some $i_0 \in \mathbb{D}$.

Proceeding by induction assume that $m \in \omega$ and we have constructed infinite sets $N_0 \supset \ldots \supset N_m$ and a sequence $i_0, \ldots, i_m$ of elements of $\mathbb{D}$ such that $\mathcal{A}^{i_k}_k | N_k$ is an infinite maximal almost disjoint family on $N_k$ for every $k \leq m$. It follows from the equality $\mathcal{A}^{i_m}_m | N_m = (\mathcal{A}^0_{m+1} \cap \mathcal{A}^{i_m}_m) | N_m \cup (\mathcal{A}^1_{m+1} \cap \mathcal{A}^{i_m}_m) | N_m$ that there exists an infinite set $N_{m+1} \subset N_m$ and $i_{m+1} \in \mathbb{D}$ such that $(\mathcal{A}^{i_{m+1}}_{m+1} \cap \mathcal{A}^{i_m}_m) | N_{m+1}$ is an infinite maximal almost disjoint family on $N_{m+1}$. It is immediate that $\mathcal{A}^{i_{m+1}}_{m+1} | N_{m+1}$ is also an infinite maximal almost disjoint family on $N_{m+1}$, so our inductive procedure can be continued to construct a decreasing family $\{N_k : k \in \omega\}$ of infinite subsets of $N$ and a sequence $\{i_k : k \in \omega\} \subset \mathbb{D}$ such that $\mathcal{A}^{i_k}_k | N_k$ is an infinite maximal almost disjoint family on $N_k$ for all $k \in \omega$.

Let $f(k) = i_k$ for every $k \in \omega$; apply Fact 3 to find an infinite set $H \subset N$ such that $H \backslash N_k$ is finite for all $k \in \omega$ and $\mathcal{A}|H$ is an infinite family. In particular, there exists $g \in \mathbb{D}^{\omega} \setminus \{f\}$ for which $A_g \cap H$ is infinite. Fix $k \in \omega$ with $i_k = f(k) \neq g(k)$; it follows from $|H \setminus N_k| < \omega$ that $A = A_g \cap N_k$ is infinite. Since $A_g \notin \mathcal{A}_k^{i_k}$, the set $A \cap B$ is finite for any $B \in \mathcal{A}_k^{i_k} | N_k$. Therefore the set $A$ witnesses that the family $\mathcal{A}_k^{i_k} | N_k$ is not maximal; this contradiction shows that Fact 4 is proved.

Returning to our solution apply Fact 1, Fact 2, and Fact 4 to find compact spaces $K$ and $L$ with the following properties:

(4)  there are points $p \in K$ and $q \in L$ such that both spaces $K \setminus \{p\}$ and $L \setminus \{q\}$ are first countable;
(5)  $K$ and $L$ are Fréchet–Urysohn spaces;
(6)  the space $Y = K \times L$ is not Fréchet–Urysohn.

Let $\pi_K : Y \to K$ and $\pi_L : Y \to L$ be the natural projections. The set $F = (\{p\} \times L) \cup (K \times \{q\})$ is closed in the space $Y$. It follows from Problem 262 that the space $Y$ is $l$-equivalent to an $\mathbb{R}$-quotient image $X$ of the space $Y_F \oplus K \oplus L$.

Observe that $Y_F \setminus \{a_F\}$ is homeomorphic to $Y \setminus F \simeq (K \setminus \{p\}) \times (L \setminus \{q\})$, so the space $Y_F \setminus \{a_F\}$ is first countable by (4). Therefore for any point $x \in Y_F \setminus \{a_F\}$ if $x \in \overline{A}$ then some sequence from $A$ converges to $x$, i.e., $Y_F$ is Fréchet–Urysohn at the point $x$.

Suppose that $A \subset Y_F \setminus \{a_F\}$ and $a_F \in \overline{A}$. If the set $G = \mathrm{cl}_Y(A)$ does not meet $F$ then $W = \{a_F\} \cup (Y \setminus (F \cup G))$ is an open neighborhood of $a_F$ in $Y_F$ (see Problem 254) such that $W \cap A = \emptyset$ which is a contradiction. Therefore $\mathrm{cl}_Y(A)$ intersects the set $F = \pi_K^{-1}(p) \cup \pi_L^{-1}(q)$; the situation being symmetric, we can assume, without loss of generality, that $\mathrm{cl}_Y(A) \cap \pi_K^{-1}(p) \neq \emptyset$ and hence $p \in \overline{p_K(A)}$. The space $K$ being Fréchet–Urysohn, we can choose a set $S = \{a_n : n \in \omega\} \subset A$ such that the sequence $\{\pi_K(a_n) : n \in \omega\}$ converges to $p$.

Take any set $U \in \tau(a_F, Y_F)$ then $V = F \cup (U \setminus \{a_F\})$ is an open neighborhood of $F$ and, in particular, $\pi_K^{-1}(p) \subset V$. By Fact 1 of S.226 we can find a set $H \in \tau(p, K)$ such that $\pi_K^{-1}(H) \subset V$. There is $m \in \omega$ such that $\pi_K(a_n) \in H$ for all $n \geq m$, so $a_n \in V$ and hence $a_n \in U$ for all $n \geq m$; this proves that the sequence $S$ converges to $a_F$ and hence the space $Y_F$ is Fréchet–Urysohn. Therefore the space $X' = Y_F \oplus K \oplus L$ is Fréchet–Urysohn as well; if $\varphi : X' \to X$ is the respective $\mathbb{R}$-quotient map then it is closed because the space $X'$ is compact. This, together with TFS-225, shows that $X$ is a Fréchet–Urysohn space. Thus the spaces $X$ and $Y$ witness that the Fréchet–Urysohn property is not $l$-invariant, i.e., our solution is complete.

**V.308.** *Let $Y$ be a space in which every closed subspace has the Baire property. Suppose that $Y$ is $l$-equivalent to a space $X$ and a nonempty set $Z \subset X$ also has the Baire property. Prove that there is a nonempty $W \subset Z$ which is open in $Z$ and homeomorphic to a subspace of $Y$.*

**Solution.** The expression $P \simeq Q$ says that the spaces $P$ and $Q$ are homeomorphic. Fix a linear homeomorphism $\varphi : C_p(Y) \to C_p(X)$ and let $\varphi^*(a) = a \circ \varphi$ for any $a \in L_p(X)$; it is an easy consequence of Fact 4 of S.489 that $\varphi^* : L_p(X) \to L_p(Y)$ is a linear isomorphism. The set $X' = \varphi^*(X)$ is a Hamel basis in the space $L_p(Y)$

such that $X' \simeq X$ and every $f \in C_p(X')$ extends to a continuous linear functional on $L_p(Y)$. Given nonempty sets $A_1, \ldots, A_n \subset L_p(Y)$ and $O_1, \ldots, O_n \subset \mathbb{R}$ let $O_1 A_1 + \ldots + O_n A_n = \{\lambda_1 x_1 + \ldots + \lambda_n x_n : \lambda_i \in O_i \text{ and } x_i \in A_i \text{ for all } i \leq n\}$.

For any $x \in X'$ we can find a uniquely determined $n \in \mathbb{N}$ and distinct points $y_1, \ldots, y_n \in Y$ such that $x = \lambda_1 y_1 + \ldots + \lambda_n y_n$ for some $\lambda_1, \ldots, \lambda_n \in \mathbb{R} \backslash \{0\}$; let $\xi_Y(x) = \{y_1, \ldots, y_n\}$ and $l_Y(x) = n$. Analogously, for any point $y \in Y$ it is possible to find a uniquely determined $m \in \mathbb{N}$ and distinct points $x_1, \ldots, x_m \in X'$ such that $y = \mu_1 x_1 + \ldots + \mu_m x_m$ for some $\mu_1, \ldots, \mu_m \in \mathbb{R} \backslash \{0\}$; let $l_X(y) = m$ and $\xi_X(y) = \{x_1, \ldots, x_m\}$. It follows from Problem 279 that $\xi_Y : X' \to \exp(Y)$ and $\xi_X : Y \to \exp(X')$ are the support maps for $\varphi$ and $\varphi^{-1}$ respectively. Applying Problem 280 we infer that both $\xi_Y$ and $\xi_X$ are lower semicontinuous.

The set $Y_n = \{y \in Y : l_X(y) \leq n\}$ is closed in $Y$ for any $n \in \mathbb{N}$ by Fact 1 of U.485; we will also need the sets $H_1 = Y_1$ and $H_n = Y_n \backslash Y_{n-1}$ for all $n > 1$. Let $Z' = \varphi^*(Z)$; then $Z' \simeq Z$ and, in particular, the space $Z'$ has the Baire property. Every set $Z_n = \{a \in Z' : l_Y(a) \leq n\}$ is closed in $Z'$. Let $G_1 = Z_1$ and $G_n = Z_n \backslash Z_{n-1}$ for all $n > 1$.

Since $Z' = \bigcup \{Z_n : n \in \mathbb{N}\}$, an easy consequence of the Baire property of $Z'$ is that there exists a nonempty open set $U$ in the space $Z'$ such that $U \subset G_n$ for some $n \in \mathbb{N}$. Pick a point $a \in U$; then $\xi_Y(a) = \{y_1, \ldots, y_n\}$ and we can choose a set $W_i \in \tau(y_i, Y)$ for every $i \leq n$ such that the family $\{W_1, \ldots, W_n\}$ is disjoint. The set $U' = \{b \in Z' : \xi_Y(b) \cap W_i \neq \emptyset \text{ for all } i \leq n\}$ is open in $Z'$ and $a \in U'$. Therefore $V = U \cap U'$ is an open neighborhood of $a$ in $Z'$.

Given any $b \in V$ it follows from $l_Y(b) = n$ that for every $i \in \{1, \ldots, n\}$ there is a unique element $x_i \in \xi_X(b) \cap W_i$; let $p_i(b) = x_i$. This gives us a mapping $p_i : V \to W_i$ for each $i \leq n$.

Let $O = \mathbb{R} \backslash \{0\}$; if $W$ is a nonempty open subset of $W_i$ then it is easy to see that $p_i^{-1}(W) = (OW_1 + \ldots + OW_{i-1} + OW + OW_{i+1} + \ldots + OW_n) \cap V$; it follows from Fact 2 of U.485 that $p_i^{-1}(W)$ is an open subset of $V$, so the map $p_i$ is continuous for any $i \leq n$.

Let $V_0 = V$; we will inductively construct nonempty open sets $V_0, \ldots, V_n$ of the space $Z'$ and $k_1, \ldots, k_n \in \mathbb{N}$ such that

(1) $V_i \subset V_{i-1}$ and $p_i(V_i) \subset H_{k_i}$ for all $i = 1, \ldots, n$.

Suppose that $j < n$ and we have $V_0, \ldots, V_j \in \tau^*(Z')$ and $k_1, \ldots, k_j \in \mathbb{N}$ such that the condition (1) is satisfied for all $i = 1, \ldots, j$. The set $\overline{p_{j+1}(V_j)}$ is closed in $Y$, so it has the Baire property; this, together with $\overline{p_{j+1}(V_j)} \subset \bigcup \{Y_n : n \in \mathbb{N}\}$ implies that there is a nonempty open subset $E$ of the space $\overline{p_{j+1}(V_j)}$ such that $E \subset H_{k_{j+1}}$ for some $k_{j+1} \in \mathbb{N}$. The set $E' = E \cap p_{j+1}(V_j)$ is nonempty and open in $p_{j+1}(V_j)$, so $V_{j+1} = p_{j+1}^{-1}(E') \cap V_j$ is open in $V_j$ and hence in $Z'$. It is straightforward that (1) now holds for all $i \leq j+1$, so our inductive procedure gives us the promised sets $V_0, \ldots, V_n$ and numbers $k_1, \ldots, k_n \in \mathbb{N}$. As a consequence,

(2) $l_X(p_i(b)) = k_i$ for any $b \in V_n$ and $i = 1, \ldots, n$.

Fix a point $b \in V_n$ and $i \in \{1, \ldots, n\}$. Let $y_i = p_i(b)$ and consider the set $\xi_X(y_i) = \{a_1^i, \ldots, a_{k_i}^i\}$; choose a disjoint family $\{U_1^i, \ldots, U_{k_i}^i\} \subset \tau^*(X')$ such that

$a_j^i \in U_j^i$ for all $j \leq k_i$. The set $Q_i = \{y \in X' : \xi_X(y) \cap U_j^i \neq \emptyset$ for all $j \leq k_i\}$ is open in $X'$, so $W_i' = Q_i \cap W_i$ is open in $X'$ and nonempty because $y_i \in W_i'$. If $y \in W_i'$ then, for every $j \leq k_i$, there is a unique point $b_j \in U_j^i \cap \xi_X(y)$; let $q_j^i(y) = b_j$. Then every $q_j^i : W_i' \to U_j^i$ is a continuous map; this is proved in the same way as we verified continuity of $p_i$.

The set $E = \bigcap\{p_i^{-1}(W_i') : i \leq n\} \cap V_n$ is open in the space $Z'$ and nonempty due to the fact that $b \in E$. For any $i \in \{1, \ldots, n\}$ and $j \in \{1, \ldots, k_i\}$ the set $S_j^i = \{c \in E : q_j^i(p_i(c)) = c\}$ is closed in $E$. If $c \in E$ and $\xi_Y(c) = \{z_1, \ldots, z_n\}$ then there are $\lambda_1, \ldots, \lambda_n \in O$ for which $c = \sum_{i=1}^n \lambda_i z_i$. For every $i \leq n$ let $\xi_X(z_i) = \{d_j^i : j \leq k_i\}$; there are $\mu_1^i, \ldots, \mu_{k_i}^i \in O$ such that $z_i = \sum_{j=1}^{k_i} \mu_j^i d_j^i$. The equality $c = \sum_{i=1}^n \lambda_i (\sum_{j=1}^{k_i} \mu_j^i d_j^i)$ shows that $c = d_j^i$ for some $i \in \{1, \ldots, n\}$ and $j \leq k_i$. By definition of the maps $p_i$ and $q_j^i$ we have $c = q_j^i(p_i(c))$, i.e., $c \in S_j^i$.

Therefore $E \subset \bigcup\{S_j^i : i \leq n, \ j \leq k_i\}$; an immediate consequence is that we can choose $i \leq n$ and $j \leq k_i$ for which there is a set $Q \in \tau^*(Z')$ with $Q \subset S_j^i$. Then $q_j^i(p_i(c)) = c$ for every $c \in Q$ which shows that the map $p_i|Q : Q \to p_i(Q)$ is a homeomorphism because $q_j^i|p_i(Q)$ is its continuous inverse.

We proved that a nonempty open set $Q$ of the space $Z' \simeq Z$ embeds in $X$. Therefore a nonempty subspace $W$ of the space $Z$ also embeds in $X$. This settles necessity and makes our solution complete.

**V.309.** *Let $X$ and $Y$ be Čech-complete $l$-equivalent spaces. Prove that every nonempty subspace of $X$ has a $\pi$-base whose elements are embeddable in $Y$. Deduce from this fact that if $X$ and $Y$ are nonempty Čech-complete $l$-equivalent spaces and $X$ is scattered then $Y$ is also scattered.*

**Solution.** Take a nonempty set $P \subset X$ and denote by $\mathcal{U}$ the family of all nonempty open subsets of $P$ which are embeddable in $Y$. Given any $U \in \tau^*(P)$ the set $Z = \overline{U}$ is closed in $X$ and hence Čech-complete. Observe that every closed subspace of $Y$ is Čech-complete and hence has the Baire property, so we can apply Problem 308 to see that some nonempty open set $W$ in the space $Z$ is embeddable in $Y$. The set $V = W \cap U$ is nonempty and open in $U$ and hence in $P$; besides, $V$ is embeddable in $Y$, so $V \in \mathcal{U}$. We proved that any $U \in \tau^*(P)$ contains an element of $\mathcal{U}$, so the family $\mathcal{U}$ is a $\pi$-base of $P$.

Finally, assume that $X \overset{l}{\sim} Y$, the spaces $X$ and $Y$ are nonempty, Čech-complete and $X$ is scattered. If $Y$ is not scattered then there exists a dense-in-itself nonempty set $Z \subset Y$. By what we proved in the previous paragraph, there exists a nonempty set $U \in \tau(Z)$ which embeds in $X$; this implies that $U$ is scattered. However, no open subspace of a dense-in-itself space can have isolated points, so $U$ is dense-in-itself; this contradiction shows that $Y$ is also scattered.

**V.310.** *Let $X$ and $Y$ be Čech-complete $l$-equivalent spaces such that $\chi(X) \leq \kappa$. Prove that $Y$ has a dense open subspace $D$ such that $\chi(x, Y) \leq \kappa$ for each $x \in D$. In particular, $\chi(D) \leq \kappa$.*

**Solution.** Apply Problem 309 to find a $\pi$-base $\mathcal{B}$ in the space $Y$ such that every $B \in \mathcal{B}$ is embeddable in $X$ and hence $\chi(B) \leq \kappa$. If $B \in \mathcal{B}$ then $B$ is open in $X$, so we have $\chi(x, Y) = \chi(x, B) \leq \kappa$ for any $x \in B$. The set $D = \bigcup \mathcal{B}$ is open and dense in $X$; every $x \in D$ belongs to some $B \in \mathcal{B}$, so $\chi(x, Y) \leq \kappa$. Therefore $D$ is the promised dense subspace of $Y$.

**V.311.** *Prove that $X$ is a closed Hamel basis of $L(X)$ for any space $X$.*

**Solution.** Let $\mathcal{L}$ be the family of all continuous maps of $X$ into locally convex spaces of cardinality not exceeding $|X| \cdot 2^\omega$. For every $\varphi \in \mathcal{L}$ let $L_\varphi$ be the locally convex space such that $\varphi : X \to L_\varphi$. If $i = \Delta\{\varphi : \varphi \in \mathcal{L}\}$ then $i$ maps $X$ in the space $L = \prod\{L_\varphi : \varphi \in \mathcal{L}\}$ and $L(X)$ is the linear hull of $i(X)$ in $L$. Let $p_\varphi : L(X) \to L_\varphi$ be the natural projection for each $\varphi \in \mathcal{L}$.

There exists $\varphi \in \mathcal{L}$ such that $\varphi$ is the canonical embedding of $X$ into $L_p(X)$. By the definition of the diagonal map, we have $(p_\varphi | i(X)) \circ i = \varphi$. It is an easy consequence of Fact 2 of S.337 that $i : X \to i(X)$ is a homeomorphism, i.e., $i$ embeds $X$ into $L(X)$ and therefore we can identify $i(x)$ with $x$ for any $x \in X$. Observe that $p_\varphi : L(X) \to L_p(X)$ is a linear map such that $p_\varphi(x) = x$ for all $x \in X$; since $X$ is linearly independent in $L_p(X)$, it has to be linearly independent in $L(X)$ as well, i.e., $X$ is a Hamel basis of $L(X)$. As an immediate consequence, the map $p_\varphi$ is an isomorphism and hence $p_\varphi^{-1}(X) = X$; since $X$ is closed in $L_\varphi = L_p(X)$ (see TFS-167), the set $X$ is closed in $L(X)$.

**V.312.** *Prove that, for any space $X$ and any continuous map $f : X \to L$ of $X$ to a locally convex space $L$, there exists a unique continuous linear map $\xi_f : L(X) \to L$ such that $\xi_f | X = f$. Observe that this makes it possible to consider that $L(X)$, as a linear space, coincides with $L_p(X) \subset C(C_p(X))$ while the topology of $L(X)$ is stronger than $\tau(L_p(X))$. In all problems that follow we use this observation identifying the underlying set of $L(X)$ with $L_p(X)$.*

**Solution.** Let $\mathcal{L}$ be the family of all continuous maps of $X$ into locally convex spaces of cardinality not exceeding $|X| \cdot 2^\omega$. For every $\varphi \in \mathcal{L}$ let $L_\varphi$ be the locally convex space such that $\varphi : X \to L_\varphi$. If $i = \Delta\{\varphi : \varphi \in \mathcal{L}\}$ then $i$ maps $X$ in the space $L = \prod\{L_\varphi : \varphi \in \mathcal{L}\}$ and $L(X)$ is the linear hull of $i(X)$ in $L$. Let $p_\varphi : L(X) \to L_\varphi$ be the natural projection for each $\varphi \in \mathcal{L}$. Since $i : X \to i(X)$ is a homeomorphism (see Problem 311), we will identify $i(x)$ with $x$ for every $x \in X$.

Let $M$ be the linear hull of the set $f(X)$ in $L$. Then $|M| \leq |X| \cdot 2^\omega$ and hence there exists $\varphi \in \mathcal{L}$ such that $\varphi : X \to L_\varphi$ coincides with $f : X \to M$, i.e., $M = L_\varphi$ and $\varphi(x) = f(x)$ for each $x \in X$. The definition of the diagonal product implies that $f = p_\varphi \circ i$. Recall that we identify $X$ with $i(X)$, so the map $p_\varphi : L(X) \to L_\varphi = M$ coincides with $f$ on $X$. Since $p_\varphi$ is linear, continuous and $M$ is a linear subspace of $L$, we can consider that $p_\varphi : L(X) \to L$ and hence $\xi_f = p_\varphi$ is the promised continuous linear extension of $f$. This extension is unique because two linear functionals on a linear space coincide whenever they coincide on a Hamel basis of the space.

Finally, observe that, for the canonical embedding $id : X \to L_p(X)$ there exists a continuous linear map $\xi : L(X) \to L_p(X)$ such that $\xi | X = id$. The set $X$

is a Hamel basis in both $L(X)$ and $L_p(X)$, so $\xi$ has to be a linear isomorphism. Therefore we can consider that the underlying set of $L(X)$ coincides with $L_p(X)$. After we identify the sets $L(X)$ and $L_p(X)$, it follows from continuity of $\xi$ that the identity map of $L(X)$ onto $L_p(X)$ is continuous, i.e., the topology of $L(X)$ contains the topology of $L_p(X)$.

**V.313.** *Suppose that $L$ is a locally convex space such that $X$ is embedded as a Hamel basis in $L$. Prove that the following conditions are equivalent:*

(i) *there exists a linear homeomorphism $i : L \to L(X)$ such that $i(x) = x$ for all $x \in X$;*

(ii) *every continuous function $f : X \to M$ of the space $X$ to a locally convex space $M$, can be extended to a continuous linear functional $\xi_f : L \to M$.*

**Solution.** Suppose that there exists a linear homeomorphism $i : L \to L(X)$ such that $i(x) = x$ for each $x \in X$. If $M$ is a locally convex space and $f : X \to M$ is a continuous function then we can apply Problem 312 to find a continuous linear functional $\mu : L(X) \to M$ such that $\mu(x) = f(x)$ for every $x \in X$. The map $\xi_f = \mu \circ i : L \to M$ is continuous and linear; since also $\xi_f(x) = \mu(i(x)) = \mu(x) = f(x)$ for all $x \in X$, we conclude that $\xi_f | X = f$, i.e., $\xi_f$ is the required extension of $f$ and hence we proved that (i)$\Longrightarrow$(ii).

Now, if (ii) holds then let $id : X \to X$ be the identity map. Since $X \subset L(X)$, we can consider that $id : X \to L(X)$, so there exists a continuous linear map $i : L \to L(X)$ such that $i | X = id$. The set $X$ is a Hamel basis in both $L$ and $L(X)$ (see Problem 311), so it follows from linearity of $i$ that $i$ is an algebraic isomorphism. By Problem 312 there exists a unique linear continuous map $j : L(X) \to L$ such that $j | X = id$. As a consequence, $j | X = id = id^{-1} = i^{-1} | X$, so $j$ and $i^{-1}$ are linear maps on $L(X)$ which coincide on $X$. The set $X$ being a Hamel basis of $L(X)$ we have the equality $i^{-1} = j$ and hence the map $i^{-1}$ is continuous. Therefore $i$ is a linear homeomorphism, i.e., we established that (ii)$\Longrightarrow$(i).

**V.314.** *Prove that, for any space $X$, the set $(L(X))^*$ coincides with the set $(L_p(X))^*$. Deduce from this fact that the weak topology of the space $L(X)$ coincides with the topology of $L_p(X)$.*

**Solution.** Recall that we consider that the spaces $L(X)$ and $L_p(X)$ have the same underlying set and operations while the topology of $L(X)$ is stronger than the topology of $L_p(X)$ (see Problem 312). Therefore $C_p(L_p(X)) \subset C_p(L(X))$ and, in particular, $(L_p(X))^* \subset (L(X))^*$. The weak topology $v$ of the space $L_p(X)$ is generated by $(L_p(X))^*$ while the weak topology $\mu$ of the space $L(X)$ is generated by the set $(L(X))^*$.

Now take an arbitrary functional $\varphi \in (L(X))^*$. The function $f = \varphi | X$ is continuous on $X$, so there exists a continuous linear functional $\xi : L_p(X) \to \mathbb{R}$ such that $\xi | X = f = \varphi | X$ (see Fact 6 of S.489). If two linear functionals coincide on a Hamel basis of a linear space then they coincide on the whole space; therefore $\varphi = \xi \in (L_p(X))^*$ which shows that $(L(X))^* = (L_p(X))^*$, i.e., the topologies

$v$ and $\mu$ are generated by the same set of linear functionals and hence $v = \mu$. Recalling that $v = \tau(L_p(X))$ (see Problem 233), we conclude that $\mu = \tau(L_p(X))$, i.e., the weak topology of $L(X)$ coincides with the topology of $L_p(X)$.

**V.315.** *Given a space $X$ let $E$ be the weak dual of $L(X)$, i.e., $E = (L(X))^*$ and the topology of $E$ is induced from $C_p(L(X))$. For every $f \in E$ let $\pi(f) = f|X$, i.e., $\pi : E \to C_p(X)$ is a restriction map. Prove that $\pi$ is a linear homeomorphism and hence $E$ is linearly homeomorphic to $C_p(X)$.*

**Solution.** Recall that we consider that the spaces $L(X)$ and $L_p(X)$ have the same underlying set and operations while the topology of $L(X)$ is stronger than the topology of $L_p(X)$ (see Problem 312). Therefore $C_p(L_p(X)) \subset C_p(L(X))$ while the map $\pi : (L_p(X))^* \to C_p(X)$ is a linear homeomorphism by Problem 235. Since $(L_p(X))^* = E$ by Problem 314, we conclude that $\pi : E \to C_p(X)$ is a linear homeomorphism as well.

**V.316.** *Observe that $l$-equivalence implies $u$-equivalence, i.e., for any spaces $X$ and $Y$, if $X \overset{l}{\sim} Y$ then $X \overset{u}{\sim} Y$. Prove that $L$-equivalence implies $l$-equivalence.*

**Solution.** If $X$ is $l$-equivalent to $Y$ then there exists a linear homeomorphism $\varphi : C_p(X) \to C_p(Y)$. By Problem 134, the maps $\varphi$ and $\varphi^{-1}$ are uniformly continuous, i.e., $\varphi$ is also a uniform homeomorphism between $C_p(X)$ and $C_p(Y)$. Therefore the space $X$ is $u$-equivalent to $Y$. Now, if $X$ is $L$-equivalent to $Y$ then $L(X)$ is linearly homeomorphic to $L(Y)$, so the weak dual $E_X$ of the space $L(X)$ is linearly homeomorphic to the weak dual $E_Y$ of the space $L(Y)$. By Problem 315, $E_X$ is linearly homeomorphic to $C_p(X)$ and $E_Y$ is linearly homeomorphic to $C_p(Y)$. Therefore $C_p(X)$ is linearly homeomorphic to $C_p(Y)$, i.e., the spaces $X$ and $Y$ are $l$-equivalent.

**V.317.** *Prove that $C_b(X)$ is complete (as a uniform space with its linear uniformity) if and only if $X$ is a $b_f$-space.*

**Solution.** For any space $Z$, if $A \subset Z$ then $\pi_A : C(Z) \to C(A)$ is the restriction map; for each $f \in C^*(A)$ let $||f||_A = \sup\{|f(x)| : x \in A\}$. If the set $A$ is bounded in $Z$ then we consider that $\pi_A : C(Z) \to C^*(A)$.

*Fact 1.* If $Z$ is a space and $A \subset Z$ is bounded in $Z$ then the set $C_A = \pi_A(C(Z))$ is closed in $C_u^*(A)$.

*Proof.* Recall that the topology of the space $C_u^*(A)$ is generated by the metric given by the formula $\rho(f, g) = ||f - g||_A$ for any $f, g \in C^*(A)$ (see Fact 1 of T.357). The following property is crucial for the proof.

(1) For any $f \in C_A$ there exists $g \in C^*(Z)$ such that $||g||_Z = ||f||_A$ and $\pi_A(g) = f$.

Fix a function $h \in C(Z)$ with $h|A = f$ and let $M = ||f||_A$. Define a function $g \in \mathbb{R}^Z$ as follows: $g(x) = h(x)$ whenever $|h(x)| \le M$; if $h(x) > M$ then $g(x) = M$ and if $h(x) < -M$ then $g(x) = -M$. It if immediate that $g \in C^*(Z)$, $g|A = f$ and $|g(z)| \le M$ for all $z \in Z$. Therefore $||g||_Z = ||f||_A$, i.e., (1) is proved.

Take any function $f \in \overline{C}_A$; there exists a sequence $S = \{f_n : n \in \omega\} \subset C_A$ which converges uniformly to $f$. Passing to a subsequence of $S$ if necessary we can assume, without loss of generality, that $\|f_n - f_{n+1}\|_A \leq 2^{-n}$ for all $n \in \omega$. Apply (1) to fix an arbitrary $g_0 \in C^*(Z)$ such that $\pi_A(g_0) = f_0$. Proceeding by induction assume that $n \in \omega$ and we have functions $g_0, \ldots, g_n \in C^*(Z)$ such that $g_i|A = f_i$ and $\|g_i - g_{i+1}\|_Z \leq 2^{-i}$ for all $i < n$. Since $\|f_n - f_{n+1}\|_A \leq 2^{-n}$, we can apply (1) to find a function $p \in C^*(Z)$ such that $p|A = f_{n+1} - f_n$ and $\|p\|_Z \leq 2^{-n}$. Then $g_{n+1} = g_n + p \in C^*(Z)$ and $\|g_n - g_{n+1}\|_Z = \|p\|_Z \leq 2^{-n}$ while $g_{n+1}|A = f_{n+1}$. Thus our inductive procedure can be continued to construct a sequence $\{g_n : n \in \omega\} \subset C^*(Z)$ such that $g_n|A = f_n$ and $\|g_n - g_{n+1}\|_Z \leq 2^{-n}$ for all $n \in \omega$. It is easy to see that $\{g_n : n \in \omega\}$ is a Cauchy sequence in a complete metric space $C_u^*(Z)$, so it converges uniformly to a function $g \in C^*(Z)$. Therefore the sequence $S' = \{g_n|A : n \in \omega\}$ converges uniformly to the function $g|A$. Since $g_n|A = f_n$ for all $n \in \omega$, the sequence $S'$ coincides with $S$, so $f = g|A \in C_A$. Therefore $\overline{C}_A = C_A$, i.e., $C_A$ is closed in $C_u^*(A)$ and Fact 1 is proved.

Assume first that $C_b(X)$ is complete and $X$ is not a $b_f$-space, i.e., there exists a function $p \in \mathbb{R}^X \backslash C_b(X)$ such that, for any bounded set $B \subset X$ we can find a function $f \in C_b(X)$ with $p|B = f|B$. It is obvious that for any $A \subset X$ the set $F_A = \{f \in C_b(X) : f|A = p|A\}$ is closed in $C_b(X)$ and $F_A \neq \emptyset$ if $A$ is bounded and nonempty. If $\mathcal{B} = \{F_B : B \neq \emptyset$ is a bounded subset of $X\}$ then it follows from $F_{A \cup B} = F_A \cap F_B$ that $\mathcal{B}$ is a filterbase in $C_b(X)$; let $\mathcal{F}$ be a filter which contains $\mathcal{B}$.

To see that the family $\mathcal{F}$ is a Cauchy filter denote by $\mathcal{U}$ the linear uniformity of $C_b(X)$ and fix any set $U \in \mathcal{U}$; there exists $W \in \tau(0, C_b(X))$ such that the set $G = \{(f, g) \in (C_b(X))^2 : f - g \in W\}$ is contained in $U$. There exists a bounded set $B \subset X$ and $\varepsilon > 0$ for which $O = \{f \in C_b(X) : f(B) \subset (-\varepsilon, \varepsilon)\} \subset W$. Then $F_B \in \mathcal{F}$ and $f, g \in F_B$ implies that $f|B = p|B = g|B$ and hence $(f - g)(x) = 0$ for any $x \in B$, i.e., $f - g \in O \subset W$. Therefore $(f, g) \in U$ for any $f, g \in F_B$; as a consequence, $F_B \times F_B \subset U$ and hence $\mathcal{F}$ is a Cauchy filter in $C_b(X)$. The uniform space $C_b(X)$ being complete there is a function $q \in \bigcap\{\overline{Q} : Q \in \mathcal{F}\} \subset \bigcap \mathcal{B}$. Since $\{x\}$ is a bounded subset of $X$, we must have $q \in F_{\{x\}}$ and therefore $q(x) = p(x)$ for any $x \in X$; this shows that $p = q \in C_b(X)$, so we obtained a contradiction which proves necessity.

Now assume that $X$ is a $b_f$-space and fix a Cauchy filter $\mathcal{F}$ in $C_b(X)$. Given a point $x \in X$ let $P_F^x = \{f(x) : f \in F\}$ for any $F \in \mathcal{F}$. Since $P_{F \cap G}^x \subset P_F^x \cap P_G^x$ for any $F, G \in \mathcal{F}$, the family $\mathcal{P}_x = \{\overline{P_F^x} : F \in \mathcal{F}\}$ is a filterbase of closed subsets of $\mathbb{R}$ and it follows from the Cauchy property of $\mathcal{F}$ that $\mathcal{P}_x$ is a Cauchy filterbase in $\mathbb{R}$. Therefore there is a unique point $c \in \bigcap \mathcal{P}_x$; let $p(x) = c$. This gives us a function $p : X \to \mathbb{R}$.

Fix an arbitrary bounded subset $B \subset X$; it is easy to see that the family $\{\pi_B(F) : F \in \mathcal{F}\}$ is a Cauchy filter in $C_B = \pi_B(C(X))$. The metric space $C_u^*(B)$ (with the metric induced by the norm $\| \cdot \|_B$) is complete; the set $C_B$ is closed in $C_u^*(B)$ by Fact 1, so it is also complete and hence $C_B$ is complete as a uniform space (see Problem 118). Consequently, there exists a unique function $h \in C_B$ such that $h \in \bigcap\{\overline{\pi_B(F)} : F \in \mathcal{F}\}$.

If $\varepsilon > 0$ then the set $W = \{f \in C_B : \|f\|_B < \frac{\varepsilon}{2}\}$ is an open neighborhood of $\mathbf{0}$ in $C_B$, so, by the Cauchy property of $\pi_B(\mathcal{F})$, there exists $F \in \mathcal{F}$ such that $\pi_B(f) - \pi_B(g) \in W$, i.e., $\|\pi_B(f) - \pi_B(g)\|_B < \frac{\varepsilon}{2}$ for any $f, g \in F$. It follows from $h \in \overline{\pi_B(F)}$ that $\|h - \pi_B(f)\|_B \leq \frac{\varepsilon}{2} < \varepsilon$ for all $f \in F$. This shows that

(2) for any $\varepsilon > 0$ there exists $F \in \mathcal{F}$ such that $\|h - \pi_B(f)\|_B < \varepsilon$ for all $f \in F$.

Now take any $x \in B$; given $\varepsilon > 0$ apply (2) to find a set $F \in \mathcal{F}$ such that $\|h - \pi_B(f)\|_B < \frac{\varepsilon}{3}$. We also have $p(x) \in \overline{\{f(x) : f \in F\}}$ by the definition of $p(x)$, so there exists $f \in F$ with $|p(x) - f(x)| < \frac{\varepsilon}{3}$. Therefore $|p(x) - h(x)| \leq \frac{2}{3}\varepsilon < \varepsilon$; since $\varepsilon > 0$ was chosen arbitrarily, we proved that $p(x) = h(x)$ for all $x \in B$. Since $h \in C_B$, there exists $g \in C_b(X)$ such that $g|B = h = p|B$. Thus the function $p$ is $b$-continuous; recalling that $X$ is a $b_f$-space we conclude that $p$ is continuous. Besides, it follows from the property (2) and the equality $p|B = h$ that

(3) for any bounded set $B \subset X$ and $\varepsilon > 0$ there exists a set $F \in \mathcal{F}$ such that $\|\pi_B(p) - \pi_B(f)\|_B < \varepsilon$ for all $f \in F$.

Finally fix any set $F \in \mathcal{F}$ and a neighborhood $U$ of the function $p$ in $C_b(X)$. We can choose a bounded set $B \subset X$ and a number $\varepsilon > 0$ such that the set $G = \{f \in C_p(X) : \|\pi_B(p) - \pi_B(f)\|_B < \varepsilon\}$ is contained in $U$. By the property (3) there exists $F' \in \mathcal{F}$ such that $\|\pi_B(p) - \pi_B(f)\|_B < \varepsilon$ for any $f \in F'$. Take any $f \in F' \cap F$; then $\|\pi_B(p) - \pi_B(f)\|_B < \varepsilon$ and hence $f \in U \cap F$. This shows that $U \cap F \neq \emptyset$ for any $U \in \tau(p, C_b(X))$, so $p \in \overline{F}$ for all $F \in \mathcal{F}$, i.e., $p \in \bigcap\{\overline{F} : F \in \mathcal{F}\}$. Therefore the uniform space $C_b(X)$ is complete and hence we settled sufficiency.

**V.318.** *Given a space $X$ call a set $P \subset C(X)$ equicontinuous at a point $x \in X$ if, for any $\varepsilon > 0$ there exists $U \in \tau(x, X)$ such that $f(U) \subset (f(x) - \varepsilon, f(x) + \varepsilon)$ for any $f \in P$. The family $P$ is called equicontinuous if it is equicontinuous at every point $x \in X$. Say that $P$ is pointwise bounded if the set $\{f(x) : f \in P\}$ is bounded in $\mathbb{R}$ for any $x \in X$. Prove that, for any equicontinuous pointwise bounded set $P \subset C(X)$, the closure of $P$ in the space $C_b(X)$ is compact. In particular, if $X$ is pseudocompact and $P \subset C(X)$ is equicontinuous and pointwise bounded then the closure of $P$ in $C_u(X)$ is compact.*

**Solution.** For any space $Z$, if $A \subset Z$ then $\pi_A : C(Z) \to C(A)$ is the restriction map; for each $f \in C^*(A)$ let $\|f\|_A = \sup\{|f(x)| : x \in A\}$. If the set $A$ is bounded in $Z$ then we consider that $\pi_A : C(Z) \to C^*(A)$. Say that a set $Q \subset C(Z)$ is adequate if it is pointwise bounded and equicontinuous. The set $\mathrm{cl}_p(Q)$ is the closure of $Q$ in the space $C_p(Z)$; analogously, $\mathrm{cl}_k(Q)$ and $\mathrm{cl}_b(Q)$ are the closures of $Q$ in the spaces $C_k(Z)$ and $C_b(Z)$ respectively.

*Fact 1.* For any space $Z$ if $Q \subset C(Z)$ is an adequate set then $Q' = \pi_A(Q) \subset C(A)$ is adequate for any $A \subset Z$.

*Proof.* Since $\{\pi_A(f)(x) : f \in Q\} = \{f(x) : f \in Q\}$ for any $x \in A$, the set $Q'$ is pointwise bounded. Given any $x \in A$ and $\varepsilon > 0$ take a set $U \in \tau(x, Z)$ such that

$f(U) \subset (f(x) - \varepsilon, f(x) + \varepsilon)$ for all $f \in Q$. Then $U' = U \cap A \in \tau(x, A)$ and $f(U') \subset (f(x) - \varepsilon, f(x) + \varepsilon)$ for all $f \in Q$ which shows that $Q'$ is equicontinuous and hence Fact 1 is proved.

**Fact 2.** For any space $Z$ if $Q \subset C(Z)$ is a pointwise bounded set then $\mathrm{cl}_p(Q)$ is also pointwise bounded. Therefore the sets $\mathrm{cl}_k(Q)$ and $\mathrm{cl}_b(Q)$ are pointwise bounded as well.

*Proof.* Let $F = \mathrm{cl}_p(Q)$ and fix a point $x \in Z$. If $e_x(f) = f(x)$ for every $f \in C(Z)$ then $e_x : C_p(Z) \to \mathbb{R}$ is a continuous function and $e_x(Q) = \{f(x) : f \in Q\}$ is a bounded subset of $\mathbb{R}$. Therefore $e_x(F) \subset \overline{e_x(Q)}$ is also bounded in $\mathbb{R}$. Since $e_x(F) = \{f(x) : f \in F\}$, we proved that the set $F$ is pointwise bounded. It follows from $\mathrm{cl}_b(Q) \subset \mathrm{cl}_k(Q) \subset \mathrm{cl}_p(Q)$ that the sets $\mathrm{cl}_b(Q)$ and $\mathrm{cl}_k(Q)$ are also pointwise bounded, so Fact 2 is proved.

**Fact 3.** For any space $Z$ if $Q \subset C(Z)$ is an equicontinuous set then $\mathrm{cl}_p(Q)$ is equicontinuous and coincides with the closure $Q'$ of the set $Q$ in $\mathbb{R}^Z$. In particular, the sets $\mathrm{cl}_k(Q)$ and $\mathrm{cl}_k(Q)$ are also equicontinuous.

*Proof.* It is evident that $\mathrm{cl}_p(Q) \subset Q'$; to prove the opposite inclusion it suffices to show that $Q' \subset C(Z)$. Every element of an equicontinuous family must be continuous, so it is sufficient to show that $Q'$ is equicontinuous. Take a point $x \in Z$ and $\varepsilon > 0$. There exists $U \in \tau(x, Z)$ such that $f(U) \subset (f(x) - \frac{\varepsilon}{2}, f(x) + \frac{\varepsilon}{2})$ for all $f \in Q$. If $f \in Q'$ and $|f(x) - f(y)| > \frac{\varepsilon}{2}$ for some $y \in U$ then the set $O = \{g \in \mathbb{R}^Z : |g(x) - g(y)| > \frac{\varepsilon}{2}\}$ is open in $\mathbb{R}^Z$ and contains $f$. Therefore there is $g \in O \cap Q$ and hence $|g(x) - g(y)| > \frac{\varepsilon}{2}$ which is a contradiction with $g(U) \subset (g(x) - \frac{\varepsilon}{2}, g(x) + \frac{\varepsilon}{2})$. Thus $f(U) \subset [f(x) - \frac{\varepsilon}{2}, f(x) + \frac{\varepsilon}{2}] \subset (f(x) - \varepsilon, f(x) + \varepsilon)$ for all $f \in Q'$ which shows that $Q'$ is equicontinuous. Consequently, $Q' \subset C(Z)$ which implies that $Q' = \mathrm{cl}_p(Q)$. Finally observe that $\mathrm{cl}_b(Q) \subset \mathrm{cl}_k(Q) \subset \mathrm{cl}_p(Q)$, so the sets $\mathrm{cl}_b(Q)$ and $\mathrm{cl}_k(Q)$ are also equicontinuous and hence Fact 3 is proved.

**Fact 4.** If $Z$ is a space and $Q \subset C(Z)$ is an adequate set then $\mathrm{cl}_p(Q)$ is compact.

*Proof.* For any $x \in Z$ fix a compact subset $K_x \subset \mathbb{R}$ such that $\{f(x) : f \in Q\} \subset K_x$ and denote by $Q'$ the closure of $Q$ in $\mathbb{R}^Z$. It follows from $Q \subset K = \prod_{x \in Z} K_x$ that $Q'$ is compact being a closed subset of $K$. Now it follows from Fact 3 that $\mathrm{cl}_p(Q) = Q'$ is also compact, so Fact 4 is proved.

**Fact 5.** If $Z$ is a space and $Q \subset C(Z)$ is an equicontinuous set then the topology $\tau_k$ induced on $Q$ from $C_k(Z)$ coincides with the topology $\tau_p$ induced on $Q$ from $C_p(Z)$.

*Proof.* It is clear that $\tau_p \subset \tau_k$, so fix a point $q \in Q$ and a set $U \in \tau_k$. Choose a compact set $K \subset Z$ and $\varepsilon > 0$ such that the set $V = \{f \in Q : |f(x) - q(x)| < \varepsilon$ for all $x \in K\}$ is contained in $U$. For every $x \in K$ there exists a set $O_x \in \tau(x, Z)$ such that $f(O_x) \subset (f(x) - \frac{\varepsilon}{3}, f(x) + \frac{\varepsilon}{3})$ for all $f \in Q$. Choose a finite set $A \subset Z$ for which $K \subset \bigcup\{O_x : x \in A\}$ and observe that the set $W = \{f \in Q : |f(x) - q(x)| < \frac{\varepsilon}{3}$ for all $x \in A\}$ belongs to $\tau_p$.

Fix any $f \in W$ and $y \in K$; there exists $x \in A$ such that $y \in O_x$ and hence $|f(y) - f(x)| < \frac{\varepsilon}{3}$. By the definition of $O_x$ we also have $|q(x) - q(y)| < \frac{\varepsilon}{3}$, so it follows from the inequality $|f(y) - q(y)| \leq |f(y) - f(x)| + |f(x) - q(x)| + |q(x) - q(y)|$ that $|f(y) - q(y)| < \frac{\varepsilon}{3} + \frac{\varepsilon}{3} + \frac{\varepsilon}{3} = \varepsilon$. The point $y \in K$ was chosen arbitrarily, so we proved that $|f(y) - q(y)| < \varepsilon$ for all $y \in K$ and hence $f \in V$. As a consequence, $W \subset V \subset U$, i.e., we proved that, for any $q \in Q$ and $U \in \tau_k$ with $q \in U$ there is $W \in \tau_p$ such that $q \in W \subset U$. Now it is easy to see that $\tau_k \subset \tau_p$, i.e., $\tau_k = \tau_p$ and hence Fact 5 is proved.

*Fact 6.* If $Z$ is a $\mu$-space, i.e., the closure of every bounded subset of $Z$ is compact then the topologies of $C_k(Z)$ and $C_b(Z)$ coincide.

*Proof.* A set $U \subset C(Z)$ belongs to $\tau(C_k(Z))$ (or to $C_b(Z)$ respectively) if and only if for any $f \in U$ there exists a compact (bounded) set $K \subset Z$ and $\varepsilon > 0$ such that the set $[f, K, \varepsilon] = \{g \in C(Z) : |f(z) - g(z)| < \varepsilon$ for all $z \in K\}$ is contained in $U$. Since every compact subset of $Z$ is bounded in $Z$, every $U \in \tau(C_k(Z))$ belongs to $\tau(C_b(Z))$, i.e., $\tau(C_k(Z)) \subset \tau(C_b(Z))$.

Now if $U \in \tau(C_b(Z))$ and $f \in U$ then there is a bounded set $B \subset Z$ and $\varepsilon > 0$ such that $[f, B, \varepsilon] \subset U$. Since $Z$ is a $\mu$-space, the set $K = \overline{B}$ is compact, so $[f, K, \varepsilon] \subset [f, B, \varepsilon] \subset U$ which shows that $U \in \tau(C_k(Z))$ and hence we proved that $\tau(C_b(Z)) \subset \tau(C_k(Z))$, i.e., $\tau(C_b(Z)) = \tau(C_k(Z))$ and hence Fact 6 is proved.

*Fact 7.* Given a space $Z$, if $p \in C(Z)$, the set $B \subset Z$ is bounded (or compact) and $\varepsilon > 0$ then the interior of the set $[p, B, \varepsilon] = \{f \in C(Z) : |f(z) - p(z)| < \varepsilon$ for all $z \in B\}$ in the space $C_b(Z)$ (or in the space $C_k(Z)$ respectively) contains the function $p$.

*Proof.* Let $U_1 = [p, B, \frac{\varepsilon}{3}]$ and $U_{n+1} = \bigcup\{[f, B, 3^{-n-1}\varepsilon] : f \in U_n\}$ for all $n \in \mathbb{N}$. The set $U = \bigcup\{U_n : n \in \mathbb{N}\}$ is open in $C_b(Z)$ (or in $C_k(Z)$ respectively) because, if $f \in U$ then $f \in U_n$ for some $n \in \mathbb{N}$ and hence $[f, B, 3^{-n-1}\varepsilon] \subset U_{n+1} \subset U$. It is evident that $p \in U$. Let $\delta_n = \frac{\varepsilon}{3} + \ldots + \frac{\varepsilon}{3^n}$ for all $n \in \mathbb{N}$. We will prove by induction that

(1) $U_n \subset [p, B, \delta_n]$ for all $n \in \mathbb{N}$.

This is clear for $n = 1$, so assume that $n \in \mathbb{N}$, $U_n \subset [p, B, \delta_n]$ and take any function $f \in U_{n+1}$. There exists $g \in U_n$ such that $f \in [g, B, 3^{-n-1}\varepsilon]$. By the induction hypothesis, we have $g \in [p, B, \delta_n]$ which implies that

$$|f(x) - p(x)| \leq |f(x) - g(x)| + |g(x) - p(x)| < 3^{-n-1}\varepsilon + \delta_n = \delta_{n+1}$$

for every $x \in B$, so $f \in [p, B, \delta_{n+1}]$ for all $f \in U_{n+1}$, i.e., $U_{n+1} \subset [p, B, \delta_{n+1}]$ and hence our inductive procedure can be carried out for all $n \in \mathbb{N}$ which shows that (1) is true. Since $\delta_n < \varepsilon$, it follows from the property (1) that $U_n \subset [p, B, \delta_n] \subset [p, B, \varepsilon]$ for all $n \in \mathbb{N}$; an immediate consequence is that $U \subset [p, B, \varepsilon]$. The set $U$ being open, we have $p \in U \subset \text{Int}([p, B, \varepsilon])$ and hence Fact 7 is proved.

*Fact 8.* Given a continuous map $\varphi : Z \to T$ let $\varphi^*(f) = f \circ \varphi$ for any $f \in C(T)$. Then the dual map $\varphi^* : C_b(T) \to C_b(Z)$ is continuous; besides, if we consider that $\varphi^* : C_k(T) \to C_k(Z)$ then $\varphi^*$ is continuous as well.

*Proof.* Fix a function $q \in C(T)$ and let $p = \varphi^*(q)$; take any set $U \ni p$ which is open in $C_b(Z)$ (or in $C_k(Z)$ respectively). There exists a bounded (compact) set $B \subset Z$ such that the set $[p, B, \varepsilon] = \{f \in C(Z) : |f(z) - p(z)| < \varepsilon$ for all $z \in B\}$ is contained in $U$ for some $\varepsilon > 0$.

The set $B' = \varphi(B)$ is bounded (compact) in the space $T$; consider the set $W' = \{g \in C(T) : |g(x) - q(x)| < \varepsilon$ for all $x \in B'\}$. If $W$ is the interior of $W'$ in the space $C_b(T)$ (or in $C_k(T)$ respectively) then $q \in W$ by Fact 7. If $g \in W$ then, for any $z \in B$, we have $\varphi^*(g)(z) = g(\varphi(z))$ and $x = \varphi(z) \in B'$. As a consequence, $|\varphi^*(g)(z) - p(z)| = |g(x) - q(x)| < \varepsilon$ for every $z \in B$, i.e., $\varphi^*(g) \in [p, B, \varepsilon] \subset U$ for all $g \in W$, so the set $W$ witnesses continuity of $\varphi^*$ at the point $q$ and hence Fact 8 is proved.

*Fact 9.* If $Z$ is a space and $Q \subset C_b(Z)$ is an adequate set then let $\varphi(z)(f) = f(z)$ for any $f \in Q$ and $z \in Z$. This defines a function $\varphi(z) \in C^*(Q)$ for any $z \in Z$ and the map $\varphi : Z \to C_u^*(Q)$ is continuous.

*Proof.* Observe first that $\varphi(z)(Q) = \{f(z) : f \in Q\}$ is a bounded subset of $\mathbb{R}$, so $\varphi(z)$ is a bounded function on $Q$ for every $z \in Z$. Every $\varphi(z)$ is continuous on the set $Q$ endowed with the pointwise convergence topology by TFS-166; therefore $\varphi(z)$ is continuous on $Q$ with the topology induced from $C_b(Z)$. This shows that $\varphi : Z \to C_u^*(Q)$.

To see that $\varphi$ is continuous take any $y \in Z$ and $U \in \tau(C_u^*(Q))$ with $\varphi(y) \in U$. There is $\varepsilon > 0$ such that the set $W = \{g \in C_u^*(Q) : \|g - \varphi(y)\|_Q < \varepsilon\}$ is contained in $U$. By equicontinuity of $Q$ there exists $V \in \tau(y, Z)$ such that $|f(z) - f(y)| < \frac{\varepsilon}{2}$ for all $z \in V$ and $f \in Q$. This is the same as saying that $|\varphi(z)(f) - \varphi(y)(f)| < \frac{\varepsilon}{2}$ for all $f \in Q$ which shows that $\|\varphi(z) - \varphi(y)\|_Q \leq \frac{\varepsilon}{2} < \varepsilon$ and hence $\varphi(z) \in W$ for every $z \in V$. In other words, $\varphi(V) \subset W \subset U$, i.e., the set $V$ witnesses that $\varphi$ is continuous at the point $y$, so Fact 9 is proved.

*Fact 10.* Given a space $Q$ and $Y \subset C_u^*(Q)$ let $e(z)(f) = f(z)$ for every $f \in Y$ and $z \in Q$. Then $e(z) \in C(Y)$ for all $z \in Z$ and the set $E = \{e(z) : z \in Q\} \subset C(Y)$ is adequate.

*Proof.* If $z \in Q$ then $e(z)$ is continuous on $Y$ with the topology induced from $C_p(Q)$ by TFS-166; since $C_u^*(Q)$ induces a stronger topology on $Y$, every $e(z)$ is continuous. Given any $f \in Y$ it follows from $f \in C^*(Q)$ that the set $f(Q)$ is bounded in $\mathbb{R}$. Therefore $\{e(z)(f) : z \in Q\} = f(Q)$ is bounded in $\mathbb{R}$ and hence the set $E$ is pointwise bounded.

Next, fix a function $f \in Y$ and $\varepsilon > 0$. The set $U = \{g \in Y : \|g - f\|_Q < \varepsilon\}$ is open in $Y$ and $f \in U$. For any $z \in Q$ and $g \in U$ we have

$$|e(z)(g) - e(z)(f)| = |g(z) - f(z)| \leq \|g - f\|_Q < \varepsilon,$$

i.e., $e(z)(U) \subset (e(z)(f) - \varepsilon, e(z)(f) + \varepsilon)$ for any $z \in Q$ and hence $U$ witnesses equicontinuity of $E$ at the point $f$. This shows that $E$ is adequate, so Fact 10 is proved.

Returning to our solution fix an adequate set $P$ in the space $C_b(X)$ and let $\varphi(x)(f) = f(x)$ for all $x \in X$ and $f \in P$. Then $\varphi(x) \in C^*(P)$ for every $x \in X$ and the map $\varphi : X \to C_u^*(P)$ is continuous by Fact 9; let $Y = \varphi(X)$. For each $f \in P$ let $e(f)(y) = y(f)$ for any $y \in Y$; then $e(f) \in C(Y)$ for every $f \in P$ and the set $Q = e(P) \subset C(Y)$ is adequate by Fact 10.

The closure $F$ of the set $Q$ in the space $C_p(Y)$ is compact by Fact 4. Furthermore, the set $F$ has to be adequate by Fact 2 and Fact 3. By Fact 5, the topology induced on $F$ from $C_k(Y)$ coincides with the topology induced on $F$ from $C_p(Y)$. Therefore the set $F$ is compact as a subspace of $C_k(Y)$. The space $Y$ is metrizable because so is $C_u^*(P)$; therefore the topologies of $C_b(Y)$ and $C_k(Y)$ coincide by Fact 6; consequently, $F$ is a compact subspace of $C_b(Y)$.

Since the dual map $\varphi^* : C_b(Y) \to C_b(X)$ is continuous by Fact 8, the set $G = \varphi^*(F)$ is a compact subspace of $C_b(X)$. Our last step is to prove that $P \subset G$, so fix an arbitrary function $f \in P$. Then $g = e(f) \in Q$; given $x \in X$ we have $\varphi^*(g)(x) = g(\varphi(x))$. If $y = \varphi(x)$ then $g(\varphi(x)) = e(f)(y) = y(f) = \varphi(x)(f) = f(x)$. Therefore $g(\varphi(x)) = f(x)$ for all $x \in X$ which shows that $\varphi^*(g) = f$ and hence $f \in \varphi^*(Q) \subset \varphi^*(F) = G$. We proved that the set $P$ is contained in a compact set $G$, so the closure of $P$ in $C_b(X)$ is compact, i.e., our solution is complete.

**V.319.** *Prove that, for any $b_f$-space $X$, a set $P \subset C(X)$ is equicontinuous and pointwise bounded if and only if the closure of $P$ in the space $C_b(X)$ is compact.*

**Solution.** Given a space $Z$ and $A \subset Z$ let $\pi_A : C(Z) \to C(A)$ be the restriction map. If $A$ is bounded in $Z$ then $||f||_A = \sup\{|f(x)| : x \in A\}$.

*Fact 1.* If $Z$ is a space and $B \subset Z$ is a bounded subspace of $Z$ then the restriction map $\pi_B : C_b(Z) \to C_u^*(B)$ is continuous.

*Proof.* Fix a function $p \in C_b(Z)$, a set $O \in \tau(\pi_B(p), C_u^*(B))$ and let $q = \pi_B(p)$. There exists $\varepsilon > 0$ such that $V = \{f \in C_u^*(B) : ||f - q||_B < \varepsilon\} \subset O$. Consider the set $W' = \{f \in C_b(Z) : |f(x) - p(x)| < \frac{\varepsilon}{2}$ for all $x \in B\}$; by Fact 7 of V.318, we have $p \in W = \mathrm{Int}(W')$. If $f \in W$ then $||f - p||_B \leq \frac{\varepsilon}{2} < \varepsilon$ which shows that $||\pi_B(f) - q||_B < \varepsilon$ and therefore $\pi_B(f) \in V$. Thus $\pi_B(W) \subset V \subset O$, so the set $W$ witnesses continuity of $\pi_B$ at the point $p$ and hence Fact 1 is proved.

Returning to our solution observe that necessity was established in Problem 318, so assume that the closure $K$ of the set $P$ in the space $C_b(X)$ is compact. It suffices to show that $K$ pointwise bounded and equicontinuous. For any point $x \in X$ let $e_x(f) = f(x)$ for every $f \in K$. Then $e_x : K \to \mathbb{R}$ is continuous if the topology $\tau_p$ is induced on $K$ from $C_p(X)$ (see TFS-166). Since the topology $\tau_b$ induced on $K$ from $C_b(X)$ contains $\tau_p$ (actually, $\tau_b = \tau_p$ but we don't need that), the map $e_x : (K, \tau_b) \to \mathbb{R}$ is continuous. Therefore the set $e_x(K) = \{f(x) : f \in K\}$ is compact and hence bounded in $\mathbb{R}$, i.e., we proved that $K$ is pointwise bounded.

To see that the set $K$ is equicontinuous fix a point $a \in X$; for any $x \in X$ the function $u_x = |e_x - e_a|$ is continuous on $C_b(X)$, so the set $u_x(K)$ is compact. Therefore the number $\varphi(x) = \sup u_x(K)$ is consistently defined; it is easy to see that $\varphi(x) = \sup\{|f(x) - f(a)| : f \in K\}$ for any $x \in X$. Our next step is to prove that the function $\varphi : X \to \mathbb{R}$ is continuous. Since $X$ is a $b_f$-space, it suffices to show that $\varphi|B$ can be extended to a continuous function on $X$ for any bounded set $B \subset X$.

So, fix a bounded set $B \subset X$; there is no loss of generality to assume that $a \in B$. The restriction map $\pi_B : C_b(X) \to C_u^*(B)$ is continuous by Fact 1, so $L = \pi_B(K)$ is a compact subset of $C_u^*(B)$. Fix an arbitrary $\varepsilon > 0$; since $L$ is metrizable, there exists a finite $Q \subset K$ such that, for any $f \in K$ there is $q \in Q$ with $\|q - f\|_B < \frac{\varepsilon}{3}$. The function $\xi : X \to \mathbb{R}$ defined by the equality $\xi(y) = \sup\{|r(y) - r(a)| : r \in Q\}$ for every $y \in X$, is continuous on $X$.

If $x \in B$ then there exists $f \in K$ such that $|f(x) - f(a)| > \varphi(x) - \frac{\varepsilon}{3}$. There is $q \in Q$ with $\|q - f\|_B < \frac{\varepsilon}{3}$ and, in particular, $|q(x) - f(x)| < \frac{\varepsilon}{3}$ and $|q(a) - f(a)| < \frac{\varepsilon}{3}$. Thus, $|f(x) - f(a)| \leq |f(x) - q(x)| + |q(x) - q(a)| + |q(a) - f(a)|$ which shows that $\varphi(x) - \frac{\varepsilon}{3} < \frac{2}{3}\varepsilon + |q(x) - q(a)|$ and hence $\varphi(x) \geq \xi(x) \geq |q(x) - q(a)| > \varphi(x) - \varepsilon$. The point $x \in B$ was chosen arbitrarily, so we established that

(1) for every $\varepsilon > 0$ there is $\xi \in C(X)$ such that $|\xi(x) - \varphi(x)| < \varepsilon$ for all $x \in B$.

Applying the property (1) we can choose, for each $n \in \omega$ a continuous function $\xi_n : X \to \mathbb{R}$ such that $|\xi_n(x) - \varphi(x)| < 2^{-n}$ for all $x \in B$. An immediate consequence is that the sequence $\{\pi_B(\xi_n) : n \in \omega\}$ converges uniformly to $\varphi|B$ and hence $\varphi|B$ belongs to the closure of the set $\pi_B(C(X))$ in $C_u^*(B)$. However, $\pi_B(C(X))$ is closed in $C_u^*(B)$ by Fact 1 of V.317, so $\varphi|B \in \pi_B(C(X))$ for any bounded $B \subset X$. The $b_f$-property of $X$ implies that the function $\varphi$ is continuous on $X$.

Finally, take any $\varepsilon > 0$; by continuity of the function $\varphi$ there exists a set $U \in \tau(a, X)$ such that $|\varphi(x) - \varphi(a)| = |\varphi(x)| < \varepsilon$ for all $x \in U$. Given $f \in K$ we have $|f(x) - f(a)| \leq |\varphi(x)| < \varepsilon$ for every $x \in U$ and hence $f(U) \subset (f(a) - \varepsilon, f(a) + \varepsilon)$. This shows that $K$ is equicontinuous and makes our solution complete.

**V.320.** *Given a space $X$, let $[P, \varepsilon] = \{\varphi \in L(X) : \varphi(P) \subset (-\varepsilon, +\varepsilon)\}$ for every $P \subset C(X)$ and $\varepsilon > 0$. Prove that a set $U \subset L(X)$ is open in $L(X)$ if and only if, for any $\eta \in U$ there exists an equicontinuous pointwise bounded set $P \subset C(X)$ and $\varepsilon > 0$ such that $\eta + [P, \varepsilon] \subset U$. In other words, the topology of $L(X)$ coincides with the topology of uniform convergence on equicontinuous pointwise bounded subsets of $C(X)$.*

**Solution.** As before, we consider that $X \subset L(X)$ and $X$ is a Hamel basis of $L(X)$. Call a set $P \subset C(X)$ *adequate* if $P$ is equicontinuous and pointwise bounded; let $\mathcal{O} = \{U \subset L(X) :$ for each $\eta \in U$ there exists an adequate set $P \subset C(X)$ and $\varepsilon > 0$ such that $\eta + [P, \varepsilon] \subset U\}$. We have to prove that $\mathcal{O} = \tau(L(X))$.

We omit an easy checking that $\mathcal{O}$ is a topology on $L(X)$. Let us prove, however, that $L_*(X) = (L(X), \mathcal{O})$ is a locally convex space. Given a set $A \subset L(X)$ we denote by $\text{Int}_*(A)$ the interior of $A$ in the space $L_*(X)$. Let us first prove that

(1) if $\eta \in L(X)$ then $\eta \in \mathrm{Int}_*(\eta + [P, \varepsilon])$ for any adequate set $P \subset C(X)$ and $\varepsilon > 0$.

Consider the set $H_1 = \eta + [P, \frac{\varepsilon}{3}]$ and let $H_{n+1} = \bigcup\{\varphi + [P, 3^{-n-1}\varepsilon] : \varphi \in H_n\}$ for all $n \in \mathbb{N}$. The set $H = \bigcup\{H_n : n \in \mathbb{N}\}$ is open in $L_*(X)$; indeed, if $\eta \in H$ then there exists $n \in \mathbb{N}$ such that $\eta \in H_n$ and hence $\eta + [P, 3^{-n-1}\varepsilon] \subset H_{n+1} \subset H$. If $n \in \mathbb{N}$ and $\delta_n = 3^{-1}\varepsilon + \ldots + 3^{-n}\varepsilon$ then it takes a straightforward induction to show that $H_n \subset \eta + [P, \delta_n]$. Since $\delta_n < \varepsilon$, we have $H_n \subset [P, \varepsilon]$ for all $n \in \mathbb{N}$ and hence $H \subset \eta + [P, \varepsilon]$. Therefore $\eta \in H \subset \mathrm{Int}_*(\eta + [P, \varepsilon])$, i.e., (1) is proved.

Let $\mathcal{B} = \{B \in \mathcal{O} : \mathbf{0} \in B\}$; it is evident that $B_0, B_1 \in \mathcal{B}$ implies $B_0 \cap B_1 \in \mathcal{B}$. If $B \in \mathcal{B}$ and $\eta \in L(X)$ then we can find an adequate set $P \subset C(X)$ and $\varepsilon > 0$ with $[P, \varepsilon] \subset B$. There are $x_1, \ldots, x_n \in X$ and $\lambda_1, \ldots, \lambda_n \in \mathbb{R}$ such that $\eta = \sum_{i=1}^n \lambda_i x_i$. Take any $i \leq n$; the set $P$ being pointwise bounded there is $q_i > 0$ such that $|f(x_i)| \leq q_i$ for all $f \in P$. If $q = \sum_{i=1}^n (|\lambda_i| + 1)q_i$ then $|\eta(f)| = |\sum_{i=1}^n \lambda_i f(x_i)| \leq q$ for all $f \in P$. Now if $t > \frac{q}{\varepsilon}$ then $|\eta(\frac{1}{t}f)| = \frac{1}{t}|\eta(f)| \leq \frac{q}{t} < \varepsilon$ for all $f \in P$ which shows that $\frac{1}{t}\eta \in [P, \varepsilon]$ and hence $\eta \in t[P, \varepsilon] \subset tB$, i.e., we proved that every set $B \in \mathcal{B}$ is absorbing.

Now take any $\eta \in L(X)\setminus\{\mathbf{0}\}$. There are distinct points $x_1, \ldots, x_n \in X$ and $\lambda_1, \ldots, \lambda_n \in \mathbb{R}\setminus\{0\}$ such that $\eta = \lambda_1 x_1 + \ldots + \lambda_n x_n$. Take a function $f \in C(X)$ such that $f(x_1) = \frac{1}{\lambda_1}$ and $f(x_i) = 0$ for all $i = 2, \ldots, n$. The set $P = \{f\}$ is adequate and $\eta \notin [P, 1]$; it follows from (1) that the set $B = \mathrm{Int}_*([P, 1])$ belongs to $\mathcal{B}$. Since $\eta \notin B$, we proved that $\bigcap \mathcal{B} = \{\mathbf{0}\}$.

Given any $B \in \mathcal{B}$ there is an adequate set $P \subset C(X)$ such that $[P, \varepsilon] \subset B$ for some $\varepsilon > 0$. Apply the property (1), to see that the set $V = \mathrm{Int}_*([P, \frac{\varepsilon}{2}])$ belongs to $\mathcal{B}$; it is straightforward that $V + V \subset [P, \varepsilon] \subset B$, so, for any $B \in \mathcal{B}$ there is $V \in \mathcal{B}$ with $V + V \subset B$.

Next assume that $B \in \mathcal{B}$ and $\eta \in B$; choose an adequate set $P \subset C(X)$ such that $\eta + [P, \varepsilon] \subset B$ for some $\varepsilon > 0$. Apply the property (1) once more to conclude that $V = \mathrm{Int}_*([P, \varepsilon])$ belongs to $\mathcal{B}$; since also $\eta + V \subset B$, we proved that, for any $B \in \mathcal{B}$ and $\eta \in B$ there exists $V \in \mathcal{B}$ with $\eta + V \subset B$.

Suppose that $B \in \mathcal{B}$ and $\varepsilon > 0$; choose an adequate set $P \subset C(X)$ such that $[P, \delta] \subset B$ for some $\delta > 0$. The set $V = \mathrm{Int}_*([P, \frac{\delta}{\varepsilon}])$ belongs to $\mathcal{B}$ by (1); if $\lambda \in (-\varepsilon, \varepsilon)$ and $\eta \in V$ then $|\lambda\eta(f)| \leq \varepsilon|\eta(f)| < \varepsilon\frac{\delta}{\varepsilon} = \delta$ for each $f \in P$ and hence $\lambda\eta \in [P, \delta] \subset B$. Therefore we can apply Problem 202 to the family $\mathcal{B}$ to see that there exists a topology $\tau$ on the set $L(X)$ such that $(L(X), \tau)$ is a linear topological space for which $\mathcal{B}$ is a local base at $\mathbf{0}$ in $(L(X), \tau)$. If $U \in \tau$ and $\eta \in U$ then $-\eta + U \in \tau$ and $\mathbf{0} \in -\eta + U$, so there is $B \in \mathcal{B}$ such that $B \subset -\eta + U$ and hence $\eta + B \subset U$. Since $B \in \mathcal{O}$, there is an adequate set $P \subset C(X)$ for which $[P, \varepsilon] \subset B$ for some $\varepsilon > 0$. Consequently, $\eta + [P, \varepsilon] \subset U$; this proves that $U \in \mathcal{O}$ and therefore $\tau \subset \mathcal{O}$.

If $U \in \mathcal{O}$ and $\eta \in U$ then it is easy to see that $B = -\eta + U \in \mathcal{O}$; since $\mathbf{0} \in B$, the set $B$ belongs to $\tau$ and hence $U = \eta + B \in \tau$. Therefore $\mathcal{O} \subset \tau$, i.e., $\mathcal{O} = \tau$, so we proved that $L_*(X)$ is a linear topological space. If $B \in \mathcal{B}$ then there exists an adequate set $P \subset C(X)$ such that $[P, \varepsilon] \subset B$ for some $\varepsilon > 0$. It is easy to see that

$[P, \varepsilon]$ is a convex set so $W = \mathrm{Int}_*([P, \varepsilon])$ is also convex (see Problem 203). Since $\mathbf{0} \in W \subset B$, we proved that $L_*(X)$ is a locally convex space.

Let $\mathcal{O}' = \{O \cap X : O \in \mathcal{O}\}$ be the topology induced on $X$ from $L_*(X)$. If $O \in \mathcal{O}'$ and $x \in O$ then there exists an adequate set $P \subset C(X)$ such that $(x + [P, \varepsilon]) \cap X \subset O$ for some $\varepsilon > 0$. By equicontinuity of $P$ there exists a set $U \in \tau(x, X)$ such that $|f(y) - f(x)| < \varepsilon$ for each $f \in P$ and $y \in U$. If $y \in U$ then $(y - x)(f) = f(y) - f(x) \in (-\varepsilon, \varepsilon)$ for every $f \in P$ which shows that $y - x \in [P, \varepsilon]$ and hence $y \in x + [P, \varepsilon] \subset O$. Thus, for every set $O \in \mathcal{O}'$ and $x \in O$ there is $U \in \tau(x, X)$ such that $U \subset O$. An immediate consequence is that $\mathcal{O}' \subset \tau(X)$ and hence the identity map $i : X \to (X, \mathcal{O}')$ is continuous. We can consider that $i : X \to L_*(X)$, so Problem 312 is applicable to find a linear continuous map $\xi : L(X) \to L_*(X)$ such that $\xi|X = i$. The set $X$ is a Hamel basis of $L(X)$, so the linear extension of $i$ is unique and hence $\xi$ has to be the identity map. It follows from continuity of $\xi$ that $\mathcal{O} \subset \tau(L(X))$.

To prove that $\tau(L(X)) \subset \mathcal{O}$ fix a set $U \in \tau(L(X))$ such that $\mathbf{0} \in U$ and choose a convex balanced set $V \in \tau(\mathbf{0}, L(X))$ with $V \subset U$. Consider the set $Q = \{\alpha \in (L(X))^* : \alpha(V) \subset \mathbb{I}\} \subset C(L(X))$. Given $\eta \in L(X)$ there is $r > 0$ such that $\eta \in tV$ for all $t > r$. Therefore $\frac{\eta}{r+1} \in V$ which implies that $|\alpha(\frac{\eta}{r+1})| \leq 1$, i.e., $|\alpha(\eta)| \leq r + 1$ for all $\alpha \in Q$. Thus $\{\alpha(\eta) : \alpha \in Q\} \subset [-r - 1, r + 1]$ which shows that the set $Q$ is pointwise bounded.

Given $\varepsilon > 0$ and $\eta \in L(X)$ the set $W = \frac{\varepsilon}{2}V$ is an open neighborhood of $\mathbf{0}$ in $L(X)$, so $\eta + W \in \tau(\eta, L(X))$. If $\alpha \in Q$ and $w \in W$ then $\alpha(\eta + w) = \alpha(\eta) + \alpha(w)$. There is $v \in V$ such that $w = \frac{\varepsilon}{2}v$ and hence $|\alpha(w)| = \frac{\varepsilon}{2}|\alpha(v)| \leq \frac{\varepsilon}{2} < \varepsilon$. This shows that $\alpha(\eta + W) \subset (\alpha(\eta) - \varepsilon, \alpha(\eta) + \varepsilon)$ for all $\alpha \in Q$ and hence the set $Q$ is equicontinuous, i.e., $Q$ is an adequate set. The set $P = \{\alpha|X : \alpha \in Q\} \subset C(X)$ is also adequate by Fact 1 of V.318. It is easy to check that

(2)  for any $\alpha \in Q$ if $f = \alpha|X$ then $\eta(f) = \alpha(\eta)$ for every $\eta \in L(X)$.

It follows from (1) that the set $O = \mathrm{Int}_*([P, 1])$ belongs to $\mathcal{B}$. If there exists a point $\eta \in O \backslash \overline{V}$ then we can apply Problem 223 to the set $\overline{V}$ to find a functional $\alpha \in (L(X))^*$ such that $\alpha(\eta) > 1$ while $\alpha(\overline{V}) \subset \mathbb{I}$ and hence $\alpha(V) \subset \mathbb{I}$, i.e., $\alpha \in Q$. If $f = \alpha|X$ then $f \in P$, so we can apply (2) to see that $\eta(f) = \alpha(\eta) > 1$ which is a contradiction with $\eta \in O \subset [P, 1]$. This contradiction shows that $O \subset \overline{V}$ and hence $O \subset U$. Therefore every neighborhood of $\mathbf{0}$ in $L(X)$ contains an element of $\mathcal{B} \subset \mathcal{O}$, so the identity map $\xi : L(X) \to L_*(X)$ is open by Fact 3 of S.496. Consequently, $\xi$ is a homeomorphism and hence $\tau(L(X)) = \mathcal{O}$, i.e., our solution is complete.

**V.321.** *Given a space $X$ and $P \subset X$ let $I_P = \{f \in C(X) : f(P) \subset \{0\}\}$. Prove that for any linear continuous functional $\varphi : C_k(X) \to \mathbb{R}$ which is not identically zero on $C_k(X)$, there exists a compact subspace $K \subset X$ (called the support of $\varphi$ and denoted by $supp(\varphi)$) such that $\varphi(I_K) = \{0\}$ and $\varphi(I_{K'}) \neq \{0\}$ whenever $K'$ is a proper compact subset of $K$.*

**Solution.** If $Z$ is a space and $A \subset Z$ then $[A, \varepsilon] = \{f \in C(Z) : f(A) \subset (-\varepsilon, \varepsilon)\}$ for any $\varepsilon > 0$. If $A \neq \emptyset$ and a function $f \in C(Z)$ is bounded on the set $A$ then $\|f\|_A = \sup\{|f(x)| : x \in A\}$.

*Fact 1.* Suppose that $Z$ is a space and $A$ is a nonempty subset of $Z$ while some $f \in C(Z)$ is bounded on $A$. Then there exists a function $g \in C^*(Z)$ such that $g|A = f|A$ and $\|g\|_Z = \|f\|_A$.

*Proof.* Let $M = \|f\|_A$ and define a function $g \in \mathbb{R}^Z$ as follows: $g(x) = f(x)$ whenever $|f(x)| \leq M$; if $f(x) > M$ then $g(x) = M$ and if $f(x) < -M$ then $g(x) = -M$. It if immediate that $g \in C^*(Z)$, $g|A = f|A$ and $|g(z)| \leq M$ for all $z \in Z$. Therefore $\|g\|_Z = \|f\|_A$, i.e., Fact 1 is proved.

Fix a nontrivial continuous linear functional $\varphi : C_k(X) \to \mathbb{R}$ and denote by $\mathcal{K}$ the family of all nonempty compact subsets of $X$ such that $\varphi(I_K) = \{0\}$. By continuity of $\varphi$ at the point $\mathbf{0}$ there exists a compact set $K \subset X$ such that $\varphi([K, \varepsilon]) \subset (-1, 1)$ for some $\varepsilon > 0$; observe that $I_K \subset [K, \varepsilon]$. If $K = \emptyset$ then $[K, \varepsilon] = C(X)$ and hence we have the inclusion $\varphi(C(X)) \subset (-1, 1)$. However, there exists $f \in C(X)$ such that $\varphi(f) \neq 0$ and hence, for the function $g = \frac{1}{\varphi(f)} \cdot f$, we have $\varphi(g) = 1$ which is a contradiction. Therefore $K$ is a nonempty set.

If $f \in I_K$ and $\varphi(f) \neq 0$ then, for $r = \frac{1}{\varphi(f)}$ we have $\varphi(rf) = 1$. Now it follows from $rf \in I_K$ that $rf \in [K, \varepsilon]$ which implies $|\varphi(rf)| < 1$; this contradiction shows that $\varphi(I_K) = \{0\}$, i.e., $K \in \mathcal{K}$ and hence the family $\mathcal{K}$ is nonempty.

If $K \in \mathcal{K}$, $f, g \in C(X)$ and $f|K = g|K$ then $f - g \in I_K$ and hence $\varphi(f - g) = 0$ which implies that $\varphi(f) = \varphi(g)$. Therefore we proved that

(1) if $K \in \mathcal{K}$ and $f, g \in C(X)$ then $f|K = g|K$ implies that $\varphi(f) = \varphi(g)$; in particular, if $\varphi(f) \neq 0$ then $f(x) \neq 0$ for some $x \in K$.

Next assume that $K, L \in \mathcal{K}$ and $K \cap L = \emptyset$. There exists a function $f \in C(X)$ such that $\varphi(f) \neq 0$ and hence $f|L$ is not identically zero by (1). Let $h(x) = 0$ for all $x \in K$ and $h(x) = f(x)$ for all $x \in L$. It is evident that $h : K \cup L \to \mathbb{R}$ is a continuous function, so there exists $g \in C(X)$ such that $g|(K \cup L) = h$ (see Fact 1 of T.218). Since $g|L = h|L = f|L$, we can apply (1) to see that $\varphi(g) = \varphi(f) \neq 0$. On the other hand, $g \in I_K$, so $\varphi(g) = 0$. This contradiction shows that

(2) $K \cap L \neq \emptyset$ for any $K, L \in \mathcal{K}$.

Now fix any $K, L \in \mathcal{K}$; by the property (2) the set $F = K \cap L$ is nonempty. If $F \notin \mathcal{K}$ then there exists a function $f \in I_F$ such that $\varphi(f) \neq 0$. The property (1) shows that $f(a) \neq 0$ for some $a \in K$. Let $g(x) = f(x)$ for all $x \in K$ and $g(x) = 0$ whenever $x \in L \backslash K$. It follows from Fact 2 of T.354 that the function $g$ is continuous on $K \cup L$, so we can apply Fact 1 of T.218 again to find a function $h \in C(X)$ such that $h|(K \cup L) = g$. Since $h|K = f|K$, the property (1) shows that $\varphi(h) = \varphi(f) \neq 0$. However, $h(L) = \{0\}$, so $\varphi(h) = 0$; the obtained contradiction proves that

(3) $K \cap L \in \mathcal{K}$ for any $K, L \in \mathcal{K}$ and hence $\mathcal{K}$ is a centered family.

It follows from (3) that the set $K = \bigcap \mathcal{K}$ is compact and nonempty. If $K'$ is a proper compact subset of $K$ then $K' \notin \mathcal{K}$ and hence $\varphi(I_{K'}) \neq \{0\}$, so it suffices to show that $K \in \mathcal{K}$. To do that we will need the following statement.

(4) Assume that $F_n \subset X$ is compact and $F_{n+1} \subset F_n$ for all $n \in \omega$. If $F = \bigcap_{n \in \omega} F_n$ then the set $\bigcup_{n \in \omega} I_{F_n}$ is dense in $I_F$.

Fix a function $f \in I_F$ and $U \in \tau(f, C_k(X))$; there exists a compact set $Q \subset X$ such that $f + [Q, \varepsilon] \subset U$ for some $\varepsilon > 0$. If $W = f^{-1}((-\varepsilon, \varepsilon))$ then $W$ is an open subset of $X$ with $F \subset U$, so we can find $n \in \omega$ such that $F_n \subset W$ (see Fact 1 of S.326). It follows from compactness of the set $F_n$ that $r = \|f\|_{F_n} < \varepsilon$. Apply Fact 1 to find a function $h \in C(X)$ such that $\|h\|_X \leq r$ and $h|F_n = f|F_n$. Then $q = f - h \in I_{F_n}$ and $\|q - f\|_X \leq r < \varepsilon$. In particular, $q - f \in [Q, \varepsilon]$, so $q \in f + [Q, \varepsilon] \subset U$. Thus every neighborhood of $f$ in $C_k(X)$ intersects the set $\bigcup_{n \in \omega} I_{F_n}$, so (4) is proved. Now we can show that

(5) if $F_n \in \mathcal{K}$ for every $n \in \omega$ then $F = \bigcap_{n \in \omega} F_n$ also belongs to $\mathcal{K}$.

It follows from (3) that there is no loss of generality to assume that $F_{n+1} \subset F_n$ for all $n \in \omega$. By the property (4) the set $H = \bigcup_{n \in \omega} I_{F_n}$ is dense in $I_F$, so the set $\{0\} = \varphi(H)$ is dense in $\varphi(I_F)$ and hence $\varphi(I_F) = \{0\}$, i.e., (5) is proved.

Finally assume that $K \notin \mathcal{K}$, i.e., there exists $f \in I_K$ such that $\varphi(f) \neq 0$. Given any $n \in \omega$ if $W_n = f^{-1}((-2^{-n}, 2^{-n}))$ then $K \subset W_n$, so there exists $F_n \in \mathcal{K}$ such that $F_n \subset W_n$. The set $F = \bigcap_{n \in \omega} F_n$ belongs to $\mathcal{K}$ by (5) and $F \subset \bigcap_{n \in \omega} W_n = f^{-1}(0)$ which shows that $f \in I_F$ and hence $\varphi(f) = 0$, a contradiction. Therefore $K$ is the minimal element of $\mathcal{K}$, i.e., our solution is complete.

**V.322.** *Recall that a set $B$ is a barrel in a locally convex space $L$ if $B$ is closed, convex, balanced and absorbing in $L$. The space $L$ is barreled if any barrel in $L$ is a neighborhood of $\mathbf{0}$. Prove that a locally convex space $L$ is barreled whenever it has the Baire property.*

**Solution.** Assume that a locally convex space $L$ has the Baire property and fix a barrel $B \subset L$. It is straightforward that the set $rB = \{rx : x \in B\}$ is also a barrel for any $r > 0$. Since $B$ is absorbing, we have the equality $L = \bigcup\{nB : n \in \mathbb{N}\}$, so it follows from the Baire property of the space $L$ that there exists $m \in \mathbb{N}$ such that $W = \text{Int}(mB) \neq \emptyset$. The set $V = \frac{1}{2m} W$ is open in $L$ and $V \subset \frac{1}{2}B$; since $\frac{1}{2}B$ is balanced, the set $U = -V$ is also contained in $\frac{1}{2}B$. It is clear that $G = U + V$ is an open neighborhood of $\mathbf{0}$. If $x \in G$ then there are $u \in U$ and $v \in V$ such that $x = u + v$. There exist $b, b' \in B$ with $u = \frac{1}{2}b$ and $v = \frac{1}{2}b'$. It follows from convexity of $B$ that $x = \frac{1}{2}b + \frac{1}{2}b' \in B$. The point $x \in G$ was chosen arbitrarily, so $G \subset B$ which shows that every barrel $B \subset L$ is a neighborhood of $\mathbf{0}$, i.e., the space $L$ is barreled.

**V.323.** *Prove that $C_k(X)$ is a barreled space if and only if $X$ is a $\mu$-space, i.e., the closure of any bounded subspace of $X$ is compact. Deduce from this fact that $C_k(X)$ is barreled for any realcompact space $X$.*

**Solution.** For any $Q \subset X$ and $\varepsilon > 0$ let $[Q, \varepsilon] = \{f \in C(X) : f(Q) \subset (-\varepsilon, \varepsilon)\}$ and $I_Q = \{f \in C(X) : f(Q) \subset \{0\}\}$; if a function $f \in C(X)$ is bounded on $Q$ then $||f||_Q = \sup\{|f(x)| : x \in Q\}$.

Assume that $C_k(X)$ is barreled and take a bounded set $P$ in the space $X$. It is easy to see that the set $B = \{f \in C(X) : f(P) \subset \mathbb{I}\}$ is a barrel in $C_k(X)$, so we can find a set $U \in \tau(C_k(X))$ for which $\mathbf{0} \in U \subset B$. There exists a compact set $K \subset X$ such that $[K, \varepsilon] \subset U$ for some $\varepsilon > 0$.

If $P \backslash K \neq \emptyset$ then fix a point $x \in P \backslash K$; there exists a function $f \in C(X)$ such that $f(x) = 2$ and $f(K) \subset \{0\}$. Therefore $f \in [K, \varepsilon] \subset B$ while $f(x) \in f(P)$ does not belong to $\mathbb{I}$. This contradiction shows that $P \subset K$ and hence the closure of any bounded set $P \subset X$ is compact, i.e., we proved necessity.

Now assume that $X$ is a $\mu$-space and fix a barrel $B$ in the space $C_k(X)$. We omit an easy checking that the set $B' = B \cap C^*(X)$ is a barrel in the space $C_u^*(X)$; the space $C_u^*(X)$ is completely metrizable, so it has the Baire property. Therefore $C_u^*(X)$ is barreled by Problem 322 and hence the set $B'$ is a neighborhood of $\mathbf{0}$ in $C_u^*(X)$; this makes it possible to pick a number $\delta > 0$ such that $[X, \delta] \subset B' \subset B$.

Denote by $\Phi$ the set of all continuous linear functionals $\xi : C_k(X) \to \mathbb{R}$ such that $\xi(B) \subset \mathbb{I}$. For any $\xi \in \Phi \backslash \{\mathbf{0}\}$ let $S_\xi$ be the support of $\xi$, i.e., $S_\xi$ is a compact set such that $\xi(I_{S_\xi}) = \{0\}$ while $\xi(I_F) \neq \{0\}$ for any proper compact set $F \subset S_\xi$ (see Problem 321). We will need the following property of the supports of the functionals from $\Phi \backslash \{\mathbf{0}\}$.

(1) if $\xi \in \Phi \backslash \{\mathbf{0}\}$ and $U \cap S_\xi \neq \emptyset$ for some $U \in \tau(X)$ then there exists a function $f \in C(X)$ such that $f(X \backslash U) \subset \{0\}$ and $\varphi(f) \neq 0$.

By minimality of the set $K = S_\xi$ there exists a function $g \in C(X)$ such that $\xi(g) \neq 0$ and $g(K \backslash U) \subset \{0\}$. Since $K$ is compact, the set $g(K)$ is bounded in $\mathbb{R}$, so there exists $r > 0$ such that, for the function $h = rg$, we have $h(K) \subset \mathbb{I}$. It is clear that $h(K \backslash U) \subset \{0\}$ and $\xi(h) \neq 0$. Let us establish that

(2) there exists $\varepsilon > 0$ such that $||f - h||_K < \varepsilon$ implies $\xi(f) \neq 0$.

Indeed, the set $O = C_k(X) \backslash \xi^{-1}(0)$ is open in $C_k(X)$, so we can find a compact set $L \subset X$ and $\varepsilon > 0$ such that $K \subset L$ and $||f - h||_L < \varepsilon$ implies $f \in O$, i.e., $\xi(f) \neq 0$. Now, if $||f - h||_K < \varepsilon$ then we can apply Fact 1 of V.321 to find a function $q \in C(X)$ such that $q|K = (f - h)|K$ and $||q||_X < \varepsilon$. Then $||q + h - h||_X = ||q||_X < \varepsilon$ which implies $||q + h - h||_L < \varepsilon$ and hence $\xi(q + h) \neq 0$. We also have $(q + h)|K = f|K$, so it follows from the definition of support that $\xi(f) = \xi(q + h) \neq 0$ and hence (2) is proved.

The set $M = \{x \in K : |h(x)| \geq \frac{\varepsilon}{2}\}$ is compact and $M \subset U$. It is easy to see that the closure $F$ of the set $X \backslash U$ in the space $\beta X$ does not meet $M$, so there exists a continuous function $d : \beta X \to [0, 1]$ such that $d(F) \subset \{0\}$ and $d(M) \subset \{1\}$. Then $u = d|X \in C(X)$ and $u : X \to [0, 1]$ while $u(X \backslash U) \subset \{0\}$ and $u(M) \subset \{1\}$. If $f = uh$ then $f(x) = h(x)$ whenever $x \in (K \backslash U) \cup M$. If $x \in (K \backslash M) \cap U$ then $|h(x)| < \frac{\varepsilon}{2}$ and hence $|f(x) - h(x)| = |h(x)| \cdot |u(x) - 1| \leq |h(x)| < \frac{\varepsilon}{2}$. This shows that $||f - h||_K \leq \frac{\varepsilon}{2} < \varepsilon$, so we can apply (2) to conclude that $\xi(f) \neq 0$; it is clear that $f(X \backslash U) \subset \{0\}$, so (1) is proved. Our next step is to show that

(3) the set $P = \bigcup\{S_\xi : \xi \in \Phi \setminus \{0\}\}$ is bounded in $X$.

If (3) is not true then there exists a sequence $S = \{\xi_n : n \in \omega\} \subset \Phi \setminus \{0\}$ and a discrete family $\mathcal{U} = \{U_n : n \in \omega\}$ of open subsets of $X$ such that $U_n \cap S_{\xi_n} \neq \emptyset$ for all $n \in \omega$ (see Fact 1 of V.245). We can assume without loss of generality (passing to an appropriate subsequence of $S$ if necessary) that $U_n \cap S_{\xi_m} = \emptyset$ whenever $m < n$. Apply the property (1) to find a function $f_n \in C(X)$ such that $\xi_n(f_n) \neq 0$ and $f_n(X \setminus U_n) \subset \{0\}$ for each $n \in \omega$.

Let $c_0 = 0$; then $\xi_0(c_0 f_0) = 0$. Proceeding inductively, assume that $n \in \omega$ and we have $c_0, \ldots, c_n \in \mathbb{R}$ such that $\xi_i(c_0 f_0 + \ldots + c_i f_i) = i$ for all $i \leq n$. If $c_{n+1} = \frac{n+1-\sum_{i=0}^n \xi_{n+1}(c_i f_i)}{\xi_{n+1}(f_{n+1})}$ then $\xi_{n+1}(c_{n+1} f_{n+1}) = n + 1 - \sum_{i=0}^n \xi_{n+1}(c_i f_i)$ and hence $\xi_{n+1}(\sum_{i=0}^{n+1} c_i f_i) = n + 1$, so our inductive procedure can be continued to obtain a sequence $\{c_n : n \in \omega\} \subset \mathbb{R}$ such that $\xi_n(c_0 f_0 + \ldots + c_n f_n) = n$ for all $n \in \omega$.

The family $\mathcal{U}$ being discrete, the function $f = \sum_{i=0}^\infty c_i f_i$ is continuous on $X$. If $g_n = \sum_{i=0}^n c_i f_i$ then it follows from $S_{\xi_n} \cap (\bigcup_{m>n} U_m) = \emptyset$ that $f|S_{\xi_n} = g_n|S_{\xi_n}$ and hence $\xi_n(f) = \xi_n(g_n) = n$ for all $n \in \omega$. The set $B$ is absorbing, so there exists $r > 0$ such that $\frac{1}{r} \cdot f \in B$ and hence $|\xi_n(\frac{1}{r} \cdot f)| \leq 1$ which implies that $n = |\xi_n(f)| \leq r$ for all $n \in \omega$. This contradiction shows that the set $P$ is bounded in $X$ and hence (3) is proved.

Since $X$ is a $\mu$-space, the set $K = \overline{P}$ is compact; given any $f \in I_K$ we have $f(S_\xi) = \{0\}$ and hence $\xi(f) = 0$ for all $\xi \in \Phi \setminus \{0\}$. If $f \notin B$ then we can apply Problem 223 to find a continuous linear functional $\xi : C_k(X) \to \mathbb{R}$ such that $\xi(B) \subset \mathbb{I}$ and $\xi(f) > 1$. Such a functional $\xi$ must belong to $\Phi \setminus \{0\}$, so $\xi(f) = 0$; this contradiction shows that $f \in B$. The function $f \in I_K$ was chosen arbitrarily, so we proved that $I_K \subset B$.

Now take any $f \in [K, \frac{8}{3}]$; then $2f \in [K, \frac{2}{3}\delta]$ and we can apply Fact 1 of V.321 to find a function $g \in C(X)$ such that $g|K = f|K$ and $\|g\|_X \leq \|f\|_K \leq \frac{1}{3}\delta$. This implies that $\|2g\|_X \leq \frac{2}{3}\delta < \delta$ and hence $2g \in [X, \delta] \subset B$. Furthermore, $2f - 2g \in I_K$, so again $2g - 2f \in B$. By convexity of $B$ we have $f = \frac{1}{2} \cdot 2g + \frac{1}{2}(2f - 2g) \in B$, i.e., we established that $[K, \frac{8}{3}] \subset B$. Now apply Fact 7 of V.318 to conclude that there is a set $W \in \tau(C_k(X))$ such that $\mathbf{0} \in W \subset [K, \frac{8}{3}] \subset B$. Thus $B$ is a neighborhood of $\mathbf{0}$, so we proved that the space $C_k(X)$ is barreled; this settles sufficiency.

Finally observe that if $X$ is a realcompact space then it is a $\mu$-space by Fact 1 of V.246, so $C_k(X)$ is barreled and hence our solution is complete.

**V.324.** *Prove that $C_p(X)$ is barreled if and only if all bounded subspaces of $X$ are finite.*

**Solution.** For any $Q \subset X$ and $\varepsilon > 0$ let $[Q, \varepsilon] = \{f \in C(X) : f(Q) \subset (-\varepsilon, \varepsilon)\}$. Assume that $C_p(X)$ is barreled and take a bounded set $P$ in the space $X$. It is easy to see that the set $B = \{f \in C_p(X) : f(P) \subset \mathbb{I}\}$ is a barrel in $C_p(X)$, so we can find a set $U \in \tau(C_p(X))$ for which $\mathbf{0} \in U \subset B$. There exists a finite set $K \subset X$ such that $[K, \varepsilon] \subset U$ for some $\varepsilon > 0$.

If $P \setminus K \neq \emptyset$ then fix a point $x \in P \setminus K$; there exists a function $f \in C_p(X)$ such that $f(x) = 2$ and $f(K) \subset \{0\}$. Therefore $f \in [K, \varepsilon] \subset B$ while $f(x) \in f(P)$ does not belong to $\mathbb{I}$. This contradiction shows that $P \subset K$ and hence every bounded set $P \subset X$ is finite, i.e., we proved necessity.

Now if every bounded subset of $X$ is finite then $X$ is a $\mu$-space, so $C_k(X)$ is barreled by Problem 323. However, every compact subset of $X$ is finite, so it is immediate from the definitions that $\tau(C_k(X)) = \tau(C_p(X))$. Therefore the space $C_p(X)$ is also barreled and hence we settled sufficiency.

**V.325.** *Give an example of a space $X$ such that $C_p(X)$ barreled but does have the Baire property.*

**Solution.** It was proved in TFS-285 that there exists a space $X$ in which all bounded subsets are finite while $C_p(X)$ does not have the Baire property. By Problem 324, the space $C_p(X)$ is barreled so $X$ is the required example.

**V.326.** *Given a point $z$ in a space $Z$, say that a family $\mathcal{B}$ of subsets of $Z$ is a local base of neighborhoods of $Z$ at the point $z$, if $z \in Int(U)$ for any $U \in \mathcal{B}$, and $z \in V \in \tau(Z)$ implies $U \subset V$ for some $U \in \mathcal{B}$. Prove that the family of all barrels in $C_k(X)$ constitutes a local base of neighborhoods of $0$ in $C_b(X)$. Deduce from this fact that the family of all barrels in $C_p(X)$ is also a local base of neighborhoods of $0$ in $C_b(X)$.*

**Solution.** We are going to need the sets $O(A, \varepsilon) = \{f \in C(X) : f(A) \subset (-\varepsilon, \varepsilon)\}$ and $Q(A, \varepsilon) = \{f \in C(X) : f(A) \subset [-\varepsilon, \varepsilon]\}$ for any $A \subset X$ and $\varepsilon > 0$. Let $\mathcal{B}_p$ be the family of all barrels in $C_p(X)$ and denote by $\mathcal{B}_k$ the family of all barrels in $C_k(X)$; it is straightforward that $\mathcal{B}_p \subset \mathcal{B}_k$.

*Fact 1.* If $Z$ is a space and $A \subset Z$ then the restriction map $\pi : C(Z) \to C(A)$ is continuous as a map between $C_k(Z)$ and $C_k(A)$. If we consider that $\pi$ maps $C_b(X)$ to $C_b(A)$ then $\pi$ is also continuous.

*Proof.* The map $\pi$ is linear, so it suffices to show its continuity at $0$ (see Fact 2 of S.496). Assume that $U$ is a neighborhood of $0$ in the space $C_k(A)$ (or in the space $C_b(A)$ respectively). There exists a compact (bounded) set $K \subset A$ such that $B = \{f \in C(A) : f(K) \subset (-\varepsilon, \varepsilon)\} \subset U$ for some $\varepsilon > 0$. Consider the set $V = \{f \in C(Z) : f(K) \subset (-\varepsilon, \varepsilon)\}$; if $W$ is the interior of $V$ in the space $C_k(Z)$ (or in $C_b(Z)$ respectively) then $0 \in W$ (see Fact 7 of V.318). Since also $\pi(W) \subset B \subset U$, the set $W$ witnesses continuity of the map $\pi : C_k(Z) \to C_k(A)$ (or continuity of the map $\pi : C_b(Z) \to C_b(A)$ respectively), so Fact 1 is proved.

Returning to our solution observe that

(1) $Q(A, \varepsilon)$ is a barrel in $C_p(X)$ and hence in $C_k(X)$ for every bounded set $A \subset X$.

Assume first that $U$ is a neighborhood of $0$ in the space $C_b(X)$. There exists a bounded set $A \subset X$ such that $O(A, \varepsilon) \subset U$ for some $\varepsilon > 0$. Then $B = Q(A, \frac{\varepsilon}{2})$ is a barrel in $C_p(X)$ such that $B \subset O(A, \varepsilon) \subset U$. Since also $B \in \mathcal{B}_k$, we proved that both families $\mathcal{B}_p$ and $\mathcal{B}_k$ are networks at the point $0$ in the space $C_b(X)$; therefore it suffices to show that $0$ belongs to the interior in $C_b(X)$ of every barrel in $C_k(X)$.

Fix any $B \in \mathcal{B}_k$; by Fact 1, the restriction map $\pi : C_k(\upsilon X) \to C_k(X)$ is continuous. Since $\pi$ is also linear, the set $B' = \pi^{-1}(B)$ is a barrel in $C_k(\upsilon X)$. Now, $\upsilon X$ is a $\mu$-space by Fact 1 of V.246, so we can apply Problem 323 to see that $C_k(\upsilon X)$ is barreled and therefore the set $B'$ is a neighborhood of $\mathbf{0}$ in $C_k(\upsilon X)$. Consequently, we can find a compact set $K \subset \upsilon X$ and a number $\varepsilon > 0$ such that $H = \{f \in C(\upsilon X) : f(K) \subset (-\varepsilon, \varepsilon)\} \subset B'$. The set $A = K \cap X$ is bounded in $X$; let us show that

(2)  $Q(A, \frac{\varepsilon}{2})$ is contained in the closure $E$ of the set $\pi(H)$ in the space $C_k(X)$.

Fix a function $f \in Q(A, \frac{\varepsilon}{2})$ and a set $U \in \tau(f, C_k(X))$. There exists a compact set $M \subset X$ such that $f + [M, \delta] \subset U$ for some $\delta > 0$. The function $f$ can be continuously extended over the space $\upsilon X$; let $g \in C(\upsilon X)$ be the unique extension of $f$. The set $F = \mathrm{cl}_{\upsilon X}(A) \subset K$ is compact and $g(F) \subset [-\frac{\varepsilon}{2}, \frac{\varepsilon}{2}]$. The space $K$ is compact and hence normal, so there exists a function $h : K \to [-\frac{\varepsilon}{2}, \frac{\varepsilon}{2}]$ such that $h|F = g|F$. Let $d(x) = h(x)$ for each $x \in K$; if $x \in M \setminus K$ then $d(x) = g(x)$. Observe that $P = K \cap M = A \cap M$ is compact and $h|P = g|P$. This, together with Fact 2 of T.354 shows that the function $d : K \cup M \to \mathbb{R}$ is continuous and hence we can find a function $q \in C(\upsilon X)$ for which $q|(K \cup M) = d$.

It follows from $q|M = f|M$ that $\pi(q) \in f + [M, \delta] \subset U$. Besides, we have the inclusions $q(K) \subset [-\frac{\varepsilon}{2}, \frac{\varepsilon}{2}] \subset (-\varepsilon, \varepsilon)$, so $q \in H$ and hence $\pi(q) \in \pi(H) \cap U$. Thus every neighborhood in $C_k(X)$ of each function $f \in Q(A, \frac{\varepsilon}{2})$ intersects the set $\pi(H)$, so (2) is proved.

The set $B$ is closed in the space $C_k(X)$ and $\pi(H) \subset B$, so $E \subset B$ and hence $O(A, \frac{\varepsilon}{2}) \subset Q(A, \frac{\varepsilon}{2}) \subset E \subset B$. Finally, apply Fact 7 of V.318 to see that $\mathbf{0}$ belongs to the interior of $O(A, \frac{\varepsilon}{2})$ in the space $C_b(X)$ and hence the interior of the set $B$ in $C_b(X)$ also contains $\mathbf{0}$. The set $B \in \mathcal{B}_k$ was chosen arbitrarily, so we checked that both families $\mathcal{B}_k$ and $\mathcal{B}_p$ are local bases of neighborhoods of $\mathbf{0}$ in the space $C_b(X)$.

**V.327.** *Let $\varphi : C_k(X) \to C_p(Y)$ be a linear continuous map. Prove that $\varphi$ is continuous considered as a map from $C_b(X)$ to $C_b(Y)$.*

**Solution.** Take a set $U \in \tau(\mathbf{0}, C_b(Y))$ and apply Problem 326 to find a barrel $B$ in the space $C_p(Y)$ such that $B \subset U$. The set $B' = \varphi^{-1}(B)$ is easily seen to be a barrel in $C_k(X)$, so we can apply Problem 326 again to find a set $W \in \tau(\mathbf{0}, C_b(X))$ such that $W \subset B'$. Then $\varphi(W) \subset B \subset U$, so the set $W$ witnesses continuity of the map $\varphi : C_b(X) \to C_b(Y)$ at the point $\mathbf{0}$. Applying Fact 2 of S.496 we conclude that the map $\varphi : C_b(X) \to C_b(Y)$ is continuous.

**V.328.** *Assuming that a space $Y$ is $l$-equivalent to a $b_f$-space $X$ prove that $Y$ is also a $b_f$-space. In other words, $b_f$-property is $l$-invariant.*

**Solution.** Fix a linear homeomorphism $\varphi : C_p(X) \to C_p(Y)$. It follows from Problem 327 that $\varphi : C_b(X) \to C_b(Y)$ is also a linear homeomorphism. The space $C_b(X)$ is complete as a uniform space with its linear uniformity (see Problem 317), so $C_b(Y)$ is also complete with respect to its linear uniformity. Applying Problem 317 again we conclude that $Y$ is a $b_f$-space.

**V.329.** *Let $\varphi : L_p(X) \to L_p(Y)$ be a linear homeomorphism. Prove that, if $X$ is a $b_f$-space then $\varphi$ is a linear homeomorphism of $L(X)$ onto $L(Y)$.*

**Solution.** For any space $Z$ let $C_Z = \{\alpha \in C_p(L_p(Z)) : \alpha$ is a linear functional on $L_p(Z)\}$. For each $f \in C_p(Z)$ there exists a unique linear functional $e_Z(f) \in C_Z$ such that $e_Z(f)|Z = f$. The map $e_Z : C_p(Z) \to C_Z$ is a linear homeomorphism whose inverse is the restriction map $\pi_Z : C_Z \to C_p(Z)$ (see Problem 235). If $P \subset C(Z)$ and $\varepsilon > 0$ then $[P, \varepsilon]_Z = \{u \in L(Z) : u(P) \subset (-\varepsilon, \varepsilon)\}$.

*Fact 1.* If $Z$ is a space, $\varepsilon > 0$ and $P \subset C(Z)$ is an equicontinuous pointwise bounded set then $[P, \varepsilon]_Z$ is a neighborhood of $\mathbf{0}$ in $L(Z)$.

*Proof.* Let $W_0 = [P, \frac{\varepsilon}{3}]_Z$ and $W_{n+1} = \bigcup\{u + [P, 3^{-n-2}\varepsilon]_Z : u \in W_n\}$ for all $n \in \omega$. If $\delta_n = \frac{\varepsilon}{3} + \ldots + \frac{\varepsilon}{3^{n+1}}$ then it takes an easy induction to show that $W_n \subset [P, \delta_n]_Z$ for every $n \in \omega$. Since $\delta_n < \varepsilon$ for all $n \in \omega$, the set $W = \bigcup_{n \in \omega} W_n$ is contained in $[P, \varepsilon]_Z$. If $u \in W$ then $u \in W_n$ for some $n \in \omega$ and hence $u + [P, 3^{-n-2}\varepsilon] \subset W_{n+1} \subset W$ which, together with Problem 320, shows that $W$ is open in $L(Z)$. Since $\mathbf{0} \in W \subset [P, \varepsilon]_Z$, the set $[P, \varepsilon]_Z$ is a neighborhood of $\mathbf{0}$ in $L(Z)$, i.e., Fact 1 is proved.

Returning to our solution observe that the spaces $X$ and $Y$ are $l$-equivalent (see Problem 237), so $Y$ is also a $b_f$-space by Problem 328. For any $\alpha \in C_Y$ let $\varphi^*(\alpha) = \alpha \circ \varphi$; it is easy to see that $\varphi^* : C_Y \to C_X$ is a linear homeomorphism, so the map $\xi = \pi_X \circ \varphi^* \circ e_Y : C_p(Y) \to C_p(X)$ is a linear homeomorphism as well. Given any $u \in L_p(X)$ let $\xi^*(u) = u \circ \xi$; this gives us a map $\xi^* : L_p(X) \to L_p(Y)$.

Fix any $u \in L_p(X)$ and $f \in C_p(Y)$; then $\xi^*(u)(f) = u(\xi(f)) = e_X(\xi(f))(u)$. Recall that $e_X(\xi(f)) = e_X(\pi_X(\varphi^*(e_Y(f)))) = \varphi^*(e_Y(f)) = e_Y(f) \circ \varphi$ and therefore $\xi^*(u)(f) = e_Y(f)(\varphi(u)) = \varphi(u)(f)$. This shows that the linear functional $\xi^*(u)$ coincides with $\varphi(u)$ for each $u \in L_p(X)$ and hence the map $\xi^*$ coincides with $\varphi$. To sum up, we proved that

(1)  there exists a linear homeomorphism $\xi : C_p(Y) \to C_p(X)$ such that $\varphi = \xi^*$, i.e., $\varphi$ is the dual map of $\xi$.

It is an easy consequence of Problem 327 that

(2)  the map $\xi : C_b(Y) \to C_b(X)$ is also a linear homeomorphism.

Fix a neighborhood $U$ of $\mathbf{0}$ in $L(Y)$; by Problem 320 there exists an equicontinuous pointwise bounded set $P \subset C(Y)$ such that $[P, \varepsilon]_Y \subset U$. If $F$ is the closure of $P$ in the space $C_b(Y)$ then $F$ is compact (see Problem 318). Therefore the set $G = \xi(F) \subset C_b(X)$ is also compact by (2). Since $X$ is a $b_f$-space, it follows from Problem 319 that $G$ is equicontinuous and pointwise bounded.

By Fact 1, there is a set $W \in \tau(L(X))$ such that $\mathbf{0} \in W \subset [G, \varepsilon]_X$. If $w \in W$ and $f \in P$ then $g = \xi(f) \in G$ and it follows from (1) that $\varphi(w) = \xi^*(w) = w \circ \xi$; as a consequence, $\varphi(w)(f) = w(\xi(f)) = w(g) \in (-\varepsilon, \varepsilon)$. This shows that we have the inclusions $\varphi(W) \subset [P, \varepsilon]_Y \subset U$, i.e., the set $W$ witnesses continuity of the map $\varphi : L(X) \to L(Y)$ at $\mathbf{0}$. By Fact 2 of S.496 the map $\varphi : L(X) \to L(Y)$ is continuous. An analogous proof shows that the map $\varphi^{-1} : L(Y) \to L(X)$ is also continuous, so $\varphi$ is a linear homeomorphism between the spaces $L(X)$ and $L(Y)$ as promised.

**V.330.** *Let X and Y be spaces one of which is a $b_f$-space. Prove that X is L-equivalent to Y if and only if X and Y are l-equivalent.*

**Solution.** If $X$ is $L$-equivalent to $Y$ then $X \overset{l}{\sim} Y$ by Problem 316, so assume that $X \overset{l}{\sim} Y$. It follows from Problem 328 that both $X$ and $Y$ are $b_f$-spaces; apply Problem 237 to see that there exists a linear homeomorphism $\varphi : L_p(X) \to L_p(Y)$. By Problem 239, the map $\varphi : L(X) \to L(Y)$ is also a linear homeomorphism, so the spaces $X$ and $Y$ are $L$-equivalent.

**V.331.** *Prove that there exist l-equivalent spaces which are not L-equivalent.*

**Solution.** Let $D$ be a discrete space of cardinality $\omega_1$; the spaces $X = L_p(D)$ and $Y = L_p(D) \oplus D$ are $l$-equivalent by Problem 278. Fix an arbitrary $n \in \mathbb{N}$; for any $x = (x_1, \ldots, x_n) \in X^n$ and $\lambda = (\lambda_1, \ldots, \lambda_n) \in \mathbb{R}^n$ let $\xi_n(x, \lambda) = \lambda_1 x_1 + \ldots + \lambda_n x_n$. This gives us a map $\xi_n : X^n \times \mathbb{R}^n \to L(X)$; since $X \subset L(X)$ and $L(X)$ is a linear topological space, every map $\xi_n$ is continuous.

The cardinal $\omega_1$ is a precaliber of $X$ by Problem 236; therefore $\omega_1$ is a precaliber of $X^n \times \mathbb{R}^n$ and hence $\omega_1$ is also a precaliber of $P_n = \xi_n(X^n \times \mathbb{R}^n)$ for all $n \in \mathbb{N}$. It follows from the equality $L(X) = \bigcup\{P_n : n \in \mathbb{N}\}$ that $\omega_1$ is a precaliber of $L(X)$ and, in particular, the space $L(X)$ has the Souslin property.

Consider the space $C_u^*(D)$ and let $||f|| = \sup\{|f(x)| : x \in D\}$ for every function $f \in C_u^*(D)$. If $d(f, g) = ||f - g||$ for all $f, g \in C_u^*(D)$ then $d$ is a metric which generates the topology of the space $C_u^*(D)$. For each $f \in C_u^*(D)$ and $r > 0$ the set $B(f, \varepsilon) = \{g \in C_u^*(D) : ||g - f|| < \varepsilon\}$ is the open ball of radius $\varepsilon$ centered at $f$.

Given $d \in D$ let $f_d(d) = 1$ and $f_d(x) = 0$ for all $x \in D\backslash\{d\}$. Since all functions are continuous on $D$, the set $E = \{f_d : d \in D\}$ is contained in $C_u^*(D)$. It is easy to see that the family $\{B(f_d, \frac{1}{2}) : d \in D\}$ is disjoint, so if $E \subset T \subset C_u^*(D)$ then the Souslin number of $T$ is uncountable.

There exists a continuous onto map $\varphi : Y \to E$ because $D$ is a retract of $Y$ which can be mapped continuously onto anything of cardinality at most $\omega_1$. We can consider that $\varphi : Y \to C_u^*(D)$; since $C_u^*(D)$ is a locally convex space, we can apply Problem 312 to find a continuous map $\zeta : L(Y) \to C_u^*(D)$ such that $\zeta|Y = \varphi$ and, in particular, $T = \zeta(L(Y)) \supset E$. By the above observation, $c(T) > \omega$, so $c(L(Y)) > \omega$; recalling that $c(L(X)) = \omega$ we conclude that the spaces $L(X)$ and $L(Y)$ are not even homeomorphic. Thus $X$ and $Y$ are $l$-equivalent spaces which fail to be $L$-equivalent.

**V.332.** *Prove that a space X has a weaker metrizable topology if and only if L(X) has a weaker metrizable topology. In particular, if X and Y are L-equivalent and X can be condensed onto a metrizable space then Y can also be condensed onto a metrizable space.*

**Solution.** Given a space $Z$ say that a set $A \subset C(Z)$ is adequate if $A$ is equicontinuous and pointwise bounded. Recall that the phrases "$Z$ has a weaker metrizable topology" and "$Z$ condenses onto a metrizable space" say the same thing.

*Fact 1.* Suppose that $Z$ is a space and we are given a metric $\rho$ on the set $Z$ such that $\rho(x, y) \leq 1$ for all $x, y \in Z$. Assume also that $\rho$ generates a topology $\tau \subset \tau(Z)$. Then the set $E(\rho) = \{f \in C(Z) : |f(x)| \leq 1 \text{ and } |f(x) - f(y)| \leq \rho(x, y) \text{ for all } x, y \in Z\}$ is adequate, closed in $C_p(Z)$ and separates the points of $Z$.

*Proof.* If $g \in C_p(Z) \backslash E(\rho)$ then there exist $x, y \in Z$ such that $|g(x) - g(y)| > \rho(x, y)$. The set $U = \{f \in C_p(Z) : |f(x) - f(y)| > \rho(x, y)\}$ is easily seen to be open in $C_p(Z)$; since also $g \in U \subset C_p(Z) \backslash E(\rho)$, the set $E(\rho)$ is closed in $C_p(Z)$.

If $x \in Z$ then $\{f(x) : f \in E(\rho)\} \subset \mathbb{I}$, so the set $E(\rho)$ is pointwise bounded. Finally, fix $x \in Z$ and $\varepsilon > 0$; the set $B = \{y \in X : \rho(x, y) < \varepsilon\}$ is an open neighborhood of $x$. If $y \in B$ and $f \in E(\rho)$ then $|f(y) - f(x)| \leq \rho(x, y) < \varepsilon$, so $f(y) \subset (f(x) - \varepsilon, f(x) + \varepsilon)$, i.e., the set $B$ witnesses equicontinuity of $E(\rho)$. Therefore the set $E(\rho)$ is adequate.

Given distinct points $a, b \in Z$ let $f(x) = \rho(a, x)$ for all $x \in Z$. It is straightforward that $f \in E(\rho)$; since also $f(a) = 0 \neq f(b)$, the set $E(\rho)$ separates the points of $Z$ and hence Fact 1 is proved.

Returning to our solution observe that $X$ embeds in $L(X)$, so if $L(X)$ has a weaker metrizable topology then $X$ also has one, i.e., we trivially have sufficiency.

Now assume that $X$ has a weaker metrizable topology and hence there exists a metric $\rho$ on the set $X$ such that $\rho(x, y) \leq 1$ for all $x, y \in X$ and the topology $\tau$ generated by $\rho$ is contained in $\tau(X)$. The set $E(\rho) = \{f \in C(X) : |f(x)| \leq 1$ and $|f(x) - f(y)| \leq \rho(x, y)$ for all $x, y \in X\}$ is adequate, closed in $C_p(X)$ and separates the points of $X$ (see Fact 1). For any $x \in X$ and $f \in E(\rho)$ let $\varphi(x)(f) = f(x)$; this gives us a map $\varphi : X \to C_u^*(E(\rho))$ which is continuous by Fact 9 of V.318. The space $C_u^*(E(\rho))$ is locally convex, so there exists a continuous linear mapping $\Phi : L(X) \to C_u^*(E(\rho))$ such that $\Phi|X = \varphi$ (see Problem 312).

If $u \in L(X) \backslash \{0\}$ then there exist distinct points $x_1, \ldots, x_n \in X$ and numbers $\lambda_1, \ldots, \lambda_n \in \mathbb{R} \backslash \{0\}$ such that $u = \lambda_1 x_1 + \ldots + \lambda_n x_n$. If $n = 1$ then let $f(x) = 1$ for all $x \in X$; it is clear that $f \in E(\rho)$ and $\Phi(u)(f) = \lambda_1 f(x_1) = \lambda_1 \neq 0$, so $\Phi(u) \neq 0$. If $n > 1$ then let $f(x) = \backslash n\{\rho(x, x_i) : 2 \leq i \leq n\}$ for every $x \in X$.

Given any points $x, y \in X$ there exist $i, j \in \{2, \ldots, n\}$ such that $f(x) = \rho(x, x_i)$ and $f(y) = \rho(x, x_j)$. The inequalities $\rho(x, x_i) \leq \rho(x, x_j) \leq \rho(x, y) + \rho(y, x_j)$ imply that $f(x) - f(y) = \rho(x, x_i) - \rho(y, x_j) \leq \rho(x, y)$. The points $x$ and $y$ are in a symmetric situation so $f(y) - f(x) \leq \rho(x, y)$ and hence $|f(x) - f(y)| \leq \rho(x, y)$ for all $x, y \in X$ which shows that $f \in E(\rho)$. It is easy to see that $\Phi(u)(f) = \lambda_1 \rho(x_1, x_i)$ for some $i \in \{2, \ldots, n\}$, so $\Phi(u)(f) \neq 0$ and hence $\Phi(u) \neq 0$. Thus $\Phi(u) \neq 0$ for all $u \in L(X) \backslash \{0\}$ which implies that $\Phi : L(X) \to C_u^*(E(\rho))$ is injective. The space $C_u^*(E(\rho))$ is metrizable, so $L(X)$ has a weaker metrizable topology, i.e., we established necessity.

Finally, if $X$ and $Y$ are $L$-equivalent spaces and $X$ condenses onto a metrizable space then so does $L(X)$. Therefore $L(Y)$ also condenses onto a metrizable space; being a subspace of $L(Y)$, the space $Y$ must be condensable onto a metrizable space.

**V.333.** *Suppose that $X$ and $Y$ are $l$-equivalent spaces. Prove that, if $X$ is metrizable, then $Y$ can be condensed onto a metrizable space.*

**Solution.** Fix a linear homeomorphism $\varphi : L_p(X) \to L_p(Y)$ (see Problem 237). Any metrizable space is easily seen to be a $b_f$-space so $X$ is a $b_f$-space and hence we can apply Problem 329 to see that $\varphi$ is a linear homeomorphism between the spaces $L(X)$ and $L(Y)$. Therefore the spaces $X$ and $Y$ are $L$-equivalent and hence $Y$ is condensable onto a metrizable space by Problem 332.

**V.334.** *Say that a space $X$ is $\sigma$-metrizable if $X$ is the countable union of its closed metrizable subspaces. Prove that a space $X$ is $\sigma$-metrizable and paracompact if and only if $L(X)$ is $\sigma$-metrizable and paracompact.*

**Solution.** If $Z$ is a space, $\mathcal{A}$ is a family of subsets of $Z$ and $Y \subset Z$ then we let $\mathcal{A}|Y = \{A \cap Y : A \in \mathcal{A}\}$. A topology $\tau$ on the set $Z$ is called a $\mu$-*approximation* for the space $Z$ if $\tau \subset \tau(Z)$ and there is a countable family $\mathcal{P}$ of closed metrizable subspaces of $(Z, \tau)$ such that $Z = \bigcup \mathcal{P}$ and $\tau|P = \tau(Z)|P$ for any $P \in \mathcal{P}$.

Let $J$ be the interval $(0, 1] \subset \mathbb{R}$. Given a nonempty set $T$ let $J_t = J \times t$ for all $t \in T$. Choose a point $\zeta \notin J \times T$ and let $H(T) = (J \times T) \cup \{\zeta\}$. For any $t \in T$ if $x = (a, t) \in J_t$ and $y = (b, t) \in J_t$ then we let $d(x, y) = |a - b|$. If $t, s \in T$, $t \neq s$ and $x = (a, t) \in J_t$, $y = (b, s) \in J_s$ then $d(x, y) = a + b$. If $t \in T$ and $x = (a, t) \in J_t$ then let $d(x, \zeta) = a$; finally, letting $d(\zeta, \zeta) = 0$ we obtain a metric space $(H(T), d)$ which is called *Kowalsky hedgehog* (see TFS-222). Consider the set $I_t = J_t \cup \{\zeta\}$ with the topology induced from $H(T)$; for any $x = (a, t) \in J_t$ let $h_t(x) = a$ and $h_t(\zeta) = 0$. It is straightforward that $h_t : I_t \to I = [0, 1]$ is a homeomorphism for each $t \in T$.

*Fact 1.* For any space $Z$, if a topology $\tau$ is a $\mu$-approximation of $Z$ then there exists a family $\{P_n : n \in \omega\}$ of metrizable closed subspaces of $(Z, \tau)$ such that $Z = \bigcup_{n \in \omega} P_n$ while $P_n \subset P_{n+1}$ and $\tau|P_n = \tau(Z)|P_n$ for every $n \in \omega$.

*Proof.* Take a family $\mathcal{Q} = \{Q_n : n \in \omega\}$ of closed metrizable subspaces of $(Z, \tau)$ such that $\bigcup \mathcal{Q} = Z$ and $\tau|Q_n = \tau(Z)|Q_n$ for all $n \in \omega$. If $P_n = Q_0 \cup \ldots \cup Q_n$ then it is easy to represent $P_n$ as a perfect image of the space $Q_0 \oplus \ldots \oplus Q_n$, so $P_n$ is metrizable for each $n \in \omega$ (see TFS-226).

If $A \subset P_n$ is a closed subset of $Z$ then $A_i = A \cap Q_i$ is closed in $Q_i$ and hence in $(Z, \tau)$ for all $i \leq n$; as a consequence, $A = \bigcup_{i \leq n} A_i$ is closed in $(Z, \tau)$ and hence $\tau(Z)|P_n = \tau|P_n$ for all $n \in \omega$. This shows that the sequence $\{P_n : n \in \omega\}$ is as promised, i.e., Fact 1 is proved.

*Fact 2.* Given a space $Z$, the following properties are equivalent:

(i)   $Z$ is $\sigma$-metrizable and paracompact;
(ii)  $Z$ has a metrizable $\mu$-approximation;
(iii) $Z$ has a paracompact $\mu$-approximation.

*Proof.* Assume that $Z$ is $\sigma$-metrizable and paracompact; then we can choose a sequence $\{M_n : n \in \omega\}$ of metrizable closed subspaces of $Z$ such that $M_n \subset M_{n+1}$ for all $n \in \omega$ and $Z = \bigcup_{n \in \omega} M_n$. If $U$ is an open subset of $Z$ then $U \cap M_n$ is an $F_\sigma$-set in $M_n$ and hence in $Z$ for every $n \in \omega$; therefore $U$ is an $F_\sigma$-subset of $Z$ and hence $Z$ is perfectly normal.

For any $n \in \omega$ the space $M_n$ is metrizable, so it has a base $\mathcal{B}_n$ such that $\mathcal{B}_n = \bigcup\{\mathcal{C}_{nm} : m \in \omega\}$ and every family $\mathcal{C}_{nm}$ is discrete in $M_n$; discarding trivial cases we can consider that $\mathcal{C}_{nm} \neq \emptyset$ and the elements of $\mathcal{C}_{nm}$ are nonempty for all $n, m \in \omega$.

Fix $n, m \in \omega$ and choose a faithful enumeration $\{C_t : t \in T_{nm}\}$ of the family $\mathcal{C}_{nm}$. It is easy to see that the family $\{\overline{C}_t : t \in T_{nm}\}$ is discrete in $Z$, so we can use paracompactness of $Z$ to find a discrete collection $\{O_t : t \in T_{nm}\} \subset \tau(Z)$ such that $\overline{C}_t \subset O_t$ for all $t \in T_{nm}$. Pick an open set $W_t \subset O_t$ such that $W_t \cap M_n = C_t$ for all $t \in T_{nm}$. By perfect normality of $Z$ we can find a function $f_t \in C(Z, I)$ such that $f_t^{-1}(J) = W_t$ for all $t \in T_{nm}$. If $x \in \overline{W}_t$ for some $t \in T_{nm}$ then let $\varphi_{nm}(x) = h_t^{-1}(f_t(x))$; if $x \in Z \setminus \bigcup\{\overline{W}_t : t \in T_{nm}\}$ then let $\varphi_{nm}(x) = \zeta$. This gives us a map $\varphi_{nm} : Z \to H(T_{nm})$. Applying Fact 2 of T.354 it is easy to see that $\varphi_{nm}$ is continuous.

For each $n \in \omega$ apply perfect normality of $Z$ once more to pick a function $q_n \in C(Z, I)$ such that $M_n = q_n^{-1}(0)$. If $\mathcal{F} = \{\varphi_{nm} : n, m \in \omega\} \cup \{q_n : n \in \omega\}$ then the diagonal product $\theta$ of the family $\mathcal{F}$ maps $Z$ continuously into the metrizable space $\prod\{H(T_{nm}) : n, m \in \omega\} \times I^\omega$; therefore the space $Y = \theta(Z)$ is metrizable. Let $\pi_{nm} : Y \to H(T_{nm})$ and $p_n : Y \to I$ be the respective natural projections for all $n, m \in \omega$.

Given distinct $x, y \in Z$ there exists $n \in \omega$ such that $\{x, y\} \subset M_n$. Since $\mathcal{B}_n$ is a base in $M_n$, we can find $m \in \omega$ and $t \in T_{nm}$ for which $x \in C_t$ and $y \notin C_t$. Then $f_t(y) = 0$ and $f_t(x) \neq 0$; therefore $\varphi_{nm}(x) \in J_t$ while $\varphi_{nm}(y) = \zeta$. Thus $\varphi_{nm}(x) \neq \varphi_{nm}(y)$ and hence $\theta(x) \neq \theta(y)$; this shows that $\theta$ is a condensation.

Let $M_n' = \theta(M_n)$; if $y \in Y \setminus M_n'$ then $y = \theta(x)$ for some $x \in Z \setminus M_n$ and hence $q_n(x) > 0$ which implies that $y \in G = p_n^{-1}(J)$, i.e., $G$ is an open neighborhood of $y$ in the space $Y$. For every $z' \in M_n'$ there is $z \in M_n$ with $z' = \theta(z)$; as a consequence, $p_n(z') = q_n(z) = 0$, so $z' \notin G$ and hence $G \cap M_n' = \emptyset$. Therefore

(1) the set $M_n'$ is closed in $Y$ for every $n \in \omega$.

Next fix $n \in \omega$, a point $y \in M_n'$ and let $x = \theta^{-1}(y)$; then $x \in M_n$. Take an arbitrary set $U \in \tau(x, M_n)$. The family $\mathcal{B}_n$ being a base in $M_n$ there exists $m \in \omega$ and $t \in T_{nm}$ such that $x \in C_t \subset U$. Therefore $\varphi_{nm}(x) \in J_t$ and hence the set $G = \pi_{nm}^{-1}(J_t)$ is an open neighborhood of $y$ in $Y$. Suppose that $z' \in M_n' \cap G$ and $z = \theta^{-1}(z')$; it follows from $\pi_{nm}(z') \in J_t$ that $\varphi_{nm}(z) \in J_t$ and therefore $f_t(z) \in J$, i.e., $z \in W_t \cap M_n = C_t \subset U$. This proves that $\theta^{-1}(G \cap M_n') \subset U$, i.e., the set $G \cap M_n'$ witnesses continuity of $\theta^{-1}|M_n'$ at the point $y$. As a consequence,

(2) the map $\theta|M_n : M_n \to M_n'$ is a homeomorphism for every $n \in \omega$.

The properties (1) and (2) show that the sequence $\{M_n : n \in \omega\}$ witnesses that the topology $\tau = \{\theta^{-1}(U) : U \in \tau(Y)\}$ is a metrizable $\mu$-approximation for the space $Z$, so we settled (i)$\Longrightarrow$(ii).

The implication (ii)$\Longrightarrow$(iii) being evident assume that $Z$ has a paracompact $\mu$-approximation and hence we can apply Fact 1 to choose a paracompact topology $\tau \subset \tau(Z)$ and a sequence $\{M_n : n \in \omega\}$ of metrizable closed subsets of the space $(Z, \tau)$ such that $\bigcup_{n \in \omega} M_n = Z$ while $M_n \subset M_{n+1}$ and $\tau|M_n = \tau(Z)|M_n$ for all $n \in \omega$. Let $\mathcal{B}_n$ be a $\sigma$-discrete base in $M_n$; observe that $\mathcal{F}_n = \{\overline{B} : B \in \mathcal{B}_n\}$ is a

$\sigma$-discrete network in $M_n$ for every $n \in \omega$. Therefore $\mathcal{F} = \bigcup_{n \in \omega} \mathcal{F}_n$ is a $\sigma$-discrete network in the space $(Z, \tau)$ as well as in the space $Z$. Since $(Z, \tau)$ is paracompact, we can choose, for each $F \in \mathcal{F}$, a set $O_F \in \tau$ such that $F \subset O_F$ and the family $\{O_F : F \in \mathcal{F}\}$ is $\sigma$-discrete in $(Z, \tau)$ and hence in $Z$.

Now take an open cover $\mathcal{U}$ of the space $Z$; say that a set $F \in \mathcal{F}$ is marked if there exists $U_F \in \mathcal{U}$ such that $F \subset U_F$. If $x \in Z$ then there exists $U \in \mathcal{U}$ with $x \in U$; since $\mathcal{F}$ is a network in $Z$, we can choose $F \in \mathcal{F}$ such that $x \in F \subset U$. The set $F$ is, evidently, marked so marked elements of $\mathcal{F}$ cover the space $Z$. It is straightforward that the family $\mathcal{V} = \{U_F \cap O_F : F$ is a marked element of $\mathcal{F}\}$ is a $\sigma$-discrete refinement of $\mathcal{U}$, so the space $Z$ is paracompact (see TFS-230); since its $\mu$-approximation witnesses that $Z$ is $\sigma$-metrizable, we established (iii)$\Longrightarrow$(i) and hence Fact 2 is proved.

*Fact 3.* Suppose that $Z$ is a space and $\tau \subset \tau(Z)$ is a metrizable $\mu$-approximation of $Z$; fix a family $\mathcal{M} = \{M_n : n \in \omega\}$ of closed subsets of $Z' = (Z, \tau)$ such that $\bigcup \mathcal{M} = Z$ and $\tau|M_n = \tau(Z)|M_n$ for all $n \in \omega$. Denote by $L_n$ the linear hull of the set $M_n$ in the linear space $L(Z)$ and let $\nu_n$ and $\nu'_n$ be the topologies on $L_n$ induced from the spaces $L(Z)$ and $L(Z')$ respectively. The set $L_n$ is closed in $L(Z')$ and the identity map $j : (L_n, \nu_n) \to L(M_n)$ is a topological isomorphism; if we consider that $j : (L_n, \nu'_n) \to L(M_n)$ then $j$ is a topological isomorphism as well. In particular, $\nu_n = \nu'_n$ for every $n \in \omega$.

*Proof.* Every set $L_n$ is closed in $L_p(Z')$ (see Fact 2 of V.246) and hence in $L(Z')$. Observe that $L_n$ and $L(M_n)$ have the same underlying set and denote by $\lambda$ the topology of $L(M_n)$ on $L_n$. The map $j$ is clearly an isomorphism, so it suffices to show that $\nu_n = \nu'_n = \lambda$. It follows from $\tau \subset \tau(Z)$ that $\tau(L(Z')) \subset \tau(L(Z))$ and hence $\nu'_n \subset \nu_n$. Since $L_n$ is a locally convex space, the identity map of $M_n$ onto $M_n$ is extendable to a continuous linear map from $L(M_n)$ onto $(L_n, \nu_n)$. The set $M_n$ being a Hamel basis in $L_n$ the mentioned linear extension is also the identity and hence it coincides with $j$. Therefore the map $j : L(M_n) \to (L_n, \nu_n)$ is continuous which shows that $\nu_n \subset \lambda$.

To see that the linear isomorphism $j : (L_n, \nu'_n) \to L(M_n)$ is also continuous apply SFFS-104 to find a continuous map $\xi : Z' \to L(M_n)$ such that $\xi|M_n = j$. By Problem 312 there exists a continuous linear map $\Phi : L(Z') \to L(M_n)$ such that $\Phi|Z' = \xi$. The set $M_n$ being a Hamel basis in the space $L_n$ we must have $\Phi|L_n = j$, so the map $j : (L_n, \nu'_n) \to L(M_n)$ is, indeed, continuous. This implies that $\lambda \subset \nu'_n$ and hence we have $\nu_n = \nu'_n = \lambda$, i.e., Fact 3 is proved.

*Fact 4.* If $Z$ is a metrizable space then $L(Z)$ is $\sigma$-metrizable and paracompact.

*Proof.* Fix a metric $\rho$ on the set $Z$ such that $\rho(x, y) \leq 1$ for all $x, y \in Z$ and $\tau(Z)$ is generated by $\rho$. The set $E = \{f \in C(X, \mathbb{I}) : |f(x) - f(y)| \leq \rho(x, y)$ for all $x, y \in Z\}$ is pointwise bounded, equicontinuous and separates the points of $X$ (see Fact 1 of V.332). Let $\xi(x)(f) = f(x)$ for any $f \in E$ and $x \in Z$; this gives us a continuous injective map $\xi : Z \to C_u^*(E)$ (see Fact 9 of V.318). Let $Y = \xi(Z)$ and denote by $H$ the linear hull of $Y$ in the space $C_u^*(Z)$. We will establish first that $E$ has sufficiently many functions for our purposes, namely, that

(3) for any closed $A \subset Z$ if $d_A(x) = \inf\{\rho(x,a) : a \in A\}$ for every $x \in Z$ (if $A = \emptyset$ then $d_A(x) = 1$ for all $x \in Z$) then $d_A \in E$.

Indeed, this is evident if $A = \emptyset$; if $A \neq \emptyset$ take any points $x, y \in Z$ and $\varepsilon > 0$. There exists a point $a \in A$ such that $d_A(y) > \rho(y,a) - \varepsilon$; it follows from the inequalities $d_A(x) \leq \rho(x,a) \leq \rho(x,y) + \rho(y,a) \leq \rho(x,y) + d_A(y) + \varepsilon$ that $d_A(x) - d_A(y) \leq \rho(x,y) + \varepsilon$. Since $\varepsilon > 0$ was chosen arbitrarily, we proved that $d_A(x) - d_A(y) \leq \rho(x,y)$. The points $x$ and $y$ are in a symmetric situation, so $d_A(y) - d_A(x) \leq \rho(y,x) = \rho(x,y)$ and hence $|d_A(x) - d_A(y)| \leq \rho(x,y)$ for all $x, y \in Z$, i.e., (3) is proved.

Fix a point $y \in Y$, let $z = \xi^{-1}(y)$ and take an arbitrary set $U \in \tau(z, Z)$. The function $f = d_{Z \setminus U}$ belongs to $E$ by (3). The set $V = \{v \in C^*(E) : v(f) > 0\}$ is open in $C_p^*(E)$ and hence in $C_u^*(E)$. Besides, $y(f) = f(z) = d(z, Z \setminus U) > 0$, so $y \in V$. If $y' \in V \cap Y$ and $z' = \xi^{-1}(y')$ then $f(z') = y'(f) > 0$ which implies $z' \in U$. This proves that $\xi^{-1}(V \cap Y) \subset U$, i.e., the set $V \cap Y$ witnesses continuity of $\xi^{-1}$ at the point $y$. Therefore $\xi^{-1} : Y \to Z$ is continuous and hence

(4) the map $\xi : Z \to Y$ is a homeomorphism.

Assume that $y_1, \ldots, y_n$ are distinct points of $Y$ and $t_1, \ldots, t_n \in \mathbb{R} \setminus \{0\}$; let $x_i = \xi^{-1}(y_i)$ for all $i \leq n$. Consider the set $A = \{x_1, \ldots, x_n\} \setminus \{x_1\}$; then the function $f = d_A$ belongs to $E$ by (3). If $v = t_1 y_1 + \ldots + t_n y_n$ then it is evident that $v(f) = t_1 f(x_1) + \ldots + t_n f(x_n) = t_1 f(x_1) \neq 0$ and hence $v \neq \mathbf{0}$. This proves that

(5) the set $Y$ is linearly independent in $C^*(E)$ and hence $Y$ is a Hamel basis of $H$.

For the point $\mathbf{0} \in H$ let $l(\mathbf{0}) = 0$. If $h \in H \setminus \{\mathbf{0}\}$ then there exists $n \in \mathbb{N}$ and uniquely determined distinct points $y_1, \ldots, y_n \in Y$ such that $h = t_1 y_1 + \ldots + t_n y_n$ for some $t_1, \ldots, t_n \in \mathbb{R} \setminus \{0\}$; let $l(h) = n$. Fix $n \in \omega$ and let $H_n = \{h \in H : l(h) \leq n\}$. If $u \in H \setminus H_n$ then there is $k > n$ such that $u = t_1 y_1 + \ldots + t_k y_k$ for some distinct $y_1, \ldots, y_k \in Y$ and $t_1, \ldots, t_k \in \mathbb{R} \setminus \{0\}$.

Let $x_i = \xi^{-1}(y_i)$ and pick a set $U_i \in \tau(x_i, Z)$ for all $i \leq k$ in such a way that the family $\{U_1, \ldots, U_k\}$ is disjoint. Consider the function $f_i = d_{Z \setminus U_i}$ for all $i \leq k$. The set $W = \{h \in H : h(f_i) \neq 0$ for all $i \leq k\}$ is open in $H$; we have $u(f_i) = t_i f_i(x_i) \neq 0$ for all $i \leq k$, so $u \in W$. If $v = s_1 z_1' + \ldots + s_m z_m' \in W$ then let $z_i = \xi^{-1}(z_i')$ for every $i \leq m$ and consider the set $A = \{z_1, \ldots, z_m\}$. It follows from $v(f_i) \neq 0$ that $A \cap U_i \neq \emptyset$ for all $i \leq k$. Since the sets $U_i$ are disjoint, there are at least $k$-many distinct elements in $A$ and hence in $\{z_1', \ldots, z_m'\}$, so $v \notin H_n$. Therefore $u \in W \subset H \setminus H_n$, i.e., every $u \in H \setminus H_n$ has a neighborhood which does not meet $H_n$; as a consequence,

(6) the set $H_n$ is closed in $H$ for every $n \in \omega$.

Fix any $n \in \mathbb{N}$ and consider the set $G_n = H_n \setminus H_{n-1}$. Take a disjoint family $\mathcal{U} = \{U_1, \ldots, U_n\}$ of nonempty open subsets of $Y$ and sets $O_1, \ldots, O_n \in \tau(\mathbb{R} \setminus \{0\})$. We claim that the set $W = O_1 U_1 + \ldots + O_n U_n$ is open in $G_n$.

To prove it take any $u \in W$; there exist $y_1, \ldots, y_n \in Z$ and $t_1, \ldots, t_n \in \mathbb{R} \setminus \{0\}$ such that $y_i \in U_i$, $t_i \in O_i$ for all $i \leq n$ and $u = t_1 y_1 + \ldots + t_n y_n$. The family

$\mathcal{U}$ being disjoint the points $y_1, \ldots, y_n$ are distinct and hence $u \in G_n$. This proves that $W \subset G_n$. Let $x_i = \xi^{-1}(y_i)$, $V_i = \xi^{-1}(U_i)$ and $f_i = d_{Z \setminus V_i}$; then $f_i \in E$ while $r_i = f_i(x_i) \neq 0$ and $f_i(Z \setminus V_i) \subset \{0\}$ for all $i \leq n$. The multiplication and taking the inverse are continuous operations in $\mathbb{R}$, so it follows from the equality $t_i r_i r_i^{-1} = t_i$ that

(7) for every $i \leq n$, there exist sets $P_i, Q_i \in \tau(\mathbb{R} \setminus \{0\})$ such that $t_i r_i \in P_i$, $r_i \in Q_i$ and $P_i \cdot (Q_i)^{-1} \subset O_i$.

It follows from (7) that $x_i \in D_i = f_i^{-1}(Q_i)$; if $g_i = d_{Z \setminus D_i}$ then $g_i \in E$ and $g_i(Z \setminus D_i) \subset \{0\}$ for all $i \leq n$. The set $N_i = \{v \in H : v(f_i) \in P_i \text{ and } v(g_i) \neq 0\}$ is open in $H$; furthermore, $u(f_i) = t_i f_i(x_i) = t_i r_i$ and $u(g_i) = t_i g_i(x_i) \neq 0$ (we used the fact that $\mathcal{U}$ is disjoint and $D_i \subset V_i$), so $u \in N_i$ for every $i \leq n$. Therefore $N = \bigcap_{i \leq n} N_i \in \tau(u, H)$.

Suppose that $v \in N \cap G_n$ and hence $l(v) = k \leq n$; then $v = s_1 z_1' + \ldots + s_k z_k'$ for distinct points $z_1', \ldots, z_k' \in Y$ and $s_1, \ldots, s_k \in \mathbb{R} \setminus \{0\}$. For each $i \leq n$ let $z_i = \xi^{-1}(z_i')$; we have $v(f_i) = s_1 f_i(z_1) + \ldots + s_k f_i(z_k) \neq 0$, so there is $m_i \leq k$ such that $f_i(z_{m_i}) \neq 0$ and hence $z_{m_i} \in V_i$. The family $\{V_1, \ldots, V_n\}$ being disjoint, the points $z_{m_1}, \ldots, z_{m_n}$ have to be distinct, so $k = n$ and we can change the order of summation if necessary to be able to assume, without loss of generality, that $v = s_1 z_1' + \ldots + s_n z_n'$ and $z_i' \in U_i$ for all $i \leq n$.

Now, $v(g_i) = s_i g_i(z_i) \neq 0$ implies that $g_i(z_i) \neq 0$ and hence $z_i \in D_i$; an immediate consequence is that $f_i(z_i) \in Q_i$. Since we also have $v(f_i) = s_i f(z_i) \in P_i$, we conclude that $s_i \in P_i \cdot Q_i^{-1} \subset O_i$ for any $i \leq n$. Thus $v \in W$ which shows that any point $u \in W$ has a neighborhood $N$ in $H$ such that $N \cap G_n \subset W$; an immediate consequence is that $W$ is open in $G_n$ and hence we proved that

(8) if a family $\{U_1, \ldots, U_n\} \subset \tau^*(Y)$ is disjoint and $O_1, \ldots, O_n \in \tau(\mathbb{R} \setminus \{0\})$ then the set $O_1 U_1 + \ldots + O_n U_n$ is open in $G_n$.

Let $i(y) = y$ for every $y \in Y$; then the map $i : Y \to H$ is continuous, so we can apply Problem 312 to find a continuous linear map $j : L(Y) \to H$ such that $j|Y = i$. It follows from (5) that $j$ is a linear isomorphism, so we can consider that the underlying set of $L(Y)$ coincides with $H$ while $\tau(H) \subset \tau(L(Y))$.

Fix $n \in \mathbb{N}$ and let $\tau_0 = \tau(H)|G_n$, $\tau_1 = \tau(L(Y))|G_n$; it is evident that $\tau_0 \subset \tau_1$. If $W \in \tau_1$ and $u \in W$ then there is a set $W' \in \tau(L(Y))$ such that $W' \cap G_n = W$. There exist distinct $y_1, \ldots, y_n \in Y$ and $t_1, \ldots, t_n \in \mathbb{R} \setminus \{0\}$ for which $u = t_1 y_1 + \ldots + t_n y_n$. Choose disjoint sets $U_1', \ldots, U_n' \in \tau(Y)$ such that $y_i \in U_i'$ for all $i \leq n$. It follows from continuity of linear operations in the space $L(Y)$ that, for each $i \leq n$ there exist $U_i'' \in \tau(y_i, Y)$ and $O_i \in \tau(t_i, \mathbb{R} \setminus \{0\})$ such that $O_1 U_1'' + \ldots + O_n U_n'' \subset W'$. Let $U_i = U_i' \cap U_i''$ for all $i \leq n$. Then the family $\{U_1, \ldots, U_n\}$ is disjoint and $u \in V = O_1 U_1 + \ldots + O_n U_n \subset W' \cap G_n = W$. The property (8) shows that the set $V$ belongs to $\tau_0$. This proves that, for any point $u \in W$ there exists $V \in \tau_0$ such that $u \in V \subset W$; an immediate consequence is that $W \in \tau_0$. In other words,

(9) the topologies $\tau(H)|G_n$ and $\tau(L(Y))|G_n$ coincide for all $n \in \mathbb{N}$.

The space $H$ is metrizable, so every $G_n$ is an $F_\sigma$-set in $H$, i.e., there exists a countable family $\mathcal{G}_n$ of closed subsets of $H$ such that $G_n = \bigcup \mathcal{G}_n$. The family

$\mathcal{G} = \{0\} \cup (\bigcup\{\mathcal{G}_n : n \in \mathbb{N}\})$ is countable, $H = \bigcup \mathcal{G}$ and it follows from (9) that $\tau(H)|P = \tau(L(Y))|P$ for every $P \in \mathcal{G}$. Therefore $H$ is a metrizable $\mu$-approximation of $L(Y)$ and hence $L(Y)$ is $\sigma$-metrizable and paracompact by Fact 2. Since $Z$ is homeomorphic to $Y$ by (4), the space $L(Z)$ is also $\sigma$-metrizable and paracompact, i.e., Fact 4 is proved.

Returning to our solution observe that if $L(X)$ is paracompact and $\sigma$-metrizable then so is $X$ because $X$ is a closed subspace of $L(X)$ (see Problem 311); this proves sufficiency.

Now if $X$ is $\sigma$-metrizable and paracompact then we can apply Fact 1 and Fact 2 to find a metrizable topology $\tau \subset \tau(X)$ such that there exists a sequence $\{M_n : n \in \omega\}$ of closed subspaces of the space $X' = (X, \tau)$ for which $\bigcup_{n \in \omega} M_n = X$ while $M_n \subset M_{n+1}$ and $\tau|M_n = \tau(X)|M_n$ for all $n \in \omega$. Apply Fact 4 to see that the space $L(X')$ is paracompact and denote by $L_n$ the linear hull of the set $M_n$ in $L(X')$. For every $n \in \omega$ apply Fact 3 to see that

(10)  the set $L_n$ is closed in $L(X')$ and $\tau(L(X'))|L_n = \tau(L(X))|L_n$ and, besides, $L_n$ is homeomorphic to $L(M_n)$.

Apply Fact 4 to find a countable family $\mathcal{L}_n$ of closed metrizable subsets of $L_n$ such that $L_n = \bigcup \mathcal{L}_n$. The family $\mathcal{L} = \bigcup_{n \in \omega} \mathcal{L}_n$ is countable and consists of closed metrizable subsets of $L(X')$. It follows from $\bigcup_{n \in \omega} L_n = L(X')$ that $\bigcup \mathcal{L} = L(X')$; apply (10) to conclude that $\tau(L(X))|A = \tau(L(X'))|A$ for any $A \in \mathcal{L}$. Thus $\tau(L(X'))$ is a paracompact $\mu$-approximation of $L(X)$, so we can apply Fact 2 to see that $L(X)$ is $\sigma$-metrizable and paracompact. This settles necessity and makes our solution complete.

**V.335.** *Suppose that a space $X$ is $l$-equivalent to a metrizable space. Prove that $X$ is $\sigma$-metrizable and paracompact.*

**Solution.** If $X$ is $l$-equivalent to a metrizable space $M$ then $X$ is $L$-equivalent to $M$ because $M$ is a $b_f$-space (see Problem 330). By Problem 334, the space $L(M)$ is $\sigma$-metrizable and paracompact and hence so is $L(X)$. Therefore $X$ is also $\sigma$-metrizable and paracompact being a closed subspace of $L(X)$.

**V.336.** *For an arbitrary space $X$, prove that $X$ is hemicompact if and only if $C_k(X)$ is first countable.*

**Solution.** For any $A \subset X$ and $O \subset \mathbb{R}$ let $[A, O] = \{f \in C_k(X) : f(A) \subset O\}$. If $X$ is hemicompact then we can find a sequence $\{K_n : n \in \omega\}$ of compact subsets of $X$ such that $X = \bigcup_{n \in \omega} K_n$ and every compact subset of $X$ is contained in some $K_n$. The set $O(n, m) = [K_n, (-2^{-m}, 2^{-m})]$ is a neighborhood of $\mathbf{0}$ for all $n, m \in \omega$ (see Fact 7 of V.318), so it suffices to prove that $\mathcal{O} = \{\text{Int}(O(n,m)) : n, m \in \omega\}$ is a local base of $C_k(X)$ at $\mathbf{0}$. To do it take any set $U \in \tau(\mathbf{0}, C_k(X))$; there exists a compact set $K \subset X$ such that $[K, (-\varepsilon, \varepsilon)] \subset U$. Choose $n, m \in \omega$ such that $K \subset K_n$ and $2^{-m} < \varepsilon$; it is straightforward that $O(n, m) \subset [K, (-\varepsilon, \varepsilon)] \subset U$, so $\mathbf{0} \in \text{Int}(O(n, m)) \subset U$ which shows that the family $\mathcal{O}$ is a countable local base of $C_k(X)$ at $\mathbf{0}$, i.e., we proved necessity.

Now assume that $\mathcal{W} = \{W_n : n \in \omega\}$ is a local base of $C_k(X)$ at $\mathbf{0}$; for each $n \in \omega$ there exists a compact set $K_n \subset X$ such that $G_n = [K_n, (-\varepsilon_n, \varepsilon_n)] \subset W_n$ for some $\varepsilon_n > 0$. If $K \subset X$ is compact then the set $W = [K, (-1, 1)]$ is a neighborhood of $\mathbf{0}$ by Fact 7 of V.318. Since $\mathcal{W}$ is a local base at $\mathbf{0}$, there exists $n \in \omega$ such that $W_n \subset W$ and hence $G_n \subset W$. If $x \in K \backslash K_n$ then we can find a function $f \in C_k(X)$ for which $f(x) = 1$ and $f(K_n) \subset \{0\}$. Then $f \in G_n \backslash W$; this contradiction shows that $K \backslash K_n = \emptyset$, i.e., $K \subset K_n$ and hence the sequence $\{K_n : n \in \omega\}$ witnesses that the space $X$ is hemicompact. This settles sufficiency.

**V.337.** *Prove that hemicompactness is preserved by l-equivalence.*

**Solution.** Suppose that a space $X$ is hemicompact and $\varphi : C_p(X) \to C_p(Y)$ is a linear homeomorphism. The space $X$ is $\sigma$-compact, so $Y$ is also $\sigma$-compact by Problem 043. Since the topology of $C_k(X)$ is stronger that the topology of $C_p(X)$, the map $\varphi : C_k(X) \to C_p(Y)$ is also continuous. Apply Problem 327 to see that the map $\varphi : C_b(X) \to C_b(Y)$ is continuous as well. Since both $X$ and $Y$ are $\mu$-spaces, we have $\tau(C_k(X)) = \tau(C_b(X))$ and $\tau(C_k(Y)) = \tau(C_b(Y))$ by Fact 6 of V.318. Thus the map $\varphi : C_k(X) \to C_k(Y)$ is continuous; an analogous proof shows that $\varphi^{-1} : C_k(X) \to C_k(Y)$ must be continuous too, so the spaces $C_k(X)$ and $C_k(Y)$ are homeomorphic. Finally apply Problem 336 to see that $C_k(X)$ is first countable; this implies that $C_k(Y)$ is first countable as well and hence we can apply Problem 336 again to conclude that $Y$ is hemicompact. Therefore hemicompactness of $X$ together with $X \overset{l}{\sim} Y$ imply that $Y$ is hemicompact, i.e., hemicompactness is preserved by $l$-equivalence.

**V.338.** *Given a space $X$ prove that*

(i) *if $X$ is a k-space then a sequence $\{f_n : n \in \omega\} \subset C_k(X)$ is convergent whenever it is linearly Cauchy;*

(ii) *for a hemicompact space $X$ the converse is true, i.e., if any linearly Cauchy sequence in $C_k(X)$ is convergent then $X$ is a k-space.*

**Solution.** If $A \subset X$ and $\varepsilon > 0$ then $[A, \varepsilon] = \{f \in C_k(X) : f(A) \subset (-\varepsilon, \varepsilon)\}$. Suppose first that $X$ is a k-space and a sequence $\{f_n : n \in \omega\} \subset C_k(X)$ is linearly Cauchy in $C_k(X)$. If $x \in X$ and $\varepsilon > 0$ then the set $W = [\{x\}, \varepsilon]$ is a neighborhood of $\mathbf{0}$ in $C_p(X)$ and hence in $C_k(X)$, so there exists $m \in \omega$ such that $f_n - f_k \in W$ for all $n, k \geq m$. This is equivalent to saying that $|f_n(x) - f_k(x)| < \varepsilon$ for all $n, k \geq m$ and therefore the sequence $\{f_n(x) : n \in \omega\}$ is convergent being Cauchy; let $f(x) = \lim f_n(x)$ for all $x \in X$.

To prove that the function $f : X \to \mathbb{R}$ is continuous fix a compact set $K \subset X$ and $\varepsilon > 0$. Apply Fact 7 of V.318 to see that $[K, \frac{\varepsilon}{2}]$ is a neighborhood of $\mathbf{0}$, so there exists $m \in \omega$ such that $f_n - f_k \in [K, \frac{\varepsilon}{2}]$ for all $n, k \geq m$. Given any $x \in K$ we have $|f_n(x) - f_k(x)| < \frac{\varepsilon}{2}$ for all $k \geq m$. Passing to the limit we conclude that $|f_n(x) - f(x)| \leq \frac{\varepsilon}{2} < \varepsilon$ for all $x \in K$ and hence

(1) the sequence $\{f_n|K : n \in \omega\}$ converges uniformly to $f|K$ for every compact set $K \subset X$.

As a consequence, the function $f|K$ is continuous for any compact $K \subset X$; since $X$ is a $k$-space, the function $f$ must be continuous. Now take any neighborhood $U$ of the function $f$ in $C_k(X)$; there exists a compact $K \subset X$ such that $f + [K, \varepsilon] \subset U$ for some $\varepsilon > 0$. The property (1) implies that there exists $m \in \omega$ such that $|f_n(x) - f(x)| < \varepsilon$ for all $n \geq m$ and $x \in K$. This is the same as saying that $f_n - f \in [K, \varepsilon]$ and hence $f_n \in f + [K, \varepsilon] \subset U$ for all $n \geq m$. Thus the sequence $\{f_n : n \in \omega\}$ converges to $f$ in $C_k(X)$ and hence we proved (i).

Now assume that $X$ is hemicompact and every linearly Cauchy sequence of $C_k(X)$ is convergent. To prove that $X$ is a $k$-space assume that $A \subset X$ is a non-closed set such that $A \cap K$ is closed in $X$ for any compact $K \subset X$; fix a point $y \in \overline{A} \backslash A$. There exists a sequence $\mathcal{K} = \{K_n : n \in \omega\}$ of compact subsets of $X$ such that $K_n \subset K_{n+1}$ for all $n \in \omega$ and every compact subset of $X$ is contained in an element of $\mathcal{K}$; there is no loss of generality to assume that $y \in K_0$. By our choice of $A$ the set $A_n = A \cap K_n$ is compact for all $n \in \omega$ and hence the set $A = \bigcup_{n \in \omega} A_n$ is $\sigma$-compact.

Take a function $f_0 \in C(X)$ such that $f_0(y) = 1$ and $f_0(A_0) \subset \{0\}$. Suppose that $n \in \omega$ and we have functions $f_0, \ldots, f_n \in C(X)$ such that $f_i(y) = 1$, $f_i(A_i) \subset \{0\}$ and $f_{i+1}|K_i = f_i$ for all $i < n$.

Let $g_{n+1}(x) = f_n(x)$ for all $x \in K_n$ and $g_{n+1}(x) = 0$ whenever $x \in A_{n+1} \backslash K_n$. It is easy to prove, using Fact 2 of T.354, that the function $g_{n+1} : K_n \cup A_{n+1} \to \mathbb{R}$ is continuous and hence there exists a continuous function $f_{n+1} : X \to \mathbb{R}$ such that $f_{n+1}|(K_n \cup A_{n+1}) = g_{n+1}|(K_n \cup A_{n+1})$. This shows that our inductive procedure can be continued to construct a sequence $\{f_n : n \in \omega\} \subset C_k(X)$ such that

(2)  $f_n(y) = 1$, $f_n(A_n) \subset \{0\}$ and $f_{n+1}|K_n = f_n|K_n$ for all $n \in \omega$.

If $x \in X$ then $x \in K_m$ for some $m \in \omega$; it follows from (2) that $f_n(x) = f_k(x)$ for all $n, k \geq m$ and hence there exists $f(x) = \lim f_n(x)$ for all $x \in X$. The property (2) also implies that $f_n|K_m = f|K_m$ for all $n \geq m$ and hence the sequence $\{f_n|K_m : n \in \omega\}$ uniformly converges to $f|K_m$ on for every $m \in \omega$.

Take an arbitrary set $U \in \tau(\mathbf{0}, C_k(X))$; there exists a compact $K \subset X$ such that $[K, \varepsilon] \subset U$ for some $\varepsilon > 0$. Recalling that the sequence $\mathcal{K}$ witnesses hemicompactness of the space $X$ we can find $m \in \omega$ such that $K \subset K_m$. Now if $n, k \geq m$ then $f_n|K_m = f_k|K_m$ and hence $f_n|K = f_k|K$ which, in turn, shows that $f_n - f_k \in [K, \varepsilon] \subset U$. Thus the sequence $\{f_n : n \in \omega\}$ is linearly Cauchy, so $f_n \to f$ in the space $C_k(X)$ which implies that the function $f$ is continuous. However, it follows from (2) that $f(y) = 1$ while $f(A) \subset \{0\}$; this contradiction with $y \in \overline{A}$ shows that $X$ is a $k$-space, i.e., we settled (ii) and completed our solution.

**V.339.** *Let $X$ be an arbitrary space. Prove that $X$ is a hemicompact space with $k$-property if and only if $C_k(X)$ is metrizable by a complete metric.*

**Solution.** Suppose first that $C_k(X)$ is metrizable by a complete metric and hence Čech-complete (see TFS-269). Since $C_k(X)$ is a linear topological space, its topology is generated by its linear uniformity $\mathcal{L}$ (see Problem 134). Apply Problem 121 to find a complete uniform space $(L, \mathcal{U})$ such that $(C_k(X), \mathcal{L})$ is a dense uniform subspace of $(L, \mathcal{U})$. For any pair $(f, g) \in C_k(X) \times C_k(X)$ let $s(f, g) = f + g$ and $i(f) = -f$.

The map $s : C_k(X) \times C_k(X) \to C_k(X)$ is continuous and linear, so we can apply Problem 134 to see that it is uniformly continuous. Since $C_k(X) \times C_k(X)$ is a dense uniform subspace of $L \times L$ and the space $L \times L$ is complete, we can apply Problem 119 to find a uniformly continuous map $s_1 : L \times L \to L$ such that $s_1|C_k(X) \times C_k(X) = s$. Analogously, the map $i : C_k(X) \to C_k(X)$ is linear and hence uniformly continuous, so there exists a uniformly continuous map $i_1 : L \to L$ such that $i_1|C_k(X) = i$.

Let $\tilde{s}(u, v) = s_1(v, u)$ for any $u, v \in L \times L$. Since addition is commutative in $C_k(X)$, we have the equality $\tilde{s}|C_k(X) \times C_k(X) = s = s_1|C_k(X) \times C_k(X)$; the space $C_k(X) \times C_k(X)$ being dense in $L \times L$ we conclude that $\tilde{s} = s_1$ (see Fact 0 of S.351), i.e., the operation $s_1$ is commutative. Analogously, consider the operations $\mu, \nu : L^3 \to L$ defined by $\mu(u, v, w) = s_1(s_1(u, v), v)$ and $\nu(u, v, w) = s_1(u, s_1(v, w))$ for each $(u, v, w) \in L^3$. The addition is an associative operation, so $\mu|(C_k(X))^3 = \nu|(C_k(X))^3$; applying Fact 0 of S.351 again we conclude that $\mu = \nu$ and hence the operation $s_1$ is associative.

Now let $s_2(u) = s_1(0, u)$ for any $u \in L$; since $s_2(f) = 0 + f = f$ for all $f \in C_k(X)$, we can apply Fact 0 of S.351 once more to conclude that $s_2(u) = u$ for all $u \in L$, so $s_2 : L \to L$ is an identity map and hence $0$ is a zero element in $L$. Furthermore, let $\lambda(u) = s_1(u, i(u))$ for all $u \in L$; then $\lambda : L \to L$ is a continuous map such that $\lambda(f) = f + (-f) = 0$ for all $f \in C_k(X)$; thus $s_1(u, i(u)) = 0$ for every $u \in L$.

All these considerations show that the operation $s_1 : L \times L \to L$ makes $L$ a commutative group (actually, $L$ is a linear topological space but we won't need that). To simplify the notation we will write $u + v$ instead of $s_1(u, v)$ and $-u$ in place of $i_1(u)$.

For each $u \in L$ let $\varphi_u(v) = u + v$ for all $v \in L$; then $\varphi_u : L \to L$ is a continuous map. It is clear that $\varphi_{-u}$ is the inverse of $\varphi_u$, so $\varphi_u$ is a homeomorphism for all $u \in L$. Suppose that $L \neq C_k(X)$ and fix a point $u \in L \backslash C_k(X)$. If $\varphi_u(C_k(X)) \cap C_k(X) \neq \emptyset$ then there exist $f, g \in C_k(X)$ such that $g = f + u$ and hence $u = g + (-f) \in C_k(X)$ which is a contradiction. Therefore $\varphi(C_k(X))$ and $C_k(X)$ are disjoint dense Čech-complete subspaces of $L$; this contradicts TFS-264 and shows that $C_k(X) = L$. As a consequence,

(1) the space $C_k(X)$ is complete with respect to its linear uniformity.

Take a sequence $\{f_n : n \in \omega\} \subset C_k(X)$ which is linearly Cauchy in $C_k(X)$ and let $P_m = \{f_n : n \geq m\}$ for all $m \in \omega$. The family $\{P_m : m \in \omega\}$ is, evidently, a Cauchy filterbase, so it follows from completeness of $C_k(X)$ that there exists a function $f \in \bigcap\{\overline{P}_m : m \in \omega\}$. Fix an arbitrary set $U \in \tau(0, C_k(X))$ and choose $V \in \tau(0, C_k(X))$ such that $\overline{V} \subset U$. There exists $m \in \omega$ for which $f_n - f_k \in V$ for all $n, k \geq m$.

Take any $n \geq m$ and consider the set $f_n - P_m \subset V$. The shift operation is a homeomorphism, so $\overline{f_n - P_m} = f_n - \overline{P}_m \subset \overline{V} \subset U$. This, together with $f \in \overline{P}_m$

shows that $f_n - f \in \overline{V} \subset U$ and hence $f_n \in f + U$ for all $n \geq m$. Therefore the sequence $\{f_n : n \in \omega\}$ converges to $f$, i.e., we proved that a sequence in $C_k(X)$ is convergent whenever it is linearly Cauchy. Applying Problem 338 we conclude that $X$ is a $k$-space and hence we established sufficiency.

Finally, assume that $X$ is a hemicompact $k$-space; it follows from Problem 336 that $C_k(X)$ is first countable. By Problem 206 there exists an invariant metric $d$ on the set $C_k(X)$ which generates the topology of $C_k(X)$. Suppose that $S = \{f_n : n \in \omega\}$ is a Cauchy sequence with respect to the metric $d$. Given any $U \in \tau(0, C_k(X))$ there exists $\varepsilon > 0$ such that $B = \{f \in C_k(X) : d(f, 0) < \varepsilon\} \subset U$. Take $m \in \omega$ such that $d(f_n, f_k) < \varepsilon$ for all $n, k \geq m$; since $d$ is invariant, we have $d(f_n - f_k, 0) < \varepsilon$ and hence $f_n - f_k \in B \subset U$ for all $n, k \geq m$. This shows that the sequence $S$ is linearly Cauchy, so we can apply Problem 338 to see that $S$ is convergent. Thus the metric $d$ has to be complete, i.e., we settled necessity.

**V.340.** *Let $X$ and $Y$ be $l$-equivalent spaces. Prove that $X$ is a hemicompact $k$-space if and only if $Y$ is a hemicompact $k$-space.*

**Solution.** Suppose that $X$ is a hemicompact $k$-space and fix a linear homeomorphism $\varphi : C_p(X) \to C_p(Y)$. The map $\varphi : C_b(X) \to C_b(Y)$ is also a linear homeomorphism by Problem 327. The space $Y$ is hemicompact by Problem 337, so both $X$ and $Y$ are $\mu$-spaces and hence we have the equalities $\tau(C_k(X)) = \tau(C_b(X))$ and $\tau(C_k(Y)) = \tau(C_b(Y))$ by Fact 6 of V.318. Therefore the map $\varphi : C_k(X) \to C_k(Y)$ is also a linear homeomorphism.

By Problem 339 the space $C_k(X)$ is metrizable and Čech-complete, so $C_k(Y)$ is metrizable and Čech-complete too. Applying Problem 339 again we conclude that $Y$ has to be a hemicompact $k$-space. Since $X$ and $Y$ are in a symmetric situation, an analogous proof shows that if $Y$ is a hemicompact $k$-space then so is $X$.

**V.341.** *Prove that any subspace of an $\aleph_0$-space is an $\aleph_0$-space and any countable product of $\aleph_0$-spaces is an $\aleph_0$-space.*

**Solution.** Suppose that $X$ is an $\aleph_0$-space and fix a countable network $\mathcal{N}$ for all compact subsets of $X$. Given $Y \subset X$ let $\mathcal{N}_Y = \{N \cap Y : N \in \mathcal{N}\}$; it is clear that $\mathcal{N}_Y$ is countable. If $K$ is a compact subset of $Y$ and $K \subset U \in \tau(Y)$ then take a set $U' \in \tau(X)$ such that $U' \cap Y = U$. There exists $N \in \mathcal{N}$ for which $K \subset N \subset U'$; then $N' = N \cap Y \in \mathcal{N}_Y$ while $K \subset N' \subset U$. This proves that $\mathcal{N}_Y$ is a network for all compact subsets of $Y$ and hence every $Y \subset X$ is an $\aleph_0$-space.

Our next step is to prove that the product of two $\aleph_0$-spaces is an $\aleph_0$-space, so fix $\aleph_0$-spaces $X$ and $Y$ and their respective networks $\mathcal{N}_X$ and $\mathcal{N}_Y$ for the families $\mathcal{K}(X)$ and $\mathcal{K}(Y)$. Let $\pi_X : X \times Y \to X$ and $\pi_Y : X \times Y \to Y$ be the respective natural projections. Call a compact set $K \subset X \times Y$ *rectangular* if $K = K_1 \times K_2$ for some compact sets $K_1 \subset X$ and $K_2 \subset Y$; a set $K \subset X \times Y$ is *standard* if it is a finite union of rectangular sets.

Let $\mathcal{N}_0 = \{P \times Q : P \in \mathcal{N}_X \text{ and } Q \in \mathcal{N}_Y\}$ and denote by $\mathcal{N}$ the family of all finite unions of the elements of $\mathcal{N}_0$. Suppose first that $K = K_1 \times K_2$ is a rectangular compact subset of $X \times Y$. If $K \subset U \in \tau(X \times Y)$ then there exist $V \in \tau(K_1, X)$

and $W \in \tau(K_2, Y)$ such that $V \times W \subset U$ (see Fact 3 of S.271). Pick $P \in \mathcal{N}_X$ and $Q \in \mathcal{N}_Y$ for which $K_1 \subset P \subset V$ and $K_2 \subset Q \subset W$. Then $R = P \times Q \in \mathcal{N}_0$ and $K \subset R \subset U$; this shows that

(1) the family $\mathcal{N}_0$ is a network for all rectangular compact subsets of $X \times Y$.

It is a trivial consequence of the property (1) that

(2) the family $\mathcal{N}$ is a network for all standard compact subsets of $X \times Y$.

Now suppose that $K \subset X \times Y$ is an arbitrary compact subset of $X \times Y$ and $K \subset U \in \tau(X \times Y)$. For any $z \in K$ we can find $V_z \in \tau(X)$ and $W_z \in \tau(Y)$ such that $z \in V_z \times W_z \subset \overline{V}_z \times \overline{W}_z \subset U$. By compactness of $K$ we can pick a finite set $A \subset K$ for which $K \subset \bigcup \{V_z \times W_z : z \in A\}$. The set $D_z = (\overline{V}_z \times \overline{W}_z) \cap K$ is compact; besides, $D_z \subset E_z = \pi_X(D_z) \times \pi_Y(D_z) \subset \overline{V}_z \times \overline{W}_z \subset U$ for every $z \in A$. Clearly, $E = \bigcup \{E_z : z \in A\}$ is a standard compact subset of $X \times Y$ and $K \subset E \subset U$. Apply (2) to find a set $N \in \mathcal{N}$ such that $E \subset N \subset U$; then $K \subset N \subset U$ and hence $\mathcal{N}$ is a network for all compact subsets of $X \times Y$, i.e., $X \times Y$ is an $\aleph_0$-space. A trivial induction shows that

(3) any finite product of $\aleph_0$-spaces is an $\aleph_0$-space.

Finally, assume that $X_n$ is an $\aleph_0$-space for all $n \in \omega$ and let $X = \prod_{n \in \omega} X_n$. For each $n \in \omega$ let $Y_n = X_0 \times \ldots \times X_n$ and apply (3) to find a countable network $\mathcal{P}_n$ in the space $Y_n$ for all compact subsets of $Y_n$; we will also need the natural projection $\pi_n : X \to Y_n$.

The family $\mathcal{N} = \bigcup \{\pi_n^{-1}(P) : P \in \mathcal{P}_n, n \in \omega\}$ is countable, so it suffices to show that $\mathcal{N}$ is a network for $\mathcal{K}(X)$. Take any compact set $K \subset X$ and $U \in \tau(K, X)$. For every point $x \in K$ there exists $n(x) \in \omega$ and a set $U_x \in \tau(Y_n)$ such that $x \in V_x = \pi_{n(x)}^{-1}(U_x) \subset U$. Take a finite set $A \subset K$ such that $K \subset \bigcup \{V_x : x \in A\}$ and let $n = \max\{n(x) : x \in A\}$. If $W_x = \pi_n(V_x)$ then $\pi_n^{-1}(W_x) = V_x$ for all $x \in A$.

The set $K' = \pi_n(K) \subset Y_n$ is compact and $K' \subset W = \bigcup \{W_x : x \in A\}$. Pick $P \in \mathcal{P}_n$ with $K' \subset P \subset W$. Then $Q = \pi_n^{-1}(P) \in \mathcal{N}$ and it follows from the inclusions $K \subset \pi_n^{-1}(K') \subset Q \subset \pi_n^{-1}(W) = \bigcup \{V_x : x \in A\} \subset U$ that $K \subset Q \subset U$ and hence $\mathcal{N}$ is, indeed, a network for all compact subsets of $X$, so we proved that any countable product of $\aleph_0$-spaces is an $\aleph_0$-space.

**V.342.** *Observe that a compact-covering continuous image of an $\aleph_0$-space is an $\aleph_0$-space. Prove that a space $X$ is an $\aleph_0$-space if and only if $X$ is a compact-covering continuous image of a second countable space.*

**Solution.** Suppose that $X$ is an $\aleph_0$-space and $f : X \to Y$ is a continuous compact-covering map. Fix a countable network $\mathcal{N}$ for all compact subsets of $X$ and let $\mathcal{N}' = \{f(N) : N \in \mathcal{N}\}$. Suppose that $K' \subset Y$ is compact and $U' \in \tau(K', Y)$; there exists a compact $K \subset X$ such that $f(K) = K'$. The set $U = f^{-1}(U')$ is open in $X$ and $K \subset U$, so there is $N \in \mathcal{N}$ with $K \subset N \subset U$. Then $N' = f(N) \in \mathcal{N}'$ and $K' \subset N' \subset U'$; this shows that $\mathcal{N}'$ is a countable network for all compact subsets of $Y$ and hence $Y$ is an $\aleph_0$-space. Therefore

(1)  any compact-covering continuous image of an $\aleph_0$-space is an $\aleph_0$-space.

If $M$ is a second countable space then take a countable base $\mathcal{B}$ in $X$ and let $\mathcal{N}$ be the family of all finite unions of the elements of $\mathcal{B}$. It is an easy exercise that $\mathcal{N}$ is a network for all compact subsets of $M$; this proves that

(2)  any second countable space is an $\aleph_0$-space.

The properties (1) and (2) show that any compact-covering continuous image of a second countable space is an $\aleph_0$-space.

To prove the converse fix an $\aleph_0$-space $X$ and let $\mathcal{P} = \{P_n : n \in \omega\}$ be a network for all compact subsets of $X$. It is straightforward that the family $\{\overline{P}_n : n \in \omega\}$ is also a network for $\mathcal{K}(X)$, so we can assume, without loss of generality, that all elements of $\mathcal{P}$ are closed in $X$.

Say that a function $f \in \omega^\omega$ is *appropriate* if there exists a point $x \in X$ such that $\mathcal{Q}_f = \{P_{f(n)} : n \in \omega\}$ is a network for the set $\{x\}$, i.e., $x \in \bigcap\{P_{f(n)} : n \in \omega\}$ and, for any $U \in \tau(x, X)$, there is $n \in \omega$ with $P_{f(n)} \subset U$; observe that the point $x$ is uniquely determined by $f$ because $\{x\} = \bigcap \mathcal{Q}_f$; let $\varphi(f) = x$. The space $M = \{f \in \omega^\omega : f \text{ is appropriate}\}$ is second countable and we have defined a map $\varphi : M \to X$.

If $x \in X$ then the family $\mathcal{P}_x = \{P \in \mathcal{P} : x \in P\}$ is a network for the set $\{x\}$. It is easy to find a function $f \in \omega^\omega$ such that $\mathcal{P}_x = \{P_{f(n)} : n \in \omega\}$ and hence $\varphi(f) = x$; this proves that $\varphi$ is a surjective map.

Now if $f \in M$ and $x = \varphi(f) \in U \in \tau(X)$ then pick a number $n \in \omega$ such that $P_{f(n)} \subset U$; this is possible because $\mathcal{Q}_f$ is a network for the set $\{x\}$. The set $W = \{g \in M : g(n) = f(n)\}$ is open in $M$ and $f \in W$. If $g \in W$ and $y = \varphi(g)$ then it follows from $P_{f(n)} \in \mathcal{Q}_g$ and $\{y\} = \bigcap \mathcal{Q}_g$ that $y \in P_{f(n)} \subset U$. Thus $\varphi(W) \subset U$ and hence $\varphi$ is a continuous map.

Next fix a compact set $K \subset X$ and let $\{\mathcal{F}_n : n \in \omega\}$ be an enumeration of all finite subfamilies of $\mathcal{P}$ which cover the set $K$; fix a finite set $D_n \subset \omega$ such that $\mathcal{F}_n = \{P_i : i \in D_n\}$ for every $n \in \omega$. The set $D = \prod_{n\in\omega} D_n \subset \omega^\omega$ is compact; suppose that $f \in D$ and $x \in (\bigcap \mathcal{Q}_f) \cap K$. Given any $U \in \tau(x, X)$ take $U' \in \tau(x, K)$ with $\overline{U'} \subset U$; there exists $P \in \mathcal{P}$ such that $\overline{U'} \subset P \subset U$. For each $y \in K\backslash\{x\}$ take a set $O_y \in \tau(y, K)$ such that $x \notin \overline{O}_y$ and choose $P_y \in \mathcal{P}$ for which $\overline{O}_y \subset P_y \subset X\backslash\{x\}$. The cover $\{O_y : y \in K\backslash\{x\}\} \cup \{U'\}$ of the compact space $K$ has a finite subcover, so there exists a finite $A \subset K$ such that $K \subset (\bigcup\{P_y : y \in A\}) \cup P$. There is $n \in \omega$ for which $\{P\} \cup \{P_y : y \in A\} = \mathcal{F}_n$. Since $P$ is the unique element of $\mathcal{F}_n$ which contains $x$, we must have $P_{f(n)} = P \subset U$. This proves that $\mathcal{Q}_f$ is a network for $\{x\}$, so $x = \varphi(f)$, i.e., we have established that

(3)  if $f \in D$ and $x \in (\bigcap \mathcal{Q}_f) \cap K$ then $\varphi(f) = x$.

Given any $x \in K$ recall that every family $\mathcal{F}_n$ covers $K$, so there is $i \in D_n$ such that $x \in P_i$; let $f(n) = i$. For the function $f \in D$ we have $x \in \bigcap \mathcal{Q}_f$, so $x = \varphi(f)$ by (3). Therefore $K \subset \varphi(M \cap D)$. If $f \in D\backslash M$ then it follows from (3) that $(\bigcap\{P_{f(n)} : n \in \omega\}) \cap K = \emptyset$ and hence there exists a finite set $Q \subset \omega$ such that $(\bigcap\{P_{f(n)} : n \in Q\}) \cap K = \emptyset$. The set $W = \{g \in \omega^\omega : g(n) = f(n)$

for all $n \in Q\}$ is an open neighborhood of $f$ in $\omega^\omega$ and it is clear that $f \in W$ while $W \cap (M \cap D) = \emptyset$. Thus the set $M' = M \cap D$ is compact being closed in $D$. The set $K' = \varphi^{-1}(K) \cap M'$ is also compact and $\varphi(K') = K$; this shows that $\varphi$ is compact-covering and hence every $\aleph_0$-space is a compact-covering continuous image of a second countable space.

**V.343.** *Prove that a space $X$ is an $\aleph_0$-space with the $k$-property if and only if it is a quotient image of a second countable space.*

**Solution.** Let us prove a couple of useful facts about the $k$-property.

*Fact 1.* Suppose that $Z$ is a $k$-space and every compact subspace of $Z$ is sequential. Then $Z$ is also sequential.

*Proof.* If $A \subset Z$ is not closed in $Z$ then there exists a compact $K \subset Z$ such that $K \cap A$ is not closed in $K$. By sequentiality of $K$ there exists a sequence $S \subset K \cap A$ which converges to a point $x \in K \backslash (K \cap A) = K \backslash A$; therefore $x \notin A$, i.e., $S \subset A$ witnesses sequentiality of $Z$ and Fact 1 is proved.

*Fact 2.* If $Z$ is a $k$-space and $f : Z \to Y$ is a quotient map then $Y$ is also a $k$-space.

*Proof.* Suppose that $A \subset Y$ is not closed in $Y$; then $B = f^{-1}(A)$ is not closed in $Z$, so there exists a compact $K' \subset Z$ such that $B' = K' \cap B$ is not closed in $K'$. Choose a point $z \in \overline{B'} \backslash B'$; then $z \in K'$ and hence $y = f(z) \in K = f(K')$. Besides, $z \in Z \backslash B$ implies that $y \notin A$; however, $y \in \overline{f(B')} \subset \overline{A \cap K}$ which shows that $y \in \overline{A \cap K} \backslash (A \cap K)$ and hence the set $A \cap K$ is not closed in $K$, i.e., the compact set $K$ witnesses that $Y$ is a $k$-space so Fact 2 is proved.

Returning to our solution assume that $M$ is a second countable space and $f : M \to X$ is a quotient map. Since $M$ is a $k$-space, it follows from Fact 2, that $X$ is also a $k$-space. Fix a countable base $\mathcal{B}$ in the space $M$ which is closed under finite unions and let $\mathcal{N} = \{f(B) : B \in \mathcal{B}\}$; observe that $\mathcal{N}$ is also closed under finite unions. Suppose that $K \subset X$ is compact and $U \in \tau(K, X)$; let $\mathcal{N}' = \{N_k : k \in \omega\}$ be the family of all elements of $\mathcal{N}$ contained in $U$. Since $f^{-1}(U)$ is a union of a subfamily of $\mathcal{B}$, we have $U = \bigcup \mathcal{N}'$ and, in particular, $K \subset \bigcup \mathcal{N}'$.

Assume that $L_k = K \backslash (N_0 \cup \ldots \cup N_k) \neq \emptyset$ for all $k \in \omega$. Recalling that $K \subset \bigcup_{k \in \omega} N_k$ we can choose a sequence $\{y_k : k \in \omega\}$ in such a way that $y_k \in L_k$ for all $k \in \omega$ and $l \neq k$ implies $y_l \neq y_k$. The space $K$ is easily seen to be metrizable, so we can find a subsequence $S = \{y_{k_i} : i \in \omega\}$ of the sequence $\{y_k : k \in \omega\}$ which converges to a point $y \in K \backslash S$. By our choice of $S$,

(1) the set $N \cap S$ is finite for each $N \in \mathcal{N}'$.

Since $y \in \overline{S} \backslash S$, the set $S$ is not closed in $X$, so the set $T = f^{-1}(S)$ is not closed in $M$. Pick a sequence $C = \{b_n : n \in \omega\} \subset T$ which converges to a point $b \notin T$. It is evident that $C \cap f^{-1}(y_{k_i})$ is finite for any $i \in \omega$, so $f(C)$ is an infinite subsequence of $S$. Therefore $f(C)$ converges to $y$ and hence $y = f(b)$. Take a set $B \in \mathcal{B}$ such that $b \in B \subset f^{-1}(U)$. Then $N = f(B) \in \mathcal{N}'$; since $B$ contains infinitely many points of $T$, the set $N$ contains infinitely many points of $S$ which contradicts (1) and shows that $K \subset N' = N_0 \cup \ldots \cup N_k$ for some $k \in \omega$. Since $N' \in \mathcal{N}'$, we have $K \subset N' \subset U$ and hence $\mathcal{N}$ is a network for all compact subsets of $X$, i.e., $X$ is an $\aleph_0$-space. Thus every quotient image of a second countable space is an $\aleph_0$-space with the $k$-property.

Finally assume that $X$ is an $\aleph_0$-space with the $k$-property and apply Problem 342 to find a second countable space $M$ such that there exists a compact-covering map $f : M \to X$. Since $X$ has a countable network, all compact subspaces of $X$ are metrizable, so $X$ is sequential by Fact 2. We claim that the mapping $f$ is quotient.

To prove this assume that $A \subset X$ is not closed while $B = f^{-1}(A)$ is closed in $M$. By sequentiality of $X$ there is a sequence $S = \{a_n : n \in \omega\} \subset A$ which converges to a point $x \in X \setminus A$. The set $K = S \cup \{x\}$ is compact, so we can find a compact $L \subset M$ such that $f(L) = K$. The set $B \cap L$ is also compact being closed in $L$; therefore the set $f(B \cap L) = S$ must be compact and hence closed in $X$, a contradiction. Thus the map $f$ is quotient, so every $\aleph_0$-space with the $k$-property is a quotient image of a second countable space.

**V.344.** *Prove that any $\aleph_0$-space of countable character is second countable.*

**Solution.** Suppose that $X$ is an $\aleph_0$-space and $\chi(X) \leq \omega$. Fix a countable network $\mathcal{N}$ for all compact subsets of $X$ and let $\mathcal{B} = \{\mathrm{Int}(P) : P$ is a finite union of elements of $\mathcal{N}\}$. The family $\mathcal{B}$ being countable it suffices to show that $\mathcal{B}$ is a base of $X$, so take a point $x \in X$ and any set $U \in \tau(x, X)$; consider the family $\mathcal{N}' = \{N_k : k \in \omega\}$ of all elements of $\mathcal{N}$ contained in $U$. Suppose that, for any $k \in \omega$, the point $x$ does not belong to the interior of the set $M_k = N_0 \cup \ldots \cup N_k$. Let $\{O_n : n \in \omega\}$ be a countable local base of $X$ at the point $x$ such that $\overline{O}_n \subset U$ for all $n \in \omega$. By our assumption about the sets $M_k$, we can choose a point $x_k \in O_k \setminus M_k$ for all $k \in \omega$. As a consequence, the sequence $S = \{x_k : k \in \omega\}$ converges to $x$ and

(1) the set $\{k \in \omega : x_k \in N\}$ is finite for every $N \in \mathcal{N}'$.

However, $K = S \cup \{x\}$ is a compact subset of $U$, so there is $N' \in \mathcal{N}$ for which $K \subset N' \subset U$; therefore $N' \in \mathcal{N}'$ and $S \subset N'$ which contradicts (1). Thus $x \in V = \mathrm{Int}(M_k)$ for some $k \in \omega$ and we have $V \subset U$; since also $V \in \mathcal{B}$, we proved that $\mathcal{B}$ is a countable base of the space $X$.

**V.345.** *Suppose that $\varphi : X \to Y$ is a continuous map; recall that the dual map $\varphi^* : C_k(Y) \to C_k(X)$ is defined by the formula $\varphi^*(f) = f \circ \varphi$ for every $f \in C_k(Y)$. Assuming that $\varphi$ is compact-covering, prove that $\varphi^*$ is an embedding.*

**Solution.** Given a space $Z$ and a function $f \in C(Z)$ we will need the set

$$[f, K, \varepsilon]_Z = \{g \in C(Z) : |g(z) - f(z)| < \varepsilon\}$$

for every $K \subset Z$ and $\varepsilon > 0$.

The map $\varphi^*$ is continuous by Fact 8 of V.318; consider the set $E = \varphi^*(C_k(Y))$. If $f, g \in C_k(Y)$ and $f \neq g$ then $f(y) \neq g(y)$ for some $y \in Y$. Any compact-covering map is surjective, so there is a point $x \in X$ such that $\varphi(x) = y$. It follows from $\varphi^*(f)(x) = f(\varphi(x)) = f(y) \neq g(y) = \varphi^*(g)(x)$ that $\varphi^*(f) \neq \varphi^*(g)$, so the mapping $\varphi^* : C_k(Y) \to E$ is a condensation.

To see that the map $\xi = (\varphi^*)^{-1} : E \to C_k(Y)$ is continuous fix a function $f \in E$, let $g = \xi(f)$ and take any $U \in \tau(g, C_k(Y))$. There exists a compact set $K \subset Y$ such that $[g, K, \varepsilon]_Y \subset U$ for some $\varepsilon > 0$. The map $\varphi$ being compact-covering, we can find a compact set $L \subset X$ for which $\varphi(L) = K$. Apply Fact 7 of V.318 to see that there is an open set $W$ in the space $C_k(X)$ such that $f \in W \subset [f, L, \varepsilon]_X$. Thus $V = W \cap E$ is an open neighborhood of $f$ in the space $E$. If $h \in V$ and $u = \xi(h)$ then $h = \varphi^*(u)$. Fix a point $y \in K$ and pick $x \in L$ with $\varphi(x) = y$. We have $|u(y) - g(y)| = |h(x) - f(x)| < \varepsilon$, so $u \in [g, K, \varepsilon]_Y$. This shows that $\xi(h) \in [g, K, \varepsilon]_Y \subset U$ for all $h \in V$, i.e., $\xi(V) \subset U$ and hence the set $V$ witnesses continuity of $\xi$ at the point $f$. Consequently, the map $\xi = (\varphi^*)^{-1}$ is continuous and hence $\varphi^* : C_k(Y) \to E$ is a homeomorphism.

**V.346.** *Given a compact subspace $K$ of a space $X$ let $v(f, x) = f(x)$ for every $f \in C_k(X)$ and $x \in K$. Prove that the map $v : C_k(X) \times K \to \mathbb{R}$ is continuous.*

**Solution.** For any function $f \in C_k(X)$ and $L \subset X$ we will need the set

$$[f, L, \varepsilon] = \{g \in C_k(X) : |g(x) - f(x)| < \varepsilon \text{ for all } x \in L\}$$

for every $\varepsilon > 0$. To prove continuity of the map $v$ fix $f_0 \in C_k(X)$, $x_0 \in K$ and $\varepsilon > 0$; let $r = f_0(x_0) = v(f_0, x_0)$. Apply Fact 7 of V.318 to see that there is an open set $W$ in the space $C_k(X)$ such that $f_0 \in W \subset [f_0, K, \frac{\varepsilon}{2}]$. The set $V = \{x \in K : f_0(x) \in (r - \frac{\varepsilon}{2}, r + \frac{\varepsilon}{2})\}$ is an open neighborhood of the point $x_0$ in the space $K$, so $O = W \times V$ is an open neighborhood of the point $(f_0, x_0)$ in $C_k(X) \times K$. If $(f, x) \in O$ then $x \in V$ and hence $|f_0(x) - f_0(x_0)| < \frac{\varepsilon}{2}$; since also $f \in [f_0, K, \frac{\varepsilon}{2}]$, we have $|f(x) - f_0(x)| < \frac{\varepsilon}{2}$ which implies that $|v(f, x) - v(f_0, x_0)| = |f(x) - f_0(x_0)| < \varepsilon$, so the set $O$ witnesses that the function $v$ is continuous at the point $(f_0, x_0)$.

**V.347.** *Prove that the following properties are equivalent for any space $X$:*

*(i) $X$ is an $\aleph_0$-space;*
*(ii) $C_k(X)$ is an $\aleph_0$-space.*
*(iii) $C_k(X)$ has a countable network.*

**Solution.** Given a space $Z$, a set $K \subset Z$ and a function $f \in C(Z)$ consider the set $[f, K, \varepsilon] = \{g \in C(Z) : |g(x) - f(x)| < \varepsilon \text{ for all } x \in K\}$ for every $\varepsilon > 0$. Furthermore, if $Q \subset \mathbb{R}$ then $\langle K, Q \rangle = \{f \in C(Z) : f(K) \subset Q\}$. Given a family $\mathcal{S}$ of open subsets of $Z$ say that a family $\mathcal{B} \subset \exp(Z)$ is an $\mathcal{S}$-network for compact subsets of $Z$ if for any compact $K \subset Z$ and $S \in \mathcal{S}$ with $K \subset S$ there exists $B \in \mathcal{B}$ for which $K \subset B \subset S$.

*Fact 1.* A space $Z$ is an $\aleph_0$-space if and only if it has a subbase $\mathcal{S}$ such that some countable family $\mathcal{B}$ of subsets of $Z$ is an $\mathcal{S}$-network for compact subsets of $Z$.

*Proof.* If $Z$ is an $\aleph_0$-space then it has a network $\mathcal{B}$ for all compact subsets of $Z$; the family $\mathcal{S} = \tau(Z)$ is a subbase of $Z$ and $\mathcal{B}$ is an $\mathcal{S}$-network for compact subsets of $Z$, so we proved necessity.

Now assume that $\mathcal{S}$ is a subbase of $Z$ and $\mathcal{B}$ is a countable $\mathcal{S}$-network for compact subsets of $Z$. Let $\mathcal{C}$ be the family of all finite intersections of the elements of $\mathcal{B}$ and denote by $\mathcal{D}$ the family of all finite unions of the elements of $\mathcal{C}$. It is clear that $|\mathcal{D}| \leq \omega$; let us prove that $\mathcal{D}$ is a network for all compact subsets of $Z$.

Fix a compact set $K \subset Z$ and $U \in \tau(K, Z)$; say that a set $G \in \tau(K)$ is *adequate* if there exists $C \in \mathcal{C}$ such that $G \subset C \subset U$. Given any $x \in K$ there exists a finite $\mathcal{S}' \subset \mathcal{S}$ such that $x \in \bigcap \mathcal{S}' \subset U$. Take a set $V \in \tau(x, Z)$ for which $\overline{V} \subset \bigcap \mathcal{S}'$; if $S \in \mathcal{S}'$ then the set $F = \overline{V} \cap K$ is contained in $S$, so we can find $P_S \in \mathcal{B}$ with $F \subset P_S \subset S$. If $C = \bigcap \{P_S : S \in \mathcal{S}'\}$ then $F \subset C \subset U$ and $C \in \mathcal{C}$; an immediate consequence is that $G = V \cap K \ni x$ is an adequate set. Thus, for any $x \in K$ there exists an adequate set $G_x$ such that $x \in G_x$. By compactness of $K$ there is a finite $A \subset K$ such that $K = \bigcup \{G_x : x \in A\}$.

For every $x \in A$ pick $C_x \in \mathcal{C}$ for which $G_x \subset C_x \subset U$; then $D = \bigcup \{C_x : x \in A\}$ is an element of $\mathcal{D}$ and $K \subset D \subset U$. This shows that $\mathcal{D}$ is a countable network for all compact subsets of $Z$, i.e., $Z$ is an $\aleph_0$-space, so we settled sufficiency and hence Fact 1 is proved.

*Fact 2.* For any space $Z$ the family $\mathcal{S} = \{\langle K, O \rangle : K$ is a compact subset of $Z$ and $O \in \tau(\mathbb{R})\}$ is a subbase of the space $C_k(Z)$.

*Proof.* Suppose that $K \subset Z$ is compact, $O \in \tau(\mathbb{R})$ and $f \in \langle K, O \rangle$. Then $f(K) \subset O$ and it follows from compactness of the set $f(K)$ that there exists $\varepsilon > 0$ such that $|f(x) - r| > \varepsilon$ whenever $x \in K$ and $r \in \mathbb{R} \backslash O$. Take any $g \in [f, K, \varepsilon]$; if $x \in K$ and $r = g(x) \in \mathbb{R} \backslash O$ then $\varepsilon < |r - f(x)| = |g(x) - f(x)| < \varepsilon$; this contradiction shows that $g \in \langle K, O \rangle$ and hence $[f, K, \varepsilon] \subset \langle K, O \rangle$. By definition of the topology of $C_k(Z)$ every set $\langle K, O \rangle$ is open in $C_k(Z)$, i.e., $\mathcal{S} \subset \tau(C_k(Z))$.

Now assume that $U \in \tau(C_k(Z))$ and $f \in U$; there exists a compact set $K \subset Z$ such that $[f, K, \varepsilon] \subset U$ for some $\varepsilon > 0$. It is easy to find a finite set $A \subset K$ for which $f(A)$ is an $\frac{\varepsilon}{3}$-net for $f(K)$, i.e., for every $x \in K$ there is $a \in A$ such that $|f(x) - f(a)| < \frac{\varepsilon}{3}$. The set $K_a = \{x \in K : |f(x) - f(a)| \leq \frac{\varepsilon}{3}\}$ is compact for every $a \in A$ and $K = \bigcup \{K_a : a \in A\}$. If $Q_a = (f(a) - \frac{\varepsilon}{2}, f(a) + \frac{\varepsilon}{2})$ then $S_a = \langle K_a, Q_a \rangle \in \mathcal{S}$ and $f \in S_a$ for all $a \in A$; therefore $f \in S = \bigcap \{S_a : a \in A\}$.

If $g \in S$ and $x \in K$ then take $a \in A$ such that $|f(x) - f(a)| < \frac{\varepsilon}{3}$; then $x \in K_a$ and it follows from $g \in S_a$ that $|g(x) - f(a)| < \frac{\varepsilon}{2}$. An immediate consequence is that $|g(x) - f(x)| \leq \frac{\varepsilon}{2} + \frac{\varepsilon}{3} < \varepsilon$ and hence $g \in [f, K, \varepsilon] \subset U$ for every $g \in S$. Thus finite intersections of the elements of $\mathcal{S}$ form a base in $C_k(Z)$, so $\mathcal{S}$ is a subbase of $C_k(Z)$, i.e., Fact 2 is proved.

Returning to our solution suppose that $X$ is an $\aleph_0$-space. By Problem 342 there exists a second countable space $M$ and a compact-covering map $\varphi : M \to X$. The dual map $\varphi^* : C_k(X) \to C_k(M)$ is an embedding by Problem 345, so it suffices to show that $C_k(M)$ is an $\aleph_0$-space (see Problem 341).

Let $\mathcal{B}$ be a countable base of $M$ closed under finite unions and finite intersections. It is easy to see that

(1) the family $\mathcal{B}$ is a network for all compact subsets of $M$.

The family $\mathcal{S} = \{\langle C, U \rangle : C \subset M \text{ is compact and } U \text{ is an open subset of } \mathbb{R}\}$ is a subbase of $C_k(M)$ by Fact 2. Denote by $\mathcal{Q}$ the family of all finite unions of intervals with rational endpoints; it is easy to see that $\mathcal{Q}$ is a network for all compact subsets of $\mathbb{R}$. Consider the countable family $\mathcal{O} = \{\langle B, Q \rangle : B \in \mathcal{B} \text{ and } Q \in \mathcal{Q}\}$.

Suppose that $K \subset \langle C, U \rangle \in \mathcal{S}$ and $K$ is a compact subset of $C_k(M)$. For any $f \in C_k(M)$ and $x \in C$ let $v(f, x) = f(x)$; then the map $v : C_k(M) \times C \to \mathbb{R}$ is continuous by Problem 346. For every $x \in C$ the set $K(x) = v(K, x)$ is compact being a continuous image of the compact set $K \times \{x\}$. We will need the following property of the sets $K(x)$.

(2) if $\{x_n : n \in \omega\} \subset C$ and $x_n \to x$ then for any $G \in \tau(K(x), \mathbb{R})$ there exists $m \in \omega$ such that $K(x_n) \subset G$ for all $n \geq m$.

If (2) does not hold then we can find a subsequence $\{x_{n_i} : i \in \omega\}$ of the sequence $\{x_n : n \in \omega\}$ and a sequence $S = \{f_i : i \in \omega\} \subset K$ such that $f_i(x_{n_i}) \notin G$ for all $i \in \omega$. Pick an accumulation point $f \in K$ for the sequence $S$; it follows from $f(x) \in K(x) \subset G$ that $v(f, x) = f(x) \in G$, so continuity of $v$ at the point $(f, x)$ implies that we can find $i \in \omega$ such that $f_i(x_{n_i}) = v(f_i, x_{n_i}) \in G$. This contradiction shows that (2) is true.

Our next step is to show that

(3) there is $B \in \mathcal{B}$ such that $C \subset B$ and $v(K, B) \subset Q \subset U$ for some $Q \in \mathcal{Q}$.

Let $\{Q_n : n \in \omega\}$ be an enumeration of all elements of $\mathcal{Q}$ contained in $U$. It is easy to see that $\chi(C, M) \leq \omega$; recalling that $\mathcal{B}$ is closed under finite unions and intersections we can choose a countable outer base $\mathcal{B}' = \{B_n : n \in \omega\} \subset \mathcal{B}$ of the set $C$ in $M$ such that $B_{n+1} \subset B_n$ for all $n \in \omega$. If (3) does not hold then $v(K, B_n) \setminus (Q_0 \cup \ldots \cup Q_n) \neq \emptyset$ and hence we can find a function $g_n \in K$ and $y_n \in B_n$ such that $g_n(y_n) \notin Q_0 \cup \ldots Q_n$ for all $n \in \omega$.

It is straightforward that the set $C \cup \{y_n : n \in \omega\}$ is compact, so the sequence $\{y_n : n \in \omega\}$ contains a subsequence $\{y_{n_i} : i \in \omega\}$ which converges to a point $y \in C$. It follows from $K \subset \langle C, U \rangle$ that $K(y) \subset U$; the set $K(y)$ is compact and $K(y) \subset \bigcup_{n \in \omega} Q_n$, so there is $n \in \omega$ such that $K(y) \subset Q = Q_0 \cup \ldots \cup Q_n$. Apply (2) to see that there exists $m \in \omega$ for which $K(y_{n_i}) \subset Q$ for all $i \geq m$. For $i = m + n$ we have $K(y_{n_i}) \subset Q \subset Q' = Q_0 \cup \ldots \cup Q_{n_i}$ and hence $g_{n_i}(y_{n_i}) \in Q'$ which is a contradiction, i.e., (3) is proved.

An immediate consequence of (3) is that $K \subset \langle B, Q \rangle \subset \langle C, U \rangle$ which proves that for the subbase $\mathcal{S}$ of the space $C_k(M)$ the family $\mathcal{O}$ is an $\mathcal{S}$-network for compact subsets of $C_k(M)$. This, together with Fact 1 shows that $C_k(M)$ is an $\aleph_0$-space and hence $C_k(X)$ is an $\aleph_0$-space, so we established that (i)$\Longrightarrow$(ii).

The implication (ii)$\Longrightarrow$(iii) is trivial, so assume that $C_k(X)$ has a countable network $\mathcal{N}$ and consider the set $P_N = \bigcap\{f^{-1}((0, +\infty)) : f \in N\}$ for each $N \in \mathcal{N}$. The family $\mathcal{P} = \{P_N : N \in \mathcal{N}\}$ is countable; let us prove that $\mathcal{P}$ is a network for all compact subsets of $X$.

Observe first that $C_k(X)$ has a stronger topology than $C_p(X)$, so we have the inequalities $nw(X) = nw(C_p(X)) \leq nw(C_k(X)) \leq \omega$ and hence $X$ is normal. Given a compact set $K \subset X$ and $U \in \tau(K, X)$ the closed sets $K$ and $X \setminus U$ are disjoint, so there exists a continuous function $f : X \to [0, 1]$ such that $f(K) \subset \{1\}$ and $f(X \setminus U) \subset \{0\}$. The set $O = (0, +\infty)$ is open in $\mathbb{R}$ and $f(K) \subset O$, so we have $f \in \langle K, O \rangle$; the family $\mathcal{N}$ being a network in $C_k(X)$ there exists $N \in \mathcal{N}$ such that $f \in N \subset \langle K, O \rangle$ (see Fact 2). If $x \in K$ then $g(x) \in O$ and hence $x \in g^{-1}(O)$ for every $g \in N$; therefore $x \in P_N$ for all $x \in K$ which shows that $K \subset P_N$. Finally, if $x \in P_N$ then $g(x) > 0$ for all $g \in N$; in particular, $f(x) > 0$ and hence $x \in U$. As a consequence, $P_N \subset U$, so $\mathcal{P}$ is a countable network for all compact subsets of $X$. Thus $X$ is an $\aleph_0$-space, i.e., we settled (iii)$\Longrightarrow$(i) and hence our solution is complete.

**V.348.** *Let $X$ and $Y$ be $l$-equivalent spaces. Prove that $X$ is an $\aleph_0$-space if and only if so is $Y$. In particular, if some space $Z$ is $l$-equivalent to a second countable space then $Z$ is an $\aleph_0$-space. Deduce from this fact that any first countable space $l$-equivalent to a second countable space must be second countable.*

**Solution.** Fix a linear homeomorphism $\varphi : C_p(X) \to C_p(Y)$ and assume that $X$ is an $\aleph_0$-space; we have the equalities $nw(Y) = nw(X) = \omega$ by Problem 001, so both $X$ and $Y$ are $\mu$-spaces. The map $\varphi : C_b(X) \to C_b(Y)$ is continuous by Problem 327; since the topologies of $C_k(X)$ and $C_b(X)$ coincide as well as the topologies of $C_k(Y)$ and $C_b(Y)$ (see Fact 6 of V.318), the map $\varphi : C_k(X) \to C_k(Y)$ is also continuous. By Problem 347 the space $C_k(X)$ has a countable network; network weight is invariant under continuous images so $nw(C_k(Y)) \leq \omega$ and therefore we can apply Problem 347 again to convince ourselves that $Y$ is an $\aleph_0$-space. Since $X$ and $Y$ are in a symmetric situation, $\aleph_0$-property of $Y$ also implies that $X$ is an $\aleph_0$-space.

Finally, observe that any second countable space is an $\aleph_0$-space, so if $Z \overset{l}{\sim} M$ and $w(M) \leq \omega$ then $Z$ is an $\aleph_0$-space. If, additionally, $\chi(Z) \leq \omega$ then $w(Z) \leq \omega$ because every first countable $\aleph_0$-space must be second countable (see Problem 344).

**V.349.** *Suppose that a space $X$ has a countable network and $Y$ is an $\aleph_0$-space. Prove that $C_p(X, Y)$ has a countable network.*

**Solution.** If $Z$ and $T$ are spaces, $z_1, \ldots, z_n \in Z$ and $O_1, \ldots, O_n \in \tau(T)$ then we have the equality $[z_1, \ldots, z_n, O_1, \ldots, O_n]_Z^T = \{f \in C_p(Z, T) : f(z_i) \in O_i$ for all $i = 1, \ldots, n\}$. It is evident that the family $\{[z_1, \ldots, z_n, O_1, \ldots, O_n]_Z^T : n \in \mathbb{N}, z_1, \ldots, z_n \in Z$ and $O_1, \ldots, O_n \in \tau(T)\}$ is a base of the space $C_p(Z, T)$.

*Fact 1.* Given spaces $Z, Z'$ and $T$ suppose that $\varphi : Z \to Z'$ is a continuous onto map. For each $f \in C_p(Z', T)$ let $\varphi^*(f) = f \circ \varphi$. Then $\varphi^* : C_p(Z', T) \to C_p(Z, T)$ is an embedding.

*Proof.* Fix an arbitrary function $h_0 \in C_p(Z', T)$ and an open neighborhood $U$ of the function $g_0 = \varphi^*(h_0)$ in $C_p(Z, T)$. There exist point $z_1, \ldots, z_n \in Z$ and

$O_1, \ldots, O_n \in \tau(T)$ such that $g_0 \in U' = [z_1, \ldots, z_n, O_1, \ldots, O_n]_Z^T \subset U$. If $y_i = \varphi(z_i)$ for all $i \leq n$ then $W = [y_1, \ldots, y_n, O_1, \ldots, O_n]_{Z'}^T$ is an open neighborhood of the function $h_0$ in the space $C_p(Z', T)$. It is straightforward that $\varphi^*(W) \subset U' \subset U$, so the map $\varphi^*$ is continuous at the point $h_0$.

To see that $\varphi^*$ is an injection take distinct $f, g \in C_p(Z', T)$. Then $f(y) \neq g(y)$ for some $y \in Z'$. If $z \in \varphi^{-1}(y)$ then $\varphi^*(f)(z) = f(y) \neq g(y) = \varphi^*(g)(z)$ and hence $\varphi^*(f) \neq \varphi^*(g)$.

Let $E = \varphi^*(C_p(Z', T))$; to see that the mapping $(\varphi^*)^{-1} : E \to C_p(Z', T)$ is continuous take any $f \in E$ and let $g = (\varphi^*)^{-1}(f)$. Fix a set $U \in \tau(g, C_p(Z', T))$; there are $y_1, \ldots, y_n \in Y$ and $O_1, \ldots, O_n \in \tau(T)$ such that

$$g \in U' = [y_1, \ldots, y_n, O_1, \ldots, O_n]_{Z'}^T \subset U.$$

Choose $x_i \in \varphi^{-1}(y_i)$ for all $i \leq n$; the set $V = [x_1, \ldots, x_n, O_1, \ldots, O_n] \cap E$ is an open neighborhood of $f$ in the space $E$. If $h \in V$ and $u = (\varphi^*)^{-1}(h)$ then $h(x_i) = \varphi^*(u)(x_i) = u(\varphi(x_i)) = u(y_i) \in O_i$ for all $i \leq n$. Thus $u = (\varphi^*)^{-1}(h) \in U'$ for any $h \in V$ and hence $(\varphi^*)^{-1}(V) \subset U' \subset U$ which proves continuity of the map $(\varphi^*)^{-1}$ at an arbitrary point $f$. Hence the mapping $\varphi^* : C_p(Z', T) \to E$ is a homeomorphism, so Fact 1 is proved.

Returning to our solution fix a countable network $\mathcal{N}$ for all compact subsets of $Y$ which is closed under finite unions; there exists a second countable space $M$ and a continuous onto map $\varphi : M \to X$ (see Fact 6 of T.250). The dual map $\varphi^* : C_p(X, Y) \to C_p(M, Y)$ is an embedding by Fact 1, so it suffices to show that $C_p(M, Y)$ has a countable network.

Let $\mathcal{B}$ be a countable base of $M$ closed under finite unions. Given $B \in \mathcal{B}$ and $N \in \mathcal{N}$ let $\langle B, N \rangle = \{f \in C_p(M, Y) : f(B) \subset N\}$. Consider the family $\mathcal{S}' = \{\langle B, N \rangle : B \in \mathcal{B} \text{ and } N \in \mathcal{N}\}$; we will show that the family $\mathcal{S}$ of all finite intersections of the elements of $\mathcal{S}'$ is a network in $C_p(M, Y)$. Our main tool will be the following property.

(1) For any $f \in C_p(M, Y)$ if $x \in M$ and $U \in \tau(f(x), Y)$ then there exists $N \in \mathcal{N}$ such that $N \subset U$ and $x \in \text{Int}(f^{-1}(N))$.

Let $\{N_k : k \in \omega\}$ be an enumeration of all elements of $\mathcal{N}$ contained in $U$ and suppose that (1) does not hold. Since $\mathcal{N}$ is closed under finite unions, the point $x$ does not belong to the interior of the set $P_k = f^{-1}(N_0) \cup \ldots f^{-1}(N_k)$ for all $k \in \omega$. Take a decreasing local base $\{W_k : k \in \omega\}$ at the point $x$ such that $W_0 \subset f^{-1}(U)$ and choose a point $y_k \in W_k \backslash P_k$ for every $k \in \omega$. It is clear that the set $F = \{y\} \cup \{y_k : k \in \omega\}$ is compact and the set $\{k \in \omega : f(y_k) \in N_i\}$ is finite for every $i \in \omega$. However, $G = f(F)$ is a compact subset of $U$, so there is $N \in \mathcal{N}$ with $G \subset N \subset U$. We have $N = N_i$ for some $i \in \omega$, so $\{f(y_k) : k \in \omega\} \subset N_i$; this contradiction shows that (1) is proved.

Fix $f \in C_p(M, Y)$ and $U \in \tau(f, C_p(M, Y))$. There are points $x_1, \ldots, x_n \in M$ and $O_1, \ldots, O_n \in \tau(Y)$ for which $f \in U' = [x_1, \ldots, x_n, O_1, \ldots, O_n]_M^Y \subset U$.

Apply (1) to find a set $N_i \in \mathcal{N}$ such that $N_i \subset O_i$ and $x_i \in V_i = \mathrm{Int}(f^{-1}(N_i))$ for all $i = 1, \ldots, n$. Pick $B_i \in \mathcal{B}$ such that $x_i \in B_i \subset V_i$; then $f \in \langle B_i, N_i \rangle$ for all $i \leq n$. If $S = \bigcap \{ \langle B_i, N_i \rangle : i \leq n \}$ then it is straightforward that $f \in S \subset U' \subset U$, so $\mathcal{S}$ is a countable network of $C_p(M, Y)$. We already observed that this implies that $nw(C_p(X, Y)) \leq \omega$ and hence our solution is complete.

**V.350.** *Given spaces $X, Y$ and a function $u : X \times Y \to \mathbb{R}$ let $u_x(y) = u(x, y)$ for all $y \in Y$; then $u_x : Y \to \mathbb{R}$ for every $x \in X$. Analogously, let $u^y(x) = u(x, y)$ for all $x \in X$; then $u^y : X \to \mathbb{R}$ for every $y \in Y$. Say that the function $u$ is separately continuous if the functions $u_x$ and $u^y$ are continuous (on $Y$ and $X$ respectively) for all $x \in X$ and $y \in Y$. Let $C_p^s(X \times Y)$ be the set of all separately continuous functions on $X \times Y$ with the topology induced from $\mathbb{R}^{X \times Y}$. Observe that $C_p^s(X \times Y)$ is a locally convex space and let $\xi(\varphi)(x, y) = \varphi(x)(y)$ for any continuous function $\varphi : X \to C_p(Y)$. Prove that $\xi(\varphi) \in C_p^s(X \times Y)$ for every $\varphi \in C_p(X, C_p(Y))$ and $\xi : C_p(X, C_p(Y)) \to C_p^s(X \times Y)$ is a linear homeomorphism.*

**Solution.** It is easy to see that $C_p^s(X \times Y)$ is a linear subspace of $\mathbb{R}^{X \times Y}$, so $C_p^s(X \times Y)$ is a locally convex space. Besides, $C_p(X, C_p(Y))$ is a linear subspace of $C_p(Y)^X$, so $C_p(X, C_p(Y))$ is also a locally convex space.

For each $x \in X$ and $\varphi \in C_p(X, C_p(Y))$ the function $\xi(\varphi)_x$ is continuous because it coincides with $\varphi(x) \in C_p(Y)$. If $y \in Y$ then let $\pi_y : C_p(Y) \to \mathbb{R}$ be the projection onto the coordinate determined by $y$; then $\xi(\varphi)^y = \pi_y \circ \varphi$ is a continuous map, so $\xi(\varphi)$ is separately continuous, i.e., $\xi(\varphi) \in C_p^s(X \times Y)$ for every $\varphi \in C_p(X, C_p(Y))$.

If $\varphi$ and $\eta$ are distinct elements of $C_p(X, C_p(Y))$ then $\varphi(x) \neq \eta(x)$ for some $x \in X$; consequently, $\varphi(x)(y) \neq \eta(x)(y)$ for a point $y \in Y$. This shows that $\xi(\varphi)(x, y) \neq \xi(\eta)(x, y)$ and hence $\xi(\varphi) \neq \xi(\eta)$, i.e., $\xi$ is an injective map.

Given any $u \in C_p^s(X \times Y)$ let $\varphi(x) = u_x$; then $\varphi(x) \in C_p(Y)$ for every $x \in X$ and therefore $\varphi : X \to C_p(Y)$. For each $y \in Y$ the map $\pi_y \circ \varphi$ is continuous because it coincides with $u^y$; applying TFS-102 we conclude that $\varphi : X \to C_p(Y)$ is continuous. We have $\xi(\varphi)(x, y) = \varphi(x)(y) = u_x(y) = u(x, y)$ for all $(x, y) \in X \times Y$; this shows that $\xi(\varphi) = u$, so $\xi$ is also surjective, i.e., $\xi : C_p(X, C_p(Y)) \to C_p^s(X \times Y)$ is a bijective map. We omit a simple verification of linearity of $\xi$; once it is done we can see that $\xi$ is a linear isomorphism.

Given $(x, y) \in X \times Y$ let $p_{(x,y)}(u) = u(x, y)$ for each $u \in C_p^s(X \times Y)$; then $p_{(x,y)} : C_p^s(X \times Y) \to \mathbb{R}$ is the projection onto the factor determined by the point $(x, y)$. For every $x \in X$ let $q_x(\varphi) = \varphi(x)$ for each function $\varphi \in C_p(X, C_p(Y))$; then $q_x : C_p(X, C_p(Y)) \to C_p(Y)$ is the projection onto the factor determined by $x$.

Fix any point $(x, y) \in X \times Y$; then $(p_{(x,y)} \circ \xi)(\varphi) = \varphi(x)(y)$ for every function $\varphi \in C_p(X, C_p(Y))$. Therefore $p_{(x,y)} \circ \xi = \pi_y \circ q_x$ is a continuous map for every $(x, y) \in X \times Y$, so we can apply TFS-102 once again to conclude that $\xi$ is continuous.

Finally take any $x \in X$ and $u \in C_p^s(X \times Y)$; if $\varphi = \xi^{-1}(u)$ then $q_x(\varphi) = u_x$; to see that the map $q_x \circ \xi^{-1} : C_p^s(X \times Y) \to C_p(Y)$ is continuous observe that

$\pi_y \circ q_x \circ \xi^{-1}(u) = u(x, y) = p_{(x,y)}(u)$ for any $y \in Y$. The map $\pi_y \circ (q_x \circ \xi^{-1}) = p_{(x,y)}$ being continuous for all $y \in Y$, the mapping $q_x \circ \xi^{-1}$ is continuous for all $x \in X$, so $\xi^{-1}$ is continuous (observe that we have just applied TFS-102 twice) and hence $\xi$ is, indeed, a linear homeomorphism.

**V.351.** *Prove that the space $C_p(X, C_p(X))$ has a countable network if and only if $X$ is countable. Deduce from this fact that $C_p(X)$ is an $\aleph_0$-space if and only if $X$ is countable.*

**Solution.** If $X$ is countable then $w(C_p(X)) = \omega$ and hence $w(C_p(X)^X) = \omega$, so it follows from $C_p(X, C_p(X)) \subset C_p(X)^X$ that $w(C_p(X, C_p(X))) = \omega$; in particular, $C_p(X, C_p(X))$ has a countable network, i.e., we proved sufficiency.

Now assume that the space $C_p(X, C_p(X))$ has a countable network; the space $C_p(X)$ is a continuous image of $C_p(X, C_p(X))$, so $nw(X) = nw(C_p(X)) = \omega$. Thus $nw(X \times X) = \omega$ and hence $X \times X$ is perfectly normal; this makes it possible to find a continuous function $u : X \times X \to [0, 1]$ such that $u^{-1}(0) = \Delta = \{(x, x) : x \in X\}$. For any $z \in X$ let $f_z(z, z) = 1$ and

$$f_z(x, y) = \frac{2u(x, y)}{u(x, y) + u(x, z) + u(z, y)} \text{ whenever } (x, y) \neq (z, z).$$

Then $f_z : X \times X \to \mathbb{R}$ and the function $f_z$ is continuous at all points of $X \times X$ except $(z, z)$. However, $f_z(z, y) = 1$ and $f_z(x, z) = 1$ for all $x, y \in X$; this proves that the function $f_z$ is separately continuous for every $z \in X$. The space $C_p^s(X \times X)$ of separately continuous functions on $X \times X$ with the topology of pointwise convergence is linearly homeomorphic to $C_p(X, C_p(X))$ by Problem 350, so $nw(C_p^s(X \times X)) = \omega$. We already saw that the set $D = \{f_z : z \in X\}$ is contained in $C_p^s(X \times X)$. Observe that, for every $z \in X$ we have $f_z(z, z) = 1$ and $f_z(x, x) = 0$ for all $x \neq z$. Therefore $z \neq z'$ implies that $f_z \neq f_{z'}$ and hence $|D| = |X|$.

The set $U_z = \{f \in C_p^s(X \times X) : f(z, z) > 0\}$ is open in $C_p^s(X \times X)$ and it is immediate that $U_z \cap D = \{f_z\}$ for every $z \in X$. Therefore $D$ is a discrete subspace of $C_p^s(X \times X)$, so it follows from $nw(C_p^s(X \times X)) = \omega$ that $|D| \leq \omega$ and hence $|X| = |D| \leq \omega$. This proves necessity and shows that

(1)  for any space $X$ we have $nw(C_p(X, C_p(X))) = \omega$ if and only if $|X| = \omega$.

Finally, observe that if $X$ is countable then $C_p(X)$ is an $\aleph_0$-space being second countable. Now if $C_p(X)$ is an $\aleph_0$-space then $nw(X) = nw(C_p(X)) \leq \omega$ and hence the space $C_p(X, C_p(X))$ has a countable network (see Problem 349), so (1) implies that $X$ is countable.

**V.352.** *Prove that a space $X$ is of second category in itself if and only if the player $E$ has no winning strategy in the Banach–Mazur NE-game on $X$.*

**Solution.** Suppose that the player $E$ does not have a winning strategy on $X$ in the Banach–Mazur $NE$-game. If $X$ is of first category in itself then we can fix a family $\mathcal{K} = \{K_n : n \in \mathbb{N}\}$ of closed nowhere dense subsets of $X$ such that

$X = \bigcup \mathcal{K}$. Now, if $n \in \mathbb{N}$ and the moves $V_1, U_1, \ldots, V_{n-1}, U_{n-1}, V_n$ are made then let $U_n = \sigma(V_1, \ldots, V_n) = V_n \backslash K_n$; it is clear that $V_n$ is a nonempty open subset of $U_n$, so we defined a strategy $\sigma$ for the player $E$ on the space $X$.

Suppose that $\mathcal{P} = \{U_n, V_n : n \in \mathbb{N}\}$ is a play in which $E$ applies the strategy $\sigma$; if $P = \bigcap \mathcal{P}$ then $P \cap K_n = \emptyset$ for all $n \in \mathbb{N}$ and hence $P = \emptyset$. Therefore the strategy $\sigma$ is winning for the player $E$; this contradiction shows that $X$ is of second category in itself, i.e., we proved sufficiency.

Now assume that $X$ is of second category in itself and $s$ is a strategy of the player $E$ in the $NE$-game on the space $X$. The family $\mathcal{B} = \{s(V) : V \in \tau^*(X)\}$ is a $\pi$-base of the space $X$, so we can choose a maximal disjoint (with respect to inclusion) subfamily $\mathcal{U}(1)$ of the family $\mathcal{B}$. It is evident that $\bigcup \mathcal{U}(1)$ is dense in $X$. For any $U \in \mathcal{U}(1)$ fix a set $V = \delta_1(U) \in \tau^*(X)$ such that $U = s(V)$ and let $\mathcal{V}(1) = \{\delta_1(U) : U \in \mathcal{U}(1)\}$. It is clear that the map $\delta_1 : \mathcal{U}(1) \to \mathcal{V}(1)$ is a bijection.

Proceeding by induction assume that $n \in \mathbb{N}$ and we have constructed families $\mathcal{U}(1), \mathcal{V}(1) \ldots, \mathcal{U}(n), \mathcal{V}(n)$ of nonempty open subsets of $X$ with the following properties:

(1) the family $\mathcal{U}(i)$ is disjoint and $\bigcup \mathcal{U}(i)$ is dense in $X$ for every $i \le n$;
(2) for every $i \le n$ a bijection $\delta_i : \mathcal{U}(i) \to \mathcal{V}(i)$ is fixed in such a way that $U \subset \delta_i(U)$ for each $U \in \mathcal{U}(i)$;
(3) if $i < n$ and $V \in \mathcal{V}(i+1)$ then there is $U \in \mathcal{U}(i)$ with $V \subset U$;
(4) if $k \le n$ and $U_1 \supset \ldots \supset U_k$ where $U_i \in \mathcal{U}(i)$ for all $i \le k$ then letting $V_i = \delta_i(U_i)$ we obtain an initial segment $V_1, U_1, \ldots, V_k, U_k$ of a play in the $NE$-game such that $U_i = s(V_1, U_1, \ldots, V_{i-1}, U_{i-1}, V_i)$ for all $i \le k$.

Fix a set $U_n \in \mathcal{U}(n)$ and let $V_n = \delta_n(U_n)$; if $1 < k \le n$ and we have the sets $V_k, U_k, \ldots, V_n, U_n$, then we can apply (3) to find a set $U_{k-1} \in \mathcal{U}(k-1)$ (which is evidently, unique) with $V_k \subset U_{k-1}$; let $V_{k-1} = \delta_{k-1}(U_{k-1})$. This inductive construction gives us the sets $U_1 \supset \ldots \supset U_n$, so the property (4) implies that $V_1, U_1, \ldots, V_n, U_n$ is an initial segment of a play in which $E$ applies the strategy $s$. Therefore we can consider the family $\mathcal{B} = \{s(V_1, \ldots, V_n, V) : V \in \tau^*(U_n)\}$ which is, evidently, a $\pi$-base in $U_n$; choose a maximal disjoint subfamily $\mathcal{U}(n+1)[U_n]$ of the family $\mathcal{B}$. It is evident that $\bigcup \mathcal{U}(n+1)$ is dense in $U_n$. For each $U \in \mathcal{U}(n+1)$ there exists $V = \delta_{n+1}(U) \in \tau^*(U_n)$ such that $U = s(V_1, \ldots, V_n, V)$; it is immediate that the family $\mathcal{V}(n+1)[U_n] = \{\delta_{n+1}(U) : U \in \mathcal{U}(n+1)\}$ consists of nonempty open subsets of $U_n$ and $\delta_{n+1} : \mathcal{U}(n+1)[U_n] \to \mathcal{V}(n+1)[U_n]$ is a bijection. If $\mathcal{U}(n+1) = \bigcup \{\mathcal{U}(n+1)[U_n] : U_n \in \mathcal{U}(n)\}$ and $\mathcal{V}(n+1) = \delta_{n+1}(\mathcal{U}(n+1))$ then the map $\delta_{n+1} : \mathcal{U}(n+1) \to \mathcal{V}(n+1)$ is a bijection while the family $\mathcal{U}(n+1)$ is disjoint and its union is dense in $X$.

It is straightforward to verify that the properties (1)–(4) now hold if we replace $n$ with $n+1$. Thus our inductive procedure can be continued to construct a sequence of families $\{\mathcal{U}(i), \mathcal{V}(i) : i \in \mathbb{N}\}$ with the conditions (1)–(4) satisfied for all $n \in \mathbb{N}$.

It follows from (1) that the set $F_i = X \backslash (\bigcup \mathcal{U}(i))$ is closed and nowhere dense in $X$ for every $i \in \mathbb{N}$; since $X$ is of second category in itself, there exists a point $x \in X \backslash (\bigcup \{F_i : i \in \mathbb{N}\})$. Thus $x \in \bigcap \{\bigcup \mathcal{U}(i) : i \in \mathbb{N}\}$, so we can choose a sequence $\{U_i : i \in \mathbb{N}\}$ such that $U_i \in \mathcal{U}(i)$ for each $i \in \mathbb{N}$ and $x \in \bigcap\{U_i : i \in \mathbb{N}\}$. For every $i \in \mathbb{N}$ it is an easy consequence of (1)–(3) that $U_{i+1} \subset U_i$; let $V_i = \delta_i(U_i)$. The property (4) shows that $\mathcal{P} = \{V_i, U_i : i \in \mathbb{N}\}$ is a play in which $E$ applies the strategy $s$. Now, $x \in \bigcap \mathcal{P}$, so $s$ is not winning for $E$. Since a strategy $s$ of the player $E$ was taken arbitrarily, we proved that no strategy of $E$ is winning; this settles necessity and makes our solution complete.

**V.353.** *Prove that a space $X$ has the Baire property if and only if the player $E$ has no winning strategy in the Banach–Mazur $E$-game on $X$.*

**Solution.** Suppose that the player $E$ does not have a winning strategy on $X$ in the Banach–Mazur $E$-game. Fix a nonempty open set $U \subset X$ and assume that $E$ has a winning strategy $s$ in the $NE$-game on the space $U$. To define a strategy $\sigma$ of $E$ in the $E$-game on $X$ let $\sigma(\emptyset) = U$. For any $V_1 \in \tau^*(U)$ we can consider $V_1$ to be the first move of $NE$ in the $NE$-game on $V$, so we can define $\sigma(V_1) = s(V_1)$. It is evident that $\tau^*(U)$ is the family of all admissible sets for $\sigma$.

Proceeding by induction assume that $n \in \mathbb{N}$ and we proved that, for each $i \leq n$, an $i$-tuple $(V_1, \ldots, V_i)$ is admissible for the strategy $s$ if and only if it is admissible for the strategy $\sigma$ and $\sigma(V_1, \ldots, V_i) = s(V_1, \ldots, V_i)$. Observe that $\{V_1, U_1, \ldots, V_n, U_n, V_{n+1}\}$ is an initial segment of a play in the $NE$-game on the space $U$ if and only if $\{U, V_1, U_1, V_2, \ldots, U_{n-1}, V_n, U_n, V_{n+1}\}$ is an initial segment of a play in the $E$-game on $X$. By the induction hypothesis we have the equalities $U_i = s(V_1, \ldots, V_i) = \sigma(V_1, \ldots, V_i)$ for all $i \leq n$. The $(n+1)$-tuple $(V_1, \ldots, V_{n+1})$ is admissible for $\sigma$ if and only if $V_{n+1} \subset U_n$ and this happens precisely when it is admissible for $s$, so we can let $\sigma(V_1, \ldots, V_{n+1}) = s(V_1, \ldots, V_{n+1})$ completing our construction of a strategy $\sigma$ for the player $E$ in the $E$-game on $X$.

The strategy $s$ is winning, so $\sigma$ is also winning because, apart from the set $U$, all plays where $E$ applies $\sigma$ coincide with the respective plays on $E$ where $E$ uses the strategy $s$. This contradiction shows that $E$ does not have a winning strategy on the $NE$-game on $U$ and hence $U$ is of second category in itself by Problem 352. Thus every $U \in \tau^*(X)$ is of second category, i.e., $X$ has the Baire property which shows that we proved sufficiency.

Now suppose that $X$ has the Baire property and $\sigma$ is a strategy of the player $E$ in the $E$-game on $X$; let $U = \sigma(\emptyset)$. Every $V_1 \in \tau^*(U)$ is admissible for $\sigma$, so we can let $s(V_1) = \sigma(V_1)$. Proceeding inductively as above it is easy to see that letting $s(V_1, \ldots, V_n) = \sigma(V_1, \ldots, V_n)$ for every admissible $n$-tuple $(V_1, \ldots, V_n)$ for the strategy $\sigma$ we consistently define a strategy $s$ for the player $E$ in the $NE$-game on the space $U$.

The set $U$ is of second category in itself, so the strategy $s$ cannot be winning by Problem 352. Since any play where $E$ applies $\sigma$ coincides (apart from the set $U$) with the respective play on $E$ where $E$ uses the strategy $s$, the strategy $\sigma$ is not winning either. Thus $E$ does not have a winning strategy in the $E$-game on the space $X$ and hence we settled necessity.

**V.354.** *Prove that*

*(i) any pseudocompact space with the moving off property is compact.*

*(ii) any paracompact locally compact space has the moving off property.*

**Solution.** If $X$ is a pseudocompact non-compact space with the moving off property then $\mathcal{S} = \{\{x\} : x \in X\}$ is a moving off family so there is an infinite $\mathcal{A} \subset \mathcal{S}$ which has a discrete open expansion. However, a pseudocompact space cannot have an infinite discrete family of nonempty open sets (see TFS-136); this contradiction shows that (i) is proved.

To prove (ii) assume that $X$ is a nonempty paracompact locally compact space. Fix a locally finite open cover $\mathcal{U}$ of the space $X$ such that $\overline{U}$ is compact for each $U \in \mathcal{U}$. Let $\mathcal{U}[A] = \{U \in \mathcal{U} : U \cap A \neq \emptyset\}$ and $\mathrm{St}(A, \mathcal{U}) = \bigcup \mathcal{U}[A]$ for any $A \subset X$. Suppose that $\mathcal{K}$ is a moving off family of compact subsets of $X$; observe that $\mathcal{K}$ has to be nonempty because, for any $x \in X$ there is $K \in \mathcal{K}$ with $x \notin K$.

Take any set $K_0 \in \mathcal{K}$; proceeding inductively assume that $n \in \omega$ and we have chosen sets $K_0, \ldots, K_n \in \mathcal{K}$ in such a way that

(1) $K_{i+1} \cap \mathrm{St}(K_0 \cup \ldots \cup K_i, \mathcal{U}) = \emptyset$ for each $i < n$.

The set $K' = K_0 \cup \ldots \cup K_n$ is compact, so the family $\mathcal{U}[K']$ is finite and hence the set $\mathrm{St}(K', \mathcal{U})$ has compact closure $F$. Using the fact that $\mathcal{K}$ is a moving off collection, we can find a set $K_{n+1} \in \mathcal{K}$ such that $K_{n+1} \cap F = \emptyset$. It is evident that the property (1) still holds for all $i \leq n$, so our inductive procedure can be continued to construct a family $\mathcal{K}' = \{K_i : i \in \omega\} \subset \mathcal{K}$ such that (1) is fulfilled for all $n \in \omega$ and, in particular, the family $\mathcal{K}'$ is infinite.

Fix a point $x \in X$ and $U \in \mathcal{U}$ with $x \in U$. If $i < j$ and $U \cap K_i \neq \emptyset$ then $U \subset \mathrm{St}(K_i, \mathcal{U})$, so it follows from (1) that $U \cap K_j = \emptyset$. Therefore, every $x \in X$ has a neighborhood which meets at most one element of $\mathcal{K}'$, i.e., the family $\mathcal{K}'$ is discrete. By paracompactness of $X$ (or, more exactly, by its collectionwise normality), the family $\mathcal{K}' \subset \mathcal{K}$ has a discrete open expansion. This shows that the space $X$ has the moving off property and hence we settled (ii).

**V.355.** *Let $X$ be a $q$-space. Prove that, if $X$ has the moving off property then it is locally compact.*

**Solution.** Fix a point $a \in X$ and say that a family $\{U_n : n \in \omega\}$ of open neighborhoods of $a$ is a $q$-sequence if any sequence $\{x_n : n \in \omega\} \subset X$ such that $x_n \in U_n$ for every $n \in \omega$, has an accumulation point. Since $X$ is a $q$-space, there exists at least one $q$-sequence. Observe that if $\{U_n : n \in \omega\}$ is a $q$-sequence, $V_n \in \tau(a, X)$ and $a \in V_n \subset U_n$ for all $n \in \omega$ then the family $\{V_n : n \in \omega\}$ is a $q$-sequence as well. Therefore we can find a $q$-sequence $\mathcal{O} = \{O_n : n \in \omega\}$ such that $\overline{O}_{n+1} \subset O_n$ for every $n \in \omega$. Besides,

(1) if a sequence $\{n_i : i \in \omega\} \subset \omega$ is strictly increasing and $y_i \in O_{n_i}$ for all $i \in \omega$ then the sequence $S = \{y_i : i \in \omega\}$ has an accumulation point.

Indeed, $i \leq n_i$, so $O_{n_i} \subset O_i$ for every $i \in \omega$, so $S$ has an accumulation point because $\mathcal{O}$ is a $q$-sequence.

Suppose that $X$ is not locally compact at the point $a$. Since the moving off property is closed-hereditary, any closed neighborhood of $a$ is a non-compact space with the moving off property; therefore it is not countably compact by Problem 354. Thus we can choose an infinite closed discrete set $D_n \subset \overline{O}_{n+1}$; observe that $D_n \subset O_n$ for every $n \in \omega$. Take a faithful enumeration $\{x_n : n \in \omega\}$ of some countably infinite set $D \subset D_0$ and consider the family $\mathcal{D} = \{\{x_n, d\} : d \in D_n, n \in \omega\}$.

If $K \subset X$ is compact then $D \cap K$ is finite, so there exists $n \in \omega$ such that $x_n \notin K$. The set $D_n \cap K$ is also finite, so we can find $d \in D_n$ with $d \notin K$. Now $\{x_n, d\} \in \mathcal{D}$ and $\{x_n, d\} \cap K = \emptyset$; this proves that $\mathcal{D}$ is a moving off collection. Consequently, there is an infinite $\mathcal{D}' \subset \mathcal{D}$ which has a discrete open expansion. Observe that if $\{x_n, d\}$ and $\{x_m, e\}$ are distinct elements of $\mathcal{D}'$ then $m \neq n$, so we can choose an increasing sequence $\{n_i : i \in \omega\} \subset \omega$ and $d_{n_i} \in D_{n_i}$ for all $i \in \omega$ in such a way that $\{\{x_{n_i}, d_{n_i}\} : i \in \omega\} \subset \mathcal{D}'$ and, in particular, $\{d_{n_i} : i \in \omega\}$ is a faithfully indexed closed discrete subset of $X$. However, $d_{n_i} \in O_{n_i}$ for all $i \in \omega$, so we can apply (1) to conclude that the sequence $\{d_{n_i} : i \in \omega\}$ has an accumulation point; this contradiction shows that $X$ is locally compact at any point $a \in X$.

**V.356.** *Prove that the following conditions are equivalent for any space $X \neq \emptyset$:*

(i) *$X$ has the moving off property;*

(ii) *given a sequence $\{\mathcal{K}_i : i \in \omega\}$ of moving off collections in $X$, we can choose $K_i \in \mathcal{K}_i$ for each $i \in \omega$, such that the family $\{K_i : i \in \omega\}$ has a discrete open expansion;*

(iii) *the player $II$ has no winning strategy in the Gruenhage–Ma game on $X$.*

**Solution.** Suppose that the space $X$ has the moving off property and we are given a sequence $\{\mathcal{K}_i : i \in \omega\}$ of moving off collections in $X$. Since nonempty compact spaces do not admit moving off collections, the space $X$ is not compact and hence not pseudocompact by Problem 354, so we can fix a faithfully indexed closed discrete set $D = \{d_n : n \in \omega\} \subset X$.

For every $n \in \omega$ consider the family $\mathcal{L}_n = \{\{d_n\} \cup L_0 \cup \ldots \cup L_n : L_i \in \mathcal{K}_i$ for all $i \leq n\}$ and let $\mathcal{L} = \bigcup_{n \in \omega} \mathcal{L}_n$. Given a compact set $F \subset X$, the set $D \cap F$ is finite, so there is $n \in \omega$ such that $d_n \notin F$. Pick $L_i \in \mathcal{K}_i$ with $L_i \cap F = \emptyset$ for all $i \leq n$; this is possible because every $\mathcal{K}_i$ is a moving off collection. Then the set $L = \{d_n\} \cup L_0 \cup \ldots \cup L_n$ belongs to $\mathcal{L}$ and $L \cap F = \emptyset$; this shows that $\mathcal{L}$ is a moving off family.

Thus there exists an infinite family $\mathcal{L}' \subset \mathcal{L}$ which has a discrete open expansion. If $L = \{d_n\} \cup L_0 \cup \ldots \cup L_n$ and $L' = \{d_m\} \cup P_0 \cup \ldots \cup P_m$ are distinct elements of $\mathcal{L}'$ then they are disjoint, so $n \neq m$. Therefore we can find an increasing sequence $\{n_i : i \in \omega\} \subset \omega$ and a set $Q_i = \{d_{n_i}\} \cup L_0^i \cup \ldots \cup L_{n_i}^{n_i} \in \mathcal{L}_{n_i}$ such that $Q_i \in \mathcal{L}'$ for each $i \in \omega$. Observe that we have $i \leq n_i$, so $L_i^{n_i} \in \mathcal{K}_i$; thus, letting $K_i = L_i^{n_i}$ for every $i \in \omega$ we obtain the promised sequence $\{K_i : i \in \omega\}$ which has a discrete open expansion. This shows that (i)$\Longrightarrow$(ii).

For each $k \in \mathbb{N}$ denote by $\mathcal{S}_k$ the family of all finite sequences of the elements of $\omega$ of length $k$ and let $\mathcal{S} = \bigcup\{\mathcal{S}_k : k \in \mathbb{N}\}$. If $s = (n_1, \ldots, n_k) \in \mathcal{S}$ then $s[i] = (s_1, \ldots, s_i)$ for every $i = 1, \ldots, k$ and $s^\frown n = (n_1, \ldots, n_k, n)$ for all $n \in \omega$.

Next, assume that the property (ii) holds for $X$ and $\sigma$ is a strategy of the player $II$ in the Gruenhage–Ma game on the space $X$. The family $\{\sigma(K) : K \in \mathcal{K}(X)\}$ is clearly moving off, so we can choose a sequence $\{K_n : n \in \omega\} \subset \mathcal{K}(X)$ such that $\{\sigma(K_n) : n \in \omega\}$ has a discrete open expansion; let $K_{(n)} = K_n$ and $L_{(n)} = \sigma(K_n)$ for every $n \in \omega$.

Proceeding by induction assume that $n \in \mathbb{N}$ and, for any nonempty finite sequence $s$ of elements of $\omega$ of length at most $n$, we have defined sets $K_s$ and $L_s$ is such a way that

(1) if $k \leq n$ and $s \in \mathcal{S}_k$ then the family $K_{s[1]}, L_{s[1]}, \ldots, K_{s[k]}, L_{s[k]}$ is an initial segment of a play in which $II$ applies the strategy $\sigma$.

(2) the family $\{L_{s^\frown n} : n \in \omega\}$ has a discrete open expansion for every sequence $s$ of length at most $n - 1$.

Fix any sequence $s \in \mathcal{S}_n$; then $K_{s[1]}, L_{s[1]}, \ldots, K_{s[n]}, L_{s[n]}$ is an initial segment of a play on $X$ in which $II$ applies the strategy $\sigma$. It is clear that the family $\{P(K) = \sigma(K_{s[1]}, \ldots, K_{s[n]}, K) : K \in \mathcal{K}(X)\}$ is moving off, so we can choose a faithfully indexed sequence $\{K_n : n \in \omega\} \subset \mathcal{K}(X)$ such that $\{P(K_n) : n \in \omega\}$ has a discrete open expansion. Let $K_{s^\frown n} = K_n$ and $L_{s^\frown n} = P(K_n)$ for all $n \in \omega$. Having done this construction for all $s \in \mathcal{S}_n$ we obtain the sets $K_t$ and $L_t$ for all $t \in \mathcal{S}_{n+1}$ and it is straightforward that the conditions (1) and (2) are satisfied if $n$ is replaced with $n + 1$. Therefore our inductive procedure can be continued to obtain a family $\{K_s, L_s : s \in \mathcal{S}\}$ for which (1) and (2) are fulfilled for all $n \in \mathbb{N}$.

Any family which has a discrete open expansion is moving off so the family $\mathcal{Q}_s = \{L_{s^\frown n} : n \in \omega\}$ is moving off for each $s \in \mathcal{S} \cup \{\emptyset\}$. Apply (2) to pick $Q_s \in \mathcal{Q}_s$ for all $s \in \mathcal{S} \cup \{\emptyset\}$ in such a way that the family $\mathcal{P} = \{Q_s : s \in \mathcal{S} \cup \{\emptyset\}\}$ has a discrete open expansion.

Thus there is $n_1 \in \omega$ such that $L_{(n_1)} \in \mathcal{P}$. Proceeding inductively assume that $k \in \mathbb{N}$ and we have a sequence $s = (n_1, \ldots, n_k)$ such that $L_{s[i]} \in \mathcal{P}$ for all $i \leq k$. By our choice of $\mathcal{P}$ there exists $n_{k+1} \in \omega$ such that $L_{s^\frown n_{k+1}} \in \mathcal{P}$. Therefore this inductive procedure can be continued to construct a sequence $\{n_i : i \in \mathbb{N}\} \subset \omega$ such that $L_{(n_1, \ldots, n_k)} \in \mathcal{P}$ for all $k \in \mathbb{N}$.

Let $P_k = K_{(n_1, \ldots, n_k)}$ and $Q_k = L_{(n_1, \ldots, n_k)}$ for all $k \in \mathbb{N}$; it follows from (1) and (2) that $\{P_k, Q_k : k \in \mathbb{N}\}$ is a play in which $II$ applies the strategy $\sigma$. It follows from $\{Q_k : k \in \mathbb{N}\} \subset \mathcal{P}$ that the family $\{Q_k : k \in \mathbb{N}\}$ has a discrete open expansion and hence the strategy $\sigma$ is not winning. Recall that a strategy $\sigma$ of the second player was taken arbitrarily so we showed that $II$ has no winning strategy if (ii) is assumed, i.e., (ii)$\Longrightarrow$(iii).

Finally, assume that $II$ has no winning strategy in the Gruenhage–Ma game on the space $X$ while $X$ does not have the moving off property. This means that there exists a moving off family $\mathcal{P}$ in the space $X$ such that no infinite subfamily of $\mathcal{P}$ has a discrete open expansion. If moves $K_1, L_1, \ldots, K_{n-1}, L_{n-1}, K_n$ are made in the game of Gruenhage–Ma on $X$ then let $\sigma(K_1, \ldots, K_n)$ be an element of $\mathcal{P}$ which does not meet $K_n \cup (\bigcup\{K_i \cup L_i : 1 \leq i < n\})$; such an element exists because $\mathcal{P}$ is a moving off family. In any play where $II$ applies $\sigma$ the set of choices of $II$ is an infinite subfamily of $\mathcal{P}$, so it does not have a discrete open expansion, i.e., $\sigma$ is a winning strategy; this contradiction shows that we settled (iii)$\Longrightarrow$(i) and thereby completed our solution.

**V.357.** *Prove that, if $C_k(X)$ has the Baire property then the space $X$ has the moving off property.*

**Solution.** If $K$ is a compact subset of the space $X$ and $O \in \tau(\mathbb{R})$ then the set $\langle K, O \rangle = \{f \in C_k(X) : f(K) \subset O\}$ is open in $C_k(X)$ by Fact 2 of V.347. Suppose that $\mathcal{P}$ is a moving off family in the space $X$ and consider, for any $n \in \omega$, the set $U_n = \bigcup\{\langle K, (n, +\infty)\rangle : K \in \mathcal{P}\}$. By our above observation the set $U_n$ is open in $C_k(X)$ for each $n \in \omega$. If $W$ is a nonempty open subset of $C_k(X)$ then there exists a function $f \in W$ and a compact set $L \subset X$ such that the set

$$H = \{g \in C_k(X) : |g(x) - f(x)| < \varepsilon \text{ for all } x \in L\}$$

is contained in $W$ for some $\varepsilon > 0$. Pick $K \in \mathcal{P}$ with $K \cap L = \emptyset$ and define a function $h : K \cup L \to \mathbb{R}$ by the equalities $h|L = f$ and $h(x) = n + 1$ for every $x \in K$. It is evident that $h$ is continuous on $K \cup L$, so there exists $g \in C_k(X)$ such that $g|(K \cup L) = h$ (see Fact 1 of T.218). It is clear that $g \in H \cap U_n \subset W \cap U_n$, i.e., $W \cap U_n \neq \emptyset$ for any nonempty open set $W \subset C_k(X)$.

This shows that every set $U_n$ is dense in $C_k(X)$, so the Baire property of $C_k(X)$ implies that we can find a function $f \in \bigcap_{n\in\omega} U_n$. Pick a set $K_0 \in \mathcal{P}$ such that $f \in \langle K_0, (0, +\infty)\rangle$ and let $a_0 = 0$; since $f(K_0)$ is compact, we can find $b_0 > a_0$ such that $f(K_0) \subset (a_0, b_0)$.

Proceeding inductively, assume that $k \in \omega$ and we have chosen $K_0, \ldots, K_k \in \mathcal{P}$ and $a_0, b_0, \ldots, a_k, b_k \in \mathbb{R}$ such that

(1) $a_i < b_i$ and $f(K_i) \subset (a_i, b_i)$ for all $i \leq k$;
(2) $b_i + 1 < a_{i+1}$ for all $i = 0, \ldots, k - 1$.

Take $n \in \omega$ with $n > b_k + 1$ and let $a_{k+1} = n$; there exists $K_{k+1} \in \mathcal{P}$ such that $f(K_{k+1}) \subset (a_{k+1}, +\infty)$. The set $f(K_{k+1})$ being compact we can find $b_{k+1} > a_{k+1}$ for which $f(K_{k+1}) \subset (a_{k+1}, b_{k+1})$. It is clear that the properties (1) and (2) now hold if we replace $k$ with $k + 1$. Thus our inductive procedure can be continued to construct families $\{K_i : i \in \omega\} \subset \mathcal{P}$ and $\{a_i, b_i : i \in \omega\} \subset \mathbb{R}$ such that the conditions (1) and (2) are satisfied for all $k \in \omega$.

It is immediate from (1) and (2) that $\{(a_i, b_i) : i \in \omega\}$ is a discrete open expansion of the family $\{f(K_i) : i \in \omega\}$, so $\{f^{-1}((a_i, b_i)) : i \in \omega\}$ is a discrete open expansion of an infinite family $\{K_i : i \in \omega\} \subset \mathcal{P}$. Therefore the space $X$ has the moving off property.

**V.358.** *Prove that, for any q-space $X$, the following conditions are equivalent:*

(i) *$C_k(X)$ has the Baire property;*
(ii) *$X$ has the moving off property;*
(iii) *the player $II$ has no winning strategy in Gruenhage–Ma game on the space $X$.*

**Solution.** If $Z$ is a space, $A \subset Z$, $f \in C_k(Z)$ and $\varepsilon > 0$ then, as usual, we let $[f, A, \varepsilon] = \{g \in C_k(Z) : |f(x) - g(x)| < \varepsilon \text{ for all } x \in A\}$. Given sets $A, B \subset Z$ we write $A \prec B$ of $B \succ A$ if $A \subset \text{Int}(B)$.

*Fact 1.* If $Z$ is a space then for any compact set $K \subset Z$ and $\varepsilon > 0$ the set $[f, K, \varepsilon]$ is open in $C_k(Z)$ for every $f \in C_k(Z)$.

*Proof.* Fix any function $g \in [f, K, \varepsilon]$ and let $h = g - f$; then $h(K) \subset (-\varepsilon, \varepsilon)$ and it follows from compactness of $h(K)$ that there exists $\delta \in (0, \varepsilon)$ such that $h(K) \subset (-\varepsilon + \delta, \varepsilon - \delta)$. Take any function $u \in [g, K, \delta]$; for every $x \in K$ we have $|u(x) - f(x)| \leq |u(x) - g(x)| + |g(x) - f(x)| < \delta + (\varepsilon - \delta) = \varepsilon$, so $u \in [f, K, \varepsilon]$ and hence we showed that for every $g \in [f, K, \varepsilon]$ there exists $\delta > 0$ for which $[g, K, \delta] \subset [f, K, \varepsilon]$. Therefore the set $[f, K, \varepsilon]$ is open in $C_k(Z)$, i.e., Fact 1 is proved.

Returning to our solution recall that it was proved in Problem 356 that (ii) $\Longleftrightarrow$ (iii) and it follows from Problem 357 that (i)$\Longrightarrow$(ii), so it suffices to establish that (iii)$\Longrightarrow$(i). To do it, assume that the second player does not have a winning strategy in the Gruenhage–Ma game on the space $X$ while $C_k(X)$ does not have the Baire property. It follows from Problem 355 that $X$ is locally compact and hence, for every compact set $K \subset X$ there exists a compact $L \subset X$ such that $K \prec L$.

It is an easy consequence of Fact 3 of T.371 that the space $C_k(X)$ is of the first category in itself, so we can find a family $\mathcal{Q} = \{Q_n : n \in \mathbb{N}\}$ of closed nowhere dense subsets of $C_k(X)$ such that $\bigcup \mathcal{Q} = C_k(X)$. Given any set $U \in \tau^*(C_k(X))$ and $n \in \mathbb{N}$, the set $\sigma_n(U) = U \backslash Q_n$ is also open and nonempty.

We are going to describe a strategy $s$ for the player $II$ in the Gruenhage–Ma game on $X$ as follows. Suppose that the first move of $I$ is a compact set $K_1 \subset X$. Choose a compact set $A_1 \subset X$ with $K_1 \prec A_1$ and let $g_1(x) = 0$ for all $x \in X$. The set $W_1 = [g_1, A_1, \frac{1}{2}]$ is open in $C_k(X)$ (see Fact 1) and nonempty, so we can fix a function $f_1 \in C_k(X)$ together with a compact set $E_1 \subset X$ and $\varepsilon_1 \subset (0, \frac{1}{2})$ such that $A_1 \subset E_1$ and $\overline{[f_1, E_1, \varepsilon_1]} \subset \sigma_1(W_1)$. Take a compact set $D_1 \succ E_1$ and let $L_1 = s(K_1) = D_1 \backslash \text{Int}(A_1)$.

Proceeding inductively assume that $n \in \mathbb{N}$ and we defined the strategy $s$ for all $i$-tuples $(K_1, \ldots, K_i) \in (\mathcal{K}(X))^i$ with $i \leq n$ in such a way that if $K_1, L_1, \ldots, K_n, L_n$ is an initial segment of a play in which $II$ applies $s$ then we also have compact sets $\{A_i, D_i, E_i : 1 \leq i \leq n\}$, functions $f_1, g_1, \ldots, f_n, g_n \in C_k(X)$ and numbers $\varepsilon_1, \ldots, \varepsilon_n > 0$ which satisfy the following conditions:

(1) $\varepsilon_i < 2^{-i}$ and $K_i \prec A_i \subset E_i \prec D_i$ for all $i = 1, \ldots, n$;
(2) $\varepsilon_{i+1} < \varepsilon_i$ and $\overline{[f_{i+1}, E_{i+1}, \varepsilon_{i+1}]} \subset \sigma_{i+1}([g_{i+1}, A_{i+1}, \varepsilon_i])$ for all $i = 1, \ldots, n-1$;
(3) if $i < n$ then $D_i \prec A_{i+1}$, $g_{i+1}|E_i = f_i$ and $g_{i+1}(A_{i+1} \backslash \text{Int}(D_i)) \subset \{0\}$;
(4) $L_i = D_i \backslash \text{Int}(A_i)$ for all $i = 1, \ldots, n$.

If the first player chooses a compact set $K_{n+1} \subset X$ then we can find a compact set $A_{n+1} \succ D_n \cup K_{n+1}$. It is easy to see that there exists a function $g_{n+1} \in C_k(X)$ such that $g_{n+1}|E_n = f_n|E_n$ and $g_{n+1}(x) = 0$ for all $x \in A_{n+1} \backslash \text{Int}(D_n)$.

Take $\varepsilon_{n+1} \in (0, 2^{-n-1})$, a compact set $E_{n+1} \supset A_{n+1}$ and $f_{n+1} \in C_k(X)$ such that $\varepsilon_{n+1} < \varepsilon_n$ and $\overline{[f_{n+1}, E_{n+1}, \varepsilon_{n+1}]} \subset \sigma_{n+1}([g_{n+1}, A_{n+1}, \varepsilon_n])$.

Take a set $D_{n+1} \succ E_{n+1}$ and let $L_{n+1} = s(K_1, \ldots, K_n) = D_{n+1} \backslash \text{Int}(A_{n+1})$. This defines a strategy $s$ for the second player on the space $X$ and if $\{K_i, L_i : i \in \mathbb{N}\}$ is a play in which $II$ applies $s$ then we also have families $\{A_i, D_i, E_i : i \in \omega\}$ together with the functions $\{f_i, g_i : i \in \mathbb{N}\}$ and positive numbers $\{\varepsilon_i : i \in \mathbb{N}\}$ with the properties (1)–(4) fulfilled for all $n \in \mathbb{N}$.

By our assumption the strategy $s$ cannot be winning, so we can fix a play $\{K_i, L_i : i \in \mathbb{N}\}$ in which $II$ applies $s$ and loses; we also have our overhead consisting of the families $\{A_i, D_i, E_i : i \in \omega\}$ together with the functions $\{f_i, g_i : i \in \mathbb{N}\}$ and positive numbers $\{\varepsilon_i : i \in \mathbb{N}\}$ with the properties (1)–(4).

It follows from (1) and (3) that $A = \bigcup\{A_n : n \in \mathbb{N}\}$ is an open subset of $X$. The properties (2) and (3) imply the inequality $|f_{i+1}(x) - f_i(x)| < 2^{-i}$ for every $x \in E_i$ and $i \in \mathbb{N}$. Since the sequence $\{E_i : i \in \mathbb{N}\}$ is increasing, we have $|f_{n+1}(x) - f_n(x)| < 2^{-n}$ for all $x \in E_i$ and $n \geq i$; it is an easy exercise that this implies that the sequence $\{f_n : n \in \mathbb{N}\}$ converges uniformly on every set $E_i$. Let $f(x) = \lim f_n(x)$ for every $x \in A$; this definition is consistent because $A = \bigcup\{E_i : i \in \mathbb{N}\}$. Letting $f(x) = 0$ for all $x \in X \backslash A$ we obtain a function $f : X \to \mathbb{R}$.

Let $W_n = [f_n, E_n, \varepsilon_n]$ for all $n \in \mathbb{N}$. It follows from (2) and (3) that $W_{n+1} \subset W_n$, so $\{f_i : i \geq n\} \subset W_n$ for all $n \in \mathbb{N}$. As a consequence,

(5) $|f_i(y) - f_n(y)| < \varepsilon_n$ for all $i \geq n$ and therefore $|f(y) - f_n(y)| \leq \varepsilon_n < 2^{-n}$ for every $y \in E_n$ and $n \in \mathbb{N}$.

Observe that $f|E_i$ is continuous on $E_i$ being a uniform limit of continuous functions. If $x \in E_i$ then $x \in \text{Int}(E_{i+1})$ by the property (1), so it follows from continuity of $f|E_{i+1}$ that the function $f$ is continuous at the point $x$.

Now, fix a point $x \in X \backslash A$ and $\varepsilon > 0$; pick $m \in \mathbb{N}$ with $2^{-m+1} < \varepsilon$. Since the play $\{K_i, L_i : i \in \mathbb{N}\}$ is a loss for $II$, the family $\{L_i : i \in \mathbb{N}\}$ is discrete and hence the set $L = \bigcup\{L_i : i \in \mathbb{N}\} \subset A$ is closed in $X$, so the set $U = X \backslash (A_m \cup L)$ is an open neighborhood of the point $x$.

If $y \in U \backslash A$ then $f(y) = 0$ and hence $|f(y) - f(x)| = 0 < \varepsilon$. If $y \in A$ then let $p = \backslash n\{n \in \mathbb{N} : y \in A_n\}$. If $y \in D_{p-1}$ then it follows from $y \notin A_{p-1}$ that $y \in L_{p-1} \subset L$ which is a contradiction. Therefore $y \in A_p \backslash D_{p-1}$, so the conditions (2) and (3) imply that $|f_p(y)| < 2^{-p}$. The property (5) shows that $|f(y) - f_p(y)| \leq 2^{-p}$ and hence $|f(y)| \leq 2^{-p+1} \leq 2^{-m+1} < \varepsilon$. Thus $f(U) \subset (-\varepsilon, \varepsilon)$ and hence $U$ witnesses continuity of the function $f$ at the point $x$.

Thus the function $f$ is continuous on the space $X$; fix any $n \in \mathbb{N}$ and a set $W \in \tau(f, C_k(X))$. There exists a compact set $P \subset X$ and $\delta > 0$ such that $E_n \subset P$ and $[f, P, \delta] \subset W$. Since the sequence $\{f_i : i \in \mathbb{N}\}$ converges to $f$ uniformly on $E_n$, there exists $i \geq n$ such that $|f(x) - f_i(x)| < \delta$ for all $x \in E_n$. If $h = f_i - f$ then $h(E_n) \subset (-\delta, \delta)$, so it follows from compactness of $E_n$ that there is $\delta' \in (0, \delta)$ such that $h(E_n) \subset [-\delta', \delta']$. By Fact 1 of V.321 there exists a function $h_0 \in C(X)$ such that $h_0|E_n = h|E_n$ and $h_0(X) \subset [-\delta', \delta']$.

If $g = f + h_0$ then $|f(x) - h_0(x)| \leq \delta' < \delta$ for all $x \in X$, so $g \in [f, P, \delta]$. On the other hand, if $x \in E_n$ then $g(x) = f_i(x)$, so $|g(x) - f_n(x)| = |f_i(x) - f_n(x)| < \varepsilon_n$

(here we used again the property (5)) and hence $g \in [f_n, E_n, \varepsilon_n] \cap [f, P, \delta] \subset W_n \cap W$. This proves that every neighborhood of $f$ in $C_k(X)$ intersects the set $W_n$, i.e., $f \in \overline{W}_n$ for all $n \in \mathbb{N}$.

Thus, $f \in \bigcap\{\overline{W}_n : n \in \mathbb{N}\}$; however, it follows from (2) and the definition of $\sigma_n$ that $\overline{W}_n \cap Q_n = \emptyset$ for all $n \in \mathbb{N}$ and therefore $\bigcap\{\overline{W}_n : n \in \mathbb{N}\} = \emptyset$. This contradiction shows that $C_k(X)$ has the Baire property, i.e., we settled (iii)$\Longrightarrow$(i) and hence our solution is complete.

**V.359.** *Let $X$ be a paracompact $q$-space. Prove that $C_k(X)$ has the Baire property if and only if $X$ is locally compact.*

**Solution.** If $X$ is locally compact then $X$ has the moving off property by Problem 354 and hence $C_k(X)$ is a Baire space by Problem 358. If, on the other hand, the space $C_k(X)$ has the Baire property then $X$ has the moving off property by Problem 357, so $X$ is locally compact by Problem 355.

**V.360.** *Let $X$ be a paracompact $q$-space. Prove that, if $X$ is $l$-equivalent to a locally compact paracompact space then $X$ is also locally compact. In particular, any first countable paracompact space $l$-equivalent to a locally compact paracompact space, is locally compact. Deduce from this fact that*

  (i) *if $X$ and $Y$ are metrizable $l$-equivalent spaces then $X$ is locally compact if and only if $Y$ is locally compact;*
  (ii) *if a first countable space $X$ is $l$-equivalent to a second countable locally compact space then $X$ is also locally compact and second countable.*

**Solution.** Suppose that $Y$ is a locally compact paracompact space and we are given a linear homeomorphism $\varphi : C_p(X) \to C_p(Y)$. Apply Problem 327 to see that $\varphi : C_b(X) \to C_b(Y)$ is also a linear homeomorphism. It is straightforward that every paracompact space is a $\mu$-space, i.e., if $Z$ is paracompact then the closure of every bounded subset of $Z$ is compact. Therefore $X$ and $Y$ are $\mu$-spaces, so we have the equalities $\tau(C_b(X)) = \tau(C_k(X))$ and $\tau(C_b(Y)) = \tau(C_k(Y))$ by Fact 6 of V.318. Therefore $\varphi : C_k(X) \to C_k(Y)$ is a linear homeomorphism as well.

It is evident that every locally compact space is a $q$-space so we can apply Problem 359 to convince ourselves that $C_k(Y)$ has the Baire property. Therefore $C_k(X)$ also has the Baire property and hence $X$ has the moving off property by Problem 357. This, together with Problem 355, implies that $X$ is locally compact. The statement (i) follows from the fact that every metrizable space is first countable (and hence a $q$-space) and paracompact.

Now if $\chi(X) = \omega$ and $X$ is $l$-equivalent to a second countable locally compact space then $X$ is second countable by Problem 348, so (i) is applicable to conclude that $X$ is locally compact.

**V.361.** *Suppose that a $q$-space $X$ is $l$-equivalent to a locally compact metrizable space. Prove that $X$ is metrizable and locally compact.*

**Solution.** The space $X$ is paracompact by Problem 335, so we can apply Problem 360 to see that $X$ is locally compact and hence Čech-complete. It follows from

Problem 333 that $X$ has a weaker metrizable topology, so the diagonal of $X$ is a $G_\delta$-set in $X \times X$. This makes it possible to apply Fact 7 of U.421 to conclude that $X$ is metrizable.

**V.362.** *Suppose that $X$ is $l$-equivalent to a metrizable space, $Y \subset X$ and $Y$ is Čech-complete. Prove that $Y$ is metrizable. In particular, if a Čech-complete space $X$ is $l$-equivalent to a metrizable space then $X$ is metrizable.*

**Solution.** If $Z$ is a space and $\mathcal{A} \subset \exp(Z)$ then $\mathcal{A}|P = \{A \cap P : A \in \mathcal{A}\}$ for any $P \subset Z$.

*Fact 1.* If a space $Z$ is $\sigma$-metrizable and paracompact then every subspace of $Z$ is $\sigma$-metrizable and paracompact.

*Proof.* Fix an arbitrary subspace $T \subset Z$; by Fact 2 of V.334 the space $Z$ has a metrizable $\mu$-approximation, i.e., there exists a metrizable topology $\nu$ on $Z$ and a countable family $\mathcal{F}$ of closed subsets of $Z$ such that $\nu \subset \tau(Z)$ while $\bigcup \mathcal{F} = Z$ and $\tau(Z)|F = \nu|F$ for every $F \in \mathcal{F}$. If $\mathcal{F}' = \mathcal{F}|T$ and $\nu' = \nu|T$ then $\mathcal{F}'$ is a family of closed subsets of $T$ and $\nu'$ is a metrizable topology on $T$ such that $\bigcup \mathcal{F}' = T$ and $\nu'|P = \tau(T)|P$ for every $P \in \mathcal{F}'$. In other words, $\nu'$ is a metrizable $\mu$-approximation for the space $T$, so we can apply Fact 2 of V.334 again to conclude that $T$ is paracompact and $\sigma$-metrizable and hence Fact 1 is proved.

Returning to our solution observe that $X$ has a weaker metrizable topology by Problem 333; therefore $Y$ also has a weaker metrizable topology and, in particular, the diagonal of $Y$ is a $G_\delta$-subset of $Y \times Y$. The space $X$ is paracompact and $\sigma$-metrizable by Problem 335, so we can apply Fact 1 to see that $Y$ is also paracompact. Finally, it follows from Fact 7 of U.421 that $Y$ is metrizable.

**V.363.** *Suppose that $X$ is $l$-equivalent to a metrizable space. Prove that $\overline{A}$ is an $\aleph_0$-space for any countable set $A \subset X$.*

**Solution.** If $Z$ is a space and $B \subset Z$ then $\langle B \rangle$ is the linear hull of $B$ in $L_p(Z)$.

*Fact 1.* If $Z$ is a space and $Y \subset Z$ then $\mathrm{cl}_{L_p(Z)}(\langle Y \rangle) = \langle \mathrm{cl}_Z(Y) \rangle$. In other words, the closure of the linear hull of $Y$ in $L_p(Z)$ coincides with the linear hull of the closure of $Y$ in $Z$.

*Proof.* The set $P = \langle \mathrm{cl}_Z(Y) \rangle$ is closed in $L_p(Z)$ by Fact 2 of V.246, so it follows from $\langle Y \rangle \subset P$ that $Q = \mathrm{cl}_{L_p(Z)}(\langle Y \rangle) \subset P$. For every $n \in \mathbb{N}$ consider the map $\varphi_n : \mathbb{R}^n \times Z^n \to L_p(Z)$ defined by the formula $\varphi_n(t, z) = t_1 z_1 + \ldots + t_n z_n$ for any $t = (t_1, \ldots, t_n) \in \mathbb{R}^n$ and $z = (z_1, \ldots, z_n) \in Z^n$.

Let $F = \mathrm{cl}_Z(Y)$; it follows from continuity of $\varphi_n$ that the set $H_n = \varphi_n(\mathbb{R}^n \times Y^n)$ is dense in the set $G_n = \varphi_n(\mathbb{R}^n \times F^n)$ for every $n \in \mathbb{N}$. As a consequence, the set $\langle Y \rangle = \bigcup\{H_n : n \in \mathbb{N}\}$ is dense in $\langle F \rangle = \bigcup\{G_n : n \in \mathbb{N}\}$, so $Q = P$, i.e., Fact 1 is proved.

Returning to our solution fix a metrizable space $M$ such that $X \overset{l}{\sim} M$ and let $A_0 = A$. We can consider that $M \subset L_p(X)$ is a Hamel basis of $L_p(X)$ such that every $f \in C(M)$ can be extended to a continuous linear functional on $L_p(X)$ (see Problem 238).

Since $M$ is a Hamel basis of the space $L_p(Z)$, there exists a countable set $B_0 \subset M$ such that $A_0 \subset \langle B_0 \rangle$. Proceeding by induction, it is easy to construct sequences $\{A_n : n \in \omega\}$ and $\{B_n : n \in \omega\}$ such that $A_n \subset X$, $B_n \subset M$ while we have the inclusions $A_n \subset \langle B_n \rangle$, $A_n \subset A_{n+1}$ and $B_n \subset \langle A_{n+1} \rangle$, $B_n \subset B_{n+1}$ for all $n \in \omega$. If $P = \bigcup_{n \in \omega} A_n$ and $B = \bigcup_{n \in \omega} B_n$ then it is straightforward that $\langle P \rangle = \langle B \rangle$, so $\langle \overline{P} \rangle = \langle \overline{B} \rangle$ by Fact 1.

By paracompactness of $X$ (see Problem 335) and $M$ the sets $\overline{P}$ and $\overline{B}$ are $C$-embedded in $X$ and in $M$ respectively. By Fact 3 of V.246, the space $\langle \overline{P} \rangle$ is linearly homeomorphic to $L_p(\overline{P})$ while $\langle \overline{B} \rangle$ is linearly homeomorphic to $L_p(\overline{B})$. Therefore the spaces $\overline{P}$ and $\overline{B}$ are $l$-equivalent; by metrizability of $M$, the space $\overline{B}$ is second countable, so $\overline{P}$ is an $\aleph_0$-space by Problem 348. It follows from $\overline{A} \subset \overline{P}$ that $\overline{A}$ is also an $\aleph_0$-space (see Problem 341), i.e., our solution is complete.

**V.364.** *For any space $X$ let $dc(X) = \sup\{|\mathcal{U}| : \mathcal{U} \subset \tau^*(X) \text{ and } \mathcal{U} \text{ is a discrete family}\}$. Prove that if $X$ is $l$-equivalent to a metrizable space then $nw(X) = dc(X)$. In particular, if $X$ is $l$-equivalent to a metrizable space then both the Souslin property of $X$ and $ext(X) \leq \omega$ imply that $X$ has a countable network.*

**Solution.** It is trivial that $dc(X) \leq c(X) \leq nw(X)$ and $dc(X) \leq ext(X)$ for any space $X$. Now, if $X$ is $l$-equivalent to a metrizable space then it is paracompact and there exists a countable family $\mathcal{F}$ of metrizable closed subsets of $X$ such that $X = \bigcup \mathcal{F}$ (see Problem 335).

Suppose that $\kappa$ is an infinite cardinal while $dc(X) \leq \kappa$ and $ext(X) > \kappa$; then there is a closed discrete set $D \subset X$ such that $|D| = \kappa^+$. By collectionwise normality of $X$ there is a discrete family $\mathcal{U} = \{U_d : d \in D\}$ such that $d \in U_d$ for every $d \in D$. Since $\mathcal{U} \subset \tau^*(X)$ and $|\mathcal{U}| > \kappa$, we obtained a contradiction which shows that $ext(X) \leq \kappa$ and hence $ext(X) \leq dc(X)$. Given $F \in \mathcal{F}$ we have $ext(F) \leq ext(X) \leq \kappa$, so it follows from metrizability of $F$ that $nw(F) \leq \kappa$ (see TFS-214). Since $\mathcal{F}$ is countable and $\bigcup \mathcal{F} = X$, we can apply SFFS-405 to conclude that $nw(X) \leq \kappa$. As a consequence, $nw(X) \leq dc(X)$ and hence $nw(X) = dc(X)$ as promised.

Finally observe that if $c(X) \leq \omega$ then $nw(X) = dc(X) \leq c(X) \leq \omega$. If $ext(X) \leq \omega$ then $nw(X) = dc(X) \leq ext(X) \leq \omega$, so both countable extent and the Souslin property of $X$ imply that $nw(X) \leq \omega$.

**V.365.** *Given a space $X$ and a first countable space $Y$ assume that there exists a continuous linear surjection $\varphi : C_p(X) \to C_p(Y)$. For any $y \in Y$ there exist $x_1, \ldots, x_n \in X$ and $\lambda_1, \ldots, \lambda_n \in \mathbb{R} \setminus \{0\}$ such that $\varphi(f)(y) = \sum_{i=1}^n \lambda_i f(x_i)$ for any $f \in C_p(X)$; denote the set $\{x_1, \ldots, x_n\}$ by $supp(y)$. Suppose that $\mathcal{U}$ is a locally finite open cover of $X$ and let $T(U) = \{y \in Y : supp(y) \cap U \neq \emptyset\}$ for every $U \in \mathcal{U}$. Prove that the family $\{T(U) : U \in \mathcal{U}\}$ is a locally finite open cover of $Y$.*

**Solution.** Observe first that, for any $y \in Y$, the definition of $supp(y)$ is consistent by Problem 279.

*Fact 1.* If $Z$ is a space then a set $A \subset Z$ is not bounded in $Z$ if and only if there exists an infinite locally finite family $\mathcal{V} \subset \tau^*(Z)$ such that $V \cap A \neq \emptyset$ for every $V \in \mathcal{V}$.

*Proof.* If $A$ is not bounded then there exists an infinite discrete family $\mathcal{V} \subset \tau^*(Z)$ such that $V \cap A \neq \emptyset$ for each $V \in \mathcal{V}$ (see Fact 1 of V.245); it is evident that $\mathcal{V}$ is locally finite, so we proved necessity.

Now assume that there is an infinite locally finite family $\mathcal{V} \subset \tau^*(Z)$ such that $V \cap A \neq \emptyset$ for every $V \in \mathcal{V}$. Fix $V_0 \in \mathcal{V}$ and a point $x_0 \in V_0 \cap A$; there exists a set $W_0 \in \tau(x_0, Z)$ such that $\overline{W}_0 \subset V_0$ and $\overline{W}_0$ meets only finitely many elements of $\mathcal{V}$. Proceeding inductively assume that $n \in \omega$ and we have chosen $V_0, \ldots, V_n \in \mathcal{V}$, points $x_0, \ldots, x_n \in A$ and sets $W_0, \ldots, W_n \in \tau(Z)$ such that

(1) $x_i \in W_i \cap A$ and $\overline{W}_i \subset V_i$ for all $i \leq n$;
(2) $V_i \cap \overline{W}_j = \emptyset$ whenever $0 \leq j < i \leq n$;
(3) for every $i \leq n$, the set $\overline{W}_i$ meets only finitely many elements of $\mathcal{V}$.

By the inductive assumption, the set $W = \overline{W}_0 \cup \ldots \cup \overline{W}_n$ meets only finitely many elements of $\mathcal{V}$, so there exists $V_{n+1} \in \mathcal{V}$ such that $V_{n+1} \cap W = \emptyset$. Pick a point $x_{n+1} \in V_{n+1} \cap A$ and choose a set $W_{n+1} \in \tau(x_{n+1}, Z)$ such that $\overline{W}_{n+1}$ meets only finitely many elements of $\mathcal{V}$ and $\overline{W}_{n+1} \subset V_{n+1}$. It is straightforward that the conditions (1)–(3) are now satisfied if we replace $n$ with $n + 1$. Therefore we can construct a set $\{x_i : i \in \omega\}$ and families $\{V_i : i \in \omega\}$ and $\{W_i : i \in \omega\}$ such that the properties (1)–(3) hold for all $n \in \omega$.

The family $\{V_i : i \in \omega\}$ is locally finite, so it follows from (1) that the family $\mathcal{W} = \{\overline{W}_i : i \in \omega\}$ is also locally finite and hence the set $F_i = \bigcup\{\overline{W}_j : j \in \omega \backslash \{i\}\}$ is closed in $Z$ for every $i \in \omega$. Observe that (1) and (2) imply that the family $\mathcal{W}$ is disjoint.

Fix a point $x \in Z$; if $x \in \overline{W}_i$ for some $i \in \omega$ then $G = Z \backslash F_i \in \tau(x, Z)$ and $G$ intersects only one element of $\mathcal{W}$. If $x \notin \bigcup \mathcal{W}$ then $X \backslash (\bigcup \mathcal{W})$ is an open neighborhood of $x$ which meets no elements of $\mathcal{W}$. This proves that the family $\mathcal{W}$ is discrete and hence $\{W_i : i \in \omega\}$ is discrete as well. Since $x_i \in W_i \cap A$ for all $i \in \omega$, we can apply Fact 1 of V.245 to conclude that $A$ is not bounded in $Z$. This settles sufficiency and shows that Fact 1 is proved.

Returning to our solution let $\xi(y) = \text{supp}(y)$ for every $y \in Y$ and observe that the map $\xi : Y \to \exp(X)$ is lower semicontinuous by Problem 280; an immediate consequence is that the set $T(U)$ is open in $Y$ for each $U \in \mathcal{U}$. If $y \in Y$ then $\xi(y)$ is a nonempty subset of $X$; since $\mathcal{U}$ is a cover of $X$, there exists $U \in \mathcal{U}$ with $U \cap \xi(y) \neq \emptyset$ and hence $y \in T(U)$. This shows that the family $\mathcal{T} = \{T(U) : U \in \mathcal{U}\}$ is a cover of $Y$.

Finally fix a point $y \in Y$ and a local base $\{O_n : n \in \omega\}$ of $Y$ at the point $y$ such that $O_{n+1} \subset O_n$ for every $n \in \omega$. If $\mathcal{T}$ is not locally finite at $y$ then every set $O_n$

intersects infinitely many elements of $\mathcal{T}$, so we can choose a faithfully indexed (and hence infinite) family $\mathcal{U}' = \{U_n : n \in \omega\} \subset \mathcal{U}$ and a set $\{y_n : n \in \omega\} \subset Y$ such that $y_n \in O_n \cap T(U_n)$ for all $n \in \omega$.

The set $K = \{y\} \cup \{y_n : n \in \omega\}$ is easily seen to be a convergent sequence, so $K$ is compact and hence the set $L = \bigcup\{\mathrm{supp}(z) : z \in K\}$ is bounded in $X$ by Problem 281. For every $n \in \omega$ the set $\mathrm{supp}(y_n)$ is contained in $L$, so it follows from $y_n \in T(U_n)$ that $U_n \cap L \neq \emptyset$. Thus $\mathcal{U}'$ is an infinite locally finite family and $U \cap L \neq \emptyset$ for every $U \in \mathcal{U}'$; therefore $L$ is not bounded by Fact 1. This contradiction shows that the cover $\mathcal{T}$ is locally finite at any $y \in Y$ and hence our solution is complete.

**V.366.** *Let $X$ and $Y$ be metrizable spaces for which there exists a continuous linear surjection of $C_p(X)$ onto $C_p(Y)$. Prove that, if $X$ is Čech-complete then $Y$ is also Čech-complete.*

**Solution.** If $Z$ is a space and $\mathcal{A}$ is a family of subsets of $Z$ then, as usual, we let $\mathrm{St}(Y, \mathcal{A}) = \bigcup\{A \in \mathcal{A} : A \cap Y \neq \emptyset\}$ for any $Y \subset Z$.

*Fact 1.* Suppose that $(M, \rho)$ is a metric space which is not Čech-complete and let $\{\mathcal{U}_n : n \in \omega\}$ be a sequence of locally finite open covers of $M$ with the following properties:

  (i)   $\mathcal{U}_{n+1}$ is a star refinement of $\mathcal{U}_n$ for all $n \in \omega$;
  (ii)  $\mathrm{diam}_\rho(U) \leq 2^{-n}$ for all $n \in \omega$ and $U \in \mathcal{U}_n$;
  (iii) if $n \in \omega$ and $U \in \mathcal{U}_{n+1}$ then $U$ meets only finitely many elements of $\mathcal{U}_n$.

Then we can choose a set $U_n \in \mathcal{U}_n$ in such a way that $\overline{U}_{n+1} \subset U_n$ for all $n \in \omega$ and $\bigcap\{U_n : n \in \omega\} = \emptyset$.

*Proof.* By TFS-237, there is a complete metric space $(N, \rho')$ such that $(M, \rho)$ is isometric to a dense subspace of $(N, \rho')$. To simplify the notation we will consider that $(N, \rho)$ is a complete metric space and $M$ is a dense subspace of $N$. If $U$ is an open subset of $M$ let $O(U) = \bigcup\{G \in \tau(N) : G \cap M \subset U\}$. It is immediate that

(1) if $U, V \in \tau(M)$ then $O(U) \cap M = U$ and $U \subset V$ implies $O(U) \subset O(V)$.

For every $n \in \omega$ let $\mathcal{V}_n = \{O(U) : U \in \mathcal{U}_n\}$. If $V_n = \bigcup \mathcal{V}_n$ then $V_n$ is an open subset of $N$ such that $M \subset V_n$ for every $n \in \omega$. If $V = \bigcap_{n \in \omega} V_n$ then $V$ is Čech-complete, so there exists a point $z \in V \setminus M$. Let us prove that

(2) the family $\mathcal{W}_n = \{W \in \mathcal{U}_n : z \in O(W)\}$ is finite for every $n \in \omega$.

Indeed, if some $\mathcal{W}_n$ is infinite then take any $U \in \mathcal{U}_{n+1}$ such that $z \in O(U)$; then, for any $W \in \mathcal{W}_n$ we have $z \in O(W) \cap O(U)$, so $O(W) \cap O(U)$ is a nonempty open subset of $N$ and hence $U \cap W = O(U) \cap O(W) \cap M \neq \emptyset$. However, this contradicts (iii) and shows that (2) is proved.

If $W \in \mathcal{U}_n$ and $W' \in \mathcal{U}_{n+1}$ say that $W \prec W'$ if $\mathrm{St}(W', \mathcal{U}_{n+1}) \subset W$. If $n \in \omega$, $k > 1$ and we are given sets $W \in \mathcal{U}_n$ and $W' \in \mathcal{U}_{n+k}$ say that $W \prec W'$ if there exist sets $W_1, \ldots, W_{k-1}$ such that $W_i \in \mathcal{U}_{n+i}$ for all $i = 1, \ldots, k-1$ and

$W \prec W_1 \prec \ldots \prec W_{k-1} \prec W'$. It is clear that $W \prec W' \prec W''$ implies $W \prec W''$; besides, it follows from (1) that for every $n \in \omega$, $m > n$ and $W \in \mathcal{U}_n$ if $W' \in \mathcal{W}_m$ and $W \prec W'$ then $W \in \mathcal{W}_n$.

The condition (i) implies that,

(3) if $n \in \omega$, $n < m$ and $W' \in \mathcal{W}_m$ then there is $W \in \mathcal{W}_n$ such that $W \prec W'$.

For every $W \in \mathcal{W} = \bigcup_{n\in\omega} \mathcal{W}_n$ let $S(W) = \{W' \in \mathcal{W} : W \prec W'\}$ and $A(W) = \{n \in \omega : S(W) \cap \mathcal{W}_n \neq \emptyset\}$.

It follows from (3) that $\bigcup\{\mathcal{W}_n : n > 0\} = \bigcup\{S(W) : W \in \mathcal{W}_0\}$, so we can choose $U_0 \in \mathcal{W}_0$ such that the set $A(U_0)$ is infinite.

Proceeding by induction assume that $k \in \omega$ and we have sets $U_0, \ldots, U_k$ such that

(4) $U_i \in \mathcal{W}_i$ and the set $A(U_i)$ is infinite for all $i \leq k$;

(5) $U_i \prec U_{i+1}$ for all $i < k$.

Let $\mathcal{H} = \{W \in \mathcal{W}_{k+1} : U_k \prec W\}$; it follows from (3) and (4) that $\mathcal{H}$ is nonempty. If $i > k + 1$ and $W \in S(U_k) \cap \mathcal{W}_i$ then there exists a set $W' \in \mathcal{H}$ such that $W' \prec W$. As a consequence, the set $\bigcup\{S(U_k) \cap \mathcal{W}_i : i > k + 1\}$ is contained in the set $\bigcup_{W\in\mathcal{H}} \bigcup_{i>k+1} S(W) \cap \mathcal{W}_i$. The family $\mathcal{H}$ being finite there exists $U_{k+1} \in \mathcal{H}$ such that $A(U_{k+1})$ is infinite; it is clear that the statements (4) and (5) remain true if we replace $k$ with $k + 1$. Therefore our inductive procedure can be continued to construct a family $\{U_i : i \in \omega\}$ such that the conditions (4) and (5) are satisfied for all $k \in \omega$. In particular, $U_n \in \mathcal{U}_n$ and $St(U_{n+1}, \mathcal{U}_{n+1}) \subset U_n$; it is an easy exercise that this implies $\overline{U}_{n+1} \subset U_n$ for all $n \in \omega$.

The condition (ii) shows that $\text{diam}_\rho(O(U_n)) = \text{diam}_\rho(U_n) \to 0$, so $\bigcap_{n\in\omega} O(U_n)$ can contain at most one point; therefore $\bigcap_{n\in\omega} O(U_n) = \{z\}$ and hence we have the equality $\bigcap_{n\in\omega} U_n = \{z\} \cap M = \emptyset$ which shows that Fact 1 is proved.

*Fact 2.* Given a metric space $(Z, d)$ a set $A \subset Z$ is totally bounded in $Z$ if and only if, for any $\varepsilon > 0$, there exists a finite cover $\mathcal{P} \subset \exp(Z)$ of the set $A$ such that $\text{diam}_d(P) < \varepsilon$ for each $P \in \mathcal{P}$.

*Proof.* For any $r > 0$ and $z \in Z$ let $B(z, r) = \{y \in Z : d(z, y) < r\}$ be the ball of radius $r$ centered at $z$. If $A$ is totally bounded and $\varepsilon > 0$ then there exists a finite set $F \subset Z$ such that $A \subset \bigcup\{B(a, \frac{\varepsilon}{3}) : a \in F\}$. If $\mathcal{P} = \{B(a, \frac{\varepsilon}{3}) : a \in F\}$ then $\text{diam}_d(P) \leq \frac{2}{3}\varepsilon < \varepsilon$ for every $P \in \mathcal{P}$ and $A \subset \bigcup \mathcal{P}$, so we proved necessity.

Now assume that $A$ can be covered by a finite number of sets of arbitrarily small diameter and fix $\varepsilon > 0$. There exists a finite family $\mathcal{P} \subset \exp(Z)$ such that $A \subset \bigcup \mathcal{P}$ and $\text{diam}_d(P) < \varepsilon$ for each $P \in \mathcal{P}$; we can assume that all elements of $\mathcal{P}$ are nonempty. For every $P \in \mathcal{P}$ choose a point $x_P \in P$; then $P \subset B(x_P, \varepsilon)$ and hence $A \subset \bigcup \mathcal{P} \subset \bigcup\{B(x_P, \varepsilon) : P \in \mathcal{P}\}$, so the finite set $F = \{x_P : P \in \mathcal{P}\}$ witnesses that $A$ is totally bounded. Fact 2 is proved.

Returning to our solution suppose that $\varphi : C_p(X) \to C_p(Y)$ is a linear surjection and $X$ is Čech-complete; fix a complete metric $\rho$ on $X$ which generates the topology of $X$. Suppose that $Y$ is not Čech-complete and take any metric $d$ on $Y$ such that

$\tau(d) = \tau(Y)$. Using paracompactness of $X$ it is easy to construct a sequence $\{\mathcal{U}_n : n \in \omega\}$ of locally finite covers of $X$ such that, for any $n \in \omega$ and $U \in \mathcal{U}_n$ we have $\mathrm{diam}_\rho(U) \leq 2^{-n}$ and $\mathcal{U}_{n+1}$ is a refinement of $\mathcal{U}_n$.

For every $y \in Y$ there exist $x_1, \ldots, x_k \in X$ and $\lambda_1, \ldots, \lambda_k \in \mathbb{R}\backslash\{0\}$ such that $\varphi(f)(y) = \sum_{i=1}^{k} \lambda_i f(x_i)$ for each $f \in C_p(X)$ (see Problem 279); let $\mathrm{supp}(y) = \{x_1, \ldots, x_k\}$. For every $A \subset X$ consider the set $T(A) = \{y \in Y : \mathrm{supp}(y) \cap A \neq \emptyset\}$; then $\mathcal{T}_n = \{T(U) : U \in \mathcal{U}_n\}$ is a locally finite open cover of $Y$ for each $n \in \omega$ (see Problem 365). This makes it possible to construct a sequence $\{\mathcal{V}_n : n \in \omega\}$ of locally finite open covers of $Y$ with the following properties:

(6) if $n \in \omega$ then $V \neq \emptyset$ and $\mathrm{diam}_d(V) \leq 2^{-n}$ for any $V \in \mathcal{V}_n$;
(7) $\mathcal{V}_{n+1}$ is a star refinement of $\mathcal{V}_n$ for each $n \in \omega$;
(8) if $n \in \omega$ then every $V \in \mathcal{V}_{n+1}$ intersects only finitely many elements of $\mathcal{V}_n$;
(9) for any $n \in \omega$ each $V \in \mathcal{V}_n$ intersects only finitely many elements of $\mathcal{T}_n$.

By Fact 1 we can choose $V_n \in \mathcal{V}_n$ in such a way that $\overline{V}_{n+1} \subset V_n$ for every $n \in \omega$ and $\bigcap\{V_n : n \in \omega\} = \emptyset$. Pick a point $y_n \in V_n$ for every $n \in \omega$; it is easy to deduce from $\bigcap_{n\in\omega} \overline{V}_n = \emptyset$ that

(10) the set $D = \{y_n : n \in \omega\}$ is infinite, closed and discrete in $Y$.

The family $\mathcal{W}_n = \{U \in \mathcal{U}_n : V_n \cap T(U) \neq \emptyset\}$ is finite by (9) and we have the inclusion $\bigcup\{\mathrm{supp}(y) : y \in V_n\} \subset \bigcup \mathcal{W}_n$ for every $n \in \omega$. Consider the set $K = \bigcup\{\mathrm{supp}(y_n) : n \in \omega\}$; given any $\varepsilon > 0$ there exists $n \in \omega$ such that $2^{-n} < \varepsilon$. Then $\{y_i : i \geq n\} \subset V_n$, so $\mathcal{P} = \{\{y_i\} : i < n\} \cup \mathcal{W}_n$ is a finite cover of $K$ such that $\mathrm{diam}_\rho(P) \leq 2^{-n} < \varepsilon$ for every $P \in \mathcal{P}$. Since $\varepsilon > 0$ was taken arbitrarily, Fact 2 shows that $K$ is totally bounded and hence $\overline{K}$ is a compact subset of $X$ (see TFS-212). By Problem 282, the set $C = \{y \in Y : \mathrm{supp}(y) \subset \overline{K}\}$ is bounded in $Y$ and hence so is $D \subset C$. Since $D$ is closed in $Y$, it must be compact (recall that $Y$ is a metric space); this contradiction shows that $Y$ is Čech-complete as promised.

**V.367.** *Prove that if a metrizable space $X$ is $l$-equivalent to a Čech-complete space then $X$ is also Čech-complete. As a consequence, if $X$ and $Y$ are $l$-equivalent metrizable spaces then $X$ is metrizable by a complete metric if and only if so is $Y$.*

**Solution.** Suppose that $Y$ is a Čech-complete space and $X \overset{l}{\sim} Y$. By Problem 362 the space $Y$ is metrizable, so we can apply Problem 366 to see that $X$ is Čech-complete. Now, if $X$ and $Y$ are $l$-equivalent metrizable spaces and $X$ is metrizable by a complete metric then $X$ is Čech-complete by TFS-269, so $Y$ is Čech-complete as well. Applying TFS-269 again we conclude that $Y$ is metrizable by a complete metric. Since $X$ and $Y$ are in a symmetric situation, metrizability of $Y$ by a complete metric also implies metrizability of $X$ by a complete metric.

**V.368.** *Show that there exist first countable $l$-equivalent spaces $X$ and $Y$ such that $X$ is locally compact and $Y$ is not locally compact.*

**Solution.** In the space $P = \omega_1 \times (\omega + 1)$ consider the set $F = \omega_1 \times \{\omega\}$; it is evident that $F$ is a retract of $P$. The space $\omega_1$ is locally compact and $\omega + 1$ is compact; it is an easy exercise that the product of two locally compact spaces is locally compact, so $P$ is locally compact and hence so is the space $X = P^+$. It is also clear that the space $X$ is first countable.

Apply Problem 258 to see that $X$ is $l$-equivalent to the space $Y = P_F \oplus F$ where $P_F$ is the $\mathbb{R}$-quotient space obtained from $P$ by contracting the set $F$ to a point. The subspace $F$ is homeomorphic to $\omega_1$ which is first countable; besides, $F$ is clopen in $Y$, so $Y$ is first countable at every point of $F$. Let $a_F$ be the point represented by the set $F$ in the space $P_F$. Observe that $P_F \setminus \{a_F\}$ is homeomorphic to $\omega_1 \times \omega$ (see Problem 252), so $P_F$ is first countable at every point of $P_F \setminus \{a_F\}$. To see that $P_F$ is also first countable at $a_F$, let $O_n = \omega_1 \times ((\omega + 1) \setminus n)$ for every $n \in \omega$ (recall that we identify every $n \in \omega$ with the set $\{0, \dots, n - 1\}$). It turns out that

(1) for any $U \in \tau(F, P)$ there exists $n \in \omega$ such that $O_n \subset U$.

To prove (1) denote by $L$ the set of all limit ordinals in $\omega_1$ and take, for any ordinal $\alpha \in L$, an ordinal $\mu(\alpha)$ and a number $m(\alpha) \in \omega$ such that the set $(\mu(\alpha), \alpha] \times ((\omega + 1) \setminus m(\alpha))$ is contained in $U$. Apply SFFS-065 to find a stationary set $A \subset \omega_1$ such that there exists $k \in \omega$ for which $m(\alpha) = k$ for all $\alpha \in A$. By SFFS-067, there exists $\beta < \omega_1$ and an uncountable set $B \subset A$ such that $\mu(\alpha) = \beta$ for every $\alpha \in B$. An immediate consequence is that $(\beta, \omega_1) \times ((\omega + 1) \setminus k) \subset U$. It is easy to deduce from compactness of $\beta + 1$ that there exists $l \in \omega$ such that $(\beta + 1) \times ((\omega + 1) \setminus l)) \subset U$; now if $n = \max\{k, l\}$ then $O_n \subset U$, i.e., (1) is proved.

It is easy to see that $F$ is a zero-set in $P$ and every $O_n$ is a clopen subset of $P$. This, together with Problem 253, shows that the set $G_n = \{a_F\} \cup (O_n \setminus F)$ is an open neighborhood of $a_F$ in $P$ for each $n \in \omega$. Next, apply (1) to conclude that the family $\{G_n : n \in \omega\}$ is a local base at the point $a_F$ in $P_F$, so $P_F$ is first countable at $a_F$ and hence the space $Y$ is first countable.

Finally, take any $U \in \tau(a_F, Y)$; by (1) there exists $n \in \omega$ such that $G_n \subset U$ and hence $Q = \omega_1 \times \{n + 1\} \subset G_n$ is a non-compact subset of $U$ which is closed in $Y$. Therefore $Q$ is also closed in $\overline{U}$, so $\overline{U}$ cannot be compact. This shows that $Y$ is not locally compact at the point $a_F$ and hence our spaces $X$ and $Y$ are first countable and $l$-equivalent while $X$ is locally compact and $Y$ is not.

**V.369.** *Given a space $Z$ let $Z'$ be the set of non-isolated points of $Z$. Suppose that $X$ and $Y$ are normal first countable $l$-equivalent spaces. Prove that if $X'$ is countably compact then $Y'$ is also countably compact. Show that this statement can be false if we omit first countability of $X$ and $Y$.*

**Solution.** It follows from Problem 263 that the space $(\omega + 1) \times \omega$ is $l$-equivalent to the Fréchet–Urysohn $\omega$-fan $V(\omega)$. The spaces $V(\omega)$ and $(\omega + 1) \times \omega$ are countable and hence normal; besides, the set of non-isolated points of $(\omega + 1) \times \omega$ is infinite and discrete while $V(\omega)$ has a unique non-isolated point. Therefore even compactness of $X'$ does not need to imply that $Y'$ is countably compact if we omit first countability of $X$.

Now assume that $X$ and $Y$ are first countable $l$-equivalent normal spaces and $X'$ is countably compact. If $Y'$ is not countably compact then we can fix a faithfully indexed closed discrete subset $D = \{d_n : n \in \omega\}$ of the subspace $Y'$. It is clear that $D$ is also closed in $Y$; it follows from normality of $Y$ that

(1) the set $D$ is not bounded in $Y$.

Choose a disjoint family $\{O_n : n \in \omega\} \subset \tau(Y)$ such that $d_n \in O_n$ for each $n \in \omega$. Fix a decreasing local base $\{B_m^n : m \in \omega\}$ at every point $d_n$ in such a way that $B_0^n \subset O_n$; pick a function $f_m^n \in C_p(Y, [0, 1])$ such that $f_m^n(d_n) = 1$ and $f_m^n(Y \setminus B_m^n) \subset \{0\}$ for every $m \in \omega$. If $\chi_n$ is the characteristic function of the point $d_n$ in $Y$ then $\chi_n$ is discontinuous and the sequence $P_n = \{f_m^n : m \in \omega\}$ converges to $\chi_n$ for all $n \in \omega$.

By Fact 1 of V.272 there exists a linear homeomorphism $\varphi : \mathbb{R}^Y \to \mathbb{R}^X$ such that $\varphi(C_p(Y)) = C_p(X)$. If $Q_n = \varphi(P_n)$ then $Q_n \subset C_p(X)$ is a sequence converging to a function $\delta_n = \varphi(\chi_n)$ which is discontinuous on $X$. Therefore we can choose a point $a_n \in X'$ such that the function $\delta_n$ is discontinuous at $a_n$ for every $n \in \omega$. The set $X'$ being countably compact and first countable, the sequence $A = \{a_n : n \in \omega\}$ has a convergent subsequence. Passing, if necessary, to an appropriate subsequence of $D$ we can assume, without loss of generality, that there is a point $a \in X'$ such that $a_n \to a$.

Choose a decreasing local base $\{W_n : n \in \omega\}$ at the point $a$ in $X$. Passing to the relevant subsequences, we can assure that $a_n \in W_n$ for all $n \in \omega$. Since $\delta_n$ is discontinuous at $a_n$, we can find a sequence $S_n = \{x_m^n : m \in \omega\} \subset W_n$ which converges to $a_n$ while

(2) the sequence $\{\delta_n(x_m^n) : m \in \omega\}$ does not converge to $\delta_n(a_n)$ for every $n \in \omega$.

The set $K_n = S_n \cup \{a_n\}$ is compact for each $n \in \omega$ and every open neighborhood of $a$ contains all but finitely many of the sets $K_n$. An immediate consequence is that the set $K = \{a\} \cup (\bigcup\{K_n : n \in \omega\})$ is compact.

For every $x \in X$ there exist distinct points $y_1, \ldots, y_n \in Y$ and numbers $\lambda_1, \ldots, \lambda_n \in \mathbb{R} \setminus \{0\}$ such that $\varphi(f)(x) = \sum_{i=1}^n \lambda_i f(y_i)$ for all $f \in C_p(Y)$ (see Problem 279); let $\operatorname{supp}(x) = \{y_1, \ldots, y_n\}$. As a trivial consequence,

(3) for every $x \in X$, if $f \in C_p(Y)$ and $f(\operatorname{supp}(x)) = \{0\}$ then $\varphi(f)(x) = 0$.

For every set $E \subset X$ let $\underline{\operatorname{supp}(E)} = \bigcup\{\operatorname{supp}(x) : x \in E\}$; from compactness of $K$ it follows that the set $\overline{\operatorname{supp}(K)}$ is bounded in $Y$ (see Problem 281 and Fact 2 of S.398). Apply (1) to conclude that $D$ is not contained in $\overline{\operatorname{supp}(K)}$ and fix $n \in \omega$ such that $d_n \in D \setminus \overline{\operatorname{supp}(K)}$. There exists $k \in \omega$ for which $B_m^n \cap \operatorname{supp}(K) = \emptyset$ and hence $f_m^n(\operatorname{supp}(K)) = \{0\}$ for all $m \geq k$. Apply (2) to see that $\varphi(f_m^n)(x) = 0$ for all $x \in K$ and $m \geq k$. The sequence $\{\varphi(f_m^n) : m \geq k\}$ converges to $\delta_n$, so $\delta_n(x) = 0$ for all $x \in K$. In particular, $\delta_n(x_m^n) = 0 = \delta_n(a_n)$ for all $m \geq k$; this contradiction with (2) shows that $Y'$ is countably compact and hence our solution is complete.

**V.370.** *Given a nonempty closed subspace $F$ of a normal space $X$ suppose that $F$ is a retract of some neighborhood of $F$. Prove that $F$ is $l$-embedded in $X$ and hence $C_p(X) \approx C_p(F) \times I$, where $I = \{f \in C_p(X) : f(F) = \{0\}\}$.*

**Solution.** Fix a set $O \in \tau(F, X)$ such that there exists a retraction $r : O \to F$. By normality of $X$ we can find a set $G \in \tau(F, X)$ for which $\overline{G} \subset O$; choose a continuous function $p : X \to [0, 1]$ with $p(F) \subset \{1\}$ and $p(X \backslash G) \subset \{0\}$. Given any $f \in C_p(F)$ define a function $\varphi(f) \in \mathbb{R}^X$ as follows: $\varphi(f)(x) = 0$ for all $x \in X \backslash G$ and $\varphi(f)(x) = f(r(x)) \cdot p(x)$ for each $x \in G$.

Observe first that, for any point $x \in F$, we have $r(x) = x$ and $p(x) = 1$, so $\varphi(f)(x) = f(x)$, i.e., $\varphi(f)|F = f$ and hence $\varphi : C_p(F) \to \mathbb{R}^X$ is an extension operator. The function $\varphi(f)$ is constant on $X \backslash G$, so it is continuous at all points of the open set $X \backslash \overline{G}$. On the set $O$ the function $\varphi(f)$ coincides with the continuous function $(f \circ r) \cdot p$, so $\varphi(f)$ is continuous on $O$ and hence $\varphi(f) \in C_p(X)$ (see Fact 1 of S.472) for every $f \in F$ which shows that $\varphi : C_p(F) \to C_p(X)$. It is straightforward that $\varphi$ is a linear map, so let us check that $\varphi$ is continuous.

Given a point $x \in X$ let $\pi_x(f) = f(x)$ for all $f \in C_p(X)$. If $x \in X \backslash G$ then $(\pi_x \circ \varphi)(f) = \varphi(f)(x) = 0$ for any $f \in C_p(F)$, so the map $\pi_x \circ \varphi$ is continuous. If $x \in G$ then $(\pi_x \circ \varphi)(f) = p(x) \cdot f(r(x)) = p(x)\pi_{r(x)}(f)$, so $\pi_x \circ \varphi : C_p(F) \to \mathbb{R}$ coincides with the continuous function $p(x) \cdot \pi_{r(x)}$ which shows that $\pi_x \circ \varphi$ is continuous for all $x \in X$ and hence $\varphi$ is continuous by TFS-102. Thus $\varphi$ is a continuous linear extender for $F$ and hence $F$ is $l$-embedded in $X$. Applying CFS-448 we conclude that $C_p(X)$ is linearly homeomorphic to $C_p(F) \times I$ as promised.

**V.371.** *Given a space $X$ assume that $X_0$ and $X_1$ are closed subspaces of $X$ such that $X = X_0 \cup X_1$ and the set $F = X_0 \cap X_1$ is $l$-embedded in $X$; suppose additionally that $C_p(F) \approx C_p(F) \times C_p(F)$. Prove that $C_p(X) \approx C_p(X_0) \times C_p(X_1)$.*

**Solution.** If $Z$ is a space and $A \subset Z$ then $I_A^Z = \{f \in C(Z) : f(A) \subset \{0\}\}$ and $\pi_A : C_p(Z) \to C_p(A)$ is the restriction map; if $z \in Z$ and $A = \{z\}$ then we write $\pi_z$ instead of $\pi_A$.

**Fact 1.** Suppose that $Z$ is a space, $Z_0$ and $Z_1$ are closed in $Z$ and $Z = Z_0 \cup Z_1$. If $Y = Z_0 \cap Z_1$ then the space $I_Y^Z$ is linearly homeomorphic to $I_Y^{Z_0} \times I_Y^{Z_1}$.

**Proof.** For any function $f \in I_Y^Z$ let $\varphi(f) = (\pi_{Z_0}(f), \pi_{Z_1}(f))$; it is immediate that $\varphi : I_Y^Z \to I_Y^{Z_0} \times I_Y^{Z_1}$ is a continuous linear map. If $f, g \in I_Y^Z$ and $f \neq g$ then there is a point $z \in Z$ such that $f(z) \neq g(z)$. Take $i \in \{0, 1\}$ with $z \in Z_i$; then $\pi_{Z_i}(f) \neq \pi_{Z_i}(g)$ and hence $\varphi(f) \neq \varphi(g)$ which shows that the map $\varphi$ is injective.

Given any $(f_0, f_1) \in I_Y^{Z_0} \times I_Y^{Z_1}$ let $f(x) = f_0(x)$ for all $x \in Z_0$ and $f(x) = f_1(x)$ whenever $x \in Z \backslash Z_0$. It is an easy consequence of Fact 2 of T.354 that $f$ is continuous on $X$; since also $\varphi(f) = (f_0, f_1)$, the map $\varphi$ is surjective.

Given $i \in \{0, 1\}$ and a point $z \in Z_i$ let $q_z^i(f) = f(z)$ for each $f \in I_Y^{Z_i}$ and let $\mu_i : I_Y^{Z_0} \times I_Y^{Z_1} \to I_Y^{Z_i}$ be the natural projection. To show that the map $\varphi^{-1}$ is continuous fix a point $z \in Z$. Then $z \in Z_i$ for some $i \in \{0, 1\}$; now, if we are given $p = (f_0, f_1) \in I_Y^{Z_0} \times I_Y^{Z_1}$ then $(\pi_z \circ \varphi^{-1})(p) = f_i(z) = q_z^i(\mu_i(p))$, so the map $\pi_z \circ \varphi^{-1}$ is continuous because it coincides with the continuous map $q_z^i \circ \mu_i$. By TFS-102 the map $\varphi^{-1}$ is continuous, so $\varphi$ is a linear homeomorphism and hence Fact 1 is proved.

Returning to our solution observe that $C_p(X) \approx C_p(F) \times I_F^X$ by CFS-448; by Fact 1, we have $I_F^X \approx I_F^{X_0} \times I_F^{X_1}$, so $C_p(X) \approx C_p(F) \times I_F^{X_0} \times I_F^{X_1}$. Recalling that $C_p(F) \approx C_p(F) \times C_p(F)$ we conclude that

(1) $C_p(X) \approx (C_p(F) \times I_F^{X_0}) \times (C_p(F) \times I_F^{X_1})$.

It is easy to see that $F$ is $l$-embedded in both $X_0$ and $X_1$, so we can apply CFS-448 again to convince ourselves that $C_p(F) \times I^{X_i} \approx C_p(X_i)$ for each $i \in \{0, 1\}$; this, together with (1), shows that $C_p(X) \approx C_p(X_0) \times C_p(X_1)$ and completes our solution.

**V.372.** *Suppose that a space $X$ has a nontrivial convergent sequence. Prove that $X$ is $l$-equivalent to $X \oplus (\omega + 1)$ and $X^+ \overset{l}{\sim} X$. Deduce from this fact that $X^+$ is $l$-equivalent to $X$ for every infinite metrizable space $X$.*

**Solution.** Since $X$ has a nontrivial convergent sequence, we can find a faithfully indexed set $A = \{a_n : n \in \omega\} \subset X$ and a point $a \in X \backslash A$ such that $A$ converges to $a$. It is clear that $S = A \cup \{a\}$ is homeomorphic to $\omega + 1$. Consider the set $I = \{f \in C_p(X) : f(S) = \{0\}\}$; it follows from CFS-482 that $S$ is $l$-embedded in $X$, so $C_p(X) \approx I \times C_p(S)$ by CFS-448. Apply Problem 276 to see that $C_p(S) \times C_p(S) \approx C_p(S)$ and therefore $C_p(X) \approx I \times C_p(S) \times C_p(S) \approx C_p(X) \times C_p(S) \approx C_p(X \oplus (\omega + 1))$ which shows that the spaces $X$ and $X \oplus (\omega + 1)$ are $l$-equivalent.

Next observe that the spaces $S$ and $S^+$ are homeomorphic so it follows from $C_p(X) \approx I \times C_p(S^+)$ that $C_p(X) \approx I \times C_p(S) \times \mathbb{R} \approx C_p(X) \times \mathbb{R} \approx C_p(X^+)$ and hence $X \overset{l}{\sim} X^+$. Finally assume that $X$ is infinite and metrizable; if $\omega + 1$ embeds in $X$ then $X^+ \overset{l}{\sim} X$ by what we proved above. If $X$ does not have nontrivial convergent sequences then it is an infinite discrete space so $X^+$ is even homeomorphic to $X$ and hence again $X^+ \overset{l}{\sim} X$.

**V.373.** *Suppose that a space $X$ has a nontrivial convergent sequence and $Y$ is $l$-embedded in $X$. Prove that $X$ is $l$-equivalent to $X_Y \oplus Y$. Consequently, for any infinite metrizable space $X$, if $Y$ is closed in $X$ then $X \overset{l}{\sim} X_Y \oplus Y$. Here $X_Y$ is the $\mathbb{R}$-quotient space obtained from $X$ by contracting $Y$ to a point.*

**Solution.** Assume first that $X$ has a nontrivial convergent sequence. It follows from Problem 258 that $X^+$ is $l$-equivalent to $X_Y \oplus Y$; thus, $X \overset{l}{\sim} X^+ \overset{l}{\sim} X_Y \oplus Y$ (see Problem 372). If $X$ is an infinite metrizable space then $X^+$ is $l$-equivalent to $X$ by Problem 372 and any closed $Y \subset X$ is $l$-embedded in $X$ by CFS-469. Therefore $X \overset{l}{\sim} X^+ \overset{l}{\sim} X_Y \oplus Y$ as promised.

**V.374.** *Suppose that $X$ is a compact space and $F$ is $l$-embedded in $X$. Prove that $X^+$ is $l$-equivalent to $F \oplus \alpha(X \backslash F)$. In particular, if $X$ is an infinite metrizable compact space then $X \overset{l}{\sim} F \oplus \alpha(X \backslash F)$ for any closed $F \subset X$.*

**Solution.** If $Z$ is a space and $Y$ is a closed subset of $Z$ then, as usual, $Z_Y$ is the $\mathbb{R}$-quotient space obtained from $Z$ by contracting $Y$ to a point.

**Fact 1.** Suppose that $K$ and $L$ are compact spaces, $a$ is a non-isolated point of $K$ and $b$ is a non-isolated point of $L$. If $K\setminus\{a\}$ is homeomorphic to $L\setminus\{b\}$ then there exists a homeomorphism $\varphi : K \to L$ such that $\varphi(a) = b$. In particular, $K$ is homeomorphic to $\alpha(K\setminus\{a\})$.

*Proof.* Fix a homeomorphism $h : K\setminus\{a\} \to L\setminus\{b\}$; let $\varphi(a) = b$ and $\varphi(x) = h(x)$ for any $x \in K\setminus\{a\}$. It is clear that $\varphi : K \to L$ is a bijection which is continuous at all points of $K\setminus\{a\}$. Analogously, $\varphi^{-1}$ is continuous at all points of $L\setminus\{b\}$. If $U \in \tau(a, K)$ then $P = K\setminus U$ is compact so $P' = h(P)$ is also compact and hence $\varphi(U) = L\setminus P'$ is open in $L$. An analogous proof shows that $\varphi^{-1}(V)$ is open in $K$ for any $V \in \tau(b, L)$. Therefore $\varphi$ is continuous at $a$ and $\varphi^{-1}$ is continuous at $b$, so $\varphi$ is a homeomorphism and hence Fact 1 is proved.

Returning to our solution apply Problem 258 to see that $X^+ \overset{l}{\sim} X_F \oplus F$. If $a_F$ is the point of $X_F$ represented by $F$ then $X_F\setminus\{a_F\}$ is homeomorphic to $X\setminus F$ by Problem 252. Applying Fact 1 we conclude that $X_F$ is homeomorphic to $\alpha(X\setminus F)$ and therefore $X^+ \overset{l}{\sim} F \oplus \alpha(X\setminus F)$. If $X$ is an infinite metrizable compact space then the space $X^+$ is $l$-equivalent to $X$ by Problem 372; besides, every closed subset of $X$ is $l$-embedded in $X$ by CFS-469, so $X \overset{l}{\sim} X^+ \overset{l}{\sim} X_F \oplus F \overset{l}{\sim} \alpha(X\setminus F) \oplus F$. The last equivalence takes place because $X_F$ is homeomorphic to $\alpha(X\setminus F)$ by Fact 1.

**V.375.** *Let $X$ and $Y$ be metrizable compact spaces. Suppose that $F$ and $G$ are closed subspaces of $X$ and $Y$ respectively such that $F \overset{l}{\sim} G$ and $X\setminus F$ is homeomorphic to $Y\setminus G$. Prove that $X$ is $l$-equivalent to $Y$.*

**Solution.** If $X$ is finite then $F$ is finite and it follows from $F \overset{l}{\sim} G$ that $|F| = |G|$ (see Problem 159); the spaces $X\setminus F$ and $Y\setminus G$ being homeomorphic, we have $|X\setminus F| = |Y\setminus G|$, so $Y$ is also finite and $|Y| = |X|$ which shows that $X$ and $Y$ are even homeomorphic.

If $X$ is infinite then either $F$ of $X\setminus F$ is infinite. In the first case $F \overset{l}{\sim} G$ implies that $G$ is infinite (see TFS-186); if $X\setminus F$ is infinite then $Y\setminus G$ is infinite, so in both cases the space $Y$ has to be infinite. Now we can apply Problem 374 to see that we have the equivalencies $X \overset{l}{\sim} F \oplus \alpha(X\setminus F)$ and $Y \overset{l}{\sim} G \oplus \alpha(Y\setminus G)$. Since $F \overset{l}{\sim} G$ and $\alpha(X\setminus F)$ is homeomorphic (and hence $l$-equivalent) to $\alpha(Y\setminus G)$, we can apply Problem 265 to conclude that $X \overset{l}{\sim} Y$.

**V.376.** *Let $X$ and $Y$ be nonempty compact metrizable spaces such that either $Y \times (\omega + 1)$ or $\alpha(Y \times \omega)$ embeds in $X$. Prove that $C_p(X) \approx C_p(X) \times (C_p(Y))^n$ for every $n \in \mathbb{N}$.*

**Solution.** As usual, we identify every $m \in \omega$ with the set $\{0, \dots, m-1\}$. Fix $n \in \mathbb{N}$ and let $F \subset X$ be homeomorphic either to $Y \times (\omega + 1)$ or to $\alpha(Y \times \omega)$. It is an easy exercise that

(1) $F \oplus (Y \times n)$ is homeomorphic to $F$ and hence $C_p(F) \approx C_p(F) \times (C_p(Y))^n$.

If $I = \{f \in C_p(X) : f(F) = \{0\}\}$ then the space $C_p(X)$ is linearly homeomorphic to $I \times C_p(F)$ (see CFS-448 and CFS-482), so we can apply the property (1) to conclude that $C_p(X) \approx I \times C_p(F) \approx I \times C_p(F) \times (C_p(Y))^n \approx C_p(X) \times (C_p(Y))^n$.

**V.377.** *Say that a metrizable compact space $K$ is universal in the dimension $n \in \omega$ if* $\dim K = n$ *and any metrizable compact space of dimension at most $n$ embeds in $K$. Prove that if $X$ and $Y$ are metrizable compact spaces universal in the dimension $n$, then $X \overset{l}{\sim} Y$.*

**Solution.** If a compact space $K$ is universal in the dimension $n$ then the space $K' = K \times (\omega + 1)$ embeds in $K$ because $K'$ is compact, metrizable and $\dim K' = n$ due to the fact that $K'$ is the countable union of its subspaces homeomorphic to $K$ (see Problem 150). It follows from Problem 376 that $C_p(K) \approx C_p(K) \times C_p(K)$. In particular, $C_p(X)$ is linearly homeomorphic to $(C_p(X))^2$ and $C_p(Y)$ is linearly homeomorphic to $(C_p(Y))^2$. By universality of $Y$ in the dimension $n$, the space $X$ embeds in $Y$ and the space $Y$ embeds in $X$ by universality of $X$ in the dimension $n$, so we can apply Problem 292 to conclude that $X \overset{l}{\sim} Y$.

**V.378.** *Suppose that a space $X$ is $l$-equivalent to $Y$ and $Y$ is a metrizable compact space universal in the dimension $n$. Prove that $X$ is also a metrizable compact space universal in the dimension $n$.*

**Solution.** Say that a compact space $K$ is *strongly universal in the dimension $n$* if $K$ is universal in the dimension $n$ and can be embedded into any nonempty open subspace of $K$.

*Fact 1.* If a compact space $K$ is universal in the dimension $n$ then it contains a compact subspace which is strongly universal in the dimension $n$.

*Proof.* Fix an $n$-dimensional compact space $P$ such that any second countable space of dimension at most $n$ embeds in $P$ and, in particular, $P$ is universal in the dimension $n$; such a space exists by Problem 162. The space $P^\omega$ is second countable and $P$ embeds in any set $U \in \tau^*(P^\omega)$. Fix a base $\mathcal{B} = \{B_n : n \in \omega\} \subset \tau^*(P^\omega)$ in the space $P^\omega$ and choose a set $P_n \subset B_n$ homeomorphic to $P$ for every $n \in \omega$. The space $G = \bigcup_{n \in \omega} P_n$ is second countable and $\dim G \leq n$ by Problem 150. By the choice of $P$, there is a set $G' \subset P$ homeomorphic to $G$; let $H = \overline{G'}$.

The space $H$ is compact, metrizable and $\dim H \leq n$; if $U \in \tau^*(H)$ then $U' = U \cap G'$ is homeomorphic to a nonempty open set $W$ of the space $G$. Take $W' \in \tau(P^\omega)$ such that $W' \cap G = W$ and pick $n \in \omega$ such that $B_n \subset W'$. Then $P_n \subset W' \cap G = W$ and hence $P_n$ embeds in $U' \subset U$. Since $P_n$ is universal in the dimension $n$, the space $H$ embeds in $P_n$ and hence in $U$; this shows that $H$ is strongly universal in the dimension $n$. Since also $\dim H \leq n$, the space $H$ embeds in $K$, so $K$ has a compact subspace which is strongly universal in the dimension $n$, i.e., Fact 1 is proved.

Returning to our solution observe first that the space $X$ is compact by Problem 138 and $\dim X = n$ by Problem 180. Apply Fact 1 to find a strongly universal subspace $K \subset Y$. By Problem 309, there exists $U \in \tau^*(K)$ which is embeddable in $X$. Being strongly universal in the dimension $n$, the space $K$ embeds in $U$ and hence in $X$. It is trivial that every compact $n$-dimensional space which contains a subspace universal in the dimension $n$ is universal in the dimension $n$, so $X$ is universal in the dimension $n$ as required.

**V.379.** *Prove that, for any $n_1, \ldots, n_k \in \mathbb{N}$, the space $C_p(\mathbb{I}^{n_1}) \times \ldots \times C_p(\mathbb{I}^{n_k})$ is linearly homeomorphic to $C_p(\mathbb{I}^n)$ where $n = \max\{n_1, \ldots, n_k\}$.*

**Solution.** For any points $x = (x_1, \ldots, x_m)$ and $y = (y_1, \ldots, y_m)$ of the space $\mathbb{R}^m$ let $d_m(x, y) = \sqrt{(x_1 - y_1)^2 + \ldots + (x_m - y_m)^2}$; as usual, for each $r > 0$, the set $B_m(x, r) = \{y \in \mathbb{R}^m : d_m(x, y) < r\}$ is the open ball of radius $r$ centered at $x$. The expression $X \simeq Y$ says that the spaces $X$ and $Y$ are homeomorphic. It is easy to see that $B_m(x, r) \simeq B_m(y, s)$ for any $x, y \in \mathbb{R}^m$ and $r, s > 0$. Let $u_m \in \mathbb{R}^m$ be the zero point of $\mathbb{R}^m$.

*Fact 1.* For any $m \in \mathbb{N}$ and $U \in \tau^*(\mathbb{R}^m)$ the space $\alpha(\mathbb{I}^m \times \omega)$ embeds in $U$.

*Proof.* If $V$ is a nonempty open subset of $\mathbb{R}^m$ then take a point $a \in V$. There exists $r > 0$ such that $B_m(a, r) \subset V$; since $B_m(a, r)$ is homeomorphic to $B_m(u_m, 2m)$, the space $B_m(u_m, 2m)$ embeds in $V$. It is straightforward that $\mathbb{I}^m \subset B_m(u_m, 2m)$, so we proved that

(1)  the space $\mathbb{I}^m$ embeds into any nonempty open subspace of $\mathbb{R}^m$.

Let $W_i = B_m(u_m, 2^{-i})$; it is immediate that $W_i \setminus \overline{W}_{i+1} \neq \emptyset$ for every $i \in \omega$ and $\{W_i : i \in \omega\}$ is a decreasing local base at $u_m$. Take an open ball $Q_i \subset W_i \setminus \overline{W}_{i+1}$ for every $i \in \omega$; then the family $\{Q_i : i \in \omega\}$ is disjoint and convergent to $u_m$ in the sense that any $U \in \tau(u_m, \mathbb{R}^m)$ contains all but finitely many $Q_i$'s.

The property (1) shows that we can find a set $J_i \subset Q_i$ with $J_i \simeq \mathbb{I}^m$ for each $i \in \omega$. It is standard that the space $K = \{u_m\} \cup \{J_i : i \in \omega\}$ is homeomorphic to $\alpha(\mathbb{I}^m \times \omega)$. Moreover, $K \subset W_0$ and $W_0$ is an open ball which is homeomorphic to any open ball contained in $U$. Therefore $W_0$ embeds in $U$ and hence $K \simeq \alpha(\mathbb{I}^m \times \omega)$ also embeds in $U$, i.e., Fact 1 is proved.

Returning to our solution, let us prove, by induction on $k \in \mathbb{N}$ that

(2)  for any $n_1, \ldots, n_k \in \mathbb{N}$ the space $C_p(\mathbb{I}^{n_1}) \times \ldots C_p(\mathbb{I}^{n_k})$ is linearly homeomorphic to $C_p(\mathbb{I}^n)$ for $n = \max\{n_1, \ldots, n_k\}$.

If $k = 1$ then (2) is clearly true, so assume that we proved (2) for all $k \leq l$ and take any $n_1, \ldots, n_l, n_{l+1} \in \mathbb{N}$; let $n = \max\{n_1, \ldots, n_{l+1}\}$. We can assume, without loss of generality, that $n_{l+1} = n$. It follows from Fact 1 that the space $\alpha(\mathbb{I}^n \times \omega)$ embeds in $\mathbb{I}^n$, so we can apply Problem 376 to see that $C_p(\mathbb{I}^n) \times C_p(\mathbb{I}^{n_1}) \approx C_p(\mathbb{I}^n)$. Therefore $\prod_{i=1}^{l+1} C_p(\mathbb{I}^{n_i}) \approx C_p(\mathbb{I}^n) \times C_p(\mathbb{I}^{n_1}) \times \prod_{i=2}^{l} C_p(\mathbb{I}^{n_i}) \approx C_p(\mathbb{I}^n) \times \prod_{i=2}^{l} C_p(\mathbb{I}^{n_i})$, so we can apply the induction hypothesis to see that $C_p(\mathbb{I}^n) \times \prod_{i=2}^{l} C_p(\mathbb{I}^{n_i}) \approx C_p(\mathbb{I}^n)$. This completes the induction step and shows that (2) is true for all $k \in \mathbb{N}$.

**V.380.** *Given* $n \in \mathbb{N}$ *and a closed subset* $F$ *of the space* $\mathbb{I}^n$ *such that* $\emptyset \neq F \neq \mathbb{I}^n$ *prove that* $(\mathbb{I}^n)_F$ *is* $l$*-equivalent to* $\mathbb{I}^n$. *Here* $(\mathbb{I}^n)_F$ *is the* $\mathbb{R}$*-quotient image of* $\mathbb{I}^n$ *obtained by contracting* $F$ *to a point.*

**Solution.** Denote by $a_F$ the point of the space $(\mathbb{I}^n)_F$ represented by the set $F$ and apply Problem 373 to convince ourselves that $\mathbb{I}^n$ is $l$-equivalent to $(\mathbb{I}^n)_F \oplus F$ and hence $C_p(\mathbb{I}^n) \approx C_p((\mathbb{I}^n)_F) \times C_p(F)$. Observe that $(\mathbb{I}^n)_F \backslash \{a_F\}$ is homeomorphic to $\mathbb{I}^n \backslash F$ (see Problem 252); since $\mathbb{I}^n \backslash F$ contains a nonempty open subspace of $V$ of the space $\mathbb{R}^n$, the set $\alpha(\mathbb{I}^n \times \omega)$ embeds in $V$ by Fact 1 of V.379. It follows from $F \subset \mathbb{I}^n$ that the set $\alpha(F \times \omega)$ embeds in $\alpha(\mathbb{I}^n \times \omega)$ and hence $\alpha(F \times \omega)$ embeds in $(\mathbb{I}^n)_F$. This, together with Problem 376, shows that $C_p(\mathbb{I}^n) \approx C_p((\mathbb{I}^n)_F) \times C_p(F) \approx C_p((\mathbb{I}^n)_F)$, i.e., $(\mathbb{I}^n)_F \overset{l}{\sim} \mathbb{I}^n$ as promised.

**V.381.** *Prove that, for any* $n \in \mathbb{N}$, *if* $U$ *is a nonempty open subset of the space* $\mathbb{R}^n$, *then the space* $\alpha(U)$ *is* $l$*-equivalent to* $\mathbb{I}^n$.

**Solution.** The set $O = (-1, 1)^n$ is homeomorphic to $\mathbb{R}^n$ and open in $\mathbb{I}^n$. Therefore $U$ is homeomorphic to an open subspace of $O$ and hence there is no loss of generality to consider that $U \subset O$. Then $F = \mathbb{I}^n \backslash U$ is a nonempty closed subset of $\mathbb{I}^n$ and $F \neq \mathbb{I}^n$. Thus $\mathbb{I}^n \overset{l}{\sim} (\mathbb{I}^n)_F$ where $(\mathbb{I}^n)_F$ is the $\mathbb{R}$-quotient space obtained from $\mathbb{I}^n$ by contracting $F$ to a point $a_F$ (see Problem 380). If $\alpha(U) = U \cup \{\xi\}$ then $U = \alpha(U) \backslash \{\xi\}$ is homeomorphic to $(\mathbb{I}^n)_F \backslash \{a_F\}$ by Problem 252. Therefore we can apply Fact 1 of V.374 to see that $(\mathbb{I}^n)_F$ is homeomorphic to $\alpha(U)$ and hence $(\mathbb{I}^n)_F \overset{l}{\sim} \alpha(U)$. Consequently, $\mathbb{I}^n \overset{l}{\sim} (\mathbb{I}^n)_F \overset{l}{\sim} \alpha(U)$, so $\alpha(U)$ is $l$-equivalent to $\mathbb{I}^n$.

**V.382.** *Given a space* $X$ *with* $\dim X = n \in \mathbb{N}$ *assume that* $X$ *is homeomorphic to a finite union of Euclidean cubes, i.e., there is a finite family* $\mathcal{F}$ *of subsets of* $X$ *such that* $\bigcup \mathcal{F} = X$ *and every* $F \in \mathcal{F}$ *is homeomorphic to* $\mathbb{I}^k$ *for some* $k \in \mathbb{N}$. *Prove that* $X$ *is* $l$*-equivalent to* $\mathbb{I}^n$.

**Solution.** The expression $Y \simeq Z$ says that the spaces $Y$ are $Z$ are homeomorphic. If $Z$ is a space and $Y$ is a closed subset of $Z$ then $Z_Y$ is the $\mathbb{R}$-quotient image of $Z$ obtained by contracting $Y$ to a point; this point will be denoted by $a_Y$.

It suffices to prove, by induction on $m$, that

(1) if $X = F_1 \cup \ldots \cup F_m$ and $F_i \simeq \mathbb{I}^{n_i}$ for all $i \leq m$ then $X \overset{l}{\sim} \mathbb{I}^n$.

If $m = 1$ then it follows from $\dim X = n$ that $n_1 = n$ (see Problem 159), so $X \simeq \mathbb{I}^n$ and hence $X \overset{l}{\sim} \mathbb{I}^n$. Assume that (1) has been proved for all $m < l$ and $X = F_1 \cup \ldots \cup F_l$ where every $F_i$ is homeomorphic to $\mathbb{I}^{n_i}$ for some $n_i \in \mathbb{N}$. It follows from Problem 155 that $n_i \leq n$ for all $i \leq l$; if $n_i < n$ for each $i \leq l$ then $\dim X < n$ by Problem 151. This contradiction shows that there exists $i \leq l$ for which $n_i = n$; there is no loss of generality to assume that $n_1 = n$.

If $F_1 \subset F = F_2 \cup \ldots \cup F_l$ then $X = F_2 \cup \ldots \cup F_l$, so $X \overset{l}{\sim} \mathbb{I}^n$ by the induction hypothesis. Therefore we can assume that $F_1 \backslash F \neq \emptyset$. If $F \cap F_1 = \emptyset$ then $X \simeq F_1 \oplus F$, so $C_p(X) \approx C_p(F_1) \times C_p(F)$; by the induction hypothesis, $C_p(F) \approx C_p(\mathbb{I}^k)$ for $k = \dim F \leq n$. Thus, $C_p(X) \approx C_p(\mathbb{I}^n) \times C_p(\mathbb{I}^k) \approx C_p(\mathbb{I}^n)$ (see Problem 379), i.e., $X \overset{l}{\sim} \mathbb{I}^n$.

If $G = F_1 \cap F \neq \emptyset$ then $G$ is a nonempty closed subset of $F_1$ with $G \neq F_1$. Consequently, $(F_1)_G$ is $l$-equivalent to $\mathbb{I}^n$ by Problem 380. The space $X_F$ is compact and $X_F \backslash \{a_F\}$ is homeomorphic to $X \backslash F$ (see Problem 252); besides, $(F_1)_G \backslash \{a_G\}$ is homeomorphic to $F_1 \backslash G = X \backslash F$. Thus, Fact 1 of V.374 is applicable to conclude that $(F_1)_G \simeq X_F$. It follows from Problem 380 that $(F_1)_G \overset{l}{\sim} \mathbb{I}^n$, so $X_F \overset{l}{\sim} \mathbb{I}^n$. Apply Problem 373 to see that $X \overset{l}{\sim} X_F \oplus F \overset{l}{\sim} \mathbb{I}^n \oplus F$. The induction hypothesis shows that $F \overset{l}{\sim} \mathbb{I}^n$, so $X \overset{l}{\sim} \mathbb{I}^n \oplus \mathbb{I}^n \overset{l}{\sim} \mathbb{I}^n$ (see Problem 379); this completes the inductive step and shows that (1) is true for any $m \in \mathbb{N}$.

**V.383.** *For any $n \in \mathbb{N}$ prove that both spaces $\mathbb{I}^n \times (\omega + 1)$ and $\alpha(\mathbb{I}^n \times \omega)$ are $l$-equivalent to $\mathbb{I}^n$.*

**Solution.** Since the interior of $\mathbb{I}^n$ in $\mathbb{R}^n$ is nonempty, it follows from Fact 1 of V.379 that the space $L = \alpha(\mathbb{I}^n \times \omega)$ embeds in $\mathbb{I}^n$. It is clear that $\mathbb{I}^n$ embeds in $L$; besides, it follows from Problem 379 that $C_p(\mathbb{I}^n) \times C_p(\mathbb{I}^n) \approx C_p(\mathbb{I}^n)$.

Let $s$ be the unique point of $L \backslash (\mathbb{I}^n \times \omega)$ and apply Fact 1 of V.379 once more to see that there is a set $K_n \subset \mathbb{I}^n \times \{n\}$ homeomorphic to $L$ for each $n \in \omega$. It is immediate to verify that the space $K = \{s\} \cup (\bigcup \{K_n : n \in \omega\})$ is homeomorphic to $\alpha(L \times \omega)$, i.e., $\alpha(L \times \omega)$ embeds in $L$, so we can apply Problem 376 to convince ourselves that $C_p(L) \approx C_p(L) \times C_p(L)$. Now, it follows from Problem 292 that the space $L$ is $l$-equivalent to $\mathbb{I}^n$.

In the space $Y = \mathbb{I}^n \times (\omega + 1)$ consider the set $F = \mathbb{I}^n \times \{\omega\}$; it follows from Problem 374 that $Y$ is $l$-equivalent to $F \oplus \alpha(Y \backslash F) = F \oplus \alpha(\mathbb{I}^n \times \omega)$. Since $F$ is homeomorphic to $\mathbb{I}^n$ and $\alpha(\mathbb{I}^n \times \omega) \overset{l}{\sim} \mathbb{I}^n$, we conclude that $Y \overset{l}{\sim} \mathbb{I}^n \oplus \mathbb{I}^n \overset{l}{\sim} \mathbb{I}^n$ (see Problem 379).

**V.384.** *Prove that $\mathbb{I}^n \times \mathbb{D}^\omega$ is not $l$-equivalent to $\mathbb{I}^n$ for any $n \in \mathbb{N}$.*

**Solution.** If the space $\mathbb{I}^n \times \mathbb{D}^\omega$ is $l$-equivalent to $\mathbb{I}^n$ for some $n \in \mathbb{N}$ then there is a nonempty open set $U \subset \mathbb{I}^n \times \mathbb{D}^\omega$ which embeds in $\mathbb{I}^n$ (see Problem 309). Take any sets $V \in \tau^*(\mathbb{I}^n)$ and $W \in \tau^*(\mathbb{D}^\omega)$ such that $V \times W \subset U$. By Fact 1 of V.379, the space $\mathbb{I}^n$ embeds in $V$, so $\mathbb{I}^n \times W$ embeds in $V \times W$ and hence in $\mathbb{I}^n$; let $Y$ be a homeomorphic copy of $\mathbb{I}^n \times W$ in $\mathbb{I}^n$. Pick any $w \in W$; then $\mathbb{I}^n \times \{w\}$ is an $n$-dimensional subset of $\mathbb{I}^n \times W$ which is nowhere dense in $\mathbb{I}^n \times W$ because $W$ has no isolated points. Therefore some $n$-dimensional subspace $K$ of the space $Y$ is nowhere dense in $Y$ and hence in $\mathbb{I}^n$. This, however, contradicts Problem 160 and shows that $\mathbb{I}^n \times \mathbb{D}^\omega$ is not $l$-equivalent to $\mathbb{I}^n$.

**V.385.** *Suppose that $K$ is a compact space and there exists a continuous bijective map of $[0, +\infty)$ onto $K$. Prove that $K$ is $l$-equivalent to $\mathbb{I}$. Deduce from this fact that if there is a continuous bijection of $\mathbb{R}$ onto a compact space $L$ then $L$ is also $l$-equivalent to $\mathbb{I}$.*

**Solution.** Recall that a continuous bijection is called *condensation* and a space $X$ *condenses* onto a space $Y$ if there exists a condensation $f : X \to Y$.

*Fact 1.* Suppose that $Z$ is a space and $\{C_t : t \in T\}$ is a centered family of compact connected subsets of $Z$. Then the set $C = \bigcap_{t \in T} C_t$ is connected.

*Proof.* If $C$ is disconnected then we can find nonempty disjoint closed sets $D, E \subset C$ such that $C = D \cup E$. In any space disjoint compact sets can be separated by open sets (see Fact 4 of T.309), so we can find disjoint $U, V \in \tau(Z)$ such that $D \subset U$ and $E \subset V$.

Apply Fact 1 of S.326 to see that there is $t \in T$ such that $C_t \subset U \cup V$. It is immediate that $U' = U \cap C_t$ and $V' = V \cap C_t$ are nonempty clopen subsets of $C_t$ with $C_t = U' \cup V'$; this contradiction with connectedness of $C_t$ shows that Fact 1 is proved.

Returning to our solution suppose that $K$ is a compact space and there exists a condensation $\varphi : [0, +\infty) \to K$; consider, for every $r > 0$, the set $R_r = \varphi([r, +\infty))$ and let $R = \bigcap\{\overline{R_r} : r > 0\}$. Observe that every $R_r$ is connected being a continuous image of a connected space $[r, +\infty)$ (see Fact 1 of T.309, Fact 1 of V.290 and Fact 2 of U.493); therefore $\overline{R_r}$ is also connected by Fact 1 of T.312, so we can apply Fact 1 to see that $R$ is a connected compact subset of $K$. Note also that $K$ is metrizable because it has a countable network. Let us prove first that

(1) a point $x$ belongs to $R$ if and only if there is a sequence $\{t_n : n \in \omega\} \subset [0, +\infty)$ such that $t_n \to +\infty$ and $\varphi(t_n) \to x$.

If we have a sequence $S = \{t_n : n \in \omega\}$ as in (1) then, for every $r > 0$, infinitely many terms of the sequence $\{\varphi(t_n) : n \in \omega\}$ belong to $R_r$ and hence $x \in \overline{R_r}$; consequently, $x \in R$. If, on the other hand, $x \in R$ then fix a local decreasing base $\{B_n : n \in \omega\}$ of the space $K$ at the point $x$. It is easy to find an increasing sequence $\{t_n : n \in \omega\} \subset [0, +\infty)$ such that $\varphi(t_n) \in R_{t_n} \cap B_n$ and $t_n > n$ for each $n \in \omega$. Then $t_n \to +\infty$ and $\varphi(t_n) \to x$, i.e., (1) is proved.

Our next step is to show that

(2) the set $[r, +\infty)$ is not contained in $\varphi^{-1}(R)$ for any $r > 0$.

Assume for contradiction that $[r, +\infty) \subset \varphi^{-1}(R)$ for some $r > 0$. The set $P_m = \varphi([m, m+1])$ is compact for every $m \in \omega$ and $R \subset \bigcup\{P_m : m \in \omega\}$. By the Baire property of $R$, the set $H = \text{Int}_R(P_m \cap R)$ is nonempty for some $m \in \omega$; fix a point $x \in H$. By (1), there is a sequence $\{t_n : n \in \omega\} \subset [0, +\infty)$ such that $t_n \to +\infty$ and $\varphi(t_n) \to x$; there is no loss of generality to consider that $t_n > r$ for every $n \in \omega$. The set $H$ being a neighborhood of $x$ in $R$, infinitely many terms of the sequence $\{\varphi(t_n) : n \in \omega\} \subset R$ belong to $H$; since $\varphi$ is a bijection, infinitely many terms of the sequence $\{t_n : n \in \omega\}$ belong to $[m, m+1]$ which is a contradiction with $t_n \to +\infty$, so (2) is proved.

It turns out that the set $\varphi^{-1}(R)$ is even bounded, i.e.,

(3) there is $a > 0$ such that $\varphi^{-1}(R) \subset [0, a]$.

If (3) does not hold then apply (2) to find an increasing sequence $\{t_n : n \in \mathbb{N}\}$ of points of $(0, +\infty) \setminus \varphi^{-1}(R)$ such that $t_n \to +\infty$ and $[t_n, t_{n+1}] \cap \varphi^{-1}(R) \neq \emptyset$ for all $n \in \mathbb{N}$. Letting $t_0 = 0$ and $Q_i = \varphi([t_i, t_i + 1]) \cap R$ for all $i \in \omega$ we obtain a sequence $\mathcal{Q} = \{Q_i : i \in \omega\}$ of disjoint closed subsets of $R$ such that $R = \bigcup \mathcal{Q}$ and $Q_i \neq \emptyset$ for all $i \in \mathbb{N}$. By Fact 5 of V.290 at most one element of $\mathcal{Q}$ is nonempty; this contradiction shows that (3) is proved.

If $E = \varphi((a, +\infty))$ then

(4) the map $\varphi : (a, +\infty) \to E$ is a homeomorphism.

Indeed, if the statement (4) is false then there is a point $t \in (a, +\infty)$ and a sequence $S = \{t_n : n \in \omega\} \subset (a, +\infty)$ which is does not converge to $t$ while the sequence $\{\varphi(t_n) : n \in \omega\}$ converges to the point $x = \varphi(t) \in E$. Passing to a subsequence of $S$ if necessary we can assume, without loss of generality, that there is $r > 0$ for which $|t_n - t| \geq r$ for all $n \in \omega$. An immediate consequence is that the set $S$ is closed and discrete in $[a, +\infty)$ and hence $t_n \to +\infty$. Therefore $x \in R$ which implies that $t \in \varphi^{-1}(R) \cap (a, +\infty)$ which is a contradiction with (3) and hence (4) is proved.

Observe that $F = \varphi([0, a])$ is a compact subspace of $K$ homeomorphic to $\mathbb{I}$, so $K$ is $l$-equivalent to $F \oplus \alpha(K \setminus F)$ by Problem 374. It follows from (4) that $K \setminus F$ is homeomorphic to $(a, +\infty)$, so $K \overset{l}{\sim} \mathbb{I} \oplus \alpha((a, +\infty)) \overset{l}{\sim} \mathbb{I} \oplus \mathbb{I} \overset{l}{\sim} \mathbb{I}$ (see Problems 381 and 379). Therefore we established that

(5) if $[0, +\infty)$ condenses onto a compact space $K$ then $K \overset{l}{\sim} \mathbb{I}$.

Now assume that $L$ is a compact space and $\varphi : \mathbb{R} \to L$ is a condensation. The proofs of some properties of $L$ will very similar to what we have proved for $K$. Let $A_r = \varphi([r, +\infty))$ and $B_r = \varphi((-\infty, r])$ for every $r \in \mathbb{R}$ and consider the sets $A = \bigcap\{\overline{A}_r : r \in \mathbb{R}\}$ and $B = \bigcap\{\overline{B}_r : r \in \mathbb{R}\}$. Mimicking the proof of (1) we convince ourselves that

(6) $A = \{x \in L :$ there exists a sequence $\{t_n : n \in \omega\} \subset \mathbb{R}$ such that $t_n \to +\infty$ and $\varphi(t_n) \to x\}$ and $B = \{x \in L :$ there exists a sequence $\{t_n : n \in \omega\} \subset \mathbb{R}$ such that $t_n \to -\infty$ and $\varphi(t_n) \to x\}$.

The proof of the following property is also analogous to the case of $K$ but we carry it out anyway.

(7) The set $[r, +\infty)$ is not contained in $\varphi^{-1}(A)$ and the set $(-\infty, r]$ is not contained in $\varphi^{-1}(B)$ for any $r \in \mathbb{R}$.

Assume for contradiction that $[r, +\infty) \subset \varphi^{-1}(A)$ (or $(-\infty, r] \subset \varphi^{-1}(B)$) for some $r \in \mathbb{R}$. The set $P_n = \varphi([m, m + 1])$ is compact for every $m \in \mathbb{Z}$ and we have the inclusion $A \cup B \subset \bigcup\{P_m : m \in \mathbb{Z}\}$. By the Baire property of $A$ (or $B$ respectively), the set $H = \mathrm{Int}_A(P_m \cap A)$ (or the set $H = \mathrm{Int}_B(P_m \cap B)$ respectively) is nonempty for some $m \in \mathbb{Z}$; fix a point $x \in H$. By (6), there is a sequence $\{t_n : n \in \omega\} \subset \mathbb{R}$ such that $t_n \to +\infty$ $(t_n \to -\infty)$ and $\varphi(t_n) \to x$; there is no loss of generality to consider that $t_n > r$ $(t_n < r)$ for every $n \in \omega$. The set $H$ being a neighborhood of $x$ in $A$ (or in $B$ respectively), infinitely many terms of the sequence $\{\varphi(t_n) : n \in \omega\} \subset A$ ($\{\varphi(t_n) : n \in \omega\} \subset B$) belong to $H$; since $\varphi$ is a bijection, infinitely many terms of the sequence $\{t_n : n \in \omega\}$ belong to $[m, m + 1]$ which is a contradiction with $t_n \to +\infty$ (or $t_n \to -\infty$ respectively), so (7) is proved.

(8) If the set $(-\infty, r]$ ($[r, +\infty)$) is not contained in $\varphi^{-1}(A)$ (or in $\varphi^{-1}(B)$ respectively) for any $r \in \mathbb{R}$ then $\varphi^{-1}(A)$ ($\varphi^{-1}(B)$ respectively) is bounded in $\mathbb{R}$.

To prove (8) apply the property (7) to find a sequence $S = \{s_n : n \in \mathbb{Z}\}$ such that $m < n$ implies $s_m + 1 < s_n$ and $S \cap \varphi^{-1}(A) = \emptyset$ (or $S \cap \varphi^{-1}(B) = \emptyset$ respectively). The set $P_n = [s_n, s_{n+1}] \cap A$ (or $P_n = [s_n, s_{n+1}] \cap B$ respectively) is closed in $A$ (in $B$) for any $n \in \mathbb{Z}$, the family $\mathcal{P} = \{P_n : n \in \mathbb{Z}\}$ is disjoint and $A = \bigcup \mathcal{P}$ (or $B = \bigcup \mathcal{P}$ respectively). By Fact 5 of V.290 at most one element of the family $\mathcal{P}$ is nonempty and hence the set $\varphi^{-1}(A)$ (or $\varphi^{-1}(B)$ respectively) is bounded in $\mathbb{R}$, i.e., (8) is proved.

(9) One of the sets $\varphi^{-1}(A)$ and $\varphi^{-1}(B)$ has to be bounded in $\mathbb{R}$.

Indeed, if $A$ and $B$ are unbounded then it follows from (7) and (8) that there are $r, s \in \mathbb{R}$ such that $(-\infty, r] \subset \varphi^{-1}(A)$ and $[s, +\infty) \subset \varphi^{-1}(B)$. The set $A$ is closed in $L$, so $B \subset \overline{B}_r \subset A$ and hence $B \subset A$ which shows that $[s, +\infty) \subset \varphi^{-1}(B) \subset \varphi^{-1}(A)$; this contradicts (7) and shows that (9) is proved.

(10) if $\varphi^{-1}(A) \subset (-\infty, r)$ (or $\varphi^{-1}(B) \subset (r, +\infty)$) then $\varphi|[r, +\infty) : [r, +\infty) \to A_r$ (or $\varphi|(-\infty, r] : (-\infty, r] \to B_r$ respectively) is a homeomorphism.

Indeed, if this is not true then there is a point $t \geq r$ ($t \leq r$) and a sequence $S = \{t_n : n \in \omega\} \subset [r, +\infty)$ (or $S = \{t_n : n \in \omega\} \subset (-\infty, r]$ respectively) which is does not converge to $t$ while the sequence $\{\varphi(t_n) : n \in \omega\}$ converges to the point $x = \varphi(t)$. Passing to a subsequence of $S$ if necessary we can assume, without loss of generality, that there is $\delta > 0$ for which $|t_n - t| \geq \delta$ for all $n \in \omega$. An immediate consequence is that the set $S$ is closed and discrete in $[r, +\infty)$ (in $(-\infty, r]$) and hence $t_n \to +\infty$ (or $t_n \to -\infty$ respectively). Therefore $x \in A$ (or $x \in B$ respectively) by the property (6) which implies that $t \in \varphi^{-1}(A) \cap [r, +\infty)$ (or $t \in \varphi^{-1}(B) \cap (-\infty, r]$ respectively) which is a contradiction and hence (10) is proved.

(11) If $\varphi^{-1}(A)$ (or $\varphi^{-1}(B)$) is bounded in $\mathbb{R}$ then there exists $r \in \mathbb{R}$ such that the set $A_s$ (or $B_s$ respectively) is compact for any $s < t$ (or $s > t$ respectively).

To prove (11) assume that $\varphi^{-1}(A) \subset (-r, r)$ (or $\varphi^{-1}(B) \subset (-r, r)$ respectively) and take any $s < r$ ($s > r$). If $x \in \overline{A}_s \setminus A_s$ (or $x \in \overline{B}_s \setminus B_s$) then there is a sequence $\{x_n : n \in \omega\} \subset A_s$ ($\{x_n : n \in \omega\} \subset B_s$) which converges to $x$. Let $t_n = \varphi^{-1}(x_n)$ for each $n \in \omega$; if the sequence $\{t_n : n \in \omega\}$ has a bounded subsequence then it will have an accumulation point in $[s, +\infty)$ (in $(-\infty, s]$) and hence $\{x_n : n \in \omega\}$ will have a accumulation point in $A_s$ (or in $B_s$ respectively) which is a contradiction. Therefore $t_n \to +\infty$ ($t_n \to -\infty$) and hence $x \in A$ ($x \in B$). As a consequence, $\overline{A}_s \subset A_s \cup A \subset A_s \cup \varphi([-r, r]) = A_s$ (or $\overline{B}_s \subset B_s \cup B \subset B_s \cup \varphi([-r, r]) = B_s$), so the set $A_s$ (or $B_s$ respectively) is closed in $L$, i.e., (11) is proved.

(12) If $\varphi^{-1}(A)$ (or $\varphi^{-1}(B)$) is bounded in $\mathbb{R}$ then there exists $r \in \mathbb{R}$ such that
$\varphi^{-1}(B) \subset [r, +\infty)$ (or $\varphi^{-1}(A) \subset (-\infty, r]$ respectively).

If $\varphi^{-1}(B)$ does not have a lower bound (or $\varphi^{-1}(A)$ does not have an upper bound respectively) then apply (11) and (7) to find a sequence $\{t_n : n \in \omega\} \subset \mathbb{R}\backslash\varphi^{-1}(B)$ (or $\{t_n : n \in \omega\} \subset \mathbb{R}\backslash\varphi^{-1}(A)$) such that $A_{t_0}$ is compact ($B_{t_0}$ is compact) and $t_{n+1} < t_n - 1$ ($t_{n+1} > t_n + 1$) while $[t_{n+1}, t_n] \cap \varphi^{-1}(B) \neq \emptyset$ (or $\varphi^{-1}(A) \cap [t_n, t_{n+1}] \neq \emptyset$) for every $n \in \omega$. Let $P_0 = A_{t_0} \cap B$ (or $P_0 = B_{t_0} \cap A$ respectively) and consider the set $P_n = \varphi([t_{n+1}, t_n]) \cap B$ (or $P_n = \varphi([t_n, t_{n+1}]) \cap A$) for all $n \in \mathbb{N}$.

Then $\mathcal{P} = \{P_n : n \in \omega\}$ is a disjoint family of compact subsets of $B$ (of $A$) such that infinitely many elements of $\mathcal{P}$ are nonempty and $B = \bigcup\mathcal{P}$ (or $A = \bigcup\mathcal{P}$ respectively). Since $B$ is connected ($A$ is connected) this is a contradiction with Fact 5 of V.290, i.e., (12) is proved.

By (9) we can assume that $\varphi^{-1}(A)$ (or $\varphi^{-1}(B)$) is bounded in $\mathbb{R}$. The property (12) implies that there exists a number $r \in \mathbb{R}$ such that $\varphi^{-1}(B) \subset (r, +\infty)$ (or $\varphi^{-1}(A) \subset (-\infty, r)$ respectively). Apply the properties (10) and (11) to find $s \in \mathbb{R}$ such that the set $A_s$ is compact (or the set $B_s$ is compact respectively) and $\varphi|(-\infty, s) : (-\infty, s) \to L\backslash A_s$ (or $\varphi|(s, +\infty) : (s, +\infty) \to L\backslash B_s$) is a homeomorphism. The set $[s, +\infty)$ $((-\infty, s])$ condenses onto $A_s$ (onto $B_s$ respectively) which is compact, so we can apply the property (5) to conclude that $A_s \overset{l}{\sim} \mathbb{I}$ (or $B_s \overset{l}{\sim} \mathbb{I}$). This proves that

(13) there is a closed set $F \subset L$ such that $F \overset{l}{\sim} \mathbb{I}$ and $L\backslash F$ is homeomorphic to an open subset of $\mathbb{R}$.

Finally, apply Problem 374 to the set $F$ from (13) to see that $L \overset{l}{\sim} F \oplus \alpha(L\backslash F)$. By Problem 381, the space $\alpha(L\backslash F)$ has to be $l$-equivalent to the space $\mathbb{I}$, so $L \overset{l}{\sim} \mathbb{I} \oplus \mathbb{I} \overset{l}{\sim} \mathbb{I}$ (see Problem 379) and hence our solution is complete.

**V.386.** *Assume that $X$ is a second countable space and* $\dim X = n \in \omega$. *Let* $\mathcal{O} = \{U \in \tau(X) : \dim U < n\}$ *and* $O = \bigcup\mathcal{O}$. *The set* $K(X) = X\backslash O$ *is called the dimensional kernel of $X$. Prove that* $\dim O < n$ *and* $\dim W = n$ *for any nonempty open subset $W$ of the space $K(X)$.*

**Solution.** The space $X$ is hereditarily Lindelöf, so there exists a countable family $\mathcal{U} \subset \mathcal{O}$ such that $O = \bigcup\mathcal{U}$. For each $U \in \mathcal{U}$ fix a countable family $\mathcal{F}_U$ of closed subsets of $X$ such that $U = \bigcup\mathcal{F}_U$; the family $\mathcal{F} = \bigcup\{\mathcal{F}_U : U \in \mathcal{U}\}$ is countable and $\bigcup\mathcal{F} = O$. Since $\dim F \leq n - 1$ for each $F \in \mathcal{F}$, we can apply Problem 150 to convince ourselves that $\dim O \leq n - 1 < n$.

Assume that $\dim W \leq n - 1$ for some nonempty open set $W$ of the space $K(X)$. Using again second countability of the space $X$ we can find a countable family $\mathcal{H}$ of closed subsets of $K(X)$ such that $\bigcup\mathcal{H} = W$. The set $G = O \cup W$ is open in $X$; besides, the elements of $\mathcal{H}$ are closed in $X$, so it follows from $G = \bigcup(\mathcal{F} \cup \mathcal{H})$ that $\dim G \leq n - 1$ (see Problem 150) and hence $G \in \mathcal{O}$. Since also $G\backslash O = G\backslash(\bigcup\mathcal{O}) = W \neq \emptyset$, we obtained a contradiction which shows that $\dim W = n$ for any $W \in \tau^*(K(X))$.

**V.387.** *Prove that if $n \in \mathbb{N}$ and a space $X$ is $l$-equivalent to $\mathbb{I}^n$ then the dimensional kernel $K(X)$ of the space $X$ is also $l$-equivalent to $\mathbb{I}^n$.*

**Solution.** The space $X$ is compact by Problem 138 and $\dim X = n$ by Problem 180. Since network weight is preserved even by $t$-equivalence, the space $X$ is metrizable. Apply Problem 273 to see that $X \times (\omega + 1)$ is $l$-equivalent to $\mathbb{I}^n \times (\omega + 1) \overset{l}{\sim} \mathbb{I}^n \overset{l}{\sim} X$ (see Problem 383). Therefore $X \times (\omega + 1)$ is $l$-equivalent to $X$.

By Problem 309, some nonempty open subspace of $\mathbb{I}^n$ embeds in $X$; since $\mathbb{I}^n$ embeds in every open subspace of $\mathbb{R}^n$, there exists a set $F \subset X$ which is homeomorphic to $\mathbb{I}^n$. If $U = F \setminus K(X) \neq \emptyset$ then $U$ is an $n$-dimensional open subspace of $F$ and $U \subset X \setminus K(X)$ while $\dim(X \setminus K(X)) < n$ (see Problem 386); this contradiction shows that $F \subset K(X)$. Apply Problem 373 to see that there exists a linear topological space $I$ such that $C_p(K(X)) \approx I \times C_p(F) \approx I \times C_p(\mathbb{I}^n) \approx I \times C_p(\mathbb{I}^n) \times C_p(\mathbb{I}^n)$ (see Problem 379). This shows that

(1)  $C_p(K(X)) \approx C_p(K(X)) \times C_p(\mathbb{I}^n)$.

Since $K(X) \times (\omega + 1)$ embeds in $X \times (\omega + 1)$, it follows from Problem 373 that there exists a linear topological space $E$ such that $C_p(X \times (\omega + 1)) \approx E \times C_p(K(X) \times (\omega + 1))$. Observe that $(K(X) \times (\omega + 1)) \oplus K(X)$ is homeomorphic to $K(X) \times (\omega + 1)$, so we have $C_p(\mathbb{I}^n) \approx C_p(X) \approx C_p(X \times (\omega + 1)) \approx E \times C_p(K(X) \times (\omega + 1))$ and it follows from $C_p(K(X) \times (\omega + 1)) \approx C_p(K(X) \times (\omega + 1)) \times C_p(K(X))$ that the space $C_p(\mathbb{I}^n)$ is linearly homeomorphic to $E \times C_p(K(X) \times (\omega + 1)) \times C_p(K(X))$ which in turn is linearly homeomorphic to $C_p(X \times (\omega + 1)) \times C_p(K(X)) \approx C_p(X) \times C_p(K(X))$. This, together with (1), shows that $C_p(\mathbb{I}^n) \approx C_p(\mathbb{I}^n) \times C_p(K(X)) \approx C_p(K(X))$ and hence $C_p(\mathbb{I}^n) \overset{l}{\approx} C_p(K(X))$, i.e., $K(X)$ is $l$-equivalent to $\mathbb{I}^n$ as promised.

**V.388.** *Call a second countable space $Y$ weakly $n$-Euclidean if $\dim Y = n$ and every $n$-dimensional subspace of $Y$ has nonempty interior and contains a homeomorphic copy of $\mathbb{I}^n$. Prove that a compact space $X$ is $l$-equivalent to $\mathbb{I}^n$ if and only if its dimensional kernel $K(X)$ has a nonempty open weakly $n$-Euclidean subspace and every $U \in \tau^*(K(X))$ contains a subset which is $l$-equivalent to $X$.*

**Solution.** Suppose that $X$ is $l$-equivalent to $\mathbb{I}^n$ and hence $\dim X = n$. By Problem 309, some $W \in \tau^*(K(X))$ embeds in $\mathbb{I}^n$; let $W' \subset \mathbb{I}^n$ be homeomorphic to $W$. Since $\dim W' = n$ (see Problem 386), the interior $V'$ of the set $W'$ in $\mathbb{I}^n$ is nonempty (see Problem 160) and hence there exists $V \in \tau^*(W)$ which is homeomorphic to $V'$. Thus $V$ is a nonempty open subspace of $K(X)$ and $V$ is a homeomorphic to an open subspace of $\mathbb{I}^n$. It follows from Problem 160 and Fact 1 of V.379 that $V$ is weakly $n$-Euclidean.

Now fix any $U \in \tau^*(K(X))$; by Problem 309 some set $G \in \tau^*(U)$ embeds in $\mathbb{I}^n$. Let $G' \subset \mathbb{I}^n$ be homeomorphic to $G$. Since $\dim G' = n$ (see Problem 386), the interior of $G'$ in $\mathbb{I}^n$ is nonempty by Problem 160; this, together with Fact 1 of V.379 shows that $\mathbb{I}^n$ embeds in $G'$ and hence in $G \subset U$. If $F \subset U$ is a homeomorphic copy of $\mathbb{I}^n$ then $F$ is $l$-equivalent to $X$ and hence we proved necessity.

Now assume that some $W \in \tau^*(K(X))$ is weakly $n$-Euclidean and every subspace $U \in \tau^*(K(X))$ contains a subset which is $l$-equivalent to $X$. The space $\mathbb{I}^n$ embeds in $W$ and hence in $X$; by Problem 373, there exists a linear topological space $E$ such that $C_p(X) \approx E \times C_p(\mathbb{I}^n) \approx E \times C_p(\mathbb{I}^n) \times C_p(\mathbb{I}^n) \approx C_p(X) \times C_p(\mathbb{I}^n)$, i.e., we proved that

(1) $C_p(X) \approx C_p(X) \times C_p(\mathbb{I}^n)$.

Let $F \subset W$ be a homeomorphic copy of $\mathbb{I}^n$; since $W$ is weakly $n$-Euclidean, we can find a set $U \in \tau^*(W)$ such that $U \subset F$. The set $U$ is also open in $K(X)$, so there exists $Y \subset U$ with $Y \overset{l}{\sim} X$. The space $Y$ embeds in $\mathbb{I}^n$; it is an easy exercise to see that $\alpha(Y \times \omega)$ embeds in $\alpha(\mathbb{I}^n \times \omega)$ which, in turn, embeds in $\mathbb{I}^n$ (see Fact 1 of V.379). Therefore $\alpha(Y \times \omega)$ embeds in $\mathbb{I}^n$ and hence we can apply Problem 376 to convince ourselves that $C_p(\mathbb{I}^n) \approx C_p(\mathbb{I}^n) \times C_p(Y)$. This, together with the property (1), shows that $C_p(X) \approx C_p(\mathbb{I}^n) \times C_p(X) \approx C_p(\mathbb{I}^n) \times C_p(Y) \approx C_p(\mathbb{I}^n)$ and hence $X \overset{l}{\sim} \mathbb{I}^n$, i.e., we settled sufficiency.

**V.389.** *Given (linear) topological spaces $X$ and $Y$ the expression $X \sim Y$ says that they are (linearly) homeomorphic. Suppose that $X^\omega \sim X$ and there exist (linear) topological spaces $E$ and $F$ such that $X \sim Y \times F$ and $Y \sim X \times E$. Prove that $X \sim Y$.*

**Solution.** Observe first that the usual properties of the Tychonoff product imply that $Z \times T \sim T \times Z$ for any (linear) topological spaces $Z$ and $T$.

It follows from $Y \sim X \times E$ that $Y \sim X^\omega \times E \sim X \times X^\omega \times E \sim X \times Y$. On the other hand, $X^\omega \sim (Y \times F)^\omega \sim Y^\omega \times F^\omega \sim Y \times Y^\omega \times F^\omega \sim Y \times X^\omega$. Recalling that $X^\omega \sim X$ we conclude that $X \sim Y \times X^\omega \sim Y \times X \sim X \times Y \sim Y$, so $X \sim Y$.

**V.390.** *Suppose that $L$ is a linear topological space, $M$ is a linear subspace of $L$ and there exists a linear retraction $r : L \to M$. Prove that $L \approx M \times r^{-1}(0)$. Deduce from this fact that for any linear topological spaces $L$ and $E$ there exists a linear topological space $N$ such that $L \approx E \times N$ (i.e., $E$ is a linear topological factor of $L$) if and only if there exists a linear retract $E'$ of the space $L$ such that $E' \approx E$.*

**Solution.** Denote the space $r^{-1}(0)$ by $N$ and let $\xi(u, v) = u + v$ for any point $(u, v) \in M \times N$. It is evident that $\xi : M \times N \to L$ is a linear map. From continuity of operations in $L$ it follows that $\xi$ is continuous. Let $\eta(x) = (r(x), x - r(x))$ for every $x \in L$; it is easy to see that $\eta : L \to M \times N$ is also a linear map. Besides, $\eta$ is continuous being the diagonal product of continuous maps. It is straightforward that $(\eta \circ \xi)(u, v) = (u, v)$ for any $(u, v) \in M \times N$ and $(\xi \circ \eta)(x) = x$ for each $x \in L$, so $\eta$ and $\xi$ are mutually inverse linear homeomorphisms. As a consequence, $L \approx M \times N = M \times r^{-1}(0)$. This implies, in particular, that any linear topological space which is linearly homeomorphic to a linear retract of $L$ is a linear topological factor of $L$.

Now assume that $E$ is a linear topological factor of $L$, i.e., $L$ is linearly homeomorphic to $E \times N$ for some linear topological space $N$. There is no loss of generality to consider that $L = E \times N$. The linear subspace $E' = \{(a, \mathbf{0}_N) : a \in E\}$ of the space $L$ is easily seen to be linearly homeomorphic to $E$. Given any point $x = (a, b) \in L$ let $r(x) = (a, \mathbf{0}_N)$; then $r : L \to E'$ is a linear continuous retraction (the relevant verification can be left to the reader).

**V.391.** *Given linear topological spaces $L, M$ and $N$ prove that $L \approx M \times N$ if and only if there exist linear subspaces $M'$ and $N'$ of the space $L$ for which $M' \approx M$, $N' \approx N$ and there exist linear retractions $r : L \to M'$ and $s : L \to N'$ such that $r(x) + s(x) = x$ for any $x \in L$.*

**Solution.** Suppose first that $L \approx M \times N$ and hence we can identify $L$ with $M \times N$. If $M' = \{(a, \mathbf{0}_N) : a \in M\}$ then $M'$ is a linear subspace of $L$ which is easily seen to be linearly homeomorphic to $M$. Analogously, $N' = \{(\mathbf{0}_M, b) : b \in N\}$ is a linear subspace of $L$ which is linearly homeomorphic $N$. For any $x = (a, b) \in L$ let $r(x) = (a, \mathbf{0}_N)$ and $s(x) = (\mathbf{0}_M, b)$. Then $r : L \to M'$ and $s : L \to N'$ are linear retractions such that $r(x) + s(x) = x$ for every $x \in L$, so we settled necessity.

Now assume that we have linear subspaces $M'$ and $N'$ of the space $L$ for which $M' \approx M$, $N' \approx N$ and there exist linear retractions $r : L \to M$ and $s : L \to N$ such that $r(x) + s(x) = x$ for every $x \in L$. Let $\xi(x) = (r(x), s(x))$ for each $x \in L$; then $\xi : L \to M' \times N'$ is a linear continuous map. For any point $x = (a, b) \in M' \times N'$ let $\eta(x) = a + b$; by continuity of the operations in $L$ the map $\eta : M' \times N' \to L$ is continuous. We omit a trivial verification of the fact that $\xi$ and $\eta$ are mutually inverse maps and hence each one of them is a linear homeomorphism. Therefore $L \approx M' \times N' \approx M \times N$, i.e., we proved sufficiency.

**V.392.** *For any linear topological space $L$ denote by $L^*$ the set of all continuous linear functionals on $L$ with the topology inherited from $C_p(L)$. Prove that, for any locally convex spaces $M$ and $N$, we have $(M \times N)^* \approx M^* \times N^*$. In particular, $C_p(Y)$ is a linear topological factor of $C_p(X)$ if and only if $L_p(Y)$ is a linear topological factor of $L_p(X)$.*

**Solution.** Denote the space $M \times N$ by $L$; by Problem 391 we can consider that $M$ and $N$ are linear subspaces of $L$ for which there exist linear retractions $r : L \to M$ and $s : L \to N$ such that $r(x) + s(x) = x$ for every $x \in L$. Let $\pi_M : C_p(L) \to C_p(M)$ and $\pi_N : C_p(L) \to C_p(N)$ be the relevant restriction maps.

If $f \in M^*$ then $\xi(f) = f \circ r$ is a continuous linear functional on $L$, so the map $\xi : M^* \to L^*$ is a linear embedding (see TFS-163) and hence the space $M' = \xi(M^*)$ is linearly homeomorphic to $M^*$. Analogously, let $\eta(f) = f \circ s$ for each $f \in N^*$; then $\eta : N^* \to L^*$ is also a linear embedding and therefore $N' = \eta(N^*) \approx N^*$.

Let $r_0 = \xi \circ \pi_M$ and $s_0 = \eta \circ \pi_N$; the restriction maps being linear and continuous, the maps $r_0 : L^* \to L^*$ and $s_0 : L^* \to L^*$ are linear and continuous as well. It follows from $M^* = \pi_M(L^*)$ and $N^* = \pi_N(L^*)$ (see Problem 224) that $r_0(L^*) = M'$ and $s_0(L^*) = N'$, so we can consider that $r_0 : L^* \to M'$ and $s_0 : L^* \to N'$.

If $h \in M'$ then $h = f \circ r$ for some $f \in M^*$; the map $r$ being a retraction, we have $f(r(x)) = f(x)$ for any $x \in M$ and hence $\pi_M(h) = f$. An immediate consequence is that $r_0(h) = f \circ r = h$, i.e., $r_0(h) = h$ for any $h \in M'$ and hence $r_0$ is a linear retraction. Analogously, the map $s_0$ is also a linear retraction.

Take any functional $f \in L^*$ and $x \in L$; since $r(x) + s(x) = x$, we conclude that $f(r(x) + s(x)) = f(x) = f(r(x)) + f(s(x))$. This shows that the equality $r_0(f)(x) + s_0(f)(x) = f(x)$ holds for any $x \in L$ and therefore $r_0(f) + s_0(f) = f$ for every $f \in L^*$. Applying Problem 391 to the retractions $r_0$ and $s_0$ we conclude that $L^* \approx M' \times N' \approx M^* \times N^*$ as promised.

Finally, if $C_p(X) \approx C_p(Y) \times L$ for some locally convex space $L$ then apply TFS-197 to see that we have the equalities $(C_p(Y))^* = L_p(Y)$ and $(C_p(X))^* = L_p(X)$, so $L_p(X) = (C_p(X))^* \approx (C_p(Y))^* \times L^* \approx L_p(Y) \times L^*$ and hence $L_p(Y)$ is a linear topological factor of $C_p(X)$. If $L_p(Y)$ is a linear topological factor of $L_p(X)$ then there is a locally convex space $M$ such that $L_p(X) \approx L_p(Y) \times M$. Recall that $(L_p(X))^* \approx C_p(X)$ and $(L_p(Y))^* \approx C_p(Y)$ (see Problem 235); as a consequence, $C_p(X) = (L_p(X))^* \approx (L_p(Y))^* \times M^* \approx C_p(Y) \times M^*$ which shows that $C_p(Y)$ is a linear topological factor of $C_p(X)$.

**V.393.** *Suppose that $X$ is a second countable non-compact space, $n \in \mathbb{N}$ and there exists a locally finite cover $\mathcal{I}$ of the space $X$ such that $\bigcup\{Int(I) : I \in \mathcal{I}\} = X$ and every $I \in \mathcal{I}$ is homeomorphic to $\mathbb{I}^n$. Prove that $X$ is l-equivalent to $\mathbb{I}^n \times \omega$.*

**Solution.** As usual, the expression $Z \simeq T$ says that the spaces $Z$ and $T$ are homeomorphic. If $L$ and $M$ are linear topological spaces then $M$ *is a linear topological factor of $L$* if there exists a linear topological space $E$ such that $L \approx M \times E$.

Let $Y = \mathbb{I}^n \times \omega$; then $Y \times \omega \simeq Y$ and hence

(1) the space $(C_p(Y))^\omega$ is linearly homeomorphic to $C_p(Y)$.

The space $X$ is non-compact and second countable, so it is not pseudocompact; take a discrete family $\{U_n : n \in \omega\}$ of nonempty open subsets of $X$ and fix a point $x_n \in U_n$ for each $n \in \omega$. Choose a set $P_n \in \mathcal{I}$ such that $x_n \in P_n$ for each $n \in \omega$. The set $P_n \cap U_n$ is open in $P_n$ and nonempty, so it contains a set $W_n$ which is homeomorphic to a nonempty open subset of $\mathbb{R}^n$; an immediate consequence is that $\mathbb{I}^n$ embeds in $W_n$ (see Fact 1 of V.379) for every $n \in \omega$. Take a set $F_n \subset W_n$ such that $F_n \simeq \mathbb{I}^n$ for each $n \in \omega$. The family $\{F_n : n \in \omega\}$ being discrete, the set $F = \bigcup\{F_n : n \in \omega\}$ is closed in $X$ and homeomorphic to $\mathbb{I}^n \times \omega = Y$. This shows that the space $Y$ embeds in $X$ as a closed subspace and hence we can apply Problem 373 to see that

(2) $C_p(Y)$ is a linear topological factor of $C_p(X)$.

The family $\mathcal{I}$ is easily seen to be countably infinite; let $\{I_n : n \in \omega\}$ be a faithful enumeration of $\mathcal{I}$. It is easy to find for each $n \in \omega$, a function $p_n \in C_p(X, [0, 1])$ such that $p_n^{-1}(0) = X \setminus Int(I_n)$. Let $p(x) = \sum_{n \in \omega} p_n(x)$ for every $x \in X$. Since $\mathcal{I}$ is locally finite, the number of summands in the definition of $p(x)$ is finite, so $p(x)$ makes sense for every $x \in X$.

Given any point $x \in X$ there exists a set $G \in \tau(x, X)$ such that the set $A = \{n \in \omega : I_n \cap G \neq \emptyset\}$ is finite. Then $p(y) = \sum\{p_n(y) : n \in A\}$ for all $y \in G$, so $f$ is continuous on the set $G$; applying Fact 1 of S.472 we convince ourselves that $p$ is a continuous function. Since $\bigcup\{Int(I_n) : n \in \omega\} = X$, we have $p(x) > 0$ for any

$x \in X$, so $q_n = \frac{p_n}{p}$ is a continuous function on $X$ for each $n \in \omega$. Observe also that $\sum_{n \in \omega} q_n(x) = 1$ and the set $B(x) = \{n \in \omega : x \in I_n\}$ is nonempty and finite for every $x \in X$.

Let $I = \bigcup \{I_n \times \{n\} : n \in \omega\}$ where the topology of $I$ is inherited from $X \times \omega$; evidently, the space $I$ is homeomorphic to $Y$. For any $a = (x, n) \in I$ let $\xi(a) = x$. This gives us a continuous onto map $\xi : I \to X$. Apply Problem 234 to find a continuous linear map $\Xi : L_p(I) \to L_p(X)$ such that $\Xi | I = \xi$.

Given $x \in X$ let $\varphi_n(x) = (x, n)$ for each $n \in B(x)$ and consider the point $\varphi(x) = \sum \{q_n(x)\varphi_n(x) : n \in B(x)\} \in L_p(I)$. Our next step is to prove that the map $\varphi : X \to L_p(I)$ is continuous; let $\pi_f(u) = u(f)$ for all $f \in C_p(I)$ and $u \in L_p(I)$.

Fix any $f \in C_p(I)$; since $L_p(I) \subset C_p(C_p(I))$, it suffices to show that $\pi_f \circ \varphi$ is continuous (see TFS-102). Let $f_n(x) = f(x, n)$ for each $x \in I_n$; it is immediate that $f_n \in C_p(I_n)$, so we can find a function $g_n \in C_p(X)$ such that $g_n | I_n = f_n$ for every $n \in \omega$. If $x \in X$ then there exists $W \in \tau(x, X)$ such that the set $A = \{n \in \omega : I_n \cap W \neq \emptyset\}$ is finite. For every point $y \in W$ we have the equalities $\pi_f(\varphi(y)) = \sum \{q_n(y)f(\varphi_n(y)) : n \in B(y)\} = \sum \{q_n(y)g_n(y) : n \in A\}$ because $B(y) \subset A$ and $q_n(y) = 0$ whenever $n \in A \backslash B(y)$.

Therefore $\pi_f \circ \varphi$ is continuous on $W$ because it coincides on $W$ with the continuous function $\sum_{n \in A} q_n g_n$. By Fact 1 of S.472 the function $\pi_f \circ \varphi$ is continuous for each $f \in C_p(I)$, so $\varphi$ is a continuous map.

By Problem 234 there exists a linear continuous map $\Phi : L_p(X) \to L_p(I)$ such that $\Phi | X = \varphi$; then $r = \Phi \circ \Xi : L_p(I) \to L_p(I)$ is also a linear continuous map. Observe that $\Xi(\varphi(x)) = \sum_{n \in B(x)} q_n(x)\Xi(\varphi_n(x)) = (\sum_{n \in B(x)} q_n(x)) \cdot x = x$ for any $x \in X$. If $X' = \varphi(X)$ then $\varphi : X \to X'$ and $\Xi | X' : X' \to X$ are mutually inverse continuous maps, so $\varphi$ is an embedding and it follows from the definition of $r$ that $r(\varphi(x)) = \varphi(x)$ for every $x \in X$. In other words,

(3) $r(y) = y$ for all $y \in X'$.

If $M$ is the linear hull of $X'$ in $L_p(I)$ then it follows from (3) and linearity of $r$ that $r(y) = y$ for all $y \in M$; observe also that $r(L_p(I)) = \Phi(L_p(X)) = M$, so $M$ is a linear retract of $L_p(I)$.

Take any nonempty finite set $E \subset X$; the family $\mathcal{P}_E = \{\xi^{-1}(x) : x \in E\}$ is disjoint so $P_E = \bigcup \mathcal{P}_E$ is linearly independent in $L_p(I)$. If $u$ is a nontrivial linear combination of the set $\varphi(E)$ then $u$ is a nontrivial linear combination of $P_E$, so $u \neq \mathbf{0}$ because $P_E$ is linearly independent. This shows that

(4) $X'$ is linearly independent and hence it is a linear basis in $M$.

It follows from the property (4) that $\Phi$ is an injection. We already saw that the map $\Xi | X' : X' \to X$ is the inverse of $\varphi$. Therefore $\Xi | M : M \to L_p(X)$ is the inverse of $\Phi : L_p(X) \to M$ which shows that $\Phi$ is an embedding. In particular, $M \approx L_p(X)$, i.e., $L_p(X)$ is linearly homeomorphic to a linear retract of $L_p(I)$. This implies that $L_p(X)$ is a linear topological factor of $L_p(I)$ (see Problem 390);

applying Problem 392 we conclude that $C_p(X)$ is a linear topological factor of $C_p(I)$. Since $I \simeq Y$, we proved that

(5) $C_p(X)$ is a linear topological factor of $C_p(Y)$.

Finally observe that the properties (1)–(2) and (5) together with Problem 389 imply that $C_p(X)$ is linearly homeomorphic to $C_p(Y) = C_p(\mathbb{I}^n \times \omega)$, so our solution is complete.

**V.394.** *Prove that, for any $n \in \mathbb{N}$, every nonempty open subspace of $\mathbb{R}^n$ is $l$-equivalent to $\mathbb{I}^n \times \omega$. In particular, if $U \in \tau^*(\mathbb{R}^n)$ then $U \overset{l}{\sim} \mathbb{R}^n$.*

**Solution.** The set $H_n = (-2^{-n}, 2^{-n})^n$ is open in $\mathbb{R}^n$ for every $n \in \omega$ and it is easy to see that the family $\{H_n : n \in \omega\}$ is a local base of $\mathbb{R}^n$ at $\mathbf{0}$. Consequently, $\mathcal{B}(x) = \{x + H_n : n \in \omega\}$ is a local base of $\mathbb{R}^n$ at the point $x$ for any $x \in \mathbb{R}^n$. Therefore the family $\mathcal{B} = \bigcup\{\mathcal{B}(x) : x \in \mathbb{R}^n\}$ is a base in $\mathbb{R}^n$; if $\mathcal{H} = \{B \in \mathcal{B} : \overline{B} \subset U\}$ then $\mathcal{H}$ is a base in $U$. It is straightforward that $\overline{H}$ is homeomorphic to $\mathbb{I}^n$ for any $B \in \mathcal{H}$.

The space $U$ is $\sigma$-compact and hence Hurewicz, so we can apply CFS-050 to find a family $\mathcal{U} \subset \mathcal{H}$ such that $\mathcal{U}$ is locally finite in $U$ and $\bigcup \mathcal{U} = U$. The family $\mathcal{I} = \{\overline{H} : H \in \mathcal{U}\}$ is also locally finite in $U$ and every $I \in \mathcal{I}$ is homeomorphic to $\mathbb{I}^n$. Observe that $U$ is not compact (see Fact 1 of CFS-493) and the interiors of the elements of $\mathcal{I}$ cover the space $U$, so we can apply Problem 393 to conclude that $U$ is $l$-equivalent to $\mathbb{I}^n \times \omega$. Therefore $U \overset{l}{\sim} \mathbb{I}^n \times \omega \overset{l}{\sim} \mathbb{R}^n$ for any $U \in \tau^*(\mathbb{R}^n)$.

**V.395.** *Given spaces $X$ and $Y$ assume that $nw(X) = \omega$ and $C_p(Y)$ is a linear topological factor of $C_p(X)$. Prove that $\dim Y \le \dim X$.*

**Solution.** Say that the *local dimension of a space $Z$ is at most $k$* if, for any $z \in Z$ there is a set $U \in \tau(z, Z)$ such that $\dim U \le k$. We will also need the subspace $R_* = \mathbb{R}\backslash\{0\}$ of the space $\mathbb{R}$.

*Fact 1.* If $Z$ is a Lindelöf space then $\dim Z \le k$ if and only if the local dimension of $Z$ is at most $k$.

*Proof.* If $\dim Z \le k$ then $U = Z$ is the required neighborhood for any $z \in Z$. Now, suppose that for any $z \in Z$ there a set $U_z \in \tau(z, Z)$ such that $\dim U_z \le k$. Choose a set $V_z \in \tau(z, Z)$ such that $\overline{V}_z \subset U_z$; then $\dim(\overline{V}_z) \le k$ for every $z \in Z$. By the Lindelöf property of $Z$ there is a countable $A \subset Z$ such that $Z = \bigcup\{V_z : z \in A\}$; then $Z = \bigcup\{\overline{V}_z : z \in A\}$, so it follows from the countable sum theorem (see Problem 150) that $\dim Z \le k$, i.e., Fact 1 is proved.

Returning to our solution observe that if $\dim X = \infty$ then there is nothing to prove, so assume that $\dim X = n \in \omega$. Since $C_p(Y)$ embeds in $C_p(X)$, we have $nw(Y) = nw(C_p(Y)) \le nw(C_p(X)) \le \omega$, so $Y$ also has a countable network.

By Problem 392, the space $L_p(Y)$ has to be a linear topological factor of $L_p(X)$, so we can consider that $L_p(Y)$ is a linear subspace of $L_p(X)$ and there exists a linear retraction $r : L_p(X) \to L_p(Y)$ (see Problem 390). For any point $y \in Y$ there exists a unique number $k \in \mathbb{N}$ such that for some uniquely determined distinct points $x_1, \ldots, x_k \in X$ and numbers $\lambda_1, \ldots, \lambda_k \in R_*$ we have $y = \lambda_1 x_1 + \ldots + \lambda_k x_k$;

let $l_X(y) = k$ and $\xi_X(y) = \{x_1, \ldots, x_k\}$. For any $u \in L_p(X)$ there exists a unique number $m \in \omega$ such that for some $\mu_1, \ldots, \mu_m \in R_*$ and uniquely determined distinct points $y_1, \ldots, y_m \in Y$ we have $r(u) = \mu_1 y_1 + \ldots + \mu_m y_m$; let $l_Y(u) = m$ and $\xi_Y(u) = \{y_1, \ldots, y_m\}$. Note that $m = 0$ means that $r(u) = \mathbf{0}$ and $\xi_Y(u) = \emptyset$. The set $L_m = \{u \in L_p(Y) : l_Y(u) \le m\}$ is closed in $L_p(Y)$ (see Fact 1 of U.485); let $L_{-1} = \emptyset$ and $L'_m = L_m \backslash L_{m-1}$ for every $m \in \omega$.

The set $Y_k = \{y \in Y : l_X(y) \le k\}$ is closed in $Y$ for any $k \le \omega$ by Fact 1 of U.485. Since $Y = \bigcup_{k \in \omega} Y_k$, it suffices to show that dim $Y_k \le n$ for every $k \in \omega$ (see Problem 150). The space $Y$ being perfect, $Y'_k = Y_k \backslash Y_{k-1}$ is an $F_\sigma$-set in $Y$; since also $Y_0 = \emptyset$, it suffices to show that dim $Y'_k \le n$ for all $k \in \mathbb{N}$.

Fix any number $k \in \mathbb{N}$; for any $\eta = (m_1, \ldots, m_k) \in \omega^k$ consider the set $O_\eta = \{y \in Y'_k :$ there exists an enumeration $\{x_1, \ldots, x_k\}$ of the set $\xi_X(y)$ such that $|\xi_Y(x_i)| \le m_i$ for all $i \le k\}$ and let $[\eta] = \sum_{i=1}^k m_i$. It is clear that

(1) $Y'_k = \bigcup\{O_\eta : \eta \in \omega^k\}$.

Given any $k$-tuple $\eta = (m_1, \ldots, m_k) \in \omega^k$ take a point $y \in Y'_k \backslash O_\eta$. Then $\xi_X(y) = \{x_1, \ldots, x_k\}$ and $|\xi_Y(x_i)| = j_i$ for each $i \le k$ while $j_1 + \ldots + j_k > [\eta]$. Since $L_{j_i - 1}$ is closed in $L_p(Y)$ for all $i \le k$, the set $U_i = \{v \in L_p(X) : r(v) \notin L_{j_i - 1}\}$ is open in $L_p(X)$ and $x_i \in U_i$ for all $i \le k$. Take a disjoint family $W_1, \ldots, W_k$ of open subsets of $X$ such that $x_i \in W_i \subset U_i$ for all $i \le k$.

The set $G = (R_* W_1 + \ldots + R_* W_k) \cap Y'_k$ is an open neighborhood of $y$ in $Y'_k$ (see Fact 2 of U.485). If $z \in G$ then $z = \lambda_1 a_1 + \ldots + \lambda_k a_k$ where $\lambda_i \in R_*$ and $a_i \in W_i$ for all $i \le k$. An immediate consequence is that $l_X(z) = \{a_1, \ldots, a_k\}$ and $l_Y(a_i) \ge j_i$ for all $i \le k$ which shows that $z \notin O_\eta$ and hence we proved that $y \in G \subset Y'_k \backslash O_\eta$, i.e., every point of $Y'_k \backslash O_\eta$ has a neighborhood in $Y'_k$ contained in $Y'_k \backslash O_\eta$. Therefore

(2) the set $O_\eta$ is closed in $Y'_k$ for any $\eta \in \omega^k$.

The properties (1) and (2) show that it suffices to prove that dim $O_\eta \le n$ for every $\eta \in \omega^k$. We will do that by induction on $m = [\eta]$. If $m = 0$ and $y \in O_\eta$ then $y$ is a linear combination of the set $\xi_X(y)$. Therefore $y = r(y)$ is a linear combination of $r(\xi_X(y)) = \{\mathbf{0}\}$, i.e., $y = \mathbf{0}$ which is a contradiction. Therefore $O_\eta = \emptyset$ and hence dim $O_\eta = -1 \le n$.

Now assume that $m > 0$ and dim $O_\eta \le n$ whenever $\eta \in \omega^k$ and $[\eta] < m$; fix a $k$-tuple $\eta = (m_1, \ldots, m_k)$ such that $[\eta] = m$. It follows from the property (2) that the set $F = \bigcup\{O_\nu : [\nu] < m\}$ is closed in $Y'_k$ and dim $F \le n$ (see Problem 150). The set $O_\eta \backslash F$ is open in $O_\eta$; since $Y$ has a countable network, $O_\eta \backslash F$ is an $F_\sigma$-subset of $O_\eta$, so it suffices to prove that $O' = O_\eta \backslash F$ has dimension at most $n$.

By Fact 1 it is sufficient to show that $O'$ has local dimension at most $n$, so fix a point $b \in O'$ and $a_1, \ldots, a_k \in X$ such that $\xi_X(b) = \{a_1, \ldots, a_k\}$. There is no loss of generality to assume that $|\xi_Y(a_i)| \le m_i$ for every $i \le k$, so it follows from $\sum_{i=1}^k |\xi_Y(a_i)| = m$ that $|\xi_Y(a_i)| = m_i$ for all $i \le k$.

Choose a disjoint family $\{G_1, \ldots, G_k\}$ of open subsets of $X$ such that $a_i \in G_i$ and $r(G_i) \cap L_{m_i - 1} = \emptyset$ for every $i \le k$. The set $G = (R_* G_1 + \ldots + R_* G_k) \cap O'$ is open in $O'$ by Fact 2 of U.485. If $y \in G$ then there is a unique point $x_i \in \xi_X(y) \cap G_i$; let $p_i(y) = x_i$ for every $i \leq k$.

This gives us a map $p_i : G \to G_i$; if $W$ is open in $G_i$ then it is straightforward that $p_i^{-1}(W) = (R_* G_1 + \ldots R_* G_{i-1} + R_* W + R_* G_{i+1} + \ldots + R_* G_k) \cap G$, so the map $p_i$ is continuous for any $i \leq k$.

Given $i \leq k$ let $\xi_Y(r(a_i)) = \{b_j^i : 1 \leq j \leq m_i\}$ and choose a disjoint family $\{H_j^i : j \leq m_i\}$ of open subsets of $Y$ such that $b_j^i \in H_j^i$ for all $j \leq m_i$. The set $H_i = \sum_{j=1}^{m_i} R_* H_j^i \subset L_{m_i}$ is open in $L_{m_i}$ and $r(a_i) \in H_i$. Besides, $r(p_i(G)) \subset L'_{m_i}$ and hence $H_i' = r^{-1}(H_i) \cap p_i(G)$ is open in $p_i(G)$ for each $i \leq k$. Therefore the set $V = \bigcap\{p_i^{-1}(H_i') : i \leq k\}$ is open in $G$ and hence in $O'$; it is easy to see that $b \in V$.

For any point $u \in H_i$ there is a unique point $y_j^i \in \xi_Y(u) \cap H_j^i$; let $q_j^i(u) = y_j^i$ for every $j \leq m_i$. The map $q_j^i : H_i \to H_j^i$ is continuous for every $j \leq m_i$; this can be proved in the same way as continuity of every map $p_i$. By our choice of the set $V$ the map $q_j^i \circ r \circ p_i$ is well defined on $V$; it is an easy exercise that the set $E_{ij} = \{c \in V : q_j^i(r(p_i(c))) = c\}$ is closed in $V$ for any $i \leq k$ and $j \leq m_i$.

If $y \in V$ then there are $x_1, \ldots, x_k \in X$ and $\lambda_1, \ldots, \lambda_k \in R_*$ such that $x_i \in H_i'$ for each $i \leq k$ and $y = \sum_{i=1}^k \lambda_i x_i$. Given $i \leq k$ there exist $y_1^i, \ldots, y_{m_i}^i \in Y$ such that $y_j^i \in H_j^i$ for all $j \leq m_i$ and $r(x_i) = \sum_{j=1}^{m_i} \mu_j^i y_j^i$. It follows from the equalities $y = r(y) = \sum_{i=1}^k \lambda_i r(x_i) = \sum_{i=1}^k \lambda_i (\sum_{j=1}^{m_i} \mu_j^i y_j^i)$ that $y = y_j^i$ for some $i \leq k$ and $i \leq m_i$; this shows that $y \in E_{ij}$ and hence

(3) $V = \bigcup\{E_{ij} : 1 \leq i \leq k, \ 1 \leq j \leq m_i\}$.

Observe that the map $(q_j^i \circ r)|p_i(E_{ij})$ is the inverse of the map $p_i|E_{ij}$; an immediate consequence is that $p_i$ embeds $E_{ij}$ in the space $X$ for any $i \leq k$ and $j \leq m_i$. It follows from Problem 155 that $\dim E_{ij} \leq \dim X = n$ for any $i \leq k$ and $j \leq m_i$. Applying Problem 150 and (3) we conclude that $\dim V \leq n$. Therefore $V$ witnesses that $O'$ has local dimension at most $n$ at any point $b \in O'$, so $\dim O' \leq n$ by Fact 1. This shows that $\dim O_\eta \leq n$ and completes our inductive proof. As a consequence, $\dim Y_k' \leq n$ for all $k \in \mathbb{N}$ and hence $\dim Y \leq n = \dim X$ as promised.

**V.396.** *Assuming that $X$ is a compact space and $C_p(Y)$ is a linear topological factor of $C_p(X)$ prove that* $\dim Y \leq \dim X$.

**Solution.** Given spaces $Z, Z'$ and a continuous onto map $\varphi : Z \to Z'$ the dual map $\varphi^* : C_p(Z') \to C_p(Z)$ is defined by $\varphi^*(f) = f \circ \varphi$ for any $f \in C_p(Z')$. Say that a set $L \subset C_p(Z)$ is *adequate* if there is a space $Z'$ and a continuous onto map $\varphi : Z \to Z'$ such that $L = \varphi^*(C_p(Z'))$. Given $n \in \omega$, a set $L \subset C_p(Z)$ is called *dimensionally $n$-adequate* if there exists a space $Z'$ and a continuous onto map $\varphi : Z \to Z'$ such that $\dim Z' \leq n$ and $L = \varphi^*(C_p(Z'))$.

*Fact 1.* Suppose that a space $K$ is compact, $L_i \subset C_p(K)$ and $L_i \subset L_{i+1}$ for every $i \in \omega$. Then

(a) if every $L_i$ is adequate then $L = \overline{\bigcup_{i \in \omega} L_i}$ is adequate;
(b) if every $L_i$ is dimensionally $n$-adequate then $L$ is also dimensionally $n$-adequate.

*Proof.* We will give a simultaneous proof for (a) and (b). Take a space $Z_i$ (with $\dim Z_i \leq n$ if we are proving (b)) for which there exists a continuous onto map $\varphi_i : K \to Z_i$ such that $L_i = \varphi_i^*(C_p(Z_i))$ for each $i \in \omega$. It follows from $L_i \subset L_{i+1}$ that there exists a continuous map $h_i : Z_{i+1} \to Z_i$ such that $h_i \circ \varphi_{i+1} = \varphi_i$ for all $i \in \omega$ (see TFS-163). For any $i, j \in \omega$ with $i < j$ let $\pi_i^j = h_i \circ \ldots \circ h_{j-1} : Z_j \to Z_i$; then $S = \{Z_i; \pi_i^j, i, j \in \omega\}$ is an inverse sequence of compact spaces. If $Z$ is the inverse limit of $S$ then $Z$ is compact (and $\dim Z \leq n$ (see Problem 174) if we are proving (b)). Let $\pi_i : Z \to Z_i$ be the limit projection for any $i \in \omega$.

It is easy to see that $\pi_i^j \circ \varphi_j = \varphi_i$ for any $i, j \in \omega$ with $i < j$. This, together with Problem 167, shows that the diagonal product $\varphi = \Delta_{i \in \omega} \varphi_i$ maps the space $K$ in $Z$.

Suppose that a point $a = (a_i : i \in \omega)$ belongs to $Z$ and let $F_i = \varphi_i^{-1}(a_i)$ for every $i \in \omega$. If $j > i$ and $x \in F_j$ then $\varphi_i(x) = \pi_i^j(\varphi_j(x)) = \pi_i^j(a_j) = a_i$ and hence $x \in F_i$. This proves that $\{F_i : i \in \omega\}$ is a decreasing family of nonempty compact sets; if $x \in \bigcap_{i \in \omega} F_i$ then $\varphi_i(x) = a_i$ for all $i \in \omega$, i.e., $\varphi(x) = a$. Consequently, $\varphi(K) = Z$; since also $\pi_i \circ \varphi = \varphi_i$, we can apply TFS-163 once more to see that $L_i \subset G = \varphi^*(C_p(Z))$ for any $i \in \omega$. The set $G$ is closed in $C_p(K)$, so $L \subset G$.

If $x$ and $y$ are distinct points of the space $Z$ then it follows from $Z \subset \prod_{i \in \omega} Z_i$ that there is $i \in \omega$ such that $\pi_i(x) \neq \pi_i(y)$. Take a function $f \in C_p(Z_i)$ with $f(\pi_i(x)) \neq f(\pi_i(y))$; then $g = f \circ \pi_i \in \pi_i^*(C_p(Z_i))$ and $g(x) \neq g(y)$. This proves that the set $A = \bigcup_{i \in \omega} \pi_i^*(C_p(Z_i))$ separates the points of $Z$; it is straightforward that $A$ is an algebra in $C_p(Z)$, so $A$ is dense in $C_p(Z)$ by TFS-192. As a consequence, $\varphi^*(A)$ is dense in $\varphi^*(C_p(Z)) = G$.

It follows from $\pi_i \circ \varphi = \varphi_i$ that $\varphi_i^* = \varphi^* \circ \pi_i^*$ and hence $L_i = \varphi^*(\pi_i^*(C_p(Z_i)))$ for any $i \in \omega$. Therefore the set $\bigcup_{i \in \omega} L_i = \varphi^*(A)$ is dense in $G$, so $L = G$ is an adequate set. In the case of (b) the set $G = \varphi^*(C_p(Z))$ is dimensionally $n$-adequate because $\dim Z \leq n$, so Fact 1 is proved.

Returning to our solution let $n = \dim X$; if $n = \infty$ then there is nothing to prove, so we can consider that $n \in \omega$. Note that the space $C_p(Y)$ embeds in $C_p(X)$, so $t(C_p(Y)) \leq t(C_p(X)) = \omega$ and hence $Y$ is Lindelöf by TFS-149. There exists a continuous linear surjection of $C_p(X)$ onto $C_p(Y)$ (see Problem 390), so $C_p(Y)$ is pseudocompact by Problem 193; as a consequence, $Y$ is compact. It follows from Problem 390 that we can assume that $C_p(Y)$ is a linear subspace of $C_p(X)$ and there exists a linear retraction $r : C_p(X) \to C_p(Y)$.

Fix any continuous onto map $\varphi : Y \to K$ of $Y$ onto a second countable space $K$. Then $L_0 = \varphi^*(C_p(K))$ is an adequate subspace of $C_p(Y)$ and $nw(L_0) = \omega$; it follows from Fact 5 of U.086 that there is an adequate subset $M_0'$ of the space $C_p(X)$ such that $L_0 \subset M_0'$ and $nw(M_0') = \omega$. Apply Problem 161 to find a set $M_0$ which is dimensionally $n$-adequate in $C_p(X)$ while $M_0' \subset M_0$ and $nw(M_0) \leq \omega$.

Proceeding inductively assume that $k \in \omega$ and we have sets $L_0, M_0, \ldots, L_k, M_k$ with the following properties:

(1) the set $L_i$ is adequate in $C_p(Y)$ and the set $M_i$ is dimensionally $n$-adequate in $C_p(X)$ for all $i \leq k$;
(2) $L_i \subset M_i$ and $nw(L_i) = nw(M_i) = \omega$ for all $i \leq k$;
(3) $M_i \subset M_{i+1}$ and $r(M_i) \subset L_{i+1}$ for every $i < k$.

Since $nw(r(M_k)) \leq nw(M_k) \leq \omega$, we can apply Fact 5 of U.086 once more to find an adequate subset $L_{k+1}$ of the space $C_p(Y)$ such that $r(M_k) \subset L_{k+1}$ and $nw(L_{k+1}) = \omega$. Analogously, we can find an adequate subset $M'_{k+1}$ of the space $C_p(X)$ such that $M_k \cup L_{k+1} \subset M'_{k+1}$ and $nw(M'_{k+1}) \leq \omega$. Apply Problem 161 to find a set $M_{k+1}$ which is dimensionally $n$-adequate in $C_p(X)$ while $M'_{k+1} \subset M_{k+1}$ and $nw(M_{k+1}) \leq \omega$. This completes the inductive step and shows that we can construct families $\{L_i : i \in \omega\}$ and $\{M_i : i \in \omega\}$ such that the properties (1)–(3) hold for all $k \in \omega$. Observe that it follows from (2) and (3) that $L_i = r(L_i) \subset r(M_i) \subset L_{i+1}$ for all $i \in \omega$.

By Fact 1 the set $L = \overline{\bigcup_{i \in \omega} L_i}$ is adequate in $C_p(Y)$ and the set $M = \overline{\bigcup_{i \in \omega} M_i}$ is dimensionally $n$-adequate in $C_p(X)$. It follows from (3) and continuity of $r$ that $r(M) \subset L$. However, the set $M$ is closed in $C_p(X)$ and $\bigcup_{i \in \omega} L_i \subset M$ by (2); this implies $L \subset M$ and hence $r(M) \supset r(L) = L$, i.e., $r(M) = L$.

There exist a compact space $Z$ and a continuous onto map $\xi : X \to Z$ such that $\dim Z \leq n$ and $M = \xi^*(C_p(Z))$. It follows from (2) that $w(Z) \leq \omega$. Analogously, there is a compact space $T$ and a continuous onto map $\eta : Y \to T$ such that $w(T) \leq \omega$ and $L = \eta^*(C_p(T))$. Since $L$ is a linear retract of $M$, the space $C_p(T)$ is linearly homeomorphic to a linear retract of $C_p(Z)$. By Problem 390, the space $C_p(T)$ is a topological linear factor of $C_p(Z)$; applying Problem 395 we obtain the inequalities $\dim T \leq \dim Z \leq n$.

Finally observe that $\varphi^*(C_p(K)) = L_0 \subset L = \eta^*(C_p(T))$. By TFS-163, there exists a continuous map $\mu : T \to K$ such that $\varphi = \mu \circ \eta$. Therefore, for any continuous map $\varphi$ of $Y$ onto a second countable space $K$ we found a space $T$ together with maps $\eta : Y \to T$ and $\mu : T \to K$ such that $\dim T \leq n$ and $\varphi = \mu \circ \eta$. Therefore $\dim Y \leq n = \dim X$ by Problem 161, so our solution is complete.

**V.397.** *(Open Mapping Theorem) Suppose that $L$ is a Banach space, $M$ is a linear topological space which is of second category in itself and $f : L \to M$ is a surjective continuous linear map. Prove that $f$ is open. In particular, any continuous linear onto map between two Banach spaces is open.*

**Solution.** There exists a norm $|| \cdot ||$ on the space $L$ which generates a complete metric on $L$. Let $B(r) = \{x \in L : ||X|| < r\}$ be the open $r$-ball centered at $\mathbf{0}_L$ for any $r > 0$. Let us establish first that

(1) $\mathbf{0}_M$ belongs to the interior of $\overline{f(B(r))}$ for any $r > 0$.

By linearity of the map $f$ we have $M = \bigcup\{n \cdot \overline{f(B(\frac{r}{2}))} : n \in \mathbb{N}\}$; the space $M$ being of second category in itself, the set $\overline{n \cdot f(B(\frac{r}{2}))}$ has nonempty interior for some $n \in \mathbb{N}$. The map $\varphi : M \to M$ defined by the equality $\varphi(x) = nx$ for any

$x \in M$ is a homeomorphism so the set $\overline{f(B(\frac{r}{2}))}$ also has nonempty interior $U$. Besides, we have the inclusions $-B(\frac{r}{2}) = B(\frac{r}{2})$ and $B(\frac{r}{2}) + B(\frac{r}{2}) \subset B(r)$ which imply that $f\left(B(\frac{r}{2})\right) - f\left(B(\frac{r}{2})\right) \subset f(B(r))$ by linearity of the map $f$. Apply Problem 203 to see that $\overline{f\left(B(\frac{r}{2})\right)} - \overline{f\left(B(\frac{r}{2})\right)} \subset \overline{f\left(B(\frac{r}{2})\right) - f\left(B(\frac{r}{2})\right)} \subset \overline{f(B(r))}$. The set $U - U$ is an open neighborhood of $\mathbf{0}_M$ and $U - U \subset \overline{f\left(B(\frac{r}{2})\right)} - \overline{f\left(B(\frac{r}{2})\right)} \subset \overline{f(B(r))}$ which shows that (1) is proved.

Our next step is to show that

(2) if $r_n > 0$ and $y_n \in \overline{f(B(r_n))}$ for any $n \in \omega$ then $r_n \to 0$ implies $y_n \to \mathbf{0}_M$.

Indeed, if $U \in \tau(\mathbf{0}_M, M)$ then there is $V \in \tau(\mathbf{0}_M, M)$ such that $\overline{V} \subset U$. By continuity of $f$, the exists $W \in \tau(\mathbf{0}_L, L)$ for which $f(W) \subset V$. Choose a number $r > 0$ with $B(r) \subset W$ and $m \in \omega$ such that $r_n < r$ for all $n \geq m$. Then $y_n \in \overline{f(B(r_n))} \subset \overline{f(B(r))} \subset \overline{f(W)} \subset \overline{V} \subset U$ for all $n \geq m$ which shows that $y_n \to \mathbf{0}_M$, i.e., (2) is proved.

The following property is crucial for our proof.

(3) $\overline{f(B(r))} \subset f(B(3r))$ for any $r > 0$.

Fix a point $y \in \overline{f(B(r))}$ and let $r_n = 2^{-n}r$ for all $n \in \omega$. It follows from (1) that $P = y - \overline{f(B(r_1))}$ is a neighborhood of $y$, so $P \cap f(B(r_0)) \neq \emptyset$ and hence we can find a point $x_0 \in B(r_0)$ in such a way that $f(x_0) \in P$ and therefore $y_0 = y - f(x_0) \in \overline{f(B(r_1))}$. Proceeding inductively, assume that $m \in \omega$ and we have chosen points $x_0, \ldots, x_m$ and $y_0, \ldots, y_m$ such that

(4) $x_i \in B(r_i)$ and $y_i \in \overline{f(B(r_{i+1}))}$ for all $i \leq m$;
(5) $y_i = y_{i-1} - f(x_i)$ for $1 < i \leq m$.

Since $y_m \in \overline{f(B(r_{m+1}))}$ and the set $P = y_m - \overline{f(B(r_{m+2}))}$ is a neighborhood of the point $y_m$ by the property (1), we have $P \cap f(B(r_{m+1})) \neq \emptyset$ and hence we can find a point $x_{m+1} \in B(r_{m+1})$ in such a way that $f(x_{m+1}) \in P$ and therefore $y_{m+1} = y_m - f(x_{m+1}) \in \overline{f(B(r_{m+2}))}$. It is easy to see that the properties (4) and (5) now hold if $m$ is replaced with $m + 1$. Thus our inductive procedure can be continued to construct sequences $\{x_i : i \in \omega\}$ and $\{y_i : i \in \omega\}$ such that the conditions (4) and (5) are satisfied for all $m \in \omega$.

Let $z_i = x_0 + \ldots + x_i$ for all $i \in \omega$. Given any numbers $n, m \in \omega$ such that $m < n$ we have $\|z_n - z_m\| \leq \sum_{i=1}^{n-m} \|x_{m+i}\| \leq \sum_{i=1}^{n-m} r_{m+i} \leq 2^{-m}r$ which easily implies that $\{z_i : i \in \omega\}$ is a Cauchy sequence. Therefore $z_i \to z$ for some $z \in L$; for any $i \in \omega$ we have $\|z_i\| \leq \sum_{j=1}^{i} \|x_j\| \leq r_0 + \ldots + r_i \leq 2r$, so $\|z\| \leq 2r < 3r$ and hence $z \in B(3r)$.

Given any number $n \in \omega$ it follows from the property (5) that we have the equalities $f(x_0) = y - y_0$, $f(x_1) = y_0 - y_1, \ldots, f(x_n) = y_{n-1} - y_n$ and therefore $f(z_n) = \sum_{i=1}^{n} f(x_i) = y - y_n$. The properties (2) and (4) show that $y_n \to \mathbf{0}_M$ and hence $f(z_n) \to y$. The function $f$ being continuous we have $f(z_n) \to f(z)$ which implies that $f(z) = y$. We showed that $y \in f(B(3r))$ for any point $y \in \overline{f(B(r))}$, i.e., $\overline{f(B(r))} \subset f(B(3r))$ and hence (3) is proved.

Finally, take an arbitrary set $U \in \tau(\mathbf{0}_L, L)$ and $s > 0$ such that $B(s) \subset U$. It follows from (1) that $V \subset \overline{f(B(\frac{s}{3}))}$ for some $V \in \tau(\mathbf{0}_M, M)$. Now apply (3) to convince ourselves that $V \subset \overline{f(B(\frac{s}{3}))} \subset f(B(s)) \subset f(U)$ and therefore we can apply Fact 3 of S.496 to conclude that $f$ is an open map.

**V.398.** *(Closed Graph Theorem) Suppose that $L$ and $M$ are Banach spaces and $f : L \to M$ is a linear map such that its graph $G = \{(x, f(x)) : x \in L\}$ is closed in $L \times M$. Prove that the map $f$ is continuous.*

**Solution.** Say that a norm $|| \cdot ||$ on a linear space $G$ is *complete* if the metric generated by this norm is complete.

*Fact 1.* Suppose that $|| \cdot ||_G$, $|| \cdot ||_H$ are complete norms on linear spaces $G$ and $H$ respectively. For any point $z = (x, y) \in G \times H$ let $||z|| = ||x||_G + ||y||_H$. Then $|| \cdot ||$ is a complete norm on $G \times H$ and the topology of the space $(G \times H, || \cdot ||)$ coincides with the topology of the product $(G, || \cdot ||_G) \times (H, || \cdot ||_H)$. In particular, the product of any two Banach spaces is a Banach space.

*Proof.* It is clear that $||z|| \geq 0$ for any $z \in E = G \times H$. If $z = (x, y) \in E$ and $||z|| = 0$ then $||x||_G = 0$ and $||y||_H = 0$ which shows that $x = \mathbf{0}_G$ and $y = \mathbf{0}_H$ and hence $z = (\mathbf{0}_G, \mathbf{0}_H) = \mathbf{0}_E$. If $t \in \mathbb{R}$ and $z = (x, y) \in E$ then the equalities $||tz|| = ||tx||_G + ||ty||_H = |t|(||x||_G + ||y||_H) = |t| \cdot ||z||$ show that the second axiom of the norm also holds.

Take any $a, b \in E$ with $a = (a_1, a_2)$ and $b = (b_1, b_2)$. By our definition of $|| \cdot ||$ we have $||a + b|| = ||a_1 + b_1||_G + ||a_2 + b_2||_H \leq ||a_1||_G + ||b_1||_G + ||a_2||_H + ||b_2||_H = ||a|| + ||b||$, so the axiom of triangle is fulfilled and hence $|| \cdot ||$ is, indeed, a norm on $E$.

Let us temporarily denote by $E'$ the space $E$ with the norm $||\cdot||$. For any $r > 0$ let $B_G(r) = \{x \in G : ||x||_G < r\}$ and $B_H(r) = \{y \in H : ||y||_H < r\}$; we will also need the $r$-ball $B(r) = \{z \in E : ||z|| < r\}$. If $U \in \tau(\mathbf{0}_E, E)$ then there exists $\varepsilon > 0$ such that $B_G(\varepsilon) \times B_H(\varepsilon) \subset U$. It is straightforward that $B(\varepsilon) \subset B_G(\varepsilon) \times B_H(\varepsilon) \subset U$, so the identity map $i : E' \to E$ is continuous by Fact 2 of S.496.

Now, if $U \in \tau(\mathbf{0}_E, E')$ then there exists $r > 0$ for which $B(r) \subset U$. It is easy to see that $B_G(\frac{r}{2}) \times B_H(\frac{r}{2}) \subset B(r)$, so $i(U)$ is a neighborhood of $\mathbf{0}_E$; this, together with Fact 3 of S.496, shows that the map $i$ is open. Thus $i : E' \to E$ is a homeomorphism, which is the same as saying that the topology of $E$ coincides which $\tau(E')$.

To prove completeness of $|| \cdot ||$ assume that $z_n = (x_n, y_n) \in E$ for all $n \in \omega$ and $\{z_n : n \in \omega\}$ is a Cauchy sequence with respect to the metric generated by $|| \cdot ||$. Given any $\varepsilon > 0$ there is $m \in \omega$ such that $||z_n - z_k|| < \varepsilon$ whenever $n, k \geq m$. Then $||x_n - x_k||_G \leq ||z_n - z_k|| < \varepsilon$ and $||y_n - y_k||_H \leq ||z_n - z_k|| < \varepsilon$ for all $n, k \geq m$ which shows that $\{x_n : n \in \omega\}$ and $\{y_n : n \in \omega\}$ are Cauchy sequences in their respective spaces. By completeness of the norms $|| \cdot ||_G$ and $|| \cdot ||_H$, we can find $x \in G$ and $y \in H$ such that $x_n \to x$ and $y_n \to y$. Therefore $z_n \to z = (x, y)$ in the product space $E$; since $\tau(E') = \tau(E)$, the sequence $\{z_n : n \in \omega\}$ converges to $z$ in $E'$, so $|| \cdot ||$ is complete and hence Fact 1 is proved.

Returning to our solution apply Fact 1 to see that $E = L \times M$ is a Banach space; let $\pi_L : E \to L$ and $\pi_M : E \to M$ be the natural projections. It is easy to see that any closed linear subspace of a Banach space is a Banach space; we omit a simple proof that $G$ is a linear subspace of $L \times M$ and hence $G$ is a Banach space. The map $p = \pi_L | G : G \to L$ is continuous, linear and surjective, so it is open by Problem 397. Since $p$ is also an injection, it has to be a homeomorphism so $f = (\pi_M | G) \circ p^{-1}$ is a continuous map, i.e., our solution is complete.

**V.399.** *Prove that, for any nonempty space $X$, there exists a continuous surjection of the space $C_p(X)$ onto $C_p(X) \times \mathbb{R}$.*

**Solution.** The expression $Y \simeq Z$ says that the spaces $Y$ and $Z$ are homeomorphic.

*Fact 1.* If $Z$ is a normal space and $\mathbb{D}^\omega \times \omega$ embeds in $Z$ as a closed subspace then $Z$ can be continuously mapped onto $\mathbb{R}^n$ for any $n \in \mathbb{N}$.

*Proof.* Fix any $n \in \mathbb{N}$ and observe that there exists a discrete family $\{F_i : i \in \omega\}$ of closed subsets of $Z$ such that $F_i \simeq \mathbb{D}^\omega$ for each $i \in \omega$. The space $K_i = [-i, i]^n \subset \mathbb{R}^n$ is compact and metrizable for any $i \in \omega$, so we can apply TFS-128 to find a continuous onto map $\varphi_i : F_i \to K_i$. If $x \in F = \bigcup_{i \in \omega} F_i$ then there is a unique $i \in \omega$ with $x \in F_i$; let $\varphi(x) = \varphi_i(x)$. This gives us a continuous map $\varphi : F \to \mathbb{R}^n$ and it is clear that $\varphi(F) = \bigcup_{i \in \omega} K_i = \mathbb{R}^n$.

For any $k \in \{1, \ldots, n\}$ let $\pi_k : \mathbb{R}^n \to \mathbb{R}$ be the projection onto the $k$-th factor. The map $\pi_k \circ \varphi : F \to \mathbb{R}$ is continuous, so there exists a continuous map $\mu_k : Z \to \mathbb{R}$ such that $\mu_k | F = \pi_k \circ \varphi$ for every $k \leq n$. If $\mu$ is the diagonal product of the family $\{\mu_k : 1 \leq k \leq n\}$ then it is straightforward that $\mu : Z \to \mathbb{R}^n$ is a continuous map such that $\mu | F = \varphi$ and therefore $\mu(Z) \supset \mu(F) = \varphi(F) = \mathbb{R}^n$, i.e., $\mu$ is a continuous map of $Z$ onto $\mathbb{R}^n$, so Fact 1 is proved.

Returning to our solution take any point $a \in X$; then $C_p(X) \simeq I \times \mathbb{R}$ where $I = \{f \in C_p(X) : f(a) = 0\}$ (see Fact 1 of S.409). Since $\mathbb{D}^\omega \times \omega$ embeds in $\mathbb{R}$ as a closed subspace, we can apply Fact 1 to see that there exists a continuous onto map $\varphi : \mathbb{R} \to \mathbb{R}^2$. If $f \in I$ and $t \in \mathbb{R}$ then let $\mu(f, t) \doteq (f, \varphi(t))$; this gives us a continuous map $\mu : I \times \mathbb{R} \to I \times \mathbb{R}^2$ and it is immediate that $\mu$ is surjective. Now it follows from $C_p(X) \simeq I \times \mathbb{R}$ and $C_p(X) \times \mathbb{R} \simeq I \times \mathbb{R}^2$ that there exists a continuous surjective map of $C_p(X)$ onto $C_p(X) \times \mathbb{R}$.

**V.400.** *Prove that there exists an infinite compact space $K$ for which there is no linear continuous surjection of $C_p(K)$ onto $C_p(K) \times \mathbb{R}$. In particular, $K^+$ is not l-equivalent to the space $K$.*

**Solution.** If $X$ is a space and $A \subset X$ then $\pi_A : C_p(X) \to C_p(A)$ is the restriction map; let $C_p(A|X) = \pi_A(C_p(X))$. For any nonempty $D$ consider the set $\mathbb{R}_*^D = \{f \in \mathbb{R}^D : f(D) \subset [-r, r] \text{ for some } r > 0\}$; in other words, $\mathbb{R}_*^D$ is the set of all bounded functions on $D$. If $f \in \mathbb{R}_*^D$ then $\|f\|_D = \sup\{|f(x)| : x \in D\}$. Given infinite sets $D$ and $E$ say that a linear continuous map $\varphi : \mathbb{R}^D \to \mathbb{R}^E$ is *adequate* if $\varphi(\mathbb{R}_*^D) = \mathbb{R}_*^E$ and the map $\varphi | \mathbb{R}_*^D : (\mathbb{R}_*^D, \| \cdot \|_D) \to (\mathbb{R}_*^E, \| \cdot \|_E)$ is continuous.

If $\varphi : \mathbb{R}^D \to \mathbb{R}$ is a continuous nontrivial linear functional then $\varphi \in L_p(D)$ and hence we can apply TFS-197 to see that

(1) there exist uniquely determined distinct $d_1, \ldots, d_n \in D$ and $\lambda_1, \ldots, \lambda_n \in \mathbb{R}\backslash\{0\}$ such that $\varphi(f) = \sum_{i=1}^n \lambda_i f(d_i)$ for any $f \in \mathbb{R}^D$.

Now, if $L$ is a dense linear subspace of $\mathbb{R}^D$ and $\varphi : L \to \mathbb{R}$ is a continuous linear functional then there exists a unique continuous linear extension of $\varphi$ over the whole space $\mathbb{R}^D$ (see Problem 224), so $\varphi$ has the same representation as in (1).

Given nonempty sets $D$ and $E$ assume that $L$ is a dense linear subspace of $\mathbb{R}^D$ and $M$ is a dense linear subspace of $\mathbb{R}^E$. If $\xi : L \to M$ is a linear continuous map and we are given a point $y \in E$ then let $\varphi(f) = f(y)$ for all $f \in M$; this defines a continuous linear functional $\varphi : M \to \mathbb{R}$. Since $\varphi \circ \xi$ is a continuous linear functional on $L$, by our above observation there exist uniquely determined distinct points $x_1, \ldots, x_n \in D$ and numbers $\lambda_1, \ldots, \lambda_n \in \mathbb{R}\backslash\{0\}$ such that $\varphi(f)(y) = \sum_{i=1}^n \lambda_i f(x_i)$ for any $f \in L$. If $\mathrm{supp}(y) = \{x_1, \ldots, x_n\}$ then for any $x \in \mathrm{supp}(y)$ there is a unique $\lambda_{yx} \in \mathbb{R}\backslash\{0\}$ such that $\varphi(f)(y) = \sum_{x \in \mathrm{supp}(y)} \lambda_{yx} f(x)$ for each $f \in L$. For any $A \subset D$ and $y \in E$ let $\mu(y, A) = \sum\{|\lambda_{yx}| : x \in \mathrm{supp}(y) \cap A\}$. Of course, $\mu(y, A)$ depends on $D, E, L, M$ and $\xi$ but we omit that to simplify notation.

As usual, $\mathbb{D} = \{0, 1\}$ is the two-point discrete space. For any map $f \in \mathbb{R}_*^\omega$ there exists a unique continuous function $[f] : \beta\omega \to \mathbb{R}$ such that $[f]|\omega = f$. If $X$ is a set and $A \subset X$ then $\chi_A : X \to \mathbb{D}$ is the characteristic function of $A$, i.e., $\chi_A(A) \subset \{1\}$ and $\chi_A(X\backslash A) \subset \{0\}$. The function $\chi_A$ formally depends on $X$ but we don't mention this because $X$ will always be clear. If $\alpha$ and $\beta$ are ordinals and $\alpha < \beta$ then $[\alpha, \beta) = \{\gamma : \alpha \leq \gamma < \beta\}$.

*Fact 1.* Suppose that $(B, ||\cdot||)$ is a normed space and $L$ is a proper closed linear subspace of $B$. For any $r > 0$ let $S(r) = \{x \in B : ||x|| \leq r\}$. Then for any positive numbers $r$ and $R$ such that $r < R$ the set $S(R)$ is not contained in $S(r) + L$.

*Proof.* Apply Problem 222 to find a nontrivial continuous linear functional $\varphi_0 : B \to \mathbb{R}$ such that $\varphi_0(L) = \{0\}$. There exists $\varepsilon > 0$ such that $\varphi_0(S(\varepsilon)) \subset (-1, 1)$; as an immediate consequence we obtain the inclusion $\varphi_0(S(1)) \subset (-\frac{1}{\varepsilon}, \frac{1}{\varepsilon})$. Therefore the number $q = \sup\{|\varphi_0(x)| : x \in S(1)\}$ is positive and consistently defined; let $\varphi = \frac{1}{q}\varphi_0$.

We have $\varphi(S(1)) \subset [-1, 1]$ and $\sup\{|\varphi(x)| : x \in S(1)\} = 1$. Take a sequence $\{x_n : n \in \omega\} \subset S(1)$ such that $|\varphi(x_n)| \to 1$. Given any $x \in B\backslash\{0\}$ we have $|\varphi(x)| = ||x||\cdot\varphi(\frac{x}{||x||}) \leq ||x||$ because $\frac{x}{||x||} \in S(1)$. If $x \in S(r)+L$ then $x = x'+y$ for some $x' \in S(r)$ and $y \in L$. Then $|\varphi(x)| = |\varphi(x') + \varphi(y)| = |\varphi(x')| \leq r$ and hence $\varphi(S(r) + L) \subset [-r, r]$.

For any $n \in \omega$ let $y_n = Rx_n$; then $y_n \in S(R)$ and it follows from $|\varphi(x_n)| \to 1$ that $|\varphi(y_n)| \to R$. Consequently, there exists $n \in \omega$ such that $|\varphi(y_n)| > r$. This implies that $y_n \in S(R)\backslash(S(r) + L)$ and hence Fact 1 is proved.

*Fact 2.* Assume that $(B_0, ||\cdot||_0)$ and $(B_1, ||\cdot||_1)$ are normed spaces and $f : B_0 \to B_1$ is a linear map. For any positive number $r$ let $S_0(r) = \{x \in B_0 : ||x||_0 \leq r\}$ and

$S_1(r) = \{y \in B_1 : ||y||_1 \le r\}$ be the closed $r$-balls in the respective spaces. Then the following conditions are equivalent:

(a) the map $f$ is continuous;
(b) there exists a number $R > 0$ such that $f(S_0(1)) \subset S_1(R)$.

If, additionally, the map $f$ is surjective and the spaces $B_1$ and $B_2$ are Banach then the conditions (a) and (b) are also equivalent to the following statement.

(c) There exist positive numbers $r$ and $R$ such that $S_1(r) \subset f(S_0(1)) \subset S_1(R)$.

*Proof.* If $f$ is continuous then there exists $\varepsilon > 0$ such that $f(S_0(\varepsilon)) \subset S_1(1)$. By linearity of $f$ we have $f(S_0(1)) \subset S_1(\frac{1}{\varepsilon})$, so letting $R = \frac{1}{\varepsilon}$ we obtain the inclusion $f(S_0(1)) \subset S_1(R)$; this shows that (a)$\Longrightarrow$(b).

Now assume that $f(S_0(1)) \subset S_1(R)$ for some positive number $R$ and take any set $U \in \tau(\mathbf{0}_1, B_1)$ where $\mathbf{0}_1$ is the zero element of $B_1$. There exists $\varepsilon > 0$ such that $S_1(\varepsilon) \subset U$; the interior $V$ of the set $S_0(\frac{\varepsilon}{R})$ contains the zero element $\mathbf{0}_0$ of the space $B_0$. If $x \in V\backslash\{\mathbf{0}_0\}$ then $\frac{x}{||x||_0} \in S_0(1)$, so $\frac{1}{||x||_0} \cdot f(x) \in S_1(R)$ and hence $\frac{||f(x)||_1}{||x||_0} \le R$, i.e., $||f(x)||_1 \le R||x||_0 \le \frac{\varepsilon}{R} \cdot R = \varepsilon$ which shows that $f(V) \subset S_1(\varepsilon) \subset U$. Therefore the map $f$ is continuous at the point $\mathbf{0}_0$, so it is continuous by Fact 2 of S.496 and hence (b)$\Longrightarrow$(a).

Now assume that $f$ is surjective. If $f$ is continuous then it is open by Problem 397; since $\mathbf{0}_0$ belongs to the interior of $S_0(1)$, the point $\mathbf{0}_1$ belongs to the interior of $f(S_0(1))$. Therefore $S_1(r) \subset f(S_0(1))$ for some $r > 0$ and hence we proved that (a)$\Longrightarrow$(c). It is evident that (c)$\Longrightarrow$(b), so (a) $\Longleftrightarrow$ (b) $\Longleftrightarrow$ (c), i.e., Fact 2 is proved.

*Fact 3.* Let $i : \mathbb{R} \to \mathbb{R}$ be the identity map, i.e., $i(x) = x$ for any $x \in \mathbb{R}$. Suppose that $X$ is a separable space and $\varphi : C_p(X) \to C_p(X) \times \mathbb{R}$ is a continuous linear surjection. Then there exists a countable dense set $D \subset X$ such that the map $\xi_D = (\pi_D \times i) \circ \varphi \circ \pi_D^{-1} : C_p(D|X) \to C_p(D|X) \times \mathbb{R}$ is a continuous linear surjection.

*Proof.* If $D$ is dense in $X$ then $\pi_D : C_p(X) \to C_p(D|X)$ is a linear bijection. Therefore $\xi_D$ is also linear; the product of any number of surjections is also a surjection, so $\xi_D$ is a linear surjection for any dense $D \subset X$. To find a set $D$ for which $\xi_D$ is continuous fix a point $a \notin X$ and identify $C_p(X) \times \mathbb{R}$ with the space $X^+ = X \oplus \{a\}$. Therefore $\varphi : C_p(X) \to C_p(X^+)$ and we can use the notion of support for the map $\varphi$.

Let $D_0$ be any dense subset of $X^+$; if $n \in \omega$ and we have a set $D_n$ then let $D_{n+1} = \bigcup\{\text{supp}(y) : y \in D_n\}$. This gives us a sequence $\{D_n : n \in \omega\}$ of subsets of $X^+$; let $D = (\bigcup_{n\in\omega} D_n)\backslash\{a\}$.

Observe first that $D$ is dense in $X$ because it contains $D_0 \cap X$ which is dense in $X$. To see that $\xi_D$ is continuous note that $C_p(D|X) \times \mathbb{R}$ is a subspace of the product $\mathbb{R}^D \times \mathbb{R} = \mathbb{R}^{D\cup\{a\}}$ and let $p_y : \mathbb{R}^{D\cup\{a\}} \to \mathbb{R}$ be the projection onto the factor determined by $y$ for any $y \in D \cup \{a\}$.

Fix any $y \in D \cup \{a\}$ and consider the map $p_y \circ \xi_D : C_p(D|X) \to \mathbb{R}$. Given any $f \in C_p(D|X)$ let $g = \pi_D^{-1}(f)$. Then $p_y(\xi_D(f)) = \varphi(g)(y) = \sum_{x\in\text{supp}(y)} \lambda_{yx} g(x)$. However, $\text{supp}(y) \subset D$ and $g(x) = f(x)$ for any $x \in D$

(recall that $f$ is the restriction of $g$ to $D$), so $p_y(\xi_D(f)) = \sum_{x \in \text{supp}(y)} \lambda_{yx} f(x)$. For any $x \in D$ let $e_x(f) = f(x)$ for any $f \in C_p(D)$. Then $e_x$ is continuous on $C_p(D)$ and hence on $C_p(D|X)$. Therefore $p_y \circ \xi_D = \sum\{\lambda_{yx} \cdot e_x : x \in \text{supp}(y)\}$ is continuous being a linear combination of continuous functions. This shows that $p_y \circ \xi_D$ is continuous for any $y \in D \cup \{a\}$, so $\xi_D$ is continuous by TFS-102 and hence Fact 3 is proved.

*Fact 4.* Suppose that $X$ is a separable pseudocompact space and we have a linear continuous surjection $\varphi : C_p(D|X) \to C_p(D|X) \times \mathbb{R}$ is for some countably infinite sets $D \subset X = \overline{D}$. Then there exists a unique adequate map $q : \mathbb{R}^D \to \mathbb{R}^D \times \mathbb{R}$ such that $q|C_p(D|X) = \varphi$.

*Proof.* It suffices to prove existence of $q$ because its uniqueness follows from the fact that $C_p(D|X)$ is dense in $\mathbb{R}^D$. Take a point $a \notin X$ and let $D^+ = D \oplus \{a\}$; we will identify $\mathbb{R}^D \times \mathbb{R}$ with $\mathbb{R}^{D^+}$.

For any $y \in D^+$ let $p_y : \mathbb{R}^{D^+} \to \mathbb{R}$ be the projection on the factor determined by $y$. The map $p_y \circ \varphi$ is a continuous linear functional on $C_p(D|X)$ and hence there exists a continuous linear functional $q_y : \mathbb{R}^D \to \mathbb{R}$ such that $q_y|C_p(D|X) = p_y \circ \varphi$ for any $y \in D^+$ (see Problem 224). The diagonal product $q = \Delta\{q_y : y \in D^+\}$ is a continuous linear map and $q : \mathbb{R}^D \to \mathbb{R}^{D^+}$; it is easy to see that $q|C_p(D|X) = \varphi$.

Let $Y = X \oplus \{a\}$; the space $X$ being pseudocompact, the norms $||\cdot||_X$ and $||\cdot||_Y$ are consistently defined on $C_p(X)$ and $C_p(Y)$ respectively; besides, it is easy to see that $B_X = (C_p(X), ||\cdot||_X)$ and $B_Y = (C_p(Y), ||\cdot||_Y)$ are Banach spaces. For any $r > 0$ let $S_X(r) = \{x \in B_X : ||x||_X \le r\}$ and $S_Y(r) = \{y \in B_Y : ||y||_Y \le r\}$; we will also need the balls $T_D(r) = \{x \in \mathbb{R}_*^D : ||x||_D \le r\}$ and $T_{D^+}(r) = \{y \in \mathbb{R}_*^{D^+} : ||y||_{D^+} \le r\}$.

Consider the identity maps $i_X : B_X \to C_p(X)$ and $i_Y : B_Y \to C_p(Y)$ and let $\Gamma_\varphi$ be the graph of the map $\varphi$. The maps $\pi_D \circ i_X$ and $\pi_{D^+} \circ i_Y$ are continuous isomorphisms, so the map $\xi = (\pi_{D^+} \circ i_Y)^{-1} \circ \varphi \circ (\pi_D \circ i_X) : B_X \to B_Y$ is consistently defined, surjective and linear. If $\Gamma_\xi$ is the graph of $\xi$ then $h = (\pi_D \circ i_X) \times (\pi_{D^+} \circ i_Y)$ is a condensation of $B_X \times B_Y$ onto $C_p(D|X) \times C_p(D^+|Y)$ and $\Gamma_\xi = h^{-1}(\Gamma_\varphi)$ which shows that $\Gamma_\xi$ is closed in $B_X \times B_Y$, so the map $\xi$ is continuous by Problem 398. Apply Fact 2 to see that

$(\bullet)$ there exist positive numbers $r$ and $R$ such that $S_Y(r) \subset \xi(S_X(1)) \subset S_Y(R)$.

Observe that $Q = (\pi_D \circ i_X)(S_X(1))$ is a dense subset of $T_D(1)$, so $\varphi(Q)$ is a dense subset of $\varphi(T_D(1))$ in the topology of pointwise convergence on $\mathbb{R}_*^{D^+}$. Besides, $\varphi(Q) = (\pi_{D^+} \circ i_Y)(\xi(S_X(1))) \subset (\pi_{D^+} \circ i_Y)(S_Y(R)) \subset T_{D^+}(R)$ by $(\bullet)$). The set $T_{D^+}(R)$ is compact under the topology of pointwise convergence on $\mathbb{R}^{D^+}$ because it coincides with the set $[-R, R]^{D^+}$. Therefore $\varphi(T_D(1)) \subset \overline{\varphi(Q)} \subset T_{D^+}(R)$ (the bar denotes the pointwise convergence closure). As a consequence, we have the formula $\varphi(T_D(n)) = n\varphi(T_D(1)) \subset nT_{D^+}(R) \subset \mathbb{R}_*^{D^+}$. Since $\mathbb{R}_*^D = \bigcup\{T_D(n) : n \in \mathbb{N}\}$, we conclude that $\varphi(\mathbb{R}_*^D) \subset \mathbb{R}_*^{D^+}$; the inclusion $\varphi(T_D(1)) \subset T_{D^+}(R)$ together with Fact 2 shows that $\varphi|\mathbb{R}_*^D$ is continuous with respect to the norm topologies on $\mathbb{R}_*^D$ and $\mathbb{R}_*^{D^+}$.

The inclusion $\varphi(Q) = (\pi_{D+} \circ i_Y)(\xi(S_X(1))) \supset (\pi_{D+} \circ i_Y)(S_Y(r))$ together with $Q \subset T_D(1)$ and compactness of $T_D(1)$ under the topology of pointwise convergence show that $\varphi(T_D(1)) \supset \overline{(\pi_{D+} \circ i_Y)(S_Y(r))} = T_{D+}(r)$. This implies the inclusion $\varphi(T_D(n)) \supset T_{D+}(nr)$ for any $n \in \mathbb{N}$, so it follows from $\mathbb{R}_*^{D^+} = \bigcup\{T_{D+}(nr) : n \in \mathbb{N}\}$ that $\varphi(\mathbb{R}_*^D) \supset \mathbb{R}_*^{D^+}$, so $\varphi(\mathbb{R}_*^D) = \mathbb{R}_*^{D^+}$ and hence the map $f$ is adequate, i.e., Fact 4 is proved.

*Fact 5.* Suppose that $D$ is a countably infinite set and $\varphi : \mathbb{R}^D \to \mathbb{R}^D \times \mathbb{R}$ is an adequate map. For an infinite set $A \subset D$ let $\{S_a : a \in A\}$ be a disjoint family of subsets of $D$. Then, for every $\delta > 0$ there exists an infinite set $B \subset A$ such that $\mu(a, \bigcup\{S_b : b \in B\backslash\{a\}\}) < \delta$ for any $a \in B$.

*Proof.* Fix any point $c \notin D$ and consider the set $D^+ = D \cup \{c\}$; we will identify $\mathbb{R}^D \times \mathbb{R}$ with the set $\mathbb{R}^{D^+}$. For any $r > 0$ let $S(r) = \{x \in \mathbb{R}_*^D : ||x||_D \le r\}$ and $S_+(r) = \{y \in \mathbb{R}^{D^+} : ||y||_{D+} \le r\}$. Let us prove first that

(2) for any $\varepsilon > 0$ and for any infinite $A' \subset A$ there exists $b \in A'$ and an infinite $A_0 \subset A'$ such that $\mu(a, S_b) < \varepsilon$ for any $a \in A_0$.

If (2) is not true then there exists $\varepsilon > 0$ and an infinite set $A' \subset A$ such that for each $b \in A'$ the set $Q_b = \{a \in A' : \mu(a, S_b) < \varepsilon\}$ is finite. By Fact 2 there exists $R > 0$ such that $\varphi(S(1)) \subset S_+(R)$. Pick a number $j > \frac{R}{\varepsilon}$ and choose distinct $b_1, \ldots, b_j \in A'$; by our assumption there exists a point $a \in A'\backslash(Q_{b_1} \cup \ldots \cup Q_{b_j})$.

We have $\mu(a, D) \ge \mu(a, \bigcup\{S_{b_i} : 1 \le i \le j\}) = \sum_{i=1}^j \mu(a, S_{b_i}) \ge j\varepsilon > R$. There exists a function $f \in [-1, 1]^D = S(1)$ such that $f(d) = \frac{\lambda_{ad}}{|\lambda_{ad}|}$ for any $d \in \text{supp}(a)$. Then $\varphi(f)(a) = \sum\{\lambda_{ad} f(d) : d \in \text{supp}(a)\} = \mu(a, D) > R$. However, $f \in S(1)$, so $\varphi(f) \in S_+(R)$ and hence we must have $|\varphi(f)(b)| \le R$ for any $b \in D^+$. This contradiction shows that (2) is proved.

Apply (2) to find a point $a_0 \in A$ and an infinite set $A_0 \subset A$ such that $\mu(a, S_{a_0}) < \frac{\delta}{2}$ for any $a \in A_0$. Proceeding by induction assume that $m \in \omega$ and we have infinite sets $A_0, \ldots, A_m$ and points $a_0, \ldots, a_m$ with the following properties:

(3) $a_{i+1} \in A_i\backslash\{a_0, \ldots, a_i\}$ and $A_{i+1} \subset A_i$ whenever $i < m$;
(4) if $i \le m$ and $a \in A_i$ then $\mu(a, S_{a_i}) < 2^{-i-1} \cdot \delta$;
(5) $S_{a_{i+1}} \cap (\bigcup\{\text{supp}(a_j) : j \le i\}) = \emptyset$ for all $i < m$.

The set $A_m$ being infinite we can find an infinite set $A' \subset A_m\backslash\{a_0, \ldots, a_m\}$ such that $S_a \cap (\bigcup\{\text{supp}(a_i) : i \le m\}) = \emptyset$ for any $a \in A'$. Apply (2) to pick an element $a_{m+1} \in A'$ and an infinite set $A_{m+1} \subset A'$ such that $\mu(a, S_{a_{m+1}}) < 2^{-m-2} \cdot \delta$ for any $a \in A_{m+1}$. It is immediate that the conditions (3)–(5) are now satisfied if we replace $m$ with $m + 1$, so our inductive procedure can be continued to construct sequences $\{A_i : i \in \omega\}$ and $\{a_i : i \in \omega\}$ for which the properties (3)–(5) hold for all $m \in \omega$.

To check that the set $B = \{a_i : i \in \omega\}$ is as promised observe that it follows from (3) that the indexation of $B$ is faithful and hence $B$ is an infinite subset of $A$. Now if $n \in \omega$ then $\text{supp}(a_n) \cap S_i = \emptyset$ for all $i > n$, so $\mu(a_n, \bigcup\{S_b : b \in B\backslash\{a_n\}\})$ is equal to $\mu(a_n, S_{a_0} \cup \ldots \cup S_{a_{n-1}}) = \sum_{i=1}^{n-1} \mu(a_n, S_{a_i}) < \delta(2^{-1} + \ldots + 2^{-n}) < \delta$ and hence Fact 5 is proved.

**Fact 6.** Given spaces $X$ and $Y$ assume that $D \subset X = \overline{D}$, $E \subset Y = \overline{E}$ and $\varphi : C_p(D|X) \to C_p(E|Y)$ is a continuous linear surjection. Suppose that $y \in E$ and $S \subset \text{supp}(y)$ is a nonempty set. Then for any $U \in \tau(S, X)$ there exists a set $V \in \tau(y, Y)$ such that $\mu(x, U \cap D) > \frac{1}{2}\mu(y, S)$ for any $x \in V \cap E$.

*Proof.* Let $r = \mu(y, S)$; it is easy to find a set $U' \in \tau(X)$ such that $U' \subset U$ and $U' \cap \text{supp}(y) = S$. There exists a function $f \in C_p(X, \mathbb{I})$ such that $f(x) = \frac{\lambda_{yx}}{|\lambda_{yx}|}$ for any $x \in S$ and $f(X \setminus U') \subset \{0\}$; let $g = f|D$. It is straightforward that $\varphi(g)(y) = r$; by continuity of the function $h = \varphi(g)$ there exists a set $V \in \tau(y, Y)$ such that $h(V \cap E) \subset (\frac{r}{2}, +\infty)$.

Take any point $x \in V \cap E$ and consider the set $T = \text{supp}(x) \cap U'$. Then

$$\mu(x, U \cap D) \geq \sum \{|\lambda_{xt}| : t \in T\} \geq \sum \{|\lambda_{xt} f(t)| : t \in T\}$$

$$\geq \sum \{\lambda_{xt} f(t) : t \in T\} = \varphi(f)(x) > \frac{r}{2}$$

and hence Fact 6 is proved.

**Fact 7.** Suppose that $D = \{d_i : i \in \omega\}$ is a faithfully indexed set and consider the set $D_n = \{d_i : i \leq n\}$ for any $n \in \omega$; assume additionally that $\varphi : \mathbb{R}^D \to \mathbb{R}^D \times \mathbb{R}$ is an adequate map. Then there exists $\varepsilon > 0$ and an infinite set $A \subset \omega$ such that $\mu(d_n, D \setminus D_n) > \varepsilon$ whenever $n \in A$.

*Proof.* Choose a point $e \notin D$ and let $D^+ = D \cup \{e\}$; we will identify $\mathbb{R}^D \times \mathbb{R}$ with $\mathbb{R}^{D^+}$. Given any $r > 0$ we will need the closed balls $S(r) = \{x \in \mathbb{R}^D_* : ||x||_D \leq r\}$ and $S_+(r) = \{y \in \mathbb{R}^{D^+}_* : ||y||_{D^+} \leq r\}$. The map $\varphi|\mathbb{R}^D_* : \mathbb{R}^D_* \to \mathbb{R}^{D^+}_*$ is continuous and onto, so there exists $r > 0$ such that $S_+(r) \subset \varphi(S(1))$ (see Fact 2). Let $\varepsilon = \frac{r}{2}$; we claim that the set $A = \{n \in \omega : \mu(d_n, D \setminus D_n) > \varepsilon\}$ is infinite.

Striving for contradiction assume that $A$ is finite and hence there exists $m \in \omega$ such that $\bigcup \{\text{supp}(d_n) : n \in A\} \cup \text{supp}(e) \subset D_m$. It turns out that

(6) if $d \in D_m \cup \{e\}$ then $\mu(d, D \setminus D_m) \leq \varepsilon$.

To see that (6) holds observe first that if $d = d_n$ for some number $n \in A$ then $\text{supp}(d) \subset D_m$ and therefore $\mu(d, D \setminus D_m) = 0$. If $d = d_n$ and $n \notin A$ then the inequalities $\mu(d, D \setminus D_m) \leq \mu(d, D \setminus D_n) \leq \varepsilon$ show that (6) is true.

If $P = \{x \in S(1) : x(D \setminus D_m) = \{0\}\}$ and $Q = \{x \in S(1) : x(D_m) = \{0\}\}$ then $P + Q = S(1)$ and hence $\varphi(P) + \varphi(Q) \supset S_+(r)$. Let $\pi : \mathbb{R}^{D^+}_* \to \mathbb{R}^{D_m \cup \{e\}}$ be the natural projection; if $T = \{x \in \mathbb{R}^{D_m \cup \{e\}} : ||x||_{D_m \cup \{e\}} \leq r\}$ then $T \subset \pi(S^+(r))$ and therefore $T \subset \pi((\varphi(P)) + \pi(\varphi(Q)))$. It is easy to see that the set $P$ is contained in an $(m + 1)$-dimensional linear subspace of $\mathbb{R}^D_*$; since the linear dimension of $\mathbb{R}^{D_m \cup \{e\}}$ is equal to $m + 2$, the set $\pi(\varphi(P))$ is contained in a proper linear subspace $L$ of $\mathbb{R}^{D_m \cup \{e\}}$ (which is automatically closed by Problem 228).

If $x \in Q$ and $d \in D_m \cup \{e\}$ then $\varphi(x)(d) = \sum \{\lambda_{dt} x(t) : t \in \text{supp}(d) \setminus D_m\}$ and therefore $|\pi(\varphi(x))(d)| \leq \sum \{|\lambda_{dt}| : t \in \text{supp}(d) \setminus D_m\} = \mu(d, D \setminus D_m) \leq \varepsilon$ by the property (6). As a consequence, the set $\pi(\varphi(Q))$ is contained in the $\varepsilon$-ball $T' = \{x \in \mathbb{R}^{D_m \cup \{e\}} : ||x||_{D_m \cup \{e\}} \le \varepsilon\}$. Now it follows from $T \subset \pi(\varphi(P)) + \pi(\varphi(Q))$ that $T \subset L + T'$ which is a contradiction with Fact 1. Thus the set $A$ is infinite and hence Fact 7 is proved.

*Fact 8.* Suppose that $X$ is a pseudocompact space, $D$ is a countable dense subset of $X$ and there exists a continuous linear surjection $\varphi : C_p(D|X) \to C_p(D|X) \times \mathbb{R}$. Then for any dense set $E \subset D$ there exists $\varepsilon > 0$ such that, for some infinite set $A \subset E$ and a family $\{S_a : a \in A\}$ of pairwise disjoint finite subsets of $D$ we have the following properties:

(a) $S_a \subset \mathrm{supp}(a)$ and $\mu(a, S_a) > \varepsilon$ for any $a \in A$;
(b) $A \cap (\bigcup_{a \in A} S_a) = \emptyset$.
(c) either $\lambda_{ad} > 0$ for all $a \in A$ and $d \in S_a$ or $\lambda_{ad} < 0$ for all $a \in A$ and $d \in S_a$.

*Proof.* By Fact 4 we can consider that $\varphi$ is defined on the whole space $\mathbb{R}^D$ and $\varphi : \mathbb{R}^D \to \mathbb{R}^D \times \mathbb{R}$ is an adequate map. Choose a faithful enumeration $\{d_n : n \in \omega\}$ of the set $D$ and let $D_n = \{d_0, \dots, d_n\}$ for any $n \in \omega$. By Fact 7 there exists an infinite set $N \subset \omega$ such that, for some $\delta > 0$ we have $\mu(d_n, D \setminus D_n) > \delta$ for each $n \in N$.

Let $n_0 = \backslash n(N)$ and $P_0 = \mathrm{supp}(d_{n_0}) \setminus D_{n_0}$. Proceeding inductively we can find an increasing sequence $\{n_i : i \in \omega\} \subset N$ such that

(7) for any $i, j \in \omega$ if $j < i$ then $\mathrm{supp}(d_{n_j}) \cup \{d_{n_j}\} \subset D_{n_i - 1}$.

Let $P_i = \mathrm{supp}(d_{n_i}) \setminus D_{n_i}$ for each $i \in \omega$; the property (7) implies that the family $\mathcal{P} = \{P_i : i \in \omega\}$ is disjoint and $\{d_{n_i} : i \in \omega\} \cap (\bigcup \mathcal{P}) = \emptyset$. For the set $P_0 \subset \mathrm{supp}(d_{n_0})$ take $U \in \tau(P_0, X)$ and $V' \in \tau(d_{n_0}, X)$ such that $U \cap V' = \emptyset$ and apply Fact 6 to find a set $V \in \tau(d_{n_0}, X)$ such that $V \subset V'$ and $\mu(a, U \cap D) > \frac{\delta}{2}$ for any $a \in V$. The set $E$ being dense in $X$ there exists $a_0 \in V \cap E$; consider the set $Q_0 = \mathrm{supp}(a_0) \cap U$.

Proceeding inductively assume that $k \in \omega$ and we have chosen distinct points $a_0, \dots, a_k \in E$ and disjoint sets $Q_0, \dots, Q_k$ such that

(8) $Q_i \subset \mathrm{supp}(a_i)$ and $\mu(a_i, Q_i) > \frac{\delta}{2}$ for all $i \le k$;
(9) $\{a_0, \dots, a_k\} \cap (\bigcup_{i \le k} Q_i) = \emptyset$.

Let $H = \{a_0, \dots, a_k\} \cup Q_0 \cup \dots \cup Q_k$ and find $i \in \omega$ such that $d_{n_i} \cup P_i$ does not meet $H$. Take a set $U \in \tau(P_i, X)$ such that $\overline{U} \cap (H \cup \{d_{n_i}\}) = \emptyset$ and apply Fact 6 to find a set $V \in \tau(d_{n_i}, X)$ for which $V \cap U = \emptyset$ and $\mu(a, U \cap D) > \frac{\delta}{2}$ for all $a \in V$. By density of $E$ we can find a point $a_{k+1} \in E \cap V$ and a set $Q_{k+1} \subset \mathrm{supp}(a_{k+1}) \cap U$ such that $\mu(a_{k+1}, Q_{k+1}) > \frac{\delta}{2}$. It is clear that the properties (7) and (8) are now fulfilled if we replace $k$ with $k + 1$, so we can pursue our inductive procedure to construct a set $\{a_i : i \in \omega\} \subset E$ and a family $\{Q_i : i \in \omega\}$ such that (8) and (9) hold for all $k \in \omega$.

For any $i \in \omega$ let $Q_i^+ = \{d \in Q_i : \lambda_{a_i d} > 0\}$ and $Q_i^- = \{d \in Q_i : \lambda_{a_i d} < 0\}$; it follows from $\mu(a_i, Q_i) = \mu(a_i, Q_i^+) + \mu(a_i, Q_i^-)$ that either $\mu(a_i, Q_i^+) > \frac{\delta}{4}$ or $\mu(a_i, Q_i^-) > \frac{\delta}{4}$. This shows that there exists an infinite set $N \subset \omega$ such that either $\mu(a_i, Q_i^+) > \frac{\delta}{4}$ or $\mu(a_i, Q_i^-) > \frac{\delta}{4}$ for all $i \in N$.

Finally, let $\varepsilon = \frac{\delta}{4}$ and $A = \{a_i : i \in N\}$. If $a \in A$ then $a = a_i$ for a unique $i \in \omega$; let $S_a = Q_i$. It is immediate that the conditions (a)–(c) are satisfied, so Fact 8 is proved.

**Fact 9.** Suppose that $\{S_n : n \in \omega\}$ is a disjoint family of subsets of $\omega$ and we have a set $\{a_t, b_t\} \subset \beta\omega$ for every $t \in T$. Assume additionally that $|T| < \mathfrak{c}$ and $[\chi_{S_n}](a_t) = [\chi_{S_n}](b_t)$ for all $t \in T$ and $n \in \omega$. Then there exists an infinite set $A \subset \omega$ such that, for any $A' \subset A$, if $S = \bigcup\{S_n : n \in A'\}$ then $[\chi_S](a_t) = [\chi_S](b_t)$ for each $t \in T$.

*Proof.* Fix a family $\mathcal{A}$ of infinite subsets of $\omega$ such that $|\mathcal{A}| = \mathfrak{c}$ and $A \cap B$ is finite for any distinct $A, B \in \mathcal{A}$ (see TFS-141). For any $A \subset \omega$ consider the set $H(A) = \mathrm{cl}_{\beta\omega}(\bigcup_{n \in A} S_n) \setminus (\bigcup_{n \in A} \mathrm{cl}_{\beta\omega}(S_n))$. Let us prove first that

(10) $\mathcal{H} = \{H(A) : A \in \mathcal{A}\}$ is a disjoint family.

Indeed, if $A, B \in \mathcal{A}$ then consider the sets $C = A \cap B$ and $A' = A \setminus B$. Observe that $\mathrm{cl}_{\beta\omega}(\bigcup_{n \in A} S_n) = \mathrm{cl}_{\beta\omega}(\bigcup_{n \in A'} S_n) \cup (\bigcup\{\mathrm{cl}_{\beta\omega}(S_n) : n \in C\})$ because the set $C$ is finite; as an immediate consequence, $H(A) = H(A')$. Analogously, if $B' = B \setminus A$ then $H(B) = H(B')$

The sets $P = \bigcup_{n \in A'} S_n$ and $Q = \bigcup_{n \in B'} S_n$ are disjoint, so the sets $\mathrm{cl}_{\beta\omega}(P)$ and $\mathrm{cl}_{\beta\omega}(Q)$ are disjoint as well by Fact 1 of S.371. Since $H(A) \subset P$ and $H(B) \subset Q$, we have $H(A) \cap H(B) = \emptyset$, i.e., (9) is proved.

It follows from $|T| < \mathfrak{c}$ and the property (9) that there exists a set $A \in \mathcal{A}$ such that $H(A) \cap \{a_t, b_t\} = \emptyset$ for any $t \in T$; fix any $A' \subset A$, $t \in T$ and let $S = \bigcup_{n \in A'} S_n$. If $\{a_t, b_t\} \cap \mathrm{cl}_{\beta\omega}(S) = \emptyset$ then $[\chi_S](a_t) = [\chi_S](b_t) = 0$. The sets $S$ and $F = \bigcup_{n \in A \setminus A'} S_n$ are disjoint so $\mathrm{cl}_{\beta\omega}(S) \cap \mathrm{cl}_{\beta\omega}(F) = \emptyset$ and, in particular, $H(A') \cap (\bigcup_{n \in A \setminus A'} \mathrm{cl}_{\beta\omega}(S_n)) = \emptyset$. Since also $H(A') \cap (\bigcup_{n \in A'} \mathrm{cl}_{\beta\omega}(S_n)) = \emptyset$, we conclude that $H(A') \cap (\bigcup_{n \in A} \mathrm{cl}_{\beta\omega}(S_n)) = \emptyset$ and therefore $H(A') \subset H(A)$.

Finally, assume that the intersection of $\{a_t, b_t\}$ and the set $\mathrm{cl}_{\beta\omega}(S)$ is nonempty. However, $\{a_t, b_t\} \cap H(A) = \emptyset$ and hence $\{a_t, b_t\} \cap H(A') = \emptyset$, so it follows from the equality $\mathrm{cl}_{\beta\omega}(S) = H(A') \cup (\bigcup\{\mathrm{cl}_{\beta\omega}(S_n) : n \in A'\})$ that $\{a_t, b_t\} \cap \mathrm{cl}_{\beta\omega}(S_n) \neq \emptyset$ for some $n \in A'$. Recalling that $[\chi_{S_n}](a_t) = [\chi_{S_n}](b_t)$ we convince ourselves that $\{a_t, b_t\} \subset \mathrm{cl}_{\beta\omega}(S_n) \subset \mathrm{cl}_{\beta\omega}(S)$ and therefore $[\chi_S](a_t) = [\chi_S](b_t) = 1$, i.e., Fact 9 is proved.

**Fact 10.** Suppose that $f_t : \omega \to \mathbb{D}$ for any $t \in T$ and $|T| < \mathfrak{c}$. Then, for any infinite set $A \subset \omega$ there exist distinct points $a, b \in \mathrm{cl}_{\beta\omega}(A)$ such that $f_t(a) = f_t(b)$ for any $t \in T$.

*Proof.* If the family $\mathcal{F} = \{f_t : t \in T\}$ separates the points of $F = \mathrm{cl}_{\beta\omega}(A)$ then the diagonal product of $\mathcal{F}$ embeds $F$ in the space $\mathbb{D}^T$ and hence $w(F) \leq w(\mathbb{D}^T) < \mathfrak{c}$. However, the space $F$ is homeomorphic to $\beta A$ (see Fact 2 of S.451) which, in turn, is homeomorphic to $\beta\omega$. Therefore $w(\beta\omega) < \mathfrak{c}$; this contradiction with TFS-368 shows that the family $\mathcal{F}$ cannot separate the points of $\mathrm{cl}_{\beta\omega}(A)$, i.e., there are distinct $a, b \in \mathrm{cl}_{\beta\omega}(A)$ such that $f_t(a) = f_t(b)$ for any $t \in T$, so Fact 10 is proved.

Returning to our solution let us introduce some notation. We will construct inductively a set of functions $\{u_\alpha : \alpha < \mathfrak{c}\} \subset \mathbb{D}^\omega$ such that its diagonal product $u = \Delta\{u_\alpha : \alpha < \mathfrak{c}\} : \omega \to \mathbb{D}^\mathfrak{c}$ will give us the promised space $K$ as the closure of the set $u(\omega)$ in $\mathbb{D}^\mathfrak{c}$.

Letting $u_n = \chi_{\{n\}}$ for each $n \in \omega$ we obtain the family $\{u_n : n \in \omega\}$, so the rest of the inductive procedure will show us how to construct the set $\{u_\alpha : \omega \le \alpha < \mathfrak{c}\}$. For any $A, B \subset \mathfrak{c}$ with $A \subset B$ the map $\pi_A : \mathbb{D}^B \to \mathbb{D}^A$ is the natural projection and $p_\alpha : \mathbb{D}^B \to \mathbb{D}$ is the projection onto the factor determined by $\alpha$, i.e., $p_\alpha = \pi_{\{\alpha\}}$ for all $\alpha \in B$. The maps $\pi_A$ and $p_\alpha$ formally depend on $B$ but the set $B$ will always be clear from the context, so we don't mention it to simplify the notation.

If $A \subset \mathfrak{c}$ and the set $\{u_\alpha : \alpha \in A\}$ is already constructed then define the map $u_A : \omega \to \mathbb{D}^A$ to be the diagonal product $\Delta\{u_\alpha : \alpha \in A\}$ and let $K_A = \overline{u_A(\omega)}$.

Say that a pair $(D, \varphi)$ is *admissible* if the following conditions are satisfied:

(11)  there is a countable set $A_D \subset \mathfrak{c}$ such that $\omega \subset A_D$ and $D \subset \mathbb{D}^{A_D}$
(12)  there is a map $\xi_D : \omega \to D$ such that $\xi_D(n)|\omega = u_n$ for each $n \in \omega$;
(13)  $\varphi : \mathbb{R}^D \to \mathbb{R}^D \times \mathbb{R}$ is an adequate map;
(14)  there exists a number $\varepsilon > 0$ such that, for some infinite set $E \subset \omega$ and a family $\{S_n : n \in E\}$ of finite pairwise disjoint subsets of $D$, we have

    (14.1)  $\xi_D(E) \cap (\bigcup\{S_n : n \in E\}) = \emptyset$;
    (14.2)  if the function $\mu(\cdot, \cdot)$ is determined by the map $\varphi$ then $S_n \subset$ $\mathrm{supp}(\xi_D(n))$ and $\mu(\xi_D(n), S_n) > \varepsilon$ for any $n \in E$;
    (14.2)  either $\lambda_{\xi_D(n)d} > 0$ for any $d \in S_n$ and $n \in E$ or $\lambda_{\xi_D(n)d} < 0$ for any $d \in S_n$ and $n \in E$.

Take a countably infinite set $Q$, a point $a \notin Q$ and compactify the discrete space $Q$ with the point $a$, i.e., all points of $Q$ are isolated in $X = Q \cup \{a\}$ and $U \in \tau(a, X)$ if and only if $a \in U$ and $X \setminus U$ is finite. Fix a point $q \in Q$ and let $L = C_p(Q|X)$; in the space $L$ consider the subspaces $M = \{f \in L : f(q) = 0\}$ and $M' = \{f \in L : f(Q \setminus \{q\}) = \{0\}\}$. For any $f \in L$ let $r(f)(x) = f(x)$ for any $x \in Q \setminus \{q\}$ and $r(f)(q) = 0$. It is easy to check that $r : L \to M$ is a linear retraction. If we let $r'(f)(x) = 0$ for all $x \ne q$ and $r'(f)(q) = f(q)$ then $r' : L \to M'$ is also a linear retraction and $x = r(x) + r'(x)$ for any $x \in Q$. Applying Problem 391 we convince ourselves that $L \approx M \times M' \approx M \times \mathbb{R}$. It is an easy exercise to see that $M \approx L$, so we proved that there exists a linear homeomorphism of $C_p(Q|X)$ onto $C_p(Q|X) \times \mathbb{R}$.

Now let $D = \{u_n : n \in \omega\} \subset \mathbb{D}^\omega$ and $A_D = \omega$. If $\xi_D(n) = u_n$ for all $n \in \omega$ then $\xi_D : \omega \to D$ is a bijection. Let $\theta(n) = 0$ for all $n \in \omega$; it is straightforward that $X = D \cup \{\theta\}$ is a one-point compactification of the discrete space $D$. By what we proved in the preceding paragraph there exists a linear homeomorphism $\varphi : C_p(D|X) \to C_p(D|X) \times \mathbb{R}$. By Fact 4 we can consider that $\varphi$ is defined on $\mathbb{R}^D$ and $\varphi : \mathbb{R}^D \to \mathbb{R}^D \times \mathbb{R}$ is an adequate map. Applying Fact 8 we conclude that there exists an infinite set $E \subset \omega$ such for which the properties (14.1)–(14.3) hold and hence $(D, \varphi)$ is an admissible pair.

It is easy to see that the cardinality of the family $\mathcal{A}$ of all admissible pairs does not exceed $\mathfrak{c}$, so we can find a set $Q \subset \mathfrak{c}\backslash\omega$ such that $\mathcal{A} = \{(D_\alpha, \varphi_\alpha) : \alpha \in Q\}$ and $A_{D_\alpha} \subset \alpha$ for any $\alpha \in Q$. Letting $D_\alpha = D$ and $\varphi_\alpha = \varphi$ for all $\alpha \in (\mathfrak{c}\backslash\omega)\backslash Q$ we obtain an enumeration $\{(D_\alpha, \varphi_\alpha) : \omega \le \alpha < \mathfrak{c}\}$ of the family $\mathcal{A}$ such that $A_{D_\alpha} \subset \alpha$ for any $\alpha \in \mathfrak{c}\backslash\omega$. To simplify the notation let $\xi_\alpha = \xi_{D_\alpha}$ and $A_{D_\alpha} = A_\alpha$ for any $\alpha \in \mathfrak{c}\backslash\omega$; observe that $A_\omega = \omega$.

*Fact 11.* There exists a family of functions $\{u_\alpha : \omega \le \alpha < \mathfrak{c}\} \subset \mathbb{D}^\omega$ and a set of pairs $\{(a_\alpha, b_\alpha) : \omega \le \alpha < \mathfrak{c}\} \subset \beta\omega \times \beta\omega$ such that, for any $\beta \in [\omega, \mathfrak{c})$ we have the following properties:

(a) $[u_\gamma](a_{\gamma'}) = [u_\gamma](b_{\gamma'})$ for all $\gamma, \gamma' \in [\omega, \beta)$;
(b) if $u_{A_\beta}(\omega) \subset D_\beta \subset K_{A_\beta}$ then for any $Q \subset \mathbb{D}^\mathfrak{c}$ such that $\pi_{A_\beta \cup \{\beta\}}(Q) \subset K_{A_\beta \cup \{\beta\}}$ and $\pi_{A_\beta}|Q : Q \to D_\beta$ is a bijection, for the function $g = \varphi_\beta(p_\beta \circ (\pi_{A_\beta}|Q)^{-1})$ we have $[g \circ \xi_\beta](a_\beta) \neq [g \circ \xi_\beta](b_\beta)$.

*Proof.* Assume that $\omega \le \alpha < \mathfrak{c}$ and we have functions $\{u_\beta : \beta < \alpha\}$ and pairs $\{(a_\beta, b_\beta) : \beta < \alpha\}$ such that the conditions (a) and (b) are satisfied for all $\beta \in [\omega, \alpha)$. If the statement $u_{A_\alpha}(\omega) \subset D_\alpha \subset K_{A_\alpha}$ is not true then let $u_\alpha = \theta$, choose any point $a_\alpha \in \beta\omega$ and let $b_\alpha = a_\alpha$; it is clear that the properties (a) and (b) now hold for all $\beta \le \alpha$.

Now assume that $u_{A_\alpha}(\omega) \subset D_\alpha \subset K_{A_\alpha}$. The properties (11) and (12) show that $\pi_{A_\omega}(\xi_\alpha(n)) = u_n$; since the set $\pi_{A_\omega}^{-1}(u_n) \cap K_{A_\alpha} \ni \xi_\alpha(n)$ is clopen in $K_\alpha$ for every $n \in \omega$, it follows from density of $u_{A_\alpha}(\omega)$ in $K_{A_\alpha}$ that $\xi_\alpha(n) = u_{A_\alpha}(n)$ for all $n \in \omega$, i.e., $\xi_\alpha = u_{A_\alpha}$.

There exist $E \subset \omega$, $\varepsilon > 0$ and a family $\mathcal{S} = \{S_n : n \in E\}$ as in the condition (14) for the set $D_\alpha$. Observe that all statements of (14) hold if we replace $E$ with any infinite subset of $E$. We will have to pass several times to an infinite subset of $E$; to simplify notation, we will denote the respective smaller set by $E$ again. Choose an increasing enumeration $\{n_i : i \in \omega\}$ of the set $E$.

Let $\tilde{n} = \xi_\alpha(n)$ for each $n \in \omega$; recalling that $K_{A_\alpha}$ is a metrizable compact space we can pass to an infinite subset of $E$ if necessary to guarantee that the sequence $\{\tilde{n}_i : i \in \omega\}$ converges to a point $s \in K_{A_\alpha}$. The family $\mathcal{S}$ is disjoint, so we can pass once more to an infinite subset of $E$ to assure that $s \notin S_n$ for all $n \in E$. As a consequence,

(15)  $S_n \cap \mathrm{cl}_{K_{A_\alpha}}(\{\tilde{n} : n \in E\}) = \emptyset$ for any $n \in E$.

**Case 1.** There exist a point $t \in K_{A_\alpha}$ and a number $\delta > 0$ such that, for any $U \in \tau(t, K_{A_\alpha})$ the set $\{n \in E : \mu(\tilde{n}, U \cap S_n) > \delta\}$ is infinite.

Choose a decreasing local base $\{U_k : k \in \omega\}$ of $t$ in $K_{A_\alpha}$. The assumption of our Case shows that we can find an increasing sequence $\{i_k : k \in \omega\} \subset \omega$ such that $\mu(\tilde{n}_{i_k}, U_k \cap S_{n_{i_k}}) > \delta$; let $m_k = n_{i_k}$ and $P_{m_k} = U_k \cap S_{n_{i_k}}$ for any $k \in \omega$. Passing to a smaller infinite subset of $E$ if necessary we can assume that $E = \{m_k : k \in \omega\}$. Therefore

(16)  $\mu(\tilde{n}, P_n) > \delta$ for any $n \in E$.

The sequence $\{P_{m_k} : k \in \omega\}$ being disjoint we can refine the set $E$ once more to be able to consider that $t \notin P_{m_k}$ for all $k \in \omega$. It is easy to find a disjoint family $\{V_{m_k} : k \in \omega\}$ of clopen subsets of $K_{A_\alpha}$ such that $P_{m_k} \subset V_{m_k} \subset U_k \backslash \{t\}$ while $V_{m_k} \cap \{\tilde{n} : n \in E\} = \emptyset$ and $\text{supp}(\tilde{m}_k) \cap V_{m_k} = P_{m_k}$ for each $k \in \omega$.

Assume that $t \in \text{supp}(\tilde{n})$ for infinitely many $n \in E$ and hence we can pursue our practice of refining $E$ and consider that $t \in \text{supp}(\tilde{n})$ for all $n \in E$. Let $q(d) = 1$ for all $d \in D_\alpha$. Then $q \in \mathbb{R}_*^{D_\alpha}$; the map $\varphi_\alpha$ being adequate, the function $q' = \varphi_\alpha(q)$ is bounded on $D_\alpha$ and hence there exists $r > 0$ such that $|q'(\tilde{n})| \le r$ for all $n \in E$. Recall that $q'(\tilde{n}) = \sum\{\lambda_{\tilde{n}d} : d \in \text{supp}(\tilde{n})\}$; since the numbers $\lambda_{\tilde{n}d}$ are of the same sign, we have $|\lambda_{\tilde{n}t}| \le r$ for all $n \in E$. Therefore the sequence $\{\lambda_{\tilde{n}t} : n \in E\}$ is bounded, so we can pass to smaller infinite subset of $E$ again to consider that it is convergent. Refining $E$ once again we can assume that

(17) either $t \notin \text{supp}(\tilde{n})$ for all $n \in E$ or there exists $\lambda \in \mathbb{R}$ such that $t \in \text{supp}(\tilde{n})$ and $\lambda_{\tilde{n}t} \in (\lambda, \lambda + \frac{\delta}{4})$ for all $n \in E$.

We still need to refine the set $E$ once more to apply Fact 5 to the family $\{V_n \cap D_\alpha : n \in \omega\}$ to be able to consider that

(18) $\mu(\tilde{n}, \bigcup\{V_k \cap D_\alpha : k \in E\backslash\{n\}\}) < \frac{\delta}{4}$ for any $n \in E$.

Let $R_n = \{k \in \omega : \tilde{k} \in V_n\}$ for all $n \in E$. It is easy to see that the family $\mathcal{B} = \{p_\beta^{-1}(0) \cap K_{A_\alpha}, p_\beta^{-1}(1) \cap K_{A_\alpha} : \beta \in A_\alpha\}$ is a clopen subbase in the space $K_{A_\alpha}$. Every set $V_n$ is compact, so it can be represented as a finite union of some finite intersections of the elements of $\mathcal{B}$. Thus the characteristic function of $R_n$ is a finite sum of finite products of the elements of the family $\{u_\beta, 1 - u_\beta : \beta \in A_\alpha\}$. This, together with the property (a) shows that $[\chi_{R_n}](a_\beta) = [\chi_{R_n}](b_\beta)$ for all $\beta < \alpha$ and $n \in E$. By Fact 9 we can refine the set $E$ once again to consider that

(19) $[\chi_{\bigcup\{R_n : n \in E'\}}](a_\beta) = [\chi_{\bigcup\{R_n : n \in E'\}}](b_\beta)$ for any $E' \subset E$ and $\beta \in [\omega, \alpha)$.

Next, apply Fact 10 to find distinct points $a_\alpha, b_\alpha \in \text{cl}_{\beta\omega}(E)$ for which we have the equality $[u_\beta](a_\alpha) = [u_\beta](b_\alpha)$ for all $\beta < \alpha$. Choose a set $E' \subset E$ such that $a_\alpha \in \text{cl}_{\beta\omega}(E')$ and $b_\alpha \notin \text{cl}_{\beta\omega}(E')$; then $b_\alpha \in \text{cl}_{\beta\omega}(E\backslash E')$. Let us check that for the function $u_\alpha = \chi_{\bigcup\{R_n : n \in E'\}}$ and the points $a_\alpha, b_\alpha$, the conditions (a) and (b) are satisfied.

We already saw that $V_n \cap \xi_\alpha(E) = \emptyset$ for any $n \in E$; this implies the equality $(\bigcup\{R_n : n \in E\}) \cap E = \emptyset$, so $u_\alpha(E) = \{0\}$ and hence $[u_\alpha](a_\alpha) = [u_\alpha](b_\alpha) = 0$. This, together with (19), shows that (a) is fulfilled for all $\beta \le \alpha$.

Assume that $Q \subset \mathbb{D}^c$, the map $\pi_{A_\alpha}|Q : Q \to D_\alpha$ is a bijection and we have the inclusion $\pi_{A_\alpha \cup \{\alpha\}}(Q) \subset K_{A_\alpha \cup \{\alpha\}}$; let $h = p_\alpha \circ ((\pi_{A_\alpha}|Q)^{-1})$. Given any $n \in \omega$ there is a unique $x \in Q$ such that $\pi_{A_\alpha}(x) = \tilde{n}$; since also $y = \pi_{A_\alpha \cup \{\alpha\}}(x) \in K_{A_\alpha \cup \{\alpha\}}$ and $y|A_\alpha = x|A_\alpha = \tilde{n}$, we conclude that $y|\omega = u_n$, so it follows from density of $\xi_{A_\alpha \cup \{\alpha\}}(\omega)$ in $K_{A_\alpha \cup \{\alpha\}}$ that $y$ is the unique point of $K_{A_\alpha \cup \{\alpha\}}$ such that $y|A_\alpha = \tilde{n}$. Consequently, $y$ coincides with $\xi_{A_\alpha \cup \{\alpha\}}(n)$ and hence $p_\alpha(x) = p_\alpha(y) = u_\alpha(n)$. This proves that $h(\tilde{n}) = u_\alpha(n)$ for any $n \in \omega$.

If $W$ is a neighborhood of $t$ then all but finitely many elements of the family $\{V_n : n \in E\}$ are contained in $W$. Therefore the function $v = \chi_{\bigcup\{V_n : n \in E'\}}$ is locally constant at every point of $K_{A_\alpha} \setminus \{t\}$. Given $d \in K_{A_\alpha} \setminus \{t\}$ fix a clopen set $G \in \tau(d, K_{A_\alpha})$ such that $v(G) = \{i\} \subset \mathbb{D}$. Then the sets $G' = G \times \{i\} \subset K_{A_\alpha \cup \{\alpha\}}$ and $G'' = K_{A_\alpha \cup \{\alpha\}} \setminus G'$ are clopen in $K_{A_\alpha \cup \{\alpha\}}$, disjoint and $G' \cup G'' = K_{A_\alpha \cup \{\alpha\}}$. Therefore, for every $d \in G$, the point $x = (d, i)$ is the unique element of $K_{A_\alpha \cup \{\alpha\}}$ such that $x|A_\alpha = d$.

Now if $d \in D_\alpha \setminus \{t\}$ and $x \in Q$ is the unique point such that $x|A_\alpha = d$ then it follows from $x' = x|(A_\alpha \cup \{\alpha\}) \in K_{A_\alpha \cup \{\alpha\}}$ that $x'$ coincides with the unique extension of $d$ over $A_\alpha \cup \{\alpha\}$. This proves that

(20) $h(d) = \chi_{\bigcup \{V_n : n \in E'\}}(d)$ for any $d \in D_\alpha \setminus \{t\}$.

For any $n \in E$ let $l_n = \lambda_{\tilde{n}t} h(t)$ if $t \in \text{supp}(\tilde{n})$ or $l_n = 0$ otherwise. Since $h(t) \in \mathbb{D}$, it follows from (17) that

(21) there exists $l \in \mathbb{R}$ such that $l_n \in (l, l + \frac{\delta}{4})$ for any $n \in E$.

Let $g = \varphi_\alpha(h)$; observe first that

(22) $g(\tilde{n}) = \sum \{\lambda_{\tilde{n}d} : d \in \text{supp}(\tilde{n}) \cap (\bigcup \{V_k : k \in E'\})\} + l_n$ for any $n \in E$.

By the property (14.2) all numbers $\lambda_{\tilde{n}d}$ are of the same sign; assume first that they are positive and let $V' = \bigcup_{k \in E'} V_k$. If $n \in E'$ then we have the equality

(23) $g(\tilde{n}) = \sum \{\lambda_{\tilde{n}d} : d \in \text{supp}(\tilde{n}) \cap V_n\} + \sum \{\lambda_{\tilde{n}d} : d \in \text{supp}(\tilde{n}) \cap (V' \setminus V_n)\} + l_n$.

Thus $g(\tilde{n}) \geq \sum \{\lambda_{\tilde{n}d} : d \in \text{supp}(\tilde{n}) \cap V_n)\} + l_n = \mu(\tilde{n}, P_n) + l_n$, so we can apply (16) and (21) to see that $g(\tilde{n}) > l + \delta$.

If $n \in E \setminus E'$ then $V' \setminus V_n = V'$, so $g(\tilde{n}) = \sum \{\lambda_{\tilde{n}d} : d \in \text{supp}(\tilde{n}) \cap (V' \setminus V_n)\} + l_n$, so it follows from (18) that $g(\tilde{n}) < \frac{\delta}{4} + l_n < l + \frac{\delta}{2}$. Recalling that $a_\alpha \in \text{cl}_{\beta\omega}(E')$ and $b_\alpha \in \text{cl}_{\beta\omega}(E \setminus E')$ we conclude that $[g \circ \xi_\alpha](a_\alpha) \geq l + \delta$ while $[g \circ \xi_\alpha](b_\alpha) \leq l + \frac{\delta}{2}$ and therefore $[g \circ \xi_\alpha](a_\alpha) \neq [g \circ \xi_\alpha](b_\alpha)$.

Now if $\lambda_{\tilde{n}d} < 0$ for any $n \in E$ and $d \in \text{supp}(\tilde{n})$ then fix any $n \in E'$. We still have (22) and hence $g(\tilde{n}) = -\mu(\tilde{n}, P_n) - \mu(\tilde{n}, V' \setminus V_n) + l_n \leq -\mu(\tilde{n}, P_n) + l_n$, so we can apply (16) and (21) to see that $g(\tilde{n}) < -\delta + l + \frac{\delta}{4} = l - \frac{3}{4}\delta$.

If $n \in E \setminus E'$ then $V' \setminus V_n = V'$, so $g(\tilde{n}) = -\mu(\tilde{n}, V' \setminus V_n) + l_n \geq l_n - \frac{\delta}{4} > l - \frac{\delta}{4}$. Recalling again that $a_\alpha \in \text{cl}_{\beta\omega}(E')$ and $b_\alpha \in \text{cl}_{\beta\omega}(E \setminus E')$ we obtain the inequality $[g \circ \xi_\alpha](a_\alpha) \leq l - \frac{3}{4}\delta$ while $[g \circ \xi_\alpha](b_\alpha) \geq l - \frac{\delta}{4}$ and therefore $[g \circ \xi_\alpha](a_\alpha) \neq [g \circ \xi_\alpha](b_\alpha)$.

Thus we carried out the induction step in Case 1.

**Case 2.** For any point $t \in K_{A_\alpha}$ and $\delta > 0$ there exists $U \in \tau(t, K_{A_\alpha})$ such that the set $N(U, \delta) = \{n \in E : \mu(\tilde{n}, U \cap S_n) > \delta\}$ is finite.

Observe first that we can replace "point" by "finite set" in the formulation of Case 2, i.e., for any finite set $P \subset K_{A_\alpha}$ there exist $U \in \tau(P, K_{A_\alpha})$ such that the set $N(U, \delta)$ is finite. To see this take an enumeration $\{x_1, \ldots, x_k\}$ of the set $P$ and pick a set $U_i \in \tau(x_i, K_{A_\alpha})$ such that the set $N(U_i, \frac{\delta}{k})$ is finite for any $i = 1, \ldots, k$; then $U = \bigcup \{U_i : 1 \leq i \leq k\}$ is an open neighborhood of the set $P$. It is immediate that $N(U, \delta) \subset \bigcup \{N(U_i, \frac{\delta}{k}) : 1 \leq i \leq k\}$, so $N(U, \delta)$ is a finite set as promised.

Take any number $n_0 \in E$ and let $\delta = \frac{\varepsilon}{8}$; by the property (15), there exists $U \in \tau(\text{supp}(\tilde{n}_0), K_{A_\alpha})$ such that $N(U, \frac{\varepsilon}{8})$ is finite and $U \cap \xi_\alpha(E) = \emptyset$. Let $i_0 = 0$ and pick disjoint clopen sets $U_{n_0}$ and $V_{n_0}$ such that $S_{n_0} \subset V_{n_0}$, $\text{supp}(\tilde{n}_0) \backslash S_{n_0} \subset U_{n_0}$ and $U_{n_0} \cup V_{n_0} \subset U$.

Proceeding inductively assume that $k \in \omega$ and we have an increasing sequence $\{i_0, \ldots, i_k\} \subset E$ together with a disjoint family $\{U_{n_{i_j}}, V_{n_{i_j}} : j \leq k\}$ of clopen subsets of $K_{A_\alpha}$ with the following properties:

(24) the set $E_j = N(U_{n_{i_j}} \cup V_{n_{i_j}}, 2^{-j-3}\varepsilon)$ is finite for any $j \leq k$;

(25) $(U_{n_{i_j}} \cup V_{n_{i_j}}) \cap \xi_\alpha(E) = \emptyset$ for any $j \leq k$;

(26) $n_{i_j} \in E \backslash (\bigcup\{E_l : l < j\})$ for all $j \leq k$;

(27) $S_{n_{i_j}} \backslash (\bigcup\{U_{n_{i_l}} \cup V_{n_{i_l}} : l < j\}) \subset V_{n_{i_j}}$ for all $j \leq k$;

(28) $\text{supp}(\tilde{n}_{i_j}) \backslash (S_{n_{i_j}} \cup (\bigcup\{U_{n_{i_l}} \cup V_{n_{i_l}} : l < j\})) \subset U_{n_{i_j}}$ whenever $j \leq k$.

Take $i_{k+1} > i_k$ such that $n_{i_{k+1}} \in E \backslash \bigcup\{E_l : l \leq k\}$ and apply (15) to find an open neighborhood $U$ of the set $P = \text{supp}(\tilde{n}_{i_{k+1}}) \backslash (\bigcup\{U_{n_{i_j}} \cup V_{n_{i_j}} : j \leq k\})$ such that the set $N(U, 2^{-k-4}\varepsilon)$ is finite and $U \cap (\xi_\alpha(E) \cup (\bigcup\{U_{n_{i_j}} \cup V_{n_{i_j}} : j \leq k\})) = \emptyset$. It is easy to find disjoint clopen sets $U_{n_{i_{k+1}}}$ and $V_{n_{i_{k+1}}}$ in the space $K_{A_\alpha}$ such that $S_{n_{i_{k+1}}} \cap P \subset V_{n_{i_{k+1}}}$, $P \backslash S_{n_{i_{k+1}}} \subset U_{n_{i_{k+1}}}$ and $U_{n_{k+1}} \cup V_{n_{k+1}} \subset U$. It is clear that the conditions (24)–(28) are now satisfied if we replace $k$ with $k+1$, so our inductive procedure can be continued to construct an increasing sequence $\{i_j : j \in \omega\} \subset \omega$ and a disjoint family $\{U_{n_{i_j}}, V_{n_{i_j}} : j \leq k\}$ such that the properties (24)–(28) hold for any $k \in \omega$.

Let $m_k = n_{i_k}$ and $P_{m_k} = S_{m_k} \backslash (\bigcup\{U_{m_j} \cup V_{m_j} : j < k\})$ for any $k \in \omega$. Observe that $\mu(\tilde{m}_k, P_{m_k}) = \mu(\tilde{m}_k, S_{m_k}) - \mu(\tilde{m}_k, (\bigcup\{U_{m_j} \cup V_{m_j} : j < k\}) \cap S_{m_k})$. Recalling that $\mu(\tilde{m}_k, S_{m_k}) > \varepsilon$ and the number $\mu(\tilde{m}_k, (\bigcup\{U_{m_j} \cup V_{m_j} : j < k\}) \cap S_{m_k})$ does not exceed $\sum\{\mu(\tilde{m}_k, (U_{m_j} \cup V_{m_j}) \cap S_{m_k}) : j < k\} \leq \sum_{j<k} 2^{-j-3}\varepsilon \leq \frac{\varepsilon}{4}$ we conclude that

(29) $\mu(\tilde{m}_k, P_{m_k}) > \varepsilon - \frac{\varepsilon}{4} = \frac{3}{4}\varepsilon$ for every $k \in \omega$.

Observe that $\{m_k : k \in \omega\} \subset E$ is an infinite set; as before, we will apply Fact 5 to the family $\{V_k \cap D_\alpha : k \in \omega\}$ to find an infinite subset of $\{m_k : k \in \omega\}$ (denoting it by $E$ again) to be able to consider that

(30) $\mu(\tilde{n}, \bigcup\{V_k \cap D_\alpha : k \in E \backslash \{n\}\}) < \frac{\varepsilon}{4}$ for any $n \in E$.

Let $R_n = \{k \in \omega : \tilde{k} \in V_n\}$ for all $n \in E$. It is easy to see that the family $\mathcal{B} = \{p_\beta^{-1}(0) \cap K_{A_\alpha}, p_\beta^{-1}(1) \cap K_{A_\alpha} : \beta \in A_\alpha\}$ is a clopen subbase in the space $K_{A_\alpha}$. Every set $V_n$ is compact, so it can be represented as a finite union of some finite intersections of the elements of $\mathcal{B}$. Thus the characteristic function of $R_n$ is a finite sum of finite products of the elements of the family $\{u_\beta, 1 - u_\beta : \beta \in A_\alpha\}$. This, together with the property (a) shows that $[\chi_{R_n}](a_\beta) = [\chi_{R_n}](b_\beta)$ for all $\beta < \alpha$ and $n \in E$. By Fact 9 we can refine the set $E$ once again to consider that

(31) $[\chi_{\bigcup\{R_n : n \in E'\}}](a_\beta) = [\chi_{\bigcup\{R_n : n \in E'\}}](b_\beta)$ for any $E' \subset E$ and $\beta \in [\omega, \alpha)$.

Next, apply Fact 10 to find distinct points $a_\alpha, b_\alpha \in \mathrm{cl}_{\beta\omega}(E)$ for which we have the equality $[u_\beta](a_\alpha) = [u_\beta](b_\alpha)$ for all $\beta < \alpha$. Choose a set $E' \subset E$ such that $a_\alpha \in \mathrm{cl}_{\beta\omega}(E')$ and $b_\alpha \notin \mathrm{cl}_{\beta\omega}(E')$; then $b_\alpha \in \mathrm{cl}_{\beta\omega}(E \backslash E')$. Let us check that for the function $u_\alpha = \chi_{\bigcup\{R_n : n \in E'\}}$ and the points $a_\alpha, b_\alpha$, the conditions (a) and (b) are satisfied.

We already saw that $V_n \cap \xi_\alpha(E) = \emptyset$ for any $n \in E$; this implies the equality $(\bigcup\{R_n : n \in E\}) \cap E = \emptyset$, so $u_\alpha(E) = \{0\}$ and hence $[u_\alpha](a_\alpha) = [u_\alpha](b_\alpha) = 0$. This, together with (31), shows that (a) is fulfilled for all $\beta \leq \alpha$.

Assume that $Q \subset \mathbb{D}^c$, the map $\pi_{A_\alpha}|Q : Q \to D_\alpha$ is a bijection and we have the inclusion $\pi_{A_\alpha \cup \{\alpha\}}(Q) \subset K_{A_\alpha \cup \{\alpha\}}$; let $h = p_\alpha \circ ((\pi_{A_\alpha}|Q)^{-1})$. Given any $n \in \omega$ there is a unique $x \in Q$ such that $\pi_{A_\alpha}(x) = \tilde{n}$; since also $y = \pi_{A_\alpha \cup \{\alpha\}}(x) \in K_{A_\alpha \cup \{\alpha\}}$ and $y|A_\alpha = x|A_\alpha = \tilde{n}$, we conclude that $y|\omega = u_n$, so it follows from density of $\xi_{A_\alpha \cup \{\alpha\}}(\omega)$ in $K_{A_\alpha \cup \{\alpha\}}$ that $y$ is the unique point of $K_{A_\alpha \cup \{\alpha\}}$ such that $y|A_\alpha = \tilde{n}$. Consequently, $y$ coincides with $\xi_{A_\alpha \cup \{\alpha\}}(n)$ and hence $p_\alpha(x) = p_\alpha(y) = u_\alpha(n)$. This proves that $h(\tilde{n}) = u_\alpha(n)$ for any $n \in \omega$.

Observe that, for each $n \in E$ and $d \in \mathrm{supp}(\tilde{n})$ the function $v = \chi_{\bigcup\{V_n : n \in E'\}}$ is locally constant at $d$, i.e., there exists a clopen set $G \in \tau(d, K_{A_\alpha})$ such that $v(G) = \{i\} \subset \mathbb{D}$. Indeed, the function $v$ is not locally constant only at the points of $\overline{V'} \backslash V'$ where $V' = \bigcup\{V_k : k \in E'\}$. However, $\mathrm{supp}(\tilde{m}_k) \subset U_{m_k} \cup V_{m_k}$, so our statement follows from the fact that $(U_{m_k} \cup V_{m_k}) \cap (\overline{V'} \backslash V') = \emptyset$ for any $k \in \omega$.

Then the sets $G' = G \times \{i\} \subset K_{A_\alpha \cup \{\alpha\}}$ and $G'' = K_{A_\alpha \cup \{\alpha\}} \backslash G'$ are clopen in $K_{A_\alpha \cup \{\alpha\}}$, disjoint and $G' \cup G'' = K_{A_\alpha \cup \{\alpha\}}$. Therefore, for every $d \in G$, the point $x = (d, i)$ is the unique element of $K_{A_\alpha \cup \{\alpha\}}$ such that $x|A_\alpha = d$.

Now if $d \in \mathrm{supp}(\tilde{n})$ for some $n \in E$ and $x \in Q$ is the unique point such that $x|A_\alpha = d$ then it follows from $x' = x|(A_\alpha \cup \{\alpha\}) \in K_{A_\alpha \cup \{\alpha\}}$ that $x'$ coincides with the unique extension of $d$ over $A_\alpha \cup \{\alpha\}$. This proves that

(32)  $h(d) = \chi_{\bigcup\{V_n : n \in E'\}}(d)$ for any $d \in \bigcup\{\mathrm{supp}(\tilde{n}) : n \in E\}$.

Let $g = \varphi_\alpha(h)$; observe first that

(33)  $g(\tilde{n}) = \sum\{\lambda_{\tilde{n}d} : d \in \mathrm{supp}(\tilde{n}) \cap (\bigcup\{V_k : k \in E'\})\}$ for any $n \in E$.

By the property (14.2) all numbers $\lambda_{\tilde{n}d}$ are of the same sign; assume first that they are positive. If $n \in E'$ then we have the equality

(34)  $g(\tilde{n}) = \sum\{\lambda_{\tilde{n}d} : d \in \mathrm{supp}(\tilde{n}) \cap V_n)\} + \sum\{\lambda_{\tilde{n}d} : d \in \mathrm{supp}(\tilde{n}) \cap (V' \backslash V_n)\}$.

Thus $g(\tilde{n}) \geq \sum\{\lambda_{\tilde{n}d} : d \in \mathrm{supp}(\tilde{n}) \cap V_n)\} = \mu(\tilde{n}, P_n) \geq \frac{3}{4}\varepsilon$ by (29).

If $n \in E \backslash E'$ then $V' \backslash V_n = V'$, so $g(\tilde{n}) = \sum\{\lambda_{\tilde{n}d} : d \in \mathrm{supp}(\tilde{n}) \cap (V' \backslash V_n)\}$ and hence it follows from (30) that $g(\tilde{n}) < \frac{\varepsilon}{4}$. Recalling that $a_\alpha \in \mathrm{cl}_{\beta\omega}(E')$ and $b_\alpha \in \mathrm{cl}_{\beta\omega}(E \backslash E')$ we conclude that $[g \circ \xi_\alpha](a_\alpha) \geq \frac{3}{4}\varepsilon$ while $[g \circ \xi_\alpha](b_\alpha) \leq \frac{\varepsilon}{4}$ and therefore we have the inequality $[g \circ \xi_\alpha](a_\alpha) > [g \circ \xi_\alpha](b_\alpha)$.

Now if $\lambda_{\tilde{n}d} < 0$ for any $n \in E$ and $d \in \mathrm{supp}(\tilde{n})$ then fix any $n \in E'$. We still have (34) and hence $g(\tilde{n}) = -\mu(\tilde{n}, P_n) - \mu(\tilde{n}, V' \backslash V_n) \leq -\mu(\tilde{n}, P_n) < -\frac{3}{4}\varepsilon$ by the property (29).

If $n \in E \setminus E'$ then $V' \setminus V_n = V'$, so $g(\tilde{n}) = -\mu(\tilde{n}, V' \setminus V_n) \geq -\frac{\varepsilon}{4}$. Recalling again that $a_\alpha \in \mathrm{cl}_{\beta\omega}(E')$ and $b_\alpha \in \mathrm{cl}_{\beta\omega}(E \setminus E')$ we obtain the inequality $[g \circ \xi_\alpha](a_\alpha) \leq -\frac{3}{4}\varepsilon$ while $[g \circ \xi_\alpha](b_\alpha) \geq -\frac{\varepsilon}{4}$ and therefore $[g \circ \xi_\alpha](a_\alpha) < [g \circ \xi_\alpha](b_\alpha)$.

Thus our induction step is also carried out in Case 2, so in all possible cases our inductive procedure can be continued to obtain the family $\{u_\alpha : \omega \leq \alpha < \mathfrak{c}\}$ and a set $\{(a_\alpha, b_\alpha) : \omega \leq \alpha < \mathfrak{c}\} \subset \beta\omega \times \beta\omega$ with the properties (a) and (b), i.e., Fact 11 is proved.

Returning to our solution again take the family $\{u_\alpha : \omega \leq \alpha < \mathfrak{c}\}$ and the set $\{(a_\alpha, b_\alpha) : \omega \leq \alpha < \mathfrak{c}\} \subset \beta\omega \times \beta\omega$ from Fact 11 and consider the infinite compact subspace $K = \overline{u(\omega)}$ of the space $\mathbb{D}^\mathfrak{c}$; recall that $u = \Delta\{u_\alpha : \alpha < \mathfrak{c}\}$ and the family $\{u_n : n \in \omega\}$ was constructed when we introduced the notation for Fact 11.

Assume that there exists a continuous linear surjection $\varphi : C_p(K) \to C_p(K) \times \mathbb{R}$. By Fact 3 there exists a countable dense subset $E$ of the space $K$ such that the map $\xi = (\pi_E \times i) \circ \varphi \circ \pi_E^{-1} : C_p(E|X) \to C_p(E|X) \times \mathbb{R}$ is a continuous linear surjection; here $i(r) = r$ for any $r \in \mathbb{R}$, i.e., $i : \mathbb{R} \to \mathbb{R}$ is the identity. By Fact 4 we can consider that $\xi$ is defined on the whole space $\mathbb{R}^E$ and $\xi : \mathbb{R}^E \to \mathbb{R}^E \times \mathbb{R}$ is an adequate map.

Observe that $u(\omega)$ is a dense set of isolated points of $K$, so $u(\omega) \subset E$. By Fact 8, we can find $\varepsilon > 0$, an infinite subset $M \subset \omega$ and a family $\{P_n : n \in M\}$ of pairwise disjoint finite subsets of $E$ such that

$(\alpha)$  $u(M) \cap (\bigcup\{P_n : n \in M\}) = \emptyset$;
$(\beta)$  $P_n \subset \mathrm{supp}(u(n))$ and $\mu(u(n), P_n) > \varepsilon$ for any $n \in M$;
$(\gamma)$  either $\lambda_{u(n)d} > 0$ for any $d \in P_n$ and $n \in M$ or $\lambda_{u(n)d} < 0$ for any $d \in P_n$ and $n \in M$.

It is easy to find a countable set $A \subset \mathfrak{c}$ such that $\omega \subset A$ and $\pi_A | E$ is injective; let $D = \pi_A(E)$, $\xi_A = u_A$ and $S_n = \pi_A(P_n)$ for any $n \in \omega$. Since $\pi = \pi_A | E : E \to D$ is a bijection, the map $\pi^* : \mathbb{R}^D \to \mathbb{R}^E$ defined by $\pi^*(f) = f \circ \pi$ for any $f \in \mathbb{R}^D$, is a linear homeomorphism such that $\pi^* | \mathbb{R}_*^D$ is a linear homeomorphism between $\mathbb{R}_*^D$ and $\mathbb{R}_*^E$ in the norm topology. Take a point $a \notin D \cup E$ and let $D^+ = D \oplus \{a\}$, $E^+ = E \oplus \{a\}$. Letting $\pi(a) = a$ we can consider that $\pi : E^+ \to D^+$ and $\xi : \mathbb{R}^E \to \mathbb{R}^{E^+}$.

Given any $f \in \mathbb{R}^D$, let $\varphi(f) = \xi(f \circ \pi) \circ \pi^{-1}$. It is easy to see that $\varphi : \mathbb{R}^D \to \mathbb{R}^{D^+}$ and $\varphi = (\pi^*)^{-1} \circ \xi \circ \pi^*$, so $\varphi$ is an adequate map.

It follows from the properties $(\alpha)$–$(\gamma)$ that the pair $(D, \varphi)$ belongs to the family $\mathcal{A}$ of all admissible pairs and hence there is $\beta \in [\omega, \mathfrak{c})$ such that $D = D_\beta$, $\varphi = \varphi_\beta$ and $A = A_\beta$. The hypothesis of the statement (b) of Fact 11 is, evidently, satisfied so, for the function $g = \varphi(p_\beta \circ \pi^{-1})$ we have $[g \circ \xi_\beta](a_\beta) \neq [g \circ \xi_\beta](b_\beta)$.

Recalling that $\xi_\beta = u_{A_\beta} = \pi_{A_\beta} \circ u$ we convince ourselves that $g \circ \xi_\beta = g \circ \pi_{A_\beta} \circ u$; by definition of $\varphi$ we have $g \circ \xi_\beta = \xi(p_\beta \circ \pi^{-1} \circ \pi) \circ \pi^{-1} \circ \pi_{A_\beta} \circ u = \xi(p_\beta | E) \circ u$.

It is clear that $p_\beta | E \in C_p(E|K)$, so $\xi(p_\beta | E) \in C_p(E|K) \times \mathbb{R}$ and hence there exists a function $q \in C_p(K)$ such that $q | E = \xi(p_\beta | E) | E$. This gives us the formula $[q \circ u](a_\beta) \neq [q \circ u](b_\beta)$.

Let $\tilde{u} : \beta\omega \to K$ be the continuous extension of $u$. It is a consequence of the property (a) of Fact 11 that

(35) $[u_\alpha](a_\gamma) = [u_\alpha](b_\gamma)$ for any $\alpha, \gamma \in [\omega, \mathfrak{c})$.

It is an easy exercise that $[u_\alpha] = p_\alpha \circ \tilde{u}$ for any $\alpha \in [\omega, \mathfrak{c})$ and hence we have the equality $\tilde{u} = \Delta\{[u_\alpha] : \alpha \in [\omega, \mathfrak{c})\}$ which implies, together with (35), that $\tilde{u}(a_\alpha) = \tilde{u}(b_\alpha)$ for any $\alpha \in [\omega, \mathfrak{c})$. In particular, $\tilde{u}(a_\beta) = \tilde{u}(b_\beta)$ and hence $(q \circ \tilde{u})(a_\beta) = (q \circ \tilde{u})(b_\beta)$. Since $[q \circ u]|\omega = q \circ u = (q \circ \tilde{u})|\omega$ and $\omega$ is dense in $\beta\omega$, the maps $[q \circ u]$ and $q \circ \tilde{u}$ coincide (see Fact 0 of S.351) and hence $[q \circ u](a_\beta) = [q \circ u](b_\beta)$; this contradiction shows that there is no linear continuous surjection of $C_p(K)$ onto $C_p(K) \times \mathbb{R}$, i.e., our solution is complete.

**V.401.** *Suppose that $\kappa$ is an infinite cardinal and $X$ is a Lindelöf $\Sigma$-space such that $t(X) \leq \kappa$ and $\pi\chi(X) \leq \kappa$. Prove that $X$ has a $\pi$-base of order $\leq \kappa$. In particular, if $X$ is a Lindelöf $\Sigma$-space with $t(X) = \pi\chi(X) = \omega$ then $X$ has a point-countable $\pi$-base. Deduce from this fact that any Lindelöf $\Sigma$-space $X$ with $\chi(X) \leq \kappa$ has a $\pi$-base of order $\leq \kappa$. In particular, any first countable Lindelöf $\Sigma$-space has a point-countable $\pi$-base.*

**Solution.** Given a family $\mathcal{A}$ of subsets of $X$ by $\bigwedge \mathcal{A}$ we denote the family of all finite intersections of the elements of $\mathcal{A}$. Recall that $\mathrm{ord}(\mathcal{A}) \leq \kappa$ if every point of $X$ belongs to at most $\kappa$-many elements of $\mathcal{A}$.

If $J$ is the set of isolated points of $X$ then the family $\mathcal{J} = \{\{x\} : x \in J\}$ is a disjoint $\pi$-base at every point of $\overline{J}$, so it suffices to construct the promised $\pi$-base in the space $Y = X \setminus \overline{J}$ which has no isolated points. Therefore we can assume, without loss of generality, that $Y = X$, i.e., $X$ has no isolated points and hence we can fix, for any $x \in X$, a $\pi$-base $\mathcal{B}_x$ at the point $x$ such that $|\mathcal{B}_x| \leq \kappa$ and $x \notin \overline{U}$ for any $U \in \mathcal{B}_x$.

Fix a countable family $\mathcal{F}$ of closed subsets of $X$ which is a network with respect to a compact cover $\mathcal{C}$ of the space $X$; there is no loss of generality to assume that $\mathcal{F}$ is closed under finite intersections. We will show by induction over all cardinals $\lambda \leq d(X)$ that

(1) if $A \subset X$ is a set with $|A| = \lambda$ then there is a family $\mathcal{B} \subset \tau^*(X)$ such that $|\mathcal{B}| \leq \lambda \cdot \kappa$ while $\mathcal{B}$ is a $\pi$-base at every point of $A$ and $\mathrm{ord}(\mathcal{B}) \leq \kappa$.

If $\lambda \leq \kappa$ then the family $\mathcal{B} = \bigcup\{\mathcal{B}_x : x \in A\}$ has cardinality at most $\kappa$, so (1) is true. Now assume that we are given a cardinal $\nu > \kappa$ and (1) is proved for all $\lambda < \nu$; fix a set $B \subset X$ with $|B| = \nu$. Let $\Lambda = \{\alpha + 1 : \alpha < \nu\}$ be the set of all successor ordinals smaller than $\nu$ and choose an enumeration $\{b_\alpha : \alpha \in \Lambda\}$ of the set $B$. To start our inductive construction let $G_0 = \emptyset$ and $\mathcal{D}_0 = \emptyset$. Now assume that, for some ordinal $\beta \in (0, \nu]$, we have constructed a family $\{G_\alpha : \alpha < \beta\}$ and a collection $\{\mathcal{D}_\alpha : \alpha < \beta\}$ with the following properties:

(2) $G_\alpha$ is a closed subset of $X$ and $\mathcal{D}_\alpha \subset \tau^*(X)$ for any $\alpha < \beta$;
(3) $|\mathcal{D}_\alpha| \leq |\alpha| \cdot \kappa$, $\mathrm{ord}(\mathcal{D}_\alpha) \leq \kappa$ and the family $\mathcal{D}_\alpha$ is a $\pi$-base at every point of $G_\alpha$ for any $\alpha < \beta$;

(4) $\gamma < \alpha < \beta$ implies $G_\gamma \subset G_\alpha$ and $\mathcal{D}_\gamma \subset \mathcal{D}_\alpha$;

(5) if $\alpha < \beta$ is a limit ordinal then $G_\alpha = \bigcup\{G_\gamma : \gamma < \alpha\}$ and $\mathcal{D}_\alpha = \bigcup\{\mathcal{D}_\gamma : \gamma < \alpha\}$;

(6) if $\gamma < \alpha < \beta$, $F \in \mathcal{F}$, $U \in \bigwedge \mathcal{D}_\gamma$ and $\overline{U} \cap F \neq \emptyset$ then $G_\alpha \cap (\overline{U} \cap F) \neq \emptyset$;

(7) if $\gamma < \alpha < \beta$ then $\overline{U} \cap G_\gamma = \emptyset$ for any $U \in \mathcal{D}_\alpha \backslash \mathcal{D}_\gamma$;

(8) if $\alpha < \beta$ and $\alpha \in \Lambda$ then $b_\alpha \in G_\alpha$.

Suppose first that $\beta$ is a successor ordinal, say $\beta = \gamma + 1$ and consider the family $\mathcal{H} = \{H : \text{there exists } U \in \bigwedge \mathcal{D}_\gamma \text{ and } F \in \mathcal{F} \text{ such that } \emptyset \neq H = \overline{U} \cap F \subset X \backslash G_\gamma\}$. For any $H \in \mathcal{H}$ pick a point $x_H \in H$; then the set $Z = (\{b_\beta\} \cup \{x_H : H \in \mathcal{H}\}) \backslash G_\gamma$ has cardinality at most $|\mathcal{H}| \leq |\beta| \cdot \kappa < \nu$, so the induction hypothesis is applicable to fix a family $\mathcal{D}'_\beta \subset \tau^*(X)$ such that $\mathcal{D}'_\beta$ is a $\pi$-base at every point of $Z$ while $|\mathcal{D}'_\beta| \leq |\beta| \cdot \kappa$, $\text{ord}(\mathcal{D}'_\beta) \leq \kappa$ and $\overline{U} \cap G_\gamma = \emptyset$ for any $U \in \mathcal{D}'_\beta$. Letting $\mathcal{D}_\beta = \mathcal{D}_\gamma \cup \mathcal{D}'_\beta$ and $G_\beta = G_\gamma \cup Z$ we obtain a family $\{G_\alpha : \alpha \leq \beta\}$ and a collection $\{\mathcal{D}_\alpha : \alpha \leq \beta\}$ such that the properties (2)–(8) are fulfilled for all $\alpha \leq \beta$.

If $\beta$ is a limit ordinal then let $G_\beta = \bigcup\{G_\alpha : \alpha < \beta\}$ and $\mathcal{D}_\beta = \bigcup\{\mathcal{D}_\alpha : \alpha < \beta\}$. It is immediate that the properties (2),(4)–(8) and the second part of (3) are fulfilled for all $\alpha \leq \beta$. To see that the first part of (3) is also fulfilled consider two cases.

a) The cofinality of $\beta$ does not exceed $\kappa$. Then it follows from (4) that $\mathcal{D}_\beta$ is a union of $\leq \kappa$-many families of order $\leq \kappa$, so $\text{ord}(\mathcal{D}_\beta) \leq \kappa$.

b) If $\text{cf}(\beta) > \kappa$ then it follows from (4) and $t(X) \leq \kappa$ that $G_\beta = \bigcup\{G_\alpha : \alpha < \beta\}$. Assume that there is a family $\mathcal{U} \subset \mathcal{D}_\beta$ such that $|\mathcal{U}| = \kappa^+$ and $P = \bigcap \mathcal{U} \neq \emptyset$. Pick a set $C \in \mathcal{C}$ such that $H = C \cap P \neq \emptyset$. We claim that

(9) the family $\mathcal{U}_C = \{\overline{U} \cap C \cap G_\beta : U \in \mathcal{U}\}$ is centered.

Indeed, fix a finite family $\mathcal{V} \subset \mathcal{U}$ and let $V = \bigcap \mathcal{V}$; it follows from (4) that there is $\alpha < \beta$ such that $\mathcal{V} \subset \mathcal{D}_\alpha$. If $F \in \mathcal{F}$ and $C \subset F$ then $F \cap V \supset H \neq \emptyset$, so it follows from (6) that $G_{\alpha+1} \cap F \cap \overline{V} \neq \emptyset$ which implies that $(G_\beta \cap \overline{V}) \cap F \neq \emptyset$ for any $F \in \mathcal{F}$ with $C \subset F$. Since the family $\{F \in \mathcal{F} : C \subset F\}$ is a network at $C$, the closed set $G_\beta \cap \overline{V}$ has to meet $C$. As a consequence, $\emptyset \neq G_\beta \cap \overline{V} \cap C \subset \bigcap\{\overline{U} \cap C \cap G_\beta : U \in \mathcal{V}\}$ which settles (9).

Since the family $\mathcal{U}_C$ consists of compact subsets of $X$, it follows from (9) that there is a point $x \in G_\beta \cap (\bigcap\{\overline{U} : U \in \mathcal{U}\})$. Consequently, there is $\alpha < \beta$ such that $x \in G_\alpha$. The property (3) shows that the family $\mathcal{U}' = \mathcal{U} \cap \mathcal{D}_\alpha$ has cardinality at most $\kappa$, so the collection $\mathcal{W} = \mathcal{U} \backslash \mathcal{U}' \subset \mathcal{D}_\beta \backslash \mathcal{D}_\alpha$ is nonempty. Besides, $\overline{U} \cap G_\alpha = \emptyset$ for any $U \in \mathcal{W}$ by (7). This contradiction with $x \in \overline{U} \cap G_\alpha$ for any $U \in \mathcal{U}$ shows that $\text{ord}(\mathcal{D}_\beta) \leq \kappa$ and hence our inductive procedure gives us a family $\mathcal{D}_\nu$ for which the properties (2)–(8) are fulfilled and, in particular, $\mathcal{B} = \mathcal{D}_\nu$ satisfies (1).

Next, apply (1) to a dense set $A \subset X$ with $|A| = d(X)$; the respective family $\mathcal{B}$ will be a $\pi$-base of $X$ of order at most $\kappa$. Finally, if $X$ is a Lindelöf $\Sigma$-space with $\chi(X) \leq \kappa$ then $t(X) \leq \kappa$ and $\pi\chi(X) \leq \kappa$, so $X$ has a $\pi$-base of order at most $\kappa$.

**V.402.** *Given a space $X$ and an infinite cardinal $\kappa$ suppose that $\pi\chi(X) \leq \kappa$ and $d(X) \leq \kappa^+$. Prove that $X$ has a $\pi$-base of order $\leq \kappa$. In particular, if $\pi\chi(X) \leq \omega$ and $d(X) \leq \omega_1$ then $X$ has a point-countable $\pi$-base.*

**Solution.** For any $x \in X$ choose a $\pi$-base $\mathcal{B}_x$ at the point $x$ such that $|\mathcal{B}_x| \leq \kappa$. If $X$ has a dense subspace $D$ of cardinality $\leq \kappa$ then $\mathcal{B} = \bigcup\{\mathcal{B}_x : x \in D\}$ is easily seen to be a $\pi$-base of $X$ of cardinality at most $\kappa$ and hence $\operatorname{ord}(\mathcal{B}) \leq \kappa$. Therefore we can assume, without loss of generality, that $d(X) = \kappa^+$.

Fix a set $D = \{d_\alpha : \alpha < \kappa^+\}$ which is dense in $X$ and let $x_0 = d_0$. Proceeding inductively, assume that $\beta < \kappa^+$ and we have chosen points $\{x_\gamma : \gamma < \beta\} \subset D$ in such a way that

(1) if $\alpha' < \alpha < \beta$ then $d_{\alpha'} \in \overline{\{x_\gamma : \gamma \leq \alpha\}}$;

(2) $x_\alpha \notin \overline{\{x_\gamma : \gamma < \alpha\}}$ for any $\alpha < \beta$.

It follows from $d(X) > \kappa$, that $Y = \{x_\gamma : \gamma < \beta\}$ cannot be dense in $X$, so the set $A = \{\gamma < \kappa^+ : d_\gamma \in X \backslash \overline{Y}\}$ is nonempty; let $\mu = \backslash n A$ and $x_\beta = d_\mu$. It is clear that the condition (2) now holds for all $\alpha \leq \beta$.

To check (1) suppose that $\alpha' < \alpha \leq \beta$. If $\alpha < \beta$ then $d_{\alpha'} \in \overline{\{x_\gamma : \gamma \leq \alpha\}}$ by the induction hypothesis. If $\alpha = \beta$ and there exists $\nu < \beta$ with $\alpha' < \nu$ then we have $d_{\alpha'} \in \overline{\{x_\gamma : \gamma \leq \nu\}} \subset \overline{\{x_\gamma : \gamma \leq \alpha\}}$, so (1) holds in this case too. Finally, if $\alpha = \beta = \alpha' + 1$ then $d_\nu \in \overline{\{x_\gamma : \gamma \leq \alpha'\}} = \overline{\{x_\gamma : \gamma < \beta\}}$ for any $\nu < \alpha'$ which shows that either $d_{\alpha'} \in \overline{\{x_\gamma : \gamma < \beta\}}$ or $\alpha' = \mu$ and hence $d_{\alpha'} = d_\mu = x_\beta \in \overline{\{x_\gamma : \gamma \leq \beta\}}$, so we proved that the conditions (1) and (2) are satisfied for all $\alpha \leq \beta$.

This shows that our inductive procedure can be continued to construct a set $E = \{x_\gamma : \gamma < \kappa^+\}$ such that the properties (1) and (2) hold for all $\beta < \kappa^+$. It follows from (1) that $E$ is dense in $X$. It is an easy consequence of (2) that the family $\mathcal{C}_\alpha = \{B \in \mathcal{B}_{x_\alpha} : B \cap \{x_\gamma : \gamma < \alpha\} = \emptyset\}$ is a $\pi$-base at the point $x_\alpha$ for every $\alpha < \kappa^+$. By density of $E$, the family $\mathcal{C} = \bigcup\{\mathcal{C}_\alpha : \alpha < \kappa^+\}$ is a $\pi$-base in $X$.

If $\operatorname{ord}(\mathcal{C}) \geq \kappa^+$ then fix a point $x \in X$ and a family $\mathcal{C}' \subset \mathcal{C}$ such that $|\mathcal{C}'| = \kappa^+$ and $x \in \bigcap \mathcal{C}'$. For each $C \in \mathcal{C}'$ there is $B_C \in \mathcal{B}_x$ with $B_C \subset C$; since $|\mathcal{B}_x| \leq \kappa$, we can find a set $B \in \mathcal{B}_x$ for which the family $\mathcal{C}'' = \{C \in \mathcal{C}' : B_C = B\}$ has cardinality $\kappa^+$. The set $E$ being dense in $X$ there is $\alpha < \kappa$ such that $x_\alpha \in B$ and therefore $x_\alpha \in C$ for all $C \in \mathcal{C}''$. The family $\mathcal{B} = \bigcup\{\mathcal{B}_{x_\beta} : \beta \leq \alpha\}$ has cardinality not exceeding $\kappa$, so there is $C \in \mathcal{C}'' \backslash \mathcal{B}$. Recalling the definition of $\mathcal{C}$ we convince ourselves that $C \in \mathcal{C}_\gamma$ for some $\gamma > \alpha$ and hence $x_\alpha \notin C$; this contradiction shows that $\mathcal{C}$ is a $\pi$-base in $X$ with $\operatorname{ord}(\mathcal{C}) \leq \kappa$.

**V.403.** *Assuming CH prove that*

(i) *any Lindelöf first countable space has a point-countable $\pi$-base;*

(ii) *any space $X$ with $\chi(X) = c(X) = \omega$ has a point-countable $\pi$-base;*

(iii) *if $\omega_1$ is a caliber of $X$ and $\chi(X) \leq \omega$ then $X$ is separable.*

**Solution.** Given a set $A$ and an infinite cardinal $\kappa$ let $\exp_\kappa(A) = \{B : B \subset A$ and $|B| \leq \kappa\}$.

**Fact 1.** For any space $X$ we have $|X| \leq d(X)^{t(X) \cdot \psi(X)}$. In particular, the inequality $|X| \leq d(X)^{\chi(X)}$ holds for every space $X$.

**Proof.** Suppose that $d(X) = \lambda$ and $t(X) \cdot \psi(X) \leq \kappa$; choose a dense subset $D$ of the space $X$ such that $|D| = \lambda$. By $\psi(X) \leq \kappa$ we can choose, for any point

$x \in X$, a family $\mathcal{B}_x$ of open neighborhoods of $x$ in $X$ such that $|\mathcal{B}_x| \leq \kappa$ and $\{x\} = \bigcap\{\overline{B} : B \in \mathcal{B}_x\}$; it follows from $t(X) \leq \kappa$ that we can find, for any $B \in \mathcal{B}_x$, a set $S_B^x \subset D \cap B$ such that $|S_B^x| \leq \kappa$ and $x \in \overline{S_B^x}$.

Let $\varphi(x) = \{S_B^x : B \in \mathcal{B}_x\}$ for every $x \in X$; then $\varphi : X \to \exp_\kappa(\exp_\kappa(D))$. Since $|\exp_\kappa(\exp_\kappa(D))| = (|D|^\kappa)^\kappa = \lambda^{\kappa \cdot \kappa} = \lambda^\kappa$, it suffices to show that $\varphi$ is an injective map. Take any distinct points $x, y \in X$ and pick $B \in \mathcal{B}_x$ such that $y \notin \overline{B}$. Then $y \notin \overline{S_B^x}$ and hence $S_B^x \notin \varphi(y)$ because all elements of $\varphi(y)$ contain $y$ in their closure. Therefore $\varphi(y) \neq \varphi(x)$ for any distinct $x, y \in X$, so the map $\varphi$ is injective and hence $|X| \leq |\exp_\kappa(\exp_\kappa(D))| = \lambda^\kappa$, i.e., we established that $|X| \leq d(X)^{t(X) \cdot \psi(X)}$ for any space $X$; observe that $t(X) \cdot \psi(X) \leq \chi(X)$ and hence also $|X| \leq d(X)^{\chi(X)}$, so Fact 1 is proved.

*Fact 2.* If $X$ is an arbitrary space then $|X| \leq 2^{l(X) \cdot \psi(X) \cdot t(X)}$. In particular, we have the inequality $|X| \leq 2^{l(X) \cdot \chi(X)}$ for any space $X$.

*Proof.* Let $\kappa = l(X) \cdot \psi(X) \cdot t(X)$ and fix a family $\mathcal{B}_x \subset \tau(x, X)$ such that $|\mathcal{B}_x| \leq \kappa$ and $\{x\} = \bigcap \mathcal{B}_x$ for each $x \in X$. For any set $A \subset X$ we will need the family $\mathcal{B}(A) = \bigcup\{\mathcal{B}_x : x \in A\}$; take any point $x_0 \in X$ and let $F_0 = \{x_0\}$. Suppose that $\alpha < \kappa^+$ and we have a family $\{F_\beta : \beta < \alpha\}$ of closed subsets of $X$ with the following properties:

(1) $|F_\beta| \leq 2^\kappa$ for each $\beta < \alpha$;
(2) $F_{\alpha'} \subset F_\alpha$ whenever $\alpha' \leq \alpha < \beta$;
(3) if $\alpha < \beta$ and $\mathcal{U}$ is a subfamily of $\mathcal{B}(\bigcup_{\gamma < \alpha} F_\gamma)$ such that $|\mathcal{U}| \leq \kappa$ and $X \setminus \bigcup \mathcal{U} \neq \emptyset$ then $F_\alpha \setminus \bigcup \mathcal{U} \neq \emptyset$.

Consider the collection $\mathbb{A} = \{\mathcal{U} : \mathcal{U}$ is a subfamily of $\mathcal{B}(\bigcup_{\alpha < \beta} F_\alpha)$ such that $|\mathcal{U}| \leq \kappa$ and $\bigcup \mathcal{U} \neq X\}$. Observe first that $|\bigcup_{\alpha < \beta} F_\alpha| \leq \kappa \cdot 2^\kappa = 2^\kappa$ and therefore the family $\mathcal{B}(\bigcup_{\alpha < \beta} F_\alpha)$ has cardinality at most $2^\kappa \cdot \kappa = 2^\kappa$. As a consequence, $|\mathbb{A}| \leq (2^\kappa)^\kappa = 2^\kappa$. For any $\mathcal{U} \in \mathbb{A}$ choose a point $x(\mathcal{U}) \in X \setminus \bigcup \mathcal{U}$ and consider the set $P = \{x(\mathcal{U}) : \mathcal{U} \in \mathbb{A}\} \cup (\bigcup_{\alpha < \beta} F_\alpha)$.

If $F_\beta = \overline{P}$ then $d(F_\beta) \leq |P| \leq 2^\kappa + 2^\kappa = 2^\kappa$, so it follows from Fact 1 and $t(X) \cdot \psi(X) \leq \kappa$ that $|F_\beta| \leq |P|^\kappa \leq (2^\kappa)^\kappa = 2^\kappa$. It is straightforward that the conditions (1)–(3) are now satisfied for all $\alpha \leq \beta$ and hence our inductive procedure can be continued to obtain a family $\{F_\alpha : \alpha < \kappa^+\}$ of closed subsets of $X$ for which the properties (1)–(3) hold for all $\beta < \kappa^+$.

If $F = \bigcup\{F_\alpha : \alpha < \kappa^+\}$ then $|F| \leq 2^\kappa \cdot \kappa^+ = 2^\kappa$, so it suffices to show that $F = X$. Note first that it follows from $t(X) \leq \kappa$ that the set $F$ is closed in $X$. If $F \neq X$ then fix a point $a \in X \setminus F$ and choose, for any $x \in F$ a set $V_x \in \mathcal{B}_x$ such that $a \notin V_x$. The family $\mathcal{V} = \{V_x : x \in F\}$ is an open cover of $F$, so it follows from $l(F) \leq l(X) \leq \kappa$ that we can find a family $\mathcal{U} \subset \mathcal{V}$ such that $|\mathcal{U}| \leq \kappa$ and $F \subset \bigcup \mathcal{U}$.

It is easy to see that there exists an ordinal $\alpha < \kappa^+$ such that $\mathcal{U} \subset \mathcal{B}(\bigcup_{\gamma < \alpha} F_\gamma)$; since $a \notin \bigcup \mathcal{U}$, the set $X \setminus \bigcup \mathcal{U}$ is nonempty, so we can apply (3) to see that $F_\alpha \setminus (\bigcup \mathcal{U}) \neq \emptyset$. However, $F_\alpha \subset F \subset \bigcup \mathcal{U}$; this contradiction shows that $F = X$ and hence we checked that $|X| \leq 2^{l(X) \cdot t(X) \cdot \psi(X)}$ for any space $X$; finally it follows from $t(X) \cdot \psi(X) \leq \chi(X)$ that $|X| \leq 2^{l(X) \cdot \chi(X)}$ and hence Fact 2 is proved.

*Fact 3.* For any space $X$ we have the inequality $|X| \leq 2^{c(X) \cdot \chi(X)}$.

*Proof.* Apply Fact 1 to see that $|X| \leq d(X)^{\chi(X)} \leq w(X)^{\chi(X)}$; we also have the inequalities $w(X) \leq \pi\chi(X)^{c(X)} \leq \chi(X)^{c(X)} \leq (2^{\chi(X)})^{c(X)} = 2^{\chi(X) \cdot c(X)}$ (see Fact 4 of U.083) and hence $|X| \leq (2^{\chi(X) \cdot c(X)})^{\chi(X)} = 2^{\chi(X) \cdot c(X)}$, i.e., Fact 3 is proved.

Returning to our solution observe that if $X$ is a Lindelöf first countable space then $|X| \leq 2^{l(X) \cdot \chi(X)} = 2^{\omega}$ by Fact 2. Recalling that we are also assuming CH, we conclude that $d(X) \leq |X| \leq \omega_1$. If $\chi(X) = c(X) = \omega$ then $|X| \leq 2^{\chi(X) \cdot c(X)} \leq 2^{\omega}$ (see Fact 3), so again $d(X) \leq |X| \leq \omega_1$. If $\omega_1$ is a caliber of $X$ then $c(X) \leq \omega$, so in cases (i)–(iii) we have a space $X$ such that $\chi(X) \leq \omega$ and $d(X) \leq \omega_1$. Therefore we can apply Problem 402 to see that $X$ has a point-countable $\pi$-base; this settles (i) and (ii). In the case (iii) the cardinal $\omega_1$ is a caliber of $X$, so the relevant point-countable $\pi$-base of $X$ has to be countable and hence $d(X) \leq \pi w(X) = \omega$ which shows that our solution is complete.

**V.404.** *Give an example of a first countable space which has no point-countable $\pi$-base.*

**Solution.** Let $\omega_i$ be the $i$-th uncountable ordinal with its usual well-ordering and the discrete topology for any $n \in \mathbb{N}$ and consider the set $X = \prod\{\omega_i : i \in \mathbb{N}\}$. Given $x, y \in X$ we say that $x \leq y$ if $x(n) \leq y(n)$ for all $n \in \mathbb{N}$. We will also need the set $R_x = \{y \in X : x \leq y\}$ for every $x \in X$. If $x \in X$ and $n \in \mathbb{N}$ then $U_n(x) = \{y \in X : y(i) = x(i)$ for all $i \leq n\}$. It is evident that the family $\mathcal{S} = \{R_x : x \in X\} \cup \{U_n(x) : x \in X, n \in \mathbb{N}\}$ covers $X$, so there exists a unique topology $\tau$ on the set $X$ such that $\mathcal{S}$ is a subbase of $(X, \tau)$; from this moment on we will consider that $X$ is the space $(X, \tau)$.

It is easy to see that the family $\{U_n(x) : x \in X, n \in \mathbb{N}\}$ is a base for the product topology $\tau_p$ on the set $X$. Since $(X, \tau_p)$ is a metrizable space (being a countable product of discrete and hence metrizable spaces), it follows from $\tau_p \subset \tau$ that $X$ is a Hausdorff space.

Let $O_n(x) = U_n(x) \cap R_x$ for any $x \in X$ and $n \in \mathbb{N}$. We claim that

(1) for any $x \in X$ the family $\{O_n(x) : n \in \mathbb{N}\}$ is a base of $X$ at the point $x$; therefore the space $X$ is first countable.

To prove (1) note first that $y \in U_n(x)$ implies that $U_n(y) \subset U_n(x)$; besides, $y \in R_x$ implies $R_y \subset R_x$. Therefore $O_n(y) \subset O_n(x)$ whenever $y \in O_n(x)$. Now assume that $x \in U \in \tau$. Since the family $\mathcal{S}$ is a subbase of the space $X$, we can find points $x_1, \ldots, x_k \in X$, a finite set $A \subset X$ and $n_1, \ldots, n_k \in \mathbb{N}$ such that $x \in W = (\bigcap\{U_{n_i}(x_i) : i \leq k\}) \cap (\bigcap\{R_y : y \in A\}) \subset U$. If $n = \max\{n_i : i \leq k\}$ then $U_n(x) \subset U_{n_i}(x) \subset U_{n_i}(x_i)$ for all $i \leq k$; besides, it follows from $x \in R_y$ that $R_x \subset R_y$ for all $y \in A$. An immediate consequence is that $x \in O_n(x) \subset W \subset U$ and hence (1) is proved.

Given $x \in X$ and $n \in \mathbb{N}$ take any point $y \in X \setminus O_n(x)$; if $y \notin U_n(x)$ then $U_n(y) \cap U_n(x) = \emptyset$ and hence $U_n(y) \cap O_n(x) = \emptyset$. If $y \in U_n(x)$ then $y \notin R_x$ and hence there exists $k > n$ such that $y(k) < x(k)$. It is clear that $U_k(y) \cap R_x = \emptyset$ and hence $U_k(y) \cap O_n(x) = \emptyset$. This shows that every $y \in X \setminus O_n(x)$ has a neighborhood contained in $X \setminus O_n(x)$, i.e., the set $X \setminus O_n(x)$ is open and hence

(2) the set $O_n(x)$ is clopen in $X$ for any $x \in X$ and $n \in \mathbb{N}$.

An immediate consequence of (1) and (2) is that the space $X$ is zero-dimensional and hence Tychonoff (see Fact 1 of S.232). Let $\omega_\omega = \sup\{\omega_i : i \in \mathbb{N}\}$; our next step is to prove that $d(X) > \omega_\omega$. Assume that $D$ is a dense subset of $X$. If $|D| \leq \omega_\omega$ then $D = \bigcup\{D_i : i \in \mathbb{N}\}$ where $|D_i| \leq \omega_i$ for any $i \in \mathbb{N}$. For each $i \in \mathbb{N}$ the cofinality of $\omega_{i+1}$ is strictly greater than the cardinality of $D_i$, so there is an ordinal $\alpha_i \in \omega_{i+1}$ such that $x(i+1) < \alpha_i$ for any $x \in D_i$. Let $x(1) = 0$ and $x(i+1) = \alpha_i$ for any $i \in \mathbb{N}$.

Take any $d \in D$; then $d \in D_i$ for some $i \in \mathbb{N}$ and hence $d(i+1) < \alpha_i = x(i+1)$. This shows that $d \notin R_x$ for any $d \in D$, i.e., $D \cap R_x = \emptyset$. Since $R_x$ is a nonempty open subset of $X$, this contradicts density of $D$ and proves that

(3) $\pi w(X) \geq d(X) > \omega_\omega$.

Finally assume that $X$ has a point-countable $\pi$-base $\mathcal{B}$ and let $\kappa = (\omega_\omega)^+$, i.e., $\kappa$ is the cardinal that follows $\omega_\omega$. It follows from (1) that we can assume, without loss of generality, that every element of $\mathcal{B}$ is the set $O_n(x)$ for some $x \in X$ and $n \in \mathbb{N}$. Therefore $\mathcal{B} = \bigcup\{\mathcal{B}_n : n \in \mathbb{N}\}$ where $\mathcal{B}_n = \{O_n(x) : x \in A_n\}$ for every $n \in \mathbb{N}$.

The property (3) implies that $|\mathcal{B}| \geq \kappa$; since the cardinal $\kappa$ is regular, there is $n \in \mathbb{N}$ for which $|A_n| = \kappa$. Since $\kappa > \omega_n = |\omega_1 \times \ldots \times \omega_n|$, we can apply regularity of $\kappa$ once more to find a set $A \subset A_n$ and a sequence $(\alpha_1, \ldots, \alpha_n) \in \omega_1 \times \ldots \times \omega_n$ such that $|A| = \kappa$ and $x(i) = \alpha_i$ for all $x \in A$.

Choose a set $A' \subset A$ with $|A'| = \omega_1$ and fix any $i > n$. Since the cofinality of $\omega_i$ is strictly greater than $\omega_1$, we can choose an ordinal $\beta_i \in \omega_i$ such that $x(i) < \beta_i$ for all $x \in A'$. Define a point $y \in X$ as follows: $y(i) = \alpha_i$ for all $i \leq n$ and $y(i) = \beta_i$ whenever $i > n$. It is immediate that $y \in R_x$ and hence $y \in O_n(x)$ for any $x \in A'$. Therefore $y \in \bigcap\{O_n(x) : x \in A'\}$, so the order of the family $\mathcal{B}$ at the point $y$ is at least $\omega_1$, i.e., $\mathcal{B}$ is not point-countable. This contradiction shows that $X$ is a first countable space which has no point-countable $\pi$-base.

**V.405.** *Let $X$ be a space for which we can find a family of sets $\{A_m : m \in \omega\}$ and a sequence $\{k_m : m \in \omega\} \subset \mathbb{N} \setminus \{1\}$ such that $\sup\{|A_m| : m \in \omega\} = |X|$ while $A_m \subset X^{k_m} \setminus \Delta_{k_m}(X)$ and $A_m$ is concentrated around $\Delta_{k_m}(X)$ for every $m \in \omega$. Prove that the space $C_p(X)$ has a point-countable $\pi$-base. In particular, the space $C_p(X)$ has a point-countable $\pi$-base if there is a set $A$ with $|A| = |X|$ such that either $A \subset X^n \setminus \Delta_n(X)$ and $A$ is concentrated around $\Delta_n(X)$ or $A \subset X$ and $A$ is concentrated around some point of $X$.*

**Solution.** Given a space $Z$ call a family $\mathcal{U} \subset \tau^*(Z)$ a $\pi$-base of a family $\mathcal{V} \subset \tau^*(Z)$ if any element of $\mathcal{V}$ contains some $U \in \mathcal{U}$. It is evident that if $\mathcal{V}$ is a $\pi$-base in $Z$ and $\mathcal{U}$ is a $\pi$-base of $\mathcal{V}$ then $\mathcal{U}$ is also a $\pi$-base in $Z$.

Denote by $\mathcal{I}$ the family of all nontrivial open intervals of $\mathbb{R}$ with rational endpoints. The collection $\mathcal{Q} = \{(Q_0, \ldots, Q_{n-1}) : n \in \mathbb{N} \setminus \{1\}, Q_i \in \mathcal{I}$ for any

$i < n$ and the family $\{\overline{Q}_0, \ldots, \overline{Q}_{n-1}\}$ is disjoint$\}$ is countable. If $x = (x_0, \ldots, x_{n-1}) \in X^n$ and $Q = (Q_0, \ldots, Q_{n-1}) \in \mathcal{Q}$ then let $[x, Q] = \{f \in C_p(X) : f(x_i) \in Q_i \text{ for all } i < n\}$. It is straightforward that the family $\mathcal{B} = \{[x, Q] : \text{there is a number } n \in \mathbb{N}\setminus\{1\} \text{ such that } x \in X^n\setminus\Delta_n(X) \text{ and } Q \in \mathcal{Q} \cap \mathcal{I}^n\}$ is a $\pi$-base in $C_p(X)$. For any $Q = (Q_0, \ldots, Q_{n-1}) \in \mathcal{Q}$ let $\mathcal{B}_Q = \{[x, Q] : x \in X^n\setminus\Delta_n(X)\}$; it is evident that $\mathcal{B} = \bigcup\{\mathcal{B}_Q : Q \in \mathcal{Q}\}$. The family $\mathcal{Q}$ being countable it suffices to find a point-countable $\pi$-base for every $\mathcal{B}_Q$, so fix any $Q = (Q_0, \ldots, Q_{n-1}) \in \mathcal{Q}$.

Our assumptions about the sequence $\{A_m : m \in \omega\}$ make it possible to choose a family $\{Y_m : m \in \omega\}$ of subsets of $X^n\setminus\Delta_n(X)$ and a sequence $\{\varphi_m : m \in \omega\}$ such that every $\varphi_m : Y_m \to A_m$ is an injection and $\bigcup\{Y_m : m \in \omega\} = X^n\setminus\Delta_n(X)$. Then $\mathcal{B}_Q = \bigcup\{\mathcal{B}_Q^m : m \in \omega\}$ where $\mathcal{B}_Q^m = \{[x, Q] : x \in Y_m\}$ for any $m \in \omega$. Thus it suffices to find a point-countable $\pi$-base for every $\mathcal{B}_Q^m$, so fix $m \in \omega$.

We still have to split $\mathcal{B}_Q^m$ into finitely many subfamilies. Given any point $x \in Y_m$ let $y = \varphi_m(x)$ and $A_x = \{i < n : x(i) \in \{y(0), \ldots, y(k_m-1)\}\}$. For any set $A \subset n$ consider the family $\mathcal{B}_Q^m(A) = \{[x, Q] : x \in Y_m \text{ and } A_x = A\}$; it is evident that we have the equality $\mathcal{B}_Q^m = \bigcup\{B_Q^m(A) : A \subset n\}$, so it suffices to find a point-countable $\pi$-base for every $\mathcal{B}_Q^m(A)$. To do the last splitting, for any $E \subset k_m$ consider the family $\mathcal{B}_Q^m(A, E) = \{[x, Q] \in \mathcal{B}_Q^m(A) : \{x(i) : i \in A_x\} = \{\varphi_m(x)(j) : j \in E\}\}$. It is immediate that $\mathcal{B}_Q^m(A) = \bigcup\{\mathcal{B}_Q^m(A, E) : E \subset k_m\}$, so it suffices to construct a point-countable $\pi$-base for every family $\mathcal{B}_Q^m(A, E)$.

To do this, for $r = k_m - |E|$ take an $r$-tuple $S = (S_0, \ldots, S_{r-1}) \in \mathcal{Q}$ such that the family $\{\overline{S}_0, \ldots, \overline{S}_r\} \cup \{\overline{Q}_i : i < n\}$ is disjoint and fix a bijection $b : k_m\setminus E \to r$. Given an arbitrary point $x = (x_0, \ldots, x_{n-1}) \in Y_m$ such that $[x, Q] \in \mathcal{B}_Q^m(A, E)$ and $\varphi_m(x) = (y_0, \ldots, y_{k_m-1})$, for any $i \in E$ there is a unique $j(i) \in A$ such that $x_{j(i)} = y_i$. Let $H(x) = \{f \in C_p(X) : f(x_i) \in Q_i \text{ for any } i < n \text{ and } f(y_i) \in S_{b(i)} \text{ if } i \in k_m\setminus E\}$. It is evident that $H(x) \subset [x, Q]$ is a nonempty open set, so the family $\mathcal{H} = \{H(x) : [x, Q] \in \mathcal{B}_Q^m(A, E)\}$ is a $\pi$-base for $\mathcal{B}_Q^m(A, E)$.

Now, take any function $f \in C_p(X)$. If $[x, Q] \in \mathcal{B}_Q^m(A, E)$ and $f \in H(x)$ then $\varphi_m(x)$ belongs to $W(x) = \prod\{f^{-1}(Q_{j(i)}) : i \in E\} \times \prod\{f^{-1}(S_{b(i)}) : i \in k_m\setminus E\}$. Since the family $\{\overline{Q}_{j(i)} : i \in E\} \cup \{\overline{S}_{b(i)} : i \in k_m\setminus E\}$ is disjoint, we conclude that $\overline{W(x)} \cap \Delta_{k_m}(X) = \emptyset$. The set $A_m$ is concentrated around the $k_m$-diagonal $\Delta_{k_m}(X)$, so $|W(x) \cap A_m| \le \omega$. The set $P = \{x \in Y_m : [x, Q] \in \mathcal{B}_Q^m(A, E) \text{ and } f \in H(x)\}$ is contained in the set $R = \{x \in Y_m : \varphi_m(x) \in W(x)\}$. The set $R$ is countable because $\varphi_m(Y_m) \subset A_m$ while $W(x) \cap A_m$ is countable and $\varphi_m$ is injective. Thus $P$ is also countable, i.e., the family $\mathcal{H}$ is point-countable at any $f \in C_p(X)$. Therefore $C_p(X)$ indeed, has a point-countable $\pi$-base.

Observe that if there is $n \in \mathbb{N}\setminus\{1\}$ and a set $A \subset X^n\setminus\Delta_n(X)$ such that $|A| = |X|$ and $A$ is concentrated around $\Delta_n(X)$ then we can let $A_m = A$ and $k_m = n$ for all $m \in \omega$ obtaining thus a sequence $\{A_m : m \in \omega\}$ which guarantees that $C_p(X)$ has a point-countable $\pi$-base.

Finally, if $a \in X$ and a set $A \subset X$ with $|A| = |X|$ is concentrated around the point $a$ then the set $A' = (A\setminus\{a\}) \times \{a\} \subset X^2\setminus\Delta_2(X)$ has the same cardinality as $X$ and $A'$ is concentrated around the diagonal $\Delta_2(X)$, i.e., in this case $C_p(X)$ also has a point-countable $\pi$-base.

**V.406.** *Prove that*

(a) $C_p(\alpha)$ *has a point-countable $\pi$-base whenever $\alpha$ is an ordinal with its order topology;*

(b) $C_p(AD(X))$ *has a point-countable $\pi$-base for any countably compact space X. Here $AD(X)$ is the Alexandroff double of X.*

**Solution.** (a) Let $(\beta, \gamma] = \{\mu : \beta < \mu \le \gamma\}$ for any ordinals $\beta, \gamma \in \alpha$ such that $\beta < \gamma$. If $\alpha$ is a countable ordinal then $C_p(\alpha)$ has a countable base, so we can assume, without loss of generality, that $|\alpha| > \omega$. Let $A = \{(\beta, \beta + 1) : \beta + 1 < \alpha\} \subset (\alpha \times \alpha) \backslash \Delta_2(\alpha)$. It is clear that $|A| = |\alpha|$; let us prove that $A$ is concentrated around the set $\Delta = \Delta_2(\alpha)$.

Assume toward a contradiction that $A$ is not concentrated around $\Delta$ and fix a set $U \in \tau(\Delta, \alpha \times \alpha)$ for which there exists an uncountable set $B \subset \alpha$ such that

(1) $\{(\beta, \beta + 1) : \beta \in B\} \subset A \backslash U$.

Let $\beta_0 = \backslash n B$ and $\beta_{n+1} = \backslash n (B \backslash \{\beta_0, \ldots, \beta_n\})$ for any $n \in \omega$. It is immediate from the construction of the sequence $S = \{\beta_n : n \in \omega\}$ that the set $P_n = \{\beta \in B : \beta < \beta_n\}$ is finite for all $n \in \omega$. Since $B$ is uncountable, there exists $\gamma \in B$ such that $\beta_n < \gamma$ for all $n \in \omega$. Consequently, the ordinal $\mu = \sup\{\beta_n : n \in \omega\}$ belongs to $\alpha$.

Since $(\mu, \mu) \in U$, there exists $\mu' < \mu$ such that $(\mu', \mu] \times (\mu', \mu] \subset U$. Take $n \in \omega$ such that $\beta_n > \mu'$; then $(\beta_n, \beta_n + 1) \in (\mu', \mu] \times (\mu', \mu] \subset U$; this contradicts (1) and proves that $A$ is concentrated around $\Delta_2(\alpha)$, so we can apply Problem 405 to conclude that $C_p(\alpha)$ has a point-countable $\pi$-base.

(b) Recall that $AD(X)$ has the underlying set $X \times \{0, 1\}$ and the topology generated by the family $\{\{(x, 1)\} : x \in X\} \cup \{(U \times \{0, 1\}) \backslash K : U \in \tau(X)$ and $K \subset X \times \{1\}$ is a finite set$\}$. If $X$ is finite then $AD(X)$ is also finite and hence there is nothing to prove, so we assume that $X$ is infinite; denote by $\Delta$ the set $\Delta_2(AD(X))$.

For every $x \in X$ consider the points $a_x = (x, 0)$ and $b_x = (x, 1)$ of the space $AD(X)$ and let $A = \{(a_x, b_x) : x \in X\} \subset (AD(X) \times AD(X)) \backslash \Delta$. It is clear that $|A| = |X| = |AD(X)|$; take any open neighborhood $U$ of the diagonal $\Delta$ in the space $AD(X) \times AD(X)$. If $A \backslash U$ is infinite then there exists an infinite set $B \subset X$ such that $A' = \{(a_x, b_x) : x \in B\} \subset A \backslash U$. It follows from countable compactness of $X$ that some $z \in X$ is an accumulation point for $B$; then $w = (a_z, a_z) \in \Delta$ and hence $w \in U$, so there exists a set $O \in \tau(z, X)$ and a finite $K \subset X$ such that, for the set $W = (O \times \{0, 1\}) \backslash (K \times \{1\})$ we have $w \in W \times W \subset U$.

Since $z$ is an accumulation point of $B$, we can pick a point $y \in (O \backslash K) \cap B$. Then $(a_y, b_y) \in W \times W$ and hence $(a_y, b_y) \in U \cap A'$ which is a contradiction. Thus the set $A \backslash U$ is finite for any $U \in \tau(\Delta, AD(X) \times AD(X))$ and hence $A$ is concentrated around $\Delta = \Delta_2(AD(X))$. Finally, apply Problem 405 to conclude that the space $C_p(AD(X))$ has a point-countable $\pi$-base.

**V.407.** *Prove that the following conditions are equivalent for any infinite space X with $l^*(X) \le \omega$:*

(i) $C_p(X)$ has a point-countable $\pi$-base;

(ii) there is a family of sets $\{A_m : m \in \omega\}$ and a sequence $\{k_m : m \in \omega\} \subset \mathbb{N}\backslash\{1\}$ such that $\sup\{|A_m| : m \in \omega\} = |X|$ while $A_m \subset X^{k_m}\backslash\Delta_{k_m}(X)$ and $A_m$ is concentrated around $\Delta_{k_m}(X)$ for every $m \in \omega$.

**Solution.** By Problem 405, it suffices to prove the implication (i)$\Longrightarrow$(ii). Denote by $\mathcal{I}$ the family of all nontrivial open intervals of $\mathbb{R}$ with rational endpoints. The collection $\mathcal{Q} = \{(Q_0, \ldots, Q_{n-1}) : n \in \mathbb{N},\ Q_i \in \mathcal{I}$ for any $i < n$ and the family $\{\overline{Q}_0, \ldots, \overline{Q}_{n-1}\}$ is disjoint$\}$ is countable. For any point $x = (x_0, \ldots, x_{n-1}) \in X^n\backslash\Delta_n(X)$ and $Q = (Q_0, \ldots, Q_{n-1}) \in \mathcal{Q}$ let $[x, Q] = \{f \in C_p(X) : f(x_i) \in Q_i$ for all $i < n\}$.

Fix a point-countable $\pi$-base $\mathcal{B} \subset \tau^*(C_p(X))$. Making the elements of $\mathcal{B}$ smaller if necessary, we can assume that, for each $B \in \mathcal{B}$ there is $n \in \mathbb{N}\backslash\{1\}$, a point $x \in X^n\backslash\Delta_n(X)$ and $Q = (Q_0, \ldots, Q_{n-1}) \in \mathcal{Q}$ such that $B = [x, Q]$. Choose an enumeration $\{Q^m : m \in \omega\}$ of the family $\mathcal{Q}$; we have $Q^m = (Q_0^m, \ldots, Q_{k_m-1}^m)$ for every $m \in \omega$. Consider the set $A_m = \{x \in X^{k_m}\backslash\Delta_{k_m}(X) : [x, Q^m] \in \mathcal{B}\}$ and let $\mathcal{B}_m = \{[x, Q^m] : x \in A_m\}$ for any $m \in \omega$; it is clear that $\mathcal{B} = \bigcup\{\mathcal{B}_m : m \in \omega\}$.

Since $w(C_p(X)) = \pi w(C_p(X)) = |X|$ (see TFS-169 and Fact 2 of T.187), we have $|\mathcal{B}| \geq |X|$ and hence $|\mathcal{B}| = |X|$ because there are at most $|X|$-many sets of the form $[x, Q]$ where $x \in X^n$ and $Q \in \mathcal{Q} \cap \mathcal{I}^n$ for some $n \in \mathbb{N}\backslash\{1\}$. Therefore $|X| = |\mathcal{B}| = \sup\{|\mathcal{B}_m| : m \in \omega\} = \sup\{|A_m| : m \in \omega\}$.

Fix a number $m \in \mathbb{N}\backslash\{1\}$ and take a set $U \in \tau(\Delta_{k_m}(X), X^{k_m})$. For any point $x = (x_0, \ldots, x_{k_m-1}) \in F = X^{k_m}\backslash U$ choose a function $f_x \in C_p(X)$ such that $f_x(x_i) \in Q_i^m$ for all $i < k_m$. Then $x \in W_x = f_x^{-1}(Q_0^m) \times \ldots \times f_x^{-1}(Q_{k_m-1}^m)$. Besides, $f_x \in [y, Q^m]$ if and only if $y \in W_x$; the family $\mathcal{B}$ being point-countable, the set $\{y \in A_m : y \in W_x\}$ is countable for any $x \in F$.

The space $F = X^{k_m}\backslash U$ is Lindelöf, so there is a countable $H \subset F$ such that $F \subset \bigcup\{W_x : x \in H\}$. Thus the set $A_m\backslash U = A_m \cap F \subset (A_m \cap (\bigcup\{W_x : x \in H\}))$ is countable because $A_m \cap W_x$ is countable for any point $x \in H$. Therefore the set $A_m \subset X^{k_m}\backslash\Delta_{k_m}(X)$ is concentrated around $\Delta_{k_m}(X)$ for any $m \in \omega$.

**V.408.** *Given a space $X$ suppose that the cardinality of $X$ is regular and uncountable while $l^*(X) = \omega$, i.e., all finite powers of $X$ are Lindelöf. Prove that the space $C_p(X)$ has a point-countable $\pi$-base if and only if there exists a natural number $n > 1$ such that some set $A \subset X^n\backslash\Delta_n(X)$ is concentrated around $\Delta_n(X)$ and $|A| = |X|$.*

**Solution.** We have sufficiency by Problem 405, so assume that the space $C_p(X)$ has a point-countable $\pi$-base. By Problem 407, we can find a family $\{A_m : m \in \omega\}$ and a sequence $\{k_m : m \in \omega\} \subset \mathbb{N}\backslash\{1\}$ such that $\sup\{|A_m| : m \in \omega\} = |X|$ while the set $A_m \subset X^{k_m}\backslash\Delta_{k_m}(X)$ is concentrated around $\Delta_{k_m}(X)$ for each $m \in \omega$. Since $|X|$ is a regular uncountable cardinal, there exists $m \in \omega$ such that $|A_m| = |X|$. Letting $A = A_m$ and $n = k_m$ we conclude the proof of necessity.

**V.409.** *Given a metrizable space $X$ prove that $C_p(X)$ has a point-countable $\pi$-base if and only if $X$ is countable.*

**Solution.** If the space $X$ is countable then $C_p(X)$ even has a countable base, so we have sufficiency. To prove necessity assume that $X$ is a metrizable space and $C_p(X)$ has a point-countable $\pi$-base $\mathcal{B}$. By SFFS-285, the cardinal $\omega_1$ is a caliber of $C_p(X)$, so $\mathcal{B}$ is countable and hence we can apply Fact 2 of T.187 to conclude that $|X| = \pi w(C_p(X)) = \omega$.

**V.410.** *Suppose that $X$ is an infinite space with $l^*(X) = \omega$. Prove that if $C_p(X)$ has a point-countable $\pi$-base then $|X| = \Delta(X)$. Here $\Delta(X) = \backslash n\{\kappa : \Delta_2(X)$ is a $G_\kappa$-subset of $X \times X\}$ is the diagonal number of the space $X$. Deduce from this fact that if $X$ is compact and $C_p(X)$ has a point-countable $\pi$-base then $w(X) = |X|$.*

**Solution.** It is evident that $\Delta(X) \leq |X|$, so assume that $\Delta(X) = \kappa < |X|$. We can find a family of sets $\{A_m : m \in \omega\}$ and a sequence $\{k_m : m \in \omega\} \subset \mathbb{N}\backslash\{1\}$ such that $\sup\{|A_m| : m \in \omega\} = |X|$ while $A_m \subset X^{k_m}\backslash\Delta_{k_m}(X)$ and $A_m$ is concentrated around $\Delta_{k_m}(X)$ for every $m \in \omega$ (see Problem 407). If $|A_m| \leq \kappa$ for all $m \in \omega$ then $|X| = \sup\{|A_m| : m \in \omega\} \leq \kappa$ which is a contradiction. Therefore there is $m \in \omega$ for which $|A_{k_m}| \geq \kappa^+$; let $n = k_m$ and $A = A_{k_m}$.

It follows from Fact 2 of T.078 that $\Delta_n(X)$ is a $G_\kappa$-subset of the space $X^n$. Choose a family $\mathcal{U} \subset \tau(\Delta_n(X), X^n)$ such that $|\mathcal{U}| \leq \kappa$ and $\bigcap \mathcal{U} = \Delta_n(X)$. Since $A$ is concentrated around $\Delta_n(X)$, the set $A\backslash U$ is countable for any $U \in \mathcal{U}$. It follows from the equality $A = \bigcup\{A\backslash U : U \in \mathcal{U}\}$ that $|A| \leq \kappa \cdot \sup\{|A\backslash U| : U \in \mathcal{U}\} \leq \kappa \cdot \omega = \kappa$; this contradiction with $|A| \geq \kappa^+$ shows that $|X| \leq \Delta(X)$ and hence $|X| = \Delta(X)$.

Finally observe that if $X$ is compact then $l^*(X) = \omega$ and $w(X) = \Delta(X)$ (see SFFS-091), so if $C_p(X)$ has a point-countable $\pi$-base then $w(X) = \Delta(X) = |X|$ as promised.

**V.411.** *Prove that if $K$ is a scattered Corson compact space then $C_p(K)$ has a point-countable $\pi$-base.*

**Solution.** If $K$ is countable then $C_p(K)$ even has a countable base, so we assume from now on that the space $K$ is uncountable. Let $\kappa = w(K)$; since $K$ is scattered, we have $hl(K) \geq |K|$ (see SFFS-005 and SFFS-006). Besides, $hl(K) \leq w(K) = nw(K) \leq |K|$ (see Fact 4 of S.307) and hence $|K| = \kappa$.

Apply CFS-188 to see that the diagonal $\Delta = \{(x,x) : x \in K\} \subset K \times K$ of the space $K$ is a $W$-set in $K \times K$. By CFS-186 the space $P = (K \times K)\backslash\Delta$ is metacompact, so there exists a point-finite open cover $\mathcal{U}$ of the space $P$ such that $\overline{U} \cap \Delta = \emptyset$ for every $U \in \mathcal{U}$ (the bar denotes the closure in $K \times K$). Since $hl(K \times K) \leq \kappa$, we can choose a subfamily $\mathcal{V} \subset \mathcal{U}$ such that $|\mathcal{V}| \leq \kappa$ and $\bigcup \mathcal{V} = P$. We also have $\Delta(K) = w(K) = \kappa$ by SFFS-091, so it follows from $\Delta = \bigcap\{(K \times K)\backslash\overline{U} : U \in \mathcal{V}\}$ that the cardinality of $\mathcal{V}$ cannot be less than $\kappa$ and hence $|\mathcal{V}| = \kappa$.

For any $z \in P$ let $\mathcal{V}_z = \{V \in \mathcal{V} : z \in V\}$ and take an arbitrary point $z_0 \in P$. Suppose that $\alpha < \kappa$ and we have a set $\{z_\beta : \beta < \alpha\}$ such that

(1) $z_\gamma \notin \bigcup\{\bigcup \mathcal{V}_{z_\beta} : \beta < \gamma\}$ for any $\gamma < \alpha$.

Since $\mathcal{V}_{z_\beta}$ is finite for every $\beta < \alpha$, the family $\mathcal{W} = \bigcup\{\mathcal{V}_{z_\beta} : \beta < \alpha\}$ has cardinality at most $\omega \cdot |\alpha| = \max\{|\alpha|, \omega\} < \kappa$. Therefore $\mathcal{W}$ cannot cover $P$ because otherwise $\Delta = \bigcap\{(K \times K)\backslash\overline{W} : W \in \mathcal{W}\}$ and hence $\psi(\Delta, K \times K) \leq |\mathcal{W}| < \kappa$ which is a contradiction. Pick any point $z_\alpha \in P\backslash(\bigcup\mathcal{W})$; it is straightforward that (1) now holds for all $\gamma \leq \alpha$ and hence our inductive procedure can be continued to construct a set $D = \{z_\beta : \beta < \kappa\} \subset P$ such that (1) holds for all $\alpha < \kappa$. Since $z_\beta \in \bigcup\mathcal{V}_{z_\beta}$ for any $\beta < \kappa$, it follows from (1) that $z_{\beta'} \neq z_\beta$ whenever $\beta \neq \beta'$ and hence $|D| = \kappa = |K|$.

If $z \in P$ then there is $U \in \mathcal{V}$ with $z \in U$. If $z_\alpha \in U$ then $U \in \mathcal{V}_{z_\alpha}$ and hence (1) shows that $z_\beta \notin U$ for all $\beta > \alpha$. Now, if $\beta < \alpha$ then $z_\alpha \notin \mathcal{V}_{z_\beta}$, so $U \notin \mathcal{V}_{z_\beta}$, i.e., $z_\beta \notin U$. Therefore every $z \in P$ has a neighborhood which contains at most one point of $D$, so $D$ is closed and discrete in $P$.

Now take any set $W \in \tau(\Delta, K \times K)$. The set $L = (K \times K)\backslash W = P\backslash W \subset P$ is compact. Therefore $D \cap L = D\backslash W$ is compact being closed in $L$; however, $D \cap L$ is a discrete space, so it must be finite. This proves that $D\backslash W$ is finite for any $W \in \tau(\Delta, K \times K)$ and hence $D$ is concentrated around $\Delta$. Finally, apply Problem 405 to conclude that $C_p(K)$ has a point-countable $\pi$-base.

**V.412.** *Prove that if $C_p(X)$ embeds in a $\Sigma$-product of first countable spaces then it has a point-countable $\pi$-base.*

**Solution.** Suppose that $N_t$ is a space for every $t \in T$ and $N = \prod_{t \in T} N_t$. Given a point $a \in N$ let $\Sigma(N, a) = \{x \in N : |\{t \in T : x(t) \neq a(t)\}| \leq \omega\}$. If $x \in \Sigma(N, a)$ then $\text{supp}(x) = \{t \in T : x(t) \neq a(t)\}$. The spaces with a countable network are called *cosmic*.

*Fact 1.* Assume that $nw(N_t) = \chi(N_t) = \omega$ for all $t \in T$, and $N = \prod_{t \in T} N_t$. Then, for any $a \in N$, every subspace of $\Sigma(N, a)$ has a point-countable $\pi$-base.

*Proof.* For each $S \subset T$ consider the set $N_S = \prod_{t \in S} N_t$ and the natural projection $\pi_S : N \to N_S$; let $a_S = \pi_S(a)$. It is easy to see that $\pi_S(\Sigma(N, a)) = \Sigma(N_S, a_S)$ for any $S \subset T$. For every $Y \subset \Sigma(N, a)$ let $\mu(Y)$ be the minimal cardinality of a set $S \subset T$ such that $Y$ embeds in $\Sigma(N_S, a_S)$. Observe first that

(1) if $Y \subset \Sigma(N, a)$ and $\mu(Y) \leq \omega$ then $Y$ has a countable $\pi$-base.

Indeed, $\mu(Y) \leq \omega$ implies that $Y$ embeds in $N_S$ for some countable $S \subset T$. The space $N_S$ is cosmic and first countable and hence so is $Y$. Take any countable dense $D \subset Y$ and fix a countable local base $\mathcal{B}_d$ in the space $Y$ at every point $d \in D$. It is an easy exercise that the family $\mathcal{B} = \bigcup_{d \in D} \mathcal{B}_d$ is a countable $\pi$-base of $Y$, so (1) is proved. Our next step is to show that

(2) if $Y \subset \Sigma(N, a)$ and $\mu(Y) = \mu(V)$ for any nonempty open set $V$ of the space $Y$, then $Y$ has a point-countable $\pi$-base.

If (2) is not true then the family $\mathcal{Y} = \{Y \subset \Sigma(N, a) : \mu(Y) = \mu(V)$ for any $V \in \tau^*(Y)$ and $Y$ does not have a point-countable $\pi$-base$\}$ is nonempty. Let $\kappa = \bigwedge\{\mu(Y) : Y \in \mathcal{Y}\}$ and choose a set $Y \in \mathcal{Y}$ with $\mu(Y) = \kappa$. It follows from (1) that $\kappa > \omega$. We can assume, without loss of generality, that $Y \subset \Sigma(N_S, a_S)$ for some $S \subset T$ with $|S| = \kappa$. Choose a faithful enumeration $\{s_\alpha : \alpha < \kappa\}$ of the set

$S$; we will need the set $S_\alpha = \{s_\beta : \beta \leq \alpha\}$ for any $\alpha < \kappa$. Let $p_\alpha(y) = y|S_\alpha$ for any $y \in Y$ and $\alpha < \kappa$; then $p_\alpha : Y \to Y_\alpha = p_\alpha(Y)$ is the restriction to the set $Y$ of the relevant natural projection.

Given any ordinal $\alpha < \kappa$, we have $Y_\alpha \subset \Sigma(N_{S_\alpha}, a_{S_\alpha})$ and $|S_\alpha| < \kappa$, so our choice of $\kappa$ guarantees that $Y_\alpha$ has a point-countable $\pi$-base $\mathcal{B}'_\alpha$; consider the set $F_\alpha = \{x \in Y_\alpha : x(s_\alpha) = a(s_\alpha)\}$ and the family $\mathcal{B}_\alpha = \{B \in \mathcal{B}'_\alpha : B \cap F_\alpha = \emptyset\}$. The family $\mathcal{E}_\alpha = \{p_\alpha^{-1}(B) : B \in \mathcal{B}_\alpha\} \subset \tau^*(Y)$ is point-countable for each $\alpha < \kappa$; let $\mathcal{E} = \bigcup\{\mathcal{E}_\alpha : \alpha < \kappa\}$.

Take any $U \in \tau^*(Y)$; there exist $\alpha_1, \ldots, \alpha_n \in \kappa$ and sets $O_1, \ldots, O_n$ such that $O_i \in \tau(N_{s_{\alpha_i}})$ for all $i \leq n$ while the set $U' = \{y \in Y : y(s_{\alpha_i}) \in O_i$ for each $i \leq n\}$ is nonempty and contained in $U$. Let $\alpha = \max\{\alpha_1, \ldots, \alpha_n\}$; if $y(s_\beta) = a(s_\beta)$ for any $\beta > \alpha$ and $y \in U'$ then $U'$ embeds in $\Sigma(N_{S_\alpha}, a_{S_\alpha})$ and hence $\mu(U') \leq |S_\alpha| < \kappa$ which is a contradiction.

Therefore there exists an ordinal $\beta > \alpha$ such that $y(s_\beta) \neq a(s_\beta)$ for some point $y \in U'$. The set $V = p_\beta(U')$ is open in $Y_\beta$ and $U' = p_\beta^{-1}(V)$ (see Fact 1 of S.298). Since $p_\beta(y) \in V \setminus F_\beta$, the open set $V \setminus F_\beta$ is nonempty and hence there is $B \in \mathcal{B}'_\beta$ such that $B \subset V \setminus F_\beta$, i.e., $B \in \mathcal{B}_\beta$. Since also $B' = p_\beta^{-1}(B) \subset p_\beta^{-1}(V) = U' \subset U$ and $B' \in \mathcal{E}$, we proved that every nonempty open subset of $Y$ contains an element of $\mathcal{E}$, i.e., the family $\mathcal{E}$ is a $\pi$-base of $Y$.

Now assume that there exists a family $\mathcal{E}' \subset \mathcal{E}$ such that $|\mathcal{E}'| = \omega_1$ and $\bigcap \mathcal{E}' \neq \emptyset$. Since $\mathcal{E}_\alpha$ is point-countable, the family $\mathcal{E}' \cap \mathcal{E}_\alpha$ is countable for every $\alpha < \kappa$. As a consequence, there exists an uncountable set $A \subset \kappa$ such that $\mathcal{E}' \cap \mathcal{E}_\alpha \neq \emptyset$ for any $\alpha \in A$. Pick a point $y \in \bigcap \mathcal{E}'$ and an element $B_\alpha \in \mathcal{B}_\alpha$ for any $\alpha \in A$ in such a way that $y \in \bigcap\{p_\alpha^{-1}(B_\alpha) : \alpha \in A\}$.

It follows from the equality $B_\alpha \cap F_\alpha = \emptyset$ that $p_\alpha(y) \notin F_\alpha$ and hence $y(\alpha) \neq a(\alpha)$ for any $\alpha \in A$. Thus $A \subset \text{supp}(y)$, so $\text{supp}(y)$ is uncountable which is a contradiction with $y \in \Sigma(N_S, a_S)$. Therefore the family $\mathcal{E}$ is a point-countable $\pi$-base of $Y$; this contradiction with the choice of $Y$ shows that (2) is proved.

Now take an arbitrary subspace $Y$ of the space $\Sigma(N, a)$ and consider the family $\mathcal{B} = \{B \in \tau^*(Y) : \mu(B) = \mu(B')$ for any $B' \in \tau^*(Y)$ with $B' \subset B\}$. If $U \in \tau^*(Y)$ then let $\kappa = \bigwedge n\{\mu(U') : U' \in \tau^*(Y)$ and $U' \subset U\}$ and choose a set $B \in \tau^*(Y)$ such that $B \subset U$ and $\mu(B) = \kappa$. It is immediate that $B \in \mathcal{B}$, so the family $\mathcal{B}$ is a $\pi$-base of $Y$. Take a maximal disjoint subfamily $\mathcal{B}'$ of the family $\mathcal{B}$; then the set $\bigcup \mathcal{B}'$ is dense in the space $Y$. The property (2) shows that, for every set $B \in \mathcal{B}'$ we can choose a point-countable $\pi$-base $\mathcal{C}_B$ for the space $B$. It is easy to see that $\mathcal{C} = \bigcup\{\mathcal{C}_B : B \in \mathcal{B}'\}$ is a point-countable $\pi$-base of $Y$, so Fact 1 is proved.

Returning to our solution assume that $M_t$ is a first countable space for any $t \in T$ and we have a point $a \in M = \prod_{t \in T} M_t$ such that $C_p(X) \subset \Sigma(M, a)$. Let $p_t : M \to M_t$ be the natural projection and consider the set $N'_t = p_t(C_p(X))$ for any $t \in T$. It follows from TFS-299 that $N'_t$ is a cosmic space and hence $N_t = N'_t \cup \{a(t)\}$ is cosmic as well for any $t \in T$; let $N = \prod_{t \in T} N_t$. It follows from $C_p(X) \subset \Sigma(N, a)$ and Fact 1 that $C_p(X)$ has a point-countable $\pi$-base, so our solution is complete.

**V.413.** *Say that a space $X$ is $P$-favorable (for the point-open game) if the player $P$ has a winning strategy in the point-open game on $X$. Prove that*

(i) *any countable space is $P$-favorable;*
(ii) *a continuous image of a $P$-favorable space is $P$-favorable;*
(iii) *the countable union of $P$-favorable space is $P$-favorable;*
(iv) *any nonempty closed subspace of a $P$-favorable space is $P$-favorable;*
(v) *any nonempty Lindelöf scattered space is $P$-favorable.*

**Solution.** (i) Suppose that $X$ is a countable space and choose an enumeration $\{x_n : n \in \omega\}$ of the set $X$. Let $s(\emptyset) = x_0$ and, if $n \in \omega$ and moves $x_0, U_0, \ldots, x_n, U_n$ are made then let $s(U_0, \ldots, U_n) = x_{n+1}$. If $\{x_n, U_n : n \in \omega\}$ is a play where $P$ applies the strategy $s$ then $\bigcup_{n \in \omega} U_n \supset \{x_n : n \in \omega\} = X$, so $s$ is a winning strategy for the player $P$, i.e., $X$ is $P$-favorable.

(ii) Assume that $s$ is a winning strategy for the player $P$ on a space $X$ and $f : X \to Y$ is a continuous onto map. If $x_0 = s(\emptyset)$ then let $y_0 = f(x_0)$ and $t(\emptyset) = y_0$. Suppose that moves $y_0, V_0, \ldots, y_n, V_n$ are made in the point-open game on $Y$ and we have points $x_0, \ldots, x_n \in X$ such that $f(x_i) = y_i$ for all $i \leq n$ and $x_0, f^{-1}(V_0), \ldots, x_n, f^{-1}(V_n)$ is an initial segment of a play on $X$ where $P$ applies the strategy $s$. If $x_{n+1} = s(f^{-1}(V_0), \ldots, f^{-1}(V_n))$ then let $y_{n+1} = f(x_{n+1})$ and $t(V_0, \ldots, V_n) = y_{n+1}$. This defines a strategy $t$ on the space $Y$.

If $\{y_n, V_n : n \in \omega\}$ is a play in which $P$ applies $t$ then there exists a sequence $\{x_n : n \in \omega\} \subset X$ such that $f(x_n) = y_n$ for each $n \in \omega$ and $\{x_n, f^{-1}(V_n) : n \in \omega\}$ is a play on $X$ in which $P$ applies the strategy $s$. Since $s$ is a winning strategy, we have $X = \bigcup_{n \in \omega} f^{-1}(V_n)$ and hence $Y = \bigcup_{n \in \omega} V_n$, so $t$ is a winning strategy for the player $P$ on $Y$, i.e., $Y$ is $P$-favorable.

(iii) Assume that $X = \bigcup_{n \in \omega} X_n$ and there exists a winning strategy $s_n$ on the space $X_n$ for every $n \in \omega$. Choose a disjoint family $\{A_n : n \in \omega\}$ of infinite subsets of $\omega$ such that $\omega = \bigcup_{n \in \omega} A_n$. If $m \in \omega$ and $S = \{x_0, U_0, \ldots, x_n, U_n\}$ is an initial segment of a play on $X$ then let $S|X_m = \{x_{i_0}, U_{i_0} \cap X_m, \ldots, x_{i_k}, U_{i_k} \cap X_m\}$ where $i_0 < \ldots < i_k$ and $\{x_{i_0}, \ldots, x_{i_k}\} = \{x_0, \ldots, x_n\} \cap X_m$. The set $S|X_m$ will be called the restriction of the segment $S$ to $X_m$.

There is a unique $m \in \omega$ such that $0 \in A_m$; let $x_0 = s_m(\emptyset)$ and $s(\emptyset) = x_0$. Proceeding inductively assume that moves $x_0, U_0, \ldots, x_n, U_n$ are made in such a way that, for any $m \in \omega$, if $S = \{x_0, U_0, \ldots, x_n, U_n\}$ then the set $S|X_m$ is an initial segment (maybe empty) of a play on $X_m$ in which $P$ applies the strategy $s_m$. There is a unique $m \in \omega$ such that $n + 1 \in A_m$; if $S|X_m = \{x_{i_0}, U_{i_0} \cap X_m, \ldots, x_{i_k}, U_{i_k} \cap X_m\}$ then the strategy $s_m$ is applicable to the segment $S|X_m$ by our induction hypothesis. Let $x_{n+1} = s_m(U_{i_0} \cap X_m, \ldots, U_{i_k} \cap X_m)$ and $s(U_0, \ldots, U_n) = x_{n+1}$. This gives a strategy $s$ for the player $P$ on the space $X$.

Now, assume that $\{x_n, U_n : n \in \omega\}$ is a play on $X$ where $P$ applies the strategy $s$ and fix an arbitrary $m \in \omega$. There is a unique enumeration $\{i_j : j \in \omega\}$ of the set $A_m$ such that $i_j < i_k$ whenever $j < k$. By our definition of the strategy $s$, the family $\{x_{i_j}, U_{i_j} \cap X_m : j \in \omega\}$ is a play on $X_m$ in which $P$ applies the strategy $s_m$. Therefore $X_m = \bigcup_{j \in \omega} U_{i_j} \cap X_m = \bigcup_{n \in A_m} U_n \cap X_m$ and hence $X_m \subset \bigcup_{n \in A_m} U_n \subset \bigcup_{n \in \omega} U_n$ for any $m \in \omega$. Consequently, $X = \bigcup_{m \in \omega} X_m \subset \bigcup_{n \in \omega} U_n$, so $X = \bigcup_{n \in \omega} U_n$ which shows that the strategy $s$ is winning for $P$, i.e., $X$ is $P$-favorable.

(iv) Suppose that $X$ is $P$-favorable and $F$ is a nonempty closed subspace of $X$. Fix a winning strategy $s$ for the player $P$ on the space $X$. If we have an $n$-th move $\{x_n, U_n\}$ of a play on $X$ say that the move $U_n$ of the player $O$ is *adequate* if $x_n \in X \setminus F$ implies $U_n = X \setminus F$. In this item we will only consider the initial segments of a play on $X$ where $P$ applies the strategy $s$ and all moves of $O$ are adequate. A play (or an initial segment of a play) is called *adequate* if all moves of $O$ are adequate. Observe that

(1) if we have an adequate play $\{x_n, U_n : n \in \omega\}$ and $F \not\subset U_0 \cup \ldots \cup U_k$ for some $k \in \omega$ then there exists $m > k$ such that $x_m \in F$.

Indeed, the strategy $s$ is winning, so $\bigcup_{n \in \omega} U_n = X$; if $x_n \notin F$ then $U_n = X \setminus F$ for all $n > k$ and hence $\bigcup_{n \in \omega} U_n = U_0 \cup \ldots \cup U_k \neq X$ which is a contradiction.

Apply the property (1) to see that there is a uniquely determined $m_0 \in \omega$ and an adequate segment $\{x_0, U_0, \ldots, x_{m_0-1}, U_{m_0-1}\}$ such that $x_i \notin F$ for all $i < m_0$ and $x_{m_0} = s(U_0, \ldots, U_{m_0-1}) \in F$. Let $y_0 = x_{m_0}$ and $t(\emptyset) = y_0$.

Proceeding inductively assume that $\{y_0, V_0, \ldots, y_k, V_k\}$ is an initial segment of a play on $F$ such that there exist natural numbers $m_0 < \ldots < m_k$ and an adequate initial segment $\{x_0, U_0, \ldots, x_{m_k}, U_{m_k}\}$ such that $y_i = x_{m_i}$ and $V_i = U_{m_i} \cap F$ for all $i \leq k$ while $\{x_0, \ldots, x_{m_k}\} \setminus \{y_0, \ldots, y_k\} \subset X \setminus F$.

If $F \subset W_k = U_0 \cup \ldots \cup U_{m_k}$ then $V_0 \cup \ldots \cup V_k = F$, so take any point $y_{k+1} \in F$ and let $t(V_0, \ldots, V_k) = y_{k+1}$. If $F \setminus W_k \neq \emptyset$ then apply the property (1) to see that there exists a uniquely determined $m_{k+1} \in \omega$ and adequate initial segment $\{x_0, U_0, \ldots, x_{m_k}, U_{m_k}, x_{m_k+1}, U_{m_k+1}, \ldots, x_{m_{k+1}-1}, U_{m_{k+1}-1}, x_{m_{k+1}}\}$ such that $x_{m_{k+1}} = s(U_0, \ldots, U_{m_{k+1}-1}) \in F$ and $x_i \notin F$ whenever $m_k < i < m_{k+1}$. Let $y_{k+1} = x_{m_{k+1}}$ and $t(V_0, \ldots, V_k) = y_{k+1}$. This inductive procedure gives us a strategy $t$ for the player $P$ on the space $F$.

If $\{y_k, V_k : k \in \omega\}$ is a play where $P$ applies the strategy $t$ then we have two possibilities.

*Case 1.* There exists $k \in \omega$ such that $F \subset V_0 \cup \ldots \cup V_k$. Then it is evident that the player $P$ is the winner.

*Case 2.* $F \not\subset V_0 \cup \ldots \cup V_k$ for all $k \in \omega$. Then there exists an adequate play $\mathcal{E} = \{x_n, U_n : n \in \omega\}$ and an increasing sequence $S = \{m_k : k \in \omega\} \subset \omega$ such that $y_k = x_{m_k}$ and $V_k = U_{m_k} \cap F$ for all $k \in \omega$. Since $P$ applied the strategy $s$ in the play $\mathcal{E}$, we have $X = \bigcup_{n \in \omega} U_n$. Besides, $U_n = X \setminus F$ for all $n \in \omega \setminus S$ and therefore $\bigcup_{n \in S} U_n \supset F$. As a consequence, $F = \bigcup_{n \in S} (U_n \cap F) = \bigcup_{k \in \omega} V_k$ and hence the player $P$ is the winner.

Therefore $t$ is a winning strategy for the player $P$ on the space $F$, i.e., $F$ is $P$-favorable.

(v) Given a scattered space $Y$ let $I(Y)$ be the set of all isolated points of $Y$. Fix any scattered space $X$ and let $X_0 = X$. Proceeding inductively assume that $\beta > 0$ is an ordinal and we have a family $\{X_\alpha : \alpha < \beta\}$ of closed subsets of $X$ with the following properties:

(2) if $\alpha + 1 < \beta$ then $X_{\alpha+1} = X_\alpha \backslash I(X_\alpha)$;
(3) if $\alpha < \beta$ is a limit ordinal then $X_\alpha = \bigcap_{\gamma < \alpha} X_\gamma$.

If $\beta = \alpha + 1$ for some ordinal $\alpha$ then let $X_\beta = X_\alpha \backslash I(X_\alpha)$; if $\beta$ is a limit ordinal then let $X_\beta = \bigcap_{\alpha < \beta} X_\alpha$. It is clear that the properties (2) and (3) still hold for all $\alpha \le \beta$, so our inductive procedure can be continued indefinitely. However, if $X_\alpha \neq \emptyset$ then $X_{\alpha+1}$ is strictly contained in $X_\alpha$ because the space $X$ is scattered. Therefore the ordinal $\mu(X) = \backslash n\{\alpha : X_\alpha = \emptyset\}$ (called the dispersion index of $X$) is well defined. The family $\{X_\alpha : \alpha < \mu(X)\}$ is called the *canonical decomposition* of $X$.

Assume that $X$ is a scattered space and $\{X_\alpha : \alpha < \mu(X)\}$ is a canonical decomposition of $X$. Suppose that $Y \subset X$ and $Y \cap X_\gamma = \emptyset$ for some ordinal $\gamma < \mu(X)$. Consider the canonical decomposition $\{Y_\alpha : \alpha < \lambda\}$ of the space $Y$. We have $Y_0 = Y \subset X = X_0$; proceeding inductively assume that $\beta$ is an ordinal and we proved that $Y_\alpha \subset X_\alpha$ for all $\alpha < \beta$. If $\beta$ is a limit ordinal then $Y_\beta = \bigcap_{\alpha < \beta} Y_\alpha \subset \bigcap_{\alpha < \beta} X_\alpha = X_\beta$. If $\beta = \alpha + 1$ then $Y_\alpha \subset X_\alpha$ and hence any non-isolated point of $Y_\alpha$ is not isolated in $X_\alpha$, i.e., $Y_\beta \subset X_\beta$. Therefore our inductive procedure can be continued to prove that $Y_\alpha \subset X_\alpha$ for any ordinal $\alpha$. In particular, $Y_\gamma \subset X_\gamma$, so it follows from $Y_\gamma \cap X_\gamma \subset Y \cap X_\gamma = \emptyset$ that $Y_\gamma = \emptyset$ and hence we established that

(4) if $\{X_\alpha : \alpha < \mu(X)\}$ is a canonical decomposition of $X$ and $Y \cap X_\gamma = \emptyset$ for some $\gamma < \mu(X)$ and $Y \subset X$ then $\mu(Y) \le \gamma$.

Say that a point $a$ of a space $X$ is *central* in $X$ if $X \backslash U$ is $P$-favorable for any $U \in \tau(a, X)$. Observe that

(5) if a space $X$ has a central point then $X$ is $P$-favorable.

Indeed, if $x_0$ is a central point of $X$ then let $s(\emptyset) = x_0$; if $U_0$ is the move of the player $O$ then there is a winning strategy $t$ for the player $P$ on the space $F = X \backslash U_0$; let $x_1 = t(\emptyset)$ and $s(U_0) = x_1$. If moves $x_0, U_0, \ldots, x_n, U_n$ are made for some $n > 0$ then let $s(U_0, \ldots, U_n) = t(U_1 \cap F, \ldots, U_n \cap F)$. This gives us a strategy $s$ on the space $X$. If $\{x_n, U_n : n \in \omega\}$ is a play where $P$ applies $s$ then, for the set $F = X \backslash U_0$, the sequence $\{x_{n+1}, U_{n+1} \cap F : n \in \omega\}$ is a play in which $P$ applies a winning strategy $t$. Therefore $F = \bigcup_{n \in \omega}(U_{n+1} \cap F)$ and hence $F \subset \bigcup_{n \in \omega} U_{n+1}$. This implies in turn that $X = \bigcup_{n \in \omega} U_n$, so $s$ is a winning strategy, i.e., $X$ is $P$-favorable and hence (5) is proved.

Now assume that $X$ is a nonempty scattered Lindelöf space. We will prove that $X$ is $P$-favorable by induction on the dispersion index of $X$. Observe that $X \neq \emptyset$ implies that $\mu(X) > 0$. If $\mu(X) = 1$ then $X$ is discrete and hence countable, so it is $P$-favorable by (i). Now assume that $\beta > 1$ is an ordinal and we proved that any nonempty Lindelöf scattered space $Y$ with $\mu(Y) < \beta$ is $P$-favorable.

Take a Lindelöf scattered space $X$ with $\mu(X) = \beta$ and let $\{X_\alpha : \alpha < \beta\}$ be the canonical decomposition of $X$. We have two cases.

*Case 1.* $\beta$ is a limit ordinal. Then $\bigcap_{\alpha < \beta} X_\alpha = \emptyset$. For any point $x \in X$ there exists $\alpha < \beta$ such that $x \notin X_\alpha$ and hence we can choose a set $U_x \in \tau(x, X)$ such that $\overline{U}_x \cap X_\alpha = \emptyset$. It follows from (4) that $\mu(\overline{U}_x) \leq \alpha < \beta$, so we can apply the induction hypothesis to see that $\overline{U}_x$ is a $P$-favorable space for any $x \in X$. The space $X$ being Lindelöf there exists a countable $A \subset X$ such that $X = \bigcup\{U_x : x \in A\}$ and hence $X = \bigcup\{\overline{U}_x : x \in A\}$. Now it follows from (iii) that $X$ is $P$-favorable.

*Case 2.* $\beta = \alpha + 1$. Then $X_\alpha$ is countable closed discrete subset of $X$. Use collectionwise normality of $X$ to find a family $\{U_x : x \in X_\alpha\}$ such that $U_x \in \tau(x, X)$ for any $x \in X_\alpha$ and the collection $\{\overline{U}_x : x \in X_\alpha\}$ is disjoint. Consider the set $F = X \setminus (\bigcup\{U_x : x \in X_\alpha\})$; it follows from $F \cap X_\alpha = \emptyset$ that we can apply (4) again to convince ourselves that $\mu(F) \leq \alpha < \beta$ and hence $F$ is $P$-favorable.

Assume that $W \in \tau(x, \overline{U}_x)$; then $F_x = \overline{U}_x \setminus W$ is a closed subset of $X$ such that $F_x \cap X_\alpha = \emptyset$ and hence we can apply the property (4) once more to see that $F_x$ is $P$-favorable by the induction hypothesis. This proves that $x$ is a central point of $\overline{U}_x$, so the property (5) guarantees that $\overline{U}_x$ is $P$-favorable for all $x \in X_\alpha$. The equality $X = F \cup (\bigcup\{\overline{U}_x : x \in X_\alpha\})$ shows that $X$ is the union of countably many $P$-favorable spaces, so it is $P$-favorable by (iii). This shows that every Lindelöf scattered space is $P$-favorable and completes our solution.

**V.414.** *Prove that*

(a) *if $X$ is $P$-favorable for the point-open game and $\psi(X) \leq \omega$ then $X$ is countable;*

(b) *a compact space $X$ is $P$-favorable for the point-open game if and only if $X$ is scattered.*

**Solution.** (a) Fix a winning strategy $s$ for the player $P$ on the space $X$. For any $x \in X$ let $\mathcal{U}_x \subset \tau(x, X)$ be a countable family such that $\{x\} = \bigcap \mathcal{U}_x$. Say that an initial segment $S = \{x_0, U_0, \dots, x_n, U_n\}$ of a play on $X$ is *s-legal* if $P$ applies the strategy $s$ in the segment $S$.

Let $x_0 = s(\emptyset)$ and $A_0 = \{x_0\}$. Proceeding inductively assume that we have a countable set $A_n \subset X$ and consider the set $A'_{n+1} = \{y \in X : \text{there exists } k \in \omega \text{ and an } s\text{-legal initial segment } \{x_0, U_0, \dots, x_k, U_k\} \text{ such that } U_i \in \bigcup\{\mathcal{U}_x : x \in A_n\} \text{ for all } i \leq k \text{ and } y = s(U_0, \dots, U_k)\}$; let $A_{n+1} = A_n \cup A'_{n+1}$ and observe that the family $\mathcal{V} = \bigcup\{\mathcal{U}_x : x \in A_n\}$ is countable, so we only have countably many finite subfamilies of $\mathcal{V}$. If $S = \{x_0, U_0, \dots, x_k, U_k\}$ is an $s$-legal initial segment then the family $\{U_0, \dots, U_k\}$ uniquely determines $S$ because we have the equalities $x_0 = s(\emptyset), x_1 = s(U_0), \dots, x_n = s(U_0, \dots, U_{n-1})$. Therefore $|A'_{n+1}| \leq |\mathcal{V}| \leq \omega$ and hence $A_{n+1}$ is countable.

Once we have the sequence $\{A_n : n \in \omega\}$ let $A = \bigcup_{n \in \omega} A_n$ and consider the family $\mathcal{W} = \bigcup\{\mathcal{U}_x : x \in A\}$. Let us show that

(1) if we have an $s$-legal segment $\{x_0, U_0, \ldots, x_n, U_n\}$ such that $U_0, \ldots,$
$U_n \in \mathcal{W}$ then $s(U_0, \ldots, U_n) \in A$.

Indeed, it is easy to find a number $m \in \omega$ such that $\{x_0, \ldots, x_n\} \subset A_m$ and
$\{U_0, \ldots, U_n\} \subset \bigcup \{\mathcal{U}_x : x \in A_m\}$. Then $s(U_0, \ldots, U_n) \in A'_{m+1} \subset A$ and hence
(1) is proved.

The set $A$ is countable, so it suffices to show that $X = A$. Assume toward a
contradiction that $X \setminus A \neq \emptyset$ and pick a point $y \in X \setminus A$.

Since $x_0 \in A$, we have $x_0 \neq y$ and hence there exists $U_0 \in \mathcal{U}_{x_0}$
such that $y \notin U_0$. Assume that $n \in \omega$ and we have an $s$-legal segment
$\{x_0, U_0, \ldots, x_n, U_n\}$ such that $y \notin U_0 \cup \ldots \cup U_n$ and $U_i \in \mathcal{W}$ for all $i \leq n$. If
$x_{n+1} = s(U_0, \ldots, U_n)$ then it follows from (1) that $x_{n+1} \in A$; thus $x_{n+1} \neq y$
and hence we can choose a set $U_{n+1} \in \mathcal{U}_{x_{n+1}}$ with $y \notin U_{n+1}$. It follows from
$\mathcal{U}_{x_{n+1}} \subset \mathcal{W}$ that $U_{n+1} \in \mathcal{W}$.

Thus we carried out the induction step, so our procedure can be continued to
construct a play $\{x_n, U_n : n \in \omega\}$ in which $P$ applies the strategy $s$ while
$y \notin U_n$ for each $n \in \omega$. As a consequence, $y \notin \bigcup_{n \in \omega} U_n$; since $s$ is a winning
strategy, we obtained a contradiction which shows that $X = A$ and hence $X$ is
countable as promised.

(b) If $X$ is a compact scattered space then $X$ must be $P$-favorable for the point-
open game (see Problem 413), so we have sufficiency. Now assume that $X$ is a
compact $P$-favorable space. If $X$ is not scattered then there exists a continuous
onto map $f : X \to \mathbb{I}$ (see SFFS-133). Any continuous image of a $P$-favorable
space is $P$-favorable by Problem 413, so $\mathbb{I}$ is $P$-favorable. Now, it follows
from (a) and $\psi(\mathbb{I}) = \omega$ that $\mathbb{I}$ is countable. This contradiction shows that $X$
is scattered and settles necessity.

**V.415.** *Given a nonempty space $X$ define a game $\mathcal{PO}'$ on the space $X$ as follows:
at the $n$-th move the player $P$ chooses a finite set $F_n \subset X$ and the player $O$ takes
a set $U_n \in \tau(X)$ such that $F_n \subset U_n$. The game ends after the moves $F_n, U_n$ are
made for all $n \in \omega$. The player $P$ wins if $\bigcup_{n \in \omega} U_n = X$; otherwise the victory is
assigned to $O$. Prove that the player $P$ has a winning strategy in the game $\mathcal{PO}'$ on
a space $X$ if and only if $X$ is $P$-favorable for the point-open game.*

**Solution.** It is evident that every winning strategy of the player $P$ in the point-open
game is a winning strategy of $P$ in the game $\mathcal{PO}'$, so we have sufficiency.

Now assume that $s$ is a winning strategy of $P$ on a space $X$ in the game $\mathcal{PO}'$. An
initial segment $S = \{F_0, U_0, \ldots, F_n, U_n\}$ of a play in $\mathcal{PO}'$ will be called $s$-legal if
$P$ applies the strategy $s$ in the segment $S$ and, besides, $F_i = \emptyset$ implies $U_i = \emptyset$ for
any $i \leq n$.

There exists a number $m \in \omega$ such that, for a uniquely determined $s$-legal initial
segment $\{F_0, U_0, \ldots, F_{m-1}, U_{m-1}\}$ we have the equalities $F_0 = \ldots = F_{m-1} = \emptyset$
while $F_m = s(U_0, \ldots, U_{m-1}) \neq \emptyset$. Choose a faithful enumeration $\{x_0, \ldots, x_k\}$ of
the set $F_m$, let $y_0 = x_0$ and $t(\emptyset) = y_0$. If $1 \leq i \leq k$ and we have an initial segment
$\{y_0, V_0, \ldots, y_{i-1}, V_{i-1}\}$ in the point-open game on $X$ then let $t(V_0, \ldots, V_{i-1}) = x_i$.

After the moves $y_i, V_i$ are made for all $i \leq k$ let $U_m = V_0 \cup \ldots \cup V_k$ to obtain an initial segment $S = \{F_0, U_0, \ldots, F_m, U_m\}$ in the game $\mathcal{PO}'$. It is evident that $S$ is $s$-legal and $\bigcup_{i \leq k} V_i = \bigcup_{i \leq m} U_m$.

Proceeding inductively assume that $\{y_0, V_0, \ldots, y_n, V_n\}$ is an initial segment in the point-open game on the space $X$ such that for some initial $s$-legal segment $\{F_0, U_0, \ldots, F_m, U_m\}$, we have $V = \bigcup_{i \leq n} V_i = \bigcup_{i \leq m} U_i$. If $V = X$ then choose any $y_{n+1} \in X$ and let $t(V_0, \ldots, V_n) = y_{n+1}$. If $V \neq X$ then there is a uniquely determined $s$-legal segment $\{F_0, U_0, \ldots, F_m, U_m, F_{m+1}, U_{m+1}, \ldots, F_{m'-1}, U_{m'-1}\}$ such that $F_{m'} = s(U_0, \ldots, U_{m'-1}) \neq \emptyset$ and $F_i = \emptyset$ whenever $m < i < m'$. Choose a faithful enumeration $\{z_0, \ldots, z_{k'}\}$ of the set $F_{m'}$; if $0 \leq i \leq k'$ and we have an initial segment $\{y_0, V_0, \ldots, y_n, V_n, y_{n+1}, V_{n+1}, \ldots y_{n+i}, V_{n+i}\}$ in the point-open game on $X$ then let $y_{n+i+1} = z_i$ and $t(V_0, \ldots, V_{n+i}) = y_{n+i+1}$.

After the moves $y_i, V_i$ are made for all $i = n+1, \ldots, n+k'+1$ consider the set $U_{m'} = \bigcup\{V_i : n+1 \leq i \leq n+k'+1\}$. For the $s$-legal segment $\{F_0, U_0, \ldots, F_{m'}, U_{m'}\}$ we have $\bigcup_{i \leq m'} U_i = \bigcup_{i \leq n+k'+1} V_i$ and hence our inductive procedure can be continued to construct a strategy $t$ for the player $P$ on the space $X$ in the point-open game. Given a play $\{y_i, V_i : i \in \omega\}$ where $P$ applies $t$, we have two possibilities.

*Case 1.* There exists $n \in \omega$ such that $V_0 \cup \ldots \cup V_n = X$. It clear that $P$ wins this play.

*Case 2.* $V_0 \cup \ldots, \cup V_n \neq X$ for any $n \in \omega$ and hence there exists an $s$-legal play $\{F_i, U_i : i \in \omega\}$ for which we can choose increasing sequences $\{m_k : k \in \omega\}$ and $\{n_k : k \in \omega\}$ of natural numbers such that $\bigcup_{i \leq n_k} V_i = \bigcup_{i \leq m_k} U_i$ for all $k \in \omega$. As a consequence, $\bigcup_{i \in \omega} V_i = \bigcup_{i \in \omega} U_i = X$, so the player $P$ wins again and hence the strategy $t$ is winning, i.e., the space $X$ is $P$-favorable. This settles necessity and makes our solution complete.

**V.416.** *Given a nonempty space $X$ define a game $\mathcal{PO}''$ on the space $X$ as follows: at the $n$-th move the player $P$ chooses a finite set $F_n \subset X$ and the player $O$ takes a set $U_n \in \tau(X)$ such that $F_n \subset U_n$. After all moves $\{F_n, U_n : n \in \omega\}$ are made let $\mathcal{U} = \{U_n : n \in \omega\}$. The player $P$ wins if $\underline{\lim}\, \mathcal{U} = X$; otherwise the victory is assigned to $O$. Prove that the player $P$ has a winning strategy in the game $\mathcal{PO}''$ on a space $X$ if and only if $X$ is $P$-favorable for the point-open game.*

**Solution.** Recall that the moves in the game $\mathcal{PO}'$ are the same as in $\mathcal{PO}''$ but the player $P$ wins if the sets chosen by $O$ cover the whole space. Suppose that $s$ is a winning strategy for the player $P$ in the game $\mathcal{PO}''$ on a space $X$. If $\{F_n, U_n : n \in \omega\}$ is a play on $X$ where $P$ applies $s$ and $\mathcal{U} = \{U_n : n \in \omega\}$ then $\underline{\lim}\, \mathcal{U} = X$ and hence $\bigcup \mathcal{U} = X$ which shows that $s$ is also a winning strategy for the player $P$ in the game $\mathcal{PO}'$. By Problem 417 the space $X$ is $P$-favorable, so we proved necessity.

Now assume that a space $X$ is $P$-favorable and apply Problem 417 once more to find a winning strategy $s$ for the player $P$ on the space $X$ in the game $\mathcal{PO}'$. Say that a sequence $\{W_0, \ldots, W_n\}$ of open subsets of $X$ is *adequate* if there exist finite sets $G_0, \ldots, G_n$ such that $\{G_0, W_0, \ldots, G_n, W_n\}$ is an initial segment of play where $P$ uses the strategy $s$. Say that a sequence $\{W_n : n \in \omega\}$ of open subsets of $X$

is adequate if its initial segment $\{W_0, \ldots, W_n\}$ is adequate for any $n \in \omega$. Given an adequate family $\{W_n : n \in \omega\}$ the set $G_{n+1} = s(W_0, \ldots, W_n)$ is consistently defined for any $n \in \omega$. If $G_0 = s(\emptyset)$ then $\{G_n, W_n : n \in \omega\}$ is a play in which $P$ applies the strategy $s$. The strategy $s$ is winning, so we have the following property.

(1) If $\{W_n : n \in \omega\} \subset \tau(X)$ is an adequate family then $\bigcup_{n \in \omega} W_n = X$.

Let $F_0 = s(\emptyset)$; it is easy to see that

(2) for any $n \in \omega$, a sequence $\{W_0, \ldots, W_n\}$ of open subsets of $X$ is adequate if and only if $F_0 \subset W_0$ and $s(W_0, \ldots, W_i) \subset W_{i+1}$ for any $i = 0, \ldots, n - 1$.

To construct a strategy $t$ for the player $P$ in the game $\mathcal{P}\mathcal{O}''$ on the space $X$ let $t(\emptyset) = F_0$ and $t(U_0) = F_0 \cup s(U_0)$ for any $U_0 \in \tau(F_0, X)$. Suppose that we have an initial segment $\{F_0, U_0, \ldots, F_n, U_n\}$ of a play on $X$ such that

(3) if $0 \le j_0 < \ldots < j_k \le n$ then $\{U_{j_0}, \ldots, U_{j_k}\}$ is an adequate sequence;
(4) if $i < n$ and $0 \le j_0 < \ldots < j_k \le i$ then $F_0 \cup s(U_{j_0}, \ldots, U_{j_k}) \subset F_{i+1}$.

Given any sequence $\{j_0, \ldots, j_k\} \subset \{0, \ldots, n\}$ such that $j_0 < \ldots < j_k$ consider the set $F(j_0, \ldots, j_k) = F_0 \cup s(U_{j_0}, \ldots, U_{j_k})$ and let

$$F_{n+1} = \bigcup \{F(j_0, \ldots, j_k) : \{j_0, \ldots, j_k\} \subset \{0, \ldots, n\} \text{ and } j_0 < \ldots < j_k\};$$

let $t(U_0, \ldots, U_n) = F_{n+1}$. It is immediate from (2) that for any $U_{n+1} \in \tau(F_{n+1}, X)$ the properties (3) and (4) hold if we replace $n$ by $n + 1$, so our inductive procedure can be continued to construct a strategy $t$ for the player $P$ on the space $X$.

Now suppose that $\{F_n, U_n : n \in \omega\}$ is a play where $P$ applies the strategy $t$ and let $\mathcal{U} = \{U_n : n \in \omega\}$. Assume for a moment that $\underline{\lim}\, \mathcal{U} \ne X$. Then there exists $x \in X$ such that $x \notin \bigcap\{U_i : i \ge n\}$ for any $n \in \omega$. Therefore we can find an increasing sequence $\{j_k : k \in \omega\} \subset \omega$ such that $x \notin U_{j_k}$ for every $k \in \omega$ and hence $x \notin \bigcup_{k \in \omega} U_{j_k}$. However, it follows from (3) that $\{U_{j_k} : k \in \omega\}$ is an adequate family, so we can apply (1) to conclude that $\bigcup_{k \in \omega} U_{j_k} = X$. This contradiction shows that $t$ is a winning strategy for the player $P$ in the game $\mathcal{P}\mathcal{O}''$, i.e., we settled sufficiency and completed our solution.

**V.417.** *Prove that a space $X$ is $P$-favorable for the point-open game if and only if $C_p(X)$ is a $W$-space.*

**Solution.** Suppose that $X$ is $P$-favorable for the point-open game. By Problem 416 there exists a strategy $s$ on the space $X$ which is winning for $P$ in the game $\mathcal{P}\mathcal{O}''$. Denote by $u$ the function which is identically zero on $X$. For any finite set $A \subset X$ and $\varepsilon > 0$ let $[A, \varepsilon] = \{f \in C_p(X) : f(A) \subset (-\varepsilon, \varepsilon)\}$.

If $F_0 = s(\emptyset)$ then $H_0 = [F_0, 1]$ is an open neighborhood of the point $u$; let $t(\emptyset) = H_0$. If the second player in the $W$-game chooses a function $f_0 \in H_0$ then let us consider that the player $O$ picks the set $U_0 = f_0^{-1}((-1, 1))$; it is easy to see that $F_0 \subset U_0$.

Proceeding inductively assume that moves $\{H_0, f_0, \ldots, H_n, f_n\}$ have been made in the Gruenhage $W$-game on $C_p(X)$ at the point $u$, i.e., $H_i \in \tau(u, C_p(X))$,

$f_i \in H_i$ for all $i \leq n$ and, besides, we have an initial segment $S = \{F_0, U_0, \ldots, F_n, U_n\}$ of a play in the game $\mathcal{PO}''$ on $X$ such that $P$ applies the strategy $s$ in the segment $S$ while

(1)  $H_i = [F_i, 2^{-i}]$ and $U_i = f_i^{-1}((-2^{-i}, 2^{-i}))$ for every $i \leq n$.

If $F_{n+1} = s(U_0, \ldots, U_n)$ then the set $H_{n+1} = [F_{n+1}, 2^{-n-1}]$ is open in $C_p(X)$ and contains $u$; let $t(f_0, \ldots, f_n) = H_{n+1}$. If the answer of the second player in the $W$-game is a function $f_{n+1} \in H_{n+1}$ then letting $U_{n+1} = f_{n+1}^{-1}((-2^{-n-1}, 2^{-n-1}))$ we obtain an open neighborhood $U_{n+1}$ of the set $F_{n+1}$ and hence the family $\{F_0, U_0, \ldots, F_n, U_n, F_{n+1}, U_{n+1}\}$ is an initial segment of a play in the game $\mathcal{PO}''$ on $X$ where $P$ applies the strategy $s$. Evidently, the property (1) now holds for all $i \leq n + 1$ and hence our inductive procedure defines a strategy $t$ of the first player in the Gruenhage $W$-game on $C_p(X)$ at the point $u$ for which the condition (1) is satisfied.

Now, assume that $\{H_n, f_n : n \in \omega\}$ is a play in the $W$-game in $C_p(X)$ at the point $u$ in which the first player applies the strategy $t$. By our construction we have a play $\{F_n, U_n : n \in \omega\}$ in the game $\mathcal{PO}''$ on $X$ in which $P$ applies the strategy $s$ and the property (1) holds for all $n \in \omega$. Let $\mathcal{U} = \{U_n : n \in \omega\}$; since the strategy $s$ is winning, we have $X = \lim \mathcal{U}$.

Given any neighborhood $G$ of the point $u$ in $C_p(X)$ we can find a finite set $P \subset X$ and $\varepsilon > 0$ such that $[P, \varepsilon] \subset G$. Fix $m \in \omega$ for which $2^{-m} < \varepsilon$; for any $x \in P$ there exists $l_x \in \omega$ such that $x \in U_n$ for all $n \geq l_x$. If $l = m + \sum\{l_x : x \in P\}$ then $x \in U_n$ whenever $x \in P$ and $n \geq l$. Furthermore, for any point $x \in P$ we have $f_n(x) \in (-2^{-n}, 2^{-n}) \subset (-2^{-m}, 2^{-m}) \subset (-\varepsilon, \varepsilon)$ which shows that $f_n \in [P, \varepsilon] \subset G$ for all $n \geq l$. Thus the sequence $\{f_n : n \in \omega\}$ converges to $u$, so the strategy $t$ is winning and hence we have the $W$-property at the point $u$. The space $C_p(X)$ being homogeneous, every $f \in C_p(X)$ is a $W$-point, i.e., we established necessity.

To prove sufficiency suppose that $u$ is a $W$-point in $C_p(X)$ and fix a winning strategy $t$ for the first player in the Gruenhage game on $C_p(X)$ at the point $u$. If $H_0 = t(\emptyset)$ then there exists a finite set $F_0 \subset X$ such that $[G_0, \varepsilon_0] \subset H_0$ for some $\varepsilon_0 > 0$; let $s(\emptyset) = F_0$. If the player $O$ chooses an open set $U_0 \supset F_0$ then we can find a function $f_0 \in C_p(X)$ such that $f_0(F_0) \subset \{0\}$ and $f_0(X \setminus U_0) \subset \{1\}$. It is clear that $f_0 \in H_0$, so $\{H_0, f_0\}$ is an initial segment of a play in the $W$-game on $C_p(X)$ at the point $u$.

Proceeding inductively assume that $n \in \omega$ and we have an initial segment $\{F_0, U_0, \ldots, F_n, U_n\}$ in the game $\mathcal{PO}'$ (see Problem 415) and, besides, there is an initial segment $\{H_0, f_0, \ldots, H_n, f_n\}$ of a play in the $W$-game on $C_p(X)$ at the point $u$ in which the first player applies the strategy $t$ while

(2)  $f_i(F_i) \subset \{0\}$ and $f_i(X \setminus U_i) \subset \{1\}$ for every $i \leq n$.

The set $H_{n+1} = t(f_0, \ldots, f_n)$ is an open neighborhood of $u$ in $C_p(X)$, so there exists a finite set $F_{n+1} \subset X$ such that $[F_{n+1}, \varepsilon_{n+1}] \subset H_{n+1}$ for some $\varepsilon_{n+1} > 0$. Let $s(U_0, \ldots, U_n) = F_{n+1}$. If $U_{n+1} \supset F_{n+1}$ is the move of the player $O$ then we can find a function $f_{n+1} \in C_p(X)$ such that $f_{n+1}(F_{n+1}) \subset \{0\}$ and $f_{n+1}(X \setminus U_{n+1}) \subset \{1\}$. Then $f_{n+1} \in H_{n+1}$ and hence we have an initial segment

$\{H_0, f_0, \ldots, H_{n+1}, f_{n+1}\}$ of a play in the $W$-game on $C_p(X)$ at the point $u$ in which the first player applies the strategy $t$ and the condition (2) is satisfied for all $i \le n + 1$. Therefore our inductive procedure can be continued to give us a strategy $s$ in the game $\mathcal{PO}'$ for the player $P$ on the space $X$ with the condition (2) is satisfied for all $n \in \omega$.

Finally assume that $\{F_n, U_n : n \in \omega\}$ is a play in which $P$ applies the strategy $s$. By our construction, there exists a play $\{H_n, f_n : n \in \omega\}$ in the $W$-game on $C_p(X)$ at the point $u$ in which the first player applies the strategy $t$ and the property (2) holds. Given any $x \in X$ the set $[\{x\}, 1]$ is an open neighborhood of $u$; since the strategy $t$ is winning, the sequence $\{f_n : n \in \omega\}$ converges to $u$ and hence there exists $i \in \omega$ such that $f_i \in [\{x\}, 1]$, i.e., $|f_i(x)| < 1$. This, together with (2) implies that $x \in U_i$; the point $x \in X$ was taken arbitrarily, so we proved that $\bigcup_{n\in\omega} U_n = X$ and hence $s$ is a winning strategy for the player $P$ in the game $\mathcal{PO}'$. Applying Problem 415 we conclude that $X$ is $P$-favorable for the point-open game so we settled sufficiency and completed our solution.

**V.418.** *Assuming that $2^\omega = \omega_1$ and $2^{\omega_1} = \omega_2$ prove that there exists a scattered Lindelöf $P$-space $X$ such that $hd^*(X) = \omega_1$ and $|X| = \omega_2$. Show that $C_p(X)$ is a $W$-space with no point-countable $\pi$-base. In particular, $C_p(X)$ is a $W$-space which cannot be embedded into a $\Sigma$-product of first countable spaces.*

**Solution.** Given a space $Z$ let $\mathcal{G}$ be the family of all $G_\delta$-subsets of $Z$. If $\tau$ is the topology on $Z$ generated by $\mathcal{G}$ as a base then $(Z, \tau)$ is called *the $\omega$-modification of $Z$.* We denote by $CH^+$ the statement $2^{\omega_1} = \omega_2$; given ordinals $\alpha$ and $\beta$ with $\alpha < \beta$ let $[\alpha, \beta) = \{\gamma : \alpha \le \gamma < \beta\}$.

*Fact 1.* Given spaces $Z$ and $P$, if $A \subset Z \times P$ and $u = (z, p) \in Z \times P$ then $u \in \overline{A}$ if and only if, for any $U \in \tau(z, Z)$ we have $p \in \pi(A \cap (U \times P))$ where $\pi : Z \times P \to P$ is the projection.

*Proof.* If $u \in \overline{A}$ and $U \in \tau(z, Z)$ then $u \in \overline{A \cap (U \times P)}$ because $U \times P$ is a neighborhood of $u$. The projection $\pi$ being continuous, it follows from $\pi(u) = p$ that $p \in \overline{\pi(A \cap (U \times P))}$ and hence we proved necessity.

Now assume that $p \in \overline{\pi(A \cap (U \times P))}$ for any $U \in \tau(z, Z)$. If $u \notin \overline{A}$ then there are $U \in \tau(z, Z)$ and $V \in \tau(p, P)$ such that $(U \times V) \cap A = \emptyset$. It follows from $p \in \overline{\pi(A \cap (U \times P))}$ that $V \cap \pi(A \cap (U \times P)) \ne \emptyset$ and hence there is a point $(x, y) \in A \cap (U \times P)$ such that $y \in V$. Since also $x \in U$, we have $(x, y) \in A \cap (U \times V)$ which is a contradiction. Therefore $u \in \overline{A}$, i.e., we settled sufficiency and hence Fact 1 is proved.

*Fact 2.* If $Z$ is a Hausdorff $P$-space and $Y$ is a Lindelöf subspace of $Z$ the $Y$ is closed in $Z$.

*Proof.* Take an arbitrary point $z \in Z \backslash Y$. By the Hausdorff property of $Z$, for each $y \in Y$ we can find $U_y, V_y \in \tau(Z)$ such that $y \in U_y$, $z \in V_y$ and $U_y \cap V_y = \emptyset$. The set $Y$ being Lindelöf there exists a countable $A \subset Y$ such that $Y \subset U = \bigcup\{U_y : y \in A\}$. The set $V = \bigcap\{V_y : y \in A\}$ is an open neighborhood of $z$ in $Z$ (because

$Z$ is a $P$-space) and $V \cap U = \emptyset$. Therefore $V \cap Y = \emptyset$ and hence any $z \in Z \backslash Y$ has an open neighborhood contained in $Z \backslash Y$. This shows that $Z \backslash Y$ is open in $Z$ whence $Y$ is closed in $Z$, i.e., Fact 2 is proved.

Returning to our solution apply $CH^+$ to conclude that the family $\mathcal{S} = \{S :$ there is $n \in \mathbb{N}$ such that $S \subset \omega_2^n$ and $|S| = \omega_1\}$ has cardinality $\omega_2^{\omega_1} = (2^{\omega_1})^{\omega_1} = 2^{\omega_1} = \omega_2$. Take an enumeration $\{P_\alpha : \alpha < \omega_2\}$ of the family $\mathcal{S}$; given any $\alpha < \omega_2$ let $n_\alpha \in \mathbb{N}$ be the natural number for which $P_\alpha \subset \omega_2^{n_\alpha}$. For an arbitrary $\alpha < \omega_2$ it follows from $|S| = \omega_1$ that there exists an ordinal $\lambda_\alpha \in [\omega_1, \omega_2)$ such that $s(i) < \lambda_\alpha$ for any $s \in P_\alpha$ and $i < n_\alpha$. Let $\mu_\alpha = \sup\{\lambda_\beta : \beta < \alpha\} + 1$ for any $\alpha < \omega_2$. It is clear that $P_\alpha \subset \mu_\alpha^{n_\alpha}$ for each $\alpha < \omega_2$ and $\alpha < \beta$ implies $\mu_\alpha < \mu_\beta$. Let $S_{\mu_\alpha} = P_\alpha$ and $n(\mu_\alpha) = n_\alpha$ for all $\alpha < \omega_2$; if $\mu \in [\omega_1, \omega_2) \backslash \{\mu_\alpha : \alpha < \omega_2\}$ then let $S_\mu = \omega_1$ and $n(\mu) = 1$. This gives us an enumeration $\{S_\mu : \omega_1 \leq \mu < \omega_2\}$ of the family $\mathcal{S}$ such that

(∗)  $S_\mu \subset \mu^{n(\mu)}$ for every $\mu \in [\omega_1, \omega_2)$.

Let $D$ be a discrete space of cardinality $\omega_1$. It is clear that $w(D^{\omega_1}) = \omega_1$; fix a base $\mathcal{B}$ in $D^{\omega_1}$ with $|\mathcal{B}| = \omega_1$. Consider the family $\mathcal{G}$ of all $G_\delta$-subsets of $D^{\omega_1}$ and let $\tau'$ be the topology on $D^{\omega_1}$ generated by $\mathcal{G}$ as a base, i.e., $E = (D^{\omega_1}, \tau')$ is the $\omega$-modification of $D^{\omega_1}$. Suppose that $x \in W \in \tau'$; there exists $G \in \mathcal{G}$ such that $x \in G \subset W$. Since $G = \bigcap_{n \in \omega} U_n$ for some family $\{U_n : n \in \omega\} \subset \tau(D^{\omega_1})$, we can pick a set $B_n \in \mathcal{B}$ such that $x \in B_n \subset U_n$ for every $n \in \omega$. Then $x \in \bigcap_{n \in \omega} B_n \subset G$. Therefore the family $\mathcal{G}'$ of all countable intersections of the elements of $\mathcal{B}$ is a base of the space $E$. Observe that $|\mathcal{G}'| \leq |\mathcal{B}|^\omega \leq \omega_1^\omega$ and apply CH to see that we have the equalities $\omega_1^\omega = (2^\omega)^\omega = 2^{\omega \cdot \omega} = 2^\omega = \omega_1$. Consequently, $|\mathcal{G}'| \leq \omega_1$ and hence $w(E) = \omega_1$.

It follows from CH and $CH^+$ that $|E| = \omega_1^{\omega_1} = (2^\omega)^{\omega_1} = 2^{\omega_1} = \omega_2$, so we can identify the underlying set of $E$ with $\omega_2$ and consider that we have a topology $\rho$ on the set $\omega_2$ such that $E = (\omega_2, \rho)$ is a $P$-space with $w(E) = \omega_1$. For every $\beta < \omega_2$ and natural $n > 1$ we will need the natural projection $\pi_\beta^n : \beta^{n-1} \times (\beta+1) \to (\beta+1)$. For any $\alpha < \omega_2$, denote by $\rho_\alpha$ the topology on $\alpha$ induced from $E$.

To start our inductive construction let $\tau_\alpha = \exp(\alpha)$ for each $\alpha < \omega_1$ and assume that, for some $\alpha \in [\omega_1, \omega_2]$ we have a collection $\{\tau_\beta : \beta < \alpha\}$ of topologies with the following properties:

(1)  $\tau_\beta$ is a topology on $\beta$ for all $\beta < \alpha$;
(2)  for any $\beta < \alpha$, the space $Y_\beta = (\beta, \tau_\beta)$ is Tychonoff, locally Lindelöf and every $G_\delta$-subset of $Y_\beta$ is open in $Y_\beta$ (i.e., $Y_\beta$ is a $P$-space); besides, for any $n \in \mathbb{N}$, a base $\mathcal{B}_\beta^n$ with $|\mathcal{B}_\beta^n| \leq \omega_1$ is chosen in $(Y_\beta)^n$;
(3)  if $\gamma < \beta < \alpha$ then $\tau_\beta \cap \exp(\gamma) = \tau_\gamma$;
(4)  $\rho_\beta \subset \tau_\beta$ for any $\beta < \alpha$;
(5)  if $\mu < \beta < \alpha$, $n = n(\mu) > 1$ and the point $\beta$ belongs to the $\rho$-closure of the set $\pi_\beta^n(S_\mu \cap (U \times (\beta+1)))$ for some $U \in \mathcal{B}_\beta^{n-1}$ then $\beta$ belongs to the $\tau_{\beta+1}$-closure of $\pi_\beta^n(S_\mu \cap (U \times (\beta+1)))$;
(6)  if $\mu < \beta < \alpha$, $n = n(\mu) = 1$ and the point $\beta$ belongs to the $\rho$-closure of $S_\mu$ then $\beta$ also belongs to the $\tau_{\beta+1}$-closure of $S_\mu$.

If $\alpha$ is a limit ordinal then let $\tau_\alpha$ be the $\tilde\omega$-modification of the topology generated by the family $\bigcup\{\tau_\beta : \beta < \alpha\}$ as a subbase. It follows from (3) that every set $\beta$ is open in $Y_\alpha = (\alpha, \tau_\alpha)$, so (2) and (3) imply that the property (3) still holds for all $\beta \le \alpha$. If $U \in \rho_\alpha$ then $U_\beta = U \cap \beta \in \tau_\beta$ for any $\beta < \alpha$ by (4). Therefore $U = \bigcup\{U_\beta : \beta < \alpha\} \in \tau_\alpha$, i.e., (4) also holds for all $\beta \le \alpha$. The properties (5) and (6) bring no new obligations for the collection $\{\tau_\beta : \beta \le \alpha\}$, so they are also fulfilled for all $\beta \le \alpha$.

As to the property (2), observe first that $Y_\alpha$ is Hausdorff its topology being stronger than the Tychonoff topology $\rho_\alpha$. Given a point $\beta < \alpha$ there is a local base $\mathcal{C}$ of Lindelöf open neighborhoods of the point $\beta$ in the space $Y_{\beta+1}$. It follows from (2) and (3) that $\mathcal{C}$ is also a local base in $Y_\alpha$ at the point $\beta$. Besides, any Lindelöf subspace is closed in a Hausdorff $P$-space (see Fact 2), so $\mathcal{C}$ is a clopen local base in $Y_\alpha$ at $\beta$. Thus $Y_\alpha$ is zero-dimensional and hence Tychonoff. Choosing the respective bases in all finite powers of $Y_\alpha$, we conclude that (2) is also fulfilled for all $\beta \le \alpha$, i.e., our construction can be carried out for any limit ordinal $\alpha$.

Now assume that $\alpha$ is a successor ordinal, say, $\alpha = \xi + 1$ and consider the set $\mathcal{M}_\xi = \{\mu < \xi$: either $n = n(\mu) > 1$ and $\xi$ is in the $\rho$-closure of the set $T_\mu^U = \pi_\beta^n(S_\mu \cap (U \times (\xi + 1)))$ for some $U \in \mathcal{B}_\xi^{n-1}$ or $n = n(\mu) = 1$ and $\xi$ is in the $\rho$-closure of $S_\mu\}$. If $\mathcal{M}_\xi = \emptyset$ then let $\tau_\alpha$ be the topology generated by $\tau_\xi \cup \{\{\xi\}\}$ as a subbase. Choosing the respective bases in all finite powers of $Y_\alpha$ we can easily see that the conditions (1)–(6) are satisfied for all $\beta \le \alpha$, so this case is easy.

Now, if the set $\mathcal{M}_\xi$ is nonempty then the family $\mathcal{H}_\xi = \{S_\mu : \mu \in \mathcal{M}_\xi$ and $n(\mu) = 1\} \cup \{T_\mu^U : \mu \in \mathcal{M}_\xi,\ n(\mu) > 1$ and $U \in \mathcal{B}_\xi^{n-1}\}$ has cardinality $\le \omega_1$, so we can choose an enumeration $\{H_\beta : \beta < \omega_1\}$ of the family $\mathcal{H}_\xi$ in which every $H \in \mathcal{H}_\xi$ occurs $\omega_1$-many times. Since $E_\alpha = (\alpha, \rho_\alpha)$ is a $P$-space of weight $\omega_1$, we can choose a local clopen base $\{B_\beta : \beta < \omega_1\}$ at the point $\xi$ in $E_\alpha$ such that $\gamma < \beta < \omega_1$ implies $B_\beta \subset B_\gamma$.

The point $\xi$ belongs to the $\rho$-closure of every element of $\mathcal{H}_\xi$, so we can pick a point $z_\beta \in H_\beta \cap B_\beta$ for every $\beta < \omega_1$. It follows from $(*)$ that $z_\beta \ne \xi$ for any $\beta < \omega_1$. The transfinite sequence $S = \{z_\beta : \beta < \omega_1\}$ converges to the point $\xi$ in $E_\alpha$, which, together with the $P$-property of $E_\alpha$ implies that $S$ is a discrete subspace of $E_\alpha$ and $\xi$ is the unique cluster point of $S$.

Given an ordinal $\beta < \omega_1$ the set $B_\beta \cap \xi$ is a clopen subspace of $Y_\xi$, so there exists a clopen Lindelöf subspace $L_\beta$ of the space $Y_\xi$ such that $z_\beta \in L_\beta \subset B_\beta$. For $\mathcal{C}_\xi = \{\{\xi\} \cup (\bigcup\{L_\gamma : \gamma \ge \beta\}) : \beta < \omega_1\}$ let $\tau_\alpha$ be the topology generated by the family $\tau_\xi \cup \mathcal{C}_\xi$ as a subbase.

An immediate consequence of the definition of $\tau_\alpha$ is that $\tau_\alpha \cap \exp(\xi) = \tau_\xi$; i.e., we have (1) and (3) for the collection $\{\tau_\beta : \beta \le \alpha\}$. The family $\mathcal{C}_\xi$ is a local clopen base at the point $\xi$, so the space $Y_\alpha$ is Tychonoff being $T_1$ and zero-dimensional. Any countable intersection of the elements of $\mathcal{C}_\xi$ belongs to $\mathcal{C}_\xi$ whence $\xi$ is a $P$-point in $Y_\alpha$; thus $Y_\alpha$ is a $P$-space. Furthermore, every $C \in \mathcal{C}_\xi$ is Lindelöf because, for any $U \in \tau(\xi, Y_\alpha)$ there is $\beta < \omega_1$ such that the set $C \setminus U$ is closed in the Lindelöf space $\bigcup\{L_\gamma : \gamma < \beta\}$. After we choose the relevant bases in all finite powers of $Y_\alpha$ we conclude that (2) also holds for the family $\{\tau_\beta : \beta \le \alpha\}$. The property (4) is clear and the property (5) must only be checked for $\beta = \xi$.

If $\mu < \xi$ and $n(\mu) > 1$ assume that the point $\xi$ is in the $\rho$-closure of the set $T_\mu^U$ for some $U \in \mathcal{B}_\xi^{n-1}$. Then $H = T_\mu^U \in \mathcal{H}_\xi$ and therefore $|H \cap S| = \omega_1$ which shows that $C \cap H \neq \emptyset$ for any $C \in \mathcal{C}_\xi$, i.e., $\xi$ is in the $\tau_\alpha$-closure of $H$. Therefore the condition (5) is satisfied for the family $\mathcal{T} = \{\tau_\beta : \beta \leq \alpha\}$. The proof that (6) is also satisfied for $\mathcal{T}$ is analogous, so our inductive construction is complete; let $\tau = \tau_{\omega_2}$ and $Y = (\omega_2, \tau)$.

It follows from (2) and (3) that every $Y_\beta$ is open in $Y$, so $Y$ is right-separated and hence scattered. The properties (2) and (3) imply that $Y$ is locally Lindelöf, Tychonoff and every $\beta \in Y$ is a $P$-point in $Y$, i.e., $Y$ is a locally Lindelöf $P$-space.

We will prove by induction that $hd(Y^n) \leq \omega_1$ for every $n \in \mathbb{N}$. To start the induction let $Y^0 = \{\emptyset\}$; then $hd(Y^0) = 1 < \omega_1$. Now assume that $n \geq 1$ and we have proved that $hd(Y^k) \leq \omega_1$ for all $k < n$. We will need the sets $\Delta_1 = \emptyset$ and $\Delta_n = \{x \in Y^n : \text{there are distinct } i, j < n \text{ with } x(i) = x(j)\}$ for any $n \in \mathbb{N}\setminus\{1\}$. By Fact 0 of T.020 it suffices to show that we have the inequality $hd(Y^n\setminus\Delta_n) \leq \omega_1$. Consider the set $G_n = \{x \in Y^n : x(0) < \ldots < x(n-1)\}$; since any permutation of the set $n$ generates a homeomorphism of $Y^n\setminus\Delta_n$ onto itself, the space $Y^n\setminus\Delta_n$ is a finite union of spaces homeomorphic to $G_n$. Therefore it suffices to establish that $hd(G_n) \leq \omega_1$ (evidently, $G_1 = Y$).

Take an arbitrary set $A \subset G_n$; there is no loss of generality to consider that $|A| = \omega_2$. We have $hd(Y^{n-1}) \leq \omega_1$ by the induction hypothesis; since $w(E) \leq \omega_1$, we can apply Fact 1 of V.023, to see that $hd(Y^{n-1} \times E) \leq \omega_1$. Therefore the set $A$ has density at most $\omega_1$ in the space $Y^{n-1} \times E$, so we can find a set $B \subset A$ such that $|B| = \omega_1$ and $B$ is $\nu$-dense in $A$ where $\nu = \tau(Y^{n-1} \times E)$; there is an ordinal $\mu < \omega_2$ such that $B = S_\mu$. Let $\pi : Y^{n-1} \times Y \to Y$ be the projection.

The set $A_0 = \{x \in A : x(n-1) \leq \mu\}$ has cardinality at most $\omega_1$. If $x \in A\setminus A_0$ then $\beta = x(n-1) > \mu$ and $x = (z, \beta)$ where $z \in \beta^{n-1}$. Given any $V \in \tau(z, Y^{n-1})$, there is $U \in \mathcal{B}_\beta^{n-1}$ such that $z \in U \subset V$. The point $x$ being in the $\nu$-closure of $S_\mu$, we can apply Fact 1 to see that the point $\beta$ is in the $\rho$-closure of the set $T_\mu^U = \pi_n(S_\mu \cap (U \times (\beta + 1)))$. The property (5) for $n > 1$ (or the property (6) if $n = 1$) shows that $\beta$ belongs to the $\tau_{\beta+1}$-closure of $T_\mu^U$.

An immediate consequence of (3) is that $\beta$ belongs to the $\tau$-closure of $T_\mu^U$. Since $U \subset V$, the point $\beta$ belongs to the $\tau$-closure of the set $\pi_n(S_\mu \cap (V \times (\beta + 1)))$; therefore $\beta$ belongs to the $\tau$-closure of the set $\pi(S_\mu \cap (V \times Y))$. Now, Fact 1 is applicable to the product $P \times Z$ where $Z = Y^{n-1}$ and $P = Y$ to conclude that $x = (z, \beta)$ is in the $\tau^n$-closure of $S_\mu$ where $\tau^n = \tau(Y^n)$. This proves that $A_0 \cup B$ is a $\tau^n$-dense subset of $A$ of cardinality at most $\omega_1$. Therefore $hd(G_n) \leq \omega_1$ and hence $hd(Y^n) \leq \omega_1$.

To finally construct the promised space $X$ take a point $a \notin Y$. To introduce a topology $\tau_X$ on the set $X = Y \cup \{a\}$ consider the family $\mathcal{L} = \{U \subset Y : Y\setminus U \text{ is a clopen Lindelöf subspace of } Y\}$ and let $\tau_X$ be the topology generated by the family $\tau \cup \{\{a\} \cup L : L \in \mathcal{L}\}$. It is straightforward that $(X, \tau_X)$ is a Lindelöf scattered $P$-space. It follows from $hd^*(Y) \leq \omega_1$ and $|X\setminus Y| = 1$ that $hd^*(X) \leq \omega_1$ (see Fact 1 of T.099).

Next, apply Problem 413 to see that $X$ is $P$-favorable for the point-open game and hence $C_p(X)$ is a $W$-space by Problem 417. Assume that $C_p(X)$ has a point-countable $\pi$-base $\mathcal{B}$. By Fact 2 of U.271 there exists a family $\{D_\alpha : \alpha < \omega_1\}$ of discrete subspaces of $C_p(X)$ such that the set $D = \bigcup\{D_\alpha : \alpha < \omega_1\}$ meets every element of the family $\mathcal{B}$; in particular, $D$ is dense in $C_p(X)$.

Apply SFFS-027 to see that $s(C_p(X)) \leq hl(C_p(X)) \leq hl^*(C_p(X)) = hd^*(X) \leq \omega_1$ and hence $|D_\alpha| \leq \omega_1$ for each $\alpha < \omega_1$; therefore $|D| \leq \omega_1$. As a consequence, $\psi(X) \leq iw(X) = d(C_p(X)) \leq \omega_1$ which shows that $\psi(X) \leq \omega_1$. In particular, $\psi(a, X) \leq \omega_1$ and hence there is a family $\{W_\alpha : \alpha < \omega_1\} \subset \tau(a, X)$ such that $\bigcap\{W_\alpha : \alpha < \omega_1\} = \{a\}$. If $F_\alpha = X \backslash W_\alpha$ then $F_\alpha$ is a Lindelöf subspace of $Y$ for any $\alpha < \omega_1$. It follows from $Y = \bigcup\{F_\alpha : \alpha < \omega_1\}$ that $l(Y) \leq \omega_1$. However, $\{Y_\alpha : \alpha < \omega_2\}$ is an open cover of $Y$ which does not have any subcover of cardinality $\omega_1$. This contradiction shows that $C_p(X)$ does not have a countable $\pi$-base and hence it cannot be embedded in a $\Sigma$-product of first countable spaces (see Problem 412), so our solution is complete.

**V.419.** *Prove that if $X_t$ is a $d$-separable space for each $t \in T$ then the product space $\prod_{t \in T} X_t$ is $d$-separable. In other words, any product of $d$-separable spaces must be $d$-separable.*

**Solution.** The expression $X \simeq Y$ says that the spaces $X$ and $Y$ are homeomorphic. Given a space $N_t$ for each $t \in T$ suppose that $a \in N = \prod_{t \in T} N_t$. The space $\sigma(N, a) = \{x \in N : \text{the set } \{t \in T : x(t) \neq a(t)\} \text{ is finite}\}$ is the $\sigma$-product of $N$ with the center $a$. For any $x \in \sigma(N, a)$ let $\operatorname{supp}(x) = \{t \in T : x(t) \neq a(t)\}$. We will say that a topological property $\mathcal{P}$ is $d$-*adequate*, if

(1) $\mathcal{P}$ is preserved by finite products;
(2) if $X$ has $\mathcal{P}$ and $U$ is open in $X$ then $U$ has $\mathcal{P}$;
(3) $\mathcal{P}$ is preserved by free unions, i.e., if $X_t$ has $\mathcal{P}$ for every $t \in T$ then the space $\bigoplus_{t \in T} X_t$ also has $\mathcal{P}$;
(4) If $X$ is a space, $\overline{Y} = X$ and $Y$ has $\mathcal{P}$ then $X$ also has $\mathcal{P}$;
(5) if $X = \bigcup_{n \in \omega} X_n$ and every $X_n$ has $\mathcal{P}$ then $X$ also has $\mathcal{P}$.

*Fact 1.* Suppose that a topological property $\mathcal{P}$ is $d$-adequate and $N_t$ has $\mathcal{P}$ for any $t \in T$; let $N = \prod_{t \in T} N_t$. Then the space $\sigma(N, a)$ has $\mathcal{P}$ for any $a \in N$ and hence the space $N$ also has $\mathcal{P}$.

*Proof.* Fix any point $a \in N$ and let $N_t^* = N_t \backslash \{a(t)\}$ for any $t \in T$; given a set $S \subset T$ let $a_S = a|S$. For every $n \in \omega$ consider the set $L_n = \{x \in \sigma(N, a) : |\operatorname{supp}(x)| = n\}$; observe that $L_0 = \{a\}$ and $\bigcup_{n \in \omega} L_n = \sigma(N, a)$.

Fix any $n \in \mathbb{N}$ and consider the family $\mathcal{F}_n = \{A \subset T : |A| = n\}$. Observe that for every $A \in \mathcal{F}_n$ the set $H_A = \prod_{t \in A} N_t^* \times \{a_{T \backslash A}\}$ is contained in $L_n$ and $L_n = \bigcup\{H_A : A \in \mathcal{F}_n\}$. If $A, A' \in \mathcal{F}_n$ and $A \neq A'$ then $A \backslash A' \neq \emptyset$ or $A' \backslash A \neq \emptyset$. If $t \in A \backslash A'$ then $x(t) \neq a(t)$ for any $x \in H_A$ while $y(t) = a(t)$ whenever $y \in H_{A'}$; an analogous reasoning for the case of $A' \backslash A \neq \emptyset$ shows that $H_A \cap H_{A'} = \emptyset$ for distinct $A, A' \in \mathcal{F}_n$, i.e.,

(6) the family $\{H_A : A \in \mathcal{F}_n\}$ is disjoint for any $n \in \mathbb{N}$.

Let $W_A = \prod_{t \in A} N_t^* \times \prod_{t \in T \setminus A} N_t$ for every $A \in \mathcal{F}_n$. It is clear that $W_A$ is an open subset of $N$ and a moment's reflection shows that $W_A \cap L_n = H_A$. Therefore every $H_A$ is open in $L_n$, so it follows from (6) that $L_n$ is homeomorphic to $\bigoplus \{H_A : A \in \mathcal{F}_n\}$. It is an easy exercise that $H_A \simeq \prod_{t \in A} N_t^*$, so it follows from the properties (1) and (2) that $H_A$ has $\mathcal{P}$ for every $A \in \mathcal{F}_n$. Applying (3) we can see that $L_n$ has the property $\mathcal{P}$. Now the property (5) implies that $\sigma(N, a)$ has $\mathcal{P}$. Finally observe that $\sigma(N, a)$ is dense in $N$ and apply (4) to conclude that $N$ has $\mathcal{P}$ as well, so Fact 1 is proved.

Returning to our solution let $\mathcal{P}$ stand for $d$-separability. Given $d$-separable spaces $X$ and $Y$ we can choose discrete subspaces $D_n^X \subset X$ and $D_n^Y \subset Y$ for each $n \in \omega$ in such a way that the set $D_X = \bigcup \{D_n^X : n \in \omega\}$ is dense in $X$ and $D_Y = \bigcup \{D_n^Y : n \in \omega\}$ is dense in $Y$. It is evident that $D_{m,n} = D_m^X \times D_n^Y$ is a discrete subspace of $X \times Y$ for any $m, n \in \omega$. Since $D = \bigcup \{D_{m,n} : m, n \in \omega\} = D_X \times D_Y$, the set $D$ is dense in $X \times Y$ and hence $X \times Y$ is $d$-separable. Therefore $\mathcal{P}$ has the property (1).

Given a $d$-separable space $X$ fix a family $\{D_n : n \in \omega\}$ of discrete subspaces of $X$ such that $D = \bigcup_{n \in \omega} D_n$ is dense in $X$. If $U$ is open in $X$ then $D_n' = D_n \cap U$ is a discrete subspace of $U$ and $D' = \bigcup_{n \in \omega} D_n' = D \cap U$ is dense in $U$, so $U$ is $d$-separable and hence $\mathcal{P}$ satisfies the condition (2).

Now assume that $X_t$ is $d$-separable and choose a family $\{D_n^t : n \in \omega\}$ of discrete subspaces of $X_t$ such that $D_t = \bigcup_{n \in \omega} D_n^t$ is dense in $X_t$ for every $t \in T$. We will identify every $X_t$ with the respective clopen subspace of $X = \bigoplus_{t \in T} X_t$. It is straightforward that the set $D_n = \bigcup \{D_n^t : t \in T\}$ is discrete for any $n \in \omega$; since also $D = \bigcup_{n \in \omega} D_n = \bigcup_{t \in T} D_t$ is dense in $X$, we conclude that $d$-separability is preserved by free unions, i.e., $\mathcal{P}$ has (3).

If $Y$ has a dense $\sigma$-discrete subspace $D$ and $\overline{Y} = X$ then $D$ is a dense $\sigma$-discrete subspace of $X$, so $X$ is also $d$-separable and hence $\mathcal{P}$ has (4). Finally, assume that $X = \bigcup_{n \in \omega} X_n$ and every $X_n$ is $d$-separable; fix a family $\{D_m^n : m \in \omega\}$ of discrete subspaces of $X_n$ such that $D_n = \bigcup \{D_m^n : m \in \omega\}$ is dense in $X_n$. Then the set $D = \bigcup \{D_m^n : n, m \in \omega\}$ is a dense $\sigma$-discrete subspace of $X$ and hence $\mathcal{P}$ also has the property (5), so $d$-separability is a $d$-adequate property. Applying Fact 1 we conclude that $\mathcal{P}$ is preserved by arbitrary products, i.e., any product of $d$-separable spaces is $d$-separable and hence our solution is complete.

**V.420.** *Given an infinite cardinal $\kappa$ and a space $X$ prove that $X^\kappa$ is $d$-separable if and only if there exists a family $\mathcal{D} = \{D_n : n \in \omega\}$ of discrete subspaces of $X^\kappa$ such that $\sup\{|D_n| : n \in \omega\} \geq d(X)$. In particular, if $X^\kappa$ has a discrete subspace of cardinality $d(X)$ then $X^\kappa$ is $d$-separable. Deduce from this fact that $X^{d(X)}$ is $d$-separable for any space $X$.*

**Solution.** If the space $X^\kappa$ is $d$-separable then there exists a family $\{D_n : n \in \omega\}$ of discrete subsets of $X^\kappa$ such that $D = \bigcup_{n \in \omega} D_n$ is dense in $X^\kappa$. Therefore $|D| = \sup\{|D_n| : n \in \omega\} \geq d(X^\kappa) \geq d(X)$, so we have necessity.

Let $\lambda = d(X)$ and assume that we have a family $\{D_n : n \in \omega\}$ of discrete subspaces of $X^\kappa$ such that $\sup\{|D_n| : n \in \omega\} \geq \lambda$. Consider the cardinal $\lambda_n = |D_n|$;

there is no loss of generality to assume that $\omega \le \lambda_n \le \lambda_{n+1}$ for all $n \in \omega$. If $X$ is discrete then $X^\kappa$ is $d$-separable by Problem 421, so we can assume that $X$ is not discrete; fix a non-isolated point $p \in X$.

It is evident that there exists a set $S \subset X \backslash \{p\}$ such that $\overline{S} = X$ and $|S| = \lambda$. It follows from $\sup_{n \in \omega} \lambda_n \ge \lambda$ that we can find a family $\{S_n : n \in \omega\}$ of subsets of $S$ such that $|S_n| \le \lambda_n$, $S_n \subset S_{n+1}$ for all $n \in \omega$ and $S = \bigcup_{n \in \omega} S_n$. For each nonempty finite set $A \subset \kappa$ and $n \in \omega$ the cardinality of the set $(S_n)^A$ does not exceed $\lambda_n$; let $\{s_\xi^{A,n} : \xi < \lambda_n\}$ be an enumeration of $(S_n)^A$.

Take a family $\{I_n : n \in \omega\}$ of subsets of $\kappa$ such that $\kappa = \bigcup_{n \in \omega} I_n$ while $I_n \subset I_{n+1}$ and $|I_n| = |\kappa \backslash I_n| = \kappa$ for every $n \in \omega$. Since $X^{\kappa \backslash I_n}$ is homeomorphic to $X^\kappa$, we can find a discrete subspace $E_n \subset X^{\kappa \backslash I_n}$ such that $|E_n| = \lambda_n$. Let $\{e_\xi^n : \xi < \lambda_n\}$ be an enumeration of $E_n$ and choose a set $U_\xi^n \in \tau(X^{\kappa \backslash I_n})$ such that $U_\xi^n \cap E_n = \{e_\xi^n\}$ for all $\xi < \lambda_n$.

For any $n \in \omega$ and $k \in \mathbb{N}$ let $\mathcal{F}_k^n = \{A \subset I_n : |A| = k\}$. Given a set $A \in \mathcal{F}_k^n$ and $\xi < \lambda_n$ define a point $x_\xi^{A,n} \in X^\kappa$ as follows: $x_\xi^{A,n}(\alpha) = s_\xi^{A,n}(\alpha)$ for all $\alpha \in A$; if $\alpha \in I_n \backslash A$ then $x_\xi^{A,n}(\alpha) = p$ and $x_\xi^{A,n}(\alpha) = e_\xi^n(\alpha)$ for all $\alpha \in \kappa \backslash I_n$.

For arbitrary $n \in \omega$ and $k \in \mathbb{N}$ consider the set $Q_k^n = \{x_\xi^{A,n} : \xi < \lambda_n,\ A \in \mathcal{F}_k^n\}$ and take any point $x_\xi^{A,n} \in Q_k^n$. The set $W = \{x \in X^\kappa : x(\alpha) \ne p$ for all $\alpha \in A$ and $x|(\kappa \backslash I_n) \in U_\xi^n\}$ is, evidently, open in $X^\kappa$ and $x_\xi^{A,n} \in W$. If $y = x_\eta^{A',n} \in Q_k^n \backslash \{x_\xi^{A,n}\}$ then we have two possibilities.

*Case 1.* If $\eta \ne \xi$ then $y|(\kappa \backslash I_n) = e_\eta^n \notin U_\xi^n$, so $y \notin W$.

*Case 2.* If $\eta = \xi$ then $A' \ne A$; if $y \in W$ then $y(\alpha) \ne p$ for all $\alpha \in A$. However, $A' = \{\alpha \in I_n : y(\alpha) \ne p\}$, so $A \subset A'$ and it follows from $|A| = |A'| = k$ that $A = A'$ and hence $y = x_\xi^{A,n}$ which is a contradiction.

Therefore $W \cap Q_k^n = \{x_\xi^{A,n}\}$ and hence we proved that

(1)  $Q_k^n$ is a discrete subspace of $X^\kappa$ for any $n \in \omega$ and $k \in \mathbb{N}$.

Thus the set $Q = \bigcup \{Q_k^n : n \in \omega,\ k \in \mathbb{N}\}$ is a $\sigma$-discrete subspace of $X^\kappa$. Take an arbitrary nonempty set $U \in \tau(X^\kappa)$. There is a finite set $A = \{\alpha_1, \ldots, \alpha_k\} \subset \kappa$ such that for some $V_1, \ldots, V_k \in \tau^*(X)$ the set $W = \{x \in X^\kappa : x(\alpha_i) \in V_i$ for all $i \le k\}$ is contained in $U$. The set $S$ being dense in $X$ we can find $t_1, \ldots, t_k \in S$ such that $t_i \in V_i$ for all $i \le k$. There exists $n \in \omega$ such that $A \subset I_n$ and $\{t_1, \ldots, t_k\} \subset S_n$; it is evident that $A \in \mathcal{F}_k^n$. Letting $s(\alpha_i) = t_i$ for all $i \le n$ we obtain a point $s \in (S_n)^A$, so there exists an ordinal $\xi < \lambda_n$ such that $s = s_\xi^{A,n}$. Now it follows from $x_\xi^{A,n}|A = s$ that $x_\xi^{A,n} \in W \cap Q \subset U \cap Q$ and hence $Q \cap U \ne \emptyset$ for any $U \in \tau^*(X^\kappa)$, i.e., $Q$ is dense in $X^\kappa$, so the space $X^\kappa$ is $d$-separable which shows that we proved sufficiency.

Finally, take any space $X$ and let $\kappa = d(X)$. If $X$ is discrete then the space $X^\kappa$ is $d$-separable by Problem 419. If $X$ is not discrete then the two-point space $\mathbb{D}$ embeds in $X$ and hence $\mathbb{D}^\kappa$ embeds in $X^\kappa$. It is an easy exercise that $\mathbb{D}^\kappa$ has a discrete subspace of cardinality $\kappa$, so $X^\kappa$ also has a discrete subspace of cardinality $\kappa$ and hence we can apply our above result to convince ourselves that $X^\kappa$ is $d$-separable.

**V.421.** *Prove that*

(a) *if $K$ is a compact space then $K^{\omega}$ is $d$-separable;*
(b) *there exists a compact space $K$ such that $K^n$ is not $d$-separable for any $n \in \mathbb{N}$. Thus $K^{\omega}$ is $d$-separable but no finite power of $K$ is $d$-separable.*

**Solution.** Given a space $X$ let $\Delta_X = \{(x,x) : x \in X\} \subset X \times X$ be the diagonal of $X$. Say that a compact space $X$ is *deep* if any nonempty $G_{\delta}$-subset of $X$ has nonempty interior.

*Fact 1.* For any compact space $X$ there exists a discrete subspace $D \subset (X \times X) \backslash \Delta_X$ such that $|D| = d(X)$.

*Proof.* If $X$ is finite then there is nothing to prove, so we assume from now on that $X$ is infinite. Call a compact space $X$ *adequate* if

(1) $d(U) = d(X)$ for any $U \in \tau^*(X)$.

We will show first that our Fact is true for all adequate compact spaces, so assume that $X$ is adequate and let $\kappa = d(X)$; then $w(U) \geq d(U) = \kappa$ for any nonempty open set $U \subset X$. Call a set $U \subset X \times X$ *standard* if there exist $V, W \in \tau^*(X)$ such that $\overline{V} \cap \overline{W} = \emptyset$ and $U = V \times W$. Observe that $\overline{U} \cap \Delta_X = \emptyset$ for any standard set $U$ and the union of the family $\mathcal{S}$ of all standard sets coincides with $(X \times X) \backslash \Delta_X$. We will need the following property of $\mathcal{S}$.

(2) If $\mathcal{U} \subset \mathcal{S}$ and $|\mathcal{U}| < \kappa$ then $((U \times U) \backslash \Delta_X) \backslash (\bigcup \mathcal{U}) \neq \emptyset$ for any $U \in \tau^*(X)$.

Assume toward a contradiction, that $(U \times U) \backslash \Delta_X$ is contained in $\bigcup \mathcal{U}$ for some family $\mathcal{U} \subset \mathcal{S}$ with $|U| < \kappa$. Choose a set $V \in \tau^*(X)$ such that $\overline{V} \subset U$ and observe that $Y = \overline{V}$ is a compact space for which $w(Y) \geq w(V) \geq \kappa$. It is easy to see that $(Y \times Y) \backslash \Delta_Y = (Y \times Y) \backslash \Delta_X \subset (U \times U) \backslash \Delta_X$ and therefore $(Y \times Y) \backslash \Delta_X \subset \bigcup \mathcal{U}$. Furthermore, $\overline{H} \cap \Delta_X = \emptyset$ for any $H \in \mathcal{U}$, so $Q_H = \overline{H} \cap (Y \times Y)$ is a compact subset of $(Y \times Y) \backslash \Delta_Y$.

It follows from the equality $(Y \times Y) \backslash \Delta_Y = \bigcup \{Q_H : H \in \mathcal{U}\}$ that $(Y \times Y) \backslash \Delta_Y$ is a union of $< \kappa$-many compact sets and hence $\Delta_Y$ is the intersection of $< \kappa$-many open subsets of $Y \times Y$, i.e., $\Delta(Y) < \kappa$. However, $w(Y) = \Delta(Y)$ (see SFFS-091) and hence $w(Y) < \kappa$; this contradiction shows that (2) is proved.

Take an arbitrary point $z_0 = (x_0, y_0) \in (X \times X) \backslash \Delta_X$; there exists a standard set $H_0$ such that $z_0 \in H_0$. Proceeding inductively assume that $\alpha < \kappa$ and we have a set $\{z_{\beta} = (x_{\beta}, y_{\beta}) : \beta < \alpha\} \subset (X \times X) \backslash \Delta_X$ and a family $\{H_{\beta} : \beta < \alpha\}$ of standard subsets of $X \times X$ with the following properties:

(3) $z_{\beta} \in H_{\beta}$ for all $\beta < \alpha$;
(4) $z_{\beta} \notin \bigcup \{H_{\gamma} : \gamma < \beta\}$ for every $\beta < \alpha$;
(5) $H_{\beta} \cap \{z_{\gamma} : \gamma < \beta\} = \emptyset$ for all $\beta < \alpha$.

The set $P = \{x_{\beta}, y_{\beta} : \beta < \alpha\}$ has cardinality strictly less than $\kappa$, so $\overline{P} \neq X$ and hence we can find a set $U \in \tau^*(X)$ with $\overline{U} \cap \overline{P} = \emptyset$. By the property (2) there are distinct points $x_{\alpha}, y_{\alpha} \in U$ such that $z_{\alpha} = (x_{\alpha}, y_{\alpha})$ does not belong to the set

$\bigcup\{H_\beta : \beta < \alpha\}$. Choose $V \in \tau(x_\alpha, X)$ and $W \in \tau(y_\alpha, X)$ such that $V \cup W \subset U$ and $\overline{V} \cap \overline{W} = \emptyset$; then $H_\alpha = V \times W$ is a standard set such that $\{z_\beta : \beta < \alpha\} \cap H_\alpha = \emptyset$ and hence the properties (3)–(5) hold for all $\beta \leq \alpha$.

This proves that our inductive procedure can be continued to construct a set $D = \{z_\alpha : \alpha < \kappa\} \subset (X \times X) \backslash \Delta_X$ and a family $\mathcal{H} = \{H_\alpha : \alpha < \kappa\}$ of elements of $\mathcal{S}$ such that the conditions (3)–(5) are satisfied for all $\alpha < \kappa$. It follows from (3)–(5) that $D \cap H_\alpha = \{z_\alpha\}$, so the set $D$ is discrete; the properties (3) and (4) imply that $\{z_\alpha : \alpha < \kappa\}$ is a faithful enumeration of $D$, so $|D| = \kappa = d(X)$ and hence our Fact is proved for all adequate compact spaces $X$.

Now take an arbitrary compact space $X$. Given a set $U \in \tau^*(X)$ consider the cardinal $\mu = \backslash n\{d(V) : V \in \tau^*(U)\}$ and pick a set $V \in \tau^*(U)$ with $d(V) = \mu$. It is evident that $d(W) = \mu$ for any nonempty open set $W \subset V$. If $W \in \tau^*(X)$ and $\overline{W} \subset V$ then it is easy to see that $\overline{W}$ is an adequate compact space. Therefore

(6)  the family $\mathcal{W} = \{W \in \tau^*(X) : \overline{W}$ is adequate$\}$ is a $\pi$-base of the space $X$.

Let $\mathcal{W}'$ be a maximal disjoint subfamily of $\mathcal{W}$; it follows from the property (6) that the set $\bigcup \mathcal{W}'$ is dense in $X$ and hence $d(X) = \sup\{d(W) : W \in \mathcal{W}'\}$. For every $W \in \mathcal{W}'$ take a set $H_W \in \tau^*(X)$ with $\overline{H}_W \subset W$; it is easy to see that the compact space $\overline{H}_W$ is adequate and $d(\overline{H}_W) = d(W)$. By what we proved for adequate compact spaces, we can find a discrete set $D_W \subset (\overline{H}_W \times \overline{H}_W) \backslash \Delta_X$ with $|D_W| = d(W)$ for each $W \in \mathcal{W}'$. It is straightforward that $D = \bigcup\{D_W : W \in \mathcal{W}'\}$ is a discrete subset of the space $(X \times X) \backslash \Delta_X$. Finally we can apply the equalities $|D| = \sup\{|D_W| : W \in \mathcal{W}'\} = \sup\{d(W) : W \in \mathcal{W}'\} = d(X)$ to see that $D$ is the promised discrete subspace of $(X \times X) \backslash \Delta_X$, so Fact 1 is proved.

*Fact 2.* If $X$ is a deep space, then $X^n$ is also deep for any $n \in \mathbb{N}$.

*Proof.* Suppose that $H$ is a $G_\delta$-subset of $X^n$ and $x = (x_1, \ldots, x_n) \in H$. Fix a family $\mathcal{U} = \{U_m : m \in \omega\} \subset \tau(X^n)$ such that $H = \bigcap \mathcal{U}$. For every $m \in \omega$ we can find sets $W_1^m, \ldots, W_n^m \in \tau(X)$ such that $x_i \in W_i^m$ for all $i \leq n$ and $W_m = W_1^m \times \ldots \times W_n^m \subset U_m$.

Given any $i \in \{1, \ldots, n\}$ observe that $x_i \in \bigcap\{W_i^m : m \in \omega\}$; the space $X$ being deep we can find a set $V_i \in \tau(x_i, X)$ with $V_i \subset \bigcap\{W_i^m : m \in \omega\}$. It is clear that $V = V_1 \times \ldots \times V_n$ is an open neighborhood of $x$ and $V \subset H$, so $H$ has nonempty interior, i.e., $X^n$ is deep and hence Fact 2 is proved.

*Fact 3.* If $X$ is a deep compact space without isolated points then $X$ cannot be $d$-separable.

*Proof.* Assume that $D_n$ is a discrete subspace of $X$ and $D = \bigcup_{n \in \omega} D_n$. Since $X$ is dense-in-itself, the set $\overline{D}_n$ is nowhere dense in $X$ for every $n \in \omega$. The space $X$ has the Baire property, so $X \backslash (\bigcup_{n \in \omega} \overline{D}_n) \neq \emptyset$ and hence $H = X \backslash (\bigcup_{n \in \omega} \overline{D}_n)$ is a nonempty $G_\delta$-subset of $X$. The space $X$ being deep, we can find a set $U \in \tau^*(X)$ such that $U \subset H$ and hence $U \cap D = \emptyset$. Therefore the set $D$ is not dense in $X$ and hence $X$ is not $d$-separable, i.e., Fact 3 is proved.

Returning to our solution observe that if $K$ is a compact space then $K \times K$ has a discrete subspace of size $d(K)$ by Fact 1. Since $K \times K$ embeds in $K^\omega$, we conclude that $K^\omega$ has a discrete subspace of size $d(K)$, so we can apply Problem 420 to see that $K^\omega$ is $d$-separable; this settles (a).

To prove (b) consider the compact space $K = \beta\omega \setminus \omega$. Given any nonempty open set $U \subset \beta\omega \setminus \omega$ there exists an infinite set $A \subset \omega$ such that $\overline{A} \cap (\beta\omega \setminus \omega) \subset U$ (see Fact 2 of S.370). The set $\overline{A}$ is uncountable by Fact 1 of S.483, so $\overline{A} \cap (\beta\omega \setminus \omega) \subset U$ is uncountable as well. Therefore $U$ is uncountable and hence $K$ has no isolated points. The space $K$ is deep by TFS-370, so $K^n$ is dense-in-itself and deep by Fact 2. Applying Fact 3 we conclude that $K^n$ is not $d$-separable for any $n \in \omega$. However, $K^\omega$ is $d$-separable by (a), so our solution is complete.

**V.422.** *Prove that if the space $C_p(X)$ is $d$-separable then there is a discrete subspace $D \subset C_p(X)$ such that $|D| = d(C_p(X))$.*

**Solution.** For any $x_1, \ldots, x_n \in X$ and any nontrivial open intervals $O_1, \ldots, O_n$ of $\mathbb{R}$, the set $[x_1, \ldots, x_n, O_1, \ldots, O_n] = \{f \in C_p(X) : f(x_i) \in O_i \text{ for all } i \leq n\}$ is open in $C_p(X)$. Call a set $U \subset C_p(X)$ standard open if there exist points $x_1, \ldots, x_n \in X$ and nontrivial open intervals $O_1, \ldots, O_n$ of $\mathbb{R}$ such that $U = [x_1, \ldots, x_n, O_1, \ldots, O_n]$. It is evident that standard open sets form a base of $C_p(X)$. Besides, it follows from Fact 1 of S.409 and Fact 1 of S.494 that

(1) every standard open subset of $C_p(X)$ is homeomorphic to $C_p(X)$.

Fix a family $\mathcal{D} = \{D_n : n \in \omega\}$ of discrete subspaces of $C_p(X)$ such that $\bigcup \mathcal{D}$ is dense in $C_p(X)$ and hence the cardinal $\lambda = \sup\{|D_n| : n \in \omega\}$ is greater than or equal to $\kappa = d(C_p(X))$; let $\lambda_n = |D_n|$ for any $n \in \omega$. Take a disjoint family $\{U_n : n \in \omega\}$ of nonempty open subspaces of $C_p(X)$ and pick a standard open set $V_n \subset U_n$ for each $n \in \omega$. It follows from (1) that $C_p(X)$ is homeomorphic to $V_n$ and therefore we can fix a discrete subspace $E_n \subset V_n$ such that $|E_n| = \lambda_n$ for every $n \in \omega$. It is evident that $E = \bigcup_{n\in\omega} E_n$ is a discrete subspace of $C_p(X)$ and $|E| = \lambda \geq \kappa$. Passing, if necessary, to an appropriate subspace of $E$ we can obtain a discrete subspace of $C_p(X)$ of cardinality $\kappa$.

**V.423.** *Given a space $X$ and $n \in \mathbb{N}$ say that a discrete subspace $D \subset X^n$ is essential if $\overline{D} \cap \Delta_n(X) = \emptyset$ and $|D| = iw(X)$. Prove that if, for some $n \in \mathbb{N}$, there exists an essential discrete set $D \subset X^n$ then $C_p(X)$ is $d$-separable. In particular, if there exists a discrete subspace of $X$ of cardinality $iw(X)$ then $C_p(X)$ is $d$-separable.*

**Solution.** For any $n \in \mathbb{N}$ let $o_n \in \mathbb{R}^n$ be the point whose all coordinates are equal to zero; for any $r > 0$, the set $\mathbb{B}_r^n = \{(x_1, \ldots, x_n) \in \mathbb{R}^n : x_1^2 + \ldots + x_n^2 \leq r^2\}$ is the closed $r$-ball in $\mathbb{R}^n$ centered at $o_n$. It is clear that $\mathbb{B}_r^1 = [-r, r] \subset \mathbb{R}$. Given a space $X$ and a set $A \subset X$ let $C_A^n(X) = \{f \in C_p(X, \mathbb{R}^n) : f(A) \subset \{o_n\}\}$ for any $n \in \mathbb{N}$. For any points $a, b \in \mathbb{R}^n$ the set $[a, b] = \{t \cdot a + (1 - t) \cdot b : t \in [0, 1]\}$ is the line segment in the space $\mathbb{R}^n$ which connects $a$ and $b$. Given a space $Z$ say that sets $P, Q \subset Z$ are *functionally separated* if there exists a function $f \in C(Z, [0, 1])$ such that $f(P) \subset \{0\}$ and $f(Q) \subset \{1\}$.

*Fact 1.* Given a space $X$ and an infinite cardinal $\kappa$ suppose that a family $\mathcal{F}$ consists of finite subsets of $X$, the cardinality of $\mathcal{F}$ is equal to $\kappa$ and there is $m \in \mathbb{N}$ such that $|F| \leq m$ for all $F \in \mathcal{F}$. Then there exists a finite set $A \subset X$ (called the core of $\mathcal{F}$) such that for any finite $B \subset X \backslash A$ there exists $U \in \tau(B, X)$ such that the family $\{F \in \mathcal{F} : F \cap U = \emptyset\}$ has cardinality $\kappa$.

*Proof.* If the family $\mathcal{F}$ has no finite core then we can choose disjoint nonempty sets $A_0, \ldots, A_m \subset X$ such that for any $i \leq m$ and $U \in \tau(A_i, X)$ there are strictly less than $\kappa$ elements of $\mathcal{F}$ which miss $U$. Now let $U_0, \ldots, U_m$ be disjoint open subsets of $X$ such that $A_i \subset U_i$ for all $i \leq m$. Every family $\{F \in \mathcal{F} : F \cap U_i = \emptyset\}$ has cardinality strictly less than $\kappa$, so there is $F \in \mathcal{F}$ such that $F \cap U_i \neq \emptyset$ for any $i \leq m$. This implies, however, that $|F| \geq m + 1$ and hence we obtained a contradiction which shows that Fact 1 is proved.

*Fact 2.* Suppose that $n \in \mathbb{N}$ and $X$ is a space for which there exists a finite set $A \subset X$ such that, for some $r > 0$ and some $\sigma$-discrete subspace $E \subset C_A^n(X)$, the set $K_n = C_p(X, \mathbb{B}_r^n) \cap C_A^n(X)$ is contained in the closure of $E$. Then the space $(C_p(X))^n$ is $d$-separable.

*Proof.* We identify $(C_p(X))^n$ with $C_p(X, \mathbb{R}^n)$. For every $k \in \mathbb{N}$ let $\varphi_k(f) = k \cdot f$ for any $f \in C_p(X, \mathbb{R}^n)$; it is immediate that $\varphi_k : C_p(X, \mathbb{R}^n) \to C_p(X, \mathbb{R}^n)$ is a homeomorphism. Therefore $E_k = \varphi_k(E)$ is a $\sigma$-discrete subspace of $C_p(X, \mathbb{R}^n)$ and $F_k = \varphi_k(K_n) \subset \overline{E_k}$ for any $k \in \mathbb{N}$. It is an easy exercise that $\bigcup\{F_k : k \in \mathbb{N}\}$ is dense in $C_A^n(X)$, so $\bigcup\{E_k : k \in \mathbb{N}\}$ is a $\sigma$-discrete dense subspace of $C_A^n(X)$, i.e., $C_A^n(X)$ is $d$-separable. If $A = \emptyset$ then $C_p(X, \mathbb{R}^n) = C_A^n(X)$ is $d$-separable. If $A \neq \emptyset$ then $C_p(X, \mathbb{R}^n)$ is homeomorphic to $C_A^n(X) \times (\mathbb{R}^n)^A$ by Fact 1 of S.409. Any product of $d$-separable spaces is $d$-separable (see Problem 421), so the space $(C_p(X))^n \simeq C_p(X, \mathbb{R}^n)$ is $d$-separable and hence Fact 2 is proved.

*Fact 3.* If $Z$ is a space and $P, Q \subset Z$ then $P$ and $Q$ are functionally separated if and only if there exist zero-sets $P'$ and $Q'$ in the space $Z$ such that $P \subset P'$, $Q \subset Q'$
and $P' \cap Q' = \emptyset$.

*Proof.* If $P$ and $Q$ are functionally separated then take a function $f \in C(Z, [0, 1])$ such that $f(P) \subset \{0\}$ and $f(Q) \subset \{1\}$; then $P' = f^{-1}(0)$ and $Q' = f^{-1}(1)$ are as promised, so we have necessity.

If, on the other hand, there exist disjoint zero-sets $P', Q'$ such that $P \subset P'$ and $Q \subset Q'$ then we can apply Fact 1 of V.140 to see that there exists a function $f \in C(Z, [0, 1])$ such that $f(P') \subset \{0\}$ and $f(Q') \subset \{1\}$. It is clear that $f$ also separates $P$ and $Q$, so Fact 3 is proved.

*Fact 4.* Suppose that $Z$ is a space and we have functionally separated nonempty sets $P, Q \subset Z$. Then, for any points $a, b \in \mathbb{R}^n$ there exists a continuous function $f : Z \to [a, b]$ such that $f(P) = \{a\}$ and $f(Q) = \{b\}$.

*Proof.* Since $\mathbb{R}^n = C_p(\{1,\ldots,n\})$, it follows from Fact 1 of S.301 that there exists a homeomorphism $\varphi : [0,1] \to [a,b]$ such that $\varphi(0) = a$ and $\varphi(1) = b$. Take a function $g \in C(Z,[0,1])$ such that $g(P) \subset \{0\}$ and $g(Q) \subset \{1\}$. Then $f = \varphi \circ g$ is as promised, so Fact 4 is proved.

**Fact 5.** Suppose that $Z$ is a space, a set $F \subset Z$ is compact and $F \subset U \in \tau(Z)$. Then there exists $V \in \tau(F,Z)$ such that $V$ is functionally separated from $X \setminus U$. In particular, $F$ is functionally separated from $Z \setminus U$.

*Proof.* The sets $F$ and $G = \mathrm{cl}_{\beta Z}(Z \setminus U)$ are disjoint and closed in $\beta Z$. By normality of $\beta Z$ there exists a continuous function $f : \beta Z \to [0,1]$ such that $f(F) \subset \{1\}$ and $f(G) \subset \{0\}$. If $g = f|Z$ then $P = g^{-1}([\frac{1}{2},1])$ and $Q = g^{-1}(0)$ are zero-sets in $Z$. Since $Z \setminus U \subset Q$ and $V = f^{-1}((\frac{1}{2},1]) \subset P$, we can apply Fact 3 to see that $V$ is functionally separated from $Z \setminus U$; besides, $F \subset V \in \tau(Z)$, so Fact 5 is proved.

**Fact 6.** Suppose that $w_1,\ldots,w_k \in \mathbb{R}^n$ and we have a disjoint family $U_1,\ldots,U_k$ of open subsets of $Z$. If $\emptyset \neq F_i \subset U_i$ and $F_i$ is functionally separated from $Z \setminus U_i$ (in particular, if $F_i$ is a compact subset of $U_i$) for each $i \leq k$ then there exists a continuous function $\varphi : Z \to \mathbb{R}^n$ such that $f(\overline{U}_i) \subset [o_n, w_i]$ and $f(F_i) = \{w_i\}$ for any $i \leq k$ while $f(z) = o_n$ for all $z \notin \bigcup_{i \leq k} U_i$.

*Proof.* Apply Fact 4 to find a continuous function $\varphi_i : Z \to [o_n, w_i]$ such that $\varphi_i(F_i) = \{w_i\}$ and $\varphi_i(Z \setminus U_i) \subset \{o_n\}$ for every $i \leq k$. Let $\varphi(z) = \varphi_i(z)$ if $z \in U_i$ for some $i \leq k$; if $z \in Z \setminus (\bigcup_{i \leq k} U_i)$ then let $\varphi(z) = o_n$. It is easy to see, applying Fact 2 of T.254, that $\varphi$ is a continuous function with the required properties. Finally observe that if $F_i$ is compact then $F_i$ is functionally separated from $Z \setminus U_i$ (see Fact 5), so Fact 6 is proved.

Returning to our solution observe that if the cardinal $\kappa = iw(X)$ is countable then $C_p(X)$ is even separable, so we can assume that $\kappa > \omega$. For any $k \in \mathbb{N}$ denote by $M_k$ the set $\{1,\ldots,k\}$. In this proof we will pass several times to a subset $D' \subset D$ with $|D'| = \kappa$. To simplify the notation we will assume each time that $D' = D$ which means that all previous reasoning can be repeated for our smaller set $D'$.

Let $\Delta_n = \Delta_n(X)$ and denote by $S_n$ the set of all bijections from the set $M_n$ onto itself. Every $\sigma \in S_n$ determines a map $p_\sigma : X^n \to X^n$ defined by the formula $p_\sigma(x) = (x_{\sigma(1)},\ldots,x_{\sigma(n)})$ for any $x = (x_1,\ldots,x_n) \in X^n$. It is clear that every $p_\sigma$ is a homeomorphism such that $p_\sigma(\Delta_n) = \Delta_n$. The set $P_d = \{p_\sigma(d) : \sigma \in S_n\}$ is finite for all $d \in D$, so we can pass, if necessary, to a subset of $D$ of cardinality $\kappa$, to assure that $D \cap P_d = \{d\}$ for any $d \in D$.

We can consider that $n$ is the minimal number for which an essential discrete subspace exists in $X^n$. For any $d = (d_1,\ldots,d_n) \in D$ let $K_d = \{d_1,\ldots,d_n\}$; if some $x \in X$ belongs to $\kappa$-many distinct elements of the family $\{K_d : d \in D\}$ then, passing to an appropriate subset of $D$ of cardinality $\kappa$ we can consider that there is $j \in M_n$ such that $d_j = x$ for all $d = (d_1,\ldots,d_n) \in D$; if $x_d = (d_1,\ldots,d_{j-1},d_{j+1},\ldots,d_n)$ then the set $E = \{x_d : d \in D\} \subset X^{n-1} \setminus \Delta_{n-1}$ is also discrete and essential which is a contradiction with the choice of $n$. Thus

(1) $|\{d \in D : x \in K_d\}| < \kappa$ for every $x \in X$.

It is easy to find a set $O \in \tau(\Delta_n, X^n)$ such that $p_\sigma(O) = O$ for any $\sigma \in S_n$ and $D \cap O = \emptyset$; if $F = \overline{D}\backslash D$ then, evidently, $F \cap O = \emptyset$. Besides, $P_d \cap O = \emptyset$ for any $d \in D$ and the set $P_d\backslash F \ni d$ is nonempty; let $m_d = |P_d\backslash F|$. Passing, if necessary, to a relevant subset of $D$ of cardinality $\kappa$ we can assume, without loss of generality, that there is $m \in \mathbb{N}$ such that $m_d = m$ for any $d \in D$.

Choose, for any $d = (d_1, \ldots, d_n) \in D$ a family $\{U_1^d, \ldots, U_n^d\}$ of open subsets of $X$ with the following properties:

(2) the family $\{\overline{U_1^d}, \ldots, \overline{U_n^d}\}$ is disjoint;
(3) if $U^d = U_1^d \times \ldots \times U_n^d$ then $U^d \cap D = \{d\}$;
(4) $U_{i_1}^d \times \ldots \times U_{i_n}^d \subset O$ whenever $i_1, \ldots, i_n \in M_n$ are not all distinct;
(5) $(U_{\sigma(1)}^d \times \ldots \times U_{\sigma(n)}^d) \cap D = \emptyset$ for any $d \in D$ and $\sigma \in S_n$ such that $p_\sigma(d) \notin F \cup \{d\}$.

Since $iw(X) = \kappa$, we can choose a base $\mathcal{B}$ of cardinality $\kappa$ of some Tychonoff topology $\mu$ on $X$ weaker than $\tau(X)$; let $X' = (X, \mu)$. There is no loss of generality to consider that $B \neq \emptyset$ for any $B \in \mathcal{B}$. From now on the bar denotes the closure in $X$ and all topological properties in which the space is not mentioned are meant to hold in the space $X$.

Apply Fact 1 to the space $X'$ to find a finite set $A \subset X$ such that, for any finite $B \subset X\backslash A$, there is $U \in \tau(B, X')$ for which the set $\{d \in D : K_d \cap \mathrm{cl}_{X'}(U) = \emptyset\}$ has cardinality $\kappa$. It follows from (1) that only $< \kappa$-many elements of the family $\mathcal{D} = \{K_d : d \in D\}$ meet $A$, so, passing if necessary, to a subset of $D$ of cardinality $\kappa$, we can assume, without loss of generality, that $K_d \cap A = \emptyset$ for any $d \in D$.

Our next step is to consider, for every $k \in \mathbb{N}$, the family $\mathcal{W}_k$ of all $3k$-tuples $(W_1, \ldots, W_k, V_1, \ldots, V_k, r_1, \ldots, r_k) \in \mathcal{B}^{2k} \times \mathbb{Q}^k$ such that

(6) $W_i \in \mathcal{B}$, $V_i \in \mathcal{B}$, $r_i \in \mathbb{I} \cap \mathbb{Q}$ for all $i \in M_k$;
(7) $\overline{V}_i \subset W_i$ and $\overline{V}_i$ is functionally separated from $X\backslash W_i$ for all $i \in M_k$;
(8) if $W = \bigcup_{i \in M_k} W_i$ then $\overline{W} \cap A = \emptyset$ and $|\{d \in D : \overline{W} \cap K_d = \emptyset\}| = \kappa$;
(9) the family $\{\overline{W}_i : i \leq k\}$ is disjoint.

It is straightforward that $|\mathcal{W}_k| \leq \kappa$ for any $k \in \omega$, so if $\mathcal{W} = \bigcup\{\mathcal{W}_k : k \in \mathbb{N}\}$ then $|\mathcal{W}| \leq \kappa$. For any element $\xi = (W_1, \ldots, W_k, V_1, \ldots, V_k, r_1, \ldots, r_k)$ of the family $\mathcal{W}$ let $k_\xi = k$, $W[\xi] = \bigcup_{i \leq k} W_i$ and $R_i(\xi) = r_i$ for all $i \leq k$. Using the property (8) it is easy to construct an injection $\varphi : \mathcal{W} \to D$ such that $\overline{W[\xi]} \cap K_{\varphi(\xi)} = \emptyset$ for any $\xi \in \mathcal{W}$.

Fix $\xi \in \mathcal{W}$; if $\varphi(\xi) = d = (d_1, \ldots, d_n)$ then we can apply Fact 5 and Fact 6 to choose a continuous function $f_\xi : X \to [-1, 2]$ and a set $H_i \in \tau(d_i, X)$ for any $i \leq n$ with the following properties:

(10) $H_i \subset U_i^d$ for all $i \leq n$ and $(\bigcup_{i \leq n} \overline{H}_i) \cap (W[\xi] \cup A) = \emptyset$;
(11) $f_\xi(d_i) = 2$ for all $i \leq n$ and $f_\xi(x) = 0$ for any $x \in X\backslash(W[\xi] \cup (\bigcup_{i \leq n} H_i))$.
(12) $f_\xi(X\backslash(\bigcup_{i \leq n} H_i)) \subset \mathbb{I}$ and $f_\xi(V_i) = \{R_i(\xi)\}$ for all $i \leq k_\xi$.

We claim that the set $I = \{f \in C_p(X, \mathbb{I}) : f(A) \subset \{0\}\}$ is contained in the closure in $C_p(X)$ of the set $\Omega = \{f_\xi : \xi \in \mathcal{W}\}$. It suffices to show that, for any finite $B = \{x_1, \ldots, x_k\} \subset X \setminus A$ and $G_1, \ldots, G_k \in \tau^*(\mathbb{I})$, there is $f \in \Omega$ such that $f(x_i) \in G_i$ for any $i \leq k$. First choose $r_1, \ldots, r_k \in \mathbb{Q}$ such that $r_i \in G_i$ for al $i \leq k$. Since $A$ is the core of the family $\mathcal{D}$ in the space $X'$, there exist $W_1, \ldots, W_k \in \mathcal{B}$ such that $x_i \in W_i$ for all $i \leq k$, the family $\mathcal{A}' = \{\mathrm{cl}_{X'}(W_i) : i \leq k\}$ is disjoint (and hence the collection $\mathcal{A} = \{\overline{W}_i : i \leq k\}$ is disjoint as well) and there $\kappa$-many elements $d \in D$ such that $K_d \cap (\bigcup \mathcal{A}) = \emptyset$.

It is easy to choose $V_i \in \tau(x_i, X')$ such that $V_i \in \mathcal{B}$ and the set $V_i$ is functionally separated from $X \setminus W_i$ in $X'$ (and hence in $X$) for all $i \in M_k$. An immediate consequence is that the $3k$-tuple $\xi = (W_1, \ldots, W_k, V_1, \ldots, V_k, r_1, \ldots, r_k)$ belongs to $\mathcal{W}$. Thus $f_\xi(x_i) = r_i \in G_i$ for all $i \leq k$ which proves that $I \subset \overline{\Omega}$.

Let us show that $\Omega$ is a discrete subspace of $C_A = \{f \in C_p(X) : f(A) \subset \{0\}\}$. It follows from (10) and (11) that $\Omega \subset C_A$. To see that $\Omega$ is discrete, fix any $\xi = (W_1, \ldots, W_k, V_1, \ldots, V_k, r_1, \ldots, r_k) \in \mathcal{W}$ and let $d = (d_1, \ldots, d_n) = \varphi(\xi)$; the set $B_\xi = \{f \in C_p(X) : f(d_i) > 1$ for all $i \in M_n\}$ is open in $C_p(X)$ and contains $f_\xi$, so it suffices to establish that $B_\xi \cap \Omega = \{f_\xi\}$. Assume toward a contradiction, that $f_\eta \in B_\xi$ for some $\eta \in \mathcal{W} \setminus \{\xi\}$; then $a = (a_1, \ldots, a_n) = \varphi(\eta) \neq d$.

For any $i \in M_n$ we have $f_\eta(d_i) > 1$ which, together with (10)–(12), implies that $d_i \in U_{j_i}^a$ for some $j_i \in M_n$. If $j_k = j_l$ for some distinct $k, l \in M_n$ then it follows from (4) that $d \in U_{j_1}^a \times \ldots \times U_{j_n}^a \subset O$ which is a contradiction with $D \cap O = \emptyset$. Thus the function $\sigma : M_n \to M_n$ defined by $\sigma(i.) = j_i$ for all $i \in M_n$, is a bijection, i.e., $\sigma \in S_n$; let $\nu = \sigma^{-1}$. If $b = (a_{k_1}, \ldots, a_{k_n}) \in P_a \setminus F$ then $d(b) = (d_{\nu(k_1)}, \ldots, d_{\nu(k_n)}) \in P_d \setminus F$ because $d(b) \in U_{k_1}^a \times \ldots \times U_{k_n}^a \subset X^n \setminus F$ (see (5)). The property (5) also implies that $d \neq d(b)$ for any $b \in P_a \setminus F$. Therefore the set $\{d\} \cup \{d(b) : b \in P_a \setminus F\}$ which has $m + 1$ elements, is contained in $P_d \setminus F$ whereas $|P_d \setminus F| = m$, a contradiction. Consequently, $B_\xi \cap \Omega = \{f_\xi\}$ which proves that $\Omega \subset C_A$ is a discrete subspace such that $I \subset \overline{\Omega}$. Finally apply Fact 2 to see that $C_p(X)$ is $d$-separable.

**V.424.** *Assume that $X$ is a space and there exists a discrete subspace $D \subset X \times X$ such that $|D| = iw(X)$. Prove that $C_p(X)$ is $d$-separable.*

**Solution.** If the cardinal $\kappa = iw(X)$ is countable then $C_p(X)$ is even separable, so we can assume that $\kappa > \omega$. For any $k \in \mathbb{N}$ denote by $M_k$ the set $\{1, \ldots, k\}$. In this proof we will pass several times to a subset $D' \subset D$ with $|D'| = \kappa$. To simplify the notation we will assume each time that $D' = D$ which means that all previous reasoning can be repeated for our smaller set $D'$. Given a space $Z$ say that sets $P, Q \subset Z$ are *functionally separated* if there exists a function $f \in C(Z, [0, 1])$ such that $f(P) \subset \{0\}$ and $f(Q) \subset \{1\}$.

Let $\Delta = \Delta_2(X)$; we can consider that there is no discrete subspace of $X$ of cardinality $\kappa$ because this case was already settled in Problem 423. In particular, $D \cap \Delta$ has cardinality $< \kappa$, so we can pass to an appropriate subset of $D$ of cardinality $\kappa$ to see that we can assume, without loss of generality, that $D \subset X^2 \setminus \Delta$. For any element $d = (d_1, d_2) \in D$ let $K_d = \{d_1, d_2\}$. If some $x \in X$ belongs to $\kappa$-many elements of $\mathcal{D} = \{K_d : d \in D\}$ then, according to our policy, we can assume that $x \in K_d$ for all $d \in D$ and therefore $D \subset (\{x\} \times X) \cup (X \times \{x\})$. This shows that either $|D \cap (\{x\} \times X)| = \kappa$ or $|D \cap (X \times \{x\})| = \kappa$. Since both sets $X \times \{x\}$ and $\{x\} \times X$ are homeomorphic to $X$, a discrete space of cardinality $\kappa$ embeds in $X$ which is a contradiction. As a consequence,

(1) $|\{d \in D : x \in K_d\}| < \kappa$ for every $x \in X$.

Choose, for any $d = (d_1, d_2) \in D$ a pair $\{U_1^d, U_2^d\}$ of open subsets of $X$ such that $\overline{U_1^d} \cap \overline{U_2^d} = \emptyset$ and, for the set $U^d = U_1^d \times U_2^d$, we have $U^d \cap D = \{d\}$.

Since $iw(X) = \kappa$, we can choose a base $\mathcal{B}$ of cardinality $\kappa$ of some Tychonoff topology $\mu$ on $X$ weaker than $\tau(X)$; let $X' = (X, \mu)$. There is no loss of generality to consider that $B \neq \emptyset$ for any $B \in \mathcal{B}$. From now on the bar denotes the closure in $X$ and all topological properties in which the space is not mentioned are meant to hold in the space $X$.

We can apply Fact 1 of V.423 to the space $X'$ to find a finite set $A \subset X$ such that, for any finite $B \subset X \backslash A$, there exists a set $U \in \tau(B, X')$ for which the cardinality of the set $\{d \in D : K_d \cap \mathrm{cl}_{X'}(U) = \emptyset\}$ is equal to $\kappa$. It follows from (1) that only $< \kappa$-many elements of the family $\mathcal{D}$ meet $A$, so, passing if necessary, to a subset of $D$ of cardinality $\kappa$, we can assume, without loss of generality, that $K_d \cap A = \emptyset$ for any $d \in D$.

Our next step is to consider for every $k \in \mathbb{N}$, the family $\mathcal{W}_k$ of all $3k$-tuples $(W_1, \ldots, W_k, V_1, \ldots, V_k, r_1, \ldots, r_k) \in \mathcal{B}^{2k} \times \mathbb{Q}^k$ such that

(2) $W_i \in \mathcal{B}$, $V_i \in \mathcal{B}$, $r_i \in \mathbb{I} \cap \mathbb{Q}$ for all $i \in M_k$;
(3) $\overline{V}_i \subset W_i$ and $\overline{V}_i$ is functionally separated from $X \backslash W_i$ for all $i \in M_k$;
(4) if $W = \bigcup_{i \in M_k} W_i$ then $\overline{W} \cap A = \emptyset$ and $|\{d \in D : \overline{W} \cap K_d = \emptyset\}| = \kappa$;
(5) the family $\{\overline{W}_i : i \leq k\}$ is disjoint.

It is straightforward that $|\mathcal{W}_k| \leq \kappa$ for any $k \in \omega$, so if $\mathcal{W} = \bigcup\{\mathcal{W}_k : k \in \mathbb{N}\}$ then $|\mathcal{W}| \leq \kappa$. For any element $\xi = (W_1, \ldots, W_k, V_1, \ldots, V_k, r_1, \ldots, r_k)$ of the family $\mathcal{W}$ let $k_\xi = k$, $W[\xi] = \bigcup_{i \leq k} W_i$ and $R_i(\xi) = r_i$ for all $i \leq k$. Using the property (4) it is easy to construct an injection $\varphi : \mathcal{W} \to D$ such that $\overline{W[\xi]} \cap K_{\varphi(\xi)} = \emptyset$ for any $\xi \in \mathcal{W}$.

Fix $\xi \in \mathcal{W}$; if $\varphi(\xi) = d = (d_1, d_2)$ then we can apply Fact 5 and Fact 6 of V.426 to choose a continuous function $f_\xi : X \to [-2, 2]$ and a set $H_i \in \tau(d_i, X)$ for any $i = 1, 2$ with the following properties:

(6) $H_i \subset U_i^d$ for all $i = 1, 2$ and $(\overline{H}_1 \cup \overline{H}_2) \cap (W[\xi] \cup A) = \emptyset$;
(7) $f_\xi(d_1) = 2$ and $f_\xi(d_2) = -2$ while $f_\xi(x) = 0$ for any $x \in X \backslash (W[\xi] \cup (H_1 \cup H_2))$.
(8) $f_\xi(X \backslash H_1) \subset [-2, 1]$, $f_\xi(X \backslash H_2) \subset [-1, 2]$ and $f_\xi(V_i) = \{R_i(\xi)\}$ for all $i \leq k_\xi$.

We claim that the set $I = \{f \in C_p(X, \mathbb{I}) : f(A) \subset \{0\}\}$ is contained in the closure in $C_p(X)$ of the set $\Omega = \{f_\xi : \xi \in \mathcal{W}\}$. It suffices to show that, for any finite $B = \{x_1, \ldots, x_k\} \subset X \backslash A$ and $G_1, \ldots, G_k \in \tau^*(\mathbb{I})$, there is $f \in \Omega$ such that $f(x_i) \in G_i$ for any $i \leq k$. First choose $r_1, \ldots, r_k \in \mathbb{Q}$ such that $r_i \in G_i$ for al $i \leq k$. Since $A$ is the core of the family $\mathcal{D}$ in the space $X'$, there exist $W_1, \ldots, W_k \in \mathcal{B}$ such that $x_i \in W_i$ for all $i \leq k$, the family $\mathcal{A}' = \{\mathrm{cl}_{X'}(W_i) : i \leq k\}$ is disjoint (and hence the collection $\mathcal{A} = \{\overline{W}_i : i \leq k\}$ is disjoint as well) and there $\kappa$-many elements $d \in D$ such that $K_d \cap (\bigcup \mathcal{A}) = \emptyset$.

It is easy to choose $V_i \in \tau(x_i, X')$ such that $V_i \in \mathcal{B}$ and the set $V_i$ is functionally separated from $X \backslash W_i$ in $X'$ (and hence in $X$) for all $i \in M_k$. An immediate consequence is that the $3k$-tuple $\xi = (W_1, \ldots, W_k, V_1, \ldots, V_k, r_1, \ldots, r_k)$ belongs to $\mathcal{W}$. Thus $f_\xi(x_i) = r_i \in G_i$ for all $i \leq k$ which proves that $I \subset \overline{\Omega}$. It is immediate from the definition that $\Omega \subset C_A = \{f \in C_p(X) : f(A) \subset \{0\}\}$.

Fix any $\xi = (W_1, \ldots, W_k, V_1, \ldots, V_k, r_1, \ldots, r_k) \in \mathcal{W}$ and consider the point $d = (d_1, d_2) = \varphi(\xi)$; the set $B_\xi = \{f \in C_p(X) : f(d_1) > 1 \text{ and } f(d_2) < -1\}$ is open in $C_p(X)$ and contains the function $f_\xi$, so it suffices to establish the equality $B_\xi \cap \Omega = \{f_\xi\}$. Assume toward a contradiction, that $f_\eta \in B_\xi$ for some $\eta \in \mathcal{W} \backslash \{\xi\}$; then $a = (a_1, a_2) = \varphi(\eta) \neq d$.

We have $f_\eta(d_1) > 1$ and $f_\eta(d_2) < -1$ which, together with (6)–(8), implies that $d_1 \in U_1^a$ and $d_2 \in U_2^a$; this shows that $d \in U^a$ contradicting $U^a \cap \Omega = \{a\}$. Consequently, $B_\xi \cap \Omega = \{f_\xi\}$ and hence $\Omega \subset C_A$ is a discrete subspace such that $I \subset \overline{\Omega}$. Finally apply Fact 2 of V.423 to see that $C_p(X)$ is $d$-separable.

**V.425.** *Let $X$ be a space such that the cardinal $\kappa = iw(X)$ has uncountable cofinality. Prove that the following conditions are equivalent:*

*(i) $(C_p(X))^n$ is $d$-separable for all $n \geq 2$;*
*(ii) $(C_p(X))^n$ is $d$-separable for some $n \geq 2$;*
*(iii) $C_p(X) \times C_p(X)$ is $d$-separable;*
*(iv) for some $m \in \mathbb{N}$ there is a discrete set $D \subset X^m$ with $|D| = \kappa$.*

**Solution.** For any $n \in \mathbb{N}$ let $\mathbb{B}_n = \{(x_1, \ldots, x_n) \in \mathbb{R}^n : x_1^2 + \ldots + x_n^2 \leq 1\}$ be the closed ball of radius 1 centered at the point $o_n \in \mathbb{R}^n$ whose all coordinates are equal to zero. Given a point $x = (x_1, \ldots, x_n) \in \mathbb{R}^n$ let $|x|_n = \sqrt{x_1^2 + \ldots + x_n^2}$; for any $k \in \mathbb{N}$ denote by $M_k$ the set $\{1, \ldots, k\}$. For any points $a, b \in \mathbb{R}^n$ the set $[a, b] = \{t \cdot a + (1 - t) \cdot b : t \in [0, 1]\}$ is the line segment in the space $\mathbb{R}^n$ which connects $a$ and $b$. If $Z$ is a space then we say that sets $P, Q \subset Z$ are *functionally separated* if there exists a function $f \in C(Z, [0, 1])$ such that $f(P) \subset \{0\}$ and $f(Q) \subset \{1\}$.

*Fact 1.* If $Z$ is a space such that the cardinal $\kappa = iw(Z)$ has uncountable cofinality and $C_p(Z)$ is $d$-separable then, for some $n \in \mathbb{N}$, there is a discrete $E \subset Z^n$ with $|E| = \kappa$.

*Proof.* Apply Problem 422 to see that there exists a discrete subspace $D \subset C_p(Z)$ such that $|D| = d(C_p(Z)) = iw(Z) = \kappa$. Let $\mathcal{B}$ be the family of all nonempty open intervals of $\mathbb{R}$ with rational endpoints. For any $m \in \mathbb{N}$ if $x = (x_1, \ldots, x_m) \in Z^m$ and $B = (B_1, \ldots, B_m) \in \mathcal{B}^m$ then the set $[x, B] = \{f \in C_p(Z) : f(x_i) \in B_i$ for all $i \leq m\}$ is open in $C_p(Z)$ and the family $\mathcal{C} = \{[x, B] : x \in Z^m$ and $B \in \mathcal{B}^m$ for some $m \in \mathbb{N}\}$ is a base in $C_p(Z)$.

For any $f \in D$ fix a set $U_f \in \mathcal{C}$ such that $U_f \cap D = \{f\}$ and $U_f = [x_f, B_f]$ where $x_f = (x_1^f, \ldots, x_{m_f}^f) \in Z^{m_f}$, $B_f = (B_1^f, \ldots, B_{m_f}^f) \in \mathcal{B}^{m_f}$, the family

$\{B_1^f, \ldots, B_{m_f}^f\}$ is disjoint and $f(x_i^f) \in B_i^f$ for all $i \leq m_f$. Since $\mathcal{B}$ is countable and $\mathrm{cf}(\kappa) > \omega$, there is $n \in \mathbb{N}$ such that, for some $D' \subset D$ of cardinality $\kappa$ and $B = (B_1, \ldots, B_n) \in \mathcal{B}^n$, we have $B_f = B$ for any $f \in D'$.

Given distinct $f, g \in D'$ if $x_f = x_g$ then $U_f = [x_f, B] = [x_g, B] = U_g$ and hence $f \in U_g$ which is a contradiction with the choice of $U_g$. Therefore the correspondence $f \to x_f$ is an injection and hence the set $E = \{x_f : f \in D'\} \subset X^n$ has cardinality $\kappa$. Fix a function $f \in D'$; the set $W = f^{-1}(B_1) \times \ldots \times f^{-1}(B_n)$ is open in $Z^n$ and $x_f \in W$. If $x_g \in W$ for some $g \in D' \backslash \{f\}$ then $f(x_i^g) \in B_i$ for all $i \leq n$ and hence $f \in [x_g, B] = U_g$ which is a contradiction. Therefore $W \cap E = \{x_f\}$ and hence $E$ is a discrete subset of $X^n$ with $|E| = \kappa$, i.e., Fact 1 is proved.

Returning to our solution observe that the implications (i)$\Longrightarrow$(ii) and (i)$\Longrightarrow$(iii) are trivial. Suppose that, for some $n \in \mathbb{N}$, the space $(C_p(X))^n \simeq C_p(X \times \{1, \ldots, n\})$ is $d$-separable and let $X_n = X \times \{1, \ldots, n\}$. Fact 1 shows that there is $m \in \mathbb{N}$ and a discrete subset $E \subset (X_n)^m$ such that $|E| = \kappa$. The space $(X_n)^m$ being a union of finitely many homeomorphic copies of $X^m$, there is a discrete $D \subset X^m$ with $|D| = \kappa$. This shows that (ii)$\Longrightarrow$(iv) and (iii)$\Longrightarrow$(iv), so all that is left is to prove that (iv)$\Longrightarrow$(i).

Fix any $n \in \mathbb{N} \backslash \{1\}$ and let $m \in \mathbb{N}$ be the minimal number for which there is a discrete subspace in $X^m$ of cardinality $\kappa$. It follows from Problems 424 and 419 that we can assume, without loss of generality, that $m > 2$.

Choose distinct points $w_1, \ldots, w_m \in \mathbb{R}^n$ such that $|w_i|_n = 2$ for all $i \leq m$ (it is precisely at this step where we apply the fact that $n \geq 2$). Consider the segment $J_i = [o_n, w_i]$ for any $i \leq m$ and let $J = \bigcup_{i \leq m} J_i$. For every $i \leq m$ fix a set $Q_i \in \tau(w_i, \mathbb{R}^n)$ such that $Q_i \cap (\mathbb{B}_n \cup (J \backslash J_i)) = \emptyset$ and the family $\{Q_1, \ldots, Q_m\}$ is disjoint.

Let $D$ be a discrete subspace of $X^m$ with $|D| = \kappa$; in this proof we will pass several times to a subset $D' \subset D$ with $|D'| = \kappa$. To simplify the notation we will assume each time that $D' = D$ which means that all previous reasoning can be repeated for our smaller set $D'$. Observe first that $\Delta_m(X)$ is the finite union of spaces homeomorphic to $X^{m-1}$, so if $|D \cap \Delta_m(X)| = \kappa$ then a discrete space of cardinality $\kappa$ embeds in $X^{m-1}$ which is a contradiction. Therefore we can assume, without loss of generality, that $D \subset X^m \backslash \Delta_m(X)$.

For any point $d = (d_1, \ldots, d_m) \in D$ let $K_d = \{d_1, \ldots, d_m\}$; if some $x \in X$ belongs to $\kappa$-many distinct elements of the family $\{K_d : d \in D\}$ then, passing to an appropriate subset of $D$ of cardinality $\kappa$ we can consider that there is $j \in M_m$ such that $d_j = x$ for all $d = (d_1, \ldots, d_m) \in D$; if $x_d = (d_1, \ldots, d_{j-1}, d_{j+1}, \ldots, d_m)$ then the set $E = \{x_d : d \in D\} \subset X^{m-1}$ is also discrete and $|E| = \kappa$ which is a contradiction with the choice of $m$. Thus

(1)  $|\{d \in D : x \in K_d\}| < \kappa$ for every $x \in X$.

Choose, for any $d = (d_1, \ldots, d_m) \in D$ a family $\{U_1^d, \ldots, U_m^d\}$ of open subsets of $X$ such that the family $\{\overline{U_1^d}, \ldots, \overline{U_m^d}\}$ is disjoint and for the set $U^d = U_1^d \times \ldots \times U_m^d$, we have $U^d \cap D = \{d\}$.

Since $iw(X) = \kappa$, we can choose a base $\mathcal{B}$ of cardinality $\kappa$ of some Tychonoff topology $\mu$ on $X$ weaker than $\tau(X)$; let $X' = (X, \mu)$. From now on the bar denotes the closure in $X$ and all topological properties in which the space is not mentioned are meant to hold in the space $X$.

Apply Fact 1 of V.423 to the space $X'$ to see that there exists a finite set $A \subset X$ such that, for any finite $B \subset X \backslash A$, we can find $U \in \tau(B, X')$ for which the set $\{d \in D : K_d \cap \mathrm{cl}_{X'}(U) = \emptyset\}$ has cardinality $\kappa$. It follows from (1) that only $< \kappa$-many elements of the family $\mathcal{D} = \{K_d : d \in D\}$ meet $A$, so, passing if necessary, to a subset of $D$ of cardinality $\kappa$, we can assume, without loss of generality, that $K_d \cap A = \emptyset$ for any $d \in D$.

Our next step is to consider, for every $k \in \mathbb{N}$, the family $\mathcal{W}_k$ of all $3k$-tuples $(W_1, \ldots, W_k, V_1, \ldots, V_k, r_1, \ldots, r_k) \in \mathcal{B}^{2k} \times (\mathbb{Q}^n)^k$ such that

(2) $W_i \in \mathcal{B}$, $V_i \in \mathcal{B}$, $r_i \in \mathbb{B}_n \cap \mathbb{Q}^n$ for all $i \in M_k$;
(3) $\overline{V}_i \subset W_i$ and $\overline{V}_i$ is functionally separated from $X \backslash W_i$ for all $i \in M_k$;
(4) if $W = \bigcup_{i \in M_k} W_i$ then $\overline{W} \cap A = \emptyset$ and $|\{d \in D : \overline{W} \cap K_d = \emptyset\}| = \kappa$;
(5) the family $\{\overline{W}_i : i \leq k\}$ is disjoint.

It is straightforward that $|\mathcal{W}_k| \leq \kappa$ for any $k \in \omega$, so if $\mathcal{W} = \bigcup\{\mathcal{W}_k : k \in \mathbb{N}\}$ then $|\mathcal{W}| \leq \kappa$. For any element $\xi = (W_1, \ldots, W_k, V_1, \ldots, V_k, r_1, \ldots, r_k)$ of the family $\mathcal{W}$ let $k_\xi = k$, $W[\xi] = \bigcup_{i \leq k} W_i$ and $R_i(\xi) = r_i$ for all $i \leq k$. Using the property (4) it is easy to construct an injection $\varphi : \mathcal{W} \to D$ such that $\overline{W[\xi]} \cap K_{\varphi(\xi)} = \emptyset$ for any $\xi \in \mathcal{W}$.

Fix $\xi \in \mathcal{W}$; if $\varphi(\xi) = d = (d_1, \ldots, d_m)$ then we can apply Fact 5 and Fact 6 of V.423 to choose a continuous function $f_\xi : X \to \mathbb{B}_n \cup J$ and a set $H_i \in \tau(d_i, X)$ for any $i \leq m$ with the following properties:

(6) $H_i \subset U_i^d$ for all $i \leq m$ and $(\bigcup_{i \leq m} \overline{H}_i) \cap (W[\xi] \cup A) = \emptyset$;
(7) $f_\xi(d_i) = w_i$ for all $i \leq m$ and $f_\xi(x) = o_n$ for any $x \in X \backslash (W[\xi] \cup (\bigcup_{i \leq m} H_i))$.
(8) $f_\xi^{-1}(Q_i) \subset H_i$ and $f_\xi(V_i) = \{R_i(\xi)\}$ for all $i \leq k_\xi$.

The set $I = \{f \in C_p(X, \mathbb{B}_n) : f(A) \subset \{o_n\}\}$ is contained in the closure in $C_p(X, \mathbb{R}^n)$ of the set $\Omega = \{f_\xi : \xi \in \mathcal{W}\}$. To prove that, it suffices to show that, for any finite $B = \{x_1, \ldots, x_k\} \subset X \backslash A$ and $G_1, \ldots, G_k \in \tau^*(\mathbb{B}_n)$, there is $f \in \Omega$ such that $f(x_i) \in G_i$ for any $i \leq k$. First choose $r_1, \ldots, r_k \in \mathbb{Q}^n$ such that $r_i \in G_i$ for al $i \leq k$. Since $A$ is the core of the family $\mathcal{D}$ in the space $X'$, there exist $W_1, \ldots, W_k \in \mathcal{B}$ such that $x_i \in W_i$ for all $i \leq k$, the family $\mathcal{A}' = \{\mathrm{cl}_{X'}(W_i) : i \leq k\}$ is disjoint (and hence the collection $\mathcal{A} = \{\overline{W}_i : i \leq k\}$ is disjoint as well) and there $\kappa$-many elements $d \in D$ such that $K_d \cap (\bigcup \mathcal{A}) = \emptyset$.

By Fact 5 of V.423 we can choose $V_i \in \tau(x_i, X')$ such that $V_i \in \mathcal{B}$ and the set $V_i$ is functionally separated from $X \backslash W_i$ in $X'$ (and hence in $X$) for all $i \in M_k$. An immediate consequence is that the $3k$-tuple $\xi = (W_1, \ldots, W_k, V_1, \ldots, V_k, r_1, \ldots, r_k)$ belongs to $\mathcal{W}$. Thus $f_\xi(x_i) = r_i \in G_i$ for all $i \leq k$ which proves that $I \subset \overline{\Omega}$.

It follows from (6) and (7) that $\Omega \subset C_A = \{f \in C_p(X, \mathbb{R}^n) : f(A) \subset \{o_n\}\}$; to see that $\Omega$ is discrete, fix any $\xi = (W_1, \ldots, W_k, V_1, \ldots, V_k, r_1, \ldots, r_k) \in \mathcal{W}$ and let $d = (d_1, \ldots, d_m) = \varphi(\xi)$; the set $B_\xi = \{f \in C_p(X) : f(d_i) \in Q_i$ for all $i \in M_m$} is open in $C_p(X, \mathbb{R}^n)$ and contains $f_\xi$, so it suffices to establish that $B_\xi \cap \Omega = \{f_\xi\}$. Assume toward a contradiction, that $f_\eta \in B_\xi$ for some $\eta \in \mathcal{W} \backslash \{\xi\}$; then $a = (a_1, \ldots, a_m) = \varphi(\eta) \neq d$.

For any $i \in M_m$, it follows from $f_\eta(d_i) \in Q_i$, together with (6)–(8), that $d_i \in U_i^a$. An immediate consequence is that $d \in U^a$ which contradicts the fact that $U^a \cap D = \{a\}$. Therefore $B_\xi \cap \Omega = \{f_\xi\}$ which proves that $\Omega \subset C_A$ is a discrete subspace such that $I \subset \overline{\Omega}$. Finally apply Fact 2 of V.423 to see that the space $(C_p(X))^n \simeq C_p(X, \mathbb{R}^n)$ is $d$-separable. This proves the implication (iv)$\Longrightarrow$(i) and makes our solution complete.

**V.426.** *Prove that*

(a) *if* $\sup\{s(X^n) : n \in \mathbb{N}\} > iw(X)$ *then* $C_p(X)$ *is $d$-separable;*
(b) *if $K$ is a Corson compact space then* $C_p(K)$ *is $d$-separable;*
(c) *if $X$ is a metrizable space then* $C_p(X)$ *is $d$-separable.*

**Solution.** (a) Let $\lambda = iw(X)$ and fix a condensation $\varphi : X \to Y$ of $X$ onto a space $Y$ with $w(Y) \leq \lambda$. Then $w(Y^n) \leq \lambda$ and hence $\Delta_n(Y)$ is a $G_\lambda$-subset of $Y^n$ for every $n \in \mathbb{N}$. For any $x = (x_1, \ldots, x_n) \in X^n$ let $\varphi^n(x) = (\varphi(x_1), \ldots, \varphi(x_n))$; it is easy to see that $\varphi^n : X^n \to Y^n$ is a condensation such that $\varphi^n(\Delta_n(X)) = \Delta_n(Y)$. Therefore $\Delta_n(X)$ is a $G_\lambda$-set in $X^n$ for each $n \in \mathbb{N}$.
Take the minimal $n \in \mathbb{N}$ such that there is a discrete $D \subset X^n$ with $|D| > \lambda$. Since $\Delta_n = \Delta_n(X)$ is the finite union of subspaces homeomorphic to $X^{n-1}$, the set $D \cap \Delta_n$ has cardinality at most $\lambda$, so, passing if necessary to the relevant subset of $D$ of cardinality $> \lambda$, we can consider that $D \subset X^n \backslash \Delta_n$. Choose a family $\mathcal{U} \subset \tau(X \times X)$ such that $|\mathcal{U}| \leq \lambda$ and $\bigcap \mathcal{U} = \Delta_n$. It follows form the equality $D = \bigcup\{D \backslash U : U \in \mathcal{U}\}$ that $|D \backslash U| > \lambda$ for some $U \in \mathcal{U}$. Letting $E' = D \backslash U$ we obtain a set $E' \subset X^n \backslash \Delta_n$ such that $\overline{E'} \cap \Delta_n = \emptyset$ and $|E'| > \lambda$. Take any set $E \subset E'$ with $|E| = \lambda$; then $E \subset X^n \backslash \Delta_n$ while $|E| = iw(X)$ and $\overline{E} \cap \Delta_n = \emptyset$. Finally, apply Problem 423 to conclude that $C_p(X)$ is $d$-separable.
(b) Apply Fact 1 of V.421 to find a discrete set $D \subset K \times K$ such that $|D| = d(K)$. It follows from CFS-121 that $w(K) = d(K)$, so we have a discrete subspace $D \subset K \times K$ such that $|D| = w(K) = iw(K)$. Now, Problem 424 shows that $C_p(K)$ is $d$-separable.
(c) It will be easy to complete our solution once we prove the following fact.

*Fact 1.* If $Y$ is a metrizable space then there exists a disjoint family $\mathcal{U} \subset \tau^*(Y)$ such that $|\mathcal{U}| = w(Y)$. Consequently, there exists a discrete subspace $D \subset Y$ such that $|D| = w(Y)$.

*Proof.* Our Fact is an easy exercise if $w(Y) \leq \omega$, so we assume that $w(Y) > \omega$. Fix a base $\mathcal{B} = \bigcup_{n \in \omega} \mathcal{B}_n \subset \tau^*(Y)$ of the space $Y$ such that the family $\mathcal{B}_n$ is discrete for every $n \in \omega$. If $\kappa = w(Y)$ has uncountable cofinality then it follows from $|\mathcal{B}| \geq \kappa$ that $|\mathcal{B}_n| \geq \kappa$ for some $n \in \omega$. Choosing any $\mathcal{U} \subset \mathcal{B}_n$ with $|\mathcal{U}| = \kappa$ we obtain the promised family $\mathcal{U}$.

Now assume that $\{\kappa_n : n \in \omega\}$ is an increasing sequence of regular uncountable cardinals such that $\kappa = \sup\{\kappa_n : n \in \omega\}$. Say that a set $U \in \tau^*(Y)$ is *adequate* if $w(V) = w(U)$ for any $V \in \tau^*(U)$. For any $U \in \tau^*(Y)$ consider the cardinal $\mu = \setminus n\{w(V) : V \in \tau^*(U)\}$ and choose $V \in \tau^*(U)$ such that $w(V) = \mu$. It is clear that $V \subset U$ is an adequate set so the family $\mathcal{W}$ of all adequate subsets of $Y$ forms a $\pi$-base in $Y$.

Suppose that $U$ is an adequate subset of $Y$ with $w(U) = \kappa$. Since $U$ is infinite, we can find a disjoint faithfully indexed family $\{U_n : n \in \omega\}$ of nonempty open subsets of $U$. It follows from $c(U_n) = w(U_n) = \kappa > \kappa_n$ that there exists a disjoint family $\mathcal{U}_n \subset \tau^*(U_n)$ such that $|\mathcal{U}_n| = \kappa_n$. The family $\mathcal{U} = \bigcup_{n \in \omega} \mathcal{U}_n \subset \tau^*(Y)$ is disjoint and $|\mathcal{U}| = \kappa$. Therefore we can assume that $w(U) < \kappa$ for any adequate set $U \subset Y$.

Let $\mathcal{V}$ be a maximal disjoint subfamily of $\mathcal{W}$. It is straightforward that $G = \bigcup \mathcal{V}$ is dense in $Y$ and hence $w(G) = w(Y) = \kappa$ (recall that $Y$ is a metrizable space). If $|\mathcal{V}| = \kappa$ then letting $\mathcal{U} = \mathcal{V}$ we obtain the promised family $\mathcal{U}$. If not then there exists $n \in \omega$ such that $|\mathcal{V}| \leq \kappa_n$.

Assume first that there exists $m \in \omega$ such that $w(V) \leq \kappa_m$ for all $V \in \mathcal{V}$. Then $w(G) = nw(G) \leq \kappa_m \cdot \kappa_n < \kappa$ which is a contradiction. Therefore we can choose a faithfully indexed family $\{V_n : n \in \omega\} \subset \mathcal{V}$ such that $w(V_n) > \kappa_n$ for any $n \in \omega$. Since also $c(V_n) = w(V_n) > \kappa_n$, we can choose a disjoint family $\mathcal{U}_n$ of nonempty open subsets of $V_n$ such that $|\mathcal{U}_n| = \kappa_n$ for every $n \in \omega$. Then $\mathcal{U} = \bigcup_{n \in \omega} \mathcal{U}_n$ is a disjoint family of nonempty open subsets of $Y$ such that $|\mathcal{U}| = \kappa$, so we settled the first part of our Fact. If we take a point $x_U \in U$ for every $U \in \mathcal{U}$ then the set $D = \{x_U : U \in \mathcal{U}\} \subset Y$ is discrete and $|D| = \kappa = w(Y)$, so Fact 1 is proved.

Finally observe that if $X$ is a metrizable space then we can apply Fact 1 to find a discrete subspace $D' \subset X$ such that $|D'| = w(X) \geq iw(X)$. Therefore we can extract a set $D \subset D'$ such that $|D| = iw(X)$, so Problem 423 is applicable to conclude that $C_p(X)$ is $d$-separable.

**V.427.** *Prove that if $C_p(X)$ is a Lindelöf $\Sigma$-space then it is $d$-separable.*

**Solution.** Say that a space is *simple* if it has at most one non-isolated point. A nonempty space will be called 0-concentrated if it is countable. If $n \in \mathbb{N}$ and we have defined $m$-concentrated spaces for any natural number $m < n$ say that a space $Z$ is $n$-concentrated if there is a point $a \in Z$ such that, for any $U \in \tau(a, Z)$, the set $Z \setminus U$ is the union of a countable family of $(n-1)$-concentrated spaces. This defines a class $\mathcal{C}_n$ of $n$-concentrated spaces for all $n \in \omega$; let $\mathcal{C} = \bigcup\{\mathcal{C}_n : n \in \omega\}$. Let us prove that

(1) for every $n \in \omega$ the class $\mathcal{C}_n$ is closed-hereditary and, for any space $Z$ such that $Z = \bigcup_{k \in \omega} Z_k$ and $Z_k \in \mathcal{C}_n$ for each $k \in \omega$, we have $Z \in \mathcal{C}_{n+1}$.

Let $\mathcal{P}_n$ be the statement that $\mathcal{C}_n$ is closed-hereditary and denote by $\mathcal{Q}_n$ the statement which says that any countable union of elements of $\mathcal{C}_n$ belongs to $\mathcal{C}_{n+1}$. Since every subspace of a countable space is countable, the statement $\mathcal{P}_0$ is true. Now if $\{Z_k : k \in \omega\} \subset \mathcal{C}_0$ and $Z = \bigcup_{k \in \omega} Z_k$ then $Z$ is countable and nonempty. Take any point $a \in Z$; then $Z \setminus U$ is countable for any $U \in \tau(a, Z)$, so $Z \in \mathcal{C}_1$. This shows that $\mathcal{Q}_0$ also holds.

Proceeding inductively assume that $n \in \mathbb{N}$ and we proved that $\mathcal{P}_i$ and $\mathcal{Q}_i$ are fulfilled for all $i < n$. Suppose that $Z \in \mathcal{C}_n$ and $F$ is a closed subspace of $Z$. Pick a point $a \in A$ such that $Z \setminus U$ is the countable union of elements of $\mathcal{C}_{n-1}$ for any $U \in \tau(a, Z)$. If $a \notin F$ then $F$ is the countable union of elements of $\mathcal{C}_{n-1}$, so $F \in \mathcal{C}_n$ by the induction hypothesis.

Now if $a \in F$ then take any $V \in \tau(a, F)$ and pick a set $U \in \tau(a, Z)$ such that $U \cap F = V$. There exists a family $\{Y_k : k \in \omega\} \subset \mathcal{C}_{n-1}$ such that $Z \setminus U = \bigcup_{k \in \omega} Y_k$. For every $k \in \omega$ the set $F'_k = F \cap Y_k$ belongs to $\mathcal{C}_{n-1}$ by the induction hypothesis; since $F \setminus V = \bigcup_{k \in \omega} F'_k$, we conclude that the point $a$ witnesses that $F \in \mathcal{C}_n$ and hence $\mathcal{P}_n$ is proved.

Assume that $Z = \bigcup_{k \in \omega} Z_k$ and $Z_k \in \mathcal{C}_n$ for every $k \in \omega$. Fix any point $a \in Z$; if $U \in \tau(a, Z)$ then $Z \setminus U = \bigcup_{k \in \omega}(Z_k \setminus U)$. It follows from $\mathcal{P}_n$ that $Z_k \setminus U \in \mathcal{C}_n$ for all $k \in \omega$, so the point $a$ witnesses that $Z \in \mathcal{C}_{n+1}$ and hence we settled $\mathcal{Q}_n$. Thus our induction procedure can be continued to show that $\mathcal{P}_n$ and $\mathcal{Q}_n$ are true for all $n \in \omega$, i.e., (1) is proved.

If $Z \in \mathcal{C}_n$ then letting $Z_k = Z$ for all $k \in \omega$ we can apply (1) to see that $Z \in \mathcal{C}_{n+1}$, i.e., $\mathcal{C}_n \subset \mathcal{C}_{n+1}$ for all $n \in \omega$. Now it takes a trivial induction to establish that

(2) $\mathcal{C}_m \subset \mathcal{C}_n$ whenever $m \leq n$.

Let us show that every class $\mathcal{C}_n$ behaves properly under continuous images, i.e.,

(3) for any $n \in \omega$, if $Z \in \mathcal{C}_n$ and $Y$ is a continuous image of $Z$ then $Y \in \mathcal{C}_n$.

Since any image of a countable set is countable, the property (3) is true for $n = 0$. Proceeding inductively assume that $m \in \mathbb{N}$ and we have (3) for all $n < m$. If $Z \in \mathcal{C}_m$ and $f : Z \to Y$ is a continuous onto map then fix a point $a \in Z$ such that $Z \setminus U$ is the countable union of elements of $\mathcal{C}_{m-1}$ for any $U \in \tau(a, Z)$. Let $b = f(a)$ and take any $V \in \tau(b, Y)$.

There exists a family $\{Z_k : k \in \omega\} \subset \mathcal{C}_{m-1}$ such that $Z \setminus f^{-1}(V) = \bigcup_{k \in \omega} Z_k$. By the induction hypothesis the set $Y_k = f(Z_k)$ belongs to $\mathcal{C}_{m-1}$ for all $k \in \omega$, so it follows from $Y \setminus V = \bigcup_{k \in \omega} Y_k$ that the point $b$ witnesses that $Y \in \mathcal{C}_m$ and hence (3) is proved.

It turns out that the family $\mathcal{C}$ is well behaved under products, namely,

(4) if $Y \in \mathcal{C}_n$ and $Z \in \mathcal{C}_m$ then $Y \times Z \in \mathcal{C}_{m+n+1}$.

We will prove (4) by induction on $k = \setminus n\{m, n\}$. If $m = 0$ then $Z$ is countable and hence $Y \times Z$ is the countable union of spaces homeomorphic to $Y$. Therefore it follows from (1) that $Y \times Z \in \mathcal{C}_{n+1} = \mathcal{C}_{n+m+1}$. An analogous argument shows that (4) holds for any $m \in \omega$ whenever $n = 0$.

Assume that $l \in \mathbb{N}$ and (4) is verified for any $k < l$. If $m = \setminus n\{m, n\} = l$ then $n \geq l$ and hence we can choose points $a \in Y$ and $b \in Z$ which witness that $Y \in \mathcal{C}_n$ and $Z \in \mathcal{C}_m$. Take an arbitrary open neighborhood $U$ of the point $z = (a, b)$ in the space $Y \times Z$. Choose $V \in \tau(a, Y)$ and $W \in \tau(b, Z)$ such that $V \times W \subset U$; there exist families $\{Y_p : p \in \omega\} \subset \mathcal{C}_{n-1}$ and $\{Z_q : q \in \omega\} \subset \mathcal{C}_{m-1}$ such that $\bigcup_{p \in \omega} Y_p = Y \setminus V$ and $\bigcup_{q \in \omega} Z_q = Z \setminus W$. By the induction hypothesis we have

$Y'_p = Y_p \times Z \in C_{m+n}$ and $Z'_q = Y \times Z_q \in C_{m+n}$ for all $p, q \in \omega$. If $F = (Y \times Z) \setminus U$ then it follows from (1) and the equality $F = (\bigcup_{p \in \omega} (Y'_p \cap F)) \cup (\bigcup_{q \in \omega} (Z'_q \cap F))$ that $(Y \times Z) \setminus U$ is the countable union of elements of $C_{m+n}$ for any $U \in \tau(z, Y \times Z)$. Therefore the point $z$ witnesses that $Y \times Z \in C_{m+n+1}$ as promised. The case of $n = l$ is symmetric, so we carried out the induction step and hence (4) is proved.

It turns out that

(5) for any $n \in \omega$, if $Y \in C_n$ then $Y$ is $d$-separable;

To establish (5) observe that it is true for $n = 0$ and assume that $m \in \mathbb{N}$ and we proved it for all $n < m$. Take a space $Z \in C_m$ and fix a point $a \in Z$ such that $Z \setminus U$ can be covered by a countable family from $C_{m-1}$ for any $U \in \tau(a, Z)$. Let $\mathcal{U}$ be a maximal disjoint family of open subsets of $Z$ such that $a \notin \overline{U}$ for any $U \in \mathcal{U}$. Every $U \in \mathcal{U}$ can be covered by a countable family of spaces from $C_{m-1}$ which are $d$-separable by the induction hypothesis; an easy consequence is that $U$ is also $d$-separable and hence $V = \bigcup \mathcal{U} \simeq \bigoplus \{U : U \in \mathcal{U}\}$ is $d$-separable as well. Thus $Z$ is $d$-separable because $V$ is a dense $d$-separable subspace of $Z$; this shows that (5) is proved.

*Fact 1.* If both $Z$ and $C_p(Z)$ are Lindelöf $\Sigma$ then there is a closed simple subspace $L \subset C_p(Z)$ which separates the points of $Z$.

*Proof.* By CFS-285, there is a condensation $\varphi : Z \to Y \subset \Sigma(A)$ for some infinite set $A$; choosing a larger set $A$ if necessary, we can assume, without loss of generality, that $|A| > \omega$. For any $f \in C_p(Y)$ let $\varphi^*(f) = f \circ \varphi$; then $\varphi^* : C_p(Y) \to C_p(Z)$ is a homeomorphism (see TFS-163).

The space $\Sigma = \Sigma(A)$ being homeomorphic to $C_p(L(\kappa))$ for $\kappa = |A|$ (see CFS-106), the space $L(\kappa)$ can be embedded in $C_p(\Sigma)$ in such a way that it generates the topology of $\Sigma$. The restriction $\pi : C_p(\Sigma) \to C_p(Y)$ maps $L(\kappa)$ onto a set $M \subset C_p(Y)$ which generates the topology of $Y$ (see Fact 1 of U.285); in particular, $M$ separates the points of $Y$ and therefore the set $N' = \varphi^*(M)$ separates the points of $Z$. The space $N'$ is a continuous image of $L(\kappa)$, so it is concentrated around a point, i.e.,

(6) there is a set $N' \subset C_p(Z)$ which is concentrated around a point and separates the points of $Z$.

It is easy to see that we can consider that the set $N'$ is concentrated around the function $u$ which is identically zero on $Z$. Take a homeomorphism $\mu : \mathbb{R} \to (-1, 1)$ such that $\mu(0) = 0$ and let $N = \{\mu \circ f : f \in N'\}$. Then $N \subset C_p(Z, (-1, 1))$ while $N$ is concentrated around the point $u$ (see TFS-091) and separates the points of $Z$.

Applying Fact 9 of U.285 we can find a condensation $\xi$ of $C_p(Z)$ into some $\Sigma(B)$. It is evident that $\xi(N) \subset \Sigma(B)$ is concentrated around the point $\xi(u)$. By Fact 1 of U.289, there exists a family $\{K_n : n \in \omega\}$ of simple compact spaces such that $\xi(u)$ is the unique point which can be non-isolated in some $K_n$ and $\xi(N) = \bigcup \{K_n : n \in \omega\}$. As a consequence, the set $K'_n = \xi^{-1}(K_n)$ is closed in $C_p(Z)$ and all points of $K'_n \setminus \{u\}$ are isolated in $K'_n$ for each $n \in \omega$. Let $L_n = 2^{-n} \cdot K'_n$ for every $n \in \omega$; it is immediate that the set $L = \bigcup_{n \in \omega} L_n$ still separates the points of the space $Z$.

Fix a function $f \in C_p(Z) \setminus \{u\}$ and a point $x \in Z$ such that $|f(x)| > 2^{-n}$ for some $n \in \omega$. The set $U = \{g \in C_p(Z) : |g(x)| > 2^{-n}\}$ is an open neighborhood of $f$ which does not meet the set $L_m$ for every $m \geq n$.

If $f \notin L$ then the set $U' = C_p(Z) \setminus (\bigcup \{L_m : m < n\})$ is also a neighborhood of $f$ in $C_p(Z)$ and hence $V = U \cap U' \in \tau(f, C_p(Z))$ while $V \cap L = \emptyset$. This proves that $L$ is closed in $C_p(Z)$.

If $f \in L$ then $U \cap L$ is closed and discrete in $U$ because $U \cap L \subset \bigcup \{L_m : m < n\}$ while every $L_m$ is closed and discrete in $C_p(Z) \setminus \{u\}$. Therefore we can find a set $U' \in \tau(f, C_p(Z))$ for which $U' \cap (\bigcup \{L_m : m < n\}) = \{f\}$. Then, for the set $V = U \cap U'$ we have $V \cap L = \{f\}$, so all points of $L \setminus \{u\}$ are isolated in $L$ and hence Fact 1 is proved.

*Fact 2.* If $C_p(Z)$ is a Lindelöf $\Sigma$-space then there exists a closed simple subspace of $C_p(Z)$ which separates the points of $Z$. In other words, Lindelöf $\Sigma$-property of $Z$ can be omitted in Fact 1.

*Proof.* Observe that $\upsilon Z$ and $C_p(\upsilon Z)$ have to be Lindelöf $\Sigma$-spaces (see CFS-206 and CFS-234); besides, the restriction map $\pi : C_p(\upsilon Z) \to C_p(Z)$ is a condensation such that $\pi | A : A \to \pi(A)$ is a homeomorphism for any countable set $A \subset C_p(\upsilon Z)$ (TFS-437). By Fact 1 there is a simple subspace $T' \subset C_p(\upsilon Z)$ which separates the points of $\upsilon Z$; therefore $T = \pi(T')$ separates the points of $Z$.

Any simple Lindelöf space is concentrated around its non-isolated point, so $T'$ is concentrated around a point $f \in T'$. It is evident that $T$ is concentrated around the point $g = \pi(f)$. If $h \in C_p(Z) \setminus \{g\}$ then there is a neighborhood $V$ of $g$ in $C_p(Z)$ such that $h \notin \overline{V}$. It is clear that $W = C_p(Z) \setminus \overline{V}$ is an open neighborhood of $h$ which contains at most countably many elements of $T$; let $E = T \cap W$ and $E' = \pi^{-1}(E)$. If $h \in \overline{E \setminus \{h\}}$ then $h' = \pi^{-1}(h)$ is in the closure of $E' \setminus \{h'\}$ because $\pi | (E' \cup \{h'\}) : E' \cup \{h'\} \to E \cup \{h\}$ is a homeomorphism. However, this contradicts the fact that $T' \setminus \{f\}$ is closed and discrete in $C_p(\upsilon Z) \setminus \{f\}$.

This contradiction shows that $T \setminus \{g\}$ is closed and discrete in $C_p(Z) \setminus \{g\}$ and hence $T$ is a simple closed subspace of $C_p(Z)$ which separates the points of $Z$, i.e., Fact 2 is proved.

Returning to our solution apply Fact 2 to find a simple closed set $T \subset C_p(X)$ which separates the points of $X$. Since $T$ is Lindelöf $\Sigma$, it is concentrated around a point $f \in T$ and hence $T \in \mathcal{C}_1$. If $A(T)$ is the minimal subalgebra of $C_p(X)$ that contains $T$ then $A(T)$ is dense in $C_p(X)$ (see TFS-192) and $A(T) = \bigcup_{n \in \mathbb{N}} A_n$ where, for every $n \in \mathbb{N}$, there is a continuous onto map $\varphi_n : T^{k_n} \times \mathbb{R}^{k_n} \to A_n$ for some $k_n \in \mathbb{N}$. The set $B_n = \varphi_n(T^{k_n} \times \mathbb{Q}^{k_n})$ is a countable union of elements of $\mathcal{C}$ by (3) and (4); besides, it is dense in $A_n$ for any $n \in \mathbb{N}$. Thus $B = \bigcup_{n \in \omega} B_n$ is a dense subspace of $A(T)$ which can be represented as the countable union of spaces from $\mathcal{C}$. It follows from (5) that $B$ is $d$-separable, so $C_p(X)$ is also $d$-separable because it contains a dense $d$-separable subspace.

**V.428.** *Prove that if $C_p(X)$ is a Lindelöf $\Sigma$-space then the space $X$ must be hereditarily d-separable.*

**Solution.** Recall that a family $\mathcal{U} \subset \exp(Z)$ is called *weakly $\sigma$-point-finite* if there exists a collection $\{\mathcal{U}_n : n \in \omega\} \subset \exp(\mathcal{U})$ such that, for any point $x \in Z$, if $N_x = \{n \in \omega : \mathcal{U}_n$ is point-finite at $x\}$ then $\mathcal{U} = \bigcup\{\mathcal{U}_n : n \in N_x\}$. Say that a space $Z$ is *simple* if $Z$ has at most one non-isolated point.

Given a set $A$ and a family $s = \{A_n : n \in \omega\} \subset \exp(A)$ consider the subspace $\Sigma_s(A) = \{x \in \mathbb{R}^A : A = \bigcup\{A_n : |x^{-1}(\mathbb{R}\backslash\{0\}) \cap A_n| < \omega\}\}$ of the space $\mathbb{R}^A$; it is easy to see that $\Sigma_s(A) \subset \Sigma(A)$. We say that a space $Y$ is a $\Sigma_s$-product of real lines if there exists a set $A$ and a family $s = \{A_n : n \in \omega\} \subset \exp(A)$ such that $Y = \Sigma_s(A)$. For any $x \in \Sigma(A)$ let $\mathrm{supp}(x) = x^{-1}(\mathbb{R}\backslash\{0\})$.

A family $\mathcal{U} \subset \exp(Z)$ is $T_0$-*separating* if, for any distinct points $x, y \in Z$ there exists $U \in \mathcal{U}$ such that $U \cap \{x, y\}$ is a singleton. If $Z$ is a space then $\mathcal{U} \subset \exp(Z)$ is called *a Gul'ko family* on $Z$ if $\mathcal{U}$ is $T_0$-separating, weakly $\sigma$-point-finite and consists of cozero subsets of $Z$.

*Fact 1.* A space $Z$ has a Gul'ko family if and only if $Z$ can be condensed into some $\Sigma_s$-product of real lines.

*Proof.* Suppose that $\mathcal{U}$ is a weakly $\sigma$-point-finite $T_0$-separating family of cozero subsets of $Z$ and fix a sequence $\{\mathcal{U}_n : n \in \omega\}$ of subfamilies of $\mathcal{U}$ which witnesses that $\mathcal{U}$ is weakly $\sigma$-point-finite. For any $U \in \mathcal{U}$ fix a function $f_U \in C_p(Z)$ such that $f_U^{-1}(0) = Z\backslash U$ and consider the diagonal product $f = \Delta\{f_U : U \in \mathcal{U}\}$. Let $Y = f(Z)$ and $A = \{f_U : U \in \mathcal{U}\}$; then $Y \subset \mathbb{R}^A$ and $f : Z \to Y$ is a condensation: this easily follows from the fact that $\mathcal{U}$ is a $T_0$-separating family.

The family $\mathcal{U}$ being point-countable (see Fact 2 of U.290), the set $Y$ is contained in $\Sigma(A)$. Let $A_n = \{f_U : U \in \mathcal{U}_n\}$ for every $n \in \omega$ and consider the family $s = \{A_n : n \in \omega\}$. Given any $y \in \Sigma(A)$ let $M_y = \{n \in \omega :$ the set $\mathrm{supp}(y) \cap A_n$ is finite$\}$.

Take any $y \in Y$ and $f_U \in A$; there is a unique $x \in Z$ such that $f(x) = y$. The family $\mathcal{U}$ is weakly $\sigma$-point-finite, so there exists $n \in \omega$ such that $\mathcal{U}_n$ is point-finite at $x$ and $U \in \mathcal{U}_n$. Then $f_U \in A_n$ and $n \in M_y$ because $\mathrm{supp}(y) \cap A_n$ coincides with the finite set $\{f_V : x \in V \in \mathcal{U}_n\}$. Therefore $A = \bigcup\{A_n : n \in M_y\}$ for any $y \in Y$ and hence $Y \subset \Sigma_s(A)$, i.e., we proved necessity.

Now assume that $A$ is a set, $s = \{A_n : n \in \omega\}$ is a family of subsets of $A$ and $f : Z \to Y$ is a condensation for some $Y \subset \Sigma_s(A)$. Fact 1 of U.290 implies that there exists a Gul'ko family on the space $\Sigma_s(A)$. It is trivial that existence of a Gul'ko family is hereditary, so $Y$ also has a Gul'ko family $\mathcal{V}$. Letting $\mathcal{U} = \{f^{-1}(V) : V \in \mathcal{V}\}$ we obtain a Gul'ko family on $Z$, so we settled sufficiency and hence Fact 1 is proved.

*Fact 2.* If $C_p(Z)$ is a Lindelöf $\Sigma$-space then the space $Z$ has a Gul'ko family and hence $Z$ can be condensed into a $\Sigma_s$-product of real lines.

*Proof.* Apply Fact 2 of V.427 to find a closed simple subspace $T \subset C_p(Z)$ which separates the points of $Z$. Let $e(x)(f) = f(x)$ for any $x \in Z$ and $f \in T$; then

$e(x) \in C_p(T)$ for any $x \in Z$ and $e : Z \to C_p(T)$ is a condensation by TFS-166. Let $Y = e(Z)$; since $T$ is a simple Lindelöf $\Sigma$-space, we can apply CFS-274 to see that $Y$ has a Gul'ko family $\mathcal{V}$. Then $\mathcal{U} = \{e^{-1}(V) : V \in \mathcal{V}\}$ is a Gul'ko family in $Z$, so Fact 2 is proved.

*Fact 3.* If $Z$ is a space and $\mathcal{U}$ is a weakly $\sigma$-point-finite family of nonempty open subsets of $Z$ then there is a $\sigma$-discrete subspace $D \subset Z$ such that $D \cap U \neq \emptyset$ for any $U \in \mathcal{U}$.

*Proof.* Assume that $\mathcal{U} \subset \tau^*(X)$ and the collection $\{\mathcal{U}_n : n \in \omega\} \subset \exp(\mathcal{U})$ witnesses that $\mathcal{U}$ is a weakly $\sigma$-point-finite family. Denote by $A_n$ the set of points at which $\mathcal{U}_n$ has finite order and let $\mathcal{V}_n = \{U \in \mathcal{U}_n : U \cap A_n \neq \emptyset\}$ for each $n \in \omega$. We are going to show that $\bigcup\{\mathcal{V}_n : n \in \omega\} = \mathcal{U}$.

Take any $U \in \mathcal{U}$ and $x \in U$. According to the definition of weakly $\sigma$-point-finite family, there exists $n \in \omega$ such that $U \in \mathcal{U}_n$ and $\mathcal{U}_n$ has finite order at $x$. Then $x \in A_n$ and $x \in U$ implies $U \in \mathcal{V}_n$.

For each $n \in \omega$ the family $\mathcal{W}_n = \{U \cap A_n : U \in \mathcal{V}_n\} \subset \tau^*(A_n)$ is point-finite; by Fact 2 of U.271 there exists a $\sigma$-discrete subset $B_n \subset A_n$ such that $B_n \cap V \neq \emptyset$ for every $V \in \mathcal{W}_n$. Therefore $B_n \cap U \neq \emptyset$ for every $U \in \mathcal{V}_n$ and $n \in \omega$; since $\bigcup_{n \in \omega} \mathcal{V}_n = \mathcal{U}$, the $\sigma$-discrete set $D = \bigcup_{n \in \omega} B_n$ intersects every element of $\mathcal{U}$, i.e., Fact 3 is proved.

*Fact 4.* Every subspace of a $\Sigma_s$-product of real lines has a weakly $\sigma$-point-finite $\pi$-base and hence the space $\Sigma_s(A)$ is hereditarily $d$-separable for any set $A$ and any family $s = \{A_n : n \in \omega\} \subset \exp(A)$.

*Proof.* Let $\mathcal{B} = \{(p,q) : p, q \in \mathbb{Q}, \ p < q \text{ and } pq > 0\}$; in other words, $\mathcal{B}$ is the family of all nonempty rational intervals of $\mathbb{R}$, which do not contain zero. It is evident that any $U \in \tau^*(\mathbb{R})$ contains an element of $\mathcal{B}$, i.e., the family $\mathcal{B}$ is a $\pi$-base in $\mathbb{R}$. In fact, $\mathcal{B}$ has even stronger property: for any $a \in \mathbb{R}\backslash\{0\}$ and any $U \in \tau(a, \mathbb{R})$, there exists $B \in \mathcal{B}$ such that $a \in B \subset U$, i.e., $\mathcal{B}$ is a base in $\mathbb{R}\backslash\{0\}$.

Given any points $a_1, \ldots, a_n \in A$, and $O_1, \ldots, O_n \in \tau(\mathbb{R})$, consider the set

$$[a_1, \ldots, a_n; O_1, \ldots, O_n] = \{x \in \Sigma_s(A) : x(a_i) \in O_i \text{ for each } i \leq n\}.$$

It is clear that $\mathcal{U} = \{[a_1, \ldots, a_n; O_1, \ldots, O_n] : n \in \mathbb{N}, \ a_i \in A$ and $O_i \in \tau(\mathbb{R})$ for all $i \leq n\}$ is a base in $\Sigma_s(A)$. If $U = [a_1, \ldots, a_n; O_1, \ldots, O_n] \in \mathcal{U}$ and $B_i \subset O_i, \ B_i \in \mathcal{B}$ for all $i \leq n$ then we have the inclusion $V = [a_1, \ldots, a_n; B_1, \ldots, B_n] \subset U$; i.e., the family $\mathcal{V} = \{[a_1, \ldots, a_n; B_1, \ldots, B_n] : n \in \mathbb{N}, \ a_i \in A$ and $B_i \in \mathcal{B}$ for all $i \leq n\}$ is a $\pi$-base of $\Sigma_s(A)$.

Denote by $u$ the element of $\Sigma_s(A)$ for which $u(a) = 0$ for all $a \in A$ and take any $Z \subset \Sigma_s(A)$. It suffices to prove our Fact for the space $Z \backslash \{u\}$. Indeed, the case when $u \notin Z$ is clear; if $u \in Z$ and $u$ is an isolated point of $Z$ then for any weakly $\sigma$-point-finite $\pi$-base $\mathcal{C}$ in the space $Z \backslash \{u\}$, the family $\{\{u\}\} \cup \mathcal{C}$ is a weakly $\sigma$-point-finite $\pi$-base in the space $Z$. If $u$ is not isolated in $Z$ then any $\pi$-base for $Z \backslash \{u\}$ is also a $\pi$-base for $Z$, so again it suffices to find a weakly $\sigma$-point-finite $\pi$-base for the space $Z \backslash \{u\}$. To simplify the notation we will assume, without loss of generality, that $u \notin Z$.

It turns out that

(1)  the family $\mathcal{W}_n(B) = \{[a_1, \ldots, a_n; B_1, \ldots, B_n] : a_i \in A$ for all $i \leq n\}$ is weakly $\sigma$-point-finite for any $n \in \mathbb{N}$ and $B = (B_1, \ldots, B_n) \in \mathcal{B}^n$

Given any numbers $i_1, \ldots, i_n \in \omega$ let $\mathcal{H}(i_1, \ldots, i_n) = \{[a_1, \ldots, a_n; B_1, \ldots, B_n] :$ the set $\{a_1, \ldots, a_n\}$ is contained in $A_{i_1} \cup \ldots \cup A_{i_n}\}$. The collection $\mathbb{H} = \{\mathcal{H}(i_1, \ldots, i_n) : i_j \in \omega$ for all $j \leq n\}$ is countable, so we can choose an enumeration $\{\mathcal{W}_i : i \in \omega\}$ of the family $\mathbb{H}$.

If $x \in \Sigma_s(A)$ and $W = [a_1, \ldots, a_n; B_1, \ldots, B_n] \in \mathcal{W}_n(B)$ then we can choose $i_1, \ldots, i_n \in \omega$ such that $a_j \in A_{i_j}$ and the set $Q_j = \text{supp}(x) \cap A_{i_j}$ is finite for any $j \leq n$. There exists $m \in \omega$ such that $\mathcal{H}(i_1, \ldots, i_n) = \mathcal{W}_m$; then $W \in \mathcal{W}_m$ and we claim that $\mathcal{W}_m$ is point-finite at $x$.

Indeed, if $x \in W' = [b_1, \ldots, b_n; B_1, \ldots, B_n] \in \mathcal{W}_m$ and $b_l \notin Q = \bigcup\{Q_j : j \leq n\}$ for some $l \leq n$ then there exists $j \leq n$ such that $b_l \in A_{i_j} \setminus Q \subset A_{i_j} \setminus Q_j$ and therefore $x(b_l) = 0 \notin B_l$. As a consequence, $b_l \in Q$ for all $l \leq n$, so $(b_1, \ldots, b_n) \in Q^n$. We proved that all elements $[b_1, \ldots, b_n; B_1, \ldots, B_n]$ from $\mathcal{W}_m$ that can contain $x$ are among those for which $(b_1, \ldots, b_n) \in Q^n$. The set $Q^n$ being finite, we proved that $\mathcal{W}_m$ is point-finite at $x$ and hence the family $\{\mathcal{W}_m : m \in \omega\}$ witnesses that $\mathcal{W}_n(B)$ is weakly $\sigma$-point-finite, i.e., (1) is proved.

It is immediate that $\mathcal{V} = \bigcup\{\mathcal{W}_n(B) : n \in \mathbb{N}$ and $B \in \mathcal{B}^n\}$, so we can apply Fact 1 of U.293 to convince ourselves that $\mathcal{V}$ is weakly $\sigma$-point-finite. Therefore the family $\mathcal{V}|Z = \{V \cap Z : V \in \mathcal{V}$ and $V \cap Z \neq \emptyset\}$ is weakly $\sigma$-point-finite as well. We claim that $\mathcal{V}|Z$ is a $\pi$-base in $Z$.

To prove it take any $U \in \tau^*(Z)$ and $x \in U$. Then $x(\alpha) \neq 0$ for some $\alpha < \kappa$; there exist ordinals $\alpha_1, \ldots, \alpha_n \in \kappa$ and $O_1, \ldots, O_n \in \tau^*(\mathbb{R})$ such that $\alpha_1 = \alpha$ and $x \in V = [\alpha_1, \ldots, \alpha_n; O_1, \ldots, O_n] \cap Z \subset U$. If $x(\alpha_i) \neq 0$ then there is $B_i \in \mathcal{B}$ such that $x(\alpha_i) \in B_i \subset O_i$; if $x(\alpha_i) = 0$ then $0 \in O_i$. This shows that we do not lose generality if we assume that there is $k \in \{1, \ldots, n\}$ such that $O_i = B_i \in \mathcal{B}$ for all $i \leq k$ and $x(\alpha_i) = 0 \in O_i$ for $i = k + 1, \ldots, n$.

Call a set $K' = \{\alpha_{i_1}, \ldots, \alpha_{i_m}\} \subset K = \{\alpha_{k+1}, \ldots, \alpha_n\}$ marked if there exists a point $y \in Z$ such that $y(\alpha_i) \in B_i$ for all $i \leq k$ and $y(\alpha_{i_j}) \in O_{i_j} \setminus \{0\}$ for every $j \leq m$. Since the set $K$ is finite, there exists a maximal marked set $M = \{\alpha_{i_1} \ldots, \alpha_{i_m}\} \subset K$ (which is possibly empty). This means that there is $y \in Z$ with $y(\alpha_i) \in B_i$ for all $i \leq k$ and $y(\alpha_{i_j}) \in O_{i_j} \setminus \{0\}$ for all $j \leq m$ while for any $z \in Z$ such that $z(\alpha_i) \in B_i$ for all $i \leq k$ and $z(\alpha_{i_j}) \in O_{i_j} \setminus \{0\}$ for all $j \leq m$, we have $z(\beta) = 0$ for every $\beta \in K \setminus M$.

Changing the enumeration of $K$ if necessary, we can restrict ourselves to the case when $M = \{\alpha_{k+1}, \ldots, \alpha_m\}$ for some $m \leq n$. Since $0 \neq y(\alpha_i) \in O_i$ for all $i \leq m$, we can choose $B_i \in \mathcal{B}$ such that $y(\alpha_i) \in B_i \subset O_i$ for all $i \in \{k+1, \ldots, m\}$. Then $W = [\alpha_1, \ldots, \alpha_m; B_1, \ldots, B_m] \cap Z \in \mathcal{V}|Z$ because $y \in W$ and hence $W \neq \emptyset$; besides, for any $z \in W$ we have $z(\alpha_i) = 0 \in O_i$ for all $i \in \{m+1, \ldots, n\}$. As a consequence, any $z \in W$ belongs to $V$, i.e., $W \subset V \subset U$. This shows that, for any $U \in \tau^*(Z)$ we can find a set $W \in \mathcal{V}|Z$ with $W \subset U$; therefore $\mathcal{V}|Z$ is a $\pi$-base in $Z$.

Finally observe that if $Z \subset \Sigma_s(A)$ then there exists a weakly $\sigma$-point-finite $\pi$-base $\mathcal{B}$ in the space $Z$. By Fact 3, there exists a $\sigma$-discrete set $D \subset Z$ such that $D \cap B \neq \emptyset$ for any $B \in \mathcal{B}$. It is immediate that $D$ is dense in $Z$, so $D$ is $d$-separable and hence Fact 4 is proved.

*Fact 5.* Suppose that $Z$ is a Lindelöf $\Sigma$-space and $\mathcal{F}$ is a fixed countable network with respect to a compact cover $\mathcal{C}$ of the space $Z$. Assume additionally that $\mathcal{F}$ is closed under finite intersections, and we have a condensation $f : Z \to Z'$. If $Y \subset Z$ and $A \subset Y$ is a set such that $f(A \cap F)$ is dense in $f(Y \cap F)$ for any $F \in \mathcal{F}$ then $Y \subset \overline{A}$.

*Proof.* Assume, toward a contradiction that there is a point $z \in Y \backslash \overline{A}$ and fix a set $C \in \mathcal{C}$ such that $z \in C$. The set $K = \overline{A} \cap C$ is compact, so $U = Z' \backslash f(K)$ is an open neighborhood of $y = f(z)$; take $V \in \tau(y, Z')$ such that $\overline{V} \subset U$. Since $\mathcal{F}$ is closed under finite intersections, we can choose a sequence $\mathcal{S} = \{F_n : n \in \omega\} \subset \mathcal{F}$ such that $C \subset F_n$ and $F_n \supset F_{n+1}$ for any $n \in \omega$ while $\mathcal{S}$ is a network at $C$, i.e., for any $O \in \tau(C, Z)$ there is $n \in \omega$ with $F_n \subset O$. We have $y \in f(Y \cap F_n)$; the set $f(A \cap F_n)$ being dense in $f(Y \cap F_n)$, we can pick a point $a_n \in A \cap F_n$ such that $f(a_n) \in V$ for every $n \in \omega$.

The sequence $S = \{a_n : n \in \omega\}$ must have an accumulation point in $C$. Indeed, if every point $z \in C$ has an open neighborhood $O_z$ such that the set $\{n \in \omega : a_n \in O_z\}$ is finite then, by compactness of $C$, there is a finite $D \subset C$ with $C \subset O = \bigcup\{O_z : z \in D\}$. Then there are only finitely many $n \in \omega$ such that $a_n \in O$ while there exists $m \in \omega$ with $F_m \subset O$ and therefore $a_n \in O$ for all $n \geq m$. This contradiction shows that there is an accumulation point $a \in C$ for the sequence $S$. Then $a \in \overline{A} \cap C = K$; since $f(a_n) \in V$ for all $n \in \omega$, we have $f(a) \in \overline{V}$ by continuity of $f$. Thus $f(a) \in U \cap f(K)$; this contradiction shows that Fact 5 is proved.

*Fact 6.* If a Lindelöf $\Sigma$-space $Z$ condenses onto a hereditarily $d$-separable space then $Z$ is hereditarily $d$-separable.

*Proof.* Fix a condensation $\varphi : Z \to Z'$ of the space $Z$ onto a hereditarily $d$-separable space $Z'$ and let $\mathcal{F}$ be a countable network of $Z$ with respect to a compact cover of $Z$. There is no loss of generality to assume that $\mathcal{F}$ is closed with respect to finite intersections. Given any $Y \subset Z$ and $F \in \mathcal{F}$ we can find a $\sigma$-discrete set $B_F \subset \varphi(Y \cap F)$ which is dense in $\varphi(Y \cap F)$. The set $A_F = \varphi^{-1}(B_F) \subset Y$ is $\sigma$-discrete for any $F \in \mathcal{F}$ and hence $A = \bigcup\{A_F : F \in \mathcal{F}\}$ is also $\sigma$-discrete.

It is immediate that $\varphi(A \cap F)$ is dense in $\varphi(Y \cap F)$ for any $F \in \mathcal{F}$, so we can apply Fact 5 to conclude that $A$ is dense in $Y$. Therefore every $Y \subset Z$ is $d$-separable, i.e., Fact 6 is proved.

Returning to our solution observe that $\upsilon X$ is a Lindelöf $\Sigma$-space as well as the space $C_p(\upsilon X)$ (see CFS-206 and CFS-234). By Fact 2, the space $\upsilon X$ can be condensed onto a subset $Y$ of a $\Sigma_s$-product of real lines. Apply Fact 4 to see the space $Y$ is hereditarily $d$-separable, so we can apply Fact 6 to see that $\upsilon X$ is also hereditarily $d$-separable. This implies that $X \subset \upsilon X$ is hereditarily $d$-separable and completes our solution.

**V.429.** *Prove that under CH,*

(a) *there exists a compact space $K$ such that $C_p(K)$ is not $d$-separable;*
(b) *there exists a space $X$ such that $X \times X$ is $d$-separable while $X$ is not $d$-separable.*

**Solution.** (a) Apply SFFS-099 to see that, under CH, there exists a non-metrizable compact space $K$ such that $hl(C_p(K)) = \omega$ and hence $s(C_p(K)) = \omega$. This implies that every $\sigma$-discrete subspace of $C_p(K)$ is countable; if $C_p(K)$ is $d$-separable then it is separable and hence $K$ is metrizable which is a contradiction. Thus $K$ is a compact space such that $C_p(K)$ is not $d$-separable.

(b) By $\mathbb{D}$ we denote the set $\{0, 1\}$ with the discrete topology. Recall that a space $Z$ is called *Luzin* if it has no isolated points and any nowhere dense subspace of $Z$ is countable. If $A$ is a set then $[A]^{<\omega}$ is the family of all finite subsets of $A$. An open set $U \subset \mathbb{D}^{\omega_1}$ is called standard if there is a finite $A \subset \omega_1$ and a point $a \in \mathbb{D}^A$ such that $U = \{x \in \mathbb{D}^{\omega_1} : x|A = a\}$. It is clear that standard open subsets of $\mathbb{D}^{\omega_1}$ constitute a base in $\mathbb{D}^{\omega_1}$.

*Fact 1.* Given a countably infinite set $A$ suppose that $B \subset A$ is nonempty, finite and $H_n$ is a nowhere dense subspace of $\mathbb{D}^A$ for any $n \in \omega$. If $H = \bigcup_{n \in \omega} \overline{H}_n$ then, for any $s, t \in \mathbb{D}^B$ there exist $h_s, h_t \in \mathbb{D}^A \backslash H$ such that $h_s|B = s$, $h_t|B = t$ and $h_s|(A \backslash B) = h_t|(A \backslash B)$.

*Proof.* Let $\pi : \mathbb{D}^A \to \mathbb{D}^{A \backslash B}$ be the projection. The sets $U = \{x \in \mathbb{D}^A : x|B = s\}$ and $V = \{x \in \mathbb{D}^A : x|B = t\}$ are open in $\mathbb{D}^A$ and both maps $\pi_U = \pi|U : U \to \mathbb{D}^{A \backslash B}$ and $\pi_V = \pi|V : V \to \mathbb{D}^{A \backslash B}$ are homeomorphisms. The sets $P = U \backslash H$ and $Q = V \backslash H$ are dense Čech-complete subspaces of $U$ and $V$ respectively, so $P' = \pi_U(P)$ and $Q' = \pi_V(Q)$ are dense Čech-complete subspaces of $\mathbb{D}^{A \backslash B}$. Any pair of dense Čech-complete subspaces of any space have nonempty intersection by TFS-264; if $h \in P' \cap Q'$ then it is straightforward that $h_s = (\pi_U)^{-1}(h)$ and $h_t = (\pi_V)^{-1}(h)$ are as required, so Fact 1 is proved.

*Fact 2.* For any infinite cardinal $\kappa$ the space $\sigma = \{x \in \mathbb{D}^\kappa : |x^{-1}(1)| < \omega\}$ is $\sigma$-discrete.

*Proof.* If $\sigma_n = \{x \in \sigma : |x^{-1}(1)| = n\}$ for each $n \in \omega$ then $\sigma = \bigcup_{n \in \omega} \sigma_n$. Given any point $x \in \sigma_n$ let $A = x^{-1}(1)$. The set $V = \{y \in \mathbb{D}^\kappa : y|A = x|A\}$ is open in $\mathbb{D}^{\omega_1}$. If $y \in V \cap \sigma_n$ then $y(\alpha) = 1$ for any $\alpha \in A$ and hence $A \subset y^{-1}(1)$. However, $|A| = |y^{-1}(1)| = n$, so $y^{-1}(1) = A$ and hence $y = x$; this proves that $V \cap \sigma_n = \{x\}$ and therefore every $\sigma_n$ is a discrete subspace of $\sigma$. Consequently, $\sigma = \bigcup_{n \in \omega} \sigma_n$ is a $\sigma$-discrete space so Fact 2 is proved.

*Fact 3.* If CH holds then there exists a dense Luzin subspace $Z$ of the $\Sigma$-product $\Sigma = \{x \in \mathbb{D}^{\omega_1} : |x^{-1}(1)| \leq \omega\}$ of the Cantor cube $\mathbb{D}^{\omega_1}$ such that the space $Z \times Z$ is $d$-separable.

*Proof.* For any set $A \subset \omega_1$ let $\pi_A : \mathbb{D}^{\omega_1} \to \mathbb{D}^A$ be the projection map defined by $\pi_A(x) = x|A$ for any $x \in \mathbb{D}^{\omega_1}$. Say that a set $N \subset \mathbb{D}^{\omega_1}$ is *canonical* if there exists a nonempty countable set $A \subset \omega_1$ and a closed nowhere dense subset $M$ of $\mathbb{D}^A$ such that $N = \pi_A^{-1}(M)$. It is immediate that every canonical set is nowhere dense in $\mathbb{D}^{\omega_1}$.

Fix a countable dense set $P \subset \mathbb{D}^{\omega_1}$ and let $\{p_n : n \in \omega\}$ be a faithful enumeration of $P$. It is easy to find a family $\mathcal{A} = \{A_{mn} : m, n \in \omega\}$ of uncountable disjoint subsets of $\omega_1$ such that $\bigcup \mathcal{A} = \omega_1$. Choose a family $\{K_\alpha : \alpha < \omega_1\}$ of finite subsets of $\omega_1$ for which we have the equality $\{K_\alpha : \alpha \in A_{mn}\} = [\omega_1]^{<\omega}$ for any $m, n \in \omega$.

Suppose that $N$ is a nowhere dense subset of $\mathbb{D}^{\omega_1}$ and choose maximal disjoint family $\mathcal{U}$ of standard open subsets of $\mathbb{D}^{\omega_1}$ which are contained in $\mathbb{D}^{\omega_1} \setminus \overline{N}$. The family $\mathcal{U}$ is countable and $U = \bigcup \mathcal{U}$ is dense in $\mathbb{D}^{\omega_1}$. There exists a countable set $A \subset \omega_1$ such that $U = \pi_A^{-1}\pi_A(U)$. The set $V = \pi_A(U)$ is open and dense in $\mathbb{D}^A$, so $M = \mathbb{D}^A \setminus V$ is nowhere dense in $\mathbb{D}^A$; the set $N' = \pi_A^{-1}(M)$ is canonical and $N \subset N'$. Therefore

(1) canonical sets are cofinal in the family of all nowhere dense subsets of $\mathbb{D}^{\omega_1}$,

i.e., for any nowhere dense set $N \subset \mathbb{D}^{\omega_1}$ there is a canonical set $N'$ such that $N \subset N'$. It follows from CH that the family of all countable subsets of $\omega_1$ has cardinality $\omega_1$; besides, the family of all closed nowhere dense subsets of $\mathbb{D}^A$ does not exceed $\omega_1$ for any countable $A \subset \omega_1$. This shows that the family of all canonical subsets of $\mathbb{D}^{\omega_1}$ has cardinality $\omega_1$.

Therefore we can find an $\omega_1$-sequence $\{B_\alpha : \alpha < \omega_1\}$ of countable subsets of $\omega_1$ such that, for any $\alpha < \omega_1$ there is a closed nowhere dense $N_\alpha$ in the space $\mathbb{D}^{B_\alpha}$ such that $\{M_\alpha = \pi_{B_\alpha}^{-1}(N_\alpha) : \alpha < \omega_1\}$ is an enumeration of all canonical subsets of $\mathbb{D}^{\omega_1}$.

Our promised space $Z$ will have the form $\{x_\alpha, y_\alpha : \alpha < \omega_1\}$. To force the space $Z$ to be Luzin it suffices to make it dense in $\mathbb{D}^{\omega_1}$ and guarantee that, for any $\alpha < \omega_1$ the points $x_\beta, y_\beta$ are outside $M_\alpha$ for all $\beta > \alpha$.

To obtain $d$-separability of the space $Z \times Z$ we will choose the points $x_\alpha, y_\alpha$ in such a way that the set $D_{mn} = \{(x_\alpha, y_\alpha) : \alpha \in A_{mn}\}$ is $\sigma$-discrete for any $m, n \in \omega$ and $D = \bigcup_{m,n < \omega} D_{mn}$ is dense in $Z \times Z$. To get $\sigma$-discreteness of $D_{mn}$ we will show that the "sum modulo 2" map $(x_\alpha, y_\alpha) \to x_\alpha * y_\alpha = x_\alpha + y_\alpha - 2x_\alpha y_\alpha$ takes $D_{mn}$ injectively onto a subspace of $\sigma = \{x \in \mathbb{D}^{\omega_1} : |x^{-1}(1)| < \omega\}$. The density of $Z$ in $\Sigma$ will be proved establishing density of $Z \times Z$ in $\Sigma \times \Sigma$ which, in turn, will be obtained by assuring that the point $(p_m, p_n)$ belongs to the closure of the set $D_{mn}$ for all $m, n \in \omega$.

To start carrying out the outlined program take any distinct points $x_0, y_0 \in \sigma$ such that $\pi_{K_0}(x_0) = \pi_{K_0}(p_m)$ and $\pi_{K_0}(y_0) = \pi_{K_0}(p_n)$ where $(m, n) \in \omega \times \omega$ is the unique pair with $0 \in A_{mn}$; let $E_0 = \omega \cup K_0 \cup x_0^{-1}(1) \cup y_0^{-1}(1)$ and fix an ordinal $\mu_0 \in E_0$ such that $x_0(\mu_0) \neq y_0(\mu_0)$.

To make the inductive step assume that $\alpha < \omega_1$ and we have chosen a family $\{E_\beta : \beta < \alpha\}$ of countably infinite subsets of $\omega_1$ and a set $\{x_\beta, y_\beta : \beta < \alpha\} \subset \Sigma$ with the following properties:

(2) if $\gamma < \beta < \alpha$ then $E_\gamma \subset E_\beta$;
(3) if $\beta < \alpha$ and $\beta \in A_{mn}$ then $\pi_{K_\beta}(x_\beta) = \pi_{K_\beta}(p_m)$ and $\pi_{K_\beta}(y_\beta) = \pi_{K_\beta}(p_n)$;

(4) $\bigcup\{B_\gamma : \gamma < \beta\} \cup (\bigcup\{K_\gamma : \gamma \le \beta\}) \subset E_\beta$ for any $\beta < \alpha$;

(5) $x_\beta^{-1}(1) \cup y_\beta^{-1}(1) \subset E_\beta$ for any $\beta < \alpha$;

(6) if $\beta < \alpha$ then there is $\mu_\beta \in E_\beta \backslash (\bigcup\{E_\gamma : \gamma < \beta\})$ such that $x_\beta(\mu_\beta) \ne y_\beta(\mu_\beta)$;

(7) $\{\pi_{E_\beta}(x_\beta), \pi_{E_\beta}(y_\beta)\} \cap \pi_{E_\beta}(M_\gamma) = \emptyset$ whenever $\gamma < \beta < \alpha$;

(8) $x_\beta * y_\beta \in \sigma$ for any $\beta < \alpha$.

Consider the set $E = \bigcup\{E_\beta : \beta < \alpha\} \cup (\bigcup\{B_\beta : \beta < \alpha\}) \cup K_\alpha$ and let $(m, n) \in \omega \times \omega$ be the unique pair such that $\alpha \in A_{mn}$. Since $B_\beta \subset E$, the set $H_\beta = \pi_E(M_\beta)$ is nowhere dense in $\mathbb{D}^E$ for any $\beta < \alpha$, so we can apply Fact 1 to find points $s, t \in \mathbb{D}^E \backslash (\bigcup\{H_\beta : \beta < \alpha\})$ such that $s|K_\alpha = p_m|K_\alpha$, $t|K_\alpha = p_n|K_\alpha$ while $s|(E\backslash K_\alpha) = t|(E\backslash K_\alpha)$. Take an ordinal $\mu_\alpha \in \omega_1 \backslash E$ and let $E_\alpha = E \cup \{\mu_\alpha\}$; define $x_\alpha, y_\alpha \in \Sigma$ by requiring that $x_\alpha|E = s$, $y_\alpha|E = t$ while $x_\alpha(\mu_\alpha) = 0 = 1 - y_\alpha(\mu_\alpha)$ and $x_\alpha(\beta) = y_\alpha(\beta) = 0$ for any $\beta \in \omega_1 \backslash E_\alpha$. It is straightforward that the properties (2)–(7) are fulfilled now for all $\beta \le \alpha$. The condition (8) is also satisfied because $\pi_{E\backslash K_\alpha}(x_\alpha) = \pi_{E\backslash K_\alpha}(y_\alpha)$ and hence the points $x_\alpha$ and $y_\alpha$ are distinct only on finitely many coordinates. Thus our inductive procedure can be continued to obtain a set $Z = \{x_\alpha, y_\alpha : \alpha < \omega_1\} \subset \Sigma$ such that the properties (2)–(8) hold for all $\alpha < \omega_1$.

Given $m, n \in \omega$, if $K \subset \omega$ is a finite set then there is $\alpha \in A_{mn}$ such that $K = K_\alpha$ and hence $\pi_K(x_\alpha) = \pi_K(p_m)$, $\pi_K(y_\alpha) = \pi_K(p_n)$ by the property (3). This proves that the point $(p_m, p_n)$ belongs to the closure of the set $D_{mn} = \{(x_\alpha, y_\alpha) : \alpha \in A_{mn}\}$ for any $m, n \in \omega$. As a consequence, the set $P \times P$ is contained in the closure of the set $D = \bigcup\{D_{mn} : m, n \in \omega\} \subset Z \times Z$ which shows that $Z \times Z$ is dense in $\mathbb{D}^{\omega_1} \times \mathbb{D}^{\omega_1}$ and hence in $\Sigma \times \Sigma$.

Now, if $N \subset Z$ is nowhere dense then $N$ is nowhere dense in $\mathbb{D}^{\omega_1}$ and hence there is $\alpha < \omega_1$ such that $N \subset M_\alpha$. An immediate consequence of (7) is that $x_\beta, y_\beta \notin M_\alpha$ for any $\beta > \alpha$ and hence the set $N \subset \{x_\beta, y_\beta : \beta \le \alpha\}$ is countable. This shows that $Z$ is a Luzin space.

We have already established that $D$ contains the set $P \times P$ in its closure, so it is dense in $Z \times Z$. The map $\varphi : D \to \mathbb{D}^{\omega_1}$ defined by $\varphi(x_\alpha, y_\alpha) = x_\alpha * y_\alpha$ is continuous and $\varphi(D) \subset \sigma$ by (8). The space $\sigma$ is $\sigma$-discrete by Fact 2, so $D' = \varphi(D)$ is $\sigma$-discrete as well. If $\alpha < \beta < \omega_1$ then it follows from (5) and (6) that $(x_\alpha * y_\alpha)(\mu_\beta) = 0$ and $(x_\beta * y_\beta)(\mu_\beta) = 1$ which shows that $\varphi((x_\alpha, y_\alpha)) \ne \varphi((x_\beta, y_\beta))$, i.e., the map $\varphi$ is a continuous bijection of $D$ onto a $\sigma$-discrete space $D'$. Thus $D$ is a $\sigma$-discrete dense subspace of $Z \times Z$ and hence $Z \times Z$ is $d$-separable, so Fact 3 is proved.

Returning to our solution apply Fact 3 to see that, under CH, there exists a dense Luzin subspace $X$ of the set $\Sigma = \{x \in \mathbb{D}^{\omega_1} : |x^{-1}(1)| \le \omega\}$ such that $X \times X$ is $d$-separable. If $Y$ is a countable subset of $\Sigma$ then the set $A = \bigcup\{x^{-1}(1) : x \in Y\}$ is countable; take any $\alpha \in \omega_1 \backslash A$ and observe that $V = \{x \in \Sigma : x(\alpha) = 1\}$ is a nonempty open subset of $\Sigma$ such that $V \cap Y = \emptyset$. In particular, $Y$ is not dense in $\Sigma$; this shows that $d(\Sigma) > \omega$, so the space $X$ is not separable being dense in $\Sigma$. Furthermore, $s(X) \le hl(X) \le \omega$ (see SFFS-043) and hence every $\sigma$-discrete subspace of $X$ is countable. Therefore $X$ is not $d$-separable while $X \times X$ is $d$-separable.

**V.430.** *Suppose that $\kappa$ is an infinite cardinal, $T \neq \emptyset$ is a set and $N_t$ is a space such that $nw(N_t) \leq \kappa$ for all $t \in T$. Assume that $D$ is a dense subspace of the product $N = \prod_{t \in T} N_t$ and $f : D \to K$ is a continuous map of $D$ onto a compact space $K$. Prove that if $\rho(K) \leq \kappa$ then $w(K) \leq \kappa$. Deduce from this fact that if a compact space $K$ is a continuous image of a dense subspace of a product of cosmic spaces then $w(K) = t(K) = \rho(K)$.*

**Solution.** Recall that a space is *cosmic* if it has a countable network. If $nw(X) \leq \lambda$ for some infinite cardinal $\lambda$ then $X$ is called $\lambda$-cosmic. If $X$ is a space then a family $\mathcal{E} \subset \exp(X)$ *separates the points of a set* $F \subset X$ *from the points of* $F' \subset X$ if for any $x \in F$ and $y \in F'$ there exists $E \in \mathcal{E}$ such that $x \in E$ and $y \notin E$.

Given a product $X = \prod_{a \in A} X_a$ and $B \subset A$ let $p_B : X \to X_B = \prod_{a \in B} X_a$ be the natural projection. Suppose that $\lambda$ is an infinite cardinal and $nw(X_a) \leq \lambda$ for all $a \in A$. Say that $F \subset X$ is an *adequate $G_\lambda$-subset* of $X$ if there exists a set $B \subset A$ (called the support of $F$ and denoted by $\text{supp}(F)$) such that $|B| \leq \lambda$ and $F = p_B^{-1}(z)$ for some $z \in X_B$. It is evident that any adequate $G_\lambda$-subset of $X$ is a $G_\lambda$-subset of $X$. An open subset $U$ of the space $X$ is called *standard* if there is a finite $B \subset A$ (which is also called the support of $U$ and denoted by $\text{supp}(U)$) and $U_a \in \tau(X_a)$ for each $a \in B$ such that $U = \prod_{a \in B} U_a \times \prod_{a \in A \setminus B} X_a$. It is clear that standard open subsets of $X$ constitute a base of $X$.

If $X$ is an arbitrary space and $D \subset X$ is a dense subspace of $X$ then we will need an extension operator $e_D^X : \tau(D) \to \tau(X)$ which is defined by the formula $e_D^X(U) = \bigcup \{V \in \tau(X) : V \cap D \subset U\}$ for any $U \in \tau(D)$. It is easy to see that, for any space $X$ and any dense $D \subset X$, we have the following properties:

(1) $e_D^X(U) \cap D = U$ for any $U \in \tau(D)$;
(2) if $U, V \in \tau(D)$ and $U \subset V$ then $e_D^X(U) \subset e_D^X(V)$;
(3) if $U, V \in \tau(D)$ and $U \cap V = \emptyset$ then $e_D^X(U) \cap e_D^X(V) = \emptyset$.

Assume that $X$ is a space, $D$ is a dense subspace of $X$ and let $e = e_D^X$. Given a map $f : D \to Y$, the set $Q(f, y) = \bigcap \{\text{cl}_X(e(f^{-1}(U))) : U \in \tau(y, Y)\}$ is a useful extension of the set $f^{-1}(y)$ (it is an easy exercise that $Q(f, y) \cap D = f^{-1}(y)$ for any $y \in Y$).

*Fact 1.* Suppose that $nw(X_a) \leq \lambda$ for any $a \in A$. Then any $G_\lambda$-subset of the space $X = \prod_{a \in A} X_a$ is a union of adequate $G_\lambda$-subsets of $X$.

*Proof.* Take any $G_\lambda$-set $Q \subset X$ and fix a family $\mathcal{U} \subset \tau(X)$ such that $|\mathcal{U}| \leq \lambda$ and $Q = \bigcap \mathcal{U}$. Given $x \in Q$, for every $U \in \mathcal{U}$ we can choose a finite $B_U \subset A$ and a set $W_a \in \tau(X_a)$ for every $a \in B_U$ such that $x \in H_x^U = \prod_{a \in B_U} W_a \times \prod_{a \in A \setminus B_U} X_a \subset U$. The set $B = \bigcup \{B_U : U \in \mathcal{U}\}$ has cardinality not exceeding $\lambda$; let $y = \pi_B(x)$. If $z \in p_B^{-1}(y)$ then $p_{B_U}(z) = y|B_U = p_{B_U}(x) \in \prod_{a \in B_U} W_a$ which shows that $z \in H_x^U$ for any $U \in \mathcal{U}$. Therefore $z \in \bigcap \mathcal{U}$ for any $z \in H_x = p_B^{-1}(y)$, i.e., $H_x$ is an adequate $G_\lambda$-set such that $x \in H_x \subset Q$ for each $x \in Q$. Thus $Q = \bigcup \{H_x : x \in Q\}$, so Fact 1 is proved.

*Fact 2.* Assume that $nw(X_a) \leq \lambda$ for any $a \in A$. If $\mathcal{E}$ is an arbitrary family of $G_\lambda$-subsets of $X = \prod_{a \in A} X_a$ then there exists $\mathcal{E}' \subset \mathcal{E}$ such that $|\mathcal{E}'| \leq \lambda$ and $\bigcup \mathcal{E}' = \bigcup \mathcal{E}$.

*Proof.* If $|A| \leq \lambda$ or $|\mathcal{E}| \leq \lambda$ then there is nothing to prove, so we assume that $|A| > \lambda$ and $|\mathcal{E}| > \lambda$. Let $Y = \bigcup \mathcal{E}$; for every set $E \in \mathcal{E}$ we can choose a family $\mathcal{S}_E$ of adequate $G_\lambda$-subsets of the space $X$ such that $\bigcup \mathcal{S}_E = E$ (see Fact 1). Then, for the family $\mathcal{S} = \bigcup \{\mathcal{S}_E : E \in \mathcal{E}\}$ we have $\bigcup \mathcal{S} = Y$.

Take an arbitrary $a_0 \in A$ and let $A_0 = \{a_0\}$, $\mathcal{G}_0 = \emptyset$. Proceeding inductively, assume that $n \in \omega$ and we have families $\{A_i : i \leq n\}$ and $\{\mathcal{G}_i : i \leq n\}$ with the following properties:

(4) $A_i \subset A_{i+1} \subset A$ and $\mathcal{G}_i \subset \mathcal{G}_{i+1} \subset \mathcal{S}$ for any $i < n$;
(5) $|A_i| \leq \lambda$ and $|\mathcal{G}_i| \leq \lambda$ for all $i \leq n$;
(6) $\bigcup \{\text{supp}(F) : F \in \mathcal{G}_i\} \subset A_{i+1}$ whenever $i < n$;
(7) $p_{A_i}(\bigcup \mathcal{G}_{i+1})$ is dense in $p_{A_i}(Y)$ for all $i < n$.

Observe that $nw(X_{A_n}) \leq \lambda$, so $nw(p_{A_n}(Y)) \leq \lambda$ and hence we can pick a dense set $D$ in the space $p_{A_n}(Y)$ with $|D| \leq \lambda$. It is easy to find a family $\mathcal{G}'_{n+1} \subset \mathcal{S}$ with $|\mathcal{G}'_{n+1}| \leq \lambda$ such that $p_{A_n}(\bigcup \mathcal{G}'_{n+1}) \supset D$ and hence $p_{A_n}(\bigcup \mathcal{G}'_{n+1})$ is dense in $p_{A_n}(Y)$. Letting $\mathcal{G}_{n+1} = \mathcal{G}'_{n+1} \cup \mathcal{G}_n$ and $A_{n+1} = A_n \cup (\bigcup \{\text{supp}(F) : F \in \mathcal{G}_{n+1}\})$ we obtain the families $\{A_i : i \leq n+1\}$ and $\{\mathcal{G}_i : i \leq n+1\}$ such that the properties (4)–(7) hold if we replace $n$ with $n + 1$. Therefore our inductive procedure can be continued to construct families $\{A_i : i \in \omega\}$ and $\{\mathcal{G}_i : i \in \omega\}$ for which the conditions (4)–(7) are satisfied for all $n \in \omega$.

Let $\mathcal{G} = \bigcup \{\mathcal{G}_n : n \in \omega\}$ and $A' = \bigcup_{n \in \omega} A_n$. To see that $\bigcup \mathcal{G}$ is dense in $Y$ take any point $y \in Y$ and $U \in \tau(y, X)$. There exists a finite set $B \subset A$ and $W_a \in \tau(y(a), X_a)$ for all $a \in B$ such that the set $V = \prod_{a \in B} W_a \times \prod_{a \in A \setminus B} X_a$ is contained in $U$.

Pick $n \in \omega$ for which $B' = B \cap A_n = B \cap A'$ and observe that $V' = p_{A_n}(V)$ is an open neighborhood of $p_{A_n}(y)$ in $X_{A_n}$. The property (7) shows that there exists $F \in \mathcal{G}_{n+1}$ such that $p_{A_{n+1}}(F) \cap V' \neq \emptyset$. Therefore there is a point $x \in F$ such that $x(a) \in W_a$ for all $a \in B'$.

Let $z(a) = x(a)$ for all $a \in A \setminus (B \setminus B')$ and $z(a) = y(a)$ whenever $a \in B \setminus B'$. The points $z$ and $x$ are distinct only on the set $B \setminus B' \subset A \setminus A'$. It follows from (6) that $\text{supp}(F) \subset A'$ and hence $p_{A'}^{-1}(p_{A'}(F)) = F$; since $p_{A'}(z) = p_{A'}(x) \in p_{A'}(F)$, we conclude that $z \in F \cap V$ and hence $\emptyset \neq F \cap V \subset F \cap U$. This shows that $U \cap (\bigcup \mathcal{G}) \neq \emptyset$ for every $U \in \tau(y, X)$ and hence $y \in \overline{\bigcup \mathcal{G}}$ for any $y \in Y$, i.e., $\bigcup \mathcal{G}$ is dense in $Y$.

Finally observe that, for any $F \in \mathcal{G}$ there exists $E_F \in \mathcal{E}$ such that $F \subset E_F$. Then the family $\mathcal{E}' = \{E_F : F \in \mathcal{G}\} \subset \mathcal{E}$ has cardinality at most $\lambda$ and the inclusions $\bigcup \mathcal{E}' \supset \bigcup \mathcal{G} \supset Y$ show that $\bigcup \mathcal{E}'$ is dense in $Y$, so Fact 2 is proved.

*Fact 3.* Assume that $nw(X_a) \leq \lambda$ for any $a \in A$. If $\mathcal{E}$ is an arbitrary family of $G_\lambda$-subsets of $X = \prod_{a \in A} X_a$ and $E = \bigcup \mathcal{E}$ then there exists a set $B \subset A$ with $|B| \leq \lambda$ and a closed subset $F$ of the space $X_B$ such that $E = p_B^{-1}(F)$.

*Proof.* It follows easily from Fact 1 and Fact 2 that there exists a family $\mathcal{S}$ of adequate $G_\lambda$-subsets of $X$ such that $|\mathcal{S}| \leq \lambda$ and $\bigcup \mathcal{S} \subset \bigcup \mathcal{E} \subset \overline{\bigcup \mathcal{S}}$ which implies that $\overline{\bigcup \mathcal{S}} = E$. The set $B = \bigcup\{\operatorname{supp}(P) : P \in \mathcal{S}\}$ has cardinality not exceeding $\lambda$ and it is easy to see that $\bigcup \mathcal{S} = p_B^{-1} p_B(\bigcup \mathcal{S})$.

Let $H = p_B(\bigcup \mathcal{S})$ and $F = \overline{H}$. The map $p_B : X \to X_B$ is open and we have the equality $\bigcup \mathcal{S} = p_B^{-1}(H)$, so we can apply Fact 1 of S.424 to see that $E = \overline{\bigcup \mathcal{S}} = p_B^{-1}(\overline{H}) = p_B^{-1}(F)$ and hence Fact 3 is proved.

*Fact 4.* Given a product $X = \prod_{a \in A} X_a$ assume that $c(X) \leq \lambda$. Then, for any $U \in \tau(X)$ there exists a set $B \subset A$ such that $|B| \leq \lambda$, the set $p_B(\overline{U})$ is closed in $X_B$ and $\overline{U} = p_B^{-1} p_B(\overline{U})$.

*Proof.* Take a maximal disjoint family $\mathcal{V}$ of standard open subsets of $X$ contained in $U$. It is immediate that $V = \bigcup \mathcal{V}$ is dense in $U$, so $\overline{V} = \overline{U}$. It follows from $c(X) \leq \lambda$ that $|\mathcal{V}| \leq \lambda$ and therefore the set $B = \bigcup\{\operatorname{supp}(W) : W \in \mathcal{V}\}$ has cardinality $\leq \lambda$. The set $V' = p_B(V)$ is dense in $U' = p_B(U)$, so $\overline{V'} = \overline{U'}$; it is easy to see that $V = p_B^{-1}(V')$. Since the map $p_B : X \to X_B$ is open, we can apply Fact 1 of S.424 to conclude that $\overline{U} = \overline{V} = p_B^{-1}(\overline{V'}) = p_B^{-1}(\overline{U'})$. As a consequence, $p_B(\overline{U}) = \overline{U'}$ is a closed subset of $X_B$ and $\overline{U} = p_B^{-1} p_B(\overline{U})$, so Fact 4 is proved.

*Fact 5.* Suppose that $X$ is a compact space and $\mathcal{F}$ is a family of closed subsets of $X$ with $|\mathcal{F}| \leq \lambda$. Then for any nonempty $G_\lambda$-subset $H$ of the space $X$ there exists a nonempty closed $G_\lambda$-set $H' \subset H$ such that, for any $F \in \mathcal{F}$, either $H' \cap F = \emptyset$ or $H' \subset F$.

*Proof.* Say that a closed set $P \subset H$ is *small with respect to a set* $F \in \mathcal{F}$ if either $P \subset F$ or $P \cap F = \emptyset$. Therefore we must find a nonempty $G_\lambda$-set $H' \subset H$ which is small with respect to every $F \in \mathcal{F}$. Let $\{F_\alpha : \alpha < \lambda\}$ be an enumeration of $\mathcal{F}$. Fix a point $x \in H$ and take a closed $G_\lambda$-set $H'_0$ such that $x \in H'_0 \subset H$ (see Fact 2 of S.328). If $H'_0 \subset F_0$ then let $H_0 = H'_0$; if $H'_0 \backslash F_0 \neq \emptyset$ then $H'_0 \backslash F_0$ is a nonempty $G_\lambda$-subset of $X$, so we can apply Fact 2 of S.328 again to find a closed nonempty $G_\lambda$-set $H_0 \subset H'_0 \backslash F_0$; it is evident that in both cases $H_0$ is small with respect to $F_0$.

Proceeding inductively assume that $\alpha < \lambda$ and we have a decreasing family $\{H_\beta : \beta < \alpha\}$ of closed nonempty $G_\lambda$-subsets of $X$ such that $H_\beta$ is small with respect to $F_\beta$ for any $\beta < \alpha$. The closed $G_\lambda$-set $H'_\alpha = \bigcap_{\beta < \alpha} H_\beta$ is still nonempty; if $H'_\alpha \subset F_\alpha$ then let $H_\alpha = H'_\alpha$. If $Q = H'_\alpha \backslash F_\alpha \neq \emptyset$ then $Q$ is a nonempty $G_\lambda$-subset of $X$, so we can apply Fact 2 of S.328 to see that there exists a nonempty closed $G_\lambda$-set $H_\alpha \subset Q$.

It is evident that, in both cases, $H_\alpha$ is small with respect to $F_\alpha$ and $H_\alpha \subset H_\beta$ for any $\beta \leq \alpha$, so our inductive procedure can be continued to construct a decreasing family $\{H_\beta : \beta < \lambda\}$ of closed nonempty $G_\lambda$-subsets of $X$ such that $H_\beta$ is small with respect to $F_\beta$ for every $\beta < \lambda$. It is immediate that $H' = \bigcap\{H_\beta : \beta < \lambda\} \subset H$ is a nonempty closed $G_\lambda$-subset of $X$ which is small with respect to every $F \in \mathcal{F}$, so Fact 5 is proved.

*Fact 6.* Suppose that $X$ is a space, $D$ is a dense subspace of $X$ and denote by $e$ the extension operator $e_D^X$. Assume that $f : D \to Y$ is a continuous map and

$\varphi : X \to Z$ is an open continuous map. For any $U \in \tau(Y)$ let $U^* = e(f^{-1}(U))$. Consider the set $N = \{z \in Z : \varphi^{-1}(z)$ is contained in $Q(f, y_z)$ for some $y_z \in Y\}$ and let $M = \mathrm{cl}_X(\varphi^{-1}(N)) \cap D$. Then

(a) $\varphi(U^*) \cap \varphi(V^*) \cap \varphi(M) = \emptyset$ for any $U, V \in \tau(Y)$ such that $\overline{U} \cap \overline{V} = \emptyset$;

(b) there exists a continuous map $g : \varphi(M) \to Y$ such that $g \circ (\varphi|M) = f|M$.

*Proof.* (a) Assume that $\varphi(U^*) \cap \varphi(V^*) \cap \varphi(M) \neq \emptyset$. The set $W = \varphi(U^*) \cap \varphi(V^*)$ is open in $Z$. By continuity of $\varphi$ the set $\varphi(M)$ is contained in $\mathrm{cl}_Z(N)$, so it follows from $W \cap \varphi(M) \neq \emptyset$ that $W \cap N \neq \emptyset$; fix a point $z \in W \cap N$. We have the inclusion $\varphi^{-1}(z) \subset Q(f, y_z)$ and it follows from $z \in \varphi(U^*)$ that $\varphi^{-1}(z) \cap U^* \neq \emptyset$, so $U^* \cap Q(f, y_z) \neq \emptyset$. Analogously, it follows from $z \in \varphi(V^*)$ that $V^* \cap Q(f, y_z) \neq \emptyset$.

To obtain the desired contradiction we will prove that $y_z \in \overline{U} \cap \overline{V}$. Indeed, suppose that $y_z \notin \overline{U} \cap \overline{V}$. Since $U$ and $V$ are in a symmetric situation, we can assume, without loss of generality, that $y_z \notin \overline{U}$. If $G = Y \backslash \overline{U}$ then it follows from (3) and the definition of $U^*$ that $G^* \cap U^* = \emptyset$. Therefore $\mathrm{cl}_X(G^*) \cap U^* = \emptyset$; since $y_z \in G \in \tau(y_z, Y)$, the set $Q(f, y_z)$ is contained in $\mathrm{cl}_X(G^*)$, so $Q(f, y_z) \cap U^* = \emptyset$; this contradiction proves that $y_z \in \overline{U} \cap \overline{V}$ which, in turn, contradicts $\overline{V} \cap \overline{U} = \emptyset$, so (a) is proved.

(b) Suppose first that there exist points $x, y \in M$ such that $\varphi(x) = \varphi(y)$ but $f(x) \neq f(y)$. Take sets $U \in \tau(f(x), Y)$ and $V \in \tau(f(y), Y)$ such that $\overline{U} \cap \overline{V} = \emptyset$ and observe that $x \in U^*$ and $y \in V^*$ which implies $\varphi(x) \in \varphi(U^*) \cap \varphi(V^*) \cap \varphi(M)$; this contradiction with (a) shows that there exists a map $g : \varphi(M) \to Y$ such that $g \circ (\varphi|M) = f|M$.

To see that $g$ is a continuous map fix any point $z \in \varphi(M)$ and $U \in \tau(g(z), Y)$; let $y = g(z)$ and choose a set $V \in \tau(y, Y)$ such that $\overline{V} \subset U$. Take a point $x \in M$ such that $\varphi(x) = z$. Then $f(x) = g(\varphi(x)) = y$. The set $V^*$ is an open neighborhood of $x$ in $X$, so $W' = \varphi(V^*) \in \tau(z, Z)$. Let us show that $W = W' \cap \varphi(M)$ witnesses continuity of $g$ at the point $z$.

Assume that $w \in W$ and $g(w) \notin \overline{V}$; fix a point $u \in M$ such that $\varphi(u) = w$. Choose a set $H \in \tau(g(w), Y)$ such that $\overline{H} \cap \overline{V} = \emptyset$. We have $f(u) = g(w) \in H$ and therefore $u \in H^*$. As a consequence, $\varphi(u) = w \in \varphi(H^*) \cap \varphi(V^*) \cap \varphi(M)$ which is again a contradiction with (a). Therefore $g(w) \in \overline{V}$ for any $w \in W$ and hence $g(W) \subset \overline{V} \subset U$, i.e., the map $g$ is continuous at every point of $\varphi(M)$. This settles (b) and shows that Fact 6 is proved.

**Fact 7.** Given a space $X$ and an infinite cardinal $\lambda$ suppose that $F$ is a closed subset of $X$ such that $\chi(F, X) \leq \lambda$. If $G$ is a compact subset of $X$ such that $G \subset F$ and $\chi(G, F) \leq \lambda$ then $\chi(G, X) \leq \lambda$.

*Proof.* Fix an outer base $\mathcal{B}$ of the set $F$ in $X$ such that $|\mathcal{B}| \leq \lambda$ and let $\mathcal{C}$ be an outer base of $G$ in $F$ such that $|\mathcal{C}| \leq \lambda$. For any $B \in \mathcal{B}$ and $C \in \mathcal{C}$ the set $B \backslash (F \backslash C)$ is an open neighborhood of $G$ in $X$. Choose a set $O(B, C) \in \tau(G, X)$ with $\overline{O(B, C)} \subset B \backslash (F \backslash C)$ (see Fact 4 of T.309). It is straightforward that the family $\mathcal{O}' = \{O(B,C) : B \in \mathcal{B}$ and $C \in \mathcal{C}\}$ has cardinality not exceeding $\lambda$. Therefore the family $\mathcal{O}$ of all finite intersections of the elements of $\mathcal{O}'$ also has cardinality at most $\lambda$.

It is evident that $\bigcap \mathcal{O} \supset G$; let us prove that $\mathcal{O}$ is an outer base of $G$ in $X$. Given any set $U \in \tau(G, X)$ there exists a set $C \in \mathcal{C}$ such that $C \subset U \cap F$; take any set $B \in \mathcal{B}$ and observe that $\overline{O(B,C)} \cap F \subset (B \backslash (F \backslash C)) \cap F = C \subset U$. Therefore the closed set $P = \overline{O(B,C)} \backslash U$ is disjoint from $F$, so we can find a set $B' \in \mathcal{B}$ such that $B' \cap P = \emptyset$. It is immediate that $O(B,C) \cap O(B',C) \subset U$, so $\mathcal{O}$ is an outer base of $G$ in $X$ and hence $\chi(G, X) \leq |\mathcal{O}| \leq \lambda$, so Fact 7 is proved.

*Fact 8.* Suppose that $X$ is a compact space and $\lambda$ is an infinite cardinal. Assume that $F$ is a $G_\lambda$-subset of $X$ and $\pi \chi(x, F) \leq \lambda$ for some $x \in F$. Then there exists a $\pi$-base $\mathcal{B}$ of $X$ at the point $x$ such that $|\mathcal{B}| \leq \lambda$ and $B \cap F \neq \emptyset$ for any $B \in \mathcal{B}$. In particular, $\pi \chi(x, X) \leq \lambda$.

*Proof.* Choose a $\pi$-base $\mathcal{C}$ of the space $F$ at the point $x$ with $|\mathcal{C}| \leq \lambda$. By Fact 2 of S.328, for every $C \in \mathcal{C}$ we can find a nonempty compact set $P_C \subset C$ such that $P_C$ is a $G_\delta$-subset of $F$. Then $\chi(P_C, X) = \psi(P_C, X) \leq \lambda$ (see Fact 2 of S.358), so we can fix an outer base $\mathcal{B}_C$ of the set $C$ in the space $X$ with $|\mathcal{B}_C| \leq \lambda$. If $\mathcal{B} = \bigcup \{\mathcal{B}_C : C \in \mathcal{C}\}$ then $|\mathcal{B}| \leq \lambda$ and $B \cap F \neq \emptyset$ for any $B \in \mathcal{B}$.

Given any set $U \in \tau(x, X)$ there exists $C \in \mathcal{C}$ with $C \subset U \cap F$ and hence $P_C \subset U$. The family $\mathcal{B}_C$ being an outer base of $C$ in $X$ there is $B \in \mathcal{B}_C$ such that $B \subset U$. This shows that $\mathcal{B}$ is a $\pi$-base of $X$ at the point $x$ and hence Fact 8 is proved.

*Fact 9.* Suppose that $nw(X_a) \leq \lambda$ for every $a \in A$ and $D$ is a dense subspace of the product $X = \prod_{a \in A} X_a$. Assume that $Y$ is a compact space with $\rho(Y) \leq \lambda$ and $f : D \to Y$ is a continuous onto map. Denote by $e$ the extension operator $e_D^X$ and let $\mathcal{P}$ be the family of all $G_\lambda$-subsets $P$ of the space $X$ such that $P \subset Q(f, y)$ for some $y \in Y$. If $M = \mathrm{cl}_X(\bigcup \mathcal{P}) \cap D$ then $f(M)$ intersects every nonempty $G_\lambda$-subset of the space $Y$.

*Proof.* Let $K_a = \beta X_a$ for every $a \in A$ and consider the space $K = \prod_{a \in A} K_a$. For any $B \subset A$ let $p_B : X \to X_B = \prod_{a \in B} X_a$ be the natural projection; we will also need the natural projection $q_B : K \to K_B = \prod_{a \in B} K_a$. It follows from $d(K_a) \leq d(X_a) \leq nw(X_a) \leq \lambda$ that $\lambda^+$ is a caliber of $K_a$ for any $a \in A$ and hence $\lambda^+$ is a caliber of $K$ (see SFFS-281); in particular, $c(K) \leq \lambda$.

For every set $U \in \tau(Y)$ let $U^* = e(f^{-1}(U))$ and $[U] = \mathrm{cl}_K(U^*)$; it is easy to see that $[U] = \mathrm{cl}_K(e_D^K(f^{-1}(U)))$, so we can apply Fact 4 to see that there exists a set $E(U) \subset A$ such that $|E(U)| \leq \lambda$ and $[U] = q_{E(U)}^{-1} q_{E(U)}([U])$.

Assume that there exists a nonempty $G_\lambda$-set $\Phi_0'$ of the space $Y$ which does not meet $f(M)$. By Fact 2 of S.328 we can pass to a smaller set if necessary to be able to assume, without loss of generality, that $\Phi_0'$ is compact.

Apply Fact 3 to the set $N = \mathrm{cl}_X(\bigcup \mathcal{P})$ to find $B(0) \subset A$ such that $|B(0)| \leq \lambda$ and $N = p_{B(0)}^{-1} p_{B(0)}(N)$. If $O = X \backslash N$ then $O = p_{B(0)}^{-1} p_{B(0)}(O)$

and $f^{-1}(\Phi_0') \subset O$. We have the inclusion $p_{B(0)}(O) \subset X_{B(0)}$ which, together with $nw(X_{B(0)}) \leq \lambda$ implies that we can find a network $\mathcal{N}_0$ in the space $p_{B(0)}(O)$ such that $|\mathcal{N}_0| \leq \lambda$.

Let $\mathcal{N}_0' = \{\mathrm{cl}_{K_{B(0)}}(N) : N \in \mathcal{N}_0\}$ and denote by $\mathcal{N}_0''$ the family of all finite intersections of the elements of $\mathcal{N}_0'$. Observe that $|\mathcal{N}_0''| \leq \lambda$ and the family $\mathcal{N}_0''$ separates the points of $p_{B(0)}(O)$ from the points of $K_{B(0)} \backslash p_{B(0)}(O)$. Consider the family $\mathcal{F}_0 = \{q_{B(0)}^{-1}(F) : F \in \mathcal{N}_0''\}$ and let $\mathcal{G}_0 = \{\mathrm{cl}_Y(f(F \cap D)) : F \in \mathcal{F}_0\}$; clearly, $\mathcal{G}_0$ consists of compact subsets of $Y$. It follows from $O \subset \bigcup \mathcal{F}_0$ that $\Phi_0' \subset \bigcup \mathcal{G}_0$. By Fact 5 there exists a nonempty closed $G_\lambda$-subset $\Phi_0$ of the space $Y$ such that $\Phi_0 \subset \Phi_0'$ and, for any $G \in \mathcal{G}_0$, either $\Phi_0 \subset G$ or $\Phi_0 \cap G = \emptyset$.

It follows from the inequality $\rho(\Phi_0) \leq \rho(Y) \leq \lambda$ that there exists $y_0 \in \Phi_0$ such that $\pi\chi(y_0, \Phi_0) \leq \lambda$ (see Fact 1 of U.086); apply Fact 8 to find a $\pi$-base $\mathcal{B}_0$ of $Y$ at the point $y_0$ such that $|\mathcal{B}_0| \leq \lambda$ and $U \cap \Phi_0 \neq \emptyset$ for all $U \in \mathcal{B}_0$.

Proceeding inductively assume that $n \in \omega$ and we have constructed, for every $i \leq n$, a nonempty compact $G_\lambda$-subset $\Phi_i$ of the space $Y$, a point $y_i \in \Phi_i$, a set $B(i) \subset A$ together with the families $\mathcal{F}_i$, $\mathcal{G}_i$, and $\mathcal{B}_i$ with the following properties:

(8) $\mathcal{F}_i$ is a family of compact subsets of $K$ and $\{p_{B(i)}(F) : F \in \mathcal{F}_i\}$ separates the points of $p_{B(i)}(O)$ from the points of $K_{B(i)} \backslash p_{B(i)}(O)$ for any $i \leq n$;
(9) $p_{B(i)}^{-1} p_{B(i)}(F) = F$ for any $F \in \mathcal{F}_i$ and $i \leq n$;
(10) $\mathcal{F}_i$ is closed under finite intersections for all $i \leq n$;
(11) $\mathcal{G}_i = \{\mathrm{cl}_Y(f(F \cap D)) : F \in \mathcal{F}_i\}$ for all $i \leq n$;
(12) $\max\{|B(i)|, |\mathcal{F}_i|, |\mathcal{G}_i|\} \leq \lambda$ for all $i \leq n$;
(13) if $i \leq n$ then $\Phi_i$ is a compact $G_\lambda$-subset of $Y$ such that, for any $G \in \mathcal{G}_i$ either $\Phi_i \cap G = \emptyset$ or $\Phi_i \subset G$;
(14) for each $i \leq n$ the family $\mathcal{B}_i$ is a $\pi$-base of $Y$ at the point $y_i$ such that $|\mathcal{B}_i| \leq \lambda$ and $U \cap \Phi_i \neq \emptyset$ for all $U \in \mathcal{B}_i$;
(15) $B(i) \subset B(i + 1)$, $\mathcal{F}_i \subset \mathcal{F}_{i+1}$ and $\Phi_{i+1} \subset \Phi_i$ for all $i < n$;
(16) for any $i < n$, if $U \in \mathcal{B}_i$ then $E(U) \subset B(i + 1)$.

The family $\mathcal{E} = \{E(U) : U \in \mathcal{B}_0 \cup \ldots \cup \mathcal{B}_n\}$ has cardinality not exceeding $\lambda$, so we can choose a set $B(n + 1) \subset A$ such that $|B(n + 1)| \leq \lambda$ and $B(n) \cup (\bigcup \mathcal{E}) \subset B(n + 1)$. It is evident that $O = p_{B(n+1)}^{-1} p_{B(n+1)}(O)$. We have $p_{B(n+1)}(O) \subset X_{B(n+1)}$ and $nw(X_{B(n+1)}) \leq \lambda$ which shows that we can find a network $\mathcal{N}_{n+1}$ in the space $p_{B(n+1)}(O)$ such that $|\mathcal{N}_{n+1}| \leq \lambda$.

Let $\mathcal{N}_{n+1}' = \{\mathrm{cl}_{K_{B(n+1)}}(N) : N \in \mathcal{N}_{n+1}\}$ and denote by $\mathcal{N}_{n+1}''$ the family of all finite intersections of the elements of $\mathcal{N}_{n+1}'$. Observe that $|\mathcal{N}_{n+1}''| \leq \lambda$ and the family $\mathcal{N}_{n+1}''$ separates the points of $p_{B(n+1)}(O)$ from the points of $K_{B(n+1)} \backslash p_{B(n+1)}(O)$. Let $\mathcal{F}_{n+1} = \{q_{B(n+1)}^{-1}(F) : F \in \mathcal{N}_{n+1}''\} \cup \mathcal{F}_n$ and $\mathcal{G}_{n+1} = \{\mathrm{cl}_Y(f(F \cap D)) : F \in \mathcal{F}_{n+1}\}$; clearly, $\mathcal{G}_{n+1}$ consists of compact subsets of $Y$. It follows from $O \subset \bigcup \mathcal{F}_{n+1}$ that $\Phi_n \subset \bigcup \mathcal{G}_{n+1}$. By Fact 5 there exists a nonempty closed $G_\lambda$-subset $\Phi_{n+1}$ of the space $Y$ such that $\Phi_{n+1} \subset \Phi_n$ and, for any set $G \in \mathcal{G}_{n+1}$, either $\Phi_{n+1} \subset G$ or $\Phi_{n+1} \cap G = \emptyset$.

It follows from the inequality $\rho(\Phi_{n+1}) \leq \rho(Y) \leq \lambda$ that there exists a point $y_{n+1} \in \Phi_{n+1}$ such that $\pi\chi(y_{n+1}, \Phi_{n+1}) \leq \lambda$ (see Fact 1 of U.086); apply Fact 8 to find a $\pi$-base $\mathcal{B}_{n+1}$ of $Y$ at the point $y_{n+1}$ such that $|\mathcal{B}_{n+1}| \leq \lambda$ and $U \cap \Phi_{n+1} \neq \emptyset$

for all $U \in \mathcal{B}_{n+1}$. It is straightforward that the conditions (8)–(16) are now satisfied if we replace $n$ with $n + 1$, so our inductive procedure can be continued to construct for every $i \in \omega$, a nonempty compact $G_\lambda$-subset $\Phi_i$ of the space $Y$, a point $y_i \in \Phi_i$, a set $B(i) \subset A$ together with the families $\mathcal{F}_i$, $\mathcal{G}_i$, and $\mathcal{B}_i$ such that the properties (8)–(16) hold for each $n \in \omega$.

Consider the sets $\Phi = \bigcap_{n \in \omega} \Phi_n$ and $B = \bigcup_{n \in \omega} B(n)$ together with the families $\mathcal{B} = \bigcup_{n \in \omega} \mathcal{B}_n$, $\mathcal{F} = \bigcup_{n \in \omega} \mathcal{F}_n$ and $\mathcal{G} = \bigcup_{n \in \omega} \mathcal{G}_n$. Choose an accumulation point $y$ for the sequence $\{y_n : n \in \omega\}$; it is easy to see that $y \in \Phi$ and $\mathcal{B}$ is a $\pi$-base of $Y$ at the point $y$.

Let $\tau_y$ be the family of all open neighborhoods of $y$ in $Y$, i.e., $\tau_y = \tau(y, Y)$. For every $U \in \tau_y$ we will need the set $H(U) = \bigcup\{V^* : V \in \mathcal{B} \text{ and } V \subset U\}$. Now let $P = \bigcap\{\mathrm{cl}_K(H(U)) : U \in \tau_y\}$. It follows from (16) that $q_B^{-1}q_B([U]) = [U]$ for any $U \in \mathcal{B}$; as an immediate consequence, we have $q_B^{-1}q_B(P) = P$. Since $P$ is compact, the set $q_B(P)$ is a compact subspace of $K_B$ and hence $R = q_B(P) \cap X_B$ is a closed subspace of $X_B$. It follows from $nw(X_B) \leq \lambda$ that $R$ is a $G_\lambda$-subset of $X_B$ and hence $P' = P \cap X = p_B^{-1}(R)$ is a $G_\lambda$-subset of $X$.

Given any set $U \in \tau_y$ it follows from the inclusion $H(U) \subset U^*$ that we have the formula $\mathrm{cl}_K(H(U)) \cap X = \mathrm{cl}_X(H(U)) \subset \mathrm{cl}_X(U^*)$. As an immediate consequence, $P' \subset \bigcap\{\mathrm{cl}_X(U^*) : U \in \tau_y\} = Q(f, y)$.

The property (9) implies that $q_B^{-1}q_B(F) = F$ for any $F \in \mathcal{F}$. Besides, the family $\mathcal{F}_B = \{q_B(F) : F \in \mathcal{F}\}$ consists of compact subsets of $K_B$ and it follows from (8), (9), and (15) that

(17) $\mathcal{F}_B$ is closed under finite intersections and separates the points of $p_B(O)$ from the points of $K_B \setminus p_B(O)$.

Take a point $x \in D$ with $f(x) = y$; let $\mathcal{F}_x = \{F \in \mathcal{F} : x \in F\}$ and consider the set $L = \bigcap \mathcal{F}_x$. Then $x \in L = q_B^{-1}q_B(L)$ and $q_B(L) \subset q_B(O) = p_B(O)$; this is an easy consequence of (17). Take an arbitrary $U \in \tau_y$ and any finite subfamily $\mathcal{F}'$ of the family $\mathcal{F}_x$. There exists $n \in \omega$ such that $\mathcal{F}' \subset \mathcal{F}_n$ and hence $F = \bigcap \mathcal{F}' \in \mathcal{F}_n$. Therefore the set $G = \mathrm{cl}_Y(f(F \cap D))$ belongs to $\mathcal{G}_n$; it follows from $y = f(x) \in G \cap \Phi_n$ that $G \cap \Phi_n \neq \emptyset$ and hence $\Phi_n \subset G$ by the property (13).

There exists $W \in \mathcal{B}_n$ such that $W \subset U$ and it follows from the property (14) that $W \cap \Phi_n \neq \emptyset$. Thus $f^{-1}(W) \cap F \neq \emptyset$ which shows that $F \cap W^* \neq \emptyset$. Consequently, $H(U) \cap (\bigcap \mathcal{F}') \neq \emptyset$ for any finite family $\mathcal{F}' \subset \mathcal{F}_x$ and $U \in \tau_y$. It is easy to see, using Fact 1 of S.326, that this implies $L' = P \cap L \neq \emptyset$. Taking in consideration that $q_B^{-1}q_B(L') = L'$ we conclude that $q_B(L') \subset q_B(P) \cap q_B(O)$, so $q_B(O) \cap q_B(P) \neq \emptyset$. Therefore $\emptyset \neq L' \subset P \cap O$ and hence $P_0 = P \cap O = P' \cap O$ is a nonempty $G_\lambda$-subset of $X$. Besides, $P_0 \subset P' \subset Q(f, y)$ and hence $P_0 \in \mathcal{P}$; this contradiction with $P_0 \subset O \subset X \setminus (\bigcup \mathcal{P})$ shows that $f(M)$ intersects every nonempty $G_\lambda$-subset of $Y$, i.e., Fact 9 is proved.

Returning to our solution assume that $\rho(K) \leq \kappa$ and denote by $\mathcal{P}$ the family of all nonempty $G_\kappa$-subsets $P$ of the product $N$ such that $P \subset Q(f, y)$ for some $y \in K$. It follows from Fact 1 and Fact 2 that we can find a family $\mathcal{A} \subset \mathcal{P}$ of adequate $G_\kappa$-subsets of $N$ such that $|\mathcal{A}| \leq \kappa$ and $\bigcup \mathcal{A}$ is dense in $\bigcup \mathcal{P}$. Let $S = \bigcup\{\mathrm{supp}(P) : P \in \mathcal{A}\}$; then we have the equality $p_S^{-1}p_S(P) = P$ for any $P \in \mathcal{A}$

and $F = cl_N(\bigcup \mathcal{P}) = cl_N(\bigcup \mathcal{A})$. Let $A = \bigcup \mathcal{A}$ and take any $z \in p_S(A)$. There is $P \in \mathcal{A}$ with $z \in p_S(P)$ and hence $p_S^{-1}(z) \subset P \subset Q(f, y)$ for some $y \in K$.

Let $D' = F \cap D$ and apply Fact 6 to the open map $\varphi = p_S$ and the map $f$ to see that there exists a continuous map $g : p_S(D') \to K$ such that $g \circ (p_S|D') = f|D'$. This implies that $f(D') = g(p_S(D'))$ and hence $nw(f(D')) \leq nw(p_S(D')) \leq \kappa$. Next apply Fact 9 to see that

(18) the set $E = f(D')$ intersects every nonempty $G_\kappa$-subset of $K$.

If $K \backslash E \neq \emptyset$ then pick a point $y \in K \backslash E$ and take a set $O_z \in \tau(z, K)$ such that $y \notin \overline{O}_z$ for any $z \in E$. Since $l(E) \leq nw(E) \leq \kappa$, there is a set $E' \subset E$ such that $|E'| \leq \kappa$ and $E \subset \bigcup \{O_z : z \in E'\}$.

Then $y \in R = \bigcap \{K \backslash \overline{O}_z : z \in E'\} \subset K \backslash E$, so $R$ is a nonempty $G_\kappa$-subset of $K$ contained in $K \backslash E$. This contradiction with (18) shows that $K = E$ and hence $w(K) = nw(K) = nw(E) \leq \kappa$ (see Fact 4 of S.307).

Finally assume that $K$ is a continuous image of a dense subspace of a product of cosmic spaces. It is trivial that $\kappa = t(K) \leq w(K)$. Since $t(\mathbb{I}^{\kappa^+}) > \kappa$, it follows from TFS-162 that the space $K$ cannot be mapped onto $\mathbb{I}^{\kappa^+}$ and hence $\rho(K) \leq \kappa$, i.e., $\rho(K) \leq t(K) \leq w(K)$. If $\rho(K) = \lambda$ then $w(K) \leq \lambda$ because $K$ is a continuous image of a dense subspace of a product of cosmic and hence $\lambda$-cosmic spaces. Thus $w(K) \leq \rho(K) \leq t(K) \leq w(K)$ which shows that $w(K) = t(K) = \rho(K)$ and hence our solution is complete.

**V.431.** *Suppose that $\kappa$ is a cardinal of uncountable cofinality, $T \neq \emptyset$ is a set and $N_t$ is a space such that $nw(N_t) \leq \omega$ for all $t \in T$. Assume that $D$ is a dense subspace of $N = \prod_{t \in T} N_t$ and $f : D \to K$ is a continuous map of $D$ onto a compact space $K$ with $w(K) = \kappa$. Prove that $K$ maps continuously onto $\mathbb{I}^\kappa$.*

**Solution.** If $X$ is a space then a family $\mathcal{E} \subset \exp(X)$ *separates the points of a set* $F \subset X$ *from the points of* $F' \subset X$ if for any $x \in F$ and $y \in F'$ there exists $E \in \mathcal{E}$ such that $x \in E$ and $y \notin E$.

Given a product $X = \prod_{a \in A} X_a$ and $B \subset A$ let $p_B : X \to X_B = \prod_{a \in B} X_a$ be the natural projection. Suppose that $\lambda$ is an infinite cardinal and $nw(X_a) \leq \omega$ for all $a \in A$. Say that $F \subset X$ is an *adequate $G_\lambda$-subset* of $X$ if there exists a set $B \subset A$ (called the support of $F$ and denoted by $\text{supp}(F)$) such that $|B| \leq \lambda$ and $F = p_B^{-1}(z)$ for some $z \in X_B$. It is evident that any adequate $G_\lambda$-subset of $X$ is a $G_\lambda$-subset of $X$. Say that $F \subset X$ is a *$G_{<\lambda}$-subset* of $X$ if $F$ is a $G_{\lambda'}$-subset of $X$ for some $\lambda' < \lambda$. An open subset $U$ of the space $X$ is called *standard* if there is a finite $B \subset A$ (which is also called the support of $U$ and denoted by $\text{supp}(U)$) and $U_a \in \tau(X_a)$ for each $a \in B$ such that $U = \prod_{a \in B} U_a \times \prod_{a \in A \backslash B} X_a$. It is clear that standard open subsets of $X$ constitute a base of $X$.

If $X$ is an arbitrary space and $D \subset X$ is a dense subspace of $X$ then we will need an extension operator $e_D^X : \tau(D) \to \tau(X)$ which is defined by the formula $e_D^X(U) = \bigcup \{V \in \tau(X) : V \cap D \subset U\}$ for any $U \in \tau(D)$. It is easy to see that, for any space $X$ and any dense $D \subset X$, we have the following properties:

(1) $e_D^X(U) \cap D = U$ for any $U \in \tau(D)$;
(2) if $U, V \in \tau(D)$ and $U \subset V$ then $e_D^X(U) \subset e_D^X(V)$;
(3) if $U, V \in \tau(D)$ and $U \cap V = \emptyset$ then $e_D^X(U) \cap e_D^X(V) = \emptyset$.

Assume that $X$ is a space, $D$ is a dense subspace of $X$ and let $e = e_D^X$. Given a map $f : D \to Y$, the set $Q(f, y) = \bigcap\{\mathrm{cl}_X(e(f^{-1}(U))) : U \in \tau(y, Y)\}$ is a useful extension of the set $f^{-1}(y)$ (it is an easy exercise that $Q(f, y) \cap D = f^{-1}(y)$ for any $y \in Y$).

*Fact 1.* Suppose that $\lambda$ is a cardinal of uncountable cofinality and $nw(X_a) \le \omega$ for any $a \in A$. Then, for any family $\mathcal{E}$ of adequate $G_{<\lambda}$-subsets of $X = \prod_{a \in A} X_a$ there exists a set $B \subset A$ and a family $\mathcal{G} \subset \mathcal{E}$ such that $|B| < \lambda$ and $|\mathcal{G}| < \lambda$ while $\bigcup\{\mathrm{supp}(P) : P \in \mathcal{G}\} \subset B$ and $\bigcup \mathcal{E} \subset \overline{\bigcup \mathcal{G}}$. In particular, $E = \bigcup \mathcal{E} = \overline{\bigcup \mathcal{G}}$ and $p_B^{-1} p_B(E) = E$.

*Proof.* Let $Y = \bigcup \mathcal{E}$; take an arbitrary $b_0 \in A$ and let $B_0 = \{b_0\}$, $\mathcal{G}_0 = \emptyset$. Proceeding inductively, assume that $n \in \omega$ and we have families $\{B_i : i \le n\}$ and $\{\mathcal{G}_i : i \le n\}$ with the following properties:

(4) $B_i \subset B_{i+1} \subset A$ and $\mathcal{G}_i \subset \mathcal{G}_{i+1} \subset \mathcal{E}$ for any $i < n$;
(5) $|B_i| < \lambda$ and $|\mathcal{G}_i| < \lambda$ for all $i \le n$;
(6) $\bigcup\{\mathrm{supp}(P) : P \in \mathcal{G}_i\} \subset B_{i+1}$ whenever $i < n$;
(7) $p_{B_i}(\bigcup \mathcal{G}_{i+1})$ is dense in $p_{B_i}(Y)$ for all $i < n$.

Observe that $nw(X_{B_n}) \le |B_n| \cdot \omega < \lambda$, so $nw(p_{B_n}(Y)) < \lambda$ and hence we can pick a dense set $D$ in the space $p_{B_n}(Y)$ such that $|D| < \lambda$. It is easy to find a family $\mathcal{G}'_{n+1} \subset \mathcal{E}$ with $|\mathcal{G}'_{n+1}| < \lambda$ such that $p_{B_n}(\bigcup \mathcal{G}'_{n+1}) \supset D$ and hence $p_{B_n}(\bigcup \mathcal{G}'_{n+1})$ is dense in the set $p_{B_n}(Y)$. Consider the family $\mathcal{G}_{n+1} = \mathcal{G}'_{n+1} \cup \mathcal{G}_n$ and the set $B_{n+1} = B_n \cup (\bigcup\{\mathrm{supp}(P) : P \in \mathcal{G}_{n+1}\})$; it is clear that for the families $\{B_i : i \le n+1\}$ and $\{\mathcal{G}_i : i \le n+1\}$ the properties (4)–(7) hold if we replace $n$ with $n+1$. Therefore our inductive procedure can be continued to construct families $\{B_i : i \in \omega\}$ and $\{\mathcal{G}_i : i \in \omega\}$ for which the conditions (4)–(7) are satisfied for all $n \in \omega$.

Let $\mathcal{G} = \bigcup\{\mathcal{G}_n : n \in \omega\}$ and $B = \bigcup_{n \in \omega} B_n$; it follows from (5) and $\mathrm{cf}(\lambda) > \omega$ that $|\mathcal{G}| < \lambda$ and $|B| < \lambda$. To see that $\bigcup \mathcal{G}$ is dense in $Y$ take any point $y \in Y$ and $U \in \tau(y, X)$. There exists a finite set $H \subset A$ and $W_a \in \tau(y(a), X_a)$ for all $a \in H$ such that the set $V = \prod_{a \in H} W_a \times \prod_{a \in A \backslash H} X_a$ is contained in $U$.

Pick $n \in \omega$ for which $H' = H \cap B_n = H \cap B$ and observe that $V' = p_{B_n}(V)$ is an open neighborhood of $p_{B_n}(y)$ in $X_{B_n}$. The property (7) shows that there exists $P \in \mathcal{G}_{n+1}$ such that $p_{B_{n+1}}(P) \cap V' \ne \emptyset$. Therefore there is a point $x \in P$ such that $x(a) \in W_a$ for all $a \in H'$.

Let $z(a) = x(a)$ for all $a \in A \backslash (H \backslash H')$ and $z(a) = y(a)$ whenever $a \in H \backslash H'$. The points $z$ and $x$ are distinct only on the set $H \backslash H' \subset A \backslash B$. It follows from (6) that $\mathrm{supp}(P) \subset B$ and hence $p_B^{-1}(p_B(P)) = P$; since $p_B(z) = p_B(x) \in p_B(P)$, we conclude that $z \in P \cap V$ and hence $\emptyset \ne P \cap V \subset P \cap U$. This shows that $U \cap (\bigcup \mathcal{G}) \ne \emptyset$ for every $U \in \tau(y, X)$ and hence $y \in \overline{\bigcup \mathcal{G}}$ for any $y \in Y$, i.e., $\bigcup \mathcal{G}$ is dense in $Y$. It follows from (6) that $\bigcup\{\mathrm{supp}(P) : P \in \mathcal{G}\} \subset B$ and hence $\overline{\bigcup \mathcal{G}} = p_B^{-1} p_B(\bigcup \mathcal{G})$.

Finally observe that it follows from $\bigcup \mathcal{G} \subset \bigcup \mathcal{E} \subset \overline{\bigcup \mathcal{G}}$ that $E = \overline{\bigcup \mathcal{E}} = \overline{\bigcup \mathcal{G}}$. The map $p_B : X \to X_B$ being open we have the equality $\overline{\bigcup \mathcal{G}} = p_B^{-1}(p_B(\bigcup \mathcal{G}))$ (see Fact 1 of S.424). As an immediate consequence, $E = p_B^{-1} p_B(E)$, so Fact 1 is proved.

*Fact 2.* Suppose that $nw(X_a) \leq \omega$ for every $a \in A$ and $D$ is a dense subspace of the product $X = \prod_{a \in A} X_a$. Assume that $\lambda$ is a cardinal of uncountable cofinality and $Y$ is a compact space which cannot be continuously mapped onto $\mathbb{I}^\lambda$ while there exists a continuous onto map $f : D \to Y$. Denote by $e$ the extension operator $e_D^X$ and let $\mathcal{P}$ be the family of all adequate $G_{<\lambda}$-subsets $P$ of the space $X$ such that $P \subset Q(f, y)$ for some $y \in Y$. If $M = \mathrm{cl}_X(\bigcup \mathcal{P}) \cap D$ then $f(M)$ intersects every nonempty $G_{<\lambda}$-subset of the space $Y$.

*Proof.* Let $K_a = \beta X_a$ for every $a \in A$ and consider the space $K = \prod_{a \in A} K_a$. For any $B \subset A$ let $p_B : X \to X_B = \prod_{a \in B} X_a$ be the natural projection; we will also need the natural projection $q_B : K \to K_B = \prod_{a \in B} K_a$. It follows from $d(K_a) \leq d(X_a) \leq nw(X_a) \leq \omega$ that $\omega_1$ is a caliber of $K_a$ for any $a \in A$ and hence $\omega_1$ is a caliber of $K$ (see SFFS-281); in particular, $c(K) \leq \omega$.

For every set $U \in \tau(Y)$ let $U^* = e(f^{-1}(U))$ and $[U] = \mathrm{cl}_K(U^*)$; it is easy to see that $[U] = \mathrm{cl}_K(e_D^K(f^{-1}(U)))$, so we can apply Fact 4 of V.430 to see that there exists a set $E(U) \subset A$ such that $|E(U)| \leq \omega < \lambda$ and $[U] = q_{E(U)}^{-1} q_{E(U)}([U])$.

Assume that there exists a nonempty $G_{<\lambda}$-set $\Phi_0'$ of the space $Y$ which does not meet $f(M)$. By Fact 2 of S.328 we can pass to a smaller set if necessary to be able to assume, without loss of generality, that $\Phi_0'$ is compact.

Apply Fact 1 to the set $N = \mathrm{cl}_X(\bigcup \mathcal{P})$ to find $B(0) \subset A$ such that $|B(0)| < \lambda$ and $N = p_{B(0)}^{-1} p_{B(0)}(N)$. If $O = X \setminus N$ then $O = p_{B(0)}^{-1} p_{B(0)}(O)$ and $f^{-1}(\Phi_0') \subset O$. We have the inclusion $p_{B(0)}(O) \subset X_{B(0)}$ which, together with $nw(X_{B(0)}) < \lambda$ implies that we can find a network $\mathcal{N}_0$ in the space $p_{B(0)}(O)$ such that $|\mathcal{N}_0| < \lambda$.

Let $\mathcal{N}_0' = \{\mathrm{cl}_{K_{B(0)}}(N) : N \in \mathcal{N}_0\}$ and denote by $\mathcal{N}_0''$ the family of all finite intersections of the elements of $\mathcal{N}_0'$. Observe that $|\mathcal{N}_0''| < \lambda$ and the family $\mathcal{N}_0''$ separates the points of $p_{B(0)}(O)$ from the points of $K_{B(0)} \setminus p_{B(0)}(O)$. Consider the family $\mathcal{F}_0 = \{q_{B(0)}^{-1}(F) : F \in \mathcal{N}_0''\}$ and let $\mathcal{G}_0 = \{\mathrm{cl}_Y(f(F \cap D)) : F \in \mathcal{F}_0\}$; clearly, $\mathcal{G}_0$ consists of compact subsets of $Y$. It follows from $O \subset \bigcup \mathcal{F}_0$ that $\Phi_0' \subset \bigcup \mathcal{G}_0$. By Fact 5 of V.430 there exists a nonempty closed $G_{<\lambda}$-subset $\Phi_0$ of the space $Y$ such that $\Phi_0 \subset \Phi_0'$ and, for any $G \in \mathcal{G}_0$, either $\Phi_0 \subset G$ or $\Phi_0 \cap G = \emptyset$.

It follows from the fact that $Y$ cannot be mapped onto $\mathbb{I}^\lambda$ that $\Phi_0$ cannot be continuously mapped onto $\mathbb{I}^\lambda$ and hence there exists a point $y_0 \in \Phi_0$ such that $\pi \chi(y_0, \Phi_0) < \lambda$ (see Fact 1 of U.086); apply Fact 8 of V.430 to find a $\pi$-base $\mathcal{B}_0$ of $Y$ at the point $y_0$ such that $|\mathcal{B}_0| < \lambda$ and $U \cap \Phi_0 \neq \emptyset$ for all $U \in \mathcal{B}_0$.

Proceeding inductively assume that $n \in \omega$ and we have constructed, for every $i \leq n$, a nonempty compact $G_{<\lambda}$-subset $\Phi_i$ of the space $Y$, a point $y_i \in \Phi_i$, a set $B(i) \subset A$ together with the families $\mathcal{F}_i$, $\mathcal{G}_i$, and $\mathcal{B}_i$ with the following properties:

(8) $\mathcal{F}_i$ is a family of compact subsets of $K$ and $\{p_{B(i)}(F) : F \in \mathcal{F}_i\}$ separates the points of $p'_{B(i)}(O)$ from the points of $K_{B(i)} \backslash p_{B(i)}(O)$ for any $i \leq n$;

(9) $p_{B(i)}^{-1} p_{B(i)}(F) = F$ for any $F \in \mathcal{F}_i$ and $i \leq n$;

(10) $\mathcal{F}_i$ is closed under finite intersections for all $i \leq n$;

(11) $\mathcal{G}_i = \{\mathrm{cl}_Y (f(F \cap D)) : F \in \mathcal{F}_i\}$ for all $i \leq n$;

(12) $\max\{|B(i)|, |\mathcal{F}_i|, |\mathcal{G}_i|\} < \lambda$ for all $i \leq n$;

(13) if $i \leq n$ then $\Phi_i$ is a compact $G_{<\lambda}$-subset of $Y$ such that, for any $G \in \mathcal{G}_i$ either $\Phi_i \cap G = \emptyset$ or $\Phi_i \subset G$;

(14) for each $i \leq n$ the family $\mathcal{B}_i$ is a $\pi$-base of $Y$ at the point $y_i$ such that $|\mathcal{B}_i| < \lambda$ and $U \cap \Phi_i \neq \emptyset$ for all $U \in \mathcal{B}_i$;

(15) $B(i) \subset B(i+1)$, $\mathcal{F}_i \subset \mathcal{F}_{i+1}$ and $\Phi_{i+1} \subset \Phi_i$ for all $i < n$;

(16) for any $i < n$, if $U \in \mathcal{B}_i$ then $E(U) \subset B(i+1)$.

The family $\mathcal{E} = \{E(U) : U \in \mathcal{B}_0 \cup \ldots \cup \mathcal{B}_n\}$ has cardinality strictly less than $\lambda$, so we can choose $B(n+1) \subset A$ such that $|B(n+1)| < \lambda$ and $B(n) \cup (\bigcup \mathcal{E}) \subset B(n+1)$. It is evident that $O = p_{B(n+1)}^{-1} p_{B(n+1)}(O)$. We have $p_{B(n+1)}(O) \subset X_{B(n+1)}$ and $nw(X_{B(n+1)}) < \lambda$ which shows that we can find a network $\mathcal{N}_{n+1}$ in the space $p_{B(n+1)}(O)$ such that $|\mathcal{N}_{n+1}| < \lambda$.

Let $\mathcal{N}'_{n+1} = \{\mathrm{cl}_{K_{B(n+1)}}(N) : N \in \mathcal{N}_{n+1}\}$ and denote by $\mathcal{N}''_{n+1}$ the family of all finite intersections of the elements of $\mathcal{N}'_{n+1}$. Observe that $|\mathcal{N}''_{n+1}| < \lambda$ and the family $\mathcal{N}''_{n+1}$ separates the points of $p_{B(n+1)}(O)$ from the points of $K_{B(n+1)} \backslash p_{B(n+1)}(O)$. Let $\mathcal{F}_{n+1} = \{q_{B(n+1)}^{-1}(F) : F \in \mathcal{N}''_{n+1}\} \cup \mathcal{F}_n$ and $\mathcal{G}_{n+1} = \{\mathrm{cl}_Y (f(F \cap D)) : F \in \mathcal{F}_{n+1}\}$; clearly, $\mathcal{G}_{n+1}$ consists of compact subsets of $Y$. It follows from $O \subset \bigcup \mathcal{F}_{n+1}$ that $\Phi_n \subset \bigcup \mathcal{G}_{n+1}$. By Fact 5 of V.430 there exists a nonempty closed $G_{<\lambda}$-subset $\Phi_{n+1}$ of the space $Y$ such that $\Phi_{n+1} \subset \Phi_n$ and, for any set $G \in \mathcal{G}_{n+1}$, either $\Phi_{n+1} \subset G$ or $\Phi_{n+1} \cap G = \emptyset$.

It follows from the fact that $Y$ cannot be mapped onto $\mathbb{I}^\lambda$ that $\Phi_{n+1}$ cannot be continuously mapped onto $\mathbb{I}^\lambda$ and hence there exists a point $y_{n+1} \in \Phi_{n+1}$ such that $\pi\chi(y_{n+1}, \Phi_{n+1}) < \lambda$ (see Fact 1 of U.086); apply Fact 8 of V.430 to find a $\pi$-base $\mathcal{B}_{n+1}$ of $Y$ at the point $y_{n+1}$ such that $|\mathcal{B}_{n+1}| < \lambda$ and $U \cap \Phi_{n+1} \neq \emptyset$ for all $U \in \mathcal{B}_{n+1}$. It is straightforward that the conditions (8)–(16) are now satisfied if we replace $n$ with $n+1$, so our inductive procedure can be continued to construct for every $i \in \omega$, a nonempty compact $G_{<\lambda}$-subset $\Phi_i$ of the space $Y$, a point $y_i \in \Phi_i$, a set $B(i) \subset A$ together with the families $\mathcal{F}_i$, $\mathcal{G}_i$, and $\mathcal{B}_i$ such that the properties (8)–(16) hold for each $n \in \omega$.

Consider the sets $\Phi = \bigcap_{n \in \omega} \Phi_n$ and $B = \bigcup_{n \in \omega} B(n)$ together with the families $\mathcal{B} = \bigcup_{n \in \omega} \mathcal{B}_n$, $\mathcal{F} = \bigcup_{n \in \omega} \mathcal{F}_n$ and $\mathcal{G} = \bigcup_{n \in \omega} \mathcal{G}_n$. It follows from $\mathrm{cf}(\lambda) > \omega$ that $|B| < \lambda$ and $|\mathcal{B}| < \lambda$. Choose an accumulation point $y$ for the sequence $\{y_n : n \in \omega\}$; it is easy to see that $y \in \Phi$ and $\mathcal{B}$ is a $\pi$-base of $Y$ at the point $y$.

Let $\tau_y$ be the family of all open neighborhoods of $y$ in $Y$, i.e., $\tau_y = \tau(y, Y)$. For every $U \in \tau_y$ we will need the set $H(U) = \bigcup\{V^* : V \in \mathcal{B}$ and $V \subset U\}$. Now let $P = \bigcap\{\mathrm{cl}_K(H(U)) : U \in \tau_y\}$. It follows from (16) that $q_B^{-1} q_B([V]) = [V]$ for any $V \in \mathcal{B}$; as an immediate consequence, we have $q_B^{-1} q_B(P) = P$. Since $P$ is compact, the set $q_B(P)$ is a compact subspace of $K_B$ and hence $R = q_B(P) \cap X_B$ is a closed subspace of $X_B$. It follows from $nw(X_B) < \lambda$ that $R$ is a $G_{<\lambda}$-subset of $X_B$ and hence $P' = P \cap X = p_B^{-1}(R)$ is a $G_{<\lambda}$-subset of $X$.

Given any set $U \in \tau_y$ it follows from the inclusion $H(U) \subset U^*$ that we have $\mathrm{cl}_K(H(U)) \cap X = \mathrm{cl}_X(H(U)) \subset \mathrm{cl}_X(U^*)$, so $P' \subset \bigcap\{\mathrm{cl}_X(U^*) : U \in \tau_y\} = Q(f, y)$.

The property (9) implies that $q_B^{-1}q_B(F) = F$ for any $F \in \mathcal{F}$. Besides, the family $\mathcal{F}_B = \{q_B(F) : F \in \mathcal{F}\}$ consists of compact subsets of $K_B$ and it follows from (8),(9) and (15) that

(17) $\mathcal{F}_B$ is closed under finite intersections and separates the points of $p_B(O)$ from the points of $K_B \backslash p_B(O)$.

Take a point $x \in D$ with $f(x) = y$; let $\mathcal{F}_x = \{F \in \mathcal{F} : x \in F\}$ and consider the set $L = \bigcap \mathcal{F}_x$. Then $x \in L = q_B^{-1}q_B(L)$ and $q_B(L) \subset q_B(O) = p_B(O)$; this is an easy consequence of (17). Take an arbitrary $U \in \tau_y$ and any finite subfamily $\mathcal{F}'$ of the family $\mathcal{F}_x$. There exists $n \in \omega$ such that $\mathcal{F}' \subset \mathcal{F}_n$ and hence $F = \bigcap \mathcal{F}' \in \mathcal{F}_n$. Therefore the set $G = \mathrm{cl}_Y(f(F \cap D))$ belongs to $\mathcal{G}_n$; it follows from $y = f(x) \in G \cap \Phi_n$ that $G \cap \Phi_n \neq \emptyset$ and hence $\Phi_n \subset G$ by the property (13).

There exists $W \in \mathcal{B}_n$ such that $W \subset U$ and it follows from the property (14) that $W \cap \Phi_n \neq \emptyset$. Thus $f^{-1}(W) \cap F \neq \emptyset$ which shows that $F \cap W^* \neq \emptyset$. Consequently, $H(U) \cap (\bigcap \mathcal{F}') \neq \emptyset$ for any finite family $\mathcal{F}' \subset \mathcal{F}_x$ and $U \in \tau_y$. It is easy to see, using Fact 1 of S.326, that this implies $L' = P \cap L \neq \emptyset$. Taking in consideration that $q_B^{-1}q_B(L') = L'$ we conclude that $q_B(L') \subset q_B(P) \cap q_B(O)$, so $q_B(O) \cap q_B(P) \neq \emptyset$. Therefore $\emptyset \neq L' \subset P \cap O$ and hence $P \cap O = P' \cap O$ is a nonempty $G_{<\lambda}$-subset of $X$.

Apply Fact 1 of V.430 to find an adequate $G_{<\lambda}$-subset $P_0$ of the space $X$ such that $P_0 \subset P' \cap O$. It follows from $P_0 \subset P' \subset Q(f, y)$ that $P_0 \in \mathcal{P}$; this contradiction with $P_0 \subset O \subset X \backslash (\bigcup \mathcal{P})$ shows that $f(M)$ intersects every nonempty $G_{<\lambda}$-subset of $Y$, i.e., Fact 2 is proved.

Returning to our solution assume that $K$ does not map continuously onto $\mathbb{I}^\kappa$ and denote by $\mathcal{P}$ the family of all nonempty adequate $G_{<\kappa}$-subsets $P$ of the product $N$ such that $P \subset Q(f, y)$ for some $y \in K$. It follows from Fact 1 that we can find a family $\mathcal{A} \subset \mathcal{P}$ and a set $S \subset T$ such that $|\mathcal{A}| < \kappa$, $|S| < \kappa$ while $\bigcup \mathcal{P} \subset \overline{\bigcup \mathcal{A}}$ and $p_S^{-1}p_S(P) = P$ for any $P \in \mathcal{A}$.

Observe that $F = \overline{\bigcup \mathcal{P}} = \overline{\bigcup \mathcal{A}}$; let $A = \bigcup \mathcal{A}$ and take any $z \in p_S(A)$. There is $P \in \mathcal{A}$ with $z \in p_S(P)$ and hence $p_S^{-1}(z) \subset P \subset Q(f, y)$ for some $y \in K$. Let $D' = F \cap D$ and apply Fact 6 of V.430 to the open map $\varphi = p_S$ and the map $f$ to see that there exists a continuous map $g : p_S(D') \to K$ such that $g \circ (p_S|D') = f|D'$. This implies that $f(D') = g(p_S(D'))$ and hence $nw(f(D')) \leq nw(p_S(D')) < \kappa$. Next apply Fact 2 to see that

(18) the set $E = f(D')$ intersects every nonempty $G_{<\kappa}$-subset of $K$.

If $K \backslash E \neq \emptyset$ then pick a point $y \in K \backslash E$ and take a set $O_z \in \tau(z, K)$ such that $y \notin \overline{O}_z$ for any $z \in E$. Since $l(E) \leq nw(E) < \kappa$, there is a set $E' \subset E$ such that $|E'| < \kappa$ and $E \subset \bigcup \{O_z : z \in E'\}$.

Then $y \in R = \bigcap \{K \backslash \overline{O}_z : z \in E'\} \subset K \backslash E$, so $R$ is a nonempty $G_{<\kappa}$-subset of $K$ contained in $K \backslash E$. This, however, contradicts (18) and implies that $K = E$, so $w(K) = nw(K) = nw(E) < \kappa$ (see Fact 4 of S.307). This final contradiction with $w(K) = \kappa$ shows that $K$ can be continuously mapped onto $\mathbb{I}^\kappa$ and hence our solution is complete.

**V.432.** *Given an infinite cardinal $\kappa$ suppose that $nw(N_t) \leq \kappa$ for any $t \in T$ and $C \subset N = \prod_{t \in T} N_t$ is a dense subspace of $N$. Assume additionally that we have a continuous (not necessarily surjective) map $\varphi : C \to L$ of $C$ into a compact space $L$. Prove that if $y \in C' = \varphi(C)$ and $h\pi\chi(y, L) \leq \kappa$ then $\psi(y, C') \leq \kappa$. Here the cardinal $h\pi\chi(y, L)$ is the hereditary $\pi$-character of the space $L$ at the point $y$, i.e., $h\pi\chi(y, L) = \sup\{\pi\chi(y, Z) : y \in Z \subset L\}$.*

**Solution.** If we have a product $Z = \prod_{t \in T} Z_t$ and $A \subset T$ then $Z_A = \prod_{t \in A} Z_t$ is the $A$-face of $Z$ and $\pi_A : Z \to Z_A$ is the natural projection. A set $F \subset Z$ *depends on* $A \subset T$ if $\pi_A^{-1}\pi_A(F) = F$; if $F$ depends on a set of cardinality $\leq \kappa$ then we say that $F$ *depends on at most $\kappa$-many coordinates*. A set $E \subset Z$ *covers a face* $Z_A$ if $\pi_A(E) = Z_A$. Suppose that, for every $t \in T$ we have a family $\mathcal{N}_t$ of subsets of $Z_t$ and let $\mathcal{N} = \{\mathcal{N}_t : t \in T\}$. If we have a faithfully indexed set $A = \{t_1, \ldots, t_n\} \subset T$ and $N_i \in \mathcal{N}_{t_i}$ for each $t \leq n$ then let $[t_1, \ldots, t_n, N_1, \ldots, N_n] = \{x \in Z : x(t_i) \in N_i$ for all $i = 1, \ldots, n\}$. A set $H \subset Z$ is called $\mathcal{N}$-*standard* (or standard if $\mathcal{N}$ is clear) if $H = [t_1, \ldots, t_n, N_1, \ldots, N_n]$ for some $t_1, \ldots, t_n \in T$ and $N_i \in \mathcal{N}_{t_i}$ for all $i \leq n$. In this case we let $\text{supp}(H) = A$ and $r(H) = n$. We also consider that $H = Z$ is the unique standard subset of $Z$ such that $r(H) = 0$. Given any point $x \in Z$ and $A \subset T$ the set $\langle x, A \rangle = \{y \in Z : y(t) = x(t)$ for any $t \in A\}$ is closed in $Z$. If $A \subset T$ then the face $Z_A$ is called $\kappa$-*residual* if $|T \backslash A| \leq \kappa$. Say that a nonempty closed set $F \subset Z$ is $\kappa$-*large* if, for any $x \in F$ and any finite $A \subset T$, the set $\langle x, A \rangle \cap F$ covers a $\kappa$-residual face of $Z$.

If $Z$ is a space and $\mathcal{H}$ is a family of subsets of $Z$ then $\mathcal{H}$ is called *a network of $Z$ at a point* $z \in Z$ if for any $U \in \tau(z, Z)$ there exists $H \in \mathcal{H}$ such that $x \in H \subset U$.

*Fact 1.* Given an infinite cardinal $\kappa$ suppose that $nw(N_t) \leq \kappa$ for any $t \in T$ and $N = \prod_{t \in T} N_t$. Assume that $C \subset N$ is dense in $N$, and we have a compact extension $K_t$ of the space $N_t$ for any $t \in T$. If a set $F \subset K = \prod_{t \in T} K_t$ is $\kappa$-large then there exists a $G_\kappa$-set $G$ in the space $K$ such that $F \subset G$ and $F \cap C = G \cap C$. In particular, $F \cap C$ is a $G_\kappa$-subset of $C$.

*Proof.* We can assume, without loss of generality, that $K \backslash F \neq \emptyset$. For every $t \in T$ fix a network $\mathcal{N}_t$ in the space $N_t$ such that $|\mathcal{N}_t| \leq \kappa$; we will need the family $\mathcal{M}_t = \{\text{cl}_{K_t}(N) : N \in \mathcal{N}_t\}$. If $\mathcal{M} = \{\mathcal{M}_t : t \in T\}$ then the $\mathcal{M}$-standard subsets of $K$ will be called *standard*. It is easy to see that

(1) the family $\mathcal{H}$ of all standard subsets of $K$ is a network in $K$ at every $x \in C$.

Given standard sets $P$ and $P'$ say that $P' \preceq P$ if $P = [t_1, \ldots, t_n, M_1, \ldots, M_n]$ and there exists a natural $k \leq n$ such that $P' = [t_{i_1}, \ldots, t_{i_k}, M_{i_1}, \ldots, M_{i_k}]$ for some distinct $i_1, \ldots, i_k \in \{1, \ldots, n\}$; if $k < n$ then we write $P' \prec P$. We also include here the case when $k = 0$, so $P' = K \preceq P$ for any standard set $P$. Say that a standard set $P$ is *minimal* if $P \cap F = \emptyset$ but $P' \cap F \neq \emptyset$ whenever $P' \prec P$. It follows from (1) that

(2) for any $x \in C \backslash F$ there exists a minimal standard set $P$ such that $x \in P$.

It will be easy to finish our proof if we establish that

(3) the family $\mathcal{S}$ of minimal standard sets has cardinality not exceeding $\kappa$.

Assume, toward a contradiction that $|\mathcal{S}| > \kappa$. Then we can choose $\mathcal{S}_0 \subset \mathcal{S}$ such that $|\mathcal{S}_0| = \kappa^+$ and there exists $n \in \omega$ with $r(P) = n$ for all $P \in \mathcal{S}_0$. Observe first that

(4) if $A \subset T$, a set $D \subset K$ covers the face $K_{T \backslash A}$ and a standard set $P$ is disjoint from $D$ then $\text{supp}(P) \cap A \neq \emptyset$.

Indeed, if $\text{supp}(P) = \{t_1, \ldots, t_k\} \subset T \backslash A$ and $P = [t_1, \ldots, t_k, M_1, \ldots, M_k]$ then it follows from $\pi_{T \backslash A}(D) = K_{T \backslash A}$ that there exists a point $x \in D$ such that $x(t_i) \in M_i$ for all $i \leq k$. Therefore $x \in D \cap P$ which is a contradiction.

The set $F$ being $\kappa$-large, there exists $A_1 \subset T$ with $|A_1| \leq \kappa$ such that $F$ covers the face $K_{T \backslash A_1}$. The property (4) shows that $\text{supp}(P) \cap A_1 \neq \emptyset$ for any $P \in \mathcal{S}_0$. There exists a point $t_1 \in A_1$ such that the family $\mathcal{S}_0' = \{P \in \mathcal{S}_0 : t_1 \in P\}$ has cardinality $\kappa^+$. Since $|\mathcal{M}_{t_1}| \leq \kappa$, we can find a family $\mathcal{S}_1 \subset \mathcal{S}_0'$ and $M_1 \in \mathcal{M}_{t_1}$ such that $|\mathcal{S}_1| = \kappa^+$ and $[t_1, M_1] \preceq P$ for any $P \in \mathcal{S}_1$.

Proceeding by induction assume that $k < n$ and we have a set $A_k \subset T$ with $|A_k| \leq \kappa$ and a family $\mathcal{S}_k$ such that $|\mathcal{S}_k| = \kappa^+$ and, for some $t_1, \ldots, t_k \in A_k$ and $M_i \in \mathcal{M}_{t_i}$ $(i = 1, \ldots, k)$, we have $[t_1, \ldots, t_k, M_1, \ldots, M_k] \preceq P$ for every $P \in \mathcal{S}_k$. Therefore $P = [t_1, \ldots, t_k, s_1, \ldots, s_{n-k}, M_1, \ldots, M_k, E_1, \ldots, E_{n-k}]$ for every $P \in \mathcal{S}_k$; let $Q(P)$ be the set in which $s_1$ and $E_1$ are omitted from the definition of $P$, i.e., $Q(P) = [t_1, \ldots, t_k, s_2, \ldots, s_{n-k}, M_1, \ldots, M_k, E_2, \ldots, E_{n-k}]$. It is clear that $Q(P) \prec P$; since $P$ is minimal, the set $Q(P)$ intersects $F$ for each $P \in \mathcal{S}_k$.

Fix a set $R \in \mathcal{S}_k$ and let $F' = F \cap Q(R)$. The set $F$ being $\kappa$-large, we can find $A \subset T$ with $|A| \leq \kappa$ such that $F'$ covers the face $K_{T \backslash A}$ and hence the face $K_{T \backslash (A \cup A_k)}$ as well. Let $A_{k+1} = A \cup A_k$ and observe that every set $P \in \mathcal{S}_k$ is disjoint from $F'$; this, together with (4) shows that $\text{supp}(P) \cap A_{k+1} \neq \emptyset$. Suppose for a moment that $P = [t_1, \ldots, t_k, s_1, \ldots, s_{n-k}, M_1, \ldots, M_k, E_1, \ldots, E_{n-k}] \in \mathcal{S}_k$ and $\{s_1, \ldots, s_{n-k}\} \cap A_{k+1} = \emptyset$. Since $F'$ covers the face $K_{T \backslash A_{k+1}}$, we can find a point $x \in F'$ such that $x(s_i) \in E_i$ for all $i \leq n - k$; since also $x(t_i) \in M_i$ for all $i \leq k$ because $x \in Q(R)$, we conclude that $x \in F' \cap P$. This contradiction implies that $\{s_1, \ldots, s_{n-k}\} \cap A_{k+1} \neq \emptyset$ and hence the set $\text{supp}(P) \backslash \{t_1, \ldots, t_k\}$ intersects the set $A_{k+1}$ for any $P \in \mathcal{S}_k \backslash \{R\}$.

Therefore we can find a family $\mathcal{S}_{k+1} \subset \mathcal{S}_k$ of cardinality $\kappa^+$ together with a point $t_{k+1} \in A_{k+1} \backslash \{t_1, \ldots, t_k\}$ and a set $M_{k+1} \in \mathcal{M}_{t_{k+1}}$ such that for any $P \in \mathcal{S}_{k+1}$ we have $[t_1, \ldots, t_{k+1}, M_1, \ldots, M_{k+1}] \preceq P$. As a consequence, our inductive procedure can be continued to construct a family $\mathcal{S}_n \subset \mathcal{S}$ such that $|\mathcal{S}_n| = \kappa^+$ while $[t_1, \ldots, t_n, M_1, \ldots, M_n] \preceq P$ for any $P \in \mathcal{S}_n$. Recalling that $r(P) = n$, we conclude that we have the equality $P = [t_1, \ldots, t_n, M_1, \ldots, M_n]$ for each $P \in \mathcal{S}_n$; this contradiction shows that $|\mathcal{S}| \leq \kappa$, i.e., (3) is proved.

It is straightforward that $G = K \backslash (\bigcup \mathcal{S})$ is a $G_\kappa$-subset of $K$ such that $F \subset G$ and $F \cap C = G \cap C$, i.e., Fact 1 is proved.

Returning to our solution observe that there is no loss of generality to assume that $C'$ is dense in $L$. Choose a compact extension $K_t$ of the space $N_t$ for any $t \in T$; then $K = \prod_{t \in T} K_t$ is a compact extension of both $N$ and $C$. There exist continuous maps $\Phi : \beta C \to L$ and $\xi : \beta C \to K$ such that $\Phi | C = \varphi$ and $\xi(x) = x$ for any $x \in C$. It is clear that both $\Phi$ and $\xi$ are surjective.

For every $t \in T$ fix a network $\mathcal{N}_t$ in the space $N_t$ such that $|\mathcal{N}_t| \leq \kappa$ and let $\mathcal{M}_t = \{\mathrm{cl}_{K_t}(N) : N \in \mathcal{N}_t\}$. If $\mathcal{M} = \{\mathcal{M}_t : t \in T\}$ then the $\mathcal{M}$-standard subsets of $K$ will be called *standard*. Our first step is to prove that

(5) the set $F_y = \xi(\Phi^{-1}(y))$ is $\kappa$-large.

Fix a point $x \in F_y$, a finite $A \subset T$ and consider the set $P = \langle x, A \rangle = \{x' \in K : x'(t) = x(t) \text{ for all } t \in A\}$. It follows from $P \cap F_y \neq \emptyset$ that $\xi^{-1}(P) \cap \Phi^{-1}(y) \neq \emptyset$ and hence $y \in Q = \Phi(\xi^{-1}(P))$. The set $Q$ is compact and it follows from $h\pi\chi(y, L) \leq \kappa$ that we can choose a $\pi$-base $\mathcal{B}$ of the space $Q$ at the point $y$ such that $|\mathcal{B}| \leq \kappa$. For every $B \in \mathcal{B}$ pick a set $O_B \in \tau(L)$ such that $\emptyset \neq O_B \cap Q \subset \overline{O}_B \cap Q \subset B$. It follows from $c(K) \leq \kappa$ and Fact 4 of V.430 that

(6) for any $U \in \tau^*(L)$, the set $\mathrm{cl}_K(\varphi^{-1}(U))$ depends on at most $\kappa$-many coordinates and coincides with the set $\xi(\mathrm{cl}_{\beta C}(\Phi^{-1}(U)))$.

Apply (6) to find a set $S \subset T$ of cardinality at most $\kappa$ for which $A \subset S$ and the set $D_B = \xi(\mathrm{cl}_{\beta C}(\Phi^{-1}(O_B)))$ depends on $S$ for any $B \in \mathcal{B}$. The face $K_{T \setminus S}$ is residual; to show that $P \cap F_y$ covers $K_{T \setminus S}$ fix any point $w \in K_{T \setminus S}$ and consider the set $E = \{z \in K : \pi_{T \setminus S}(z) = w \text{ and } \pi_S(z) \in \pi_S(P)\}$. Clearly, $E$ is a nonempty compact subset of $P$.

Fix any $B \in \mathcal{B}$; it follows from $O_B \cap Q \neq \emptyset$ that there is a point $u \in \xi^{-1}(P)$ such that $\Phi(u) \in O_B$; thus $u \in \Phi^{-1}(O_B)$ which shows that $\xi(u) \in D_B \cap P$. Define a point $u' \in K$ by the equalities $\pi_{T \setminus S}(u') = w$ and $\pi_S(u') = \pi_S(\xi(u))$. Since the sets $D_B$ and $P$ depend on $S$, we conclude that $u' \in D_B \cap P$. On the other hand, $\pi_S(u') \in \pi_S(P)$, so $u' \in E$, and therefore $E \cap D_B \neq \emptyset$.

As a consequence, $\Phi(\xi^{-1}(E)) \cap \overline{O}_B \neq \emptyset$ and hence $\Phi(\xi^{-1}(E)) \cap B \neq \emptyset$ for any $B \in \mathcal{B}$; since $\Phi(\xi^{-1}(E))$ is a closed subset of $Q$ and $\mathcal{B}$ is a $\pi$-base of $Q$ at $y$, we must have $y \in \Phi(\xi^{-1}(E))$ which implies that $\xi^{-1}(E) \cap \Phi^{-1}(y) \neq \emptyset$ and hence $E \cap F_y \neq \emptyset$. If $v \in E \cap F_y$ then $w = \pi_{T \setminus S}(v) \in \pi_{T \setminus S}(P \cap F_y)$; the point $w \in K_{T \setminus S}$ was chosen arbitrarily, so $P \cap F_y$ covers $K_{T \setminus S}$ and hence (5) is proved.

By Fact 1 there exists a $G_\kappa$-set $G$ in the space $K$ such that $F_y \subset G$ and $G \cap C = F_y \cap C = \varphi^{-1}(y)$. Therefore we can choose a family $\mathcal{F}$ of compact subsets of $K$ such that $|\mathcal{F}| \leq \kappa$ and $C \setminus F_y \subset \bigcup \mathcal{F} \subset K \setminus F_y$. For any $F \in \mathcal{F}$ the set $W_F = L \setminus \Phi(\xi^{-1}(F))$ is an open neighborhood of $y$ in $L$ and it is straightforward that $H = \bigcap \{W_F : F \in \mathcal{F}\}$ is a $G_\kappa$-subset of $L$ such that $H \cap C' = \{y\}$. This shows that $\psi(y, C') \leq \kappa$ and hence our solution is complete.

**V.433.** *Suppose that $C$ is a dense subspace of a product $N = \prod_{t \in T} N_t$ such that $nw(N_t) \leq \kappa$ for each $t \in T$. Assume that $K$ is a compact space with $t(K) \leq \kappa$ and $\varphi : C \to K$ is a continuous (not necessarily surjective) map; let $C' = \varphi(C)$. Prove that every closed subspace of $C'$ is a $G_\kappa$-set; in particular, $\psi(C') \leq \kappa$.*

**Solution.** Fix a nonempty closed set $F'$ in the space $C'$ and let $F = \mathrm{cl}_K(F')$. Consider the quotient map $p : K \to K_F$ obtained by contracting the set $F$ to a point and let $q = p|C'$. It is easy to see, applying TFS-162, that we have the inequalities $t(K_F) \leq t(K) \leq \kappa$; denote by $y$ the point of the space $K_F$ represented by $F$ and let $C'' = p(C')$.

*Fact 1.* If $X$ is a compact space then $h\pi\chi(x, X) \leq t(X)$ for any $x \in X$.

*Proof.* Fix any set $Z \subset X$ such that $x \in Z$ and let $P = \overline{Z}$; it is an easy exercise that $\pi\chi(x, Z) = \pi\chi(x, P)$. The space $P$ is compact and $t(P) \leq t(X)$, so we can apply TFS-331 to see that $\pi\chi(P) \leq t(P) \leq t(X)$ and hence $\pi\chi(x, Z) = \pi\chi(x, P) \leq t(X)$ which shows that Fact 1 is proved.

Returning to our solution observe that $h\pi\chi(y, K_F) \leq \kappa$ by Fact 1; therefore we can apply Problem 432 to the map $p \circ \varphi$ to see that $\psi(y, C'') \leq \kappa$. Since $F' = q^{-1}(y)$, we conclude that $F'$ is a $G_\kappa$-subset of $C'$.

**V.434.** *Suppose that $C$ is a dense subspace of a product $N = \prod_{t \in T} N_t$ such that $nw(N_t) \leq \kappa$ for each $t \in T$. Assume additionally that $l(C) \leq \kappa$ and $K$ is a compact space with $t(K) \leq \kappa$ such that there exists a continuous (not necessarily surjective) map $\varphi : C \to K$. Prove that if $C' = \varphi(C)$ then $hl(C') \leq \kappa$.*

**Solution.** We have $l(C') \leq \kappa$ while every closed subspace of the space $C'$ is a $G_\kappa$-set by Problem 433. Now, SFFS-001 shows that $hl(C') \leq \kappa$.

**V.435.** *Prove that if $C$ is a dense subspace of a product of cosmic spaces and $K$ is a compact space then, for any continuous map $\varphi : C \to K$, we have $\Psi(\varphi(C)) \leq t(K)$.*

**Solution.** Let $\kappa = t(K)$ and $Y = \varphi(C)$. The space $C$ is dense in the product of cosmic and hence $\kappa$-cosmic spaces, so we can apply Problem 433 to see that every closed subspace of $Y$ is a $G_\kappa$-set. Therefore $\Psi(Y) \leq \kappa$ as promised.

**V.436.** *Suppose that $C$ is a dense subspace of a product of cosmic spaces and $\varphi : C \to K$ is a continuous (not necessarily surjective) map into a compact space $K$ of countable tightness; let $Y = \varphi(C)$. Prove that $Y$ is a perfect space and $\pi w(Y) \leq \omega$. In particular, if $\varphi : C_p(X) \to K$ is a continuous map of $C_p(X)$ into a compact space $K$ with $t(K) \leq \omega$ then $\varphi(C_p(X))$ is a perfect space of countable $\pi$-weight.*

**Solution.** There is no loss of generality to assume that $Y$ is dense in $K$. Apply Problem 435 to see that $\Psi(Y) \leq t(K) \leq \omega$, so $Y$ is a perfect space.

The cardinal $\omega_1$ is a precaliber of $C$, so it also a precaliber of $Y$ (see SFFS-277, SFFS-278 and SFFS-280). Therefore $\omega_1$ is a caliber of $K$ (see SFFS-279). By TFS-332, we can find a point-countable $\pi$-base $\mathcal{B}$ in the space $K$; this, together with $\omega_1$ being a caliber of $K$ implies that $\mathcal{B}$ is countable. Therefore the family $\mathcal{B}_Y = \{B \cap Y : B \in \mathcal{B}\}$ is a countable $\pi$-base in the space $Y$ and hence $\pi w(Y) \leq \omega$. Finally observe that our solution is also applicable to any space $C = C_p(X)$ because $C_p(X)$ is a dense subspace of $\mathbb{R}^X$.

**V.437.** *Suppose that $C$ is a dense Lindelöf $\Sigma$-subspace of a product of cosmic spaces and $\varphi : C \to K$ is a continuous (not necessarily surjective) map into a compact space $K$ of countable tightness. Prove that $nw(\varphi(C)) \leq \omega$. In particular, if $C_p(X)$ is a Lindelöf $\Sigma$-space and $\varphi : C_p(X) \to K$ is a continuous map of $C_p(X)$ into a compact space $K$ of countable tightness then $\varphi(C_p(X))$ is cosmic.*

**Solution.** Let $Y = \varphi(C)$; observe that $C \times C$ is a Lindelöf $\Sigma$-space and the space $Y \times Y \subset K \times K$ is a continuous image of $C \times C$. It is clear that $C \times C$ also embeds as a dense subspace in a product of cosmic spaces; besides, $t(K \times K) \leq \omega$ (see Fact 1 of V.274), so we can apply Problem 434 to convince ourselves that $Y \times Y$ is hereditarily Lindelöf and hence $\Delta(Y) \leq \omega$. Since $Y$ is a Lindelöf $\Sigma$-space, we can apply SFFS-300 to conclude that $nw(Y) \leq \omega$. Finally observe that our solution is also applicable to any space $C = C_p(X)$ because $C_p(X)$ is a dense subspace of $\mathbb{R}^X$.

**V.438.** *Given an infinite cardinal $\kappa$ observe that the property $\mathcal{P}_\kappa$ of being strongly $\kappa$-cosmic is stronger than being $\kappa$-cosmic. Besides, $\mathcal{P}_\kappa$ is preserved by subspaces and continuous images. Prove that if $\mu < cf(\kappa)$ and $X_\alpha$ is strongly $\kappa$-cosmic for any $\alpha < \mu$ then $X = \prod_{\alpha < \mu} X_\alpha$ is also strongly $\kappa$-cosmic.*

**Solution.** Suppose that $Z = \bigcup\{Z_\alpha : \alpha < \kappa\}$ and $nw(Z_\alpha) < \kappa$ for every $\alpha < \kappa$. Fix a network $\mathcal{N}_\alpha$ in the space $Z_\alpha$ for each $\alpha < \kappa$ and observe that $\mathcal{N} = \bigcup_{\alpha < \kappa} \mathcal{N}_\alpha$ is a network in $Z$ with $|\mathcal{N}| \leq \kappa$. Therefore every strongly $\kappa$-cosmic space is $\kappa$-cosmic.

If $Y \subset Z$ then let $Y_\alpha = Y \cap Z_\alpha$ for every $\alpha < \kappa$. Then $Y = \bigcup_{\alpha < \kappa} Y_\alpha$ and $nw(Y_\alpha) \leq nw(Z_\alpha) < \kappa$ for each $\alpha < \kappa$. This shows that every subspace of a strongly $\kappa$-cosmic space is strongly $\kappa$-cosmic.

Now assume that $f : Z \to T$ is a continuous onto map and let $T_\alpha = f(Z_\alpha)$ for all $\alpha < \kappa$. Then $T = \bigcup_{\alpha < \kappa} T_\alpha$ and $nw(T_\alpha) \leq nw(Z_\alpha) < \kappa$ for each $\alpha < \kappa$. Therefore every continuous image of a strongly $\kappa$-cosmic space is strongly $\kappa$-cosmic.

*Fact 1.* If $Z$ is a strongly $\kappa$-cosmic space and $\lambda = cf(\kappa)$ then there exists a family $\{Z_\alpha : \alpha < \lambda\}$ of subspaces of $Z$ such that $\alpha < \beta < \lambda$ implies that $Z_\alpha \subset Z_\beta$ while $\bigcup_{\alpha < \lambda} Z_\alpha = Z$ and $nw(Z_\alpha) < \kappa$ for every $\alpha < \lambda$.

*Proof.* Choose a family $\{Y_\alpha : \alpha < \kappa\}$ of subspaces of $Z$ such that $Z = \bigcup_{\alpha < \kappa} Y_\alpha$ and $nw(Y_\alpha) < \kappa$ for each $\alpha < \kappa$. If $Y'_\alpha = \bigcup\{Y_\gamma : \gamma \leq \alpha$ and $nw(Y_\gamma) \leq |\alpha|\}$ then $nw(Y'_\alpha) \leq |\alpha| \cdot |\alpha| \leq |\alpha| \cdot \omega < \kappa$ for any $\alpha < \kappa$. It is immediate that $\alpha < \beta$ implies $Y'_\alpha \subset Y'_\beta$. Given a point $z \in Z$ there is an ordinal $\alpha < \kappa$ such that $z \in Y_\alpha$; if $\beta = \max\{\alpha, nw(Y_\alpha)\}$ then $\beta < \kappa$ and $z \in Y_\alpha \subset Y'_\beta$. This shows that $Z = \bigcup_{\alpha < \kappa} Y'_\alpha$.

Observe that it follows from $cf(\kappa) = \lambda$ that there is a function $\xi : \lambda \to \kappa$ such that $\alpha < \beta < \lambda$ implies $\xi(\alpha) < \xi(\beta)$ and for any $\gamma < \kappa$ there is $\alpha < \lambda$ such that $\xi(\alpha) > \gamma$. Letting $Z_\alpha = Y'_{\xi(\alpha)}$ for any $\alpha < \lambda$ it is easy to see that we obtain the promised family $\{Z_\alpha : \alpha < \lambda\}$, so Fact 1 is proved.

Returning to our solution assume that $\mu < \lambda = \mathrm{cf}(\kappa)$ and the space $X_\alpha$ is strongly $\kappa$-cosmic for any $\alpha < \mu$; let $X = \prod_{\alpha < \mu} X_\alpha$. By Fact 1 we can choose a family $\{Z_\beta^\alpha : \beta < \lambda\}$ such that $\beta < \beta'$ implies $Z_\beta^\alpha \subset Z_{\beta'}^\alpha$, while $nw(Z_\beta^\alpha) < \kappa$ for each $\beta < \lambda$ and $\bigcup_{\beta < \lambda} Z_\beta^\alpha = X_\alpha$ for every $\alpha < \mu$.

Letting $Y_\beta = \prod_{\alpha < \mu} Z_\beta^\alpha$ we obtain a family $\{Y_\beta : \beta < \lambda\}$ of subsets of $X$. Given any $\beta < \lambda$ note that $nw(Y_\beta) \leq \sup\{nw(Z_\beta^\alpha) : \alpha < \mu\} < \kappa$ because $\mu < \lambda = \mathrm{cf}(\kappa)$ and $nw(Z_\beta^\alpha) < \kappa$ for each $\alpha < \mu$. Thus $nw(Y_\beta) < \kappa$ for each $\beta < \lambda$.

For any $x \in X$ and $\alpha < \mu$ there exists $\beta_\alpha < \lambda$ such that $x(\alpha) \in Z_{\beta_\alpha}^\alpha$. Take any $\beta < \lambda$ with $\{\beta_\alpha : \alpha < \mu\} \subset \beta$; it is immediate that $x \in Y_\beta$, so we proved that $\bigcup_{\beta < \lambda} Y_\beta = X$ and hence $X$ is strongly $\kappa$-cosmic, i.e., our solution is complete.

**V.439.** *Given an infinite cardinal $\kappa$ prove that $\mathbb{D}^\kappa$ is not strongly $\kappa$-cosmic and hence no strongly $\kappa$-cosmic space can be continuously mapped onto $\mathbb{I}^\kappa$.*

**Solution.** It is easy to find a disjoint family $\{A_\alpha : \alpha < \kappa\}$ of subsets of $\kappa$ such that $|A_\alpha| = \kappa$ for any $\alpha < \kappa$ and $\bigcup_{\alpha < \kappa} A_\alpha = \kappa$. If $\mathbb{D}^\kappa$ is strongly $\kappa$-cosmic then fix a family $\{Y_\alpha : \alpha < \kappa\}$ such that $nw(Y_\alpha) < \kappa$ for every $\alpha < \kappa$ and $\bigcup_{\alpha < \kappa} Y_\alpha = \mathbb{D}^\kappa$.

Let $\pi_\alpha : \mathbb{D}^\kappa \to \mathbb{D}^{A_\alpha}$ be the natural projection; since $nw(\pi_\alpha(Y_\alpha)) < \kappa$, it is impossible that $\pi_\alpha(Y_\alpha) = \mathbb{D}^{A_\alpha}$, so we can choose a point $x_\alpha \in \mathbb{D}^{A_\alpha} \backslash \pi_\alpha(Y_\alpha)$ for each $\alpha < \kappa$. For every $\beta < \kappa$ there is a unique $\alpha < \kappa$ such that $\beta \in A_\alpha$; let $x(\beta) = x_\alpha(\beta)$. This gives us a point $x \in \mathbb{D}^\kappa$. For each $\alpha < \kappa$ it follows from $\pi_\alpha(x) = x_\alpha \notin \pi_\alpha(Y_\alpha)$ that $x \notin Y_\alpha$. Consequently, $x \notin \bigcup_{\alpha < \kappa} Y_\alpha$; this contradiction shows that $\mathbb{D}^\kappa$ is not strongly $\kappa$-cosmic.

Finally observe that if a strongly $\kappa$-cosmic space $X$ maps continuously onto $\mathbb{I}^\kappa$ then $\mathbb{I}^\kappa$ and $\mathbb{D}^\kappa \subset \mathbb{I}^\kappa$ must be strongly $\kappa$-cosmic by Problem 438. This final contradiction proves that no strongly $\kappa$-cosmic space can be continuously mapped onto $\mathbb{I}^\kappa$.

**V.440.** *Assume that $\kappa$ is an infinite cardinal and $K$ is a strongly $\kappa$-cosmic compact space. Prove that there exists $x \in K$ such that $\chi(x, K) < \kappa$.*

**Solution.** Say that $F \subset K$ is a $G_{<\kappa}$-set if $F$ is a $G_\lambda$-subset of $K$ for some $\lambda < \kappa$. Assume, toward a contradiction, that $\chi(x, K) \geq \kappa$ for any $x \in K$ and the space $K$ is strongly $\kappa$-cosmic. If $\mu = \mathrm{cf}(\kappa)$ then there exists a $\mu$-sequence $\{Y_\alpha : \alpha < \mu\}$ such that $\alpha < \beta$ implies $Y_\alpha \subset Y_\beta$ and $K = \bigcup\{Y_\alpha : \alpha < \mu\}$ while $nw(Y_\alpha) < \kappa$ for every $\alpha < \mu$ (see Fact 1 of V.438). It follows from TFS-327 and Fact 7 of V.430 that

(1) For any $\lambda < \kappa$, if $F$ is a closed $G_\lambda$-subset of $K$ then $\chi(x, F) \geq \kappa$ for any $x \in F$.

Since $nw(Y_0) < \kappa$, it is impossible that $Y_0 = K$, so we can pick a point $x \in K \backslash Y_0$. Take a set $O_z \in \tau(z, K)$ such that $x \notin \overline{O}_z$ for any $z \in Y_0$. Since $l(Y_0) \leq nw(Y_0) < \kappa$, there is a set $E_0 \subset Y_0$ such that $|E_0| < \kappa$ and $Y_0 \subset \bigcup\{O_z : z \in E_0\}$. Then $x \in R_0 = \bigcap\{K \backslash \overline{O}_z : z \in E_0\} \subset K \backslash Y_0$; since $R_0$ is a $G_{<\kappa}$-subset of $K$, we can apply Fact 2 of S.328 to find a closed $G_{<\kappa}$-subset $F_0$ of the space $K$ such that $x \in F_0 \subset R_0 \subset K \backslash Y_0$.

Proceeding inductively assume that $\alpha < \mu$ and we have constructed a family $\{F_\beta : \beta < \alpha\}$ of nonempty closed $G_{<\kappa}$-subsets of $K$ with the following properties:

(2) $\beta < \gamma < \alpha$ implies $F_\gamma \subset F_\beta$;

(3) $F_\beta \cap Y_\beta = \emptyset$ for any $\beta < \alpha$.

It follows from $\alpha < \mu$ that $F = \bigcap_{\beta < \alpha} F_\beta$ is a nonempty closed $G_{<\kappa}$-subset of $K$, so we can derive from the property (1) that $\chi(x, F) \geq \kappa$ for any $x \in F$. As a consequence, $w(F) = nw(F) \geq \kappa > nw(Y_\alpha)$ and therefore $Y_\alpha$ cannot cover $F$.

Pick any point $x \in F \backslash Y_\alpha$; there exists a set $O_z \in \tau(z, K)$ such that $x \notin \overline{O}_z$ for any $z \in Y_\alpha$. Since $l(Y_\alpha) \leq nw(Y_\alpha) < \kappa$, there is a set $E_\alpha \subset Y_\alpha$ such that $|E_\alpha| < \kappa$ and $Y_\alpha \subset \bigcup \{O_z : z \in E_\alpha\}$. Then $x \in R_\alpha = \bigcap \{K \backslash \overline{O}_z : z \in E_\alpha\} \subset K \backslash Y_\alpha$; since $R_\alpha$ is a $G_{<\kappa}$-subset of $K$, we can apply Fact 2 of S.328 to find a closed $G_{<\kappa}$-subset $H$ of the space $K$ such that $x \in H \subset R_\alpha \subset K \backslash Y_\alpha$. It is evident that if $F_\alpha = H \cap F$ then (2) and (3) are fulfilled for all $\beta \leq \alpha$, so we can construct a family $\{F_\alpha : \alpha < \mu\}$ with the properties (2) and (3) satisfied for all $\alpha < \mu$. The space $K$ being compact, the set $P = \bigcap \{F_\alpha : \alpha < \mu\}$ is nonempty; it follows form (3) that $P \subset K \backslash (\bigcup \{Y_\alpha : \alpha < \mu\})$, so we obtained a contradiction which shows that $K$ must have a point $x$ such that $\chi(x, K) < \kappa$.

**V.441.** *Suppose that $\kappa$ is an uncountable cardinal and $X$ is a space such that $w(X) \leq \kappa$ and $l(X) < cf(\kappa)$. Prove that $C_p(X)$ is strongly $\kappa$-cosmic. In particular, if $cf(\kappa) > \omega$ and $X$ is a Lindelöf space with $w(X) = \kappa$ then $C_p(X)$ is strongly $\kappa$-cosmic.*

**Solution.** It follows from $w(X) \leq \kappa$ that we can assume, without loss of generality, that $X \subset \mathbb{I}^\kappa$. For any $A \subset \kappa$ let $p_A : \mathbb{I}^\kappa \to \mathbb{I}^A$ be the natural projection. If $\alpha < \kappa$ and $A = \{\beta : \beta < \alpha\}$ then $p_\alpha = p_A|X$ and $Y_\alpha = p_\alpha(X)$.

Let $p_\alpha^*(g) = g \circ p_\alpha$ for any $g \in C_p(Y_\alpha)$; then the map $p_\alpha^* : C_p(Y_\alpha) \to C_p(X)$ is an embedding for any $\alpha < \kappa$ and for $Z_\alpha = p_\alpha^*(C_p(Y_\alpha))$ we have $nw(Z_\alpha) = nw(Y_\alpha) < \kappa$. It follows from TFS-298 that any continuous function $f : X \to \mathbb{R}$ factorizes through a $\leq l(X)$-face in $\mathbb{I}^\kappa$, i.e., there is a set $A \subset \kappa$ such that $|A| \leq l(X)$ for which there is a continuous map $g : p_A(X) \to \mathbb{R}$ such that $g \circ (p_A|X) = f$. It follows from $|A| < cf(\kappa)$ that we can find $\alpha < \kappa$ with $A \subset \alpha$. It is easy to see that there exists a continuous map $h : Y_\alpha \to \mathbb{R}$ for which $h \circ p_\alpha = f$ and hence $f = p_\alpha^*(h)$, i.e., $f \in Z_\alpha$. This proves that $\bigcup \{Z_\alpha : \alpha < \kappa\} = C_p(X)$, so $C_p(X)$ is strongly $\kappa$-cosmic.

**V.442.** *Let $\kappa$ be a cardinal of uncountable cofinality. Prove that if $X$ is a Lindelöf $\Sigma$-space with $nw(X) \leq \kappa$ then $C_p(X)$ is strongly $\kappa$-cosmic.*

**Solution.** We have $d(C_p(X)) \leq nw(C_p(X)) = nw(X) \leq \kappa$, so there exists a set $\{g_\alpha : \alpha < \kappa\}$ which is dense in $C_p(X)$; let $P_\alpha = \{g_\beta : \beta < \alpha\}$ for every $\alpha < \kappa$. Any Lindelöf $\Sigma$-space is stable (see SFFS-266), so $C_p(X)$ is monolithic and hence the space $F_\alpha = \overline{P}_\alpha$ is $|\alpha|$-cosmic, i.e., $nw(F_\alpha) \leq |\alpha| < \kappa$ for any $\alpha < \kappa$. Besides, it follows from $t(C_p(X)) = \omega$ and $cf(\kappa) > \omega$ that $\bigcup \{F_\alpha : \alpha < \kappa\} = C_p(X)$, so the space $C_p(X)$ is strongly $\kappa$-cosmic.

**V.443.** *Given a cardinal $\kappa > \omega$ suppose that a space $X$ is strongly $\kappa$-monolithic, i.e., $w(\overline{A}) \leq \kappa$ for every $A \subset X$ with $|A| \leq \kappa$ and, additionally, $l(X) < cf(\kappa)$. Prove that $w(K) < \kappa$ for any compact continuous image $K$ of the space $C_p(X)$.*

*Deduce from this fact that if $X$ is a Lindelöf strongly $\omega_1$-monolithic space (in particular, if $l(X) = \omega$ and $w(X) \leq \omega_1$) then every compact continuous image of $C_p(X)$ is metrizable.*

**Solution.** For any $A \subset X$ let $\pi_A : C_p(X) \to C_p(A) \subset \mathbb{R}^A$ be the restriction map. Suppose that $K$ is a compact space, $\varphi : C_p(X) \to K$ is a continuous onto map and $w(K) \geq \kappa$. Since $C_p(X)$ is a dense subspace of $\mathbb{R}^X$, we can apply Problem 430 to see that $\rho(K) = w(K)$. If $w(K) = \kappa$ then $\mathrm{cf}(\kappa) > \omega$ together with Problem 431 show that $K$ can be continuously mapped onto $\mathbb{I}^\kappa$. On the other hand, if $w(K) > \kappa$ then it follows from $\rho(K) = w(K)$ that $K$ can be continuously mapped onto $\mathbb{I}^{\kappa^+}$ and hence $K$ also maps continuously onto $\mathbb{I}^\kappa$.

Therefore, in all possible cases there is a continuous onto map $q : K \to \mathbb{I}^\kappa$. Since $C_p(X)$ is a dense subspace of $\mathbb{R}^X$, it follows from TFS-299 that we can find a set $A \subset X$ with $|A| \leq \kappa$ and a continuous map $g : \pi_A(C_p(X)) \to \mathbb{I}^\kappa$ for which $g \circ \pi_A = q \circ \varphi$. If $B = \overline{A}$ then it is easy to see that there exists a continuous map $h : \pi_B(C_p(X)) \to K$ such that $h \circ \pi_B = q \circ \varphi$.

By strong $\kappa$-monolithity of $X$ we have $w(B) \leq \kappa$; since $B$ is closed in $X$, we conclude that $l(B) \leq l(X) < \mathrm{cf}(\kappa)$. This makes it possible to apply Problem 441 to see that the space $C_p(B)$ is strongly $\kappa$-cosmic and hence so is $\pi_B(C_p(X)) \subset C_p(B)$. Since the space $\mathbb{I}^\kappa$ is a continuous image of $\pi_B(C_p(X))$, we obtained a contradiction with Problem 439 and hence $w(K) < \kappa$.

Finally, observe that the above result is applicable in the case when $\kappa = \omega_1$ and $X$ is a Lindelöf strongly $\omega_1$-monolithic space. Thus, for every compact continuous image $K$ of the space $C_p(X)$ we have $w(K) < \omega_1$, i.e., $w(K) \leq \omega$ and hence $K$ is metrizable.

**V.444.** *Given a cardinal $\kappa$ with $\mathrm{cf}(\kappa) > \omega$ assume that $X$ is a $\kappa$-monolithic Lindelöf $\Sigma$-space. Prove that if $K$ is a compact continuous image of $C_p(X)$ then $w(K) < \kappa$. In particular, if $X$ is an $\omega_1$-monolithic Lindelöf $\Sigma$-space then every compact continuous image of $C_p(X)$ is metrizable.*

**Solution.** For any $A \subset X$ let $\pi_A : C_p(X) \to C_p(A) \subset \mathbb{R}^A$ be the restriction map. Suppose that $K$ is a compact space, $\varphi : C_p(X) \to K$ is a continuous onto map and $w(K) \geq \kappa$. Since $C_p(X)$ is a dense subspace of $\mathbb{R}^X$, we can apply Problem 430 to see that $\rho(K) = w(K)$. If $w(K) = \kappa$ then $\mathrm{cf}(\kappa) > \omega$ together with Problem 431 show that $K$ can be continuously mapped onto $\mathbb{I}^\kappa$. On the other hand, if $w(K) > \kappa$ then it follows from $\rho(K) = w(K)$ that $K$ can be continuously mapped onto $\mathbb{I}^{\kappa^+}$ and hence $K$ also maps continuously onto $\mathbb{I}^\kappa$. Therefore, in all possible cases there is a continuous onto map $q : K \to \mathbb{I}^\kappa$.

Since $C_p(X)$ is a dense subspace of $\mathbb{R}^X$, it follows from TFS-299 that we can find a set $A \subset X$ with $|A| \leq \kappa$ and a continuous map $g : \pi_A(C_p(X)) \to \mathbb{I}^\kappa$ for which $g \circ \pi_A = q \circ \varphi$. If $B = \overline{A}$ then it is easy to see that there exists a continuous map $h : \pi_B(C_p(X)) \to K$ such that $h \circ \pi_B = q \circ \varphi$.

In particular, the space $C_p(B) = \pi_B(C_p(X))$ maps continuously onto $\mathbb{I}^\kappa$. By $\kappa$-monolithity of the space $X$ we have $nw(B) \leq \kappa$; since $B$ is closed in $X$, it is a Lindelöf $\Sigma$-space, so we can apply Problem 442 to see that $C_p(B)$ is strongly $\kappa$-cosmic; this contradiction with Problem 439 shows that $w(K) < \kappa$.

Finally, observe that the above result is applicable in the case when $\kappa = \omega_1$ and $X$ is an $\omega_1$-monolithic Lindelöf $\Sigma$-space. Thus, for every compact continuous image $K$ of the space $C_p(X)$ we have $w(K) < \omega_1$, i.e., $w(K) \leq \omega$ and hence $K$ is metrizable.

**V.445.** *Let $\kappa$ be a cardinal and denote by $D_\kappa$ a discrete space of cardinality $\kappa$. For the compact space $K = \beta D_\kappa$ prove that $C_p(K)$ maps continuously onto $\mathbb{I}^\kappa$.*

**Solution.** There exists a continuous surjective map $\varphi : C_p(K) \to C_p(K, \mathbb{I})$ (see TFS-092). If $\pi : C_p(K, \mathbb{I}) \to \mathbb{I}^{D_\kappa}$ is the restriction map then we have the equality $\pi(C_p(K, \mathbb{I})) = \mathbb{I}^{D_\kappa}$; therefore $\pi \circ \varphi$ maps $C_p(K)$ continuously onto the space $\mathbb{I}^{D_\kappa}$. It is easy to find a homeomorphism $\xi : \mathbb{I}^{D_\kappa} \to \mathbb{I}^\kappa$; then $\xi \circ \pi \circ \varphi$ maps $C_p(K)$ continuously onto $\mathbb{I}^\kappa$.

**V.446.** *Prove that every one of the following statements is equivalent to Luzin's axiom $(2^{\omega_1} > \mathfrak{c})$:*

 (i) *for every separable compact space $K$ any compact continuous image of $C_p(K)$ is metrizable;*
(ii) *any compact continuous image of $C_p(\beta\omega)$ is metrizable;*
(iii) *any compact continuous image of $C_p(\mathbb{I}^\mathfrak{c})$ is metrizable;*
(iv) *for every compact space $K$ with $w(K) \leq \mathfrak{c}$, any compact continuous image of $C_p(K)$ is metrizable.*

**Solution.** A space $X$ is called *extremally disconnected* if $\overline{U}$ is open for any $U \in \tau(X)$. It is an easy exercise that a space $X$ is extremally disconnected if and only if for any disjoint sets $U, V \in \tau(X)$ we have $\overline{U} \cap \overline{V} = \emptyset$. For any infinite cardinal $\kappa$ let $D_\kappa$ be a discrete space of cardinality $\kappa$.

*Fact 1.* Suppose that $f : X \to Y$ is a closed irreducible onto map. If $Y$ is extremally disconnected then $f$ is a homeomorphism.

*Proof.* For any $U \in \tau(X)$ let $f^\sharp(U) = Y \setminus f(X \setminus U)$; it is evident that $f^\sharp(U) \subset f(U)$ and $f^\sharp(U)$ is open in $Y$ while $f^{-1}(f^\sharp(U)) \subset U$; besides, $f^\sharp(U)$ is dense in $f(U)$ by Fact 1 of S.383.

Suppose that $x, y \in X$ and $x \neq y$. Take disjoint sets $U, V \in \tau(X)$ such that $x \in U$, $y \in V$ and $U \cap V = \emptyset$. It is straightforward that $f^\sharp(U) \cap f^\sharp(V) = \emptyset$. The space $Y$ being extremally disconnected, we have $\overline{f^\sharp(U)} \cap \overline{f^\sharp(V)} = \emptyset$. Furthermore, $\overline{f(U)} = \overline{f^\sharp(U)}$ and $\overline{f(V)} = \overline{f^\sharp(V)}$ which shows that $\overline{f(U)} \cap \overline{f(V)} = \emptyset$. Since $x \in f(U)$ and $y \in f(V)$, we conclude that $f(x) \neq f(y)$ whenever $x$ and $y$ are distinct points of $X$. Therefore $f$ is a bijection, so it is a homeomorphism by TFS-155 and hence Fact 1 is proved.

*Fact 2.* For any cardinal $\kappa$ the space $\beta D_\kappa$ is extremally disconnected.

*Proof.* If $\kappa$ is finite then $\beta D_\kappa = D_\kappa$ is a finite space, so it is discrete and hence extremally disconnected. Now if $\kappa \geq \omega$ then take any set $U \in \tau(\beta D_\kappa)$ and let $V = U \cap D_\kappa$. Then $\overline{U} = \overline{V}$ because $D_\kappa$ is dense in $\beta D_\kappa$. Besides, $\overline{V}$ is clopen in

$\beta D_\kappa$ by Fact 2 of T.371, so $\overline{U}$ is clopen in $\beta D_\kappa$ for any $U \in \tau(\beta D_\kappa)$ and hence the space $\beta D_\kappa$ is extremally disconnected, i.e., Fact 2 is proved.

*Fact 3.* If $2^{\omega_1} = \mathfrak{c}$ and $D$ is a discrete space of cardinality $\omega_1$ then the space $\beta D$ embeds in $\beta \omega$.

*Proof.* Since $d(\beta(D)) \leq |D| = \omega_1$, we can apply Fact 2 of S.368 to see that $w(\beta D) \leq 2^{\omega_1} = \mathfrak{c}$ and hence we can choose a set $F \subset \mathbb{I}^\mathfrak{c}$ which is homeomorphic to $\beta D$. Take a countable set $E \subset \mathbb{I}^\mathfrak{c}$ which is dense in $\mathbb{I}^\mathfrak{c}$ and choose a surjection $\varphi : \omega \to E$. The map $\varphi$ is continuous because $\omega$ is discrete, so there exists a continuous map $\Phi : \beta \omega \to \mathbb{I}^\mathfrak{c}$ such that $\Phi|\omega = \varphi$ (see TFS-257). The set $\Phi(\beta \omega)$ contains the dense set $E$ of the space $\mathbb{I}^\mathfrak{c}$, so $\Phi(\beta \omega) = \mathbb{I}^\mathfrak{c}$.

Consider the set $G = \Phi^{-1}(F)$ and let $\xi = \Phi|G$; then $\xi : G \to F$ is a perfect map. Therefore there exists a closed set $G' \subset G$ such that $\xi(G') = F$ and $\xi|G'$ is irreducible (see TFS-366). Therefore $\xi|G' : G' \to F$ is a homeomorphism by Fact 1, so $F$ embeds in $\beta \omega$ and hence $\beta D$ embeds in $\beta \omega$, i.e., Fact 3 is proved.

Returning to our solution observe that (iv)$\Longrightarrow$(iii) because $w(\mathbb{I}^\mathfrak{c}) \leq \mathfrak{c}$. If a compact space $K$ has weight at most $\mathfrak{c}$ then $K$ embeds in $\mathbb{I}^\mathfrak{c}$ and hence we can consider that $K \subset \mathbb{I}^\mathfrak{c}$. The restriction $\pi_K : C_p(\mathbb{I}^\mathfrak{c}) \to C_p(K)$ is a continuous onto map and hence any continuous image of $C_p(K)$ is also a continuous image of $C_p(\mathbb{I}^\mathfrak{c})$; thus (iii)$\Longrightarrow$(iv), i.e., the statements (iii) and (iv) are equivalent.

It is evident that (i)$\Longrightarrow$(ii); since $\mathbb{I}^\mathfrak{c}$ is separable, we also have (i)$\Longrightarrow$(iii). Any separable Tychonoff space has weight at most $\mathfrak{c}$ (see Fact 2 of S.368), so (iv)$\Longrightarrow$(i) which shows that the properties (i),(iii) and (iv) are all equivalent and each one of them implies (ii).

Now, if Luzin's Axiom holds then $|C_p(\mathbb{I}^\mathfrak{c})| = \mathfrak{c} < 2^{\omega_1} = |\mathbb{I}^{\omega_1}|$, so the space $C_p(\mathbb{I}^\mathfrak{c})$ cannot be mapped (even discontinuously) onto $\mathbb{I}^{\omega_1}$. If $K$ is a compact continuous image of $C_p(\mathbb{I}^\mathfrak{c})$ then $K$ cannot be mapped onto $\mathbb{I}^{\omega_1}$ either so $w(K) = \rho(K) \leq \omega$ (see Problem 430) and hence $K$ is metrizable.

Thus, Luzin Axiom implies (iii); since (iii) implies all the statements (i)–(iv), Luzin's Axiom implies (i)–(iv).

Now, assume that $2^{\omega_1} = \mathfrak{c}$ (i.e., Luzin's Axiom does not hold) and let $D$ be a discrete space of cardinality $\omega_1$. By Fact 3 we can consider that $\beta D$ is a subspace of $\beta \omega$. The restriction $\pi_{\beta D} : C_p(\beta \omega) \to C_p(\beta D)$ is a continuous onto map and the space $C_p(\beta D)$ can be mapped on $\mathbb{I}^{\omega_1}$ by Problem 445, so $C_p(\beta \omega)$ also maps continuously onto $\mathbb{I}^{\omega_1}$. Therefore (ii) does not hold and hence (ii) implies Luzin's Axiom together with all statements (i)–(iv). Thus Luzin's Axiom is equivalent to every one of the statements (i)–(iv), i.e., our solution is complete.

**V.447.** *Under Luzin's Axiom prove that*

(a) *if $X$ is Lindelöf and $w(X) \leq \mathfrak{c}$ then every compact continuous image of $C_p(X)$ is metrizable.*

(b) *if $X$ is Lindelöf and first countable then every compact continuous image of $C_p(X)$ is metrizable.*

**Solution.** (a) We have $|C_p(X)| \leq w(X)^{l(X)} \leq 2^\omega$ (see Fact 7 of U.074), so it follows from Luzin's Axiom that $|C_p(X)| < 2^{\omega_1} = |\mathbb{I}^{\omega_1}|$ and hence $C_p(X)$ cannot be (continuously) mapped onto $\mathbb{I}^{\omega_1}$. If $K$ is a compact continuous image of $C_p(X)$ then $K$ does not map onto $\mathbb{I}^{\omega_1}$ either and hence $\rho(K) \leq \omega$. Applying Problem 430 we conclude that $w(K) = \rho(K) \leq \omega$ and hence $K$ is metrizable.
(b) It follows from Fact 2 of V.403 that $|X| \leq 2^{l(X) \cdot \chi(X)} \leq 2^\omega$. Take a countable local base $\mathcal{B}_x$ at every point $x \in X$. Then $\mathcal{B} = \bigcup \{\mathcal{B}_x : x \in X\}$ is a base in $X$ such that $|\mathcal{B}| \leq \mathfrak{c} \cdot \omega = \mathfrak{c}$. Therefore $w(X) \leq \mathfrak{c}$ and hence we can apply (a) to conclude that every compact continuous image of $C_p(X)$ is metrizable.

**V.448.** *Prove that, for every hereditarily Lindelöf first countable space $X$, any compact continuous image of $C_p(X)$ is metrizable. In particular, for each perfectly normal compact space $X$, any compact continuous image of $C_p(X)$ is metrizable.*

**Solution.** For any $A \subset X$ let $\pi_A : C_p(X) \to C_p(A)$ be the restriction map. If some compact continuous image $K$ of the space $C_p(X)$ is not metrizable then we have $\rho(K) = w(K) > \omega$ (see Problem 430) and hence $K$ maps continuously onto $\mathbb{I}^{\omega_1}$. Therefore there exists a continuous surjective map $\varphi : C_p(X) \to \mathbb{I}^{\omega_1}$. By TFS-299 we can find a set $A \subset X$ such that $|A| \leq \omega_1$ and there is a continuous map $q : \pi_A(C_p(X)) \to \mathbb{I}^{\omega_1}$ for which $q \circ \pi_A = \varphi$.

Fix a countable local base $\mathcal{B}_x$ at every point $x \in X$ and consider the family $\mathcal{B} = \bigcup \{\mathcal{B}_x : x \in A\}$; it is clear that $|\mathcal{B}| \leq \omega_1$ and $\{B \cap A : B \in \mathcal{B}\}$ is a base in $A$. Thus $w(A) \leq \omega_1$ and hence $A$ is a Lindelöf space of weight $\leq \omega_1$ which implies that $C_p(A)$ is strongly $\omega_1$-cosmic by Problem 441. The space $\pi_A(C_p(X)) \subset C_p(A)$ is also strongly $\omega_1$-cosmic and hence so is $\mathbb{I}^{\omega_1}$ (see Problem 438) which is a contradiction with Problem 439.

Finally observe that if $X$ is a perfectly normal compact space then $\chi(X) \leq \omega$ and $X$ is hereditarily Lindelöf by SFFS-001. Therefore every compact continuous image of $C_p(X)$ is metrizable.

**V.449.** *Suppose that $X$ is a space such that $l(C_p(X)) = t(C_p(X)) = \omega$. Prove that every compact continuous image of $C_p(X)$ is metrizable.*

**Solution.** Given a space $Z$ recall that a set $S = \{x_\alpha : \alpha < \kappa\} \subset Z$ is called *a free sequence in $Z$* if $\overline{\{x_\beta : \beta < \alpha\}} \cap \overline{\{x_\beta : \beta \geq \alpha\}} = \emptyset$ for any $\alpha < \kappa$. The cardinal $\kappa$ is called the length of the free sequence $S$.

*Fact 1.* Suppose that $\kappa$ is an infinite cardinal and $Z$ is a space such that $l(Z) \leq \kappa$ and $t(Z) \leq \kappa$. Then $Z$ has no free sequences of length $\kappa^+$.

*Proof.* Assume that $S = \{x_\alpha : \alpha < \kappa^+\}$ is a free sequence in $Z$ and consider the sets $F_\alpha = \overline{\{x_\beta : \beta \geq \alpha\}}$ and $G_\alpha = \overline{\{x_\beta : \beta < \alpha\}}$ for any $\alpha < \kappa^+$. Let $U_\alpha = Z \backslash F_\alpha$ for each $\alpha < \kappa^+$; if $\bigcap\{F_\alpha : \alpha < \kappa^+\} = \emptyset$ then the family $\{U_\alpha : \alpha < \kappa^+\}$ is an open cover of the space $Z$ such that $\alpha < \alpha'$ implies $U_\alpha \subset U_{\alpha'}$. It follows from $l(Z) \leq \kappa$ that there exists $\beta < \kappa^+$ such that $U = \bigcup\{U_\alpha : \alpha < \beta\} = Z$. However, $x_\beta \notin U$ which is a contradiction. Consequently, there exists a point $x \in F = \bigcap_{\alpha < \kappa^+} F_\alpha$.

It follows from $x \in \overline{\{x_\alpha : \alpha < \kappa^+\}} = F_0$ that we can choose a set $A \subset \kappa^+$ such that $|A| \leq \kappa$ and $x \in \overline{\{x_\alpha : \alpha \in A\}}$; pick an ordinal $\beta < \kappa^+$ with $A \subset \beta$; then $x \in \overline{\{x_\alpha : \alpha \in A\}} \subset G_\beta$. It follows from $x \in F$ that $x \in F_\beta$ and hence $x \in F_\beta \cap G_\beta$ which is a contradiction with the definition of free sequence. Thus $Z$ has no free sequences of length $\kappa^+$, i.e., Fact 1 is proved.

Returning to our solution suppose that $\varphi : C_p(X) \to K$ is a continuous onto map of $C_p(X)$ onto a compact space $K$. If $\{x_\alpha : \alpha < \omega_1\}$ is a free sequence in $K$ then pick a point $f_\alpha \in \varphi^{-1}(x_\alpha)$ for every $\alpha < \omega_1$. It is straightforward that $\{f_\alpha : \alpha < \omega_1\}$ is a free sequence in $C_p(X)$ which contradicts $t(C_p(X)) = l(C_p(X)) = \omega$ (see Fact 1).

Therefore the space $K$ has no uncountable free sequences and hence $t(K) \leq \omega$ by TFS-328. Applying Problem 430 we conclude that $w(K) = t(K) \leq \omega$, so $K$ is metrizable and hence our solution is complete.

**V.450.** *Prove that if $X$ is a space such that $C_p(X)$ is a Lindelöf $\Sigma$-space then every compact continuous image of $C_p(X)$ is metrizable.*

**Solution.** Denote by $\upsilon X$ the Hewitt realcompactification of the space $X$. It is evident that $C_p(X)$ is a continuous image of the space $C_p(\upsilon X)$. Besides, $Y = \upsilon X$ is a Lindelöf $\Sigma$-space (see CFS-206) and hence $t(C_p(Y)) = \omega$; since $C_p(Y)$ is also a Lindelöf $\Sigma$-space (see CFS-234), we can apply Problem 449 to see that every compact continuous image of $C_p(Y)$ is metrizable. Therefore every compact continuous image of $C_p(X)$ is also metrizable.

**V.451.** *For the double arrow space $K$ prove that the space $C_p(K)$ can be condensed onto $\mathbb{I}^\omega$.*

**Solution.** Suppose that $Z$ is a space, $f : Z \to \mathbb{R}$ and $z \in Z$. Then the number $osc(f,z) = \inf\{diam(f(U)) : U \in \tau(z,Z)\}$ is called the *oscillation* of $f$ at the point $z$. Recall that $f$ is continuous at the point $z$ if and only if $osc(f,z) = 0$ (Fact U.347).

As usual, $\mathbb{D} = \{0,1\}$; given $a,b \in \mathbb{R}$ we will have to distinguish the point $(a,b) \in \mathbb{R}^2$ from the interval $(a,b) = \{t \in \mathbb{R} : a < t < b\} \subset \mathbb{R}$. To avoid confusion, the point in the plane whose coordinates are $a$ and $b$ will be denoted by $\langle a,b \rangle$. Recall that $K = K_0 \cup K_1 \subset \mathbb{R}^2$ where $K_0 = (0,1] \times \{0\}$ and $K_1 = [0,1) \times \{1\}$. For any $x = \langle t,i \rangle \in K$ let $p(x) = t$; thus $p : K \to [0,1]$ is the restriction to $K$ of the respective projection of the plane.

If $x = \langle t,0 \rangle \in K_0$ and $0 < a < t$ then $O_a(x) = \{x\} \cup ((a,t) \times \mathbb{D})$. The family $\{O_a(x) : 0 < a < p(x)\}$ is a local base of $K$ at the point $x$. If $x = \langle t,1 \rangle \in K_1$ and $t < a < 1$ then $O_a(x) = \{x\} \cup ((t,a) \times \mathbb{D})$. The family $\{O_a(x) : p(x) < a < 1\}$ is a local base of $K$ at the point $x$.

We will also need the lexicographical order $\prec$ on the set $K$. Given two points $x = \langle t,i \rangle$ and $y = \langle s,j \rangle$ of the set $K$ say that $x \prec y$ if either $t < s$ or $t = s$ and $i < j$. It is an easy exercise that $(K, \prec)$ is a linearly ordered set and the topology of $K$ is generated by the order $\prec$.

The space $K$ is first countable (TFS-384), so we can fix, for any $x \in K$, a local base $\{U_n(x) : n \in \omega\}$ of the space $K$ at the point $x$ such that $U_{n+1} \subset U_n$ for any $n \in \omega$. For any $A \subset K$ let $\pi_A : C_p(K) \to C_p(A)$ be the restriction map and let $C_p(A|K) = \pi_A(C_p(K))$.

*Fact 1.* If $A$ is a countable dense subset of the double arrow space $K$ then $C_p(A|K)$ is an $F_{\sigma\delta}$-subset of $\mathbb{R}^A$ and, in particular, $C_p(A|K)$ is a Borel set.

*Proof.* Given any points $x, y \in A$ and $m \in \omega$ it is easy to see that the set $D(x, y, m) = \{f \in \mathbb{R}^A : |f(x) - f(y)| \le 2^{-m}\}$ is closed in $\mathbb{R}^A$. The oscillation of a function $f \in \mathbb{R}^A$ at a point $a \in A$ is equal to zero if and only if for any $m \in \omega$ there exists $n \in \omega$ such that $|f(x) - f(y)| \le 2^{-m}$ for all $x, y \in U_n(a) \cap A$. The set $E(m, n, a) = \bigcap\{D(x, y, m) : x, y \in U_n(a) \cap A\}$ is also closed in $\mathbb{R}^A$ for any $m, n \in \omega$ and $a \in A$; it is immediate that a function $f \in \mathbb{R}^A$ is continuous at $a$ if and only if $f \in \bigcap_{m \in \omega} \bigcup_{n \in \omega} E(m, n, a)$. Therefore a function $f \in \mathbb{R}^A$ is continuous on $A$ if and only if $f \in E = \bigcap_{a \in A} \bigcap_{m \in \omega} \bigcup_{n \in \omega} E(m, n, a)$, i.e., $C_p(A) = E$. It is clear that $E$ is an $F_{\sigma\delta}$-subset of $\mathbb{R}^A$, so we proved that

(1)  $C_p(A)$ is an $F_{\sigma\delta}$-subset of $\mathbb{R}^A$.

However, the set $C_p(A|K)$ is strictly smaller than $C_p(A)$, so we need something else to characterize $C_p(A|K)$. Consider the following property of a function $f \in \mathbb{R}^A$.

($\theta$)  for any sequence $S = \{a_n : n \in \omega\} \subset A$ if $S \to x$ and $x \in K \backslash S$ then the sequence $\{f(a_n) : n \in \omega\}$ is convergent.

Let $P$ be the set of all functions from $\mathbb{R}^A$ which have the property ($\theta$); it is clear that $C_p(A|K) \subset P \cap C_p(A)$. To prove the reverse inclusion take any function $f \in C_p(A)$ which has the property ($\theta$). For any point $x \in K$ there is a sequence $\{a_n : n \in \omega\} \subset A \backslash \{x\}$ which converges to $x$; let $g(x)$ be the limit of the sequence $\{f(a_n) : n \in \omega\}$ which exists by ($\theta$).

If $\{b_n : n \in \omega\} \subset A \backslash \{x\}$ is any other sequence that converges to $x$ and $\{f(b_n) : n \in \omega\}$ does not converge to $g(x)$ then let $z_{2n} = a_n$ and $z_{2n+1} = b_n$ for any $n \in \omega$. The sequence $\{z_n : n \in \omega\} \subset A \backslash \{x\}$ still converges to $x$ but $\{f(z_n) : n \in \omega\}$ is not convergent; this contradiction with ($\theta$) shows that we have consistently defined a function $g : K \to \mathbb{R}$ such that $g|A = f$.

Assume that the function $g$ is discontinuous at some point $x \in K$ and take $\varepsilon > 0$ such that $\text{diam}(g(U)) > \varepsilon$ for any $U \in \tau(x, K)$. For every $n \in \omega$ fix points $y_n, z_n \in U_n(x)$ such that $|g(y_n) - g(z_n)| > \varepsilon$. It is easy to find sequences $\{a_k^n : k \in \omega\} \subset (A \cap U_n(x)) \backslash \{x, y_n, z_n\}$ and $\{b_k^n : k \in \omega\} \subset (A \cap U_n(x)) \backslash \{x, y_n, z_n\}$ such that $a_k^n \to y_n$ and $b_k^n \to z_n$ (when $k \to \infty$). By our construction of the function $g$ we have $g(a_k^n) \to g(y_n)$ and $g(b_k^n) \to g(z_n)$ when $k \to \infty$. Therefore we can choose $k_n, l_n \in \omega$ such that $|g(a_{k_n}^n) - g(y_n)| < \frac{\varepsilon}{3}$ and $|g(b_{l_n}^n) - g(z_n)| < \frac{\varepsilon}{3}$.

It is easy to see that $\{a_{k_n}^n, b_{l_n}^n : n \in \omega\} \subset A \backslash \{x\}$ is a sequence which converges to $x$. Therefore the sequence $S = \{g(a_{k_n}^n), g(b_{l_n}^n) : n \in \omega\}$ must be convergent. If, for some $n \in \omega$ we have $|g(a_{k_n}^n) - g(b_{l_n}^n)| < \frac{\varepsilon}{3}$ then

$$|g(y_n) - g(z_n)| \le |g(y_n) - g(a_{k_n}^n)| + |g(a_{k_n}^n) - g(b_{l_n}^n)| + |g(b_{l_n}^n) - g(z_n)| < \varepsilon,$$

which is a contradiction. Therefore $|g(a_{k_n}^n) - g(b_{l_n}^n)| \ge \frac{\varepsilon}{3}$ for all $n \in \omega$ and hence the sequence $S$ cannot be convergent. This contradiction shows that $g$ is continuous and hence we proved that $C_p(A|K)$ consists of the functions from $C_p(A)$ which satisfy the condition $(\theta)$.

Let us also consider the following condition for a function $f \in \mathbb{R}^A$.

$(\Theta)$ For each $m \in \omega$ we can find $n \in \omega$ such that for any $a_0, \ldots, a_n \in A$ with
$a_0 \prec \ldots \prec a_n$, there exists $i < n$ for which $|f(a_i) - f(a_{i+1})| \le 2^{-m}$.

Denote by $Q$ the set of functions from $\mathbb{R}^A$ which satisfy the condition $(\Theta)$. To show that $C_p(A|K) \subset Q$ take any $f \in C_p(A|K)$; then $f = g|A$ for some $g \in C_p(K)$. Given any $m \in \omega$ it follows from compactness of $K$ that there exists an open cover $\{U_0, \ldots, U_k\}$ of the space $K$ such that $\text{diam}(g(U_i)) < 2^{-m}$ for every $i \le k$. Making the sets $U_i$ smaller if necessary we can assume that every $U_i$ is an interval with respect to the order $\prec$.

Let $n = k + 1$ and take any points $a_0, \ldots, a_n \in A$ such that $a_0 \prec \ldots \prec a_n$. It follows from $n > k$ that there are $i, j \le n$ such that $i < j$ and $a_i, a_j \in U_l$ for some $l \le k$. Since the set $U_l$ is an interval, it follows from $a_i \prec a_{i+1} \preceq a_j$ that $a_{i+1} \in U_l$, so $|f(a_i) - f(a_{i+1})| = |g(a_i) - g(a_{i+1})| \le \text{diam}(U_l) \le 2^{-m}$ which shows that $f$ has the property $(\Theta)$ and hence we established that $C_p(A|K) \subset Q$.

It turns out that $Q \subset P$ or, equivalently, the property $(\Theta)$ for a function $f \in \mathbb{R}^A$ implies that $f$ satisfies $(\theta)$. To see this, assume that a function $f \in \mathbb{R}^A$ has $(\Theta)$ and take any sequence $S = \{a_i : i \in \omega\} \subset A$ which converges to some point $x \notin S$. If the sequence $\{f(a_i) : i \in \omega\}$ is not convergent then it is not a Cauchy sequence and hence there exists $\varepsilon > 0$ and a sequence $\{i_k, j_k : k \in \omega\} \subset \omega$ such that $i_k < j_k < i_{k+1}$ for all $k \in \omega$ while $|f(a_{i_k}) - f(a_{j_k})| > \varepsilon$ for all $k \in \omega$.

From now on we assume that $x = \langle t, 0 \rangle$ for some $t \in (0, 1]$; it is an exercise for the reader to carry out the respective adjustments to make the proof work in the case when $x \in K_1$. It follows from $S \to x$ that the set $\{i \in \omega : a_i \succ x\}$ is finite and the set $\{i \in \omega : a_i \prec y\}$ is finite for any $y \prec x$.

Let $b_0 = a_{i_0}$; proceeding inductively assume that $n \in \omega$ and we have points $b_0, \ldots, b_n \in S$ such that $b_0 \prec \ldots \prec b_n$ and $|f(b_i) - f(b_{i+1})| > \frac{\varepsilon}{2}$ for any $i < n$. Since only finitely many elements of $S$ are below $b_n$ under the order $\prec$, we can find $k \in \omega$ such that $a_{i_k} > b_n$ and $a_{j_k} > b_n$. If $|f(a_{i_k}) - f(b_n)| \le \frac{\varepsilon}{2}$ and $|f(a_{j_k}) - f(b_n)| \le \frac{\varepsilon}{2}$ then $|f(a_{i_k}) - f(a_{j_k})| \le \varepsilon$ which is a contradiction. Therefore we can find a point $b_{n+1} \in \{a_{i_k}, a_{j_k}\}$ such that $|f(b_n) - f(b_{n+1})| > \frac{\varepsilon}{2}$.

Therefore our inductive procedure can be continued to construct a sequence $S' = \{b_i : i \in \omega\} \subset S$ such that $b_i \prec b_{i+1}$ and $|f(b_i) - f(b_{i+1})| > \frac{\varepsilon}{2}$ for all $i \in \omega$. Take $m \in \omega$ such that $2^{-m} < \frac{\varepsilon}{2}$ and observe that the sequence $S'$ is a counterexample to the existence of the number $n$ promised in $(\Theta)$. This contradiction shows that $Q \subset P$ and hence $C_p(A|K)$ consists of functions from $C_p(A)$ with the property $(\Theta)$, i.e., $C_p(A|K) = Q \cap C_p(A)$.

For arbitrary points $a_0, \ldots, a_n \in A$ and any number $m \in \omega$ consider the set $H(a_0, \ldots, a_n) = \{f \in \mathbb{R}^A : |f(a_i) - f(a_{i+1})| \le 2^{-m} \text{ for some } i < n\}$; it follows from the equality $H(a_0, \ldots, a_n) = \bigcup \{ D(a_i, a_{i+1}, m) : i < n \}$ that $H(a_0, \ldots, a_n)$ is closed in $\mathbb{R}^A$. It is straightforward that $Q = \bigcap_{m \in \omega} \bigcup_{n \in \omega} \bigcap \{ H(a_0, \ldots, a_n) : a_0, \ldots, a_n \in A$ and $a_0 \prec \ldots \prec a_n \}$, so $Q$ is and $F_{\sigma\delta}$-subset of $\mathbb{R}^A$. Applying (1) we conclude that $C_p(A|K) = Q \cap C_p(A)$ is an $F_{\sigma\delta}$-subset of $\mathbb{R}^A$ and hence Fact 1 is proved.

Returning to our solution take a countable dense subspace $A$ of the space $K$. It is clear that $\pi_A : C_p(K) \to C_p(A|K)$ is a condensation. By Fact 1 the space $C_p(A|K)$ is a Borel set. If $C_p(A|K)$ is $\sigma$-compact then $A$ is a $P$-space (see Fact 4 of S.186); any countable $P$-space is discrete while $A$ has no isolated points because $K$ has no isolated points. This contradiction shows that $C_p(A|K)$ is not $\sigma$-compact, so we can apply SFFS-354 to see that $C_p(A|K)$ can be condensed onto $\mathbb{I}^\omega$. Therefore $C_p(K)$ also condenses onto $\mathbb{I}^\omega$ and hence our solution is complete.

**V.452.** *Prove that $C_p(\mathbb{D}^c)$ condenses onto $\mathbb{I}^\omega$.*

**Solution.** Suppose that $Z$ is a space and $A$ is a dense subspace of $Z$. Given a continuous function $\varphi : A \to \mathbb{R}$ let $\mathrm{osc}(\varphi, z) = \inf \{ \mathrm{diam}(\varphi(U \cap A)) : U \in \tau(z, Z) \}$ for any $z \in Z$; the number $\mathrm{osc}(\varphi, z)$ is called the *oscillation* of $\varphi$ at the point $z$. If $Z$ is a space and $D \subset Z$ then $\pi_D : C_p(Z) \to C_p(D)$ is the restriction map and $C_p(D|Z) = \pi_D(C_p(Z))$.

**Fact 1.** Suppose that $M$ is a compact space and $D$ is a countable subset of $C_p(M)$ such that the closure $K$ of the set $D$ in the space $\mathbb{R}^M$ is compact. Then $C_p(D|K)$ is an $F_{\sigma\delta}$-subset of $\mathbb{R}^D$.

*Proof.* To characterize the set $C_p(D|K)$ we must describe what functions from $C_p(D)$ can be extended over $K$. Given any $k \in \mathbb{N}$ and $t = (t_1, \ldots, t_k) \in M^k$ let $\mathrm{coord}(t) = \{ t_1, \ldots, t_k \}$. For an arbitrary finite set $A \subset M$ and $f, g \in D$ let $\rho_A(f, g) = \sup \{ |f(a) - g(a)| : a \in A \}$; clearly, $O_A(f, \delta) = \{ g \in \mathbb{R}^M : \rho_F(f, g) < \delta \}$ is an open subset of $\mathbb{R}^M$ for any $\delta > 0$, $f \in \mathbb{R}^M$ and finite $A \subset M$.

The following condition on a function $\varphi \in \mathbb{R}^D$ turns out to be a characterization of functions from $C_p(D|K)$.

$(\theta)$ for any $m \in \omega$ there exists a number $n \in \omega$ and a finite set $A \subset M$ such that $|\varphi(f) - \varphi(g)| \leq 2^{-m}$ whenever $f, g \in D$ and $\rho_A(f, g) < 2^{-n}$.

Denote by $P$ the set of all functions from $\mathbb{R}^D$ which have the property $(\theta)$ and fix a function $\varphi \in P$. Given any $h \in K$ and $\varepsilon > 0$ take $m \in \omega$ with $2^{-m} < \varepsilon$ and apply $(\theta)$ to find a number $n \in \omega$ and a finite set $A \subset M$ such that $\rho_A(f, g) < 2^{-n}$ implies $|\varphi(f) - \varphi(g)| < 2^{-m}$ for any $f, g \in D$. The set $U = \{ f \in K : \rho_A(f, h) < 2^{-n-1} \}$ is an open neighborhood of $h$ in $K$. If $f, g \in U \cap D$ then it is easy to see that $\rho_A(f, g) \leq \rho_A(f, h) + \rho_A(h, g) < 2^{-n}$ and hence $|\varphi(f) - \varphi(g)| \leq 2^{-m} < \varepsilon$; this shows that $\mathrm{diam}(\varphi(U \cap D)) \leq \varepsilon$. Since $\varepsilon > 0$ was chosen arbitrarily, we proved that, for any $\varepsilon > 0$, there exists a set $U \in \tau(h, K)$ such that $\mathrm{diam}(\varphi(U \cap D)) \leq \varepsilon$; therefore $\mathrm{osc}(\varphi, h) = 0$ for any $h \in K$. Applying Fact 3 of T.368 we conclude that the function $\varphi$ can be continuously extended over $K$, i.e., $\varphi \in C_p(D|K)$ and therefore $P \subset C_p(D|K)$.

To prove the inclusion $C_p(D|K) \subset P$ take any function $\xi \in C_p(K)$ and let $\varphi = \xi|D$. Fix any $m \in \omega$; by continuity of $\xi$, for any $h \in K$ we can choose a finite set $A_h \subset M$ and $\delta_h > 0$ such that the diameter of the set $\xi(O_{A_h}(h, 2\delta_h) \cap K)$ does not exceed $2^{-m}$.

The space $K$ being compact we can extract a finite subcover of the cover $\{O_{A_h}(h, \delta_h) : h \in K\}$, so fix a finite $F \subset K$ such that $K \subset \bigcup\{O_{A_h}(h, \delta_h) : h \in F\}$. Let $A = \bigcup\{A_h : h \in F\}$, $\delta = \backslash n\{\delta_h : h \in F\}$ and pick a number $n \in \omega$ such that $2^{-n} < \delta$.

Given any $f, g \in D$ with $\rho_A(f, g) < 2^{-n}$ take a function $h \in F$ such that $f \in O_{A_h}(h, \delta_h)$. We have $|f(x) - g(x)| < 2^{-n} < \delta \leq \delta_h$ for any $x \in A$ and hence these inequalities also hold for any $x \in A_h$. It is easy to see that this implies $g \in O_{A_h}(h, 2\delta_h)$, so $|\varphi(f) - \varphi(g)| \leq \operatorname{diam}(\xi(O_{A_h}(h, 2\delta_h) \cap K)) \leq 2^{-m}$. This proves that any function from $C_p(D|K)$ has $(\theta)$ and hence $C_p(D|K) \subset P$, so we established that $P = C_p(D|K)$.

For any $m, n \in \omega$ and $k \in \mathbb{N}$ consider the set $F(m, n, k) = \{(\varphi, t) \in \mathbb{R}^D \times M^k :$ for any $f, g \in D$ either $\rho_{\operatorname{coord}(t)}(f, g) \geq 2^{-n}$ or $|\varphi(f) - \varphi(g)| \leq 2^{-m}\}$.

Take any $(\xi, s) \in (\mathbb{R}^D \times M^k) \backslash F(m, n, k)$; there exist functions $f, g \in D$ such that $\rho_{\operatorname{coord}(s)}(f, g) < 2^{-n}$ and $|\xi(f) - \xi(g)| > 2^{-m}$. Let $s = (s_1, \ldots, s_k)$; the functions $f$ and $g$ being continuous on $M$, the set

$$W = \{t = (t_1, \ldots, t_k) \in M^k : |f(t_i) - g(t_i)| < 2^{-n} \text{ for all } i \leq k\}$$

is open in $M^k$ and $s \in W$. It is clear that $G = \{\varphi \in \mathbb{R}^D : |\varphi(f) - \varphi(g)| > 2^{-m}\}$ is open in $\mathbb{R}^D$ and $\xi \in G$. We have $(\xi, s) \in G \times W$ and $(G \times W) \cap F(m, n, k) = \emptyset$, so every point of $(\mathbb{R}^D \times M^k) \backslash F(m, n, k)$ has a neighborhood contained in the set $(\mathbb{R}^D \times M^k) \backslash F(m, n, k)$. Therefore the complement of $F(m, n, k)$ is open in $\mathbb{R}^D \times M^k$ and hence $F(m, n, k)$ is closed in $\mathbb{R}^D \times M^k$ for any $m, n \in \omega$ and $k \in \mathbb{N}$.

Let $\pi : \mathbb{R}^D \times M^k \to \mathbb{R}^D$ be the natural projection. The space $M^k$ being compact, the map $\pi$ is closed (see Fact 3 of S.288), so the set $E(m, n, k) = \pi(F(m, n, k))$ is closed in $\mathbb{R}^D$ for all $m, n \in \omega$ and $k \in \mathbb{N}$. Now it is easy to see that a function $\varphi \in \mathbb{R}^D$ has $(\theta)$ if and only if $\varphi \in \bigcap_{m \in \omega} \bigcup_{n \in \omega} \bigcup_{k \in \mathbb{N}} E(m, n, k)$.

Thus we have the equality $P = \bigcap_{m \in \omega} \bigcup_{n \in \omega} \bigcup_{k \in \mathbb{N}} E(m, n, k) = C_p(D|K)$ which shows that $C_p(D|K)$ is an $F_{\sigma\delta}$-subset of $\mathbb{R}^D$ and hence Fact 1 is proved.

Returning to our solution consider the compact space $M = \mathbb{D}^\omega$; then the space $K = \mathbb{D}^M$ is compact. The space $D = C_p(M, \mathbb{D}) \subset C_p(M)$ is countable and dense in $K$ (see Fact 1 of U.077 and Fact 1 of S.390). Therefore Fact 1 can be applied to see that $C = C_p(D|K)$ is an $F_{\sigma\delta}$-subset of $\mathbb{R}^D$, so $C$ is a Borel set. If $C$ is $\sigma$-compact then $D$ has to be a $P$-space by Fact 4 of S.186. Any countable $P$-space is discrete; since $D$ has no isolated points, we obtained a contradiction which shows that $C$ is not $\sigma$-compact. Thus we can apply SFFS-354 to see that $C$ can be condensed onto $\mathbb{I}^\omega$. The space $C_p(K)$ condenses onto $C$, so $C_p(K)$ can also be condensed onto $\mathbb{I}^\omega$. Finally observe that $K$ is homeomorphic to $\mathbb{D}^{\mathfrak{c}}$ and hence $C_p(\mathbb{D}^{\mathfrak{c}})$ condenses onto the space $\mathbb{I}^\omega$.

**V.453.** *Suppose that $X$ is a nonempty $\sigma$-compact second countable space. Prove that $C_p(X)$ condenses onto a compact space.*

**Solution.** Observe first that $\mathbb{R}$ is a locally compact space, so it condenses onto a compact space $K$ by Fact 3 of T.357. Therefore $\mathbb{R}^X$ condenses onto the compact space $K^X$. This shows that if $X$ is discrete then $C_p(X) = \mathbb{R}^X$ condenses onto a compact space.

Now assume that $X$ is not discrete and fix a non-isolated point $a \in X$. It is easy to find a countable dense set $A \subset X$ such that $a \in A$ and hence the point $a$ is not isolated in $A$, which implies that the space $A$ is not discrete. The restriction map $\pi_A : C_p(X) \to C_p(A)$ is a condensation of $C_p(X)$ onto $C_p(A|X) = \pi_A(C_p(X))$.

It follows from SFFS-368 that $C_p(A|X)$ is a Borel set in $\mathbb{R}^A$. If the space $C_p(A|X)$ is $\sigma$-compact then $A$ has to be a $P$-space by Fact 4 of S.186; since any countable $P$-space must be discrete, $A$ is discrete which is a contradiction. Therefore $C_p(A|X)$ is not $\sigma$-compact so there exists a condensation $\varphi : C_p(A|X) \to \mathbb{I}^\omega$ by SFFS-354. The map $\varphi \circ \pi_A$ is a condensation of $C_p(X)$ onto $\mathbb{I}^\omega$, so $C_p(X)$ condenses onto a compact space in all possible cases.

**V.454.** *Prove that there exists a set $X \subset \mathbb{R}$ such that $C_p(X)$ does not condense onto an analytic space.*

**Solution.** Given a space $Z$ and $A \subset Z$ let $\pi_A : C_p(Z) \to C_p(A)$ be the restriction map and $C_p(A|Z) = \pi_A(C_p(Z))$.

*Fact 1.* Suppose that $Z$ is a metrizable space and $D \subset Z$ is a dense subspace of $Z$. If $D \subset Y \subset Z$ and $Y \neq Z$ then $C_p(D|Z) \subset C_p(D|Y)$ and $C_p(D|Z) \neq C_p(D|Y)$.

*Proof.* It is evident that $C_p(D|Z) \subset C_p(D|Y)$; pick a point $y \in Z \setminus Y$ and fix a metric $d$ which generates the topology of $Z$. Letting $f(x) = \frac{1}{d(x,y)}$ for any $x \in Y$ we obtain a function $f \in C_p(Y)$ and hence $\pi_D(f) \in C_p(D|Y)$.

Assume that there exists a function $h \in C_p(Z)$ such that $h|D = f|D$ and take a sequence $\{a_n : n \in \omega\} \subset D$ which converges to $y$. Then $S = \{h(a_n) : n \in \omega\}$ must converge to $h(y)$. However $h(a_n) = f(a_n) = \frac{1}{d(a_n,y)} \to \infty$ and hence $S$ is not convergent. This contradiction shows that $\pi_D(f) \in C_p(D|Y) \setminus C_p(D|Z)$, so Fact 1 is proved.

Returning to our solution denote by $\mathbb{K}$ the Cantor set and fix a countable dense set $Q \subset \mathbb{K}$; we consider that $\mathbb{K} \subset \mathbb{R}$. Say that a set $E \subset \mathbb{K}$ has the *Bernstein property* if $E \cap F \neq \emptyset$ for any uncountable closed set $F \subset \mathbb{K}$. Observe that $\mathcal{A} = \{\{x\} \times \mathbb{K} : x \in \mathbb{K}\}$ is a disjoint family of uncountable closed subsets of $\mathbb{K} \times \mathbb{K}$ such that $|\mathcal{A}| = \mathfrak{c}$. Since $\mathbb{K} \times \mathbb{K}$ is homeomorphic to $\mathbb{K}$ by SFFS-348, there exists a disjoint family $\mathcal{A}'$ of uncountable closed subsets of $\mathbb{K}$ such that $|\mathcal{A}'| = \mathfrak{c}$. As an immediate consequence,

(1) $|E| = \mathfrak{c}$ for any set $E \subset \mathbb{K}$ with the Bernstein property.

Apply Fact 5 of S.151 to fix a set $E \subset \mathbb{K}$ such that $Q \subset E$ and both sets $E$ and $\mathbb{K} \setminus E$ have the Bernstein property. Consider the family $\mathcal{F}$ of all continuous functions

$\varphi : B \to \mathbb{R}^{\omega}$ where $B$ is an analytic subset of $\mathbb{R}^D$ for some countable $D \subset \mathbb{K}$ with $Q \subset D$. It is easy to see that $|\mathcal{F}| = \mathfrak{c}$, so we can take an enumeration $\{\varphi_{\alpha} : \alpha < \mathfrak{c}\}$ of the family $\mathcal{F}$. Thus, for every $\alpha < \mathfrak{c}$ we have a countable set $D_{\alpha}$ with $Q \subset D_{\alpha} \subset \mathbb{K}$ and an analytic set $B_{\alpha} \subset \mathbb{R}^{D_{\alpha}}$ such that $\varphi_{\alpha} : B_{\alpha} \to \mathbb{R}^{\omega}$.

Denote by $\mathcal{G}$ the family of all $G_{\delta}$-subsets of $\mathbb{K}$ which contain $E$. Observe first that

(2)  $\mathbb{K} \backslash G$ is countable for any $G \in \mathcal{G}$.

Indeed, $\mathbb{K} \backslash G = \bigcup \{K_n : n \in \omega\}$ where every $K_n$ is closed in $\mathbb{K}$. If $\mathbb{K} \backslash G$ is uncountable, then the set $K_n$ is uncountable for some $n \in \omega$; since $K_n \cap E = \emptyset$, we obtain a contradiction with the Bernstein property of $E$. It follows from (1) that $|\mathbb{K} \backslash E| = \mathfrak{c}$, so the following condition is an immediate consequence of (2):

(3)  if $\mathcal{G}' \subset \mathcal{G}$ and $|\mathcal{G}'| < \mathfrak{c}$, then the set $(\bigcap \mathcal{G}') \backslash E$ has cardinality $\mathfrak{c}$.

Let $u(t) = 0$ for every $t \in \mathbb{K}$ and denote by $X_0$ the set $E$. We have to consider 4 cases:

*Case 1.* $D_0 \backslash X_0 \neq \emptyset$. Pick $y_0 \in D_0 \backslash X_0$ and fix an arbitrary point $x_0 \in \mathbb{K} \backslash E$ distinct from $y_0$. Take $A_0 = \mathbb{K}$ and $f_0 = g_0 = u$.

*Case 2.* $D_0 \subset X_0$ and $C_p(D_0|X_0) \backslash B_0 \neq \emptyset$. Take any $h_0 \in C_p(X_0)$ such that $h_0|D_0 \notin B_0$ and apply Fact 3 of T.368 to find a set $A_0 \in \mathcal{G}$ and a continuous function $f_0 : A_0 \to \mathbb{R}$ such that $X_0 \subset A_0$ and $f_0|X_0 = h_0$. Let $g_0 = f_0$ and apply (3) to pick $x_0 \in A_0 \backslash X_0$ and $y_0 \in \mathbb{K} \backslash (X_0 \cup \{x_0\})$.

*Case 3.* $D_0 \subset X_0$, $C_p(D_0|X_0) \subset B_0$ and $\varphi_0|C_p(D_0|X_0)$ is not injective. Take distinct functions $p_0, q_0 \in C_p(X_0)$ such that $\varphi_0(p_0|D_0) = \varphi_0(q_0|D_0)$ and apply Fact 3 of T.368 once more to find a set $A_0 \in \mathcal{G}$ and continuous functions $f_0, g_0 : A_0 \to \mathbb{R}$ such that $X_0 \subset A_0$ while $f_0|X_0 = p_0$ and $g_0|X_0 = q_0$. Pick a point $x_0 \in A_0 \backslash X_0$ and take any point $y_0 \in \mathbb{K} \backslash (X_0 \cup \{x_0\})$.

*Case 4.* $D_0 \subset X_0$, $C_p(D_0|X_0) \subset B_0$ and $\varphi_0|C_p(D_0|X_0)$ is injective. Then choose a point $x_0 \in \mathbb{K} \backslash X_0$ and take any point $y_0 \in \mathbb{K} \backslash (X_0 \cup \{x_0\})$. Finally, let $A_0 = \mathbb{K}$ and $f_0 = g_0 = u$.

Proceeding inductively assume that $0 < \alpha < \mathfrak{c}$ and we have constructed a set $\{x_{\beta}, y_{\beta} : \beta < \alpha\} \subset \mathbb{K} \backslash E$, a set of functions $\{f_{\beta}, g_{\beta} : \beta < \alpha\}$ and a family of sets $\{A_{\beta} : \beta < \alpha\} \subset \mathcal{G}$ with the following properties:

(4)  $\{f_{\beta}, g_{\beta}\} \subset C_p(A_{\beta})$ for any $\beta < \alpha$;
(5)  if $\beta < \gamma < \alpha$ then $x_{\beta} \neq x_{\gamma}$;
(6)  $x_{\beta} \neq y_{\gamma}$ for any $\beta, \gamma < \alpha$;
(7)  if $\beta < \alpha$, $X_{\beta} = E \cup \{x_{\gamma} : \gamma < \beta\}$ and $D_{\beta} \backslash X_{\beta} \neq \emptyset$ then $y_{\beta} \in D_{\beta} \backslash X_{\beta}$;
(8)  $x_{\beta} \in \bigcap \{A_{\gamma} : \gamma \leq \beta\}$ for any $\beta < \alpha$;
(9)  for each $\beta < \alpha$, if $D_{\beta} \subset X_{\beta}$ and $C_p(D_{\beta}|X_{\beta}) \backslash B_{\beta} \neq \emptyset$, then $X_{\beta} \subset A_{\beta}$ and $f_{\beta}|D_{\beta} \notin B_{\beta}$;
(10)  if $\beta < \alpha$, $D_{\beta} \subset X_{\beta}$, $C_p(D_{\beta}|X_{\beta}) \subset B_{\beta}$ and $\varphi_{\beta}|C_p(D_{\beta}|X_{\beta})$ is not injective, then $X_{\beta} \subset A_{\beta}$, $f_{\beta} \neq g_{\beta}$ and $\varphi_{\beta}(f_{\beta}|D_{\beta}) = \varphi_{\beta}(g_{\beta}|D_{\beta})$.

Let $X_{\alpha} = E \cup \{x_{\beta} : \beta < \alpha\}$. If $D_{\alpha} \backslash X_{\alpha} \neq \emptyset$ then pick any point $y_{\alpha} \in D_{\alpha} \backslash X_{\alpha}$; let $A_{\alpha} = \mathbb{K}$ and $f_{\alpha} = g_{\alpha} = u$. The set $A = (\bigcap \{A_{\beta} : \beta < \alpha\}) \backslash E$ is has cardinality c by (3), so we can choose a point $x_\alpha \in A \backslash (X_\alpha \cup \{y_\beta : \beta \leq \alpha\})$. It is evident that the conditions (4)–(10) are still satisfied for all $\beta \leq \alpha$.

Now if $D_\alpha \subset X_\alpha$ and $C_p(D_\alpha | X_\alpha) \backslash B_\alpha \neq \emptyset$ then take any $h_\alpha \in C_p(X_\alpha)$ such that $h_\alpha | D_\alpha \notin B_\alpha$ and apply Fact 3 of T.368 to find a set $A_\alpha \in \mathcal{G}$ and a continuous function $f_\alpha : A_\alpha \rightarrow \mathbb{R}$ such that $X_\alpha \subset A_\alpha$ and $f_\alpha | X_\alpha = h_\alpha$. Let $g_\alpha = f_\alpha$ and apply (3) to pick $x_\alpha \in (\bigcap \{A_\beta : \beta \leq \alpha\}) \backslash (X_\alpha \cup \{y_\beta : \beta < \alpha\})$. Take any point $y_\alpha \in \mathbb{K} \backslash (X_\alpha \cup \{x_\alpha\})$ and observe that the properties (4)–(10) now hold for all $\beta \leq \alpha$.

Suppose that we have the case when $D_\alpha \subset X_\alpha$ and $C_p(D_\alpha | X_\alpha) \subset B_\alpha$ while the map $\varphi_\alpha | C_p(D_\alpha | X_\alpha)$ is not injective. Take distinct functions $p_\alpha, q_\alpha \in C_p(X_\alpha)$ such that $\varphi_\alpha(p_\alpha | D_\alpha) = \varphi_\alpha(q_\alpha | D_\alpha)$ and apply Fact 3 of T.368 once more to find a set $A_\alpha \in \mathcal{G}$ and continuous functions $f_\alpha, g_\alpha : A_\alpha \rightarrow \mathbb{R}$ such that $X_\alpha \subset A_\alpha$ while $f_\alpha | X_\alpha = p_\alpha$ and $g_\alpha | X_\alpha = q_\alpha$. Pick $x_\alpha \in (\bigcap \{A_\beta : \beta \leq \alpha\}) \backslash (X_\alpha \cup \{y_\beta : \beta < \alpha\})$ and take any $y_\alpha \in \mathbb{K} \backslash (X_\alpha \cup \{x_\alpha\})$; observe that the properties (4)–(10) still hold for all $\beta \leq \alpha$.

The remaining case is when $D_\alpha \subset X_\alpha$ and $C_p(D_\alpha | X_\alpha) \subset B_\alpha$ while the map $\varphi_\alpha | C_p(D_\alpha | X_\alpha)$ is injective. Then apply the property (3) once more to choose a point $x_\alpha \in (\bigcap \{A_\beta : \beta \leq \alpha\}) \backslash (X_\alpha \cup \{y_\beta : \beta < \alpha\})$ and take any point $y_\alpha \in \mathbb{K} \backslash (X_\alpha \cup \{x_\alpha\})$. Let $A_\alpha = \mathbb{K}$, $f_\alpha = g_\alpha = u$ and observe that the properties (4)–(10) keep holding for all $\beta \leq \alpha$.

Therefore we can carry out the c-many steps in our inductive procedure to obtain a set $\{x_\beta, y_\beta : \beta < \mathfrak{c}\} \subset \mathbb{K} \backslash E$, a set of functions $\{f_\beta, g_\beta : \beta < \mathfrak{c}\}$ and a family $\{A_\beta : \beta < \mathfrak{c}\} \subset \mathcal{G}$ such that the properties (4)–(10) hold for all $\alpha < \mathfrak{c}$. Our promised set will be $X = E \cup \{x_\alpha : \alpha < \mathfrak{c}\}$. Suppose that $C_p(X)$ condenses onto an analytic space $Y$. Then $nw(Y) \leq nw(C_p(X)) \leq \omega$ and hence $Y$ is condensable onto a second countable space $Z$; it is clear that $Z$ has to be analytic. Thus $C_p(X)$ condenses onto a second countable analytic space $Z$ and we can consider that $Z \subset \mathbb{R}^\omega$.

Let $\mu : C_p(X) \rightarrow Z$ be a condensation and apply TFS-299 to see that we can find a countable set $D \subset X$ such that $Q \subset D$ and there is a continuous map $\xi : C_p(D | X) \rightarrow Z$ for which $\xi \circ \pi_D = \mu$; it is clear that $\xi$ is a condensation. By Fact 3 of T.368 there exists a $G_\delta$-subset $H$ of the space $\mathbb{R}^D$ such that $C_p(D | X) \subset H$ and there is a continuous map $\theta : H \rightarrow \mathbb{R}^\omega$ with $\theta | C_p(D | X) = \xi$. If $B = \theta^{-1}(Z)$ and $\varphi = \theta | B$ then $B$ is an analytic subset of $\mathbb{R}^D$ (see SFFS-338) such that $C_p(D | X) \subset B$ while $\varphi : B \rightarrow Z$ is a continuous map and $\varphi | C_p(D | X)$ is injective.

The map $\varphi$ belongs to the family $\mathcal{F}$, so there exists an ordinal $\alpha < \mathfrak{c}$ such that $\varphi = \varphi_\alpha$, $B = B_\alpha$ and $D = D_\alpha$. If $D_\alpha \backslash X_\alpha \neq \emptyset$ then it follows from (7) that $y_\alpha \in D$. However, the property (6) shows that $y_\alpha \notin X$ while $y_\alpha \in D \subset X$ which is a contradiction.

Therefore $D_\alpha \subset X_\alpha$; assume for a moment that $C_p(D_\alpha | X_\alpha) \backslash B_\alpha \neq \emptyset$. Then we can apply the property (9) to see that $X_\alpha \subset A_\alpha$ and $f_\alpha | D_\alpha \notin B_\alpha$. Since $x_\beta \in A_\alpha$ for all $\beta \geq \alpha$ by (8), we conclude that $X \subset A_\alpha$ and hence $g = f_\alpha | X$ is a function from $C_p(X)$ such that $g | D = f_\alpha | D \notin B_\alpha$; this is a contradiction with $C_p(D | X) \subset B_\alpha$.

Thus $D_\alpha \subset X_\alpha$ and $C_p(D_\alpha | X_\alpha) \subset B_\alpha$. Assume that the map $\varphi | C_p(D_\alpha | X_\alpha)$ is not injective. Then the property (10) is applicable and hence $X_\alpha \subset A_\alpha$ while

$f_\alpha \neq g_\alpha$ and $\varphi(f_\alpha|D) = \varphi(g_\alpha|D)$. Since $x_\beta \in A_\alpha$ for all $\beta \geq \alpha$ by (8), we conclude that $X \subset A_\alpha$ and hence $f = f_\alpha|X$ and $g = g_\alpha|X$ are distinct functions from $C_p(X)$, so the functions $f|D$ and $g|D$ are distinct as well by Fact 0 of S.351. However, $\varphi(f|D) = \varphi(f_\alpha|D) = \varphi(g_\alpha|D) = \varphi(g|D)$ which contradicts the fact that $\varphi|C_p(D|X)$ is injective.

Finally assume that we have $D_\alpha \subset X_\alpha$ while $C_p(D_\alpha|X_\alpha) \subset B_\alpha$ and the map $\varphi|C_p(D|X_\alpha)$ is injective. It follows from $C_p(D|X_\alpha) \subset B$ that $\varphi(C_p(D|X_\alpha)) \subset Z$. By Fact 1, there exists a function $h \in C_p(D|X_\alpha)\backslash C_p(D|X)$. By injectivity of the map $\varphi|C_p(D|X_\alpha)$ we have $h \in Z\backslash\varphi(C_p(D|X))$; this final contradiction with the equality $\varphi(C_p(D|X)) = Z$ shows that $C_p(X)$ cannot be condensed onto an analytic space, so our solution is complete.

**V.455.** *Given a nonempty space $X$ such that $|C_p(X)| < 2^{d(X)}$ prove that $C_p(X)$ cannot be condensed onto a compact space. Deduce from this fact that neither one of the spaces $C_p(\beta\omega\backslash\omega)$ and $C_p(\Sigma(\mathbb{D}^c))$ condenses onto a compact space. Here $\Sigma(\mathbb{D}^c) = \{x \in \mathbb{D}^c : |x^{-1}(1)| \leq \omega\}$ is the $\Sigma$-product of $\mathbb{D}^c$.*

**Solution.** It is an easy consequence of the inequality $|C_p(X)| < 2^{d(X)}$ that $X$ is an infinite space. Suppose that $\varphi : C_p(X) \to K$ is a condensation of $C_p(X)$ onto a compact space $K$ and let $\kappa = d(X)$. If $\chi(x, K) \geq \kappa$ for any $x \in K$ then $|K| \geq 2^\kappa$ (see TFS-330). However, $|K| \leq |C_p(X)| < 2^\kappa$ which is a contradiction. Therefore we can find a point $x \in K$ such that $\psi(x, K) \leq \chi(x, K) < \kappa$.

Fix a family $\mathcal{U} \subset \tau(x, K)$ such that $|\mathcal{U}| < \kappa$ and $\{x\} = \bigcap\mathcal{U}$. If $\varphi(f) = x$ then $\{f\} = \varphi^{-1}(x)$ and hence $\{f\} = \bigcap\{\varphi^{-1}(U) : U \in \mathcal{U}\}$. This proves that $\psi(f, C_p(X)) \leq |\mathcal{U}| < \kappa$. The space $C_p(X)$ being homogeneous, we must have $\psi(C_p(X)) < \kappa$ and hence $d(X) = \psi(C_p(X)) < \kappa = d(X)$ (see TFS-173) which is a contradiction. Thus $C_p(X)$ cannot be condensed onto a compact space.

Now let $X = \beta\omega\backslash\omega$; it follows from separability of $\beta\omega$ that $|C_p(\beta\omega)| \leq \mathfrak{c}$. Since $C_p(\beta\omega)$ maps continuously onto $C_p(X)$ (see TFS-380), we have $|C_p(X)| \leq \mathfrak{c}$. Next apply TFS-371 to see that $d(X) \geq c(X) = \mathfrak{c}$ and hence $|C_p(X)| = \mathfrak{c} < 2^\mathfrak{c} \leq 2^{d(X)}$, so $C_p(X)$ cannot be condensed onto a compact space.

Finally assume that $X = \Sigma(\mathbb{D}^c)$. The space $\mathbb{D}^c$ is separable, so $|C_p(\mathbb{D}^c)| \leq \mathfrak{c}$. For any $A \subset \mathfrak{c}$ let $p_A : \mathbb{D}^c \to \mathbb{D}^A$ be the natural projection. It is easy to see that $p_A(X) = \mathbb{D}^A$ for any countable $A \subset \mathfrak{c}$, so we can apply Fact 2 of S.433 to see that the space $X$ is pseudocompact (in fact, it is countably compact) and hence $\mathbb{D}^c = \beta X$ (see Fact 2 of S.309). This implies that the restriction map $\pi_X : C_p(\mathbb{D}^c) \to C_p(X)$ is surjective and hence $|C_p(X)| \leq |C_p(\mathbb{D}^c)| \leq \mathfrak{c}$. If $D \subset X$ and $|D| < \mathfrak{c}$ then the set $A = \bigcup\{x^{-1}(1) : x \in D\}$ has cardinality not exceeding $|D| \cdot \omega < \mathfrak{c}$, so we can choose an ordinal $\alpha \in \mathfrak{c}\backslash A$. Then $W = \{x \in X : x(\alpha) = 1\}$ is nonempty and open in $X$ while $W \cap D = \emptyset$. Therefore $D$ is not dense in $X$; this proves that $d(X) \geq \mathfrak{c}$ and hence $|C_p(X)| \leq \mathfrak{c} < 2^\mathfrak{c} \leq 2^{d(X)}$ which shows that $C_p(X)$ cannot be condensed onto a compact space.

**V.456.** *Prove that if $2^\omega = \omega_1$ and $2^{\omega_1} > \omega_2$ then $C_p(\mathbb{D}^{\omega_2})$ does not condense onto a compact space.*

**Solution.** The cardinal $\omega_2$ is not $\omega$-cofinal; thus, for any function $f : \omega \to \omega_2$ there exists an ordinal $\alpha < \omega_2$ such that $f(\omega) \subset \alpha$ and hence $f \in \alpha^\omega$. Therefore $\omega_2^\omega = |\bigcup\{\alpha^\omega : \alpha < \omega_2\}| \leq \omega_1^\omega \cdot \omega_2$. Besides, $\omega_1^\omega \leq (2^\omega)^\omega = 2^\omega = \omega_1$. As a consequence, $\omega_2^\omega \leq \omega_1 \cdot \omega_2 = \omega_2$.

Apply Fact 7 of U.074 to see that $|C_p(\mathbb{D}^{\omega_2})| \leq w(\mathbb{D}^{\omega_2})^\omega = \omega_2^\omega = \omega_2$. If the space $\mathbb{D}^{\omega_2}$ is separable then $w(\mathbb{D}^{\omega_2}) \leq 2^{d(\mathbb{D}^{\omega_2})} \leq 2^\omega = \omega_1$ (see Fact 2 of S.368) which is a contradiction. Therefore $d(\mathbb{D}^{\omega_2}) \geq \omega_1$ and hence we can conclude that $|C_p(\mathbb{D}^{\omega_2})| = \omega_2 < 2^{\omega_1} \leq 2^{d(\mathbb{D}^{\omega_2})}$, so it follows from Problem 455 that $C_p(\mathbb{D}^{\omega_2})$ does not condense onto a compact space.

**V.457.** *Suppose that $2^\omega < 2^{\omega_1}$ and $C_p(X)$ condenses onto a compact space. Prove that $X$ is separable if and only if $|C_p(X)| \leq \mathfrak{c}$.*

**Solution.** If $X$ is separable then fix a countable dense set $D \subset X$; the restriction map condenses $C_p(X)$ onto a subspace of $C_p(D)$. This implies the inequalities $|C_p(X)| \leq |C_p(D)| \leq |\mathbb{R}^D| \leq \mathfrak{c}^\omega = \mathfrak{c}$, so $|C_p(X)| \leq \mathfrak{c}$. Observe that we need neither Luzin's Axiom nor condensations of $C_p(X)$ onto a compact space.

Now assume that $C_p(X)$ condenses onto a compact space $K$ while $|C_p(X)| \leq \mathfrak{c}$ and Luzin's Axiom holds. Fix a condensation $\varphi : C_p(X) \to K$. If $\chi(x, K) \geq \omega_1$ for any $x \in K$ then $|K| \geq 2^{\omega_1}$ (see TFS-330) which, together with $|K| = |C_p(X)| \leq \mathfrak{c} < 2^{\omega_1}$ gives a contradiction. Therefore $\chi(x, K) \leq \omega$ for some $x \in K$, so we can fix a countable family $\mathcal{U} \subset \tau(x, K)$ such that $\{x\} = \bigcap \mathcal{U}$. Take a function $f \in C_p(X)$ such that $\varphi(f) = x$. It is straightforward that $\{f\} = \bigcap\{\varphi^{-1}(U) : U \in \mathcal{U}\}$ and hence $\psi(f, C_p(X)) \leq \omega$. The space $C_p(X)$ is homogeneous and therefore $\psi(C_p(X)) = \omega$, so we can apply TFS-173 to see that $d(X) = \psi(C_p(X)) = \omega$, i.e., $X$ is separable.

**V.458.** *Suppose that $l(X^n) = \omega$ for all $n \in \mathbb{N}$ and $C_p(X)$ is Lindelöf. Observe that if $C_p(X)$ condenses onto a compact space $K$ then $K$ is metrizable and $X$ is separable. Prove that if $C_p(X)$ condenses onto a $\sigma$-compact space $Y$ then $X$ is separable and $\psi(Y) = \omega$. Deduce from this fact that if $X$ is a non-metrizable Corson compact space then $C_p(X)$ does not condense onto a $\sigma$-compact space.*

**Solution.** Suppose first that $C_p(X)$ condenses onto a compact space $K$. It follows from $l^*(X) = \omega$ that $t(C_p(X)) = \omega$ and hence we can apply Problem 449 to convince ourselves that $K$ is metrizable. The existence of a condensation onto $K$ together with $w(K) \leq \omega$ shows that $iw(C_p(X)) \leq \omega$ and hence $d(X) \leq \omega$ (see TFS-173).

Now assume that $C_p(X)$ condenses onto a $\sigma$-compact space and fix a condensation $\varphi : C_p(X) \to Y$ and a family $\{K_n : n \in \omega\}$ of compact subsets of $Y$ such that $Y = \bigcup_{n \in \omega} K_n$. The set $F_n = \varphi^{-1}(K_n)$ is closed in $C_p(X)$ for every $n \in \omega$. If $n \in \omega$ and $S$ is an uncountable free sequence in $K_n$ then the set $S' = \varphi^{-1}(S)$ is an uncountable free sequence in $F_n$ which is impossible because $l(F_n) \leq l(C_p(X)) = \omega$ and $t(F_n) \leq t(C_p(X)) = \omega$ (see Fact 1 of V.449). This contradiction shows that the space $K_n$ has no uncountable free sequences and therefore $t(K_n) \leq \omega$ for any $n \in \omega$ (see TFS-328).

Apply SFFS-432 to see that there exists $n \in \omega$ such that the space $C_p(X)$ is homeomorphic to $C$ for some $C \subset F_n$. Since $\varphi|C$ maps $C$ into $K_n$, it follows from Problem 433 that $\psi(\varphi(C)) \leq \omega$. Since the map $\varphi|C$ is a condensation, we must have $\psi(C_p(X)) = \psi(C) \leq \psi(\varphi(C)) = \omega$ and hence $d(X) = \psi(C_p(X)) = \omega$, i.e., $X$ is separable as promised.

It is a consequence of $\psi(C_p(X)) = \omega$ that $C_p(X)\backslash\{f\}$ is an $F_\sigma$-set for any $f \in C_p(X)$. The space $C_p(X)$ being Lindelöf, $C_p(X)\backslash\{f\}$ is Lindelöf as well. Therefore $Y\backslash\{y\}$ is Lindelöf for any $y \in Y$; this implies that $\psi(Y) \leq \omega$.

Finally note that $l(C_p(X)) = t(C_p(X)) = \omega$ for any Corson compact space $X$ (see CFS-150), so if $C_p(X)$ condenses onto a $\sigma$-compact space then $X$ is separable and hence metrizable by CFS-121.

**V.459.** *Given a space $X$ prove that the space $C_p(X)$ condenses onto a space embeddable in a compact space of countable tightness if and only if $C_p(X)$ condenses onto a second countable space.*

**Solution.** If $C_p(X)$ condenses onto a second countable space $M$ then $M$ is embeddable in a metrizable compact space $K$ (see TFS-209); of course then $t(K) \leq \omega$ and hence we proved sufficiency.

Now if $\varphi : C_p(X) \to Y$ is a condensation of $C_p(X)$ onto a space $Y$ such that $Y \subset K$ for some compact space $K$ of countable tightness then $\psi(Y) \leq \omega$ by Problem 433. It is an easy exercise that $\psi(C_p(X)) \leq \psi(Y) \leq \omega$, so $iw(C_p(X)) = \omega$ (see TFS-173), i.e., $C_p(X)$ condenses onto a second countable space.

**V.460.** *Assuming $MA+\neg CH$ prove that if $K$ is a compact space such that $C_p(K)$ is Lindelöf and condensable onto a $\sigma$-compact space then $K$ is metrizable.*

**Solution.** Apply Problem 458 to see that $K$ is separable; it follows from CFS-080 that $K$ is $\omega$-monolithic, so $w(K) = nw(K) \leq \omega$ by Fact 4 of S.307.

**V.461.** *Assume that $C_p(X)$ is a Lindelöf $\Sigma$-space and there exists a condensation of $C_p(X)$ onto a $\sigma$-compact space $Y$. Prove that $nw(X) \leq \omega$ and $nw(Y) \leq \omega$.*

**Solution.** Denote by $\upsilon X$ the Hewitt realcompactification of the space $X$. Then $Z = \upsilon X$ is a Lindelöf $\Sigma$-space by CFS-206; since $C_p(Z)$ is also a Lindelöf $\Sigma$-space (see CFS-234), we conclude that $t(C_p(Z)) = l(C_p(Z)) = \omega$. The space $C_p(Z)$ condenses onto $C_p(X)$ and hence $C_p(Z)$ is condensable onto $Y$; this shows that the space $Z$ is separable (see Problem 458). It follows from stability of $C_p(Z)$ (see SFFS-266) that $Z$ is monolithic and hence $nw(Z) \leq \omega$. Since $X \subset Z$, we have $nw(X) \leq nw(Z) = \omega$. Finally observe that $nw(Y) \leq nw(C_p(X)) = nw(X) \leq \omega$.

**V.462.** *Prove that, for any compact space $X$ such that $w(X) \leq \omega_1$, the space $C_p(X)$ is hereditarily metalindelöf.*

**Solution.** Recall that a space $Z$ is called *metalindelöf* if every open cover of $Z$ has a point-countable refinement. If $\mathcal{A}$ is a family of subsets of $Z$ and $Y \subset Z$ then $\mathcal{A}|Y = \{A \cap Y : A \in \mathcal{A}\}$. If $d$ is a metric on a set $Z$ then $\tau(d)$ is the topology generated by $d$. Given a continuous map $\varphi : Z \to T$ let $\varphi^*(f) = f \circ \varphi$ for any $f \in C(T)$; then $\varphi^* : C(T) \to C(Z)$ is the dual map of the map $\varphi$.

*Fact 1.* A space $Z$ is hereditarily metalindelöf if and only if every open subspace of $Z$ is metalindelöf.

*Proof.* Since necessity is evident, assume that every open subspace of the space $Z$ is metalindelöf and fix an arbitrary set $Y \subset Z$. If $\mathcal{U} \subset \tau(Y)$ is a cover of $Y$ then choose a set $O_U \in \tau(Z)$ such that $O_U \cap Y = U$ for every $U \in \mathcal{U}$. The set $G = \bigcup\{O_U : U \in \mathcal{U}\}$ is open in $Z$ and $\mathcal{V} = \{O_U : U \in \mathcal{U}\}$ is an open cover of $G$. By our assumption $G$ is metalindelöf, so there exists a point-countable open refinement $\mathcal{W}$ of the family $\mathcal{V}$. It is straightforward that the family $\mathcal{W}' = \mathcal{W}|Y$ is a point-countable open refinement of $\mathcal{U}$, so $Y$ is metalindelöf and hence Fact 1 is proved.

*Fact 2.* Suppose that $(Z, d)$ is a metric space and $\mu$ is a topology on $Z$ with $\mu \subset \tau(d)$. Assume additionally that we have a family $\{F_\alpha : \alpha < \omega_1\}$ of closed subsets of $(Z, \mu)$ with the following properties:

(a) $\alpha < \beta < \omega_1$ implies $F_\alpha \subset F_\beta$;
(b) $\bigcup\{F_\alpha : \alpha < \omega_1\} = Z$;
(c) for any $\alpha < \omega_1$ the set $F_\alpha$ is second countable if considered with the topology induced from $(Z, d)$.
(d) if $\alpha < \omega_1$ is a limit ordinal then the set $\bigcup\{F_\beta : \beta < \alpha\}$ is dense in $F_\alpha$ in the topology induced from $(Z, d)$.

Then the space $(Z, \mu)$ is hereditarily metalindelöf.

*Proof.* Take any set $U \in \mu$ and let $F'_\alpha = F_\alpha \cap U$ for any $\alpha < \omega_1$. If we consider the metric $d' = d|(U \times U)$ and the topology $\mu' = \mu|U$ then the space $(U, d')$ together with the family $\{F'_\alpha : \alpha < \omega_1\}$ trivially satisfies the conditions (a)–(c).

Now if $\alpha < \omega_1$ is a limit ordinal then take any nonempty open subset $W$ of the space $F'_\alpha$. The set $F'_\alpha$ is open in $F_\alpha$, so $W$ is also open in $F_\alpha$ and hence it follows from the property (d) that $W \cap F_\beta \neq \emptyset$ for some $\beta < \alpha$. Since $W \subset U$, we have $W \cap F'_\beta = W \cap F_\beta \cap U = W \cap F_\beta \neq \emptyset$ and hence the set $\bigcup\{F'_\beta : \beta < \alpha\}$ is dense in $F'_\alpha$, i.e., the space $U$ also has the property (d). Thus the properties (a)–(d) are open-hereditary, so we can apply Fact 1 to see that it suffices to establish that $(Z, \mu)$ is metalindelöf.

Let $B(x, r) = \{y \in Z : d(y, x) < r\}$ be the $r$-ball centered at $x$ for any $x \in Z$. Take an arbitrary open cover $\mathcal{O}$ of the space $(Z, \mu)$ and let $r_x = \sup\{r :$ there exists $O \in \mathcal{O}$ such that $B(x, r) \subset O\}$; every element of $\mathcal{O}$ is open in $(Z, d)$, so $r_x > 0$ for any $x \in Z$. For each $x \in Z$ fix a set $O_x \in \mathcal{O}$ with $B(x, \frac{3}{4}r_x) \subset O_x$.

Let $F_{-1} = \emptyset$ and fix, for each $\alpha \in \omega_1 \cup \{-1\}$ a countable dense subset $C_\alpha$ of the set $F_{\alpha+1} \backslash F_\alpha$ with the topology induced from $(Z, d)$. Consider the family $\mathcal{U}_\alpha = \{O_z \backslash F_\alpha : z \in C_\alpha\}$ for each $\alpha \in \omega_1 \cup \{-1\}$. It is clear that every element of $\mathcal{U}_\alpha$ is open in $(Z, \mu)$ and contained in some element of $\mathcal{O}$; let $\mathcal{U} = \bigcup\{\mathcal{U}_\alpha : \alpha \in \omega_1 \cup \{-1\}\}$.

Given a point $x \in Z$ there is $\alpha < \omega_1$ such that $x \in F_\alpha$, so it follows from the property (a) that $x \notin \bigcup \mathcal{U}_\beta$ for any $\beta \geq \alpha$. Since $\bigcup\{\mathcal{U}_\beta : \beta < \alpha\}$ is countable, we proved that every $x \in Z$ belongs to at most countably many elements of $\mathcal{U}$, i.e., the family $\mathcal{U}$ is point-countable.

To see that $\mathcal{U}$ covers $Z$ fix any point $x \in Z$ and denote by $\eta_x$ the least ordinal $\eta < \omega_1$ such that $x \in F_\eta$. It turns out that

(1)  there exists $\alpha < \eta_x$ such that $(F_{\alpha+1} \backslash F_\alpha) \cap B(x, \frac{1}{4}r_x) \neq \emptyset$.

Indeed, if $\eta_x = \alpha+1$ for some $\alpha \in \omega_1 \cup \{-1\}$ then $x \in (F_{\alpha+1} \backslash F_\alpha) \cap B(x, \frac{1}{4}r_x)$. If $\eta_x$ is a limit ordinal then it follows from the property (d) that we can find a minimal $\alpha \in \omega_1 \cup \{-1\}$ such that $F_{\alpha+1} \cap B(x, \frac{1}{4}r_x) \neq \emptyset$; it easily follows from minimality of $\alpha$ that $(F_{\alpha+1} \backslash F_\alpha) \cap B(x, \frac{1}{4}r_x) \neq \emptyset$, so (1) is proved.

The set $C_\alpha$ being dense in $F_{\alpha+1} \backslash F_\alpha$ there exists a point $z \in C_\alpha \cap B(x, \frac{1}{4}r_x)$. It follows form the inclusions $B(z, \frac{1}{2}r_x) \subset B(x, \frac{3}{4}r_x) \subset O_x$ that $r_z \geq \frac{1}{2}r_x$. Furthermore, $B(z, \frac{3}{8}r_x) \subset B(z, \frac{3}{4}r_z) \subset O_z$. We have $d(z, x) < \frac{1}{4}r_x < \frac{3}{8}r_x$ which shows that $x \in B(z, \frac{3}{8}r_x) \subset O_z$. It follows from $\alpha < \eta_x$ that $x \notin F_\alpha$, so $x \in O_z \backslash F_\alpha \in \mathcal{U}$ and therefore the family $\mathcal{U}$ covers $Z$ which shows that $\mathcal{U}$ is a point-countable refinement of the cover $\mathcal{O}$. Thus $(Z, \mu)$ is metalindelöf; we already saw that this implies that every subspace of $Z$ is metalindelöf and hence Fact 2 is proved.

Returning to our solution fix any compact space $X$ with $w(X) \leq \omega_1$; we can consider that $X \subset \mathbb{I}^{\omega_1}$. For every countable ordinal $\alpha > 0$ let $\pi_\alpha : X \to \mathbb{I}^\alpha$ be the restriction to $X$ of the natural projection; if $X_\alpha = \pi_\alpha(X)$ then we can consider that $\pi_\alpha : X \to X_\alpha$. If $0 < \beta < \alpha < \omega_1$ then we will also need the projection $p_\beta^\alpha : X_\alpha \to X_\beta$.

It is immediate that the set $F_\alpha = \pi_\alpha^*(C(X_\alpha))$ is an algebra in $C(X)$ for every $\alpha > 0$; let $F_0 = \{u\}$ where $u(x) = 0$ for all $x \in X$ and observe that $F_\alpha$ is closed in $C_p(X)$ for each $\alpha < \omega_1$. It follows from TFS-298 that $\bigcup \{F_\alpha : \alpha < \omega_1\} = C(X)$. If $0 < \beta < \alpha < \omega_1$ then it is an easy consequence of the equality $\pi_\beta = p_\beta^\alpha \circ \pi_\alpha$ that $F_\beta \subset F_\alpha$; clearly $F_0 \subset F_\alpha$ for every $\alpha < \omega_1$.

Given any $\alpha \in \omega_1 \backslash \{0\}$ the space $C_u(X_\alpha)$ is second countable by Fact 2 of T.357; since the map $\pi_\alpha^* : C_u(X_\alpha) \to C_u(X)$ is continuous by Fact 8 of V.318, the set $F_\alpha$ is separable and hence second countable with the topology induced from $C_u(X)$.

Next assume that $\alpha < \omega_1$ is a limit ordinal and let $G_\beta = (p_\beta^\alpha)^*(C(X_\beta))$ for every $\beta < \alpha$. The family of maps $\{p_\beta^\alpha : 0 < \beta < \alpha\}$ separates the points of $X_\alpha$ and hence the algebra $G = \bigcup \{G_\beta : \beta < \alpha\}$ separates the points of $X_\alpha$. Applying TFS-191 we conclude that $G$ is dense in $X_\alpha$. Note that $F_\beta = \pi_\alpha^*(G_\beta)$ for every $\beta < \alpha$ and hence $\bigcup \{F_\beta : \beta < \alpha\} = \pi_\alpha^*(G)$ is dense in $\pi_\alpha^*(C(X_\alpha)) = F_\alpha$ in the topology induced from $C_u(X)$.

Let $d(f, g) = \sup\{|f(x) - g(x)| : x \in X\}$ for any $f, g \in C(X)$; then $d$ is a metric on $C(X)$ which generates the topology of $C_u(X)$. If $\mu$ is the topology of $C_p(X)$ then the space $Z = C(X)$ with the metric $d$ and the family $\{F_\alpha : \alpha < \omega_1\}$ satisfies the conditions (a)–(d) of Fact 2 which shows that the space $(Z, \mu) = C_p(X)$ is hereditarily metalindelöf and hence our solution is complete.

**V.463.** *Given an uncountable discrete space $X$ prove that the space $C_p(\beta X, \mathbb{D})$ is not metalindelöf; therefore $C_p(X)$ is not metalindelöf either.*

**Solution.** Given a space $Z$ and a set $A \subset Z$ suppose that $U_a \in \tau(a, Z)$ for every $a \in A$; then the family $\mathcal{U} = \{U_a : a \in A\}$ is called *an open expansion* of the set $A$. The expansion $\mathcal{U}$ is called *point-countable* if the set $\{a \in A : x \in U_a\}$ is countable for any $x \in Z$.

*Fact 1.* If $Z$ is a metalindelöf space then every closed discrete subspace of $Z$ has a point-countable open expansion.

*Proof.* Let $D$ be a closed discrete subspace of $Z$. For every $x \in Z$ take a set $U_x \in \tau(x, Z)$ such that $|U_x \cap D| \le 1$ and choose a point-countable open refinement $\mathcal{V}$ of the open cover $\mathcal{U} = \{U_x : x \in Z\}$ of the space $Z$. For every $d \in D$ pick a set $O_d \in \mathcal{V}$ with $d \in O_d$. Then $\mathcal{O} = \{O_d : d \in D\}$ is an open expansion of $D$. If $d$ and $d'$ are distinct points of $D$ and $O_d \ni d'$ then pick $U \in \mathcal{U}$ with $O_d \subset U$ and observe that $\{d, d'\} \subset U$ which is a contradiction with the choice of $\mathcal{U}$. Therefore $O_d \cap D = \{d\}$ for all $d \in D$ and hence

(1)  $O_d \ne O_{d'}$ for distinct $d, d' \in D$.

Now if $\mathcal{O}$ is not point-countable then there is an uncountable $D' \subset D$ such that $\bigcap \{O_d : d \in D'\} \ne \emptyset$. It follows from (1) that the family $\{O_d : d \in D'\} \subset \mathcal{V}$ is uncountable, so $\mathcal{V}$ is not point-countable which is a contradiction. Therefore $\mathcal{O}$ is an open point-countable expansion of $D$ and hence Fact 1 is proved.

*Fact 2.* Suppose that $\kappa$ is an infinite regular cardinal, and $E$ is a set with $|E| = \kappa$. Assume that $\mathcal{A}$ is a family of subsets of $E$ such that $|\mathcal{A}| = \kappa$ and $|A| = \kappa$ for any $A \in \mathcal{A}$. If, additionally, $|A \cap A'| < \kappa$ for any distinct $A, A' \in \mathcal{A}$ then there exists a set $B \subset \kappa$ such that $|B| = \kappa$ and $|B \cap A| < \kappa$ for any $A \in \mathcal{A}$.

*Proof.* Take a faithful enumeration $\{A_\alpha : \alpha < \kappa\}$ of the family $\mathcal{A}$. Choose an arbitrary point $x_0 \in A_0$; proceeding inductively, assume that $\beta < \kappa$ and we have a set $\{x_\alpha : \alpha < \beta\}$ such that

(2)  $x_\alpha \ne x_{\alpha'}$ for any distinct $\alpha, \alpha'$ and $x_\alpha \notin A_\gamma$ whenever $\gamma < \alpha < \beta$.

The set $A_\beta$ has cardinality $\kappa$ while $|A_\alpha \cap A_\beta| < \kappa$ for any $\alpha < \beta$. By regularity of $\kappa$ we can choose a point $x_\beta \in A_\beta \backslash (\{x_\alpha : \alpha < \beta\} \cup \bigcup \{A_\alpha : \alpha < \beta\})$. It is clear that (2) now holds for all $\alpha \le \beta$, so we can construct a set $B = \{x_\alpha : \alpha < \kappa\}$ such that (2) is satisfied for all $\alpha < \kappa$ and, in particular, $|B| = \kappa$. Now, if $\beta < \kappa$ then it follows from (2) that $B \cap A_\beta \subset \{x_\alpha : \alpha \le \beta\}$ and hence $|B \cap A_\beta| < \kappa$ for all $\beta < \kappa$, i.e., Fact 2 is proved.

*Fact 3.* Suppose that $\kappa$ is an infinite regular cardinal and $E$ is a set with $|E| = \kappa$. Then there exists a family $\mathcal{A}$ of subsets of $E$ such that $|\mathcal{A}| = \kappa^+$ while $|A| = \kappa$ for any $A \in \mathcal{A}$ and $|A \cap A'| < \kappa$ for any distinct $A, A' \in \mathcal{A}$.

*Proof.* Say that a family $\mathcal{B}$ of subsets of $E$ is almost disjoint if $|\mathcal{B}| \ge \kappa$ while $|B| = \kappa$ for any $B \in \mathcal{B}$ and $|B \cap B'| < \kappa$ for any distinct $B, B' \in \mathcal{B}$. It is easy to find a disjoint family $\mathcal{A}_\kappa = \{A_\alpha : \alpha < \kappa\}$ of subsets of $E$ such that $|A_\alpha| = \kappa$

for any $\alpha < \kappa$; of course, $\mathcal{A}_\kappa$ is almost disjoint. Proceeding inductively assume that $\kappa \le \beta < \kappa^+$ and we have a family $\{A_\alpha : \alpha < \beta\}$ such that $\mathcal{A}_\gamma = \{A_\alpha : \alpha \le \gamma\}$ is almost disjoint for any $\gamma < \beta$.

It is an easy exercise that the family $\{A_\alpha : \alpha < \beta\}$ is also almost disjoint. By Fact 2 there exists a set $A_\beta \subset E$ such that $|A_\beta| = \kappa$ and $|A_\beta \cap A_\alpha| < \kappa$ for any $\alpha < \beta$, i.e., the family $\mathcal{A}_\beta = \{A_\alpha : \alpha \le \beta\}$ is almost disjoint. Therefore our inductive procedure can be continued to construct a family $\mathcal{A} = \{A_\alpha : \alpha < \kappa^+\}$ such that $\mathcal{A}_\beta$ is almost disjoint for every $\beta < \kappa^+$. If $A, A'$ are distinct elements of $\mathcal{A}$ then there is $\beta < \kappa^+$ such that $A, A' \in \mathcal{A}_\beta$ and hence $|A \cap A'| < \kappa$ because $\mathcal{A}_\beta$ is almost disjoint. This shows that $\mathcal{A}$ is almost disjoint and hence Fact 3 is proved.

*Fact 4.* Let $Z$ be a discrete space of cardinality $\omega_1$. Then there exists a closed discrete subset $D \subset C_p(\beta Z, \mathbb{D})$ such that $|D| = \omega_1$ and $D$ has no point-countable open expansion.

*Proof.* Recall that for any $A \subset Z$ the set $\overline{A}$ is clopen in $\beta Z$ by Fact 2 of T.371 (the bar denotes the closure in $\beta Z$). Let $\{z_\alpha : \alpha < \omega_1\}$ be a faithful enumeration of $Z$ and denote by $Z_\alpha$ the set $\{z_\beta : \beta < \alpha\}$ for all $\alpha < \omega_1$. Given any $f \in \mathbb{D}^Z$ there exists a unique function $e(f) \in C_p(\beta Z, \mathbb{D})$ such that $e(f)|Z = f$. For any $\alpha < \omega_1$ let $f_\alpha(z_\beta) = 1$ for all $\beta < \alpha$ and $f_\alpha(z_\beta) = 0$ whenever $\beta \ge \alpha$, i.e., $f_\alpha$ is the characteristic function of the set $Z_\alpha$; let $g_\alpha = e(f_\alpha)$. We will show that $D = \{g_\alpha : \alpha < \omega_1\}$ is the promised set.

Given any function $f \in C_p(\beta Z, \mathbb{D})$ and a finite subset $K$ of the space $\beta Z$ let $[f, K] = \{g \in C_p(\beta Z, \mathbb{D}) : g|K = f|K\}$; it is clear that the family $\{[f, K] : K$ is a finite subset of $\beta Z\}$ is a local base of the space $C_p(\beta Z, \mathbb{D})$ at the point $f$.

If $\alpha = \beta + 1 < \omega_1$ and $O = [g_\alpha, \{z_\beta, z_\alpha\}]$ then $g_\gamma(z_\beta) = 0$ and hence $g_\gamma \notin O$ for any $\gamma < \alpha$. If $\gamma > \alpha$ then $g_\gamma(z_\alpha) = 1$ and hence $g_\gamma \notin O$. This shows that $O \cap D = \{g_\alpha\}$.

Next, assume that $\alpha \le \omega_1$ is a limit ordinal. It is straightforward that the family $\mathcal{E}_\alpha = \{\overline{Z_\alpha \backslash Z_\beta} : \beta < \alpha\}$ is centered and consists of nonempty compact subsets of $\beta Z$. Therefore $E_\alpha = \bigcap \mathcal{E}_\alpha \ne \emptyset$. Furthermore $\overline{Z_\beta} \cap \overline{Z_\alpha \backslash Z_\beta} = \emptyset$ (see Fact 1 of S.382) and hence $\overline{Z_\beta} \cap E_\alpha = \emptyset$ for any $\beta < \alpha$; an immediate consequence is that

(3) for any limit ordinal $\alpha \le \omega_1$ the set $E_\alpha$ is nonempty and $g_\beta(E_\alpha) = \{0\}$ for each $\beta < \alpha$.

Now if $\alpha < \omega_1$ is a limit ordinal then pick a point $x \in E_\alpha$ and consider the set $O = [g_\alpha, \{x, z_\alpha\}]$. If $\gamma > \alpha$ then $g_\gamma(z_\alpha) = 1 \ne g_\alpha(z_\alpha)$ and hence $g_\gamma \notin O$. If $\gamma < \alpha$ then $g_\gamma(x) = 0$ by (3). However, $x \in \overline{Z_\alpha}$ and $g_\alpha(\overline{Z_\alpha}) = \{1\}$; this shows that $g_\alpha(x) = 1$ and hence $g_\gamma \notin O$ for all $\gamma < \alpha$. Thus $O \cap D = \{g_\alpha\}$ and hence $D$ is a discrete subset of $C_p(\beta Z, \mathbb{D})$.

To prove that $D$ is closed in $C_p(\beta Z, \mathbb{D})$ take any function $f \in C_p(\beta Z, \mathbb{D}) \backslash D$ and assume that $f(z_\alpha) < f(z_\beta)$ for some ordinals $\alpha, \beta \in \omega_1$ with $\alpha < \beta$. It is immediate that $U = [f, \{z_\alpha, z_\beta\}]$ is an open neighborhood of $f$ such that $U \cap D = \emptyset$. Therefore, to prove that $f \notin \overline{D}$, we can consider that

(4) $f(z_\alpha) \ge f(z_\beta)$ whenever $\alpha < \beta < \omega_1$.

The function $f$ cannot be identically zero on $Z$ because otherwise it would coincide with the function $g_0 \in D$. This shows that the set $f^{-1}(1)$ is nonempty and hence $f^{-1}(1) \cap Z \neq \emptyset$. Consider the ordinal $\alpha = \sup\{\beta : z_\beta \in f^{-1}(1)\}$. If $\alpha < \omega_1$ and $z_\alpha \in f^{-1}(1)$ then it follows from (4) that $f^{-1}(1) \cap Z = Z_{\alpha+1}$ and hence $f = g_{\alpha+1}$ which is a contradiction. If $\alpha < \omega_1$ and $z_\alpha \notin f^{-1}(1)$ then $\alpha$ must be a limit ordinal and it is easy to deduce from (4) that $f|Z = f_\alpha$ and hence $f = g_\alpha$ which is again a contradiction.

Finally, assume that $\alpha = \omega_1$ and hence we can apply (4) once more to see that $f(z_\alpha) = 1$ for all $\alpha < \omega_1$. Thus $f(z) = 1$ for all $z \in \beta Z$. Apply (3) to take a point $x \in E_\alpha$; then $g_\beta(x) = 0$ for all $\beta < \omega_1$ while $f(x) = 1$ and hence $U = [f, \{x\}]$ is an open neighborhood of $f$ such that $U \cap D = \emptyset$. We established that any $f \in C_p(\beta Z, \mathbb{D}) \setminus D$ has an open neighborhood which does not meet $D$ and hence $D$ is closed in $C_p(\beta Z, \mathbb{D})$.

Suppose that $U_\alpha \in \tau(g_\alpha, C_p(\beta Z, \mathbb{D}))$ for any $\alpha < \omega_1$ and $\{U_\alpha : \alpha < \omega_1\}$ is a point-countable extension of $D$. For any $\alpha < \omega_1$ there exists a finite set $K_\alpha \subset \beta Z$ such that $V_\alpha = [g_\alpha, K_\alpha] \subset U_\alpha$; it is evident that the family $\{V_\alpha : \alpha < \omega_1\}$ is also a point-countable open extension of the set $D$.

The set $Y = \bigcup\{\overline{Z}_\alpha : \alpha < \omega_1\}$ is open in $\beta Z$; let $L_\alpha = K_\alpha \cap Y$ for all $\alpha < \omega_1$. Given any $x \in Y$ let $\xi(x) = \setminus n\{\alpha : x \in \overline{Z}_\alpha\}$ and $q(x) = z_{\xi(x)}$. Since $M_\alpha = q(L_\alpha)$ is a finite subset of $Z$ for every $\alpha < \omega_1$, we can apply SFFS-038 to find a finite set $R \subset Z$ and an uncountable $A \subset \omega_1$ such that $M_\alpha \cap M_\beta = R$ for any distinct $\alpha, \beta \in A$. Let $M'_\alpha = M_\alpha \setminus R$ and $L'_\alpha = \{x \in L_\alpha : q(x) \notin R\}$ for each $\alpha \in A$.

Given any finite set $P \subset Z$ consider the ordinals $m(P) = \setminus n\{\alpha : z_\alpha \in P\}$ and $j(P) = \max\{\alpha : z_\alpha \in P\}$ if $P \neq \emptyset$; if $P = \emptyset$ then let $j(P) = m(P) = 0$. The family $\mathcal{M} = \{M'_\alpha : \alpha \in A\}$ is disjoint and hence, for any $\alpha < \omega_1$ only countably many elements of $\mathcal{M}$ intersect $Z_\alpha$. Therefore we can pass to an appropriate uncountable subset of $A$ to be able to consider, without loss of generality, that for the ordinal $\alpha^* = j(R)$ we have $m(M'_\alpha) > \alpha^*$ and $\alpha > \alpha^*$ for any $\alpha \in A$. Let $L''_\alpha = L_\alpha \setminus L'_\alpha$ for any $\alpha \in A$; then $q(L''_\alpha) = R \subset Z_{\alpha^*}$; since $g_\beta(Z_{\alpha^*}) = \{1\}$ and $L''_\alpha \subset \overline{Z}_{\alpha^*}$, we convince ourselves that

(5) $g_\beta(L''_\alpha) \subset \{1\}$ for any $\alpha, \beta \in A$.

If $E = \beta Z \setminus Y$ then it turns out that

(6) for any subsets $P$ and $Q$ of the set $Z$ if $|P \cap Q| < \omega_1$ then $\overline{P} \cap \overline{Q} \cap E = \emptyset$.

To prove (6) choose $\alpha < \omega_1$ such that $P \cap Q \subset Z_\alpha$ and assume that $z \in \overline{P} \cap \overline{Q} \cap E$. Consider the family $\mathcal{C}_z = \{U \cap Z : U \in \tau(z, \beta Z)\}$ and apply Fact 2 of T.371 to see that $P \in \mathcal{C}_z$ and $Q \in \mathcal{C}_z$. The family $\mathcal{C}_z$ is a filter, so $P \cap Q \in \mathcal{C}_z$ and hence $z \in \overline{P \cap Q} \subset \overline{Z}_\alpha$; this contradiction shows that (6) is proved.

Furthermore, $g_\alpha(\beta Z \setminus \overline{Z}_\alpha) = \{0\}$; since $E \cap \overline{Z}_\alpha = \emptyset$, we have the equality

(7) $g_\alpha(E) = \{0\}$ for any $\alpha < \omega_1$.

Given a countable set $P \subset Z$, only countably many elements of $\mathcal{M}$ intersect $P$, so it is easy to construct by transfinite induction $\omega_1$-sequences $\{\mu_\alpha : \alpha < \omega_1\} \subset A$ and $\{\nu_\alpha : \alpha < \omega_1\} \subset \omega_1$ with the following properties:

(8)  $v_0 = \alpha^*$ and $v_\alpha + 1 < m(M'_{\mu_\alpha})$ for any $\alpha < \omega_1$;

(9)  $j(M'_{\mu_\alpha}) \leq v_{\alpha+1}$ for every $\alpha < \omega_1$;

(10)  $v_\alpha > \sup\{v_\beta : \beta < \alpha\}$ for any $\alpha < \omega_1$.

Consider the set $H_\alpha = Z_{v_{\alpha+1}} \backslash Z_{v_\alpha+1}$; it is straightforward from the definitions that $L'_{\mu_\alpha} \subset \overline{H}_\alpha$ for every $\alpha < \omega_1$.

Apply Fact 3 to find a family $\Omega$ of subsets of $\omega_1$ such that $|\Omega| = \omega_2$ while $|S| = \omega_1$ and $\backslash n(S) > \alpha^*$ for any $S \in \Omega$ and $S \cap S'$ is countable whenever $S$ and $S'$ are distinct elements of $\Omega$. The family $\{H_\alpha : \alpha < \omega_1\}$ is disjoint, so if we let $Q_S = \bigcup\{H_\alpha : \alpha \in S\}$ for any $S \in \Omega$ then the family $\mathcal{Q} = \{Q_S : S \in \Omega\}$ will be almost disjoint in the sense that $Q_S \cap Q_{S'}$ is countable for any distinct $S, S' \in \Omega$.

It follows from (6) that $\overline{Q}_S \cap \overline{Q}_{S'} \cap E = \emptyset$ for any distinct $S, S' \in \Omega$. The set $K = \bigcup\{K_\alpha \cap E : \alpha < \omega_1\}$ has cardinality $\leq \omega_1$ while the family $\{\overline{Q}_S \cap E : S \in \Omega\}$ is disjoint and has cardinality $\omega_2$. Therefore we can find a set $S \in \Omega$ such that $\overline{Q}_S \cap K = \emptyset$.

To define a function $h : Z \to \mathbb{D}$ let $h(z_\alpha) = 1$ for all $\alpha \leq \alpha^*$ and $h(z_\alpha) = 0$ whenever $\alpha > \alpha^*$ and $z_\alpha \notin \bigcup\{H_\alpha : \alpha \in S\}$. Given any $\beta \in S$ observe that $\overline{H}_\beta$ is an open set which contains $L'_{\mu_\beta}$, so we can choose a disjoint family $\{O_d : d \in L'_{\mu_\beta}\}$ of open subsets of $\beta Z$ in such a way that $d \in O_d \subset \overline{H}_\beta$ for every $d \in L'_{\mu_\beta}$. The family $\mathcal{O} = \{O_d \cap Z : d \in L'_{\mu_\beta}\}$ is disjoint; let $h(z) = 0$ if $z \in H_\beta \backslash (\bigcup \mathcal{O})$ and $h(z) = g_{\mu_\beta}(d)$ for any $z \in O_d \cap Z$ and $d \in L'_{\mu_\beta}$.

This gives us a function $h : Z \to \mathbb{D}$; let $w = e(h)$. Take any $\beta \in S$ and observe that it follows from $d \in \overline{O_d \cap Z}$ that $w(d) = g_{\mu_\beta}(d)$ for each $d \in L'_{\mu_\beta}$. Furthermore, if $d \in L''_{\mu_\beta}$ then $d \in \overline{Z}_{\alpha^*+1}$ and hence $w(d) = 1$, so it follows from (5) that $w(d) = g_{\mu_\beta}(d)$. If $d \in K_{\mu_\beta} \cap E$ then observe that $h^{-1}(1) \subset Z_{\alpha^*+1} \cup Q_S$ while $\overline{Z}_{\alpha^*+1} \cap E = \emptyset$ and $\overline{Q}_S \cap K_{\mu_\beta} = \emptyset$. Therefore $\overline{h^{-1}(1)} \cap K_{\mu_\beta} = \emptyset$ and hence we can apply (7) to see that $w(x) = 0 = g_{\mu_\beta}(x)$ for any $x \in K_{\mu_\beta} \cap E$.

We proved that $w|K_{\mu_\beta} = g_{\mu_\beta}|K_{\mu_\beta}$ which shows that $w \in V_{\mu_\beta} \subset U_{\mu_\beta}$ for any $\beta \in S$ and hence $w$ witnesses that the family $\{U_\alpha : \alpha < \omega_1\}$ is not point-countable. Thus $D$ is a closed discrete subset of $C_p(\beta Z, \mathbb{D})$ which has no point-countable open expansion, i.e., Fact 4 is proved.

Returning to our solution assume that $X$ is an uncountable discrete space and hence there exists a surjective map $\varphi : X \to Z$ of $X$ onto a discrete space $Z$ such that $|Z| = \omega_1$; clearly, $\varphi$ is continuous. There exists a continuous map $\xi : \beta X \to \beta Z$ such that $\xi|X = \varphi$; let $\xi^*(f) = f \circ \xi$ for any $f \in C_p(\beta Z)$. The dual map $\xi^* : C_p(\beta Z) \to C_p(\beta X)$ is a closed embedding and it is straightforward that $\xi^*(C_p(\beta Z, \mathbb{D})) \subset C_p(\beta X, \mathbb{D})$. Apply Fact 4 to find a closed discrete subspace $D \subset C_p(\beta Z, \mathbb{D})$ which has no point-countable open expansion in $C_p(\beta Z, \mathbb{D})$. It is immediate that $\xi^*(D)$ is a closed discrete subset of the space $C_p(\beta X, \mathbb{D})$ which has no point-countable open expansion in $C_p(\beta X, \mathbb{D})$ and hence the space $C_p(\beta X, \mathbb{D})$ is not metalindelöf by Fact 1. The property of being metalindelöf is easily seen to be closed-hereditary; since $C_p(\beta X, \mathbb{D})$ is a closed subspace of $C_p(\beta X)$, the space $C_p(\beta X)$ is not metalindelöf either, i.e., our solution is complete.

**V.464.** *Prove that*

(i) *every strong $\Sigma$-space is a D-space. In particular, any Lindelöf $\Sigma$-space is a D-space;*

(ii) *every space with a point-countable base is a D-space.*

**Solution.** (i) Fix a strong $\Sigma$-space $X$ and take its compact cover $\mathcal{K}$ for which there exists a network $\mathcal{F}$ modulo $\mathcal{K}$ such that $\mathcal{F} = \bigcup_{i \in \omega} \mathcal{F}_i$ and every $\mathcal{F}_i$ is discrete. For each $i \in \omega$ the family $\mathcal{G}_i = \bigcup \{\mathcal{F}_j : j \leq i\}$ is locally finite; choose an ordinal $\mu_i$ for which there exists a faithful enumeration $\{G^i_\alpha : \alpha < \mu_i\}$ of the family $\mathcal{G}_i$.

Take an arbitrary neighborhood assignment $N$ on the space $X$. Given a set $U \in \tau(X)$ say that an element $F \in \mathcal{F}$ is *$U$-special* if there exists a set $K \in \mathcal{K}$ such that $K \subset F$, $K \backslash U \neq \emptyset$ and we can find a finite set $E \subset K \backslash U$ for which $N(E) \supset F \backslash U$; such a set $E$ will be called *a $U$-kernel* of the set $F$.

If the set $G^0_0$ is $\emptyset$-special then choose a $\emptyset$-kernel $E^0_0$ for $G^0_0$; if $G^0_0$ is not $\emptyset$-special then let $E^0_0 = \emptyset$. Proceeding inductively assume that $\beta < \mu_0$ and we have chosen a set $E^0_\alpha \subset G^0_\alpha$ for any $\alpha < \beta$ in such a way that

(1) if $\alpha < \beta$ and $G^0_\alpha$ is $N(\bigcup_{\gamma < \alpha} E^0_\gamma)$-special then $E^0_\alpha$ is an $N(\bigcup_{\gamma < \alpha} E^0_\gamma)$-kernel of $G^0_\alpha$; otherwise, $E^0_\alpha = \emptyset$.

If $G^0_\beta$ is $N(\bigcup_{\alpha < \beta} E^0_\alpha)$-special then let $E^0_\beta$ be an $N(\bigcup_{\alpha < \beta} E^0_\alpha)$-kernel of $G^0_\beta$; otherwise, let $E^0_\beta = \emptyset$. It is trivial that the property (1) now holds for all $\alpha \leq \beta$, so we can construct a family $\mathcal{E}_0 = \{E^0_\alpha : \alpha < \mu_0\}$ such that (1) holds for all $\beta < \mu_0$.

Proceeding inductively assume that $n \in \omega$ and we have constructed a family $\mathcal{E}_i = \{E^i_\alpha : \alpha < \mu_i\}$ for each $i \leq n$ in such a way that

(2) if $i < n$, $\beta < \mu_{i+1}$, $A(i, \beta) = \bigcup\{E^j_\alpha : j \leq i$ and $\alpha < \mu_j\} \cup \{E^{i+1}_\alpha : \alpha < \beta\}$ and the set $G^{i+1}_\beta$ is $N(A(i, \beta))$-special then $E^{i+1}_\beta$ is an $N(A(i, \beta))$-kernel of $G^{i+1}_\beta$.

Let $A(n, 0) = \bigcup\{E^j_\alpha : j \leq n$ and $\alpha < \mu_j\}$; if $G^{n+1}_0$ is $N(A(n, 0))$-special then choose an $N(A(n, 0))$-kernel $E^{n+1}_0$ of the set $G^{n+1}_0$; otherwise, let $E^{n+1}_0 = \emptyset$. Assume that $\beta < \mu_{n+1}$ and we have a family $\{E^{n+1}_\alpha : \alpha < \beta\}$ such that

(3) for any ordinal $\alpha < \beta$, if $A(n, \alpha) = A(n, 0) \cup \bigcup_{\gamma < \alpha} E^{n+1}_\gamma$ and the set $G^{n+1}_\alpha$ is $N(A(n, \alpha))$-special then $E^{n+1}_\alpha$ is an $N(A(n, \alpha))$-kernel of $G^{n+1}_\alpha$.

Let $A(n, \beta) = A(n, 0) \cup \bigcup_{\alpha < \beta} E^{n+1}_\alpha$; if $G^{n+1}_\beta$ is $N(A(n, \beta))$-special then take any $N(A(n, \beta))$-kernel $E^{n+1}_\beta$ for the set $G^{n+1}_\beta$; otherwise let $E^{n+1}_\beta = \emptyset$. This shows that our construction can be continued to obtain a family $\{E^{n+1}_\alpha : \alpha < \mu_{n+1}\}$ such that the condition (3) is satisfied for every $\beta < \mu_{n+1}$. Letting $\mathcal{E}_{n+1} = \{E^{n+1}_\alpha : \alpha < \mu_{n+1}\}$ we obtain a family $\mathcal{E}_{n+1}$ for which the property (2) holds for $i = n$ and all $\beta < \mu_{n+1}$ and hence we can construct the family $\mathcal{E}_i = \{E^i_\alpha : \alpha < \mu_i\}$ for each $i \in \omega$ in such a way that the condition (2) is satisfied for all $n \in \omega$ and $\beta < \mu_{n+1}$.

Observe first that it follows from (1) and (2) that $E_\alpha^i \subset G_\alpha^i$ for any $\alpha < \mu_i$, so every family $\mathcal{E}_i$ is locally finite. Any locally finite union of finite sets is closed and discrete, so the set $D_i = \bigcup \mathcal{E}_i$ is closed and discrete in $X$ for every $i \in \omega$.

Let $D = \bigcup_{i \in \omega} D_i$ and assume that $N(D) \neq X$; then there exists a set $K \in \mathcal{K}$ such that $K' = K \backslash N(D) \neq \emptyset$. By compactness of $K'$ we can find a finite set $Q \subset K'$ such that $K' \subset N(Q)$. The compact space $K'' = K \backslash N(Q)$ is covered by the family $\{N(D_i) : i \in \omega\}$, so there exists $n \in \omega$ such that $K'' \subset N(\bigcup_{i \leq n} D_i)$. Take a set $F \in \mathcal{F}$ for which $K \subset F \subset N(\bigcup_{i \leq n} D_i) \cup N(Q)$; there exists $m > n$ such that $F = G_\alpha^{m+1}$ for some $\alpha < \mu_{m+1}$. It is easy to see that the sets $K$ and $Q$ witness that $G_\alpha^{m+1}$ is $N(A(m, \alpha))$-special and hence $G_\alpha^{m+1} \backslash N(A(m, \alpha)) \subset N(E_\alpha^{m+1})$. As an immediate consequence, $F \backslash N(D) \subset G_\alpha^{m+1} \backslash N(A(m, \alpha)) \subset N(E_\alpha^{m+1}) \subset N(D)$; this contradiction proves that $N(D) = X$.

Finally take any point $x \in X$; since $N(D) = X$, we can choose a number $n \in \omega$ such that $x \in N(D_n)$. The property (2) implies that $N(D_n)$ is an open neighborhood of $x$ which does not meet $\bigcup_{i > n} D_i$. Since every $D_i$ is closed and discrete, the set $\bigcup_{i \leq n} D_i$ is also closed and discrete, so there exists a set $V \in \tau(x, X)$ which contains at most one element of $\bigcup_{i \leq n} D_i$. Thus the set $V \cap N(D_n)$ contains at most one element of $D$. The point $x \in X$ was chosen arbitrarily, so we proved that every $x \in X$ has a neighborhood which contains at most one element of $D$, i.e., $D$ is closed an discrete. Since also $N(D) = X$, we established that $X$ is a $D$-space.

(ii) Take a space $X$ with a point-countable base $\mathcal{B}$; this means that the family $\mathcal{B}_x = \{B \in \mathcal{B} : x \in B\}$ must be countable for any $x \in X$. It is easy to find infinite disjoint subsets $\{A_n : n \in \omega\}$ of the set $\omega$ such that $\{0, \dots, n + 1\} \subset A_0 \cup \dots \cup A_n$ for every $n \in \omega$ and $\bigcup_{n \in \omega} A_n = \omega$. Given a neighborhood assignment $N$ on the space $X$ observe first that we can assume, without loss of generality, that $N(x) \in \mathcal{B}$ for any $x \in X$. Indeed, if we proved (ii) for all neighborhood assignments $N$ such that $\{N(x) : x \in X\} \subset \mathcal{B}$, assume that $M$ is an arbitrary neighborhood assignment on $X$. For each $x \in X$ choose a set $B_x \in \mathcal{B}$ such that $x \in B_x \subset M(x)$ and let $N(x) = B_x$. For the assignment $N$, there exists a closed discrete set $D \subset X$ with $N(D) = X$; since $M(D) \supset N(D) = X$, we have $M(D) = X$, i.e., $D$ witnesses the fact that $X$ is a $D$-space.

We will construct by transfinite induction countable sets $D_\alpha$ whose union will be a closed discrete set $D$ such that $N(D) = X$. Let $D_0 = \emptyset$ and assume that, for some ordinal $\alpha$ we have the family $\{D_\beta : \beta < \alpha\}$. If $N(\bigcup\{D_\beta : \beta < \alpha\}) = X$, then our construction stops. If not, pick a point $x_0 \in Q = X \backslash N(\bigcup\{D_\beta : \beta < \alpha\})$ and let $H_0 = \{x_0\}$. Choose an enumeration $\{U_n : n \in A_0\}$ of the family $\mathcal{B}_{x_0}$. If we have a finite set $H_n = \{x_0, \dots, x_n\} \subset Q$ and an enumeration $\{U_n : n \in A_i\}$ of the family $\mathcal{B}_{x_i}$ for any $i \leq n$, observe that there is a unique $k \leq n$ such that $n + 1 \in A_k$ and consider the following condition

$E(\alpha, n)$: the set $R(\alpha, n) = \{x \in Q \backslash N(H_n) : N(x) \in \{U_m : m \in A_k\}\}$ is nonempty.

If $E(\alpha, n)$ holds, then choose the minimal $m \in A_k$ such that $N(x) = U_m$ for some $x \in Q \backslash N(H_n)$ and take $x_{n+1}$ to be any point in $Q \backslash N(H_n)$ such that $N(x_{n+1}) = U_m$; let $H_{n+1} = H_n \cup \{x_{n+1}\}$. If $R(\alpha, n) = \emptyset$, then let $x_{n+1} = x_0$ and $H_{n+1} = H_n$. After we have the sets $\{H_n : n \in \omega\}$ let $D_\alpha = \bigcup_{n \in \omega} H_n$.

Note that the sets $D_\alpha$ are disjoint and $D_\alpha \neq \emptyset$ for any ordinal $\alpha > 0$ if $D_\alpha$ is defined. Therefore, there exists an ordinal $\mu < |X|^+$ at which our inductive construction stops; let $D = \bigcup \{D_\alpha : \alpha < \mu\}$. By definition of our inductive procedure we must have $N(D) = X$, so the only thing left is to prove that $D$ is a closed discrete subset of $X$.

Fix any $x \in X$ and let $\alpha$ be the minimal ordinal such that $x \in N(D_\alpha)$. If $\beta > \alpha$, then $D_\beta \subset X \setminus N(D_\alpha)$, so $N(D_\alpha)$ is a neighborhood of $x$ that does not meet the set $\bigcup \{D_\beta : \alpha < \beta < \mu\}$. There exists $n \in \omega$ such that $x \in N(H_n)$ where $\{H_n : n \in \omega\}$ is the family used for the construction of $D_\alpha$. The condition $E(\alpha, n)$ shows that $W = N(H_n)$ is a neighborhood of $x$ that does not meet $D_\alpha \setminus H_n$ and hence $W \cap D_\alpha$ is finite.

Finally assume that $N(x) \cap D_\beta \neq \emptyset$ for some ordinal $\beta < \alpha$ and pick a point $y \in D_\beta \cap N(x)$. Let $\{G_n : n \in \omega\}$ be the family used for the construction of $D_\beta$. Then $y \in G_n$ for some $n \in \omega$ and there is $k \in \omega$ such that $\mathcal{B}_y = \{U_n : n \in A_k\}$; observe that $N(x) \in \mathcal{B}_y$ and therefore $N(x) = U_m$ for some $m \in \omega$. Besides, the point $x$ witnesses the fact that $E(\beta, n)$ holds for any $n \in A_k$. Therefore at the inductive step $n$ a point $x_n \in D_\beta$ will be chosen in such a way that $N(x_n) = U_p$ for some $p \leq m$. If $l > n$ and $l \in A_k$, then $x_l \notin N(x_n) = U_p$ and hence $N(x_l) \neq U_p$. This shows that there will be infinitely many steps of the induction procedure at which *distinct* sets are chosen from the finite family $\mathcal{U} = \{U_n : n \in A_k, n \leq m\}$; this contradiction shows that $N(x) \cap D_\beta = \emptyset$ for any $\beta < \alpha$. Consequently, the set $N(x) \cap N(D_\alpha) \cap W$ is a neighborhood of $x$ that meets finitely many elements of $D$. Thus $D$ is a closed discrete set that witnesses the $D$-property of $X$.

**V.465.** *Prove that,*

(i) *for any infinite cardinal $\kappa$, if a space is monotonically $\kappa$-monolithic then it is $\kappa$-monolithic;*

(ii) *for any infinite cardinal $\kappa$, if $X$ is a monotonically $\kappa$-monolithic space and $Y \subset X$ then $Y$ is also monotonically $\kappa$-monolithic;*

(iii) *if $\kappa$ is an infinite cardinal, then any countable product of monotonically $\kappa$-monolithic spaces is monotonically $\kappa$-monolithic;*

(iv) *for any infinite cardinal $\kappa$, every closed continuous image of a monotonically $\kappa$-monolithic space is monotonically $\kappa$-monolithic;*

(v) *a space $X$ is monotonically monolithic if and only if $X$ is monotonically $\kappa$-monolithic for any infinite cardinal $\kappa$;*

(vi) *for any infinite cardinal $\kappa$, a space $X$ is monotonically $\kappa$-monolithic if and only if, to any finite set $F \subset X$ we can assign a countable family $\mathcal{O}(F) \subset exp(X)$ in such a way that for any set $A \subset X$ with $|A| \leq \kappa$, the family $\bigcup \{\mathcal{O}(F) : F \in [A]^{<\omega}\}$ is an external network at all points of $\overline{A}$;*

(vii) *a space $X$ is monotonically monolithic if and only if, to any finite set $F \subset X$ we can assign a countable family $\mathcal{O}(F) \subset exp(X)$ in such a way that for any $A \subset X$, the family $\bigcup \{\mathcal{O}(F) : F \in [A]^{<\omega}\}$ is an external network at all points of $\overline{A}$;*

(viii) *if $X$ is a space and the set $X'$ of non-isolated points of $X$ has a countable network then $X$ is monotonically monolithic. In particular, if a space $X$ is cosmic or has a unique non-isolated point then $X$ is monotonically monolithic;*

(ix) *given an infinite cardinal $\kappa$, if a space $X$ is monotonically $\kappa$-monolithic and $t(X) \leq \kappa$, then $X$ is monotonically monolithic; .*

(x) *for any infinite cardinal $\kappa$, every $\sigma$-product of monotonically $\kappa$-monolithic spaces is monotonically $\kappa$-monolithic. In particular, any $\sigma$-product of monotonically monolithic spaces is monotonically monolithic.*

**Solution.** (i) Suppose that a space $X$ is monotonically $\kappa$-monolithic and take the respective operator $\mathcal{O}$ on $X$. If $A \subset X$ and $|A| \leq \kappa$ then $\mathcal{N} = \{P \cap \overline{A} : P \in \mathcal{O}(A)\}$ is a network in $\overline{A}$ and $|\mathcal{N}| \leq |\mathcal{O}(A)| \leq \max\{|A|, \omega\} \leq \kappa$, so $X$ is $\kappa$-monolithic.

(ii) Suppose that $X$ is monotonically $\kappa$-monolithic and take the respective operator $\mathcal{O}$ on the space $X$. If $Y \subset X$ and $A \subset Y$ then consider the family $\mathcal{O}_Y(A) = \{P \cap Y : P \in \mathcal{O}(A)\}$. It is straightforward that $\mathcal{O}_Y$ is an operator that witnesses monotone $\kappa$-monolithity of $Y$.

(iii) Let $X_i$ be a monotonically $\kappa$-monolithic space for any $i \in \omega$ and consider the space $X = \prod_{i \in \omega} X_i$. Fix an operator $\mathcal{O}_i$ which witnesses monotone $\kappa$-monolithity of $X_i$ and consider the projection $\pi_i : X \to X_i$ for each $i \in \omega$. Given a collection $\mathbf{A} = \{\mathcal{A}_i : i \in \omega\}$ of families such that $\mathcal{A}_i \subset \exp(X_i)$ for all $i \in \omega$, the family $\mathcal{W}(\mathbf{A}) = \{(\prod_{i \in Q} A_i) \times (\prod_{i \in \omega \setminus Q} X_i) : Q$ is a finite subset of $\omega$ and $A_i \in \mathcal{A}_i$ for each $i \in Q\}$ consists of subsets of $X$.

Given any set $A \subset X$ with $|A| \leq \kappa$ let $A_i = \pi_i(A)$ and $\mathcal{A}_i = \mathcal{O}_i(A_i)$ for every $i \in \omega$. If $\mathbf{A} = \{\mathcal{A}_i : i \in \omega\}$ then it is easy to check that the family $\mathcal{O}(A) = \mathcal{W}(\mathbf{A})$ is an external network for the set $\overline{A}$ in $X$. We omit a simple verification that the operator $\mathcal{O}$ witnesses monotone $\kappa$-monolithity of $X$.

(iv) Assume that $X$ is monotonically $\kappa$-monolithic and $f : X \to Y$ is a closed map. Fix a point $x_y \in f^{-1}(y)$ for any $y \in Y$ and take an operator $\mathcal{O}$ which witnesses monotonic $\kappa$-monolithity of $X$. Given a set $B \subset Y$ with $|B| \leq \kappa$ let $A = \{x_y : y \in B\}$ and $\mathcal{N}(B) = \{f(P) : P \in \mathcal{O}(A)\}$. To see that $\mathcal{N}$ is an external network of $\overline{B}$ in $Y$ take any $y \in \overline{B}$ and $U \in \tau(y, Y)$. The map $f$ being closed we have $f(\overline{A}) = \overline{B}$, so there is a point $x \in \overline{A}$ with $f(x) = y$. If $P \in \mathcal{O}(A)$ and $x \in P \subset f^{-1}(U)$ then $f(P) \in \mathcal{N}(B)$ and $y \in f(P) \subset U$. It is easy to see that $\mathcal{N}$ satisfies all other conditions of the definition of monotone $\kappa$-monolithity.

(v) If $X$ is monotonically monolithic and an operator $\mathcal{O}$ witnesses this, then $\mathcal{O}$ witnesses that $X$ is monotonically $\kappa$-monolithic for any infinite cardinal $\kappa$, i.e., we proved necessity.

Now if $X$ is monotonically $\kappa$-monolithic for any infinite cardinal $\kappa$ then take an operator $\mathcal{O}$ that witnesses this for $\kappa = \max\{|X|, \omega\}$. Since $|A| \leq \kappa$ for any $A \subset X$, the operator $\mathcal{O}$ satisfies all conditions of the definition of (strong) monotone monolithity and hence we settled sufficiency.

(vi) We will need the following fact.

*Fact 1.* If $\kappa$ is an infinite cardinal and $Z$ is a monotonically $\kappa$-monolithic space with a respective operator $\mathcal{O}$, then for any set $A \subset Z$ with $|A| \leq \kappa$, we have the equality $\mathcal{O}(A) = \bigcup \{\mathcal{O}(F) : F \in [A]^{<\omega}\}$.

*Proof.* If $A$ is finite, then there is nothing to prove, so assume, proceeding by induction, that $\mu \leq \kappa$ is a cardinal and we proved our equality for any set $B \subset Z$ with $|B| < \mu$. If $A \subset Z$ and $|A| = \mu$, then it is easy to find an increasing family $\{B_\alpha : \alpha < \mu\}$ such that $\bigcup \{B_\alpha : \alpha < \mu\} = A$ and $|B_\alpha| < \mu$ for any $\alpha < \mu$. By the induction hypothesis, we have $\mathcal{O}(B_\alpha) = \bigcup \{\mathcal{O}(F) : F \in [B_\alpha]^{<\omega}\}$ for any $\alpha < \mu$. Since the operator $\mathcal{O}$ is $\kappa$-monotone, we have $\mathcal{O}(A) = \bigcup \{\mathcal{O}(B_\alpha) : \alpha < \mu\}$, so it follows from the equalities $\bigcup \{\mathcal{O}(B_\alpha) : \alpha < \mu\} = \bigcup \{\bigcup \{\mathcal{O}(F) : F \in [B_\alpha]^{<\omega}\} : \alpha < \mu\} = \bigcup \{\mathcal{O}(F) : F \in [A]^{<\omega}\}$ that $\mathcal{O}(A) = \bigcup \{\mathcal{O}(F) : F \in [A]^{<\omega}\}$, i.e., Fact 1 is proved.

Returning to our solution assume that in a space $X$ we have an operator $\mathcal{O}$ on all finite subsets of $X$ as in (vi). Letting $\mathcal{O}'(A) = \bigcup \{\mathcal{O}(F) : F \in [A]^{<\omega}\}$ for every set $A \subset X$ with $|A| \leq \kappa$, it is immediate that we obtain a monotone $\kappa$-monolithity operator $\mathcal{O}'$ on $X$. This proves sufficiency.

Now, if $\mathcal{O}$ is a monotone $\kappa$-monolithity operator on $X$, then let $\mathcal{O}'(F) = \mathcal{O}(F)$ for any finite subset of $X$. If $A \subset X$ and $|A| \leq \kappa$, then

$$\bigcup \{\mathcal{O}'(F) : F \in [A]^{<\omega}\} = \bigcup \{\mathcal{O}(F) : F \in [A]^{<\omega}\} = \mathcal{O}(A)$$

by Fact 1, so the family $\bigcup \{\mathcal{O}'(F) : F \in [A]^{<\omega}\}$ is an external network at all points of $\overline{A}$, i.e., we proved necessity.

(vii) Assume that in a space $X$ we have an operator $\mathcal{O}$ on all finite subsets of $X$ as in (vii). Letting $\mathcal{O}'(A) = \bigcup \{\mathcal{O}(F) : F \in [A]^{<\omega}\}$ for any $A \subset X$, it is immediate that we obtain a monotone monolithity operator $\mathcal{O}'$ on $X$. This proves sufficiency.

Now, if $\mathcal{O}$ is a monotone monolithity operator on $X$, then let $\mathcal{O}'(F) = \mathcal{O}(F)$ for any finite subset of $X$. If $A \subset X$, then

$$\bigcup \{\mathcal{O}'(F) : F \in [A]^{<\omega}\} = \bigcup \{\mathcal{O}(F) : F \in [A]^{<\omega}\} = \mathcal{O}(A)$$

by Fact 1, so the family $\bigcup \{\mathcal{O}'(F) : F \in [A]^{<\omega}\}$ is an external network at all points of $\overline{A}$, i.e., we proved necessity.

(viii) Let $\mathcal{N}$ be a countable network in the space $X'$. Given an arbitrary set $A \subset X$ let $\mathcal{O}(A) = \{\{a\} : a \in A \backslash X'\} \cup \mathcal{N}$. It is trivial that the operator $\mathcal{O}$ witnesses monotonic monolithity of $X$.

(ix) Let $\mathcal{O}'$ be an operator of monotonic $\kappa$-monolithity on $X$. Consider the family $\mathcal{O}(F) = \mathcal{O}'(F)$ for any finite set $F \subset X$. Given an arbitrary set $A \subset X$, take any point $x \in \overline{A}$ and $U \in \tau(x, X)$. It follows from $t(X) \leq \kappa$ that we can find a set $B \subset A$ such that $|B| \leq \kappa$ and $x \in \overline{B}$. The operator $\mathcal{O}'$ is applicable to $B$, so there exists $N \in \mathcal{O}'(B)$ such that $x \in N \subset U$. By Fact 1, there is a finite set $F \subset B$ such that $N \in \mathcal{O}'(F) = \mathcal{O}(F)$. Therefore the family $\bigcup\{\mathcal{O}(F) : F \in [A]^{<\omega}\}$ contains an external network at every point of $\overline{A}$ which, together with (vi), implies that $X$ is monotonically monolithic.

(x) Suppose that $X_t$ is monotonically $\kappa$-monolithic and fix the respective operator $\mathcal{O}_t$ on the space $X_t$ for any $t \in T$. Let $X = \prod_{t \in T} X_t$ and fix a point $a \in X$. We must prove that the space $\sigma(X, a) = \{x \in X : |\{t \in T : x(t) \neq a(t)\}| < \omega\}$ is monotonically $\kappa$-monolithic. For any $S \subset T$ we will need the point $a_S \in \prod_{t \in S} X_t$ such that $a_S(t) = a(t)$ for any $t \in S$. Let $p_t : X \to X_t$ be the projection for each $t \in T$.

For every point $x \in \sigma(X, a)$ let $\text{supp}(x) = \{t \in T : x(t) \neq a(t)\}$. Suppose that we have a finite set $K \subset T$ and a family $\mathcal{A}_t$ of subsets of the space $X_t$ for any $t \in K$. Let $\mathcal{A} = \{\mathcal{A}_t : t \in K\}$ and $\mathcal{H}(K, \mathcal{A}) = \{\prod_{t \in K} P_t \times \{a_{T \setminus K}\} : P_t \in \mathcal{A}_t$ for every $t \in K\}$. Now we are ready to construct a monotonic $\kappa$-monolithity operator on $\sigma(X, a)$.

To do so, fix a set $A \subset \sigma(X, a)$ with $|A| \leq \kappa$ and let $\lambda = \max\{|A|, \omega\}$. It is evident that the family $\mathcal{E}(A) = \bigcup\{\text{supp}(x) : x \in A\}$ has cardinality at most $\lambda$; for every finite $K \subset \mathcal{E}(A)$ let $\mathcal{A}_K(A) = \{\mathcal{O}_t(p_t(A)) : t \in K\}$. Then the cardinality of the family $\mathcal{H}(K, \mathcal{A}_K(A))$ does not exceed $\lambda$. Therefore the family $\mathcal{O}(A) = \bigcup\{\mathcal{H}(K, \mathcal{A}_K(A)) : K$ is a finite subset of $\mathcal{E}(A)\}$ also has cardinality at most $\lambda$.

To check that the operator $\mathcal{O}$ is as promised, observe that if $A \subset B \subset \sigma(X, a)$ then $p_t(A) \subset p_t(B)$ for any $t \in T$. Therefore $\mathcal{H}(K, \mathcal{A}_K(A)) \subset \mathcal{H}(K, \mathcal{A}_K(B))$ for every finite $K \subset \mathcal{E}(A)$ which in turn implies that $\mathcal{O}(A) \subset \mathcal{O}(B)$. To see that the increasing union property holds for $\mathcal{O}$ observe that $\mathcal{O}(A)$ is defined by using finitary operations with families which have this property. To finally see that $\mathcal{O}(A)$ is an outer network at all points of $\overline{A}$ take any $x \in \overline{A}$ and any $U \in \tau(x, \sigma(X, a))$. There exists a finite set $K \subset T$ and a family $\{W_t : t \in K\}$ such that $\text{supp}(x) \subset K$ and $W_t \in \tau(x(t), X_t)$ for any $t \in K$ while $W = \{y \in \sigma(X, a) : y(t) \in W_t$ for each $t \in K\} \subset U$.

Since $x(t) \in \overline{p_t(A)}$, there exists $P_t \in \mathcal{O}(p_t(A))$ such that $x(t) \in P_t \subset W_t$ for every $t \in K$. The set $P = \prod_{t \in K} P_t \times \{a_{T \setminus K}\}$ belongs to $\mathcal{O}(A)$ and $x \in P \subset U$, so $\mathcal{O}$ is, indeed, a monotone $\kappa$-monolithity operator on $\sigma(X, a)$.

**V.466.** *Prove that,*

*(i)  a space $X$ is strongly monotonically monolithic if and only if $X$ is strongly monotonically $\omega$-monolithic;*

*(ii)  every space with a point-countable base must be strongly monotonically monolithic;*

*(iii)  if a space $X$ is strongly monotonically monolithic, then it must be strongly monolithic and monotonically monolithic;*

*(iv)  every subspace of a strongly monotonically monolithic space is strongly monotonically monolithic;*

*(v)  any countable product of strongly monotonically monolithic spaces must be strongly monotonically monolithic;*

*(vi) if $X$ is strongly monotonically monolithic and $f : X \to Y$ is an open map such that $d(f^{-1}(y)) \leq \omega$ for any $y \in Y$ then $Y$ is strongly monotonically monolithic;*

*(vii) every countably compact strongly monotonically monolithic space is compact and metrizable.*

**Solution.** (i) We must only prove sufficiency, so assume that a space $X$ is strongly monotonically $\omega$-monolithic and fix a respective operator $\mathcal{O}$ on $X$. Take any set $A \subset X$, a point $x \in \overline{A}$ and $U \in \tau(x, X)$. Note first that $\mathcal{O}(\{x\})$ contains a countable local base at the point $x$, so there is a countable set $B \subset A$ such that $x \in \overline{B}$. Take an increasing family $\{B_n : n \in \omega\}$ of finite sets such that $B = \bigcup\{B_n : n \in \omega\}$. The operator $\mathcal{O}$ is applicable to $B$, so there exists $N \in \mathcal{O}(B)$ for which $x \in N \subset U$. By $\omega$-monotonicity of $\mathcal{O}$, we can find $n \in \omega$ such that $N \in \mathcal{O}(B_n)$. This proves that for any set $A \subset X$, the family $\mathcal{O}'(A) = \bigcup\{\mathcal{O}(F) : F \in [A]^{<\omega}\}$ contains a local base at every point of $\overline{A}$. It is immediate that the operator $\mathcal{O}'$ is monotone, so it witnesses that $X$ strongly monotonically monolithic.

(ii) Assume that a space $X$ has a point-countable base $\mathcal{B}$. For any $A \subset X$ let $\mathcal{O}(A) = \{B \in \mathcal{B} : B \cap A \neq \emptyset\}$. If $x \in \overline{A}$ and $U \in \tau(x, X)$ then there is $B \in \mathcal{B}$ such that $x \in B \subset U$; we also have $B \cap A \neq \emptyset$, so $B \in \mathcal{O}(A)$ and hence the family $\mathcal{O}(A)$ is an external base of $\overline{A}$ in $X$. It is evident that $\mathcal{O}(A)$ has all other properties required by the definition of the operator of strong monotone $\kappa$-monolithity.

(iii) Since any external base of a set is also an external network of the same set, any operator of strong monotone monolithity is also an operator of monotone monolithity. If $X$ is strongly monotonically monolithic and $A \subset X$, then there is an external base $\mathcal{B}$ for the set $\overline{A}$ such that $|\mathcal{B}| \leq \max\{|A|, \omega\}$. It is clear that the family $\mathcal{B}' = \{B \cap \overline{A} : B \in \mathcal{B}\}$ is a base in $\overline{A}$ such that $|\mathcal{B}'| \leq \max\{|A|, \omega\}$; this proves that $X$ is strongly monolithic.

(iv) Suppose that $X$ is strongly monotonically monolithic and take the respective operator $\mathcal{O}$ on the space $X$. If $Y \subset X$ and $A \subset Y$ then consider the family $\mathcal{O}_Y(A) = \{P \cap Y : P \in \mathcal{O}(A)\}$. It is straightforward that $\mathcal{O}_Y$ is an operator that witnesses strong monotone monolithity of $Y$.

(v) Let $X_i$ be a strongly monotonically $\kappa$-monolithic space for any $i \in \omega$ and consider the space $X = \prod_{i \in \omega} X_i$. Fix an operator $\mathcal{O}_i$ which witnesses strong monotone $\kappa$-monolithity of $X_i$ and consider the projection $\pi_i : X \to X_i$ for each $i \in \omega$. Given a collection $\mathbf{A} = \{\mathcal{A}_i : i \in \omega\}$ of families such that $\mathcal{A}_i \subset \tau(X_i)$ for all $i \in \omega$ it is easy to see that the family $\mathcal{W}(\mathbf{A}) = \{(\prod_{i \in Q} A_i) \times (\prod_{i \in \omega \setminus Q} X_i) : Q$ is a finite subset of $\omega$ and $A_i \in \mathcal{A}_i$ for each $i \in Q\}$ consists of open subsets of $X$.

Given any set $A \subset X$ with $|A| \leq \kappa$ let $A_i = \pi_i(A)$ and $\mathcal{A}_i = \mathcal{O}_i(A_i)$ for every $i \in \omega$. If $\mathbf{A} = \{\mathcal{A}_i : i \in \omega\}$ then it is easy to check that the family $\mathcal{O}(A) = \mathcal{W}(\mathbf{A})$ is an external base for the set $\overline{A}$ in $X$. We omit a simple verification that the operator $\mathcal{O}$ witnesses strong monotone $\kappa$-monolithity of $X$.

(vi) Take a countable dense set $A_y$ in the space $f^{-1}(y)$ for any $y \in Y$ and let
$\mathcal{O}$ be the strong monotone monolithity operator in $X$. If $B \subset Y$ then let
$Q_B = \bigcup\{A_y : y \in B\}$ and $\mathcal{N}(B) = \{f(E) : E \in \mathcal{O}(Q_B)\}$.
To see that $\mathcal{N}(B)$ is an external base for $\overline{B}$ take a point $y \in \overline{B}$ and a set
$U \in \tau(y, Y)$; observe that $\overline{f^{-1}(B)} \cap f^{-1}(y) \neq \emptyset$ because the map $f$ is
open. Since $Q_B$ is dense in $f^{-1}(B)$, there exists a point $x \in f^{-1}(y)$ with
$x \in \overline{Q}_B$. Take a set $V \in \mathcal{O}(Q_B)$ such that $x \in V \subset f^{-1}(U)$; then $W =$
$f(V) \in \mathcal{N}(B)$ and $y \in W \subset U$, i.e., $\mathcal{N}(B)$ is an external base for $B$. An easy
proof that $\mathcal{N}$ satisfies all other conditions of the definition of strong monotone
monolithity can be left to the reader.

(vii) Given a strongly monotonically $\omega$-monolithic countably compact space $X$
take an operator $\mathcal{O}$ which witnesses strong monotone $\omega$-monolithity of $X$ and
fix a point $a \in X$; let $A_0 = \{a\}$. Proceeding inductively assume that $n \in \omega$
and we have countable subsets $A_0, \ldots, A_n$ of the space $X$ with the following
properties:

(1) $A_i \subset A_{i+1}$ for any $i < n$;
(2) if $i < n$ and $\mathcal{U}$ is a finite subfamily of $\mathcal{O}(A_i)$ such that $\bigcup \mathcal{U} \neq X$ then
there is a point $x \in A_{i+1} \backslash (\bigcup \mathcal{U})$.

The set $A_n$ being countable the family $\mathcal{V}$ of all finite unions of the elements
of $\mathcal{O}(A_n)$ is also countable. Let $\mathcal{V}' = \{V \in \mathcal{V} : V \neq X\}$ and choose a point
$x_V \in X \backslash V$ for any $V \in \mathcal{V}'$. The set $A_{n+1} = A_n \cup \{x_V : V \in \mathcal{V}'\}$ is countable and
it is immediate that the properties (1) and (2) still hold if we replace $n$ with $n + 1$.

Therefore our inductive procedure can be continued to construct a sequence $\{A_n :$
$n \in \omega\}$ such that (1) and (2) are true for all $n \in \omega$. If $A = \bigcup_{n \in \omega} A_n$ is not dense
in $X$ then pick a set $W \in \tau^*(X)$ such that $\overline{W} \cap \overline{A} = \emptyset$. The family $\mathcal{O}(A)$ being an
external base for $\overline{A}$ we can choose a set $V_x \in \mathcal{O}(A)$ such that $x \in V_x \subset X \backslash \overline{W}$ for
any $x \in \overline{A}$.

The space $\overline{A}$ is countably compact and second countable because $\mathcal{O}(A)$ is a
countable external base of $\overline{A}$. Thus the set $\overline{A}$ is compact and hence there exists a
finite $Q \subset \overline{A}$ such that $\overline{A} \subset \bigcup\{V_x : x \in Q\}$. It follows from the definition of
strong monotone $\omega$-monolithity that $\mathcal{V} = \{V_x : x \in Q\} \subset \mathcal{O}(A_n)$ for some number
$n \in \omega$. The property (2) implies that $A_{n+1} \backslash (\bigcup \mathcal{V}) \neq \emptyset$ while $A_{n+1} \subset A \subset \bigcup \mathcal{V}$;
this contradiction shows that $A$ is dense in $X$, so $X$ is compact and metrizable
because $\mathcal{O}(A)$ is a countable base in $X$.

**V.467.** *For any infinite cardinal $\kappa$ prove that*

(a) *a space $X$ is monotonically $\kappa$-stable if and only if $C_p(X)$ is monotonically
$\kappa$-monolithic.*
(b) *a space $X$ is monotonically $\kappa$-monolithic if and only if $C_p(X)$ is monotonically
$\kappa$-stable.*

**Solution.** Fix a countable base $\mathcal{B}$ in the space $\mathbb{R}$; given a family $\mathcal{A}$ of subsets of a
space $Z$ let $\mathcal{W}(\mathcal{A}) = \{[A_1, \ldots, A_n; B_1, \ldots, B_n] : n \in \mathbb{N}, A_i \in \mathcal{A}$ and $B_i \in \mathcal{B}$ for

every $i \leq n$}; here, as usual, $[A_1, \ldots, A_n; B_1, \ldots, B_n] = \{f \in C_p(Z) : f(A_i) \subset B_i$ for any $i \leq n\}$. If $\varphi : Z \to Z'$ is an onto map, then its dual map $\varphi^* : C_p(Z') \to C_p(Z)$ is defined by the formula $\varphi(f) = f \circ \varphi$ for any $f \in C_p(Z')$.

**Fact 1.** If $Z$ is a space, then the assignment $\mathcal{A} \to \mathcal{W}(\mathcal{A})$ is $\mu$-monotone on $\exp(Z)$ for any infinite cardinal $\mu$.

*Proof.* The family $\mathcal{B}$ being countable, we have $|\mathcal{W}(\mathcal{A})| \leq \max\{|\mathcal{A}|, \omega\}$ for any set $A \subset Z$. It is evident that $A \subset A'$ implies $\mathcal{W}(\mathcal{A}) \subset \mathcal{W}(\mathcal{A}')$; if $\lambda \leq \mu$ and $\{\mathcal{A}_\alpha : \alpha < \lambda\}$ is an increasing family of subsets of $Z$, then take any set $Q = [A_1, \ldots, A_n; B_1, \ldots, B_n] \in \mathcal{A} = \bigcup\{\mathcal{A}_\alpha : \alpha < \lambda\}$. There exists an ordinal $\alpha < \lambda$ such that $\{A_1, \ldots, A_n\} \subset \mathcal{A}_\alpha$ and hence $Q \in \mathcal{W}(\mathcal{A}_\alpha)$. Therefore $\mathcal{W}(\bigcup_{\alpha<\lambda} \mathcal{A}_\alpha) = \mathcal{W}(\mathcal{A})$, i.e., Fact 1 is proved.

**Fact 2.** Suppose that $Z$ is a space and $\varphi : Z \to Z'$ is a continuous onto map. If $\mathcal{N}$ is a network for the map $\varphi$, then $\mathcal{W}(\mathcal{N})$ is an external network of $\varphi^*(C_p(Z'))$ in $C_p(Z)$.

*Proof.* Take a function $f \in \varphi^*(C_p(Z'))$ and a set $W \in \tau(f, C_p(Z))$. There exist points $x_1, \ldots, x_n \in Z$ and sets $B_1, \ldots, B_n \in \mathcal{B}$ such that $f \in U = \{g \in C_p(Z) : g(x_i) \in B_i$ for any $i \leq n\} \subset W$. Fix a function $h \in C_p(Z')$ such that $f = h \circ \varphi$. Let $y_i = \varphi(x_i)$; it follows from $f(x_i) = h(y_i) \in B_i$ that $y_i \in G_i = h^{-1}(B_i)$ for any $i \leq n$. The family $\mathcal{N}$ being a network for $\varphi$, we can find sets $N_1, \ldots, N_n \in \mathcal{N}$ such that $x_i \in N_i$ and $\varphi(N_i) \subset G_i$ for each $i \leq n$. Then $Q = [N_1, \ldots, N_n; B_1, \ldots, B_n] \in \mathcal{W}(\mathcal{N})$ and $f \in Q \subset U \subset W$, i.e., Fact 2 is proved.

**Fact 3.** If $Z$ is a space and $Y \subset Z$, then let $\pi_Y : C_p(Z) \to C_p(Y)$ be the restriction map. If $\mathcal{N}$ is an external network of the set $Y$ in $Z$, then $\mathcal{W}(\mathcal{N})$ is a network for the map $\pi_Y$.

*Proof.* Take any $f \in C_p(Z)$ and $U \in \tau(\pi_Y(f), C_p(Y))$. Choose $y_1, \ldots, y_n \in Y$ and $B_1, \ldots, B_n \in \mathcal{B}$ such that $\pi_Y(f) \in V = \{g \in C_p(Y) : g(y_i) \in B_i$ for all $i \leq n\} \subset U$. There exists $N_i \in \mathcal{N}$ such that $y_i \in N_i \subset f^{-1}(B_i)$ for every $i \leq n$. The set $H = [N_1, \ldots, N_n; B_1, \ldots, B_n]$ belongs to $\mathcal{W}(\mathcal{N})$ and $f \in H$. Now, if $h \in H$, then we have the inclusions $h(y_i) \in h(N_i) \subset B_i$ for each $i \leq n$; since $\pi_Y(h)(y_i) = h(y_i)$, we conclude that $\pi_Y(h)(y_i) \in B_i$ for every $i \leq n$ which shows that $\pi_Y(h) \in V \subset U$ for any $h \in H$, i.e., $\pi_Y(H) \subset U$ and hence Fact 3 is proved.

**Fact 4.** If a space $Z$ is monotonically $\mu$-monolithic for some infinite cardinal $\mu$, then $C_p(C_p(Z))$ is also monotonically $\mu$-monolithic.

*Proof.* Recall that for any space $Y$, if $A \subset Y$, then $\pi_A : C_p(Y) \to C_p(A)$ is the restriction map and $C_p(A|Y) = \{f|A : f \in C_p(Y)\} = \pi_A(C_p(Y))$. Apply TFS-300 to fix, for any $f \in C_p(C_p(Z))$ a countable set $S(f) \subset Z$ and a continuous function $p(f) : C_p(S(f)|Z) \to \mathbb{R}$ such that $p(f) \circ \pi_{S(f)} = f$. If $B \subset Z$ and $S(f) \subset B$, then let $q_{S(f)}^B : C_p(B|Z) \to C_p(S(f)|Z)$ be the restriction map. It is clear that $f = p(f) \circ \pi_{S(f)} = p(f) \circ q_{S(f)}^B \circ \pi_B$ which shows that $f \in \pi_B^*(C_p(C_p(B|Z)))$. This proves that

(1) if $f \in C_p(C_p(Z))$ and $S(f) \subset B \subset Z$, then $f \in \pi_B^*(C_p(C_p(B|Z)))$.

Now assume that $Z$ is monotonically $\mu$-monolithic and fix the respective operator $\mathcal{O}$ on the space $Z$. For any set $A \subset C_p(C_p(Z))$ with $|A| \leq \mu$ consider the set $S(A) = \bigcup \{S(f) : f \in A\}$. It is easy to check that

(2) the assignment $A \to S(A)$ is $\mu$-monotone.

Now let $\mathcal{N}(A) = \mathcal{W}(\mathcal{W}(\mathcal{O}(S(A))))$ for any set $A \subset C_p(C_p(Z))$ with $|A| \leq \mu$. Apply Fact 1 and the property (2) to see that $\mathcal{N}$ is the composition of four $\mu$-monotone maps, so it is $\mu$-monotone and hence we only have to prove that $\mathcal{N}(A)$ is an external network for $\overline{A}$.

Let $B = \overline{S(A)}$ and $Y = C_p(B|Z)$. The map $\pi_B : C_p(Z) \to Y$ is open by TFS-152, so the set $E = \pi_B^*(C_p(Y))$ is closed in $C_p(C_p(Z))$ by TFS-163. The property (1) shows that $A \subset E$ and hence $\overline{A} \subset E$. By the choice of the operator $\mathcal{O}$, the family $\mathcal{O}(S(A))$ is an external network for $B$, so $\mathcal{W}(\mathcal{O}(S(A)))$ is a network for the map $\pi_B$ by Fact 3. Now, apply Fact 2 to see that $\mathcal{N}(A) = \mathcal{W}(\mathcal{W}(\mathcal{O}(S(A))))$ is an external network for the set $E$; since $\overline{A} \subset E$, the family $\mathcal{N}(A)$ is also an external network for $\overline{A}$, i.e., the operator $\mathcal{N}$ witnesses that $C_p(C_p(Z))$ is monotonically $\mu$-monolithic and hence Fact 4 is proved.

Returning to our solution assume that $X$ is monotonically $\kappa$-stable and fix the respective operator $\mathcal{O}$. Given a set $A \subset C_p(X)$ with $|A| \leq \kappa$, denote by $\varphi$ the reflection map with respect to $\overline{A}$, i.e., $\varphi : X \to C_p(\overline{A})$ and $\varphi(x)(f) = f(x)$ for any $x \in X$ and $f \in \overline{A}$. The family $\mathcal{O}(A)$ is a network for the map $\varphi$; let $\mathcal{N}(A) = \mathcal{W}(\mathcal{O}(A))$. The assignment $A \to \mathcal{N}(A)$ is $\kappa$-monotone being the composition of two $\kappa$-monotone mappings. By Fact 2, the family $\mathcal{N}(A)$ is an external network in $C_p(X)$ for the set $\varphi^*(C_p(Y))$ where $Y = \varphi(X)$. Observe that $\overline{A} \subset \varphi^*(C_p(Y))$ by Fact 5 of U.086 and therefore $\mathcal{N}(A)$ is an external network in $C_p(X)$ for the set $\overline{A}$. Thus, the operator $\mathcal{N}$ witnesses that $C_p(X)$ is monotonically $\kappa$-monolithic, i.e., we proved necessity in (a).

Now assume that $C_p(X)$ is monotonically $\kappa$-monolithic and let $\mathcal{N}$ be the respective operator on $C_p(X)$. Let $e : X \to C_p(C_p(X))$ be the natural embedding defined by $e(x)(f) = f(x)$ for any $x \in X$ and $f \in C_p(X)$ (see TFS-167). Take a set $A \subset C_p(X)$ with $|A| \leq \kappa$ and apply Fact 3 to the restriction map $\pi : C_p(C_p(X)) \to C_p(\overline{A})$ to note that $\mathcal{W}(\mathcal{N}(A))$ is a network for $\pi$ in $C_p(C_p(X))$. Letting $\mathcal{O}(A) = \{e^{-1}(H \cap e(X)) : H \in \mathcal{W}(\mathcal{N}(A))\}$ we obtain a $\kappa$-monotone operator, so it suffices to show that $\mathcal{O}$ witnesses monotone $\kappa$-stability of $X$, i.e., $\mathcal{O}(A)$ is a network for the reflexion map $\varphi : X \to C_p(\overline{A})$ defined by $\varphi(x)(f) = f(x)$ for any $x \in X$ and $f \in \overline{A}$.

To do this, observe first that we have the equality $\varphi = \pi \circ e$; take any point $x \in X$ and a set $U \in \tau(\varphi(x), C_p(\overline{A}))$. Since $\mathcal{W}(\mathcal{N}(A))$ is a network for the map $\pi$ and $\varphi(x) = \pi(e(x)) \in U$, we can find a set $Q \in \mathcal{W}(\mathcal{N}(A))$ such that $e(x) \in Q$ and $\pi(Q) \subset U$. Then the set $P = e^{-1}(Q \cap e(X))$ belongs to $\mathcal{O}(A)$ and $x \in P$. Since also $\varphi(P) = \pi(e(P)) \subset \pi(Q) \subset U$, the set $P$ witnesses the fact that $\mathcal{O}(A)$ is a network for $\varphi$ and hence the space $X$ is monotonically $\kappa$-stable, i.e., we completed the proof of (a).

(b) If $X$ is monotonically $\kappa$-monolithic, then so is $C_p(C_p(X))$ by Fact 4 and hence $C_p(X)$ is monotonically $\kappa$-stable by (a). If $C_p(X)$ is monotonically $\kappa$-stable, then $C_p(C_p(X))$ is monotonically $\kappa$-monolithic by (a) and therefore $X$ is monotonically $\kappa$-monolithic being embeddable en $C_p(C_p(X))$. This settles (b) and makes our solution complete.

**V.468.** *Prove that if $X$ is a Lindelöf $\Sigma$-space then the space $C_p(X)$ must be monotonically monolithic. In particular, if $C_p(Y)$ is a Lindelöf $\Sigma$-space then $Y$ is monotonically monolithic.*

**Solution.** For any $E_1, \ldots, E_n \subset X$ and $U_1, \ldots, U_n \in \tau(\mathbb{R})$ we will need the set $[E_1, \ldots, E_n, U_1, \ldots U_n] = \{f \in C_p(X) : f(E_i) \subset U_i \text{ for every } i \le n\}$. Fix a countable base $\mathcal{B}$ in $\mathbb{R}$ and take a compact cover $\mathcal{K}$ of the space $X$ such that there exists a countable network $\mathcal{N}$ modulo $\mathcal{K}$. For every family $\mathcal{E}$ of subsets of $X$ let $\mathcal{C}(\mathcal{E}) = \{N \backslash E : E \in \mathcal{E}, \ N \in \mathcal{N}\}$ and consider the family $\mathcal{F}(\mathcal{E})$ of all finite unions of the elements of $\mathcal{E}$; besides, let $\mathcal{W}(\mathcal{E}) = \{[E_1, \ldots, E_n, U_1, \ldots, U_n] : E_i \in \mathcal{E} \text{ and } U_i \in \mathcal{B} \text{ for all } i \le n\}$.

Given a set $A \subset C_p(X)$ let $\mathcal{E} = \{f^{-1}(B) : f \in A \text{ and } B \in \mathcal{B}\}$; if $\mathcal{H} = \mathcal{C}(\mathcal{F}(\mathcal{E}))$ and $\mathcal{O}(A) = \mathcal{W}(\mathcal{H})$ then $\mathcal{O}(A)$ is an external network for $\overline{A}$. To see it take any function $f \in \overline{A}$; we claim that

(1) for any $x \in X$ and $B \in \mathcal{B}$ such that $f(x) \in B$ there exists $P \in \mathcal{H}$ for which $f \in [P, B] \subset [\{x\}, B]$.

To prove (1), take a set $K \in \mathcal{K}$ with $x \in K$. For any $y \in X \backslash f^{-1}(B)$ pick a set $B_y \in \mathcal{B}$ such that $f(y) \in B_y$ and $f(x) \notin \overline{B}_y$; there exists a function $a_y \in A$ such that $a_y(x) \in B \backslash \overline{B}_y$ and $a_y(y) \in B_y$. The family $\mathcal{A} = \{a_y^{-1}(B_y) : y \in X \backslash f^{-1}(B)\}$ covers the set $X \backslash f^{-1}(B)$ and hence we can choose a finite set $Q \subset X \backslash f^{-1}(B)$ such that $K \subset G = \bigcup\{a_y^{-1}(B_y) : y \in Q\} \cup f^{-1}(B)$. There exists a set $N \in \mathcal{N}$ such that $K \subset N \subset G$; observe that we have $x \notin a_y^{-1}(B_y)$ for any point $y \in Q$, so $x \in P = N \backslash (\bigcup\{a_y^{-1}(B_y) : y \in Q\}) \subset f^{-1}(B)$. This shows that $f(P) \subset B$ and hence $f \in [P, B]$ and it follows from $x \in P$ that $[P, B] \subset [\{x\}, B]$; since $P \in \mathcal{H}$, the property (1) is proved.

Now, take any $V \in \tau(f, C_p(X))$; there exist $x_1, \ldots, x_n \in X$ and sets $B_1, \ldots, B_n \in \mathcal{B}$ such that $V' = [\{x_1\}, \ldots, \{x_n\}, B_1, \ldots, B_n] \subset V$. Observe that we have the equality $V' = \bigcap\{[\{x_i\}, B_i] : i \le n\}$ and apply the property (1) to find sets $P_1, \ldots, P_n \in \mathcal{H}$ such that $f \in [P_i, B_i] \subset [\{x_i\}, B_i]$ for each $i \le n$. Then the set $L = [P_1, \ldots, P_n, B_1, \ldots, B_n]$ belongs to $\mathcal{O}(A)$ and $f \in L \subset V' \subset V$, i.e., $\mathcal{O}(A)$ is, indeed an external network for $\overline{A}$. It is straightforward that the operator $\mathcal{O}$ satisfies all other conditions of the definition of monotonic monolithity, so $C_p(X)$ is monotonically monolithic.

Finally observe that if $C_p(Y)$ is a Lindelöf $\Sigma$-space then $C_p(C_p(Y))$ is monotonically monolithic; the space $Y$ embeds in $C_p(C_p(Y))$, so it must be monotonically monolithic by Problem 465.

**V.469.** *Prove that if $\upsilon X$ is a Lindelöf $\Sigma$-space then the space $C_p(X)$ must be monotonically $\omega$-monolithic. In particular, if $\upsilon(C_p(X))$ is a Lindelöf $\Sigma$-space then $C_p(X)$ is monotonically $\omega$-monolithic. Deduce from these facts that if $X$ is pseudocompact then $C_p(X)$ is monotonically $\omega$-monolithic.*

**Solution.** Let us first formulate the following easy fact for future references.

*Fact 1.* Given an infinite cardinal $\kappa$, suppose that $X$ is a monotonically $\kappa$-monolithic space and $f : X \to Y$ is a continuous bijection such that $f|A$ is a homeomorphism for any set $A \subset X$ with $|A| \leq \kappa$. Then $Y$ is monotonically $\kappa$-monolithic.

*Proof.* It is easy to see that $f(\overline{A}) = \overline{f(A)}$ and $f|\overline{A}$ is a homeomorphism for any set $A \subset X$ with $|A| \leq \kappa$. Let $\mathcal{O}$ be an operator which witnesses monotone $\kappa$-monolithity of $X$. For any set $B \subset Y$ with $|B| \leq \kappa$ let $\mathcal{A}(B) = \{f(P) : P \in \mathcal{O}(f^{-1}(B))\}$; it follows from $f\left(\overline{f^{-1}(B)}\right) = \overline{B}$ that $\mathcal{A}(B)$ is an external network for $\overline{B}$. The map $f$ being a bijection, the rest of conditions for $\mathcal{A}$ to be a monotone $\kappa$-monolithity operator on $Y$ are trivially satisfied, so Fact 1 is proved.

Returning to our solution note that the set $X$ is dense in $\upsilon X$ and the restriction map $\pi : C_p(\upsilon X) \to C_p(X)$ is a continuous bijection such that $\pi|A$ is a homeomorphism for any $A \subset C_p(\upsilon X)$ with $|A| \leq \omega$ (see TFS-436). The space $C_p(\upsilon X)$ is monotonically monolithic by Problem 468, so Fact 1 is applicable to convince ourselves that $C_p(X)$ is monotonically $\omega$-monolithic. Furthermore, if $\upsilon(C_p(X))$ is a Lindelöf $\Sigma$-space then apply CFS-206 to see that $\upsilon X$ is Lindelöf $\Sigma$ and therefore $C_p(X)$ is monotonically $\omega$-monolithic. Finally, if $X$ is pseudocompact then $\upsilon X = \beta X$ (see TFS-417 and Fact 3 of S.309), so $C_p(X)$ is monotonically $\omega$-monolithic.

**V.470.** *Let $X$ be a monotonically $\kappa$-monolithic space with $ext(X) = \lambda \leq \kappa$. Prove that*

*(i) if $\lambda < \kappa$ then $l(X) \leq \lambda$;*
*(ii) if $\lambda = \kappa$ and $t(X) \leq \kappa$ then $l(X) \leq \kappa$.*

**Solution.** Assume that $Z$ is a monotonically $\nu$-monolithic space and fix an operator $\mathcal{O}$ which witnesses that; let $N$ be a neighborhood assignment on $Z$. For every $P \subset Z$ say that $x \in P$ is a central point of $P$ if $P \subset N(x)$; denote by $K(P)$ the set of all central points of $P$. For an open set $U \subset Z$ say that a set $A \subset Z$ is $U$-saturated if $|A| \leq \nu$ and $K(P) \subset N(A) \cup U$ for any $P \in \mathcal{O}(A)$. If $U = \emptyset$ then $U$-saturated sets will be called saturated.

*Fact 1.* Assume that $Z$ is a monotonically $\nu$-monolithic space with the respective operator $\mathcal{O}$; let $N$ be a neighborhood assignment on $Z$ and fix a set $U \in \tau(Z)$. Suppose that $\beta \leq \nu$ is an ordinal and we are given a $\beta$-sequence $\{A_\alpha : \alpha < \beta\}$ of $U$-saturated sets such that $\alpha < \alpha'$ implies $A_\alpha \subset A_{\alpha'}$. Then $A = \bigcup\{A_\alpha : \alpha < \beta\}$ is $U$-saturated.

*Proof.* Take any $P \in \mathcal{O}(A)$; there exists $\alpha < \beta$ such that $P \in \mathcal{O}(A_\alpha)$. The set $A_\alpha$ being $U$-saturated, we have $K(P) \subset N(A_\alpha) \cup U \subset N(A) \cup U$ which shows that $A$ is $U$-saturated, i.e., Fact 1 is proved.

**Fact 2.** Suppose that $Z$ is a monotonically $\nu$-monolithic space; fix the respective operator $\mathcal{O}$ and let $N$ be a neighborhood assignment on $Z$. Then for any $A \subset Z$ with $|A| \leq \nu$ and $U \in \tau(Z)$ such that $N(A) \subset U$ there exists a closed discrete set $D \subset Z \backslash U$ such that $|D| \leq \max\{|A|, \omega\}$ and the set $A \cup D$ is $U$-saturated.

*Proof.* We will use induction on the cardinal $\mu = |A|$. To start off, assume that $|A| \leq \omega$ and take a family $\{\Omega_n : n \in \omega\}$ of infinite disjoint subsets of $\omega$ such that $\bigcup\{\Omega_n : n \in \omega\} = \omega$ and $\{0, \ldots, n\} \subset \Omega_0 \cup \ldots \cup \Omega_n$ for all $n \in \omega$. Enumerate the family $\mathcal{O}(A)$ as $\{P_n : n \in \Omega_0\}$ and let $B_0 = A$. Proceeding inductively, assume that $k \in \omega$ and we have sets $B_0, \ldots, B_k$ with the following properties:

(1) $B_i \subset B_{i+1}$ and $|B_{i+1} \backslash B_i| \leq 1$ for any $i < k$;
(2) an enumeration $\{P_n : n \in \Omega_j\}$ for the family $\mathcal{O}(B_j)$ is chosen for each $j \leq k$;
(3) if $i < k$, $i \in \Omega_j$ and $\bigcup\{K(P_n) : n \in \Omega_j\} \backslash (N(B_i) \cup U) \neq \emptyset$ then, for the number $m = \backslash n\{n \in \Omega_j : K(P_n) \backslash (N(B_i) \cup U) \neq \emptyset\}$ we have $B_{i+1} = B_i \cup \{d\}$ for some $d \in K(P_m) \backslash (N(B_i) \cup U)$.

There is a unique $j \in \omega$ such that $k \in \Omega_j$ and hence $j \leq k$. If we have the inclusion $\bigcup\{K(P_n) : n \in \Omega_j\} \subset N(B_k) \cup U$ then let $B_{k+1} = B_k$; if not, then consider the number $m = \backslash n\{n \in \Omega_j : K(P_n) \backslash (N(B_k) \cup U) \neq \emptyset\}$, choose a point $d \in K(P_m) \backslash (N(B_k) \cup U)$ and let $B_{k+1} = B_k \cup \{d\}$. Choose an enumeration $\{P_n : n \in \Omega_{k+1}\}$ of the family $\mathcal{O}(B_{k+1})$ and observe that now the conditions (1)–(3) are satisfied if we replace $k$ with $k + 1$. Therefore our inductive procedure can be continued to construct a family $\{B_i : i \in \omega\}$ such that the properties (1)–(3) hold for all $k \in \omega$.

Let $B = \bigcup_{i \in \omega} B_i$ and consider the set $D = B \backslash A$. It follows from (1) that $|D| \leq \omega$; the condition (3) shows that $D \subset Z \backslash U$. To see that the set $A \cup D = B$ is $U$-saturated take any $P \in \mathcal{O}(B)$; there exists $j \in \omega$ such that $P \in \mathcal{O}(B_j)$ and hence $P = P_n$ for some $n \in \Omega_j$. If $K(P) \backslash (N(B) \cup U) \neq \emptyset$ then $K(P_n) \backslash (N(B_i) \cup U) \neq \emptyset$ for all $i \in \Omega_j$ which is impossible because the condition (3) implies that after at most $n$ inductive steps a point $d \in K(P_n) \backslash (N(B_i) \cup U)$ has to be chosen for some $i \in \Omega_j$ and hence $P_n \subset N(d) \subset N(B)$; this contradiction shows that $A \cup D$ is $U$-saturated.

Now fix any point $x \in Z$; if $x \in U$ then $U$ is a neighborhood of $x$ with $U \cap D = \emptyset$. If $x \notin N(B) \cup U$ and $x \in \overline{D}$ then there exists $P \in \mathcal{O}(D)$ such that $x \in P \subset N(x)$ and hence $x \in K(P) \backslash (N(B) \cup U)$. We have $\mathcal{O}(D) \subset \mathcal{O}(B)$, so $P \in \mathcal{O}(B)$ which contradicts the fact that $B$ is $U$-saturated. Thus $x \notin \overline{D}$ for any $x \notin N(B) \cup U$.

Finally, if $x \in N(B) \backslash U$ then let $m$ be the minimal $i$ for which $x \in N(B_i)$. It follows from (3) that $N(B_i)$ is an open neighborhood of $x$ which does not meet the set $D \backslash B_i$; this, together with (1) shows that $N(B_i) \cap D$ is finite. Thus every $x \in Z$ has a neighborhood whose intersection with $D$ is finite and hence $D$ is closed and discrete. This completes our proof for a countable $A$.

Now assume that $\mu \le \nu$ is an uncountable cardinal and our lemma is proved whenever $A \subset Z$, $|A| < \mu$ and $U \in \tau(N(A), Z)$. Take an arbitrary set $A \subset Z$ such that $|A| = \mu$, let $\{a_\alpha : \alpha < \mu\}$ be an enumeration of $A$ and take any $U \in \tau(N(A), Z)$; we will also need the set $A_\alpha = \{a_\beta : \beta < \alpha\}$ for all $\alpha < \mu$.

By the induction hypothesis there is a countable closed discrete set $D_\omega \subset Z \backslash U$ such that the set $A_\omega \cup D_\omega$ is $U$-saturated. Proceeding inductively assume that $\omega < \alpha < \mu$ and we have a family $\{D_\beta : \omega \le \beta < \alpha\}$ with the following properties:

(4) if $U_\beta = U \cup N(\bigcup_{\omega \le \gamma < \beta} D_\gamma)$ then $D_\beta$ is a closed discrete subspace of $Z \backslash U_\beta$ and $|D_\beta| \le |\beta|$ for all $\beta \in [\omega, \alpha)$;

(5) if $B_\beta = A_\beta \cup \bigcup_{\omega \le \gamma < \beta} D_\gamma$ then $B_\beta \cup D_\beta$ is $U$-saturated whenever $\omega \le \beta < \alpha$;

Let $U_\alpha = U \cup N(\bigcup_{\omega \le \beta < \alpha} D_\beta)$; it follows from (4) that the cardinality of the set $B_\alpha = A_\alpha \cup \bigcup_{\omega \le \beta < \alpha} D_\beta$ does not exceed $|\alpha|$; besides, $N(B_\alpha) \subset U_\alpha$, so we can apply our induction hypothesis to find a closed discrete subset $D_\alpha$ of the set $Z \backslash U_\alpha$ such that $B_\alpha \cup D_\alpha$ is $U_\alpha$-saturated. The equality $U_\alpha \cup N(B_\alpha \cup D_\alpha) = U \cup N(B_\alpha \cup D_\alpha)$ shows that $B_\alpha \cup D_\alpha$ is also $U$-saturated.

It is clear that the conditions (4) and (5) are satisfied for all $\beta \in [\omega, \alpha]$, so our induction procedure can be continued to construct a family $\{D_\beta : \omega \le \beta < \mu\}$ for which the conditions (4) and (5) are satisfied whenever $\omega < \alpha < \mu$.

If $D = \bigcup_{\omega \le \beta < \mu} D_\beta$ then it follows from (4) that $D \subset Z \backslash U$. Take any set $P \in \mathcal{O}(A \cup D)$. There exists $\beta < \mu$ such that $P \in \mathcal{O}(B_\beta)$, so the condition (5) implies that $K(P) \subset U \cup N(B_\beta \cup D_\beta) \subset U \cup N(D)$; this proves that $A \cup D$ is $U$-saturated.

Take any $x \in Z$; if $x \in U$ then $U$ is a neighborhood of $x$ which does not meet $D$. If $x \notin U \cup N(D)$ and $x \in \overline{D}$ then there exists $P \in \mathcal{O}(D) \subset \mathcal{O}(A \cup D)$ such that $x \in P \subset N(x)$ and hence $x \in K(P) \backslash (U \cup N(D))$ which is a contradiction with the fact that $A \cup D$ is $U$-saturated. Therefore $x \notin \overline{D}$. If $x \in N(D) \backslash U$ then let $\alpha$ be the minimal ordinal such that $x \in N(D_\alpha)$. If $\alpha = \gamma + 1$ for some ordinal $\gamma$ then the set $E_\alpha = \bigcup_{\omega \le \beta < \alpha} D_\beta$ is contained in $B_\gamma \cup D_\gamma$. If $x \in \overline{E_\alpha}$ then $x \in P \subset N(x)$ for some $P \in \mathcal{O}(E_\alpha) \subset \mathcal{O}(B_\gamma \cup D_\gamma)$. The set $B_\gamma \cup D_\gamma$ being $U$-saturated, we have $x \in K(P) \subset U \cup N(B_\gamma \cup D_\gamma)$ and hence $x \in N(\bigcup_{\omega \le \beta \le \gamma} D_\beta)$ which contradicts the choice of $\alpha$. This proves that $x \notin \overline{E_\alpha}$.

If $\alpha$ is a limit ordinal then $B_\alpha = \bigcup_{\omega \le \beta < \alpha}(B_\beta \cup D_\beta)$, so it follows from (5) and Fact 1 that $B_\alpha$ is $U$-saturated. If $x \in \overline{E_\alpha}$ then $x \in P \subset N(x)$ and hence $x \in K(P)$ for some $P \in \mathcal{O}(E_\alpha) \subset \mathcal{O}(B_\alpha)$. The set $B_\alpha$ being $U$-saturated, we have $K(P) \subset U \cup N(B_\alpha)$, since $x \notin U$, we conclude that $x \in N(E_\alpha)$ which is again a contradiction with the choice of $\alpha$. Therefore $x \notin \overline{E_\alpha}$ in all cases. The property (4) shows that $N(D_\alpha)$ is a neighborhood of $x$ which does not meet the set $\bigcup_{\alpha < \beta < \mu} D_\beta$. Consequently, the point $x$ has a neighborhood which does not meet the set $\bigcup \{D_\beta : \beta \in [\omega, \mu) \backslash \{\alpha\}\}$. The set $D_\alpha$ is closed and discrete, so $x$ has a neighborhood which contains at most one point of $D$. Therefore the set $D$ is closed and discrete in $Z$ and it follows from the property (4) that $|D| \le \mu$; thus, we carried out the inductive step and hence Fact 2 is proved

Returning to our solution observe that there is no loss of generality to assume that $\lambda \geq \omega$. Fix an operator $\mathcal{O}$ which witnesses monotone $\kappa$-monolithity of $X$. Since in both cases we have to show that $l(X) \leq \lambda$, we are going to give a common reasoning for (i) and (ii) indicating the places in which the proofs differ.

Assume that $l(X) > \lambda$ and take an arbitrary neighborhood assignment $N$ on the space $X$ such that $N(A) \neq X$ for any set $A \subset X$ with $|A| \leq \lambda$.

Fix a point $a \in X$ and apply Fact 2 to the sets $\{a\}$ and $U = N(a)$ to find a closed discrete set $D_0' \subset X \backslash U$ such that the set $D_0 = \{a\} \cup D_0'$ is $U$-saturated; it is easy to see that $D_0$ is closed, discrete and saturated.

Proceeding by induction assume that $0 < \alpha < \lambda^+$ and we have chosen a family $\{D_\beta : \beta < \alpha\}$ of closed nonempty discrete subsets of $X$ with the following properties:

(6) the set $\bigcup_{\gamma \leq \beta} D_\gamma$ is saturated for any $\beta < \alpha$;
(7) $D_\beta \subset X \backslash N(\bigcup_{\gamma < \beta} D_\gamma)$ for each $\beta < \alpha$.

Observe that $|D_\beta| \leq ext(X) \leq \lambda$ for any $\beta < \alpha$ and therefore $|\bigcup_{\beta < \alpha} D_\beta| \leq \lambda$ which shows that $N(\bigcup_{\beta < \alpha} D_\beta) \neq X$ by our choice of $N$. Let $Q = \bigcup_{\beta < \alpha} D_\beta$ and choose a point $b \in X \backslash N(Q)$. Apply Fact 2 to the sets $Q \cup \{b\}$ and $V = N(Q \cup \{b\})$ to find a closed discrete set $D_\alpha' \subset X \backslash N(\{b\} \cup Q)$ such that $Q \cup \{b\} \cup D_\alpha'$ is $V$-saturated and let $D_\alpha = \{b\} \cup D_\alpha'$. It is easy to see that the set $D_\alpha \cup Q = \bigcup_{\beta \leq \alpha} D_\beta$ is saturated, so (6) and (7) still hold for all $\beta \leq \alpha$.

Therefore this inductive construction can be continued to construct a family $\mathcal{D} = \{D_\alpha : \alpha < \lambda^+\}$ with the properties (6) and (7). It follows from (7) that the sets $N(\bigcup_{\beta < \alpha} D_\beta)$ are strictly increasing and the family $\mathcal{D}$ is disjoint, so the cardinality of the set $D = \bigcup_{\beta < \lambda^+} D_\beta$ is equal to $\lambda^+$.

Fix any point $x \in X$; we are going to prove that $x$ has a neighborhood which contains at most one point of $D$. If $x \notin \overline{D}$ then there is nothing to prove, so we assume from now on that $x \in \overline{D}$.

If we are proving (i), observe that $|D| = \lambda^+ \leq \kappa$ which shows that operator $\mathcal{O}$ is applicable to $D$. If $x \in \overline{D} \backslash N(D)$ then $x \in P \subset N(x)$ for some $P \in \mathcal{O}(D)$ and hence the set $D$ is not saturated which is a contradiction with Fact 1.

In the case (ii) if $x \in \overline{D} \backslash N(D)$ then $x \in \overline{H} \backslash N(D)$ for some set $H \subset D$ with $|H| \leq \kappa$. The operator $\mathcal{O}$ is applicable to $H$, so there is $P \in \mathcal{O}(H)$ such that $x \in P \subset N(x)$. There exists $\alpha < \lambda^+$ such that $H \subset \bigcup_{\beta \leq \alpha} D_\beta$ and hence $P \in \mathcal{O}(\bigcup_{\beta \leq \alpha} D_\beta)$ which is a contradiction with the fact that $\bigcup_{\beta \leq \alpha} D_\beta$ is a saturated set by (6).

Therefore in both cases (i) and (ii) we must have $x \in N(D)$. Denote by $\gamma$ the least ordinal such that $x \in N(D_\gamma)$. It follows from (7) that $N(D_\gamma)$ is a neighborhood of $x$ which does not meet the set $\bigcup_{\gamma < \beta < \lambda^+} D_\beta$. Apply the property (6) and Fact 1 to see that the set $E = \bigcup_{\beta < \gamma} D_\beta$ is saturated. If $x \in \overline{E}$ then $x \in P \subset N(x)$ for some $P \in \mathcal{O}(E)$. Therefore $x \in K(P) \subset N(E)$ which is a contradiction with the choice of $\gamma$. Thus $x \notin \overline{E}$ and hence the point $x$ has a neighborhood which meets only the set $D_\gamma$. Since $D_\gamma$ is closed and discrete, the point $x$ has a neighborhood which contains at most one point of $D$. As a consequence, $D$ is a closed discrete subset of $X$ with $|D| = \lambda^+$; this contradiction makes our solution complete.

**V.471.** *Prove that every subspace of a monotonically monolithic space is a D-space. As a consequence,*

(i) *if $X$ is a Lindelöf $\Sigma$-space then every subspace of $C_p(X)$ is a D-space.*

(ii) *Observe that $ext(Y) = l(Y)$ for any D-space $Y$. Therefore (i) generalizes Baturov's theorem (SFFS-269).*

**Solution.** Assume that $X$ is a monotonically monolithic space and fix an operator $\mathcal{O}$ which witnesses that; let $N$ be a neighborhood assignment on $X$. For every $P \subset X$ say that $x \in P$ is a central point of $P$ if $P \subset N(x)$; denote by $K(P)$ the set of all central points of $P$. For an open set $U \subset X$ say that a set $A \subset X$ is $U$-*saturated* if $K(P) \subset N(A) \cup U$ for any $P \in \mathcal{O}(A)$. If $U = \emptyset$ then $U$-saturated sets will be called *saturated*.

Observe first that monotone monolithity is a hereditary property by Problem 465, so it suffices to show that every monotonically monolithic space $X$ is a D-space. Let $\kappa = |X|$; there is no loss of generality to assume that $\kappa \geq \omega$. Fix an operator $\mathcal{O}$ which witnesses monotone monolithity of $X$. Take an arbitrary point $a \in X$ and apply Fact 2 of V.470 to the sets $\{a\}$ and $U = N(a)$ to find a closed discrete set $D_0' \subset X \backslash U$ such that the set $D_0 = \{a\} \cup D_0'$ is $U$-saturated; it is easy to see that $D_0$ is closed, discrete and saturated.

Proceeding by induction assume that $0 < \alpha < \kappa^+$ and we have chosen a family $\{D_\beta : \beta < \alpha\}$ of closed nonempty discrete subsets of $X$ with the following properties:

(1) the set $\bigcup_{\gamma \leq \beta} D_\gamma$ is saturated for any $\beta < \alpha$;
(2) $D_\beta \subset X \backslash N(\bigcup_{\gamma < \beta} D_\gamma)$ for each $\beta < \alpha$.

If $N(\bigcup_{\beta < \alpha} D_\beta) \neq X$ then let $Q = \bigcup_{\beta < \alpha} D_\beta$ and choose a point $b \in X \backslash N(Q)$. Apply Fact 2 of V.470 to the sets $Q \cup \{b\}$ and $V = N(Q \cup \{b\})$ to find a closed discrete set $D_\alpha' \subset X \backslash N(\{b\} \cup Q)$ such that $Q \cup \{b\} \cup D_\alpha'$ is $V$-saturated and let $D_\alpha = \{b\} \cup D_\alpha'$. It is easy to see that the set $D_\alpha \cup Q = \bigcup_{\beta \leq \alpha} D_\beta$ is saturated, so the properties (1) and (2) still hold for all $\beta \leq \alpha$.

Therefore this inductive construction can be continued to construct a set $D_\alpha$ as soon as $N(\bigcup_{\beta < \alpha} D_\beta) \neq X$. It follows from (2) that the sets $N(\bigcup_{\beta < \alpha} D_\beta)$ are strictly increasing, so it is impossible to obtain the family $\{D_\alpha : \alpha < \kappa^+\}$ and hence there exists $\alpha < \kappa^+$ such that $N(\bigcup_{\beta < \alpha} D_\beta) = X$. Let $D = \bigcup_{\beta < \alpha} D_\beta$ and fix any point $x \in X$.

Denote by $\gamma$ the least ordinal such that $x \in N(D_\gamma)$. It follows from (2) that $N(D_\gamma)$ is a neighborhood of $x$ which does not meet the set $\bigcup_{\gamma < \beta < \alpha} D_\beta$. Apply the property (1) and Fact 1 of V.470 to see that the set $E = \bigcup_{\beta < \gamma} D_\beta$ is saturated. If $x \in \overline{E}$ then $x \in P \subset N(x)$ for some $P \in \mathcal{O}(E)$. Therefore $x \in K(P) \subset N(E)$ which is a contradiction with the choice of $\gamma$. Thus $x \notin \overline{E}$ and hence the point $x$ has a neighborhood which meets only the set $D_\gamma$. Since $D_\gamma$ is closed and discrete, the point $x$ has a neighborhood which contains at most one point of $D$. This proves that $D$ is a closed discrete subset of $X$ and hence $X$ is a D-space.

Finally apply Problem 468 to see that if $X$ is a Lindelöf $\Sigma$-space then $C_p(X)$ is monotonically monolithic and hence every subspace of $C_p(X)$ is a $D$-space; this proves (i).

Now, if $Y$ is a $D$-space and $ext(Y) \leq \kappa$, take any open cover $\mathcal{U}$ of the space $Y$. For any $x \in Y$ choose a set $U_x \in \mathcal{U}$ such that $x \in U_x$. There exists a closed discrete set $D \subset X$ such that $\bigcup\{U_x : x \in D\} = Y$. Therefore $\{U_x : x \in D\}$ is a subcover of $\mathcal{U}$ of cardinality $\leq ext(Y) \leq \kappa$, i.e., we proved that $l(Y) \leq ext(Y)$ which implies $l(Y) = ext(Y)$. In particular, if $X$ is a Lindelöf $\Sigma$-space and $Y \subset C_p(X)$, then $Y$ is a $D$-space by (i) and hence $ext(Y) = l(Y)$. This settles (ii) and makes our solution complete.

**V.472.** *Given an infinite cardinal $\kappa$, suppose that a space $X$ is monotonically $\kappa$-monolithic and $nw(X) \leq \kappa^+$. Prove that every subspace of $X$ is a $D$-space. Deduce from this fact that for any pseudocompact space $Z$ with $nw(Z) \leq \omega_1$, every subspace of the space $C_p(Z)$ is a $D$-space.*

**Solution.** Observe first that both monotone $\kappa$-monolithity and network weight not exceeding $\kappa^+$ are hereditary properties, so it suffices to show that every monotonically $\kappa$-monolithic space of network weight at most $\kappa^+$ is a $D$-space.

Fix an operator $\mathcal{O}$ of monotone $\kappa$-monolithity in $X$ and let $\mathcal{E} = \{E_\alpha : \alpha < \kappa^+\}$ be a network in $X$. Take an arbitrary neighborhood assignment $N$ on $X$. For every $P \subset X$ say that $x \in P$ is a central point of $P$ if $P \subset N(x)$; denote by $K(P)$ the set of all central points of $P$. For an open set $U \subset X$ say that a set $A \subset X$ of cardinality $\leq \kappa$ is $U$-saturated if $K(P) \subset N(A) \cup U$ for any $P \in \mathcal{O}(A)$. If $U = \emptyset$ then $U$-saturated sets will be called saturated.

Take an arbitrary point $a \in X$ and apply Fact 2 of V.470 to the sets $\{a\}$ and $U = N(a)$ to find a countable closed discrete set $D_0' \subset X \setminus U$ such that the set $D_0 = \{a\} \cup D_0'$ is $U$-saturated; it is easy to see that $D_0$ is closed, discrete and saturated.

Proceeding by induction assume that $0 < \alpha < \kappa^+$ and we have chosen a family $\{D_\beta : \beta < \alpha\}$ of closed nonempty discrete subsets of $X$ with the following properties:

(1) the set $\bigcup_{\gamma \leq \beta} D_\gamma$ is saturated for any $\beta < \alpha$;
(2) $|D_\beta| \leq \kappa$ and $D_\beta \subset X \setminus N(\bigcup_{\gamma < \beta} D_\gamma)$ for each $\beta < \alpha$.
(3) if $\beta < \alpha$ and $M = \{\eta < \kappa^+ : K(E_\eta) \setminus N(\bigcup_{\gamma < \beta} D_\gamma) \neq \emptyset\}$ then, for the ordinal $\xi = \setminus n M$ we have $D_\beta \cap K(E_\xi) \neq \emptyset$.

Let $Q = \bigcup_{\beta < \alpha} D_\beta$; if $N(Q) = X$ then our inductive construction ends. If $N(Q) \neq X$ then consider the set $M = \{\eta < \kappa^+ : K(E_\eta) \setminus N(\bigcup_{\gamma < \alpha} D_\gamma) \neq \emptyset\}$. If $M = \emptyset$ then choose a point $b \in X \setminus N(Q)$ arbitrarily. If $M \neq \emptyset$ then let $\xi = \setminus n M$ and take a point $b \in K(E_\xi) \setminus N(Q)$. Apply Fact 2 of V.470 to the sets $Q \cup \{b\}$ and $V = N(Q \cup \{b\})$ to find a closed discrete set $D_\alpha' \subset X \setminus N(\{b\} \cup Q)$ such that $|D_\alpha'| \leq \kappa$ and $Q \cup \{b\} \cup D_\alpha'$ is $V$-saturated; let $D_\alpha = \{b\} \cup D_\alpha'$. It is easy to see that the set $D_\alpha \cup Q = \bigcup_{\beta \leq \alpha} D_\beta$ is saturated, so the properties (1)–(3) still hold for all $\beta \leq \alpha$.

Therefore this inductive construction can be continued to construct a set $D_\alpha$ whenever $\alpha < \kappa^+$ and $N(\bigcup_{\beta<\alpha} D_\beta) \neq X$. Suppose that our construction did not stop at any step $\alpha < \kappa^+$. If $\bigcup\{N(D_\beta) : \beta < \kappa^+\} \neq X$ then pick a point $x \in X \setminus \bigcup\{N(D_\beta) : \beta < \kappa^+\}$ and $E \in \mathcal{E}$ such that $x \in E \subset N(X)$. There is $\xi < \kappa^+$ such that $E = E_\xi$; since $x \in K(E_\xi)$, we have $K(E_\xi) \setminus \bigcup\{N(D_\beta) : \beta < \alpha\} \neq \emptyset$ for any $\alpha < \kappa^+$. However, this is impossible because the ordinal $\xi$ must be the minimal element of the set $M = \{\eta < \kappa^+ : K(E_\eta) \setminus N(\bigcup_{\gamma<\beta} D_\gamma) \neq \emptyset\}$ for some $\beta \leq \xi$ and hence $K(E_\xi)$ has to meet the set $\bigcup\{D_\beta : \beta \leq \xi + 1\}$ which in turn implies that $x \in E_\xi \subset \bigcup\{N(D_\beta) : \beta \leq \xi + 1\} \subset \bigcup\{N(D_\beta) : \beta < \kappa^+\}$, a contradiction.

This proves that there exists an ordinal $\alpha \leq \kappa^+$ such that $N(\bigcup_{\beta<\alpha} D_\beta) = X$. Let $D = \bigcup_{\beta<\alpha} D_\beta$ and fix any point $x \in X$. Denote by $\gamma$ the least ordinal such that $x \in N(D_\gamma)$. It follows from (2) that $N(D_\gamma)$ is a neighborhood of $x$ which does not meet the set $\bigcup_{\gamma<\beta<\alpha} D_\beta$. Apply the property (1) and Fact 1 of V.470 to see that the set $E = \bigcup_{\beta<\gamma} D_\beta$ is saturated. If $x \in \overline{E}$ then $x \in P \subset N(x)$ for some $P \in \mathcal{O}(E)$ (the operator $\mathcal{O}$ is applicable to $E$ because $|E| \leq \kappa$ by (2)). Therefore $x \in K(P) \subset N(E)$ which is a contradiction with the choice of $\gamma$. Thus $x \notin \overline{E}$ and hence the point $x$ has a neighborhood which meets only the set $D_\gamma$. Since $D_\gamma$ is closed and discrete, the point $x$ has a neighborhood which contains at most one point of $D$. This proves that $D$ is a closed discrete subset of $X$ and hence $X$ is a $D$-space.

Finally observe that if $Z$ is pseudocompact and $nw(Z) \leq \omega_1$ then $C_p(Z)$ is monotonically $\omega$-monolithic (see Problem 469) and $nw(C_p(Z)) = nw(Z) \leq \omega_1$, so $C_p(X)$ is a hereditary $D$-space.

**V.473.** *Prove that any countably compact monotonically $\omega$-monolithic space must be Corson compact.*

**Solution.** Let $Z$ be a space; given any $A \subset Z$ let $[A]_\omega = \bigcup\{\overline{B} : B \subset A$ and $|B| \leq \omega\}$ and $[A]^\omega = \{x \in X : $ if $H$ is a $G_\delta$-subset of $Z$ and $x \in H$ then $H \cap A \neq \emptyset\}$. Recall that the Gruenhage's $W$-game is played on a nonempty closed subset $F$ of a space $Z$ as follows: at the $n$-th move the first player called $O$ chooses an open set $U_n \in \tau(F, Z)$ and the player $P$ responds by taking a point $x_n \in U_n$. After $\omega$-many moves have been made, the player $O$ wins if the sequence $S = \{x_n : n \in \omega\}$ converges to $F$, i.e., the set $\{n \in \omega : x_n \notin V\}$ is finite for any $V \in \tau(F, Z)$; otherwise $P$ is the winner.

*Fact 1.* Suppose that $X$ is a monotonically $\omega$-monolithic space and there exists a family $\{X_\alpha : \alpha < \omega_1\}$ of cosmic subsets of $X$ such that $X = \bigcup_{\alpha<\omega_1} X_\alpha$. Then $X$ is monotonically monolithic.

*Proof.* It is easy to see that $d(X) \leq \omega_1$; if $d(X) \leq \omega$, then the space $X$ is cosmic and hence monotonically monolithic. Now assume that we have $d(X) = \omega_1$, take a countable dense subspace $M_\alpha$ in the space $X_\alpha$ for each $\alpha < \omega_1$ and let $\{d_\alpha : \alpha < \omega_1\}$ be a faithful enumeration of the set $D = \bigcup_{\alpha<\omega_1} M_\alpha$. We will also need the sets $D_\alpha = \{d_\beta : \beta < \alpha\}$ and $F_\alpha = \overline{D}_\alpha$ for every $\alpha < \omega_1$.

Given any $x \in X$, there is $\beta < \omega_1$ such that $x \in X_\beta$ and hence $x \in \overline{M}_\beta$. There exists $\alpha < \omega_1$ such that $M_\beta \subset D_\alpha$ and hence $x \in \overline{D}_\alpha = F_\alpha$. This proves that $\bigcup_{\alpha < \omega_1} F_\alpha = X$. Fix an operator $\mathcal{A}$ which witnesses monotone $\omega$-monolithity of $X$ and call a set $Y \subset X$ *small* if $Y \subset F_\alpha$ for some $\alpha < \omega_1$.

For every small set $Y \subset X$ denote by $\gamma(Y)$ the least ordinal $\alpha < \omega_1$ such that $Y \subset F_\alpha$ and let $\mathcal{O}(Y) = \mathcal{A}(D_{\gamma(Y)})$. If a set $Y \subset X$ is not small then let $\mathcal{O}(Y) = \bigcup\{\mathcal{A}(D_\alpha) : \alpha < \omega_1\}$. It is immediate that $\mathcal{O}(Y)$ is an external network of $Y$ in $X$ such that $|\mathcal{O}(Y)| \leq \max\{|Y|, \omega\}$ for any $Y \subset X$ and $Y \subset Z \subset X$ implies $\mathcal{O}(Y) \subset \mathcal{O}(Z)$.

Next, assume that $Y = \bigcup\{Y_\xi : \xi < \beta\} \subset X$ for an ordinal $\beta$ and $Y_\xi \subset Y_\eta$ whenever $\xi < \eta < \beta$. If some $Y_\xi$ is not small then $\mathcal{O}(Y_\xi) = \mathcal{O}(X) = \mathcal{O}(Y)$. If every $Y_\xi$ is small then let $\alpha_\xi = \gamma(Y_\xi)$ and observe that if $\alpha = \sup\{\alpha_\xi : \xi < \beta\} = \omega_1$ then $\bigcup_{\xi < \beta} \mathcal{O}(Y_\xi) = \bigcup_{\xi < \beta} \mathcal{A}(D_{\alpha_\xi}) = \bigcup_{\xi < \omega_1} \mathcal{A}(D_\xi) = \mathcal{O}(X) = \mathcal{O}(Y)$.

Finally, if $\alpha < \omega_1$ then $Y_\xi \subset \overline{D}_\alpha$ for every $\xi < \beta$ and hence $Y \subset \overline{D}_\alpha$ which shows that $\mathcal{O}(Y) \subset \mathcal{A}(D_\alpha) = \bigcup_{\xi < \beta} \mathcal{A}(D_{\alpha_\xi}) = \bigcup_{\xi < \beta} \mathcal{O}(Y_\xi)$, so Fact 1 is proved.

*Fact 2.* Suppose that $X$ is countably compact and $F$ is a closed subset of $X$ for which there exists a family $\mathcal{U} = \{U_n : n \in \omega\} \subset \tau(F, X)$ such that $F = \bigcap \mathcal{U}$ and $\overline{U}_{n+1} \subset U_n$ for every $n \in \omega$. Then $\mathcal{U}$ is an outer base of $F$ in $X$ and hence $\chi(F, X) \leq \omega$.

*Proof.* Take any set $U \in \tau(F, X)$ and consider the set $G_n = \overline{U}_n \cap (X \setminus U)$ for each $n \in \omega$. It follows from $\bigcap_{n \in \omega} \overline{U}_n = F$ that $\bigcap_{n \in \omega} G_n = \emptyset$. Since $\{G_n : n \in \omega\}$ is a decreasing family of closed subsets of a countably compact space $X$, it is impossible that $G_n \neq \emptyset$ for all $n \in \omega$, so there exists $n \in \omega$ with $G_n = \emptyset$ and hence the inclusions $F \subset U_n \subset \overline{U}_n \subset U$ show that $\mathcal{U}$ is an outer base of $F$ in $X$, i.e., Fact 2 is proved.

*Fact 3.* Let $Z$ be a countably compact space. Then $[[A]_\omega]^\omega = \overline{A}$ for any $A \subset Z$.

*Proof.* Since it is evident that $[[A]_\omega]^\omega \subset \overline{A}$, let us prove that $\overline{A} \subset D = [[A]_\omega]^\omega$. If $x \in \overline{A} \setminus D$ then it is easy to find a sequence $\mathcal{U} = \{U_n : n \in \omega\} \subset \tau(x, Z)$ such that $\overline{U}_{n+1} \subset U_n$ for every $n \in \omega$ and $F = \bigcap_{n \in \omega} U_n \subset Z \setminus [A]_\omega$.

By Fact 2, the family $\mathcal{U}$ is an outer base for the set $F$. Since $x \in F$, for every $n \in \omega$ we can choose $x_n \in U_n \cap A$ because $x \in \overline{A}$ and $U_n$ a neighborhood of $x$. The set $B = \{x_n : n \in \omega\} \subset A$ has cardinality $\leq \omega$, so $F \cap \overline{B} \subset (\bigcap \mathcal{U}) \cap [A]_\omega = \emptyset$. Since $\mathcal{U}$ is an outer base of $F$ in $Z$, there is $n \in \omega$ with $U_n \cap \overline{B} = \emptyset$ which is a contradiction with the fact that $x_n \in U_n \cap B$, so Fact 3 is proved.

*Fact 4.* If $X$ is a countably compact space and $t(X) > \omega$ then $X$ has an uncountable free sequence.

*Proof.* There exists an $\omega$-closed non-closed set $A \subset X$ by Fact 1 of S.328. Then $[A]_\omega = A$ and hence $\overline{A} = [A]^\omega$ by Fact 3. Fix any $x \in \overline{A} \setminus A$; then $x \in [A]^\omega$, so

$(*)$  $H \cap A \neq \emptyset$ for any $G_\delta$-set $H \ni x$.

Take $a_0 \in A$ arbitrarily and let $H_0 = X$. Suppose that $\alpha < \omega_1$ and we have constructed points $\{a_\beta : \beta < \alpha\} \subset A$ together with closed $G_\delta$-sets $\{H_\beta : \beta < \alpha\}$ with the following properties:

(1) $\{x, a_\beta\} \subset H_\beta$ for all $\beta < \alpha$;
(2) $H_\beta \subset H_{\beta'}$ if $\beta' < \beta < \alpha$;
(3) $\overline{\{a_\gamma : \gamma < \beta\}} \cap H_\beta = \emptyset$ for all $\beta < \alpha$.

Since $x \notin P = \overline{\{a_\gamma : \gamma < \alpha\}}$, there exists a closed $G_\delta$-set $H \ni x$ such that $H \cap P = \emptyset$. If we let $H_\alpha = H \cap (\bigcap\{H_\beta : \beta < \alpha\})$ and take any $a_\alpha \in H_\alpha \cap A$ (this choice is possible because of $(*)$), then the same conditions are fulfilled for all $\beta \leq \alpha$ and hence the inductive construction can be continued to provide a set $S = \{a_\alpha : \alpha < \omega_1\} \subset A$.

We claim that $S$ is a free sequence. Indeed, if $\beta < \omega_1$ then $\overline{\{a_\gamma : \gamma < \beta\}} \cap H_\beta = \emptyset$ while $\{a_\gamma : \gamma \geq \beta\} \subset H_\beta$ by (1) and (2). The set $H_\beta$ being closed, we have $\overline{\{a_\gamma : \gamma < \beta\}} \cap \overline{\{a_\gamma : \gamma \geq \beta\}} \subset \overline{\{a_\gamma : \gamma < \beta\}} \cap H_\beta = \emptyset$. As a consequence, $S$ is a free sequence of length $\omega_1$, so Fact 4 is proved.

*Fact 5.* If $X$ is a monotonically $\omega$-monolithic countably compact space, then $X$ is compact.

If tightness of $X$ is uncountable then we can apply Fact 4 to find a free sequence $S = \{s_\alpha : \alpha < \omega_1\} \subset X$. Let $S_\alpha = \{s_\beta : \beta < \alpha\}$ for any $\alpha < \omega_1$; it is straightforward that the set $F = \bigcup_{\alpha < \omega_1} \overline{S}_\alpha$ is countably compact. Since every set $F_\alpha = \overline{S}_\alpha$ is cosmic, it follows from Fact 1 that $F$ is monotonically monolithic. Therefore $F$ is compact being a countably compact $D$-space (see Problem 471).

For the compact space $F$ we have $F = \overline{S}$; since $F$ can be continuously mapped onto $\omega_1 + 1$ (see Fact 8 of T.298), we can apply Problem 465 to see that $\omega_1 + 1$ is monotonically $\omega$-monolithic and hence the space $\omega_1$ is monotonically $\omega$-monolithic as well. However, $ext(\omega_1) = t(\omega_1) = \omega$, so $\omega_1$ must be Lindelöf by Problem 470; this contradiction shows that $t(X) \leq \omega$. By Problem 465(ix), the space $X$ is monotonically monolithic, so it has $D$-property by Problem 471. As a consequence, $X$ is compact being a countably compact $D$-space (see Problem 471(ii)), i.e., Fact 5 is proved.

*Fact 6.* If $X$ is a countably compact monotonically $\omega$-monolithic space, then for every nonempty closed set $F \subset X$, the player $O$ has a winning strategy at $F$.

*Proof.* The space $X$ is compact by Fact 5; fix an operator $\mathcal{N}$ that witnesses $\omega$-monolithity of $X$ and let $\sigma(\emptyset) = X$. Suppose that $n \in \omega$ and $U_0, \ldots, U_n$, $p_0, \ldots, p_n$ are the moves of the players $O$ and $P$ respectively such that $U_0 = \sigma(\emptyset)$ and $U_{i+1} \subset U_i$ for any $i < n$. Choose an enumeration $\{N_{ij} : j \in \omega\}$ for the family $\mathcal{N}(\{p_0, \ldots, p_i\})$ for any $i = 0, \ldots, n$. The family $\mathcal{F}_n = \{N_{ij} : i \leq n, \ j \leq n \text{ and } \overline{N}_{ij} \cap F = \emptyset\}$ is finite, so we can find an open set $U_{n+1} \subset U_n$ such that $F \subset U_{n+1}$ and $\overline{U}_{n+1} \cap (\bigcup \mathcal{F}_n) = \emptyset$; let $\sigma(p_0, \ldots, p_n) = U_{n+1}$.

To see that the strategy $\sigma$ is winning for the player $O$ suppose that a family $\{U_n, p_n : n \in \omega\}$ is a play in which $O$ applies $\sigma$. If the sequence $S = \{p_n : n \in \omega\}$ does not converge to the set $F$, then, by compactness of $X$, there exists a cluster point $x$ for the sequence $S$ such that $x \notin F$. Take a set $V \in \tau(x, X)$ such that $\overline{V} \cap F = \emptyset$; since $\mathcal{N}(S)$ is an external network for $\overline{S} \ni x$, we can find a set $N \in \mathcal{N}(S)$ such that $x \in N \subset V$. It follows from the equality $S = \bigcup\{\{p_0, \dots, p_n\} : n \in \omega\}$ that $N \in \mathcal{N}(\{p_0, \dots, p_n\})$ for some $n \in \omega$ and hence $N = N_{ij}$ for some $i \leq n$ and $j \in \omega$.

By our choice of the strategy $\sigma$, the set $\overline{U}_k$ does not meet $N$ for any $k > m = \max\{n, j\}$, so for the neighborhood $W = X \backslash \overline{U}_{m+1}$ of the point $x$ we have $W \cap S \subset \{p_0, \dots, p_m\}$, i.e., $x$ is not a cluster point of $S$ which is a contradiction. Therefore $\sigma$ is a winning strategy for the player $O$ and hence Fact 6 is proved.

Returning to our solution suppose that $X$ is a countably compact $\omega$-monolithic space. The space $X$ is compact by Fact 5; apply Problem 465(iii) to see that $X \times X$ is also monotonically $\omega$-monolithic, so we can apply Fact 6 to conclude that for every nonempty closed subset $F$ of $X \times X$, the player $O$ has a winning strategy in the Gruenhage's game on $F$. Therefore $X$ is Corson compact by Problem CFS-188 and hence our solution is complete.

**V.474.** *Give an example of a Corson compact space that is not monotonically $\omega$-monolithic.*

**Solution.** We will consider several trees identifying any tree $(T, \leq)$ with the underlying set $T$. Recall that a set $C \subset T$ is *a chain* if $(C, \leq)$ is a linearly ordered set. Maximal chains are called *branches*. To simplify the notation, we denote the order on $\omega_1$ and all trees we use by the same letter; the set on which the order is taken will be always clear from the context. Given a point $x \in T$, the set $L_x = \{y \in T : y < x\}$ is well-ordered; its order type will be denoted by $ht(x)$. It $\alpha$ is an ordinal, then the set $\{x \in T : ht(x) = \alpha\}$ is called the $\alpha$-th level of $T$.

A tree $T$ is *continuous* if for any $x \in T$ such that $ht(x)$ is a limit ordinal, if $z \in T$ and $y < z$ for any $y \in L_x$, then $x \leq z$. We consider that zero is a limit ordinal, so continuity of a tree $T$ implies that there is only one element $x \in T$ such that $ht(x) = 0$. We say that a tree $T$ is *complete* if every branch of $T$ has a maximal element. For any $x \in T$ let $V(x) = \{y \in T : x \leq y\}$ and $\hat{x} = \{y \in T : y \leq x\}$. A set $A \subset T$ is *dense* in $T$ if $V(x) \cap A \neq \emptyset$ for any $x \in T$.

Two elements $x, y$ of a tree $T$ are called *comparable* if either $x \leq y$ or $y \leq x$; if $x$ and $y$ are not comparable, they are *incomparable*. Say that $x \in T$ is an *immediate successor of* $y \in T$ if $y$ is the maximal element of $L_x$. The point $y$ is called *the immediate predecessor of* $x$. A point $x \in T$ is *an immediate successor* if it has an immediate predecessor. It is an easy exercise that $x \in T$ is an immediate successor if and only if $ht(x)$ is a successor ordinal.

Apply SFFS-065 and SFFS-066 to fix a stationary set $S \subset \omega_1$ such that $\omega_1 \backslash S$ is also stationary and denote by $\mathbb{T}_0$ the family of all subsets of $S$ closed in $\omega_1$. It is immediate that all elements of $\mathbb{T}_0$ are countable. Given $x, y \in \mathbb{T}_0$, say that $x \leq y$ if $x \subset y$ and $\alpha < \beta$ whenever $\alpha \in x$ and $\beta \in y \backslash x$. The set $\mathbb{T}_0$ considered with the above order is a tree: this was proved in Fact 1 of U.176. Every $x \in \mathbb{T}_0 \setminus \{\emptyset\}$ is compact, so it has a maximal element which will denoted by $\max(x)$. For notational purposes it is convenient to consider that $\max(\emptyset) = 0$.

The set $\mathbb{D}$ is the doubleton $\{0, 1\}$ with the discrete topology. We let $\mathbb{D}^0 = \{\emptyset\}$ and $\mathbb{D}^n$ is the set of all functions $f : \{0, \dots, n-1\} \to \mathbb{D}$ for any $n \in \mathbb{N}$. We will also need the set $\mathbb{D}^{<\omega} = \bigcup \{\mathbb{D}^n : n \in \omega\}$. The set $\mathbb{D}^{<\omega}$ is easily seen to be a tree if we define its order by the condition that $f \leq g$ if and only if $\mathrm{dom}(f) \subset \mathrm{dom}(g)$ and $g|\mathrm{dom}(f) = f$. The expression $\mathrm{Lim}(\omega_1)$ stands for the set of all limit ordinals in $\omega_1$.

*Fact 1.* Given a tree $T$ suppose that $D$ is a dense subset of $T$ and $C \subset D$ is a maximal antichain in $D$. Then $C$ is a maximal antichain in $T$.

*Proof.* Take any $x \in T$; by density of $D$ there is $d \in D$ such that $x \leq d$. It follows from maximality of $C$ in $D$ that there exists $y \in C$ which is comparable with $d$. If $y \leq d$ then $y$ is comparable with $x$ because $\{x, y\} \subset \hat{d}$ and the set $\hat{d}$ is linearly ordered. If $d \leq y$, then $x \leq d \leq y$ implies that $x \leq y$, so any point of $T$ is comparable with an element of $C$, i.e., $C$ is a maximal antichain in $T$ and hence Fact 1 is proved.

*Fact 2.* Given a complete continuous tree $T$ consider the set $T' = \{x \in T : \mathrm{ht}(x)$ is a successor ordinal$\}$ and denote by $\tau_T$ the topology generated by the family $S = \{V(x) : x \in T'\} \cup \{T \setminus V(x) : x \in T'\}$ as a subbase. Then

(a) the family $S_0 = \{V(x) : x \in T'\}$ is $T_0$-separating, i.e., for any distinct points $x, y \in T$, there exists $z \in T'$ such that $V(z) \cap \{x, y\}$ is a singleton;
(b) $(T, \tau_T)$ is a compact Hausdorff space.

*Proof.* (a) Take any distinct points $x, y \in T$. If $x$ and $y$ are comparable say, $x < y$, then $y \in V(y)$ and $x \notin V(y)$. If $x$ and $y$ are incomparable and the ordinal $\mathrm{ht}(x)$ is a successor, then $x \in V(x)$ and $y \notin V(x)$. Now, if both $\mathrm{ht}(x)$ and $\mathrm{ht}(y)$ are limit ordinals, then by continuity of $T$ there exists $z \in T$ such that $z < x$ and $y \notin V(z)$. Since $x \in V(z)$, we proved that the family $S_0$ is $T_0$-separating.
(b) It follows from (a) that $\{x\} = \bigcap \{S \in S : x \in S\}$; since every element of $S$ is clopen in $(T, \tau_T)$, the set $\{x\}$ is closed in $(T, \tau_T)$ for any $x \in T$, i.e., $(T, \tau_T)$ is Tychonoff being a zero-dimensional $T_1$-space (see Fact 1 of S.232).

To prove that $X = (T, \tau_T)$ is compact it suffices to show that any cover $\mathcal{U} \subset S$ of the space $X$ has a finite subcover (see TFS-118). So take any $\mathcal{U} \subset S$ such that $\bigcup \mathcal{U} = T$ and let $\mathcal{U}_0 = \{U : U \in \mathcal{U} \cap S_0\}$. It is easy to see that

(1) if $x, y \in T$ are incomparable, then $(T \setminus V(x)) \cup (T \setminus V(y)) = T$,

and therefore we can assume, without loss of generality, that the set $C = \{x \in T : T \setminus V(x) \in \mathcal{U}\}$ is a chain. It is an easy exercise to see that there exists a branch $B \subset T$ such that $C \subset B$; let $a$ be the maximal element of $B$. Observe that if $x \notin B$, then $a \notin V(x)$ and $a \notin (T \setminus V(y))$ for any $y \in B$. Therefore there exists a minimal $b \in B$ such that $V(b) \in \mathcal{U}$; since the ordinal $\mathrm{ht}(b)$ is a successor, there exists $c \in B$ which is an immediate predecessor of $b$, i.e., $b$ is the minimal element of $B$ greater than $c$. If $x \notin B$, then $c \notin V(x)$; besides, $c \notin V(x)$ for any $x \in B$ such that $V(x) \in \mathcal{U}$. As a consequence, there exists $x \in B$ for which $c \in (T \backslash V(x))$; this implies that $b \leq x$ and therefore $V(b) \cup (T \backslash V(x)) = T$, i.e., $\{V(b), T \backslash V(x)\}$ is a finite subcover of $\mathcal{U}$, so $(T, \tau_T)$ is compact and hence Fact 2 is proved.

*Fact 3.* Given a continuous tree $T$ denote by $\mathcal{B}$ the family of all branches of $T$ which have no maximal element. Take a faithfully indexed set $Q = \{t_b : b \in \mathcal{B}\}$ such that $Q \cap T = \emptyset$ and let $\tilde{T} = T \cup Q$. For each $b \in \mathcal{B}$, declare that $x < t_b$ for any $x \in b$ and $t_b$ is incomparable with all elements of $(T \backslash b) \cup (Q \backslash \{t_b\})$. Then $\tilde{T}$ with the original order on $T$ together with the order for the elements of $Q$ is a complete continuous tree called the completion of $T$.

*Proof.* It is straightforward that $\leq$ is a partial order on $\tilde{T}$. If $b \in \mathcal{B}$, then no element of $\tilde{T}$ is strictly greater than $t_b$; this shows that $\{y \in \tilde{T} : y < x\} = \{y \in T : y < x\}$ is a well-ordered set for any $x \in T$. Now, if $b \in \mathcal{B}$, then $\{y \in \tilde{T} : y < t_b\} = b$ is also a well-ordered set, so $\tilde{T}$ is, indeed, a tree.

To see that $\tilde{T}$ is continuous take any $x \in \tilde{T}$ such that $\mathrm{ht}(x)$ is a limit ordinal. We will assume first that $x \in T$. Suppose that $z \in \tilde{T}$ and $y < z$ whenever $y < x$. If $z \in T$ then $x \leq z$ by continuity of $T$. If $b \in \mathcal{B}$ and $z = t_b$, then observe that set $L_x = \{y \in T : y < x\}$ is not a branch, so $L_x \subset b$ and $L_x \neq b$; take any $t \in b \backslash L_x$. Then $y < t$ for any $y \in L_x$, so $x \leq t$ by continuity of $T$; it follows from $x \leq t < t_b$ than $x < t_b = z$.

Finally take any $t_b \in Q$ and assume that $z \in \tilde{T}$ and $y < t_b$ for any $y \in b = L_{t_b}$. It follows from maximality of $b$ that $z \notin T$. If $z = t_{b'}$ for some $b' \neq b$, then the branches $b$ and $b'$ are distinct, so $b \backslash b' \neq \emptyset$ by maximality of $b$. If $y \in b \backslash b'$, then $t_{b'}$ is not comparable with $y$; this contradiction shows that $z = t_b$, so the tree $\tilde{T}$ is continuous and hence Fact 3 is proved.

*Fact 4.* Assume that $A_n$ is a maximal antichain in $\mathbb{T}_0$ and let $O_n = \bigcup \{V(a) : a \in A_n\}$ for every $n \in \omega$. Then the set $O = \bigcap_{n \in \omega} O_n$ is dense in $\mathbb{T}_0$, i.e., for any $x \in \mathbb{T}_0$ there exists $y \in O$ such that $x \leq y$.

*Proof.* Fix an arbitrary point $x \in \mathbb{T}_0$; it follows from maximality of $A_n$ that

(2)  for any $t \in \mathbb{T}_0$ and $n \in \omega$ there exist $u \in A_n$ and $s \in \mathbb{T}_0$ such that $\{u, t\} \subset \hat{s}$.

We are going to construct by transfinite induction a family $\{E_\alpha : \alpha < \omega_1\}$ and an $\omega_1$-sequence $\{\delta_\alpha : \alpha < \omega_1\} \subset \omega_1$ with the following properties:

(3)  $\delta_0 = \max(x)$ and $E_0 = \{x\}$;
(4)  $E_\alpha \subset \mathbb{T}_0$ is countable and $\max(p) \leq \delta_\alpha$ for any $p \in E_\alpha$ and $\alpha < \omega_1$;
(5)  if $\alpha \in \mathrm{Lim}(\omega_1)$ then $\delta_\alpha = \sup\{\delta_\beta : \beta < \alpha\}$ and $E_\alpha = \bigcup\{E_\beta : \beta < \alpha\}$;
(6)  if $\alpha < \beta < \omega_1$ then $\alpha \leq \delta_\alpha < \delta_\beta$, $E_\alpha \subset E_\beta$ and, for any $p \in E_\alpha$ and $n \in \omega$ there is $u \in A_n$ and $q \in E_\beta$ such that $\{p, u\} \subset \hat{q}$ and $\max(q) > \delta_\alpha$.

To satisfy the condition (3) we must start with $\delta_0 = \max(x)$, $E_0 = \{x\}$; if $\alpha < \omega_1$ is a limit ordinal and we have the set $\{\delta_\beta : \beta < \alpha\}$ and the family $\{E_\beta : \beta < \alpha\}$ then let $\delta_\alpha = \sup\{\delta_\beta : \beta < \alpha\}$ and $E_\alpha = \bigcup\{E_\beta : \beta < \alpha\}$. This guarantees (5).

Now, suppose that, for some ordinal $\nu < \omega_1$, we have the set $\{\delta_\alpha : \alpha \le \nu\}$ and the family $\{E_\alpha : \alpha \le \nu\}$ with the properties (3)–(6) fulfilled for all $\alpha, \beta \le \nu$. For every element $t \in E_\nu$ and $n \in \omega$ fix $u(t,n) \in A_n$ and $s(t,n) \in \mathbb{T}_0$ such that $t \le s(t,n)$, $u(t,n) \le s(t,n)$ and $\max(s(t,n)) > \max\{\nu, \delta_\nu\}$ (this is possible by (2)).

Let $E'_{\nu+1} = E_\nu \cup \{s(t,n) : t \in E_\nu, n \in \omega\}$, $E_{\nu+1} = \bigcup\{\hat{p} : p \in E'_{\nu+1}\}$ and $\delta_{\nu+1} = \sup\{\max(p) : p \in E_{\nu+1}\}$. It is straightforward that the properties (3)–(6) hold for the set $\{\delta_\alpha : \alpha \le \nu + 1\}$ and the family $\{E_\alpha : \alpha \le \nu + 1\}$, so our inductive procedure can be continued to construct the promised $\omega_1$-sequence $\{\delta_\alpha : \alpha < \omega_1\}$ and the family $\{E_\alpha : \alpha < \omega_1\}$ with the properties (3)–(6).

Observe that

(7) the set $H = \{\alpha < \omega_1 : \alpha = \delta_\alpha\}$ is closed and unbounded in $\omega_1$.

Indeed, if $\{\alpha_n\}_{n\in\omega} \subset H$ is an increasing sequence and $\alpha_n \to \alpha$ then it follows from (5) that $\delta_\alpha = \sup\{\delta_{\alpha_n} : n \in \omega\} = \sup\{\alpha_n : n \in \omega\} = \alpha$ because $\delta_{\alpha_n} = \alpha_n$ for any $n \in \omega$. This proves that the set $H$ is closed. Given any $\beta < \omega_1$, let $\alpha_0 = \beta$ and $\alpha_{n+1} = \delta_{\alpha_n} + 1$ for any $n \in \omega$. A consequence of (6) is that $\alpha_n < \alpha_{n+1}$ for any $n \in \omega$; if $\alpha = \sup_{n\in\omega} \alpha_n$ then $\alpha \in H$ and $\alpha > \beta$ which shows that $H$ is cofinal in $\omega_1$ and (7) is proved.

Our set $S$ being stationary, it follows from the property (7) that there exists a limit ordinal $\alpha \in H \cap S$; fix an increasing sequence $\{\alpha_n : n \in \omega\}$ such that $\sup_{n\in\omega} \alpha_n = \alpha$. Applying the property (7) once more we conclude that $\sup\{\delta_{\alpha_n} : n \in \omega\} = \alpha$.

Observe that $E_0 \subset E_{\alpha_0}$, so we can take $p_0 = x \in E_{\alpha_0}$. Suppose that $n \in \omega$ and we have sets $\{p_i : i \le n\} \subset \mathbb{T}_0$ and $\{u_i : i < n\} \subset \mathbb{T}_0$ with the following properties:

(8)  $p_i \in E_{\alpha_i}$ for all $i \le n$;
(9)  $p_i \le p_{i+1}$ and $\max(p_{i+1}) > \alpha_i$ for any $i < n$;
(10)  $u_i \in A_i$ and $u_i \le p_{i+1}$ for every $i < n$.

The property (6) implies that we can choose $p_{n+1} \in E_{\alpha_{n+1}}$ and $u_n \in A_n$ such that $u_n \le p_{n+1}$, $p_n \le p_{n+1}$ and $\max(p_{n+1}) > \delta_{\alpha_n} \ge \alpha_n$. It is immediate that (8)–(10) are fulfilled for the sets $\{p_i : i \le n+1\}$ and $\{u_i : i < n+1\}$, so our inductive procedure gives us sequences $\{p_n : n \in \omega\}$ and $\{u_n : n \in \omega\}$ with the properties (8)–(10).

It follows from (4), (6) and (8) that $\alpha \notin p_n$ for any $n \in \omega$. It is easy to see that the set $p = (\bigcup_{n\in\omega} p_n) \cup \{\alpha\}$ belongs to $\mathbb{T}_0$ and $u_n \le p_{n+1} < p$; in particular, $p \in V(u_n) \subset O_n$ for any $n \in \omega$. Therefore $p \in O$ and $x = p_0 < p$, so $O$ is dense in $\mathbb{T}_0$ and hence Fact 4 is proved.

*Fact 5.* An element $x \in \mathbb{T}_0$ is an immediate successor if and only if $\alpha = \max(x)$ is an isolated point of the set $x$.

*Proof.* Suppose that $y$ is an immediate predecessor of $x$ and let $\beta = \max(y)$. All points of $x \setminus y$ are greater than $\beta$. If $\gamma = \setminus n(x \setminus y)$, then $y < y \cup \{\gamma\} \le x$ which shows that $y \cup \{\gamma\} = x$ and hence $\gamma = \alpha$, so the set $\{\alpha\} = (\beta, \omega_1) \cap x$ is open in $x$; this proves necessity.

If the point $\alpha$ is isolated in $x$, then $\{\alpha\}$ is an open subset of $x$, so $y = x \backslash \{\alpha\} \in \mathbb{T}_0$ is an immediate predecessor of $x$ and hence Fact 5 is proved.

*Fact 6.* The tree $\mathbb{T}_0$ is continuous.

*Proof.* Suppose that $x \in \mathbb{T}_0$ and $\mathrm{ht}(x)$ is a limit ordinal. Then $x$ has no immediate predecessor and hence $\alpha = \max(x)$ is not isolated in $x$ by Fact 5. If $z < y$ for any $z$ with $z < x$, then $[0, \beta] \cap x \subset y$ for any $\beta < \alpha$ and therefore $x \backslash \{\alpha\} \subset y$. Since $x \backslash \{\alpha\}$ is dense in $x$, the set $x$ is contained in $y$.

If $\gamma \in y \backslash x$ and $\gamma < \alpha$, then pick an ordinal $\beta \in (\gamma, \alpha) \cap x$ and observe that the ordinal $\gamma \in y \backslash ([0, \beta] \cap x)$ witnesses the fact that $y$ is not a successor of $[0, \beta] \cap x < x$; this contradiction shows that $x \leq y$ and hence Fact 6 is proved.

Returning to our solution consider the set $\mathbb{T}_1 = \mathbb{T}_0 \times \mathbb{D}^{<\omega}$. For every $t \in \mathbb{T}_0$ choose a faithfully indexed set $\{b_{t,\alpha} : \alpha \in S, \ \alpha > \max(t)\}$ of branches in the tree $\mathbb{D}^{<\omega}$ such that $|\{\alpha : s \in b_{t,\alpha}\}| = \omega_1$ for any $s \in \mathbb{D}^{<\omega}$. Given $x, y \in \mathbb{T}_1$ such that $x = (t, s)$ and $y = (t', s')$ we declare that $x < y$ if and only if

(11) $t = t'$ and $s < s'$; or
(12) there exists $\alpha > \max(t)$ such that $t \cup \{\alpha\} \leq t'$ and $s \in b_{t,\alpha}$.

Let us check that $\leq$ is a partial order on $\mathbb{T}_1$. It is evident that the inequalities $x < y$ and $y < x$ cannot hold together, so we only have to check transitivity. Suppose that $x, y, z \in \mathbb{T}_1$, $x = (t, s)$, $y = (t', s')$, $z = (t'', s'')$ and $x < y < z$. It follows from (11) and (12) that $t \leq t' \leq t''$; assume first that $t = t'$ and hence $s < s'$. If $t' = t''$, then $s < s' < s''$, so $x = (t, s) < z = (t, s'')$. If $t' < t''$, then there exists $\alpha > \max(t')$ such that $t' \cup \{\alpha\} \leq t''$ and $s' \in b_{t',\alpha}$. The set $b_{t',\alpha}$ being a branch, it follows from $s < s'$ that $s \in b_{t',\alpha} = b_{t,\alpha}$; this, together with the fact that $t \cup \{\alpha\} = t' \cup \{\alpha\} \leq t''$ shows that $x < z$ by (12). Finally, assume that $t < t'$ and fix $\alpha > \max(t)$ such that $t \cup \{\alpha\} \leq t'$ and $s \in b_{t,\alpha}$; since also $t' \cup \{\alpha\} \leq t' \leq t''$, the property (12) implies that $x < z$.

To see that $\mathbb{T}_1$ is a tree, take any $x = (t, s) \in \mathbb{T}_1$. We will first prove that the set $L_x = \{y \in \mathbb{T}_1 : y < x\}$ is linearly ordered by $<$, i.e., any two elements of $L_x$ are comparable. Take any $y = (t', s') \in L_x$ and $z = (t'', s'') \in L_x$. Observe that $\mathbb{T}_0$ is a tree and $t' \leq t$ and $t'' \leq t$ which implies that $t'$ and $t''$ are comparable, so we can assume, without loss of generality, that $t' \leq t''$. If $t' = t''$ then there exist ordinals $\alpha$ and $\beta$ greater than $\max(t') = \max(t'')$ such that $t' \cup \{\alpha\} \leq t$ and $t'' \cup \{\beta\} \leq t$; it is easy to see that this implies $\alpha = \beta$. Therefore $s' \in b_{t',\alpha}$ and $s'' \in b_{t',\alpha}$; since $b_{t',\alpha}$ is a chain, the elements $s'$ and $s''$ must be comparable, so the property (11) shows that $y = (t', s')$ and $z = (t', s'')$ are comparable.

Next, assume that $t' < t''$ and let $\alpha = \backslash n (t'' \backslash t')$; then $t' \cup \{\alpha\} \leq t''$. There exists $\beta < \omega_1$ such that $t' \cup \{\beta\} \leq t$ and $s' \in b_{t',\beta}$; it follows from $t' \cup \{\alpha\} \leq t'' \leq t$ that $t' \cup \{\alpha\} \leq t$ and therefore $\alpha = \beta$. Finally, note that $s' \in b_{t',\beta} = b_{t',\alpha}$ and therefore $(t', s') < (t'', s'')$, i.e., $y$ and $z$ are comparable by (12).

To see that the set $L_x$ is well-ordered, take any nonempty set $A \subset L_x$ and consider the element $t_0 = \backslash n \{t' : (t', s') \in A$ for some $s' \in \mathbb{D}^{<\omega}\}$. If $A' = \{(t', s') \in A : t' = t_0\}$, then it immediate from (11) and (12) that $y < z$ whenever

$y \in A'$ and $z \in A \setminus A'$, so it suffices to find the minimal element in $A'$. We already know that all elements of $A'$ are comparable and $A' = \{(t_0, s') : S' \in D_0\}$ for some $D_0 \subset \mathbb{D}^{<\omega}$, so it follows from (11) that all elements of $D_0$ are comparable; since $\mathbb{D}^{<\omega}$ is a tree, there is a minimal element $d \in D_0$. It is immediate that $(t_0, d)$ is a minimal element of $A'$ and hence by our above observation, $(t_0, d)$ is the minimal element of $A$, i.e., we proved that $\mathbb{T}_1$ is, indeed, a tree.

Let us establish the following relationship between the trees $\mathbb{T}_0$ and $\mathbb{T}_1$:

(13)  for any elements $t, t' \in \mathbb{T}_0$ we have $t \leq t'$ if and only if $(t, \emptyset) \leq (t', \emptyset)$.

Note that if $(t, \emptyset) \leq (t', \emptyset)$ then $t \leq t'$: this is an immediate consequence of (11) and (12), so we have sufficiency in (13). Now, assume that $t < t'$ and let $\alpha = \setminus n(t' \setminus t)$. Then $t \cup \{\alpha\} \leq t'$; the set $b_{t,\alpha}$ being a branch, we have $\emptyset \in b_{t,\alpha}$ and hence $(t, \emptyset) < (t', \emptyset)$ by (12). This settles necessity and shows that (13) is proved.

The following property characterizes the elements of $\mathbb{T}_1$ that have no immediate successor.

(14)  given any $x = (t, s) \in \mathbb{T}_1$ its height $\mathrm{ht}(x)$ is a limit ordinal if and only if $s = \emptyset$.

If $s \neq \emptyset$ then it has an immediate predecessor $s'$ in the tree $\mathbb{D}^{<\omega}$. It is straightforward that $(t, s')$ is an immediate predecessor of $x$, i.e., we proved necessity. Now, if $y = (t', s') < x = (t, \emptyset)$, then there is an ordinal $\alpha < \omega_1$ such that $t' \cup \{\alpha\} \leq x$ and $s' \in b_{t',\alpha}$. If $s'' \in b_{t',\alpha}$ and $s' < s''$, then $(t', s') < (t', s'') < x$, i.e., $(t', s')$ is not an immediate predecessor of $x$ and hence (14) is proved.

Our next step is to show that $\mathbb{T}_1$ is continuous, so fix any element $x \in \mathbb{T}_1$ such that $\mathrm{ht}(x)$ is a limit ordinal and assume that $y = (t_0, s_0)$ is greater than any element of $L_x = \{y \in \mathbb{T}_1 : y < x\}$. It follows from (14) that $x = (t, \emptyset)$ for some $t \in \mathbb{T}_0$. Consider first the case when the element $t$ has an immediate predecessor $t'$. Then $t = t' \cup \{\alpha\}$ for some $\alpha > \max(t')$; for any $s \in b_{t',\alpha}$ we can apply (12) to see that $(t', s) < x$ and hence $(t', s) < y$ by our choice of $y$. There exists an ordinal $\beta(s) > \max(t')$ such that $t' \cup \{\beta(s)\} \leq t_0$ and $s \in b_{t',\beta(s)}$. Observe that $\beta(s) \neq \beta(s')$ implies that the elements $t' \cup \{\beta(s)\}$ and $t' \cup \{\beta(s')\}$ are incomparable in the tree $\mathbb{T}_0$; this contradiction with the inclusion $\{t' \cup \{\beta(s')\}, t' \cup \{\beta(s)\}\} \subset \hat{t}_0$ shows that there exists $\beta < \omega_1$ such that $\beta(s) = \beta$ for any $s \in b_{t',\alpha}$ and hence $b_{t',\alpha} \subset b_{t',\beta}$. Since $b_{t',\alpha}$ and $b_{t',\beta}$ are branches of the tree $\mathbb{D}^{<\omega}$, we have $b_{t',\alpha} = b_{t',\beta}$ and hence $\alpha = \beta$ by faithfulness of the indexation of the family $\{b_{t',\gamma} : \gamma > \max(t')\}$. Consequently, $t = t' \cup \{\alpha\} = t' \cup \{\beta\} \leq t_0$ and therefore $x = (t, \emptyset) \leq (t_0, s_0) = y$ as promised.

Finally assume that $\mathrm{ht}(t)$ is s limit ordinal. If $t' < t$, then it follows from (13) that $(t', \emptyset) < (t, \emptyset) = x$ and hence $(t', \emptyset) < (t_0, s_0)$ which in turn implies that $t' < t_0$. By continuity of the tree $\mathbb{T}_0$ (see Fact 5) we have $t \leq t_0$ and therefore $x = (t, \emptyset) \leq (t_0, \emptyset) \leq (t_0, s_0) = y$ and hence we proved that

(15)  the tree $\mathbb{T}_1$ is continuous.

Let $T' = \{(t, s) \in \mathbb{T}_1 : s \neq \emptyset\}$; it follows from (14) that $T'$ is precisely the set of the elements of $\mathbb{T}$ that have an immediate predecessor. Consider the family $\mathcal{B}$ of the branches of $\mathbb{T}_1$. Given any $(t, s) \in \mathbb{T}_1$ it is easy to find $s' \in \mathbb{D}^{<\omega}$ such that $s < s'$

and hence $(t, s) < (t, s')$. This proves that every element of $\mathbb{T}_1$ has a successor and hence $\mathcal{B}$ coincides with the set of the branches of $\mathbb{T}_1$ with no maximal element.

Take a faithfully indexed set $Q = \{t_b : b \in \mathcal{B}\}$ such that $Q \cap T = \emptyset$ and let $\mathbb{T} = \mathbb{T}_1 \cup Q$. For each $b \in \mathcal{B}$, declare that $x < t_b$ for any $x \in b$ and $t_b$ is incomparable with all elements of $(\mathbb{T}_1 \backslash b) \cup (Q \backslash \{t_b\})$. Then the set $\mathbb{T}$ with the respective order is a complete continuous tree by Fact 3. Since no element of $Q$ has an immediate predecessor, also for the tree $\mathbb{T}$, the set $T'$ coincides with the set of all elements that have an immediate predecessor. Denote by $\tau$ the topology on the tree $\mathbb{T}$ generated by the family $\mathcal{S} = \{V(x) : x \in T'\} \cup \{\mathbb{T} \backslash V(x) : x \in T'\}$ as a subbase. By Fact 2, the space $K = (\mathbb{T}, \tau)$ is compact and Hausdorff.

Suppose that $C \subset \mathbb{T}$ is an infinite chain. Since $Q$ is an antichain, the set $C$ has at most one element of $Q$ and hence the chain $C' = C \backslash Q \subset \mathbb{T}_1$ has the same cardinality as $C$. Given any elements $(t, s) \in C'$ and $(t', s') \in C'$, it follows from their comparability, that $t$ and $t'$ are comparable in $\mathbb{T}_0$ by (11) and (12). Therefore the set $E = \{t \in \mathbb{T}_0 : \text{there exists } s \in \mathbb{D}^{<\omega} \text{ such that } (t, s) \in C'\}$ is a chain in $\mathbb{T}_0$. Since all chains in $\mathbb{T}_0$ are countable by Fact 1 of U.176, we conclude that $|E| \leq \omega$. As a consequence, $|C| = |C'| \leq |E \times \mathbb{D}^{<\omega}| \leq \omega$ and hence we proved that

(16)  every chain in the tree $\mathbb{T}$ is countable.

Observe that $\mathcal{U} = \{V(x) : x \in T'\}$ is a $T_0$-separating family of clopen subsets of $K$ by Fact 2. If $x \in \mathbb{T}$, then it follows from (16) that the set $\{y \in T' : x \in V(y)\} \subset \hat{x}$ is countable because $\hat{x}$ is a chain in $\mathbb{T}$. This shows that $K$ has a $T_0$-separating point-countable family of clopen sets and therefore $K$ is Corson compact (see CFS-118).

Since monotone $\omega$-monolithity is a hereditary property, to show that $K$ is not monotonically $\omega$-monolithic, it suffices to disprove monotone monolithity of $\mathbb{T}_1$ with the topology inherited from $K$.

*Fact 7.* Suppose that $A_n$ is an antichain in $\mathbb{T}_1$ for each $n \in \omega$. Then there exists a maximal antichain $B$ in $\mathbb{T}_1$ such that $B \cap \hat{x} = \emptyset$ for any $x \in A = \bigcup_{n \in \omega} A_n$.

*Proof.* For any $a = (t, s) \in A$ consider the set $C(a) = \{t \cup \{\alpha\} : \alpha \in S, \alpha > \max(t) \text{ and } s \in b_{t,\alpha}\}$. We will prove first that

(17)  $E_n = \bigcup \{C(a) : a \in A_n\}$ is an antichain in $\mathbb{T}_0$.

Take distinct elements $a = (t, s)$ and $a' = (t', s')$ of the set $A_n$ and assume that some set $t \cup \{\alpha\} \in C(a)$ is comparable with a set $t' \cup \{\beta\} \in C(a')$. There is no loss of generality to assume that $t \cup \{\alpha\} \leq t' \cup \{\beta\}$. If $t \cup \{\alpha\} \neq t' \cup \{\beta\}$, then $\beta = \max(t' \cup \{\beta\}) > \alpha = \max(t \cup \{\alpha\})$ which implies that $t \cup \{\alpha\} \leq t'$ and hence $(t, s) \leq (t', s')$ by (12), i.e., we obtained a contradiction. Therefore $t \cup \{\alpha\} = t' \cup \{\beta\}$ and hence $\beta = \max(t' \cup \{\beta\}) = \max(t \cup \{\alpha\}) = \alpha$ which shows that $t = t'$. We also have $s \in b_{t,\alpha}$ and $s' \in b_{t,\alpha}$; the set $b_{t,\alpha}$ being a branch, the elements $s$ and $s'$ are comparable and hence $(t, s)$ is comparable with $(t, s') = (t', s')$ by (11). This contradiction shows that no element of $C(a)$ is comparable with an element of $C(a')$ if $a \neq a'$. Since it is evident that every $C(a)$ is an antichain, the set $E_n$ is also an antichain, so (17) is proved.

Take a maximal antichain $D_n$ in the tree $\mathbb{T}_0$ such that $E_n \subset D_n$; it turns out that

(18) for any $a = (t, s) \in A_n$, the set $\hat{t}$ does not meet $D_n$.

To see that (18) is true, assume that $d \leq t$ and $(t, s) \in A_n$ for some $s \in \mathbb{D}^{<\omega}$ and $d \in D_n$. If $d \in D_n \backslash E_n$, then take $\alpha > \max(t)$ such that $s \in b_{t,\alpha}$ and observe that $d \leq t \leq t \cup \{\alpha\} \in E_n$ which is a contradiction with the fact that $d$ and $t \cup \{\alpha\}$ belong to the same antichain. Now, if $d \in E_n$, then $d = t' \cup \{\beta\}$ where $(t', s') \in A_n$ and $s' \in b_{t',\beta}$ for some $s' \in \mathbb{D}^{<\omega}$. This, together with (12) implies that $(t', s') \leq (t, s)$, so $t' = t$ and hence $t \cup \{\beta\} \leq t$ which is again a contradiction, i.e., we proved (18).

Let $H_n = \bigcup \{V(t) : t \in D_n\}$ for every $n \in \omega$; the set $H = \bigcap_{n \in \omega} H_n$ is dense in $\mathbb{T}_0$ by Fact 4. Take a maximal antichain $B'$ in the set $H$; then $B'$ is a maximal antichain in $\mathbb{T}_0$ by Fact 1. To see that the set $B = \{(t, \emptyset) : t \in B'\}$ is as promised observe first that distinct elements of $B$ are not comparable by (13), i.e., $B$ is an antichain. Now, if $(t', s') \in \mathbb{T}_1$, then take $\alpha > \max(t')$ such that $s' \in b_{t',\alpha}$. The set $t' \cup \{\alpha\}$ is comparable with some $t \in B'$ by maximality of $B'$. If $t' \cup \{\alpha\} \leq t$, then $(t', s') \leq (t, \emptyset)$ by (12). If $t < t' \cup \{\alpha\}$ then $\alpha = \max(t' \cup \{\alpha\})$ does belong to $t$ and hence $t \leq t'$. Then $(t, \emptyset) \leq (t', s')$, so in all possible cases $(t', s')$ is comparable with an element of $B$; this proves that $B$ is a maximal antichain in $\mathbb{T}_1$.

Finally assume that $(t', \emptyset) \in B$ and there exists $n \in \omega$ such that $(t', \emptyset) \leq (t, s)$ for some $x = (t, s) \in A_n$; this evidently implies that $t' \leq t$. However, $t' \in H_n$ and hence there is $d \in D_n$ such that $d \leq t'$, so it follows from $d \leq t' \leq t$ that $d \leq t$ which is a contradiction with (18). Consequently, $\hat{x} \cap B = \emptyset$ for any $x \in A$, i.e., Fact 7 is proved.

Finally, assume that an operator $\mathcal{N}$ witnesses that the space $\mathbb{T}_1$ is monotonically $\omega$-monolithic. For every $x \in \mathbb{T}_1$ let $\mathcal{N}^*(x) = \mathcal{N}(\hat{x})$; observe that the operator $\mathcal{N}$ is applicable to the set $\hat{x}$ because $\hat{x}$ is countable by (16). Given any $N \in \mathcal{N}^*(x)$ let $M(N, x) = \{y \in N \cap V(x) : V(y) \cap N = \{y\}\}$ and observe that the set $M(N, x)$ is an antichain for any $N \in \mathcal{N}^*(x)$.

Let $A_0$ be a maximal antichain in the tree $\mathbb{T}_1$; proceeding by induction assume that $n \in \omega$ and we have maximal antichains $A_0, \ldots, A_n$ in the tree $\mathbb{T}_1$ with the following properties:

(20) if $i < n$, then $A_{i+1} = \bigcup \{A_{i+1}(a) : a \in A_i\}$ where $A_{i+1}(a) \subset V(a) \backslash \{a\}$ is a maximal antichain in $V(a)$ for any $a \in A_i$;

(21) if $i < n$ and $a \in A_i$, then $A_{i+1}(a) \cap \hat{u} = \emptyset$ whenever $u \in M(N, a)$ and $N \in \mathcal{N}^*(a)$.

For every $a \in A_n$ let $\mathcal{M}_a = \{\{a\}\} \cup \{M(N, a) : N \in \mathcal{N}^*(a)\}$; since $\mathcal{M}_a$ is a countable family of antichains of $\mathbb{T}_1$, we can apply Fact 7 to find a maximal antichain $B_a$ in the tree $\mathbb{T}_1$ such that $\hat{x} \cap B_a = \emptyset$ whenever $x \in \bigcup \mathcal{M}_a$. Then $A_{n+1}(a) = B_a \cap V(a)$ is a maximal antichain in $V(a)$ and $A_{n+1}(a) \subset V(a) \backslash \{a\}$. To see it, take any $x \in V(a)$; by maximality of $B_a$ there exists $y \in B_a$ comparable with $x$. If $x \leq y$, then $a \leq x \leq y$ implies that $y \in V(a)$. If $y < x$, then it follows from $a \leq x$, that $y$ is comparable with $a$. Since the inequality $y \leq a$ is impossible because of the choice of $B_a$ and the fact that $a \in \bigcup \mathcal{M}_a$, we have $a < y$, i.e., $y \in V(a) \cap B_a = A_{n+1}(a)$ and hence $A_{n+1}(a) \subset V(a) \backslash \{a\}$ is a maximal antichain in $V(a)$.

Let $A_{n+1} = \bigcup \{A_{n+1}(a) : a \in A_n\}$. Take any distinct $x, y \in A_{n+1}$. If $x, y \in A_{n+1}(a)$ for some $a \in A_n$, then $x$ and $y$ are incomparable because $A_{n+1}(a)$ is an antichain. Now, if there are distinct $a, a' \in A_n$ such that $x \in A_{n+1}(a)$ and $y \in A_{n+1}(a')$, then $V(a) \cap V(a') = \emptyset$ because $A_n$ is an antichain; since $V(x) \subset V(a)$ and $V(y) \subset V(a')$, we have the inclusion $V(x) \cap V(y) \subset V(a) \cap V(a') = \emptyset$ and therefore the elements $x$ and $y$ are incompatible. This proves that $A_{n+1}$ is an antichain and hence the property (20) holds for all $i \leq n$.

Take any $x \in \mathbb{T}_1$. Since the antichain $A_n$ is maximal, there exists $a \in A_n$ which is comparable with $x$. If $x \leq a$, then any element of the nonempty set $A_{n+1}(a)$ is comparable with $x$. If $x \geq a$, then we can apply maximality of $A_{n+1}(a)$ to find an element $y \in A_{n+1}(a)$ comparable with $x$. This shows that $A_{n+1}$ is a maximal antichain in $\mathbb{T}_1$. Finally, observe that the property (21) holds by our choice of $A_{n+1}(a)$ for any $a \in A_n$ and hence our inductive procedure can be continued to obtain a sequence $\{A_n : n \in \omega\}$ of maximal antichains in $\mathbb{T}_1$ that satisfy the conditions (20) and (21) for any $n \in \omega$.

Apply Fact 7 once more to find an element $x \in \mathbb{T}_1$ such that $x \notin \hat{a}$ for any point $a \in \bigcup_{n \in \omega} A_n$. By maximality of $A_n$ we can find $a_n \in A_n$ such that $a_n < x$ for any $n \in \omega$. The set $Q = \{a_n : n \in \omega\} \subset \hat{x}$ is linearly ordered, so it follows from (20) that $a_n < a_{n+1}$ for any $n \in \omega$. Let $y$ be the minimal element of $\hat{x}$ such that $a_n < y$ for any $n \in \omega$.

It is easy to see that $y$ has no immediate predecessor and hence $\text{ht}(y)$ is a limit ordinal. Take any open set $U$ in the space $\mathbb{T}_1$ such that $y \in U$. There exist points $p_1, \ldots, p_m, q_1, \ldots, q_r \in \mathbb{T}_1$ such that

$$y \in W = (\mathbb{T} \backslash V(p_1)) \cap \ldots \cap (\mathbb{T}_1 \backslash V(p_m)) \cap V(q_1) \cap \ldots \cap V(q_r) \subset U.$$

Observe first that the inequality $a_n < y$ implies that $a_n \in \mathbb{T}_1 \backslash V(p_i)$ for any $i \leq m$ and $n \in \omega$ and hence $Q \subset W_0 = (\mathbb{T} \backslash V(p_1)) \cap \ldots \cap (\mathbb{T}_1 \backslash V(p_m))$. Every element $q_i$ has an immediate predecessor, so $q_i < y$ and hence we can pick a point $a_{n(i)} > q_i$ for any $i = 1, \ldots, r$. If $k > \max\{n(1), \ldots, n(r)\}$, then $a_k \in W_1 = V(q_1) \cap \ldots \cap V(q_r)$ and therefore $a_k \in W_0 \cap W_1 = W \subset U$. This proves that $U \cap Q \neq \emptyset$ for any set $U \in \tau(y, \mathbb{T}_1)$, i.e., $y \in \overline{Q}$.

Take $t \in \mathbb{T}_0$ and $s \in \mathbb{D}^{<\omega}$ such that $y = (t, s)$. There is $n \in \omega$ such that $s \in \mathbb{D}^n$; let $s_i | n = s$ and $s_i(n) = i$ for each $i \in \mathbb{D}$. We leave it to the reader to check that $s_0$ and $s_1$ are the only immediate successors of $s$, i.e.,

(22) for any $s' \in \mathbb{D}^{<\omega}$, if $s < s'$, then $s_i \leq s'$ for some $i \in \mathbb{D}$.

Let $y_0 = (t, s_0)$ and $y_1 = (t, s_1)$; it is clear that $y < y_i$ for every $i \in \mathbb{D}$. We will prove that $y_0$ and $y_1$ are the only immediate successors of $y$, i.e.,

(23) for any $z \in \mathbb{T}_1$, if $y < z$, then $y_i \leq z$ for some $i \in \mathbb{D}$.

To see that (23) holds take any $z = (t', s') \in \mathbb{T}_1$ such that $y < z$. If $t' = t$, then $s < s'$ and hence $s_i \leq s'$ for some $i \in \mathbb{D}$ by (22). Then $y_i \leq z$, so the property (23) is true for this case. If $t' \neq t$, then there exists $\alpha > \max(t)$ such that $t \cup \{\alpha\} \leq t'$ and $s \in b_{t,\alpha}$. Since $b_{t,\alpha}$ has no maximal element, we can pick $s'' \in b_{t,\alpha}$ with $s < s''$. Apply (22) once more to see that $s_i \leq s''$ for some $i \in \mathbb{D}$ and hence $s_i \in b_{t,\alpha}$. This shows that $y_i = (t, s_i) \leq (t', s') = z$ by (12), so (23) is proved.

It is an immediate consequence of (23) that $V(y) \backslash \{y\} = V(y_0) \cup V(y_1)$, so the set $G = (\mathbb{T}_1 \backslash V(y_0)) \cap (\mathbb{T}_1 \backslash V(y_1))$ is an open neighborhood of the point $y$. Therefore we can find $N \in \mathcal{N}(Q)$ such that $y \in N \subset G$. Since $\mathcal{N}$ is $\omega$-monotone, there exists $n \in \omega$ such that $N \in \mathcal{N}(\{a_0, \ldots, a_n\}) \subset \mathcal{N}^*(a_n)$. By our choice of the set $G$ we have $y \in M(N, a_n)$, so the property (21) shows that $A_{n+1}(a_n) \cap \hat{y} = \emptyset$ which is a contradiction with $a_{n+1} \in A_{n+1}(a_n) \cap \hat{y}$. Therefore $\mathbb{T}_1$ is not monotonically $\omega$-monolithic and hence our Corson compact space $K \supset \mathbb{T}_1$ is not monotonically monolithic either, i.e., our solution is complete.

**V.475.** *Observe that if $X$ is a monotonically $\omega$-monolithic compact space and $\omega_1$ is a caliber of $X$, then the space $X$ is metrizable. Prove that there exists a monotonically $\omega_1$-monolithic (and hence monotonically $\omega$-monolithic) non-compact pseudocompact space $X$ such that $\omega_1$ is a caliber of $X$.*

**Solution.** If $X$ is a monotonically $\omega$-monolithic compact space, then $X$ is Corson compact by Problem 473, so it is metrizable if $\omega_1$ is a caliber of $X$ by CFS-132.

The following statement gives a method of construction of monotonically $\kappa$-monolithic spaces.

*Fact 1.* If $X$ is a space and every subspace of $X$ of cardinality $\leq \kappa$ is closed in $X$ then $X$ is monotonically $\kappa$-monolithic.

*Proof.* If $X'$ is the set $X$ with the discrete topology and $f : X' \to X$ is the identity map then $f|A$ is homeomorphism for any $A \subset X'$ with $|A| \leq \kappa$. The space $X'$ being monotonically monolithic, Fact 1 of V.469 guarantees that $X$ is monotonically $\kappa$-monolithic, so Fact 1 is proved.

*Fact 2.* Given an uncountable regular cardinal $\kappa$ assume that $\kappa$ is a caliber of a space $X$ and $Y \subset X$ is $G_\kappa$-dense in $X$, i.e., every nonempty $G_\kappa$-subset of $X$ intersects $Y$. Then $\kappa$ is also a caliber of $Y$.

*Proof.* Take any family $\mathcal{U} \subset \tau^*(Y)$ with $|\mathcal{U}| = \kappa$. For each $U \in \mathcal{U}$ choose a set $O_U \in \tau(X)$ such that $O_U \cap Y = U$; then $U \neq V$ implies $O_U \neq O_V$, so the family $\mathcal{O} = \{O_U : U \in \mathcal{U}\} \subset \tau^*(X)$ has cardinality $\kappa$. Since $\kappa$ is a caliber of $X$, there exists a family $\mathcal{V} \subset \mathcal{U}$ such that $|\mathcal{V}| = \kappa$ and $G = \bigcap \{O_U : U \in \mathcal{V}\} \neq \emptyset$. Now, $G$ is a nonempty $G_\kappa$-subset of $X$, so $\bigcap \mathcal{V} = G \cap Y \neq \emptyset$ and hence the family $\mathcal{V}$ witnesses that $\kappa$ is a caliber of $Y$, so Fact 2 is proved.

Returning to our solution let $\nu = 2^{\omega_1}$ and take a set $A$ of cardinality $\nu$. Choose a disjoint family $\{A_\alpha : \alpha < \nu\}$ of subsets of $A$ such that $A = \bigcup_{\alpha < \nu} A_\alpha$ and $|A_\alpha| = \nu$ for every $\alpha < \nu$. If $B \subset A$ and $|B| \leq \omega_1$ then $|\mathbb{I}^B| \leq (2^\omega)^{\omega_1} = \nu$. Since the family of all subsets of $A$ of cardinality $\omega_1$ of the set $A$ has cardinality $\nu$, we can choose an enumeration $\{f_\alpha : \alpha < \nu\}$ of the set $\bigcup\{\mathbb{I}^B : B \subset A \text{ and } |B| \leq \omega_1\}$. For any $\alpha < \nu$ there is a unique $B_\alpha \subset A$ such that $|B_\alpha| \leq \omega_1$ and $f_\alpha \in \mathbb{I}^{B_\alpha}$.

Fix $\alpha < \nu$ and define a point $x_\alpha \in \mathbb{I}^A$ as follows: $x_\alpha(a) = f_\alpha(a)$ for any $a \in B_\alpha$; if $a \in A_\alpha \backslash B_\alpha$ then $x_\alpha(a) = 1$ and $x_\alpha(a) = 0$ for all $a \in A \backslash (A_\alpha \cup B_\alpha)$.

Let us show that $X = \{x_\alpha : \alpha < \nu\}$ is the promised space. If $B \subset A$, $|B| \leq \omega_1$ and $y \in \mathbb{I}^B$ then $y = f_\alpha$ and $B = B_\alpha$ for some ordinal $\alpha < \nu$. As a consequence, $x_\alpha | B = f_\alpha | B = y$ and hence $X$ covers all $\omega_1$-faces of $\mathbb{I}^A$. Therefore $X$ is pseudocompact by Fact 2 of S.433. It is easy to see that $X$ is $G_{\omega_1}$-dense in $\mathbb{I}^A$; since $\kappa = \omega_1$ is a caliber of $\mathbb{I}^A$ (see SFFS-282), we can apply Fact 2 to convince ourselves that $\omega_1$ is a caliber of $X$.

If $Y \subset X$ and $|Y| \leq \omega_1$ then fix any point $x_\alpha \in X$. Choose a set $E \subset \nu$ such that $|E| \leq \omega_1$ and $Y \subset \{x_\beta : \beta \in E\}$. The set $G = \bigcup\{B_\beta : \beta \in E \cup \{\alpha\}\}$ has cardinality at most $\omega_1$, so $A_\alpha \backslash G \neq \emptyset$; pick a point $a \in A_\alpha \backslash G$ and note that $x_\alpha(a) = 1$ while $x_\beta(a) = 0$ for any $\beta \in E \backslash \{\alpha\}$. This shows that the set $\{y \in X : y(a) > 0\}$ is an open neighborhood of $x_\alpha$ whose intersection with $Y$ has at most one point. This proves that every set $Y \subset X$ of cardinality at most $\omega_1$ is closed and discrete in $X$, so $X$ is monotonically $\omega_1$-monolithic by Fact 1. The space $X$ is not compact because it is dense in $\mathbb{I}^A$ and the point whose all coordinates are equal to zero does not belong to $X$.

**V.476.** *Suppose that $X$ is a monotonically Sokolov space. Prove that*

(a) $X$ *is monotonically $\omega$-monolithic;*

(b) *any $F_\sigma$-subset of $X$ is monotonically Sokolov.*

**Solution.** (a) For any countable family $\mathcal{F}$ of closed subsets of the space $X$ fix a retraction $r_\mathcal{F} : X \to X$ and a countable family $\mathcal{N}(\mathcal{F}) \subset \exp(X)$ that witness the monotone Sokolov property of $X$. Given a countable set $A \subset X$ let $\mathcal{F}(A) = \{\{x\} : x \in A\}$ and $\mathcal{O}(A) = \mathcal{N}(\mathcal{F}(A))$. It is immediate that the assignment $A \to \mathcal{F}(A)$ is $\omega$-monotone, so the operator $\mathcal{O}$ is also $\omega$-monotone. Since $r_{\mathcal{F}(A)}(F) \subset F$ for any $F \in \mathcal{F}(A)$, we have the equality $r_{\mathcal{F}(A)}(x) = x$ for any $x \in A$. The continuity of the map $r_{\mathcal{F}(A)}$ implies that $r_{\mathcal{F}(A)}(x) = x$ for any $x \in \overline{A}$ and therefore $\overline{A} \subset r_{\mathcal{F}(A)}(X)$. The family $\mathcal{N}(\mathcal{F}(A))$ being a countable external network for $r_{\mathcal{F}(A)}(X)$ in $X$, it must be a countable external network for $\overline{A}$ in $X$, so $X$ is monotonically $\omega$-monolithic.

(b) For any countable family $\mathcal{G}$ of closed subsets of $X$ fix a continuous retraction $r_\mathcal{G} : X \to X$ and a countable collection $\mathcal{N}(\mathcal{G}) \subset \exp(X)$ that witness the monotone Sokolov property of $X$. Take any $F_\sigma$-subset $Y$ of the space $X$ and let $\mathcal{H}$ be a countable family of closed subsets of $X$ with $Y = \bigcup \mathcal{H}$.

Given a countable family $\mathcal{G}$ of closed subsets of the space $Y$ it is easy to see that $\mathcal{F}(\mathcal{G}) = \mathcal{H} \cup \{G \cap H : G \in \mathcal{G} \text{ and } H \in \mathcal{H}\}$ is a family of closed subsets of $X$ and the assignment $\mathcal{G} \to \mathcal{F}(\mathcal{G})$ is $\omega$-monotone. We will prove that the map $s_\mathcal{G} = r_{\mathcal{F}(\mathcal{G})} | Y$ and the family $\mathcal{O}(\mathcal{G}) = \{N \cap Y : N \in \mathcal{N}(\mathcal{F}(\mathcal{G}))\}$ witness the monotone Sokolov property of the space $Y$.

If $y \in Y$, then $y \in H$ for some $H \in \mathcal{H}$. Since $H \in \mathcal{H} \subset \mathcal{F}(\mathcal{G})$, by the choice of $r_{\mathcal{F}(\mathcal{G})}$, we have $r_{\mathcal{F}(\mathcal{G})}(H) \subset H$. Hence $s_{\mathcal{G}}(y) = r_{\mathcal{F}(\mathcal{G})}(y) \in H \subset Y$, so $s_{\mathcal{G}} : Y \to Y$ is a continuous retraction. The family $\mathcal{N}(\mathcal{F}(\mathcal{G}))$ being an external network for $r_{\mathcal{F}(\mathcal{G})}(X)$ in $X$, it follows that $\mathcal{O}(\mathcal{G})$ is an external network for $s_{\mathcal{G}}(Y)$ in $Y$. The assignments $\mathcal{F} \to \mathcal{F}(\mathcal{G})$ and $\mathcal{F}(\mathcal{G}) \to \mathcal{N}(\mathcal{F}(\mathcal{G}))$ are $\omega$-monotone, so the operator $\mathcal{O}$ must also be $\omega$-monotone. Finally, we have the inclusion $s_{\mathcal{G}}(G) \subset G$ for each $G \in \mathcal{G}$. Indeed, take any $y \in G$ for some $G \in \mathcal{G}$. Since $y \in Y = \bigcup \mathcal{H}$, we can find $H \in \mathcal{H}$ with $y \in H$. Then $y \in G \cap H \in \mathcal{F}(\mathcal{G})$. It follows from the choice of $r_{\mathcal{F}(\mathcal{G})}$ that $r_{\mathcal{F}(\mathcal{G})}(G \cap H) \subset G \cap H$. Thus $s_{\mathcal{G}}(y) = r_{\mathcal{F}(\mathcal{G})}(y) \in G \cap H \subset G$.

**V.477.** *Suppose that a space $X$ is monotonically retractable and fix, for any countable set $A \subset X$, a retraction $r_A : X \to X$ and a network $\mathcal{N}(A)$ for the map $r_A$ that witness monotone retractability of $X$. Prove that, for any countable family $\mathcal{G}$ of closed subsets of $X$ we can find a countable set $P(\mathcal{G}) \subset X$ such that $r_{P(\mathcal{G})}(G) \subset G$ for any $G \in \mathcal{G}$ and the assignment $\mathcal{G} \to P(\mathcal{G})$ is $\omega$-monotone.*

**Solution.** Let $\mathcal{N}(X) = \bigcup\{\mathcal{N}(A) : A$ is a countable subset of $X\}$. For any closed set $G \subset X$ and $H \in \mathcal{N}(X)$, if $G \cap H \neq \emptyset$ then pick a point $p_{G,H} \in G \cap H$. Given a countable family $\mathcal{G}$ of closed subsets of $X$ and a countable subfamily $\mathcal{H}$ of $\mathcal{N}(X)$, let $E(\mathcal{G}, \mathcal{H}) = \{p_{G,H} : G \in \mathcal{G}, H \in \mathcal{H}, \text{ and } G \cap H \neq \emptyset\}$. If $G$ is a nonempty closed subset of $X$, then pick a point $p_G \in G$.

**Fact 1.** Suppose that we have families $\mathcal{A} \subset \exp(X)$, $\mathcal{B} \subset \exp(Y)$ and $\mathcal{C} \subset \exp(Z)$ for some sets $X, Y$ and $Z$.

(a) If the maps $p : \mathcal{A} \to \mathcal{B}$ and $q : \mathcal{B} \to \mathcal{C}$ are $\omega$-monotone then the map $q \circ p : \mathcal{A} \to \mathcal{C}$ is also $\omega$-monotone;

(b) if $\mathcal{B}$ is invariant under countable unions, $|T| \leq \omega$ and $p_t : \mathcal{A} \to \mathcal{B}$ is $\omega$-monotone for any $t \in T$ then the map $p : \mathcal{A} \to \mathcal{B}$ defined by $p(A) = \bigcup\{p_t(A) : t \in T\}$ for any $A \in \mathcal{A}$ is also $\omega$-monotone.

*Proof.* (a) If $A \in \mathcal{A}$ is countable, then $B = p(A)$ is countable and hence $q(B) = (q \circ p)(A)$ is countable. If $A \subset A'$, then $B = p(A) \subset p(A') = B'$ because the map $p$ is $\omega$-monotone. Therefore $(q \circ p)(A) = q(B) \subset q(B') = (q \circ p)(A')$. Finally, if we have a set $A = \bigcup_{n \in \omega} A_n$ such that $A_n \subset A_{n+1}$ for any $n \in \omega$, then $(q \circ p)(A) = q(p(A)) = q(\bigcup_{n \in \omega} p(A_n))$ by $\omega$-monotonicity of the map $p$. Since also $p(A_n) \subset p(A_{n+1})$ for any $n \in \omega$, by monotonicity of the map $q$, we have $q(\bigcup_{n \in \omega} p(A_n)) = \bigcup_{n \in \omega} q(p(A_n))$; this settles (a).

(b) If $A \in \mathcal{A}$ is a countable, then $p(A)$ is countable being a countable union of countable sets. If $A \subset A'$, then $p_t(A) \subset p_t(A')$ for every $t \in T$, so $p(A) \subset p(A')$. Finally assume that $A = \bigcup_{n \in \omega} A_n$ and $A_n \subset A_{n+1}$ for each $n \in \omega$. We only have to show that $p(A) \subset \bigcup_{n \in \omega} p(A_n)$, so take any point $x \in p(A)$. There exists $t \in T$ such that $x \in p_t(A)$; by $\omega$-monotonicity of $p_t$ we can find $k \in \omega$ for which $x \in p_t(A_k)$. Now it follows from $p_t(A_k) \subset p(A_k)$ that $x \in p(A_k) \subset \bigcup_{n \in \omega} p(A_n)$; this settles (b) and shows that Fact 1 is proved.

Returning to our solution, for any countable family $\mathcal{G}$ of closed subsets of $X$ let $Q(\mathcal{G}) = \{p_G : G \in \mathcal{G} \setminus \{\emptyset\}\}$. The assignment $\mathcal{G} \to Q(\mathcal{G})$ is easily seen to be

$\omega$-monotone. Define $P_0(\mathcal{G}) = Q(\mathcal{G})$ and $P_{n+1}(\mathcal{G}) = P_n(\mathcal{G}) \cup E(\mathcal{G}, \mathcal{N}(P_n(\mathcal{G})))$ for any $n \in \omega$. Then the set $P(\mathcal{G}) = \bigcup\{P_n(\mathcal{G}) : n \in \omega\}$ is as promised.

It takes an easy induction and Fact 1 to show that each of the assignments $P_n$ is $\omega$-monotone and hence $P$ is also $\omega$-monotone. Let us show that $r_{P(\mathcal{G})}(G) \subset G$ for each $G \in \mathcal{G}$.

If this is not true, then $r_{P(\mathcal{G})}(x) \notin G$ for some $G \in \mathcal{G}$ and $x \in G$. Then $U = X \setminus G$ is a neighborhood of $r_{P(\mathcal{G})}(x)$ which does not intersect $G$. Since $\mathcal{N}(P(\mathcal{G}))$ is a network for $r_{P(\mathcal{G})}$, there exists a set $H \in \mathcal{N}(P(\mathcal{G}))$ such that $x \in H$ and $r_{P(\mathcal{G})}(H) \subset U$. The equality $\mathcal{N}(P(\mathcal{G})) = \mathcal{N}(\bigcup\{P_n(\mathcal{G}) : n \in \omega\}) = \bigcup\{\mathcal{N}(P_n(\mathcal{G})) : n \in \omega\}$ implies that $H \in \mathcal{N}(P_n(\mathcal{G}))$ for some $n \in \omega$. It follows from $x \in G \cap H$ that $G \cap H \neq \emptyset$, so a point $p_{G,H} \in G \cap H$ has been chosen and

$$p_{G,H} \in E(\mathcal{G}, \mathcal{N}(P_n(\mathcal{G}))) \subset P_{n+1}(\mathcal{G}) \subset P(\mathcal{G}) \subset r_{P(\mathcal{G})}(X).$$

Therefore $p_{G,H} = r_{P(\mathcal{G})}(p_{G,H}) \in r_{P(\mathcal{G})}(H) \subset U \subset X \setminus G$. This contradiction shows that $r_{P(\mathcal{G})}(G) \subset G$ for any $G \in \mathcal{G}$.

**V.478.** *Given a space $Y$ denote by $\mathcal{CL}(Y)$ the family of all closed subsets of $Y$ and let $\mathcal{CL}^*(Y) = \bigcup\{\mathcal{CL}(Y^n) : n \in \mathbb{N}\}$. Prove that, for any space $X$, the following conditions are equivalent:*

*(a) $X$ is monotonically retractable:*

*(b) for any countable family $\mathcal{G}$ of closed subsets of $X$, there exists a retraction $s_{\mathcal{G}} : X \to X$ and a countable network $\mathcal{O}(\mathcal{G})$ for the map $s_{\mathcal{G}}$ such that $s(G) \subset G$ for any $G \in \mathcal{G}$ and the assignment $\mathcal{O}$ is $\omega$-monotone;*

*(c) for any countable family $\mathcal{G} \subset \mathcal{CL}^*(X)$ there exists a retraction $t_{\mathcal{G}} : X \to X$ and a countable network $\mathcal{P}(\mathcal{G})$ for the map $t_{\mathcal{G}}$ such that, for any $n \in \mathbb{N}$, we have the inclusion $t_{\mathcal{G}}^n(G) \subset G$ for any $G \in \mathcal{G} \cap \mathcal{CL}(X^n)$ and the assignment $\mathcal{P}$ is $\omega$-monotone.*

**Solution.** Recall that if $Y$ is a space and $f : Y \to Y$, then for any $n \in \mathbb{N}$, the function $f^n : Y^n \to Y^n$ is defined by the equality $f^n(y) = (f(y_1), \dots, f(y_n))$ for any $y = (y_1, \dots, y_n) \in Y$.

It is immediate that (c)$\Longrightarrow$(b). If (b) holds and $A$ is a countable subset of $X$ then consider the family $\mathcal{G}(A) = \{\{x\} : x \in A\}$, the retraction $r_A = s_{\mathcal{G}(A)}$ and the family $\mathcal{N}(A) = \mathcal{O}(\mathcal{G}(A))$. It is easy to check that $r_A$ and $\mathcal{N}(A)$ witness that $X$ is monotonically retractable, so we have (b)$\Longrightarrow$(a).

Next assume that (a) holds and fix, for any countable $B \subset X$, a continuous retraction $s_B : X \to X$ and a countable network $\mathcal{O}(B)$ for the map $s_B$ such that $B \subset s_B(X)$ and the assignment $\mathcal{O}$ is $\omega$-monotone.

Take an arbitrary point $a \in Y = X^{\mathbb{N}}$ and let $p_n : Y \to X$ be the projection onto the $n$-th factor for any $n \in \mathbb{N}$. Observe that the set $E(A) = \bigcup\{p_n(A \cup \{a\}) : n \in \mathbb{N}\} \subset X$ is countable for any countable $A \subset Y$. For each $y \in Y$ let $(s_{E(A)})^{\mathbb{N}}(y)(n) = s_{E(A)}(y(n))$ for any $n \in \mathbb{N}$. We leave it to the reader to check that the map $r_A = (s_{E(A)})^{\mathbb{N}} : Y \to Y$ is a retraction such that $A \subset r_A(Y)$. Denote by

$\mathcal{N}(A)$ the family of all sets $U \subset Y$ such that $U = \prod\{N_n : n \in \mathbb{N}\}$ where, for some finite set $F \subset \mathbb{N}$, we have $N_n \in \mathcal{O}(E(A))$ for all $n \in F$ and $N_n = X$ if $n \notin F$.

It is straightforward to verify that the assignment $A \to \mathcal{N}(A)$ is $\omega$-monotone. To see that the family $\mathcal{N}(A)$ is a network for the map $r_A$ take any point $y \in Y$ and a set $W \in \tau(r_A(y), Y)$. There exists a number $n \in \mathbb{N}$ and sets $V_1, \ldots, V_n \in \tau(X)$ such that $r_A(y) \in V \subset W$ where $V = V_1 \times \ldots \times V_n \times X^{\mathbb{N}\setminus\{1,\ldots,n\}}$. The family $\mathcal{O}(E(A))$ being a network for the map $s_{E(A)}$, it follows from $r_A(y)(i) = s_{E(A)}(y(i)) \in V_i$ that we can find $F_i \in \mathcal{O}(E(A))$ such that $y(i) \in F_i$ and $s_{E(A)}(F_i) \subset V_i$ for each $i \le n$. Then $y \in F = F_1 \times \ldots \times F_n \times X^{\mathbb{N}\setminus\{1,\ldots,n\}} \in \mathcal{N}(A)$ and $r_A(F) \subset V \subset W$ which shows that $\mathcal{N}(A)$ is a network for the map $r_A$ and hence $r_A$ and $\mathcal{N}(A)$ witness that the space $X^{\mathbb{N}}$ is monotonically retractable.

Now we can apply Problem 477 to convince ourselves that for any countable family $\mathcal{H}$ of closed subsets of $Y$ we can choose a countable set $P(\mathcal{H}) \subset Y$ such that $r_{P(\mathcal{H})}(H) \subset H$ for any $H \in \mathcal{H}$ and the assignment $P$ is $\omega$-monotone.

Observe that the set $Z_n = \{y \in Y : \{k \in \mathbb{N} : y(k) \ne a(k)\} \subset \{1, \ldots, n\}\} \subset Y$ is closed in $Y$ and the map $\varphi_n = p_n|Z_n : Z_n \to X^n$ is a homeomorphism for which we have the equality $s_{E(A)}^n = \varphi_n \circ r_A \circ \varphi_n^{-1}$ for any $n \in \mathbb{N}$. If $\mathcal{G}$ is a countable subfamily of $\mathcal{CL}^*(X)$, then fix $n(G) \in \mathbb{N}$ such that $G \subset X^{n(G)}$ for every $G \in \mathcal{G}$ and consider the family $\mathcal{G}' = \{\varphi_{n(G)}^{-1}(G) : G \in \mathcal{G}\} \subset \mathcal{CL}(X^{\mathbb{N}})$.

Let $t_{\mathcal{G}} = s_{E(P(\mathcal{G}'))} : X \to X$ and $\mathcal{P}(\mathcal{G}) = \mathcal{O}(E(P(\mathcal{G}')))$. Then $t_{\mathcal{G}}$ is a retraction and $\mathcal{P}(\mathcal{G})$ is a network for $t_{\mathcal{G}}$. Since the assignments $P, E, \mathcal{O}$ and $\mathcal{G} \to \mathcal{G}'$ are $\omega$-monotone, the assignment $\mathcal{P}$ is also $\omega$-monotone. Finally, for any $G \in \mathcal{G} \cap \mathcal{CL}(X^n)$ we have $n(G) = n$ and therefore

$$(t_{\mathcal{G}})^n(G) = (s_{E(P(\mathcal{G}))})^n(G) = \varphi_n((s_{E(P(\mathcal{G}))})^{\mathbb{N}}(\varphi_n^{-1}(G))) \subset \varphi_n(\varphi_n^{-1}(G)) = G,$$

i.e., the map $t_{\mathcal{G}}$ and the family $\mathcal{P}(\mathcal{G})$ witness that (c) holds, so we proved that (a)$\Longrightarrow$(c) and hence our solution is complete.

**V.479.** *Given a monotonically retractable space $X$, prove that*

*(a) every $F_\sigma$-subset of $X$ is monotonically retractable;*
*(b) the space $X$ is Sokolov; in particular, $X$ is $\omega$-stable, collectionwise normal and $ext(X) \le \omega$;*
*(c) $C_p(X)$ is a Lindelöf D-space.*

**Solution.** (a) Suppose that $Y$ is an $F_\sigma$-subset of $X$ and let $\mathcal{F}$ be a countable family of closed subsets of $X$ such that $Y = \bigcup \mathcal{F}$. Apply Problem 478 to assign to any countable family $\mathcal{G}$ of closed subsets of $X$ a retraction $s_{\mathcal{G}} : X \to X$ and a countable network $\mathcal{O}(\mathcal{G})$ for $s_{\mathcal{G}}$ such that $s_{\mathcal{G}}(G) \subset G$ for any $G \in \mathcal{G}$ and the assignment $\mathcal{O}$ is $\omega$-monotone.

For any countable set $A \subset Y$ let $\mathcal{G}(A) = \mathcal{F} \cup \{\{x\} : x \in A\}$. It is clear that the assignment $A \to \mathcal{G}(A)$ is $\omega$-monotone; observe first that $s_{\mathcal{G}(A)}(Y) \subset Y$. Indeed, if $y \in Y$ then $y \in F$ for some $F \in \mathcal{F} \subset \mathcal{G}(A)$, so $s_{\mathcal{G}(A)}(F) \subset F$ which implies that $s_{\mathcal{G}(A)}(y) \in F \subset Y$. Therefore $r_A = s_{\mathcal{G}(A)}|Y : Y \to Y$ is a retraction.

Since $\mathcal{O}(\mathcal{G}(A))$ is a network for $s_{\mathcal{G}(A)}$, the family $\mathcal{N}(A) = \{P \cap Y : P \in \mathcal{O}(\mathcal{G}(A))\}$ is a network for $r_A$ and it is immediate that the assignment $\mathcal{N}$ is $\omega$-monotone.

For any $x \in A$ it follows from $r_A(\{x\}) \subset \{x\}$ that $r_A(x) = x$; this shows that we have the inclusion $A \subset r_A(Y)$, so the map $r_A$ and the family $\mathcal{N}(A)$ witness that $Y$ is monotonically retractable.

(b) Suppose that $F_n$ is a closed subset of $X^n$ for any $n \in \mathbb{N}$. Apply Problem 478(c) to find a retraction $r : X \to X$ and a countable network $\mathcal{N}$ for the map $r$ such that $r^n(F_n) \subset F_n$ for any $n \in \mathbb{N}$. It is straightforward that the family $\{f(N) : N \in \mathcal{N}\}$ is a countable network in $r(X)$, so the map $r$ witnesses that $X$ is a Sokolov space. The rest of the statements of (b) follow from the fact that every Sokolov space is $\omega$-stable, collectionwise normal and has countable extent by CFS-161 and CFS-163.

(c) Fix a countable base $\mathcal{B}$ in the space $\mathbb{R}$; given a family $\mathcal{A}$ of subsets of a space $Z$ let $\mathcal{W}(\mathcal{A}) = \{[A_1, \ldots, A_n; B_1, \ldots, B_n] : n \in \mathbb{N},\ A_i \in \mathcal{A} \text{ and } B_i \in \mathcal{B} \text{ for every } i \leq n\}$; here, as usual, $[A_1, \ldots, A_n; B_1, \ldots, B_n] = \{f \in C_p(Z) : f(A_i) \subset B_i \text{ for any } i \leq n\}$. If $\varphi : Z \to Z'$ is an onto map, then its dual map $\varphi^* : C_p(Z') \to C_p(Z)$ is defined by the formula $\varphi(f) = f \circ \varphi$ for any $f \in C_p(Z')$.

For any countable set $A \subset X$ fix a retraction $r_A : X \to X$ and a network $\mathcal{N}(A)$ for the map $r_A$ that witness monotone retractability of $X$. Given a set $S \subset X$, let $\pi_S : C_p(X) \to C_p(S)$ be the restriction map. Let $\mathcal{O}(A) = \mathcal{W}(\mathcal{N}(A))$ for any countable set $A \subset X$. By Fact 1 of V.467 and Fact 1 of V.477, the assignment $\mathcal{O}$ is $\omega$-monotone.

We will prove the following statement which will imply everything we need:

(1) if $N : C_p(X) \to \tau(C_p(X))$ is a neighborhood assignment on $C_p(X)$, then there exists a countable closed discrete $D \subset X$ such that $\bigcup\{N(f) : f \in D\} = C_p(X)$.

Take an arbitrary neighborhood assignment $N$ on $C_p(X)$; there is no loss of generality to assume that $N(f)$ is a standard open subset of $C_p(X)$ and, in particular, there exists a finite set $S_f \subset X$ such that $N(f) = \pi_{S_f}^{-1}\pi_{S_f}(N(f))$ for any $f \in C_p(X)$. For any set $B \subset C_p(X)$ say that $f \in B$ is a central point of $B$ if $B \subset N(f)$; denote by $Z(B)$ the set of all central points of $B$. It is easy to find disjoint infinite sets $\{\Omega_i : i \in \omega\}$ such that $\{0, \ldots, n\} \subset \Omega_0 \cup \ldots \cup \Omega_n$ for every $n \in \omega$ and $\omega = \bigcup_{n \in \omega} \Omega_n$.

Pick a function $f_0 \in C_p(X)$ arbitrarily; let $A_0 = S_{f_0}$ and take an enumeration $\{P_k : k \in \Omega_0\}$ of the family $\mathcal{O}(A_0)$. Proceeding inductively, assume that we have functions $f_0, \ldots, f_n \in C_p(X)$ and countable subsets $A_0, \ldots, A_n$ of the space $X$ with the following properties:

(2) $A_i \subset A_{i+1}$ for all $i < n$;
(3) $S_{f_i} \subset A_i$ and $\{P_k : k \in \Omega_i\}$ is an enumeration of $\mathcal{O}(A_i)$ for every $i \leq n$;
(4) $\bigcup\{Z(P_j) : j < i\} \subset \bigcup\{N(f_j) : j \leq i\}$ for all $i \leq n$;
(5) $f_{i+1} \in C_p(X) \setminus \bigcup\{N(f_j) : j \leq i\}$ or $f_{i+1} = f_i$ for each $i < n$.

To construct $f_{n+1}$ let $U_n = \bigcup\{N(f_i) : i \leq n\}$. It follows from (3) and $n \in \bigcup_{i \leq n} \Omega_i$ that the set $P_n$ is defined; if $Z(P_n) \subset U_n$ then let $f_{n+1} = f_n$. This ensures that (4) holds for all $i \leq n + 1$. If $Z(P_n)\backslash U_n \neq \emptyset$ then choose $f_{n+1} \in Z(P_n)\backslash U_n$. This implies that $Z(P_n) \subset P_n \subset N(f_{n+1})$, so (4) now holds for all $i \leq n + 1$.

If $A_{n+1} = A_n \cup S_{f_{n+1}}$ and $\{P_k : k \in \Omega_{n+1}\}$ is an enumeration of $\mathcal{O}(A_{n+1})$, then the properties (2)–(5) still hold if we replace $n$ by $n + 1$, so we can construct a set $D = \{f_i : i \in \omega\}$ and a family $\mathcal{A} = \{A_i : i \in \omega\}$ such that the conditions (2)–(5) are satisfied for all $n \in \omega$; let $A = \bigcup_{i \in \omega} A_i$.

We will show that $\bigcup\{N(f_i) : i \in \omega\} = C_p(X)$, so fix any function $f \in C_p(X)$. Observe that $g = f \circ r_A \in r_A^*(C_p(Y))$ where $Y = r_A(X)$. Since $\mathcal{N}(A)$ is a network for $r_A$, we can apply Fact 2 of V.467 to see that $\mathcal{O}(A)$ is an external network for $r_A^*(C_p(Y))$ and hence we can find a set $P \in \mathcal{O}(A)$ such that $g \in P \subset N(g)$ and in particular, $g \in Z(P)$. By (2) and $\omega$-monotonicity of $\mathcal{O}$, there exists a number $n \in \omega$ such that $P \in \mathcal{O}(A_n)$ and hence $P = P_k$ for some $k \in \Omega_n$. The property (4) shows that $g \in Z(P_k) \subset \bigcup\{N(f_i) : i \in \omega\}$, so there exists $m \in \omega$ such that $g \in N(f_m)$. It follows from (2) and (3) that $S_{f_m} \subset A_m \subset A \subset Y$. Since $r_A(x) = x$ for any $x \in A$, we have the equality $f|S_{f_m} = g|S_{f_m}$, so it follows from $g \in N(f_m)$ that we have the inclusions $f \in N(f_m) \subset \bigcup\{N(f_i) : i \in \omega\}$. The point $f \in C_p(X)$ was taken arbitrarily, so we proved that $\bigcup\{N(f_i) : i \in \omega\} = C_p(X)$.

Given any $f \in C_p(X)$ take $n \in \omega$ such that $f \in N(f_n)$; by the property (5) we have $N(f_n) \cap D \subset \{f_0, \dots, f_n\}$ and hence every point of $C_p(X)$ has a neighborhood whose intersection with $D$ is finite. This shows that $D$ is a closed discrete subset of $C_p(X)$ and finishes the proof of (1).

It is clear that (1) implies that $C_p(X)$ is a $D$-space. Given any open cover $\mathcal{U}$ of the space $C_p(X)$ take a set $U_f \in \mathcal{U}$ such that $f \in U_f$ for any $f \in C_p(X)$. Apply (1) to the neighborhood assignment $\{U_f : f \in C_p(X)\}$ to find a countable set $D \subset C_p(X)$ such that $\bigcup\{U_f : f \in D\} = C_p(X)$. Since $\{U_f : f \in D\}$ is a countable subcover of $\mathcal{U}$, we proved that $C_p(X)$ is Lindelöf. This settles (c) and makes our solution complete.

**V.480.** *Prove that*

(a) *any $\sigma$-product of monotonically retractable spaces is monotonically retractable;*

(b) *any $\Sigma$-product of monotonically retractable spaces is monotonically retractable and hence every countable product of monotonically retractable spaces is monotonically retractable;*

(c) *any closed subspace of a $\Sigma$-product of cosmic spaces is monotonically retractable.*

**Solution.** If $Z$ is a set, then $[Z]^{\leq \omega}$ is the family of all countable subsets of $Z$. Suppose that in a space $X_t$, for any countable set $A \subset X_t$ we have a retraction $r_A^t : X_t \to X_t$ and a network $\mathcal{N}_t(A)$ for the map $r_A^t$ that witness monotone retractability of $X_t$ for any $t \in T$. Fix a point $a \in X = \prod_{t \in T} X_t$. We must prove that the spaces

$\Sigma(X, a) = \{x \in X : \text{the set supp}(x) = \{t \in T : x(t) \neq a(t)\} \text{ is countable}\}$ and $\sigma(X, a) = \{x \in X : \text{supp}(x) \text{ is finite}\}$ are both monotonically retractable.

*Fact 1.* Suppose that $h_t : X_t \to X_t$ is a continuous retraction such that $h_t(a(t)) = a(t)$ for every $t \in T$ and we have a set $E \subset T$. For any $x \in X$ let $s(x)(t) = h_t(x(t))$ for each $t \in E$ and $s(x)(t) = a(t)$ whenever $t \in T \backslash E$. Then

(i)  $s : X \to X$ is a continuous retraction;
(ii)  $s(\Sigma(X, a)) \subset \Sigma(X, a)$;
(iii)  $s(\sigma(X, a)) \subset \sigma(X, a)$.

*Proof.* Let $p_t : X \to X_t$ be the natural projection for every $t \in T$. Assume first that $t \in T \backslash E$; then $p_t(s(x)) = s(x)(t) = a(t)$ for any $x \in X$ and hence $p_t \circ s$ is continuous being a constant function. If $t \in E$, then $p_t(s(x)) = s(x)(t) = h_t(x(t))$ for any $x \in X$, so $p_t \circ s = h_t \circ p_t$ is also a continuous function. This proves that the map $s$ is continuous (see TFS-102). To see that $s$ is a retraction, take any $y \in s(X)$ and fix $x \in X$ such that $y = s(x)$. If $t \in E$, then $s(y)(t) = h_t(y(t)) = h_t(h_t(x(t))) = h_t(x(t)) = y(t)$. If $t \in T \backslash E$, then $s(y)(t) = a(t) = s(x)(t) = y(t)$. Therefore $s(y) = y$ for any $y \in s(X)$ and hence $s$ is a continuous retraction; this settles (i).

Now take any point $x \in X$ and $t \in T \backslash \text{supp}(x)$. If $t \in E$, then the equalities $s(x)(t) = h_t(x(t)) = h_t(a(t)) = a(t)$ show that $t \notin \text{supp}(s(x))$. If $t \notin E$, then $s(x)(t) = a(t)$ and again $t \notin \text{supp}(s(x))$. This shows that $\text{supp}(s(x)) \subset \text{supp}(x)$ for any $x \in X$; this immediately implies (ii) and (iii) and shows that Fact 1 is proved.

Suppose that $A \subset X$ is a countable set and $E \subset T$. Define a map $r[A, E] : X \to X$ by the formula $r[A, E](x)(t) = r^t_{A_t}(x(t))$ if $t \in E$; here $A_t = p_t(A \cup \{a\})$. If $t \in T \backslash E$, then let $r[A, E](x)(t) = a(t)$. Observe that $a(t) \in A_t$ and hence $r^t_{A_t}(a(t)) = a(t)$ for any $t \in T$, so it follows from Fact 1 that $r[A, E]$ is a continuous retraction such that $r[A, E](\Sigma(X, a)) \subset \Sigma(X, a)$. Call a set $Y \subset \Sigma(X, a)$ *monotonically invariant* if we can assign to any countable set $A \subset Y$ a subset $E(A) \supset \bigcup \{\text{supp}(x) : x \in A\}$ of the set $T$ such that $r[A, E(A)](Y) \subset Y$ and the assignment $E$ is $\omega$-monotone. Observe that it follows from Fact 1 that $r[A, E(A)]|Y : Y \to Y$ is a continuous retraction whenever $Y$ is a monotonically invariant subset of $\Sigma(X, a)$ and $A$ is a countable subset of $Y$.

*Fact 2.* If $Y$ is a monotonically invariant subset of $\Sigma(X, a)$, then $Y$ is monotonically retractable.

*Proof.* Fix a map $E : [Y]^{\leq \omega} \to [T]^{\leq \omega}$ that witnesses monotonical invariance of $Y$. Fix a countable set $A \subset Y$ and let $s_A = r[A, E(A)]|Y$; the map $s_A : Y \to Y$ is a continuous retraction by Fact 1. If $x \in A$, then $x(t) \in A_t$ and hence $s_A(x)(t) = r^t_{A_t}(x(t)) = x(t)$ for any $t \in E(A)$. If $t \in T \backslash E(A)$ then $x(t) = a(t)$ because $\text{supp}(x) \subset E(A)$. This implies that $s_A(x)(t) = a(t) = x(t)$, so $s_A(x)(t) = x(t)$ for any $t \in T$, i.e., $s_A(x) = x$ for any $x \in A$.

Say that a set $Q \subset X$ is $A$-standard if $Q = \prod_{t \in T} Q_t$ and there is a finite $F \subset E(A)$ such that $Q_t \in \mathcal{N}_t(A_t)$ for any $t \in F$ and $Q_t = X_t$ for all $t \in T \setminus F$. It is easy to see that the family $\mathcal{N}(A) = \{Q \cap Y : Q$ is an $A$-standard set$\}$ is countable; we will show that $\mathcal{N}(A)$ is a network for $s_A$.

Suppose that $x \in Y$ and $y = s_A(x) \in U \in \tau(Y)$. There exists a set $V = \prod_{t \in T} V_t$ such that $V_t \in \tau(X_t)$ for all $t \in T$ while $V_t = X_t$ if $t \in T \setminus F$ for some finite set $F \subset T$ and $y \in V \cap Y \subset U$. If $t \in K = F \cap E(A)$, then it follows from $y(t) = r^t_{A_t}(x(t)) \in V_t$ that we can find $Q_t \in \mathcal{N}_t(A_t)$ such that $x(t) \in Q_t$ and $r^t_{A_t}(Q_t) \subset V_t$. Letting $Q_t = X_t$ for any $t \in T \setminus K$, we obtain an $A$-standard set $Q = \prod_{t \in T} Q_t$. It follows from $x \in Q$ that $x \in Q' = Q \cap Y \in \mathcal{N}(A)$.

Take any $z \in Q'$; if $t \in K$, then $s_A(z)(t) = r^t_{A_t}(z(t)) \subset r^t_{A_t}(Q_t) \subset V_t$. If $t \in F \setminus K$, then $s_A(z)(t) = a(t) = s_A(x)(t) = y(t) \in V_t$. Therefore $s_A(z)(t) \in V_t$ for all $t \in F$ which implies $s_A(z) \in V \cap Y \subset U$ for each $z \in Q'$, i.e., $s_A(Q') \subset U$. This proves that $\mathcal{N}(A)$ is a network for the map $s_A$.

Let us show that the operator $\mathcal{N}$ is $\omega$-monotone. It is evident that the family $\mathcal{N}(A)$ is countable for any countable $A \subset Y$. If $A \subset A'$, then $A_t \subset A'_t$ and hence $\mathcal{N}_t(A_t) \subset \mathcal{N}_t(A'_t)$ for any $t \in T$. Since also $E(A) \subset E(A')$, every $A$-standard set is $A'$-standard, so $\mathcal{N}(A) \subset \mathcal{N}(A')$. Finally, if $\{A_n : n \in \omega\}$ is a non-decreasing family of countable subsets of $Y$ and $A = \bigcup_{n \in \omega} A_n$, then it suffices to show that $\mathcal{N}(A) \subset \bigcup_{n \in \omega} \mathcal{N}(A_n)$, so take any set $Q' \in \mathcal{N}(A)$. There exists an $A$-standard set $Q = \prod_{t \in T} Q_t$ such that $Q' = Q \cap Y$. Let $F \subset T$ be a finite set such that $Q_t \in \mathcal{N}_t(A_t)$ for any $t \in F$ and $Q_t = X_t$ whenever $t \in T \setminus F$. By monotonicity of every $\mathcal{N}_t$, we can find $n \in \omega$ such that $Q_t \in \mathcal{N}_t((A_n)_t)$ for every $t \in F$ and hence $Q = \prod_{t \in T} Q_t$ is an $A_n$-standard set, i.e., $Q' \in \mathcal{N}(A_n)$. Therefore we have verified that the retractions $s_A$ and the families $\mathcal{N}(A)$ witness monotone retractability of the space $Y$ and hence Fact 2 is proved.

Returning to our solution let $Y$ be one of the spaces $\Sigma(X, a)$ or $\sigma(X, a)$. Observe that if we let $E(A) = \bigcup\{\text{supp}(x) : x \in A\}$ for any countable set $A \subset Y$, then the assignment $E$ is $\omega$-monotone. Since also $r[A, E(A)](Y) \subset Y$ by Fact 1, we proved that both $\Sigma(X, a)$ and $\sigma(X, a)$ are monotonically invariant in $\Sigma(X, a)$, so they are monotonically retractable by Fact 2. This settles (a) and (b).

(c) Assume that $nw(X_t) = \omega$ for all $t \in T$ and let $Y$ be a closed subset of $\Sigma(X, a)$. Fix a countable network $\mathcal{E}_t$ in the space $X_t$ for every $t \in T$. If, for any countable set $A \subset X$, we let $\mathcal{N}_t(A) = \mathcal{E}_t$ and $r^t_A(x) = x$ for all $x \in X_t$, then the maps $r^t_A$ and the families $\mathcal{N}_t(A)$ witness monotone retractability of the space $X_t$ for every $t \in T$.

Taking in consideration the structure of the maps $r^t_A$, we can see that for any $A \subset Y$ and $B \subset T$, the map $r[A, B] : X \to X$ acts as follows: $r[A, B](x)(t) = x(t)$ if $x \in B$ and $r[A, B](x)(t) = a(t)$ whenever $t \in T \setminus B$. For every countable set $A \subset Y$ let $S(A) = \bigcup\{\text{supp}(x) : x \in A\}$. It is immediate that the assignment $S$ is $\omega$-monotone.

Say that a set $Q = \prod_{t \in T} Q_t$ is canonical, if $Q \cap Y \neq \emptyset$, the set $H(Q) = \{t \in T : Q_t \neq X_t\}$ is finite and $Q_t \in \mathcal{E}_t$ for every $t \in H(Q)$. For each canonical set

$Q$ pick a point $x_Q \in Q \cap Y$ and let $D(B) = \{x_Q : Q$ is a canonical set and $H(Q) \subset B\}$ for any countable $B \subset T$. Again, it is an easy exercise to prove that the operator $D$ is $\omega$-monotone.

For any countable set $A \subset Y$, consider the sets $E_0(A) = S(A)$ and $E_{n+1}(A) = E_n(A) \cup S(D(E_n(A)))$ for any $n \in \omega$; let $E(A) = \bigcup_{n \in \omega} E_n(A)$. Using Fact 1(a) of V.477 it takes a simple induction to prove that every operator $E_n$ is $\omega$-monotone. Therefore the assignment $A \to E(A)$ is also $\omega$-monotone by Fact 1(b) of V.477.

Fix any countable set $A \subset Y$ and let $B = E(A)$. If $y \in D(B) = D(\bigcup_{n \in \omega} E_n(A))$, then $y \in D(E_n(A))$ for some $n \in \omega$ by monotonicity of the operator $D$. This implies that $\text{supp}(y) \subset S(D(E_n(A))) \subset E_{n+1}(A) \subset E(A) = B$; by definition of the map $r[A, B]$ we have $r[A, B](y) = y$ and hence we proved that

(1) $r[A, B](y) = y$ for any $y \in D(B)$.

Now take any point $y \in r[A, B](Y)$ and a set $U \in \tau(y, X)$; pick $x \in Y$ such that $y = r[A, B](x)$. There exists a set $V = \prod_{t \in T} V_t$ such that $V_t \in \tau(X_t)$ for all $t \in T$ while $V_t = X_t$ if $t \in T \setminus F$ for some finite set $F \subset T$ and $y \in V \subset U$. Let $K = F \cap B$; the family $\mathcal{E}_t$ being a network in $X_t$, we can find a set $Q_t \in \mathcal{E}_t$ such that $y(t) \in Q_t \subset V_t$ for any $t \in K$. Let $Q_t = X_t$ for all $t \in T \setminus K$ and consider the set $Q = \prod_{t \in T} Q_t$. Since $K \subset B$, it follows from the definition of $r[A, B]$ that $x(t) = y(t) \in Q_t$ for every $t \in K$; therefore $x \in Q \cap Y$ and hence $Q$ is a canonical set. The sequence $\{E_n(A) : n \in \omega\}$ is increasing and $\bigcup_{n \in \omega} E_n(A) = B \supset K$, so there exists a number $n \in \omega$ such that $H(Q) \subset K \subset E_n(A)$. As an immediate consequence, $x_Q \in D(E_n(A)) \subset D(B)$. Since also $x_Q \in Q \subset V \subset U$, we proved that $D(B) \cap U \neq \emptyset$ for any set $U \in \tau(y, X)$ and hence $y \in \overline{D(B)}$ for any point $y \in r[A, B](Y)$. Therefore we can apply the property (1) to see that $r[A, B](D(B)) = D(B) \subset Y$ is dense in $r[A, B](Y)$. Since the set $Y$ is closed in $\Sigma(X, a)$ and we have the inclusion $r[A, B](Y) \subset \Sigma(X, a)$ by Fact 1(ii), we conclude that $r[A, B](Y) \subset \text{cl}_{\Sigma(X,a)}(D(B)) \subset Y$ and hence we proved that $Y$ is a monotonically invariant subset of $\Sigma(X, a)$. Finally apply Fact 2 to conclude that $Y$ is monotonically retractable; this finishes the proof of (c) and makes our solution complete.

**V.481.** *For any space $X$ prove that*

*(a) $X$ is monotonically retractable if and only if $C_p(X)$ is monotonically Sokolov.*
*(b) $X$ is monotonically Sokolov if and only if $C_p(X)$ is monotonically retractable.*

**Solution.** Fix a countable base $\mathcal{B}$ in the space $\mathbb{R}$; given a family $\mathcal{A}$ of subsets of a space $Z$, let $\mathcal{W}(\mathcal{A}) = \{[A_1, \ldots, A_n; B_1, \ldots, B_n] : n \in \mathbb{N}, A_i \in \mathcal{A}$ and $B_i \in \mathcal{B}$ for every $i \leq n\}$; here, as usual, $[A_1, \ldots, A_n; B_1, \ldots, B_n] = \{f \in C_p(Z) : f(A_i) \subset B_i$ for any $i \leq n\}$. If $\varphi : Z \to Z'$ is an onto map, then its dual map $\varphi^* : C_p(Z') \to C_p(Z)$ is defined by the formula $\varphi(f) = f \circ \varphi$ for any $f \in C_p(Z')$. For any space $Z$ denote by $\mathcal{CL}(Z)$ the family of all closed subsets of $Z$ and let $\mathcal{CL}^*(Z) = \bigcup\{\mathcal{CL}(Z^n) : n \in \mathbb{N}\}$.

Recall that a subset $Y$ of a space $Z$ is a retract of $Z$ if there exists a continuous onto map $r : Z \to Y$ (called a retraction) such that $r(z) = z$ for any $z \in Y$.

However, it is equivalent to say that a continuous map $r : Z \to Z$ is a retraction if $r \circ r = r$, so we will often consider that a retraction on $Z$ is a map from $Z$ to $Z$. To make the perception easier for the readers used to another definitions of retraction, we will often say that a retraction is continuous neglecting the fact that in this book, continuity is a part of the definition of retraction.

**Fact 1.** Suppose that $Z$ is a space and $F$ is a retract of $Z$. If $r : Z \to F$ is a continuous retraction, then $s = r^* \circ \pi_F : C_p(Z) \to r^*(C_p(F))$ is also a continuous retraction.

*Proof.* It is clear that the map $s$ is continuous. The set $F$ must be $C$-embedded in $Z$ by Fact 1 of S.398, so $\pi_F(C_p(Z)) = C_p(F)$ and hence $s(C_p(Z)) = r^*(C_p(F))$. Take any function $f \in s(C_p(Z))$; there exists $g \in C_p(F)$ such that $f = g \circ r$. It is easy to see that $f|F = g$ and hence $s(f) = (f|F) \circ r = g \circ r = f$, so $s(f) = f$ for any $f \in r^*(C_p(F))$, i.e., $s$ is a retraction and hence Fact 1 is proved.

**Fact 2.** Suppose that $f : Z \to Y$ is a continuous onto map and $\mathcal{U}$ is a base of $Y$. Assume that, for the family $\mathcal{F} = \{f^{-1}(\overline{U}) : U \in \mathcal{U}\}$, there exists a continuous map $r : Z \to Z$ such that $r(F) \subset F$ for any $F \in \mathcal{F}$. Then $f = (f|r(Z)) \circ r$.

*Proof.* Take any point $x \in Z$. Given a set $U \in \mathcal{U}$ such that $f(x) \in \overline{U}$ observe that $x \in F = f^{-1}(\overline{U}) \in \mathcal{F}$. By our hypothesis we have $r(x) \in F$, so $f(r(x)) \in \overline{U}$. Thus $f(r(x)) \in \overline{U}$ for any $U \in \mathcal{U}$ with $f(x) \in U$ and hence $f(x) = f(r(x))$ for any $x \in X$, i.e., Fact 2 is proved.

**Fact 3.** If a space $Z$ is monotonically retractable, then $C_p(Z)$ is monotonically Sokolov.

*Proof.* Apply Problem 478 to assign to any countable family $\mathcal{G} \subset \mathcal{CL}^*(Z)$ a continuous retraction $t_{\mathcal{G}} : Z \to Z$ and a countable network $\mathcal{P}(\mathcal{G})$ for the map $t_{\mathcal{G}}$ such that, for any $n \in \mathbb{N}$ and $G \in \mathcal{G} \cap \mathcal{CL}(Z^n)$ we have $(t_{\mathcal{G}})^n(G) \subset G$ and the assignment $\mathcal{P}$ is $\omega$-monotone.

Let $\mathcal{F}$ be a countable family of closed subsets of $C_p(Z)$. Given any $F \in \mathcal{F}$, for any $n \in \mathbb{N}$ and $B_1, \ldots, B_n \in \mathcal{B}$ consider the subset

$$G(F, B_1, \ldots, B_n) = \{(x_1, \ldots, x_n) \in Z^n : [x_1, \ldots, x_n; B_1, \ldots, B_n] \cap F = \emptyset\}$$

of the space $Z^n$. Take any $z = (z_1, \ldots, z_n) \in Z^n \backslash G(F, B_1, \ldots, B_n)$. There exists a function $f \in [z_1, \ldots, z_n; B_1, \ldots, B_n] \cap F$; since $W_i = f^{-1}(B_i)$ is an open neighborhood of $z_i$ for any $i \leq n$, the set $W = W_1 \times \ldots \times W_n$ is an open neighborhood of $z$ such that $W \cap G(F, B_1, \ldots, B_n) = \emptyset$. This proves that every set $G(F, B_1, \ldots, B_n)$ is closed in $Z^n$ and therefore $\mathcal{G}(\mathcal{F}) = \{G(F, B_1, \ldots, B_n) : F \in \mathcal{F}, B_1, \ldots, B_n \in \mathcal{B}, \text{and } n \in \mathbb{N}\}$ is a countable subfamily of $\mathcal{CL}^*(Z)$. The assignment $\mathcal{F} \to \mathcal{G}(\mathcal{F})$ is easily seen to be $\omega$-monotone. Consider the set $M(\mathcal{F}) = t_{\mathcal{G}(\mathcal{F})}(Z)$; we will prove that the retraction $r_{\mathcal{F}} = (t_{\mathcal{G}(\mathcal{F})})^* \circ \pi_{M(\mathcal{F})}$ and the family $\mathcal{N}(\mathcal{F}) = \mathcal{W}(\mathcal{P}(\mathcal{G}(\mathcal{F})))$ witness that $C_p(Z)$ is monotonically Sokolov.

By Fact 1 the map $r_{\mathcal{F}} : C_p(Z) \to C_p(Z)$ is a retraction. The family $\mathcal{P}(\mathcal{G}(\mathcal{F}))$ being a network for the map $t_{\mathcal{G}(\mathcal{F})}$, we can apply Fact 2 of V.467 to see that the family $\mathcal{W}(\mathcal{P}(\mathcal{G}(\mathcal{F})))$ is an external network for $(t_{\mathcal{G}(\mathcal{F})})^*(C_p(M(\mathcal{F})))$ in $C_p(Z)$. This is the same as saying that the family $\mathcal{W}(\mathcal{P}(\mathcal{G}(\mathcal{F})))$ is an external network for the set $(t_{\mathcal{G}(\mathcal{F})})^* \circ \pi_{M(\mathcal{F})}(C_p(Z))$ in $C_p(Z)$, i.e., $\mathcal{N}(\mathcal{F})$ is an external network for $r_{\mathcal{F}}(C_p(Z))$ in $C_p(Z)$. Since the assignments $\mathcal{W}$, $\mathcal{P}$ and $\mathcal{G}$ are $\omega$-monotone (see Fact 1 of V. 467), the assignment $\mathcal{N}$ is $\omega$-monotone by Fact 1 of V.477. Now, to finish the proof, we only need to show that $r_{\mathcal{F}}(F) \subset F$ for each $F \in \mathcal{F}$.

Suppose that $r_{\mathcal{F}}(f) \notin F$ for some set $F \in \mathcal{F}$ and $f \in F$. Since $F$ is closed in the space $C_p(Z)$, we can choose $x_1, \ldots, x_n \in Z$ and $B_1, \ldots, B_n \in \mathcal{B}$ such that the set $U = [x_1, \ldots, x_n; B_1, \ldots, B_n]$ contains $r_{\mathcal{F}}(f)$ and $U \cap F = \emptyset$. This shows that $(x_1, \ldots, x_n) \in G(F, B_1, \ldots, B_n) \in \mathcal{G}(\mathcal{F}) \cap \mathcal{CL}(Z^n)$. Since $(t_{\mathcal{G}(\mathcal{F})})^n(G) \subset G$ for any $G \in \mathcal{G}(\mathcal{F}) \cap \mathcal{CL}(Z^n)$, we have $(t_{\mathcal{G}(\mathcal{F})})^n(x_1, \ldots, x_n) \in G(F, B_1, \ldots, B_n)$. As a consequence, $V \cap F = \emptyset$ if $V = [t_{\mathcal{G}(\mathcal{F})}(x_1), \ldots, t_{\mathcal{G}(\mathcal{F})}(x_n); B_1, \ldots, B_n]$. On the other hand, $r_{\mathcal{F}}(f) \in U$ implies $r_{\mathcal{F}}(f)(x_k) = f(t_{\mathcal{G}(\mathcal{F})}(x_k)) \in B_k$ for $k = 1, \ldots, n$, i.e., $f \in V \cap F$; this contradiction shows that $r_{\mathcal{F}}(F) \subset F$ for each $F \in \mathcal{F}$ and hence Fact 3 is proved.

*Fact 4.* If a space $Z$ is monotonically Sokolov, then $C_p(Z)$ is monotonically retractable

*Proof.* We can assign to any countable $\mathcal{F} \subset \mathcal{CL}(Z)$ a retraction $r_{\mathcal{F}} : Z \to Z$ and a countable external network $\mathcal{N}(\mathcal{F})$ for $r_{\mathcal{F}}(Z)$, such that $r_{\mathcal{F}}(F) \subset F$ for every $F \in \mathcal{F}$ and the assignment $\mathcal{N}$ is $\omega$-monotone. For any countable set $E \subset C_p(Z)$ consider the family $\mathcal{F}(E) = \{f^{-1}(\overline{B}) : f \in E$ and $B \in \mathcal{B}\}$. The assignment $E \to \mathcal{F}(E)$ is easily seen to be $\omega$-monotone; denote by $K(E)$ the set $r_{\mathcal{F}(E)}(Z)$. It turns out that the map $s_E = (r_{\mathcal{F}(E)})^* \circ \pi_{K(E)}$ and the family $\mathcal{O}(E) = \mathcal{W}(\mathcal{N}(\mathcal{F}(E)))$ witness that $C_p(Z)$ is monotonically retractable.

It follows from Fact 1 that $s_E = (r_{\mathcal{F}(E)})^* \circ \pi_{K(E)} : C_p(Z) \to C_p(Z)$ is a retraction. Since the family $\mathcal{N}(\mathcal{F}(E))$ is an external network for $K(E)$ in $Z$, Fact 3 of V.467 shows that the family $\mathcal{W}(\mathcal{N}(\mathcal{F}(E)))$ is a network for the map $\pi_{K(E)}$. It follows from continuity of $(r_{\mathcal{F}(E)})^*$ that $\mathcal{W}(\mathcal{N}(\mathcal{F}(E)))$ is a network of $(r_{\mathcal{F}(E)})^* \circ \pi_{K(E)}$, i.e., the family $\mathcal{O}(E)$ is a network for the map $s_E$. All assignments $\mathcal{W}$, $\mathcal{N}$, and $\mathcal{F}$ being $\omega$-monotone, the assignment $\mathcal{O}$ is $\omega$-monotone as well.

To convince ourselves that $E \subset s_E(C_p(Z))$ take any $f \in E$ and observe that $\{f^{-1}(\overline{B}) : B \in \mathcal{B}\} \subset \mathcal{F}(E)$. By the choice of $r_{\mathcal{F}(E)}$, we have $r_{\mathcal{F}(E)}(F) \subset F$ for any $F \in \mathcal{F}(E)$. Therefore we can apply Fact 2 to conclude that

$$f = (f \mid K(E)) \circ r_{\mathcal{F}(E)} = (r_{\mathcal{F}(E)})^* \circ \pi_{K(E)}(f) = s_E(f) \in s_E(C_p(Z)),$$

i.e., the space $C_p(Z)$ is monotonically retractable and hence Fact 4 is proved.

Returning to our solution observe that if $X$ is a monotonically retractable space, then $C_p(X)$ is monotonically Sokolov by Fact 3. Now, if $C_p(X)$ is monotonically Sokolov, then $C_p(C_p(X))$ is monotonically retractable by Fact 4. Being embeddable in $C_p(C_p(X))$ as a closed subspace (see TFS-167), the space $X$ must be monotonically retractable by Problem 479, so we proved (a).

If $X$ is monotonically Sokolov, then $C_p(X)$ is monotonically retractable by Fact 4. If, on the other hand, the space $C_p(X)$ is monotonically retractable, then $C_p(C_p(X))$ is monotonically Sokolov by Fact 3. By TFS-167 the space $X$ embeds in $C_p(C_p(X))$ as a closed subspace; applying Problem 476 we can see that $X$ is monotonically Sokolov. This settles (b) and makes our solution complete.

**V.482.** *Prove that*

(a) *any monotonically Sokolov space is Lindelöf and Sokolov;*

(b) *any countable product of monotonically Sokolov spaces is monotonically Sokolov;*

(c) *every $\mathbb{R}$-quotient image of a monotonically Sokolov space is monotonically Sokolov;*

(d) *every $\mathbb{R}$-quotient image of a monotonically retractable space $X$ must be monotonically retractable;*

(e) *A monotonically retractable space must be $\omega$-monolithic but it can fail to be monotonically $\omega$-monolithic.*

**Solution.** (a) If $X$ is monotonically Sokolov, then $C_p(X)$ is monotonically retractable by Problem 481, so it is Sokolov by Problem 479. Therefore $X$ is a Sokolov space by CFS-156. Besides, $C_p(C_p(X))$ must be Lindelöf by Problem 479, so $X$ is Lindelöf being homeomorphic to a closed subset of $C_p(C_p(X))$.

(b) We will first prove that $X^\omega$ is monotonically Sokolov if $X$ is monotonically Sokolov; it is clear that we can assume that $X$ is infinite. In this case the space $C_p(X)$ is monotonically retractable and hence $C_p(C_p(X))$ is monotonically Sokolov by Problem 481. It follows from TFS-177 that the space $Y = C_p(C_p(X))$ is homeomorphic to $Y^\omega$. The space $X$ is homeomorphic to a closed subspace of $Y$ and therefore $X^\omega$ is also homeomorphic to a closed subspace of $Y \simeq Y^\omega$, so it follows from Problem 476 that $X^\omega$ is monotonically Sokolov.

Now assume that $X_i$ is monotonically Sokolov for any $i \in \omega$. Then the space $X = \bigoplus\{X_i : i \in \omega\}$ is also monotonically Sokolov because every $C_p(X_i)$ is monotonically retractable by Problem 481 and hence $C_p(X) \simeq \prod\{C_p(X_i) : i \in \omega\}$ is also monotonically retractable by Problem 480. Thus $X^\omega$ is monotonically Sokolov and hence so is $\prod\{X_i : i \in \omega\}$ being homeomorphic to a closed subspace of $X^\omega$.

(c) If $Y$ is an $\mathbb{R}$-quotient image of a monotonically Sokolov space $X$ then $C_p(Y)$ embeds in $C_p(X)$ as a closed subspace by TFS-163. The space $C_p(X)$ is monotonically retractable by Problem 481 and hence so is $C_p(Y)$ by Problem 479. Applying Problem 481 once more we conclude that $Y$ is monotonically Sokolov.

(d) If $Y$ is an $\mathbb{R}$-quotient image of a monotonically retractable space $X$ then $C_p(Y)$ embeds in $C_p(X)$ as a closed subspace by TFS-163. The space $C_p(X)$ is monotonically Sokolov by Problem 481 and hence so is $C_p(Y)$ by Problem 476. Applying Problem 481 again we conclude that $Y$ is monotonically retractable.

(e) If $X$ is monotonically retractable, then it is Sokolov by Problem 479. Every
Sokolov space is $\omega$-monolithic by CFS-163 so $X$ must be $\omega$-monolithic. Now,
take a Corson compact space $K$ which is not monotonically $\omega$-monolithic.
Such a space exists by Problem 474. Since $K$ embeds in a $\Sigma$-product of real
lines, it is monotonically retractable by Problem 480. Therefore $K$ is a compact
monotonically retractable space which is not monotonically $\omega$-monolithic.

**V.483.** *Prove that a compact space $X$ is monotonically retractable if and only if $X$
is Corson compact.*

**Solution.** For any space $Z$ we denote by $\mathcal{CL}(Z)$ the family of all closed subsets of
$Z$. Recall that the Gruenhage's $W$-game is played on a nonempty closed subset $F$
of a space $Z$ as follows: at the $n$-th move the first player, called $O$, chooses an open
set $U_n \in \tau(F, Z)$ and the player $P$ responds by taking a point $x_n \in U_n$. After $\omega$-
many moves have been made, the player $O$ wins if the sequence $S = \{x_n : n \in \omega\}$
converges to $F$, i.e., the set $\{n \in \omega : x_n \notin V\}$ is finite for any $V \in \tau(F, Z)$;
otherwise $P$ is the winner.

*Fact 1.* If $Z$ is a compact monotonically retractable space and $F \neq \emptyset$ is a closed
subset of $Z$, then the player $O$ has a winning strategy in the $W$-game on $F$.

*Proof.* Apply Problem 478 to assign to any countable family $\mathcal{F} \subset \mathcal{CL}(Z)$ a
continuous retraction $r_{\mathcal{F}} : Z \to Z$ and a network $\mathcal{N}(\mathcal{F})$ for the map $r_{\mathcal{F}}$ such
that $r_{\mathcal{F}}(G) \subset G$ for any $G \in \mathcal{F}$ and the assignment $\mathcal{N}$ is $\omega$-monotone. Observe
first that the family $\mathcal{N}'(\mathcal{F}) = \{\overline{P} : P \in \mathcal{N}(\mathcal{F})\}$ is still a network for the map $r_{\mathcal{F}}$
and the assignment $\mathcal{N}'$ is also $\omega$-monotone so, passing to $\mathcal{N}'$ if necessary, we can
consider, without loss of generality, that every element of $\mathcal{N}(\mathcal{F})$ is closed in $Z$.

To start defining a strategy $s$ for the player $O$ in the $W$-game on $F$ let $s(\emptyset) = Z$
and $\mathcal{F}_0 = \{F\}$. Proceeding by induction assume that $n \in \omega$ and we have moves
$U_0, a_0, \ldots, U_{n-1}, a_{n-1}, U_n$ in the $W$-game on $F$ and countable families $\mathcal{F}_0, \ldots, \mathcal{F}_n$
of closed subsets of $Z$ with the following properties:

(1) an enumeration $\{Q_{i,k} : k \in \omega\}$ is chosen for the family $\mathcal{F}_i$ for each $i \leq n$;
(2) $U_0 = s(\emptyset) = Z$ and $a_i \in U_i$ for all $i < n$;
(3) $\overline{U}_{i+1} \subset U_i$ for all $i < n$;
(4) $\mathcal{F}_{i+1} = \mathcal{F}_i \cup \mathcal{N}(\mathcal{F}_i \cup \{\{a_0\}, \ldots, \{a_i\}\})$ whenever $i < n$;
(5) if $\max(j, k) \leq i < n$ and $Q_{j,k} \cap F = \emptyset$, then $\overline{U}_{i+1} \cap Q_{j,k} = \emptyset$.

After the player $P$ picks a point $a_n \in U_n$, define a family $\mathcal{F}_{n+1}$ by the formula
$\mathcal{F}_{n+1} = \mathcal{F}_n \cup \mathcal{N}(\mathcal{F}_n \cup \{\{a_0\}, \ldots, \{a_n\}\})$ and choose an enumeration $\{Q_{n+1,k} :
k \in \omega\}$ of $\mathcal{F}_{n+1}$. The family $\mathcal{G} = \{Q_{j,k} : \max(j, k) \leq n$ and $Q_{j,k} \cap F = \emptyset\}$ is
finite, so the set $\bigcup \mathcal{G}$ is closed and does not meet $F$. By normality of the space $Z$
we can find a set $U_{n+1} \in \tau(F, Z)$ such that $\overline{U}_{n+1} \subset U_n$ and $\overline{U}_{n+1} \cap (\bigcup \mathcal{G}) = \emptyset$.
Let $s(a_0, \ldots, a_n) = U_{n+1}$ and observe that the conditions (1)–(5) still hold if we
replace $n$ with $n + 1$, so we completed the definition of a strategy $s$ such that (1)–(5)
hold for all $n \in \omega$ whenever the player $O$ applies $s$.

Suppose that $\mathcal{P} = \{U_n, a_n : n \in \omega\}$ is a play on $F$ where $O$ applies $s$ and hence
we have a collection $\{\mathcal{F}_n : n \in \omega\}$ of families with the properties (1)–(5). To see

that the $O$ wins in the play $\mathcal{P}$ assume that the sequence $S = \{a_n : n \in \omega\}$ does not converge to $F$; by compactness of $Z$ there exists a cluster point $p \notin F$ for the sequence $S$. Observe that the family $\{\{a_n\} : n \in \omega\}$ is contained in $\mathcal{F} = \bigcup_{n \in \omega} \mathcal{F}_n$ by (4), and hence $r_{\mathcal{F}}(a_n) = a_n$ for each $n \in \omega$. It follows from $p \in \overline{S}$ and continuity of the retraction $r_{\mathcal{F}}$ that $r_{\mathcal{F}}(p) = p \in Z \backslash F$. Since $\mathcal{N}(\mathcal{F})$ is a network for $r_{\mathcal{F}}$, there exists $N \in \mathcal{N}(\mathcal{F})$ such that $p \in N$ and $r_{\mathcal{F}}(N) \cap F = \emptyset$. The operator $\mathcal{N}$ being $\omega$-monotone, we can find $n \in \omega$ such that $N \in \mathcal{F}_n$ and hence $N = Q_{n,k}$ for some $k \in \omega$. Besides, $N \in \mathcal{F}_{n+1}$ by (4) and hence $N \in \mathcal{F}$ which implies that $r_{\mathcal{F}}(N) \subset N$. Since also $F \in \mathcal{F}$, we have $r_{\mathcal{F}}(F) \subset F$; if $x \in F \cap N$, then $r_{\mathcal{F}}(x) \in r_{\mathcal{F}}(N) \cap F = \emptyset$ which is a contradiction. Therefore $N \cap F = \emptyset$.

Now, if we consider the number $m = \max(n, k) + 1$, then (5) can be applied to the set $N = Q_{n,k}$ to see that $N \cap \overline{U}_m = \emptyset$. By the properties (2) and (3) we have $\{a_i : i \geq m\} \subset U_m$ and therefore $G = (Z \backslash \overline{U}_m)$ is an open neighborhood of the point $p \in N$ such that $G \cap S \subset \{a_0, \ldots, a_{m-1}\}$ which implies that $p$ is not a cluster point of $S$; this contradiction shows that the sequence $S$ converges to $F$, so $s$ is a winning strategy for the player $O$ and hence Fact 1 is proved.

Returning to our solution observe that if $X$ is Corson compact, then it embeds in a $\Sigma$-product of real lines, so it is monotonically retractable by Problem 480(c); this proves sufficiency.

Finally, assume that $X$ is a compact monotonically retractable space. By Problem 480(b) the space $X \times X$ is also monotonically retractable, so we can apply Fact 1 to conclude that for every nonempty closed subset $F$ of $X \times X$, the player $O$ has a winning strategy in the Gruenhage's $W$-game on $F$. Therefore $X$ is Corson compact by Problem CFS-188 and hence our solution is complete.

**V.484.** *Prove that a monotonically retractable space $X$ is monotonically Sokolov if and only if $X$ is monotonically $\omega$-monolithic. In particular, a compact space is monotonically Sokolov if and only if it is monotonically $\omega$-monolithic.*

**Solution.** For any space $Z$ we denote by $\mathcal{CL}(Z)$ the family of all closed subsets of $Z$. Any monotonically Sokolov space is monotonically $\omega$-monolithic by Problem 476, so necessity holds trivially.

To prove sufficiency assume that $X$ is a monotonically retractable monotonically $\omega$-monolithic space and apply Problem 478 to assign to any countable family $\mathcal{F}$ of closed subsets of $X$ a continuous retraction $r_{\mathcal{F}} : X \to X$ and a network $\mathcal{N}(\mathcal{F})$ for the map $r_{\mathcal{F}}$ such that $r_{\mathcal{F}}(G) \subset G$ for any $G \in \mathcal{F}$ and the assignment $\mathcal{N}$ is $\omega$-monotone. Observe first that the family $\mathcal{N}'(\mathcal{F}) = \{\overline{P} : P \in \mathcal{N}(\mathcal{F})\}$ is still a network for the map $r_{\mathcal{F}}$ and the assignment $\mathcal{N}'$ is also $\omega$-monotone so, passing to $\mathcal{N}'$ if necessary, we can consider, without loss of generality, that every element of $\mathcal{N}(\mathcal{F})$ is closed in $X$. By monotone $\omega$-monolithity of $X$, to any countable set $A \subset X$ we can assign an external network $\mathcal{M}(A)$ for the set $\overline{A}$ in such a way that the operator $\mathcal{M}$ is $\omega$-monotone. For any nonempty closed subset $F$ of the space $X$ fix a point $x_F \in F$.

Given a countable $\mathcal{G} \subset \mathcal{CL}(X)$ consider the family $\mathcal{P}(\mathcal{G}) = \{\{x_G\} : G \in \mathcal{G}\setminus\{\emptyset\}\}$; it is straightforward that the operator $\mathcal{P}$ is $\omega$-monotone. For any countable family $\mathcal{F} \subset \mathcal{CL}(X)$ let $\mathcal{O}_0(\mathcal{F}) = \mathcal{F} \cup \mathcal{P}(\mathcal{F})$ and

(1) $\mathcal{O}_{n+1}(\mathcal{F}) = \mathcal{O}_n(\mathcal{F}) \cup \mathcal{P}(\mathcal{O}_n(\mathcal{F})) \cup \mathcal{N}(\mathcal{O}_n(\mathcal{F}) \cup \mathcal{P}(\mathcal{O}_n(\mathcal{F})))$

for any $n \in \omega$. It takes a trivial induction together with Fact 1(a) of V.477 to see that every operator $\mathcal{O}_n$ is $\omega$-monotone. Let $\mathcal{O}(\mathcal{F}) = \bigcup_{n\in\omega} \mathcal{O}_n(\mathcal{F})$ and apply Fact 1(b) of V.477 to see that the operator $\mathcal{O}$ is also $\omega$-monotone.

Letting $M(\mathcal{F}) = \{x_F : F \in \mathcal{O}(\mathcal{F})\setminus\{\emptyset\}\}$ we also obtain an $\omega$-monotone operator which assigns a countable set to any countable family of closed subsets of $X$. Given a nonempty set $F \in \mathcal{O}(\mathcal{F})$, there exists a number $n \in \omega$ such that $F \in \mathcal{O}_n(\mathcal{F})$ and hence we have the inclusions $\{x_F\} \in \mathcal{O}_{n+1}(\mathcal{F}) \subset \mathcal{O}(\mathcal{F})$. This proves that

(2) $\{x_F\} \in \mathcal{O}(\mathcal{F})$ and hence $r_{\mathcal{O}(\mathcal{F})}(x_F) = x_F$ for any $F \in \mathcal{O}(\mathcal{F})\setminus\{\emptyset\}$.

Now, if $F \in \mathcal{N}(\mathcal{O}(\mathcal{F}))$ then $F \in \mathcal{N}(\mathcal{O}_n(\mathcal{F}))$ by $\omega$-monotonicity of $\mathcal{N}$. Then $F \in \mathcal{O}_{n+1}(\mathcal{F}) \subset \mathcal{O}(\mathcal{F})$ which implies that

(3) $\mathcal{N}(\mathcal{O}(\mathcal{F})) \subset \mathcal{O}(\mathcal{F})$ and hence $M(\mathcal{F})$ is dense in $r_{\mathcal{O}(\mathcal{F})}(X)$.

Indeed, we must only prove the second part of the statement (3), so observe that the family $\mathcal{N}(\mathcal{O}(\mathcal{F}))$ is a network for the map $r_{\mathcal{O}(\mathcal{F})}$ and hence for any set $Q \subset X$ such that $Q \cap N \neq \emptyset$ for any nonempty set $N \in \mathcal{N}(\mathcal{O}(\mathcal{F}))$, the set $r_{\mathcal{O}(\mathcal{F})}(Q)$ is dense in $r_{\mathcal{O}(\mathcal{F})}(X)$. Now it follows from (2) and (3) that $M(\mathcal{F})$ is dense in $r_{\mathcal{O}(\mathcal{F})}(X)$. Therefore the family $\mathcal{E}(\mathcal{F}) = \mathcal{M}(M(\mathcal{F}))$ is an external network of the set $r_{\mathcal{O}(\mathcal{F})}(X)$ and it follows from $\mathcal{F} \subset \mathcal{O}(\mathcal{F})$ that $r_{\mathcal{O}(\mathcal{F})}(F) \subset F$ for any $F \in \mathcal{F}$. This shows that the $\omega$-monotone operator $\mathcal{E}$ and the retraction $s_{\mathcal{F}} = r_{\mathcal{O}(\mathcal{F})}$ witness the monotone Sokolov property of $X$, so we proved sufficiency.

Finally observe that if $X$ is an $\omega$-monolithic compact space, then $X$ is Corson compact by Problem 473. Therefore $X$ is monotonically retractable by Problem 483 and hence $X$ is monotonically Sokolov, i.e., our solution is complete.

**V.485.** *Suppose that $X$ is a first countable countably compact subspace of an ordinal with its order topology. Prove that $X$ is monotonically retractable and hence $C_p(X)$ is a Lindelöf D-space.*

**Solution.** Suppose that $\alpha$ is an ordinal and $Y$ is a nonempty subset of $\alpha$. Say that a set $A \subset Y$ is *saturated in* $Y$ if $\setminus nY \in A$ and every isolated point of $A$ is also isolated in $Y$. If $Y$ is clear, the set $A$ will be called *saturated*.

**Fact 1.** Suppose that $\alpha$ is an ordinal and $Y \subset \alpha$. If a set $A \subset Y$ is saturated in $Y$, then $B = \overline{A}$ is also saturated in $Y$.

*Proof.* It follows from $\setminus n(Y) \in A \subset B$ that $\setminus n(Y) \in B$. Now, if $y \in B$ is isolated in $B$, then $\{y\}$ is open in $B$, so density of $A$ in $B$ implies that $\{y\} \cap A \neq \emptyset$, i.e., $y \in A$ and hence $y$ is isolated in $A$. The set $A$ being saturated, the point $y$ must be isolated in $Y$, so Fact 1 is proved.

**Fact 2.** Suppose that $\alpha$ is an ordinal and $A \subset \alpha$ is countable. Then $\overline{A}$ is also countable.

*Proof.* If $\overline{A}$ is uncountable, then the set $P = \overline{A}\backslash A$ is also uncountable; consider the ordinal $\alpha_0 = \backslash n P$. Proceeding by induction assume that $\gamma < \omega_1$ and we have chosen ordinals $\{\alpha_\beta : \beta < \gamma\} \subset P$ such that

(1) $\beta < \beta' < \gamma$ implies $\alpha_\beta < \alpha_{\beta'}$ and the set $P \cap \alpha_\beta$ is countable for any $\beta < \gamma$.

If $\mu = \sup\{\alpha_\beta : \beta < \gamma\}$ then $P \cap (\mu + 1)$ is also countable so, for the ordinal $\alpha_\gamma = \backslash n(P\backslash(\mu + 1))$, the property (1) still holds for all $\beta \leq \gamma$. Therefore our construction can be continued to construct a set $\{\alpha_\beta : \beta < \omega_1\} \subset P$ for which the condition (1) is satisfied for all $\gamma < \omega_1$. The set $O_\beta = (\alpha_\beta, \alpha_{\beta+1}]$ is open in $\alpha$ for any $\beta < \omega_1$ and the family $\{O_\beta : \beta < \omega_1\}$ is disjoint. It follows from $\alpha_{\beta+1} \in \overline{A}$ that $O_\beta \cap A \neq \emptyset$ for any $\beta < \omega_1$, so the set $A$ cannot be countable; this contradiction shows that Fact 2 is proved.

*Fact 3.* Suppose that $\alpha$ is an ordinal and $Y$ is a countably compact subset of $\alpha$. Then $Y$ is $\omega$-closed in $\alpha$, i.e., $\overline{A} \subset Y$ for any countable $A \subset Y$ (the bar denotes the closure in $\alpha$). As a consequence, $\mathrm{cl}_Y(A)$ is compact and countable for any countable set $A \subset Y$.

*Proof.* The set $B = \overline{A}$ is countable by Fact 2; since $B$ is the closure of a countably compact set $B \cap Y$, it has to be pseudocompact, so $B$ is second countable and hence compact. If $x \in B\backslash Y$ then it follows from $x \in \overline{A}\backslash A$ that there exists a nontrivial sequence $S \subset A$ which converges to $x$. It is straightforward that $S$ is an infinite closed discrete subset of $Y$; this contradiction with countable compactness of $Y$ shows that $B \subset Y$, i.e., $\mathrm{cl}_Y(A) = \overline{A}$ is a countable compact subset of $Y$ and hence Fact 3 is proved.

*Fact 4.* Suppose that $\alpha$ is an ordinal, $Y$ is a countably compact subset of $\alpha$ and $F$ is a compact saturated subset of $Y$. For every $x \in Y$ let $r_F(x) = \max\{y \in F : y \leq x\}$. Then $r_F : Y \to F$ is a continuous retraction.

*Proof.* Observe first that $r_F(x)$ is consistently defined for every $x \in Y$ because the set $\{y \in F : y \leq x\}$ is nonempty, closed in $F$ and hence compact. It is clear from the definition that $r_F(x) = x$ for any $x \in F$, so we only need to prove that the map $r_F$ is continuous.

To do that, fix a point $x \in Y$ and an open neighborhood $U$ of the point $y = r_F(x)$ in $F$. Assume first that $x \notin F$; then $y < x$ and, for the set $V = (y, x] \cap Y$, we have $r_F(V) = \{y\} \subset U$, so the set $V \in \tau(x, Y)$ witnesses that $r_F$ is continuous at the point $x$. Now, if $x \in F$ then $y = x$. If $x$ is isolated in $F$ then it is also isolated in $Y$, so the map $r_F$ is trivially continuous at $x$. If $x$ is not isolated in $F$ then there exists $a < y$ such that $(a, y] \cap F \subset U$ and hence we can find a point $z \in (a, x) \cap F$. The set $V = (z, x] \cap Y$ is an open neighborhood of $x$ in $Y$ and $r_F(V) \subset [z, x] \cap F \subset U$, so again $r_F$ is continuous at $x$ and hence Fact 4 is proved.

Returning to our solution fix an ordinal $\mu$ and a first countable countably compact subspace $X \subset \mu$; there is no loss of generality to assume that $X$ is infinite. From now on, any interval is considered only for the points of $X$; in particular,

$[x, y] = \{z \in X : x \le z \le y\}$ while $[x, y) = \{z \in X : x \le z < y\}$ and $[x, \rightarrow) = \{z \in X : x \le z\}$ whenever $x, y \in X$ and $x < y$. Let $I(X)$ be the set of isolated points of $X$ and $z_0 = \backslash n(X)$.

For any countable compact saturated set $K \subset X$, the map $r_K : X \rightarrow K$ is the retraction defined in Fact 4. If $x \in X \backslash I(X)$, then fix a set $S_x \subset I(X)$ such that $\sup(S_x) = x$ and $S_x \backslash (z, x]$ is finite for any $z < x$. In other words, $S_x \subset (\leftarrow, x)$ is a sequence of isolated points of $X$ that converges to $x$. Given a countable set $A \subset X$, let $E(A) = \{z_0\} \cup A \cup \bigcup \{S_x : x \in A \backslash I(X)\}$. It turns out that

(2) The set $E(A)$ is saturated for any countable $A \subset X$.

To prove the property (2) we must only check that every isolated point of the set $E(A)$ is isolated in $X$, so take any $y \in E(A)$ which is isolated in $E(A)$. If $y = z_0$ or $y \in \bigcup \{S_x : x \in A \backslash I(X)\} \subset I(X)$, then $y$ is certainly isolated in $X$. The same is true if $y \in A \cap I(X)$. Since it follows from $x \in \overline{S_x \backslash \{x\}}$ for every $x \in A \backslash I(X)$, that the points of $A \backslash I(X)$ are not isolated in $E(A)$, we have no other possibilities for $y$ and hence (2) is proved.

We leave it to the reader to verify that

(3) the assignment $A \rightarrow E(A)$ is $\omega$-monotone.

For any countable set $A \subset X$ the family

$$\mathcal{O}(A) = \{\{x\} : x \in A\} \cup \{[x, \rightarrow) : x \in A\} \cup \{[x, y) : x, y \in A \text{ and } x < y\}$$

is countable and it is again easy to check that the assignment $\mathcal{O}$ is $\omega$-monotone. This shows that letting $\mathcal{N}(A) = \mathcal{O}(E(A))$ for any countable set $A \subset X$, we obtain an $\omega$-monotone assignment $A \rightarrow \mathcal{N}(A)$ by Fact 1(a) of V.477. The set $K(A) = \overline{E(A)}$ is compact and saturated by the property (2), Fact 1 and Fact 3, so $r_{K(A)} : X \rightarrow K(A)$ is a continuous retraction by Fact 4. Since $A \subset K(A)$, it suffices to show that

(4) the family $\mathcal{N}(A)$ is a network for the map $r_{K(A)}$ for any countable set $A \subset X$.

To prove (4) take any point $x \in X$ and a set $U \in \tau(r_{K(A)}(x), X)$. By density of $E(A)$ in $K(A)$, there exists a point $y \in E(A)$ such that $[y, r_{K(A)}(x)] \subset U$.

*Case 1.* We have $z \le x$ for any $z \in E(A)$. Then $x \in Q = [y, \rightarrow) \in \mathcal{N}(A)$ and the inclusion $r_{K(A)}(Q) \subset [y, r_{K(A)}(x)] \subset U$ shows that $Q \in \mathcal{N}(A)$ witnesses that $\mathcal{N}(A)$ is a network for $r_{K(A)}$.

*Case 2.* There exists $z \in E(A)$ such that $x < z$. Then the point $u = \backslash n\{t \in E(A) : x < t\}$ belongs to $E(A)$ and $x \in Q = [y, u) \in \mathcal{N}(A)$; it is easy to check that again $r_{K(A)}(Q) \subset [y, r_{K(A)}(x)] \subset U$; this finishes the proof of (4) and shows that our space $X$ is indeed, monotonically retractable.

Finally, observe that $C_p(X)$ must be a Lindelöf $D$-space by Problem 479, so our solution is complete.

**V.486.** *Prove that*

(a) *the set $E$ of all ordinals $\alpha < \omega_2$ of uncountable cofinality is monotonically $\omega$-monolithic but is neither a $D$-space nor monotonically $\omega_1$-monolithic;*

*(b) for set $Y$ of all ordinals $\alpha < \omega_2$ of countable cofinality, the space $X = Y \cup \{\omega_2\}$ is countably compact and $ext(C_p(X)) = \omega$ while $l(C_p(X)) = \omega_2$. In particular, $X$ is a countably compact space such that $C_p(X)$ is not a D-space.*

**Solution.** (a) It is easy to see that if $A \subset E$ and $|A| = \omega_1$ then $A$ has an accumulation point; as a consequence, $ext(X) \leq \omega$.

*Fact 1.* If $Z$ is a space and every subspace of $Z$ of cardinality $\leq \kappa$ is closed in $Z$ then $Z$ is monotonically $\kappa$-monolithic.

*Proof.* If $Z'$ is the set $Z$ with the discrete topology and $f : Z' \to Z$ is the identity map then $f|A$ is homeomorphism for any $A \subset Z'$ with $|A| \leq \kappa$. The space $Z'$ being monotonically monolithic, Fact 1 of V.469 guarantees that $Z$ is monotonically $\kappa$-monolithic, so Fact 1 is proved.

Returning to our solution observe that every countable subset of $E$ is closed in $E$, so $E$ is monotonically $\omega$-monolithic by Fact 1. The family $\{\alpha \cap E : \alpha < \omega_2\}$ is an open cover of $E$ which has no countable subcover, so $E$ is not Lindelöf; applying Problem 470 we conclude that $E$ is not monotonically $\omega_1$-monolithic.

Finally let $N(\alpha) = \{\beta \in E : \beta \leq \alpha\}$ for any $\alpha \in E$; then $N$ is a neighborhood assignment on $E$. If $D$ is a closed discrete subset of $E$ then $|D| \leq \omega$ and hence the set $Z = N(D)$ is not cofinal in $\omega_2$. The set $E$ being cofinal in $\omega_2$, we can find $\alpha \in E$ such that $\alpha \notin Z$, so $N(D) \neq E$ for any closed discrete $D \subset E$ which shows that $E$ is not a D-space.

(b) If $A \subset Y$ is a countably infinite set then we can choose a strictly increasing sequence $\{\alpha_n : n \in \omega\} \subset A$. The ordinal $\alpha = \sup\{\alpha_n : n \in \omega\}$ is countably cofinal and hence $\alpha \in Y$; it is easy to check that $\alpha$ is an accumulation point of $A$. This shows that $X$ is a countably compact space.

By the definition of the order topology on $\omega_2 + 1$, the point $\omega_2$ cannot belong to the closure of any subset of $Y$ of cardinality less than $\omega_2$. Therefore $t(X) \geq \omega_2$ and hence $l(C_p(X)) \geq \omega_2$ by TFS-189; besides, $l(C_p(X)) \leq nw(C_p(X)) = nw(X) = \omega_2$, so $l(C_p(X)) = \omega_2$.

*Fact 2.* For any $f \in C_p(Y)$ there exists $\alpha \in Y$ such that $f(\beta) = f(\alpha)$ for all $\beta \in Y$ with $\beta \geq \alpha$.

*Proof.* Take any continuous function $f : Y \to \mathbb{R}$. Assume that, for any $\alpha \in Y$, there is an ordinal $\beta = \beta(\alpha) > \alpha$ such that $\beta \in Y$ and $f(\beta) \neq f(\alpha)$; let us also fix $n = n(\alpha) \in \mathbb{N}$ with $|f(\beta) - f(\alpha)| \geq \frac{1}{n}$. There exists $m \in \mathbb{N}$ and a set $A \subset Y$ such that $|A| = \omega_2$ and $n(\alpha) = m$ for each $\alpha \in A$. Take any $\alpha_0 \in A$ and $\beta_0 = \beta(\alpha_0)$; if we have $\alpha_i, \beta_i$ for all $i \leq n$, find $\alpha_{n+1} \in A$ with $\alpha_{n+1} > \max\{\alpha_i, \beta_i : i \leq n\}$ and let $\beta_{n+1} = \beta(\alpha_{n+1})$. This inductive construction gives us sequences $\{\alpha_i : i \in \omega\}$ and $\{\beta_i : i \in \omega\}$ such that $\beta_i = \beta(\alpha_i)$ and $\beta_i < \alpha_{i+1}$ for all $i \in \omega$.

The ordinal $\alpha = \sup\{\alpha_i : i \in \omega\}$ belongs to $Y$; the function $f$ being continuous at the point $\alpha$, there is $\beta < \alpha$ such that $|f(\gamma) - f(\alpha)| < \frac{1}{2m}$ for all $\gamma \in (\beta, \alpha) \cap Y$. Choose $n \in \omega$ with $\beta < \alpha_n < \beta_n$; then

$$|f(\alpha_n) - f(\beta_n)| \leq |f(\alpha_n) - f(\alpha)| + |f(\beta_n) - f(\alpha)| < \frac{1}{2m} + \frac{1}{2m} = \frac{1}{m}.$$

This contradiction with $|f(\alpha_n) - f(\beta_n)| \geq \frac{1}{m}$ proves that, for some $\alpha \in Y$, there will be no $\beta > \alpha$, $\beta \in Y$ with $f(\beta) \neq f(\alpha)$, i.e., $f(\beta) = f(\alpha)$ for all $\beta \geq \alpha$, $\beta \in Y$ and hence Fact 2 is proved.

Returning to our solution let $I = \{f \in C_p(X) : f(\omega_2) = 0\}$; then $C_p(X) \simeq I \times \mathbb{R}$, so it suffices to show that $ext(I) \leq \omega$ (see Fact 1 of V.242). To do it, fix any $\alpha \in Y$ and consider the set $F_\alpha = \{y \in X : y > \alpha\}$. Observe that both sets $F_\alpha$ and $Y_\alpha = X \setminus F_\alpha$ are clopen in $X$ and let $P_\alpha = \{f \in C_p(X) : f(F_\alpha) = \{0\}\}$; if $\pi_\alpha : C_p(X) \to C_p(Y_\alpha)$ is the restriction map then it is easy to see that $p_\alpha = \pi_\alpha | P_\alpha : P_\alpha \to C_p(Y_\alpha)$ is an embedding. It is evident that $\pi_\alpha(P_\alpha) = C_p(Y_\alpha)$, so $P_\alpha$ is homeomorphic to $C_p(Y_\alpha)$. Since $Y_\alpha \subset \omega_2$ is a countably compact first countable subset of $\omega_2$, we can apply Problem 485 to convince ourselves that $C_p(Y_\alpha)$ is Lindelöf. This proves that

(1) the space $P_\alpha$ is Lindelöf for any $\alpha \in Y$.

Finally, assume that a set $D \subset I$ is closed, discrete and $|D| = \omega_1$. By Fact 2, for any $f \in D$ there exists $\alpha_f \in Y$ such that $f(\beta) = f(\alpha_f) = 0$ for any $\beta \in Y$ with $\beta \geq \alpha_f$. The set $Y$ is cofinal in $\omega_2$, so there exists an ordinal $\alpha \in Y$ such that $\alpha_f < \alpha$ for all $f \in D$. As a consequence, $D \subset P_\alpha$ and hence $D$ is an uncountable closed discrete subset of $P_\alpha$; this contradiction with the property (1) shows that $ext(C_p(X)) = ext(I) = \omega$. Finally observe that $ext(Z) = l(Z)$ for any $D$-space $Z$, so it follows from $ext(C_p(X)) = \omega < l(C_p(X))$ that $C_p(X)$ is not a $D$-space.

**V.487.** *Suppose that $X$ and $C_p(X)$ are Lindelöf $\Sigma$-spaces. Prove that both $X$ and $C_p(X)$ must be monotonically retractable and monotonically Sokolov. In particular, $X$ and $C_p(X)$ have the Sokolov property*

**Solution.** Let $\mathcal{B}$ be a fixed countable base in $\mathbb{R}$. If $Z$ is a space and $x_1, \ldots, x_k \in Z$ then as usual, $[x_1, \ldots, x_k; B_1, \ldots, B_k] = \{f \in C_p(Z) : f(x_i) \in B_i$ for all $i = 1, \ldots, k\}$ whenever $B_1, \ldots, B_k \in \mathcal{B}$. If $U = [x_1, \ldots, x_k; B_1, \ldots, B_k]$, then $supp(U) = \{x_1, \ldots, x_k\}$.

If $r : Z \to Y$ is a continuous map then it dual map $r^* : C_p(Y) \to C_p(Z)$ is defined by the equality $r^*(f) = f \circ r$ for any $f \in C_p(Y)$. Given a space $Z$ recall that $\pi_M : C_p(Z) \to C_p(M)$ is the restriction map for any $M \subset Z$; if $L \subset C_p(Z)$, then $e_L : Z \to C_p(L)$ is the reflection map defined by $e_L(x)(f) = f(x)$ for any $f \in L$. If we have a fixed set $Y \subset C_p(Z)$ suppose that $P \subset Z$ and $Q \subset Y$; we will say that sets $M \subset P$ and $L \subset Q$ are $(P, Q)$-*conjugate* if $\pi_M(L) = \pi_M(Q)$ and $e_L(M) = e_L(P)$; the sets $M$ and $L$ are $(P, Q)$-*preconjugate* if $\pi_M(L)$ is dense in $\pi_M(Q)$ and $e_L(M)$ is dense in $e_L(P)$. If no confusion is possible, then $(Z, Y)$-conjugate sets will be called conjugate and $(Z, Y)$-preconjugate sets will be simply called preconjugate. The following statement was proved as Fact 2 in U.285.

**Lemma 1.** *Suppose that $Z$ is a space and a set $Y \subset C_p(Z)$ generates the topology of $Z$. Assume also that $M \subset X$ and $L \subset Y$ are conjugate sets. Then the mappings*

$u = e_L|M : M \rightarrow e_L(M)$ and $v = \pi_M|L : L \rightarrow \pi_M(L)$ are homeomorphisms; besides, $r = u^{-1} \circ e_L : X \rightarrow M$ and $q = v^{-1} \circ \pi_M : Y \rightarrow L$ are continuous retractions such that $q = r^* \circ (\pi_M|Y)$. The maps $r$ and $q$ are called the pair of retractions corresponding to the conjugate pair $(M, L)$.

We will also need the lemma given below; it was proved as Fact 7 in U. 285.

**Lemma 2.** *Suppose that $X$ is a Lindelöf $\Sigma$-space and a Lindelöf $\Sigma$-space $Y \subset C_p(X)$ generates the topology of $X$. Assume additionally that some countable families $\mathcal{P} \subset exp(X)$ and $\mathcal{Q} \subset exp(Y)$ are closed under finite intersections and finite unions and there exist compact covers $\mathcal{K}$ and $\mathcal{C}$ of the spaces $X$ and $Y$ respectively such that $\mathcal{P}$ is a network with respect to $\mathcal{K}$ and $\mathcal{Q}$ is a network with respect to $\mathcal{C}$. Assume that, for some sets $M \subset X$ and $L \subset Y$ the pair $(M \cap P, L \cap Q)$ is $(P, Q)$-preconjugate for any $(P, Q) \in \mathcal{P} \times \mathcal{Q}$. Then the pair $(cl_X(M), cl_Y(L))$ is conjugate.*

The following simple fact is often useful when we consider networks for mappings.

*Fact 1.* Suppose that $f : Z \rightarrow Y$ is a continuous map and $\mathcal{N} \subset exp(Z)$ is a network for $f$. Then, for any continuous map $g : Y \rightarrow T$, the family $\mathcal{N}$ is also a network for $g \circ f$.

*Proof.* Take any $x \in Z$ and $U \in \tau(g(f(x)), T)$; by continuity of $g$ there exists a set $V \in \tau(f(x), Y)$ such that $g(V) \subset U$. Since $\mathcal{N}$ is a network for $f$, we can find a set $N \in \mathcal{N}$ such that $x \in N$ and $f(N) \subset V$. Then $(g \circ f)(N) \subset g(V) \subset U$, so $\mathcal{N}$ is a network for $g \circ f$ and hence Fact 1 is proved.

Our next step is to prove that any countable subset of a space $Z$ can be extended to a countable set that generates many preconjugate sets.

*Fact 2.* Suppose that $Z$ is a space and a set $Y \subset C_p(Z)$ generates the topology of $Z$. Assume also that $\mathcal{P} \subset exp(Z)$ and $\mathcal{Q} \subset exp(Y)$ are countable families. Then for any countable set $A \subset Z$, it is possible to find countable sets $M(A) \subset Z$ and $L(A) \subset Y$ such that $A \subset M(A)$, the pair $(M(A) \cap P, L(A) \cap Q)$ is $(P, Q)$-preconjugate whenever $(P, Q) \in \mathcal{P} \times \mathcal{Q}$ and the assignments $A \rightarrow M(A)$ and $A \rightarrow L(A)$ are $\omega$-monotone.

*Proof.* Choose first an enumeration $\{(P_n, Q_n) : n \in \omega\}$ of the set $\mathcal{P} \times \mathcal{Q}$ in which every pair $(P, Q) \in \mathcal{P} \times \mathcal{Q}$ occurs infinitely many times. For any $n \in \omega$, denote by $\mathcal{U}_n$ the family of all nonempty sets $[x_1, \ldots, x_k; B_1, \ldots, B_k] \cap Q_n$, where $x_1, \ldots, x_k \in Z$, $B_1, \ldots, B_k \in \mathcal{B}$ and $k \in \mathbb{N}$; we will also need the family $\mathcal{V}_n$ of all nonempty sets $[f_1, \ldots, f_k; B_1, \ldots, B_k] \cap e_Y(P_n)$, where $f_1, \ldots, f_k \in Y$, $B_1, \ldots, B_k \in \mathcal{B}$ and $k \in \mathbb{N}$. For any $n \in \omega$ and $U \in \mathcal{U}_n$ pick a function $f_U \in U$; for each $V \in \mathcal{V}_n$ fix a point $x_V \in P_n$ such that $e_Y(x_V) \in V$.

Take a countable set $A \subset Z$. We will recursively construct $M(A)$ and $L(A)$. Let $M_0(A) = A$ and $L_0(A) = \emptyset$. Assume that $n \in \omega$ and we have countable sets $M_0(A), \ldots, M_n(A) \subset Z$ and $L_0(A), \ldots, L_n(A) \subset Y$. Let

$$L_{n+1}(A) = L_n(A) \cup \{f_U : U \in \mathcal{U}_n \text{ and supp}(U) \subset M_n(A)\}, \text{and}$$

$$M_{n+1}(A) = M_n(A) \cup \{x_V : V \in \mathcal{V}_n \text{ and supp}(V) \subset L_{n+1}(A)\}.$$

Note that $M_{n+1}(A) \subset Z$ and $L_{n+1}(A) \subset Y$ are countable sets because $M_n(A)$ and $L_n(A)$ are countable. So our inductive procedure can be continued to construct sequences $\{M_i(A) : i \in \omega\}$ and $\{L_i(A) : i \in \omega\}$. We will prove that the sets $M(A) = \bigcup\{M_n(A) : n \in \omega\}$ and $L(A) = \bigcup\{L_n(A) : n \in \omega\}$ are as promised. Let us show first that

(1) the assignments $A \to M(A)$ and $A \to L(A)$ are $\omega$-monotone.

It takes a straightforward induction to see that the sets $M(A)$ and $L(A)$ are countable for any countable set $A \subset X$ and the assignments $A \to M_n(A)$ and $A \to L_n(A)$ are $\omega$-monotone for every $n \in \omega$. Applying Fact 1(b) of V.477 we can see that the assignments $A \to M(A) = \bigcup\{M_n(A) : n \in \omega\}$ and $A \to L(A) = \bigcup\{L_n(A) : n \in \omega\}$ are $\omega$-monotone and hence we proved (1).

Our last step is to establish that

(2) the pair $(M(A) \cap P, L(A) \cap Q)$ is $(P, Q)$-preconjugate for any $P \in \mathcal{P}$ and $Q \in \mathcal{Q}$.

To prove (2), take any pair $(P, Q) \in \mathcal{P} \times \mathcal{Q}$ and consider the sets $M = M(A) \cap P$ and $L = L(A) \cap Q$; we have to establish that the pair $(M, L)$ is $(P, Q)$-preconjugate. Take any function $f \in Q$ and $W \in \tau(\pi_M(f), \pi_M(Q))$. By the definition of the topology of pointwise convergence there are points $x_1, \ldots, x_k \in M \subset M(A)$ and sets $B_1, \ldots, B_k \in \mathcal{B}$ such that $f \in U = [x_1, \ldots, x_k; B_1, \ldots, B_k] \cap Q \subset Y$ and we have $\pi_M(f) \in \pi_M(U) \subset W$.

It follows from the choice of our enumeration of $\mathcal{P} \times \mathcal{Q}$ and the fact that the sequence $\{M_i(A) : i \in \omega\}$ is increasing, that there exists a number $n \in \omega$ such that $(P_n, Q_n) = (P, Q)$ and $\{x_1, \ldots, x_k\} \subset M_n(A)$. Since $U \in \mathcal{U}_n$, it easily follows from the inclusions $f_U \in L_{n+1}(A) \cap U \subset L(A) \cap Q = L$ that $\pi_M(f_U) \in \pi_M(U) \subset W$. Therefore $\pi_M(L)$ is dense in $\pi_M(Q)$.

To show that the set $e_L(M)$ is dense in the space $e_L(P)$ fix any point $x \in P$ and a set $W \in \tau(e_L(x), e_L(P))$. There exist functions $f_1, \ldots, f_k \in L \subset L(A)$ and sets $B_1, \ldots, B_k \in \mathcal{B}$ such that $e_Y(x) \in V = [f_1, \ldots, f_k; B_1, \ldots, B_k] \cap e_Y(P)$ and we have the inclusion $e_L(x) \in \pi_L(V) \subset W$. It follows from the choice of our enumeration of $\mathcal{P} \times \mathcal{Q}$ and the fact that the sequence $\{L_i(A) : i \in \omega\}$ is increasing, that there is $n \in \omega$ for which $(P_n, Q_n) = (P, Q)$ and $\{f_1, \ldots, f_k\} \subset L_n(A)$. Then $V \in \mathcal{V}_n$ and it follows from $x_V \in M_{n+1}(A) \cap P_n \subset M(A) \cap P = M$ that $e_L(x_V) = \pi_L(e_Y(x_V)) \in \pi_L(V) \subset W$. Therefore $e_L(M)$ is dense in the space $e_L(P)$, i.e., we settled (2) and hence Fact 2 is proved.

*Fact 3.* For any Lindelöf $\Sigma$-space $Z$, if $C_p(Z)$ is also Lindelöf $\Sigma$, then $Z$ is monotonically retractable.

*Proof.* Observe that the set $Y = C_p(Z)$ generates the topology of $Z$; choose countable families $\mathcal{P} \subset \exp(Z)$ and $\mathcal{Q} \subset \exp(Y)$ closed under finite intersections and finite unions which are networks with respect to a compact cover of $X$ and $Y$ respectively.

It follows from Fact 2 that for any countable $A \subset Z$, there exist countable sets $M(A) \subset Z$ and $L(A) \subset Y$ such that $A \subset M(A)$ and the pair $(M(A) \cap P, L(A) \cap Q)$ is $(P, Q)$-preconjugate for any $(P, Q) \in \mathcal{P} \times \mathcal{Q}$ while the assignment $A \to L(A)$ is $\omega$-monotone. If $F = \mathrm{cl}_Z(M(A))$ and $G = \mathrm{cl}_Y(L(A))$ then the pair $(F, G)$ has to be conjugate by Lemma 2. Applying Lemma 1 we can convince ourselves that the map $u_A = e_G | F : F \to e_G(F)$ is a homeomorphism and the map $r_A = (u_A)^{-1} \circ e_G : Z \to F$ is a continuous retraction.

Since $Z$ is a Lindelöf $\Sigma$-space, the space $C_p(Z)$ must be monotonically monolithic by Problem 468; applying Problem 467 we conclude that $Z$ is monotonically stable and, in particular, it is monotonically $\omega$-stable. Therefore, to each countable set $L \subset Y$ we can assign a network $\mathcal{O}(L) \subset \exp(Z)$ for the map $e_{\overline{L}}$ in such a way that the operator $\mathcal{O}$ is $\omega$-monotone. If $A \subset Z$ is a countable set, then $(u_A)^{-1}$ is a homeomorphism, so the family $\mathcal{O}(L(A))$ is a network for $r_A$ by Fact 1. Finally, let $K(A) = \mathrm{cl}_Z(M(A))$ and $\mathcal{N}(A) = \mathcal{O}(L(A))$. Then $A \subset K(A) \subset Z$ and $r_A : Z \to K(A)$ is a continuous retraction. Besides, $\mathcal{N}(A)$ is a countable network for $r_A$ and the assignment $\mathcal{N}$ is $\omega$-monotone; this shows that $Z$ is monotonically retractable, so Fact 3 is proved.

Returning to our solution, assume that $X$ and $C_p(X)$ are Lindelöf $\Sigma$-spaces. By Fact 3, the space $X$ is monotonically retractable and hence $C_p(X)$ is monotonically Sokolov by Problem 481. Since $C_p(C_p(X))$ is also a Lindelöf $\Sigma$-space by CFS-219, we can apply Fact 3 to the space $Z = C_p(X)$ to conclude that $C_p(X)$ is monotonically retractable and hence $X$ is monotonically Sokolov by Problem 481. Finally, apply Problem 479 to see that both spaces $X$ and $C_p(X)$ are Sokolov.

**V.488.** *Prove that*

(a) *any Eberlein compact space $X$ has a $\sigma$-closure-preserving local base at every point, i.e., for any $x \in X$ there exists a local base $\mathcal{B}_x$ at the point $x$ such that $\mathcal{B}_x = \bigcup_{n \in \omega} \mathcal{B}_x^n$ and every $\mathcal{B}_x^n$ is closure-preserving;*

(b) *there exists an Eberlein compact space $K$ that does not have a closure-preserving local base at some point;*

(c) *there exists a non-metrizable compact space $Y$ such that $C_p(Y)$ has a closure-preserving local base at every point.*

**Solution.** (a) Given a space $Z$ and a closed set $F \subset Z$ recall that a family $\mathcal{U} \subset \tau(F, Z)$ is *an outer base* of $F$ in $Z$ if for any $V \in \tau(F, X)$ there is $U \in \mathcal{U}$ such that $U \subset V$. A family $\mathcal{A} \subset \exp(Z)$ is *$\sigma$-closure-preserving* if $\mathcal{A} = \bigcup_{n \in \omega} \mathcal{A}_n$ and every $\mathcal{A}_n$ is closure-preserving.

*Fact 1.* If $Z$ is a zero-dimensional Eberlein compact space and $F$ is a closed subset of $Z$, then $F$ has a $\sigma$-closure-preserving clopen outer base in $Z$.

*Proof.* The space $Z \setminus F$ must be $\sigma$-metacompact, i.e., every open cover of $Z \setminus F$ has a $\sigma$-point-finite refinement (see CFS-363). For any $x \in Z \setminus F$ choose a set $U_x \in \tau(x, Z)$ with $F \cap \overline{U}_x = \emptyset$. There exists a refinement $\mathcal{U}$ of the cover $\{U_x : x \in Z \setminus F\}$ such that $\mathcal{U} = \bigcup_{n \in \omega} \mathcal{U}_n$ while $\mathcal{U}_n \subset \mathcal{U}_{n+1}$ and $\mathcal{U}_n$ is point-finite for any $n \in \omega$.

Observe that $\overline{U}$ is compact for any $U \in \mathcal{U}$, so the cover $\mathcal{U}$ can be shrunk, i.e., we can find, for any $U \in \mathcal{U}$, a compact set $Q_U \subset U$ such that $\{Q_U : U \in \mathcal{U}\}$ is still a cover of $Z \setminus F$ (see Fact 1 of U.188). Using normality of $Z$ it is easy to find a clopen set $G_U$ such that $Q_U \subset G_U \subset U$ for any $U \in \mathcal{U}$. Given a compact $Q \subset Z \setminus F$ there is a finite $\mathcal{V} \subset \mathcal{U}$ such that $Q \subset \bigcup \{G_V : V \in \mathcal{V}\}$. Consequently, the family $\mathcal{W}$ of all finite intersections of the family $\{Z \setminus G_U : U \in \mathcal{U}\}$ is an outer base for the set $F$ in $Z$; observe that all elements of $\mathcal{W}$ are clopen in $Z$.

To see that $\mathcal{W}$ is $\sigma$-closure-preserving denote by $\mathcal{W}_n$ the family of all finite intersections of the elements of the family $\{Z \setminus G_U : U \in \mathcal{U}_n\}$. Using the fact that the sequence $\{\mathcal{U}_n : n \in \omega\}$ is increasing, it is easy to prove that $\mathcal{W} = \bigcup_{n \in \omega} \mathcal{W}_n$, so it suffices to show that each $\mathcal{W}_n$ is closure-preserving.

Suppose that $n \in \omega$ and there exists a point $x \in Z$ together with a family $\mathcal{W}' \subset \mathcal{W}_n$ such that $x \in \overline{\bigcup \mathcal{W}'} \setminus \bigcup \mathcal{W}'$. The family $\{G_U : U \in \mathcal{U}_n\}$ being point-finite, the set $H = \bigcap \{G_U : x \in G_U\}$ is open in $X$. If $W \in \mathcal{W}'$ and $W = \bigcap \{X \setminus G_{U_i} : 1 \leq i \leq n\}$ then it follows from $x \notin W$ that $x \in G_{U_i}$ for some $i \leq n$ and hence $W \cap H = \emptyset$. Consequently, $(\bigcup \mathcal{W}') \cap H = \emptyset$; this contradiction with $x \in \overline{\bigcup \mathcal{W}'}$ shows that every $\mathcal{W}_n$ is closure-preserving and hence Fact 1 is proved.

Now assume $X$ is an arbitrary Eberlein compact space and $x \in X$. Take a zero-dimensional Eberlein compact space $Z$ for which there exists a continuous onto map $f : Z \to X$ (see CFS-336). Passing to an appropriate closed subset of the space $Z$ if necessary we can assume, without loss of generality, that the map $f$ is irreducible, i.e., $f(E) \neq X$ for any proper closed set $E \subset Z$ (see TFS-366).

Apply Fact 1 to find a clopen outer base $\mathcal{U}$ of the set $G = f^{-1}(x)$ in the space $Z$ such that $\mathcal{U} = \bigcup_{n \in \omega} \mathcal{U}_n$ and each $\mathcal{U}_n$ is closure-preserving in $Z$. For every set $U \in \mathcal{U}_n$ let $H(U) = X \setminus f(Z \setminus U)$; observe that the set $H(U)$ is open in $X$ and $x \in H(U)$. The map $f$ being irreducible, the set $H(U)$ must be dense in $f(U)$ for any $U \in \mathcal{U}_n$ by Fact 1 of S.383.

Since every $U \in \mathcal{U}_n$ is compact, the family $\mathcal{H}_n = \{f(U) : U \in \mathcal{U}_n\}$ consists of compact subsets of $X$; the fact that the map $f$ is closed easily implies that $\mathcal{H}_n$ is closure-preserving in $X$. Therefore the family $\mathcal{V}_n = \{H(U) : U \in \mathcal{U}_n\} \in \tau(x, X)$ is also closure-preserving in $X$ for every $n \in \omega$. It is an easy exercise that the family $\mathcal{V} = \{H(U) : U \in \mathcal{U}\}$ is a local base at $x$ in $X$; since $\mathcal{V} = \bigcup_{n \in \omega} \mathcal{V}_n$, the family $\mathcal{V}$ is $\sigma$-closure-preserving and hence we settled (a).

(b) Recall that for any set $D$, the expression $[D]^{<\omega}$ stands for the family of all finite subsets of $D$.

*Fact 2.* Assume that we are given a map $\xi : [\omega_1]^{<\omega} \to [\omega_1]^{<\omega}$ such that $A \cap \xi(A) = \emptyset$ for any $A \in [\omega_1]^{<\omega}$. Then there exists a disjoint family $\{A_n : n \in \omega\} \subset [\omega_1]^{<\omega}$ such that $|A_n| \to \infty$ and the sets $\bigcup_{n \in \omega} A_n$ and $\bigcup_{n \in \omega} \xi(A_n)$ are disjoint.

*Proof.* It is easy to find a sequence $\{\mathcal{U}_n : n \in \omega\}$ of disjoint families of finite subsets of $\omega_1$ such that the family $\bigcup_{n \in \omega} \mathcal{U}_n$ is disjoint and, for any $n \in \omega$ we have $|\mathcal{U}_n| = \omega_1$ and $|U| > n$ for each $U \in \mathcal{U}_n$.

Passing to a smaller family $\mathcal{U}_n$ if necessary, we can assume, without loss of generality, that, for every $n \in \omega$, the family $\{\xi(U) : U \in \mathcal{U}_n\}$ is a $\Delta$-system with a root $D_n$, i.e., we have $U \cap V = D_n$ for any distinct $U, V \in \mathcal{U}_n$ (see SFFS-038).

The set $D = \bigcup_{n \in \omega} D_n$ being countable we can pass again to a smaller family $\mathcal{U}_n$ to guarantee that $(\bigcup \mathcal{U}_n) \cap D = \emptyset$ for any $n \in \omega$. Take an arbitrary $A_0 \in \mathcal{U}_0$ and assume, proceeding inductively, that $n \in \omega$ and we have chosen a set $A_i \in \mathcal{U}_i$ for every $i \leq n$ in such a way that the sets $B_n = A_0 \cup \ldots \cup A_n$ and $S_n = \xi(A_0) \cup \ldots \cup \xi(A_n)$ are disjoint.

The set $B_n \cup (S_n \setminus D) \subset \omega_1 \setminus D$ is finite while both families $\{\xi(U) \setminus D : U \in \mathcal{U}_{n+1}\}$ and $\mathcal{U}_{n+1}$ are disjoint, so we can find $A_{n+1} \in \mathcal{U}_{n+1}$ such that the sets $A_{n+1} \cup \xi(A_{n+1})$ and $B_n \cup (S_n \setminus D)$ are disjoint. This shows that our inductive procedure can be continued to construct a family $\{A_n : n \in \omega\}$ such that the sets $\bigcup_{n \in \omega} A_n$ and $\bigcup_{n \in \omega} \xi(A_n)$ are disjoint and $A_n \in \mathcal{U}_n$ for every $n \in \omega$. This implies that $|A_n| \to \infty$, so $\{A_n : n \in \omega\}$ is the promised family, i.e., Fact 2 is proved.

Returning to our solution, recall that $A(\omega_1)$ is the one-point compactification of a discrete set of cardinality $\omega_1$. Therefore $A(\omega_1) = D \cup \{a\}$ where $|D| = \omega_1$ and $a \notin D$ is the unique non-isolated point of $A(\omega_1)$. The space $K = A(\omega_1)^\omega$ is Eberlein compact (see CFS-107, CFS-301 and CFS-307). Denote by $p$ the point of $K$ all coordinates of which are equal to $a$; we will prove that $K$ does not have a local closure-preserving base at $p$.

To do it, take an arbitrary local base $\mathcal{B}$ at the point $p$. Given any finite $E \subset D$ let $n = |E|$ and consider the set $O(E) = \{x \in K : x(i) \notin E \text{ for any } i \leq n\}$. It is clear that every $O(E)$ is an open subset of $K$ and, besides, the family $\{O(E) : E \in [D]^{<\omega}\}$ is a local base of $K$ at $p$. As a consequence, for any $E \in [D]^{<\omega}$ there exist sets $B(E) \in \mathcal{B}$ and $\xi(E) \in [D]^{<\omega}$ such that $\xi(E) \cap E = \emptyset$ and $O(E \cup \xi(E)) \subset B(E) \subset O(E)$.

Apply Fact 2 to find a disjoint family $\{E_n : n \in \omega\}$ of finite subsets of $D$ such that $|E_n| \to \infty$ while the sets $E = \bigcup_{n \in \omega} E_n$ and $S = \bigcup_{n \in \omega} \xi(E_n)$ are disjoint. We can consider that $|E_n| > n$ for any $n \in \omega$.

For each $n \in \omega$ pick a point $z_n \in E_n$ and define a point $q \in K$ by $q(n) = z_n$ for any $n \in \omega$. We will also need the point $q_n \in K$ such that $q_n(i) = z_i$ for all $i \leq n$ and $q_n(i) = a$ whenever $i > n$. It is straightforward that $q_n \to q$. It follows from $q(n) = z_n \in E_n$ that $q \notin O(E_n)$ and hence $q \notin B(E_n)$ for all $n \in \omega$.

However, $z_i \notin E_n \cup \xi(E_n)$ for all $i < n$ which shows that $q_{n-1} \in O(E_n \cup \xi(E_n))$ and hence $q_{n-1} \in B(E_n)$ for every $n \in \omega \setminus \{0\}$. Therefore $q \in \overline{\bigcup_{n \in \omega} B(E_n)}$ which shows that the family $\mathcal{B}$ is not closure-preserving and finishes the proof of (b).

(c) Given a set $D$ say that a family $\mathcal{A} \subset [D]^{<\omega}$ is cofinal in $[D]^{<\omega}$ if for any $F \in [D]^{<\omega}$ there is a set $A \in \mathcal{A}$ such that $F \subset A$.

*Fact 3.* Given spaces $Z$ and $Z'$ suppose that $Z$ has a closure-preserving local base $\mathcal{B}$ at a point $z$ and $Z'$ has a closure-preserving local base $\mathcal{B}'$ at a point $z'$. Then the family $\mathcal{C} = \{B \times B' : B \in \mathcal{B}$ and $B' \in \mathcal{B}'\}$ is a closure-preserving local base at the point $(z, z')$ in the space $Z \times Z'$.

*Proof.* We leave it to the reader to verify that $\mathcal{C}$ is a local base at $(z, z')$. Take any family $\mathcal{D} \subset \mathcal{C}$ and assume that $u \in \overline{\bigcup \mathcal{D}} \backslash \bigcup \{\overline{C} : C \in \mathcal{D}\}$ for some $u = (x, y) \in Z \times Z'$. For every $C = B \times B' \in \mathcal{D}$ let $C_0 = B$ and $C_1 = B'$; observe that $\overline{C} = \overline{C}_0 \times \overline{C}_1$. Consider the families $\mathcal{D}_0 = \{C \in \mathcal{D} : x \notin \overline{C}_0\}$ and $\mathcal{D}_1 = \{C \in \mathcal{D} : y \notin \overline{C}_1\}$.

The family $\mathcal{B}$ being closure-preserving, it follows from $x \notin \bigcup \{\overline{C}_0 : C \in \mathcal{D}_0\}$ that we can find a set $U \in \tau(x, Z)$ such that

(1) $U \cap C_0 = \emptyset$ for any $C \in \mathcal{D}_0$.

Analogously, there exists a set $V \in \tau(y, Z')$ such that

(2) $V \cap C_1 = \emptyset$ for any $C \in \mathcal{D}_1$.

The set $W = U \times V$ is an open neighborhood of the point $u$ in $Z \times Z'$. If $C \in \mathcal{D}$, then $u \notin \overline{C}_0 \times \overline{C}_1$ and hence we have two cases:

*Case 1.* $x \notin \overline{C}_0$. Then $C \in \mathcal{D}_0$, so $U \cap C_0 = \emptyset$ and therefore $(U \times V) \cap (C_0 \times C_1) = \emptyset$.
*Case 2.* $y \notin \overline{C}_1$. Then $C \in \mathcal{D}_1$, so $V \cap C_1 = \emptyset$ and therefore $(U \times V) \cap (C_0 \times C_1) = \emptyset$.

Thus, in all possible cases, we have $W \cap C = \emptyset$ for any $C \in \mathcal{D}$, i.e., $W \cap (\bigcup \mathcal{D}) = \emptyset$; this contradiction shows that $\mathcal{C}$ is closure-preserving and hence Fact 3 is proved.

*Fact 4.* Suppose that $Z$ is a space with a countable dense set $D$ of isolated points. Assume that there exists a point $a \in Z \backslash D$ and a family $\mathcal{F} \subset [Z]^{<\omega} \backslash \{\emptyset\}$ cofinal in $[Z]^{<\omega}$ such that for any closed neighborhood $C$ of the point $a$, we can find a finite set $E \subset Z \backslash C$ such that $E \cap F \neq \emptyset$ whenever $F \in \mathcal{F}$ and $F \backslash C \neq \emptyset$. Then $C_p(Z)$ has a closure-preserving local base at every point.

*Proof.* Let $u \in C_p(Z)$ be the function equal to zero at all points of the space $Z$. If $I = \{f \in C_p(Z) : f(a) = 0\}$, then $C_p(Z)$ is homeomorphic to $I \times \mathbb{R}$ (see Fact 1 of S.409). It is trivial that $\mathbb{R}$ has a closure-preserving local base at every point, so if we prove that $I$ has a closure-preserving local base at $u$, then Fact 3 can be applied to see that $C_p(Z)$ has a closure-preserving local base at some point. This implies that $C_p(Z)$ has a closure-preserving base at every point (see e.g., TFS-062, TFS-115, TFS-116 and Fact 1 of S.496). Therefore it suffices to show that $I$ has a closure-preserving local base at $u$.

Given a function $f \in I$, a finite set $A \subset Z$ and a number $\varepsilon > 0$ consider the set $M(f, A, \varepsilon) = \{g \in I : |g(x) - f(x)| < \varepsilon$ for all $x \in A\}$. It is clear that the family $\{M(f, A, \varepsilon) : A \in [Z]^{<\omega}$ and $\varepsilon > 0\}$ is a local base at $f$ in $I$ for any $f \in I$.

Observe first that if $D$ is finite then $Z$ is finite and hence our Fact is trivially true, so we can assume, without loss of generality, that $D$ is infinite and hence we can fix a faithful enumeration $\{d_n : n \in \mathbb{N}\}$ of the set $D$. Let $D_n = \{d_1, \ldots, d_n\}$

for any $n \in \mathbb{N}$. For every $F \in \mathcal{F}$ let $m_F = |F|$ and $A(F) = F \cup D_{m_F}$; then $W(F) = M(u, A(F), \frac{1}{2m_F})$ is an open neighborhood of $f$. Let us establish first that

(3) the family $\mathcal{W} = \{W(F) : F \in \mathcal{F}\}$ is a local base at $u$.

Take any $U \in \tau(u, I)$; there exists a finite set $K \subset Z$ and $\varepsilon > 0$ such that $M(u, K, \varepsilon) \subset U$. Since $\mathcal{F}$ is cofinal in $[Z]^{<\omega}$, we can find $F \in \mathcal{F}$ such that $K \subset F$ and $m_F > \frac{1}{\varepsilon}$. It is clear that $u \in W(F) \subset M(u, K, \varepsilon) \subset U$, so (3) is proved.

To finish the proof of our Fact, it suffices to establish that $\mathcal{W}$ is closure-preserving, so take any family $\mathcal{G} \subset \mathcal{F}$ and assume that $f \notin \bigcup\{\overline{W(F)} : F \in \mathcal{G}\}$. The set $H = f^{-1}(\mathbb{R}\backslash\{0\})$ is open and nonempty because $f \neq u$. Therefore $H \cap D \neq \emptyset$, so we can find $k, n \in \mathbb{N}$ such that $|f(d_n)| > \frac{1}{2k}$ and hence $r = |f(d_n)| - \frac{1}{2k} > 0$.

Consider the set $U = M(f, \{d_n\}, r)$. If $F \in \mathcal{G}$ and $|F| \geq l = \max\{k, n\}$, then the point $d_n$ belongs to $A(F)$ and $|g(d_n)| < \frac{1}{2m_F} \leq \frac{1}{2l} \leq \frac{1}{2k}$ for any $g \in W(F)$. On the other hand, if $h \in U$, then $|h(d_n) - f(d_n)| < r$ and hence $|h(d_n)| > |f(d_n)| - r = \frac{1}{2k}$. This shows that $h \neq g$ and hence

(4) $U \cap W(F) = \emptyset$ for any $F \in \mathcal{G}_l = \{F \in \mathcal{G} : m_F \geq l\}$.

For any $i < l$ let $\mathcal{G}_i = \{F \in \mathcal{G} : m_F = i\}$; the set $C_i = \{x \in Z : |f(x)| \leq \frac{1}{2i}\}$ is a closed neighborhood of the point $a$, so we can find a finite set $E_i \subset Z\backslash C_i$ such that $E_i \cap F \neq \emptyset$ whenever $F \in \mathcal{F}$ and $F\backslash C_i \neq \emptyset$. There is no loss of generality to assume that $D_i\backslash C_i \subset E_i$. Observe that we have the inequality $|f(x)| - \frac{1}{2i} > 0$ for any $x \in E_i$, so the number $\varepsilon_i = \backslash n\{|f(x)| - \frac{1}{2i} : x \in E_i\}$ is positive and hence the set $U_i = M(f, E_i, \varepsilon_i)$ is an open neighborhood of $f$ for any $i < l$.

Fix any $F \in \mathcal{G}_i$ and $g \in W(F)$; it follows from $f \notin \overline{W(F)}$, that $|f(x)| > \frac{1}{2i}$ for some $x \in F \cup D_i$. Given any function $h \in U_i$ we have two cases.

*Case 1.* $F \subset C_i$; then $x \in D_i\backslash C_i \subset E_i$, so it follows from the inequalities

$$|h(x)| > |f(x)| - \varepsilon_i \geq |f(x)| - (|f(x)| - \frac{1}{2i}) = \frac{1}{2i}$$

and the fact that $|g(x)| < \frac{1}{2i}$ that $h \neq g$.

*Case 2.* $F\backslash C_i \neq \emptyset$; by the choice of $E_i$ we can find a point $y \in E_i \cap F$, so the inequalities $|h(y)| > |f(y)| - \varepsilon_i \geq |f(y)| - (|f(y)| - \frac{1}{2i}) = \frac{1}{2i}$ together with the fact that $|g(y)| < \frac{1}{2i}$ show that $h \neq g$. This proves that

(5) if $1 \leq i < l$, then $U_i \cap W(F) = \emptyset$ for all $F \in \mathcal{G}_i$.

As an immediate consequence of (4) and (5), the set $V = U \cap U_1 \cap \ldots \cap U_{l-1}$ is an open neighborhood of the function $f$ that does not meet the set $\bigcup\{W(F) : F \in \mathcal{G}\}$, so $f \notin \overline{\bigcup\{W(F) : F \in \mathcal{G}\}}$ and hence the family $\mathcal{W}$ is a closure-preserving local base at $u$, so Fact 4 is proved.

Returning to our solution let $Y$ be the compact space $M_0$ described in TFS-387. Recall that $Y = D \cup A \cup \{p\}$ where the sets $D, A$ and $\{p\}$ are disjoint, $D$ is a countably infinite dense set of isolated points of $Y$, the set $A$ is uncountable and $A \cup \{p\}$ is homeomorphic to the one-point compactification of a discrete space while $p$ is the unique non-isolated point of $A \cup \{p\}$. Let $\{d_n : n \in \mathbb{N}\}$ be a faithful enumeration of the set $D$. For any $F \in [A]^{<\omega} \setminus \{\emptyset\}$ let $m_F = |A|$ and $Q_F = F \cup \{p, d_1, \ldots, d_{m_F}\}$. We claim that the family $\mathcal{F} = \{Q_F : F \in [A]^{<\omega} \setminus \{\emptyset\}\}$ and the point $p$ satisfy the hypotheses of Fact 4.

It is immediate that $\mathcal{F}$ is cofinal in $[Y]^{<\omega}$, so take any closed neighborhood $C$ of the point $p$. There is nothing to prove if $C = Y$; if $C \neq Y$ then $(Y \setminus C) \cap D \neq \emptyset$ because $D$ is dense in $Y$. Therefore $d_n \in D \setminus C$ for some $n \in \mathbb{N}$; we claim that the set $E = (A \cup \{d_1, \ldots, d_n\}) \setminus C$ witnesses that the hypothesis of Fact 4 holds for $Y$. It is easy to see that $E$ is finite; assume that $Q_F \in \mathcal{F}$ and $Q_F \setminus C \neq \emptyset$. Since $A \setminus C \subset E$, there is no loss of generality to consider that $F \subset C$. If $|F| \geq n$, then $d_n \in Q_F \cap E$. If $|F| < n$, then $\emptyset \neq Q_F \setminus C \subset \{d_1, \ldots, d_n\} \setminus C \subset E$, so $d_i \in Q_F \cap E$ for some $i \leq n$.

We proved that $E \cap G \neq \emptyset$ whenever $G \in \mathcal{F}$ and $G \setminus C \neq \emptyset$, so Fact 4 can be applied to convince ourselves that the space $C_p(Y)$ has a closure-preserving local base at every point. Finally observe that the compact space $Y$ is not metrizable because $A$ is an uncountable discrete subset of $Y$; this settles (c) and completes our solution.

**V.489.** *Given a compact space $K$ let $\|f\| = \sup\{|f(x)| : x \in K\}$ for any $f \in C(K)$; if $f, g \in C(K)$ then $\rho_K(f, g) = \|f - g\|$. As usual, $C_u(K)$ is the set $C(K)$ with the topology generated by the metric $\rho_K$. If $A \subset C_p(K)$ and $A \neq \emptyset$ then $\mathrm{diam}(A) = \sup\{\|f - g\| : f, g \in A\}$. For each $\varepsilon > 0$ say that a family $\mathcal{A}$ of subsets of $C(K)$ is $\varepsilon$-small if $\mathrm{diam}(A) < \varepsilon$ for any $A \in \mathcal{A}$. The space $C_p(K)$ is said to have the property JNR if for every $\varepsilon > 0$ we can find a family $\{M_n : n \in \omega\}$ of closed subsets of $C_p(K)$ such that $C_p(K) = \bigcup_{n \in \omega} M_n$ and, for every $n \in \omega$ there exists an $\varepsilon$-small family $\mathcal{U}_n \subset \tau(M_n)$ such that $\bigcup \mathcal{U}_n = M_n$. Prove that the following conditions are equivalent:*

*(i) the space $C_p(K)$ has the property JNR;*
*(ii) there exists a $\sigma$-discrete family $\mathcal{N}$ in the space $C_p(K)$ such that $\mathcal{N}$ is a network in both spaces $C_p(K)$ and $C_u(K)$.*

*In particular, if $C_p(K)$ has the property JNR then it has a $\sigma$-discrete network.*

**Solution.** Assume that $C_p(K)$ has the property JNR. For any $f \in C(K)$ and $r > 0$ the set $B(f, r) = \{g \in C(K) : \rho_K(f, g) < r\}$ is the $r$-ball centered at $f$ in the space $C_u(K)$. Given $\delta > 0$ say that a family $\mathcal{A}$ of subsets of $C(K)$ is $\delta$-uniformly discrete if, for any $f \in C(K)$ the set $B(f, \delta)$ intersects at most one element of $\mathcal{A}$.

For each $n \in \omega$ consider the open cover $\mathcal{U}_n = \{B(f, 2^{-n}) : f \in C(K)\}$ of the space $C_u(K)$; by Fact 1 of T.373 we can find a refinement $\mathcal{V}_n \subset \tau(C_u(K))$ of the cover $\mathcal{U}_n$ such that $\mathcal{V}_n = \bigcup_{m \in \omega} \mathcal{B}_{nm}$ and, for every $m \in \omega$, the family $\mathcal{B}_{nm}$ is $\delta_{nm}$-uniformly discrete for some $\delta_{nm} > 0$. Applying Fact 1 of U.050 it is easy to see that the family $\mathcal{B} = \bigcup \{\mathcal{B}_{nm} : n, m \in \omega\}$ is a base in the space $C_u(K)$.

For every pair $(n, m) \in \omega \times \omega$ apply the property JNR to obtain a family $\{E_{nmk} : k \in \omega\}$ of closed subspaces of $C_p(K)$ such that $C_p(K) = \bigcup_{k \in \omega} E_{nmk}$ and, for every $k \in \omega$, there exists a $\delta_{nm}$-small cover $\mathcal{G}_{nmk} \subset \tau(E_{nmk})$ of the space $E_{nmk}$. Let $\mathcal{N}_{nmk} = \{B \cap E_{nmk} : B \in \mathcal{B}_{nm}\}$ for any $n, m, k \in \omega$. Given any point $f \in E_{nmk}$

take $G \in \mathcal{G}_{nmk}$ with $f \in G$ and observe that it follows from $\operatorname{diam}(G) < \delta_{nm}$ that $G \subset B(f, \delta_{nm})$ and hence $G$ intersects at most one element of $\mathcal{B}_{nm}$. Therefore $G$ intersects at most one element of $\mathcal{N}_{nmk}$; since $f \in E_{nmk}$ was chosen arbitrarily and $G$ is open in $E_{nmk}$, we conclude that the family $\mathcal{N}_{nmk}$ is discrete in $E_{nmk}$; the set $E_{nmk}$ being closed in $C_p(K)$, the family $\mathcal{N}_{nmk}$ is discrete in $C_p(K)$ for any $n, m, k \in \omega$.

Thus $\mathcal{N} = \bigcup \{\mathcal{N}_{nmk} : n, m, k \in \omega\}$ is a $\sigma$-discrete family in $C_p(K)$. Take an arbitrary $f \in C_u(K)$ and $O \in \tau(f, C_u(K))$. Since $\mathcal{B}$ is a base in $C_u(K)$, we can find $n, m \in \omega$ such that $f \in B \subset O$ for some $B \in \mathcal{B}_{nm}$. Choose $k \in \omega$ for which $f \in E_{nmk}$ and observe that $B' = B \cap E_{nmk} \in \mathcal{N}$. Since also $f \in B' \subset O$, we proved that $\mathcal{N}$ is a network in $C_u(K)$; it is evident that this implies that $\mathcal{N}$ is also a network in $C_p(K)$ and hence we established that (i)$\Longrightarrow$(ii).

*Fact 1.* Suppose that $Z$ is a compact space; let $\rho_Z(f, g) = \sup\{|f(x) - g(x)| : x \in Z\}$ for any $f, g \in C(Z)$. As usual, $\operatorname{diam}_{\rho_Z}(B) = \sup\{\rho_Z(f, g) : f, g \in B\}$ for any nonempty set $B \subset C(Z)$. Given a set $A \subset C(Z)$ let $\overline{A}$ be the closure of $A$ in the space $C_p(Z)$. Then $\operatorname{diam}_{\rho_Z}(A) = \operatorname{diam}_{\rho_Z}(\overline{A})$ for any nonempty set $A \subset C(Z)$.

*Proof.* Let $r = \operatorname{diam}_{\rho_Z}(A)$; it suffices to show that $\operatorname{diam}_{\rho_Z}(\overline{A}) \leq r$. If this is false then we can find $f, g \in \overline{A}$ for which $\rho_Z(f, g) > r + \varepsilon$ for some $\varepsilon > 0$. Therefore there exists a point $x \in Z$ such that $|f(x) - g(x)| > r + \varepsilon$. Take $f_0, g_0 \in A$ for which $|f_0(x) - f(x)| < \frac{\varepsilon}{3}$ and $|g(x) - g_0(x)| < \frac{\varepsilon}{3}$. Then

$$|f_0(x) - g_0(x)| = |f_0(x) - f(x) + f(x) - g(x) + g(x) - g_0(x)| \geq$$

$$|f(x) - g(x)| - |f_0(x) - f(x)| - |g_0(x) - g(x)| \geq r + \varepsilon - \frac{2}{3}\varepsilon = r + \frac{\varepsilon}{3} > r,$$

which shows that $\operatorname{diam}_{\rho_Z}(A) \geq \rho_Z(f_0, g_0) > r$ which is a contradiction. Therefore $\operatorname{diam}_{\rho_Z}(A) = \operatorname{diam}_{\rho_Z}(\overline{A})$, i.e., Fact 1 is proved.

Returning to our solution assume that $\mathcal{N}$ is a network in the space $C_u(K)$ such that $\mathcal{N} = \bigcup_{n \in \omega} \mathcal{N}_n$ and every family $\mathcal{N}_n$ is discrete in the space $C_p(K)$. The family $\mathcal{M}_n = \{\overline{N} : N \in \mathcal{N}_n\}$ is still discrete in $C_p(K)$ for every $n \in \omega$; let $\mathcal{M} = \bigcup_{n \in \omega} \mathcal{M}_n$. Take a point $f \in C_u(K)$ and $U \in \tau(f, C_u(K))$; there exists $r > 0$ such that $B(f, r) \subset U$. The family $\mathcal{N}$ being a network of $C_u(K)$ we can find $N \in \mathcal{N}$ with $f \in N \subset B(f, \frac{r}{3})$. Apply Fact 1 to see that if $M = \overline{N}$ then $\operatorname{diam}(M) = \operatorname{diam}(N) \leq \frac{2}{3}r$. Furthermore, $f \in M$, so $M \subset B(f, \frac{3}{4}r) \subset B(f, r) \subset U$. This shows that $\mathcal{M}$ is also a network in the space $C_u(K)$.

Given any $\varepsilon > 0$ let $M_n = \bigcup \{M \in \mathcal{M}_n : \operatorname{diam}(M) < \varepsilon\}$ for each $n \in \omega$. It is easy to see that every set $M \in \mathcal{M}_n$ is open in $\bigcup \mathcal{M}_n$ (in the topology on $\bigcup \mathcal{M}_n$ induced from $C_p(K)$); therefore the family $\mathcal{U}_n = \{M \in \mathcal{M}_n : \operatorname{diam}(M) < \varepsilon\}$ is an $\varepsilon$-small open cover of the set $M_n$. Take any point $f \in C_p(K)$; since $\mathcal{M}$ is a network in $C_u(K)$, we can find $n \in \omega$ and $M \in \mathcal{M}_n$ such that $f \in M \subset B(f, \frac{\varepsilon}{3})$. This implies that $\operatorname{diam}(M) \leq \operatorname{diam}(B(f, \frac{\varepsilon}{3})) \leq \frac{2}{3}\varepsilon < \varepsilon$, so $M \in \mathcal{U}_n$ and hence $f \in M_n$.

Thus $C_p(K) = \bigcup M_n$ which shows that $C_p(K)$ has the property JNR; this settles (ii)$\Longrightarrow$(i) and makes our solution complete.

**V.490.** *Prove that*

(a) *if $L$ is a linearly ordered separable compact space then $C_p(L)$ is a $\sigma$-space, i.e., it has a $\sigma$-discrete network;*

(b) *for any separable dyadic compact space $K$, the space $C_p(K)$ has a $\sigma$-discrete network.*

**Solution.** (a) For any set $A \subset \mathbb{R}$ let $\mathrm{dm}(A) = \sup\{|t - s| : s, t \in A\}$, i.e., $\mathrm{dm}(A)$ is the diameter of $A$ in $\mathbb{R}$. Given a space $X$ and a continuous function $f$ on the space $X$ let $||f||_Y = \sup\{|f(x)| : x \in Y\}$ for any $Y \subset X$. If $P \subset C_p^*(X)$ then let $\mathrm{diam}_X(P) = \sup\{||f - g||_X : f, g \in P\}$. A family $\mathcal{P}$ of subsets of $C_p^*(X)$ is $\varepsilon$-small if $\mathrm{diam}_X(P) < \varepsilon$ for any $P \in \mathcal{P}$.

Denote by $<$ the order that generates the topology of $L$. Let $a$ be the smallest element of $L$ and denote by $b$ its largest element (see TFS-305). For any $x, y \in L$ the expression $x \leq y$ says that either $x < y$ or $x = y$. For any $x, y \in L$ let $[x, y] = \{z \in L : x \leq z \leq y\}$ and $(x, y) = \{z \in L : x < z < y\}$. We will also need the intervals $[x, y) = \{z \in L : x \leq z < y\}$ and $(x, y] = \{z \in L : x < z \leq y\}$. Fix a countable dense set $S \subset L$ such that $\{a, b\} \subset S$.

*Fact 1.* Suppose that $X$ is a space, $\varepsilon > 0$ and $\mathcal{A}$ is a family of subsets of $X$ such that $X = \bigcup \mathcal{A}$; let $F = \{f \in C(X) : \mathrm{dm}(f(A)) \leq \varepsilon$ for any $A \in \mathcal{A}\}$. Assume additionally, that $D \subset X$ and $D \cap A \neq \emptyset$ for any $A \in \mathcal{A}$. Then $||f - g||_X \leq 3\varepsilon$ for any pair of functions $f, g \in F$ such that $||f - g||_D \leq \varepsilon$.

*Proof.* Take any $t \in X$ and choose $A \in \mathcal{A}$ such that $t \in A$. There exists $d \in D \cap A$; it follows from $||f - g||_D \leq \varepsilon$ that $|f(d) - g(d)| \leq \varepsilon$. Besides, we have the inequalities $|f(t) - f(d)| \leq \mathrm{dm}(f(A)) \leq \varepsilon$ and $|g(t) - g(d)| \leq \mathrm{dm}(g(A)) \leq \varepsilon$. As an immediate consequence, $|f(t) - g(t)| \leq |f(t) - f(d)| + |f(d) - g(d)| + |g(d) - g(t)| \leq 3\varepsilon$ for any $t \in X$, i.e., $||f - g||_X \leq 3\varepsilon$ and hence Fact 1 is proved.

Returning to our solution consider, for any $x, y \in L$ with $x \leq y$ and any $\varepsilon > 0$ the set $A(x, y, \varepsilon) = \{f \in C_p(L) :$ there exist $p, q \in [x, y]$ such that $p \leq q$ and $[x, y] = [x, p] \cup [q, y]$ while $\mathrm{dm}(f([x, p])) \leq \varepsilon$ and $\mathrm{dm}(f([q, y])) \leq \varepsilon\}$. Take any function $f \in C_p(X) \backslash A(x, y, \varepsilon)$; to construct a neighborhood of $f$ which does not meet $A(x, y, \varepsilon)$ we will need the points $u = \sup\{t \in [x, y] : \mathrm{dm}(f([x, t])) \leq \varepsilon\}$ and $v = \inf\{t \in [x, y] : \mathrm{dm}(f([t, y])) \leq \varepsilon\}$.

It is easy to see that $\mathrm{dm}(f([x, u])) \leq \varepsilon$ and $\mathrm{dm}(f([v, y])) \leq \varepsilon$; this, together with $f \notin A(x, y, \varepsilon)$ implies that $[x, y] \neq [x, u] \cup [v, y]$ and hence there exists a point $z \in (u, v)$. By our choice of $u$ and $v$ there exist $r, s \in [x, z]$ and $t, w \in [z, y]$ such that $|f(r) - f(s)| > \varepsilon$ and $|f(t) - f(w)| > \varepsilon$.

The set $W = \{g \in C_p(L) : |g(r) - g(s)| > \varepsilon$ and $|g(t) - g(w)| > \varepsilon\}$ is open in $C_p(L)$ and $f \in W$. If $g \in W \cap A(x, y, \varepsilon)$ then there exist $p, q \in L$ such that $p \leq q$ and $[x, y] = [x, p] \cup [q, y]$ while $\mathrm{dm}(g([x, p])) \leq \varepsilon$ and $\mathrm{dm}(g([q, y])) \leq \varepsilon$. Therefore the segment $[x, p]$ cannot contain the points $r$ and $s$ and hence $p < z$. Analogously, the set $\{t, w\}$ cannot be contained in $[q, y]$, so $z < q$.

Thus $z \in [x, y] \backslash ([x, p] \cup [q, y])$ which is a contradiction. Therefore $W \cap A(x, y, \varepsilon) = \emptyset$, so we proved that every $f \in C_p(L) \backslash A(x, y, \varepsilon)$ has a neighborhood contained in $C_p(L) \backslash A(x, y, \varepsilon)$. As a consequence,

(1) for every $\varepsilon > 0$ the set $A(x, y, \varepsilon)$ is closed in $C_p(L)$ for any $x, y \in L$ with $x \leq y$.

Our next step is to prove that

(2) for any $\varepsilon > 0$ and $x, y \in L$ with $x \leq y$, if $f \in A(x, y, \varepsilon)$ then there exists $U_f \in \tau(f, C_p(L))$ such that $\|g - f\|_{[xy]} \leq 3\varepsilon$ for each $g \in U_f \cap A(x, y, \varepsilon)$.

Fix points $p, q \in [x, y]$ such that $p \leq q$ and $[x, y] = [x, p] \cup [q, y]$ while we have the inequalities $\text{dm}(f([x, p])) \leq \varepsilon$ and $\text{dm}(f([q, y])) \leq \varepsilon$. Let $D = \{x, y, p, q\}$ and observe that the set $U_f = \{f \in C_p(L) : |g(t) - f(t)| < \varepsilon$ for all $t \in D\}$ is an open neighborhood of $f$ in $C_p(L)$ such that $\|f - g\|_D \leq \varepsilon$ for any $g \in U_f$. Take any function $g \in U_f \cap A(x, y, \varepsilon)$ and choose points $r, s \in [x, y]$ such that $r \leq s$ and $[x, y] = [x, r] \cup [s, y]$ while $\text{dm}(g([x, r])) \leq \varepsilon$ and $\text{dm}(g([s, y])) \leq \varepsilon$. There are several cases to consider.

*Case 1.* $r = p = q$ and hence $q \leq s$. Then the family $\mathcal{A} = \{[x, r], [s, y]\}$ in the space $X = [x, y]$ and the set $D$ together with the functions $f|X$ and $g|X$ satisfy the assumptions of Fact 1, so $\|f - g\|_X \leq 3\varepsilon$.

*Case 2.* $r = p < q$; it is an easy exercise that this implies $s \leq q$. Then the family $\mathcal{A} = \{[x, p], [q, y]\}$ in the space $X = [x, y]$ and the set $D$ together with the functions $f|X$ and $g|X$ satisfy the assumptions of Fact 1, so $\|f - g\|_X \leq 3\varepsilon$.

*Case 3.* $r < p$; it is easy to see that this implies $s \leq p$. Then the family $\mathcal{A} = \{[x, r], [s, p], [q, y]\}$ and the set $D$ together with the functions $f|X$ and $g|X$ satisfy the assumptions of Fact 1, so $\|f - g\|_X \leq 3\varepsilon$.

*Case 4.* $r > p$; it is straightforward that this implies $r \geq q$. Then the family $\mathcal{A} = \{[x, p], [q, r], [s, y]\}$ and the set $D$ together with the functions $f|X$ and $g|X$ satisfy the assumptions of Fact 1, so $\|f - g\|_X \leq 3\varepsilon$.

Thus, in all possible cases we have $\|f - g\|_X \leq 3\varepsilon$, i.e., (2) is proved. Now let us establish that

(3) for any $f \in C_p(L)$ and $\varepsilon > 0$ there exists a sequence $\{s_0, \ldots, s_n\} \subset S$ such that $s_0 = a < s_1 < \ldots < s_n = b$ and $f \in A(s_i, s_{i+1}, \varepsilon)$ for any $i < n$.

Recall that a set $P \subset L$ is called *convex* if $[x, y] \subset P$ for any $x, y \in P$. It is easy to see that the family $\mathcal{C}$ of convex open subsets of $L$ is a base of $L$. By continuity of $f$ for any $x \in L$ there exists $C_x \in \mathcal{C}$ such that $\text{dm}(f(C_x)) \leq \varepsilon$. Choose a finite $K \subset L$ such that $L = \bigcup\{C_x : x \in K\}$; it is easy to find a sequence $Q = \{s_0, \ldots, s_n\} \subset S$ such that $a = s_0 < \ldots < s_n = b$ and $Q \cap C_x \neq \emptyset$ for any $x \in K$.

Fix any $i < n$ and consider the families $\mathcal{E}_0 = \{C_x : x \in K$ and $s_i \in C_x\}$ and $\mathcal{E}_1 = \{C_x : x \in K$ and $s_{i+1} \in C_x\}$. Observe first that $[s_i, s_{i+1}] \subset \bigcup \mathcal{E}_0 \cup \bigcup \mathcal{E}_1$. Indeed, if $t \in [s_i, s_{i+1}] \backslash (\bigcup \mathcal{E}_0 \cup \bigcup \mathcal{E}_1)$ then $t \in C_x$ for some $x \in K$; by convexity of $C_x$, it follows from $C_x \cap \{s_i, s_{i+1}\} = \emptyset$ that $C_x \subset (s_i, s_{i+1})$. There exists a point $q \in Q \cap C_x$; then $q \in (s_i, s_{i+1})$ which is a contradiction. This proves that $[s_i, s_{i+1}] \subset \bigcup \mathcal{E}_0 \cup \bigcup \mathcal{E}_1$.

If $E, E' \in \mathcal{E}_0$ then it is easy to deduce from convexity of $E$ and $E'$ that either $E \cap [s_i, s_{i+1}] \subset E' \cap [s_i, s_{i+1}]$ or $E' \cap [s_i, s_{i+1}] \subset E \cap [s_i, s_{i+1}]$. Consequently, there exists $E_0 \in \mathcal{E}_0$ such that $E_0 \cap [s_i, s_{i+1}] = (\bigcup \mathcal{E}_0) \cap [s_i, s_{i+1}]$. Analogously, we can find $E_1 \in \mathcal{E}_1$ such that $E_1 \cap [s_i, s_{i+1}] = (\bigcup \mathcal{E}_1) \cap [s_i, s_{i+1}]$. Therefore $[s_i, s_{i+1}] \subset E_0 \cup E_1$.

If $x \in E_0 \cap E_1 \cap [s_i, s_{i+1}]$ then it is easy to see that the points $p = q = x$ witness that $f \in A(s_i, s_{i+1}, \varepsilon)$. If $E_0 \cap E_1 \cap [s_i, s_{i+1}] = \emptyset$ then, for the points $p = \sup(E_0 \cap [s_i, s_{i+1}])$ and $q = \inf(E_1 \cap [s_i, s_{i+1}])$, we have $p \le q$ and it is straightforward that $p$ and $q$ witness that $f \in A(s_i, s_{i+1}, \varepsilon)$, so (3) is proved.

Fix an arbitrary $\varepsilon > 0$ and observe that for any points $s_0, \ldots, s_n \in S$ such that $a = s_0 < \ldots < s_n = b$, the set $M(s_1, \ldots, s_n) = \bigcap \{A(s_i, s_{i+1}, \frac{\varepsilon}{9}) : 0 \le i < n\}$ is closed in $C_p(L)$. The family $\mathcal{M} = \{M(s_0, \ldots, s_n) : n \in \mathbb{N}, \; s_0, \ldots, s_n \in S$ and $a = s_0 < \ldots < s_n = b\}$ is countable and it follows from (1) that $\mathcal{M}$ consists of closed subsets of $C_p(L)$. The property (3) shows that $\bigcup \mathcal{M} = C_p(L)$. Now fix any $s_0, \ldots, s_n \in S$ such that $a = s_0 < \ldots < s_n = b$ and take any function $f \in H = M(s_0, \ldots, s_n)$.

The property (2) makes it possible to find sets $U_0^f, \ldots, U_{n-1}^f \in \tau(f, C_p(L))$ such that $\|g - f\|_{[s_i, s_{i+1}]} \le \frac{\varepsilon}{3}$ for any $g \in U_i^f \cap H$; if $U^f = \bigcap_{i<n} U_i^f$ then $\|g - f\|_L \le \frac{\varepsilon}{3}$ for any $g \in U^f \cap H$ and hence $G_f = U^f \cap H$ is an open subset of $H$ such that $\text{diam}_L(G^f) \le \frac{2\varepsilon}{3} < \varepsilon$. Therefore $\{G^f : f \in H\}$ is an $\varepsilon$-small open cover of $H$; this demonstrates that $C_p(L)$ has the property JNR and hence we can apply Problem 489 to conclude that $C_p(L)$ has a $\sigma$-discrete network.

(b) We will first prove that the space $C_p(\mathbb{D}^{\mathfrak{c}})$ has a $\sigma$-discrete network. To that end we will identify $\mathbb{D}^{\mathfrak{c}}$ with $\mathbb{D}^S$ where $S = \mathbb{D}^\omega$ is the Cantor set. For every $x \in \mathbb{D}^S$ let $[x, E] = \{y \in \mathbb{D}^S : y|E = x|E\}$ for any finite set $E \subset S$. If $f \in C_p(\mathbb{D}^S)$ then $\|f\| = \sup\{|f(x)| : x \in \mathbb{D}^S\}$. For any bounded set $B \subset \mathbb{R}$ let $\text{dm}(B) = \sup\{|t - s| : s, t \in B\}$, i.e., $\text{dm}(B)$ is the diameter of $B$ in $\mathbb{R}$. Given a set $N \subset C_p(\mathbb{D}^S)$ let $\text{diam}(N) = \sup\{\|f - g\| : f, g \in N\}$. A family $\mathcal{N}$ of subsets of $C_p(\mathbb{D}^S)$ is called $\varepsilon$-small if $\text{diam}(N) < \varepsilon$ for any $N \in \mathcal{N}$.

Say that a family $\mathcal{F}$ of subsets of $S$ is *adequate* if $\mathcal{F}$ is nonempty, disjoint, finite and $F \ne \emptyset$ for any $F \in \mathcal{F}$. Given an adequate family $\mathcal{F}$ say that $L \subset S$ is *selection set for* $\mathcal{F}$ if $|L| = |\mathcal{F}|$ and $L \cap F \ne \emptyset$ for any $F \in \mathcal{F}$; besides, let $k(F) = |\{F \in \mathcal{F} : |F| > 1\}|$.

For any adequate family $\mathcal{F}$ and $\varepsilon > 0$ let $A(\mathcal{F}, \varepsilon) = \{f \in C_p(\mathbb{D}^S) :$ there exists a selection set $L$ for $\mathcal{F}$ such that $|f(x) - f(y)| \le \varepsilon$ whenever $x, y \in \mathbb{D}^S$ and $x|L = y|L\}$. If $\mathcal{F}$ is an adequate family, $\varepsilon > 0$ and $f \in C_p(\mathbb{D}^S) \setminus A(\mathcal{F}, \varepsilon)$ then a set $Q \subset \mathbb{D}^S$ is *representative for the function* $f$ if for any selection set $L$ for the family $\mathcal{F}$ there exist $x, y \in Q$ such that $x|L = y|L$ and $|f(x) - f(y)| > \varepsilon$. It turns out that

(4) given an adequate family $\mathcal{F}$ and $\varepsilon > 0$, if $f \in C_p(\mathbb{D}^S) \backslash A(\mathcal{F}, \varepsilon)$ then there exists a finite representative set $Q \subset \mathbb{D}^S$ for the function $f$.

We will prove (4) by induction on $k(\mathcal{F})$. If $k(\mathcal{F}) = 0$ then all elements of $\mathcal{F}$ are singletons and hence there exists a unique selection set $L$ for $\mathcal{F}$. Since $f \notin A(\mathcal{F}, \varepsilon)$, there must exist $x, y \in \mathbb{D}^S$ such that $x|L = y|L$ and $|f(x) - f(y)| > \varepsilon$. It is clear that $Q = \{x, y\}$ is a representative set for $f$.

Now assume that $n \in \mathbb{N}$ and we proved (4) for all families $\mathcal{F}$ such that $k(\mathcal{F}) < n$. If $\mathcal{F}$ is an adequate family and $k(\mathcal{F}) = n$ then take any selection set $M$ for the family $\mathcal{F}$ and fix $a, b \in \mathbb{D}^S$ such that $a|M = b|M$ and $|f(a) - f(b)| > \varepsilon$. The set $O = \{(x, y) \in \mathbb{D}^S \times \mathbb{D}^S : |f(x) - f(y)| > \varepsilon\}$ is an open neighborhood of the point $(a, b)$, so there exist $G \in \tau(a, \mathbb{D}^S)$ and $H \in \tau(b, \mathbb{D}^S)$ such that $G \times H \subset O$. We can find a finite set $E \subset S$ such that $M \subset E$ while $[a, E] \subset G$ and $[b, E] \subset H$. As a consequence,

(5)  for any $x, y \in \mathbb{D}^S$, if $x|E = a|E$ and $y|E = b|E$ then $|f(x) - f(y)| > \varepsilon$.

Pick $u, v \in \mathbb{D}^S$ such that $u|E = a|E$, $v|E = b|E$ and $u|(S \backslash E) = v|(S \backslash E)$. For each $F \in \mathcal{F}$ let $s_F$ be the unique point of $M \cap F$ and denote the set $E \cap F$ by $E_F$. For any $s \in E_F \backslash \{s_F\}$ consider the family $\mathcal{D}(F, s) = \{\{s\}\} \cup (\mathcal{F} \backslash \{F\})$; it is evident that $\mathcal{D}(F, s)$ is adequate and $k(\mathcal{D}(F, s)) = n - 1$. Since also $f \in C_p(\mathbb{D}^S) \backslash A(D(\mathcal{F}, s), \varepsilon)$, the induction hypothesis is applicable to find a finite representative set $Q(F, s)$ for the family $\mathcal{D}(F, s)$.

The set $Q = \{u, v\} \cup \bigcup \{Q(F, s) : F \in \mathcal{F}$ and $s \in E_F \backslash \{s_F\}\}$ is finite; we claim that it is representative for $f$. Indeed, take any selection set $L$ for the family $\mathcal{F}$ and denote by $t_F$ the unique point of $L \cap F$ for any $F \in \mathcal{F}$. If $t_F \in E_F \backslash \{s_F\}$ for some $F \in \mathcal{F}$ then $L$ is a selection set for the family $\mathcal{D}(F, t_F)$; since $Q(F, t_F)$ is a representative set for $f$ we can find $x, y \in Q(F, t_F) \subset Q$ such that $x|L = y|L$ and $|f(x) - f(y)| > \varepsilon$. If $t_F \notin E_F \backslash \{s_F\}$ then it is easy to see that $u(t_F) = v(t_F)$ for each $F \in \mathcal{F}$ and hence $u|L = v|L$; since $|f(u) - f(v)| > \varepsilon$ by (5), the functions $u, v \in Q$ witness that $Q$ is representative for $f$. This completes our inductive step so (4) is proved.

Now take any adequate family $\mathcal{F}$ and $\varepsilon > 0$; if $f \in C_p(\mathbb{D}^S) \backslash A(\mathcal{F}, \varepsilon)$ then apply (4) to find a finite representative set $Q$ for the function $f$ and consider the set $W = \{g \in C_p(\mathbb{D}^S) : |g(x) - g(y)| > \varepsilon$ whenever $x, y \in Q$ and $|f(x) - f(y)| > \varepsilon\}$; it is immediate that $W \in \tau(f, C_p(\mathbb{D}^S))$. If $g \in W$ and $L$ is a selection set for $\mathcal{F}$ then there exist $x, y \in Q$ such that $x|L = y|L$ and $|f(x) - f(y)| > \varepsilon$; by the definition of $W$ we have $|g(x) - g(y)| > \varepsilon$ and hence $g \notin A(\mathcal{F}, \varepsilon)$. This proves that every $f \in C_p(\mathbb{D}^S) \backslash A(\mathcal{F}, \varepsilon)$ has a neighborhood contained in $C_p(\mathbb{D}^S) \backslash A(\mathcal{F}, \varepsilon)$ and therefore

(6)  the set $A(\mathcal{F}, \varepsilon)$ is closed in $C_p(\mathbb{D}^S)$ for any adequate family $\mathcal{F}$ and $\varepsilon > 0$.

Fix $\varepsilon > 0$ and an adequate family $\mathcal{F}$ of subsets of $S$. Given $f \in A(\mathcal{F}, \varepsilon)$ take a selection set $L$ for the family $\mathcal{F}$ such that $|f(x) - f(y)| \le \varepsilon$ whenever $x|L = y|L$. The set $R = \{x \in \mathbb{D}^S : x$ is constant on $S \backslash \bigcup \mathcal{F}$ and on $F \backslash L$ for any $F \in \mathcal{F}\}$ is easily seen to be finite. The set $V = \{g \in C_p(\mathbb{D}^S) : |g(x) - f(x)| < \varepsilon$ for any

$x \in R\}$ is open in $C_p(\mathbb{D}^S)$; take any $g \in V \cap A(\mathcal{F}, \varepsilon)$. There exists a selection set $M$ for $\mathcal{F}$ such that $|g(x) - g(y)| \leq \varepsilon$ for all $x, y \in \mathbb{D}^S$ such that $x|M = y|M$. Take an arbitrary $x \in \mathbb{D}^S$; it is easy to find $y \in R$ such that $x|(M \cup L) = y|(M \cup L)$.

The choice of $y$ guarantees that $|f(x) - f(y)| \leq \varepsilon$ and $|g(x) - g(y)| \leq \varepsilon$; it follows from $y \in R$ that $|f(y) - g(y)| < \varepsilon$. Since $x \in \mathbb{D}^S$ was chosen arbitrarily, it follows from $|f(x) - g(x)| \leq |f(x) - f(y)| + |f(y) - g(y)| + |g(y) - g(x)| \leq 3\varepsilon$ that $|f(x) - g(x)| \leq 3\varepsilon$ for any $x \in \mathbb{D}^S$, i.e., $||f - g|| \leq 3\varepsilon$. As a consequence,

(7) if $\mathcal{F}$ is an adequate family and $\varepsilon > 0$ then for any $f \in A(\mathcal{F}, \varepsilon)$ there exists a set $V \in \tau(f, C_p(\mathbb{D}^S))$ such that $||f - g|| \leq 3\varepsilon$ for any $g \in V \cap A(\mathcal{F}, \varepsilon)$.

Take any function $f \in C_p(\mathbb{D}^S)$ and $\varepsilon > 0$. By continuity of $f$, for every $x \in \mathbb{D}^S$ there is a finite set $E_x \subset S$ such that $\mathrm{dm}(f([x, E_x])) < \varepsilon$. By compactness of $\mathbb{D}^S$ we can choose a finite set $P \subset \mathbb{D}^S$ such that $\mathbb{D}^S = \bigcup\{[x, E_x] : x \in P\}$; let $E = \bigcup\{E_x : x \in P\}$. Now if $x, y \in \mathbb{D}^S$ and $x|E = y|E$ then pick $a \in P$ for which $x \in [a, E_a]$. Then $y|E_a = x|E_a = a|E_a$ which shows that $y \in [a, E_a]$ and hence $|f(x) - f(y)| \leq \mathrm{dm}(f([a, E_a])) < \varepsilon$. This proves that

(8) for any $f \in \mathbb{D}^S$ and $\varepsilon > 0$ there is a finite set $E \subset S$ such that $|f(x) - f(y)| < \varepsilon$ whenever $x|E = y|E$.

Denote by $\mathcal{B}$ the family of all nonempty clopen subsets of $S$ (recall that $S = \mathbb{D}^\omega$ is the Cantor set). Then $\mathcal{B}$ is countable, so the collection $\mathbb{F}$ of all adequate families of elements of $\mathcal{B}$ is countable. Fix an arbitrary $\varepsilon > 0$ and consider the countable family $\mathcal{A} = \{A(\mathcal{F}, \frac{\varepsilon}{9}) : \mathcal{F} \in \mathbb{F}\}$. It follows from (6) that all elements of $\mathcal{A}$ are closed in $C_p(\mathbb{D}^S)$. Given any $A(\mathcal{F}, \frac{\varepsilon}{9}) \in \mathcal{A}$ apply (7) to find a set $V_f \in \tau(f, C_p(\mathbb{D}^S))$ such that $||f - g|| \leq \frac{\varepsilon}{3}$ for any $f \in A(\mathcal{F}, \varepsilon)$ and $g \in G_f = V_f \cap A(\mathcal{F}, \frac{\varepsilon}{9})$. An immediate consequence is that $\mathrm{diam}(G_f) \leq \frac{2}{3}\varepsilon < \varepsilon$ for any $f \in A(\mathcal{F}, \frac{\varepsilon}{9})$ and hence $\{G_f : f \in A(\mathcal{F}, \frac{\varepsilon}{9})\}$ is an open $\varepsilon$-small cover of $A(\mathcal{F}, \frac{\varepsilon}{9})$. If $f \in C_p(\mathbb{D}^S)$ then we can apply (8) to find a set $E = \{s_1, \ldots, s_n\} \subset S$ such that $|f(x) - f(y)| < \frac{\varepsilon}{9}$ whenever $x|E = y|E$. Take disjoint sets $B_1, \ldots, B_n \in \mathcal{B}$ such that $s_i \in B_i$ for each $i \leq n$. Then $\mathcal{F} = \{B_1, \ldots, B_n\}$ is an adequate family which, together with the set $E$, witnesses that $f \in A(\mathcal{F}, \frac{\varepsilon}{9})$. Therefore $\bigcup \mathcal{A} = C_p(\mathbb{D}^S)$, so we proved that $C_p(\mathbb{D}^S)$ has the property JNR and hence $C_p(\mathbb{D}^S) \simeq C_p(\mathbb{D}^c)$ has a $\sigma$-discrete network by Problem 489.

Finally assume that $K$ is a separable dyadic compact space; by Fact 2 of S.368, we have $w(K) \leq 2^{d(K)} \leq \mathfrak{c}$. Suppose that $A$ is a set and $\varphi : \mathbb{D}^A \to K$ is a continuous onto map. It follows from TFS-299 that we can find a set $B \subset A$ such that $|B| \leq \mathfrak{c}$ and there exists a continuous map $\xi : \mathbb{D}^B \to K$ for which $\xi \circ \pi = \varphi$ where $\pi : \mathbb{D}^A \to \mathbb{D}^B$ is the natural projection. In particular, $\mathbb{D}^B$ maps continuously onto $K$ and hence $\mathbb{D}^c$ also maps continuously onto $K$. Therefore $C_p(K)$ embeds in $C_p(\mathbb{D}^c)$, so $C_p(K)$ has a $\sigma$-discrete network, i.e., we proved (b) and hence our solution is complete.

**V.491.** *Prove that there exists a scattered compact space $K$ with a countable dense set of isolated points such that $C_p(K, \mathbb{D})$ is not perfect. Deduce from this fact that $C_p(\beta\omega, \mathbb{D})$ is not perfect and, in particular, the space $C_p(\beta\omega)$ does not have a $\sigma$-discrete network.*

**Solution.** Recall that a space $Z$ is *perfect* if every $U \in \tau(Z)$ is an $F_\sigma$-subset of $Z$. Let $S$ be the set of all finite sequences of natural numbers, i.e., $S = \bigcup\{\omega^n : n \in \omega\}$ where $\omega^0 = \{\emptyset\}$. We consider that the set $\omega^\omega$ carries the usual product topology. For any $s \in S$ let $[s] = \{x \in \omega^\omega : x|\mathrm{dom}(s) = s\}$; it is clear that $\{[s] : s \in S\}$ is a base in $\omega^\omega$. If $s \in S$, $\mathrm{dom}(s) = n$ and $i \in \omega$ then $t = s^\frown i \in \omega^{n+1}$ is defined by the conditions $t|n = s$ and $t(n) = i$.

If $x \in \omega^\omega$ then let $B_x^m = \{x|n : n \geq m\}$ for any $m \in \omega$. To introduce a topology on the set $L = S \cup \omega^\omega$ declare all points of $S$ to be isolated; if $x \in \omega^\omega$ then let the family $\mathcal{B}_x = \{\{x\} \cup B_x^m : m \in \omega\}$ be a local base at the point $x$. It is straightforward that $L$ is a locally compact Tychonoff space; denote by $K$ the one-point compactification of $L$ and let $\theta$ be the unique point of $K\backslash L$.

It is immediate that $K$ is a scattered separable compact space and $S$ is a countable dense set of isolated points of $K$. To show that $C_p(K, \mathbb{D})$ is not a perfect space consider the set $F = \{f \in C_p(K, \mathbb{D}) : f^{-1}(1) \subset S\}$. It follows from the equality $F = \{f \in C_p(K, \mathbb{D}) : f(\omega^\omega \cup \{\theta\}) = \{0\}\}$ that the set $F$ is closed in $C_p(K, \mathbb{D})$. If $C_p(K, \mathbb{D})$ is perfect then we can find a family $\{P_n : n \in \omega\}$ of closed subsets of $C_p(K, \mathbb{D})$ such that $C_p(K, \mathbb{D})\backslash F = \bigcup_{n\in\omega} P_n$.

Given any $x \in \omega^\omega$ the set $B_x = \{x\} \cup B_x^0$ is easily seen to be clopen in $K$; let $g_x$ be the characteristic function of $B_x$, i.e., $g_x(y) = 1$ for all $y \in B_x$ and $g_x(y) = 0$ whenever $y \in K\backslash B_x$. Then $\{g_x : x \in \omega^\omega\} \subset \bigcup_{n\in\omega} P_n$; the Baire property of $\omega^\omega$ implies that there exists $n \in \omega$ such that the set $Q = \{x \in \omega^\omega : g_x \in P_n\}$ is dense in $[s]$ for some $s \in S$; let $m = \mathrm{dom}(s)$. Choose a point $x_i \in Q$ such that $x_i \in [s^\frown i]$ for every $i \in \omega$.

For the set $A = \{s|i : i \leq m\} \subset K$ let $g \in C_p(K, \mathbb{D})$ be the characteristic function of $A$; then $A = g^{-1}(1)$ and hence $g \in F$. It is straightforward that $B_{x_i} \cap B_{x_j} = A$ for any distinct $i, j \in \omega$ which easily implies that the sequence $\{g_{x_i} : i \in \omega\} \subset P_n$ converges to $g$. However, $P_n$ is closed in $C_p(K, \mathbb{D})$ and $g \notin P_n$; this contradiction shows that $C_p(K, \mathbb{D})\backslash F$ is not an $F_\sigma$-subset of $C_p(K, \mathbb{D})$ and hence $C_p(K, \mathbb{D})$ is not a perfect space.

Finally, observe that there exists a continuous surjective map $\varphi : \beta\omega \to K$ because $\beta\omega$ can be continuously mapped onto any separable compact space. Let $\varphi^*(f) = f \circ \varphi$ for any $f \in C_p(K)$; then $\varphi^* : C_p(K) \to C_p(\beta\omega)$ is an embedding such that $\varphi^*(C_p(K, \mathbb{D})) \subset C_p(\beta\omega, \mathbb{D})$. The property of being perfect is easily seen to be hereditary so $C_p(\beta\omega, \mathbb{D})$ is not perfect. Any space with a $\sigma$-discrete network is perfect and hence $C_p(\beta\omega) \supset C_p(\beta\omega, \mathbb{D})$ has no $\sigma$-discrete network.

**V.492.** *Prove that, for any ultrafilter $\xi \in \beta\omega\backslash\omega$, the following conditions are equivalent:*

(i) *$\xi$ is not a $P$-point in $\beta\omega\backslash\omega$;*
(ii) *$\sigma(\mathbb{R}^\omega)$ is homeomorphic to a closed subspace of $C_p(\omega_\xi)$;*
(iii) *$C_p(\omega_\xi)$ is not hereditarily Baire.*

**Solution.** Let $\omega^0 = \{\emptyset\}$ and consider the set $\Omega = \bigcup\{\omega^n : n \in \omega\}$ of all finite sequences of natural numbers. If $s \in \Omega$ then let $l(s) = |\mathrm{dom}(s)|$ be the length of $s$. We follow the usual practice of identifying any ordinal with the set of its predecessors. In particular, $0 = \emptyset$ and $i = \{0, \ldots, i-1\}$ for any $i \in \mathbb{N}$. Thus, for any $s \in \Omega$, if $l(s) = n \geq i \in \omega$ then $s|i$ is the restriction of $s$ to the set $\{0, \ldots, i-1\}$.

For any $s \in \Omega$ and $n \in \omega$ the sequence $t = s^\frown n$ is defined by the equalities $t(i) = s(i)$ for any $i < k = \mathrm{dom}(s)$ and $t(k) = n$. Given $s, t \in \Omega$ the expression $s \prec t$ says that $t$ extends $s$, i.e., $\mathrm{dom}(s) \subset \mathrm{dom}(t)$ and $t(i) = s(i)$ for any $i \in \mathrm{dom}(s)$. Analogously, $u \in \omega^\omega$ extends $s$ (we also denote this by $s \prec u$) if $u(i) = s(i)$ for any $i \in \mathrm{dom}(s)$. If a metric $\rho$ is considered on the space $X$ then, for any point $x \in X$ and $r > 0$ the set $B(x, r) = \{y \in X : \rho(x, y) < r\}$ is the open ball of radius $r$ centered at $x$ and $\mathrm{diam}(A) = \sup\{\rho(x, y) : x, y \in A\}$ for any $A \subset X$.

A family $\{V(s) : s \in \Omega\} \subset \tau(X)$ is an $A$-system on $X$ if

(A1) $V(\emptyset) = X$ and $V(s) = \bigcup\{V(s^\frown i) : i \in \omega\}$ for any $s \in \Omega$;
(A2) $\mathrm{diam}(V(s)) < \frac{1}{l(s)}$ for any $s \in \Omega \backslash \{\emptyset\}$;
(A3) $\bigcap\{V(u|i) : i \in \omega\} \neq \emptyset$ for any $u \in \omega^\omega$.

For linear spaces $L$ and $M$ the expression $L \approx M$ says that $L$ is linearly homeomorphic to $M$. Given an ultrafilter $\xi \in \beta\omega \backslash \omega$, the Gul'ko–Sokolov game on the space $\omega_\xi$ is played as follows: at the $n$-th move player $NU$ takes a set $S_n \in \exp(\omega) \backslash \xi$ and player $U$ replies by choosing a finite set $T_n \subset \omega$ in such a way that the family $\{S_i : i \leq n\} \cup \{T_i : i \leq n\}$ is disjoint. The game ends after the moves $S_n, T_n$ are made for all $n \in \omega$ and player $U$ wins if $\bigcup\{T_n : n \in \omega\} \in \xi$. Otherwise, the player $NU$ is the winner.

*Fact 1.* A separable metrizable space $X$ is hereditarily Baire if and only if there is no closed subset of $X$ homeomorphic to $\mathbb{Q}$.

*Proof.* If some closed set $F \subset X$ is homeomorphic to $\mathbb{Q}$ then $F$ is not Baire, so $X$ is not hereditarily Baire; this proves necessity. Now if $X$ is not hereditarily Baire then we can find a closed set $F \subset X$ which is not Baire. Therefore there exists a set $U \in \tau^*(F)$ which is of first category in $F$. The space $X$ being second countable we can find a family $\{P_n : n \in \omega\}$ of closed nowhere dense subsets of $F$ such that $U = \bigcup_{n \in \omega} P_n$. It is easy to see that every $P_n$ is nowhere dense in $U$ and hence in $\overline{U}$. Since $P = \overline{U} \backslash U$ is also closed and nowhere dense in $\overline{U}$, the set $\overline{U} = U \cup P$ is of first category in itself. Observe that no space of first category in itself can have isolated points, so we can apply Fact 4 of T.351 to conclude that some closed subset $Q$ of the space $\overline{U}$ is homeomorphic to $\mathbb{Q}$. The set $Q$ is also closed in $X$, so we settled sufficiency, i.e., Fact 1 is proved.

*Fact 2.* If $X$ is a separable metrizable space, then player $E$ has a winning strategy in the Choquet game on $X$ if and only if $X$ is not a hereditarily Baire space.

*Proof.* Suppose that $X$ is not a hereditarily Baire space and hence we can find a closed set $F \subset X$ which is homeomorphic to $\mathbb{Q}$; let $\{q_n : n \in \omega\}$ be a faithful enumeration of the set $F$ and fix a metric $\rho$ which generates the topology of $X$. For every number $i \in \omega$ let $Q_i = \{q_0 \ldots, q_i\}$.

Choose a point $x_0 \in F \backslash \{q_0\}$ and $r_0 \in (0, 1)$ such that the closure of the set $U_0 = B(x_0, r_0)$ does not contain $q_0$ and let $\sigma(\emptyset) = (x_0, U_0)$.

Proceeding inductively assume that $n \in \omega$ and we have moves $(x_0, U_0), V_0, \ldots,$ $(x_{n-1}, U_{n-1}), V_{n-1}, (x_n, U_n)$ in the Choquet game on $X$ such that

(1) $x_i \in F$ for all $i \leq n$;
(2) $U_i = B(x_i, r_i)$ where $0 < r_i < 2^{-i}$ for all $i \leq n$;
(3) $\overline{U}_i \cap Q_i = \emptyset$ for each $i \leq n$;
(4) $\overline{U}_{i+1} \subset V_i$ for all $i < n$.

If the player $NE$ makes a move $V_n$ then we can choose a point $x_{n+1} \in V_n \backslash Q_{n+1}$ because the space $F$ has no isolated points. Pick a number $r_{n+1} \in (0, 2^{-n-1})$ such that the closure of the set $U_{n+1} = B(x_{n+1}, r_{n+1})$ is contained in $V_n \backslash Q_{n+1}$ and let $(x_{n+1}, U_{n+1}) = \sigma(V_0, \ldots, V_n)$. It is easy to see that this defines a strategy $\sigma$ of the player $E$ on the space $X$. To see that $\sigma$ is winning assume that we have a play $\{(x_i, U_i), V_i : i \in \omega\}$ in which the strategy $\sigma$ is applied. The property (2) shows that $\text{diam}(U_i) \to 0$, so if $\bigcap_{i \in \omega} U_i \neq \emptyset$ then this intersection is a singleton, i.e., $\bigcap_{i \in \omega} U_i = \{x\}$ for some $x \in X$.

Furthermore, it follows from $x \in U_i$ and (2) that $\rho(x_i, x) < 2^{-i}$ for every $i \in \omega$ and hence $x_i \to x$. The set $F$ being closed in $X$, the point $x$ has to belong to $F$ and hence $x = q_n$ for some $n \in \omega$. The property (3) implies that $q_n \notin \overline{U}_n$; however, $S = \{x_i : i \geq n\} \subset \overline{U}_n$ and hence $q_n = x \in \overline{S} \subset \overline{U}_n$; this contradiction shows that $\bigcap_{i \in \omega} U_i = \emptyset$ and hence $\sigma$ is a winning strategy for $E$, i.e., we proved sufficiency.

Now assume that $\sigma$ is a winning strategy for the player $E$; for any $U \in \tau(X)$ and $x \in U$ take a number $r_U^x > 0$ such that $\overline{B(x, r_U^x)} \subset U$. If $(x_0, U_0) = \sigma(\emptyset)$ then let $A(\emptyset) = \{(x_0, U_0)\}$ and $q(\emptyset) = x_0$. For any $s \in \Omega$ with $l(s) = n > 0$ let $A(s) = \{M_0, V_0, \ldots, M_{n-1}, V_{n-1}, M_n\}$ if the following conditions are satisfied:

(5) $A(s)$ is an initial segment of a play in the Choquet game on $X$, i.e., $M_i = (x_i, U_i)$ while $x_i \in V_i \subset U_i$ for all $i < n$ and $U_i \subset V_{i-1}$ whenever $0 < i \leq n$;
(6) the moves of the player $E$ are made applying the strategy $\sigma$, i.e., $M_0 = \sigma(\emptyset)$ and $M_{i+1} = \sigma(V_0, \ldots, V_i)$ for any $i < n$;
(7) $V_i = B(x_i, 2^{-s(i)} r_{U_i}^{x_i})$ for each $i < n$.

It is evident that the family $A(s)$ is uniquely determined by the sequence $s$; let $q(s) = x_n$. We are going to show that the set $F = \{q(s) : s \in \Omega\}$ witnesses that the space $X$ is not hereditarily Baire.

It is clear that the set $F$ is countable; fix an arbitrary $s \in \Omega$ and $\varepsilon > 0$. If $l(s) = n$ then $A(s) = \{M_0, V_0, \ldots, M_{n-1}, V_{n-1}, M_n\}$ where $M_i = (x_i, U_i)$ for every $i \leq n$. Then $q(s) = x_n$ and we can choose $k \in \omega$ such that $2^{-k} r_{U_n}^{x_n} < \varepsilon$. Let $t = s^\frown k$ and fix any $u \in \omega^\omega$ such that $t \prec u$. The function $u$ defines a play $P = \{N_i, W_i : i \in \omega\}$ such that $A(u|m) = \{N_0, W_0, \ldots, N_{m-1}, W_{m-1}, N_m\}$ for each $m \in \omega$. As an immediate consequence, $E$ played in $P$ according to the strategy $\sigma$, so the play $P$ is a win for $E$ and hence $\bigcap_{i \in \omega} W_i = \emptyset$. It follows from $t \prec u$ that $N_i = M_i$ and $W_i = V_i$ for all $i < n$ while $M_n = N_n$ and $W_n = B(x_n, 2^{-k} r_{U_n}^{x_n})$.

If the set $\{x_j : j \in \omega\}$ is finite then some $x_j$ belongs to infinitely many $W_i$'s and hence $\bigcap_{i \in \omega} W_i \neq \emptyset$ which is a contradiction. Therefore there exists $m > n + 1$ such that $x_m \neq x_n$. The point $x_m$ has to belong to the set $W_n$ and hence we have the inequality $\rho(x_m, x_n) < 2^{-k} r_{U_n}^{x_n} < \varepsilon$ which shows that $x_m = q(u|m) \in F \backslash q(s)$ is an element of $F \backslash \{q(s)\}$ which belongs to the set $B(q(s), \varepsilon)$. Therefore every $\varepsilon$-ball centered at $q(s)$ contains a point of $F \backslash \{q(s)\}$ and hence $q(s)$ is not isolated in $F$ for any $s \in \Omega$, i.e., the set $F$ has no isolated points. By SFFS-349, the space $F$ is homeomorphic to $\mathbb{Q}$ and hence $F$ is of first category in itself.

To see that $F$ is a closed subset of the space $X$ assume that there exists a point $z \in \overline{F} \backslash F$ and choose a faithfully enumerated sequence $Z = \{z_n : n \in \omega\} \subset F$ which converges to $z$. For every $s \in \Omega$ consider the set $H(s) = \{q(t) : t \in \Omega \backslash \{s\}$ and $s \prec t\}$ and let $H_i(s) = \{q(t) : t \in \Omega$ and $s^\frown i \prec t\}$ for any $i \in \omega$; it is evident that $H(s) = \bigcup_{i \in \omega} H_i(s)$.

We claim that

(8) given any $s \in \Omega$ and $D \subset \omega$ the set $Z_D = \{z_n : n \in D\}$ intersects only finitely many sets $H_i(s)$.

To prove (8) assume that the set $Q = \{i \in \omega : Z_D \cap H_i(s) \neq \emptyset\}$ is infinite and fix $\varepsilon > 0$. We have $A(s) = \{M_0, V_0, \ldots, M_{m-1}, V_{m-1}, M_m\}$ for some $m \in \omega$ where $M_i = (x_i, U_i)$ for all $i \leq m$.

There exists $i \in Q$ such that $2^{-i} r_{U_m}^{x_m} < \varepsilon$; take $n \in D$ such that $z_n \in H_i(s)$ and hence $z_n = q(t)$ for some $t \in \Omega$ such that $t|m = s$ and $t(m) = i$. This implies that $z_n \in B(q(s), 2^{-i} r_{U_m}^{x_m})$ and hence $\rho(z_n, q(s)) < \varepsilon$. Since $\varepsilon > 0$ was taken arbitrarily, we established that $q(s)$ is an accumulation point of $Z_D$ which is impossible because $Z_D$ is a nontrivial subsequence of $Z$, so it has to converge to $z \neq q(s)$. This contradiction shows that (8) is proved. It is easy to deduce from (8) that

(9) for any $s \in \Omega$ if the set $D = \{n \in \omega : z_n \in H(s)\}$ is infinite then there exists $m \in \omega$ such that $D' = \{n \in \omega : z_n \in H_m(s)\}$ is also infinite.

It is easy to construct by induction, using the property (9), a function $u \in \omega^\omega$ such that the set $D_n = \{i \in \omega : z_i \in H(u|n)\}$ is infinite for any $n \in \omega$. There exists a play $\mathcal{M} = \{M_i, V_i : i \in \omega\}$ in which $E$ applies the strategy $\sigma$ and, for every $n \in \omega$, we have $A(u|n) = \{M_0, V_0, \ldots, M_{n-1}, V_{n-1}, M_n\}$ where $M_i = (x_i, U_i)$ for all $i \leq n$. Fix any $n \in \omega$; then the set $J = \{i \in \omega : z_i = q(t_i)$ for some $t_i \succ u|n\}$ is infinite. If $i \in J$ and $l(t_i) = k > n$ then $A(t_i) = \{M_0, V_0, \ldots, M_n, W_n, \ldots, N_{k-1}, W_{k-1}, N_k\}$; since $A(t_i)$ is an initial segment of a play, we must have $z_i = q(t_i) \in W_{k-1} \subset U_n$ and hence $\{z_i : i \in J\} \subset U_n$. As a consequence, $z \in \overline{\{z_i : i \in J\}} \subset \overline{U}_n$ for each $n \in \omega$ which shows that $z \in \bigcap_{n \in \omega} \overline{U}_n = \bigcap_{n \in \omega} U_n$ which is a contradiction with the fact that the play $\mathcal{M}$ is favorable for $E$ and hence $\bigcap_{n \in \omega} U_n = \emptyset$. Therefore $F$ is a closed subspace of $X$ which is of first category in itself so $X$ is not hereditarily Baire, i.e., we settled necessity and hence Fact 2 is proved.

*Fact 3.* Given a separable metrizable space $X$, the player $NE$ has a winning strategy in the Choquet game on $X$ if and only if $X$ is Čech-complete.

*Proof.* Assume that $X$ is Čech-complete and hence we can find a complete metric $\rho$ on the set $X$ with $\tau(X) = \tau(\rho)$. If $n \in \omega$ and the player $E$ chooses a pair $(x_n, U_n)$ where $x_n \in U_n \in \tau(X)$ then the player $NE$ takes a number $r \in (0, 1)$ such that

$\overline{B(x_n, r)} \subset U_n$ and declares his move to be the set $V_n = B(x, 2^{-n}r)$. This gives a strategy $\sigma$ for the player $NE$ in the Choquet game on $X$. If $\mathcal{P} = \{(x_i, U_i), V_i : i \in \omega\}$ is a play in which $NE$ applies $\sigma$ then we have a decreasing family $\{\overline{V}_i : i \in \omega\}$ of closed subsets of $X$ such that $\mathrm{diam}(\overline{V}_i) \leq 2^{-i+1}$ for every $i \in \omega$ and hence $\mathrm{diam}(\overline{V}_i) \to 0$. The space $(X, \rho)$ being complete we can apply TFS-236 to see that $\bigcap_{i \in \omega} \overline{V}_i \neq \emptyset$ and hence $\bigcap_{i \in \omega} U_i = \bigcap_{i \in \omega} \overline{V}_i \neq \emptyset$ which shows that $\sigma$ is a winning strategy of $NE$ and hence we proved sufficiency.

Now suppose that the player $NE$ has a winning strategy $\sigma$ in the Choquet game on the space $X$. Take a metric $\rho$ which generates the topology of $X$. For every $x \in X$ let the move of the player $E$ be the pair $(x, B(x, \frac{1}{3}))$; the strategy $\sigma$ gives us a set $V_x = \sigma((x, B(x, \frac{1}{3})))$. Choose a countable subcover $\{V_{x_n} : n \in \omega\}$ of the open cover $\{V_x : x \in X\}$ of the space $X$; let $q(s) = x_{s(0)}$, $U(s) = B(q(s), \frac{1}{3})$ and $V(s) = V_{q(s)} = \sigma((q(s), U(s)))$ for any $s \in \omega^1$. We will also need the set $V(\emptyset) = X$.

Suppose that, for some $n \in \mathbb{N}$ we have defined sets $U(s), V(s) \in \tau(X)$ and a point $q(s) \in X$ for any $s \in \Omega \backslash \{\emptyset\}$ with $l(s) \leq n$ in such a way that

(10) if $0 \leq l(s) < n$ then $V(s) = \bigcup \{V(s^\frown i) : i \in \omega\}$;
(11) if $1 \leq l(s) \leq n$ then $\mathrm{diam}(U(s)) \leq 2 \cdot 3^{-l(s)}$;
(12) if $1 \leq l(s) \leq n$ then $\{(q(s|i), U(s|i)), V(s|i) : 1 \leq i \leq l(s)\}$ is an initial segment of a play on $X$ in which $E$ applies the strategy $\sigma$

Fix any $s \in \Omega$ with $l(s) = n$ and consider an open cover $\{B(y, r_y) : y \in V(s)\}$ such that $r_y \in (0, 3^{-n-1})$ and $B(y, r_y) \subset V(s)$ for any $y \in V(s)$. Pick a set $\{y_n : n \in \omega\} \subset V(s)$ such that $\bigcup_{i \in \omega} B(y_i, r_{y_i}) = V(s)$; let $q(s^\frown i) = y_i$ and $U(s^\frown i) = B(y_i, r_{y_i})$ for any $i \in \omega$. If $i \in \omega$ then the sequence

$$(q(s|1), U(s|1)), V(s|1), \ldots, (q(s), U(s)), V(s), (q(s^\frown i), U(s^\frown i))$$

is an initial segment of a play in $X$ is which $E$ applies the strategy $\sigma$, so this strategy gives us the set $V(s^\frown i) = \sigma((q(s|1), U(s|1)), \ldots, (q(s), U(s)), (q(s^\frown i), U(s^\frown i)))$. After we construct the point $q(s^\frown i)$ and sets $U(s^\frown i), V(s^\frown i)$ for all $s \in \omega^n$ and $i \in \omega$, we will have a point $q(s)$ and sets $U(s), V(s)$ for all $s \in \Omega \backslash \{\emptyset\}$ with $l(s) \leq n + 1$. It is straightforward that the conditions (10)–(12) are satisfied if we replace $n$ with $n + 1$, so our inductive procedure can be continued to construct a point $q(s)$ and sets $U(s), V(s)$ for all $s \in \Omega \backslash \{\emptyset\}$ in such a way that (10)–(12) hold for all $n \in \mathbb{N}$.

It follows from the properties (11) and (12) that if $s \in \Omega$ and $l(s) = n > 0$ then $\mathrm{diam}(V(s)) \leq \mathrm{diam}(U(s)) \leq 2 \cdot 3^{-n} < \frac{1}{n}$. Given any $u \in \omega^\omega$, it follows from (12) that $\{(q(u|i), U(u|i)), V(u|i) : i \in \mathbb{N}\}$ is a play in the Choquet game on $X$ in which $E$ applies the strategy $\sigma$ and hence $\bigcap_{i \in \mathbb{N}} V(u|i) = \bigcap_{i \in \mathbb{N}} U(u|i) \neq \emptyset$. This, together with the property (10) shows that the family $\mathcal{V} = \{V(s) : s \in \Omega\}$ is an $A$-system on $X$ and hence $X$ is Čech-complete by Fact 3 of S.491. This settles necessity and finishes the proof of Fact 3.

To simplify the notation, let $\omega^* = \beta\omega \backslash \omega$.

*Fact 4.* The following conditions are equivalent for any ultrafilter $\eta \in \omega^*$:

(i) $\eta$ is a $P$-point in $\omega^*$;
(ii) for any countable family $\mathcal{E} \subset \exp(\omega) \backslash \eta$, there exists $A \in \eta$ such that $A \cap E$ is finite for any $E \in \mathcal{E}$;
(iii) for any countable disjoint family $\mathcal{E} \subset \exp(\omega) \backslash \eta$, there exists $A \in \eta$ such that $A \cap E$ is finite for any $E \in \mathcal{E}$.

*Proof.* It is evident that (ii)$\Longrightarrow$(iii). Assume that (iii) holds and $\eta$ is not a $P$-point in $\omega^*$; then there exists a $\sigma$-compact set $G \subset \omega^* \backslash \{\eta\}$ such that $\eta \in \overline{G}$. Choose a family $\{K_n : n \in \omega\}$ of compact subsets of $\omega^*$ such that $\bigcup_{n \in \omega} K_n = G$. It is easy to find a clopen set $U_n$ in the space $\omega^*$ such that $K_n \subset U_n \subset \omega^* \backslash \{\eta\}$. If $U = \bigcup_{n \in \omega} U_n$ then $\eta \in \overline{U} \backslash U$. Let $V_0 = U_0$ and $V_{n+1} = U_{n+1} \backslash (\bigcup \{U_i : i \leq n\})$ for any $n \in \omega$. It is easy to see that $\{V_n : n \in \omega\}$ is a disjoint family of clopen subsets of $\omega^*$ such that $\bigcup_{n \in \omega} V_n = U$.

It is an easy exercise that there exists a clopen subset $W_n$ of the space $\beta\omega$ such that $W_n \cap \omega^* = V_n$; if $H_n = W_n \cap \omega$ then $\mathrm{cl}_{\beta\omega}(H_n) \cap \omega^* = V_n$ for every $n \in \omega$. Assume for a moment that $n \neq m$ and the set $B = H_n \cap H_m$ is infinite. Then $F = \mathrm{cl}_{\beta\omega}(B) \cap \omega^* \neq \emptyset$ and $F \subset V_n \cap V_m = \emptyset$ which is a contradiction. Therefore the set $H_n \cap H_m$ is finite for any distinct $n, m \in \omega$ and hence we can find a set $E_n \subset H_n$ such that $|H_n \backslash E_n| < \omega$ for every $n \in \omega$ and the family $\{E_n : n \in \omega\}$ is disjoint. It follows from Fact 1 of S.370 that $\mathrm{cl}_{\beta\omega}(E_n) \cap \omega^* = V_n$; in particular, $\eta \notin \mathrm{cl}_{\beta\omega}(E_n)$ and hence $E_n \notin \eta$ for each $n \in \omega$.

Thus the disjoint family $\mathcal{E} = \{E_n : n \in \omega\} \subset \exp(\omega)$ contains no elements of $\eta$, so we can apply (iii) to find a set $A \in \eta$ such that $A \cap E_n$ is finite for any $n \in \omega$. Apply Fact 1 of S.371 to see that the set $Q = \mathrm{cl}_{\beta\omega}(A) \cap \omega^*$ does not meet $\mathrm{cl}_{\beta\omega}(E_n) \cap \omega^* = V_n$ for any $n \in \omega$. However, $Q \in \tau(\eta, \omega^*)$ by Fact 1 of S.370 so it follows from $Q \cap U = \emptyset$ that $\eta \notin \overline{U}$; this contradiction shows that (iii)$\Longrightarrow$(i).

To prove the implication (i)$\Longrightarrow$(ii) assume that the ultrafilter $\eta$ is a $P$-point in $\omega^*$ and fix a countable family $\mathcal{E} \subset \exp(\omega) \backslash \eta$. Then $\eta \notin \bigcup \{\mathrm{cl}_{\beta\omega}(E) : E \in \mathcal{E}\}$ and therefore $G = \bigcup \{\mathrm{cl}_{\beta\omega}(E) \cap \omega^* : E \in \mathcal{E}\}$ is an $F_\sigma$-subset of $\omega^*$ such that $\eta \notin G$. The ultrafilter $\eta$ being a $P$-point in $\omega^*$ we can find a set $U \in \tau(\eta, \omega^*)$ such that $U \cap G = \emptyset$. By Fact 2 of S.370 there exists $A \subset \omega$ with $\eta \in \mathrm{cl}_{\beta\omega}(A)$ such that $\mathrm{cl}_{\beta\omega}(A) \cap \omega^* \subset U$. Then $A \in \eta$ and it follows from $\mathrm{cl}_{\beta\omega}(A) \cap \mathrm{cl}_{\beta\omega}(E) \cap \omega^* = \emptyset$ that $A \cap E$ is finite for any $E \in \mathcal{E}$ (see Fact 1 of S.371); this settles (i)$\Longrightarrow$(ii) and shows that Fact 4 is proved.

Given an ultrafilter $\xi \in \beta\omega \backslash \omega$, the Gul'ko–Sokolov game on $\omega_\xi$ is played as follows: at the $n$-th move player $NU$ takes a set $S_n \in \exp(\omega) \backslash \xi$ and player $U$ replies by choosing a finite set $T_n \subset \omega$ in such a way that the family $\{S_i : i \leq n\} \cup \{T_i : i \leq n\}$ is disjoint. The game ends after the moves $S_n, T_n$ are made for all $n \in \omega$ and player $U$ wins if $\bigcup \{T_n : n \in \omega\} \in \xi$. Otherwise, the player $NU$ is the winner.

*Fact 5.* Player $NU$ has a winning strategy in the Gul'ko–Sokolov game on $\omega_\xi$ if and only if $\xi$ is not a $P$-point in $\beta\omega \backslash \omega$.

*Proof.* Assume first that $\xi$ is not a $P$-point in $\omega^*$ and apply Fact 1 to find a disjoint family $\mathcal{E} = \{E_n : n \in \omega\} \subset \exp(\omega)\backslash\xi$ such that no element of $\xi$ has a finite intersection with all elements of $\mathcal{E}$. Let $S_0 = E_0$ and $\sigma(\emptyset) = S_0$. If $n \in \omega$ and moves $S_0, T_0, \ldots, S_{n-1}, T_{n-1}$ are made in the Gul'ko–Sokolov game at $\omega_\xi$ then let $S_n = E_n\backslash(\bigcup\{T_i : i < n\})$ and $\sigma(T_0, \ldots, T_{n-1}) = S_n$. This defines a strategy $\sigma$ of the player $NU$ on the space $\omega_\xi$. If $\{S_n, T_n : n \in \omega\}$ is a play in which $NU$ applies the strategy $\sigma$ then $S_n \subset E_n$ and $E_n\backslash S_n$ is finite for any $n \in \omega$. If $T = \bigcup_{n\in\omega} T_n \in \xi$ then, for some $n \in \omega$, the set $T \cap E_n$ is infinite by our choice of the family $\mathcal{E}$. This implies that $T \cap S_n \neq \emptyset$ which is a contradiction. Therefore $T \notin \xi$, i.e., $NU$ wins every play in which he applies $\sigma$, so $\sigma$ is a winning strategy of $NU$ and hence we proved sufficiency.

To establish necessity, assume that $\sigma$ is a winning strategy of the player $NU$ on the space $\omega_\xi$ and the ultrafilter $\xi$ is a $P$-point of the space $\omega^*$. Consider the family $\mathcal{E}$ of all possible moves of the player $NU$ made according to the strategy $\sigma$. In other words, $\mathcal{E} = \{\sigma(\emptyset)\} \cup \{\sigma(T_0, \ldots, T_n) : n \in \omega$ and the $(n + 1)$-tuple $(T_0, \ldots, T_n)$ constitutes the moves of $U$ in some play $\{S_i, T_i : i \in \omega\}$ in which $NU$ applies the strategy $\sigma\}$. Since we only have countably many finite sequences of finite subsets of $\omega$, the family $\mathcal{E}$ is countable. Furthermore, $\mathcal{E} \subset \exp(\omega)\backslash\xi$, so we can apply Fact 1 to see that there exists an element $A \in \xi$ such that $A \cap E$ is finite for every $E \in \mathcal{E}$. Let $\{a_i : i \in \omega\}$ be a faithful enumeration of the set $A$.

Let $S_0 = S_0' = \sigma(\emptyset)$; we will also need the set $T_{-1}' = \emptyset$. Passing to the set $A\backslash\sigma(\emptyset)$ if necessary, we can assume, without loss of generality, that $S_0 \cap A = \emptyset$. Proceeding inductively assume that $n \in \omega$ and we have families $\{S_i : 0 \le i \le n\}$ and $\{S_i' : 0 \le i \le n\}$ together with the families $\{T_i : 0 \le i < n\}$ and $\{T_i' : -1 \le i < n\}$ such that

(13)  $S_n = \{S_0, T_0, \ldots, S_{n-1}, T_{n-1}, S_n\}$ is an initial segment of a play in which $NU$ applies the strategy $\sigma$;

(14)  $S_n' = \{S_0', T_0', \ldots, S_{n-1}', T_{n-1}', S_n'\}$ is also an initial segment of a play in which $NU$ applies the strategy $\sigma$;

(15)  if $0 \le i < n$ then $\{a_0, \ldots, a_i\} \subset M_i = (T_0 \cup T_0') \cup \ldots \cup (T_i \cup T_i')$;

(16)  if $0 \le i \le n$ then $(S_0 \cup \ldots \cup S_i) \cap A \subset L_i = \bigcup\{T_j' : -1 \le j < i\}$;

(17)  if $0 \le i < n$ then $(S_0' \cup \ldots \cup S_i') \cap A \subset K_i = \bigcup\{T_j : 0 \le j \le i\}$.

The finite set $F = S_n' \cap A$ is disjoint from the set $L_n$ and we have the inclusion $(S_0 \cup \ldots \cup S_n) \cap A \subset L_n$ by the property (16), so $F$ is disjoint from $S_0 \cup \ldots \cup S_n$. If $a_n \notin M_{n-1}$ then let $T_n = \{a_n\} \cup (F\backslash K_{n-1})$; if $a \in M_{n-1}$ then define the set $T_n$ to be equal to $F\backslash K_{n-1}$. The above observations show that, in both cases, $T_n$ is an admissible move for $U$ to continue the play started in $S_n$. Besides, the condition (17) is now satisfied for all $i \in \{0, \ldots, n\}$.

The strategy $\sigma$ gives us the set $S_{n+1} = \sigma(T_0, \ldots, T_n)$; observe that $S_{n+1}$ is disjoint from $K_n = T_0 \cup \ldots \cup T_n$. Besides, $(S_0' \cup \ldots \cup S_n') \cap A \subset K_n$, so $G = S_{n+1} \cap A$ is disjoint from $(S_0' \cup \ldots \cup S_n')$. Therefore the set $T_n' = G\backslash L_n$ is a valid move for $U$ to continue the play started in $S_n'$. The property (16) now holds for all $i \in \{0, \ldots, n+1\}$ and it is evident that we also have (15) for $i = n$. Letting $S_{n+1}' = \sigma(T_0', \ldots, T_n')$ we obtain families $S_{n+1} = S_n \cup \{T_n, S_{n+1}\}$ and $S_{n+1}' = S_n' \cup \{T_n', S_{n+1}'\}$ together with the families $\{T_i : 0 \le i \le n\}$ and $\{T_i' : -1 \le i \le n\}$ for which the properties (13)–(17) hold if we replace $n$ with $n + 1$.

Therefore our inductive procedure can be carried out for every $n \in \omega$ to construct families $S = \{S_n, T_n : n \in \omega\}$ and $S' = \{S_n', T_n' : n \in \omega\}$ such that the conditions (13)–(17) are satisfied for all $n \in \omega$. It is clear that $S$ and $S'$ are plays in which $NU$ applies the strategy $\sigma$, so neither of the sets $T = \bigcup_{n\in\omega} T_n$ and $T' = \bigcup_{n\in\omega} T_n'$ belongs to the ultrafilter $\xi$. Therefore $\xi \notin \mathrm{cl}_{\beta\omega}(T)$ and $\xi \notin \mathrm{cl}_{\beta\omega}(T')$ which shows that $\xi \notin \mathrm{cl}_{\beta\omega}(T \cup T')$ and hence $T \cup T' \notin \xi$. However, it follows from (15) that $A \subset T \cup T'$; since $A \in \xi$, we have $T \cup T' \in \xi$. This contradiction shows that $\xi$ cannot be a $P$-point, so we settled necessity and hence Fact 5 is proved.

*Fact 6.* Suppose that $X$ is a space and $F$ is closed subspace of $X$. Consider the linear subspace $I = \{f \in C_p(X) : f(F) \subset \{0\}\}$ of the space $C_p(X)$ and let $\pi : C_p(X) \to C_p(X \setminus F)$ be the restriction map. Then the map $p = \pi|I$ is a linear embedding of $I$ in $C_p(X \setminus F)$.

*Proof.* Let $J = p(I)$; we must prove that $p : I \to J$ is a homeomorphism. The map $p$ is linear and continuous because so is the map $\pi$. If $f, g \in I$ and $f \neq g$ then there exists $x \in X \setminus F$ such that $f(x) \neq g(x)$; therefore the point $x$ witnesses that $p(f) \neq p(g)$ and hence $p$ is a condensation.

To see that the map $q = p^{-1} : J \to I$ is continuous, fix a function $g_0 \in J$ and let $f_0 = q(g_0)$; then $g_0 = p(f_0)$. If $U \in \tau(f_0, I)$ then we can find a finite set $A \subset X$ and $\varepsilon > 0$ such that the set $V = \{f \in I : |f(x) - f_0(x)| < \varepsilon$ for all $x \in A\}$ is contained in $U$. If $B = A \setminus F$ then the set $W = \{g \in J : |g(x) - g_0(x)| < \varepsilon$ for all $x \in B\}$ is an open neighborhood of $g_0$ in $J$.

Given any function $g \in W$ let $f = q(g)$; then $f|(X \setminus F) = g$ and, in particular, $|f(x) - f_0(x)| = |g(x) - g_0(x)| < \varepsilon$ for any $x \in B$. Besides, $f(x) = f_0(x) = 0$ for any $x \in A \setminus B$ which shows that $|f(x) - f_0(x)| < \varepsilon$ for any $x \in A$, i.e., $f \in V$. This proves that $q(W) \subset V \subset U$, so the set $W$ witnesses continuity of the map $q$ at the point $g_0$. Thus $p$ is a homeomorphism and hence Fact 6 is proved.

*Fact 7.* Suppose that $\eta \in \beta\omega \setminus \omega$ and consider the set $I = \{f \in C_p(\omega_\eta) : f(\eta) = 0\}$; let $\pi : C_p(\omega_\eta) \to C_p(\omega) = \mathbb{R}^\omega$ be the restriction map. Then $C_p(\omega_\eta) \approx \pi(I)$.

*Proof.* By Fact 1 it suffices to show that $C_p(\omega_\eta)$ is linearly homeomorphic to $I$. Take an infinite set $A \subset \omega$ such that $\omega \setminus A$ is also infinite; then exactly one of the sets $A$ and $\omega \setminus A$ belongs to $\eta$. Renaming the relevant set if necessary, we can assume, without loss of generality, that $A \notin \eta$.

It is clear that $F = \{f \in I : f(\omega_\eta \setminus A) = \{0\}\}$ is a linear subspace of $I$; for any $f \in I$ let $r(f)(x) = 0$ if $x \in \omega_\eta \setminus A$ and $r(f)(x) = f(x)$ for every $x \in A$. Then $r : I \to F$ is a linear continuous map such that $r(f) = f$ for any $f \in F$, i.e., $r$ is a retraction. It follows from Problem 390 that $I \approx F \times M$ for some linear topological space $M$. Since also, $F \approx \mathbb{R}^A \approx \mathbb{R}^\omega$, we conclude that $I \approx \mathbb{R}^\omega \times M$ and therefore $\mathbb{R} \times I \approx (\mathbb{R} \times \mathbb{R}^\omega) \times M \approx \mathbb{R}^\omega \times M \approx I$. Recalling that $\mathbb{R} \times I \approx C_p(\omega_\eta)$ (see Fact 1 of S.409), we conclude that $I \approx C_p(\omega_\eta)$ and hence Fact 7 is proved.

Returning to our solution let $F_n = \{x \in \mathbb{R}^\omega : |x^{-1}(\mathbb{R}\setminus\{0\})| \leq n\}$ for any $n \in \omega$. We omit a simple proof that $F_n$ is a closed nowhere dense subspace of $\sigma(\mathbb{R}^\omega)$; it follows from the equality $\sigma(\mathbb{R}^\omega) = \bigcup_{n\in\omega} F_n$ that the space $\sigma(\mathbb{R}^\omega)$ is of the first category in itself. This shows that if $\sigma(\mathbb{R}^\omega)$ is homeomorphic to a closed subspace of $C_p(\omega_\xi)$ then $C_p(\omega_\xi)$ is not hereditarily Baire, i.e., we proved that (ii)$\Longrightarrow$(iii).

Now assume that $\xi$ is not a $P$-point in $\beta\omega\setminus\omega$ and apply Fact 4 to find a disjoint family $\{A_n : n \in \omega\}$ of infinite subsets of $\omega$ such that $A_n \notin \xi$ for any $n \in \omega$ and, for every $U \in \xi$, the set $U \cap A_n$ is infinite for some $n \in \omega$. Choose a faithful enumeration $\{a_k^n : k \in \omega\}$ of the set $A_n$ for each $n \in \omega$. Since $\omega\setminus(\bigcup_{n\in\omega} A_n)$ cannot belong to $\xi$, there is no loss of generality to assume that $\bigcup_{n\in\omega} A_n = \omega$. For any $f \in \mathbb{R}^\omega$ let $e(f)$ be the function from $\mathbb{R}^{\omega_\xi}$ such that $e(f)|\omega = f$ and $e(f)(\xi) = 0$. It follows from Fact 7 that the set $C_\xi = \{f \in \mathbb{R}^\omega : e(f) \in \mathbb{R}^{\omega_\xi}$ is continuous$\}$ is homeomorphic to $C_p(\omega_\xi)$.

Let $B = \bigcup\{A_n : n \in \mathbb{N}\}$. For any function $f \in \mathbb{R}^{A_0}$ consider a function $\varphi(f) \in \mathbb{R}^B$ defined as follows: $\varphi(f)(a_k^n) = 0$ whenever $k \in \omega$, $n \in \mathbb{N}$ and $k < n$; if $k \geq n$ then

$$(18) \quad \varphi(f)(a_k^n) = |f(a_k^0)| \cdot \sum\{|f(a_{i_1}^0) \cdot \ldots \cdot f(a_{i_n}^0)| : 0 \leq i_1 < \ldots < i_n \leq k\}.$$

This gives us a map $\varphi : \mathbb{R}^{A_0} \to \mathbb{R}^B$ and it is easy to see, using TFS-102, that $\varphi$ is continuous. Given a function $f \in \mathbb{R}^B$ let

$$(19) \quad \mu(f)(a_k^n) = k \cdot \sum\{|f(a_i^n)| : i \leq k\}$$

for any $n \in \mathbb{N}$ and $k \in \omega$; this gives us a map $\mu : \mathbb{R}^B \to \mathbb{R}^B$ and again it is straightforward that $\mu$ is continuous. Therefore the map $\nu = \mu \circ \varphi : \mathbb{R}^{A_0} \to \mathbb{R}^B$ is continuous.

For each $f \in \mathbb{R}^{A_0}$ there exists a unique function $u_f \in \mathbb{R}^\omega$ such that $u_f|A_0 = f$ and $u_f|B = \nu(f)$. An evident homeomorphism between $\mathbb{R}^{A_0} \times \mathbb{R}^B$ and $\mathbb{R}^\omega$ shows that we can identify the set $\Gamma = \{u_f : f \in \mathbb{R}^{A_0}\}$ with the graph of the map $\nu$, so $\Gamma$ is closed in $\mathbb{R}^\omega$. Furthermore, if we let $q(f) = u_f$ for any $f \in \mathbb{R}^{A_0}$ then $q : \mathbb{R}^{A_0} \to \Gamma$ is a homeomorphism (see Fact 4 of S.390).

The set $E = \Gamma \cap C_\xi$ is closed in $C_\xi$; we claim that $E = q(\sigma(\mathbb{R}^{A_0}))$. Indeed, take any $f \in \sigma(\mathbb{R}^{A_0})$; there exists $m \in \omega$ such that $f(a_k^0) = 0$ for any $k \geq m$. Now, if $n > m$ and $k \geq n$ then every summand in the expression for.$\varphi(f)(a_k^n)$ involves a product of values of $f$ at $(m + 1)$ distinct coordinates. Since one of those values is bound to be zero, we conclude that $\varphi(f)(a_k^n) = 0$ for all $n > m$ and $k \in \omega$, i.e., $\varphi(f)|A_n$ is identically zero for all $n > m$. As an immediate consequence, $\nu(f)|A_n$ is identically zero for all $n > m$.

Therefore the function $u_f$ is identically zero on the set $A' = \bigcup\{A_i : i > m\}$; since $A_i \notin \xi$ for all $i \leq m$, the set $A'$ belongs to $\xi$ and hence the function $e(u_f)$ is continuous on $\omega_\xi$, i.e., $q(f) \in E$. This shows that $q(\sigma(\mathbb{R}^{A_0})) \subset E$.

Now, if $f \in \mathbb{R}^{A_0}\setminus\sigma(\mathbb{R}^{A_0})$ then, for any $n \in \mathbb{N}$ we can find positive integers $k_1 < \ldots < k_n$ such that $f(a_{k_i}^0) \neq 0$ for all $i \leq n$. Then it follows from (18) that $\varphi(f)(a_{k_n}^n) \geq r = |f(a_{k_n}^0)| \cdot |f(a_{k_1}^0)| \cdot \ldots \cdot |f(a_{k_n}^0)| > 0$.

Apply (19) to see that $\mu(\varphi(f))(a_k^n) \geq kr$ for any $k > k_n$; this proves that,

(20) for any $n \in \mathbb{N}$, we have $\nu(f)(a_k^n) \to +\infty$ as $k \to \infty$.

If $u_f \in C_\xi$ then there exists $U \in \xi$ such that $u_f(U) \subset (-1, 1)$. Choose a number $n \in \mathbb{N}$ such that $U \cap A_n$ is infinite; then $|\nu(f)(a_k^n)| < 1$ for infinitely many $k$. This, however, gives a contradiction with (20) and proves that $q(\mathbb{R}^{A_0} \setminus \sigma(\mathbb{R}^{A_0})) \cap C_\xi = \emptyset$ and hence $q(\sigma(\mathbb{R}^{A_0})) = E$ which shows that $E$ is a closed subspace of $C_\xi$ homeomorphic to $\sigma(\mathbb{R}^{A_0})$. Since $C_\xi \approx C_p(\omega_\xi)$, the space $\sigma(\mathbb{R}^{A_0})$ embeds in $C_p(\omega_\xi)$ as a closed subspace, so we settled (i)$\Longrightarrow$(ii).

To show that (iii)$\Longrightarrow$(i) assume that $C_p(\omega_\xi)$ is not hereditarily Baire and hence $C_\xi$ is not hereditarily Baire either. If $f \in C_\xi$, $\varepsilon > 0$ and $A$ is a finite subset of $\omega$ then let $[f, A, \varepsilon] = \{g \in C_\xi : |f(n) - g(n)| < \varepsilon$ for each $n \in A\}$. It is clear that the family $\{[f, A, \varepsilon] : A$ is a finite subset of $\omega$ and $\varepsilon > 0\}$ is a local base of $C_\xi$ at $f$.

Apply Fact 2 to see that we can choose a winning strategy $\sigma$ for the player $E$ in the Choquet game on the space $C_\xi$. Our plan is to construct a winning strategy $s$ for the player $NU$ on the space $\omega_\xi$.

If $(f_0, U_0) = \sigma(\emptyset)$ then let $S_0 = \{n \in \omega : |f_0(n)| \geq 1\}$. Since $e(f_0)$ is continuous and equals zero at $\xi$, the set $S_0$ does not belong to $\xi$, so letting $s(\emptyset) = S_0$ we obtain a strategy $s$ for the first move of the player $NU$.

If the player $U$ chooses a finite set $T_0 \subset \omega \setminus S_0$ then $|f_0(n)| < 1$ for any $n \in T_0$ and therefore we can find $\varepsilon_0 \in (0, 1)$ together with a finite set $A_0 \subset \omega$ such that $T_0 \subset A_0$ and $[f_0, A_0, \varepsilon_0] \subset U$ while $[f_0(n) - \varepsilon_0, f_0(n) + \varepsilon_0] \subset (-1, 1)$ for any $n \in A_0 \setminus S_0$. If $V_0 = [f_0, A_0, \varepsilon_0]$ then the strategy $\sigma$ gives us a pair $(f_1, U_1) = \sigma(V_0)$, so we can define the set $S_1 = \{n \in \omega : |f_1(n)| \geq 2^{-1}\} \setminus (S_0 \cup T_0)$; let $S_1 = s(T_0)$.

Proceeding inductively, assume that $m \in \mathbb{N}$ and we have an initial segment $\mathcal{S}_m = \{S_0, T_0, \ldots, S_{m-1}, T_{m-1}, S_m\}$ of a play in the Gul'ko–Sokolov game on $\omega_\xi$ and an initial segment $\mathcal{U}_m = \{(f_0, U_0), V_0, \ldots, (f_{m-1}, U_{m-1}), V_{m-1}, (f_m, U_m)\}$ of a play in the Choquet game on $C_\xi$ with the following properties:

(21) the player $E$ applies the strategy $\sigma$ in $\mathcal{U}_m$;
(22) the player $NU$ applies the strategy $s$ in $\mathcal{S}_m$;
(23) $S_i = \{n \in \omega : |f_i(n)| \geq 2^{-i}\} \setminus \bigcup_{j < i}(S_j \cup T_j)$ for any $i \in \{1, \ldots, m\}$;
(24) $V_i = [f_i, A_i, \varepsilon_i]$ and $\varepsilon_i \in (0, 2^{-i})$ for any $i < m$;
(25) $T_i \cup A_{i-1} \subset A_i$ for any $i \in \{1, \ldots, m-1\}$;
(26) $[f_i(n) - \varepsilon_i, f_i(n) + \varepsilon_i] \subset (f_{i-1}(n) - \varepsilon_{i-1}, f_{i-1}(n) + \varepsilon_{i-1})$ for any $n \in A_{i-1}$ and $i \in \{1, \ldots, m-1\}$;
(27) $[f_i(n) - \varepsilon_i, f_i(n) + \varepsilon_i] \subset (-2^{-i}, 2^{-i})$ for any $n \in T_i$ and $i < m$.

The properties (22) and (24) show that $(f_m, U_m) = \sigma(V_0, \ldots, V_{m-1})$ and, in particular, $f_m(n) \in (f_{m-1}(n) - \varepsilon_{m-1}, f_{m-1}(n) + \varepsilon_{m-1})$ for any $n \in A_{m-1}$. If $T_m$ is the move of the player $U$ in the Gul'ko–Sokolov game then it follows from (23) that we have the inequality $|f_m(n)| < 2^{-m}$ for any $n \in T_m$. This makes it possible to find a number $\varepsilon_m \in (0, 2^{-m})$ and a finite set $A_m \subset \omega$ such that $A_{m-1} \cup T_m \subset A_m$ and $[f_m, A_m, \varepsilon_m] \subset U_m$ while the properties (26) and (27) are fulfilled for $i = m$. Letting $V_m = [f_m, A_m, \varepsilon_m]$ we can apply the strategy $\sigma$ to obtain the move $(f_{m+1}, U_{m+1}) = \sigma(V_0, \dots, V_m)$ of the player $E$. Consider the set $S_{m+1} = \{n \in \omega : |f_{m+1}(n)| \geq 2^{-m-1}\} \backslash \bigcup_{j \leq m}(S_j \cup T_j)$; it is easy to see that if we let $s(T_0, \dots, T_m) = S_{m+1}$ then the properties (21)–(27) are now fulfilled for $m$ replaced with $m + 1$.

Therefore our inductive procedure can be continued to define a strategy $s$ for the player $NU$ on the space $\omega_\xi$. Assume that $\mathcal{S} = \{(S_m, T_m) : m \in \omega\}$ is a play on $\omega_\xi$ in which $NU$ applies $s$. The properties (21)–(27) show that there is a play $\{(f_m, U_m), V_m : m \in \omega\}$ on the space $C_\xi$ in which $E$ applies the strategy $\sigma$ and the conditions (21)-(27) are satisfied for any $m \in \omega$.

Let $A = \bigcup_{i \in \omega} A_i$; it follows from (25) that $T = \bigcup_{i \in \omega} T_i \subset A$. Given any $n \in A$ the family $\mathcal{E}(n) = \{[f_i(n) - \varepsilon_i, f_i(n) + \varepsilon_i] : i \in \omega\}$ consists of compact decreasing sets whose diameters tend to zero; therefore there is a unique number $f(n) \in \bigcap \mathcal{E}(n)$. Letting $f(n) = 1$ for any $n \in \omega \backslash A$ we obtain a function $f : \omega \to \mathbb{R}$. If $f \in C_\xi$ then it is easy to see that $f \in \bigcap_{n \in \omega} V_n = \bigcap_{n \in \omega} U_n$ which is a contradiction with the fact that $\sigma$ is a winning strategy for the player $E$. Therefore $f \notin C_\xi$, i.e., the function $e(f)$ is discontinuous on $\omega_\xi$.

It follows from (27) that the function $f$ tends to zero on $T$; this easily implies that if $T \in \xi$ then $e(f)$ is continuous on $\omega_\xi$ which is a contradiction. Therefore $T \notin \xi$ and hence $NU$ is the winner in the play $\mathcal{S}$. Thus $s$ is a winning strategy for $NU$ on $\omega_\xi$, so we can apply Fact 5 to see that $\xi$ is not a $P$-point in $\beta\omega \backslash \omega$. This proves that (iii)$\Longrightarrow$(i) and makes our solution complete.

**V.493.** Let $\xi, \eta \in \beta\omega \backslash \omega$. Prove that the spaces $\omega_\xi$ and $\omega_\eta$ are $l$-equivalent if and only if there exists a bijection $b : \omega \to \omega$ such that $b(\xi) = \{\{b(U) : U \in \xi\} = \eta$. In particular, $\omega_\xi$ and $\omega_\eta$ are $l$-equivalent if and only if they are homeomorphic.

**Solution.** We will need the following set-theoretic fact known as Katetov's three sets lemma.

*Fact 1.* Suppose that $D$ is a nonempty set and $f : D \to D$ is a map such that $f(x) \neq x$ for any $x \in D$. Then there exist disjoint subsets $A_0, A_1, A_2$ of the set $D$ such that $D = A_0 \cup A_1 \cup A_2$ and $f(A_i) \cap A_i = \emptyset$ for any $i = 0, 1, 2$.

*Proof.* Say that a set $U \subset D$ is *invariant* if $f(U) \subset U$; let us call $U$ *strongly invariant* if it is invariant and $f^{-1}(U) \subset U$. A set $U \subset D$ will be called *adequate* if there exist disjoint sets $U_0, U_1$ and $U_2$ such that $U = U_0 \cup U_1 \cup U_2$ and $f(U_i) \cap U_i = \emptyset$ for each $i \leq 2$; we will call the family $\{U_0, U_1, U_2\}$ an adequate decomposition of $U$.

Suppose that $U$ is an adequate set and $\{U_0, U_1, U_2\}$ is its adequate decomposition. Say that an adequate set $V$ *properly contains* $U$ if there exists an adequate decomposition $\{V_0, V_1, V_2\}$ of the set $V$ such that $U_i \subset V_i$ for all $i \leq 2$.

Given any point $x \in D$ let $x_0 = x$ and, if $x_i$ is defined for some $i \in \omega$ then let $x_{i+1} = f(x_i)$. The set $F(x) = \{x_i : i \in \omega\}$ is easily seen to be invariant; besides, $|F(x)| \geq 2$ for any $x \in D$.

Fix a point $x \in D$ and assume first that $F(x)$ is finite, say $F(x) = \{x_0, \dots, x_k\}$. Let $U_0 = \{x_0\}$, $U_1 = \{x_i : i \leq k \text{ and } i \text{ is odd}\}$ and $U_2 = \{x_i : 0 < i \leq k \text{ and } i \text{ is}$

even} and observe that $f(U_0) \subset U_1$ while $f(U_1) \subset U_0 \cup U_2$ and $f(U_2) \subset U_0 \cup U_1$, i.e., the sets $U_0$, $U_1$ and $U_2$ witness that $F(x)$ is an adequate set.

If $F(x)$ is infinite and hence $\{x_i : i \in \omega\}$ is a faithful enumeration of the set $F(x)$ then letting $U_0 = \emptyset$, $U_1 = \{x_{2i} : i \in \omega\}$ and $U_2 = \{x_{2i+1} : i \in \omega\}$ we convince ourselves that $F(x)$ is also an adequate set. This proves that

(1) for any point $x \in D$ there exists an adequate invariant set $F \subset D$ such that $x \in F$.

Suppose that $U$ is an invariant adequate set and $\{U_0, U_1, U_2\}$ is its adequate decomposition. Consider the map $s : \{0, 1, 2\} \to \{0, 1, 2\}$ defined by $s(0) = 1$, $s(1) = 2$ and $s(2) = 0$. If $V_i = U_i \cup (f^{-1}(U_{s(i)}) \setminus U)$ then $U_i \subset V_i$ for any $i \leq 2$. Given distinct $i, j \leq 2$ we have $U_i \cap U_j = \emptyset$ and $U_i \cap (f^{-1}(U_{s(j)}) \setminus U) = \emptyset$. This, together with the equalities $U_j \cap (f^{-1}(U_{s(i)}) \setminus U) = \emptyset$ and $f^{-1}(U_{s(i)}) \cap f^{-1}(U_{s(j)}) = \emptyset$ implies that $V_i \cap V_j = \emptyset$ for any distinct $i, j \leq 2$.

It is immediate from the definition that the set $V_i$ is contained in $f^{-1}(U)$ for every $i \leq 2$, so $V = V_0 \cup V_1 \cup V_2 \subset f^{-1}(U)$. It is evident that $U \subset V$; if $x \in f^{-1}(U) \setminus U$ then $f(x) \in U_{s(i)}$ and hence $x \in V_i$ for some $i \leq 2$. This proves that $V = f^{-1}(U)$ and, in particular, $V$ is invariant. Furthermore, $f(V_i) \subset f(U_i) \cup U_{s(i)}$ which, together with the equalities $f(U_i) \cap U_i = \emptyset$ and $U_{s(i)} \cap U_i = \emptyset$ shows that $f(V_i) \cap V_i = \emptyset$ for every $i \leq 2$ and hence the set $V$ is adequate. Therefore,

(2) for any invariant adequate set $U$ the set $V = f^{-1}(U)$ is also invariant and adequate and $U$ is properly contained in $V$.

Again, fix an invariant adequate set $U \subset D$ and its adequate decomposition $\{U_0, U_1, U_2\}$. Apply the property (2) inductively to obtain a sequence $\{V_i^n : n \in \omega\}$ for any $i \in \{0, 1, 2\}$ in such a way that

(3) $V^n = V_0^n \cup V_1^n \cup V_2^n$ is an invariant adequate set and $\{V_0^n, V_1^n, V_2^n\}$ is its adequate decomposition for any $n \in \omega$;
(4) $f^{-1}(U) = V_0$ and $f^{-1}(V^n) = V^{n+1}$ for any $n \in \omega$;
(5) $U_i \subset V_i^0$ and $V_i^n \subset V_i^{n+1}$ for any $n \in \omega$ and $i \in \{0, 1, 2\}$.

Let $V_i = \bigcup_{n \in \omega} V_i^n$ for each $i \in \{0, 1, 2\}$; it is straightforward to see that $V = \bigcup_{n \in \omega} V^n$ is a strongly invariant adequate set and $\{V_0, V_1, V_2\}$ is its adequate decomposition. Therefore it follows from (1) that

(6) for any $x \in D$ there exists a strongly invariant adequate set $U \subset D$ such that $x \in U$.

Apply the property (6) to construct a maximal disjoint family $\mathcal{U}$ of nonempty strongly invariant adequate subsets of $D$. We omit an easy proof that $U = \bigcup \mathcal{U}$ is a strongly invariant adequate subset of $D$. If $D' = D \setminus U \neq \emptyset$ then the property (6) applied to the set $D'$ and the map $f|D' : D' \to D'$ gives us a nonempty set $V \subset D'$ which is strongly invariant and adequate with respect to the map $f|D'$. Since $D'$ is strongly invariant, the set $V$ is strongly invariant with respect to the map $f$; this implies that $\mathcal{U} \cup \{V\}$ is a disjoint family of strongly invariant adequate subsets of $D$ which is a contradiction with maximality of $\mathcal{U}$. Therefore $\bigcup \mathcal{U} = D$ and hence $D$ is an adequate set, i.e., Fact 1 is proved.

Returning to our solution say that ultrafilters $\xi$ and $\eta$ are *equivalent* if there exists a bijection $b : \omega \to \omega$ such that $b(\xi) = \eta$. Observe first that we have the equalities $\tau(\xi, \omega_\xi) = \{A \cup \{\xi\} : A \in \xi\}$ and $\tau(\eta, \omega_\eta) = \{B \cup \{\eta\} : B \in \eta\}$. Therefore, for any bijection $b : \omega \to \omega$ such that $b(\xi) = \eta$ we can define a map $h : \omega_\xi \to \omega_\eta$ letting $h(\xi) = \eta$ and $h(n) = b(n)$ for any $n \in \omega$; then $h(\tau(\xi, \omega_\xi)) = \tau(\eta, \omega_\eta)$ and hence $h$ is a homeomorphism.

On the other hand, if $\omega_\xi$ is homeomorphic to $\omega_\eta$ then $\omega_\xi \overset{l}{\sim} \omega_\eta$; take any homeomorphism $h : \omega_\xi \to \omega_\eta$ and observe that we must have $h(\xi) = \eta$, so $b = h|\omega$ is easily seen to be a bijection which takes the elements of $\xi$ onto the elements of $\eta$. Therefore the ultrafilters $\xi$ and $\eta$ are equivalent if and only if $\omega_\xi$ is homeomorphic to $\omega_\eta$.

From now on we assume that $\omega_\xi$ is $l$-equivalent to $\omega_\eta$; we already saw that it suffices to prove that the ultrafilters $\xi$ and $\eta$ are equivalent. For any function $f \in \mathbb{R}^\omega$ let $e_\xi(f)(n) = f(n)$ for each $n \in \omega$ and $e_\xi(f)(\xi) = 0$. It follows from Fact 2 of V.489 that the set $C_\xi = \{f \in \mathbb{R}^\omega : e_\xi(f) \text{ is continuous on } \omega_\xi\}$ is linearly homeomorphic to $C_p(\omega_\xi)$. Analogously, let $e_\eta(f)(n) = f(n)$ for every $n \in \omega$ and $e_\eta(f)(\eta) = 0$ for any $f \in \mathbb{R}^\omega$. Applying Fact 2 of V.489 again we conclude that the space $C_\eta = \{f \in \mathbb{R}^\omega : e_\eta(f) \text{ is continuous on } \omega_\eta\}$ is linearly homeomorphic to $C_p(\omega_\eta)$.

Given any $n \in \omega$ let $\delta_n(f) = f(n)$ for any $f \in \mathbb{R}^\omega$; then $\delta_n : \mathbb{R}^\omega \to \mathbb{R}$ is a continuous linear functional on $\mathbb{R}^\omega$. Now assume that $\varphi : C_\xi \to \mathbb{R}$ is a nontrivial continuous linear functional. By Problem 224, there exists a continuous linear functional $\Phi : \mathbb{R}^\omega \to \mathbb{R}$ with $\Phi|C_\xi = \varphi$. We can apply TFS-197 to see that $\Phi = r_1 \delta_{k_1} + \ldots + r_n \delta_{k_n}$ for some $n \in \mathbb{N}$, $r_1, \ldots, r_n \in \mathbb{R} \setminus \{0\}$ and distinct $k_1, \ldots, k_n \in \omega$. The set $C_\xi$ being dense in $\mathbb{R}^\omega$, the functional $\delta_i$ is the unique extension of every $\delta_i|C_\xi$ over $\mathbb{R}^\omega$. This makes it possible to identify $\delta_i|C_\xi$ with $\delta_i$, so, to simplify our notation, we will write that $\varphi = r_1 \delta_{k_1} + \ldots + r_n \delta_{k_n}$. Analogously, if $\mu : C_\eta \to \mathbb{R}$ is a nontrivial continuous linear functional then there exists $m \in \mathbb{N}$ such that $\mu = s_1 \delta_{l_1} + \ldots + s_m \delta_{l_m}$ for some $s_1, \ldots, s_m \in \mathbb{R} \setminus \{0\}$ and distinct $l_1, \ldots, l_m \in \omega$ (here we identify $\delta_{l_i}|C_\eta$ with $\delta_{l_i}$ for any $i \leq m$).

Fix a linear homeomorphism $T : C_\xi \to C_\eta$. For any $n \in \omega$ the map $\delta_n \circ T$ is a nontrivial continuous linear functional on $C_\xi$, so there exists a nonempty finite set $P(n) \subset \omega$ such that $\delta_n \circ T = \sum_{m \in P(n)} r_{mn} \delta_m$ where $r_{mn} \neq 0$ for any $m \in P(n)$. Analogously, there exists a nonempty finite set $Q(n) \subset \omega$ such that $\delta_n \circ T^{-1} = \sum_{m \in Q(n)} s_{mn} \delta_m$ where $s_{mn} \neq 0$ for all $m \in Q(n)$. The following property is a key to our proof.

(7)  Suppose that $A \in \eta$, $B \subset \omega$ and $P(n) \cap B \neq \emptyset$ for any $n \in A$; if, additionally, $R_K = \{n \in A : P(n) \cap B \subset K\}$ is finite for any finite $K \subset B$ then $B \in \xi$.

Assume that $B \notin \xi$; we are going to construct a function $f \in C_\xi$. Let $f(n) = 0$ for any $n \in \omega \setminus B$; it follows from $\omega \setminus B \in \xi$ that $f$ will belong to $C_\xi$ no matter what values it will take on $B$. It follows from (7) that $B$ is infinite, so we can choose a faithful enumeration $\{n_i : i \in \omega\}$ of the set $B$; let $B_i = \{n_0, \ldots, n_i\}$ for every $i \in \omega$.

If $R_{B_0} = \emptyset$ then let $f(n_0) = 0$; otherwise choose a number $f(n_0)$ such that $|r_{n_0 n} f(n_0)| \geq 1$ for any $n \in R_{B_0}$. Observe that it is possible because the set $R_{B_0}$ is finite by (7) and $r_{n_0 n} \neq 0$ for each $n \in R_{B_0}$.

Proceeding inductively assume that $j \in \omega$ and we have defined $f(n_i)$ for any number $i \leq j$. If the set $R_{B_{j+1}} \setminus R_{B_j}$ is empty then let $f(n_{j+1}) = 0$. Otherwise, choose $f(n_{j+1})$ in such a way that

(∗) $|\sum_{m \in P(n) \cap B} r_{mn} f(m)| \geq 1$ for all $n \in R_{B_{j+1}} \setminus R_{B_j}$.

This is possible because $n_{j+1} \in P(n) \cap B$ while there are only finitely many elements in $R_{B_{j+1}} \setminus R_{B_j}$ and $r_{n_{j+1} n} \neq 0$ for any $n \in R_{B_{j+1}} \setminus R_{B_j}$ which shows that $f(n_{j+1})$ can be chosen sufficiently big to guarantee that the absolute value of all the sums in (∗) is greater than or equal to 1.

Therefore our inductive procedure can be continued to construct a function $f \in \mathbb{R}^\omega$ such that $f(\omega \setminus B) = \{0\}$ and the property (∗) holds for all $j \in \omega$. We already saw that $f \in C_\xi$ and hence $g = T(f) \in C_\eta$.

Take any $n \in A$; we have $g(n) = \delta_n(g) = (\delta_n \circ T)(f) = \sum_{m \in P(n)} r_{mn} \delta_m(f)$. However, the function $f$ is identically zero on $\omega \setminus B$, so $g(n) = \sum_{m \in P(n) \cap B} r_{mn} f(m)$. If $j = \max\{k : n_k \in P(n) \cap B\}$ then $n \in R_{B_j} \setminus R_{B_{j-1}}$ if $j > 0$ and $n \in R_{B_0}$ if $j = 0$. Therefore it follows from (∗) or the choice of $f(n_0)$ respectively, that $|g(n)| \geq 1$; since we checked this for any $n \in A$, the function $g$ cannot belong to $C_\eta$. This contradiction shows that (7) is proved. It is important to see that an analogous proof shows that we have the counterpart of (7) if we interchange $\eta$ and $\xi$, namely,

(7′) suppose that $B \in \xi$, $A \subset \omega$ and $Q(n) \cap A \neq \emptyset$ for any $n \in B$; if, additionally, the set $\{n \in B : Q(n) \cap A \subset K\}$ is finite for any finite $K \subset A$ then $A \in \eta$.

Next assume that $A \in \eta$ and we have a nonempty set $P_0(n) \subset P(n)$ for each $n \in A$. Given a set $B \subset \omega$ with $P_0(n) \cap B \neq \emptyset$ for every $n \in A$, it is an immediate consequence of (7) that

(8) if the set $\{n \in A : m \in P_0(n)\}$ is finite for any $m \in B$ then $B \in \xi$.

We will also need the version of (8) obtained by interchanging $\eta$ and $\xi$. That is, assume that $B \in \xi$ and we have a nonempty set $Q_0(n) \subset Q(n)$ for each $n \in B$. Given a set $A \subset \omega$ with $Q_0(n) \cap A \neq \emptyset$ for every $n \in B$, it is an immediate consequence of (7′) that

(8′) if the set $\{n \in A : m \in Q_0(n)\}$ is finite for any $m \in A$ then $A \in \eta$.

We will apply the property (8) to show that our task can be reduced to establishing the following statement.

(9) If there exists a set $E \in \eta$ for which we can find an injection $\varphi : E \to \omega$ such that $\varphi(n) \in P(n)$ for any $n \in E$ then the ultrafilters $\xi$ and $\eta$ are equivalent.

Choosing a smaller element of $\eta$ if necessary, we can assume, without loss of generality, that the sets $\omega \setminus E$ and $\omega \setminus \varphi(E)$ are infinite. Take an arbitrary bijection $\varphi_0 : \omega \setminus E \to \omega \setminus \varphi(E)$ and define a map $h : \omega \to \omega$ by the equalities $h(n) = \varphi(n)$ if $n \in E$ and $h(n) = \varphi_0(n)$ for any $n \in \omega \setminus E$.

It is clear that the map $h$ is a bijection; fix an arbitrary element $C \in \eta$ and let $P_0(n) = \{\varphi(n)\}$ for any $n \in A = C \cap E$. The map $\varphi$ being injective it is easy to see that (8) is applicable to the sets $A$ and $B = \varphi(A) = h(A)$, so $B \in \xi$. It follows from $h(C) \supset B$ that $h(C) \in \xi$. This shows that $h(C) \in \xi$ for any $C \in \eta$, i.e., $h(\eta) \subset \xi$. To see that $h(\eta) = \xi$ take any element $B \in \xi$; if $A = h^{-1}(B) \notin \eta$ then $A' = \omega \backslash A \in \eta$ and hence $\omega \backslash B = h(A') \in \xi$ which is a contradiction. Therefore $A \in \eta$ which shows that $B = h(A) \in h(\eta)$. This proves that $h(\eta) = \xi$ and hence the ultrafilters $\xi$ and $\eta$ are equivalent, i.e., (9) is proved.

For each $n \in \omega$ consider the sets $P_0(n) = \{m \in P(n) : n \in Q(m)\}$ and $Q_0(n) = \{m \in Q(n) : n \in P(m)\}$. It follows from the equality $\delta_n = \delta_n \circ T \circ T^{-1}$ that $\delta_n = \sum_{m \in P(n)} r_{mn} \sum_{k \in Q(m)} s_{km} \delta_k$; the family $\{\delta_i : i \in \omega\}$ being linearly independent, there exists $m \in P(n)$ such that $n \in Q(m)$ and hence $m \in P_0(n)$. Therefore the set $P_0(n)$ is nonempty for any $n \in \omega$. Analogously, $Q_0(n) \neq \emptyset$ for every $n \in \omega$.

Next observe that the family $\mathcal{P} = \{P_0(n) : n \in \omega\}$ is point-finite because for any $m \in \omega$ if $m \in P_0(n)$ then $n \in Q(m)$ and the set $Q(m)$ is finite. An analogous reasoning shows that the family $\mathcal{Q} = \{Q_0(m) : m \in \omega\}$ is also point-finite.

It will be easy to finish our proof after we establish the following property of the families $\mathcal{P}$ and $\mathcal{Q}$.

(10) The set $E = \{n \in \omega : Q_0(m) = \{n\}$ for any $m \in P_0(n)\}$ belongs to the ultrafilter $\eta$.

Note first that it is straightforward from the definition that $n \in Q_0(m)$ for any $m \in P_0(n)$. If the property (10) does not hold then there exists an element $A \in \eta$ such that, for any $n \in A$ we can choose a number $l_n \in P_0(n)$ such that there exists $\Phi(n) \in Q_0(l_n) \backslash \{n\}$. For every $n \in \omega \backslash A$ choose a number $\Phi(n) \neq n$ arbitrarily. This gives a map $\Phi : \omega \to \omega$ such that $\Phi(n) \neq n$ for all $n \in \omega$.

By Fact 1 we can find disjoint sets $A_0, A_1, A_2$ such that $\omega = A_0 \cup A_1 \cup A_2$ and $\Phi(A_i) \cap A_i = \emptyset$ for any $i \in \{0, 1, 2\}$. It is easy to see that one of the sets $A_0 \cap E, A_1 \cap E, A_2 \cap E$ belongs to $\eta$; denote this set by $W$. It follows from the choice of the sets $A_i$ that $\Phi(W) \cap W = \emptyset$.

Consider the set $W' = \{l_n : n \in W\}$; then $l_n \in P_0(n) \cap W'$, so $P_0(n) \cap W' \neq \emptyset$ for any $n \in W$ and hence we can apply the property (8) to see that $W' \in \xi$. If $W'' = \Phi(W)$ then $\Phi(n) \in Q_0(l_n) \cap W''$ and hence $Q_0(l_n) \cap W'' \neq \emptyset$ for any $n \in \omega$, so we can apply (8') to conclude that $W'' \in \eta$. Thus, $W \in \eta$, $W'' \in \eta$ and $W \cap W'' = \emptyset$ which is a contradiction, so the property (10) is proved.

Finally, choose an element $\varphi(n) \in P_0(n)$ for any $n \in E$. Given distinct $n, m \in E$ if $\varphi(n) = \varphi(m) = k$ then $\{m, n\} \subset Q_0(k)$ which is a contradiction with the choice of $E$. Therefore $\varphi(n) \neq \varphi(m)$ for any distinct $m, n \in E$, i.e., the map $\varphi : E \to \omega$ is an injection, so we can apply (9) to see that the ultrafilters $\xi$ and $\eta$ are equivalent and hence our solution is complete.

**V.494.** *Suppose that $\alpha$ and $\beta$ are ordinals such that $\omega \leq \alpha \leq \beta < \omega_1$. Prove that $(\alpha + 1)$ is $l$-equivalent to $(\beta + 1)$ if and only if $\beta < \alpha^\omega$. Deduce from this fact that there exist countable compact u-equivalent spaces which are not $l$-equivalent.*

**Solution.** The expression $Y \simeq Z$ says that the spaces $Y$ and $Z$ are homeomorphic. If $L$ and $M$ are linear topological spaces then we write $L \approx M$ if $L$ and $M$ are linearly homeomorphic. Given a space $Z$ let $I(Z)$ be the set of isolated points of $Z$ and $D_0(Z) = Z$. If $\xi$ is an ordinal and we have $D_\xi(Z)$ then $D_{\xi+1}(Z) = D_\xi(Z) \setminus I(D_\xi(Z))$; if $\xi$ is a limit ordinal and we have $D_\eta(Z)$ for any $\eta < \xi$ then $D_\xi(Z) = \bigcap_{\eta < \xi} D_\eta(Z)$. The set $D_\xi(Z)$ will be called the $\xi$-th derivative of the space $Z$. If $Y$ is a space and $A \subset Y$ is a closed subspace of $Y$ then $Y_A$ is the space obtained from $Y$ by contracting the set $A$ to a point. If $\varphi : X \to Y$ is a continuous map then let $\varphi^*(f) = f \circ \varphi$ for any $f \in C(Y)$; then $\varphi^* : C(Y) \to C(X)$ is called the dual map of the map $\varphi$.

Given two linearly ordered spaces $L$ and $M$ we say that they are *canonically homeomorphic* if there exists a homeomorphism $f : L \to M$ which is also an order isomorphism between $L$ and $M$. If $<_L$ and $<_M$ are the orders on $L$ and $M$ respectively then the *lexicographic order* $<$ on $L \times M$ is defined as follows: for any $a, b \in L \times M$ with $a = (a_1, a_2)$ and $b = (b_1, b_2)$ we declare that $a < b$ if either $a_1 <_L a_2$ or $a_1 = a_2$ and $b_1 <_M b_2$. It is an easy exercise that the lexicographic order on the product of two well-ordered sets is also a well-order.

If $K$ is a nonempty countable compact space then $K$ is scattered. Therefore, if $\xi$ is an ordinal and $D_\xi(K) \neq \emptyset$ then $D_{\xi+1}(K)$ is a proper closed subset of $D_\xi(K)$. Therefore $D_\xi(K)$ has to be empty for some countable ordinal $\xi$; it is easy to see that $\mu = \setminus n\{\eta < \omega_1 : D_\eta(K) = \emptyset\}$ is a successor ordinal; the predecessor of $\mu$ is called the *dispersion index* of $K$ and is denoted by $di(K)$. In other words, an ordinal $\delta$ is the dispersion index of a countable compact space $K$ if $\delta = \setminus n\{\eta < \omega_1 : D_\eta(K)$ is finite$\}$. An easy proof by transfinite induction shows that, for any countable scattered compact space $K$,

(1)   if $F$ is a closed subset of $K$ then $D_\xi(F) \subset D_\xi(K) \cap F$; if the set $F$ is clopen in $K$ then $D_\xi(F) = D_\xi(K) \cap F$ for any ordinal $\xi$.

Say that an infinite ordinal $\xi$ is a *prime component* if $\eta + \mu < \xi$ for any ordinals $\eta$ and $\mu$ such that $\eta < \xi$ and $\mu < \xi$. Given ordinals $\eta$ and $\xi$ with $\eta \leq \xi$ let $[\eta, \xi] = \{\mu : \eta \leq \mu \leq \xi\}$; all other ordinal intervals are defined analogously. In particular, for an ordinal $\xi$ we will often denote the space $\xi + 1$ by $[0, \xi]$.

*Fact 1.* Suppose that a well-ordered set $A$ is isomorphic to an ordinal $\xi$ and a well-ordered set $B$ is isomorphic to an ordinal $\eta$. Consider the set

$$C = C(A, B) = (\{0\} \times A) \cup (\{1\} \times B)$$

with the following order: if $c_1, c_2 \in C$ and $c_i = (a_i, b_i)$ for each $i \in \mathbb{D}$ then say that $c_1 < c_2$ if either $a_1 < a_2$ (which actually means that $a_1 = 0$ and $a_2 = 1$) or $a_1 = a_2$ and $b_1 < b_2$ (observe that $b_1 < b_2$ makes sense because $a_1 = a_2$ implies that either $\{b_1, b_2\} \subset A$ or $\{b_1, b_2\} \subset B$). Then $C$ with the described order is isomorphic to $\xi + \eta$. In other words, to obtain the sum of two well-ordered sets $A$ and $B$ we must place all elements of $B$ after all elements of $A$.

*Proof.* There is no loss of generality to consider that $A = \xi$ and $B = \eta$. Therefore $C = C(\xi, \eta) = (\{0\} \times \xi) \cup (\{1\} \times \eta)$. We will prove our Fact by induction on $\eta$. If $\eta = 0$ then $C = \{0\} \times \xi$ is isomorphic to $\xi = \xi + \eta$.

Assume that $\mu$ is an ordinal and we proved that $C(\xi, \eta)$ is isomorphic to $\xi + \eta$ for all $\eta < \mu$. If $\mu = \eta + 1$ for some ordinal $\eta$ then $C(\xi, \mu) = C(\xi, \eta) \cup \{(1, \eta)\}$ and $(1, \eta)$ is the greatest element of $C(\xi, \mu)$. By the induction hypothesis, $C(\xi, \eta)$ is isomorphic to $\xi + \eta$, so $C(\xi, \mu)$ has the same order type as $(\xi + \eta) + 1 = \xi + (\eta + 1)$, so $C(\xi, \mu)$ is isomorphic to $\xi + \mu$.

If $\mu$ is a limit ordinal then we have the equalities $\xi + \mu = \sup\{\xi + \eta : \eta < \mu\}$ and $C(\xi, \mu) = \bigcup\{C(\xi, \eta) : \eta < \mu\}$. By the induction hypothesis, $C(\xi, \eta)$ is isomorphic to $\xi + \eta$ and hence $C(\xi, \eta) < \xi + \mu$ for any $\eta < \mu$. As a consequence, the order type of $C(\xi, \mu)$ does not exceed $\xi + \mu$. Besides, if the order type of $C(\xi, \mu)$ is strictly less than $\xi + \mu$ then it is strictly less than $\xi + \eta$ for some $\eta < \mu$. However, $\xi + \eta$ is isomorphic to $C(\xi, \eta) \subset C(\xi, \mu)$, so the order type of $\xi + \eta$ does not exceed the order type of $C(\xi, \mu)$; this contradiction shows that $C(\xi, \mu)$ is isomorphic to $\xi + \mu$, so we carried out the induction step and hence Fact 1 is proved.

*Fact 2.* If $\xi$ is a prime component then $\xi$ is a limit ordinal; besides, if $\eta < \xi$ then $\eta + \xi = \xi$ and hence the space $[0, \eta] \oplus [0, \xi]$ is homeomorphic to $[0, \xi]$.

*Proof.* If $\xi = \eta + 1$ then the ordinals $\eta$ and $\mu = 1$ witness that $\xi$ is not a prime component; this contradiction shows that $\xi$ has to be a limit ordinal. Therefore $\xi \leq \eta + \xi = \sup\{\eta + \mu : \mu < \xi\} \leq \xi$ because $\eta + \mu < \xi$ for any $\mu < \xi$. Consequently, $\eta + \xi = \xi$; it is easy to deduce from Fact 1 that $[0, \eta + \xi]$ is homeomorphic to $[0, \eta] \oplus [1, \xi] \simeq [0, \eta] \oplus [0, \xi]$, so $[0, \xi] \simeq [0, \eta] \oplus [0, \xi]$, i.e., Fact 2 is proved.

*Fact 3.* The ordinal $\omega^\xi$ is a prime component for any countable ordinal $\xi > 0$.

*Proof.* If $\xi = 1$ then $\omega^\xi = \omega$ is a prime component because $m + n < \omega$ whenever $m < \omega$ and $n < \omega$. Suppose that $\xi > 1$ is an ordinal and we proved that $\omega^\eta$ is a prime component for any $\eta < \xi$. Assume first that $\xi$ is a limit ordinal and we have ordinals $\mu, \mu'$ such that $\delta = \max\{\mu, \mu'\} < \omega^\xi$. There exists $\eta < \xi$ for which $\delta < \omega^\eta$; by the induction hypothesis, $\mu + \mu' < \omega^\eta \leq \omega^\xi$, so $\mu + \mu' < \omega^\xi$.

Now, if $\xi = \eta + 1$ then take any ordinals $\mu, \mu'$ such that $\delta = \max\{\mu, \mu'\} < \omega^\xi$ and observe that $\omega^\xi = \omega^\eta \cdot \omega = \sup\{\omega^\eta \cdot n : n \in \omega\}$. Pick $n \in \omega$ such that $\delta < \omega^\eta \cdot n$; then $\mu + \mu' \leq \omega^\eta \cdot n + \omega^\eta \cdot n = \omega^\eta \cdot (2n) < \omega^\xi$. Thus $\omega^\xi$ is also a prime component, so our inductive procedure shows that $\omega^\xi$ is a prime component for any $\xi \in [1, \omega_1)$, i.e., Fact 3 is proved.

*Fact 4.* For any countably infinite ordinal $\eta < \omega_1$ there exists a unique ordinal $\xi > 0$ such that $\omega^\xi \leq \eta < \omega^{\xi+1}$.

*Proof.* It is easy to prove by transfinite induction that $\omega^\mu \geq \mu$ for any ordinal $\mu$. In particular, $\omega^{\eta+1} \geq \eta + 1 > \eta$, so we can define the ordinal $\xi' = \backslash n\{\mu : \omega^\mu > \eta\}$. If $\xi'$ is a limit ordinal then $\omega^{\xi'} = \sup\{\omega^\mu : \mu < \xi'\}$, so it follows from $\eta < \omega^{\xi'}$ that $\eta < \omega^\mu$ for some $\mu < \xi'$ which is a contradiction. Therefore $\xi' = \xi + 1$ for some ordinal $\xi$, so, by our choice of $\xi'$, we have $\omega^\xi \leq \eta < \omega^{\xi+1}$ as promised.

Suppose that $\omega^\delta \leq \eta < \omega^{\delta+1}$ and $\delta \neq \xi$; if $\delta > \xi$ then $\delta \geq \xi+1$, so $\omega^\delta \geq \omega^{\xi+1} > \eta$ which is a contradiction. If $\delta < \xi$ then $\delta + 1 \leq \xi$ and therefore $\omega^{\delta+1} \leq \omega^\xi \leq \eta$; this contradiction shows that $\delta = \xi$, so $\xi$ is uniquely determined and hence Fact 4 is proved.

*Fact 5.* If $\xi$ is a countable ordinal then $D_\xi([1, \omega^\xi]) = \{\omega^\xi\}$.

*Proof.* If $\xi = 0$ then $\omega^\xi = 1$, so $D_0([1, \omega^\xi]) = [1, 1] = \{\omega^\xi\}$. Proceeding by induction assume that $0 < \xi < \omega_1$ and $D_\eta([1, \omega^\eta]) = \{\omega^\eta\}$ for any $\eta < \xi$. If $\xi$ is a limit ordinal then $[1, \omega^\xi] = \bigcup\{[1, \omega^\eta] : \eta < \xi\} \cup \{\omega^\xi\}$. Given any $\eta < \xi$, the set $[1, \omega^\eta]$ is clopen in $[1, \omega^\xi]$, so $D_\eta([1, \omega^\xi]) \cap [1, \omega^\eta] = D_\eta([1, \omega^\eta]) = \{\omega^\eta\}$ by the induction hypothesis. As a consequence, $D_\xi([1, \omega^\xi]) \cap [1, \omega^\eta] = D_\xi([1, \omega^\eta]) = \emptyset$ for any $\eta < \xi$ and hence $D_\xi([1, \omega^\xi]) \subset \{\omega^\xi\}$. On the other hand, we have the equality $D_\xi([1, \omega^\xi]) = \bigcap\{D_\eta([1, \omega^\xi]) : \eta < \xi\}$; the family $\mathcal{D} = \{D_\eta([1, \omega^\xi]) : \eta < \xi\}$ is decreasing and we saw that $\omega^\eta \in D_\eta([1, \omega^\eta]) \subset D_\eta([1, \omega^\xi])$ and hence all elements of $\mathcal{D}$ are nonempty. By compactness of the space $[1, \omega^\xi]$ we have $\bigcap \mathcal{D} \neq \emptyset$, so $D_\xi([1, \omega^\xi]) = \bigcap \mathcal{D} = \{\omega^\xi\}$.

If $\xi = \eta + 1$ for some countable ordinal $\eta$ then it follows from the equalities $[1, \omega^\xi] = [1, \omega^\eta \cdot \omega] = \bigcup\{[1, \omega^\eta \cdot n] : n \in \omega\} \cup \{\omega^\xi\}$ that there is a family $\{B_n : n \in \omega\}$ of disjoint clopen subsets of $[1, \omega^\xi]$ such that $[1, \omega^\xi] = \{\omega^\xi\} \cup \bigcup\{B_n : n \in \omega\}$ and $B_n \simeq [1, \omega^\eta]$ for every $n \in \omega$. The set $D_\eta([1, \omega^\xi]) \cap B_n = D_\eta(B_n)$ is a singleton by the induction hypothesis; fix a point $y_n$ such that $D_\eta([1, \omega^\xi]) \cap B_n = \{y_n\}$ for every $n \in \omega$. As a consequence, the set $D_\eta([1, \omega^\xi]) = \{\omega^\xi\} \cup \{y_n : n \in \omega\}$ is a convergent sequence with limit $\omega^\xi$. Therefore $D_\xi([1, \omega^\xi]) = D_1(D_\eta([1, \omega^\xi])) = \{\omega^\xi\}$, so our inductive procedure shows that $D_\xi([1, \omega^\xi]) = \{\omega^\xi\}$ for any $\xi < \omega_1$ and hence Fact 5 is proved.

*Fact 6.* Given a countable ordinal $\xi$, a countable compact space $K$ is homeomorphic to $[1, \omega^\xi]$ if and only if $|D_\xi(K)| = 1$.

*Proof.* Since necessity is immediate from Fact 5, let us prove sufficiency by induction on $\xi$. If $\xi = 0$ and $D_\xi(K) = K$ is a singleton then $\omega^\xi = 1$ and hence $K$ is homeomorphic to $\{1\} = [1, \omega^\xi]$.

Assume that $\xi > 0$ is a countable ordinal and we proved that, for any countable compact space $K$, if $\eta < \xi$ and $|D_\eta(K)| = 1$ then $K \simeq [1, \omega^\eta]$. Suppose first that $\xi$ is a limit ordinal and fix a countable compact space $K$ such that $D_\xi(K) = \{a\}$ for some point $a \in K$.

Choose a family $\{O_n : n \in \omega\}$ of clopen subsets of $K$ such that $O_0 = K$ while $U_{n+1} \subset U_n$ for each $n \in \omega$ and $\bigcap_{n \in \omega} U_n = \{a\}$. If $P_n = O_n \setminus O_{n+1}$ then $P_n$ is a clopen compact subset of $K \setminus \{a\}$ for any $n \in \omega$.

Apply (1) to see that $D_\xi(P_n) = D_\xi(K) \cap P_n = \emptyset$ and hence $\xi_n = di(P_n) < \xi$ for each $n \in \omega$. Given any $n \in \omega$ the set $A_n = D_{\xi_n}(P_n)$ is finite, so it is easy to find a clopen disjoint finite cover $\mathcal{O}_n$ of the space $P_n$ such that $|D_{\xi_n}(O)| = 1$ for every $O \in \mathcal{O}_n$. By the induction hypothesis, every $O \in \mathcal{O}_n$ is homeomorphic to $[1, \omega^{\xi_n}]$; if we choose an enumeration $\{U_k : k \in \omega\}$ of the family $\bigcup_{n \in \omega} \mathcal{O}_n$ then $U_k \simeq [1, \omega^{\mu_k}]$ for every $k \in \omega$.

Using Fact 1 of V.199 it is easy to construct a sequence $\{v_k : k \in \omega\}$ of countable ordinals with the following properties:

(2)  $v_i < v_{i+1}$ for each $i \in \omega$;

(3)  $v_0 = \omega^{\mu_0}$ and the interval $B_{i+1} = [v_i + 1, v_{i+1}]$ is canonically homeomorphic to $[1, \omega^{\mu_i+1}]$ for every $i \in \omega$.

Let $B_0 = [1, \omega^{\mu_0}]$; the ordinal $v_i$ being a finite sum of ordinals which are all strictly smaller than $\omega^\xi$, it follows from Fact 2 and Fact 3 that $v_i < \omega^\xi$ for all $i \in \omega$. Therefore the ordinal $v = \sup\{v_i : i \in \omega\}$ does not exceed $\omega^\xi$. On the other hand, if $\eta < \xi$ then $D_{\eta+1}(K)$ is an infinite set, so $D_{\eta+1}(K) \cap U_k \neq \emptyset$ for some $k \in \omega$. Consequently, $di(U_k) = \mu_k \geq \eta + 1 > \eta$ which shows that the order type of $[1, v_{k+1}]$ is greater than or equal to $\omega^{\mu_k} \geq \omega^{\eta+1} > \omega^\eta$. Thus $v \geq \omega^\eta$ for any $\eta < \xi$; since $\omega^\xi = \sup\{\omega^\eta : \eta < \xi\}$, we conclude that $v \geq \omega^\xi$ and hence $v = \omega^\xi$.

It follows from (2) and (3) that $\{B_i : i \in \omega\}$ is a clopen compact cover of the space $[1, \omega^\xi)$ and therefore $[1, \omega^\xi]$ is the one-point compactification of the space $\bigoplus_{i \in \omega} B_i$. The family $\{U_i : i \in \omega\}$ is a clopen compact cover of $K \backslash \{a\}$, so $K$ is a one-point compactification of the space $\bigoplus_{i \in \omega} U_i \simeq \bigoplus_{i \in \omega} B_i$. This shows that $K \simeq [1, \omega^\xi]$ and hence we carried out the induction step for any limit ordinal $\xi$.

Now assume that $\xi$ is a successor and fix an ordinal $\eta$ such that $\xi = \eta + 1$. For the space $Y = D_\eta(K)$ the set $D_1(Y) = D_\xi(K) = \{a\}$ is a singleton, so $Y$ is a compact space with a unique non-isolated point $a$, i.e., $Y$ is a sequence which converges to $a$. It is easy to find a decreasing sequence $\mathcal{O} = \{O_n : n \in \omega\}$ of clopen subsets of $K$ such that $O_0 = K$ while $\bigcap \mathcal{O} = \{a\}$ and $(O_n \backslash O_{n+1}) \cap (Y \backslash \{a\}) \neq \emptyset$ for any $n \in \omega$. Splitting every $O_n \backslash O_{n+1}$ into a finitely many disjoint clopen subsets if necessary, we obtain a disjoint clopen compact cover $\{W_n : n \in \omega\}$ of the space $Y \backslash \{a\}$ such that $W_n \cap Y$ is a singleton; let $a_n \in Y \backslash \{a\}$ be the point for which $W_n \cap Y = \{a_n\}$ for each $n \in \omega$.

For each $n \in \omega$ we have the equalities $D_\eta(W_n) = D_\eta(K) \cap W_n = \{a_n\}$, so $W_n \simeq [1, \omega^\eta]$ by the induction hypothesis. Therefore the space $K$ is the one-point compactification of $\bigoplus_{n \in \omega} W_n$ which is homeomorphic to $[1, \omega^\eta] \times \omega$. Observe that $[1, \omega^\xi) = \bigcup\{[1, \omega^\eta \cdot n] : n \in \omega\} = \bigcup_{n \in \omega} [\omega^\eta \cdot n + 1, \omega^\eta \cdot (n + 1)]$ and the set $[\omega^\eta \cdot n + 1, \omega^\eta \cdot (n + 1)]$ is clopen in $[1, \omega^\xi]$ and homeomorphic to $[1, \omega^\eta]$, so $[1, \omega^\xi)$ is homeomorphic to $[1, \omega^\eta] \times \omega$ and hence $[1, \omega^\xi]$ is the one-point compactification of $[1, \omega^\eta] \times \omega$. Therefore $K$ is homeomorphic to $[1, \omega^\xi]$, i.e., we completed the induction step and hence Fact 6 is proved.

*Fact 7.* Suppose that $K$ is a countable compact space and $A = D_\xi(K) \neq \emptyset$ for some countable ordinal $\xi$. If $a$ is the point represented by $A$ in the space $K_A$ then $D_\xi(K_A) = \{a\}$.

*Proof.* If $x \in K \backslash A$ then take a clopen set $U \in \tau(x, K)$ such that $U \cap A = \emptyset$. It follows from (1) and $D_\xi(K) \cap U = \emptyset$ that $D_\xi(U) = \emptyset$ and hence $D_\xi(K_A) \cap U = \emptyset$; in particular, $x \notin D_\xi(K_A)$. Therefore $D_\xi(K_A) \subset \{a\}$; to see that $a \in D_\xi(K_A)$ observe that $K$ is scattered, so we can find an isolated point $b$ in the space $A$ and a clopen set $V$ such that $V \cap A = \{b\}$. The set $V' = (V \setminus \{A\}) \cup \{a\} \subset K_A$ is the one-point compactification of the space $V \setminus A = V \setminus \{b\}$ which shows that $V$ is also the one-point compactification of $V \setminus A$.

Consequently, $V \simeq V'$; since $b \in D_\xi(K) \cap V = D_\xi(V)$ we must have $D_\xi(V') \neq \emptyset$ and hence $D_\xi(K_A) \neq \emptyset$. Therefore $D_\xi(K_A) = \{a\}$, i.e., Fact 7 is proved.

*Fact 8.* If $K$ and $L$ are countable compact spaces such that $di(L) < di(K)$ then $K \oplus L \simeq K$. In particular, if $K$ is infinite and $L$ is finite then $K \oplus L \simeq K$.

*Proof.* Let $\xi = di(K)$; since $D_\xi(K) \neq \emptyset$, we can find $n \in \mathbb{N}$ and distinct points $a_1, \dots, a_n \in K$ such that $D_\xi(K) = \{a_1, \dots, a_n\}$. There exist disjoint clopen sets $U_1, \dots, U_n$ for which $K = U_1 \cup \dots \cup U_n$ and $a_i \in U_i$ for every $i \leq n$. We have the equality $D_\xi(U_i) = \{a_i\}$ and hence $U_i \simeq [1, \omega^\xi]$ for each $i \leq n$ by Fact 6. Observe that $K \simeq U_1 \oplus \dots \oplus U_n$ and $D_\xi(U_1 \oplus L) = D_\xi(U_1) \oplus D_\xi(L) = D_\xi(U_1) = \{a_1\}$ which shows that $U_1 \oplus L \simeq [1, \omega^\xi] \simeq U_1$ by Fact 6. As an immediate consequence, $K \oplus L \simeq (U_1 \oplus L) \oplus \dots \oplus U_n \simeq U_1 \oplus \dots \oplus U_n \simeq K$, i.e., Fact 8 is proved.

*Fact 9.* If $K$ and $L$ are countably infinite compact spaces and $di(K) = di(L)$ then $K \overset{l}{\sim} L$.

*Proof.* If $\xi = di(K) = di(L)$ then $D_\xi(K)$ and $D_\xi(L)$ are nonempty finite sets. We can find $n \in \mathbb{N}$ and distinct points $a_1, \dots, a_n \in K$ such that $D_\xi(K) = \{a_1, \dots, a_n\}$. There exist disjoint clopen sets $U_1, \dots, U_n$ for which $K = U_1 \cup \dots \cup U_n$ and $a_i \in U_i$ for every $i \leq n$. Then $D_\xi(U_i) = \{a_i\}$ and hence $U_i \simeq [1, \omega^\xi]$ for each $i \leq n$ by Fact 6.

The set $F = \{a_1, \dots, a_n\}$ is easily seen to be a retract of $K$, so $K$ is $l$-equivalent to the space $K_F \oplus F$ (see Problem 373). Since $K_F$ is infinite and $F$ is finite, we can apply Fact 8 to convince ourselves that $K_F \oplus F \simeq K_F$. It follows from Fact 6 and Fact 7 that $D_\xi(K_F)$ is a singleton and therefore $K_F \simeq [1, \omega^\xi]$. This shows that the space $K$ is $l$-equivalent to $[1, \omega^\xi]$. Since we can repeat the same reasoning for $L$, we conclude that $K \overset{l}{\sim} [1, \omega^\xi] \overset{l}{\sim} L$, so $K \overset{l}{\sim} L$ and hence Fact 9 is proved.

*Fact 10.* If $\omega \leq \mu < \omega_1$ and $\mu$ is a prime component then $[1, \mu]$ is $l$-equivalent to $[1, \nu]$ for any ordinal $\nu$ such that $\mu \leq \nu < \mu^\omega$.

*Proof.* Take the ordinal $\xi > 0$ such that $\mu = \omega^\xi$ and let $K = [1, \nu]$. The set $F = D_\xi(K)$ is nonempty because $D_\xi([1, \omega^\xi]) = \{\mu\} \subset F$ (see Fact 5). Since $\mu^\omega = \sup\{\mu^n : n \in \omega\}$, there exists a number $n \in \omega$ such that $\mu^n \leq \nu < \mu^{n+1}$. We will prove our Fact by induction on $n$.

If $n = 1$ then $\omega^\xi \leq \nu < (\omega^\xi)^2 = \omega^{\xi \cdot 2}$. Observe that $K$ is a clopen subspace of the space $[1, \omega^{\xi \cdot 2}]$ and hence we can apply (1) together with Fact 5 to see that $D_{\xi \cdot 2}(K) = D_{\xi \cdot 2}([1, \omega^{\xi \cdot 2}]) \cap K = \{\omega^{\xi \cdot 2}\} \cap K = \emptyset$. Therefore $D_\xi(F) = D_{\xi \cdot 2}(K) = \emptyset$; this shows that $di(F) < \xi = di([1, \omega^\xi])$ and hence $[1, \omega^\xi] \oplus F \simeq [1, \omega^\xi]$ by Fact 8.

The space $K_F \oplus F$ is $l$-equivalent to $K$ by Problem 373; besides, $K_F \simeq [1, \omega^\xi]$ by Fact 7 and Fact 6. Therefore $K \overset{l}{\sim} K_F \oplus F \overset{l}{\sim} [1, \omega^\xi] \oplus F \simeq [1, \omega^\xi]$ and hence we settled the case of $n = 1$.

Now assume that $\mu^n \le \nu < \mu^{n+1}$ for some number $n > 1$ and we have proved that $[1, \delta] \overset{l}{\sim} [1, \mu]$ for any ordinal $\delta$ such that $\mu^k \le \delta < \mu^{k+1}$ for some $k < n$. Since $K$ is a clopen subspace of $[1, \omega^{\xi \cdot (n+1)}]$, we can apply the property (1) together with Fact 5 to see that $D_{\xi \cdot (n+1)}(K) = D_{\xi \cdot (n+1)}([1, \omega^{\xi \cdot (n+1)}]) \cap K = \{\omega^{\xi \cdot (n+1)}\} \cap K = \emptyset$. Therefore $D_{\xi \cdot n}(F) = D_{\xi \cdot (n+1)}(K) = \emptyset$; this shows that $di(F) < \xi \cdot n$.

The space $K_F \oplus F$ is $l$-equivalent to $K$ by Problem 373; besides, $K_F \simeq [1, \omega^\xi]$ by Fact 7 and Fact 6. There exists an ordinal $\delta$ such that $F \simeq [0, \delta]$ (see Problem 199). If $\delta < \mu$ then $di(F) < \xi$ and hence $F \oplus [1, \omega^\xi] \simeq [1, \omega^\xi]$ by Fact 8. Therefore $K \overset{l}{\sim} F \oplus [1, \omega^\xi] \simeq [1, \omega^\xi]$ as promised.

If $di(F) \ge \xi$ then $\delta \ge \mu$ and it follows from $di(F) < \xi \cdot n$ that $\delta < \omega^{\xi \cdot n}$. Observe also that $[1, \delta] \simeq [1, \delta] \oplus \{0\} \simeq [0, \delta] \simeq F$ by Fact 8. Since $\delta < \omega^{\xi \cdot n}$, there exists $k < n$ such that $\omega^{\xi \cdot k} \le \delta < \omega^{\xi \cdot (k+1)}$, i.e., $\mu^k \le \delta < \mu^{k+1}$, so our induction hypothesis can be applied to see that $F \simeq [1, \delta] \overset{l}{\sim} [1, \mu]$. As a consequence, $K \overset{l}{\sim} K_F \oplus F \simeq [1, \mu] \oplus [1, \delta] \overset{l}{\sim} [1, \mu] \oplus [1, \mu]$. Besides, $di([1, \mu] \oplus [1, \mu]) = \xi = di([1, \mu])$, so we can apply Fact 9 to conclude that $K \overset{l}{\sim} [1, \mu] \oplus [1, \mu] \overset{l}{\sim} [1, \mu]$; this completes our induction step and shows that Fact 10 is proved.

*Fact 11.* If $(B_1, \|\cdot\|_1)$ and $(B_2, \|\cdot\|_2)$ are normed spaces then $B_1 \approx B_2$ if and only if there exists a surjective linear map $\varphi : B_1 \to B_2$ such that, for some number $K > 0$, we have $\|x\|_1 \le \|\varphi(x)\|_2 \le K\|x\|_1$ for any $x \in B_1$.

*Proof.* Let $\mathbf{0}_1$ and $\mathbf{0}_2$ be the zero vectors in the spaces $B_1$ and $B_2$ respectively. Suppose first that $\varphi : B_1 \to B_2$ is a surjective linear map as in the hypothesis of our Fact. If $x \in B_1$ and $\varphi(x) = \mathbf{0}_1$ then $\|x\|_1 \le \|\varphi(x)\|_2 = 0$, so $\|x\|_1 = 0$ and hence $\varphi$ is injective. For any $r > 0$ let $S_1(r) = \{x \in B_1 : \|x\|_1 \le r\}$ and $S_2(r) = \{x \in B_2 : \|x\|_2 \le r\}$.

The second inequality of our hypothesis implies that $\varphi(S_1(1)) \subset S_2(K)$, so the map $\varphi$ is continuous by Fact 2 of V.400. If $\xi : B_2 \to B_1$ is the inverse of $\varphi$ then the first inequality of our hypothesis shows that $\|\xi(y)\|_1 \le \|y\|_2$ because, for $x = \xi(y)$ we have $y = \varphi(x)$ for any $y \in B_2$. Therefore $\xi(S_2(1)) \subset S_1(1)$ and hence $\xi$ is also continuous by Fact 2 of V.400. It turns out that $\varphi$ is a homeomorphism, so we settled sufficiency.

Now assume that $\xi : B_1 \to B_2$ is a linear homeomorphism and let $\eta : B_2 \to B_1$ be its inverse. By Fact 2 of V.400 there exists $R > 0$ such that $\eta(S_2(1)) \subset S_1(R)$. Let $\varphi(x) = R \cdot \xi(x)$ for any $x \in B_1$. It is immediate that $\varphi : B_1 \to B_2$ is a surjective linear map. If $x \in B_1 \setminus \{\mathbf{0}_1\}$ and $y = \varphi(x) = R \cdot \xi(x)$ then $\frac{1}{\|y\|_2} \cdot y$ belongs to $S_2(1)$ and therefore $\frac{1}{\|y\|_2} \cdot \|\eta(y)\|_1 \le R$. Observe that $\eta(y) = \eta(R \cdot \xi(x)) = R \cdot x$, so we have $\frac{1}{\|y\|_2} \cdot R \cdot \|x\|_1 \le R$, so $\|x\|_1 \le \|y\|_2 = \|\varphi(x)\|_2$ and hence we have the first of the promised inequalities for $\varphi$.

The map $\varphi$ being continuous, we can apply Fact 2 of V.400 once more to see that $\varphi(S_1(1)) \subset S_2(K)$ for some $K > 0$. If $x \in B_1 \setminus \{\mathbf{0}_1\}$ then $y = \frac{1}{\|x\|_1} \cdot x \in S_1(1)$

and hence $||\varphi(y)||_2 \leq K$, i.e., $\frac{1}{||x||_1} \cdot ||\varphi(x)||_2 \leq K$ which shows that we have the inequality $||\varphi(x)||_2 \leq K||x||_1$ for every $x \in B_1$ and hence we established necessity, so Fact 11 is proved.

**Fact 12.** Suppose that $(B_1, || \cdot ||_1)$ and $(B_2, || \cdot ||_2)$ are normed spaces and a linear continuous map $\varphi : B_1 \to B_2$ is not a topological embedding (i.e., $\varphi$ is not a homeomorphism between $B_1$ and $\varphi(B_1)$). Then, for any $\varepsilon > 0$ there exists $x \in B_1$ such that $||x||_1 = 1$ and $||\varphi(x)||_2 < \varepsilon$.

*Proof.* If $\varphi$ is not injective then there exists $x' \in B_1 \backslash \{0_1\}$ such that $\varphi(x') = 0_2$. Then $x = \frac{1}{||x'||_1} \cdot x'$ is as promised. If $\varphi$ is injective then let $C = \varphi(B_1)$ and consider the inverse map $\xi : C \to B_1$ for the map $\varphi$. The map $\xi$ cannot be continuous, so there exists a sequence $\{y_n : n \in \omega\} \subset C \backslash \{0_2\}$ such that $y_n \to 0_2$ and $\{\xi(y_n) : n \in \omega\}$ does not converge to $0_1$.

Passing to a subsequence if necessary, we can assume that there is $\delta > 0$ such that $||\xi(y_n)||_1 \geq \delta$; let $x_n = \frac{1}{||\xi(y_n)||_1} \cdot \xi(y_n)$ for all $n \in \omega$. Then, for every $n \in \omega$ we have $||x_n||_1 = 1$, so it follows from $||\varphi(x_n)||_2 \leq \frac{1}{\delta} \cdot ||y_n||_2$ that the sequence $\{||\varphi(x_n)||_2 : n \in \omega\}$ converges to zero. Thus there exists $n \in \omega$ for which $||\varphi(x_n)||_2 < \varepsilon$, i.e., $x = x_n$ is as promised and hence Fact 12 is proved.

**Fact 13.** Suppose that $K$ be an infinite countable compact space and $a \in K$ is a non-isolated point of $K$. Let $I = \{f \in C_u(K) : f(a) = 0\}$. Then the Banach space $I$ (with the norm inherited from $C_u(K)$) is linearly homeomorphic to $C_u(K)$.

*Proof.* Recall that $||f|| = \sup\{|f(x)| : x \in K\}$ for any $f \in C_u(K)$ and the topology of $C_u(K)$ is generated by the norm $|| \cdot ||$. Given a point $b \notin K$, the space $\{b\} \oplus K$ is homeomorphic to $K$ by Fact 8. Therefore we can find an isolated point $z \in K$ such that $Q = K \backslash \{z\} \simeq K$. To simplify the notation we denote the norms in $I$ and $C_u(Q)$ by the same symbol $|| \cdot ||$.

For every $f \in I$ let $\delta(f)(x) = f(x) + f(z)$ for any $x \in Q$. This defines a linear map $\delta : I \to C_u(Q)$. If $g \in C_u(Q)$ then let $f(x) = g(x) - g(a)$ for all $x \in Q$ and $f(z) = g(a)$. It is straightforward that $\delta(f) = g$, so the map $\delta$ is surjective. It is immediate that $||\delta(f)|| \leq 2||f||$ for any $f \in I$. We claim that also

(4) $||f|| \leq 2||\delta(f)||$ for every $f \in I$.

Indeed, fix a function $f \in I$ and let $r = ||f||$. If $f = 0$ then the promised inequality clearly holds, so assume that $f \neq 0$ and hence $r > 0$. If $|f(z)| \geq \frac{r}{2}$ then $|\delta(f)(a)| = |f(z)| \geq \frac{r}{2}$ and hence $||\delta(f)|| \geq |f(z)| \geq \frac{r}{2}$ as required. If $|f(z)| < \frac{r}{2}$ then there is a point $x \in Q$ such that $|f(x)| = r$; since $|\delta(f)(x)| = |f(x) + f(z)| > \frac{r}{2}$, we conclude that $||\delta(f)|| \geq |f(x) + f(z)| > \frac{r}{2}$, so (4) is proved.

Finally, let $\varphi(f) = 2\delta(f)$ for any $f \in I$. Then $\varphi : I \to C_u(Q)$ is a surjective linear map and it follows from (4) that $||f|| \leq ||\varphi(f)|| \leq 4||f||$ for any $f \in I$, so we can apply Fact 11 to convince ourselves that $I \approx C_u(Q) \approx C_u(K)$ and hence Fact 13 is proved.

**Fact 14.** Suppose that $K$ is a compact space and $F$ is a closed subset of $K$. Assume additionally that there exists a retraction $r : K \to F$. For every function $f \in C_u(F)$

let $\varphi(f) = f \circ r$. Then $\varphi : C_u(F) \to C_u(K)$ is a linear isometric embedding, i.e., $\varphi : C_u(F) \to L = \varphi(C_u(F))$ is a linear homeomorphism and $||f|| = ||\varphi(f)||$ for any $f \in C_u(F)$.

*Proof.* It is evident that $\varphi$ is a linear map; recall that $||f|| = \sup\{|f(x)| : x \in F\}$ for any function $f \in C_u(F)$. If $g \in C_u(K)$ then $||g|| = \sup\{|g(x)| : x \in K\}$. It follows from $\varphi(f)|F = f$ that $||\varphi(f)|| \geq ||f||$ for any $f \in C_u(F)$. We also have $|\varphi(f)(x)| = |f(r(x))| \leq ||f||$ for any $x \in K$, so $||\varphi(f)|| \leq ||f||$ which shows that $||f|| = ||\varphi(f)||$ for all $f \in C_u(F)$, so Fact 11 is applicable to conclude that $\varphi : C_u(F) \to L$ is a linear homeomorphism and hence Fact 14 is proved.

*Fact 15.* For any ordinals $\xi$ and $\eta$ the order type of $\xi \cdot \eta$ coincides with the order type of the set $\eta \times \xi$ with the lexicographic order.

*Proof.* If $\eta = 0$ then $\xi \cdot \eta = 0$ is the empty set as well as $\eta \times \xi$, so the order types of $\xi \cdot \eta$ and $\eta \times \xi$ coincide for $\eta = 0$. Proceeding by induction assume that $\mu > 0$ is an ordinal and our Fact is proved for all ordinals $\eta < \mu$.

Assume first that $\mu$ is a successor ordinal, i.e., $\mu = \eta + 1$ for some ordinal $\eta$. Then $\xi \cdot \mu = \xi \cdot \eta + \xi$. Observe that $\xi \times \mu = A \cup B$ where $A = \xi \times \eta$ and $B = \xi \times \{\eta\}$; we have $x < y$ whenever $x \in A$ and $y \in B$, so $\xi \times \mu$ has the order type of $A + B$ (see Fact 1). Furthermore, $A$ is isomorphic to $\xi \cdot \eta$ by the induction hypothesis and $B$ is isomorphic to $\xi$. Thus $\xi \times \mu$ has the order type of $\xi \cdot \eta + \xi$ as promised.

Now if $\mu$ is a limit ordinal then $\xi \cdot \mu = \sup\{\xi \cdot \eta : \eta < \mu\}$. It follows from the equality $\xi \times \mu = \bigcup\{\xi \times \eta : \eta < \mu\}$ and the induction hypothesis that the order type of $\xi \cdot \eta$ is the same as the order type of $\xi \times \eta$, so it does not exceed the order type of $\xi \times \mu$ for any $\eta < \mu$. This proves that the order type of $\xi \cdot \mu$ does not exceed the order type of $\xi \times \mu$. If, on the other hand, the order type of $\xi \times \mu$ is strictly less than $\xi \cdot \mu$ then the order type of $\xi \times \mu$ is strictly less than $\xi \cdot \eta$ for some $\eta < \mu$ which is impossible because the ordinal $\xi \cdot \eta$ has the same type as the subset $\xi \times \eta$ of the set $\xi \times \mu$. This contradiction shows that the order type of $\xi \times \mu$ coincides with $\xi \cdot \mu$, so we completed our induction step and hence Fact 15 is proved.

*Fact 16.* Suppose that $\xi$ and $\eta$ are infinite ordinals. Then there exists a closed set $A \subset [1, \xi \cdot \eta]$ with the following properties:

(a)  $A$ is canonically homeomorphic to $[1, \eta]$;
(b)  there is a retraction $r : [1, \xi \cdot \eta] \to A$ and a set $B \subset A$ such that $r^{-1}(\nu)$ is canonically homeomorphic to $[1, \xi]$ for each $\nu \in B$ and $r^{-1}(B)$ is dense in $[1, \xi \cdot \eta]$.

*Proof.* By Fact 15 we can identify $[0, \xi \cdot \eta)$ with the set $\eta \times \xi$ with the lexicographic order. Take a point $s \notin \eta \times \xi$ and let $Z = ((\eta \times \xi) \backslash \{(0,0)\}) \cup \{s\}$; the order $\prec$ in $Z$ is the same as in $\eta \times \xi$ for the points of $\eta \times \xi$ and $s > t$ for any $t \in Z \backslash \{s\}$. Then $Z$ is canonically homeomorphic to $[1, \xi \cdot \eta]$.

Consider the set $A = \{(\nu, 0) : 1 \leq \nu < \eta\} \cup \{s\}$; let $\varphi(s) = \eta$ and $\varphi((\nu, 0)) = \nu$ for each $\nu \in [1, \eta)$. This gives us a bijection $\varphi : A \to [1, \eta]$ and it is straightforward that $\varphi$ is a canonical homeomorphism.

Let $r(a) = a$ for any $a \in A$. If $z \in Z \setminus A$ then $z = (\nu, \delta)$ for some $\nu < \eta$ and $\delta > 0$; let $r(z) = (\nu + 1, 0)$. For any $\nu < \eta$ the set $H_\nu = (\{\nu\} \times [1, \xi)) \cup \{(\nu + 1, 0)\}$ is easily seen to be clopen in $Z$ and canonically homeomorphic to $[1, \xi]$; it follows from the equality $H_\nu = r^{-1}((\nu + 1, 0))$ that the map $r$ is continuous at all points of $H_\nu$. The family $\mathcal{H} = \{H_\nu : \nu < \eta\}$ is disjoint, so $r$ is continuous at all points of $H = \bigcup \mathcal{H}$. Furthermore, $Z \setminus A \subset H$ and hence $H$ is dense in $Z$. Therefore, for the set $B = \{(\nu + 1, 0) : \nu < \eta\}$ we have $r^{-1}(B) = H$, so $r^{-1}(B)$ is dense in $Z$.

If $\nu < \eta$ is a limit ordinal then $r^{-1}((\nu, 0)) = \{(\nu, 0)\}$. Given any ordinal $\mu < \nu$ the set $W = \{z \in Z : (\mu + 1, 0) \prec z \prec (\nu, 1)\}$ is an open neighborhood of $(\nu, 0)$ such that $r(W) \subset (\mu, \nu]$. This shows that $r$ is also continuous at all points of $Z \setminus H$, so $r$ has all promised properties, i.e., Fact 16 is proved.

Returning to our solution take any countable infinite ordinals $\alpha$ and $\beta$ such that $\alpha \leq \beta < \alpha^\omega$. By Fact 4 there exists an ordinal $\xi > 0$ such that $\omega^\xi \leq \alpha < \omega^{\xi+1}$. Since $\omega^{\xi+1} = \omega^\xi \cdot \omega = \sup\{\omega^\xi \cdot n : n \in \omega\}$, we can find a number $n \in \omega$ for which $\omega^\xi \cdot n \leq \alpha < \omega^\xi \cdot (n + 1)$. It is easy to see that $[1, \omega^\xi \cdot (n + 1)] \simeq Y_0 \oplus \ldots \oplus Y_n$ where $Y_i \simeq [1, \omega^\xi]$ for every $i \leq n$. Therefore $di([1, \omega^\xi \cdot (n + 1)]) = \xi = di([1, \omega^\xi])$, so it follows from $[1, \omega^\xi] \subset [1, \alpha] \subset [1, \omega^\xi \cdot (n + 1)]$ that $di([1, \alpha]) = \xi$ and hence $[1, \alpha] \overset{l}{\sim} [1, \omega^\xi]$ by Fact 9.

The ordinal $\mu = \omega^\xi$ is a prime component and $\mu \leq \alpha \leq \beta < \alpha^\omega$; it follows from $\alpha^\omega = \sup\{\alpha^m : m \in \omega\}$ that there exists $m \in \omega$ such that $\beta < \alpha^m \leq (\mu \cdot \omega)^m$. The ordinal $\mu$ is infinite, so $\mu \cdot \omega \leq \mu \cdot \mu$ and therefore $\beta \leq (\mu^2)^m = \mu^{2m} < \mu^\omega$; thus $\mu \leq \beta < \mu^\omega$, so we can apply Fact 10 to convince ourselves that $[1, \beta] \overset{l}{\sim} [1, \mu]$ and hence $[1, \alpha] \overset{l}{\sim} [1, \beta]$. The ordinals $\alpha$ and $\beta$ being infinite, we can apply Fact 8 to see that $\alpha + 1 = [0, \alpha] \simeq [1, \alpha]$ and $\beta + 1 = [0, \beta] \simeq [1, \beta]$. This implies that $(\alpha + 1) \overset{l}{\sim} (\beta + 1)$ and hence we proved sufficiency.

Now suppose that $\alpha$ and $\beta$ are infinite ordinals such that $\alpha \leq \beta$ and $C_p(\alpha + 1)$ is linearly homeomorphic to $C_p(\beta + 1)$. Observe that $[1, \alpha] \simeq \alpha + 1$ and $[1, \beta] \simeq \beta + 1$ by Fact 8. It follows from Problem 327 that the spaces $C_u([1, \alpha])$ and $C_u([1, \beta])$ are linearly homeomorphic. Let $\mu = \setminus n\{\xi$ : there exists a linear homeomorphic embedding of $C_u([1, \alpha])$ into $C_u([1, \xi])\}$. It is clear that $\mu$ is an infinite ordinal; if $\mu$ is a successor then $\mu = \mu' + n$ for some limit ordinal $\mu'$ and $n \in \mathbb{N}$. Apply Fact 8 once more to see that $[1, \mu] \simeq [1, \mu'] \oplus [1, n] \simeq [1, \mu']$ and hence $C_u([1, \alpha])$ also linearly embeds in $C_u([1, \mu'])$ which is a contradiction with the choice of $\mu$. Therefore $\mu$ is a limit ordinal.

Assume that $\beta \geq \alpha^\omega \geq \mu^\omega$. Then the set $[1, \mu^\omega] \subset [1, \beta]$ is a retract of $[1, \beta]$ (see SFFS-316), so $C_u([1, \mu^\omega])$ linearly embeds in $C_u([1, \beta]) \approx C_u([1, \alpha])$ (see Fact 14) and hence in $C_u([1, \mu])$. The space $I = \{f \in C_u([1, \mu]) : f(\mu) = 0\}$ is linearly homeomorphic to $C_u([1, \mu])$ by Fact 13 so $C_u([1, \mu^\omega])$ linearly embeds in $I$.

Fix a linear subspace $L$ of the space $I$ which is linearly homeomorphic to $C_u([1, \mu^\omega])$. We denote the norm in $I$ and in $C_u([1, \mu^\omega])$ by the same symbol $\| \cdot \|$. Let $\varphi : L \to C_u([1, \mu^\omega])$ be a linear homeomorphism; by Fact 11, there exists

$K > 0$ such that $||f|| \le ||\varphi(f)|| \le K||f||$ for any $f \in L$. Pick a number $n \in \omega$ with $n > 4K$; the space $[1, \mu^n]$ is a retract of $[1, \mu^\omega]$ by SFFS-316 so $C_u([1, \mu^n])$ is isometric to a subspace of $C_u([1, \mu^\omega])$ by Fact 14.

Therefore there exists a linear subspace $M$ of the space $L$ and a surjective map $\theta : M \to C_u([1, \mu^n])$ such that $||f|| \le ||\theta(f)|| \le K||f||$ for any $f \in M$. Let $g_0 \in C_u([1, \mu^n])$ be the function which is identically equal to 1. The map $\theta$ being a linear homeomorphism, it has an inverse $\zeta : C_u([1, \mu^n]) \to M$. Let $f_0 = \zeta(g_0)$; it follows from $f_0(\mu) = 0$ that we can find an ordinal $\gamma_0 < \mu$ such that $|f_0(\gamma)| < \frac{1}{n+1}$ for all $\gamma \ge \gamma_0$.

Apply Fact 16 to find a closed set $Y_0 \subset [1, \mu^n]$ canonically homeomorphic to $[1, \mu]$ and a retraction $r_0 : [1, \mu^n] \to Y_0$ such that, for some $B_0 \subset Y_0$, the set $r_0^{-1}(B_0)$ is dense in $[1, \mu^n]$ and $r_0^{-1}(x)$ is canonically homeomorphic to $[1, \mu^{n-1}]$ for all $x \in B_0$. The dual map $r_0^* : C_u(Y_0) \to C_u([1, \mu^n])$ is a linear embedding by Fact 14. Let $\pi_0 : I \to C_u([1, \gamma_0])$ be the restriction map. Since $Y_0 \simeq [1, \mu]$, the linear map $\pi_0 \circ \zeta \circ r_0^*$ cannot be an embedding by the choice of $\mu$, so we can apply Fact 12 to find a function $g_1 \in r_0^*(C_u(Y_0))$ such that $||g_1|| = 1$ and $|\zeta(g_1)(\gamma)| < \frac{1}{n+1}$ for all $\gamma \le \gamma_0$. Let $f_1 = \zeta(g_1)$; it follows from $f_1(\mu) = 0$ that we can find $\gamma_1 > \gamma_0$ such that $|f_1(\gamma)| < \frac{1}{n+1}$ for all $\gamma \ge \gamma_1$.

The map $g_1$ is constant on every set $r_0^{-1}(x)$; since $r_0^{-1}(B_0)$ is dense in $[1, \mu^n]$ and $||g_1|| = 1$ we can choose $x \in B_0$ such that $|g_1(z)| \ge \frac{1}{2}$ for any $z \in Z_1 = r_0^{-1}(x)$.

Proceeding inductively assume that $1 \le k < n$ and we have constructed functions $f_0, g_0, f_1, g_1, \ldots, f_k, g_k$, ordinals $\gamma_0 < \gamma_1 < \ldots < \gamma_k < \mu$ and sets $Z_0 = [1, \mu^n] \supset Z_1 \supset \ldots \supset Z_k$ with the following properties:

(5)  $f_i \in I$ and $g_i = \theta(f_i)$ for all $i \le k$;
(6)  $||g_i|| = 1$ and $|f_i(\gamma)| < \frac{1}{n+1}$ for all $\gamma \le \gamma_{i-1}$ and $i \in \{1, \ldots, k\}$;
(7)  $|f_i(\gamma)| < \frac{1}{n+1}$ for all $\gamma \ge \gamma_i$ and $i \le k$;
(8)  $g_i(Z_i) = \{s_i\}$ and $|s_i| \ge \frac{1}{2}$ for all $i \le k$;
(9)  $Z_i$ is canonically homeomorphic to $[1, \mu^{n-i}]$ for each $i \le k$.

Apply Fact 16 to find a closed set $Y_k \subset Z_k$ canonically homeomorphic to $[1, \mu]$ and a retraction $r_k : Z_k \to Y_k$ such that, for some $B_k \subset Y_k$, the set $r_k^{-1}(B_k)$ is dense in $Z_k$ and $r_k^{-1}(x)$ is canonically homeomorphic to $[1, \mu^{n-k-1}]$ for all $x \in B_k$. The dual map $r_k^* : C_u(Y_k) \to C_u(Z_k)$ is a linear embedding by Fact 14. Since $Z_k$ is a retract of $[1, \mu^n]$, the space $C_u(Z_k)$ embeds in $C_u([1, \mu^n])$ as a linear subspace (see Fact 14); to simplify the notation, we consider that $r_k^*(C_u(Y_k))$ is a linear subspace of $C_u([1, \mu^n])$.

Let $\pi_k : I \to C_u([1, \gamma_k])$ be the restriction map. Since $Y_k \simeq [1, \mu]$, the linear map $\pi_k \circ \zeta \circ r_k^*$ cannot be an embedding by the choice of $\mu$, so we can apply Fact 12 to find a function $g_{k+1} \in r_k^*(C_u(Y_k))$ such that $||g_{k+1}|| = 1$ and $|\zeta(g_{k+1})(\gamma)| < \frac{1}{n+1}$ for all $\gamma \le \gamma_k$. Let $f_{k+1} = \zeta(g_{k+1})$; it follows from $f_{k+1}(\mu) = 0$ that we can find $\gamma_{k+1} > \gamma_k$ such that $|f_{k+1}(\gamma)| < \frac{1}{n+1}$ for all $\gamma \ge \gamma_{k+1}$.

The map $g_{k+1}$ is constant on every set $r_k^{-1}(x)$; since $r_k^{-1}(B_k)$ is dense in $[1, \mu^{n-k}]$ and $||g_{k+1}|| = 1$, we can choose $x \in B_k$ such that $|g_{k+1}(z)| \ge \frac{1}{2}$ for any $z \in Z_{k+1} = r_k^{-1}(x)$. It is clear that the conditions (5)–(9) are now satisfied if we replace

$k$ with $k + \bar{1}$, so our inductive procedure can be continued to construct functions $\{f_i, g_i : i \leq n\}$, ordinals $\gamma_0 < \ldots < \gamma_n$ and sets $Z_0 \supset \ldots \supset Z_n$ such that the properties (5)–(9) hold for all $k \leq n$. Choose a number $\varepsilon_i \in \{-1, 1\}$ such that $\varepsilon_i s_i > 0$ for every $i \leq n$ and consider the function $f = \sum_{i=0}^{n} \varepsilon_i f_i$; then $g = \theta(f) = \sum_{i=0}^{n} \varepsilon_i g_i$.

Take a point $z \in Z_n$; it follows from the property (8) that $g_i(z) = s_i$ for all $i \leq n$. Recall that $||f_i|| \leq ||\theta(f_i)|| = ||g_i|| = 1$ for all $i \leq n$ and observe that $||\theta(f)|| = ||\sum_{i=0}^{n} \varepsilon_i \theta(f_i)|| \geq \sum_{i=0}^{n} \varepsilon_i f_i(z) = \sum_{i=0}^{n} |s_i| \geq \frac{n+1}{2}$.

Let $E_0 = [1, \gamma_0)$, $E_{n+1} = [\gamma_n, \mu]$ and $E_{i+1} = [\gamma_i, \gamma_{i+1})$ for every $i < n$. Given any $g \in [1, \mu]$, there is a unique $j \leq n + 1$ such that $\gamma \in E_j$. It follows from (6) and (7) that $|f_i(\gamma)| < \frac{1}{n+1}$ for all $i \neq j$; since $||f_j|| \leq ||\theta(f_j)|| = ||g_j|| = 1$, we have $|f_j(\gamma)| \leq 1$. Therefore $|f(\gamma)| \leq 1 + n \cdot \frac{1}{n+1} < 2$ for any $\gamma \in [1, \mu]$ and hence $||f|| = \max\{|f(\gamma)| : \gamma \in [1, \mu]\} < 2$.

We must have the inequalities $||\theta(f)|| \leq K||f|| < 2K$. However, $||\theta(f)|| \geq \frac{n+1}{2}$ which shows that $2K > \frac{n+1}{2}$, i.e., $n + 1 < 4K$; this contradiction with the choice of $n$ shows that the inequality $\beta \geq \alpha^\omega$ is impossible, so we settled necessity.

Finally observe that if $K = [1, \omega]$ and $K' = [1, \omega^\omega]$ then the countable compact spaces $K$ and $K'$ are not $l$-equivalent by our above results; however $K$ is $u$-equivalent to $K'$ by Problem 200, so our solution is complete.

**V.495.** *Suppose that $X$ is a metrizable zero-dimensional compact space and there exists a continuous linear surjection of $C_p(X)$ onto $C_p(Y)$. Prove that $Y$ is a metrizable compact zero-dimensional space. In particular, there exists no continuous linear surjection of $C_p(\mathbb{K})$ onto $C_p(\mathbb{I})$. Here, as usual, $\mathbb{K}$ is the Cantor set and $\mathbb{I} = [-1, 1] \subset \mathbb{R}$.*

**Solution.** Recall that "closed map" means "continuous closed onto map." A map $\varphi : T \to Z$ is *finite-to-one* if $\varphi^{-1}(z)$ is a finite set for any $z \in Z$.

*Fact 1.* Suppose that $Z$ is a zero-dimensional space and $T$ is a space for which there exists a closed finite-to-one map $\varphi : T \to Z$. Then $T$ is also zero-dimensional.

*Proof.* Fix an arbitrary point $x \in T$ and a set $U \in \tau(x, T)$. Choose a faithful enumeration $\{x_1, \ldots, x_n\}$ of the set $\varphi^{-1}\varphi(x)$; there is no loss of generality to assume that $x = x_1$. Take a disjoint family $\{V_1, \ldots, V_n\} \subset \tau(T)$ such that $V_1 \subset U$ and $x_i \in V_i$ for all $i \in \{1, \ldots, n\}$. If $V = V_1 \cup \ldots \cup V_n$ then $\varphi^{-1}\varphi(x) \subset V$, so we can apply Fact 1 of S.226 to find a set $W \in \tau(\varphi(x), Z)$ such that $\varphi^{-1}(W) \subset V$.

The space $Z$ being zero-dimensional, there exists a clopen set $G$ such that $\varphi(x) \in G \subset W$ and hence $H = \varphi^{-1}(G) \subset V$. The set $H$ is clopen in $T$; since $V_1$ is clopen in $V$, the set $H' = H \cap V_1$ is clopen in $H$ and hence in $T$. We have $x \in H' \subset V_1 \subset U$, so clopen subsets of $T$ constitute a base in $T$, i.e., $T$ is zero-dimensional and hence Fact 1 is proved.

Returning to our solution observe that $nw(C_p(X)) = nw(X) = w(X) \leq \omega$; this implies that $nw(C_p(Y)) \leq \omega$ and hence $nw(Y) \leq \omega$, so $Y$ is a $\mu$-space. Therefore we can apply Problem 285 to see that $Y$ is compact; since $w(Y) = nw(Y) \leq \omega$, the space $Y$ must be metrizable.

Apply Problem 237 to see that $L_p(Y)$ embeds in $L_p(X)$ as a linear subspace; in particular, we can consider that $Y$ is a closed linearly independent subspace of $L_p(X)$. For any $n \in \mathbb{N}$, $a = (a_1, \ldots, a_n) \in \mathbb{R}^n$ and $x = (x_1, \ldots, x_n) \in X^n$ let $\theta_n(a, x) = a_1 x_1 + \ldots + a_n x_n$; this gives a continuous map $\theta_n : \mathbb{R}^n \times X^n \to L_p(X)$ for any $n \in \mathbb{N}$. Let $M_0 = \{\mathbf{0}\}$ where $\mathbf{0}$ is the zero vector of $L_p(X)$; for any $n \in \mathbb{N}$ the set $M_n = \theta_n(\mathbb{R}^n \times X^n)$ is closed in $L_p(X)$ by Fact 1 of U.485. Therefore the set $A_n = M_n \backslash M_{n-1}$ is open in $M_n$; it follows from $nw(L_p(X)) = \omega$ that $A_n$ is an $F_\sigma$-subset of $L_p(X)$ for every $n \in \mathbb{N}$. The set $Y$ is linearly independent in $L_p(X)$, so $Y \subset \bigcup\{A_n : n \in \mathbb{N}\}$. Therefore we can find a family $\{Y_n : n \in \omega\}$ of closed subsets of $Y$ and a sequence $\{k_n : n \in \omega\} \subset \mathbb{N}$ such that $Y = \bigcup_{n \in \omega} Y_n$ and $Y_n \subset A_{k_n}$ for each $n \in \mathbb{N}$. For every point $u \in L_p(X) \backslash \{\mathbf{0}\}$ we can find a uniquely determined $n \in \mathbb{N}$ and $a_1, \ldots, a_n \in \mathbb{R} \backslash \{0\}$ such that $u = a_1 x_1 + \ldots + a_n x_n$ for some distinct $x_1, \ldots, x_n \in X$; let $\mathrm{supp}(u) = \{x_1, \ldots, x_n\}$.

Say that a set $G \subset L_p(X)$ is *adequate* if there exists $n \in \mathbb{N}$ such that, for some disjoint family $\{U_1, \ldots, U_n\} \subset \tau(X)$ and a family $\{O_1, \ldots, O_n\} \subset \tau(\mathbb{R} \backslash \{0\})$ we have $G = O_1 \cdot U_1 + \ldots + O_n \cdot U_n = \theta_n(O_1 \times \ldots \times O_n, U_1 \times \ldots \times U_n)$. Adequate sets form a base in every $A_n$ by Fact 3 of U.485. Therefore every $Y_n$ can be covered by finitely many adequate subsets of $A_{k_n}$.

Consequently, there exists a family $\{P_n : n \in \omega\}$ of compact subsets of $Y$ and a sequence $\{m_n : n \in \omega\} \subset \mathbb{N}$ such that $Y = \bigcup_{n \in \omega} P_n$ and the set $P_n$ is contained in an adequate subset $G_n$ of $A_{m_n}$ for every $n \in \omega$.

Fix any $n \in \omega$ and choose the sets $O_1, \ldots, O_{m_n} \in \tau(\mathbb{R} \backslash \{0\})$ and disjoint sets $U_1, \ldots, U_{m_n} \in \tau(X)$ such that $G_n = O_1 \cdot U_1 + \ldots + O_{m_n} \cdot U_{m_n}$; let $O = O_1 \times \ldots \times O_{m_n}$ and $U = U_1 \times \ldots \times U_{m_n}$. The map $\theta_{m_n} | (O \times U) : O \times U \to G_n$ is a homeomorphism (see Fact 4 of U.485), so we can find a set $P'_n \subset O \times U$ such that $P'_n \simeq P_n$ and $\theta_{m_n}(P'_n) = P_n$.

Let $\pi_n : P'_n \to X^{m_n}$ be the natural projection and denote by $K$ the compact set $\pi_n(P'_n)$. Fix any point $x = (x_1, \ldots, x_{m_n}) \in K$; if $u \in \pi_n^{-1}(x)$ then $u = (a, x) \in P'_n$ for some $a = (a_1, \ldots, a_{m_n}) \in O$ and hence $\theta_n(u) = a_1 x_1 + \ldots + a_{m_n} x_{m_n} \in Y$. If the set $\pi_n^{-1}(x)$ is infinite, then there exists an infinite set $Q \subset Y$ such that $\bigcup\{\mathrm{supp}(w) : w \in Q\} \subset \{x_1, \ldots, x_{m_n}\}$. Therefore $Q$ is contained in the linear span of the finite set $\{x_1, \ldots, x_{m_n}\}$; this contradiction with linear independence of $Q$ shows that $\pi_n^{-1}(x)$ is finite for any $x \in K$.

The spaces $P'_n$ and $K$ being compact, the map $\pi_n : P'_n \to K$ is closed; the space $K \subset X^{m_n}$ is zero-dimensional and we already saw that $\pi_n$ is finite-to-one, so the space $P'_n$ is zero-dimensional by Fact 1. It follows from $P'_n \simeq P_n$ and SFFS-306 that $\dim(P_n) = 0$ for all $n \in \omega$, so we can apply the countable sum theorem (see Problem 150) to conclude that $\dim(Y) = 0$ and hence $Y$ is zero-dimensional. Finally, observe that the Cantor set $\mathbb{K}$ is zero-dimensional and $\mathbb{I}$ isn't, so there exists no continuous linear surjection of $C_p(\mathbb{K})$ onto $C_p(\mathbb{I})$.

**V.496.** *Suppose that a space $X$ is metrizable, compact and $\sigma$-zero-dimensional. Prove that if there exists a continuous linear surjection of $C_p(X)$ onto $C_p(Y)$ then $Y$ is also a compact metrizable $\sigma$-zero-dimensional space. In particular, if $X$ is compact, metrizable and finite-dimensional then there is no continuous linear surjection of $C_p(X)$ onto $C_p(\mathbb{I}^\omega)$.*

**Solution.** Recall that "closed map" means "continuous closed onto map." A map $\varphi : T \to Z$ is *finite-to-one* if $\varphi^{-1}(z)$ is a finite set for any $z \in Z$. Observe first that $nw(C_p(X)) = nw(X) = w(X) \leq \omega$; this implies that $nw(C_p(Y)) \leq \omega$ and hence $nw(Y) \leq \omega$, so $Y$ is a $\mu$-space. Therefore we can apply Problem 285 to see that $Y$ is compact; since $w(Y) = nw(Y) \leq \omega$, the space $Y$ must be metrizable.

Apply Problem 237 to see that $L_p(Y)$ embeds in $L_p(X)$ as a linear subspace; in particular, we can consider that $Y$ is a closed linearly independent subspace of $L_p(X)$. For any $n \in \mathbb{N}$, $a = (a_1, \ldots, a_n) \in \mathbb{R}^n$ and $x = (x_1, \ldots, x_n) \in X^n$ let $\theta_n(a, x) = a_1 x_1 + \ldots + a_n x_n$; this gives a continuous map $\theta_n : \mathbb{R}^n \times X^n \to L_p(X)$ for any $n \in \mathbb{N}$. Let $M_0 = \{\mathbf{0}\}$ where $\mathbf{0}$ is the zero vector of $L_p(X)$; for any $n \in \mathbb{N}$ the set $M_n = \theta_n(\mathbb{R}^n \times X^n)$ is closed in $L_p(X)$ by Fact 1 of U.485. Therefore the set $A_n = M_n \backslash M_{n-1}$ is open in $M_n$; it follows from $nw(L_p(X)) = \omega$ that $A_n$ is an $F_\sigma$-subset of $L_p(X)$ for every $n \in \mathbb{N}$. The set $Y$ is linearly independent in $L_p(X)$, so $Y \subset \bigcup\{A_n : n \in \mathbb{N}\}$. Therefore there exists a family $\{Y_n : n \in \omega\}$ of closed subsets of $Y$ and a sequence $\{k_n : n \in \omega\} \subset \mathbb{N}$ such that $Y = \bigcup_{n \in \omega} Y_n$ and $Y_n \subset A_{k_n}$ for each $n \in \mathbb{N}$. For every point $u \in L_p(X)\backslash\{\mathbf{0}\}$ we can find a uniquely determined $n \in \mathbb{N}$ and $a_1, \ldots, a_n \in \mathbb{R}\backslash\{0\}$ such that $u = a_1 x_1 + \ldots + a_n x_n$ for some distinct $x_1, \ldots, x_n \in X$; let $\text{supp}(u) = \{x_1, \ldots, x_n\}$.

Say that a set $G \subset L_p(X)$ is *adequate* if there exists $n \in \mathbb{N}$ such that, for some disjoint family $\{U_1, \ldots, U_n\} \subset \tau(X)$ and a family $\{O_1, \ldots, O_n\} \subset \tau(\mathbb{R}\backslash\{0\})$ we have $G = O_1 \cdot U_1 + \ldots + O_n \cdot U_n = \theta_n(O_1 \times \ldots \times O_n, U_1 \times \ldots \times U_n)$. Adequate sets form a base in every $A_n$ by Fact 3 of U.485. Therefore every $Y_n$ can be covered by finitely many adequate subsets of $A_{k_n}$.

Consequently, there exists a family $\{P_n : n \in \omega\}$ of compact subsets of $Y$ and a sequence $\{m_n : n \in \omega\} \subset \mathbb{N}$ such that $Y = \bigcup_{n \in \omega} P_n$ and the set $P_n$ is contained in an adequate subset $G_n$ of $A_{m_n}$ for every $n \in \omega$.

Fix any $n \in \omega$ and choose the sets $O_1, \ldots, O_{m_n} \in \tau(\mathbb{R}\backslash\{0\})$ and disjoint sets $U_1, \ldots, U_{m_n} \in \tau(X)$ such that $G_n = O_1 \cdot U_1 + \ldots + O_{m_n} \cdot U_{m_n}$; let $O = O_1 \times \ldots \times O_{m_n}$ and $U = U_1 \times \ldots \times U_{m_n}$. The map $\theta_{m_n}|(O \times U) : O \times U \to G_n$ is a homeomorphism (see Fact 4 of U.485), so we can find a set $P'_n \subset O \times U$ such that $P'_n \simeq P_n$ and $\theta_{m_n}(P'_n) = P_n$.

Let $\pi_n : P'_n \to X^{m_n}$ be the natural projection and denote by $K$ the compact set $\pi_n(P'_n)$. Fix any point $x = (x_1, \ldots, x_{m_n}) \in K$; if $u \in \pi_n^{-1}(x)$ then $u = (a, x) \in P'_n$ for some $a = (a_1, \ldots, a_{m_n}) \in O$ and hence $\theta_n(u) = a_1 x_1 + \ldots + a_{m_n} x_{m_n} \in Y$. If the set $\pi_n^{-1}(x)$ is infinite, then there exists an infinite set $Q \subset Y$ such that $\bigcup\{\text{supp}(w) : w \in Q\} \subset \{x_1, \ldots, x_{m_n}\}$. Therefore $Q$ is contained in the linear span of the finite set $\{x_1, \ldots, x_{m_n}\}$; this contradiction with linear independence of $Q$ shows that $\pi_n^{-1}(x)$ is finite for any $x \in K$.

The spaces $P'_n$ and $K$ being compact, the map $\pi_n : P'_n \to K$ is closed. The space $X^{m_n}$ is easily seen to be $\sigma$-zero-dimensional, so the space $K \subset X^{m_n}$ is also $\sigma$-zero-dimensional. Take a family $\{E_i : i \in \omega\}$ of zero-dimensional subspaces of $K$ such that $K = \bigcup_{i \in \omega} E_i$.

If $E'_i = \pi_n^{-1}(E_i)$ then the map $\pi_n|E'_i : E'_i \to E_i$ is closed for every $i \in \omega$ by Fact 1 of S.261; it is clearly finite-to-one so the space $E'_i$ is zero-dimensional by Fact 1 of V.495. We have $P'_n = \bigcup_{i \in \omega} E'_i$, so the space $P'_n$ is $\sigma$-zero-dimensional.

It follows from $P_n' \simeq P_n$ that the space $P_n$ is $\sigma$-zero-dimensional for all $n \in \omega$, so the space $Y = \bigcup_{n \in \omega} P_n$ is $\sigma$-zero-dimensional as well.

Finally, observe that if $X$ is finite-dimensional then it representable as the finite union of its zero-dimensional subspaces (see Problem 164); in particular, the space $X$ must be $\sigma$-zero-dimensional. If there exists a continuous linear surjection of $C_p(X)$ onto $C_p(\mathbb{I}^\omega)$ then $\mathbb{I}^\omega$ must be $\sigma$-zero-dimensional; this contradiction with Problem 098 shows that there exists no continuous linear surjection of $C_p(X)$ onto $C_p(\mathbb{I}^\omega)$, so our solution is complete.

**V.497.** *Prove that a space $K$ is $l$-equivalent to the Cantor set $\mathbb{K}$ if and only if there exists a continuous linear surjection of $C_p(K)$ onto $C_p(\mathbb{K})$ as well as a continuous linear surjection of $C_p(\mathbb{K})$ onto $C_p(K)$.*

**Solution.** Necessity being trivial, assume that there exists a continuous linear surjection of $C_p(\mathbb{K})$ onto $C_p(K)$ and vice versa. We can apply Problem 495 to see that $K$ is a metrizable zero-dimensional compact space. It follows from existence of a continuous linear surjection of $C_p(K)$ onto $C_p(\mathbb{K})$ that $L_p(\mathbb{K})$ embeds in $L_p(K)$ as a linear subspace (see Problem 237). In particular, there exists a linearly independent set $F \subset L_p(K)$ which is homeomorphic to $\mathbb{K}$.

For any $u \in F$ there exists a nonempty finite set $\varphi(u) \subset K$ such that $u$ belongs to the linear hull $L_u$ of the set $\varphi(u)$. The linear space $L_u$ is finite-dimensional; since $F$ is linearly independent, the set $L_u \cap F$ has to finite. Therefore we have a map $\varphi : F \to [K]^{<\omega} = \{A \subset K : |A| < \omega\}$ such that $\varphi^{-1}(A)$ is finite for any $A \in [K]^{<\omega}$. If the space $K$ is countable then $[K]^{<\omega}$ is also countable and hence $F = \bigcup\{\varphi^{-1}(A) : A \in [K]^{<\omega}\}$ is countable as well; this contradiction shows that $K$ is uncountable and hence $K \overset{l}{\sim} \mathbb{K}$ by Problem 297.

**V.498.** *Let $I$ be the closed interval $[0, 1] \subset \mathbb{R}$. Given any $n \in \omega$ suppose that $K$ is a metrizable compact space and $\dim K \leq n$. Prove that there exists a compact subspace $K' \subset I^{2n+1}$ with the following properties:*

*(i)  $K'$ is homeomorphic to $K$;*
*(ii) for any function $\varphi \in C(K')$, we can choose $\varphi^1, \ldots, \varphi^{2n+1} \in C(I)$ such that*
$$\varphi(y) = \varphi^1(y_1) + \ldots + \varphi^{2n+1}(y_{2n+1}) \text{ for any } y = (y_1 \ldots, y_{2n+1}) \in K'.$$

*Deduce from this fact that if $X$ is a second countable space and $\dim X \leq n$ then $X$ embeds in $\mathbb{I}^{2n+1}$.*

**Solution.** Recall that an open cover $\mathcal{U}$ of a space $Z$ is *shrinkable* if, for each $U \in \mathcal{U}$, we can choose a closed set $F_U \subset U$ such that $\bigcup\{F_U : U \in \mathcal{U}\} = Z$. Say that a family $\mathcal{V}$ of subsets of $Z$ is *inscribed* in $\mathcal{U}$ if, for any $V \in \mathcal{V}$ there exists $U \in \mathcal{U}$ such that $V \subset U$. We will often use the following simple observation:

(1) if $Z$ is a space then a finite family $\mathcal{A}$ of subsets of $Z$ is discrete if and only if the family $\{\overline{A} : A \in \mathcal{A}\}$ is disjoint.

If $Z$ is a space and $A, B \subset Z$ then the sets $A$ and $B$ are called *separated* if $\overline{A} \cap B = \overline{B} \cap A = \emptyset$; say that $A$ and $B$ are *open-separated* if there exist disjoint sets $U, V \in \tau(Z)$ such that $A \subset U$ and $B \subset V$.

*Fact 1.* Suppose that $Z$ is a hereditarily normal space and $Y \subset Z$. If $\mathcal{G} \subset \tau(Y)$ is a finite disjoint family then there exists a disjoint family $\{U_G : G \in \mathcal{G}\} \subset \tau(Z)$ such that $G \subset U_G$ for any $G \in \mathcal{G}$.

*Proof.* Take a faithful enumeration $\mathcal{G} = \{G_1, \ldots, G_m\}$ of the family $\mathcal{G}$. Our Fact is trivially true if $m = 1$. Proceeding by induction assume that we proved it for $m = k$ and $G = \{G_1, \ldots, G_{k+1}\}$ is a disjoint family of open subsets of $Y$. It is immediate that the sets $G' = G_1 \cup \ldots \cup G_k$ and $G_{k+1}$ are separated in $Z$, so we can apply Fact 1 of S.291 to see that $G'$ and $G_{k+1}$ are open-separated. Fix disjoint sets $U', U_{k+1} \in \tau(Z)$ such that $G' \subset U'$ and $G_{k+1} \subset U_{k+1}$.

Our induction hypothesis is now applicable to the family $\{G_1, \ldots, G_k\}$ and the space $U'$, so we can find disjoint sets $U_1, \ldots, U_k \in \tau(U')$ such that $G_i \subset U_i$ for all $i \leq k$. Then $\{U_1, \ldots, U_{k+1}\}$ is a disjoint family of open subsets of $Z$ such that $G_i \subset U_i$ for every $i \leq k + 1$. This completes the induction step and shows that Fact 1 is proved.

*Fact 2.* If $Z$ is a metrizable compact space and $m \in \omega$ then $\dim Z \leq m$ if and only if, for any open cover $\mathcal{U}$ of the space $Z$ and any $k \geq m + 1$ there exists a collection $\mathcal{C} = \{\mathcal{V}_1, \ldots, \mathcal{V}_k\}$ of discrete families of open subsets of $Z$ such that every $\mathcal{V}_i$ is inscribed in $\mathcal{U}$ and for any set $A \subset \{1, \ldots k\}$ with $|A| = m + 1$, the family $\bigcup\{\mathcal{V}_i : i \in A\}$ is a cover of $Z$.

*Proof.* Given a number $k \geq m + 1$ let $\mathcal{E}(k)$ be the statement which says that for any open cover $\mathcal{U}$ of the space $Z$ there exists a collection $\mathcal{C} = \{\mathcal{V}_1, \ldots, \mathcal{V}_k\}$ of discrete families of open subsets of $Z$ such that every $\mathcal{V}_i$ is inscribed in $\mathcal{U}$ and for any set $A \subset \{1, \ldots k\}$ with $|A| = m + 1$, the family $\bigcup\{\mathcal{V}_i : i \in A\}$ is a cover of $Z$. If $\mathcal{E}(m + 1)$ holds then every open cover $\mathcal{U}$ of the space $Z$ has a refinement $\mathcal{V}$ which is a union of $(m + 1)$-many discrete subfamilies. This implies $\mathrm{ord}(\mathcal{V}) \leq m + 1$ and hence $\dim Z \leq m$, i.e., we proved sufficiency.

Now assume that $\dim Z \leq m$; we will prove, by induction on $k$ that $\mathcal{E}(k)$ holds for each $k \geq m + 1$. Apply Problem 164 to find sets $Z_0, \ldots, Z_m \subset Z$ such that $Z = Z_0 \cup \ldots \cup Z_m$ and $\dim Z_i = 0$ for all $i \leq m$. Let $\mathcal{U}$ be an open cover of $Z$; since $Z$ is compact, we can assume, without loss of generality, that $\mathcal{U} = \{U_1, \ldots, U_l\}$ for some $l \in \mathbb{N}$. For any $j \leq m$, apply Fact 2 of T.311 to the open cover $\{U_i \cap Z_j : i \leq l\}$ of the space $Z_j$, to find a disjoint family $\mathcal{V}_j = \{V_1^j, \ldots, V_l^j\}$ of clopen subsets of $Z_j$ such that $V_i^j \subset U_i$ for every $i \leq l$ and $\bigcup \mathcal{V}_j = Z_j$.

By Fact 1 there exists a disjoint family $\mathcal{W}_j = \{W_1^j, \ldots, W_l^j\} \subset \tau(Z)$ such that $V_i^j \subset W_i^j$ for all $i \leq l$. If $H_i^j = W_i^j \cap U_i$ for every $i \leq l$ then the family $\mathcal{H}^j = \{H_1^j, \ldots, H_l^j\}$ is disjoint, inscribed in $\mathcal{U}$ and $Z_j \subset \bigcup \mathcal{H}_j$ for each $j \leq m$. Therefore the family $\mathcal{H} = \bigcup\{\mathcal{H}_j : 0 \leq j \leq m\}$ is an open cover of the space $Z$. By Fact 1 of U.188 the cover $\mathcal{H}$ is shrinkable and hence we can choose a closed set $F_i^j$ in the space $Z$ such that $F_i^j \subset H_i^j$ for every $i \leq l$ and $j \leq m$ in such a way that the family $\{F_i^j : i \leq l \text{ and } j \leq m\}$ covers $Z$.

By normality of the space $Z$ we can choose, for every $j \leq m$ and $i \leq l$, a set $G_i^j \in \tau(Z)$ such that $F_i^j \subset G_i^j \subset \mathrm{cl}_Z(G_i^j) \subset H_i^j$. Since the family

$\{cl_Z(G_i^j) : i \leq l\}$ is disjoint, we can apply (1) to see that the family $\mathcal{V}_j = \{G_i^j : i \leq l\}$ is discrete for each $j \leq m$. It is clear that every $\mathcal{V}_j$ is inscribed in $\mathcal{U}$ and $\bigcup\{\mathcal{V}_j : 0 \leq j \leq m\}$ is a cover of $Z$, so we proved $\mathcal{E}(m + 1)$.

Now assume that we established the statement $\mathcal{E}(k)$ for some $k \geq m + 1$ and fix an open cover $\mathcal{U}$ of the space $Z$. By the induction hypothesis there exists a collection $\mathcal{C} = \{\mathcal{V}_1, \ldots, \mathcal{V}_k\}$ of discrete families of open subsets of $Z$ such that every $\mathcal{V}_i$ is inscribed in $\mathcal{U}$ and $\bigcup\{\mathcal{V}_i : i \in A\}$ is a cover of $Z$ for any set $A \subset \{1, \ldots, k\}$ of cardinality $m + 1$.

Let $\mathcal{A} = \{A \subset \{1, \ldots, k\} : |A| = m\}$; consider the set $P_A = Z \backslash \bigcup\{\bigcup \mathcal{V}_i : i \in A\}$ for every $A \in \mathcal{A}$. If $A$ and $B$ are distinct elements of $\mathcal{A}$ then $|A \cup B| \geq m + 1$ and hence $\bigcup_{i \in A} \mathcal{V}_i \cup \bigcup_{i \in B} \mathcal{V}_i$ is a cover of $X$; as a consequence, $P_A \cap P_B = \emptyset$. Therefore $\{P_A : A \in \mathcal{A}\}$ is a disjoint family of closed subsets of $Z$, so we can find a discrete family $\{O_A : A \in \mathcal{A}\} \subset \tau(Z)$ such that $P_A \subset O_A$ for each $A \in \mathcal{A}$.

For every $A \in \mathcal{A}$ fix an index $j_A \in \{1, \ldots, k\} \backslash A$; since the family $\bigcup_{i \in A} \mathcal{V}_i \cup \mathcal{V}_{j_A}$ is a cover of $Z$, the family $\mathcal{V}_{j_A}$ covers $P_A$. Thus the family $\mathcal{D}_A = \{V \cap O_A : V \in \mathcal{V}_{j_A}\}$ is discrete, inscribed in $\mathcal{V}_{j_A}$ (and hence in $\mathcal{U}$) and covers $P_A$. This easily implies that the family $\mathcal{V}_{k+1} = \bigcup\{\mathcal{D}_A : A \in \mathcal{A}\}$ is discrete and inscribed in $\mathcal{U}$. To see that the family $\{\mathcal{V}_1, \ldots, \mathcal{V}_{k+1}\}$ is as promised take any set $B \subset \{1, \ldots, k, k + 1\}$ of cardinality $m + 1$. If $B \subset \{1, \ldots, k\}$ then $\bigcup_{i \in B} \mathcal{V}_i$ is a cover of $Z$ by the induction hypothesis. If $k + 1 \in B$ then $A = B \cap \{1, \ldots, k\} \in \mathcal{A}$ and hence $\mathcal{V}_{k+1}$ covers the set $P_A = Z \backslash \bigcup_{i \in A} \bigcup \mathcal{V}_i$. This shows that $\bigcup\{\mathcal{V}_i : i \in B\}$ is a cover of $Z$, so we carried out the induction step and hence $\mathcal{E}(k)$ holds for any $k \geq m + 1$, i.e., Fact 2 is proved.

Returning to our solution, let $Q = \{1, \ldots, 2n + 1\}$ and fix a metric $\rho$ on the space $K$ which generates the topology of $K$. For any $k \in \omega$ we will need the set $I_k = [0, 1 - 2^{-k-1}] \subset I$. Let $\text{diam}(Z) = \sup\{\rho(x, y) : x, y \in Z\}$ for any $Z \subset K$. Apply Fact 2 to construct, for any $q \in Q$, a discrete family $\mathcal{C}_{q,0}$ of open subsets of $K$ such that $\text{diam}(U) \leq 1$ for every $U \in \bigcup\{\mathcal{C}_{q,0} : q \in Q\}$ and, for any set $A \subset Q$ with $|A| = n + 1$, the family $\bigcup_{q \in A} \mathcal{C}_{q,0}$ is a cover of $K$.

Let $E(r) = \{i \cdot r^{-1} : i \in \{0, \ldots, r\}\}$ for any prime number $r \in \mathbb{N}$. Take distinct prime numbers $r_0^1, \ldots, r_0^{2n+1} \in \mathbb{N}$ such that $r_0^q > 2|\mathcal{C}_{q,0}|$ for each $q \in Q$ and choose a number $\varepsilon_0 > 0$ such that $\varepsilon_0 < \backslash n\{\frac{1}{2}, (r_0^1)^{-1}, \ldots, (r_0^{2n+1})^{-1}\}$. Using normality of the space $K$ it is easy to construct, for any $q \in Q$, a function $f_0^q \in C(K, I_0)$ such that, for each $U \in \mathcal{C}_{q,0}$ we have $f_0^q(U) = \{a_U\}$ for some number $a_U \in E(r_0^q) \cap I_0$ and $a_U \neq a_V$ for distinct $U, V \in \mathcal{C}_{q,0}$.

Proceeding inductively assume that $m \in \omega$ and we have constructed, for any $k \leq m$, a collection $\{\mathcal{C}_{q,k} : q \in Q\}$ of discrete families of open subsets of $K$, prime numbers $r_k^1, \ldots, r_k^{2n+1} \in \mathbb{N}$, a set of functions $\{f_k^q : q \in Q\} \subset C(K, I_k)$ and a number $\varepsilon_k > 0$ with the following properties:

(2) if $1 \leq k \leq m$ then $\varepsilon_k < \frac{1}{2}\varepsilon_{k-1}$;
(3) $\varepsilon_k < \backslash n\{2^{-k-1}, (r_k^1)^{-1}, \ldots, (r_k^{2n+1})^{-1}\}$ for any $k \leq m$;
(4) if $k \leq m$ then $\text{diam}(U) \leq 2^{-k}$ for any $U \in \bigcup\{\mathcal{C}_{q,k} : q \in Q\}$;
(5) if $p, q \in Q$ and $k, l \leq m$ then $r_k^q = r_l^p$ if and only if $p = q$ and $k = l$;

(6) if $k \leq m$ then for any set $A \subset Q$ with $|A| = n + 1$, the family $\bigcup_{q \in A} C_{q,k}$ is a cover of $K$;

(7) if $k \leq m$ and $U \in \mathcal{C}_{q,k}$ then $f_k^q(U) = \{a_U\}$ for some number $a_U \in E(r_k^q) \cap I_k$ and $a_U \neq a_V$ for distinct $U, V \in \mathcal{C}_{q,k}$;

(8) if $j < k \leq m$ and $q \in Q$ then $f_j^q(x) \leq f_k^q(x) \leq f_j^q(x) + \varepsilon_j - \varepsilon_k$ for all $x \in K$.

There exists a number $\delta > 0$ such that $|f_m^q(x) - f_m^q(y)| < \frac{1}{4}\varepsilon_m$ for any $q \in Q$ whenever $\rho(x, y) \leq \delta$. Take a collection $\{\mathcal{C}_{q,m+1} : q \in Q\}$ of discrete families of open subsets of $K$ such that $\text{diam}(U) < \backslash n\{\delta, 2^{-m-1}\}$ for any $U \in \bigcup\{\mathcal{C}_{q,m+1} : q \in Q\}$ and $\bigcup\{\mathcal{C}_{q,m+1} : q \in A\}$ is a cover of $K$ for any $A \subset Q$ with $|A| = n + 1$.

Choose distinct prime numbers $r_{m+1}^1, \ldots, r_{m+1}^{2n+1} \in \mathbb{N} \backslash \{r_k^q : k \leq m \text{ and } q \in Q\}$ such that $r_{m+1}^q > (|\mathcal{C}_{m+1}^q| + 2) \cdot \frac{4}{\varepsilon_m}$ for every $q \in Q$. Fix a number $q \in Q$ and $U \in \mathcal{C}_{q,m+1}$; observe that the diameter of the set $f_m^q(U)$ does not exceed $\frac{1}{4}\varepsilon_m$. If $d_U = \inf f_m^q(U)$ then $f_m^q(U) \subset [d_U, d_U + \frac{1}{4}\varepsilon_m]$ and the interval $[d_U + \frac{1}{4}\varepsilon_m, d_U + \frac{1}{2}\varepsilon_m]$ contains at least $|\mathcal{C}_{m+1}^q|$-many points of the set $E(r_{m+1}^q)$. Therefore it is possible to choose a number $a_U \in [d_U + \frac{1}{4}\varepsilon_m, d_U + \frac{1}{2}\varepsilon_m] \cap E(r_{m+1}^q)$ in such a way that $a_U \neq a_V$ for any distinct $U, V \in \mathcal{C}_{m+1}^q$. It is immediate that $a_U \in I_{m+1}$ for all $U \in \mathcal{C}_{q,m+1}$.

Given any set $U \in \mathcal{C}_{m+1}^q$ let $g_q(x) = a_U - f_m^q(x)$ for every $x \in \overline{U}$. Then $g_q$ is a continuous function on the closed set $F = \bigcup\{\overline{U} : U \in \mathcal{C}_{q,m+1}\}$ and, besides, $0 \leq g_q(x) \leq \frac{1}{2}\varepsilon_m$ for any point $x \in F$. Therefore we can find a continuous function $h_q : K \to [0, \frac{1}{2}\varepsilon_m]$ such that $h_q|F = g_q$. Letting $f_{m+1}^q = h_q + f_m^q$ we obtain a function $f_{m+1}^q \in C(K, I_{m+1})$ such that $f_{m+1}^q(U) = \{a_U\}$ and $a_U \in E(r_{m+1}^q) \cap I_{m+1}$ for any set $U \in \mathcal{C}_{q,m+1}$. Choose a number $\varepsilon_{m+1} > 0$ such that $\varepsilon_{m+1} < \frac{1}{2}\varepsilon_m$ and $\varepsilon_{m+1} < \backslash n\{2^{-m-1}, (r_{m+1}^1)^{-1}, \ldots, (r_{m+1}^{2n+1})^{-1}\}$; it is straightforward that the conditions (2)–(7) are now satisfied for all $k \leq m + 1$.

To prove (8), fix any index $q \in Q$ and observe that we have the inequalities $f_m^q(x) \leq f_{m+1}^q(x) \leq f_m^q(x) + \frac{1}{2}\varepsilon_m < \varepsilon_m - \varepsilon_{m+1}$ for every $x \in K$, i.e., the property (8) holds for $j = m$ and $k = m + 1$. Now, if $j < m$ then $f_m^q(x) \leq f_j^q(x) + \varepsilon_j - \varepsilon_m$ for all $x \in K$ and therefore $f_{m+1}^q(x) \leq f_j^q(x) + \varepsilon_j - \varepsilon_m + \frac{1}{2}\varepsilon_m < f_j^q(x) + \varepsilon_j - \varepsilon_{m+1}$ for each $x \in K$; this shows that (8) is true for $k = m + 1$ and any $j < k$. If $k \leq m$ and $j < k$ then (8) is true by the induction hypothesis, so we checked that (8) holds if we replace $m$ with $m + 1$.

Thus we carried out the induction step, so our procedure guarantees that we can construct, for any $k \in \omega$, a collection $\{\mathcal{C}_{q,k} : q \in Q\}$ of discrete families of open subsets of the space $K$, prime numbers $r_k^1, \ldots, r_k^{2n+1} \in \mathbb{N}$, a set of functions $\{f_k^q : q \in Q\} \subset C(K, I_k)$ and a number $\varepsilon_k > 0$ such that the conditions (2)–(8) are satisfied for all $m \in \omega$.

The properties (3) and (8) show that the sequence $\{f_k^q(x) : k \in \omega\}$ is increasing and bounded by 1; therefore there exists $f^q(x) = \lim f_k^q(x)$ for any $x \in K$ and hence we have a function $f^q : K \to I$ for every $q \in Q$. The property (8) implies that $|f^q(x) - f_j^q(x)| \leq \varepsilon_j$ for all $x \in K$ and $j \in \omega$; therefore the sequence

$\{f_k^q : k \in \omega\}$ converges uniformly to the function $f^q$ and hence $f^q \in C(K, I)$ for each $q \in Q$.

Consider the family $\mathcal{D}_{q,k} = \{f^q(U) : U \in \mathcal{C}_{q,k}\}$ for any numbers $q \in Q$ and $k \in \omega$. Recall that $f_k^q(U) = \{a_U\}$ for some $a_U \in E(r_k^q)$. The property (8) implies that $f_k^q(x) \leq f^q(x) \leq f_k^q(x) + \varepsilon_k$ for all $x \in K$ and therefore $f^q(U) \subset [a_U, a_U + \varepsilon_k]$ for every $U \in \mathcal{C}_{q,k}$. The property (7) shows that if $U$ and $V$ are distinct elements of $\mathcal{C}_{q,k}$ then $a_U \neq a_V$ and hence $|a_U - a_V| \geq (r_k^q)^{-1}$; by the property (3) we have $[a_U, a_U + \varepsilon_k] \cap [a_V, a_V + \varepsilon_k] = \emptyset$, so the family $\{[a_U, a_U + \varepsilon_k] : U \in \mathcal{C}_{q,k}\}$ is disjoint and hence $\mathcal{D}_{q,k}$ is discrete for any $q \in Q$ and $k \in \omega$.

Fix an arbitrary continuous function $h : K \to \mathbb{R}$; let $k_0 = 0$ and $\chi_0^q(t) = 0$ for all $q \in Q$ and $t \in \mathbb{R}$. Proceeding by induction assume that $m \in \omega$ and we have a set of numbers $\{k_i : i \leq m\} \subset \omega$ and functions $\chi_0^q, \ldots, \chi_m^q \in C(\mathbb{R})$ for every $q \in Q$ with the following properties:

(9)  $k_i < k_{i+1}$ whenever $i < m$;

(10)  if $h_i = \sum_{q \in Q} \sum_{j=0}^{i} \chi_j^q \circ f^q$ and $M_i = \sup\{|h(x) - h_i(x)| : x \in K\}$ for each $i < m$, then $|\chi_{i+1}^q(t)| \leq (n+1)^{-1} M_i$ for any $q \in Q$ and $t \in \mathbb{R}$;

(11)  for every $i < m$ and $U \in \mathcal{C}_{q,k_{i+1}}$ if $D = f^q(U) \in \mathcal{D}_{q,k_{i+1}}$ then $\chi_{i+1}^q(D) = \{s_D\}$ where $s_D = (n+1)^{-1}(h(y_D) - h_i(y_D))$ for some $y_D \in U$.

(12)  if $i < m$ then $|(h - h_i)(x) - (h - h_i)(y)| < (2n+2)^{-1} M_i$ for any $x, y \in K$ such that $\rho(x, y) \leq 2^{-k_{i+1}}$.

Let $h_m = \sum_{q \in Q} \sum_{j=0}^{m} \chi_j^q \circ f^q$; if $M_m = \sup\{|(h - h_m)(x)| : x \in K\} = 0$ then $h = h_m$ and our construction stops. If $M_m > 0$ then, by uniform continuity of $h - h_m$, there exists $\delta > 0$ such that $|(h - h_m)(x) - (h - h_m)(y)| < (2n+2)^{-1} M_m$ for any $x, y \in K$ with $\rho(x, y) < \delta$. Pick a number $k_{m+1} \in \omega$ such that $k_{m+1} > k_m$ and $2^{-k_{m+1}} < \delta$.

Fix an arbitrary $q \in Q$; for every $U \in \mathcal{C}_{q,k_{m+1}}$ take a point $z_U \in U$ and let $\chi(t) = (n+1)^{-1}(h(z_U) - h_m(z_U))$ for any $t \in \overline{f^q(U)}$. This gives a continuous function $\chi : Z = \bigcup\{\overline{D} : D \in \mathcal{D}_{q,k_{m+1}}\} \to [-(n+1)^{-1} M_m, (n+1)^{-1} M_m]$. The family $\mathcal{D}_{q,k_{m+1}}$ is discrete, so there exists $\chi_{m+1}^q \in C(\mathbb{R})$ such that $\chi_{m+1}^q | Z = \chi$ and $|\chi_{m+1}^q(t)| \leq (n+1)^{-1} M_m$ for all $t \in \mathbb{R}$.

It is straightforward that the conditions (9)–(12) are now satisfied if we replace $m$ by $m' + 1$, so our inductive procedure can be continued. If our induction stops at a step $m$ then we obtain a family of functions $\{\chi_i^q : i \leq m, \, q \in Q\} \subset C(\mathbb{R})$ such that $h = h_m = \sum_{q \in Q} \sum_{j=0}^{m} \chi_j^q \circ f^q$. Letting $\varphi^q = \sum_{j=0}^{m} \chi_j^q$ for every $q \in Q$ we obtain a family $\{\varphi^q : q \in Q\} \subset C(\mathbb{R})$ such that $h = \sum_{q \in Q} \varphi^q \circ f^q$.

Now, if our inductive construction did not stop at any finite step then we have a sequence $\{k_i : i \in \omega\} \subset \omega$ and a family of functions $\{\chi_i^q : q \in Q, \, i \in \omega\} \subset C(\mathbb{R})$ such that the conditions (9)–(12) are satisfied for all $m \in \omega$.

Fix any $i \in \omega$, $q \in Q$ and a point $x \in \bigcup \mathcal{C}_{q,k_{i+1}}$; if $x \in U \in \mathcal{C}_{q,k_{i+1}}$ then there is $y \in U$ such that $\chi_{i+1}^q(f^q(x)) = \chi_{i+1}^q(f^q(y)) = (n+1)^{-1}(h - h_i)(y)$ for any $x \in U$. Furthermore, $\text{diam}(U) \leq 2^{-k_{i+1}}$, so we can apply (12) to see that $|(h - h_i)(x) - (h - h_i)(y)| < (2n+2)^{-1} M_i$ and therefore

(13) $|(n+1)^{-1}(h-h_i)(x) - \chi^q_{i+1}(f^q(x))| < (n+1)^{-1}(2n+2)^{-1} M_i$ for any point $x \in \bigcup C_{q,k_{i+1}}$

Take an arbitrary point $x \in X$. If the set $A = \{q \in Q : x \in \bigcup C_{q,k_{i+1}}\}$ has cardinality less than $n+1$ then $|Q \backslash A| \geq n+1$ and $\bigcup_{i \in Q \backslash A} C_{q,k_{i+1}}$ does not cover the point $x$ which is a contradiction with (6). Thus $|A| \geq n+1$, so we can choose a set $B \subset A$ with $|B| = n+1$. It follows from (10) and (13) that

$$|(h-h_{i+1})(x)| = |(h-h_i)(x) - \textstyle\sum_{q \in Q} \chi^q_{i+1}(f^q(x))| \leq$$
(14) $|(h-h_i)(x) - \textstyle\sum_{q \in B} \chi^q_{i+1}(f^q(x))| + |\textstyle\sum_{q \in Q \backslash B} \chi^q_{i+1}(f^q(x))| <$
$\frac{1}{2n+2} M_i + \frac{n}{n+1} M_i = \frac{2n+1}{2n+2} M_i.$

It follows from (14) that $M_{i+1} < \frac{2n+1}{2n+2} M_i$ for every $i \in \omega$ and hence

(15) $M_i < (\frac{2n+1}{2n+2})^i M_0$ for all $i \in \omega$,

so $|(h-h_i)(x)| < (\frac{2n+1}{2n+2})^i \cdot M_0$ for any $i \in \omega$ and $x \in K$. As a consequence, for every point $x \in K$ the sequence $\{h_i(x) : i \in \omega\}$ converges to $h(x)$.

Fix an index $q \in Q$; if $d_i = \frac{1}{n+1} \cdot (\frac{2n+1}{2n+2})^i M_0$ then it follows from (10) and (15) that $|\chi^q_i(t)| \leq d_i$ for any $i \in \omega$ and $t \in \mathbb{R}$. It is evident that the series $\sum_{i=0}^{\infty} d_i$ converges, so we can apply TFS-030 to see that the sequence $\{\chi^q_0 + \ldots + \chi^q_i : i \in \omega\}$ converges uniformly to a function $\varphi^q \in C(\mathbb{R})$.

Let $\xi^q_i = \chi^q_0 + \ldots + \chi^q_i$ for every $q \in Q$ and $i \in \omega$. Given any point $x \in K$ the sequence $\{\xi^q_i(f^q(x)) : i \in \omega\}$ converges to the point $\varphi^q(f^q(x))$. Therefore the sequence $\mathcal{S} = \{\sum_{q \in Q} \xi^q_i(f^q(x)) : i \in \omega\}$ converges to $\sum_{q \in Q} \varphi^q(f^q(x))$; recalling that $\mathcal{S} = \{h_i(x) : i \in \omega\}$, we conclude that $\mathcal{S}$ also converges to $h(x)$, so we have the equality $h(x) = \sum_{q \in Q} \varphi^q(f^q(x))$ for each $x \in K$. Thus, in all cases, we obtained the equality $h = \sum_{q \in Q} \varphi^q \circ f^q$ which shows that

(16)  for any continuous function $h : K \to \mathbb{R}$ there exist $\varphi^1, \ldots, \varphi^{2n+1} \in C(\mathbb{R})$ such that $h = \sum_{q \in Q} \varphi^q \circ f^q$.

If $x$ and $y$ are distinct points of $K$ and $f^q(x) = f^q(y)$ for all $q \in Q$ then fix a function $h \in C(K)$ such that $h(x) = 0$ and $h(y) = 1$. If $\{\varphi^q : q \in Q\} \subset C(\mathbb{R})$ is a family as in (16) then $\varphi^q(f^q(x)) = \varphi^q(f^q(y))$ for all $q \in Q$ which, together with (16), implies that $h(x) = h(y)$ which is a contradiction.

Therefore the family $\{f^q : q \in Q\}$ separates the points of $K$. Define a map $f : K \to I^{2n+1}$ by the equality $f(x) = (f^1(x), \ldots, f^{2n+1}(x))$ for any $x \in K$. It is immediate that $f$ is an embedding; let $K' = f(K)$.

Given any continuous function $\varphi : K' \to \mathbb{R}$ the function $h = \varphi \circ f$ is continuous and hence we can apply (16) to find a family $\{\varphi^q : q \in Q\} \subset C(\mathbb{R})$ such that $h = \sum_{q \in Q} \varphi^q \circ f^q$. Since $f^q(K) \subset I$, we can restrict every function $\varphi^q$ to $I$ to obtain the same equality. Therefore we can consider, without loss of generality, that $\varphi^q \in C(I)$ for every $q \in Q$.

If $y = (y_1, \ldots, y_{2n+1}) \in K'$ then there is a unique $x \in K$ with $y_q = f^q(x)$ for any $q \in Q$. We have $\varphi(y) = \varphi(f(x)) = \sum_{q \in Q} \varphi^q(f^q(x))$; in other words, $\varphi(y) = \varphi^1(y_1) + \ldots + \varphi^{2n+1}(y_{2n+1})$ for any point $y = (y_1, \ldots, y_{2n+1}) \in K'$, so the set $K' \subset I^{2n+1}$ is as promised.

Finally observe that we proved, in particular, that any compact space $K$ with dim $K \leq n$ embeds in $I^{2n+1}$ and hence in $\mathbb{I}^{2n+1}$. If $X$ is a second countable space and dim $X \leq n$ then there exists a second countable compact space $K$ such that $X \subset K$ and dim $K \leq n$ (see Problem 162). If $\mu : K \to \mathbb{I}^{2n+1}$ is an embedding then $\mu|X$ embeds $X$ in $\mathbb{I}^{2n+1}$ and hence our solution is complete.

**V.499.** *Prove that, for any finite-dimensional metrizable compact space $K$, there exists a surjective continuous linear mapping of $C_p(\mathbb{I})$ onto $C_p(K)$.*

**Solution.** The expression $L \approx M$ says that the linear topological spaces $L$ and $M$ are linearly homeomorphic. Let $I$ be the closed interval $[0, 1] \subset \mathbb{R}$ and fix $n \in \omega$ such that dim $K \leq n$. By Problem 498, we can consider that $K \subset I^{2n+1}$ and,

(1) for any continuous function $f : K \to \mathbb{R}$ there exist $h_1, \ldots, h_{2n+1} \in C(I)$ such that $f(x) = \sum_{i=1}^{2n+1} h_i(x_i)$ for every point $x = (x_1, \ldots, x_{2n+1}) \in K$.

Let $\pi_i : I^{2n+1} \to I$ be the natural projection of $I^{2n+1}$ onto its $i$-th factor; the function $p_i = \pi_i|K : K \to I$ is continuous for each $i \leq 2n + 1$. The dual map $p_i^* : C_p(I) \to C_p(K)$ is a linear embedding by TFS-163; thus $Z_i = p_i^*(C_p(I))$ is a linear subspace of $C_p(K)$ homeomorphic to $C_p(I)$ for every $i \leq 2n+1$. Therefore the space $Z = Z_1 \times \ldots \times Z_{2n+1}$ is linearly homeomorphic to $(C_p(I))^{2n+1}$.

For any point $g = (g_1, \ldots, g_{2n+1}) \in Z$ let $\mu(g) = g_1 + \ldots + g_{2n+1}$; then $\mu : Z \to C_p(K)$ is a linear continuous map. Given any $f \in C_p(K)$ apply (1) to find functions $h_1, \ldots, h_{2n+1} \in C(I)$ such that $f(x) = \sum_{i=1}^{2n+1} h_i(x_i)$ for every point $x = (x_1, \ldots, x_{2n+1}) \in K$.

Therefore $f(x) = \sum_{i=1}^{2n+1} h_i(p_i(x))$ for every $x \in K$ and hence we have the equality $f = \sum_{i=1}^{2n+1} h_i \circ p_i$. Letting $g_i = p_i^*(h_i) = h_i \circ p_i$ for all $i \leq 2n + 1$ we obtain a point $g = (g_1, \ldots, g_{2n+1}) \in Z$ such that $\mu(g) = f$. Therefore $\mu : Z \to C_p(K)$ is a linear continuous surjective map. It follows from $Z \approx (C_p(I))^{2n+1} \approx (C_p(\mathbb{I}))^{2n+1}$ that $Z$ is linearly homeomorphic to $(C_p(\mathbb{I}))^{2n+1}$. Take a space $J_i$ homeomorphic to $\mathbb{I}$ for every $i = 1, \ldots, 2n + 1$ and let $J = J_1 \oplus \ldots \oplus J_{2n+1}$; it follows from Problem 265 that $C_p(J) \approx \prod_{i=1}^{2n+1} C_p(J_i) \approx (C_p(\mathbb{I}))^{2n+1}$. Apply Problem 266 to conclude that $C_p(J) \approx C_p(\mathbb{I})$ and hence $C_p(\mathbb{I}) \approx Z$, so there exists a continuous linear surjection of $C_p(\mathbb{I})$ onto $C_p(K)$.

**V.500.** *Prove that $l(X) = l(Y)$ for any $l$-equivalent spaces $X$ and $Y$. In particular, if $X \overset{l}{\sim} Y$ then $X$ is Lindelöf if and only if so is $Y$.*

**Solution.** Given a set $A$ we denote by Fin$(A)$ the family of all finite subsets of $A$. A family $\{U_n : n \in \omega\}$ of subsets of a space $Z$ is called *increasing* if $U_n \subset U_{n+1}$ for each $n \in \omega$. Suppose that we have spaces $Z$ and $T$ and a map $\varphi : Z \to \exp(T)$; let $\varphi_l^{-1}(U) = \{z \in Z : \varphi(z) \cap U \neq \emptyset\}$ and $\varphi_u^{-1}(U) = \{z \in Z : \varphi(z) \subset U\}$ for any $U \subset T$. A map $\varphi : Z \to \exp(T)$ is called *lower semicontinuous* if $\varphi_l^{-1}(U)$ is open in $Z$ for any $U \in \tau(T)$.

If $Z$ is a set and $\mathcal{A} = \{A_n : n \in \omega\}$ is a sequence of subsets of $Z$ then the expression $\mathcal{A} \to Z$ says that $Z = \bigcup_{n \in \omega} \bigcap_{k \geq n} A_k$. For any $\mathcal{A} \subset \exp(Z)$ we denote by $\bigwedge \mathcal{A}$ the family of all finite intersections of elements of $\mathcal{A}$.

If $Z$ and $T$ are spaces and $\varphi : Z \to \mathrm{Fin}(T)$ say that a map $e : \tau(T) \to \exp(Z)$ is an *extractor for $\varphi$ (or $\varphi$-extractor)* if the following conditions are satisfied:

(1) $\varphi_u^{-1}(U) \subset e(U)$ for any $U \in \tau(T)$;
(2) if $U, V \in \tau(T)$ and $U \subset V$ then $\varphi(x) \cap (V \backslash U) \neq \emptyset$ whenever $x \in e(V) \backslash e(U)$;
(3) if $\mathcal{U} = \{U_n : n \in \omega\} \subset \tau(T)$ is an increasing sequence and $\{e(U_n) : n \in \omega\} \to Z$ then $\bigcup \mathcal{U} = T$.

Given an extractor $e : \tau(T) \to \exp(Z)$ let $e^*(U) = Z \backslash e(U)$ for any $U \in \tau(T)$. If $Z$ is a space and $\mathcal{A}$ is a family of subsets of $Z$ then we write $l(\mathcal{A}) \leq \kappa$ if $l(A) \leq \kappa$ for any $A \in \bigwedge \mathcal{A}$. For every family $\mathcal{U}$ of subsets of $Z$ let $[\mathcal{U}]_\kappa = \{\bigcup \mathcal{V} : \mathcal{V} \subset \mathcal{U}$ and $|\mathcal{V}| \leq \kappa\}$. Say that a cover $\mathcal{U}$ of the space $Z$ is *$\kappa$-trivial* if $Z \in [\mathcal{U}]_\kappa$; if $\mathcal{U}$ is not $\kappa$-trivial, we call it *$\kappa$-nontrivial*.

If $\varphi : Z \to \mathrm{Fin}(T)$ say that an extractor $e : \tau(T) \to \exp(Z)$ is *$\kappa$-synchronous* if there exists a base $\mathcal{B}$ in the space $T$ and a family $\mathcal{L}$ of subsets of $Z$ such that $l(\mathcal{L}) \leq \kappa$ and, for any $\kappa$-nontrivial cover $\mathcal{U} \subset \mathcal{B}$ of the space $T$, if $U \in [\mathcal{U}]_\kappa$ then there exists $V \in [\mathcal{U}]_\kappa$ such that $U \subset V$ and $e^*(V) \in \mathcal{L}$.

*Fact 1.* Given spaces $Z$ and $T$, let $\varphi : Z \to \exp(T)$ be a lower semicontinuous map. Then, for any $U \in \tau(T)$ and $n \in \mathbb{N}$, the set $V = \{x \in Z : |\varphi(x) \cap U| \geq n\}$ is open in $Z$.

*Proof.* Fix a point $x \in V$ and choose distinct points $t_1, \ldots, t_n \in \varphi(x) \cap U$. There exist disjoint sets $W_1, \ldots, W_n \in \tau(T)$ such that $t_i \in W_i \subset U$ for any $i \leq n$. The set $G = \bigcap_{i \leq n} \varphi_l^{-1}(W_i)$ is open in $Z$ and it is straightforward that $x \in G \subset V$, so every point of $V$ is in the interior of $V$, i.e., $V$ is open in $Z$ and hence Fact 1 is proved.

*Fact 2.* Suppose that $Z$ and $T$ are spaces and $\varphi : Z \to \mathrm{Fin}(T)$ is a lower semicontinuous map. If $e : \tau(T) \to \exp(Z)$ is an extractor for $\varphi$ and $\{U_n : n \in \omega\} \subset \tau(T)$ is an increasing sequence then, for any point $x \in Z$, either $I = \{n \in \omega : x \in e(U_n)\}$ or $\omega \backslash I$ is finite.

*Proof.* If both sets $I$ and $\omega \backslash I$ are infinite then it is easy to see that we can find a sequence $\{k_n : n \in \omega\} \subset \omega$ such that $k_{n+1} > k_n + 1$ while $k_n \in \omega \backslash I$ and $k_n + 1 \in I$ for any $n \in \omega$. Therefore $x \in e(U_{k_n + 1}) \backslash e(U_{k_n})$ and hence $\varphi(x) \cap (U_{k_n + 1} \backslash U_{k_n}) \neq \emptyset$ for any $n \in \omega$ by the property (2). The family $\{U_{k_n + 1} \backslash U_{k_n} : n \in \omega\}$ being disjoint, the set $\varphi(x)$ has to be infinite; this contradiction shows that Fact 2 is proved.

*Fact 3.* Suppose that $Z$ and $T$ are spaces, $\varphi : Z \to \mathrm{Fin}(T)$ is a lower semicontinuous map and $e : \tau(T) \to \exp(Z)$ is an extractor for $\varphi$. Assume additionally that $\{U_n : n \in \omega\} \subset \tau(T)$ is an increasing sequence such that for any $k, n \in \omega$ with $k \leq n$ we have $|\varphi(x)| \geq n - k$ for each $x \in \bigcap_{i=k}^n e^*(U_i)$. Then $\{e(U_n) : n \in \omega\} \to Z$ and hence $T = \bigcup_{n \in \omega} U_n$.

*Proof.* Fix an arbitrary point $x \in Z$; we must prove that there exists $m \in \omega$ such that $x \in e(U_n)$ for any $n \geq m$. By Fact 2, it suffices to show that the set $I = \{n \in \omega : x \in e(U_n)\}$ is infinite. However, if $I$ is finite then there exists $k \in \omega$ such that $x \in \bigcap_{i=k}^n e^*(U_i)$ and hence $|\varphi(x)| \geq n - k$ for any $n \geq k$. This implies that $\varphi(x)$ is infinite, so we obtained a contradiction which shows that Fact 3 is proved.

*Fact 4.* Suppose that $Z$ and $T$ are spaces, $\varphi : Z \to \text{Fin}(T)\backslash\{\emptyset\}$ is a lower semicontinuous map and $\kappa$ is an infinite cardinal. If $e : \tau(T) \to \exp(Z)$ is a $\kappa$-synchronous extractor for $\varphi$ and $l(Z) \leq \kappa$ then $l(T) \leq \kappa$.

*Proof.* There exists a base $\mathcal{B}$ in the space $T$ and a family $\mathcal{L}$ of subsets of $Z$ such that $l(\mathcal{L}) \leq \kappa$ and, for any $\kappa$-nontrivial cover $\mathcal{U} \subset \mathcal{B}$ of the space $T$, if $U \in [\mathcal{U}]_\kappa$ then there exists $V \in [\mathcal{U}]_\kappa$ such that $U \subset V$ and $e^*(V) \in \mathcal{L}$.

Assume that $l(T) > \kappa$; then we can find a $\kappa$-nontrivial open cover $\mathcal{U} \subset \mathcal{B}$ of the space $T$. For any $x \in Z$ there exists a set $U_x \in \mathcal{U}$ such that $\varphi(x) \cap U_x \neq \emptyset$ and hence $x \in \varphi_l^{-1}(U_x)$. The cover $\{\varphi_l^{-1}(U_x) : x \in Z\}$ has a subcover of cardinality $\leq \kappa$ and hence there exists $A \subset Z$ such that $|A| \leq \kappa$ and $Z = \bigcup\{\varphi_l^{-1}(U_x) : x \in A\}$. If $\mathcal{U}_0 = \{U_x : x \in A\} \subset \mathcal{U}$ then $|\mathcal{U}_0| \leq \kappa$ and $\varphi(x) \cap (\bigcup\mathcal{U}_0) \neq \emptyset$ for any $x \in Z$.

By the choice of $\mathcal{B}$ and $\mathcal{L}$ there exists $U_0 \in [\mathcal{U}]_\kappa$ such that $\bigcup\mathcal{U}_0 \subset U_0$ and $e^*(U_0) \in \mathcal{L}$. If $x \in e^*(U_0)$ then the property (1) implies that $\varphi(x)$ is not contained in $U_0$, so it follows from $\varphi(x) \cap U_0 \neq \emptyset$ that $|\varphi(x)| \geq 2$.

Proceeding inductively, assume that $n \in \omega$ and we have sets $U_0, \ldots, U_n \in [\mathcal{U}]_\kappa$ such that

(4) $e^*(U_i) \in \mathcal{L}$ for any $i \leq n$;
(5) $|\varphi(x)| \geq m - k + 2$ for any $x \in F_{k,m} = \bigcap_{i=k}^{m} e^*(U_i)$ and $k \leq m \leq n$;
(6) $|\varphi(x) \cap U_{m+1}| \geq m - k + 2$ for any $x \in F_{k,m}$ and $k \leq m < n$;
(7) $U_m \subset U_{m+1}$ for any $m < n$.

Fix any $k \leq n$; the set $F_{k,n}$ belongs to $\bigwedge\mathcal{L}$, so $l(F_{k,n}) \leq \kappa$. By (5), we can choose, for any $x \in F_{k,n}$ a finite family $\mathcal{W}_x \subset \mathcal{U}$ such that $|\varphi(x) \cap \bigcup\mathcal{W}_x| \geq n - k + 2$. The set $H_x = \{y \in Z : |\varphi(y) \cap \bigcup\mathcal{W}_x| \geq n - k + 2\}$ is an open neighborhood of $x$ by Fact 1, so the cover $\{H_x : x \in F_{k,n}\}$ of the set $F_{k,n}$ has a subcover of cardinality not exceeding $\kappa$.

Therefore there exists a family $\mathcal{U}_{n+1} \subset \mathcal{U}$ such that $|\mathcal{U}_{n+1}| \leq \kappa$ and we have $|\varphi(x) \cap \bigcup\mathcal{U}_{n+1}| \geq n - k + 2$ for any $x \in F_{k,n}$ and $k \leq n$. By the choice of $\mathcal{B}$ we can find a set $U_{n+1} \in [\mathcal{U}]_\kappa$ such that $U_n \cup \bigcup\mathcal{U}_{n+1} \subset U_{n+1}$ and $e^*(U_{n+1}) \in \mathcal{L}$. It is immediate that the conditions (4), (6) and (7) are satisfied if we replace $n$ with $n+1$. To see that (5) is also true it suffices to check it for all sets $F_{k,n+1} = \bigcap_{i=k}^{n+1} e^*(U_i)$, so fix $k \leq n+1$ and a point $x \in F_{k,n+1}$.

If $k = n+1$ then it follows from the inclusion $U_0 \subset U_{n+1}$ that $\varphi(x) \cap U_{n+1} \neq \emptyset$. Since $x \notin e(U_{n+1})$, by the property (1) it is impossible that $\varphi(x) \subset U_{n+1}$, so $\varphi(x)\backslash U_{n+1} \neq \emptyset$ and hence $|\varphi(x)| \geq 2 = (n+1) - (n+1) + 2$.

If $k \leq n$ then $F_{k,n+1} \subset F_{k,n}$, so $x \in F_{k,n}$. Applying (6) for $m = n$ we can see that $|\varphi(x) \cap U_{n+1}| \geq n - k + 2$; since also $x \in e^*(U_{n+1})$, we must have $\varphi(x)\backslash U_{n+1} \neq \emptyset$ by the property (1), so $|\varphi(x)| \geq n - k + 2 + 1 = (n+1) - k + 2$ and hence (5) also holds if we replace $n$ with $n + 1$.

Therefore our induction procedure can be continued to construct an increasing sequence $\{U_n : n \in \omega\} \subset [\mathcal{U}]_\kappa$ for which the conditions (4)–(7) are satisfied for all $n \in \omega$.

It follows from the property (5) and Fact 3 that $\{e(U_n) : n \in \omega\} \to Z$ and hence $T = \bigcup_{n \in \omega} U_n$. As a consequence, $T \in [\mathcal{U}]_\kappa$, i.e., $\mathcal{U}$ is $\kappa$-trivial; this contradiction shows that $l(T) \leq \kappa$ and hence Fact 4 is proved.

Returning to our solution observe that the spaces $X$ and $Y$ are in a symmetrical situation, so it is sufficient to prove that $l(Y) \leq l(X)$. There is no loss of generality to assume that $l(X)$ is an infinite cardinal; let $\xi : C_p(Y) \to C_p(X)$ be a linear homeomorphism. For any point $x \in X$ and $f \in C_p(X)$ let $\delta_x(f) = f(x)$. Then $\delta_x : C_p(X) \to \mathbb{R}$ is a nontrivial continuous linear functional and hence the map $\delta_x \circ \xi : C_p(Y) \to \mathbb{R}$ is a nontrivial continuous linear functional on $C_p(Y)$. By TFS-197, there exists a nonempty finite set $A_x \subset Y$ and a set $\{\lambda_y(x) : y \in A_x\} \subset \mathbb{R}\backslash\{0\}$ such that $\xi(g)(x) = \sum_{y \in A_x} \lambda_y(x) \cdot g(y)$ for any $g \in C_p(Y)$. Since the point $x$ will always be clear from the context, in the expression for $\xi(g)(x)$ we will write $\lambda_y$ instead of $\lambda_y(x)$. For technical reasons, we will consider that any sum with the empty set of summands is equal to zero.

Letting $\varphi(x) = A_x$ for any point $x \in X$ we obtain a finite-valued lower semicontinuous map $\varphi : X \to \mathrm{Fin}(Y)\backslash\{\emptyset\}$ (see Problem 280). Fix an arbitrary point $y \in Y$ and consider a continuous nontrivial linear functional $\mu : C_p(Y) \to \mathbb{R}$ defined by $\mu(g) = g(y)$ for every $g \in C_p(Y)$. Then $\mu \circ \xi^{-1}$ is a continuous nontrivial linear functional on $C_p(X)$, so there exists a nonempty finite set $B \subset X$ and a set $\{t_x : x \in B\} \subset \mathbb{R}\backslash\{0\}$ such that $\xi^{-1}(f)(y) = \sum_{x \in B} t_x \cdot f(x)$ for any $f \in C_p(X)$.

If $y$ does not belong to the finite set $F = \bigcup\{A_x : x \in B\}$ then pick a function $g \in C_p(Y)$ such that $g(y) = 1$ and $g(F) = \{0\}$. For the function $f = \xi(g)$ we have $f(x) = \sum_{y \in A_x} \lambda_y \cdot g(y) = 0$ because $g(A_x) \subset g(F) = \{0\}$. Therefore $f(x) = 0$ for any $x \in B$ and hence $g(y) = \xi^{-1}(f)(y) = \sum_{x \in B} t_x \cdot f(x) = 0$ which is a contradiction. Thus $y \in F$, i.e., there exists $x \in X$ such that $y \in A_x$. Since the point $y \in Y$ was chosen arbitrarily, we proved that

(8) the map $\varphi$ is surjective, i.e., $\bigcup\{\varphi(x) : x \in X\} = Y$.

Given an open set $V \subset Y$ and $x \in X$ let $r_V(x) = \sum\{\lambda_y : y \in \varphi(x)\backslash V\}$ and consider the set $e(V) = \{x \in X : r_V(x) = 0\}$. It turns out that

(9) the map $e : \tau(Y) \to \exp(X)$ is a $\varphi$-extractor.

If $V \in \tau(Y)$ and $\varphi(x) \subset V$ then $\varphi(x)\backslash V = \emptyset$ and hence $r_V(x) = 0$ by our agreement on sums with the empty set of summands; therefore $\varphi_u^{-1}(V) \subset e(V)$, i.e., the property (1) holds for $e$.

Now assume that $U, V \in \tau(Y)$ and $U \subset V$; if $x \in e(V)\backslash e(U)$ then $r_V(x) = 0$ and $r_U(x) \neq 0$, so it follows from $r_U(x) = r_V(x) + \sum\{\lambda_y : y \in \varphi(x) \cap (V\backslash U)\}$ that $\sum\{\lambda_y : y \in \varphi(x) \cap (V\backslash U)\} \neq 0$ and hence $\varphi(x) \cap (V\backslash U) \neq \emptyset$, i.e., we proved that $e$ satisfies the condition (2).

To prove that the map $e$ has (3), fix an increasing sequence $\{U_n : n \in \omega\} \subset \tau(Y)$ such that $\{e(U_n) : n \in \omega\} \to X$. If $U = \bigcup_{n \in \omega} U_n \neq Y$ then fix a point $y \in Y\backslash U$ and consider the set $G = \{g \in C_p(Y) : |g(y)| < 1\}$. The map $\xi$ being open, the set $\xi(G)$ is an open neighborhood of the zero function of $C_p(X)$; in particular, there exists a finite set $F \subset X$ such that $\xi(G)$ contains the set $P = \{f \in C_p(X) : f(F) \subset \{0\}\}$. The set $Q = \bigcup\{\varphi(x) : x \in F\}$ is finite, so we can find a number $k \in \omega$ such that $Q \cap U = Q \cap U_k$ and $F \subset e(U_k)$.

Choose $g \in C_p(Y)$ such that $g(Q \cap U_k) \subset \{0\}$ and $g(\{y\} \cup (Q \backslash U_k)) = \{1\}$. For each $x \in F$ we have the equality

$$\xi(g)(x) = \sum_{z \in \varphi(x) \cap U_k} \lambda_z \cdot g(z) + \sum_{z \in \varphi(x) \backslash U_k} \lambda_z \cdot g(z) = r_{U_k}(x) = 0,$$

and hence $\xi(g) \in \xi(G)$. Now, $\xi$ is a bijection, so $g \in G$ whence $|g(y)| < 1$; this contradiction shows that $Y = U$ and hence $e$ has the property (3), i.e., (9) is proved.

Recall that a set $U \subset Y$ is called *functionally open* if we can find a set $O \in \tau(\mathbb{R})$ and $g \in C_p(Y)$ such that $U = g^{-1}(O)$. Functionally open sets are also called *cozero sets*. The complements of functionally open sets are called *functionally closed*. The family $\mathcal{B}$ of all functionally open subsets of $Y$ is a base in $Y$ (see Fact 1 of T.252). Say that a set $U \in \mathcal{B}$ is *adequate* if there exists an increasing sequence $\{F_n : n \in \omega\}$ of functionally closed subsets of $Y$ such that $U = \bigcup_{n \in \omega} F_n$ and $\varphi_u^{-1}(U) \backslash \varphi_u^{-1}(F_n) \neq \emptyset$ for any $n \in \omega$. The following lemma shows that there are sufficiently many adequate sets for our purposes.

**Lemma 1.** *Suppose that $\kappa$ is an infinite cardinal and $\mathcal{U} \subset \mathcal{B}$ is a $\kappa$-nontrivial cover of $Y$. Then, for every $V \subset \mathcal{U}$ with $|V| \leq \kappa$ there exists a family $\mathcal{W} \subset [\mathcal{U}]_\omega$ such that*

(a) $\mathcal{W}$ *is closed under finite unions and* $|\mathcal{W}| \leq \kappa$;
(b) *every* $W \in \mathcal{W}$ *is adequate and* $\bigcup V \subset \bigcup \mathcal{W}$.

*Proof.* Let $V = \bigcup \mathcal{V}$; the cover $\mathcal{U}$ being $\kappa$-nontrivial, the set $Y \backslash V$ is nonempty. The map $\varphi$ is surjective by (8), so there exists $x_0 \in X$ such that $\varphi(x_0) \backslash V \neq \emptyset$. Pick $U_0 \in [\mathcal{U}]_\omega$ with $\varphi(x_0) \subset U_0$. Proceeding inductively, assume that $n \in \omega$ and we have points $x_0, \ldots, x_n \in X$ and sets $U_0, \ldots, U_n \in [\mathcal{U}]_\omega$ such that

(10) $\varphi(x_i) \subset U_i$ for all $i \leq n$;
(11) $\varphi(x_0) \backslash V \neq \emptyset$ and $\varphi(x_{i+1}) \backslash (V \cup U_0 \cup \ldots \cup U_i) \neq \emptyset$ for any $i < n$.

Since $V' = V \cup U_0 \cup \ldots \cup U_n \in [\mathcal{U}]_\kappa$, we can pick a point $x_{n+1} \in X$ such that $\varphi(x_{n+1}) \backslash V' \neq \emptyset$. Taking a set $U_{n+1} \in [\mathcal{U}]_\omega$ with $\varphi(x_{n+1}) \subset U_{n+1}$ we obtain points $x_0, \ldots, x_n, x_{n+1} \in X$ and sets $U_0, \ldots, U_n, U_{n+1} \in [\mathcal{U}]_\omega$ such that the conditions (10) and (11) are satisfied if we replace $n$ with $n + 1$. Therefore our inductive procedure can be continued to obtain a sequence $\{x_n : n \in \omega\} \subset X$ and a family $\{U_n : n \in \omega\} \subset [\mathcal{U}]_\omega$ such that (10) and (11) are true for every $n \in \omega$.

It is clear that $U = \bigcup_{n \in \omega} U_n \in [\mathcal{U}]_\omega$. Denote by $\mathcal{V}_{fu}$ the family of all finite unions of elements of $\mathcal{V}$. For every $H \in \mathcal{V}_{fu}$ let $O_H = H \cup U$ and consider the family $\mathcal{W} = \{O_H : H \in \mathcal{V}_{fu}\}$. It is easy to see that the family $\mathcal{W}$ is closed under finite unions; besides, $|\mathcal{W}| \leq \kappa$ and $V \subset \bigcup \mathcal{W}$.

To see that every $W \in \mathcal{W}$ is adequate observe first that a countable union of functionally open sets is a functionally open set (see Fact 1 of T.252), so we have the inclusion $\mathcal{W} \subset \mathcal{B}$. If $H \in \mathcal{V}_{fu}$ and $W = O_H$ then choose an increasing sequence $\{F_n : n \in \omega\}$ of functionally closed sets such that $H = \bigcup_{n \in \omega} F_n$. We can also find, for every $n \in \omega$, an increasing sequence $\{F_n^k : k \in \omega\}$ of functionally closed sets such that $\bigcup_{k \in \omega} F_n^k = U_n$. Let $G_n = F_n \cup F_n^0 \cup \ldots \cup F_n^n$ for every $n \in \omega$. It is immediate that $\{G_n : n \in \omega\}$ is an increasing sequence of functionally closed subsets of $Y$ and $O_H = \bigcup_{n \in \omega} G_n$. Since also $\{x_n : n \in \omega\} \subset \varphi_u^{-1}(O_H)$ but $x_{n+1} \notin \varphi_u^{-1}(G_n)$ for every $n \in \omega$, we conclude that $W = O_H$ is an adequate set and hence Lemma 1 is proved.

**Lemma 2.** *Given an infinite cardinal $\kappa$ suppose that $\mathcal{V} \subset \tau(Y)$ is a family of adequate sets with $|\mathcal{V}| \leq \kappa$ which is closed under finite unions. Then $e^*(\bigcup \mathcal{V})$ is an $F_\kappa$-subset of $X$.*

*Proof.* Let $V = \bigcup \mathcal{V}$ and choose an indexation $\{V_s : s \in S\}$ of the family $\mathcal{V}$. For every $s \in S$ there exists an increasing sequence $\{F_n^s : n \in \omega\}$ of functionally closed subsets of $Y$ such that $V_s = \bigcup_{n \in \omega} F_n^s$ and $\varphi_u^{-1}(V_s) \backslash \varphi_u^{-1}(F_n^s) \neq \emptyset$ for each $n \in \omega$.

For any $s \in S$ and $n \in \omega$ the set $F_n^s$ is functionally separated from $Y \backslash V_s$ (see Fact 1 of V.140), so we can find a function $g_n^s \in C_p(Y)$ such that $g_n^s(F_n^s) \subset \{0\}$ and $g_n^s(Y \backslash V_s) \subset \{n\}$; let $f_n^s = \xi(g_n^s)$. For any $s \in S$ and $x \in Q_s = \varphi^{-1}(V_s)$ denote by $n_x^s$ the first integer $m \in \omega$ such that $\varphi(x) \subset F_m^s$. It follows from the inclusion $\varphi(x) \subset F_{i+n_x^s}^s$ that we have $f_{i+n_x^s}^s(x) = 0$ for any $i \in \omega$ and therefore the set $U_k^s(x) = \bigcap_{i \leq k} \{y \in X : |f_{i+n_x^s}^s(y)| < 1\}$ is an open neighborhood of the point $x$ for any $k \in \omega$.

As a consequence, $A_s = \bigcap_{k \in \omega} \bigcup_{x \in Q_s} U_k^s(x)$ is a $G_\delta$-subset of $X$ for each $s \in S$. The map $\varphi$ being lower semicontinuous, $B_s = \varphi_l^{-1}(V \backslash V_s) = \bigcap_{n \in \omega} \varphi_l^{-1}(V \backslash F_n^s)$ is also a $G_\delta$-subset of $X$ for any $s \in S$. Therefore $A = \bigcap_{s \in S}(A_s \cup B_s)$ is a $G_\kappa$-subset of $X$, so it suffices to show that $e^*(V) = X \backslash A$.

Take any $y \in e^*(V)$; the family $\mathcal{V}$ being invariant under finite unions, we can find $s \in S$ such that $\varphi(y) \cap V = \varphi(y) \cap V_s$ and hence $y \notin B_s$. Pick $k \in \omega$ such that $\varphi(y) \cap V \subset F_k^s$ and $k |r_V(y)| \geq 1$; this is possible because $r_V(y) \neq 0$.

To see that $y \notin \bigcup \{U_k^s(x) : x \in Q_s\}$ take any $x \in Q_s$ and observe that $f_{k+n_x^s}^s(y) = (k + n_x^s) r_V(y)$, so it follows from $(k + n_x^s)|r_V(y)| \geq k |r_V(y)| \geq 1$ that $y \notin U_k^s(x)$. Thus $y \notin \bigcup \{U_k^s(x) : x \in Q_s\}$, so $y \notin (A_s \cup B_s)$ and hence $y \notin A$. This proves that $e^*(V) \subset X \backslash A$.

To prove that $X \backslash A \subset e^*(V)$, take any $y \in X \backslash A$; there exists $s \in S$ such that $y \notin A_s \cup B_s$. It follows from $y \notin B_s$ that $\varphi(y) \cap V = \varphi(y) \cap V_s$, so we can choose a number $p \in \omega$ such that $\varphi(y) \cap V \subset F_p^s$. There exists $k \in \omega$ such that $y \notin \bigcup \{U_k^s(x) : x \in Q_s\}$; besides we can find $x \in Q_s$ for which $\varphi(x) \backslash F_p^s \neq \emptyset$ and hence $n_x^s > p$. It follows from $y \notin U_k^s(x)$ that $|f_{i+n_x^s}^s(y)| \geq 1$ for some $i \leq k$. We have the inclusions $\varphi(y) \cap V \subset F_p^s \subset F_{i+n_x^s}^s$ which imply that $f_{i+n_x^s}^s(y) = (i + n_x^s) r_V(y)$. Therefore $(i + n_x^s)|r_V(y)| \geq 1$; in particular, $r_V(y) \neq 0$ and hence $y \in e^*(V)$.

This proves that $X \backslash A \subset e^*(V)$, so $X \backslash A = e^*(V)$ and hence $e^*(V)$ is, indeed, an $F_\kappa$-set in $X$, so Lemma 2 is proved.

Finally, let $\kappa = l(X)$ and consider the family $\mathcal{L}$ of all $F_\kappa$-subsets of $X$. We have $l(L) \leq \kappa$ because $\bigwedge \mathcal{L} = \mathcal{L}$. Therefore Lemma 1 and Lemma 2 say that the families $\mathcal{L}$ and $\mathcal{B}$ witness that $e$ is a $\kappa$-synchronous extractor for $\varphi$, so $l(Y) \leq \kappa$ by Fact 4 and hence our solution is complete.

# Chapter 3
# Bonus results: Some Hidden Statements

The reader has, evidently, noticed that an essential percentage of the problems of the main text is formed by purely topological statements some of which are quite famous and difficult theorems. A common saying among $C_p$-theorists is that any result on $C_p$-theory contains only 20% of $C_p$-theory; the rest is general topology.

It is evident that the author could not foresee all topology which would be needed for the development of $C_p$-theory; so a lot of material had to be dealt with in the form of auxiliary assertions. After accumulating more than seven hundred such assertions, the author decided that some deserve to be formulated together to give a "big picture" of the additional material that can be found in solutions of problems.

This section presents 80 topological statements which were proved in the solutions of problems without being formulated in the main text. In these formulations the main principle is to make them clear for an average topologist. A student could lack the knowledge of some concepts of the formulation; so the index of this book can be used to find the definitions of the necessary notions. After every statement we indicate the exact place (in this book) where it was proved.

The author considers that most of the results that follow are very useful and have many applications in topology. Some of them are folkloric statements and quite a few are published theorems, sometimes famous ones. For example, Fact 5 of V.290 is a classical theorem of Sierpiński that can be found in practically all textbooks and, in particular, in Engelking (1977). Fact 1 of V.274 is a well-known result of Arhangel'skii published in Arhangel'skii (1978b) and Fact 3 of V.403 is a celebrated theorem of Hajnal and Juhász that has an infinity of applications (see the survey of Hodel (1984)).

To help the reader find a result he/she might need, we have classified the material of this section according to the following topics: *standard spaces, compact spaces and their generalizations, properties of continuous maps, cardinal invariants and set theory, locally convex spaces and homotopies, zero-dimensional spaces, and connected spaces.* The last section is entitled *raznoie* which in Russian means "miscellaneous" and contains unclassified results. The author hopes that once we understand in which subsection a result should be, then it will be easy to find it.

## 3.1 Standard spaces

By *standard spaces* we mean the real line, its subspaces and it powers, Tychonoff and Cantor cubes as well as ordinals together with the Alexandroff and Stone–Čech compactifications of discrete spaces.

**V.022. Fact 4.** *The space $\mathbb{R}^\omega$ is not $\sigma$-totally bounded (as a linear topological space).*

**V.148. Fact 1.** *Given a cardinal $\kappa > \omega$ let $\Sigma = \{x \in \mathbb{D}^\kappa : |x^{-1}(1)| \leq \omega\}$. Then the set $P = \mathbb{D}^\kappa \backslash \Sigma$ is a pseudocompact non-countably compact (and hence non-normal) dense subspace of $\mathbb{D}^\kappa$.*

**V.160. Fact 1.** *Suppose that $A, B \subset \mathbb{R}$ are countable dense subsets and a function $f : A \to B$ is a bijection such that $a < b$ implies $f(a) < f(b)$. Then there exists a unique homeomorphism $h : \mathbb{R} \to \mathbb{R}$ such that $h|A = f$ and $x < y$ implies $h(x) < h(y)$.*

**V.160. Fact 8.** *For any $n \in \mathbb{N}$, if $A$ and $B$ are countable dense subspaces of $\mathbb{R}^n$ then there exists a homeomorphism $h : \mathbb{R}^n \to \mathbb{R}^n$ such that $h(A) = B$.*

**V.266. Fact 1.** *Given a space $Z$ suppose that $a \in Z$ and $Z_0, Z_1$ are closed subsets of $Z$ such that $Z_0 \cup Z_1 = Z$ and $Z_0 \cap Z_1 = \{a\}$. If there exist homeomorphisms $h_0 : \mathbb{I} \to Z_0$ and $h_1 : \mathbb{I} \to Z_1$ such that $h_0(1) = h_1(1) = a$ then $Z$ is homeomorphic to $\mathbb{I}$.*

**V.299. Fact 1.** *For any nonempty open set $U \subset \mathbb{R}^n$ there is an open set $V \subset U$ such that $V \simeq \mathbb{R}^n$.*

**V.302. Fact 1.** *For the space $N = \omega_1 \times (\omega_1 + 1)$ let $F = \{(\alpha, \omega_1) : \alpha < \omega_1\}$. For any $f \in C_p(N)$ with $f(F) = \{0\}$ there exists an ordinal $\gamma < \omega_1$ for which we have the inclusion $\{(\alpha, \beta) : \gamma < \alpha < \omega_1 \text{ and } \gamma < \beta \leq \omega_1\} \subset f^{-1}(0)$.*

**V.302. Fact 2.** *The space $T = ((\omega_1 + 1) \times (\omega_1 + 1)) \backslash \{(\omega_1, \omega_1)\}$ is not normal.*

**V.302. Fact 3.** *For any ordinal $\xi$ let $f(\alpha, \beta) = \backslash n\{\alpha, \beta\}$ for every $(\alpha, \beta) \in \xi \times \xi$. Then the map $f : \xi \times \xi \to \xi$ is continuous.*

**V.434. Fact 2.** *For any infinite cardinal $\kappa$ the space $\sigma = \{x \in \mathbb{D}^\kappa : |x^{-1}(1)| < \omega\}$ is $\sigma$-discrete.*

**V.434. Fact 3.** *If the Continuum Hypothesis holds then there exists a dense Luzin subspace $Z$ of the $\Sigma$-product $\Sigma = \{x \in \mathbb{D}^{\omega_1} : |x^{-1}(1)| \leq \omega\}$ of the Cantor cube $\mathbb{D}^{\omega_1}$ such that the space $Z \times Z$ is d-separable.*

**V.452. Fact 2.** *For any cardinal $\kappa$ the space $\beta D_\kappa$ is extremally disconnected.*

**V.452. Fact 3.** *If $2^{\omega_1} = \mathfrak{c}$ and $D$ is a discrete space of cardinality $\omega_1$ then the space $\beta D$ embeds in $\beta\omega$.*

**V.488. Fact 1.** *The following conditions are equivalent for any ultrafilter $\eta \in \omega^*$:*

*(i)  $\eta$ is a P-point in $\omega^*$;*

*(ii)  for any countable family $\mathcal{E} \subset \exp \omega \backslash \eta$, there exists $A \in \eta$ such that $A \cap E$ is finite for any $E \in \mathcal{E}$;*

*(iii) for any countable disjoint family $\mathcal{E} \subset \exp \omega \backslash \eta$, there exists $A \in \eta$ such that $A \cap E$ is finite for any $E \in \mathcal{E}$.*

## 3.2   Compact spaces and their generalizations

This Section contains some statements about compact and countably compact spaces.

**V.072. Fact 1.** *Any sequentially compact space is countably compact; if $Z$ is countably compact and sequential then $Z$ is sequentially compact.*

**V.072. Fact 2.** *Any finite product of sequentially compact spaces is sequentially compact and hence countably compact.*

**V.073. Fact 1.** *Suppose that $Z$ is a space and $\lambda$ is a cardinal such that $l(Z) \leq \lambda$ and $\overline{A}$ is compact for any $A \subset Z$ with $|A| \leq \lambda$. Then $Z$ is compact.*

**V.161. Fact 1.** *Given $m \in \omega$ suppose that $K$ is a compact space with $\dim K \leq m$ and $N$ is a second countable space. Then, for any continuous map $p : K \to N$ there exists a second countable space $L$ such that $\dim L \leq m$ and there are continuous maps $q : K \to L$ and $r : L \to N$ for which $r \circ q = p$.*

**V.200. Fact 1.** *If $K$ is a countably infinite compact space then for any point $a \notin K$ the space $K \oplus \{a\}$ is homeomorphic to $K$.*

**V.290. Fact 5.** *Suppose that $Z$ is a connected compact space and $\mathcal{A} = \{Z_n : n \in \omega\}$ is a disjoint family of closed subsets of $Z$ such that $Z = \bigcup_{n \in \omega} Z_n$. Then at most one element of $\mathcal{A}$ is nonempty.*

**V.296. Fact 1.** *For any infinite compact space $X$ let $\Omega(X)$ be the one-point compactification of the space $X \times \omega$. Then*

*(1) $w(X) = w(\Omega(X))$; in particular, if $X$ is second countable then $\Omega(X)$ is also second countable;*
*(2) if $X$ is zero-dimensional then $\Omega(X)$ is also zero-dimensional;*
*(3) $\Omega(X) \oplus X \simeq \Omega(X)$.*

**V.423. Fact 1.** *For any compact space $X$ there exists a discrete $D \subset (X \times X) \backslash \Delta_X$ such that $|D| = d(X)$.*

**V.480. Fact 5.** *If $X$ is a countably compact space and $t(X) > \omega$ then $X$ has an uncountable free sequence.*

## 3.3   Properties of continuous maps

We consider the most common classes of continuous maps: open, closed, perfect, and quotient. The respective results basically deal with preservation of topological properties by direct and inverse images.

**V.077. Fact 1.** *Given spaces $Z, T$ and an open continuous onto map $f : Z \to T$, the map $f_A = f | f^{-1}(A) : f^{-1}(A) \to A$ is open for any $A \subset T$.*

**V.192. Fact 3.** *Assume that $n \in \mathbb{N}$ and $f : Z \to T$ is an open map with $|f^{-1}(t)| = n$ for any $t \in T$. Then*

*(a) $f$ is perfect;*
*(b) $f$ is a local homeomorphism, i.e., for any $z \in Z$ there is $W \in \tau(z, Z)$ such that $f | W : W \to f(W)$ is a homeomorphism.*

**V.194. Fact 9.** *Given metrizable spaces $Z$ and $T$ and a continuous map $f : Z \to T$ the following conditions are equivalent:*

*(a) $f$ is an almost perfect map, i.e., $f(F)$ is closed in $T$ for any closed $F \subset Z$ and $f^{-1}(t)$ is compact for any $t \in T$ (recall that an almost perfect continuous map is perfect if and only if it is surjective);*
*(b) a sequence $S = \{z_n : n \in \omega\} \subset Z$ has a convergent subsequence if and only if the sequence $S' = \{f(z_n) : n \in \omega\} \subset T$ has a convergent subsequence.*

**V.251. Fact 2.** *For any spaces $Z$ and $T$, a map $g : Z \to T$ is $\mathbb{R}$-quotient if and only if, for any function $h : T \to \mathbb{I}$, it follows from continuity of $h \circ g$ that $h$ is continuous.*

**V.272. Fact 1.** *Given spaces $T$ and $W$ if $\varphi : C_p(T) \to C_p(W)$ is a linear homeomorphism then there exists a linear homeomorphism $\Phi : \mathbb{R}^T \to \mathbb{R}^W$ such that $\Phi | C_p(T) = \varphi$.*

**V.274. Fact 1.** *Given spaces $P, Q$ and an infinite cardinal $\kappa$ suppose that $t(Q) \leq \kappa$ and there exists a continuous closed onto map $f : P \to Q$ such that $t(f^{-1}(q)) \leq \kappa$ for any $q \in Q$. Then $t(P) \leq \kappa$. As a consequence, if $K$ is a compact space with $t(K) \leq \kappa$ then $t(Q \times K) \leq \kappa$.*

**V.304. Fact 2.** *If $A$ is a retract of a space $Z$ then the set $F = cl_{\beta Z}(A)$ is a retract of $\beta Z$.*

**V.306. Fact 1.** *For any space $Z$ its diagonal $\Delta = \{(z, z) : z \in Z\}$ is a retract of $Z \times Z$.*

**V.343. Fact 2.** *If $Z$ is a $k$-space and $f : Z \to Y$ is a quotient map then $Y$ is also a $k$-space.*

**V.452. Fact 1.** *Suppose that $f : X \to Y$ is a closed irreducible onto map. If $Y$ is extremally disconnected then $f$ is a homeomorphism.*

## 3.4 Cardinal invariants and set theory

To classify function spaces, using cardinal invariants often gives crucial information. This Section includes both basic, simple results on the topic and very difficult classical theorems.

**V.023. Fact 1.** Given an infinite cardinal $\kappa$, if $hd(Y) \leq \kappa$ and $w(Z) \leq \kappa$ then $hd(Y \times Z) \leq \kappa$. In particular, the product of a second countable space and a hereditarily separable space is hereditarily separable.

**V.274. Fact 1.** Given spaces $P, Q$ and an infinite cardinal $\kappa$ assume that $t(Q) \leq \kappa$ and there exists a continuous closed onto map $f : P \to Q$ such that $t(f^{-1}(q)) \leq \kappa$ for any $q \in Q$. Then $t(P) \leq \kappa$. As a consequence, if $K$ is a compact space with $t(K) \leq \kappa$ then $t(Q \times K) \leq \kappa$.

**V.274. Fact 2.** Given a space $P$ suppose that $\{U_a : a \in A\}$ is an open cover of $P$ such that $t(U_a) \leq \kappa$ for any $a \in A$. Then $t(P) \leq \kappa$.

**V.274. Fact 3.** The weak functional tightness of the space $P = V(\omega) \times V(\mathfrak{c})$ is uncountable.

**V.403. Fact 1.** For any space $X$ we have $|X| \leq d(X)^{t(X) \cdot \psi(X)}$. In particular, the inequality $|X| \leq d(X)^{\chi(X)}$ holds for every space $X$.

**V.403. Fact 2.** If $X$ is an arbitrary space then $|X| \leq 2^{l(X) \cdot \psi(X) \cdot t(X)}$. In particular, we have the inequality $|X| \leq 2^{l(X) \cdot \chi(X)}$ for any space $X$.

**V.403. Fact 3.** For any space $X$ we have the inequality $|X| \leq 2^{c(X) \cdot \chi(X)}$.

**V.413. Fact 1.** Assume that $N_t$ is a space with $nw(N_t) = \chi(N_t) = \omega$ for all $t \in T$, and $N = \prod_{t \in T} N_t$. Then, for any $a \in N$, every subspace of $\Sigma(N, a)$ has a point-countable $\pi$-base.

**V.439. Fact 1.** If $X$ is a compact space then $h\pi\chi(x, X) \leq t(X)$ for any $x \in X$.

**V.456. Fact 1.** Suppose that $\kappa$ is an infinite cardinal and $Z$ is a space such that $l(Z) \leq \kappa$ and $t(Z) \leq \kappa$. Then $Z$ has no free sequences of length $\kappa^+$.

**V.490. Fact 1.** Suppose that $D$ is a nonempty set and $f : D \to D$ is a map such that $f(x) \neq x$ for any $x \in D$. Then there exist disjoint subsets $A_0, A_1, A_2$ of the set $D$ such that $D = A_0 \cup A_1 \cup A_2$ and $f(A_i) \cap A_i = \emptyset$ for any $i = 0, 1, 2$.

## 3.5  Locally Convex Spaces and Homotopies

The methods of the homotopy theory were used by Gul'ko to establish that $C_p(\omega_1)$ is not homeomorphic to its square. Here we present the results he used to prove this famous theorem.

**V.024. Fact 3.** *The space $L_p(Z)$ has the Souslin property for any space $Z$.*

**V.100. Fact 1.** *Given spaces $X, Y, Z$ and maps $f, g, h$ such that $f, g \in C(X, Y)$ and $h \in C(Y, Z)$, if $f$ is homotopic to $g$ then $h \circ f$ is homotopic to $h \circ g$.*

**V.100. Fact 2.** *Given spaces $X, Y, Z$ and maps $f, g, h$ such that $f, g \in C(Y, Z)$ and $h \in C(X, Y)$, if $f$ is homotopic to $g$ then $f \circ h$ is homotopic to $g \circ h$.*

**V.100. Fact 3.** *Homotopical equivalence is an equivalence relation on the class of topological spaces.*

**V.100. Fact 9.** *If $n < m$ then any continuous function $f : S^n \to S^m$ is homotopic to a constant map.*

**V.100. Fact 10.** *For any $n \in \omega$ and $p \in \mathbb{R}^{n+1}$ the spaces $\mathbb{R}^{n+1} \backslash \{p\}$ and $S^n$ are homotopically equivalent.*

**V.100. Fact 11.** *For any distinct $n, m \in \mathbb{N}$ the spaces $S^n$ and $S^m$ are not homotopically equivalent.*

**V.100. Fact 15.** *Suppose that $L$ is a linear topological space and $K \subset L$ is a linear subspace of $L$ of codimension $n$. Then $L \backslash K$ is homotopically equivalent to $S^{n-1}$.*

**V.228. Fact 1.** *Suppose that $N$ is a linear topological space and $G \subset N$ is a locally compact linear subspace of $N$. Then $G$ is closed in $N$.*

**V.233. Fact 1.** *Given a linear space $L$ suppose that $\tau$ and $\mu$ are linear space topologies on $L$ such that, for any $U \in \tau$ with $\mathbf{0}_L \in U$, there exists $V \in \mu$ such that $0 \in V \subset U$. Then $\tau \subset \mu$.*

**V.246. Fact 2.** *For any space $Y$, if $Z$ is a closed subspace of $Y$ then the linear hull $H$ of the set $Z$ in $L_p(Y)$ is closed in $L_p(Y)$.*

**V.246. Fact 3.** *Given a space $Y$, if $Z \subset Y$ is $C$-embedded in $Y$ then, for the linear hull $H$ of the set $Z$ in $L_p(Y)$, there exists a linear homeomorphism $\varphi : L_p(Z) \to H$ such that $\varphi(z) = z$ for any $z \in Z$.*

**V.250. Fact 1.** *Suppose that $L$ is a linear topological space and $K$ is a compact subset of $L$. Then the set $K + A$ is closed in $L$ for any closed $A \subset L$.*

**V.250. Fact 2.** *Given a locally convex space $L$ suppose that $M$ is a closed linear subspace of $L$ and $G$ is a finite-dimensional linear subspace of $L$. Then $M + G$ is a closed subspace of $L$.*

## 3.6   Zero-dimensional Spaces and Connected Spaces

A selection of results on connected spaces and zero-dimensional spaces is pre-sented here. Even though every space $C_p(X)$ is connected and hence never zero-dimensional, each of the above-mentioned properties turns out to be important in $C_p$-theory.

**V.097. Fact 1.** *Suppose that $X$ is a cosmic space such that $X \setminus \{a\}$ is zero-dimensional for some point $a \in X$. Then $X$ is zero-dimensional. In other words, adding a point to a cosmic zero-dimensional space gives a zero-dimensional space.*

**V.097. Fact 2.** *Suppose that $X$ is a second countable space and $F, G \subset X$ are disjoint closed subsets of $X$. Then, for any zero-dimensional $Z \subset X$, there is a partition $C$ between the sets $F$ and $G$ such that $C \cap Z = \emptyset$.*

**V.290. Fact 1.** *Suppose that $Z$ is a space and, for any points $x, y \in Z$, there exists a connected subspace $C \subset Z$ such that $\{x, y\} \subset C$. Then $Z$ is connected.*

**V.290. Fact 2.** *Suppose that $Z$ is a space and $C = \{C_t : t \in T\}$ is a family of connected subspaces of $Z$ such that $\bigcap C \neq \emptyset$. Then the set $C = \bigcup C$ is connected. As a consequence, the component of any point of $Z$ is a closed connected subspace of $Z$.*

**V.290. Fact 3.** *Given any space $Z$ and $z \in Z$ we have $K_z \subset Q_z$, i.e., the component of any point is contained in its quasi-component. If $Z$ is compact then $K_z = Q_z$.*

**V.290. Fact 4.** *Suppose that $Z$ is a connected compact space and $F$ is a nonempty closed subset of $Z$ such that $F \neq Z$. Then, for any $z \in F$, the component $K$ of the point $z$ in the space $F$ intersects the boundary $B = F \setminus \text{Int}(F)$ of the set $F$ in $Z$.*

**V.290. Fact 5.** *Suppose that $Z$ is a connected compact space and $\mathcal{A} = \{Z_n : n \in \omega\}$ is a disjoint family of closed subsets of $Z$ such that $Z = \bigcup_{n \in \omega} Z_n$. Then at most one element of $\mathcal{A}$ is nonempty.*

**V.298. Fact 1.** *A second countable zero-dimensional space is Čech-complete if and only if it embeds in $\mathbb{P}$ as a closed subspace.*

**V.495. Fact 1.** *Suppose that $Z$ is a zero-dimensional space and $T$ is a space for which there exists a closed finite-to-one map $\varphi : T \to Z$. Then $T$ is also zero-dimensional.*

## 3.7  Raznoie (Unclassified results)

Last but not least, we placed here some interesting results which do not fit in any of the earlier sections.

**V.023. Fact 3.** *Suppose that $X$ is a space such that $C_p(X)$ is Lindelöf. Given a set $A \subset X$ assume that every countable subset of $A$ is $C$-embedded in $X$. Then $A$ is $C$-embedded in $X$.*

**V.140. Fact 1.** *Given disjoint functionally closed sets $F$ and $G$ in a space $Z$ there exists a function $f \in C(Z, [0,1])$ such that $f(F) \subset \{0\}$ and $f(G) \subset \{1\}$. In particular, there exist functionally open sets $O(F)$ and $O(G)$ and disjoint functionally closed sets $P(F)$, $P(G)$ such that $F \subset O(F) \subset P(F)$ and $G \subset O(G) \subset P(G)$.*

**V.140. Fact 2.** *Suppose that $F$ is functionally closed in a space $Z$ and $O$ is a functionally open subset of $Z$ for which $F \subset O$. Then there is a functionally open set $G \subset Z$ such that $F \subset G \subset \overline{G} \subset O$.*

**V.147. Fact 1.** *Given a Tychonoff space $Z$, suppose that $F_1, \ldots, F_m$ are functionally closed subsets of $Z$ such that $F_1 \cap \ldots \cap F_m = \emptyset$. Then $cl_{\beta Z}(F_1) \cap \ldots \cap cl_{\beta Z}(F_m) = \emptyset$.*

**V.156. Fact 1.** *Given a space $Z$ and a natural number $m$ we have $\dim Z \leq m$ if and only if any functionally open cover $\mathcal{U} = \{U_0, \ldots, U_{m+1}\}$ of the space $Z$ has a functionally open shrinking $\mathcal{V} = \{V_0, \ldots, V_{m+1}\}$ such that $\bigcap_{i \leq m+1} V_i = \emptyset$.*

**V.245. Fact 1.** *Given a space $Z$, a set $Y \subset Z$ is not bounded in $Z$ if and only if there exists a discrete family $\{U_n : n \in \omega\} \subset \tau(Z)$ such that $U_n \cap Y \neq \emptyset$ for any $n \in \omega$.*

**V.246. Fact 1.** *If $Y$ is a Dieudonné complete space and $B$ is a bounded subset of $Y$ then $\overline{B}$ is compact.*

**V.291. Fact 1.** *Any paracompact locally Čech-complete space is Čech-complete.*

**V.291. Fact 2.** *Given a paracompact space $Z$ suppose that $D$ is a closed discrete subset of $Z$ and $\chi(d, Z) \leq \omega$ for each $d \in D$. If $Z \setminus D$ is Čech-complete then $Z$ is also Čech-complete.*

**V.301. Fact 1.** *Any space with at most one non-isolated point is hereditarily paracompact.*

**V.433. Fact 3.** *If $Z$ is a space and $\mathcal{U}$ is a weakly $\sigma$-point-finite family of nonempty open subsets of $Z$ then there is a $\sigma$-discrete subspace $D \subset Z$ such that $D \cap U \neq \emptyset$ for any $U \in \mathcal{U}$.*

**V.433. Fact 5.** *Suppose that $Z$ is a Lindelöf $\Sigma$-space and $\mathcal{F}$ is a fixed countable network with respect to a compact cover $\mathcal{C}$ of the space $Z$. Assume additionally that $\mathcal{F}$ is closed under finite intersections, and we have a condensation $f : Z \to Z'$. If $Y \subset Z$ and $A \subset Y$ is a set such that $f(A \cap F)$ is dense in $f(Y \cap F)$ for any $F \in \mathcal{F}$ then $Y \subset \overline{A}$.*

**V.433. Fact 6.** *If a Lindelöf $\Sigma$-space $Z$ condenses onto a hereditarily $d$-separable space then $Z$ is hereditarily $d$-separable.*

# Chapter 4
# Open problems

This Chapter contains 100 unsolved problems which are classified by topics presented in seven sections, the names of which outline what the given group of problems is about. At the beginning of each section we define the notions *which are not defined in the main text*. Each published problem has a reference to the respective paper or book. If it is unpublished, then my opinion on who is the author is expressed. The last part of each problem is a very brief explanation of its motivation and/or comments referring to the problems of the main text or some papers for additional information. If the paper is published and the background material is presented in the main text, we mention the respective exercises. If the main text contains no background, we refer the reader to the original paper. If no paper is mentioned in the motivation part, then the reader must consult the paper/book in which the unsolved problem was published.

To do my best to assign the right author to every problem, I implemented the following simple principles:

1. If the unsolved problem is published, then I cite the publication and consider myself not to be involved in the decision about who is the author. Some problems are published many times and I have generally preferred to cite the articles in journals/books which are more available for the Western reader. Thus it may happen that I do not cite the earliest paper where the problem was formulated. Of course, I mention it explicitly, if the author of the publication attributes the problem to someone else.

2. If, to the best of my knowledge, the problem is unpublished then I mention the author according to my personal records. The information I have is based upon my personal acquaintance and communication with practically all specialists in $C_p$-theory. I am aware that it is a weak point and it might happen that the problem I attributed to someone was published (or invented) by another person. However, I did an extensive work ploughing through the literature to make sure that this does not happen.

© Springer International Publishing Switzerland 2016
V.V. Tkachuk, *A Cp-Theory Problem Book*, Problem Books in Mathematics,
DOI 10.1007/978-3-319-24385-6_4

## 4.1   Mappings which involve $C_p$-spaces

The existence of an algebraic structure compatible with the topology of $C_p(X)$ radically improves its topological properties. In particular, if $C_p(X)$ is an open continuous image of some nice space $Z$ then $C_p(X)$ might have even better properties than $Z$. For example, if $C_p(X)$ is an open image of a metrizable space then it is second countable. Also, if a space $Z$ is a continuous image of $C_p(X)$ then we can expect very strong restrictions on $Z$ if, say, $Z$ is compact. A lot of research has been done in this area and this Section contains a compilation of the respective open questions.

**4.1.1** Suppose that $C_p(Y)$ is an image of $C_p(X)$ under a continuous (an open continuous) mapping. Is it true that $s(Y) \leq s(X)$?

**Published in** Okunev (1997a)
**Motivated by** the fact that spread is $t$-invariant (Problem 068)

**4.1.2** Suppose that $C_p(Y)$ is an image of $C_p(X)$ under a continuous (an open continuous) mapping. Is it true that $hd(Y) \leq hd(X)$?

**Published in** Okunev (1997a)
**Motivated by** the fact that hereditary density is $t$-invariant (Problem 069)

**4.1.3** Suppose that $C_p(Y)$ is an image of $C_p(X)$ under a continuous (an open continuous) mapping. Is it true that $hl(Y) \leq hl(X)$?

**Published in** Okunev (1997a)
**Motivated by** $t$-invariance of hereditary Lindelöf number (Problem 070)

**4.1.4** Assume that $X$ is Eberlein compact and $S$ is a dense subspace of $C_p(X)$. Suppose that $Y$ is a continuous image of $S$ and $B \subset Y$ is compact. Is it true that $B$ has countable tightness (is sequential or monolithic)?

**Published in** Tkachenko (1985)
**Motivated by** the fact that $t(B) \leq \omega_1$.

**4.1.5** Let $K$ be a compact space with $|K| \leq \mathfrak{c}$. Is it true in ZFC that every compact continuous image of $C_p(K)$ is metrizable?

**Published in** Tkachenko and Tkachuk (2005)
**Motivated by** the fact that this is true under Luzin's Axiom (Problem 452)

**4.1.6** Let $K$ be a first countable compact space. Is it true in ZFC that every compact continuous image of $C_p(K)$ is metrizable?

**Published in** Tkachenko and Tkachuk (2005)
**Motivated by** the fact that this is true under Luzin's Axiom (Problem 452)

**4.1.7** Let $K$ be a Fréchet–Urysohn compact space. Is it true that every compact continuous image of $C_p(K)$ is metrizable?

**Published in** Tkachenko and Tkachuk (2005)
**Comment** this is true under Luzin's Axiom if $K$ has countable character (Problem 452)

**4.1.8** Let $K$ be a compact space of countable tightness. Is it true that every compact continuous image of $C_p(K)$ is metrizable?

**Published in** Tkachenko and Tkachuk (2005)
**Comment** this is true under Luzin's Axiom if $K$ has countable character (Problem 452)

**4.1.9** Let $K$ be a hereditarily separable compact space. Is it true in ZFC that every compact continuous image of $C_p(K)$ is metrizable?

**Published in** Tkachenko and Tkachuk (2005)
**Motivated by** the fact that this is true under Luzin's Axiom (Problem 452)

**4.1.10** Suppose that $K$ is a compact space and $C_p(K)$ maps continuously onto $\mathbb{I}^\kappa$ for some uncountable cardinal $\kappa$. Is it true that the space $K$ contains a discrete $C^*$-embedded subset of cardinality $\kappa$?

**Published in** Tkachenko and Tkachuk (2005)
**Related to** Problem 451

**4.1.11** Let $K$ be a scattered (or left-separated) compact space. Is it true that every compact continuous image of $C_p(K)$ is metrizable?

**Published in** Tkachenko and Tkachuk (2005)
**Comment** this is true if $K$ is Corson compact (Problem 456)

**4.1.12** Let $X$ be a space such that $C_p(X)$ is Lindelöf. Is it true that every compact continuous image of $C_p(X)$ is metrizable?

**Published in** Tkachenko and Tkachuk (2005)
**Motivated by** the fact that it is true if $t(C_p(X)) = \omega$ (Problem 456)

**4.1.13** Let $X$ be a Lindelöf $P$-space. Is it true that every compact continuous image of $C_p(X)$ is metrizable?

**Published in** Tkachenko and Tkachuk (2005)

**4.1.14** Suppose that $K$ is an Eberlein compact space and $\varphi : C_p(K) \to Y$ is a continuous surjective map of $C_p(K)$ onto a $\sigma$-compact space $Y$. Must $Y$ be cosmic?

**Published in** Tkachuk (2009b)
**Motivated by** the fact that this is true if $\varphi$ is a condensation

**4.1.15** Suppose that $K$ is a Corson compact space and $\varphi : C_p(K) \to Y$ is a continuous surjective map of $C_p(K)$ onto a $\sigma$-compact space $Y$. Must $Y$ be cosmic?

**Published in** Tkachuk (2009b)
**Motivated by** the fact that this is true if $\varphi$ is a condensation

## 4.2   Properties preserved by *t*-equivalence

The main line of study in this area is to find out which common properties the
spaces $X$ and $Y$ must have if $C_p(X)$ is homeomorphic to $C_p(Y)$. In many cases the
respective open question reflects an attempt to check whether an $l$-invariant property
is also $t$-invariant.

**4.2.1** Suppose that $X$ and $Y$ are compact scattered $t$-equivalent spaces. Must $X$
and $Y$ be $u$-equivalent?

**Published in** Arhangel'skii (1992b)
**Comment** the answer is "yes" if $X$ and $Y$ are countable (Problem 200).

**4.2.2** Let $X$ and $Y$ be $t$-equivalent spaces. Is it true that dim $X = \dim Y$?

**Published in** Arhangel'skii (1990a)
**Motivated by** the fact that this is true if the spaces $X$ and $Y$ are $u$-
equivalent (Problems 180–182).

**4.2.3** Is $C_p(\mathbb{I})$ homeomorphic to $C_p(\mathbb{K})$?

**Published in** Arhangel'skii (1998b)
**Motivated by** the fact that a positive answer would show that the dimension
dim is not $t$-invariant. It is also an open question whether $C_p(\mathbb{I})$ is
homeomorphic to $C_p(\mathbb{I} \times \mathbb{I})$ (Arhángel'skii (1992b))

**4.2.4** Let $X$ be an infinite (compact) space. Is it true that $C_p(X)$ is homeomorphic
to $C_p(X) \times \mathbb{R}$?

**Published in** Arhangel'skii (1998b)
**Motivated by** the fact that, for any $n \in \mathbb{N}$, there is a space $Y$ such that
$C_p(X)$ is homeomorphic to $Y \times \mathbb{R}^n$.
**Related to** Problems 399 and 400

**4.2.5** Suppose that there exists a homeomorphism $\varphi : \mathbb{R}^X \to \mathbb{R}^Y$ for which we
have the equality $\varphi(C_p(X)) = C_p(Y)$. Is it true that dim $X = \dim Y$?

**Published in** Arhangel'skii (1992b)
**Related to** Problems 180–182, 054

**4.2.6** Let $X$ and $Y$ be $t$-equivalent spaces. Is it true that $l(X) = l(Y)$?

**Published in** Arhangel'skii (1992b)
**Motivated by** the fact that this is true if the spaces $X$ and $Y$ are $l$-equivalent
(Problem 500)

**4.2.7** Suppose that $X$ is $t$-equivalent to $Y$ and $X$ is a $k$-separable space. Must $Y$
be $k$-separable?

**Published in** Arhangel'skii (1989a)
**Motivated by** the fact that the answer is "yes" if $X = C_p(X')$ and $Y =
C_p(Y')$ for some spaces $X'$ and $Y'$ (Problem CFS-315).

**4.2.8** Let $X$ be a metrizable $\sigma$-compact space. Does there exist a compact space $Y$ such that $X$ is $t$-equivalent to $Y$?

**Published in** Arhangel'skii (1992b)
**Related to** Problem 043

**4.2.9** Let $X$ be a first countable space which is $t$-equivalent to a second countable space. Must $X$ be second countable?

**Published in** Arhangel'skii (1992b)
**Related to** Problem 348

**4.2.10** Suppose that $X$ is $t$-equivalent to $Y$ and $Y$ is $\sigma$-pseudocompact. Must $X$ be $\sigma$-pseudocompact?

**Published in** Arhangel'skii (1989a)
**Related to** Problems 043 and 045

**4.2.11** Suppose that $X$ is $t$-equivalent to $Y$ and $Y$ is $\sigma$-countably compact. Must $X$ be $\sigma$-countably compact?

**Published in** Arhangel'skii (1992b)
**Related to** Problems 043 and 045

**4.2.12** Suppose that $X$ is $t$-equivalent to $Y$ and $Y$ is a Hurewicz space. Must $X$ be a Hurewicz space?

**Published in** Arhangel'skii (1989a)
**Related to** Problems 014 and 043

**4.2.13** Does there exist an infinite compact space $X$ such that any space, which is $t$-equivalent to $X$, is compact (homeomorphic to $X$)?

**Published in** Arhangel'skii (1992b)
**Motivated by** the fact that there exists a space $Y$ such that $C_p(X) \simeq C_p(Y)$ implies that $X$ is pseudocompact (Problem TFS-400).

**4.2.14** Is it true that every infinite compact space is $t$-equivalent to a compact space containing a nontrivial convergent sequence?

**Published in** Arhangel'skii (1990a)
**Motivated by** the fact that there are compact spaces without nontrivial convergent sequences which are $l$-equivalent to compact spaces with nontrivial convergent sequences (Problem 270)

**4.2.15** Suppose that $X$ is $t$-equivalent to $Y$ and all countable subsets of $X$ are closed. Is it true that all countable subsets of $Y$ must be closed?

**Published in** Arhangel'skii (1983a)
**Motivated by** the fact that if all countable subsets of $X$ are closed and $C$-embedded then all countable subsets of the space $Y$ also have this property (Problem TFS-485).

**4.2.16** Given a cardinal $\kappa$, a space $X$ is called *$\kappa$-scattered*, if $|\overline{A}| \leq \kappa$ for any $A \subset X$ with $|A| \leq \kappa$. Suppose that $X$ is a $\kappa$-scattered space and $C_p C_p(X)$ is homeomorphic to $C_p C_p(Y)$. Must the space $Y$ be $\kappa$-scattered?

**Published in** Arhangel'skii (1989a)

**Motivated by** the fact that $Y \overset{t}{\sim} X$ implies $Y$ is $\kappa$-scattered (Problem SFFS-187).

**4.2.17** Say that a space $X$ is *a $k_\omega$-space* if it is a hemicompact $k$-space. Is it true that the $k_\omega$-property is preserved by $t$-equivalence?

**Published in** Arhangel'skii (1988a)
**Related to** Problem 340

**4.2.18** Suppose that a space $X$ is $t$-equivalent to a compact space. Must $X$ have the $k_\omega$-property? Must $X$ be $k$-space or hemicompact space?

**Published in** Arhangel'skii (1988a)
**Related to** Problem 340

**4.2.19** Suppose that a space $X$ is $t$-equivalent to $\mathbb{R}$. Must $X$ be $k_\omega$-space?

**Published in** Arhangel'skii (1988a)
**Related to** Problem 340

**4.2.20** Suppose that $X$ is $t$-equivalent to a $K_{\sigma\delta}$-space $Y$. Must $X$ be a $K_{\sigma\delta}$-space?

**Published in** Arhangel'skii (1988a)
**Motivated by** the fact that $K$-analyticity is $t$-invariant (Problem 043).

**4.2.21** Assume that $X$ and $Y$ are $t$-equivalent spaces. Is it true that the space $X \times Z$ is $t$-equivalent to $Y \times Z$ for any compact space $Z$?

**Published in** Okunev (1995b)
**Motivated by** the fact that it is true if $X \overset{l}{\sim} Y$.

**4.2.22** Suppose that $C_p(X)$ is homeomorphic to $C_p(Y)$. Must the space $L_p(X)$ be homeomorphic to $L_p(Y)$?

**Published in** Arhangel'skii and Choban (1989)
**Motivated by** the fact that this is true if $X \overset{l}{\sim} Y$ (Problem 237).

**4.2.23** Suppose that $L_p(X)$ is homeomorphic to $L_p(Y)$. Must $C_p(X)$ be homeomorphic to $C_p(Y)$?

**Published in** Arhangel'skii and Choban (1989)
**Motivated by** the fact that this is true if $L_p(X)$ is linearly homeomorphic to $L_p(Y)$ (Problem 237).

## 4.3   Properties preserved by $u$-equivalence

The main line of study in this area is to find out which common properties the spaces $X$ and $Y$ must have if $C_p(X)$ is uniformly homeomorphic to $C_p(Y)$. In many cases the respective open question reflects an attempt to check whether an $l$-invariant property is also $u$-invariant.

**4.3.1** Say that a space $X$ is *a $k_\omega$-space* if it is a hemicompact $k$-space. Is it true that the $k_\omega$-property is preserved by $u$-equivalence?

> **Published in** Arhangel'skii (1992b)
> **Related to** Problem 340

**4.3.2** Suppose that a space $X$ is $u$-equivalent to $\mathbb{R}$. Must $X$ be $k_\omega$-space?

> **Published in** Arhangel'skii (1988a)
> **Related to** Problem 340

**4.3.3** Is the class of $\aleph_0$-spaces $u$-invariant?

> **Published in** Arhangel'skii (1992b)
> **Related to** Problem 348

**4.3.4** Is the class of perfect spaces $u$-invariant?

> **Published in** Arhangel'skii (1992b)
> **Related to** Problem 070

**4.3.5** Suppose that there exists a homeomorphism $\varphi : \mathbb{R}^X \to \mathbb{R}^Y$ for which we have the equality $\varphi(C_p(X)) = C_p(Y)$. Is it true that $X$ is $u$-equivalent to $Y$?

> **Published in** Arhangel'skii (1992b)
> **Related to** Problem 139

**4.3.6** Let $X$ and $Y$ be $u$-equivalent metrizable spaces. Is it true that $X$ is Čech-complete if and only if so is $Y$?

> **Published in** Marciszewski and Pelant (1997)
> **Motivated by** the fact that this is true if the spaces $X$ and $Y$ are $l$-equivalent (Problems 366–367).

## 4.4  Properties preserved by $l$-equivalence

The topology of a space $X$ is completely determined by the structure of topological algebra on $C_p(X)$; so it is natural to ask what is implied by linear topological properties of $C_p(X)$. In other words, the main line of research is to find out what common properties the spaces $X$ and $Y$ must have if $C_p(X)$ is linearly homeomorphic to $C_p(Y)$. Recall that a space $X$ is $l$-stable if $X \overset{l}{\sim} X \oplus A$ for any $l$-embedded set $A \subset X$. The space $X$ is $S$-stable if $X \times (\omega + 1)$ is $l$-equivalent to $X$.

**4.4.1** Suppose that $X$ is a countably compact space and $X$ is $l$-equivalent to a space $Y$. Must $Y$ be countably compact?

**Published in** Arhangel'skii (1988a)
**Motivated by** by the fact that pseudocompactness is even $u$-invariant and hence $l$-invariant (Problems 134 and 193)

**4.4.2** Say that a space $X$ is $\kappa$-*pseudocompact* for some cardinal $\kappa$, if $f(X)$ is compact for any continuous $f : X \to Y$ such that $w(Y) \leq \kappa$. Is it true that $\kappa$-pseudocompactness is preserved by $l$-equivalence?

**Published in** Arhangel'skii (1998a)
**Motivated by** the fact that $\omega$-pseudocompactness is $l$-invariant because it coincides with pseudocompactness (Problems 134 and 193).

**4.4.3** Suppose that $X$ is a $k$-separable space and $X$ is $l$-equivalent to a space $Y$. Must $Y$ be $k$-separable?

**Published in** Arhangel'skii (1988a)
**Motivated by** the fact that the answer is positive if $X = C_p(X')$ and $Y = C_p(Y')$ for some spaces $X'$ and $Y'$ (Problem CFS-315).

**4.4.4** Is the class of perfect spaces $l$-invariant?

**Published in** Arhangel'skii (1992b)
**Motivated by** the fact that the hereditary Lindelöf property is known to be even $t$-invariant (Problem 070).

**4.4.5** Suppose that $X$ and $Y$ are metrizable spaces and each one embeds into the other as a closed subspace. Is it true that $X \overset{l}{\sim} Y$?

**Published in** Arhangel'skii (1992b)
**Motivated by** the fact that this is true if one of them is $l$-stable

**4.4.6** Suppose that $X$ and $Y$ are (compact, or compact metrizable) spaces and, for each one, there exists an $l$-embedding into the other. Is it true that $X \overset{l}{\sim} Y$?

**Published in** Arhangel'skii (1991)
**Motivated by** the fact that this is true if one of them is $l$-stable

**4.4.7** Is every infinite metrizable compact space $l$-stable (or $S$-stable)?

**Published in** Arhangel'skii (1991)
**Motivated by** the fact that $X \times (\omega + 1)$ is $S$-stable for any space $X$.

**4.4.8** Is every infinite metrizable space $l$-stable (or $S$-stable)?

**Published in** Arhangel'skii (1991)
**Motivated by** the fact that $X \times (\omega + 1)$ is $S$-stable for any space $X$.

**4.4.9** Is it true that every infinite $l$-stable (compact) space is $S$-stable?

**Published in** Arhangel'skii (1991)
**Comment** for metrizable spaces, $S$-stability implies $l$-stability.

**4.4.10** Suppose that $Y$ is a nonempty space with $Y \oplus Y \overset{l}{\sim} Y$. Is it true that $C_p(Y)$ is linearly homeomorphic to $C_p(Y) \times \mathbb{R}$?

**Published in** Arhangel'skii (1991)
**Motivated by** the fact that the conclusion is true for metrizable spaces.

**4.4.11** Suppose that $X$ is a $K_{\sigma\delta}$-space which is $l$-equivalent to a space $Y$. Must $Y$ be a $K_{\sigma\delta}$-space?

**Published in** Arhangel'skii (1988a)
**Motivated by** the fact that $K$-analyticity is $t$-invariant (Problem 043).

**4.4.12** Is it true that every infinite compact space is $l$-equivalent to a compact space containing a nontrivial convergent sequence?

**Published in** Arhangel'skii (1990a)
**Motivated by** the fact that there are compact spaces without nontrivial convergent sequences which are $l$-equivalent to compact spaces with nontrivial convergent sequences (Problem 270).

**4.4.13** Is it true that every infinite compact space $K$ is $l$-equivalent to a compact space containing a point of countable character?

**Published in** Arhangel'skii (1990a)

**4.4.14** Suppose that $X$ and $Y$ are (perfectly) normal $l$-equivalent spaces. Is it true that $ext(X) = ext(Y)$?

**Published in** Baars and Gladdines (1996)
**Motivated by** $l$-invariance of the Lindelöf property (Problem 500).

## 4.5   Generalizations of functional equivalences

Considering condensations or linear embeddings instead of linear homeomorphisms gives a natural weakening of $l$-equivalence; so there were numerous attempts to strengthen respectively the positive results on $l$-invariance of topological properties. To obtain the relevant results, sometimes, instead of $l$-equivalence of $X$ and $Y$ it suffices to assume that $C_p(X) \overset{l}{\sim} C_p(Y)$ or even $C_p(C_p(X)) \overset{l}{\sim} C_p(C_p(Y))$. The problems of this Section constitute a sample of what we could expect if we try to prove the counterparts of some well-known results for a weakened notion of $l$-equivalence.

**4.5.1** Given spaces $X$ and $Y$, suppose that $X$ is compact and there exists a linear condensation of $C_p(X)$ onto $C_p(Y)$ as well as a linear condensation of $C_p(Y)$ onto $C_p(X)$. Must $Y$ be compact?

**Published in** Arhangel'skii (1992b)
**Motivated by** the fact that compactness is $l$-invariant (Problem 193).

**4.5.2** Given spaces $X$ and $Y$, suppose that $X$ is Lindelöf and there exists a (linear) condensation of $C_p(X)$ onto $C_p(Y)$ as well as a (linear) condensation of $C_p(Y)$ onto $C_p(X)$. Must $Y$ be Lindelöf?

**Published in** Arhangel'skii (1992b)
**Motivated by** $l$-invariance of the Lindelöf property (Problem 500).

**4.5.3** Given spaces $X$ and $Y$, suppose that $X$ is $\sigma$-compact and there exists a (linear) condensation of $C_p(X)$ onto $C_p(Y)$ as well as a (linear) condensation of $C_p(Y)$ onto $C_p(X)$. Must $Y$ be $\sigma$-compact?

**Published in** Arhangel'skii (1992b)
**Motivated by** the fact that $\sigma$-compactness is $t$-invariant (Problem 043).

**4.5.4** Given spaces $X$ and $Y$, suppose that $X$ is a Lindelöf $\Sigma$-space and there exists a (linear) condensation of $C_p(X)$ onto $C_p(Y)$ as well as a (linear) condensation of $C_p(Y)$ onto $C_p(X)$. Must $Y$ be a Lindelöf $\Sigma$-space?

**Published in** Arhangel'skii (1992b)
**Motivated by** $t$-invariance of the Lindelöf $\Sigma$-property (Problem 043).

**4.5.5** Given spaces $X$ and $Y$, suppose that $X$ is analytic and there exists a (linear) condensation of $C_p(X)$ onto $C_p(Y)$ as well as a (linear) condensation of $C_p(Y)$ onto $C_p(X)$. Must $Y$ be analytic?

**Published in** Arhangel'skii (1992b)
**Motivated by** the fact that analyticity is $l$-invariant (Problem 044).

**4.5.6** Suppose that $X$ is Lindelöf and $C_p(Y)$ (linearly) embeds in $C_p(X)$. Must $Y$ be Lindelöf?

**Published in** Arhangel'skii (1992b)
**Motivated by** $l$-invariance of the Lindelöf property (Problem 500).

**4.5.7** Suppose that $X$ is Lindelöf and each one of the spaces $C_p(X)$ and $C_p(Y)$ (linearly) embeds in the other. Must $Y$ be Lindelöf?

**Published in** Arhangel'skii (1992b)
**Motivated by** $l$-invariance of the Lindelöf property (Problem 500).

**4.5.8** Suppose that $C_p(C_p(X))$ is homeomorphic to $C_p(C_p(Y))$ and the space $Y$ is ($\sigma$-)compact. Must $X$ be $\sigma$-compact?

**Published in** Arhangel'skii (1988a, 1989a)
**Motivated by** the fact that $\sigma$-compactness is $t$-invariant (Problem 043).

**4.5.9** Suppose that $C_p(C_p(X))$ is homeomorphic to $C_p(C_p(Y))$ and $Y$ is a Lindelöf $\Sigma$-space. Must $X$ be Lindelöf $\Sigma$-space?

**Published in** Arhangel'skii (1988a)
**Motivated by** $t$-invariance of the Lindelöf $\Sigma$-property (Problem 043).

**4.5.10** Suppose that $C_p(C_p(X))$ is homeomorphic to $C_p(C_p(Y))$ and $Y$ is discrete. Must $X$ be discrete?

**Published in** Arhangel'skii (1989a)
**Motivated by** the fact that discreteness is $t$-invariant (Problem TFS-487).

**4.5.11** Suppose that $C_p(C_p(X))$ is homeomorphic to $C_p(C_p(Y))$ and $Y$ is countable (has cardinality $\kappa$). Must $X$ be countable (have cardinality $\kappa$)?

**Published in** Arhangel'skii (1988a)
**Motivated by** the fact that cardinality is $t$-invariant (Problem 001).

**4.5.12** Suppose that $X$ is finite and $C_p(C_p(C_p(X)))$ is (linearly) homeomorphic to $C_p(C_p(C_p(Y)))$. Must $Y$ be finite?

**Published in** Arhangel'skii (1988a)
**Motivated by** the fact that discreteness is $t$-invariant (Problem TFS-487).

**4.5.13** Suppose that $X$ is a dyadic compact space such that $C_p(X)$ embeds in $C_p(Y)$ for some Eberlein compact $Y$. Must $X$ be metrizable?

**Published in** Arhangel'skii (1992b)
**Motivated by** the fact that the Eberlein property is $t$-invariant in compact spaces (Problem 033).

**4.5.14** Suppose that $X$ is a compact space such that $C_p(X)$ embeds in $C_p(Y)$ for some Eberlein compact $Y$. Must $X$ be Eberlein compact?

**Published in** Arhangel'skii (1992b)
**Motivated by** the fact that the Eberlein property is $t$-invariant in compact spaces (Problem 033).

**4.5.15** Suppose that $X$ is a compact space such that $C_p(X)$ embeds in $C_p(Y)$ for some Corson compact $Y$. Must $X$ be Corson compact?

**Published in** Arhangel'skii (1992b)
**Motivated by** the fact that the Corson property is $t$-invariant in compact spaces (Problem 033).

**4.5.16** Suppose that $X$ and $Y$ are compact spaces such that $C_{p,n}(X)$ is homeomorphic to $C_{p,n}(Y)$ for some $n \in \mathbb{N}$. Is it true that $C_p(X)$ is $K$-analytic if and only if $C_p(Y)$ is $K$-analytic?

**Published in** Arhangel'skii (1988a)
**Motivated by** the fact that the space $X$ is Gul'ko compact if and only if so is $Y$ (Problem CFS-243).

**4.5.17** Say that spaces $X$ and $Y$ are *at-equivalent* if there exists a homeomorphism $\varphi : \mathbb{R}^X \to \mathbb{R}^Y$ for which we have the equality $\varphi(C_p(X)) = C_p(Y)$. Suppose that $X$ and $Y$ are *at*-equivalent and $X$ is an $\aleph_0$-space. Is it true that $Y$ is also $\aleph_0$-space?

**Published in** Arhangel'skii (1992b)
**Motivated by** the fact that $\aleph_0$-property is $l$-invariant (Problem 348).

**4.5.18** Suppose that $C_p(X)$ is linearly homeomorphic to $C_p(Y) \times L$ for some linear topological space $L$. Is it true that $\dim Y \leq \dim X$?

**Published in** Arhangel'skii (1990a)
**Motivated by** the fact that this is true for compact spaces (Problem 396).

## 4.6  Fuzzy questions

A question belongs to this Section if it does not say exactly what its author wants us to prove or find out but rather expresses an intuitive idea of what should be done. Such questions can have many different solutions and their inherent difficulty consists in impossibility to be sure whether a given solution is satisfactory.

**4.6.1**  Given $n \in \omega$, find a characterization of the property "dim $X \le n$" in terms of the linear topological structure of $C_p(X)$.

**Published in** Arhangel'skii (1992b)
**Related to** Problems 180–182

**4.6.2**  Given $n \in \omega$, find a characterization of the property "dim $X \le n$" in terms of the uniform structure of $C_p(X)$.

**Published in** Arhangel'skii (1992b)
**Related to** Problems 180–182

**4.6.3**  Find a method of constructing $t$-equivalent but not $l$-equivalent spaces.

**Published in** Arhangel'skii (1989a)

**4.6.4**  Does there exist a natural (linear) topological property of the space $C_p(X)$ which characterizes Lindelöf spaces $X$?

**Published in** Arhangel'skii (1989a)
**Motivated by** the fact that Lindelöf property is $l$-invariant (Problem 500)

**4.6.5**  Characterize hereditary Lindelöf degree of $X$ by a well-formulated topological property of $C_p(X)$.

**Published in** Arhangel'skii (1998b)
**Motivated by** the fact that hereditary Lindelöf degree is $t$-invariant (Problem 070)

**4.6.6**  Characterize spread of $X$ by a topological property of $C_p(X)$.

**Published in** Arhangel'skii (1998b)
**Motivated by** the fact that spread is $t$-invariant (Problem 068)

**4.6.7**  Characterize hereditary density of a space $X$ by a topological property of $C_p(X)$.

**Published in** Arhangel'skii (1998b)
**Motivated by** the fact that hereditary density is $t$-invariant (Problem 069)

**4.6.8**  Characterize the spaces $X$ such that $C_p(X)$ embeds into some Lindelöf space $C_p(Y)$.

**Published in** Arhangel'skii (1998b)
**Comment** $\mathbb{R}^{\omega_1}$ is not embeddable in a Lindelöf space $C_p(Y)$ (Problem 023)

**4.6.9** For any $n \in \mathbb{N}$, characterize compact metrizable spaces $l$-equivalent to the cube $\mathbb{I}^n$.

**Published in** Arhangel'skii (1992b)
**Related to** Problem 299

**4.6.10** Characterize analyticity of $X$ by a topological property of $C_p(X)$.

**Published in** Arhangel'skii and Calbrix (1999)
**Motivated by** the fact that analyticity is $t$-invariant (Problem 044)

**4.6.11** In which classes of spaces is the extent $l$-invariant?

**Published in** Baars and Gladdines (1996)

## 4.7 Raznoie (unclassified questions)

It is usually impossible to completely classify a complex data set such as the open problems in $C_p$-theory. This last group of problems contains the open questions which do not fit into any previous Section.

**4.7.1** Let $X$ be a perfectly normal compact space. Is it true in ZFC that $C_p(X)$ is (hereditarily) metalindelöf?

> **Published in** Arhangel'skii (1998b)
> **Comment** This is true under CH (Arhangel'skii (1998b))
> **Related to** Problems 469 and 470

**4.7.2** Is it true that $C_p(X)$ is submetalindelöf for any compact space $X$?

> **Published in** Arhangel'skii (1997)
> **Motivated by** the fact that there exist compact spaces $X$ such that $C_p(X)$ is not metalindelöf (Problem 470)

**4.7.3** Let $X$ be a scattered compact space. Must $C_p(X)$ have a point-countable $\pi$-base?

> **Published in** Tkachuk (2005d)
> **Motivated by** the fact that this is true if $X$ is Corson compact (Problem 412).

**4.7.4** Is $C_p(\beta\omega)$ homeomorphic to $C_p(\beta\omega) \times \mathbb{R}^{\omega}$?

> **Author** V.V. Tkachuk
> **Related to** Problem 025

**4.7.5** Suppose that $C_p(X)$ has a uniformly dense monotonically $\omega$-monolithic subspace. Must $C_p(X)$ be monotonically $\omega$-monolithic?

> **Author** V.V. Tkachuk

**4.7.6** Suppose that $X$ is a monotonically monolithic compact space. Must $X$ have a dense metrizable subspace?

> **Published in** Tkachuk (2013b)
> **Motivated by** the fact that this is true for Gul'ko compact spaces (Problem TFS-293)

**4.7.7** Suppose that $X$ is a monotonically monolithic compact space such that $c(X) \le \omega$. Must $X$ be metrizable?

> **Published in** Tkachuk (2013b)
> **Motivated by** the fact that this is true for Gul'ko compact spaces (Problem CFS-294)

**4.7.8** Is it true that, for any compact $X$, the space $C_p(X)$ can be embedded in a Lindelöf $C_p(Y)$?

**Published in** Arhangel'skii (1998b)
**Motivated by** the fact that if $X$ is a discrete uncountable space then $C_p(X)$ cannot be embedded in a Lindelöf $C_p(Y)$ (Problem 023)

**4.7.9** Suppose that $X$ is a compact space such that $C_p(X)$ can be embedded in a Lindelöf $C_p(Y)$. Is it true that $t(X) = \omega$?

**Published in** Arhangel'skii (1998b)
**Motivated by** the fact that if $X$ is a discrete uncountable space then $C_p(X)$ cannot be embedded in a Lindelöf $C_p(Y)$ (Problem 023)

**4.7.10** Suppose that $\overline{D}$ has countable pseudocharacter for any discrete subspace $D \subset C_p(X)$. Must the space $X$ be separable, or, equivalently, is it true that $\psi(C_p(X)) = \omega$?

**Published in** Tkachuk (2015a)

**4.7.11** Suppose that $\overline{D}$ is Lindelöf for any discrete set $D \subset C_p(X) \times C_p(X)$. Must $C_p(X)$ be Lindelöf?

**Published in** Tkachuk (2015a)

**4.7.12** Suppose that the subspace $\overline{D}$ has the Fréchet–Urysohn property for any discrete set $D \subset C_p(X)$. Must the space $C_p(X)$ be Fréchet–Urysohn?

**Published in** Tkachuk (2015a)

**4.7.13** Assume that the subspace $\overline{D}$ is realcompact for any discrete set $D \subset C_p(X)$. Must the space $C_p(X)$ be realcompact?

**Published in** Tkachuk (2015a)

# Bibliography

My original intention was to reference *all* research publications on $C_p$-theory. However, once this work was started, the impossibility of this task became evident. Not only are there hundreds of journals and institutions where the results are being published, but it is too hard to even imagine how many papers on $C_p$-theory are in the process of publication at the present time. This book has been in preparation for more than ten years; so I reluctantly decided to be content with a *reasonably complete* bibliography. The selection below has 800 items; it covers the material of this book and can claim to reflect the state of the art of modern $C_p$-theory.

AFANAS, D.N., CHOBAN, M.M.
[2000]    *Topological properties of function spaces,* Bul. Acad. Stiinte Repub. Mold. Mat. **3**(2000), 28–52.

AHARONI, I., LINDENSTRAUSS, J.
[1978]    *Uniform equivalence between Banach spaces,* Bull. Amer. Math. Soc., **84**(1978), 281–283.

ALAS, O.T.
[1978]    *Normal and function spaces,* Topology, Vol. II, Colloq. Math. Soc. János Bolyai, 23, North-Holland, Amsterdam, 1980, 29–33.

ALAS, O.T., GARCIA-FERREIRA, S., TOMITA, A.H.
[1999]    *The extraresolvability of some function spaces,* Glas. Mat. Ser. III **34(54):1**(1999), 23–35.

ALAS, O.T., TAMARIZ-MASCARÚA, A.
[2006]    *On the Čech number of $C_p(X)$ II,* Questions Answers Gen. Topology **24:1**(2006), 31–49.

© Springer International Publishing Switzerland 2016
V.V. Tkachuk, *A Cp-Theory Problem Book*, Problem Books in Mathematics,
DOI 10.1007/978-3-319-24385-6

ALAS, O.T., TKACHUK, V.V., WILSON R.G.,
[2009]   *A broader context for monotonically monolithic spaces,* Acta Math.
         Hungarica, **125:4**(2009), 369–385.

ALSTER, K.
[1979]   *Some remarks on Eberlein compacta,* Fund. Math., **104:1**(1979),
         43–46.

ALSTER, K., POL, R.
[1980]   *On function spaces of compact subspaces of $\Sigma$-products of the real
         line,* Fund. Math., **107:2**(1980), 135–143.

AMIR, D., LINDENSTRAUSS, J.
[1968]   *The structure of weakly compact sets in Banach spaces,* Annals Math.,
         **88:1**(1968), 35–46.

ANDERSON, B.D.
[1973]   *Projections and extension maps in $C(T)$,* Illinois J. Math., **17**(1973),
         513–517.

ANGOA, J., TAMARIZ-MASCARÚA, A.
[2006]   *Spaces of continuous functions, $\Sigma$-products and box topology,*
         Comment. Math. Univ. Carolin. **47:1**(2006), 69–94.

ARENS, R.
[1946]   *A topology of spaces of transformations,* Annals Math., **47**(1946),
         480–495.
[1952]   *Extensions of functions on fully normal spaces,* Pacific J. Math.,
         **2**(1952), 11–22.

ARENS, R., DUGUNDJI, J.
[1951]   *Topologies for function spaces,* Pacific J. Math., **1:1**(1951), 5–31.

ARGYROS, S., FARMAKI, V.
[1985]   *On the structure of weakly compact subsets of Hilbert spaces and
         applications to the geometry of Banach spaces,* Trans. Amer. Math.
         Soc., **289**(1985), 409–427.

ARGYROS, S., ARVANITAKIS, A.D., MERCOURAKIS, S.
[2008]   *Talagrand's $K_{\sigma\delta}$ problem* Topology Appl. **155:15**(2008), 1737–1755.

ARGYROS, S.A., DODOS, P., KANELLOPOULOS, V.
[2008]   *A classification of separable Rosenthal compacta and its applications,*
         Dissertationes Math. (Rozprawy Mat.) **449**(2008).

ARGYROS, S.A., MANOUSSAKIS, A., PETRAKIS, M.
[2003]   *Function spaces not containing $l_1$,* Israel J. Math. **135**(2003), 29–81.

ARGYROS, S., MERCOURAKIS, S.
[1993] *On weakly Lindelöf Banach spaces,* Rocky Mountain J. Math., **23**(1993), 395–446.

ARGYROS, S., MERCOURAKIS, S., NEGREPONTIS, S.
[1983] *Analytic properties of Corson compact spaces,* General Topology and Its Relations to Modern Analysis and Algebra, 5. Berlin, 1983, 12–24.

ARGYROS, S., NEGREPONTIS, S.
[1983] *On weakly K-countably determined spaces of continuous functions,* Proc. Amer. Math. Soc., **87:4**(1983), 731–736.

ARHANGEL'SKII, A.V.
[1959] *An addition theorem for the weight of sets lying in bicompacta (in Russian),* DAN SSSR, **126**(1959), 239–241.
[1976] *On some topological spaces occurring in functional analysis (in Russian),* Uspehi Mat. Nauk, **31:5**(1976), 17–32.
[1978a] *On spaces of continuous functions with the topology of pointwise convergence (in Russian),* Doklady AN SSSR, **240:3**(1978), 506–508.
[1978b] *The structure and classification of topological spaces and cardinal invariants (in Russian),* Uspehi Mat. Nauk, **33:6**(1978), 29–84.
[1980] *On the relationship between the invariants of topological groups and their subspaces (in Russian),* Uspehi Mat. Nauk, **35:3**(1980), 3–22.
[1981] *Classes of topological groups (in Russian),* Uspehi Mat. Nauk, **36:3**(1981), 127–146.
[1982a] *On linear homeomorphisms of function spaces (in Russian),* Doklady AN SSSR, **264:6** (1982), 1289–1292.
[1982b] *A theorem on $\tau$-approximation and functional duality (in Russian),* Matem. Zametki, **31:3**(1982), 421–432.
[1982c] *Factorization theorems and function spaces: stability and monolithity (in Russian),* Doklady AN SSSR, **265:5**(1982), 1039–1043.
[1982d] *On relationship between topological properties of $X$ and $C_p(X)$,* General Topology and Its Relations to Modern Analysis and Algebra, 5. Berlin, 1982, 24–36.
[1983a] *Function spaces and completeness-like conditions (in Russian),* Vestnik Mosk. Univ., Math., Mech., **38:6**(1983), 4–9.
[1983b] *Topological properties of function spaces: duality theorems (in Russian),* Doklady AN SSSR, **269:6**(1983), 1289–1292.
[1983c] *Functional tightness, Q-spaces and $\tau$-embeddings,* Comment. Math. Univ. Carolinae, **24:1**(1983), 105–120.
[1984a] *Function spaces with the topology of pointwise convergence and compact spaces (in Russian),* Uspehi Mat. Nauk, **39:5**(1984), 11–50.
[1984b] *Continuous mappings, factorization theorems and function spaces (in Russian),* Trudy Mosk. Mat. Obsch., **47**(1984), 3–21.

[1985]   *Function spaces with the topology of pointwise convergence (in Russian)*, General Topology: Function Spaces and Dimension (in Russian), Moscow University P.H., 1985, 3–66.

[1986]   *Hurewicz spaces, analytic sets and fan tightness of function spaces (in Russian)*, Doklady AN SSSR, **287:3**(1986), 525–528.

[1987]   *A survey of $C_p$-theory*, Questions and Answers in General Topology, **5**(1987), 1–109.

[1988a]  *Some results and problems in $C_p(X)$-theory*, General Topology and Its Relations to Modern Analysis and Algebra, VI, Heldermann Verlag, Berlin, 1988, 11–31.

[1988b]  *Some problems and lines of investigation in general topology*, Comment. Math. Univ. Carolinae, **29:4**(1988), 611–629.

[1989a]  *Topological Function Spaces (in Russian)*, Moscow University P.H., Moscow, 1989.

[1989b]  *Hereditarily Lindelöf spaces of continuous functions*, Moscow University Math. Bull., **44:3**(1989), 67–69.

[1989c]  *On iterated function spaces*, Bull. Acad. Sci. Georgian SSR, **134:3**(1989), 481–483.

[1989d]  *On linear topological and topological classification of spaces $C_p(X)$*, Zbornik radova Filoz. fakult. u Nišu, Mat., **3**(1989), 3–12.

[1990a]  *Problems in $C_p$-theory*, in: Open Problems in Topology, North Holland, Amsterdam, 1990, 603–615.

[1990b]  *On the Lindelöf degree of topological spaces of functions, and on embeddings into $C_p(X)$*, Moscow University Math. Bull., **45:5**(1990), 43–45.

[1991]   *On linear topological classification of spaces of continuous functions in the topology of pointwise convergence*, Math., USSR–Sbornik, **70:1**(1991), 129–142.

[1992a]  *Topological Function Spaces (translated from Russian)*, Kluwer Academic Publishers, Dordrecht, 1992.

[1992b]  *$C_p$-theory*, in: Recent Progress in General Topology, North Holland, Amsterdam, 1992, 1–56.

[1995a]  *Spaces of mappings and rings of continuous functions*, in: General Topology, 3, Encyclopedia Math. Sci., Springer, Berlin, **51**(1995), 73–156.

[1995b]  *A generic theorem in the theory of cardinal invariants of topological spaces*, Comment. Math. Univ. Carolinae, **36:2**(1995), 303–325.

[1995c]  *General Topology III*, Encycl. Math. Sciences, **51**, Springer-Verlag, Berlin Heidelberg, New York, 1995.

[1996a]  *On Lindelöf property and spread in $C_p$-theory*, Topol. and Its Appl., **74:(1-3)**(1996), 83–90.

[1996b]  *On spread and condensations*, Proc. Amer. Math. Soc., **124:11**(1996), 3519–3527.

[1997]   *On a theorem of Grothendieck in $C_p$-theory*, Topology Appl. **80**(1997), 21–41.

[1998a]  *Some observations on $C_p$-theory and bibliography,*  Topology Appl., **89**(1998), 203–221.

[1998b]  *Embeddings in $C_p$-spaces,* Topology Appl., **85**(1998), 9–33.

[2000a]  *On condensations of $C_p$-spaces onto compacta,* Proc. Amer. Math. Soc., **128:6**(2000), 1881–1883.

[2000b]  *Projective $\sigma$-compactness, $\omega_1$-caliber and $C_p$-spaces,* Topology Appl., **104**(2000), 13–26.

[2002a]  *Relative normality and dense subspaces,* Topology Appl., **123**(2002), 27–36.

[2002b]  *Topological invariants in algebraic environment,* Recent Progress in Gen. Top. II, North-Holland, Amsterdam, (2002), 1–57.

[2002c]  *Erratum to: "On condensations of $C_p$-spaces onto compacta" [Proc. Amer. Amer. Math. Soc. 128:6(2000), 1881–1883],* Proc. Amer. Math. Soc. **130:6**(2002), 1875

[2005]   *D-spaces and covering properties,* Topology Appl., **146–147**(2005), 437–449.

ARHANGEL'SKII, A.V., BUZYAKOVA, R.Z.

[1999]   *On linearly Lindelöf and strongly discretely Lindelöf spaces,* Proc. Amer. Math. Soc., **127:8**(1999), 2449–2458.

[2002]   *Addition theorems and D-spaces,* Comment. Math. Univ. Carolinae, **43:4**(2002), 653–663.

ARHANGEL'SKII, A.V., CALBRIX, J.

[1999]   *A characterization of $\sigma$-compactness of a cosmic space $X$ by means of subspaces of $\mathbf{R}^X$,* Proc. Amer. Math. Soc., **127:8**(1999), 2497–2504.

ARHANGEL'SKII, A.V., CHOBAN M.M.

[1988]   *The extension property of Tychonoff spaces and generalized retracts,* Comptes Rendus Acad. Bulg. Sci., **41:12**(1988), 5–7.

[1989]   *$C_p(X)$ and some other functors in general topology. Continuous extenders,* in: Categorical Topology, World Scientific, London, 1989, 432–445.

[1990]   *On the position of a subspace in the whole space,* Comptes Rendus Acad. Bulg. Sci., **43:4**(1990), 13–15.

[1992]   *Extenders of Kuratowski–van Douwen and classes of spaces,* Comptes Rendus Acad. Bulg. Sci., **45:1**(1992), 5–7.

[1996]   *On continuous mappings of $C_p$-spaces and extenders,* Proc. Steklov Institute Math., **212**(1996), 28–31.

ARHANGEL'SKII, A.V., OKUNEV, O.G.

[1985]   *Characterization of properties of spaces by properties of their continuous images (in Russian)* Vestnik Mosk. Univ., Math., Mech., **40:5**(1985), 28–30.

ARHANGEL'SKII, A.V., PAVLOV, O.I.
[2002] *A note on condensations of $C_p(X)$ onto compacta*, Comment. Math. Univ. Carolinae, **43:3**(2002), 485–492.

ARHANGEL'SKII, A.V., PONOMAREV, V.I.
[1974] *Basics of General Topology in Problems and Exercises (in Russian)*, Nauka, Moscow, 1974.

ARHANGEL'SKII, A.V., SHAKHMATOV, D.B.
[1988] *On pointwise approximation of arbitrary functions by countable families of continuous functions (in Russian)*, Trudy Sem. Petrovsky, **13**(1988), 206–227.

ARHANGEL'SKII, A.V., SZEPTYCKI, P.J.
[1997] *Tightness in compact subspaces of $C_p$-spaces*, Houston J. Math., **23:1**(1997), 1–7.

ARHANGEL'SKII, A.V., TKACHUK, V.V.
[1985] *Function Spaces and Topological Invariants (in Russian)*, Moscow University P.H., Moscow, 1985.
[1986] *Calibers and point-finite cellularity of the spaces $C_p(X)$ and some questions of S. Gul'ko and M. Hušek*, Topology Appl., **23:1**(1986), 65–74.

ARHANGEL'SKII, A.V., USPENSKIJ, V.V.
[1986] *On the cardinality of Lindelöf subspaces of function spaces*, Comment. Math. Univ. Carolinae, **27:4**(1986), 673–676.

ASANOV, M.O.
[1979] *On cardinal invariants of spaces of continuous functions (in Russian)*, Modern Topology and Set Theory (in Russian), Izhevsk, 1979, N 2, 8–12.
[1980] *On spaces of continuous maps*, Izvestiia Vuzov, Mat., 1980, N 4, 6–10.
[1983] *About the space of continuous functions*, Colloq. Math. Soc. Janos Bolyai, **41**(1983), 31–34.

ASANOV, M.O., SHAMGUNOV, N.K.
[1981] *The topological proof of the Nachbin–Shirota's theorem*, Comment. Math. Univ. Carolinae, **24:4**(1983), 693–699.

ASANOV, M.O., VELICHKO, M.V.
[1981] *Compact sets in $C_p(X)$ (in Russian)*, Comment. Math. Univ. Carolinae, **22:2**(1981), 255–266.

BAARS, J.
[1992] *Equivalence of certain free topological groups*, Commentationes Math. Univ. Carolin. **33:1**(1992), 125–130.
[1993a] *Topological equivalence of function spaces*, Ann. N.Y. Acad. Sci., **704**(1993), 351–352.

[1993b]  *A note on linear mappings between function spaces,* Comment. Math. Univ. Carolinae, **34:4**(1993), 711–715.

[1993c]  *On the $l_p^*$-equivalence of certain locally compact spaces,* Topology Appl. **52:1**(1993), 43–57.

[1994]  *Function spaces of first countable paracompact spaces,* Bull. Polon. Acad. Sci., Math., **42**(1994), 29–35.

BAARS, J., DE GROOT, J.

[1988]  *An isomorphical classification of function spaces of zero-dimensional locally compact separable metric spaces,* Comment. Math. Univ. Carolinae, **29**(1988), 577–595.

[1991]  *On the l-equivalence of metric spaces,* Fund. Math., **137**(1991), 25–43.

[1992]  *On Topological and Linear Equivalence of Certain Function Spaces,* Center for Mathematics and Computer Sciences, Amsterdam, CWI Tract **86**(1992).

BAARS, J., DE GROOT, J., MILL, J.

[1989]  *A theorem on function spaces,* Proc. Amer. Math. Soc., **105:4**(1989), 1020–1024.

BAARS, J., DE GROOT, J., MILL, J. VAN, PELANT, J.

[1989]  *On topological and linear homeomorphisms of certain function spaces,* Topology Appl., **32**(1989), 267–277.

[1993]  *An example of $l_p$-equivalent spaces which are not $l_p^*$-equivalent,* Proc. Amer. Math. Soc., **119**(1993), 963–969.

BAARS, J., DE GROOT, J., PELANT, J.

[1993]  *Function spaces of completely metrizable spaces,* Trans. Amer. Mathem. Soc., **340:2**(1993), 871–879.

BAARS, J., GLADDINES, H.

[1996]  *On linear invariance of Lindelöf numbers,* Canadian Math. Bull., **39:2**(1996), 129–137.

BAARS, J., GLADDINES, H., MILL, J. VAN

[1993]  *Absorbing systems in infinite-dimesional manifolds,* Topology Appl., **50:2**(1993), 147–182.

BANACH, T., CAUTY, R.

[1997]  *Universalité forte pour les sous-ensembles totalement bornés. Applications aux espaces $C_p(X)$,* Colloq. Math., **73**(1997), 25–33.

BANDLOW, I.

[1991]  *A characterization of Corson compact spaces,* Commentationes Math. Univ. Carolinae, **32:3**(1991), 545–550.

[1994]  *On function spaces of Corson–compact spaces,* Comment. Math. Univ. Carolinae, **35:2**(1994), 347–356.

BARTLE, R.G.
 [1955]  *On compactness in functional analysis,* Trans. Amer. Math. Soc.,
         **79**(1955), 35–57.

BARTLE, R.G., GRAVES, L.M.
 [1952]  *Mappings between function spaces,* Trans. Amer. Math. Soc.,
         **72**(1952), 400–413.

BASHKIROV, A.I.
 [1972]  *Normality and bicompactness of spaces of mappings (in Russian),*
         Vestnik Mosk. Univ., Math., Mech., 1972, N 3, 77–79.

BATUROV D.P.
 [1987]  *On subspaces of function spaces (in Russian),* Vestnik Moskovsk.
         Univ., Math., Mech., **42:4**(1987), 66–69.
 [1988]  *Normality of dense subsets of function spaces,* Vestnik Moskovsk.
         Univ., Math., Mech., **43:4**(1988), 63–65.
 [1990]  *Normality in dense subspaces of products,* Topology Appl., **36**(1990),
         111–116.
 [1990]  *Some properties of the weak topology of Banach spaces,* Vestnik Mosk.
         Univ., Math., Mech., **45:6**(1990), 68–70.

BELL, M., MARCISZEWSKI, W.
 [2004]  *Function spaces on t-Corson compacta and tightness of polyadic
         spaces,* Czech. Math. J., **54(129)**(2004), 899–914.
 [2007]  *On scattered Eberlein compact spaces,* Israel J. Math. **158**(2007),
         217–224.

BELLA, A., YASCHENKO, I.V.
 [1999]  *On AP and WAP spaces,* Comment. Math. Univ. Carolinae,
         **40:3**(1999), 531–536.

BENYAMINI, Y., RUDIN, M.E., WAGE, M.
 [1977]  *Continuous images of weakly compact subsets of Banach spaces,*
         Pacific J. Mathematics, **70:2**(1977), 309–324.

BENYAMINI, Y., STARBIRD, T.
 [1976]  *Embedding weakly compact sets into Hilbert space,* Israel J. Math.,
         **23**(1976), 137–141.

BESSAGA, C., PELCZINSKI, A.
 [1960]  *Spaces of continuous functions, 4,* Studia Math., **19**(1960), 53–62.
 [1975]  *Selected Topics in Infinite-Dimensional Topology,* PWN, Warszawa,
         1975.

BLASKO, J.L.
 [1977]  *On $\mu$-spaces and $k_R$-spaces,* Proc. Amer. Math. Soc., **67:1**(1977),
         179–186.
 [1990]  *The $G_\delta$-topology and K-analytic spaces without perfect compact sets,*
         Colloq. Math., **58**(1990), 189–199.

BORGES, C.J.
[1966a] *On stratifiable spaces,* Pacific J. Math., **17:1**(1966), 1–16.
[1966b] *On function spaces of stratifiable spaces and compact spaces,* Proc. Amer. Math. Soc., **17**(1966), 1074–1078.

BOURGAIN, J.
[1977] *Compact sets of the first Baire class,* Bull. Soc. Math. Belg., **29:2**(1977), 135–143.
[1978] *Some remarks on compact sets of first Baire class,* Bull. Soc. Math. Belg., **30**(1978), 3–10.

BOURGAIN, J., FREMLIN, D.H., TALAGRAND, M.
[1978] *Pointwise compact sets of Baire-measurable functions,* Amer. J. Math., **100**(1978), 845–886.

BOURGAIN, J., TALAGRAND, M.
[1980] *Compacité extremal,* Proc. Amer. Math. Soc., **80**(1980), 68–70.

BOUZIAD, A.
[1990a] *Jeux topologiques et points de continuité d'une application separement continue,* Comptes Rendus de la Académie des Sciences, Paris, Ser. I, **310**(1990), 359–362.
[1990b] *Une classe d'espaces co-Namioka,* Comptes Rendus de la Académie des Sciences, Paris, Ser. I, **310**(1990), 779–782.
[1994] *Notes sur la propriété de Namioka,* Trans. Amer. Math. Soc., **344:2**(1994), 873–883.
[1996] *The class of co-Namioka spaces is stable under product,* Proc. Amer. Math. Soc., **123**(1996), 983–986.
[2001] *Le degré de Lindelöf est l-invariant,* Proc. Amer. Math. Soc., **129:3**(2001), 913–919.

BOUZIAD, A., CALBRIX, J.
[1995] *Images usco-compactes des espaces Čech-complets de Lindelöf,* C. R. Acad. Sci. Paris Sér. I Math. **320:7**(1995), 839–842.

BROWN, R.
[1964] *Function spaces and product topologies,* Oxford J. Math., **15:2**(1964), 238–250.

BUCHWALTER, H., SCHMETS, J.
[1973] *Sur quelques propriétés de l'espace $C_s(T)$,* J. Math. Pures Appl., **52:3**(1973), 337–352.

BURKE, D.K.
[1984] *Covering properties,* Handbook of Set-Theoretic Topology, ed. by K. Kunen and J.E. Vaughan, Elsevier Science Publishers B.V., 1984, 347–422.
[2007] *Weak-bases and D-spaces,* Comment. Math. Univ. Carolin. **48:2**(2007), 281–289.

BURKE, D.K., LUTZER, D.J.
[1976]   *Recent advances in the theory of generalized metric spaces,* in:
         Topology: Proc. Memphis State University Conference, Lecture Notes
         in Pure and Applied Math., Marcel Dekker, New York, 1976, 1–70.

BURKE, D.K., POL, R.
[2003]   *On Borel sets in function spaces with the weak topology,* Journal
         London Math. Soc. **(2)68:3**(2003), 725–738.
[2005]   *Note on separate continuity and the Namioka property,* Topology
         Appl. **152:3**(2005), 258–268.

BURKE, M.R., TODORCEVIC, S.
[1996]   *Bounded sets in topological vector spaces,* Math. Ann. **305:1**(1996),
         103–125.

BUROV, YU.A.
[1984]   *On mutual decompositions of weak topological bases of a topological
         vector space (in Russian),* Uspehi Mat. Nauk, **39:5**(1984), 237–238.
[1986a]  *On the properties of (weakly) l-equivalent spaces (in Russian).* Gen-
         eral Topology. Mappings of Topological Spaces (in Russian), Moscow
         University P.H., 1986, 13–19.
[1986b]  *On l- and M-equivalences,* Topics in Geometry and Topology,
         Petrozavodsk University P.H., Petrozavodsk, 1986, 3–13.

BUZYAKOVA, R.Z.
[1993]   *On splittable spaces,* Vestnik Moskov. Univ. Mat., Mekh., **48:6**(1993),
         83–84.
[1996]   *Splittability of compacta over linearly ordered spaces and over the
         class of all linearly ordered spaces,* Vestnik Moskov. Univ. Mat.,
         Mekh., **51:3**(1996), 81–84.
[2002]   *On D-property of strong $\Sigma$-spaces,* Comment. Math. Univer. Caroli-
         nae, **43:3**(2002), 493–495.
[2004a]  *In search for Lindelöf $C_p$'s,* Comment. Math. Univ. Carolinae,
         **45:1**(2004), 145–151.
[2004b]  *On cleavability of continua over LOTS,* Proc. Amer. Math. Soc.,
         **132:7**(2004), 2171–2181.
[2004c]  *Hereditary D-property of function spaces over compacta,* Proc. Amer.
         Math. Soc., **132:11**(2004), 3433–3439.
[2005]   *Cleavability of compacta over the two arrows,* Topology Appl., **150:
         1-3**(2005), 144–156.
[2006a]  *Spaces of continuous step functions over LOTS,* Fund. Math.,
         **192**(2006), 25–35.
[2006b]  *Spaces of continuous characteristic functions,* Comment. Math. Uni-
         versitatis Carolinae, **47:4**(2006), 599–608.
[2007]   *Function spaces over GO spaces,* Topology Appl., **154:4**(2007),
         917–924.

[2008]   *How sensitive is $C_p(X, Y)$ to changes in $X$ and/or $Y$?* Comment. Math. Univ. Carolin., **49:4**(2008), 657–665.

[2010]   *Injections into function spaces over ordinals,* Topology Appl., **157**(2010), 2844–2849.

[2012]   *More on injections into function spaces over ordinals,* Topology Appl., **159**(2012), 1573–1577.

CALBRIX, J.

[1985a]   *Classes de Baire et espaces d'applications continues,* Comptes Rendus Acad. Sci. Paris, Ser I, **301:16**(1985), 759–762.

[1985b]   *Espaces $K_\sigma$ et espaces des applications continues,* Bulletin Soc. Math. France, **113**(1985), 183–203.

[1987]   *Filtres sur les entiers et ensembles analytiques,* Comptes Rendus Acad. Sci. Paris, Ser I, **305**(1987), 109–111.

[1988]   *Filtres Boreliens sur l'ensemble des entiers et espaces des applications continues,* Rev. Roumaine Math. Pures et Appl., **33**(1988), 655–661.

[1996]   *k-spaces and Borel filters on the set of integers (in French),* Trans. Amer. Math. Soc., **348**(1996), 2085–2090.

CALBRIX, J., TROALLIC, J.

[1979]   *Applications separement continues,* C. R. Acad. Sci. Paris, Ser I, **288**(1979), 647–648.

CASARRUBIAS-SEGURA, F.

[1999]   *Realcompactness and monolithity are finitely additive in $C_p(X)$,* Topology Proc., **24**(1999), 89–102.

[2001]   *On compact weaker topologies in function spaces,* Topology Appl., **115**(2001), 291–298.

CASARRUBIAS-SEGURA, F., OKUNEV, O., PANIAGUA RAMÍREZ, C.G.

[2008]   *Some results on $L\Sigma(\kappa)$-spaces,* Comment. Math. Univer. Carolinae, **49:4**(2008), 667–675.

CASARRUBIAS-SEGURA, F., ROJAS-HERNÁNDEZ, R.

[2015]   *On some monotone properties,* Topology Appl., **182**(2015), 36–52.

CASCALES, B.

[1987]   *On K-analytic locally convex spaces,* Arch. Math., **49**(1987), 232–244.

CASCALES, B., KĄKOL, J., SAXON, S.A.

[2002]   *Weight of precompact subsets and tightness,* J. Math. Anal. Appl., **269**(2002), 500–518.

CASCALES, B., MANJABACAS, G, VERA, G.

[1998]   *Fragmentability and compactness in $C(K)$-spaces,* Studia Math., **131:1**(1998), 73–87.

CASCALES, B., NAMIOKA, I.
[2003]   *The Lindelöf property and σ-fragmentability,* Fund. Math., **180**(2003),
         161–183.

CASCALES, B., NAMIOKA, I., ORIHUELA, J.
[2003]   *The Lindelöf property in Banach spaces,* Studia Math., **154:2**(2003),
         165–192.

CASCALES, B., NAMIOKA, I., VERA, G.
[2000]   *The Lindelöf property and fragmentability,* Proceedings American
         Math. Society., **128:11**(2000), 3301–3309.

CASCALES, B., ONCINA, L.
[2003]   *Compactoid filters and USCO maps,* J. Math. Anal., Appl., **282**(2003),
         826–845.

CASCALES, B., ORIHUELA, J.
[1987]   *On compactness in locally convex spaces,* Math. Z., **195**(1987),
         365–381
[1988]   *On pointwise and weak compactness in spaces of continuous functions,*
         Bull. Soc. Math. Belg., **40:2**(1988), 331–352.
[1991a]  *A sequential property of set-valued maps,* J. Math. Analysis Appl.,
         **156:1**(1991), 86–100.
[1991b]  *Countably determined locally convex spaces,* Portugal. Math.
         **48:1**(1991), 75–89.

CASCALES, B., RAJA, M.
[2004]   *Bounded tightness for weak topologies,* Arch. Math., **82**(2004),
         324–334.

CASCALES, B., VERA, G.
[1994]   *Topologies weaker than the weak topology of a Banach space,* J. Math.
         Analysis Appl., **182:1**(1994), 41–68.

CAUTY, R.
[1974]   *Rétractions dans les espaces stratifiables,* Bull. Soc. Math. France
         **102**(1974), 129–149.
[1991]   *L'espace de fonctions continues d'un espace metrique denombrable,*
         Proc. Amer. Math. Soc., **113**(1991), 493–501.
[1998]   *La classe borélienne ne détermine pas le type topologique de $C_p(X)$,*
         Serdica Math. J. **24:3-4**(1998), 307–318.

CAUTY, R., DOBROWOLSKI, T., MARCISZEWSKI, W.
[1993]   *A contribution to the topological classification of the spaces $C_p(X)$,*
         Fundam. Math., **142**(1993), 269–301.

CHOBAN, M.M.
   [1998a] *General theorems on functional equivalence of topological spaces,* Topol. Appl., **89**(1998), 223–239.
   [1998b] *Isomorphism problems for the Baire function spaces of topological spaces,* Serdica Math. J. **24:1**(1998), 5–20.
   [1998c] *Isomorphism of functional spaces,* Math. Balkanica (N.S.) **12: 1-2**(1998), 59–91.
   [2001] *Functional equivalence of topological spaces,* Topology Appl., **111**(2001), 105–134.
   [2005] *On some problems of descriptive set theory in topological spaces,* Russian Math. Surveys **60:4**(2005), 699–719.

CHRISTENSEN, J.P.R.
   [1974] *Topology and Borel Structure,* North Holland P.C., Amsterdam, 1974.
   [1981] *Joint continuity of separably continuous functions,* Proceedings of the Amer. Math. Soc., **82:3**(1981), 455–461.

CHRISTENSEN, J.P.R., KENDEROV, P.S.
   [1984] *Dense strong continuity of mappings and the Radon–Nykodym property,* Math. Scand., **54:1**(1984), 70–78.

CIESIELSKI, K.
   [1993] *Linear subspace of $\mathbb{R}^\lambda$ without dense totally disconnected subsets,* Fund. Math. **142** (1993), 85–88.

CIESIELSKI, K., POL, R.
   [1984] *A weakly Lindelöf function space $C(K)$ without any continuous injection into $c_0(\Gamma)$,* Bull. Acad. Polon. Sci., Ser. Math., Astron. y Phys., **32:(11-12)**(1984), 681–688.

COMFORT, W.W.
   [1968] *On the Hewitt realcompactification of a product space,* Transactions Amer. Math. Soc. **131**(1968), 107–118.
   [1971] *A survey of cardinal invariants,* General Topology and Appl. **1:2**(1971), 163–199.
   [1977] *Ultrafilters: some old and some new results,* Bull. Amer. Math. Soc. **83:4**(1977), 417–455.
   [1988] *Cofinal families in certain function spaces,* Commentationes Math. Univ. Carolin. **29:4**(1988), 665–675.

COMFORT, W.W., FENG, L.
   [1993] *The union of resolvable spaces is resolvable,* Math. Japon. **38:3**(1993), 413–414.

COMFORT, W.W., HAGER, A.W.
   [1970a] *Estimates for the number of real-valued continuous functions,* Trans. Amer. Math. Soc., **150**(1970), 619–631.

[1970b]  *Dense subspaces of some spaces of continuous functions*, Math. Z. **114**(1970), 373–389.

[1970c]  *Estimates for the number of real-valued continuous functions*, Trans. Amer. Math. Soc. **150**(1970), 619–631.

COMFORT, W.W., NEGREPONTIS, S.A.

[1965]   *The ring $C(X)$ determines the category of $X$*, Proceedings Amer. Math. Soc. **16**(1965), 1041–1045.

[1966]   *Extending continuous functions on $X \times Y$ to subsets of $\beta X \times \beta Y$*, Fund. Math. **59**(1966), 1–12.

[1974]   The theory of ultrafilters. Die Grundl. Math. Wiss., **211**, Springer, New York, 1974.

[1982]   *Chain Conditions in Topology*, Cambridge Tracts in Mathematics, **79**, New York, 1982.

CONTRERAS-CARRETO, A., TAMARIZ-MASCARÚA, A.

[2003]   *On some generalizations of compactness in spaces $C_p(X, 2)$ and $C_p(X, \mathbb{Z})$*, Bol. Soc. Mat. Mexicana **(3)9:2**(2003), 291–308.

CORSON, H.H.

[1959]   *Normality in subsets of product spaces*, American J. Math., **81:3**(1959), 785–796.

[1961]   *The weak topology of a Banach space*, Trans. Amer. Math. Soc., **101:1**(1961), 1–15.

CORSON, H.H., LINDENSTRAUSS, J.

[1966a]  *On function spaces which are Lindelöf spaces*, Trans. Amer. Math. Soc., **121:2**(1966), 476–491.

[1966b]  *On weakly compact subsets of Banach spaces*, Proc. Amer. Math. Soc., **17:2**(1966), 407–412.

CÚTH, M.

[2014]   *Characterization of compact monotonically $(\omega)$-monolithic spaces using system of retractions*, Topology Appl., **171**(2014), 87–90.

DEBS, G.

[1985]   *Espaces K-analytiques et espaces de Baire de fonctions continues*, Mathematika, **32**(1985), 218–228.

[1986]   *Pointwise and uniform convergence on a Corson compact space*, Topol. Appl., **23:3**(1986), 299–303.

DEVILLE, R.

[1989]   *Convergence ponctuelle et uniforme sur un espace compact*, Bull. Acad. Polon. Sci., Math., **37**(1989), 7–12.

DEVILLE, R., GODEFROY, G.
[1993]   *Some applications of projectional resolutions of identity,* Proc. London Math. Soc., Ser. (3), **67:1**(1993), 183–199.

DIESTEL, J.
[1975]   *Geometry of Banach spaces — selected topics,* Lecture Notes Math., **11:485**(1975).

DIJKSTRA, J.
[2005]   *On homeomorphism groups and the compact-open topology,* Amer. Math. Monthly **112:10**(2005), 910–912.

DIJKSTRA, J., DOBROWOLSKI, T., MARCISZEWSKI, W., MILL, J. VAN, MOGILSKI, J.
[1990]   *Recent classification and characterization results in geometric topology,* Bull. Amer. Math. Soc., **22:2**(1990), 277–283.

DIJKSTRA, J., GRILLOT, T., LUTZER, D., VAN MILL, J.
[1985]   *Function spaces of low Borel complexity,* Proc. Amer. Math. Soc., **94:4**(1985), 703–710.

DIJKSTRA, J., MOGILSKI, J.
[1996]   $C_p(X)$-*representation of certain Borel absorbers,* Topology Proc., **16**(1991), 29–39.
[1996]   *The ambient homeomorphy of certain function spaces and sequence spaces,* Comment. Math. Univ. Carolinae, **37:3**(1996), 595–611.

DI MAIO, G., HOLÁ, L., HOLÝ, D., MCCOY, R.A.
[1998]   *Topologies on the space of continuous functions,* Topology Appl. **86:2**(1998), 105–122.

DIMOV, G.
[1987]   *Espaces d'Eberlein et espaces de type voisins,* Comptes Rendus Acad. Sci., Paris, Ser. I, **304:9**(1987), 233–235.
[1988]   *Baire subspaces of $c_0(\Gamma)$ have dense $G_\delta$ metrizable subsets,* Rend. Circ. Mat. Palermo (2) Suppl. **18**(1988), 275–285.
[2014]   *An internal topological characterization of the subspaces of Eberlein compacta and related compacta-I,* Topology Appl., **169**(2014), 71–86.

DIMOV, G., TIRONI, G.
[1987]   *Some remarks on almost radiality in function spaces,* Acta Univ. Carolin. Math. Phys. **28:2**(1987), 49–58.

DOBROWOLSKI, T., GUL'KO, S.P., MOGILSKI, J.
[1990]   *Function spaces homeomorphic to the countable product of $\ell_f^2$,* Topology Appl., **34**(1990), 153–160.

DOBROWOLSKI, T., MARCISZEWSKI, W.

[1995]   *Classification of function spaces with the pointwise topology deter-mined by a countable dense set,* Fund. Math., **148**(1995), 35–62.

[2002]   *Failure of the factor theorem for Borel pre-Hilbert spaces,* Fund. Math., **175**(2002), 53–68.

[2004]   *Infinite-dimensional topology,* in: Encycl. General Topology, Elsevier Sci. Publ., Amsterdam, 2004, 497–502.

DOBROWOLSKI, T., MARCISZEWSKI, W., MOGILSKI, J.

[1991]   *Topological classification of function spaces $C_p(X)$ of low Borel complexity,* Trans. Amer. Math. Soc., **328**(1991), 307–324.

DOBROWOLSKI, T., MOGILSKI, J.

[1992]   *Certain sequence and function spaces homeomorphic to the countable product of $\ell_f^2$,* J. London Math. Soc., **45:2**(1992), 566–576.

DOUWEN, E.K. VAN

[1975a]  *Simultaneous extension of continuous functions,* PhD Dissertation, **99**(1975), Amsterdam, Free University.

[1975b]  *Simultaneous linear extensions of continuous functions,* General Topology Appl., **5**(1975), 297–319.

[1984]   *The integers and topology,* Handbook of Set–Theoretic Topology, K. Kunen and J.E. Vaughan, editors, Elsevier Science Publishers B.V., 1984, 111–167.

DOUWEN, E.K. VAN, LUTZER, D.J., PRZYMUSINSKI, T.C.

[1977]   *Some extensions of the Tietze–Urysohn theorem,* American Math. Monthly, **84**(1977), 435–441

DOUWEN, E.K. POL, R.

[1977]   *Countable spaces without extension properties,* Bull. Polon. Acad. Sci., Math., **25**(1977), 987–991.

DOW, A.

[2005a]  *Closures of discrete sets in compact spaces,* Studia Sci. Math. Hungar. **42:2**(2005), 227–234.

[2005b]  *Property D and pseudonormality in first countable spaces,* Comment. Math. Univ. Carolin. **46:2**(2005), 369–372.

DOW, A., JUNNILA H., PELANT, J.

[1997]   *Weak covering properties of weak topologies,* Proceedings of the London Math. Soc., **(3)75:2**(1997), 349–368.

[2006]   *Coverings, networks and weak topologies,* Mathematika **53:2**(2006), 287–320.

DOW, A., PAVLOV, O.
  [2006]  *More about spaces with a small diagonal,* Fund. Math. **191:1**(2006), 67–80.
  [2007]  *Perfect preimages and small diagonal,* Topology Proc. **31:1**(2007), 89–95.

DOW, A., SIMON, P.
  [2006]  *Spaces of continuous functions over a Ψ-space,* Topology Appl. **153:13**(2006), 2260–2271.

DRANISHNIKOV, A.N.
  [1986]  *Absolute F-valued retracts and function spaces with the topology of pointwise convergence (in Russian),* Sibirsk. Mat. Zhurnal, **27**(1986), 74–86.

DUGUNDJI, J.
  [1951]  *An extension of Tietze's theorem,* Pacific J. Math., **1**(1951), 353–367.

EBERLEIN, W.F.
  [1947]  *Weak compactness in Banach spaces, I,* Proc. Nat. Acad. Sci. (USA), **33**(1947), 51–53.

EFIMOV, B.A.
  [1977]  *Mappings and imbeddings of dyadic spaces, I,* Math. USSR Sbornik, **32:1**(1977), 45–57.

ELEKES, M., KUNEN, K.
  [2002]  *Transfinite sequences of continuous and Baire class 1 functions,* Proc. Amer. Math. Soc., **131:8**(2002), 2453–2457.

ENGELKING, R.
  [1977]  *General Topology,* PWN, Warszawa, 1977.
  [1978]  *Dimension Theory,* PWN, Warszawa, 1978.

ESENIN–VOL'PIN, A.S.
  [1949]  *On the existence of a universal bicompactum of arbitrary weight (in Russian),* Dokl. Acad. Nauk SSSR, **68**(1949), 649–652.

FENG, Z., GARTSIDE, P.
  [2007]  *More stratifiable function spaces,* Topology Appl. **154:12**(2007), 2457–2461.
  [2013]  *Function spaces and local properties,* Fund. Math., **223**(2013), 207–223.

FERRANDO, J.C.
  [2009]  *Some characterizations for υX to be Lindelof Σ or K-analytic in terms of $C_p(X)$,* Topology Appl., **156**(2009), 823–830.

FERRANDO, J.C., KĄKOL, J.
    [2008]    *A note on spaces $C_p(X)$ K-analytic-framed in $\mathbb{R}^X$*, Bull. Aust. Math.
              Soc., **78**(2008), 141–146.

FLORET, K.
    [1980]    *Weakly Compact Sets,* Lecture Notes in Math., **801**(1980), Springer,
              Berlin.

FORT, M.K.
    [1951]    *A note on pointwise convergence,* Proc. Amer. Math. Soc., **2**(1951),
              34–35.

FOX, R.H.
    [1945]    *On topologies for function spaces,* Bull. Amer. Math. Soc., **51**(1945),
              429–432.

FREMLIN, D.H.
    [1977]    *K-analytic spaces with metrizable compacta,* Mathematika, **24**(1977),
              257–261.
    [1994]    *Sequential convergence in $C_p(X)$,* Comment. Math. Univ. Carolin.,
              **35:2**(1994), 371–382.

FULLER, R.V.
    [1972]    *Condition for a function space to be locally compact,* Proc. Amer.
              Math. Soc., **36**(1972), 615–617.

GALE, D.
    [1950]    *Compact sets of functions and function rings,* Proc. Amer. Math. Soc.,
              **1**(1950), 303–308.

GARCÍA, F., ONCINA, L., ORIHUELA, J.
    [2004]    *Network characterization of Gul'ko compact spaces and their rela-
              tives,* J. Math. Anal. Appl. **297:2**(2004), 791–811.

GARCÍA–FERREIRA, S., TAMARIZ–MASCARÚA, A.
    [1994]    *p-Fréchet–Urysohn property of function spaces,* Topology Appl.,
              **58:2**(1994), 157–172.
    [1994]    *p-sequential like properties in function spaces,* Comment. Math.
              Universitatis Carolinae, **35:4**(1994), 753–771.

GARTSIDE, P.
    [1997]    *Cardinal invariants of monotonically normal spaces,* Topology Appl.
              **77:3**(1997), 303–314.
    [1998]    *Nonstratifiability of topological vector spaces,* Topol. Appl.
              **86:2**(1998), 133–140.

GARTSIDE, P., GLYN, A.
[2005]   *Closure preserving properties of $C_k$(metric fan)*, Topology Appl. **151:1-3**(2005), 120–131.

GARTSIDE, P., LO, J.T.H., MARSH, A.
[2003]   *Sequential density*, Topology Appl., **130**(2003), 75–86.

GARTSIDE, P.M., REZNICHENKO, E.A.
[2000]   *Near metric properties of function spaces*, Fund. Math. **164:2**(2000), 97–114.

GERLITS, J.
[1974]   *On the depth of topological spaces*, Proc. Amer. Math. Soc. **44**(1974), 507–508.
[1980]   *Continuous functions on products of topological spaces*, Fund. Math. **106:1**(1980), 67–75.
[1983]   *Some properties of $C(X)$, II,*  Topology Appl., **15:3**(1983), 255–262.

GERLITS, J., NAGY, ZS.
[1982]   *Some properties of $C(X)$, I,* Topology Appl., **14:2**(1982), 151–161.

GERLITS, J., JUHÁSZ, I., SZENTMIKLÓSSY, Z.
[2005]   *Two improvements on Tkačenko's addition theorem*, Comment. Math. Univ. Carolin. **46:4**(2005), 705–710.

GERLITS, J., NAGY, ZS., SZENTMIKLOSSY, Z.
[1988]   *Some convergence properties in function spaces*, in: General Topology and Its Relation to Modern Analysis and Algebra, Heldermann, Berlin, 1988, 211–222.

GILLESPIE, D.C, HURWITZ, W.A.
[1930]   *On sequences of continuous functions having continuous limits*, Trans. Amer. Math. Soc., **32**(1930), 527–543.

GILLMAN, L., JERISON, M.
[1960]   *Rings of Continuous Functions*, D. van Nostrand Company Inc., Princeton, 1960.

GODEFROY, D.
[1980]   *Compacts de Rosenthal*, Pacific J. Math., **91:2**(1980), 293–306.

GODEFROY, D., TALAGRAND, M.
[1982]   *Espaces de Banach representables*, Israel J. Math., **41**(1982), 321–330.

GORÁK, R.
[1977]   *Function spaces on ordinals*, Comment. Math. Univ. Carolinae, **46:1**(2005), 93–103.

GORDIENKO, I.YU.
  [1990]   *Two theorems on relative cardinal invariants in $C_p$-theory,* Zb. Rad.
           Filozofskogo Fak. Nišu, Ser. Mat., **4**(1990), 5–7.

GRAEV, M.I.
  [1950]   *Theory of topological groups, I (in Russian),* Uspehi Mat. Nauk,
           **5:2**(1950), 3–56.

GRANADO, M., GRUENHAGE, G.
  [2006]   *Baireness of $C_k(X)$ for ordered $X$,* Comment. Math. Univ. Carolin.
           **47:1**(2006), 103–111.

GROTHENDIECK, A.
  [1952]   *Critères de compacité dans les espaces fonctionnels génereaux,* Amer.
           J. Math., **74**(1952), 168–186.
  [1953]   *Sur les applications linéaires faiblement compactes d'espaces du type
           $C(K)$,* Canadian J. Math., **5:2**(1953), 129–173.

GRUENHAGE, G.
  [1976]   *Infinite games and generalizations of first-countable spaces,* General
           Topology and Appl. **6:3**(1976), 339–352.
  [1984a]  *Covering properties of $X^2 \backslash \Delta$, W-sets and compact subspaces of
           $\Sigma$-products,* Topology Appl., **17:3**(1984), 287–304.
  [1984b]  *Generalized metric spaces,* Handbook of Set-Theoretic Topology,
           North-Holland, Amsterdam, 1984, 423–501.
  [1986a]  *Games, covering properties and Eberlein compacta,* Topol. Appl.,
           **23:3**(1986), 291–298.
  [1986b]  *On a Corson compact space of Todorcevic,* Fund. Math., **126**(1986),
           261–268.
  [1987]   *A note on Gul'ko compact spaces,* Proc. Amer. Math. Soc., **100**(1987),
           371–376.
  [1997]   *A non-metrizable space whose countable power is $\sigma$-metrizable,* Proc.
           Amer. Math. Soc. **125:6**(1997), 1881–1883.
  [1998]   *Dugundji extenders and retracts of generalized ordered spaces,*
           Fundam. Math., **158**(1998), 147–164.
  [2002]   *Spaces having a small diagonal,* Topology Appl., **122**(2002), 183–200.
  [2006a]  *A note on D-spaces,* Topology Appl., **153**(2006), 2229–2240.
  [2006b]  *The story of a topological game,* Rocky Mountain J. Math. **36:6**(2006),
           1885–1914.
  [2008]   *Monotonically compact and monotonically Lindelöf spaces,* Questions
           and Answers Gen. Topology **26:2**(2008), 121–130.
  [2011]   *A survey of D-spaces,* Contemporary Math. **533**(2011), 13–28.
  [2012]   *Monotonically monolithic spaces, Corson compacts, and D-spaces,*
           Topology Appl., **159**(2012), 1559–1564.

GRUENHAGE, G., MA, D.K.
[1997]   *Bairness of $C_k(X)$ for locally compact X*, Topology Appl., **80**(1997), 131–139.

GRUENHAGE, G., MICHAEL, E.
[1983]   *A result on shrinkable open covers*, Topology Proc., **8:1**(1983), 37–43.

GRUENHAGE, G., TAMANO, K.
[2005]   *If X is σ-compact Polish, then $C_k(X)$ has a σ-closure-preserving base*, Topology Appl. **151:1-3**(2005), 99–106.

GRUENHAGE, G., TSABAN, B., ZDOMSKYY, L.
[2011]   *Sequential properties of function spaces with the compact-open topology*, Topology Appl., **158**(2011), 387–391.

GUERRERO SÁNCHEZ, D.
[2010]   *Closure-preserving covers in function spaces*, Comment. Math. Univ. Carolinae, **51:4**(2010), 693–703.

GUERRERO SÁNCHEZ, D., TKACHUK, V.V.
[2012]   *Dense subspaces vs closure-preserving covers of function spaces*, Topology Proc., **39**(2012), 219–234.

GUL'KO, S.P.
[1977]   *On properties of subsets lying in Σ-products (in Russian)*, Doklady AN SSSR, **237:3**(1977), 505–507.

[1978]   *On properties of some function spaces (in Russian)*, Doklady AN SSSR, **243:4**(1978), 839–842.

[1979]   *On the structure of spaces of continuous functions and their hereditary paracompactness (in Russian)*, Uspehi Matem. Nauk, **34:6**(1979), 33–40.

[1981]   *On properties of function spaces (in Russian)*, Seminar Gen. Topol., Moscow University P.H., Moscow, 1981, 8–41.

[1988]   *The space $C_p(X)$ for countable infinite compact X is uniformly homeomorphic to $c_0$*, Bull. Acad. Polon. Sci., **36:(5-6)**(1988), 391–396.

[1990]   *Spaces of continuous functions on ordinals and ultrafilters (in Russian)*, Matem. Zametki, **47:4**(1990), 26–34.

[1993]   *Uniformly homeomorphic spaces of functions*, Proc. Steklov Inst. Math., **3**(1993), 87–93.

GUL'KO, S.P., KHMYLEVA, T.E.
[1986]   *The compactness is not preserved by the relation of t-equivalence (in Russian)*, Matem. Zametki, **39:6**(1986), 895–903.

GUL'KO, S.P., OKUNEV, O.G.
  [1986]   *Local compactness and M-equivalence (in Russian),* in: Topics in
           Geometry and Topology, Petrozavodsk University P.H., Petrozavodsk,
           1986, 14–23.

GUL'KO, S.P., OS'KIN, A.V.
  [1975]   *Isomorphic classification of spaces of continuous functions on well-
           ordered compact spaces (in Russian),* Functional Analysis Appl.,
           **9:1**(1975), 61–62.

GUL'KO, S.P., SOKOLOV, G.A.
  [1998]   *P-points in* $\mathbb{N}^*$ *and the spaces of continuous functions,* Topology
           Appl., **85**(1998), 137–142.
  [2000]   *Compact spaces of separately continuous functions in two variables,*
           Topology and Its Appl. **107:1-2**(2000), 89–96.

GUTHRIE, J.A.
  [1973]   *Mapping spaces and cs-networks,* Pacific J. Math., **47**(1973),
           465–471.
  [1974]   *Ascoli theorems and pseudocharacter of mapping spaces,* Bull.
           Austral. Math. Soc., **10**(1974), 403–408.

HAGER, A.W.
  [1969]   *Approximation of real continuous functions on Lindelöf spaces,* Proc.
           Amer. Math. Soc., **22**(1969), 156–163.

HAGLER, J.
  [1975]   *On the structure of S and C(S) for S dyadic,* Trans. Amer. Math. Soc.,
           **214**(1975), 415–428.

HANSARD, J.D.
  [1970]   *Function space topologies,* Pacific J. Math., **35**(1970), 381–388.

HANSEL, G., TROALLIC, J.P.
  [1992]   *Quasicontinuity and Namioka's theorem,* Topology Appl., **46**(1992),
           135–149.

HAO-XUAN, Z.
  [1982]   *On the small diagonals,* Topology Appl., **13:3**(1982), 283–293.

HART, J.E., KUNEN, K.
  [2002]   *Bohr topologies and compact function spaces,* Topology Appl.,
           **125**(2002), 183–198.
  [2005]   *Limits in function spaces and compact groups,* Topology Appl. **151:
           1-2**(2005), 157–168.
  [2006]   *Inverse limits and function algebras,* Topology Proc. **30:2**(2006),
           501–521.

HAYDON, R.G.

[1972] *Compactness in $C_s(T)$ and applications,* Publ. Dép. Math. Lyon, **9:1**(1972), 105–113.

[1978] *On dual $L^1$-spaces and injective bidual Banach spaces,* Israel J. Math., **31**(1978), 142–152.

[1990] *A counterexample to several questions about scattered compact spaces,* Bull. London Math. Soc., **22**(1990), 261–268.

[1994] *Countable unions of compact sets with the Namioka property,* Mathemat., **41**(1994), 141–144.

[1995] *Baire trees, bad norms and the Namioka property,* Mathematika, **42**(1995), 30–42.

HAYDON, R.G., ROGERS, C.A.

[1990] *A locally uniformly convex renorming for certain $C(K)$,* Mathematika, **37**(1990), 1–8.

HEATH, R.W.

[1973a] *Monotonically normal spaces,* Trans. Amer. Math. Soc., **178**(1973), 481–493.

[1973b] *Some comments on simultaneous extensions of mappings on closed subsets,* in: Topology Conference, Springer, Berlin, VPI **375**(1973), 114–119.

[1977] *Extension properties of generalized metric spaces,* Univ. North Carolina at Greensboro, 1977, 1–46.

HEATH, R.W., LUTZER, D.J.

[1974a] *The Dugundji extension theorem and collectionwise normality,* Bull. Acad. Polon. Sci., Ser. Math., **22**(1974), 827–830.

[1974b] *Dugundji extension theorem for linearly ordered spaces,* Pacific J. Math., **55**(1974), 419–425.

HEATH, R.W., LUTZER, D.J., ZENOR, P.L.

[1975] *On continuous extenders,* Studies in Topology, Academic Press, New York, 1975, 203–213.

HELMER, D.

[1981] *Criteria for Eberlein compactness in spaces of continuous functions,* Manuscripta Math., **35**(1981), 27–51.

HERNÁNDEZ RENDÓN, A., OKUNEV, O.

[2010] *On $L\Sigma(n)$-spaces: $G_\delta$-points and tightness,* Topology Appl., **157**(2010), 1491–1494.

HEWITT, E.

[1948] *Rings of real-valued continuous functions, I,* Trans. Amer. Math. Soc., **64:1**(1948), 45–99.

HEWITT, E., ROSS, K.
  [1963]   *Abstract Harmonic Analysis, Volume 1,* Springer–Verlag, Berlin, 1963.

HODEL, R.
  [1984]   *Cardinal Functions I,* in: Handbook of Set-Theoretic Topology, Ed. by
           K. Kunen and J.E. Vaughan, Elsevier Science Publishers B.V., 1984,
           1–61.

HRUŠAK, M., SZEPTYCKI, P.J., TAMARIZ-MASCARÚA, A.
  [2005]   *Spaces of continuous functions defined on Mrówka spaces,* Topol.
           Appl., **148**(2005), 239–252.

HUŠEK, M.
  [1972]   *Realcompactness of function spaces and* $\upsilon(P \times Q)$, General Topology
           and Appl. **2**(1972), 165–179.
  [1977]   *Topological spaces without κ-accessible diagonal,* Comment. Math.
           Univ. Carolinae, **18:4**(1977), 777–788.
  [1979]   *Mappings from products. Topological structures, II,* Math. Centre
           Tracts, Amsterdam, **115**(1979), 131–145.
  [1997a]  *Productivity of some classes of topological linear spaces,* Topology
           Appl. **80:1-2**(1997), 141–154.
  [1997b]  *Mazur-like topological linear spaces and their products,* Comment.
           Math. Univ. Carolin. **38:1**(1997), 157–164.
  [2000]   *Counterexample on products of bornological spaces,* Arch. Math.
           (Basel) **75:3**(2000), 217–219.
  [2005]   $C_p(X)$ *in coreflective classes of locally convex spaces,* Topol. Appl.
           **146/147**(2005), 267–278.

IVANOV, A.V.
  [1978]   *On bicompacta all finite powers of which are hereditarily separable,*
           Soviet Math., Doklady, **19:6**(1978), 1470–1473.

JARDÓN, D.
  [2002]   *Weakly Eberlein compact spaces,* Topology Proc., **26**(2001–2002),
           695–707.

JARDÓN, D., TKACHUK, V.V.
  [2002]   *Ultracompleteness in Eberlein–Grothendieck spaces,* Boletín de la
           Sociedad Mat. Mex., **10:3**(2004), 209–218.
  [2015]   *Splittability over some classes of Corson compact spaces,* Topol.
           Appl., **184**(2015), 41–49.

JAYNE, J.E.
  [1974]   *Spaces of Baire functions, 1,* Ann. Inst. Fourier, **24:4**(1974), 47–76.

JAYNE, J.E., NAMIOKA, I., ROGERS, C.A.
  [1990]  *Norm fragmented weak-star compact sets,* Collect. Math., **41**(1990), 133–163.
  [1992]  *σ-fragmentable Banach spaces,* Mathematika, **39**(1992), 161–188.

JOHDAI, N.
  [1989]  *Note on equivalent properties relating with $L_p(X)$,* Kobe J. Math., **6**(1989), 201–206.

JUHÁSZ, I.
  [1971]  *Cardinal functions in topology,* Mathematical Centre Tracts, **34**, Amsterdam, 1971.
  [1980]  *Cardinal Functions in Topology—Ten Years Later,* Mathematical Centre Tracts, North Holland P.C., Amsterdam, 1980.
  [1991]  *Cardinal functions,* Recent Progress in General Topology, North-Holland, Amsterdam, 1992, 417–441.
  [1992]  *The cardinality and weight-spectrum of a compact space,* Recent Devel. Gen. Topol. and Appl., Math. Research Berlin **67**(1992), 170–175.

JUHÁSZ, I., MILL, J. VAN
  [1981]  *Countably compact spaces all countable subsets of which are scattered,* Comment. Math. Univ. Carolin. **22:4**(1981), 851–855.

JUHÁSZ, I., SOUKUP, L., SZENTMIKLÓSSY, Z.
  [2007]  *First countable spaces without point-countable π-bases,* Fund. Math. **196:2**(2007), 139–149.

JUHÁSZ, I., SZENTMIKLÓSSY, Z.
  [1992]  *Convergent free sequences in compact spaces,* Proc. Amer. Math. Soc., **116:4**(1992), 1153–1160.
  [1995]  *Spaces with no smaller normal or compact topologies,* 1993), Bolyai Soc. Math. Stud., **4**(1995), 267–274.
  [2002]  *Calibers, free sequences and density,* Topology Appl. **119:3**(2002), 315–324.
  [2008]  *On d-separability of powers and $C_p(X)$,* Topology Appl. **155:4**(2008), 277–281.

JUHÁSZ, I., SZENTMIKLO'SSY, Z., SZYMANSKI, A.
  [2007]  *Eberlein spaces of finite metrizability number,* Comment. Mathem. Univ. Carolinae **48:2**(2007), 291–301.

JUST, W., SIPACHEVA, O.V., SZEPTYCKI, P.J.
  [1996]  *Non-normal spaces $C_p(X)$ with countable extent,* Proc. American Math. Society, **124:4**(1996), 1227–1235.

KADEC, M.I.
   [1967]   *A proof of topological equivalence of all separable infinite-dimensional Banach spaces (in Russian)*, Functional Analysis Appl., **1**(1967), 53–62.

KĄKOL, J.
   [2000]   *Remarks on linear homeomorphisms of function spaces*, Acta Math. Hungarica, **89**(2000), 315–319.

KĄKOL, J., KUBIŚ, W., LÓPEZ-PELLICER, M.
   [2011]   *Descriptive Topology in Selected Topics of Functional Analysis*, Developments in Mathematics, **24**, Springer, New York, 2011.

KĄKOL, J., LÓPEZ-PELLICER, M.
   [2003]   *On countable bounded tightness for spaces $C_p(X)$*, J. Math. Anal. Appl., **280**(2003), 155–162.

KĄKOL, J., LÓPEZ-PELLICER, M., OKUNEV, O.
   [2014]   *Compact covers and function spaces*, J. Math. Anal. Appl., **411**(2014), 372–380.

KALAMIDAS, N.D.
   [1985]   *Functional properties of $C(X)$ and chain conditions on $X$*, Bull. Soc. Math. Grèce **26**(1985), 53–64.
   [1992]   *Chain condition and continuous mappings on $C_p(X)$*, Rendiconti Sem. Mat. Univ. Padova, **87**(1992), 19–27.

KALAMIDAS, N.D., SPILIOPOULOS, G.D.
   [1992]   *Compact sets in $C_p(X)$ and calibers,* Canadian Math. Bull., **35:4**(1992), 497–502.

KALENDA, F.K.O.
   [2004]   *Note on countable unions of Corson countably compact spaces*, Comment. Math. Univ. Carolin., **45:3**(2004), 499–507.

KAUL, S.K.
   [1969]   *Compact subsets in function spaces*, Canadian Math. Bull., **12**(1969), 461–466.

KAWAMURA, K., MORISHITA, K.
   [1996]   *Linear topological classification of certain function spaces on manifolds and CW*, Topology Appl., **69:3**(1996), 265–282.

KENDEROV, P.S.
   [1980]   *Dense strong continuity of pointwise continuous mappings*, Pacific Journal Math., **89**(1980), 111–130.
   [1987]   *$C(T)$ is weak Asplund for every Gul'ko compact $T$*, Comptes Rendus Acad. Bulg. Sci., **40:2**(1987), 17–19.

KHAN, L.A., MORISHITA, K.
[1996]    *The minimal support for a functional on $C^b(X)$*, Topology Appl., **73**(1996), 285–294.

KHMYLEVA, T.E.
[1979]    *Classification of function spaces on segments of ordinals (in Russian)*, Siberian Math. Journal, **20:3**(1979), 624–631.

KHURANA, S.S.
[1981]    *Pointwise compactness on extreme points*, Proc. Amer. Math. Soc., **83**(1981), 347–348.

KISLYAKOV, S.V.
[1975]    *Isomorphic classification of spaces of continuous functions on ordinals (in Russian)*, Sibirsk. Mat. Zhurnal, **16:3**(1975), 293–300.

KOČINAC, L.D., SCHEEPERS, M.
[1999]    *Function spaces and strong measure zero sets*, Acta Math. Hungar. **82:4**(1999), 341–351.
[2002]    *Function spaces and a property of Reznichenko*, Topology Appl. **123:1**(2002), 135–143.

KOLMOGOROV, A.N.
[1957]    *On representation of continuous functions of several variables as a superposition of continuous functions of one variable and summing (in Russian)*, DAN SSSR, **114:5**(1957), 953–956.

KOROVIN, A.V.
[1992]    *Continuous actions of pseudocompact groups and axioms of topological groups*, Comment. Math. Univ. Carolinae, **33**(1992), 335–343.

KOYAMA, A., OKADA, T.
[1987]    *On compacta which are l-equivalent to $I^n$*, Tsukuba J. Math., **11:1**(1987), 147–156.

KRAWCZYK, A., MARCISZEWSKI, W., MICHALEWSKI, H.
[2009]    *Remarks on the set of $G_\delta$-points in Eberlein and Corson compact spaces*, Topology Appl., **156**(2009), 1746–1748.

KRIVORUCHKO, A.I.
[1972]    *On the cardinality of the set of continuous functions*, Soviet Math., Dokl., **13**(1972), 1364–1367.
[1973]    *On cardinal invariants of spaces and mappings*, Soviet Math., Doklady, **14**(1973), 1642–1647.
[1975]    *The cardinality and density of spaces of mappings*, Soviet Math., Doklady, **16**(1975), 281–285.

KRUPSKI, M.
  [2015]   *A note on condensations of function spaces onto σ-compact and analytic spaces,* Proc. Amer. Math. Soc., **143**(2015), 2263–2268.

KUBIS, W., LEIDERMAN, A.
  [2004]   *Semi-Eberlein spaces,* Topology Proc. **28:2**(2004), 603–616.

KUBIS, W., OKUNEV, O., SZEPTYCKI, P.J.
  [2006]   *On some classes of Lindelof Σ-spaces,* Topology Appl., **153**(2006), 2574–2590.

KUNDU, S., MCCOY, R.A.
  [1993]   *Topologies between compact and uniform convergence on function spaces,* Internat. J. Math. Math. Sci. **16**(1993), 101–110.
  [1995]   *Weak and support-open topologies on $C(X)$,* Rocky Mountain J. Math. **25:2**(1995), 715–732.

KUNDU, S., MCCOY, R.A., OKUYAMA, A.
  [1989]   *Spaces of continuous linear functionals on $C_k(X)$,* Math. Japon. **34:5**(1989), 775–787.

KUNDU, S., MCCOY, R.A., RAHA, A.B.
  [1993]   *Topologies between compact and uniform convergence on function spaces II,* Real Anal. Exchange **18:1**(1992/93), 176–189.

KUNDU, S., OKUYAMA, A.
  [1993]   *Complete duals of $C^*(X)$,* Math. Scand. **72:1**(1993), 33–46.

KUNEN, K.
  [1980]   *Set Theory. An Introduction to Independence Proofs,* Studies Logic Found. Mathematics, **102**(1980), North Holland P.C., Amsterdam, 1980
  [1981]   *A compact L-space under CH,* Topology Appl., **12**(1981), 283–287.
  [1998]   *Bohr topologies and partition theorems for vector spaces,* Topology Appl. **90:1-3**(1998), 97–107.

KUNEN, K., MILL, J. VAN
  [1995]   *Measures on Corson compact spaces,* Fund. Math., **147**(1995), 61–72.

KUNEN, K., DE LA VEGA, R.
  [2004]   *A compact homogeneous S-space,* Topology Appl., **136**(2004), 123–127.

KURATOWSKI, C.
  [1966]   *Topology, vol. 1,* Academic Press Inc., London, 1966.

LAMBRINOS, P.
[1981] *The bounded-open topology on function spaces,* Manuscripta Math., **36**(1981), 47–66.

LEE, J.P., PIOTROWSKI, Z.
[1985] *A note on spaces related to Namioka spaces,* Bull. Austral. Math. Soc., **31**(1985), 285–292.

LEIDERMAN, A.G.
[1984] *On properties of spaces of continuous functions (in Russian),* Cardinal Invariants and Mappings of Topological Spaces (in Russian), Izhevsk, 1984, 50–54.
[1985] *On dense metrizable subspaces of Corson compact spaces (in Russian),* Matem. Zametki, **38:3**(1985), 440–449.
[1988] *Adequate families of sets and function spaces,* Commentationes Math. Univ. Carolinae, **29:1**(1988), 31–39.

LEIDERMAN, A., LEVIN, M., PESTOV, V.
[1997] *On linear continuous open surjections of the spaces $C_p(X)$,* Topol. Appl., **81**(1997), 269–279.

LEIDERMAN, A.G., MALYKHIN, V.I.
[1988] *On nonpreservation of final compactness by products of spaces $C_p(X)$ (in Russian),* Sibirsk. Mat. Zhurnal, **29:1**(1988), 84–93.

LEIDERMAN, A.G., MORRIS, S.A., PESTOV, V.G.
[1997] *The free Abelian topological group and the free locally convex space on the unit interval,* J. London Math. Soc., **56**(1997), 529–538.

LEIDERMAN, A.G., SOKOLOV, G.A.
[1984] *Adequate families of sets and Corson compacts,* Commentat. Math. Univ. Carolinae, **25:2**(1984), 233–246.

LELEK, A.
[1969] *Some cover properties of spaces,* Fund. Math., **64:2**(1969), 209–218.

LINDENSTRAUSS, J.
[1972] *Weakly compact sets — their topological properties and the Banach spaces they generate,* Annals Math. Studies, **69**(1972), 235–273.

LINDENSTRAUSS, J., STEGALL, C.
[1975] *Examples of separable spaces which do not contain $\ell^1$ and whose duals are nonseparable,* Studia Math., **54**(1975), 81–105.

LINDENSTRAUSS, J., TZAFRIRI, L.
[1977] *Classical Banach Spaces I,* Springer, Berlin, 1977.

LUTZER, D.J., MCCOY, R.A.
[1980]   *Category in function spaces I,* Pacific J. Math., **90:1**(1980), 145–168.

LUTZER, D.J., MILL, J. VAN, POL, R.
[1985]   *Descriptive complexity of function spaces,* Trans. Amer. Math. Soc., **291**(1985), 121–128.

LUTZER, D. J., MILL, J. VAN, TKACHUK, V.V.
[2008]   *Amsterdam properties of $C_p(X)$ imply discreteness of $X$,* Canadian Math. Bull. **51:4**(2008), 570–578.

MA, D.K.
[1993]   *The Cantor tree, the $\gamma$-property, and Baire function spaces,* Proc. Amer. Math. Soc., **119:3**(1993), 903–913.

MALYKHIN, V.I.
[1987]   *Spaces of continuous functions in simplest generic extensions,* Math. Notes, **41**(1987), 301–304.
[1994]   *A non-hereditarily separable space with separable closed subspaces,* Q & A in General Topology, **12**(1994), 209–214.
[1998]   *On subspaces of sequential spaces (in Russian),* Matem. Zametki, **64:3**(1998), 407–413.
[1999]   *$C_p(I)$ is not subsequential,* Comment. Math. Univ. Carolinae, **40:4**(1999), 785–788.

MALYKHIN, V.I., SHAKHMATOV, D.B.
[1992]   *Cartesian products of Fréchet topological groups and function spaces,* Acta Math. Hungarica, **60**(1992), 207–215.

MARCISZEWSKI, W.
[1983]   *A pre-Hilbert space without any continuous map onto its own square,* Bull. Acad. Polon. Sci., **31:(9-12)**(1983), 393–397.
[1987]   *Lindelöf property in function spaces and a related selection theorem,* Proc. Amer. Math. Soc., **101**(1987), 545–550.
[1988a]  *A remark on the space of functions of first Baire class,* Bull. Polish Acad. Sci., Math., **36:(1-2)**(1997), 65–67.
[1988b]  *A function space $C(K)$ not weakly homeomorphic to $C(K) \times C(K)$,* Studia Mathematica, **88**(1988), 129–137.
[1989]   *On classification of pointwise compact sets of the first Baire class functions,* Fund. Math., **133**(1989), 195–209.
[1991]   *Order types, calibers and spread of Corson compacta,* Topology Appl., **42**(1991), 291–299.
[1992]   *On properties of Rosenthal compacta,* Proc. Amer. Math. Soc., **115**(1992), 797–805.
[1993]   *On analytic and coanalytic function spaces $C_p(X)$,* Topology and Its Appl., **50**(1993), 241–248.

[1995a] *On universal Borel and projective filters,* Bull. Acad. Polon. Sci., Math., **43:1**(1995), 41–45.

[1995b] *A countable X having a closed subspace A with $C_p(A)$ not a factor of $C_p(X)$,* Topology Appl., **64**(1995), 141–147.

[1995c] *On sequential convergence in weakly compact subsets of Banach spaces,* Studia Math., **112:2**(1995), 189–194.

[1997a] *A function space $C_p(X)$ not linearly homeomorphic to $C_p(X) \times R$,* Fundamenta Math., **153:2**(1997), 125–140.

[1997b] *On hereditary Baire products,* Bull. Polish Acad. Sci., Math., **45:3**(1997), 247–250.

[1997c] *On topological embeddings of linear metric spaces,* Math. Ann., **308:1**(1997), 21–30.

[1998a] *P-filters and hereditary Baire function spaces,* Topology Appl., **89**(1998), 241–247.

[1998b] *Some recent results on function spaces $C_p(X)$,* Recent Progress in Function Spaces., Quad. Mat. **3**(1998), Aracne, Rome, 221–239.

[1998c] *On van Mill's example of a normed X with $X \not\approx X \times R$,* Proc. Amer. Math. Society, **126:1**(1998), 319–321.

[2000] *On properties of metrizable spaces X preserved by t-equivalence,* Mathematika, **47**(2000), 273–279.

[2002] *Function Spaces,* in: Recent Progress in General Topology II, Ed. by M. Hušek and J. van Mill, Elsevier Sci. B.V., Amsterdam, 2002, 345–369.

[2003a] *A function space $C_p(X)$ without a condensation onto a σ-compact space,* Proc. Amer. Math. Society, **131:6**(2003), 1965–1969.

[2003b] *On Banach spaces $C(K)$ isomorphic to $c_0(\Gamma)$,* Studia Math., **156:3**(2003), 295–302.

[2004] *Rosenthal compacta,* in: Encycl. General Topology, Elsevier Sci. Publ., Amsterdam, 2004, 142–144.

MARCISZEWSKI, W., MILL, J. VAN
[1998] *An example of $t_p^*$-equivalent spaces which are not $t_p$-equivalent,* Topology Appl., **85**(1998), 281–285.

MARCISZEWSKI, W., PELANT, J.
[1997] *Absolute Borel sets and function spaces,* Trans. Amer. Math. Soc., **349:9**(1997), 3585–3596.

MARCISZEWSKI, W., POL, R.
[2009] *On Banach spaces whose norm-open sets are $F_\sigma$-sets in the weak topology,* J. Math. Anal. Appl., **350:2**(2009), 708–722.

[2010] *On some problems concerning Borel structures in function spaces,* Revista de la Real Acad. Cien. (RACSAM), **104:2**(2010), 327–335.

MÁTRAI, T.

[2004]    *A characterization of spaces l-equivalent to the unit interval,* Topology
          Appl., **138**(2004), 299–314.

MCCOY, R.A.

[1975]    *First category function spaces under the topology of pointwise conver-*
          *gence,* Proc. Amer. Math. Soc., **50**(1975), 431–434.

[1978a]   *Characterization of pseudocompactness by the topology of uniform*
          *convergence on function spaces,* J. Austral. Math. Soc., **26**(1978),
          251–256.

[1978b]   *Submetrizable spaces and almost σ-compact function spaces,* Proc.
          Amer. Math. Soc., **71**(1978), 138–142.

[1978c]   *Second countable and separable function spaces,* Amer. Math.
          Monthly, **85:6**(1978), 487–489.

[1980a]   *Countability properties of function spaces,* Rocky Mountain J. Math.,
          **10**(1980), 717–730.

[1980b]   *A K-space function space,* Int. J. Math. Sci., **3**(1980), 701–711.

[1980c]   *Necessary conditions for function spaces to be Lindelöf,* Glas. Mat.,
          **15**(1980), 163–168.

[1980d]   *Function spaces which are k-spaces,* Topology Proc., **5**(1980),
          139–146.

[1983]    *Complete function spaces,* Internat. J. Math. Math. Sci. **6:2**(1983),
          271–277.

[1986]    *Fine topology on function spaces,* Internat. J. Math. Math. Sci.
          **9:3**(1986), 417–424.

MCCOY, R.A., NTANTU, I.

[1985]    *Countability properties of function spaces with set-open topologies,*
          Topology Proceedings **10:2**(1985), 329–345.

[1986]    *Completeness properties of function spaces,* Topology Appl.,
          **22:2**(1986), 191–206.

[1988]    *Topological Properties of Spaces of Continuous Functions,* Lecture
          Notes in Math., 1315, Springer, Berlin, 1988.

[1992]    *Properties of C(X) with the epi-topology,* Boll. Un. Mat. Ital. B
          **(7)6:3**(1992), 507–532.

[1995]    *Properties of C(X) with the weak epi-topology,* Q&A General Topol.
          **13:2**(1995), 139–152.

MERKCOURAKIS, S., NEGREPONTIS, S.

[1992]    *Banach spaces and Topology, 2,* Recent Progress in Topology,
          Elsevier, N.Y., 1992.

MEYER, P.R.

[1970]    *Function spaces and the Alexandroff–Urysohn conjecture,* Estratto
          degli Annali di Matematica Pura ed Aplicata, Ser. 4, **86**(1970), 25–29.

MICHAEL, E.

[1961]   *On a theorem of Rudin and Klee,* Proc. Amer. Math. Soc., **12**(1961), 921.

[1966]   $\aleph_0$-*spaces,* J. Math. and Mech., **15:6**(1966), 983–1002.

[1971]   *Paracompactness and the Lindelöf property in finite and countable Cartesian products,* Compositio Math., **23**(1971), 179–214.

[1973]   *On k-spaces, $k_R$-spaces and k(X),* Pacific J. Math., **47:2**(1973), 487–498.

[1977]   $\aleph_0$-*spaces and a function space theorem of R. Pol,* Indiana Univ. Math. J., **26**(1977), 299–306.

MICHAEL, E., RUDIN, M.E.

[1977a]  *A note on Eberlein compacta,* Pacific J. Math., **72:2**(1977), 487–495.

[1977b]  *Another note on Eberlein compacta,* Pacific J. Math., **72:2**(1977), 497–499.

MICHALEWSKI, H.

[2001]   *An answer to a question of Arhangel'skii,* Commentat. Math. Univ. Carolinae, **42:3**(2001), 545–550.

MILL, J. VAN

[1982]   *A homogeneous Eberlein compact space which is not metrizable,* Pacific J. Mathematics, **101:1**(1982), 141–146.

[1984]   *An introduction to $\beta\omega$,* Handbook of Set-Theoretic Topology, North-Holland, Amsterdam, 1984, 503–567.

[1987]   *Topological equivalence of certain function spaces,* Compositio Math., **63**(1987), 159–188.

[1989]   *Infinite-Dimensional Topology.* Prerequisites and Introduction, North Holland, Amsterdam, 1989.

[1999]   $C_p(X)$ *is not $G_{\delta\sigma}$: a simple proof,* Bull. Polon. Acad. Sci., Ser. Math., **47**(1999), 319–323.

[2002]   *The Infinite-Dimensional Topology of Function Spaces,* North Holland Math. Library **64**, Elsevier, Amsterdam, 2002.

MILL, J. VAN, PELANT, J., POL, R.

[2003]   *Note on function spaces with the topology of pointwise convergence,* Arch. Math., **80**(2003), 655–663.

MILL, J. VAN, POL, R.

[1986]   *The Baire Category Theorem in products of linear spaces and topological groups,* Topology Appl., **22**(1986), 267–282.

[1993]   *A countable space with a closed subspace without measurable extender,* Bull. Acad. Polon. Sci., Math., **41**(1993), 279–283.

MISRA, P.R.

[1982]   *On isomorphism theorems for C(X),* Acta Mathematica Hungarica, **39:4**(1982), 389–390.

MOLINA LARA, I., OKUNEV, O.

[2010]   *$L\Sigma(\leq \omega)$-spaces and spaces of continuous functions*, Cent. Eur. J. Math., **8:4**(2010), 754–762.

MORISHITA, K.

[xxxx]   *The $k_R$-property of function spaces*, Preprint.

[1992a]  *The minimal support for a continuous functional on a function space*, Proc. Amer. Math. Soc. **114:2**(1992), 585–587.

[1992b]  *The minimal support for a continuous functional on a function space. II*, Tsukuba J. Math. **16:2**(1992), 495–501.

[1999]   *On spaces that are l-equivalent to a disk*, Topology Appl., **99**(1999), 111–116.

MORITA, K.

[1956]   *Note on mapping spaces*, Proc. Japan Acad., **32**(1956), 671–675.

MYCIELSKI, J.

[1973]   *Almost every function is independent*, Fund. Math., **81**(1973), 43–48.

MYKHAYLYUK V.V.

[2006]   *Namioka spaces and topological games*, Bull. Austral. Math. Soc. **73:2**(2006), 263–272.

[2007]   *Metrizable compacta in the space of continuous functions with the topology of pointwise convergence*, Acta Math. Hungar. **117:4**(2007), 315–323.

NACHBIN, L.

[1954]   *Topological vector spaces of continuous functions*, Proceedings Nat. Acad. Sci., (USA), **40:6**(1954), 471–474.

NAGAMI, K.

[1969]   *$\Sigma$-spaces*, Fund. Math., **65:2**(1969), 169–192.

NAGATA J.

[1949]   *On lattices of functions on topological spaces and of functions on uniform spaces*, Osaka Math. J., **1:2**(1949), 166–181.

NAKHMANSON, L. B.

[1982]   *On continuous images of σ-products (in Russian)*, Topology and Set Theory, Udmurtia Universty P.H., Izhevsk, 1982, 11–15.

[1984]   *The Souslin number and calibers of the ring of continuous functions (in Russian)*, Izv. Vuzov, Matematika, 1984, N 3, 49–55.

[1985]   *On Lindelöf property of function spaces (in Russian)*, Mappings and Extensions of Topological Spaces, Udmurtia University P.H., Ustinov, 1985, 8–12.

NAKHMANSON, L. B., YAKOVLEV, N.N.
  [1982]  *On compacta, lying in σ-products (in Russian),* Comment. Math. Univ. Carolinae, **22:4**(1981), 705–719.

NAMIOKA, I.
  [1974]  *Separate continuity and joint continuity,* Pacific J. Math., **51:2**(1974), 515–531.
  [2002]  *On generalizations of Radon-Nikodym compact spaces,* Topol. Proc. **26:2**(2001/02), 741–750.

NAMIOKA, I., PHELPS, R.R.
  [1975]  *Banach spaces which are Asplund spaces,* Duke Math. J., **42**(1975), 735–750.

NAMIOKA, I., POL, R.
  [1992]  *Mappings of Baire spaces into function spaces and Kadec renorming,* Israel J. Math., **78**(1992), 1–20.
  [1998]  *σ-fragmentability of mappings into $C_p(K)$,* Topology Appl., **89**(1998), 249–263.
  [1996]  *σ-fragmentability and analyticity,* Mathematika **43:1**(1996), 172–181.

NEGREPONTIS, S.
  [1984]  *Banach spaces and topology,* Handbook of Set-Theoretic Topology, North Holland, Amsterdam, 1984, 1001–1142.

NOBLE, N.
  [1969a]  *Products with closed projections,* Trans Amer. Math. Soc., **140**(1969), 381–391.
  [1969b]  *Ascoli theorems and the exponential map,* Trans Amer. Math. Soc., **143**(1969), 393–411.
  [1970]  *The continuity of functions on Cartesian products,* Transactions Amer. Math. Soc., **149**(1970), 187–198.
  [1971]  *Products with closed projections II,* Trans Amer. Math. Soc., **160**(1971), 169–183.
  [1974]  *The density character of function spaces,* Proc. Amer. Math. Soc., **42:1**(1974), 228–233.

NOBLE, N., ULMER, M.
  [1972]  *Factoring functions on Cartesian products,* Trans Amer. Math. Soc., **163**(1972), 329–339.

NYIKOS, P.
  [1981]  *Metrizability and the Fréchet–Urysohn property in topological groups,* Proc. Amer. Math. Soc., **83:4**(1981), 793–801.
  [1989]  *Classes of compact sequential spaces,* Set Theory Appl., Lecture Notes in Math., 1401, Springer, Berlin, 1989, 135–159.

OHTA, H., YAMADA, K.

[1998]   *On spaces with linearly homeomorphic function spaces in the compact open topology,* Tsukuba Math. J., **22**(1998), 39–48.

OKADA, T.

[1985]   *A remark on Eberlein compacta,* Kobe J. Math., **2**(1985), 65–66.

OKADA, T., OKUYAMA, A.

[1987]   *Note on topologies of linear continuous functionals and real measure functions,* Kobe J. Math. **4:1**(1987), 73–77.

OKUNEV, O.G.

[1984]   *Hewitt extensions and function spaces (in Russian),* Cardinal Invariants and Mappings of Topological Spaces (in Russian), Izhvsk, 1984, 77–78.

[1985]   *Spaces of functions in the topology of pointwise convergence: Hewitt extension and τ-continuous functions,* Moscow Univ. Math. Bull., **40:4**(1985), 84–87.

[1986]   *On non-preserving a property by the relation of M-equivalence (in Russian),* Continuous Functions on Topological Spaces (in Russian), Riga, 1986, 123–125.

[1989]   *On compact-like properties of spaces of continuous functions in the topology of pointwise convergence,* Bull. Acad. Sci. Georgian SSR, **134:3**(1989), 473–476.

[1990]   *Weak topology of an associated space, and t-equivalence,* Math. Notes, **46:(1-2)**(1990), 534–538.

[1993a]  *On Lindelöf Σ-spaces of functions in the pointwise topology,* Topol. Appl., **49**(1993), 149–166.

[1993b]  *On analyticity in cosmic spaces,* Comment. Math. Univ. Carolinae, **34**(1993), 185–190.

[1995a]  *On Lindelöf sets of continuous functions,* Topology Appl., **63**(1995), 91–96.

[1995b]  *M-equivalence of products,* Trans. Moscow Math. Soc., **56**(1995), 149–158.

[1996]   *A remark on the tightness of products,* Comment. Math. Univ. Carolinae, **37:2**(1996), 397–399.

[1997a]  *Homeomorphisms of function spaces and hereditary cardinal invariants,* Topology Appl., **80**(1997), 177–188.

[1997b]  *On the Lindelöf property and tightness of products,* Topology Proc., **22**(1997), 363–371.

[2002]   *Tightness of compact spaces is preserved by the t-equivalence relation,* Comment. Math. Univ. Carolin. **43:2**(2002), 335–342.

[2005a]  *Fréchet property in compact spaces is not preserved by M-equivalence,* Comment. Math. Univ. Carolin. **46:4**(2005), 747–749.

[2005b]  *A σ-compact space without uncountable free sequences can have arbitrary tightness,* Questions Answers Gen. Topology **23:2**(2005), 107–108.

[2009]   *On the Lindelöf property of spaces of continuous functions over a Tychonoff space and its subspaces,* Comment. Math. Univ. Carolin. **50:4**(2009), 629–635.

[2011a]  *A relation between spaces implied by their t-equivalence,* Topology Appl. **158**(2011), 2158–2164.

[2011b]  *The Lindelöf number of $C_p(X) \times C_p(X)$ for strongly zero-dimensional X,* Cent. Eur. J. Math., **9:5**(2011), 978–983.

[2013]   *The one-point Lindelöfication of an uncountable discrete space can be surlindelöf,* Cent. Eur. J. Math., **11:10**(2013), 1750–1754.

OKUNEV, O.G., RAMÍREZ, H.D.

[2013]   *Linear homeomorphisms of function spaces and properties close to countable compactness,* Topology Proc., **40**(2013), 1–8.

OKUNEV, O., REZNICHENKO, E.

[2013]   *A note on surlindelöf spaces,* Topology Proc., **31**(2007), 667–675.

OKUNEV, O.G., SHAKHMATOV, D.B.

[1987]   *The Baire property and linear isomorphisms of continuous function spaces (in Russian),* Topological Structures and Their Maps, Latvian State University P.H., Riga, 1987, 89–92.

OKUNEV, O.G., TAMANO, K.

[1996]   *Lindelöf powers and products of function spaces,* Proceedings Amer. Math. Soc., **124:9**(1996), 2905–2916.

OKUNEV, O.G., TAMARIZ-MASCARÚA, A.

[2004]   *On the Čech number of $C_p(X)$,* Topology Appl., **137**(2004), 237–249.

OKUNEV, O.G., TKACHUK, V.V.

[2001]   *Lindelöf $\Sigma$-property in $C_p(X)$ and $p(C_p(X)) = \omega$ do not imply countable network weight in X,* Acta Mathematica Hungarica, **90:(1–2)**(2001), 119–132.

[2002]   *Density properties and points of uncountable order for families of open sets in function spaces,* Topology Appl., **122**(2002), 397–406.

[2014]   *Calibers, $\omega$-continuous maps and function spaces,* RACSAM, **108:2**(2014), 419–430.

OKUYAMA, A.

[1981]   *Some relationships between function spaces and hyperspaces of compact sets,* General Topology and Its Relations to Modern Analysis and Algebra, V, Fifth Prague Topolol. Symposium, 1981, 527–535.

[1983]   *Questions on angelic and strictly angelic spaces,* Questions Answ. Gen. Topol. **1:1**(1983), 51–53.

[1987a]  *On a topology of the set of real measure functions,* Kobe J. Math. **3:2**(1987), 237–242.

[1987b]   *On a topology of the set of linear continuous functionals,* Kobe Journal
          Math. **3:2**(1987), 213–217.

[1990]    *Note on spaces of continuous functionals on* $C^*(X)$*,* Questions
          Answers Gen. Topology **8:1**(1990), 79–82.

OKUYAMA, A., TERADA, T.

[1989]    *Function spaces,* Topics in General Topology, 1989, 411–458.

O'MEARA, P.

[1971]    *On paracompactness in function spaces with the compact-open topol-
          ogy,* Trans. Amer. Math. Soc., **29:1**(1971), 183–189.

ORIHUELA, J.

[1987]    *Pointwise compactness in spaces of continuous functions,* Journal
          London Math. Soc., **36:2**(1987), 143–152.

[1992]    *On weakly Lindelöf Banach spaces,* Progress in Functional Analysis,
          Elsevier S.P., 1992, 279–291.

ORIHUELA, J., SCHACHERMAYER, W., VALDIVIA, M.

[1991]    *Every Radon–Nykodym Corson compact space is Eberlein compact,*
          Studia Mathematica, **98:2**(1991), 157–174.

OSTRAND, PH. A.

[1965]    *Dimension of metric spaces and Hilbert's problem 13,* Bull. Amer.
          Math. Soc., **71:4**(1965), 619–622.

OXTOBY, J.C.

[1961]    *Cartesian products of Baire spaces,* Fund. Math., **49:2**(1961),
          157–166.

PASYNKOV, B.A.

[1967]    *On open mappings,* Soviet Math. Dokl., **8**(1967), 853–856.

PAVLOVSKII, D.S.

[1979]    *The spaces of open sets and spaces of continuous functions (in
          Russian),* Doklady AN SSSR, **246:4**(1979), 815–818.

[1980]    *On spaces of continuous functions (in Russian),* Doklady AN SSSR,
          **253:1**(1979), 38–41.

[1982]    *On spaces which have linearly homeomorphic spaces of continuous
          functions endowed with the topology of pointwise convergence (in
          Russian),* Uspehi Matem. Nauk, **37:2**(1982), 185–186.

PELANT, J.

[1988]    *A remark on spaces of bounded continuous functions,* Indag. Math.,
          **91**(1988), 335–338.

PELCZIŃSKI, A.
[1960]   *Projections in certain Banach spaces,* Studia Math., **19**(1960),
         209–228.
[1968]   *Linear extensions, linear averaging, and their applications to linear
         topological classification of spaces of continuous functions,* Disserta-
         tiones Math., (Rozprawy Mat.), **58**(1968).

PELCZIŃSKI, A., SEMADENI, Z.
[1959]   *Spaces of continuous functions (3) (spaces $C(\Omega)$ for $\Omega$) without
         perfect subsets,* Studia Math., **18**(1959), 211–222.

PESTOV, V.G.
[1982]   *Coincidence of the dimension* dim *of l-equivalent topological spaces
         (in Russian),* Doklady AN SSSR, **266:3**(1982), 553–556.
[1984]   *Some topological properties preserved by the relation of M -equivalence
         (in Russian),* Uspehi Mat. Nauk, **39:6**(1984), 203–204.

PIOTROWSKI, Z.
[1985]   *Separate and joint continuity,* Real Analysis Exchange, **11**(1985/86),
         293–322.

PLEBANEK, G.
[1995]   *Compact spaces that result from adequate families of sets,* Topol.
         Appl., **65:3**(1995), 257–270.

POL, R.
[1972]   *On the position of the set of monotone mappings in function spaces,*
         Fund. Math., **75**(1972), 75–84.
[1974]   *Normality in function spaces,* Fund. Math., **84:2**(1974), 145–155.
[1977]   *Concerning function spaces on separable compact spaces,* Bull. Acad.
         Polon. Sci., Sér. Math., Astron. et Phys., **25:10**(1977), 993–997.
[1978]   *The Lindelöf property and its analogue in function spaces with weak
         topology,* Topology. 4-th Colloq. Budapest, **2**(1978), Amsterdam,
         1980, 965–969.
[1979]   *A function space $C(X)$ which is weakly Lindelöf but not weakly
         compactly generated,* Studia Math., **64:3**(1979), 279–285.
[1980a]  *A theorem on the weak topology of $C(X)$ for compact scattered $X$,*
         Fundam. Math., **106:2**(1980), 135–140.
[1980b]  *On a question of H.H. Corson and some related matters,* Fund. Math.,
         **109:2**(1980), 143–154.
[1982]   *Note on the spaces $P(S)$ of regular probability measures whose topol-
         ogy is determined by countable subsets,* Pacific J. Math., **100**(1982),
         185–201.
[1984a]  *An infinite-dimensional pre-Hilbert space not homeomorphic to its
         own square,* Proc. Amer. Math. Soc., **90:3**(1984), 450–454.

[1984b]  *On pointwise and weak topology in function spaces,* Univ. Warsza-
         wski, Inst. Matematiki, Preprint 4/84, Warszawa, 1984.
[1986]   *Note on compact sets of first Baire class functions,* Proceedings Amer.
         Math. Soc., **96:1**(1986), 152–154.
[1989]   *Note on pointwise convergence of sequences of analytic sets,* Mathem.,
         **36**(1989), 290–300.
[1995†]  *For a metrizable X, $C_p(X)$ need not be linearly homeomorphic to
         $C_p(X) \times C_p(X)$,* Mathematika, **42**(1995), 49–55.

PONTRIAGIN, L.S.
[1984]   *Continuous Groups (in Russian),* Nauka, Moscow, 1984.

POPOV, V.V.
[xxxx]   *The space $C_p(\beta\mathbb{N})$ has not a projectively Baire property,* Preprint.

PREISS, D., SIMON, P.
[1974]   *A weakly pseudocompact subspace of Banach space is weakly com-
         pact,* Comment. Math. Univ. Carolinae, **15:4**(1974), 603–609.

PRYCE, J.D.
[1971]   *A device of R.J. Whitley's applied to pointwise compactness in
         spaces of continuous functions,* Proc. Amer. Math. Soc., **23:3**(1971),
         532–546.

PTÁK, V.
[1954a]  *On a theorem of W.F. Eberlein,* Studia Math., **14:2**(1954), 276–284.
[1954b]  *Weak compactness in locally convex topological vector spaces,*
         Czechoslovak Math. J., **79**(1954), 175–186.
[1955]   *Two remarks on weak compactness,* Czechoslovak Math. J., **80**(1955),
         532–545.
[1963]   *A combinatorial lemma on the existence of convex means and its
         application to weak compactness,* Proc. Sympos. Pure Math., 7,
         Convexity, 1963, 437–450.

PYTKEEV, E.G.
[1976]   *Upper bounds of topologies,* Math. Notes, **20:4**(1976), 831–837.
[1982a]  *On the tightness of spaces of continuous functions (in Russian),* Uspehi
         Mat. Nauk, **37:1**(1982), 157–158.
[1982b]  *Sequentiality of spaces of continuous functions (in Russian),* Uspehi
         Matematicheskih Nauk, **37:5**(1982), 197–198.
[1985]   *The Baire property of spaces of continuous functions, (in Russian),*
         Matem. Zametki, **38:5**(1985), 726–740.
[1990]   *A note on Baire isomorphism,* Comment. Math. Univ. Carolin.
         **31:1**(1990), 109–112.
[1992a]  *On Fréchet–Urysohn property of spaces of continuous functions,
         (in Russian),* Trudy Math. Inst. RAN, **193**(1992), 156–161.

[1992b] *Spaces of functions of the first Baire class over K-analytic spaces (in Russian)*, Mat. Zametki **52:3**(1992), 108–116.

[2003] *Baire functions and spaces of Baire functions*, Journal Math. Sci. (N.Y.) **136:5**(2006), 4131–4155

PYTKEEV, E.G., YAKOVLEV, N.N.

[1980] *On bicompacta which are unions of spaces defined by means of coverings*, Comment. Math. Univ. Carolinae, **21:2**(1980), 247–261.

RAJAGOPALAN, M., WHEELER, R.F.

[1976] *Sequential compactness of X implies a completeness property for C(X)*, Canadian J. Math., **28**(1976), 207–210.

REYNOV, O.I.

[1981] *On a class of compacts and GCG Banach spaces*, Studia Math., **71**(1981), 113–126.

REZNICHENKO, E.A.

[1987] *Functional and weak functional tightness (in Russian)*, Topological Structures and Their Maps, Latvian State University P.H., Riga, 1987, 105–110.

[1989a] *Paracompactness is not preserved by l-equivalence*, General Topology, Spaces and Mappings, Mosk. Univ., Moscow, 1989, 155–156.

[1989b] *A pseudocompact space in which only the subsets of not full cardinality are not closed and not discrete*, Moscow Univ. Math. Bull. **44:6**(1989), 70–71.

[1990a] *Normality and collectionwise normality in function spaces*, Moscow Univ. Math. Bull., **45:6**(1990), 25–26.

[1990b] *Convex compact spaces and their maps*, Topology Appl., **36**(1990), 117–141.

[1994] *Extension of functions defined on products of pseudocompact spaces and continuity of the inverse in pseudocompact groups*, Topology Appl., **59:3**(1994), 233–244.

[2008] *Stratifiability of $C_k(X)$ for a class of separable metrizable X*, Topology Appl. **155:17-18**(2008), 2060–2062.

[xxxxa] *Groups with separately continuous and jointly continuous multiplication*, Preprint.

[xxxxb] *Talagrand compact spaces: $\pi$-points and Stone–Čech extensions*, Preprint.

REZNICHENKO, E. A., SHKARIN, S.A.

[2002] *On linear extension operators*, Russ. J. Math. Phys. **9:2**(2002), 188–197.

RIBARSKA, N.K.

[1987] *Internal characterization of fragmentable spaces*, Mathematika, **34**(1987), 243–257.

ROJAS-HERNÁNDEZ, R.
  [2014a]  *On monotone stability,* Tooplogy Appl., **165**(2014), 50–57.
  [2014b]  *Function spaces and D-property,* Topology Proc., **43**(2014), 301–317.

ROJAS-HERNÁNDEZ, R., TAMARIZ-MASCARÚA, A.,
  [2012]  *D-property, monotone monolithicity and function spaces,* Topology
          Appl., **159**(2012), 3379–3391.

ROJAS-HERNÁNDEZ, R., TKACHUK, V.V.
  [2014]  *A monotone version of the Sokolov property and monotone retractabil-
          ity in function spaces,* J. Math. Anal. Appl., **412:1**(2014), 125–137.

ROSENTHAL, H.P.
  [1974]  *The heredity problem for weakly compactly generated Banach spaces,*
          Compositio Math., **28:1**(1974), 88–111.
  [1977]  *Pointwise compact subsets of the first Baire class,* Amer. J. Math.,
          **99:2**(1977), 362–378.
  [1978]  *Some recent discoveries in the isomorphic theory of Banach spaces,*
          Bull. Amer. Math. Soc., **84**(1978), 803–831.

RUDIN, M.E.
  [1956]  *A note on certain function spaces,* Arch. Math., **7**(1956), 469–470.

RUDIN, W.
  [1973]  *Functional Analysis,* McGraw-Hill Book Company, New York, 1973.

SAINT RAYMOND, J.
  [1976]  *Fonctions boreliennes sur quotients,* Bull. Sci. Math., **100:2**(1976),
          141–147.
  [1984]  *Jeux topologiques et espaces de Namioka,* Proc. Amer. Math. Soc.,
          **87**(1984), 499–504.

SAKAI, M.
  [1988a]  *On supertightness and function spaces,* Comment. Math. Univ.
           Carolin., **29:2**(1988), 249–251.
  [1988b]  *Property $C''$ and function spaces,* Proc. Amer. Math. Soc.,
           **104:3**(1988), 917–919.
  [1992]   *On embeddings into $C_p(X)$ where $X$ is Lindelöf,* Comment. Math.
           Univ. Carolinae, **33:1**(1992), 165–171.
  [1995]   *Embeddings of $\kappa$-metrizable spaces into function spaces,* Topol. Appl.,
           **65**(1995), 155–165.
  [2000]   *Variations on tightness in function spaces,* Topology Appl., **101**(2000),
           273–280.
  [2003]   *The Pytkeev property and the Reznichenko property in function spaces,*
           Note di Matem. **22:2**(2003/04), 43–52

[2006] *Special subsets of reals characterizing local properties of function spaces,* Selection Principles and Covering Properties in Topology, Dept. Math., Seconda Univ. Napoli, Caserta, Quad. Mat., **18**(2006), 195–225,

[2007] *The sequence selection properties of* $C_p(X)$, Topology Appl., **154**(2007), 552–560.

[2008] *Function spaces with a countable $cs^*$-network at a point,* Topol. Appl., **156:1**(2008), 117–123.

[xxxx] *On calibers of function spaces,* Preprint.

SCHACHENMAYER, W.

[1977] *Eberlein compacts et espaces de Radon,* C. R. Acad. Sci. Paris, **284**(1977), 405–477.

SCHEEPERS, M.

[1997a] *Combinatorics of open covers III. Games,* $C_p(X)$, Fund. Math. **152:3**(1997), 231–254.

[1997b] *A sequential property of $C_p(X)$ and a covering property of Hurewicz,* Proc. Amer. Math. Soc., **125:9**(1997), 2789–2795.

[1998a] $C_p(X)$ *and Arhangel'skii's $\alpha_i$-spaces,* Topology Appl., **89**(1998), 265–275.

[1998b] *Strong measure zero sets, filters, and pointwise convergence,* East-West Journal Math. **1:1**(1998), 109–116.

[1999a] *Sequential convergence in $C_p(X)$ and a covering property,* East-West J. Math., **1**(1999), 207–214.

[1999b] *Combinatorics of open covers. VI. Selectors for sequences of dense sets,* Quaest. Math. **22:1**(1999), 109–130.

[2008] *Rothberger's property in all finite powers,* Topology Appl. **156:1**(2008), 93–103.

SCHEPIN, E.V.

[1979] *On $\kappa$-metrizable spaces (in Russian),* Izvestiya Acad. Nauk SSSR, Ser. Math., **43:2**(1979), 442–478.

SCHMETS, J.

[1976] *Espaces de fonctions continues,* Lecture Notes in Math., **519**(1976), N XII.

SEMADENI, Z.

[1960] *Banach spaces non-isomorphic to their Cartesian squares,* Bull. Acad. Polon. Sci., Math., **8**(1960), 81–84.

[1971] *Banach spaces of continuous functions,* Monog. Mat., **1:55**(1971).

SHAKHMATOV, D.B.

[1986] *A pseudocompact Tychonoff space all countable subsets of which are closed and $C^*$-embedded,* Topology Appl., **22:2**(1986), 139–144.

SHAPIROVSKY, B.E.
  [1974]  *Canonical sets and character. Density and weight in compact spaces,*
          Soviet Math. Dokl., **15**(1974), 1282–1287.
  [1978]  *Special types of embeddings in Tychonoff cubes, subspaces of*
          *Σ-products and cardinal invariants,* Colloquia Mathematica Soc.
          Janos Bolyai, **23**(1978), 1055–1086.
  [1981]  *Cardinal invariants in bicompacta (in Russian),* Seminar on General
          Topology, Moscow University P.H., Moscow, 1981, 162–187.

SHIROTA, T.
  [1954]  *On locally convex vector spaces of continuous functions,* Proceedings
          Japan Acad., **30:4**(1954), 294–298.

SIMON, P.
  [1976]  *On continuous images of Eberlein compacts,* Commentationes Math.
          Univ. Carolinae, **17:1**(1976), 179–194.

SIPACHEVA, O.V.
  [1986]  *Description of the topology of free topological groups without using*
          *structures of universal uniformities (in Russian),* General Topology.
          Mappings of Topological Spaces, Moscow University P.H., Moscow,
          1986, 122–130.
  [1988]  *Lindelöf Σ-spaces of functions and the Souslin number,* Moscow Univ.
          Math. Bull., **43:3**(1988), 21–24.
  [1989]  *On Lindelöf subspaces of function spaces over linearly ordered sep-*
          *arable compacta,* General Topology, Spaces and Mappings, Mosk.
          Univ., Moscow, 1989, 143–148.
  [1990]  *The structure of iterated function spaces in the topology of pointwise*
          *convergence for Eberlein compacta (in Russian),* Matem. Zametki,
          **47:3**(1990), 91–99.
  [1992]  *On surlindelöf compacta (in Russian),* General Topology. Spaces,
          Mappings and Functors, Moscow University P.H., Moscow, 1992,
          132–140.

SOKOLOV, G.A.
  [1984]  *On some classes of compact spaces lying in Σ-products,* Comment.
          Math. Univ. Carolinae, **25:2**(1984), 219–231.
  [1986]  *On Lindelöf spaces of continuous functions (in Russian),* Mat.
          Zametki, **39:6**(1986), 887–894.
  [1993]  *Lindelöf property and the iterated continuous function spaces,*
          Fundam. Math., **143**(1993), 87–95.

STEGALL, C.
  [1975]  *The Radon–Nikodym property in conjugate Banach spaces,* Trans.
          Amer. Math. Soc., **206**(1975), 213–223.

[1987a] *Applications of the Descriptive Topology in Functional Analysis*, Linz Univ., 1985–1987.

[1987b] *Topological spaces with dense subspaces that are homeomorphic to complete metric spaces and the classification of $C(K)$ Banach spaces*, Mathematika, **34**(1987), 101–107.

[1988a] *Generalizations of a theorem of Namioka*, Proc. Amer. Math. Soc., **102**(1988), 559–564.

[1988b] *The topology of certain spaces of measures*, Univ. de Murcia, Notas Mat.,**3**(1988).

STOKKE, R.

[1997] *Closed ideals in $C(X)$ and $\Phi$-algebras*, Topology Proc., **22**(1997), 501–528.

STONE, A.H.

[1963] *A note on paracompactness and normality of mapping spaces*, Proc. Amer. Math. Soc., **14**(1963), 81–83.

TALAGRAND, M.

[1975] *Sur une conjeture de H.H. Corson*, Bull. Sci. Math., **99:4**(1975), 211–212.

[1977] *Espaces de Banach faiblement $K$-analytiques*, Comptes Rendus de la Academie de Sciences, Paris, **284:13**(1977), 745–748.

[1978] *Comparaison des boreliens d'un espace de Banach pour les topologies fortes et faibles*, Indiana Univ. Math. J., **27**(1978), 1001–1004.

[1979a] *Espaces de Banach faiblement $K$-analytiques*, Annals Math., **110:3**(1979), 407–438.

[1979b] *Sur la $K$-analyticité des certains espaces d'operadeurs*, Israel J. Math., **32**(1979), 124–130.

[1979c] *Deux généralisations d'un théorème de I. Namioka*, Pacific J. Math., **81**(1979), 239–251.

[1980] *Compacts de fonctions mesurables et filtres non mesurables*, Studia Math., **67**(1980), 13–43.

[1984] *A new countably determined Banach space*, Israel J. Math., **47:1**(1984), 75–80.

[1985] *Espaces de Baire et espaces de Namioka*, Math. Ann., **270**(1985), 159–164.

TAMANO, K., TODORCEVIC, S.

[2005] *Cosmic spaces which are not $\mu$-spaces among function spaces with the topology of pointwise convergence*, Topology Appl. **146/147**(2005), 611–616.

TAMARIZ–MASCARÚA, A.

[1996] *Countable product of function spaces having p-Frechet-Urysohn like properties*, Tsukuba J. Math. **20:2**(1996), 291–319.

[1998]   $\alpha$-pseudocompactness in $C_p$-spaces, Topology Proc., **23**(1998), 349–362.

[2006]   *Continuous selections on spaces of continuous functions,* Comment. Math. Univ. Carolin. **47:4**(2006), 641–660.

TANI, T.

[1979]   *Perfectly finally compact spaces are hard,* Math. Japonica, **24**(1979), 323–326.

[1986]   *On the tightness of $C_p(X)$,* Memoirs Numazu College Technology, **21**(1986), 217–220.

TKACHENKO, M.G.

[1978]   *On the behaviour of cardinal invariants under the union of chains of spaces (in Russian),* Vestnik Mosk. Univ., Math., Mech., **33:4**(1978), 50–58.

[1979]   *On continuous images of dense subspaces of topological products (in Russian),* Uspehi Mat. Nauk, **34:6**(1979), 199–202.

[1981a]  *On a result of E. Michael and M.E. Rudin (in Russian),* Vestnik Mosk. Univ., Math., Mech., **36:5**(1981), 47–50.

[1981b]  *On one property of bicompacta (in Russian),* Seminar on General Topology, Moscow University P.H., Moscow, 1981, 149–156.

[1982]   *On continuous images of dense subspaces of $\Sigma$-products of compacta (in Russian),* Sibirsk. Math. J., **23:3**(1982), 198–207.

[1985]   *On continuous images of spaces of functions,* Sibirsk. Mat. Zhurnal, **26:5**(1985), 159–167.

TKACHENKO, M.G., TKACHUK, V.V.

[2005]   *Dyadicity index and metrizability of compact continuous images of function spaces,* Topol. Appl., **149:(1–3)**(2005), 243–257.

TKACHUK, V.V.

[1983a]  *On a method of constructing examples of $M$-equivalent spaces (in Russian),* Uspehi Mat. Nauk, **38:6**(1983), 127–128.

[1983b]  *On cardinal functions of Souslin number type (in Russian),* Dokl. Akad. Nauk SSSR, **270:4**(1983), 795–798.

[1984a]  *On multiplicativity of some properties of function spaces with the topology of pointwise convergence (in Russian),* Vestnik Mosk. Univ., Math., Mech., **39:6**(1984), 36–38.

[1984b]  *Characterization of the Baire property in $C_p(X)$ in terms of the properties of the space $X$ (in Russian),* Cardinal Invariants and Mappings of Topological Spaces (in Russian), Izhevsk, 1984, 76–77.

[1984c]  *On a supertopological cardinal invariant,* Vestn. Mosk. Univ., Matem., Mech., **39:4**(1984), 26–29.

[1985a]  *Duality with respect to the $C_p$-functor and cardinal invariants of Souslin number type (in Russian),* Matem. Zametki, **37:3**(1985), 441–451.

[1985b] *A characterization of the Baire property of $C_p(X)$ by a property of $X$,* *(in Russian)* Mappings and Extensions of Topological Spaces, Ustinov, 1985, 21–27.

[1986a] *The spaces $C_p(X)$: decomposition into a countable union of bounded subspaces and completeness properties,* Topology Appl., **22:3**(1986), 241–254.

[1986b] *Approximation of $\mathbf{R}^X$ with countable subsets of $C_p(X)$ and calibers of the space $C_p(X)$,* Comment. Math. Univ. Carolinae, **27:2**(1986), 267–276.

[1987a] *The smallest subring of the ring $C_p(C_p(X))$ which contains $X \cup \{1\}$ is dense in $C_p(C_p(X))$ (in Russian),* Vestnik Mosk. Univ., Math., Mech., **42:1**(1987), 20–22.

[1987b] *Spaces that are projective with respect to classes of mappings,* Trans. Moscow Math. Soc., **50**(1987), 139–156.

[1988] *Calibers of spaces of functions and metrization problem for compact subsets of $C_p(X)$ (in Russian),* Vestnik Mosk. Univ., Matem., Mech., **43:3**(1988), 21–24.

[1991] *Methods of the theory of cardinal invariants and the theory of mappings applied to the spaces of functions (in Russian),* Sibirsk. Mat. Zhurnal, **32:1**(1991), 116–130.

[1992] *A note on splittable spaces,* Comment. Math. Univ. Carolinae, **33:3**(1992), 551–555.

[1994] *Decomposition of $C_p(X)$ into a countable union of subspaces with "good" properties implies "good" properties of $C_p(X)$,* Trans. Moscow Math. Soc., **55**(1994), 239–248.

[1995] *What if $C_p(X)$ is perfectly normal?* Topology Appl., **65**(1995), 57–67.

[1997] *Some non-multiplicative properties are l-invariant,* Comment. Math. Univ. Carolinae, **38:1**(1997), 169–175.

[1998] *Mapping metric spaces and their products onto $C_p(X)$,* New Zealand J. Math., **27:1**(1998), 113–122.

[2000] *Behaviour of the Lindelöf $\Sigma$-property in iterated function spaces,* Topology Appl., **107:3-4**(2000), 297–305.

[2001] *Lindelöf $\Sigma$-property in $C_p(X)$ together with countable spread of $X$ implies $X$ is cosmic,* New Zealand J. Math., **30**(2001), 93–101.

[2003] *Properties of function spaces reflected by uniformly dense subspaces,* Topology Appl., **132**(2003), 183–193.

[2005a] *A space $C_p(X)$ is dominated by irrationals if and only if it is $K$-analytic,* Acta Math. Hungarica, **107:4**(2005), 253–265.

[2005b] *Function spaces and d-separability,* Quaestiones Math., **28:4**(2005), 409–424.

[2005c] *A nice class extracted from $C_p$-theory,* Comment. Math. Univ. Carolinae, **46:3**(2005), 503–513.

[2005d] *Point-countable $\pi$-bases in first countable and similar spaces,* Fundam. Math., **186**(2005), 55–69.

[2007a] *Condensing function spaces into $\Sigma$-products of real lines,* Houston Journ. Math., **33:1**(2007), 209–228.

[2007b] *Twenty questions on metacompactness in function spaces,* Open Problems in Topology II, ed. by E. Pearl, Elsevier B.V., Amsterdam, 2007, 595–598.

[2007c] *A selection of recent results and problems in $C_p$-theory,* Topol. Appl., **154:12**(2007), 2465–2493.

[2009a] *Monolithic spaces and $D$-spaces revisited,* Topology Appl., **156:4**(2009), 840–846.

[2009b] *Condensations of $C_p(X)$ onto $\sigma$-compact spaces,* Appl. Gen. Topology, **10:1**(2009), 39–48.

[2009c] *Embeddings and frames of function spaces,* J. Math. Anal. Appl., **357**(2009), 237–243.

[2010] *Lindelöf $\Sigma$-spaces: an omnipresent class,* RACSAM, **104:2**(2010), 221–244.

[2011] *Countably compact first countable subspaces of ordinals have the Sokolov property,* Quaestiones Math., **34:2**(2011), 225–234.

[2012] *The Collins-Roscoe property and its applications in the theory of function spaces,* Topology Appl., **159:6**(2012), 1529–1535.

[2013a] *Some criteria for $C_p(X)$ to be an $L\Sigma(\leq \omega)$-space,* Rocky Mountain J. Math., **43:1**(2013), 373–384.

[2013b] *Lifting the Collins-Roscoe property by condensations,* Topol. Proc., **42**(2013), 1–15.

[2015a] *Discrete reflexivity in function spaces,* Bull. Belgian Math. Soc., **21:5**(2014), 1–14.

[2015b] *Lindelöf $P$-spaces need not be Sokolov,* Math. Slovaca, **65**(2015), to appear.

TKACHUK, V.V., SHAKHMATOV, D.B.
[1986] *When the space $C_p(X)$ is $\sigma$-countably compact? (in Russian),* Vestnik Mosk. Univ., Math., Mech., **41:1**(1986), 70–72.

TKACHUK, V.V., YASCHENKO, I.V.
[2001] *Almost closed sets and topologies they determine,* Comment. Math. Univ. Carolinae, **42:2**(2001), 395–405.

TODD, A.R.
[1991] *Pseudocompact sets, absolutely Warner bounded sets and continuous function spaces,* Arch. Math., **56**(1991), 474–481.

TODORCEVIC, S.
[1989] *Partition Problems in Topology,* Contemporary Mathematics, American Mathematical Society, **84**(1989). Providence, Rhode Island, 1989.

[1993] *Some applications of $S$ and $L$ combinatorics,* Ann. New York Acad. Sci., **705**(1993), 130–167.

[1999]   *Compact sets of the first Baire class*, J. Amer. Math. Soc., **12**:4(1999),
         1179–1212.
[2000]   *Chain-condition methods in topology*, Topology Appl. **101**:1(2000),
         45–82.

TROALLIC, J.P.
[1979]   *Espaces fonctionnels et théorèmes de Namioka*, Bull. Soc. Math.
         France, **107**(1979), 127–137.
[1981]   *Une approche nouvelle de la presque-périodicité faible*, Semigr.
         Forum, **22**(1981), 247–255.
[1990]   *Quasi-continuité, continuité separée et topologie extremale*, Proc.
         Amer. Math. Soc., **110**(1990), 819–827.

TULCEA, A.I.
[1974]   *On pointwise convergence, compactness and equicontinuity*, Adv.
         Math., **12**(1974), 171–177.

TYCHONOFF A.N.
[1935]   *Über einer Funktionenraum*, Math. Ann., **111**(1935), 762–766.

USPENSKIJ, V.V.
[1978]   *On embeddings in functional spaces (in Russian)*, Dokl. AN SSSR,
         **242**:3(1978), 545–546.
[1982a]  *On frequency spectrum of functional spaces (in Russian)*, Vestnik
         Mosk. Univ., Math., Mech., **37**:1(1982), 31–35.
[1982b]  *A characterization of compactness in terms of the uniform structure in
         function space (in Russian)*, Uspehi Mat. Nauk, **37**:4(1982), 183–184.
[1983a]  *On the topology of free locally convex space (in Russian)*, Doklady AN
         SSSR, **270**:6(1983), 1334–1337.
[1983b]  *A characterization of realcompactness in terms of the topology of
         pointwise convergence on the function space*, Comment. Math. Univ.
         Carolinae, **24**:1(1983), 121–126.

VALOV, V.M.
[1986]   *Some properties of $C_p(X)$*, Comment. Math. Univ. Carolinae,
         **27**:4(1986), 665–672.
[1991a]  *Linear topological classification of certain function spaces*, Trans.
         Amer. Math. Soc., **327**:2(1991), 583–600.
[1991b]  *Linear topological classifications of certain function spaces*, Trans.
         Amer. Math. Soc. **327**:2(1991), 583–600.
[1993]   *Linear topological classifications of certain function spaces II*,
         Mathem. Pannon. **4**:1(1993), 137–144.
[1997a]  *Function spaces*, Topology Appl. **81**:1(1997), 1–22.
[1997b]  *Spaces of bounded functions with the compact-open topology*, Bull.
         Polish Acad. Sci. Math. **45**:2(1997), 171–179.
[1999]   *Spaces of bounded functions*, Houston J. Math. **25**:3(1999), 501–521.

VALOV, V., VUMA, D,
[1996] *Lindelöf degree and function spaces,* Papers in honour of Bernhard Banaschewski, Kluwer Acad. Publ., Dordrecht, 2000, 475–483.
[1998] *Function spaces and Dieudonné completeness,* Quaest. Math. **21:** 3-4(1998), 303–309.

VAŠAK, L.
[1981] *On one generalization of weakly compactly generated Banach spaces,* Studia Mathematica, **70:1**(1981), 11–19.

DE LA VEGA, R., KUNEN, K.
[2004] *A compact homogeneous S-space,* Topology Appl., **136**(2004), 123–127.

VELICHKO, N.V.
[1981] *On weak topology of spaces of continuous functions (in Russian),* Matematich. Zametki, **30:5**(1981), 703–712.
[1982] *Regarding the theory of spaces of continuous functions (in Russian),* Uspehi Matem. Nauk, **37:4**(1982), 149–150.
[1985] *Networks in spaces of mappings (in Russian),* Mappings and Extensions of Topological Spaces, Udmurtia University P.H., Ustinov, 1985, 3–6.
[1993] *On the theory of spaces of continuous functions,* Proc. Stekl. Inst. Math. **193:3**(1993), 57–63.
[1995a] $C(X)$ *in the weak topology,* Georgian Math. J. **2:6**(1995), 653–657.
[1995b] *On normality in function spaces,* Math. Notes, **56:5-6**(1995), 1116–1124.
[1996] *On the sequential completeness of* $C_p(X)$ *(in Russian),* Proc. Institute Math. Mech. **4**(1996), 118–123.
[1998a] *The Lindelöf property is l-invariant,* Topology Appl., **89**(1998), 277–283.
[1998b] *On the sequential completeness of* $C_\lambda(X)$ *(in Russian),* Fund. Prikl. Mat. **4:1**(1998), 39–47.
[1998c] *On the family of λ-topologies in a space of functions,* Siber. Math. J., **39:3**(1998), 422–430.
[2000] *On spaces* $C_\lambda(X)$, Topology Appl. **107:1-2**(2000), 191–195.
[2001] *On subspaces of functional spaces,* Proc. Steklov Inst. Math. **2**(2001), 234–240.
[2002] *Remarks on* $C_p$*-theory,* Proc. Steklov Inst. Math., **2**(2002), 190–192.
[2004] *Cardinal invariants of λ-topologies,* Siberian Math. J., **45:2**(2004), 241–247.

VIDOSSICH, G.
[1969a] *On topological spaces whose function space is of second category,* Invent. Math., **8:2**(1969), 111–113.

[1969b] *A remark on the density character of function spaces,* Proc. Amer. Math. Soc., **22**(1969), 618–619.

[1970] *Characterizing separability of function spaces,* Invent. Math., **10:3**(1970), 205–208.

[1971] *On a theorem of Corson and Lindenstrauss on Lindelöf function spaces,* Israel J. Math., **9**(1971), 271–278.

[1972] *On compactness in function spaces,* Proc. Amer. Math. Soc., **33**(1972), 594–598.

WAGE, M.L.

[1980] *Weakly compact subsets of Banach spaces,* Surveys in General Topology, New York, Academic Press, 1980, 479–494.

WARNER, S.

[1959] *The topology of compact convergence on continuous function spaces,* Duke Math. J., **25:2**(1959), 265–282.

WHITE, H.E., JR.

[1978] *First countable spaces that have special pseudo-bases,* Canadian Mathem. Bull., **21:1**(1978), 103–112.

WILDE, M. DE

[1974] *Pointwise compactness in space of functions and R.C. James theorem,* Math. Ann., **208**(1974), 33–47.

YAKOVLEV, N.N.

[1980] *On bicompacta in $\Sigma$-products and related spaces,* Comment. Math. Univ. Carolinae, **21:2**(1980), 263–283.

YASCHENKO, I.V.

[1989] *Baire functions as restrictions of continuous ones (in Russian),* Vestnik Mosk. Univ., Math., Mech., **44:6**(1989), 80–82.

[1991] *On rings of Baire functions (in Russian),* Vestnik Mosk. Univ., Math., **46:2**(1991), 88.

[1992a] *On the extent of function spaces,* Comptes Rendus Acad. Bulg. Sci., **45:1**(1992), 9–10.

[1992b] *Embeddings in $R^n$ and in $R^\omega$ and splittability,* Vestnik Mosk. Univ., Math., Mech., **47:2**(1992), 107.

[1992c] *On fixed points of mappings of topological spaces,* Vestnik Mosk. Univ., Math., Mech., **47:5**(1992), 93.

[1992d] *Cardinality of discrete families of open sets and one-to-one continuous mappings,* Questions and Answers in General Topology, **2**(1992), 24–26.

[1994] *On the monotone normality of functional spaces,* Moscow University Math. Bull., **49:3**(1994), 62–63.

YOSHIOKA, I.
[1980]   *Note on topologies for function spaces,* Math. Japonica, **25**(1980),
373–377.

YOUNG, N.J.
[1973]   *Compactness in function spaces; another proof of a theorem of
J.D. Pryce,* J. London Math. Soc., **6**(1973), 739–740.

ZENOR, PH.
[1980]   *Hereditary m-separability and hereditary m-Lindelöf property in
product spaces and function spaces,* Fund. Math., **106**(1980),
175–180.

# List of special symbols

For every symbol of this list we refer the reader to a place where it was defined. There could be many such places but we only mention one here. Note that a symbol is often defined in one of the previous volumes of this book. The first volume entitled "Topology and Function Spaces" is denoted by *TFS*. The second volume has the title "Special Features of Function Spaces"; it is denoted by *SFFS*, and the third volume whose title is "Compactness in Function Spaces" is referred to as *CFS*. We *never* use page numbers; instead, we have the following types of references:

(a) *A reference to an introductory part of a section.*
For example,
$\mathcal{A}_\xi$ · · · · · · · · · · · · · · · · · · **1.1**
says that $\mathcal{A}_\xi$ is defined in the Introductory Part of Section 1.1 of this volume. Of course,
$C_p(X)$ · · · · · · · · · · · · · · · **TFS-1.1**
shows that $C_p(X)$ is defined in the Introductory Part of Section 1.1 of the book TFS. Analogously,
$\mathbb{K}$ · · · · · · · · · · · · · · · · · **SFFS-1.4**
says that $\mathbb{K}$ was defined in the Introductory Part of Section 1.4 of the book SFFS and
$C_{p,n}(X)$ · · · · · · · · · · · · · · · **CFS-1.2**
means that $C_{p,n}(X)$ was defined in the Introductory Part of Section 1.2 of the book CFS.

(b) *A reference to a problem.*
For example,
$h\pi\chi(y, L)$ · · · · · · · · · · · · · · **432**
says that $h\pi\chi(y, L)$ was defined in Problem 432 of this volume, while
$C_u(X)$ · · · · · · · · · · · · · · · **TFS-084**
indicates that the expression $C_u(X)$ is defined in Problem 084 of the book TFS and, naturally,

© Springer International Publishing Switzerland 2016
V.V. Tkachuk, *A Cp-Theory Problem Book*, Problem Books in Mathematics,
DOI 10.1007/978-3-319-24385-6

$X[A]$ $\cdots\cdots\cdots\cdots\cdots$ **CFS-090**
shows that $X[A]$ was defined in Problem 090 of the book CFS.
(c) *A reference to a solution.*
For example,
$O(f, K, \varepsilon)$ $\cdots\cdots\cdots\cdots\cdots$ **S.321**
says that the definition of $O(f, K, \varepsilon)$ can be found in the Solution of Problem 321
of the book TFS.
Analogously, we can infer from
$\Delta_n(Z)$ $\cdots\cdots\cdots\cdots\cdots$ **T.019**
that the definition of $\Delta_n(Z)$ can be found in the Solution of Problem 019 of the
book SFFS.
From the expression
$B_1(X)$ $\cdots\cdots\cdots\cdots\cdots$ **U.463**
we can understand that the definition of $B_1(X)$ can be found in the Solution of
Problem 463 of the book CFS and, finally,
$\Omega(K)$ $\cdots\cdots\cdots\cdots\cdots$ **V.300**
says that the definition of $\Omega(K)$ can be found in the Solution of Problem 300 of this
volume.

Every problem is short; so it won't be difficult to find a reference in it. An
introductory part *is never longer than two pages*; so, hopefully, it is not hard to
find a reference in it either. Please, keep in mind that a solution of a problem can be
pretty long but its definitions *are always given in the beginning.*

The symbols are arranged in alphabetical order; this makes it easy to find the
expressions $B(x, r)$ and $\beta X$ but it is not immediate what to do if we are looking
for $\bigoplus_{t \in T} X_t$. I hope that the placement of the expressions which start with Greek
letters or mathematical symbols is intuitive enough to be of help to the reader. Even
if it is not, then there are only three pages to plough through. The alphabetic order is
*by line* and not by column. For example, the first three lines contain symbols which
start with "A" or something similar and lines 3–5 are for the expressions beginning
with "B," "$\beta$," or "$\mathbb{B}$."

| | | | |
|---|---|---|---|
| $A(\kappa)$ $\cdots$ **TFS-1.2** | | $a(X)$ $\cdots$ **TFS-1.5** | |
| $AD(X)$ $\cdots$ **TFS-1.4** | | $\alpha(X)$ $\cdots$ **1.4** | |
| $\bigwedge \mathcal{A}$ $\cdots$ **T.300** | | $\bigvee \mathcal{A}$ $\cdots$ **U.031** | |
| $\mathcal{A}|Y$ $\cdots$ **T.092** | | $[a_0, \ldots, a_m]$ $\cdots$ **1.1** | |
| $\mathcal{A}_\xi$ $\cdots$ **1.1** | | $\alpha + \beta$ $\cdots$ **1.5** | |
| $\alpha \cdot \beta$ $\cdots$ **1.5** | | $\alpha^\beta$ $\cdots$ **1.5** | |
| $B(x, r)$ $\cdots$ **TFS-1.3** | | $B_d(x, r)$ $\cdots$ **TFS-1.3** | |
| $B_1(X)$ $\cdots$ **U.463** | | (B1)–(B2) $\cdots$ **TFS-006** | |
| $\beta X$ $\cdots$ **TFS-1.3** | | $\mathbb{B}(X)$ $\cdots$ **SFFS-1.4** | |
| $\text{cl}_X(A)$ $\cdots$ **TFS-1.1** | | $\text{cl}_\tau(A)$ $\cdots$ **TFS-1.1** | |
| $C(X)$ $\cdots$ **TFS-1.1** | | $C^*(X)$ $\cdots$ **TFS-1.1** | |
| $C(X, Y)$ $\cdots$ **TFS-1.1** | | $C_p(X, Y)$ $\cdots$ **TFS-1.1** | |
| $C_u(X)$ $\cdots$ **TFS-084** | | $C_k(X)$ $\cdots$ **1.3** | |
| $C_p(X)$ $\cdots$ **TFS-1.1** | | $C_p^*(X)$ $\cdots$ **TFS-1.1** | |

# Index

© Springer International Publishing Switzerland 2016                                                      721
V.V. Tkachuk, *A Cp-Theory Problem Book*, Problem Books in Mathematics,
DOI 10.1007/978-3-319-24385-6

Printed in the United States
By Bookmasters